ISBN 978-0-331-03509-4
PIBN 11104623

FB &c Ltd, Dalton House, 60 Windsor Avenue, London, SW19 2RR.
Company number 08720141. Registered in England and Wales.

Supplement, July [illegible], 1890.

THE

CHEMICAL TRADE JOURNAL:

A WEEKLY NEWSPAPER DEVOTED TO

The Commercial Aspect of the Chemical and Allied Industries.

EDITED BY

GEORGE E. DAVIS,

FORMERLY ONE OF HER MAJESTY'S INSPECTORS OF ALKALI WORKS
CHEMICAL ENGINEER AND CONSULTING CHEMIST.

VOLUME VI.

[JANUARY TO JUNE, 1890.]

MANCHESTER:
DAVIS BROS., 32, BLACKFRIARS STREET.

1890.

f Chemi
t Supplem

MANCHESTER:
PRINTED BY DAVIS BROS., 32, BLACKFRIARS STREET.

THE

CHEMICAL TRADE JOURNAL.

Publishing Offices: 32, BLACKFRIARS STREET, MANCHESTER.

No. 137. **SATURDAY, JANUARY 4, 1890.** **Vol. VI.**

Contents.

Notices.

All communications for the *Chemical Trade Journal* should be addressed, and Cheques and Post Office Orders made payable to—

DAVIS BROS., 32, Blackfriars Street, MANCHESTER.

Our Registered Telegraphic Address is—
"Expert, Manchester."

The Terms of Subscription, commencing at any date, to the *Chemical Trade Journal*,—payable in advance,—including postage to any part of the world, are as follow:—

Yearly	(52 numbers)	**12s. 6d.**
Half-Yearly	(26 numbers)	**6s. 6d.**
Quarterly	(13 numbers)	**3s. 6d.**

Readers will oblige by making their remittances for subscriptions by Postal or Post Office Order, crossed.

Communications for the Editor, if intended for insertion in the current week's issue, should reach the office not later than **Tuesday Morning.**

Articles, reports, and correspondence on all matters of interest to the Chemical and allied industries, home and foreign, are solicited. Correspondents should condense their matter as much as possible, write on one side only of the paper, and in all cases give their names and addresses, not necessarily for publication. Sketches should be sent on separate sheets.

We cannot undertake to return rejected manuscripts or drawings, unless accompanied by a stamped directed envelope.

Readers are invited to forward items of intelligence, or cuttings from local newspapers, of interest to the trades concerned.

As it is one of the special features of the *Chemical Trade Journal* to give the earliest information respecting new processes, improvements, inventions, etc., bearing upon the Chemical and allied industries, or which may be of interest to our readers, the Editor invites particulars of such—when in working order—from the originators; and if the subject is deemed of sufficient importance, an expert will visit and report upon the same in the columns of the *Journal*. There is no fee charged for visits of this kind.

We shall esteem it a favour if any of our readers, in making inquiries of, or opening accounts with advertisers in this paper, will kindly mention the *Chemical Trade Journal* as the source of their information.

Advertisements intended for insertion in the current week's issue, should reach the office by **Wednesday morning** at the latest.

Advertisements.

Prepaid Advertisements of Situations Vacant, Premises on Sale or To be Let, Miscellaneous Wants, and Specific Articles for Sale by Private Contract, are inserted in the *Chemical Trade Journal* at the following rates:—

20 Words and under	2s. 0d.	per insertion.
Each additional 10 words	0s. 6d.	,,

Advertisements of Situations Wanted are inserted at one-half the above rates when prepaid, viz:—

20 Words and under	1s. 0d.	per insertion.
Each additional 10 words	0s. 3d.	,,

[illegible] Announcements in the Directory Columns, and all Adver[illegible] are charged at the Tariff rates which will be forwarded on [illegible]

THE CHEMICAL AND ALLIED TRADES IN 1889.

"Was I deceived or did a sable cloud
Turn forth her silver lining on the night."—*Milton.*

THE dark cloud has for a long time past o'ershadowed many branches of the chemical trade. Increased competition, a continual lowering of prices, and in some instances a diminished consumption, have all combined to make the chemical manufacturer's life an uneasy one. Many, if not most, of the difficulties that have arisen may be traced directly to the commercial surroundings of the undertaking, as a manufacturer can only consider his business successful or non-successful according to whether his revenue exceeds his expenditure or *vice-versa*. It will thus be seen that even with chemical operations in *statu quo*, the various alternations in price, either of raw materials or finished products, will banish certain processes from the arena of practicability, or bring up others that have been devised before their time. The present year has seen all of these things in degrees of greater or lesser magnitude, and perhaps

THE ALKALI TRADE

is one that has had a keener watch set upon it than most other branches of industry. The year will be remembered as being the breaking up of all conventions regulating the sales of alkali products; hydrochloric acid, bleach, chlorate caustic, and ash: all the paper bonds of union have been scattered to the four winds of heaven, and each manufacturer is now free to make and sell just as much as he pleases. It is difficult to say at this early date what will be the effect of abrogating the many conventions that have of late existed in the alkali trade, but the immediate result has been to cause a fall in bleach values of considerable amount, for while at this time last year the price of this commodity was £7. 7s. 6d. per ton, it has now fallen to £5. and even less is whispered.

There is no doubt the introduction of sulphite wood pulp into papermaking has had a considerable effect upon the alkali trade, as very fair paper, useful for a large number of purposes, can be made without the intervention of a particle of either caustic soda or bleaching powder, and the use and manufacture of sulphite pulp is extending.

The anxious times passed through by the Leblanc makers have served to stimulate them to an extraordinary degree. It is true they have not yet learned the lesson of the folly of making low strength ash and caustic, giving casks and salt to the consumer, who, nevertheless, has to pay the carriage upon it, but no doubt the time will come when but little else than 70 per cent. will be made by the Leblanc process.

In most processes the gradual expansion of prices for raw material has necessitated the strictest economy in all operations connected therewith, and whereas 55 to 60 cwts. of salt were required twelve years ago to make a ton of "bleach" by the Weldon process, it is now being made from 42 cwts., a number, we feel sure, cannot be reduced by the said process without material modification. At the same time the Deacon process has been so improved, that a ton of bleach is now being made from a little more than a ton of salt, a quantity yet leaving room for improvement.

Alkali values have improved somewhat, and we hope this is to be a permanent feature of the trade; refined 48 % ash having risen during the year fr m £4. 17s. 6d. to £5. 5s., and 70 % caustic soda from £7. to £7. 5s., 60's caustic having also advanced 5s.

Coal and salt, two very important raw materials, have advanced in price, 4s. 6d. slack has risen to about 8s., and salt from 5s. 9d. to 10s., hard facts that will no doubt tend to the observance of stricter economy.

But by far the most notable event in the alkali trade during the current year has been the setting to work of the Chance-Claus sulpher recovery process, which is now working successfully in several places, so far as the operations themselves are concerned. The plant is, however, costly, though easy to mani ulate, and to our minds the future of the process taken as a whole is a little bit uncertain. In fact it has already been discovered that even with the best arranged plant, sulphur, as sulphuretted hydrogen from vat-waste, cannot compete in cost with the sulphur from Spanish pyrites, and that in order to secure success, sulphur must be the marketable product. But again there lies a stumbling block in the way, far greater than any other with which the commercial departments of the alkali works have eve had to deal. Invention succeeds invention, the new displaces the old, and the great market for "flowers" or flour of sulphur—the gunpowder manufacture—is slowly but surely becoming closed. The new powders, and especially the picric compounds, are attaining such a degree of perfection as to have sooner or later a most decided effect upon the flour of sulphur trade. In fact, we may say the reaction has now set in and those Leblanc makers producing flour of sulphur have already discovered it is no easy task to sell their product. One works in which the process has been recently started, is endeavouring to secure the maximum quantity of "roll" sulphur and the minimum of "flour," but whether this will aid them in the long run remains to be seen.

The ammonia soda works still continue in the even tenour of their way. What the Cheshire Alkali Co. are doing we are unable to say, but a third works of this character have now placed a high-class quality of ammonia-soda in the market. After many trials and tribulations, Messrs. Mathieson and Co. have so far completed their work as to offer considerable quantities for sale. Messrs. Brunner, Mond, and Co, have again paid substantial dividends to their shareholders, and though they seem no nearer the solution of the problem as regards the manufacture of bleaching powder, yet they have, it is rumoured, succeeded in getting more latent value from their fuel than any other firm has yet been able to accomplish.

Intimately connected with the chemical trade, bound up with it by ties of no common order, is the

PAPER MANUFACTURE,

which is one of the first to feel any improvement in the general condition of trade, the consumption of paper being so widely distributed over the civilised world.

This has just been the case during the year that has just passed away. Paper makers have been busy, and in many cases extensive alterations and additions to existing plant have been made to meet the growing requirements of the demands of the market, and to keep pace with the ceaseless tide of invention which seems to flood every industry. It is confidently expected that the healthy tone which pervades the trade at the present time will continue during the next 12 months, if it does not materially improve.

The paper trade has not escaped the syndicate fever which has attacked every description of business enterprise—for we have witnessed the total collapse of the projected "News Syndicate," and the successful conversion of large publishing and stationery firms into limited companies. The development [illegible] wood-pulp manufacture in this country during [illegible]ar, has taken place to a greater extent than [illegible]anufacturers of paper are becoming more ac[illegible] the qualities and capabilities of the different [illegible] pulp manufactured for the most part in Norway and Sweden and in Germany, and have now fully recognised the vast influence which these fibres will have upon the industry. Several of the largest pulp manufacturing concerns have been successfully floated with British capital, and in notable instances the operations of these companies will be carried out in England. The Kellner Partington Co., at Barrow, the Hull Co., at Hull, and the very recently projected company at the same place—a reference to which we had in our last but one issue—have for their object the manufacture of wood pulp by what is known as the bisulphite process, and the eagerness with which the public subscribed towards the formation of them is a sure indication of the further development of this industry in our midst. We hear further that the large installation of Dahl's system for the preparation of wood-pulp by what is known as the "*Sulphate* process" is being adopted by Messrs. Dickinson, of Croxley Mills; and that the Inverkeithing Co., on the shores of the Firth of Forth, has decided to extend its pulp manufacture by the old caustic soda process, first introduced into this country by Lee and Houghton, beyond the mere experimental output of the present year. But the extension of these manufactures has been carried out on a much wider scale in America, Germany, Austria, Norway and Sweden, than at home. These countries possess abundant supplies of timber suitable for pulp manufacture, and thus are supposed to have advantages which are wanting in England. All these countries have received extensive additions to their wood-pulp producing powers.

Owing to the collapse of the Bleach Association the value of Bleaching Powder has materially decreased, giving a specific advantage to the paper maker in the cost of bleaching the various raw fibres which he uses, whilst the action of the Salt Union in raising the price of chemical salt, and the influence of the increasing cost of fuel upon Leblanc soda makers have had a tendency to increase the cost of alkalies. What, therefore, may be saved by the diminishing value of bleach, will only neutralize the increased cost of the alkali, and the outlook in the immediate future for the further developement and introduction of electrolytic methods of bleaching seems somewhat gloomy. The promoters of the electrolytic methods of bleaching have been somewhat silent during the year, probably owing to to the fact that rival processes are said to exist, and the instability of the future cost of bleaching powder is very apparent. It was confidently hoped that the present year would see the thorough development of M. Hermite's process, but with the exception of a little information of questionable value, nothing has been said with respect to the advantages, first, as a practical process, and second, as a question of economy, which this invention offers, nor, indeed, that its success in either of these respects has been assured.

The year 1889 has not been marked by the introduction of any new invention of a revolutionary tendency, but rather by the steady introduction of improvements on existing machinery, and by the more extended use of processes already widely known, and which are being more highly appreciated as factors in the advancement of the times. There have, also, been plain indications that old rule of thumb methods of manufacture are now looked upon with uncertainty, and that technical education of a high class is necessary to understand the nature and possibilities of the numerous inventions which flood the market, and to successfully lutroduce them into our paper mills.

Perhaps next in interest, and also, we may say, in importance, is the

SCOTTISH MINERAL OIL INDUSTRY,

which to-day looks back on a year of steady, though not very swift, recuperation from the serious exhaustion of the foregoing season. The various companies—formerly destructively at war with each other—were found to work wonderfully well in combination, much better than the most hopeful had expected, and although some little jostling and individual interests now and again cropped up, there had arisen nothing of very grave import in that connection. The remembrance of the general peril of 1887 was still fresh, and mutual forbearance in the last resort continued a ruling influence. The Mineral Oil Association, which may be regarded as a select committee of the whole industry with executive powers,

. in hand the price and market position of
:ther with the rate of production (restricted)
ach of the makers ; and the same authority,
'ith the Standard Oil Co. of America, had
regulation in like manner, of paraffin scale
The other varieties of the paraffin manu.
lubricating oils, naphtha and sulphate of
ieanwhile left to find their natural levels in
ket. In January, 1889, the paraffin scale
ent fell in for renewal, a representative
producers attending the meeting held
hat purpose. The Oakbank Oil Co. at first
inclination to become a party for another
itlook for a little seemed risky ; but, finally,
emoved so far, and the compact entered its
h the original members intact and a few
andle making firms added. On that occasion
e and wax, and of the paraffin-candles manu.
m, were left as originally fixed, although
rongly advocated a slight advance. These,
at 2d. and 2¹⁄₁₆d. per lb. for scale, soft and
; 2⅞d. for wax 120° semi-refined ; and 4d.
as the standard for household candles
dealers, while burning oil by agreement was
Scotch buyers, at not less than 5¾d. to 6d. a
to quality. A heavy penalty is by mutual
ed to infringement of this rule, but it has to
makers are at liberty to sell to *bona-fide*
foreign consumers at what they can get and
In February proposals were seriously ad.
ociation of financiers in London to buy up
eral Oil Company in Scotland, in order to
dustry as one single close concern ; but
ject agitated the various managements for
ue reconciliation of individual interests
too huge an undertaking, and it faded out
practical finance without result. The
were holding surprisingly well together
n of the Association, and in short the time
Had the offer come, in almost any shape
months of 1887, it would have been snapped
rage was for selling out of "oil" then, and
e thought of buying in. At the annual
s year—April and May—the dividends an.
eek to week gave indication of a vastly
condition. Be it noted that in 1888,
refining companies were doing no more
om the brink of collapse, there were no
ccount. That is, the Broxburn Oil Com.
ears had paid 25 per cent, managed with
le for 5 ; a second company paid 3, and all
anks—some of them worse, for there were
s to declare. This year the show was
able. The Broxburn Company came up
the Hermand (crude oil only, crude works
bject to the same vicissitudes), 17½ ; the
the Oakbank, 8 ; Young's Paraffin Com.
Clippens, 5 ; and the Holmes, 4. The
ompany had nothing to divide, but that was
l causes, well understood and previously
Burntisland works had just engaged in the
ndles at a heavy outlay for plant, and their
et new and unknown, had to a large extent
he market unplaced. During the season
opportunities of this concern for disposing
es should be much more favourable. The
ny has been unfortunate from the begin.
ew years ago, owing to the shale field to
having come up in productiveness to the
the vendors, and, as yet, there has been
shareholders, although the concern in its
as confessedly been well managed. The
mpany made a small profit, but not sufficient
he shareholders had, therefore, to repeat
k of the two foregoing years. Prospects
mpany are not, however, by any means
agement. There has been much trouble
is now amended at considerable outlay,

and a fine new field of shale contiguous to the works has j
been opened out, so that with decent values ruling in the p
ducts market, the balance of next year ought to leave sor
thing for division. A very uncertain factor in the lot of th
shale oil works in Scotland is the attitude of the shale min
employed, of whom there are several thousands at work wh
the industry is in full blast. A strike in 1887 inflicted
disastrous injury, irrespective of the loss in wages to t
strikers themselves, the latter variously estimated fr
£50,000., upwards. The strikers were worsted, and since
turning to work, better values being had for most of t
paraffin products, wages have been increased withc
pressure applied. The men, or their agents for the
have been clamouring in a small way for a furth
rise and holding out threats ; but prices really do n
permit of that, and were the miners so mad as to co
out in force again, as in 1887, nothing but general disast
could be the result. The fact is, that towards the close
the year the position of burning oil (the chief product, a
that on which dividend prospects depend) became less sat
factory, owing partly to the phenomenal mildness of t
winter, and the Association prices (raised in July to 6¼d.
6¾d. at the works), were maintained with some difficult
Just at the close, the gas workers' agitation in England h
somewhat stimulated the demand for Scotch burning oil, a
also for the improved oil lamps which one or two of t
Paraffin companies manufacture ; but this is mainly
transient influence, although it is certainly true that son
householders, after being thus compelled to try oil *versus* co
gas, are delighted with the perfected results of the forme
and disposed to stick to it as a permanency. During the ye
one new works has been added to the list, or rather, it shou
be said, one of the old shipwrecked concerns has been r
vived, in new hands and under more hopeful auspices. Th
is the Caledonian Oil Company, the shalefields and refine
works acquired by them being those of the old Lanark Com
pany, which went down a few years ago when the oil outloo
in Scotland first began to be an anxious one, and prices we
unremunerative. The proprietors, that is to say the share
holding body, are chiefly of London, where the head offic
at present is, and the place also of the statutory meeting
The direction of the works is in the hands of Mr. Sutherlan
of Bathgate, formerly the managing director of the We
Lothian Oil Company, and Mr. A. C. Thomson, late of th
Pumpherston Company, has been acting as chemical enginee
and consulting chemist. The oils and solid paraffins of thi
new works are expected to be ready for offer to consume
early in 1890. The Hermand Oil Company, of Mid Lothia
till now confined to the production of crude oil, resolved a
its annual meeting in May to begin refining, and at an earl
date to consider the advisability of starting candle-makin
also for these purposes £60,000. of additional capital ha
been called for from original shareholders and new investor
and this, it is believed, has been coming in to fair satisfactio
under the circumstances. These represent the only addition
of the year to the bulk of the paraffin industry in Scotlan
and the old concerns are all in life and all likely to liv
although one or two have still formidable difficulties to co
tend with. They form a section of their own on the Exchang
and speculative buying and selling go on more or less active
from year's end to year's end. The stock market is "torpedoed
(to use an American oil-well technicality) from time to time b
speculative operators with sometimes surprising results. Du
ing September, quotations took a big jump under a report th
both America and Russia, as oil producers, were fallin
behind in available stocks, but the reactive drop in Novemb
more than discounted September's rise, on the impressio
regaining footing that both these foreign fields still possesse
abundant supplies.
These Oil Companies of Scotland number 15 in all, e
cluding one or two private affairs of quite small compass. Th
capital involved amounts, roughly speaking, to two and a ha
millions, of which Young's Paraffin Co., which is the largest
an aggregate investment, appropriates nearly £600,000. Th
Broxburn Co. comes next with its £335,000, followed close
by the Burntisland and Clippens Companies. The Linlith
gow also is heavily weighted with capital, and the next
order of amount is the Pumpherston Co. with £130,000. T

smallest is the Dalmeny Co. with £27,000, a concern of handsome dividends throughout, but confined to the production of crude oil only, and so liable to fewer risks. Four of the companies are fully paid up; the rest are for the most part called to within one-fifth.

Passing from the Mineral Oil Companies, we turn our attention to the processes of

TAR DISTILLING AND REFINING,

in which industries, however, there is nothing revolutionary to chronicle. It is the outside of this industry that affects the inside, and tar distillers as a rule have to go plodding on, and accept as the result of their labours whatever prices the dealers may see fit to offer them. Benzol has kept a remarkably steady value during the whole year, as 90's came in at 3s. 2d., and went out at 3s. 5d., while 50/90's entered the year at 2s. 6d., and found 2s. 7d. at the end of December. Carbolic Acid (crude) and Anthracene have declined, but the latter is now very firm; in fact, with an exception here and there, all tar products have become enhanced in value, to the sole gain, in many instances, of the gas works situated in positions populated with keen competitors.

Taking all in all, tar distillers as a rule cannot look back with regret upon the year 1889, after having passed through years such as '85, since which date there is scarcely a product that has not doubled itself in value, and several become trebled and quadrupled. Yet the conditions are practically the same as before. There is more tar undergoing distillation now than in 1885, and though the yield of benzol from tar may be less, there is much more carbonization benzol in the market than there was four years ago.

During the year there have been considerable developments in the use of certain bye-products of the coal tar industry; creosote is now very largely employed for purposes of illumination in large open spaces, and the "Wells" light is perhaps the most portable and useful form of apparatus, as the oil is converted into vapour before combustion, and hence there is no spray of oil thrown about as with most other lamps. This outlet for creosote (cooled and filtered) is one which all tar-distillers should foster as, if the price is not inordinately raised, there should be a steady business passing for many years to come, that is, of course, supposing the tar distillation industry continues to exist. But there is no doubt that the longevity of this industry is contingent upon the non-success of the "Dinsmore process," which we have noticed in another column, and it is most probable that tar distillers of the keener perception are watching that process with more than ordinary interest. It was at one time thought that a big fillip would be given to the pyridine industry, as an agent for odorizing "water-gas," but as this gas is now under a cloud, these expectations are not being realised. Neither as a medium for "denaturing" spirit has much pyridine been required.

The year has not been marked with any invention or discovery of note, unless we mention the introduction of synthetical carbolic acid by the Badische Anilin and Soda Fabrik. The tar distillation industry is too uncertain as regards its supply of raw material for manufacturers to expend much capital in experimental plant, or to trouble themselves in developing business which might not be theirs to-morrow. It is strange, yet interesting, how one industry may beget another, and the new one grow into immense importance. Such has been the case, however, with

THE GAS INDUSTRY,

which is a manufacturing operation requiring for its full technical development the highest scientific knowledge and skill, both in engineering and chemistry. Twenty years ago this was not recognised, but it is pleasing to observe that in all works of standing the chemistry of the process is to-day receiving due attention. Gas managers have of late turned their attention more deeply to the working up, on the premises, of their residual products, this having, in a great measure, been brought about by the action of the railway companies. We have already called attention to many anomalous rates and charges, but there has never been anything more iniquitous than the charges for carriage of ammoniacal liquor, which have been extracted from the pockets of the gas companies, until it has well nigh extinguished the carrying trade of gas water, the gas works being compelled to work up the liquor themselves, in order to escape the fearful tax imposed upon them, often double the value of the substance carried.

In the sulphate of ammonia manufacture the old form of still is becoming a *rara avis*, and no one with any knowledge would even keep them working if they possessed them. The column still using 7 cwt. of fuel per ton of sulphate when worked with 5° Tw. liquor (ten ounce) is the one that is now being used almost universally, and of which there are several patterns to choose from. In this connection, especially when the gas liquor made in the works is used up therein, it is surprising that no serious attempt should be made to reduce the water used in scrubbing, so strengthening the liquor, say to 16 or 18 ounce, and avoiding the inevitable loss of illuminants by excessive water. If 18 ounce liquor were used with a column still of the most approved construction, the fuel consumption should be brought down to 4 cwts. per ton of sulphate.

During the past year, which has not been an uneventful one for several gas undertakings, we have published in our columns "A Study of Coal Gas," being the results of a research emanating from the laboratories of the Paris Gas Co. These experiments deserve the careful consideration of the Board of Management of every gas undertaking in the United Kingdom, showing as they do the necessity of a careful study of the subject of gas-making and the highly technical character of the various operations.

Those responsible for the due carrying on of the gas industry have noticed perhaps the steady progress made by practical electricity, and though we may be some distance from the time when gas will be superseded for purposes of illumination, yet the recent strikes have taught the lesson how uncertain the gas supply of the future may be, and how much may be done with paraffin oil better and at less cost than gas. These are lessons which, if not learned, had better not have been read us at all: who can say that he has profited by them?

In the

ANILINE COLOUR TRADE

there is not much to report in the scientific department, nevertheless the outlook of the industry in England looks very hopeful. In the latter months of the year, the old established colour works of Messrs. Dan Dawson Bros. was converted into a Limited Liability Company, and is, we understand, to be carried on with renewed vigour. The works are well situated and well stocked, both with chemical and colour plant, for the manufacture of Magenta, Blue, Saffranine, Naphthalamine, Bismarck Brown, Chrysoidine, and the Naphthalene colours. When it is remembered that the German colour makers alone export goods to the value of over £5,000,000 per annum, it does seem strange that no great effort has been made to increase the make at home. But perchance the tide which has for so many years been running in one direction, will soon be restored into its old channels, and perhaps we may be the first to announce that the well-known colour firm of Mr. Ivan Levinstein has been converted into a private limited company, with a capital of £150,000, of which the partners are two of the most important German companies, *viz.*, the Actiengesellschaft für Anilin Fabrikation, in Berlin, and the Farbwerke, vormals, Friedrich Bayer and Cie, of Elberfeld, who will, with Mr. Levinstein himself, trade as I. Levinstein and Co., Limited. Mr. Ivan Levinstein will be the managing director. It is, we believe, the intention to very considerably extend and enlarge the Crumpsall Vale Chemical Works, and there is no doubt that the new enterprise will give a greater impetus to a re-transfer of the coal tar colour industry to England than all the clamour for Technical Education.

Another trade that has aroused from lethargy, having been awakened, as is usually the case, from the outside, is the

WOOLLEN INDUSTRY,

especially that branch devoted to the scouring of the raw wool. Fortunes in the bye-products from certain operations have been sent down the Aire and Calder simply for the want of a little enterprise on the part of the manufacturer, a little

are fairly well described by the lines of one of Spenser's poems:—

"Full little knowest thou that hast not tried
What hell it is in suing long to bide;
To loose good dayes that might be better spent,
To waste long nights in pensive discontent.
To speed to-day, to be put back to-morrow;
To feed on hope, to pine with feare and sorrow.

* * * * * *

To fret thy soule with crosses and with cares,
To eate thy heart through comfortlesse dispaires;
To fawn, to crowche, to waite, to ryde, to ronne,
To spend, to give, to want, to be undonne."—*Spenser.*

Perhaps the year '89 has been most propitious to

THE MERCHANT

who has conducted his business with due regard for the future. In our last New Year's number we pointed out that there seemed every hope for a revival of trade, especially in the chemical and allied industries, and in many instances this has been so where the free course of trade had not been previously fettered by protectionist conventions; these industries which have been hindered in their free expansion will have to find their normal level before they commence to improve. Iron and coal and salt have assumed values far in excess of what even the most sanguine seller could have dreamed of two years ago, and the time has come when more economy will have to be practised in the use of all these things. It is notorious that excessive waste has taken place in the use of these two last commodities for many years past, with the result of excessive air pollution. The merchant will in many cases benefit, but the lesson to be learned is the necessity for the manufacturer to study the commercial aspect of the situation more deeply, and to delegate a little more of the works management to technical and trustworthy hands. Early on in the year '89 it was seen that a rise in the price of iron was inevitable; the rise in the price of coal was also a foregone conclusion, yet all the while makers of chemical products were wrangling away over conventions instead of improving the yield of their processes. Sulphuric acid, especially in Yorkshire, has experienced an immense rise in price; B.O.V. was early in the year advanced 2s. 6d., then shortly after, another 2s. 6d., and finally 7s. 6d., so that if there was a profit before the rise, there must be a substantial one now. Whether this prosperity will result in other works being built, remains to be seen, but it has had the effect of causing large users to erect chambers for manufacturing their own requirements.

In glancing over the history of the past year, we observe an almost uninterrupted improvement in

MANURES AND MANURE MATERIALS

of all kinds, if we except sulphate of ammonia, which has just maintained the price of this time last year, and nitrate of soda, which, owing to causes we shall presently discuss, has reached a lower point than we ever remember at this period of the year.

The rapid advance in Charleston freights 18 months ago, causing an equivalent advance in the c.f. and i. price of phosphate rock, having seriously curtailed purchases for United Kingdom during the latter half of 1888, it was anticipated that a large stock of rock would have accumulated at the shipping ports, and that with easier freights after the cotton shipping season easier prices might be expected, but when it transpired that the United States had taken all the available supplies of rock, the United Kingdom buyers began to realise that only a very material drop in freights could afford them any hope of a return to anything like the former level of prices. 1889 opened with somewhat easier freights, but, having no surplus stocks of rock, shippers were still very firm and were still able to obtain 9d. per unit c.i.f. to United Kingdom—in some cases 9¼d. for prompt shipment. Up to April buyers still held off as far as possible, but with their Autumn requirements staring them in the face they were then compelled to make their purchases, and a large business was done at 9¼d. c.i.f. to United Kingdom. In May freights fell to within 5s. per ton of the lowest point of 1888, but only for a short time, and with the aid of the huge United States demand, shippers were able to maintain their price, and, by the end of July, to advance it to 10d. per unit c.i.f. Since then the market has steadily hardened, and at the close of the year shippers almost refuse to quote ahead—the value being fully 11d. per unit for cargoes c.i.f. to United Kingdom ports.

The shipping season for Canadian phosphate opened at 9d. for 70 %, 10½d. for 75 %, and 1s. for 80 %, with a ⅛th rise, delivered Thames, Mersey, or Clyde, and continued at those prices until in April, in sympathy with other phosphates, it was advanced to 9½d., 11d., and 13d., and in May to 10d., 11½d., and 13d., for 70, 75, and 80 % respectively. In August, considerable business was done at 10¼d. and 10½d. for 70 %, 1s. for 75 %, 13½d. for 80 %., and more money has since been paid. Very full prices have been bid for 1890 shipment, but so far shippers refuse to commit themselves, looking for something near 1s. for 70 % and relative prices for 75 and 80 % strength later on.

Large shipments of Somme and Belgian have been made during the year, Somme at prices ranging from 8½d. and 10d. for 55/60 %, and 9½d. and 11d. for 60/65 %, ⅛th rise; Belgian at ½d. to 1d. per unit more for the same strengths. Aruba has kept on about the same level of prices as Canadian for 75 %, with ⅛th rise.

Bone Ash early in the year was somewhat neglected, and could easily be bought at about £4. 2s. 6d. on 70 per cent. in cargoes, but in March £4. 5s. had been paid, and at the end of May the whole of the 1889 South American production was quietly bought up for German account at from £4. 5s. to £4. 7s. 6d. per ton on 70 per cent., Hamburg terms. In August re-selling commenced at £4. 13s. 9d., Hamburg terms, and a large portion of the River Plate supply has since changed hands at up to £5. 2s. 6d. on 70 per cent. Closing quotation is £5. 5s.

The year opened with a very firm market for River Plate cargoes of bones, £5. being required for United Kingdom or Continent, but in February the value fell to £4. 15s.; and in April a large arrived cargo was sold at £4. 10s. In May, however, the United States having come in as a large buyer for summer and autumn shipment, price was again advanced to £5. for shipment, though an April cargo was subsequently sold at £4. 16s. 3d. Since then values have steadily advanced, to £5. in July, £5. 5s. in August, £5. 7s. 6d. in September, £5. 8s. 9d. in October, and £5. 10s. in November, for United Kingdom or Continent, the United States all the time paying rather more and securing nearly the whole of the River Plate supply as well as several thousand tons Mediterranean. United Kingdom has had to depend largely upon East Indian crushed bones and bone meal. Prices of crushed bones have ranged from £4. 6s. 3d. in March, for autumn shipment, up to £5. 10s. on spot in December, the latter price being now nearest value in any position. East Indian bone meal opened at £5. in January last, falling as low as £4. 12s. 6d. on spot in March, under pressure of heavy arrivals. For shipment, however, the lowest price touched was £4. 17s. 6d. in April, since when it has gradually advanced to £5. 12s. 6d. delivered Thames, Mersey, or Clyde—the closing value. It should be remarked that speculators have got hold of the bulk of the supplies for shipment December forward, but still we do not think that prices have been so far unduly forced, having regard to the present values of mineral phosphates and the very large U.S. requirements of phosphatic manure material.

Turning to nitrogenous material, nitrate of soda claims first attention. A year ago this article stood at 11s., which price was practically maintained up to the end of March, thanks to the skilful manipulation of shippers and speculators who coutrived to delude buyers into the belief that there was bound to be a scarcity of supply for the spring demand, though by the end of June the visible supply had increased within the twelve months no less than 100,000 tons. The movement, however, collapsed at the end of March under pressure of figures, and by the end of May 8s. 4½d. was touched, a price quite as high as was warranted by the figures. Since the spot price has varied from 8s. 3d. end of October, 8s. 9d. end of November, when it was stated that three-fourths of the nitrate works would be stopped throughout December There has been no stoppage, however, and the close of the year finds us again at 8s. 4½d., and with an increase of visible supply within the twelve months of something like 150,000 tons. So far as can be seen the prospects of nitrate were never more hopeless, and only an actual stoppage of production over a considerable period, or a material curtailment of output all round, can save the market from a further serious decline. There can be no question that the capacity to produce has been enormously increased within the last few years, owing to the formation of limited companies out of the old private concerns and the consequent introduction of larger capital and better appliances. It recently transpired that one of the leading companies had within the last two years increased its output capacity from 150,000 qtls. per six months to 280,000 qtls.; and if the other companies have increased theirs in anything like the same proportion, what has been said above only becomes the more evident. No doubt a low price for nitrate will stimulate the consumption, but it must take a long time before it can meet the present output capacity.

Sulphate of ammonia has apparently not been influenced by nitrate of soda at all, but actually, we think, it has, and that the influence of nitrate has been counterbalanced by that of extreme scarcity and relatively high prices of all other kinds of nitrogenous material. The present price of sulphate works out only to 10s. per unit of ammonia, whereas dried blood is fetching quite 11s. 6d. per unit, and Liebig's and other similar meat guanos would find ready buyers at over 11s. per unit of ammonia. So that where sulphate ammonia can be substituted it is substituted, and we are of opinion that only this cause has saved sulphate from a very considerable decline in value. Superphosphate has maintained a very firm position throughout the year, excepting for a short time in May and June, when, to clear their surplus stock, manufacturers submitted to lower prices, opening with 45s. per ton in bulk, f.o.r. at works, for 26 per cent., that price was maintained until late in May, when as low as 40s. was accepted. The market, however, recovered in August, along with advanced prices for materials, and 45s. again became the current quotation. Closing value cannot be put under 46s. 3d. in bulk f.o.r. at works.

There are three facts which we think manure manufacturers would do well to take into account in making provision for the future—the increased buying capacity of the British farmer, the enormously increasing demand for all kinds of manure material for the United States, and the inadequacy of new sources of supply. We think these facts point to a higher range of prices over 1890 than we have had in 1889, excepting for nitrate of soda, and sulphate ammonia in so far as it may be affected by nitrate.

And now we must say a few words

TO OUR READERS

who have appreciated the value of accurate trade information disseminated amongst them at short intervals. This is clearly shown us by the support accorded to the *Chemical Trade Journal* by the members of the chemical and allied industries both at home and abroad. We have now a very fair circulation all over the civilised world, and in the current year we hope to materially improve the Journal and make it even more interesting to subscribers outside the general body of manufacturers of heavy chemicals. The number of industries in the United Kingdom depending more or less upon commercial chemistry is legion, and we shall not be satisfied until we have gathered them all into our fold.

We are now entering into our sixth volume, and the experiences of the past five have not been thrown away upon us; we can see now, even more clearly than before, the great need of a Journal carried on upon the lines we have followed up to to-day, not only to be written for the makers of chemicals pure and simple, but to be a bond of union between the producer and consumer, to be the medium of announcing new products and criticising new processes in a manner that can only be done by private enterprise.

We have endeavoured to do all our work in connection with the Journal absolutely free from bias; we hope we have given praise where praise has been due, and if we have blamed where blame was considered deserving we hope to be credited with purity of intention; but withal, the fact remains, that if our remarks have any weight with our readers, as we hope they have, it will make us cautious in our dispensation of the Attic salt, of which all editors have more or less for disposal.

In many industries the new year bodes much good, but there are others that will not look so healthy at the close of '90 as at the end of '89. It would be invidious to particularise, we only hope that we may be mistaken in our forecast, and that all our readers may have

A Happy and Prosperous New Year.

AMMONIA STILLS.

THE old-fashioned method of placing ammonia water in a boiler and blowing steam into it until all the carbonate and sulphide was driven over into sulphuric acid is very properly coming to a close. It has seen its day, may have rendered good service to its employers, but it cannot hold its own against newer and improved processes and machinery. Ten years ago the foregoing method was pretty generally adopted, but in many works dealing with ammonia salts, the gases and steam passing away from the last boiling were utilised in a tall tower filled with bricks, and down which the gas liquor was made to flow. In this form of apparatus the quantity of steam used was enormous, and we have known of instances where 30 cwts. of fuel have been required to each ton of sulphate of ammonia produced. Moreover, the amount of lime employed with the old form of apparatus was such that one never went into a sulphate works without seeing an enormous heap of it piled up somewhere, and there were always constant complaints of the irregular character, and the large quantity of ammonia left in the spent water. Nevertheless the old process was for a long time in much favour, and manufacturers did not look with approbation upon the apparent complications of the new form of still, which in 1878 was presented to their notice. We allude now to the Grüneberg still, of which there are many in use abroad, but very few in this country.

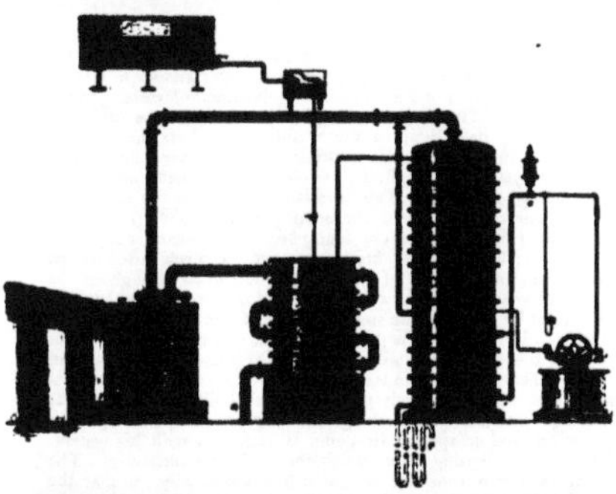

[illegible] still has been much improved of late years, and perhaps [illegible] is that made by Messrs. Ashmore, Benson, Pease and Co., [illegible] of Stockton-on-Tees. The accompanying illus-[illegible] a complete apparatus guaranteed to produce twenty [illegible] of ammonia per week from 5° Tw. liquor of good [illegible] the liquor is 5½° Tw. it will produce 25 tons, and [illegible] Tw. [illegible] weekly.

The [illegible] consists, as shown, of an overhead tank, in which [illegible] of gaswater is constantly kept. From this tank it [illegible] regulating tank, fitted with a lever and float, so [illegible] constant level, and so ensure the regular and [illegible] through the apparatus. The remainder of the [illegible] a tall column still, a heater of peculiar construction, [illegible] joints, the ordinary and well-known saturator [illegible] is fished, and a vessel from which the lime is [illegible] into the still. The saturator being of the ordinary [illegible] any description, and the same may be said of the [illegible] exception that it is not used for making up the lime in, but only for containing a supply of finely-sieved milk for pumping into the still.

The heater into which the gas-water first runs from the regulating tank above, receives the waste steam and gases from the saturator, and, passing forwards to the outlet, heats up the gas liquor in its passage, so that when the still is properly at work the gaswater enters the tall still at 160° F., while the gases escape at 100°-120° F. If the heater is overworked the exit gases will be hotter than the liquor going into the still, and this is not desirable. The heater is so constructed that all the joints are visible, and outside., *i.e.*, round the periphery, so that if a leakage takes place it is at once detected. The heater is the weakest point in all other sulphate apparatus, but the one illustrated gets over every difficulty which has been urged against them. In the earlier forms of this still the tall column consisted of a series of plates perforated with a large number of quarter-inch holes through which the steam passed, the liquor passing down from plate to plate by means of overflow pipes. This form is now made for special work, but it possesses the disadvantage that it will only work the exact quantity it is designed for. It does its work admirably, but when the supply of liquor is decreased from the normal, the reduced pressure lets the liquor fall from tray to tray, and when the steam-pressure is increased to meet a larger liquor supply, the steam cannot get quickly enough through the plates.

The new still, such as is illustrated above, and as made by Messrs. Ashmore, Benson, Pease and Co, overcomes all the foregoing difficulties. A four-feet still in experienced hands can be made to work any quantity of sulphate, or liquor representing it, from 50 to 500 gallons per hour. When working 300 gallons per hour, the amount of ammonia left in the spent water is so small that it has to be Nesslerised, whilst, when working 500 gallons per hour, the quantity need not exceed 0·006 per cent., if due care be exercised in the supply of milk of lime.

The amount of fuel employed naturally varies according to the strength of the liquor operated upon, the degree of perfection of the non-conducting covering with which the still is closed, and the amount worked through the still, the point of greatest economy is that which the still is guaranteed to do. Thus a four feet still is at its greatest perfection when working so as to produce 20 tons of sulphate per week from 9° Tw. liquor, and when covered with two inches in thickness of Haack's fossil meal covering, this, together with the heating of the boiler feed-water by means of the heat in the spent liquor, will require a consumption of coal equal to four cwts. per ton of sulphate: with 6½° Tw. liquor, but without the heating of the boiler feed-water, the coal consumption is 6¼ cwts. per ton of sulphate, while with 5½° Tw. liquor the consumption is 7 cwts. The cost of making sulphate by means of these stills as follows:—

	£	s.	d.
Vitriol (ordinary chamber acid)	1	15	0
Fuel 7 cwt. at 6s.	0	2	1
Lime...	0	1	6
Bags...	0	3	6
Labour......................................	0	4	0
	£2.	6	1

Of course in some places acid will be dearer than this, and in other places cheaper; but we presume the price quoted will be the general average. The amount of lime will vary according to the whim of the workman; it is of no use putting a large quantity of thick lime into the still. In ordinary liquors the fixed ammonia requires about 2½ cwts. to the ton of sulphate, and this may be obtained as a thin milk from about 3 cwts. of lump lime, when sieved through a screen (in the wet state) having 30 wires to the inch. This still enables the foul gases to be dealt with easily. As a rule these gases pass away from the heater much below the temperature of boiling water, so that they require but little cooling and can be burned with ease, and the products of combustion sent into the vitriol chambers where they make a substantial amount of sulphuric acid. Theoretically, the sulphuretted hydrogen evolved from the distillation of gaswater should give back about one quarter of the vitriol needed for the conversion of the ammonia into sulphate, and this enables the foul gases to be readily and economically dealt with.

The Condition of the Salford Sewage Works was discussed at a recent meeting of the Town Council. A resolution, moved by Mr. Alderman Walmsley, in favour of removing the sewage works and the treatment of sewage from the control of the Building Committee, was rejected. The Council appointed a committee to consider and report as to the establishment of a technical school for Salford; and a step was taken, by the adoption of a report presented by the Town Clerk, in the direction of improving the condition of artisans' dwellings in the borough.

EVAPORATION BY MULTIPLE EFFECT.

THE BERRYMAN MULTIPLE STILL.

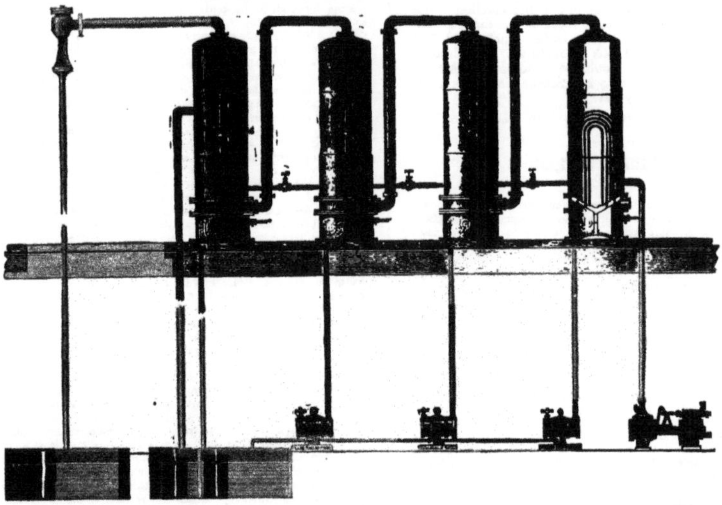

MANY years ago a patent was taken out by John Dale, for evaporating soda liquors by multiple effect; but, like most specifications of that date, when chemical industry was young, the instructions, if duly carried out, could have led to nothing but failure. It is not clear whether Dale recognised that he was utilising the latent heat of the steam in each successive effect, but anyhow, nothing came of the project, and soda liquors continued and continue to the present day to be boiled up in open cast-iron or wrought-iron vessels. In these evaporating pans about four pounds of water (not more) are evaporated per pound of fuel, so it will be seen there is a great room for improvement. Evaporation, by means of steam coils placed in the liquor to be evaporated, is much more economical, as, if a boiler is not overworked, 10 pounds of water can be converted into steam of 30 pounds pressure, by the combustion of a pound of coal, and every pound of steam condensed in an evaporating coil, will evaporate nearly a pound of water if the liquid is already at the boiling point. But if we consider that 10 per cent. is utilised in heating up the liquor to its boiling point, then 10 pounds of water, condensed in the coil, should evaporate nine pounds of water from the surrounding medium, and while the coil is kept full of steam and connected with a good steam trap, there cannot be any waste of the heating agent. Moreover, in many factories hot clean water is a desideratum, and after the steam has done its work there it remains as condensed water, useful for boiler feeding or for a multitude of other purposes.

Nine pounds of water evaporated per pound of fuel may be reckoned on with a single effect where ten pounds is the work of the boiler, but it is when several stills are combined in multiple effect that the economy comes in. The latent heat of the steam from the first effect instead of escaping unutilised passes into the second effect and yields us another nine pounds there, and so on for the addition of each still or effect as the French engineers have styled them.

Upon the supposition that ten pounds of water is the duty of one pound of fuel, in the steam boilers, two effects will evaporate 18 lbs. of water per pound of fuel; three effects, 27; four effects, 36; five effects, 45; six effects, 54; seven effects, 63; eight effects, 72; nine effects, 81; and ten effects, 90 lbs. of water per pound of fuel.

Most of the apparatus working on this principle is very intricate and expensive, and therefore, they have not come into general use, even in situations where they would have been exceedingly economical, but with a lower cost, and when they become better known it is very probable we shall find a more extended application. Messrs. Joseph Wright and Co., of Tipton, have to a great extent solved the difficulty as to cost. They have adapted the well-known "Berryman" feed-water heater to the purposes of a multiple still and have fitted up at their works a system of four effects, in order to show them working to their intending customers. The illustration will give the reader an idea of their construction, and of the small space they take up.

The woodcut shows a system of four effects, the steam is supplied from the steam boiler to the first effect, and the steam raised in the first still passes into the coil of the second, and so on throughout the series, all the water formed by condensation of the steam in the various coils is let out by means of the steam traps placed on the ground level. The liquor to be evaporated is pumped into the first effect in a constant stream, and the flow from still to still is regulated by means of an ordinary graduated tap. The last effect is worked under a vacuum of 26 inches of mercury, at which pressure water boils at 120°F. The vacuum is produced while absorbing the heat given off by the vapour in the last effect, and no special air pump is necessary with this system, the fall of the condensing water doing the work very efficiently. The concentrated liquid runs out of the still automatically, so that the degree of concentration depends entirely upon the rate at which the liquor is pumped through the apparatus. The stills are placed about 26 feet or more above the ground level, that is about equivalent to the second floor of a building, but the liquor supply pump and the tanks for weak and concentrated liquors are placed upon the ground floor. In this position the apparatus is as well nigh automatic as it can be, the steam is drawn away and condensed by means of a jet condenser, and a sixteen feet fall of water can be made to produce a 27 inch vacuum, when supplying sufficient water to condense the whole of the steam from the last still. The liquor being pumped into the first still at a given rate, evaporation goes on during its passage from still to still, until it reaches the overflow from the last still from whence it flows into the tank placed to receive it. If space will not permit of this arrangement, an air-pump of the ordinary pattern is used to extract the liquor from the last effect, and in such case the whole of the apparatus may be placed a few feet above the ground level.

Industrial Celebrities.

V.

PETER SPENCE.

LORD Macaulay in his essay on Milton describes the Puritans as:—"Men whose minds had derived a peculiar character from the daily contemplation of superior beings and eternal interests. Not content with acknowledging in general terms an over-ruling Providence, they habitually ascribed every event to the will of the Great Being, for whose power nothing was too vast, for whose inspection nothing was too minute. To know Him, to serve Him, to enjoy Him, was with them the great end of existence. If they were unacquainted with the works of philosophers and poets, they were deeply read in the works of God. If their names were not found on the register of heralds, they were recorded in the Book of Life." It was from families of this class that Peter Spence was descended. His mother's family had for generations farmed lands, situated on the Grampians; his father was a hand-loom weaver in the burgh of Brechin. Though of lowly station, his parents were distinguished for their genuine piety and lofty character; his father commanded the greatest veneration, he was a man of wisdom and sound judgment, of great purity and uprightness.

Peter Spence was born at Brechin, on the 19th February, 1806, and in the parish school of that town he acquired the rudiments of knowledge; he was trained to a life of hardihood and industry, and from his earliest years was made familiar with a religious life and doctrine that had in them much that was intensely stern and narrow, but which, nevertheless, produced characters of great strength and beauty.

At an early age he was sent from home to the city of Perth, where he was apprenticed to a grocer. During these years he manifested a great love of reading, works on science having a peculiar attraction for him. We shall see how in after years Peter Spence became a fertile inventor, but in his earlier years when he was engaged in the prosaic duties of a grocer's store he manifested the gifts of the poet.

There lies before us a volume of his poems "written in early life." His thoughts are revealed to us in the subjects he has selected:

"In the Primeval Forest,"
"The Destruction of Pompeii,"
"Ode to Palestine,"
"The Death of Bolivar,"
"Navarino,"
"Napoleon's Russian Campaign,"

and others similar. But it is not merely in efforts to sing the glories of nature, or the great tragedies of history, or the struggles of patriots for freedom that he invokes his muse, he can delight us with a pretty ballad, "The Tale of a Minstrel," or amuse us with a clever, witty parody on Gray's Elegy:—

"Beneath yon rugged elms, that beech tree's shade,
Where rests the river o'er its pebbly bed,
Each in his place the finny tribe is laid
Till tide shall help them o'er the ford to sped.

Ah, soon for them, the angler with his rod,
With wormy bait, or fly hook feathery clad,
Or fisher's net shall lay them on the sod
To writhe with pain and make their captors glad.

Ah, soon for them the blazing hearth shall burn,
And cook shall then the oily sauce prepare,
While gourmand Cockneys to the dish return
And gobble each his more than glutton share.

Let not ambition mock the humble theme
That now expands obedient to my wish,
Nor cast aside, as 'twere an idle dream
The short and simple annals of a fish."

We have selected these few verses just to illustrate his humour. There are sonnets, and there is a pretty little poem he terms "A Lyric," of which we also venture to quote a verse or two, to show the man.

"I love the land where my fathers have dwelt.
The rock, and the stream, and the plain,
I love to read how the southern felt
The might of these patriot men.

I love the stream where
And the slippery slid
And the big stepping st
To run our little mill

I love to look on a han
(And many a one I se
When a smile on her ru
And I love when she

I love to engage in a fri
With one who will ar
I love to hear others my
Who does not, may so

The industrious apprentice, the sc
youth is, we believe, a faithful pictu
he spent in the city of Perth and its
few helps in his chemical studies,
on books; his daily occupations we
the contrary. Several men, who h
have been won to the science by hav
amidst surroundings that invited
opportunities to experiment and for
advantages. It is a mystery to
works on chemistry after he had spe
the shop. When work was done, c
down in the morning, the grocer's ap
his studies
and month
elementary
crude appa
seed time o
when, in p
himself the
panions, fit
more influe
wise have
work that
Dundee Ga
apprentices
sight into s
connected v
cation of g
the princip
turing caree

In the ye
ceeded to
himself as a
earliest pate
address bei
road, in the
essay in in
Peter Spence
zation of W
in the prese
facture Prus
and Plaster
the refuse li

PETER SPENCE.

The London venture was a failure a
no outlet for some eight or nine ye
Spence able to do anything that he
although he had been incessantly e
of November in that year, he sought p
manufacture of copperas and alum. T
accident; he had been seeking in every
alumina, and was at this time expe
experiment had been a failure, and he
but after he had one day completed his
left the materials in the basin which w
next morning to his delight and surprise
crystals. This accident was the founda

The scene of his labours is now tran
in Cumberland. He here manufacture
and alum from burnt or calcined coal
designed to obtain liquors, sufficiently c
crystallise without any supplementary
the initiation of Spence's alum process,
pletely to revolutionise the alum trad
product, as very greatly to increase its

But Cumberland was not the best
Spence soon saw that Manchester and it
advantageously situated; coal shale c
obtained, close at hand, and the m
were equally convenient.

The next patented invention came from Pendleton, on the 12th November 1850, and this was also for alum, and with it cement. The previous patent was very crude and simple, but it contained the germ, which during the succeeding five years had developed until the one now under consideration appeared, which was in every way very much more complete.

Shale from the coal measures is the raw material which he principally relies on. After being calcined in large heaps over flues, the burnt shales are cast, warm, into leaden tanks in which there is sulphuric acid; the shales in time absorb up the acid and become quite dry. The vat is then emptied of these saturated shales, which are put into another vat and covered with boiling liquor; the sulphate of alumina is hereby dissolved out, and crystalised alum obtained. If it is found the sulphate of alumina is not sufficiently soluble, that the shales have not been sufficiently acted on by the sulphuric acid, the saturated shales are heated in the furnace and then put into the dissolving vat. But one point Spence kept in view was that no expense must be incurred by any process of evaporation prior to crystallisation—"the clear solutions of sulphate of alumina must be of sufficient strength for the crystallisation of alum without evaporation."

The same patent specifies that the mother liquors obtained after the alum is crystallised out, and which contain a considerable quantity of free sulphuric acid, shall be stored in a covered tank, this tank shall have a pipe at the top leading to the flue or chimney, and another, a lead pipe, leading into it at the bottom to convey ammoniacal compounds and vapour of water, obtained when ammoniacal gas liquor is being boiled or distilled into it. Sulphate of ammonia is formed, quantities of sulphuretted are given off, this gas is conveyed away to the chimney and the sulphate liquor is used in mixing with the sulphate of alumina, and so forming ammonia alum. The same specification claims a process for distilling ammoniacal gas liquor and other volatile, or partially volatile, liquors by an arrangement of two or more boilers or vessels, by which the liquid to be distilled can be run from one to the other through the whole range, and the passing of steam charged with volatile matter from one to the other through the whole range in an opposite direction.

The formation of cement, which he calls Patent Zinc Cement, from the spent shale after the sulphate of alumina was washed out, was another item of this most inclusive patent.

He took two parts of refuse lime from the gas works, and one part of his spent shale, and to that he added a little sulphate of zinc solution; these are mixed to about the consistence of mortar made into bricks, and dried, afterwards calcined or burnt at a moderate red heat, after cooling it is finely ground and constitutes cement.

The zinc was used for two purposes: to prevent the oxydisation of any iron, and so causing the colour of the stone to be degraded, and also to prevent the stone being attacked by mosses and lichens.

The last point he claimed is one of much interest; it related to the manufacture of the carbonates of the alkalies from the sulphates by means of sulphate of baryta.

Powdered sulphate of baryta was mixed with carbonaceous matter and heated. After this mixture is burnt it is lixiviated, and a solution of sulphide of barium is obtained; to this the sulphate of the alkalies (in solution) is added, sulphate of baryta is thrown down, and the sulphide of the alkali remains in solution. This solution is pumped into a tank, in which it is heated to boiling, and then run through a tower packed with coke, through an atmosphere of carbonic acid gas; carbonate of the alkali is formed and sulphuretted hydrogen is given off, which is collected and burnt to produce sulphuric acid; the liquor which runs through the tower is collected in a tank and evaporated to soda ash, or is made into soda crystals. Spence says "I am aware that a process similar to this has been previously described, but the temperature was not raised, my point being that the liquor must be acted on at an elevated temperature."

Spence's alum process was a very great success, both chemically and commercially, and at Pendleton his business rapidly grew, until he was the chief alum manufacturer in the world.

The prosperity that attended him enabled him to prosecute various researches, and from this time forth the records of the Patent Office reveal how varied were the subjects he investigated, and how fertile he was in resource.

It is impossible for us in this short article to follow his labours and ideas as they are unfolded in his various patents; we must content ourselves with classifying the patents and summarising their results.

The patent last described, and worked out at the Pendleton Alum Works, was his main achievement. On it he built up a vast and successful business; and the other patents to which we shall advert are quite subsidiary as compared to it.

[illegible]wing are the subjects to which his patents refer:—

[illegible]an Blue and Prussiate of Potash.
[illegible]facture of Sulphate of Alumina and Alum.
[illegible]facture of Copperas.
[illegible]facture of Sulphuric Acid, and obtaining Sulphur from [illegible]ites.
5. Calcination of Copper and other ores.
6. Separation of Copper from its ores.
7. Separation of Zinc from its ores.
8. Separation of various metals from their ores.
9. Purification of Gas.
10. Production of Salammoniac.
11. Manufacture of Sulphocyanide of Ammonium.
12. Manufacture of White Lead.
13. Treatment of Phosphate of Alumina and Iron.
14. Treatment of Sewage.
15. Manufacture of Manure.
16. Treatment of Bauxite, and the Manufacture of Aluminoferric Cake.
17. Purification of Water.
18. Treatment of Gases arising from the manufacture of Salts of Ammonia from Gas Liquors.
19. The Use of Oxide of Manganese in Freeing Sulphate of Alumina Solutions from Iron.
20. Obtaining Power by Steam.
21. Repairing Fractured Bells.

Most of the above are the subjects of more than one patent, some of several, and the same subject is treated after the interval of several years. Some of the patents are modifications and improvements of preceding patents, others put forth entirely new ideas. Some had little or no practical value, others revealed results and methods of great and permanent interest; and nearly all assist in marking the progress of chemical discovery in the sphere to which they apply.

One of his patents for producing Prussian blue and prussiate of potash, illustrates his tendency to try and utilise refuse. He employed the alkaline sulphates, fused them in a pot, and, when in a fused state, added refuse leather, and calcined them together; then lixiviated and crystallised. To the mother liquors he added sulphuric acid, and converted the alkalies into the sulphates. This being done, he added sulphate of iron, and so precipitated Prussian blue.

In the Island of Anglesey, at the Parys Mine, there are enormous heaps of refuse ore, called bluestone. It contains of zinc, lead, copper, and a little silver, but the percentages of the various metals are so low that metallurigists have failed to devise a method to extract any or all the metals, so as to make the process a commercial success.

To use up this abundant refuse, which is found in numerous localities in many parts of the world, has been one of the problems which metallurgists have failed to solve. Peter Spence tried his hand at it. He ground the ore into a fine powder, and treated it with hydrochloric acid, so dissolving out the sulphide of lead, the sulphide of zinc remaining unacted on; the undissolved blend he calcined, and extracted the zinc by volatilising it in the usual way, and the residue containing the copper and silver he melted, and separated the metals by one of the well-known processes. He succeeded no better than others who have undertaken the same work. Then his treatment of copper ores was, by two or three methods, to bring the copper into solution, and out of the solution to precipitate the copper as sulphide with sulphuretted hydrogen obtained from chemical waste.

In his smelting processes, his patents refer to the construction of calciners. His object was to utilise, as much as possible, waste heat, and to bring the calcined ore, while hot, direct from the calciners into the melters. Some of his furnaces were built directly over the smelting furnaces, and others had a series of floors and the ore was drawn from one to the other; then, again, he invented a mechanical rabble which worked backwards and forwards at intervals, and he designed a method for cooling this rabble so as to prevent its destruction.

But the "Henderson" process excelled all his attempts to treat copper ores by the wet way. His calciners are still used in certain districts, where "smalls" are plentiful.

The discovery of immense quantities of phosphate of alumina and iron in the West Indies, on the island of Redonda, aroused him to seek to discover a method of utilising this mineral, so as to produce alum from the alumina, and use the phosphoric acid in the manufacture of manures. He patented two processes, the first in 1870, which was incomplete, and was perfected by a second patent in 1873. The first idea was to dissolve the phosphate of alumina in sulphuric acid, and to the solution add ammonia or potash to form an alum; then crystallise. To the mother liquor add ammoniacal gas water, iron and any residue of alumina were precipitated, and the clear liquor was boiled down and dried to a state fit for manure; the amendment of this patent was that after the iron, etc., had been thrown down by the ammoniacal gas water and phosphate of ammonia obtained, pure ammonia was added to the monophosphate, and a tribasic phosphate of ammonia was formed.

The second process consisted in mixing the mineral phosphate with sulphate of soda (salt cake) coal and a little oxide of iron, and heating these together until decomposition took place; then lixiviating out the phosphate of soda that is formed; to this solution caustic lime is added, phosphate of lime is precipitated, and caustic soda remains in the liquor, this is boiled down in the usual way; the precipitated [illegible] is either used as it is, or it is converted into superphosphate.

Spence was so confident of the success of this process that he decided to erect plant on a large scale, and to alter the arrangement of his works. He was too sanguine, and his conclusions were at least premature, for after incurring heavy expense, the process had to be abandoned, and large quantities of the mineral were thrown on his hands.

His manufacture of sulpho-cyanide of ammonium was also an invention worthy of note. He patented this first in 1863, and completed it in [illegible]. In this process he utilised the ammoniacal liquor from gas works, drove off the volatile salts of ammonia, and then crystallised the chloride of ammonium.

The mother liquor he then diluted so as to precipitate extraneous matters, and the clear solution was taken for use. To this was added a mixed solution of sulphate of iron and copper, by which sulpho-cyanide of copper is produced, which, being treated with a solution of sulphide of ammonium the copper was separated, and sulphocyanide of ammonium formed and obtained in crystals.

In 1875 he took out in connection with his son Frank his first patent for the treatment of Bauxite and the production of Alumino-ferric cake. These salts he found to be of value in the purification of sewage, and also in the purification of water; valuable also in paper-making and dyeing.

His last patent, taken out just a year before he died, is the one [illegible] associated with the memorable law suit between A. G. Kurtz and Co., of St. Helens, and his firm. The invention is that in solutions of sulphate of alumina, which contain iron, the iron is precipitated by black oxide of manganese being added to the liquor when hot, before the sediment has settled out. The iron goes down with the manganese.

The same tastes that induced Peter Spence, as a youth, to be an active member of the Debating Society, at Brechin, caused him to take a lively interest in the proceedings of the Manchester Literary and Philosophical Society.

Several papers which he read before that society are published; they have principally a scientific, rather than a literary, interest.

A favourite theme of his was embodied in a paper read before the society and published in pamphlet form in the year 1857, it is entitled "Coal, Smoke and Sewage scientifically and practically considered, with suggestions for the sanitary improvement of the drainage of towns, and the beneficial application of the sewage." He divided the contamination of the atmosphere proceeding from drains and chimneys into evils visible and invisible, and the latter he regarded as much more pernicious than the former. The removal of solid filth from drains and cesspools, and of black smoke and soot from chimneys was, he considered, a most imperfect, crude, and unscientific method of dealing with the evils.

He writes: "The invisible evil is completely ignored, the smoke nuisance is held as *the* evil of our atmospheric condition. 'Burn your smoke,' is the united cry of the Sanitary Association, the public, and the law." He calls the attempt to do away entirely with black smoke "the sanitary smoke-consuming mania."

"Perfect freedom from smoke would, if accomplished, only increase the evil arising from the purely gaseous results of combustion;" and adds: "While demonstrating that to the mere economist, there is great inducement to get rid of all visible smoke, as a consequence of his obtaining perfect combustion of fuel, I would at the same time say to the sanitary smoke consumer, that he had better not interfere further, as every step he takes in enforcing the consumption of smoke, will only tend to deteriorate the atmosphere, and that every cloud of [illegible] smoke he is successful in dispelling, is only making way for a [illegible] baneful, though invisible agent." He regarded carbonic acid, [illegible] oxide, sulphurous and sulphuric acids, as far more prejudicial to health than visible smoke; indeed, he believed soot to have an [illegible] influence, and to be beneficial rather than prejudicial.

From the economical standpoint he did not advocate permitting imperfect combustion, but he would construct apparatus that all ammonia would be condensed and converted into sulphate of ammonia, [illegible] gases as could not be condensed and utilised should be carried up lofty stacks, and diffused before they could alight among human [illegible], and so become innocuous.

The nitrogen, phosphates, and alkalies in sewage he would also [illegible]; he regarded the system of town drainage into watercourses [illegible] and injurious in the extreme. In fact, he would have all [illegible] connected with drains; the drains should be smoke culverts as well as water conduits; the gases resulting from combustion [illegible] destructively on the sewage gases, and both should be carried [illegible] by means of a very strong draft up very lofty chimneys; his [illegible] feet high, at least.

[illegible] his projects presented to his mind, he thus described:— "[illegible] from the interior of all our dwellings, from the mansion [illegible], and even to the cellar, of the slightest trace of those [illegible] which feed our fevers, consumption, and cholera, [illegible], more than any other physical cause, serve to lower [illegible] of our town populations, making them an easy prey [illegible]

Instead of the atmosphere of our towns being, as now, a dim, and dull and murky compound of black and yellow and grey smoke, blended into a haze, and containing besides its proper elements a mixture of carbon, carbonic oxide, carbonic acid, and sulphurous acid gases, with the condensible hydro-carbon compounds of our domestic fires, we should then continually revel in an atmosphere transparent, pure, and salubrious as that which encircles the mountain's side, or reclines on the ocean's bosom; and instead of vegetation, with its charming green being banished from the interior of our towns, and even in their suburbs holding only a stunted, miserable, and withering struggle for life, every nook and corner where light could descend might then have its shrubs or its tree; our most bustling streets might be enlivened by the evergreen leaves of the climbing ivy, or with other parasitical plants, and might be made fragrant to the smell, while the eye was delighted with their beautiful and variegated flowers."

He felt assured these anticipations were not extravagant, that this vision was no chimera; but "a most desirable and practicable reality," which he was convinced would in time be realised.

It is mournful to contemplate, that in this very year, 1857, when he published this pamphlet, he was himself being assailed by a number of the residents of Pendleton, for permitting gases to issue from his works, which they asserted were noxious and injurious to health and vegetation.

The trial of Regina *v.* Spence, which occupied three days of August, 1857, when the firm was proceeded against for having so conducted their works as to be a public nuisance, is one of those notorious cases long to be remembered. Baron Channell was the Judge; Mr. Wilde, Q.C., was leading counsel for the plaintiff, and Sir Frederick Thesiger, Q.C., for the defendant. A large number of witnesses were called on both sides, and eminent experts in chemistry, horticulture, and botany gave evidence for and against.

Dr. E. Frankland, who, at that time was a professor at Owen's College, and who had been employed by Mr. Spence to superintend his processes, so as to be able to suggest, if possible, any apparatus or arrangement that might prevent any nuisance arising, was induced to give evidence for the plaintiff.

Mr. Spence felt this very much. Having retained Dr. Frankland's services, and being prepared, as far as possible, to carry out any improvements he could suggest, he considered Dr. Frankland had no right to side with the prosecutors against him. To say the least, it was an unfortunate circumstance, and we are not suprised that it gave rise to very bitter reproaches from Sir Frederick Thesiger. Dr. Angus Smith, and Professor Crace Calvert were called by Mr. Spence.

Scientific witnesses have too often justified a severe judgment that has been frequently pronounced against them; it is with satisfaction we find Baron Channell, referring to the scientific witnesses that gave evidence in this case, and saying: "I think, when you come to sift and examine for yourselves the evidence of those whom I have called scientific witnesses, you will not find a great discrepancy between them. There are a great many points upon which they agree, and the difference will really be with reference to those whom I have ventured to call the non-scientific witnesses."

The verdict went against Spence, but only on one count out of three; the jury found that a disagreeable smell emanated at times from the works, but that it was not proved that they were injurious either to health or vegetation. He was compelled to leave Pendleton and remove to the site of his present works at Miles Platting. He had anticipated that this would be the only way out of his troubles, and with that energy and foresight, which were so characteristic of him, had made his arrangements beforehand.

In reviewing this celebrated trial, we believe Peter Spence had to contend against a great amount of prejudice, and that the social position of those who instigated the proceedings, told heavily against him; we think also it was probably a mistake that the special jury were not brought over to view the works and the neighbourhood.

The Literary and Philosophical Society afforded him a constant and agreeable retreat from the worries of business; he frequently had interesting communications to make, and ably assisted in the discussion of scientific subjects, introduced by others.

He was much inclined to humour, and thoroughly enjoyed a joke; indeed, it was noticed that on the very day he died his bright humour did not forsake him.

One trait of his character used to exhibit itself at these meetings, his aversion to take anything for granted; he would try to test theories and to prove facts.

The question of the effect of cold on iron was raised at one of the meetings of the Society. A railway accident had just occurred, occasioned by the breaking of the tyres of the carriage wheels. It was generally advanced that cold and frost made iron and steel more brittle, but Dr. Joule stated that, however general the impression might be, he knew of no experiments that tended to prove that impression to be a correct one.

Peter Spence at once determined to make a series of thorough and conclusive experiments, the results of which were that he proved "that a specimen of cast iron having at 70° Fahr. a given power of resistance to transverse strain, will, on its temperature being reduced to zero, have that power increased by 3 per cent."

One of the discoveries of great scientific interest and practical utility which he made, was communicated by him to the chemical section of the British Association, at Exeter, in August, 1862. He desired to obtain a temperature of 228° Fahr., to extract alumina in the form of sulphate from minerals containing that earth. His apparatus was heated by steam and fire, and he was able to digest for a long period the solutions at the necessary temperature. It happened that, by accident, the fire was neglected and the steam alone operated, nevertheless, to his surprise, he observed the high temperature was preserved; he was led to investigate, and experiment on this phenomenon. Being convinced that the high boiling point of his liquors had something to do with the phenomenon, he selected a solution of a salt (nitrate of soda) having a high boiling point—about 250° Fahr. The nitrate of soda was placed in a vessel surrounded by a jacket; steam was let into the intervening space, until a temperature of nearly 212° Fahr. was obtained; the steam was then shut off, and an open pipe immersed in the solution, and steam from the same source was thrown directly into the liquor; in a few seconds the thermometer slowly, but steadily, moved, and minute after minute progressed until it touched 250° Fahr.

This thoroughly confirmed the results obtained in the digesting vessels and became to the author of immense practical value.

As a corroboration of the theory, which seems to explain the apparent paradox, the author found that the temperatures of his solutions were in the exact ratios of their specific gravities, and had no connexion with the temperature of the steam which never exceeded 212° Fahr. The greater specific gravity of the acid solutions, the higher the boiling-point; and, therefore, whatever the boiling-point of the solution in water of any salt, to that point, or nearly, will steam of 212° Fahr. raise it.

When Peter Spence made this statement it was received with general surprise, amounting almost to incredulity; in fact, Professor Williamson declined to accept the fact, until he had himself seen the experiment performed by Spence, and had personally examined the thermometer. The explanation was then appparent, that the latent heat evolved by the condensing steam had become sensible heat, measurable by the thermometer.

In 1881-2 Peter Spence gave evidence before a House of Commons Committee, on Railway Rates; and his evidence is reprinted in pamphlet form, with the title:—"How the Railway Companies are crippling British Industry and destroying the Canals; with suggestions for reforming the whole system of railway charges and for rescuing the water-ways permanently for the nation."

This pamphlet is still worthy of careful perusal, as it abounds in well-authenticated facts, and contains many valuable and ingenious suggestions. When the scheme of the Manchester Ship Canal was started it had his enthusiastic support, and he was one of those who first subscribed £1,000 towards the needful funds.

Peter Spence never felt that his duties were limited by his business. He never regarded his workmen as mere "hands," destined to toil and poverty, that by their labours the few might rule in affluence and care. He threw his activities into the church in Oxford-road, under the pastoral care of Dr. McLaren, and was elected one of its deacons, in which capacity his intelligence, enterprise, and generosity were most valuable.

It is a pleasure to us to record here an incident that occurred at a very critical period of Peter Spence's business. After the great losses which he incurred by the too hasty alteration of his plant, to work the Redonda Phosphate process, which proved to be a great mistake, he was so crippled financially that there appeared to be no alternative but to liquidate the business, and get it converted into a limited company.

This was a crushing sorrow and disappointment, after the years of persevering industry and successful invention. The wife of his friend and pastor becoming acquainted with the circumstances, privately and totally without Spence's cognisance, mentioned the matter to one of the worthy deacons of the chapel, who at once asked Spence to call on him, and at the interview, inquired what amount would enable him to extricate himself from his difficulties. Spence said it would take five thousand pounds; his friend at once replied, he should instruct his banker to place the amount needed, to his credit. Spence was overwhelmed at the splendid generosity of his brother deacon, and was unspeakably thankful for the unexpected succour thus accorded to him, and which he ever regarded as a most merciful Providence. The prompt and generous aid thus afforded, enabled him to re-arrange works, and revert to his former processes. The months that red brought with them a period of very great prosperity, and the year he was enabled to repay the total sum lent to him. se days such an incident would, perhaps, be regarded as an and unscientific interference with that gospel, so dear to strong lists, that the "fittest must survive," the "weak must go to the

by the terrible and wide-spread evils of drunkenness, he became at an early age an adherent of the great temperance movement, and throughout the remainder of his life was one of its most earnest advocates. He was a teetotaler, a Good Templar, a Blue Ribbon-man, a member of the United Kingdom Alliance, patron of Bands of Hope, and President of the Manchester Temperance Union. Although all these, and similar movements had his hearty support, he was unable to approve of the policy of the Church of England Temperance Society. He believed total abstinence was absolutely necessary in combating the evils of drunkenness, and that temperance reformers made a great mistake in seeking to sanction and secure mere moderate drinking.

He stamped out drunkenness amongst those in his employ, and so secured a staff of sober and industrious men, to whom he proved himself a kind and generous employer.

He was an active member of the Manchester Chamber of Commerce, and a very attentive Justice of the Peace, also a director of the Mechanics' Institution (now the Technical School.) In fact, Peter Spence took a keen interest in all public movements: in politics he was a Radical, in Church polity a Nonconformist; a Parliamentary position was offered him, but this he declined.

Although he inherited a delicate constitution, with a tendency to consumption, yet his habits through life were so temperate, and the measures he took, to preserve his health, so wise, that he enjoyed excellent health, and was, after he had passed threescore years and ten, a picture of happy, bright, vigorous old age. Although short in stature, his appearance was most attractive, there was always a merry twinkle in his eye, and a cheerful, hopeful look on his countenance.

The death of his wife, in February, 1883, was a blow from which he could not rally; without any attack of disease, his strength gradually failed, until on the 5th July, the same year, he passed peacefully to his rest, at the age of seventy-eight.

He was another illustrious example of men, who have climbed the ladder of success from very insignificant beginnings, who have shown that the noblest heritage men can receive, is the inspiring power of sound training and of noble lives, that the diligent and persistent pursuit of knowledge is the way to wealth and honour, and that the man who cares for others has discovered the secret of a happy life.

We conclude this notice with the beautiful words which Dr. McLaren applied to him—"one has gone from us who was as well-known in other circles as in the Church, and was everywhere the same man. Clear and active in intellect, prompt in deed, generous and conscientiously liberal, fervently and always passionately attached to such principles and causes that had won his allegiance, and possessing in a very remarkable manner a force of will, tenacious, adherence to his convictions, with large sympathies and warm affections. His devout Christianity was of an unfortunately rare type; his conscientious liberality of rarer type still. To a green old age he kept much of the interest, the vigour, the buoyancy of youth, and he died in peace, calm of mind, and clear of heart, not eager to go, but satisfied with what God had given him, leaving behind him in many a good cause a great blank, and in many of our hearts a green and perpetual memory."

The Water in Paris.—It is clear that the Parisians will have to bestow serious attention on the question of their supply of water. It has never been of the best, and tourists are well aware that they have always been cautioned to drink as sparingly as possible of it. Now, it seems that the difficulties connected with the supply of fairly potable water are greater than ever. The Vanne Canal, whence the bulk of the water was drawn, has broken down; the scheme to turn on the limpid stream of the Arve in the Eure department has not been carried out; and for the present there is nothing for it but to blend nasty Seine water with that of the springs so as to provide for the needs of the great city.

A New Electric Tramcar.—An electric car on the new principle was tried on the Bristol-road tramway route at Birmingham recently with entire success. The car is of the same size as the cars on the Birmingham cable tramway. It is lighted with electricity, and it is to have an electric bell. The motor is in the front bogie, and the accumulators in which the electrical energy is carried are under the seats on both sides. It is expected that one of these cars will be able to run 70 miles with a single charge. This is a full days work of the present cars. They are, however, only reckoned as half-day service cars, but the whole of the receptacle containing the power driving a car can be taken out at the depôt and recharged in the space of five minutes. The present car is the outcome of a recent union of interests between the Julien, the Sprague, and the Electrical Power Storage systems. It has enabled the Central Company to meet the hopes recently expressed that the ultimate form in which electricity might be applied as the motive power of tramways in Birmingham should be the self-contained car and not a separate engine. Until recently the great difficulty in the employment of large cars with self-contained power has been the use of the driving chain, which has been found in practice to be noisy and liable to break. But this difficulty has been quite overcome by an ingenious arrangement which absolutely dispenses with the chain, and which is practically noiseless.

simply by having at my command an excess of boiler power over and above our daily or hourly requirements. It is those people who are always pressing their boilers to get as much work out of them as possible that make the black smoke, and for my part I attribute much of the nuisance to the intense desire of many manufacturers to cheapen the cost of production by turning out as much steam as possible in a given space. There is a limit that cannot be exceeded, and herein lies all the difficulty. A 30 ft. × 7 ft. 6 in. Lancashire Boiler, with two separate flues and working at 70 lbs. pressure, can consume, even with careless hand-firing, 25 or 30 tons of fuel per week of 132 hours, without the chimney showing more than a dim haze above it. Increase the quantity of slack to 35 or 40 tons per week, and there will be a light brown smoke, while more than 50 tons are consumed the smoke becomes blacker in hue until at 60 tons per boiler per week there is nothing but a continual volume of black smoke.

I am much surprised to find the Chief Inspector of Alkali Works leading the van in what is called the smoke nuisance agitation. The Alkali Acts have all along specially exempted products of the combustion of coal from the operation of these Acts, and there are many people like myself who think the Inspector should devote a little more attention to those things he was appointed to do, and which anyone with a nose, living within the vicinity of a chemical works, will say are not yet complete, and let those things alone which he cannot do better than anyone else.—I am, &c.,

A MANUFACTURER OF TWENTY YEARS' STANDING.

HOUSEHOLD SMOKE.

To the Editor of the Chemical Trade Journal.

SIR,—In your leader of the 28th December you describe the evils of town fogs to be due to household smoke; surely a demonstration of this fact is not necessary. I am a traveller into town by the S. W. Ry. every morning to business, and when coming into Waterloo almost the whole way from Clapham Junction we are over the housetops and can easily observe the vast area of chimneys reeking with domestic coal smoke. Cannot anything be done by modern science to prevent the greater portion of this evil. I know it is a difficult matter, but surely there must be some remedy. I went many times to the Smoke Abatement Exhition in order to discover how smokeless chimneys may be brought about, but I must say I returned no wiser than I went, and I do not find that the result of that Exhibition has been any the less smoke, or that the world of coal consumers has been enlightened in any way. In my humble opinion these exhibitions and trials are all rubbish, and this is borne out by the barren results of them.

A great deal, if not the whole, of the blame really lies with the property owner or the house builder, fire-grates are not selected as a rule on any account save that of cheapness, and especially is this so with cottage property. Then again, in our mansions it is of no use putting in any arrangement with the slightest trouble attached to it, as the servants complain and out it has to come, the architect being blamed for inserting any such arrangement.—I am, &c.,

AN ARCHITECT.

Our Book Shelf.

THE GAS MANAGER'S HANDBOOK. By Thomas Newbigging, M.I.C.E., Fifth Edition, 524 pp., with 193 figures in the text. London: Walter King, 11, Bolt Court, Fleet Street.

It is very seldom that a Handbook written for any one special branch of industry is of use to many others; this work is an exception to the rule. Written ostensibly for the use of gas engineers and managers, it is a handbook of great importance to every worker in the chemical and allied industries. And the cause of this is not far to seek, gas-making is essentially a chemical operation; in the earlier days of chemistry the worker in the laboratory was seldom illustrated without his retort, and one of the earliest experiments recommended to the tyro in chemistry is the preparation of coal-gas by heating coal in an old tobacco pipe, and storing the gas so produced in a bell jar over water. The chemistry of gas-making has not advanced so rapidly as other industries until within the last few years, but considerable attention is now being paid to it, and when we know that every operation is essentially a chemical one, from the carbonizing of the coal to the measuring of the gas at the station meter, it is not suprising to find the most important gasworks employing chemists of experience. The handbook before us has been very considerably enlarged from former editions and copiously illustrated, and takes the reader not only through the whole series of operations carried on in a gasworks, but treats of the construction of the plant besides; in fact it is now arranged in such a readable form that it will soon find favour and a ready sale amongst many other readers than gas engineers.

The book is up to date with regard to the information contained therein, for we find described good accounts of Generator furnaces, t[illegible]

use of pure oxygen in purification, Claus' process, the Dinsmore process, and others comparatively recent.

Amidst the mass of information of such a useful character, there is a section that should be revised—p. 400—403. Letheby's average composition of London gas, and the comparative illuminating power of other light-giving materials, may have been very interesting before it was discovered that the principal illuminating power was due to benzol vapour; but it is out of date now, and there exists plenty of material for a due correction.

Similarly the relative cost of the magnesium light and coal-gas, which was as 52s. 6d. to 1s. 9½d. in 1865, might also be revised with advantage, seeing that magnesium has fallen from 21s. to 2s. 6d per ounce.

Respecting the relative illuminating power of various materials, as set forth on paper, it is interesting to note the difference between theory and practice. In ordinary use a room is lighted with a 3-light gasalier, each light burning four cubic feet per hour, or for six hours 72 cubic feet; which, at 2s. 6d. per 1,000 cubic feet=2·16 pence, and the occupants say there is only just enough light to see by. Suddenly, through the action of the gas stokers, a duplex lamp is brought into operation; the occupants say the light is as good as with gas, and as only a pint of oil is burned in six hours, costing 8d. per gallon, the light from paraffin oil costs 1d. as against 2·16 pence for gas. The light may not be theoretically as powerful, but practically it is the same, and is sufficient, whereas less than the three burners of the gasalier would not be sufficient.

We do not find any reference in this handbook to the Wenham, Fourness, or other regenerative lamps, at which we are much surprised, seeing that they are in such general use.

The few foregoing remarks must not be taken in terms of disparagement, they relate more to the general aspect of the question than to the work itself. It is a handbook we can confidently recommend to every chemical engineer and manager throughout the world, they will earn a very great deal from it, and though written for the gas industry, is equally applicable to many other branches of the chemical trade.

HANDBOOK OF QUANTITATIVE ANALYSIS, by John Mills and Barker North, 208 pp., and 34 illustrations in the text, 1889. London: Chapman and Hall, Limited.

There are many works on Quantitative Analysis, but the majority of them are not sufficiently clear to be placed in the hands of beginners, especially those who happen to be situated at a distance from teaching centres. The explanation of the preliminary operations of analysis are well though shortly explained in the work before us, and when we look further and find "simple examples in gravimetric analsyis," we are quite pleased with the lucidity of description. Chapter III. contains exercises in volumetric analysis, which at one time was sadly neglected in most elementary treatises on practical analysis. The book is one that can be strongly recommended to the student upon first commencing quantitative work, or as explained in the preface, to those entering for Honours at the Chemistry Examinations of the Science and Art Department, and for the Associateship of the Institute of Chemstry.

CATALOGUE of new and second-hand books, English and Foreign, on Chemistry and the Allied Sciences. W. F. Clay, 2, Teviot Place, Edinburgh, 1890.

Mr. Clay, who has established his reputation as a collector and vendor of chemical literature, sends us his catalogue of new and second-hand chemical books, and as it is the only complete list, to our knowledge, published in England, we hasten to advise our readers of its appearance. To those who cull old chemical literature a perusal of the catalogue will be interesting, for there we find such works as Boerhaave's new method of Chemistry, 1753; Dalton's new system of Chemical Philosophy, 1808; Lavoisier's Elements of Chemistry, 1790; Rumford's Essays, 1802; Thomson's system of Chemistry, 1804, and many others equally rare. To those requiring the latest editions the present catalogue will be welcome, as we find it contains C. E. Grove's Chemical Technology, 1889; Watt's New Dictionary of Chemistry, and all others that have been published of recent date. Mr. Clay has a case in our Permanent Chemical Exhibition, so that intending purchasers should call and see specimens of his stock.

INDIA-RUBBER STREETS.—So pleased are they with the caoutchouc pavement laid down, by way of experiment, on a bridge at Hanover, that the corporation of the town has ordered an extension of this kind of pavement over a distance of 1,500 metres. A street in Berlin has been laid in a similar way, and now Hamburg is going to try it. It is as durable as stone, they say, is not slippery like asphalte, and suffers neither from heat nor cold. It is silent we may be sure. Safe for the horse and pleasant for the driver, how will it be with the man on the street? To avoid knocking down pedestrians like nine-pins the horses and vehicles that fly over India-rubber streets must be well provided with warning bells.

THE "WELLS LIGHT."

THERE is no place where the want of a light of considerable intensity is for space illumination felt so much, as in a chemical works. Those who have had charge of processes in the night time will agree with us that night work is never so good as day work and that the reason for this is mainly want of light.

Up to within the last few years the arc lamp in an electric current was the only source of concentrated illumination, if we may except, however, the Siemens high power gas lamp that we have seen in use in one or two establishments. But gas is not everywhere available, and it cannot easily be shifted from place to place as progress of work often demands. The "Wells" light which has now established itself as a practical source of illumination for large open spaces, overcomes all the objections that can be urged against electric light and coal-gas, and the fact that over 2,000 lamps have been sold since their introduction speaks volumes as to their usefulness. Certainly a better source of light for chemical and allied works we have never seen. We were at a chemical works the other night near Manchester where night was turned into day by the aid of a "Wells' Light," which the manager assured us gave no trouble whatever.

To illustrate these remarks we show a small hand lamp of 500 candle power, possessing a cylinder 10 inches in diameter and 16 inches deep, the weight empty is but 40 lbs., and when filled with oil, 70 lbs., and when burning at the above named candle power throws a flame 12 inches long, consuming half-a-gallon of oil per hour.

The principle upon which the lamp acts is by generating vapour from the oil employed, in contradistinction to those lamps driven by continuous supplies of air or steam. This makes the arrangement portable and self-contained. The oil is forced into the cylinder by means of a pump and hosepipe, and compresses the air already in the cylinder to about 20 lbs. to the square inch pressure. In ordinary cases the burner is then heated by the cup and chimney provided with the lamp, and, the valve being opened, the oil is forced up by the air pressure into the heated burner where it is converted into gas, and is there ignited, burning with a large and brilliant flame. The heat of this flame passing through the generating tubes continually converts the ascending oil into gas. In order to increase the facility of lighting a self-lighter has been added, doing away with all smoke, and the need of waste and oil. This apparatus consists of a small oil vessel filled with oil, fitted with a nozzle, and a hook for hanging to the burner, and is connected with a flexible tube to the top of the tank, with a simple "push on" coupling. When air is admitted the blast sprays the oil, and forms a flame that heats the burner sufficiently in about four minutes. When the burner is hot and running well, the lighter is detached, and may be used on another lamp. The old method of heating was undoubtedly a drawback, especially where the lamp was used in the interior of buildings, but all smoke and smell is now obviated.

The "Ship Canal" pattern is, perhaps, the most useful size made, and ranges from 2,000 to 2,500 candle power, using about a gallon

the stock of liquors, when they have attained equilibrium, will be equal to a solution made from 72 per cent. caustic soda.

Many paper-makers place implicit confidence in the tests of the percentage of alkali in the recovered ash, but we beg to point out here that this is not a very good criterion to go by. A 35 per cent. recovered ash may, and often does, give a much better liquor than one testing over 40% as experience has shown over and over again.

For many years now in our laboratory we have discarded the plan of testing recovered soda in the same manner as is employed for soda-ash, and as the process we now use may be of service in the paper-makers laboratory, we give the outline of it here.

First, the sample is roughly ground in an iron mortar, and 100 grms. of the ash placed to digest with 400 c.c. of water at a gentle heat. The settled solution is then poured through a coarse filter, and the residue treated with successive portions of hot water until the whole of the filtrate is brought to 20° Tw. The filtrate is then allowed to cool to 60° F. (15° 5C.) made exactly to 20° Tw., and carefully measured. This gives us a relative test of the quantity of liquor yielded by any kind of ash. Now in order to test the quality of the ash it is far better to judge by inference from an examination of the 20° Tw. liquor produced as already described. Certain quantities can be pipetted off, and the quantities of the various constituents determined. The following analyses have been done by this method, and show extreme conditions which cannot be good for the manufacture :—

No.	C.C. of 20° Tw. liquor from 100 grms. of ash.	% of Insoluble residue.	Grms. per litre in 20° Tw. liquor.				
			Alkali.	Sulphate.	Sulphate after Oxid.	Sodium Chloride.	Silica.
1	698	30·0	40·0	10.5	26·5	14·8	7·1
2	862	12·0	37·7	24·6	25·3	10·6	12·2
3	910	1·2	48·0	7·2	7·6	6·7	10·4
4	700	15·3	50·0	3·1	3·9	9·0	1·9
5	940	11·5	44·0	5·6	6·1	4·9	26·9

Another point worth mentioning in this connection is the mode of treating recovered ash. Some manufacturers draw the ash from the calciner in a green or flaming condition and wheel it straightway into "dens" to "burn off." This is by no means necessary, nay, it is even injurious to the quality of the liquors, as we have always found that there is far more silica in liquors made from "burned off" ash, than from that vatted in the green state.

On the other hand, it is injurious both to the quality of the ash and to the life of the furnace, to carry the calcination process too far. If carried to incipient fusion, the alumina of the brick lining is strongly attacked, and this alumina subsequently causes a loss of soda. The foregoing statements are not mere opinions, they are inferences drawn from the results of many analyses made in our laboratory from samples received from many forms of evaporators ; the arguments hold good for all forms.

BLEACHING POWDER.

THE recent collapse of all agreements and conventions connected with the various products of the alkali trade has naturally caused much attention to be paid to bleaching powder. In years gone by, and not many years either, bleach has been sold at £3. 15s. per ton ; and in the absence of all conventions, and with manufacturers striving to turn out all their plant will possibly produce, there are many consumers expecting a return to the foregoing low price. Our readers may possibly expect us to say something upon the subject in this our New Year's number, so that it would be as well to review the various processes that have passed before the showman's pointer. Few outsiders have properly grasped the changes brought about by modern chlorine processes. It has often been urged, by those who are only partially acquainted with manufacturing details, that the manganese process resulted in producing a greater weight of bleaching powder from a ton of salt than was the general rule before. This was brought about by better condensation. The manufacture of hydrochloric acid is now quite a different matter to what it was twenty years ago, when wash towers were to be found sending acid away into the drains at 10° and 12° Tw. as unfit for use.

This is clearly shown from some figures culled from an old manufacturing record of 1871. In this work, which shall be nameless, the decomposition of 1,220 tons of salt produced 194 tons of bleaching powder of 36% strength, or about 6 tons of salt being consumed per ton of bleach made. That the acid was really being wasted in the wash towers may be seen from the fact that the maganese stills were charged with 600 lbs. of 70% manganese each, and 50 cubic feet acid testing 25% of real H Cl, ten of these stills used regularly to duce a chamber of 36% bleaching powder, weighing on an aver hree months exactly four tons. This means 14 cwts. of 70% manganese ınd four tons of liquid muriatic acid at 28°Tw. per ton of bleach or a on of bleach from about 45 cwts. of salt. This amount was not for nany years reached by the Weldon process. In 1876 the general con- ıumption of salt per ton of bleach was 60 cwts., and it was considered :xtremely good working if ever 55 cwts. was reached. How then did he Weldon process aid the manufacturer? Simply by cheapening his nanganese, for whilst at the present time the native maganese neces- ary for a ton of bleach would cost at least £2. 8s., the Weldon man- ganese costs but the odd eight shillings. Even up to 1884 a great deal of acid was lost as free acid in the still liquors, and in that year, or the year before, Mr. George E. Davis read, before the Manchester section of the Society of Chemical Industry, a paper, the results of some experi- nents carried out in 1875, with a view of utilising this acid and making t into bleach. Like most other improvements the paper was attacked by men who should have known better; they used all their arts and wiles to prove the impossibility of attempting such a thing, but the proper answer is that the process is being employed in the year of Grace 1890, with the result of making a ton of bleach from 45 cwts. of salt, a quantity equal to that used in the old native stills* Mr. Davis found in practice the free acid in the still liquor was equivalent to the "base" of the Weldon mud, and based his process on this information.

In 1873 the bleaching powder made by the Weldon process was sup- posed to cost as follows, and the second column exhibits the cost to-day:—

	1879. £	s.	d.	1890. £	s.	d.
Coal	0	6	10	0	6	0
Lime	1	1	4	0	14	0
Limestone	0	3	3	0	1	0
Manganese	0	8	1	0	4	2
Wages	0	17	8	0	13	4
Casks	0	17	1	0	13	0
	3	14	3	2	11	6

Nothing is here reckoned for the cost of the muriatic acid of which, as already seen, four tons are required, neither is there anything set down for repairs or depreciation.

The necessary hydrochloric acid at sixpence per carboy, comes to £2. per ton on the bleach, so that with bleach at £5. free on rails at makers' works there is not much profit for the manufacturer, and if prices continue at this low level, there will be not a few balance sheets that will look ugly at this time next year. To the foregoing costs of production must be added:—

	s.	d.
Depreciation at 10%	5	0
Repairs	4	3
Material for Repairs and General Expenses	6	8
	15	11

o that it will be seen the manufacturer does not net sixpence per bottle for his acid. With the cost price of bleach at £5. made up as follows:

	£	s.	d.
Coal		6	0
Lime		14	0
Limestone dust		1	0
Manganese		4	2
Wages		13	4
Casks		13	0
Depreciation		5	0
Repairs		4	3
General Expenses, &c.		6	8
Acid (4 tons)	1	12	7
	£5	0	0

Four tons of acid at £1. 12s. 7d. is fourpence and thirteen-twentieths of a penny per carboy.

But there are other processes by which bleaching powder is being manufactured, Deacon's classic process introduced in 1871 is still at work in several places, and if it had been properly pushed and worked it might have produced a ton of bleach from a ton of salt. But it has not yet done so. We believe no accurate working figures have ever been published in this country in connection with this process, which has always seemed under a cloud from its very initiation. In actual practice it has only been found convenient to decompose the gas coming from the pot; we cannot theorise on what might have been done, or at least attempted, we must stick to actual facts. If we take t that 75 per cent. of the total gas evolved from salt is sent away from the pot, we shall not be far from the truth. The average decomposition, in practice, that we have found, is 60%; 75 × 0·6 = 45 per cent., of which about five is lost in the wash towers, &c., leaving, say 40 for conversion into bleach. At this rate a ton of bleach would be made from 30 cwts. of salt.

*See Chemical Trade Journal, Vol. 1, p. 132.

We have often wondered why those manufacturers who have this process in use have not endeavoured ere this to strengthen up their weak acid, and to dehydrate it for further use in the process. Two processes were at use in Manchester in 1877, which might profitably be employed in the Deacon process for the preparation of substantially dry muriatic acid gas, and by this means it would be possible to make a ton of bleaching powder for every ton of salt decomposed. We believe some such process has been worked by Hasenclever on the Continent, and by Solway in his experimental works, and that they have found it successful. When liquid muriatic acid is mixed either with strong sulphuric acid or a saturated solution of chloride of calcium, or with solid chloride of calcium, magnesium or zinc chlorides, the water stays with the deliquescent salts, while the acid gas escapes. The present arrangement of acid making towers, wash towers, weak acid towers, and so forth, would remain almost unchanged, it would only be necessary to concentrate the deliquescent material from time to time as it became too weak for evolving gas.

But we have to deal with practical results that have been attained on the manufacturing scale. We believe the most successful run on the Deacon process, for any length of time, has been when the salt-cake pot, working 30 tons of salt per week of 132 hours, has turned out 21 tons of bleach per week, with a consumption of 23 cwts. of copper scales per 1,000 tons of bleach produced, and the cost of production may be accurately stated as follows:—

	£	s.	d.
Labour	0	15	1
Fuel, at 6s.	0	2	8
Lime	0	7	0
Copper scales	0	1	2
Casks	0	13	0
Repairs	0	4	0
Depreciation	1	13	4
General expenses	0	13	4
Acid	0	10	5
	£5	0	0

One item deserves explanation, and that is "Depreciation." In the Weldon process this amounts to 5s. per ton, but in the Deacon to 33s. 4d. We base our figures on the cost of a plant erected to take the gas from one pot and furnace, and the longest run it ever had was one of 37 weeks, during which time 588 tons of bleach were made in it. This plant cost £13,000. whereas a Weldon plant, to produce 100 tons of bleach weekly, would cost but £12,000.

But in this cost sheet of the Deacon process it must not be forgotten that nearly 60 per cent. of the acid is turned into another channel. In the Weldon process all is used up, so that where a weak acid can be utilised there is no great difference between the two methods financially, the more especially as a Deacon plant to do 15 tons of bleach per week could be put up to-day for less than £13,000. Herein lies the solution of the riddle thrown out in an early number of this journal, the production of a ton of bleach from a ton of salt; but science moves very rapidly now-a-days, and this ratio has been largely exceeded, the next problem we shall attempt to unfold to our readers, in a few weeks time, is *a ton and a half of bleach from a ton of salt*, with a comparatively inexpensive plant, one simple and easy to work, requiring the minimum of labour, and free from liability to escapes of gas, so frequent under existing systems.

THE LAMP TRADE.—The Central News says:—In consequence of the strikes and agitation in the gas trade in London and the provinces, there has been an enormous demand for lamps, especially of the cheaper kind. Many of the large wholesale houses have been cleared of their stocks, and an extremely good business is being done in lamp glasses. Although the stock in the hands of retailers is still very large, the shopkeepers have been for some days past laying in large stocks of petroleum, oil, and candles, and householders appear to have been buying largely of the latter. Strange to say, however, tallow in the market has not been affected to any appreciable extent, for prices at the last sales were practically unchanged save for an occasional 3d. to 6d. drop. A similar phenomenon is observable in the petroleum market. Retailers have been largely increasing their stocks, but the wholesale market is flat, and within the last few days petroleum has declined a farthing per gallon, and Russian is one-eighth of a penny lower. Colza is firm, owing to natural causes, and other oils are not affected. Coals, however, have shown much movement. Best coals, which at the end of November were quoted on the exchange at 19s., are now 21s. 6d. per ton; and seconds, which were 18s. have been run up to 20s. 6d. Many firms are now retailing house coals at 28s. per ton.

THE DINSMORE PROCESS AND ITS PROBABLE EFFECT ON THE BENZOLE MARKET.

THE Dinsmore process of gas making, with material modifications, patented by Mr. Isaac Carr, is now receiving the careful attention of gas engineers, and there is no doubt we shall see it introduced into many works. How much patent there is about the whole thing is a matter which no doubt will have to be decided before long, seeing that many specifications have been published for the destructive distillations of tar, and to our own knowledge at least one patent claims the superheating of the gas after leaving the primary retort and before its entry into the hydraulic main. It was the offer of this process by the patentee which induced the writer to enter into a series of experiments, some results of which were read before the Society of Chemical Industry (London Section) in 1885. It has never been denied to our knowledge that a gas of good illuminating power could be obtained by the destructive distillation of ordinary gas tar, but no one, before the date of the experiments alluded to, had attempted to satisfy himself which of the constituents of the tar yielded the illuminants and which the diluents, or how much gas could be produced of a certain illuminating power, from a given weight of various tar products. Shortly after this paper was read, Mr. Lewis T. Wright, then gas engineer to the Nottingham Corporation, attempted to show the absurdity of attempting to enrich gas by the aid of tar products, but his experiments were faulty, his inferences incorrect and his logic bad, and, moreover, his paper exhibited the fact that he must have been unacquainted at that date with the experiments that had been made at the works of the Paris Gas Company, and which we have now published in these columns.

In the paper to be found in the J. S. Ch. Ind., January, 1886, when treating of the gasification of tar, the following paragraph appears:—"Supposing for a moment that these crude attempts, and others of a similar nature which followed them, had been successful in the matter of gas-making, pure and simple, it is doubtful how far they would have conduced to the enrichment of the gas, as I have found by experiment that the pitch which constitutes about three-fifths of the total weight of the tar produces a gas absolutely devoid of illuminating power, in quantity about 25 per cent. of its weight."

And again:—"To attempt the carburation of gas by employing any compound of the benzene series seems to me the height of absurdity, seeing that it is from coal-tar that these products are produced, and the cost of extraction must be added to the original cost of the tar. If these compounds are to be used at all, it is the crude tar which should be operated upon in the gasworks, the tar being made to replace cannel."

Experiments were made by the reader of the paper, from which it was argued that the seven tons of tar produced from 100 tons of coal is of value to the gas maker, equal to nearly ten tons of cannel coal, and there would be produced besides, pitch and anthracene, which would yield a very respectable revenue. In the experiments cited, ordinary coal gave 10,000 cubic feet of 17 candle gas, or in all 170,000 candles. This is perhaps above the average for ordinary coal, and 160,000 candles is nearer the mark. The tar oils operated upon gave 16,000 cubic feet of 50 candle gas per ton, or a total of 800,000 candles per ton, so that a ton of tar oils for this purpose would be five times the value of a ton of ordinary coal. This is exactly what the ratio is to-day, when coal is at 9s. per ton, and oils at 3d. per gallon.

Mr. Carr has read a very interesting paper before the Manchester District Association of Gas Engineers, and in it gave the result of an eight weeks trial of the Dinsmore process while using Abram Arley mine slack, in which 130 tons of coal produced 1,545,000 cubic feet of 21·32 candle gas, without the use of a particle of cannel. These results are much in excess of those obtained by the writer when working in a somewhat different way, but if they can be repeated the process has a great future before it, and must in time become generally adopted.

And now we may turn our attention to another phase of the question; the Dinsmore tar will be entirely denuded of its benzol, toluol and xylol, the carbolic acid and light creosote oils; of what value will it be to the tar distiller? Practice can only answer the question properly. It may be that the tar will be so thick as not to run through pipes in the cold weather, if so what is to become of it? Perhaps it will in such a case be burned in the works wherein it is produced. If this be so then the anthracene and pitch will go as well into thin air. From whence then will the colour-makers procure the raw material?

It must be remembered that the make of benzol from coal-tar has not increased in late years in proportion to the coal carbonised; higher heats have been employed in gas works, yielding more gas per ton; that this poor gas has robbed the tar of a portion of its lighter products: no one knows this better than the modern tar distiller. At the present time about two millions of gallons of benzol are annually produced from gas-tar, so that if the Dinsmore process becomes general, this quantity must be made up by new carbonising works. The year 1890 will most probably see about 385,000 gallons of 90% benzol turned out from the carbonising works of
millions of gallons procured from or

The Dinsmore process is then en
tar-distiller, quite as much as that of
only sell their *surplus* tar, so that it t
the wording of their contracts, and
article that may be next to worthles
should endeavour to act honestly in t
illuminating power from the tar, this
endeavour to persuade themselves or t
good as before.

We shall await with great interest
are to be conducted at the Liverpool
results as Mr. Carr has obtained at V
system all over England is certain, a
are prepared we may have a return
astonished the world in 1883. Do
cycles? If so, 1890 may not be an un
not oversold themselves.

TENSILE TESTS AT VAR

M. ANDRE le Chatelier has recei
ments on the tensil strengths
The specimens tested were annealed
heated in an air bath. The results ob
table, the figures being the breakin
inch:—

	Te	
	15 deg.	100 deg.
Copper	16·0	14·5
Aluminum	11·7	9·5
Nickel	35·0	35·0
Silver	10·9	10·1
Aluminum bronze	33·8	33·3
Copper, iron, nickel	26·9	26·8
Zinc	7·9	1·5

THE PATENT OFFIC

"EXPERTS?" said a prominent
recently. "Why, they are as
summer, and you can hardly enter a r
meeting one or more. The work here is
embracing has it does the entire domain
unless the men who decide upon the ne
thousands of applications for patents, whi
not thoroughly understand the state of th
be crowded with the cases of litigious inv

Many of the examiners in the differe
learning. Especially is this true of t
applications are out of the usual run,
inventions, such as the telephone and
lighting. The experts who are called upo
Edison and others, in regard to new electr
be men who thoroughly understand the
been almost a life study, and there is no b
not thoroughly familiar with. There is
examiner is busy in the laboratory.

But the experts of this office are by
divisions of electricity and chemistry.
treated of here, and the examiners must,
be thoroughly familiar with their subjects.

The inventive faculties of the country se
at present the subject of naval projectiles
attention at their hands. This requires th
their claims should know his business thor

Mr. P. B. Pierce, who has charge of th
another man who can be considered an ex
sewing machines than any man in the cou
ject a life study.

And so it is throughout the office. Ev
his particular line, and some of their p
difficult to fill. The pay of these men is e
invaluable services they render. Every o
deal more out of the office as an attorney,
this is, nobody seems to know, unless it is
and they are so wrapped up in it as to
other interests for the sake of the opportu
which their position so abundantly gives t

A STUDY OF COAL GAS,

By M. EM. SAINTE-CLAIRE DEVILLE.

(Engineer at the experimental works of the Paris Company.)
(Continued from 337, vol. v.)

The following are the averages of the results obtained, the extreme limits of variations being given in each case :—

1. Gas entirely deprived of benzene, but still containing 4·4 litres of non-aromatic hydrocarbons per 100 litres.

The gas burner was placed at one metre and the standard light at 1·40 metres from the photometric screen.

The amount of gas burned varied from 133-151 litres (mean 141·9 litres).

From this it follows :—

Illuminating power (volume per unit standard light) 293·7 litres, or illuminating power of the gas per 100 litres per hour, 0·334 candles. This makes 0·076 candles per 1% of olefines and acetylenes, or 0·056 per gramme of these same hydrocarbons.

2. Gas cooled to 22° ; that is the formed gas entering in addition 23·151 grms. of benzenes (92% of which is pure benzene) per cubic metre, or 2·315 grms. per 100 litres.

This gas burnt at a Bengel burner placed at one metre from the screen, at the rate of 124·4 litres (limits 150-127) gives the same light as a standard candle, at the same distance. The number 124·4, therefore expresses the effective volume per standard candle. Intensity per 100 litres of gas burnt is then 0·804 candles, *i.e.* 0·470 candles more than the former gas.

We may therefore say that the 2·315 grms. of benzene have supplied the illuminating power corresponding to 0·470 candles, or 0·203 candles per grm. of benzene.

3. Ordinary gas. That is to say the second sample, containing in addition 16·245 grms. of benzenes (52% of which is pure benzene), per cubic metre, or 1·624 grms. per 100 litres. This gas burnt at one metre from the screen at the rate of 104·6 litres (limits 103-107) gives a light equal to that of a standard candle at the same distance.

Its illuminating power per 100 litres of gas is, therefore, 0·956 candles, *i.e.*, 0·152 more than the preceding.

The additional 1·624 grms. of benzene have, therefore, increased the illuminating power by 0·152 candles, or 0·093 per gramme of benzene.

We have, therefore, for the analysis of the illuminating power of our ordinary gas per 100 litres :

(1.) 0·334 candles due to 4·4 litres, or approximately 6 grammes of various hydrocarbons, making 0·076 candles per litre, or 0·056 per grm. of these.

(2.) 0·470 candles due to 2·315 grms. of benzene, containing 92% of pure benzene, or 0·203 per gramme.

(3.) 0·152 candles due to 1·624 grms. of benzene containing 52% of pure benzene, or 0·093 per grm.

These give a total of 0·622 candles produced by 3·939 grms. of benzene containing 77% of pure benzene, or 0·157 per gramme.

The total illuminating power must, therefore, be appointed in the following manner:

Due to various hydrocarbons	34·9	34·9
" 92% benzene ..	49·2	65·1
" 52% " ..	15·9	
	100·0	100·0

Experiments made since the printing of this article, show that the three 3-4% of ethylene present in coal-gas do not contribute more than 10-13% (*i.e.* 3·25% per 1% of ethylene) to the illuminating power. About 22-25% of the illuminating power, therefore, remains to be accounted for by acetylene and propylene. In view of this result, it necessary to fall in with the opinion held by Lewis T. Wright and Percy Frankland, who maintain that Marsh gas and its homologues play a much more important part than is usually believed. The following table is borrowed from the report of the 23rd November, 1886 (No. 229), in which are given the results of experiments on the influence of carbonic acid, several other gases and ethylene on the illuminating power of coal-gas. The table shows the probable share of each class of hydrocarbons in the production of 100 units of light.

	Rich Gas Types III. IV. & V.	Poor Gas. I. & II.	Average Gas.
Benzene .... (Aromatic series)........	40-50	70-85	65
Other hydrocarbons...... Absorbable by bromine.. Ethylene, propylene & acetylene	30-15	15-7	10-20
Marsh Gas	12-10	10-8	10
Higher members of Marsh Gas series	18-25	5·0	15·5

THE ILLUMINATING POWER OF A GAS DOES NOT INCREASE IN PROPORTION TO ITS RICHNESS.

A superficial examination of these numbers might lead to the conclusion that pure benzene has a much greater illuminating power than an equal weight of the other hydrocarbons of the series.

1 grm. of benzene gives (92% of pure benzene), 0·203 candles.
1 " " (52% "), 0·093 "

or less than half.

This effect would, to some extent, be expected, for benzene is richer in carbon than the other hydrocarbons of the series. I am, however, convinced that the enormous difference which we have shown to exist between the illuminating power due to the benzene condensed between 22° and 70°, and that due to benzene condensed at 22°, is due to another more important cause.

It appears very probable that the luminosity of one gramme of a hydrocarbon varies considerably with the richness of the gas in which it is contained, and with which it is burnt.

The addition of one gramme of benzene to a poor gas, that is to say, produces a much greater increase in its illuminating power than the addition of the same amount to a gas already rich.

It must, however, be borne in mind that this variation in the specific luminosity of a hydrocarbon may perhaps only be apparent, and that it will not be shown, or only in a less degree if each gas be tested under the most favourable conditions, that is, with a burner adapted to its composition, as the Bengel burner is to the average composition of Paris gas.

Whatever may be the correct interpretation of the above figures, from the point of view of the theory of combustion, they suffice to enable us to answer the following question :

What is, measured under the usual conditions of the city of Paris, the increase in illuminating power which can be expected if, by the application of the process known as "hot condensation," or by any other means, it becomes possible to retain in the gas, all or part of the benzene which is now found dissolved in the tar?

This addition would mean continuing the series of successive enrichments of the gas which we have already considered.

We have seen that, setting out with gas entirely freed from benzene, the addition of 2·315 grams. of this hydrocarbon to 100 litres of gas, produced an improvement of 0·203 candles per gramme, and that a further addition of 1·524 grms. only produces an improvement of 0·003 candles per gramme.

It is obvious that a third addition would make still less improvement, and that in estimating it at 0·08 candles per gramme of benzene added to 100 litres of gas, we are well within the truth.

Now, all the benzene contained in the tar would not enrich the gas by more than 2·9 grms. per cubic metre, as we have already seen, which amounts to 0·29 grms. per 100 litres.

The extreme limit of enrichment of the gas, which can never be actually attained, is, therefore,

$$0·08 \times 0·29 = 0·023 \text{ candles.}$$

In short, the benzene left in the tar represents less than 2 per cent. of the illuminating power of the gas, and is, therefore, not worth retaining in the gas. (It must be remembered that this applies to the average gas of the Parisian Company, in the works of which 65—70% of the tar is condensed hot, both in the syphons and collector. Such are the facts as to the part played by benzene in average gas).

Other investigations have also been made, the gas being drawn directly from the retort and condensed, etc., by means of laboratory apparatus, so arranged as to avoid the formation of a stock of tar such as accumulates in the coke columns of the scrubbers. The experiments were not continued long, owing to the trouble caused by them in the retort houses.

The results are expressed in the accompanying table, the numbers in which represent the amount of benzene in the gas, the position of the the figures showing the period of the distillation at which it was collected.

Weight of benzene in grammes, per cubic metre of gas :—

Age of charge	1	2	3	4	5	6	7
4th hour	25·300	24·008	25·166	—	24·833	—	—
	—	—	—	24·250	—	31·875	—
	—	—	—	—	—	—	27·700
3rd hour	—	—	—	31·147	27·000	—	—
	35·086	30·886	34 029	—	—	39·017	—
	—	—	—	—	—	—	31·953
2nd hour	—	—	—	37·184	—	—	—
	34·075	33·935	42·300	—	29·100	39·125	—
	—	—	—	39·808	—	—	41·592
1st hour	—	—	—	32·798	—	37·100	—
	25·278	28·806	28·444	—	36·815	34·700	39·086
	—	—	—	—	—	—	40·205
	—	—	—	—	—	28·642	—

It appears from these numbers that benzene is present in moderate quantity at the commencement of the distillation, and that its quantity increases gradually to a maximum at the end of 30 or 45 minutes, this being maintained until the distillation is half over. It then slowly decreases till the close of the operation.

These results admit of easy explanation by Berthelot's theory, that benzene is one of the chief products formed when a mixture of hydrocarbons is submitted to a high temperature.

(To be continued.)

Market Reports.

THE LIVERPOOL COLOUR MARKET.

COLOURS are unaltered. Ochres: Oxfordshire quoted at £10. £12., £14., and £16.; Derbyshire, 50s. to 55s.; Welsh, best, 50s. to 55s.; seconds, 47s. 6d.; and common, 18s.; Irish, Devonshire, 40s. to 45s.; French, J.C., 55s., 45s. to 60s.; M.C., 65s. to 67s. 6d. Umber: Turkish, cargoes to arrive, 40s. to 50s.; Devonshire, 50s. to 55s. White lead, £22. 10s. Red lead, £19. 10s. Oxide of zinc: V.M. No. 1, £24. 10s. to £25.; V.M. No. 2, £23. Venetian red, £6. 10s. Cobalt: Prepared oxide, 10s. 6d.; black, 9s. 9d.; blue, 6s. 6d. Zaffres: No. 1, 3s. 6d.; No. 2, 2s. 6d. Terra Alba: Finest white, 60s.; good, 40s. to 50s. Rouge: Best, £24.; ditto for jewellers, 9d. per lb. Drop black, 25s. to 28s. 6d. Oxide of iron, prime quality, £10. to £15. Paris white, 60s. Emerald green, 10d. per lb. Derbyshire red 60s. Vermillionette, 5d. to 7d. per lb.

TAR AND AMMONIA PRODUCTS.

TUESDAY.

The holiday season has not been without its effect upon the tar products market, but so far only as regards the demand, values have not suffered to any degree, nor from what we can see are they likely to if producers will be circumspect. Benzoles may be quoted at old rates, 3s. 5d. for 90's and 2s. 7d. for 50/90's; solvent naphtha and carbolic acid as before. Creosote of good quality for lighting purposes has risen a halfpenny, and good fluid makes are difficult to purchase ahead. Anthracenes are very firm, and rates already quoted are applicable to-day. Pitch is still in good demand, and 32s. Garston and 34s. at East Coast ports is the ruling figure.

The sulphate of ammonia market is quiet on account of the holidays, and though buyers are not disposed to pay the values now asked by the sellers, yet it is the general opinion that the highest figures paid before Christmas will be again reached as soon as the New Year is well in. To-day's prices may be quoted as £12. 5s. Hull and Leith, London £12. 6s. 3d. to £12. 7s. 6d., while Beckton is variously spoken of at both these latter prices. There is not much sulphate offering, so that producers need not feel any alarm at business being a little less brisk during Christmas week and the New Year. The year 1890 commences better than 1889, and there is no reason why prices should not be maintained.

WEST OF SCOTLAND CHEMICALS.

GLASGOW, Tuesday.

This afternoon the Scotch works close for the New Year's holiday, and some of them will not resume till Monday. The feeling for fully a week has been more lively, and the New Year is to begin with a larger share of immediate promise in it for the chemical industry than could have been expected two or three weeks ago. Bleaching powder and soda crystals appear to have called a halt from the late downward run, and there has even been a recovery—very partial, however, as yet. Caustics have further gained. Lubricating oil retains advantage recently acquired. Perhaps sulphate of ammonia is scarcely so strong, but the shrinkage is trifling. The article is on offer to-day at £12. 5s., that price, however, being at the moment just in excess of buyers' views, and true Leith figures therefore must be stated at £12. 3s. 9d. to £12. 5s. Bichromate of potash and bichromate of soda have been moving off in moderate parcels to the Continent farther into the winter season than usual, owing to the persistent openness of the temperature, and consequent postponed closing of North Sea ports. In other respects the demand for these articles is not up to the average, and works could do with bigger order lists. Chief prices current are:—Soda crystals, 42s. 6d. net Tyne; alum in lump, £4. 17s. 6d., less 2½% Glasgow; borax, English refined £30., and boracic acid, £37. 10s. net Glasgow; soda ash, 48/52°, 1⅛d. less 5% Tyne; caustic soda, white

76°, £9., 70/72°, £7. 7s. 6d., 60/62°, £
less 2½% Liverpool; bicarbonate of
casks, £5. 5s. net Tyne; refined
Tyne; saltcake, 26s. to 28s.; bleach
less 5% f.o.r. Glasgow; bichromate
less 5 and 6% to Scotch and English
potash, 4½d., less 5% any port; nit
sulphate of ammonia, £12. 3s. 9d. to
1st and 2nd white, £36. and £34., less
£23. 10s., less 5% Liverpool; paraf
paraffin wax, 120°, semi-refined, 2⅞d
paraffin oil (burning), 6¾d. to 7d.,
865°, £5. to £5. 10s.; 885°, £5.
£7. to £7. 10s. Week's imports c
cargoes.

MISCELLANEOUS CH

In all branches of the chemical trad
tendency to higher prices, and prospe
fairly encouraging. In the alkali trad
in values during the past week, but
firmly held by makers for future deliv
rails is the current quotation, but for
ding to make. Caustic soda, £7. 5s.
for prompt and forward at 25s. per
4½d. to 4¾d. per lb. Sulphate of
being done at £23. Acetates of lime
good demand. Acetic acids are incl
material. Other acetates without mov
prices unchanged. Acetic still maintai
both for home and export trade. Anilin
is a fair amount of enquiry. Prussiates
the spot, but the high price quoted for
forward business being done. Vitriol st
for all purposes.

THE LIVERPOOL MINI

Notwithstanding the holidays and the
the year a good business has been done
Manganese: Arrivals have increased, bu
tion prices continue strong. Magnesite:
prices in buyers' favour; raw ground, £
£10. to £11. Bauxite (Irish Hill Bran
that the supply is scarcely able to keep
firm—lump, 20s; seconds. 16s; thirds, 1
7s. 6d. per ton at the mine; arrivals conti
have been large, but prices have been fu
G.G.B. "Angel-White" brand—90s. to
superfine. Barytes (carbonate) steady; s
£6.; No. 1 lumps, 90s.; best, 80s.; second
50s.; best ground, £6., and selected cry
unchanged; best lump, 35s. 6d.; good
6d. to 27s. 6d.; common, 18s. 6d. to 20s.
brand, 60s.; common, 45s.; grey, 32s.
unchanged, ground at £10., and espec
quality, £13. Iron ore is brisk. Bilbao
to 10s. 6d. f.o.b.; Irish, 11s. to 12s. 6d
Purple ore continues scarce at last quotati
ore in strong demand at 27s. 6d. for 20 p
brands in good demand, and bring full pri
at £5. 10s. to £6. and smalls, £5. to £5.
45s. to 50s. for best blue and yellow; fir
Scheelite, wolfram, tungstate of soda, a
inquired for. Chrome metal, 5s. 6d. per
lb. Chrome ore, best qualities are sought af
ore and antimony metal firm. Uranium
Best rock £17. to £18.; brown grades,
ore dearer; Smalls £12. to £13.; se
Calamine: Best qualities scarce—60s.
sulphate (celestine) steady, 16s. 6d. to 1
to £16.; powered (manufaetured), £11.
manufactured, old G.G.B. brand brings
English. Plumbago: More offering; bes
tions; Italian and Bohemian, £4, to £22
cargoes, scarce on spot—20s. to 22. 6d.
scarce, especially high brands. Ground m
offering—common, 18s. 6d.; good mediu
to 35s. (at Runcorn.)

New Companies.

BENGAL INDIGO MANUFACTURING COMPANY, LIMITED.—This company was egistered on the 12th inst., with a capital of £150,000., in £10. shares, to carry on usiness as indigo planters, and manufacturers of, and dealers in indigo, and also s farmers, graziers, growers, manufacturers, and dealers in all kinds of tropical roducts. The subscribers are :— Shares.

	Shares.
W. E. Medland, Withington, near Manchester, merchant and manufacturer	1
M. A. Herrman, Hamburg, merchant	1
E. C. Schretkky, Dresden	1
J. Leigh, Knutsford	1
S. Mosley, 104, King-street, Manchester, chartered accountant	1
J. B. Lonsdale, 9, Fennel-street, Manchester, merchant	1
J. W. Radcliffe, Werneth-park, Oldham, merchant, &c.	1
F. Karuth, 58, Perham-road, West Kensington	1

DAVID MARTINEAU AND SONS, LIMITED.—This company was registered on the 2th inst., with a capital of £250,000., in £10. shares, to acquire as a going concern he business of David Martineau and Sons, of Christian-street, sugar refiners. The subscribers are :—

	Shares.
David Martineau, 21, Mincing-lane, sugar refiner	1
G. Martineau, 21, Mincing-lane, sugar refiner	1
W. S. Hodgson, Esher, Surrey	1
C. H. Gill, Staines, sugar refiner	1
E. S. Hobson, 21, Mincing-lane, clerk	1
R. W. Brown, 41, Kirkstall-road, Streatham-hill, clerk	1
C. F. Worters, 125, Sunderland-road, Forest-hill	1

WITTY AND WYATT, LIMITED.—This company was registered on the 12th inst., vith a capital of £30,000, in £1. shares, to manufacture and deal in asbestos and sbestos goods, and to enter into an agreement with Messrs. Witty and Wyatt. The ubscribers are :—

	Shares
H. H. Witty, 88, Leadenhall-street, merchant	7,499
R. C. Wyatt, 88, Leadenhall-street, merchant	7,499
Torre Diaz, 41, Moorgate-street, merchant	7,500
J. M. de Zulueta, 41, Moorgate-street, merchant	7,499
P. Pate, 88, Leadenhall-street, secretary to a company	1
F. H. Lovegrove, 88, Leadenhall-street, clerk	1
J. T. Hancock, 13, Goodwin-road, Forest-gate, secretary to a company	1

ENGLISH PORTLAND CEMENT COMPANY, LIMITED.—This company was regis ered on the 18th inst., with a capital of £25,000., in £1. shares, to manufacture ement and similar substances, and to adopt an agreement with Messrs. Irwin, Hopper, and Co. The subscribers are :—

	Shares.
R. Morris, J.P., Doncaster	500
Mrs. G. Morris, Doncaster	120
Miss H. E. Morris, Doncaster	10
Miss M. Morris, Doncaster	10
Mrs. M. A. Morris, Doncaster	100
W. R. Morris, Doncaster, merchant	250
Miss C. Woodhouse, Doncaster	10

KUHN PATENT PASTEURIZATION SYNDICATE, LIMITED.—This syndicate was egistered on the 16th inst., with a capital of £20,000., in £1. shares, to acquire the atent rights of Wm. Kuhn, relating to the Pasteurization of beer, wines, and other quids. The subscribers are :—

	Shares.
E. W. Owles, 14, Ordnance-road, St. John's Wood, clerk	1
A. E. Pridmore, 2, Broad-street-buildings, architect	1
T. J. Sturgeon, Broad-street-buildings, commission agent	1
O. Vernede, 10, New Broad-street, solicitor	1
J. Wakefield, 4, Keppel-street, Russell-square, commission agent	1
J. M. Sharp, 1, Finsbury-square, solicitor	1
R. J. Jeffes, 4, Westbourne-terrace, Lower Clapton, clerk	1

HEATHLEY, JANISCH, AND COMPANY, LIMITED.—This company was regis ered on the 14th inst., with a capital of £3,000., in £1. shares, to trade as chemists, ruggists, and dealers in proprietary articles, and in all kinds of electrical and cientific apparatus. The subscribers are :—

	Shares.
J. Heathley, Blyth, grocer	
G. A. Heathley, Newcastle-on-Tyne, chemist's assistant	1
Ralph Carr, Newcastle-on-Tyne, insurance agent	1
W. Turnbull, Newcastle-on-Tyne, draper	1
Mrs. Turnbull, Newcastle-on-Tyne	1
Mrs. E. Janisch, Newcastle-on-Tyne	1
E. Towers, Tynemouth, accountant	1

AUDENSHAW PAINT AND COLOUR COMPANY, LIMITED.—This company was registered on the 18th inst., with a capital of £11,000., in £1. shares, to acquire the business of the Andenshaw Paint and Colour Co., at Palatine Works, Guide-lane, Guide-bridge, near Manchester, and at 11, Great Ducie-street, Strangeways, Manchester. The subscribers are :—

	Shares.
Wm. Hellewell, Sowerby-bridge, York	1
J. R. Meadows, Guide-bridge, Manchester, colour-maker	1
B. Huggins, Guide-bridge, Manchester, bookkeeper	1
Wm. Penny, 10, Adela-street, Belfast, commercial traveller	1
J. Hudson, Upper-mill, Oldham, plumber and painter	1
L. Pugh, 87, Broad-street, Manchester, colourman	1
A. Howorth, 176, Milnrow-road, Rochdale, clerk	1

Gazette Notices.

PARTNERSHIPS DISSOLVED.

D. S. HAYNES AND H. P. FINEMORE, Whitfield-street, Finsbury, London, manufacturing chemists and commission agents.

J. AND E. THOMAS AND GREEN, Wooburn, paper makers, as far as regards E. Thomas.

L. ASHWORTH, J. ASHWORTH, AND J. WALSH, Cheadle and Manchester, under the style of Demming's Printing Co., calico printers.

Adjudications.

WALTER CUNNINGHAM, Grange Park-road, LEYTON AND DAY, W. REUBEN, Vicarage-road, Leyton (trading as Cunningham and Co.), Leyton, timber, lime, and cement merchants.

Notices of Dividends.

JOHN BROWN, of Rochdale, mineral water manufacturer, first and final dividend of 3s. 10¼d., December 30. Official Receiver's Office, Oldham.

The Patent List.

This list is compiled from Official sources in the Manchester Technical Laboratory, under the immediate supervision of George E. Davis and Alfred R. Davis.

APPLICATIONS FOR LETTERS PATENT.

Measurement of Liquids. J. Pocock and S. Johnson. 20,198. Dec. 16.
Manufacture of Alloys of Aluminium. S. Pearson and J. H. Pratt. 20,214. December 16.
Manufacture of Metallic Electrodes.—(Complete Specification). F. Marx. 20,217. December 16.
Manufacture of Azoxy Aniline and of Colouring Matters derived therefrom. J. Imray. 20,219. December 16.
Disinfectants A. Artmann and H. W. Kufeke. 20,231. December 16.
Electricity Meters. A Reckenzaun. 20,318. December 17.
Field-Magnets of Dynamo-Electric Machines. S. Kapp. 20,326. Dec. 18.
Manufacture of Hypophosphites. J. A Kendall. 20,392. December 18.
Method and Apparatus for Extracting the Tannin from Bark used for Tanning Purposes. J. Hutchings and W. N. Hutchings. 20,403. Dec. 18.
Production of Ozone. A. Muirhead 20,426. December 19.
Secondary Batteries. G. Philippart. 20,516. December 20.
Charging Gas Retorts. J. West. 20,551. December 21.
Electric Meters. J. Einstein and S. Kornprobst. 20,582. December 21.

IMPORTS OF CHEMICAL PRODUCTS

AT

THE PRINCIPAL PORTS OF THE UNITED KINGDOM.

LONDON.

Week ending Dec. 14th.

myl Acetate,	£23 F.	Stahlschmidt & Co.
ntimony—		
apan,	139 t.	H. R. Merton & Co.
ntimony Ore—		
France,	40 t.	Pillow, Jones & Co.
A. Turkey,	30	Quirk, Barton & Co.
Japan,	2	H. R. Merton
Holland,	8 c.	H. Boyce
Portugal,	9	H. Emanuel
,,	3	Redfern, Alexander & Co.
cetic Acid—		
Holland,	199 pkgs.	Beresford & Co.
ntimony Regulus—		
Portugal,	10 t.	Knowles & Co
arytes—		
Germany,	44 cks.	W. Harrison & Co.
,,	19	A. Boltz
,,	101	J. Matton
,,	23	W. Harrison & Co.
Holland,	20	J. Owen
,,	20	Prprs. of Scott's Whf.
Barium Peroxide—		
Germany,	£109	Domeier & Co.
Bone Meal—		
E. Indies,	200 t.	F. A. Hodgkinson & Co.
Borax—		
Germany,	£49	F. Stahlschmidt & Co.
Bromine—		
Germany,	£225	A. & M. Zimmermann
Caoutchouc—		
Madagascar,	120 c.	Kebbel, Son & Co.
,,	193	Carey & Browne
E. Indies,	10	J. H. & G. Scovell
E. Africa,	720	W. Johnson & Co.
C. America,	33	R. G. Hall & Co.
E. Africa,	6	L. & I. D. Jt. Co.
,,	86	,,
Madagascar,	13	Matheson & Co.
E. Indies,	159	Anderson, Weber & Smith
B. Honduras,	7	Belsize Estate & Prod. Co.
Holland,	8	W. Balchin
E. Africa,	5	L. & I. D. Jt. Co.
,,	45	Kebble, Son & Co.
,,	108	Clarke & Smith
,,	22	Forbes, Forbes & Co.
Aden,	24	L. & I. D. Jt. Co.
,,	12	Clarke & Smith
E. Indies,	132	L. & I. D. Jt. Co.
Chemicals *(otherwise undescribed)*		
Germany,	£60	T. H. Lee
,,	1,480	A. & M. Zimmermann
,,	12	T. H. Lee
,,	275	A. & M. Zimmermann
,,	12	J. Paysen
,,	18	Phillips & Graves
Cream Tartar—		
France,	10 pkgs.	M. Ashby
Holland,	14	Middleton & Co.
France,	5	Webb, Warter & Co.
Spain,	5	,,
Holland,	70	N. Smith
Camphor—		
Germany,	2 tubs.	L. & I. D. Jt Co.
,,	4 cs.	J. W. Drysdale & Co.
Caustic Potash—		
Holland,	21 pkgs.	Spies Bros. & Co.
France,	8	F. W. Berk & Co.
Carbonic Acid—		
Germany,	25 pkgs.	Elkan & Co.
Dyestuffs—		
Saffron		
Spain,	1 pkg.	P. W. Hore
,,	3	Major & Field
,,	2	Mancha & Co.
,,	1	Burgoyne, Burbidge & Co.
,,	1	Hodgkinsons & Co.
,,	3	Price, Hickman & Co.
,,	1	R. Quincey & Son
,,	1	Price, Hickman & Co
,,	1	J. Knill & Co.
Extracts		
France,	25 pkgs.	Miller & Co.
Austria,	84	L. J. Levinstein & Sons
Denmark,	1	Fullwood & Bland
France,	175	L. J. Levinstein & Sons
Cutch		
Burmah,	500 pkgs.	H. C. Schmidt
,,	600	Ross & Deering
Gambier		
E. Indies,	183	W. H. Cole & Co.
,,	84	P. & O. Co.

„ 510 J. H. & G. Scovell
„ 233 W. H. Cole & Co.
Sumac
Italy, 100 pkgs. F. Roberts
Orchella
Portugal, 70 pkgs. N. Smith
Aniline
Holland, 1 pkg. Hooton & Yates
U.S. 3 W. H. Cole & Co.
Cochineal
Canaries, 2 pkgs. Swanston & Co.
„ 22 Marshall & French
„ 17 A. Levy & Co.
„ 6 Beriro & Co.
„ 20 Bruce & White
Indigo
Greytown, 26 srns. Chalmers, Guthrie & Co.
E. Indies, 37 chts. L. & I. D. Jt. Co.
„ 98 T. H. Allan & Co.
„ 5 Phillips & Graves
Annatto
U. S. 2 pkgs. Lewis & Peat.
Ceylon, 25 Hoare, Wilson & Co.
Myrabolams
E. Indies, 1,094 pkgs. J. Forsey
Madder
Holland, 2 pkgs. J. Owen
„ 2 Union Lighterage Co.
Tanners' Bark
Queensland, 1 t. L. & I. D. Jt. Co.

Glucose—
Germany, 19 pkgs./154 c. L. Sutro & Co.
„ 200 200 Anderson, Weber & Smith
„ 26 233 Union Lighterage Co.
„ 100 100 J. Barber & Co.
„ 25 220 Hyde, Nash & Co.
„ 4 33 F. Stahlschmidt & Co.
„ 700 700 J. Barber & Co
„ 54 180 L. & I. D. Jt. Co.
„ 20 180 Dutrulle, Solomon & Co.
„ 300 300 J. F. Ohlmann
„ 40 236 Proprs. Chamberlains Wf.
„ 20 160 T. M. Duche & Sons
„ 2,000 2,000 L. & I. D. Jt. Co.
„ 500 500 Olyett & Francis
„ 4 35 J. Cooper
„ 40 326 Proprs Chamberlains Wf.
„ 100 200 J. Forsey.
„ 21 180 C. Tennant, Sons & Co.
France, 232 232 Dutrulle, Solomon & Co.
Germany, 4 32 T. H. Lee
„ 125 990 T. J. Darren
„ 50 415 Anderson, Weber & Smith
„ 21 175 C. Tennant, Sons & Co.

Gutta Percha—
E. Indies, 585 c. J. H. & G. Scovell
„ 4 L. & I. D. Jt. Co.
„ 380 Kaltenbach & Schmitz

Iron Bromide—
Germany, £500 A. & M. Zimmermann

Isinglass—
B.W. Indies, 1 cs. Malcolm, Kearton & Co.
E. „ 16 pkgs. Clarke & Smith
„ 3 L. & I. D. Jt. Co.
„ 22 „
B.W.I. 1 W. M. Smith & Sons
Sts. Settlements, 63 Hale & Son

Manure—
Phosphate Rock
France, 120 t. T. Farmer & Co.
„ 180 Hunter, Waters & Co.
„ 472 Moller, Graetz & Co.
Holland, 100 Lowes Manure Co.
Waste Salt
Germany, 30 t. Petri Bros.
Manure (*otherwise undescribed*)
U. S. 2 t. C. A. & H. Nichols

Mineral Salt—
Germany, £100 N. S. S. Co.

Magnesia—
Holland, £140 Beresford & Co.

Naphtha—
Holland, 7 brls. Domeier & Co.
Germany, 7 „
Belgium, 10 H. Lorenz

Potassium Sulphate—
Germany, 7 pkgs. Elkan & Co.
Holland, 11 Beresford & Co.
„ 10 A. Henry & Co.

Potassium Muriate—
Holland, 24 pkgs. H. Johnson & Sons
Germany, 500 F. W. Berk & Co.

Potassium Bicarbonate—
Holland, £25 A. & M. Zimmermann

Pitch—
France, 88 brls. H. Hill & Sons
N. Russia, 350 Sieveking, Droop & Co.
Holland, 5 Rosenbery, Loewe & Co.
„ 5 J. Owen
Germany. 49 Berlandina Brs. & Co.
Sweden, 85 „

Plumbago—
Holland, 90 cks. Brown & Elmslie
Ceylon, 157 E. Barber & Co
„ 586 J. Thredder, Sons & Co.

Pyrites—
Spain, 1,428 t. Societe Commerciale, &c.
Pomaron, 1,080 Mason & Barry

Potassium Carbonate—
Holland, 11 pkgs. Spies Brs. & Co.
France, 9 E. W. Carling & Co.

Paraffin Wax—
U. S. 257 brls. H. Hill & Sons
„ 5,054 J. H. Usmar & Co.
„ 160 H. Hill & Sons

Sugar of Lead—
Germany, 39 pkgs. Beresford & Co.
„ 16 C. J. Capes & Co.

Salammoniac—
Holland, £120 Barr, Moering & Co.

Stearine—
Holland, 53 bgs. H. Hill & Sons
„ 183 brls. J. H. Usmar & Co.
France, 76 sks. J. Goddard & Co
„ 144 „
U. S. 12 brls. Croager Brs.

Sodium Hyposulphite—
Holland, £70 Beresford & Co

Salicylic Acid—
Germany, 8 pkgs. L. & I. D. Jt. Co.
„ 11 Burgoyne, Burbidge & Co.

Saltpetre—
Holland, 18 cks. Augspurg, Hopf & Co
Germany, 60 kegs. P. Hecker & Co.
E. Indies, 841 bgs. Ralli Brs.
„ 1,195 C. Wimble & Co.

Sodium Sulphate—
Holland, £35 A. Henry & Co.
„ 80 Reddaway, Martin & Co.

Tartaric Acid—
Holland, 9 pkgs. Webb, Warter & Co.
„ 70 H. S. Smith & Co.
„ 11 Webb, Warter & Co.
„ 8 „
„ 20 Middleton & Co.
„ 4 Augspurg, Hopf & Co.
„ 14 N. Smith
„ 6 S. Smith
„ 3 Webb, Warter & Co.
„ 14 Middleton & Co.

Tar—
Libau, 24 brls. C. Arendt

Ultramarine—
Holland, 15 cs. A. Henry & Co.
Germany, 6 pkgs. H. Brooks & Co.
Holland, 12 cs. Ohlenschlager Brs.
„ 12 „
„ 1 pkg. Glaeser & Vogeler

Zinc Oxide—
Germany, 130 cks. M. Ashby
„ 13 Beresford & Co
„ 50 W. A. Rose & Co.

LIVERPOOL.

Week ending Dec. 18th.

Acetate of Lime—
St. Nazaire, 7 cks.
New York, 1829 bgs

Bones—
Boston, 165 cks. J. Gordon & Co.
Parnahiba, 4 t. R. Singlehurst & Co.

Barytes—
Antwerp, 27 cks.

Cream Tartar—
Bordeaux, 9 cks.
Rotterdam, 7

Copper Precipitate—
Huelva, 159 t Beauford & Co.

Copper Ore—
Hamburg, 297 bgs.
Lisbon, 19 t. Harrington & Co.
New York, 175 brls. Am. Metal Co.

Charcoal—
Hamburg, 1 pkg.

Chemicals (*otherwise undescribed*)
Rotterdam, 2 cks.

Caoutchouc—
Philadelphia, 50½ brls.
Gaboon, 128 cks. Hatton & Cookson
Eloby, 62 „

Dyestuffs—
Saffron
Valencia, 2 cs. Okell & Co.
Valonia
Patras, 677 bgs. H. Theakstone & Co.
„ 85 Barff & Co.
Constantinople, 106 bgs. A. Paparoglu
Smyrna, 34 bgs.
Argols
Oporto, 22 cks. F. Leyland & Co
Sumac
Bordeaux, 50 bls.
Annatto
Bordeaux, 2 hds. 3 cks.

Glycerine—
St. Nazaire, 4 cks.
Rotterdam, 2 dms.
Hamburg, 7 cbys. 3 cs.
Bordeaux, 50 cks.

Glucose—
Hamburg, 50 cks.
Philadelphia, 200 brls.
New York, 250

Litharge—
Rotterdam, 12 cks.

Manganese Ore—
Bordeaux, 203 t. G. G. Blackwell
Poti, 1920

Potash—
Rouen, 19 cks. R. J. Francis & Co.
„ 29 21 dms. Co-op. W'sale Society

Pyrites—
Huelva, 1357 t. Beauford & Co.

Phosphate—
St. Valery, 200 t.

Saltpetre—
Hamburg, 100 cks.
„ 5

Sodium Phosphate—
Rotterdam, 16 cks.

Salt—
Antwerp, 1 bg.

Stearine—
Philadelphia, 5 tcs. 29 hds.
New York, 25

Sulphur—
Catania, 400 t.

Tartaric Acid—
Rotterdam, 40 cks.
„ 7

Tartar—
Bordeaux, 54 cks.

Ultramarine—
Hamburg, 20 cs.
Rotterdam, 3 cks. 5
„ 2 8

Verdigris—
Bordeaux, 2 cks.

Waste Salt—
Hamburg, 326 bgs.

Zinc Oxide—
St. Nazaire, 25 brls.
Rotterdam, 100 cks. O. & H. Marcus
Antwerp, 40 brls. J. T. Fletcher & Co.
Antwerp 140

TYNE.

Week ending December 19th.

Alum Earth—
Hamburg, 18 cks Tyne S.S. Co.

Glycerine—
Hamburg, 1 drm.

Sulphur Ore—
Drontheim, 400 t. R. S. Story & Co.

Ultramarine—
Rotterdam, 20 cs. Tyne S.S. Co.

GLASGOW.

Week ending Dec. 19th.

Acetate of Lime—
New York, 354 bgs

Brimstone—
Girgenti, 600 t.

Dyestuffs—
Sumac
Palermo, 1,800 brs. 119 bales.
Alizarine
Rotterdam, 70 cks. J. Rankine & Son
„ 1 cs. 65 „

New Companies.

Bengal Indigo Manufacturing Company, Limited.—This company was registered on the 12th inst., with a capital of £150,000., in £10. shares, to carry on business as indigo planters, and manufacturers of, and dealers in indigo, and also as farmers, graziers, growers, manufacturers, and dealers in all kinds of tropical products. The subscribers are:—

	Shares.
W. E. Medland, Withington, near Manchester, merchant and manufacturer	1
M. A. Herrman, Hamburg, merchant	1
E. C. [illegible], [illegible]	1
J. Leigh, Knutsford	1
S. Bradley, 104, King-street, Manchester, chartered accountant	1
J. B. Lonsdale, 4, [illegible]-street, Manchester, merchant	1
J. W. Radcliffe, Werneth-park, Oldham, merchant, &c.	1
F. Karath, 58, Perham-road, West Kensington	1

David Martineau and Sons, Limited.—This company was registered on the 12th inst., with a capital of £250,000, in £10. shares, to acquire as a going concern the business of David Martineau and Sons, of Christian-street, sugar refiners. The subscribers are:—

	Shares.
David Martineau, 21, Mincing-lane, sugar refiner	1
G. Martineau, 21, Mincing-lane, sugar refiner	1
W. S. Hodgson, Esher, Surrey	1
C. H. Gill, Staines, sugar refiner	1
E. S. Hobson, 21, Mincing-lane, clerk	1
R. W. Brown, 41, Kirkstall-road, Streatham-hill, clerk	1
C. F. Worters, 105, Sunderland-road, Forest-hill	1

Witty and Wyatt, Limited.—This company was registered on the 12th inst., with a capital of £30,000, in £1. shares, to manufacture and deal in asbestos and asbestos goods, and to enter into an agreement with Messrs. Witty and Wyatt. The subscribers are:—

	Shares
H. H. Witty, 88, Leadenhall-street, merchant	7,499
R. C. Wyatt, 88, Leadenhall-street, merchant	7,499
Torre Diaz, 41, Moorgate-street, merchant	7,500
J. M. de Zulueta, 41, Moorgate-street, merchant	7,499
P. Pate, 88, Leadenhall-street, secretary to a company	1
F. H. Lovegrove, 88, Leadenhall-street, clerk	1
J. T. Hancock, 13, Godwin-road, Forest-gate, secretary to a company	1

English Portland Cement Company, Limited.—This company was registered on the 18th inst., with a capital of £25,000., in £1. shares, to manufacture cement and similar substances, and to adopt an agreement with Messrs. Irwin, Hopper, and Co. The subscribers are:—

	Shares.
R. Morris, J.P., Doncaster	500
Mrs. G. Morris, Doncaster	100
Miss H. E. Morris, Doncaster	10
Miss M. Morris, Doncaster	10
Mrs. M. A. Morris, Doncaster	100
W. R. Morris, Doncaster, merchant	250
Miss C. Woodhouse, Doncaster	10

Kuhn Patent Pasteurisation Syndicate, Limited.—This syndicate was registered on the 16th inst., with a capital of £20,000., in £1 shares, to acquire the patent rights of Wm. Kuhn, relating to the Pasteurisation of beer, wines, and other liquids. The subscribers are:—

	Shares.
E. W. Owles, 14, Ordnance-road, St. John's Wood, clerk	1
A. E. Friskmore, 2, Broad-street-buildings, architect	1
T. J. Sturgeon, Broad-street-buildings, commission agent	1
O. Vernede, 10, New Broad-street, solicitor	1
J. Wakefield, 4, Keppel-street, Russell-square, commission agent	1
J. M. Sharp, 1, Finsbury-square, solicitor	1
R. J. Jeffes, 4, Westbourne-terrace, Lower Clapton, clerk	1

Heathley, Janisch, and Company, Limited.—This company was registered on the 14th inst., with a capital of £3,000., in £1. shares, to trade as chemists, druggists, and dealers in proprietary articles, and in all kinds of electrical and scientific apparatus. The subscribers are:—

	Shares.
J. Heathley, Blyth, grocer	
G. A. Heathley, Newcastle-on-Tyne, chemist's assistant	1
Ralph Carr, Newcastle-on-Tyne, insurance agent	1
W. Turnbull, Newcastle-on-Tyne, draper	1
Mrs. Turnbull, Newcastle-on-Tyne	1
Mrs. E. Janisch, Newcastle-on-Tyne	1
E. Towers, Tynemouth, accountant	1

Audenshaw Paint and Colour Company, Limited.—This company was registered on the 12th inst., with a capital of £11,000., in £1. shares, to acquire the business of the Audenshaw Paint and Colour Co., at Palatine Works, [illegible], Guide-bridge, near Manchester, and at 11, Great Ducie-street, [illegible], Manchester. The subscribers are:—

	Shares.
Wm. Hallewell, Sowerby-bridge, York	1
J. R. Meadows, Guide-bridge, Manchester, colour-maker	1
B. Huggins, Guide-bridge, Manchester, bookkeeper	1
Wm. Fenny, 10, Adela-street, Belfast, commercial traveller	1
J. Hudson, Upper-mill, Oldham, plumber and painter	1
L. Pugh, 67, Broad-street, Manchester, colourman	1
A. Howorth, 176, Milnrow-road, Rochdale, clerk	1

Gazette Notices.

PARTNERSHIPS DISSOLVED.

D. S. Haynes and E. P. Finemore, Whitfield-street, Finsbury, London, manufacturing chemists and commission agents.

J. and E. Thomas and Green, Wooburn, paper makers, as far as regards E. Thomas.

L. Ashworth, J. Ashworth, and J. Walsh, Cheadle and Manchester, under the style of Demming's Printing Co., calico printers.

Adjudications.

Walter Cunningham, Grange Park-road, Leyton and Day, W. Rauben, Vicarage-road, Leyton (trading as Cunningham and Co.), Leyton, timber, lime, and cement merchants.

Notices of Dividends.

John Brown, of Rochdale, mineral water manufacturer, first and final dividend of 3s. 10¼d., December 30. Official Receiver's Office, Oldham.

The Patent List.

This list is compiled from Official sources in the Manchester Technical Laboratory, under the immediate supervision of George E. Davis and Alfred R. Davis.

APPLICATIONS FOR LETTERS PATENT.

Measurement of Liquids. J. Pocock and S. Johnson. 20,198. Dec. 16.

Manufacture of Alloys of Aluminium. S. Pearson and J. H. Pratt. 20,21[illegible]. December 16.

Manufacture of Metallic Electrodes.—(Complete Specification). F. Mar[illegible]. 20,217. December 16.

Manufacture of Azoxy Aniline and of Colouring Matters derived therefrom. J. Imray. 20,219. December 16.

Disinfectants. A. Artmann and H. W. Kufeke. 20,231. December 16.

Electricity Meters. A. Reckenzaun. 20,318. December 17.

Field-Magnets of Dynamo-Electric Machines. S. Kapp. 20,326. Dec. 18.

Manufacture of Hypophosphites. J. A. Kendall. 20,392. December 18.

Method and Apparatus for Extracting the Tannin from Bark used for Tanning Purposes. J. Hutchings and W. N. Hutchings. 20,403. Dec. 18.

Production of Ozone. A. Muirhead. 20,416. December 19.

Secondary Batteries. G. Philippart. 20,516. December 20.

Charging Gas Retorts. J. West. 20,551. December 21.

Electric Meters. J. Einstein and S. Kornprobst. 20,581. December 21.

IMPORTS OF CHEMICAL PRODUCTS

AT

THE PRINCIPAL PORTS OF THE UNITED KINGDOM.

Calcium Chloride—
Auckland 2 t. 10 c. £5
Carbolic Acid—
New York 9 t. 3 c. 2 q. £886
Barcelona 2 3 19
Amsterdam 10 52
Citric Acid—
New Orleans 2 c. £19
Dunging Salts—
New York 5 t. £49
Dried Blood—
Valencia 6 t. £56
Epsom Salts—
Rio Janeiro 100 kegs.
Glycerine—
New York 3 t. 21 cks. £206
Glauber Salt—
Bombay 1 t. £5
Iron Oxide—
Genoa 2 t. £10
Magnesia—
Hamburg 10 cs.
Rio Janeiro 10
Mexico 5
Santander 3
Manure—
Lisbon 4 t. £10
Havana 1 40
Havre 98 1 c. 2 q. 294
Oxalic Acid—
Genoa 1 t. £46
Pitch—
Naples 210 t.
New York 35 cks.
Potassium Chlorate—
Venice 6 c. £16
Rotterdam 5 15
New York 22 t. 10 1132
Genoa 1 44
Naples 1 42
Gothenburg 29 1 1368
Hamburg 1 48
Rotterdam 5 10 263
Copenhagen 5 6 242
Havana 1 15 83
Shanghai 1 10 77
Potassium Bichromate—
Boston 4 c. £14
Ancona 11 1 q. 22
Havre 4 t. 15 140
Odessa 9 14
Potassium Chromate—
Naples 10 c. 2 q. £15
Leghorn 2 t. 15 112
Potassium Carbonate—
Chicago 2 t. 10 c. £45
Potash—
Philadelphia 12 c. 1 q. £15
Phosphate Rock—
Hamburg 310 t. 1 c. 2 q. £1496
Potassium Prussiate—
Barcelona 3 t. 10 c. £227
Salt—
Axim 25 t.
Amsterdam 150
Antwerp 255
Barbadoes 106
Bonny 200
Calcutta 1042
Chicago 130
Chama 10
Copenhagen 105
Christiana 500
Grand Bassa 10
Halifax 400
Havana 20 cs.
Kingston 40
Kobe 1
Livingstone 200
Loango 27
M. Video 12
New York 291
Opofo 750
Port Natal 40
Porsgrund 20
Portland M. 8[illegible]0
Ponce 50 cs.
Rosario 12
Rotterdam 98
Rio Grande do Sul 70 cs.
Rouen 10
Salt Pond 7
San Francisco 72
Sierra Leone 200
Wellington 463
Soda Ash—
Barcelona 32 cks. 12 brls.
Baltimore 764 tcs. 355
Bahia 10
Bari 4
Boston, 200 bags. 296
Bilbao 35
B Ayres 100
Boston, 266 tcs. 46 cks. 604 bgs.
B. Ayres 250 brls.
Buffalo 60
Calcutta 7
Ceara 50
Ghent 14 cks.
Genoa 5
Leghoan 1
Malaga 10
Marhrham 30
Marseilles 20
New Orleans 6
New York 155
Naples 35
New York 460 tcs.
Oporto 6
Philadelphia 899 tcs. 220 cks.
Odessa 100
Salonica 28
San Francisco 67
Syra 220
Rio Janeiro 135
Santos 15
Sherbrooke 17 tcs.
Vola 15
Wellington 20
Sodium Silicate—
Barcelona 16 cks.
Constantinople 20
Malaga 6 brls.
Seville 6
Smyrna 20
Valencia 18
Venice 32
Sodium Carbonate—
Poston 196 brls.
San Luis Potosi 20 cks.
Gothenburg 28
Leghorn 1
Soda Crystals—
Calcutta 85 brls.
Constantinople 50
Gibraltar 10
Halifax 103
London, Ont. 25
New Orleans 100
New York 38 cks.
San Sebastian 30 bgs.
Smyrna 90
Philadelphia 125
Singapore 8
Sodium Bicarbonate—
Antwerp 20 pkgs.
B. Ayres 80 cks.
Melbourne 320 kgs. 54
San Juan P.R. 20
Seville 20
Sydney 200 42
Sulphur—
Savanilla 13 c. 3 q. £9
Crenfuegos 17 3 7
Calcutta 10 t. 10 61
Vera Cruz 9 16 1 62
Iquique 7 10 50
Boston 40 176
New York 25 10 115
Portland M. 30 9 145
Mexico 4 18 2 31
Sodium Nitrate—
Cartagena 5 2 £45
Sodium Biborate
Antwerp 7 t. 14 c. 3 q. £239
Bordeaux 3 2 2 93
Portland M. 1 13 2 35
Rotterdam 4 13 125
Salammoniac—
Algiers 3 c. £5
Alexandria 16 28
Naples 18 2 q. 30
Smyrna 2 t. 11 2 87
Palermo 9 15
Trieste 1 34
Genoa 9 1 16
Constantinople 14 23
Bombay 294 499
Syra 3 5
Superphosphate—
New Orleans 300 t. 9 c. 1 q. £601
Stettin 25 1 1 78
Sodium Sulphate—
Portland 5 t. 2 c. £10
Saltpetre—
Valparaiso 11 t. 15 c. 1 q. £251
Syra 3 9 1 74
Monaos 5 3 7
Ceara 1 4 2 30
Sodium Bichromate—
Yokohama 3 11 2 £107
Saltcake—
Baltimore 199 t. 19 c. £353
Philadelphia 25 10 47
Sheep Dip—
E. London 1 t. 1 q. £43
Tar—
Ambriz 10 brls.
Santander 6 cks.
Tartar—
New York 30 t. £2850
Zinc Muriate—
Bombay 11 cks. 7 t. 3 c. 3 q. £159

TYNE.

Week ending Dec. 19th.

Alkali—
Copenhagen, 11 t. 17 c.
Rotterdam, 10
Christiania, 15
Anthracene—
Rotterdam, 58 brls.
Ammonia Carbonate—
Odense, 1 t. 5 c.
Ammonia Sulphate—
Odense, 2 t.
Antwerp, 30
Antimony—
Hamburg, 2 t. 10 c.
New York, 15
Bleaching Powder—
Copenhagen, 106 t. 8 c.
Libau, 33 5
Rotterdam, 15 4
" 5 13
Horsens, 1 3
Antwerp, 9 6
" 40 7
Hamburg, 5 13
Gothenburg, 5
Riga, 60 12
New York, 120 18
Stockholm, 1 11
Copenhagen, 9 5
Bone Ash—
Antofagasta, 2 c.
Ergastiria, 17
Barytes Carbonate—
Hamburg, 8 t.
Caustic Soda—
Hamburg, 16 c.
Christiania, 6 6
Philadelphia, 29
New York, 94 6
Stockholm, 4 7
Copenhagen, 2 16
Flint Ground—
Antwerp, 50 t.
Gypsum—
New York, 35 t.
Litharge—
Antwerp, 1 t. 13 c.
Hamburg, 1 12
Stockholm, 1
Magnesia—
Antwerp, 2 c.
Hamburg, 7
Pitch—
Copenhagen, 40 t.
Antwerp, 103 5 c.
Plumbago Crucibles—
Ergastiria 4 c.
Plumbago—
Ergastiria, 1 c.
Pearl Ash—
Ergastiria, 1 c.
Potassium Chlorate—
Hamburg, 1 t.
Philadelphia, 2
Paraffin Wax—
Stockholm, 5 t. 5 c.
Soda—
Copenhagen, 21 t. 8 c.
Rotterdam, 17 5
" 21 11
Aalborg, 9 1
" 25 15
Odense, 10 13
Nykjobing 6 14
Svendborg, 14 8
Christiania, 10
Philadelphia, 257 16
Stockholm, 43 14
" 16 3
Copenhagen, 25
Soda Ash—
Antwerp, 16 t.
Riga, 15 6 c.
New York, 100 4
Copenhagen, 1 19
Libau, 6 t.
Soda Salts—
Philadelphia, 5 t.

Alka
Lib
Mar
Ode
Benz
Dun
Bone
Han
Caus
Ant
Cop
Han
Liba
Stet
Chem
Antv
Gen
Ams
Han
Bren
Ham
Berg
Liba
Napl
Chri
Cope
Dron
Dunk
Goth
New
Odes
Roue
Rotte
Stetti
Stava
Dyest
Antw
Chris
Hamb
Reval
Rotte
Manu
Hamb
Napht
Dunki
Pitch-
Antwe
Naple
Slag—
Amste
Turpe
Dront
Tar—
Genoa
Hamb

Wee

Chemi
Diepp
Hamb
Dyest
Diepp

W

Benzo
Rotte
Terne
Bleac
Rotte
Chemi
Boulo
Haml
Rotte
Coal
Cala
"
Rotte
Dyest
Ghen
"
Haml
Rotte
Manu
Boulc
Ghen
Ham
Pitch-
Antw

Bilbao 35
B. Ayres 100
Boston, 266 tcs. 46 cks. 604 bgs.
B. Ayres 250 brls.
Buffalo 60
Calcutta 7
Ceara 50
Ghent 14 cks.
Genoa 5
Leghoan 1
Malaga 10
Marhrham 30
Marseilles 20
New Orleans 6
New York 155
Naples 35
New York 460 tcs.
Oporto 6
Philadelphia 899 tcs. 220 cks.
Odessa 100
Salonica 28
San Francisco 67
Syra 220
Rio Janeiro 135
Santos 15
Sherbrooke 17 tcs.
Vola 15
Wellington 20

Sodium Silicate—
Barcelona 16 cks.
Constantinople 20
Malaga 6 brls.
Seville 6
Smyrna 20
Valencia 18
Venice 32

Sodium Carbonate—
Boston 196 brls.
San Luis Potosi 20 cks.
Gothenburg 28
Leghorn 1

Soda Crystals—
Calcutta 85 brls.
Constantinople 50
Gibraltar 10
Halifax 103
London, Ont. 25
New Orleans 100
New York 38 cks.
San Sebastian 30 bgs.
Smyrna 90
Philadelphia 125
Singapore 8

Sodium Bicarbonate—
Antwerp 20 pkgs.
B. Ayres 80 cks.
Melbourne 320 kgs. 54
San Juan P.R. 20
Seville 20
Sydney 200 42

Sulphur—
Savanilla 13 c 3 q. £69
Cienfuegos 17 3 7
Calcutta 10 t. 10 61
Vera Cruz 9 16 1 62
Iquique 7 10 50
Boston 40 176
New York 25 10 115
Portland M. 30 9 145
Mexico 4 18 2 31

Sodium Nitrate—
Cartagena 5 2 £43

Sodium Biborate
Antwerp 7 t. 14 c. 3 q. £239
Bordeaux 3 2 2 93
Portland M. 1 13 2 35
Rotterdam 4 13 125

Salammoniac—
Algiers 3 c. £5
Alexandria 16 28
Naples 18 2 q. 30
Smyrna 2 t. 11 2 87
Palermo 9 15
Trieste 1 34
Genoa 9 1 16
Constantinople 14 23
Bombay 294 499
Syra 3 5

Superphosphate—
New Orleans 300 t. 9 c. 1 q. £601
Stettin 25 1 1 78

Sodium Sulphate—
Portland 5 t. 2 c. £10

Saltpetre—
Valparaiso 11 t. 15 c. 1 q. £251
Syra 3 9 1 74
Monaco 5 3 7
Ceara 1 4 0 30

Sodium Bichromate—
Yokohama 3 11 2 £107

Saltcake—
Baltimore 199 t. 19 c. £353
Philadelphia 25 10 47

Sheep Dip—
E. London 1 t. 1 q. £43

Tar—
Ambriz 10 brls.
Santander 6 cks.

Tartar—
New York 30 t. £2850

Zinc Muriate—
Bombay 11 cks. 7 t. 3 c. 3 q. £159

TYNE.

Week ending Dec. 19th.

Alkali—
Copenhagen, 11 t. 17 c.
Rotterdam, 10
Christiania, 15

Anthracene—
Rotterdam, 58 brls.

Ammonia Carbonate—
Odense, 1 t. 5 c.

Ammonia Sulphate—
Odense, 2 t.
Antwerp, 30

Antimony—
Hamburg, 2 t. 10 c.
New York, 15

Bleaching Powder—
Copenhagen, 106 t. 8 c.
Libau, 33 5
Rotterdam, 15 4
" 5 13
Horsens, 1 3
Antwerp, 9 6
" 40 7
Hamburg, 5 13
Gothenburg, 5
Riga, 60 12
New York, 120 18
Stockholm, 1 11
Copenhagen, 9 5

Bone Ash—
Antofagasta, 2 c.
Ergastiria, 17

Barytes Carbonate—
Hamburg, 8 t.

Caustic Soda—
Hamburg, 16 c.
Christiania, 6 6
Philadelphia, 29
New York, 94 6
Stockholm, 4 7
Copenhagen, 2 16

Flint Ground—
Antwerp, 50 t.

Gypsum—
New York, 35 t.

Litharge—
Antwerp, 1 t. 13 c.
Hamburg, 1 12
Stockholm, 1

Magnesia—
Antwerp, 2 c.
Hamburg, 7

Pitch—
Copenhagen, 40 t.
Antwerp, 103 5 c.

Plumbago Crucibles—
Ergastiria 4 c.

Plumbago—
Ergastiria, 1 c.

Pearl Ash—
Ergastiria, 1 c.

Potassium Chlorate—
Hamburg, 1 t.
Philadelphia, 2

Paraffin Wax—
Stockholm, 5 t. 5 c.

Soda—
Copenhagen, 21 t. 8 c.
Rotterdam, 17 5
" 21 11
Aalborg, 9 1
" 25 15
Odense, 10 13
Nykjobing 6 14
Svendborg, 14 8
Christiania, 10
Philadelphia, 257 16
Stockholm 43 14
" 16 3
Copenhagen, 25

Soda Ash—
Antwerp, 15 t.
Riga, 15 6 c.
New York, 100 4
Copenhagen, 1 19
Libau, 6 t.

Soda Salts—
Philadelphia, 5 t.

HULL.

Week ending Dec. 17th.

Alkali—
Libau, 205 cks.
Marseilles, 350 drms.
Odessa, 371

Benzol—
Dunkirk, 23 cks.

Bone Size—
Hamburg, 5 cks.

Caustic Soda—
Antwerp, 19 drms.
Copenhagen, 40
Hango, 11 cks.
Libau, 67
Stettin, 10

Chemicals (*otherwise undescribed*)
Antwerp, 4 cs.
Genoa, 49 5 cks. 60 pkgs.
Amsterdam, 6 10 1
Hango, 3 4
Bremen, 2
Hamburg, 36 15 10
Bergen, 21 10
Libau, 20
Naples, 2
Christiania, 7 pkgs. 203 slabs
Copenhagen, 2 115
Drontheim, 82
Dunkirk, 1 cks
Gothenburg, 13 40 56
New York, 75 cs.
Odessa, 194 cks 96 pkgs.
Rouen, 40
Rotterdam, 27 cks. 40 cs.
Stettin, 4 1 pkg
Stavanger 20

Dyestuffs (*otherwise undescribed*)
Antwerp, 1 ck.
Christiana, 2 cks.
Hamburg, 13 1 keg.
Reval, 68
Rotterdam, 1 cs. 6 cks.

Manure—
Hamburg, 2,856 bgs.

Naphtha—
Dunkirk, 29 cks.

Pitch—
Antwerp, 539 t.
Naples, 255

Slag—
Amsterdam, 75

Turpentine—
Drontheim, 3 c.

Tar—
Genoa, 210 cks.
Hamburg, 8 cs.

GRIMSBY.

Week ending December 14th.

Chemicals (*otherwise undescribed*)
Dieppe, 40 pkgs.
Hamburg, 15

Dyestuffs (*otherwise undescribed*).
Dieppe, 2 cks.

GOOLE.

Week ending Dec. 11th.

Benzol—
Rotterdam, 24 cks.
Terneuzen, 56 drms.

Bleaching Preparations—
Rotterdam, 76 c s.

Chemicals (*otherwise undescribed*)
Boulogne, 2 c s
Hamburg, 62
Rotterdam, 1 cs.
" 160

Coal Products—
Calais, 3 c s.
" 10 puns.
Rotterdam, 92 cks.

Dyestuffs (*otherwise undescribed*)
Ghent, 2 cases.
" 7 c s.
" 1 drm.
Hamburg, 105 c s.
Rotterdam, 2

Manure—
Boulogne, 42 bgs.
Ghent, 216
Hamburg, 304

Pitch—
Antwerp, 232 t.

beverage for 19. Phosphorus is credited with 11 deaths, and hydrochloric acid or spirits of salt 14. Strychnia or nux vomica killed 17. Those poisons in least request seem, however, to have been the most deadly—arsenic, mercury, and corrosive sublimate. If we may use the arguments recently urged with regard to the deaths from water gas, we should agitate for the total restriction of sale of all poisonous substances, but here we are confronted with a difficulty, as we find acetic acid or vinegar is credited with one death, alcohol with six, chlorodyne with 10, various fungi with 21, fish with four, improper food with one, decomposed food with four, so that it is quite clear our food stuffs, or those substances known as food stuffs, are as liable to cause death as the veritable poisons. Perhaps those who are so anxious to place all poisons under the especial care of chemists and druggists, will be good enough to tell us what poisons really are, and whether it is not true that nearly all the poisons which people have purchased for homicidal purposes are not, in by far the majority of cases, purchased from chemists and druggists.

THE JARROW CHEMICAL COMPANY.

An announcement, says the *Shields Daily Gazette* will be found in our news columns to-day which, while it can occasion no great astonishment to those who are acquainted with the chemical trade—for something of this character must have been anticipated for some time past—must be painful to a large number of the inhabitants of South Shields, and will be received with a feeling of regret throughout the whole of Tyneside. The Jarrow Chemical Company has to-day announced to its workmen that it finds itself compelled to close its extensive works at Tyne Dock. The decision, we are happy to say, is not to take immediate effect. There is to be an interval during which the whole of the men engaged at the works will have an opportunity of obtaining new employment. There will be no sudden dispersion. This is not one of those cases in which a large number of men are thrown out of work at once, on a labour market already over crowded. With one exception, all the industries of Tyneside are, happily in a most prosperous condition, and men who find it necessary to seek new employment are not likely to be for any long time out of work. In this instance, and under the circumstances in which the works of the Jarrow Chemical Company are to be closed, the whole of the workmen will be exceptionally well situated for making a new provision for themselves before their present engagement expires. It has happened frequently in the history of our manufactures that by new inventions, or by the discovery of new processes, or by unexpected diversions in the channels of trade, large classes of the population have been driven in some sudden manner to seek new employments. In the present case there are elements neither of suddenness nor of unexpectedness. Those conditions which have been so much to the advantage of trade generally have operated only to the disadvantage of the chemical industry, which has been carried on, if not always at a loss, certainly at a profit which was far from yielding a fair percentage on the capital employed. Quite recently the successful prosecution of chemical manufacture has been hampered by a new condition. It is no longer possible to carry on the other branches of the trade without adding a department for the making of bleaching powder. At the Friar's Goose works of the Jarrow Chemical Company, bleaching powder has been one of the branches of production. Bleaching powder has not been produced at the works at Tyne Dock. The directors were therefore confronted with a dilemma. They must either, at very considerable cost, and with no certainty of remuneration, make an expensive addition to their works, or take, however reluctantly, the step which is announced to-day. We have good grounds for stating that this step has been taken with the very utmost unwillingness; not until some time after it had seemed to become unavoidable, and with a most generous consideration for the men whose services would have to be dispensed with. The works will not be finally closed for at least some three or four months to come. This is the hopeful feature of the situation. The directors kept the works on throughout the whole of the late period of depression, and they have chosen such a time for bringing them to an end as would be most to the advantage of the workmen in their employment. All classes of the population will witness with sorrow the disappearance of an old Shields industry. It is sixty-six years since the chemical works at Tyne Dock were commenced. For twenty years they were in the hands of Messrs. Cookson, and for the remaining forty-six years they have been in the hands of the Jarrow Chemical Co. The relations of employers and workmen have throughout that long period been of a uniformly agreeable character, and there are men in the works whose fathers and grandfathers were there before them. We feel certain that the directors will not less regret the necessary breaking of these relationships than will the men whom they have so long kept in their employment. To one aspect of the matter, as affecting the welfare of the borough of South Shields generally, a merely passing reference may be made. The land occupied by the Jarrow Chemical Co. at Tyne Dock is among the most valuable land on Tyneside, and there is no doubt that its position and its other advantages will soon cause it to be occupied by other and more profitable industries, to the benefit of the town and of the surrounding district.

A STUDY OF COAL GAS,

By M. Em. Sainte-Claire Deville.

(Engineer at the experimental works of the Paris Company.)

(Concluded from page 19.)

DIRECT EXPERIMENTS ON THE INFLUENCE OF HEAT ON THE PRODUCTION OF BENZENE.

I must, however, in this connection, point out that numerous experiments made for the purpose of ascertaining the effect of a high temperature on the production of benzene, have not led to quite such simple conclusions.

I have found that if gas taken at the exit of the meter at the commencement of the charge, *i.e.*, rich, but not tarry gas, be superheated, it requires a very intense heat applied to a slow stream of gas to produce any perceptible difference between the heated and unheated gas. The difference thus produced consists in the loss of about 12 per cent. of its benzene, and of 35 per cent. of its other hydrocarbons, absorbable by bromine, accompanied by the deposition of carbon in the porcelain tube, and production of naphthalene and anthracene.

If, instead of superheating the gas after purification from tar, it be heated together with the volatile products formed during the distillation, the results are the same as regards the benzene and rich hydrocarbons, so that all that can be concluded is that benzene resists a high temperature better than other hydrocarbons. This question as to the relative amounts of benzene formed during the successive periods of the distillation, which is of great theoretical interest, requires further investigation.

However this may be, we know sufficient to be able to affirm that, among all the hydrocarbons which contribute to the illuminating power, benzene is the only one which exists in considerable amount throughout the whole of the distillation.

Experiments carried out on the laboratory scale show that, speaking generally, a high temperature favours its production.

It remains to be learned whether the proportion of benzene depends on the temperature of the coal at the precise moment of decomposition, or on the temperature to which the volatile products are submitted after their formation, during their passage along the red hot walls of the retorts.

RECAPITULATION AND CONCLUSIONS.

The two chief subjects treated in this article are:—

(*a*) The classification of gas coals according to their elementary composition.

(*b*) A study of the amount of benzene contained in the gas, and on the part played by this hydrocarbon in affecting the illuminating power of the gas.

I. At the experimental works of the Paris Company, during the last twelve years, 1,012 coal tests have been carried out, 36,000 kilos being employed for each test.

Of these 1,012 tests, 898 have been made with coal from 23 sources, that from each source having been tested on the average 39 times.

The average results of this large number of experiments are therefore free from variations due to accidental deviations from the normal character of the coal, and consequently clearly show the influence of the composition of the coal upon the various properties of the gas, which are of interest to the gas manufacturer.

The basis selected for this classification is the amount of oxygen contained in the coal, which is considered to represent the combined water, and thus to be the best indication of the geological age of the coal and of its degree of carbonization.

By dividing the coals tested into five classes of types, containing increasing amounts of oxygen, from 5·5—12 per cent., and by calculating the average results for each type, the following conclusions have been arrived at:—

As the coal becomes richer in oxygen, the gas produced by its distillation becomes gradually richer in heavy constituents—carbonic acid, carbonic oxide, marsh gas, and illuminating hydrocarbons, and poorer in pure hydrogen.

Its density and illuminating power both increase. It is also to be observed that the proportion of volatile matter, determined by calcination in a crucible, also increases, even when the amount of combined water is deducted.

ing in the retorts becomes larger as the
. The products of condensation (tar
with the amount of oxygen.
coal, it is found that type III. com-
namely which give a rich gas and yet
of good quality. Types I. or II. give

and V.—give a rich gas, but their coke
quality.
in gas may be estimated by simply
at —22°, and adding to the weight
moist at 15°) a constant equal to 23·5

d with a known weight of pure benzene,
70°, and that the amount which con-
0°, is exactly equal to 23·5 grms. per
in.); this number agrees with that cal-
s for the vapour pressure of benzene.
ame manner, deposits, whatever be its
d between the same temperatures, from
at 0° r760 min., which corresponds to
re, moist at 15°.
posited consists of almost exactly 23·5
ith a little toluene, xylene, etc.
termination under very varied circum-
llowing conclusions:—
ity of Paris contains 39–40 grms. of
sed of:—

....	77
of the aromatic series	23
	100

, and from 1–1·1% of the volume of
benzene vapour.
he gas contains 4–4·5% by volume of
o°.
gas is found to be:—

....	65
drocarbons	35
	100

gas may be classed in two symmetrical

h, gases.
mated coals.
on, at a low temperature.
ist stages of the distillation.

oor, gases.
oor in oxygen.
on at high temperatures.
ast stages of the distillation.
ntained in a cubic metre of gas is
ties.
on gases of the second group is a little
consequently, a little poorer in toluene
n those of the first group. The differ-
d.
oup at the moment of leaving the retort
an those of the second group, or, to
emperature of the coal at the exact
gher for gases of the second group than

f temperature appears to have the effect
hydrocarbons of the olefine and acety-
he aromatic series, with the exception
at least, of not destroying benzene; of
nd, finally, of forming naphthalene.

, be considered as an industrial confir-
n the subject.
power, benzene plays a secondary part
part in gases of average richness and is
st given by very poor gases. If the whole
a gas be removed by refrigeration to
ing power thus occasioned divided by the
ost constant for all gases), it is found
as the candle power of one gramme of
gas, is much higher for poor than for

se 0·040 candles and 0·250 candles.
f devoid of illuminating power, be pro-
ocarbon, that the illuminating power of

the mixture increases much less rapidly than its richness in illuminating material.

It is necessary to remark that these facts have only been proved by applying to all gases, rich or poor, the photometric method founded on the use of the Bengel burner.

It would be interesting to study the variation of illuminating power when each gas was burned at a burner suited to its composition, as the Bengel burner is adapted to the average Paris gas.

MANURES FOR POTATOES.

(Experiments by Dr. J. H. Gilbert.)

In a former article we shewed that twelve years' experiments, carried out on the same land by Dr. J. H. Gilbert, the renowned English agriculturist, prove that, contrary to the view usually taken, nitrate of soda and sulphate of ammonia produce the best results with potatoes, provided that they are employed in conjunction with superphosphate and salts of potassium.

According to these data, the best fertiliser for potatoes seems to be composed of:—

Sulphate of ammonia	120	kilos.
Nitrate of soda	200	kilos.
Superphosphate	600-800	kilos.
Potassium	100-200	kilos.

In clayey soils a certain proportion of the superphosphate may be replaced by phosphate. In a second communication, Dr. Gilbert, summarising experiments also carried out during 12 years, shows that ordinary manure employed by itself never gives results equal to those obtained by the use of mineral fertilisers or of manure, to which mineral ingredients have been added.

As a mean of the 12 years, manure alone produced only 4,200 kilos. per acre, while 5,150 kilos. were got with manure mixed with mineral fertilisers, and 6,742 kilos. by the aid of nitrate of soda and sulphate of ammonium, mixed with phosphate and potassium fertilisers.

These facts are due to the rapid absorption of mineral fertilisers, while ordinary farm manure is not sufficiently rapidly decomposed to produce its effect on the crop of potatoes. In discussing the results of his experiments, Dr. Gilbert shows that the potato is unable to avail itself of the nitrogen supplied to it unless there be present in the soil a sufficient quantity of mineral matter, and hence proceeds the necessity of employing large amounts of phosphoric acid. Phosphoric acid also facilitates the absorption of potassium by the plant. One of the effects of superphosphate applied to spring crops is to increase, to a large extent, the development of superficial nourishing roots. It also appears that under the influence of superphosphate, which is exerted both on the plant and the soil, the potato becomes better able to absorb a large amount of potassium.

When the potato has mineral matter at its disposal, it is able to do with less nitrogen than any other vegetable. The main factor in its cultivation seems, therefore, to be phosphoric acid. The farmers of Merville, a district in the neighbourhood of the Nord, where the potato is the chief article of cultivation, have always obtained good results by the use of Peru guano, on account of its richness in phosphoric acid and nitrogen. The fertiliser now employed with good results in this district contains 9% of nitrogen, 6% of phosphoric acid, and 5% of potassium. The proportion of phosphoric acid here is hardly sufficient, but it must be remembered that the lands are now rich in phosphoric acid, guano having been constantly employed for 25 years.

Many countries, however, maintain the practice of using large quantities of ordinary manure for potatoes, and even of spreading this manure among the tubercles themselves. It seems probable that, apart from its richness in constituents of every kind, the beneficial effect produced by manure is chiefly due to the alteration in the physical character of the soil introduced by its use, for it renders the earth porous, and thus enables it to be penetrated by the superficial roots on whose development the success of the crop so largely depends. We ought, perhaps, also to attribute some part of its good effects to the rise of temperature produced by the decomposition of the organic substances contained in it, and to the carbonic acid which it evolves, and which, thanks to the moisture present, renders soluble the mineral constituents of the soil.

Tendencies to disease are naturally more pronounced with very rich, nitrogenous manures, the potatoes obtained being less dense, and more watery, than those grown in land which has only received moderate manuring. The character of the season always plays a part of the utmost importance with respect to the quality of the crop, and the diseases to which it is liable. Dry years, as everybody knows, are the most favourable, damp years the most disastrous to the potato harvest.
—*L'Engrais.*

McDOUGALL'S PATENT MECHANICAL STOKERS.

McDOUGALL'S PATENT MECHANICAL STOKERS.

WE do not think it necessary to apologise to our readers for making a few remarks on boiler-firing in general, and a mechanical stoker in particular. The steam raiser is at last beginning to open his eyes to the fact that, if he wishes to keep abreast with the times, and live in peace and good-will with his neighbours, he must discard the old haphazard system of slovenliness, and avail himself of the resources that modern progress puts within his reach. To dwell upon the folly and wastefulness of employing men whose only qualification for their post, is an unfitness for anything else—to point out the loss in evaporative power obtained from the fuel employed—to lament over the clouds of black smoke reeking from the chimney of the mismanaged factory, to do all this would be to tell an already too oft told tale, and we turn with pleasure to the brighter side of the picture.

Very little study of the subject is necessary to show that the chief points to be attained is a complete combustion of the cheapest quality

Fig. 1.

of fuel. This is a comprehensive object to aim at, and includes all those minor features which enable the manufacturer of the mechanical stoker to make out such a good case for his apparatus. For, with complete combustion, black smoke is a thing unknown, and unburnt coal is not to be found among the ash and clinker.

The McDougall Patent Mechanical Stoker is made by the Chadderton Iron Works Co., Ltd., of Chadderton, near Manchester. The stoker proper is made in two forms, the variation being intended to meet difficulties arising from differences in the burning quality of the coal, as Mr. McDougall, after twenty years' experience in the matter, is strongly of opinion that no one form of mechanical stoker is adapted to deal with all the varied qualities of coal. This is only what could reasonably be expected, and failure to realise the fact is responsible for much disappointment and vexation.

The form most strongly recommended, where a soft and clinkering coal is used, is that styled by the makers "The Level Reciprocating Bar Arrangement" (fig. 1). Combined with their patent air blast it represents their latest and most improved system of stoking, promotes a most vigorous combustion and effectually prevents the formation of smoke. In this arrangement a ram, situated at the bottom of the hopper, distributes the fuel evenly and regularly over the front part of the grate, and the bars are kept in constant motion by a series of cams fixed on a shaft beneath the front of the grate, the construction being such that they are pushed in altogether, and are drawn out again in sections. By this means the fuel is slowly carried towards the rear end of the grate, and, as combustion is completed by the time the bridge is reached, only clinkers and ashes remain to be here delivered on to the bottom of the flue, whence a door, specially fitted beneath the bridge, allows them to be raked out at convenience. It will be at once seen that by this arrangement the clinker has very little chance of "settling" on the bars; and as the rich vapours distilled from the newly-charged fuel, mixed with air, have to pass over the incandescent mass at the back of the grate, the chances of their passing away unconsumed are reduced to a minimum. Where very heavy work is required from the boiler the advantage of the air-blast is very pronounced, and its low cost enables the makers to recommend its addition to the stoker in all cases. In this, as in the other form of the apparatus, the mechanism is of the simplest possible description, and no springs or tappets are employed.

The advantage of the mechanical stoker was recently shewn by a comparative test made at Millwall Docks by the engineer, Mr. F. E. Duckham. It reduced the coal bill by a trifle over 25 per cent., by allowing "small" coal at 11s. 3d. per ton to be used instead of "hard steam," at 15s. 3d., the work done per pound of coal being practically the same in both cases, but, if anything, slightly in favour of the cheaper quality.

It should be mentioned that if at any time it is desirable not to use the stoker, the disposition of parts is such that the fires can be charged by hand in the ordinary way.

The second form of the mechanical stoker is designed for cases where the load is fairly constant, and the fuel a hard one, which does not clinker. It is somewhat cheaper than the first type, but has already stood the test of several years, with the most satisfactory results. Instead of a ram, an inclined step introduces the fuel on to the coking plate, and the firebars, instead of being moved on the level and actuated by cams, are driven from an eccentric shaft, and, in addition to a reciprocating motion, have an alternate rising and falling motion. The effect is the same as in the No. 1 arrangement, the fuel being carried steadily rearwards, and the clinker, being prevented from attaching itself to the bars, is finally deposited with the ash at the bottom of the flue for removal, as above described.

In this case, also, hand-firing can be resorted to, if circumstances at any time render it necessary.

Recognising the fact that a mechanical stoker might, under certain circumstances, be undesirable, or its use impracticable, Mr. McDougall has also given his attention to hand-firing, and has brought out a system of continuous moving bars, shown in figs. 2 and 3, which will go far towards compensating for the faults and failings of the fireman. These automatic moving bars obtain their motion in a very simple and effective manner from a series of eccentrics working on a shaft placed

FIG. 2.

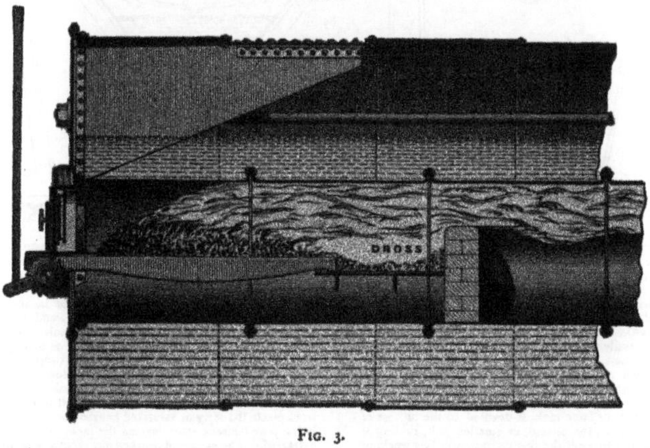

FIG. 3.

beneath the front of the fire-grate, and move in suitable jaws provided for their reception in the ends of the fire-bars. By these means the bars receive a continuous and even motion, and the arrangement of the eccentrics being only alternately uniform, each bar moves in a direction exactly opposite to that taken by the bars next adjacent. In this way the adhesion of clinker is effectually prevented, the bars are protected from burning, and the air spaces kept clean and effective. Behind the fire-grate, and at the same level (*see fig.* 3), is situated an open space or chamber, bounded on the side farthest from the front by the fire-bridge. This space forms a kind of combustion chamber, and can also be utilised for the putting aside of a mass of incandescent fuel to re-start the fires after clinkering or cleaning. The price of this apparatus is very moderate, compared with the advantages in economy and convenience to be derived from its use.

We think, with Mr. McDougall, that before one or other of the three mechanical arrangements above described all ordinary difficulties connected with the consumption of solid fuel should vanish, and we have no doubt that the genius of the inventor, with the skill of the makers, can successfully cope with any of those special obstacles which do not come under the head of everyday experience.

We are very pleased to learn that these mechanical devices for securing a more scientific and economical use of feul are coming more and more into fashion day by day. The period of depression through which we have just passed has done much to minimise wastefulness of every kind and degree, and in few cases has this beneficial result been more pronounced than where the combustion of fuel is concerned. But as yet, we stand only on the threshold. Practice is never perfect till it is made to harmonise with true theory, and how far we are from attaining this is only too deplorably manifest to those who know what a pound of fuel *might* do, and what it *does* do. Every nearer approximation to the attainment of this harmony is a step in the right direction, and we hail the appliances of which we have been treating as a means towards this end. The boiler shed is the starting point of the factory or workshop. Here is *the* place of all others where the principles of economy should be studied and put into force. We hope that the year 1890 will see a yet wider introduction of the McDougall Patent Mechanical Stoker, and that twelve months hence we shall be able to chronicle still greater triumphs for its energetic patentees.

A NEW WOOL-SCOURING MACHINE.

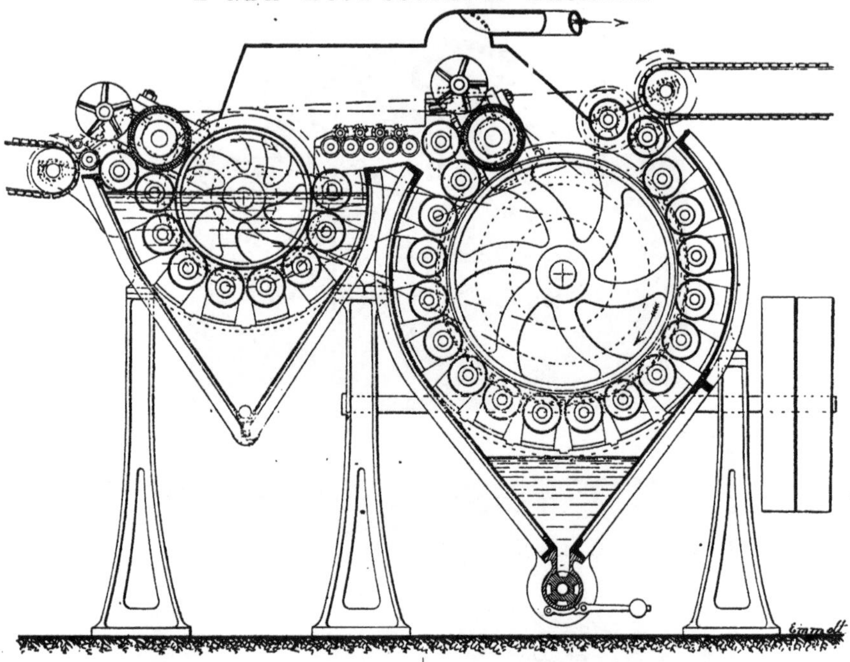

Messrs. Pullan, Tuke, and Gill, engineers, of Leeds, have just constructed a machine for wool-scouring on the solvent process, which promises to have a marked effect on an important local industry. The machine is the invention and patent of Mr. George Burnell, of Leeds. Mr. Burnell is a native of Otley, and has had some 40 years' experience as a wool-scourer in the colonies. The emulsion method of wool-scouring, by which the grease is washed out with an alkali, is just now almost universally used, and a high stage of perfection has been reached in the process, but it is a system that possesses many disadvantages, and much attention is being given just now to the solvent method, by which the fat is dissolved in a liquid. Mr. Burnell's machine, roughly speaking, is divided into two parts, or V-shaped tanks. The first, or large tank, is devoted to the process of freeing the wool from grease, the second, or smaller tank, ridding it of the potash soaps that remain after the first operation. The working parts of the large tank consist of a central revolving drum, 40in. in diameter, surrounded by 16 smaller compressing rollers, so arranged as to press the wool against the large central roller, or, if necessary, to yield and permit different thicknesses of wool to pass through. The solvent used is petroleum-benzine, and the first tank is filled with water (in continuous flow) up to the bottom of the rollers, the remainder of the tank, up to the nip of the highest small roller containing the benzine. The wool to be treated is fed in at the top of the tank, on the right-hand side. It passes into and is saturated by the benzine, and is then compressed between one of the small rollers and the large drum. On escaping from the rollers it again becomes impregnated by the benzene (which does not, like the water, render it clammy, but allows it to expand), and passes on to the next roller, passing around the large drum in this way. The action may, in fact, be likened to holding the wool in the hand, dipping it into the benzene, and then squeezing it out, and repeating the action a succession of times. The benzene dissolves the wool fat, which it thus removes from the wool, and the dirt attaching to the wool by means of the fat is also set at liberty, and drops through the solvent and water on to a grooved roller at the bottom of the tank, and thus escapes. While the wool is fed into the tank from right to left, a continuous flow of pure benzene is fed from left to right, so that the crude wool meets the dirty or saturated benzene on entering, and is immersed in the pure benzene when leaving the tank, after having had the fat extracted. The saturated benzene passes out at a point just below where the wool enters to a receiver, is distilled, and returned to the tank in its pure state to be used again, a continuous supply thus being kept up. Just before emerging from the first tank the wool passes through wringing rollers, is squeezed, and then conveyed to the second tank, which contains water at a suitable temperature. It is there subjected to a similar process to that of the first tank; the potash and other impurities are thus removed, and the wool, after receiving a final compression, is conveyed by an endless band to the drying chamber. The potash soap liquor that remains is led away to a vapour tank, and the carbonate of potash recovered in the ordinary manner. The machine can also be

goods from oil or grease left in after manu- of danger in the process is the volatile and petroleum-benzine employed as a solvent, but d by a galvanized iron hood, which enables it is condensed and used over again. It is ine, unlike disulphide of carbon, which is largely is not poisonous, and further, that by the pro- perfectly, and is not torn or tangled in any is built by Messrs. Pullan, Tuke and Gill, but to the Neville Works, Elland-road, the new ary Tuke and Co., which are more convenient ition of the machine, which will be ready in

OL-SCOURING BY PROFESSOR HUMMELL.

ult., Prof. Hummell, of the Yorkshire College, bers of the Society of Dyers and Colourists, in the ool, on the subject of "Wool-scouring with vola- ht, head master of the school, presided, and there The lecturer said that no operation to which l had to be submitted by the manufacturer was an that of scouring. It was the first operation to, and if inefficiently performed, it gave rise to the manufacturer and the dyer. The ulti- bly a clammy or harsh feel, and were void of ere poor and irregular, and lacked brilliancy. of yarn or cloth, wool had invariably to be o remove the oil introduced by the spinner. It refore, that the scouring process should be as ras, while it should efficiently remove the yolk wise injure or felt the fibres. At the present two distinct methods of scouring wool—the e solvent method. The first, and most usual, shing the wool in a warm and dilute solution alkaline salt; as, for example, carbonate of tty matter was broken up under the influence d an emulsion, and in this manner was removed. onsisted in treating the wool with some liquid e fatty matter from the wool, and it was well the case of greasy wool the potash soaps must y a separate treatment with water, since these ood solvents for potash soap. Contrasting the ntages of the two methods, he remarked that d in its favour the prestige of long-established he agents employed were inexpensive, and no Further, by a long process of evolution, the applied a form of apparatus which was all but l which enabled the scouring to be performed ner. The disadvantages of the emulsion s:—First, the agents used had an injurious slight though it was, it might be materially on the part of the workmen to the tration of the alkaline solutions employed; agents were wholly or partially lost, and red by tedious processes, and in altered sity for invariably employing water, and the calcareous, thus giving rise to the precipita- ps on the wool. The advantages accruing to couring wool seemed to be as follows:—The no agent which had not a deleterious action upon which the operation might be made automatic carelessness of workmen, the facility with be recovered in its original form with but oo, of recovering the waste products of the necessity of employing water, especially in the ogether with the possibility of lessening the d in the case of scouring raw wool. The ap- e solvent method had to contend with were actual practice was concerned, attending the absence of any apparatus which had usage to answer every requirement. Any ered all the pros and cons of the two methods could not but conclude that if they only s the solvent method would, in the end, be and would be universally employed. The plied a choice of more or less suitable solvents, the engineer to invent and construct some hich to apply them. He then gave a *resume* arrangements which had been proposed from ne 1856, in order to apply volatile liquids as ol, including carbon-disulphide, alcohol and and petroleum-benzine. The chief portion of to a description, with the aid of lantern sides, rranged and constructed to work the solvent" ting Deiss, Leyfferth, Moison, Lunge, Heyl, lesson, Mullings, Thomas, Cox, C. W. Smith, Patry (E. De Pass), Owen, Singer and Judell, and lastly that Mr. G. Burnell. In the course of his description of the vario machines, he mentioned that one fatal objection to many of the pr cesses was the use of carbon-disulphide, which, under certain conditio coloured the wool yellow. From the fact, he continued, that fo machines for scouring wool by volatile liquids were brough out during 1888, it seemed as though the time was now ripe f making a vigorous attempt to introduce the solvent method scouring. They lived in a day of the utilisation of waste product and it ought certainly to be taken into account in connection with t wool-scouring industry. He shewed that if the potash liquor, whic was one of the residuals, was treated in an ordinary gas retort, the would get a certain amount of illuminating gas. There then remaine the coke, which would furnish crude carbonate of potash, and th being free from soda would be especially useful to glass manufacturer From wool fat they got lanoline, which was used as a salve when mixe with medicaments, and it seemed it was very readily absorbed in the skin. Then wool fat was used in a crude condition for lubricati heavy machinery, and might be used for other purposes. Its futu uses really had to be found out. They tried at the Yorkshi College whether it was suitable for candles or night-lights, but it ga a smoky flame, and did not burn very brilliantly. He then shewe that under the present system the waste soap liquors had to through several processes by different manufacturers before comi back to the wool-scourer. By the solvent method the solvent w recovered at once, and made use of again in the process.

Mr. Slatter alluded to the danger attending the solvent methods account of the inflammable nature of the agents, and suggested th the colouring of the wool in the carbon disulphide process was perha due to the presence of free sulphur. He suggested that if the by products were utilised as mentioned their value would be very great diminished, for a time at least.

Mr. Beckett thought that, although in the first tank the wool woul not wrap round the rollers because of the benzine, still, in the seco portion of the machine, when it went through water, it would be liab to wrap round.

Mr. Burnell said that, with regard to the second part of the machin the grease had been extracted when it reached there, and the wo did not adhere to the rollers.

Mr. Wilkinson said that while the liberating of the fat caused t dirt to drop from the wool, it would drop on to other rollers beneat and particularly so on the ascending side of the tank, and that therefo the wool must retain some dirt.

Mr. Burnell observed that all dirt that remained after passing throu the second tank could be got rid of by simply shaking the wool. reply to further remarks, he said that the machine could be worked fo halfpenny per pound. He pointed out that by the other processes t agents had to be constantly renewed, but that in his machine the age was used over and over again. The loss in condensing the benzi was only 5 per cent., and the cost of condensing was 2d. a gallon, that if they took the bye-products into account they had the scouri done for nothing, and would get a profit from the products.

It was pointed out that the Burnell process cleansed the wool at o operation. In the Singer and Judell process the wool was treat twelve times.

The Chairman said the discrepancy might be accounted f by the fact that in the Burnell machine, in passing through t sixteen rollers, the wool got sixteen dips and sixteen squeezes. course, in the Singer and Judell the wool went through twelve succe sive tanks containing the pure agent, while in the Burnell machine merely went through successive strata which might or might not be in state of purity.

THE SUBSIDENCES IN CHESHIRE.—The Salt Union directors, the result of their conference at Crewe with the members of the Nort wich and Winsford Local Boards, have forwarded to the two Boards list of questions, to which answers have been appended, and the doc ment returned to the Salt Union. The questions are framed on t assumption that the land subsidences in Cheshire have been caused the pumping of brine by the Salt Union and their predecessors, an that compensation should be paid in respect of damage resulting ther from. The Boards propose that compensation should be paid to no receivers of brine royalties (but not to royalty receivers or their tenant within the area covered by the bill of 1881. Any works using brin are to be excluded, as well as railways, rivers, canals, and bridge Compensation to the extent of 1d. a ton or 1d. per 1,000 gallons brine pumped is only demanded for future damage, the distinction b tween that and past damage to be determined by a general survey; n further applications to Parliament for increase of compensation to allowed for ten years. The Boards, in answer to the question wheth since 1881 anything has occurred to alter the right to compensation fo damage, assert that prices have been largely increased, and that pub lic opinion has greatly altered, the right being now admitted by mo of the opponents of the Bill of that year.

Correspondence.

ANSWERS TO CORRESPONDENTS.

MAGDEBURG.—We shall be pleased to receive essays from every part of the world.

B.—Thanks for your communication. It will, no doubt, be interesting to some of our Yorkshire readers.

R. C.—We do not believe any such process possible.

J. F. S.—The varnish might be useful for rigid steel goods, but scarcely for wire. Try it.

D. R.—The numbers were posted as usual. Your copy should have been delivered on Saturday morning.

J. D., AQUA, KNENO, AND H. J.—See reply to D. R. You should make some allowance for New Year's week.

CONFECTIONER.—The one named, and all similar essences, are artificial. 2. Whether harmless or not, we cannot say.

WATER GAS.—You are quite correct, there have been more deaths from coal gas than from water gas. 2. True, but people should not be so fond of handing over their money to these company mongers, without they wish to lose it.

S.—We are much obliged for the information contained in your favour of the 2nd inst.

PLANT.—Iron borings wetted with a weak solution of salammoniac.

BENZOL.—Yes, there are six carbonising plants now at full work.

THE JARROW CHEMICAL COMPANY.

PROPOSED CLOSING OF THE WORKS AT TYNE DOCKS.

THE Directors of the Jarrow Chemical Company, Limited, have announced to their officials and workmen their intention of gradually closing their Alkali Works, at Tyne Dock, South Shields, as soon as their existing engagements and contracts will permit. They will continue to carry on their works at Friars Gorse, Gateshead, as these are fully equipped for the manufacture of bleaching powder, and the recovery from soda waste by Chance's new patent process. The alkali manufacture cannot now be profitably carried on, except in connection with the production of bleaching powder; and the Directors did not feel warranted in embarking the capital necessary for the erection of bleach and sulphur recovery plant at the Jarrow Chemical Works. These works occupy one of the most valuable sites for commercial and industrial undertakings in the North of England, adjoining the quays of the Tyne Dock, with all the facilities for landing and shipping of materials and goods, and exemption from dock dues secured by the Tyne Dock Act. The directors believe that the premises will not remain unoccupied, but will still contribute to the trade and industry of the town and neighbourhood.

The following intimation of the intention of the directors has been issued to the workmen to-day:—

The directors of this company think it their duty to give the earliest intimation to the various persons in their employment of a resolution at which they have arrived with regard to the South Shields works of the company.

Formerly their industry depended on the manufacture of alkali and soda, and, until the last few years, it was possible to carry on the works without the manufacture of bleaching powder. This is no longer the case, and the directors have had before them the alternatives either of embarking a large amount of fresh capital in bleaching powder plant, as they have done at the Friars Gorse Works, or of discontinuing the manufacture.

The present prospects of the chemical trade do not seem to them to warrant such an outlay, and they have with much regret come to the resolution to close the works at South Shields when they have completed their existing engagements and worked up their stocks in process of manufacture.

The directors look back upon the very harmonious and agreeable relations which have subsisted between the company and the work people, and which, along with the skill and steadiness of their workmen, the intelligence and attention of their foremen, and the zeal and loyalty of all their officials, have contributed so much to the success of their business.

It is a considerable mitigation of this change that it takes place during a period when the demand for labour is larger than for many years past, and when new employment is more easily to be found.

In a public point of view there is also a mitigation of the closing of these works in the circumstance that they occupy an admirable site, one of the best in the country, available for any kind of commercial and manufacturing industry, and possessing the immense advantage of contiguity of the quays of the Tyne Docks, with valuable rights of landing and shipping raw materials and goods, and exemption from dock dues secured by the Tyne Dock Act. The directors believe that the premises will not remain unoccupied, but will still contribute to the trade and industry of the town and neighbourhood.

PRIZE COMPETITION.

A LITTLE more than twelve months ago, one of our frien[ds] suggested that it might be well to invite articles [for] competition, on some subject of special interest to [our] readers; giving a prize for the best contribution, he at th[e] same time offering to defray the cost of the first prize, [and] suggesting a subject in which he was personally interested.

At that date we were totally unprepared for such a sug[gestion], but on thinking over the matter at intervals [of] leisure, we have come to think that the elaboration of such [a] scheme would not only prove of benefit to the manufactur[ers], but would go some way towards defraying the holid[ay] expenses of the prize winner.

We are therefore prepared to receive competitions in th[e] following subjects:—

SUBJECT I.

The best method of pumping or otherwise lifting or forci[ng] water aqueous hydrochloric of 30 deg. Tw.

SUBJECT II.

The best method of separating or determining the relati[ve] quantities of tin and antimony when present together in com[mercial] samples. On each subject, the following prizes a[re] offered:—

One first prize of five pounds, one second prize of [one] pound, five additional prizes to those next in order of m[erit], consisting of a free copy of the *Chemical Trade Journal* f[or] twelve months.

The competition is open to all nationalities residing in a[ny] part of the world. Essays must reach us on or before Ap[ril] 30th, for Subject II, and on or before May 31st, in Subject [I]. The prizes will be announced in our issue of June 28th.

We reserve to ourselves the right of publishing the co[n]tributions of any competitor.

Essays may be written in English, German, French, Spani[sh] or Italian.

TIN PLATE WORKS IN FRANCE.

SOUTH Wales has a rival in France—in a district which may [be] described as one of its own kin. In the valley which is cal[led] "Kerglau" which in the Breton language signifies the "Villag[e of] Rain," and which is about two miles above Hennebont, a little sea[port] four and a half miles from l'Orient roadstead, there is a manufac[tory] of tin plates. The *Australian Mining Standard* has a descripti[on of] them. During the first year the production of sheet iron and tin pl[ates] was about 900 tons per year. In 1885 it reached 10,000 tons, and [at] present it exceeds 12,000 tons. At the same time, the little por[t of] Hennebont, formerly almost completely deserted, frequented only [by] some fishermen, became very important. All the crude mater[ials] necessary for the manufacture come by ship to the locality. [The] materials are discharged in the port, unloaded in barges which asc[end] the Blavet, and towed by steam or horse-power to the quays of [the] factory. The chief crude materials employed are coal, which co[mes] direct from England, there being consumed on an average 70 to [...] tons per day, pig iron, scrap iron, carbonate of lime, magnesia, c[...] clay, tin, grease, chloride of zinc, and the acids used in the manu[fac]ture of tin plates. About 700 workers of both sexes are employe[d in] the works, and the motive power used is about 1,000 horses, [a] quarter of which is supplied by a turbine fed by the Blavet, [the] remainder by different steam engines. The work of the factory ma[y be] classified into five principal parts—(1) the manufacture of the pig [and] scrap iron into steel ingots; (2) the manufacture of the ingots into b[ars] and thin sheets; (3) the preparation of the sheets for the process; [(4)] the manufacture of the tin plates; (5) the decoration and stampin[g of] the tin plates.

The manufacture into bars and sheets employs about 300 workmen, [and] brings into play almost the whole of the motive force. The steel i[...] is first drawn into bars, and then into thin sheets or black iron. [The] steel ingots are heated in re-heating furnaces, and then taken [to a] steam hammer, which shapes them and cuts them into two. [The] pieces are heated over again for half-an-hour, then passed to the [...] trains, of which there are three. The bar train is an ordinary ro[lling] mill, which draws the iron out, making bars from 20ft. to 23f[t. in] length 4in. wide, and 0·4in. in thickness. These bars are at [...] cooled by immersion in water, and cut into small lengths of from [...] to 12in. in length, according to the length of the sheets to be m[ade]. These steel strips are taken red-hot to annealing furnaces, and [...]

Market Reports.

MISCELLANEOUS CHEMICAL MARKET.

In the alkali trade there is an upward movement in prices both of soda ash and caustic soda, the latter being particularly firm, and makers as a rule are indisposed to bind themselves on sales for forward delivery even at a premium upon present values; 70's on the spot are worth £7. 10s. at makers' works. Soda ash is firm at 1⅛ f.o.b. for good brands of carbonated ash and $1\frac{1}{16}$ for caustic ash. Some of the leading brands both of ammonia and Leblanc soda ash are now quoted ⅛d. per deg. above these prices. Bleaching powder is firm for prompt deliveries, at £5. 5s. f.o.r. makers' works, and maintains a strong position for forward deliveries, £7. 7s. 6d. to £7. 10s. having been paid during the past week. For export, £7 15s. is nearest price to-day f.o.b. Chlorate of potash rather firmer, and more business doing at 4⅛d. to 4¼d. Sulphur in all grades as now produced from the "Chance" recovery process is well sold. Vitriol still continues in good demand, and prices are without change. Sulphate of copper is moving freely at £23. 10s. Lead acetate and nitrate are only in moderate request, though prices continue firm and without likelihood of giving way owing to the firmness of the lead market. Acetates of all kinds are in slightly better demand for spot delivery, but very little doing beyond this except in grey acetate of lime, which has been sold freely, and is value for £14. per ton both for prompt and forward delivery. Brown acetate of lime £8. per ton. Acetic acids in all qualities and acetate of soda are slightly improved. White powdered arsenic is in good request for home and export trade. Citric and tartaric acids are quiet. Oxalic acid irregular; prices varying from 3d. to 3⅛d. There is a firmer tendency in potash, carbonate, and caustic, and demand is good. There is some uncertainty at present as to the future of iodine. Negotiations are on foot for renewal of the combination, but it is yet uncertain as to the outcome and, meantime, price is nominally 8d. to 9d. per oz.

WEST OF SCOTLAND CHEMICALS.

GLASGOW, Tuesday.

Makers have been in a greater hurry to resume after the holiday than for many a New Year past, and some of the works closed for the two first days only. All were restarted by yesterday morning, and the transaction of business on the market is by to-day back to about its normal swing. The opening is very cheering in character, demand for most articles being on the mount, and prices rising as a general rule. The alkalies continue to mend, and bleaching powder is now getting some shillings over the recent very deep bottom. Caustics, alum, soda crystals, and chlorate of potash, are also all on the rise. Sulphate of ammonia is a trifle easier for prompt Leith, value to-day being no more than £12. 2s. 6d., and there is nothing doing for forward. In the paraffin section there is a further firmness to note in lubricating oils, partly owing to reports of American makers having added another 10s. a ton to their prices. Scale, wax and candles are also pretty firm, and the interest of makers of these is at present on the acute stage over the meeting in London appointed for the 14th inst, of English, Scotch, and American delegates to settle the convention for another year. Mr. Bedford, of the Standard Oil Company, New York, is on his way across, and will be present on behalf of the American producers. Chief prices current are :—Soda crystals, 43s. 6d. net Tyne; alum in lump, £5. 0s. 0d., less 2½% Glasgow; borax, English refined £30., and boracic acid, £37. 10s. net Glasgow; soda ash, 48/52°, 1⅛d. less 5% Tyne; caustic soda, white, 76°, £9., 70/72°, £7. 12s. 6d., 60/62°, £6. 15s., and 60/62° cream, £6. 7s. 6d, all less 2½% Liverpool; bicarbonate of soda, 5 cwt. casks, £5., and 1 cwt. casks, £5. 5s. net Tyne; refined alkali, 48/52°, 1⅛d., less 7½% Tyne; saltcake, 26s. to 28s.; bleaching powder, £5. 15s. to £6. 0s.; less 5% f.o.r. Glasgow; bichromate of potash, 4d., and of soda 3d., less 5 and 6% to Scotch and English buyers respectively; chlorate of potash, 4⅛d., less 5% any port; nitrate of soda 8s. 4½d. to 8s. 6d., sulphate of ammonia, £12. 2s. 6d. f.o.b. Leith; salammoniac, 1st and 2nd white, £36. and £34., less 2½% any port; sulphate of copper; £23. 0s., less 5% Liverpool; paraffin scale, hard, $2\frac{1}{16}$d., soft, 2d.; paraffin wax, 120°, semi-refined, $2\frac{11}{16}$d.; paraffin spirit (naphtha), 10d.; paraffin oil (burning), 6¾d. to 7d., at Glasgow; ditto (lubricating), 865°, £5. to £5. 10s.; 885°, £5. 10s. to £6. 0s.; and 890/895°, £7. to £7. 10s. Week's imports of sugar at Greenock were 28,669 bags.

THE LIVERPOOL MINERAL MARKET.

Our market still continues to improve, and a further advance has taken place in most minerals. Manganese: Arrivals have continued small, and prices have further advanced. Magnesite: Raw lump, unchanged, raw ground, £6. 10s., and calcined ground, £10. to £11. Bauxite (Irish Hill Brand): is in very strong demand, bringing full prices. Lump done at 20s.; seconds, 16s.; thirds, 12s.; ground, 35s. Dolomite, 7s. 6d. per ton at the mine; arrivals continue. French chalk: arrivals have been rather small this week, and prices continue strong, especially for G.G.B. "Angel-White" brand—90s. to 95s. medium, 100s. to 105s. superfine. Barytes (carbonate) steady; selected crystal lump scarce at £6; No. 1 lumps, 90s.; best, 80s.; seconds and good nuts, 70s.; smalls, 50s.; best ground, £6., and selected crystal ground, £8. Sulphate unchanged; best lump, 35s. 6d.; good medium, 30s.; medium, 25s. 6d. to 27s. 6d.; common, 18s. 6d. to 20s.; ground, best white, G.G.B brand, 60s; common, 45s.; grey, 32s. 6d. to 40s. Pumicestone is unchanged, ground at £10., and specially selected lump, finest quality, £13. Iron ore has still improved. Bilbao and Santander firmer at 9s. to 10s. 6d. f.o.b.; Irish, 11s. to 12s. 6d.; Cumberland, 14s. to 18s. Purple ore, 12s. 6d. Spanish manganiferous are fully sold, but orders are being taken for forward delivery at 27s. 6d. for 20 per cent. Emery-stone: best brands scarce, and bringing full prices. No. 1 lump is quoted at £5. 10s. to £6., and smalls, £5. to £5. 10s. Fullers' earth steady; 45s. to 50s. for best blue and yellow; fine impalpable ground, £7. Scheelite, wolfram, tungstate of soda, and tungsten metal have somewhat improved. Chrome metal, 5s. 6d. per lb. Tungsten alloys 2s. per lb. Chrome ore, best qualities scarce and demand; prices firmer. Antimony ore: Prices firmer. Antimony metal, £73. to £75. Uranium oxide 24s. to 26s. Asbestos: Best rock £17. to £18.; brown grades, £14. to £15. Potter's lead ore dearer; Smalls £12. to £13.; selected lump, £14. to £15. Calamine: Best qualities scarce—60s. to 80s. Strontia steady; sulphate (celestine) steady, 16s. 6d. to 17s. Carbonate (native) £15. to £16.; powered (manufactured), £11. to £12. Limespar: English manufactured, old G.G.B. brand, brings full prices; 50s. for ground English. Plumbago: More offering; best Ceylon lump at last quotations; Italian and Bohemian, £4. to £12. French sand, in cargoes, scarce on spot—20s. to 22s. 6d. Ferro-manganese selling at very much higher prices. Ground mica, £50. China clay freely offering—common, 18s. 6d.; good medium, 22s. 6d. to 25s.; best 30s. to 35s. (at Runcorn).

THE LIVERPOOL COLOUR MARKET.

COLOURS remain unchanged. Ochres: Oxfordshire quoted at £10., £12., £14., and £16.; Derbyshire, 50s. to 55s.; Welsh, best, 50s. to 55s.; seconds, 47s. 6d.; and common, 18s.; Irish, Devonshire, 40s. to 45s.; French, J.C., 55s., 45s. to 60s.; M.C., 65s. to 67s. 6d. Umber: Turkish, cargoes to arrive, 40s. to 50s.; Devonshire, 50s. to 55s. White lead, £21. 10s. to £22. Red lead, £18. 10s. Oxide of zinc: V.M. No. 1, £24. 10s. to £25.; V.M. No. 2, £23. Venetian red, £6. 10s. Cobalt: Prepared oxide, 10s. 6d.; black, 9s. 9d.; blue, 6s. 6d. Zaffres: No. 1, 3s. 6d.; No. 2, 2s. 6d. Terra Alba: Finest white, 60s.; good, 40s. to 50s. Rouge: Best, £24.; ditto for jewellers, 9d. per lb. Drop black, 25s. to 28s. 6d. Oxide of iron, prime quality, £10. to £15. Paris white, 60s. Emerald green, 10d. per lb. Derbyshire red 60s. Vermillionette, 5d. to 7d. per lb.

REPORT OF MANURE MATERIAL.

The new year opens with a good inquiry for all kinds of phosphatic material, but scarcity of supplies stands in the way of a large business. Nitrogenous material, other than nitrate of soda, is also inquired for, and anything offering is readily placed at a full price.

The value of Charleston River phosphate rock is nominally about 11d. per unit, c.f. and i., to U. K., but shippers are very shy about offering at all in quantity for shipment ahead, anticipating still higher prices later on. For 55/60% Somme, 10¾d., for 60/65%, 1s., and for 65/70%, 13½d., c.i f. in bulk, in cargoes to U. K., would have to be paid, ⅛th rise in each case. We have no alteration to report in values, of Belgian, the higher qualities, for the most part, being withheld for the present. Neither have we any business in Canadian to report, the extreme prices demanded by shippers at present precluding business.

For bone ash, £5. 5s. on 70%, Hamburg terms, now asked, but £5. 3s. 9d. might possibly still be business, this being the price of last sale reported.

Sale of an October cargo of River Plate bones is reported at £5. 10s. to U. K., and the same price would be submitted for a December shipment. No further business reported in crushed India bones, but £5. 10s. would be paid for fine parcels near at hand, or £5. 7s. 6d. later shipment. £5. 12s. 6d. required for East Indian bone meal in any position, and sales have been made at this price to an outport.

Nitrate of soda continues very quiet at 8s. 4½d. per cwt. on spot. Nearest value of due cargoes is 8s. 4½d. to 8s. 6d., and November shipments, 8s. 7½d., but there is very liitle disposition to speculate at all in the face of the very heavy visible supplies.

Further sales of River Plate blood are reported to have been made at full prices, but exact figures do not transpire. Prices for ground horn and hoof and other similar nitrogenous material unchanged.

Superphosphates much inquired for, and 26 per cent. is now worth fully 47s. 6d. per ton, in bulk, f.o.r. at works, with every prospect of a further advance.

Since the turn of the year there seems to be a greater disposition on the part of manure manufacturers to complete their purchases, and we anticipate that a large business will be concluded within the next week or two.

TAR AND AMMONIA PRODUCTS.

The demand for tar products is fairly good all round. Benzols are firmer, and have advanced slightly in price, 90's being 3s. 6d., and 50/90's 2s. 8d. Solvent naphtha continues in very good request, and creosote has moved off well enough to keep producers low in stock. The demand for anthracene has been very good, and much has changed hands, though it does not seem that it has gone to the ultimate user. There are indications that speculators have absorbed most of that which has found its way into the market, and the times is coming when it will have to be re-sold. Selling second-hand anthracene is, however, not such an easy matter, so that the future of this article is a little bit uncertain. Pitch is slightly easier, though only a matter of 1s. or so, but at present values a considerable business has been done.

The sulphate of ammonia market is quiet, though it has preserved its steadiness so far as prices are concerned. Business has been reported from Hull at prices ranging from £12. 2s. 6d. to £12. 3s. 9d., and at Liverpool, £12. to £12. 2s. 6d. The price at Leith may be quoted at £12. 2s. 6d., though it is rumoured that £12. 1s. 3d. has been accepted. The Continental demand is hardly up to the mark at present, though there is no doubt it will presently improve. In the meanwhile there are no heavy stocks weighing on the market, and there is a fair home demand. The nitrate market is exceedingly dull, and large arrivals and heavy shipments have not tended to improve the position.

New Companies.

CAMBORNE GAS COMPANY, LIMITED.—This company was registered on the 12th ult., with a capital of £10,000., in £5. shares, to supply gas, electricity, or other illuminating agent, at Camborne, Cornwall. The subscribers are:—

	Shares
F. W. Thomas, Camborne, accountant	1
W. Bailey, Camborne, bank manager	1
T. S. Lowry, Camborne, bank manager	1
R. H. Williams, Camborne, secretary	1
J. R. Daniell, Camborne, solicitor	1
A. G. Richards, Camborne, clerk	1
W. Pike, Camborne, mine purser	1

GEMMING AND MINING COMPANY OF CEYLON, LIMITED.—This company was registered on the 12th ult., with a capital of £100,000., in £2. shares, 50 being founders' shares, to acquire land in Ceylon and elsewhere, supposed to contain sapphires, cat's eyes, rubies, tourmalines, amethysts, topaz, star stones, and other precious stones and other minerals; and in particular the estate of Everton (Kabragallakelle) and Aberfoyle (Kalkanda), containing respectively about 760 and 520 acres, and also the mining and gemming rights over the Rangweltenne Estate of 404 acres, and the Springwood and Barro Estate of 1,100 acres, all situate in Ceylon. The subscribers are:—

	Ord. Shares.
Neil Morris, Wallwood-road, Leytonstone, clerk	1
Lionel Walsh, Dancer-road, Fulham clerk	1
Gordon Hunter, 6, Handley-road, N.E., clerk	1
F. Vernon Ball, 126, Whitehorse-road, Stepney, commission agent	1
Charles Dixon, 57, Moorgnte-street, accountant	1
Wm. Bowman, 31, Gilmore-street, Lewisham, accountant	1
A. Lightoller, 12, Argyle-square, W.C., clerk	1

MILWALL LEAD COMPANY, LIMITED.—This company was registered on the 12th ult., with a capital of £75,000., in £100. shares, to acquire all or part of the business of Pontifex and Wood, Limited, of Millwall and Shoe-lane, relating to lead, antimony, white lead, oils, and colours. The subscribers are:—

	Shares
A. E Walker, 102, Lenham Gardens	1
A. H. Lancaster, St. Peter's Chambers, Cornhill, lead merchant	1
F. J. Walker, 7, Longridge-road, S.W.	1
J. H. Enthoven, 17, Gracechurch-street, lead merchant	1
J. Enthoven, 17, Gracechurch-street, lead merchant	1
F. Enthoven, 17, Gracechurch-street, lead merchant	1
A. L. Enthoven, 14, Connaugh-place	1

NEW BALLESWIDDEN SYNDICATE, LIMITED.—This company was registered on the 13th ult., with a capital of £15 000., in £1. shares, to acquire the lease of the New Balleswinden Mine, parish of Saint Just, Cornwall, The subscribers are:—

	Shares
Commander The Hon. W. Grimston, R.N., St. Albans	1
C E. Holdsworth, Woodside-grange, Wimbledon, miner	1
F. B. Henderson, C.E, 62, St. Clement's-house	1
J. G. Dalzell, 12, Clement's-inn, solicitor	1
E. Bennett, Harlesden	1
E. Mackay, 27, Clement's-lane, clerk	1
T. Clay, Suffolk House, E.C., engineer	1

PRECIOUS STONES AND MINERAL EXPLORATION SYNDICATE OF CEYLON, LIMITED —This company was registered on the 12th ult., with a capital of £10,200, in £1. shares, 200 being deferred or founder's shares, to acquire mines and mining rights in Ceylon or elsewhere. The subscribers are :— Shares

N. T. Treveen, Gate House, Hampton Court 1
N. P. Miles Tronson, 8, Drapers' Gardens, financial agent 1
R. Atwood, 8, Drapers' Gardens, accountant 1
Lieut. Col. F. W. Wheler, 18, Lawn-terrace, Blackheath 1
G. H. Homan, 16, Hartham-road, N , secretary to a company.......... 1
J. Hodgkyns, 8, Drapers' Gardens 1
H. G. Revell Brade, 10, Herbert-crescent, Han's-place, S.W., journalist 1

CORONA HILL SILVER MINING COMPANY, LIMITED.—This company was registered on the 17th ult., with a capital of £150,000., in £1. shares, to acquire mineral and other property in New South Wales, and in any other of the Australian Calonies, or elsewhere. The subscribers are :— Shares.

John Proffitt, 32, Great George-street, solicitor.......... 1
A. O. Scott, 32, Great George-street, solicitor 1
J. McDonnell, 114, Christchurch-road, Tulse-hill, solicitor 1
H. Allan, 12, Delahay-street, S.W., chartered accountant.......... 1
A. G. Weddell, Dorking, solicitor 1
H. T. Bird, 4, Great George-street, chartered accountant 1
F. C. Potter, 4, Great George-street, chartered accountant 1

GULF COAST MINING COMPANY, LIMITED.—This company was registered on the 16th ult., with a capital of £20,000., in £1. shares, to carry on mining operations (locality not stated). The subscribers are :— Shares.

A. Charles, 14, Baalbec-road, Highbury, clerk 1
A. H. Foster, 65, Albyn-road, St. John's, S.E., clerk 1
A. W. Langton, 67, Manor-road, Brockley, clerk 1
W. S. Gay, 249, Farringdon-road-buildings, clerk 1
T. Hancock, 4, The Terrace, Barn-street, N , clerk 1
A. Hawkins, 15, Lander-terraee, Wood-green 1
A. G. Blackman, 72, Alscot-road, Bermondsey, clerk 1

KLIPFONTEIN MINING AND ESTATE COMPANY, LIMITED.—This company was registered on the 16th ult., with a capital of £70,000 , in £1. shares, to carry on mining operations in Africa. An agreement will be entered into with Wolf Carlis, and Ferdinand Eugene Schuler The subscribers are :— Shares.

S. B. Garcia, 18, Henrietta-street, Covent Garden 1
C. Graham Bennett, 7, Wakefield-road, Tottenham 1
W. R. Mercer, 14, Anhalt-road, Battersea-park, clerk 1
W. Vize. 64, Craford-road, Tufnell-park, clerk 1
H. W. Moore, 29, Versailles-road, Anerley, clerk 1
T. J. Seel, 5, Copthall-buildings, accountant 1
W. H. Adams, 168, Friern-road, S.E. 1

NORTH AFRICAN GAS SYNDICATE, LIMITED.—This company was registered on the 13th inst., with a capital of £30,000., in £1. shares, to manufacture and supply gas or other luminant in Algeria and other parts of Northern Africa. The subscribers are :— Shares.

T. M. McGeagh, 21, Bedford-place, engineer.......... 1
R. Huichinson, C.E , M.E., 2, Clyde-street, Kensington 1
W. McLachlah, M.E., College-hill-chambers, E.C.......... 1
J. G. Pigott. College-hill-chambers. E.C , accountant.......... 1
H. M. Dunstan, 50, Avenue-road, West Kensington 1
H Schallehn, 13, Stockwell-park-road 1
C. F. Bentley, 53, Gloucester-road, N., accountant 1

RUGBY AND NEWBOLD CEMENT COMPANY, LIMITED.—This company was registered on the 19th ult , with a capital of £12,000., in £5. shares, to acquire certain freehold property at Newbold, Warwick, and the goodwill of the business of manufacturers of lime and cement carried on there. The subscribers are :— Shares.

Frank Creke, 8, Newhall-street, Birmingham, chartered accountant 1
J. Hopewell, Rugby, stationer 1
A. H. Singleton, Balsall Heath, Birmingham, clerk.......... 1
D. Parkes, Wellington-street, Birmingham, clerk 1
D. Daly, Jun., Handsworth, Birmingham, clerk 1
J. Garrick, 92, Sloane-street, Chelsea, contractor 1
R. Brown, 19, Exeter-place, Chelsea, clerk.......... 1

TRANSVAAL PETROLEUM, LIMITED.—This company was registered on the 21st ult., with a capital of £5,000., in £1. shares, 2,000 being founders' shares, to acquire concessions, oil wells, lands, estates, and properties in Burmah, India, Canada, Australia, New Zealand, Borneo, Beluchistan, South Africa, and elsewhere, and to develop the resources thereof; and to carry on business as oil prospectors, refiners, and merchants. The subscribers are :— Shares.

T. W. Watson, Pailton, near Rugby.......... 1
A. T. Angus, 15, Fielding-road, W., engineer.......... 1
H. Haycroft, 4, Farnham Royal, Kennington, shorthand writer.......... 1
C. E. Harvey, 110, Cannon-street, secretary to a company 1
W. G Hampton, 77, King William-street, manager.......... 1
E E. Young, 12, Cranbourne-road, Stratford, clerk 1
A. S. Macnaughton, 31, Lombard-street, accountant 1

AUTOMATIC TAP SYNDICATE LIMITED.—This syndicate was registered on the 24th ult., with a capital of £2,000., in £10. shares, to acquire and work patent rights granted to James Rigg, Thomas Meacock, and Angus William Cuthbert Ward for improvements in valves, taps, or cocks, for domestic, sanitary, and other uses. The subscribers are :— Shares.

C. W. Duncan, 41, Linden-gardens, solicitor 5
Lucian de Rin, Ethelburga-house, ship-broker 5
W. H. C. Mahon, 33, Ely-place, solicitor 5
H. A. Jackson, Burgess-hill, Sussex 5
J. Howard Calls, Moorgate-street, builder 5
J. McMillan, 4, Fenchurch-avenue. shipbroker.......... 2
C. Kemp Wild, 113, Cheapside, artists' colourman 3

ELECTRICAL STANDARDISING, TESTING AND TRAINING INSTITUTION, LIMITED.—This company was registered on the 31st ult., with a capital of £10,000., in £5. shares, to establish, maintain, and carry on any standardising, testing, or calibrating institution or institutions, where all or any electric measuring instruments can be tested calibrated, or standardised. The subscribers are :— Shares.

W. A. Pittman, 7, St. Helen's Gardens, North Kensington 1
J. Whitehead, Heycot, Crouch-end, electrical engineer 1
T. E. Toweison, F.C S., 1, Portland-road, Finsbury-park, analytical chemist 1
H. Linklater, 117, Bishopsgate-street, accountant.......... 1
R. Acock, 117. Bishopsgate-street, secretary 1
J. A. Roxburgh, 117, Bishopsgate-street, clerk 1
O. W. Seligman, 3, Moreton-gardens, South Kensington 1

ELLIS, DREW, AND COMPANY.—This company was registered on the 30th ult., with a capital of £10,000 , in £5. shares, to
are :—

K. Francis, Burgess-hill, Sussex, wine
J. W. Miles, Burgess-hill, Sussex, nurse
D Parsons, St. John's-common, enginee
C. Holland, Burgess-hill, shop assistant
W. Pollington, 61, Queen's-road, Bright
E. A. Gravett, Burgess-hill, potter ....
J. T. Martin, Buckingham-street, Stran

FOLKESTONE SANITARY STEAM LAUNDRY
was registered on the 27th ult., with a capital
indicated in title. The subscribers are :—

T. G Heron, Folkestone, grocer ......
H. W. Taylor, East Folkestone, laundry.
D. Baker, Folkestone, contractor ......
W. D. Fagg, Folkestone ....
H. Marchant, Folkestone, hotelkeeper...
G. C Conley. Cheriton, Sandgate, contra
H. H. Barton. Folkestone, auctioneer ...

GOLD KOPPE MINING COMPANY, LIMITED.
31st ult. with a capital of £333,007.. in £1 sh
rights in Austria and elsewhere and to adopt
between Andrew Wm. Gill and Arthur Wm. V

T. W. Rudderforth, 12 Prideaux-road, Cl
C R. A. Proctor, 31, Lothian-road, North
W. D. Turner, Mowbray-road, Upper No
C. K. Greenwell, 28. Upper Hamilton-terr
C. W. Cuthbert 3, Upper John street, Kin
C. E. Green, 208, Ashmore-road, W., secre
J. Hobbs, 30, Union-street, Hackney-road,

HONDURAS CORPORATION, LIMITED.—This
ult.. with a capital of £62,500., in £25. shares
duras, and to develop and turn the same to acc
ing, agricultural, trading, and other operations

J. N. McAdam, C.E , Surrey House, Victo
G. T. Jellicoe, 328, Mansion-house chamber
J. H. Hamilton, 21, Bedford-street, Strand,
F L Marshall, 83. Queen-street, secretary
H C. Eaton, 42, Doughty-street, W.C. ...
T. Goddard, 11, Queen Victoria-street. finar
G. H. De Lasaux, 30, High-street, Canterbu

LEEDS AND WAKEFIELD BREWERIES, LIMIT
conversion to a company of the brewery busines
J. Pilling Kirk, of Leeds, and of Messrs. H. M
of Wakefield. It was registered on the 31st ult ,
shares, 10,000 being six per cent. cumulative p
are :—

W. Barrett, Forest-road, E. accountant ....
E. R. Porter, 47, Lupton-road, N.W. ......
J. Clarke, 12, Gordon-road, Hornsey ......
H. J. Bass, 177, King's-road, N.W , factory
G. E. Friend, 8, Larcom-street, S.E., mercha
J. Kelly, 24w, Peabody-square, Jomes-street,
W. Glynn 14, Little Albany-street, N.W., cle

PEAT UTILISATION COMPANY, LIMITED.—Th
24th ult., with a capital of £200,000., in £5. shar
George Henry Bérand, for the treatment and uti
transformation into peat wool and its application f
are :—

Frederick Wm. Bird, Chiswick ....
E. H. Cannot, Temple, barrister ....
Cyril Leslie Johnson, Ealing ....
F. Ernest Bird, 26, Budge-row, solicitor ....
H. Richards, Courtenay-road, Penge, accountar
G. Benezet, 32, Regent-square ....
C. T. Plant, 30, Harewood-square ....

SANITATION COMPANY, LIMITED.—This company
with a capital of £10,000., in £5. shares, 300 of which
on business as sanitary engineers. The subscribers a

R. J. Arbon, 14, Victoria-street ....
G. Gill, Warwick-lane, publisher ....
W. E Garstin, 24, Marlbro'-road, N.W..........
S. Call, 85, Baker-street, W., Accountant ....
A. Garstin, Ivy Walls, Barnes-common ....
H. C. Burt, 42A, Bowe-lane, silk agent..........
W. C. Watts, 110, Upland-road, Dulwich, clerk

SEPOY COMPANY, LIMITED.—This company was re
a capital of £1,000., in £5. shares, to acquire the busi
facturers, carried on by J Hudspeth Pearsons of Pe
Sepoy Rubbing oils, Indian Balm and Morning Liver
Long Causeway, Peterborough.

SKELSEY'S ADAMANT CEMENT COMPANY, LIM
registered on the 1st inst., with a capital of £50,000
the business of cement manufacturer, carried on by
subscribers are :—

G. H. Skelsey, Hull ....
R. Brearby, Batley, cloth manufacturer ....
A. Brearby, Batley, cloth manufacturer ....
W. Skelsey, Batley, cloth manufacturer ....
A. Shelsey, Batley, cloth manufacturer ....
F. Reynolds, Knottingley, lime merchant ....
Hy. Skelsey, Bridlington-quay ....

SODEN MINERAL PRODUCE COMPANY, LIMITED.—
on the 27th ult., with a capital of £20,000., in £1
mineral pastilles and waters from the mineral springs o
and to carry on business as wholesale and retail chemist

J. Mammelsdorff, 52, Bread-street, merchant .....
Jean Bach, 52, Bread-street, merchant ....
R. Morgenstern, Frankfort-on-Maine ....
M. Frank, Frankfort-on-Maine, merchant ....
L. A. Ricard Abenheimer, Frankfort-on-Maine .
J. E. Caher, Frankfort-on-Maine..........
N. Marx, Frankfort-on-Maine ....

TRANSVAAL MINERALS, LIMITED.—This company w
with a capital of £300,000., in £1. shares, to acquire
1882, granted by the South African Republic to Alois

the smelting of iron ores, and the manufacture and sale of iron goods of all kinds. To explore, work, and develop mines. The subscribers are :— Shares.

Tom Donald, 50, Antill-road, Bow, clerk 1
W. Gordon Hughes, 18, Winchester-road, Belsize-park, clerk 1
C. P. Beale, 33, Wakehurst-road, Wandsworth-common, clerk 1
J. Eustace, 12, St. Donatt's-road, accountant, Newcross 1
T. Brown, 85, Hornsey-park-road, accountant 1
J. W. Nolan, 31, De Laune-street, Kennington-park, clerk 1
T. Hodge, 2, Park-villas, Albion-road, N., clerk 1

HIGGOTT'S DRUG STORES, LIMITED.—This company was registered on the 3rd inst., with a capital of £800., in £1. shares, to acquire the business of H. Higgott, of 23, Market-place, Great Bridge, Stafford, chemist. The subscribers are :— Shares.

Eliza Higgott, Stapenhill, Burton-on-Trent 1
H. Higgott, Great Bridge, chemist 1
Sarah Jane Higgott, Stapenhill 1
John Higgott, Burton-on-Trent, clerk 1
W. Higgott, Burton-on-Trent, farmer 1
S. Higgott, Burton-on-Trent 1
Mrs Jane Higgott, Burton-on-Trent 1

HONITON GAS AND COKE, LIMITED.—This is the application of limited liability to this company, which was originally constituted by Deed of Settlement on the 20th November, 1834. It was registered on the 4th inst., with a capital of £4,900., in £10. shares, the whole of which are taken up, and are fully paid.

MIDLAND COAL, COKE, AND IRON COMPANY, LIMITED.—This company was registered on the 3rd inst., with a capital of £375,000., in £10. shares, to acquire collieries and other mines, ironstone works, blast furnaces, brickworks, coke ovens, &c., in the parishes of Audley and Wolstanton, County of Stafford, known as Podmore Hall, Hayeswood, Apedale and Chesterton. The subscribers are :— Shares.

J. T. Smith, Stratford-on-Avon, bank director 1
J. H. E. Heathcote, M.P., Apedale Hall, Newcastle-under-Lyne 1
B. Gibbons, 70, Gt. Bridgewater-street, Manchester, merchant 1
J. E. Huxtable, Holland-park-gardens, W., solicitor 1
W. F. Pollock, 12, Cumberland-terrace, Regent's-park 1
W. B. Blyth, Rockbourne-road, Forest-hill 1

Gazette Notices.

PARTNERSHIPS DISSOLVED.

CHESTER AND GIBB, Bucklersbury, London, and JOHANNESBURG AND BARBERTON, South Africa, consulting mining engineers and sole agents for vending mining and other machinery manufactured by Fraser and Chalmers, of Chicago, and by the Sprague Electric Railway and Motor Co., New York.

J. HARTLEY AND J. SPENCE, under the style of Paton and Charles, Wapping, soap manufacturers.

B. AND H. WADDINGTON, Bradford, chemists, druggists, and dentists.

J.C. PAIN AND SONS, Walworth-road, London, and Mitcham, and elsewhere, pyrotechnists, as far as regards J. C. Pain.

E. BROWNLOW AND F. W. BROWNLOW, under the style of Slack and Brownlow, Manchester, filter manufacturers.

RAWLINS AND WILSON, Trentham, brick and tile manufacturers.

R. H. MONRO, SIR F. G. MILNER, THE HON. R. PARKER, THE HON. G. N. DAWNAY, and the HON. C. T. DUNDAS, under the style of the Tadcaster Tower Brewery Co., York and Tadcaster, brewers and maltsters, and wine and spirit merchants, as far as regards R. H. Monro.

W. G. HARRISON AND W. WALKER, Junr., under the style of W. G. Harrison and Co., Liverpool, manufacturing chemists.

THE BANKRUPTCY ACT, 1883.

Receiving Orders.

WM. WILLIAMS (trading as Williams and Co.), Ramsgate, brewer and mineral water manufacturer.

First Meetings, and Public Examinations.

THOS. HART, Heaton Norris, pharmaceutical chemist, January 10, Official Receiver's Offices, Stockport, January 15, Court House, Stockport.

CHAS. POULTON, Ipswich, mineral water manufacturer. January 10, Official Receiver's Office, Ipswich, January 16, Shire Hall, Ipswich.

Adjudications.

CHAS. POULTON, Ipswich, mineral water manufacturer.

Notices of Dividends.

HERBERT REES TREHEARNE, Gt. Chapel-street, Soho, London, also [illegible] Wharf, Lombard-road, Battersea, London, late of Lavender Gardens, [illegible] Hill, London, stone, lime and cement merchant, first div. of 7s. 6d. any da[illegible] Official Receiver's Offices, 33, Carey-street, Lincoln's Inn, London.

JOHN ANTHONY ENRIGHT (trading as J. A. Enright & Co.), Llanerchy[illegible] mineral water manufacturer, first and final dividend of 1s. 8¼d., January [illegible] Brynalaw, Menai Bridge, Anglesey.

The Patent List.

This list is compiled from Official sources in the Manchester Technical Laboratory, under the immediate supervision of George E. Davis and Alfred R. Davis.

APPLICATIONS FOR LETTERS PATENT.

Method of Dividing Glass Cylinders. J. G. Sowerby. 20,619, Dec. 23.

Sprinklers for the Automatic Extinction of Fires. W. Whitehead. 20,622, December 23.

Apparatus for Raising Liquids. W. P. Theermann. 20,624, December 23.

Rendering Textile and other Fabrics fire proof. W. S. Somers. 20,625, December 23.

Electric Distribution. W. M. Mordey. 20,627, December 23.

Collecting for Further Utilisation the Excess of Carbonic Acid Generated During the Manufacture of Beer. L. Haas. 20,630, December 23.

Heel-ball Composition for Boots. H. W. Farrer and E. Young. 20,637, December 23.

Sensitizing and Developing Chloride of Silver for Photographic Purposes. W. H. Caldwell. 20,662, December 23.

Apparatus for Grinding and Amalgamating Ore Containing Gold and Silver. M. Crawford. 20,663, December 23.

Treatment of Textile Fabrics by means of Ammoniacal Oxide of Copper to Increase the Waterproof and Non-Combustible Character of the Fabrics (complete specification). C. Baswitz. 20,665, December 23.

Manufacture of Colouring Matters Suitable for Dyeing. J. Y. Johnson. 20,668, December 23.

Anti-Friction Bearings. H. W. Libbey. 20,670, December 23.

Apparatus for Treating Finely-divided Metalliferous Material for the Purposes of Separation and Amalgamation. W. L. Wise. 20,677, December 23.

Production of Compounds of the Diphenylmethan Group and the Rosaniline Series. O. Imray. 20,678, December 23.

Smokeless and Gasless Fuel. A. Myall. 20,679, December 23.

Materials for Covering Roofs. A. Myall. 20,680, December 23.

Tanks for Melting or Founding Glass. S. Washington. 20,687, Dec. 24.

Drying and Heating Chambers. J. H. R. Dinsmore. 20,705, Dec. 24.

Dyeing and Scouring Machines.—(Complete Specification). J. P. Delahunty. 20,715, December 24.

Boiling and Precipitating Tower.—(Complete Specification). G. E. Pennycook. 20,720, December 24.

Obtaining Ammonia from Atmospheric Nitrogen and the Hydrogen of Decomposed Steam. J. C. Fell. 20,722, Dec. 24.

Apparatus for Drying Minerals. R. Morris and J. Wood. 20,730, Dec. 24.

Magneto-electric Machines. — (Complete Specification). R. Haddan 20,734, December 24.

Bunsen Burners. G. Reimann. 20,744, December 24.

Electrical Accumulators. F. Marx. 20,751, December 24.

Manufacture of Electrodes for Electric Batteries. F. Marx. 20,752, December 24.

Process for Determining the Amount of Water Contained in Steam. M. Gehre. 20,755, December 24.

Process of Producing Mixed Coloured Silk for Obtaining Mixed Coloured Silk Fibre. R. J. MacLean. 20,771, December 27.

Apparatus for Measuring the Flow of Liquids. H. Sutcliffe. 20,775, December 27.

Apparatus for Evaporating Liquids. R. A. Robertson and W. J. Mirrlees. 20,784, December 27.

Treatment of Spent Acid. J. Longshaw. December 27.

Ventilating Waterproof Garments. G. C. Mandleberg, H. L. Rothband, and S. L. Mandleberg. 20,804, December 27.

Refrigerating Apparatus. T. B. Hill and J. Sinclair. 20,811, Dec 27.

IMPORTS OF CHEMICAL PRODUCTS

AT

THE PRINCIPAL PORTS OF THE UNITED KINGDOM.

LONDON.

Week ending Dec. 21st.

Alkali—		
Holland,	200 c.	H. Wallace & Co.
France,	135	Arnati & Harrison
Holland,	100	Beresford & Co.
,,	900	W. G. Taylor & Co.
Acetate of Soda—		
France,	£85.	W. G. Blagden
,,	100	Lister & Biggs
,,	23	Barr, Moering & Co.
,,	75	Hunter, Waters, & Co.
Acetic Acid—		
Holland,	15 pkgs.	A. & M. Zimmermann
,,	50	,,
Antimony Ore—		
France,	20 t.	Pillow, Jones, & Co.
Portugal,	77	H. Emanuel
Alum—		
Holland,	£50.	Ohlenschlager Bros.
Bones—		
Holland,	13 t.	J. Gibb & Co.
Egypt,	25	M. D. Co.
Barium Chloride—		
Germany,	£80.	Petri Bros.
Barytes—		
Germany,	22 cks.	O. Hara & Hoar
,,	22	W. Harrison & Co.
,,	22	O. Hara & Hoar
Belgium,	22	W. Harrison & Co.
Holland,	40	Burrell & Co.
,,	45	D. Storer & Sons
,,	41	Prop. Scott's Whf.
Germany,	19	W. G. Taylor & Co.
Brimstone—		
Italy,	5 t.	G. Boor & Co.
,,	115	Johnson & Hooper
,,	15	W. C. Bacon & Co.
,,	500	Seager, White, & Co.
Boracic Acid—		
Italy,	24 pkgs.	Howards & Son
Borax—		
Italy,	£277.	A. Faber & Co.
Chemicals *(otherwise undescribed)*		
Germany,	£5.	Phillipps & Graves
,,	743	A. & M. Zimmermann
Holland,	14	R. Morrison & Co.
,,	10	Phillipps & Graves
Germany,	90	T. Farmer & Co.
Belgium,	10	T. H. Lee
France,	620	A. Henry & Co.
Germany,	141	T. H. Lee
,,	5	Grosscurth & Luboldt
Caoutchouc—		
Natal,	12 c.	H. Rooke, Sons, & Co.
Pt. Louis,	4	Arbuthnot, Latham, & Co.
E. Indies,	112	Wallace Bros.
,,	20	L. & I. D. Jt. Co.
France,	560	W. Balchin
W. Africa,	17	Matheson & Co.
Mauritius,	6	Blyth, Green, & Co.
E. Indies,	20	L. & I. D. Jt. Co.
Holland,	1	Norris & Joyner
Pt. Louis,	5	L. & I. D. Jt. Co.
U. States,	5	Williams, Torrey, & Co.
E. Indies,	53	Henderson Bros.
Copper Ore—		
Norway	10 c.	W. E. Bott & Co.
Cream Tartar—		
Holland,	8 pkgs.	W. C. Bacon & Co.
Italy,	28	B. Jacob & Son
France,	5	C. F. Gerhardt
,,	5	Webb, Warter, & Co.
Holland,	50	A. & M. Zimmerman
Italy,	9	W. C. Bacon & Co.
Holland,	6	Middleton & Co.
Copperas—		
Germany,	£35	Burgoyne, Burbidge, & Co.

Campher—
Germany, 2 pkgs. L. & I. D. Jt. Co.

Dyestuffs—
Extracts
Germany, 1 pkg. H. Johnson & Sons
France, 31 T. H. Lee
" 150 L. J. Levinstein & Sons
Germany, 20 Hermann, Keller, & Co.
" 2 L. & I. D. Jt. Co.
U. States, 50 W. Burton & Sons
Holland, 40 Burt, Boulton, & Co.
" 4 pkgs. T. H. Lee
Austria, 600 A. J. Humphery
Orchella
E. Indies, 38 pkgs. Beresford & Co.
Ceylon, 42 Kebbel, Son, & Co.
" 16 L. & I. D. Jt. Co.
Natal, 200 H. Rooke, Sons, & Co.
Ceylon, 16 Lambert & Strong
Gambier
E. Indies, 403 pkgs. R. & J. Henderson
" 403 "
" 89 Brinkmann & Co.
Argols
Cape, 15 pkgs. Brown & Elmslie
Annatto
Ceylon, 50 pkgs. Hoare, Wilson, & Co.
France, 5 Fullwood & Bland
Ceylon, 19 J W. Cater, Sons & Co.
Cochineal
Canaries, 80 pkgs. W. M. Smith & Sons
" 10 Swanston & Co.
" 12 Johnson, Rolls, & Co.
" 7 H. R. Toby & Co.
" 3 Kuhner, Hendschel, & Co.
Myrabolams
E. Indies, 366 pkgs. J. Graves
" 1818 Beresford & Co.
" 1000 Marshall & French
" 1366 L. & I. D. Jt. Co.
Dyestuffs *(otherwise undescribed)*
E. Indies, 750 bgs. Ralli Bros.
Sweden, 2 cs. M. D. Co.
Indigo
E. Indies, 18 chts. W. Brandt, Sons, & Co.
" 6 Caldwell, Watson, & Co.
" 6 L. & I. D. Jt Co.
" 12 "
" 19 "
Natal, 16 Arbuthnot, Latham, & Co.
" 16 Parsons & Keith
E. Indies, 7 Benecke, Souchay, & Co.
Greytown, 35 sons A. Jiminez & Sons
" 7 Lewis & Peat
Saffron
Spain, 1 pkg. C. Brumlen
" 1 Major & Field
" 1 Pickford & Co.
Sumac
Italy, 400 pkgs. J. Kitchin
" 100 W. France
" 100 Oakley
" 1000 Beresford & Co.
" 100 J. Knill & Co.
Tanners' Bark
Natal, 7 t. W. Dunn & Co.

Glycerine—
Germany, £86 Becker & Ulrich
" 140 H. Lambert
" 8 Craven & Co.
" 100 Prop. Scott's Whf.
" 35 Beresford & Co.
" 15 R. W. Greef & Co.
Holland, 68 Knight & Morris
" 380 Spies Bros. & Co.
" 150 F. W, Heilgers & Co.
Germany, 320 T. H. Lee

Gutta Percha—
Germany, 30 c. Kaltenbach & Schmitz
Holland, 22 Kleinwort, Sons, & Co.
E. Indies, 870 H. W. Jewesbury & Co.

Glucose—
Germany, 100 pkgs. 100 c. J. Barber & Co.
" 77 616 M. D. Co.
" 113 204 Barrett, Tagant & Co.
" 21 185 C. Tennant, Sons, & Co.
" 285 300 H.A. Litchfield & Co.
U. States, 500 500 Olyett & Francis
Germany, 300 300 J. F. Ohlmann
" 300 300 J. Barber & Co.
" 200 200 Anderson, Weber, & Smith
" 200 200 Horne & Co.
" 25 200 Props. Scott's Whf.
" 15 123 J. Cooper
France, 100 100 Barrett, Tagant, & Co.
Germany, 100 200 J. Forsey
" 25 210 Fellows, Norton, & Co.
" 100 100 Trier, Meyer, & Co.
" 25 200 Anderson, Weber, & Smith
" 25 175 W. Bush & Co.
" 100 200 J. Forsey
" 3 26 L. & I. D. Jt. Co.
" 500 500 J. F. Ohlmann
" 200 200 J. Cooper
U. States, 240 240 L. Sutro & Co.
Germany, 200 400 J. Forsey
" 25 210 L. Sutro & Co.
" 60 498 Anderson, Weber, & Smith
" 100 955 Seaward Bros.
" 200 200 Howell & Co.
" 60 480 T. M. Duche & Son
" 100 100 M. D. Co.
" 100 100 Barrett, Tagant, & Co.

Isinglass—
E. Indies, 7 pkgs. L. & I. D. Jt. Co.

Insect Powder—
U. States, 28 £. Davies, Turner, & Co.

Manure—
Phosphate Rock
France, 120 t. T. Farmer & Co
" 236 Moeller, Graetz, & Co.
Manure *(otherwise undescribed)*
Germany, 2 cs. Craven & Co
France, 5 t. J. Burnett & Sons
Germany, 20 F. W. Berk & Co.

Naphtha—
Germany, 18 brls. Stein Bros.
Holland, 8 Domeier & Co.

Plumbago—
Holland, 42 cks. Brown & Elmslie
Belgium, 9 W. Brandts, Sons, & Co.
Ceylon, 43 Hoare, Wilson, & Co.
Germany, 18 Fellows & Co.
" 2 Pokorney, Fielder, & Co.
" 20 W. W. Lupton
Ceylon, 437 Prop. Chamb. Whf.
Germany, 20 cs. O. Scholzig
" 3 Beresford & Co.

Potassium Sulphate—
Germany, 17 pkgs. Ross & Deering

Pyrites—
La Laja, 700 t. G. Ward & Sons
" 500 Forbes, Abbott, & Co.

Paraffin Wax—
U. States, 325 brls. H. Hill & Sons
" 3658 J. H. Usmar & Co.

Potassium Muriate—
Germany, 529 pkgs. Bessler, Walchter, & Co.

Saltpetre—
Germany, 120 brls. P. Hecker & Co.
Holland, 135 G. Meyer & Co.
Germany, 4 cks. Craven & Co.
" 35 P. Hecker & Co.
E. Indies, 594 bgs. Ralli Bros.
Germany, 16 cks. P. Hecker & Co.
" 100 T. Merry & Son
E. Indies, 209 L. & I. D. Jt. Co.

Stearine—
Holland, 50 bgs. H. Hill & Sons
France, 90 sks. J. Goddard & Co.
" 293 "
Belgium, 9 cks. H. Hill & Sons

Tartars—
Italy, 36 pkgs. Howards & Son
France, 22 Thames Steam Tug Co.
Italy, 7 " "
" 24 " "
" 22 Fellows, Norton, & Co.

Tartaric Acid—
France, 4 pkgs. Webb, Warter, & Co.
Holland, 12 "
" 10 N. Smith

Ultramarine—
Germany, 4 cks. G. Steinhoff
Belgium, 5 pkgs. Ohlenshlager Bros.
" 2 W. Harrison & Co.
Holland, 3 cks. "
Belgium, 2 Leach & Co.
Germany, 5 cs. G. Steinhoff

Zinc Oxide—
Holland, 334 cks. M. Ashby

LIVERPOOL.

Week ending Dec. 25th.

Albumen—
Marseilles, 10 cs.

Ammonia—
Bordeaux, 1 cs.

Borax—
Leghorn, 100 cs. 169 cks.

Brimstone—
Catania, 150
" 50

Barytes—
Rotterdam, 23 cks.
Antwerp, 20 bgs.

Boracic Acid—
Leghorn, 40 cks.

Bone Meal—
Calcutta, 667 bgs.
" 2,000

Bones—
Karachi, 4,000 bgs. Ralli Bros.
Boston, 4 cs.

Copperas—
Hamburg, 31 brls.

Chemicals *(otherwise undescribed)*
Boston, 36 bxs. J. Cassels & Co.
New York, 5 cs. Robinson & Allen

Carbon—
New York, 10 bxs. T. H. Boyce & Co.

Caoutchouc—
Pernambuco, 6 bls.
Ceara, 117 Bieber & Co.
Bageida, 6 pns. 31 brls. F. & A. Swanzy
Quittah, 3 25 "
" 2 J. H. Rayner & Co.
Accra, 5 F. & A. Swanzy
" 1 1 cs. C. Lane & Co.
" 28 cks. 12 J. P. Werner
" 14 5 1 J. J. Fischer & Co.
Salt Pond, 3 1 A. Millerson & Co.
Cape Coast, 23 1 L. Hart & Co.
" 37 1 Fletcher & Fraser
" 1 C L. Clare & Co.
" 1 S. E. Nordlinger
" 3 Havard & Co.
" 3 Whimsted & Watson
" 1 Grimwade, Ridley & Co.
" 2 Ihlers & Bell
" 8 "
Attaoboe, 2 bgs. Radcliffe & Durant
" 1 Pickering & Berthoud
Grand Bassam, 15 cks. F. & A. Swanzy
" 3 bgs. Edwards Bros.
" 2 Pickering & Berthoud
" 1 4 brls. W. D. Woodin
" 1 J. Sladelmann
" 6 cks. A. Verdier
" 3
Sierra Leone, 14 brls. Pickering & Berthoud
" 7 3 pns.
" 4 N. Waterhouse & Son
Isles de Los, 266 F. Colin & Co.

Caustic Potash—
Dunkirk, 9 dms.

Copper Ore—
Leghorn, 126 t.
Galveston, 3,480 sks.

Copper Precipitate—
Huelva, 6,947 bgs. Matheson & Co.
" 3,069

Dyestuffs—
Valonia
Constantinople, 133 bgs.
Smyrna, 196 t. Barry Bros.
Syra, 268 bgs. C. G. Nicolaidi
" 34 C. Nicolaides
" 277 bls.

Divi Divi—
Curacao, 225 bgs. D. Midgley & Sons
" 139
Rio Hache, 135 t. E. Brownbill & Co.
Dextrine
Bremerhaven, 160 bgs.
Sumac
Palermo, 200 bgs. J. Kitchin & Co.
" 260 bls. 675 bgs.
" 150
Venice, 500 bgs.
Argols
Bordeaux, 50 bgs. 1 ck.
Extracts
Antwerp, 1 ck. J. T. Fletcher & Co.
Cochineal
Grand Canary, 40 bgs. Beach & Co.
" 40 Lathbury & Co.
Teneriffe, 50 Intnl. Bank of London

Glu
H.
Gly
R
Hor
Br
A
M
Iodi
Va
Lith
Ro
Pyri
Hu
"
"
Pots
Ha
Pots
Ha
Phos
Dur
Pota
Har
Pitch
Ams
Sodi
Iqui
Rott
Saltp
Han
Calc
Sulphu
Catani
"
Stearin
Antwe
"
Tartar
Naples
"
"
Tartar
Messin
Rotterd
Verdig
Bordea
Utram
Rotter
Waste
Hambu
Zinc Ox
Antwerp

Week

Alkali—
Bremen,
Albume
Riga,
Barytes
Bremen,
Chemica
Stettin,
Danzig,
Antwerp,
Hamburg
Genoa,
Marseille
Dyestuff
Sumac
Trieste,
Palermo,
Alizari
Rotterda
"
"
Dextri
Hamburg
Dyest
Hamburg
Glucose—
New Yor
Hamburg
"
Stettin,
Hamburg
Glycerin
Venice,
Hamburg
Horn P
Hamburg

Manure
Hamburg, 17 cks. Wilson, Sons, & Co.

Pitch—
Bordeaux, 24 cks.
Antwerp, 9
Amsterdam, 14 H. F. Pudsey

Sugar of Lead—
Hamburg, 21 cks. Wilson, Sons, & Co.

Verdigris—
Bordeaux, 2 cks. Rawson & Robinson

GLASGOW.

Week ending December 26th.

Alum—
Hamburg, 117 cks.
Rotterdam, 13

Bone Meal—
Bombay, 4,104 bgs.

Barytes—
Rotterdam, 20 cks. J. Rankine & Son

Chemicals *(otherwise undescribed)*
Hamburg, 1 cs.
Rotterdam, 2

Cream Tartar—
Marseilles, 10 cks.

Chlorate of Manganese—
Rotterdam, 1 ck. J. Rankine & Son

Dyestuffs—
Tannin
Hamburg, 1 ck.

Madder
Rotterdam, 5 cks.
Bordeaux, 10
Rotterdam, 1

Alizarine
Rotterdam, 120 cks. J. Rankine & Son
" 220

Cutch
Halifax, 52 bxs.

Glucose—
Hamburg, 42 cks.

Phosphate Rock—
Antwerp, 500 bgs.

Potassium Prussiate—
Hamburg, 2 cks.

Potash—
Hamburg, 20 cks.

Sulphur Ore—
Drontheim, 150 t.

Sugar of Lead—
Hamburg, 20 cks.

Saltpetre—
Rotterdam, 7 cks.
Hamburg, 23 J. Poynter & Son

Stearine—
Bordeaux, 50 bgs.

Tar—
Marseilles, 63 brls.

Tartar—
Bordeaux, 2 cks.
" 2 J. & P. Hutchinson

Tartaric Acid—
Marseilles, 43 brls.

Ultramarine—
Rotterdam, 4 cks. 3 cs.

Zinc Oxide—
Rotterdam, 50 cks.

TYNE.

Week ending December 12th.

Alum Earth—
Hamburg, 18 cks. Tyne S. S. Co.

Copper Precipitate—
Huelva, 5,552 bgs. A. Guthrie

Glucose—
Hamburg, 15 cks. Tyne S. S. Co.
Rotterdam, 100 bgs.

Nitre Saltcake—
Harburg, 131 t.

Saltpetre—
Hamburg, 14 cks.

Sulphur Ore—
Huelva, 202 t. A. Guthrie

GOOLE.

Week ending Dec. 18th.

Alkali—
Ghent, 6 cbys.

Dyestuffs—
Alizarine
Rotterdam, 29 cks.
" 24

Madder
Rotterdam, 4 cks.

Dyestuffs *(otherwise undescribed)*
Boulogne, 72 cks.
" 2 cs.

Glucose—
Hamburg, 25 cks.
" 200 bgs.
Calais, 100

Potash—
Calais, 81 drms.
" 13 cks.

Saltpetre—
Rotterdam, 85 bgs.

Tartar—
Rotterdam, 5 cks.

Waste Salt—
Hamburg, 200 cks.
Hamburg, 508 bgs.

Waste Alum—
Rotterdam, 8 cks.

GRIMSBY.

Week ending December 21st.

Albumen—
Hamburg, 4 pkgs.

Chemicals *(otherwise undescribed)*
Hamburg, 5 pkgs.

Dyestuffs *(otherwise undescribed)*
Antwerp, 5 pkgs.

Glucose—
Hamburg, 25 cks.
" 90 bgs.
" 20

Zinc Ashes—
Antwerp, 26 cks.

EXPORTS OF CHEMICAL PRODUCTS

FROM

THE PRINCIPAL PORTS OF THE UNITED KINGDOM.

LONDON.

Week ending December 21st.

Ammonia Carbonate—
Adelaide, 5 c. £11

Arsenic—
Yokohama, 1 t. 10 c. £21
Napier, 2 27

Alum—
Salonica, 20 t. 2 c. £110

Acids *(otherwise undescribed)*
Sydney, 12 cs. £31

Ammonia Sulphate—
Berbice, 20 t. £250
Martinique, 30 17 c. 392
Cologne, 35 422
Hamburg, 28 9 335
Valencia, 500 6,000

Acetic Acid—
Auckland, 8 cks. £15

Ammonia Liquid—
B. Ayres, 20 cs. £40

Boracic Acid
Melbourne, 1 t. 2 c. £42

Benzoic Acid—
Philadelphia, 1 q. £10

Bone Meal—
Martinique, 144 t. 16 c. £470

Borax—
Melbourne, 17 c. £20
Keppel Bay, 20 cs. 83

Brimstone—
M. Video, 1 t. 4 c. £17

Copper Sulphate
Adelaide, [illegible] £111
Odessa, 6 [illegible] 111
B. Ayres, 5 [illegible] 111
Cape Town, 1 72
Brisbane, [illegible] 117
Dunedin, [illegible] 17

Cream Tartar
Adelaide, [illegible] £171
Sydney, 5 [illegible]
Melbourne, [illegible] 47

Caustic Soda
[illegible] lbs. £[illegible]
[illegible] 74
[illegible] £12
[illegible]
41

Citric Acid—
Barcelona, 2 cks. £75
Hamburg, 5 c. 17
Amsterdam, 2 kegs. 13
Revel, 1 t. 137
Genoa, 1 15 c. 283
Trieste, 5 40

Chemicals *(otherwise undescribed)*
Auckland, 40 cs. £30
Cadiz, 8 pkgs. 48
Sydney, 4 t. 12 c. 103

Disinfectants—
Melbourne, 21 cs. £40

Guano—
Jersey, 83 t. £800

Manure—
Singapore, 31 t. 7 c. £240
Puerto Angel, 10 87
Rotterdam, 30 375
Boston, 2 cs. 11
Copenhagen, 160 t. 736
Alicante, 110 16 c. 836
Penang, 675 5,400
St Lucia, 103 996
Bordeaux, 43 4 193
Barbadoes, 75 750
Demerara, 81 9 927

Oxalic Acid—
Venice, 26 t. £1,090
Paris, 5 175
Melbourne, 12 c. 25

Potassium Chlorate—
Lisbon, 10 c. £21

Phosphorus—
Las Palmas, 2 c. £21
Dunedin, 2,300 lbs. 215
Otago, 500 50

Phosphoric Acid Solid—
Demerara 2 t. 4 c. £322

Saltpetre—
Adelaide, 1 t. 15 c. £31

Sulphuric Acid—
Bombay, 9 c. £9
Jersey, 10 4
Algoa Bay, 5 5
Vigo, 10 cs. 15
Natal 1 t. 13 c. 9

Sheep Dip
N[illegible] £[illegible]
Calcutta [illegible] 115
Patagonia [illegible] 400
Algoa Bay, 200 560
Gallegos, 50 100

Sodium Silicate—
Auckland, 4 t. 18 c. £17

Salicylic Acid—
Brisbane, 1 cs. £30

Sodium Sulphate—
Ghent, 90 t. £90

Tartaric Acid—
Adelaide, 1 t. £131
Kurrachee, 10 c. 69
Sydney, 58 kgs. 300
Gibraltar, 2 c. 18

LIVERPOOL.

Week ending Dec. 21st.

Alkali—
Liverpool, 68 tcs.

Ammonia Carbonate—
Hamburg, 6 c. 1 q. £8
New York, 14 t. 10 1 470

Alum—
Alexandria, 6 t. £31
Beyrout, 8 1 c. 41
Smyrna, 19 8 1 q. 99
Volo, 1 9 2 8
Callao, 12 2 10
Kurrachee, 81 7 368
Buenos Ayres, 3 t. 15

Ammonia Sulphate—
Bordeaux, 59 t. £723
Hamburg, 20 7 c. 1 q. 244
Valencia, 193 17 2 2,410
Rotterdam, 52 3 3 626

Ammonia Muriate—
Boston, 3 t. 15 c. £95
Philadelphia, 6 3 127

Acetate of Lime—
Odessa, 15 t. 9 c. 1 q. £165

Bleaching Powder—
Baltimore, 113 cks. 5 cs.
Boston, 385 192 tcs.
Bilbao, 41 kgs.
Boston, 89
Calcutta, 92
Genoa, 34
Montreal, 4
Malaga, 35 brls.
Nantes, 104
New York, 107
New Orleans, 5 78
Philadelphia, 322 brls.
Vigo, 6

Borax—
Amsterdam, 5 c. 3 q. £8
Rouen, 10 t. 14 301
Halifax, 9 14

Borax Salts—
Havana, 2 t. 7 c. 2 q. £76

Brimstone—
Calcutta, 20 t. 12 c. 2 q. £98

Caustic Soda—
Antwerp, 8 bxs.
Astoria, 20 dms.
Barcelona, 200 90
Carthagena, 5
Constantinople, 10
Fiume, 45
Genoa, 80
Ghent, 12
Havana, 50
Honolulu, 10
Kobe, 50 cks.
La Guayra, 6
Lisbon, 150
Leghorn, 100
Madrid, 57
Mauritius, 4
Messina, 60
New Orleans, 200
New York, 850
Norrkoping, 35
Oporto 15
Pasages, 25
Pernambuco, 30
Portland O. 75
Puerto Cabello, 31
Rouen, 15
Terneuzen, 18
Talcahuano, 40
Trieste, 32
Valencia, 20 brls. 14
Valparaiso, 30
Yokohama, 300

Copper Sulphate—
Genoa, 5 t. 2 q. £90
Venice, 4 17 c. 1 88
Barcelona, 23 18 2 493
St Nazaire, 5 1 2 100
Bilbao, 66 1 1,213
Bordeaux 10 180
Rouen, 20 7 441
Tarragona, 6 127
Alexandria, 9 c. 11
Barcelona, 67 t. 17 c. £1,297
Antwerp, 5 3 2 q. 93
Leghorn, 1 1 3 21
Marseilles, 11 14 2 250
Bilbao, 10 6 1 185
Genoa, 63 6 2 1,171
Nantes, 60 3 3 1,320
San Sebastian, 10 4 185
Trieste, 1 1 1 24

Carbolic Acid—
Genoa, 14 c. 61
New York, 5 t. 8 3 q. 369
Rouen, 2 185

Copperas—
Malaga, 6 c. £14
Melbourne, 11 t. 20
Syra, 7 18 14
Smyrna, 9 8 20

Charcoal—
Rouen, 50 t. £223

Citric Acid—
Vera Cruz, 5 c. £37

Caustic Potash—
Calcutta, 1 t. 1 c. £21
New York, 12 11 252
Brantford, 4 6 86
Havana, 5 1 q. 9

Chemicals (*otherwise undescribed*)
New York, 1 cs, 4 t. 14 c. 1 q. £447
East London, 2 8
Colon, 1 keg. 5
Vera Cruz, 2 8 c. 3 q. 102
Santos, 8 pkgs. 160
Rio de Janeiro, 99 cs. 129

Copper Precipitate—
Rotterdam, 47 t. 10 c. £1 660
Antwerp, 27 14 1 q. 831

Calcium Chloride—
Rouen, 5 t 6 c. £9

Epsom Salts—
Rio Janeiro, 50 kgs.

Glycerine—
Madeira, 1 cs. £5
Vera Cruz, 2 drs. 46
Hamburg, 5t. 19 c. 170

Iron Oxide—
Rio Grande do Sul, 14 t. 2 c 3 q. £35

Magnesia—
Rio Janeiro, 40 cs.

Manure—
Genoa 56 t. 3 c. 2 q. £196

Magnesium Chloride—
Halifax, 4 t. £16

Muriatic Acid—
Callao, 10 c. £6

Oxalic Acid—
Barcelona, 1 t. 8 c. 3 q. £60
Oporto, 3 6

Oxolate Potash—
Genoa, 12 t. 8 c. £409

Potassium Chlorate—
Philadelphia, 10 t. £560
Rio de Janeiro, 5 c 10
Antwerp, 1 42
New York, 18 11 746
Valparaiso, 16 40
Yokohama, 5 210
Genoa, 1 55
Hiogo, 15 800
Hamburg, 13 624
Santander, 10 12
Santos, 3 163

Phosphorus—
Buenos Ayres, 1 t. 8 c. 3 q. £246
Havana, 18 3 219

Potassium Bichromate—
Havre, 19 c. £30
Yokohama, 11 t. 8 404
Rio de Janeiro, 1 3 2 q. 44

Potassium Carbonate—
New York, 7 t. 4 c. £108

Picric Crystals—
New York, 10 c. £85

Potash—
Amsterdam, 1 t 7 c. £29
Monte Video, 1 6 3 q. 40

Potassium Muriate—
St John's, 1 t. 19 c. 1 q. £16

Phosphate—
Hamburg, 54 t. £220

Potassium Prussiate—
Rio de Janeiro, 10 c. £36

Salt—
Antwerp, 46 t.
Boston, 100
Calcutta 3,022
Cape Town, 74
Chicago, 190
Demerara, 40
Lagos, 107
Lavannah, 18
Manoh, 45
New Orleans 1695
Philadelphia, 80
Portland O., 100
San Francisco, 230
Sydney 300
Trinidad, 12

Soda Ash—
Baltimore, 343 cks.
Barcelona 10
Beyrout, 50 brls.
Boston, 387 128 tcs 1031 bgs.
Constantinople, 200 brls.
Corfu, 50
Dundas, 18 tcs.
Malta, 20 cks.
M. Video, 50
Naples, 50
New York, 220 851 tcs.
Oporto, 21
Philadelphia, 610
Puerto Cabello, 16 brls.
Savona 33 cks.
Sherbrooke, 18 tcs.
Tuticorin, 14 brls
Rio Janeiro, 20
Smyrna, 460

Sodium Bicarbonate—
Barcelona, 155 kgs.
Batoum, 60
Bombay, 20
Copenhagen, 330
Catania, 50
Genoa, 10 cks.
Hamilton, 50 cks.
Halifax, 120
Karachi, 200
Montreal 9
New Brunswick 50
Odessa, 100 kgs.
Talcahuano 10

Sodium Silicate—
Corunna, 10 brls

Soda Crystals—
Beyrout, 27 brls.
Boston, 280
Constantinople, 197
Halifax, 25
Mauritius, 13
Malta, 50
New Orleans, 25
Rotterdam, 20
Smyrna, 245
Valparaiso, 50

Sheep Dip—
Cape Town, 2 t 5 c. £49
B. Ayres, 1 13 1 q. 41
M. Video, 200 drs. 47
East London, 10 23

Salammoniac—
Beyrout, 3 t. 11 c 2 q. £119
Marseilles, 2 2 71
Alexandria, 1 34
Havana, 3 5
Malta, 5 8

Sodium Sulphate—
Philadelphia, 5 t. 2 c. 2 q. £26

Saltpetre—
Pernambuco, 1 t. 5 c. £30
Valparaiso, 8 5 1 q. 176

Saltcake—
Baltimore, 231 t. 12 c. £402
Philadelphia, 5 12
New York, 25 50

Sodium Potash—
Bari, 2 q. £10

Superphosphate—
Stettin, 100 t. 1 c. 1 q £312
Hamburg, 210 1 1 707

Sodium Nitrate—
Malaga, 3 t. £26
Shanghai, 5 2 c. £46
Ancona, 12 2 1 q. 103
Cartagena, 3 2 35

Sodium Chlorate—
Vera Cruz, 1 ck. £8

Sodium Biborate
Rotterdam, 10 c. 1 q. £15
Shanghai, 1 t. 3 3 34
Antwerp, 5 17 3 163

Sulphur—
St. John's N. F., 1 t. 9 c. £12
Perambuco, 1 2 3 q. 7
Boston 58 1 259
Halifax, 30 2 135
Calcutta, 1 7

Sulphuric Acid—
Santos, 10 pkgs. £15

Sodium Phosphate—
Pasages, 10 c. £14

Tartar—
New York, 12 t. £1,140

Wood Acid—
Vera Cruz, 6 cs. £16

Zinc Oxide—
Santos 1 t. £22

Zinc Chloride—
Halifax, 1 t. 3 c. 1 q £9.

HULL.

Week ending Dec. 24th.

Alkali—
Fiume, 18 cks
Harlingen, 75
Hango, 4 drms.
Venice, 31
Trieste, 10

Bleaching Powder—
Fiume, 1 ck.
Trieste, 21 cks.
Venice, 10

Bone Size—
Harlingen, 10 cks.

Chemicals (*otherwise undescribed*)
Amsterdam, 200 cks.
Antwerp, 1
Bremen, 30
Copenhagen, 3 pkgs.
Christiania, 82 5 cs.
Dunkirk 33 19 drms.
Fiume, 22
Gothenburg, 2 4 12
Hamburg, 13 4 1
Lbao, 5
Riga 50
Rotterdam, 60 50 40
Reval, 37 100
Stettin, 24 100
Trieste, 73
Venice, 45

Caustic Soda—
Copenhagen, 130 drms.
Riga, 810
Reval, 66
Trieste, 113 cks.

Dyestuffs (*otherwise undescribed*)
Antwerp, 1 drm.
Hamburg, 1 1 cs 4 cks.
Rouen, 3 bls.
Rotterdam, 30 20 pkgs.

Manure—
Antwerp, 30 t.
Esbjerg, 1,700 bgs.
Hamburg, 188

Pitch—
Antwerp, 107 t.

Slag—
Hamburg, 328 t.

TYNE.

Week ending Dec. 25th.

Alum—
Newfairwater, 1 t.

Alkali—
Amsterdam, 5 t. 10 c.
Reval, 9 15

Bleaching Powder—
Newfairwater, 57 t. 19 c.
" 20 1
Amsterdam, 18 3
" 26 3
Pasages, 5 4
Reval, 70 7
Antwerp, 21 7
Hamburg, 21 14
Gothenburg 15 2
Baltimore, 50 5
Christiania, 10 4
" 1 4

Caustic Soda—
Antwerp, 1 q.
" 7 t. 17 c.
Christiana, 6 11

Litharge—
Hamburg, 12 c.

Manure—
Hamburg, 50 t. 8 c.

Magnesia—
Hamburg, 3 c.

Paraffin Wax—
Danzig, 10 t

Pitch—
Antwerp, 1 q.
" 110 t.

Potassium Chlorate—
Hamburg, 5 c.
Riga, 4 19 c.

Plumbago—
Bilbao 2 t. 9 c.

Soda Ash—
Newfairwater, 50 t. 14 c.
Antwerp, 13

Soda—
Newfairwater, 4 t. 19 c.
Amsterdam, 2 10
" 23 6
Launceston, 43 14
Antwerp, 2
" 1
Aarhuus, 23 3
Christiania, 10

Salt—
Laun

Sulph
Gothe

Sodiu
Haml

Sodiu
Baltin
Bilbac
Christ

Sodiu
Bilbac

Turpe
Antwe
"
Christ

Tar—
Antwe
"
Hambu

Tar Oil
Bilbao,

G

We

Benzol—
Dieppe,
Rotterda

Charcoa
Rouen,

Dyestuff
Anilir
Calcutta,

Epsom S
Melbour
Christian

Gutta P
Amsterda

Mangane
Rouen,

Potassiu
Christiani
Rouen,
Rotterdar

Pitch—
Christiani
Santande
Dieppe,

Paraffin
Rouen,
Dieppe,
Santander

Sodium B
Rouen,
New York

G

Week e

Coal Prod
Dieppe,

Chemicals
Hamburg,

C

Week e

Bone Size
Boulogne,

Benzol—
Dunkirk,
Hamburg,
Ternenzen,

Chemicals
Boulogne,
Hamburg,
"
Rotterdam

Coal Prod
Dunkirk,
Ghent,
Hamburg,
Rotterdam

Dyestuffs
Dunkirk,
Ghent, 1
Hamburg,

Manure—
Boulogne,
Hamburg,

Pitch—
Dunkirk,

PRICES CURRENT.

THURSDAY, JANUARY 9, 1890.

PREPARED BY HIGGINBOTTOM AND CO., 116, PORTLAND STREET, MANCHESTER.

The values stated are F.O.R. *at maker's works, or at usual ports of shipment in U.K. The price in different localities may vary*

			£	s.	d.
Acids:—					
Acetic, 25 °/₀ and 40 °/₀		per cwt	7/6 & 12/-		
,, Glacial		,,	2	10	0
Arsenic, S.G., 2000°		,,	0	12	0
Chromic 82 °/₀	nett	per lb.	0	0	6½
Fluoric		,,	0	0	3¼
Muriatic (Tower Salts), 30° Tw.		per bottle	0	0	6½
,, (Cylinder), 30° Tw.		,,	0	2	11
Nitric 80° Tw.		per lb.	0	0	2
Nitrous		,,	0	0	1¼
Oxalic	nett	,,	0	0	3¼
Picric		,,	0	1	5
Sulphuric (fuming 50 °/₀)		per ton	15	10	0
,, (monohydrate)		,,	5	10	0
,, (Pyrites, 168°)		,,	3	2	6
,, (,, 150°)		,,	1	6	0
,, (free from arsenic, 140/145°)		,,	1	10	0
Sulphurous (solution)		,,	2	15	0
Tannic (pure)		per lb.	0	1	0½
Tartaric		,,	0	1	4
Arsenic, white powered		per ton	12	10	0
Alum (looselump)		,,	4	12	6
Alumina Sulphate (pure)		,,	5	0	0
,, ,, (medium qualities)		,,	4	10	0
Aluminium		per lb.	0	15	0
Ammonia, ·880=28°		per lb.	0	0	3
,, =24°		,,	0	0	1⅛
,, Carbonate	nett	,,	0	0	3½
,, Muriate		per ton	24	0	0
,, ,, (sal-ammoniac) 1st & 2nd		per cwt.	36/-& 34/-		
,, Nitrate		per ton	40	0	0
,, Phosphate		per lb.	0	0	10½
,, Sulphate (grey), London		per ton	12	5	0
,, ,, (grey), Hull		,,	12	2	6
Aniline Oil (pure)		per lb.	0	0	10
Aniline Salt ,,		,,	0	0	9
Antimony		per ton	75	0	0
,, (tartar emetic)		per lb.	0	1	1
,, (golden sulphide)		,,	0	0	10
Barium Chloride		per ton	7	0	0
,, Carbonate (native)		,,	3	15	0
,, Sulphate (native levigated)		,,	45/- to 75/-		
Bleaching Powder, 35 %		,,	5	5	0
,, Liquor, 7 %		,,	2	10	0
Bisulphide of Carbon		,,	12	15	0
Chromium Acetate (crystal		per lb.	0	0	6
Calcium Chloride		per ton	2	0	0
China Clay (at Runcorn) in bulk		,,	16/- to 32/-		
Coal Tar Dyes:—					
Alizarine (artificial) 20 %		per lb.	0	0	9
Magenta		,,	0	3	9
Coal Tar Products					
Anthracene, 30 %, f.o.b. London		per unit per cwt.	0	1	1
Benzol, 90 % nominal		per gallon	0	3	6
,, 50/90 ,,		,,	0	2	8
Carbolic Acid (crystallised 35°)		per lb.	0	0	11¼
,, (crude 60°)		per gallon	0	2	10
Creosote (ordinary)		,,	0	0	2
,, (filtered for Lucigen light)		,,	0	0	3
Crude Naphtha 30 % @ 120° C.		,,	0	1	2
Grease Oils, 22° Tw.		per ton	3	0	0
Pitch, f.o.b. Liverpool or Garston		,,	1	12	0
Solvent Naphtha, 90 % at 160°		per gallon	0	1	6
Coke-oven Oils (crude)		,,	0	0	2½
Copper (Chili Bars)		per ton	51	7	6
,, Sulphate		,,	23	10	0
,, Oxide (copper scales)		,,	52	10	0
Glycerine (crude)		,,	30	0	0
,, (distilled S.G. 1250°)		,,	55	0	0
Iodine		nett, per oz.	8d. to 9d.		
Iron Sulphate (copperas)		per ton	1	7	6
,, Sulphide (antimony slag)		,,	2	0	0
Lead (sheet) for cash		,,	17	0	0
,, Litharge Flake (ex ship)		,,	18	0	0
,, Acetate (white ,,)		,,	23	10	0
,, ,, (brown ,,)		,,	19	0	0
,, Carbonate (white lead) pure		,,	19	0	0
,, Nitrate		,,	22	5	0
Lime, Acetate (brown)		,,	8	0	0
,, ,, (grey)		per ton	14	0	0
Magnesium (ribbon and wire)		per oz.	0	2	3
,, Chloride (ex ship)		per ton	2	7	6
,, Carbonate		per cwt.	1	17	6
,, Hydrate		,,	0	10	0
,, Sulphate (Epsom Salts)		per ton	2	15	0
Manganese, Sulphate		,,	18	0	0
,, Borate (1st and 2nd)		per cwt.	60/- & 42/6		
,, Ore, 70 %		per ton	4	5	0
Methylated Spirit, 61° O.P.		per gallon	0	2	2
Naphtha (Wood), Solvent		,,	0	3	6
,, ,, Miscible, 60° O.P.		,,	0	4	0
Oils:—					
Cotton-seed		per ton	22	10	0
Linseed		,,	22	10	0
Lubricating, Scotch, 890°—895°		,,	7	5	0
Petroleum, Russian		per gallon	0	0	5¼
,, American		,,	0	0	6⅛
Potassium (metal)		per oz.	0	3	10
,, Bichromate		per lb.	0	0	4
,, Carbonate, 90% (ex ship)		per ton	18	5	0
,, Chlorate		per lb.	0	0	4⅝
,, Cyanide, 98%		,,	0	2	0
,, Hydrate (Caustic Potash) 80/85 %		per ton	23	0	0
,, ,, (Caustic Potash) 75/80 %		,,	21	15	0
,, ,, (Caustic Potash) 70/75 %		,,	20	15	0
,, Nitrate (refined)		per cwt.	1	3	6
,, Permanganate		,,	4	5	0
Prussiate Yellow		per lb.	0	0	9½
,, Sulphate, 90 %		per ton	9	10	0
,, Muriate, 80 %		,,	7	15	0
Silver (metal)		per oz.	0	3	7⅛
,, Nitrate		,,	0	2	5½
Sodium (metal)		per lb.	0	4	0
,, Carb. (refined Soda-ash) 48 %		per ton	5	5	0
,, ,, (Caustic Soda-ash) 48 %		,,	4	2	6
,, ,, (Carb. Soda-ash) 48%		,,	4	12	6
,, ,, (Carb. Soda-ash) 58 %		,,	5	12	6
,, ,, (Soda Crystals)		,,	2	12	0
,, Acetate (ex-ship)		,,	15	10	0
,, Arseniate, 45 %		,,	10	0	0
,, Chlorate		per lb.	0	0	6¼
,, Borate (Borax)		nett, per ton.	29	10	0
,, Bichromate		per lb.	0	0	3
,, Hydrate (77 % Caustic Soda)		per ton.	8	5	0
,, ,, (74 % Caustic Soda)		,,	8	0	0
,, ,, (70 % Caustic Soda)		,,	7	10	0
,, ,, (60 % Caustic Soda, white)		,,	6	10	0
,, ,, (60 % Caustic Soda, cream)		,,	6	2	6
,, Bicarbonate		,,	5	0	0
,, Hyposulphite		,,	5	0	0
,, Manganate, 25%		,,	8	5	0
,, Nitrate (95 %) ex-ship Liverpool		per cwt.	0	8	4½
,, Nitrite, 98 %		per ton	27	0	0
,, Phosphate		,,	15	15	0
,, Prussiate		per lb.	0	0	6
,, Silicate (glass)		per ton	5	7	6
,, ,, (liquid, 100° Tw.)		,,	3	17	6
,, Stannate, 40 %		per cwt.	2	0	0
,, Sulphate (Salt-cake)		per ton.	1	5	0
,, ,, (Glauber's Salts)		,,	1	10	0
,, Sulphide		,,	7	15	0
,, Sulphite		,,	5	5	0
Strontium Hydrate, 98 %		,,	8	0	0
Sulphocyanide Ammonium, 95 %		per lb.	0	0	7¼
,, Barium, 95 %		,	0	0	5½
,, Potassium		,,	0	0	7¼
Sulphur (Flowers)		per ton.	7	0	0
,, (Roll Brimstone)		,,	6	0	0
,, Brimstone: Best Quality		,,	4	15	0
Superphosphate of Lime (26 %)		,,	2	10	0
Tallow		,,	25	10	0
Tin (English Ingots)		,,	101	0	0
,, Crystals		per lb.	0	0	6¼
Zinc (Spelter		per ton.	24	7	6
,, Chloride (solution, 96° Tw.		,,	6	0	0

THE

CHEMICAL TRADE JOUI

Publishing Offices: 32, BLACKFRIARS STREET, MANCHESTE.

No. 139. **SATURDAY, JANUARY 18, 1890.**

Contents.

Notices.

All communications for the *Chemical Trade Journal* should be addressed, and Cheques and Post Office Orders made payable to—

DAVIS BROS., 32, Blackfriars Street, MANCHESTER.

Our Registered Telegraphic Address is—

"Expert, Manchester."

The Terms of Subscription, commencing at any date, to the *Chemical Trade Journal*,—payable in advance,—including postage to any part of the world, are as follow:—

Yearly (52 numbers) .... **12s. 6d.**
Half-Yearly (26 numbers) .... **6s. 6d.**
Quarterly (13 numbers) .... **3s. 6d.**

Readers will oblige by making their remittances for subscriptions by Postal or Post Office Order, crossed.

Communications for the Editor, if intended for insertion in the current week's issue, should reach the office not later than **Tuesday Morning.**

Articles, reports, and correspondence on all matters of interest to the Chemical and allied industries, home and foreign, are solicited. Correspondents should condense their matter as much as possible, write on one side only of the paper, and in all cases give their names and addresses, not necessarily for publication. Sketches should be sent on separate sheets.

We cannot undertake to return rejected manuscripts or drawings, unless accompanied by a stamped directed envelope.

Readers are invited to forward items of intelligence, or cuttings from local newspapers, of interest to the trades concerned.

As it is one of the special features of the *Chemical Trade Journal* to give the earliest information respecting new processes, improvements, inventions, etc., bearing upon the Chemical and allied industries, or which may be of interest to our readers, the Editor invites particulars of such—when in working order—from the originators; and if the subject is deemed of sufficient importance, an expert will visit and report upon the same in the columns of the *Journal*. There is no fee required for visits of this kind.

We shall esteem it a favour if any of our readers, in making inquiries of, or opening accounts with advertisers in this paper, will kindly mention the *Chemical Trade Journal* as the source of their information.

Advertisements intended for insertion in the current week's issue, should reach the office by **Wednesday morning** at the latest.

Advertisements.

Prepaid Advertisements of Situations Vacant, Premises on Sale or To be Let, Miscellaneous Wants, and Specific Articles for Sale by Private Contract, are inserted in the *Chemical Trade Journal* at the following rates:—

30 Words and under .... 2s. 0d. per insertion.
Each additional 10 words .... 0s. 6d. „

Advertisements of Situations Wanted are inserted at one-half the above rates when prepaid, viz:—

30 Words and under .... 1s. 0d. per insertion.
Each additional 10 words .... 0s. 3d. „

Trade Advertisements, Announcements in the Directory Columns, and all Advertisements not prepaid, are charged at the Tariff rates which will be forwarded on application.

CURRENT TO

MORE FINANC

It is rumoured that negotiations b purpose of amalgamating the a prietary, somewhat similar to the S proposal has already been welco sidered by some of the leading man

It would take a very wise man to p a wild proposal; but we hear the Promoters are generally hopeful, they selves, hope in their schemes, and the lous British public, but in this instanc them to sit down and count the cost, profit nowhere. Manufacturers will a tion in the improvement of their proc engineering.

SMOKE AND COAL

How times have changed since the d one Edmund Rowe was tried, convicte crime of burning sea-coal in London— was issued forbidding its use in Lon when prelates of the Established Chu town on account of the noisome smell by the burning of sea-coal"—when London would not come into any hous coals were burned, nor eat of the meat t with sea-coal." London alone burnt last year.

The question now is: with all our m coal be burned without causing so mu which we reply: most assuredly, pro wish this object attained go about the in the right way. It is now an open se project of a large fund for testing cert appliances is bound to end in failure t aptly styled the apathy of the stea chester appeal has been like the mo has brought forth nothing but the m mistake to presume that the steam us subject of economy in the combustion c economy that produces black smoke. more in steam raising than the mere f heat units, but this is not always patent loudest on the smoke question.

Many times in these columns have we that the ordinary householder is as m manufacturer, and that it is the invi which causes so much destruction to ve of our large towns. Has any serious at force the householder to feel his respon

PETROLEUM DANGE

When the petroleum steamer blew the public was placed in a state of grea of a new danger. When the petroleum took fire in the river Wear, at Sunderla tocsin was sounded again in many quarters. When will our capitalists learn that the handling of hazardous goods should only be entrusted to those who possess the necessary technical qualifications? We shall await the Board of Trade enquiry with interest, but at the same time the thought strikes us that many new industries have been nipped in the bud by the short sightedness of the introducers; if a little more of the profits were spent in providing good technical advice, so much better would it be for all parties concerned.

ELECTRIC LIGHTING DANGERS.

A paper, by Mr. Chas. W. Vincent, in the current issue of the *Nineteenth Century*, is well worth reading. The subject is "The Dangers of Electric Lighting," and, though we by no means share his alarmist views, there is a great deal we should all bear in mind. There is no doubt that the use of electricity is extending, and that very rapidly, so that it is imperative everyone should have a good general knowledge respecting it. There seems a desire in many quarters to intensify the dangers that might arise from the introduction of electricity, but this can in most instances be traced to interested sources.

A writer in the *Newcastle Daily Chronicle* tells us he has "personally experienced shocks of 500 volts. and upwards without receiving the slightest injury," but he should consider that we are not all electrical salamanders.

SOIL.

SINCE the system of artificial fertilisation is based upon the addition to the soil of the materials wanting in it, it is of the greatest importance to know the composition of the land which is to be treated, and the analysis of arable soils has therefore come to be of much importance in scientific agriculture.

M. Gosselet, professor of geology to the faculty of science at Lille, has pointed out in his recent lectures on the geology of the department of the Nord, the interest attached not only to the study of the soil itself, but also to that of the sub-soil, which, it is obvious, must exert considerable influence on the properties of the arable soil. M. Gosselet defines the soil and sub-soil as follows:—

To farmers, the soil is that part of the earth which they work, which is impregnated with organic detritus and which furnishes support and nourishment to plants. The sub-soil is the layer immediately underlying this, which is not disturbed by the ploughshare. The soil is merely a portion of the sub-soil which has been repeatedly broken up and enriched by manure and fertilisers. It follows that a knowledge of the sub-soil is necessary to enable the nature of the soil itself to be comprehended. Many of our agriculturists do not appear to suspect this. Led away by the wonders of chemistry, they imagine that they know all about their fields when they have taken up a shovelful of earth here and there and had it analysed. This soil, however, taken from the surface, alters every year according to the culture to which it is submitted, and the particular material thus removed, according to the various manners with which it is treated, according to the action of the rain in washing out soluble matter, or bringing to it substances from its neighbourhood. But what can chemical analysis tell as to the physical properties of the soil, its permeability, its density, its structure?

The various samples taken from the same property are still frequently mixed and the mixture analysed, which is very like ascertaining the value of an orchard, by taking a fruit from each tree and making a preserve from the collection. Hitherto the French farmer has been indifferent to geology. He will wait to learn it until he has suffered from the applications of it made by foreign competitors.

The idea entertained by geologists of the soil and sub-soil is quite opposed to the preceding. For them the arable portion has no interest; it is an artificial product, and is neglected. They call soil the superficial layer formed under the influence of denudation and deposition, which have fashioned the surface of the earth during the last geological periods. The sub-soil is composed of more ancient layers, which, in our country at least, are of aqueous origin, and which have contributed their fragments to constitute the present soil.

The terms soil and sub-soil on French geological maps have to be interpreted in this way. Those constructed to a small scale only show the soil; others, on a larger scale, such as the new geological map of Belgium, show both soil and sub-soil; whilst others, again, show the soil occasionally, but are mainly intended to represent the sub-soil. Of this kind are maps on a moderate scale, such as that of France on the scale of one-eighty thousandth.

Legal.

THE ALLEGED NUISANCE AT MASSEY BROOK BONE WORKS.

A CASE of considerable interest to manufacturers on the one hand, and to property owners on the other, was tried in the Nisi Prius Court at the Liverpool Assizes on the 18th and 19th ult., when Mr. Justice Charles, sitting without a jury, heard the Chancery Action—Rylands and another v. Gibson & Co. The plaintiffs were represented by Mr. Gully, Q.C., Mr. Joseph Walton, and Mr. Rylands, and the defendants by Mr. Bigham, Q.C., Mr. Pickford, and Mr. Maberly.

The Massey Brook Bone Works are situated about 500 yards from Massey Hall, Thelwall, a handsome mansion built by the late Mr. Rylands upon an estate of 38 acres purchased by him in 1882. When Mr. Rylands died in February, 1887, he left the house and estate to his wife, Mrs. Caroline Rylands, and the reversion to Mr. Lewis Gordon Rylands. These are the plaintiffs in present action.

Mrs. Rylands did not continue in the house long after her husband's death. It was put into the hands of an agent for sale, but various negotiations for its purchase fell through, on account, it is said, of the unpleasant odours emanating from the bone mill adjacent.

The works in question have been in existence some 40 years, and have, during that time, been used for various purposes. The present defendants came into possession of them in 1885, and commenced the business of bone boiling, erecting amongst other plant a chimney about 100 feet high. From this date the nuisance seems to have made itself felt, and from this chimney were evolved the odours that have given rise to the action.

The process of boiling has been carried on in three pans, holding 2, 3, and 7 tons of bones respectively. It lasts about 48 hours, and the liquor is subsequently drawn off into an evaporator, where it is concentrated to a thicker consistency. The steam and organic vapours driven off from the pans and evaporator were taken into the boiler flue, and, for the defence, it was submitted that this treatment was sufficient to ensure the combustion of all the offensive volatile organic matters, and effectually prevent the escape of unpleasant odours from the chimney top. It was further suggested that the smells complained of arose from nightsoil being conveyed along the canal or spread upon the surrounding fields, from the Bridgewater canal itself; or even from a pond in the plaintiffs' own grounds. The case for the defendants was also supported by the evidence of Dr. Burghardt and Mr. Charles Estcourt, both analytical chemists, of Manchester, who spoke to the excellence of the arrangements at the works for dealing with the matter, and to the impossibility of offensive smells proceeding from the chimney. Witnesses on the other side, however, were unanimous as to the existence of the evil, and in their belief that it proceeded from the bone mill chimney. What might be called the low level nuisance, arising from the "green" and boiled bones in stock, would not travel nearly so far as Massey Hall. Letters of complaint had been addressed by Mrs. Rylands to the Lymm Local Board in 1886, and again in 1888, but to no purpose.

His Lordship gave judgment on Friday. He said that in that case Mrs. Caroline Rylands and another sought an injunction to restrain the defendant, who traded under the name of Gibson & Co., from causing certain offensive odours which the plaintiffs alleged proceeded from his works. The claim originally was for damages for injury to the plaintiffs' premises in consequence of the emission of those odours, but that was not insisted upon at the hearing, and the plaintiffs confined themselves to an application for an injunction to restrain the defendant from continuance, or repetition, of the nuisance alleged to have been committed by him. The plaintiffs' premises were originally bought by the late Mr. Peter Rylands, somewhere about the year 1882. They consisted of a house, which he enlarged when he purchased the property, together with a considerable number of acres surrounding it. The late Mr. Rylands resided there until his death in 1887; not all the year round, because during a part of the year he was in London. He died in the early part of 1887, and Mrs. Rylands continued to reside there until October, 1888. Between the beginning of 1887 and October, 1888, however, she placed the property in the hands of a house-agent, with a view to it being let; but from then until the present time the property had remained unlet, in consequence, it was alleged, of the offensive odours which were given off from the defendant's works and reached the garden and windows of the plaintiff's house. The property was a valuable one. It cost a good many thousand pounds when it was bought, but according to the valuation now placed upon it by the agent who was called it had depreciated considerably, owing to the alleged nuisance from defendant's works. The house was in a neighbourhood which was distinctly residential, and that was a very material matter to be considered in a case of that kind, for there could be no doubt that the old doctrine of what was called going to a nuisance was no longer law. A person must not be too particular if he planted him-men, he had very little doubt that his decision would not necessarily result in the stoppage of the defendant's trade. There seemed to be a high probability that more efficient precautions might be taken whereby no nuisance would be occasioned, but it was his duty to grant an injunction restraining the defendant from repetition of the nuisance, with costs.

On the application of Mr. Maberley, execution was stayed for a month, pending appeal, on defendant paying costs to plaintiffs' solicitors, the latter giving an undertaking to return the money if judgment were reversed.

ALUMINIUM.

IT has frequently happened in the history of chemical industry that by some improvement in the process of manufacture the cost of a valuable substance is considerably lessened, so that its useful properties are placed within the reach of those who may find employment for them. Such is notably the case with the metallic element aluminium, which, until quite recently, has partaken more of the nature of a chemical curiosity than of a substance which might, under any circumstances, come to be generally used by the manufacturer or industrial artist. A few years ago aluminium sold at 40s. *per ounce*, and was manufactured by the pound, but recently the process for its production has been so simplified that it now sells at 20s. *per pound*, and is manufactured by the ton. This revolution in its manufacture is due to the investigations of Mr. Castner. Large quantities of sodium are used in the process, and it has been more in the direction of cheapening the cost of production of sodium that the improvement in question has been effected; in truth, the original process remains practically the same as a chemical operation. And by this means the valuable properties of aluminium are placed within the reach of the engineer, the sculptor, the instrument-maker, and the mechanician generally, in whose hands they promise a wide application. What these properties are, and how they may be applied, may not be altogether devoid of interest at the present time. To begin with, aluminium is widely distributed among the materials forming the earth's crust, being present (in a combined form, of course) in large quantities in clay; so that the raw material from which it is extracted may be obtained in abundance. The properties which chiefly recommend it for wider use are its lightness, tensile strength, durability, and malleability. Its specific gravity is about 2·6, while iron is 7·8; in other words, a given volume of aluminium will only weigh one-third of the same volume of iron. Again, its tensile strength is about 14 tons per square inch, thus comparing favourably with wrought iron. Its ductility and malleability are quite as good as gold or silver; observing certain precautions it may be drawn out into fine wire, beaten out as thin as gold leaf, hammered into thin sheets, wrought into plates, or cast into bars. Its melting point is much lower than iron, being about 700° centigrade; recent experiments have shown that it is the best conductor of heat yet known, and it conducts electricity almost as well as copper. It can be alloyed with silver, copper, iron, and most other metals. A finished bar or sheet is capable of taking on a very high polish, equal to that of silver, being, however, superior to silver in respect of its not being tarnished by sulphurous fumes. Lastly, it is not attacked by air or water, so that it cannot rust or wear by any exposure to the elements as in the case of iron. These then are a few of the chief properties of this somewhat remarkable metal, and naturally it will be asked, How are they to be employed?

Combining as it does lightness with strength, aluminium could with advantage be employed in many circumstances where iron can now alone be used. At present its cost precludes its use in large amount, but there are many objects which might be constructed of it. Many of these might be mentioned, but, for example, we may name optical instruments, scientific instruments, cooking utensils, reading lamps, chandeliers, spoons, forks, surgical and dental instruments. In many of these its lightness, and more especially its capability of taking on a high polish which would resist tarnishing, aluminium would be beneficial, and effect a great improvement. Further, we may reasonably look forward with some degree of confidence to the cost of its production being again considerably lessened, and may contemplate the effect of constructing ships, bridges, and many kinds of exposed metal work with aluminium. Its chief advantage in the first of these would be its lightness. It has been estimated that if an Atlantic liner were built of aluminium it would have only one-third the draught, and be propelled with the same engine power at double the speed that it is at present when built of iron—a previsionary advance in ocean travelling which would place New York within three days of Liverpool. It would also enable vessels of large tonnage to sail up rivers which are at present only navigable to those of light draught. Bridges built of this metal would be quite as strong as built of iron, would only have one-third the weight, and the natural non-tarnishable brightness of the metal would obviate the necessity for its being made hideous with red lead paint to prevent corrosion in a moist atmosphere.

But perhaps the sphere for the most promising employment of aluminium is in connection with its alloys. The result of alloying 10 % of aluminium with 90 % of copper is to produce an alloy of golden colour, capable of taking a high polish, and with a tensile strength exceeding that of wrought iron, and nearly equal to that of steel. This alloy, known as aluminium bronze, is one concerning which great hopes are entertained. It is, of course, much cheaper than aluminium, and hence there may be expected of it a wider use. It makes good castings, and is easily worked under the hammer, its malleability increasing with repeated melting or tempering. It is thus suitable for large castings, such as statues and other large works of art. It is hard enough to be substituted for steel in the bearings and other parts of machines, and also for the manufacture of gun barrels and cannon; in short, aluminium bronze might be substituted for steel in nearly all its applications. When aluminium is added to steel it appears to raise its tensile strength and to lower its melting point. The latter effect is of some advantage in the manufacture of steel castings. It is well known that cast metal often contains blow-holes, due to air or other gases being imprisoned in the molten metal just before solidification takes place. By adding aluminium to molten steel or iron, the melting point of the latter is lowered; it therefore is rendered more fluid, and consequently the possibility of the formation of blow-holes is much lessened. Besides this advantageous use of the metal, experiments at present going on seem to show that more valuable discoveries may yet be made as to the effect of alloying aluminium and steel. Many other possible uses for this remarkable substance could be mentioned; but enough has been said to show that, provided its price comes within the reach of manufacturing industry, there will probably open up many opportunities for its employment, and thus hasten the advent of what has in a prophetic sense been termed the "Age of Aluminium."

NORTHWICH LOCAL BOARD AND THE WIDNES BRINE SCHEME.

THE projected scheme for the conveyance of brine from Wincham to Widnes occupied the attention of the Northwich Local Board on Tuesday night.

The clerk submitted a copy of an intended Act for incorporating and conferring powers upon the Widnes Brine Supply Company to erect a pumping station in a field in the township of Wincham (adjoining the Board's district) for the purpose of pumping brine to be conveyed therefrom through pipes terminating in the township of Widnes, in the county of Lancaster, and also giving the company power from or by their works to supply brine and salt to any person, company or corporation on such terms and conditions as may be agreed upon. He had been asked to make a report upon the Bill. The only report he could make was that the Bill was drawn in the usual form, and that the terms of it were against the best interests of the inhabitants of the Board's district.

Mr. Worsley, with great pleasure, moved the following resolution:—"That the conveying of brine from the township of Wincham to the township of Widnes, as proposed by the intended Act of the Widnes Brine Supply Company, is against the best interests of the inhabitants of the Board's district; and further, the Board by this minute of resolution authorises and directs the Clerk to do all acts and things as may be necessary for preparing and presenting to Parliament a petition against the intended Act."

Mr. Moreton seconded.

Mr. Leicester: What *locus standi* have we, if it is outside the Board's district?

The Clerk asked that no discussion should take place before the petition came up for sealing.

Mr. Leicester failed to see the wisdom of preparing a petition if they had no *locus standi*. Wincham was out of their district, and it would not affect them if they started pumping.

Mr. Worsley's impression was that they had a *locus standi*.

The Chairman (Mr. Clough): It is very desirable that we should stop it, if possible.

Mr. Leicester: Of course; but we shall get stopped on the threshold and told we have no *locus standi*.

Mr. Moreton thought it was worth while making the effort.

Mr. Leicester said it was thought that the pumping of brine at Marbury by the Mersey Salt Company would be prejudicial to Northwich; but they could not stop it.

Mr. Williams: The passing of this resolution does not bind the Board's hands in any shape or form.

The Chairman: We might find ourselves in an awkward position if we did not pass the resolution.

Mr. Leicester: It does not empower the Clerk to incur any expense?

The Chairman: No.

The resolution was then carried *nem con.*

PRIZE COMPETITION.

A LITTLE more than twelve months ago, one of our friends suggested that it might be well to invite articles for competition, on some subject of special interest to our readers; giving a prize for the best contribution, he at the same time offering to defray the cost of the first prize, and suggesting a subject in which he was personally interested.

At that date we were totally unprepared for such a suggestion, but on thinking over the matter at intervals of leisure, we have come to think that the elaboration of such a scheme would not only prove of benefit to the manufacturer, but would go some way towards defraying the holiday expenses of the prize winner.

We are therefore prepared to receive competitions in the following subjects:—

SUBJECT I.

The best method of pumping or otherwise lifting or forcing water aqueous hydrochloric of 30 deg. Tw.

SUBJECT II.

The best method of separating or determining the relative quantities of tin and antimony when present together in commercial samples. On each subject, the following prizes are offered:—

One first prize of five pounds, one second prize of one pound, five additional prizes to those next in order of merit, consisting of a free copy of the *Chemical Trade Journal* for twelve months.

The competition is open to all nationalities residing in any part of the world. Essays must reach us on or before April 30th, for Subject II, and on or before May 31st, in Subject I. The prizes will be announced in our issue of June 28th.

We reserve to ourselves the right of publishing the contributions of any competitor.

Essays may be written in English, German, French, Spanish, or Italian.

RAILWAY RATES AND CHARGES.

WE have from time to time directed attention to revision of railway rates, says the *South Wales Daily News*, a subject of so intricate a nature that only those directly affected have taken interest in it sufficient to induce consideration of present circumstances. By the act of last year, the railway companies had to submit to the Board of Trade new schedules of rates. This they have done; and the traders of the country having objected to the proposed new rates, it has become the duty of the Board of Trade to make inquiry, and to fix what the new charges shall be. It is now clearly apparent that the railway companies have an intention to increase their charges all round, and to levy terminals where they have never levied them before. They are seeking powers to increase the charges now actually paid by 50 per cent. Last year the railways carried 201,576,616 tons of minerals, at a cost of £16,158,881., and the companies now propose a terminal charge upon minerals ranging from 1s. to 1s. 8d. a ton, which, at the minimum rate of 1s., would add £10,078,830. to the rate on minerals. They carried last year 80,171,823 tons of general merchandise, at a cost of £21,239,841. They now propose to add terminals to this charge, which, upon an average, will yield 3s. 6d. a ton, making £14,015,032. addition to the present charge. These two items of increase amount together to £24,093,862. a year, and as the total of gross receipts last year was £72,894,665., it is nearly one-third of new additional charge, and that will be wholly an increase of dividends to shareholders, as the working expenses will not be increased by a penny. In short, the railway companies are seeking to tax the whole community for their own advantage. Their position is that of landowners exacting an enormous rent by means of protection. The shareholder holds a monopoly, which is protection, and the whole country have to pay him tribute. The enormous traffic upon British railways would enable the companies to pay large dividends if the directors adopted the policy which has proved so completely successful elsewhere: that policy is one directly opposite to existing methods. At present it is the practice to put on traffic the heaviest rates it will bear, instead of developing business by carrying at lowest remunerative charges. In countries which differ as widely as Belgium, Prussia, and Ireland, the policy of reducing rates has been attended with most gratifying results. Lower rates have yielded higher dividends, because of the greatly-increased traffic which has been promoted; and the experience of other countries ·ld be repeated here if their wiser policy were adopted.

the retort, and, being volatile, not requiring any lixiviation. In addition to the above, there was also a short descriptive sketch of the salt mines at Khewra, Hindostan, read by the author, Mr. E. Rodger, Glasgow, from notes made during a recent visit.

ELECTRIC LIGHTING IN BIRMINGHAM.

A COMPANY has been formed, with a nominal capital of £100,000, for lighting the Midland metropolis, and it is expected that the system will be completed before next winter. The area to be dealt with includes the central portion of the town, with one or two spurs, and the new illuminant will be available for offices, shops, business premises, public buildings, educational institutions, and hotels. But one station will be employed. The generating apparatus will consist of steam-driven dynamos, with special arrangements for securing continuity and uniformity day and night. The current will at once be sufficient to actuate 5,000 16-candle power or 10,000 8-candle power incandescent lamps, but the mains will be ample for carrying a much larger current. The initial outlay is estimated at £30,000. Arrangements for the public safety have been carefully made. The leads or conductors will lie in V-shaped conduits underground, overhead wires being prohibited; and the pressure of the current will be low, averaging 150 volts. This is a plan which cannot endanger life, the lethal currents, about which so much has been heard, being something like 2,000 volts. The current will also actuate arc lamps should these be desired. The wires of supply will enter the consumer's premises, and the user will have a choice of fittings and appliances. The maximum charge under the provisional order is 10d. per Board of Trade unit—that is, sufficient to run 34 eight-candle glow lamps for one hour; but there may be a sliding scale giving advantage to the large consumer. The current will also be available for motive power.

THE TIN-PLATE TRADE.

EXPORTS FOR 1889.

THE export of tin plates and sheets for the year just ended exceeds that of any previous year in the history of the trade. Over 430,000 tons were exported during the twelve months, representing an aggregate value of over £6,000,000. For the month of December the quantity of exports was below the average for the year, reaching only 29,232 tons; but with the holidays and what not, the last month of a year is seldom a typical one. We append statistics showing the volume of trade for the 12 months:—

Where to.	Tons.	Value. £
Holland	3,795	56,234
Germany	4,179	62,174
France	4,322	62,696
Australasia	6,620	97,726
British North America	15,385	214,338
United States	336,692	4,674,455
Other countries	59,630	862,873
Total	430,623	£6,030,496

It will be instructive to compare the figures for the year just ended with those of the two years which preceded it:

	Tons.	Value. £
1887	353,506	4,792,854
1888	391,361	5,546,228
1889	430,623	6,030,496

No better testimony to the vitality and the development of the tin-plate industry need be sought after than is supplied by the statistics given above, which may be left to tell their own story.

In consequence of the development of the exports of tin-plates from Swansea, the harbour trustees are taking steps to extend the already extensive storage accommodation at the East Dock.

A NEW IDEA.—A cargo of pitch brought by the steamship *Tynedale* has been unloaded at the Reading railroad coal wharves in Philadelphia. The company has a process by which the refuse coal dust, with the addition of pitch, can be made into fuel. The idea is patented, and has already been tried, it is said, with success, but not on so extensive a scale.

GLYCERINE JELLY.—A new cement is described in the *Monde de la Science*, which is attacked neither by petroleum, benzine, turpentine oil, nor any similar liquids. It consists of a mixture of glycerine with gelatine or glue. The product is quite liquid when hot, and resembles caoutchouc when cold, alike in appearance as in its elastic properties. Applied hot to the interior of a barrel or any wooden vessel, the latter is thereby, according to the *Monde de la Science*, made safe for holding any of the above-mentioned substances.

VANADIUM IN CAUSTIC POTASH.*

BY EDGAR F. SMITH.

THE occurrence of vanadium in commercial caustic soda has been noticed. As far as I am aware, it has not been observed in caustic potash, hence the following lines:—

While engaged in making certain decompositions, for which I employed the ordinary stick potash, I was rather surprised, after saturating the alkaline solutions with hydrogen sulphide, acidulating with hydrochloric acid, and heating for several hours, to discover that the separated sulphur showed a decidedly chocolate brown colour. This occurred repeatedly. The quantity of dark material was never very great, yet it was present, even in potash from alcohol. Some preliminary experiments were made which pointed to vanadium. I therefore saturated the warm aqueous solution of three pounds of stick potash with hydrogen sulphide. Heat was applied for several hours longer, and during this period the liquid gradually assumed a yellow to deep red colour. Hydrochloric acid was added to distinct acid reaction. The separated sulphur was quite dark in colour. After filtration and washing, the residue was dried and treated with carbon disulphide to extract the free sulphur. The chocolate coloured mass remaining after this treatment dissolved, with exception of a slight quantity, in yellow ammonium sulphide, from which it was reprecipitated by dilute acid. Again washed, treated with carbon disulphide and carefully ignited, there remained a crystalline mass. With a small quantity of this last product I made the phosphorous bead test—yellow, passing through dark to green when cool. Another special test consisted in dissolving a portion of the ignited residue in a drop of concentrated sulphuric acid, free from iron. One cc. of water was next added, and to this colourless solution from one to three drops of a dilute potassium ferrocyanide solution, whereupon the liquid acquired a fine green colouration. This latter test Dr. Walz (*American Chemist*, vol. vi., p. 453) employed in detecting the minute quantities of vanadium in American magnetites. He considers it conclusive evidence of the presence of vanadium. The most important test, and that which I applied to the remainder of the ignited residue, after effecting its solution in nitric acid, was to add a large piece of ammonium chloride to the ammoniacal solution. The morning following, crystals of ammonium vanadate had separated. These gave the true vanadium bead, and when ignited, moistened with pure, strong nitric acid and evaporated, left the deep red-coloured residue characteristic of vanadium compounds.

The vanadium sulphide, as first obtained, was impure, consequently the reactions were at times masked, and it was only after eliminating some silver and iron that the reactions were unquestionable.

The impure sulphide from the three pounds of caustic potash weighed about one-half gram.

THE NITRATE OF SODA TRADE IN 1889.

THE year's statistics as to the nitrate of soda trade, which are just out, possess considerable interest. They show that during the year there has been a production and export from the producing countries with which consumption in this country and in Europe generally have not kept pace. The total production is estimated at 900,000 tons—about 500,000 by British joint-stock companies, and about 400,000 by private firms and Chilian companies. From all the nitrate ports, however, there have been exported 930,000 tons, as against 750,000 tons in 1888, 400,000 tons in 1886, and 535,000 tons in 1884. Of the exports the United Kingdom has received about 121,000 tons (the quantity in 1888 being 103,000 tons); the Continent, 673,000 tons (the quantity in 1888 being 547,400 tons); and the United States and California, 90,000 tons, other places taking up the balance. Then as regards consumption, the United Kingdom (by deliveries and exports), has disposed of 104,000 tons, the same as in 1888; the Continent has absorbed 575,000 tons, as compared with 532,000 tons last year; and the United States have consumed 80,000 tons, as against 65,000 in the year preceding. It thus appears that while the United Kingdom has received 18,000 tons more than in 1888, and the Continent 126,000 tons additional, there has only been an increased consumption under these two headings of 43,000 tons, and stocks, therefore, have grown substantially. The stocks in this country now stand at 30,000 tons, having risen from 17,900 tons in June last, and they compare with 14,600 tons a year ago. On the Continent the stocks have accumulated from 64,000 tons at the close of 1888, and 93,000 tons six months since to a total of 162,000, which gives a gross increase for the Continent and the United Kingdom on the year of 110,000 tons. The visible supply for the United States (including stock and nitrate afloat) is about 40,000 tons, 12,000 tons in excess of that for last year. Then there are 380,000 tons afloat for Europe, the largest probably on record; the amount for this time 1888 being 325,000 tons, and 1886, 178,000. It is believed that larger stocks exist in the hands of dealers and consumers in the interior of Germany than at date last year, the increased deliveries having been chiefly since the close of the Spring consuming season. Slightly smaller stocks are believed to exist in producers' hands than at the end of 1888. The visible supply for Europe during the next four months consists of stocks at ports 192,000 tons, and afloat 380,000 tons (or a total quantity (exclusive of stocks in interior or in transit) of about 572,000 tons, against 404,000 tons in 1888. Deliveries for consumption during same period have been, in 1889, 334,000 tons, with an average spot price of 10s. 9d.; in 1888, 351,000 tons, with price 10s. 3d.; in 1887, 246,000 tons, with price 10s. 9d.; in 1886, 212,000 tons, with price 11s.; in 1885, 260,000 tons, with price 9s. 6d.; in 1884, 270,000 tons, with average spot price of 10s. per cwt. The value on spot at date last year was 10s. 9d. to 11s. per cwt., and to-day it is 8s. 4½d. to 8s. 6d. per cwt. The highest point was in January, when 11s. 6d. was reached for small due cargoes, but values gradually receded until 10s. was reached at the close of March, 8s. 6d. at the close of April, and 8s. 1d. in the middle of May for arrived cargoes. The lowest points during the year for distant arrival were about 8s. 9d. in May, and 8s. 3d. per cwt. in November, and the highest 10s. in February and March, and 9s. 3d. in July and August. Holders of cargoes for Spring arrival are firm in their possessions, in the belief that some improvement in the present very low range of value, notwithstanding the large excess in visible supply, is propable during the consuming season. Buyers having already, to a considerable extent, contracted for their requirements for the early Spring, meantime abstain from operating freely, pending some assured action on the part of the producers to restrict production and shipments during 1890. Freights from the nitrate ports have fluctuated from 25s. in February and March, 35s. in June, 37s. 6d. in August and September, to 32s. 6d. in November and December, which is the closing quotation on spot; while for distant loading 35s. per ton has been paid. Last year's range was from 25s. to 35s., the closing rate being 30s. per ton. Exchange has fluctuated from 30d. in January to 25⅜d. in June, 24⅞d. in July, 26d. in August, since ruling at 25d. to 25½d. which is the closing rate. Last year's fluctuations were from 26d. in June to 29d. in December. Cost of nitrate, free on board, has ruled from 2dols. 75c. in spring to 2dols. 55c. in June, 2dols. 65c. in July and August, to 2dols. 45c. in November, 2dols. 50c. per quintal being the closing quotation. The shipment duty remains at the equivalent of slightly over 2s. 6d. per cwt.

PAPER MILLS IN INDIA.

THE paper mills industry in India is one of yearly growing importance in the considerable addition it makes to the extent of local manufactures, and the consequently diminished requirements from foreign markets. It is only under the present outlook of the relations of silver that any economic calculations can be projected from any data which can be made to show that the people of this country will need less of the manufactures and productions of other nations owing to enlarged indigenous productions, throwing a large weight in adjusting the balance trade on silver imports and India Council's drawings.

Paper manufacture by steam machinery was first commenced in India by the late John Marshman, at Serampore. Working by a sort of rule-of-thumb process with anything but the most modern machinery—and no scientifically tested process—the highest success scored by the mills was the well-known "Serampore paper." Even that was attained with great difficulty, owing to the effect of climate on the bleaching medium used. After various vicissitudes the mill was broken up, and the King of Burmah became the happy possessor of the machinery, lock, stock, and barrel. There are now nine paper mills—five in the Bombay Presidency, two in Bengal, one at Lucknow, and one at Gwalior. Three of these are under private, and the remainder under Joint Stock Association auspices. These mills manufacture blotting, brown, and white cartridges, white country paper, writing and foolscap papers, and coloured coarse papers. The materials used are wheat and rice straw, rags, various kinds of grasses, old jute, and hemp rope, or bagging, wood-pulp and waste paper. It is needless to say that the supply of rags is of too precarious a character and limited extent to admit of any extension of the quantity of any of the finer class of papers these mills can manufacture. The invested capital in these mills totals up to nearly £500,000. The largest mill is at Bally, and produces yearly about 3,100 tons of paper. The value of the yearly production of all the mills collectively is R29,00,000, and they give employment to about 3,500 people. The sale of the Indian-made paper is yearly increasing, and under more skilled superintendence, and the improved machinery which has recently been imported, the class of paper turned out is of a better quality, and commands a larger sale than hitherto. In the last five years, the value of the paper turned out from the Indian Mills has risen from R18,00,000 to R24,40,000. There is no apparent effect from the local industry on the value of the imports into India from foreign markets; the expansion of the trade has been very great, and the consumption is now very large. In 1867-68 the value of the paper imports was R28,50,000; in 1877-78 R30,50,000, and in 1888-89, R40,26,921.—*Indian Daily News.*

*Read at recent meeting Chemical Section Franklin Institute.

...HAM BRINE SCHEME.

...s Brine Scheme was recently brought
...thwich Local Board. From minutes
... a meeting of the Board, as a com-
..., when there were present—Messrs.
..., Woodcock, Worsley, Rogerson,
...eting was held for the purpose of
...pplication intended to be made to
...for a bill to incorporate a company
...f works for a pumping station in
...r, to the township of Widnes, in the
...liamentary notice and plans having
...was moved by Mr. Clough, seconded
... the clerk be authorised to obtain a
...rt thereon to the board.
...tten to the solicitors of the Widnes
...king for a printed copy of the plan
...otice. In reply, Messrs. Bridgman,
...said, "Before agreeing to send them
... behalf you apply for them, and if
...re them will enter into a guarantee,
...ot be used for opposition on standing
...o send them." In reply to that letter
... the plans were for the use of his
...sh them without the guarantee and
... withdraw the board's application.
...ncise as possible, and detain you only
...ceived from our clerk a Christmas
... place, to wish him and all around this
... pleasurable Christmas rejoicings. I
...ur town at heart more than our clerk.
...ving in respect to this greeting, I hope
... to himself. On the Christmas card
...d the future of poor old Northwich—
... through a bridge or locks, with an
...und in sight, but a few birds, whether
..., I know not; but if this comes to
... scheme created, it will be a very
...years hence for a good many of the
... respect to this new brine scheme, we
... doing an immense amount of damage,
...made here going directly out of our
...body to deal out more destruction, and
...utside? It brings to my mind a very
...n of our great and intrepid traveller,
...overcome all dangers and opposition
... his way through 400 miles of forrest
...ated faithfully his orders, relieved
...them safe out of their difficulties,
... like a labourer returning home on a
... good week's work, with well-earned
...ly for the Sabbath of to-morrow."
...hese words than is represented in their
...r has well earned his fortune; he will
...ur laws; he will be an honoured and
...nobles of our land, and undoubtedly
...us Majesty, our Empress and Queen.
...tunity of making ready for that eternal
...ost people are wishful to attain. Well,
...anleys in Northwich. We have our
...ly morn till late at night; we have our
...apour heat of mid-day and midnight;
...ough their daily routine of life—a love,
...s hazardous and dangerous to life,
... lost their lives in these avocations.
... people are in a much different
... Sabbath than Mr. Stanley. Take
... Mr. Gough, before Mr. Ritchie, on
...The subsidences did away with a
...se of it belonging to men who
...es, whose property represented their
... were thus ruined." On the one side
...e, contentment of mind, a desire to
...ly, and contemplate calmly the coming
...ople may be so constituted that they,
...and dissatisfaction with life, at variance
... state of opposition to an all-wise Provi-
... where men have honestly obtained, by
...ich should be a permanent income for
...o-called estates are just as worthy of
...as those of the greatest noblemen of
... place in juxtaposition the estate of his

Grace the Duke of Westminster and Lord Egerton's by the side of Mr. Pickering's, of Dane Bridge, and Mr. Lamb's, of Leicester-street. Had either of these noblemen gone through the same anxiety and distraction of mind, and the same losses, I venture to say you would have a law made to-morrow, in the Houses of Lords and Commons, to meet difficulties in connection with the same. These are only instances of many others. Then, sir, as to excuses in denying justice, look at the huge amount of wealth going out of our districts. May I ask you how many vendors to the Salt Union there were in the salt trade? You agree that about forty will represent the whole lot. Now, we will threw over the three-quarter million of money, entirely, a portion capitalised to expenses, purchases, &c., and take simply the three millions. In ordinary business transactions of life you can always have a return of 2½ to 5 per cent. What is 2½ per cent. on this three millions? Only £25,000. per annum. Then take the profits of Messrs Brunner, Mond and Co., where entire strangers come into your district, making a limited company amongst strangers, and within ten years pay off their original capital nearly three times over, at the present time paying 25 per cent. What does one-fifth of their profits mean, still leaving a dividend of 20 per cent? It means over £40,000. per annum, together £115,000. per annum. And, gentlemen, £8,000., which you, as a Board, have asked for, is denied. We are to have a definite reply from the Salt Union meeting, I presume, early in January, and I do hope in the meantime every endeavour will be made to have all our details ready. I say our present position is a disgrace to civilisation; it is a disgrace to our noblemen of Cheshire to quietly see and know this still to exist; and it is a disgrace to ourselves as a community. I do hope we shall be prepared and able to have some mite of equity accorded us.

RELATIVE VALUE OF VARIETIES OF LOGWOOD.

L. BRUHE, who has examined a large number of specimens of the logwood of commerce, maintains that the statement occurring in many text books, even such as have been published by prominent authorities, that the name of the wood is at once the criterion of its value, the brands, San Domingo, Yucatan, Monte Christo, and Laguna being generally considered the best, and the others as of inferior value, is incorrect. The author gives his grounds for this assertion in the following analyses of samples:—

Name and mark of woods.	Water. %	Ash. %	Combustible matter. %	Alcoholic extract. %	Ethereal extract. %	Residue. %	Yield of extract. %
Yucatan logwood	13·00	1·09	65·71	37·46	60·12	2·42	20·20
Laguna logwood	12·38	0·96	65·66	47·95	1·37	0·68	21·00
Domingo logwood	13·19	1·88	70·73	53·47	44·95	1·58	14·02
Monte Christo logwood, 1884 ..	13·20	2·94	65·11	60·32	32·00	7·78	18·75
Monte Christo logwood, 1887 ..	14·70	1·03	70·27	54·10	34·72	11·18	14·00
Fort Liberte logwood, 1886-87	13·10	0·82	65·75	54·11	1·89	4·00	20·33
Fort Liberte logwood, 1887 ..	10·12	0·88	73·00	47·92	50·00	2·18	16·00
Fort Liberte logwood, 1885-86	12·11	2·03	68·41	45·17	59·72	5·21	17·45
Fort Liberte logwood, J. B., 1887	11·84	1·03	69·13	34·81	59·24	5·95	18·00
Yucatan logwood, E. J.	14·51	1·20	66·95	38·51	58·34	3·15	17·34
Domingo logwood, D.	13·71	2·14	64·85	50·32	43·81	5·81	19·30
Jamaica wood ..	14·10	1·14	66·06	50·50	43·20	6·30	18·70
Jamaica wood ..	12·10	1·30	68·60	50·71	43·05	6·24	18·00
Jamaica wood, roots	15·30	2·30	71·70	30·12	52·99	16·89	10·70

THE LONGEST TELEPHONE LINE.—The longest one in Europe is that now laid between Pesth and Prague. It is 600 kilometres in length, or about 375 miles. The trials just made on this line of communication are reported highly sasisfactory. It is, we may note, nearly double the distance between Paris and Brussels.

MR. CHRISTIAN ALLHUSEN, founder of the Newcastle Chemical Works, and for many years chairman of the Newcastle and Gateshead Chamber of Commerce, died yesterday morning at his residence, near Windsor. The deceased gentleman, who was 83 years of age, was a recognised authority on trade matters, and he rendered material assistance to the late Mr. Cobden in concluding the commercial treaty with France.

Correspondence.

ANSWERS TO CORRESPONDENTS.

PETER.—The number is out of print, but we will publish the table again as it has had a large sale.

ENQUIRER.—The journal circulates very largely amongst the Allied Industries.

Q. E. D.—By no means. 2. We cannot advise you.

SUBSCRIBER.—If you will send us your full name and address we will insert your query as you have written it.

L——s.—Don't try to appear funny.

A READER.—We trust you have received our almanac by this time.

C. T. B.—Thanks for the newspaper. The local news is always a source of interest.

OILMAN.—Certainly; the oil and colour trade cannot be conducted without some knowledge of technology.

C.—We will send a representative over to see the apparatus if you will make an appointment.

K. B.—We have tried the process in our laboratory, and have met with nothing but failure.

MANUFACTURER.—Call at our offices; we shall be pleased to talk over the matter with you.

COAL.—The matter would take too much space to explain here. 2. Yes. 3. Be careful how you play with fire.

CHEMICAL.—We have received many similar complaints, but have not printed them, as our solicitors call them libels.

PLATE GLASS.

ALTHOUGH the foreign manufacturers of plate glass have lost their hold on this market through the enterprise and ability of American producers, the condition of trade in their home markets more than makes amends for the loss. They are united in a strong combination, and with an unusually large consumption in all parts of Europe, and no surplus stock anywhere, they are getting good prices.

The foreign manufacturers of plate glass are encountering the same difficulty that is said to obtain in this country, namely a scarcity of skilled labour. Particular difficulty is experienced by them in finding men who are qualified to fill the office of manager, and on this account the building of several new works, that have been projected, is at a standstill.

A plate glass factory is to be built at Dusseldorf, Germany, and another is reported for Monstier, Belgium. Plans have also been drawn for a works at Auvelais, which is but a few miles from Monstier.

A plate glass works, which is the first of its kind in that country, is now being built at Pisa, Italy, and is almost completed. It is a branch establishment of a French firm of manufacturers, whose principal factory is one of the largest of that kind in the world.

THE LISHMAN SYSTEM OF SMOKE PREVENTION.

THE Lishman system of perfecting combustion and preventing smoke has been submitted to a further test at the works of Messrs. Inglis and Wakefield, Busby, near Glasgow, with satisfactory results. The apparatus, in the instance referred to, was attached to two Lancashire boilers, 28 feet long by 7 feet 6 inches in diameter, and Scotch coal was used as fuel. A Scotch newspaper thus describes the test:—"By invitation of Mr. Thomas Lishman, the inventor of the Lishman steam generator, we had the pleasure of seeing one of his improved methods of perfecting combustion and preventing smoke as applied to the ordinary Lancashire type of steam boilers at the works of Messrs. Inglis and Wakefield, Busby. In this instance the apparatus was attached to two boilers, to each of which there are two furnaces. There are other four boilers of recent construction, of similar dimensions, connected with the same chimney stack. For the purpose of efficiently testing the merits of the invention for preventing smoke, advantage was taken of the opportunity afforded for so doing when the other boilers had finished working, so that no smoke could be emitted from the chimney but from the two boilers to which the patent was applied. After work had ceased at one o'clock operations began, all four fires being cleared and clinked previously. The fuel was levelled in the furnace, then tried in the usual way, each furnace in succession being similarly treated. The chimney top was anxiously watched by the spectators. From their firings no smoke was perceptible. The fires were then stoked, and fresh fuel to each furnace supplied, when nothing more than a light stream of thin vapour issued from the chimney. To prove the difference of the two systems three of the other boilers of similar dimensions delivering into the same chimney were fired up in the ordinary way, when, immediately afterwards, dense volumes of black smoke issued from the chimney. The trial, which lasted over an hour, clearly demonstrated that smoke from Scotch coal can practically be prevented by the application of this method. The main f[illegible] the apparatus consist of a blower, through which the air at a pressure is introduced. The air for sustaining combustion is conducted by a conduit to the front of the boilers where it is delivered into a receiver placed below the floor plates. A cast or wrought iron pipe is placed on the floor of each furnace, connected at the one end to the air receiver, and at the other end to an internal air-box. This pipe is perforated on each side, from which air is equally distributed to the fuel and the bars. Air is also, in regulated quantities, admitted to the air-box behind the bridge, wherein it is highly heated and distributed by means of about 100 small orifices into the escaping gases from the fuel chamber. Air is also admitted in a number of rectangular streams at the front of the fuel. There are 226 thin streams of air operating upon the fuel and gases at right angles to the currents, and thereby the gases become thoroughly mixed and the combustion perfected. The apparatus is of a very simple nature, and easy to deal with."

SUPPLIES FOR THE NEW YORK ARSENAL.

THE following is a list of bids for furnishing supplies to the New York Arsenal, which we take from the Government *Advertiser*:

CHEMICALS, SALTS, GUMS, ETC.

10 gals. acid, carbolic, 45c. per gal., bidder, Charles H. Pleasant.
25 lbs. acid, muriatic, 6½c. per lb., do.
10 lbs. acid, hydrochloric, 8½c. per lb., do.
50 lbs. acid, nitric, 42° specific gravity, 10½c. per lb., do.
25 lbs. acid, oxalic, 11c. per lb., do.
8 ozs. acid, citric, 4c. per oz., McKesson and Robbins.
100 lbs. acid, sulphuric, 66°, 5½c. per lb., Chas. H. Pleasant.
1 lb. acid, acetic, 14c. per lb., do.
100 gals. alcohol, grain, 95 per cent., $2. 18 per gal., do.
2 lbs. alum, powdered, 4⅜c. per lb., Geo. Gabb's Sons.
25 lbs. ammonia, aqua, FFFF, 9c. per lb., Chas. H. Pleasant.
5 lbs. ammonia, sulphide, 50c. per lb., McKesson and Robbins.
20 lbs. borax, pulverised, 10c. per lb., Charles H. Pleasant.
5 lbs. chalk, red, lump, 8c. per lb., McKesson and Robbins.
50 lbs. chalk, white, lump, 1¼c. per lb., do.
100 lbs. copper, sulphate (blue stone), blue vitriol, 7c. per lb., Chas. H. Pleasant.
500 lbs. electrophone fluid, in carboys, net weight (carboys to be returned to pier 6, N.R.), $3 95, bids rejected.
30 grs. gold, chloride of, 3⅛c. per gr., McKesson and Robbins.
100 lbs. gum arabic, 75c. to 45c. per lb., do.
100 lbs. gum camphor, bids rejected.
5 lbs. of shellac, gum, 25c. per lb., McKesson and Robins.
5 lbs. shellac, white, 27c. per lb., do.
100 lbs. glue, common, No. 1¼, 11c. per lb., Geo. Gabb's Sons.
50 lbs. glue, white, best quality, 21c. per lb., do.
5 lbs. glue, cabinetmakers' best, 12c. per lb., McKesson and Robbins.
48 boxes lye, concentrated, 1 lb. boxes, 7c. per box, Charles H. Pleasant.
2 lbs. iron, photosulphate of, 8c. per lb., do.
75 lbs. of mercury, for battery purposes, 54c. per lb., McKesson and Robbins.
1 oz. mercury, bichloride of, C.P., 25c. per oz., C. H. Pleasant.
50 lbs. potassa, chlorate of, 15⅜c. per lb., McKesson and Robbins.
30 lbs. of potassa, bichromate of, 14c. per ih., do.
2 lbs. potash, carbonate of, 16c. per lb., C. H. Pleasant.
2 lbs. potash, neutral oxalate of, 24c. per lb., McKesson and Robbins.
1 lb. potash, yellow prussiate of, 26c. per lb., C. H. Pleasant.
2 ozs. potassium, bromide of, 4c. per oz., McKesson and Robbins.
2 lbs. potassium, nitrate of, C.P., 12c. per lb., C. H. Pleasant.
30 lbs. salammoniac, for Leclanche battery, 10c. per lb., do.
50 lbs. sal soda, 2c. per lb., do.
12 ozs. silver, nitrate of, 63c. per oz., do.
25 lbs. soda, hyposulphite of, 4c. per lb., do.
2 ozs. soda, acetate of, 12c. per oz., do.
2 lbs. soda, sulphate of, crystals, 4c. per lb, do.
5 lbs. of soda, carbonate of, 4c. per lb., McKesson and Robbins.
50 lbs. strontia, nitrate of, 12c. per lb., do.

WILL OF THE LATE MR. WOOLLEY.—Probate of the will, dated August 18th, 1885, of the late Mr. Harold Woolley, of Harefield, Holland-road, Crumpsall, and of Manchester, pharmaceutical chemist, who died on October 16th last, leaving personalty valued at £30,225. 19s. 7d., has been granted to the executors, Mrs. Woolley, Mr. Ewar Christensen, of Christiana, and Mr. James Lowndes, of Manchester, to each of whom, as executor, the testator bequeaths £50.; to Mrs. Woolley £100., and his plate, pictures, furniture, and household effects. He devises all his real estate, and bequeaths the residue of his personal estate in trust to pay the income thereof to Mrs. Wooley for her life, and on her death to distribute the estate in equal shares amongst all his children.

ELECTRIC TRANSMISSION OF POWER.

THE electric transmission of energy for mining operations has been making rapid progress during the past few years both at home and abroad; and now we may safely assume electricity to be one of the recognised means of distributing energy below ground. Indeed, it is probable that no mining engineer would nowadays put in new plant without first taking into serious consideration the comparative cost and advantage to be derived from its use. Electricity is coming into favour for several very important reasons, chief among these being the cheap and efficient manner in which power can be transmitted through long distances. When electricity is more extensively used, the best part of the underground work of collieries will be done by this means. The ease with which energy can be carried by the copper conductors will commend electricity, and by its means power will be taken into parts now inaccessible, except to hand and horse labour. Indeed, it is not improbable that, in some of the colliery centres, where a number of pits is worked within a comparatively small area, large power-stations will be erected on similar lines to the central lighting stations. From these centres power will be distributed to motors below ground. The cables in many cases will be run on the surface and carried below through special boreholes of small diameter. The cost of insulation in the cables will be by this means much reduced. Mechanical coal-cutting is likely to receive a new impulse now energy can be cheaply supplied at the "faces," and assuming such machines are used in non-gaseous seams, the cost of coal-getting should be reduced very considerably. In America there are already some four or five different types of electric coal-getters working successfully. In England Messrs. Blackman and Goolden, at Allerton Main and other collieries have made considerable progress. Their cutter is a high-speed rotary-bar. Messrs. Immisch are also giving this important branch their consideration, not only with reference to getting coal, but also for drift-driving and rock-drilling generally. The electric work hitherto done in England has been so successful that we look to a large increase in the present year.—*Engineer.*

CHINESE PRODUCTS.

THE British Consul of the Ichang district reports that White wax ("insect wax," produced by the Coccus Pe-la), shows a great and continued increase in quantity, but the value is less than in 1887, the price having gone down.

The customs tables show an export of vegetable tallow, 400,341 pounds, valued at £4,233. Besides the ordinary vegetable tallow from the stillingia, there is a kind obtained from the seeds of the varnish tree, used for candles for the hot weather. The export of varnish was 65,227 pounds, value £2,918. Varnish is produced in the highlands westward of Ichang: some, perhaps most, of it appears to find its way to the Yang-tsze lower down than Ichang.

The exports described under the general name of medicines are more than in any previous year. The same is the case with nut-galls. So also with musk, the export of which, more than two tons, seems enormous for such an article. Safflower shows an increase on the preceding year, but is less than in two of the former years; so also with rhubarb.

Of native opium, large quantities are brought down from Szech'wan by porters travelling overland; but probably not much comes into Ichang except for local consumption, the route striking the Yang-tsze someway lower down. In the latter part of 1887, a station was established inland, westward from Ichang, to collect a tax on opium brought down overland, instead of collecting it at Sha-sze, alias Sha-shih on the Yang-tsze below Ichang). I fear however, that this expedient has not been very successful.

COAL: HOW ACQUIRED, HOW SPENT.—It is calculated that the space of time required for the production of the South Wales coalfield was about 640,000 years. The coal measures consist of various seams, and between these seams occur layers of sandstone and shale beds which must have evidently been formed under water. The land has alternately sunk and risen. Whilst it was above water huge tropical forests sprang up, and then they were submerged, to be covered with the deposits of rivers in estuaries. The combined thickness of coal in South Wales is about 120 feet, and it is reckoned that a bed three feet in thickness took a thousand years to form, but as this coal is interspersed with about 120,000 feet of sandstone, the time for the deposition of this strata must be added, which runs to the 640,000 years already mentioned. At the rate it is being extracted now a few hundred years will observe the last of it. Talk about a miser and a spendthrift son, not one of the former ever hoarded his wealth with the care or patience of Nature, nor did one of the latter ever squander his paternal inheritance with the extravagance of mankind in regard to the coal.

Trade Notes.

AN ELECTRO-TECHNICAL EXHIBITION.—A committee has been formed at Frankfort to arrange for the holding of an International Electrical Technical Exhibition from June to October this year.

OTTO VON GUERICKE'S AIRPUMP.—The Mayor of Magdeburg's air-pump has been presented to the Berlin Physical Society, the President of which has just been showing that this ancient instrument is still in good working order. But as it was made in 1675, it cannot be Otto Von Guericke's first air-pump, inasmuch as it was in 1650 that he invented the machine.

AN ELECTRIC TYPE WRITER.—Mr. F. Higgins, a well-known electrician, has given the title of electric type writer to an instrument which he has invented and improved for printing news, transmitted with electricity, on a sheet of paper, in manner similar to that of the ordinary type writer. A single wire transmits the news from a distance to the new machine, which, by electrical power, prints the words, as they are transmitted, in clear type, on a sheet of paper—not a tape—with proper spaces between the words and between the lines.

THE BIOGRAPHY OF COLONEL NORTH.—It was stated, when Col. North went to Chili, that Dr. W. H. Russell accompanied him in the capacity of his biographer. Nothing has been heard of the book since the return of the party, but the idea has not been abandoned. Dr. Russell is, in fact, now at work upon its production. It will be illustrated by Mr. Melton Prior, and will be produced in what the Nitrate King regards as a manner worthy of the subject.

LOCH KATRINE WATER.—The monthly report of the quality of Loch Katrine water, prepared by Professor E. J. Mills, D.Sc., F.R.S., Glasgow and West of Scotland Technical College, has been issued. The results are returned in parts per 100,000:—Total solid impurity, 2·6; organic carbon, ·129; organic nitrogen, ·029; nitric nitrogen, ·003; ammonia, ·000; total combined nitrogen, ·032; hardness, 1·11; chlorine, ·75. Temperature, 5°·8 C. = 42°·44 Fahr. The water was sampled on December 16th. It was brown in light colour, and contained a little suspended matter.

AMERICANS MANUFACTURING PATENT FUEL.—It has been told how the Canadians had determined to convert their mountains of sawdust into patent fuel, and now news comes from Philadelphia that the vast quantities of coal dust annually produced throughout the Pennsylvanian coal fields are to be treated in the same manner as is done in South Wales. The plant has been constructed in the most substantial manner, and the Reading Anthracite Compressed Fuel Company is already in operation, turning out blocks which burn readily, giving an intense heat, with an entire absence of clinkers.

PAPER FROM SAWDUST.—Ottawa—a great timber emporium—has for years been puzzled by the problem associated with the disposal of the vast quantities of sawdust. At last the happy idea has been hit upon of converting it into paper. A mill has been constrncted capable of treating 12,00 tons of sawdust annually. The paper from this material can be made so thick and strong that when tarred and dried it can be used for building purposes, while paper of a finer quality can be obtained by adding one-fourth of wastepaper to the sawdust pulp.

WATER GAS.—The North British Water Gas Syndicate seem to have had a lively meeting in Glasgow on Friday, the 10th inst. A resolution was proposed in favour of the appointment of a committee to investigate the position and prospects of the company, including all facts relating to the patents and the promotion of the undertaking. The resolution was rejected by 35 votes against 27, the directors opposing the appointment of a committee of investigation.

* * * *

We don't want to work,
But by jingoes if we do,
We've got the men, we've got the fools,
We've got their bawbees too.

MR. PICKARD, M.P., ON THE PRICES OF COAL.—Yesterday, Mr. B. Pickard, M.P., who has for the last fortnight been confined to bed, left Barnsley for Southport. Speaking to the representatives of the press, the hon. gentlemen said he was surprised at the continued advance of prices of coal. House coal had gone up 4s. or 5s. per ton since 1888, gas coal 2s. a ton since last July on contracts, and they did not know how much otherwse. Engine fuel had gone up from 3s. to 8s. 6d. per ton, and coking fuel 166 per cent. The men were getting 30 per cent. on the getting price of 1s. 6d. per ton, or 5d. per ton more money. If the owners considered the men would be content with that, they were making a great mistake. Referring to the movements of the men in the north of England for a 15 per cent. advance, Mr. Pickard said that coke had risen in value from 17s. to 37s. per ton, and the owners could give 190 per cent. if they chose. Half the coal got in Durham was coking coal, and the men wer[illegible] from 6½d. to 9½d. per ton for getting it.

NEWCASTLE GAS SUPPLY.—Mr. John Pattinson, city analyst, reports to the Corporation:—I have the honour to report to you the following results of my testing of the illuminating power and purity of the gas supplied to this city by the Newcastle-upon-Tyne and Gateshead Gas Company;—Date of testing, 9th January. Illuminating power in sperm candles, 15·6; grains of sulphur in 100 cubic feet of gas, 8·46; sulphuretted hydrogen nil. According to Act of Parliament, the gas, when tested, should not be of less illuminating power than 15½ standard sprem candles, nor contain more than 20 grains of sulphur in 100 cubic feet. The gas should also be quite free from sulphuretted hydrogen.

CREMATION.—Cremation is making rapid advance in public favour. The first cremation at Woking was conducted in the spring of the year 1885. About 50 cremations have taken place since last May, when operations were resumed after the entire reconstruction of the building hitherto existing there, and the addition of the handsome little chapel. On several occasions there have been two bodies cremated on the same day. Not only is hostile sentiment giving way, but it is beginning to be found that for middle and upper class funerals cremation is really cheaper than burial. The total cost of it, including everything ordinarily implied when we speak of a funeral, is stated to be £15.

ARSENIC IN THE OFFSPRING OF A POISONED MOTHER.—In connection with the foregoing there is brought to mind Roussin's experiments, whereby it was found that arsenic is capable of replacing, to a partial extent, the phosphorous of the calcium phosphates in bone. Having injected calcium arsenate into a rabbit, M. Roussin found that the bones of the young rabbits, born of the poisoned animal, contained a considerable quantity of arsenic. Also, that their urine deposited in noteworthy proportions ammoniacal-magnesian arsenate. But M. Pouchet's experiment does more than confirm M. Roussin's observation. For it proves that not only is calcium arsenate fixed in the bone, but arsenic in the condition of arsenious acid (white arsenic), or sodium arsenate, which, in the substance of the bone, combines with the lime to form calcium arsenate, after all traces of the poison have vanished from the viscera.

CANADIAN PHOSPHATES.—Returns which are now appearing in Canada indicate that the past season has been a fairly prosperous one in the phosphate industry. The exports of rock phosphate from Montreal to Europe during the season of navigation amounted to 23,540 tons, an increase of no less than 9,000 tons as compared with the season of 1888. In addition 2,565 tons of ground phosphate went to the United States from the Ottawa Valley, and 361 tons from the Kingston mines, making a total export to the United States of 3,026 tons. The quantity held over is estimated at 5,000 tons. This season's production of 32,466 tons is an increase of about 9,000 tons as compared with either 1888 or 1887. It is noteworthy, too, that the exports to Europe—that is, substantially to Great Britain—which were in 1882 only 6 per cent. of the total imports, and in 1886 12 per cent., are now over 70 per cent. of the total imports.—*Canadian Gazette.*

THE DEVELOPMENT OF THE PETROLEUM TRADE.—Russian petroleum is rapidly overhauling the American article. From 1883 the exports of the latter have remained stationery, whereas those of the former have jumped with wonderful bounds. In the first of these years America sent us 1,329,004 barrels, while Russia supplied us with only 502. In 1889 America shipped to this country 1,465,265 barrels, and Russia 701,924. But the output shows a more remarkable development. In 1882 the daily quantity in the United States was 82,303 barrels, and this declined in 1888 to 46,707 barrels, whereas the daily output from Baku rose to 14,850 barrels in 1882 to 56,601 barrels in 1888. In the matter of transport Russia has proved to be quite as fertile in resource as her American rival. There are now 62 tank steamers employed in the transport of oils from the United States and Batoum. By these means the oil has become very much cheaper to the consumer. At Cardiff and Swansea some endeavours have been made to afford accommodation for the trade, and more should certainly be done, for in six years the imports have been nearly doubled.

THE NEW ASBESTOS COMPANY, LIMITED.—The New Asbestos Company, Limited, has been formed to acquire and carry on as a going concern, and to develop and extend a very important asbestos industry, thoroughly established and in complete working order, known as the "Société Francaise des Amiantes," of Tarascon. The Tarascon Company was incorporated in France in the latter half of last year, and its business proved a great success. No doubt can be entertained as to the fact that it is capable of enormous development. It has been found that further capital is necessary to keep pace with and adequately meet the increasing volume of business. It is with this view, and in order to secure the benefits of English capital and co-operation, that the present company has been formed. Judging by the commanding position occupied by similar companies, notably Bell's Asbestos Company, the present undertaking offers excellent opportunities for investment. The capital is £120,000., divided into 30,000 shares of £4 each, and there is a present issue of 25,000 shares, whereof the French proprietary of the Société Francaise des Amiantes take 8,332 shares.

selling at very much higher prices. Ground mica, £50. China clay freely offering—common, 18s. 6d.; good medium, 22s. 6d. to 25s.; best 30s. to 35s. (at Runcorn).

REPORT OF MANURE MATERIAL.

The increased activity indicated in our last report continues, and a very considerable business has been concluded during the week in phosphatic material.

There would be buyers of Charleston River phosphate rock at 11d. per unit, in cargoes c.f. and i. to U.K., but in the absence of offers from the other side we do not hear of any business. For Somme phosphates the prices quoted in our last still hold for early shipment, but they would not be accepted far ahead. Owing to strikes there is difficulty in shipping Belgian, and contracts will only be accepted with ample margin in time for shipment. The quotation is 7½d. c.f. and i. for 40/45%, and as much as 10½d. is asked for 48/53%. We do not hear of any transactions in Canadian, producers still refusing to name any possible prices. £5. 12s. 6d. bid and refused for a small November cargo of River Plate bones, but it would be accepted for a December shipment of larger size. There are several cargoes wanted for U.K., and it looks as though buyers would have to come up to sellers' ideas of prices. Several parcels of crushed East Indian bones arrived have been sold at £5. 12s. 6d., and there would be further buyers for immediate delivery. There are buyers for shipment at £5. 7s. 6d. and sellers at £5. 10s. Owing to arrivals East Indian bone meal has been somewhat easier, but the quay having been cleared at £5. 10s., there is nothing now to be had on spot under £5. 12s. 6d., and that price would have to be submitted "to arrive."

Nitrate of soda remains very dull; 8s. 4½d. is full value on spot for fine quality, and 8s. 3d. would buy ordinary. Due cargoes are worth about 8s. 4½d., and November shipments 8s. 7½d., but there is still much indisposition to speculate in this article, and consumers are buying from hand to mouth. The quotations for dried blood, ground hoof, &c., remain without change, and there are no important transactions to report.

The tendency of supers is to harden, and value of 26% is now 47s. 6d. and 50s. per ton in bulk, f.o.r., according to position of works.

Manufacturers have still large purchases to make, but before completing they would like to see their stocks of manures moving into consumption.

WEST OF SCOTLAND CHEMICALS.

GLASGOW, Tuesday.

In general chemical saltcake and alum are slightly easier, and the bichromate works are perhaps less well supplied with orders than is desirable; but nearly all the other lines are healthy, having further added to the firmness announced a week ago. Bleaching powder experiences a steady recovery so far, and soda crystals, partly because of certain important trade changes on the Tyne, are run up higher than for a long time. Caustic sodas (except in the highest strength, which is steady at £9.) are again dearer and in enhanced enquiry; soda ash, refined alkali, and chlorate of potash, are all also the turning of the scales to the good. Water gas circles here are much agitated, and the North British Directorate had a bad quarter of an hour at last week's shareholding meeting, although victorious in the final voting. The meeting of the paraffin scale and wax British-American delegates has been put off some few days without a definite fixture; but the representative of the Americans is now here, and the important business on hand will be consummated one way or other within very brief space. Paraffins are meanwhile all firm except burning oil and naptha, the latter having lost part of recent gain. Lubricating oils still harden. The sulphate of ammonia market is idle for the moment, but stocks are not large and a rally is probable. To-day £12. 2s. 6d. Leith cannot be got, but on the other hand, it does not appear that less has been taken. Chief prices current are:—Soda crystals, 47s. 6d. and 50s. net Tyne; alum in lump, £4. 17s. 0d., less 2½% Glasgow; borax, English refined £30., and boracic acid, £37. 10s. net Glasgow; soda ash, 48/52°, 1⅛d. less 2½% Tyne; caustic soda, white, 76°, £9., 70/72°, £7. 15s, 60/62°, £7., and 60/62° cream, £6. 10s., all less 2½% Liverpool; bicarbonate of soda, 5 cwt. casks, £5., and 1 cwt. casks, £5. 5s. net Tyne; refined alkali, 48/52°, 1⅛d., less 5% Tyne; saltcake, 26s. to 27s.; bleaching powder, £6. 2s. 6d. to £6. 10s.; less 5% f.o.r. Glasgow; bichromate of potash, 4d., and of soda 3d., less 5 and 6% to Scotch and English buyers respectively; chlorate of potash, 4¾d., less 5% any port; nitrate of soda 8s. 4½d. to 8s. 6d., sulphate of ammonia, £12. to £12. 2s. 6d. f.o.b. Leith; salammoniac, 1st and 2nd white, £36. and £34., less 2½% any port; sulphate of copper; £23. 10s., less 5% Liverpool; paraffin scale, hard, 2⅛d., soft, 2d.; paraffin wax, 120°, semi-refined, 2⅞d.; paraffin spirit (naphtha), 9d.; paraffin oil (burning), 6¾d. to 7d., at Glasgow; ditto (lubricating), 865°, £5. to £5. 10s.; 885°, £5. 10s. to £6. 0s.; and 890/895°, £7. to £7. 10s. Week's imports of sugar at Greenock were 57,332 bags.

TAR AND AMMONIA PRODUCTS.

TUESDAY.

Benzoles are still very firm, and it is expected they will rise higher yet; by some dealers the price of 90's is set at 3s. 8d., and 50/90's at 2s. 10d., but as a rule purchasers do not rise to such high figures. One point is certain that there are no stocks of benzol awaiting sale, and we have news of more carbonising works being projected, though it is very probable that with coal at the present price we may find the uninitiated burning their 'prentice fingers. Solvent naphtha and creosote continue in good demand, and though carbolic acid is in good request for the manufacture of explosives, the price is low. Anthracene has been moving off well of late, and prices quoted in our last report hold good for to-day. Pitch is rather easier, but former prices are still current at all ports.

The sulphate of ammonia market is neglected, and prices have drooped to £12. 1s. 3d., or even £12. f.o.b. Hull, and the same values for Leith. Beckton price is now down to £12., and outside London makes find buyers with difficulty at £12. 2s. 6d. Enquiries have reached us asking how this is, in face of dear nitrogen in manures. The reply is simple, the "bear" speculators have led foreign buyers to believe that if they only hold off the market, they will be able to get sulphate at their own price, but it is quite probable that in holding off too long they may have to pay a largely advanced price in the Spring. Sulphate makers would do well to note this and act accordingly.

MISCELLANEOUS CHEMICAL MARKET.

During the past week there has been a strong demand for soda ash, and the prices have advanced; spot values may be taken as 1$\frac{1}{16}$d. for caustic ash and 1$\frac{5}{16}$d. for carbonate. Bleaching powder remains in the same position as last advised both for spot and forward deliveries. Chlorate of potash decidedly firmer at 4¾d. White caustic soda has also advanced, and is extremely scarce on the spot, most makers being sold for some time to come. Values to-day, for 70's., £8. per ton, and for 60's., £7. per ton f.o.b. Liverpool. Sulphate of copper still in good request at £24. per ton, and there is no sign of weakness in this article. Acetates and acetic acids in all grades continue to receive more inquiry, and prices for early deliveries are firm, as per list in prices current. Carbolic acids, crude and crystal, are still in uncertain position as to the future. Spot prices are, for crude, 60's., 2s. 11d., and for crystal, 34's. 11¾d. Prussiate of potash firm, 9¾d. and 10d. Bichromates of potash and soda are in abundant supply, and without change in prices. There is still a general good demand for all heavy chemicals, and the influence of higher prices in wages and fuel is manifest in the upward tendency of values in the same.

Gazette Notices.

PARTNERSHIPS DISSOLVED.

MARSHALL, KINDER AND CO., Denton, hat manufacturers.
RITSON AND CO., Gateshead, electrical and mechanical engineers.
W. W. LIDCOMBE AND J. A FIELDING, under the style of Thomas Ryder and Co., Manchester, brewers, engineers and coppersmiths.
JAMES SINCLAIR AND SON, Southwark-street, Southwark, London, and at Liverpool, manufacturers of and dealers in soap and other articles.

THE BANKRUPTCY ACT, 1883.

Receiving Orders.

F. ASHE Carnaby-street, Soho, London, oil and colourman.
TOM MARSDEN (trading as John Marsden and Sons), Wyke, near Bradford, manufacturing chemist and logwood cutter.

First Meetings, and Public Examinations.

J. HUNTER WATTS, Seething-lane, London, E.C., colour manufacturer and merchant, January 22, Bankruptcy Buildings, Portugal-street, Lincoln's-Inn-Fields, London, February 11, 34, Lincoln's-Inn-Fields.
TOM MARSDEN (trading as John Marsden and Sons), Wyke, near Bradford, manufacturing chemist and logwood cutter, January 20, Official Receiver's Chambers, Bradford, February 7, County Court, Bradford.

Adjudications.

TOM MARSDEN (trading as John Marsden and Sons), Wyke, near Bradford, manufacturing chemist and logwood cutter.

Application for Debtor's Discharge.

E. WINDSBANK, Manchester, paper merchant, February 8, Court House, Manchester.

The Patent List.

This list is compiled from Official sources in the Manchester Technical Laboratory, under the immediate supervision of George E. Davis and Alfred R. Davis.

APPLICATIONS FOR LETTERS PATENT.

Pumps for Medical and Surgical Laboratories. C. Breuillard. 20,879. December 28.
Elements for Voltaic Batteries. D. G. Fitzgerald and A. H. Hough. 20,896 December 28.
Manufacture of Nitrate of Ammonia and Bicarbonate of Soda in conjunction therewith. Watson Smith and C. G. Creswell. 20,869. Dec. 30.
Apparatus for Recovering Chemical from Solutions. M. B. Mason, S. D. Warren, and F. Warren. 20,908. December 31.
Battery Plates for Storage Batteries.—(Complete Specification). A. E. Woolf. 20,911. December 31.
Apparatus for generating and maintaining Electric Currents of Constant Quantity in Circuits of Varying Resistance. W. P. Thompson. 20,928. December 31.
Exposing Liquids to the Action of Gases. J. E. Bedford. 20,925. Dec. 31.
Apparatus for Distilling, Evaporating, or Concentrating Liquids. J. Wright. 20,954. December 31.
Dynamo-Electric Machines. J. J. Wood. 20,955. December 31.
A System of Hydro-Electric Distribution. F. Rysselberghe. 20,974. December 31
Process of and Apparatus for Manufacturing Gas. H. E. Newton. 20,985 December 31.

IMPORTS OF CHEMICAL PRODUCTS

AT

THE PRINCIPAL PORTS OF THE UNITED KINGDOM.

LONDON.

Week ending Dec. 20th.

Antimony Ore—		
Austria,	5 t.	Bailey & Leetham
Sydney,	7	Saddington & Co.
Japan,	90	H. R. Merton & Co.
Bastia,	298	H. Emanuel
"	124	Johnson, Matthey & Co.
Portugal,	18	H. Emanuel
Spain,	40	B. Kuhn
Austria,	40	H. R. Merton & Co.
Alkali—		
France,	100 c.	Arnati & Harrison
Acetate of Lime—		
U.S.,	£180	Johnson & Hooper
Bones—		
U. S.,	31 t. 10 c.	H. A. Lane & Co.
E. Indies,	240 t.	Cornwell, Son & Co.
Holland,	10	J. Lucey & Sons
Bone Meal—		
E. Indies,	100 t.	F. A. Hodgkinson & Co.
Barytes—		
Holland,	23 cks.	W. Harrison & Co.
Germany,	35	O'Hara & Hoar
Belgium,	154	"
Holland,	39	Purrell & Co
Germany,	46	W. Harrison & Co.
"	80	D. Storer & Sons
"	17	O'Hara & Hoar
Chemicals *(otherwise undescribed)*		
Germany,	£13	T. H. Lee
"	260	A. & M. Zimmermann
Holland,	190	Webb, Warter & Co.
Germany,	93	T. H. Lee
"	380	A. & M. Zimmermann
"	20	L. & I. D. Jt. Com
Cream Tartar—		
Holland,	7 pkgs.	Middleton & Co.
Spain,	3	Webb, Warter & Co.
Italy,	64	Howards & Son
"	5	C. F. Gerhardt
Spain,	2	Evans, Lescher & Co.
"	10	Webb, Warter & Co.
France,	52	W. C. Bacon & Co.
Holland	14	J. Thornton
Spain	9	C. Christopherson & Co.
Italy,	10	Webb, Warter & Co.
Caustic Soda—		
Holland,	80 c.	Beresford & Co.
Caoutchouc—		
E. Indies,	62 c.	Jackson & Till
Madagascar,	21	Kebbel, Sons & Co.
E. Indies,	58	Kleinwort, Sons & Co.
Straits Settlements,		T. Merry & Sons
E. Africa,	19	Lewis & Peat
"	20	Kebbel, Son & Co.
France,	26	"
E. Indies,	44	Wallace Bros.
"	13	L. & I. D. Jt. Com.
Copperas—		
Germany,	£45	Charles & Fox
Camphor—		
Germany,	27 pkgs.	L. & I. D. Jt. Co
Caustic Potash—		
France,	13 pkgs.	F. W. Berks & Co.
Cobalt Ore—		
New Caledonia,	7 t.	Gellatly & Co.
Dyestuffs—		
Cochineal		
Canaries,	4 pkgs.	Swanston & Co.
"	4	Beriro & Co.
"	8	Kuhner, Hendschel, & Co.
Alizarine		
Holland,	60 pkgs.	Union Lighterage Co.
Cutch		
E. Indies,	500 pkgs.	Hoare, Wilson & Co.
Tanners' Bark		
Belgium,	11 t.	J. Graves
"	29 10 c.	Schlosser Bros.
Germany,	25 10	Hicks, Nash & Co.
Belgium,	8 10	"
"	40	Schlosser Bros.
Indigo		
E. Indies,	74 chts.	T. Ronaldson & Co.
"	36	L. & I. D. Jt. Co.
"	10	Parsons & Keith
"	69	T. H. Allan & Co.
"	32	L. & I. D. Jt. Co.
"	2	Benecke, Souchay, & Co.
"	169	Arbuthnot, Latham & Co.
"	3	Parsons & Keith
"	68	L. & I. D. Jt. Co.
"	14	W. Brandt, Sons, & Co.
Sumac		
Italy,	100 pkgs.	F. Roberts
Extracts		
Austria,	41 pkgs.	L. J. Levinstein & Sons
"	500	A. J. Humphery
Canada,	840	"
U. S.	250	T. Ronaldson & Co.
France,	103	Burt, Boulton, & Co.
Denmark,	1	G. Scrutton & Co.
France,	103	Burt, Boulton, & Co.
"	180	L. J. Levenstein & Sons
U. S.	40	W. A. Bowditch
"	100	McCracken & Co.
Valonia		
A. Turkey,	33 t.	A. J. Humphery

Ceylon, 189 Prop. Chamb. Whf.
" 42 Hoare, Wilson & Co.

Phosphoric Acid—
Germany, 14 pkgs. Phillips & Graves

Potassium Bicarbonate—
Holland, £20 A. & M. Zimmermann

Potash Salts—
Germany £110 F. J. Puts & Co.

Pyrites—
Spain, 1,932 t. C. Tennant, Sons & Co.

Saltpetre—
Germany, 35 cks. P. Hecker & Co.
E. Indies, 580 bgs. G. Ward & Sons
" 703 L. & I. D. Jt. Com.
Germany 3 cks Craven & Co.
E. Indies, 694 bgs. Henckell & Co.
" 367 L. & I. D. Jt. Co.

Tartaric Acid—
Holland, 2 pkgs. Middleton & Co.
" 27 Webb, Warter, & Co.

Tar—
Bilbao, 554 brls. Burt, Boulton & Co.

Ultramarine—
Holland, 10 cs. W. Harrison & Co
" 20 pkgs. Haeffner, Hilpert & Co
" 20 cks. Seaward Bros.

Zinc Oxide—
Holland, 100 cks. Props. Dowgate Dock
Germany, 50 brls. Julius Matton

LIVERPOOL.

Week ending Jan. 1st.

Albumen—
Antwerp, 4 cs.

Bone Meal—
Bombay, 926 bgs.

Bones—
Glasgow, 211 bgs.
Cadiz, 627 M. Greenwood

Bone Ash—
Rio Grande, 315 t.

Caoutchouc—
New York, 3 brls.
Philadelphia, 50 pkgs.
New York, 4 cs. Farnworth & Jardine
Manaos, 16 Carden & Co.
" 50 Bard & Co.
" 670
Saltpond, 1 brl. 2 bxs. Pickering & Berthoud
" 1 hd. 3 brls. 2 cs. W. Griffiths & Co.
" 1 cs. C. Rathman
" 1 brl.
Cape Coast Castle, 2 brls. Grimwade, Ridley, & Co.
" 2 Edwards Bros.
Axim, 2 bgs. Radcliffe & Durant
" 1 Edwards Bros.
Grand Bassa, 1 brl. Com. Franc
" 2 H. F. Attia
Sierra Leone, 4 Pickering & Berthoud
" 22 22 pns. "
Bathurst 5 Hutton & Co.
" 5 cks. Cie Francaise
" 4 A. Cros & Co.
" 1

Chemicals (*otherwise undescribed*)
Rouen, 85 cks. Co-op. W'sale Society, Ltd.

Copper Ore—
Lisbon, a quantity, Harrington & Co.
" 215 t.
Sestri Levante, 1,030 t. H. Bath & Sons

Camphor—
Rouen, 6 cks. R. J. Francis & Co.

Cream Tartar—
Barcelona, 6 cks. Sachse & Klemm
Tarragona, 5 "
" 26

Copper Precipitate—
Huelva, 148 t.

Charcoal—
St. Nazaire, 726 sks.

Dyestuffs—
Myrabolams
Bombay, 346 bgs. D. Sassoon & Co.
" 333 "
" 400 "
Hamburg, 1 brl.
Indigo
Calcutta, 48 chts.
Argols
Oporto, 23 cks. German Bank
" 23 "
Sumac
Venice, 50 bgs.
Saffron
Valencia, 2 cks. 1 bg.
Cutch
Rangoon, 1,300 bxs.

Guano—
Huanillos, 972 t. Norman & Pigott

Glucose—
Rouen, 5 cks. Co-op. W'sale Society, Ltd.
Hamburg, 19 cks.
" 57

Glycerine—
Marseilles, 7 brls. R. J. Francis & Co.

Hypophosphite—
Rouen, 1 cs. R. J. Francis & Co.

Horn Piths—
Rio Grande, 5 t.

Isinglass—
Para, 16 pkgs. W. Hooton & Yates

Manganese Ore—
Gothenburg, 120 t. Macqueen Bros.

Nitrate—
Iquique, 620 t. N. Waterhouse & Sons

Oxalic Acid—
Rotterdam, 60 cks. O. & H. Marcus

Potash—
Portland, 15 brls.
Rouen, 21 drms. 83 cks.

Phosphate—
Ghent, 1,000 bgs.
Antwerp, 200 t.

Pitch—
Rotterdam, 24 cks.

Pyrites—
Huelva, 1,350 t. Matheson & Co.
" 1,221 Tennants & Co.

Potash Salts—
Hamburg, 1,827 bgs.

Paraffin Wax—
Rangoon, 188 bxs.

Salt—
Calcutta, 1 bg.

Sulphur Ore—
Huelva, 1,430 t. Matheson & Co.
" 1,118 "
" 1,492 "

Sugar of Lead—
Rouen, 87 cks. R. J. Francis & Co.

Stearine—
Antwerp, 1 bg.

Sulphur—
Catania, 582 bgs. 67 brls.

Soda—
Hamburg, 15 cks.

Saltpetre—
Hamburg, 174 cks.

Tartaric Acid—
Rotterdam, 28 cks.

Tar—
Marseilles, 144 brls.

Ultramarine—
Rotterdam, 10 cks.

Waste Salt—
Hamburg, 200 t.

Zinc Oxide—
Rotterdam, 5 cks.
Antwerp, 5 brls.
" 183
Hamburg, 10

Zinc Ashes—
Boston, 9 brls.

HULL.

Week ending Jan. 2nd.

Aluminium—
Rotterdam, 41 cks. W. & C. L. Ringrose

Ammonia—
Rotterdam, 14 cks. W. & C. L. Ringrose
Hamburg, Wilson, Sons & Co.

Acetic Acid—
Rotterdam, 30 cbys. Geo. Lawson & Sons

Barytes—
Bremen, 5 cks. Veltmann & Co.
" 12 J. & T. Seddon
Antwerp, 100 bgs. Bailey & Leetham

Chemicals (*otherwise undescribed*)
Bremen, 2 cs. 2 cks. Veltmann & Co.
Dunkirk, 72 pkgs. Wilson, Sons & Co.
" 16 cks. "
Antwerp, 39 13 drms. "
Hamburg, 90 "

Dyestuffs—
Alizarine
Rotterdam, 24 pkgs. W. & C. L. Ringrose
" 133 cks. Hutchinson & Son
" 87 pkgs. W. & C. L. Ringrose
" 13 cks. Hutchinson & Son
" 27 W. & C. L. Ringrose
" 1 "
Madder
Rotterdam, 4 cks. J. Pyefinch & Co.
" 4 Geo. Lawson & Sons
" 5 Wilson, Sons & Co.
Extracts
Dunkirk, 20 cks. Wilson, Sons & Co.
Rouen, 108 Rawson & Robinson
Dextrine
Hamburg, 100 bgs. Wilson, Sons & Co.

Glucose—
Hamburg, 10 cks. Rawson & Robinson
" 15 Woodhouse & Co.
" 400 bgs. C. M. Lofthouse & Co.
Stettin, 24 cks. Wilson, Sons & Co.
Hamburg, 28 "
" 12 Bailey & Leetham
" 50

Glycerine—
Amsterdam, 9 pkgs.

Manure—
Danzig, 4 cks.
Ghent, 240 t. Wilson, Sons & Co.
" 100
" 344 "

Naphthol—
Rotterdam, 5 brls. Hutchinson & Son

Pitch—
Ghent, 97 cks. Wilson, Sons & Co.

Pyrites—
Huelva, 1,437 t. J. Dalton, Holmes & Co.

Potash—
Rotterdam, 3 cks. Geo. Lawson & Sons

Plumbago—
Hamburg, 19 cks. Wilson, Sons & Co.
" 9 Furley & Co.

Soda—
Rotterdam, 43 cks. G. Lawson & Sons

Stearine—
Antwerp, 3 bgs. Bailey & Leetham

Tartaric Acid—
Rotterdam, 7 cks W. & C. L. Ringrose
" 4

Turpentine—
Libau, 50 brls.
Konigsberg, 40

Waste Salt—
Hamburg, a quantity. Geo. Lawson & Sons

GLASGOW.

Week ending Jan. 2nd.

Alum—
Rotterdam, 40 cks.

Barytes—
Rotterdam, 46 cks.

Chemicals (*otherwise undescribed*)
Rouen, 12 pkgs. J. & P. Hutchison

Carbon—
New York, 15 cs.

Copper Ore—
Huelva, 1,222 t. Tharsis Co.

Dyestuffs—
Sumac
Rouen, 100 bgs. J. & P. Hutchison
Alizarine
Rotterdam, 58 cks. 2 cs.

Pitch—
Antwerp, 9 cks.

Stearine—
Antwerp, 13 bgs.
New York, 20 hhds.
Baltimore, 20

Sodium Nitrate—
Pisagua, 1,413 t. A. Cross & Sons

Sulphur Ore—
Drontheim, 155 t.

Tartar—
Bordeaux, 8 cks.

Tartaric Acid—
Rotterdam, 3 cks. J. Rankine & Sons

TYNE.

eek ending January 1st.

Earth—
aburg, 18 cks. Tyne S.S. Co.
er Precipitate—
lva, 419 t.
icals *(otherwise undescribed)*
aburg, 1 cs. Tyne S S. Co.
ose—
erdam, 20 bgs. Tyne S.S. Co.
aburg. 45 cks. "
arine—
aburg, 11 cs. 3 cbys. Tyne S S. Co.
um —
en, 150 t.
es—
lva, 1 321 t.
etre—
aburg, 250 t. Clephans & Wiencke

Ultramarine—
Rotterdam, 1 ck. Tyne S.S. Co.
Waste Salt—
Hamburg, 50 t. Tyne S.S. Co.

GRIMSBY.

Week ending December 28th.

Albumen—
Antwerp, 13 pkgs.
" 7 cks.
Chemicals *(otherwise undescribed)*
Hamburg, 24 cks.
Extract of Tan—
Dieppe, 20 pkgs.
Glucose—
Hamburg, 12 cks.
" 40
" 12
Zinc Ashes—
Antwerp, 24 pkgs.

GOOLE.

Week ending December 25th.

Antimony Ore—
Boulogne, 9 cks.
Charcoal—
Antwerp, 75 bgs.
Copper Ore—
Huelva, 660 tons
Dyestuffs—
Alizarine
Rotterdam, 34 cks.
Dyestuffs *(otherwise undescribed)*
Boulogne, 30 cks.
" 46 cs.
Glucose—
Dunkirk, 120 bgs.
Hamburg, 21 cks.
" 161 bgs.
Calais, 200

Glycerine—
Hamburg, 3 cs.
Plumbago—
Boulogne, 7 pkgs.
" 8 cks.
" 6 cs.
Potash—
Calais, 19 pkgs.
" 69 drmns.
Antwerp, 16 cks.
Dunkirk, 35
Sodium Chlorate—
Rouen, 3 cks.
Saltpetre—
Hamburg, 43 cks.
Rotterdam, 100 bgs.
" 270
Tartaric Acid—
Rotterdam, 1 ck.
Waste Salt—
Hamburg, 35 cks.

EXPORTS OF CHEMICAL PRODUCTS

FROM

THE PRINCIPAL PORTS OF THE UNITED KINGDOM.

LONDON.

ek ending December 28th.

—*(otherwise undescribed)*
gne, 300 btles. £457
—
nica, 6 t. £47
onia—
hama, 5 cwts. £11
o, 25 cs. £17
onia Sulphate—
aburg, 341 t. 19 cwts. £4 127
inique, 202 10 2,500
onia Carbonate—
gne, 1 t. 2 cwts. £29
eaux, 1 2 33
ington, 5 cs. 12
c Acid—
hama, 4 cwts. £11
ey, 8 cs 27
phite of Lime—
to, 1 t. 13 cwt. £23
stone—
l, 3 t. 10 cwt. £27
x—
aburg, 1 t. 1 cwt. £60
n Tartar—
bourne, 3 t. 5 cwts. 28 pkgs. £480
ey, 1 t. £114
eras—
ador, 10 t. 14 cwts. £27
Acid—
gne, 4 cks. £147
aburg, 1 t. 13 c. 1 ck. 254
o, 3 cwts. 23
lle, 18 cwts. 131
bourne, 15 cwts. 105
l, 2 t. 10 cwts. 326
ey, 10 75
sels, 12 95
er Sulphate—
stchurch, 1 t. 10 cwts. £33
ey, 2 47
olic Acid—
Town, 16 cwts. £43
ogne, 4 t. 14 cwts. £173
seilles, 15 1,680
icals *(otherwise undescribed)*
cland, 27 cs. £191
York, 11 pkgs. 106
tic Potash—
ier, 5 t. 5 cwts. £102
fectants—
ey, 20 cks. £33
aburg, 25 100
bourne, 42 cs. 43
o—
ritius, 2,119 bgs. 172 cks. £1,000
Carbon—
sels, 3 t. 15 cwts. £68
ire—
bourg, 152 t. 7 cwts. £595
aburg, 50 400

Jersey, 152 12 1,492
Guernsey, 6 10 41
Antwerp, 50 3 350
Launceston, 5 54
Hobart, 15 162
Oxalic Acid—
Dunkirk, 8 t. 5 cwts. £270
Potassium Bichromate—
M. Video, 10 cwts. £23
Phosphorus—
Wellington, 19 cs. 4½ cwts. £162
Otago, 18 1,000 lbs. £266
Phosphoric Acid—
Berbice, 5 t. 14 cwts £88
Saltpetre—
Oporto, 11 t. 1 cwt. 227
Lisbon, 40 15 867
Sydney, 2 10 55
Santos, 7 16 174
N. Orleans, 25 525
Adelaide, 1 5 27
Saccharum Cake—
Dunkirk, 50 t. £75
Sulphuric Acid—
Natal, 1 t. 15 cwts. £9
Kurrachee, 6 3 60
Sheep Dip—
Pt. Howard, 300 cs. £600
Melbourne, 41 pt. 52
Natal, 6 t. 3 cwts. 280
Sodium Sulphate—
Lyttelton, 18 cks. £30
Tartaric Acid—
Melbourne, 95 kegs. £764
Sydney, 54 174

LIVERPOOL.

Week ending Dec. 28th.

Alkali—
Antwerp, 10 cks.
Genoa, 30
Ammonia Carbonate—
Oporto, 5 c. £8
Ammonia Muriate—
Philadelphia, 32 t. 3 c. 3 q. £736
San Francisco, 1 17
Antimony Regulus—
Portland, M., 1 t. £74
Rotterdam, 3 213
Antwerp, 10 c. 33
Alum—
Bombay, 82 t. 19 c. £386
Seville, 1 5
Syra, 9 19 1 q. 50
Oporto, 1 4 7
Sydney, 1 5
Constantinople, 2 t. 9 c. 1 q. 12
Smyrna, 3 15
Ammonia Sulphate—
Boston, 2 t. 4 c. 1 q. £15
Demerara, 33 3 431
Ammonia—
Monte Video, 5 c. £14

Bleaching Powder—
Amsterdam, 7 cks.
Batoum, 26
Boston, 639
" 10 kgs. 257 tcs. 303 cks.
Genoa, 45 cks.
Ghent, 141
Hamburg, 54
Havana, 8
Malaga, 4 brls.
Messina, 35
Naples, 4
New York, 191
" 78 t. 8 cks. £428
Oporto, 30 brls.
Palermo, 7
Pasages, 42 cks.
Rouen, 130
Santos, 4 cs.
Trieste, 29
Venice, 54
Vera Cruz, 25
Borax—
Rouen, 42 t. 15 c. £1,232
Borate of Lime—
Havre, 50 t. 19 c. 1 q. £612
Caustic Soda—
Algiers, 35 bxs.
Amsterdam, 183 dms.
Antwerp, 44 10 brls.
Baltimore, 100
Barcelona, 100
Barbadoes, 120
Batoum, 200
Beyrout, 25
Bombay, 15
Bordeaux, 8
Boston, 500
Calcutta, 30
Carthagena, 75 25 cks 5 brls.
Catania, 115
Cienfuegos, 10
Constantinople, 25
Genoa, 55
Ghent, 5
Guantanamo, 10
Hamburg, 35
Halifax, 25
Havana, 40
Hiogo, 200
Lisbon, 20
Messina, 30
Maceio, 30
Manilla, 100
Malaga, 256 126 kgs. 10 brls.
Maracaibo, 10
Malta, 5
Matanzas, 50
Nantes, 192
New Orleans, 200
Naples, 650
New York, 668
Odessa, 98
Oporto, 89
Pasages, 169
Pernambuco, 10
Philadelphia, 135 brls.
Port Natal, 22 cs.

Portland, M. 400 dms.
" O. 20
Rotterdam, 40
Rouen, 161
Rio Janeiro, 15
Santo Domingo, 40 kgs.
Santander, 12
Seville, 45
Shanghai, 80
Sydney, 46
Tampico, 300 kgs
Talcahuano, 40
Tarragona, 113 30 cks.
Tacuma, 50
Trebizonde, 7
Valencia, 25
Vera Cruz, 15
Yokohama, 300
Chemicals *(otherwise undescribed)*
Bahia, 7 pkgs. £50
Antwerp, 16 t. 18 c. 514
Hamburg, 1 14 2 q. 85
New York, 10 30
Copper Sulphate—
Bilbao, 5 t. £115
Pasages, 10 218
Rouen, 11 12 c. 136
Barcelona, 10 1 q. 180
Bordeaux, 10 3 220
Venice, 20 2 294
B. Ayres, 40 kgs. 47
Constantinople, 6 c 6
Carbolic Acid—
Santander, 6 c. 3 q. £27
New York, 2 t. 112
Paris, 1 5 2 146
Hamburg, 1 10 192
Amsterdam, 17 2 12
Caustic Potash—
Tuticorin, 19 c. £21
Melbourne, 3 t. 2 62
Calcutta, 2 7 47
Charcoal—
Nantes, 14 t. £70
Copperas—
Syra, 2 t. 13 c. £6
Calcium Chloride—
B. Ayres, 5 t. £10
Disinfectants—
Bombay, 3 c. 3 q. £7
Epsom Salts—
Rio Janeiro, 100 kgs.
Glycerine—
Portland, 10 t. 13 c. £446
Iron Oxide—
Boca, 179 t. 9 c. 2 q. £449
Barcelona, 15 2 70
Rosario, 20 t. 59
Lime Carbonate—
Lisbon, 1 t. 14 c. £9
Magnesia—
Trieste, 1 cs.

Rangoon, 9 brls.
Rio Janeiro, 10 cks.
Syra, 250 brls.
Smyrna, 75 cks.
Valparaiso, 100 brls.
Yokohama, 120

Sodium Silicate—
Bilbao, 58 brls. 16 cks.
Beyrout, 5
Barcelona, 30
Constantinople, 5
Mollendo, 10
Oporto, 32 cks.

Soda Crystals—
Bombay, 400 kgs.
Boston, 280 brls.
Constantinople, 180
Calcutta, 35
New York, 438 cks.
Portland, O., 300
San Francisco, 360 brls.
Tacoma, 430
Valparaiso, 30 100 dms.

Sodium Bichromate—
Sydney, 120 kgs.

Sheep Dip—
Algoa Bay, 2 c. 2 q. £5
Buenos Ayres, 90 t. 3 c. 2 q. 2,085

Sodium Sulphate—
Seville, 18 c. £5

Salammoniac—
Syra, 4 c. £7
Constantinople, 2 t. 5 76
Naples, 18 31
Kurrachee, 1 9 49
Genoa, 6 3 q. 11
Pernambuco, 1 5
Bilbao, 8 1 13

Sodium Nitrate—
Demerara, 19 t. 19 c. £197
Portland, 1 7 3 q. 12
New York, 8 11
Batoum, 1 11
Seville, 1 5 11
Fiume, 6 2 2 51

Sulphur—
Guantanamo, 12 c. £5
Batoum, 1 t. 15 39
Halifax, 7 12 2 q. 45
Tampico, 10 1 5
Vera Cruz, 1 6 3 9
Iquique, 41 17 1 272
Boston, 67 2 293

Saltcake—
New York, 49 t. 10 c. £95

Superphosphate—
Hamburg, 100 t. 1 c. 1 q. £112

Tar—
Bay Beach, 12 brls.
Carra Azul 8
New York, 8
Guayaquil, 50

Tartaric Acid—
Montreal, 2 t. 6 c. £351

Waste Salt—
Opobo, 731 t.
Portland, 99
Port Townsend, 115
Portland, O. 885
Port Natal, 28
Rangoon, 1,928
Salt Pond, 7 17 c.
Savannah, 500
Seattle, 250
Sierra Leone, 40
Talcahuano, 44 10
Tacoma, 400

Zinc Chloride—
New York, 4 t. £68

GLASGOW.

Week ending January 2nd.

Ammonia Sulphate—
Valencia, £1,083

Ammonia—
Hamburg, 10 t. £116

Bleaching Powder—
Karachi, 2 t. 8 c. £23

Benzol—
Rotterdam, 58 puns. £690

Copperas—
Karachi, 79 t. 1 c. £212

Chemicals *(otherwise undescribed)*
Karachi, 5 t. 10 c. £74

Charcoal—
Hamburg, 462 bgs. 41 t. 12 c. £190

Epsom Salts—
Sydney, 3 t. 2 c. £24

Litharge—
Karachi, 1 t. 10 c. £26

Manure—
Valencia, 165 t. 24 c. £1,308

Manganese—
Rotterdam, 18 cks. £88

Potassium Bichromate—
Karachi, 3 t. 1 c. £109

Pitch—
Antwerp, 640 t.

Paraffin Wax—
Bilbao, 5 t. 12 c. £80

Sheep Dip—
Melbourne, 6 t. 10 c. £200

Sulphuric Acid—
Karachi, 15 t. 10 c. £77

Sodium Bichromate—
Karachi, 1 t. £26

Sugar of Lead—
Karachi, 4 t. 8 c. £78

Soda Ash—
Baltimore, 31 t. 15 c. £390

Tar Oil—
Bilbao, 10 t. £40

TYNE.

Week ending Jan. 1st.

Alkali—
Christiania, 1 t. 2 c.
Gothenburg, 5 8
Leghorn, 29 7
Antwerp, 5 2
Rotterdam, 1 1

Alum Cake—
Copenhagen, 7 t. 9 c.
Bandholm, 6 2

Arsenic—
Genoa, 1 t. 15 c.

Antimony—
Rotterdam, 2 t.
Rouen, 6

Ammonia Sulphate—
Genoa, 30 t.
Helsingborg, 30 t. 9 c.
Hamburg, 15

Bleaching Powder—
Hamburg, 13 t. 10 c.
Lisbon, 8
Gothenburg, 14 8
Christiania, 19
Copenhagen, 2 4
Rotterdam, 17 17
Rouen, 54 2
Antwerp, 100 3

Barytes Carbonate—
Antwerp, 9 t. 18 c.
Rouen, 57

Barytes—
Rouen, 35 t.

Caustic Soda—
Lisbon 1 t. 19 c.
Genoa, 18 15
Antwerp, 170 1
Dunkirk, 17
Syra, 110 6

Copperas—
Christiania, 3 t. 1 c.
Odense, 5 7

Glauber Salts—
Christiania, 1 t. 7 c.

Litharge—
Genoa, 1 t. 2 c.
Rouen, 7 17

Magnesia—
Antwerp, 6 c.
Barcelona, 5 t. 16
Rouen, 11

Nitrate of Baryta—
Rouen, 5 t. 3 c.

Potassium Chlorate—
Genoa, 1 t.
Leghorn, 1 12 c.
Naples, 10
Antwerp, 2
Copenhagen, 5
Malmo, 2 10

Pitch—
Leghorn, 200 t.
Antwerp, 100
Valencia, 25

Soda Ash—
Lisbon, 12 t. 9 c.
Gothenburg, 6 11
Dunkirk, 6 10
Copenhagen, 19 5
Genoa, 3 1
Leghorn, 2 3

Soda—
Nykjobing, 21 t. 14 c.
Christiania, 5 6
Malmo, 11 t. 27 c.
Genoa, 11 22
Catania, 2
Huelva, 2 3
Horsens 13 17
Odense, 28 7
Antwerp, 6 9
" 5 7
Syra, 29 19
Nakskov, 19 17
Rotterdam, 23 6

Sodium Bicarbonate—
Christiania, 1 c.

HULL.

Week ending December 31st.

Alum—
Gothenburg, 670 slbs.

Alkali—
Riga, 255 ckr.

Borax—
Danzig, 26 cks.

Bone Size—
Harlingen, 1 ck.

Bleaching Powder—
Stettin, 110 cks.

Chemicals *(otherwise undescribed)*
Antwerp, 1 cs. 39 cks
Amsterdam, 7 20
Alexandria, 12
Bergen 1
Bremen, 6
Bordeaux, 36
Christiania, 16 15 drms.
Copenhagen, 13 200 pkgs.
Danzig, 24 88
Dunkirk, 44
Gothenburg, 6 50 13
Hamburg, 234 6 35
Konigsberg, 4 10
Libau, 1
New York, 100 88
Rotterdam, 30 123
Rouen, 9
Stavanger, 7
Stettin, 15 123

Caustic Soda—
Danzig, 86 drms.
Libau, 35
Riga, 363

Dyestuffs *(otherwise undescribed)*
Hamburg, 1 cs.
Reval, 20 cks.
Rotterdam, 4

Manure—
Gothenberg, 10 cks.
Hamburg, 1055 bgs.
St. Michael's 5 t.

Pitch—
Antwerp, 342 t.

Slag—
Amsterdam, 155 t.
Danzig, 407

GOOLE.

Week ending Dec. 25th.

Benzol—
Antwerp, 26 cks.
Rotterdam, 118
Terneuzen, 32 drs.

Chemicals *(otherwise undescribed)*
Antwerp, 20 cs.
Ghent, 10 cks.
Rotterdam, 25

Coal Products—
Dunkirk, 45 cks.

Dyestuffs *(otherwise undescribed)*
Boulogne, 1 cs.
" 2 cks.
Ghent, 20

Manure—
Boulogne, 24 bgs.
Ghent, 680

Pitch—
Antwerp, 650 tons
Dunkirk, 56

GRIMSBY.

Week ending Dec. 28th.

Chemicals *(otherwise undescribed)*
Dieppe, 34 cks.
Hamburg, 8 pkgs.
Rotterdam, 24 drms.

Coal Products—
Dieppe, 10 cks.

Dyestuffs *(otherwise undescribed)*
Dieppe, 1 ck.

PRICES CURRENT.

THURSDAY, JANUARY 16, 1890.

PREPARED BY HIGGINBOTTOM AND CO., 116, PORTLAND STREET, MANCHESTER.

[Th]e values stated are F.O.R. at maker's works, or at usual ports of shipment in U.K. The price in different localities may vary

		£	s.	d.
[Ac]ids :—				
Acetic, 25 °/₀ and 40 °/₀	per cwt	7/6 & 12/6		
,, Glacial	,,	2	10	0
Arsenic, S.G., 2000°	,,	0	12	0
Chromic 82 °/₀	nett per lb.	0	0	6½
Fluoric	,,	0	0	3¼
Muriatic (Tower Salts), 30° Tw.	per bottle	0	0	8
,, (Cylinder), 30° Tw.	,,	0	2	11
Nitric 80° Tw.	per lb.	0	0	2
Nitrous	,,	0	0	1¼
Oxalic	nett ,,	0	0	3¼
Picric	,,	0	1	5
Sulphuric (fuming 50 °/₀)	per ton	15	10	0
,, (monohydrate)	,,	5	10	0
,, (Pyrites, 168°)	,,	3	2	6
,, (,, 150°)	,,	1	6	0
,, (free from arsenic, 140/145°)	,,	1	10	0
Sulphurous (solution)	,,	2	15	0
Tannic (pure)	per lb.	0	1	0½
Tartaric	,,	0	1	4
[Ar]senic, white powered	per ton	12	10	0
[Al]um (looselump)	,,	4	12	6
[Al]umina Sulphate (pure)	,,	5	0	0
,, ,, (medium qualities)	,,	4	10	0
[Al]uminium	per lb.	0	15	0
[Am]monia, ·880=28°	per lb.	0	0	3
,, =24°	,,	0	0	1⅞
,, Carbonate	nett ,,	0	0	3½
,, Muriate	per ton	24	0	0
,, ,, (sal-ammoniac) 1st & 2nd	per cwt.	36/- & 34/-		
,, Nitrate	per ton	40	0	0
,, Phosphate	per lb.	0	0	10½
,, Sulphate (grey), London	per ton	12	5	0
,, ,, (grey), Hull	,,	12	2	6
[An]iline Oil (pure)	per lb.	0	0	10
[An]iline Salt ,,	,,	0	0	9
[An]timony	per ton	75	0	0
,, (tartar emetic)	per lb.	0	1	1
,, (golden sulphide)	,,	0	0	10
[Ba]rium Chloride	per ton	7	0	0
,, Carbonate (native)	,,	3	15	0
,, Sulphate (native levigated)	,,	45/- to 75/-		
[Bl]eaching Powder, 35 %	,,	5	5	0
,, Liquor, 7 %	,,	2	10	0
[Bi]sulphide of Carbon	,,	12	15	0
[Ch]romium Acetate (crystal	per lb.	0	0	6
[Ca]lcium Chloride	per ton	2	0	0
[Ch]ina Clay (at Runcorn) in bulk	,,	16/- to 32/-		
[Co]al Tar Dyes :—				
Alizarine (artificial) 20 %	per lb.	0	0	9
Magenta	,,	0	3	9
[Co]al Tar Products				
Anthracene, 30 %, f.o.b. London	per unit per cwt.	0	1	1
Benzol, 90 % nominal	per gallon	0	3	6
,, 50/90 ,,	,,	0	2	8
Carbolic Acid (crystallised 35°)	per lb.	0	0	11¼
,, (crude 60°)	per gallon	0	2	10
Creosote (ordinary)	,,	0	0	2
,, (filtered for Lucigen light)	,,	0	0	3
Crude Naphtha 30 % @ 120° C.	,,	0	1	2
Grease Oils, 22° Tw.	per ton	3	0	0
Pitch, f.o.b. Liverpool or Garston	,,	1	12	0
Solvent Naphtha, 90 % at 160°	per gallon	0	1	6
[Co]ke-oven Oils (crude)	,,	0	0	2½
[Co]pper (Chili Bars)	per ton	50	5	0
,, Sulphate	,,	24	0	0
,, Oxide (copper scales)	,,	52	10	0
[Gl]ycerine (crude)	,,	30	0	0
,, (distilled S.G. 1250°)	,,	55	0	0
[Io]line	nett, per oz.	8d. to 9d.		
[Iro]n Sulphate (copperas)	per ton	1	7	6
Sulphide (antimony slag)	,,	2	0	0
[Le]ad (sheet) for cash	,,	17	0	0
,, Litharge Flake (ex ship)	,,	17	10	0
,, Acetate (white ,,)	,,	23	10	0
,, ,, (brown ,,)	,,	19	0	0
,, Carbonate (white lead) pure	,,	19	0	0
,, Nitrate	,,	22	5	0

		£	s.	d.
Lime, Acetate (brown)	,,	8	5	0
,, ,, (grey)	per ton	14	0	0
Magnesium (ribbon and wire)	per oz.	0	2	3
,, Chloride (ex ship)	per ton	2	7	6
,, Carbonate	per cwt.	1	17	6
,, Hydrate	,,	0	10	0
,, Sulphate (Epsom Salts)	per ton	2	15	0
Manganese, Sulphate	,,	18	0	0
,, Borate (1st and 2nd)	per cwt.	60/- & 42/6		
,, Ore, 70 %	per ton	4	5	0
Methylated Spirit, 61° O.P.	per gallon	0	2	2
Naphtha (Wood), Solvent	,,	0	3	6
,, ,, Miscible, 60° O.P.	,,	0	4	0
Oils :—				
Cotton-seed	per ton	22	10	0
Linseed	,,	22	10	0
Lubricating, Scotch, 890°—895°	,,	7	5	0
Petroleum, Russian	per gallon	0	0	5¼
,, American	,,	0	0	6⅛
Potassium (metal)	per oz.	0	3	10
,, Bichromate	per lb.	0	0	4
,, Carbonate, 90% (ex ship)	per ton	18	5	0
,, Chlorate	per lb.	0	0	4¼
,, Cyanide, 98%	,,	0	2	0
,, Hydrate (Caustic Potash) 80/85 %	per ton	23	0	0
,, ,, (Caustic Potash) 75/80 %	,,	21	15	0
,, ,, (Caustic Potash) 70/75 %	,,	20	15	0
,, Nitrate (refined)	per cwt.	1	3	6
,, Permanganate	,,	4	5	0
Prussiate Yellow	per lb.	0	0	9¼
,, Sulphate, 90 %	per ton	9	10	0
,, Muriate, 80 %	,,	7	15	0
Silver (metal)	per oz.	0	3	8¼
,, Nitrate	,,	0	2	5½
Sodium (metal)	per lb.	0	4	0
,, Carb. (refined Soda-ash) 48 %	per ton	5	5	0
,, ,, (Caustic Soda-ash) 48 %	,,	4	2	6
,, ,, (Carb. Soda-ash) 48%	,,	4	15	0
,, ,, (Carb. Soda-ash) 58 %	,,	5	12	6
,, ,, (Soda Crystals)	,,	2	12	0
,, Acetate (ex-ship)	,,	16	0	0
,, Arseniate, 45 %	,,	10	0	0
,, Chlorate	per lb.	0	0	6¼
,, Borate (Borax)	nett, per ton.	29	10	0
,, Bichromate	per lb.	0	0	3
,, Hydrate (77 % Caustic Soda)	per ton.	8	15	0
,, ,, (74 % Caustic Soda)	,,	8	5	0
,, ,, (70 % Caustic Soda)	,,	7	17	6
,, ,, (60 % Caustic Soda, white)	,,	6	17	6
,, ,, (60 % Caustic Soda, cream)	,,	6	10	0
,, Bicarbonate	,,	5	0	0
,, Hyposulphite	,,	5	0	0
,, Manganate, 25%	,,	8	10	0
,, Nitrate (95 %) ex-ship Liverpool	per cwt.	0	8	3
,, Nitrite, 98 %	per ton	27	10	0
,, Phosphate	,,	15	15	0
,, Prussiate	per lb.	0	0	7
,, Silicate (glass)	per ton	5	7	6
,, ,, (liquid, 100° Tw.)	,,	3	17	6
,, Stannate, 40 %	per cwt.	2	0	0
,, Sulphate (Salt-cake)	per ton.	1	6	0
,, ,, (Glauber's Salts)	,,	1	10	0
,, Sulphide	,,	7	15	0
,, Sulphite	,,	5	5	0
Strontium Hydrate, 98 %	,,	8	0	0
Sulphocyanide Ammonium, 95 %	per lb.	0	0	7¼
,, Barium, 95 %	,,	0	0	5½
,, Potassium	,,	0	0	7¼
Sulphur (Flowers)	per ton.	7	0	0
,, (Roll Brimstone)	,,	6	0	0
,, Brimstone : Best Quality	,,	4	15	0
Superphosphate of Lime (26 %)	,,	2	10	0
Tallow	,,	25	10	0
Tin (English Ingots)	,,	99	0	0
,, Crystals	per lb.	0	0	6¼
Zinc (Spelter	per ton.	24	10	0
,, Chloride (solution, 96° Tw.	,,	6	0	0

THE

[C]HEMICAL TRADE JOURNAL.

Publishing Offices: 32, *BLACKFRIARS STREET, MANCHESTER.*

... SATURDAY, JANUARY 25, 1890. Vol. VI.

Contents.

Notices.

...nications for the *Chemical Trade Journal* should be ... Cheques and Post Office Orders made payable to—

...., 32, Blackfriars Street, MANCHESTER.

Our Registered Telegraphic Address is—

"Expert, Manchester."

... of Subscription, commencing at any date, to the ...de *Journal*,—payable in advance,—including postage ... the world, are as follow:—

...(52 numbers)	12s. 6d.
...ly (26 numbers)	6s. 6d.
...y (13 numbers)	3s. 6d.

... oblige by making their remittances for subscriptions by ... Office Order, crossed.

...tions for the Editor, if intended for insertion in the ... issue, should reach the office not later than Tuesday ...

...orts, and correspondence on all matters of interest to the ... allied industries, home and foreign, are solicited. ...s should condense their matter as much as possible, ...de only of the paper, and in all cases give their names ..., not necessarily for publication. Sketches should be ...ate sheets.

...ndertake to return rejected manuscripts or drawings, ...anied by a stamped directed envelope.

... invited to forward items of intelligence, or cuttings from ...rs, of interest to the trades concerned.

... of the special features of the *Chemical Trade Journal* ...liest information respecting new processes, improvements, ..., bearing upon the Chemical and allied industries, or ... of interest to our readers, the Editor invites particulars ... in working order—from the originators; and if the ...sed of sufficient importance, an expert will visit and ... same in the columns of the *Journal*. There is no fee ...its of this kind.

...steem it a favour if any of our readers, in making ... opening accounts with advertisers in this paper, will ... the *Chemical Trade Journal* as the source of their ...

...nts intended for insertion in the current week's issue, ...he office by Wednesday morning at the latest.

Advertisements.

...tisements of Situations Vacant, Premises on Sale or ... Miscellaneous Wants, and Specific Articles for Sale by ..., are inserted in the *Chemical Trade Journal* at the ...:—

...s and under	2s.	0d.	per insertion.
...dditional 10 words	0s.	6d.	,,

...ents of Situations Wanted are inserted at one-...er rates when prepaid, viz:—

... and under	1s.	0d.	per insertion.
...dditional 10 words	0s.	3d.	,,

...nouncements in the Directory Columns, and all Adver-... are charged at the Tariff rates which will be forwarded on ...

FOREIGN LABOUR IN THE AMERICAN GLASS TRADE.

THE American glass trade is going through an experience which, though by no means peculiar, is yet replete with interest to those who study the intricate relations subsisting between labour and capital in the trades and manufactures. The Glass Workers' Association, with a view to keeping up rates of wages, and otherwise protecting the interests of its members, has used its power, after a very arbitrary fashion, to limit apprentices to a number out of all proportion to the requirements of a growing trade. The laws of supply and demand—mightier than any Glass Workers' Association—were not to be interfered with, and the American workmen have now the mortification of seeing foreigners making their way in to fill positions which an over-strained policy has withheld from their own countrymen. Nor is their vexation likely to be diminished by the fact that the invocation of the alien contract labour law promises to be of little avail to them.

We are far from being devoid of sympathy for the American, as for all other working men, but we cannot help dwelling on this as another of the many cases where an uncurbed militant trade unionism defeats its own ends, and ruins the best interests of those who combine to give it their practical advocacy. A trades union is a labour syndicate, and trades unionists too often fail to realise the fact that the foundations of all syndicates are exceedingly unstable. The experience of the great copper gamble—so recently gone to take its place among the great failures of the past—was not needed to teach us that moderation is the first element of success in any syndicate or combination. Exaggerated demands and one-sided restrictions always tend, directly or indirectly, to compass the downfall of the undertaking in which their sway is too freely allowed. Had the American glass workers taken care to put, so to say, their own house in order first, and secured their position by a more liberal regulation of their own numbers, so that they could at any time have dealt with the demands of an increased trade, then things would have been better for them. The result of their dog-in-the-manger policy may be a reduction, through increased competition, in the rates of wages that they have been seeking to raise. We hope the lesson will act as a warning to our own countrymen. It is quite evident to all that our national welfare for the next few years, at least, will depend very largely upon the position taken up by the working man as represented in the trades unions. If these bodies act wisely and study the true interests of their members, we see no reason to anticipate the decadence of England's greatness, but if, on the other hand, the working man allows himself to be robbed of his sound common sense by the blatant mob orator or the irresponsible sower of agitation, and blocks the course of trade and industry because he is unable to obtain his not always reasonable demands, then, for himself and his country, we fear the time may be a disastrous one, and may see such a deviation of our trade to other quarters that our industrial pre-eminence will cease to be recognised, and our commercial prestige irretrievably lost. We repeat, with emphasis, that in our opinion, almost everything depends upon the moderation displayed by employers and employed in their mutual relations, and we trust that both parties may have so profited by past experience, that no serious check may be given to the flow of our industrial enterprise. It is a repetition of the old story of the "Belly and the Members." The prosperity of the one cannot be permanently dissociated from that of the other. In the long run they must stand or fall together.

Legal.

IMPORTANT ACTION AGAINST A STYAL FARMER.

CHESHIRE FARMERS AND A MANCHESTER MANURE COMPANY.

AT the Altrincham County Court, on the 8th inst., before his Honour Judge Ffoulkes, the Manchester Phosquane Company sued John Walton, a farmer, of Styal, near Altrincham, for the sum of £6. 0s. 9d., for manures sold and delivered. Mr. Sutton, barrister, (instructed by Messrs. Addleshaw and Warburton, of Manchester), appeared for the company, and Mr. Gray, of Stockport, defended.

Mr. Sutton, in stating the case, said when this manure was sold to the defendant no complaints were made about it; and all Mr. Smith, the company's agent, had said was that he had sold it previously at a higher price; but when the bill was applied for he said there was no fertilising power in the manure. He mentioned the case at Stockport which had been taken against the company, and the result of the analysis there was that there were no fertilising ingredients at all in the manure; and his Honour gave judgment for defendant. This surprised the company.

His Honour said the issue was for the defendant.

Mr. Sutton thought it well that he should mention this.

Mr. Gray submitted that his case did not rest solely upon the manure; here were direct misrepresentations made at the selling of this manure. It would be more convenient for the plaintiffs' witnesses to be called first.

His Honour held that defendant's witnesses should be called first.

Mr. Gray, continuing, said this was not a case which affected the defendant alone, but a large class of men in Cheshire. For about a year a man in the employ of the company had been going about selling to the farmers a certain manure as being phosphate manure, and stating that it was bone phosphate—made of blood and bones. This agent had said that his employers had been obliged to get rid of this stuff because of the offensive smell, and that it caused a fever, and also that he had sold some of the same kind to Mr. Rogerson and Mr. Booth, both well-known farmers in the neighbourhood, and that that they had done great things with it. He went on puffing this manure up, and the result was defendant eventually bought three tons. The stuff was invoiced at £6 0s. 9d., including 10s. 6d. for bags, and defendant wrote back stating that it was to be bags free, which the plaintiffs denied. About two or three days afterwards defendant opened the bags and saw nothing but very large lumps. He put this stuff on two halves of a meadow field, and left the other half bare in order to test it; and he found it had no good effect on the land. Samples of the manure were sent to Mr. Thomson, a well-known analyst in Manchester, and he would tell them that instead of containing 14% of phosphate of lime, as plaintiffs guaranteed in their letter, it only contained 3·19, and it was such a small quantity that it was practically worthless. He then called

The defendant, who stated that on the 12th November he took a sample of this manure to Manchester, out of the the three or four bags which were left unused. He delivered the sample to a lad named Garner, who took it to the Royal Institution, and who further took it to someone in the laboratory.—Cross-examined: He took it from three different bags which had not been used at all.

James Garner, clerk to Messrs. Brown and Ainsworth, solicitors, Stockport, proved taking the sample.

Mr. William Thomson said he was a member of the firm of Crace, Calvert, and Thomson, Manchester, and that in November, he received a sample of manure for analysis from his assistant, Mr. Hart. He examined the contents of the bag, and found it contained phosphoric acid equal to tribasic phosphate of lime of 3·93 per cent. Manures were estimated for their fertilising worth by the quantity of tribasic phosphate of lime or their equivalent. One of the chief constituents of this manure was plaster of Paris, and, assuming that it was in its best condition, it would be worthless to throw on the land. The manure in this condition was of no value; and there was just as much chance of it doing harm as good if put on the land. Cross-examined: Any manure which contained Prussian blue would injure plant life. The manure might contain some appreciable amount of bone, but no blood. The sample he had produced was practically not soluble; theoretically it was, but it might take 20, 30, or 50 years.

The defendant, recalled, stated that in April of last year a man named Smith called at his house and said he had a manure to offer him which would be of very great advantage. He described the manure as made from blood and bone, suitable for present crops, as meadows, grass, and oats, and that it was soluble. He said the price was £3 per ton, but owing to a fever breaking out near to the works, he was offering it at 35s. per ton, carriage and bags free. Smith (the agent) had told him that he had sold some to Mr. Rogerson, of Handforth Hall, and also to Mr. Booth, of Dairy House Farm, Handforth. Defendant knew these gentlemen as being practical farmers, and he gave the agent an order for three tons. When the manure arrived, he found the bulk was in hard, dry lumps. He had the manure put on two halves of meadow and also on some oats, in most favourable weather. After it had been on the fields some ten or twelve days they had to pick some of it off again, as it was not soluble. When they came to cut the grass they found the lumps, and it was with great difficulty that they cut the grass, as the manure got in the knives; and in consequence they had to cut it higher. He afterwards wrote to the company complaining of the quality of the manure.—Cross-examined: The reason he took it off the land was because the lumps would not break. One of the representations made by the agent Smith, was that it would grow clover up to the waist.

Evidence was then called by the plaintiff company.

John Heaton, manager to the company, gave evidence as to what the manure was composed of, and also spoke as to taking a sample to Mr. Estcourt. Cross-examined, he said a great many farmers had refused to pay since an action was taken at Stockport.

George Smith, agent to the company, said the manure produced was not a fair sample. He remembered calling on the defendant and sending a sample to him. They usually carried samples about with them, but he could not say whether he had one on this occasion. Defendant asked him if he had been selling anything in the neighbourhood, and he replied that he had been selling some fish and bone manure to others in the district. He did not tell defendant it was blood and bone manure. He admitted that he wrote to Mr. Rogerson, of Handforth Hall, stating that he was selling bone phosphate at 35s. per ton, and asking him to speak well of it. Since the case at Stockport a great many of the Cheshire farmers had refused to pay.

Samuel Arnold, clerk to the company, said defendant called at the company's office, and offered to pay for the manure, less the price of the bags.

Mr. C. Estcourt, analyst to the city of Manchester, said he had analysed the sample of manure sent him, and found it was of a reasonable merchantable value. Cross-examined: He could not agree with Mr. Thomson that it was worthless.

His Honour, in summing up, said the only point left for him to determine was whether what was sold to the defendant was worthless as a marketable commercial commodity; if it had any value in it at all the defence failed. The defendant took what was offered him believing that he would get value in the manure, but when he finds out that he does not get the advantages he expected the proper course for defendant to pursue would have been to have charged the plaintiff company with the amount of the labour involved in order to get the article into a proper condition. This case was one which ought to be decided upon scientific evidence which was very good on both sides. Mr. Thomson had given his evidence for the defendant in an extremely able manner, but in cross-examination he admitted that there were things in this manure of fertilising properties, and also that there might be something else in it which neutralised the effect. The evidence went to prove that it was not worthless. He felt sorry that defendant did not get the benefits he expected from it, but if he had allowed it to stop in the ground it might have been better. He should give judgment for the plaintiffs with costs.—The case lasted about five hours, and there was a good deal of interest taken in it by farmers.

A NEW RED DYE.

THE Clayton Aniline Co., Limited, are introducing a new red, patented and manufactured by them under the name of *Stanley Red for Silk*.

This red is a new azo colour, and possesses several advantages for silk dyeing over the azo colours now employed; it does not bleed by lying in water, is fast to acids and alkalis, fairly fast to light and soap and does not mark off by warm pressing between white paper.

The ungummed silk (hanks or pieces) is dyed in a bath acidulated with sulphuric acid (¼ to ½lb. for 10lbs. silk) with or without the addition of boiled off liquor.

For 10lbs. silk use 27 gallons of water.

2½ to 3 ,, boiled off liquor,

and a sufficient quantity of sulphuric acid to render the bath slightly acid; then add a solution of 4 to 6ozs. of Stanley red in water, enter the silk and raise the temperature slowly up to 170 degs.—190 degs. F., work for ½ to ¾ hour, wash in water, wring off and brighten in a weak solution of either acetic acid or tartaric acid (1½—3ozs. for 30 gallons of water), wring off and dry. Stanley red can be dyed in the same bath with every acid colour, and is suitable for mixed goods of wool and silk, as it dyes equally well with sulphuric acid or with a mixture of sulphuric acid and Glauber's salt.

EXTRACTS FROM BOARD OF TRADE JOURNAL.

DISCOVERY OF SPONGES ON THE SICILIAN COAST.

The *Perseveranza* of Milan states that important sponge-banks have lately been discovered close to the island of Lampedusa, on the southern coast of Sicily. These deposits of sponges extend for over a surface of from 15 to 18 marine leagues, and are situated about an equal distance from the south-eastern extremity of the island. The smallest depths above these banks is 20 ells; the greatest depth is from 30 to 31 ells. At the lesser depths rock is met with, on which the sponge grows; at greater depths a sandy soil is found. All varieties of sponge are discovered here, including those which are in the greatest commercial request, and they are easy to obtain. Greek and Italian vessels have already proceeded to Lampedusa to take advantage of this discovery.

SULPHATE OF COPPER FOR VINEYARDS IN BARCELONA.

Her Majesty's Consul at Barcelona reports to the Secretary of State for Foreign Affairs that notices have been published in the *Official Gazette* of the Province by the "Provincial Deputation," signed by "E. Maluquer, President," to the effect that the deputation have decided to obtain in foreign markets, where it is manufactured, a quantity of pure sulphate of copper for distribution to vine growers in the province, whose vineyards are attacked by mildew.

Manufacturers are, therefore, invited to send samples with quotation of prices to the "Deputation" at Barcelona either direct or through their agent.

THE ITALIAN VINTAGE OF 1889.

"According to the returns furnished from the various wine-growing districts of Italy, early in September the grape crop was estimated to produce 22,308,366 hectolitres of wine, or 490,784,052 imperial gallons, being, 61·13 per cent. of an average yield. Storms, cold weather, excessive damp, and insect ravages have, however, done more serious damage to the vines than had been expected, and according to the reports received by the Ministry of Agriculture, the actual production of wine in Italy this year has been limited to 21,139,100 hectolitres, or 465,060,200 imperial gallons, being only 57·77 per cent. of an average yield.

"In no part of Italy has the crop approached an average, and it has been excessively light in Venetia, in Lombardy, in Piedmont, and in Liguria. The quality of the wine is stated to be $\frac{1}{12}$th excellent, $\frac{2}{12}$th good, $\frac{8}{12}$th moderate, and $\frac{1}{12}$th inferior."

TRADE BETWEEN SPAIN AND THE UNITED KINGDOM.

I.—IMPORTS INTO THE UNITED KINGDOM FROM SPAIN.

Principal Articles.		Quantity. Three Months ended December. 1888.	Quantity. Three Months ended December. 1889.	Value. Three Months ended December. 1888.	Value. Three Months ended December. 1889.
				£	£
Chemical products, unenumerated	Value	—	—	12,811	12,364
Copper, ore and regulus	Tons	25,528	19,327	573,974	424,201
" unwrought and part wrought	"	5	—	3 00	—
Cork, unmanufactured	"	84	357	1,952	3,546
" manufactured	Lbs.	164,018	292,350	11,210	20,582
Iron Ore	Tons	615,476	890,693	414,268	659,932
Lead, ore	"	596	10	3,374	90
" pig and sheet	"	20,868	20,881	285,552	274,097
Manganese ore	"	1,840	3,004	5,520	7,423
Oil, olive	Tuns	34	1,483	1,151	51,186
Pyrites of iron or copper	Tons	136,588	120,050	282,582	223,235
Quicksilver	Lbs.	403,784	412,631	50,669	43,561
Rags, esparto	Tons	12,723	16,316	78,346	95,967
Silver ore	Value	—	—	62,129	50,284
Wine	Galls.	1,023,110	1,152,654	228,538	252,864
Wool, sheep and lambs'	Lbs.	282,262	1,027,430	7,790	34,474
Zinc ore	Tons	220	—	560	—

2.—EXPORTS OF BRITISH AND IRISH PR KINGDOM TO SP

Principal Articles.		Quar Three Mon Decen 1888.	
Alkali	Cwts.	76,910	
Bags and sacks, empty	Dozs.	7,788	
Caoutchouc, manufactures of	Value	—	
Cement	Tons	2,715	
Chemical products and preparations (including dye stuffs)	Value	—	
Clay, and manufactures of	"	—	
Coals, cinders and fuel	Tons	360,160	3
Coal, products of, &c., including naphtha, paraffine, paraffine oil, and petroleum	Value	—	
Cotton yarn	Lbs.	81,100	[illegible]
Cottons, entered by the yard	Yards	660,300	94
Cottons, entered at value	Value	—	
Glass manufactures	"	—	
Grease, tallow, and animal fat	Cwts.	4,671	3
Hardware and cutlery, unenumerated	Value	—	
Implements and tools	"	—	
Jute yarn	Lbs.	953,800	2,176
Linen yarn	"	963,000	990
Linens, entered by the yard	Yards	268,400	305
Linens, entered at value.	Value	—	
Machinery, steam engines	"	—	
" all other sorts	"	—	
Manure	"	—	
Metals, iron, wrought and unwrought	Tons	15,347	18,
Metals, brass, manufactures of	Cwts.	380	1,
Metals, copper, wrought and unwrought	"	717	7
Metals, tin, unwrought	"	747	1,4
Oil, seed	Tons	67	1
Oil, other sorts	Value	—	-
Oil and floorcloth	Sq. yds	76,500	67,
Painters' colours and materials	Value	—	-
Paper of all sorts	Cwts.	748	1,6
Provisions (including meat)	Value	—	-
Silk manufactures	"	—	-
Soap	Cwts.	679	4
Telegraphic wires and apparatus	Value	—	-
Wood, hewn and sawn, and manufactures of	"	—	-
Wool, foreign, dressed in the United Kingdom	Lbs.	47,900	164,8
Wool, flocks and rag wool	"	6,600	14,2
Woollens, entered by the yard	Yards	216,500	479,1
Woollens, entered at value	Value	—	—

CUSTOMS TARIFF OF NICARAGU

The following is a statement of the rates of im on the articles specified under the Customs tariff duty in each case is leviable per Spanish *libra* or pound avoirdupois are equal to 101·6 Spanish pou

Note.—Peso 4s. 2d. (nominal val

ARTICLES.

MEDICINES, DRUGS, &C.

Oils, olive, linseed, almond, castor, cocoa-nut, coc any other similar oils.......................

Acids, chlorhydric, muriatic, sulphuric, and nitric

Do. carbolic, acetic, and oxalic ....

	Pes. Cts.
rpentine oil, gaseous mineral oil, acidulated waters......	0·02
¡ua-fortis, orange and rose water........................	0·05
u de Cologne, lavender, Florida, holy, Kananga, and other similar toilette waters................................	0·07
hite lead or carbonate of lead............................	0·04
um	0·02
quid ammonia or volatile alkali	0·04
·een copperas or sulphate of iron........................	0·01
ılphur of any kind	0·03
ıgar candy	0·08
carbonate of soda	0·01
›rax or borate of sodium	0·05
ernary bandages of any kinds	0·20
›xes of wood or cardboard for the use of chemists..........	0·02
ırbonate of soda crystallised and chloride of lime..........	0·01
edicinal sweetmeats or pastiles	0·15
›rks for stoppers of bottles and casks.....................	0·10
edicinal barks	0·15
ıalk, clay, Tripoli..	0·02
ases of glass and articles of any kind for chemists and druggists..	0·05
rtificial teeth and dental mastic	0·50
erfumed and medicinal essences	1·00
ponges of any kind	1·00
iquid amber	0·05
um arabic or gum lac	0·10
lycerine	0·08
int	0·15
enna leaves and rosemary.................................	0·06
oaps and syrups, medicinal	0·15
yringes, syphons, and any other articles of caoutchouc and gutta-percha	0·30
›o., and any other articles of metal, except gold and silver ..	0·10
›o., and any other articles of glass	0·05
.inseed or linseed flour	0·03
›ocoa butter	0·15
'at for ointments or pomades........	0·05
ledicines in gelatine capsules, patented medicines, as well as balms and mixtures of oils in drugs, pastes, powders, in a liquid state, tragacanth, and medicines in any form and prepared in any way not specially mentioned	0·15
ledicines and drugs, in pastes, powders, liquids, tragacanth in any form, and prepared in any way, not specially mentioned	0·15
Mustard, in seed or flour	0·08
;all nuts	0·05
ırtificial eyes of any material	0·13
)xide of zinc	0·04
'iltering paper	0·05
'itch, resinous	0·02
'ill makers' and other utensils, and instruments of metal for chemists and druggists....................................	0·15
Epsom and Glauber Salts	0·02
;altpetre, or nitrate of potash	0·03
;alammoniac	0·05
Caustic soda	0·02
Medicinal berries	0·15
;ulphate of iron	0·01
;ulphate of zinc, or white vitriol	0·02
;ulphate of copper, or blue vitriol	0·03
;ulphate of quinine	1·00
'ersian sherbet...	0·05
;uspenders, waistbands, bandages, and similar articles	0·20
Taffetas, cerecloth, &c.	0·25
Turpentine..	0·05
Utensils for chemists and druggists, of faïence, stone, and composition, such as mortars, basins, &c..............	0·03
'oisons for tanning, insects, &c.	0·10
Medicinal wines, such as quinine, extract of meat, &c.	0·15
Vaseline, casmoline, and petrolate	0·10
Bristol sarsaparilla and other similar kinds	0·08

The following Articles pay specific duties as under :—

Articles.	Rates of Duty.	Pes. Cts.
Foreign Spirits more than 12 degrees up to 25 degrees inclusive by the Cartier alcoholmeter..	Bottle	0·40
Do., more than 25 degrees by the Cartier alcoholmeter (with special authority of the Government) pay same duty as the preceding articles with an increase per degree in excess of		0·03
Tobacco, in rolls or snuff	Spanish Pound	0·40

The following Articles are free from import duty :—

Manure for agricultural use.	Lime and cement.
Stills, under special authority from the Government,	Tubing of iron, galvanised or not, and the corresponding plugs.
Metallic fencing wire, with posts, rails, and other accessories.	Coal and animal charcoal.
Stone breakers.	Crucibles for melting metals.
Mercury for mines.	Ice.
Asphalt.	Gold in lumps, bars, dust, or coined.
Sorting and ventilating machines for coffees and seeds.	Silver in lumps, bars, dust, or coined.
Water pumps of metal of any kind.	Fire bricks and tiles for smelting furnaces.
Hydraulic pumps of metal of any kind.	Hydraulic presses.

Articles not mentioned in the tariff pay the duties leviable on similar articles; those not mentioned in the composition of which there are several materials pay the duty leviable on the chief material; finally, for those not mentioned which cannot be classified, there is levied 50 per cent. on the invoice value. If the original invoice is missing a valuation of the goods will be made by experts.

The import duties are fixed on the gross weight without any deduction for cases and packages.

THE COPPER TRADE IN 1889.

MESSRS. James Lewis and Son, in their review of the Copper Trade for 1889, write :—Whereas the year 1888 will be memorable in the history of the copper trade for the rise of the French syndicate formed to control all the supplies of this metal, the year 1889 will be still more memorable for its fall. The original conception of this speculation contained elements of success, but the manner in which its details were carried out doomed it to failure. The level of prices at which contracts were made with the different producing companies, varying from £61. 10s. to £70. per ton, was far too high, as there is little doubt that most of them would at one time have gladly contracted for their production at about £50. per ton—a price that would have left a fair profit not only to the mining companies, but also to the syndicate themselves, without materially contracting the consumption. Moreover, very much less capital would have been required by the syndicate, the want of which was the primary cause of its collapse. Had the syndicate been able to finance their large stock a little longer, and met consumers by a moderate reduction in price, they would have obtained the benefit of the exhaustion of all outside stocks, and of the great revival in the metal trade generally which sprang up shortly after their downfall, and have thus enabled them to dispose of part of their holding. Moreover, American producers were willing to accept a reduction on their contract prices, and were on their way to France to arrange the details at the time of the collapse of the syndicate, and the English mining companies would probably have followed their example. The effect of the syndicate's operations, extending over a period of one year and five months, was an increase in the stocks of copper in public and private warehouses in England, France, and the United States—from the minimum reached on the 31st December, 1887, of 58,000 tons, to the maximum attained on the 1st May, 1889, of 179,000 tons—of 121,000 tons, about one-half of which was due to increased supplies and the other half to diminished consumption. During this period the value of copper advanced from £40. on the 21st October, 1887, to £107. on the 9th September, 1888, and receded to £35. per ton on the 18th March, 1889. The statistics show that the direct import of copper into England and France in 1889 was 14,077 tons less than in 1888; that, exclusive of the Chili bars transferred from England to France, the export of copper from England exceded that of 1888 by 26,118 tons; and that the apparent consumption of England was 23,197 tons greater than in 1885, while the apparent consumption of France was 3,338 tons less. The figures with regard to France are, however, misleading, in consequence of the large quantity of English copper shipped to France in 1888 and not returned in the public stocks, chief portion of which appeared to. have gone into consumption in 1888, whereas it has really been consumed during 1889. Taking the average English and French consumption and the English export for the two past years, 123,640 tons, it is 2,700 tons per annum less than that of the previous two years, nearly 11,000 tons less than that of the years 1885 and 1884, and nearly 7,000 tons per annum less than the average of the four years 1884 to 1887. It is, therefore, evident that the large deliveries of the past nine months have hardly made good the great depletion of stocks all over the world, without in any way supplying the greatly increased demand due to the present revival in trade, and to the special demand arising from the extended use of electricity and of sulphate of copper. The value of telegraphic wires and apparatus exported in 1889 was £1,040,082. against £521,055. in 1888, or more than double, as the cost of the copper used in 1888 was higher than in 1889, and the value of machinery and mill-work exported in 1889 was £15,254,658. against £12,939,267. in 1888; in this case, however, the value of the iron used was greater in 1889 than in 1888. The consumption of the United States has exceeded that of 1888 by 27,500 tons. The impetus given to production by the high prices paid by the syndicate increased the import into England and France from 117,000 tons in 1887 to 160,000 tons in 1888, but during the past year it has fallen to 146,000 tons under the influence of the low prices which followed the collapse of the syndicate. The most notable decrease has been in the shipments from Chili, 8,500 tons, and from "other countries" nearly 8,000 tons, while from the United States it is 500 tons, from Australia 500 tons, and from Japan nearly 2,000 tons. The increase from Spain and Portugal is, however, 1,500 tons, from the Cape of Good Hope 2,700 tons, from Quebrada 700 tons, and from Mexico 1,800 tons. The total production of the world for the past year we estimate at 263,000 tons, against 260,000 tons in 1888. During 1889 the public stocks in England and France decreased 1,319 tons, and private stocks in France and elsewhere have been very largely reduced, while in the United States they have decreased 5,800 tons—the total reduction during the year being probably over 25,000 tons. On the 1st January, 1889, the total stock was probably 150,000 tons; by the 1st of April it had increased to about 175,000 tons; and we now estimate it at about 125,000 tons. The quantity of copper produced during 1890 will mainly depend upon the level at which the value is maintained. At £50. for good merchantable copper there is little doubt that most, if not all, of the large producers can work to a fair profit, while this price will in no way interfere with consumption. This latter promises to be very large with the great extension of the use of electric light and power, the increasing demand for sulphate of copper, the brass required for the numerous war and other steamships in course of construction, and the locomotives and machinery for which makers are full of orders up to nearly the end of the year. The production of the United States is estimated as follows (tons fine):—

	1885.	1886.	1887.	1888.	*1889.
Lake Superior	32,209	35,593	33,331	38,574	38,300
Arizona	10,137	6,990	8,036	14,195	14,300
Montana	30,267	25,720	35,223	43,704	46,900
Other States and Territories	1,439	1,507	2,517	4,194	6,000
Total	74,052	69,810	79,107	100,667	105,500
Imported as Pyrites	2,271	2,009	2,367	2,192	2,200
Total	76,323	71,819	81,474	102,859	107,700
Stock, 31st December	15,600	13,400	12,000	34,800	29,000
Estimated consumption	40,000	50,000	62,000	48,000	75,500

*Estimated.

THE GOVERNMENT AND THE TELEPHONE.

THE present position of the telephone question in this country is peculiar, and we are certain to see very considerable changes in connection with it during the year 1890. It will be recollected that when the telephone first came to be introduced—namely, in 1879 and 1880—it was thought that the new industry could be carried on independently of the Post Office telegraphic monopoly, and telephone exchanges in the principal cities of the kingdom were started upon that footing. At that time, also, the rival patents of Graham, Bell, and Edison were represented by separate organisations, and entered into competition with each other. The Post Office, however, quickly asserted its claim to control the use of the new invention wherever it was employed for the purpose of communicating between two persons, or, in fact, for any other purpose than upon the exclusively private affairs of one individual or firm. In the result it was held that the Department was entitled to take this ground, in virtue of the Telegraph Acts, and in consequence it was arranged to grant licences to the companies then in the field, empowering them to continue their business subject to certain restrictions, and in particular to a payment to the Department, by way of royalty, of 10 per cent. of the gross takings. These licences were granted as from the 1st January, 1881, and for a period of thirty-one years from that date, but it was a condition that the Postmaster-General might at certain intervals signify his intention to purchase the business of any company, the price to be paid being fixed by an arbitrator in the usual way. The first date upon which this option can be declared is 30th June, 1890; and, if exercised, the purchase would take effect as from the 31st December following. Should the Goverment elect not to buy at present, the next break in the licence will not occur until 1897. As between the companies themselves, competition soon became to be regarded as suicidal, and an amalgamation of the interests of Bell and Edison was arranged early, so that it was to a combination of these two groups that the

licences referred to were granted. The ir
the so-called "receiver," or instrument wh
hear the message; that of Edison to the "
into which the words are spoken. Each in
for his invention. For instance, it was o
Bell instrument was equally available for
transmitting. But this was speedily foun
able. Finally, there have been minor inven
the efficiency of the original discoveries, in
but the patents for these, also, are mainly e
the National Telephone Company, now
organisation in the kingdom. So much as
dustry in its relation to the Post Office Dep
of the business as presently founded upon
of the instruments employed. These pa
approaching their expiry, when, of course, i
construct exchanges free from the control o
patent of Bell will terminate with the yea
middle of 1891, and minor patents later.
practically as soon as 1890 is out new and riv
operations, and, as a matter of fact, a prop
under discussion, to establish what are called
somewhat on the lines of a club. It appear
year one of two things may happen, either c
plete change in the existing position as re
Government elects to purchase, the telepho
ment of the Post Office, and all future develo
through that medium. If on the other hand,
not to interfere, and is willing—as in such ci
doubt be willing—to grant further licences,
competition next year between the existing
cerns which are certain to spring into exis
into any discussion at present as to what t
under either alternative, there will be a dec
part of the public to lean to the view that to a
and open competition to have full scope will
tension of use and a cheap service to the publi
the interest of the Department, which has ex
public money in acquiring the telegraph mo
sight of.—*Scotsman.*

THE MANCHESTER ASSO
ENGINEERS.

MR. John West, M.Inst.C.E., gave his pre
the above association on the 11th inst.
or two local matters, Mr. West went on to con
of electric and gas lighting. There were ele
formed for lighting London, with a united capi
sterling, and these companies were now erec
ting down steam boilers, engines, electrical pla..
plying the various districts. Electricity was su
companies on the following different systems,
continuous current; (2) the direct continuous c
and (3) the alternate current transformer system.
the various systems were stated, after which Mr.
with the question of the cost of the supply.

The charges for electricity ranged from 7d. to
unit of 1,000 watts, and some companies would
consumer under a minimum charge of £14 per
admitted that electricians had made some p
machinery for securing a better and more regular
it must not be forgotten that improved lighting b
very rapid strides. Having in 1882 dealt with th
tive burners, he pointed out that since that perio
burners have been perfected. These were the re
Siemens, Sugg, Wenham, Fourness, and others,
burners of Welsbach, Clamond, and others, bot
come into general use, particularly the regenerat
them a better light can be obtained with from
quantity of gas. He had been favoured with
experiments made by Mr. Alexander Smith, of Ab
rator burners of suitable size for lighting halls, sho
as follows:—

By consuming 10 cubic feet of gas per hour in o
light equal to 155·2 candles was obtained, or 15·
cubic foot of gas consumed per hour. This resu
gas, which, when tested according to the method
Work Clauses Act, gives a light equal to 5·6 c
foot, thus showing a gain of 9·92 candles for every
when used by these lamps, or a gain of over 177 p

Similar tests have been made with London
Wenham and other regenerative lamps, which s

actory results. A photometric test of the Welsbach lamp made by Professor Carlton J. Lambert showed that an illuminating power of 18·8 candles was obtained with a consumption of three cubic feet per hour of London 18-candle gas. The relative cost of gas and electricity with equal light from each was then compared. Eminent electricians stated that a 16-candle-power incandescent lamp was calculated to last about 1,000 hours, and would absorb 60 Board of Trade units of electricity in that time. The cost of 60 units at 8d. per unit, the usual Board of Trade limit of charge, would be 40s., to which must be added 3s. 9d., the cost of the lamp, which made a total of 43s. 9d. per 60 units. To make the comparison clear as to the cost of gas in London, Manchester, and Aberdeen to produce same amount of light, both with ordinary burners and regenerator burners Mr. West prepared the following table.

COMPARISON OF COST BETWEEN GAS AND ELECTRICITY.

THE ILLUMINATING POWER AND PRICE OF GAS PER 1,000 CUBIC FEET AS SUPPLIED IN LONDON, MANCHESTER, AND ABERDEEN.

ELECTRICITY.	One incandescent lamp of 16 candle-power requires 60 watts per hour, or 60 Board of Trade units every 1,000 hours, which at 8d. per unit (the price now charged by the House to House Electric Supply Co., Kensington, and the Electric Supply Co., Liverpool) amounts to 40s. 0d. To which must be added the cost of lamp, 3s. 9d.				Equivalent light in candles for one hour.	Comparative cost.	Ratio of cost, showing the number of times cheaper gas is than electricity.
					16,000	s. d. 43 9	1
WITH ORDINARY GAS BURNERS.	Quality of gas.	No of cubic feet required to produce 16 candle-power per hour.	No. of cubic feet required for 1,000 hours.	Price of gas.			
	candles	cubic feet	cubic feet	s. d.			
London	16	5	5,000	2 6	16,000	12 6	3·5
Manchester	20	4	4,000	2 6	16,000	10 0	4·4
Aberdeen	28	2·88	2,880	3 6	16,000	10 1	4·3
WITH REGENERATIVE BURNERS.							
London	16	2½	2,500	2 6	16,000	6 3	7
Manchester	20	2	2,000	2 6	16,000	5 0	8·7
Aberdeen	28	1·44	1,440	3 6	16,000	5 0½	8·7

The comparison is made with the best type of burners, as these are in the reach of everyone, and the gas is compared with an electric lamp which is assumed to give the same light throughout the 1,000 hours, but such is not the case, as it is found, after they have been used for a time, that the carbon filaments get volatliised and deposited on the glass, consequently the size of the filament is diminished, and the surface of illumination reduced; the resistance of the filament would also be increased, and the intensity of the incandesence decreased for a given power, the carbon on the glass further reducing the light. These facts are confirmed by the experiments of Mr. W. H. Preece, chief electrician to the Post Office, who states that the ordinary commercial lamps consuming 3·54 watts per candle-power when new, require 6·1 watts per candle-power after burning 900 hours, the mean absorption being 5.25 per lamp. Stated the other way, it appears that the candle power of the lamps after burning 900 hours is only 58·5 per cent. of the initial duty, and the mean value of a lamp during its life is only 67 per cent. of that which it possesses when new. Assuming an average use of four hours per day tor a lamp, this means not only that all incandescent lamps must be renewed before 225 days, when their luminosity is little more than half what it was when new; but taking all the lamps in an establishment together, three should be provided to do the work of two if the intended average lighting effect is to be maintained. The advantage of electric lighting in a low-pitched room where ventilation is defective was admitted, but this could be obtained with regenerative burners. Mr. West then went on to give his views as to the cheapest and best way of producing and delivering electricity to the consumer under varying circumstances. He was informed that in one town a complete electric lighting plant driven by steam power, and supplying from a central station 3,000 incandescent lamps of 16 candle-power, with cables laid along the streets, would cost about £20,000. Mr. Ferguson Bell, of Stafford, stated that in their town it would cost £14,500. for a like installation, but he did not include the cost of land in his estimate, details of which were given.

If the number of lights was sufficient, there was the plan of a consumer providing himself with electric light by means of a complete plant, driven either by a gas engine or steam engine, on his own premises. A friend of his had given him the cost of a complete electric installation, including fittings for 320 incandescent lamps of 16 candle-power each, which had been working for about two years at one of the banks in London, and was driven by Crossley Bros.' gas engines. The plant included two 9-horse-power Otto engines, two dynamos, and two sets of accumulators, and cost £3,000. The working expenses—for 1,800 hours per annum, the building being dark—were, including interest on capital at 10 per cent., £1,012. 10s. per annum. The current supplied was equal to about 38,500 Board of Trade units. The analysis of the above figures shows that—

	d.				
The cost for generating electricity, including wear and tear and interest on capital, is	5·58	per	Board	of	Trade unit.
Interest on fittings	·56		,,		,,
Cost for renewal of lamps........	·7				
The total cost per Board of Trade unit on the complete installation	6·84		,,		,,

From these figures it will be seen that if the bank had obtained electricity from a central station electric lighting company, at 8d. per unit, the total cost to them would have been:

	d.				
Electricity........................	8·0	per	Board	of	Trade unit.
Interest on fittings	·56		,,		,,
Cost for renewal of lamps	·7		,,		,,
Total	9·26		,,		,,

So that by generating electricity on their own premises they saved 2·42d. per unit, or £357. 19s. 2d. per annum. He thought it would be found in practice that the most economical and satisfactory way of producing and supplying electricity to the consumers would be by gas engines on their own premises, or by a system of very small central stations, where a few houses or shops were in close proximity and worked by the direct system of distribution. By this plan we would avoid expensive outlay on buildings, at the same time be advantageously using the existing gas mains, besides being able to dispense with the cost and maintenance of an electrical meter on the premises. The cost of generating plant for 400 incandescent 16 c.p. lamps on this plan would be as follows:—Plant, £900.; working expenses for 1,800 hours per annum, £660; current supplied, 43,200 Board of Trade units; cost per unit, 3·66d. In the case of an installation of 50 lamps of the same power, the cost of plant would be £275.; the working expenses, £110. 19s. the current, 5,400 units; and the cost per unit, 4·93d.

By this plan electricity could be supplied, as shown by these examples, at prices varying from 3·66d. to 4·93d. per Board of Trade unit, according to the size of the installation, inclusive of working expenses, repairs, depreciation, and interest on capital. The usual charge by the London electrical companies, whose supply was generated for large districts in large stations driven by steam power, was 8d. per unit, showing a saving of 4·34d. and 3·07d. per unit respectively in favour of the gas engine scheme. To explain how the charge for working by gas was obtained, he took first the quantity of gas required per indicated horse-power per hour at 20 cubic feet; with this amount of gas a maximum of 8 lamps of 16 candle-power each could be maintained per hour, and by allowing a little extra for irregularities, and taking 50 cubic feet of gas to produce one Board of Trade unit, the gas would cost 1¼d., to which must be added the other figures in the table of working expenses just given. The reason why electricity could be distributed to customers by gas engines cheaper than from a large central station, was because the very expensive cables for distributing the electricity (which in most cases take one-third of the capital employed) could be dispensed with, and the further expenses for renewal after they have been in use for a time. It was stated that Edison's cables in Berlin only lasted three years. These cables cost £90,000. for 36,000 lamps of 16 candle-power and 144 arc lamps. From these facts it was quite clear that the interest on the first cost and maintenance and renewals of distributing plant alone would be more than the cost of gas delivered on the consumers' premises from the existing mains. While approving of the action of the Corporation of Manchester in applying for a provisional order, he thought they should not put down a central station, but rather encourage separate installations in which gas engines would be used, Messrs. Crossley Brothers having sold 550 for this purpose. Gas engines and dynamos might be hired to consumers in the same way as stoves are. In many cases where arc lights of high power were employed on out-door work, they had been superseded by the "Lucigen" and

PRIZE COMPETITION.

A LITTLE more than twelve months ago, one of our friends suggested that it might be well to invite articles for competition, on some subject of special interest to our readers; giving a prize for the best contribution, he at the same time offering to defray the cost of the first prize, and suggesting a subject in which he was personally interested.

At that date we were totally unprepared for such a suggestion, but on thinking over the matter at intervals of leisure, we have come to think that the elaboration of such a scheme would not only prove of benefit to the manufacturer, but would go some way towards defraying the holiday expenses of the prize winner.

We are therefore prepared to receive competitions in the following subjects:—

SUBJECT I.

The best method of pumping or otherwise lifting or forcing water aqueous hydrochloric of 30 deg. Tw.

SUBJECT II.

The best method of separating or determining the relative quantities of tin and antimony when present together in commercial samples. On each subject, the following prizes are offered:—

One first prize of five pounds, one second prize of one pound, five additional prizes to those next in order of merit, consisting of a free copy of the *Chemical Trade Journal* for twelve months.

The competition is open to all nationalities residing in any part of the world. Essays must reach us on or before April 30th, for Subject II, and on or before May 31st, in Subject I. The prizes will be announced in our issue of June 28th.

We reserve to ourselves the right of publishing the contributions of any competitor.

Essays may be written in English, German, French, Spanish, or Italian.

THE COMPOSITION OF AMERICAN LARD.—It will interest consumers to learn that the adulteration of this product, which has always been more or less doctored, has now assumed shocking proportions. Formerly, says *Iron*, the American manufacturers were satisfied to add merely from 15 to 20 per cent. of water, thus obtaining the value of lard for water. Now "lard" is made of cheap cottonseed oil, to which is added sufficient stearine, obtained as a refuse in the manufacture of margarine, to give the substance the consistency and hardness of natural lard. It is estimated that the annual American production of lard amounts to 2,700,000 cwt., of which quantity 35 per cent. is adulterated. A well-known American firm produces annually 270,000 cwt., and uses for the manufacture 120,000 cwt. of cottonseed oil and 30,000 cwt. of stearine. This "lard" is put into the market under the high sounding name of "refined lard," or "pure refined lard." A sanitary commission of the canton of St. Gallen, in Switzerland, has recently bestowed special attention upon the matter, and its investigations have fully confirmed the large extent to which adulteration is carried on. In the face of such facts, Dr. Ambuhl, a chemist of St. Gallen, has urged the Federal Government to either forbid the importation of the adulterated article altogether, or to impose such a heavy duty as to cause the product to disappear from the Swiss market.

DISSOLUTION OF PARTNERSHIP.—We notice that the partnership between Messrs. G. Dyson, F. Dyson, H. Dyson, and E. A. Brotherton, manufacturers of sulphate of ammonia, etc., at Leeds and Wakefield, has been dissolved as from June 30th, 1889. Since this date Mr. Brotherton has been continuing the business on his own account, and we have no doubt that under his able management the reputation of the old firm will be maintained, and the success of the new one assured.

ICE ON TELEGRAPH WIRES.—The man is not yet born who can tell what really happens in a wire when an electric current passes along it. That being so, he is, of course, not yet born who can explain why the silicium bronze telephone wires between Vienna and Buda Pesth have, during the recent frosts, been covered with ice from five to eight centimetres thick, whereas the adjoining iron telegraph wires have been covered to a thickness of only three centimetres.

REDUCTION IN TELEPHONE CHARGES.—The National Telephone Company formally announced yesterday to their subscribers to their Liverpool and district system that the rent for the use of telephones in that centre has been reduced from £20. to £15. per annum; and, further, that arrangements are now being rapidly completed for the establishment of telephonic communication between Liverpool and London.

Correspondence.

ANSWERS TO CORRESPONDENTS.

., C.—Re Prize Competition—The methods should be applicable not only to loys, but to any class of material containing the metals in question.
:RY.—We are not surprised that the fact has struck you. We shall have some-ing to say about it before long.
.—Look up the back numbers to which you refer. You cannot possibly have ss time at your disposal than we have.
.—You have based your remarks on copper continuing at £50. per ton. What if rises or falls?
N.—Perhaps not; but we do not deal in prophecy.
I.—The process requires development before a sensible man would venture to cpress an opinion.
.—To do as you wish is impossible. We are surprised at you asking it.
OURMAN.—The uses that have been found for the oil are simply legion. 2. You ve not looked through our advertising columns.
RCHANT.—Yes; they are making hay while the sun shines. They know it will ot shine long.
. S.—Thanks. We have just heard of another firm who propose to treat their quors in exactly the same way, so you are not alone in the matter.
INBER.—If the combination should break down, but not otherwise.

We do not hold ourselves responsible for the opinions expressed by our correspondents.

To the Editor of the Chemical Trade Journal.

DEAR SIR,—I do not wish to find any fault with the general tone the article on Multiple Evaporation, and I am prepared to accept result that can be shown in actual practice, but it seems to me t in the sentence quoted your writer strings on his "effects" without slightest regard to the conditions under which they can be added. I e it that if the initial pressure and temperature of the steam is such t four effects suffice to reduce it to 26" of mercury and 120° Fahr., addition of a fifth effect will bring no further economy, unless the tial pressure and temperature be correspondingly raised.

As regards the evaporation of caustic liquors by this method, I don't nk it will be possible to utilise high pressures, even with the most eful management, if rivetted boilers are used. We have evaporated uor of from 15° to 20° Tw. in Lancashire boilers for several years, but leakage is almost uncontrollable, even when the steam is escaping ely, and the cocks only partially closed when blowing off the liquor, ich only requires 3 or 4 lbs. of pressure.—Yours truly,

JAMES KAY HILL.

Fullarton Cottage, Irvine, 10th Jan., 1890.

THE COAL SUPPLY OF FRANCE.

A very new and original question (says the Paris correspondent of the *Daily Telegraph*) is beginning to occupy the attention of the ar Office authorities, and as it is one on which M. de Freycinet, ing to his training at the Ecole Polytechnique and his engineering perience, is peculiarly qualified to pronounce an opinion, it is by no eans unlikely that it may meet with a practical solution while he is t the guiding spirit in the Rue Saint Dominique. It is a noteworthy t that during the past ten years France has become more and more lebted to Belgium, England, and Germany, for a substantial pro-rtion of her supply of coal. English fuel finds its way into upwards forty-nine departments, Belgian into twenty-four, and German into eteen. Now it is argued that if hostilities were to break out tween France and her Eastern neighbour coal would in all proba-ity be regarded "as contraband of war," in which case England and lgium, at once, would cease to furnish this country with that very im-rtant commodity. Not only would the foreign supply fail at a very tical moment, but the department of the Nord, owing to its position ar the frontier, might be menaced with invasion, and it is precisely this quarter that Frenchmen turn for a large proportion of their own al. When, moreover, the fact is borne in mind that, with the ception of the coal in possession of the railway, gas, and other mpanies, which are always bound to have a certain quantity at the sposal of the Minister of War, the stock in hand at the present date very small, it will at once be seen that if serious complications were arise suddenly the outlook would be anything but reassuring. It is lculated that about 105,000 men are employed in the different coal nes. About one-half of these would have to leave and join their giments in the event of a "mobilisation," to say nothing of the horses iich would also be requisitioned. Under these adverse circumstances al would only be forthcoming at the rate of 5,500,000 tons per num for a country which consumes yearly 32,000,000 tons. Even suming, for the sake of argument, that coal could be dispensed with domestic and manufacturing purposes throughout the length and eadth of the land, this would by no means suffice. It is estimated at more than 14,000,000 tons would be required for the army, the navy, the mercantile marine, railways, and other indispensible purposes. It is with a view to remedying—if only in part—this un-satisfactory state of things that the War Office had just turned its attention to a plan for the military organisation of the coal mines which may ere long be laid before the Chamber of Deputies. If this project be adopted only a fraction of the coal miners will be called out when-ever war breaks out; the great majority continuing to work in their respective pits under the direction of the military authorities. All the miners will spend a twelvemonth with the military engineers, in order to be initiated in all the drill of that branch of the service: those among them who may be summoned to the colours in the event of hostilities rejoining the engineers, partly with a view to their working the enemy's coal mines, if their are successful in pushing to some distance beyond the frontier. Such are the broad outlines of a scheme which bids fair to find favour with the War office and the public at large.

Trade Notes.

A RISE IN QUININE.—The price of quinine has risen this last week in consequence of the inroads made upon the available stock by the consumption of influenza patients.

PURCHASE OF AN ANILINE COLOUR WORKS.—WE hear that the Berlin Aniline Dye Company, Limited, have bought up the Aniline Colour Works of George Carl Zimmer, of Manheim on the Rhine. With this they take over the manufacture of methylene blue, and enter the convention that exists for the sale of this colour.

A SUBSTITUTE FOR GLASS.—The new translucent substance in-tended as a substitute for glass has been adopted for some months in some of the public buildings of London, and various advantages are claimed for it, among these being such a degree of pliancy that it may be bent backwards and forwards like leather, and be subjected to very considerable tensile strains with impunity; it is almost as transparent as glass, and of a pleasant amber colour, varying in shade from very light golden to pale brown. The basis of the material is a web of fine iron wire, with warp and weft threads about one-twelfth of an inch apart, this being enclosed like a fly in amber, in a sheet of translucent varnish, of which the base is linseed oil. There is no resin or gum in this varnish, and, once having become dry, it is capable of standing heat and damp without undergoing any change, neither hardening nor becoming sticky. Briefly, the manufacture is accomplished by dipping the sheets edgeways into deep tanks of varnish and then allowing the coating, which they thus receive, to dry in a warm atmosphere.

SNOW-SWEEPING BY ELECTRICITY.—An electric snow-sweeper is used on the West End-street Railway of Boston. It consists of a low truck or car fitted with electric motors, drawing current from an overhead conductor. These motors not only propel the car, but drive the snow brushes, which are cylindrical, like the rotary hair brushes of a barber, one for each rail, before and behind the car, so that it can sweep forwards or backwards, as may be required.

THE SUFFOCATION OF A GASWORKS MANAGER.—A coroner's inquest was held on the 16th inst. at Festiniog on Mr. Morgan, manager of Festiniog Gasworks, who was suffocated on Tuesday by an escape of gas. Evidence was given that Mr. Morgan, assisted by a workman named Hughes, was endeavouring to increase the pressure of the gas in the street pipes by introducing it direct from the gasometer, when there was an escape, which rendered both men in-sensible. The medical testimony showed that Mr. Morgan was asphyxiated, and Hughes had a narrow escape, all efforts to resusci-tate him proving for a time futile. The jury returned a verdict of accidental death.

A NEW ELECTRIC TRAMCAR.—An electric car on the new principle was tried on the Bristol-road tramway route at Birmingham recently with entire success. The car is of the same size as the cars on the Birmingham cable tramway. It is lighted with electricity, and it is to have an electric bell. The motor is in the front bogie, and the accumulators in which the electrical energy is carried are under the seats on both sides. It is expected that one of these cars will be able to run 70 miles with a single charge. This is a full days work of the present cars. They are, however, only reckoned as half-day service cars, but the whole of the receptacle containing the power driving a car can be taken out at the depôt and recharged in the space of five minutes. The present car is the outcome of a recent union of interests between the Julien, the Sprague, and the Electrical Power Storage systems. It has enabled the Central Company to meet the hopes recently expressed that the ultimate form in which electricity might be applied as the motive power of tramways in Birmingham should be the self-contained car and not a separate engine. Until recently the great difficulty in the employment of large cars with self-contained power has been the use of the driving chain, which has been found in practice to be noisy and liable to break. But this difficulty has been quite overcome by an ingenious arrangement which absolutely dispenses with the chain, and which is practically noiseless.

DECLINE OF THE CHESHIRE SALT TRADE.—The returns of salt exported from Cheshire show that during the past year there was a greatly diminished volume of business transacted, represented by a total of 745,111 tons, against 1,022,334 tons in 1888, and 994,726 tons in 1887. There was a falling off of something like 25,000 tons to the United States, and 15,000 tons to British North America. The exports to Calcutta were 221,616 tons, against 314,332 in 1888, or a drop of 92,000 tons. There is now a more active demand for salt, and a revival of trade is expected.

SEWAGE DISPOSAL AT SALE.—At a special meeting of the Sale Local Board, held last Wednesday, Mr. H. Burgess (chairman) presiding, the surveyor (Mr. A. G. M'Beath) submitted details of a scheme for the disposal of the sewage of the Board's district. The system to be adopted is known as the international process, and recommended as the most suitable for it, as the effluent can be turned into the river Mersey without fear of incurring complaint. The scheme will deal with 2,220,000 gallons of sewage per day, and pumping machinery will be constructed to deal with eight millions per day. The site of the works is at Dr. White's bridge. It is a little over 5½ acres, and the cost of impounding tanks, filter beds, and pumping apparatus is estimated at £14,733. The cost of maintenance would be equal to a rate of 3d. in the pound per annum. It was decided that the plans should be thoroughly investigated with a view to their adoption.

THE WATER IN PARIS.—It is clear that the Parisians will have to bestow serious attention on the question of their supply of water. I has never been of the best, and tourists are well aware that they have always been cautioned to drink as sparingly as possible of it. Now, it seems that the difficulties connected with the supply of fairly potable water are greater than ever. The Vanne Canal, whence the bulk of the water was drawn, has broken down; the scheme to turn on the limpid stream of the Arve in the Eure department has not been carried out; and for the present there is nothing for it but to blend nasty Seine water with that of the springs so as to provide for the needs of the great city.

ARSENIC IN DEAD MEN'S BONES.—About the time that the public opinion of this country was being agitated by the Maybrick case, there was a corresponding feeling astir beyond the English Channel in consequence of the poisonings at Havre. The inquiries instituted in connection with the latter case are reported to have fully confirmed the results of the recent researches by M. G. Pouchet. The most important is that arsenic, if administered during life, will be found in the bones of a dead body when every trace of the poison shall have disappeared from the intestines, and even from the liver, where the arsenic is especially apt to be fixed. The arsenic is found to accumulate in the spongy tissue of all the bones, and especially when it has been absorbed in small doses over a long period. When the dose is sufficient to cause death in a few hours, then it is in the bones rich in compact tissue where the poison mostly accumulates.

FRAUDULENT DAMAGE ALLOWANCES.—In reviewing this subject the Secretary of the United States Treasury says:—"The law under which rebate of duties is allowed on imported merchandise for damage on the voyage of importation was passed in 1799, when water transportation was confined to sailing vessels. Owing to long voyages and incidental exposure to weather and water, merchandise was liable to damage and deterioration, from which the owner could not then, as now, protect himself by marine insurance. Within recent years almost all merchandise, subject to damage, is transported in steamers making quick transit, and there is but little liability to actual damage, so that the causes which led to the enactment of the original law have largely disappeared. The law has now become a convenient means for the perpetration of frauds of the most scandalous character, is demoralising to customs officials, and operates so uniformly and largely to the advantage of the unscrupulous, that its repeal is generally demanded by honest merchants throughout the country.

GERMAN VEGETARIANISM receives a severe blow just at the turn of the year. Its most zealous scientific partisan, its most-quoted learned authority, the writer of so many leaflets and polemical pamphlets; Dr. Alanus, sends the vegetarians his farewell. —"Warum ich nicht mehr vegetarisch lebe" (Why I no longer live as a vegetarian), such is the title of an article sent to the Rhenish *Courier* by Dr. Alanus. The former preacher of the vegetable diet writes:—"Having lived for a long time as a vegetarian without feeling any better or worse than formerly with mixed food, I made one day the disagreeable discovery that my arteries began to show signs of atheromatous degeneration. Particularly in the temporal and radical arteries this morbid process was unmistakable. Being still under 40 years of age, I could not interpret this symptom as a manifestation of old age, and being furthermore not addicted to drinking, I was utterly unable to explain the matter. I turned it over and over in my mind without finding a solution of the enigma. I, however, found the explanation which I had sought so long quite accidently in a work of that excellent physician Dr. E. Monin, of Paris. The following is the verbal translation of the passage in question:—'In order to continue

the criticism of vegetarianism we dare no
lamented Gubler on the influence of the
degeneration of the arteries. Vegetab
salts than that of animal origin, introduces
blood. Raymond has observed numero
monastery of vegetarian friars, amongst o
man scarcely 32 years old, whose arteries
indurated, The naval surgeon Treille h
atheromatous degeneration in Bombay a
people live exclusively on rice. The veg
the blood vessels and makes prematurely ol
as old as his arteries.' It must produce
senile arch of the cornea, and phosphaturia.'
these newest results of medical investigation
I have, as a matter of course, returned
longer consider purely vegetable food as t
only as a curative method which is of the great
states. Some patients may follow this diet
it is not adapted for everybody's contined u
the starving cure, which cures some patien
continually by the healthy. I have become
which has shown me that one single bruta
most beautiful theoretical building."—*Cologn*

THE BELLITE COMPANY.—The Isle of M
new industry. A company called the Bel
works in Sweden, have purchased abou
Ballaskeig, Cornah, in the parish of Mougl
having works constructed for the purpose o
explosive called bellite. The reason for st
is that the port of Stockholm is practically
months, and the cost of fuel and materia
great as in Sweden. The company desire
goods at all times during the year, and this
now from their new works. About 50 hands
but when the manufacture is in full swing man

THE SCOTCH OIL TRADE.—The *Economi*
of a few days a conference, fraughtwith importa
Scotch oil trade, will be held in London. Mr.
representative of the Standard Oil Company, of
over the Atlantic, in order to discuss certain
cropped up amongst the producers of paraffin sc
prices ruling in 1889, when compared with 1886
Scotch trade of about £200,000., contrasted wi
of not less than £520,000., and compared w
losses must be something enormous, as will be
calculated that in the case of Young's Paraffin C
represent a difference in income of over £500,00
for the skill and management of those associa
that, notwithstanding these great losses, they h
maintaining the trade in its present by no means

BRIQUETTES.—Patent fuel has had a fashio
upon it, viz., "coal briquettes." Why not coal
N. Proctor, of Hunslet, New-road, has pate
claims to possess the necessary qualifications
"briquettes" at a much less cost than the a
Apart from the question of driving power, the c
not be more than one-third that of machines a
the small amount of pressure placed upon the
render these articles much more suitable for comb
present system. The introduction of steam into
the coal dust and pitch dust are mixed, rende
plastic, and great pressure is therefore found to
machine turning out 16 blocks of 5 lbs. each per
21 tons per day, requires a driving power of abou
sizes may be made if required. Briefly, the
machine is fed with coal and pitch dust thro
material then falls into a cylinder, where it i
steamed, and made adhesive. By a "worm" arr
the material is forced into a mould, and is then
mittent movement sends the brick into position
operation of the same intermittent motion takes pl
comes down and forces the brick out of the machi
and the next brick also into position for being pre

A PROSPEROUS COAL-CARRYING RAILWAY.
Trade inquiry as to railway rates some interesting
given in the last few days. A witness connected
Railway—one of the great coal-carrying railways
that the dividend paid was 15 per cent. for the
even higher rates had been paid. Beside such a
North-Eastern Railway becomes very small in c
dividend is paid out of a low charge for the ca
per ton per mile; then it became ⅞d. per ton pe
0·55d. per ton per mile for full train loads. Poss
that form one of the reasons why South Wa
quantities of coal.

DUST EXPLOSIONS.—Dr. Marcet has been contending, before the [Roy]al Meteorological Society, for the inflammability and explosive [cha]racter of dust when mixed with air. He gives instances of ex[plos]ions due to the presence of fine dust in coal mines and flour mills. [Th]e same opinion, as a frequent cause of explosions in mines, is gaining [grou]nd among those appointed by several European Governments to [inve]stigate the subject. It is a doctrine that should be familiar to [Nort]h-country readers, as having been long and ably advocated by Mr. [Atk]inson, one of the local mine inspectors.

Market Reports.

MISCELLANEOUS CHEMICAL MARKET.

[I]n the alkali trade, the firmness of prices continues, and all the [pro]ducts are sharing in the improvement. Bleaching powder, while [unc]hanged at £5. 2s 6d. to £5 5s. for soft wood casks on rails at [ma]kers' works, is moving off freely, and makers are not willing to [sell] for future deliveries, except at a premium on these figures. [Sod]a ash and caustic soda are still in good demand, and the higher [pric]es are freely paid. Chlorate of potash at 6¼d. per lb. is without material change in its position. Soda crystals are more scarce [and] dearer, advances of 2s. 6d. to 5s per ton being now quoted for leading makes. Sulphate, vitriol, and muriatic acid are all sell[ing] freely at current rates. Sulphate of copper still disappoints those [wh]o look for a fall in price, and is in brisk demand, more particularly [for] export trade, at £24. to £24. 10s. Sulphate of iron, more scarce [and] dearer generally. Acetates of lime and acetic acids firm on the [spo]t. Acetate of soda without change. Lead nitrate and acetate [qui]et with very little inquiry, but the higher values of raw material [kee]p prices firm. Tartaric and citric acids generally disappointing [h]olders, and only slow of sale at 1s. 3d. and 1s. 4d. respectively. [Pot]ash, caustic, and carbonate firmer. Prussiate of potash scarce and [firm] at 9½d. to 9¾d. per lb, according to make. Picric acid is find[ing] ready outlet at 1s. 3½d. to 1s. 5d. per lb. Carbolic acid [cry]stals, dull at 11d. to 11¼d. per lb. for 34°, and there is a dis[pos]ition to reduce prices for forward contracts. Iodine still un[set]tled, 8½d. to 9d. per oz. The state of trade generally in all [che]micals for the use of dyers, calico printers, and other textile [tra]des is still quiet. Aniline oil and salt are still firmly held by makers [at] the improved prices which have been ruling recently.

REPORT ON MANURE MATERIAL.

[T]HE demand for mineral phosphates still continues, and any-thing afloat or for immediate steamer shipment is fetching [full]y full prices. The market for other material seems scarcely so [stro]ng, though prices do not indicate any great variation from those [quo]ted a week ago.

[C]harleston River phosphate rock is worth fully 11d. per unit in [car]goes, c f. and i. to United Kingdom, afloat or for immediate ship[me]nt, and it is doubtful whether less would be accepted for shipment [furt]her ahead. The strikes which were interfering with the shipment [of] Belgian are at an end, but quotations for this description of phos[ph]ate have advanced to about 8d. per unit, cost, freight, and insurance, [for] 40/45%, and fully 10½d. would have to be paid for 48 to 53 %. [Qu]otations for Somme remain without change, and for Canadian the [pri]ces must be considered as nominal in the absence of any business [rep]orted, say at about 1s. for 70%, 13½d. for 75%, and 15d. for 80 %, [del]ivered Thames, Mersey, or Clyde. One-fifth rise in each case.

[A] small cargo of River Plate bones, November sailing, sold at [£5]. 15s., but this price must be considered as exceptional, sellers being [un]able to get even £5. 12s. 6d. for a 500 ton mixed December cargo. [Th]e market for East Indian bones remains without material change, [sell]ers requiring £5. for shipment ahead, and buyers being unwilling to [giv]e more than £5. 7s. 6d. East Indian bone meal is a shade easier, [the]re being now sellers at £5. 10s., Liverpool, Hull, or Glasgow, and [buy]ers meantime holding off in expectation of doing a shade better.

[N]itrate of soda continues very dull, and 8s. 3d. must be considered [the] closing spot value. Due cargoes are worth barely 8s. 4½d., and [8s.] 6d. would have to be taken for November shipments. Buyers are [pur]suing a hand-to-mouth policy in purchasing, believing that the [ma]rket must further succumb under the very large visible supply.

[T]he tendency of supers is still to harden, and value of 26% can now [sca]rcely be quoted under 50s. per ton, in bulk, f.o.r. at works.

[W]e have no further business to report in ground hoof, but shipments [of] River Plate dried blood are now on the way, and we shall presently, [no] doubt, see more doing in this article.

WEST OF SCOTLAND CHEMICALS.

GLASGOW, Tuesday.

In the paraffin section there is a good rise to note as regards lubricating oils, these being dearer in all the grades, so far as the better qualities of makes are concerned. The mount is a substantial one, ranging from 5s. to 10s. a ton, but dealers are reckoning that it is not to hold for any considerable time, being due mainly to passing influences, chief of these being a higher quotation by the American producers. The British-American paraffin scale delegates met in London last week, but adjourned without issue, decision to be arrived at next meeting, which will probably have place in the first week of February. Here there is much dubiety as to the result, but some makers are sanguine that the Americans will consent to a fractional rise in scale at least. Burning oil remains very weak, and uncalled-for stocks attain serious bulk in the hands of makers, who will probably feel compelled, ere all is over, to carry forward portion of their customers' contracts to next season. Sulphate of ammonia is quite idle at about £11. 17s. 6d., Leith nominal value, although reserves by no means bulk largely. Demand for general chemicals has slackened during the week, and recovery in caustic and bleach has been checked. Soda crystals, however, are firmer, forming about the only other exception, unless sulphate of copper be included, with its slightly enhanced quotation. Chief prices current are:—Soda crystals, 50s. net Tyne; alum in lump, £5., less 2½% Glasgow; borax, English refined £30., and boracic acid, £37. 10s. net Glasgow; soda ash, 1⅛d. less 2½% Tyne; caustic soda, white, 76°, £9., 70/72°, £7. 15s., 60/62°, £7., and 60/62° cream, £6. 10s., all less 2½% Liverpool; bicarbonate of soda, 5 cwt. casks, £5., and 1 cwt. casks, £5. 5s. net Tyne; refined alkali, 48/52°, 1⅜d., less 2½% Tyne; saltcake, 25s. 6d. to 26s.; bleaching powder, £6. 2s. 6d. to £6. 10s.; less 5% f.o.r. Glasgow; bichromate of potash, 4d., and of soda 3d., less 5 and 6% to Scotch and English buyers respectively; chlorate of potash, 4¾d., less 5% any port; nitrate of soda 8s. 3d. to 8s. 4½d., sulphate of ammonia, £11. 17s. 6d. f.o.b. Leith; salammoniac, 1st and 2nd white, £36. and £34., less 2½% any port; sulphate of copper, £24. less 5% Liverpool; paraffin scale, hard, 2$\frac{1}{16}$d., soft, 2d.; paraffin wax, 120°, semi-refined, 2$\frac{11}{16}$d.; paraffin spirit (naphtha), 9d.; paraffin oil (burning), 6¾d. to 7d., at Glasgow; ditto (lubricating), 865°, £5. 10s. 885°, £6. to £6. 10s.; and 890/895°, £7. 10s. to £8. Week's imports of sugar at Greenock were 7,284 bags only.

THE LIVERPOOL COLOUR MARKET.

COLOURS without material alteration. Ochres: Oxfordshire quoted at £10., £12., £14., and £16.; Derbyshire, 50s. to 55s.; Welsh, best, 50s. to 55s.; seconds, 47s. 6d.; and common, 18s.; Irish, Devonshire, 40s. to 45s.; French, J.C., 55s., 45s. to 60s.; M.C., 65s. to 67s. 6d. Umber: Turkish, cargoes to arrive, 40s. to 50s.; Devonshire, 50s. to 55s. White lead, £21. 10s. to £22. Red lead, £18. 10s. Oxide of zinc: V.M. No. 1, £24. 10s. to £25.; V.M. No. 2, £23. Venetian red, £6. 10s. Cobalt: Prepared oxide, 10s. 6d.; black, 9s. 9d.; blue, 6s. 6d. Zaffres: No. 1, 3s. 6d.; No. 2, 2s. 6d. Terra Alba: Finest white, 60s.; good, 40s. to 50s. Rouge: Best, £24.; ditto for jewellers', 9d. per lb. Drop black, 25s. to 28s. 6d. Oxide of iron, prime quality, £10. to £15. Paris white, 60s. Emerald green, 10d. per lb. Derbyshire red 60s. Vermillionette, 5d. to 7d. per lb.

TAR AND AMMONIA PRODUCTS.

TUESDAY.

There is a good inquiry for Benzoles of all description, and prices have risen considerably; 90's are now worth 4s. per gallon, and 50/90's 3s. 2d. to 3s. 3d. Solvent naphtha is in good demand, and the better qualities are moving upwards in price. Creosote is moving off well, there being no stocks in the makers' hands. Crude carbolic acid remains in much the same state as quoted in our few last reports, the position of this article for the future being rather uncertain. In anthracene, speculators continue to be active, and sales of a quality are reported at 1s. 6d. per unit, while for B quality 1s. 3d. to 1s. 3½d. is to-day's market value. Consumers generally appear to be holding back, as the prices now reached are out of all proportion to the price of alizarine, which is easily obtainable at 8½d. to 9d. per lb.

The sulphate market continues dull, but still there is no fear that prices will continue so low as are quoted to-day. The Hull market has fallen to £11. 17s. 6d., though we believe sales have been made at £11. 18s. 9d. London price is £12., but Beckton has, we understand, been offered at £11. 17s, 6d, and this is the price also f.o.b. Leith. It is strange that sulphate makers should show such little faith in the strength of their position, and, as a rule, give way at times when their ideas should be stronger, and we would advise them to carefully peruse that portion of the article on nitrogenous manures which appeared in our leader commencing the New Year's number.

Notices of Dividends.

WHITHAM, JOSEPH, Miles Platting, chemical manufacturer, trading as Joseph Whitham and Company, Pendleton, dyer and printer, and as the Whitworth Printing Company, Harwarden, galvaniser; also trading with T. Brayshaw, as Whitham and Brayshaw, at Manchester, and residing at Nelson. Third and final dividend of ¾d. any day, Trustees' Office, 2, Clarence Buildings, Booth-street, Manchester.

New Companies.

ELECTRICAL ENGINEERING CORPORATION, LIMITED.—This company was registered on the 15th inst., with a capital of £150,000., in £5. shares, 400 of which are founders' shares, to carry out an agreement (unregistered) of the 9th inst., between R. B. Rennie (for the United Electrical Engineering Company, Limited) of one part, and J. G. Statter (for this company) of the other part; and another agreement of even date between J. G. Statter and Company, Limited, of one part, and E. Manville (for this company) of the other part, for the transfer to this company of the businesses of the United Electrical Engineering Company, Limited, and J. G Statter and Company, Limited. The subscribers are:—

	Ord. Shares.
C. Samuel, 176, Sutherland-avenue	1
W. L. Madgen, 3, Prince's-mansions, electrical engineer	1
W. G. Ainslie, M P, 21, Ennismore-gardens	1
J. G. Statter, C.E., Engineering Works, West Drayton	1
H. B. Statter, Snapethorpe, near Wakefield, Associate Royal School of Mines	1
P. E. Scrutton, Alliance Engineering Works, West Drayton, secretary	1
E. Manville, 3, Prince's mansions, Victoria-street, electrical engineer	1

JESSE FISHER AND SON, LIMITED.—This company was registered on the 15th inst., with a capital of £10,000., in £1. shares, to trade as chemical manufacturers, and as makers of surgical instruments, appliances, and requisites. An unregistered agreement of the 13th inst., between W. S. Rothband and G. W. Fox will be adopted. The subscribers are:—

	Shares.
W. S. Rothband, 61, Elizabeth-street, Cheetham, Manchester, manufacturing chemist	1
M. Kerab, 7, Dantzic-street, Manchester, wholesale clothier	1
L. King, 20, Thomas-street, Manchester, cigar manufacturer	1
A. Barder, 20, Thomas-street, Manchester, fancy goods dealer	1
J. Bayley, 44, Kennedy-street, Manchester, accountant	1
J. Crompton, Chorlton-on-Medlock, Manchester, clerk	1
T. Ashton, 10, Buckingham-street, Manchester, cashier	1

MINERALS AND FULLERS' EARTH SYNDICATE, LIMITED.—This syndicate was registered on the 13th inst., with a capital of £2,000., in £100. shares, to purchase mines, quarries, beds of fullers' earth, and other minerals in the United Kingdom. The subscribers are:—

	Shares.
Joseph Day, Bath, engineer	1
H. N. Garrett, Bath	1
C. W. Thickle, Bath, clerk	1
A. J. King, Bath, solicitor	1
L. S Cocking, Huddersfield, drysalter	1
A. G. D. Moger, Bath	1
W. B. Hallett, 11, Queen Victoria-street, auctioneer	1

PURNELL HIGH-SPEED GAS ENGINE AND MANUFACTURING COMPANY, LIMITED.—This company was registered on the 13th inst., with a capital of £50,000., in £1. shares, to acquire certain patents referred to in an unregistered agreement of 2nd inst., between John James Purnell and Joseph White; and to carry on business as mechanical, gas electrical, and general engineers. The subscribers are:—

	Shares.
Joseph White, 2, Goldsmith-street, Gough-square, E.C., publisher, &c.	1
A. W. Mantle, 52, Queen Victoria-street, engineer	1
Sidney Smith, 2, Robin Hood-court. Shoe-lane, contractor	1
J. E. Jackson, 3, Eton-road, Hampstead, engineer	1
A. H. Skan, 40, Algernon-road, Lewisham, shorthanq writer	1
J. H. Jack, 41, Chesilton-road, Fulham, secretary to a company	1
H. C. Fowler, 8, Plevna-villas, St. Paul's-road, Tottenham, shorthand writer	1

GLENDON LAND AND BRICK COMPANY, LIMITED.—This company was registered on the 10th inst., with a capital of £15,000., in £5. shares, to carry on at Kettering, Northampton, the business of brick, tile, and patent fuel manufacturers. The subscribers are:—

	Shares.
G. H. Hoyle, Victoria-street, Westminster, solicitor	1
R. H. Hepburn, Parliament-mansions, engineer	1
J. Spence, Kettering	1
R. Waddington, Kettering, stationer	1
A. D. Studd, Kettering, engineer	1
T. Morris, Kettering, engineer	1
E. J. Deacon, Kettering, engineer	1

LONDON PAPER MILLS COMPANY, LIMITED.—This company was registered on the 10th inst, with a capital of £75,000., divided into 4,990 ordinary shares of £10. each, and 100 founders' shares of £1. each, to carry on business as paper manufacturers in all branches. The subscribers are:—

	Ord. Shares.	Founders' Shares.
T. L. Roberts, Bedford-park, Croydon	1	
R. D. Wilkinson, 2 Elmwood-road, Croydon	1	
A. L. Poulter, 6, Arthur-street West E.C., wholesale stationer	1	60
A. E. Reed, Highdene, Sidcup, paper maker	1	40
E. C. D Poulter, 6, Arthur-street West, wholesale stationer	1	
G. Griffiths, J.P., 24, Fitzjohn's-avenue, N.W.	1	
A. J. Rhodes, Chislehurst	1	

NEW ASBESTOS COMPANY, LIMITED.—This company was registered on the 10th inst., with a capital of £120,000., in £4. shares, to acquire the business of asbestos manufacturers and merchants, carried on by the Société Francaise des Amiantes at Tarascon, France, and elsewhere, including certain asbestos properties in France and Italy. The subscribers are:—

	Shares.
John Pound, 81, Leadenhall-street, manufacturer	1
H. W. Maynard, 34, Gracechurch-street	1
W. B. Wright, E.C., Rosslyn, Ealing	1
H. H. Hyde Clarke, 32, St. George's-square, S.W.	1
J. Simpson, Hampton Wick, merchant	1
W. Field, 22, Buckingham-street, Strand, engineer	1
W. M. Donelly, 37, Lombard-street, secretary	1

FOWLER, LANCASTER, AND COMPANY, LIMITED.—This company was registered on the 8th inst., with a capital of £22,000., in £10. shares, to carry on business as electrical engineers, manufacturers, and contractors. The subscribers are :—

	Shares.
Arthur Mackenzie, Temple Chambers, E.C., agent	1
W. M. Pyke, 2, Metal Exchange-buildings, solicitor	1
A. E. Shill, 30, Dayton-grove, Peckham, architect	1
H. L. Collins, 67, Lombard-street, clerk	1
C. W. Rawlinson, 11, Lincoln's-inn-fields, solicitor	1
R. W. Bonder, 11, Lincoln's-inn-fields, solicitor	1
C. A. E. Good, 48, Lincoln's-inn-fields, solicitor	1

UNION DRUG COMPANY, LIMITED.—This company was registered on the 7th inst., with a capital of £1,000., in 5s. shares, 500 being founders' shares, to take over a business carried on under the style of the Union Drug Company, and to manufacture certain cures and pills. The subscribers are :—

	Shares.
F. C. Peachey, 241, Compton-buildings, Goswell-road, jeweller	1
G. C. Cardnell, 20, Canetowers-road, N.W., correspondent	1
Sydney H. Monckton, 34, Great Queen-street, W.C., clerk	1
H. T. Richard, 8, Shepherd's-place, Kennington	1
H. Davies, 73, St. Augustine-road, N.W., clerk	1
A. Twallin, 12, Salisbury-villas, Stamford-hill, clerk	1
W. B. Perkin, 38, The Grove, Hammersmith, architect	1

IMPORTS OF CHEMICAL PRODUCTS

AT

THE PRINCIPAL PORTS OF THE UNITED KINGDOM.

LONDON.

Week ending Jan. 4th.

Acetic Acid—
Holland, 98 pkgs. A. & M. Zimmermann

Acetate of Lime—
U. States, £350 W. Thatcher
" 184 Johnson & Hooper
" 120 M. D. Co.
" 86 Johnson & Hooper
" 165 "

Antimony Ore—
N. Zealand, 51 t. N. Z. Antimony Co.
Sydney, 11 H. Bath & Son
A. Turkey, 43 t. Quirk, Barton & Co.
Sydney, 4 E. Hawkins
Portugal, 4 J. Ford & Co.

Ammonia—
U. S., £270 Davies, Turner, & Co.

Alum—
Holland, £55 Ohlenschlager Bros.

Acetate of Soda—
France, £40 Lister & Biggs
" 85 Hunter, Waters, & Co.

Antimony—
Adelaide, 4 t. S. Lambert & Co.
Sydney, 10 E. Hawkins

Bones—
E. Indies, 10 t. R. & J. Henderson
N. S. Wales, 14 Goad, Rigg, & Co.
Queensland, 8 Dyster, Nalder, & Co.
Spain, 12 Goad, Rigg, & Co.

Borate Manganese—
Germany, £85 T. H. Lee

Bone Meal—
E. Indies, 200 t. Hunter, Waters, & Co.

Barytes—
Belgium, 111 cks. O'Hara & Hoar
" 130 A. Zumbeck & Co.
Germany, 22 O'Hara & Hoar
" 23 W. Harrison & Co.

Brimstone—
Italy, [illegible] t. Brandram Bros.
" 100 W. C. Bacon & Co.
" [illegible] Soundy & Son

Borax—
Germany, £28 F. Stahlschmidt & Co.

Barium—
Germany, £[illegible] Holm

Barium Chloride—
Germany, £70 Petri Bros.

Caoutchouc—
Denmark, [illegible] Davies, Turner, & Co.
Portugal, [illegible] Kleinwort, Sons, & Co.
France, [illegible] Pickford & Co.
" [illegible] T. H. Lee
E. Indies, [illegible] L. & I. D. Jt. Co.
U. States, [illegible] "
Cape, [illegible] "
Germany, [illegible] J. E. Wilson

Cream Tartar
France, [illegible] C. F. Gerhardt
Spain, 5 Webb, Warter, & Co.
Holland, 4 pkgs. A. Hopf & Co.
" 15 Middleton & Co.
Spain, 5 C. Christopherson & Co.

Chemicals *(otherwise undescribed)*
£15 P. Morrison & Co.
5 Phillipps & Graves
30 T. H. Lee
30 L. & I. D. Jt. Co.
5 Beresford & Co.
W. K. Bott & Co.
T. H. Lee
"
L. Zimmermann
H. Lambert
[illegible]mmermann
[illegible]D. Jt. Co.

Belgium, 18 T. H. Lee

Caustic Soda—
Belgium, 45 c. Beresford & Co.

Copper Regulus—
Holland, 99 t. Johnson, Matthey, & Co.

Dyestuffs—

Extracts
France, 125 pkgs. L. J. Levinstein & Sons
Holland, 80 Burt, Boulton, & Co.
U. States, 200 M. Dk. C.
" 100 W. Burton & Sons
" 61 A. J. Humphery

Myrabolams
E. Indies, 67 pkgs. L. & I. D. Jt. Co.
" 67 "

Gambier
E. Indies, 177 pkgs. S. Barrow & Bro.
" 1,030 L. & I. D. Jt. Co.
" 214 "

Indigo
E. Indies, 227 chts. Arbuthnot, Latham, & Co.
" 5 bxs. 85 G. Ward & Sons
" 14 Benecke, Souchay & Co.
" 24 L. & I. D. Jt. Co.
" 65 Ernsthauser & Co.
" 32 Elkan & Co
" 17 L. & I. D. Jt. Co.
" 20 A. Harvey
" 16 Elkan & Co.
" 19 T. Barlow & Bros.
" 144 Patry & Pasteur
" 24 Hoare, Miller & Co.
" 4 Walker, Munsie, & Co.
" 49 L. & I. D. Jt. Co.
" 55 P. & O. Co.
" 12 H. Grey, jun.
" 18 G. Dards
" 7 Elkan & Co.
" 17 L. & I. D. Jt. Co.
" 8 F. W. Heilgers & Co.
" 122 Stansbury & Co.
" 341 Arbuthnot, Latham, & Co.
" 6 Phillipps & Graves
" 11 G. Dards
" 24 L. & I. D. Jt. Co.
" 16 F. Huth & Co.
" 13 Ernsthausen & Co.

Saffron
Spain, 1 pkg Pickford & Co.
" 1 A. Ruffer & Sons
" 1 C. Brumlen
France, 3 R. G. Clements & Son

Tanners' Bark
Belgium, 9 t. T. H. Lee
" 5 Brown & Elmslie
" 3 Hicks, Nash & Co.
U. States, 15 A. J. Humphery

Sumac
Italy, 1,126 pkgs. C. E. Fosbrooke
" 350 J. Kitchen
" 650 W. France

Valonia
A. Turkey, 25 t. A. & W. Nesbitt
" 25 Anning & Cobb
" 90 J. Graves

Aniline
U. States, 5 pkgs. W. W. Young

Argols
Greece, 1 pkg. A. M. Messinesi

Madder
France, 2 pkgs. W. France
Holland, 6 Phillips & Graves

Safflower
E. Indies, 30 pkgs. Ripley, Roberts, & Co.

Glycerine—
Germany, £160 Becker & Ulrich
Holland, 440 Spies, Bros. & Co.
Germany, 21 Becker & Ulrich
" 58 Craven & Co.

Glucose—
Germany, 100 pkgs. 200 cwts. J. Forsey
" 25 203 Pillow, Jones & Co.
" 200 200 L. Sutro & Co.
" 40 321 Prop. Chamb. Wf.
" 300 300 J. Barber & Co.
" 300 300 Horne & Co.
U. States, 600 600 Humphery & Co.
Germany, 100 200 J. Forsey
" 200 200 Horne & Co.
" 25 222 J. Knill & Co.
" 37 290 P. W. Hore
" 22 185 G. Tennant, Sons, & Co.
" 5 40 Perkins & Homer
" 200 200 Horne & Co.
" 8 70 L. & I. D. Jt. Co.
" 4 24 T. H. Lee
" 25 200 W. Balchin
" 100 200 J. Knill & Co.
" 200 200 J. Barber & Co.
" 390 385 J. Cooper
" 200 400 J. Forsey
U. States, 500 500 W. Parkinson
Germany, 200 400 J. Forsey
" 25 202 Page, Son, & East
" 200 200 T. E. Howell
" 400 400 J. F. Ohlmann
" 25 150 C. Tennant, Sons, & Co.
U. States, 600 1,056 W. Parkinson

Guano—
Belgium, 37 t. 10 c. Soundy & Son
Chili, 72 M. D. Co.

Isinglass—
St. Settlements, 21 pkgs. L. & I. D. Jt. Co.
China, 3 W. M. Smith & Sons
B. W. Indies, 1 "
" 1 Malcolm, Kearton, & Co.
E. Indies, 3 L. & I. D. Jt. Co.

Manure—

Phosphate Rock
U. States, 2,652 t. Wyllie & Gordon
France, 50 Anglo Contn'l Guano Works Agency
Belgium, 190 Millar & Johnson
B. W. I., 750 Pickford & Winkfield
France, 171 T. Farmer & Co.
" 170 Lawes Manure Co.
Belgium, 135 Anglo Contn'l Guano Works Agency
France, 140 F. C. Hills & Co.
" 150 T. Farmer & Co.
Belgium, 395 Lawes Manure Co.
" 170 Odam's Cheml. Manure Co.
" 257 T. Farmer & Co.
" 400 W. H. Carey & Sons
" 105 Moeller, Graetz, & Co.
" 100 Lawes Manure Co.

Manure *(otherwise undescribed)*
U. S., 88 t. T. D. Charlesworth & Co.
" 4 C. A. & H. Nichols

Naphtha—
Belgium, 10 brls E. Wood
Holland 8 Domeier & Co.

Potassium Carbonate—
Holland, 6 pkgs. Spies, Bros., & Co.

Potash Salts—
Holland, £70 R. Morrison & Co.

Plumbago—
Holland, 50 cks. Brown & Elmslie
Germany, 58 Beresford & Co.
" 31 Wright & Ashbridge
Holland, 25 Brown & Elmslie
Germany, 152 cs. Props. Scott's Wharf
E. Indies, 35 cks. L. & I. D. Jt. Co.
Ceylon, 164 "

Pyrites—
Sydney, 10t. H. Bath & Son

Pitch—
Holland, 22 brls Rosenburg, Lowe, & Co.

Paraffin Wax—
U. States, 7,098 brls. J. H. Usmar & Co.

Sugar of Lead—
Holland, 34 pkgs. Beresford & Co.
" 19 J. Kitchen

Saltpetre—
Germany, 50 cks. T. Merry & Son

Stearine—
N. Zealand, 10 bgs. Miles, Bros., & Co.
Holland, 76 T. H. Lee

Soda—
France, £57 Props. Hays Whf.
" 57 "

Tartaric Acid—
Holland, 1 pkg. Middleton & Co.
" 31 Webb, Warter, & Co.
" 4 "
" 7 J. Smidt
" 4 Middleton & Co.
France, 39 Webb, Warter, & Co.

Tar—
Pillau, 56 brls. Linch, Moeller, & Co.
France, 500 Burt, Boulton, & Co.
Reval, 3 cks. Chope & Co.

Tincal—
E. Indies, £40 W. W. & R. Johnson & Sons

Tartar—
Italy, 81 pkgs. Thames S. Tug Co.

Ultramarine—
Belgium, 12 pkgs. Leach & Co.
Germany, 27 G. Steinhoff
Belgium, 2 cs. W. Harrison & Co.
Germany, 8 cks. A. Henry & Co.

Zinc Oxide—
Holland, 20 cks. Beresford & Co.
" 50 H. Hodson & Co.
" 128 M. Ashby
" 12 Beresford & Co.
" 2 H. Lambert
" 25 brls. M. Ashby

LIVERPOOL.

Week ending Jan. 8th.

Acetic Acid—
Rotterdam, 50 bskts.

Albumen—
Antwerp, 10 cks. J. T. Fletcher & Co.
Marseilles, 4 cs.

Ammonia Skimmings—
Philadelphia, 16 cks.

Acetate of Lime—
New York, 630 bgs.

Barytes—
Antwerp, 7 ck

Bleaching Powder—
New York, 6 crts.

& Roberts
L. Graham
Balli Bros.
Graham & Co.
bell & Co.
Union Co.
L. Graham & Co.
Cockbain, Allardice
Niger Co.
P. Werner
ats & Co.
A. Swanny
ickering & Berthoud
A. Swanny
Robbin, & Co.
Iver & Co.
& Durant
Nordlinger
mlinson & Co.
ibbs & Co.
n, Bros., & Co.
her & Co.
Iver & Co.
& Fraser
nson & Co.
T. Crouch
lert & Co.
urs & Bell
mlinson & Co.
ard & Co.
& Watson
aunt Bros.
rson & Co.
ckering & Berthoud
tadelmann
L. Verdiers
A. Swanny
3 brls.
ther & Co.
ane & Co.
A. Swanny
allie & Co.
lson & Co.
aselle, Son, & Co.
a Trading Co.
Francaise
Palma
ading Co.
Cie Francaise
Paterson,
pais & Co.
ickering & Berthoud
urst & Co.
ann & Co.
aber & Co.
, Bood, & Co.
gton & Co.
l S. S. Co.
ury & Co.
Daranty, & Son
olp & Co.
gson & Co.
Cole & Co.
e, Sons, & Co.
L. Kitchen

Gambier
Singapore, 6,594 bls.
" 897 E. Boustead & Co.
Extracts
New York, 17 brls.
Philadelphia, 100 bxs.
Bordeaux, 394 cks.
Orchella
Colon, 26 bls. Mathison & Bausire
Argols
Bordeaux, 52 cks.
Oporto, 23 Ger. Bank
Madder
Bordeaux, 10 cks.
Divi Divi
Hamburg, 444 bgs.
Glycerine—
Antwerp, 1 brl.
Bordeaux, 14 cks.
Glucose—
Stettin, 25 cks. W. Slater
" 8 J. Brooks
Baltic, 11
Hamburg, 185 bgs. 69 cks.
" 10
Horn Piths—
Valparaiso, 661 bgs.
Hydrated Potash—
Dunkirk, 7 cks.
Iodine—
Valparaiso, 176 brls. A. Gibbs & Son
" 7 W. & J. Lockett
Isinglass—
Maranham, 2 cks Gunston, Sons, & Co.
" 1 R. Singlehurst & Co.
Para, 14 cs. Bieher & Co.
Manganese Ore—
Gothenburg, 105 t. Macqueen, Bros., & Co.
Phosphates—
Antwerp, 150 t.
Pyrites—
Huelva, 1,044 t. Tennants & Co.
" 1,330 "
" 1,427 "
Plumbago—
Genoa, 11 cks.
Phosphate Rock—
Coosaw, 2,000 t. J. Adger & Co.
Potash—
Dunkirk, 9 cks.
Pitch—
Bordeaux, 50 cks.
Potash Salts—
Hamburg, 494 bgs.
Salt—
New York, 18 crts. L. Rapp & Co.
Antwerp, 20 bgs.
Saltpetre—
Rotterdam, 135 bgs.
Sodium Silicate—
Rotterdam, 7 cks.
Soda—
Rotterdam, 40 brls.
" 1 cs.
Sodium Nitrate—
Pisagua, 9,864 bgs.
Sulphur—
Catania, 140 bgs.
Palermo, 20
Stearine—
New York, 150 brls.
Antwerp, 13 bgs.
Saltpetre—
Hamburg, 50 cks.
Tartaric Acid—
Rotterdam, 39 cks.
Hamburg, 8
Tartar—
Rotterdam, 2 cks.
Valparaiso, 9 bls. Bates, Stokes & Co.
Bordeaux, 53 cks.
Tar—
New York, 531 brls. Cayser, Irvine & Co.
Tartar Emetic—
Hamburg, 13 cks.
Ultramarine—
Rotterdam, 33 cks.
Verdigris—
Marseilles, 10 cks.
Zinc Ashes—
New York, 43 brls. R. Crooks & Co.
Zinc Oxide—
Antwerp, 200 brls. 25 cks.
Stettin, 150
Antwerp, 75 J. T. Fletcher & Co.

GLASGOW.

Week ending Jan. 19th.

Alum—
Rotterdam, 79 cks.
Brimstone—
Sicily— 800 t. T. Lawrie
Boracite Ore—
Panderma, 518 t. Jos. Townsend, Ltd.
Barytes—
Rotterdam, 44 cks. Jas. Rankine & Son
" 111
Charcoal—
Antwerp, 210 bgs.
Chemicals *(otherwise undescribed)*
Hamburg, 4 cs.
Dyestuffs—
Valonia
Smyrna, 500 t. Martin & Miller
Extracts—
New York, 6 brls.
Antwerp, 27 cks. 1 cs.
Alizarine
Rotterdam, 121 cks. Jas. Rankine & Son
" 23 "
" 36 "
Cutch
Boston, 10 cks. 1 kg.
Gypsum—
Rouen, 140 t.
Glucose—
Hamburg, 200 bgs.
Manganese—
Rotterdam, 4 cks. Jas. Rankine & Son
Pyrites—
Huelva, 1,124 t. Tharsis Co.
Potash—
Dunkirk, 20 drms. M. Perls & Co.
Soda—
Rouen, 10 cks.
Sulphur Ore—
Huelva, 1,279 t. The Rio Tinto Co., London.
Salt—
Rotterdam, 3 cks. Jas. Rankine & Son
Spirits of Turpentine—
Charleston, 1,500 cks. L. M'Lellan
Stearine—
Rotterdam, 33 cks. J. Rankine & Son
Tar—
Wilmington, 1,750 brls. John Parker
Turpentine—
Wilmington, 400 brls. John Parker
Tartar—
Rotterdam, 4 cks. Jas. Rankine & Son
Verdigris—
Bombay, 1 brl.
Ultramarine—
Rotterdam, 5 cs. Jas. Rankine & Son
" 4 cks. "
Waste Salt—
Hamburg, 300 t.
Zinc Oxide—
Rotterdam, 50 cks. Jas. Rankine & Son
Antwerp, 85

HULL.

Week ending Jan. 9th.

Acid—*(otherwise undescribed)*
Rotterdam, 1 ck. Geo. Lawson & Sons
Amsterdam, 24 W. & C. L. Ringrose
Alumina—
Rotterdam, 40 cks. W. & C. L. Ringruse
Barytes—
Rotterdam 22 cks.
Bremen, 23 Tudor & Son
" 19 Blundell, Spence & Co.
Chemicals *(otherwise undescribed)*
Antwerp, 6 brls. 54 pkgs. Wilson, Sons & Co
" 4 cks. Bailey & Leetham
Hamburg, 61 brls. 11 pkgs. "
Bremen, 1 ck. J. Pyefinch & Co.
Bari, 94 pkgs. 40 cks. Wilson, Sons & Co.
Dyestuffs—
Alizarine
Rotterdam, 16 cks. Hutchinson & Son
" 34 W. & C. L. Ringrose
" 35 2 cs. Hutchinson & Son
" 41 pkgs. W. & C. L. Ringrose
Sumac
Trieste, 109 bgs. Wilson, Sons & Co.
Palermo, 325
" 1,850

Glucose—
Hamburg, 200 bgs. Wilson, Sons & Co
" 200 C. M. Lofthouse & Co.
" 100 A. C. Hornstead & Co.
" 17 cks. Rawson & Robinson
" 100 cs. 450 bgs. 59 cks.
" 35 cks. Woodhouse & Co.
" 30 Rawson & Robinson
Stettin, 37 Wilson, Sons & Co.
Guano—
Bretthausen, 1,667 sks.
Nitrate—
Rotterdam, 2 brls. Wilson, Sons & Co.
Plumbago—
Hamburg, 20 cs. Wilson, Sons & Co.
" 18 cks. G. Buckton & Son
Pitch—
Antwerp, 7 cks. H. F. Pudsey
Soda—
Rotterdam, 6 cks. Geo. Lawson & Sons
Stearine—
Rotterdam, 18 bgs. Wilson, Sons & Co.
Saltpetre—
Hamburg, 5 cks.
Sulphur—
Catania, 116 cks. 1,685 bgs. Wilson, Sons & Co.
Turpentine—
Wilmington, 150 brls.
Savannah, 1,721

TYNE.

Week ending Jan. 8th.

Alum Earth—
Hamburg, 18 cks. Tyne S. S. Co.
Barytes—
Rotterdam, 20 cks. Tyne S. S. Co.
Glucose—
Hamburg, 40 bgs. Tyne S. S. Co.
Pyrites—
Huelva, 1,799 t. Scott Bros.
" 1,547 "
Phosphate—
Valle, 405 t.
Sodium Nitrate—
Taltal, 6,480 bgs. Scott Bros.
Zinc Oxide—
Antwerp, 85 brls.

GOOLE.

Week ending Jan. 1st.

Antimony—
Boulogne, 2 cks.
Caoutchouc—
Boulogne, 3 cks.
" 4
Copper Ore—
Huelva, 700 t.
Dyestuffs—
Alizarine
Rotterdam, 6 cks.
Dyestuffs *(otherwise undescribed)*
Boulogne, 6 cs.
" 34 cks.
Glucose—
Hamburg, 200 bgs.
Plumbago—
Boulogne, 5 cs.
Potash—
Calais, 10 cks.
" 145 drms.
Saltpetre—
Rotterdam, 100 bgs.
" 100
Sulphur Ore—
Seville, 1,040 t.
Tartaric Acid—
Boulogne, 5 cs.
Waste Salt—
Hamburg, 610 bgs.
" 508

GRIMSBY.

Week ending Jan. 4th.

Albumen—
Hamburg, 6 cks.
Caoutchouc—
Hamburg, 5 pkgs.
Dyestuffs *(otherwise undescribed)*
Antwerp, 2 cs.
Glucose—
Hamburg, 100 bgs.

EXPORTS OF CHEMICAL PRODUCTS

FROM

THE PRINCIPAL PORTS OF THE UNITED KINGDOM.

LONDON.

Week ending Jan. 4th.

Ammonia Sulphate—
Mauritius, 10 t. £160
Hamburg, 309 2 c. 3,717
Jersey, 46 650
Genoa, 100 1,250
Ghent, 150 1,842
Madeira, 2 1 c. 25

Ammonia Carbonate—
Cologne, 1 t. 2 c. £30

Ammonia—
Singapore, 3 c. 2 q. £20

Acetic Acid—
Melbourne, 1 t. 4 c. £33

Acids (*otherwise undescribed*)
Pt. Natal, 6 cs. £16

Alumina Sulphate—
Bordeaux, 7 t. 6 c. £25

Arsenic—
Wellington 13 c. £12

Acetic Acid—
Wellington 20 cs. £35

Bone Superphosphate—
Las Palmas, 6 t. £26

Bisulphite of Lime—
Bombay, 2 t. 13 c. £26

Carbolic Acid—
Rotterdam, 36 brls. £273

Copper Sulphtate—
Brussels, 6 t. 4 c. £130
Alexandria, 2 cks. 11
Wellington, 1 24
Antwerp, 9 19 220

Citric Acid—
Libau, 10 c. £74
Hamburg, 2 kegs. 15
Brussels, 15 111
Karachi, 2 q. 1
Genoa, 2 15

Cream Tartar—
Brisbane, 17 c. £[illegible]

Carbonate of Lime—
New Orleans, 5 t. 19 c. £83

Caustic Potash—
Algoa Bay, 12 drs. £13
Adelaide, 18 c. 31

Carbolic Solid—
Rotterdam, [illegible] cks. £[illegible]

Caustic Soda—
Hobart, 15 t. 5 c. £115

Chemicals (*otherwise undescribed*)
Sydney, 9 cks. £137

Disinfectants—
Natal, [illegible] 1 c. £17
Hamburg, 10 cks. 40

Dried Blood—
Las Palmas, [illegible] t. 1 c. £[illegible]

Manure—
Hamburg, [illegible] t. £2,517
Amsterdam, [illegible] [illegible]
[illegible] 41 17 c. [illegible]
New York, 1 15
Mauritius, 45 [illegible]
[illegible] [illegible] [illegible]
[illegible] [illegible] [illegible]
[illegible] [illegible] 217
[illegible] [illegible] 13 [illegible]
Natal [illegible] [illegible]

Magnesium Sulphate—
Sydney, [illegible] [illegible] £[illegible]

Oxalate Potash
[illegible] [illegible] [illegible] £[illegible]

[illegible]sium Chlorate—
[illegible] [illegible] 17 [illegible] £[illegible]
[illegible] [illegible] kegs. [illegible]
[illegible] 2 [illegible] [illegible]
[illegible] 2 [illegible] [illegible]

Phosphorous—
Wellington, 4 cs. £40

Sheep Dip—
E. London, 25 cs. 12 c. £103
Pt. Natal, 1 t 40
Cape Town, 3 8 140
Gallegos, 5 cks. £32

Sulphuric Acid—
E. London, 15 c. £10
Natal, 5 t. 16 23

Salammoniac—
Alexandria, 5 cks. £49

Soda Ash—
Hobart, 13 t. 2 c. £69

Saltpetre—
Oporto, 7 t. 15 c. £160

Sodium Bichromate—
San Francisco, 1 c. £2

Sodium Sulphate—
Ghent, 50 t. £50

Sulphur Flowers—
Algoa Bay, 50 pkgs. £32

Superphosphate of Lime—
Oporto, 9t. 17 c. £45
Las Palmas, 6 18

Tartaric Acid—
Brisbane, 15 c. £118
Sydney, 5 30
Bombay, 15 cs. 90

LIVERPOOL.

Week ending Jan. 4th.

Alkali—
Boston, 5 cks.
Havre, 105
Philadelphia, 38 tcs.

Ammonia Muriate—
Havre, 1 t. 5 c. £49
Antwerp, 8 2 q. 11
Philadelphia, 11 t. 256
Manilla, 5 c. 8
Santander, 10 15

Alum Sulphate—
M. Video, 22 t. 0 c. 3 q. £88

Ammonia Sulphate—
Antwerp, 30 t. £360
Genoa, 30 10 c 365
Bordeaux, 48 10 570
Valencia, 100 1,297
Venice, 20 240
Hamburg, 90 10 2 q. 1,086
Ghent, 55 15 672
New York, 1 35
Demerara, 167 14 2 2,098

Alum—
Portland, M. 6 t. 1 c. £30
Smyrna, 16 16 1 q. 80
Melbourne, 8 1 30
Malta, 4 1 19
Algiers, 2 9
Messina, 2 10
Piraeus, 14 15 72
Salonica, 5 2 25
Santos, 1 8
Damas, 5 1 25

Antimony Regulus—
Hamburg, 5 t. 1 c. £334

Ammonia Carbonate—
Antwerp, 3 t. 4 c. £90
Hamburg, 3 1 q. 7
Havana, 1 10 1 45
Barcelona, 10 2 17

Acetic Acid—
Messina, 12 t. 2 c. 3 q. £150
Vera Cruz, 23 cks. 110

Bleaching Powder—
Alexandria, 5 cs.
Bari, 2 cks.
Bombay, 10 cs.
Boston, 910 cks. 362 tcs.
Bilboa, 19 brls.
Baltimore, 47
Barcelona, 2
Genoa, 203
Havre, 50 cks.
Corunna, 8 brls.
Havana, 5 cks.
Leghorn, 20 cks.
Lisbon, 15
Nantes, 128
New York, 531
Philadelphia, 169
Seville, 18 brls.
Terneuzen, 25 cks.
Vigo, 10 brls.

Blood Albumen—
Vera Cruz, 3 t. 1 c. £114

Brimstone—
Vera Cruz, 50 t. £225
Ponce, 4 9 c. 1 q. 32
Boston, 25 140

Borax—
Alexandria, 19 c. 3 q. £28
Hamburg, 13 19
M. Video, 1 t. 28
Adelaide, 1 3 c. 33
San Juan 4 3 q 6
New York, 55 3 1,696
Batoum, 10 20
Malaga, 3 5
Syra, 17 25

Caustic Soda—
Algiers, 20 dms.
Algoa Bay, 10
Amsterdam, 35
Ancona, 40
" 20 brls. 10
Antwerp, 7 bxs.
Bahia, 30
Baltimore 300
Barcelona, 40 cks. 323 dms.
Bari, 221
Beyrout, 10
Boston, 200
Calcutta, 204
Canada, 33
Carril, 6 brls. 6 6 kgs.
Galveston, 68
Genoa, 148
Gothenburg, 193 kgs
Hamburg, 23
Hiogo, 100
La Guayra, 10
La Libertad 40
Malaga, 55 brls.
Nantes, 82
New Orleans, 60
New York, 715
Oporto, 670
Palermo, 10
Pernambuco, 10
Pasages, 5
Philadelphia, 27 brls. 8
Portland, M. 600
Port Alegre, 20
Puerto Cabello, 20
Rio Janeiro, 25
San Francisco, 20 bxs 125 dms. 20 brls.
Seville, 50 kgs. 446 dms. 97 brls.
Sydney, 25
Tarragona, 48
Trieste, 15
Vigo, 30
Varna, 10
Venice, 18
Yokohama, 500

Chemicals (*otherwise undescribed*).
Colon, 7 pkgs. £41
Vera Cruz, 5 cks. 30
Lisbon, 18 c. 3 q. 31
Colombo, 2 t. 16 c. 10

Copper Sulphate—
Pasages, 29 t. 19 c 3 q. £527
Ancona, 35 7 645
Antwerp, 10 3 200
Barcelona, 115 8 1 2,74[illegible]
Fiume, 5 [illegible]
Rotterdam, 1 t. [illegible]
Rouen, 34 18 [illegible]
Bordeaux, 92 1 3 1,89[illegible]
Lisbon, 14 19 [illegible]
Sydney, 4 8[illegible]
Tarragona, 8 2 14[illegible]
Trieste, 5 3 2 10[illegible]
Venice, 75 5 1,245
Leghorn, 1 13 1 30
Genoa, 37 5 1 698
San Sebastian 16 2 1 304
Nantes, 53 5 969

Carbolic Acid—
Rotterdam, 18 t. 10 c. £2,536
Vera Cruz, 1 7
Montreal, 5 1 q. 19
Calcutta, 4 16
New York, 2 3 116

Copper Precipitate—
Rouen, 9 c. £14

Copperas—
Nantes, 5 t. 10 c. £115
Calcutta, 43 4 2 q. 76
Bombay, 12 17 2 29
Constantinople, 2 9 10

Charcoal—
Nantes, 203 t. 13 c. £820

Citric Acid—
Havana, 1 c. £8
Montreal, 10 7

Disinfectants—
Callao, 40 glns. £15

Epsom Salts—
Catania, 16 brls.

Glycerine—
Hamburg, 5 t. 11 c. £140

Magnesia—
Genoa, 2 cs.
Havana, 12

Manure—
Grand Canary, 1 t 19 c. £9
Valencia, 10 30

Magnesium Chloride—
Bombay, 10 t £25

Nitric Acid—
Santos, 5 cbys. £20

Pitch—
Genoa, 1,605 t.
Marseilles, 1,401
Sierra Leone, 30 kgs.

Potassium Chlorate—
Genoa, 1 t. £42
Lisbon, 2 10 c. 145
New York, 15 695
Rotterdam, 5 241
Hiogo, 5 210
Terneuzen, 1 46
Philadelphia, 10 56
Boston, 23 17 600
Mexico, 2 5
Alexandria, 1 49
Rio Janeiro, 3 163
Gothenburg, 9 13 481
Hong Kong, 1 10 70
M. Video, 3 2 q. 11
Bari, 1 41

Phosphorus—
Kingston, 2 q. £7
San Juan, 5 c. 1 107
Vera Cruz, 15 100
Buenos Ayres, 1 t. 19 c. 115

Picric Acid—
Boston, 5 c. £28

Potashes—
Antwerp, 1 t. 15 c. 2 q. £38

Potassium Bichromate—
Havana, 4 c. 2 q. £7

Saltpetre—
Sydney, 12 t. 10 c. £275
Maceio, 1 12 40
Maranham, 21 1 q. 53
Paita, 16 18
Las Palmas, 2 3 2 42

Sodium Sulphate—
San Francisco, 13 c. 1 q. £5
Catania, 2 t. 7
Antwerp, 10 brls. 11

Sulphur—
Boston, 73 t. 8 c. £332
Portland, 35 5 3 q. 170
Madras, 4 19 31
Rio de Janeiro, 9 6 59
Bahia, 1 7
Iquique, 29 4 3 172
Monte Video, 76 16 388

Sodium Biborate—
Rotterdam, 10 c. 1 q. £16
Havre, 1 t. 2 2 34
Bordeaux, 3 2 2 94

Tar—
Habana, 10 brls.
Maranham, 5
Calcutta, 50

Tannic Acid—
Rouen, 7 c. £42
Magog, 4 2 q. 32
Vera Cruz, 1 cs. 28

Tartar—
New York, 12 c. £1,140

Zinc Muriate—
Bombay, 5 t. 19 c. £60

Zinc Chloride—
Rotterdam, 5 t. 4 c £80
Bombay, 10 8

Zinc Oxide—
Maranham, 8 c. £10

GLASGOW.

Week ending January 9th.

Ammonia Sulphate—
Antwerp, 30 t. £360
Hamburg, 10 132

Ammonia—
Buenos Ayres, 1 t. 2 c. £36

Bones—
Antwerp, 10 t. £32

Bleaching Powder—
Calcutta, 5 t. 4 c. £40
Melbourne, 6 40

Benzol—
Rotterdam, 4 drms. 30 pns. £320

Caustic Soda—
Japan, 29 t. £205

Litharge—
Antwerp, a quantity

Manure—
Antigua, 22 t. 19 c. £151

Manganese—
Rotterdam, 3 cks. £12

Naphtha— S
Rouen, 19 t. £342
" 86

Potassium Bichromate—
Japan, 38 t. 9 c. £1,300
Antwerp, 23 10 721
Italy, 9 9 273
Barcelona, 16 12 600
Rouen, 9 6 260
Boston, 3 13 140
Bergen, 1 ck. 14
Amsterdam, 12 175

Paraffin Wax—
Antwerp, 4 t. 1 c. £91
Italy, 11 10 251
Cape Colony, 6 3 125
Bordeaux, 6 11 157
Barcelona, 34 894

Pitch—
Rochfort, a quantity
Amsterdam, 99 t. £30
Calcutta, 3 30
Rouen, 130 130
Bordeaux, 350 350

Potassium Prussiate—
Italy, 2 t. 8 c. £132
New York, 19 £76

Salt—
Brisbane, 150 t.

Sulphuric Acid—
Bombay, 7 t. 10 c. £37

Sodium Bichromate—
Italy, 19 t. 5 c. £501
Rouen, 10 9 325
Boston, 3 15 100

Sulphur—
Madras, 7 t. 13 c. £42

Salammoniac—
Melbourne, 2 t. 5 c. £73

Soda Ash—
Boston, 4 t. 16 c. £30

Tar—
Singapore, 31 t. £105
Madras, 5 18 c. 21
Cacutta, 14 18 47

HULL.

Week ending Jan. 7th.

Alkali—
Venice, 66 cks.

Benzol—
Rotterdam, 24 cks.

Chemicals *(otherwise undescribed)*
Amsterdam, 25 cks.
Antwerp, 6 cs. 28 cks. 19 pkgs.
Bergen, 10
Bombay, 5 2,208 5
Copenhagen, 40
Christiania, 4 2 20
Dunkirk, 1
Genoa, 18 25
Ghent, 2
Gothenburg, 40 74 25
Hamburg, 99 22 6
Harlingen, 57
Naples, 10
Reval 1
Rotterdam, 1 50
Stettin, 10
Trieste, 22
Venice, 35

Caustic Soda—
Bergen, 4 drms.

Dyestuffs *(otherwise undescribed)*
Antwerp, 1 pkg.
Bombay, 182 cks.
Genoa, 50 t.
Hamburg, 2 drms. 5 cks.
Reval, 1 cs.
Rotterdam, 1 pkg. 16

Horn Piths—
Hamburg, 54 pkgs.

Manure—
Hamburg, 458 t

Naphtha—
Dunkirk, 26 cks.

Pitch—
Antwerp, 522 t. 20 bls.
Dunkirk, 125
Genoa, 214

TYNE.

Week ending January 8th.

Alum Lump—
San Francisco, 20 t. 11 c.

Alum Cake—
Copenhagen, 12 t. 8 c.

Alkali—
Rotterdam, 5 t. 12 c.
New York, 50 3
Rotterdam, 1 7
" 1

Ammonia Sulphate—
Rotterdam, 20 t.

Antimony—
Rotterdam, 10 t.
Hamburg, 1 4 c.
New York, 31
Barcelona, 12
Rotterdam, 5

Aluminium—
Hamburg, 1 cs.

Alum—
Rotterdam, 10 t. 6 c.

Bleaching Powder—
San Francisco, 30 t. 4 c.
Antwerp, 49 18
Copenhagen, 32 9
Antwerp, 31 4
Rotterdam, 13 13
Antwerp, 9 2
Rotterdam, 10 6
Antwerp, 28 4
Rotterdam, 6 12
Hamburg, 1 19
Christiania, 33 11
Dunkirk, 5
New York, 235 t. 12 c.
Rotterdam, 5
" 2 11
" 15 1
Copenhagen, 22 9

Barytes Carbonate—
Hamburg, 9 t. 14 c.
New York, 50

Caustic Soda—
Antwerp, 2 t.
Hamburg, 2 16 c.
Christiania, 9 11
Barcelona, 33
New York, 101
" 15 7
Rotterdam, 2

Gypsum—
New York, 105 t. 4 c.

Glycerine—
Rotterdam, 8 t. 7 c.

Hydrate of Baryta—
Hamburg, 30 t. 10 c.

Litharge—
Copenhagen, 11 c.

Magnesia—
Rotterdam, 1 c.
Barcelona, 6 t. 5
New York, 1 12

Manure—
St. Heliers, 200 t.

Potassium Chlorate—
Rotterdam, 20 c.
Antwerp, 2
" 3 t.
Hamburg, 5
Barcelona, 1 2
Randers, 1
Rotterdam, 5

Plumbago Crucibles—
Aarhuus, 5 c.

Soda Ash—
San Francisco, 53 t. 1 c.
Antwerp, 1 6
Christiania, 5 12
Gothenburg, 46 11
New York, 115 10
Copenhagen, 19 14

Soda—
San Francisco, 109 t. 19 c.
Aarhuus, 24 3
Frederikshavn, 10 8
Copenhagen, 20 5
Rotterdam, 5 1
Svendborg, 28 8
Ergastiria, 10
Aarhuus, 23 10
Aalborg, 47 9
Barcelona, 16 8
Randers, 54 3

Sodium Suphate—
Copenhagen, 14 t. 16 c.

Sodium Bicarbonate
Rotterdam, 10 c.

GRIMSBY.

Week ending Jan. 4th.

Coal Products—
Dieppe, 10 cks.

Chemicals *(otherwise undescribed)*
Dieppe, 24 cks.
" 12 drms.
Hamburg, 26 cks.
Rotterdam, 12

Dyestuffs *(otherwise undescribed)*
Rotterdam, 2 cks.

GOOLE.

Week ending Jan. 1st.

Alkali—
Dunkirk, 20 cks.

Benzol—
Dunkirk, 48 cks.
Hamburg, 18 drms.
Rotterdam, 83 cks.

Coal Products—
Calais, 11 cks.
" 162
Ghent, 4 cs.

Chemicals *(otherwise undescribed)*
Dunkirk, 17 cks.
Rotterdam, 136
Rouen, 50 bgs.

Dyestuffs *(otherwise undescribed)*
Ghent, 15 cks.
" 6 drms.
Rouen, 4 cks.

Manure—
Boulogne, 102 bgs.
Dunkirk, 430
Rotterdam, 296
" 1 ck.

PRICES CURRENT.

THURSDAY, JANUARY 23, 1890.

PREPARED BY HIGGINBOTTOM AND CO., 116, PORTLAND STREET, MANCHESTER.

:e values stated are F.O.R. at maker's works, or at usual ports of shipment in U.K. The price in different localities may vary

		£	s.	d.
:ids:—				
Acetic, 25 % and 40 %	per cwt	7/6 & 12/6		
,, Glacial	,,	2	10	0
Arsenic, S.G., 2000°	,,	0	12	0
Chromic 82 % nett	per lb.	0	0	6½
Fluoric	,,	0	0	3¾
Muriatic (Tower Salts), 30° Tw.	per bottle	0	0	8
,, (Cylinder), 30° Tw.	,,	0	2	11
Nitric 80° Tw.	per lb.	0	0	2
Nitrous	,,	0	0	1¾
Oxalic nett	,,	0	0	3
Picric	,,	0	1	5
Sulphuric (fuming 50 %)	per ton	15	10	0
,, (monohydrate)	,,	5	10	0
,, (Pyrites, 168°)	,,	3	2	6
,, (,, 150°)	,,	1	8	0
,, (free from arsenic, 140/145°)	,,	1	10	0
Sulphurous (solution)	,,	2	15	0
Tannic (pure)	per lb.	0	1	0½
Tartaric	,,	0	1	3
Arsenic, white powered	per ton	12	15	0
Alum (looselump)	,,	4	12	6
Alumina Sulphate (pure)	,,	5	0	0
,, ,, (medium qualities)	,,	4	10	0
Aluminium	per lb.	0	15	0
Ammonia, ·880=28°	per lb.	0	0	3
,, =24°	,,	0	0	1⅞
,, Carbonate nett	,,	0	0	3½
,, Muriate	per ton	24	0	0
,, ,, (sal-ammoniac) 1st & 2nd	per cwt.	36/-& 34/-		
,, Nitrate	per ton	37	10	0
,, Phosphate	per lb.	0	0	10½
,, Sulphate (grey), London	per ton	12	0	0
,, ,, (grey), Hull	,,	11	17	6
Aniline Oil (pure)	per lb.	0	0	11½
Aniline Salt ,,	,,	0	0	11
Antimony	per ton	75	0	0
,, (tartar emetic)	per lb.	0	1	1
,, (golden sulphide)	,,	0	0	10
Barium Chloride	per ton	7	0	0
,, Carbonate (native)	,,	3	15	0
,, Sulphate (native levigated)	,,	45/- to 75/-		
Bleaching Powder, 35 %	,,	5	2	6
,, Liquor, 7 %	,,	2	10	0
Bisulphide of Carbon	,,	12	15	0
Chromium Acetate (crystal	per lb.	0	0	6
Calcium Chloride	per ton	2	0	0
China Clay (at Runcorn) in bulk	,,	16/- to 32/-		
Coal Tar Dyes:—				
Alizarine (artificial) 20 %	per lb.	0	0	9
Magenta	,,	0	3	9
Coal Tar Products				
Anthracene, 30 % A, f.o.b. London	per unit per cwt.	0	1	6
Benzol, 90 % nominal	per gallon	0	4	0
,, 50/90 ,,	,,	0	3	3
Carbolic Acid (crystallised 35°)	per lb.	0	0	11
,, (crude 60°)	per gallon	0	2	10
Creosote (ordinary)	,,	0	0	2½
,, (filtered for Lucigen light)	,,	0	0	3
Crude Naphtha 30 % @ 120° C.	,,	0	1	2
Grease Oils, 22° Tw.	per ton	3	0	0
Pitch, f.o.b. Liverpool or Garston	,,	1	12	0
Solvent Naphtha, 90 % at 160°	per gallon	0	1	6
Coke-oven Oils (crude)	,,	0	0	2½
Copper (Chili Bars)	per ton	49	0	0
,, Sulphate	,,	24	0	0
,, Oxide (copper scales)	,,	52	10	0
Glycerine (crude)	,,	30	0	0
,, (distilled S.G. 1250°)	,,	55	0	0
Iodine	nett, per oz.	8d. to 9d.		
Iron Sulphate (copperas)	per ton	1	10	0
,, Sulphide (antimony slag)	,,	2	0	0
Lead (sheet) for cash	,,	16	0	0
,, Litharge Flake (ex ship)	,,	17	10	0
,, Acetate (white ,,)	,,	23	10	0
,, ,, (brown ,,)	,,	19	0	0
,, Carbonate (white lead) pure	,,	19	0	0
,, Nitrate	,,	22	5	0

		£	s.	d.
Lime, Acetate (brown)	,,	8	5	0
,, ,, (grey)	per ton	14	0	0
Magnesium (ribbon and wire)	per oz.	0	2	3
,, Chloride (ex ship)	per ton	2	7	6
,, Carbonate	per cwt.	1	17	6
,, Hydrate	,,	0	10	0
,, Sulphate (Epsom Salts)	per ton	2	15	0
Manganese, Sulphate	,,	18	0	0
,, Borate (1st and 2nd)	per cwt.	60/- & 42/6		
,, Ore, 70 %	per ton	4	5	0
Methylated Spirit, 61° O.P.	per gallon	0	2	2
Naphtha (Wood), Solvent	,,	0	3	9
,, ,, Miscible, 60° O.P.	,,	0	4	0
Oils:—				
Cotton-seed	per ton	22	10	0
Linseed	,,	22	10	0
Lubricating, Scotch, 890°—895°	,,	7	5	0
Petroleum, Russian	per gallon	0	0	5¾
,, American	,,	0	0	6⅛
Potassium (metal)	per oz.	0	3	10
,, Bichromate	per lb.	0	0	4
,, Carbonate, 90% (ex ship)	per ton	18	5	0
,, Chlorate	per lb.	0	0	4¾
,, Cyanide, 98%	,,	0	2	0
,, Hydrate (Caustic Potash) 80/85 %	per ton	23	0	0
,, ,, (Caustic Potash) 75/80 %	,,	21	15	0
,, ,, (Caustic Potash) 70/75 %	,,	20	15	0
,, Nitrate (refined)	per cwt.	1	3	6
,, Permanganate	,,	4	5	0
Prussiate Yellow	per lb.	0	0	9¾
,, Sulphate, 90 %	per ton	9	10	0
,, Muriate, 80 %	,,	7	15	0
Silver (metal)	per oz.	0	3	8⅛
,, Nitrate	,,	0	2	5½
Sodium (metal)	per lb.	0	4	0
,, Carb. (refined Soda-ash) 48 %	per ton	5	5	0
,, ,, (Caustic Soda-ash) 48 %	,,	4	2	6
,, ,, (Carb. Soda-ash) 48%	,,	4	15	0
,, ,, (Carb. Soda-ash) 58 %	,,	5	12	6
,, ,, (Soda Crystals)	,,	2	13	6
,, Acetate (ex-ship)	,,	16	0	0
,, Arseniate, 45 %	,,	10	0	0
,, Chlorate	per lb.	0	0	6¾
,, Borate (Borax)	nett, per ton.	29	10	0
,, Bichromate	per lb.	0	0	3
,, Hydrate (77 % Caustic Soda)	per ton.	9	0	0
,, ,, (74 % Caustic Soda)	,,	8	10	0
,, ,, (70 % Caustic Soda)	,,	7	17	6
,, ,, (60 % Caustic Soda, white)	,,	6	17	6
,, ,, (60 % Caustic Soda, cream)	,,	6	10	0
,, Bicarbonate	,,	5	0	0
,, Hyposulphite	,,	5	0	0
,, Manganate, 25%	,,	8	10	0
,, Nitrate (95 %) ex-ship Liverpool	per cwt.	0	8	3
,, Nitrite, 98 %	per ton	27	10	0
,, Phosphate	,,	15	15	0
,, Prussiate	per lb.	0	0	7
,, Silicate (glass)	per ton	5	7	6
,, ,, (liquid, 100° Tw.)	,,	3	17	6
,, Stannate, 40 %	per cwt.	2	0	0
,, Sulphate (Salt-cake)	per ton.	1	6	0
,, ,, (Glauber's Salts)	,,	1	10	0
,, Sulphide	,,	7	15	0
,, Sulphite	,,	5	5	0
Strontium Hydrate, 98 %	,,	8	0	0
Sulphocyanide Ammonium, 95 %	per lb.	0	0	7¾
,, Barium, 95 %	,	0	0	5½
,, Potassium	,,	0	0	7¾
Sulphur (Flowers)	per ton.	7	0	0
,, (Roll Brimstone)	,,	6	0	0
,, Brimstone: Best Quality	,,	4	15	0
Superphosphate of Lime (26 %)	,,	2	10	0
Tallow	,,	25	10	0
Tin (English Ingots)	,,	99	0	0
,, Crystals	per lb.	0	0	6¾
Zinc (Spelter	per ton.	24	10	0
,, Chloride (solution, 96° Tw.	,	6	0	0

THE

CHEMICAL TRADE JOUR

Publishing Offices: 32, *BLACKFRIARS STREET, MANCHESTER.*

No. 141. **SATURDAY, FEBRUARY 1, 1890.**

Contents.

Notices.

All communications for the *Chemical Trade Journal* should be addressed, and Cheques and Post Office Orders made payable to—

DAVIS BROS., 32, Blackfriars Street, MANCHESTER.

Our Registered Telegraphic Address is—

"Expert, Manchester."

The Terms of Subscription, commencing at any date, to the *Chemical Trade Journal*,—payable in advance,—including postage to any part of the world, are as follow:—

Yearly (52 numbers) .... **12s. 6d.**
Half-Yearly (26 numbers) .... **6s. 6d.**
Quarterly (13 numbers) .... **3s. 6d.**

Readers will oblige by making their remittances for subscriptions by Postal or Post Office Order, crossed.

Communications for the Editor, if intended for insertion in the current week's issue, should reach the office not later than **Tuesday Morning.**

Articles, reports, and correspondence on all matters of interest to the Chemical and allied industries, home and foreign, are solicited. Correspondents should condense their matter as much as possible, write on one side only of the paper, and in all cases give their names and addresses, not necessarily for publication. Sketches should be sent on separate sheets.

We cannot undertake to return rejected manuscripts or drawings, unless accompanied by a stamped directed envelope.

Readers are invited to forward items of intelligence, or cuttings from local newspapers, of interest to the trades concerned.

As it is one of the special features of the *Chemical Trade Journal* to give the earliest information respecting new processes, improvements, inventions, etc., bearing upon the Chemical and allied industries, or which may be of interest to our readers, the Editor invites particulars of such—when in working order—from the originators; and if the subject is deemed of sufficient importance, an expert will visit and report upon the same in the columns of the *Journal*. There is no fee required for visits of this kind.

We shall esteem it a favour if any of our readers, in making inquiries of, or opening accounts with advertisers in this paper, will kindly mention the *Chemical Trade Journal* as the source of their information.

Advertisements intended for insertion in the current week's issue, should reach the office by **Wednesday morning** at the latest.

Advertisements.

Prepaid Advertisements of Situations Vacant, Premises on Sale or To be Let, Miscellaneous Wants, and Specific Articles for Sale by Private Contract, are inserted in the *Chemical Trade Journal* at the following rates:—

30 Words and under.......... 2s. 0d. per insertion.
Each additional 10 words .... 0s. 6d. ,,

Advertisements of Situations Wanted are inserted at one-half the above rates when prepaid, viz:—

30 Words and under.......... 1s. 0d. per insertion.
Each additional 10 words .... 0s. 3d. ,,

Trade Advertisements, Announcements in the Directory Columns, and all Advertisements not prepaid, are charged at the Tariff rates which will be forwarded on application.

THE GOVERNMENT TAX-GAT WAR PATH.

WE have often stated in these colum enemy to the expansion of Eng merce is the Englishman. By this we do Englishman, though even he often spite his face, but the species of the gen particularly refer is the official Britisher Government tax-gatherer. The chemi enough, and more than enough, to put u the result of being legislated for by a set sons, totally ignorant of the ways and patent laws have been designed, ostensi of protecting the native inventor, but stopped English progress and placed mc of the foreigner; our spirit laws have be the use of alcohol in home industries that and have been driven abroad. In this la trade, it is a cause for chagrin to find tl have been and are created by the Goveri who should, whatever else may happen, to place the interests of their own countr of departmental aims and ends. The chie ment departments should be men of und who can swallow the gnat, even if they h occasional camel, and who, moreover, w much of that intermeddling so natural to whose only chance of promotion lies having at some period of their existence d nest.

The Customs are, perhaps, more sinnec ning, but the Inland Revenue department close its zeal to the world by playing th and interfering with native manufactur warrantable manner, and in a way not believe, by the framers of the Acts which of action of this department.

It will be known to many of our readers been made to render tar distillers liable t imposed upon all alcohol distillers, and th of some meddling officer have even endea of vitriol rectifying retorts into the revenu and some vitriol rectifiers have even paid test, in order to escape further annoyance

There is, no doubt, the intention of th keep a sharp look out upon all appliance sibly be used to produce illicit alcohol, I emphatically, that either a tar still or a r be used to produce alcohol, if employed fc which it is ordinarily intended. If the C of Inland Revenue thinks otherwise, we st to have the "Dew of Ben Nevis" run ove from a mellow old tar still, and send t friends as a drop of tasty "Old Scotch."

But the subject has even a more ridiculous side from the orts which have lately been made to induce sulphate of ımonia makers to pay a tax on their sulphate of ammonia ills. No doubt the exciseman has thought that he has reseen a grand haul here, but in getting the little fishes to his net he must take care that the big ones do not burst ie meshes and let all out together. A sulphate of ammonia till is about as unlike an apparatus contemplated in the ıct as anything can possibly be. In it a true distillation oes not take place, but by means of a rapid current of team the ammonia gas, the carbonic acid, and the sulphurıtted hydrogen are driven out of the liquor undergoing treatnent, and are passed through oil of vitriol to arrest the ımmonia and allow the foul gases to escape. But the lynx ye of the exciseman sees a similarity, and so he "goes for" the poor manufacturer.

What is distillation? Nothing more nor less than the process of evaporation in a closed vessel, so that to be consistent the exciseman should pounce down upon all owners of kettles and saucepans, to say nothing of the vacuum boilers of the sugar refiners and the drying ovens of a multitude of other trades.

But we will now give the Board of Inland Revenue a line which they have hitherto neglected, and which may bring some grist to their mill, for which they seem ever on the alert. They have shamefully neglected what appears to us, looking at the comic side of the question, a source of revenue, easy to collect, extensive in its ramifications, and with a *locale* easily identified—we refer to the steam-boiler. What is this but a huge still; in it the evaporation is carried on without contact with the air, the distillate is collected and utilised; what more could the exciseman want to prove his case? Why not make the owner of a steam-boiler pay a still-licence and throw his works open to the prying inspection of a troop of tub-gaugers. Suppose some day an enterprising steam user were to put fermented mash in his boiler and work his engines with the weak alcohol vapour distilled over, finally condensing the exhaust alcohol and so defrauding the revenue, what a serious loss it would be to the department!

We hear the official mind exclaiming—Oh! that can never be. It is quite as probable as that whisky will ever be produced from a tar still, or that "fine Hollands" will be made in a sulphate of ammonia still, and we hope that all sulphate makers and tar distillers will combine and stand solid against the innovation of the still tax.

THE DEMAND FOR ALUMINUM.

THE growing favour with which aluminum is regarded is shown by the rapidly-increasing demand for the new metal. This demand has grown to such an extent that the producing capacity has become inadequate, and manufacturers are being pushed. The Pittsburgh Reduction Company has, since its recent cut in prices, been far behind its orders, and has decided on a step which will insure a temporary relief at least from the pressure.

On Monday they placed orders for three 208 h.p. Babcock and Wilcox boilers, two 250 h.p. Westinghouse Compound engines, and two 125,000 light dynamos.

The dynamos are to be used in the separation of the metal from the clay, and were ordered from the United States Electric Light Company. This step is the initial one in a movement to increase the output of their plant to six times the present amount. Their capacity will then be two or three times larger than that of any other plant in world. They will at once begin the erection of an iron building, 70 by 120 feet, adjoining their present works on Smallman-street, between 32nd and 33rd, and will have the new works in operation inside of 90 days. The urgent necessity of the step on the part of the company is readily understood when it is stated that for the month of November orders equalling twice the present output were received, and the same is true of December.

SUPPLY AND USE OF NICKEL AND ITS ALLOYS WITH STEEL.

The most interesting paper and the most instructive excursion of the recent meeting of the American Institute of Mining Engineers at Ottawa related to the Sudbury, Ontario, copper-nickel deposits. The paper was read by Dr. E. D. Peters, manager of the Canadian Copper Company at Sudbury, and it covered an exhaustive description of the deposits, which were originally thought to be of such importance as a source of copper supply that apprehension was felt in some quarters that they would affect the price of the metal. Such, so far at least, has not been the case, though the workings have proved immense bodies of nickel-bearing pyrrhotite, with occasional pockets of copper pyrites. In places this bed has been proved to be 100 feet thick, and its limits have not yet been ascertained. The three mines in the district belonging to the Canadian Copper Company are not uniform in character, and vary considerably in the amount of nickel contained in the ore. The Stobie mine, which possesses the largest bodies of ore, and is worked by open cast, as much as 560 tons being thrown down by one blast recently, is low in nickel, but is valuable from its iron contents, after roasting, as a flux for the oars containing a higher percentage of nickel and copper, but more mixed with gangue.

Mining on this system means cheap production, and we can quite believe that Dr. Peters is correct in his estimate that he can produce from this mine 80 tons a day, at 30 to 35 cents a ton. In the Copper Clift mine the ore occurs in irregular masses, but is very rich in nickel, and large bodies are developed, carrying from 8 to 10 per cent. in that metal. The Evans mine also has a large body of pyrrhotite, but is more highly nickeliferous than the Stobie. This mine produces about 60 tons of first-class ore a day.

The roasting and smelting arrangements, as might be expected under Dr. Peters' management, are models of ingenuity and efficiency, and the result is that about 40 tons of matte are produced a day, averaging about 27 per cent. of copper and 15 to 18 per cent. of nickel. The furnace work is worth recording, one smelter averaging for months of continuous work 125 tons of ore for 24 hours, and having gone as high as 156 tons. Fuel seems to be the only disadvantage, Connellsville coke being used at the somewhat high cost of about 7 dols. 25 cents a ton, but against this is to be set the judicious handling of the ore and its fluxing qualities, which enables the fuel to carry a burden of 8 to 1.

The result of these operations at Sudbury will be an enormous increase in the world's supply of nickel. The supply hitherto has been principally from the mines of the French Company in New Caledonia, and this supply has been regulated to a great extent by the demand, at about 1,000 tons a year, maintaining the price at what the company considered a profitable basis, or rather as high a figure as it could without decreasing consumption, for it has never shown very great profits. The Sudbury production already exceeds the world's consumption, and Dr. Peters has no doubt that he can produce 2,000 tons of nickel a year.

The important question is, will there be a market for this increased supply of the metal even at considerably lower prices than those at present ruling? Mr. James Riley, the well-known metallurgist and manager of the Steel Company of Scotland, in a paper prepared by him at the request of the Council of the Iron and Steel Institute of Great Britain on tests made by him of alloys of nickel and steel, furnishes data which convince him that there will be such a market. It appears that in France a patent has been taken out for these alloys, and Mr. Riley visited the works at which the process was carried on, and continued his tests at his own works in Scotland with most remarkable and satisfactory results. His data, as usual, are clear, and the results are conclusive, although, as he says himself, several series of tests involving a very large number of separate experiments are necessary to a full investigation. We have not space to give in detail here the actual tests carried out, but some of the conclusions arrived at will be sufficient for our steel makers to appreciate the importance of the subject.

The alloy can be made in any good open-hearth furnace working at a fairly good heat. The charge can be made in as short a time as an ordinary "scrap" charge of steel—say about seven hours. Its working demands no extraordinary care; in fact, not so much as is required in working many other kinds of charges, the composition being easily and definitely controlled. If the charge is properly worked, nearly all the nickel will be found in the steel—almost none is lost in the slag, in this respect being widely different from charges of chrome steel. Any scrap produced in the subsequent operations of hammering, rolling, shearing, etc., can be re-melted in making another charge without loss of nickel.

The addition of 4·7 per cent. of nickel raises the elastic limit from 16 up to 28 tons and breaking strain from 30 up to 40 tons per square inch without impairing the elongation or contraction of area to any noticeable extent. With only 3 per cent. of nickel somewhat similar

OZOKERITE.

THIS peculiar mineral has for some years past occupied a very important position in its useful application to the improvement and preservation of insulated conductors. Being an insulator itself of a high quality, it readily lends itself, from its wax-like nature, to a combination with other insulators, or with textile fabrics.

Hitherto it has not been found in any considerable quantities, except in Galicia and Moldavia, in Austria, from which places our supplies have been received; but latterly an important discovery has been made of some mines in Utah, in the United States, which are now being opened out with very great success. The supplies, which hitherto were only obtainable from Austria, will now, fortunately for America, be obtained there. This is a favourable circumstance, as very large quantities of ozokerite are used in the States for purposes other than those electrical. It is employed for the numerous needs for which "wax" is required, and in the electrical industry it is used either by itself, or in combination with other materials. Certain well-known insulating products rest upon ozokerite as their principal support.

This mineral, which is fairly well-known, is obtained by mining, but from the careless character of these operations its production has been attended with some risk to life. Small quantities have been found in the United Kingdom, but our supplies commercially have come from Austria, and the shipments to this country have been very large, as ozokerite has been extensively used in what may be termed the "wax," as well as the electrical, industry.

Ozokerite, or "ozokerit," as it is indifferently spelt, is composed of about—

Hydrogen	13·75
Carbon	86·25

And it melts at a rather low temperature, but slightly higher than "paraffin" wax, to which, in many ways, it bears a faint relation. The character of the mineral slightly varies, and its quality or degree of hardness undergoes variations according to the different methods of purifying and refining. These qualities differ to the extent that the melting point extends from as low as 140 degs. to about 170 degs., which is about the finest quality that is produced.

The fact that ozokerite was an insulating substance was known soon after its introduction commercially into this country, for we find, by referring to our patent records (where the history of many a subject may be traced), that a patent was taken out on the 31st December, 1869, by Augustus Matthiessen, for an "insulating substance for the covering of electric telegraph conducting wires, and for electric telegraph apparatus and insulators." This invention, we learn, consists in the application of "ozokerite" or "earth wax." "Ozokerite may be applied in its natural state, or its residues or products may be incorporated with gutta-percha, India-rubber, or other known insulating substances, with or without the admixture of fibrous substances."

In 1872, on the 29th April, the late Henry Highton applied for a patent—for which provisional protection only was granted—for the application to the conducting wires of cables which are covered with gutta-percha, caoutchouc, or similar materials, of a solution of paraffin, *ozokerite*, or shellac, to fill up the minute pores and improve the insulation.

In 1875, on the 27th May, a patent was applied for and granted to Messrs. Field and Talling, which is interesting, as it relates to the first patent in 1869. "The compounds made according to Specification 3,778-1869, viz., by mixing together ozokerite and gutta-percha, or like substances, by heat, having been found to be brittle and therefore useless as coatings for telegraph wires, the present invention consists in using solvents to dissolve the 'ozokerit' and elastic gums, or in masticating with or without the use of solvents."

In the same year, on November 25th, a patent was taken out by the late W. T. Henley, for improvements in insulating conductors. "The improvement in the method of insulating the conductor consists in curing the India-rubber covered conductors in ozokerite, paraffin, or other similar hydro-carbon, instead of in high-pressure steam." "A double vessel, or steam-jacketted cylinder, is used, the outer vessel or jacket only being charged with steam; the inner vessel is charged with hot hydro-carbon." "According to another plan, the core is cured by high-pressure steam, dried, and treated with ozokerite or paraffin under heat and pressure." In the following year Mr. Henley took out a further patent for passing the insulated conductors through ozokerite. Subsequently, patents were taken out by others, but these have no particular interest.

Previous to this date, and through many subsequent years, the Messrs. Field, who were the sole importers of "ozokerite," were investigating with much skill and trouble, and at considerable expense, the various qualities and peculiarities of this material: and they succeeded in producing the various very fine qualities of ozokerite for electrical and other purposes which are generally so well known. It may be mentioned, however, that they were successful in obtaining a compound of this material and India-rubber, which they termed "nigrite," which has been used for insulating wire in a manner similar to gutta-percha. We are not aware that this wire has come into any extensive use, but that it has been tried and found capable of being manufactured is a matter of fact. We have ourselves tried some experiment with this kind of wire, extending over a very considerable period, and have found it to stand many severe tests, remaining after the lapse of several years in an unaltered condition.

Amongst the users of ozokerite will be found the Postal Telegraph Department, who have for a period exceeding eleven years used it almost entirely for the protection of their underground gutta-percha, insulated wires, as a substitute for the Stockholm tar hitherto almost universally used, but abandoned on account of its action upon the quality of the gutta-percha wire at that time employed. The action of Stockholm tar upon gutta-percha was well known as tending to considerably reduce the insulation, but it had been, and still is, in certain quarters, used in combination with some fabric, as the best preservative for gutta-percha wires underground or in tunnels. The gutta-percha covered wires were protected with tape well-tarred, and with the best quality of material this had been found to answer remarkably well; but about the year 1878, the use of tar was abandoned, and some experiments having been for a length of time tried with ozokerite, it was found suitable, and since that date it has been almost continuously used. The use of ozokerite was tried experimentally several years before tar was finally abandoned. As a preservative by itself it was not found to be of the practical value anticipated, as it did not tend to preserve the fabric it was used in connection with to the extent expected. A certain admixture of Stockholm tar—whose antiseptic qualities have been so well proved—was found to give the desired results, and since the abandonment of pure tar the underground gutta-percha wires of the Post Office have been protected by tape well soaked in a mixture of ozokerite and tar. As a practical result it may be worth while mentioning that the use of ozokerite has been found to improve the insulation.

The preparation of the cables for the underground service of London and provincial towns is an important work, and has invariably been carried out, formerly by the Electric and International Telegraph Company, and subsequently by the Post Office at Gloucester Road Factory. Where the insulated wires have been received direct from the manufacturers, and first having been thoroughly covered with the tape prepared with ozokerite compound, have been made into the necessary cables. The tape used for the purpose is of a special class, and is first passed through a bath of compound and then cooled by passing through water. The compound consists of about three parts by weight of ozokerite to one part of Stockholm tar; it is kept heated in a boiler, and maintained at the lowest temperature at which it can be kept fluid; the water cools the prepared tape, and prevents it sticking on being wound off on to a bobbin. The wires are taped in the ordinary manner; gutta-percha wire thus treated is found to last in a very satisfactory manner. The quality of the ozokerite used by the Post Office is different to the harder sort mostly in use, being softer, and possessing somewhat different characteristics, it melts also at a lower temperature, somewhere about 156° Fahrenheit, and, as may be supposed, mixes thoroughly with the tar.

It will be seen, therefore, from what we have said, that this mineral or "earth-wax" is an important industry, and that a very large supply is in demand not only for our underground wires, but also for our electric light wires, for in their insulation and in their external preservation all cable and wire manufacturers use "ozokerite" very largely. It would be very difficult to supply its place for the manufacture of electric light wires, for whilst ozokerite has many of the qualities of paraffin wax, it has others which that material does not possess. The continuous supply of this material is therefore much to be desired, and the recent discoveries in the United States are consequently welcome.—*Electrical Review.*

Hematite and the Scotch Bears.—The sudden drop in hematite during the past fortnight is attributed to the action of some Glasgow iron merchants who are under engagements to complete old contracts at low prices, and are, therefore, doing their best to bear prices to their benefit. There is no reason whatever that warrant holders should allow this, as the present prices are abnormally low for this class of iron. A fortnight ago, when it touched 82s. 6d., makers were complaining that they could not make it for the money in the face of scarcity of coke, and the advanced price of ore and wages. How much more possible is it for them to manufacture more cheaply now when coke is almost unobtainable, and ore shippers ask more for future deliveries? This movement of the "bears" has resulted in frightening weak men out of their warrants, but it has transpired that there are not many of this class, the bulk of the holders being stronger than the "bears" anticipated. The consequence is that prices are firmer, and under the present condition of affairs hematite is expected by impartial authorities to reach 95s. or 100s.

NATIVE WYOMING SOAP.

BY HERMAN WESTPHAL, PH. G.

THIS mineral, which is known locally under the [illegible] of "Native Wyoming Soap," occurs in the Blue Ridge [illegible]o miles west and south-west of Soudance, Wyoming Territo[illegible] elevation of about 5,000 feet.

It is found in two distinct forms. 1. The [illegible] which occurs in sink holes, or in the neighbourhood of [illegible] covers, probably, an area of several hundred acres. It f[illegible]ns a thick very tenacious pasty mass, about the consistency of butter. Its colour is light yellowish grey. Taste slightly saline, clay-like. Odour argillaceous. Moistened between the fingers with a little water it feels like soap or some greasy substance, hence its name. This soapy feel is probably due to the extreme fineness of the silicates which it contains. In hot weather the edges of the sink holes become hard and brittle, and on some places show an efflorescence of fine crystals of magnesium sulphate.

These holes, which seem to be almost bottomless, at times become very annoying to the ranchers, as cattle frequently get into them, and unless discovered in time and pulled out are sure to perish, as it is impossible for them to free themselves.

2. The dry variety occurs underground in veins like coal. It is hard and dry, and looks very much like chalk; the colour is somewhat darker, varying from yellowish white to dirty greenish yellow. On addition of a little water, however, it is converted into the soft variety as found on the surface of the earth. It appears probable that the wet variety is formed by springs running over beds of dry soap, washing it up and in time accumulating it in large quantities.

It is used by cowboys and ranchers who live in the vicinity as a substitute for soap, and for removing grease by absorption. They also use it for making "hard water."

Five grams of the soap exposed to the atmosphere for several weeks, at a temperature of about 24° C. lost 41·20 % of moisture. It had become a very hard and brittle mass, varying from dirty white, greyish green to orange yellow. Taste and odour were unaltered; it adhered to the tongue, and when cut with a knife exhibited a very smooth and shining surface. On ignition the wet soap lost 53·30 %, while the air-dry soap lost 12·10 %, corresponding to 41·20 % on exposure. The air-dry soap reduced to an impalpable powder gave the following composition on analysis:—

	Per cent.
SiO_2	61·08
Fe_2O_3	3·91
Al_2O_3	17·12
MnO	traces.
CaO	2·76
MgO	1·82
Na_2O	0·20
SO_3	0·88
H_2O	12·10
Total	99·87

—*Am. Jour. Pharm.*

OLIVE OIL AND OLIVES.

POSSIBLY no food product was more extensively shown at the Paris Exhibition than olives and olive oil. In the French official catalogue 606 exhibitors of olive oil are specially named, besides numerous collective exhibits, and many others which are included under the general term "comestible" or edible oils; 448 of these exhibitors are from Portugal, 128 from California, and only 12 from France. One French exhibit, however, is made by 67 associated producers.

The Mediterranean has from time immemorial been the seat of the olive culture, and according to the *Journal of the Society of Arts* Spain has about 3,000,000 acres under olives, Italy 2,250,000, and France about 330,000. Tunis has over 4,000,000 trees, Algeria 3,000,000, Nice 1,000,000, where olive oil forms four-fifths of the agricultural produce, and Syria several million. The number of trees in other countries is unknown. Tuscany first exported olive oil, hence its old name "Florence oil."

Forty-five distinct species of the olive tree have been described, and in countries where it is indigenous the tree sometimes reaches a height of 60 feet with a trunk circumference of 12 feet. Besides the difference in the nature of the wood, foliage and habit of growth, there are large olives and small olives, pointed, oval, round and curved fruit, and of all colours, ranging from white to black and from green to red. The flavour of the fruit is mild, sharp, or bitter, and according to the variety there is obtained sweet oil light coloured and of exquisite flavour, up to dark green, thick and of a bitter taste, strong and very unpleasant to the taste. Hence it follows that olive oil may be perfectly pure and also quite unfit for food or culinary purposes, only suitable for greasing machines and making soap.

...E FOR GUM ARABIC.

...acacia has led Trojanowsky to seek for a ...substitute, and this he believes may be found ...seed. By boiling the seed with water and pre...decoction with twice its volume of alcohol, he ... which, after drying, consisted of opaque, ...fragments, somewhat brittle, but not easily ...dissolving in water to a turbid mucilaginous solu...were sufficient to emulsionize an ounce of cod ...quantity of alcohol, however, required for the ...difficulty of drying the adhesive product being ... further experiments were made, and, by still ... the source of the mucilage, and treating with ... more closely resembling acacia was obtained. ...one part of flaxseed with eight of dilute sulphuric ... of water, until the mixture, which at first ...fluid; this is then strained through muslin, ...liquid is added four times its volume of strong ... being collected on a filter, washed with alcohol ... is in the form of translucent, grayish-brown, ...y pulverised, and without odour or taste, and ...ise an ounce of cod liver oil.

...EAD, AND SPELTER IN 1889.

...metal markets for 1889, the *Engineering and* ...states that the prospects for the copper trade must ...dly encouraging. It may be that consumption ... present enormous rate, but everything indicates ... 1890. Some increase in production is probable ...Michigan, but no anxiety need be felt in this ...emand will absorb any likely increase. Neither ...spectre of the warehoused stocks, which exerted ...ence on the market in the early part of 1889. ...is still piled up in warehouse in New York some ...t be forgotten that hardly one of the smelting ... whatever, and our more interested readers will ...e fact that often during the past ten years the ...ompany alone have held about as large a stock ... held of lake copper by the foreign bankers in

...tions favour a continuance of prices quite as high, ...e now ruling for the next year, and this prospect ...nished by the reports of a serious fire in the ... which reliable information cannot at present be ...ust result in some reduction of output. Stocks ... and at present the total stock of copper in this ...0,000 pounds.

...art of this is held by the banks which took over ...yndicate, yet quite an important quantity of this ...port, and is only awaiting shipment.

...e history of the trade have the mining companies ...and were it not for the syndicate's accumulation ...ay be absolutely bare of copper, and a copper ...tened.

...f electrolytic copper steadily increases, and in ... less than 21,000,000 pounds, as compared with ... 1888. The coming year will still further ...oduction.

...PER PRODUCTION BY STATES.

	1888. Pounds.	1889. Pounds.	1889. Long tons.
....	86,584,124	86,000,000	38,393
....	34,497,300	32,000,000	14,286
....	97,897,968	105,000,000	46,875
....	1,631,271	3,400,000	1,518
....	1,570,021	1,700,000	759
....	3,300,000	3,000,000	1,339
....	2,131,047	2,400,000	1,071
......	3,241,725	3,000,000	1,339
...ction..	227,853,456	236,500,000	105,580
...ores ..	5,000,000	5,100,000	2,277
....	232,853,453	241,600,000	107,857
...f year..	40,000,000	75,000,000	33,482
....	272,853,456	316,600,000	141,337
...in ore, ...	78,000,000	82,000,000	36,607
....	119,853,453	169,600,000	75,714
...d of year	75,000,000	65,000,000	29,016

The consumption of lead for pipes used in protecting underground electric wires has assumed an enormous expansion, and the year closes with the smelters in full operation, and the market practically bare of the metal.

With regard to the general question of production and consumption, it would appear that these two factors now nearly balance each other, and there is reason to believe that of the heavy stocks accumulated during the reckless speculation of the previous year not more than 10,000 tons are left. Part of this is still the subject of litigation, and part is strongly held for higher prices. The production of lead in the United States during the last three years is shown in the following table:—

	Arizona and California.	Colorado.	Idaho and Montana.	Mo., Kan. Ill. and Wiscon.
1887	1,000	63,000	27,000	28,000
1888	1,000	65,442	34,875	30,000
1889	1,500	70,000	30,000	34,000
	Nevada.	Utah.	Other States.	Total production.
1887	3,400	22,000	16,300	160,700
1888	2,400	22,283	*30,000	186,000
1889	1,500	22,000	*31,000	190,000

*Including 28,636 tons from Mexico.

*Including 19,000 tons imported in Mexican ores.

The consumption of spelter has steadily increased, and the greater part of such increase may be attributed to the great activity in the galvanizing trade, which has been in a very prosperous condition throughout the whole of the year. So large has been the demand that on several occasions during the year some of the galvanizing works have been compelled to shut down for a time for want of the necessary supplies of metal. Without doubt an increase in demand has also been attributed to the requirements for brass making, and the manufacture of sheets was also very good up to the last six weeks or so, when the prices for the manufactured article did not respond to the advance in raw material.

Although spelter prices in this country may now be regarded as at a moderately high level, they are still comparatively much lower than values now ruling in Europe, where this metal has been exceedingly scarce for some months past. Imports have now almost entirely ceased, and the total quantity of foreign spelter imported during the past year only amounted to about 800 tons, the whole of which was undoubtedly exported again in a manufactured state, by which means the duty originally paid would be returned again as "drawback." Unless all signs prove deceptive, the consumptive demand for spelter is likely to increase still further.

PRODUCTION OF SPELTER IN THE UNITED STATES, 1882 TO 1889, INCLUSIVE, BY STATES.

Year.	Illinois.	Kansas.	Missouri.	Eastern and Southern States.
1882........	18,201	7,366	2,500	5,698
1883........	16,792	9,010	5,730	5,340
1884........	17,594	7,859	5,230	7,861
1885........	19,427	8,502	4,677	8,082
1886........	21,077	8,932	5,870	6,762
1887........	22,279	11,955	8,600	7,446
1888........	22,445	10,442	13,465	9,561
1889........	24,000	11,000	12,500	12,000

AMERICAN v. RUSSIAN KEROSENE.

U. S. Consul Pettus, in a recent report from China, gives the following regarding the consumption of oil in that country:—

One of the most striking features of recent customs returns is the immense and rapid growth in the import of Russian kerosene oil into this country. Indeed, judging by the leaps and bounds by which the new trade has advanced, one might, with reason, conclude that the American oil was giving way before that of the Muscovites. Brother Jonathan had surely needs bestir himself or the oil wells of Pennsylvania and other states will cease to supply this illuminant to the far East, as they have done for years past. It is not hard to explain the reason of the progress of Russian oil here. It is somewhat cheaper than its older rival, and its sources of production at Baku are simply unlimited. There is nothing a Chinaman appreciates more than being able to save the most trifling sum, and it is because of its being a trifle under the cost of American oil that Russian kerosene is finding such demand in Shanghai. The trade is not yet two years old, yet the stock of Russian kerosene held in Shanghai to-day is larger than the other. There are at present some 277,000 cases of Russian oil held here, against 185,000 of American. The prices ruling, as given in Mr. Bielfeld's last report, are somewhat about 1·37½ to 1.40 taels for Russian and 1·50 taels for American. The arrivals from January [illegible] July this year were about 225,000 cases Russian, against 353,000 ca[illegible] American, or, roughly speaking, the American oil is represented [illegible]

:r cent. and the Russian by 40. These figures speak for thems. Five steamers of different nationalities brought full)es of Russian oil from Batoum to Shanghai since the beginning of /ear, and two more are on their way. The carrying capacity of : vessels is enormous, and most of them have been specially built he trade. They fly almost every flag, but principally the German Italian. The Italian steamer Palestro and the German steamer enfels, which arrived here during the past month, brought, respecly, 99,800 and 84,000 cases, the first named being consigned to ssrs. Jardine, Matheson, & Co., and the second to the China and an Trading Company. The Italian steamer's cargo was the biggest signment of oil ever brought by any vessel to Shanghai. Looking k over previous customs returns, we find that the import of American osene has dropped from 5,995,710 gallons in the three months April June, 1886 (when, no doubt, there was an oil boom which many of readers will remember, perhaps regretfully), to 1,857,200 gallons, per last return, while the Russian oil grew from nothing in 1887 to :50,440 gallons for the same period, according to the same authority. this quantity about half of the American and a third of the Russian : re-exported. The first attempt to place Russian oil in the local ırket was made early last year, Messrs. Jardine, Matheson, & Co. ing the pioneers in the trade. They soon had imitators, and now ere are two other large firms engaged in the business, and two aucons of the Russian oil are at present held daily. The Chinese, as they ways are of innovations were at first distrustful, but the superior ıality of the article soon established its reputation, and the prices corspondingly improved. The principal, and, indeed, the only objection hich the Chinese have to the Russian oil is the inferior workmanship f the cases, which give great trouble and entail heavy loss ı leakage, while the American tins are very strong and well nished. We would be sorry to see the American trade crushed y its new rival, but we suppose that it is a contest of the survial of the fittest, or rather the cheapest, in this as in every ther struggle in human affairs. We imagine, too, that with the resent energetic governor of Formosa at the head of affairs we shall ome day hear of the development of the oil deposits of that wonderful sland, and the discovery of the petroleum wells in the interior of Formosa which Mr. Dodd made years ago will prove another source of revenue to His Excellency Liu Ming-Chuan. Mr. Charles Marvin, in that highly-interesting volume, "The Region of Eternal Fire," writing of the oil supply at Baku, which he terms "the real base of Russian operations against India," and, contrasting the resources of the Russian with those of the American oil supply, says that he saw 400 wells around Baku, all at different depths, and therefore apparently springing from independent reservoirs. Some of these wells varied in depth from 259 to 580 feet, while one well which had been worked for generations was only 70 feet below the surface. The famous Droobja fountain well, spouting oil 300 feet high at the rate of 2,000,000 gallons per diem, came from a depth of 574 feet, which shows the enormous natural forces which must have been at work below, while all around were small wells of 300 feet deep, throwing up their spouts of the valuable oil quite unaffected by the giant well of Droobja. Many of the pumpings had been worked from remote times without any apparent diminution of their resources. The whole peninsula of Apsheron is honey-combed with thousands of oil wells, one of which had given 1,500,000 barrels, and yet the pumps were drawing the oil as freely as when it was first tapped years ago. The wells in America are much deeper, and a man thinks nothing of boring 1,000 feet for oil, but 300 feet appears to be the average in the Baku region. In 1883 two flowing wells at Baku threw up 30,000,000 gallons apiece of oil in two months, and they were finally plugged to "cork up" for future use. Nobel Brothers, the Russian oil kings, and rivals of the Devoes, have 14 such gigantic reservoirs corked up, because the crude petroleum will not fetch more than a few pence a ton at Baku, and the deepest of these basins is only 800 feet from the surface, while in the Bradford region in America there are numbers of wells 2,000 or 3,000 feet deep, and one in West Virginia which is over 5,000 feet deep. The pipe line to Batoum, too, is what gives the Russian oil the great pull, while the American has nearly all to be sent, with consequent extra freight, to New York for refinement, whence it is almost exclusively shipped.

MANCHESTER SHIP CANAL.—To the purely commercial reader as to the skilled engineer there are divers ways of giving a true conception of the magnitude of the works at the Manchester Ship Canal. Mr. Leader Williams, the engineer of the great undertaking, is master of all the ways, and not least so of the way befitting a popular audience. Here is a quotation from that gentleman's lecture in the Salford Town Hall:—"There are 11,489 men and boys, 182 horses, 5,900 waggons, and 169 locomotives employed at the Canal works. There have been constructed 213 miles of temporary railway; and 10,000 tons of coal and 8,0000 tons of Portland cememt are used in the excavations each month."

NOTES FROM KUHLOW.

ARTIFICIAL MANURES.—The German agriculturatists have had the disadvantages of the convention among the artifici manufacturers brought home to them in an exceedingly u manner. One of the principal articles affected is Thomas sla steel slag, which a few years ago was regarded as hardly cany , has, since being ground to a dust fineness, become a v valuable manure. To enable farmers to obtain the article a erate price the German Agricultural Society concluded a contract for three years with the Thomas meal manufacturers at the end of 1887. In accordance with this contract the price per 1000 kilos. of 20 per cent. stuff was 310 mks. at Peine. Since the expiration of the contract the price has rapidly risen. In the autumn of 1888 it stood at 380 mks., in the course of the present year it has advanced to 510 mks., and a further increase is said to be in prospect. To obtain increased prices the Thomas slag meal manfacturers during the summer sold some 50,000 tons abroad. Much lower prices had to be accepted for the foreign sales than what was received for the inland ones. It has been proved that in many cases the price of this 17 per cent. export article at the works was only 290 mks., and even 200 mks. per 1,000 kilos. For reasons readily comprehensible it was stipulated that the stuff should not be resold to Germany under penalty of a considerable sum. The manufucturers were thus selling their slag meal to foreign countries at abnormally low prices, seeking by this means to keep up the high prices in the home market. It is natural that German consumers should not regard this state of things with equanimity. In our last issue we mentioned that the persons interested in agriculture were about to adopt vigorous measures of self-defence. At the same time complaints are heard of the advance in the price of phosphoric acid, which result is equally to be attributed to the activity of a coalition of the manufacturers. At the last meeting of the Pomeranian Economical Society at Goeslin, it was shown that in the last year the price of that article had been forced up 80 per cent. The manager of a chemical works endeavoured to prove that the advance was justified by the increased cost of the raw materials and higher wages. Notwithstanding, the resistance against the operations of the manufacturers' coalition remained general. The advice was given to farmers to cease using the article or to restrict its quantity; it was further recommended to purchase phosphoric acid from England and Belgium, in which countries it is considerably cheaper. In such an event foreign competition will certainly be of advantage to German consumers. Attempts have, it is true, been made in the Reichstag to get a protective duty imposed upon artificial manures, but they have always failed. In the interests of a sound state of business it is to be hoped that the attempts of the artificial manure consumers to bring about reasonable prices will be attended with success.

THE IMPORT DUTY ON LUBRICATING MINERAL OILS.—A modification of the duty on lubricating mineral oils is advocated by the Leipsic Chamber of Commerce. This article came in duty free from 1865 to 1879, but after the latter year a duty of 6 mks. per 100 kilos. was imposed, and since 1885 the duty has been 10 mks. per 100 kilos. In passing a judgment upon the oil duties the Chamber of Commerce remarks—"Of what importance these oils are to the whole German industry is seen by the weight upon which duty is paid. In the last three years, notwithstanding the high duty, equalling 50 per cent. of the value, the quantities imported were 307,165, 226,354, and 330,115 dble. ctrs. In the reports received by us the production in the inland of these lubricating oils is described as extremely limited, and as hardly capable of increasing. The increase in the duty seems to have been of little benefit to the rape-seed oil, which is being more and more pressed by the mineral oils, because consumers believe that the latter merit the preference. The Russian lubricating oils are not only used as such, but they are also employed in printers' black manufactories, whereas for that purpose the native oils are not to be used, no matter how cheap they might be. The printers' black manufacturers, whose markets have seriously suffered through the increase in foreign duties, therefore earnestly desire a reduction in the duty on lubricating mineral oils." In the official trade statistics the average value of the lubricating mineral oils in 1888 is estimated at 18·50 mks. per 100 kilos. The existing duty is consequently considerably more than half the value of the article.

THE UPPER SILESIAN ZINC MARKET IN 1889.—The Upper Silesian zinc industry can look back with satisfaction on the second half of 1889. Prices steadily pursued an upward course, and every quantity brought into the market found an immediate purchaser. No such prices as are reached at present have been attained since 1841. The strike agitation exercised no serious influence on the Upper Silesian zinc industry. It was only in a few cases and for a short time that work was suspended in some of the works. At the whole of the works higher wages were conceded, representing an appreciable improvement in the wages conditions. The leading brand, W. H.,

menced at the beginning of January with 18 mks., then rising to 5 mks., and was quoted at the end of February at 18·25 mks. er special brands stood at 17·70—17·90 mks. towards the end of ruary. In March, both actual buyers and speculators were very erved, the tone consequently becoming weaker. After a fall of 75 s.—1 mk. the tendency again became firmer. W. H. was quoted 17·90 mks, "Godulla" 16·70 mks, and the Vereinsmarke 17 mks. The upward direction then went steadily forward to the end of the year, business everywhere being very lively. During the last few months it was everywhere difficult to obtain goods for immediate delivery; the whole of the stocks were cleared out. The sales in the last few months of the year were almost all for 1890 delivery, and some works are even engaged for the second quarter of this year. At the close of the year "Godulla" was quoted at 23 65 mks., and W.H. at 24 mks. per 50 kilos free on rails Breslau; the latter brand could not be had. The advance above the lowest price during the year amounted to 7 mks per 50 kilos. Twenty-three works were running.

At a meeting of the sixteen South German cement manufacturers, held at Frankfort-on-the-Maine, it was decided to increase the price by at least 35 pfgs. per 50 kilos. The increase to come into force on the 1st of January.

Kali Market.—According to the report of the Sales Syndicate of the Kali works the deliveries of chloride of potassium in November amounted to 282,248 ctrs. (against 225,000 ctrs. in 1888). During the first half of this year the deliveries were 1,145,000 ctrs., and from the 1st of July to the end of November 1,149,000 ctrs., so that the whole of the December deliveries may be regarded as a surplus.

German Sugar Industry.—The *Reichsanzeiger* publishes a preliminary review of the results of the beet sugar manufacture in the season 1889-90, from which it appears that in the German Empire up to the 1st of December, in 400 beet sugar manufactories, 65,050,467 dble. ctrs. of beet were consumed, yielding 9,921,516 dble. ctrs. full measure. Probably a further quantity of 30,732,806 dble. ctrs. beet will still be used, so that the total for the season would then be 95,783,273 dble. ctrs., against 78,961,830 dble. ctrs. last season.

Probable Advance in the Price of Type Metal.—It is reported from type founding circles that in consequence of the enormous increase in the price of antimony, a very important metal in type production, caused by its recent large employment for military purposes, a considerable advance in the price of printing materials, will probably take place shortly. As a matter of fact the price of antimony has risen more than 100 per cent. in 10 months, and people in the branch are expecting a further heavy rise.

Advance in the Price of Lime.—The whole of the limekiln proprietors of Upper Silesia have intimated that from the 1st of January the prices will be increased as follows, unslacked piece lime to 44 pfgs., fresh lime ash to 14 pfgs. per 50 kilos. at Gogolin. The reason for this step is the considerably increased expenses for wages and coal. Should coal still further rise, the foregoing prices will be again increased. The difference in the carriage between Gogolin and other stations will be equalised bs reducing the price when there is a higher carriage, and increasing it when the carriage is lower.

Cement Prices.—The Portland Cement-fabrik Hemmoor has issued a circular in which it is stated that in consequence of the enduring and considerable rises in the price of all raw stuffs, especially coal, the company, like the other manufacturers, is compelled to advance its prices for Portland cement, more particularly in view of the lively demand for that article. In the first instance, by reason of some coal contracts that are still running, the advance will only be 25 pfg. (3d.), but very shortly it is to be increased to 50 pfgs. (6d.) per cask of 180 kilos. or per two sacks of 85 kilos. each without packing.

Cement Products.—Mr. A. Kutschbach, cement goods manufacturer, Eutritzscher Strasse, Leipsic, affords surprising evidence in his works of the progress attained in the application of cement for building purposes. As an imitation of sandstone, cement has made great progress, and this fact in itself is a sufficient evidence of its merits in th: particular branch. The numerous new discoveries in the practical ap lication of cement made in the factory in question deserve special att ntion. Among these may be mentioned the cement floor plates, wi ch in their variegated colours testify to the progress made in colouring and when perforated and rimmed are very valuable in works, br weries, paper mills and sidewalks. They are said to have a consic rable sale. Mr. Kutschbach has opened out a new field in his im ation of Rochlitzer sandstone, and also in his coating-stone, which la er had previously been made of clay, and consequently did not by fa possess the endurance of the cement stone.

Thomas' Phosphate Meal.—In the latest number of a German agricultural periodical an appeal is made to the German Agricultural Society, under reference to a report of the Manure Section, in which German farms were requested to restrict their consumption of Thomas meal in 1890, "in order thereby to put a increase of the price brought about by Thomas phosphate meal manufacturers," manure until "the Thomas phosphate meal price at present fixed by the convention of at Wanne, thus at 18—19 pfgs." The abo sumers of Thomas' phosphate meal to the When the moral effect is taken into accou which is signed by the leading German ag on the German and foreign consumption, it Hamburg paper, represent less than a furth ctrs., so that the total diminution for 1890 w ctrs., or half the production of 1889. It is movement is attended with prospects of succ

Coke Production.—The prevailing high a great increase in the production, which is reaction. In France many new coke wor and many more are nearing their comple works that were stopped last year are now ag places new works are even being constructed be observed in Germany. Thus, for instan hundred and sixty new furnaces are being ere structed for the extraction of the accessory pr On all sides the newly constructed furnaces a of the derivatives. This is a noteworthy poin result. A slight difference is caused betwe coke on account of the reduced cost of pro accessory product, sulphate of ammonia, ca and, further, a large and rapid expansion industry is thereby accomplished. The quanti in 1888 amounted to 122,785 tons, of whic was won in the manufacture of coke.

Lutzen.—The sugar manufactory here i Germany; no less than 10,000 ctrs. of beet bei the course of last summer new diffusers we things, and they have turned out excellently. it took half an hour to empty each diffuser, emptied instantaneously. Instead of the empt side, the new diffusers have openings at the locked. When the 70 ctrs. of beet chippings i got out the locks are opened, steam is directed the vessel is immediately emptied.

Horse-Power.—With the dissemination of term horse-power has come to be very largely employment is so frequent and general, very ma or ignorance of what the expression really mea that an engine requires a certain power or forc however, according to the nature of the engine. this power it is necessary to adopt a certain stan goes by the name of horse-power. One horse- seconds-kilogramme-metre, that is to say that in of 1 kilo. can be lifted 1 metre high, or inversel can be lifted 75 metres high. Take, for example impelled by steam, gas, or compressed air, of 6 h above rule such and engine must possess a powe 6×75 kilos.=450 kilos. exactly 1 metre high, o form, a working machine which requires 6 h.p. to be worked by a falling weight of 6×75 kilos.=45 second will have fallen exactly 1 metre. If th second be greater than 1 metre the weight must b raises in 1 second 75 kilos. 6 metres high, or 6 k kilos. 18 metres high, or, putting it in the oth machine is driven by a falling weight which in 1 and weighs 75 kilos., or falls 18 metres with a we whole of these performances would represent 6 power of a man is reckoned at 13—18 kilogramme the work required. In turning a wheel round six as equalling 1 h.p. From this it will be perceive is the employment of mechanical power, even w force is required.

Loch Katrine Water.—The monthly repo Loch Katrine water, prepared by Professor E. J. M Glasgow and West of Scotland Technical Colle The results are returned in parts per 100,000.—T 2·7; organic carbon, ·104; organic nitrogen, ·0 ·003; ammonia, ·000; total combined nitrogen, ·0 chlorine, ·75, Temperature, 6°·0 C.=42°·8 Fah sampled on January 15th. It was light brown in c a little suspended matter. The temperature, hardn matter are unusually high. The organic carbon i

Correspondence.

ANSWERS TO CORRESPONDENTS.

MANUFACTURES.—We are afraid we cannot help you.
D. C.—You must give us fuller information.
K. K.—We are glad to have your independent testimony to the value of the apparatus.
OZONE.—We shall probably have something to say about this before long.
DUBIOUS.—Your question is really too absurd for our serious consideration.
R. B.—You will find an account of it in a recent number of this journal.
J. M.—Your informant is wrong. There *is* a point above 40 per cent. anhydrous at which the fuming acid is liquid at the ordinary temperature.

NATURAL GAS IN INDIANA.

A TOUR through the Indiana gas belt and its pipe line cities and towns shows up several facts worth remembering. First, that the Indiana field is larger than any other yet discovered, and the area of territory fed from it is greater than that supplied from any other natural gas source. Second, judging from the location of the field, the lower initial pressure of the gas and slower consumption of the supply, it is likely to last longer than any other gas territory thus far found. Third, comparing it with any other field thus far developed, it has brought more wealth, a larger access of population and greater comfort and convenience to this commonwealth than have come to all the other states in the union combined from the finding of natural gas inside or outside of their borders.

Four hundred miles of pipe lines have been completed, including those supplying the city of Indianopolis from the Hamilton county field. Of the Indianaopolis lines the Consumer's Trust, with the recent purchase of the Broad Ripple plant, has about seventy-four miles of pipe line, and with its 9,000 connections, supplies about thirty-five thousand people. Its pipe line includes thirty miles of eight-inch, fourteen miles of ten-inch, twelve miles of twelve-inch and eighteen miles of sixteen-inch pipe, and it has about one hundred and fifteen miles of street mains, varying from sixteen inches to three inches in diameter. The Indianopolis Natural Gas Company has thirty-five miles of pipe line, and with 5,000 connections, supplies 20,000 people. Its pipe line includes six miles of six-inch, eight miles of eight-inch and twenty-one miles of twelve-inch pipe, and its street mains extend over sixty miles. The two companies combined supply 55,000 domestic consumers at a saving over former coal bills of 250,000 dols. annually, and up to a comparatively recent period each supplied about one hundred and fifty factories and public buildings, with an annual saving over former coal bills of more than 100,000 dols. At this time the Consumers' Trust Company has shut off its factory supply for the winter. The other completed pipe lines in the state vary in length from five to forty-eight miles, and in diameter from two to twelve inches, the majority of them averaging eight inches. Upward of twenty cities and towns of the state, with an aggregate population of 260,000, are supplied with pipe line service, and if the cities and towns of the state supplied directly from the well are added, the list will be increased to seventy-one cities and towns, with an aggregate population of 411,000 in round numbers.

Placing the calculation on the basis that one-fourth of the population in cities and towns supplied with natural gas are consumers, it would give the State upwards of 100,000 municipal consumers, with an approximate annual saving over former fuel of 3,000,000 dols. The saving to manufacturers in the four cities of Muncie, Anderson, Marion, and Kokomo alone foots up 1,045,000 dols. a year, and if the savings of all the other State manufacturers using gas fuel did not exceed another million it would make up a total of 5,000,000 dols. a year as the amount of the saving effected by the new fuel to the people of the State, saying nothing about the increased comfort and incidental advantages which attend its use.

A good many explanations have been given concerning the difficulties found in forcing natural gas through long pipe lines, and many theories have been put forward based on atmospheric changes; heat, cold, etc. The latter have come to be rejected by practical operators who are engaged in the business. They say that neither heat nor cold in the atmosphere can affect the gas in the ground at the depth from which it is taken; that when moving through the pipe line it carries warmth sufficient to melt snow, if a hole be dug and the snow laid on the outside of the pipe in the coldest weather; and that even if the temperature of the gas were lower than the pipe line, the gas would only suffer a slight condensation that would make it move the faster. Initial pressure and friction in the pipe are the only factors admitted into the calculation.

When, by lengthening the pipe, enough friction is caused to overcome the initial or rock pressure at the first end of the pipe, no gas is delivered at the other end. When the pipe is shortened the gas is delivered at the farther end, with a pressure reduced in exact proportion to the distance travelled and the resistance encountered from friction. Starting with 600 pounds pressure at Sheffield, gas is delivered at Buffalo, ninety miles distant, with a pressure of less than one hundred pounds. Starting with gas at 300 pounds pressure at Chesterfield, Ind., it gets to Richmond, Ind., forty-five miles away with less than one hundred pounds pressure. This is the argume[illegible] the operators, briefly stated, and it looks reasonable. If co[illegible] scheme of carrying Indiana natural gas to Chicago and C[illegible] impossible, except through some device for pumping o[illegible] and this is not believed to be practicable.

CANADIAN OIL STATISTICS.

THE quantity of Canadian refined oil consumed in the Dominion of Canada is ascertained from the amount of Canadian refined oil inspected each month by the Inland Revenue Department at each of the different refineries in Canada, where the oil is manufactured, and as no refined oil can be offered for sale before it is inspected for safety, a perfect record of the consumption can therefore be obtained. The official returns for the first eleven months of last year are therefore as follows:—

1889.	Quantity inspected Gallons refined oil.	Crude equiv. in barrels.
January	729,452	52,003
February	737,523	52,644
March	605,963	43,583
April	361,082	25,792
May	451,142	32,224
June	487,233	33,953
July	633,886	45,278
August	649,604	46,400
September	1,361,069	97,219
October	1,311,752	93,696
November	1,153,476	82,391

The total quantity of Canadian refined oil inspected from the 30th June to the 30th of June in each year by the Inland Revenue Department since 30th June, 1881, and therefore consumed by the people was as follows:—

	Gallons.	Crude equiv.
1882	6,169,353	440,668
1883	7,135,580	509,684
1884	7,836,949	559,782
1885	7,843,043	560,217
1886	8,351,203	595,800
1887	8,436,938	602,638
1888	9,769,265	697,805
1889	9,684,336	691,738

The total quantity of Canadian and American refined oil consumed in Canada each year from 1882 to 1889 has been as follows:—

	Can. oil.	Americ'n oil.	Total.
1882	6,109,353	3,026,186	9,195,539
1883	7,135,589	3,088,414	10,223,994
1884	7,836,949	3,148,920	10,985,869
1885	7,843,033	3,813,379	11,656,412
1886	8,341,203	3,803,724	12,144,927
1887	8,436,938	4,309,397	12,746,335
1888	9,769,265	4,493,924	14,263,189
1889	9,684,336	4,723,698	14,408,034

By the above return it will be seen that there is a steady increase in the consumption of American oil in Canada. Last year the consumption of American oil increased, while that of the Canadian product decreased. America furnishes two-thirds of the total consumption in Canada.

The shipments for 1889 of crude and refined oil over the Grand Trunk and the Michigan Central railways, together with the crude equivalent, are as follows;—

GRAND TRUNK.

	Crude.	Refined.	Crude equivalent.
January	9,035	8,508	34,479
February	8,155	7,333	32,904
March	10,080	6,048	26,849
April	7,939	5,427	24,608
May	8,946	8,007	28,963
June	11,470	8,788	33,440
July	9,900	9,064	32,560
August	7,240	11,011	34,767
September	11,326	19,881	56,528
October	21,272	21,177	81,715
November	17,380	17,888	62,109
December	26,652	18,800	73,552

	MICHIGAN CENTRAL.		
	[illegible]	4,637	17,451
	2,800	5,268	19,644
	5,110	4,175	16,955
	3,290	3,163	9,716
	3,245	3,620	12,865
	3,210	6,215	19,347
	4,890	5,210	17,715
	5,292	6,240	20,992
	[illegible]	8,239	25,518
	4,415	9,195	27,403
	5,405	10,680	31,959
	4,550	8,595	23,037
	TOTAL.		
	12,455	13,145	51,930
	11,965	12,601	52,545
	15,190	10,223	43,745
	11,759	8,590	25,324
	12,761	11,627	32,828
	15,280	15,003	33,787
	11,590	14,274	45,275
	12,632	17,251	46,759
	19,147	28,120	98,046
	25,687	33,372	93,059
	22,795	28,508	83,064
	29,212	26,995	96,689

rtiser, a journal specially devoted to the Canadian mmenting upon the increased consumption of ada, says :—"We have been asked by some of sh again our views about the flash test of our oil, e matter prominently before both the crude and of this place as such a change, we think would e oil interest of Canada, and we invite discussion to bring out the views of the intelligent thinkers question. Some people think the gravity clause d from the Petroleum Act, so as to enable the uality of oil they like on the market so long as it y think this would satisfy the crude oil interest, as e would be more than ever disgusted with bad t buy it at all. Others think the gravity should be compel the refiners to make good oil equal to p our surplus crude. Some of our refiners say y of ·790 Canadian oil is now being manufactured on the Canadian market in competition with hat it gives just as good satisfaction to those who merican oil. We think, however, that the flash anada is too high, even higher than in any other , 95° by the Abel instrument is our test, and it is ls who think that 73° oil is safe enough, which is instrument in England. We do not agree with we do not think it suitable for our climate, but we ° would be perfectly safe, and this would enable us ight burning percentage out of our Canadian crude the high class of American oil coming into this it we only get 7 gallons of 790° refined oil out of a of crude, imperial measure, which is fit to com- merican oil of the finest quality, and it is thought was reduced to 88° or 90° we could get 9 gallons rrel of crude, of a quality as good as the best that the importation of American oil would be lly, and in consequence this would give a greater le oil.

ANUAL FOR 1889-90, BY WALTER R. SKINNER. n Lane.]—The third annual publication of this es, by its bulk and fulness of detail, to the great mining enterprise, as well as to the industry of ires additional importance in the eyes of investors, cent establishment of the Mining Exchange, which transactions in mining securities equal facilities and as are afforded for the sale and purchase of ordinary the Stock Exchange. Some idea of the expansion last year, more particularly in the department of , may be formed from the fact that whereas in companies, with 84 millions of capital, and in es with 123 millions of capital, the record for 1899 mpanies with 157 millions of capital. Hence the ity pages in the present issue of the manual. A of the British South Africa Company, which is the book, will be found useful for reference, and ventors will be almost indispensible to investors in p.

PRIZE COMPETITION.

A LITTLE more than twelve months ago, one of our friends suggested that it might be well to invite articles for competition, on some subject of special interest to our readers; giving a prize for the best contribution, he at the same time offering to defray the cost of the first prize, and suggesting a subject in which he was personally interested.

At that date we were totally unprepared for such a suggestion, but on thinking over the matter at intervals of leisure, we have come to think that the elaboration of such a scheme would not only prove of benefit to the manufacturer, but would go some way towards defraying the holiday expenses of the prize winner.

We are therefore prepared to receive competitions in the following subjects :—

SUBJECT I.

The best method of pumping or otherwise lifting or forcing warm aqueous hydrochloric of 30 deg. Tw.

SUBJECT II.

The best method of separating or determining the relative quantities of tin and antimony when present together in commercial samples.

On each subject, the following prizes are offered :—

One first prize of five pounds, one second prize of one pound, five additional prizes to those next in order of merit, consisting of a free copy of the *Chemical Trade Journal* for twelve months.

The competition is open to all nationalities residing in any part of the world. Essays must reach us on or before April 30th, for Subject II, and on or before May 31st, in Subject I. The prizes will be announced in our issue of June 28th.

We reserve to ourselves the right of publishing the contributions of any competitor.

Essays may be written in English, German, French, Spanish, or Italian.

Trade Notes.

OUR INDUSTRIAL CELEBRITIES.—We are requested to state that the memoir of the late Mr. Peter Spence, that recently appeared in our columns, has been reprinted in pamphlet form ; various additional facts of interest having been embodied in it by Mr. Frank Spence (Manchester Alum Works, Manchester), who will be pleased to forward a copy to anyone desirous of having it.

ELECTRICITY FOR LONDON TRAMWAYS.—The London Tramway Company have introduced a bill which was considered by Mr. Campion, examiner for private bills, yesterday, for the purpose of applying electricity as a motive power for their various systems of tramways, by means of wire ropes, chains, or other electrical apparatus placed underground.

ELECTRIC LAUNCHES ON THE THAMES.—Next season there will be 24 electric launches on the Upper Thames, and in a few years steam launches will be extinct. An electric launch dispenses with boiler, furnace, and engines, which take up so much room on steam launches. No stoker or engineer is needed. The charged motors are laid along the keel beneath the flooring of the hold, and the launch need not slow up for 60 miles, and then it can replenish its storage batteries at one of the many stations en route. The largest electric ship on the Upper Thames is the Viscountess Bury. It cost £2,000, is 65 feet long, carries 60 passengers, and can be hired at 60 guineas a week.

THE tallest smoke-stack in the United States, and, in fact, says the *Boston Globe*, the tallest in the world designed solely for the purpose of providing a draught for boilers, is receiving its final courses in Fall River, Mass. It is intended to meet the requirements of the entire steam plant of the four new mills of the Fall River Iron Company. Some idea of its size can be had from the following figures, furnished by the contractor. From the top of the granite foundation to the cap is 350 ft., the diameter at the base is 30 ft., at the top 21 ft., the flue is 11 ft. throughout, and the entire structure rests on a solid granite foundation 55 ft. by 30 ft., 16 ft. deep. In its construction there were used 1,700,000 bricks, 2,000 tons of stone, 2,000 barrels of mortar, 1,000 loads of sand, 1,000 barrels of Portland cement, and the estimated cost is 40,000 dols. It is arranged for two flues 9 ft. 6 in. b[illegible] 6 ft., connecting with forty boilers, which are to be run in connect[illegible] with four triple expansion engines of 1,350 horse power each.

ELECTRICITY IN STREET LIGHTING.—The Newcastle Corporation's lighting committee will not carry out their implied threat of lighting the street lamps by means of electricity, for they have agreed with the Gas Company, and renewed the contract. Nevertheless, they mean to try electric lighting on a small and experimental scale. They propose to put the new light into the big lamps in the large open spaces at the top of Westgate Hill, the Central Station, the Cattle Market, and Barras Bridge. This experiment ought to yield satisfactory results.

OTTO OF ROSE.—The St. Petersburg ladies are complaining of the costliness of absolutely pure otto (or attar, as they call it) of rose. But may they not consider themselves happy if they can get it pure at any price. The best authorities say that there is no such thing as pure otto of rose imported into Britain. It is always adulterated, mostly with Indan oil of geranium, otherwise known as Rusa oil and oil of ginger-grass. It takes 20,000 roses (*Rosa damascena)* to yield otto of rose equal in weight to a two-shilling piece.

ANOTHER REFINERY NUISANCE.—Richard R. Quay, acting for himself and others, made information before Squire Singleton, of Beaver, against Jesse Dubbs, manager of the Gas City Oil Works, charging him with maintaining a common nuisance. The works are located on the Cleveland and Pittsburgh-road, near the residence of Senator Quay, and the complaint is made that the stench arising therefrom is unbearable.

FREE ALCOHOL IN GERMANY.—An order has just been issued by which German soap-makers are allowed to use alcohol in their manufactures free of duty, provided they conform to certain regulations. These rules, says the *Chemist and Druggist*, are very simple, and consist simply in the manufacturer giving notice on an official form of the quantity of spirit which he intends to use, and the time when the manufacturing process is to take place. The alcohol is then forwarded to the works in bond, and in officially sealed packages, and at the appointed hour an excise officer attends at the works, breaks the seal, and watches the incorporation of the alcohol into the soap stock until the spirit has been fully mixed with the caustic soda lye. The concession has been made specially in the interests of the manufactures of transparent toilet soaps. For the manufacture of perfumery with duty-free alcohol facilities are already given.

BREAKING AWAY FROM THE CONTINENTAL.—On the 12th inst. the Florence Oil and Refining Company, of Florence, Col., cut loose from the Continental Oil Company, so says the *Refiner* of that place. It seems that several years ago the refiners of Florence entered into an agreement whereby the Continental agreed to take all the refined oil produced by the other companies at a price which allowed a small profit to those companies. This arrangement worked well enough while the supply was limited, but during the last eighteen months the supply has greatly increased, and the Continental has failed to take the oil from the companies as they produced it, because, the *Refiner* says, "neither the Continental or Standard wish to see the Colorado oil industry become of so much importance as to rival their own industry in the East, from whence is brought millions of gallons of oil each year to supply territory which belongs to Colorado." The Florence Company has secured a line of tank cars, and with fair treatment from the railroads will ship oil as far east as Omaha and Kansas City, south to Galveston and into Old Mexico, north to the British possessions and west to the Pacific coast.

A NEW OUTLET FOR COPPER.—E. A. Pontifex, chairman of the famous Cape Copper Company, in an address to the stockholders called attention in the following words to a matter of the greatest interest to the copper trade. Said he: There is one new outlet for copper, which, in addition to the large use of the metal for electrical purposes, promises to be a very important feature. I allude to its use in the form of sulphate of copper in vine culture. It has been found to be the only panacea for the phylloxera. We have ourselves, together with our friends of the Briton Ferry Copper Company, erected works capable of turning out a large quantity of sulphate of copper. The sales which we have already effected for present and future delivery will absorb 1,000 tons of the metal, and we have advices from France that works are in course of construction there capable of turning out 40,000 tons of sulphate of copper annually, which is equivalent to using up 10,000 tons of copper a year. And it must be remembered that copper thus used is lost for ever. It does not return to the maker as old copper for manufacture, with only some 8 or 10 per cent. of waste, as copper does which has been used for locomotive-plates or boiling-pans, or even pots and kettles. An increased consumption of over 10,000 tons a year for this one purpose—in addition to its use in the ever-swelling volume of trade and manufactures—must go a long way toward absorbing the greater output which higher prices may stimulate. I think, therefore, that we may fairly look forward to better prices than ruled before the French combination entered upon their disastrous campaign, and I am in a measure confirmed in this hope by the very prosperous condition of the iron trade, for it is an axiom with metal merchants and miners that iron carries nearly all other metals up with it.

Market Reports.

THE LIVERPOOL MINERAL MARKET.

The improvement reported in our market last week has be[illegible]more than maintained. Manganese: Arrivals are nil, whil[illegible] have been very much drawn upon, and prices are firmer. [illegible]: Raw lump without alteration, raw ground, £6. 10s., and calcined ground, £10. to £11. Bauxite (Irish Hill brand): The demand keeps more than pace with production, and prices are very firm. Lump done at 20s.; seconds, 16s.; thirds, 12s.; ground, 35s. Dolomite, 7s. 6d. per ton at the mine. French Chalk: Arrivals have been small, whilst stocks have been further drawn upon, and prices are firmer, especially for G.G.B. "Angel-White" brand—90s. to 95s. medium, 100s. to 105s. superfine. Barytes (carbonate), steady; selected crystal lump scarce at £6.; No. 1 lumps, 90s.; best, 80s.; seconds and good nuts, 70s.; smalls, 50s.; best ground, £6.; and selected crystal ground, £8. Sulphate unchanged; best lump, 35s. 6d.; good medium, 30s.; medium 25s. 6d. to 27s. 6d.; common, 18s. 6d. to 20s.; ground best white, G.G.B. brand, 60s.; common, 45s.; grey, 32s. 6d. to 40s. Pumicestone is unaltered; ground at £10., and specially selected lump, finest quality, £13. Iron ore continues in good demand at full prices. Bilbao and Santander firmer at 9s. to 10s. 6d. f.o.b.; Irish, 11s. to 12s. 6d.; Cumberland, 14s to 18s. Purple ore, 12s. 6d. Spanish manganiferous ore commands ready sale at full prices. Emery-stone: Best brands rather scarce, and bringing full prices. No. 1 lump is quoted at £5. 10s. to £6., and smalls £5. to £5. 10s. Fullers' earth steady; 45s. to 50s. for best blue and yellow; fine impalpable ground, £7. Scheelite, wolfram, tungstate of soda and tungsten metal more inquired for. Chrome metal, 5s. 6d. per lb. Tungsten alloys 2s. per lb. Chrome ore inquired for, especially the higher grades. Antimony ore and metal have further advanced. Uranium oxide, 24s. to 26s. Asbestos: Best rock, £17. to £18.; brown grades, £14. to £15. Potter's lead ore dearer on account of the suspension of the production through strike; smalls, £14. to £15.; selected lump, £16. to £17. Calamine: Best qualities scarce—60s. to 80s. Strontia steady; sulphate (celestine) steady, 16s. 6d. to 17s. Carbonate (native) £15. to £16.; powdered (manufactured), £11. to £12. Limespar: English manufactured, old G.G.B. brand sought after, and brings full prices; 50s. for ground English. Felspar, 40s. to 50s.; fluorspar, 20s. to £6. Bog ore more inquired for—22s. to 25s. Plumbago: More offering; best Ceylon lump at last quotations; Italian and Bohemian, £4. to £12. per ton. French sand, in cargoes, continues scarce on spot—20s. to 22s. 6d. Ferro-manganese sells readily at higher figures. Ground mica, £50. China clay freely offering—common, 18s. 6d.; good medium, 22s. 6d. to 25s.; best, 30s. to 35s. (at Runcorn).

METAL MARKET REPORT.

	Last week.		This week.
Iron	60/4	....	56/1½
Tin	£94 0 0	....	£94 0 0
Copper	48 17 6	....	48 17 6
Spelter	24 10 0	....	24 2 6
Lead	13 10 0	....	13 0 0
Silver	44⅜	....	44½

COPPER MINING SHARES.

	Last week.			This week.	
Rio Tinto	16 7/16	16½	....	16 5/16	16 7/16
Mason & Barry	6 13/16	6 15/16	....	6¾	6⅞
Tharsis	4⅝	4¾	....	4⅝	4¾
Cape Copper Co.	3 7/16	3 9/16	....	3 7/16	3 9/16
Namaqua	2 7/16	2 9/16	....	2 5/16	2 7/16
Copiapo	2 7/16	2 9/16	....	2 7/16	2 9/16
Panulcillo	1	1⅛	....	1 1/16	1 3/16
New Quebrada	⅞	1	....	⅞	1⅛
Libiola	3½	3⅝	....	3 9/16	3 11/16
Tocopilla	1/9	2/3	....	1/9	2/3
Argentella	6d.	1/-	....	6d.	1/-

REPORT OF MANURE MATERIAL.

The demand for mineral phosphates continues, there being very little stuff available near at hand to satisfy it, but there has been less doing in all other kinds of manure material, and for these prices on the week are somewhat in favour of buyers.

Charleston River Phosphate Rock is still quoted at 11d. per unit, but we do not hear of anything to be had even at that price near at hand or for early shipment. The syndicate movement in Carolina is still in progress, and meantime the raisers do not show much willingness to make contracts for shipment ahead.

There is a somewhat greater disposition to sell Somme description, but we have no alteration in values to note. Belgian continues as last quoted, and prices for Canadian must still be considered as nominally 1s. for 70%, 13½d. for 75%, and 15d. for 80%, ⅛th rise, delivered Thames, Mersey, or Clyde, the shippers still refusing to commit themselves for the coming season's output.

Bones are easier in all positions, freights permitting shipments from different ports in the Mediterranean, and exchange leading shippers of East Indian to modify their recent quotations. For a day or two the market has been somewhat demoralized, but at the close it is firm, at £5. 7s. 6d. for crushed East Indian, and £5. 2s. 6d. for mixed; and £5. 5s. for good grinding bones, ex-quay Liverpool. Nearest value of River Plate cargoes we should consider to be £5. 10s., though 2s. 6d. per ton more is asked for December shipments.

There have been sales of East Indian bone meal at as low as £5. 5s. February and March shipment, but the market quickly recovered, and that price has since been refused for 1,000 tons in a line, £5. 7s. 6d. being the lowest that sellers will now entertain ex-ship Glasgow or Liverpool.

Nitrate of soda remains dull at 8s. 3d. per cwt. on the spot. Due cargoes might fetch a fraction over 8s. 3d., and November shipments are worth about 8s. 6d. There is very little demand in any position, buyers just taking what they require to meet their current needs.

We have no change to note in value of supers, though there is a good damand, price remaining at about 50s. per ton, in bulk, f.o.r. at works.

There have been sales of home prepared dried blood during the week on private terms. Nearest value is 11s. 6d. per unit, f.o.r. at works, which price would be accepted for River Plate now on the way and shortly due.

Manure manufacturers seem to be suspending further purchases, until they see their manufactured goods going out into the country.

MISCELLANEOUS CHEMICAL MARKET.

The prices of all products in the alkali trade continue firm and makers are amply supplied with orders for some time to come. Supplies of caustic soda are now more easily obtainable, but the current prices are firmly adhered to, viz., 70%, £7. 17s. 6d. to £8, and 60% £7. per ton f.o.b. The demand for soda ash continues brisk, and the tendency is rather to higher prices than otherwise. For soda crystals £2. 12s. 6d. to £2. 15s. are now the lowest figures, according to package, &c. Bleaching powder is in fairly good demand, and nearest value to-day f.o.r. makers' works is £5. 2s. 6d. per ton. Hardwood packages, f.o.b., firm at £5. 12s. 6d. per ton. Chlorate of potash, 5¾d. to 6d. per lb. Lump sulphur, £5. per ton. Vitriol, 145%, free from arsenic, is in demand at 29s. 6d. to 30s. Sulphate of copper still scarce and firm, £24. 10s. per ton. White powdered arsenic is dearer and selling at £13. net per ton at usual points. Acetates of lime unchanged in prices, but only a moderate amount of business passing at current rates. Acetic acids rather dearer. Yellow prussiate of potash advanced to 10½d. per lb. Alum and sulphate of alumina are realising higher prices generally. Lead acetates and nitrate continue to be neglected, but prices are stationary. Aniline oil and salt are dearer, and there is more general inquiry now from those who require these products, and who missed the opportunity of covering when prices were low. Potash caustic and carbonate are in good demand and prices are firm. Sulpho-cyanides are more inquired for, and makers are now quoting advanced prices. Carbolic acid crystals and liquid show no signs of improvement at present.

THE LIVERPOOL COLOUR MARKET.

COLOURS.—Demand quieter; prices unaltered. Ochres: Oxfordshire
ioted at £10., £12., £14., and £16.; Derbyshire, 50s. to 55s.; Welsh,
st, 50s. to 55s.; seconds, 47s. 6d.; and common, 18s.; Irish, Devonshire,
s. to 45s.; French, J.C., 55s., 45s. to 60s.; M.C., 65s. to 67s. 6d.
mber: Turkish, cargoes to arrive, 40s. to 50s.; Devonshire, 50s.
55s. White lead, £21. 10s. to £22. Red lead, £18. 10s. Oxide
zinc: V.M. No. 1, £24. 10s. to £25.; V.M. No. 2, £23. Venetian
d, £6. 10s. Cobalt: Prepared oxide, 10s. 6d.; black, 9s. 9d.; blue,
s. 6d. Zaffres: No. 1, 3s. 6d.; No. 2, 2s. 6d. Terra Alba: Finest
iite, 60s.; good, 40s. to 50s. Rouge: Best, £24.; ditto for jewellers',
per lb. Drop black, 25s. to 28s. 6d. Oxide of iron, prime quality,
10. to £15. Paris white, 60s. Emerald green, 10d. per lb. Derby-
ire red 60s. Vermillionette, 5d. to 7d. per lb.

WEST OF SCOTLAND CH

C

Although as a rule holders' quotations keep
give way, enquiry is not of satisfactory volum
some stimulant. Sulphate of ammonia, as abc
the chemical list, here and elsewhere, has gon
attentions of buyers, and as low as £11. 15s.
been taken. Operators feel distinctly nonplus
to make of present unfavourable developme
forward. Soda crystals are easier; there is
caustics, and even in bleach, but the only unq
substantial rises are salammoniac and lubric
having been put up 20s. and the other in some
improved on late gains. The three local makers c
contrary to the general rule, have of late been
for orders at combination prices, and this equal
of soda also. The Scotch paper makers have
to their men with a view to some basis of co-op
the future, so as to avoid mutually ruinous strik
considering just now how this novel proposal
in a spirit of faith, or with suspicion. Chief
Soda crystals, 48s. 6d. to 50s. net Tyne; alum i
Glasgow; borax, English refined £30., and b
net Glasgow; soda ash, 1⅛d. less 2½% Tyne;
£9. 70s., 70/72°, £7. 17s. 6d., 60/62°, £7
cream, £6. 15s., all less 2½% Liverpool; bica
casks, £5., and 1 cwt. casks, £5. 5s. net
48/52°, 1⅜d., net Tyne; saltcake, 25s. 6d. to
£5. 15s., less 5% f.o.r. Glasgow; bichromate
soda 3d., less 5 and 6% to Scotch and Englis
chlorate of potash, 4¾d., less 5% any port;
sulphate of ammonia, £11. 15s. f.o.b. Leith;
2nd white, £37. and £35., less 2½% any po
£24. 10s., less 5% Liverpool; paraffin scale,
paraffin wax, 120°, semi-refined, 2⅛d.; par
9d.; paraffin oil (burning), 6¾d. to 7d., at Gla
ing), 865°, £5. 10s. to £6., 885°, £6. to £
£7. 10s. to £8. Week's imports of sugar at
bags.

TAR AND AMMONIA PR

The benzol market remains without change,
90's benzol still finding a market at 4s., and 50
Crude naphtha 20% at 120° is fetching 1s. 2d.
qualities at a corresponding increase. Crude c
tinues unsatisfactory for the sellers, but the de
solvent naphtha is very good. The Anthracene
out change, there being, apparently, a large sp
on. Pitch is weak, and though 33s. 6d. may
some cases on the East coast, and 32s. Garston
small, and purchasers seem to have covered the
The sulphate of ammonia market may be cor
than last week, and, to a certain degree, price
recover. To-day's value at Hull may certainly b
and £11. 15s. to £11. 16s. 3d. at Leith.
£ , while outside London makes are
rumoured that several brokers have large deli
month, having considerably oversold themselve
why such an effort has been made to "bear" the
If sellers will only refer to the Exports they wi
able quantity of sulphate has been shipped dur
which does not accord well with the state
demand.

New Compan

STANDARD PORTLAND CEMENT COMPANY, LIMITE
of the Standard Portland Cement Company, in liquid
the 22nd inst., with a capital of £100,000., divided into
preference shares, 12,280 ordinary shares of £5. each, a
each. The property of the company is situate at Ba
known as the Barrington Cement, Brick, and Lime Wo
Portland Cement Works. The subscribers are:—

J. H. Smith, Wallington, Surrey ....
W. Morris, Hitchin ....
F. J. Lee Smith, Newlands Park, Sydenham ....
Col. A. Hamilton, 41, Lennox-gardens ....
G. Stanford, 17, Benland-road Clapham, secretary
F. W. Toole, 34, St. Mary's-square, S.E., clerk ..
W. S Park, 20, New North-road, secretary to a
F. Hunter, 3, Vicarage-terrace, Leyton, private

Gazette Notices.

PARTNERSHIPS DISSOLVED.

CHAPPELL AND Co., Heathfield, near Bovey Tracey, electric carbon manufacturers.
HEY AND SMITH, Bradford, wholesale druggists and drysalters.

Notices of Dividends.

LIVERSIDGE ELLIOT, Huddersfield, electric engineer. First dividend of 1s. 3d., February 3rd, Messrs. Haigh and Sons, New-street, Huddersfield.

The Patent List.

This list is compiled from Official sources in the Manchester Technical Laboratory, under the immediate supervision of George E. Davis and Alfred R. Davis.

APPLICATIONS FOR LETTERS PATENT.

Coating Surfaces with Paints. Kenneth Henry Cornish. 5. January 1.
Preventing Action of Water on Lead. James Smillie. 26. January 1.
Apparatus for Washing Ores.—(Complete Specification.)—Robert Angus. 27. January 1.
Distilling Heavy Petroleum Oil. 27. A. M. Clark. January 1.
Electrical Coating of Metals. S. V. Dardier. 29. January 1.
Hydrocarbon Lights. F. W. S. Stokes. 34. January 1.
Manufacture of Raw Materials for Dyestuffs. R. J. Friswell. 39. January 1.
Decorating Wood. C. Hettwig and F. Heckner. 40. January 1.
Scoops for Charging Gas Retorts. S. J. Ball. 44. January 1.
Automatic Lubricator. J. McDonnell. 49. January 2.
Distillation of Coal or Shale. T. Parker. 67. January 2.
Separating Tin from other Metals. J. Bang and A. Ruffin. 70. January 2.
Manufacture of White Lead. G. Bischof. 71. January 2.
Supplying Vessels with Hydro-Carbons. R. Bruno. 72. January 2
Steam Boilers. J. G. Williams. 83. January 2.
Oil Filler for Lamps and Stoves. E. Banfield. 91. January 3.
Electric Meters. W. Mosley. 112. January 3.
Burning Fuel Economically. W. Potts. 116. January 3.
Production of Azo Colouring Matters. R. J. Friswell and A. G. Green. 134. January 3.
Apparatus for Delivering Regular Quantities of Liquid. R. [illegible] and D. Winterbottom. 148. January 4.
Paraffin Lamps. J. G. Patterson. 153. January 4.
Feeding Furnaces with Fuel. J. W. Claridge. 164. January 4.
Hot Blast Fuel Heater. J. Judge and W. Dix. 165. January 4.
Drawing Metal Tubes. H. Isray. 174. January 4.
Steam Jet Pumps.—(Complete Specification). E. O. Johnson. 217. Jan. 4.
Separating Phenol from Creosote. P. Richm. 236. January 4.
Carbonising Fabrics. J. Walker and W. Brown. 220. January 5.
Electro Motors and Dynamos. J. Hopkinson, E. Hopkinson, and G. A. Grindle. 230. January 5.
Extracting Gold from Quartz. R. Brown and G. H. Irvine. 237. Jan. 5.
Dynamos. W. F. King. 240. January 7.
Prevention of Choked Ascension Pipes. J. Stears. 260. January 7.
Electro Magnetic Despatch Apparatus.—(Complete Specification). R. Hadden. 272. January 7.
Treatment of Ammonium Chloride, Ammonium Sulphate, and Calcium Chloride for obtaining Hydrochloric Acid. A. T. Smith, jun. 289. January 7.
Plates for Storage Batteries.—(Complete Specification). H. H. Lake. 300. January 7.
Secondary Batteries.—(Complete Specification). H. H. Lake. 309. Jan. 7.
Treatment of Sewage. 309. T. Kay. January 7.
Making Coal Gas in Vertical Retorts. W. T. Cotton and E. F. B. Crowther. 343. January 8.
Electrodes for Batteries. L. Epstein. 350. January 8.
Dextrine. S. Wohle, A. C. Irwin, and G. O. Jacob. 357. January 8.
Smoke Economiser. G. H. Hunt. 373. January 9.
Dyeing Yarn. J. B. Whiteley, E. Whitely, and J. B. McKay. 385. January 9.
Manufacture of Nitrate of Potash. O. C. Townsend. 381. January 9.
Evaporation of Brine. R. C. Wilson. 393. January 9.
Evaporation of Liquids. W. Wild. 398. January 9.
Manufacture of Sulphate of Copper. H. J. K. Hamesbottom. 411. Jan. 9.
Fixing Atmospheric Nitrogen. P. R. de Lambilly and E. L. Chaleter. 419. January 9.
Ore Concentrating Machinery. W. J. Smith. 424. January 9.
Improvements in the Manufacture of Alum. P. M. Spence, D. D. Spence, and A. Keilman. 448. January 10.
Treating Liquids and Solids with Gases. C. Blagburn. 474. Jan. 10.
Purifying Water. E. Devonshire. 480 January 10.
Primary Batteries. M. Sappey. 488. January 10.
Concentrating Liquids by Evaporation. T. E. Wilson. 510. January 11
Separating Iron from Bones. F. Christy and J. H. Carter. 530. Jan. 11
Gas Producers. W. C. Sellar. 551. January 11.
Manufacture of Copper. T. Troynam. 560. January 11.
Blue Colouring Matters. Kern and Sandoz. 569. January 11.

IMPORTS OF CHEMICAL PRODUCTS

AT

THE PRINCIPAL PORTS OF THE UNITED KINGDOM.

LONDON.

Week ending Jan. 11th.

Alkali—		
Holland,	33 t.	H. Wallace & Co.
France,	240	Arnati & Harrison
Antimony Ore—		
Melbourne,	5 t.	Knight & Morris
Portugal,	12	H. Emanuel
Acetic Acid—		
Holland,	175 pkgs.	Beresford & Co.
"	30	A. & M. Zimmermann
Arsenic—		
Germany,	£73	Williams, Farrey & Co.
Bones—		
U.S.	16 t.	H. A. Lane & Co.
N.S. Wales,	4½	Good, Rigg & Co.
France,	70	Darling Bros.
N.S. Wales,	3 7 c.	Dalgety & Co
N. Zealand,	3 12	Johnson & Allsop
"	1½	Flack, Chandler & Co.
Germany,	10	W. M. Smith & Sons
Barium Chloride—		
Holland,	£90	Petri Bros.
Barytes—		
Holland,	38 cks.	O'Hara & Hoar
"	82	Prop. Scott's Whf.
"	44	W. Harrison & Co.
"	89	D. Storer & Sons
Boracic Acid—		
Italy,	49 pkgs.	Howards & Son
"	9	J. Puddy & Co.
Caoutchouc—		
[illegible]	£11	T. H. Lee
Madagascar,	5 t. 17 c.	Keddie, Son & Co.
S. Africa,	4	L. & I. D. Jt. Co.
E. Indies,	10 7	Aachenheim & Fuhrken
[illegible],	1	D. C. Thomas & Sons
[illegible]	16	L. & I. D. Jt. Co.
[illegible]	80	Lewis & Peat
[illegible]	[illegible]	Arnold, Cheney & Co.
E. Africa,	17	Matheson & Co.
"	1 16	Clarke & Smith
Aden,	1 2	Anderson, Weber & Smith
E. Indies,	1 14	Clarke & Smith
"	1 10	F. Stahlschmidt & Co.
E. Africa,	19	Price, Boustead & Co.
Sts. Settlements,	1 t.	L. & I. D. Jt. Co.
A. Turkey,	£12	M. D. Co.
Germany,	£100	L. Mendiney
Chemicals *(otherwise undescribed)*		
Germany,	£10	F. Stahlschmidt & Co.
Holland,	15	W. Harrison & Co.
Germany,	10	Carroll & Frost
Holland,	10	T. H. Lee
Germany,	33	L. & I. D. Jt. Co.
"	254	A. & M. Zimmermann
"	113	W. Burton & Sons
"	34	T. H. Lee
Holland,	100	A. & M. Zimmermann
Germany,	70	
"	19	T. H. Lee
Caustic Potash—		
Holland,	21 pkgs.	Francis & Co.
France,	13 drms.	J. Knight & Sons
"	22 pkgs.	Northcott & Sons
Camphor—		
Germany,	2 tubs.	L. & I. D. Jt. Co.
Japan,	191	Beresford & Co.
"	79	Devitt & Hett
"	30	L. & I. D. Jt. Co.
Caustic Soda—		
Holland,	27 c.	Beresford & Co.
Cream Tartar—		
France,	16 pkgs.	E. W. Carling & Co.
Copper Regulus—		
Sydney,	30 t.	J. A. Drew
Dolomit—		
Germany,	£72	J. B. Orr & Co.
Dyestuffs—		
Argols		
Newcastle,	33 pkgs.	Hoare, Wilson & Co.
Dyestuffs (otherwise undescribed)		
Germany,	5 pkgs.	H. Kohnstamm
Indigo		
E. Indies,	60 chts.	L. & I. D. Jt. Co.
"	79	Arbuthnot, Latham & Co.
"	7	A. Harvey
"	2	Parsons & Keith
"	24	Langstaff & Co.
"	76	F. Stahlschmidt & Co.
"	71	G. Dards
"	48	L. & I. D. Jt. Co.
"	41	Elkan & Co.
"	18	J. Forsey
"	8	H. Johnson & Sons
"	7	Walker, Munsie & Co.
"	19	T. H. Allan & Co.
"	4	Parsons & Keith
"	70	Patry & Pasteur
"	6	G. Dards
"	12	H. Grey, jun.
"	163 pkgs.	T. Ronaldson & Co.
"	79 chts.	L. & I. D. Jt. Co.
"	5	W. Brandt, Sons & Co.
"	5	Seton, Laing & Co.
"	3	Benecke, Souchay & Co.
"	47	Arbuthnot, Latham & Co.
"	30	A. Harvey
"	1	L. & I. D. Jt. Co.
"	10	Ernsthausen & Co.
"	20	L. & I. D. Jt. Co.
"	11	Langstaff & Co.
"	25	L. & I. D. Jt. Co.
Gambier		
E. Indies,	14 pkgs.	S. Barrow & Bro.
China,	501	L. I. D. Jt. Co.
E. Indies,	1,074	Brinkmann & Co.
"	393	J. H. & G. Scovell
"	249	Beresford & Co.
"	315	L. & I. D. Jt. Co.
Valonia		
A. Turkey,	205 t.	J. Greene
"	10	Baxter & Hoare
"	200	A. & W. Nesbitt
"	100	J. Kitchin
"	68 16 c.	Boutcher, Mortimore & Co.
"	33 18	Beresford & Co.
Sumac		
Italy,	500 pkgs.	W. Shelcott
"	100	Union Lighterage Co.
Myrabolams		
E. Indies,	217 pkgs.	Beresford & Co.
"	560	Marshall & France
"	2,124	L. & I. D. Jt. Co.
Cutch		
E. Indies,	11 pkgs.	L. & I. D. Jt. Co.
Extracts		
France,	125 pkgs.	L. J. Levinstein & Sons
Holland,	71	Burt, Boulton & Co.
France,	40	Prop. Chamble's Wf.
Orchella		
	24 pkgs.	N. Smith
Cochineal		
Canaries,	16 pkgs.	A. Levy & Co.
"	18	Marshall & France
"	8	Barrie & Co.
"	6	Johnson, Rolls & Co.
"	1	Kuhner, Henderson & Co.
Tanners' Bark		
Belgium,	50 t.	J. Kitchin
"	24	S. Barrow & Bro.
"	13	Bevington & Sons
Cape,	10½	A. & W. Nesbitt
Dextrine—		
A. Turkey,	£45	Anderson, Weber & Co.
France,	£160	R. Warner & Co
Gutta Percha—		
E. Indies,	3 t. 10 c.	L. & I. D. Jt. Co.
"	26 7	Kaltenbach & Schmitz
Germany,	9 9	
E. Indies,	14 10	J. H. & G. Scovell

Glucose—
Germany, 200 pkgs. 200 c. J. Cooper
" 200 200 Fellows, Morton & Co.
" 35 283 Proprs. Chamberlain's Wf.
" 210 280 M. D. Co.
" 50 50 Hyde, Nash & Co.
" 100 100 Barrett, Tagant & Co.
U.S. 500 500 Francis & Co. Co.
Germany, 100 890 L. & I. D Jt.
" 25 210 J. Knill & Co.
U.S. 600 600 Humphery & Co.
" 600 600 A. Dawson & Co
Germany, 200 200 J. F. Ohlmann
" 100 200 T. M. Duche & Sons
" 200 100 Fellows, Morton & Co.
" 400 400 J. Barber & Co.
" 24 215 Andorsen, Becker & Co.
" 21 180 C. Tennant, Sons & Co.
" 25 200 Fellows, Morton & Co.
" 25 129 Proprs. Scott's Whf.
" 100 150 H. A. Litchfield & Co.
" 200 200 Becker & Ulrich
" 25 200 P. Windmuller
" 37 350 J. Cooper
" 160 160 C. Tennant, Sons & Co.
France, 100 100 J. Barber & Co.

Glycerine—
Germany, £86 Craven & Co.
" 120 G. Haller & Co.
" 35 A. & M Zimmermann

Isinglass—
Sts. Settlements, 2 pkgs. Hale & Son
B. Guiana, 1 pkgs. W. Baird & Co
E. Indies, 10 L. & I. D. Jt. Co.
" 34 Clarke & Smith
Aden, 7 L. & I. D. Jt. Co.
China, 5 "
B. W. I. 1 W. M. Smith & Sons
E. Indies, 12 Hale & Son
" 2 J. P. Alpe & Co.
Sts. Settlements, 26 pkgs. L. & I. D. Jt. Co.

Manure—
Phosphate Rock
863 t. Lawes Manure Co.
140 T. Farmer & Co.

Manganese Ore—
N. Zealand, 4 t. L. & I. D. Jt. Co.

Naphtha—
Holland, 190 brls. Stein Brs.
Belgium, 10 Domeier & Co.

Potassium Carbonate—
Holland, 11 pkgs. Francis & Co.
" 6 Charles & Fox

Potassium Muriate—
Holland, £10 H. Johnson & Son

Plumbago—
Holland, 32 cf. J. Barber & Co.
Germany, 12 cks. Beresford & Co.
Ceylon, 90 Prop. Chamb. Wf.
" 50 Doulton & Co.
" 523 Thredder, Son & Co.
France, 4 cs. A. Henry & Co.
Holland, 17 cks. Brown & Elmslie
Ceylon, 97 Thredder, Son & Co.
" 20 H. Johnson & Sons
E. Indies, 43 L. & I. D. Jt. Co.

Pitch—
Germany, 26 brl. Berlandina Bros. & Co.
Italy, 30 "

Potassium Oxy Muriate—
Belgium, 100 kgs. G. Boor & Co.

Pyrites—
Spain, 2,048 t. C. Tennant, Sons & Co.

Paraffin Wax—
U. S. 999 brls. J. H. Usmar & Co.

Pearl Ash—
France, 120 c. F. W. Berk & Co.
" 100 Petri Bros.

Potassium Hydrate—
France, 3 pkgs. J. A. Reid
" 4 Spies Brs. & Co.

Quinine—
Germany, £1,000 L. & I. D. Jt. Co.

Saltpetre—
E. Indies, 953 bgs 570 pkgs. C. Wimble & Co.
" 1,597 Clift, Nicholson & Co.
Germany, 58 pkgs. Craven & Co.
E. Indies, 349 100 cks. L. & I. D. Jt, Co.
" 349 Tulloch & Co.
Germany, 35 cks. P. Hecker & Co.

Stearine—
Holland, 10 cks. Perkins & Homer
Belgium, 112 bgs. H. Hill & Sons
France, 150 G. Boor & Co.
" 148 sks. J. Goddard & Co.

Salicylic Acid—
Germany, 2 pkgs. Burgoyne, Burbidge & Co.

Sodium Hyposulphite—
Holland, £23 J. James

Sodium—
Holland, £85 Beresford & Co.

Tartar—
Italy, 39 pkgs. Howards & Son
" 28 B. Jacob & Sons

Tartaric Acid—
Germany, 2 pkgs. L. & I. D. Jt. Co.
Italy, 80 Webb, Warter & Co.
Holland, 52 "
Germany, 12 Middleton & Co.
Holland, 2 W. C. Bacon & Co.

Tin Ore—
France, 48 c. Hughes, Chemery & Co.

Verdigris—
France, £110 Webb, Warter & Co.

Ultramarine—
Holland, 5 pkgs. H. Peron & Co.
" 4 cks. Spies, Brs. & Co.
" 4 Ohlenschlager Brs.
Germany, 40 cs. G. Steinkoff

Zinc Oxide—
Holland, 3 brls. H. Lambert
Austria, 7 cks. Braithwaite & Co.
Holland, 120 M. Ashby

LIVERPOOL.

Week ending Jan. 15th.

Albumen—
Hamburg, 1 ck.

Bone Meal—
Calcutta, 5,832 bgs.

Cream Tartar—
Rotterdam, 22 cks.
Barcelona, 17 hds.
Moncofar, 4 cks.
" 5 Sachse & Klemm

Charcoal—
Rouen, 100 bgs. Co-op. W'sale Society

Copper Precipitate—
Pomaron, 20,318 bgs. Mason & Barry

Caoutchouc—
Mussera, 2 cks. Taylor, Laughland & Co.
" 5 bgs. Samson & de Liagre
Ambrizette, 92 "
" 22 cks. Taylor, Laughland & Co.
Landana, 1 Valle, Azevedo & Co.
Black Point, 8 brls. 1 ck. 2 cs. Pinto, Leite & Nephew
" 1 ck. Edwards Bros.
Loango, 2 brls. 8 cks. "
N'gore, 9 cks. R. W. Roulston
Cape Lopez, 26 J. Holt & Co.
Gaboon, 12 A. V. Heyder
" 7 Daumas & Co.
" 3 J. Holt & Co.
Beta, 7 "
Addah, 2 brls. 3 cks. F. & A Swanzy
" 7 cks. 1 cs. J. P. Werner
" 1 Radcliffe & Durant
Accra, 7 cks. F. & A. Swanzy
" 8 Hutton & Co.
" 73 J. P. Werner
" 9 Lane & Co.
" 3 Taylor, Laughland & Co.
" 3 cs. 12 cks. Fischer & Co.
Salt Pond, 4 brls. Edwards Bros.
" 4 5 cs. Fletcher & Fraser
" 1 2 Havard & Co
" 1 A. Reis
" 1 Radcliffe & Durant
" 2 Griffiths & Co
" 1
Cape Coast, 2 brls. F. & A. Swanzy
" 2 T. Crouch
" 22 [illegible]
" 5 Ihlers & Bell
" 1 C. L. Clare & Co.
" 1 R. Davies & Co.
" 9 Millerson & Co.
" 1 Hobson & Co.
" 4 Radcliffe & Durant
" 1 cs. 6 brls. Havard & Co.
" 1 2 Edwards Bros.
" 2 Tomlinson & Co.
" 2 cs. 24 brls.
Elmina, 3 cks. F. & A. Swanzy
" 2 cs. Radcliffe & Durant
Axim, 5 bgs. Edwards Bros.
" 2 Pickering & Berthoud
Sierra Leone, 6 brls. Millington & Co.
Isle de Los, 235 Colin & Co.
Bathurst, 2 pkgs. French & Co.

Dyestuffs—
Orchella
Colon, 13 bls. M. Beausire
Bay Beach, 7 bgs.
Quittah, 7
Cochineal
Grand Canary, 16 bgs. Levy & Co.
Teneriffe, 15 bgs. H. Pool
" 12 M. Proda
" 5 T. Taylor & Co.
Las Palmas, 60 Lathbury & Co.
Valonia
Syra, 79 bgs. C. G. Nicolaidis
Divi Divi
Smyrna, 2 brls.
Sumac
Barcelona, 100 bgs. A. Ruffer & Sons
Argols
Leghorn, 5 cks.
Indigo
Calcutta, 29 chts. 2 bxs. J. S. Morgan & Co.
" 108 1 Baring Bros. & Co.
" 59 1 Amos, Keay Mfg. Co.
Myrabolams
Bombay, 1 pol. Okell & Owen
" 2,248 bgs. D. Sassoon & Co.
" 1,334 W. & R. Graham & Co.
" 901
" 1,675 Betry, Craig & Co

Glycerine—
Rotterdam, 10 cks. 5 carboys.
Venice, 21

Glucose—
Hamburg, 15 bgs. 87 cks.

Magnesia—
Rotterdam, 64 cs.

Manganese—
Poti, 1,450 t.

Potash—
Rotterdam, 41 cks.
Rouen, 73 Co.-op. Wls. Society

Pyrites—
Huelva, 1,652 t. Matheson & Co.
" 1,500 Peauford & Co.

Pitch—
Rotterdam, 22 cks.

Phosphate Lime—
Dieppe, 200 t Phosphate Guano Co.

Phosphates—
Dieppe, 350 t. Phosphate Guano Co.

Salt—
Rotterdam, 12 t.

Saltpetre—
Hamburg, 77 cks.

Sodium Nitrate—
Caleta Buena, 6,353 bgs.

Tartaric Acid—
Rotterdam, 2 cks.

Tincal—
Calcutta, 205 bgs.

Tartar—
Constantinople, 5 bgs.

Turepntine—
Trieste, 2 brls.

Tar—
New York, 233 brls.

Ultramarine—
Rotterdam, 40 brls. 2 cks, 5 cs.

Zinc Ashes—
Rotterdam, 16 cks.

TYNE.

Week ending Jan. 16th.

Alum Earth—
Hamburg, 18 cks.

Glucose—
Hamburg, 100 bgs. 24 cks.

G
Wee
Boracic
Leghorn
Barytes
Rotterd
Cream T
Genoa,
Dyestuf
Extra
Antwerp
Alizar
Rotterda
Gutta-pe
Amsterd
Guano—
Huanic lo
Oxalic A
Rotterda
Pyrites—
Huelva,
Phospha
Antwerp,
Charlesto
Sulphur-
Antwerp,
Catania,
Salt—
Rotterdar
Turpenti
Savannah
Tartaric
Rotterdan
Zinc Oxi
Antwerp,
New York
Rotterdam
Antwerp,

Week
Chemical
Rotterdam
Dyestuffs
Alizari
Rotterdam
Dyestu
Boulogne
Glucose—
Hamburg,
Dunkirk,
Calais,
Potash—
Dunkirk,
Calais,
Saltpetre
Rotterdam
Hamburg,
Sodium S
Rotterdam
Sodium M
Rotterdar
Zinc Oxid
Antwerp,

GI
Wee
Albumen-
Dieppe,
Chemical
Hamburg,

Weel
Aluminiu
Stettin,
Acid—(ot
Rotterdan
Bones—
Gothenbu
Amsterda
Bone Mea
Bombay,
Barytes—
Antwerp,

Rotterdam, 35 brls. Hutchinson & Son
Bremen, 18 cks. T. W. Flint & Co.

Chemicals *(otherwise undescribed)*
Stettin, 14 cks. Wilson, Sons, & Co.
Antwerp, 25 pks. "
Dunkirk, 6 cks. "
Hamburg, 3 "
Bremen, 5 kgs. Veltmann & Co.
" 1 ck. "

Carbon—
New York, 80 brls. Wilson, Sons, & Co.

Cobalt Oxide—
Hamburg, 9 cs. G. Malcolm, & Son

Dyestuffs—
Myrabolams
Bombay, 3,759 bgs. Wilson, Sons, & Co
Alizarine
Rotterdam, 53 cks. W. & C. L. Ringrose
" 35 Hutchinson & Son
" 22 1 cs. W. & C. L. Ringrose

Madder
Rotterdam, 4 cks. Hutchinson & Sons
" 2 W. & C. L. Ringrose

Glucose—
Stettin, 50 cks. Wilson, Sons, & Co
Rotterdam, 20 bgs. Hutchinson & Son
Hamburg, 46 cks. 20 bgs. Wilson, Sons, & Co.

Manure—
Hamburg, 170 t. Wilson, Sons, & Co

Naphthol—
Rotterdam, 5 brls. Wilson, Sons,

Pitch—
Alexandria, 102 cks. Wilson, S

Turpentine—
Libau, 90 cks. Morley, Clarke,

Tartaric Acid—
Rotterdam, 12 cks. G. Lawson &

Tar—
Reval, 2 cks. R. G.

EXPORTS OF CHEMICAL PRODUCTS

FROM

THE PRINCIPAL PORTS OF THE UNITED KINGDOM.

LONDON.

Week ending Jan. 11th.

Ammonia Sulphate—
Ghent, 40 t. £476
Penang, 1 13
Cologne, 5 60
Yokohama, 2 26
Hamburg, 151 10 c. 3,027

Arsenic—
Lyttleton, 20 cks. £56
Antwerp, 9 t. 15 cwts. 140
Napier, N. Z'land, 1 3 c. 18
Melbourne, 1 14

Ammonia Carbonate—
Adelaide, 14 c. £11
Calcutta, 4 11
Melbourne, 1 t. 2 45
Yokohama, 30 cs. 60

Acetic Acid—
Canterbury, 15 cs. 2 cks. £42
Launceston, 6 12
Melbourne, 12 cwts. 32

Alum Cake—
Calcutta, 19 t. 18 c. £75

Antimony Sulphate—
Philadelphia, 15 c. £63

Ammonia—
Rosario, 25 cs. £26

Bisulphite of Lime—
Calcutta, 4 t. 12 c. £89

Bisulphide of Carbon—
Cape Town, 20 t. 3 c. £240
M. Video, 12 cs. 15
Rosario, 100 drs. £52

Boracic Acid—
Yokohama, 20 kegs. £38
Melbourne, 1 t. 11 c. 64

Bluestone—
Hobart, 1 t. 17 c. £46

Chemicals *(otherwise undescribed)*
Hamburg, 20 pkgs. £78
Natal, 9 cs. 53
Yokohama, 80 pkgs. 115 kegs. 631
Hiogo, 50 cs. 1,313

Caustic Soda—
San Diego, 55 t. 19 c. £376
Auckland, 2 t. 20
Canterbury, 8 t. 2 c. 60
Algoa Bay, 1¾ c. £15

Copper Sulphate—
Libau, 2 cks. £11
Lisbon, 5 t. 115

Cream Tartar—
Melbourne, 2 t. 10 c. £278
Canterbury, 5 28
Auckland, 5 28

Citric Acid—
Rotterdam, 1 t. 10 c. £215
Hamburg, 2 cks. 20 kegs. 212
Genoa, 1 t. 14 c. 217
Durban, 1 8
Odessa, 1 keg. 7
Yokohama, 5 35

Carbolic Solid—
Antwerp, 24 drs. £362
Adelaide, 4 cs. 30
Yokohama, 30 cs. 188

Copperas—
Alexandria, 5,050 lbs. £28
Mogador, 4 t. 15 c. £13

Carbolic Acid—
Algoa Bay, 2 c. £40
Cape Town, 200 gals. 30
Melbourne, 5 c. 7
Yokohama, 4,000 lbs. 82

Caustic Potash—
Algoa Bay, 5½ c. £12

Disinfectants—
Colombo, 4 cs. £25
Sydney, 12 cs. 51

Guano—
Penang, 500 t. £5,000
Jamaica, 25 250
Demerara, 75 825
Jersey, 56 519

Manure—
Penang, 475 t. £3,800
Hamburg, 30 375
Jamaica, 25 t. 1 c. 64
Demerara, 50 500
Jersey, 102 779
Martinique, 287 11 1,200
Valencia, 419 4 4,220
E. London, 1 42

Oxalic Acid—
Reval, 5 t. 7 c. £160
Melbourne, 7 12

Phosphorus—
Oporto, 6 c. £40

Potassium Bichromate—
Melbourne, 3 c. £6

Potassium Bicarbonate—
Philadelphia, 10 c. £14

Potassium Chlorate—
Yokohama, 400 kegs. £809

Potassium Cyanide—
Yokohama, 4 cs. £20

Sheep Dip—
B. Ayres, 67 t. 17 c. £3,140
Natal, 3 8 140
Algoa Bay, 1 5 70
Faulkland Is., 83 pkgs. 179
Natal, 100 92
Melbourne, 200 drs. 119

Salommoniac—
Libau, 2 cks. £19

Saccharum—
Sydney, 34 t. 2 c. £600

Sulphur—
Trinidad, 2 t. 16 c. £30
E. London, 8 18 94

Sodium Sulphate—
Ghent, 105 t. £105

Soda Crystals—
Sydney, 10 t. 14 c. £29
Hong Kong, 10 t. 3 c. £27

Saltpetre—
Rio Janeiro, 1 t. 9 c. £32
Algoa Bay, 15 18
Oporto, 1 2 24
Santos, 1 5 27

Sulphuric Acid—
Algoa Bay, 25 cs. £17

Sulphur Di-oxide—
Melbourne, 4 cs. £48

Sodium Tartrate—
Melbourne, 6½ c. £23

Tartaric Acid—
Algoa Bay, 1 cs. £19
Melbourne, 20 kegs. 5 cks. 179
Rangoon, 2 cwt. 14
Brisbane, 3 c. 22
Sydney, 5 cks. 21 kegs. 113
Durban, 168 lbs. 12
M. Video, 2 cks. 70

LIVERPOOL.

Week ending Jan. 11th.

Alkali—
Boston, 76 tcs.
Harve, 100 cks.

Ammonia Carbonate—
New York, 13 t. 12 . £435
Ancona, 5 2 q. 10
Harve, 19 239
Santander, 1 3 34
Venice, 10 21
Genoa, 10 21

Alum—
San Francisco, 20 t. £95
Havana, 8 39
Cape Town, 3 8 c. 15
Seville, 2 6 12
Santander, 4 19 1 q. 24
Palermo, 8 6 4
Callao, 3 6 18
Patras, 1 10 8

Ammonia Sulphate—
Barcelona, 5 t. 15 c. 3 q. £70
Ghent, 20 240
Bordeaux, 26 13 1 319
Hamburg, 66 18 800
Rotterdam, 5 62
Morlain, 19 15 240

Antimony Regulus—
Canada, 1 t. £74

Ammonia Muriate—
San Francisco, 1 t. 4 c. 3 q. 18
Philadelphia, 42 4 3 994

Antimony—
New York, 6 t. 10 c. £450

Bleaching Powder—
Amsterdam, 11 cks.
Antwerp, 3
Bombay, 5 cs.
Boston, 583
Hamburg, 42
Genoa, 30
Havana, 9
Monte Video, 14 brls.
Naples, 35 cks.
New York, 247
Oporto, 14 brls.
Palermo, 6 cks.
Philadelphia, 50
Portland, M. 20
Rouen, 24
Trieste, 39
Venice, 12

Brimstone—
Boston, 25 t. £136
Calcutta, 10 13 c. 2 q. 54

Boracic Acid—
Antwerp, 5 t. 12 c. £116

Bone Phosphate—
Hamburg, 20 t. 1 c. £96

Caustic Soda—
Adelaide, 100 drms.
Alexandria, 20
Algoa Bay, 32 bxs. 15 cs.
Alicante, 18 drms.
Amsterdam, 105
Antwerp, 34
Bahia, 15
Baltimore, 200
Barcelona, 110
Batoum, 200
Beyrout, 8
Bilbao, 100
Boston, 400
Cape Town, 5 bxs.
Catania, 61
Calcutta, 160
Ceara, 30
Chicago, 100
Durban, 19
Ferrol, 6 brls.
Genoa, 382
Havana, 50
Kobe, 50
La Guayra, 5
Leghorn, 145
Lisbon, 131
Madeira, 2
Melbourne, 163
Madrid, 100
Mexico, 100 kegs.
Malaga, 15 100
Nantes, 53
New York, 749 50 brls.
Pernambuco, 30
Philadelphia, 25
Piræus, 4
Port Natal, 10 bxs.
Rotterdam, 6
Rouen, 144
Salonica, 4
Santander, 10
Seville, 100
Valencia, 41
Venice, 100
Vigo, 20 brls.
Yokohama, 450

Copper Sulphate—
Barcelona, 55 t. 1 c. 3 q.
Bordeaux, 32 13 3
Genoa, 14 1 2
San Sebastian, 20
Nantes, 30 3 1
Ancona, 12 2 3
Leghorn, 45 3 2
Cape Town, 6 14 0
Palma, 50 4
Santos, 10
Lima, 3 2
Adelaide, 10
Antwerp, 10 3 0
Bilbao, 20 6 3
Barcelona, 14 6 1
Saigon, 1
San Thomé, 1
Singapore, 5 10
Tarragona, 5 4 2
Venice, 1 19 2
Genoa, 1
Leghorn, 10
Melbourne, 2 10
Bordeaux, 50 1 1

Caustic Potash—
Beyrout, 5 c.
Antwerp, 1 t. 15

Chemicals *(otherwise undescr*
East London, 6 c. 1 q.
Philadelphia, 1 t.

Carbolic Acid—
Rotterdam, 12 c. £67
Hamburg, 5 3 q. 36
Gibraltar, 8 1 25
Cobalt Oxide—
Gijon, 1 c. £45
Charcoal—
Hamburg, 45 t. £225
St. Nazaire, 16 15 c. 50
Copperas—
Talcahuano, 4 t. 8 c. 1 q. £10
Salaora, 4 0 2 8
Disinfectants—
Bombay, 6 t. 5 c £42
Rio de Janeiro, 5 3 q. 14
Epsom Salts—
Rio de Janeirio, 200 kegs.
Santos, 20 brls.
Glucose—
Malta, 40 bgs.
Manure—
Bordeaux, 200 t. 11 c. 2 q £779
Genoa, 399 7 1 1,389
Havre 50 0 3 202
Las Palmas, 1 10 2 19
Grand Canary, 12 t. 99
Magnesium Oxide—
Shanghai, 12 c. £8
Magnesium Carbonate—
Mexico, 8 c. £18
Oxalic Acid—
Trieste, 14 c. £20
Pitch—
Ancona, 1,647 t.
Port de Bouc, 1,002 2 c.
Sherbro, 5 brls.
Potassium Bichromate—
Barcelona, 11 t. 2 c. £389
Yokohama, 7 10 1 q. 254
Seville. 4 7
New York, 10 1 15
Varna, 6 3 13
Leghorn, 2 45
Potassium Chlorate—
Genoa, 2 t. 5 c. £134
Antwerp, 2 102
Pasages, 1 10 77
Hamburg, 1 3 52
Havana, 2 88
Leghorn, 1 5 64
Rotterdam, 10 21
Kobe, 12 4 598
Monte Video, 1 53
Nagasaki, 1 5 61
New York, 10 560
Shanghai, 2 10 105
Palermo, 1 47
Gothenburg, 5 242
Montreal, 10 20
Naples, 1 10 70
New York, 10 396
Naples, 10 25
Catania, 10 25
Potassium Carbonate—
Boston, 9 t. £161
Bombay, 1 13 c. 30
Potassium Chromate—
Havre, 2 t. £60
Genoa, 9 c. 1 q. 15
Phosphorus—
Bangkok, 1 q. £8
Vera Cruz, 3 3 40
Salt—
Adelaide, 11 c.
Akassa, 8½ t.
Amsterdam, 188
Antwerp, 247
Baltimore, 400
Bordeaux, 100
Boston, 325
Buguma, 60
Chicago, 65
Gopenhagen, 82
Coquimbo, 91
Drammen, 450
East London, 62½
Ghent, 120
Isle de Los, 50
Manaos, 19 13
Moss, 450
Newcastle, 100
New York, 138
Norfolk, 1,270
Old Calabar, 100 10
Opobo, 654 15
Portland, M. 80
Sodium Bicarbonate—
Adelaide, 20 kgs.
Brisbane, 30
B. Ayres, 10 130 cks.
Carthagenia, 120
Genoa, 9
Hamburg, 44 brls.

Melbourne, 57
Monte Video, 50
Newcastle, 50
New York, 196
Rotterdam, 122
Rouen, 60 4
Salonica, 20
Santander, 23
San Jose de Guatimala, 20 kgs.
Sydney, 67 kgs. 50
Trieste, 10
Vera Cruz, 5 2 cks.
Yokohama, 2,500
Soda Ash—
Adelaide, 42 brls.
Baltimore, 141 cks. 420 tcs.
Barcelona, 60
Boca, 100
Bombay, 18
Boston, 202 cks 1,408 bgs.
B. Ayres, 23 brls.
Buffalo, 60
Calcutta, 5
Callao, 30
Ghent, 11
Hamilton, 80 tcs.
Havre, 17
Madeira, 4
Malta, 150 cks.
Melbourne, 243 cks. 18 brls.
Montreal. 50
New York, 707 929 tcs.
Naples, 65
Piræus, 30 brls.
Philadelphia, 290 tcs. 252 cks.
Portland, 82
Rio de Janeiro, 75 brls.
Rotterdam, 545 t.
San Francisco, 110 cks. 218 tcs.
Santos, 30 brls.
Smyrna, 155 50 bxs. 145 cks.
Sydney, 101
Valencia, 4 cks.
Sodium Silicate—
Barcelona, 10 cks.
Genoa, 35 brls.
Naples, 20
Salonica, 8
Soda Crystals—
Boston, 280 brls.
B. Ayres, 10 cks.
Calcutta, 15
Callao, 140
New Orleans, 84
New York, 140
Philadelphia, 76
St. Johns, N'fld., 60
Sydney, 15 kegs.
Saltcake—
Philadelphia, 25 t. 7 c. £48
New York, 100 10 c. 217
Sodium Nitrate—
Shanghai, 5 t. £45
Trieste, 12 6 c. 1 q. 107
Carthagena, 1 5 1 10
Malaga, 10 0 1 87
Salammoniac—
Constantinople, 6 c. £10
Ghent, 18
Samsoun, 7 1 q. 12
Venice, 10 17
Mexico, 8 3 16
Genoa, 30 1 55
Durazzo, 9 15
Philadelphia, 5 t. 1 3 173
Patras, 5 1 9
Smyrna, 1 34
Batoum, 11 1 30
Syra, 3 5
Ancona, 9 1 16
Alexandria, 1 18 66
Sulphur
Boston, 75 t. 2 c. £318
Barbadoes, 20 3 3 q. 6
Sodium Biborate—
New York, 1 t. 3 c. 1 q. £50
Sheep Dip—
Bahia Blanca, 18 t. £945
Sulphuric Acid—
Para, 9 c. £6
Sodium Sulphate—
Lisbon, 2 t. 14 c. 3 q. £7
Saltpetre—
Callao, 1 t. 5 c. £28
Syra, 1 3 3 q. 27
Tar—
Cabinda, 10 cks.
Gaboon, 15 brls.
Loando, 10
Sherbro, 10
Tartaric Acid—
Havana, 1 c. £8
Tin Oxymuriate—
Rio Janeiro, 2 c. 2 q. £9

Waste Salt—
Oporbo, 181 t.
Sherbro, 350
Sierra Leone, 300
Sinoe, 25
Sydney, 500

HULL.

Week ending Jan. 14th.

Bleaching Powder—
Drontheim, 31 cks.
Danzig, 15
Stettin, 19
Bone Size—
Harlingen, 11 cks.
Chemicals (*otherwise undescribed*)
Antwerp, 45 cks.
Boston, 50
Christiania, 10 pks.
Copenhagen, 5
Drontheim, 10 pkgs.
Danzig, 5 30
Dunkirk, 173
Ghent, 5
Gothenberg, 13 cs. 40 pkgs.
Hamburg, 71 7 6
Konigsberg, 2 10 pks.
Libau, 27 60
New York, 100
Rouen, 65
Reval, 34
Rotterdam, 46 14
Stettin, 17 20 pkgs.
Caustic Soda—
Copenhagen, 40 drms.
Drontheim, 34 drums.
Dunkirk, 19
Hango, 80 drms. 6 cks.
Hamburg, 1 cs
Libau, 250 drums.
Reval, 100 cks.
Dyestuffs (*otherwise undescribed*)
Boston, 15 cks.
Christiania, 1 keg.
Hamburg, 2
Konigsberg, 1
Rotterdam, 46
Manure—
Copenhagen, 16 t.
Esbjerg, 1,400 bgs.
Ghent, 1,199
Hamburg, 1,575
Pitch—
Antwerp, 474 t.
Dunkirk, 75
Slag—
Danzig, 350 t.
Stettin, 46
Tar—
Hamburg, 10 cks.

GLASGOW.

Week ending Jan. 16th.

Ammonium Sulphate—
Calcutta, 2 t. 10 c. £32
Trinidad, 224 15½ £2,646
Ammonia—
Trinidad, 452 c. £258
Ammonium Carbonate—
Boston, 20 c. £31
Bones—
Trinidad, 714 c. £171
Benzol—
Rotterdam, 10 puns £320
Chemicals—
Rangoon, 4 t. 16 c. £50
Caustic Soda—
Dunedin, 199¾ c. £99
Copper Sulphate—
Marseilles, 5 t. 11¾ c. £105
Dyestuffs—
Sumac
Dunedin, 35 c. £23
Manure—
Trinidad, 162 t. 19¾ c. £604
Magnesium Chloride—
Penang, £25
Potassium Sulphate—
Trinidad, 84¼ c. £31
Pitch—
Bordeaux, 720 t. £1,080
Calcutta, 28½ 55
Potassium Bichromate—
Rotterdam, 12 cks. £175
Dunedin, 14¾ c. 8 lbs. 27
Marseilles, 1 t. 17¾ c. 60
Potassium Prussiate—
Rotterdam, 4 cks. £57

Paraffin
Oporto,
Sulphuric
Rangoon,
Sodium B
Rotterdam
Tartar Cr
Dunedin,
Tar—
Cape Town
Natal
Turpentin
Penang,
Waste sal
Bombay,

Week e

Ammonia
Hamburg,
Alum—
Rotterdam,
Aalborg,
Alkali—
Libau, 9
Ammoniac
Huelva,
Bleaching
Hamburg,
Rotterdam,
"
Libau,
Antwerp,
"
Caustic So
Antwerp,
Flint Grou
Antwerp,
Hydrochlo
Huelva,
Litharge—
Hamburg,
Lisbon,
Magnesia—
Hamburg,
Naphthalin
Antwerp,
Potassium
Hamburg,
Libau,
Antwerp,
Pitch—
Antwerp,
Soda—
Rotterdam,
Horsens,
Sodium Bi
Malmo,

G

Week

Bleaching
Antwerp, 3
Bone Size—
Boulogne,
Benzol—
Rotterdam,
Terneuzen,
Chemicals
Antwerp,
Ghent,
Hamburg,
Rotterdam,
Coal Produ
Rotterdam,
Dyestuffs (
Boulogne,
Ghent,
Manure—
Boulogne,
Hamburg,
Pitch—
Antwerp,

GE

Week

Chemicals
Antwerp, 8
Dieppe, 1
Coal Produ
Antwerp, 7
Dieppe, 24

PRICES CURRENT.

THURSDAY, JANUARY 30, 1890.

PREPARED BY HIGGINBOTTOM AND CO., 116, PORTLAND STREET, MANCHESTER.

'he values stated are F.O.R. *at maker's works, or at usual ports of shipment in U.K. The price in different localities may vary*

		£	s.	d.
eids :—				
Acetic, 25 °/o and 40 °/o	per cwt	7/6 & 12/6		
,, Glacial	,,	2	10	0
Arsenic, S.G., 2000°	,,	0	12	0
Chromic 82 °/o	nett per lb.	0	0	6½
Fluoric	,,	0	0	3¾
Muriatic (Tower Salts), 30° Tw.	per bottle	0	0	8
,, (Cylinder), 30° Tw.	,,	0	2	11
Nitric 80° Tw.	per lb.	0	0	2
Nitrous	,,	0	0	1¾
Oxalic	nett ,,	0	0	3
Picric	,,	0	1	5
Sulphuric (fuming 50 °/o)	per ton	15	10	0
,, (monohydrate)	,,	5	10	0
,, (Pyrites, 168°)	,,	3	2	6
,, (,, 150°)	,,	1	8	0
,, (free from arsenic, 140/145°)	,,	1	10	0
Sulphurous (solution)	,,	2	15	0
Tannic (pure)	per lb.	0	1	0½
Tartaric	,,	0	1	3
rsenic, white powered	per ton	13	0	0
lum (looselump)	,,	4	12	6
lumina Sulphate (pure)	,,	5	0	0
,, ,, (medium qualities)	,,	4	10	0
luminium	per lb.	0	15	0
mmonia, ·880=28°	per lb.	0	0	3
,, =24°	,,	0	0	1⅞
,, Carbonate	nett ,,	0	0	3½
,, Muriate	per ton	24	0	0
,, ,, (sal-ammoniac) 1st & 2nd	per cwt.	36/- & 34/-		
,, Nitrate	per ton	37	10	0
,, Phosphate	per lb.	0	0	10½
,, Sulphate (grey), London	per ton	12	0	0
,, ,, (grey), Hull	,,	11	17	6
niline Oil (pure)	per lb.	0	0	11¾
niline Salt ,,	,,	0	0	11
ntimony	per ton	75	0	0
,, (tartar emetic)	per lb.	0	1	1
,, (golden sulphide)	,,	0	0	10
arium Chloride	per ton	7	10	0
,, Carbonate (native)	,,	3	15	0
,, Sulphate (native levigated)	,,	45/- to 75/-		
leaching Powder, 35%	,,	5	2	6
,, Liquor, 7%	,,	2	10	0
isulphide of Carbon	,,	12	15	0
hromium Acetate (crystal	per lb.	0	0	6
alcium Chloride	per ton	2	0	0
hina Clay (at Runcorn) in bulk	,,	16/- to 32/-		
al Tar Dyes :—				
Alizarine (artificial) 20%	per lb.		0	9
Magenta	,,	8	3	9
al Tar Products				
Anthracene, 30% A, f.o.b. London	per unit per cwt.	0	1	6
Benzol, 90% nominal	per gallon	0	4	0
,, 50/90 ,,	,,	0	3	3
Carbolic Acid (crystallised 35°)	per lb.	0	0	10¾
,, (crude 60°)	per gallon	0	2	10
Creosote (ordinary)	,,	0	0	2½
,, (filtered for Lucigen light)	,,	0	0	3
Crude Naphtha 30% @ 120° C.	,,	0	1	2
Grease Oils, 22° Tw.	per ton	3	0	0
Pitch, f.o.b. Liverpool or Garston	,,	1	12	0
Solvent Naphtha, 90% at 160°	per gallon	0	1	8
oke-oven Oils (crude)	,,	0	0	2½
opper (Chili Bars)	per ton	48	15	0
,, Sulphate	,,	24	10	0
,, Oxide (copper scales)	,,	52	10	0
ycerine (crude)	,,	30	0	0
,, (distilled S.G. 1250°)	,,	55	0	0
dine	nett, per oz.	8d. to 9d.		
on Sulphate (copperas)	per ton	1	10	0
, Sulphide (antimony slag)	,,	2	0	0
ead (sheet) for cash	,,	16	0	0
,, Litharge Flake (ex ship)	,,	17	10	0
,, Acetate (white ,,)	,,	23	10	0
,, ,, (brown ,,)	,,	19	0	0
,, Carbonate (white lead) pure	,,	19	0	0
,, Nitrate	,,	22	5	0

		£	s.	d.
Lime, Acetate (brown)	,,	8	10	0
,, ,, (grey)	per ton	14	0	0
Magnesium (ribbon and wire)	per oz.	0	2	3
,, Chloride (ex ship)	per ton	2	7	6
,, Carbonate	per cwt.	1	17	6
,, Hydrate	,,	0	10	0
,, Sulphate (Epsom Salts)	per ton	3	0	0
Manganese, Sulphate	,,	18	0	0
,, Borate (1st and 2nd)	per cwt.	60/- & 42/6		
,, Ore, 70%	per ton	4	5	0
Methylated Spirit, 61° O.P.	per gallon	0	2	2
Naphtha (Wood), Solvent	,,	0	3	9
,, ,, Miscible, 60° O.P.	,,	0	4	0
Oils :—				
Cotton-seed	per ton	22	0	0
Linseed	,,	22	10	0
Lubricating, Scotch, 890°—895°	,,	7	5	0
Petroleum, Russian	per gallon	0	0	5¾
,, American	,,	0	0	6
Potassium (metal)	per oz.	0	3	10
,, Bichromate	per lb.	0	0	4
,, Carbonate, 90% (ex ship)	per ton	18	5	0
,, Chlorate	per lb.	0	0	4¾
,, Cyanide, 98%	,,	0	2	0
,, Hydrate (Caustic Potash) 80/85%	per ton	23	0	0
,, ,, (Caustic Potash) 75/80%	,,	21	15	0
,, ,, (Caustic Potash) 70/75%	,,	20	15	0
,, Nitrate (refined)	per cwt.	1	3	6
,, Permanganate	,,	4	5	0
Prussiate Yellow	per lb.	0	0	10½
,, Sulphate, 90%	per ton	9	10	0
,, Muriate, 80%	,,	7	15	0
Silver (metal)	per oz.	0	3	8½
,, Nitrate	,,	0	2	5½
Sodium (metal)	per lb.	0	4	0
,, Carb. (refined Soda-ash) 48%	per ton	5	7	6
,, ,, (Caustic Soda-ash) 48%	,,	4	7	6
,, ,, (Carb. Soda-ash) 48%	,,	4	15	0
,, ,, (Carb. Soda-ash) 58%	,,	5	12	6
,, ,, (Soda Crystals)	,,	2	13	6
,, Acetate (ex-ship)	,,	16	0	0
,, Arseniate, 45%	,,	10	0	0
,, Chlorate	per lb.	0	0	6¼
,, Borate (Borax)	nett, per ton.	29	10	0
,, Bichromate	per lb.	0	0	3
,, Hydrate (77% Caustic Soda)	per ton.	9	10	0
,, ,, (74% Caustic Soda)	,,	8	15	0
,, ,, (70% Caustic Soda)	,,	7	17	6
,, ,, (60% Caustic Soda, white)	,,	6	17	6
,, ,, (60% Caustic Soda, cream)	,,	6	10	0
,, Bicarbonate	,,	5	0	0
,, Hyposulphite	,,	5	0	0
,, Manganate, 25%	,,	8	10	0
,, Nitrate (95%) ex-ship Liverpool	per cwt.	0	8	3
,, Nitrite, 98%	per ton	27	10	0
,, Phosphate	,,	15	15	0
,, Prussiate	per lb.	0	0	7¾
,, Silicate (glass)	per ton	5	7	6
,, ,, (liquid, 100° Tw.)	,,	3	17	6
,, Stannate, 40%	per cwt.	2	0	0
,, Sulphate (Salt-cake)	per ton.	1	6	0
,, ,, (Glauber's Salts)	,,	1	10	0
,, Sulphide	,,	7	15	0
,, Sulphite	,,	5	5	0
Strontium Hydrate, 98%	,,	8	0	0
Sulphocyanide Ammonium, 95%	per lb.	0	0	7¾
,, Barium, 95%	,	0	0	5½
,, Potassium	,,	0	0	7¾
Sulphur (Flowers)	per ton.	7	0	0
,, (Roll Brimstone)	,,	6	0	0
,, Brimstone : Best Quality	,,	4	17	6
Superphosphate of Lime (26%)	,,	2	10	0
Tallow	,,	25	10	0
Tin (English Ingots)	,,	99	0	0
,, Crystals	per lb.	0	0	6¾
Zinc (Spelter	per ton.	24	0	0
,, Chloride (solution, 96° Tw.	,	6	0	0

THE

CHEMICAL TRADE JOURI

Publishing Offices: 32, *BLACKFRIARS STREET, MANCHESTER.*

No. 142. **SATURDAY, FEBRUARY 8, 1890.**

Contents.

Notices.

All communications for the *Chemical Trade Journal* should be addressed, and Cheques and Post Office Orders made payable to—

DAVIS BROS., 32, Blackfriars Street, MANCHESTER.

Our Registered Telegraphic Address is—

"Expert, Manchester."

The Terms of Subscription, commencing at any date, to the *Chemical Trade Journal*,—payable in advance,—including postage to any part of the world, are as follow:—

Yearly (52 numbers) .... **12s. 6d.**
Half-Yearly (26 numbers) .... **6s. 6d.**
Quarterly (13 numbers) .... **3s. 6d.**

Readers will oblige by making their remittances for subscriptions by Postal or Post Office Order, crossed.

Communications for the Editor, if intended for insertion in the current week's issue, should reach the office not later than **Tuesday Morning**.

Articles, reports, and correspondence on all matters of interest to the Chemical and allied industries, home and foreign, are solicited. Correspondents should condense their matter as much as possible, write on one side only of the paper, and in all cases give their names and addresses, not necessarily for publication. Sketches should be sent on separate sheets.

We cannot undertake to return rejected manuscripts or drawings, unless accompanied by a stamped directed envelope.

Readers are invited to forward items of intelligence, or cuttings from local newspapers, of interest to the trades concerned.

As it is one of the special features of the *Chemical Trade Journal* to give the earliest information respecting new processes, improvements, inventions, etc., bearing upon the Chemical and allied industries, or which may be of interest to our readers, the Editor invites particulars of such—when in working order—from the originators; and if the subject is deemed of sufficient importance, an expert will visit and report upon the same in the columns of the *Journal*. There is no fee required for visits of this kind.

We shall esteem it a favour if any of our readers, in making inquiries of, or opening accounts with advertisers in this paper, will kindly mention the *Chemical Trade Journal* as the source of their information.

Advertisements intended for insertion in the current week's issue, should reach the office by **Wednesday morning** at the latest.

Advertisements.

Prepaid Advertisements of Situations Vacant, Premises on Sale or To be Let, Miscellaneous Wants, and Specific Articles for Sale by Private Contract, are inserted in the *Chemical Trade Journal* at the following rates:—

30 Words and under.......... 2s. 0d. per insertion.
Each additional 10 words .... 0s. 6d. ,,

Advertisements of Situations Wanted are inserted at one-half the above rates when prepaid, viz:—

30 Words and under.......... 1s. 0d. per insertion.
Each additional 10 words .... 0s. 3d. ,,

Trade Advertisements, Announcements in the Directory Columns, and all Adver. ements not prepaid, are charged at the Tariff rates which will be forwarded on plication.

CENTRAL versus LOCAL ADI

WE note with deep concern the Tuesday's meeting of the Manch Society of Chemical Industry. After th of papers by Mr. W. Younger, the subj Instruction Act was brought forward by the Owens College, and our attention, l meeting that appeared in the followi *Guardian*, will be principally devoted his utterances.

In the first place, we wish to remind have all along withheld our cordial ass visions of the Act and to the proposed m tration, so far as this can be foreshadov that the measure was being hurried alo allow its various details to be worked out and precision to render it permanently u once passed, it would be a barrier to reason of the finality that always attac a legal instrument.

But it is perfectly incomprehensible to Bailey's position should have remained ing last under the erroneous impressions the meeting. Surely it is only in the things that a large and wealthy city like provide the wherewithal for its own edu and enterprises, and should bear its pr of extending the operations of any nati to districts less favoured in their resou trouble appears to be that the Science of South Kensington intends to have a fi ministration of the Act. To us this trou a blessing in disguise. We are far from s no evils attendant upon a centralisatio for this is a fact only too patent to all. this, that local management brings in innumerable multitude of evils that the comes insignificant in comparison. Loc jealousies, and the various other local infl be so easy to recapitulate, would all pl parts in the misuse of the enormous sum ratepayer would be invited to place at the tral administration avoids much of this. tial though sometimes unfeeling. One m dealt out to all, and the interest of the fa the interpretation of the wants of the ma

But we were most concerned by the an Bailey that he wiped his hands of any a matter before the City Council. This s hope it does not mean that he will rel cause of education, or withdraw his supp cause to which he has rendered such Manchester is still in want of "more ligt discouraging commentary on the actior parties if they force Dr. Bailey to p execution.

The Paper Makers' Column.

'HE ELECTRIC DECOMPOSITION OF CHLORIDES IN SOLUTION.

MANY years ago the patent records told us how we might prepare caustic soda and chlorine from a solution of common salt by assing a current of electricity through it; but those were the days of rimary batteries, when the cost of the electric current put all such lecompositions out of the question. The improvements in dynamos luring the last two decades have entirely altered the situation, though, n our opinion, the matter is still unsolved—yet the problem is becoming nearer of solution day by day.

There are three steps by which practical electrolysis must be ffected; the first is an economical supply of power; the second, an conomical transformation of that power into electricity; and, hirdly, the discovery of practical methods of decomposition. The irst step will probably be solved by the Hargreaves thermo-motor; he second may be regarded as practically solved, though students of electrolysis on the large scale will possibly prefer to blunder along naking their own discoveries than to accept the results of contemporary workers.

It is the third step, however, which remains a fertile field for the inventor, and practical processes of decomposition are alone needed, as it is a well-known fact that nearly every chemical compound can be decomposed when sufficient electric force is applied to it. The decomposition of chemical compounds by this means demands very careful study and attention, as the results are liable to be influenced to a very large extent by small changes in the physical conditions, such as intensity of current, surface of electrodes and their distances asunder, the nature of the electrodes, and, perhaps most of all, the manner in which the electricity is applied.

We have now in our mind's eye two processes in practical operation, depending for their success upon the economical application of electricity. The Hermite process of electric bleaching is one that has been frequently alluded to in these columns, and depends upon the decomposition of magnesium chloride, while the other is the Webster process for the electric treatment of sewage, which doubtless owes its efficacy to the decomposition of the chlorides that all sewage contains. It is easy to show theoretically how, under one set of conditions, the process could end in nothing but failure, while under more favourable reatment success may follow, and, as an illustration, we offer a few emarks upon the electrical decomposition of a solution of magnesium chloride.

Everyone knows that the theoretical maximum yields of the laboratory are not to be attained in practice; but combining them with knowledge gained by practice on the large scale very truthful figures can be obtained.

The reactions which occur during the electrolysis of an aqueous olution of magnesium chloride may be stated for our purpose here, is:—

$$MgCl_2 + H_2O + H_2O = MgH_2O_2 + H_2 + Cl_2$$

It will be seen that this is not a mere splitting up of the chloride into magnesium metal and chlorine. If it were so we should have to overcome affinities represented by 93·5 calories.

The chemical equivalent of chloride of magnesium is made up of:—

Magnesium	12·0
Chlorine	35·5
Magnesium Chloride	47·5
Heat developed in calories	93·5

om which we can easily calculate the number of volts required to vercome the chemical affinity. The formula

$$\frac{He}{23} = E = \frac{93·5}{23} = 4·06$$

pplies this information, showing that 5 volts would be required in actice, and that 4 volts would not be sufficient.

The gases, however, appear at the electrodes and not the metal, so at we have to overcome the heat of formation of dry magnesium hloride and water, less the heat of formation of magnesium hydrate; is is:—

$$75·5 \times 34·5 - 74·9 = 35·1 \text{ calories.}$$

If we substitute this value for 9·35 calories above mentioned, we ve:—

$$\frac{He}{23} = E = \frac{35·1}{23} = 1·53 \text{ volts.}$$

We do not claim rigid mathematical accuracy for this figure, but the practical proof of its being very near the mark lies in the fact that a strong solution of magnesium chloride can be decomposed by the current from a single Bunsen cell.

The next equation shows us that to produce one kilogramme (2·2 lbs. of chlorine per hour, we shall require theoretically 1·54 H.P.:—

$$\frac{35·1 \times 1000 \times 424}{35·5 \times 3600 \times 75} = 1·54 \text{ H.P.,}$$

and a current of 782 amperes is required, as shown by the following:—

$$\frac{1000}{0·000355 \times 3600} = 782.$$

This is all fair sailing, but the foregoing figures refer only to the actual work required for theoretically overcoming the chemical affinity of the chlorine for the magnesium; the actual work in practice can only be known beforehand when the resistances are known. In practice the resistances can be easily measured and kept within reasonable limits.

If we allow 50 square centimetres for each ampere of current, a surface of 4 square metres will be required to yield one kilogramme of chlorine per hour, and knowing the specific resistance of a solution of chloride of magnesium, and allowing the distance of one centimetre between each electrode, we can fairly well ascertain what the resistance of the electrolysing tanks will be, per ampere. The resistance of the conductors can be easily ascertained from their length and section.

The counter E.M.F. being $= \frac{\text{Volts}}{\text{Amperes}}$ is also readily calculated.

The internal resistance of the dynamo; resistance due to the polarization of the electrodes and other causes, bring up a total of not less in practice than 0·01 ohm,

$$0·01 \times 782 = 7·82 \text{ volts.}$$

We have already seen that the actual decomposition requires 1·53 volts so that the work there realised is but:—

$$\frac{100 \times 1·53}{7·82} = 20\%$$

of the total, showing clearly that in practice 7·9 brake horse power is required to furnish one kilo. of chlorine per hour, which is equal to 2·8 kilos. per hour of a bleaching powder of 35 per cent. strength.

In order, then, to replace the equivalent of bleaching powder, equal to the employment of 2·8 kilos. (6·2 lbs.) per hour, the installation would have to consist of:—

A steam engine of 8 brake horse power and a dynamo capable of yielding 6,115 watts of effective current.

This can hardly be called a practical installation, let us multiply it by 8 times, it will then be equal to 50 lbs. of bleaching powder per hour, or three tons per working week, and there will have to be installed, one steam engine of 64 brake horse power, and a dynamo or dynamos yielding, say, 50,000 watts.

We have now to determine on the size and style of dynamos. We may have one passing a large current of low intensity, say 10,000 amperes at 5 volts; or eight separate dynamos each giving 1,250 amperes at 5 volts; or we may have one dynamo producing 1,250 amperes at a pressure of 40 volts.

Although but 1·53 volts are required theoretically to decompose the chloride in solution, yet in practice, 5 volts are reached, partly because of the magnesia, which, attaching itself to the cathode, increases the resistance very considerably, and partly on account of the inferior conductivity of dilute solutions, as well as the fact that a considerable current density is necessary, as otherwise the cost of electrodes becomes a most serious question. Five volts may therefore be calculated upon as being absolutely necessary in an ordinary electrolyser.

But though we have any one of the foregoing alternatives before us, it by no means follows that they are equally efficacious in action. The first proposal to send a large current of 5 volt intensity through what is practically a pair of electrodes of immense surface, is so much at variance with modern experience as to be at once discarded, and the second, to send a weak current from eight smaller machines through eight different tanks, is known to result in such a loss of electric efficiency as to be utterly uneconomical. M. Gramme, in his experiments upon the decomposition of water by a current of low intensity, obtained an efficiency of barely 35 per cent., so that in this ratio it is more than probable that in the case of the first problem an engine of 120 brake horse power would be required to drive the machine necessary to produce the equivalent of 3 tons of bleaching powder per week, and there are besides other grave disadvantages into which it is not necessary to enter here. The second installation may also be dismissed from our minds in a similar line of argument, so that we now come to the third proposition of the dynamo yielding 1,250 amperes at 40 volts, and worked in series through eight electrolysers. It is in this connection that our calculations lead us to believe that 64 horse power will be con-

In most cases merchandise is exported by some party other than the manufacturer, and in very many cases the exportation of the article on which drawback is allowed, is only incidental to the exportation of some other article, as in the case of bags exported filled with grain, flour, etc., and tin cans exported filled with petroleum, turpentine, fish, fruits, and meats. The application of the drawback law reduces the cost of placing domestic products in foreign markets to what that cost would be in case there was no import duty on the materials used in the manufacture of the packages in which such products are exported.

Certain manufacturing interests have undertaken to persuade the Secretary of the Treasury so to administer existing law that exporters of domestic products might be forced to buy export packages from them, and these manufacturers are now instructing Congress so to revise and amend the drawback law that exporters will be obliged to purchase packages of their make instead of using other packages equally serviceable and costing less. These manufacturers have substantial reasons for trying to make the department and Congress forget that the object of the drawback system is not to give to any manufacturer the right indirectly to tax or throttle the export trade of the country.

Since the right to claim drawback clearly belongs to the party who places the exported article on the foregin market, it is important that the right of that party only should be recognised in the law. The conditions of exportation are such as to make it difficult in many cases to determine who is the exporter of merchandise shipped. Ownership of much of that merchandise changes on the deck of the exporting vessel. Where merchandise is shipped from interior points on "through" bills of lading, ownership often changes while the goods are in transit from point of shipment to port of exportation. These and other conditions make it practically impossible for an executive officer to secure the information which would make him competent to decide how and where to apply the term exporter. The bill of lading, which in all business transactions is recognised as the commercial title to the merchandise covered thereby, always reveals a party who is either the exporter or his agent. These facts clearly indicate the confusion which would result from a legal requirement that the Government should in allowing drawback deal directly with the exporter only, and as clearly indicate that the Government may safely, and should, deal with the exporter, either directly or through his agent.

Some opponents claim that the present drawback system gives an opportunity to use domestic material in the manufacture of an article to be exported, and gives an opportunity to swear that such article was made from imported material, thus laying the foundation for a claim for drawback by means of which the Government would be defrauded. A manufacturer of tin cans might start the manufacture of tin plates in this country, and at a cost considerably above that of imported plates procure plates from which to manufacture cans for export. A manufacturer of bags made from burlaps or bagging, might, by paying a little more than the imported article would cost, procure a like article of domestic manufacture and from that material make bags to be exported, using a like quantity and kind of imported material to make bags for domestic use. What has been said of the manufacturers named may be said of all manufacturers excepting such as are made from material the like of which is not produced in this country. The opportunity for perjury and fraud on the part of the manufacturer and exporter is, in all the cases noted, clearly manifest. To get the measure of the danger to which the revenue is exposed because of these opportunities, it is only necessary to ascertain the strength of the motive or impulse to which manufacturers and exporters are subject. Thorough search discovers no reason why a refiner of sugar should set apart for export the product of raw sugars grown in the United States or the Sandwich Islands, and put on the home market the product of raw sugars on which import duty has been paid. Absolutely nothing could be gained by perjury and fraud in the case considered. In case of tin cans, considerably less than nothing could be gained by the use of like means. Perjury and fraud on the part of manufacturers of bags would, in the present condition of the market, result in a gain for the criminals of a trifle less than nothing. In no case could the use for export of domestic material in place of like imported material benefit either manufacturer or exporter, because the present system would necessitate the setting apart for domestic consumption of a like quantity of imported material, so, unless it be assumed that manufacturers and exporters will commit perjury just for the fun of the thing, and defraud the Government at considerable expense to themselves, the fears expressed relative to danger of the revenue from frauds of the kind considered are groundless, and it seems worse than absurd to propose to cripple the export trade by laws to prevent motiveless crime and profitless fraud.

Free traders claim that duties on raw material prevent the exportation of articles made from such material. The drawback system gives all the advantages which free trade can offer to manufacturers and exporters by practically making material for exports free.—*Oil, Paint, and Drug Reporter.*

Legal.

SIR WILLIAM THOMSON'S COMPASS.

'N the Chancery Division of the High Court of Justice, the case of · Thomson v. Hughes and Son has been heard. Lord Justice lorth was moved by the plaintiff to grant an injunction that the efendants, their servants, agents, and workmen might be restrained ntil the trial, or further order, from making or selling any compass ards similar to the card now being manufactured and sold by them as ie "Paget" card, and from making and selling any other cards similar).the cards described in the plaintiff's specification, except only cards ie same as the compass cards sent by the defendants' solicitor to the laintiff's solicitor on the 1st January, 1889, and referred to in the rder of the Court of Appeal made in this action on the 6th March, 889, and that the defendants be ordered to pay the cost of this notion. The Attorney-General (Sir Richard Webster, Q.C., M.P.), Ir. Aston, Q.C., and Mr. Bousfield were retained to argue the case or the plaintiff, while the defendants were represented by Mr. 'letcher Moulton, Q.C., and Mr. Chadwyck Healey. The Attorney. ieneral, in opening the case for the plaintiff, said the action was rought on Thomson's patent of 1876. The defendants were making imething which was in no way a "Paget" card, but which was in all ssential particulars the same as the "Moore," in respect of which the efendants had submitted to an injunction. This the plaintiff had iscovered in November, 1889, whereupon he at once wrote to the olicitors of the defendants, requiring an undertaking that they would ot sell any more "Paget" cards. The defendants wrote back that ney had kept an account of the cards made under the "Paget" pecification, and would continue to do so. The "Paget" patent, how. ver, had expired, and if the defendants were really making a "Paget's" ard there was not the slightest reason to keep an account. If his the Attorney-General's) affidavits were correct, there was no doubt hat the defendants were not making a card acccording to Paget's speci. cation, but making a card taking all the merit of Sir Wm. Thomson's nvention, and selling it as a "Paget" card. Prior to Sir Wm. 'homson's card, the best compass known was the Admiralty Compass. Compasses were affected by a number of errors, the chief being the ermanent magnetism of the ship, the magnetism induced by the earth n the soft iron of the ship, and what was known as the "heeling" rror. Sir William Thomson for the first time practically showed eople how to use short magnets, which corrected these errors. He, or the first time, combined a light card with short needles so placed hat the result was the obtaining of a very slow period of oscillation, vhile at the same time a steady compass, and one to which these cor. ections could be very readily applied was secured. Sir William Thomson was led to the line of thought which resulted in his inven. ion by having to write the biographical notice of Mr. Archibald Smith, member of the Bar. He found such difficulty in pressing the matter hat he had to take up the manufacture of his discovery, and had to lemonstrate the efficiency of his compass cards on his own yacht, and n vessels on which he fixed them gratuitously, with the result that here had been a number of attempts to infringe his patent. The alidity of the patent was fought before Mr. Justice Kekewich, and udgment was given in his favour. The same thing happened n Ireland and, practically, in Scotland. Mr. Moulton, Q.C., on be. alf of the defendants, submitted that the matter should stand over in rder to give the defendants an opportunity of cross-examining the laintiff's witnesses. He further contended that what his clients were nanufacturing was substantially the "Paget" compass card, with egard to which no undertaking was given in the Court of Appeal. Ir. Justice North said that it seemed to him that the defendants had nade a card which, for all practical purposes, was the same as that vhich he had undertaken not to make. Looking to the balance of onvenience and inconvenience, he was not, however, in favour of ranting an injunction. It seemed to him that the plaintiff could lose ttle, if anything, if he did not grant an injunction, and the plaintiff vas successful at the trial, while the defendants might lose a great deal he granted an injunction, and the defendant turned out successful. Jnder these circumstances, on the authority of a case in the Irish Court of Appeal, he must refuse to grant an injunction, asking the efendants to give an undertaking, by counsel, that they will keep an ccount of all the cards they make during the continuance of the atent. Mr. Chadwyck Healey signified that he would give the ndertaking asked for. The question of the costs of this motion was eserved.

BEEF EXTRACT.—As a source of "beef extract" the horse has found formidable rival. The chemist at the Russian whaling station of 'ort Wladimir, on the Murman coast, announces his ability to produce he article from whale flesh. The new product is said to be equal to enuine beef-extract in appearance and taste.

THE ESTIMATION OF THE RELATIVE AMOUNTS OF CHROMATE AND BICHROMATE IN A MIXTURE.

By J. ARTHUR WILSON.

THERE are a few operations in which, not only is it requisite to determine the actual percentage of chromic acid, but also to know the state in which it exists, and the amount, for instance, in mordanting with bichromate and sulphuric acid, in woollen dyeing, and in the manufacture of bichromates themselves.

The new indicator of Traube and Hock, viz., lacmoid, reacts neutral to bichromate, and alkaline to chromate, so that it may be used. The method is very simple; the solution is titrated with N/2 sulphuric acid, till paper painted over with neutral lacmoid solution shows a faint red. The following shows the results obtained from my notebook, in some test experiments :—

Substance taken.	c.c. Standard Acid.	Found.
1. 1grm. pure K_2CrO_4	5·15c.c. of Normal H_2SO_4	1·0012
2. ,, ,, ,,	,, ,, ,,	,,
3. 1grm. pure K_2CrO_4		
0·5 ,, ,, $K_2Cr_2O_7$	5·15 ,, N H_2SO_4	1·0012

There is another method which does not seem to be sufficiently well known, and I was somewhat surprised to find no mention of it in looking over the latest edition of Sutton's Volumetric Analysis. It depends on the fact that the blue colour of chrominm heptoxide, imparted to ether, is not produced till slight excess of sulphuric acid is present. The moment this excess is produced, the blue colour is imparted to ether. The method applied to mixtures of chromic and bichromic acids is as follows: 5 grammes of the substance (say a mixture of $K_2Cr_2O_7$, and K_2CrO_4) is dissolved in about 50 c.c. of water, and 3 c.c. of hydrogen peroxide added (the acidity of the hydric peroxide must be slight, or if more than usual it must be estimated and allowed for), and the liquid covered with a layer of ether, a few centimetres deep. The spit of the burette should be long, and pass through an indiarubber stopper which fits loosely into the test tube or small flask holding the mixture. Normal sulphuric acid is then added, with agitation, till the well-known blue colour is produced in the ethereal layer; Results are a little too high, but good enough for all practical purposes. Test trials from my laboratory notebook :—

Weight of Substance, taken in grammes.	c.c. of Normal Sulphuric Acid.	Amount, from the Acid consumed.
1. 2·5017	12·9	2·507
2. 2·5012	12·9	2·517
3. 2·508 K_2CrO_4		
1·034 $K_2Cr_2O_7$	13·0	2·527

THE FREIGHT MARKET.

MESSRS. Angier Bros., in their last circular, say freights during the past fortnight have been dull, not much doing, and rates weaker generally. In China waters a few coasting orders were filled at 14 to 16 cents. per picul, Saigon to Hong Kong, but homewards employment is difficult to find. From Burmah and India the late improvements in rates was barely maintained, and orders are scarce since the last few fixtures. Black Sea business has been very dull for both prompt and spring loading at lower prices. Mediterranean employment is poor, with less demand for tonnage at a decline in rates. American business keeps fair, and the fixtures show little falling off, except that the demand is more limited. Nearly all the steamers bound out have experienced unusually severe weather and long passages, materially interfering with shipment and contract dates. The River Plate offers fair rates for March to May loading. Very little business offers for outside boats from the Brazils. In time charters orders are scarce, rates 9s. to 10s. Outward coal rates keep very low all round, with considerable delay in loading. Iron and berth cargoes follow the coal rate. Bunker coals are slightly cheaper and easier to get; wages are maintained at the late advance. Under present circumstances old boats are making little or no real profit, and the cheaply bought new boats are giving seriously diminished returns.

BARBADOES.—Messrs. Laurie and Co. in their circular say:—During the past month we have been amply supplied with seeking tonnage suitable for all inquiries, and quite a number of vessels have been taken up. The sugar season is near at hand, and we have already made some engagements from Martinique and Guadeloupe to French ports. We anticipate a large demand for tonnage during the reaping season, as the crops all through the West Indies are reported as being very large.

CARDIFF.—Messrs. Henderson Brothers advise us of the following fixtures yesterday:—Gibraltar, 7s., Ulleswater; Smyrna, 10s. 6d., Acheen; 19s., Badsworth; Tarragona, 9s. Dunkirk; 9s., clean copper, Blackwater.

DEMERARA.—Messrs. Sandbach, Parker, & Co. advise us of the following charters and freights:—Charters—A. C. Wade, sugar, U.S.A., 15c direct port, 16c Delaware Breakwater; Acadia, sugar, 14c New York, 16c Philadelphia; Arecuna, sugar, 15c Philadelphia; Greve Frijs, molasses, Copenhagen, 30s. Freights—Steam to London, sugar, 1s. 6d per cwt.; rum, 2¼d. per gallon. Steam to Liverpool, sugar, 1s.6d per cwt.; rum, 2¼d. per gallon. Sail to London, sugar, 1s. per cwt.; rum, 1½d. per gallon. Sail to Liverpool, sugar, 1s. 3d. per cwt.; rum, 2d. per gallon.

DUNDEE.—Our correspondent writes that for the Baltic trade (wood) vessels are being fixed to load timber, at Baltic ports, at 30s. for Dundee. Last year the season opened with rates at 35s.

GLASGOW, Monday.—Our correspondent telegraphs that there was a quiet feeling in the Glasgow Freight Market to-day, and little business was completed. Ships reported fixed:—San Francisco, to United Kingdom, Havre, or Antwerp, 36s. 3d.; West Coast of South America to United Kingdom, 31s. 3d.; Newcastle, New South Wales, to San Francisco, 21s. In steam tonnage a vessel is fixed—Leith to Genoa, 9s. 6d. Bombay open charters:—25s. is quoted Feb. loading.

MAURITIUS.—Messrs. Chadwick and Co. have received the following telegram from Messrs. Scott and Co., dated January 23:—Freights steady—to United Kingdom or Colonies, 25s. and 5 per cent. per ton; to Bombay, six to eight annas per bag.

NEWCASTLE, N.S.W.—Mr. Wallace's circular contains the following fixtures:—For San Francisco—The export is 12,901 tons coal; the Glencairn, Crown of Italy, Harvester, and Celtic Chief are here to load about 9,000 tons, under home charter. For San Diego—Five vessels have gone forward, taking 7,359 tons; there is nothing here to follow. For Hong Kong—The Altcar chartered 16s., Ringleader at 15. 6d., and the steamer Tannadice, on ship's account, have sailed with 4,403; 15s. 6d. is offering. Freights—Hong Kong, 15s. 6d.; Manilla, 15s.; Ilo Ilo, 15s. 6d.; Java, 17s.; Singapore, 16s.; Saigon, 18s.; Cheefoo, 18s.; Amoy, 18s.; Padang, 21s. (nominal); Madras, 18s.; Bombay, 18s.; Colombo, 18s.; San Francisco, 17s.; San Diego, 18s.; Wilmington, 17s. 6d.; Mauritius, 16s.; Honolulu, 14s.(nominal); Valparaiso for orders, 22s. to 25s.

ODESSA.—Mr. Marshall Little has received the following telegram from Messrs. M'Nabb, Rougier and Co., of Odessa, dated February 2:—Market quiet. Berth—Odessa or Sebastopol—January-February loading, London or Hull, 17s.; Antwerp or Rotterdam, 17s.; March-April loading, London or Hull, 18s. 9d.; Antwerp or Rotterdam, 18s. 9d.; Nicolaieff, 1s. 3d. extra; Mediterranean, fcs., 1·55. Charters—Rates for opening navigation unchanged. Require two 9,000 qrs., to load at Odessa 23s. late March, early April, less 1s. 3d. reduction direct port U.K. or Continent; Nicolaieff 1s. 3d. extra, option of a safe port between Bergen, Copenhagen 3s. over direct rates.

VALPARAISO.—Messrs. Cockbain, Allardice and Co. have received the following telegram from their Valparaiso house:—Nitrate and wheat, for orders, 35s. U.K. or Continent, less 1s. 3d. direct port, for ready vessels; guano, for orders, 31s. 3d. U.K. and 36s. 3d. Continent, less 1s. 3d. direct port, for ready vessels; copper or manganese ores, 31s. 3d. U.K. and 36s. 3d. Continent, less 1s. 3d. direct port, for ready vessels.

THE USE OF PLUMBAGO IN COTTON GINS.—The following circular, signed by a large number of New England cotton manufacturers, has been sent to the New Orleans Cotton Exchange:—"Graphite, or plumbago, appears to be coming into common use in the South as a lubricant for the bearings of cotton gins. Numerous pieces of black, greasy cotton have been found in bales when opened at the mills. In one factory alone some sixty bales have been found to contain small pieces of this very dirty substance. We suppose that in cleaning off the journals of the gins a piece of cotton is used, and the dirty cotton, covered with the greasy plumbago, carelessly thrown into the lint pile, gets packed in the bales of cotton, or that the shaft which carries the belt drips the grease into the lint. The loss, if the plumbago is found, is trifling, but if a small piece is carried into the machinery, a mass of cotton is blackened and ruined by it. There is no known solvent for plumbago, and it smuts everything it touches. We beg you to draw the attention of all cotton receivers to this matter, and to ask them to notify the planters of the trouble, and get them to put a stop to it as soon as possible."

LINSEED OIL IN THE ENGL

IN their annual market report just to hand, M
Rose of London, have the following to s
product:—

Linseed.—A very appreciable improvement
advance in the value of most articles in wl
were marked features of the year just clos
product under review, sales were more readily
time past, and the demand being fully equal
prices with few exceptions, gradually improve
though checked in some degree by the serious
business and consequent loss of trade resulting
during August and September. The total imp
the United Kingdom was 2,270,000 qrs., ag
1888, 2,341,175 in 1887, 2,081,283 qrs. in 188
1885. The Indian crop was reported to be a
ments up to the present, probably on account
ruling there lately bringing out all available st
previous years, a largely increased quantity ha
though this country has received rather less
Russia, through official and private sources, re
in the year of a disastrous failure of the crops,
out to be totally unfounded, as the supply
unusually large; values, however, did not give
becoming apparent. The crop from the Rive
poor one and the quality inferior, owing to a
time of harvesting. Many cargoes were shippe
and arrived heated and damaged to such an ext
for manufacturing purposes, and giving great dis
cerned. America helped to supply her increa
largely of Indian seeds in our markets, over 10
transhipped from London, about 21,000 qrs.
11,000 qrs. from Liverpool.

We start the year with fair supplies on hand, b
on passage, and shipments from India for the
likely to be exceptionally small. Some parcels
lately unsold on importer's account have been ho
of a marked improvement in rates before the arr
Satisfactory rains having fallen in India, pros
(sales of which up to the present have been very
far as can be judged now, considered favourab
Plate come good reports of the crop, both as
quality. The market opens steady with price
Spot Calcutta 41s. 6d. ex-ship; December-Janua
January-February, 42s. 3d. @ 42s. 6d.; March, 4
April-June, 39s. 6d. @ 40s. Bombay, 43s. sp
spot Calcutta, 42s.; April-June, 40s. @ 40s.
43s. 6d. Azov, spot 41s. Reval, 95 per cent., 3
Riga, 7½ measure, 37s. La Plata, sailer Janua
39s. Bombay, January-February Dunkirk, 44s.; F
April-June, Amsterdam, 42s.

The import into London was from the following

East Indies........................
Black Sea ....
Baltic and White Sea ....
Other ports........................

	1889.	1888.	18
	qrs.	qrs.	q
Total imports into U.K....	2,270,000	2,542,027	

Afloat by last mail advices:—

	1890.	1889.	18
	qrs.	qrs.	qr
From E. Indies to U.K.	134,880	177,23	

The stocks in warehouse and vessels dischargi
are:—

	1890.	1889.	1
	qrs.	qrs.	(
London	45,000	25,000	3
Hull	140,000	164,000	12
Liverpool	35,000	10,500	1

Linseed Oil.—The total production in the Uni
a large falling off compared with the previous tw
can be estimated, was as much as some 23,000 to
commenced the year with a stock of about 7,000
in Hull and most of the principal markets on the
exceptionally heavy, and these rather increased t
the early months, but by the autumn they were a
The article met with considerable activity at time
whole showed a market improvement upon 1888
months, the low prices, both on the spot and for

ortness of the Indian seed crop, created a good speculative business, hich, with the decided improvement in trade generally, continued to e end of the year. Hull oil in this market is not dealt in to any rge extent; the exports show an increase of 1,500 tons upon that of 388.

	1889.	1888.	1887.	1886.
	Tons.	Tons.	Tons.	Tons.
xport from Hull for the year	6,770	5,200	8,060	10,244
roduction in the U.K., we estimate	105,000	128,000	116,000	98,000

Oil Cakes.—Values of London-made Linseed ruled higher than in 388, and, except at the close of the year, generally met with a fair ile at current rates. January saw the highest figure of the year, up to 9. being paid for best qualities. In June £7. 5s. and £7. 10s. was ccepted, the market improving later to £8. and £8. 5s. present value.

As a comparison to the foregoing report, we present the following emarks from the annual circular of Game, Bowers, and Co., of ondon.

Linseed.—The imports from India have again been very considerable, nd a large quantity of North Russian seed was imported into Hull uring the summer and autumn. Prices of India seed have ruled or the most part in sellers' favour, regular supplies having been taken om this market to supply the American demand. Although in the arly spring large sales of forward shipment Calcutta were made as low s 38s., the value steadily advanced until in the autumn 44s.@44s. 6d. as reached for seed in all positions. Owing to heavy arrivals at the nd of the year, and small demand, the price of spot seed has fallen to 1s. 6d. at which several parcels have changed hands. To-day the narket closes firmer, with buyers at 42s. January-February shipment quoted 42s. 6d., and spring shipments 41s.@39s. 6d. according to osition. The business in Bombay to the United Kingdom has been maller than usual, this class of seed having been mostly sold to the Continent direct. We quote December-January shipment 44s. Black ea is 41s. for parcels afloat to Hull. Petersburg 95 per cent. is 40s. or shipment, 39s. is asked for La Plata cargoes.

The export from Calcutta to the United Kingdom for the second alf of December is telegraphed as 2,400 tons against 7,100 tons in 888. The total import into London during 1889 has been 871,788 qrs. against 954,093 qrs. corresponding period last year. The quantity float from the East Indies, per last advices is:—

	Calcutta		Bombay	
	1890	1889	1890	1889
	Qrs.	Qrs.	Qrs.	Qrs.
To London	114,853	92,296	....	4,369
Liverpool	33,183	28,855	5,844	5,462
Hull	....	64,931	....	....
Other ports	....	1,369	....	....
Coast for orders	.....	....	....	....
Continent	....	....	18,098	7,928
Total	148,036	187,451	23,942	17,751

Linseed oil.—The make has been smaller than the two previous ears, and the large stock held in warehouse having been consumed, here has been an upward tendency in this article during the past welve months. The market closes steady.

A STORY OF EARLY PETROLEUM DAYS.

QUINCY Robinson related an incident of the early history of the oil regions recently, which may give the children of the resent generation a vague idea of the magnitude of the transactions which took place when oil was $8 and $9 a barrel, and poor people gained a competency by scooping it off the surface of creeks, or gathered t from pools around the tanks which had overflowed. The story as old by Mr. Robinson was as follows:—"Within a month after Colonel Drake had struck the first petroleum ever brought to the surface in America by means of drilling, my father and the father of my relatives here brought a tract of land comprising 1,280 acres adjoining the farm n which the Drake well was located, for $350,000. Not long afterwards I was sitting in their office one day—I remember it as distinctly s though it happened only yesterday—when an agent for an eastern yndicate walked in and offered $500,000 for the 1,280 acres. The owners looked at him rather incredulously for a moment, but before hey could speak he had counted out on the table $500,000 in cash nd drafts which he offered for a deed of the tract. I was appalled by he size of the pile, but my father and the father of these gentlemen etired for consultation, and decided that if the property was worth $500,000 it was worth $1,000,000, and the offer was refused. Their heirs still own the land, and now it is valued at $20,000. Where they ould have got dollars we could scarcely get nickels. Thus you can ee what seemingly fairy stories could be told of those days. They are lmost incomprehensible to the present generation, but they were red-hot facts," and a sigh of regret that the offer had not been accepted went round the circle.—*Pittsburg Dispatch.*

A NEW INDIGO VAT.

WE are pleased to notice the introduction of a Patent Indigo Powder by Mr. J. Cowan of Glasgow, which promises to create a revolution in this branch of dyeing.

In five minutes anyone may now have a genuine Indigo vat, by simply adding to hot water in an iron or zinc vessel heated by fire, "dumb" or unperforated steam pipes, or a steam jacket or casing, equal quantities—by weight—of Patent Indigo in the form of prepared powder, and liquid bi-sulphite of soda. After boiling for a minute or two, put off boil, and let dye-liquor settle for several minutes. It will then have a pure yellowish green appearance, with a coppery film on surface, and dyeing is conducted in the clear liquor, procedure as regards handling, squeezing, and airing, being practically same as by old processes.

A dye-liquor, prepared as above directed, but subject to modifications of temperature as detailed in the next paragraph, will dye wool, silk, cotton, and linen, to any desired shade, and in one or more operations, according to concentration. In place of about 120° Fahr. as hitherto for woollen work, and cold for cotton and linen, the following heat may be employed in dyeing with this new vat:—For unspun wool, 180° to 200° Fahr.; for woollen piece goods and light serges, 150° to 180° Fahr.; for woollen yarns, 130° to 160° Fahr.; for silk, 120° to 160° Fahr.; and for cotton and linen, 120° to 150° Fahr.

As there is no lime in this Patent Indigo Powder, no necessity arises for those vexatious delays between dyeings, at present required when fresh stuffs are added, five minutes sufficing with this new vat for dye-liquor to clear, after fresh indigo and bi-sulphite have been stirred in. In large dyeworks such stuff may be conveniently added to vat as required, from a stock kept ready boiled up in separate boiler, and prepared as instructed in paragraph 1, but in a concentrated form. This arrangement allows vat liquor to be kept at the heat suited to the class of work in hand. It follows, therefore, that the total absence of lime from this new vat, permits of continuous dyeing in it, the importance of which fact will be at once apparent.

In dyeworks where a large daily output is imperative, great economy in plant—and consequently in space—is effected by the use of this patent indigo, as its power of continuous dyeing enables one vat to equal, or even exceed the production of several, under old methods. In the case of cold dyed cotton and linen yarns, a series of vats is at present indispensable, but by this new process one vat gives any shade of blue wished, and in one or more "runs" or "dips" in dyer's option. Unspun wool dyed in this vat is beautifully soft, elastic, and uniform in shade, while piece goods and yarns are better penetrated by the colour, owing to the higher dyeing heat available. For the same reason they are much more easily and cheaply cleaned.

It may be stated here with regard to scouring aud sulphuring, that such blues are in these respects entirely satisfactory, and, also, that under prolonged exposure to light, and the severest weather conditions, they remain absolutely unchanged. Cotton cloth, yarn, and thread treated in this vat—as also all linen products—are unusually level in shade, and well dyed through. Such blues are very fast to soaping, and far excel cold-dyed blues in resistance to bleaching. Warps or "chains" are dyed to shade in one vat, and in one continuous operation, thus saving time and labour, and keeping the fibre agreeably soft. Being extremely clean, these blues, when woven into white goods, can be calendered direct from loom without staining the fabric, unlike cold-dyed blues, which are so imperfectly fixed on the fibre as to necessitate elaborate precautions being taken ere calendering can proceed, to protect goods into which they are woven from being soiled during the process. Fast greens of all shades for shipment are also readily obtained in this vat.

When prepared as a paste, it will be found that the patent indigo powder may be used for printing calico, or other tissues and fibres, either by block or cylinder, and in conjunction with colours for "steaming styles." These indigo prints stand the strongest steaming and soaping, without blurring or "running" in the slightest degree on the fabric, which requires no preparation beyond that given for steaming colours. Pure indigo blues, dyed in this new vat, give the genuine indigo re-action with nitric acid, and stand all "government" or other usual tests, whether acid or alkaline. The customary "bottomings" and "toppings" used for indigo work on wool, woollen goods, cotton, and linen, may all be employed with this patent indigo.

NORTH-EASTERN RAILWAY.—The return of the North-Eastern Railway Company's receipts for the past week is the best this year, though the gross total of the earnings is far below the summer average. Compared, however, with a year ago, passengers yield an increase of £2,064.; merchandise and cattle give £5,109. more; minerals have increased by £4,454.; and dock dues, &c., by £97. There is thus, for the week, the increase of £11,814.; and for the half year, so far, the increase is £31,208. It seems as if the current half-year would be a better one than has been the past.

covered fully two centuries ago, by Brand, an alchemist, of Hamburg, who prepared it by distilling evaporated urine and sand along with carbon, and condensed the distillate. About a century later the existence of phosphate of lime in bones was discovered, and a few years subsequently the Swedish chemist, Scheele, announced a process by which phosphorus could be obtained from bone ash. From the days of Scheele till quite recently the latter was used for the preparation of phosphorus (held as a compound body until Lavoisier proved it to be an element), but now, however, native mineral phosphate of lime has taken the place of the other. In selecting a mineral phosphate for purposes of this manufacture, it is of importance to obtain one with a high percentage of phosphorus and a minimum of iron and alumina. As a typical example of suitable phosphates met with in the market may be taken the Canadian, which frequently contains as high as 17 per cent. of phosphorus. The phophate is first ground to a very fine powder, and then decomposed with sufficient sulphuric acid to convert all the lime there present into sulphate of lime, this operation being performed in large wooden tuns, with agitators, and heated by steam. After decomposition the contents of the tun are run out into filters to separate the sparingly soluble sulphate of lime from the phosphoric acid liquor. This effluent is then evaporated down to a syrupy consistence in lead-lined evaporators, and is afterwards mixed with carbonaceous matter, such as wood, charcoal, or coke, and carefully dessicated either in iron pots or other suitable furnaces. So prepared, the mixture is distilled in short retorts of best Stourbridge clay placed in a furnace similar to the Belgian furnaces used in the distillation of zinc, with a capacity of about 30 retorts each. This distillation occupies about 16 hours, crude phosphorus of a dull mahogany-brown colour being obtained. Refining is effected by fusing under water and agitating along with bichromate of potash and sulphuric acid, the result being the pale yellow phosphorus of commerce. Dr. Readman concluded this instalment of his paper by showing how small, really, was the quantity of phosphorus used in the production of matches, ordinary boxes of these containing each only from half a grain to two-and-a-half grains, according to make and brand. Highly appreciative remarks were made by several members, and attention called to the errors found in the text-books as touching the manufacture of phosphorus and other chemicals. Mr. G. H. Gemmell read a short paper on "Moisture in Pulp," referring to the variations in straw and wood paper pulp. For his own part he thought 10 per cent. was a fair average of moisture in air-dried pulp, although paper makers themselves average at 8 or 9, and others contend for as high a figure as 12.

PRIZE COMPETITION.

A LITTLE more than twelve months ago, one of our friends suggested that it might be well to invite articles for competition, on some subject of special interest to our readers; giving a prize for the best contribution, he at the same time offering to defray the cost of the first prize, and suggesting a subject in which he was personally interested.

At that date we were totally unprepared for such a suggestion, but on thinking over the matter at intervals of leisure, we have come to think that the elaboration of such a scheme would not only prove of benefit to the manufacturer, but would go some way towards defraying the holiday expenses of the prize winner.

We are therefore prepared to receive competitions in the following subjects :—

SUBJECT I.

The best method of pumping or otherwise lifting or forcing warm aqueous hydrochloric acid of 30 deg. Tw.

SUBJECT II.

The best method of separating or determining the relative quantities of tin and antimony when present together in commercial samples.

On each subject, the following prizes are offered :—

One first prize of five pounds, one second prize of one pound, five additional prizes to those next in order of merit, consisting of a free copy of the *Chemical Trade Journal* for twelve months.

The competition is open to all nationalities residing in any part of the world. Essays must reach us on or before April 30th, for Subject II, and on or before May 31st, in Subject I. The prizes will be announced in our issue of June 28th.

We reserve to ourselves the right of publishing the contributions of any competitor.

Essays may be written in English, German, French, Spanish or Italian.

Correspondence.

ANSWERS TO CORRESPONDENTS.

STILL TAX.—We shall be pleased to open our columns to a discussion of the question, but please stick to facts.

COMPETITOR.—We do not wish to receive any of the prize competitions before the time specified.

JULIUS.—We cannot vouch for the correctness of your views when you cynically state that all the talent of the chemical trade is centered at 32, Blackfriars-street, but we can tell you that the bulk of the information is centered there.

J. B.—Purely an agricultural question.

JOHN.—We fear you have not read the article too carefully. The process you mention has been tried in several different ways, and all have failed.

R. C. S.—We will see; but Bristol is not a port of any great importance.

OILMAN.—It is our intention—as we have been requested by so many—to pay more attention to the oil, paint, and colour trades.

CHLORINE.—Bleaching liquor is a lime salt, and not a soda salt, as you infer. It is made (1) by passing chlorine into milk of lime, or (2) by dissolving bleaching powder in water.

T. R.—We are not aware of any such process.

BETA.—A sulphuret of antimony produced in the wet way. 2. ree sulphur.

EX-OFFICIO.—We are not aware of the exact conditions. Apply to the Secretary, who will doubtless supply the information.

ANALYST.—We are biassed in favour of litmus, it is a wel tried, true, and trusty friend; but some of the newer indicators have their special uses.

C. C. C.—Yes. Many years ago.

A RIVER POLLUTOR.—We would rather not offer any opinion.

B.—A correspondent wishes to know whether a still holding a gallon, and used for producing distilled water, is liable to the still tax. We do not know. If it is so, a great many photographers are liable.

B. J. R.—We should regard it as an adulteration, and we think a court of law would come to the same conclusion.

P. J. A.—See an article on the subject in this week's number.

AMMONIA.—No doubt you are in some respects correct. We have come to the same conclusion. It is a pity some men should travel about the country endeavouring to corrupt the business morals of all they come in contact with.

T. P.—The substance you mention is sold in this country as hot neck grease sometimes alone, and sometimes melted with a portion of tallow, for use in lubricating the rolls in iron works.

H. Cl.—(1) It matters not whether the methods sent are old or new. (2) We shall judge them by their intrinsic merit only. (3) Yes; we should prefer drawings, either coloured or shaded.

*** *We do not hold ourselves responsible for the opinions expressed by our correspondents.*

THE TAX ON STILLS.

To the Editor of the Chemical Trade Journal.

SIR,—You will have done good service to the chemical trade by calling public attention to those absurd visits of excisemen so well known to all entrusted with the management of chemical works.

When we first started to make sulphate of ammonia, it was a new thing in this district, and our stills were old boilers placed horizontally and worked into closed saturators. We had only been going a few weeks when in walks Mr. Exciseman. Oh! says he, you are working a still here without a licence. That's where you make a mistake says I, we've no stills here. To be sure, I must see says he; so you can says I, and he went wandering all about the yard and nearly burned his nose at the ammonia boiler without ever being able to catch a glimpse of even a ghost of a still. We never worked *stills*, they are too dangerous to be trusted, so we always bought old boilers, which the exciseman thought were steam boilers, and so let us alone.—I am, &c., WILL O' TH' WISP.

EXPERT EVIDENCE.

To the Editor of the Chemical Trade Journal.

SIR,—It is morally certain that no sensible man can take up your journal of the last few weeks and read the Legal intelligence therein recorded without coming to the opinion that some radical change is needed in the method of receiving and giving scientific evidence in our courts of law.

In the Massey-Brook bone works nuisance, two experts gave evidence that they had examined the works, and although it is presumed they carried their noses with them, they could not discover any nuisance, plainly smellable by anyone residing in the neighbourhood. Of what value is such evidence as this?

The next case is the action against a Styal farmer, and with all due deference to the comparative ability of the experts engaged in the case, it does not argue very well for the reputation of chemistry as an exact science. One chemist says the manure is "worthless to throw on the land," another says "it was of reasonable merchantable value." Pray who is to be believed?—I am, &c., A CHESHIRE FARMER.

THE STILL TAX.

To the Editor of the Chemical Trade Journal.

SIR,—Your leader in last week's issue as to the manner in which the officials of the Board of Inland Revenue do their work, and the petty annoyances they heap on the small trader, is very much to the point, and, it is to be hoped, will serve a useful purpose, even if only to awaken the trade to a sense of the injustice shown to many chemists by the subordinate officials of the Board of Inland Revenue. It is not many years ago that I was manager and part owner of a small vitriol works, where vitriol was rectified and sold, and, to the best of my belief, we paid a tax of five shillings a year for the privilege of using a still, though I could never see why we should have paid it. I am very strong in the belief that the interested branches of the trade should form a committee to sift the question, and take a test case to the law courts if found necessary. I feel sure there would be no difficulty in raising the wind for so laudable and necessary an object, and it might result in the Inland Revenue department being obliged to refund the monies it has wrongfully taken from the chemical trade for so many years.—I am, &c.,

EX-MANUFACTURER.

IMPURITIES IN GLYCERINE.

UNDER the title of "Adulteration of Glycerine," F. Jean contributes an article to the *Journal de Pharmacie d'Alsace Lorraine*, in which he considers not merely adulterations intentionally added, but impurities due to carelessness in its manufacture or purification. Among them are oxide of lead, lime, and butyric acid. French perfumers and manufacturers of cosmetics test their glycerine with nitrate of silver. If no turbidity or change of colour takes place in 24 hours, it is considered good.

The chloroform test for glycerine consists in mixing equal volumes of chloroform and glycerine, shaking thoroughly, and then letting them stand. The upper strata is pure glycerine, while the lower one is chloroform containing all the impurities. If there are no impurities in the glycerine the chloroform remains unchanged, otherwise there will be a turbid layer just beneath the glycerine.

On adding a few drops of dilute sulphuric acid to a mixture of equal parts of glycerine and distilled water, and then a little alcohol, the presence of lime or lead will be shown by a white precipitate. The latter is recognised by sulphydric acid, which turns the precipitate black.

Butyric acid is detected by mixing the glycerine with absolute alcohol and sulphuric acid of 66° B. On gently heating the mixture, the butyric ether is easily recognised by its agreeable odour.

Formic and oxalic acids are also found in glycerine, impurities which are of special importance to pharmacists.

They are detected as follows:—Equal volumes of glycerine and sulphuric acid, specific gravity 1.83, are mixed together. Pure glycerine does not give off any carbonic oxide gas; but if either of the acids mentioned is present, an evolution of that gas will be observed. To decide whether both acids are present, and if not, which one, some alcohol of 40° B. and 1 drop of sulphuric acid are added, and then gently heated. Formic ether (used in making essence of peaches) will be recognised at once by its characteristic odour, and proves the presence of formic acid. To another sample of the glycerine add a little solution of chloride of calcium (free from carbonate), when it will give a precipitate of oxalate of lime, if oxalic acid is present.

Sugar, glucose, dextrin, and gum are often used as intentional adulterations of glycerine, and are tested for as follows:—The glycerine is mixed with 150 or 200 drops of distilled water, and 3 or 4 centigrammes of molybdate of ammonia are added, and one drop of pure nitric acid. It is boiled about 30 seconds. If sugar or dextrin is present, the mixture will be blue.

Glycerine adulterated with cane sugar or syrup acquires a brownish-black colour when boiled with sulphuric acid. Glucose is detected by boiling it with caustic soda, which turns it brown.

If detected qualitatively, the quantity may be estimated by the following method:—5 grams of glycerine are weighed out and mixed with 5 cubic centimetres of distilled water. It is boiled in a little flask, with Barreswil's alkaline solution of tartrate of copper. The sub-oxide of copper is precipitated, and the precipitate disolved again in hydrochloric acid. An excess of ammonia is added, and it is poured into a vessel containing an excess of nitrate of silver. A precipitate of metallic silver is formed and filtered out. It is washed with warm water and ammonia, calcined at a red heat, and weighed; 109.6 parts of metallic silver represent 100 of glucose.

If cane sugar or dextrin is found, it is boiled for half-an-hour with acidified water to convert these substances into glucose.

If none of these impurities are present, the amount of water is found by Vogel's well-known method.—*National Druggist.*

one on the head and the other on the right flank. At the end of each wire was a water-soaked sponge. At a given signal Mr. Brown turned on a 900-volt current, and at the same instant the animal dropped dead. There was no appreciable interval between the turning on of the current and the animal's evidently painless death. Further tests of the machine with a Westinghouse dynamo and an alternating current were also satisfactory. This is the way the preparations for the execution will be made: An electro, covered with a wet sponge, will be placed on the top of the condemned man's head, and another in a large shoe on one of his feet. His arms will then be strapped across his breast, and a similar strap placed around his ankles. He will then be placed on the chair, the straps will be attached to hooks provided for the purpose, a button will be touched, and then all will be over.

THE TREATMENT OF SEWAGE.—In 1888 Sir Henry Roscoe issued a report based upon a number of careful experiments upon the action of the usual deodorants upon sewage. Since that date some further important experiments have been made in reference to Mr. William Webster's process of electrical treatment of sewage carried out at Crossness pumping station, and by which it is found that the process is capable of removing practically the whole of the matters in suspension, and an average of 22 per cent. of the the oxidisable organic matters in solution at a cost of £1. 15s. per million gallons of sewage, or £102,200. per annum for the whole of the daily flow of the 160 millions gallons of London sewage, exclusive of labour, interest on capital, wear and tear, etc., and the cost of disposing of the sludge.

SOUTH AFRICAN GOLD.—The returns of the export of raw gold from South Africa *via* Natal and Cape Colony during December, compiled by the Capetown Chamber of Commerce, have now reached us. The amount shipped *via* Natal during the month was £43,901., and the amount *via* Cape Colony was £87,870. The total for 1889 *via* Natal is £584,405., against £402,018. for 1888, and the total *via* Cape Colony is £857,366., against £516,886. for 1888. The total for both colonies for 1889 is £1,441,771., against £918,904. for 1888. The yield is beyond doubt increasing, though the total for December is not the largest monthly record, being only £131,771., against £144,338. in June last and £136,902. in July; but with these exceptions it beats any other month. The total value of the gold export for 1889 is, however, still only about one-third the value of the diamonds annually exported from the Kimberley mines.

THE "BAILEY-FRIEDERICH" STEAM MOTOR.—By invitation of Mr. Alderman W. H. Bailey, the head of the firm, a large number of members of the Arts Club and Brazennose Club and other gentlemen interested, paid a visit to the works of Messrs. W. H. Bailey and Co., in Oldfield-road, Salford, for the purpose of making an inspection of the "Bailey-Friederich" steam motor. The invention of a gentleman of Vienna, this motor is a steam engine and boiler in combination. It is an exceedingly compact piece of mechanism, and works, as was shown yesterday, automatically and noiselessly. The furnace enables coal, coke, tanners' bark, or any other kind of fuel to be used, and it is claimed for it that it requires less fuel and a great deal less water than any other form of steam motor. This economy in the consumption of fuel is secured by the heat of the exhaust steam being utilised by the condenser, and by the very great heating surface which is obtained in consequence of the peculiar construction of the furnace. The inventor and Mr. Bailey—by whom some minor improvements have been introduced—are hopeful that the use of the motor may solve the question of the introduction of the electric light into moderate-sized houses, as it is estimated that 40 16-candle power incandescent lamps can be kept supplied at a cost of about 6d. per night for fuel. In Continental countries, where coal is less abundant and therefore more costly than with us, the economy of fuel which the motor effects has already brought it into extensive use, and now it is being brought under notice in this country. At Messrs. Bailey's works, by the use of one of the motors, 40 16-candle power lamps are kept going at a cost of 1d. per hour.—*Manchester Guardian.*

MAKING GLASSWARE BY MACHINERY.—The manufacture of glassware by machinery on a permanent scale has been undertaken at the long-idle Ellenville (N. N.) Glass Works in the village of that name. When it was reported that a machine for blowing glass bottles had been invented and successfully worked in England, a company of American glass manufacturers was formed with a view of introducing the machines in this country, and one of the members was sent over there to examine and report upon the merits of the invention. The machine as now fitted up will blow quart bottles only. It is operated by a man and boy, and is very simple of construction. It consists of an iron upright, around which revolve arms fitted with moulds for shaping the glass. A pipe supplied with a current of air and readily manipulated by the operator does the work of blowing. The machine is operated with astonishing celerity, and is said to be capable of turning off 100 dozen of perfect bottles a day. Three glass workers at Martin's Ferry, O., have invented a machine which they claim will radically change the method heretofore employed in pressing glassware. It is claimed that by the use of this machine pressed ware can be made to look as well as blown ware.

Market Reports.

THE LIVERPOOL MINERAL MARKET.

Our market has well maintained its previous advance, and for some minerals still higher prices have been paid. Manganese: Arrivals have been very small, whilst stocks have been further drawn upon, and sales have been made at much higher prices, both for prompt and forward delivery. Magnesite: Raw lump without change, raw ground, £6. 10s., and calcined ground, £10. to £11. Bauxite (Irish Hill brand): In increasing demand, selling freely at full prices. Lump, 20s.; seconds, 16s.; thirds, 12s.; ground, 35s. Dolomite, 7s. 6d. per ton at the mine. French Chalk: Arrivals have improved a little, but prices remain firm and unaltered, especially for G.G.B. "Angel-White" brand — 90s. to 95s. medium, 100s. to 105s. superfine. Barytes (carbonate), steady; selected crystal lump scarce at £6.; No. 1 lumps, 90s.; best, 80s.; seconds and good nuts, 70s.; smalls, 50s.; best ground, £6.; and selected crystal ground, £8. Sulphate quiet; best lump, 35s. 6d.; good medium, 30s.; medium 25s. 6d. to 27s. 6d.; common, 18s. 6d. to 20s.; ground best white, G.G.B. brand, 60s.; common, 45s.; grey, 32s. 6d. to 40s. Pumicestone steady; ground at £10., and specially selected lump, finest quality, £13. Iron ore continues in good demand at full prices. Bilbao and Santander firmer at 9s. to 10s. 6d. f.o.b.; Irish, 11s. to 12s. 6d.; Cumberland, 16s. to 18s. Purple ore, unchanged. Spanish manganiferous ore commands ready sale at full prices. Emery-stone: Best brands in good demand, bringing full prices. No. 1 lump is quoted at £5. 10s. to £6., and smalls £5. to £5. 10s. Fullers' earth steady; 45s. to 50s. for best blue and yellow; fine impalpable ground, £7. Scheelite, wolfram, tungstate of soda and tungsten metal more inquired for. Chrome metal, 5s. 6d. per lb. Tungsten alloys 2s. per lb. Chrome ore: High percentage inquired for at fair prices. Antimony ore and metal have further advanced. Uranium oxide, 24s. to 26s. Asbestos: Best rock, £17. to £18.; brown grades, £14. to £15. Potter's lead ore: The miners' strike has terminated, but prices are still firm; smalls, £14. to £15.; selected lump, £16. to £17. Calamine: Best qualities scarce—60s. to 80s. Strontia steady; sulphate (celestine) steady, 16s. 6d. to 17s. Carbonate (native) £15. to £16.; powdered (manufactured), £11. to £12. Limespar: English manufactured, old G.G.B. brand sought after, and brings full prices; 50s. for ground English. Felspar, 40s. to 50s.; fluorspar, 20s. to £6. Bog ore more inquired for—22s. to 25s. Plumbago: More offering; best Ceylon lump at last quotations; Italian and Bohemian, £4. to £12. per ton. French sand, in cargoes, continues scarce on spot—20s. to 22s. 6d. Ferro-manganese sells easily at advanced figures. Ground mica, £50. China clay freely offering—common, 18s. 6d.; good medium, 22s. 6d. to 25s.; best, 30s. to 35s. (at Runcorn).

WEST OF SCOTLAND CHEMICALS.

GLASGOW, Tuesday.

In the paraffin section a rather important arrangement has just been concluded (taking effect from February 1st.), whereby the Scotch and American producers of scale and wax conjointly are to charge about a halfpenny a pound more for their output, the compact to last as thus readjusted till the end of January 1891. The advances thus artifically brought about consist of ⅝ths of a penny for soft scale (commonly termed "match paraffin" in the manufacture), for hard scale 11/16ths, and for paraffin wax 120° semi-refined 13/16nds. At the same time, also by agreement, the Associated Candle Makers of England and Scotland (amongst which most of the Scotch oil companies are classed), advance their finished products one halfpenny per pound. This agreement will add largely to the gross revenue of these Scotch Companies, but, already a good part of it is in reality discounted by recent rises in the costs of production. The general market for chemicals has improved slightly, and most quotations are firmer. Sulphate of ammonia is an exception, being easier, and very idle, buyers offering about £11. 12s. 6d. at which low mark, however, no sales are on report. The local feeling in bleaching powder is stronger, and holders look for better prices. During the week bichromates have been rather less in demand. Chief prices current are:—Soda crystals, 49s. net Tyne; alum in lump £4. 17s. 6d., less 2½% Glasgow; borax, English refined £30., and boracic acid, £37. 10s. net Glasgow; soda ash, 1⅛d. less 2½% Tyne; caustic soda, white, 76°, £10., 70/72°, £8. 5s., 60/62°, £7. 7s. 6d., and 60/62° cream, £6. 15s., all less 2½% Liverpool; bicarbonate of soda, 5 cwt. casks, £5., and 1 cwt. casks, £5. 5s. net Tyne; refined alkali 48/52°, 1½d., less 5% Tyne; saltcake, 25s. 6d. to 26s. 6d; bleaching powder, £5. 15s. to £6. less 5% f.o.r. Glasgow; bichromate of potash, 4d., and of soda 3d., less 5 and 6% to Scotch and English buyers respectively; chlorate of potash, 4¼d., less 5% any port; nitrate of soda 8s. 3d. to 8s. 6d.; sulphate of ammonia, £11. 12s. 6d. to £11. 15s. f.o.b. Leith; salammoniac, 1st and 2nd white, £37. and £35., less 2½% any port; sulphate of copper, £25. less 5% Liverpool; paraffin scale, hard, 2½d., soft, 2⅛d.; paraffin wax, 120°, semi-refined, 3 5/32d.; paraffin spirit (naphtha), 9d.; paraffin oil (burning), 6¼d. to 7d., at Glasgow; ditto (lubricating), 865°, £5. 10s. to £6., 885°, £6. to £6. 10s.; and 890/895°, £7. 10s. to £8 Week's imports of sugar at Greenock were 27,295 bags.

MISCELLANEOUS CHEMICAL MARKET.

There has been an excited market during the past week in caustic soda, which on the spot is now worth for 70 % £9. per ton, and for 60% £8. per ton, f.o.b., Liverpool. The future of this article is somewhat uncertain, and meanwhile the makers are not willing to book far ahead. It is not unreasonable to expect an increase in the prices of this product, inasmuch as it is most directly affected by the advances in salt and fuel. Recent curtailments of production have brought about the present rise in values, and if makers continue to work upon the same lines, it is not at all unlikely that even the present prices will be maintained. Soda ash is holding a firm position at 1⅛d. to 1¼d. per deg., and soda crystals are naturally in sympathy with the latter and firm at £2. 12s. 6d. on rails. Bleaching powder continues steady in price at £5. 2s. 6d. to £5. 5s., on rails at works, and £5. 12s. 6d. to £5. 15s. f.o.b. in export casks Chlorate of potash, steady at 6¼d. Sulphur in all grades is finding ready outlet at current rates. Sulphate of copper is scarce and firm in price at £24. 10s. to £25. per ton. Lead salts of all kinds are still slow of sale, but prices are unaltered. There is brisk demand in all quarters for green copperas, which is scarce and higher in price generally. Ammonia salts are rather quiet, with the exception of salammoniac, which is in brisk demand at the advanced prices. Carbolic acid crystals are only moving slowly, and present quotations are nominally 10½d. to 10¾d. per lb. for 34/35°. There is a fair ready sale for cresylic acid of well-rectified quality, at 1s. to 1s. 3d. per gallon. Potash, caustic and carbonate, are firm, and there is moderate demand. Prussiate of potash still scarce and firm at 10d. to 10½d. per lb.

THE LIVERPOOL COLOUR MARKET.

COLOURS quiet; prices unaltered. Ochres: Oxfordshire quoted at £10., £12., £14., and £16.; Derbyshire, 50s. to 55s.; Welsh, best, 50s. to 55s.; seconds, 47s. 6d.; and common, 18s.; Irish, Devonshire, 40s. to 45s.; French, J.C., 55s., 45s. to 60s.; M.C., 65s. to 67s. 6d. Umber: Turkish, cargoes to arrive, 40s. to 50s.; Devonshire, 50s. to 55s. White lead, £21. 10s. to £22. Red lead, £18. 10s. Oxide of zinc: V.M. No. 1, £24. 10s. to £25.; V.M. No. 2, £23. Venetian red, £6. 10s. Cobalt: Prepared oxide, 10s. 6d.; black, 9s. 9d.; blue, 6s. 6d. Zaffres: No. 1, 3s. 6d.; No. 2, 2s. 6d. Terra Alba: Finest white, 60s.; good, 40s. to 50s. Rouge: Best, £24.; ditto for jewellers', 9d. per lb. Drop black, 25s. to 28s. 6d. Oxide of iron, prime quality, £10. to £15. Paris white, 60s. Emerald green, 10d. per lb. Derbyshire red 60s. Vermillionette, 5d. to 7d. per lb.

TAR AND AMMONIA PRODUCTS.

TUESDAY.

The Benzol market has not maintained its firmness, and prices today are just a trifle weaker, 90's being value for 3s. 11d. to 4s., and 50/90's 3s. 2d. to 3s. 3d. There is, however, not any very great change, and sellers will do well to watch the market carefully. The demand for solvent naphtha continues to increase, and prices are still advancing. We hear that consumers who have not covered their requirements are being asked 2s. for good solvent for prompt delivery. Creosote is still moving off well, and though prices have not advanced from last week's quotations, yet there seems to be some difficulty in consumers getting regular supplies. The crude carbolic acid market still continues in an unsatisfactory state, though many attribute this to the "bear" attitude of several purchasers. It is quite certain that carbolic acid is being used as largely as ever, and hence the difficulty of accounting for the flat state of the market. Anthracene is perhaps a trifle weaker, and 1s. 2½d. to 1s. 3d. may be considered to-day's value. Pitch is weak, and there is not much business doing, but prices are maintained in the meantime.

The sulphate of ammonia market remains *in statu quo*, though attempts are being made to "bear" prices to an extraordinary extent. Hull is quoted at £11. 13s. 9d., and Leith at £11. 12s. 6d., but we have not been able to hear of any actual transactions at these prices. London is also quoted £11. 17s. 6d., and Beckton £11. 15s., but sellers will require to be very cautious in their transactions, as these prices are not warranted either by the business which is being done, or the opinions of purchasers on the Continent. It is well known that several speculators have largely oversold themselves for this month's delivery, and the "bear" operations have been devised for shifting their losses on to the shoulders of the manufacturer. The demand for sulphate all the world over has not abated one iota, and there is no reason at all why prices should be any lower than they were at this time last year.

REPORT ON MANURE MATERIAL.

There has been a moderate business throughout the week in manure material, with no great alteration in values, except that nitrate of soda has recovered somewhat, owing to further negotiations among producers to curtail their output or to stop their works altogether for a time.

Charleston River phosphate rock continues to be quoted 11d. per unit, cost, freight, and insurance to United Kingdom in cargoes, but it is not freely offered, and we do not hear of any important business passing. Somme 55/60 % is worth 10d., 60/65 % 11½d., 65/70 % 13d., per unit, ⅛th rise, in cargoes, c.i.f. to United Kingdom in bulk, and Belgian 40/45 % 8d., and 48/53 % 10d. per unit, c.i.f., in bulk, in cargoes to United Kingdom. There is a moderate business passing in these descriptions. We do not hear of anything doing in Canadian Rock, values being nominally as quoted in our last report.

A moderate amount of business is passing in bones and bone meal, though the attitude of manure manufacturers is rather to hold off until they see their manures going out, before they complete their purchases of material for the coming season. River Plate cargoes of bones, December shipment, are worth about £5. 10s., and later shipments would hardly fetch as much. Crushed East Indian bones have been sold during the week at £5. 5s. to arrive, and that price has, we understand, been accepted for meal, but at the close the nearest value is £5. 7s. 6d. for meal, and £5. 5s. for crushed bones, shipment ahead. From store, £5. 10s. to £5. 12s. 6d. required. There has been more enquiry within the past few days, and there are prospects of an improvement in values.

Nitrate of soda cannot now be had under 8s. 6d. on spot. Due cargoes are worth 8s. 4½d. to 8s. 6d., and November shipment might fetch a little more though the attitude of both buyers and sellers is rather in the direction of waiting to see what comes of the producers' present negotiations for the reduction of the output.

The demand for superphosphates continues, and 26 % may still be quoted 50s. in bulk, f.o.r. at works.

Dried blood is somewhat easier, and 11s. per unit would probably buy River Plate shipments now on the way and shortly due. Home prepared is quoted 11s. 6d. per unit f.o.r. at works.

METAL MARKET REPORT.

	Last week.		This week.
Iron	56/1½	....	52/10½
Tin	£94 0 0	....	£93 12 6
Copper	48 17 6	....	48 17 6
Spelter	24 2 6	....	23 0 0
Lead	13 0 0	....	12 17 6
Silver	44½	....	$44\frac{5}{16}$

COPPER MINING SHARES.

	Last week.			This week.	
Rio Tinto	$16\frac{5}{16}$	$16\frac{7}{16}$	....	$16\frac{9}{16}$	$16\frac{11}{16}$
Mason & Barry	6¼	6⅞	....	$6\frac{9}{16}$	$6\frac{11}{16}$
Tharsis	4⅝	4¾	....	4½	4⅝
Cape Copper Co.	$3\frac{7}{16}$	$3\frac{9}{16}$	....	3⅛	3⅜
Namaqua	$2\frac{5}{16}$	$2\frac{7}{16}$	....	2⅛	2⅜
Copiapo	$2\frac{7}{16}$	$2\frac{9}{16}$	....	$2\frac{9}{16}$	$2\frac{11}{16}$
Panulcillo	$1\frac{1}{16}$	$1\frac{3}{16}$	....	1	1¼
New Quebrada	⅞	1⅛	....	⅞	1⅛
Libiola	$3\frac{9}{16}$	$3\frac{11}{16}$	....	$3\frac{7}{16}$	$3\frac{11}{16}$
Tocopilla	1/9	2/3	....	1/6	2/-
Argentella	6d.	1/-	....	6d.	1/-

New Compani

GRINDLEY AND COMPANY, LIMITED.—This company
ult., with a capital af £20,000., in £5. shares, to take o
business of tar, rosin, and naphtha distillers, and of pit
varnishes, disinfectant, sheep dip, creosote, asphaltum
jelly manufacturers, carried on at Upper North-street,
Company. The subscribers are:

W. F. Rolfe, Barchester-street, Poplar, felt manufac
J. Smart, Stevenage, Herts., asphalte manufacturer.
W. S. Duff, 32, Etchingham-road, Leyton-road, E.,
J. Grindley, Millfield, Highgate, merchant........
F. Ramel, 36, Graham-road, Dalston ....
J. J. Patrick, 174, Burdett-road, E., commercial trav
T. W. Holmes, 9, Westbourne-road, Barnsbury, cler

NORWICH ELECTRICITY COMPANY, LIMITED.—This
the 25th ult., with a capital of £50,000., in £10. shares,
elsewhere the business of an electricity company in all
are:—

G. Fowell Buxton, Thorpe, Norwich ....
A. R. Chamberlin, Ipswich-road, Norwich ....
J. Bugg Coaks, Thorpe, Norwich ....
R. J. Colman, Beacondale-woods, Norwich ....
C. R. Gilman, Eaton, Norwich ....
F. W. Harmer, Cringleford, Norfolk ....
G. B. Kennett, Norwich ....

PARAGON SMOKE PREVENTOR COMPANY, LIMITE
registered on the 29th ult., with a capital of £3,000.,
patents for improvements in the manufacture of mechani
the prevention or consumption of smoke. The subscriber

W. P. White, 29, Bold-street, Manchester, engineer
C. H. Sidebottom, 15, Cromford-court, Manchester, b
W. H. Cottrell, Bolton, commercial traveller ....
J. Whittaker, Accrington, engineer ....
J. Haythornthwaite, Church, potato merchant ......
W. Kitchen, Accrington, licensed victualler ....
A. Dewhurst, Church, boiler maker ....

Gazette Notice

PARTNERSHIPS DISSOL

J. J. THOMAS and SON, Rochdale, chemists and dru
manufacturers.
SALTER and SQUIRE, Washington, Durham, and Wylam-
and slag merchants.
BLAKEY and PETERS, Leeds, oil refiners, soap manuf
chants.

The Patent Li

This list is compiled from Official sources in the Laboratory, under the immediate supervision of Alfred R. Davis.

APPLICATIONS FOR LETTERS PATE

Rotary Motor Engines, etc. J. Roots. 570. Janua
Manures. M. C. Ginster. 575. January 13.
Dry Gas Meters.—(Complete Specification.)—J. T. Wy
598. January 13.
Brushes for Dynamos. J. C. Mewburn. 600. Janu
Filter Presses. J. Brock and T. Minton. 606. Janua
Electric Motors. R. Peacock and H. L. Lange. 607.
Manufacture of Glass Bottles. W. C. Bennett. 612.
Regulators for Electric Currents. R. B. Evered
January 13.
Preservation of Wood. F. T. P. Wells. 617. Janua
Steam Generators. J. T. Thornycroft. 645. January
Pulverizing Mills.—(Complete Specification). M. B.
Filters.—(Complete Specification). E. M. Knight. 648.
Rock Breakers.—(Complete Specification). M. B. Dod
Soaps. J. Templeman. 656. January 14.
Reversible Electric Motors. P. J. R. Crampton. 66
Ageing Liquors. W. P. Thompson. 667. January 14
Separators. J. S. Rigby and J. R. Wylde. 668. Janu
Ascertaining the Requisite Time of Exposure i
Ballard. 669. January 14.
Purifying Salt.—(Complete Specification). A. Domei
Elevators. L. Apostoloff. 676. January 14.
Accelerating Fermentation. E. Edwards. 678. Ja
Gas or Oil Motor Engines. F. W. Crossley. 684.
Electric Motors.—(Complete Specification). S. C. C.
Illuminants for Electric and other Lamps. J. Cl
" " " "
" " " "
Gas Burners.—(Complete Specification). H. H. Lake.
Pulverizing Mills.—(Complete Specification). H. H.
Hydrocarbon Burners.—(Complete Specification). J.
Addition of Smoke Consuming Chambers to Furn
712. January 15.

Explosives. F. Y. Wolseley and H. R. Punchon. 715. January 15.
Red Dyes. J. J. Hummel 724. January 15.
Pressure Gauges. W. Clifford. 731. January 15.
Boiler Flues. B Brown. 732. January 15.
Dioxynaphthaline-mono-sulpho Acid. A. Bang. 735. January 15.
Bridge Protector for Furnaces. W. McG. Greaves. 745. January 15.
Generation of Electricity by Gas Batteries.—(Complete Specification) O. Dahl. 761. January 15.
Heating, Cooling, or Condensing. J. H. Selwyn. 762. January 15.
Telephones. W. Oesterreich and W. Genest 771. January 15
Obtaining Alumina and Phosphoric Acid from Phosphates of Alumina. H. H. Leigh. 789. January 15.
Machinery for Rolling Glass. Chance Bros. and Co, Limited, and E. F. Chance. 785. January 15.
Apparatus for Washing Gas. W. W. Horn. 788. January 15.
Manufacture of Alcohol. G. Guignard and A. Hédouin. 792. January 15.
Carburetting Gas. H. S. Maxim and G. S. Sedgwick. 794. Jan. 15.
The Utilisation of Peat, Sawdust, etc. R. Stone. 795. January 15.
Steam Boiler Furnaces W. Truswell. 811. January 16.
Preservative Composition. W. H. Ness and W. C. Folkard. 812. Jan. 16.
Filters A. Smith. 828. January 16.
Manufacture of Glass Vessels. H. L Phillips. 830. January 16.
Manufacture of Ammonium Nitrate, and Sulphate, or Chloride of Sodium, and of Potassium.
Manufacture of Aluminium. W. White. 859. January 16.
Gas or Vapour Burners. T. W. Johnson. 863. January 16.
Fire Extinguishers. J. G Lorrain. 864. January 16.
Vapour Generators. G. W. Garrett. 870. January 17.
Dynamos. J. Perry. 872. January 17.
Dynamos. J. Swinburne. 878 January 17.
Sheep Dip. G Craig. 889. January 17.
Insulating and Waterproofing Composition. A. N. Ford. 893. Jan 17.
Air Compressing Machines. J. C. Mewburn. 894. January 17.
Dyeing Vats. E. Charles. 903. January 17.
The Electro-deposition of Metallic Alloys. W. A. Thoms. 907. Jan. 17.
The Distillation of Coal. E. Freund. 910. January 17.
Enclosed Fireplaces. G. C. Kendal. 931. January 17.
Manufacture of Caustic Soda. W. Feld. 931 January 18.
Dynamos.—(Complete Specification). W. Thompson. 950. January 18.
Oil Burners for Furnaces. G. S. Grimston. 952. January 18.

IMPORTS OF CHEMICAL PRODUCTS
AT
THE PRINCIPAL PORTS OF THE UNITED KINGDOM.

LONDON.
Week ending Jan. 18th.

Acetic Acid—
Holland, £21 Phillips & Graves
Germany, 40 pkgs. Pickford & Co
Holland, 10 A. & M. Zimmermann
,, 190 Beresford & Co.

Antimony Ore—
Spain, 14 t. H. Bath & Son
U. S., 16 F. Haerberlin
France, 10 Pillow, Jones, & Co.
Spain, 30 F. Williams
Sydney, 11 W. Caudery & Co.

Bones—
U. S., 27 t. H. A. Lane & Co.
Victoria, 8 Culverwell, Brooks, & Co.
E. Indies, 50 Goad, Rigg & Co.
N. S. Wales, 1 t. 7 c. ,,
,, 7 17 ,,
N. Zealand, 7 Flack, Chandler, & Co.

Bone Ash—
France, 5 c. Arnati & Harrison

Barytes—
Germany, 22 cks. O'Hara & Hoar
,, 57 T. & W. Farmiloe
,, 23 O'Hara & Hoar
,, 56 W. Harrison & Co.
Belgium, 12 J. L. Lyon & Co.
,, 134 A. Zumbeck & Co.
,, 153 pkgs. W. Harrison & Co.
Germany, 19 cks. W. G. Taylor & Co.

Brimstone—
Italy, 250 t. Brandram, Bros., & Co.
,, 200 W. C. Bacon & Co.
,, 150 G. Ward & Sons

Caoutchouc—
E. Indies, 2 t. 5 c. Matheson & Co.
E. Africa, 45 11 L. & I. D. Jt. Com.
E. Indies, 16 Knighton & Co.
,, 8 1 L. & I. D. Jt. Com.
E. Africa, 10 Clarke & Smith
U. S., 3 J. Patton

Chemicals (*otherwise undescribed*)
Germany, £23 T. H. Lee
,, 93 ,,
,, 15 L. & I. D. Jt. Com.
France, 14 Carey & Sons
Germany, 23 Phillips & Graves
,, 357 A. & M. Zimmermann
,, 138 T. H. Lee

Cream Tartar—
Spain, 4 pkgs. C. Christopherson & Co.
Holland, 11 Middleton & Co.

Camphor—
Japan, 66 tubs. Lewis & Peat
China, 35 L. & I. D. Jt. Com.
Germany, 3 ,,
Japan, 90 May & Baker

Copper Precipitate—
Spain, 66 t. G. Ward & Sons

Copper Regulus—
Sydney, 78 t. H. Rogers, Sons, & Co.
Adelaide, 26 Johnson, Matthey, & Co

Caustic Potash—
France, 8 pkgs. F. W. Berk & Co.

Carbonic Gas—
Holland, 110 pkgs. Phillips & Graves

Copperas—
Germany, £30 Burgoyne, Burbidge, & Co.

Cobalt—
New Caledonia, 1 t. Gellatly & Co.

Dyestuffs—

Cochineal
Canaries, 10 pkgs. W. M. Smith & Sons
,, 13 Johnson, Rolls, & Co.
,, 15 H. R. Toby & Co.
,, 1 Kuhner, Henschel, & Co.

Tanners' Bark
Belgium, 50 t. Hicks, Nash, & Co.
N. S. Wales, 94 J. Swire & Sons
Victoria, 17 Newcomb & Son
New Zealand, 92 W. H. Boult

Indigo
E. Indies, 57 chts. L. & I. D Jt. Co.
,, 31 Arbuthnot, Latham, & Co.
,, 10 Benecke, Souchay, & Co.
,, 5 W. Brandt, Sons, & Co.
,, 18 T. D. Brand
,, 13 Elkan & Co.
,, 5 Darling Bros.
,, 12 P. & O. Co.
,, 8 G. Dards
,, 7 J. Owen
,, 131 1 bx. Elkan & Co.
,, 125 P. & O. Co.
,, 160 L. & I. D. Jt. Co.
,, 8 Frauling & Gaskell
,, 38 T. Barlow & Bros.
,, 84 T. Ronaldson & Co.
,, 10 Henderson Bros
,, 15 Elkan & Co.
,, 1 Anderson Bros.
,, 9 Williams, Torrey, & Co.
,, 47 Phillips & Graves
,, 113 L. & I. D. Jt. Com.
,, 4 E. Bower & Co.
,, 25 Patry & Pasteur
,, 10 Parsons & Keith
,, 61 Elkan & Co.
,, 12 F. Huth & Co.
,, 23 H. Grey, junr.
,, 31 J. Owen
,, 10 Elkan & Co.
,, 22 Langstaff & Co

Extracts
U. S., 20 pkgs. H. Johnson & Co.
E. Indies, 350 T. Ronaldson & Co.

Orchella
E. Africa, 228 pkgs. Phillips & Graves

Annatto
Ceylon, 30 pkgs. Hoare, Wilson, & Co.
,, 5 ,,

Sumac
Italy, 500 pkgs. C. E. Fosbrooke
,, 1,000 I. Kitchen
,, 750 C. E. Fosbrooke
,, 350 W. France

Saffron
Spain, 1 pkg. Mancha & Co.
,, 2 W. Balchin

Safflower
E. Indies, 30 pkgs. L. & I. D. Jt. Co.

Myrabolams
E. Indies, 67 pkgs. Arbuthnot, Latham, & Co.

Gambier
E. Iudies, 95 pkgs. L. & I. D. Jt. Co.
,, 214 Lewis & Peat

Rennet
Denmark, 13 cs. A. F. White & Co.

Valonia
A. Turkey, 100 t. J. Graves
,, 68 A. & W. Nesbitt

Glycerine—
S. Australia, £90 H. Hill & Sons
Germany, 350 Prop. Scott's Wf.
,, 360 Beresford & Co.
Victoria 226 ,,
Germany, 60 T. H Lee
,, 17 Craven & Co.
,, 64 A. & M. Zimmermann
N. Z., 70 Hicks, Nash, & Co.

Guano—
Sweden, 70 t. Matthews & Luff
,, 170 ,,
Lobos de Afuna, 1,200 t. Anglo Cont. Guano Wks.
Sweden, 72 t. Anglo Swedish Co.

Glucose—
Germany, 40 pkgs. 313 cwts. T. J. Warier
,, 120 120 Barrett, Tagant, & Co.
,, 100 100 J. Barber & Co.
,, 10 82 Props. Chamb. Whf.
U. S., 2,500 2,500 J. R. Francis & Co.
,, 500 500 M. D. Co.
Germany, 20 161 Leach & Co.
,, 2 15 L. & I. D. Jt. Co.
,, 200 200 L. Sutro & Co.
,, 15 120 Hyde, Nash, & Co.

Glucose (*continued*).
Germany 200 pkgs. 200 cwts. J. F. Ohlmann
,, 600 600 F. Barber & Co.
,, 50 400 T. M. Duche & Sons
,, 60 60 H. Lorenz
,, 10 55 Props. Chamb. Whf.
,, 60 69 Perkins & Homer
,, 17 140 H. Lorenz
U. S., 200 200 Barrett, Tagant, & Co.
,, 530 550 ,,

Gutta Percha—
B. W. I., 49 c. D. Baird & Co.
E. Indies, 565 Kleinwort, Sons, & Co.

Isinglass—
E. Indies, 12 pkgs. L. & I. D. Jt. Co.
Vancouver's Isld., 8 pkgs. Hudson's Bay Co.
N. Russia, 8 J. Vickers
China, 3 W. M. Smith & Sons

Manure—

Phosphate Rock
France, 120 t. T. Farmer & Co.
Belgium, 380 Anglo Contl. Guano Works Agency.
,, 100 Sawer, Mead, & Co.
Holland, 162 Lawes Manure Co.
Aruba, 487 G. M. Bauer
Germany, 4 Petri Bros.

Superphosphate
Holland, 100 t. Lawes Manure Co.

Manure (*otherwise undescribed*)
Germany, 5 t. Union Lighterage Co.
,, 6 N. S. S. Co.
,, 50 Hunter, Waters, & Co.
,, 256 J. Gibbs

Naphtha—
10 brls. Kister & Biggs

Nitre—
Chili, 8,985 bgs. Anglo Contl. Guano Works

Potassium Prussiate—
Germany, 16 pkgs. A. & M. Zimmermann.
,, £23 ,,

Potassium Carbonate—
France, £23 Beresford & Co

Paraffin Wax—
U. S., 470 brls. H. Hill & Sons

Pitch—
Germany, 54 brls. Berlandina, Bros., & Co.
San Sebastian, 29 brls. ,,

Potassium Sulphate—
France, 21 pkgs. Fredrichstadt
,, 7 Carey & Sons
Germany, 305 G. M. Bauer
,, 21 W. V. Robinson & Co.
France, 27 C. Pawson

Plumbago—
Holland, 92 cks. Brown & Elmslie
Ceylon, 701 J. Thresser, Son, & Co.
,, 713 H. W. Ison
,, 394 Marshall & French
,, 125 H. J. Perlbach & Co.
,, 1,210 H. W. Ison

Pyrites—
Spain, 1,412 t. G. Ward & Sons
,, 400 F. C. Hills & Co.
,, 200 Anderson, Anderson, & Co.

Potash Salts—
Germany, £59 Howard & Sons

Potassium Bicarbonate—
Holland, 14 pkgs. A. & M. Zimmermann

Potassium Muriate—
Germany, 305 pkgs. Odam's Cheml. M. Co.

Quinine—
Holland, £1,250 F. W. Heilgers & Co.

Saltpetre—
Germany, 25 btls. 120 cks. P. Hecker & Co.
E. Indies, 35 cks. 1,122 bgs. L. & I. D. Jt. Com.
,, 982 bgs. Henckell du Buisson
,, 286 ,,
,, 1,106 Henderson Bros.
,, 400 Ralli Bros.
,, 181 C. Wimble & Co.
Germany, 50 cks. T. Merry & Son
,, 41 J. Hall & Son

Stearine—
Germany, 33 cks. W. Balchin
France, 145 sks. J. Goddard & Co

Sodium Acetate—
France, £85 Hunter, Waters, & Co.

Sodium Prussiate—
Holland, £46 Spies, Bros., & Co.

Salammoniac—
Germany, £17 Hening & Co

Tartaric Acid—
Italy, 40 pkgs. Webb, Warter, & Co.
France, 19 ,,
Holland, 7 Middleton & Co.
,, 10 Webb, Warter, & Co.

Tar—
Germany, 46 brls. Linck, Moeller, & Co.

Tartar—
Italy, 25 pkgs. Thames S. Tug Co.

Turpentine—
U.S., 504 brls. Nicholl & Knight
,, 250 Game, Bowes, & Co.

Ultramarine—
Holland, 6 cs. W. Harrison & Co.
,, 3 H. Peron & Co
Belgium, 8 cks. Leach & Co.

Zinc Oxide—
Holland, 2 cks. W. Bennett
,, 250 N. Smith
,, 175 M. Ashby
Germany, 10 D. Storer & Sons

Zinc Sulphate—
Germany, 43 pkgs. Howard & Sons
,, £70 F. Smith

LIVERPOOL.

Week ending Jan. 22nd.

Boracic Acid—
Leghorn, 40 cks.

Borax—
Leghorn, 10 cs.

Bones—
Constantinople, 173 bgs. Jivan, Tekeian, & Co.
Monte Video, 2 bgs.
Karachi, 4,000 Ralli Bros.
Maraham, 128 R. Singlehurst & Co.
Ceara, 1,252 ,,

Bone Ash—
Rio Grande do Sul, 181 t. A. Tesdorpf & Co

Borate of Lime—
Valparaiso, 1,430 cks. Cockbain, Allardice, & Co.
,, 705

Barytes—
Ghent, 43 brls.
Antwerp, 40 bgs.

Carbonate—
New York, 50 brls. Meade-King, & Robinson

Copper Ore—
Salaverry, 78 bgs. Bates, Stokes & Co.

Cream Tartar—
Tarragona, 13 cks.

Caoutchouc—
Ceara, 21 bgs. Gunston, Sons, & Co.
,, 58 Leech, Harrison, & Co.
,, 111 Rosing, Bros., & Co.
,, 42 Bieber & Co.
,, 1
Accra, 5 F. & A. Swanzy
,, 1 L. Hart & Co.
,, 3 Hutton & Co.
C. C. Castle, 8 brls. 1 ck. Fletcher & Fraser
,, 1 hd. 6 brls. Ellis & Co.
,, 20
,, 1 cs. 1 A. Reis
,, 1 C. L. Clare & Co.
Monravia, 1 brl. Hart & Co.
,, 2 cks. Brierley & Co.
Sierra Leone, 1 brl. N. Waterhouse & Co.
,, 5 A. Herschell
Sherbro, 1 brl. 1 bg. Comp. Tran.
,, 2 Paterson Zochoms

Dyestuffs—

Extracts
Havre, 110 cks. Cunard S S. Co., Ltd.
New York, 5 brls.

Annatto
Bordeaux, 25 cks.

Myrabolams
Bombay, 2,974 bgs.
,, 1,906 Beyts, Craig & Co.
,, 700 E. D. Sassoon & Co.
,, 667 Arbuthnot, Ewart & Co.

Sumac
Palermo, 100 bgs. J. Kitchen
,, 900

Valonia
Syra, 298 bgs. A. C. Cozzifachi
,, 543 C. G. Nicolaidis
Smyrna, 115 bgs. G. T. Scrini
,, 410

Indigo
Calcutta, 51 chts.
,, 139 1 bx.

Cochineal
Gd. Canary, 9 bgs. Lathbury & Co.
,, 20 J. Walker & Co.
Teneriffe, 8 sks. E. Posselt & Co.

Glycerine—
Antwerp, 24 cks.

Glucose—
New York, 200 brls Schoellkoph & Hartford
,, 26

Horn Piths—
Rio Grande do Sul, 5,100 t. A. Tesdorpf & Co
Ceara, 148 bgs. R. Singlehurst & Co.

Iodine—
Valparaiso, 19 brls. A. Gibbs & Son
,, 36 W. & J. Lockett

Pitch—
Amsterdam, 7 cks.

Paraffin Wax—
New York, 100 brls. Makin & Bancroft

Pyrites—
Huelva, 1,404 t. Matheson & Co.

Phosphates—
Ghent, 600 bgs.

Quinine—
Havre, 47 pkgs. Cunard S.S. Co., Ld

Stearine—
Boston, 36 hds.

Tartar—
Naples, 3 cks.
Valparaiso, 3 bgs.

Tartaric Acid—
Bilbao, 16 brls.

Verdigris—
Bordeaux, 2 cks.

Zinc Oxide—
Antwerp, 80 brls.

Zinc Ashes—
Genoa, 1 brl.
Antwerp, 95

GLASGOW.

Week ending Jan. 23rd.

Acetate of Lime—
New York, 200 bgs.

Blood Albumen—
Rotterdam, 7 cks. J. Rankine & Son

Barytes—
Rotterdam, 136 cks.

Chemicals (*otherwise undescribed*)
Hamburg, 2 cs.

Copper Ore—
Hommelvik, 100 t.

Dyestuffs—

Alizarine
Rotterdam, 32 cks. J. Rankine & Son

Dextrine
Hamburg, 30 bgs.

Glucose—
Hamburg, 10 cks.

Manganese—
Rotterdam, 4 cks. J. Rankine & Son

Pyrites—
Huelva, 1,625 t. Tharsis Co., Ltd.

Saltpetre—
Rotterdam, 7 cks. J. Rankine & Son

Ultramarine—
Rotterdam, 5 cs. J. Rankine & Son

GRIMSBY.

Week ending Jan. 18th.

Albumen—
Antwerp, 16 pkgs.
Hamburg, 13 cks. 2 cs.

Dyestuffs (*otherwise undescribed*)
Antwerp, 5 pkgs.

HULL.

Week ending Jan. 21st.

Acid (*otherwise undescribed*)
Rotterdam, 1 ck.

Chemicals (*otherwise undescribed*)
Hamburg, 68 pkgs. 100 bgs. Wilson, Sons, & Co.
Antwerp, 94 ,,
Rotterdam, 6 7 cks. W. & C. L. Ringrose
Stettin, 1 cs. Wilson, Sons, & Co.
Leghorn, 1 ck. ,,
Genoa, 7 ,,

Caoutchouc—
Hamburg, 1 cs. C. M. Lofthouse & Co.

Dyestuffs—

Alizarine
Rotterdam, 69 pkgs. W. & C. L. Ringrose
Rotterdam, 56 cks. Hutchinson & Son

Valonia
Rotterdam, 12

Dextrine
Stettin, 90 b

Rennet
Copenhagen,

Glucose—
Hamburg, 60
,, 700
,, 200
Stettin,

Manure—
Ghent, 270

Potash—
Dunkirk, 30
Rotterdam, 24

Phosphate—
Dunkirk, 1,000

Saltpetre—
Hamburg, 6 br

Sulphur Ore—
Seville, 1,020 t.

Stearine—
Rotterdam, 15

Tar—
Pillau, 10
Marseilles, 25

Turpentine—
Libau, 50

Ultramarine—
Rotterdam, 5 c
,, 2 c

Waste Salt—
Hamburg, 200 b

Zinc Oxide—
Rotterdam, 25 c

GO

Week end

Alum—
Rotterdam, 31

Acetic Acid—
Rotterdam, 10

Dyestuffs—

Alizarine
Rotterdam, 30

Dyestuffs (*ot*
Boulogne, 30

Glucose—
Hamburg, 72

Potash—
Hamburg, 13
Calais, 34
Dunkirk, 25

Phosphate—
Ghent, 80

Potassium—
Boulogne, 2 ck

Sugar of Lea
Hamburg, 19 ck

Sodium Silica
Hamburg, 9

Saltpetre—
Rotterdam, 200

Waste Salt—
Hamburg, 525

Zinc Oxide—
Antwerp, 20

TYNE.

Week ending Jan. 23rd.

Alum Earth—
Hamburg, 18 cks. Tyne S. S. Co.
Barytes—
Rotterdam, 86 cks. J. Robinson & Sons
Bones—
Randers, 7 cks.
Copper Precipitate—
Pomaron, 2,750 bgs. Mason & Barry
Huelva, 1,906
" 4,452
Copper Regulus—
Huelva, 45 t.
Glucose—
Hamburg, 7 cks. 100 bgs. Tyne S. S. Co.
Pyrites—
Huelva, 1,391 t. Scott Bros.
" 660
Sulphur Ore—
Pomaron, 2,100 t. Mason & Barry
" 1,160
Salt—
Hamburg, 101 bgs. Tyne S. S. Co

EXPORTS OF CHEMICAL PRODUCTS

FROM

THE PRINCIPAL PORTS OF THE UNITED KINGDOM.

LONDON.

Week ending Jan. 15th.

Acids—(*otherwise undescribed*)
Natal, 5 t. 10 c. £22
New York, 30 cs. 85
Jersey, 4,140 lbs. 52
Alum Cake—
Calcutta, 175 cks. £113
Ammonium Sulphate—
Hamburg, 100 t. 600 bgs. £1,625
Cologne, 18 10 c. 222
Havre, 35 3 390
Ammonium Carbonate—
Odessa, 2 t. 2 c. £61
Acetic Acid—
Buenos Ayres, 16 c. £13
Ammonium Muriate—
Auckland, 2 t. 4 c. £45
Arsenic—
Amsterdam, 6 cks. £17
Brimstone—
Madras, 10 t. 1 c. £59
Bisulphide of Carbon—
Rotterdam, 4 t. 19 c. £59
Copper Sulphate—
Barcelona, 10 cs. £81
Salonica, 1 t. 2 c. 26
Madras, 10 11
Samsoun, 10 12
Tarragona, 10 19 229
Barcelona, 12 10 251
Bombay, 5 114
Treport, 36 cks. 214
Chemicals (*otherwise undescribed*)
Barcelona, 10 cs. £81
Monte Video, 10 50
Auckland, 23 pkgs. 175
Calcutta, 6 38
Trinidad, 4 63
Citric Acid—
Seville, 6 c. £44
Barcelona, 1 t. 140
Cadiz, ½ 4
Lisbon, 11 72
Brussels, 10 74
Genoa, 20 148
Odessa, 5 kegs. 37
Cream Tartar—
Cape Town, 2 c. £15
Sydney, 20 cks. 110
Wellington, 6 36
Algoa Bay, 65 cs. 71
Caustic Soda—
Cadiz, 3 t. £25
Melbourne, 3 30
Algoa Bay, 1 14 c. 69
Carbolic Acid—
Yokohama, 2,000 lbs. £41
Falkland, 1 t. 11 c. 28
Colombo, 4 21
Disinfectants—
Madras, 2 t. 11 c. £20
Guano—
Morlaix, 189 t. £1,890
Hydro Carbon—
Brussels, 3 t. 14 c. £66
Manure—
Hamburg, 48 t. 16 c. £210
Rotterdam, 20 67
Jersey, 158 12 2,073
Oporto, 71 380
Valencia, 20 265
St. Lucia, 70 700
Jamaica, 5 55
Dominica, 5 65
Demerara, 50 500
Magnesia—
Lisbon, 10 c. £18
Nitric Acid—
Jersey, 2 t. 6 c. £43
Oxalic Acid—
Moscow, 5 t. 6 c. £175
Odessa, 7½ 12
Libau, 28 cks. 268
Stettin, 12 14
Brussels, 23 30
Genoa, 2 6 58
Antwerp, 5 11 120
Potassium Chlorate—
Oporto, 2 t. £90
Sydney, 10 520
Picric Acid—
Gothenburg, 1 cs. £10
Potash—
Toronto, 10 c. 35
Potassium Bichromate—
Marseilles, 7 c. 13
Salammoniac—
Varna, 1 t. 1 c. £34
Sulphuric Acid—
Buenos Ayres, 11 cs. £12
Port Natal, 3 t. 6 c. 19
Algoa Bay, 25 cs. 96 pkgs. 118
Sulphur Flowers—
Algoa Bay, 140 kgs. 100 cks. £93
Lisbon, 20 brls. 6 20
Saltpetre—
Oporto, 3 t. 14 c. £80
Wellington, 20 24
Seville, 4 cks. 47
Adelaide, 1 23
Sheep Dip Powder—
Algoa Bay, 55 c. 25 cs. £224
San Salvador, 30 60
Hobart, 5 t. 13 120
E. London, 1 10 125 428
Superphosphate—
Cherbourg, 94 t. 19 c. £280
Tartaric Acid—
Wellington, 5½ c. £42
Lisbon, 5½ 36
Sydney, 20 cks. 136

LIVERPOOL.

Week ending Jan. 18th.

Alkali—
Cape Town, 10 bxs. 1 pkg.
Alum—
Callao, 1 t. 1 q. £5
Santander, 5 26
Boston, 2 15 c. 17
Calcutta, 30 3 14
Valencia, 2 9
Malaga, 15 10 91
Talcahuano, 1 t. 5 6
Havana, 19 5
Ammonium Sulphate—
Ghent, 15 t. 7 c. £179
Rotterdam, 55 10 685
Hamburg, 10 17 3 q. 130
New York, 4 17 165
Bordeaux, 70 840
Hamburg, 110 13 3 1,487
Valencia, 112 16 3 1,465
Antwerp, 30 360
Ammonium Carbonate—
St. Johns, N. B., 9 c. £14
Santander, 1 t. 2 3 q. 33
Madeira, 4 7
Bahia, 13 17
New York, 13 3 1 440
Havana, 6 3 12
Naples, 12 2 15
Antimony Regulus—
Nantes, 2 t. 3 c. 1 q. £135
Ammonium Muriate—
New York, 13 t. 11 c. £330
Ammonia—
B. Ayres, 19 c. 2 q. £25
Bleaching Powder—
Alexandria, 3 brls.
Ancona, 7 cks.
Antwerp, 6
Baltimore, 112
Bordeaux, 35
Boston, 389 85 tcs.
Chicago, 50 bxs.
Christiania, 19
Genoa, 484
Hamburg, 25
Havana, 5
Havre, 37
Lisbon, 70
Nantes, 107
New York, 155
Rotterdam, 14
Venice, 5
Barium Sulphate—
Ferrol, 2 t. 1 q. £5
Borax—
Rouen, 42 t. 17 c. £1,200
Vera Cruz, 8 12
Matanzas, 5 2 q. £8
Savanilla, 1 6
Cardenas, 3 3 5
Havana, 2 3 3 61
Barium Chlorate—
Vigo, 5 c. £19
Boracic Acid—
Havana, 4 c. 1 q. £8
Brimstone—
Boston, 50 t. £290
New York, 15 c. 6
Caustic Soda—
Alicante, 80 dms. 20 brls.
Amsterdam, 25
Antwerp, 100 51
Bahia, 10
Baltimore, 50
Barcelona, 225 25
Bari, 73
Batoum, 150
Bombay, 15
Bordeaux, 100
Brisbane, 36
Calcutta, 30
Callao, 30
Catania, 55
Chicago, 50 bxs.
Constantinople, 15
Dunedin, 29
Genoa, 80
Ghent, 30
Hamburg, 26
Havana, 40
Havre, 9
Leghorn, 14
Malaga, 62 20 kgs. 10 brls.
Matanzas, 49
Melbourne, 10 pkgs. 15 bxs.
Naples, 10 4 brls.
Nantes, 35
New York, 360 10 kgs.
Oporto, 55
Palermo, 10
Passages, 35
Philadelphia, 200 31 cks. 135 brls.
Piræus, 180
Portland, M. 310
Port Natal, 40 cs. 2 bxs.
Rotterdam, 22
San Francisco, 105 10 brls. 20 bxs.
San Vanilla, 20
Seville, 186
San Sebastian, 30
Talcahuano, 25
Tampico, 10
Valparaiso, 70
Vera Cruz, 10
Yokohama, 100
Chemicals (*otherwise undescribed*)
Beyrout, 2 t. 7 c. 2 q. £79
Colon, 1 cs. 2 kgs. 21
Puerto Cabello, 5 pkgs. 8
Hiogo, 3 t. 19 c. 130
New Orleans, 4 cks. 5
Algoa Bay, 2 q. 9
Copper Sulphate—
Bordeaux, 5 t. 2 q. £110
Rouen, 45 7 c. 1,088
Marseilles, 7 1 1 q. 166
Pachuras, 20 1 2 471
Barcelona, 25 10 1 546
Beyrout, 9 2 11
Genoa, 29 9 2 556
Hamburg, 5 2 2 90
Nantes, 10 1 2 186
Salonica, 11 10
Santander, 25 5 451
Carbolic Acid—
Rouen, 1 t. 2 c. £121
Shangai, 1 12 65
Havre, 20 238
Paris, 4 7 480
Hamburg, 6 0 2 q. 756
Piræus, 1 8
Havana, 3 3 12
Chloride of Lime—
Boston, 50 t. 15 c. £325
New York, 50 272
Charcoal—
Hamburg, 146 t. 8 c. £435
Caustic Potash—
New York, 8 t. 1 c. £161

Copperas—
Rio de Janeiro, 2 t. 8 c. £5
Alexandria, 6 19
Oporto, 2 16 6
Piræus, 3 1 6
Copper Precipitate—
Halifax, 6 t. £245
Disinfectants—
Bahia, 1 c 1 q. £8
Epsom Salts—
Santander, 5 brls.
Glycerine—
Rouen, 18 t. 15 c. £450
New York, 23 16 720
Monte Video, 2 cs. 7
Rotterdam, 10 440
Guano—
Barcelona, 73 t. 16 c 1 q. £719
Iron Oxide—
Havana, 4 c. £9
Magnesia—
Rio Janeiro, 5 cs.
Porto Alegre, 5
Magnesium Sulphate—
Santander, 2 t. 1 q. £7
Manure—
Honolulu, 20 t. £200
Gr. Canary, 20 147
Rochefort, 610 10 c. 2,005
Oxalic Acid—
Syra, 3 c. £56
Trieste, 2 t. 10 2 q. 60
Messina, 3 3 13
Pitch—
Cette, 1,967 t.
Coquimbo, 40 brls.
Potassium Bichromate—
Rouen, 5 t. 12 c. 2q. £200
Rio de Janeiro, 1 cks. 15
Calcutta, 3 c. 5
Alexandria, 6 12
Potassium Chlorate—
Hamburg, 5 c. £12
Yokohama, 20 t. 856
Coruna, 10 20
Hong Kong, 1 114 85
Kobe, 16 5 706
New York, 1 5 73
Portland, M. 12 15
Gothenburg, 17 3 807
Boston, 15 42
Santos, 3 148
Potassium Carbonate—
Mozambique, 1 t. 5 c. £24
New York, 7 10 120
Phosphate—
Cape Town, 2 t. £5
Pearl Ashes—
Calcutta, 14 c. £21
Potassium Prussiate—
Lisbon, 1 t. 3 q. £105
Salt—
Adelaide, 125 t.
Baltimore, 180
Barbadoes, 55
Bibundi, 50
Boston, 310
Brisbane, 160
Calcutta, 12,498
Cameroons, 36 3 c.
Cape Coast, 7½
Chicago, 130
Copenhagen, 482
Coquimbo, 220
East London, 11
Elobey, 218¾
Ghent, 30
Grand Bassam, 20
Lagos, 21
Lavannah, 26¾
Manoh, 13 8
Monrovia, 50
Monte Video, 13 14
New Orleans, 2,531 16
Newfoundland, 5
New York, 130
Porland, M. 22
Rio Nunez, 20
Ronne, 210
Rotterdam, 292
Saltpond, 6¾
San Francisco, 839
Soda Crystals—
Alexandria, 30 brls.
Calcutta, 7 cks.
Malta, 50
New York, 386
Philadelphia, 60
Salonica, 44
Trieste, 3
Sodium Silicate—
Alicante, 10 brls.
Almeria, 10

Sodium Bicarbonate—
Almeria, 10 cks.
Auckland, 40 kgs.
Bombay, 200
Calcutta, 500
Constantinople, 20
Dunedin, 50
Gothenburg, 40
New York, 100
San Francisco, 30
Smyrna, 40
Wellington, 20
Soda Ash—
Baltimore, 709 cks. 328 trs.
Bombay, 5 kgs.
Boston, 878 bgs. 87 cks.
Buenos Ayres, 50 brls.
Calcutta, 42
Constantinople, 20
Lisbon, 10
Malaga,
Melbourne, 295
Monte Video, 25
Montreal, 10
New York, 123 1,558 trs.
Oporto, 5 15
Para,
Philadelphia, 21 354
Portland, M. 170
Rosario, 116
Salonica, 325
San Francisco, 65
Smyrna, 400
Venice, 15
Sodium Carbonate—
New York, 140 brls.
Sulphosite—
Bombay, 11 c. 3 q. £47
Salommoniac—
Samsoun, 16 c. £26
Vigo, 2 2 q. 5
Alexandria, 2 t. 69
Salonica, 1 1 39
Trieste, 1 2 35
Constantinople, 5 2 175
Sodium Biborate—
Antwerp, 7 t. 7 c. 1 q. £221
Hamburg, 14 19 1 450
Portland, M., 12 14 3 394
Buenos Ayres, 4 1 6
Rotterdam, 12 17
Saltcake—
New York, 149 t. 15 c. £326
Sodium Sulphate—
Valparaiso, 10 c. £9
Sheep Dip—
Melbourne, 151 galls. £26
Cape Town, 25 cs. 50
Sodium Nitrate—
Valencia, 3 t. 6 c. 1 q. £27
Venice, 7 11 1 66
Sulphur
Boston, 60 t. 5 c. £269
Trinidad, 5 16 35
Superphosphate—
Barcelona, 10 t. 3 q. £240
Gr. Canary, 11 33
Hamburg, 100 1 c. 1 350
Sodium Bichromate—
Piræus, 1 t. £29
Sulphuric Acid—
Madras, 3 t. 10 c. £15
Tar—
Bahia, 6 brls.
Calcutta, 50
Callao, 10
Genoa, 3 cks.
Lima, 20 20 dms.
Sierra Leone, 20
Tin Oxylate—
Vera Cruz, 5 cks. £56
Tartar—
New York, 20 t. £1,900
Vitriol—
Adelaide, 5 t. £115
Barcelona, 14 10 c. 2 q. 90
Pasages, 4 19 3 87
Bordeaux, 50 4 2 1,200
Montreal, 3 5 54
Nantes, 131 7 2 3,142
Oporto, 1 1 17
Calcutta, 32 670
Constantinople, 1 15 38
Genoa, 96 18 1 2,105
Smyrna, 8 3 8
Talcahuano, 1 2 25
Trieste, 1 3 21
Venice, 40 1 2 829
Leghorn, 17 3 3 335
Naples, 2 7 2 40
Zinc Chloride—
Bombay, 3 t. 15 c. 2 q. £55

TYNE.

Week ending January 22nd.

Alkali—
Amsterdam, 24 t. 14 c.
Savona, 62 18
Leghorn, 29
Rotterdam, 11
Barcelona, 15 6
Hamburg, 3 9
Copenhagen, 5 11
Anthracene—
Rotterdam, 16 t.
Ammonia Sulphate—
Rotterdam, 20 t.
Antimony—
Rotterdam, 10 t.
Hamburg, 3 10 c.
Bleaching Powder—
Copenhagen, 1 t. 9 c.
" 9 19
Rotterdam, 22 5
Hamburg, 34 9
Antwerp, 93 5
Lisben, 20 3
Libau, 25
Amsterdam, 1,443
Bergen, 5 3
Odense, 9 5
Barytes Carborate—
Rotterdam, 5 t.
Caustic Soda—
Copenhagen, 3 t. 12 c.
Antwerp, 68 2
Amsterdam, 5
Copperas—
Esbjerg, 6 t. 7 c.
Copper, Sulphate—
Leghorn, 4 c.
Flint Ground—
Rotterdam, 10 t.
Iron Oxide—
Lisbon, 2
Litharge—
Palermo, 11
Amsterdam, 5 t. 9
Magnesia—
Rotterdam, 15 c.
Hamburg, 2
Lisbon, 10
Barcelona, 7 t. 13
" 8 15
Potassium Chlorate—
Rotterdam, 11 c.
Antwerp, 3 t. 5
Pitch—
Antwerp, 100 t.
Soda—
Gothenburg, 10 t. 10 c.
Drontheim, 45 3
Copenhagen, 25 5
Rotterdam, 10 15
Svendborg, 5 4
Nakskov, 10 15
Soda Ash—
Lisbon, 30 t. 17 c.
Savona, 44 10
Leghorn, 6 9
Antwerp, 71
Naples, 30 2
Sodium—
Hamburg, 2 cs.
Sodium Chlorate—
Hamburg, 1 t.
Superphosphate—
Venice, 300 t.
Sodium Bicarbonate—
Amsterdam, 1 t. 1 c.

HULL.

Week ending Jan. 21st.

Alkali—
Bari, 4 cks.
Fiume, 56
Marseilles, 250 drms
Odessa, 60
Benzol—
Dunkirk, 24 cks.
Bleaching Powder—
Danzig, 55 cks.
Venice, 30
Bone Size—
Ghent, 4 cks.
Chemicals *(otherwise undescribed)*
Antwerp, 30 pks.
Amsterdam, 1 cs. 50 cks.
Bergen, 20
Bremen, 75
Bordeaux, 30

Copenhage
Christiania
Dunkirk,
Genoa,
Gothenburg
Hamburg,
Libau,
Naples,
Odessa,
Rouen,
Rotterdam,
Caustic So
Libau,
Dyestuffs (
Hamburg,
Rouen,
Rotterdam,
Reval,
Ferro Man
Genoa, 1
Manure—
Copenhagen
Ghent,
Hamburg,
Rotterdam,
Naptholine
Antwerp,
Libau,
Pitch—
Antwerp,
Dunkirk,
Genoa,
Slag—
Danzig,
Stettin,
Tar—
Christiansan
Hamburg,
Turpentine
Malta,

GL

Week en

Ammonium
Demerara,
Bordeaux,
Bleaching P
Sydney, 7 t
Bones—
Demerara, 1
Copper Sul
Bordeaux, 1
Guano—
Demerara, 2
Manure—
Trinidad, 2
Manganese
Antwerp, 6
Potassium
Boston,
Antwerp, 3
Rotterdam, 2
" 1
Paraffin W
Sydney, 7
Bordeaux, 5
Pitch—
Bordeaux, 2
Sodium Bic
Boston, 3 t.

G

Week

Benzol—
Antwerp,
Hamburg,
Rotterdam,
Chemicals (
Antwerp,
Hamburg,
Coal Produ
Calais,
Rotterdam
Dyestuffs (
Boulogne,
Manure—
Ghent, 4
Hamburg, 9
Pitch—
Antwerp,
Dunkirk,

GR

Week

Chemicals (
Dieppe,

PRICES CURRENT.

THURSDAY, February 5, 1890.

Prepared by Higginbottom and Co., 116, Portland Street, Manchester.

The values stated are F.O.R.* *at maker's works, or at usual ports of shipment in U.K. The price in different localities may vary.*

		£	s.	d.
Acids:—				
Acetic, 25 °/o and 40 °/o	per cwt	7/6 & 12/6		
,, Glacial	,,	2	10	0
Arsenic, S.G., 2000°	,,	0	12	0
Chromic 82 °/o .. nett	per lb.	0	0	6½
Fluoric	,,	0	0	3¾
Muriatic (Tower Salts), 30° Tw.	per bottle	0	0	8
,, (Cylinder), 30° Tw.	,,	0	2	11
Nitric 80° Tw.	per lb.	0	0	2
Nitrous	,,	0	0	1¾
Oxalic .. nett	,,	0	0	3
Picric	,,	0	1	5
Sulphuric (fuming 50 °/o)	per ton	15	10	0
,, (monohydrate)	,,	5	10	0
,, (Pyrites, 168°)	,,	3	2	6
,, (,, 150°)	,,	1	8	0
,, (free from arsenic, 140/145°)	,,	1	10	0
Sulphurous (solution)	,,	2	15	0
Tannic (pure)	per lb.	0	1	0½
Tartaric	,,	0	1	3
Arsenic, white powered	per ton	13	0	0
Alum (looselump)	,,	4	12	6
Alumina Sulphate (pure)	,,	5	0	0
,, ,, (medium qualities)	,,	4	10	0
Aluminium	per lb.	0	15	0
Ammonia, ·880=28°	per lb.	0	0	3
,, =24°	,,	0	0	1⅞
,, Carbonate .. nett	,,	0	0	3½
,, Muriate	per ton	24	0	0
,, ,, (sal-ammoniac) 1st & 2nd	per cwt.	37/- & 35/-		
,, Nitrate	per ton	37	10	0
,, Phosphate	per lb.	0	0	10⅝
,, Sulphate (grey), London	per ton	12	0	0
,, ,, (grey), Hull	,,	11	17	6
Aniline Oil (pure)	per lb.	0	0	11¾
Aniline Salt ,,	,,	0	0	11
Antimony	per ton	75	0	0
,, (tartar emetic)	per lb.	0	1	1
,, (golden sulphide)	,,	0	0	10
Barium Chloride	per ton	7	10	0
,, Carbonate (native)	,,	3	15	0
,, Sulphate (native levigated)	,,	45/- to 75/-		
Bleaching Powder, 35%	,,	5	2	6
,, Liquor, 7%	,,	2	10	0
Bisulphide of Carbon	,,	12	15	0
Chromium Acetate (crystal	per lb.	0	0	6
Calcium Chloride	per ton	2	0	0
China Clay (at Runcorn) in bulk	,,	16/- to 32/-		
Coal Tar Dyes:—				
Alizarine (artificial) 20%	per lb.	0	0	9
Magenta	,,	0	3	9
Coal Tar Products				
Anthracene, 30% A, f.o.b. London	per unit per cwt.	0	1	6
Benzol, 90% nominal	per gallon	0	4	0
,, 50/90 ,,	,,	0	3	3
Carbolic Acid (crystallised 35°)	per lb.	0	0	10½
,, (crude 60°)	per gallon	0	2	10
Creosote (ordinary)	,,	0	0	2½
,, (filtered for Lucigen light)	,,	0	0	3
Crude Naphtha 30% @ 120° C.	,,	0	1	3
Grease Oils, 22° Tw.	per ton	3	0	0
Pitch, f.o.b. Liverpool or Garston	,,	1	12	0
Solvent Naphtha, 90% at 160°	per gallon	0	1	10
Coke-oven Oils (crude)	,,	0	0	2½
Copper (Chili Bars)	per ton	48	15	0
,, Sulphate	,,	24	10	0
,, Oxide (copper scales)	,,	52	10	0
Glycerine (crude)	,,	30	0	0
,, (distilled S.G. 1250°)	,,	55	0	0
Iodine	nett, per oz.	8d. to 9d.		
Iron Sulphate (copperas)	per ton	1	10	0
,, Sulphide (antimony slag)	,,	2	0	0
Lead (sheet) for cash	,,	15	0	0
,, Litharge Flake (ex ship)	,,	17	10	0
,, Acetate (white ,,)	,,	23	10	0
,, ,, (brown ,,)	,,	19	0	0
,, Carbonate (white lead) pure	,,	19	0	0
,, Nitrate	,,	22	5	0

		£	s.	d.
Lime, Acetate (brown)	,,	8	10	0
,, ,, (grey)	per ton	14	0	0
Magnesium (ribbon and wire)	per oz.	0	2	3
,, Chloride (ex ship)	per ton	2	7	6
,, Carbonate	per cwt.	1	17	6
,, Hydrate	,,	0	10	0
,, Sulphate (Epsom Salts)	per ton	3	0	0
Manganese, Sulphate	,,	18	0	0
,, Borate (1st and 2nd)	per cwt.	60/- & 42/6		
,, Ore, 70%	per ton	4	10	0
Methylated Spirit, 61° O.P.	per gallon	0	2	2
Naphtha (Wood), Solvent	,,	0	3	9
,, ,, Miscible, 60° O.P.	,,	0	4	4½
Oils:—				
Cotton-seed	per ton	22	0	0
Linseed	,,	22	10	0
Lubricating, Scotch, 890°—895°	,,	7	5	0
Petroleum, Russian	per gallon	0	0	5¾
,, American	,,	0	0	6
Potassium (metal)	per oz.	0	3	10
,, Bichromate	per lb.	0	0	4
,, Carbonate, 90% (ex ship)	per ton	18	5	0
,, Chlorate	per lb.	0	0	4¾
,, Cyanide, 98%	,,	0	2	0
,, Hydrate (Caustic Potash) 80/85%	per ton	23	0	0
,, ,, (Caustic Potash) 75/80%	,,	21	15	0
,, ,, (Caustic Potash) 70/75%	,,	20	15	0
,, Nitrate (refined)	per cwt.	1	3	6
,, Permanganate	,,	4	5	0
Prussiate Yellow	per lb.	0	0	10¾
,, Sulphate, 90%	per ton	9	10	0
,, Muriate, 80%	,,	7	15	0
Silver (metal)	per oz.	0	3	8½
,, Nitrate	,,	0	2	5½
Sodium (metal)	per lb.	0	4	0
,, Carb. (refined Soda-ash) 48%	per ton	5	10	0
,, ,, (Caustic Soda-ash) 48%	,,	4	7	6
,, ,, (Carb. Soda-ash) 48%	,,	4	15	0
,, ,, (Carb. Soda-ash) 58%	,,	5	12	6
,, ,, (Soda Crystals)	,,	2	13	6
,, Acetate (ex-ship)	,,	16	0	0
,, Arseniate, 45%	,,	10	0	0
,, Chlorate	per lb.	0	0	6¼
,, Borate (Borax)	nett, per ton.	29	10	0
,, Bichromate	per lb.	0	0	3
,, Hydrate (77% Caustic Soda)	per ton.	10	10	0
,, ,, (74% Caustic Soda)	,,	10	0	0
,, ,, (70% Caustic Soda)	,,	8	17	6
,, ,, (60% Caustic Soda, white)	,,	7	17	6
,, ,, (60% Caustic Soda, cream)	,,	7	10	0
,, Bicarbonate	,,	5	5	0
,, Hyposulphite	,,	5	2	6
,, Manganate, 25%	,,	8	10	0
,, Nitrate (95%) ex-ship Liverpool	per cwt.	0	8	3
,, Nitrite, 98%	per ton	27	10	0
,, Phosphate	,,	15	15	0
,, Prussiate	per lb.	0	0	7¾
,, Silicate (glass)	per ton	5	7	6
,, ,, (liquid, 100° Tw.)	,,	3	17	6
,, Stannate, 40%	per cwt.	2	0	0
,, Sulphate (Salt-cake)	per ton.	1	7	6
,, ,, (Glauber's Salts)	,,	1	10	0
,, Sulphide	,,	7	15	0
,, Sulphite	,,	5	5	0
Strontium Hydrate, 98%	,,	8	0	0
Sulphocyanide Ammonium, 95%	per lb.	0	0	8½
,, Barium, 95%	,,	0	0	5½
,, Potassium	,,	0	0	7¾
Sulphur (Flowers)	per ton.	7	0	0
,, (Roll Brimstone)	,,	6	0	0
,, Brimstone: Best Quality	,,	4	17	6
Superphosphate of Lime (26%)	,,	2	10	0
Tallow	,,	25	10	0
Tin (English Ingots)	,,	99	0	0
,, Crystals	per lb.	0	0	6¾
Zinc (Spelter	per ton.	22	10	0
,, Chloride (solution, 96° Tw.	,,	6	0	0

THE

CHEMICAL TRADE JOURNAL.

Publishing Offices: 32, BLACKFRIARS STREET, MANCHESTER.

No. 1[illegible]. SATURDAY, FEBRUARY 15, 1890. Vol. VI.

Contents.

Notices.

All communications for the *Chemical Trade Journal* should be addressed, and Cheques and Post Office Orders made payable to—

[illegible] BROS., 32, Blackfriars Street, MANCHESTER.

Our Registered Telegraphic Address is—

"Expert, Manchester."

The Terms of Subscription, commencing at any date, to the *Chemical Trade Journal*,—payable in advance,—including postage to any part of the world, are as follow:—

Yearly (52 numbers)	**12s. 6d.**
Half-Yearly (26 numbers)	**6s. 6d.**
Quarterly (13 numbers)	**3s. 6d.**

Readers will oblige by making their remittances for subscriptions by Postal or Post Office Order, crossed.

Communications for the Editor, if intended for insertion in the current week's issue, should reach the office not later than **Tuesday Morning.**

Articles, reports, and correspondence on all matters of interest to the Chemical and allied industries, home and foreign, are solicited. Correspondents should condense their matter as much as possible, write on one side only of the paper, and in all cases give their names and addresses, not necessarily for publication. Sketches should be sent on separate sheets.

We cannot undertake to return rejected manuscripts or drawings, unless accompanied by a stamped directed envelope.

Readers are invited to forward items of intelligence, or cuttings from local newspapers, of interest to the trades concerned.

As it is one of the special features of the *Chemical Trade Journal* to give the earliest information respecting new processes, improvements, inventions, etc., bearing upon the Chemical and allied industries, or which may be of interest to our readers, the Editor invites particulars of such—when in working order—from the originators; and if the subject is deemed of sufficient importance, an expert will visit and report upon the same in the columns of the *Journal*. There is no fee [illegible] visits of this kind.

We shall esteem it a favour if any of our readers, in making [illegible] or opening accounts with advertisers in this paper, will kindly mention the *Chemical Trade Journal* as the source of their information.

Advertisements intended for insertion in the current week's issue, should reach the office by **Wednesday morning** at the latest.

Advertisements.

[illegible] Advertisements of Situations Vacant, Premises on Sale or [illegible], Miscellaneous Wants, and Specific Articles for Sale by [illegible], are inserted in the *Chemical Trade Journal* at the [illegible]:—

[illegible] Words and under	2s. 0d.	per insertion.
[illegible] additional 10 words	0s. 6d.	,,

[illegible] of Situations Wanted are inserted at one-[illegible] when prepaid, viz:—

[illegible] under	1s. 0d.	per insertion.
[illegible] Words	0s. 3d.	,,

[illegible] in the Directory Columns, and all Adver-[illegible] at the Tariff rates which will be forwarded on [illegible]

ON THE DETERMINATION OF MOISTURE IN WOOD AND STRAW PULP.

BY GEORGE E. DAVIS AND ALFRED R. DAVIS.

The determination of the actual quantity of paper-making material in pulp as it comes from the factory or department where it is made is attended with considerable difficulties. In the first place the task of sampling is no light one, as almost every bale varies in its percentage of moisture, and again, one portion of a bale may differ considerably from another. In the case of this fibrous material it is impossible to treat it by the ordinary methods for reducing large samples to a workable size, and the custom has obtained of cutting a piece here and there off bales, supposed to be representative of the bulk, and of estimating the moisture in each piece separately, striking an average for the final result. No doubt this method will give fairly accurate results, if the number of bales sampled be large enough in proportion to the total number; but those interested should bear in mind that to draw two or three samples from as many hundreds of bales is quite a haphazard proceeding, which can only furnish satisfactory results by the merest chance. But supposing a fairly representative sample to have been drawn; we are soon confronted with another source of discrepancy. The pieces of wet pulp are too often wrapped up in paper or some other absorbent material, and part with varying percentages of moisture according to the time that elapses before they eventually reach the laboratory and are weighed. The next process is the drying in the water or steam bath at 212° F., and from this point the custom of the trade diverges in two directions. Some firms take the weight dried at 212° F., and add on a certain definite percentage to obtain the weight of merchantable pulp at ordinary temperature; while others take the dried pulp and expose it to the atmosphere in an ordinary room for a certain length of time in order that it may re-absorb the normal amount of moisture. The weight after exposure is supposed to indicate the proportion of "air-dried pulp" in the sample.

It will be at once seen that the latter method introduces possibilities of conflict. If the dried pulp is not exposed for a uniform length of time, and under uniform conditions as regards temperature, air-currents, atmospheric moisture and the like, the results are certain to be discordant, and the difficulty of obtaining this uniformity of conditions is well known to all who have had anything to do with the subject. Where the "exposure" method is in vogue the time adopted for re-absorption is generally 24 hours. As a rule this is about sufficient, but to be on the safe side we prefer 36 hours.

It must be borne in mind that when the pulp has absorbed its full quantum of moisture, its weight will vary according to the hygrometrical conditions of the room, and, supposing draughts to be excluded, will fall with a rise of temperature, and *vice versa*.

The experience of the Manchester Technical Laboratory is not in favour of the exposure method. As the outcome of a large number of investigations upon this class of material, it is recommended to take the weight, dried, at 212°, and add a definite percentage, to obtain the weight of "air-dried" pulp. In the very great majority of cases, 10 per cent. expresses, with almost absolute accuracy, the amount to be so added, and if this figure were universally adopted, it would give much plainer sailing to all parties concerned. Even if a certain kind of pulp were known to absorb considerably more or less than this amount, it would be quite easy to regulate the price accordingly, but the necessity for this seldom arises.

To surmount the difficulty of the possibility of loss of weight between the taking of the sample and its delivery to the chemist, the following precautions are recommended. The buyers' and sellers' representatives should superintend the weighing off, on the spot, say of 100 grammes or 1,000 grains of each agreed sample, and immediately wrap up in thin rubber tissue. If this is quickly and carefully done, and the wrapped up samples are guarded from heat or pressure, they can be delivered to the chemist without undergoing any practical loss of weight. He will of course check the figures handed him with the samples, but if the operation has been properly performed, will never find a deviation of more than two or three tenths of a per cent.

If this course were universally adopted, with the 10 per cent. addition to convert the weight at 212° into "air dried" weight, we feel certain that the disputes, now so numerous between buyer and seller, would become much less frequent, and that in the long run the sacrifice of accuracy would be comparatively nil.

THE MANCHESTER TECHNICAL LABORATORY,

32, Blackfriars Street.

Feb. 11th, 1890.

POTASSIUM CARBONATE AS A FERTILISING AGENT.

CARBONATE of potash is, according to M. Georges Ville, about to play a very important part as a potash fertiliser. Much better results are to be obtained by its use than are given by the other salts of potassium, such as the sulphate and chloride. The use of carbonate of potash in agriculture, is indeed already well known, and its high price alone has deterred agriculturists from recommending it. Too much importance is usually attached to this question of the price of the material employed, but it should be borne in mind that the effect produced by a given quantity of a product, must be considered before a fair judgment can be formed of its merits. Now, without going into the details of agricultural experiments, we may say in general, that the action of carbonate of potash is rapid, and that it produces very marked results.

Everyone has noticed how luxuriantly vegetation springs up on hillocks and slopes, the hard and dry herbage of which has been lately burned. A small plant of a dark green colour rapidly covers these formerly arid spots, and this effect is due to the potassium salts contained in the ash of the incinerated plants. The chief of these salts is carbonate of potash, and it is probably the chief agent in producing the rapid and striking alteration just described. The wandering herdsmen in Algeria, America, and even in Greece, are well aware of the improvement produced in pasturage by the incineration of the vegetation of a district, and they do not hesitate to fire vast prairies and even forests in order to ensure good pasturage for coming years.

As long as five or six years ago carbonate of potash was employed with success in the preparation of artificial compound manures. The author of this article recollects that Messrs. Guionnet and Co., of Orleans, a firm deservedly esteemed in the trade, obtained marvellous results by means of a manure made up for liguminous plants on a basis of potash, supplied by the refined carbonate of commerce.

We are not aware whether this firm continues the use of carbonate of potash in the manufacture of its manures, but, speaking generally, we may say that, owing to excessive competition and the low price to which all manures have fallen, carbonate of potash has had to be given up. Cheapness was essential, the potassium had to be supplied at the lowest possible price, and therefore almost everyone made use of the chloride and sulphate as fulfilling this condition.

Since the statements of M. Georges Ville, whose vigorous personality, eclipsed for a short period, seems about to resume the authority and popularity to which his agricultural labours entitle him, there has been a very pronounced tendency in favour of the use of carbonate of potash as a manure. We, therefore, believe that at the present moment our readers will be glad to receive any commercial information which will guide them in their purchase of this material.

SOURCE.

Plants require for their proper growth and development mineral constituents which they take from the soil, and organic constituents which they find in the atmosphere. A large proportion of this mineral matter is potassium, and certain plants in particular have a special power of appropriating its salts. If, then, the organic portion of the plant be destroyed the mineral portion remains, and this is the most fruitful source of the potash salts now produced.

POTASHES.

In former times all the potash of commerce was extracted from the ashes of wood. American and Russian *potashes*, which are now disappearing from the market, have always been obtained in this way. Large quantities of ash were required for the production of even a small amount of potash, and entire forests had to be destroyed to supply the needs of industry.

This barbarous process had to disappear as civilisation gradually penetrated into the countries in which it was in use, and science following in its footsteps discovered other sources, both more abundant and less wasteful.

POTASHES FROM THE SUGAR BEET.

The root of the sugar beet is very rich in mineral matter, which it absorbs from the soil, and contains potash salts in quantities which depend on the condition of the ground, becoming smaller if the land has been impoverished by successive cultivations of the same root. The sugar refiners remove from the beet the greater part of its organic matter, the sugar and the pulp, whilst, on the other hand, the mineral salts accumulate in the mother liquors, along with a large amount of sugar, which they render uncrystallizable, and which cannot be removed. This mixture of sugar and mineral salts constitutes the molasses of the sugar works. The sugar of the molasses is converted into alcohol by fermentation, and the whole distilled, the alcohol being condensed, and the final residue obtained in the form of the *vinasse*, which necessarily contains all the soluble mineral constituents of the beet-root, together with a small quantity of organic matter.

BEET-ROOT SALTS.

The evaporated vinasse yields a black, syrupy substance, which is burnt in kilns, and thus converted into a greyish black flaky mass, resembling the ash left by coal. This is the crude beet-root potash or salt.

The composition of this material varies according to the nature of the soil in which the beet-root from which it comes has been grown, and also according to the process by which it has been worked up.

Virgin soils yield salts which sometimes contain as much as 50% of carbonate of potash. This is the case with the Austrian and German products, which are largely imported into France. These foreign salts are advantageous, as they furnish carbonate of potash at a low price, when allowance is made for the value of the sulphate and chloride in them. The following analysis shows the composition of this class of material, the sample being of German origin :—

Potassium carbonate	51·22
Sodium carbonate	10·40
Potassium sulphate	8·10
Potassium chloride	6·60
Insoluble residue	18·05
Water	5·63
	100·00

The price is calculated on the carbonate, the other salts not being reckoned. The above sample would, for example, be paid for as containing 51·22 units of carbonate, nothing being charged for the 8·10 of sulphate, and 6·60 of chloride. To express these amounts as oxide of potassium, K_2O, it must be borne in mind that

1 of carbonate = 0·6811 of potassium oxide.
1 of sulphate = 0·5407 ,, ,,
1 of chloride = 0·6314 ,, ,, ,,

These numbers may be employed for converting the amounts of potassium salts in all the analysis given into oxide of potassium, and thus ascertaining the price paid in terms of this. The actual price paid for the material is from 46-47 centimes per unit of carbonate for large sales. This price is higher than that of refined carbonate of potash, because the refiner who buys those salts separates the constituents and extracts carbonate of soda, and sulphate, and chloride of potassium which he is able to sell, although he only pays for the carbonate.

The following tables contain the analysis of various samples of beet-root salts :—

	Aisne.	Belgium.	District of the Nord.	Lille.	Nord.
Potassium carbonate..	41·02	40·25	22·14	26·70	11·70
Sodium carbonate	12·41	22·75	23·10	21·10	34·10
Potassium sulphate ..	10·35	7·40	15·01	17·42	15·22
Potassium chloride....	14·64	20·00	19.90	15·25	19·40
Insoluble	14·54	7·00	14·85	15·10	13·58
Water..............	7·0	2·60	5·0	4·40	6·00
	100·00	100·00	100·00	100·00	100·00

REFINED CARBONATE OF POTASH.

These salts are produced by all distillers of molasses; they are bought by the potash refiners, all established in the district *du Nord*, whose object is to separate from them their various constituents and sell them separately. We shall not enter upon a description of this process which has previously been described in our columns, but will simply state that by the process the refiner obtains refined potashes, carbonate os soda, sulphate of potash, and chloride of potash. The purity of the refined potashes varies considerably, but, speaking generally, does not transcend the limits of 70 and 93%. The usual products contain 75-80, 78-82, 80-85, 88-92 per cent.

It takes the form of pure white, small, round granules, mixed with other larger pieces, varying between the size of a hazel nut and a walnut. If the calcination has been well conducted, all the pieces are of a dull, pure white, even to the centre, rounded like pebbles, very hard, and sometimes tinted faintly blue. The actual price of refined potash is 43 centimes per unit for large sales, 70-85 % quality; richer qualities being priced at 45-46 centimes per unit.

The following table gives analyses of various types of refined potashes:—

	70/75	75/80	78/82	80/85	90
Carbonate of potassium....	72·57	77·31	80·94	84·21	91·94
Carbonate of sodium......	15·36	14·34	11·50	8·55	2·25
Chloride of potassium	5·51	3·50	3·19	3·02	2·10
Sulphate of potassium	3·46	2·90	2·50	2 20	2·40
Insoluble residue	0·43	0·40	0·42	0·43	0·73
Water	0·70	0·42	0·43	0·50	0·58
Phosphates..............	1·97	1·13	1·02	1·09	—
	100·00	100·00	100·00	100·00	100·00

MANUFACTURED REFINED POTASH.

We have said that the refining of the beet salts consists in separating out the sulphate and chloride of potassium; the salts can themselves be converted into carbonate of potash by the process employed in the manufacture of soda ash from salt. The Leblanc process, in short, may be applied to the potassium salts obtained from Stassfurt.

This process did not at first give such satisfactory results with potash as with soda salts, but it has now been perfected, and gives a product which may be used in all the industries which require pure carbonate. The product obtained contains 94%.

The material now turned out is a magnificent product consisting of very white small grains or fine powder, which contains 88-94%. Its actual price is 46 centimes per unit of potassium carbonate.

Analysis of a sample of manufactured carbonate of potassium:—

Potassium carbonate..........	91·25
Sodium carbonate............	2·22
Potassium sulphate	2·40
Potassium chloride	2·21
Water	1·00
Insoluble and not determined..	0·92
	100·00

POTASH DERIVED FROM SUINT.

Sheep's wool is impregnated with a fatty substance, which amounts to 15·20% of the total weight, and is known as suint. This substance is excreted from the sheep's skins by the aid of small glands, situated at the base of the hairs. It increases in amount with the fineness of the wool, amounting to as much as 35% of the weight in fine fleeces, whereas it may not be more than 8-10% in coarser wool.

The sunit consists of the potassium salt of an organic acid, together with fatty matter, either free or combined with earthy basis, such as lime or alumina. The potassium compounds are very soluble, the simple process of washing the wool in water sufficing to dissolve them. One hundred kilos. of wool contain from 4-8% of potassium salts, according to the amount of sunit.

The wash waters are evaporated, the residue calcined, and then sent into the market as sunit potash. It is produced in all the great wool districts, Roubaix, Tourcoing, Dorignies (Nord) Reims, Verviers, etc.

It is particularly suitable for agricultural use since it contains very little sodium.

The actual price is 0·40—0·42 francs per unit of carbonate of potassium.

The following analysis show the usual composition of this material:

	1	2	3	4	Steeping Water.
Carbonate of potassium..	79·25	76·28	82·05	70·25	48·26
" " soda......	2·00	3·02	2·07	3·29	4·01
Chloride of potassium ..	4·85	4·81	3·85	6·40	6·40
Sulphate " ..	3·90	4·24	3·10	5·82	5·90
Silicate " ..	—	1·80	—	1·22	4·80
Insoluble	5·22	5·81	4·01	7·42	27·53
Sulphides and Sulphites	—	—	1·25	—	—
Water	4·78	3·50	3·67	5·60	3·10
	100·00	100·00	100·00	100·00	100·00

PRODUCTION.

The following table gives the actual p
potassium in France:—

Beet salts, going from the distillers to the
Worked up by the distillers themselves ..
Imported from Germany, Russia, and Bel
Suint potash ....
Refined potash.......................
Manufactured potash ....

The refining industry extracts from the
chloride, 10% of sulphate, and 25% of soda.

The production of beet salts during 1888
being dear, the molasses were all utilised for

The approaching spring should give much
because sugar being very low, much more
conversion into sugar at the present prices,
The salts produced by the sugar makers
composition:—

Carbonate of potassium..........
Carbonate of sodium ....
Sulphate of potassium ....
Chloride of potassium ....
Residue ....

THE USE OF INDUSTRIAL MANURES.

WOOLLEN WASTE.

THE waste from the woollen industries is ge
and sometimes in potash. Crude wool
consisting of two parts:—1, the true wool, cont
2, the suint, which contains up to 33% of p
understood that the first of these leaves residues
fertilizers.

All the waste coming from wool, comprisi
combing, and sifting the sweepings of the rooms,
etc., contains nitrogen in the same proportion a
the admixture of earthy and inert matter is ofte
percentage of nitrogen considerably, and this
the greatest extent with material which is fine
residues consist chiefly of the fine pile their ri
amount to 13%. Those which come into the marl
3-5% of nitrogen, together with 0·3 and 0·18%
sell at about the same price as other fertilizers c
tracts should never be entered upon without a
of the value of the goods by analysis.

Dust from sifting is usually richer and contain
by separating the finest parts from the pile by
the former only contains 2% of nitrogen, whil
contain as much as 5 or 6%. These residues ar
directly, since their decomposition in the soil is
is therefore necessary to submit them to a prepa

WOOL DUST.

The treatment to which the wool is subjected
duct a very fine powder, called wool dust or
contains:—

	Petermann.	A
Nitrogen	3·00	[illegible]
Phosphoric Acid ..	0·85	[illegible]
Potash	0·87	[illegible]

RESIDUES FROM WASHIN

During the washing of crude wool a deposit c
found. A sample of this, containing half its
found to contain:—

Organic Nitrogen ....
Phosphoric Acid ....
Potash ....

The chemical treatment of the wool also yi
small quantities of phosphoric acid, and consi
of potash. These solutions may be employed
but as they are acid they must first be neutrali
done most economically by means of mineral ph

DIRECT APPLICATION AS FERTI

In utilising these various products for agricul
to submit them to previous treatment, either by

compose in heaps, watering them periodically, or mixing them with lime, or making them up with other materials into composts. In this way their fertilising action is rendered more efficacious, and their decomposition, which would otherwise only proceed very slowly, is considerably accelerated.

INDUSTRIAL TREATMENT.

The wool waste is in some cases boiled for several hours with alkalies, such as carbonate of soda or lime, which break up and dissolve the fabrics; or it may be dissolved in sulphuric or hydrochloric acid. Another process often employed is that of dry distillation. Recently a process has been introduced by which the waste is treated with steam at 150°, and under a pressure of 5·6 atmospheres. After about 7·8 hours of this treatment a solution is obtained, which, on evaporation, yields a brittle brown substance, which has a cinchoidal fracture, and has been called *azoline*. In all of these processes a certain amount of ammonia is formed which ought of course to be collected.

M. Petermann gives the following analytical details:—

	Organic Nitrogen.	Nitrogen as Ammonia.
Waste heated in closed vessels..	4·18	1.09
Rags treated with steam	7·5—8·5	0·75—1·00
Clippings, with steam	8·52	0·74

—*L'Engrais.*

Our Book Shelf.

THE PAPER TRADE DIRECTORY for 1890. Published at the offices of "Paper-making," 66, Ludgate Hill.

This exceedingly-interesting annual directory of the Paper Trade contains a concise account of all the paper mills of the United Kingdom. The English mills are classified as hand-made mills, machine mills, and board mills; and the information given concerning each of them is of such a character as to prove of extreme utility. The chapter on "the chief centres and paper-making districts" will be found useful for those fresh on the road; and one of the first lessons in travelling is aptly put on p. 66 as follows:—"Always enquire at the station the nearest way to the mill, and valuable information can always be obtained, particularly if a small silver coin change ownership."

A DICTIONARY OF APPLIED CHEMISTRY. Edited by T. E. Thorpe, B. Sc., Ph. D., F.R.S. Vol. I., price 42s. London: Longmans, Green & Co., 1890.

It must always be a difficult task to produce a treatise on technological chemistry; and even to edit the writings of more practical men is a work that can only be accomplished by a well-trained mind. There are many drawbacks to contend against, but no doubt they are minimised by the system of employing simultaneous specialists; and such method has, we are pleased to say, been adopted in the volume before us. By this means descriptions of processes are brought down to as recent a date as it is well-nigh possible to achieve; and, as now-a-days, technical knowledge is too vast a subject for one man to know completely, it stands to reason that when the various processes are explained by men who have had the technical supervision of them, their description is to that of the mere theorist as the substance to the shadow. One omission, however, we cannot well pass over. In the list of abbreviated titles of books and journals, we do not find the *C. T. J.* It may be that the Editor does not intend to draw any information from our columns, in which case of course the introduction of our initial letters would be superfluous.

In this volume, Mr. A. H. Allen writes the article on disinfectants, of which subject he has had considerable experience; Mr. C. F. Cross, cellulose; Mr. L. Field, candles; Mr. R. J. Friswell, aniline products; Professor J. J. Hummel, dyeing; Professor G. Lunge, ammonia, bromine, and chlorine; Professor R. Meldola, azo-colouring matters; Professor W. C. Roberts Austin, alloys and assaying; Mr. C. O'Sullivan, carbohydrates; Dr. Otto N. Witt, azines. We do not know of any other specialists more capable of treating these subjects than the names we have selected, and the editor is to be congratulated on securing their valuable services.

With regard to the matter itself, there is the fact of a difference in quality of the various articles; the palm must be given to the true technologist. We need only cite the articles on copper and chlorine to prove our point. The first is very commonplace, and does not contain anything not to be found elsewhere, while that on chlorine by Dr. Lunge is a specimen of what technical articles should be. We are much pleased with the work, mainly on account of its freedom from mere copying so often found in treatises of this kind, where the traditions of yesterday, with much fiction, are served up anew as an intellectual food for to-day. We would advise the article on alums to be re-written for the next edition.

The chapter on azo-colouring matters is of a very valuable kind, and the extent of this industry may be seen by the 12 pages of patents referring to them, all of which have been taken out during the last 12 years. The benzene article will furnish much information to our English benzol refiners, who have not usually the most scientific appliances; an illustration is given of Ader's latest rectifying apparatus. The article on the manufacture of Portland cement does not come up to our ideal of technical information, neither does the article on chromium. To the writer of the matter on colorimeters we commend the form in C. N., p. 299, vol xxvii.

A very valuable table is to be found on page 586 *et seq.*—the detection of colours on dyed fabrics. This table occupies twelve pages, and will be of the utmost service in all dye works laboratories.

Turning to matter of a physical rather than of a technological character, there are several articles deserving of special attention, the balance, by Professor Dittmar, and on distillation by Professor Sydney Young, which latter should form the basis for instruction in this very important department of laboratory practice. Unfortunately too many students pass through their laboratory career without having been made thoroughly acquainted with such processes as these, and when they go forth into the world, are thrown upon their own resources, and find they have much to learn. The dictionary of applied chemistry will furnish us with some very interesting reading, and we have no hesitation in pronouncing it the best of its kind in the English language, so far as it has gone yet; it will form a handy book of reference for years to come.

THE GALLOWAY BOILER. Published by Galloways, Limited, Manchester.

How each generation managed to acquit itself so well with the appliances at its disposal, is always a matter of wonder to the generation that succeeds it. With feelings of this kind we have turned over the pages of the small publication with which we have just been favoured by Messrs. Galloways, Ltd. Not only for the individual interested directly in steam-boilers, but for everybody else into whose hands the book may fall, an exceedingly interesting half-hour is provided. A brief sketch of the history of the development of the modern steam boiler introduces a very complete and detailed description of the process of manufacture, and the machinery employed in the various operations. The text is supplemented by a series of well designed and well executed cuts.

Most people standing in front of a Galloway boiler of the latest type are conscious that the huge aggregation of tubes and plates before them, represents the expenditure of no little force and skill, but few realise the fact in anything like its entirety. Every detail in the construction, is of sufficient importance to occupy the inventive genius of the machine maker, from bending the plates, to fastening the rivets or welding the furnaces and flues. We note that Messrs. Galloways have turned out 338 boilers in one year, weighing in the aggregate 3,770 tons, and as another proof of the magnitude of their operations, it may be mentioned that their consumption of rivets amounts to about 250 tons per annum. Exclusive of the Knott Mill Works, which is now used solely for engine construction, the boiler-works proper at Ardwick, occupies eight acres, and finds employment for 600 men.

RUNCORN SOAP AND ALKALI COMPANY.

THE annual meeting of the shareholders of the Runcorn Soap and Alkali Company was held at the Law Association Rooms, Liverpool, on Monday last; Mr. A. P. Fletcher, the chairman, presiding. The report of the directors for the past year expressed regret at the result of the working, which, however, had been carried on under exceptional circumstances. The net profit for the year had been £963. 16s. 9d., and the balance from last year £5,864. 16s. 7d. being added to this made £6,823. 13s. 4d. as the present available surplus. It was proposed not to pay a dividend, but to carry forward the whole sum. The past year had been a very disappointing one for the alkali trade, for, with the exception of bleaching powder, nearly every article manufactured had failed to realise cost.—The report was adopted, Messrs. Chas. Langton and E. W. Rayner were re-elected directors of the company, and a motion for the payment of £750. as directors' fees (being £250. less than last year) was carried *nem. con.*

BURMAN PETROLEUM.—The report is that Sir Lepel Griffin and others, having constituted themselves a Syndicate, have received a concession of four square miles of oil-producing land in Upper Burmah. Burmah, we have not to add, has long been known for its mineral oil. It indeed furnished the chief supply before the development of the Trans-Caucasian and U.S. oil-fields. Burman Petroleum is said to resemble Canadian as being especially rich in solid paraffins.

Legal.

[illegible] COURT OF JUSTICE.—CHANCERY DIVISION.—Feb. 4th.

(BEFORE MR. JUSTICE NORTH.)

[illegible]TT AND ELERS (LIMITED) V. DAY—DAY V. FOSTER.

[illegible]ship delivered an elaborate judgment recently upon two [illegible] in these actions. Some important questions arose [illegible]peration of section 32 of the Patents Act of 1883, which [illegible] "when any person claiming to be the patentee of an in- [illegible], advertisements, or otherwise, threatens any other [illegible] any legal proceedings or liabilities in respect of any alleged [illegible], use, sale, or purchase of the invention, any person or [illegible]ggrieved thereby may bring an action against him, [illegible] obtain an injunction against the continuance of such [illegible] may recover such damage (if any) as may have [illegible] thereby if the alleged manufacture, use, sale, or [illegible] to which the threats related, was not in fact an infringe- [illegible] legal rights of the person making such threats. Provided [illegible]ection shall not apply if the person making such threats, with [illegible]ce, commences and prosecutes an action for infringement of [illegible]"

[illegible]ens-Hardy, Q.C., and Chadwyck Healey were for Barrett [illegible] and for Foster; Mr. Moulton, Q.C., and Mr. Thomas [illegible] for Day.

[illegible]ice North concluded his elaborate judgment as follows:—In [illegible], the prosecution of an action is vexatious when it is clear [illegible]ief can be granted at the trial. The only relief sought in this [illegible] and Elers is an injunction to restrain Day from threatening [illegible] licensees or assigns, with reference to alleged infringements [illegible] and Varley's patent, and damages and costs. As regards the [illegible] the action of "Day v. Foster," no injunction to restrain [illegible] could be granted at the trial, nor could have been at any time [illegible] in "Barrett and Elers v. Day," as no such threat was any [illegible] after the action had been brought, and no damages could [illegible] in respect of a threat which did not give rise to any cause of [illegible] as regards any other threats, none are proved or alleged, nor [illegible] from which any intention on the part of Day to make [illegible] could be inferred, and no injunction can be granted to restrain [illegible] from doing an act which it is not proved or even alleged that [illegible]ended or threatened to do (see "Stannard v. Vestry of [illegible] Ch. D., 190; "Proctor v. Bayley," 42 Ch. D., 390): [illegible], of course, be any damages for an act which was never [illegible]. What Barrett and Elers avowedly desire to do is to try [illegible]y's patent is valid or not. But no such question can arise [illegible], under the circumstances I have mentioned, as there is [illegible] it can be revelant, although, no doubt, according to [illegible] v. Royle" and "Kurtz v. Spence" (36 Ch. D., 770), [illegible] might have arisen if any threatening or intention to [illegible] Day's part was alleged and proved, and there was an issue [illegible] at the trial whether such threatening was justifiable or [illegible] circumstances I am of opinion that there is not, and [illegible] cause of action in "Barrett and Elers v. Day" in [illegible] relief can be granted at the trial, and that, this [illegible] action is vexatious, and the proceedings therein [illegible] stayed, and I accordingly direct such stay, and that the [illegible] Day's costs should be paid by the plaintiffs. In deciding [illegible] I do not mean to impute to the plaintiffs therein that [illegible] for the purpose of annoyance. I have no doubt it [illegible] legitimate move in the war game of litigation between [illegible] but this is no reason why the defendant should be harassed [illegible] of proceedings which, in my opinion, cannot result [illegible] in the plaintiffs' favour. Foster's summons in "Day v. [illegible] to have all proceedings in that action stayed until after [illegible] other action. I see no reason whatever for this. No [illegible] or as to the validity of Day's patent can be tried [illegible] Foster, but that is the proper action in which to [illegible] which Foster has made, are, or are not, in- [illegible] patent. The questions in the two actions are [illegible] Foster's affidavit in support of his summons says that [illegible] he is willing to be bound by the decision in [illegible] action, so far as the issues as to validity and in- [illegible]. Inadvertently he lets out that what he is [illegible] the validity of Day's patent—a thing which [illegible] for valuable consideration that he will not [illegible] attempt not creditable to him and vexatious in [illegible] his summons with costs.

FERTILISERS FOR PLANTS.

FOR no one manure can the same amount of infallibility be asserted, for none can so universal an application be proclaimed as for So-and-So's cosmopolitan pills. Every gardener, too, knows, what the public appears not to realise in the case of physic, that a manure which may be excellent at one time or for one plant, may be useless or even injurious at another. Gardeners, too, are often made to pay an exorbitant price for an article otherwise good. This was made apparent some years since, when we published a series of analyses of popular manures and insecticides. All of these proved good in their way, but the prices charged were in most instances extremely high. Any village chemist capable of compounding a prescription for a cow can mix a preparation suitable for plants under different circumstances, but as an aid to gardeners and others, we cite the following from a recent number of the *Jardin*. They or something equivalent have often been cited before, but the proportions of the several ingredients have not always been stated in so convenient a fashion. Dr. Jeannel's prescription, containing, as it does, notable proportions of nitrogen, phosphoric acid, and potash, is of general utility, though, indeed, the amount of potash in the soil is rarely deficient:—

Nitrate of ammonia	380	parts.
Biphosphate of ammonia	300	"
Nitrate of potash	260	"
Biphosphate of lime	50	"
Sulphate of iron	10	"
	1,000	

For flowering plants in pots the following mixture is recommended:—

Superphosphate of lime	4	parts.
Sulphate of lime	2	"
Nitrate of soda	½	"
Sulphate of ammonia	½	"
Chloride of potassium	½	"

For foliage plants in beds:—

Superphosphate of lime	4	parts.
Nitrate of soda	3	"
Chlorine of potassium	1	"
Sulphate of lime	1	"

300 grammes (say ½ lb.) to be used per square yard at the time of planting. By citing it in this fashion the quantity to be made may be large or small. M. le Marquis de Paris recommends for foliage plants in pots:—

Nitrate of soda	1	part.
Sulphate of ammonia	1	"
Superphosphate of lime	2	"
Sulphate of lime		"
Chloride of potassium	½	"

For flowering plants in beds:—

Nitrate of soda	2	parts.
Superphosphate of lime	10	"
Chloride of potassium	2	"
Sulphate of lime	4	"

Of these mixtures a teaspoonful should be used in a gallon of water once a week.—*Gardeners' Chronicle.*

ELECTRICAL.

1. How strong a current is used to send a message over an Atlantic cable? A. Thirty cells of battery only, equal to thirty volts.

2. What is the longest distance over which conversation by telephone is daily maintained? A. About 750 miles, from Portland, Maine, to Buffalo, New York.

3. What is the fastest time made by an electric railway? A. A mile a minute, by a small experimental car. Twenty miles an hour on street railway system.

4. How many miles of submarine cable are there in operation? A. Over 100,000 miles, or enough to girdle the earth four times.

5. What is the maximum power generated by an electric motor? A. Seventy-five horse power. Experiments indicate that 100 horse power will soon be reached.

6. How is a break in a submarine cable located? A. By measuring the electricity needed to charge the remaining unbroken part.

7. How many miles of telegraph wire are in operation in the United States? A. Over a million, or enough to encircle the globe forty times.

8. How many messages can be transmitted over a wire at one time? A. Four, by the quadruplex system, in daily use.

. How is telegraphing from a moving train accomplished? A. ough a circuit from the car roof, inducing a current in the wire on s along the track.

o. What are the most widely separated points between which it is sible to send a telegram? A. British Columbia and New Zealand, America and Europe.

1. How many miles of telephone wire are in operation in the United tes? A. More than 170,000, over which 1,055,000 messages are t daily.

2. What is the greatest candle power of arc light used in a light-se? A. Two millions, in the lighthouse at Houstholm, Denmark.

3. How many persons in the United States are engaged in busi-s depending solely on electricity? A. Estimated 250,000.

4. How long does it take to transport a message from San ncisco to Hong Kong? A. About fifteen minutes, *via* New k, Canso, Penzance, Aden, Bombay, Madras, Penang, and gapore.

5. What is the fastest time made by an operator sending messages the Morse system? A. About forty-two words a minute.

6. How many telephones are in use in the United States? A. out 300,000.

7. What war vessel has the most complete electrical plant? A. ited States man-of-war, Chicago.

8. What is the average cost per mile of a transatlantic sub rine cable? A. About 1,000 dols.

9. How many miles of electric railway are there in operation in United States? A. About 400 miles, and much more under con-iction.

o. What strength of current is dangerous to human life? A. e hundred volts, but depending largely on physical conditions.—*r of Steel.*

PRIZE COMPETITION.

LITTLE more than twelve months ago, one of our friends suggested that it might be well to invite articles for mpetition, on some subject of special interest to our iders; giving a prize for the best contribution, he at the me time offering to defray the cost of the first prize, and ggesting a subject in which he was personally interested.

At that date we were totally unprepared for such a sug-stion, but on thinking over the matter at intervals of sure, we have come to think that the elaboration of such a heme would not only prove of benefit to the manufacturer, t would go some way towards defraying the holiday penses of the prize winner.

We are therefore prepared to receive competitions in the llowing subjects :—

SUBJECT I.

The best method of pumping or otherwise lifting or forcing irm aqueous hydrochloric acid of 30 deg. Tw.

SUBJECT II.

The best method of separating or determining the relative antities of tin and antimony when present together in com-ercial samples.

On each subject, the following prizes are offered :—

One first prize of five pounds, one second prize of one und, five additional prizes to those next in order of merit, nsisting of a free copy of the *Chemical Trade Journal* for elve months.

The competition is open to all nationalities residing in any rt of the world. Essays must reach us on or before April th, for Subject II, and on or before May 31st, in Subject I. ie prizes will be announced in our issue of June 28th.

We reserve to ourselves the right of publishing the con-butions of any competitor.

Essays may be written in English, German, French, Spanish Italian.

COCOANUT BUTTER.—The *Times'* Calcutta correspondent tells of new trade that has arisen in India and attained extraordinary iensions." It is the trade in cocoanut butter, which is being chiefly nufactured at Mannheim, in Germany. It is news that the com-dity is meeting with such a ready market in India. But, some time , we were careful to explain the character of this new industry. As ards its constituents and specific gravity, cocoa-nut oil is liker ter than any other natural fat. The insoluble fatty acids are the ie in each; and they have nearly the same percentage of total acids. . these acids are not present in the same proportion—this and the erent melting-points mark the difference.

Correspondence.

ANSWERS TO CORRESPONDENTS.

J.R.C.—We do not think the users of the apparatus would show it you in action especially as you are in the trade.

CLARA.—Write us an account, it will be interesting to some of our readers.

JOHN BULL.—We are tired of railway rates. The traders would have done better by following our lead last April.

COPPER.—We are not able to answer your query. 2. Consult some experienced analytical chemist.

JOHN.—You are sadly mistaken; there is plenty of vitriol now in the market made from coal-brasses.

P.P.—Give us a call the next time you are in Manchester.

R.P.L.—There are many reasons why we should *not* do as you suggest.

H H.—The matter is a trade secret, and we do not know where you could get the information.

C.—We endeavour to be as correct as possible, but errors will sometimes creep in, in spite of continual watchfulness.

SALTS.—Unless you wish to gain your experience at a high cost, we could not advise you to build your plant from text books.

CHLORINE.—Yes, we believe that bleach is now being made by the Deacon process, with less than a ton of salt. It was Hasenclever who first showed us how to do it on a practical scale. 2, Five plants we think.

AQUA.—See our replies on same subject to correspondents in last week's issue.

E.G.M.—Perfectly safe.

R.T.—We cannot reply to such queries in this column. It is a matter of which much may be said on both sides.

Market Reports.

THE LIVERPOOL MINERAL MARKET.

Our market has continued further to advance, and prices all round are in favour of sellers. Manganese: There have been no arrivals this week, whilst stocks have been heavily drawn upon, and a considerable advance in prices has been paid for spot and forward delivery. Magnesite: Raw lump quiet, raw ground, £6. 10s., and calcined ground, £10. to £11. Bauxite (Irish Hill brand): The supply can scarcely keep pace with the demand, and prices are strong. Lump, 20s.; seconds, 16s.; thirds, 12s.; ground, 35s. Dolomite, 7s. 6d. per ton at the mine. French Chalk: Arrivals have been considerably larger, but prices are unaltered, especially for G.G.B. "Angel-White" brand — 90s. to 95s. medium, 100s. to 105s. superfine. Barytes (carbonate), steady; selected crystal lump scarce at £6.; No. 1 lumps, 90s.; best, 80s.; seconds and good nuts, 70s.; smalls, 50s.; best ground, £6.; and selected crystal ground, £8. Sulphate quiet; best lump, 35s. 6d.; good medium, 30s.; medium 25s. 6d. to 27s. 6d.; common, 18s. 6d. to 20s.; ground best white, G.G.B. brand, 60s.; common, 45s.; grey, 32s. 6d. to 40s. Pumicestone steady; ground at £10., and specially selected lump, finest quality, £13. Iron ore continues in good demand at full prices. Bilbao and Santander firmer at 9s. to 10s. 6d. f.o.b.; Irish, 11s. to 12s. 6d.; Cumberland, 16s. to 20s. Purple ore unchanged. Spanish manganiferous ore commands ready sale at full prices. Emery-stone: Best brands continued to be enquired for, bringing full prices. No. 1 lump is quoted at £5. 10s. to £6., and smalls £5. to £5. 10s. Fullers' earth steady; 45s. to 50s. for best blue and yellow; fine impalpable ground, £7. Scheelite, wolfram, tungstate of soda and tungsten metal more inquired for. Chrome metal, 5s. 6d. per lb. Tungsten alloys 2s. per lb. Chrome ore: High percentage inquired for at fair prices. Antimony ore and metal have further advanced. Uranium oxide, 24s. to 26s. Asbestos: Best rock, £17. to £18.; brown grades, £14. to £15. Potter's lead ore: The miners' strike has terminated, but prices are still firm; smalls, £14. to £15.; selected lump, £16. to £17. Calamine: Best qualities scarce—60s. to 80s. Strontia steady; sulphate (celestine) steady, 16s. 6d. to 17s. Carbonate (native) £15. to £16.; powdered (manufactured), £11. to £12. Limespar: English manufactured, old G.G.B. brand sought after, and brings full prices; 50s. for ground English. Felspar, 40s. to 50s.; fluorspar, 20s. to £6. Bog ore in fair demand at 22s. to 25s. Plumbago: More offering; best Ceylon lump at last quotations; Italian and Bohemian, £4. to £12. per ton. French sand, in cargoes, continues scarce on spot—20s. to 22s. 6d. Ferro-manganese selling easily at advanced figures. Ground mica, £50. China clay freely offering—common, 18s. 6d.; good medium, 22s. 6d. to 25s.; best, 30s. to 35s. (at Runcorn).

MISCELLANEOUS CHEMICAL MARKET.

There has been a slight relapse during the week in the value of caustic soda. 70%, £8. 12s. 6d. to £8. 15s. per ton on the spot. 60%, £7. 17s. 6d. per ton, and cream 60%, £7. 12s. 6d., all f.o.b. Soda ash keeps steady in price, at 1 1/8 to 1¼ per deg., and there is a fair amount of inquiry. Soda crystals keep up in value, and there is at present a brisk demand. Chlorate of potash quiet but firm, at 6¼d. per lb. Sulphur is suffering from keen competition by certain sellers, determined upon forcing their product upon the market, and notwithstanding a general good demand, there is a decline in values all round. Muriatic acid and vitriol still continue to find ready outlet at full prices. Sulphate of copper firm, at £24. 10s. for early deliveries, but easier for forward. Brown acetate of lime slow of sale, and rather lower in price, at £8. 5s. to £8. 7s. 6d. at usual centres of consumption. Grey, firm, £13. 15s. to £14. Acetate of soda in fair demand at current low rates. Acetic acids are quiet. Acetates of lead meet with little enquiry, but prices keep firm. Nitrate of lead selling slowly, at £22. to £22. 10s. Chloride of magnesium plentiful on the spot, price unaltered. White powdered arsenic shows a tendency to rise in value, and sells on the spot at £13. net at usual points of delivery. Tin crystals are slightly lower in sympathy with the metal. Yellow prussiate of potash still scarce, at 10d. to 10¼d. per lb. Oxalic acid is less plentiful, and not so easy to obtain at the low figures recently ruling. Potash caustic and carbonate can be obtained on slightly easier terms for early delivery, and the demand is light. Iodine still unsettled.

METAL MARKET REPORT.

	Last week.		This week.
Iron	56/1½	....	51/10½
Tin	£94 0 0	....	£91 0 0
Copper	48 17 6	....	46 17 6
Spelter	24 2 6	....	21 10 0
Lead	13 0 0	....	12 12 6
Silver	44½	....	44d.

COPPER MINING SHARES.

	Last week.			This week.	
Rio Tinto	16 5/16	16 7/16	....	14⅞	15
Mason & Barry	6¾	6⅞	....	6¼	6½
Tharsis	4⅝	4¾	....	4 7/16	4 9/16
Cape Copper Co.	3 7/16	3 9/16	....	3 5/16	3 7/16
Namaqua	2 5/16	2 7/16	....	2	2¼
Copiapo	2 7/16	2 9/16	....	2 7/16	2 9/16
Panulcillo	1 1/16	1 3/16	....	1	1⅛
Libiola	3 9/16	3 11/16	....	3⅜	3⅝
New Quebrada	⅞	1⅛	....	⅝	⅞
Tocopilla	1/9	2/3	....	1/6	2/-
Argentella	6d.	1/-	....	6d.	1/-

WEST OF SCOTLAND CHEMICALS.

GLASGOW, Tuesday.

The market is decidedly stronger for most of the ordinary chemicals, and where changes occur these are mostly to the rise. Caustic soda has been largely dealt in, and the demand has resulted in very stiff prices for the moment, as compared with anything ruling in this section for a long time back. Bleaching powder has been much steadier, and last quotations for the local supply have been quite fully maintained. Soda crystals are very firm, as are also soda ash and refined alkali. Nitrate of soda at the same time is slightly firmer, and fetching improved conditions. Sulphate of ammonia has bettered itself to the extent of quite 2s. 6d. to 5s, and to-day for spot Leith £11. 17s. 6d. has been paid. At the same time, speculative dealers here are a good deal at sea as regards the precise character of the influences at work, and they make no certainty as to the course of the market in the immediate future. Forward section is still pretty much a blank. In paraffins there has been an improvement in the call for oils, both burning and lubricating, the former having been helped considerably by the dense fogs locally prevailing. The 885° gravity of the latter is dearer by some shillings. Chief prices current are :—Soda crystals, 50s. net Tyne; alum in lump £4. 17s. 6d., less 2½% Glasgow; borax, English refined £30., and boracic acid, £37. 10s. net Glasgow; soda ash, 1¼d. less 5% Tyne; caustic soda, white, 76°, £11., 70/72°, £8. 17s. 6d., 60/62°, £8., and 60/62° cream, £7. 10s., all less 2½% Liverpool; bicarbonate of soda, 5 cwt. casks, £5., and 1 cwt. casks, £5. 5s. net Tyne; refined alkali 48/52°, 1½d., less 2½% Tyne; saltcake, 25s. 6d. to 26s. 6d; bleaching powder, £5. 15s. to £6. less 5% f.o.r. Glasgow; bichromate of potash, 4d., and of soda 3d., less 5 and 6% to Scotch and English buyers respectively; chlorate of potash, 4⅝d. to 4¾d., less 5% any port; nitrate of soda 8s. 4½d. to 8s. 6d.; sulphate of ammonia, £11. 17s. 6d. f.o.b. Leith; salammoniac, 1st and 2nd white, £37. and £35., less 2½% any port; sulphate of copper, £25. less 5% Liverpool; paraffin sc
2⅜d.; paraffin wax, 120°, semi-refined,
(naphtha), 9d.; paraffin oil (burning), 6¾d.
ditto (lubricating), 865°, £5. 10s. to £6.,
890/895°, £7. 10s. to £8. Week's imports
were 40,404 bags.

THE LIVERPOOL COLOUR

COLOURS quiet; prices unaltered. Ochres
at £10., £12., £14., and £16.; Derbyshire,
best, 50s. to 55s.; seconds, 47s. 6d.; and common,
40s. to 45s.; French, J.C., 55s., 45s. to 60s.;
Umber: Turkish, cargoes to arrive, 40s. to 5
to 55s. White lead, £21. 10s. to £22. Red le
of zinc: V.M. No. 1, £25.; V.M. No. 2,
£6. 10s. Cobalt: Prepared oxide, 10s. 6d.;
6s. 6d. Zaffres: No. 1, 3s. 6d.; No. 2, 2s. 6d.
white, 60s.; good, 40s. to 50s. Rouge: Best, £2
9d. per lb. Drop black, 25s. to 28s. 6d. Oxide
£10. to £15. Paris white, 60s. Emerald green
shire red 60s. Vermillionette, 5d. to 7d. per lb.

TAR AND AMMONIA PRO

Benzoles are a trifle weaker, and 3s 10d,
50/90's are about to-day's values. Solvent naphth
and 1s. 9d. f.o.b. is being asked for prompt d
moving off freely, while Anthracene is, on the
weaker. The pitch market is very weak, and it
see lower prices; 32s. 6d. is the price in London, 3
generally, and 31s. 6d. Garston. At these prices
could be done, though perhaps in some cases with

Speculators have all through the week been doin
depress the sulphate of ammonia market, but it has
principally through the firm position taken up b
very little business has been done below £11. 17s.
there are buyers at this figure to-day both at I
Some Leith business is reported at £11. 16s. 3d., bu
be repeated. London values are £11. 17s. 6d.,
£11. 16s. 3d, to £11. 17s. 6d.

REPORT OF MANURE MATE

There has been less business doing during the past
both nitrogenous and phosphatic material are som
buyers. Supplies seem to have overtaken the dem
usual result.

We have no alteration to notice in the quotati
River phosphate rock, shippers refusing to entertain
unit for cargoes, cost, freight, and insurance to U.K.
land rock might perhaps be had at ¼d. to ½d. per
tions for Somme and Belgian remain as quoted a
don't hear of any important transactions having ta
small orders are being booked all the time. Quotat
rock remain nominal, there being nothing doing in t

Supplies of bones have begun to come forward from
Spain, and Valparaiso, and values on spot have giv
Common grinders would not now fetch more than £
to £5. 5s. for good clean hard bones. There is n
River Plate bones, the cargoes offering being end of
and buyers not liking to run any risk about the ar
for to meet their Spring requirements. Five pound ten
U.K. port, is nearest value for large cargoes; £5. 15s. a
There are now sellers of crushed East Indian bones
ex-quay Liverpool, February-March steamer shi
nearer at hand would probably fetch 1s. 3d. to 1s.
East Indian bone meal is freely offered at £5. 7s. 6
Liverpool, Hull, or Glasgow, and it seems likely th
presently be taken, there being pressing sellers at
spot, ex store, higher prices are demanded, but t
actual business passing.

The stronger feeling in nitrate of soda last week
8s. 4½d. is again spot price. The nearest valu
8s. 4½d., and a little more might perhaps be obtain
shipments, but the negotiations for a reduction of
hang fire, and buyers having again lost confidence, d
on at all.

The value of 26% superphosphate remains as las
per ton in bulk, f.o.r. at works.

We do not hear of any fresh business in dried
parcels of River Plate have arrived per steamers, an
auction as soon as landed. Meantime, buyers ar
anticipation of lower prices than have been recently

Gazette Notices.

PARTNERSHIPS DISSOLVED

HUNTER, WATERS, AND Co., Gracechurch-street, London, chemical brokers and agents.

First Meetings and Public Examinations.

?, ASHE, Carnaby-street, Soho, London, oil and colourman, first meeting, February 18th, Bankruptcy-buildings, Portugal-street, Lincoln's Inn Fields. February 27th, 34, Lincoln's Inn Fields.

Adjudications.

HENRY OSWALD JAMES, Wolverhampton, mineral agent.
H. V. MURRAY. Colorado Mining Syndicate, Lombard-street, London,

The Patent List.

This list is compiled from Official sources in the Manchester Technica Laboratory, under the immediate supervision of George E. Davis and Alfred R. Davis.

APPLICATIONS FOR LETTERS PATENT.

Lubricators. W. B. Sayers. 973. January 20.
Artificially Maturing Spirits. H. Grimshaw. 980. January 20.
Recovery of Tannin from Waste Leather. H. Grimshaw. 981. Jan. 20.
Soap and Soap Powders. H. Grimshaw. 983. January 20.
Generation of Carbon Dioxide. H. Grimshaw. 984. January 20.
An Improved White Pigment. H. Grimshaw. 985. January 20.
Manufacture of Artificial Fuel. W. H. Nevill. 1,001. January 20.
Lubricators.—(Complete Specification). A. J. Boult. 1,003. January 20.
Pumps for Viscous, Pulpy, or Semi-Liquid Substances. W. P. Thompson. 1,006. January 20.
Manufacture of Soda and Potash. F. Ellershausen. 1,015. January 20.
" " F. Ellershausen. 1,016.
The Manufacture of Disinfectants. W. Dammann. 1,017. January 20.
Manufacture of Water Gas. T. Walrond-Smith. 1,022. January 20.
Electrical Apparatus.—(Complete Specification). W. P. Thompson. 1,028. January 21.
Soaps. E. Ireland. 1,047. January 21.
Dynamos and Electric Motors.—(Complete Specification). C. Bollé. 1,050. January 21.
Recovery of Zinc from Galvanised Iron and Steel. H. Grimshaw. 1,055. January 21.
The Production of Azo Colours upon Cotton. S. Knowles and J. Knowles. 1,067. January 21.
Dynamos and Electric Motors. G. B. Lückhoff and E. H. Hungerbühler. 1,068. January 21.
Decolourising and Solidifying Drying Oils. A. F. St. George. 1,069. January 21.
Electrical Switches. G. Binswanger. 1,072. January 21.
Hydrometers, Saccharometers, and Lactometers. G. C. Topp. 1,082. January 21.
Dealing with Spent Soda Lyes from Paper Mills. T. Goodall. 1,087. January 21.
Secondary or Storage Batteries.—(Complete Specification). H. H. Lake. 1,094. January 21.
Electrical Distribution.—(Complete Specification). H. H. Carpenter. 1,106. January 21.
The Treatment of Residues from Oil Manufacture. H. Noerdlinger. 1,109. January 21.
Galvanic Batteries.—(Complete Specification.—C A. Hitchcock. 1,110. Jan. 21.
Manufacture of Chromium Compounds. J. Massignon and E. Watel. 1,117. January 21.
The Concentration and Distillation of Liquids. T. E. Wilson. 1,122 January 22.
Treatment of Spent Soap Lyes. C. W. Hazlehurst and S. Pope. 1,126. January 22.
Apparatus for Dyeing and Bleaching. G. Young and F. Pearn. 1,157. January 22.
Working Blast Furnaces. T. Turner. 1,161. January 22.
Boiler Feed Water Apparatus. A. E. Tavernier and E. Casper. 1,164. January 22.
Burning Tan and other Refuse Material. C. A. Brown. 1,170. Jan. 22.
Waterproofing Composition. W. Sinclair. 1,178. January 22.
Gas Burners. W. Hemingway. 1,184. January 22.
Machinery for Making Clay Goods. J. Morton. 1,194. January 23.
Silos. J. Wilson. 1,208. January 23.
Gas Burners. J. H. R. Paterson. 1,209. January 23.
Electric Lamps. J. A. McMullen. 1,226. January 23.
Alloys for Anti-friction Purposes. E. C. Miller. 1,227. January 23.
Regenerative Gas Furnaces. R. Mannesmann. 1,234. January 23.
Gas Compressers. R. Mannesmann. 1,235. January 23.
" " R. Mannesmann. 1,236. January 23.
Manufacture of Gas. R. Mannesmann. 1,237. January 23.
" " R. Mannesmann. 1,238. January 23.
Vaporizing Liquids. D. Bethmont. 1,239. January 23.
The Electrolytic Generation of Chlorine. D. G. Fitz-Gerald and A. C. Falconer. 1,240. January 23.
Manufacture of Sugar. A. Fairgrieve. 1,247. January 23.
Compressing Gypsum. A. Moseley. 1,279. January 24.
Electrical Switches. C. L. Baker. 1,280. January 24.
Manufacture of Sugar.—(Complete Specification). C. D. Abel. 1,282. Jan 24.
Cokeing Apparatus. O. Imray. 1,283. January 24.
Boiler and Furnace Fires. O. Imray. 1,284. January 24.
Gas Producers. O. Imray. 1,285. January 24.
Barium Oxides.—(Complete Specification). H. H. Leigh. 1,300. January 24.
Filtering Apparatus. A. Capillery. 1,307. January 24.
The Production of Sterilized Milk.—(Complete Specification). A. Schmidt-Mülheim. 1,326. January 25.
Steam Engine Boilers. R. Hollingdrake. 1,330. January 25.
Oil Refining. T. H. Gray and S S. Bromhead. 1,343. January 25.
Fluid for Primary Batteries. T. Coad. 1,347. January 25.
Apparatus for the Gasification of Fuels. W. L. Wise. 1,349. January 25.
Coating Metals for Prevention of Corrosion. A. E. Haswell and A. G. Haswell. 1,355. January 25.
Rough, Glazed, and Coloured Facing Stones.—(Complete Specification). A. J. Boult. 1,355. January 25.
Manufacture of Candles. G. Bouton. 1,360. January 25.
Solution of Iodine for Medical Purposes. R. J. Downes. 1,380. Jan. 27.
Rotary Apparatus for Blowing, Exhausting, or Pumping Fluids. R. Johnson. 1,383. January 27.
Separating the Isomers contained in Crude Nitrotoluol. Martin Lange. 1,407. January 27.
Manufacturing Cumarone and its Homologues.—(Complete Specification). A. A. Vale. 1,422. January 27.
Manufacture of Metallic Compounds or Alloys. M. Netto. 1,427. Jan. 27.
Reduction of Metallic Ores and Apparatus therefor. J. T. King. 1,443. January 28.
Feed Water Purifiers for Steam Boilers.—(Complete Specification). C. A Knight. 1,453. January 28.
Method and Apparatus for Treating, Scouring, and Washing Wool. Jno. Smith, I. Smith, and Jos. Smith. 1,461. January 28.
Extracting Moisture from Wool and Cotton. J. B. Whiteley and E. Whiteley. 1,462. January 28.
Improvements in the Manufacture of Oilcake. F. C. Calthrop. 1,471. January 28.
Storage Batteries or Accumulators.—(Complete Specification). H. H. Lake. 1,495. January 28.
Extracting Substances Dissolved in Waste Water of Paper Works. F. C. Alkier. 1,514. January 28.
Digesters or Boilers for Manufacture of Paper Pulp.—(Complete Specification). W. W. Key. 1,517. January 28.
Coating Hoop Iron with Tin or Terne. T. H. Johns. 1,522. January 28.
Vaporizing Solutions containing Nitrates of Manganese. G. Wischin. 1,524. January 28.
Kilns or Ovens for Burning Bricks, Earthenware, Limestone, Cement and other Articles.—(Complete Specification). I. Button, E. Peters, and J. W. Goodsell. 1,525. January 28.
Filtering or Purifying Feed Water for Steam Boilers. J. B. Edmiston. 1,571. January 29.
Manufacture of Azoamines by the Reduction of Azo-Colouring Matters. J. Imray. 1,579. January 29.
Improvements in the Manufacture of Soda and Potash. F. Ellershausen. 1,584. January 29.
Improvements in the Manufacture of Disinfectants. C. T. Kingzett. 1,589. January 29.
Mordanting and Dyeing Wool and Cotton Fibres. John Smith, Isaac Smith, and Joseph Smith. 1,607. January 30.
Apparatus for Treating Sewage. W. Warner. 1,623. January 30.
Machine for Extraction of Precious Metals by Chlorination and Amalgamation. F. K. S. Lowndes and J. C. Kaller. 1,625. January 30.
Improvements in Calcining and Refining Copper Ore.—(Complete Specification). F. W. Dahne 1,641. January 30.
Improvements in the Manufacture of Soap. M. Perls. 1,650. Jan. 30.
Improvements in the Manufacture of Paper. M. Perls. 1,651. Jan. 30.
Extracting Oil and Grease from Cotton Waste. B. D. Barnett. 1,652. January 30.
Improvements in Saturators for Sulphate of Ammonia Plant.—(Complete Specification). G. Kennedy. 1,653. January 30.
Improvements in the Manufacture of Lime. M. B. Parrington. 1,654. January 30.
Manufacture of Lubricating Grease.—(Complete Specification). P. Plisson. 1,659. January 30.
Improvements in Carburating Gases and Purifying Hydro-Carbons. C. Heyer. 1,698. January 31.
Manufacture of Colouring Matters of the Induline Series. O. Imray. 1,699. January 31.
Antifouling Compositions. V. B. Lewes. 1,757. January 31.
Feed Water Heaters.—(Complete Specification). V. F. L. Smidth. 1,761. January 31.
Yellow Dyes or Colouring Matters. T. R. Shillito. 1,771. January 31.

IMPORTS OF CHEMICAL PRODUCTS

AT

THE PRINCIPAL PORTS OF THE UNITED KINGDOM.

LONDON.

The period Jan. 19th—Feb. 5th.

Alum—

Holland	£230.	Ohlenschlager Bros.

Acetic Acid—

Holland,	88 pkgs.	A. & M. Zimmermann
Belgium,	20	Leach & Co.
Holland,	4	Beresford & Co.

Alumina Sulphate—

Holland,	£160.	C. S. Lovell

Antimony Ore—

France,	24 t.	Pillow, Jones, & Co
Austria,	10	H. R. Merton & Co.
Portugal,	5	S. Sernza
Austria,	30	H. R. Merton & Co.
Oporto,	9	Knowles & Co.
Portugal,	8	H. Emanuel
Salonica,	49	"
N. Zealand,	80	N. Z. Antimony Co.

Austria, 49 H. R. Merton & Co.
Brisbane, 20 A. Hughes
,, 4 Vivian, Younger & Co.
Austria, 4 Typke & King
A. Turkey, 35 Quirk, Barton, & Co.

Acetate of Lime—
U. States, £189. Johnson & Hooper

Alkali—
France, 100 c. Arnati & Harrison

Antimony—
Japan, 67 t. H. R. Merton & Co.

Ammonia Sulphate—
Germany, £26 Pillow, Jones, & Co.

Bismuth—
Germany, 49 c. Grosscurth & Luboldt
Pt. Jackson, 4 t. Skinner & Co.

Bones—
U. States, 26 t. H. A. Lane & Co.
France, 12 c. Arnati & Harrison
Holland, 10 t. J. Lucy & Sons
Belgium, 8 Sawer, Mead, & Co.
Germany, 12 W. Klein & Sons
N. S. Wales, 6 Goad, Rigg, & Co.
France, 6 c. Hernu, Peron, & Co.
Queensland, 8 t. Dyster, Nalder & Co.
Cape, 30 D. de Pass & Co.
E. Indies, 200 Cornwall, Son, & Co.
,, 178 ,,
, 150 A. Cross & Sons
,, 125 L. & I. D. Jt. Co.
Queensland, 1 Brown & Elmslie
N. Zealand, 1 N. Z. L. and M. A. Co.
Belgium, 7 c. D C. Thomas & Son
E. Indies, 9 t. R. & J. Henderson
Queensland, 5 t. Anning & Cobb
Egypt, 110 Dyster, Nalder, & Co.
Queensland, 2 Hicks, Nash, & Co.
E. Indies, 150 Hunter, Waters, & Co.
U. States, 18 A. & W. Nesbitt
Queensland, 9 c. Dyster, Nalder, & Co.

Barium Peroxide—
Germany, £121. W. Burston & Co.

Bone Meal—
E. Indies, 300 t. C. C. Bryden & Co.

Brimstone—
Italy, 23 t. G. Boor & Co.
,, 24 Chilworth Gunpowder Co.
,, 3 Typke & King
,, 20 A. Zumbeck & Co.
,, 300 W. C. Bacon & Co.
Phillipine Isles, 10 t. F. Parbury & Co.
Italy, 4 Soundy & Son
Belgium, 16 Bull, Bevan, & Co.
France, 1 E. W. Carling & Co.
Italy, 100 J. Stutchbutry & Sons

Barytes—
Germany, 40 cks. W. Harrison & Co.
Belgium, 21 cks. 9 pkgs. ,,
,, 210 bgs. Leach & Co.
Holland, 46 cks. W. Harrison & Co.
Germany, 22 ,,
,, 69 D. Sterer & Sons
Belgium, 11 E. W. Carling & Co.
,, 11 T. H. Lee
,, 104 A Zumbeck & Co.
Germany, 19 P. Jantzen
,, 22 Pillow, Jones, & Co.
,, 23 W. Harrison & Co.
,, 19 H. A. Litchfield & Co.
,, 69 S. Ward & Co.

Boracic Acid—
Italy, 50 pkgs. J. Batt & Co.
,, 9 J. Puddy & Co.
Germany, 6 Beresford & Co.
Italy, 25 Howard & Sons

Barium Chloride—
Holland, £90. Petri Bros
Germany, 75 ,,

Bromine—
Germany, £100, A. & M. Zimmermann
,, 163 Craven & Co.

Caoutchouc—
Aden, 50 c. L. & I D. Jt. Co.
France, 2 Hoare, Wilson, & Co.
Greytown, 14 C. de Murietta & Co.
France, 310 T. H. Lee
E. Indies, 104 Huttenbach & Co.
Spain, 20 Kebbel, Son, & Co.
Madagascar, 6 ,,
Cape, 80 Hammond & Co.
E. Indies, 23 Adamson, Gilfillan & Co.
Natal, 13 J. Owen
Cape, 4 W. M. Smith & Sons
U. States, 37 Kebbel, Son, & Co.
E. Indies, 19 L. & I. D. Jt. Co.
E. Africa, 38 ,,
E. Indies, 82 Huttenbach & Co.
Portugal, 334 W. Brandt, Sons, & Co.
E. Africa, 25 Anderson, Weber, & Smith
Natal, 56 Ross & Deering
Cape, 65 Union Lighterage Co.
Madagascar, 155 Carey & Browne
E. Africa, 5 L. & I. D. Jt. Co.
Madagascar, 374 Kebbel, Son, & Co.
N.E. Africa, 71 Union Lighterage Co.
France, 250 T. H. Lee
E. Indies, 65 J. H. & G. Scovell
E. Africa, 16 Anderson, Weber, & Smith

Cream Tartar—
France, 5 pkgs. Webb, Warter, & Co.
,, 5 C. F. Gerhardt
Holland, 20 W. C. Bacon & Co.
Spain, 5 F. Smith
Italy, 11 Webb, Warter, & Co.
,, 48 R. Jacob & Sons
,, 6 R. Tucker & Co.
,, 120 Leager, White, & Co.
,, 4 Webb, Warter, & Co.
,, 14 Credit Lyonnais
Holland, 85 N. Smith
,, 10 Middleton & Co.
,, 4 Webb, Warter, & Co.
Italy, 5 ,,
France, 4 E. W. Carling & Co.
,, 5 B. & F. Wf. Co.
,, 20 W. C. Bacon & Co.

Chemicals (*otherwise undescribed*)
Germany, £357. A. & M. Zimmermann
,, 125 T. H. Lee
Belgium, 13 Best, Ryley, & Co.
,, 8 R. E. Drummond & Co.
Spain, 26 J. Hall, Jr., & Co.
Germany, 8 Phillipps & Graves
,, 8 J. Elinguis
Holland, 10 Beresford & Co.
Germany, 679 A. & M. Zimmermann
,, 25 Phillipps & Graves
,, 114 A. & M. Zimmermann
Holland, 20 H. Boyce
,, 25 Ohlenschlager Bros.
France, 29 Carey & Sons
Germany, 18 Ross & Deering
,, 58 T. H. Lee
,, 430 A. & M. Zimmermann
Holland, 20 Phillipps & Graves
Germany, 102 T. H. Lee
,, 302 A. & M. Zimmermann
,, 24 ,,
,, 28 F. Stahlschmidt & Co.
Italy, 15 Hughes, Chemery, & Co.
Germany, 756 A. & M. Zimmermann
,, 30 L. & I D. Jt. Co.
,, 10 ,,
Holland, 20 J Parlevliet
Germany, 139 T. H. Lee
,, 93 A. & M. Zimmermann
,, 115 C. Faust & Co.

Caustic Potash—
France, 21 pkgs. Fuerst Bros.
,, 21 Carey & Sons
,, 21 F. W. Berk & Co.

Camphor—
Germany, 4 tubs. L. & I. D. Jt Co.
,, 4 pkgs. ,,
Japan, 345 tubs. A. Faber & Co.

Copper Ore—
Pt. Augusta, 23 t. Harrold Bros.
N. Zealand, 3 L. & I. D. Jt. Co.
Sydney, 54 J. A. Drew
,, 7 F. Manders
Melbourne, 15 Vivian & Sons

Cobalt Ore—
Sydney, 76 t. Fellows, Morton, & Co.

Copper Sulphate—
Germany, £60. Charles & Fox

Caustic Soda—
Holland, 90 c. Beresford & Co.
,, 200 ,,

Carbonic Acid—
Holland, 73 pkgs. G. Rahn & Co.

Dyestuffs—
Extracts
Holland, 35 pkgs. Burt, Boulton & Co.
France, 25 B. & F. Whf. Co.
,, 295 L. J. Levinstein & Sons
,, 23 Forbes, Abbott, & Co.
,, 20 L. J. Levinstein & Sons
Holland, 28 Burt, Boulton, & Co.
France, 70 G. S. N. Co.
Belgium, 35 Burt, Boulton, & Co.
U. States, 100 M. D. Co.
Belgium, 36 Burt, Boulton, & Co.
Holland, 68 Union Lighterage Co.
Denmark, 21 A. F. White & Co.
France, 245 L. J. Levinstein & Sons
,, 50 W. France & Co.
,, 100 Bailey & Leetham
Austria, 82 L. J. Levinstein & Sons
France, 25 T. H. Lee
,, 20 Levinstein & Sons
Holland, 2 J. Sinclair & Son

Tanners' Bark
Victoria, 92 t. T. J. & T. Powell
U. States, 55 S. E. Dk. Co.
,, 50 G. Meyer & Co.
,, 1 C. F. Gerhardt
Belgium, 30 H. Henle's Succ'r.

Gambier
E. Indies, 169 pkgs. P. S. Evans & Co.
,, 842 E. Boustead & Co.
,, 301 S. Barrow & Bro.
,, 564 Beresford & Co.
,, 205 Elkan & Co.
,, 590 R. & J. Hendersen
,, 413 Anderson, Weber, & Smith
,, 515 L. & I. D. Jt. Co.
,, 199 Lewis & Peat
,, 430 Johnson, Rolls, & Co.
,, 634 Adamson, Gilfillan, & Co.
,, 207 Elkan & Co.
,, 272 S. Barrow & Bro.
,, 210 Adamson, Gilfillan, & Co.
,, 283 Lewis & Peat
,, 420 L. & J. D. Jt. Co.
,, 843 ,,
,, 418 Hoare, Wilson, & Co.
,, 435 J. H. & G. Scovell
,, 477 L. & I. D. Jt. Co.
,, 416 Hoare, Wilson, & Co.

Indigo
E. Indies, 13 ckts. L. & I. D. Jt. Co.
,, 27 T. Barlow & Bros.
,, 10 T. W. Heilgers & Co.
,, 88 L. & I. D. Jt. Co.
,, 229 4 bxs. G. Ward & Sons
,, 34 A. Harvey
,, 102 T. H. Allen & Co.
,, 1 Parsons & Keith
,, 15 Benecke, Souchay & Co.
,, 78 Arbuthnot, Latham, & Co.
,, 33 J. Owen
,, 7 ,,
,, 150 T. Ronaldson & Co.
,, 69 T. Hallen & Co.
,, 15 Fruhling & Goschen
,, 13 Anderson Bros.
,, 32 L. & I. D. Jt. Co.
,, 20 Langstaffe & Co.
,, 32 F. Huth & Co.
,, 46 L. & I. D. Jt. Co.
S. W. I. 8 Patry & Pasteur
E. Indies, 1 bx. Elkan & Co.
,, ocs. Pickford & Co.
,, 12 chts. W. Brandt, Sons, & Co.
Greytown, 10 Cotesworth & Powell
E. Indies, 32 L. & I. D. Jt. Co.
,, 21 Elkan & Co.
,, 6 Gray, Dawes, & Co.
,, 58 J. Owen
,, 9 F. Huth & Co.
,, 22 L. & I. D. Jt. Co.
,, 10 Union Lighterage Co.
,, 10 Ross & Deering
,, 87 Elkan & Co.
,, 12 A. Harvey & Co.
,, 4 Ernsthausen & Co.
,, 3 Hatton, Hall, & Co.
,, 22 Vokins & Co.
,, 97 G. Dards
,, 30 Schenker & Co.
,, 7 cs. Langstaffe & Co.
,, 24 J. Owen
,, 28 T. H. Allan & Co.
,, 9 Kleinwort, Sons, & Co.
,, 27 Ldn. & Hanseatic Bank
,, 132 L. & I. D. Jt Co.
,, 3 Patry & Pasteur
,, 10 Union Lighterage Co.
,, 10 L. & I. D. Jt. Co.
,, 25 T. Ronaldson & Co.
,, 15 Williams, Torrey, & Co.
,, 55 P. & O. Co.
,, 191 Ralli Bros.
,, 7 G. Dards
,, 8 Hoare, Miller & Co.
,, 19 Stansbury & Co.
,, 6 Ldn. & Hanseatic Bank
Holland, 3 L. & I. D. Jt. Co.
E. Indies, 37 Elkan & Co.
,, 7 Darling Bros.
,, 70 L. & I. D. Jt. Co.
,, 1 Ross & Deering
,, 15 F. Huth & Co.
,, 38 Langstaff & Co.
,, 238 L. & I. D. Jt. Co.
,, 16
,, 1
,, 7
,, 10
,, 10
,, 1
,, 5
,, 3
,, 1
,, 4
,, 1
,, 2
,, 2

Rennet
Germany,
Denmark, 3

Valonia
A. Turkey, 1
,, 8
,, 5
,,

Safflower
E. Indies, 1

Myrabolan
E. Indies, 1,
,, 2,
,, 1,6
,,
,,
,,
,,
,, 1,

Orchella
Portugal,
W. Africa,
E. Africa,
Ceylon,

Sumac
Italy, 1

Aniline
Holland,

Argols
Cape,

Cochineal
Canaries,
,,
,,
,,
,,
,,
,,

Annatto
Ceylon,

Dyestuffs
France,

Saffron
Spain,

Cutch
E. Indies 4

Glycerine—
France, £44
Germany, 1
,,
,,
Holland, 7
,, 2
France, 1
Germany,
Holland,
Germany, 1

Glucose—
France, 100
,, 38c
,, 200
Germany, 63
Belgium, 200
,, 50
,, 350
Germany, 85
,, 70
,, 30
France, 100
Germany, 200
Belgium, 100
Germany, 125
,, 35
,, 13
,, 25
,, 50c
,, 30c

	800	800	J. Cooper
	200	200	L. Sutro & Co.
...ates,	100	554	W. Parkinson
...any,	12	100	Page, Son, & East
,	80	80	Barrett, Tagant, & Co.
...ce,	200	100	Carey & Son
...any,	200	205	J. Cooper
"	4	22	Henderson, Craig, & Co.
"	21	168	C. Tennant, Sons, & Co
"	100	100	Barrett, Tagant, & Co.
"	35	193	Anderson, Weber, & Co.
"	50	425	Andersen, Becker, & Co,
"	200	200	De Paiva, Norman, & Co.
"	13	100	J. Barber & Co,
"	28	240	L. & I. D. Jt. Co.
"	200	200	Howell & Co.
"	13	120	Union Lighterage Co.
"	200	203	Horne & Co.
"	100	100	Barrett, Tagant, & Co.
"	200	200	J. Barber & Co.
"	25	200	Pillow, Jones, & Co.
"	4	34	T. H. Lee
"	200	200	Horne & Co.
"	25	210	L. & I. D Jt. Co
"	400	400	H. A. Lichfield
"	19	160	C. Tennant, Sons, & Co.
"	25	200	J. Barber & Co.
"	955	984	R. E. Drummond & Co.
U. States,	800	800	Barrett, Tagant, & Co.
Germany,	23	170	C. Tennant, Sons, & Co.
"	200	200	J. Barber & Co.
"	200	200	Horne & Co.
U. States,	500	500	M. D. Co.

...tta Percha—

E. Indies,	523 c.	Kaltenbach & Schmitz
Germany,	1	A. Lyle & Son
E. Indies,	272	Huttenbach & Co.
"	556	Kleinwort, Sons, & Co.
"	730	Jewesbury & Co.
"	600	R. & J. Henderson
"	201	Kaltenbach & Schmitz
"	245	Huttenbach & Co.
"	80	Soundy & Son
"	485	J. H. & G. Scovell
"	705	H. W. Jewesbury & Co.
U. States,	59	J. H. & G. Scovell

...uano—

Chili,	787 t.	Anglo Cont. Guano Works

...inglass—

E. Indies,	103 pkgs.	L. & I. D. Jt. Co.
"	29	Clarke & Smith
Germany,	6	Williams & Co.
E. Indies,	30	L. & I. D. Jt Co.
B. W. Indies,	1 pkgs.	W. M. Smith & Sons
Holland,	3	Pearson & Co.
China,	8	L. & I. D. Jt. Co.
E. Indies,	3	Hale & Son
B. W. Indies,	2	W. M. Smith & Sons
"	1	Malcolm, Kearton, & Co.
S. Settlements,	8	L. & D. Jt. Co.
"	8	"
E. Indies,	78	"

...sect Powder—

Austria,	£30.	H. Rubeck
"	45	W. Jacques

...agnesia—

Holland,	£30.	Beresford & Co.

...anure—

Phosphate Rock

France,	150 t.	T. Farmer & Co.
"	50	W. H. Carey & Sons
Belgium,	120	Lawes Manure Co.
France,	136	Anglo. Contl. Guano Works
"	125	Moeller, Graefe, & Co.
"	540	Anglo Contl. Guano Works
"	156	Lawes Manure Co.
"	20	Anglo Contl. Guano Works
"	115	A. Hunter & Co.
"	176	Anglo Contl. Guano Works

Manure *(otherwise undescribed)*

U. States,	11 t.	Rose, Wilson, & Co.
Germany,	20	Petri Bros.
"	50	Hunter, Waters, & Co.

Manganese Ore—

Pt. Augusta,	379 t.	Harrold Bros.

Manganese Borate—

Germany,	£126.	T. H. Lee

Naphtha—

Germany,	13 brls.	Stein Bros.
Holland,	190	Lister & Biggs
Germany,	4	Stein

Potassium Carbonate—

France,	9 pkgs.	Beresford & Co.
Germany,	5	Anderson, Becker, & Co.
"	5	"

Potassium Bicarbonate—

Holland,	10 pkgs.	A. & M. Zimmermann

Potassium Sulphate—

Holland,	10 pkgs.	Carey & Sons
Germany,	102	D. de Pass & Co
Holland,	10	Hernu, Peron & Co.
Germany,	27	Webb, Warter, & Co.

Plumbago—

Ceylon,	184 cks.	Beresford & Co.
"	65	J. Swire & Sons
"	184	Props. Chamberlains Whf.
Germany,	100 cs.	Beresford & Co
Ceylon,	310 brls.	H. W. Ison
"	152 cks.	J. Thredder, Son, & Co.
Holland,	28	Brown & Elmslie
Ceylon,	85	L. & I. D. Jt. Co
"	55	Doulton & Co
"	94	J. Thredder, Son, & Co.

Pearl Ash—

France,	100 c.	F. W. Berk & Co.
"	380	"
"	100	Petri Bros.

Potash—

France,	5 pkgs.	J. A. Reid
Germany,	£30.	L. & I. D. Jt. Co.
France,	90	J. A. Reid

Pitch—

Holland,	12 brls.	Rosenberg, Lowe, & Co.

Potassium Prussiate—

Holland,	£75.	T. H. Lee

Paraffin Wax—

U. States,	126 brls.	H. Hill & Sons

Quinine—

France,	£3,740	L. & I. D Jt. Co
Holland,	600	Kebbel, Son, & Co
"	124	F. W. Heilgers & Co

Sodium Nitrate—

Holland,	£16.	C. Faust & Co.

Strontium Nitrate—

Holland,	£276.	G. Boor & Co.

Salammoniac—

Holland,	£110.	Barr, Moering, & Co.
"	90	Barr, Moering & Co.

Saltpetre—

Germany,	61 cks.	J. Hall & Son
"	50	T. Merry & Son
"	72	Soundy & Son
E. Indies,	670	L. & I. D. Jt. Co.
U. States,	4 pkgs.	H. W. Ison
Germany,	118 cks.	P. Hecker & Co.
"	321	C. Wimble & Co
"	34	Craven & Co
"	100	T. Merry & Son
"	48	38 pkgs. P. Hecker & Co.
"	78	"
"	60	C. Wimble & Co.
"	635 bgs.	Hunter, Waters, & Co.

Stearine—

France,	148 sks.	J. Goddard & Co.
"	144	"
Germany,	6 cs.	L. & I. D. Jt. Co.
Holland,	125 bgs.	H. Hill & Sons
France,	12 cks.	J. Goddard & Co.

Salicylic Acid—

Germany,	16 pkgs.	Burgoyne, Burbidge & Co.

Sodium Sulphate—

Holland,	£14.	Hernu, Peron, & Co.

Soda—

Holland,	200 c.	Henderson, Craig, & Co.

Sugar of Lead—

Germany,	1 pkg.	L. & I. Jt. Co.

Tartaric Acid—

Holland,	4 pkgs.	A. Hopf & Co.
"	5	Northcott & Sons
"	34	Webb, Warter, & Co.
"	40	"
Italy,	40	"
Holland,	11	Middleton, & Co.
"	4	Hernu, Peron, & Co.
"	34	Webb, Warter, & Co.
France,	9	"
Holland,	8	"

Tartars—

Italy,	22 pkgs.	Thames S. Tug Co.
"	40	B. Jacob & Sons
"	11	Fellows & Co.

Tartar Salts—

Germany,	£36.	Charles & Fox

Tar—

Germany,	50 brls.	W. Peters & Sons

Ultramarine—

Holland,	2 cks.	G. Rahn & Co.
"	12 pkgs.	Haeffner, Helpert, & Co.
Germany,	100 cs.	Evon Rehn
Belgium,	9 pkgs.	W. Harrison & Co.
Germany,	29 cs.	G. Steinhoff
Holland,	13 pkgs.	Ohlenschlager Bros.
Germany,	12 cs.	G. Steinhoff
"	4	H. Brooks & Co.
Holland,	9 pkgs.	Ohlenschlager Bros.
Belgium,	20	Arnati & Harrison
Germany,	9	G. Steinhoff
Belgium,	12 cks.	Leach & Co.
France,	22 cs.	Haeffner, Helper, & Co.

Zinc Oxide—

Germany,	10 cks,	R. W. Greef & Co.
Belgium,	150	J. Matton
Holland,	125	M. Ashby
"	20 brls.	Beresford & Co,
"	250	N. Smith
"	170 cks.	M. Ashby
U. States,	100	"
Holland,	65	"
"	500	Soundy & Son
U. States,	125 cks.	M. Ashby
Holland,	50	"

LIVERPOOL.

The period Jan. 23rd—Feb. 6th.

Alum—

Rotterdam,	80 cks.	

Acetic Acid—

Ghent,	510 bgs.	

Bones—

Valparaiso,	1,550 bgs.	Jones & Roberts
Karachi,	2,668	Ralli Bros.
Cadiz,	1,262 1 cs.	S. N. S. Brown
Alexandria,	683	
Kurrachee,	1,056	W. & R. Graham & Co.
Rio Grande,	191,948 kilos.	
Constantinople,	520 bgs.	

Bone Ash—

Rio Grande,	256 t.	

Bone Dust—

Kurrachee,	88 bgs.	W. & R. Graham & Co.

Bone Meal—

Kurrachee,	274 bgs.	
Bombay,	1,784	
Calcutta,	1,500	

Borate of Lime—

Antofagasta,	2,197 bgs.	Cockbain, Allardice, & Co.
Valparaiso,	680 cks.	

Brimstone—

Catania,	206 t.	R. Roberts, Son, & Co.

Copper Ore—

Chunaral,	128 bgs.	A. Gibbs & Sons
Portland,	145	
Almeria,	34 t.	
"	23	G. G. Blackwell
Leghorn,	126	
Antwerp,	1,817 bgs.	

Cream Tartar—

Bordeaux,	2 cks.	
Barcelona,	3	
Castellon,	5	Sachse & Klemm
"	2	F. & L. Pitzler

Calcium Chloride—

Antwerp,	1 brl.	

Copper Regulus—

Chunaral,	5,490 bgs.	A. Gibbs & Sons

Charcoal—

St. Nazaire,	567 sks.	

Caoutchouc—

Manaos,	1,800 brls.	
Itacoatiaria,	11 cs.	
Para,	51	Bieber & Co.
"	865	
Sierra Leone,	10 brls.	Pickering & Berthoud
Bay Beach,	4 pns. 1 brl.	F. & A. Swanzy
Quittah,	1 27	"
Salt Pond,	1	S. & C. Nordlinger
"	8 4 cs.	Pickering & Berthoud
"	5 4	W. Griffiths & Co.
"	2	Edwards Bros.
"	7	F. & A. Swanzy
"	1	Pitt, Bros. & Co.
"	2	W. B. McIver & Co.
"	3	Fletcher & Fraser
"	10	J. J. Fischer & Co.
Cape Coast,	1	A. Millerson & Co.
"	2 1	H. B. W. Russell
"	1	Pickering & Berthoud
"	2 1	C. L. Clare & Co.
"	3 1	Fletcher & Fraser
Dixcove,	1 bx.	Edwards Bros.
Axim,	1 bg.	Havard & Co.
Assinee,	3	Millward, Bradbury & Co.
"	2	F. & A. Swanzy
"	1	A. Verdier
"	1 1	W. D. Woodin
"	1	Pickering & Berthoud
"	8	A. Millerson & Co.
"	2	M. Ridyard
"	5	W. Duff & Co.
"	2	W. D. Woodin
Grand Bassam,	6 pns.	A. Verdier
"	11 brls. 3 cs.	F. & A. Swanzy
"	3 bgs.	A. Reis
"	2	J. Stadelman
"	3 1	W. D. Woodin
"	1 1	M. Levin
"	1 6	Millward, Bradbury & Co.
"	6	Edwards Bros.
"	3 1 cs.	A. Millerson & Co.
"	1 2 brls.	
"	1	A. Ridyard
"	1	Pickering & Berthoud
Grand Bassa,	3	Cie Francaise
"	1 ck.	H. S. Attia Bros.
Sierra Leone,	31	Broadhurst, Son, & Co.
"	3 pns. 3	Paterson, Zochonis, & Co.
"	4 44	
Colon,	4	A. Dobell & Co.
Ceara,	6	R. Singlehurst & Co.
"	20	Rosing, Bros., & Co.
Pernambuco,	2	
Boston,	16 cs.	T. Turner
Quittah,	7	F. & A. Swanzy
Akassa,	118 bgs.	Royal Niger Co. Lmtd.
Quittah,	1	J. H. Rayner & Co.
Accra,	2 cks. 1 cs. 18 brls.	Pickering & Berthoud
"	5	F. F. Fischer & Co.
"	5 pkgs.	F. & A. Swanzy
Cape Coast,	2 pkgs.	Whinster & Watson
"	5 brls.	Hutton & Co.
"	3	Ihlers, Bell & Co.
"	2	Havard & Co.
"	10	Fletcher & Fraser
Dixcove,	1	F. & A. Swanzy
Axim,	3 bgs.	Pickering & Berthoud
Assinee,	1 ck.	F. & A. Swanzy
Grand Bassam,	3 pn.	A. Verdier
"	8 brls.	L. Hart & Co.
"	7 cks.	F. & A. Swanzy
Manoh,	5 brls.	Cie Francaise
Sierra Leone,	6	A. Herschell
"	8 pns.	Cie Francaise
"	19	P. Zochonis & Co.
"	57 brls.	
Conakry,	291	F. Collins & Co.

Chemicals *(otherwise undescribed).*

Rotterdam,	8 cks.	

Dried Blood—

Trieste,	60 bgs.	Animal Products Co. L.

Dyestuffs—

Valonia

Syra,	494 bgs.	Barff & Co.
"	458	Valiono Bros.
Constantinople,	10	
Syra	200	Androner Bros.

Dextrine

Hamburg,	50 bgs.	
Bremerhaven,	120	

Sumac
Barcelona, 50 bgs.
Palermo, 200 J. Kitchin
" 1,660
Divi Divi
Hamburg, 247 bgs.
Extracts
St. Nazaire, 25 cks.
Boston, 5 brls. Crown Cheml. Co.
Havre, 1 ck. Cunard S. S. Co. Lmtd.
Bordeaux, 25 cks.
Philadelphia, 20 brls.
Rouen, 553 cks. Co-op. W'sale Socy., Lmtd.
Trieste, 41 brls. L. J. Levinstein & Sons
" 41
Indigo
Calcutta, 268 chts. Baring, Bros., & Co.
" 116 H. Morgan & Co.
" 39 cks. Brown, Bros., & Co.
" 66 J. L. D. & S. Riker
Argols
Bordeaux, 84 brls.
Brindisi, 22 cks.
Bordeaux, 156
Orchella
Lisbon, 73 pkgs. A. Barbosa & Co.
Annatto
St. Nazaire, 10 cks.
Cochineal
Grand Canary, 9 bgs. Kuhner, Hendschel, & Co.
" 38 H. R. Toby
" 69 Beach & Co.
Las Palmas, 30 H. M. Coly & Co.
" 50 Lathbury & Co.
Teneriffe, 11 H. R. Toby & Co.
Grand Canary, 43 Lathbury & Co.
Teneriffe, 25 German Bank, London
" 12 Bruce & White
Glycerine—
Marseilles, 3 brls. R. J. Francis & Co.
Havre, 200 cs. Cunard S. S. Co., Lmtd.
Bordeaux, 6 cks.
Rotterdam, 2 dms.
Amsterdam, 6 carboys 3 cs.
Hamburg. 1 dm.
Glucose—
Hamburg. 278 cks.
Philadelphia, 200 brls.
Hamburg, 5 cks. D. Currie & Co.
Bordeaux, 40 brls.
Guano—
Rio Grande, 537 bgs.
Horn Piths—
Rio Grande, 12,000
Iodine—
Valparaiso, 93 brls. A. Gibbs & Son
Manganese—
Hamburg, 4 cks.
Rotterdam, 8
Naphthaline—
Brrcelona, 11 brls.
Pitch—
Amsterdam, 17 cks.
Rotterdam, 63
Phosphate—
Ghent, 1,610 bgs.
" 2,550 J. T. Fletcher & Co.
Potash—
Portland, 50 brls.
Rouen, 83 cks. Co-op. W'sale Socy., Lmtd.
Potassium Prussiate—
Hamburg 1 ck.
Phosphate of Lime—
Treport, 450 t.
Paraffin Wax—
Rangoon, 657 bxs.
Pyrites—
Huelva, 3,103 t.
" 3,025 Matheson & Co.
" 1,225 Tennants & Co.
" 1,899
Stearine—
Antwerp, 13 bgs.
Sugar of Lead—
Hamburg, 46 brls. 1 ck.
Rotterdam, 30 cks.
Saltpetre—
Hamburg, 160 cks.
Calcutta, 371 bgs.
Sodium Acetate—
Rouen, 9 cks. Co-op. W'sale Socy. Lmtd.
Tartar Salts—
Rotterdam, 25 cks.

Tar—
New York, 320 brls.
Tartar—
Bordeaux, 4 cks.
Tartaric Acid—
Rotterdam, 20 kgs. 32 cks.
Ultramarine—
Rotterdam, 12 cks. 4 cs.
Verdigris—
Bordeaux, 3 cks.
Zinc Oxide—
St. Nazaire, 25 brls.
Antwerp, 50 36 cks.
Zinc Ashes—
Trieste, 9 brls.
Boston, 14

HULL.

The period Jan. 22nd—Feb. 7th.

Acid *(otherwise undescribed)*
Rotterdam, 4 cks. Hutchinson & Sons
Bark—
Antwerp, 991 bgs. Bailey & Leetham
" 305
Barytes—
Antwerp, 42 cks. Bailey & Leetham
Bremen, 22 Veltmann & Co.
" 12 Gibson Bros.
" 23 Tudor & Son
" 7 Cammell, Woolf, & Haigh
" 57 T. W. Flint & Co.
" 42 Blundell, Spence & Co.
" 22
" 40 T. W. Flint & Co.
" 124
Rotterdam, 23 brls.
Caoutchouc—
Hamburg, 1 cs. C. M. Lofthouse
Chemicals *(otherwise undescribed)*
Bremen, 1 pkg. Veltmann & Co.
Dunkirk, 77 pkts, 9 cks. Wilson, Sons & Co.
Antwerp, 5 cks. Wilson, Sons & Co.
Bari, 62 drms " "
Bremen, 3 cs. Veltmann & Co.
Rotterdam, 2 cs. Hutchinson & Son
Hamburg, 1 brl. 2 cases 92 pkgs. Wilson, Sons & Co.
Antwerp, 69 pkg. Wilson, Sons & Co.
Dunkirk, 83 pkgs. " "
Rotterdam, 1 ck. Hutchinson & Son
" 2
Colours—
Bremen, 4 cks. Veltmann & Co.
" 4 Cammell, Woolf, & Haigh
" 6 Hargreaves, Bros. & Co.
" 8 Reckett & Sons
Rotterdam, 2 pks. W. & C. L. Ringrose
Hamburg, 9 20 cases Wilson, Sons & Co.
Rotterdam, 21 cks. 2 Hutchinson & Son
" 2 Geo. Lawson & Sons
" 1 cs. Wilson, Sons & Co.
" 36 cks. 10 cs. W. & C. L. Ringrose
Hamburg, 1 cs. Bailey & Leetham
Ghent, 43 cks. Wilson, Sons & Co.
Rotterdam, 1 ck. Hutchinson & Son
" 17
Hamburg, 2 cs. C. M. Lofthouse & Co.
Bremen 5 cs. Veltmann & Co.
Rotterdam, 13 pkgs. W. & C. L. Ringrose
" 14 cs. 9 cks. Hutchinson & Son
Antwerp, 34 cks. Wilson, Sons & Co.
Rotterdam, 1 brl. Hutchinson & Son
" 25 pks. 3 cks.
Drugs—
Hamburg. 49 pkgs. Wilson, Sons & Co.
" 4 lb. Bailey & Leetham
Trieste, 42 pkgs. Wilson, Sons & Co.
Rotterdam, 1 cs. W. & C. L. Ringrose
Hamburg, 14 Wilson, Sons & Co.
Dyestuffs—
Myrabolams
Bombay, 1,599 pkgs.
Alizarine
Rotterdam, 53 cks. 3 cases Hutchinson & Son
" 10 cks.
" 9 1 case W. & C. L. Ringrose
" 25 pkgs. W. & C. L. Ringrose
" 20 cks. Hutchinson & Son
" 54 pkgs. 42 cks. W. & C. L. Ringrose
" 20 cks. Hutchinson & Son
" 1 cs. W. & C. L. Ringrose
" 16 Hutchinson & Son
Sumac
Trieste, 195 bales Wilson, Sons & Co.
Palermo, 125 bgs. " "
" 1,650 " "
Extracts
Trieste, 80 cks. Wilson, Sons & Co.
Dunkirk, 33 " "
Farina—
Harlingen, 25 bgs. W. & C. L. Ringrose
Rotterdam, 415 " "
Harlingen, 1,878
Stettin, 100 Wilson, Sons & Co.
Hamburg, 25 C M. Lofthouse & Co.
" 71 Wilson, Sons & Co.
Glucose—
Hamburg. 5 carboys 20 bags
" 16 cks. Bailey & Leetham
" 900 bgs. C. M. Lofthouse & Co.
Stettin, 44 cks. Wilson, Sons & Co.
" 15 cs. Bailey & Leetham
Hamburg, 12 " "
" 7
" 5 cks. Wilson Sons & Co.
Glue—
Hamburg, 100 bgs.
Rotterdam, 50 cs. George Lawson & Sons
Antwerp, 5 cks. Bailey & Leetham
Logwood—
Hamburg, 50 bgs. Wilson, Sons & Co.
Manure—
Antwerp, 100 tons Wilson, Sons & Co.
Muriate Potash—
Hamburg, 5 bgs. Wilson, Sons & Co.
Naphthaniline—
Rotterdam, 1 ck. W. & C. L. Ringrose
Naphthol—
Rotterdam, 1 ck. W. & C. L. Ringrose
Ochre—
Bremen, 42 cks. Hanger, Watson, & Harris
" 1 Todd & Son
Petroleum—
New York, 9,871 brls.
Phosphorus—
Antwerp, 1 cs. Wilson, Sons & Co.
Rouen, 10 cs. Rawson & Robinson
Pitch—
Antwerp, 10 cks. Wilson, Sons & Co.
Ghent, 96 " "
Potash—
Antwerp, 22 pks. Wilson, Sons & Co.
Dunkirk, 49 cks. " "
Saltcake—
Hamburg, 11 pkgs. Wilson, Sons & Co.
" 2 cks.
Soap—
Hamburg, 19 cs. Wilson, & Co.
Soda—
Rotterdam, 12 cks. Hutchinson & Sons
Sodium Nitrate—
Rotterdam, 1 ck. W. & C. L. Ringrose
Stearine—
Antwerp, 78 bgs. Wilson, Sons & Co.
Sulphur—
Catania, 380 brls. 435 bgs. 61 tons Wilson, Sons & Co.
Tallow—
New York, 175 brls. Wilson, Sons & Co.
Tannin—
Hamburg, 2 cks. Wilson, Sons & Co.
Rotterdam, 20 Hutchinson & Sons
Ultramarine—
Hamburg, 5 cks.
Varnish—
Rotterdam, 1 box 1 hamper Hutchinson & Son
White Lead—
Antwerp, 61 cks. Bailey & Leetham
" 84 " "
Wood Pulp—
Kallundborg, 30 tons

Zinc—
Antwerp, 6 pk
Zinc Oxide—
Rotterdam, 4 cs
" 2 ck
" 2
" 10

GLA[SGOW.]

The period, Ja[n.]

Aluminium Sul[phate]
Amsterdam, 11
Ammonia—
Rotterdam, 1
Bone Meal—
Bombay, 5,000 b
Barytes—
Rotterdam, 92 c
Chemicals *(othe[rwise undescribed])*
Rotterdam, 7
Hamburg, 2
Christiania, 8
Cream Tartar—
Marseilles, 5 b
Caoutchouc—
Balimore, 278
Colours—
Rotterdam, 5 cs.
Dyestuffs—
Dyewood Ext[ract]
Antwerp, 15 cks
" 1 cs.
Alizarine
Rotterdam, 120 c
" 127
" 48
" 191
Dyewood
Rouen, 100 bxs.
Bark Extract
Baltimore, 12 brls.
Logwood Ext[ract]
Rotterdam, 2 c
" 3
Logwood Root[s]
Jamaica, 620 t.
Esparto Grass—
Oran, 1,122 t.
Amsterdam, 378
Farina—
Rotterdam, 50 br
Glycerine—
Marseilles, 6 br
Manure—
Bilbao, 330 t.
" 100
Oxalic Acid—
Christiania, 20 c
Ochre—
Marseilles, 25 b
Phosphate of L[ime]
Rouen, 206
Potash—
Christiania, 31
Rotterdam, 20
Rosin—
New York, b1,491
" 507
Red Lead—
Rotterdam, 16
Stearine—
Rotterdam, 24 cks.
Tannin—
Hamburg, 1 ck.
Ultramarine—
Rotterdam, 3 cs.
Wax—
Baltimore, 300 brls
Wood Pulp—
Christiania, 3,291
" 4,495
Amsterdam, 4,683
Christiania, 4,194
White Lead—
Rotterdam, 36
" 10
" 23
" 32
" 36
Zinc Ashes—
Antwerp, 7
Zinc Oxide—
Rotterdam, 60
Antwerp, 50
Rotterdam, 15
New York, 25

Pyragallic Acid—
Yokohama, 10 cwts. £75

Phosphorous—
Melbourne, £60
Wellington, 4½ c. 51
Otago, 600 lbs. 52
Port Chalmers, 9 100

Potassium Ferrocynade—
New York, 280 lbs. £11

Potash Crystals—
Melbourne, 20 c. £100

Potassium Cloras—
Hong Kong, 30 c. £67

Prussiate of Potash—
Cordova, 2 cwts. £11
Copenhagen, 5½ 18

Raspberry Acid—
Melbourne, £27

Saltpetre—
Adelaide, 1 t. £22
Auckland, 10 cwts. 11
Santos, 139 156
Rio Janeiro, 39 44
Adelaide, 18 20
Santos, 299 c. 332
" 179 200
Melbourne, 50 54
Dunedin, 20 24
Sydney, 100 112

Salammoniac—
Marseilles, 66 t. £785
Shanghai, 18 c. 32
Hamburg, 40 475

Sulphate Alumina—
Bordeaux, 100 c. £20

Sulphate of Ammonia—
Hamburg, 150 t. £1,763
" 1,219 c. 725
Cologne, 100 t. 1,175
Ghent, 30 367
Cherbourg, 20 240
Hamburg, 1,020 c. 605
Demerara. 97 1,164
Genoa, 200 2,400
Demerara, 10 6 175
Hamburg, 50 11 650
Demarara, 5 19 13

Sulphate of Copper—
Treport, 145 c. £165
Rouen, 264 320
Salonica, 23 cwts. 27
Adelaide, 5 t. 120
Odrssa, 203 220
Antwerp, 33 42
Colombo, 10 12
Bordeaux, 9 19 230
Oporto, 5 15 150
Dunedin, 25 29
Barcelona, 198 205

Sheep Dip—
Falkland, 1 c. £50
Elizabeth Isld. 13 30
Cary Bay, 1 t. 16 80
Dinero, 18 40
Dunedin, 2,300 gns. 299
Melbourne, £143
B. Ayres, 80 gl. 30
Algoa Bay, 12 4 c. 520
Natal, 10 cs. 24
" 10 10
" 100 280
Calcutta, 310 drms. 5 cks. 221
Natal, 19
Hobart, 9,760 gls. 1,196
E. London, 200 40
Algoa Bay, 7,009 c. 203
Melbourne, £90

Strychnine—
Sydney, 200 ozs. £24

Sulphate de Baryte—
Leghorn, £75

Sulphuric Acid—
Aden, 120 glns. £15
Natal, 4 t. 4 c. 30
" 5 5 25
Zanzibar, 80 gls. 15
Singapore, 1 11
P. Elizabeth, 3 22
" 8 t. 15 40
Natal, 116 lbs. 23
" 1,008 14
" 76 c. 16

Salammoniac—
Aden, 1 t. 1 cwt. £35
Antwerp, 40 66
Bagdad, 21 38
Salonica, 21 36
Berbice, 5 70
Rosario, 6 11
Hamburg, 60 10 650
Lisbon, 1 15 53

Silicate Soda—
Barcelona, 20 t. 3 c. £103

Soda Acetate—
Sydney, 25 c. £42

Superphosphate—
Cherbourg, 165 t. £495

Sulphuric Ether—
Melbourne, 1,400 lbs. £68

Sulphate of Soda—
Ghent. 64 t. £64

Tartaric Acid—
Cape, 1 cwt. £10
Bilbao, 2 13
Melbourne, 11 70
Sydney, 35
Dunedin, 5 37
" 10 88
Melbourne, 40 274
B. Ayres, 8 c. 64

White Arsenic—
Wellington, 2 t. £31

LIVERPOOL.

The period Jan. 24th—Feb. 8th.

Alkali—
Ancona, 10 cks.
Batoum, 40 tcs.
Boston, 70 brls.
Leghorn, 3 tcs.
New York, 294

Alum Cake—
Calcutta, 12 t. 13 c. 2 q. £51
Sydney, 5 4 3 17
Montreal, 22 14 2 105

Ammonium Muriate—
New. York, 5 t. 14 c. £127
Manila, 6 9
Bilbao, 4 17 123
Melbourne, 12 1 q. 16

Ammonium Carbonate—
Buenos Ayres, 4 cs. £16
Santander, 20 kgs. 34
Genoa, 15 25
Genoa, 1 t. 10 c. 63
Genoa, 3 1 120
Madrid, 4 7
Genoa, 9¾ cwts. 16

Acids—(*otherwise undescribed*)
Vera Cruz, 1 ck. 3 cs. £42

Ammonia Sulphate—
Ghent, 50 t. 5 c. £603
Bordeaux, 48 566
Valencia, 68 8 2 q. 822
Gr. Canary, 15 6 1 183
Bordeaux, 133 bgs. 156
Demerara, 226 cks. 797
Hamburg, 403 bgs. 492
Valencia, 55 7 c. 690
Bordeaux, 20 6 3 343
Gr. Canary, 26 312
Ancona, 20 240
Valencia, 81 10 2 978
Talcahuana, 4 12 3 41
Gr. Canary, 6 3 75
Rouen, 5 61

Alum—
New York, 15 t 15 c £80
Bombay, 17 7 98
Portland, M. 3 0 1 q. 15
Alexandria, 4 1 21
Lisbon, 1 6
Madras, 3 15
Natal, 2 2 11
Valparaiso, 1 6
Patras, 10 cks. 7
Syra, 30 23
Sydney, 61 85
Alexandria, 20 t. 8 c. 99
Havana, 7 17 39
Galatz, 4 12 24
Syra, 4 10 22
Santander, 5 26
Alexandria, 5 3 1 q. 25
Capetown, 2 10
Smyrna, 3 15
Madras, 22 2 1 133
Melbourne, 2 14 14
Ancona, 10 30
Leghorn, 10 30

Arsenic Acid—
New York, 5 t. 4 c. 3 q. £48

Arsenic—
Rio Grande do Sul 10 c. £8

Aluminium Sulphate—
Monte Video, 35 cks. £30
Calcutta, 91 brls. 30
Dunedin, 12

Acetic Acid—
Lisbon, 7 cks. £30

Antimony—
Odessa, 1 t. 0 c. 1 q. £85

Bleaching Powder—
Alexandria, 20 kegs.
Ancona, 242 cks.
Antwerp, 11
Baltimore, 158
Barcelona, 72
Batoum, 50 brls.
Bombay, 10 cs.
Boston, 263 cks.
Bremerhaven, 3
Bombay, 5 cs.
Callao, 7 cs. 3 cks.
Chicago, 27
Copenhagen, 29
Calcutta, 145 cs.
Dunedin, 13 brls.
Genoa, 330 611 cks.
Ghent 55
Havana, 4
Leghorn, 12
Lisbon, 4
Naples, 18
New York, 10 bxs. 96 brls. 288 cks. 81 tcs.
Oporto, 35 brls.
Malaga, 10 kegs.
Philadelphia, 164 cks.
Rouen, 229
San Francisco, 139 brls. 88 cks.
San Sebastian, 76 cks.
Santander, 25 brls.
Terneuzen, 25 cks.
Trieste, 4

Bone Waste—
Rouen, 12 t. 2 c. £96
" 250 bgs. 160

Borax—
Rotterdam, 5 t. 2 c. 3 q. £143
Antwerp, 5 8
New York, 16 2 22
Rouen, 34 4 956
Havana, 1 13 2 50
Nantes, 5 6 150

Brimstone—
Havana, 1 t. 6 c. £9
Calcutta, 16 t. 5 c. 2 q. 76
Maccio, 13 2 q. 6

Boracic Acid—
Bordeaux, 10 t. £300
Bremerhaven, 2 13 c. 3 q. 75
Genoa, 1 2 40
Halifax, 3 3 7

Caustic Soda—
Amsterdam, 142 dms.
Algoa Bay, 20 bxs.
Antwerp, 73 dms.
Alexandria, 52
Alicante, 7
Barcelona, 25 brls.
Bombay, 69 cks. 50 bxs. 224 drs.
Batoum, 208 dms.
Bari, 50
Boston, 1,525 10 brls.
Brisbane, 20
Calcutta, 179
Catanio, 5
Corfu, 40
Canada, 10
Chicago, 90
Dunedin, 20
East London, 24 bxs.
Fiume, 50
Genoa, 445
Havana, 70
Hiogo, 500
Kobe, 50
La Guayra, 50
Leghorn, 68
Marseilles, 80
Malta, 85
Maceio, 80
Malmo, 4
Manila, 150
Monte Video, 150 50 cs.
Nagasaki, 200
Nantes, 18
New York, 2,230 177 cks. 10 bxs.
" 300 100 brls. 5
Oporto, 80
Penang, 100
Philadelphia, 445 135
Pernambuco, 30
Portland M. 235 25 cks.
Piræus, 30
Port Natal, 59 cs.
Puerto Cabello, 10
Rio Janeiro, 4
Rosario, 10
Rotterdam, 130
Rouen, 20
Rio Grande do Sul, 68 dms.
San Francisco, 200
San Sebastian, 60

Sydney,
Santander,
Seville,
Santiago de C
Tarragona,
Toronto,
Venice,
Valencia,
Valparaiso,
Wellington,
Yokohama,

Calcium Chlo
Constantinople
Sydney,

Copper Preci
Antwerp, 50 t.

Caustic Pota
Melbourne, 2
Dunedin,
New York, 6
Rosario, 1
Melbourne, 2
New York, 11

Chlorate—
Rotterdam, 2
Vigo, 5
Genoa, 5

Copperas—
Bahia,
Beyrout,
Calcutta, 23
Piræus, 3
Kurrachee, 3
Alexandria, 6
Madras, 7
Alexandria, 4
Madras, 6
" 13
Oporto 5
Patras, 2
Alexandria, 1
Calcutta,
Syra,
Calcutta,

Carbolic Solid
Genoa,
Cadiz,

Carbolic Acid
Genoa,
Rio de Janeiro,
Marseilles,
New York,
Boston,
Algoa Bay,
Rotterdam,
Leghorn,
Constantinople,
Paris,
St. Louis,
New York,
Odessa,
Cadiz,
Philadelphia,
Hiogo,
Rotterdam, 1

Copper Sulph
Pasages, 20 t
Barcelona, 25
Volo,
Leghorn, 41
Beyrout 6 c
Bari, 1 t
Smyrna, 7
Bordeaux, 50
Genoa, 44
Leghorn, 31
Nantes, 116
Barcelona, 29
Sydney, 4
Monte Video, 2
Ancona, 6
Trieste, 5
Bilbao, 17
Marseilles 111
Valencia, 9
Venice, 20
Genoa, 92
Patras,
Rouen, 22
Bilbao, 10
Tarragona, 10
Valencia, 5
Lisbon, 2
Santander, 8
Bilbao, 81
Cadiz, 122
Genoa, 25
" 140
" 21
La Union 40
Leghorn, 20
Melbourne, 40
Bilbao, 30
Bordeaux, 5
Lisbon, 5
Barcelona, 3

Bordeaux, 15 t. 3 c. [illegible] [illegible]
" 52 1 [illegible] 1,238
Dunkirk, 5 0 [illegible] 120
Iquique, [illegible] [illegible] [illegible] 470
Barcelona, 5 [illegible] [illegible] 885
Bari, 1 [illegible] [illegible] 80
Bilbao, [illegible] 27 3 185
Genoa, 13 17 [illegible] 634
" 15 3 [illegible] [illegible]
Nantes, 76 6 [illegible] [illegible]
" [illegible] 11 [illegible]
Adelaide, 30 [illegible] 229
Venice, 3 1 [illegible] 33
Havre, 5 [illegible]
Bordeaux, 64 10 [illegible]
Genoa, 8 [illegible] 3 [illegible]
Fiume, 4 [illegible] 210
Nantes, 9 18 [illegible] 930
Venice, 25 7 3 370
Sydney, 24 [illegible] 230
Tarragona, 5 [illegible] 38
Rosario, 3 [illegible] 63
Barcelona, 29 29 2 400
B. Ayres, 6 [illegible] 135
Panama, 136 cks. 448
Constantinople, 7 brls. 18
Lima, 4 6
Carril, 4 80
Barcelona, 5 26
Panama, 80 387
Barcelona, 105 cks. 453
Bordeaux, 331 brls. 160 cks. 2,652
Havre, 23 cks. 180

Chloride of Lime—
Callao, 12 c. £6
New York, 99 t. 5 c. 550

Chloride—
Trinidad, 2 t. 6 c. 3 q. £68

Chemicals (otherwise undescribed)—
Porto Alegre, 7 c. 3 q. £13
Boston, 10 cks. 156
Vera Cruz, 11 brls. 50
Acajutla, 4 cks. 17
Havre, 316 1,147
Monte Video, 1 cak. 8

Cream Tartar—
St. John, N.B., 10 c. 1 q. £52

Disinfectants—
Bombay, 2 t. 17 c. £41
Canada, 32 cks. 76

Disinfectant Powder—
Para, 2 cs. 8
Bombay, 50 cks. 13
Bombay, 5 t. 39
New York, 5 cks. 45

Dried Blood—
Bordeaux, 21 t. 3 c. £180

Epsom Salts—
Bombay, 100 kgs.
Calcutta, 100
Mauritius, 5

Glycerine—
Philadelphia, 9 t. £240

Gypsum—
New York, [illegible] £07

Guano—
Valencia, 2,387 bgs. £2,594

Iron Chromate—
Gothenburg, 11 £11

Iron Oxide—
Ensenada, 150 cks. £93
Antwerp, 10 t. 8
Philadelphia, 22 13 c. 465

Magnesia—
Carril, 1 cs.
Havre, 30
Rio Janeiro, [illegible]
Santos, 20
Trieste, 1

Magnesium Chloride—
Bombay, 5 t. 3 c. £14
Rouen, 10 9 293

Manure—
Marseilles, 750 bgs. [illegible] brls. £481
Gr. Canary, 48 t. 14 c. 406
Malaga, 5 10 45
Genoa, 197 1 q. 723
Havre, 116 [illegible] 480

Nitric Acid—
Callao, [illegible] £02

Oxalic Acid—
Ghent, 6 t. [illegible] £78
Genoa, 1 cak. 7
Barcelona, 15 [illegible] 91

Pitch—
[illegible]
[illegible]
[illegible] 20 brls.
[illegible] £430
[illegible]
[illegible] 17 c. £34

Boston, 6 [illegible] 127
" 18 17 240
New York, 13 13 [illegible]

Potashes—
Rosario, 2 dms. £2
Buenos Ayres, 2 t. 19 c. 1 q. 25
Palermo, 6 127

Potassium Bichromate—
Salonica, 11 c. 2 q. £19
Alexandria, 10 16
Havana, 5 29
Alexandria. 9 15
Smyrna, 9 2 27
Odessa, 1 t. 3 39
Venice, 0 3 17
New York, 11 t. 385
Calcutta, 3 c. 5
Marseilles, 6 cks. 80
" 2 30

Potassium Chlorate—
New York, 8 t. 15 c. £413
Havana, 9 10 118
Rotterdam, 10 21
Barcelona, 1 126
Rotterdam, 10 27
New York, 5 204
Hamburg, 1 47
Hamburg, 15 6 693
New York, 13 15 641
Kobe, 6 11 320
Yokohama, 5 198
Genoa, 1 64
Portland M., 10 80
Leghorn, 13 35
Shanghai, 3 10 188
Valencia, 1 31
Antwerp, 7 339
Barcelona, 1 51
Hamburg, 5 10 3 q. 140
Shanghai, 10 80
Hiogo, 5 210

Potassium Chromate—
Genoa, 2 t. 13 c. £75

Potassium Iodide—
Carril, 1 cs. £11

Phosphorus—
Buenos Ayres, 10 cs. £200

Potassium Stannate—
Bombay, 1 ck. £40

Potassium Chloride—
Hamburg, 44 pkgs. £93
Gothenburg, 140 kgs. £196
Hiogo, 150 £300
Genoa, 20 £40

Potash Salts—
Portland (Maine) 2 t. 1 c. £60

Salt—
Antwerp, 655 t.
Algoa Bay, 0
Bombay, 2
Boston, 719
Baltimore, 145
Barbadoes, 40
Calcutta, 690
Conakry, 25
Chicago, 230
Chittagong, 3,273
E. London, 13 t.
Grand Bassa, 7
Galveston, 569
Ghent, 180
Georges Bay, 9
Lagos, 20
Malta, 5
Marstrano, 55
Newcastle, 200
New Orleans, 4,244
Norfolk, 1,100
Old Calabar, 100
Opobo, 317
Para, 35
Pensacola, 900
Rangoon, 1,642
Rio Grande do Sul, 1 t. 100 cs.
Rotterdam, 289
Rouen, 10
Sandheads, 6,786
San Francisco, 70
Savannah, 860
Talcahuano, 100
Valparaiso, 3
Wellington, 70

Soda Crystals—
Boston, 280 cks. 500 kgs. 634 brls.
Bombay, 550
Calcutta, 50
Halifax, 50
Kurrachee, 25
Malta, 50
Malaga, 10 cks.
Otago, 14
Philadelphia, 60
Smyrna, 30 150
Valparaiso, 20 10

Sodium Silicate—
Alicante, 14 cks.
Alexandria, 16
Bombay, 10
Smyrna, 60
Seville, 35 brls.
Venice, 30
Yokohama, [illegible]

Sodium Carbonate—
Boston, 248 dms. 468 brls.
Calcutta, 34
Hamburg, 17
Melbourne, 5 brs.
New York, 498
Syra, 16

Soda Ash—
Adelaide, 40 brls.
Baltimore, 735 cks. 818 tcs.
Barcelona, 30 40
Bombay, 4
Boston, 473 393 1,360 bgs.
B. Ayres, 200 200 brls.
Canada, 10
Dundee, 17
Dunedin, 20
Ghent 4
Kingsey, 14
Hamilton, 90
Lisbon, 10 135
Macceo, 90
Malaga, 12
Maranham, 175
Melbourne, 80
M. Video, 35
New Glasgow, 80
New York, 946 137
Oporto, 60 10 pps.
Patras 30 brls.
Philadelphia, 589 cks. 709
Puerto Cabello, 12
Porto Alegre, 100
Rangoon, 8
Rio Janeiro, 100
Rosario, 100
Rithymo, 200
Sherbrooke, 47
Sydney, 235
Smyrna, 76
San Francisco, 17 cks.
Trieste, 36
Valparaiso, 50
Venice, 15
Wellington, 98 40

Sodium Bicarbonate—
Adelaide, 65 kegs
Amsterdam, 15 cks
Ancona, 20 kegs
Bombay, 195
Buenos Ayres, 100
Calcutta, 200
Copenhagen, 6 cks.
Dunedin, 90 kegs
Genoa, 20 cks.
Halifax, 25 brls. 150 kegs
Hamburg, 45 kegs
Kurrachee, 50
Melbourne, 274
Matanzas, 10 kegs 1 cs.
Marseilles, 40
Malaga, 50
Nagasaki, 200
Nantes, 10 cks.
Naples, 40 kegs
Napier, 20
Rangoon, 50
Rotterdam, 125 kegs 9 cks.
Rouen, 115
Sydney, 270
Valparaiso, 50
Wellington, 50
Yokohama, 2,000

Sugar and Lead—
Havana, 4 c. 2 q. £5

Sulphuric Acid—
Monte Video, 1 t. 10 c. £18

Sodium Biborate—
Hamburg, 5 t. 7 c. 2 q. £161
Rotterdam, 19 3 31
Santos, 6 9
Antwerp, 14 10 3 436
Dunkirk, 5 190
Boston, 2 6 1 70
Yokohama, 3 11 2 166
New York, 13 cks. 65

Sheep Dip—
Natal, 8 c. 3 q. £22
Algoa Bay, 17 3 55
Cape Town, 2 t. 4 100
Buenos Ayres, 14 t. 16 c. £237
Buenos Ayres, 400 dms. 100 cks. 750
" 250 180

Sodium Sulphate—
Wellington, 2 t. 1 c. £15

Superphosphate—
Hamburg, 100 t. 1 c. [illegible] £350

Sulphur—
[illegible], 98 [illegible]
[illegible], 24
[illegible], 1 [illegible]
[illegible], 2 20 [illegible]
[illegible], 2 10
St. John's, N.B., 22. [illegible]
[illegible]
New York, 50 [illegible]
[illegible]
New York, [illegible]
Calcutta, 60 brls.
[illegible]
[illegible]
Wellington, [illegible]
Callao, 101 bgs.
Iquique, 22 [illegible]
Boston, 335

Sodium Bichromate—
Piraeus, 4 t. 19 c. 3 q.
Boston, 17 t.
Oporto, 13 c.

Saltcake—
Philadelphia, 200 t. 7 c.
New York, 49 t. 5 c.
Philadelphia, [illegible]
Boston, 71 cs. 60 cks.
Boston, 60 cks.
Bremen, 17 brls.
Baltimore, 291 cks.

Sheep Wash—
Buenos Ayres, 12 t. 6 c. [illegible]

Salammoniac—
Smyrna, 1 t. 16 c. [illegible]
Alexandria, 4 8
Genoa, 17 0
Piraeus, 3
Marseilles, 2
Moreyra, 3
New York, 1 18
Salonica, 7
Calcutta, 3 19
Amsterdam, 4
Santander, 5
Vera Cruz, [illegible]
New York, 4 [illegible]
Alexandria, 2 1
Marseilles 10 brls.
Antwerp, 1 ck.
Naples, 4
Marseilles, 37
Calcutta, 33
Genoa, 7
New York, 15
B. Ayres, 43
Patras, 1 cak.
Philadelphia, 0
" 52
Smyrna, 10 brls.
Syra, 5
Santander, 1 t. 10 c.

Sodium Nitrate—
Almeria, 1 t. 17 c. 3 q.
Shanghai, 3 2
Fiume, 6 7 0
Bilbao, 5 1 1
Copenhagen, 3 13 3
Malaga, 6 7 1
Las Palmas, 19 19
Lisbon, 19 13 3
Rio de Janeiro, 5 t.
Carthagena, 20 1 1
Almeria, 6 8 2
Malaga, 6 6 3

Sodium Sulphate—
Philadelphia, 1 t. 10 c. 1 q.
Bombay, 1

Sodium Chlorate—
Philadelphia, 2 t. 10 c.

Saltpetre—
Maccio, 1 t. 8 c. 2 q.
Bahia, 5 t. 14 c. 2 q.
New Orleans, 100 cks.
Maccio, 10
Piraeus, 50 brls.
Syra, 10
Para, 30

Tar—
Lagos, 20 brls.
Sierra Leone, 29 brls.

Tannic Acid—
New York, 3 t. 12 c. 3 q.

Tartaric Acid—
Valparaiso, 7 c.
Grand Canary, 5 c. [illegible]

Tin Crystals—
Montreal, 5 c.

Tartar—
Halifax, 2 t. 13 c.

Zinc Sulphate—
Hamburg, 58 17 c. 2 q.

Zinc Oxide—
Bilbao, 3 c. 3 q. £5
Cephalonia, 12 14

Zinc Chloride—
Piraeus, 10 c. 1 q. £10
Rouen, 1 t. 7 c. 21
Boston, 1 t. 17 c. 3 q. 13
Marseilles, 2 cks. £11

Zinc Muriate—
Bombay, 6 t. £42

GLASGOW.

The period, Jan. 24th—Feb. 8th.

Alum—
Lisbon, 5 t. 6¾ cwt. £24

Asbestos—
Bombay, 14 cwt. £37
Bibao, 5¾ 63

Benzol—
Dieppe, 94 puns. £920
Rotterdam, 25 290
" 9 drums. 230

British Gum—
Lisbon, 35 cwt. £41

Blood Albumen—
Lisbon, £34

Cement—
Jamaica, 209¾ cwt. £22. 13s.

Copperas—
Bombay, 36 t. 19 cwt. £109

Copper Sulphate—
Italy, 5 t. 6 cwt. £98
Palermo, 5 4½ 105
Bilbao, 11 4½ 183

Chemicals (*otherwise undescribed*)
Bombay, 5 t. 15¾ cwt. £65
Rangoon, 9 16 67

Dyestuffs—
Logwood Extracts
Halifax, £129
Aniline
Melbourne, £5
Logwood
Melbourne, 15 t. £146
Dyewood Liquor
Lisbon, 6 t. 14¾ cwt. £33. 9s.
Dyestuffs (*otherwise undescribed*)
Lisbon, 16 t. 7 cwt. £345
Tannin
Lisbon, 3 cwt. £20
Dye-Colours
Calcutta, 1½ cwt. £31
Alizarine
Bombay, 2 t. 8¾ cwt. £173
Lisbon. 55 £24
Aniline
Oporto, 17 cwt. £60

Epsom Salts—
Christiania, 5 tons. £34

Farina—
Rouen, 5 t. £50

Iron Liquor—
Lisbon, £43. 10s.

Litharge—
Bombay, 1 t. 14¾ c. £28

Naphtha—
Rouen, £330

Ochre—
Calcutta, 105½ cwt. £35

Potassium Prussiate—
Italy, 2 t. 10½ cwt. £100
Rouen, 5 t. 7¾ cwt. £366

Potassium Bichromate—
Antwerp, 114¾ cwt. £195
Christiania, 4 cks. 51
Bombay, 1¾ tons 62
Rotterdam, 32 cks. 215
Italy, 234½ cwt. 440
" 11 tons 13 cwts. 363
Antwerp, 62 cwt. 110
Rotterdam, 15 cks. 218
Calcutta, 17½ cwt. 4 lbs. 21
Rouen, 8 tons 14½ cwt. 944
Amsterdam, 4 cks. 170

Pitch—
Peru, 60 t. 14½ cwt. £85
Palermo, 33 12½ 65
Calcutta, £33. 10s.
Bombay, 46 10s.
Dieppe, 640

Paraffin Wax—
Natal, 7,126 brls. £100
Italy, 403 cwt. 535
Antwerp, 158 213
Lisbon, 11¾ 16
Bilbao, 30 t. 10½ cwt. 757
Rouen, 4 6½ 105

Red Lead—
Bombay, £106
Melbourne, 30 cwt. £29 15s

Rosin—
Bombay, 20 t. 8¾ cwt. £96
" 6 9 £43. 11s.
" 85 408

Sugar of Lead (Brown)—
Bombay, 4 t. 7¾ cwt. 78

Sulphate of Ammonia—
Antwerp, 20 t. £243
Valencia, 55 t. 1 cwt. 612
Antwerp, 10 121

Sodium Bichromate—
Rotterdam, 4 cks. £28
Italy, 641½ cwt. 897
Antwerp, 148½ 240
Rouen, 22 t. 17¾ cwt. 478

Sulphuric Acid—
Calcutta, 14 t. 18¾ cwt. £65
Bombay, 143
" 6 7 £36. 15s.
Rangoon, 12 8¾ 135
Bombay, £52. 10s.

Soap—
Jamaica, 628 cwt. £511
Halifax, 447
Colombo, 24 23

Sodium Nitrate—
Singapore, 28 t. 15¾ c. £244

Salammoniac—
Melbourne. 2 t. 5½ cwt. £74

Sulphur—
Halifax, 30 cwt. £14

Tar—
Madras, 15 t. 6 cwt. £40
Bombay, 94
" £21. 10s.

Varnish—
Natal, 3¾ cwt. £22

Wood Pulp—
Melbourne, 10 t. 2 cwt. £150

White Lead—
Bombay, £127
" 8 t. 18¾ cwt. 133
" 17 170
Calcutta, 103
Melbourne 842 444

TYNE.

The period, Jan. 23rd—Feb. 7th.

Arsenic—
Copenhagen, 5 t. 13 c.
Gothenburg, 1 11
Venice, 3 3

Ammonium Sulphate—
Odense, 5 t.
Hamburg, 50

Antimony—
New York, 30 t.
Rouen, 2 2 c.
Helsingborg, 6

Alkali—
New York, 250 t. 6 c.
Rotterdam, 15 9
Antwerp, 22 12
Christiania, 15 18
Amsterdam, 45 145
Genoa, 8 4

Barytes Nitrate—
Rouen, 5 t. 1 c.
Antwerp, 7

Barytes—
Rouen, 25 t.
Copenhagen, 2

Barytes Chlorate—
Hamburg, 8 c.

Barytes Carbonate—
Rouen, 20 t.
New York, 50
Venice, 5

Bleaching Powder—
New York, 277 t. 2 c.
Boston. 33 12
Copenhagen, 132 9
Libau, 165 14
Rouen, 19 15
Rotterdam, 60 8
Bilbao, 10 19
Antwerp, 142 14
Malmo, 9 18
Hamburg, 63 18
Dunkirk, 20 7
Christiania, 15 2
Gothenburg, 15 14
New York, 257 6
Amsterdam, 35 11
Genoa, 26 9
Venice, 2 18

Copperas—
Copenhagen, 4 t.

Caustic Soda—
New York, 175 t. 11 c.
Rouen, 13 13
Antwerp, 22 1
Gothenburg, 30 5
Copenhagen, 10 11
Bilbao, 6 2
Libau, 30
Hamburg, 28 17

Dyestuffs—
Alizarine
New York, 8 t 4 c.

Ferro Manganese—
Bilbao, 35 t.

Gypsum—
New York, 108 t. 16 c.

Litharge—
Copenhagen, 1 t. 7 c.
Antwerp, 3 2
Gothenburg, 5
New York, 5 6

Magnesia—
New York, 3 t. 2 c.
Antwerp, 1
Hamburg, 3

Naptholine—
Copenhagen, 25 t.

Oxalic Acid—
New York, 11 t.

Pitch—
Antwerp 240 t.

Potassium Bichromate—
Bergen, 14 c.

Phosphorus—
Malmo, 9 c.

Potassium Chlorate—
Rotterdam, 1 t. 8 c.
Antwerp, 4 13

Superphosphate—
Cherbourg, 135 t.
Genoa, 498 19 c.
Venice, 800 1

Sodium Sulphate—
New York, 31 t. 3 c.
Rotterdam, 10 2
Copenhagen, 5 5

Soda—
New York, 105 t. 14 c.
Rotterdam, 39 7
Esbjerg, 6
Odense, 6
Aarhuus, 13 19
Copenhagen, 25 1
Svendborg, 10 19
Genoa, 8 4
Venice, 26 10
Gothenburg, 5

Soda Ash—
Gothenburg, 10 c. 7 c.
Odense, 3 5
Hamburg, 2 12
Christiania, 11 18
New York, 61 8
Antwerp, 13 19
Genoa, 42 19
Venice, 41 9

Sodium Bicarbonate—
Genoa, 5 c.

Sodium Chlorate—
Antwerp, 11 c.
Hamburg, 2 t. 22

Sodium Hyposulphite—
Amsterdam, 5 t. 11 c.

HULL.

The period, Jan. 22nd—Feb. 7th.

Alkali—
Marseilles, 185 drums.
Rotterdam, 2 cases, 6 casks.

Bleaching Powder—
Bordeaux, 15 casks
Christiania, 19
Gothenburg, 11

Caustic Soda—
Antwerp, 11 drums.
Bergen, 2
Christiania, 6
Danzig, 18

Chemicals (*otherwise undescribed*)
Antwerp, 4 cs. 1 ck. 36 pkgs. 337 drs.
Bordeaux, 20
Bergen, 27
Copenhagen, 2 1 case
Christiania, 4 cks. 20 pkgs.
Danzig, 15 72
Genoa, 16 56 cs.
Ghent, 2
Gothenburg, 40 24 5
Hamburg, 34 95 59
Konigsberg, 21
Libau, 24

Marseilles,
Naples,
New York,
Rotterdam, 10
Reval, 2
Rouen, 2
Stettin, 10
Stavanger, 1

Dyestuffs—
Antwerp,
Hamburg,
Rotterdam,
Reval, 4
Rouen,

Farina—
Rotterdam,

Glue—
New York, 1

Manure—
Ghent, 276

Pitch—
Antwerp, 1
Dunkirk, 1
Genoa, 1
Naples, 1

Painters' Col
Antwerp, 1
Bremen,
Bergen,
Copenhagen,
Christiania,
Christiansund,
Dunkirk,
Ghent,
Gothenburg,
Hamburg, 15
New York,
Rotterdam,
Rouen,

Soap—
Copenhagen,

Slag—
Danzig, 279 to

Tar—
Christiania,
Hamburg,
Rotterdam,

Varnish—
Antwerp, 1
Copenhagen, 2
Gothenburg,
Hamburg,

GO

The period, Ja

Benzole—
Antwerp, 72 c
Ghent, 31 d
Hamburg, 82 c
Rotterdam, 73 c
Terneuzen, 32 d

Bleaching Po
Ghent, 25 c
Rotterdam, 27

Chemicals (*ot*
Ghent, 14 c
Boulogne, 4
Hamburg, 44

Coal Product
Boulogne, 2 c
Hamburg, 54
Calais, 31

Dyestuffs—
Boulogne, 1 c
Ghent, 22 c

Manure—
Boulogne, 129
Dunkirk, 163
Ghent, 184
" 909
Hamburg, 218
Rotterdam, 2
Dunkirk, 105

Pitch—
Antwerp, 335
Dunkirk, 470

GRI

The period, J

Coal Product
Dieppe, 15 ck
" 77

Chemicals (*ot*
Dieppe, 14
" 3
"
Hamburg,
Rotterdam,

PRICES CURRENT.

THURSDAY, FEBRUARY 13, 1890.

PREPARED BY HIGGINBOTTOM AND CO., 116, PORTLAND STREET, MANCHESTER.

The values stated are F.O.R. at maker's works, or at usual ports of shipment in U.K. The price in different localities may vary.

			£	s.	d.
Acids:—					
Acetic, 25 °/ₒ and 40 °/ₒ		per cwt	7/6 & 12/6		
,, Glacial		,,	2	10	0
Arsenic, S.G., 2000°		,,	0	12	0
Chromic 82 °/ₒ	nett	per lb.	0	0	6½
Fluoric		,,	0	0	3¼
Muriatic (Tower Salts), 30° Tw.		per bottle	0	0	8
,, (Cylinder), 30° Tw.		,,	0	2	11
Nitric 80° Tw.		per lb.	0	0	2
Nitrous		,,	0	0	1¼
Oxalic	nett	,,	0	0	3½
Picric		,,	0	1	5
Sulphuric (fuming 50 °/ₒ)		per ton	15	10	0
,, (monohydrate)		,,	5	10	0
,, (Pyrites, 168°)		,,	3	2	6
,, (,, 150°)		,,	1	8	0
,, (free from arsenic, 140/145°)		,,	1	10	0
Sulphurous (solution)		,,	2	15	0
Tannic (pure)		per lb.	0	1	0½
Tartaric		,,	0	1	2¼
Arsenic, white powered		per ton	13	0	0
Alum (looselump)		,,	4	12	6
Alumina Sulphate (pure)		,,	5	0	0
,, ,, (medium qualities)		,,	4	10	0
Aluminium		per lb.	0	15	0
Ammonia, ·880=28°		per lb.	0	0	3
,, =24°		,,	0	0	1⅞
,, Carbonate	nett	,,	0	0	3½
,, Muriate		per ton	23	15	0
,, ,, (sal-ammoniac) 1st & 2nd		per cwt.	37/- & 35/-		
,, Nitrate		per ton	40	0	0
,, Phosphate		per lb.	0	0	10½
,, Sulphate (grey), London		per ton	11	18	9
,, ,, (grey), Hull		,,	11	17	6
Aniline Oil (pure)		per lb.	0	1	0
Aniline Salt ,,		,,	0	0	11
Antimony		per ton	75	0	0
,, (tartar emetic)		per lb.	0	1	1
,, (golden sulphide)		,,	0	0	10
Barium Chloride		per ton	7	10	0
,, Carbonate (native)		,,	3	15	0
,, Sulphate (native levigated)		,,	45/- to 75/-		
Bleaching Powder, 35 %		,,	5	7	6
,, Liquor, 7 %		,,	2	10	0
Bisulphide of Carbon		,,	12	15	0
Chromium Acetate (crystal		per lb.	0	0	6
Calcium Chloride		per ton	2	0	0
China Clay (at Runcorn) in bulk		,,	17/6 to 35/-		
Coal Tar Dyes:—					
Alizarine (artificial) 20 %		per lb.	0	0	9
Magenta		,,	0	3	9
Coal Tar Products					
Anthracene, 30 % A, f.o.b. London		per unit per cwt.	0	1	6
Benzol, 90 % nominal		per gallon	0	3	11
,, 50/90 ,,		,,	0	3	0
Carbolic Acid (crystallised 35°)		per lb.	0	0	10¼
,, (crude 60°)		per gallon	0	2	10
Creosote (ordinary)		,,	0	0	2½
,, (filtered for Lucigen light)		,,	0	0	3
Crude Naphtha 30 % @ 120° C.		,,	0	1	3
Grease Oils, 22° Tw.		per ton	3	0	0
Pitch, f.o.b. Liverpool or Garston		,,	1	12	0
Solvent Naphtha, 90 % at 160°		per gallon	0	1	11
Coke-oven Oils (crude)		,,	0	0	2½
Copper (Chili Bars)		per ton	47	0	0
,, Sulphate		,,	24	10	0
,, Oxide (copper scales)		,,	52	0	0
Glycerine (crude)		,,	30	0	0
,, (distilled S.G. 1250°)		,,	55	0	0
Iodine		nett, per oz.	8d. to 9d.		
Iron Sulphate (copperas)		per ton	1	10	0
,, Sulphide (antimony slag)		,,	2	0	0
Lead (sheet) for cash		,,	14	15	0
,, Litharge Flake (ex ship)		,,	17	0	0
,, Acetate (white ,,)		,,	23	5	0
,, ,, (brown ,,)		,,	19	0	0
,, Carbonate (white lead) pure		,,	19	0	0
,, Nitrate		,,	22	5	0

		£	s.	d.
Lime, Acetate (brown)	,,	8	7	6
,, ,, (grey)	per ton	14	0	0
Magnesium (ribbon and wire)	per oz.	0	2	3
,, Chloride (ex ship)	per ton	2	6	3
,, Carbonate	per cwt.	1	17	6
,, Hydrate	,,	0	10	0
,, Sulphate (Epsom Salts)	per ton	3	0	0
Manganese, Sulphate	,,	18	0	0
,, Borate (1st and 2nd)	per cwt.	60/- & 42/6		
,, Ore, 70 %	per ton	4	10	0
Methylated Spirit, 61° O.P.	per gallon	0	2	2
Naphtha (Wood), Solvent	,,	0	3	9
,, ,, Miscible, 60° O.P.	,,	0	4	4½
Oils:—				
Cotton-seed	per ton	22	0	0
Linseed	,,	23	0	0
Lubricating, Scotch, 890°—895°	,,	7	5	0
Petroleum, Russian	per gallon	0	0	5½
,, American	,,	0	0	5⅜
Potassium (metal)	per oz.	0	3	10
,, Bichromate	per lb.	0	0	4
,, Carbonate, 90% (ex ship)	per ton	18	5	0
,, Chlorate	per lb.	0	0	4¼
,, Cyanide, 98%	,,	0	2	0
,, Hydrate (Caustic Potash) 80/85 %	per ton	22	10	0
,, ,, (Caustic Potash) 75/80 %	,,	21	10	0
,, ,, (Caustic Potash) 70/75 %	,,	20	15	0
,, Nitrate (refined)	per cwt.	1	3	6
,, Permanganate	,,	4	5	0
Prussiate Yellow	per lb.	0	0	10
,, Sulphate, 90 %	per ton	9	10	0
,, Muriate, 80 %	,,	7	15	0
Silver (metal)	per oz.	0	3	8
,, Nitrate	,,	0	2	5½
Sodium (metal)	per lb.	0	4	0
,, Carb. (refined Soda-ash) 48 %	per ton	5	10	0
,, ,, (Caustic Soda-ash) 48 %	,,	4	7	6
,, ,, (Carb. Soda-ash) 48%	,,	4	15	0
,, ,, (Carb. Soda-ash) 58 %	,,	5	12	6
,, ,, (Soda Crystals)	,,	2	13	6
,, Acetate (ex-ship)	,,	16	0	0
,, Arseniate, 45 %	,,	10	0	0
,, Chlorate	per lb.	0	0	6¼
,, Borate (Borax)	nett, per ton.	29	10	0
,, Bichromate	per lb.	0	0	3
,, Hydrate (77 % Caustic Soda)	per ton.	10	10	0
,, ,, (74 % Caustic Soda)	,,	10	0	0
,, ,, (70 % Caustic Soda)	,,	8	15	0
,, ,, (60 % Caustic Soda, white)	,,	7	17	6
,, ,, (60 % Caustic Soda, cream)	,,	7	10	0
,, Bicarbonate	,,	5	5	0
,, Hyposulphite	,,	5	2	6
,, Manganate, 25%	,,	8	10	0
,, Nitrate (95 %) ex-ship Liverpool	per cwt.	0	8	4½
,, Nitrite, 98 %	per ton	27	10	0
,, Phosphate	,,	15	15	0
,, Prussiate	per lb.	0	0	7¾
,, Silicate (glass)	per ton	5	7	6
,, ,, (liquid, 100° Tw.)	,,	3	17	6
,, Stannate, 40 %	per cwt.	2	0	0
,, Sulphate (Salt-cake)	per ton.	1	7	6
,, ,, (Glauber's Salts)	,,	1	10	0
,, Sulphide	,,	7	15	0
,, Sulphite	,,	5	5	0
Strontium Hydrate, 98 %	,,	8	0	0
Sulphocyanide Ammonium, 95 %	per lb.	0	0	8¼
,, Barium, 95 %	,	0	0	5½
,, Potassium	,,	0	0	8
Sulphur (Flowers)	per ton.	6	15	0
,, (Roll Brimstone)	,,	5	17	6
,, Brimstone: Best Quality	,,	4	10	0
Superphosphate of Lime (26 %)	,,	2	10	0
Tallow	,,	25	10	0
Tin (English Ingots)	,,	96	0	0
,, Crystals	per lb.	0	0	6¾
Zinc (Spelter	per ton.	21	10	0
,, Chloride (solution, 96° Tw.	,	6	0	0

THE

CHEMICAL TRADE JOURI

Publishing Offices: 32, *BLACKFRIARS STREET, MANCHESTER.*

No. 144. **SATURDAY, FEBRUARY 22, 1890.**

Contents.

Notices.

All communications for the *Chemical Trade Journal* should be addressed, and Cheques and Post Office Orders made payable to—

DAVIS BROS., 32, Blackfriars Street, MANCHESTER.

Our Registered Telegraphic Address is—

"Expert, Manchester."

The Terms of Subscription, commencing at any date, to the *Chemical Trade Journal*,—payable in advance,—including postage to any part of the world, are as follow:—

Yearly (52 numbers) .... **12s 6d.**
Half-Yearly (26 numbers) .... **6s. 6d.**
Quarterly (13 numbers) .... **3s. 6d.**

Readers will oblige by making their remittances for subscriptions by Postal or Post Office Order, crossed.

Communications for the Editor, if intended for insertion in the current week's issue, should reach the office not later than **Tuesday Morning**.

Articles, reports, and correspondence on all matters of interest to the Chemical and allied industries, home and foreign, are solicited. Correspondents should condense their matter as much as possible, write on one side only of the paper, and in all cases give their names and addresses, not necessarily for publication. Sketches should be sent on separate sheets.

We cannot undertake to return rejected manuscripts or drawings, unless accompanied by a stamped directed envelope.

Readers are invited to forward items of intelligence, or cuttings from local newspapers, of interest to the trades concerned.

As it is one of the special features of the *Chemical Trade Journal* to give the earliest information respecting new processes, improvements, inventions, etc., bearing upon the Chemical and allied industries, or which may be of interest to our readers, the Editor invites particulars of such—when in working order—from the originators; and if the subject is deemed of sufficient importance, an expert will visit and report upon the same in the columns of the *Journal*. There is no fee required for visits of this kind.

We shall esteem it a favour if any of our readers, in making inquiries of, or opening accounts with advertisers in this paper, will kindly mention the *Chemical Trade Journal* as the source of their information.

Advertisements intended for insertion in the current week's issue, should reach the office by **Wednesday morning** at the latest.

Advertisements.

Prepaid Advertisements of Situations Vacant, Premises on Sale or To be Let, Miscellaneous Wants, and Specific Articles for Sale by Private Contract, are inserted in the *Chemical Trade Journal* at the following rates:—

30 Words and under.......... 2s. od. per insertion.
Each additional 10 words .... os. 6d. ,,

Advertisements of Situations Wanted are inserted at one-half the above rates when prepaid, viz:—

30 Words and under.......... 1s. od. per insertion.
Each additional 10 words .... os. 3d. ,,

Trade Advertisements, Announcements in the Directory Columns, and all Advertisements not prepaid, are charged at the Tariff rates which will be forwarded on application.

THE PERMANENT CHEMICAL

THE Permanent Chemical Exhibitio ready for a formal opening, is ad well to enable intending exhibitors to s through the scheme, and to form a good fulness of such a collection of samples a therefore invite our readers to call upon opportunity of explaining what we consi portant feature to the trade.

It may be well to describe what has b present date. The getting in of the exhibi in order after their arrival, is necessarily a one which we cannot well hasten, as the earl our control. The main exhibition room, s page in plan, is now fairly well sprinkled w interesting kind, and there is not much comers. Already the number of callers, t to see, has shown us that when the coll there will be no lack of visitors, and it is make the rooms form a comfortable lo purposes.

The rooms are heated by Musgrave's combustion stoves, and there are writing-r and lavatories, all at the service of visitors, t large and small that may be used for pur or arbitrations. When it is understood t is less than two minutes' walk from the Rc convenience of having samples and speci deposited there for reference, will be mani

Besides what our exhibitors think fit tc paring a series of trade products, that ma consulted when required. Our experienc is probably as great, or greater, than that technologists, and we shall be pleased to p at the disposal of our friends. Already t an excellent collection of valves, cocks, a useful kind in acid-resisting metal made b Newcastle-on-Tyne; there is the Brach the Bracher "Desideratum" mixer; pumping engine; one of Davis' patent a others, amongst the appliances, while dotted about for the exhibition of sa already a neat "kiosk" from the Melbour by Messrs. Brunner, Mond and Co., Limi case of the Clayton Aniline Co., Limit returned from the Paris Exhibition, wher medal; Mr. G. G. Blackwell's fine case of Irish Exhibition in London; Messrs. E. D was so admired at the Manchester Royal containing dye woods, dye wood extracts, shaw Bros., Limited, have a case that wa chester Exhibition, showing specimens chemicals. Mr. Alfred Smith, of Clayton, Messrs. Jewsbury and Brown have also ca Doulton and Co., of Lambeth, have also a on of models of their celebrated stoneware chemical appliances, and the Buckley Brick and Tile Company, of Buckley, ear Chester, have a stand showing the most important of their ares, used in the Chemical and Allied Industries. A mural iblet of the firm of Messrs. Chas. Lowe and Co., of Reddish, roclaims the nature of the various phenol products manuictured by them. A very interesting exhibit of injectors, iectors, and air-blowers is being fitted up by Messrs. Melrum Bros., which will give visitors a good idea where to fill ieir requirements, while on a wall near the case of the Ianchester Aniline Co. (Chas. Truby and Co.), is a Blackian air-propeller in course of erection. The practical side eulivened by a collection of chemical literature exhibited y the enterprising Mr. W. F. Clay, of Edinburgh, with whom, o doubt, most of our readers are acquainted.

This short notice will give an idea of the scope of the chibition; its usefulness goes without saying, and we can do iuch through the medium of the *Chemical Trade Journal*, bringing before the notice of those interested in such matters the chief objects of interest contained within the walls of the building. Some of our readers may think we have made but slow progress; if it has been slow, we hope it has been sure, and the enquiries we are now receiving day by day from a distance, without any undue advertising, has made us very hopeful of the future. There can be no doubt that such a collection as we are getting together will be of immense use to all engaged in the Chemical and Allied Industries; there need be no trade secrets brought within our walls, there are plenty of things in the ordinary daily run of trade, interesting and useful, and we do not know of any other place in the British Isles where such things could be exhibited to greater advantage to the exhibitor than under the roof of the *Chemical Trade Journal*.

Most of our readers will be unaware of the nature and number of the enquiries on trade matters, we receive by each post; it would be the advantage of everyone if, as a reply to most of these queries, we could point to an exhibit in the Permanent Chemical Exhibition.

Dr. E. Frankland, F.R.S., was examined by Mr. Addison. He said he had had an enormous experience in the treatment and purification of sewage in this and other countries. He had looked into the scheme they were now enquiring into. He believed the method that was contemplated to be adopted here, both in regard to chemical treatment and to subsequent purification was the best method at present known. As regarded the chemical treatment, it was a consecutive application of sulphate of alumina and lime. The alumina was generally added first. The best works he was acquainted with were at Coventry. After passing through the tanks and the subsidence of the precipitate, the liquid was clear and usually colourless, unless there was dye water in the sewage. It was then passed through land by what was known as downward intermittent filtration. In that process nearly the whole of the remaining organic matter in the effluent was entirely removed. It passed out of the land mainly as gases, and therefore the land did not get clogged or exhausted by the prolonged filtration of effluent of this kind. He had examined this particular plot of land for filtration, and he thought it was particularly well adapted for this process of intermittent filtration. He had not only seen the land in *situ*, but had a sample brought to his laboratory, and had analysed the soil obtained from one of the trial holes. In the first place the soil previously dried in air contained 51·9 per cent. of stones, varying in size from half-an-inch to 3in. in diameter. Nearly the whole of it was a very porous material of such a varying coarseness as one would make as an artificial filter. He should not like to say the area would be quite sufficient for a population of 15,000, but it would be quite sufficient for double the present population. In a process of this kind the sludge was pumped through underground conduits directly into presses, and a large proportion of water was pressed out, and what remained was a firm consistent cake that could be carried about and did not not readily putrify, and which only contained then about 49 or 50 per cent. of water. There would be no nuisance whatever unless the works were carried out in a slovenly way, and no possible injury to health. That had been proved over and over again. At Coventry there were villa residences as near the filtering beds as in this case, and he was informed that no complaints had been received. A very pure effluent would be discharged, which would contain less organic matter in all probability than the Irwell on entering the borough. So far as his knowledge went this was the best possible scheme that could be devised, so far as chemical treatment and purification was concerned. The commission of which he was a member was decidedly of opinion that each district ought to deal with its own sewage, and discharge the effluent into the river as soon as possible.

Mr. Sutton cross-examined Dr. Frankland at great length as to the chemical processes involved in the method proposed to be adopted, and quoted from published works of Dr. Frankland. There was no free lime left in the sludge. The sulphate of alumina took out about one-half of the polluting matter, and it was proposed to get rid of the other half by a filtration process. There might be a mere trace of sulphuretted hydrogen about the works, but he did not think it would affect the health of the people living about. He did not think there would be any difficulty in disposing of the sludge cakes. If they could not dispose of them they might store them or bury them. If the worst came to the worst they could burn them, but he should not think that would be advisable. At Coventry they had no difficulty in disposing of them. They could not make them fast enough. They were worth from 1s. 0d. to 2s. 6d. per ton. He did not think the sulphate of alumina would have any effect whatever on woollen manufacture. The harder the water the more soap must be used, but any increase of expenditure would be a vanishing quantity, perhaps ¼d. in the pound. It would have more effect on calico print works than on woollen works.

Two samples of woollen goods were handed in, which showed a considerable difference of texture, but Dr. Frankland did not think the difference had been produced by sulphate of lime. It had probably been brought about by some corrosive material. He did not think it was advisable to have the sewage carried right out of the borough. He thought it was best to return the purified water to the river as soon as possible. Any advantage to be derived from carrying the sewage away from the population was counterbalanced by depriving the river of so much water, and inasmuch as it could be successfully treated, without nuisance in the midst of the population, he did not see what advantage there would be in going to great expense in removing it from the district. He did not think it would depreciate Mr. Disley's house or the cottages adjoining. He had not seen any property near that would be depreciated at all. He would prefer these works to the existing conditions if he lived there. He did not imagine it would have any effect on the malting which was proposed to be carried on near.

Mr. Charles Estcourt, analyst to the city of Manchester, put in analyses of six samples of sewage obtained from the outlet at Acremill, the results of which were not stated, except that it was a very dilute sewage, compared with Manchester. He had heard Dr. Frankland's evidence, and he agreed with the opinions expressed by the doctor.

Mr. Mellis, C.E., was called and examined by Mr. Addison. He said he had devised and constructed the sewage works at Coventry and other places, and had several similar works on hand at present. He had looked at the plans of the borough surveyor, and generally the scheme for the sewering and sewage treatment of this part of the borough seemed to be a very suitable scheme, and a very good scheme. It was a well-devised scheme, and quite up to date. There would be no smell or nuisance from these works if they were properly conducted. The weekly amount of sludge would be about ten tons, not more. Cross-examined by Mr. Sutton: He proposed to use from seven to eight grains of sulphate of alumina, and six grains of lime per gallon. There was generally no difficulty in getting rid of sludge cakes. Cross-examined by Mr. Woodcock: There was no sewage tank at Coventry, but he had known them at other places, and they were quite unobjectionable.

Mr. E. F. Coddington, examined by Mr. Addison, said he had been manager of sewage works at Coventry for 16 years, and before that he had been manager of other works. He was well acquainted with the purification processes of other towns. Coventry had a population of 30,000, and they dealt with 2,000,000 gallons in 24 hours. There was no nuisance whatever, and no complaints from the neighbours; and they had no difficulty in disposing of the sludge cakes. Cross-examined by Mr. Sutton: They used to sell the sludge cakes at 1s. a ton, but there was such a demand for it that the Corporation had raised it to 1s. 6d. a ton. The farmers put it on their fields. At present they were using 8½ grains of sulphate of alumina to the gallon and 6·4 grains of lime. There were no works on the river below where they turned in the effluent. The works were about a mile from the centre of the town, and just outside the municipal boundry. It was in a nice open country, not a narrow gorge, as here.

This concluded the case for the Corporation.

Mr. Sutton, for the opposition, held that the case for the Corporation had not been made out, and said it was incomplete in three respects. It was incomplete because they were really intending to renew in a great measure the filtering tanks at Acremill.

The Inspector said that question was not before him, and he could not go into it.

Mr. Sutton said there were still two other points fatal to the application. In the first place in Stacksteads there was included the Cowpe valley, which must be drained, and it was admitted they had made no provision for draining it here. The surveyor said it was practically impossible to do it except at very great expense. Thus while they were applying for land to drain Stacksteads, there was a portion of the district they could not drain at all. Then it was admitted in answer to a question from the inspector that the present stone drains must be replaced.

The Inspector: That is the subsidary drains. That is no part of the sewerage.

Mr. Sutton: But it is part of their plans. When they come to borrow the money I shall have something to say about that.

The Inspector: All they are seeking now is compulsory power to acquire this land.

Mr. Sutton: On purpose for their sewage scheme. He would also prove that the scheme was a most extravagant scheme. They had omitted in the Glentop estimate the cost of pumping, which would be £259. a year, and that, capitalised at 3½ per cent., came to £7,5000. Then the surveyor said the drain to Rawtenstall would cost £21,000., but whether he meant it was the total cost of the Bacup proportion did not seem to be quite clear. But whichever was meant he would show that it was absolutely wrong. If it was meant to be the Bacup proportion, then that gave the total cost for the sewer at £42,000, which was absolutely out of the question, and a sum which even the surveyor flinched at when it was put to him. If they meant that the total cost of the sewer would be £21,000., then they had £10,000. to take off the Rawtenstall estimate. It was thus evident that the Rawtenstall scheme would be the cheaper of the two. Then with regard to the effluent, it was all very well to come and say it did no harm in a district like Coventry, but here his clients lived upon their trade, and anything that would damage the manufactures of the district was of vital importance. He should produce evidence to show that this effluent would seriously damage these works. With regard to the nuisance, it was clear there would be a nuisance, unless the works were carried out in an absolutely accurate way, if not in any case.

Mr. John Newton, C.E., examined by Mr. Sutton, said he had been in practice for 30 years, and had carried out a large number of sewage works in various parts of the country. He was well acquainted with the Bacup and Rossendale Valley, and he had surveyed the district with a view to this enquiry. He had inspected the site to be acquired, the land at Ewood Bridge, and at Acremill. The proposed site at Glen-top comprised 3a. 1r. 20p. Three-quarters of an acre, occupied by the mill, was not shown as utilised on the plan. Of the remaining 2½ acres, 1¼ acres were shown as filtering areas, and the rest for tanks and buildings. He did not consider that sufficient for the purpose. He thought that Glen-top site was an unsuitable one, it was so shut in, and because it was so enormously expensive. They would require pumping engines which would cost £750., and this would involve working expenses of at least £185., which if capitalised would amount to £5,000. By the Inspector: It could not be treated within the borough except by pumping. Cross-examined by Mr. Addison: He had designed about thirty sewage works. Some of them were nuisances, and some were not. Some of them were nuisances because they were carelessly conducted.

Charles Burghardt, Ph.D., examined by Mr. Sutton, said he had considered the Bacup sewage scheme from a chemical point of view. In his opinion it was not the most efficient method that could be devised. It would produce emanations at times of foul vapours or gases which would be a nuisance and a probable injury to health. The treatment with alumina and lime was simply a mechanical and not a chemical process. It purified the sewage to a certain extent, but not as completely as other methods would do. He considered that hydrated oxide iron was the best purifier. In the scheme proposed, the sulphate of lime in the effluent would harden the water, which would cause a serious loss in soap to the print works. Cross-examined by Mr. Addison: The iron system was in use at Buxton. The Inspector: Oh; I know Buxton. Witness did not wish to throw doubt on the Coventry scheme.

Mr. J. Carter Bell said he had made experiments on a large scale to find what quantity of organic matter was held in solution in the effluent when treated with different systems of sewerage. The amount of sulphate of lime and other solid matters in solution in this effluent would be from 70 to 80 grains to the gallon, which would have a serious effect on the manufactures of the district. He considered that in a confined space like this these works would be very objectionable.

Mr. Alderman Disley, examined by Mr. Sutton, said he was the owner of Waterbarn Mills, and the owner of some of the land in the vicinity. In case of increasing his works he would probably have to go in the direction of the land now scheduled. It was part of the land which he purchased with the mill. The rateable value of the mill was £515. He only used the river water for power. He employed about 100 hands. His house was about 40 yards from the proposed works. The prevailing winds would bring the emanations in the direction of the houses. Cross-examined by Mr. Addison: The piece of land proposed to be acquired from him was about 1,000 square yards. It previously belonged to his father-in-law. They had made no practical use of the land hitherto.

The Inspector said he should take time to consider his report.

THE "DIAMOND" PATENT OIL FILTER.

THIS apparatus, shown in the accompanying cut, is intended for the purification of refuse oil. As it is suitable for cleansing practically every kind and quality of this material, it will very readily be seen

...of application is boundless. The oil must be heated ... (regulated by its consistency) between 140°-212° F. ... by means of piping connected either with the main ... the exhaust. The waste oil is poured into the ... it is freed from the coarser impurities by the grid *b*. ... upward through the chamber *d*, which is packed with ... twist, capable of being compressed or loosened by ... screws above. The oil, after passing through this ..., trickles over the ledge *e* on to the top of another ... or wadding in the chamber *f*. Through this it filters ... *g*, and can be drawn off by the cock *i* when wanted. ... water sink to the bottom of the chamber *a*, and are ...dically by the tap *c*.

...ing the apparatus it is advisable to saturate the filter with ... otherwise the dry twist would take up so much scum and ... its efficiency much sooner than when first soaked with ...

...g material only requires changing five or six times a year, ...be done by any ordinary mechanic in a few minutes, the ...ion necessary being to see that the twist or wadding is ...in at the sides and corners, to prevent the oil from running ... crevices unfiltered.

...acturers are Messrs. Woodhouse and Rawson, Limited, ...ctoria-street, E C.

Legal.

CHANCERY OF LANCASHIRE.

(VICE-CHANCELLOR SIR H. FOX-BRISTOWE, Q.C.)

POLLUTION OF A STREAM.

... v. CHADWICK.—Messrs. Salis Schwabe and Co., of the ... Works, near Middleton, and Messrs. Lees, owners of ... which runs the Wince Brook, sought an injunction last ...ain the defendants, Messrs. J. Chadwick and Co., calico ...rs, and bleachers, at Oldham, from polluting the stream ...ay to cause injury to Messrs. Schwabe in their works. ... with Mr. Roby appeared for the plaintiffs, and Mr. ... Mr. Hopkinson for the defendants. Mr. Maberly said ... was one of the sources of water supply to which the ... a right by agreement with Messrs. Lees. The defendants ...e refuse of their dye works into the stream and polluted ...ks of the defendants were on the bank of the Spring ...tary of Stock Brook. The Stock Brook entered the ..., which in its turn was a confluent of the river Irk just ...rks of the plaintiffs. The plaintiffs took the water from ...ook near the weir, at the junction with the Irk, and con...heir lodges by artificial channels. In 1884 the plaintiffs ... water was distinctly polluted with some black sub... the following year the pollution was traced to the ...e defendants; the defendants were communicated ... plaintiffs, who pointed out the injury that was ... them. On the 13th March the defendants replied that ...ould have the immediate attention of the firm. In that ...d that there was no indication of the line of defence ...aken later, namely, that they had a right to turn their ... brook. A considerable correspondence went on during ... years, but the defendants did not then allege any right to ...ter, simply making excuses for not taking any steps to ... dye refuse from getting into the stream. In 1886 the ...ded four dye-tanks to their works, with the consequence ...t waste was increased. The defendants carried on a very ... process, having what were called "jiggers"—large ... loose dye, the fabrics being drawn through the dye ... rollers. When the vessels were to be emptied ...ed out at the bottom of the tanks, and the liquid was ... into a gutter, whence it found its way into the brook. ...le, therefore, that a large amount of colouring matter ...t from the works into the brook. The plaintiffs would ...isfied if the defendants would put down catch pans and ... to allow the solids to settle.

...Chancellor said that in order to clear the water from ...ight first be allowed to stand in settling tanks, where ...ould be precipitated; it could then be allowed to flow ...r filter beds, where the liquid would really be disinteg... portions that remained would be taken away. The ...then pass into the brook in a comparatively innocuous ...

...ly said that that process would meet all the requirements ...ffs. In fact, something less than the elaborate filter beds ...ed suggested would perfectly satisfy the plaintiffs. With ...fence now put forward by the defendants Mr. Maberly said the plaintiffs had never heard of the assertion of a right to pollute until after this action was begun a year ago. In the letters which had previously passed the defendants simply dealt in excuses for delay, and there was no claim of a right.

Mr. Edmund Salis Schwabe, senior partner of the firm of Salis Schwabe and Co., described his sources of water supply, and said his firm had been carrying on business at the same works since 1833. The water supply was of course a vital matter in their business. They had authority to use the water from the Wince Brook under a deed entered into with Messrs. Lees, the riparian owners. He could speak from personal knowledge of the supply since 1861. From 1875 to 1880 there was a gradual deterioration of the water coming from the Wince Brook. The deterioration began to get really serious after 1880.

ROTHESAY.—THE OLD MINERAL WELL.

FOR a long time the old mineral well at Bogany, once so much famed as a medicinal spring, has been lost sight of and neglected, but the Town Improvement Trust recently took up the matter, and had the spring opened up and the well cleaned out. The water has also been analysed by Mr. W. Ivison Macadam, Professor of Chemistry in Edinburgh, the result being as follows:—

	Grains.
Calcium Sulphate	137·04
Calcium Carbonate	15·72
Calcium } Magnesium } Phosphates	0·26
Magnesium Sulphate	170·18
Magnesium Carbonate	6·97
Magnesium Chloride	138·72
Sodium Sulphate	5·37
Sodium Sulphide	0·63
Sodium Chloride	1335·55
Potassium Sulphide	0·31
Potassium Chloride	27·73
Soluble Silica	0·25
Ferrous Carbonate	0·17
Organic and Volatile matter	8·14
Total solid matter per gallon, 1,847·04 grains.	
Gases	Cub. Inches.
Sulphuretted Hydrogen	22·67
Carbonic Anhydride	3·54
Oxygen	0·91
Nitrogen	5·94
Total gases per gallon, 33·06 cubic inches.	

In his report on the well Mr. Macadam says it is most nearly allied to the famed Harrogate springs, and that in many respects it compares favourably with the Strathpeffer wells. For medicinal purposes he says it will be found fully as good as the Harrogate water, and from the greater quantity of saline material superior for baths. As compared with Strathpeffer, the Bogany water will be found superior in cases requiring salts along with sulphuretted hydrogen, and equally good when sulphuretted hydrogen alone is necessary. He sums up his report by saying that in his opinion the Bogany well is of high-class quality as a medicinal water, both for internal and external use, and if properly cared for and attended to, would be a valuable addition to the many advantages Rothesay offers to those in search of health. It has been remitted to a committee to make inquiry and to report as to the best means of protecting the well and making it available for the public use.

PHOTOGRAPHIC NOTES.

A PLATINUM TONING SOLUTION FOR SILVER PRINTS.

M. GASTINE recommends the following:—

A.

Water	9¼ oz.
Sodium chloride	300 grains.
„ bitartrate	150 „

B.

Water	3½ oz.
Bichloride of platinum	150 grains.

Add two drachms of B to A and tone. "It is advisable," says the author, "to eliminate the silver salts from the paper by one or two washings before toning. If bitartrate is not at hand, take five parts of tartaric acid and mix with four and a half parts of carbonate of soda."

GELATINOUS BOTTLE WAX FOR COVERING CORKS.

In storing volatile liquids which are solvents of resinous material, the ordinary bottle wax in which bottle necks are ordinarily dipped is generally inadmissible, by reason of the solvent action of the liquids upon it. In such cases the following answers admirably, giving a ct closing; and, moreover, the top is easily pared off with a when the bottle is to be opened:—

Soft gelatine or good glue	3 parts.
Water	9 ,,
Glycerine	2 ,,

elt the gelatine in the water, and then stir in the glycerine. Any uring matter can be added, and the necks should be quite free grease when dipped. A second dip can be given if the first does give a sufficient thickness. Manufacturers, sending out photo-hic preparations containing volatile liquids, should give this aration a trial. The top can be stamped while soft with a slightly sed metal seal, or, when set, a warm stereotype (slightly oiled) or ndia rubber stamp may be used.

J. H. BIGGS' METHOD OF MAKING SILVER PRINTS ON ROUGH DRAWING PAPER.

SALTING SOLUTION.

Soft sheet gelatine	4 grains. (about ·0259 gramme)
Sodium chloride (common salt)	5 grains. (about 0·324 gramme)
Water	1 ounce. (about 28 cubic cents)

oak the gelatine and dissolve in a water bath. Whatman's drawing er is floated on this and allowed to dry spontaneously.

SENSITIZING SOLUTION.

Nitrate of silver	60 grains. (about 3·88 grammes)
Water	1 ounce. (about 28 cubic cents.)

Ammonia is added, drop by drop, till the precipitate formed is just redissolved.

the paper on a flat board, the salted side up, and apply the sensi-ng solution freely with a Buckle's brush. Pin the paper up by one ner, attaching a fragment of blotting paper to the opposite corner, prevent the accumulation of solution. When dry, print, tone, and as in the case of a paint on albumenised paper. Mr. Biggs recom-nds a carbonate of soda toning bath. For description of the Buckle ish, see next paragraph.

THE BUCKLE'S BRUSH.

This useful implement was quite common and well known in the ly days of photography, when negative paper processes were much d, and is very useful when it is important to have a clean brush, or her mop, for every operation. It is made by taking a piece of glass e, about half an inch in diameter and six or seven inches long, and wing a tuft of cotton wool partly into one end of it by a thread or ook of silver wire, an arrangement easily extemporized.—*Photo-aphic Review.*

IMPROVEMENTS IN THE MANUFACTURE OF STARCH.

N interesting paper was read before the "Vereins der Starke-Interessenten Deutschlands," by Saare, on the "Difference in uality of Moist Starches." The chief impurities found are fibre, nd, and remains of the pulp water. The water varied from 47·5 to per cent., the average being 48·5 per cent. The impurities ranged m 0·1 to 1·5 per cent., average 0·5 per cent. Sand averaged 0·22 r cent. Fibre was found from 0·01 to 0·3 per cent., average 0·15 r cent. The impurities soluble in water varied from 0·08 to 0·2 per it., mean, 0·12 per cent. To prevent the introduction of sand into finished starch, the portion which settles in the washing tank next the axis of the stirring apparatus should be removed and not mixed th the rest of the starch. Fibre gets into the finished starch owing to perfect arrangement of the sieves. Pulp water remains in the starch ing to bad washing. Spoilt and sound potatoes of the same origin re also examined. In the case of spoilt potatoes a considerable loss —40 per cent.) takes place during the preliminary washing, besides ich a large quantity of slime forms on the sieves, and a dark pulp ter, which ferments later on, is obtained. The starch settles very sely, and after draining contains 67 to 70·5 per cent. of water, ilst starch from sound potatoes contains an average of 48·5 per cent. ter. The addition of 1 grm. of sulphuric acid per liter of water ises the starch to separate somewhat cleaner. Calcium bisulphite s a similar, but superior effect. The fineness of the sieve mesh has o some influence on the purity of the starch obtained. As regards pulpy mass remaining after the starch has been washed out, not re than 5 per cent. of the starch remaining in it should be remove-le by washing, and not more than 50—60 per cent. of the dry, mpletely washed pulp should consist of chemically bound starch.—*ingl. Polyt.*

THE NATURE OF ELECTRICITY.

SOME years ago a controversy raged somewhat hotly with regard to the question as to the nature of electricity, whether it was a fluid in itself or merely a peculiar condition assumed by matter. General opinion has now settled down in favour of its being a condition some-thing like polarisation taken up by matter, and many experiments adduced in favour of that assumption are familiar to scientists. Professor S. P. Thompson, the director of the Finsbury Technical College, London, has just exhibited at the Physical Society some experiments which would furnish powerful arguments in favour of the fluid theory. He made use of the interesting Lichtenberg figures formerly employed in the lecture-room, but not much known now. They are figures pro-duced on the surface of a cake of pitch, shellac, or resin, by discharging a Leyden jar, which has been previously charged with high-tension electric-ity by means of a frictional machine. The knob of the jar is rubbed over the surface, discharging electricity upon it; then lycopodium or flower of sulphur is sprinkled over the slab or cake, and the powder adhering to the pitch forms peculiar figures resembling the twigs and branches of trees or leaves, according as the charge previously given to the Leyden jar was of positive or of negative electricity. Professor Thompson modifies the experiment by holding the Leyden jar firmly in his hand, and as it were throwing electricity from the charged knob at the disc, instead of rubbing the surface with it as is usually done. After doing this several times in front of different parts of the plate he again peppers it with the dust, and the result is seen in the presence of figures which do not at all resemble those of Lichtenberg, but rather look like the splashes from a heavily charged paint-brush loaded with colour, thrown on to the surface in the manner employed by bookbinders when pro-ducing the well-known mottled edges of books, or by decorators in imitating granite Professor Thompson finds that the forms of these figures, which he has called electric splashes, vary with the size, degree of polish, &c., given to the knob of the Leyden jar, and attributes the phenomenon to some modification of the so-called brush discharge; but the experiment is a novel one, and its exhibition created much interest on the part of the audience.

THE ECONOMY OF FUEL IN THE DISTILLATION OF AMMONIACAL LIQUOR.

IN these days of small margins of profit, the manufacturer can no longer afford to throw anything away until he is quite certain that its capacity as a revenue producer is exhausted. Applying this principle to the distillation of ammoniacal liquor in the column-still, one is struck by the fact that in the process, as ordinarily conducted, the spent liquor leaves the still at a boiling temperature, containing the heat equivalent of a considerable quantity of fuel. These few remarks are intended to shew how, in the case of the Davis still, this heat may be recovered and utilised.

In the case of the Davis still, the spent liquor escapes from the apparatus through a series of U tubes. It is proposed to cause these U tubes to pass through a water-tight box of any convenient shape and material, and the ammoniacal liquor to be treated is conducted through this box on its way to the still, so that it becomes heated by contact with the U tubes, through which the spent liquor is escaping. In one works this arrangement has been already adopted, and answers very successfully. The temperature of the ammoniacal liquor is raised to 180°F., while the spent liquor only registers 112°F. as it escapes to the drain. Where other means are adopted for the pre-heating of the ammoniacal liquor, the "Economiser" (as the apparatus is called), may be employed to heat water for the boiler supply, or for any other purpose.

In certain cases it may be desirable to substitute a coil of pipes for the series of U tubes but, if this is done, care must be taken to so arrange the coil that it forms a seal equivalent in its effect to the U tubes.

These outlet tubes or coils may be best cleaned out by connecting temporarily with a high pressure water supply, communication with the still being of course cut off while the flushing or cleansing operation is going on. A mixed jet of steam and water, where it can be got, answers this purpose very effectually.

It may be thought by some that the saving to be effected by the use of the "Economiser" is insignificant. Compared with the differences between "ancient" and modern stills, this saving may appear a small one, but it must be remembered that the field of the economist is much more restricted in its area than was the case some few decades ago. We are very much nearer the goal of theoretical perfection, and our rate of progress must be expressed in smaller units. Those manufac-turers who fail to appreciate small individual improvements, are con-siderable losers in the aggregate, and are among the first to drop out from the struggle where the fittest only survive. The proverbial sweet-ness of little fishes has its parallel in the acceptableness of small economies.

having an effective length of 15 feet for every 1½ per cent. of material picked off. The cost for labour will probably be from ¼d. to 1¼d. for every 1 per cent. picked off. Balanced screens, on which the coal is picked, are available only when the amount of material to be picked off is very small, say 1 to 1½ per cent. For all small under ¼-inch, and for dross from 1½ inches downwards, with more refuse than from 2 to 4 per cent., the wet process is most applicable. In the wet process it is desirable to have the arrangement so that the small coal can be delivered direct from the screens into the washing tanks without the intervention of wagons. In all the systems of washing the best results are obtained by sizing the small coal before it reaches the machine. This can most conveniently be done by passing it through revolving screens with meshes of varying size. The supply and degree of pulsation or agitation of the water require careful adjustment to suit the various sizes of coal to be treated, and the relative specific gravity of coal and impurities. To remove the refuse from the smaller sizes, say under ⅛ inch, the felspar washer is the most effective. The felspar system is most valuable, where the coal is crushed before washing, and is to be used for coke making. Where the coal and the refuse approach one another in specific gravity, it appears that in some cases the trough washer gives the best results. It is applicable for small quantities only, and requires a large flow of water and extra labour, but it has the recommendation of simplicity and small capital cost. It may also be sometimes utilised as a means of transport where the distance from the pit to the wagons or coke ovens is considerable. The Robinson washer is cheap as regards first cost and upkeep, and requires little water. It largely depends for its efficiency on the attention and skill of the man in charge, who may often be tempted to pass more through it than it can effectually clean. Speaking generally, more elaborate machinery is effective in avoiding waste in proportion to its cost, but the capital charges and upkeep are also high in proportion. Other things being equal, coal will be washed best with an abundant supply of clean water; but the more water used the greater the risk of fine coal being lost and the greater the difficulty of filtration. Water to wash coal for coking should not be often used over again, as dirty water dulls the coke. The particulars furnished as to settling ponds do not give sufficient data to justify any definite conclusion as to their capacity in relation to the quantity of coal washed. In most cases no record was kept of the quantity of water used; but settling ponds are a necessity, and their capacity will depend on the special circumstances of each case. There seems no better way of filtering the foul water, after it has passed through the settling ponds, than pumping it on to the rubbish heap, and allowing it to percolate through, as at Earnock. The washed gum of coal not suited for coking is meantime used almost entirely for firing colliery boilers. Briquettes are made of it to a small extent, but new outlets are required for this product The large quantity to be treated daily, and the varying nature and proportions of the coal and dirt to be separated, render washing, at most collieries, a troublesome process, and unqualified satisfaction is seldom expressed as regards any machine in use. In some cases the machine may not be quite adapted to the peculiarities of the coal treated, or it may be over-driven, or not have a sufficiency of water, or be allowed to get out of repair, all or any of these causes leading to disappointment as to results. A separate siding for each class of coal is a desirable arrangement. While the investigations of the committee have not led them to prefer and recommend any particular system as the best in all circumstances, the examples given will, it is hoped, be valuable in directing attention to the conditions in which each system reported on is likely to be most successful and profitable, leaving it to individual skill and judgment to adapt and modify as special circumstances may require."

A SATISFACTORY BALANCE SHEET.—The Directors of Brunner, Mond, and Co., Limited, Winnington, Northwich, state that the balance-sheet and profit and loss account show a balance to the credit of profit and loss account on the working of the half year ended on the 31st December, 1889, of £120,640., which, with the amount of £15,735. brought forward from the previous half year, makes a total of £136,375. The Directors propose to deal with this balance as follows:—Dividend on the preference capital at 7 per cent. per annum, £11,944.; dividend on the ordinary capital at 30 per cent. per annum, £88,125.; amount to be written off patents account, £2,500.; amount to be placed to suspense account, £20,000.; leaving a balance to be carried forward of £13,806. The Directors propose again to place to suspense account a sum of £20,000, towards meeting the cost of the re-modelling of a portion of the plant, which is in progress. The sum of £6,500. has been written off freehold land account, in order to reduce to a reasonable figure the cost of adjoining lands which the directors have felt bound in the interest of the Company to purchase during the last few months, and for which they have been compelled to pay fancy prices. The directors who retire by rotation are Mr. Henry Brunner, Mr. Holland, and Mr. Galloway, all of whom are eligible for re-election.

Correspondence.

ANSWERS TO CORRESPONDENTS.

H. S.—There is a Directory of Chemical Manufacturers of England, published by Kent and Co, London.
H. F.—Thanks for the cutting re chimney,
C. H. A.—We have endeavoured to get the information you require, but merchants are not wishful to part with it.
J. W. M.—Please send us your list occasionally.
INQUIRER.—1. No specified distance, but should be at least 50 feet. 2. Should be capable of lifting any quantity.
MARSHALL.—Your query must be a matter of business.
B. AND T.—You will see the article was from the "National Druggist of America." It is there where the method is well known, or in Alsace.
H. H.—Marsh's test is the one most applicable to your case; but you must be sure of the purity of the zinc you employ.
R. C.—From 10 to 12lbs. per ton of coal. One per cent. of volatile sulphur would be 22·4lbs., but we do not think this has been reached save in one instance.
CALC.—A table of milk of lime values has already appeared in our pages.
DYER.—It is quite possible to purify the liquid to which you refer, but not in the way you mention.

*** *We do not hold ourselves responsible for the opinions expressed by our correspondents.*

THE TREATMENT OF MINERAL OILS.

To the Editor of the Chemical Trade Journal.

SIR,—I have always understood that vaselines were products of petroleum, obtainable by filtration only, and without either distillation or treatment with acid or chemicals of any kind. Hence this purity and absolute freedom from compounds which might—and undoubtedly would—act injuriously on the skin, especially when that is in an unhealthy condition.

On account of the great development of the vaseline manufactures and the number of uses—both medicinal and for toilet purposes—to which it is now put, it is of very great importance to dealers and to the public that the former should be able to know whether the article they are offering to the latter is genuine or otherwise.

As I have always understood that pure vaselines or petroleum jellies have never been treated with chemicals—only the spurious or imitation articles being so treated, and consequently produced at a much lower price—I should be glad of a line or two in your next issue from some of your friends who can speak authoritatively on the matter, and can give some hints as to the best means of distinguishing between the spurious and the genuine article.—Yours, etc.,

ROBERT BRADSHAW, JUNR.

PRIZE COMPETITION.

A LITTLE more than twelve months ago, one of our friends suggested that it might be well to invite articles for competition, on some subject of special interest to our readers; giving a prize for the best contribution, he at the same time offering to defray the cost of the first prize, and suggesting a subject in which he was personally interested.

At that date we were totally unprepared for such a suggestion, but on thinking over the matter at intervals of leisure, we have come to think that the elaboration of such a scheme would not only prove of benefit to the manufacturer, but would go some way towards defraying the holiday expenses of the prize winner.

We are therefore prepared to receive competitions in the following subjects :—

SUBJECT I.

The best method of pumping or otherwise lifting or forcing warm aqueous hydrochloric acid of 30 deg. Tw.

SUBJECT II.

The best method of separating or determining the relative quantities of tin and antimony when present together in commercial samples.

On each subject, the following prizes are offered :—

One first prize of five pounds, one second prize of one pound, five additional prizes to those next in order of merit, consisting of a free copy of the *Chemical Trade Journal* for twelve months.

The competition is open to all nationalities residing in any part of the world. Essays must reach us on or before April 30th, for Subject II, and on or before May 31st, in Subject I. The prizes will be announced in our issue of June 28th.

We reserve to ourselves the right of publishing the contributions of any competitor.

Essays may be written in English, German, French, Spanish or Italian.

CHESHIRE COUNTY COUNCIL AND THE POLLUTION OF THE IRWELL AND MERSEY.

THE Local Government Board, having received notice from the Cheshire County Council of their intention to apply for a provisional order authorising them to constitute a joint committee representing the several administrative counties on the watersheds of the rivers Irwell and Mersey for enforcing the provisions of the Rivers Pollution Prevention Act, 1876, have requested the county authority to supply them with full information as to the grounds on which the application is based. A statement has been drawn up for the approval of the County Council at its meeting at Chester, on Thursday next, setting forth that the grounds on which the application is made are that the Irwell and Mersey and their tributaries, in consequence of their pollution by sewage and the refuse from manufactories and other places, are offensive both to smell and to sight, and are often injurious to health. Within two years it is expected, the Manchester Ship Canal will be filled with water from these rivers and their tributaries. The Ship Canal will not be a continuous stream, but a succession of ponds with a stream running through them, and the danger to health arising from those rivers will be undoubtedly increased. The Cheshire County Council are concerned directly with the state of the waters of the Ship Canal as it passes through a large number of the Cheshire townships, and indirectly as the communication between Manchester and Salford and Cheshire is so great and continuous that any epidemic in those cities would certainly spread through large portions of Cheshire. There are more than 70 local authorities in the watersheds of the Mersey and Irwell, and only about one-half of them have completed arrangements for constructing sewage works. It is therefore (in the opinion of the Council) necessary that further steps should be taken to enforce the Rivers Pollution Prevention Act, and this will best be done by the appointment of a joint committee, enabling the local authorities to act in unison. The advantage of a joint committee would also, it is conceived, be very great in dealing with the pollution of streams by manufacturers. While it is desirable that the waters of the Irwell and Mersey shall be pure, it is not desirable that the industries which have grown up on their banks should be damaged or destroyed, and the employment of labour thereby curtailed or stopped. It is also most undesirable that the manufacturers within the area of one authority should be dealt with differently from those within another. It appears to the County Council that a joint committee would be much more independent, and would treat all dwellers and manufacturers equally and justly, and would so act that the industries of this wide district should not be injured, nor the real prosperity of its inhabitants be prejudiced. The authorities affected by the proposed scheme are the County Councils of Chester, Lancaster, and Derby, and the county boroughs of Bolton, Bury, Rochdale, Oldham, Salford, Manchester, and Stockport.

THE SALT UNION.—The Directors of the Salt Union, Limited, recommend a dividend at the rate of 10 per cent. per annum for the last half year, carrying forward £41,000.

Market Reports.

REPORT ON MANURE MATERIAL.

The business during the past week has been on a very moderate scale, both in nitrogenous and phosphatic manure material, but at the close there seems to be more enquiry, and prices are perhaps the turn in favour of sellers.

Charleston River phosphate rock is still quoted 11d. per cent. for cargoes c.i.f. to U. K. Ports, but ½d. less would probably be business for land rock, early shipment. Quotations for Somme and Belgian remain without alteration, the business done for U. K. being only on a very moderate scale, but we hear of large transactions on the Continent at full prices. United Kingdom buyers do not seem inclined to purchase for next season's requirements on basis of present prices; they will wait and see what may be the probable bulk of the new sources of phosphate supply, and whether easier freights may not favour lower prices later on. Quotations for Canadian rock remain nominal, and we do not hear of any transactions.

There has been nothing done in River Plate cargoes of bones, the smaller ones offering being held for extreme prices, and the larger ones being rather late for the spring requirements. £5. 10., delivered good U. K. port, is perhaps the nearest value for December shipment.

After sales of East Indian bone meal at £5. 3s. 9d., afloat or for shipment, the market has recovered somewhat; and we don't now hear of any sellers under £5. 5s. 800 tons arrived at Liverpool will be stored without being offered at present, and further shipments about due will also be stored on arrival. So that the quantity available will be in very moderate compass, and sellers seem inclined to demand some advance on recent quotations. We do not hear of any recent business in crushed East Indian bones, but £5. 3s. 9d. would be accepted for February-March steamer shipment to Liverpool. There are buyers at more money to other ports.

Nitrate of soda is very weak in all positions. Spot price cannot be quoted above 8s. 3d. to 8s. 4½d., according to quality. Due cargoes are worth barely 8s. 4½d., and February steamer shipment may be bought at 8s. 3d. per cwt. ex quay. The combination arrangements do not seem to progress very rapidly, and, meantime, holders are getting frightened by the very heavy visible supply.

Superphosphates remain without alteration. 50s. per ton in bulk, delivered f.o.r. at works.

823 bgs. rough River Plate dried blood sold at 10s. 3d. per unit. of ammonia per ton, ex quay Liverpool, and 400 bags fine quality at 10s. 6d., which price would be accepted for 900 bags similar quality now landing at Liverpool.

METAL MARKET REPORT.

	Last week.		This week.
Iron	51/10½	....	53/8½
Tin	£91 0 0	....	£91 5 0
Copper	47 17 6	....	47 17 6
Spelter	21 10 0	....	23 0 0
Lead	12 12 6	....	13 5 0
Silver	44d.	....	43¾

COPPER MINING SHARES.

	Last week.			This week.	
Rio Tinto	$14\frac{7}{8}$	15	....	$15\frac{1}{2}$	$15\frac{5}{8}$
Mason & Barry	$6\frac{1}{4}$	$6\frac{1}{2}$	....	$6\frac{5}{16}$	$6\frac{7}{16}$
Tharsis	$4\frac{7}{16}$	$4\frac{9}{16}$	....	$4\frac{9}{16}$	$4\frac{11}{16}$
Cape Copper Co.	$3\frac{5}{16}$	$3\frac{7}{16}$	....	$3\frac{7}{16}$	$3\frac{9}{16}$
Namaqua	2	$2\frac{1}{4}$	....	2	$2\frac{1}{4}$
Copiapo	$2\frac{7}{16}$	$2\frac{9}{16}$	....	$2\frac{9}{16}$	$2\frac{11}{16}$
Panulcillo	1	$1\frac{1}{8}$	....	1	$1\frac{1}{8}$
New Quebrada	$\frac{5}{8}$	$\frac{5}{8}$	....	$\frac{5}{8}$	$\frac{7}{8}$
Libiola	$3\frac{1}{8}$	$3\frac{1}{2}$	....	3	$3\frac{1}{4}$
Tocopilla	1/6	2/-	....	1/6	2/-
Argentella	6d.	1/-	....	6d.	1/-

MISCELLANEOUS CHEMICAL MARKET.

There is a general firmness of tone in all products of the alkali trade. Manufacturers all round are busy, and with the continued scarcity and high prices of fuel the values of all products are tending upwards. Saltcake has become scarce, and is difficult to obtain for early deliveries; 32s. 6d. has been paid for the same in bulk at makers' works. Bleaching powder is firm at £5. 7s. 6d. to £5. 10s., free on rails, and manufacturers are well filled with orders. Caustic soda has again recovered in price, and is firm for spot delivery at £9. f.o.b. for 70%. Makers are still unwilling to sell forward at any concession on present values. Soda ash is firm both for prompt and future deliveries. Chlorate of potash nominally 4¾d. per lb. on the spot, but higher prices are spoken of as probable. Sulphur is in moderate request only. Vitriol still finds ready outlet at current rates. Sulphate of copper is very scarce, and for prompt delivery commands £26. 10s.; forward delivery quoted easier. Aniline oil and salt are being privately sold at easier rates. Alum generally is firmer, better prices being obtained for export trade. Brown lime acetate is rather lower in price and more freely offered; grey, scarce and firm. Potash caustic and carbonate can be had on slightly easier terms for forward shipments. Arsenic is firm, and orders are being freely placed at current rates. Trade in the finer chemicals for textile trades is still quiet, and there is consequently but little movement in values.

THE LIVERPOOL MINERAL MARKET.

The upward progress of prices has further continued, and there is every prospect of same being maintained. Manganese: Arrivals have been very limited, whilst stocks are practically exhausted; prices, therefore, for prompt and forward delivery have still further considerably advanced. Magnesite: Raw lump quiet, raw ground, £6. 10s., and calcined ground, £10. to £11. Bauxite (Irish Hill brand) continues in strong demand, and brings full prices. Lump, 20s.; seconds, 16s.; thirds, 12s.; ground, 35s. Dolomite, 7s. 6d. per ton at the mine. French Chalk: Arrivals have somewhat improved, but the largest proportion has gone into consumption, and prices remain strong, especially for G.G.B. "Angel-White" brand—90s. to 95s. medium, 100s. to 105s. superfine. Barytes (carbonate), steady; selected crystal lump scarce at £6.; No. 1 lumps, 90s.; best, 80s.; seconds and good nuts, 70s.; smalls, 50s.; best ground, £6.; and selected crystal ground, £8. Sulphate quiet; best lump, 35s. 6d.; good medium, 30s.; medium, 25s. 6d. to 27s. 6d.; common, 18s. 6d. to 20s.; ground best white, G.G.B. brand, 60s.; common, 45s.; grey, 32s. 6d. to 40s. Pumicestone steady; ground at £10., and specially selected lump, finest quality, £[...] tinues in good demand at full prices. Bilbao [...] at 9s. to 10s. 6d. f.o.b.; Irish, 11s. to 12s. [...] 16s. to 20s. Purple ore unchanged. Sp[...] ore commands ready sale at full prices. Emery [...] continued to be enquired for, bringing full pric[...] quoted at £5. 10s. to £6., and smalls £5. t[...] earth steady; 45s. to 50s. for best blue an[...] palpable ground, £7. Scheelite, wolfram, tu[...] tungsten metal continue inquired for. Chrome m[...] Tungsten alloys 2s. per lb. Chrome ore: High [...] for at fair prices. Antimony ore and metal hav[...] Uranium oxide, 24s. to 26s. Asbestos: Best r[...] brown grades, £14. to £15. Potter's lea[...] firm; smalls, £14. to £15.; selected lu[...] Calamine: Best qualities scarce—60s. to 80s. [...] sulphate (celestine) steady, 16s. 6d. to 17s. [...] £15. to £16.; powdered (manufactured), £1[...] spar: English manufactured, old G G.B. brand [...] brings full prices; 50s. for ground English. [...] 50s.; fluorspar, 20s. to £6. Bog ore in fair [...] 25s. Plumbago: Notwithstanding large new [...] unchanged; Spanish, £6.; best Ceylon lump [...] Italian and Bohemian, £4. to £12. per ton. Fre[...] continues scarce on spot—20s. to 22s. 6d. Fer[...] easily at advanced figures. Ground mica, £50. [...] offering—common, 18s. 6d.; good medium, 22s [...] 30s. to 35s. (at Runcorn).

WEST OF SCOTLAND CHEM[...]

GLA[...]

Caustic soda on the market has taken the turn d[...] quence of slackened demand, but makers are un[...] this as legitimate, and endeavour to stick to the [...] Bleaching powder, also, is slightly weaker, an[...] getting into stock on terms the least thing more [...] pared with those procurable a week ago. Crys[...] same tendency, and are easier. On the other han[...] is firmer, and rises have taken place in the marke[...] cake (which is scarce), alum, bicarbs, and sulphat[...] mates, at fixed prices, have been in somewhat impr[...] the week Sulphate of ammonia is very idle indeed, [...] tions are concerned. Eleven pounds, seventeen and si[...] and makers, being for the most part already fairly we[...] duction, refuse to do for less. Topmost offers for [...] £11. 16s. 3d. without result; and there are still [...] fixtures to note. Paraffins are generally fir[...] been put now on the same price footing as h[...] lb. Chief prices current are:—Soda crystals, [...] Tyne; alum in lump £5. less 2½% Glasgo[...] refined £30., and boracic acid, £37. 10s. net Glas[...] less 5% Tyne; caustic soda, white, 76°, £10. 15[...] 60/62°, £8., and 60/62° cream, £7. 10s., all le[...] bicarbonate of soda, 5 cwt. casks, £5. 7s. 6d., 1 cw[...] net Tyne; refined alkali 48/52°, 1½d., less 2[...] 29s. to 31s.; bleaching powder, £5. 10s. to £5. [...] Glasgow; bichromate of potash, 4d., and of sod[...] to Scotch and English buyers respectively; chlor[...] less 5% any port; nitrate of soda 8s. 4½d. to 8s. [...] monia, £11. 15s. to £11. 17s. 6d. f.o.b. Leith; [...] 2nd white, £37. and £35., less 2½% any [...] copper, £26. 10s. less 5% Liverpool; paraffin scal[...] 2½d.; paraffin wax, 120°, semi-refined, 3[...] (naphtha), 9d.; paraffin oil (burning), 6¾d. to [...] ditto (lubricating), 865°, £5. 10s. to £6., 8[...] 890/895°, £7. 10s. to £8. Last week's imports [...] were 35,906 bags.

THE LIVERPOOL COLOUR M[...]

COLOURS quiet; prices unaltered. Ochres: [...] at £10., £12., £14., and £16.; Derbyshire, 5[...] best, 50s. to 55s.; seconds, 47s. 6d.; and common, 18[...] 40s. to 45s.; French, J.C., 55s., 45s. to 60s.; M[...] Umber: Turkish, cargoes to arrive, 40s. to 50[...] to 55s. White lead, £21. 10s. to £22. Red lea[...] of zinc: V.M. No. 1, £25.; V.M. No. 2, £[...] £6. 10s. Cobalt: Prepared oxide, 10s. 6d.; [...] 6s. 6d. Zaffres: No. 1, 3s. 6d.; No. 2, 2s. 6d. [...] white, 60s.; good, 40s. to 50s. Rouge: Best, £24. [...] 9d. per lb. Drop black, 25s. to 28s. 6d. Oxide o[...] £10. to £15. Paris white, 60s. Emerald green, [...] shire red 60s. Vermillionette, 5d. to 7d. per lb.

[illegible]

TAR AND AMMONIA PRODUCTS

[illegible]

Gazette Notices.

PARTNERSHIPS DISSOLVED

[illegible]

The Bankruptcy Act, 1883.

[illegible]

Adjudications.

[illegible]

New Companies.

[illegible]—This company was registered on [illegible] shares, to acquire and work man[illegible] Auckland, New Zealand, and in par[illegible] registered agreement, certain rights and [illegible] Welby. The subscribers are:—

	Shares.
[illegible]	1
[illegible] manufacturer	1
[illegible]	1
[illegible] manufacturer	1
[illegible]	1
[illegible]facturer	1
[illegible]	1

[illegible]—This company was registered on [illegible] shares, to acquire as a going con[cern] [illegible] Red [illegible]-yard, Redcliff-street, Bristol, [illegible] Strachan, senior, the proprietor of [illegible] importers, and manufacturers of oil, [illegible] chemicals, drugs, lubricants, illuminants,

	Shares.
[illegible]-street, Bristol, oil merchant	1
[illegible]-street, Bristol, oil merchant	1
[illegible]-street, Bristol, oil merchant	1
[illegible] Bristol, carriage builder	1
[illegible] Bristol, solicitor	1
[illegible]-street, Bristol, accountant	1
[illegible] commercial traveller	1

[illegible] LIMITED—This company was registered on the [illegible] £[illegible] shares, to acquire as a going concern, [illegible] and colour manufacturer, carried on by [illegible]-street, Manchester and Crumpsall-vale, near Man[chester] [illegible] and Co. The subscribers are:—

	Shares.
[illegible] Manchester, chemical manufacturer	1
[illegible] Manchester, clerk	1
[illegible] Germany	1
[illegible]	1
[illegible] Germany	1
[illegible] Manchester, solicitor	1
[illegible]	1
[illegible]	1
[illegible]	1

[illegible] COMPANY, LIMITED.—This company was registered [illegible] £[illegible], in £1 shares to acquire from James [illegible] certain electric light machinery and plant, [illegible] an electric light company in all branches. The [subscribers are]:—

	Shares.
[illegible] Fareham	1
[illegible] Fareham	1
[illegible] Fareham	1
[illegible] Fareham	1
[illegible] Fareham	1
[illegible] Fareham	1
[illegible] New Bond-street	1

[illegible] COMPANY, LIMITED.—This company was registered on [illegible] in £1 shares, to carry on the business of elec[illegible] branches. The subscribers are:

	Shares.
[illegible] Eaton-place, electrical engineer	73
[illegible] Inverness-terrace, W., electrician	2
[illegible]	2
[illegible] Victoria-street, S.W., electrical engineer	5
[illegible] Victoria street, S.W., electrical engineer	5
[illegible] Lane Fox Pitt, [illegible] Grosvenor-gardens	12
[illegible] Hendon, barrister	1

[illegible] COMPANY, LIMITED.—This company was registered on the [illegible], with a capital of £[illegible], in £5 shares, to acquire the Leeswood [illegible] Works, Leeswood, Flint, and to carry on the business of a smelting, [illegible] and metallurgical company. The subscribers are:—

	Shares.
[illegible] Cory, Swansea, colliery owner	1
[illegible] Swansea, merchant	1
[illegible] Swansea, shipowner	1
[illegible] H. Burgess, Swansea, shipbroker	1
[illegible] R. Leaver, C.E., Swansea	1
W. [illegible] Richards, Swansea, colliery proprietor	1
[illegible] Parry, Mold, colliery proprietor	1

The Patent List.

This list is compiled from Official sources in the Manchester Technical Laboratory, under the immediate supervision of George E. Davis and Alfred R. Davis.

APPLICATIONS FOR LETTERS PATENT.

Benzoline Lamps. J. F. Barker. 1,779. February 3.
Hot Water Heater. T. Thorburn. 1,781. February 3.
The Production of Ozone. T. A. Garrett. 1,784. February 3.
Automatic Flushing Syphon. T. Thomas. 1,789. February 3.
Extraction of Gold and Silver. C. Pfeiffer. 1,794. February 3.
Mechanical Forced Draught Lamp. T. M. Thompson. 1,796. Feb. 3.
Syphons. F. Bolens. 1,801. February 3.
Glass Moulding. L. Appert. 1,803. February 3.
Heating by Means of Gas. P. Scharf. 1,804. February 3.
Manufacture of Steel. M. R. Conley and J. H. Lancaster. 1,806. Feb.
Yellow Dye. O. Imray. 1,808. February 3.
Dyeing. R. Holliday. 1,811. February 3.
Dyeing. R. Holliday. 1,812 February 3.
Dyeing. R. Holliday. 1,813 February 3.
Purifying Sewage. G. W. Ewens. 1,816. February 3.
Boiler Furnaces. A. Jobson. 1,817. February 3.
Brick Kilns.—(Complete Specification). H. Gurney. 1,820. February 3.
Thermometers for Brewers. P. S. Conron. 1,823. February 3.

Boiler Flues. S. K. Barnes. 1,827. February 3.
Azo Dyes. B. Willcox. 1,828. February 3.
Recovery of Float Gold. E. L. Mayer and J. G. Lorrain. 1,842. Feb. 4.
Filtering Apparatus. W. Holmes. 1,845. February 4.
Distillation of Ammonia. A. R. Davis. 1,848. February 4.
Construction of Boilers. C. Arnold and H. Arnold. 1,858. February 4.
Electrical Apparatus. J. A. McMullen. 1,867. February 4.
Refrigerating Machinery.—(Complete Specification). T. B. Lightfoot 1,875. February 4.
Air Compressors.—(Complete Specification). A. Riedler. 1,877. Feb. 4
Electrical Insulators.—(Complete Specification). J. B. Williams. 1,878. Feb. 4.
Generating and Utilising Combustible Vapours. B. J. B. Mills. 1,883. February 4.
Pyrotechnic Powders.—(Complete Specification). P. Jensen. 1,884. Feb. 4.
Air Pumps.—(Complete Specification). A. Berrenberg. 1,893. February 4.
Enamelling Metals. A. E. Robinson and W. H. Wheatley. 1,894. Feb. 4.
Insulating Composition. E. Fahrig. 1,897. February 4.
Treatment of Scrap Tin Plates. W. P. Thompson. 1,913. February 5.
Manufacture of Alum. F. M. Spence and D. D. Spence. 1,914. February 5.
Brewing Coppers. C. Gibbs. 1,921 February 5.
Paving and Building Stone. C. Fifield. 1,934. February 5.
Oil Lamps. T. B. Burns. 1,935. February 5.
Recuperative Lamps. E. See. 1,936 February 5.
Oil Vapour Motors. C. D. Abel. 1,943. February 5.
Water Softening Apparatus.—(Complete Specification). J. Wright. 1,946. February 5.
Explosives. H. S. Maxim. 1,951. February 5.

Lamps. T. Bass and T. Preece. 1,957. February 5.
Yellow Dyestuffs. A. Bang. 1,962. February 6
Gas Generating Apparatus. W. O. Kibble. 1,964. Febr
Pressure Filters. G. Sellars. 1,970. February 6.
Electro-plating Metals. S. Wohle. 1,987. February 6.
Lucifer Matches. W. P. Thompson 1,995. February 6.
Brewing. J. F. Littleton. 1,996 February 6.
Manufacture of Aluminium. G. Shenton. 2,002. Februa
Aerated Drinks. J. Wood. 2,005. February 6.
Air Propellers and Ventilating Fans. W. Sayer. 2,014.
Disinfecting Candles. H. Oppenheim and G. H. Elliott.
Apparatus for Heating Liquid and Gaseous Bodies. 2,055. February 7.
Smelting Tin Ores. A. J. Campion. 2,057. February 7.
Boilers Used in the Manufacture of Paper Pulp. C. February 8.
Furnaces for Burning Liquid Fuel. C. A. Sahlström and February 8.
Water and Gas Pipes. E. S. Barlow. 2,077. February 8.
Storage Batteries. J. C. Chamberlain. 2,080. February 8.
Storing Petroleum. W. P. Thompson. 2,089. February 8.
Secondary Batteries. W. P. Thompson. 2,091. February
Air Pumps. R. Haddan. 2,094. February 8.
Treatment of Margarine. P. Wild. 2,104. February 8.
Switches for Electro-Motors. C. T. Whitmore and W. February 8.
Treatment of Incandescents. C. Dellwik. 2,110. Februa

IMPORTS OF CHEMICAL PRODUCTS
AT
THE PRINCIPAL PORTS OF THE UNITED KINGDOM

LONDON.
Week ending February 14th.

Arsenic—
Germany, £227 Williams Torrey & Co.

Antimony—
Italy, 10 t. H. R. Merton & Co.
Bastia, 94 Johnson, Matthey, & Co.
Portugal, 2 J. Ford & Co.
Melbourne, 6 L. & I. D. Jt. Co.

Ammonia—
U. S. A., £134 Walsh, Lovett, & Co.

Alkali—
Pearl Ash
France, 200 c. F. W. Berk & Co.
Holland, 200 c. Taylor, Sommerville, & Co.

Acetate of Lime—
U. S. A., £202 W. Thatcher.

Acetic Acid—
Holland, 47 pkg. Beresford & Co.
Germany, 104 pkgs. A. & M. Zimmermann

Aluminite—
N. S. Wales, £32 W. H. Cole & Co

Barytes—
Germany, 24 bgs. W. Harrison & Co.
Holland, 44 ,,
Belgium, 91 cks. ,,
Germany, 99 J. Matton
,, 21 O'Hara & Hoar
Belgium, 500 bgs. A. Zumbeck & Co.

Boracic Acid—
Italy, 20 pkgs. J. Puddy & Co.
,, 35 pkgs. Howards & Sons

Bone Ash—
France, 3¼ c. Arnati & Harrison

Barium Chloride—
Germany, £55 Petri Bros.
,, 15 W. J. Cook
,, 40 pkgs. Petri Bros.

Bone Meal—
E. Indies, 100 t. Bryden & Co.

Brimstone—
France, 10 t. J. Parbury & Co.
Italy, 350 W. C. Bacon & Co.

Cream Tartar—
Holland, 4 pgs. Middleton & Co.
Spain, 9 Webb, Warter & Co.
Germany, £60 Charles & Fox
,, 189 A. & M. Zimmermann
Italy, 16 pgs. Thames S. Tug Co.
France, 10 Gellatly & Co.
Holland, 2 W. C. Bacon & Co.
Italy, 31 Thames S. Tug Co

Copper Pyrites—
Spain, 855 t. Forbes, Abbott, & Co.
La Laja, 772 G. Ward & Son

Caoutchouc—
Columbia, 36 c. Mildred, Goyeneche & Co.
E. Africa, 9 L. & I. D. Jt Co.
Strts. Settlements, T. Merry & Sons
Madagascar, 17 c. Kebbel, Son, & Co.
France, 2 ,,
Strts. Settlmts., 31 Middleton S. S. Co.
Belgium, 28 T. H. Lee
E. Africa, 36 L. & I. D. Jt. Co.
,, 50 T. Ronaldson & Co
E. Indies, 14 L. & I. D Jt. Co.
Madagascar, 6½ Townsend Bros.

Cinchona Bark—
Peru, 430 bls. Patry & Pasteur

Dyestuffs—
Indigo
E. Indies, 151 chts. L. & I. D Jt Co.
,, 10 Parsons & Keith
,, 14 W. Brandt & Sons
,, 35 London & Hauseatic Bank
,, 15 Elkan & Co.
,, 2 cs. L. & I D. Jt. Co.
,, 10 chts. M. Smith
,, 109 8 bxs. G. Ward & Sons
,, 78 2 E. Elkan & Co.
,, 9 1 F. W. Heilgers & Co.
,, 33 1 J. Owen
,, 107 F. Stahlschmidt
China, 5 chts. L. & I. D. Jt. Co.
E. Indies, 27 Anderson Bros.
,, 143 L. & I. D. Jt. Co.
,, 130 Arbuthnot, Latham, & Co.
,, 11 London & Hauseatic Bank
,, 104 T. Barlow & Co.
,, 7 Hatton, Hall, & Co.
,, 15 F. W. Heilgers & Co.
,, 94 J. Owen
,, 16 Kaul & Haeulein
,, 47 R. & J. Henderson
,, 14 Phillipps & Graves
,, 5 ,,
,, 449 T. Ronaldson
,, 16 Pickford & Co.
,, 36 L. & I. D. Jt. Co.
,, 60 F. Stahlschmidt
,, 12 Benecke, Souchay & Co.
,, 10 Stansbury & Co.
,, 8 Kleinwort & Sons
,, 20 Hoare, Millar, & Co.
,, 22 W. Brandt & Sons
,, 5 F. Huth & Co.
,, 2 J. Owen
,, 5 F. Huth & Co.

Gambier
E. Indies, 20 pkgs. Brinkmann & Co.
,, 83 T. J. & T. Powell
Italy, 630 Adamson, Gilfillan & Co.

E. Indies, 92 L. & I. D. Jt. Co.
,, 1,282 Beresford & Co.

Sumac
Italy, 50 pgs. J. Kitchin
Italy, 300 Soundy & Son
,, 210 Beresford & Co.
,, 1,050 J. Kitchin
Italy, 50 pgs. W. France & Co.
,, 100 R. Von Glehn & Sons

Valonia
A. Turkey, 41 t. J. Graves
,, 25½ J. Graves
,, 50 Hicks Nash & Co.

Extracts
France, 100 pgs. Levenstein & Sons
Germany, 50 Fleming Oil Chem. Co.
U. S. A. 80 pgs. Donaldson & Co.
France, 80 Levenstein & Sons
,, 50 G. Ward & Son
,, 117 Burt, Boulton, & Co.
Holland, 56 Burt, Boulton, & Co.
U. S. A., 25 J. Graves
France, 31 M'Cracken, Fenwick, & Co.

Tanners' Bark
Belgium, 49 t. J. Kitchin
,, 5 H. Heule's succrs.

Cochineal
Canaries, 17 pgs. Bruce & White
Germany, 10 Craven & Co.

Cutch
Strts. Settlmts, 300 pgs. Anderson, Weber, & Smith

Logs
B. W. Indies, 58 t. Anderson, Anderson, & Co.

Fustic
B. W. Indies, 96½ t. Anderson, Anderson & Co

Annatto
Ceylon, 27 pgs. Hoare, Wilson, & Co.

Orchella
Ceylon, 18 pgs. E. Barber & Co.
France, 1 L. & I. D. Jt. Co.
Portugal, 32 pgs. N. Smith

Saffron
Spain 1 pgs. Props. Chamb. Whf.
,, 1 T. Allen

Esparto—
Tripoli, 4,000 brls. E. & E. Arbib

Farina—
Holland, 286 pgs. Dunlop Bros.
,, 200 Prop. Scott's Whf.
Germany, 100 H. Lorenz
,, 70 L. Sutro & Co.
,, 20 Fellows, Morton, & Co.
,, 50 Anderson, Weber, & Smith
,, 100 J. Anderson & Co.
,, 200 C. F. Webster & Co.
,, 100 J. Anderson & Co.
,, 100 bgs. ,,

Glycerine—
Germany, £115
Holland, 100 F.
,, 82
Germany, 64
,, 860
,, 13
Holland, 15
Germany, 63
,, 56
,, 30

Glucose—
Germany, 25 pgs.
,, 25
,, 50
,, 75
,, 200
U. S. A., 350 57
Germany, 200
,, 25
,, 500
,, 200
U. S. A., 50
Germany, 25
,, 400
,, 68
,, 110
,, 32
,, 200
,, 25

Gutta Percha—
E Indies, 73 c.
,, 59
,, 952 Kalt

Galls—
A. Turkey, 2 bgs.
,, 20 K
China, 50
A. Turkey, 10
,, 28 Ke

Horns—
France, 200 c.
E. Indies, 420
U. S. A., 26
China, 68
France, 200
U. S. A., 71
E. Indies, 16½

Horn Tips—
France, 194 c.

Manure—
Phosphate Rock
France, 375 t. A. Hunter & Co.
" 250 "
Belgium, 50 B. Jacob & Sons
U.S.A., 5½ C. A. & H. Nichols
France, 105 A. Hunter & Co.
" 212 Lawes Manure Co.
" 200 Anglo Contnl. Guano Works Agency
Germany, 5 F. W. Berk & Co.
France, 110 A. Hunter & Co.

Magnesium—
Germany, 40 pgs Phillipps & Graves

Naphtha—
Germany, 19 brls. Stein Bros.
Holland, 8 drms. Domeier & Co.
" 9 "

Nitre—
Holland, 1 csk. Spies Bros.

Ochre—
France, 10 pgs. E. & G. Hirsch & Co.
" 59 W. Harrison & Co.
U. S. A., 200 brls. G. Steinhoff
France, 40 cks. W. Harrison & Co.

Petroleum—
U. S A., 500 brls. R. H. Annison
France, 118 Burt, Boulton & Co.
U. S., 3,400 J. H. Usmar & Co.
" 35 A. Palliser
Germany, 50 A. Brown & Co.
S. Russia, 769 2,807 gls. F. Huth & Co.
U. S., 1,709 t. 80,000 C. T. Bowring
Holland, 38 dms. Burt, Boulton, &
U. S. A., 150 brls. H. Hill & Sons
" 9 Major & Field

Pitch—
Russia, 28 brls. Vivian, Younger, & Co.

Plumbago—
Ceylon, 45 cks. Doulton & Co
" 87 Prop. Chamb. Whf.
" 191 Lambert & Strong
" 20 Thames S. Tug Co.
Holland, 200 Brown & Elmslie
Ceylon, 149 Prop. Chamb. Whf.

Pyrites—
Spain, 300 t. Forbes, Abbott, & Co
" 28 T S. Whitehead & Chambers
Pomaron, 1,120 t. Mason & Barry

Potassium Carbonate—
France, 5 pgs J. A Reid
Germany, 5 Andorsen, Becker, & Co.
France, 10 C. Pawson

Potassium Prussiate—
Germany, 9 pgs. A. & M. Zimmermann
" £112 T. H. Lee

Potassium Sulphate—
France, 10 pgs. Weatherly, Mead, & Co.

Paraffin Wax—
U. S. A., 265 brls. H Hill & Sons
" 170 G. H. Frank
" 2,051 J. H. Usmar & Co.

Paraffin—
Canada, 200 brls. G. H. Frank

Rosin—
U. S., 439 brls. F. & S. Chiesman & Co.

Red Lead—
Germany, 160 c. Levin Rowehl

Stearine—
Belgium, 16 cks. Perkins & Homer
Germany, 330 Craven & Co.
" 100 W. Balchin
France, 293 sks. J. Goddard & Co.
New Zealand, 5 cks. Van Geelkerken & Co.

Sodium Acetate—
France, £150 Weatherly, Mead, & Co.
" 150 "
" 210 J. Burnett & Sons

Saltpetre—
Germany, 43 cks. Craven & Co.
E. Indies, 1.736 bgs. L. & I. D. Jt. Co.
" 615 Henckett du Buisson
Germany, 60 cks. 250 bgs. C. Wimble & Co.
" 100 T. Merry & Sons
" 20 Baiss, Bros., & Co.

Sugar of Lead—
Germany, 33 pgs. C. J Capes & Co.
" 74 Beresford & Co.
Holland, 1 Webb, Warter, & Co.

Sodium Sulphate—
Holland, £60 Prop. Dowgate Dk.
Belgium, 50 Beresford & Co.

Sodium Nitrite—
Germany, £100 Kaltenbach & Schmitz

Salammoniac—
Holland. £154 Barr, Moering, & Co.

Tartaric Acid—
Holland, 12 pgs. W. C. Bacon & Co.
" 45 Webb, Warter, & Co.
France, 5 "
Holland, 31 Webb, Warter, & Co.

Tallow—
New Zealand, 171 cks. Anning & Cobb
" 31 Culverwell, Brooks, & Co.
Belgium, 23 cs. Sawer, Mead, & Co.
Portugal, 8 cks. Goad, Rigg, & Co.

Tartar—
Italy, 170 pgs. Seager, White, & Co.

Ultramarine—
Holland, 9 cks. Ohlenschlager Bros.
U. S. A., 12 hds. J. Melvin
Germany, 20 cks. Seaward Bros.
" 27 H. Lambert
" 4 R. Baelz
" 11 Spies, Bros., & Co.
" 2 Henderson, Craig, & Co.
Holland, 45 Ross & Deering
" 1 Keller & Co.
France, 7 pgs. Flageollet & Co.
Holland, 6 Ohlenschlager Bros.
" 7 cks. R. Baelz & Co.
Germany, 1 L. & I. D. Jt. Co.
" 5 T. H. Lee
France 4 Flageollet & Co.
Holland, 8 pgs. Hernu, Peron, & Co.
" 41 cs. G. Steinhoff
" 2 R. Baelz & Co.
" 37 Spies, Bros., & Co.

Varnish—
U.S.A., 15 brls. J. Stutchbury & Sons

Wood Pulp—
Sweden, 81 pgs. M. Dk. Co.
Norway, 32 F. E. Foulger
" 160 Howell & Co.
Sweden, 41 M. Dk. Co.
" 81 A. Afurdson
" 200 J. Dickinson & Co.
" 68 C. S. Lovell
" 164 Taylor, Sommerville, & Co.
Norway, 240 "
" 119 H. Seiger
Germany, 500 W. Balchin
" 122 Tough & Henderson

White Lead—
France, 9 cks. Beresford & Co.
Belgium, 20 E. & G. Hirsch & Co.
" 40 T. & W. Farmiloe
Germany, 53 "
" 107 Randall Bros.
" 8 Elkan & Co.
" 10 Petri Bros.
France, 21 Harrison & Co.
Germany, 19 D. Storer & Sons
Belgium, 59 T. & W. Farmiloe
" 7 Leach & Co.
Holland, 36 A. & M. Zimmermann

Wax—
France, 31 brls. S. Frankenstein & Co.
Morocco, 9 cks. A. Afriat
Germany, 6 brls. M. D. Co.
" 50 H. Birdseye & Co.
France, 1 Kebbel, Son, & Co
Germany, 4 pgs. L. & I. D. Jt. Co.
" 17 Grosscurth & Luboldt
Holland, 10 J. Hall, jun. & Co.
" 29 brls. Perkins & Homer

Zinc Oxide—
Germany, 82 cks. M. Ashby
" 100 J. Matton
" 3 L. & I. D. Jt. Co.
" 9 T. H. Lee
" 5 G. Steinhoff
E. Indies, 1 cs. L. & I. D. Jt. Co.
Holland, 5 G. Rahn & Co.

HULL.

Week ending Feb. 14th.

Alkali—
Dunkirk, 22 drms. Wilson, Sons, & Co.

Alumina—
Rotterdam, 40 cks. W. & C. L. Ringrose

Aluminium—
Antwerp, 1 cs. Bailey & Leetham

Albumen—
Hamburg, 14 cks. Furley & Co.

Alum Earth—
Hamburg, 19 cks. Bailey & Leetham

Barytes—
Bremen, 23 brls. Tudor & Sons
Rotterdam, 37 brls. T. W. Flint & Co.

Brimstone—
Catania, 127 bgs. 128 ckr.

Bone Meal—
Bombay, 5,757 bgs.

Camphor—
Hamburg, 2 cks Wilson, Sons, & Co.

Chemicals (*otherwise undescribed*)
Hamburg, 278 pgs. Wilson, Sons, & Co.

Chemicals—
Christiania, 12 cks Wilson, Sons, & Co.
Dunkirk, 43 "
Rotterdam, 4 1 cby. Geo. Lawson & Sons
Stettin, 23 cks. Wilson, Sons, & Co.
Danzig, 4 "
Bari, 20 "
Bremen, 6 4 cs. Veltmann & Co.
" 1 J. Pyefinch & Co.
" 5

Copperas—
Bremen, 22 cks.

Caoutchouc—
Hamburg, 3 cs. C. M. Lofthouse & Co.

Dyestuffs—
Alizarine
Rotterdam, 15 cks. W. & C. L. Ringrose
Rotterdam, 26 cks. Hutchinson & Son
" 43 3 cs. W. & C. L. Ringrose
Dyewood
Hamburg, 1 ck. 50 bgs. Wilson, Sons, & Co.
Dyestuffs (*otherwise undescribed*)
Palermo, 400 bgs. Wilson, Sons, & Co.
Extracts
Rouen, 42 cks. Rawson & Robinson
" 40 Wilson, Sons, & Co
Fiume, 750 brls. "
Dunkirk, 55 cks. "
Fiume, 912 "
Madder
Rotterdam, 6 cks. Hutchinson & Son
Myrabolams
Bombay, 8 463 bgs.
Sumac
Fiume, 20 bgs. Wilson, Sons, & Co.
Palermo, 5 bls. "
" 300 bgs. "
" 800
Tannin
Rotterdam, 20 cks. Hutchinson & Son
Valonia
Smyrna, 100 bgs.

Drugs—
Bergen, 19 pkgs. Wilson, Sons, & Co.
Trieste, 13 "
Venice, 2 cs. "

Farina—
Harlingen, 1,020 bgs.
Hamburg, 10
Stettin, 560 Wilson, Sons, & Co
Danzig, 695 "

Glycerine—
Amsterdam, 21 pkgs. W. & C. L. Ringrose
Hamburg, 3 crbys. 1 drum.

Glue—
Rouen, 21 cks. Rawson & Robinson

Glucose—
Stettin, 100 bgs. Wilson, Sons, & Co.
" 1,450

Horns—
Amsterdam, 235 bgs. Dumoulin & Goaschalk

Manure—
Ghent, 340 t. Wilson, Sons, & Co.

Mastic—
Copenhagen, 1 ck. "

Naphthol—
Hamburg, 1 ck.
Rotterdam, 20 brls. Hutchinson & Son

Naphthylamine—
Rotterdam, 1 brl. Hutchinson & Son

Potash—
Dunkirk, 20 drms. Wilson, Sons, & Co.
" 60

Pitch—
Dunkirk, 28 cks. "

Sodium Nitrate—
Rotterdam, 3 brls. "

Tar—
Dunkirk, 5 cs. "

TYNE.

Week ending Feb. 14th.

Alum Earth—
Hamburg, 19 cks. Tyne S.S. Co.

Cryolite—
Copenhagen, 5 cks.

Calcined Spar—
Rotterdam, 100 bgs. Tyne S.S. Co.

Copper Precipitate—
Huelva, 954 bgs. Bede Metal and Chemical Co.

Glycerine—
Hamburg, 1 cs. Tyne S.S. Co.

Glue—
Rotterdam, 20 bsks. "

Limespar—
Rotterdam, 250 bgs. "

Saltpetre—
Hamburg, 207 t.

Waste Salt—
Hamburg, 200 tons. Clephans & Wiencke

LIVERPOOL

Week ending Feb. 13th.

Acetate of Lime—
St. Nazaire, 60 sks.

Acetic Acid—
Ghent, 1 csk. T. Fletcher & Co.

Bone Dust—
Philadelphia, 251 bgs.

Bones—
Alexandria, 101 bgs.
Maranham, a quantity. H. Ev[illegible]
Cadiz, 494 bgs. H. S. Vidal

Cream Tartar—
Bordeaux, 2 cks.

Charcoal—
Rouen, 67 bgs. Co-op. W'sale Society, Limited

Caramel—
Hamburg, 15 bgs. 11 cks.

Copper Pyrites—
Huelva. 1,433 tons. Matheson & Co.

Chemicals (*otherwise undescribed*)
Rouen, 42 cks. Co-op. W'sale Society, Limited

Dyestuffs—
Madder Roots
Odessa, 19 bls.
Dyewood Extract
Antwerp, 7 cs. 20 cks. J. Fletcher & Co.
" 24 cks.
Gambier
Singapore, 1,258 bls. E Boustead
" 3,364
" 422
Cochineal
Grand Canary, 2 bgs. Swanston & Co.
Teneriffe, 21 Kuhner, Hendsch[illegible] & Co.
Argols
Oporto, 23 cks. Ger. Bk. Lond[illegible]

Drugs—
New York, 192 cs. W. Cuthbertson
" 26
Havre, 7 Cunard S.S. Co. Ld.
" 24 pkgs.
New York, 156 cs. W. Cuthbertson

Dried Blood—
Monte Video, 762 bgs. J. Purdo[illegible]
" 823

Esparto Grass—
Gabes, 4,174 bls.

Extracts—
Philadelphia, 25 brls.
St. Nazaire, 400 cks.
Fiume, 257 brls. Boutcher, Bacon & Co.
" 243 Millers' Tanning Ex[illegible]tract Co.
" 320 Th. Burde[illegible]
New York, 25 bxs.
Rouen, 275 cks. Co-op. W'sale Society, Limited
Havre, 82 pgs. Cunard S.S. Co., Limited

Fish Glue—
Boston, 23 kgs. 3 brls. 6 cs. Richards, Ferry & Co.

Glucose—
Rouen, 5 cks. Co-op. W'sale Society, Limited

New York, 25 brls. E. M. Duche & Son
" 200
Hamburg, 104 cks.
Philadelphia, 218 brls.

Glycerine—
Venice, 20 cks.

Glue—
Hamburg, 49 brls.

Horns—
Galveston, 29 sks.

Mercury—
Havre, 2 pgs. Cunard S.S. Co. Limited

Manganese Ore—
Odessa, 20 tons.

Manganese—
St. Jago de Cuba, 35 brls. T. Nickels & Co.

Ochre—
Rouen, 85 cks. Co-op. W'sale Society, Limited

Petroleum—
New York, 8,955 brls.

Pitch—
Bordeaux, 50 cks.

Potassium Prussiate—
Hamburg, 2 casks.

Plumbago—
Genoa, 80 bgs.

Potash—
Rouen, 24 cks 21 drms. Co-op. W'sale Society, Limited

Phosphate—
Ghent, 1,640 bgs.

Red Lead—
Rotterdam, 47 cks.

Soap—
Malta, 1 bag. E. Mahum

Saltpetre—
Hamburg, 99 cks.

Soapstone—
Bordeaux, 200 sacks. G. G. Blackwell

Sodium Acetate—
Rouen, 9 cks. Co-op. W'sale Society, Limited

Tartar Emetic—
Hamburg, 8 cks.

Tartar—
Bordeaux, 7 cs.

Ultramarine—
Trieste, 30 cs.

Verdigris—
Bordeaux, 5 cs.

Valonia—
Smyrna, 40 bgs Essayam Shahum

White Lead—
Rotterdam, 4 cks.

Wax—
Philadelphia, 100 brls,

Zinc Oxide—
Antwerp, 125 brls.

GOOLE.

Week ending Feb. 14th.

Alum—
Antwerp, 36 brls.

Antimony—
Boulogne, 8 cks.

Caoutchouc—
Boulogne, 5 cks.

Dyestuffs—
Boulogne, 93 cks.

Glucose—
Hamburg, 130 bgs.
" 15 cks.
Dunkirk, 180 bgs.

Glue—
Boulogne, 2 pgs.

Muriate of Potash—
Rotterdam, 8 cks.

Ochre—
Calais, 37 cks.
Rouen, 112

Potash—
Dunkirk, 8 drums.
" 22 cks.
Calais, 9

Potassium—
Boulogne, 1 ck.

Phosphates—
Ghent, 1,600 bgs.

Rosin—
Brunswick (Ga.), 2,320 brls.

Silicate of Soda—
Rotterdam, 1 ck.

Saltpetre—
Hamburg, 6 brls.
" 60 kgs.
Antwerp, 85 bgs.

Waste Salt—
Hamburg, 1,073 bgs.

Zinc Oxide—
Antwerp, 10 brls.

GRIMSBY.

Week ending February 8th.

Albumen—
Hamburg, 11 pgs.
Dieppe, 4 cs.

Chemicals (*otherwise undescribed*)
Hamburg, 16 pgs.
" 4 cs.

Dyestuffs—
Rotterdam, 3 cks.

Size—
Rotterdam, 27 pgs.

Zinc Ashes—
Antwerp, 9 cks.

GLASGOW.

Week ending Feb. 13th.

Alum—
Rotterdam, 82 cks.

Albumen—
Antwerp, 6 cs. 2 crates

Copper Ore—
Huelva, 1,472 t. Tharsis Co.

Dyestuffs—
Alizarine
Rotterdam, 63 cks.
Indigo
Calcutta, 33 chest
Shellac
Calcutta, 100 cases
Sumac
Marseilles, 150 cks.
Extracts
Bordeaux, 135 cks.

Farina—
Rotterdam, 390 brls.

Glucose—
New York, 50 brls.
Rotterdam, 12 cks.

Ochre—
Bordeaux, 15 cks.

Oxalic Acid—
Christiania, 10 cks.

Pitch—
Antwerp, 11 cks.

Painters' Colours—
Antwerp, 62 cases

Saltpetre—
Calcutta, 1 box.

Soda—
New York, 7 cks.

Stearine—
Bordeaux, 50 bgs.

Tar—
Marseilles, 75 cks.

Tartar—
Lisbon, 8 cks.
Bordeaux, 11

Ultramarine—
Rotterdam, 15 cs. 2

White Lead—
Antwerp, 3 casks.

Wood Pulp—
Christiania, 1,200 bls.

Zinc Oxide—
Antwerp, 75 cks.

EXPORTS OF CHEMICAL PRODUCTS

FROM

THE PRINCIPAL PORTS OF THE UNITED KINGDOM.

LONDON.

Week ending Feb. 8th.

Acids (*otherwise undescribed*)—
Bombay, 1 cs. £1
Singapore, 15 11

Alum—
Varna, 20 brls. £18
New York, 5 cks. 21

Arsenic—
Boston, 150 cs. £218

Alum Cake—
Bombay, 5 t. 7 cwt. £25
Calcutta, 156 cks. 101
Calcutta, 23 t. 15 cwt. 85

Ammonium Sulphate—
Hamburg, 20 tons. £240
Cologne, 40 475
Hamburg, 150 1,763
Bombay, 10 t. 2 cwt. 70
Dominica, 5 10 60
Demerara, 50 2 645
Marseilles, 40 11 475
Ghent, 80 10 926
" 1,000 bgs. 1,215
Dunkirk, 10 tons. 115
" 335 bgs. 365

Ammonia Liquid—
B. Ayres, 20 cs. £36
Bombay, 3 11
Calcutta, 1 2
Brisbane, 5 cks. 17
Mauritius, 8 20
Yokohama, 3 12
Calcutta, 1 22

Ammonium Muriate—
Melbourne, 21 cwt. £26
Auckland, 8 cks. 35

Ammonium Carbonate—
Sydney, 2 cs. £19

Bisulphite of Lime—
Rangoon, 23 c. £23
Calcutta, 57 54

Borax—
Melbourne, 23 c. 35
Natal, 2 kgs. 3
Rio de Janeiro, 8 c. 12
Lyttleton, 3 4
Melbourne, 5 8

Boracic Acid—
New York, 1 cs. £4

Camphor—
Bombay, 2 cs. £47

Carbolic Acid—
Yokohama, 5 cs. £32
Cape Town, 1 19
Shanghai, 1 55
New York, 20 100
Mauritius, 15 8
" 75 drms. 30
Yokohama, 30 181

Caustic Soda—
Bourgas, 64 c. £24
East London, 7 14
Townsville, 10 t. 1 c. 80
Lyttleton, 38 drms. 76
Newcastle, 96 50
Natal, 8 18

Chemicals (*otherwise undescribed*)
Durban, 3 t. £25
Singapore, 5 pg. 18
Odessa, 1 cs. 10
Whampoa, 3 6
Shanghai, 3 3
Hong Kong, 1 3
Barbadoes, 2 12
Belize, 10 73
Bombay, 26 87
Monte Video, 5 30
B. Ayres, 10 40
Hiogo, 18 497
Mauritius, 2 7
" 8 34
M. London, 60 kg. 25
Calcutta, 2 5
Natal, 5 pgs. 14
Colombo, 7 17
Calcutta, 15 cks. 6
" 3 26
Madrid, 13 94
Timaru, 3 fkns. 4
Brisbane, 34 pgs. 24
" 12 cks. 21
Durban, 1 7
Halifax, 7 85
Singapore, 2 20
Rio Janeiro, 5 kgs. 5
Mauritius, 2 cks. 5
Melbourne, 1 8
Lisbon, 80 brls. 61
Paris, 12 24
Cadiz, 2 8
Valencia, 3 8
Port Natal, 10 pgs. 40
Adelaide, 12 76
Madeira, 4 cks 24
Calcutta, 10 kgs. 57

Copper Sulphate—
B. Ayres, 10 cks. £63
Odessa, 103 cwt. 124
Varna, 9 10
Paris, 10 t. 1 cwt. 240
Lisbon, 5 1 112
Kurrachee, 36 cwt. 48
Freemantle, 40 49
Havre, 5 tons. 120
Barcelona, 10 3 cwt. 212

Cream Tartar—
Canterbury, 5 cwt. £32
E. London, 1 7
Hawkes' Bay, 2 14
Lyttleton, 10 kgs
Newcastle, 3 cwt.
E. London, 3 cs.
Natal, 1
E London, 1
Freemantle, 5
Melbourne, 40 kgs.
Adelaide, 20 cks.
Gibraltar, 1
Napier, 2 kegs,
E. Londen 7 cs.

Citric Acid—
Hamburg, 2 kgs.
Rotterdam, 3
Hamburg, 1 ck.
Cape Town, 8 cwt.
Cologne, 2 cks.
Genoa, 7 kegs.
Rotterdam, 1
Boston, 2 cks.

Carbolic Acid Cryst[illegible]
New York, 20 drms.

Chloride of Lime—
Gibraltar, 4 cks.

Copperas—
Melbourne, 12 cks.

Disinfecting Powder
Santos, 10 cwt.

Disinfectants—
Brisbane, 10 drms.
Calcutta, 25 cks.

Fertilizer—
Baltimore, 400 b[illegible]

Guano—
Sydney, 25 t.
Adelaide, 25
Trinidad, 80
Guadeloupe, 570
Jersey, 98
" 215

Glycerine—
Amsterdam, 20 drms. 271
Cologne, 17 406
Shanghai, 3 cs. 41
New York, 49 c. 102
Sydney, 15 cks. 80

Hypophosphites—
New York, 15 cs. £100

Magnesia—
Buenos Ayres, 1 cs. £68
Foochow, 1 c. 8
Monte Video, 4 22
Rio Janeiro, 10 29
Melbourne, 20 95

Manure—
Rotterdam, 1,665 bgs. £1,000
Cullera, 510 t. 13 c. 4,001
Bordeaux, 80 390
Rochefort, 114 480
Charente, 23 173
Cape Town, 20 17 110
Canary Islands, 20 200
Mauritius, 19 9 170
Hamburg, 100 1,000
Ghent, 100 1,200
Dominica, 27 12 112
Hamburg, 49 6 163
Demerara, 30 105
Hamburg, 30 600
Mauritius, 115 4 603
Orotava, 100 bgs. 47
Dunkirk, 50 600
Bordeaux, 47 165
Rouen, 91 400
Havre, 24 100
Jersey, 99 910

Nitric Acid—
Sydney, 8 cs. £15

Phosphorus—
Santa Cruz, 3 cwt. £91
Hiogo, 27 398
Yokohama, 14 219
Kobe, 2 cs. 96
Hong Kong, 2 37

Potash—
B. Ayres, 38 pgs. £97
Santos, 5 cwt. 21

Potassium Chlorate—
Hiogo, 150 kgs. £291
Yokohama, 100 190

Potassium Acetate—
Hiogo, 5 cs. £29

Potassium Cyanide—
B. Ayres, 6 cs. £15
Shanghai, 7 pgs. 29
New York, 12 cwt. 44

Potassium Bromide—
Melbourne, 1 cs. £8

Potassium Sulphate—
Mauritius, 95 cks. £315

Potassium Bichromate—
Hong Kong, 7 kegs. £16
Bombay, 16 cwt. 45

Quinine Sulphate—
New York, 20 cs. £1,184

Sodium Bicarbonate—
Sydney, 20 cwt. £7
Bombay, 10 3

Sulphuric Acid—
Madeira, 5 drms. £6
Natal, 15 c. £7
Trinidad, 4 1
Calcutta, 5 cs. 5
Algoa Bay, 25 18

Soda—
Monte Video, 13 c. £3

Salammoniac—
Varna, 2 cks. £10
Constantinople, 15 49
Bourgas, 12 43
Casa Blanca, 1 brl. 4
Port Natal, 4 fkns. 4
Kurrachee, 32 c. 56
Lisbon, 2 kgs. 5

Saltpetre—
Amsterdam, 50 cks. £210
Leghorn, 55 c. 59
Rio Grande, 20 23
Monte Video, 22 24
Rotterdam, 20 72
Natal, 5 6
Newcastle, 30 33
Melbourne, 25 27
Dominica, 5 t. 3 90
Lyttleton, 25 kgs. 38
Gothenburg, 16 cks 49
Sydney, 40 kegs. 46

Sulphate of Soda—
Ghent, 100 t. £100

Sodium Carbonate—
Natal, 4 c. £3
Hawke's Bay, 3 4

Sodium Hyposulphite—
Kurrachee, 2 cs. £2

Soda Ash—
Bombay, 17 c. £10

Sulphur—
Brisbane, 30 cks. £46
Lisbon, 20 brls. 76

Saltcake—
Antwerp, 140 t. £230

Superphosphate—
Mauritius, 126 t. £690
Harlingen, 514 bgs. £200

Soda Crystals—
Newcastle, 4 t. 18 c. £16
Mauritius, 20 cks. 10

Sodium Nitrate—
Dominica, 8 bgs. £10

Sugar of Lead—
Calcutta, 10 kegs. £13

Sodium Silicate—
Townsville, 5 t. 3 cwt. £29
Newcastle, 20 7

Sheep Dip—
Freemantle, 36 pgs. £18

Sulphide of Iron—
Martinique, 16 cwt. £3

Strychnine Crystals—
Adelaide, 2 cs. £04

Tannic Acid—
Mauritius, 1 cs. £3
Madeira, 1 cs. 2

Tartaric Acid—
Hawkes' Bay, 2 cwt. £16
Newcastle, 10 80
Natal, 1 cs. 8
Mauritius, 2 15
Lyttleton, 3 21
Sydney, 21 cks. 190
Kurrachee, 7 16
Madras, 2 14
Canterbury, 5 cwt. 40
Bombay, 6 cases. 44

LIVERPOOL.

Week ending Feb. 15th.

Antimony Regulus—
Bordeaux, 2 cks. £67
Bordeaux, 1 t.

Alum—
Constantinople, 40 cks. £31
Galatz, 5 11
Melbourne, 90 kegs. 12
Santander, 25 brls. 25
Sydney, 15 kegs. 4
Talcahuano, 10 5
Bari, 4 5
Santos, 1 t. 1 q. £8

Acids—
Santos, 50 cs. £75

Antimony—
N. York, 100 kegs. £80
Bombay, 1 ck. 38
Bombay, 1
Halifax, 1

Acetic Acid—
Santos, 1 ck. £5

Acetate of Iron—
Santos, 1 ck. £11

Ammonium Nitrate—
Calcutta, 5 cs. £20
Cartagena, 22 47

Ammonium Muriate—
Philadelphia, 45 cks. £551

Ammonium Carbonate—
Genoa, 10 kegs. £21
Rotterdam, 3 brls. 11
Toronto, 5 cks. 43
Barcelona, 20 34

Ammonium Sulphate—
Hamburg, 1,171 bgs. £1,488
Ghent, 50 61
Las Palmas, 115 117
Genoa, 665 840
Grand Canary, 20 25
Demerara, 92 t. 10 c. 1,159
Valencia, 516 bgs. 580

Aniline Colours—
Bombay, 2 pgs.

Ammonia Salts—
Porto Alegre, 1 ck.

Borax—
Havana, 19 brls.
Nantes, 40
Matanzas, 2 cks. £23
Coruna, 1 9
Cadiz, 2 14

Bleaching Powder—
Antwerp, 29 cks.
Baltimore, 54
Bombay, 28
Boston, 230 86 cs.
Calcutta, 75
Ghent, 211
New York, 900 100 tcs.
Philadelphia, 253
Santos, 8
Antwerp, 35
Calcutta, 80
Genoa, 149
Naples, 36
New York, 60
Oporto, 59
Philadelphia, 41
Portland, M., 100 brls. 31 hs. 38 pgs.
Rouen, 106 cks.
San Francisco, 175

Blood Albumen—
Vera Cruz, 12 cks.
" 12 £100
Santos, 1 10

Bone Manure—
Las Palmas, 100 bgs. £12
" " 20

Caustic Soda—
Boston, 400 drums.
Antwerp, 56 brls.
Genoa, 90
La Guayra, 10
Madras, 27
Manilla, 10
Melbourne, 18
Messina, 50
New York, 980
Rio Grande do Sul, 25 drums.
San Francisco, 50 drms.
Tampico, 200 kgs.
Terneusen, 35 drums.
Yokohama, 100
Algiers, 6
Alicante, 14
Amsterdam, 10
Antwerp, 31 26 brls.
Barcelona, 96
Bilbao, 40
Boston, 200
Calcutta, 106
Canada, 150
Cartagena, S. 50
" U.S.C. 8
Ceara, 12
Cienfuegos, 60
Corintho, 35
East London, 15 cs.
Genoa, 168
Hamburg, 96
Madrid, 50
Malaga, 70
Malta, 75
Maracaibo, 10
Maranham, 15
Melbourne, 30 5 bxs.
New York, 380
Philadelphia, 216
Portland, M. 20 10
Rio Janeiro, 5
Rotterdam, 2 brls.
Rouen, 95
San Francisco, 100
Santander, 3
Seville, 120 50
Syra, 10 cs.
Trieste, 10
Trinidad, 7
Venice, 15
Vigo, 10 cks.

Caoutchouc—
Savanilla, 5 bls.
Valparaiso, 4 cs.
Yokohama, 2

Cement—
Akassa, 21 t.
Curacoa, 25 brls.
Vera Cruz, 17 t.
Iquique, 6 cks
Manaos, 50 brls.
St. Thomas, 45 t.

Chemicals *(otherwise undescribed)*
Dunkirk, 23 brls.
Genoa, 56
" 2 cks.
Iquique, 88 brls.
Nantes, 348
New York, 4 cs. £28
Boston, 16 cks. 245
San Francisco, 10 klkns. 25

Cobalt Oxide—
New York, 5 cs.
Antwerp, 1

Caustic Potash—
Portland, 10 drms. £11

Carbolic Acid Crystals—
Malta, 2 c. £12

Carbolic Acid—
Hamburg, 14 [illegible]
Rotterdam, 1 ck.
Rouen, 4
Bombay, 8
Hamburg, 40

Carbon Bisulphide—
Bahia, 1 cs.

Copperas—
Alexandria, 100 brls.
Portland, 125
Alexandria, 40
Syra, 20

Copper Sulphate—
Barcelona, 30 cks.
Genoa, 8
" 67 [illegible]
Rouen, 85
Barcelona, 80
Genoa, 4 t. 19 c. 2 q.
" 26 19
Havre, 5 3
Barcelona, 1 17
Bordeaux, 10 7 1
Amsterdam, 6 cks.
Bilbao, 45
Valencia, 400
Bordeaux, 100
Genoa, 40
Havre, 205
Bordeaux, 28
Seville, 2
Tarragona, 20
Vigo, 8
Palermo, 21
Genoa, 20 cks.
Barcelona, 15
Bilbao, 12
Nantes, 41
Bordeaux, 216
Tarragona, 37 [illegible]
Constantinople, 5
Genoa, 72
Ancona, 110
Portland, 10
Trieste, 20
Bordeaux, 307

Citric Acid—
Coruna, 6 [illegible]

Drugs—
Colon, 5 cs.
Constantinople, 1
Corinto, 7
Kingston, 8 cks.
Monte Video, 1
New York, 10 cs.
Oporto, 20
Port Elizabeth, 1
Portland, 1 ck.
Savanilla, 16 pgs.
Sierra Leone, 9 cs.
Trinidad, 2

Dyes—
New York, 3 cs.
Vera Cruz, 3
Cape Town, 3 [illegible]
Dunedin, 6 [illegible]
New York, 20 [illegible]
Santos, 3 [illegible]

Dried Blood—
Halifax, 7 bgs.
La Palmas, 30
Grand Canary, 19

Disinfectants—
Alexandria, 10 cks.
Sierra Leone, 4

Epsom Salts—
Rio Janeiro, 50 kgs.

Glauber's Salts—
Rio Janeiro, 40 kgs.

Guano—
Valencia, 243 t.

Magnesia—
Antwerp, 12 pgs.

Manure—
Rouen, 53 t. 7 c.
Bordeaux, 445 13
Genoa, 117 24
St. Nazaire, 139 8
Havre, 157 10
La Palmas, 20
Rochefort, 517 3

Magnesium Chloride—
N. York, 25 drums.

Ochre—
N. York, 30 cks.

Oxide of Iron—
Gijon, 4 cks.
Antwerp, 13

Oxalic Acid—
Marseilles, 6 cks.

Paint—
Bahia Blanca, 15 cks.
Madeira, 1 cs.

Nassau, 6 kgs.
New York, 2
Panama, 17 cks.
St. Croix, 4
Savanilla, 4
Vera Cruz, 3

Potash Salts—
Portland, 4 cks.

Pitch—
Akassa, 6 brls.
Leghorn, 200 t.
Valparaiso, 10

Potassium Chlorate—
Hamburg, 70 kegs. £151
Boston, 100 211
N. York, 10 t. 440
Yokohama, 50 105
Philadelphia, 200 560
Coruna, 4 brls. 20
N. York, 100 210
Kobe, 25 160

Picric Acid—
N. York, 20 cs. £175
Rangoon, 20 88

Phosphorus—
Penang, 5 cs. £45

Potash—
Alexandria, 2 brls. £19
Venice, 3 17

Potassium Carbonate—
Portland, 4 t. 16 c. £76

Potassium Bichromate—
Barcelona, 30 cks. £388
Venice, 1 bx. 6
Bombay, 20 kgs. 38
Santos, 1 cs. 8
Ancona, 1 ck. 19

Quinine—
New York, 30 cs.

Salt—
Antwerp, 210 t.
Barbadoes, 100
Boston, 200
Brisbane, 100
Buenos Ayres, 25
Calcutta, 3,850
Cameroons, 50
Chicago, 199
E. London, 17¾
Galveston, 725
Halifax, 510
Hamburg, 50
Hong Kong, 1¾
Lagos, 20
Manaos, 17
Melbourne, 35
Newcastle, 302½
New York, 100
New Orleans, 200
Philadelphia, 50
Rotterdam, 125
Sydney, 50
Trinidad, 19 8 c.
San Francisco, 117
Boston, 225
Cameroons, 42¾
Christiania, 315
East London, 9
Grand Bassam, 10
Malta, 5
Santos, 4
Wellington, 51¾

Soda Ash—
Baltimore, 288 tcs.
Boston, 72 52 cks.
Ghent, 8
New York, 137
Baltimore, 350
Barcelona, 35 brls.
Boston, 268 cks. 142 tcs. 500 bgs.
Dunedin, 30 brls.
Monte Video, 100 drms.
New York, 106 tcs. 288 cks.
Philadelphia, 197
Rio Janeiro, 78 brls.
Smyrna, 550
Yokohama, 206

Soda Crystals—
Boston, 280 brls.
Malaga, 12 cks.
Philadelphia, 280
Salonica, 25
Boston, 140
Calcutta, 24 cks.
" 34 kgs.
New York, 747
Philadelphia, 167 brls.

Sodium Bicarbonate—
Monte Video, 50 kgs.
Punta Arenas, 2 cks.
Yokohama, 500
Amsterdam, 35
Antwerp, 36
Genoa, 40

Hamburg, 20 kegs
Rio Janeiro, 2
Rotterdam, 2 cks.

lag—
Nantes, 465½ t.

Sodium Silicate—
Malaga, 10 cks.
Salonica, 40
San Sebastian, 16

Spent Lyes—
Rouen, 43 dms.

Saltcake—
Canada, 63 cks. £126
Baltimore, 321 478
Philadelphia, 54 tcs. 79

Saltpetre—
Cartagena, 80 bgs. £115
Piræus, 50 brls. 162
Pernambuco, 50 cks. 64

Salammoniac—
Genoa, 2 cks. £16
Alexandria, 4 brls. 15
Philadelphia, 20 347
Alexandria, 8 40
" 10 33
Samsoun, 2 14

Sodium Biborate—
Coquimbo, 1 brl. £8
Portland, M. 30 143
Rotterdam, 10 cks. 158
" 1 16
New York, 15 41

Sheep Wash—
Galveston, 10 cs. £20
Algoa Bay, 3 t. 117
B. Ayres, 2 25
Brisbane, 10 26
B. Ayres, 50 4 c. 3 q. 1,125
S Francisco, 30 cs. 36
Algoa Bay, 150 drms. 35

Sodium Nitrate—
Shanghai, 5 t. 1 c. 1 q. £45
Leghorn, 9 9 85
Portland, 1 4 10
Cartagena, 4 10 1 43
Las Palmas, 9 7 1 78
Bari, 2 1 3 18
Pasages, 5 18 1 62

Superphosphate—
Hamburg, 772 bgs. £1,711
Las Palmas, 90 36
" 50 14

Sodium Arsenate—
Lisbon, 14 cks. £70

Sulphur—
Rouen, 80 bgs. £43
Toronto, 25 brls. 15
Boston, 667 bgs. 450
N. York, 334 220
San Juan, 10 brls. 10
Boston, 134 87
" 267 180

Tar—
Akassa, 33 brls.
Galveston, 8 cs.
Madras, 37
Ambriz, 10
Manos, 10
Rio Janeiro, 50
Sierra Leone, 30

Tannic Acid—
Santos, 1 kg. £6
Coruna, 6 brls. 43

Varnish—
Bahia Blanca, 10 dms.

Waste Salt—
Portland, 80 t.
Porte Natal, 45
Salt Pond, 27
San Francisco, 799½
Para, 124 8 c.
Sulymah, 15 12
Wellington, 125

Zinc Chloride—
Bombay, 2 cks. £8

TYNE.

Week ending Feb. 14th.

Anthracene—
Rotterdam, 248 c.

Alkali—
Rotterdam, 150 c.
Trieste, 207
Gothenburg, 50

Ammonium Sulphate—
Rotterdam, 70 t.
Hamburg, 1.905 c.

Bleaching Powder—
Copenhagen, 1,324 c.
Bergen, 106
Rotterdam, 788
Aarhuus, 23
Hamburg, 1,267
Antwerp, 3.560

Caustic Soda—
Copenhagen, 620 c.
Hamburg, 613
Antwerp, 1,800

Cement—
Copenhagen, 80 c.
Rotterdam, 44 t. 6
Huelva, 357

Copperas—
Alexandria, 418 c.

Litharge—
Hamburg, 58 c.
Copenhagen, 5

Magnesia—
Hamburg, 6 c.

Red Lead—
Copenhagen, 78 c.
Rotterdam, 1 t. 2
Hamburg, 207

Soda—
Rotterdam, 143 c.
Copenhagen, 538
Esbjerg, 552
Odense, 119
Malmo, 107
Randers, 192

Sodium Chlorate—
Rotterdam, 1 t. 2 c.

Salammoniac—
Copenhagen, 101 c.

Sodium Hyposulphite—
Helsingborg, 100 c.

Tar—
Hamburg, 20 t.

White Lead—
Hamburg, 4 c.
Copenhagen, 54

HULL.

Week ending Feb. 14th.

Alkali—
Odessa, 23 cks.

Bleaching Powder—
Bremen, 15 cks. 20 kgs.
Gothenburg, 2
Odessa, 23

Bone Size—
Harlingen, 13 cks.

Benzole—
Hamburg, 16 drms.

Caustic Soda—
Fiume, 4 drms.

Cement—
Amsterdam, 5 bgs.
Copenhagen, 8 cs.

Chemicals *(otherwise undescribed)*
Antwerp, 31 cs. 25 pgs.
Amsterdam, 3 27
Bremen, 2 cks.
Bergen, 42
Constantinople, 1
Christiania, 44
Copenhagen, 20
Dunkirk, 36
Gothenburg, 20 5 5 pgs.
Hamburg, 157 10
Konigsberg, 13
Libau, 1 20
Odessa, 388 30
Rotterdam, 30 37
Rouen, 8
Stavanger, 2
Venice, 16

Dyestuffs *(otherwise undescribed)*
Antwerp, 1 drms.
Boston, 11 cks.
Hamburg, 6 kgs. 7 cs.
Rotterdam, 2 4
Rouen, 3 cks.

Horns—
Hamburg, 51 bgs.

Manure—
Bremen, 51 t.
Hamburg, 3,417 bgs.

Ochre—
San Francisco, 71 cks.

Pitch—
Antwerp, 233 t.
St. Nazaire, 350

Tar—
Christiania, 5 cks.
Copenhagen, 10
Hamburg, 13
Rotterdam, 3
Stettin, 2

GLA

Week endi

Ammonium Sul
Demerara, 53 t.

Benzol—
Rotterdam, 25 p
" 20

Bleaching Pow
Sydney, 7 t. 4½

Boiler Coating
Sydney, 10 t. 1

Cement—
Sydney, 2 t. 6 c

Coal Tar Resid
Baltimore, 4 t.

Chemical Produ
Boston, 13 c.

Drugs—
Jamaica,

Logwood Extra
Halifax,

Litharge—
Buenos Ayres, 3 c

Manure—
Demerara, 90 t.

Oxalic Acid—
Boston, 17¾ c.

Paraffin Wax—
Sydney, 750 lbs.

Painters' Colour
New York, 5 t.

Potassium Pruss
Rotterdam, 4 ck.

Potassium Bichr
Bombay, 2 t.
Rotterdam, 35 cks
" 12
Boston, 318 c.

Potassium Iodid
Boston, 1 c.

Red Lead—
Monte Video, 1 t.
B. Ayres,

Sodium Bichrom
Rotterdam, 1 ck.

Turpentine—
Boston, 257 gls.

White Lead—
Sydney, 3 t.

Soda Ash—
Baltimore, 1,706¾

Soda—
New York, 5 t. 14

Sodium Nitrate—
Demerara, 82 c.

GOO

Week ending

Benzol—
Hamburg, 8 drun
Rotterdam, 68 cks.

Chemicals *(otherw*
Ghent, 2 cks.
Rotterdam, 101

Coal Products—
Calais, 20 c
Rotterdam, 55

Copper Precipitat
Hamburg, 200 bgs.

Dyestuffs *(otherwi*
Ghent, 100 d
Rotterdam, 20 ck

Manure—
Boulogne, 114 bg
Ghent, 814
Hamburg, 432

Pitch—
Antwerp, 324 t.
Dunkirk, 392

Tar—
Ghent, 100 cs.

GRIM

Week ending

Chemicals *(otherw*
Dieppe, 62 cs.
" 9 cks.
Rotterdam, 1
" 8 cs.

PRICES CURRENT.

THURSDAY, February 20, 1890.

Prepared by Higginbottom and Co., 116, Portland Street, Manchester.

The values stated are F.O.R. *at maker's works, or at usual ports of shipment in U.K. The price in different localities may vary.*

		£	s.	d.
Acids :—				
Acetic, 25 °/。 and 40 °/。	per cwt	7/6 & 12/6		
,, Glacial	,,	2	10	0
Arsenic, S.G., 2000°	,,	0	12	9
Chromic 82 °/。 nett	per lb.	0	0	6½
Fluoric	,,	0	0	3¼
Muriatic (Tower Salts), 30° Tw.	per bottle	0	0	8
,, (Cylinder), 30° Tw.	,,	0	2	11
Nitric 80° Tw.	per lb.	0	0	2
Nitrous	,,	0	0	1¼
Oxalic nett	,,	0	0	3¼
Picric	,,	0	1	4½
Sulphuric (fuming 50 °/。)	per ton	15	10	0
,, (monohydrate)	,,	5	10	0
,, (Pyrites, 168°)	,,	3	2	6
,, (,, 150°)	,,	1	8	0
,, (free from arsenic, 140/145°)	,,	1	10	0
Sulphurous (solution)	,,	2	15	0
Tannic (pure)	per lb.	0	1	0½
Tartaric	,,	0	1	2¼
Arsenic, white powered	per ton	13	0	0
Alum (loose lump)	,,	4	15	0
Alumina Sulphate (pure)	,,	5	0	0
,, ,, (medium qualities)	,,	4	10	0
Aluminium	per lb.	0	15	0
Ammonia, ·880=28°	per lb.	0	0	3
,, =24°	,,	0	0	1⅞
,, Carbonate nett	,,	0	0	3½
,, Muriate	per ton	23	15	0
,, ,, (sal-ammoniac) 1st & 2nd	per cwt.	37/- & 35/-		
,, Nitrate	per ton	40	0	0
,, Phosphate	per lb.	0	0	10½
,, Sulphate (grey), London	per ton	11	18	9
,, ,, (grey), Hull	,,	11	17	6
Aniline Oil (pure)	per lb.	0	0	11½
Aniline Salt ,,	,,	0	0	10
Antimony	per ton	75	0	0
,, (tartar emetic)	per lb.	0	1	1
,, (golden sulphide)	,,	0	0	10
Barium Chloride	per ton	7	10	0
,, Carbonate (native)	,,	3	15	0
,, Sulphate (native levigated)	,,	45/- to 75/-		
Bleaching Powder, 35 %	,,	5	10	0
,, Liquor, 7 %	,,	2	10	0
Bisulphide of Carbon	,,	12	15	0
Chromium Acetate (crystal	per lb.	0	0	6
Calcium Chloride	per ton	2	0	0
China Clay (at Runcorn) in bulk	,,	17/6 to 35/-		
Coal Tar Dyes :—				
Alizarine (artificial) 20 %	per lb.	0	0	9
Magenta	,,	0	3	9
Coal Tar Products				
Anthracene, 30 % A, f.o.b. London	per unit per cwt.	0	1	6
Benzol, 90 % nominal	per gallon	0	3	11
,, 50/90 ,,	,,	0	3	0
Carbolic Acid (crystallised 35°)	per lb.	0	0	10
,, (crude 60°)	per gallon	0	2	10
Creosote (ordinary)	,,	0	0	2½
,, (filtered for Lucigen light)	,,	0	0	3
Crude Naphtha 30 % @ 120° C.	,,	0	1	3
Grease Oils, 22° Tw.	per ton	3	0	0
Pitch, f.o.b. Liverpool or Garston	,,	1	12	0
Solvent Naphtha, 90 % at 160°	per gallon	0	1	11
Coke-oven Oils (crude)	,,	0	0	2½
Copper (Chili Bars)	per ton	47	17	6
,, Sulphate	,,	26	10	0
,, Oxide (copper scales)	,,	52	0	0
Glycerine (crude)	,,	30	0	0
,, (distilled S.G. 1250°)	,,	55	0	0
Iodine	nett, per oz.	8d. to 9d.		
Iron Sulphate (copperas)	per ton	1	10	
,, Sulphide (antimony slag)	,,	2	0	
Lead (sheet) for cash	,,	14	15	
,, Litharge Flake (ex ship)	,,	17	0	
,, Acetate (white ,,)	,,	23	5	
,, ,, (brown ,,)	,,	19	0	
,, Carbonate (white lead) pure	,,	19	0	
,, Nitrate	,,	22	5	8

		£	s.	d.
Lime, Acetate (brown)	,,	8	5	0
,, ,, (grey)	per ton	14	0	0
Magnesium (ribbon and wire)	per oz.	0	2	3
,, Chloride (ex ship)	per ton	2	6	3
,, Carbonate	per cwt.	1	17	6
,, Hydrate	,,	0	10	0
,, Sulphate (Epsom Salts)	per ton	3	0	0
Manganese, Sulphate	,,	18	0	0
,, Borate (1st and 2nd)	per cwt.	60/- & 42/6		
,, Ore, 70 %	per ton	4	10	0
Methylated Spirit, 61° O.P.	per gallon	0	2	2
Naphtha (Wood), Solvent	,,	0	3	11
,, ,, Miscible, 60° O.P.	,,	0	4	6
Oils :—				
Cotton-seed	per ton	22	0	0
Linseed	,,	23	0	0
Lubricating, Scotch, 890°—895°	,,	7	5	0
Petroleum, Russian	per gallon	0	0	5⅛
,, American	,,	0	0	5⅞
Potassium (metal)	per oz.	0	3	10
,, Bichromate	per lb.	0	0	4
,, Carbonate, 90% (ex ship)	per ton	18	0	0
,, Chlorate	per lb.	0	0	4¼
,, Cyanide, 98%	,,	0	2	0
,, Hydrate (Caustic Potash) 80/85 %	per ton	22	10	0
,, ,, (Caustic Potash) 75/80 %	,,	21	10	0
,, ,, (Caustic Potash) 70/75 %	,,	20	15	0
,, Nitrate (refined)	per cwt.	1	3	6
,, Permanganate	,,	4	5	0
Prussiate Yellow	per lb.	0	0	10
,, Sulphate, 90 %	per ton	9	10	0
,, Muriate, 80 %	,,	7	15	0
Silver (metal)	per oz.	0	3	7¼
,, Nitrate	,,	0	2	5½
Sodium (metal)	per lb.	0	4	0
,, Carb. (refined Soda-ash) 48 %	per ton	5	10	0
,, ,, (Caustic Soda-ash) 48 %	,,	4	10	0
,, ,, (Carb. Soda-ash) 48%	,,	4	15	0
,, ,, (Carb. Soda-ash) 58 %	,,	5	12	6
,, ,, (Soda Crystals)	,,	2	13	6
,, Acetate (ex-ship)	,,	16	0	0
,, Arseniate, 45 %	,,	10	0	0
,, Chlorate	per lb.	0	0	6¼
,, Borate (Borax)	nett, per ton.	29	10	0
,, Bichromate	per lb.	0	0	3
,, Hydrate (77 % Caustic Soda)	per ton.	10	10	0
,, ,, (74 % Caustic Soda)	,,	10	0	0
,, ,, (70 % Caustic Soda)	,,	8	17	6
,, ,, (60 % Caustic Soda, white)	,,	7	17	6
,, ,, (60 % Caustic Soda, cream)	,,	7	10	0
,, Bicarbonate	,,	5	7	6
,, Hyposulphite	,,	5	2	6
,, Manganate, 25%	,,	8	10	0
,, Nitrate (95 %) ex-ship Liverpool	per cwt.	0	8	4½
,, Nitrite, 98 %	per ton	27	10	0
,, Phosphate	,,	15	15	0
,, Prussiate	per lb.	0	0	7¼
,, Silicate (glass)	per ton	5	7	6
,, ,, (liquid, 100° Tw.)	,,	3	17	6
,, Stannate, 40 %	per cwt.	2	0	0
,, Sulphate (Salt-cake)	per ton.	1	12	6
,, ,, (Glauber's Salts)	,,	1	10	0
,, Sulphide	,,	7	15	0
,, Sulphite	,,	5	5	0
Strontium Hydrate, 98 %	,,	8	0	0
Sulphocyanide Ammonium, 95 %	per lb.	0	0	8¼
,, Barium, 95 %	,,	0	0	5½
,, Potassium	,,	0	0	8
Sulphur (Flowers)	per ton.	6	15	0
,, (Roll Brimstone)	,,	6	0	0
,, Brimstone : Best Quality	,,	4	7	6
Superphosphate of Lime (26 %)	,,	2	10	0
Tallow	,,	25	10	0
Tin (English Ingots)	,,	96	0	0
,, Crystals	per lb.	0	0	6¼
Zinc (Spelter	per ton.	23	0	0
,, Chloride (solution, 96° Tw.	,,	6	0	0

THE

CHEMICAL TRADE JOURI

Publishing Offices: 32, *BLACKFRIARS STREET, MANCHESTER.*

No. 145. SATURDAY, MARCH 1, 1890.

Contents.

Notices.

All communications for the *Chemical Trade Journal* should be addressed, and Cheques and Post Office Orders made payable to—

DAVIS BROS., 32, Blackfriars Street, MANCHESTER.

Our Registered Telegraphic Address is—

"Expert, Manchester."

The Terms of Subscription, commencing at any date, to the *Chemical Trade Journal*,—payable in advance,—including postage to any part of the world, are as follow:—

Yearly (52 numbers)	**12s. 6d.**
Half-Yearly (26 numbers)	**6s. 6d.**
Quarterly (13 numbers)	**3s. 6d.**

Readers will oblige by making their remittances for subscriptions by Postal or Post Office Order, crossed.

Communications for the Editor, if intended for insertion in the current week's issue, should reach the office not later than **Tuesday Morning.**

Articles, reports, and correspondence on all matters of interest to the Chemical and allied industries, home and foreign, are solicited. Correspondents should condense their matter as much as possible, write on one side only of the paper, and in all cases give their names and addresses, not necessarily for publication. Sketches should be sent on separate sheets.

We cannot undertake to return rejected manuscripts or drawings, unless accompanied by a stamped directed envelope.

Readers are invited to forward items of intelligence, or cuttings from local newspapers, of interest to the trades concerned.

As it is one of the special features of the *Chemical Trade Journal* to give the earliest information respecting new processes, improvements, inventions, etc., bearing upon the Chemical and allied industries, or which may be of interest to our readers, the Editor invites particulars of such—when in working order—from the originators; and if the subject is deemed of sufficient importance, an expert will visit and report upon the same in the columns of the *Journal*. There is no fee required for visits of this kind.

We shall esteem it a favour if any of our readers, in making inquiries of, or opening accounts with advertisers in this paper, will kindly mention the *Chemical Trade Journal* as the source of their information.

Advertisements intended for insertion in the current week's issue, should reach the office by **Wednesday morning** at the latest.

Advertisements.

Prepaid Advertisements of Situations Vacant, Premises on Sale or To be Let, Miscellaneous Wants, and Specific Articles for Sale by Private Contract, are inserted in the *Chemical Trade Journal* at the following rates:—

30 Words and under	2s. od.	per insertion.
Each additional 10 words	os. 6d.	"

Advertisements of Situations Wanted are inserted at one-half the above rates when prepaid, viz:—

30 Words and under	1s. od.	per insertion.
Each additional 10 words	os. 3d.	"

Trade Advertisements, Announcements in the Directory Columns, and all Advertisements not prepaid, are charged at the Tariff rates, which will be forwarded on application.

Legal.

THE CARTVALE CHEMICAL WOR NUISANCE.

THE SHERIFF PRINCIPAL'S INTERLO

Sheriff-Principal Cheyne has issued his judg
against Sheriff-Substitute Steel's decision in the
Allison, timber merchant, against Messrs. Campb
Chemical Works, in which pursuer sought interdict
from carrying on their works so as to constitute a
lowing is the Sheriff's interlocutor and note:—

PAISLEY

The Sheriff, having at avizandum considered t
the appeal, affirms the interlocutor of the Sheri
appealed against, finds the defenders liable in ad
scale 2, including a special debate fee of £5., and

NOTE.—The Sheriff-Substitute has gone so full
agreeing as I do with his conclusion and generally wit
very clear note appended to his interlocutor, I think
may be confined within the limits of a few sentences.
me, on behalf of the defenders, that the Sheriff-Sub
sufficient weight on the fact that the pursuer's pro
surrounded by public works of various descriptions
true that it is not specially dealt with in the note,
to the Sheriff-Substitute, it should be said that he
dental remarks which show that he did not overlook
ever, as it may, there can be no doubt on the auth
mining whether the operations at a particular
nuisance, locality is a circumstance, and an import
be kept in view. When it is said that a man has
air diffused over his property, fresh and free from
that means only that he is entitled to air as fresh
reasonably be expected in the locality, consistently
and legitimate use of property, and accordingly to
Lord Westbury in the case of St. Helens Smeltin
Tipping, 11 H. L. Cas. 642—"If a man lives i
subject himself to the consequences of those operat
may be carried on in his immediate locality, which
sary for trade and commerce, and also for the enjo
and for the benefit of the inhabitants of the town a
large." But while, as was remarked in the same
not in such questions stand on extreme rights, a
living in the town cannot expect complete amenit
smell, or noise, I apprehend that it is only the am
which is *necessarily* produced by the legitimate trad
neighbours, that a residenter in a town is obliged
other words, I think it stands to reason that a ma
on the ground that the atmosphere of the locality
contributions, heavily charged with smoke or v
throwing into the air of any impurities which there
of guarding against, and which materially increa
and annoyance of the neighbouring proprieto
doing so constitutes a nuisance against which the
prietor is entitled to be protected. In applying the
present case, I have come, after a careful con
evidence, to agree with the Sheriff-Substitute in
that, while the defenders may occasionally have bee
was the fault of Messrs. Robin and Houston, the
some time prior to, the commencement of these p
offenders in the matter of black smoke, and that
them, each of the scientific witnesses agree might
vented, indeed it seems to have been successfully
apparatus introduced into the works in the course o
which, owing to the lowness of the chimneys in
pursuer's property undoubtedly got, when the wind
direction, the full benefit, added, in a material deg
fort and annoyance of the pursuer and his family.
further complains of the emission from the defenders
vapours and charcoal dust, which, he says, injure

id shrubs in his garden, affected the health of his family, and interred in various other ways with his comfortable enjoyment of his property. ow, I agree with the Sheriff-Substitute that it has not been shown at the vegetation of the garden has suffered from anything traceable to e defenders' works, and I confess that I am sceptical about the lleged injury to health, but on the other hand, I have no doubt that own to the raising of this action, and indeed for some time after that ate, charcoal dust did come at intervals, and sometimes in large uantities, into the pursuer's ground from the defendants' works, and nat the pursuer was thereby subjected to considerable and sensible iscomfort, and to that extent I hold this branch of the complaint to be stablished. There remains the allegation about the offensive smells, nd as to this I am bound to say that if it had stood alone I would, ooking to the fact that so far as I can discover, the first complaint on he subject was made in the petition, have been inclined to think that here was some exaggeration in the evidence, and that the grievance vas not of such a serious nature as to require or justify the intervention of the Court. Still, the proof seems to establish that there does come rom the defenders' works, either from the evaporators or from the 85 eet chimney into which the uncondensed gases are led, or from oth, the smell which in certain states of the wind is perceptible n the pursuer's grounds and house, and is disagreeable to the members of his family, and though I attach comparatively little importance to this part of the case, I do not think that I can or ought to throw it altogether out of view. On the whole case, keeping in view what I gather to be the unanimous opinion of the scientific witnesses, that this particular business unless carefully conducted may rightly become a nuisance to the neighbourhood; and treating the question as a jury-question I am satisfied, as the Sheriff-Substitute who heard the evidence was, that the defenders' operations, as conducted at the date when this action was raised, were a nuisance to the pursuer, and if I am right in so holding, it seems to follow that the pursuer is entitled to the interdict which the Sheriff-Substitute has granted, notwithstanding that the improvements since introduced by the defenders into their works have gone far to remove, if they have not entirely removed, all reasonable ground of complaint. Indeed, in view of these improvements, one cannot but suspect that the only point of any real interest to either of the parties is now a question of expenses. (Intld.) J. C.

Agent for pursuer: Mr. John Abercrombie, writer, Moss-street.
Agent for defenders: Mr. Dugald D. Dickie, writer, High-street.

PENNSYLVANIA SALT MANUFACTURING COMPANY.

THE Pennsylvania Salt Manufacturing Company, manufacturing chemists and importers of kryolith, No. 115, Chestnut-street, Philadelphia, is one of the most substantial concerns in the United States. This representative and widely known company was chartered by the Legislature of Pennsylvania, September 25th, 1850, with ample capital. It was organised for the purpose of engaging in the manufacture of soda ash of commerce under the corporate name of the Pennsylvania Salt Manufacturing Co. This title was partially a misnomer, as the manufacture of salt was not the principal object of the organisers, but at that period there existed no law in the State under which a company could be incorporated for the production of chemicals. There was, however, a general manufacturing law containing a clause for the manufacture of salt and the products derivable therefrom, and under this clause the charter was issued, as soda is by a certain process a direct product from salt. The company's extensive works are located at Natrona, on the Pennsylvania railroad, about twenty-four miles from Pittsburg. The property purchased by the company contains coal in unlimited abundance, and a plentiful supply of salt water is obtained upon sinking wells. The works are admirably equipped with the latest improved machinery, apparatus and appliances, necessary for the systematic and successful conduct of this important and steadily increasing industry. In their works, mines, and quarries 1,400 men are employed, who earn upwards of 700,000 dols. annually. The capital and surplus of the Pennsylvania Salt Manufacturing Co. now amounts to several millions of dollars. The company manufactures extensively sulphuric acid, soda ash, caustic soda, sal soda, bicarbonate of soda, saponifer or concentrated lye, Glauber salt, alum, copperas, chloride of calcium, nitric and muriatic acids, nitrate of lead, Epsom salts, and many other chemical compounds. In consequence of the great expense of the preparation of soda compounds by the old methods, the company, in 1864, directed its attention to the mineral "kryolith" which is composed of sodium, aluminum and florine. This mineral is found on the southwest coast of Greenland, and was first discovered by the Esquimaux. Nowhere else has it been found, except in small quantities at Miask, in the Ural Mountains, between Russia and Siberia. In Greenland it is a solid mass 600 feet long, 200 feet wide, and 100 feet deep. A fleet of the company's ice fortified vessels, built expressly for this trade, bring many thousands of tons of this kryolith annually to Philadelphia, whence it is shipped by rail to the works at Natrona. To describe the company's machinery and apparatus in detail, and the various operations, would require several columns. In short, the kryolith mills, calcining houses, leaching vats, buildings for carbonising and crystalisation, massive tanks holding 2,000 tons of soda each, immense agitators, cast iron kettles weighing eleven tons each, leaden chambers containing 3,000 tons of lead, platina stills costing over 100,000 dols., saw mills, box factories, forty steam engines, and twenty-nine boilers, &c., all these constitute only a part of the valuable and extensive apparatus utilised in these immense works. Upwards of 200,000 tons of freight are handled annually by this mammoth industry. A branch of the company's works are situated in Philadelphia, where over 1,500,000 dols. is invested. At these works acids, alum, and the famous Lewis lye are manufactured. This Lewis lye is powdered and perfumed, and packed in pound cans for family use, and it is extensively useful in the household. In the company's copper and refining works at Natrona, Rio Tinto or Spanish and native ores are utilised. Natrona copper is a miracle of purity, and the silver invariably tests 999 fine. It has always been the aim of this reliable company to produce its goods on a scale and at a cost, that would increase uniform excellence, as well as cheapness. All wares. chemicals, etc., that bear "Natrona," are always recognised and appreciated by the trade and public as standard productions, possessing all the qualities claimed for them by the manufacturers. These goods have no superiors in the home or foreign market, and the reputation of the company for liberal and just dealing would be prized by the oldest commercial houses of the world. The company's central office is at No. 115, Chestnut-street, Philadelphia. At its works the company has erected several hundred substantial brick buildings for the accommodation of its operators. There are likewise several school houses and churches, and great interest is taken by the officers to make the homes of the company's workmen pleasant and attractive. The existence of this grand industry in our midst, its struggles and successes are eminently suggestive. Under practical free trade, when the United States depended upon Great Britain for its supplies of alkali and alum, the cost to the consumer was from 200 to 300 per cent. greater than at the present day. The following gentlemen, who are widely and favourably known in financial and manufacturing circles for their enterprise, ability, and honourable methods, are the officers and directors, viz.: T. Armstrong, president; F. P. Steel, vice-president; A. M. Purvis, treasurer; R. Dale Benson, B. A. Knight, J. W. McAllister, J. S. Jenks and Thos. W. Sparks, directors; P. A. Bour, general manager; R. G. Ewer, superintendent of Natrona works; G. F. Bihn, superintendent of the Greenwich works; principal chemists, G. F. Bihn, Otto Luthy, and Robert Heerlein. In conclusion, we would observe that the prosperity of the Pennsylvania Salt Manufac'uring Co. presents a forcible illustration of the material benefits arising from a federal policy affording protection to American industries, resulting in the development of the nation's wonderful resources, and in the creation of such great corporations as this one, thereby rendering the United States for ever independent of foreign manufacturers.

MESSRS. GRIMSHAW BROTHERS, LIMITED.—In our leader of last week we described this firm as exhibitors of indiarubber chemicals. This was an error, as their specialities are zinc compounds and sizing materials for cotton warps and piece goods.

ALUMINIUM DIRECT FROM CLAY.—Referring to the patented process of Julius Emmner, for extracting this metal direct from North Carolina clay, a correspondent suggests the suitability of certain Irish clays for this purpose. He quotes an analysis shewing:—

Alumina	46·64 %
Silica	40·26
Oxide of iron	3·00
Lime	2·81
Magnesia	traces.

FRENCH CHALK PENCILS.—A very useful novelty has been just brought out by Mr. G. G. Blackwell, of Liverpool, in the form of a pencil of French chalk. It is specially adapted for marking on iron, steel, and other like substances, and its convenient shape will no doubt cause it to come into very general use and give it a ready sale.

COAL EXPORTS.—The shipments of coal were very unsatisfactory last month. Most of our local ports, and all the Welsh ports, had decreases to record. In the total, the North-East ports sent out 105,000 tons less last month than a year ago; and the four ports in South Wales sent out 84,000 tons less. Higher prices are evidently greatly lessening the export coal trade—aided by the finer weather, which limits the home consumption.

Industrial Celebrities.

VI.

JOSIAS CHRISTOPHER GAMBLE.

DURING the last half century no name has been more constantly prominent amongst the public men of St. Helens than that of Gamble. Six times it has been associated with the mayoralty of that borough, five in the person of Colonel Gamble, and once in that of his eldest son; and in the present year his son-in-law bears the honours of that position.

When, in the year 1868, St. Helens was incorporated, Colonel Gamble was chosen its first Mayor, as for nearly a quarter of a century before that time he had, with indefatigable zeal, given his services to the town as a member of the Board of Commissioners; indeed, he was one of those who were instrumental in obtaining the first Local Improvement Act.

Very few towns in England have attained a position, in connection with the Volunteer movement, so distinguished as St Helens, and this success is very largely owing to the support which Colonel Gamble gave to it from its inception in 1860. He was one of the most active members of the General Committee, who raised the funds for its first establishment, and which included all the leading men of the district; he was chosen the Captain of the first company that was enrolled, and as the regiment grew in strength and thereby became successively entitled to a Major and to a Lieutenant-Colonel, he was promoted to those ranks. By his generosity the excellent drill-hall and parade-ground were provided, and during the seven and twenty years that he was commanding officer of the battalion, its organization was perfect; the 47th Lancashire was distinguished for the soldierly bearing of its members on parade, for the admirable commissariat arrangements which drew official notice when a review was held distant from home, and in musketry it won not only the highest honours in its county competitions, but also carried off, in the great national contests at Wimbledon, numerous and valuable prizes, on one occasion bearing away the blue ribbon of victory, the Queen's Prize. In yachting circles, too, for many years Colonel Gamble has been a familiar figure, having been Commodore of the Royal Mersey Yacht Club for the past ten years. Having always resided almost within the precincts of the town, he has taken a constant and intimate interest in all its concerns, not merely discharging public duties, but also filling a social position, and thus assisting to arrest wretched deterioration which too often takes place in the towns devoted to manufacturing industries, especially when the processes are noxious, polluting the atmosphere and destroying the vegetation.

JOSIAS CHRISTOPHER GAMBLE.

The bane of such places is that the swarming population of the labouring classes is left comparatively uncared for by its employers, who go to reside in more agreeable neighbourhoods, or who are only members of public companies who justify their gross neglect of social duties, by asserting that boards of directors have no right to be the dispensers of charity, and that their servants must not give their time or attention to public affairs.

The lamentable consequence of this vicious system is that such towns lack efficient local government, are not provided with those institutions and social arrangements which are essential to health, morality, and civilisation. Matthew Arnold once designated such towns "perfect hell-holes."

Colonel Gamble built his residence close to the town, and within an easy walk of his works; the scene of his labours has been constantly before him, and he has been keenly alive to the needs of the district: he has associated himself with every local movement, has spent his wealth among those who, to some extent, have contributed to its acquisition, and he has made a local residence, at least bearable if not attractive. It is not to be wondered at that his fellow-townsmen have been anxious to record their grateful appreciation of his services or that they rejoiced in the public honours which have been awarded him. But the local influence and the social position of the Gamble family today has not been reached by any royal road, it is the result of sixty years of patient plodding industry, and of great care and thrift; the extensive works of Gerard's Bridge and Hardshaw Brook are the results of slow and cautious development, not that there has been any lack of

enterprise or disregard of the progress of inventio
it was at Gerard's Bridge that the Weldon inventi
and every discovery in the manufacture of chemical
noted, but no step forward has been rashly taken
successful business is that it has been slow but sur
Josias Christopher Gamble, who was born in the y
called Graan, near Enniskillen. The Gambles
family, but were driven from Scotland, and tool
"when King James the Sixth thrust prelacy upon
were staunch and steadfast Presbyterians, and chos
conscience sake. The almost universal ambition of
that one of their sons might be a scholar and a mi
the lot of Josias Christopher Gamble; he was se
University, but to Glasgow, where he studied an
his degree of M.A. in 1797.

In December, 1799, he was ordained minister
Kirk in Enniskillen, and he continued to be the
gregation for over four years, until February, 1804.

On leaving Enniskillen he accepted a "call
Belfast.

In the present day missionaries, before proceedi
labour, frequently pass through a short course o
and medicine, and no small portion of their mi
attending to the bodies as well as to the souls of m
their gospel as being a message from "The
"healeth diseases" as
iniquities": so in the en
it was customary for theo
were being trained for
Presbyterian Church, to
medical lectures. One
Gamble attended at Gla
presided over by Dr. C
chemistry. The student
stirred by the enthusias
his teacher, and so fascin
in natural science, that
long vacations in the
studies and in making c
His friends did not at
his tasks had taken; w
fumes he created, they
heresy. But Gamble's
those of a philosophical
practical in his pursuits.
neighbourhood in which
upon their lands, and th
and spun it. Gamble ut
he had acquired in his st
prepare solutions of chl
linen which was manuf
loom weavers.

William Gossage use
cess of the Gambles wa
be attributed to their per
efforts to use up ever
Josias Christopher Gamble, even in his earliest op
bleaching solutions, did not overlook the impor
any waste; he used to work up the residues
chlorine, into Glauber's salts. These experimen
indications and suggestions of the great commerci
processes, and a vision of a Land of Promise op
in a lucrative and beneficent industry, so that du
Belfast he decided to resign his charge, and to f
mind, and, without wholly abandoning the sacre
himself to the manufacture of chemicals as a
After he had given up his church and had es
business in Dublin, he used occasionally to off
street Church in that city.

The process of the manufacture of bleaching p
tion of chlorine by slacked lime was patented
grandfather of the present Sir Charles, in 1799, an
erect a small plant in County Monaghan to work
Tennant had patented on the other side of the Chan
the plant was, may be judged from the fact that
chambers consisted of half casks inverted over a
lime; he had no sulphur, ovens or vitriol chambe
acid he required he obtained from Tennants, of G

Before the year 1815 the Monaghan Works wer
suitable premises, which Gamble built on the ban
below Island Bridge and above Dublin, on land
from Sir William Worthington, a Dublin mercha
the house with extensive garden and grounds,
works, and is at the present day occupied by Mr.

began to make his own sulphuric acid, instead of getting it from St. Rollox; he claimed to be the first manufacturer who introduced into the United Kingdom what was known as the French plan of working vitriol chambers, that was, with a constant draught through them. When his plant was being altered to this plan, his anxiety to retain the secret of the arrangement, induced him with his own hands and during the night to do the necessary plumbing work himself. The articles he manufactured were Sulphuric Acid, Bleaching Powder, Alum, and Glauber's Salts, the last of which he continued to make up from his chlorine residues. The alum he made from pipe-clay which he imported from Poole, and potash obtained from what was known as sulphur ashes. These sulphur ashes were the residues left in the brimstone burners or ovens, into which brimstone mixed with nitrate of potash used to be cast; this work was done by small boys, who with iron spoons kept regularly throwing in the charge.

The process for the manufacture of Glauber's Salts from the residues of the chlorine stills, though little mentioned in chemical books, was carried out on an extensive scale nearly down to the year 1840, about which date Messrs. Thos. Bell and Son, of Newcastle-on-Tyne, prepared sulphate of soda for the alkali manufacture by this process. We believe they found it necessary to suspend the process during the summer, as the conversion of sulphate of manganese into sulphate of soda, did not take place except at a winter temperature. While Mr. Gamble was manufacturing chemicals in Ireland, both his raw products of what is now called the soda manufacture, viz.:—Brimstone, and common salt, were subject to heavy duties. A drawback on the brimstone consumed, was refunded upon the affidavit of the manufacturer; but the exciseman attended the chlorine stills as carefully as they did the stills for making spirits, and saw the salt weighed in and gave a certificate for a return of the duty. It would thus seem that the chemical manufacturer was not subject to the duty on salt, but there were other inconveniences when it was found the duty-paid salt lost far more by so-called drainage on its voyage from Northwich to Dublin, than it did afterwards when no duty had been paid upon it. In the former case it was worth stealing, and in the latter the temptation was not so great.

At the age of forty-four, during the time that he was in business in Dublin, Gamble married Hannah Gower, the daughter of a Dublin solicitor; to them were born one son and three daughters, the daughters are all deceased, one died in Dublin the other two in St. Helens. Mrs. Gamble only survived her husband five years, and on the 16th December, 1852, died at the residence of her brother, John Gower, of Roundwood Park, County Wicklow.

During Gamble's residence in Dublin he was acquainted with James Muspratt, who had, in partnership with a Mr. Abbott, also commenced the manufacture of chemicals in that city, but in 1822 Muspratt, having perceived that Dublin was less favourably situated for his business than Lancashire, left Ireland and came to England. In the year that the Salt Tax was repealed, 1823, he started his works in Liverpool; five years afterwards Gamble also came to England, leaving Dublin and coming to reside in Liverpool in the year 1828. Several months elapsed before he could make up his mind as to which was the best locality and site for an alkali works; for a time he was inclined to select a plot of land at the head of the Birkenhead Float, where a copper ore yard is now situated; his idea was that the brine should be conveyed thither by pipes from Northwich, a scheme similar to that which the Mersey Salt and Brine Company carried out nearly fifty years afterwards, when they established their works at Runcorn. After months of search and consideration, he decided to come to St. Helens, and in this enterprise he was joined by James Muspratt. They erected their works on the banks of the St. Helens Canal, close to the double locks, on the spot which is now occupied by the Globe Alkali Company's works. Gamble was seventeen years Muspratt's senior; they were both men of impetuous natures and strong wills; their partnership only lasted two years. Muspratt, in the year 1830, commenced his Newton works, and Gamble remained in sole possession of the St. Helens property; he carried on his operations for five years without any partner, but in 1835 a works situated near to his own came into the market; it consisted of five sulphuric acid chambers, two of which had never been worked, and a plant for making alum from the blue clay called "warrant," which was obtained from the coal-pits. These works were erected about the year 1830 by Edward Rawlinson, a blind solicitor, from the north of Lancashire; they were managed by a Mr. Williams, who, it is believed, was his brother-in-law. The business was a failure. A bankrupt chemical concern is invariably sold for very much less than its value; Gamble saw that a bargain was to be obtained, and he induced Joseph and James Crosfield, the soap-boilers, of Warrington, to join him in the purchase, which they completed on such terms that the works, after being stripped of the lead they contained, were held by them at a cost of about £400. The connection thus commenced led to the Crosfields becoming partners in Gamble's business in the year 1836, and in the following year Simon Crosfield, a younger brother, who was a tobacco manufacturer in Liverpool, also joined the firm, and undertook the commercial and financial management of the business. The title of the firm was Gamble and Crosfields.

Gamble and Muspratt ran a great risk when they established their works at St. Helens, for the district at that time was a picture of rich fertility. Farmsteads, with their gardens and orchards, came right up to the streets of the small country town, the valleys were watered by brooks that had "here and there a lusty trout, and here and there a greyling" in their clear waters; game also was abundant; even after various works had began to spring up, on wintry nights pheasants would venture to seek shelter from the blast, in the warm sheds that glowed with the light of furnaces. On one occasion, about the year 1845, a stag pursued by Sir John Gerard's hounds scampered into Gamble's works, and crashed into a pile of disused carboys. The chimneys of colliery engine-houses, and cones of glass-works rose amidst farm buildings and stacks of hay and corn. Cornfields, luxuriant hedgerows, and healthy though always stunted timber, for the soil was not congenial to its growth, adorned the landscape. Families that constituted the society of that day, had their comfortable halls, or commodious and well-built residences, with ample and highly-cultivated gardens.

The names of Greenall, Cross, Caldwell, Fildes, Speakman, Orrell, Robinson, Watson, Pilkington, Gaskell, Casey, Daglish, Bromilow, Keates, West, Fincham, Cotham, Morley, the three Johnsons, Grundy, Eckersley, Haddock, and others will recall many memories of those days; a vivid picture presents itself of the pleasant social life of which St. Helens was the scene, and many a tale is still told of the "marlocks" of the men who were then young and of the convivial gatherings of their fathers.

The Plate, Crown, Flint and Bottle Glass trades were all established in St. Helens when Gamble settled there, and the Crown glass makers soon after began to use sulphate of soda; the colliery proprietors welcomed a new trade which promised to increase very largely the demand for fuel; these interests lessened the opposition which would otherwise have been fatal to a manufacture which was associated with serious injury to vegetation, and which was suspected of being prejudicial to health. From the agricultural interest, the pioneers of alkali making encountered bitter opposition. The lawsuits brought against Muspratt by the Corporation of Liverpool, and the landowners of Newton are historical; at St. Helens, Gamble, Crosfield, Clough, Darcey, Kurtz, and Morley only avoided the harass of ceaseless and ruinous litigation by the liberal payment of compensation. That the acid vapours emitted from their works did very great damage was unquestionable, but to make a fair and just assessment of the amount of injury was no easy matter, and it is to be feared that frequently the manufacturers were subjected to extravagant and exhorbitant demands, and that the loudest and least reasonable of the claimants often obtained the lion's share of the plunder. Some farmers found that harassing the chemical manufacturers paid them far better than good farming. Doubtless Gamble and Muspratt under-rated the risks they were running, and magnified the advantages of cheap fuel, ready transit, a good market, and abundant supplies of water, with convenient drainage.

How to condense the noxious and destructive gases, and take away the terrible tax which was being imposed upon them by the farmers, was a question of the first importance; and in the year 1836 William Gossage conferred an incalculable benefit on the alkali trade by the invention of his condensing towers. Gossage also patented one of his processes for the recovery of the sulphur from vat-waste, involving the using up the weak hydrochloric acid which was obtained, and which being produced from the open reverberatory furnace was too dilute to be employed in the decomposition of manganese in the production of chlorine. Gossage's plan was to place the green waste in layers on shelves in wooden stills, to run the weak muriatic acid over it, so causing sulphuretted hydrogen and carbonic acid to be given off, these mixed gases were passed through a further quantity of the waste placed on perforated shelves, such as are used in gas purifiers, and a further quantity of sulphuretted hydrogen was given off by the action of the carbonic acid. The sulphuretted hydrogen evolved in these operations, was burnt with an excess of air in an oven covered with a perforated arch, on which was placed a layer, several feet in depth, of broken bricks, and through these the sulphurous acid was passed into the ordinary vitriol chambers.

Joseph Crosfield, dismayed by the demands made upon their firm for damage to trees, hedges, and crops, insisted on the immediate adoption of Gossage's patents; to this Gamble was strongly opposed, as he believed some plans he had himself conceived, would supersede Gossage's inventions. Crosfield, however, was so impressed with the urgency of the situation, and the value of Gossage's work, that on his own responsibility, and in defiance of his partner's wishes, he concluded an agreement with Gossage to erect the necessary plant for them, and put his process into operation. Gamble was very indignant at his wishes being disregarded, and at a contract being made without his consent; a quarrel ensued between the partners, which was so serious that Joseph Crosfield never put his feet inside the works again. Notwithstanding this, the new process was brought into operation, but the use of it was discontinued about a couple of years afterwards.

Simon Crosfield avoided being drawn into the dispute, and was always on the most friendly terms with Gamble, but the differences between the partners ended in a dissolution of the partnership in 1845, some years prior to which date Gamble had acquired the sole interest in the alum works at Gerard's Bridge. At these works he first had as a partner a Mr. Marsden, the firm being known as Gamble and Marsden, but within a few years it underwent changes, first to Jos. C. Gambleand Son, then to Gamble, Son, and Sinclair, and then back again to Jos. C. Gamble and Son, which it has retained till the present day, when it is consists of Colonel Gamble, in partnership with his four sons, Josias, William, David, and George.

The patent which Mr. Gamble took out in 1839, and which he first put into operation at the Gerard's Bridge Works, claimed :—

First, iron retorts worked in connection with each other. Two decomposing furnaces are connected with a roasting furnace.

Second, the iron retorts constantly worked through a door, open or partly open while the process is going forward, the draft of the chimney drawing in a portion of the external air with the muriatic acid into the receivers. This claim was disclaimed in January, 1845, on the ground that the words have a doubtful meaning.

Third, the use of receivers so arranged that the acid can pass from one to the other or can be cut off at pleasure when strong acid is required; these are filled with glass or pebbles.

Fourth, the use of earthen stills with leaden heads, incased in iron heated by the circulation of hot water, saline solutions, or by steam.

Fifth, the mode of alternate charging of the lime receivers, by which lime already in part saturated with chlorine, is exposed to the strongest gas, and the remnant of gas is absorbed by a surface of fresh lime.

From this patent dated the introduction of the use of the "iron pot" in the decomposition of salt, which, with various modifications, has been continued up to the present day. But some manufacturers, though they adopted the apparatus so patented, maintained that its principle was not patentable, and after protracted litigation succeeded in releasing themselves from any legal liability to pay Gamble royalty for his invention, but Messrs. Tennants, of Glasgow, who produced probably three-fourths of all the bleaching powder then made in the United Kingdom, and who had bought Gamble's patent rights for the whole of Scotland, declined to avail themselves of this decision, and continued their annual payment for the whole term of 14 years, for which the patent was granted. Up to the time they bought Gamble's Scotch patent the Tennants worked the old method of making chlorine from salt, manganese, and sulphuric acid.

Advancing years and declining health told upon Gamble, and in the year 1841 he was compelled to obtain assistance in the superintendence of the works of Gamble and Crosfields. He selected James Shanks, a Scotch engineer, who was born at Johnstone, near Glasgow, and whom Gossage had brought to Lancashire to erect and superintend the plant of his various patent processes. James Shanks proved himself to be a most capable man, aud when in the year 1845 the partnership between Gamble and the Crosfields was dissolved, the Crosfields retaining the business, James Shanks was made a partner in the new firm of Crosfield Bros. and Co., his department being the practical management of the works, whilst Simon Crosfield continued to direct its financial affairs. Shanks continued to be a partner as long as he lived, and his energetic and careful management largely contributed to the repute and success of his firm; not until sometime after Shanks' death, and after the retirement of Mr. John Brock, whom he had trained did Crosfield Bros., & Co. disappear from amongst the chemical manufacturers of Lancashire.

On the 4th December, 1844, Gamble took out a patent to utilise the heat generated in pyrites burners, his idea being to apply it to the concentration of sulphuric acid by passing the hot gas under leaden pans. It is curious to know that Andrew Kurtz applied for a patent for a similar purpose, passing the heat over instead of under the pans. After the final separation from the Crosfields had taken place, Josias Christopher Gamble, on account of declining health, took the world very quietly, visiting his works at Gerard's Bridge two or three times a week, often not getting off his pony during these visits. He was troubled by chronic bronchitis, and the acid gases inseparable from an alkali works were not bearable by him; his health was permanently impaired, and the anxiety which his lawsuits entailed told upon him very heavily, indeed, it may be said that the final struggle with Andrew Kurtz killed both these enterprising and energetic pioneers of the alkali trade. Both these men had to encounter difficulties of which the present generation has no conception, their business had raised up a host of enemies among the farmers who would have annihilated them if they could, trained and capable assistants were not to be obtained, the master had to train his men in every grade of service, new inventions multiplied themselves, and competitive jealousies brought about lamentable and destructive struggles for supremacy.

Gamble did not take opposition lightly; indolence, stupidity, and neglect irritated him intensely, he threw himself h work, and he expected those in his employ to do

The striking feature in the life of Josias Chris contrast which his career presents to that of present day. They start after having received a ing in their youth. Technology was an unknown petent technologist means a man who has a fair al natural science, he must know something of geol even botany. He must be a fair mechanic, an chemist, in addition to this he must be able to rea works in French and German, but in Gamble's e struction could be obtained, the manufacturer h knowledge of the science of his subject, he had t path and feel his way in dim uncertainty.

Another point of contrast is the facilities with any apparatus, however novel or complicated can matter what process is devised, that suitable promptly be manufactured is never suggested; pr or chemical action present no difficulties, the c ready at once to make the way plain, but in Gamb be his own chemist, engineer, and architect. He nature of his reactions, to devise his tests, to co design their construction; he had no trained dra he could entrust his plans, he had to do all this

The difficulties which he encountered may be ill in which he had to make his chlorine stills. They ware, in them was placed the ground manganese charge was kept continually agitated by a mecha kept hot by the circulation of hot water. Ther manufacturers of such vessels as Gamble required a take their construction himself. They were m being built up like pots used in making gla months to build and dry them before they co first made vessels of 100 gallons but afterwards h to 400 gallons. These earthenware stills were, c when unground manganese and live steam began

It was thus throughout his whole works, insignificant as thermometers and hydrometers, wh in his own laboratory, being an expert with the bl

But not only had he to devise and construc apparatus, he had also personally to instruct as w his workmen in each department. He never empl a chemist; if he wanted assistance he taught m required them to do.

Though Gamble loved his business he was n work. He was a great reader, both of scientif with his business and of the current literatur learned French after he was married, probably of his wife, and could read French scientific w facility. Though his family never heard him spe his way in France when little of the English lang in French hotels.

Gamble, when he first came to St. Helens, resid Duke-street; he then took a farm and lived for seve house on the property which belonged to Sarah extended from Raven-street to Pocket Nook. T Bishop's Bridge; the farm is now covered by the Chemical Works, etc.

In the year 1839 he removed to Sutton House, a the close of his life.

He died at his residence, on the 27th Januar after the birth of his first grandson, and was in which his two daughters had been interred, Chapel, St. Helens.

Sutton House is now converted into the Cottage

Gamble's personal appearance was as remarka was original. He habitually wore black broa cravat, and his hat had a brim almost broad en The black cloth of his clothes was often bes showing numerous red spots. Although descende were adherents of the old-fashioned Toryism of displayed pronounced Liberal opinions, and was of the body of United Irishmen; indeed, it is agitation burst out, he hid in his house some of th of that party.

He remained a steadfast and decided Liberal, Free-Trader, and a liberal supporter of the Ant He was an advanced Financial Reformer; he adv of all Customs Duties; and, indeed, of all in much so, that he would have applied the Income even so low as twenty shillings per week, and hav taxes.

The presentment which we have of him is not th theorist or an ingenious inventor, not that of a ma days and nights in experimental research, but of one who was eminently practical, who sought to turn to useful and profitable account the knowledge he had acquired.

His University training did not unfit him for the concerns of manufactures and of commerce, and the benefits which his successors are conferring on the town in which he found his home, would indicate that the deep impressions of the work with which his own life opened are still being perpetuated in the characters and labours of his descendants.

The business was founded by courageous enterprise and self-reliance, was conducted in the face of difficulties and opposition which called forth great endurance and perseverance, and has been established, built up, and extended by unwearying industry and vigilant attention. On the qualities which this story reveals are founded the securities of permanent and increasing prosperity, and the benefits of individual success are ever enhanced when they are associated with devotion to the public weal.

ANALYSIS OF MARINE OILS.

By Russell W. Moore, A.M., M.Sc.

In connection with an extended examination of a large number of oils and fats, the results obtained in the case of some porpoise and blackfish oils were of so unusual a character as to deserve particular mention.

It has for some time been considered that the analysis of the various fats and oils occurring in nature would show a percentage of insoluble fatty acids amounting to some figure in the neighbourhood of 95 per cent. A notable exception is butter fat, in which the insoluble fatty acids range between 87·5 and 89·5%, and, in some cases, even slightly higher. The first chemical process used to distinguish butter from other fats was based solely upon this difference. Later, cocoanut oil was found to contain fatty acids soluble in large amounts of water, and sufficient washing was found to reduce the per cent. of insoluble fatty acids to a figure even lower than that given by butter. (*Chemical News*, December 5th, 1884.)

It must, however, be granted that while this opinion was founded on a very large number of analyses of butter, the number of such analyses made on other fats was extremely small when the great number and variety of fats and oils occurring in nature is taken into consideration. Oleomargarine, beef and mutton suet and lard, and a few oils, comprised nearly the whole list of fats analysed.

An examination, however, of most of the natural fats and oil serves to establish in most cases the fact that the insoluble fatty acids amount to about 95 per cent. A notable exception is found in the case of some marine oils examined by the writer.

The oils in question are known in the market as porpoise jaw and blackfish jaw oils, and are obtained from the soft fat of the head and jaw by allowing the oil to exude from the fat. The oil thus obtained is exposed to cold, and the portion remaining fluid is racked off. The resulting oil, carefully skimmed and strained, is of a straw yellow colour, thin and limpid, and by no means of an unpleasant odour. It is used for lubricating fine machinery and commands a very high price.

The oils examined were five in number, as follows:—

No. 1. Porpoise jaw oil skimmed and strained.
No. 2. Porpoise jaw oil skimmed and strained.
No. 3. Porpoise jaw oil not skimmed and strained.
No. 4. Blackfish jaw oil skimmed and strained.
No. 5. Blackfish body oil.

The oils were first examined by the wash process to determine the percentages of soluble and insoluble fatty acids. The results were as follows, the figures given being the mean of duplicates agreeing closely.

	Soluble Fatty Acids.	Insoluble Fatty Acids.
No 1.	17.18%	72.05%
No. 2.	21.44	68.41
No 3.		96.50
No. 4.	21.79	66.28
No. 5.	2.46	93.07

The oils were next examined by the Reichert process with the modification suggested at the time by Dr. Waller. The distillation was continued by adding 50 c.c. of water to the flask and distilling until a practically neutral distillate was obtained, adding 50 c.c. of water between distillations, titrating as in the Reichert process and calculating the acidity to butyric acid.

The results obtained were as follows:—

	Reichert Figure.	Total Acidity.
No. 1	47.77 c.c.	17.18%
No. 2	56.00	20.97
No. 3	2.08	1.42
No. 4	65.92	24.31
No. 5	5.60	2.34

The saponification number was obtained by the process of Koettstorfer. Considerable difficulty was experienced in this determination in obtaining concordant results, since the combination of the alkali with the fat appeared so feeble that even very dilute standard hydrochloric acid appeared to decompose the soap and liberate fatty acids. This was seen on diluting with water, when oily drops would appear while the liquid was still alkaline. By using large quantities of alcohol concordant results were obtained as follows:—

	Mgs. K O H per Grm.
No. 1	253.7
No. 2	272.3
No. 3	143.9
No. 4	290.0
No. 5	197.3

The iodine coefficient was also determined by the Hübl method and gave the following results:—

	Grms. Iodine per 100 Grms.
No. 1	49 6
No. 2	30.9
No. 3	76.8
No. 4	32.8
No. 5	99.5

It will thus be seen that the oils which had received the full treatment furnish abnormally high figures for soluble or volatile fatty acids and correspondingly low figures for the insoluble fatty acids. They constitute the most notable exception to ordinary fats in this respect. Of course the high per cent. of soluble acids is due to the treatment which the oils undergo in the refining process by which the glycerides of the lower fatty acids are concentrated in the oil that is finally strained off. This is conclusively shown by the foregoing figures, oils Numbers 1, 2, and 4, that received the full treatment, giving extraordinarily high results for soluble fatty acids and saponification equivalent and correspondingly lower iodine absorption coefficients, since the lower glycerides are of the acetic acid series and are indifferent to iodine.

The volatile acid present in these oils should be calculated to valeric acid. It was, however, calculated by butyric acid by the writer in order that the results should be comparable with other analyses.

It would be a matter of interest to conduct some experiments with butter fat by subjecting it to a similar process of freezing and straining and examining the resulting product, and the writer proposes to conduct such a series at his earliest opportunity. Valuable information regarding the nature of the composition of the molecular groupings in butter fat in this way may be obtained.

The oils examined as above were obtained by Dr. E. Waller from the manufacturer, and the analyses were made in his laboratory, New York City, and under his direction.—*J. Am. Chem. Soc.*

THE STANDARD OIL TRUST.

(From the Boston *Herald*).

It is a curious fact that while the public sentiment in this country has been turned strongly against the formation and operation of industrial trusts, and while such organisations as the sugar trust, the American cotton-seed oil trust and others of a like class have been the objects of popular denunciation, very little has been said concerning the operations and effect of the first, the largest and the most successful undertaking of this character. The Standard Oil Trust is the association which all of these other combinations have endeavoured to copy. Its managers, or lawyers, were the first to conceive the idea of welding various corporations into an impersonal and intangible entity, and it was the enormous wealth that was accumulated by the Standard oil magnates that made the sugar refiners and the bagging manufacturers desirous of following in their footsteps.

A merit that the Standard Oil Trust possesses, which is not shared by most of those that have copied its methods, is that it is in no way dependent upon the protective tariff. If the customs tax upon sugar was entirely removed, or a uniform tax was levied both upon raw and refined sugars, the sugar trust, while it might pay to the holders of its certificates a fair return, could certainly not continue to pay 10 per cent. dividends upon a capital inflated by water three times over; for the importation of foreign refined sugars would steady the American market and keep the prices down to a reasonable level. The same remark would hold true of the bagging trust. Jute bagging would be freely imported from England if the wide margin of profit that has lately existed was not kept to the American manufacturers interested in the trust by means of the customs barriers. But the Standard Oil Trust has been free from, and independent of, foreign competition, for the reason that it had possessed, and probably for a long time to come [illegible] the best field for producing, collecting, refining [illegible] petroleum oil that exists on the face of the earth.

[illegible] of the Standard Oil Trust is the shrewdness shown by [illegible] in not attempting to defiantly oppose public opinion. [illegible] had the market to itself, and yet it has not attempted to [illegible] profits it has secured by materially increasing the [illegible] oil to American consumers. If it had put up the [illegible], as it probably could have done, two or three cents a gallon, it [illegible] made an addition to the profits of its shareholders amounting [illegible] number of millions of dollars a year, but in this way [illegible] opposition would have been aroused, and legislative and con[illegible] interference would have been invited, thus imperilling the [illegible] profits already secured. In other words, the managers of this [illegible] realised from the first that they had a bountiful bonanza at their [illegible], that might be counted upon to yield great returns year after [illegible], as long as it was carefully nursed, while the eager desire for too [illegible] a profit might wreck the whole enterprise in a relatively short [illegible].

What the managers of the Standard Oil Trust have done has been to [illegible] their influence to keep down the price of crude petroleum to the lowest possible point, to improve and utilise in every way methods of [illegible] the cost of transporting and refining oil, and to keep the large [illegible] resulting from these and kindred economies to themselves, while [illegible] oil to the consuming public at no higher prices than the latter [illegible] before the Standard Oil Trust was formed. The system pursued [illegible] well illustrated by comparing it with another form of consolida[illegible]—that is, the trunk lines of railway to the West. If Mr. Vander[illegible], after he had succeeded in establishing a through line to Chicago, [illegible] able to command the field, and had put into his own pocket [illegible] of the economies brought about by consolidation and by the [illegible] price of steel rails and railroad equipments, the 100,000,000 [illegible] or more of which he died possessed would have been but a [illegible] of the accumulations that he might have laid up. But Com[illegible] Vanderbilt was compelled, by the competition of other rail[illegible], to give to the public some share in these great savings, so that [illegible] rates over the New York Central and connecting lines are only [illegible] a quarter of what they were before his plan of consolidation went [illegible] effect.

[illegible] grounds it would be no more than justice for the Standard Oil Trust to grant to the American public some share in its tremendous [illegible], but this, considering human nature, and the human nature of [illegible] who are in the management of this association, is altogether too [illegible] to expect. They possess a great natural monopoly; they pro[illegible] to work this in such a way that, without arousing public opposi[illegible], they may secure through a long series of years the largest possible [illegible] of profit. Unless foreign competition sets in, or all trusts are [illegible] interdicted by natural legislation, the chances are that in a few years more the leading men in the Standard Oil Trust will be the [illegible] capitalists in the country.

THE INCREASING DEMAND FOR COPPER.

RECENTLY there have been numerous inquiries made relative to [illegible] future of the copper market, the prospective demand for [illegible] metal, and the indications as to whether that demand will increase, [illegible] or remain as it is now. The authorities most capable of [illegible] at the exact situation agree in saying that in less than six [illegible] after the collapse of the French syndicate it became evident [illegible] current production of the mines could not meet the great [illegible] for copper in new uses, hence the only source from which the [illegible] of consumption over production could be supplied was the surplus [illegible] by the syndicate. About half of this stock has already been [illegible], and it is estimated that it will entirely disappear before the [illegible] the year. The new demand is mainly for electrical purposes, [illegible] field, the edges of which have just been touched. As an [illegible] of what the demand from this quarter is likely to be, it is [illegible] that 80,000 tons of copper will be consumed in lighting the [illegible], to say nothing of what will be taken by the other great [illegible] world. In addition it is claimed that 10,000 tons will be [illegible] sulphate of copper to destroy the vine pest phylloxera, and a [illegible] very much greater amount in the introduction of the new [illegible] powder.

[illegible] in the rise of electricity for artificial light and [illegible] power, bids fair to make a most decided change in the value of [illegible]. The old mines can hardly increase their product this year, [illegible] cannot prepare for regular production in time to meet [illegible] demand. Therefore it is evident that under the con[illegible] liable to continue and become more pronounced, [illegible] great prosperity in store for the copper interest.—[illegible].

THE ALKALIS AND THE TARIFF IN AMERICA.

NO class of products better illustrates the effects of tariff and the evil effects of free trade on prices than do the alkalies. The uses of soda ash, the caustic sodas, bleach and other similar chemical products that are known generally as the alkalies, are too well known to need any description. In paper making, in glass making, in the manufacture of textiles, and in many other of the important industries, these alkalis play a very important part.

Most of these alkalis have for many years been subject to duties that have been fairly protective. This duty has resulted in the establishment of the manufacture of these alkalis as important industries in this country. The last industry to become fixed with us is that of soda ash. Under the operation of these protective duties, and as a result of them, these manufactures have been established, and the price of the alkalis is gradually declining, until their prices are lower than they ever have been before.

To this statement there is but one exception, and that is with reference to bleach or chloride of lime, on which there is no duty, and which is not manufactured in this country. While the price of soda ash, of caustic soda and the other alkalis which are dutiable, has declined, the price of bleach, which is not dutiable and not manufactured in this country, has increased.

The importation of soda ash, on which there is a quarter of a cent per pound duty, and the invoice values of the same from 1886 to 1888 were as follows:—

Year.	Pounds imported.	Value per lb.
1886..................	279,931,929 ..	.012c.
1887..................	263,274,392 ..	.011c.
1888..................	267,896,710 ..	.010c.

The importation of chloride of lime or bleaching powder for the same years has been as follows:—

Year.	Pounds imported.	Value per lb.
1886..................	98,046,208 ..	.014c.
1887..................	103,097,847 ..	.015c.
1888..................	101,699,978 ..	.016c.

Now, chloride of lime, or bleach, as it is called, is a by-product of the manufacture of soda ash by the Le Blanc process. It is free of duty, and none is manufactured in this country, and it has been a notorious fact that the manufacturers of the Le Blanc soda ash have attempted to make up their loss on soda ash by increasing the price of bleach, and hence, as will be seen from the above table, while there has been a decrease of 1-10th of a cent a pound each year in the invoice value of the soda ash imported, there has been an increase of 1-10th of a cent a pound each year in the invoice value of bleach.

With a duty on bleach there is no reason why it should not be manufactured in this country.—*American Manufacturer*.

PRODUCTION AND SALE.—One of the leading iron circulars puts the cost of producing hematite iron at 25s. for the ore for one ton of pig iron, and 38s. 6d. for the coke, in West Cumberland. Something is to be added for wages, limestone, depreciation, making over £4 per ton. The hematite iron warrant price is now £3 12s., or 8s. less than the estimated cost of production. Sale below the cost of production cannot be long maintained, and either prices must rise or production must fall soon.

BOHEMIAN GLASS.—Gablouz, on the Neisse, is the centre of the Bohemian glass industry. When the colliers of Northumberland strike, those wanting fuel can go to Wales for it; and when the mining is paralysed in Belgium, the purchasers flock to Westphalia or elsewhere. But if the glassmakers of Bohemia abandon their work *en masse*, it will be no easy matter to supply the chemists and others in all countries who depend upon their produce—upon the Bohemian glass, which, of all glass, is that which is least acted upon by chemical reagents. Bohemian glass is a silicate of potash and lime, and is very difficultly fusible.

THE TIN-PLATE TRADE.—The advisability of the adoption of some course for the restriction of make is emphasised by the very unremunerative prices (considering the still high quotations for bars) and the stoppage of several works owing to the indisposition of their owners to continue working at a loss. From a recent official report of the Swansea Metal Exchange, we extract the following:—"It was reported that six tin-plate works have already stopped, as they could not pay the present price of steel bars and manufacture plates at a cost without incurring a positive loss; therefore they consider it the wisest course to cease making and close their works. These will be followed by a great many more works, who will cease working upon the completion of the balance of any unexpected orders on hand. In the face of this disturbed state of things buyers of tin-plates are evidently keeping back orders, and are holding off, evidently with the view of ascertaining how far the price of tin-plates will be affected, and a large majority of the makers present concurred in this decision."

Correspondence.

**** We do not hold ourselves responsible for the opinions expressed by our correspondents.*

DETERMINATION OF MOISTURE IN WOOD PULP.

To the Editor of the Chemical Trade Journal.

SIR,—In the article "On the Determination of Moisture in Wood and Straw Pulp," by Messrs. Geo. E. Davis and Alfred R. Davis, in your last week's issue, Messrs. Davis recommend that the pulp should be dried at 212° Fahr., and, in order to convert the weight thus obtained into the weight of "air-dried" pulp, that an addition of 10 per cent. be made. But why attempt to perpetuate the practice of buying and selling pulp on so indefinite and uncertain a basis as its air-dried weight must necessarily be? Surely it would be much better to buy and sell this article on the amount of pulp it contains when dried at 212° Fahr. In this case it would only be necessary to stipulate in the contracts that the pulp should contain, say 45 per cent. of pulp dried at 212° Fahr., instead of, say 50 per cent. of "air-dried" pulp. Chemists then would be able to state without doubt exactly how much of such dried pulp is contained in a given sample. This method would be much more likely to prevent disputes than the methods suggested by Messrs. Davis, which, in my opinion, is liable to be misunderstood and misinterpreted.—I am, dear sir, yours truly,

75, Side, Newcastle-upon-Tyne. JOHN PATTINSON.

To the Editor of the Chemical Trade Journal.

SIR,—Allow me to make a few remarks on the article "The Electric Decomposition of Chlorides in Solution," which appeared in No. 142 of the *Chemical Trade Journal.*

First of all, as I am a foreigner, I ask your indulgence for my bad English.

The writer of the article enters into some theoretical demonstrations which I will leave aside, because all the equations and figures show that 1·53 volt. will be sufficient for the production of chlorine by electrolysis, and that 1·54 h.p. per hour will give 1 kilo. of chlorine. The problem to be solved is the production of a substitute for hypochlorite of calcium, which is as powerful a bleaching agent, costs less, and has all the qualities, but not the drawbacks, of bleaching powder. I will not follow or criticise the article, the end of which is devoted to the demonstration of the results obtained in eight electrolysers from a 64 h.p. engine, driving a 1,250 amperes and 40 volts. dynamo, and which he represents as being the equivalent of three tons of bleaching powder per week of 132 hours.

He is evidently a chemist, and chemists sometimes make mistakes when they argue against their new enemy, electro-chemistry, as did Dr. F. Hurter and others, or as did Messrs. Raoul Pictet, Cross and Bevan, etc., who made the *Mariée trop belle*, and not only said that electrolytic Mg. (CLO)2 bleaches better than bleaching powder in the ratio of 5 to 3, but declared that they had obtained a yield of chlorine varying from 1·40 to 1·81 gramme, and on an average 1·47 gramme per ampere hour.

I do not find in the above-mentioned article any data respecting the density of the solution or the yield of chlorine. A concentrated solution is expensive, as bleaching liquor has to be applied to bulky materials, such as tons of paper pulp or of fibres and tissues.

The question of the yield of chlorine per ampere hour is still more important, as the cost of bleaching must be much higher if the yield of chlorine is only 0·60 grammes per ampere hour instead of ·70, 0·80, 0·90, or 1 gramme. Theoretically, the yield of chlorine is 1·322 gramme per ampere hour.

This result is never practically obtained, and still less have Messrs. Raoul Pictet, Cross and Bevan, etc., obtained the 1·47 gramme per ampere hour which they claimed in their report on the Hermite process.

I will take as the basis of the following estimate the yield of one gramme of chlorine per ampere hour, to show that electrolysis compares favourably with purely chemical methods.

In twenty tanks, I will produce 20 kilos. of chlorine after the passage of 1,000 amperes through them during one hour. Within 24 hours, the production of chlorine will be 480 kilos. which will represent almost 1½ ton of chloride of lime.

To obtain 5,000 watts, assuming that the brake horse power is equal to 600 watts, it requires 8·33 h.p., the price of one kilo. of chlorine is, therefore, the cost of 8·33 h.p. per hour. In a set of 20 tanks 166·66 h.p. are necessary to generate 20 kilos. of chlorine, and for the constant generation of these 20 kilos. of chlorine per hour, during 24 hours, 3,998 h.p. hours are necessary, and will have transformed a given quantity of chloride of sodium solution into a hypochlorite of sodium solution, containing 480 kilos. of chlorine.

Assuming that the consumption of coal is 2½ lbs. per h.p. hour, it will represent in round figures six tons, to the value of which, adding that of the loss in salt and other items, the production of hypochlorite of sodium containing the same quantity of chlorine as a given quantity of bleaching powder will cost about 50 per cent. less.—Yours truly, EMILE ANDREOLI.

London, February 24th, 1890.

NEWCASTLE CHEMICAL WORKS COMPANY, LIMITED.

THE following is the report of the above company:—The accounts for the year 1889 are presented herewith to the shareholders. The net profit is £7,049. 8s. 6d., out of which a dividend of 6 per cent. per annum will be paid to the holders of the preference shares, absorbing £3,600. To the balance, viz., £3,449. 8s. 6d. must be added £8,989. 4s. brought from the last account, making £12,438. 12s. 6d., from which the sum of £5,000 is written off against expenditure on caustic soda plant, leaving a balance of £7,438. 12s. 6d. to be carried forward. It is a matter of extreme regret to the directors that the Le Blanc chemical industry does not participate in the general prosperity of the country.

The association which has existed amongst the makers for many years with the object of controlling the production of bleaching powder expired on the 31st December last, all efforts to effect a re-establishment having proved ineffectual. The value of this product has, in consequence, recently suffered a serious decline, to the extent of £2 per ton or thereabouts. Raw materials (coals especially) and labour show a sharp advance, and the cost of manufacture of the company's products will be seriously enhanced during the course of the ensuing year. On the other hand, caustic soda shows an advance of about 30s. per ton as compared with the lowest price of last year.

The plant for the recovery of sulphur, which is approaching completion, will probably be put into operation early in May next, and the sale of sulphur ought materially to influence the result of the company's operations. The cost of the erections will exceed the anticipations of the directors, and financial assistance was obtained from the late chairman upon the security of mortgage debentures. The sums so lent to the company amounted up to 31st December last to £54,000.

The directors regret to have to record the decease of Mr. Christian Allhusen, which took place on 13th ult. He had been chairman of the company since its formation, and gave its business his daily attention. By substantial pecuniary support to the company in the expenditure which it has had to make, he constantly sought to establish it in a position of permanent prosperity. It is proposed to fill up the vacancy at the Board by the election of Mr. John Edward Davidson to the position of managing director, with the same remuneration as granted to the other directors. Mr. Davidson, as one of the executors of the late Mr. Allhusen, and also as a considerable holder of shares in his own right, is largely interested in the company. He has been exclusively engaged in its business for the last 24 years, and during the last 12 years has acted as secretary and general manager of the commercial department. The directors believe him to be not only entitled to, but well qualified for, the position to which he is now recommended.

The salt works give satisfactory results. A well has recently been sunk upon another part of the royalty held by the company. A salt bed, 30 feet in thickness, was met with at a depth of 778 feet, being 312 feet nearer the surface than at the other portion of the royalty already proved by the company.

The retiring director, Mr. Alfred Allhusen, is eligible for re-election. The auditors, Messrs. Monkhouse, Goddard and Co., offer themselves for re-election.

HOW DECORTICATED COTTON CAKE IS MADE.—The cotton seed mostly grown in America, and from which the decorticated cotton cake is made is not clean and free of cotton, but is tightly bound round with lint. The main bulk of the cotton is picked off by the negroes, but the seed is afterwards sent to the ginners, who give off a deal more cotton. It is now in a fit state to be marketable, a large proportion of the seed finds its way back to the land for manure, and the smaller proportion reaches the seed crushers. These gentlemen take off as much more cotton as they can by means of linters; the seed is thence passed over to the "hullers," which sculp clean off all the black husk of the cotton seed, together with any cotton remains, and leave only the kernel. The husks now proceed to the boiler fire; the kernels are ground, rolled, heated, and then put into the hydraulic presses, where the oil is extracted, and the decorticated cotton cake results. Owing to improved machinery the oil left in the cotton cake made in the new mills shows under 8% and the cake is left perfectly hard. The idea of extracting as much oil as possible from the seed is now considered to be by no means an advantage so far as the feeding value of the cake is concerned.—*Tropical Agriculturist.*

PRIZE COMPETITION.

A LITTLE more than twelve months ago, one of our friends suggested that it might be well to invite articles for competition, on some subject of special interest to our readers; giving a prize for the best contribution, he at the same time offering to defray the cost of the first prize, and suggesting a subject in which he was personally interested.

At that date we were totally unprepared for such a suggestion, but on thinking over the matter at intervals of leisure, we have come to think that the elaboration of such a scheme would not only prove of benefit to the manufacturer, but would go some way towards defraying the holiday expenses of the prize winner.

We are therefore prepared to receive competitions in the following subjects :—

SUBJECT I.

The best method of pumping or otherwise lifting or forcing warm aqueous hydrochloric acid of 30 deg. Tw.

SUBJECT II.

The best method of separating or determining the relative quantities of tin and antimony when present together in commercial samples.

On each subject, the following prizes are offered :—

One first prize of five pounds, one second prize of one pound, five additional prizes to those next in order of merit, consisting of a free copy of the *Chemical Trade Journal* for twelve months.

The competition is open to all nationalities residing in any part of the world. Essays must reach us on or before April 30th, for Subject II, and on or before May 31st, in Subject I. The prizes will be announced in our issue of June 28th.

We reserve to ourselves the right of publishing the contributions of any competitor.

Essays may be written in English, German, French, Spanish or Italian.

TECHNICAL EDUCATION IN ELEMENTARY SCHOOLS.

THE committee of the National Association for the Promotion of Technical and Secondary Education have submitted to the Education Department the following suggestions for the modification of the code as regards elementary technical education :—

DRAWING.—(1) Drawing to be introduced in infant schools, at least for boys. (2) Drawing to be made compulsory in boys' schools. (3) The minute requiring cookery to be taught in girls' schools as a condition of receiving grant for drawing to be repealed.

OBJECT LESSONS.—(4) No school to be recognised as efficient which does not provide in the three lower standards a graduated scheme of object lessons in continuation of Kindergarten instruction in the infant school.

SCIENCE.—(5) In order to encourage science as a class subject, the clause requiring English as one of the class subjects to be cancelled, and the teaching of science as a class subject to be further encouraged in the upper standards by an additional grant. (6) Scholars of any public elementary school to be allowed to attend science classes held at any place approved by the inspector, and such attendance to count as school attendance; (7) Examinations in science to be conducted orally, and not on paper, especially in the first five standards. If the inspection is satisfactory, an attendance grant of 4s. to be made for scientific specific subjects. (8) Managers to be encouraged to submit alternative courses of instruction in specific subjects under Art. 16 (Code 1888). Such subjects to receive a grant on the same principle as the subjects enumerated in Art. 15. [Art. 16. "Any other subject, other than those mentioned in Art. 15, may, if sanctioned by the Department, be taken as a specific subject provided that a graduated scheme of teaching it be submitted to and approved by the inspector." But Art. 109 (*g*), which lays down the condition for grants, says, "The specific subjects which may be taken are those enumerated in Art. 15."] (9) Grants to be made towards apparatus for science teaching and school museums.

MANUAL INSTRUCTION.—(10) Manual instruction to be introduced in boys' schools, corresponding to needlework for girls. (11) Instruction in the use of simple tools to be introduced in the higher standards as a specific subject, and grants to be paid thereon. (12) Provision to be made for the introduction of elementary modelling in connection with the teaching of drawing, and a grant to be made in connection therewith. (13) Instruction in laundry work to be e[...] schools, so far as practicable, as a part of domestic e[...]

EVENING SCHOOLS.—(14) The clause providing [...] may be presented for examination in the additional s[...] be cancelled, to enable scholars to earn grants thou[...] instruction in the standard subjects. (15) The numb[...] subjects" which may be taken to be increased from [...]

TRAINING COLLEGES.—(16) Day training college[...] of training to be recognised. The universities an[...] colleges to be utilised for the training of teacher[...] arrangements can be made.

THE SALARIES OF SCOTTISH [...] PROFESSORS.

A Parliamentary paper, just published, contains w[...] a "nominal return" of the salaries, fees, and other e[...] professors of the Scottish universities, which wo[...] account in fixing the pensions to which these gentl[...] titled after a certain period of service. The ret[...] chief prizes in connection with university teaching in[...] found in the faculty of medicine at Edinburgh. T[...] Dr. A. Crum Brown, the Professor of Chemistry[...] £3,450. ; and of Mr. William Turner, the Professo[...] £3,000. on an average ; while the following professo[...] more than £2,000., viz., Professor Rutherford, Insti[...] £2,521. ; Professor Cossar Ewart, Natural History, [...] Greenfield, General Pathology, £2,351. ; Professo[...] Medica, £2,235. ; and Professor Balfour, Botany, [...] Faculty of Arts the highest income is Professor S[...] £1,233. ; in the Faculty of Divinity the highest is [...] Biblical Criticism, £902. ; and in the Faculty [...] Rankine's, Scots Law, £1,111. In Glasgow, [...] Anatomy, heads the list with £2,233. (value of resid[...] Professor Jack, Mathematics, formerly of Owens Co[...] comes next with £1,758. (value of residence include[...] the only incomes above a thousand are Professor Ni[...] History, with £1,054., and Professor Struther's, Anat[...] At St. Andrew's, Professor Scott Lang, Mathem[...] Principal Donaldson, £572., and Principal Cunningl[...] sive of residence.

Trade Notes.

THE DISPOSAL OF RECOVERED SULPHUR.—At [...] producers of recovered sulphur under the Chance [...] held the other day in the North-Western Hotel, it [...] decided to form a limited company for the sale of [...] estimated that this decision will affect 40,000 or 50,[...] of sulphur, which is chiefly produced in Widnes and [...]

PIG IRON PRODUCTION OF GERMANY.—Accor[...] tical returns of the German Iron and Steel Manufa[...] the pig iron production of the German Empire an[...] December amounted to 391,523 tons, made up of[...] spiegel iron 184,379 tons, Bessemer pig 33,788 [...] 124,386 tons, and foundry pig 48,970 tons. In De[...] production was 354,866 tons, in November 1889, [...] total pig iron production in 1889 was 4,387,505 ton[...] tons in 1888.—*Kuhlow*.

A CARBONIC OXIDE INDICATOR.—An apparatu[...] presence of carbonic oxide gas in the atmosphere [...] by M. Rasine. Its principle depends on the prope[...] num to absorb carbonic oxide with evolution of sens[...] circuit of an electric current, two metallic plates ar[...] over each other, wh[...] l w[...]n touching close the cir[...] dent lever is placed near these plates, in the fo[...] from one end of which the upper contact plate is s[...] of an easily combustible thread. This thread is [...] containing a little cotton powder dusted over with [...] If this arrangement is exposed in an atmosphere c[...] oxide, the spongy platinum will absorb it and set [...] which will in turn burn the thread, and so cause th[...] to complete the circuit, and thereby ring a bell, [...] designed indication of the action of the instrument [...] seen whether the spongy platinum would always r[...] in this way, and also that the same effect would [...] any other gas to which the apparatus would be ex[...] *Gas Lighting*.

A) Yankee Tall (S)talk.—We are accustomed to look for big ings to America. It is a country of big lakes, big rivers, big rail- ys, big jokes, and big lies. To these must now be added big imney stalks. We have it on the authority of the *Boston Globe* that : tallest chimney in the world is receiving its finishing touches Fall River, Mass. Its height, according to our contemporary, from : top of the granite foundation to the cap is 350 feet. Would it rprise the editor of the *Boston Globe* to know that, in chimneys at ıst, his country does not lick creation; and that, in Glasgow, we ve a stalk—that of Messrs. Charles Tennant and Co., Ltd., at St. ɔllox—whose total height is 455 feet 6 inches and from surface 435 :t 6 inches. In other respects also the stalk at St. Rollox leaves its ınkee cousin behind. The outside diameter of the former being at ındation 50 feet, and at surface 40 feet; while the latter is only 30 :t in diameter at the base. A peculiarity of the St. Rollox stalk is, at it has an inner cone inverted, whose total height is 263 feet, and ɔm surface 243 feet, with an inside diameter at foundation of 12 feet. educting the usual discount from information we receive from nerican sources, we may calculate that the little fellow hiding up the ɪwser leg of St. Rollox chimney is about the height of "the tallest ɪlk in the world" at Fall River. We have only to add that Glasgow s another larger still, should brother Jonathan's (s)talk grow taller.

Market Reports.

MISCELLANEOUS CHEMICAL MARKET.

The market for alkalies has been tolerably steady during the past ek, and manufacturers find no difficulty in obtaining full market es. Caustic soda 70 % moves off steadily at £9. per ton f.o.b., and s white and cream are at corresponding values. Soda ash is very n, and makers generally are well booked with orders for some time come. Prices on the spot are now 1¼d. per deg. for ordinary ash, d 1 7/16d. to 1½d. for refined. Soda crystals are equally firm and in ɔd demand. Chlorate of potash steady at 4¾d. per lb. Bleaching wder is selling freely at £5. 5s. to £5 7s. 6d, free on rails makers' rks. There is but little fresh business doing in sulphur of various ılities, and prices on the spot are unchanged. The English pro- cers are, however, well sold, and the continuance of good demand vitriol is keeping the Chance Recovery Plant well employed. Salt te is still scarce, and worth 32s. 6d. per ton in bulk. Muriatic d is without any change. Potash caustic and carbonate are in derate request, and prices are steady. White powdered arsenic is ding ready outlet, and prices are firmly maintained. Sulphate of pper is at £26. 10s. to £27. for prompt delivery, but buyers are nerally holding off from forward purchases. There is but little vement in acetates of lime, and prices remain without change. etate of soda is in a similar position, and all other acetates and tic acid are quiet. Yellow prussiate of potash can be bought at htly lower price—9½d. to 9¾d., according to make and quantity. lphocyanides scarce on the spot and prices tending upwards. Tartaric d quiet at 1s. 2½d. to 1s. 2¾d. per lb. Carbolic acid crystals 10d. 10¼d. per lb.; liquid 1s. per gallon. Picric acid in fair demand at 3½d. to 1s. 4d. per lb.

REPORT ON MANURE MATERIAL.

Business continues on a very moderate scale both in nitrogenous l phosphatic manure material, the very low prices being realised by ners for their crops, making manure manufacturers less sanguine out the future than they were a month or two back. Still, there is pressure of supply of phosphatic material upon the market, and ces are, consequently, fairly maintained. With nitrogenous terial it is different, the excessive supply of nitrate causing a weak- s all round, and a decided tendency to lower prices.

Charleston River phosphate rock cannot be bought under 11d. per t, in cargoes, cost, freight, and insurance to U.K. ports, but offers cargoes of Lank rock are invited at ½d. per unit less. Further ge business is reported in Somme and Belgian for Continental sumption, but we do not hear of any important business done for United Kingdom. There is considerable inquiry for Canadian k, but raisers still refusing to name reasonable prices. We have , so far, heard of any contracts for next season.

There are no sales of River Plate cargoes of bones reported, though ıe concession would be made on prices recently asked for December pments £5. 8s. 9d. would be accepted for a mixed December go, and we should say that £5. 10s. was about the value of a cargo bones only.

The market for East Indian bones and bone meal is somewhat ıer, 800 tons meal, arrived at Liverpool, having been stored with- out being put on the market, and owners intimating their intention of storing further shipments now about due. Spot price is £5. 7s. 6d. ex quay; £5. 5s. has been accepted for a shipment near at hand, but it is doubtful whether this price would now do for earlier than February —March steamer shipment Further sales of crushed Kurrachee bones have been made at £5. 3s. 9d. for shipment, and the purchase could be repeated, but for parcels shortly due at least £5. 5s. would have to be paid. The tendency of this market seems decidedly firmer.

Nitrate of soda continues very weak in all positions. Spot price is 8s. 3d. for fine quality, and something less than this would have to be taken for cargoes arrived or due. There seems to be no disposition whatever to speculate for an advance, notwithstanding the reports of the combination of producers to reduce the output.

Superphosphates are still quoted 50s. in bulk delivered f.o.r. at works.

Further sales of fine River Plate dried blood have been made at 10s. 6d. per unit, ex quay; less being refused. For ground ammonite lowest price is 9s. 6d. per unit, net cash, delivered f.o.r. at works, and the ava ilable supply being in very small compass.

METAL MARKET REPORT.

	Last week.		This week.
Iron	51/10½	....	51/8
Tin	£91 0 0	....	£89 17 6
Copper	47 17 6	....	47 0 0
Spelter	21 10 0	....	22 10 0
Lead	12 12 6	....	13 0 0
Silver	44d.	....	43⅞

COPPER MINING SHARES.

	Last week.			This week.	
Rio Tinto	14⅞	15	....	15¾	15⅞
Mason & Barry	6¼	6½	....	6 9/16	6 7/16
Tharsis	4 7/16	4 9/16	....	4½	4¾
Cape Copper Co.	3 13/16	3 15/16	....	3⅜	3½
Namaqua	2	2¼	....	2	2⅛
Copiapo	2 7/16	2 9/16	....	2 7/16	2 9/16
Panulcillo	1	1⅛	....	15/16	1 1/16
New Quebrada	⅝	⅝	....	⅝	¾
Libiola	3⅜	3½	....	2⅜	8⅝
Tocopilla	1/6	2/-	....	1/6	2/-
Argentella	6d.	1/-	....	3d.	9d.

THE LIVERPOOL COLOUR MARKET.

Colours quiet; prices unaltered. Ochres: Oxfordshire quoted at £10., £12., £14., and £16.; Derbyshire, 50s. to 55s.; Welsh, best, 50s. to 55s.; seconds, 47s. 6d.; and common, 18s.; Irish, Devonshire, 40s. to 45s.; French, J.C., 55s., 45s. to 60s.; M.C., 65s. to 67s. 6d. Umber: Turkish, cargoes to arrive, 40s. to 50s.; Devonshire, 50s. to 55s. White lead, £21. 10s. to £22. Red lead, £18. 10s. Oxide of zinc: V.M. No. 1, £25.; V.M. No. 2, £23. Venetian red, £6. 10s. Cobalt: Prepared oxide, 10s. 6d.; black, 9s. 9d.; blue, 6s. 6d. Zaffres: No. 1, 3s. 6d.; No. 2, 2s. 6d. Terra Alba: Finest white, 60s.; good, 40s. to 50s. Rouge: Best, £24.; ditto for jewellers', 9d. per lb. Drop black, 25s. to 28s. 6d. Oxide of iron, prime quality, £10. to £15. Paris white, 60s. Emerald green, 10d. per lb. Derbyshire red 60s. Vermillionette, 5d. to 7d. per lb.

THE LIVERPOOL MINERAL MARKET.

Prices all round still continue to advance, and there is every prospect we shall see a further rise in figures. Manganese: Arrivals continue very limited, whilst stocks are almost nil; sales, therefore, are being made for prompt and forward delivery at advanced figures. Magnesite: Raw lump is a drug, stocks being very large; raw ground, £6. 10s.; and calcined ground, £10. to £11. Bauxite (Irish Hill brand): During the week there has been an unprecedented demand, and prices are very strong, with a certainty of further advance. Lump, 20s.; seconds, 16s.; thirds, 12s.; ground, 35s. Dolomite, 7s. 6d. per ton at the mine. French Chalk: Arrivals continue steady, most of them having gone into consumption. Prices continue firm, especially for G.G.B. "Angel-White" brand—90s. to 95s. medium, 100s. to 105s. superfine. Barytes (carbonate), steady; selected crystal lump scarce at £6.; No. 1 lumps, 90s.; best, 80s.; seconds and good nuts, 70s.; smalls, 50s.; best ground, £6.; and selected crystal ground, £8. Sulphate quiet; best lump, 35s. 6d.; good medium, 30s.; medium, 25s. 6d. to 27s. 6d.; common, 18s. 6d. to 20s.; ground best white, G.G.B. brand, 65s.; common, 45s.; grey,

32s. 6d. to 40s. Pumicestone steady; ground at £10., and specially selected lump, finest quality, £13. Iron ore continues in good demand at full prices. Bilbao and Santander firmer at 9s. to 10s. 6d. f.o.b.; Irish, 11s. to 12s. 6d.; Cumberland, 16s. to 20s. Purple ore has been sold largely for forward delivery at advanced figures. Spanish manganiferous ore commands ready sale at full prices. Emery-stone: Best brands continued to be enquired for, bringing full prices. No. 1 lump is quoted at £5. 10s. to £6., and smalls £5. to £5. 10s. Fullers, earth steady; 45s. to 50s. for best blue and yellow; fine impalpable ground, £7. Scheelite, wolfram, tungstate of soda and tungsten metal continue inquired for. Chrome metal, 5s. 6d. per lb. Tungsten alloys 2s. per lb. Chrome ore: Finest qualities in good demand, bringing full prices. Antimony ore and metal have further advanced. Uranium oxide, 24s. to 26s. Asbestos: Best rock, £17. to £18.; brown grades, £14. to £15. Potter's lead ore: Prices are firm; smalls, £14. to £15.; selected lump, £16. to £17. Calamine: Best qualities scarce—60s. to 80s. Strontia steady; sulphate (celestine) steady, 16s. 6d. to 17s. Carbonate (native), £15. to £16.; powdered (manufactured), £11. to £12. Limespar: English manufactured, old G G.B. brand sought after, and brings full prices; 50s. for ground English. Felspar, 40s. to 50s.; fluorspar, 20s. to £6. Bog ore in fair demand at 22s. to 25s. Plumbago: steady; Spanish, £6.; best Ceylon lump at last quotations; Italian and Bohemian, £4. to £12. per ton. French sand, in cargoes, continues scarce on spot—20s. to 22s. 6d. Ferro-manganese strong at advanced figures. Ground mica, £50. China clay freely offering—common, 18s. 6d.; good medium, 22s. 6d. to 25s.; best, 30s. to 35s. (at Runcorn).

WEST OF SCOTLAND CHEMICALS.

GLASGOW, Tuesday.

Scotch makers of bichromate of potash complain of undue quietness in the order department, even after allowing that this is the off season for Continental despatches. The same applies to the market for bichromate of soda, of which the sales for the week have been very small. Prices of these bichromates are unaffected, under long-standing agreement between the makers for the maintenance of quotations above a stated level. Local makers of bleach and soda crystals, also, suffer from feebleness of demand, and prices in each case have given way. There is at the same time a slight fall in the lower strengths of caustic soda. Nitrate of soda is done at unwontedly low figures, and altogether the Scotch market stands in need of some quickening at the moment. Sulphate of copper, bichromate of soda, and saltcake being scarce, are fetching slightly higher values, and the late position in paraffins is at least maintained. In sulphate of ammonia the feeling has been weaker, with a sale or two on report down to £11. 12s. 6d., but there has been partial recovery. Yesterday in Glasgow £11. 17s. 6d. was paid, but for a specially-urgent parcel wanted at once for transatlantic shipment. Leith value at present stands no higher than £11. 15s., and, perhaps, less may be taken. Chief prices current are:—Soda crystals, 48s. 6d. net Tyne; alum in lump £4. 17s. 6d., less 2½% Glasgow; borax, English refined £30., and boracic acid, £37. 10s. net Glasgow; soda ash, 1⅛d. less 5% Tyne; caustic soda, white, 76°, £10. 15s., 70/72°, £8. 15s, 60/62° £7. 17s 6d., and 60/62° cream, £7. 7s. 6d., all less 2½% Liverpool; bicarbonate of soda, 5 cwt. casks, £5. 10s., and 1 cwt. casks, £5. 17s 6d. net Tyne; refined alkali 48/52°, 1½d., less 2½% Tyne; saltcake, 32s. to 34s. 6d; bleaching powder, £5. 5s. to £5. 10s. less 5% f.o.r. Glasgow: bichromate of potash, 4d., and of soda 3d., less 5 and 6% to Scotch and English buyers respectively; chlorate of potash, 4¾d., less 5% any port; nitrate of soda 8s. 3d. to 8s. 4½d.; sulphate of ammonia, spot, £11. 15s. f.o.b. Leith; salammoniac, 1st and 2nd white, £37. and £35., less 2½% any port; sulphate of copper, £27. less 5% Liverpool; paraffin scale, hard and soft, 2½d.; paraffin wax, 120°, semi-refined, 3$\frac{5}{32}$d.; paraffin spirit (naphtha), 9d.; paraffin oil (burning), 6¾d. to 7d., at Glasgow; ditto (lubricating), 865°, £5. 10s. to £6., 885°, £6. 10s.; and 890/895°, £7. 10s. to £8. Last week's imports of sugar at Greenock were 19,534 bags.

TAR AND AMMONIA PRODUCTS.

TUESDAY.

The tar products market is decidedly easier, and 3s. and 3s. 10d. may be stated as the prices of 50/90's and 90's benzol respectively. Crude carbolic acid is flat, while solvent naphtha continues in good demand and enhanced values. Anthracene still remains as already quoted, say at 1s. 2½d. per unit for B quality. Of A quality there are no parcels offering. Pitch is in a very weak condition, and foreign purchasers will not increase their stocks.

The sulphate of ammonia market is in a somewhat peculiar for while makers have been obtaining £11. 17s. 6d., £11. 18s. 9d. f.o.b. Hull, dealers and speculators are actual ing £11. 12s. 6d. to £11. 13s. 9d. The latter evidently be a further decline is inevitable, and they keep running down tion both here and abroad. Leith is now quoted at and Liverpool at £11. 12s. 6d., but the latter price repre makes only.

Perhaps the very flat state of the nitrate market has someth with speculators' movements and sulphate.

THE TYNE CHEMICAL REPORT.

TU

With the exception of bleaching powder and soda crystals, chemicals have slightly advanced during the week. Soda good demand, and has advanced 2s. 6d. per ton. There business doing in caustic soda, but prices remain the same week. Sulphate of soda has gone up 2s 6d. a ton. Bleach dull at the moment and price easier. Crystals have dropped To-day's prices are:—Soda crystals, 47s. 6d. per ton; caus 77%, £10. per ton for prompt delivery; bleach powder, s casks, £5. 7s. 6d.; hardwood, casks, £5. 10s. per ton; soda 52%, 1⅛d., less 5%; Sulphate, ground, in casks, 42s. 6d. sulphate, in bulk, 32s. 6d. per ton; chlorate of potash, 4¾ for prompt, 5d. per lb. forward; sulphur, £4. 5s. per ton; phite of soda, casks, £4. 5s. per ton; hyposulphite of soda kegs, £4. 15s. per ton; silicate of soda, 75° Tw., £2. 10s. 100° Tw., £3. 7s. 6d., 140° Tw., £4.; pure white sulphat mina, £4. 10s. per ton; blanc fixe, £7. 10s. per ton; ch barium, £8. per ton; nitrate of baryta, crystals, £18. 10s. ground, £19.; sulphide of barium, £5. 10s. per ton. All f.o. or f.o r. makers' works.

At the seventh annual meeting of the Newcastle Chemic Co., Ltd., held at their offices on Thursday last, Mr. John Davidson (who for many has acted as secretary and general m the commercial department) was unanimously elected a n director of the company. Mr. Albert Septimus Cook has pointed secretary in the room of Mr. Davidson.

The following figures are taken from the Tyne Improveme missioner's analysis of exports for the past year:—

	1889.		1888.
Alkali and soda ash........	17,632 tons.	..	27,054 t
Bleach powder	26,953 ,,	..	29,712
Caustic soda	16,091 ,,	..	12,238
Bicarbonate soda	852 ,,	..	1,294
Manganate of soda	1,819 ,,	..	1,757
Alum and alum cake......	3,017 ,,	..	3,936
Pearl Hardening	1,748 ,,	..	883
Soda crystals...............	81,891 ,,	..	100,115
Sulphate of soda	1,041 ,,	..	927

Prices of manufacturing coals are unchanged. Northumberla small is quoted 7s. 6d. to 8s. a ton, and Durham small 9s. per is generally expected that the dispute in the Durham coal t be amicably settled. The owners have made an offer of 5 advance to all classes of workmen, or they are willing to su question of an advance to arbitration.

The masters' offer has been received with satisfaction, and that a settlement can now be arrived at without loss of either side. A vote will be taken by the miners, and the res known on Saturday next.

The Sheriff Hill Colliery Company, which is to lease and mines at Sheriff Hill and Low Fell in connection with the Chemical Works, has just been registered as a limited compa capital is set down at £30,000., in £10. shares. The first su are:—W. Allhusen, W. H. Allhusen, H. E. Allhusen, A. J. E. Davidson, J. J. Potts, A. S. Cook, one share each.

The Patent List.

This list is compiled from Official sources in the Manchester Laboratory, under the immediate supervision of George E. D Alfred R. Davis.

APPLICATIONS FOR LETTERS PATENT.

Electrodes. G A. Schoth. 2,113. February 10.

Photographic Apparatus. W. Bentley, M. H. Smith and F. Wals Feb 10.

Controlling Delivery of Fluids. J. Pollett and J. Mangnall. 2,128

Regulating the Generation of Heat.—(Complete Specification.) 2,139. Feb. 10.

Improvements in Compressed Air Supply Pipes. J. C. Mowb Feb. 10.

phone Switch Boards. D. Dewar. 2,153. Feb. 10.
uretting Gas Fittings. W. Aubert, junr. 2,155. Feb. 10.
Burners.—(Complete Specification.) F. Diemel. 2,166. Feb. 11.
tches for Electric Circuits. H. Boardman. 2,172. Feb. 11.
ro-Carbon Explosive Engines. H. A. Stuart and C. R. Binney. 2,181. Feb. 11.
ting Apparatus. C. Kendall. 2,183. Feb 11.
ts for Lead or other Pipes. W. Hughes and E. Gipprich. 2,193. Feb. 11.
go Dyeing. J. Braithwaite. 2,195. Feb. 11.
esive Cement.—(Complete Specification.) W. P. Thompson. 2,197. Feb. 11.
atures for Electro-Magnets.—(Complete Specification) S. C. C. Currie. 2,199. Feb. 11.
uous Steam Generators. R. Marshall and E. Fitzgerald. 2,217. Feb. 11.
ace Condensers. R. Marshall and E. Fitzgerald. 2,218. Feb 11.
al Exraction. A. Lebedeff. 2,226. Feb. 11.
ro-Carbon Lamps. W. L. Wise. 2,227. Feb. 11.
tro-mechanical Movements.—(Complete Specification.) S. E. Nutting. 2,234. Feb. 11.
our Burners.—(Complete Specification.) L. S Calder. 2,237. Feb. 11.
Treatment of Excreta.—(Complete Specification.) C. W. Doughty. 2,238. Feb. 11.
Manufacture of Benzol and Allied Products. F. Hlawaty. 2,242. Feb. 11.
ization of Tin-plate Scrap. F. W Harbord. 2,248. Feb. 12.
phony.—(Complete Specification.) G. Macaulay-Cruikshank. 2,255. Feb. 11.
roved Lubricant. J. Green and G. Macdonald. 2,264. Feb. 12.
ention of Phylloxera. J Pipe and H. D. Andross. 2,265. Feb. 12.
aratus for Testing Electric Currents. A. C. Cockburn and F. Teague. 2,272. Feb. 12.
nary Batteries. G. R. Postlethwaite 2,273. Feb. 12.
er Coil and Pipe Cleansing Boilers.—(Complete Specification.) The Western Coil and Pipe Cleansing Co 2,278. Feb. 12.
shing and Grinding Mills. J. C. Cole. 2,281. Feb. 12.
Extinguishers.—(Complete Specification.) A. J. Boult. 2,282. Feb. 12.
nufacture of Oil Gas. W. Hughes and E. Gipprich. 2,283. Feb. 12.
ctrolytic Decomposition of Salts. J. C. Richardson and T. J. Holland. 2,296. Feb. 12.
ctrolytic Apparatus. J. C. Richardson and T. J. Holland. 2,297. Feb. 12.
serving Eggs. D. Brown. 2,300. Feb. 12.
ctric Batteries. H. Weymersch and R. McKenzie. 2,302. Feb. 12.
stitutes for Lacquer Resins.—(Complete Specification.) H. Beck. 2,308. Feb 12.
traction of Tin. J. Teague. 2,312. Feb. 12.
ssure Gauges for Compressed Gases. W. M. Jackson. 2,314 Feb 13.
k-making Machines. S. Denison, G. H. Denison, and W. Ward. 2,320. Feb. 13.

Brick Pressing Machines. S. Denison, G. H. Denison, and W. Ward. 2,321. Feb. 13.
An Improved Motor for Steam, Hydraulic, or other Power. J. Robson. 2,334. Feb. 13.
Manufacture of Lubricating Grease. H. Harford. 2,336. Feb. 13.
Electro-magnetic Motors. G. R. Postlethwaite. 2,337. Feb. 13.
Gas Regulators. T. Thorp and T. G. March. 2,339. Feb. 13.
Odorizing Gases. F. Scudder and H. G. Colman. 2,341. Feb. 13.
Mechanical Stokers. T. Wrigley. 2,343. Feb. 13.
Glow Lamps and Portable Galvanic Batteries.—(Complete Specification) A. L. Davis. 2,364. Feb. 13.
Tanning. A. Leturque. 2,365. Feb. 13
Bricks and Fire Clay Substitutes. C. H. Edwards. 2,371. Feb. 13.
Manufacture of Paper. H. Colley. 2,375. Feb. 13.
The Dyeing of Skins or Fabrics. J. Koenigswerther. 2,381. Feb. 13.
Mechanical Clinkering Arrangement. B. H. Thwaite. 2,388. Feb. 14.
Precipitation of Gold or Silver. J. Buchanan, junr. 2,390. Feb. 14.
Stone Breakers. J. W. Lodge and B. L. Fletcher. 2,399. Feb. 14.
Electrical Switches. F. Bathurst. 2,404. Feb. 14.
Concentration of Solutions. A. Fryer. 2,427. Feb. 14.
Cleansing Tin and Terne Plates from Grease. S. B. Bowen. 2,428. Feb. 14.
Gas Pressure Regulators. H. K. Hiller. 2,429. Feb. 14.
Manufacture of Nitrate of Ammonia. A. M. Chance and H. W. Crowther. 2,432. Feb. 14.
Storage Batteries. F. S. Roberts. 2,439. Feb. 14.
Manufacture of Soap. A. des Cressonnières and E. des Cressonnières. 2,446. Feb. 14.
Steam Injectors for Steam Boilers.—(Complete Specification.) J. Desmond. 2,456. Feb. 14.
Water Motor and Sewage Pump. J. Downton 2,458. Feb. 14.
Cooling Liquids. J. W. Hartley, 2,466. Feb. 14
The Aeration of Liquids. B. J. Sharp. 2,471. Feb. 14.
Fusible Cut-outs for Electrical Circuits. A. H. Walton. 2,476. Feb. 14.
Extraction of Tin. T Teague. 2,477. Feb. 14.
Aerated Waters. O. Avedyk and E. Halot. 2,486. Feb. 14.
Electrical Insulators. W. Gritten and A. Parmley. 2,491. Feb. 14.
Hodro-Carbon Lamps. J. T. Woodward, T. Woodward, and J. Thompson. 2,493. Feb. 14.
Brick Pressing Machinery. R. H. Robinson and W. Jackson. 2,494. Feb. 14.
Electrical Conductors. J. B. King and T. E. Bickle. 2,496. Feb. 14.
Soluble Blue Cotton Dyestuff. H. H. Lake. 2,499. Feb. 14.
Heating, Cooling, and Condensing Apparatus. W. R. Renshaw. 2,502. Feb. 14.
Electric Lighting. L. Saunderson. 2,505. Feb. 14.
Refrigerators. W. Mueller. 2,505. Feb. 14.
Electric Batteries. E. C. F. Verschave and A. Baron. 2,506. Feb. 14.

IMPORTS OF CHEMICAL PRODUCTS
AT
THE PRINCIPAL PORTS OF THE UNITED KINGDOM.

LONDON.
Week ending Feb. 18th.

m—
olland, £25 Ohlenschlager Brs.
olland, 100 Beresford & Co.
timony Salt—
ermany, £148 W. Balchin
timony Ore—
pan, 4 t. H. Rogers, Sons & Co.
ewcastle, 6 E. Hawkins
state of Lime—
.S., £120 W. Thatcher
" 150 Lister & Biggs
senic—
ermany, £150 T. H. Lee
nes—
pain, 8 t. 8 c. Goad, Rigg & Co.
.S. 18½ H. A. Lane & Co.
rytes—
olland, 22 kgs. J. Skitt & Co.
elgium, 12 E. W. Carling & Co.
ermany, 19 cks. Taylor, Sommervill & Co.
elgium, 20 A. Zumbeck & Co.
mstone—
rance, 3 t. 15 c. Docks Co.
aly, 100 Johnson & Hooper
omide of Iron—
ermany, £750 A. & M. Zimmermann
omine—
ermany, £300 A. & M. Zimmermann
am Tartar—
olland, 3 pkgs. Magee, Taylor & Co.
" 23 B. & F. Wf. Co.
aly, 8 Thames S. Tug Co.
" 10 W. H. Mitchell
rance, 18 E. W. Carling & Co.
aly, 32 B. Jacob & Sons
rance, 5 W. C. Bacon & Co.
aly, 148 Seager, White & Co.

Caustic Soda—
France, 100 c. C. Pawson
Holland, 220 Beresford & Co.
Caoutchouc—
B.W.I., 25 c. L. & I. D. Jt. Co.
Belgium, 157 Kebbel, Son & Co.
E. Africa, 43 Wallace Brs.
" 68 L. & I. D. Jt. Co.
Madagascar, 4 Kebbel, Son & Co.
Cobalt Ore—
New Caledonia, 153 t. Newcomb & Son
Chemicals (*otherwise undescribed*)
Germany, £8 T. H. Lee
Holland, 5 Phillipps & Graves
N.S.Wales, 10 Henry, Peabody & Co.
Germany, 101 T. H. Lee
" 920 A. & M. Zimmermann
Caustic Potash—
France, 20 pkgs. Fuerst Brs.
Dyestuffs—
Indigo
E. Indian, 25 chts. L. & I. D. Jt. Co.
" 5 J. C. Shurn & Sons
Japan, 13 W. Isted
E. Indies, 200 L. & I. D. Jt. Co.
" 24 T. H. Allen & Co.
" 32 Patry & Pasteur
" 27 Arbuthnot, Latham & Co.
" 22 G. Dards
" 28 Lond. & Hanseatic Bank
" 7 L. & I. D. Jt. Co.
" 18 Oldemeyer & Hadenfeldt
" 15 Williams, Torrey & Co.
" 42 T. Ronaldson & Co.
" 64 P. & O. Co
" 28 Lond. & Hanseatic Bank
U.S. 51 "
E. Indies, 21 Walker, Munsie & Co.
" 70 1 box. Wrightson & Son
" 2 L. & I. D. Jt. Co.
" 18 Langstaff & Co.
" 2 Elkan & Co.
" 7 Schenker & Co.
" 4 L. & I. D. Jt. Co.
Gambier
E. Indies, 192 pkgs. Elkan & Co.
" 663 E. Boustead & Co.
" 365 S. Barrow & Bro.
" 194 Elkan & Co.
" 170 P. & S. Evans & Co.
" 636 T. J. & T. Powell
China, 218 L. & I. D Jt. Co.
Orchella
France, 30 pkgs. Kebbel, Son & Co.
Extracts
U.S., 10 pkgs. A. F. White & Co.
France, 160 Levinstein & Sons
" 50 W. France & Co.
U.S., 75 W. Burton & Sons
Denmark, 39 A. F. White & Co.
Annatto
France, 5 pkgs. R. J. Fullwood & Bland
Argols
Natal, 20 pkgs. J R. Thomas & Co.
Tanners' Bark
Belgium, 40 t. T. J. Williamson & Son
" 50 Hicks, Nash & Co
Glucose—
U.S., 50 pkgs. 305 cwts. Barrett, Tagant & Co.
France, 1 3 J. Blank
Germany, 20 20 T. H. Lee
" 70 589 Proprs. Chamberlains Wf.
" 200 200cwts. J. F. Ohlmann
U.S., 40 210 Page, Son & East
" 50 300 Proprs. Scott's Wf.
" 1 5 Petty, Wood & Co.
" 100 100 M. D. Co.
" 120 120 J. Barber & Co.

Gutta Percha—
E. Indies, 970 c. H. W. Jewesbury & Co.
" 24 Katz Brs
" 530 Huttenbach & Co.
Guano—
Holland, 17 t. 14 c. Soundy & Son
Glycerine—
Germany, £65 Beresford & Co.
" 50 "
Isinglass—
St. Settlements, 5 pkgs. L. & I. D. Jt. Co.
E. Indies, 2 pkgs. Kurtz, Stubeck & Co.
St. Settlements, 44 pkgs. L. & I. D Jt. Co.
Menthol—
Japan, £1,050 Spies, Brs. & Co.
Manure—
Phosphate Rock
Belgium, 200 t. Miller & Johnson
France, 135 D. de Pass & Co.
Manure (*otherwise undescribed*)
Germany, 6 t. W. G. Blagden
U.S., 4½ C. A. H. Nichols
Manganese Ore—
S. Australia, 377 t. Harrold Brs.
Nitre—
Chili, 161,606 cubic bgs. Montgommery & Co.
Naphtha—
Holland, 7 brls. Domeier & Co.
Pyrites—
Spain, 1,504 t. Societe Commerciale
" 1,931 C. Tennant, Sons & Co.
Plumbago—
Holland, 10 cks. Brown & Elmslie
Ceylon, 466 Thredder, Son & Co.
" 24 cks. Doulton & Co.
Germany, 65 Beresford & Co.
Pitch—
France, 70 brls. H. Hill & Son

Potassium Sulphate—
Germany, 77 pkgs. Bessler, Wachter & Co
" 102 T. Farmer & Co.
" 73 B. & F. Wf. Co
Potassium Carbonate—
France, 10 pkgs. J. A. Reid
" 6 "
" 4 W. Harloch
Stearine—
N. Z. 100 pkgs. Phœnix Whvs. Co.
" 53 Brown & Elmslie
France, 150 bgs. H. Hill & Sons
Soda—
Holland, 300 c. E. Olyett & Sons
Saltpetre—
Germany, 118 cks. P. Hecker & Co.
Belgium, 1,000 bgs. Anglo Contl. Guano Wks.
Holland, 135 cks. G. Meyer & Co.
Germany, 81 J. Hall & Son
Tartar—
Italy, 9 pkgs. Thames S. Tug Co.
Tar—
Holland, 3 brls. Burt, Boulton & Co.
Tartaric Acid—
Holland, 5 pkgs. Northcott & Son
France, 14 B. & F. Wf. Co
Ultramarine—
Holland, 4 pkgs. J. Owen
" 4 Spies, Bro. & Co.
" 2 cks. G. Rahn & Co.
Belgium, 1 Leach & Co.
Zinc Oxide—
Holland, 105 cks. M. Ashby
" 100 Proprs. of Dowgate Dk.
" 100 M. Ashby

LIVERPOOL.

Week ending Feb. 20th.

Arsenic—
Hamburg, 3 cks.
Antimony Ore—
Smyrna, 521 bgs.
Bones—
Constantinople, 451 bgs. Jivan, Tekeian, & Co.
Sur, 50 t.
Bone Meal—
Bombay, 8,193 bgs.
Barytes—
Antwerp, 18 brls. J. T. Fletcher & Co.
" 12 cks.
Borax—
Leghorn, 30 sks. Shropshire Union C. Co.
" 1 cs. 30 cks.
Borate of Lime—
Valparaiso, 720 sks.
Cream Tartar—
Genoa, 6 cks.
Caoutchouc—
Pernambuco, 9 brls.
New York, 2 cs. Robinson & Allen
Brazil, 59 brls. 5 bls. Fry, Miers, & Co.
Boston, 10 cs. T Turner & Co.
Quittah, 3 brls. 1 pn. F. & A. Swanzy
Accra, 1 1 cs. Davis, Robbin, & Co.
C. C. Castle, 11 Grimwade, Ridley, & Co.
" 1 Hobson & Co.
Monrovia, 12 cks. A. Aschee
Sierra Leone, 36 Pickering & Berthoud
" 2 N. Waterhouse & Son
" 10 cks.
Cape Coast, 5 W. B. M'Iver & Co.
" 16 2 cs. Fletcher & Fletcher
" 2 F. & A Swanzy
" 1 Edwards Bros.
" 2 W. Duff & Co.
" 1 H. B. W. Russell
" 6 A. Millerson & Co.
" 1 Havard & Co.
" 1 R. H Brotherton
" 1 L. Hart & Co.
" 1 cs. Scott, Bros., & Co.
" 4 Davis, Robbin, & Co.
" 3 Whimster & Co.
" 1 Kronig & Seigler
" 1 E. G. Gunnell
" 1 cs. A. Reis
" 2 T. Crouch
" 2 Hawkes Sommerville
Elmina, 1 ck. Tanter & Melen
" 2 1 cs. A. Reis
Axim, 1 bg. Radcliffe & Durant
Chemicals *(otherwise undescribed)*
Rotterdam, 4 cks.
Dyestuffs—
Gambier
Singapore, 2,101 bls.
Extract
Bordeaux, 15 cks.
Boston, 25 brls. L. J. Levenstein & Sons
Baltimore, 60 J. C. Bloomfield & Co.
Argols
Bordeaux, 12 cks.
Oporto, 23 German Bk., London
Cochineal
Grand Canary, 23 bgs. Widow, Duranty, & Son
" 8 K. Hindschell
" 20 Swanston & Co.
" 10 H. R. Toby & Co.
Teneriffe, 12 K. Henderson
" 12 N. P. Nathan & Son
" 15 Beriro & Co.
" 13 M. L. Tuly & Co.
" 13 Kolp & Co.
Valonia
Syra, 652 bgs. Barff & Co.
" 250
Smyrna, 842 211½ t.
Patras, 200 C. P. Papayanni & Jeremies
Indigo
Bombay, 39 cs. Forbes, Forbes, & Co.
Calcutta, 260 chts.
Dyewood
Colon, 50 cs. E. Chesney & Co.
Divi Divi
Curacoa, 193 bgs. D. Midgley & Son
Sumac
Palermo, 128 bgs.
Glycerine—
Rouen, 1 drm. 10 cks. R. J. Francis & Co.
Rotterdam, 10 cs.
Hamburg, 5 carboys
Rotterdam, 5 cs.
Glucose—
New York, 30 brls.
Hamburg, 80 cks.
Iodine—
Valparaiso, 234 brls. Gibbs & Sons
" 11 W. & J. Lockett
Manganese Ore—
Cartagena, 2,350 t. Wigan Coal & Iron Co.
Phosphates—
Ghent, 2,040 bgs. J. T. Fletcher & Co.
Antwerp, 120 t. 100 bgs. "
Potash—
Rotterdam, 28 cks.
Pitch—
Amsterdam, 12 cks.
Rotterdam, 51
Pyrites—
Huelva, 1,848 t. Tennants & Co.
" 1,630 Matheson & Co.
" 2,046 Beauford & Co.
Saltpetre—
Calcutta, 340 bgs.
Hamburg, 16 cks.
Salts—
Hamburg, 350 t. 1,000 bgs.
Stearine—
Rotterdam, 10 cs.
Antwerp, 20 bgs.
Sugar of Lead—
Rouen, 47 cks. R. J. Francis & Co.
Rotterdam, 29 cks.
Soda—
Rotterdam, 25 brls.
Tartaric Acid—
Rotterdam, 7 cks.
" 9
Tartar—
Naples, 4 cks. R. Powles
Bordeaux, 5 1 hd.
Ultramarine—
Rotterdam, 24 cks.
Verdigris—
Bordeaux, 3 cks.
Zinc Sulphate—
Rotterdam, 7 cks.
Zinc Oxide—
Hamburg, 25 cks.
Antwerp, 130 brls.

TYNE.

Week ending Feb. 20th.

Barytes—
Rotterdam, 69 cks. Tyne S.S. Co.

GLASGOW.

Week ending Feb. 20th.

Alum—
Rotterdam, 1 ck.
Alumina Sulphate—
Rotterdam, 480 cks. H. Burrell
Rotterdam, 26 csks. J. Rankine & Son
Barytes—
Rotterdam, 92 cks.
" 69
Dyestuffs—
Alizarine
Rotterdam, 31 cks. J. Rankine & Son
" 111
Extracts
Nantes, 691 cks.
Antwerp, 27
Rouen, 35
" 10 cs.
Glycerine—
Rotterdam, 13 cs. J. Rankine & Son
Amsterdam, 5 cbys. "
Phosphate—
Antwerp, 100 t.
Pyrites—
Huelva, 1,147 t. Tharsis Co.
Stearine—
New York, 20 hhds.
Rotterdam, 36 cks. J. Rankine & Son
Antwerp, 20 cks.
Sulphur Ore—
Drontheim, 220 t.
Tartar—
Bordeaux, 3 cks.
Zinc Ashes—
Antwerp, 6 cks.

HULL.

Week ending Feb. 20th.

Ammonia—
Rotterdam, 2 cks. W. & C. L. Ringrose
Ammonium Nitrate—
Rotterdam, 28 cks. Hutchinson & Son
Albumen—
Libau, 23 pkgs. Morley, Clarke, & Co.
Savannah, 2 cs. C. M Lofthouse & Co.
Bones—
Gothenburg, 3 brls. Wilson, Sons, & Co.
Barytes—
Bremen, 22 cks. Blundell, Spence, & Co.
" 60
Antwerp, 120 bgs. 42 brls. Bailey & Leetham
Brimstone—
Marseilles, 20 cks. Wilson, Sons, & Co.
Chemicals *(otherwise undescribed)*
Bremen, 3 cs. Veltmann & Co.
" 5 kgs. Storry, Smithson, & Co.
Genoa, 18 pkgs. Wilson, Sons, & Co.
Stettin, 23 cks.
Rotterdam, 2 W. & C. L. Ringrose
Caoutchouc—
Savannah, 1 cs. C. M. Lofthouse & Co.
Dyestuffs—
Argols
Bordeaux, 44 bgs. Rawson & Robinson
Alizarine
Rotterdam, 31 pkgs. Hutchinson & Son
" 60
" 7 cks. W. & C. L. Ringrose
Extracts
Boston, 20 cs. Wilson, Sons, & Co.
Antwerp, 15
69 48 cks. Rawson & Robinson
Rennet
Rotterdam, 5 cks. G. Law
Copenhagen, 2 pkgs. Wils
Indigo
Amsterdam, 9 chts.
Glucose—
Stettin, 1,400 bgs. C. M
Savannah, 5 cks.
Glycerine—
Hamburg, 10 cs. L
" 1 drum. E. &
Manure—
Ghent, 260 t. Wilson, S
Danzig, 9 pks.
Naphthol—
Rotterdam, 2 cks. V
Pitch—
Danzig, 43 cks. Wilson, S
Potash—
Rotterdam, 10 cks. Hutchin
Antwerp, 11 drums.
Salt—
Hamburg, 150 bgs. G. Laws
Soda—
Rotterdam, 7 cks. G. Law
" 4 "
Silicate—
Antwerp, 1 brl.
Stearine—
Rotterdam, 30 bgs. Wilso
Bergen, 8 brls.
Tartaric Acid—
Rotterdam, 4 cks.
Turpentine—
Savannah, 1,000 brls.
Tartar—
Bordeaux, 3 cks. Rawson &
Tar—
Marseilles, 50 cks. Wilso

GRIMSBY

Week ending Feb. 15t

Albumen—
Hamburg, 3 cks.
Chemicals *(otherwise unde*
Hamburg, 13 pkgs.
Dyestuffs—
Extracts
Dieppe, 39 cks.
Glucose—
Hamburg, 200 bgs. 12 cks.
Zinc Ashes—
Antwerp, 19 cks.

GOOLE.

Week ending Feb. 12t

Acetic Acid—
Rotterdam, 1 ck.
Chromium Acetate—
Rotterdam, 10 cks.
Caoutchouc—
Boulogne, 4 cks.
Dyestuffs—
Alizarine
Rotterdam, 78 cks.
Dyestuffs *(otherwise unde*
Boulogne, 73 cks. 4 cs.
Glucose—
Hamburg, 10 cks.
Logwood Extract—
Rotterdam, 2 cks.
Naphthylamine—
Rotterdam, 2 brls.
Phosphate—
Ghent, 1,500 bgs.
Phtalic Acid—
Rotterdam, 2 brls.
Potash—
Calais, 51 cks 69 drums.
Saltpetre—
Rotterdam, 70 bgs.
" 170
Hamburg, 4 cks.
Antwerp, 85
Ultramarine—
Rotterdam, 1 ck.
Waste Salt—
Hamburg, 194 cks. 102 b

... 90 bags.
... £50
... 24 c. 4,360
... 298
... 2,200
... 6 t. 17 c. £268
... 7
... £10
... 20 cs.
... 30
... 12 cks. 10 cs.
...ple, 2
... 12
...n Chloride—
... 1 c. 1 q. £161
...(otherwise undescribed)
... 157 t. 12 c. £544
... 110 8 478
... 22 91
... 24 c. 3 q. £31
... Bichromate—
... 5 t. 10 c. £181
... 3 2 144
... 9 3 q. 18
... 14 18
... 11 £140
... 3 brls.
... 648
...rystals—
... 23 t. 14 c. £438
... Chlorate—
... 17 t. £747
... 20 c. 21
... 93
... 403
... 117
... 44
... 30
... 564
... 685
... 46
... 40
... 71
... 42
... 22
... 397
... 2 t. 16 c. 2 q. £57
...yres, 1 t. 16 c. £200
... 11 c.
... 6 cs.
... 26 c.
... 18 cks.
... 13 brls.
... 42 brls.
... 5 cks.
... 538 tcs.

Boston, 170 cks. 780 bgs. 130 tcs.
Calcutta, 49 brls.
Genoa, 33 cks.
Ghent, 44
Maranham, 20
Mitylene, 200 brls.
Monte Video, 10
New York, 53 cks.
Patras, 30
Philadelphia, 195 tcs. 339 cks.
Porto Alegre, 37 brls.
Rio Grande do Sul, 45 cks.
Rio Janeiro, 130 brls.
Santander, 6
Santos, 10
Smyrna, 400
Sydney, 65 cks.

Soda Crystals—
Havana, 6 brls.
Malta, 50
Pernambuco, 200
Philadelphia, 50 cks.
Santander, 65
Singapore, 5
Bombay, 520
Calcutta, 34
Halifax 50
New York, 28
Philadelphia, 60

Sodium Bicarbonate—
Amsterdam, 20 kgs.
Barcelona, 150
Buenos Ayres, 25 cks. 20 kgs.
Calcutta, 298 brls.
Ghent, 15 kgs.
Hamburg, 22 54 pkts.
Kurrachee, 350
Monte Video, 90
Sydney, 200
Toronto, 20 cks.
Yokohama, 1,000
Antwerp, 16
Genoa, 80 62 bxs.
Havre, 2 cks.
Marseilles, 225
Naples, 30
Rouen, 50 brls.
Trieste, 183

Sodium Silicate—
Maranham, 10 brls.

Sodium Nitrate—

Carthagena,	2 t.	10 c.	3 q.	£22
Montreal,	1	4	2	11
New York,		8		11
Alicante,	1	19	3	16
Venice,	3	18	2	33
Boston,		5	1	15
Malaga,	3	5	3	27

Saltpetre—

Las Palmas,	3 t.	19 c.	2 q.	£82
Madeira,		16		22
Pernambuco,	2	17		64
Carril,	1	1	2	25
Carthagena,	5	1		96
Ceara,		12	2	13

Sulphuric Acid—

Bombay,		7 c.	3 q.	£5
Ceara,	53 gals.			16

Sodium Biborate—

Antwerp,	8 t.	12 c.	3 q.	£260
Rotterdam,	18			27

Sodium Bichromate—

Havre,	12 c.	2 q.	£20

Sheep Dip—

E. London,	5 t.			£250
Buenos Ayres,		3 c.	1 q.	7
Cape Town,	20 cs.			40

Salammoniac—

Philadelphia,	4 t.	2 c.		£141
Shanghai,		10	1 q.	19
Alexandria,	2	4		78
Catania,		10	2	20
Lagos,		4		9
Rotterdam,		17	1	30
Constantinople,	2			68
Montreal,		4		8
New York,	4	16		165
Patras,		8		14
Galatz,		10		17
New York,	4	16		166

Sulphur—

Porto Alegre,	2 t.	9 c.		£16
Madras,	5	5		29
Iquique,	15	1	3 q.	121
Calcutta,	3	0		26
Boston,	100	7		426

Saltcake—

Philadelphia,	36 t.	17 c.	£78
Baltimore,	185	10	324
New York,	100	3	246

Soda Salts—

Portland, M.,	1 t.	6 c.	2 q.	£38

Tar—
Bahia, 40 brls.

HULL.

Week ending Feb. 18th.

Acids (*otherwise undescribed*)—
Haugesund, 2 cks.

Bone Size—
Harlingen, 11 cks.
Hamburg, 5

Bleaching Powder—
Danzig, 232 cks.
Venice, 30

Chemicals (*otherwise undescribed*)

Antwerp,	8 cs.	5 cks.	
Bremen,	2	30	
Bombay,		2,172	135 pks.
Christiania,			5 pkgs.
Copenhagen,		108	
Danzig,	2		60 pks.
Dunkirk,		16	
Gothenburg,	15	45	100 322 olbs.
Ghent,			8 pkgs.
Hamburg,			10
New York,	100		
Reval,		10	80
Rouen,		30	
Rotterdam,	41	5	70
Stettin,	1	1	1
Trieste,		10	
Venice,		103	

Caustic Soda—
Antwerp, 34 drms. 1 ck.
Danzig, 10
Gothenburg, 36
Venice, 60
Libau, 141
Reval, 900

Dyestuffs (*otherwise undescribed*)

Bombay,		53 cks.	
Hamburg,	1 cs.		
Reval,	4	84	
Rotterdam,		3	1 keg.

Manure—
Dunkirk, 882 bgs.
Ghent, 280
Venice, 6 t.
Hamburg, 278

Pitch—
Antwerp, 194 t.

Tar—
Copenhagen, 20 cks.
Danzig, 6

GLASGOW.

Week ending Feb. 20th.

Ammonia—
Buenos Ayres, 3 t. 1 c. £70

Ammonium Sulphate—
Nantes, 60 t. 17¾ c. £727

Benzol—
Rotterdam, 78 drums £970

Bleaching Powder—
Bombay, 2 t. 18 c. £26

Boracic Acid—
Circassia, 32 c. £57

Chemicals (*otherwise undescribed*)
Rangoon, 5 t. 2¼ c. £28

Copper Sulphate—
Nantes, 15 t. 12 c. £786

Gutta Percha—
Amsterdam, 194 lbs. £8

Iodine—
Halifax, 29½ c.

Naphtha—
Rouen, £798

Potassium Bichromate—

Japan,	6 t.	19¾ c.	£253
Barcelona,		109¾	200
Rouen,	58	14	1,856
Barcelona,		330	68
Bergen,		3	40
Rotterdam,		3	25
Amsterdam,		24	352
Christiania,	1 ck.		15
Halifax,	4½ t.		142

Potassium Prussiate—
Rouen, 5 t. 6½ c. £275

Pitch—
Nantes, 442 t. £442
Barcelona, 250 375

Sodium Bichromate—

Rouen,	29 t.	1 c.	£780
Barcelona,		68¾	100
Rotterdam,	1 ck.		12

Soda Ash—
Halifax, 1,228½ c.

Sulphuric Acid—

Bombay,		10 c.	£37
Rangoon,	12 t	19½	135

Tar—
Colombo, 8 t.
Penang, 4½ c. 108

TYNE.

Week ending Feb. 20th.

Alkali—

Amsterdam,	105 t.	
Rotterdam,	2 t.	19 c.
Antwerp,	10	6

Alum Cake—
Nyborg, 22 t. 17 c.

Bleaching Powder—

Antwerp,	17 t.	6 c.
Hamburg,	33	9
Lisbon,	6	4
Baltimore,	50	7
Christiania,	4	14
Amsterdam,	31	16
Copenhagen,	116	14
Rotterdam,	15	1
Gothenburg	1	8
Malmo,	1	10
Aarhuus,	6	

Barytes Carbonate—
Copenhagen, 1

Caustic Soda—

Antwerp,	5	2
Hamburg,	7	15
Carthagena,	1	1
Baltimore,	155	1
Gothenburg,	61	17
Copenhagen,	29	

Magnesia—
Antwerp, 3 c.
Hamburg, 7

Manure—
Hamburg, 40 t.

Pitch—
Antwerp, 15 c.

Potassium Bichromate—
Nyberg, 11

Potassium Chlorate—
Rotterdam, 1 t. 10 c.

Soda—

Nyborg,	23 t.	14 c.
Esbjerg,	11	15
Frederikshavn	7 t.	9 c.

Sodium Sulphate—

Copenhagen,	1 t.	17 c.
Baltimore,	201	
Aarhuus,	10	9

Soda Ash—

Christiania,	10	1
Lisbon,	16	9
Baltimore,	29	19
Bergen,	3	10
Copenhagen,	9	19
Gothenburg,	46	4

Sodium Bicarbonate—
Amsterdam, 3 t. 10 c.

Sodium Chlorate—
Hamburg, 1 t.

GRIMSBY.

Week ending February 15th.

Chemicals (*otherwise undescribed*)
Antwerp, 2 cs. 18 pkgs.
Dieppe, 40 cks.
Hamburg, 10
Rotterdam, 4 cs. 18

Coal Products—
Antwerp, 10 cks.
Dieppe, 24 cs.

GOOLE.

Week ending Feb. 12th.

Bleaching Powder—
Ghent, 28 cks.

Bleaching Preparations—
Rotterdam, 28 cks.

Benzole—
Rotterdam, 75 cks.

Chemicals (*otherwise undescribed*)
Antwerp, 2 pkgs.
Ghent, 1 cks.
Hamburg, 5

Coal Products—
Rotterdam, 161 cks.

Dyestuffs—
Ghent, 2 cks.

Manure—
Dunkirk, 494 bgs.
Ghent, 334
Rouen, 95

Pitch—
Antwerp, 160 t.
Dunkirk, 192

PRICES CURRENT.

WEDNESDAY, February 26, 1890.

Prepared by Higginbottom and Co., 116, Portland Street, Manchester.

[Th]e values stated are F.O.R. at maker's works, or at usual ports of shipment in U.K. The price in different localities may vary.

Article		£	s.	d.
[Ac]ids:—				
Acetic, 25 °/ₒ and 40 °/ₒ	per cwt	7/6 & 12/6		
,, Glacial	,,	2	10	0
Arsenic, S.G., 2000°	,,	0	12	9
Chromic 82 °/ₒ	nett per lb.	0	0	6½
Fluoric	,,	0	0	3¼
Muriatic (Tower Salts), 30° Tw.	per bottle	0	0	8
,, (Cylinder), 30° Tw.	,,	0	2	11
Nitric 80° Tw.	per lb.	0	0	2
Nitrous	,,	0	0	1¼
Oxalic	nett ,,	0	0	3¼
Picric	,,	0	1	4½
Sulphuric (fuming 50 °/ₒ)	per ton	15	10	0
,, (monohydrate)	,,	5	10	0
,, (Pyrites, 168°)	,,	3	2	6
,, (,, 150°)	,,	1	8	0
,, (free from arsenic, 140/145°)	,,	1	10	0
Sulphurous (solution)	,,	2	15	0
Tannic (pure)	per lb.	0	1	0½
Tartaric	,,	0	1	2½
[Ar]senic, white powered	per ton	13	0	0
[Al]um (loose lump)	,,	4	15	0
[Al]umina Sulphate (pure)	,,	5	0	0
,, ,, (medium qualities)	,,	4	10	0
[Al]uminium	per lb.	0	15	0
[Am]monia, ·880=28°	per lb.	0	0	3
,, =24°	,,	0	0	1⅞
,, Carbonate	nett ,,	0	0	3½
,, Muriate	per ton	23	10	0
,, ,, (sal-ammoniac) 1st & 2nd	per cwt.	37/- & 35/-		
,, Nitrate	per ton	40	0	0
,, Phosphate	per lb.	0	0	10½
,, Sulphate (grey), London	per ton	11	18	9
,, ,, (grey), Hull	,,	11	17	6
[An]iline Oil (pure)	per lb.	0	0	11
[An]iline Salt ,,	,,	0	0	9½
[An]timony	per ton	75	0	0
,, (tartar emetic)	per lb.	0	1	1
,, (golden sulphide)	,,	0	0	10
[Ba]rium Chloride	per ton	7	10	0
,, Carbonate (native)	,,	3	15	0
,, Sulphate (native levigated)	,,	45/- to 75/-		
[Bl]eaching Powder, 35 %	,,	5	5	0
,, Liquor, 7 %	,,	2	10	0
[Bi]sulphide of Carbon	,,	12	15	0
[Ch]romium Acetate (crystal	per lb.	0	0	6
[Ca]lcium Chloride	per ton	2	0	0
[Ch]ina Clay (at Runcorn) in bulk	,,	17/6 to 35/-		
[Co]al Tar Dyes:—				
Alizarine (artificial) 20 %	per lb.	0	0	9
Magenta	,,	0	3	9
[Co]al Tar Products				
Anthracene, 30 % A, f.o.b. London	per unit per cwt.	0	1	6
Benzol, 90 % nominal	per gallon	0	3	10
,, 50/90 ,,	,,	0	3	0
Carbolic Acid (crystallised 35°)	per lb.	0	0	10
,, (crude 60°)	per gallon	0	2	10
Creosote (ordinary)	,,	0	0	2½
,, (filtered for Lucigen light)	,,	0	0	3
Crude Naphtha 30 % @ 120° C.	,,	0	1	4½
Grease Oils, 22° Tw.	per ton	3	0	0
Pitch, f.o.b. Liverpool or Garston	,,	1	12	0
Solvent Naphtha, 90 % at 160°	per gallon	0	1	11
[Co]ke-oven Oils (crude)	,,	0	0	2½
[Co]pper (Chili Bars)	per ton	47	2	6
,, Sulphate	,,	26	10	0
,, Oxide (copper scales)	,,	52	0	0
[Gl]ycerine (crude)	,,	30	0	0
,, (distilled S.G. 1250°)	,,	55	0	0
[Io]dine	nett, per oz.	8½d. to 9d.		
[Iro]n Sulphate (copperas)	per ton	1	10	0
,, Sulphide (antimony slag)	,,	2	0	0
[Le]ad (sheet) for cash	,,	14	15	0
,, Litharge Flake (ex ship)	,,	17	0	0
,, Acetate (white ,,)	,,	23	0	0
,, ,, (brown ,,)	,,	19	0	0
,, Carbonate (white lead) pure	,,	19	0	0
,, Nitrate	,,	22	5	0

Article		£	s.	d.
Lime, Acetate (brown)	,,	8	5	0
,, ,, (grey)	per ton	14	0	0
Magnesium (ribbon and wire)	per oz.	0	2	3
,, Chloride (ex ship)	per ton	2	6	3
,, Carbonate	per cwt.	1	17	6
,, Hydrate	,,	0	10	0
,, Sulphate (Epsom Salts)	per ton	3	0	0
Manganese, Sulphate	,,	18	0	0
,, Borate (1st and 2nd)	per cwt.	60/- & 42/6		
,, Ore, 70 %	per ton	4	10	0
Methylated Spirit, 61° O.P.	per gallon	0	2	2
Naphtha (Wood), Solvent	,,	0	3	11
,, ,, Miscible, 60° O.P.	,,	0	4	9
Oils:—				
Cotton-seed	per ton	22	10	0
Linseed	,,	23	10	0
Lubricating, Scotch, 890°—895°	,,	7	5	0
Petroleum, Russian	per gallon	0	0	5¼
,, American	,,	0	0	5⅞
Potassium (metal)	per oz.	0	3	10
,, Bichromate	per lb.	0	0	4
,, Carbonate, 90% (ex ship)	per ton	18	0	0
,, Chlorate	per lb.	0	0	4¼
,, Cyanide, 98%	,,	0	2	0
,, Hydrate (Caustic Potash) 80/85 %	per ton	22	10	0
,, ,, (Caustic Potash) 75/80 %	,,	21	10	0
,, ,, (Caustic Potash) 70/75 %	,,	20	15	0
,, Nitrate (refined)	per cwt.	1	3	6
,, Permanganate	,,	4	5	0
Prussiate Yellow	per lb.	0	0	9¼
,, Sulphate, 90 %	per ton	9	10	0
,, Muriate, 80 %	,,	7	15	0
Silver (metal)	per oz.	0	3	7 15/16
,, Nitrate	,,	0	2	5½
Sodium (metal)	per lb.	0	4	0
,, Carb. (refined Soda-ash) 48 %	per ton	5	12	6
,, ,, (Caustic Soda-ash) 48 %	,,	4	12	6
,, ,, (Carb. Soda-ash) 48%	,,	4	17	6
,, ,, (Carb. Soda-ash) 58 %	,,	5	15	0
,, ,, (Soda Crystals)	,,	2	13	6
,, Acetate (ex-ship)	,,	15	15	0
,, Arseniate, 45 %	,,	10	0	0
,, Chlorate	per lb.	0	0	6¼
,, Borate (Borax)	nett, per ton.	30	0	0
,, Bichromate	per lb.	0	0	3
,, Hydrate (77 % Caustic Soda)	per ton.	10	10	0
,, ,, (74 % Caustic Soda)	,,	10	0	0
,, ,, (70 % Caustic Soda)	,,	8	17	6
,, ,, (60 % Caustic Soda, white)	,,	7	17	6
,, ,, (60 % Caustic Soda, cream)	,,	7	10	0
,, Bicarbonate	,,	5	15	0
,, Hyposulphite	,,	5	2	6
,, Manganate, 25%	,,	8	10	0
,, Nitrate (95 %) ex-ship Liverpool	per cwt.	0	8	4½
,, Nitrite, 98 %	per ton	27	10	0
,, Phosphate	,,	15	15	0
,, Prussiate	per lb.	0	0	7¼
,, Silicate (glass)	per ton	5	7	6
,, ,, (liquid, 100° Tw.)	,,	3	17	6
,, Stannate, 40 %	per cwt.	2	0	0
,, Sulphate (Salt-cake)	per ton.	1	12	6
,, ,, (Glauber's Salts)	,,	1	10	0
,, Sulphide	,,	7	15	0
,, Sulphite	,,	5	5	0
Strontium Hydrate, 98 %	,,	8	0	0
Sulphocyanide Ammonium, 95 %	per lb.	0	0	8¼
,, Barium, 95 %	,,	0	0	5½
,, Potassium	,,	0	0	8
Sulphur (Flowers)	per ton.	6	15	0
,, (Roll Brimstone)	,,	6	0	0
,, Brimstone: Best Quality	,,	4	7	6
Superphosphate of Lime (26 %)	,,	2	10	0
Tallow	,,	25	10	0
Tin (English Ingots)	,,	95	0	0
,, Crystals	per lb.	0	0	6¼
Zinc (Spelter)	per ton.	22	5	0
,, Chloride (solution, 96° Tw.	,,	6	0	0

THE

CHEMICAL TRADE JOURNAI

Publishing Offices: 32, BLACKFRIARS STREET, MANCHESTER.

No. 146. **SATURDAY, MARCH 8, 1890.** **Vol.**

Contents.

Notices.

All communications for the *Chemical Trade ournal* should be addressed, and Cheques and Post Office Orders made payable to—

DAVIS BROS., 32, Blackfriars Street, MANCHESTER.

Our Registered Telegraphic Address is—

"Expert, Manchester."

The Terms of Subscription, commencing at any date, to the *Chemical Trade Journal*,—payable in advance,—including postage to any part of the world, are as follow:—

Yearly (52 numbers) .... **12s. 6d.**
Half-Yearly (26 numbers) .... **6s. 6d.**
Quarterly (13 numbers) .... **3s. 6d.**

Readers will oblige by making their remittances for subscriptions by Postal or Post Office Order, crossed.

Communications for the Editor, if intended for insertion in the current week's issue, should reach the office not later than **Tuesday Morning**.

Articles, reports, and correspondence on all matters of interest to the Chemical and allied industries, home and foreign, are solicited. Correspondents should condense their matter as much as possible, write on one side only of the paper, and in all cases give their names and addresses, not necessarily for publication. Sketches should be sent on separate sheets.

We cannot undertake to return rejected manuscripts or drawings, unless accompanied by a stamped directed envelope.

Readers are invited to forward items of intelligence, or cuttings from local newspapers, of interest to the trades concerned.

As it is one of the special features of the *Chemical Trade Journal* to give the earliest information respecting new processes, improvements, inventions, etc., bearing upon the Chemical and allied industries, or which may be of interest to our readers, the Editor invites particulars of such—when in working order—from the originators; and if the subject is deemed of sufficient importance, an expert will visit and report upon the same in the columns of the *Journal*. There is no fee required for visits of this kind.

We shall esteem it a favour if any of our readers, in making inquiries of, or opening accounts with advertisers in this paper, will kindly mention the *Chemical Trade Journal* as the source of their information.

Advertisements intended for insertion in the current week's issue, should reach the office by **Wednesday morning** at the latest.

RAILWAY RATES

WE hope the traders are so far satisfied wit
crusade against the rates and charges of the 1
Companies. From the outset we foresaw that the
would be a very unequal one, and we endeavoured t
the subject between such narrow walls, that the fi
evidence, even of such a kind as was within the reach
traders, would have made their arguments irresistible
course was not followed, and the Traders' Associatio
had the chagrin of seeing their champions, one by o
molished, without having made one point of vantage.

Time has only served to strengthen the opinions w
more than once expressed in these columns, that it
have been better to have adopted the various class
posed by the companies, and even the terminals i
bargain, rectifying any incongruity by the introducti
good system of classification. We do not see why thi
not be done for every particular branch of industry,
far as the chemical and allied industries were concern
attempted to show, and we think with a fair measure
cess, how such a classification could be establishe
rational basis.

It will be clear to those who have followed the proce
of the Board of Trade Commission, that the Commis
do not intend to allow of the introduction of any r
tionary schemes, the practical effects of which are p
matical; and it must not be forgotten that at any tim
a very difficult task to blend such a variety of inter
are represented before the court of enquiry.

It is not too late yet for the traders to take in han
question of a systematic classification; there has been
time, but the opportunity has so far been missed. W
it on good authority that the companies are willing
more, anxious to arrive at a satisfactory and lasting cl
cation, if the traders would only approach them with r

Nothing but a satisfactory classification will stop
fraudulent description of goods, brought out so well the
day by Mr. David Howard in his evidence at the en
and this is the *bête noire* of all honest traders. We fea
agitation has for the most part been fomented and encou
by those who have been labouring under private
advantages, often of their own choosing in the way o
and they have not recognised the fact that they are no
in the great majority of cases charged for carriage, et
the full extent that the companies are legally entitl
charge. We are fully aware, however, that there are
grievances requiring a remedy, but we have all along
that the traders have not gone the right way to work t
them redressed. The Lancashire and Cheshire Confe
carried on its meetings with a great flourish of trur
before the battle began; where is it now?

> "False Sextus to the mountains
> Turned first his horse's head,
> And fast fled Ferentinum,
> And fast Lanuvium fled;
> The horsemen of Nomentum
> Spurred hard out of the fray,
> The footmen of Velitræ
> Threw shield and spear away."

CHEMICAL ENGINEERING.*

XXI.

MOVING LIQUIDS.

ıVING now dealt with the conveyance of solid substances from place to place, it becomes necessary to treat of the ment of liquids. It may at first be thought that with ds, the only problem is that of raising them, as unlike bodies, they require no exertion to move them from a er to a lower level, or from place to place on the same . To a certain extent only is this true; there is more needed, perhaps, in designing machinery for elevating a d, than for allowing of its descent, but there are many ts to be considered even in this latter case, such as the of the pipes required for a certain flow in a given time, for an intermittent or a continuous supply. These subs can, however, be treated as applying also to the problem aising fluids, when the principles underlying them have n explained.

The velocity, and necessarily the quantity, of any fluid ding its way through an orifice depends in a great measure on the force employed to send it through. The shape of aperture is likewise a point necessary to be taken into nsideration, but the force is the main factor; it is a well ertained quantity and is universally described as "head." H = the height of the water above the centre of an orifice, feet, and d = the diameter of the orifice in inches, the eoretical discharge in gallons per minute may be obtained m the following formula:—

$$\sqrt{H} \times d^2 \times 16{\cdot}3$$

practical application of this formula will be given in the xt chapter, but as mentioned, the result is only an pproximation: it is the *theoretical* discharge. The nature d form of the orifice determines to a very large extent the tual discharge, a hole in a thin plate, a short tube or a long be, each giving different quantities.

What I wish to establish in your minds in the first place, is e exact quantity represented by the varying pressure, or rce, or head, which determines the velocity—cœteris paribus -at which a fluid will issue from a given orifice. H is the eight of the fluid above the centre of the orifice, and a obile fluid under the pressure of a "head" H will issue om that orifice at the same velocity, as that of a body fallg freely from the height H. If V be the velocity in feet per econd, the theoretical velocity is represented by:—

$$V = \sqrt{H} \times 8{\cdot}025$$

t will thus be seen that the "head" in feet can easily be onverted into lbs. per sq. inch. It is usual in practice to ssume that a head of water in feet is double the number of bs. per square inch, this is often sufficiently near, but the rue figures are 0·4327 lbs. per square inch, to each foot ead of water, the head in any other fluid may easily be btained when the specific gravity is known.

The following tables have been calculated for rapid comparison of these numbers:—

Lbs. per sq. inch.		Head of water in feet.		Head in inches of mercury.
1	....	2·311	....	2·046
2	....	4·622	....	4·092
3	....	6·933	....	6·138
4	....	9·244	....	8·184
5	....	11·555	....	10·230
6	....	13·866	....	12·276
7	....	16·177	....	14·322
8	....	18·488	....	16·368
9	....	20·800	....	18·414
10	....	23·110	....	20·460

Head of water in feet.		Lbs. per sq. inch		Head in inches of mercury
1	....	·4327	....	·8853
2	....	·8654	....	1·7706
3	....	1·2981	....	2·6560
4	....	1·7308	....	3·5413
5	....	2·1635	....	4·4266
6	....	2·5962	....	5·3120
7	....	3·0289	....	6·1973
8	....	3·4616	....	7·0826
9	....	3·8942	....	7·9679
10	....	4·3270	....	8,8533

Head in inches of mercury		Head of water in feet		Lbs. per sq. inch.
1	....	1·1295	....	·4887
2	....	2·2590	....	·9775
3	....	3·3885	....	1·4662
4	....	4·5181	....	1·9550
5	....	5·6476	....	2·4437
6	....	6·7771	....	2·9325
7	....	7·9066	....	3·4212
8	....	9·0361	....	3·9100
9	....	10·1656	....	4·3987
10	....	11·2952	....	4·8875

The discharge of water from orifices, and its flow in open shutes or channels, will be explained in the next lecture, though to some extent perhaps it should be mentioned now; but we shall have enough before us to consider the flow in pipes, however that flow is produced. It is well known, even in the simple operation of decanting through a funnel or "tundish" some attention must be paid to the duct by which the fluid escapes, and experiments may easily be made in the laboratory to show the practical effect of lengthening the fall-tube of the funnel, or of increasing the horizontal outflow pipe from a tank of liquid. In short, the longer the pipe the less liquid will flow through it in a given time under the same pressure or head, or, to maintain the same flow through a long pipe as through a short one, the pressure or head must be increased. We have no such practical illustrations of the truth of this argument in this country as they have in America. In the States a 4in. oil main of 60 or 70 miles long is no uncommon experience, and the friction of the main itself often amounts in such distances to nearly half a ton per square inch.

Hawksley's formula for the flow of water through pipes is:—

$$G = \sqrt{\frac{(15\ D)^5\ H}{L}} \text{ and}$$

$$D = \frac{1}{15}\sqrt[5]{\frac{G^2\ L}{H}}$$

where G = the gallons per hour,
L = length of pipe in yards,
H = head of water in feet,
D = diameter of pipe in inches.

The foregoing formula pre-supposes the pipe to be a straight one. Eytelwein gives another formula for straight pipes as:—

$$W = 4{\cdot}72\sqrt{\frac{D^5 H}{L}} \text{ and}$$

$$D = 538\sqrt{\frac{LW^2}{H}}$$

where D = diameter of pipe in inches,
H = head of water in feet,
L = length of pipe in feet,
W = discharge per minute, in cubic feet.

Taking water or an equally mobile fluid, the head necessary to overcome friction may be obtained approximately from the following formula:—

$$H = \frac{G_2 L}{(3d)^5}$$

where H = the head in feet,
L = length in yards,
d = diameter of pipe in inches,
G = gallons of discharge per minute.

When bends occur in the discharge pipe the case is often a complicated one, but thanks to the labours of a few experimentalists, we have data for the calculation of "head" necessary to overcome the friction which they cause.

If A = the angle of bend with the forward line,
V = the velocity of flow in feet per second,
K = a co-efficient for angle, as under,
H = head necessary to overcome friction.
then H = 0·0155 V^2K.

*A course of lectures delivered by Mr. George E. Davis, at the Manchester Technical School.

A	K	A	K	A	K
20°	·046	60°	·364	90°	·982
40°	·139	80°	·741	100°	1·264

All the foregoing formulæ have a very important bearing upon the problem of designing plant, but it must not by any means be thought that the subject is exhausted. The matter is often a very difficult one, but the theoretical aspects of the case are more often than not subdued in practice by making the conduit large enough. If a certain pipe is not found of sufficient size to convey a certain quantity of liquid in a given time, the rule of thumb method is to pull it down and replace it by another, the diamater of which has been selected hap-hazard. Another feature that must not be neglected is the nature of the liquid under consideration, which is often a very important factor. For instance, in calculating pipes for the flow of glycerine the foregoing water formulæ could not be employed, and with a substance like gas-tar, so much depends upon its viscosity that when the suction and delivery pipes are subjected to atmospheric influences there is a marked difference in the rate of delivery in summer and winter.

The viscosity may often be reduced by warming the pipes through which the liquid is made to flow, and a special instance is still in my memory. The pressure on a tar main was 21olbs. per square inch when the pump was running, and consequently difficulty was found in keeping the joints tight; the tar was very viscous, but by threading a half-inch steam pipe through the delivery pipe of the system a full flow was readily maintained with a pressure of only 60lbs. per square inch.

It must not be forgotten, moreover, that theoretical calculations refer only to new pipes or to those of smooth bore, the effects of corrosion, or deposits, upon the interior surface of pipes is sufficiently serious to demand special attention. Incrustations upon the interior of pipes, not only decrease the delivery capacity by constricting the area, but increase the friction to an alarming extent.

There are many interesting problems that can be solved by the application of the foregoing formulæ, and more will be said upon the subject in the lecture on weighing and measuring, but in the meantime enough has been put forward to show you that there need be no "rule of thumb" work in matters relating to the moving of liquids from one place to another.

Having our mains laid, it now remains to apply the head, or pressure, necessary to force the liquid through them. This may either be done with a natural head, or a pressure supplied by artifical means. In either case there is no difficulty in measuring it; the actual measurement above the orifice of outflow will give us the quantity we seek in the first instance, while actual measurement with a pressure gauge will determine the head in the second. The power required is also well known and calculated when we remember that 33,000 foot pounds per minute means one horse-power. In actual practice, double the theoretical power is generally required to drive pumps of ordinary construction, if indeed a much larger proportion is not found necessary, as without a pump is large enough to be called a "pumping engine," it is usually a much abused and neglected instrument. It is not my intention to give you a dissertation on pump construction, as that belongs entirely to the mechanical engineer, though it may be as well to call attention to several points that must be studied by the chemical engineer. First, there is the material of which the pump is constructed, and this is no unimportant point in connection with its use in the chemical and allied industries. Secondly, the type of pump best suited to the work in hand, and thirdly, the form and nature of the valves employed in construction. The material used for the construction of the pump must, of course, be selected on account of its chemical and physical properties, an iron pump would be of little use in dealing with aqueous hydrochloric acid, and similarly experience shows us that a brass pump is rapidly corroded by alkaline solutions. A general chemical knowledge will, in most cases, enable the proper selection of material, effect, even of slight chemical action, must never looked. And again, even when it is known that the side is perfect, the physical character of the material unsuitable, and here there is often room for a cons amount of inventive genius. As an instance, the m timony will resist, fairly well, even strong hydrochlo but this material is a very brittle and crystallisabl and requires special treatment in order to adapt it with corrosive liquids. Other materials, such a earthenware, stoneware, ebonite, etc., have been us more or less success in pump construction, but as a gen they have not found much favour, chiefly, I believe, fro design, whereby fractures too often occurred.

The second feature—the type of pump—which sh selected for varying classes of work, is often hazarde out "rhyme or reason." The types of pumps in the is legion, and it is very seldom that a pump user is give a satisfactory account of his reasons for selecti an average chemical works in England, the pump is fully treated and sorely neglected, and usually a p chosen irrespective of any intrinsic merits, but mainly it will stand a considerable amount of battering abou off and patching at the hands of the works fitter, with badly packed glands and worn out valves, absolutely breaking down. This is not what it sho but I am afraid it will be so until *foremen* in charge things are in possession of more technical informati they possess at present. When a pump absolutely ce deliver (and not till then) it is usually overhauled mechanic, and as a result it is more often than not le worse state than before. If those in charge of pumps only appreciate that 95 per cent. of pump trouble between the end of the suction pipe and the plunger pump, they will have learned their first lesson in pum Wherever it is possible, the suction pipe should be pr with an efficient foot-valve, and both suction and d pipes be large enough to reduce the friction to a min Friction in the suction pipes causes a loss of pow increases the tendency to entrance of air by the joi glands—friction in the delivery pipes is also a waste of and a cause of leaky joints.

Amongst the types of pumps there are many to from, the single acting lift pump, and the old fashion and force pattern are forms the beginner may study wi vantage, as all the forms of ram and piston pumps d for their action upon modifications of the principles th be found utilised therein. Steam pumps are single and double acting, some with fly-wheels, and some wi and the greatest diversity of opinion prevails as t suitability of this or that pattern for special work. It my intention to put myself in conflict with the mak any particular form of pump, by describing the pattern by Mr. A or Mr. X as the best, but I would enjoin eve to look well to three points before purchase: first, that all the parts are so arranged as to be readily acc for lubrication and cleaning; second, that there is pl room for packing glands, turning nuts, pet cocks, &c., w being cramped in any way; and third, that the chambers are readily accessible without undue labou the shifting of heavy parts.

There then remains the question of the suitability valves to the fluid to be pumped, and I am afra thing but actual experience will be of much service direction, except that the pump is often blamed for the comings of those who work it. A pump that will pum ders, stones, barrel-shives, or liquids equally well, is desideratum in many establishments, where no effort is to prevent solid matters entering the suction pipe.

But there is a method of scientifically testing the effi of pumps, and that is by the indicator diagram. mechanics know perfectly well how to take a diagram steam cylinder, but it has never occurred to them th same methods may be applied to a water cylinder. faults may often be revealed by taking a diagram from a cylinder, yet how many have courted this meth examination.

(To be continued.)

THE tables of the density of ammonia solutions in relation to their composition, give such discordant results that they cannot be employed even for the testing of commercial solutions. Such are those of Ure and Carius. (In Messrs. Weil Goetz and Desor's Treatise on "Ammoniacal Solutions and Waste Purifying Materials," p. 15). A comparative table is given of the results obtained by the various chemists who have endeavoured to determine the amount of NH_3 present by means of the density of the solutions, with the exception of that of Messrs. Lunge and Wiernik, which had not then been published.

The tables of Davy and Dalton were constructed without taking into account the change of volume which occurs when water is added to a concentrated solution of ammonia. Ure and Meissner corrected this error, but their determinations of ammonia are far from presenting the degree of precision which can now be obtained. Carius and Wachsmuth made new sets of determination, but, unfortunately, at different temperatures. That of Carius being for 14°, and that of Wachsmuth for 12°, so that comparison is impossible. The last table published has been constructed for 14° by L. Smith.

In the following table the results of a very careful set of determinations are embodied. The estimations of ammonia were made either by platinum chloride or by titration with normal acid, methyl-orange being used as indicator.

TABLE OF DENSITIES OF AMMONIA SOLUTIONS AT 15° (GRUNEBERG).

Density.	NH_3 % by weight.
0·880	35·50
0·885	33·40
0·890	31·40
0·895	29·50
0·900	27·70
0·905	26·00
0·910	24·40
0·915	22·85
0·920	21·30
0·925	19·80
0·930	18·35
0·935	16·90
0·940	15·45
0·945	14·00
0·950	12·60
0·955	11·20
0·960	9·80
0·965	8·40
0·970	7·05
0·975	5·75
0·980	4·50
0·990	3·30
0·995	2·15
1·000	1·05

Jour. Eclair. au Gaz.

THE FUNCTION OF AMMONIA IN THE NUTRITION OF PLANTS.

COMMUNICATED BY M. A. MUNTZ TO THE ACADEMIE DES SCIENCES.

THE question, In what form nitrogen is absorbed by the roots of plants has given rise to many investigations. It was first believed that the nitrogen of organic matter could contribute directly to the nutrition of plants. Boussingault, however, showed that nitrogen could only be taken up in the form of some salt of a mineral acid, and at a period already becoming remote, this property of being assimilated by plants was ascribed to salts of ammonia alone. Even Kuhlmann, after having confirmed the beneficial effect of nitrate of soda as a fertiliser, imagined that this salt was first converted into ammonia in the soil by reducing agents. This opinion held its own for a long time, even although Boussingault showed that nitrates are directly absorbed by plants.

At the present time, however, the inverse of this theory tends to prevail, and it is generally believed that the nitrogen of plants is assimilated in the form of nitrates, and that, in opposition to Kuhlmann's views, ammonia must pass through the form of nitrate before being [illegible]

There [illegible], a doubt as to the true part played by ammonia, [illegible] instituted do not seem to have fully [illegible] it appears easy to solve a problem which [illegible] the conditions are greatly complicated by [illegible] The good effect of ammonium

THE FIXING OF NITROGENOUS COLOURING MATTER ON VEGETABLE FIBRES.

(FROM THE SAINT DENIS SOCIETY OF COLOURING MATTERS AND CHEMICAL PRODUCTS.)

NITROGENOUS colouring matters, both simple and complex, only possess a very slight affinity for vegetable fibres. Up to the present time nitrogenous colouring matters have been almost the only ones in use, but they have nevertheless been employed in an artificial manner by combining them with sodium carbonate, sodium chloride, sodium phosphate, etc. The new process employs them directly, after first subjecting the cotton to a preliminary process, which renders it capable of being directly dyed by nitrogenous colours. A quantity of the tissue—say a hundred yards—is passed through a bath composed of

Magnesium acetate at 30° B .... 50 litres.
Aluminium acetate at 15° B .... 50 „
Water .... 50 „

After drying the material is passed into:

Sulphate of zinc .... 5 kilogrammes.
Caustic soda .... 10 „
Water ....100 „

On leaving the vat the stuff is washed, and then passed into the dyeing bath, made up to a strength of 2 or 4 per cent.

The temperature of the bath should be about 80°, and the dyeing should not last longer than half an hour. The various shades of fronceau and rocullin, almost all the oranges, the bordeaux, cerosine, the croceins, and the homologous benzidine colours can thus be well fixed on the cloth. The recipe given above may be varied according to circumstances. The substitution of the corresponding sulphates for the acetates of aluminium and magnesia is advantageous economically, and does not appreciably affect the results. The solution would then be made up in the following manner:—

Magnesium sulphate .... 15 kilos.
Alum .... 5 „
Water ....100 litres.

In addition to this change, calcium acetate may take the place of magnesium acetate, etc., etc.

Moniteur de la Teinture.

BOILER EXPLOSION IN SCOTLAND.

ABOUT ten o'clock on Saturday night last a boiler explosion took place at Pumpherston Shale Oil Works, Midlothian, with the result that three men were killed and two injured. The boiler was one of a battery of low-pressure boilers used for supplying steam to the shale distilling retorts. It was of the double-flue Lancashire type, and the working pressure was only 5lb. to the square inch. At the time of the explosion very few men were about the works. The boiler was blown some thirty yards from its site, and carried away two large tar tanks with their supports. A large area was strewn with bricks and broken timbers. The men killed were Edward Dempsey and John Scott, boiler firemen, and Andrew Taylor, retort fireman. Scott's body was found about thirty yards away, while Dempsey was blown to the top of some retorts about 25ft. high, a distance of about 40 yards. Taylor was hurled through the air a long distance over the top of the retorts and across a public road. All the bodies were shockingly mutilated, and the clothing had for the most part disappeared. Several miners' houses near where Taylor's body was found had large holes knocked in their tile roofs by falling bricks, but, though all the houses were inhabited, no one in them was injured. The two men who were hurt are recovering. The cause of the disaster is unknown. Scott and Taylor each leave a widow and three children. The accident caused great consternation in the locality.

NEW TABLE OF THE DENSITY OF AMMONIA SOLUTIONS.

BY MESSRS. G. LUNGE AND T. WIERNIK.*

THE authors point out the discrepancies which exist between the various tables which have been given for determining the strength of ammonia solutions from their densities, and give a new table, constructed with all possible accuracy.

The solutions were prepared with ammonia frequently distilled from lime. The gas was absorbed by distilled water cooled by ice, very concentrated solutions being thus obtained, which were preserved at a low temperature in well stoppered bottles.

*Translated by the Bull. Soc. Chim.

The density determinations were made by means of the p
all necessary precautions being taken to avoid loss of ammo
Three determinations were generally made, at 13°, 15
With very concentrated solutions it is impossible to make
minations above 15°. They made them at 11° and 9°, and
able to ascertain the co-efficient of dilatation of the liquid.

The solutions were subsequently titrated. For this
liquid was acidified with normal sulphuric acid, the exc
was determined by titration with methylorange as indicator.

The concentrated solutions were also analysed by means
chloride. They estimate the error of these determinations
for strong solutions, and 0·05% for weak solutions.

The table given is constructed for 15°. When the
differs from this it is necessary to make a correction which
calculated by means of the co-efficients of expansion dete
described above. The value of this correction for an
decrease of one degree is given in the table for every densit

TABLE OF DENSITY OF AMMONIA SOLUTIONS AT 1
(LUNGE AND WIERNIK.)

Density.	NH_3 %.	1 litre contains NH_3 in grammes.	Correc density
1·000	0	0	0·00
·998	0·45	4·5	0·00
·996	0·91	9·1	·00
·994	1·37	13·6	·00
·992	1·84	18·2	·00
·990	2·31	22·9	·00
·988	2·80	27·7	·00
·986	3·30	32·5	·00
·984	3·80	37·4	·000
·982	4·30	42·2	·000
·980	4·80	47·0	·000
·978	5·30	51·8	·000
·976	5·80	56·6	·000
·974	6·30	61·4	·000
·972	6·80	66·1	·000
·970	7·31	71·9	·000
·968	7·82	75·7	·000
·966	8·33	80·5	·000
·964	8·84	85·2	·000
·962	9·35	89·9	·000
·960	9·91	95·1	·000
·958	10·47	100·3	·000
·956	11·03	105·4	·000
·954	11·60	110·7	·000
·952	12·17	115·9	·000
·950	12·74	121·0	·000
·948	13·31	126·2	·000
·946	13·88	131·3	·000
·944	14·46	136·5	·000
·942	15·04	141·7	·000
·940	15·63	146·9	·000
·938	16·22	152·1	·000
·936	16·82	157·4	·000
·934	17·42	162·7	·000
·932	18·03	168·1	·000
·930	18·64	173·4	·000
·928	19·25	178·6	·000
·926	19·87	184·2	·000
·924	20·49	189·3	·000
·922	21·12	194·7	·000
·920	21·75	200·1	·000
·918	22·39	205·6	·000
·916	23·03	210·9	·000
·914	23·68	216·3	·000
·912	24·33	221·9	·000
·910	24·99	227·4	·000
·908	25·65	232·9	·000
·906	26·31	238·3	·000
·904	26·98	243·9	·000
·902	27·65	249·4	·000
·900	28·33	255·0	·000
·898	29·01	260·5	·000
·896	29·69	266·0	·000
·894	30·37	271·5	·000
·892	31·05	277·0	·00
·890	31·75	282·6	·00
·888	32·50	288·6	·00
·886	33·25	294·6	·00
·884	34·10	301·4	·00
·882	34·95	308·3	·00

—*Jour. Eclair.*

[illegible]

... otherwise known as ... take the ... is dispensed with, and ... may be set up or improvised ... were given, beginning ... of ... sand, ... this point the precipitate ... by aid of a sponge ... go with a lighted flame ... shows ... the process, and the author ... part of which he ... remarks followed the reading ... three contributors concerned ... The meeting was more largely att...

AMENDMENT OF THE COMPANIES ACTS 1862-80.

A [illegible] ... Chambers of Commerce for the ... was held a few days ago; Mr. Henry ... the chair. The invitations to the ... Manchester Chamber, and accepted ... Chamber, Halifax Chamber ... Association. These and other Chambers of ... represented at the Conference ... who opened the discussion ... agreed that some attempt ought ... Act of 1862, or at least that portion ... formed under the limited liability ... also take it that the attempt was not ... repeal the limited liability system ... entertained about that system, he ... them to appreciate it. He thought it ... the country in various ways. Many ... been brought out, or could not have ... the limited liability system. Their ... Act, and make it more effective and more ... safeguards against fraud.

... as follows:—

"... conference two registrations, a pro... are necessary."

"... capital should be subscribed and ... registration."

"... paid on application, and 10 on allot..."

"... any industrial company shall not be ... and working plant."

"... must convene a meeting of ... may appoint a committee of five or more ... nominees, or any intending shareholder ... investigate all contracts, existing or ... vendors and the intended company ... contained, nor any investigation by ... shall exempt any promoter from any ... or misrepresentation or otherwise."

"... every company shall contain a copy of the ... and be accompanied by a schedule showing ... occupations of the promoters or provisional ... contain the names and descriptions of all ... any commission, brokerage, or agency ... the same in reference to, or contingent ... the company."

"... every company shall in all cases where the ... consist of plant or property, clearly state ... which has been deducted from ... such plant or property; and shall have ... statement of cost of all items which have been ... the issuing of the last balance sheet."

"... every company shall be bound to certify ... the declared rate of depreciation has been ... that the detailed list of the additions to capital ... books and accounts correct. For every ... certificate, or for giving a false certificate, each ... a penalty of not exceeding £10, for every such ..."

"Every ... of a company who shall sign or authorise the issue ... which the rate of depreciation, or the items passed ... account are incorrectly stated, shall be liable to a ... not less than £10."

It was resolved to draft a bill embodying these resolutions.

Legal.

YOUNG, &c., v. THE HERMAND OIL COMPANY.

ON the 1st inst. Lord Kinnear closed the record in an action at the instance of William Young, consulting chemist, residing at Priersford, Peebles, and George Thomas Beilby, chemist, residing at St. Kitts, Slateford, against the Hermand Oil Company, Limited, having its registered office at 6, Shandwick-place, Edinburgh. The action is brought in the form of a claim for count and reckoning, but is understood to raise the question of validity of patents which pursuers hold relating to the distillation of shale. Pursuers sue for count and reckoning with reference to the quantity of shale distilled or manufactured by the defenders in apparatus or retorts made or worked according to the principles of the inventions set forth in the letters patent, or for payment of £1,000. Defenders, it is stated, hold licences from pursuers for the working of these patents, and the licence requires that within ten days after 31st December in each year defenders should give pursuers a statement in writing showing the number of apparatus worked, or capable of being worked, and the number of tons of shale with which the apparatus was charged. Between March, 1886, and May, 1889, defenders were required to pay a royalty of 1½d. per ton, and thereafter by a new agreement a royalty of 2d. per ton. Pursuers further allege that the annual statement is erroneous in respect that defenders have a number of the patent retorts working besides the 120 mentioned in the statement, and upon the shale distilled in which they should have paid the royalty. In reply, defenders deny this allegation, and plead, among other things, that pursuers' statements are irrelevant, wanting in specification, and unfounded in fact.

Counsel for Pursuers—Mr. Daniell. Agents—Smith and Mason, S.S.C.

Counsel for defenders—Mr. Dickson. Agents—Drummond and Reid, W.S.

THE RAILWAY RATES INQUIRY.

THE statement made by Lord Balfour of Burleigh, says the *Manchester Guardian*, with reference to the course of procedure to be adopted in connection with the inquiry into the railway classifications and schedules may not unreasonably cause anxiety as to the probable result of the whole matter. A very definite warning has been given to the traders that if they intend to put their case effectively before the representatives of the Board of Trade, they must compress and focus their statements. The Board are anxious to bring either a bill or a report before Parliament before the end of the present session; and that this should be done, if possible, is beyond question. It is desirable in the interests of both the companies and the traders that the present state of uncertainty should be terminated at the earliest possible date; and if the inquiry is to result in revisions which will remove competitive disadvantages under which the industries and trade of the country are said to be suffering—and this is the express purpose of the Act and the inquiry to which it has led—then the sooner the revisions are made the better. But it scarcely seems possible that anything very thorough can be done. The inquiry began in the middle of last October, and the Court has scarcely yet got beyond the hearing of the case for the railway companies. In presenting their case the companies have the advantage of a certain degree of unity. They are defending classifications and schedules agreed upon beforehand; the arguments that are used on behalf of one company may with very little modification be used on behalf of all; and two or three leading officials can put the case as completely as a crowd of witnesses. The interests of the traders, on the other hand, are as varied as the articles in which they deal, and even every locality in the country may be said to have its special interest in each of these articles. If the railway case has occupied since last October, it might be assumed that the traders' case, treated on the same scale, would last through many Octobers yet to come, and it is not improbable that not a few readers have already ceased to interest themselves in the inquiry as likely to prove hopelessly protracted. Lord Balfour of Burleigh tells us that the total number of traders' witnesses proposed from various sources considerably exceeds 400. Yet, after the witnesses have been heard, it will be the duty of the Board of Trade to carry on a negotiation for an agreement with the railway companies; and if this negotiation fails, then the Board will be called upon to propose what they consider to be a "just and reasonable" classification and schedule of rates and charges. Each of these duties implies that the Board should first very carefully sift, weigh, and consider the evidence brought forward during the inquiry.

A ray of hope is, however, afforded by Lord Balfour's sketch of the kind of witness and evidence which the Court wishes to hear. The witnesses, he said in affect, must be representative, and must confine themselves to the exposition of carefully thought-out opinions bearing
on certain broad principles. Firstly, the Court wishes to g
well-understood general principle on which all classifications
based. Secondly, it asks what is the principle on which
terminals are to be charged—we may add if they are to be c
all. Thirdly, the principle of service terminals is to be
Finally, some principle on which the rates for each classificati
be graduated must be enunciated. Now it may frankly be sa
far as the classifications and schedules presented by the com
concerned, they do not seem to be based on any princip
except the principle of making use of every possible pretext fo
an extra charge. From the first, the proposals of the compa
appeared like a huge joke than anything else. Far from si
the rate-books, their effect would be to make confusion w
founded. The companies really have only themselves to blan
are credited with having no more serious purpose than that o
any revision impossible, and protracting the inquiry with the h
the agitation may die out and the traders ultimately be glad
content with the arrangements hitherto existing, slightly
just to save appearances. Whether, from their point
the companies are justified in adopting this policy n
be discussed. The practical thing is that if the trader
obtain any satisfactory result at all they must give heed
Balfour's hint, and rather aim at putting a workable set of p
before the Court than at merely destroying the classificati
schedules of the companies or proposing modifications in detai
Court has plainly enough intimated that, for its part, it is bent
plification. It, for instance, considers that eight distinct cl
quite sufficient to permit of an equitable and tolerably m
grouping of rates. Then there is the question of terminals. T
ments presented by the companies really make it probable
Court will require little inducement to sweep away termin
gether, except as regards very special and extra services. If
terminals are allowed, however, there can be no justification f
except a great reduction and simplification of the rates. If term
not allowed, then the rates must be more varied in accordance
variation in the services beyond that of mere hauling rendered
Company in the case of each class. The traders must make the
tion—terminals with low and tolerably uniform rates on the on
or no terminals and a rather more complicated rate-book on th
They should very emphatically decline to accept any proposal
which are not based on these principles. Considering all thing
ever, it is a comfort to reflect that, whatever agreement is ar
between the Board of Trade and the companies, Parliament w
the right to veto it.

THE ALBION WATERPROOFING COMP (MANDLEBERG AND CO., LIMITED).

THE first annual general meeting of this company was h
morning at the Town Hall, Manchester, Mr. G. Milner
man of the company, presiding.—The Chairman moved the a
of the report, which showed a net profit for 1889 of £24,161. 1
enabling the directors to recommend a dividend of 10 per cent.
ordinary shares. He said that any anxiety which in the first y
the Company's business might have been felt had been entirely r
as they became familiar with the business, and the more they s
the more they had been satisfied that a sound and healthy l
admirably managed and conducted with remarkable ability ha
handed over to the Company. After payment of dividend ther
balance of £10,368. 4s. 8d. Of this sum the directors prop
place £2,000. to a depreciation fund and £5,000. to a reserv
carrying forward £3,368. 4s. 8d.—Mr. J. R. Hampson secon
adoption of the report, which was supported by Mr. G. C. Manc
and carried.—Mr. J. R. Hampson was re-elected a direct
Messrs. Turquand, Young and Co., auditors.—The meeting wa
wards made special, and the name of the company chang
Mandelberg and Co., Limited.

LAKE OF ACID.—In White Island, N.Z., there is a lake wh
seething, bubbling mass of muriatic acid, and as the liquid boil
bubbles have their sides in shadow, reflecting a green tinge, th
surface emitting an intensely noxious vapour. The island is of
origin, and is destitute of animal life. There are large dep
sulphur in it, the poorest yielding 62 per cent. of sulphur.

A NEW ALLOY has been discovered by Herr Reith, of Bock
Germany, which is said to practically resist the attack of most a
alkaline solutions. Its composition is as follows:—Copper, 1
tin, 2·34 parts; lead, 1·82 parts; antimony, 1 part. This
therefore, a bronze, with the addition of lead and antimony
inventor claims that it can be very advantageously used in the
tory to replace vessels or fittings of ebonite, vulcanite, or porce

Correspondence.

ANSWERS TO CORRESPONDENTS.

-We do not know any particular use for the worn rollers. We understand hey are used over again with new material.

Weak bleach always settles better than strong. A quarter of a pound of to a quart of water gives a solution of 11° Tw.

IA.—Good grey sulphate of 24½ per cent. can be made with pyrites acid. dvertise.

—Nothing but a dynamo would pay. It would be absurd to use batteries old type.

—We cannot advise on such matters. (2) Ask your patent agent.

r.—(1) All fixed. (2) The use of caustic soda would not pay. (3) When the ses in a gas works are carried out chemically, but not before.

Ve are preparing a series of articles on the subject.

four queries are only suitable for our advertising columns.

-A little learning is a dangerous thing. Read up your electrical manual h.

MAKER.—The matter is under consideration. It would be well for you to end your judgment in the meantime.

Ve do not hold ourselves responsible for the opinions expressed by our correspondents.

IE DETERMINATION OF MOISTURE IN WOOD PULP.

To the Editor of the Chemical Trade Journal.

e are quite at one with Mr. Pattinson on the subject of estimating noisture in wood pulps. We think with him that the weight as at 212° F. is the one on which all prices should be based and lations made. But in our communication of the 15th ult., we ly dealt with the two alternatives, of which one or the other is rally adopted in the trade, viz., to dry the pulp at 212° F., and a fixed percentage to obtain the air-dried weight, or to dry at 212° nd allow the re-absorption of moisture in a normal atmosphere. e advised the former course on account of the difficulty of obtain-uniform conditions for the second mode of procedure, and a little mination will show that our advice is not really at variance with Pattinson's suggestion, for the addition of a constant figure such as ecommended does not affect the value of the 212° F. dried weight as sis for the calculation of prices, etc.

Ve thank Mr. Pattinson for his friendly criticism,—And are

GEORGE E. DAVIS,
he Manchester Technical Laboratory. ALFRED. R. DAVIS.

LEAD POISONING AT SHEFFIELD.

HE Water Committee of the Sheffield Corporation have been engaged in an exhaustive inquiry into the facts connected with many complaints of lead-poisoning that have appeared in the local bers.

Mr. Alderman Gainsford, Chairman of the Water Committee, in ning the inquiry, said that they were most anxious to ascertain the e facts of the case. He supposed there was no doubt that the water n the reservoir was of such a character as to be liable to take up lead newhat freely. None of them would wish to deny that lead-poisoning sted, but they must satisfy themselves as to its extent. The next nt for inquiry was the cause of the poisoning; and, further, they nted to find out whether it was so serious a matter as to call for nedy at any cost. The third question related to the remedies that ght to be applied. He regretted that the response to the committee's uest for evidence from the general public had been almost entirely bout result, except so far as the medical profession were concerned, whom the committee were very greatly indebted.

Evidence was then given to the effect that there had been many es of sickness arising from lead poisoning which were very severe; eed, in one instance the death of the patient was attributed to this se. The evidence of several members of the medical profession, o were examined before the committee, went to show that in some of samples of water which were analysed, lead existed to the amount half a grain per gallon.

The preliminary investigations of the committee have been fruitful much startling evidence on the subject of plumbism. The dis-ures last week, were, indeed, of quite a sensational order, for we of persons been paralysed, crippled for life, and even rendered tless from lead poisoning, arising from consumption of a soft moor- water in an unfiltered condition, from the Redmires district, nce Sheffield derives part of its supply. The odd thing is that this mires high-level supply has been in use for half-a-century without estion of its injurious action on lead-pipes. The fact is, however patent that plumbism has of late years been manifesting itself in an unpleasantly rapid fashion. True it is that the percentage of consumers affected is small, but the evil, even if of comparatively small proportions, is of great intensity, and has been such as to render imperative some remedial action. The Committee find that the lead is "derived from the pipes and cisterns of the consumers," and state their preparedness to sanction practical remedial experiments. It appears that household charcoal filters are useful in removing lead, but that unless they are cleansed and recharged at frequent intervals the remedy is but temporary, and the committee are of opinion that to aim at securing the requisite conditions is to attempt the impossible. Three alternative methods are advocated—treatment of the Redmires water by the lime process; provision of filter beds; or removal of lead service pipes—and eventually the committee have resolved to give the first of these a trial, in the confident assurance that its efficacy will be realised without leading to hardening of the water, at present so much appreciated for its softness.—*Daily Graphic.*

PRIZE COMPETITION.

A LITTLE more than twelve months ago, one of our friends suggested that it might be well to invite articles for competition, on some subject of special interest to our readers; giving a prize for the best contribution, he at the same time offering to defray the cost of the first prize, and suggesting a subject in which he was personally interested.

At that date we were totally unprepared for such a suggestion, but on thinking over the matter at intervals of leisure, we have come to think that the elaboration of such a scheme would not only prove of benefit to the manufacturer, but would go some way towards defraying the holiday expenses of the prize winner.

We are therefore prepared to receive competitions in the following subjects:—

SUBJECT I.

The best method of pumping or otherwise lifting or forcing warm aqueous hydrochloric acid of 30 deg. Tw.

SUBJECT II.

The best method of separating or determining the relative quantities of tin and antimony when present together in commercial samples.

On each subject, the following prizes are offered:—

One first prize of five pounds, one second prize of one pound, five additional prizes to those next in order of merit, consisting of a free copy of the *Chemical Trade Journal* for twelve months.

The competition is open to all nationalities residing in any part of the world. Essays must reach us on or before April 30th, for Subject II, and on or before May 31st, in Subject I. The prizes will be announced in our issue of June 28th.

We reserve to ourselves the right of publishing the contributions of any competitor.

Essays may be written in English, German, French, Spanish or Italian.

THE SALT UNION.

THE first ordinary general meeting of the shareholders of the Salt Union, Limited, was held at the Cannon-street Hotel a few days since, under the presidency of Lord Thurlow. In moving the adoption of the report, the Chairman said that they had been accused of raising the price of salt to an exhorbitant extent; but he could not admit they had done so. They no doubt had raised the price; but he pointed out that the object of the Salt Union was to raise prices so as to save the trade from impending ruin. The present prices were not higher than in 1872 and 1873. The directors desired to alleviate the sufferings ariving from subsidences in the salt districts; but they had to remember that the money they were dealing with was not their own, and they were advised that no liability attached to the Salt Union. With regard to the subsidences, the saltmakers and royalty-holders would be responsible if any persons were. The directors had decided to bore in Cheshire to see if they could not find coal, petroleum, or minerals, and believed they were on the eve of a discovery greater than the discovery of coal in Kent. Mr. Verdin seconded the motion. A discussion followed, and the chairman explained the provisions made for depreciation of stock, and promised to publish a monthly return of the business. The report was adopted.

Trade Notes.

"Gassed" in a Chemical Works.—While a married man, named Michael Duffy, who lives with his wife and family in Peasley Cross-lane, was working in the chlorate house at Messrs. Kurtz's Chemical Works, on the 26th ult., he was overcome by still gas, which is known among the men as "Roger." He was picked up in an unconscious state and conveyed home, where he lies in a precarious condition.

A Submerged Staffordshire Coalfield.—At the annual meeting of the South Staffordshire and East Worcestershire Institute of Mining Engineers, at Mason College, Birmingham, on Monday, Mr. Henry Lea, in his presidential address, said the effect of the Mines Drainage Act had been to reduce the daily income of water into the mines from 48,000,000 gallons in 1873 to 17,000,000 gallons. The cost of pumping was reduced from 11d. to 4¼d. per 25,000 gallons. When the underground levels were completed he anticipated the whole area would be unwatered and the 100,000,000 tons of coal liberated.

A New Way of Filling Pots.—The E. P. Gleason Manufacturing Company, of Brooklyn, N.Y., will try a new method of filling in "batch." Over every pot is an iron tank or boiler which is constantly filled with batch and which is kept constantly warm, thereby shortening the time for melting, as the tanks are connected with each pot by a funnel-like arrangement so that it is only necessary to pull out a small trap and the batch of glass runs into the pots of its own accord. Although the idea is a good one it has not been in operation, as the mixing department is too low down in the building and difficulty is experienced in getting the batch into the tanks. To remedy this the firm is removing its sand blasting room to another portion of the building and will move the mixing rooms higher up, when the idea be then put into successful effect.—*Commoner and Glassworker.*

International Exhibition of Mining and Metallurgy, London, 1890.—An International Exhibition of Mining and Metallurgy on an extensive scale, the result of a proposal which emanated from the *Mining Journal*, will be held during the forthcoming summer at the Crystal Palace, Sydenham. One-half of the surplus of the Exhibition, as certified by the auditors, is to be paid to the exhibitors *pro rata* to the amount paid by them for space occupied, and the other half is to be disposed of by the Council, either in founding a Scholarship at the Royal School of Mines, or in helping some other institution connected with Mining and Metallurgy. The Exhibition will open on 2nd July, and close on 30th September, 1890, and, besides several features of special attraction, important collections of exhibits are expected from the Colonies and foreign countries. The Honorary Secretary is Mr. Geo. A. Ferguson, editor of the *Mining Journal*, 18, Finch-lane, London, E.C., from whom prospectuses and application forms for space may be obtained.

The Will of a Newcastle Millionaire.—Probate of the will, dated 19th January, 1888, with codicils made the 28th March, 12th April, and 26th October, 1888, and the 15th August, 1889, of the late Mr. Christian Allhusen, of Stoke Court, Bucks, J.P., and of the Newcastle Chemical Works, who died on the 13th January, aged 84 years, and was the son of Mr. Charles Christian Frederick Allhusen, of Kiel, has been granted to the executors, Mr. Wilton Allhusen, of Newcastle-on-Tyne, chemical manufacturer, and Mr. William Hutt Allhusen, merchant, sons of the testator; Mr. Edward Horatio Neville, of Skellingthorpe, Lincolnshire, his son-in-law, and Mr. John Edward Davidson, of Newcastle-on-Tyne (secretary of the Newcastle Chemical Works Company), by whom the value of the testator's personal estate has been sworn at £1,126,852. 1s. 10d., including American railway securities valued at £140,500., stocks and bonds of foreign Governments £76,950., shares of the Newcastle Chemical Works Company £47,000., and shares and debentures of other public companies £600,000.

A Novel Project.—When it was stated some weeks since in the newspapers that the building of a milk pipe line from a point in New York State to New York City was projected there was a rather general smile, and the matter was treated as a joke. The projectors were, however, it seems, in sober earnest. A company with a capital of 500,000 dols. has, it is announced, been formed at Middletown, N. Y., for the purpose of constructing such a line. The proposed method of forwarding the milk is in cylindrical tin cans, surrounded and propelled by water, and the promoters of the scheme assert that the time of transportation, for a distance of 100 miles, will not exceed an hour, while the profit will be about one cent. a gallon. *Fire and Water* thinks if this sort of thing goes on, we need not be surprised ere long to find New York the converging point, not only of oil, natural gas, and milk pipe lines, but of whisky ducts from the blue grass regions, and beer ducts from Cincinnatti, St. Louis, and Milwaukee. The pipe manufacturers may well feel cheerful at the prospect before them.—*Scientific American.*

MISCELLANEOUS CHEMICAL MARKE

Caustic soda has been in brisk demand during the past we there is a stronger market. Price for 70's is firm at £9. 5s. f. for 77 per cent., £10. 15s. net. Other strengths in proportio ash also commands full prices, and has a ready outlet at 1¼ for ordinary brands, and 1⅜d. per deg. for refined. E powder is without material change, and prices are on £5. 2s. 6d. to £5. 5s at makers' works. Chlorate of potas 4¾d. per lb. Soda crystals somewhat scarce, and price Sulphate of copper easier in price, £25. 10s. to £26. delivery, and £23. 10s. for April. Lump and flour sulphur u in position meantime, but the new company which is to deal greater part of the British production from the "Chance" process, will no doubt strengthen prices by reducing the a competition. Vitriol is still in good demand in all localities. powdered arsenic is dearer. Yellow prussiate of potash sell freely and firm at 9½d. to 9¾ per lb. Acetates of lime are i better demand, £8. 5s. to £8. 10s for brown, having been spot deliveries; prices rather easier forward. Grey scarce at £ to £14. Acetates of lead and other acetates without change. of lead slow. Carbolic acids are still quiet and in uncertain the business transacted being only for short periods as a rule acid 1s 4d. per lb. There is rather more demand for potash, carbonate, and muriate, and current rates are firmly ma Ammonia salts are generally quiet.

METAL MARKET REPORT.

	Last week.		This wee
Iron..............	51/7½	....	50/0½
Tin	£89 15 0	....	£90 0
Copper	47 2 6	....	46 15
Spelter	22 5 0	....	22 15
Lead	12 12 6	....	12 7
Silver	44d.	....	44.

COPPER MINING SHARES.

	Last week.			This wee	
Rio Tinto	15¾	15⅞	....	15⅛	15
Mason & Barry	6 15/16	6 7/16	....	6⅛	6
Tharsis	4½	4¾	....	4 7/16	4
Cape Copper Co. ..	3⅜	3½	....	3⅜	3
Namaqua	2	2⅛	....	1⅞	2
Copiapo	2 7/16	2 9/16	....	2 7/16 -	2
Panulcillo	15/16	1 1/16	....	1	1
New Quebrada	⅝	¾	....	⅞	1
Libiola	2⅝	8⅝	....	2⅜	2
Tocopilla	1/6	2/-	....	1/-	1
Argentella	3d.	9d.	....	3d.	9

Market Reports.

TAR AND AMMONIA PRODUCTS.

The benzol market remains in much the same condition a ported last week. In some quarters, it is said, a change for th has commenced, but we do not see any signs of such havir place in the meantime. Practically, prices of last week may b as the values of to-day. Solvent naphtha still continues in mand, and the higher price realised during the past few we caused the introduction into the market of many low-class q 1s. 10d. to 1s. 11d. may be considered as to-day's value for g at 160, f.o.b. Creosote is moving off very freely, and the ab stocks on the part of producers keeps prices firm. Crude ca still an enigma. It is being absorbed as fast as produced, bu cannot be raised. The pitch market continues very dull, co buyers evidently not being wishful to increase their purchases at

The sulphate of ammonia market still is in an abnormal position however, are maintained in spite of the lower prices said to h accepted by the London makers. We hear that £11. 15s. accepted at Hull, and the same price at Leith; and though Be quoted at £11. 13s. 9d, yet outside London makes have £11. 16s. 3d. We have it, however, on good authority, Wednesday a considerable quantity of sulphate was sold f delivery at Hull at £11. 17s. 6d., usual terms and conditions, those manufacturers who have been accepting lowest mon certainly been selling their production below the market pric day. Dealers still continue to depress values, but there are requirements to be filled, which is evidenced by the fact of th th which parcels offered by makers are absorbed. We consider s values to be £11. 17s. 6d. f.o.b. Hull, £11. 15s. Leith, and 2s 6d. Liverpool.

REPORT ON MANURE MATERIAL.

re has been much more inquiry, and decidedly more business both in phosphoric and a n noniacal manure material, though have not experienced a ıy great fluctuation. Proximity of consumption, no doubt, accounts for this better feeling, and we that some advance in prices may not unlikely be scored all within the next week or two.

have no alteration to report in values of mineral phosphates. of Charleston River rock have been made for summer shipment l. per unit, and in some quarters rather more than that is ed. For Charleston Land rock 10d. per unit, cost, freight, and nce to United Kingdom might, perhaps, be business. We still of large business in Somme and Belgian for Continental con- ion, but so far as is reported only retail transactions are being for United Kingdom for immediate delivery or shipment. s of Canadian remain nominal, and there seems to be no disposi- n the part of raisers to make sales.

have no transactions to report in River Plate cargoes of bones, the December shipments get nearer due, there seems to be a er desire on the part of owners to effect sales. £5. 7s. 6d. and os, delivered United Kingdom, we should now consider nearest s for large and small cargoes respectively. Shipments from erranean have begun to offer more freely, and value of these is more than £5. to £5. 2s. 6d., ex quay Liverpool.

nsiderable sales of East Indian bone meal have been made at 7s. 6d., ex quay, and that is lowest closing value. For parcels , same price is now required Sales of crushed Kurrachee bones been made from shipments arrived at £5. 5s, ex quay, and at nt there are no more to be had at the price. For shipment, 2s. 6d. nearest value.

thin the past few days nitrate of soda has shown some improve- , owing mainly to the very heavy deliveries during the first two hs of the year. We think this will turn out to be somewhat mis- ng, and that actual consumption will not be found to agree with ctual deliveries, but in the meantime an increase in the deliveries ore than 70,000 tons within two months over those within the same od last year is having its natural effect.

perphosphates are perhaps a shade easier, there being sellers in tity at 47s. 6d. per ton, in bulk, delivered f.o.r. at works.

e have nothing fresh to report in dried blood, quotations for e prepared remaining 11s. per unit, and 10s. 9d. being asked for River Plate ex-store. Ammonite is still quoted 9s. 6d. per unit, cash, in bags gross weight, delivered f.o.r. at works.

ltogether we think there is a better tone in the market than has experienced for the last month or six weeks.

THE TYNE CHEMICAL REPORT.

TUESDAY.

very quiet market to report this week. Bleaching powder and soda tals dull, with little business doing. Bleach has dropped 2s. 6d. a and crystals 6d. per ton. Prices of soda ash and sulphate of soda well maintained. Caustic soda in good demand, with a tendency igher prices. Makers generally are well supplied with orders present delivery, but not much inquiry for forward business. To- s quotations are:—Caustic soda, 77%, £10. per ton for prompt very; bleaching powder, softwood, casks, £5. 5s.; hardwood, s, £5. 7s. 6d. per ton; soda ash, 48—52%, 1¼d., less 5%; crys- 47s. per ton; sulphate of soda, in bulk, 32s. 6d. per ton; sul- e, ground, in casks, 42s. 6d. per ton; recovered sulphur, £4. 5s. ton; chlorate of potash, 4¾d. per lb. for prompt, 5d. per lb. for- d; hyposulphite of soda, casks, £4. 5s. per ton; hyposulphite of , 1 cwt. kegs, £4. 15s. per ton; silicate of soda, 140° Tw., £4. per 100° Tw., £3. 7s. 6d., 75° Tw., £2. 10s.; pure white sulphate of nina, £4. 10s. per ton; blanc fixe, £7. 10s. per ton; chloride of um, £8. per ton; nitrate of baryta, crystals, £18. 10s. per ton; nd, £19.; sulphide of barium, £5. 10s. per ton.

r. Spence Watson's decision with regard to the wages dispute veen Messrs. C. Tennant and Partners, Limited, Hebburn, and r workmen is expected to be made known in about a fortnight's .

n Friday last a deputation from Liverpool waited on the principal ufacturers here and discussed the question of a proposed syndicate, , it is said, very favourable results.

he export of chemicals from the Tyne for last month, with the ption of soda crystals, show a substantial increase in tonnage over ruary, 1889, viz.:—

	1889.	1890.
Alkali and soda ash	642 tons.	807 tons.
Bicarbonate soda	13 ,,	4 ,,
Bleach powder	1,446 ,,	2,037 ,,
Manure	50 ,,	582 ,,
Soda crystals	1,126 ,,	659 ,,
Sulphate of soda	135 ,,	342 ,,
Other chemicals	1,098 ,,	1,877 ,,
Total	4,510 ,,	6,308 ,,

Manufacturing coals in good demand and prices very firm. Northumberland steam small 8s. to 8s. 6d. per ton, and Durham small 9s. 6d. to 10s. per ton.

General satisfaction is felt at the decision of the Durham miners to accept the 5 per cent. advance offered by the masters. The result of the voting was as follows:—

For accepting 5 per cent.	17,251 votes.
For a strike	14,378 ,,
For arbitration	1,307 ,,

WEST OF SCOTLAND CHEMICALS.

GLASGOW, Tuesday.

There has been no improvement in the inquiry after potash and soda bichromates, which continue as neglected, comparatively, as they were during the preceding fortnight, leaving makers rather anxiously on the outlook for a change. Stocks have been accumulating inconveniently. In other respects the Scotch chemicals market is about equally depressed, and, excluding caustic soda, price changes have been in buyers' favour for the most part. Crystals, ash, and refined alkali are all obtainable on slightly easier terms; and if bleach maintains last week's quotation, it is not without an effort, for the signs are all towards a further fall. Caustic soda alone stands out as conspicuously reinforced, having marked a further substantial rise in three of the qualities. There has been only slight change in the paraffin section, naphtha having come down a further step from its late elevation, now quoting at from 8d. to 8½d. only. Sulphate of ammonia spot, Leith, has not improved on last quotation and still stands at £11. 15s., with no movement in forward fixtures; but at Glasgow, for prompt West India shipment, as high as £11. 18s. 9d. was paid—a special parcel. Chief prices current are:—Soda crystals, 47s. 6d. net Tyne; alum in lump £5, less 2½% Glasgow; borax, English refined £30., and boracic acid, £37. 10s. net Glasgow; soda ash, 1¼d. less 7½% Tyne; caustic soda, white, 76°, £10. 15s., 70/72°, £9. 2s. 6d., 60/62° £8. 5s, and 60/62° cream, £7. 10s., all less 2½% Liverpool; bicarbonate of soda, 5 cwt. casks, £5. 10s., and 1 cwt. casks, £5. 15s. net Tyne; refined alkali 48/52°, 1½d., less 5% Tyne; saltcake, 32s. to 34s. 6d; bleaching powder, £5. 5s. to £5. 10s. less 5% f.o.r. Glasgow; bichromate of potash, 4d., and of soda 4d., less 5 and 6% to Scotch and English buyers respectively; chlorate of potash, 4¾d., less 5% any port; nitrate of soda 8s. 1½d. to 8s. 6d.; sulphate of ammonia, spot, £11. 15s. f.o.b. Leith; salammoniac, 1st and 2nd white, £37. and £35., less 2½% any port; sulphate of copper, £26. 10s. less 5% Liverpool; paraffin scale, hard and soft, 2½d. per lb.; paraffin wax, 120°, semi-refined, 3 5/32d.; paraffin spirit (naphtha), 8½d.; paraffin oil (burning), 6¾d. to 7d. per gallon at Glasgow; ditto (lubricating), 865°, £5. 10s. to £6. per ton; 885°, £6. 10s.; and 890/895°, £7. 10s. to £8. Last week's imports of sugar at Greenock were 29,777 bags.

THE LIVERPOOL MINERAL MARKET.

Our market still remains strong, prices continue to advance, and there is every probability of seeing figures still higher. Manganese: Arrivals are nil, stocks are almost exhausted; prices, therefore, for prompt and forward delivery have further advanced. Magnesite: unaltered; raw ground, £6. 10s.; and calcined ground, £10. to £11. Bauxite (Irish Hill brand): The strong demand continues, and prices are very firm. Lump, 20s.; seconds, 16s.; thirds, 12s.; ground, 35s. Dolomite, 7s. 6d. per ton at the mine. French Chalk: Arrivals have been small this week, all of which have gone into consumption, and prices are still firm, especially for G.G.B. "Angel-White" brand—90s. to 95s. medium, 100s. to 105s. superfine. Barytes (carbonate), steady; selected crystal lump scarce at £6.; No. 1 lumps, 90s.; best, 80s.; seconds and good nuts, 70s.; smalls, 50s.; best ground, £6.; and selected crystal ground, £8. Sulphate quiet; best lump, 35s. 6d.; good medium, 30s.; medium, 25s. 6d. to 27s. 6d.; common, 18s. 6d. to 20s.; ground best white, G.G.B. brand, 65s.; common, 45s.; grey, 32s. 6d. to 40s. Pumicestone quiet; ground at £10., and specially selected lump, finest quality, £13. Iron ore continues in good demand at full prices. Bilbao and Santander firmer at 9s. to 10s. 6d. f.o.b.; Irish, 11s. to 12s. 6d.; Cumberland, 16s. to 20s. Purple ore, firm at last quotations. Spanish manganiferous ore sells freely, a full prices. Emery-stone: Best brands continued to be enquired for, bringing full prices. No. 1 lump is quoted at £5. 10s. to £6., and smalls £5. to £5. 10s. Fuller's earth, steady; 45s. to 50s. for best blue and yellow; fine impalpable ground, £7. Scheelite, wolfram, tungstate of soda and tungsten metal continue inquired for. Chrome metal, 5s. 6d. per lb. Tungsten alloys 2s. per lb. Chrome ore: High qualities in good demand, bringing full prices. Antimony ore and metal have further advanced. Uranium oxide, 24s. to 26s. Asbestos: Best rock, £17. to £18.; brown grades, £14. to £15. Potter's lead ore: Prices are firm; smalls, £14. to £15.; selected lump, £16. to £17. Calamine: Best qualities scarce—60s. to 80s. Strontia steady; sulphate (celestine) steady, 16s. 6d. to 17s. Carbonate (native), £15. to £16.; powdered (manufactured), £11. to £12. Limespar: English manufactured, old G.G.B. brand in demand, and brings full prices; 50s. for ground English. Felspar, 40s. to 50s.; fluorspar, 20s. to £6. Bog ore firm at 22s. to 25s. Plumbago: steady; Spanish, £6.; best Ceylon lump at last quotations; Italian and Bohemian, £4. to £12. per ton. French sand, in cargoes, continues scarce on spot—20s. to 22s. 6d. Ferro-manganese strong at advanced figures. Ground mica, £50. China clay freely offering—common, 18s. 6d.; good medium, 22s. 6d. to 25s.; best, 30s. to 35s. (at Runcorn).

THE LIVERPOOL COLOUR MARKET.

COLOURS.—Prices remain without alteration. Ochres: Oxfordshire quoted at £10., £12., £14., and £16.; Derbyshire, 50s. to 55s.; Welsh, best, 50s. to 55s.; seconds, 47s. 6d.; and common, 18s.; Irish, Devonshire, 40s. to 45s.; French, J.C., 55s., 45s. to 60s.; M.C., 65s. to 67s. 6d. Umber: Turkish, cargoes to arrive, 40s. to 50s.; Devonshire, 50s. to 55s. White lead, £21. 10s. to £22. Red lead, £18. 10s. Oxide of zinc: V.M. No. 1, £25.; V.M. No. 2, £23. Venetian red, £6. 10s. Cobalt: Prepared oxide, 10s. 6d.; black, 9s. 9d.; blue, 6s. 6d. Zaffres: No. 1, 3s. 6d.; No. 2, 2s. 6d. Terra Alba: Finest white, 60s.; good, 40s. to 50s. Rouge: Best, £24.; ditto for jewellers', 9d. per lb. Drop black, 25s. to 28s. 6d. Oxide of iron, prime quality, £10. to £15. Paris white, 60s. Emerald green, 10d. per lb. Derbyshire red 60s. Vermillionette, 5d. to 7d. per lb.

Gazette Notices.

The Bankruptcy Act, 1883.

Receiving Orders.

DAVID GARNER (trading as D. Garner and Son), Harrow-road, London, W., oil and colour man.

LEON HARRIS, Plasket-lane, Upton Park East, London, varnish and colour manufacturer, and oil and glass merchant.

HARRY SHAW, High-street, Poplar, London, oil and colour man.

First Meetings and Public Examinations.

HENRY WILLIAM AUSTIN (formerly trading as Austin and Son), late of Aston, near Birmingham, drysalter, first meeting March 5, 25, Colemore-row, Birmingham; public examination, March 20, County Court, Birmingham.

DAVID GARNER (trading as D. Garner and Son), Harrow-road, London, W., oil and colour man, March 12, Bankruptcy Buildings, Portugal-street, Lincoln's-Inn-Fields; March 27, 34, Lincoln's-Inn-Fields, London.

Adjudications.

DAVID GARNER (trading as D. Garner and Son), Harrow-road, London, W., oil and colour man.

H. W. AUSTIN (formerly trading as Austin and Son), late of Aston, near Birmingham, drysalter.

HARRY SHAW, High-street, Poplar, London, oil and colour man.

Notices of Dividends.

W. J. ABBOTT (trading as Abbott and Co.), Pilton, paper manufacturer, first and final dividend of 10d. any day, Official Receiver's Office, Taunton.

New Companies.

ARCHER PIPE COMPANY, LIMITED.—This company was registered on the 13th inst. with a capital of £10,000., in £1. shares, to acquire two letters patent granted in 1886 and 1889 for improvements in pipe joints, and for improvements in the method of the machinery for making earthenware pipes for drainage, water, and other purposes. The subscribers are:—

	Shares.
Col. A. C. Hamilton, 41, Lennox-gardens	1
Isaac Williams, 5, Malborough-road, Chiswick	1
H. A. Hart, 16, Ashchurch-grove, Shepherd's-bush	1
Harry Smith, The Bourne, Chiswick, contractor	1
H. Bird, 11, Queen Victoria-street, wine merchant	1
E. H. Bayley, 142, Newington-causeway	1
R. Ewing, 51, Gracechurch-street, estate agent	1

SENSITISED OPAL CARD COMPANY, LIMITED.—This company was registered on the 13th inst., with a capital of £8,000, in £1. shares, whereof 2,000 are cumulative £5. per cent. preference shares, to acquire processes invented by Friesé Green for rendering cardboard, paper, and other like substances, impervious to the action of water and photographic chemicals, and for printing photographs by light upon cards, ivory and similar substances, and to take over the [illegible] carried on under the style of the Friesé Greene Photographic Manufactu[illegible] pany. The subscribers are:—

Friesé Greene, 92, Picadilly, photographer	[illegible]
P. C. Novelli, 4, Eastcheap, merchant	[illegible]
Gordon Cox, 50, Gracechurch-street, solicitor	[illegible]
Harold Laforce, 50, Gracechurch-street, solicitor	[illegible]
Thos. Cave, 50, Gracechurch-street, solicitor	[illegible]
F. S. McMaster, 50, Gracechurch-street, solicitor	[illegible]
F. W. Darch, Heber-road, Dulwich, clerk	[illegible]

TYNE DRY SALTERY AND PACKING COMPANY, LIMITED.—This com[illegible] registered on the 26th ult., with a capital of £2,000, in £10. shares, to pu[illegible] business of drysalter and manufacturer of Tyne boiler fluid, carried on [illegible] Donald, of 5 and 7, Elswick-place, Newcastle-on-Tyne, trading as Don[illegible] and Co. The subscribers are:—

G. W. Emmorson, Newcastle, commission agent	[illegible]
W. J. Murday, Newcastle, electrical engineer	[illegible]
S. Kent, Newcastle, master mariner	[illegible]
R. Donald, Cullercoats, drysalter	[illegible]
R. Thompson, Backworth, engineer	[illegible]
G. Proud, Newcastle, agent	[illegible]
G. Dixon, Newcastle, agent	[illegible]

WILLIAMS, FRY, AND COMPANY, LIMITED.—This company was regi[illegible] the 21st ult., with a capital of £20,000., in £5. shares, to take over as a g[illegible] cern the business of Portland cement manufacturers, carried on by Art[illegible] Williams, and Stephen Henry Fry, at Greenhithe, Kent, under style of [illegible] Fry, and Co. The subscribers are:—

A. J. Williams, Stone, Greenhithe, cement manufacturer	[illegible]
S. H. Fry, Stone, Greenhithe, cement manufacturer	[illegible]
Lieut.-Col. F. C. Goad, 15, Harlington-road, Ealing	[illegible]
J. O. Collier, 59 and 60, Chancery-lane, clerk	[illegible]
J. W. Davies, 34, Guildford-street, W.C., accountant	[illegible]
R. W. Vaughan, C.E., 64, Broad-street-avenue	[illegible]
W. M. Bemister, Broad-street-avenue, clerk	[illegible]

The Patent List.

This list is compiled from Official sources in the Manchester T[illegible] Laboratory, under the immediate supervision of George E. Da[illegible] Alfred R. Davis.

APPLICATIONS FOR LETTERS PATENT.

Furnace Fronts and Fittings. J. J Meldrum and T. F. Meldrum [illegible] February 17.
Pumps. J. G Elliott. 2,521 February 17.
Borax and other Borates. The Borax Co., Limited. 2,526. Februar[illegible]
Regulation of the Carbons in Arc Lamps. O. Cornwell. 2,527. [illegible]
Low Water Indicators for Steam Boilers.—(Complete Specification[illegible] Toovey. 2,528. February 17.
Recovery of Metal from Iron Scrap. C. Thompson. 2,533 Febru[illegible]
Man-hole Covers for Sewer Ventilators. F. Bird. 2,541. Februa[illegible]
Sewer Ventilators. F. Bird. 2,542. February 17.
Water Jacketed Blast Furnaces. J. W. Chenhall. 2,544. February [illegible]
Gunpowder from Nitrocellulose. J. Y. Johnson. 2,547. February [illegible]
Gas Stoves.—(Complete Specification). C. Clamond. 2,551. February [illegible]
Ventilators for Rooms. H. Sharp. 2,557. February 17.
Carburetting Apparatus. H. S. Maxim and G. S. Sedgwick. 2,559. [illegible]
Disinfecting Sewers. S. F. Milligan and R. Barklie. 2,563. Februar[illegible]
Lubricating Bearings. E. W Cooper. 2,576. February 18.
Electrical Transmitter and Receiver. J. R. Fraser, 2,578. Febru[illegible]
Obtaining Fresh Water from Salt Water. J. Smith. 2,581. Feb[illegible]
Valve Apparatus for Gas Motor Engines. J. Fielding. 2,587. Fe[illegible]
Mechanism for Charging Furnaces.—(Complete Specification). [illegible] Haddan. 2,618. February 18.
Preservation of Food. F. Crognet, 2,620. February 18.
Electrically-driven Fans.—(Complete Specification). H. G. Watel. [illegible] February 18
Regenerative Gas Lamps.—(Complete Specification). T. Gordon. [illegible] February 18.
Opalescent Glass or Enamel. C. Huelser. 2,626. February 18. [illegible]
Apparatus for the Production of Magnesium Flash-light. A. [illegible] 2,628. February 18.
Centrifugal Filters. W. P. Thompson. 2,632. February 18.
Treatment of Ammonium Chloride with Alkali Waste. J. Leith [illegible] February 18.
Hot Air Engines. C. Wells. 2,642. February 18.
Air Engines or Motors.—(Complete Specification). H. H. Lake. Fe[illegible]
Electric Meters. S. Z de Ferranti. 2,653. February 18.
Electric Cables. J. E Kingsbury. 2,654. February 18.
Insulating Electric Wires. J. E. Kingsbury. 2,655. February 18.
An Improved Pinch Cock. P. Braham. 2,658. February 19.
Spray Apparatus. E. Embrey and E. Parry. 2,664. February 19.
Gas Motor Engines. H. Campbell. 2,670. February 19.
Photo-metric Apparatus. W. Foster. 2,671. February 19
Ventilating and Dust-catching Apparatus. G. M. Parkinson and [illegible] kinson. 2,680. February 19.
Utilising Waste Product for Electrical Purposes. A. W. Ar[illegible] 2,687. February 19.
A Steam Engine Governor.—(Complete Specification). J. W. Brown [illegible] W Sutcliffe. 2,691. February 19
Ventilating and Refrigerating. H. J. Haddan. 2,695. February 19
Smoke Consuming Apparatus. W. C. Wood and C. Whitaker [illegible] February 19.
Treating Coal and Coke with Limestone or Gypsum for im[illegible] Combustion. A. Moseley. 2,745. February 19.
Compressed Coal and Coke Block Fuel. A. Moseley. 2,746. Feb[illegible]
Lighting Railway Trains by Electricity. J. A Timmis. 2,750. [illegible]
Dyeing Textile Fibres. W. L. Wise. 2,763 February 19.
Manufacture of Projectiles. R. Cail. 2,765. February 19.
Percussion Fuses. H. A. Schlund. 2,769 February 19.
The Separation and Recovery of Gold. E. L. Mayer and J. G. [illegible] 2,772. February 19.

IMPORTS OF CHEMICAL PRODUCTS

AT

THE PRINCIPAL PORTS OF THE UNITED KINGDOM.

LONDON.

Week ending Feb. 26th.

tate of Lime—
S., £224 Johnson & Hooper
, 1812 350 Lister & Biggs

tic Acid—
rmany, 9 pkgs. G. Meyer & Co.
olland, 50 pkgs. Beresford & Co.

timony Ore—
dney, 10 t. Bank of N. S. Wales
ngapore, 10 Pillow, Jones, & Co.
" 9 c. L. & I. D. Jt Co.

ali—
olland, 200 c. Taylor, Sommerville, & Co.

nes—
ape, 3 t. Hammond & Co.
atal, 3 Flack, Chandler, & Co.
rance, 2 c. Hernu Peron, & Co.
. S., 54½ A. & W. Nesbitt
. S. Wales, 13½ t. Hicks, Nash, & Co.
rance, 12 c. Hernu, Peron, & Co.
olland, 10 t. J. Lucey & Sons

rytes—
ermany, 76 cks. D. Storer & Sons
elgium, 22 O'Hara & Hoar
ermany, 22 "
" 129 T. & W. Farmiloe
olland, 46 A Boltz
" 22 W. Harrison & Co.
elgium, 103 A. Zumbeck & Co.
" 87 W. A. Rose & Co.
" 606 A. Zumbeck & Co.
ermany, 45 O'Hara & Hoar
" 76 D. Storer & Sons
olland, 22 W. Harrison & Co.

rium Super Oxide—
ermany, £45 L. & I. D. Jt. Co

imstone—
taly, 200 t. Northcott & Sons

rium Chloride—
olland, £70 Petri Bros.

racic Acid—
taly, 30 pkgs. Howards & Son

emicals *(otherwise undescribed)*
ermany, £56 M. D. Co.
" 390 A. & M. Zimmermann
" 29 Phillipps & Graves
" 369 A. & M. Zimmermann
rance, 15 C. Brumlen
taly, 25 W. Nicolson & Co.
ermany, 80 T. H. Lee
" 693 Phillipps & Graves
" 330 A. & M. Zimmermann
" 35 T. H. Lee
" 127 A. & M. Zimmermann.

outchouc—
adagascar, 4 c. Kebbel, Son, & Co.
. Indies, 36 L. & I. D. Jt. Co.
. Africa, 56 "
" 20 Elkan & Co.
olombia, 62 Stiebel Bros.
ermany, 40 L. & I. D Jt. Co.
" 7 W. Pope & Co.
ape, 50 Hammond & Co.
. Indies, 30½ E. Boustead & Co.
" 141 Kleinwort, Sons, & Co.
. Africa, 14 Kebbel, Son, & Co.
" 26 L. & I. D. Jt. Co.

eam Tartar—
olland, 23 pkgs. Middleton & Co.
taly, 35 B. Jacob & Sons
rance, 15 W. C. Bacon & Co.
" 7 B. & F. Wf. Co.
France, 4 pkgs. W. & C. Pantin
Holland, 7 B. & F. Wf. Co.

Caustic Potash—
France, 23 pkgs. Beresford & Co.
" 21 F. W. Berk & Co.

Camphor—
Japan, 315 tubs. R. Warner & Co.
" 185 Lewis & Peat
China, 219 cs. "

Caustic Soda—
Holland, 110 c. Beresford & Co.

Dyestuffs—

Sumac
Italy, 100 pkgs. F. Roberts

Myrabolams
E. Indies, 320 pkgs. Marshall & French
" 667 L. & I. D. Jt. Co.
" 650 "
" 350 "
" 350 Baxter & Hoare

Indigo
E. Indies, 67 chts. J. Owen
" 71 N. S. S. Co.
" 6 Elkan & Co.
" 61 W. Brandt, Sons, & Co.
" 25 L. Achard & Co.
" 178 L. & I. D. Jt. Co
" 11 Hackett, Ausender, & Co.
" 11 Phillipps & Graves
W. Indies, 3 L. & I. D. Jt. Co.
Ceylon, 29 F. Huth & Co.
E. Indies, 27 "
" 68 Ross & Deering
" 83 Union Lighterage Co.
" 5 Phillipps & Graves
" 4 Langstaff & Co.
" 1 bx. Elkan & Co.
" 31 "
" 254 L. & I. D. Jt. Co.
" 19 Anderson Bros.
" 6 Parsons & Keith
" 176 T. Barlow & Bro.
" 12 Hatton, Hall, & Co.
" 17 J. Owen
" 63 F. Stahlschmidt & Co.
" 16 G. Dards
" 10 Fruhling & Goschen
" 160 L. & I. D. Jt. Co.
China, 2 cs. "
" 3 brls. "
E. Indies, 339 chts. "
" 97 Benecke, Souchay, & Co.
" 24 Stansbury & Co.
" 23 Hatton, Hall, & Co.
" 20 Parsons & Keith
" 11 C. J. Hambro
" 4 bxs. Elkan & Co.
" 44 Parsons & Keith
" 270 L. & I. D. Jt. Co.
" 25 Arbuthnot, Latham, & Co.
" 35 F. W. Heilgers & Co.
" 16 cs. H. Johnson & Sons
Holland, 2 Ripley, Roberts, & Co.
F. Indies, 18 L. & I. D. Jt. Co.
" 47 A. Harvey
" 41 Ernsthausen & Co.

Orchella
Natal, 94 pkgs. L. & I. D. Jt. Co.

Extracts
France, 70 pkgs. B. & F. Wf. Co
Austria, 720 A. J. Humphrey
Denmark, 43 Bailey & Leetham
Norway, 1 Johnson & Jorjensen
U. S., 50 M. D. Co.
Germany, 1 T. H. Lee
France, 20 "
Holland, 88 Burt, Boulton, & Co.

Annatto
France, 4 pkgs. Hurford & Middleton

Saffron
E. Indies, 6 pkgs. L & I. D. Jt. Co.

Argols
Italy, 634 pkgs. B. Jacob & Sons
Cape, 6 Brown & Elmslie

Valonia
Greece, 50 t. T. J. & T. Powell
A. Turkey, 50 Hicks, Nash, & Co.
" 279 J. Graves
" 50 Hicks, Nash, & Co.
" 50 Anning & Cobb

Cochineal
Canaries, 167 pkgs. W. M. Smith & Son
" 8 Bruce & White
" 42 Hammond & Co.
" 14 A. Jemenez & Sons
" 80 W. H. Cole & Co.
" 19 W. M. Smith & Sons
" 31 Swanston & Co.
" 10 Johnston, Rolls, & Co.
" 25 H. R. Toby & Co.
" 6 J. Owen
" 6 Kuhner, Hendschel, & Co.

Gambier
E. Indies, 414 pkgs. L. & I. D. Jt. Co.
" 429 Brinkmann & Co.
" 13 S. Barrow & Bro.
" 190 Elkan & Co.
" 170 S. Barrow & Bro.
" 347 L. & I. D. Jt. Co.
" 171 Elkan & Co.
" 251 H. Lambert
" 356 J. H. & G Scovell
" 387 R & J. Henderson
" 834 Cox, Patterson, & Co.
" 403 Beresford & Co.
Germany, 178 Von der Meden Bros.

Madder
Holland, 7 pkgs. Williams, Torrey, & Co.

Tanners' Bark
Natal, 20½ t. Baxter & Hoare

Yellow Berries
A. Turkey, 3 pkgs. F. W. Bowyer & Bartlett

Cutch
E. Indies, 500 pkgs. Anderson, Weber, & Smith

Glucose—
Germany, 410 pkgs 424 c. Herne & Co.
" 25 202 Pillow, Jones, & Co.
France, 100 100 J. Barber & Co.
U. S., 50 299 Seaward Bros.
" 20 120 M. D Co
" 600 600 A. Dawson
" 100 500 J. Keiller & Son
" 1,831 1,831 W. R. Spence
France, 400 400 J. Barber & Co.
Germany, 25 200 W. Balchin
" 40 44 Magee, Taylor, & Co.
U. S., 100 300 M. D. Co.
Germany, 10 pkgs. 81 cwt. M. D. Co.
" 800 800 J. F. Ohlmann
" 37 296 J. Knill & Co.
U. States, 500 500 T. C. Howell
" 40 240 Pillow, Jones & Co.
Germany, 30 249 Prop. Chamberlain's Wf.
U. States, 60 360 Props. Scotts' Wf.
Germany, 37 350 J. Cooper
Germany, 12 pkgs. 80 c. Dutrullo Solomon & Co.
" 25 207 L. Sutro & Co.
" 70 560 T. M. Duche & Sons
" 100 100 Barrett, Tagant, & Co.
U. States, 48 450 J. Barber & Co.

Gutta Percha—
E. Indies, 740 c. J. H. & G. Scovell
" 582½ R. & J. Henderson
" 894 Kleinwort, Sons, & Co.
France, 40 Flageollet & Co.
E. Indies, 420 Kaltenbach & Schmitz

Glycerine—
Germany, £160. Beresford & Co.
" 60 "
Holland, 60 F. W. Hielgers & Co.
" 335 Spies, Bros., & Co.
" 70 Knight & Morris

Isinglass—
Sts. Settlements, 21 pkgs. L. & I. D. Jt. Co.
Germany, 1 pkg. Williams & Co.

Manure—

Phosphate Rock
Belgium, 30 t. Leach & Co.
" 30 Fox, Roy, & Co.
" 160 J. Gibb
France, 175 A. Hunter & Co.
Belgium, 170 Lawes Manure Co.
France, 70 Arnati & Harrison
U. S., 2,551 Wyllie & Gordon

Manure *(otherwise undescribed)*
Germany, 6 t. T. H. Lee
" 50 F. J. Cornwall & Co.
" 30 F. W. Berk & Co.
" 19 Petri Bros.
" 6 T. H Lee

Naphtha—
Belgium, 10 drms. Stein Bros.
Germany, 9 "

Nitre—
Holland, 7 cks. R. Morrison & Co
" 1 cs. cubic. Spies, Bros, & Co.

Paraffin Wax—
U States, 422 brls. G. H. Frank
" 427 H. Hill & Sons
" 2,000 J H. Usmar & Co.
" 80 Rose, Wilson, & Co.

Potassium Carbonate—
France, 10 pkgs. J. A Reid
" 11 Charles & Fox

Petroleum—
U S., 4,867 brls. C. T. Bowring & Co.
Germany, 8 Hernu, Peron, & Co.
Belgium, 6 Mwll. Dk. Co.
U. S., 5 "
Germany, 50 T. H. Lee
U. S., 1,000 Mordaunt Bros.
Holland, 46 drms. Burt, Boulton. & Co.
U. S., 700 brls. Cheseboro' Mfg. Co.
" 2,283 Mordaunt Bros.
" 1,010 Rose, Wilson & Rose
E. Turkey, 10 H. French & Co.
U. S., 15 L. & I. D. Jt Co.
" 1,343 Anglo-Amer. Oil Co.

Pearl Ash—
France, 300 c. F. W. Berk & Co.
" 185 Petri Bros.

Potash—
France, 10 cks. Weatherley, Mead & Co.
Holland, £70. R. Morrison & Co.

Potassium Sulphate—
Germany, 10 pkgs H. Lambert

Potassium Bicarbonate—
Holland, 12 pkgs. A. & M. Zimmermann

Plumbago—
Ceylon, 190 cks. J. Forsey
" 66 Prop. Chamberlain's Wf.
" 81 Doulton & Co.
" 70 Thredder, Son, & Co.
" 30 L. & I. D. Jt. Co.
" 191 Barrow & Gibson
Holland, 11 Brown & Elmslie

Pyrites—
Spain, 400 t. G. Ward & Sons

Quinine—
Holland, £600. Kebbel, Son, & Co.
Austria, 3,500 F. W. Heilgers & Co.

Salicylic Acid—
Holland, £90. B. Morrison & Co.

Salammoniac—
Holland, £77. Barr, Moering, & Co.

Sodium Hyposulphite—
Germany, £38. Beresford & Co

Sodium Acetate—
France, £82. Prop. Hay's Wf.
" 80 . Barr, Moering, & Co.

Stearine—
France, 35 bgs. H. Hill & Sons
" 55 "
" 5 sks. Van Gulkerkin & Co.

Saltpetre—
Germany, 74 cks. J. Hall & Son
E. Indies, 286 bgs. Ralli Bros.
" 940 Clift, Nicholson, & Co.
Germany, 100 kgs. P. Hecker & Co.
E. Indies, 549 bgs. Hunter, Waters, & Co.
" 817 L. & I. D. Jt. Co.

Sodium Sulphate—
Holland, £25. Prop. Dowgate Dk.

Sodium Nitrate—
Germany, £33. Spies Bros & Co.

Tartars—
Italy, 97 pkgs. Thames S. Tug Co.
" 377 Seager, White, & Co.

Tartaric Acid—
Germany, 20 pkgs. O. Andrea & Co.
Holland, 8 B & F. Wf. Co.
" 4 Hernu, Peron, & Co.

Turpentine—
U.S., 2,227 brls. Nickoll & Knight

Ultramarine—
Belgium, 6 pkgs. Leach & Co.
Germany, 2 G. Steinhoff
Holland, 12 Haeffner, Hilpert, & Co.
" 13 Hernu, Peron, & Co.
" 2 cks. Ohlenschlager Bros.
Germany, 6 cs. Haeffner, Hilpert, & Co.
Holland, 20 cks. Hernu, Peron, & Co.

Zinc Oxide—
Germany, 40 cks. D. Storer & Sons
Holland, 260 M. Ashby
U.S, 100 "

LIVERPOOL

Week ending Feb. 27th.

Acids (*otherwise undescribed*)
Havre, 1 pkg. Cunard S. S. Co.

Acetic Acid—
Rotterdam, 10 botls.

Boracic Acid——
Leghorn, 5 cks.

Borax—
Leghorn, 14 cks. 100 cs

Bones—
Valparaiso, 664 sks. Jones & Roberts
Alexandria, 87 bgs.
Malta, 203 t.
Valparaiso, 231 sks. G. Batcheldor & Son
Bombay, 1,736 c.

Barytes—
Antwerp, 40 bgs.
Bremerhaven, 22 cks.
Rotterdam, 5

Chemicals (*otherwise undescribed*)
Rotterdam, 2 cks.
Frederikstad, 19
Rouen, 50
Rotterdam, 1

Cream Tartar—
Messina, 2 cs.
Genoa, 10 cks.
Patras, 3 cs. Papayanni & Jeremias
" 11 brls. A. G. Frangopulo & Co.
Spain, 5 cks. Sachse & Klemm
Constantinople, 2 bls.

Caoutchouc—
Kinsembo, 3 cks. T. Millington & Co.
" 3 bgs. Samson & de Liagre
Kinsembo, 2 cks. Taylor, Laughland, & Co.
Mussera, 3 bgs. Samson & de Liarge
Ambrizette, 1 43 "
" 10 J. Holt & Co.
" 13 Taylor, Laughland, & Co.
Muculla, 75 Samson & de Liagre
Boma, 48 Daumas & Co.
Black Point, 4 cs. 1 brl. 6 cks. Pinto, Leite & Nephew
Loango, 9 cks. Edwards Bros.
Cape Lopez, 3 Daumas & Co.
Elobey, 8 J. Holt & Co.
Old Calabar, 3 brls.
Quittah, 4 cks. F. & A. Swanzy
Salt Pond, 4 brls. A. Miller, Bro. & Co.
" 2 3 cs Fletcher & Fraser
" 1 S. & C. Nordlinger
" 1 1 Radcliffe & Durant
" 4 1 W. Griffiths & Co.
" 1 Davies, Robbins, & Co.
" 2 Edwards Bros.
Assince, 1 A. Verdier
" 1 Pitt, Bros. & Co.
" 1 bg. Radcliffe & Durant
" 1 Pickering & Berthoud
" 1 W. D. Woodin
" 1 Edwards Bros.
" 1 A. Reis
" 1 J. H. Robertson
Grand Bassam, 2 bgs.
Accra, 4 cks. 2 cs. Pickering & Berthoud
" 1 4 Davies, Robbins, & Co.
Lavannah, 9 brls. Palma Tradg. Co
Manoh, 2 "
Sulymah, 5 "
Sierra Leone, 25 Pickering & Berthoud
New York, 1 cs.
Manoas, 24 W. Curden & Co.
" 85 Fould, Freres, & Co
" 248 Bieber & Co.
" 3,402

Copper Precipitate—
Pomaron, a quantity Mason & Barry

Charcoal—
Rouen, 354 bgs.

Dyestuffs—
Sumac
Mediterranean, 50 bgs. J. Kitchin
Trieste, 25 brls.
Palermo, 650 bgs.
Extracts
Havre, 38 bxs. 70 cks. Cunard S.S. Co.
New York, 5 brls.
Nantes, 130 cks. R. J. Francis & Co.
Orchella
Colon, 8 pkgs. Matheson & Beausire
Valonia
Patras, 526 bgs. J E Sicopulo
" 500

Dextrine—
Hamburg, 45 bgs.

Glucose—
Rouen, 5 cks.
Hamburg, 37
Dunkirk, 100 bgs.
Philadelphia, 130 brls.

Glycerine—
Amsterdam, 10 cbys.
Rotterdam, 20 cs. 5 drms.
Fiume, 30

Pearl Ash—
Portland, 10 brls.

Potash—
Rouen, 18 drms. 61 cks.
Dunkirk, 7

Phosphate—
Antwerp 3,000 bgs. J T. Fletcher & Co.
Dunkirk, 2,000 sks. G. & L. Pilkington
Dunkirk, 200 t.
Belgium, 1,523 bgs.

Pyrites—
Huelva, 1,321 t. Tennants & Co.
" 1,119 Matheson & Co.
" 1,360 "
" 1,421 Matheson & Co.

Pitch—
Amsterdam, 13 cks.

Phosphoric Acid—
Bremerhaven. 6 cs.

Sugar of Lead—
Rotterdam, 20 cks.

Sodium Silicate—
Rotterdam, 1 ck.

Saltpetre—
Hamburg, 180 cks.

Tartar—
Messina, 2 cs.
Bordeaux, 4 cks.
Naples, 1 R Powles
" 1

Tartar Salts—
Rotterdam, 10 cks.

Tartaric Acid—
Rotterdam, 27 cks.

Ultramarine—
Rotterdam, 7 cs. 13 cks.
Trieste, 5

Zinc Oxide—
Antwerp, 75 brls. J. T. Fletcher & Co.
" 150
Hamburg, 25 cks.
New York, 100 brls.
Rotterdam, 180 cs. O. & H. Marcus
New York, 100 brls.

HULL.

Week ending Feb. 27th.

Acetic Acid—
Rotterdam, 56 pkgs. Furley & Co.

Acids (*otherwise undescribed*)—
Bremen, 1 ck. Steedman & Bro.

Barytes—
Bremen, 44 cks. T. W. Flint & Co
" 23 Tudor & Sons
" 13 Hurst & Cooke
Antwerp, 210 bgs. Aire & Calder N. Co.

Chemicals (*otherwise undescribed*)
Bremen, 1 keg. 2 cs. Veltmann & Co
" 1 ck. J. Pyefinch & Co.
Genoa, 5 Wilson, Sons, & Co
Gothenburg, 3 bxs. "
Antwerp, 45 pks. "
Hamburg, 2 cs. 2 brls. "

Dyestuffs—
Alizarine
Rotterdam, 48 pkgs. W. & C. L. Ringrose
" 81 Hutchinson & Sons
" 27 pks. 11 cks. W. & C. L. Ringrose
" 10 "
Extracts
Rouen, 22 cks. Rawson & Robinson
Hamburg, 20 bxs. Wilson, Sons, & Co.
Myrabolams
Bombay, 4,574 bgs.
Indigo
Amsterdam, 17 chts. W. & C. L. Ringrose
Rennet
Rotterdam, 5 cks. W. & C. L. Ringrose
Copenhagen, 2 1 cs. Wilson, Sons, & Co.
Madder
Rotterdam, 3 cks. W. & C. L. Ringrose

Glucose—
Hamburg, 10 cks. Rawson & Robinson
" 25 Woodhouse & Co.
" 4
New York, 69 brls. Wilson, Sons, & Co.
Hamburg, 26
Stettin, 15 530 "
" 35 1,200 bgs.
Hamburg, 15 Rawson & Robinson
" 13 Wilson, Sons, & Co.
Dunkirk, 10 pks. "

Litharge—
Rotterdam, 1 ck.

Manure—
Ghent, 160 t. Wilson, Sons, & Co.

Naphthol—
Rotterdam, 3 brls. G. Lawson & Sons

Potash—
Rotterdam, 2 cs. G Lawson & Sons
" 4 cks.

Stearine—
Antwerp, 3 cks. Bailey & Leetham

Salt—
Hamburg, 712 bgs.

Saltpetre—
Hamburg, 17 brls.

Soda—
Rotterdam, 10 cks. W. & C. L. Ringrose

Tartaric Acid—
Rotterdam, 7 cks. W. & C. L. Ringrose
" 6

Tar—
Marseilles, 20 cks. Wilson, Sons, & Co.

TYNE.

Week ending Feb. 27th

Alum Earth—
Hamburg, 18 cks. Tyne

Glycerine—
Hamburg, 4 cs.

Glucose—
Hamburg, 25 cks. Tyne

Pyrites—
Pomaron, 1,200 t. Mason

Ultramarine—
Hamburg, 4 cks.

Zinc Oxide—
Antwerp, 25 brls. Tyne

GLASGOW.

Week ending Feb. 27th

Alum—
Rotterdam, 44 cks. J. Rankin

Barytes—
Rotterdam, 193 cks.
" 22 J. Rankin

Chemicals (*otherwise undesc*
Hamburg, 1 cs.

Dried Blood—
Rouen, 3 cs.

Dyestuffs—
Alizarine
Rotterdam, 110 cks.
" 201 J. Rankin
Extracts
New York, 100 bxs.
Nantes, 300 cks.
Madder
Bordeaux, 10 cks. J. & P. Hu

Glycerine—
Rouen, 10 cs.

Glucose—
Philadelphia, 39 brls.

Saltpetre—
Hamburg, 120 kegs. J. Poynter

Tartar—
Oporto, 2 cks. J. & P. Hu
Bordeaux, 1

Turpentine—
Savannah, 3,442 c. Rowley

Ultramarine—
Rotterdam, 1 ck. J. Rankin
" 10 cs. "

Waste Salt—
Hamburg, 708 bgs.

GRIMSBY.

Week ending February 22

Glucose—
Hamburg, 17 cks.

GOOLE.

Week ending Feb. 19th

Acetate of Chrome—
Rotterdam, 2 cks.

Caoutchouc—
Boulogne, 4 cks.

Dyestuffs—
Extracts
Rotterdam, 6 cks.
Alizarine
Rotterdam, 15 cks.
Dyestuffs (*otherwise undescr*
Boulogne, 12 cs. 13 cks.
Dunkirk, 2

Glucose—
Hamburg, 200 bgs. 41 cks.
Dunkirk, 220
Calais, 100

Potassium Muriate—
Rotterdam, 26 cks.

Potash—
Calais, 9 cks.

Saltpetre—
Hamburg, 12 cks. 15 kegs.

Sodium Silicate—
Rotterdam, 1 ck.

Tartar—
Rotterdam, 18 cks.

Waste Salt—
Hamburg, 40 bgs.

Zinc Oxide—
Antwerp, 30 brls.

EXPORTS OF CHEMICAL PRODUCTS

FROM

THE PRINCIPAL PORTS OF THE UNITED KINGDOM.

LONDON.

Week ending February 22th.

Arsenic—
Boston, 25 t. £300
Hong Kong, 4 10 c. 61
Sydney, 1 14

Ammonium Carbonate—
Yokohama, 3 t. £107
Hong Kong, 10 c. 19
Sydney, 10 cs. 18
" 1 t. 2 75

Ammonium Sulphate—
Rotterdam, 5 t. 14 c. £65
Hamburg, 30 310
Rotterdam, 60 720
Cologne, 30 360
Ghent, 20 235
Martinique, 15 1 200
Hamburg, 50 560
" 101 19 £3,475

Acetic Acid—
Wellington, 4 t. 18 c. £113

Alumina Sulphate—
Bordeaux, 5 t. 14 c. £20

Acids (*otherwise undescribed*)
Cologne, 600 bls. £918

Brimstone—
Trinidad, 2 t. 19 c. £23

Carbolic Solid—
Hamburg, 3 c. £11
New York, 14 cs. 83

Caustic Soda—
Natal, 2 c. £11
Auckland, 5 t. 1 37

Citric Acid—
Antwerp, 2 t. £261
Cape Town, ½ c. 3
Melbourne, 20 kgs. 76
Trieste, 1 140
Genoa, 2 5 313
Marseilles, 2 13
Genoa, 1 5 177
Sydney, 11 cs. 68
Montreal, 1 35
Amsterdam, 3 c. 21
Karachi, 1½ 9
Yokohama, 10 70
Hamburg, 1 7

Copper Sulphate—
Genoa, 5 t. £121
" 14 19 c. 380
Tarragona, 10 1 210

Cream of Tartar—
Townsville, 23 c. £132
Adelaide, 1 t. 3 131
" 1 1 80
Newcastle, 7 40
Sydney, 15 79
Nelson, 10 61
Melbourne, 1 t. 10 160

Carbolic Acid—
Stettin, 25 cks. £170
Foochow, 150 gls. 17

Carbolic Disinfectant—
Sydney, 33 cs. £51

Chemicals (*otherwise undescribed*)
La Libertas, 41 pkgs. £63
Callao, 20 64
Masulipatam, 14 c. 50
Brisbane, 79 252
E. London, 95 37
Sydney, 50 319
Natal, 12 kegs. 50
Hong Kong, 9 cks. 37
Trinidad, 15 brls. 20
Sydney, 68 256
Yokohama, 35 240
" 121 179
" 100 kegs. 192

Copperas—
Salonica, 18 t. 17 c. £42

Camphor—
Melbourne, 5 c. £43

Disinfectant Liquid—
Melbourne, 4 cks. £32

Guano—
Martinique, 24 t. 11 c. £250
Barbadoes, 30 300
Trinidad, 100 1,000

Manure—
Terneuzen, 264 t. 3 c. £1,770
Bordeaux, 9 17 72
Genoa, 10 7 85
Bordeaux, 19 200
Dunkirk, 50 600
Rouen, 117 730
Penang, 5 30
Martinique, 64 1 225
Ghent, 50 400
Guadeloupe, 65 15 1,485
Timaru, 100 1,130
Hong Kong, 50 278
Rotterdam, 19 10 101
Newfairwater, 610 4,820
Gothenburg, 150 3 548
Jersey, 80 640
Valencia, 70 11 504
Jersey, 20 175
Penang, 29

Phosphorous—
Yokohama, 10 cs. 10 c. £373
Hong Kong, 3 40
Wellington, 4½ 30

Potassium Oxalate—
Antwerp, 1 t. 2 c. £90

Potassium Chlorate—
Gothenburg, 10 c. £23

Potassium Iodine—
Algoa Bay, 1 c. £63

Potassium Citrate—
Hamburg, 7 c. £48

Salammoniac—
Constantinople, 2 t. £70

Soda Crystals—
Nelson, 5 t. £14
Hong Kong, 8 t. 2 c. 21

Sulphuric Acid—
Natal, 7 t. 16 c. £60
Singapore, 2 10 86
Melbourne, 1 5 10
Sydney, 5 cs. 18
Natal, 1 10 16

Sulphur—
Calcutta, 7 t. 19 c. £45

Sheep Dip—
Natal, 50 cs. £133
E. London, 50 500 drs. 245
Sydney, 210 gls. 50
*Karachi, 1 t. 3 c. 24

Soda Ash—
Yokohama, 27 t. £138

Saltpetre—
Oporto, 5 t. 18 c. £103
Melbourne, 1 28

Salicylic Acid—
Melbourne, 1 c. £28

Sodium Nitrate—
Hong Kong, 1 t. £10

Sodium Sulphate—
Ghent, 60 t. £60

Superphosphate—
Cherbourg, 145 t. £435
Rotterdam, 30 300
Martinique, 256 1,000
Cherbourg, 300 1,180

Tannic Acid—
Yokohama, 5 cs. £31

Tartaric Acid—
Townsville, 10 c. £72
Newcastle, 7 52
Montreal, 1 t. 130

LIVERPOOL.

Week ending Feb. 26th.

Alum—
Seville, 7 t. 13 c. 3 q. £40
Calcutta, 30 9 145
Alexandretta, 1 12 8
Algiers, 1 13 7
Patras, 5 6 27
Smyrna, 6 19 35

Alum Cake—
Melbourne, 8 t. 15 c. 2 q. £26

Ammonium Muriate—
New York, 8 t. 3 c. £202

Ammonium Sulphate—
Antwerp, 54 t. 10 c. £640
Ghent, 150 1,778
Bordeaux, 2 10 29
Philadelphia, 2 kegs. 18
Antwerp, 6 1 69
Valencia, 4 19 1 57

Ammonium Carbonate—
Genoa, 2 t. £65
Naples, 15 c. 1 q. 25
Bordeaux, 2 60

Alkali—
Maranham, 175 cks.

Borate of Lime—
Rotterdam, 10 t. 1 q. £115

Bone Waste—
Rouen, 6 t. 8 c. £45

Brimstone—
Calcutta, 10 t. 11 c. 2 q. £30

Boracic Acid—
Havre, 12 c. £20

Borax—
Rotterdam, 17 c. 2 q. £40
Bilbao, 6 9

Bleaching Powder—
Antwerp, 100 cks.
Boston, 203
Havana, 5
Jarail, 38
Naples, 46
" 25
Rotterdam, 58
Trieste, 18
Vigo, 8
Antwerp, 35
Baltimore, 97
Boston, 302 83 tcs.
Ghent, 163
Montreal, 10
Nantes, 107
New York, 284

Copper Sulphate—
Barcelona, 40 t. 3 c. £436
Rouen, 46 6 1,152
Santos, 1 11 61
Barcelona, 37 15 627
Galatz, 10 9
Bordeaux, 6 3 3 q. 118
Genoa, 10 9 213
" 15 4 1 180
Marseilles, 15 2 300
Monte Video, 9 3 12
Piraeus, 8 11
Rouen, 5 2 100
Trieste, 4 19 112
Bilbao 2 2 49
Patras, 5 7
Pasages, 20 380
Tarragona, 40 2 799
Genoa, 137 3 2 2,526
Halifax, 9 2 11
Bilbao, 9 18 3 200
San Sebastian, 19 18 3 400
Bordeaux, 100 4 2,200
" 5 100
Tarragona, 24 17 1 370
Trieste, 5 2 125
Venice, 5 4 2 95
Bombay, 6 2 8
Buenos Ayres, 1 7 1 30
Barcelona, 20 2 1 392
Pasages, 25 445
Bordeaux, 39 1 947
Genoa, 49 2 934
Trieste, 5 1 3 106

Chemicals (*otherwise undescribed*)
Vera Cruz, 13 brls. £70
Antwerp, 4 t. 15 c. 3 q. 134
Hamburg, 2 8
Bombay, 7 9 1 54

Citric Acid—
Valencia, 10 c. £70

Carbon Sulphide—
Rio de Janeiro, 3 c. £5

Chrome Acetate—
Naples, 3 c. 1 q. £10

Carbolic Acid—
Genoa, 4 c. £30
Philadelphia, 17 87
Hamburg, 2 t. 3 2 q. 247
Havre, 1 4 2 140
Genoa, 9 3 48
St. Louis, 9 16 3 300
Calcutta, 9 16 2 140
New York, 6 435
Rotterdam, 11 1,515
Hong Kong, 11 8

Carbon—
Bahia, 2 cs. £7

Copperas—
Patras, 3 t. 1 c. £7
Calcutta, 15 19 1 q. 30
Volo, 3 10 7
Valparaiso, 12 19 28
Hamburg, 5 11 76

Chloride of Lime—
Baltimore, 25 t. 5 c. £132
New York, 43 15 240
Canada, 10 75
Boston, 50 10 335
New York, 99 15 537

Calcium Chloride—
Boston, 10 t. 5 c. £62

Caustic Potash—
Portland, 3 t. 15 c. £75

Buenos Ayres, 3 7
New York, 4 [illegible]
" 6 [illegible]

Cream of Tartar—
Halifax, 9 c. 2 q.

Caustic Soda—
New Orleans, 100 drums
New York, 950 [illegible]
Pernambuco, 40
Portland, [illegible]
Rotterdam, 40
Santa Cruz, 50
Santander, 5
Santos, 50
Seville, 58
Tarragona, 40
Vigo, 20
Algoa Bay, [illegible]
Amsterdam, [illegible]
Antwerp, 16
Bahia, 60
Barcelona, 30
[illegible], 200
[illegible], 360
Canada, 40
Copenhagen, 70
Galatz, 40
Genoa, 35
Maceio, 38
Maranham, 70
Melbourne, 30 [illegible]
Alexandria, [illegible]
Barcelona, 75
Boston, 322
Calcutta, 177
Galata, 505
Malaga, [illegible]
New York, 1,400
Philadelphia, 100
Santos, 20
Tunis, 6

Disinfectants—
Bombay, 2 t. 6 c.

Dried Blood—
Nantes, 10 t. 12 c. 3 q.

Epsom Salts—
Halifax, 13 brls.
Rio Janeiro, 40 kegs.

Glycerine—
Philadelphia, 9 t.

Guano—
Tarragona, 19 t. 18 c. 3 q.
Rotterdam, 6 19

Gypsum—
Calcutta, 3 t. 6 c.

Iron Oxide—
Genoa, 1 t. 7 c.

Iodine—
New York, 6 t.

Magnesia—
Rio Grande do Sul, 12 cs.

Magnesium Chloride—
Bombay, 4 t. 19 c.

Manure—
Ghent, 19 t. 4 c. 1 q.
Bordeaux, 411 8 1
St. Nazaire, 200 1 3
Genoa, 584 12 2
Havre, 400 7 2
Las Palmas, 25

Naphthaline—
Antwerp, 34 t. 10 c.

Oxalic Acid—
Rouen, 2 t. 16 c.
Piraeus, 3

Potashes—
Monte Video, 12 cs.

Potassium Nitrate—
Savanilla, 5 c. 1 q.

Phosphorus—
Havana, 5 c. 2 q.
Vera Cruz, 8 3
Havana, 6 3

Potassium Prussiate—
Shanghai, 10 c.

Potassium Chromate—
Genoa, 5 t. 5 c.
Leghorn, 3 4

Potassium Chlorate—
Valparaiso, 3 q.
Hiogo, 5 t.
Naples, 2
New York, 4 6 c.
Shanghai, 2 10
Chicago, 2 10
Boston, 5

Genoa, 5 1 1 185
Yokohama, 5 210
New York, 2 10 99
Boston, 2 10 99
Gothenburg, 5 100
Hiogo, 7 10 150

Potassium Bichromate—
Alexandretta, 5 c 2 q. £10
Kobe, 3 t. 13 2 150
Vera Cruz, 12 3 18
Coruna, 4 2 9
Bilbao, 19 1 36

Pitch—
Antwerp, 27 t.
Cette, 962

Sodium Silicate—
Callao, 10 brls.
Malta, 9 cks.
B Ayres, 7
Carthagena, 6
Rotterdam, 10

Soda Ash—
Trieste, 7 cks.
Yokohama, 127 brls.
Baltimore, 169 164 tcs.
Barcelona, 70
Bilbao, 65
Boston, 64
Canada, 10
Lisbon, 40
Madeira, 4
Madrid, 50
Maranham, 25
New York, 73
Seville, 50
Singapore, 90
Baltimore, 483
Boston, 442 158 2,060 bgs.
B. Ayres, 50
Callao, 22 drms.
Constantinople, 100
Corfu, 60
Ghent, 44 cks.
Maranham, 40 15
Monte Video, 100 100 cks.
Philadelphia, 198
Rio Grande do Sul, 50
Santos, 15
Smyrna, 300

Saltcake—
Philadelphia, 37 t. 2 c. £79
Boston, 99 8 211

Sheep Wash—
Buenos Ayres, 33 t. 17 c. 2 q. £875

Sodium Nitrate—
Seville, 12 t. 13 c. 1 q £110
Malaga, 4 7 3 38
Valencia, 3 5 1 27
Alicante, 1 16 3 15
Cartagena, 11 19 2 105

Sodium Biborate—
Hamburg, 19 t. 16 c. 3 q. £596
Rotterdam, 3 0 3 109

Sulphuric Acid—
Santos, 10 cbys. £15

Sugar of Lead—
Santos, 1 t. 2 c. 2 q. £25

Stannous Chloride—
Trinidad, 6 £275

Sulphur—
New York, 1 t. £17
Algoa Bay, 4 9 c. 1 q. 38
Boston, 15 70
Calcutta, 30 175
Boston, 25 108
Portland, 2 12
New York, 10 1 45

Salammoniac—
Patras, 3 c. 1 q. £6
Samsoun, 3 2 9
Syra, 2 3 5
Toronto, 16 28
Alexandria, 1 t. 35
Campana, 3 5
Galatz, 11 18
Smyrna, 1 12 2 56
Trieste, 2 1 70
Hiogo, 2 10 3 89

Saltpetre—
Syra, 6 t. 15 c. £144
Ceara, 1 23
Portland, 1 10 34
Syra, 2 19 3 66
Maceio, 2 13 2 65

Superhosphate—
Stettin, 200 t. 2 c. £625
Las Palmas, 20 8 2 84
Nantes, 974 bgs. 95

Sheep Dip—
Algoa Bay, 6 t. 11 c. 2 q. £305

Salt—
Savanilla, 2 t.
Antwerp, 414

Antigua, 3
Bakana, 20
Boston, 165
B. Ayres, 46
Copenhagen, 658
Detroit, 65
G. Bassam, 15
N. Orleans, 2,302
New York, 1,500
Opobo, 50
Para, 339 3 c.
Saltpond, 6 5
Bakana, 20
Bonny, 10
B. Ayres, 20
Buguma, 25
Cape Coast, 6¼
George's Bay, 9
N. Orleans 698
O. Calabar, 100
Port Natal, 5
Salt Pond, 6¼
Santos, 1 2
Valparaiso, 100

Sodium Bicarbonate—
Antwerp, 16 pkgs.
Hamburg, 20 3 brls.
Ceara, 2 cs.
Genoa, 56 brls.
Hamburg, 125 kegs.
Rotterdam, 52

Soda Crystals—
Constantinople, 100 brls.
Malta, 150
Piræus, 30
Alexandria, 30
Boston. 250
Calcutta, 34

Tar—
Bakana, 5 brls.
Curacoa,

Wood Acid—
Vera Cruz, 2 cs.

Zinc Chloride—
Bombay, 12 c. 1 q. £8
Oporto, 5 t 2 1 80
Rotterdam, 3 12 1 60

TYNE.

Week ending Feb. 27th.

Arsenic—
Naples, 1 t. 19 c.

Alkali—
Rotterdam, 18 c.
Barcelona, 20 t. 8
Ergastiria, 10
Gothenburg, 6 1
New York, 49 18
Hamburg, 13 1
Antwerp, 10 16

Antimony—
Barcelona, 6 c.

Alumina Sulphate—
Odense, 16 c.

Alum—
Rotterdam, 10 t. 4 c.

Bleaching Powder—
New York, 240 t. 13 c.
Rotterdam, 32 7
Odense, 2 4
Barcelona, 22 2
Gothenburg, 5 6
" 15 2
Libau, 79 18
New York, 151 15
Hamburg, 32 7
Antwerp, 55 14
Christiania, 30 3

Bone Ash—
Ergastiria, 1 t.

Caustic Soda—
New York, 100 t. 2 c.
Genoa, 37 11
Leghorn, 10 3
Hamburg, 16
Antwerp, 10 5
New York, 150
Horsens, 1
Barcelona, 30
Libau, 43 10

Copperas—
Odense, 10 t 14 c.
Aarhuus, 2 5
Randers, 2 5

Gypsum—
New York, 32 t. 8 c.

Litharge—
Genoa, 11 c.
Leghorn, 3 t 1
Hamburg, 3 15

Magnesia—
Barcelona, 10 t. 16 c.
" 6 19
Hamburg, 1

Naphthalene—
Libau, 25 t.

Potassium Chlorate—
New York, 2 t 10 c.
Hamburg, 11
Antwerp, 2 13

Pitch—
Malmo, 10 t.
Antwerp, 150

Phosphorus—
Barcelona, 1 t. 12 c.

Paraffin Wax—
Barcelona, 5 t 1 c.

Superphosphate—
Cherbourg, 200 t. 1 c.

Sodium Hyposulphite—
Gothenburg, 1 t. 8 c.

Soda Ash—
Gothenburg, 6 t. 13 c.
New York, 50 2
Rotterdam, 6 16
New York, 100 6
Hamburg, 9 7
Christiania, 5 5
Antwerp, 13

Soda—
Korsoer, 20 t. 16 c.
New York, 53 18
Rotterdam, 16 12
Landskrona, 50 8
New York, 59 18
Antwerp, 3 5

Sodium Chlorate—
Rotterdam, 11 c.
Antwerp, 10 c.

Sodium Sulphate—
New York, 62 t. 3 c

Tar—
Hamburg, 20 t.

GLASGOW.

Week ending Feb. 27th.

Ammonium Sulphate—
Demerara, 75 t. 6 c. £920
Antwerp, 20 3 241

Alumina Sulphate—
Melbourne, 6 t. 15¾ c. £24

Albumen—
Lisbon, £51

Bleaching Powder—
Melbourne, 7 t. 5½ c. £40

Dyestuffs—
Indigo
New York, 103 c. £1,936
Madder
Melbourne, 39½ c. £69. 11s.
Alizarine
Lisbon, £49

Dyestuffs *(otherwise undescribed)*
Melbourne, £26

Iodine—
Genoa, 10 c. £640

Manure—
Trinidad, 7½ c. £1,639

Potassium Prussiate—
Amsterdam, 4 cks. £58
Gothenburg, 1 t 1½ c. 89
Genoa, 37¾ 100

Potassium Bichromate—
Boston, 84 c. £156
Genoa, 16 t. 5¾ 540
Nantes, 15 17¾ 500
New York, 661 1,233
Antwerp, 100 125
" 206½ 295
Rotterdam, 36 cks. 490
Malta, 21¾ c. 38

Pitch—
Amsterdam, 301 t. £301

Phosphate—
Trinidad, 10 t. 18¾ c. £183

Paraffin Wax—
Genoa, 26 c. £36
Rouen, 17½ 22

Paraffin—
Genoa, 56¾ c. £77

Sodium Bichromate—
Boston, 63 c. £88
Genoa, 40 t. 5¾ 1,257
Rotterdam, 16 cks. 191

Verdigris—
Melbourne, 3¾ c. £24

HULL.

Week ending Feb. 2

Bleaching Powder—
Drontheim, 31 cks.

Bone Size—
Harlingen, 13 cks.

Chemicals *(otherwise und*
Alexandria, 12 cs.
Bremen, 71
Bordeaux, 30
Copenhagen, 4
Christiania, 5 pks. 1 8
Drontheim, 11
Genoa, 11
Gothenburg,
Ghent, 10 pks. 4
Hango, 1
Hamburg, 1
Libau, 68
Lisbon, 3
Leghorn, 8
Marseilles, 60
Stettin, 140 pkgs.
Rouen, 36
Rotterdam, 40 pks. 30 cs. 1
Antwerp, 9 4

Caustic Soda—
Antwerp, 37 cs.
Amsterdam, 20 drms.
Bergen, 4
Dunkirk, 18
Genoa, 6
Hango, 3 cks.
Konigsberg, 17

Dyestuffs *(otherwise unde*
Christiania, 20 cks.
Genoa, 5
Libau,
Reval, 2

Manure—
Aarhuus, 186 t.
Copenhagen, 16
Dunkirk, 101
Granville, 738 bgs.
Ghent, 2,408
Rotterdam, 600

Naphtha—
Dunkirk, 25 cks.

Pitch—
Antwerp, 259 t
Dunkirk, 86
Marseilles, 5 cks.

Turpentine—
Christiania, 2 cks.
Drontheim, 2 brls.
Gothenburg, 1 ck.

GRIMSBY

Week ending Feb. 22

Chemicals *(otherwise unde*
Antwerp, 2 cks.
Dieppe, 145 pkgs.
Hamburg, 2 cs. 5

GOOLE.

Week ending Feb. 19

Alkali—
Rotterdam, 4 cks.

Bleaching Preparations-
Rotterdam, 28 cks.

Benzol—
Dunkirk, 20 cs.
Ghent, 18 drums.
Rotterdam, 48 cks.

Chemicals *(otherwise und*
Hamburg, 57 cks.
Rotterdam, 26

Dyestuffs *(otherwise unde*
Boulogne, 2 cks. 3 pkgs.
Ghent, 5 6 drum
Rotterdam, 3

Manure—
Boulogne, 121 bgs.
Binges, 273
Calais, 130
Dunkirk, 536
Ghent, 555

Pitch—
Antwerp, 154 t.
Dunkirk, 79

PRICES CURRENT.

WEDNESDAY, MARCH 5, 1890.

PREPARED BY HIGGINBOTTOM AND CO., 116, PORTLAND STREET, MANCHESTER.

The values stated are F.O.R. at maker's works, or at usual ports of shipment in U.K. The price in different localities may vary.

Article		Unit	£	s.	d.
Acids :—					
Acetic, 25 °/ₒ and 40 °/ₒ		per cwt	7/6 & 12/6		
,, Glacial		,,	2	10	0
Arsenic, S.G., 2000°		,,	0	12	9
Chromic 82 °/ₒ	nett	per lb.	0	0	6½
Fluoric		,,	0	0	3¼
Muriatic (Tower Salts), 30° Tw.		per bottle	0	0	8
,, (Cylinder), 30° Tw.		,,	0	2	11
Nitric 80° Tw.		per lb.	0	0	2
Nitrous		,,	0	0	1¼
Oxalic	nett	,,	0	0	3¼
Picric		,,	0	1	4
Sulphuric (fuming 50 °/ₒ)		per ton	15	10	0
,, (monohydrate)		,,	5	10	0
,, (Pyrites, 168°)		,,	3	2	6
,, (,, 150°)		,,	1	8	0
,, (free from arsenic, 140/145°)		,,	1	10	0
Sulphurous (solution)		,,	2	15	0
Tannic (pure)		per lb.	0	1	0½
Tartaric		,,	0	1	2½
Arsenic, white powered		per ton	13	15	0
Alum (loose lump)		,,	4	15	0
Alumina Sulphate (pure)		,,	5	0	0
,, ,, (medium qualities)		,,	4	10	0
Aluminium		per lb.	0	15	0
Ammonia, ·880=28°		per lb.	0	0	3
,, =24°		,,	0	0	1⅞
,, Carbonate	nett	,,	0	0	3½
,, Muriate		per ton	23	10	0
,, ,, (sal-ammoniac) 1st & 2nd		per cwt.	37/- & 35/-		
,, Nitrate		per ton	40	0	0
,, Phosphate		per lb.	0	0	10½
,, Sulphate (grey), London		per ton	11	17	6
,, ,, (grey), Hull		,,	11	17	6
Aniline Oil (pure)		per lb.	0	0	11
Aniline Salt ,,		,,	0	0	9½
Antimony		per ton	75	0	0
,, (tartar emetic)		per lb.	0	1	1
,, (golden sulphide)		,,	0	0	10
Barium Chloride		per ton	7	10	0
,, Carbonate (native)		,,	3	15	0
,, Sulphate (native levigated)		,,	45/- to 75/-		
Bleaching Powder, 35 %		,,	5	2	6
,, Liquor, 7 %		,,	2	10	0
Bisulphide of Carbon		,,	12	15	0
Chromium Acetate (crystal		per lb.	0	0	6
Calcium Chloride		per ton	2	0	0
China Clay (at Runcorn) in bulk		,,	17/6 to 35/-		
Coal Tar Dyes :—					
Alizarine (artificial) 20 %		per lb.	0	0	9
Magenta		,,	0	3	9
Coal Tar Products					
Anthracene, 30 % A, f.o.b. London		per unit per cwt.	0	1	6
Benzol, 90 % nominal		per gallon	0	3	10
,, 50/90 ,,		,,	0	3	0
Carbolic Acid (crystallised 35°)		per lb.	0	0	10
,, (crude 60°)		per gallon	0	2	10
Creosote (ordinary)		,,	0	0	2½
,, (filtered for Lucigen light)		,,	0	0	3
Crude Naphtha 30 % @ 120° C.		,,	0	1	5
Grease Oils, 22° Tw.		per ton	3	0	0
Pitch, f.o.b. Liverpool or Garston		,,	1	12	0
Solvent Naphtha, 90 % at 160°		per gallon	0	1	11
Coke-oven Oils (crude)		,,	0	0	2½
Copper (Chili Bars)		per ton	46	15	0
,, Sulphate		,,	25	10	0
,, Oxide (copper scales)		,,	52	0	0
Glycerine (crude)		,,	30	0	0
,, (distilled S.G. 1250°)		,,	55	0	0
Iodine		nett, per oz.	8½d. to 9d.		
Iron Sulphate (copperas)		per ton	1	10	0
,, Sulphide (antimony slag)		,,	2	0	0
Lead (sheet) for cash		,,	14	0	0
,, Litharge Flake (ex ship)		,,	17	0	0
,, Acetate (white ,,)		,,	23	5	0
,, ,, (brown ,,)		,,	18	15	0
,, Carbonate (white lead) pure		,,	19	0	0
,, Nitrate		,,	22	5	0
Lime, Acetate (brown)		,,	8	5	0
,, ,, (grey)		per ton	14	0	0
Magnesium (ribbon and wire)		per oz.	0	2	3
,, Chloride (ex ship)		per ton	2	6	3
,, Carbonate		per cwt.	1	17	6
,, Hydrate		,,	0	10	0
,, Sulphate (Epsom Salts)		per ton	3	0	0
Manganese, Sulphate		,,	18	0	0
,, Borate (1st and 2nd)		per cwt.	60/- & 42/6		
,, Ore, 70 %		per ton	4	10	0
Methylated Spirit, 61° O.P.		per gallon	0	2	2
Naphtha (Wood), Solvent		,,	0	3	11
,, ,, Miscible, 60° O.P.		,,	0	4	9
Oils :—					
Cotton-seed		per ton	22	10	0
Linseed		,,	23	10	0
Lubricating, Scotch, 890°—895°		,,	7	5	0
Petroleum, Russian		per gallon	0	0	5¼
,, American		,,	0	0	5⅝
Potassium (metal)		per oz.	0	3	10
,, Bichromate		per lb.	0	0	4
,, Carbonate, 90% (ex ship)		per ton	17	15	0
,, Chlorate		per lb.	0	0	4¼
,, Cyanide, 98%		,,	0	2	0
,, Hydrate (Caustic Potash) 80/85 %		per ton	22	10	0
,, ,, (Caustic Potash) 75/80 %		,,	21	10	0
,, ,, (Caustic Potash) 70/75 %		,,	20	15	0
,, Nitrate (refined)		per cwt.	1	3	6
,, Permanganate		,,	4	5	0
Prussiate Yellow		per lb.	0	0	9¼
,, Sulphate, 90 %		per ton	9	10	0
,, Muriate, 80 %		,,	7	15	0
Silver (metal)		per oz.	0	3	8¾
,, Nitrate		,,	0	2	5½
Sodium (metal)		per lb.	0	4	0
,, Carb. (refined Soda-ash) 48 %		per ton	5	15	0
,, ,, (Caustic Soda-ash) 48 %		,,	4	15	0
,, ,, (Carb. Soda-ash) 48%		,,	4	17	6
,, ,, (Carb. Soda-ash) 58 %		,,	5	15	0
,, ,, (Soda Crystals)		,,	2	15	0
,, Acetate (ex-ship)		,,	15	15	0
,, Arseniate, 45 %		,,	10	0	0
,, Chlorate		per lb.	0	0	6¼
,, Borate (Borax)		nett, per ton.	30	0	0
,, Bichromate		per lb.	0	0	3
,, Hydrate (77 % Caustic Soda)		per ton.	10	15	0
,, ,, (74 % Caustic Soda)		,,	10	7	6
,, ,, (70 % Caustic Soda)		,,	9	2	6
,, ,, (60 % Caustic Soda, white)		,,	8	2	6
,, ,, (60 % Caustic Soda, cream)		,,	7	15	0
,, Bicarbonate		,,	5	15	0
,, Hyposulphite		,,	5	5	0
,, Manganate, 25%		,,	8	10	0
,, Nitrate (95 %) ex-ship Liverpool		per cwt.	0	8	4½
,, Nitrite, 98 %		per ton	27	10	0
,, Phosphate		,,	15	15	0
,, Prussiate		per lb.	0	0	7¼
,, Silicate (glass)		per ton	5	7	6
,, ,, (liquid, 100° Tw.)		,,	3	17	6
,, Stannate, 40 %		per cwt.	2	0	0
,, Sulphate (Salt-cake)		per ton.	1	12	6
,, ,, (Glauber's Salts)		,,	1	10	0
,, Sulphide		,,	7	15	0
,, Sulphite		,,	5	5	0
Strontium Hydrate, 98 %		,,	8	0	0
Sulphocyanide Ammonium, 95 %		per lb.	0	0	8
,, Barium, 95 %		,	0	0	5¼
,, Potassium		,,	0	0	8
Sulphur (Flowers)		per ton.	6	15	0
,, (Roll Brimstone)		,,	6	0	0
,, Brimstone : Best Quality		,,	4	7	6
Superphosphate of Lime (26 %)		,,	2	10	0
Tallow		,,	25	10	0
Tin (English Ingots)		,,	95	0	0
,, Crystals		per lb.	0	0	6¾
Zinc (Spelter)		per ton.	22	5	0
,, Chloride (solution, 96° Tw.		,,	6	0	0

THE

CHEMICAL TRADE JOURNAI

Publishing Offices: 32, BLACKFRIARS STREET, MANCHESTER.

No. 147. SATURDAY, MARCH 15, 1890. Vol.

Contents.

Notices.

All communications for the *Chemical Trade Journal* should be addressed, and Cheques and Post Office Orders made payable to—

DAVIS BROS., 32, Blackfriars Street, MANCHESTER.

Our Registered Telegraphic Address is—

"Expert, Manchester."

The Terms of Subscription, commencing at any date, to the *Chemical Trade Journal*,—payable in advance,—including postage to any part of the world, are as follow:—

Yearly (52 numbers) .... **12s. 6d.**
Half-Yearly (26 numbers) .... **6s. 6d.**
Quarterly (13 numbers) .... **3s. 6d.**

Readers will oblige by making their remittances for subscriptions by Postal or Post Office Order, crossed.

Communications for the Editor, if intended for insertion in the current week's issue, should reach the office not later than **Tuesday Morning.**

Articles, reports, and correspondence on all matters of interest to the Chemical and allied industries, home and foreign, are solicited. Correspondents should condense their matter as much as possible, write on one side only of the paper, and in all cases give their names and addresses, not necessarily for publication. Sketches should be sent on separate sheets.

We cannot undertake to return rejected manuscripts or drawings, unless accompanied by a stamped directed envelope.

Readers are invited to forward items of intelligence, or cuttings from local newspapers, of interest to the trades concerned.

As it is one of the special features of the *Chemical Trade Journal* to give the earliest information respecting new processes, improvements, inventions, etc., bearing upon the Chemical and allied industries, or which may be of interest to our readers, the Editor invites particulars of such—when in working order—from the originators; and if the subject is deemed of sufficient importance, an expert will visit and report upon the same in the columns of the *Journal*. There is no fee required for visits of this kind.

We shall esteem it a favour if any of our readers, in making inquiries of, or opening accounts with advertisers in this paper, will kindly mention the *Chemical Trade Journal* as the source of their information.

Advertisements intended for insertion in the current week's issue, should reach the office by **Wednesday morning** at the latest.

Advertisements.

TECHNICAL INSTRUCTION.

IF ever there was sufficient proof that the T Instruction Act is not a popular measure, and general public (aye, and even those whom it is supp benefit) do not care a jot for it, that proof may be f the meeting that took place on Tuesday in last weel Owens College, Manchester. It was previously anr that Mr. W. Mather, M.P., would address the meeti the question, but even the magic of Mr. Mather's na not sufficient to draw an audience. The Technical Ed propagandists may well now inquire why the pub interested public—having been piped to, have not Why have they not listened to the voice of the charm he charmed never so wisely?

The meeting at Owens College was a dismal failur we say it was not even representative we shall b correct. The press was well represented, four r being present, but when the meeting commenced the a consisted of twelve souls, all told, and by the time t ness was concluded the number had reached twenty-f spite of this want of interest, as evidenced by the pa attendance, a resolution was put and carried, as if world had been concerned in it, and a memorial p upon the instructions of that resolution for submissio Manchester City Council.

We have for a long time urged against the exte these tactics for the purpose of raising the wind by issue. To a few it is perfectly well known that th certain institutions, like many individuals, sadly in hard cash, but they have not the courage to come be public and say so in true English fashion. The m wanted, and if it can be raised compulsorily from th payers without letting them know definitely what i done with it so much the better for the memorialist have never yet seen any scheme formulated tending to us that the money so raised would be judiciously sp that the Technical Education zealots have any scheme in particular except that of supplying some pa institution with the needful. In these mercenary d are taught to believe in the doctrine of the survival fittest; we are taught to believe in ordinary daily life, thing is not supported by the mass of the population thing not required—that it must give way to others t necessary—that it is a sin to foster it by subsidies an artificial means.

The apathy shown by the general public to the Te Education problem, as it at present exists before shows that it belongs to the class which should the wall to make room for some better matured s and it is only now kept alive by means of such artifici cesses as have prompted these present remarks. by no means without experience in the needs and r ments of technical instruction in this country, and it the greatest humiliation that we regard this measure we remember that some of our most competent mind been employed in its production.

Correspondence.

ANSWERS TO CORRESPONDENTS.

S.—We do not know any particular use for the worn rollers. We understand they are used over again with new material.

—Weak bleach always settles better than strong. A quarter of a pound of ch to a quart of water gives a solution of 11° Tw.

NIA.—Good grey sulphate of 24½ per cent. can be made with pyrites acid. Advertise.

B.—Nothing but a dynamo would pay. It would be absurd to use batteries e old type.

—We cannot advise on such matters. (2) Ask your patent agent.

ST.—(1) All fixed. (2) The use of caustic soda would not pay. (3) When the esses in a gas works are carried out chemically, but not before.

-We are preparing a series of articles on the subject.

-Your queries are only suitable for our advertising columns.

—A little learning is a dangerous thing. Read up your electrical manual n.

MAKER.—The matter is under consideration. It would be well for you to end your judgment in the meantime.

We do not hold ourselves responsible for the opinions expressed by our correspondents.

IE DETERMINATION OF MOISTURE IN WOOD PULP.

To the Editor of the Chemical Trade Journal.

are quite at one with Mr. Pattinson on the subject of estimating oisture in wood pulps. We think with him that the weight as at 212° F. is the one on which all prices should be based and lations made. But in our communication of the 15th ult., we y dealt with the two alternatives, of which one or the other is ally adopted in the trade, viz., to dry the pulp at 212° F., and fixed percentage to obtain the air-dried weight, or to dry at 212° d allow the re-absorption of moisture in a normal atmosphere. advised the former course on account of the difficulty of obtain- niform conditions for the second mode of procedure, and a little ination will show that our advice is not really at variance with Pattinson's suggestion, for the addition of a constant figure such as commended does not affect the value of the 212° F. dried weight as is for the calculation of prices, etc.

e thank Mr. Pattinson for his friendly criticism,—And are

GEORGE E. DAVIS,
e Manchester Technical Laboratory. ALFRED. R. DAVIS.

LEAD POISONING AT SHEFFIELD.

IE Water Committee of the Sheffield Corporation have been engaged in an exhaustive inquiry into the facts connected with nany complaints of lead-poisoning that have appeared in the local rs.

. Alderman Gainsford, Chairman of the Water Committee, in ing the inquiry, said that they were most anxious to ascertain the facts of the case. He supposed there was no doubt that the water the reservoir was of such a character as to be liable to take up lead what freely. None of them would wish to deny that lead-poisoning d, but they must satisfy themselves as to its extent. The next for inquiry was the cause of the poisoning; and, further, they ed to find out whether it was so serious a matter as to call for dy at any cost. The third question related to the remedies that to be applied. He regretted that the response to the committee's st for evidence from the general public had been almost entirely out result, except so far as the medical profession were concerned, om the committee were very greatly indebted.

idence was then given to the effect that there had been many of sickness arising from lead poisoning which were very severe; d, in one instance the death of the patient was attributed to this . The evidence of several members of the medical profession, were examined before the committee, went to show that in some of mples of water which were analysed, lead existed to the amount lf a grain per gallon.

e preliminary investigations of the committee have been fruitful ich startling evidence on the subject of plumbism. The dis- es last week, were, indeed, of quite a sensational order, for we of persons been paralysed, crippled for life, and even rendered ess from lead poisoning, arising from consumption of a soft moor- water in an unfiltered condition, from the Redmires district, e Sheffield derives part of its supply. The odd thing is that this ires high-level supply has been in use for half-a-century without tion of its injurious action on lead-pipes. The fact is, however patent that plumbism has of late years been manifesting itself in an unpleasantly rapid fashion. True it is that the percentage of consumers affected is small, but the evil, even if of comparatively small proportions, is of great intensity, and has been such as to render imperative some remedial action. The Committee find that the lead is "derived from the pipes and cisterns of the consumers," and state their preparedness to sanction practical remedial experiments. It appears that household charcoal filters are useful in removing lead, but that unless they are cleansed and recharged at frequent intervals the remedy is but temporary, and the committee are of opinion that to aim at securing the requisite conditions is to attempt the impossible. Three alternative methods are advocated—treatment of the Redmires water by the lime process; provision of filter beds; or removal of lead service pipes—and eventually the committee have resolved to give the first of these a trial, in the confident assurance that its efficacy will be realised without leading to hardening of the water, at present so much appreciated for its softness.—*Daily Graphic.*

PRIZE COMPETITION.

A LITTLE more than twelve months ago, one of our friends suggested that it might be well to invite articles for competition, on some subject of special interest to our readers; giving a prize for the best contribution, he at the same time offering to defray the cost of the first prize, and suggesting a subject in which he was personally interested.

At that date we were totally unprepared for such a suggestion, but on thinking over the matter at intervals of leisure, we have come to think that the elaboration of such a scheme would not only prove of benefit to the manufacturer, but would go some way towards defraying the holiday expenses of the prize winner.

We are therefore prepared to receive competitions in the following subjects:—

SUBJECT I.

The best method of pumping or otherwise lifting or forcing warm aqueous hydrochloric acid of 30 deg. Tw.

SUBJECT II.

The best method of separating or determining the relative quantities of tin and antimony when present together in commercial samples.

On each subject, the following prizes are offered:—

One first prize of five pounds, one second prize of one pound, five additional prizes to those next in order of merit, consisting of a free copy of the *Chemical Trade Journal* for twelve months.

The competition is open to all nationalities residing in any part of the world. Essays must reach us on or before April 30th, for Subject II, and on or before May 31st, in Subject I. The prizes will be announced in our issue of June 28th.

We reserve to ourselves the right of publishing the contributions of any competitor.

Essays may be written in English, German, French, Spanish or Italian.

THE SALT UNION.

THE first ordinary general meeting of the shareholders of the Salt Union, Limited, was held at the Cannon-street Hotel a few days since, under the presidency of Lord Thurlow. In moving the adoption of the report, the Chairman said that they had been accused of raising the price of salt to an exhorbitant extent; but he could not admit they had done so. They no doubt had raised the price; but he pointed out that the object of the Salt Union was to raise prices so as to save the trade from impending ruin. The present prices were not higher than in 1872 and 1873. The directors desired to alleviate the sufferings ariving from subsidences in the salt districts; but they had to remember that the money they were dealing with was not their own, and they were advised that no liability attached to the Salt Union. With regard to the subsidences, the saltmakers and royalty-holders would be responsible if any persons were. The directors had decided to bore in Cheshire to see if they could not find coal, petroleum, or minerals, and believed they were on the eve of a discovery greater than the discovery of coal in Kent. Mr. Verdin seconded the motion. A discussion followed, and the chairman explained the provisions made for depreciation of stock, and promised to publish a monthly return of the business. The report was adopted.

Trade Notes.

"GASSED" IN A CHEMICAL WORKS.—While a married man, named Michael Duffy, who lives with his wife and family in Peasley Cross-lane, was working in the chlorate house at Messrs. Kurtz's Chemical Works, on the 26th ult., he was overcome by still gas, which is known among the men as "Roger." He was picked up in an unconscious state and conveyed home, where he lies in a precarious condition.

A SUBMERGED STAFFORDSHIRE COALFIELD.—At the annual meeting of the South Staffordshire and East Worcestershire Institute of Mining Engineers, at Mason College, Birmingham, on Monday, Mr. Henry Lea, in his presidential address, said the effect of the Mines Drainage Act had been to reduce the daily income of water into the mines from 48,000,000 gallons in 1873 to 17,000,000 gallons. The cost of pumping was reduced from 11d. to 4½d. per 25,000 gallons. When the underground levels were completed he anticipated the whole area would be unwatered and the 100,000,000 tons of coal liberated.

A NEW WAY OF FILLING POTS.—The E. P. Gleason Manufacturing Company, of Brooklyn, N.Y., will try a new method of filling in "batch." Over every pot is an iron tank or boiler which is constantly filled with batch and which is kept constantly warm, thereby shortening the time for melting, as the tanks are connected with each pot by a funnel-like arrangement so that it is only necessary to pull out a small trap and the batch of glass runs into the pots of its own accord. Although the idea is a good one it has not been in operation, as the mixing department is too low down in the building and difficulty is experienced in getting the batch into the tanks. To remedy this the firm is removing its sand blasting room to another portion of the building and will move the mixing rooms higher up, when the idea be then put into successful effect.—*Commoner and Glassworker.*

INTERNATIONAL EXHIBITION OF MINING AND METALLURGY, LONDON, 1890.—An International Exhibition of Mining and Metallurgy on an extensive scale, the result of a proposal which emanated from the *Mining Journal*, will be held during the forthcoming summer at the Crystal Palace, Sydenham. One-half of the surplus of the Exhibition, as certified by the auditors, is to be paid to the exhibitors *pro rata* to the amount paid by them for space occupied, and the other half is to be disposed of by the Council, either in founding a Scholarship at the Royal School of Mines, or in helping some other institution connected with Mining and Metallurgy. The Exhibition will open on 2nd July, and close on 30th September, 1890, and, besides several features of special attraction, important collections of exhibits are expected from the Colonies and foreign countries. The Honorary Secretary is Mr. Geo. A. Ferguson, editor of the *Mining Journal*, 18, Finch-lane, London, E.C., from whom prospectuses and application forms for space may be obtained.

THE WILL OF A NEWCASTLE MILLIONAIRE.—Probate of the will, dated 19th January, 1888, with codicils made the 28th March, 12th April, and 26th October, 1888, and the 15th August, 1889, of the late Mr. Christian Allhusen, of Stoke Court, Bucks, J.P., and of the Newcastle Chemical Works, who died on the 13th January, aged 84 years, and was the son of Mr. Charles Christian Frederick Allhusen, of Kiel, has been granted to the executors, Mr. Wilton Allhusen, of Newcastle-on-Tyne, chemical manufacturer, and Mr. William Hutt Allhusen, merchant, sons of the testator; Mr. Edward Horatio Neville, of Skellingthorpe, Lincolnshire, his son-in-law, and Mr. John Edward Davidson, of Newcastle-on-Tyne (secretary of the Newcastle Chemical Works Company), by whom the value of the testator's personal estate has been sworn at £1,126,852. 1s. 10d., including American railway securities valued at £140,500., stocks and bonds of foreign Governments £76,950., shares of the Newcastle Chemical Works Company £47,000., and shares and debentures of other public companies £600,000.

A NOVEL PROJECT.—When it was stated some weeks since in the newspapers that the building of a milk pipe line from a point in New York State to New York City was projected there was a rather general smile, and the matter was treated as a joke. The projectors were, however, it seems, in sober earnest. A company with a capital of 500,000 dols. has, it is announced, been formed at Middletown, N. Y., for the purpose of constructing such a line. The proposed method of forwarding the milk is in cylindrical tin cans, surrounded and propelled by water, and the promoters of the scheme assert that the time of transportation, for a distance of 100 miles, will not exceed an hour, while the profit will be about one cent. a gallon. *Fire and Water* thinks if this sort of thing goes on, we need not be surprised ere long to find New York the converging point, not only of oil, natural gas, and milk pipe lines, but of whisky ducts from the blue grass regions, and beer ducts from Cincinnatti, St. Louis, and Milwaukee. The pipe manufacturers may well feel cheerful at the prospect before them.—*Scientific American.*

MISCELLANEOUS CHEMICAL MARKE

Caustic soda has been in brisk demand during the past w
there is a stronger market. Price for 70's is firm at £9. 5s. f
for 77 per cent., £10. 15s. net Other strengths in proporti
ash also commands full prices, and has a ready outlet at 1¼
for ordinary brands, and 1½d. per deg. for refined.
powder is without material change, and prices are on
£5. 2s. 6d. to £5. 5s at makers' works. Chlorate of pota
4¾d. per lb. Soda crystals somewhat scarce, and price
Sulphate of copper easier in price, £25. 10s. to £26.
delivery, and £23. 10s. for April. Lamp and flour sulphur u
in position meantime, but the new company which is to deal
greater part of the British production from the "Chance'
process, will no doubt strengthen prices by reducing the a
competition. Vitriol is still in good demand in all localitie
powdered arsenic is dearer. Yellow prussiate of potash sel
freely and firm at 9½d. to 9¾ per lb. Acetates of lime are
better demand, £8. 5s. to £8. 10s for brown, having been
spot deliveries; prices rather easier forward. Grey scarce at
to £14. Acetates of lead and other acetates without change.
of lead slow. Carbolic acids are still quiet and in uncertain
the business transacted being only for short periods as a rul
acid 1s 4d. per lb. There is rather more demand for potash
carbonate, and muriate, and current rates are firmly m
Ammonia salts are generally quiet.

METAL MARKET REPORT.

	Last week.		This we
Iron	51/7½	....	50/0
Tin	£89 15 0	....	£90 c
Copper	47 2 6	....	46 15
Spelter	22 5 0	....	22 15
Lead	12 12 6	....	12 7
Silver	44d.	....	44

COPPER MINING SHARES.

	Last week.			This we	
Rio Tinto	15¾	15⅞	....	15⅛	15
Mason & Barry	6 15/16	6 7/16	....	6⅛	6
Tharsis	4½	4¾	....	4 7/16	4
Cape Copper Co.	3⅜	3½	....	3⅜	3
Namaqua	2	2⅛	....	1⅞	2
Copiapo	2 7/16	2 9/16	....	2 7/16	2
Panulcillo	15/16	1 1/16	....	1	1
New Quebrada	⅜	¾	....	⅞	1
Libiola	2⅜	8⅝	....	2⅜	2
Tocopilla	1/6	2/-	....	1/-	1
Argentella	3d.	9d.	....	3d.	9

Market Reports.

TAR AND AMMONIA PRODUCTS.

The benzol market remains in much the same condition
ported last week. In some quarters, it is said, a change for
has commenced, but we do not see any signs of such havi
place in the meantime. Practically, prices of last week may
as the values of to-day. Solvent naphtha still continues in
mand, and the higher price realised during the past few w
caused the introduction into the market of many low-class
1s. 10d. to 1s. 11d. may be considered as to-day's value for
at 160, f.o.b. Creosote is moving off very freely, and the a
stocks on the part of producers keeps prices firm. Crude c
still an enigma. It is being absorbed as fast as produced, b
cannot be raised. The pitch market continues very dull, c
buyers evidently not being wishful to increase their purchases a
The sulphate of ammonia market still is in an abnormal positio
however, are maintained in spite of the lower prices said to b
accepted by the London makers. We hear that £11. 15s.
accepted at Hull, and the same price at Leith; and though E
quoted at £11. 13s. 9d, yet outside London makes hav
£11. 16s. 3d. We have it, however, on good authority,
Wednesday a considerable quantity of sulphate was sold
delivery at Hull at £11. 17s. 6d., usual terms and condition
those manufacturers who have been accepting lowest mo
certainly been selling their production below the market pri
day. Dealers still continue to depress values, but there are
requirements to be filled, which is evidenced by the fact of

MEETINGS OF COMPANIES.

:e Manchester Ship Canal.—At the last meeting of the share- rs Manchester Ship Canal Company, Limited, the deputy chairman oseph Lee) stated that it was the intention of the directors to deal the remainder of the unissued ordinary shares of the company, nting to some £138,000, during the next few months, and that ought he might safely predict that the shareholders and friends of hip Canal would support the directors in taking up some of those s, thus clearing off the liability which the Board had in the matter, enabling them to issue the balance of the debentures. We are ally informed that the entire balance of the unissued ordinary shares iow been placed at par among friends of the Ship Canal without aission or discount of any kind. The whole of the share capital ig now been disposed of, the directors will be able to issue the inder of the mortgage debentures during the present year, and in issue the shareholders will have the preference of allotment.

)burite Explosives Co., Limited. — The fourth ordinary :al meeting of this company was held on Monday, at the City iinus Hotel, under the presidency of Sir John Stokes. In moving doption of the report the Chairman stated that great progress had made by the company during the past year, and the value of their its had increased by the business being extended to other countries. plant and buildings had also been improved, and the manufac- g costs had thus been reduced. The gross profit last year had £6,205., or, adding certain special items, £7,342., compared £1,642. in 1888; and the balance of the gross profit was for last £4,672, against £157. for the previous year. The general ex- iture has been reduced by £1,000. during the past year, and they d to make a further reduction of similar amount in the current Up to the 6th inst. they had received this year 161 orders and ered 38 tons, against 30 tons in the corresponding period of last The receipts in this period had increased by 28 %, while the s had fallen 24 %, owing to improvements in the manufacture. In · to extend the use of roburite they had lowered its price. They ved no complaints against the article. Mr. H. W. Maynard ıded the motion. Replying to certain criticisms respecting the ress of the company and the expenses, the chairman repeated that were going on very well in the business both in South Africa and ralia. As regarded the ammonia plant there had been great ulties, but everything that could be done to remove them would one. They had not been successful in introducing roburite into wall, but it was not an explosive very well adapted to Cornish s, the rock being very hard. He did not think the salaries could ' be regarded as too high. He felt sure that in the present year would show a remarkable amount of business done. He then put aotion, which he declared carried by a majority; and the retiring tors and auditors were afterwards re-elected.

ie Leeds Forge Company, Limited.—The first annual ordinary ral meeting of the Leeds Forge Company, Limited, was held a few ago at the Leeds Forge. There was a good attendance of share- ers, and the chair was occupied by Mr. John Scott, C.B (chairman e company), who was supported by Mr. Samson Fox, Mr. G. W. ings, M.P., Mr. J. S. Walker (secretary), and others. The :t, which recommends a dividend of 17½ per cent. on ordinary :s, has already appeared. The Chairman, in moving its adoption, ratulated the shareholders on the result of the year's operations. ' would observe that the directors had placed a sum of 114. 15s. 3d. to the reserve fund, that being the amount of profit h accrued for the four months ended December, 1888, but which were advised could not be treated as devisible profit. The reserve was established with the distinct understanding that it was to be permanent character, and not to be available in future years for urposes of division among the shareholders either in dividends or ses. (Applause.) The directors, on the same footing, proposed ld, as shown in the profit and loss account, a further sum of 000. from the transactions for 1889, these sums answering the ose of a depreciation fund, which could at any time be written n the other side of the account if thought desirable by the iholders. On the formation of the company the purchase of certain assets of the old Leeds Forge sold to the new ıny for £564,000., included stock-in-trade and work in progress, :he shareholders might think the value of that stock and work in ress, being a fluctuating asset, should be deducted from the price '564,000., and carried to the debit of the trading account. This been done to the extent of £25,920. 15s., leaving the value of the : permanent asset at £538,079. 5s. In addition to the purchase , the new company was by agreement to pay all the liabilities of ld Leeds Forge Company except the mortgages, and to receive all lebts. The carrying out of these transactions resulted in the wing figures—the liabilities were £52,222. 15s. 10d., and the debts ınted to £18,921. 2s. 1d., leaving a differenee of £33,301. 13s., which was the cost to the company of the old company's assets, in addition to the sum of £564,000. The directors would have been justified in adding the whole of that figure of £33,301. 13s. 9d. to the £538,079. 5s., and to have treated the total, £571,380. 18s. 9d., as being the cost of the new company of the permanent assets. The course they adopted was to place so much, namely £25,920. 15s., as would bring the £538,079. 5s. back again to the £564,000, and to charge the balance, £7,380. 18s. 9d., against the trading of the year. The shareholders would thus readily see that if the £7,380. 18s. 9d. had not been charged against the year's trading, the profit might with perfect justice have been stated at £84,338. 7d. 10d. (Applause.) The business of the company had continued in a high state of activity, and there was every prospect of its being considerably augmented in the current year. There was, too, continued prosperity in the production of the speciali- ties possessed by the company. The marine engineering trade required one of their principal specialities, and though a certain amount of slackening had taken place in the mercantile marine branch, that depart- ment of their business was still, in consequence of the operations of Her Majesty's Government, who had voted a large sum for the augmenta- tion of the fleet, assured for the year, and they would have a large amount of work—(applause)—and he was not unsanguine as to the amount of work which the mercantile marine would supply them with. He recognised the efficient services of the staff and workmen, and moved that the report should be adopted, and a sum of £1,500. appro- priated as remuneration for the services of the directors during 1889. Mr. Samson Fox, who was received with applause, seconded the motion, and, alluding to the erection of a special shed, covering 11,000 square yards, for the manufacture of Fox's patent pressed steel frames for rolling stock, he said that the death rate of the rolling stock, on the railways throughout the world was such as demanded the replacement, every four minutes, night and day, of at least one wagon or carriage, and they decided to put down a shed which would, equipped to the full extent, enable them to turn out a waggon at least every ten minutes. They took no unreasonable view. They were working orders for twelve different railway companies, and if they could get work from so many in so short a time there was no doubt that if they did their part more orders would come. The best railway authori- ties were agreed they had once more "struck oil," and got to the root of a great question. (Applause) The resolution was adopted.—On the motion of Mr. G. W. Hastings, M P., seconded by Mr. Joseph Craven, Mr. A S. Kirk was re-elected a director.—On the motion of Mr. Redmayne, seconded by Mr. William Fox, Mr. John Gordon, jun., was re-appointed auditor, with a remuneration of 100 guineas for the year.—In reply to Mr. Leah, the Chairman said the water-gas had been used at the works for nearly two years, that it had worked with most admirable regularity, and had effected a large saving in fuel and labour. He believed that before the next meeting the water-gas plant would be doubled, if not trebled.—A vote of thanks to the Chairman and directors was passed, and the meeting terminated.

MANUFACTURING WHITE LEAD BY ELECTRICITY.

Many attempts have been made to substitute electrolytic methods for the purely chemical ones in the manufacture of a variety of products, but thus far few have come into general use, due perhaps to lack of perseverance rather than to the impossibility of achieving the desired results. It is, therefore, with considerable satisfaction that we are able to bring to the attention of our readers a process for the electrolytic production of white lead, recently patented and put in practical operation by Mr. Turner D. Bottome, of Hoosick, N.Y., and described in a recent issue of an American contemporary.

The process devised by Mr. Bottome consists in electrolytically dis- solving a lead electrode in an electrolyte containing nascent or free carbon dioxide, whereby the lead compound formed by electrolytic action is precipitated to form hydrated carbonate of lead, or pure white lead, which is then removed, washed and dried.

The manner in which this is accomplished is as follows:—The electrolytic solution is prepared by dissolving in the proportion one half-pound each of sodium nitrate and ammonium nitrate to one gallon of water, and then saturating the solution thus formed with carbon dioxide, which can be done in various ways. Sodium carbonate and ammonium carbonate may be used in the place of the nitrates; but in that case nitric acid must be added until the bath is about neutral, which results in the larger portion of the carbon dioxide being driven off during effervescence. The electrolytic solution is then placed in a tank and electrodes of metallic lead are immersed in the same. The electrodes are then connected to the generating dynamo, and a current density of about 15 ampères per square foot of anode surface is main- tained. Upon the passage of such a current between the electrodes through the bath the white lead begins to fall very rapidly. As the carbon dioxide is taken up from the bath to form the hydrated carbonate of lead, it is, of course, necessary to have the bath replenished with additional carbon dioxide as the process continues. This can be done in several ways. A convenient way in doing this consists in burning limestone, washing the gas produced by the disassociation of the constituents of the limestone, and supplying the gas directly to the bath.

Where the electrolytic solution is made up of sodium and ammonium nitrates in water, and the solution is saturated with carbon dioxide in its free state, the reaction taking place upon the passage of the electric current, according to the inventor, are as follows:—The bath is decomposed, yielding at the anode nitrogen pentoxide, ozone and oxygen, and at the cathode sodium hydrate, ammonia and hydrogen. The lead is attacked by the powerfully-oxidising nitrogen pentoxide (N_2O_5) and ozone; but nitrogen pentoxide in the presence of water is decomposed and forms nitric acid (HNO_3). During the double decomposition which takes place nitric acid (HNO_3) and plumbic acid or hydroxide of lead, $PbO(OH)_2$, are formed. The nitric acid combines with the free ammonia and sodium hydrate to again form sodium and ammonium nitrate, while the plumbic acid is precipitated by the free carbon dioxide.

The first question raised on any invention in electro-chemistry is this: "Is it not cheaper to do the same thing by mechanical or chemical means?" But upon looking into this matter closely, one finds that the mere cost of electricity for separating or dissolving metals is comparatively slight. Mr. Bottome's calculations on this point are as follows:—

Lead is dissolved at the rate of 59·52 grains per ampère hour, and if the dissolved lead is caused to combine with some element before deposition takes place it will add to its own weight that of the element. Thus lead in combining with water and carbon dioxide gains 19·88 per cent., or nearly 20 per cent. Hence one ampère hour forms 71·35 grains of pure white lead. The counter E. M. F. of the white lead bath is from 20-100 to ¼ volt, so that the ¼ volt is required to operate one tank; one watt hour will then deposit or form 142·7 grains of white lead, or 15·2 pounds per horse power hour, or 152 pounds a day per horse power, or 27½ tons per year. Figuring power at 55 dols. a year, 50 cents. a ton for the electric current would cover the making of white lead. The cost of the water entering in combination is nil, and that of the carbon dioxide also, because 575 pounds of lime rock evolve 230 pounds of the gas at a cost of 100 pounds of hard coal, leaving a by-product of 345 pounds of quicklime, which more than pays for the coal; while the 230 pounds of gas just make one ton of white lead, 170 pounds of water also entering into combination with 1,600 pounds of lead to form 2,000 pounds, or one ton.

The inventor has favoured us with a specimen of this interesting product which is in an extremely fine state of division, the result of a simple drying by spontaneous evaporation, in marked contrast to the white lead obtained by the old acetic acid process, which has to be ground mechanically before it can be used.—*Invention.*

THE MANURIAL EFFECT OF NITRATE OF SODA COMPARED WITH THAT OF SULPHATE OF AMMONIA.

THE relative values of two of the most important fertilisers, nitrate of soda and sulphate of ammonia, have often been discussed in connection with experiments which have been made on the growth of grain. Sulphate of ammonia is always a higher price per ton than nitrate of soda, but the nitrate contains a smaller percentage of nitrogen. It does not, however, follow that the results obtained are in proportion to this percentage. There are many circumstances which have to be considered. Nitrate of soda produces more straw than sulphate of ammonia; hence, when straw is a saleable article, this point is to be considered by the farmer in the selection of his manure.

According to Mr. Warrington the nitrate apparently liberates the ash constituents of the soil, rendering them available to the growing crop to a far greater extent than is accomplished by ammonia salts. Again, nitrate of soda is more suitable for use upon the land in a very dry than in a wet season, for the simple reason that it is extremely soluble and easily carried through the soil by rain. The wheat experiment at Woburn showed this most distinctly in the year 1882, which was a wet year; the nitrate producing 26 bushels and the ammonia salts 32 bushels per acre. In the dry season of 1887 the figures were reversed, for the nitrate produced 35 bushels and the ammonia 26. A similar disproportion occurred in the wheat which was manured by the two manures, with the addition of ash constituents and also when they received double quantities of each fertiliser. Therefore, although it is important to gauge the nature of the season, the farmer will use his judgment as far as he can, and employ nitrates in the districts where the rainfall is low, and sulphate of ammonia where it is high. It has been recommended that the sulphate should be ploughed in with the crop after broadcasting, but that the nitrate should be broadcasted, preferably in two or even three dressings. The
sulphate must not be mixed either with ashes or with basic slag,
wise there may be a loss of ammonia; but it may be mixed w
earth, or even with superphosphate or gypsum. During 18 ye
periments with barley Sir John Lawes obtained an increased p
of grain on the land dressed with sulphate of ammonia to the e
86·8 per cent. as compared with what was obtained upon land
with nitrate of soda. The Royal experiments at Woburn pr
similar results, but there was a great difference in the yield of
The experiments with wheat resulted in a somewhat simila
ner, the sulphate of ammonia being equally effective wi
nitrate. With wheat the grain yielded by the use of sulpl
ammonia was equal to 74·8 per cent. of that yielded by the
quantity of nitrogen in the form of nitrate; but at the Woburn
the percentage of grain yielded was 95·7. Thus, while nitrate w
most economical at Rothamsted, the sulphate answered best
burn. When used upon grass it was found that nitrate was far s
to ammonia salts, except in those instances where phosphatic m
and potash were added, when there was little difference betwe
two fertilisers. The same may be said with regard to the use of t
manures for potato growing. Nitrate was much more satisfactor
sulphate when both were used alone, but when potash and phos
were added to both the difference was but trifling. When use
mangold, however, nitrate of soda is infinitely superior to amm
whatever connection it is given. It is now generally recomm
bearing these experiments in mind, that, whether with grain
potatoes, or grass, phosphates should at least be added, and with
and potatoes in particular potash also. It does not appear, ho
to be so necessary to use the mineral manures with nitrate as wi
phate of ammonia, for, as previously remarked, the nitrate app
have greater influence in making the natural mineral fertilisers
soil available for the use of plant life.

THE POLLUTION OF THE RIVERS MERS AND IRWELL.

A LETTER from the Local Government Board was laid befo
Council in which information was asked for respectin
memorial presented by the Council for the formation of a joint
mittee to enforce the Rivers Pollution Act, 1876, with respect
rivers Mersey and Irwell, and on the motion of the chai
seconded by Mr. T. W. Killick, it was resolved to send the foll
reply:—

"SIR,—I am now directed by the County Council for the cou
Chester to reply to your letter of the 30th November last,
subject of their application for a provisional order to be ma
constitute a joint committee, representing the several ad
trative counties on the watersheds of the rivers Irwell and M
for enforcing the provisions of the Rivers Pollution Preve
Act, 1876, and with reference to the several points in your a
mentioned letter on which you ask for information I beg to st
follows:—1. The grounds on which the application is made are th
rivers Irwell and Mersey and their tributaries, in consequence of
pollution by sewage and the refuse from manufactories and
places, are offensive both to smell and to sight, and are often inj
to health. Within two years it is expected that the Manc
Ship Canal will be filled with water from these rivers and
tributaries. The Ship Canal will not be a continuous st
but a succession of ponds with a stream running through them.
danger to health arising from those rivers will be undoubted
creased. The Cheshire County Council are concerned directly
the state of the waters of the Manchester Ship Canal, as it
through or by a large number of townships in the county of Ch
and also indirectly, as the communication between Mancheste
Salford and Cheshire is so great and continuous that the outbre
any epidemic in those cities would certainly spread through
portions of the county of Chester. There are more than 70
authorities in the watersheds of the Mersey and Irwell, and only
one-half of this number have completed any arrangements for the
struction of sewage works. From the above statement it is evic
necessary that further steps should be taken to enforce the I
Pollution Prevention Act, 1876, and it appears to the Council
this will best be done by a joint committee appointed under
section 3 of section 14 of the Local Government Act, 1888.
appointment of such a joint committee would enable local authc
to act together and in unison, and in many cases by the substituti
one large body in the place of two or three smaller ones money
be saved, and in other cases friction would be avoided, as it is u
stood several authorities have complained of want of common a
The advantage of a joint committee would also, it is conceived, be
great in dealing with the pollution of streams by manufacturers. Whil
desirable that the waters of the Mersey and Irwell shall be pure
not desirable that the industries which have grown up on their

should be damaged or destroyed, and the employment of labour be thereby curtailed or stopped. It is also most undesirable that the manufacturers within the area of one Authority should be dealt with differently than those within another, which, if each Authority acts separately, may result from local favour or local jealousy. It appears to the Cheshire County Council that a joint committee would be much more independent, and would treat all dwellers and manufacturers equally and justly, and would so act that the industries of this wide district should not be injured nor the real prosperity of its inhabitants prejudiced.

2. The map sent herewith shows precisely the limits of the area within which it is proposed that the Joint Committee shall have jurisdiction. The area covers the whole of the water-sheds of the Irwell and Mersey, until the tide is reached at Warrington.

3. A list of the County Councils affected :—Counties of Lancaster, Cheshire, and Derby. County boroughs of Bolton, Bury, Rochdale, Oldham, Salford, Manchester, and Stockport.—I am, sir, your obedient servant, THOMAS ROBERTS, Deputy Clerk to the Council.

The Secretary, Local Government Board, Whitehall, London, S.W.

THE BLAST FURNACES OF THE UNITED STATES.

The following table, showing the number of furnaces in blast and their capacities on the first of each month since January, 1886, may be of interest to our readers :—

Date.	Charcoal.		Anthracite.		Bituminous.	
	No.	Weekly Capacity.	No.	Weekly Capacity.	No.	Weekly Capacity.
1886.						
January 1	57	7,804	104	29,811	112	59,436
February 1	52	7,403	107	30,100	109	56,832
March 1	46	6,515	104	30,115	105	55,436
April 1	44	6,766	117	33,280	118	66,277
May 1	47	8,462	122	37,767	128	68,079
June 1	54	9,884	123	35,136	131	75,756
July 1	59	10,420	119	33,225	132	78,106
August 1	63	10,980	119	33,602	131	77,750
September 1	66	11,105	121	34,091	132	76,005
October 1	59	11,271	118	33,476	135	77,460
November 1	62	10,565	112	32,761	138	79,270
December 1	67	11,818	115	32,634	141	78,127 405
1887.						
January 1	66	11,895	126	35,633	140	80,132
February 1	68	12,235	137	38,099	148	85,921
March 1	61	11,572	142	39,767	146	83,834
April 1	59	11,337	143	39,477	151	86,709
May 1	54	10,819	143	40,873	149	86,822
June 1	62	11,809	145	41,288	104	54,767
July 1	77	13,969	136	37,662	101	57,355
August 1	80	14,396	127	35,278	120	70,855
September 1	79	13,900	130	36,872	143	87,953
October 1	73	15,171	122	36,044	151	93,423
November 1	80	14,145	125	36,720	148	94,982
December 1	73	13,104	120	35,361	148	93,295
1888.						
January 1	73	13,237	117	35,259	151	92,224
February 1	70	13,742	103	29,689	140	83,725
March 1	62	11,713	102	29,066	129	76,800
April 1	60	12,393	95	27,971	130	75,983
May 1	60	11,956	104	30,366	133	80,230
June 1	68	13,116	101	29,859	129	80,040
July 1	68	12,753	96	28,176	119	74,743
August 1	69	13,248	92	27,846	119	77,408
September 1	68	12,623	98	28,946	132	84,513
October 1	73	12,983	99	29,586	138	87,141
November 1	75	14,005	97	28,412	141	88,273
December 1	73	13,270	107	31,052	151	94,960
1889.						
January 1	71	13,213	108	31,837	154	97,117
February 1	66	12,291	104	30,772	146	91,752
March 1	59	11,568	100	33,383	152	97,783
April 1	63	12,150	100	33,267	149	98,835
May 1	60	11,951	103	35,078	138	91,239
June 1	66	11,577	97	34,514	129	83,591
July 1	63	11,204	90	31,848	131	89,356
August 1	66	11,712	88	32,118	132	95,595
September 1	69	11,769	94	35,497	137	92,915
October 1	66	12,672	93	36,496	149	104,378
November 1	70	13,257	98	38,181	161	111,279
December 1	69	13,226	106	40,228	162	113,501
1890.						
January 1	66	12,963	111	41,964	168	120,345
February 1	61	12,653	112	45,681	166	116,798

EXTRACTS FROM THE BOARD OF [illegible] JOURNAL.

THE MANUFACTURE OF MARGARINE IN [illegible]

A DESPATCH, dated the 7th January, has been [illegible] Foreign Office from Sir F. R. Plunkett, Her Majesty [illegible] at Stockholm, transmitting translation of a [illegible] Royal Medical Board concerning the special [illegible] be adopted with regard to the manufacture of margarine. [illegible] ing is a copy of the translation in question :—

By the fourth paragraph of the law of October 11th, 1889, [illegible] vision of the manufacture and sale of margarine, except whe[re] merely for the use of a single household, is intrusted [illegible] appointed by the Governor of the province, whose duty i[s to] see that the special directions of the Medical Board are c[arried out] and that none but perfectly pure and harmless materials are [used in the] preparation of the margarine. The special instructions [illegible] Medical Board are as follows :—

The inspector should visit the manufactory at [illegible] unknown hours, at least on four days in every week. He [illegible] on the greatest possible cleanliness being observed both in th[e] [illegible] of manufacture and in regard to the premises, vessels, [illegible]

With regard to the premises, he should make sure that [illegible] under and surrounding the building is kept clean and free fr[om] [illegible] scourings, and refuse, and that the rooms of the manufactory [illegible] clean as possible, floors, walls, and ceilings, and that they b[e] ventilated ; that all receptacles and implements, such as [illegible] cleansing the tallow, tallow cutting machines, tallow [illegible] presses, pressing cloths, churns, basins, kneading machines, [illegible] cases, etc., etc., be kept clean. He should see that no [illegible] implements that could convey any metallic poison to the [illegible] used.

As regards the workpeople, he should see that every [illegible] ployed be free from contagious diseases, sores, eruptions, [illegible] boils, and similar affections ; that they be very particular [illegible] cleanliness of their bodies, clothes, and belongings, espe[cially] forearms, hands, fingers, and nails ; that they do not take [illegible] smoke, or chew tobacco while at work. That while at work [illegible] a special blouse of thin material, that can easily be washed, [illegible] must not be worn out of the factory ; that a sufficient [illegible] stands, supplied with hot and cold water, soap, nail-brushes, [illegible] towels, be always at hand ; also that antiseptic bandages an[d] [illegible] be kept ready in case of injury.

He must see that any workpeople who may have been [illegible] with people attacked by the following diseases, viz., ty[phus] typhoid fever, scarlet fever, diphtheria, cholera, small-pox, [illegible] or have been in houses where such diseases exist, must [illegible] bodies and clothes thoroughly cleansed and disinfected be[fore being] permitted to enter the factory or assist in the making of [illegible]

As to the materials used, he must see that all be of goo[d] [illegible] that both the tallow and oleomargarine, and the vegetabl[e] [illegible] other fats used, be in appearance, smell, and taste, clean [illegible] that the milk added when churning be not taken from pl[aces where] there are any contagious diseases ; that the colouring matter [illegible] be of good quality, and that no poisonous colouring matter [illegible]

As to the manufactured article, he must see that [illegible] margarine be of proper appearance, good taste, and fre[e] [illegible] agreeable smell ; that it be packed cleanly and neatly, and [illegible] vessels formed and marked as appointed by the Law of Oct[ober] [illegible] 1889.

Should the inspector detect any irregularity in the manuf[acture] [illegible] should report the matter to the Board of Health, or [illegible] commune, and also to the Governor of the province.

PRODUCTION OF SALT IN GERMANY.

The French Ambassador at Berlin says, in a recent [illegible] statement has recently been issued by the German Government [illegible] subject of the production of salt works in Germany.

From this statement it appears that in the year 1888-89, [illegible] produced in the Empire 15,934 tons of crystallised salt, [illegible] rock-salt not crystallised, and 516,521 tons of salt produced [illegible] tion, the two last qualities showing an increase on the [illegible] last 10 years. Crystallised salt comes from the Prussian [illegible] Saxony, rock-salt from Wurtemburg, from the province [illegible] from Anhalt, and from Posen. The salt produced by [illegible] from the provinces of Saxony, Hanover, Alsace-Lorr[aine] [illegible] Thuringia.

In the course of the year 1888-89, 734,623 tons of native [illegible] been disposed of for the home trade, of which [illegible] tons for food have paid the tax of 12 marks per 100 kilos. and [illegible] utilised by industry and agriculture, have been exempt [illegible]

Exports reached 134,171 tons, distributed as follows :— [illegible] for Bremen, Hamburg, and the other parts of the empire [illegible] in the Zollverein ; 24,567 tons to Austria-Hungary, 19,638 tons to Belgium, 19,107 tons to Holland, 10,514 tons to Russia, 6,534 tons to Denmark, 6,504 tons to Sweden and Norway, 6,444 tons to the British Indies, 5,100 tons to England.

Exported salt does not pay any duty.

Imports were 28,057 tons, of which 25,265 tons were of origin.

Foreign salt introduced through the ports pay an import duty equal to the tax. That imported by land routes pays 12·80 marks per 100 kilos.

The average consumption is 7·8 kilos. per inhabitant in the whole Zollverein.

THE INFLUX OF FOREIGN CAPITAL INTO AMERICA.

ABOUT eighteen months ago, investments of foreign capital, particularly that of British capital, began to be made in American commercial enterprises.

This investment was suggested and stimulated by the organisation of a corporation which "syndicated" the sale of the Sir Edwin Guinness ale and stout breweries, of Scotland. The surplus capital of the world's metropolis having turned such a quick penny in this speculation, immediately sought for a new field, and naturally attention was directed to this country. A thorough system of exploitation began. London and other large English municipalities have a body of highly-trained and skilful men called "chartered accountants," whose reports are accepted with implicit faith by all financial institutions in England. The task which these men had to perform was to verify the book accounts of the various institutions, take an inventory of the plants examined, and substantiate statements made in regard to business. The work of investing in American enterprises in a short time became a regular organised business. A large number of properties have been examined by these chartered accountants, and negotiated through the brokers of various syndicates. It is not one but many different organised financial associations which have taken up these properties and "floated" or "placed" them.

The result, the editor of the *Architecture and Building* believes, will be highly beneficial to the industrial interests of the country, provided the vendors keep good faith with the vendees. This provision is necessary, because the system pursued generally is to buy any given property outright, and retain some one who has been prominently connected with the business to superintend its continuance under a salary. Thus the business is conducted by experienced hands. The idea involved in the purchase on the part of the vendee is that for himself and his associates he can command a business which will pay as an investment much more than the paltry two or three per cent. which the same money will command at home. The investments taken are generally sound, substantial, thoroughly established business ventures, and the direct consequence of this influx of foreign capital, so far, has been highly beneficial. The ultimate tendency, we believe, will be to secure further investment in enterprises that need additional capital to promote further progress. We do not believe in the alleged statements of the opponents of these investments, that the ultimate outcome will be a drain on our resources to enrich a foreign bond-holder, for whenever that point is reached that his interests are greater in this country than in the land of his nativity, it has generally resulted in the moving to that country of the foreign bondholder. It is a monstrous and pernicious doctrine to uphold that the foreign bondholder has been a drain upon this country's resources. On the contrary, we owe to him the means by which our resources have been brought to their present great development, and his coming should be encouraged rather than impeded.—*Scientific American.*

THREE GREAT BOILERS.—The Polson Iron Works Company, of Toronto, 28th ult., shipped to Owen Sound the last of four large boilers constructed by them for the car ferry they are now building at their shipyard there for the Canadian Pacific Railway Company. These boilers are the largest ever made in Canada, and also the largest ever carried by rail on this continent. They are of the cylindrical return multitubular type, and are 13 feet 3 inches in diameter and 14 feet long, weighing 37 tons each. The shell-plates are 11-16 of an inch in thickness, and were specially rolled in Scotland. The tubes are of German manufacture, and are 4 inches in diameter, 11 feet long, and 148 in number. There are in each boiler three of Fox's corrugated furnaces, 42 inches in diameter and 10 feet 11 inches long. The Government test showed an allowance of 94 pounds working pressure. The riveting of these boilers was done by a Tweddell hydraulic riveter, with a gap of 8 feet 4 inches, lately erected in the company's shops. The boilers when completed were lifted bodily on to the cars by a large overhead travelling crane, which has a lifting capacity of 50 tons.—*American Manufacturer.*

THE "TIPPING" SYSTEM IN TRADE.

IT is characteristic of the Corporation spirit that the Recorder Central Criminal Court should be found defending from the the "tipping" system in trade, although the Legislature has m penal to bribe and corrupt the servants of public bodies. The qu of gratuities has two sides, remarked Sir Thomas Chambers, b side which has impressed itself upon Parliament, and every on who has inquired into the subject and noted its effects, is the ab demoralisation to which it leads. If the system is so innocen the gratuity merely a reward for honest services rendered, why i so much secrecy, or why should such mystifying methods of keeping be adopted? The occasion which gave the Record opportunity of uttering his sentiments on the subject was the tr the Old Bailey of two men employed by the Bell's Asbestos Com on the charge of obtaining money by false pretences. The d was that the money had been placed at their disposal for the purp tipping, and had been so expended; and the jury, apparently ad this view, brought in a verdict of "Not guilty."

The evidence in this case has amply confirmed information w long had in our possession as to the wholesale system of ti pursued by some firms. In many instances the travellers are cons instructed to secure this or that man in particular works "and him well." From the manager to the engineer, from the high the lowest, everyone who can in any way influence an order receive a "tip" in amount and form most likely to be acceptable recipient. The heads of this particular firm do not attempt to di the fact that their business is conducted on this principle; indee admit that they "palmed" right and left. Here is a specimen letters sent from the firm to their representatives, taken fro evidence produced in court in the Bell case:—

"When the grouse season comes on we are thinking of sen brace of birds to all those men who are in a position to influence between now and Christmas. This seasonable gift would, of c only be sent to men of the better class; and we think it would amiss to include an employer here and there in our list of reci The grouse would be sent out about the latter end of Augus we propose sending you the labels, to be addressed by yourself t of the men in your district as would be likely to show their appre of the compliment in the manner desired. Please let us kno many labels you will want; the managers of sewage, gas, water and such concerns, also as surveyors of local boards."

The excuse is made that without these tips goods do not g play. But what chance of fair play would a firm have who d adopt the same tactics? Although pursued so recklessly this ping" is not done aboveboard. The man who receives thes siderations does not look upon them as a clerk might do who present from his employer, but regards them as bribes, and is c to maintain secrecy with respect to their receipt. In the case of bodies it is now illegal to give or receive bribes; in general tr may still be done so far as the law is concerned; but ev Recorder will not contend that there is any moral, although ther be a legal distinction. It is beyond all question that the ex system of tipping which is now carried on leads to wholesale de lisation and corruption, and is undermining all honest trade.—*Mall Gazette.*

EDINBURGH WATER SUPPLY.—The Works Committee of burgh Water Trust have issued a report giving the results of an i instituted in the summer of last year upon the question of add water supply to the city. They visited the Falla, Manor, a Mary's Loch waters in Peeblesshire. The report states that th sumption of water by Edinburgh and district has since 1879 in from 7,000,000 gallons per day to 15,000,000 gallons, leaving margin of 1,000,000 gallons to meet the increasing demand. T of this scheme, as estimated by the engineers, would be £664,0 £42,266. per million gallons per day. The Committee express ment with the engineers' opinions, and as seven or eight year elapse before the new supply is available, it would not, in thei be prudent for the trustees to delay much longer. The cost Falla scheme would be £97,000. for 24 million gallons a day. first instalment as regards St. Mary's Loch would be £681,0 the total £1,065,000. For the Manor scheme it is estimated t progressive surplus is sufficient to meet the increased expendi 1896-7, when an additional 1d. per £1. on the rate would be made. The Tweed scheme would create a larger defic 1896-7, and would require the raising of the rate for 1897-8 by per £1. Under the St. Mary's Loch scheme the first de would occur in 1894-5, which could be met by an increase of £1., rising in 1895-6 to 2d. per £1., and in 1896-7 and 1897-8 t per £1.

Correspondence.

ANSWERS TO CORRESPONDENTS.

P.H —True; but the increased demand for electrical purposes is absorbing enormous quantities.

J A.W.—See our article on the Porion Evaporation in this Journal of November 19th, 1888. It contains all the information you require.

J.—You seem sceptical, but we shall convince you.

W.W.—Yes. Water such as you describe can be easily dealt with.

G.Y.—From your description it appears to us that your friends have been sending you Bone Ash, instead of Bone Dust.

C.R.—By all means. Come and see it at any time.

*** *We do not hold ourselves responsible for the opinions expressed by our correspondents.*

BEESWAX.

To the Editor of the Chemical Trade Journal.

SIR,—Some time ago I had to test a sample of beeswax for a large calico printing firm, which gave bad results when made into paint, along with other materials for protecting the parts intended to be white, in the engraving of copper rollers. The sample contained only :—

8·2% Cerotic acid,

53·7% Cericin,

and 35% of hydrocarbon wax was isolated from it.

As this may possibly occur at other places, I thought it would interest the readers of the *Chemical Trade Journal*.—Faithfully yours,

Crossfield Terrace, J. ARTHUR WILSON.
Tottington, near Bury.

PRIZE COMPETITION.

A LITTLE more than twelve months ago, one of our friends suggested that it might be well to invite articles for competition, on some subject of special interest to our readers; giving a prize for the best contribution, he at the same time offering to defray the cost of the first prize, and suggesting a subject in which he was personally interested.

At that date we were totally unprepared for such a suggestion, but on thinking over the matter at intervals of leisure, we have come to think that the elaboration of such a scheme would not only prove of benefit to the manufacturer, but would go some way towards defraying the holiday expenses of the prize winner.

We are therefore prepared to receive competitions in the following subjects :—

SUBJECT I.

The best method of pumping or otherwise lifting or forcing warm aqueous hydrochloric acid of 30 deg. Tw.

SUBJECT II.

The best method of separating or determining the relative quantities of tin and antimony when present together in commercial samples.

On each subject, the following prizes are offered :—

One first prize of five pounds, one second prize of one pound, five additional prizes to those next in order of merit, consisting of a free copy of the *Chemical Trade Journal* for twelve months.

The competition is open to all nationalities residing in any part of the world. Essays must reach us on or before April 30th, for Subject II, and on or before May 31st, in Subject I. The prizes will be announced in our issue of June 28th.

We reserve to ourselves the right of publishing the contributions of any competitor.

Essays may be written in English, German, French, Spanish or Italian.

A COMPANY has been organised to utilise the power of Niagara River above the falls, not of the falls, by means of a main tunnel about two miles and a half long with certain cross tunnels, and the New York *Evening Post* says that money is already subscribed, and that one of the best-known banking houses in Wall-street is deeply interested in this undertaking. Work, it is said, will begin within a month.

THE RAILWAY RATES [illegible]

CHEMICAL MANURES.

IN evidence given at a recent sitting by Mr. G[illegible]
to the rates proposed for chemical manures, [illegible]
representative of the Berks and Oxon Chamber of C[illegible]
of 170 of the leading agriculturists and traders attending [illegible]
Reading markets. Respecting the classification, of [illegible]
original Acts of the North-Eastern, Midland, South-Western[illegible]
South-Eastern, London, Chatham, and Dover, and other [illegible]
panies, all kinds of manures were placed in the lowest class. [illegible]
now seeking to put artificial manures and chemical [illegible]
same class as grain. The manure rate on the London a[illegible]
Western Railway Company, under their old Parliamentary [illegible]
fifty miles is 5s. 2d. Their grain rate is 10s. 9d. In j[illegible]
railway companies say, first, that artificial manures w[illegible]
unknown at the time they obtained their Parliamentary [illegible]
secondly, that these artificial manures are very valuable; th[illegible]
they are in bags; and, fourthly, that in drafting their [illegible]
fication they followed the Clearing House classification, [illegible]
rates that are in force by the Clearing House classification [illegible]
been objected to hitherto, they have a right to use them [illegible]
regard to the first point, I may say that the period wh[illegible]
railway companies obtained their powers was from 18[illegible]
and during those eleven years there was an import[illegible]
1,570,000 tons of artificial manures and guano. That [illegible]
to manure at the rate of 2 cwt. to the acre, 2,000,[illegible]
every year. Is it possible that Parliament could not [illegible]
that this manure was being used? The guano of th[illegible]
was very different to the guano of these days. It contai[illegible]
per cent. of ammonia. There are about twenty artificial [illegible]
classified. Only one of these touches the figure of £10. a to[illegible]
was quoted by Mr. Findlay. Some are as low as 30s. per to[illegible]
ficial manures are almost undamageable. I have paid for[illegible]
charges, directly and indirectly, on a thousand tons since I h[illegible]
in business, and I have never had occasion to make a claim [illegible]
railway company for damage. The fact of artificial manu[illegible]
packed is an advantage to the railway company, because th[illegible]
are kept cleaner and can be loaded and unloaded with greater [illegible]
As a practical agriculturist, I say there is no limit to the fe[illegible]
the soil with a proper application of manure; the only differ[illegible]
tween our soil and that of America is that the latter pos[illegible]
ammonia, phosphates, and alkalis abundantly, whereas we h[illegible]
at all; and, therefore, it is a matter of life and death to the f[illegible]
be able to procure such commodities cheaply.

Cross-examined by Sir H. James, Q.C.: A ton of farmyard [illegible]
is worth to the sender about 6d. a ton, but a ton of sul[illegible]
ammonia is worth to-day about £11. 5s. or £11. 10s. per [illegible]
think all manures ought to be carried at the same rate.

Sir H. James: Although if farmyard manure were injured [illegible]
the railway company would only have to pay a few shillings, [illegible]
if the sulphate of ammonia were injured they would have to p[illegible]

Mr. Baylis: I never heard of ammonia being injured in [illegible]
have never had it injured by wet, nor have I known of its [illegible]
injured in railway-trucks.

Sir H. James: Would chemical manure suffer from being w[illegible]

Mr. Baylis: I scarcely think it would suffer sufficiently fo[illegible]
to claim damage. The wet would run off. It would never [illegible]
through. It is bagged in proper bags.

At the sitting of the inquiry on Wednesday, February [illegible]
course of his examination as to artificial manures, Sir John La[illegible]
that high rates on these for carriage were a burden upon agric[illegible]
whose produce now fetched a comparatively low price.

Mr. Foster, of the Anglo-Continental Guano Works, sa[illegible]
were about 1,000,000 tons of manure used in the country du[illegible]
year, mainly distributed in small parcels. The importation o[illegible]
which reached at one time 1,000,000 tons per annum, has wo[illegible]
off, and the artificial manures had risen. On the railways [illegible]
services were discharged by consignors and consignees, and [illegible]
sidered it unfair that a charge should be made by the comp[illegible]
terminal services which were not rendered by them.

The witness admitted, in cross-examination, that [illegible]
needed sheeting, and that that was a service.

Mr. Edward Packard, of Bramford, also gave evidence [illegible]
displacement of guano by artificial manures, and that fair [illegible]
these were of immense importance to agriculturists. He co[illegible]
the proposal to put the large and small station expenses toget[illegible]
charge them upon the traders as unfair. He would not object [illegible]
his manure charged as farmyard manure.

Mr. Vickers, manure manufacturer, of Manchester, Wi[illegible]
Sandbach, complained that the rates already charged by the [illegible]
and North-Western Railway were above their statutory [illegible]
objected to the proposed rates, as they would nearly double the [illegible]
charges.

PAINT PRESERVATIVES FOR IRON.

TOO much stress cannot be laid upon the condition of the surface of the iron at the time of coating; and it is perfectly essential either to have a dry surface or else a composition which is not affected by water. Prof. Lewis remarks that when an old iron structure is broken up, on the backs of the plates may often be seen the numbers painted on them in white lead and linseed oil when the work was put together, and under the paint the iron is in a perfect state of preservation, the secret being that the paint was put on while the plates were hot and dry.

Compounds prepared with boiled linseed oil are open to objection, on account of the presence of lead. The drying of boiled linseed oil is due to the fact of its containing a certain quantity of an organic compound of lead; and the drying property is, moreover, imparted by boiling it with litharge (oxide of lead), so that lead compounds are present even when the oil is not mixed with red or white lead pigment. When boiled oil dries, it does so by absorbing oxygen from the air, and becomes converted into a kind of resin, the acid properties of which also have a bad effect upon iron. Protectives of the class of tar and its derivatives, such as pitch and black varnish, and also asphalt and mineral waxes, are regarded by Professor Lewes as among the best. Certain precautions, however, must be taken in the case of tar and tar products, both of which are liable to contain small quantities of acid and ammonia salts. If care is taken to eliminate these, and if it could be contrived to always apply this class of protectives hot to warm iron, the question of protection would be practically solved; bituminous and asphaltic substances forming an enamel on the surface of iron which is free from the objections to be raised against all other protectives—that is, of being microscopically porous and therefore pervious to water. Spirit or naphtha varnishes are condemned by Professor Lewes as open to several objections. Varnishes to which a body has been given by some pigment, generally a metallic oxide, are preferable to the last class, "if the solvent used is not too rapid in its evaporation, and if care has been taken to select substances which do not themselves act injuriously upon iron, or upon the gums or resins which are to bind them together, and are also free from any impurities which could do it."

At the present time, as the author truly remarks, the favourite substance for this purpose is the red oxide of iron; but care should be taken to exclude from it free sulphuric acid and soluble sulphates, which are common impurities and extremely injurious. The finest coloured oxides are, as a rule, the worst offenders in this respect, as they are made by heating green vitriol (sulphate of iron), and in most cases the whole of the sulphuric acid is not driven off, the heat required being injurious to the colour. The acid is often neutralised by washing the oxide with dilute soda solution; but very little trouble, as a rule, is taken to wash it free from the resulting sulphate of soda, which is left in the oxide. The best form of oxide of iron to use for paint making is obtained by calcining a good specimen of hematite iron ore at a high temperature. When prepared in this way it contains no sulphates, but a proportion of clay which is harmless if it does not exceed 12 to 18%. Paint makers can easily test their red oxide for soluble sulphates by warming a little of it with pure water, filtering, and adding to the clear solution a few drops of pure hydrochloric acid and a little chloride of barium solution. If a white sediment forms in the solution, the sample should be at once rejected.

In the application of a preservative coating to iron, Prof. Lewes directs, first, thorough scraping and scrubbing from all non-adherent old paint and rust. New iron should be pickled with dilute acid to get rid of every trace of mill scale; the acid to be neutralised afterward by a slightly alkaline wash, and this again to be washed off by clean water. Under these conditions, and given a composition of good adhering properties, but little apprehension need be felt with regard to the ravages of corrosion, the chief remaining risks being from abrasion or other mechanical injury to the composition, coupled with improper constituents in itself.—*Scientific American.*

A GREAT STEEL PLANT.—The *Iron Age* prints the following report of the business of the Illinois Steel Company in 1889:—

Capital issued	17,622,600 dols.
Number of employés	9,247
Total wages paid	4,577,000 dols.
Value of product	19,000,000
Pig metal produced, tons	572,095
Rails, tons	461,147
Wire, tons	43,488
Merchant bar and nails, tons	60,230
Billets, tons	50,289
Spiegel and ferro-manganese, tons	18,031
Beams and slabs, tons	4,030

The works consumed 775,000 tons of ore, 575,000 tons of coke 140,000 tons of coal, and 200,000 tons of limestone.

Trade Notes.

THE IMPORTATION OF FOREIGN CATTLE INTO DUNDEE.—meeting of a sub-committee of the Dundee Harbour Board, the bour engineer submitted a plan prepared by him for providing ac modation for the landing, storage, and sale of foreign cattle at Di Harbour. After consideration, it was agreed to recommend tha board approve of the scheme, and instruct Mr. Cunningham engineer) to take in offers for building sheds and other necessary tions, and also that he proceed with the construction of the whar other requisite works. The committee expect that the sheds wi commodate about 600 head of cattle, and that there will be roor any future extension as circumstances may require.

THE ODESSA PETROLEUM MARKET.—Although the demar small, prices are steady—1 rouble 12 copecks, 1 rouble, 13 copecl pood in tank; and 1 rouble 32 copecks, 1 rouble 33 copecks in t barrels. The tank steamer Lux, which was in the dock at Tr left that place for Batoum for kerosine. The present winter was unfavourable for the petroleum trade. Usually with the adve autumn the price of kerosine advances in comparison with the sur prices, which advancement continues up to December and Januar all wholesale merchants make their purchases at that time; wher days lengthen the prices gradually go down. This winter, how the opposite occurred—the price was pretty high during the sur on account of the temporary difficulty of getting and taking ker to Batoum; with the advent of autumn it began to go down during the winter kerosine was cheaper than during the sum Therefore, the wholesale merchants did not hurry to make l purchases.

NITRATE OF SODA.—Messrs. Laird and Adamson's Liverpoo cular shows that there have been shipped from the West Co America to Europe this year 48,000 tons of nitrate, and that there loading on the 1st inst. a further 48,000. There were afloat a beginning of this month 329,000 tons (including cargoes between of call and destination, but deducting 8,000 tons lost at sea), ag 340,000 tons same date last year. The table of comparative p during past years is very interesting, as showing the extent o reduction in the cost of the article. Ten years ago on the 1st M the price was 18s. 9d to 19s., whereas now it stands about 8s. 1 as against 11s on March 1st, 1889. It appears that there are 8,000 tons in stock in Liverpool at present.

A NEW GERMAN GLASS.—A new red glass has been recentl vented in Germany, and appears to be attracting a good deal of a tion. Besides its use for the manufacture of bottles, goblets and of various kinds, it will be found applicable in photography an chemists' and opticians' laboratories. This glass is produced by ing in an open crucible the following ingredients: Fine sand, parts; red oxide of lead (minium), 400; carbonate of potash, lime, 100; phosphate of lime, 20; cream of tartar, 20; borax, red oxide of copper (protoxide), 9; and bioxide of tin, 13 parts. single melting a transparent red glass is thus obtained of a very quality, of which various objects can be manufactured directly, wi it being necessary to submit the glass to a second heating with the of intensifying the colour.

Market Reports.

REPORT ON MANURE MATERIAL.

The better feeling in the market, which we referred to in our week's report, continues, and a fair amount of business has been at somewhat better prices for phosphates, but without material ch for nitrogenous material.

We hear of further business in Charleston River phosphate roc 11d. per unit, which is now the current price, cost, freight, insurance to U.K. For land rock ½d. per unit less would prob be accepted. We have no alteration to report in values of Somme Belgian phosphates. The Florida phosphate rock is now defin offering in quantity, but shippers do not seem much inclined to ac prices which U. K. buyers are prepared to give. There is, how a disposition to try this new description of rock, and we don't t that any difference of idea as to price will stand very long in the of business. The Canadian shippers are now offering their rock at prices which so far preclude any extensive business.

A small cargo of River Plate bones, December shipment, has sold for U.K. on private terms, but understood to be at £5. 8s Less would be accepted for a large mixed cargo December shipn but pending arrival buyers seem shy of touching it. Further sal crushed Kurrachee bones for March shipment have been mad

PRICES CURRENT.

WEDNESDAY, March 5, 1890.

Prepared by Higginbottom and Co., 116, Portland Street, Manchester.

alues stated are F.O.R. *at maker's works, or at usual ports of shipment in U.K. The price in different localities may vary.*

		£	s.	d.
:—				
ic, 25 °/₀ and 40 °/₀	per cwt	7/6 & 12/6		
Glacial	,,	2	10	0
nic, S.G., 2000°	,,	0	12	9
omic 82 °/₀	nett per lb.	0	0	6½
oric	,,	0	0	3¾
iatic (Tower Salts), 30° Tw.	per bottle	0	0	8
(Cylinder), 30° Tw.	,,	0	2	11
ic 80° Tw.	per lb.	0	0	2
ous	,,	0	0	1¾
lic	nett ,,	0	0	3¼
ic	,,	0	1	4
phuric (fuming 50 °/₀)	per ton	15	10	0
,, (monohydrate)	,,	5	10	0
,, (Pyrites, 168°)	,,	3	2	6
,, (,, 150°)	,,	1	8	0
,, (free from arsenic, 140/145°)	,,	1	10	0
phurous (solution)	,,	2	15	0
nic (pure)	per lb.	0	1	0½
taric	,,	0	1	2½
ic, white powered	per ton	13	15	0
(loose lump)	,,	4	15	0
ina Sulphate (pure)	,,	5	0	0
,, (medium qualities)	,,	4	10	0
inium	per lb.	0	15	0
onia, ·880=28°	per lb.	0	0	3
, =24°	,,	0	0	1⅞
, Carbonate	nett ,,	0	0	3½
, Muriate	per ton	23	10	0
, ,, (sal-ammoniac) 1st & 2nd	per cwt.	37/-& 35/-		
, Nitrate	per ton	40	0	0
, Phosphate	per lb.	0	0	10½
, Sulphate (grey), London	per ton	11	17	6
, ,, (grey), Hull	,,	11	17	6
ne Oil (pure)	per lb.	0	0	11
ne Salt ,,	,,	0	0	9½
nony	per ton	75	0	0
, (tartar emetic)	per lb.	0	1	1
, (golden sulphide)	,,	0	0	10
m Chloride	per ton	7	10	0
, Carbonate (native)	,,	3	15	0
, Sulphate (native levigated)	,,	45/- to 75/-		
hing Powder, 35 %	,,	5	2	6
, Liquor, 7 %	,,	2	10	0
phide of Carbon	,,	12	15	0
nium Acetate (crystal	per lb.	0	0	6
um Chloride	per ton	2	0	0
Clay (at Runcorn) in bulk	,,	17/6 to 35/-		
Tar Dyes:—				
zarine (artificial) 20 %	per lb.	0	0	9
genta	,,	0	3	9
Tar Products				
thracene, 30 % A, f.o.b. London	per unit per cwt.	0	1	6
nzol, 90 % nominal	per gallon	0	3	10
, 50/90 ,,	,,	0	3	0
bolic Acid (crystallised 35°)	per lb.	0	0	10
,, (crude 60°)	per gallon	0	2	10
osote (ordinary)	,,	0	0	2½
,, (filtered for Lucigen light)	,,	0	0	3
ide Naphtha 30 % @ 120° C.	,,	0	1	5
ease Oils, 22° Tw.	per ton	3	0	0
ch, f.o.b. Liverpool or Garston	,,	1	12	0
vent Naphtha, 90 % at 160°	per gallon	0	1	11
oven Oils (crude)	,,	0	0	2½
er (Chili Bars)	per ton	46	15	0
Sulphate	,,	25	10	0
Oxide (copper scales)	,,	52	0	0
rine (crude)	,,	30	0	0
, (distilled S.G. 1250°)	,,	55	0	0
e	nett, per oz.	8½d. to 9d.		
Sulphate (copperas)	per ton	1	10	0
Sulphide (antimony slag)	,,	2	0	0
(sheet) for cash	,,	14	0	0
Litharge Flake (ex ship)	,,	17	0	0
Acetate (white ,,)	,,	23	5	0
,, (brown ,,)	,,	18	15	0
Carbonate (white lead) pure	,,	19	0	0
Nitrate	,,	22	5	0

		£	s.	d.
Lime, Acetate (brown)	,,	8	5	0
,, ,, (grey)	per ton	14	0	0
Magnesium (ribbon and wire)	per oz.	0	2	3
,, Chloride (ex ship)	per ton	2	6	3
,, Carbonate	per cwt.	1	17	6
,, Hydrate	,,	0	10	0
,, Sulphate (Epsom Salts)	per ton	3	0	0
Manganese, Sulphate	,,	18	0	0
,, Borate (1st and 2nd)	per cwt.	60/- & 42/6		
,, Ore, 70 %	per ton	4	10	0
Methylated Spirit, 61 O.P.	per gallon	0	2	2
Naphtha (Wood), Solvent	,,	0	3	11
,, ,, Miscible, 60° O.P.	,,	0	4	9
Oils:—				
Cotton-seed	per ton	22	10	0
Linseed	,,	23	10	0
Lubricating, Scotch, 890°—895°	,,	7	5	0
Petroleum, Russian	per gallon	0	0	5¼
,, American	,,	0	0	5⅞
Potassium (metal)	per oz.	0	3	10
,, Bichromate	per lb.	0	0	4
,, Carbonate, 90% (ex ship)	per ton	17	15	0
,, Chlorate	per lb.	0	0	4¼
,, Cyanide, 98%	,,	0	2	0
,, Hydrate (Caustic Potash) 80/85 %	per ton	22	10	0
,, ,, (Caustic Potash) 75/80 %	,,	21	10	0
,, ,, (Caustic Potash) 70/75 %	,,	20	15	0
,, Nitrate (refined)	per cwt.	1	3	6
,, Permanganate	,,	4	5	0
Prussiate Yellow	per lb.	0	0	9¼
,, Sulphate, 90 %	per ton	9	10	0
,, Muriate, 80 %	,,	7	15	0
Silver (metal)	per oz.	0	3	8⅛
,, Nitrate	,,	0	2	5½
Sodium (metal)	per lb.	0	4	0
,, Carb. (refined Soda-ash) 48 %	per ton	5	15	0
,, ,, (Caustic Soda-ash) 48 %	,,	4	15	0
,, ,, (Carb. Soda-ash) 48%	,,	4	17	6
,, ,, (Carb. Soda-ash) 58 %	,,	5	15	0
,, ,, (Soda Crystals)	,,	2	15	0
,, Acetate (ex-ship)	,,	15	15	0
,, Arseniate, 45 %	,,	10	0	0
,, Chlorate	per lb.	0	0	6¼
,, Borate (Borax)	nett, per ton.	30	0	0
,, Bichromate	per lb.	0	0	3
,, Hydrate (77 % Caustic Soda)	per ton.	10	15	0
,, ,, (74 % Caustic Soda)	,,	10	7	6
,, ,, (70 % Caustic Soda)	,,	9	2	6
,, ,, (60 % Caustic Soda, white)	,,	8	2	6
,, ,, (60 % Caustic Soda, cream)	,,	7	15	0
,, Bicarbonate	,,	5	15	0
,, Hyposulphite	,,	5	5	0
,, Manganate, 25%	,,	8	10	0
,, Nitrate (95 %) ex-ship Liverpool	per cwt.	0	8	4½
,, Nitrite, 98 %	per ton	27	10	0
,, Phosphate	,,	15	15	0
,, Prussiate	per lb.	0	0	7¼
,, Silicate (glass)	per ton	5	7	6
,, ,, (liquid, 100° Tw.)	,,	3	17	6
,, Stannate, 40 %	per cwt.	2	0	0
,, Sulphate (Salt-cake)	per ton.	1	12	6
,, ,, (Glauber's Salts)	,,	1	10	0
,, Sulphide	,,	7	15	0
,, Sulphite	,,	5	5	0
Strontium Hydrate, 98 %	,,	8	0	0
Sulphocyanide Ammonium, 95 %	per lb.	0	0	8
,, Barium, 95 %	,	0	0	5¼
,, Potassium	,,	0	0	8
Sulphur (Flowers)	per ton.	6	15	0
,, (Roll Brimstone)	,,	6	0	0
,, Brimstone: Best Quality	,,	4	7	6
Superphosphate of Lime (26 %)	,,	2	10	0
Tallow	,,	25	10	0
Tin (English Ingots)	,,	95	0	0
,, Crystals	per lb.	0	0	6⅞
Zinc (Spelter)	per ton.	22	5	0
,, Chloride (solution, 96° Tw.	,,	6	0	0

THE

CHEMICAL TRADE JOURNAL

Publishing Offices: 32, BLACKFRIARS STREET, MANCHESTER.

No. 147. **SATURDAY, MARCH 15, 1890.** **Vol. **

Contents.

Notices.

All communications for the *Chemical Trade Journal* should be addressed, and Cheques and Post Office Orders made payable to—

DAVIS BROS., 32, Blackfriars Street, MANCHESTER.

Our Registered Telegraphic Address is—

"Expert, Manchester."

The Terms of Subscription, commencing at any date, to the *Chemical Trade Journal*,—payable in advance,—including postage to any part of the world, are as follow:—

Yearly (52 numbers) .... **12s. 6d.**
Half-Yearly (26 numbers) .... **6s. 6d.**
Quarterly (13 numbers) .... **3s. 6d.**

Readers will oblige by making their remittances for subscriptions by Postal or Post Office Order, crossed.

Communications for the Editor, if intended for insertion in the current week's issue, should reach the office not later than **Tuesday Morning.**

Articles, reports, and correspondence on all matters of interest to the Chemical and allied industries, home and foreign, are solicited. Correspondents should condense their matter as much as possible, write on one side only of the paper, and in all cases give their names and addresses, not necessarily for publication. Sketches should be sent on separate sheets.

We cannot undertake to return rejected manuscripts or drawings, unless accompanied by a stamped directed envelope.

Readers are invited to forward items of intelligence, or cuttings from local newspapers, of interest to the trades concerned.

As it is one of the special features of the *Chemical Trade Journal* to give the earliest information respecting new processes, improvements, inventions, etc., bearing upon the Chemical and allied industries, or which may be of interest to our readers, the Editor invites particulars of such—when in working order—from the originators; and if the subject is deemed of sufficient importance, an expert will visit and report upon the same in the columns of the *Journal*. There is no fee required for visits of this kind.

We shall esteem it a favour if any of our readers, in making inquiries of, or opening accounts with advertisers in this paper, will kindly mention the *Chemical Trade Journal* as the source of their information.

Advertisements intended for insertion in the current week's issue, should reach the office by **Wednesday morning** at the latest.

TECHNICAL INSTRUCTION.

If ever there was sufficient proof that the Tec Instruction Act is not a popular measure, and th general public (aye, and even those whom it is suppo benefit) do not care a jot for it, that proof may be fo the meeting that took place on Tuesday in last week Owens College, Manchester. It was previously anno that Mr. W. Mather, M.P., would address the meeting the question, but even the magic of Mr. Mather's nan not sufficient to draw an audience. The Technical Edu propagandists may well now inquire why the publi interested public—having been piped to, have not d Why have they not listened to the voice of the charmer he charmed never so wisely?

The meeting at Owens College was a dismal failure, we say it was not even representative we shall be correct. The press was well represented, four rep being present, but when the meeting commenced the au consisted of twelve souls, all told, and by the time the ness was concluded the number had reached twenty-fiv spite of this want of interest, as evidenced by the pau attendance, a resolution was put and carried, as if ha world had been concerned in it, and a memorial pre upon the instructions of that resolution for submission Manchester City Council.

We have for a long time urged against the extens these tactics for the purpose of raising the wind by issue. To a few it is perfectly well known that the certain institutions, like many individuals, sadly in ne hard cash, but they have not the courage to come befo public and say so in true English fashion. The mo wanted, and if it can be raised compulsorily from the payers without letting them know definitely what is done with it so much the better for the memorialists. have never yet seen any scheme formulated tending to us that the money so raised would be judiciously spe that the Technical Education zealots have any de scheme in particular except that of supplying some part institution with the needful. In these mercenary da are taught to believe in the doctrine of the survival fittest; we are taught to believe in ordinary daily life, i thing is not supported by the mass of the population i thing not required—that it must give way to others th necessary—that it is a sin to foster it by subsidies and artificial means.

The apathy shown by the general public to the Tec Education problem, as it at present exists before shows that it belongs to the class which should the wall to make room for some better matured sc and it is only now kept alive by means of such artificia cesses as have prompted these present remarks. W by no means without experience in the needs and re ments of technical instruction in this country, and it is the greatest humiliation that we regard this measure, we remember that some of our most competent minds been employed in its production.

eters. S. Z de Ferranti. 3,096. February 26.
eters. S. Z. de Ferranti. 3,097. February 26.
Matters. H. H. Lake. 3,098. February 26.
on Burners.—(Complete Specification). J. W. B. Wright. 3,100.
y 26.
es. W. Keyworth. 3,116. February 27.
re of Aerated Waters J. P. Jackson. 3,117 February 27.
re of Glass Bottles. D Rylands. 3 121. February 27.
ents in Gas Lamps. T. C. J. Thomas 3 130. February 27.
amps. F. R. Boardman. 3,133. February 27
. J. S. McDougall and J. T. McDougall. 3,134. February 27.
re of Glass Bottles. W. Ambler. 3,138. February 27.
il Gas Furnace or Forge. The Lucigen Light Co., Ltd. and T.
hain.
stating Compound. W. Grove and L Lewis. 3,151. Feb 27.
re of Electrical Conductors. W. A. Thoms. 3,152. Feb 27.
re of Electrical Conductors. W A. Thoms. 3,153. Feb. 27.
re of Alloys. W. A. Thoms. 3,154. February 27.
hic Camera Slides. F. A. Gregory and H. F. Ainley. 3,155.
y 27.
re of Water Gas. B. von Steenbergh. 3,160. February 27.
witches. P. P. Alexander. 3,163 February 27.
eating Apparatus.—(Complete Specification). C. E. Carpenter.
February 27.
l Stokers. J. Proctor. 3,186. February 28.
Apparatus. J. Swinburne. 3,193. February 28.
Conductors. F. Cook. 3,205. February 28.
of Sand from Water. G. F. W. Hope. 3,209. February 28.
of Metallic Sodium and Chlorine Gas. J. Greenwood. 3,220.
y 28.

The Automatic Regulation of Electromotive Force. J. Shipp. 5,231. February 28.
Hot Air Gas Lamps. F. Siemens. 3,236 February 28.
Manufacture of Artificial Tartaric Acids. A. A. Brehier and B. G. Talbot 3,240. February 28.
Use of Liquid Hydro-carbons for Heating and Lighting. H. H. Doty 3,242. February 28
Extraction of Gold and Silver. W. D. Bohm. 3,245. February 28.
" " " " 3,246. February 28
Water Filtering Apparatus.—(Complete Specification). H H. Lake. 3,261. March 1.
Dynamos. C. N. Russell and R. A Scott 3,271. Merch 1.
Electric Switches. G. E Fletcher. 3,272. March 1.
The Reduction of Metallic Ores. G Simonin. 3,275. March 1.
Telephonic Apparatus. A. Whalley 3,276. March 1.
Gas Governors. E. Patterson. 3,277. March 1.
Water Heaters. W H. Skinner 3 279. March 1.
Brick Making Machinery. W Sayer. 3,285. March 1.
Manufacture of Glass. J. G. Sowerby. 3,286. March 1.
Tubular Apparatus for Heating Feed Water. A Schneider. 3,292. March 1.
Manufacture of Tanning Liquors. P Houston and C. Beakbane. 3,295. March 1.
Apparatus for Drying Steam. J. S. McDougall, J. T. McDougall, and T. D. Sugden. 3,302. March 1.
Dyestuffs. B. Willcox. 3,303. March 1.
Electrical Apparatus. M. C. Greenhill. 3,304. March 1.
A Continuous Coking Furnace.—(Complete Specification). H. Ekelund. 3,306. March 1.
Treatment of Sewage. W. E. Adeney and W K. Parry. 3.312. March 1.

IMPORTS OF CHEMICAL PRODUCTS

AT

THE PRINCIPAL PORTS OF THE UNITED KINGDOM.

LONDON.

ek ending Mar. 4th.

y Ore—
9 t. J. W. Cater, Sons, & Co.
5 H. Bath & Son
8½ B. Kuhn
cid—
53 pkgs. Little & Johnston

19 t. 2 c. Dyster, Nalder, & Co.
alia, 1½ Flack, Chandler, & Co.
1 12 C. Hicks, Nash, & Co.
17 Anning & Cobb
16 W. C. Bacon & Co.
—
21 cks. W. A. Rose & Co.
21 W. Harrison & Co.
20 A. Zumbeck & Co.
500 "
, 23 O'Hara & Hoar
22 "
46 A Boltz
—
, £800 A. & M. Zimmermann
ouc—
, 1½ c. W. M. Smith & Sons
41 Kebbel, Son, & Co.
1½ "
, 30 Union Lighterage Co
3 J. F. O'Bree & Co.
, 63 G. Ward & Sons
14 Cadwell, Watson, & Co.
, 17 L. & I. D. Jt. Co.
75 Chalmers, Guthrie & Co
, 17 Craven & Co.
27½ Kebbel, Son, & Co.
car, 4 c. "
artar—
10 pkgs. T. Ronaldson & Co
4 cks. B. & F. Wf. Co.
5 pkgs. C. F. Gerhardt
18 W. C. Bacon & Co.
ls *(otherwise undescribed)*
, £340 A. & M. Zimmermann
17 T. H. Lee
76 "
5 B. Burton & Sons
400 A. & M. Zimmermann
24 Phillipps & Graves
13 Kaul & Haenlein
21 Beresford & Co.
5 Phillipps & Graves
62 T. H. Lee
9 Evans, Lescher, & Co.

Caustic Potash—
Holland, 7 pkgs. Spies Bros., & Co.
Germany, 1 Domeier & Co.
Dyestuffs—
Gambier
E. Indies, 2 pkgs L. & I. D. Jt. Co.
" 123 Beresford & Co.
" 842 L. & I D. Jt Co.
" 429 Brinkmann & Co
" 250 H. Lambert
" 167 Elkan & Co.
" 168 P. S. Evans & Co.
Extracts
France, 1 pkg. G. Vogt & Co.
Holland, 68 J. Forsey
France, 70 Levinstein & Sons
U. S., 220 T. Ronaldson & Co.
France, 148 Prop. Chambln's Wf.
" 291 Levinstein & Sons
Canada, 160 A. J. Humphery
U. S., 40 W A. Bowditch
" 25 W. Burton & Sons
Norway, 1 Johnson & Jorjenson
Annatto
France 4 pkgs. Fullwood & Bland
Tanners' Bark
N. Zealand, 39 t. L. & I. D. Jt. Co.
S. Australia, 39 T. J. & T Powell
France, 5 c. W Jacques
Belgium, 3 T. H. Lee
N. Zealand, 10 19 W. H. Boult
Indigo
E. Indies, 115 chts. L. & I D. Jt. Co.
" 10 Phillipps & Graves
" 58 T. H. Allan & Co.
" 8 Stahlschmidt & Co
W. Indies, 3 Patry & Pasteur
E. Indies, 71 L. & I. D. Jt. Co.
" 39 P. & O. Co.
" 267 Patry & Pasteur
" 5 Phillipps & Graves
" 12 Williams, Torrey, & Co.
" 10 H. Lambert
" 19 L. & I. D. Jt. Co.
" 36 Elkan & Co.
" 1 G. Dards
" 18 F. Stahlschmidt & Co.
" 16 H. Johnson & Sons
" 10 Gordon, Woodroffe, & Co.
" 20 T. Ronaldson & Co.
" 5 J. Owen
" 10 A. Hagan
" 18 W. Brandt, Sons, & Co.
" 48 L. & I. D Jt. Co.
" 19 W. France & Co.
" 53 4 bxs. J. B. Westray & Co.
" 24 Langstaffe & Co.
S W. Indies, 20 pkgs. J. A. Haddew & Co.

E. Indies, 1 box. Elkan & Co.
" 25 chts. T Ronaldson & Co.
" 284 Arbuthnot, Latham, & Co.
" 34 Benecke, Souchay, & Co
" 19 cs. Widemann, Broicher, & Co.
" 14 F. W. Heilgers & Co.
" 6 Stansbury & Co.
" 2 bxs. 17 Ross & Deering
" 3 cs. 125 L. & I. D. Jt Co.
" 3 tubs. "
Myrabolams
E. Indies, 2,259 pkgs G. S. N. Co
" 683 Hoare, Wilson, & Co.
" 248 Beresford & Co.
" 1,340 "
" 2,000 J. Graves
Valonia
A. Turkey, 96 t. A. & W. Nesbitt
" 57 J. Graves
" 7 Zarifi Bros. & Co.
Orchella
Ceylon, 7 pkgs. Walker, Munsie, & Co.
Turmeric
E. Indies, 174 bgs. L. & I. D. Jt. Co.
Dyestuffs *(otherwise undescribed)*
—— 28 pkgs. T. Clarke

Glycerine—
Holland, £345 Prop. Scott's Wf.
Germany, 68 Beresford & Co.
France, 20 R. Baelz & Co.
S Australia, 175 H. Hill & Sons
Glucose—
Denmark, 10 pkgs. 20 c. M. D. Co.
Germany, 850 850 J. Barber & Co.
" 30 240 T. M. Duche & Sons
" 45 354 L. Sutro & Co.
" 23 180 J. Knill & Co.
" 30 240 Pillow, Jones, & Co.
" 50 403 "
" 25 120 M. D. Co.
" 5 43 "
" 10 80 C. S. Lovell
" 100 120 Burgoyne, Burbidge, & Co.
Germany, 200 pkgs 200 cwts. J. Barber & Co.
" 55 410 J. H. Epstein
" 60 43 M. D. Co.
" 200 200 J. F. Ohlmann
" 200 200 Howell & Co.
" 4 22 Knill & Co.
" 17 160 R. G. Hall & Co.

" 5 pkgs. 50 c. J. Forsey
" 37 294 L. Sutro & Co.
" 30 240 Pillow, Jones, & Co.
" 200 200 J. Barber & Co.
U. S., 600 600 A. Dawson
Gutta Percha—
E. Indies, 52 c. Cundall & Co.
" 172 Kaltenbach & Schmitz
" 135 "
" 172 Kleinwort, Sons, & Co.
Isinglass—
E. Indies, 6 pkgs. L. & I. D. Jt. Co.
B. W. I., 1 "
" 1 Hale & Sons
Manure—
Phosphate Rock
France, 140 t. Lawes' Manure Co.
Belgium, 170 "
France, 182 "
" 285 A. Hunter & Co.
Aruba, 432 G. M. Barier
Belgium, 30 Leach & Co.
" 30 D. de Pass & Co.
Manure *(otherwise undescribed)*
Germany, 51 t. Perkins & Homer
" 10 Cornwall, Son, & Co.
Naphtha—
Germany, 8 brls. Lister & Biggs
Holland, 95 "
" 95 "
Paraffin Wax—
U. S., 1,000 brls. J. H. Usmar & Co.
" 200 E & H. Holdsworth
" 20 Anglo American Oil Co.
" 245 G. H. Frank
" 145 H. Hill & Son
Petroleum—
Belgium, 4 brls. J. L. Lyons & Co.
" 10 Sawer, Mead, & Co.
" 150 Rose, Wilson, & Rose
U. S., 560 C. T. Bowring & Co.
" 50 Barrett, Tagant, & Co.
" 30 Grindley & Co.
" 7,370 Anglo American Oil Co.
" 51 A. Browne & Co.
" 40 Mwll. Dk. Co.
" 75 "
" 1,343 Anglo American Oil Co.
Belgium, 10 Sawer, Mead, & Co
" 9 "
Pitch—
France, 62 brls. H. Hill & Sons
Plumbago—
Germany, 19 cks. W. W. Lupton
Ceylon, 74 R. G. Hall & Co.
Germany, 81 Prop. Scotts' Wf.
Holland, 26 Brown & Elmslie
Ceylon, 60 L. & I. D. Jt. Co.

y Bros., & Co.
Carey & Sons
—
Charles & Fox
T. A. Reid
Cutting & Co.
, Bros., & Co.
huson & Sons
V. Berk & Co.
, Bros. & Co.
Zimmermann
therley & Co.
H. Lambert
Petri Bros.
Paris & Co.
Secker & Co.
Craven & Co.
. Hall & Son
Secker & Co.
Vimble & Co.
Contl. Guano Works
Jomar & Co.
ddard & Co.
H. Lambert
Bros., & Co.
F. Wf. Co.
Hopf, & Co.
F. Wf. Co.
each & Co.
H. Lambert
Leach & Co
hinger Bros.
Leach & Co
rrison & Co.
C. Brunlen
eron, & Co.
W. Balchin

OOL.

5th.

sher & Co.

Ralli Brs.

Ralli Brs.

Niger Co., Limited
hurst, Son & Co.
Francaise
A. Swanny
"
rner & Co.
Davies,
kin & Co
C Lane & Co.
son & Co.
Berthoud
ver & Co
, G. Gun nell
Herschell
elt & Co.
W. King
son & Co.
Brs. & Co
art & Co.
n.
nn Amen
Berthoud
fisman & Co.

—— 10 cks. Pickering & Berthoud
" 2 2 cs. Radcliffe & Durant
Accra, 2 1 bg. Perrin & Co.
" 1 Hutton & Co.
" 13 J. J. Fisher & Co.
" 1 Davies, Robbin & Co
" 15 J. P. Werner
" 2 Edwards Brs.
" 2
Axim, 3 bgs. Pickering & Berthoud
" 1 Millerson & Co
Grand Basso, 1 ck. H. S. Attia
Salt Pond, 6 brls. A. Miller Brs.
" 14 bgs. G. C. Nordlinger
" 2 brls. Pitt, Brs. & Co.
" 1 cs. 2 brls R Barbour & Brs.
" 5 Fletcher & Fraser
" 2 F. & A. Swanzy
" 1 brl.
C. C. Castle, 1 Edwards Brs.
" 3 Millerson & Co
" 2 L. Hart & Co.
" 2 Havard & Co.
" 4 Hobson & Co.
" 3 Davies, Robbins & Co.
" 2 Ihlers & Bell
" 1 Pickering & Berthoud
" 1 F. J. Eaton & Son
" 2 Radcliffe & Durant
" 2 H. B. McIver & Co.
" 1 cs. 11 brls. Fletcher & Fraser
" 2 brls. A. Reis
" 7 C. Freese
" 2 cs. F. & A. Swanzy
" 2 brls. W. Duff & Co.
" 1
Sierra Leone, 14 Pickering & Berthoud
" 1 bg. Cie Francaise
" 50 pkgs.
Bathurst, 3 cks. A. Cros & Co.
" 7 Cie Francaise
Goree, 10 cks. A. Cros & Co.
" 4 S. Johnston & Co.
Cape Coast, 2 brls. Swanzy & Co.
Appollonia, 14 bgs. Pickering & Berthoud
" 2 Edwards Brs.
" 4 Radcliffe & Durant
Assinee, 1
Grand Bassam, 1 brl. 1 cs. Millward, Bradbury & Co.
" 2 bgs. Edwards Brs.
Sierra Leone, 1 brl. 14 pns Paterson, Zachonis & Co

Cream of Tartar—
Bordeaux, 2 cks

Caustic Soda—
St. Nazaire, 19 cks.

Copper Regulus—
Valparaiso, 71 bgs Brownell, Lewis & Co

Dyestuffs—
Orchella
Lisbon, 433 bls. 57 bgs.
Bay Beach, 17 bgs.
Cochineal
Grand Canary, 7 bgs. Kolf & Co.
Teneriffe, 19 bgs. Bruce & White
" 8 bgs. Swanston & Co
Gambier
Akassa, 604 bgs. Royal Niger Co., Limited
Extracts
Havre, 7 cks. 2 cs. Cunard S.S. Co., Limited
Philadelphia, 100 bxs
New York, 9 brls.
" 10
Sumac
Barcelona, 50 bgs.

Glucose—
New York, 95 brls.
Hamburg, 15 cks.

Iodine—
Valparaiso, 12 brls. San Sebastian Nitrate Co.
" 42 W. & J. Lockett
" 38 A. Gibbs & Sons

Manganese Powder—
Hamburg, 1 cs.

Phosphate—
Ghent, 204 bgs.
Valle, 472 t. J. Halphen

Pyrites—
Huelva, 1,405 t. Tennants & Co.
" 1,728 Matheson & Co.
" 1,593 Matheson & Son
" 1,533 "

Saguaro—
Valparaiso, 210 sks.

Stearine—
Baltimore, 250 tcs.
Antwerp, 25 bgs.

Tartar—
Bordeaux, 13 cks.

Verdigris—
Marseilles, 23 cs

Waste Salt—
Hamburg, 50 t.

Zinc Oxide—
Antwerp, 180 brls.

HULL.

Week ending Mar. 6th.

Alumina—
Rotterdam, 41 cks. W. & C. L. Ringrose

Acids (*otherwise undescribed*)—
Rotterdam, 1 brl Hutchinson & Sons
Amsterdam, 25 cks. W. & C. L. Ringrose

Albumen—
Reval, 30 cs. Wilson, Sons, & Co.

Barytes—
Rotterdam, 11 cks.
Bremen, 45

Chemicals (*otherwise undescribed*)
Dunkirk, 72 pks. Wilson, Sons, & Co.
Antwerp, 80 "
Hamburg, 18 "
" 8 cks. G. Malcolm & Sons
Rotterdam, 1 W. & C. L. Ringrose
Dunkirk, 7
Christiania, 23 Wilson, Sons, & Co.
Bremen, 2 Veltmann & Co.
Rotterdam, 1 Hutchinson & Son
" 4

Dyestuffs—
Alizarine
Rotterdam, 14 cks. W. & C. L. Ringrose
" 26 pkgs. Hutchinson & Son
" 22 W. & C. L. Ringrose
" 40 Hutchinson & Son
" 10 "
" 13 W. & C. L. Ringrose
Extracts
Rouen, 105 cks. Rawson & Robinson
" 65 20 cs "
" 30 Wilson, Sons, & Co.
New York, 5 bls. "
Madder
Rotterdam, 2 cks. G. Lawson & Sons
Rennet
Copenhagen, 8 pks. Wilson, Sons, & Co.

Glucose—
Dunkirk, 40 bgs. Wilson, Sons, & Co.
Hamburg, 5 cks. "
" 24 G. Malcolm & Sons
" 33
Stettin, 57 100 bgs. Wilson, Sons, & Co.
" 1,200
Hamburg, 10 Geo. Lawson & Sons
" 20 Rawson & Robinson
" 5
" 7 C. M. Lofthouse & Co.

Guano—
Lyeskil, 3,000 bgs.

Glycerine—
Hamburg, 6 drms.

Manure—
Christiansand, 750 bgs Wilson, Sons, & Co.
Ghent, 180 t. "

Potash—
Rouen, 2 pks.
Rotterdam, 43 cks. G. Lawson & Sons

Pitch—
Rouen, 41 cks. Wilson, Sons, & Co.

Pyrites—
Huelva, 750 t. J. Dalton Holmes & Co.

Plumbago—
Hamburg, 102 cks Wilson, Sons, & Co.

Stearine—
Antwerp, 13 bgs. Wilson, Sons, & Co.

Salt—
Hamburg, 51 bgs. Furley & Co.
" 50 J. Dalton & Co.

TYNE.

Week ending March 6th.

Glycerine—
Hamburg, 1 ck.

Glucose—
Hamburg, 60 bgs. 10 cks.
New York, 25 brls.
" 50 C. Hassell

Pyrites—
Huelva, 2,006 t. Scott Bros.

Zinc Ashes—
Christiania, 62 bgs.

Zinc Oxide—
Antwerp, 25 brls. Tyne S. & Co.

GLASGOW.

Week ending March 6th.

Alum—
Rotterdam, 41 cks. J. Rankine & Son

Barytes—
Rotterdam, 46 cks.

Dyestuffs—
Alizarine
—— 12 cks. J. Rankine & Son
Rotterdam, 40 "
Madder
Rotterdam, 10 "

Glucose—
New York, 48 brls.

Oxalic Acid—
Christiania, 12 cks.

Saltpetre—
Hamburg, 120 kgs. J. Poynter & Son

Tartaric Acid—
Rotterdam, 1 ck. J. Rankine & Son

Tartar Emetic—
Hamburg, 10 cks.

Ultramarine—
Rotterdam, 2 cks. J. Rankine & Son

Waste Salt—
Hamburg, 200 bgs. J. Poynter & Son
Hamburg, 48 cks.

Zinc Oxide—
Rotterdam, 100 cks. J. Rankine & Son

GOOLE.

Week ending Feb. 26th.

Antimony—
Boulogne, 2 cks.

Caoutchouc—
Boulogne, 3 cks.

Caustic Potash—
Hamburg, 12 cks.

Dyestuffs—
Alizarine
Rotterdam, 51 cks.
Dyestuffs (*otherwise undescribed*)
Boulogne, 104 cks. 5 cs.

Glucose—
Hamburg, 270 bgs. 53 cks.

Glycerine—
Rotterdam, 10 cs.

Pyrites—
Seville, 1,150 t.

Potash—
Dunkirk, 21 drums.

Saltpetre—
Rotterdam, 325 bgs.

Sodium Nitrate—
Rotterdam, 6 cks.

Tartar—
Rotterdam, 2 cks.

Tartaric Acid—
Boulogne, 3 cs. 1 ck.

Waste Salt—
Hamburg, 508 bgs.

GRIMSBY.

Week ending March 1st.

Albumen—
Antwerp, 7 cks.
Hamburg, 5 pkgs.

Caoutchouc—
Hamburg, 3 cs

Chemicals (*otherwise undescribed*)
Hamburg, 4 pkgs.

Dyestuffs (*otherwise undescribed*)
Antwerp, 2 pkgs.

Glucose—
Hamburg, 200 bgs. 12 cks.

Manure—
Kainit
Hamburg, 500 bgs.

LONDON.

Week ending Mar. 1st.

Ammonium Sulphate—
Ghent, 25 t. £993
Cologne, 20 10 c. 318
Hamburg, 200 2,301
Ghent, 150 14 1,875
Lemvig, 4 9 35
Hamburg, 128 1,470
Demerara, 50 910
Hamburg, 50 15 600
Ghent, 60 18 725

Brimstone—
Christiania, 20 t. £95

Bisulphide of Carbon—
New Orleans, 20 drs. £19

Bone Meal—
Rotterdam, 29 t. 16 c. £130

Coal Products—
Pitch
Cette, 1,199 t. £2,000
Antwerp, 815 1,403
Christiania, 33 c. 18
Odessa, 25 t. 52
Naphtha
Melbourne, 36 drs. £37
Benzole
Calais, 80 c. £200
Yokohama, 10 47
Carbolic Acid
Mauritius, 11 c. £77
Yokohama, 27 181
Naphthaline
Stettin, 183 c. £43
New York, 205 144
Philadelphia 207 113
Treport, 183 52
Tar
Dantzic, 200 brls. £103
Rangoon, 200 rundlets. 499 cks. 273
Melbourne, 200 kgs. 30 brls. 54
Calcutta, 29 t. 40
Natal, 140 drs. 5 rnlts. 5½ brls. £18

Citric Acid—
Rotterdam, 1 c. £8
Hamburg, 20 76
M. Video, 1 ck. 14
Liban, 2 t. 10 c. 375

Carbolic Acid—
Yokohama, 1 t. 7 c. £181
Mauritius, 11 77

Chemicals (*otherwise undescribed*)
Melbourne, 2 cs. £25
Durban, 3 t. 23
Calcutta, 5 cs 40
Auckland, 24 pkgs. 32
New Orleans, 6 cs. 47

Carbolic Disinfectant—
Auckland, 10 cks. £3

Cream of Tartar—
Brisbane, 12 c. £67
Lyttleton, 15 pkgs. 79
Otago, 5 c. 26
Auckland, 2 t. 3 c. 233

Caustic Soda—
Lyttleton, 1 t. £20
New Orleans, 13 3 c. 45
Natal, 5,040 lbs. 120

Chloride of Lime—
New York, 1 t. 1 c. £25

Copper Sulphate—
Barcelona, 9 t. 18 c. £205
Galatz, 71 19

Chemical Fluid—
E. London, 810 drs. £013

Guano—
Colombo, 8 t. 8 c. £26
Demerara, 100 1,200
Guadeloupe, 420 4,740

Hydrocarbon—
Brussels, 6 t. £104

Manure—
Jersey, 15 t. 10 c. £98
Hamburg, 2,000 bgs. 1,800
Ghent, 30 t. 600
Cillers, 605 10 6,060
Jersey, 5 70
Demerara, 3 8 42
[redacted] 18 158
[redacted] 100 1,200
[redacted] 100 1,000
[redacted] 205 9,100
[redacted] 58 12 830
[redacted] 60 810
[redacted] 29 11 330
[redacted] 5 75
[redacted] 5 2,000
[redacted] 26 200

Oxalic Acid—
Auckland, 5 c. £10
Liban, 5 t. 146
" 14 cks. 145

Phosphorus—
Otago, 600 lbs. £52
Montreal, 19 cs. 111

Superphosphate—
Rotterdam, 31 t. £320

Saltpetre—
Piraeus, 1 t. 3 c. £26
Boston, 16 20
Auckland, 1 5 30

Salammoniac—
Yokohama, 3 t. 14 c. £124
Trebizonde, 4 140
Lisbon, 17 18

Sulphur Chloride—
Dunkirk, 6 cs. £95

Sheep Dip—
Punta Arenas, 1 t. £20
Sydney, 45 pkgs. 97
Boca, 1,550 gls. 195
Natal, 500 drms. 202

Soda Ash—
Yokohama, 30 t. 1 c. £203

Sulphuric Acid—
Jamaica, 9½ c. £10

Sodium Acetate—
New York, 10 t. £155

Soda Crystals—
New Orleans, 13 t. 3 c. £45
Fremantle, 3 5 10

Salammoniac—
Galatz, 1 t. 13 c. £55

Tartaric Acid—
Lyttleton, 3 pkgs. £19
Sydney, 8 c. 60
Auckland, 5 40

LIVERPOOL.

Week ending Mar. 5th.

Alkali—
Syra, 30 cks.

Acetate of Chrome—
Vera Cruz, 2 cks. £3

Ammonium Muriate—
Havre, 1 t. 4 c. £20
Melbourne, 11 3 q. 13

Ammonium Carbonate—
Valparaiso, 11 c. £28
Hamburg, 9 1 q. 16
Odessa, 3 t. 7 111
Canada, 1 t. 30
Odessa, 11 14 c. 3 q. 309
Rouen, 3 1 88

Ammonium Sulphate—
Nantes, 50 t. 15 c. £615
Dunkirk, 20 240
" 15 176
Hamburg, 24 5 293
Nantes, 35 5 403
Odessa, 10 120
Ghent, 30 369

Alum—
Buenos Ayres, 10 t. 2 c. 1 q. £51
Galatz, 1 5
Santa Catherina, 1 5 7
Buenos Ayres, 5 6 26
Havana, 8 14 44
Melbourne, 1 6
Seville, 1 5

Alum Cake—
Leghorn, 16 t. 3 c. £48
Philadelphia, 6 16 20

Bone Phosphate—
Hamburg, 16 t. £80

Brimstone—
Calcutta, 14 c. £69

Bleaching Powder—
Antwerp, 28 cks.
Bordeaux, 13
Boston, 167 tcs.
Chicago,
Demerara, 4
Dunkirk, 14
Ghent, 131
Hamburg, 16
Leghorn, 24
Montreal, 24
New York, 471 987
Antwerp, 134
Baltimore, 50
Boston 30 tcs.
Buenos Ayres, 4 cks.
Corunna, 6
Nantes, 15
Odessa, 30 brls.
Oporto, 4
Palermo, 7
Philadelphia, 82
Odessa, 9
Oporto, 14
Philadelphia, 108
Seattle, 31
Syra, 6

Boracic Acid—
Gijon, 2 t. 9 c. 3 q. £80

Borax—
Rotterdam, 4 t. 19 c. £143
Havre, 3 8
Santos, 6 20
Rouen, 33 17 1,506

Blood Albumen—
Santos, 6 c. £130

Bone Waste—
Rouen, 4 t. 12 c. £31
" 22 16 130

Caustic Soda—
Adelaide, 10 drms.
Barcelona, 55 brls.
Bilbao, 109
Bombay, 10 kegs.
Boston, 407
Carthagena, 95 25 cks.
Havana, 5
Malaga, 30
Matanzas, 6
Nantes, 35
N. Orleans, 100
New York, 30
Odessa, 83
Philadelphia, 135 brls.
Rosario, 10
Rotterdam, 10 2
San Francisco, 100 drms.
Santos, 15
Shanghai, 8
Smyrna, 20
Talcahuano, 10
Yokohama, 200
Honolulu, 20
Ibrail, 30
Leghorn, 25
Malta, 40
Melbourne, 30
Messina, 50
New Orleans, 100
New York, 940
Odessa, 142
Palermo, 6
Panama, 10
Alicante, 20 50 cks.
Amsterdam, 80
Antwerp, 8
Barcelona, 118
Batoum, 100
Bilbao, 14 27
Boston, 200
Calcutta, 106
Carril, 30
Genoa, 40
Ghent, 30
Pernambuco, 30
Rio Janeiro, 15
Rotterdam, 45
Seville, 380 160 brls.
Vigo, 20
Yokohama, 300

Copper Sulphate—
Barcelona, 25 t. 2 c. £499
Genoa, 3 10 90
Bordeaux, 14 2 q. 293
Nantes, 61 17 1,315
Tarragona, 5 19 1 120
Trieste, 5 1 124
Bordeaux, 15 2 17
Havre, 5 3 130
Barcelona, 5 3 121
Smyrna, 1 16 1 44
Coruna, 1 10 37
Venice, 12 13 1 219
Bordeaux, 45 17 1,345
Genoa, 5 128
" 35 3 3 798
Leghorn, 15 10 2 340
Montreal, 14 2 285
Nantes, 10 3 250
Naples, 2 10 1 50
Rio Grande do Sul, 5 6
Accra, 1 t. 1 28
Barcelona, 42 3 1 996
" 10 3 221
Bari, 12 7 3 260
Bilbao, 7 2 2 193
Bordeaux, 110 3 2,620
Bombay, 3 80
Tarragona, 20 178
Trieste, 9 18 199
Venice, 41 981

Sodium Nitrate—
New York, 16 c. 2 q. £22
Venice, 9 t. 17 1 85
Vera Cruz, 5 1 42
Genoa, 12 7 2 106
Leghorn, 7 15 66

Sodium Silicate—
Coruna, 17 brls.
Halifax, 25
Gothenburg, 4
Havre, 7

Sodium Bicarbonate—
Adelaide, 100 kgs.
Barcelona, 118 dms. 100 kgs.
Bombay, 250
Bordeaux, 60
Boston, 450
Buenos Ayres, 80 cks.
Canada, 100
Copenhagen, 10 2
Genoa, 30
Hamburg, 50
Kobe, 200 bgs.
Lisbon, 60
Malaga, 25
Malta, 42 dms.
New York, 200
Odessa, 200
Palermo, 20
Stettin, 40
Syra, 20
Yokohama, 250
Calcutta, 500
Genoa, 70
Kurrachee, 50
Melbourne, 20 54
Mexico City, 50
Naples, 140
New York, 7
San Francisco, 30
Trieste, 100
Valencia, 100 20
Valparaiso, 100

Soda Crystals—
Buenos Ayres, 500 brls.
Gibraltar, 100
Malta, 50
Monte Video, 32
New York, 280 cks.
Smyrna, 20
Buenos Ayres, 156
Hamburg, 80 bgs. 8 tcs.
New York, 747 kgs.
Rotterdam, 10 cks.

Soda Ash—
Adelaide, 76 brls.
Baltimore, 67 cks.
Boston, 380 197 tcs.
Canada, 258
Chicago, 400 bgs
Mytelene, 100 brls.
New York, 987
Philadelphia, 88 680
Sierra Leone, 50 bxs.
Santos, 10 brls.
Yokohama, 204
San Sebastian. 2
Smyrna, 109 brls.
Adelaide, 40
Batoum, 34
Bayrout, 50
Boston, 37
B. Ayres, 300 dms.
Carthagena, 30 brls.
Cienfuegos, 8
Ghent, 32
Kingsey, 15
Maceio, 15
New Glasgow, 28
New York, 102 1 39
Odessa, 75
Rio Janeiro, 265
San Francisco, 217
Santos, 5 brls.
Sherbro, 34
Varna, 20
Yokohama, 100

Salt—
Algoa Bay, 5 t.
Barbadoes, 17
Bilbao, 2½
Boston, 340
Buenos Ayres, 25
Cape Town, 13¼
East London, 18
Montreal, 115
Para, 321½
Pernambuco, 4
West Point, 555
Adelaide, 9
Alligator Pond, 9
Antwerp, 200
Baltimore, 500
Barbadoes, 50
Belize, 22
Boca, 19
Buenos Ayres, 30 t.
Cameroons, 109¾
Gaboon, 17½
Ghent, 300
Hamburg, 100
Manaos, 26
New Orleans, 20
New York, 10 cs.
Old Calabar, 59
Opobo, 41½
Trinidad, 63 16
Victoria, 65

Sulphur—
Boston, 25 t. 1 c. £110
Iquique, 52 16 1 q. 337
Cardenas, 15 5

Sugar of Lead—
Rio de Janeiro, 5 c. £7

Sodium Bichromate—
Havre, 6 t. 18 c. £205

Superphosphate—
Hamburg, 327 t. £887

Sodium Chlorate—
Hamburg, 2 t. £112

Sodium Biborate—
Hamburg, 1 t. 4 c. £36
Montreal, 1 30

Sheep Dip—
E. London, 16 c. 3 q. £34

Sheep Wash—
Baltimore, 2 t. £50
Odessa, 3 8 c. 1 q. 131

Saltpetre—
Syra, 1 t. 16 c. 2 q. £40

Salammoniac—
Amsterdam, 5 c. £8
Larnaca, 4 1 q. 8
New York, 4 t. 18 1 169
Genoa, 2 7 82
Alexandria, 5 8

Tar—
Sierra Leone, 10 brls.

Zinc Chloride—
Bombay, 3 t 10 c. £55
Piræus, 15 12
Rotterdam, 5 8 36

HULL.

Week ending March 4th.

Alum—
Christiansand, 621 slabs

Alkali (*otherwise undescribed*)
Libau, 40 kgs. 18 cks.

Bleaching Powder—
Danzig, 54 cks.
Stettin, 57

Bone Size—
Harlingen, 15 cks.

Caustic Soda—
Antwerp, 37 drms.
Amsterdam, 50
Konigsberg, 19
Libau, 37
Trieste, 10

Chemicals (*otherwise undescribed*)
Harlingen, 19 cks.
Konigsberg, 75 pkgs.
Libau, 1 cs. 20
Malta, 4
Messina, 22
Odessa, 202 25
Rouen, 48
Reval, 50
Rotterdam, 10 32 21
Stettin, 5
Trieste, 31
Venice, 11
Amsterdam, 1 11
Antwerp, 10 7 5
Bremen, 5
Bari, 3
Bergen, 20 1 15
Christiania, 5 2 cks. 6 pks 161 slbs.
Dunkirk, 2 6 cks.
Danzig, 21 155 pks.
Fiume, 6
Ghent, 4
Gothenburg, 10
Hamburg, 22 3 39

Dyestuffs (*otherwise undescribed*)
Hamburg, 1 drm. 2 cs. 1 ck.
Reval, 19
Rotterdam, 3 3 kegs

Manure—
Aarhuus, 190 t.
Copenhagen, 350 bgs.
Dunkirk, 99
Ghent, 1,426
Hamburg, 935

Pitch—
Antwerp, 382 t.
Cette, 1,095

Tar—
Christiania, 20 cks.
Odessa 1,220
Rotterdam, 6

TYNE.

Week ending Mar. 6th.

Alumina Sulphate—
Rouen, 4 t. 19 c.

Arsenic—
Genoa, 13 c.
Venice, 30 t. 3
Copenhagen, 1

Alkali—
San Francisco, 40 t. 18 c.
Rotterdam, 1 13
Gothenburg, 2 16

Ammonium Sulphate—
Valencia, 15 t. 2 c.

Bleaching Powder—
Lisbon, 10 t. 1 c.
Antwerp, 106 8
Hamburg, 60
Baltimore, 49 17
Lisbon, 11 6
Rotterdam, 6 4
Rouen, 20 8
Gothenburg, 5

Baryta Nitrate—
Rouen, 5 t. 1 c.

Barytes Carbonate—
Rouen, 50 t.

Caustic Soda—
Baltimore, 159 t. 11 c.
San Francisco, 161 11
Christiania, 1 14
Rotterdam, 1 3
Rouen, 2 18
Antwerp, 5 9
Valencia, 10 3
Gothenburg, 50 4

Copperas—
San Francisco, 200 brls.
Hamburg, 10 c.

Calcined Magnesia—
Lisbon, 5 c.

Ferro Manganese—
Bilbao, 37 t. 2 c.

Glycerine—
Rotterdam, 6 t. 10 c.

Litharge—
Hamburg, 4 t. 5 c.
Gothenburg, 1 19

Magnesia—
Hamburg, 10 c.
Genoa, 1 t. 16
Valencia, 3
Rouen, 3

Magnesium Carbonate—
Valencia, 11 c.

Potassium Chlorate—
Antwerp, 3 t. 10 c.
Hamburg, 10
Rotterdam, 1 12
Reval, 20
Rotterdam, 6

Pitch—
Havre, 466 t.
Antwerp, 60

Soda—
Rotterdam, 30 t. 11 c.
Aalborg, 52 30
Odense, 10 4
Ancona 22
Nakskov, 13 14
Malmo, 5 12
Kjerteminde, 7 16
Lisbon, 5 7
Rudkjobing, 10 5
San Francisco, 105 t.
Gothenburg, 5 4

Superphosphate—
Malmo, 83 t. 6 c.
Christiania, 5
Aalborg, 1 5
Esbjerg, 4 8
Valencia, 55

Soda Ash—
Genoa, 13 t. 9 c.
Ancona, 18 3
Venice, 10 9
Rotterdam, 13

Sodium Sulphate—
San Francisco, 25 t. 1 c.
Copenhagen, 5 t. 14

Sulphur—
Bilbao, 2 q.
Rouen, 26 t.

Sodium Hyposulphite—
San Francisco, 25 t.
Rouen, 2 t.

Tar—
Hamburg, 20 t.

GLASGOW

Week ending Mar.

Benzole—
Rotterdam, 43 puns.

Boracic Acid—
New York, 221 c.

Carbolic Acid—
Hong Kong, 22 c.

Caustic Soda—
Oporto, 29¾ c.

Dyestuffs—
Extracts
Boston,

Guano—
Penang, 76½ t.

Glycerine—
Rouen, 11 t. 10½ c.

Manure—
Penang, 25 t.
Oporto, 10

Naphtha—
Rouen,

Pitch—
Bombay,
Gijong, 726 t. 6½ c.
Christiania, 13¾ t
Bordeaux, 917

Potassium Bichromate—
Rouen 52 t 10 c.
Christiania, 1 ck.
Rotterdam, 145
Hong Kong, 5 t.
Boston, 72½ c.

Paraffin Wax—
Oporto, 41½ c.
Malta, 5 t. 8½ c.
Nantes, 2,204 lbs.
Rouen, 7 t.
Christiania, 24½ c.
Hong Kong, 71¾

Sodium Bichromate—
Rouen, 9 t. 13¾ c.

Sodium Silicate—
Penang, 1 t. 12¾ c.
" 27¾

Salt—
Brisbane, 100 t.

Sulphuric Acid—
Bombay,

Tar—
Bombay, 23½ t.
"

Ultramarine—
Mauritius, 1 t. 3½ c.

GOOLE.

Week ending Feb. 2

Benzole—
Ghent, 17 drs.
Hamburg, 50 cks.
Rotterdam, 20

Bleaching Preparation
Hamburg, 56 cks.

Copper Precipitate—
Rotterdam, 233 cks.

Dyestuffs (*otherwise und*
Ghent, 1 cs. 3 cks.

Chemicals (*otherwise und*
Antwerp, 2 cks.
Boulogne, 6
Rotterdam, 2

Manure—
Bruges, 232 bgs.
Ghent, 1,267
Hamburg, 439

Pitch—
Antwerp, 90 t.

GRIMSBY

Week ending Mar.

Chemicals (*otherwise und*
Dieppe, 121 pkgs.
Hamburg, 1 16 cks.

PRICES CURRENT.

WEDNESDAY, MARCH 12, 1890.

PREPARED BY HIGGINBOTTOM AND CO., 116, PORTLAND STREET, MANCHESTER.

lues stated are F.O.R. at maker's works, or at u,ual ports of shipment in U.K. The price in different localities may vary.

		£	s.	d.
c, 25 °/ₒ and 40 °/ₒ	per cwt	7/6 & 12/6		
Glacial	,,	3	2	6
hic, S.G., 2000°	,,	0	12	9
mic 82 °/ₒ	nett per lb.	0	0	6½
ric	,,	0	0	3¼
atic (Tower Salts), 30° Tw.	per bottle	0	0	8
(Cylinder), 30° Tw.	,,	0	2	11
c 80° Tw.	per lb.	0	0	2
ous	,,	0	0	1¼
ic	nett ,,	0	0	3¼
c	,,	0	1	4
huric (fuming 50 °/ₒ)	per ton	15	10	0
,, (monohydrate)	,,	5	10	0
,, (Pyrites, 168°)	,,	3	2	6
,, (,, 150°)	,,	1	8	0
,, (free from arsenic, 140/145°)	,,	1	10	0
hurous (solution)	,,	2	15	0
hic (pure)	per lb.	0	1	0½
aric	,,	0	1	2½
, white powdered	per ton	13	10	0
loose lump)	,,	4	15	0
a Sulphate (pure)	,,	5	0	0
,, (medium qualities)	,,	4	10	0
ium	per lb.	0	15	0
nia, ·880=28°	per lb.	0	0	3
=24°	,,	0	0	1⅛
Carbonate	nett ,,	0	0	3½
Muriate	per ton	23	10	0
,, (sal-ammoniac) 1st & 2nd	per cwt.	37/- & 35/-		
Nitrate	per ton	42	0	0
Phosphate	per lb.	0	0	10½
Sulphate (grey), London	per ton	11	16	3
,, (grey), Hull	,,	11	17	6
Oil (pure)	per lb.	0	0	11
Salt ,,	,,	0	0	9½
ony	per ton	75	0	0
(tartar emetic)	per lb.	0	1	1
(golden sulphide)	,,	0	0	10
Chloride	per ton	7	15	0
Carbonate (native)	,,	3	15	0
Sulphate (native levigated)	,,	45/- to 75/-		
ing Powder, 35 %	,,	5	2	6
Liquor, 7 %	,,	2	10	0
ide of Carbon	,,	12	15	0
ium Acetate (crystal	per lb.	0	0	6
n Chloride	per ton	2	0	0
Clay (at Runcorn) in bulk	,,	17/6 to 35/-		
ar Dyes:—				
rine (artificial) 20 %	per lb.	0	0	9
nta	,,	0	3	9
ar Products				
racene, 30 % A, f.o.b. London	per unit per cwt.	0	1	5
ol, 90 % nominal	per gallon	0	3	6
50/90 ,,	,,	0	2	8
olic Acid (crystallised 35°)	per lb.	0	0	10
,, (crude 60°)	per gallon	0	2	10
sote (ordinary)	,,	0	0	2½
,, (filtered for Lucigen light)	,,	0	0	3
e Naphtha 30 % @ 120° C.	,,	0	1	5
se Oils, 22° Tw.	per ton	3	0	0
, f.o.b. Liverpool or Garston	,,	1	12	0
nt Naphtha, 90 % at 160°	per gallon	0	1	11
ven Oils (crude)	,,	0	0	2½
(Chili Bars)	per ton	46	15	0
Sulphate	,,	26	0	0
Oxide (copper scales)	,,	51	0	0
ne (crude)	,,	30	0	0
(distilled S.G. 1250°)	,,	55	0	0
	nett, per oz.	8½d. to 9d.		
lphate (copperas)	per ton	1	10	0
lphide (antimony slag)	,,	2	0	0
heet) for cash	,,	14	0	0
itharge Flake (ex ship)	,,	16	0	0
cetate (white ,,)	,,	23	5	0
,, (brown ,,)	,,	18	10	0
arbonate (white lead) pure	,,	19	0	0
itrate	,,	22	0	0

		£	s.	d.
Lime, Acetate (brown)	,,	8	5	0
,, ,, (grey)	per ton	14	0	0
Magnesium (ribbon and wire)	per oz.	0	2	3
,, Chloride (ex ship)	per ton	2	6	3
,, Carbonate	per cwt.	1	17	6
,, Hydrate	,,	0	10	0
,, Sulphate (Epsom Salts)	per ton	3	0	0
Manganese, Sulphate	,,	18	0	0
,, Borate (1st and 2nd)	per cwt.	60/- & 42/6		
,, Ore, 70 %	per ton	4	10	0
Methylated Spirit, 61° O.P.	per gallon	0	2	2
Naphtha (Wood), Solvent	,,	0	3	11
,, ,, Miscible, 60° O.P.	,,	0	4	8
Oils:—				
Cotton-seed	per ton	22	10	0
Linseed	,,	23	0	0
Lubricating, Scotch, 890°—895°	,,	7	5	0
Petroleum, Russian	per gallon	0	0	5¼
,, American	,,	0	0	5⅛
Potassium (metal)	per oz.	0	3	10
,, Bichromate	per lb.	0	0	4
,, Carbonate, 90% (ex ship)	per ton	17	15	0
,, Chlorate	per lb.	0	0	5
,, Cyanide, 98%	,,	0	2	0
,, Hydrate (Caustic Potash) 80/85 %	per ton	22	10	0
,, ,, (Caustic Potash) 75/80 %	,,	21	10	0
,, ,, (Caustic Potash) 70/75 %	,,	20	15	0
,, Nitrate (refined)	per cwt.	1	3	6
,, Permanganate	,,	4	5	0
Prussiate Yellow	per lb.	0	0	9¼
,, Sulphate, 90 %	per ton	9	10	0
,, Muriate, 80 %	,,	7	15	0
Silver (metal)	per oz.	0	3	8½
,, Nitrate	,,	0	2	5½
Sodium (metal)	per lb.	0	4	0
,, Carb. (refined Soda-ash) 48 %	per ton	5	15	0
,, ,, (Caustic Soda-ash) 48 %	,,	4	17	6
,, ,, (Carb. Soda-ash) 48%	,,	5	2	6
,, ,, (Carb. Soda-ash) 58 %	,,	6	0	0
,, ,, (Soda Crystals)	,,	2	18	9
,, Acetate (ex-ship)	,,	15	15	0
,, Arseniate, 45 %	,,	10	10	0
,, Chlorate	per lb.	0	0	6¼
,, Borate (Borax)	nett, per ton.	30	0	0
,, Bichromate	per lb.	0	0	3
,, Hydrate (77 % Caustic Soda)	per ton.	11	5	0
,, ,, (74 % Caustic Soda)	,,	10	17	6
,, ,, (70 % Caustic Soda)	,,	9	17	6
,, ,, (60 % Caustic Soda, white)	,,	8	17	6
,, ,, (60 % Caustic Soda, cream)	,,	8	12	6
,, Bicarbonate	,,	5	17	6
,, Hyposulphite	,,	5	5	0
,, Manganate, 25%	,,	9	0	0
,, Nitrate (95 %) ex-ship Liverpool	per cwt.	0	8	4½
,, Nitrite, 98 %	per ton	27	10	0
,, Phosphate	,,	15	15	0
,, Prussiate	per lb.	0	0	7¼
,, Silicate (glass)	per ton	5	7	6
,, ,, (liquid, 100° Tw.)	,,	3	17	6
,, Stannate, 40 %	per cwt.	2	0	0
,, Sulphate (Salt-cake)	per ton.	1	12	6
,, ,, (Glauber's Salts)	,,	1	10	0
,, Sulphide	,,	7	15	0
,, Sulphite	,,	5	5	0
Strontium Hydrate, 98 %	,,	8	0	0
Sulphocyanide Ammonium, 95 %	per lb.	0	0	8
,, Barium, 95 %	,	0	0	5¼
,, Potassium	,,	0	0	8
Sulphur (Flowers)	per ton.	6	15	0
,, (Roll Brimstone)	,,	6	0	0
,, Brimstone: Best Quality	,,	4	7	6
Superphosphate of Lime (26 %)	,,	2	10	0
Tallow	,,	25	10	0
Tin (English Ingots)	,,	95	0	0
,, Crystals	per lb.	0	0	6⅛
Zinc (Spelter)	per ton.	22	0	0
,, Chloride (solution, 96° Tw.	,,	6	0	0

THE

CHEMICAL TRADE JOURNAL.

Publishing Offices: 32, *BLACKFRIARS STREET, MANCHESTER.*

No. 148. **SATURDAY, MARCH 22, 1890.** **Vol. VI**

Contents.

Notices.

All communications for the *Chemical Trade Journal* should be addressed, and Cheques and Post Office Orders made payable to—

DAVIS BROS., 32, Blackfriars Street, MANCHESTER.

Our Registered Telegraphic Address is—

"Expert, Manchester."

The Terms of Subscription, commencing at any date, to the *Chemical Trade Journal*,—payable in advance,—including postage to any part of the world, are as follow:—

Yearly (52 numbers) .... **12s. 6d.**
Half-Yearly (26 numbers) .... **6s. 6d.**
Quarterly (13 numbers) .... **3s. 6d.**

Readers will oblige by making their remittances for subscriptions by Postal or Post Office Order, crossed.

Communications for the Editor, if intended for insertion in the current week's issue, should reach the office not later than **Tuesday Morning**.

Articles, reports, and correspondence on all matters of interest to the Chemical and allied industries, home and foreign, are solicited. Correspondents should condense their matter as much as possible, write on one side only of the paper, and in all cases give their names and addresses, not necessarily for publication. Sketches should be sent on separate sheets.

We cannot undertake to return rejected manuscripts or drawings, unless accompanied by a stamped directed envelope.

Readers are invited to forward items of intelligence, or cuttings from local newspapers, of interest to the trades concerned.

As it is one of the special features of the *Chemical Trade Journal* to give the earliest information respecting new processes, improvements, inventions, etc., bearing upon the Chemical and allied industries, or which may be of interest to our readers, the Editor invites particulars of such—when in working order—from the originators; and if the subject is deemed of sufficient importance, an expert will visit and report upon the same in the columns of the *Journal*. There is no fee required for visits of this kind.

We shall esteem it a favour if any of our readers, in making inquiries of, or opening accounts with advertisers in this paper, will kindly mention the *Chemical Trade Journal* as the source of their information.

Advertisements intended for insertion in the current week's issue, should reach the office by **Wednesday morning** at the latest.

Advertisements.

Prepaid Advertisements of Situations Vacant, Premises on Sale or To be Let, Miscellaneous Wants, and Specific Articles for Sale by Private Contract, are inserted in the *Chemical Trade Journal* at the following rates:—

30 Words and under.......... 2s. 0d. per insertion.
Each additional 10 words .... 0s. 6d. ,,

Advertisements of Situations Wanted are inserted at one-half the above rates when prepaid, viz:—

30 Words and under.......... 1s. 0d. per insertion.
Each additional 10 words .... 0s. 3d. ,,

Trade Advertisements, Announcements in the Directory Columns, and all Advertisements not prepaid, are charged at the Tariff rates, which will be forwarded on application.

FACTORY SMOKE.

DURING the past week or two, several correspondent the Manchester papers have exercised their though ness in endeavouring to persuade the public that b smoke is preventable, and that efforts should be made to sure its total abolition. Dr. Patterson, from Oldham, ope the ball by suggesting the formation of an association prosecute the offenders. Mr. Fred Scott, the secretary of Manchester and Salford Sanitary Association, followed pointing out that his association had been working at the ject for some considerable time, and that, therefore, all eff towards Smoke Abolition should be centred in that Assc tion. These two letters led to another, by "Engineer," pointed out the possibility of preventing factory smoke, writing of an exhibition test, where 10 to 11 lbs. of water evaporated per lb. of fuel by one of the well-known mechan stoking appliances. The correspondence having reac this length, Mr. A. E. Fletcher, the chief inspe of alkali works for the Government, entered the li and adduced reasons why something should be done tow the solution of the problem. Another letter signed "Cle ness," and still another by Mr. J. W. Holden, conclude that has been said within the past few weeks, and it is ceedingly strange that all these correspondents have negle an enumeration of the evils arising from the combustio the sulphur of the coal. Why they should have so negle the sulphurous acid, which is the real damage-doing gredient, it is hard to conceive, but until this evi thoroughly appreciated by our legislators, as well as public, we are afraid the smoke problem will remain solved. It will be of but little use to the general publi see the carbon of the smoke totally consumed, and the still laden with those sulphurous vapours which destroy time, almost anything with which they come in contact. correspondent suggests that moderate fines have done good, and by this we presume he would go in for a whole slaughter with heavy fines, and a daily penalty so long the smoke continued. Nothing could be more errone than the belief that such method would lead to good resu There are so many smoke producers in the country that one with any sense would dare to tackle them all at one ti and any wholesale system of fining would certainly defea own object in the long run.

There is no doubt but that as the consumption of increases, so will the devastation produced from coal sm also increase, and this may be seen plainly enough in suburbs of all our large towns. Even within the past twe years this is only too patent around Manchester, and have also heard our friends make the same remarks regard to London, Birmingham, Leeds, Nottingham, Leicester. None of the correspondents we have alre quoted as writing to the Manchester papers have even hi what they would have us do with the sulphurous acid of coal smoke. We remember some time ago having our at tion called to an invention for washing coal smoke with li by means of which the sulphurous acid was extracted;

as at that day there was no outlet for the sulphite of lime which was produced during the operation, the process fell to the ground. To-day there is a market for sulphite of lime, and the people who were interested in the process at that time might possibly do worse than revive the practice which they tried then to introduce. We do not desire in any way to undervalue the efforts of the Manchester and Salford Noxious Vapours' Association, nor those services which have been gratuitously rendered by the chief inspector of alkali works, but we do say that they have adopted the wrong methods and the wrong arguments, and that if they are allowed to follow their bent in this particular instance, we shall be farther off from the realisation of our wish to abolish smoke than we were before. If we take Mr. Fred Scott's letter we shall see that the Manchester and Salford Noxious Vapours' Association has been in existence for a good many years, but so far as coal smoke is concerned, and the sulphurous emanations which spring from it, we are absolutely worse off than before that Association was formed, and we do not think it can be pointed out that that Association has done anything to reduce the amount of damage due to sulphurous acid. Mr. Herbert Fletcher, who is continually quoted as having taken so much interest in this smoke-consuming question, has, it is stated, six boilers producing steam at his collieries. Now, supposing these six boilers burn between them 100 tons of coal per week; these 100 tons will give not less than three tons of concentrated oil of vitriol, or say 42 carboys a week of sulphuric acid at 170° Tw. Now, then, we say "What has Mr. Fletcher done to prevent damage being caused by the dissemination of that quantity of oil of vitriol over the surrounding neighbourhood"? The emanation of a small proportion of carbon in the form of black smoke cannot possibly cause one tithe of the injury produced by this sulphur in the coal; and when Mr. Fletcher shows us that he has prevented this sulphur from escaping from his chimney—then, and then only—shall we joyfully realise that we are at the beginning of the end.

Respecting Mr. A. E. Fletcher's letter, he puts forward the proposed scope of the work upon which he has set his mind in testing appliances and methods for consuming coal smokelessly; but as we have said all along, we have no faith whatever in the methods by which he seeks to attain his end. It is all very well, no doubt, to hold exhibitions, to issue reports, and to be able to push forward some one or other of the numerous so-called smoke-preventing appliances, but that these will relieve us all our sufferings we do not believe. We may, however, turn to the printed proposed scope of the work. Under the first head we have a collation of the results of past experience. This, no doubt, would be valuable, and could be done by a committee at comparatively small expense, without the aid of any expert, and without the employment of any guarantee fund, and moreover, all experience must be of the past, but we do not see that all such information, cannot be obtained from the records of somebody's working. Under the second head we have examinations or tests, to be conducted by experts appointed by the committee, at places where the appliances for coal consumption are already at work, and this has been subdivided into:—*a*, Practical freedom from smoke; *b*, reasonable amount of duty; *c*, economy of fuel; *d*, moderate cost in wear and tear; and *e*, moderate cost of application. Surely it would be sufficient to select works that are known to be well-conducted, and to take figures, say, for the past fifteen or twenty years, which would be far more valuable than any examination or test conducted by an expert. It must be known to the committee that there are many places where coal is smokelessly burned, and if so, why cannot their practice be imitated? [illegible] or tests conducted by an [illegible] Over and [illegible] almost [illegible] [illegible]

Herbert Fletcher deserves the thanks of all interested in smoke abatement for his efforts to suppress the nuisance, and were all to follow his example, *and have plenty of boiler room*, there would be little ground for complaint." The fourth heading of Mr. Fletcher's letter is devoted to a report embodying the results of the committee's operations. What has become of the voluminous report of the Smoke Abatement Exhibition at South Kensington? Has that been forgotten, and, if so, is it not quite as likely that any report issued by another committee will be consigned to the limbo of oblivion, too? There are two reasons why the public are half-hearted in the matter of the prevention of coal smoke. Coal smoke pays to produce. The general depreciation of property and things, the killing of our trees and shrubs, the painting and cleaning of our buildings, show us that most tradesmen get something out of the dirt and dilapidation. It is true that those who can afford to do so, live outside the area of the baneful influence of the sulphurous element, but in their ordinary daily toil many of them are dependent upon the work which the removal of the dirt and dust and sulphur produces. The coal merchant wants to sell as much coal as possible, and the carter wants to cart it, and the dustman wants to keep his horse going removing the ashes. The picture dealer and picture frame maker desires work. The upholsterer wishes to supply more curtains and more furniture, which the sulphurous acid has destroyed, and in this manner work is brought all round, and, in our opinion, the evils arising from the combustion of coal will only be forgotten after the question has been tackled by central legislation, or by the enhancement of the value of coal to such an extent as will make people more economical in the future, and not only strive to burn as little as possible but will endeavour to make use of the by-products which are to be found in the gaseous products of combustion. There is another thing which may help to hasten this time forward. If coal should be found in paying quantities within a few miles of London, it will cause manufactories, mills, and works to spring up in enormous quantities within measurable distance of that city. Coal also will become cheaper there, and much larger quantities will be used for domestic purposes than has heretofore been employed. This will mean such an abominable nuisance to the inhabitants of London that the Legislature will be compelled to interfere, and perhaps when half of the fair lands of Kent have been devastated by the sulphurous fumes of the factories and works congregated round the coalpits there, we may see the time when the combustion of coal will be carried on without causing a nuisance to the inhabitants, and the unit of steam will be raised at a less cost than is now accomplished.

ADVANCE IN THE PRICE OF COAL.

THE very general stoppage of work at the collieries throughout Lancashire has naturally caused prices to be very irregular, outside customers especially paying almost any figure to secure supplies. In the Wigan district on Monday last most of the collieries advanced their pit prices 2s. 6d. and 3s. per ton best coal, which was previously selling at 12s. 6d. to 13s., being now quoted at 15s., and seconds [illegible] have gone up from 11s. 6d. and 12s. to 14s. per ton, whilst in all descriptions of fuel for manufacturing purposes the advance has been [illegible] greater, steam and forge coals having gone up from 9s. 6d. and [illegible] to 13s. and 14s., burgy from 8s. and 8s. 6d. to 11s. and 12s., and [illegible] from 7s. and 7s. 6d. to 10s. and 11s. per ton. These prices, [illegible] however, are of course only temporary, are being quoted to [illegible] customers; outside buyers have to pay in many instances considerably above these figures, and there are cases where 20s. has been paid for burgy and 15s. per ton for slack at the pit mouth. Two or three of the large colliery firms hold fairly heavy stocks, but generally [illegible] are only light. In the Manchester district the leading colliery [illegible] have not as yet made any actual advance on their list rates, but have practically ceased booking new orders, and are confining themselves to meeting as far as possible the requirements of consumers who are already customers on their books, merchants and dealers [illegible] seek supplies wherever they can obtain them. In some instances coalowners' stocks might enable them to keep the bulk of their customers, so far as really urgent requirements for consumption are [illegible], going for a fortnight or so, but this is by no means [illegible] instances, and the very great pressure for supplies at [illegible] [illegible] an indication that consumers are already feeling very [illegible] [illegible] of the strike.

CHEMICAL ENGINEERING.*

XXIII.

BEFORE concluding my remarks upon pumps generally, when used for the purpose of moving liquids, I would fain direct your attention once more to the formula (see page 154)

$$H = \frac{G^2 L}{(3d)^5}$$

for the head necessary to overcome friction. Mr. G. A. Ellis has calculated the loss in lbs. pressure from friction for a length of one hundred feet of cast iron pipe of various diameters, discharging the stated quantities per minute, and the numbers are given in the following table:—

Imperial Gallons.	Inside Diameter of Pipes.											
	¾"	1"	1¼"	1½"	2"	2½"	3"	4"	6"	8"	10"	12"
4.	3.3	0.84	.31	.12								
8.	13.	3.16	1.05	.47	.12							
12.	28.7	6.98	2.38	.97	.27							
16.	50.4	12.30	4.07	1.66	.42							
20.	78.	19.00	6.40	2.62	.67	.21	.10					
25.		27.5	9.15	3 75	.91	.30	.12					
29.		37.	12.4	5.05	1.26	.42	.14					
33.		48.	16.1	6.52	1.60	.51	.17					
37.			20.2	8.15	2.01	.62	.27					
41.			24.9	10.00	2.44	.81	.35	.09				
62.			56.1	22.40	5.32	1.80	.74	.21				
83.				39.	9.46	3.20	1.31	.33	.05			
103.				48.1	14.9	4.89	1.99	.51	.07			
124					21.2	7.00	2.85	.69	.10	.02		
145.					28.1	9.46	3.85	.95	.14	.03		
166.					37.5	12.47	5 02	1.22	.17	.05	.01	
207.					47.7	19.66	7.76	1 89	.26	.07	.03	
249.						28.06	11.20	2.66	37	.09	.04	.005
290.						33.41	15.20	3.65	.50	.11	.05	.007
332.						42.96	19.50	4.73	.65	.15	.06	.01
373.							25.00	6.01	.81	.20	.08	.02
415.							30.80	7.43	.96	.25	.09	.04
671.								14.32	2.21	.53	.18	.08
830.									3 88	.94	.32	.13
1037.										1.46	.49	.20
1245.										2.09	.70	.29
1450.											.95	.38
1660.											1.23	.49
1867.												.63
2075.												.77
2490.												1.11

The foregoing numbers will only hold good for a mobile fluid such as water or crude petroleum, and it is presumed that the pipe is a straight one and free from bend or irregularities. If it is desired to work out the friction for other lengths and quantities expressed in pounds pressure per square inch, the following formula will give the same:—

P = pounds per square inch.

$$P = \frac{G^2 L}{2{\cdot}3\,(3d)^5}$$

The numbers given by Ellis do not compare exactly with those calculated from the above formula, but nevertheless, they are sufficiently near for all practical purposes.

An inspection of the table will show its usefulness—suppose it is desired to pass 1,200 gallons per hour through a pipe one hundred feet in length, from one tank to another situated at a lower level: the table tells us, that if a pipe of one inch bore be used, the head necessary to simply overcome friction is equal to 19 pounds pressure, or 43·91 feet of water. The formula says 54 feet, and it would be safer to take the larger figure. If, however, a three-inch pipe be employed, the head necessary to overcome friction is but 0·1 pounds or 0·23 feet of water. This illustration is quite sufficient to show the importance of careful attention to the bore of pipes used in connection with pumps and other water raisers.

There is one more matter that should be mentioned in connection with the movement of liquids, and which should be more carefully studied by both workman and superin dent alike, as it has many uses in connection with chem plant and apparatus—I now allude to the employment of old time-honoured syphon—and the bearing it has upon working of many processes. It is necessary to mention as strange misconceptions regarding its principle of acti are to be found everywhere it is employed. I have only with one man, who imagined that he could syphon a liq up hill, but I have met with many who have fancied they co increase the flow by lengthening the longer limb in a h zontal direction.

The syphon is often used for drawing off a clear liquor fr a deposited sediment; for emptying tanks where no apertures are available, such as the rectifying glasses, co taining oil of vitriol and many similar operations.

The construction of the syphon scarcely requires illust tion; it usually consists of a bent metal or glass tube, short end of which dips into the liquid to be drawn off, shown in the accompanying illustration. The syphon

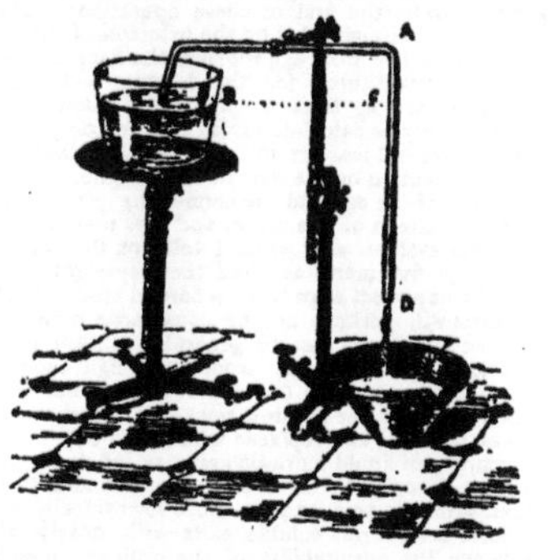

"set" in several ways; first, and most usual by filling *completely* with the fluid it is desired to withdraw, and quicl (and without allowing any to escape) inverting the sh limb into the fluid. The height of the point A, shown in t illustration, above the line C—B, must always be *less* th the head of water due to the barometric pressure at t time, as there is a certain amount of head consumed friction in passing from B to A, and from A to C; this fricti increases, of course, with the quantity of fluid passing in given time, and this quantity is regulated by the bore of t syphon, and the height from C to D, that is supposing t pipe joining these two points to be perpendicular. Anoth method of setting a syphon may be seen in the meth usually adopted in emptying or "running off" the conte of the rectifying glasses in the manufacture of D O V. this case the syphon—usually of small bore, say half-an-in —is placed with its short end dipping below the surface of vitriol in the retort, the long end is temporarily connect with a small portable leaden air-pump, one stroke of wh withdraws the air from the syphon, the fluid following i mediately, when the pump is instantly disconnected.

In filling carboys or drums by means of the syphon, end of the longer limb is furnished with a tap to stay flow at the right moment.

There are many modifications and methods of fitting the syphon in actual practice, and perhaps many more suggest themselves to the true technologist who has to w with them. Reference to the illustration will show a v common method of starting and stopping the syphon wh and as required. So long as the level of the liquor C—B above the point D, the flow is through B, A, C, D; but if any means the level E is raised to C—B the flow at o stops, though the syphon remains full of fluid, ready immediate action so soon as the level E is made to fall be

* A course of lectures delivered by Mr. George E. Davis at the Manchester Technical School.

B. This is brought about by suspending a cylinder F ınd the lower end of the longer limb of the syphon, by a in passing over a pulley; when the top of the cylinder is ught above the level C—B the syphon ceases to flow, but nediately starts again when the line E—E is brought ›w C—B. The formula already given as to loss of head friction, for velocity of efflux, and for the number of ons delivered per minute or per hour, will apply equally yphons, so that the student may find plenty of exercise vorking out problems of his own devising.

o far, we have only considered a fluid such as water, in ch the difficulties of construction of the apparatus are uced to a minimum, but in chemical operations, the ids to be dealt with are generally more or less corrosive, that the ordinary pump is utterly unsuitable. In such es special appliances have been devised for raising rosive fluids, such as, for instance, oil of vitriol, to the top ıbsorbing and denitrating towers, or weak hydrochloric d to the top of condensing towers. An ordinary pump ıld not do for these operations, so an egg and force pump n general use for the first of these operations. The egg l force pump is constructed on the principle of the labora-y wash bottle, the iron egg replaces the flask, and an air ce-pump is substituted for the human bellows. The ıssure necessary to force up a fluid of known specific vity, may easily be calculated from the foregoing formulæ, : there are several matters in this connection that require y careful attention on the part of the designer. The first he strength of the egg and the connecting pipes, and the ond is the pattern of the valves and taps used in connec-n with this system, and when I tell you that within my n knowledge five men have had their eyesight partially stroyed by imperfect apparatus, a careful study of all the ious parts will perhaps not be considered unnecessary. e eggs used in vitriol forcing generally contain 40 cubic t of acid, they are of cast iron, and weigh each about cwts.; some are cast in one piece, while others have a ad piece (in which the greatest wear and tear takes place), ıt can be removed and replaced when worn out.

The pumping of liquid muriatic acid has led to a variety inventions in order to overcome the difficulties of the eration. Muriatic acid, or to be more scientifically correct, drochloric acid, forms soluble salts with nearly all the tals, hence the unsuitability of the ordinary metals for mps having to deal with this fluid. Earthenware and glass ve on several occasions been employed for this work, but risks of fracture and consequent other damage have vays been considered too great for ordinary practice, ass is a very good material for acid pumps, and perhaps. me day, our constructive engineers will appreciate its vantages.

But in pumping such liquids as hydrochloric acid, acetic id, nitric acid, and a few other corrosive compounds, the mp is not the only difficulty, the suction and delivery pipes often a sore trouble, to some extent this has been got er by the use of ebonite pipes, and still more frequently by reading a length of india-rubber tubing through an iron pe, a little larger in internal diameter than the outside ımeter of the rubber tubing, in order to support it laterally der the necessary pressure. Weak acids of the foregoing aracter are generally transported through thick leaden pes; they wear away in time by internal corrosion, and ıen this state is reached, they are taken down and replaced new. Lead, as a rule, does not wear away evenly, and, of urse, breaks down where most corrosion has taken place, ıen this occurs, the old lead is melted up, and is a good set- against the cost of the new lead. Iron when pulled out is nost useless, except for sale as scrap.

At one time very large quantities of liquid muriatic acid of specific gravity (5° Tw.) were run away to waste, because suitable pump had been devised capable of forcing it to top of the strong or acid making towers. Mr. Hazlehurst, Runcorn, paid considerable attention to the subject, and produced a pump answering the purpose, which sounded key-note to others who have taken up and improved on invention. The diaphragm pump as now made in rmany is shown in the annexed illustration.

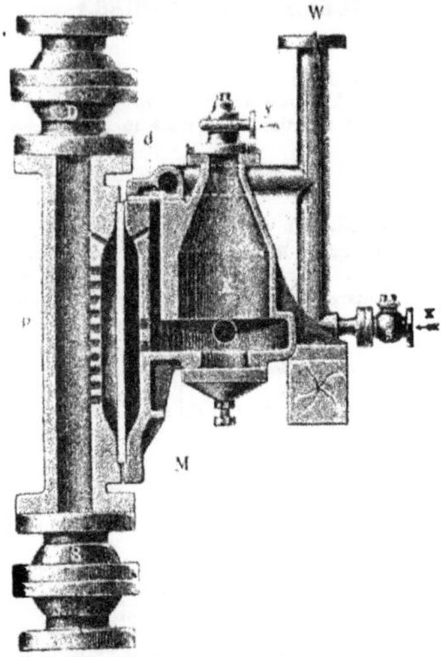

The subject of the moving of liquids is not nearly ex-hausted, but the main points have been stated, and the student will now be able to do most of his work by rule of three; we may therefore pass on to the transportation of gases.

(To be continued.)

THE SITUATION IN WIDNES.—The works are gradually closing down and the men being discharged daily, as their supply of slack gets used up. Some of the works can now only keep in their boiler fires, and by the end of this week nearly all the furnaces in Widnes will be extinguished.

URANIUM.—About six months since the discovery was announced of a lode of the rare metal uranium at the Union Mines, Grampound Road, Cornwall. This is believed to be the only known lode of that metal in the world, as it had previously been found only in isolated pockets and patches. Since the discovery was made steps have been taken to develop the lode, and to work the mine for this metal. Experimental works for producing the metal from the ore were fitted up in London, and there the ore has been submitted to treatment in order to determine the best and most e[illegible]on[illegible]l method of extracting the metal from it. The experimental [illegible]ge is now stated to have been passed, and the commercial production of the metal is being started at the works which we e recently visited. In extracting the metal the ore is first roasted and then crushed to a fine powder. It is then dissolved in acid and precipitated by the aid of certain chemicals. The precipitate is then filtered, and that precipitation is dried and pounded, the result being a yellow powder, which is the commercial oxide of uranium. There are mainly two oxides of this metal, the sesqui-oxide, which is yellow, and the protoxide, which is black. The former of these is employed in glass manufacture to produce a beautiful golden colour, and, in conjunction with other minerals, opalescent tints. The protoxide is used in the production of costly black porcelain. The uranium occurs in the lode in Cornwall as a sesqui-oxide, the yield ranging from 18 to 29 per cent. of metal. The yield at the works is stated to be about 18 per cent. from the lode near the surface which is now being worked, the products being sold to glass manufacturer[illegible] The market price of the metal is about £2,000. per ton, which is [illegible] indication of its rarity. There are other applications of the metal [illegible] contemplation by the company. Now that it can be obtained in qua[illegible]tity it is proposed to substitute it for gold in electro-plating, for whi[illegible] purposes it is well adapted. Possessing, as it does, a high electric[illegible] resistance, it is also expected to prove very useful in electric light in[illegible]stallation. The present works are capable of turning out about half [illegible] ton of uranium per week.

Industrial Celebrities.

JAMES SHANKS.

VII.

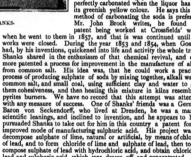

In Lancashire, James Shanks, as the managing partner of Crossfields Bros. & Co., has won for himself the name of being an exemplary works' manager. In this capacity he signalised himself by great industry, energy, tact, and uprightness. For years he made a point of being at the works before the six o'clock bell rang of a morning, he watched the workmen file in to their duties, and if any unfortunate laggard was a few minutes late, he would be greeted with ironical obsequiousness by his master taking off his hat, bowing to him, and saying, "Good morning, sir!" But Shanks was not only an admirable superintendent of men and director of works, he was an excellent practical chemist, and although possessed of but little of the marvellous inventive genius of William Gossage, he was fertile in resource, and has associated his name with one invention of great and abiding utility in the manufacture of alkali, the "Shanks' Vat."

James Shanks was born at Johnstone, in Renfrewshire, on the 24th April, 1800, his father, William Shanks, who belonged to Fife, was a practical millwright and engineer, in the year 1807 he came to Linwood, a small village in the neighbourhood of Johnstone, to erect and fit up a cotton mill, his ability was discerned, and he was made a partner in one of the large cotton mills of Johnstone. He continued his engineering work, and made machinery for the whole district. He had six sons, all of whom were trained up to a practical acquaintance with his business. James Shanks, therefore, had the sound practical training of a mechanic's shop, and the engineering knowledge and experience he there acquired proved of great value to him in after life. The routine of the shop was an admirable school for the man who was to make his mark as a capable and highly successful chemical manufacturer. The father selected James, his eldest son, to enjoy the advantages of a University training and to study for the medical profession. He remained at home, assisting his father and learning his business until he was of age, he then proceeded to Glasgow University, and in due time obtained his diploma. Returning to Johnstone, he there entered on his professional career, and, for two or three years, was the young doctor of that small manufacturing town. During his medical studies at Glasgow, Dr. Andrew Ure was the professor of chemistry, under whom it was his privilege to be placed.

About the time that Shanks went to Glasgow (1821), Ure was bringing out the first edition of his "Dictionary of Chemistry."

Ure was a native of Glasgow, having been born in that city in 1778. He was undoubtedly an admirable teacher, his works indicate great labour and research, and he was specially accurate in all he did. It is recorded, "he was remarkable for his accuracy in chemical analysis, and it is asserted none of his results have ever been upset."

JAMES SHANKS.

In these days, in almost every walk of life, men are compelled to be specialists, they have to select some subject with which they shall become thoroughly and profoundly acquainted, but a few of the great names of the past generation have taken a wide survey in the world of science, and Ure was one of these.

In 1818 appeared his "New experimental researches on some of the leading doctrines of caloric, particularly on the elasticity, temperature, and latent heat of different vapours, and on thermometric measurement and capacity." This subject was brought before the Royal Society and published in their annual transactions.

In 1829, he published his work entitled, "System of Geology," and also works on "The Philosophy of Manufactures," and "On the Cotton Manufactures of Great Britain."

Then in 1830 and 1831, he gave to the world his great book, "The Dictionary of Arts, Manufactures, and Mines," which has gone through several editions, and been translated into most of the European languages. He was the intimate friend of Sir Humphrey Davy, Dr. Wollaston, and Dr. E. D. Clarke. He was a member of the Astronomical Society, and other learned and scientific societies. At Glasgow, he succeeded Dr. Birkbeck as Andersonian Professor of Chemistry. This, then, was the master under whom Shanks studied, an enthusiastic teacher, accurate, painstaking, laborious, a man of broa culture and wide views, an author who occupies a first rank technological literature, and whose name we associate with Paye Knapp, Wagner, Watts, and Muspratt. Shanks was always of enthusiastic nature, there was a genuine glow of healthy enthusiasm everything he took up. When he returned from the University, brought with him some of his master's spirit, and in connection with th Mechanics' Institute at Johnstone, he used to give scientific lecture always to crowded audiences. Whether it was that chemistry ha greater attractions for him than medicine, or whether the pecunia advantages which the career of a chemical manufacturer appeared present to him were greater, we know not, but he decided to aband his profession and started a small Chemical Works in Paisley, whe he manufactured Alum and Chromate of Potash. This early ventu was not a success, and Paisley was left; a situation was accepted Worcester, where he remained only a very short time, taking anoth appointment at Newcastle-on-Tyne. His force of character, intelligen and shrewd common sense, as well as his scientific attainments, we not long in being discovered, and in 1836, when William Gossage w bringing out his valuable inventions, he needed thoroughly capable a intelligent men to erect the plant and carry out the processes designed. Two of these Gossage selected were Shanks and Ellic Elliot afterwards became associated with Russell, and were th inventors of the revolving black-ash furnace. Together they managed th Greenbank Alkali Works, at St. Helens, where they worked o Longmaid's Soda Process. It is said that Elliot was the first introduce into Lancashire the Scotch lixiviating vat, which Dunlop h first worked at St. Rollox. Shanks was em ployed by Gossage in putting up his condensi towers, and his plant for the recovery sulphur from vat-waste, and whilst th engaged, he was sent to St. Helens to constru these works for Gamble and Crossfields. Th firm, needing some one to assist Josi Christopher Gamble in the practical manag ment of their works, as his health was failin they retained the services of James Shank When Gamble retired from the partnership wi Crossfields, Shanks remained with them, a was made a member of the firm of Crossfield Bros., and Co., a position which he filled long as he lived.

In the spring of 1841, about the time of coming to St. Helens, he took out his fi patent, for improvements in the manufactu of carbonate of soda. These he specifies being, first, the placing of black-ash in impu carbonate of soda in fragments, on trays in chamber, making it moist, and passing ca bonic acid gas through it, until the whole the soda and lime is carbonated. Second, a l made from the black-ash is made to percola through a bed of pebbles, through whi carbonic acid is being passed. The soda perfectly carbonated when the liquor has lo its greenish yellow colour. He says this la method of carbonating the soda is preferre Mr. John Brock writes, he found th patent being worked at Crossfields' wor when he went to them in 1857, and that is was continued until th works were closed. During the year 1853 and 1854, when Gossa had, by his inventions, quickened into life and activity the whole trad Shanks shared in the enthusiasm of that chemical revival, and on more patented a process for improvement in the manufacture of alka from common salt. His idea was, that he could work a practic process of producing sulphate of soda by mixing together, alkali wast common salt, and small coal, using sufficient clay and water to gi them cohesiveness, and then heating this mixture in kilns resembli pyrites burners. We have no record that this attempt was attend with any measure of success. One of Shanks' friends was a Germa Baron von Seckendorff, who lived at Dresden, he was a man scientific leanings, and inclined to invention, and he appears to ha pursuaded Shanks to take out for him in this country a patent for improved mode of manufacturing sulphuric acid. His project was decompose sulphate of lime, natural or artificial, by means of chlori of lead, and to form chloride of lime and sulphate of lead, then to d compose sulphate of lead with hydrochloric acid, and obtain chloride lead and sulphuric acid, which was drawn off and concentrated. W fancy it must be cited as one of the instances of Shanks' good natur that he allowed himself to stand sponsor for this idea of his friend. September, 1858, Shanks patented a process of preparing chlorine the use of chromate of lime. He placed a quantity of chromate of lin in a stone or other still, and to it added hydrochloric acid, until th came of a grass green colour, when about half the chlorine elled, heat was applied by steam or otherwise, either externally nally, as is commonly practised in bleaching-powder stills. e residue, chromate of lime was recovered by running it into a lding to it a quantity of hot water and lime, by preference milk to neutralize the acid, and adding a further quantity of lime pitate the oxide of chromium, and an excess to combine with the ic acid obtained in the calcination of the precipitate, for the ate is collected in a drainer, and afterwards manipulated in a at low redness, with a free admission of air, chromate of lime is overed ready to perform its part in the repitition of the original in the production of chlorine.

his patent a medal was awarded Shanks by the Jury of the tional Exhibition, 1862, and Dr. Hofmann, the reporter, to it in his report. He says :—"The reaction between hydro- acid and chromate of potassium is well known, chloride of m, chromic chloride and water being formed with liberation of e :—

$$K_2Cr_2O_4+8HCl=2KCl+Cr_2Cl_3+4H_2O+3Cl.$$

Peligot and M. Gentéle have recommended this process for the ation of chlorine upon a large scale, but unless there be a d for the chromic chloride the reaction would probably prove stly.

analogous process, patented by Mr. Shanks, appears to be much practicable. This process consists in submitting hydrochloric the action of chromate of calcium, the products being chloride cium, chromic chloride, water and chlorine.

$$2Ca_4Cr_2O_4+16HCl=4CaCl+2Cr_2Cl_3+8H_2O+3Cl.$$

Mr. Shanks' process the chromic oxide acts as a carrier between ygen of the air and the hydrogen of the hydrochloric acid. He ot obtain, however, so much chlorine as the manganic oxide ates from the same quantity of hydrochloric acid. For whereas action of manganic oxide 16 equivalents of hydrochloric acid 8 of chlorine, Mr. Shanks' process only eliminates 6 equivalents he same quantity of hydrochloric acid.

s a question only to be solved by experience on a large scale, er the cost of roasting the mixture of chromic oxide and lime will ceed the value of an equivalent quantity of manganese.

s also doubtful whether the expense of precipitating by lime and ng be not greater than the expense involved in the regeneration oxide of manganese from the manganese residues.

king into consideration all these circumstances, it seems im- ble that this process, ingenious as it is, will be extensively ted."

now know how accurate this forecast was; still the patent is an nce of Shanks' inventive ingenuity.

e only other patent with which his name is associated is one "to ove the manufacture of caustic soda and caustic potash," taken n December, 1863.

s process is thus described :—First, the dessication and oxydation n such oxydation is required) of a mixture of carbonate of soda or tash, as the case may be, by heating the same in a reverberatory ce, or other suitable apparatus with access of air, and then pro- g a dry compound in a state favourable for the extraction of caus- da or caustic potash by lixiviation and filtration. Solutions of nate of soda or solutions ordinarily occurring in the manufacture la, and known as blackish liquors or red liquors, are employed. se solutions are not already in a state of saturation, or nearly so, are concentrated until carbonate of soda crystals begin to deposit, the quantity of carbonate of soda contained in the liquid is es- ed, and about forty-two parts of caustic lime is added for each our parts of carbonate of soda present therein, and mix them well her. A violent reaction takes place, and the mixture becomes a mass. This is treated as above until it no longer shows the nce of sulphide. In working with solutions of carbonate of potash forty-two parts of caustic lime are used for each seventy parts of nate of potash.

cond, the abstraction of solutions of caustic soda and caustic potash such dessicated compounds by lixiviation and filtration, such solu- being thereby obtained at higher gravities than similar solutions hitherto been obtained from mixtures of solutions of carbonate of or carbonating of potash with lime. The dry powder withdrawn the furnace is mixed with water or with weak solution of caustic and by injection of steam therein, together with thorough agita- the mixture is reduced to the consistence of cream; this mixture nsferred on to a filtering bed, under which a partial vacuum is ed.

hen the filtration has proceeded to such an extent that the solid er is no longer covered with the supernatant fluid, water is made w into the filtering vessel, so as to keep the solid matter con- ly covered with liquid, and continue such addition of water until olution running off is free or nearly free from caustic soda.

This patent appears never to have gone beyond the experimental stage; we cannot learn that the process was ever adopted in the works.

The one process with which Shanks' name is permanently associated is not connected with any patent. When he came to Lancashire, and for some years afterwards, he found in use the old apparatus employed for lixiviating the black-ash, but when he undertook the management of Crossfields' works, he introduced the system of vats which to the present day everywhere goes by the name of "Shanks' vats."

The lixiviation of black-ash would at first appear to be an exceedingly simple matter; it is, however, one of the stages in the production of alkali requiring great care and attention. The main points to keep in view are :—

1. To secure a good production. by washing out completely the soda contained in the black-ash.

2. To get as much as possible of the soda out as carbonate.

3. To get concentrated liquors.

To achieve these ends, it has been found that the lixiviation should be performed :—

(1). As quickly as possible.

(2). With a minimum quantity of water.

(3.) At as low a temperature as possible.

There have been three arrangements of apparatus for the purpose.

The first, the oldest, and the one which was universally used in Great Britain, consisted of tanks arranged tier above tier. The black-ash was put into the lowest tank, and after being washed with the lyes from the tanks above it, yielding a concentrated solution, the black-ash was emptied by hand labour, with spades from the lowest tank to the one next above it, and so on, but this operation was tedious, costly, and ineffective. It entailed much labour in emptying the black-ash from one vat to another, the handling of the ash caused the balls to break up and so become dense instead of retaining their porosity, and the moist residues were exposed to the air, and so the sodium carbonate was to some extent decomposed. The principle of this method was nothing more than mere filtration.

The second arrangement was one invented by Clement-Désormes. It was principally used in France, Germany, and Belgium. We believe that Muspratt was the only English manufacturer who ever adopted it. Clement-Désormes' method was to place the black-ash balls in perforated boxes that were suspended in tanks, so that the solution in the tanks, when they were full, just covered the black-ash in the boxes.

Clement-Désormes had observed that any substance dissolves more rapidly when placed near the surface of the solution than when at the bottom, the reason being manifest, the more material dissolved the greater the density of the solution, and the denser portion sinking, causing the lighter and less concentrated to rise and so come in contact with the substance, or if the tank was being regularly supplied with fresh water or weak lye, then the more dense sinking to the bottom was drawn off into another vessel by a pipe from the bottom.

In some respects this was a great improvement on the original plant; it prevented the breaking to so great an extent of the black ash balls, and they retained more of their porosity: it also prevented the exposure of the ash to the air for any length of time; but the plant took up too much room, was costly, and entailed too much labour in working, still in some places on the Continent it has only been abandoned during the latter years.

The third arrangement or "Shanks' Vats" is the one now universally adopted in Great Britain. Wagner, who was professor of technology at Wartzburg, and whose work, "Chemical Technology," which is translated and edited in this country by William Crookes, F.R.S., states :—"Mr James Shanks, of St. Helens, Lancashire, was the first to found a rational and economical plan of lixiviation, on what is termed methodical filtration, based upon the fact that a solution becomes more dense the more salnic matter it has in solution, and that a column of weak ley of a certain height equilibrates a shorter column of a stronger ley;" but Lunge, in his treatise on the manufacture of alkali, states :—"It was usual to call it Shanks' lixiviating process, after A. W. Hofmann, in his report by the Juries of Exhibits in the International Exhibition, 1862—page 22 of the report, upon the authority of Mr. Gossage, had unconditionally claimed the honour from the late Mr. James Shanks, of St. Helens.

Muspratt had previously asserted (in his Dictionary II. p. 926), that he knows it is a foreign invention which was introduced by Mr. C. T. Dunlop, into the St. Rollox Chemical Works, about the year 1843. "but Hofmann's statement was almost universally credited." Lunge then proceeds to say :—"This cannot be done now since Scheurer Kestner (in the 'Bulletin de la Société Industrielle de Mülhouse', February 28th, 1868), has completely elucidated the matter, and proved Muspratt's statement to be substantially correct." He reports, that in October 1856, he had visited Messrs. Tennants' works, on a journey to Scotland, in company with Mr. Gundelach, Mannheim.

Mr. Dunlop then showed them the apparatus working by displacement of the liquor, *just in the same form as to-day*, and stated to them that it had been at work for more than ten years (this agrees with Muspratt's year 1843); moreover, Mr. Dunlop told them that he had constructed the apparatus in consequence of advice given him by Mr. Gundelach, on a previous visit. Mr. Gundelach now informed Mr. Scheurer Kestner, that the original idea had been given to him by the well-known physicist, Professor Buff, of Giessen, who a few years before had made a few experiments with it at Kestner's works at Thann, to which no further development was given; on the strength of these, Mr. Gundelach had given that advice to Mr. Dunlop."

This explanation by Lunge would appear to prove that to Shanks did not belong the honour of priority of invention.

In the report of the Juries, page 22, there is a note which says:—"The reporter makes this statement upon the authority of his friend and brother juror, Mr. William Gossage, who had most attentively watched the development of the soda manufacture. It deserves, however, to be noticed, that the honour of invention is claimed by others." Notwithstanding Lunge's explanation, we are not prepared to deny to Shanks the originality, if not the absolute priority of his vats.

The claim of Dunlop was not unknown to Gossage when he asserted Shanks' right to the place of honour; neither is it likely that he could have been ignorant of the fact that Elliot had also introduced similar vats into the works under his management, for, as we have previously remarked, Elliott was in his employ, and he would be thoroughly acquainted with all improvements that were taking place. He would never have authorised the award of honour to Shanks had he not known—if not in the absolute originality of the idea, at least in the perfection of its application—that Shanks was fully entitled to it. On the authority of Mr. Brock we can state: "Mr. Shanks did not patent them (the vats, &c.), but I have frequently heard him express regret at not having done so."

If James Shanks had not been perfectly convinced that he was the inventor of his vats he would never have expressed regret that he did not adopt and patent as his own the invention of another. Probity and sterling uprightness were strong traits in his character: he was not the man to accept honours and awards which he knew belonged to others; and if Hofmann or Gossage were mistaken, Shanks would have corrected their error, and given the honour to whom it was due.

Professor Buff's suggestions and experiments, to which Lunge refers, do not appear to have been thought very much of by Mr. Gundelach, for they were not followed up or developed by him; and Professor Buff's own colleagues, Drs. Henry Will and Hermann Kopp, the Professors of Chemistry at Giessen, have no knowledge of Dr. Buff having considered himself as the originator of the perfected method of lixiviation. Dr. Buff is no longer with us: we cannot refer the disputed point to him for further elucidation; and we do not feel justified in reversing the verdict of the best instructed and most experienced contemporaries, and shall most certainly attribute to Shanks the honour of being one, at least, who was mainly instrumental in introducing the present perfect arrangement and method of lixiviation, and shall continue to apply to them the term "Shanks' Vats," and not, as Lunge suggests, "Buff-Dunlop Vats."

"Shanks' Vats" have proved successful in accomplishing all that was sought in perfect lixiviation;—

1st. Rapidity of solution.
2nd. Little water employed.
3rd. Low temperature,

and they have secured:—Good production; minimum of decomposition; greatest concentration.

As the arrangement is capable of extended application, we draw attention to Lunge's remarks (II., page 468), where he says, "In the soda manufacture, and following its example, in many other cases, the circulation of the liquid is caused without any mechanical assistance, simply by its hydrostatical pressure. Singularly enough, this matter, extremely simple as it is, is wrongly explained in Hofmann's Report by the Juries. There the principle of the liquor-motion is reduced to the fact that solutions become heavier as they become richer, and that any given column of a weak solution is balanced by a shorter column of a dense one. Hence, in a series of horizontally disposed lixiviating vats, the water level will be lower in each successive vat, viz., highest in that receiving pure water, lowest in that containing saturated liquor. Thus, though the vats be horizontal, a "working declivity" of from 12 to 15 inches is stated to be gained. Now, a declivity, of course, exists, but not a "working" one; the liquor cannot flow from the *weaker* tanks to the *stronger* ones, although their level may be very different if this difference is only caused by that in the specific gravities of the solutions. In the connecting tubes, &c., a certain amount of friction has to be overcome by a corresponding pressure, the tanks must be made higher, and those openings from which the strong liquor issues must be made low enough to satisfy not merely the difference of level between the water and the strong liquor, but also the pressure required for overcoming the friction. *The head of water between the top of the tank and the lateral exit opening is the real moving principle*; if it is made too low, or, if by a partial obstruction of the connecting pipes the friction is increased, the weak tank will run over before the st[...] one begins to run."

Dr. Hofmann concludes his remarks on this subject by observi[...] "It is not only in the manufacture of soda that the process of meth[...]cal lixiviation is applicable. It is available in all branches of arts [...] manufactures, for the economical collection and concentration o[...] kinds of soluble matter, whenever such matter is diffused in s[...] proportion throughout the substance of bulky, porous, insoluble ma[...]

Considered, then, in its general bearings, methodical lixiviation, [...] perfected by Shanks, in continuation of Dèsormes' ever-memor[...] improvements, is undoubtedly entitled to rank among the m[...] valuable and beautiful of the great typical processes in applied che[...]try; and it will probably be regarded by posterity as one of the m[...] important industrial bequests of our age."

Associated with Shanks in the working of the business was Si[...] Crossfield, who was a kindly, easy-going man; he attended to [...] commercial routine, but it was Shanks who put all fire and "go" [...] the concern.

He took into his employ, as a young man, in the year 1857, [...] John Brock, in whom he found a man after his own heart, and w[...] he trained to be his successor; what the character of that training w[...] is seen in the establishment and management of the British Al[...] Works at Widnes, which has been under Mr. Brock's direction [...] management from its commencement.

Shanks and Simon Crossfield were both generous men; the dev[...]ing competition of the present day, with its watchword of "[...] survival of the fittest," and its policy of crushing down into extinc[...] the weak, would have ill accorded with their natures or their p[...]ciples; certainly they lived in times when the struggle for existence [...] not appear to be so fierce as it is to-day, but for all that we can scar[...] conceive such a revulsion of ideas and sympathies in these men, [...] who, instead of crushing down a weak competition, are known to h[...] given valuable assistance to a young firm just starting in the same b[...]ness. An unselfish, chivalrous spirit manifested itself sometime[...] those days, which might well be treasured as an inheritance more [...]cious even than the most splendid inventions.

"Do as you would be done by," was not then considered [...] exploded fallacy.

Amongst his workpeople Shanks was a strict disciplinarian; t[...] was no laxity or want of vigilance in his management, he expe[...] every soul in his employ to do his duty—he was everywh[...] he looked *into*, not merely *at*, everything. Still he was very ten[...]hearted and felt deep sympathy with the labouring classes. In [...] the poor ever had a large-hearted and wise friend, and by those in [...] employ he was most highly respected, one might almost say belov[...]

James Shanks will ever be remembered by those amongst whom [...] moved as one of the cheeriest and most genial of men, full of ki[...]ness, and poesessed of a quaint bright humour; fun twinkled in [...] eye, and even in his more serious moods there was a sprightly pla[...]ness in his manner.

He paid little attention to mere conventionalities, and of appeara[...] he was utterly regardless. He might be often seen driving through [...] town with articles on the seat of his open carriage that he had [...]chased at his baker's or his grocer's.

He was no worshipper of the Sartor; the importance or dig[...] which dress could confer went for very little in his eyes; he had a supr[...] contempt for the dandy, his estimate of men was akin to that of [...] nation's bard:—

"The rank is but the guinea stamp,
The man's the gowd for a' that."

James Shanks was intensely social, he delighted in entertainin[...] numerous friends. He had acquired the art of making the gather[...] at his house extremely interesting; he himself was a centr[...] vivacity and good humour. He usually was enthusiastic over s[...] new philosophical or scientific apparatus or contrivance. The g[...]scope was a great delight to him. Then he frequently manage[...] include amongst his guests some one or two who could give sp[...] interest to the gathering—sometimes a traveller who had come h[...] from some perilous expedition, or a philanthropist full of some [...] scheme of beneficence. The evenings were seldom dull, and n[...] commonplace, Shanks himself was so delightfully unaffected [...] free from all self-consciousness.

But his sympathies were far too broad to allow him to confine [...] social gifts to a mere circle of private friends.

His connection with the Mechanics' Institute of his native t[...] when he left college, and when mechanics' institutes were a nov[...] gave him a life-long interest in those institutions, and their motto [...] truly expressed what Shanks ardently desired and ceaselessly so[...] to accomplish: "To make the man a better mechanic, and [...] mechanic a better man."

For many years he was the president of the Mechanics' Institut[...] St. Helens, and together with others, amongst whom may be [...]tioned Messrs. Watson and Wilson, of the Bridgewater Works, [...] the brothers Lacey, who, for so many years, have done such sple[...]

EXPORTS OF CHEMICAL PRODUCTS

LONDON.

Veek ending Mar. 1st.

nium Sulphate—
, 25 t. £295
ne, 20 10 c. 318
urg, 200 2,301
, 152 14 1,875
ca, 4 2 55
urg, 128 1,470
rara, 50 500
urg, 50 15 600
, 60 18 725

tone—
iania, 20 t. £95

hide of Carbon—
Orleans, 20 drs. £19

Meal—
rdam, 29 t. 16 c. £150

Products—
h
1,199 t. £2,000
rp, 815 1,425
iania, 56 c. 18
a, 25 t. 52
phtha
ourne, 36 drs. £37
zole
, 80 c. £202
nama, 10 47
bolic Acid
us, 11 c. £77
nama, 27 181
phthaline
, 103 c. £45
York, 205 144
lelphia 201 113
rt, 123 52
,
ic, 200 brls. £125
oon, 200 runlets. 299 cks. 273
ourne, 200 kgs. 30 brls. 54
tta, 22 t. 40
, 149 drs. 5 rnlts. 5½ brls. £18

Acid—
rdam, 1 c. £8
ourg, 10 76
ideo, 1 ck. 14
, 2 t. 10 c. 375

lic Acid—
hama, 1 t. 7 c. £181
itius, 11 77

icals (*otherwise undescribed*)
ourne, 2 cs. £25
an, 3 t. 23
tta, 5 cs 40
land, 24 pkgs. 32
Orleans, 6 cs. 47

lic Disinfectant—
land, 10 cks. £3

of Tartar—
ane, 12 c. £67
eton, 15 pkgs. 79
, 5 c. 26
land, 2 t. 3 c. 233

ic Soda—
eton, 1 t. £20
Orleans, 13 3 c. 45
, 5,040 lbs. 120

ide of Lime—
York, 1 t 1 c. £25

r Sulphate—
elona, 9 t. 18 c. £205
, 71 19

ical Fluid—
ondon, 810 drs. £213

—
mbo, 8 t. 8 c. £26
erara, 100 1,200
eloupe, 420 4,740

ocarbon—
sels, 6 t. £104

re—
y, 15 t. 10 c. £98
burg, 2,000 bgs. 1,800
it, 50 t. 600
a, 605 10 6,060
y, 5 70
erara, 3 8 42
15 158
kirk, 100 1,200
adoes, 100 1,000
leloupe, 205 2,100
eaux, 58 12 250
art, 60 810
ucia, 29 11 330
ras, 5 75
skov, 450 5 2,080
inique, 19 14 200

Oxalic Acid—
Auckland, 5 c. £10
Libau, 5 t. 146
" 14 cks. 145

Phosphorus—
Otago, 600 lbs. £52
Montreal, 12 cs. 111

Superhosphate—
Rotterdam, 31 t. £310

Saltpetre—
Piræus, 1 t. 5 c. £26
Boston, 16 20
Auckland, 1 5 30

Salammoniac—
Yokohama, 3 t. 14 c. £124
Trebizonde, 4 140
Lisbon, 12 18

Sulphur Chloride—
Dunkirk, 6 cs. £25

Sheep Dip—
Punta Arenas, 1 t. £40
Sydney, 45 pkgs. 97
Boca, 1,550 gls. 195
Natal, 500 dms. 101

Soda Ash—
Yokohama, 50 t. 1 c. £225

Sulphuric Acid—
Jamaica, 9½ c. £10

Sodium Acetate—
New York, 10 t. £155

Soda Crystals—
New Orleans, 13 t. 3 c. £45
Fremantle, 3 5 10

Salammoniac—
Galatz, 1 t. 13 c. £55

Tartaric Acid—
Lyttleton, 3 pkgs. £19
Sydney, 8 c. 60
Auckland, 5 40

LIVERPOOL.

Week ending Mar. 5th.

Alkali—
Syra, 30 cks.

Acetate of Chrome—
Vera Cruz, 2 cks. £5

Ammonium Muriate—
Havre, 1 t. 4 c. £20
Melbourne, 11 3 q. 13

Ammonium Carbonate—
Valparaiso, 11 c. £28
Hamburg, 9 1 q. 16
Odessa, 3 t. 7 111
Canada, 1 t. 30
Odessa, 11 14 c. 3 q. 509
Rouen, 3 1 88

Ammonium Sulphate—
Nantes, 50 t. 15 c. £615
Dunkirk, 20 240
" 15 176
Hamburg, 24 5 293
Nantes, 35 5 423
Odessa, 10 120
Ghent, 30 362

Alum—
Buenos Ayres, 10 t. 2 c. 1 q. £51
Galatz, 1 5
Santa Catherina, 1 5 7
Buenos Ayres, 5 6 26
Havana, 8 14 44
Melbourne, 1 6
Seville, 1 5

Alum Cake—
Leghorn, 16 t. 3 c. £48
Philadelphia, 6 16 20

Bone Phosphate—
Hamburg, 16 t. £80

Brimstone—
Calcutta, 14 c. £69

Bleaching Powder—
Antwerp, 28 cks.
Bordeaux, 13
Boston, 167 tcs.
Chicago,
Demerara, 4
Dunkirk, 14
Ghent, 131
Hamburg, 16
Leghorn, 24
Montreal, 24
New York, 471 987
Antwerp, 154
Baltimore, 50
Boston 92 tcs.
Buenos Ayres, 4 cks.
Corunna, 6
Nantes, 15
Odessa, 30 brls
Oporto, 4
Palermo, 7
Philadelphia, 82
Odessa, 9
Oporto, 14
Philadelphia, 168
Stettin, 31
Syra, 6

Boracic Acid—
Gijon, 2 t. 9 c. 3 q. £80

Borax—
Rotterdam, 4 t. 19 c. £142
Havre, 5 8
Santos, 6 10
Rouen, 53 12 1,526

Blood Albumen—
Santos, 6 c. £130

Bone Waste—
Rouen, 4 t. 12 c. £31
" 22 16 130

Caustic Soda—
Adelaide, 10 dms.
Barcelona, 55 brls.
Bilbao, 109
Bombay, 10 kegs.
Boston, 492
Carthagena, 95 25 cks.
Havana, 5
Malaga, 30
Matanzas, 6
Nantes, 35
N. Orleans, 100
New York, 30
Odessa, 83
Philadelphia, 135 brls.
Rosario, 10
Rotterdam, 10 3
San Francisco, 100 dms.
Santos, 15
Shanghai, 8
Smyrna, 20
Talcahuano, 10
Yokohama, 200
Honolulu, 20
Ibrail, 30
Leghorn, 25
Malta, 42
Melbourne, 30
Messina, 50
New Orleans, 100
New York, 940
Odessa, 142
Palermo, 6
Panama, 10
Alicante, 20 50 cks.
Amsterdam, 80
Antwerp, 8
Barcelona, 118
Batoum, 100
Bilbao, 14 27
Boston, 200
Calcutta, 106
Carril, 30
Genoa, 40
Ghent, 30
Pernambuco, 30
Rio Janeiro, 15
Rotterdam, 45
Seville, 380 162 brls.
Vigo, 20
Yokohama, 300

Copper Sulphate—
Barcelona, 25 t. 2 c. £499
Genoa, 3 10 90
Bordeaux, 14 2 q. 293
Nantes, 61 17 1,315
Tarragona, 5 19 1 120
Trieste, 5 1 124
Bordeaux, 15 2 17
Havre, 5 3 130
Barcelona, 5 3 121
Smyrna, 1 16 1 44
Coruna, 1 10 37
Venice, 12 13 1 219
Bordeaux, 45 17 1,345
Genoa, 5 128
" 35 3 3 798
Leghorn, 15 10 1 340
Montreal, 14 1 225
Nantes, 10 3 250
Naples, 2 10 1 50
Rio Grande do Sul, 5 6
Accra, 1 t. 1 28
Barcelona, 42 3 1 996
" 10 3 223
Bari, 12 7 3 260
Bilbao, 7 2 2 125
Bordeaux, 110 3 2,610
Bombay, 3 80
Tarragona, 10 178
Trieste, 9 18 199
Venice, 47 981
Genoa, 7 t. 12 c. 170
Leghorn, 5 2 1 135
Nantes, 5 130
Barcelona, 15 4 3 337
Rouen, 42 15 1,112
Rio Janeiro, 4 3 5

Chemicals (*otherwise undescribed*)
Gijon, 4 t. 3 c. 3 q. £100
New York, 11 100
Hamburg, 5 540
Havre, 163 10 1,143
Rio Jane'ro, 1 1 6

Caustic Potash—
Rosario, 2 c. 1 q. £5
Genoa, 6 7
Hamburg, 5 t. 4 107
New York, 14 16

Citric Acid—
Pacasmayo, 1 c. 1 q. £12

Carbolic Acid—
Philadelphia, 12 c. £132
Genoa, 3 25
Leghorn, 3 c. 3 q. £25
Havre, 3 t. 9 162
Rotterdam, 24 5 3,390
Bombay, 20 239
Rotterdam, 18 5 2,540
Yokohama, 3 15 3 500

Charcoal—
Hamburg, 51 t. £276
Nantes, 16 80
Hamburg, 49 17 c. 99

Dried Blood—
Rouen, 10 t. 3 c. £81

Disinfectants—
New York, 10 cs. 99
Montreal, 1 c. 6
Calcutta, 2 t. 5 2 q. 11

Guano—
St. John's, N. F., 19 c. 3 q. £5

Gypsum Spar—
New York, 6 t 3 c. 1 q. £39

Iodine—
Genoa, 10 c. 600

Manganese Oxide—
Shanghai, 12 c. £8

Magnesia—
New York, 55 cs. 6 crts.
Corunna, 2 cs.

Magnesium Chloride—
Bombay, 4 t. 19 c. £13
Rouen, 20 17 45

Manure—
Genoa, 59 t. 17 c. 2 q. £233
St. Malo, 327 17 1 2,070
Havre, 1 7 1 11
" 10 16 1 63

Oxalic Acid—
Portland, 15 c. 3 q. £60

Phosphorus—
Vera Cruz, 3 c. 2 q. £54

Potassium Carbonate—
Philadelphia, 7 c. £9

Potassium Chlorate—
Hamburg, 4 t. £173
New York, 2 10 c. 131
Portland, 5 11
Rotterdam, 10 21
Odessa, 6 14
Shanghai, 2 5 110
Yokohama, 2 10 125
Antwerp, 2 89
New York, 1 5 69
Yokohama, 5 186
Shanghai, 2 10 128
Mexico City, 10 22
New York, 5 222
" 5 216
Shanghai, 2 10 114

Pitch—
Vigo, 11 t.
Cette, 1,187
Monte Video, 31 brls.

Potashes—
Rosario, 3 c. 3 q. £5

Potash Salts—
New York, 13 c. 3 q. £23

Potassium Silicate—
Amsterdam, 1 t. 2 c. 1 q. £10

Potassium Bichromate—
Havre, 5 t. £140
Barcelona, 11 385
Rio de Janeiro, 8 c. 2 q. 15

Picric Acid—
Boston, 5 c. £28
New York, 2 t. 15 418

Potassium Prussiate—
Barcelona, 1 t. 2 q. £60

Saltcake—
New York, 24 t. 18 c. £36
Baltimore, 287 5 544

Sodium Nitrate—
New York, 16 c. 2 q. £22
Venice, 9 t. 17 1 85
Vera Cruz, 5 1 42
Genoa, 12 7 2 106
Leghorn, 7 15 66

Sodium Silicate—
Coruna, 17 brls.
Halifax, 25
Gothenburg, 4
Havre, 7

Sodium Bicarbonate—
Adelaide, 100 kgs.
Barcelona, 118 dms. 100 kgs.
Bombay, 250
Bordeaux, 60
Boston, 450
Buenos Ayres, 80 cks.
Canada, 100
Copenhagen, 10 2
Genoa, 30
Hamburg, 50
Kobe, 200 bgs.
Lisbon, 60
Malaga, 25
Malta, 42 dms.
New York, 200
Odessa, 200
Palermo, 20
Stettin, 40
Syra, 20
Yokohama, 250
Calcutta, 500
Genoa, 70
Kurrachee, 50
Melbourne, 20 54
Mexico City, 50
Naples, 140
New York, 7
San Francisco, 30
Trieste, 100
Valencia, 100 20
Valparaiso, 100

Soda Crystals—
Buenos Ayres, 500 brls.
Gibraltar, 100
Malta. 50
Monte Video, 32
New York, 280 cks.
Smyrna, 20
Buenos Ayres, 136
Hamburg, 80 bgs. 8 tcs.
New York, 747 kgs.
Rotterdam, 10 cks.

Soda Ash—
Adelaide, 76 brls.
Baltimore, 67 cks.
Boston, 380 197 tcs.
Canada, 258
Chicago, 400 bgs
Mytelene, 100 brls.
New York, 987
Philadelphia, 88 680
Sierra Leone, 50 bxs.
Santos, 10 brls.
Yokohama, 204
San Sebastian. 2
Smyrna, 109 brls.
Adelaide, 40
Batoum, 34
Bayrout, 50
Boston, 37
B. Ayres, 300 dms.
Carthagena, 30 brls.
Cienfuegos, 8
Ghent, 32
Kingsey, 15
Maceio, 15
New Glasgow, 28
New York, 102 1 39
Odessa, 75
Rio Janeiro, 265
San Francisco, 217
Santos, 5 brls.
Sherbro, 34
Varna, 20
Yokohama, 100

Salt—
Algoa Bay, 5 t.
Barbadoes, 17
Bilbao, 2½
Boston, 340
Buenos Ayres, 25
Cape Town, 13½
East London, 18
Montreal, 115
Para, 321½
Pernambuco, 4
West Point, 555
Adelaide, 9
Alligator Pond, 9
Antwerp, 200
Baltimore, 500
Barbadoes, 50
Belize, 22
Boca, 19

Buenos Ayres, 30 t.
Cameroons, 109¼
Gaboon, 17½
Ghent, 300
Hamburg, 100
Manaos, 26
New Orleans, 20
New York, 10 cs.
Old Calabar, 59
Opobo, 41½
Trinidad, 63 16
Victoria, 65

Sulphur—
Boston, 25 t. 1 c. £110
Iquique, 52 16 1 q. 337
Cardenas, 15 5

Sugar of Lead—
Rio de Janeiro, 5 c. £7

Sodium Bichromate—
Havre, 6 t. 18 c. £205

Superphosphate—
Hamburg, 327 t. £887

Sodium Chlorate—
Hamburg, 2 t. £112

Sodium Biborate—
Hamburg, 1 t. 4 c. £36
Montreal, 1 30

Sheep Dip—
E. London, 16 c. 3 q. £34

Sheep Wash—
Baltimore, 2 t. £50
Odessa, 3 8 c. 1 q. 131

Saltpetre—
Syra, 1 t. 16 c. 2 q. £40

Salammoniac—
Amsterdam, 5 c. £8
Larnaca, 4 1 q. 8
New York, 4 t. 18 1 169
Genoa, 2 7 82
Alexandria, 5 8

Tar—
Sierra Leone, 10 brls.

Zinc Chloride—
Bombay, 3 t 10 c. £55
Piræus, 15 12
Rotterdam, 5 8 36

HULL.

Week ending March 4th.

Alum—
Christiansand, 621 slabs

Alkali (*otherwise undescribed*)
Libau, 40 kgs. 18 cks.

Bleaching Powder—
Danzig, 54 cks.
Stettin, 57

Bone Size—
Harlingen, 15 cks.

Caustic Soda—
Antwerp, 37 drms.
Amsterdam, 50
Konigsberg, 19
Libau, 37
Trieste, 10

Chemicals (*otherwise undescribed*)
Harlingen, 19 cks.
Konigsberg, 75 pkgs.
Libau, 1 cs. 20
Malta, 4
Messina, 22
Odessa, 202 25
Rouen, 48
Reval, 50
Rotterdam, 10 32 21
Stettin, 5
Trieste, 31
Venice, 11
Amsterdam, 1 11
Antwerp, 10 7 5
Bremen, 5
Bari, 3
Bergen, 20 1 15
Christiania, 5 2 cks. 6 pks 161 slbs.
Dunkirk, 2 6 cks.
Danzig, 21 155 pks.
Fiume, 6
Ghent, 4
Gothenburg, 10
Hamburg, 22 3 39

Dyestuffs (*otherwise undescribed*)
Hamburg, 1 drm. 2 cs. 1 ck.
Reval, 19
Rotterdam, 3 3 kegs

Manure—
Aarhuus, 190 t.
Copenhagen, 350 bgs.
Dunkirk, 99
Ghent, 1,426
Hamburg, 935

Pitch—
Antwerp, 382 t.
Cette, 1,095

Tar—
Christiania, 20 cks.
Odessa 1,290
Rotterdam, 6

TYNE.

Week ending Mar. 6th.

Alumina Sulphate—
Rouen, 4 t. 19 c.

Arsenic—
Genoa, 13 c.
Venice, 30 t. 3
Copenhagen, 1

Alkali—
San Francisco, 40 t. 18 c.
Rotterdam, 1 13
Gothenburg, 2 16

Ammonium Sulphate—
Valencia, 15 t. 2 c.

Bleaching Powder—
Lisbon, 10 t. 1 c.
Antwerp, 106 8
Hamburg, 60
Baltimore, 49 17
Lisbon, 11 6
Rotterdam, 6 4
Rouen, 20 8
Gothenburg, 5

Baryta Nitrate—
Rouen, 5 t. 1 c.

Barytes Carbonate—
Rouen, 50 t.

Caustic Soda—
Baltimore, 159 t. 11 c.
San Francisco, 161 11
Christiania, 1 14
Rotterdam, 1 3
Rouen, 2 18
Antwerp, 5 9
Valencia, 10 3
Gothenburg, 50 4

Copperas—
San Francisco, 200 brls.
Hamburg, 10 c.

Calcined Magnesia—
Lisbon, 5 c.

Ferro Manganese—
Bilbao, 37 t. 2 c.

Glycerine—
Rotterdam, 6 t. 10 c.

Litharge—
Hamburg, 4 t. 5 c.
Gothenburg, 1 19

Magnesia—
Hamburg, 10 c.
Genoa, 1 t. 16
Valencia, 3
Rouen, 3

Magnesium Carbonate—
Valencia, 11 c.

Potassium Chlorate—
Antwerp, 3 t. 10 c.
Hamburg, 10
Rotterdam, 1 12
Reval, 20
Rotterdam, 6

Pitch—
Havre, 466 t.
Antwerp, 60

Soda—
Rotterdam, 30 t. 11 c.
Aalborg, 52 30
Odense, 10 4
Ancona 22
Nakskov, 13 14
Malmo, 5 12
Kjerteminde, 7 16
Lisbon, 5 7
Rudkjobing, 10 5
San Francisco, 105 t.
Gothenburg, 5 4

Superphosphate—
Malmo, 83 t. 6 c.
Christiania, 5
Aalborg, 1 5
Esbjerg, 4 8
Valencia, 55

Soda Ash—
Genoa, 13 t. 9 c.
Ancona, 18 3
Venice, 10 9
Rotterdam, 13

Sodium Sulphate—
San Francisco, 25 t. 1 c.
Copenhagen, 5 t. 14

Sulphur—
Bilbao, 2 q.
Rouen, 26 t.

Sodium Hyposulphite—
San Francisco, 25 t.
Rouen, 2 t.

Tar—
Hamburg, 20 t.

GLASGOW

Week ending Mar.

Benzole—
Rotterdam, 43 puns.

Boracic Acid—
New York, 221 c.

Carbolic Acid—
Hong Kong, 22 c.

Caustic Soda—
Oporto, 29¼ c.

Dyestuffs—
Extracts
Boston,

Guano—
Penang, 76½ t.

Glycerine—
Rouen, 11 t. 10½ c.

Manure—
Penang, 25 t.
Oporto, 10

Naphtha—
Rouen,

Pitch—
Bombay,
Gijong, 726 t. 6½ c.
Christiania, 13¼ t
Bordeaux, 917

Potassium Bichromate—
Rouen 52 t 10 c.
Christiania, 1 ck.
Rotterdam, 145
Hong Kong, 5 t.
Boston, 72½ c.

Paraffin Wax—
Oporto, 41½ c.
Malta, 5 t. 8½ c.
Nantes, 2,204 lbs.
Rouen, 7 t.
Christiania, 24½ c.
Hong Kong, 71¼

Sodium Bichromate—
Rouen, 9 t. 13¼ c.

Sodium Silicate—
Penang, 1 t. 12¼ c.
" 27¼

Salt—
Brisbane, 100 t.

Sulphuric Acid—
Bombay,

Tar—
Bombay, 23½ t.
"

Ultramarine—
Mauritius, 1 t. 3½ c.

GOOLE.

Week ending Feb. 26

Benzole—
Ghent, 17 drs.
Hamburg, 50 cks.
Rotterdam, 20

Bleaching Preparations
Hamburg, 56 cks.

Copper Precipitate—
Rotterdam, 233 cks.

Dyestuffs (*otherwise und*
Ghent, 1 cs. 3 cks.

Chemicals (*otherwise und*
Antwerp, 2 cks.
Boulogne, 6
Rotterdam, 2

Manure—
Bruges, 232 bgs.
Ghent, 1,267
Hamburg, 439

Pitch—
Antwerp, 90 t.

GRIMSBY

Week ending Mar.

Chemicals (*otherwise und*
Dieppe, 121 pkgs.
Hamburg, 2 16 cks.

t lumps, 90s.; best, 80s.; seconds and good nuts, 70s.; ; best ground, £6.; and selected crystal ground, £8. Sul- et; best lump, 35s. 6d.; good medium, 30s.; medium, to 27s. 6d.; common, 18s. 6d. to 20s.; ground e, G.G.B. brand, 65s.; common, 45s; grey, 32s. s. Pumicestone quiet; ground at £10., and specially lump, finest quality, £13. Iron ore continues in mand at full prices. Bilbao and Santander stronger at 9s. to .o.b.; Irish, 11s. to 12s. 6d.; Cumberland in strong de- 8s. to 24s. Purple ore unchanged at last quotations. Spanish rous ore in demand at full prices. Emery-stone: Best brands spot and for forward delivery, bringing full prices No. 1 lump at £5. 10s. to £6., and smalls £5. to £5. 10s. Fullers' earth s. to 50s. for best blue and yellow; fine impalpable ground, £7. wolfram, tungstate of soda, and tungsten metal more in- . Chrome metal, 5s. 6d. per lb. Tungsten alloys 2s. per lb. re: High grades continue to be inquired for; low qualities of sale. Antimony ore and metal have further advanced. oxide, 24s. to 26s. Asbestos: Best rock, £17. to £18.; brown 14. to £15. Potter's lead ore: Prices are firm; smalls, £14. to ected lump, £16. to £17. Calamine: Best qualities scarce— 80s. Strontia steady; sulphate (celestine) steady, 16s. 6d. Carbonate (native), £15. to £16.; powdered (manu- £11. to £12. Limespar: English manufactured, old brand in demand, and brings full prices; 50s. for ground Felspar, 40s. to 50s.; fluorspar, 20s. to £6. Bog ore s. to 25s. Plumbago: steady; Spanish, £6.; best Ceylon ast quotations; Italian and Bohemian, £4. to £12. per ton; £5. to £6.; Blackwell's "Mineraline," £10. French cargoes, continues scarce on spot—20s. to 22s. 6d. nganese strong at advanced figures. Ground mica, £50. ay freely offering—common, 18s. 6d.; good medium, to 25s.; best, 30s. to 35s. (at Runcorn).

THE LIVERPOOL COLOUR MARKET.

RS.—Whilst there is more doing prices are unaltered. Oxfordshire quoted at £10., £12., £14., and £16.; Derbyshire, s.; Welsh, best, 50s. to 55s.; seconds, 47s. 6d.; and common, 18s, vonshire, 40s. to 45s.; French, J.C., 55s., 45s. to 60s.; M.C., s. 6d. Umber: Turkish, cargoes to arrive, 40s. to 50s.; ire, 50s. to 55s. White lead, £21. 10s. to £22. Red lead, Oxide of zinc: V.M. No. 1, £25.; V.M. No. 2, £23, red, £6. 10s. Cobalt: Prepared oxide, 10s. 6d.; black, blue, 6s. 6d. Zaffres: No. 1, 3s. 6d.; No. 2, 2s. 6d. Terra inest white, 60s.; good, 40s. to 50s. Rouge: Best, £24.; ditto llers', 9d. per lb. Drop black, 25s. to 28s. 6d. Oxide of iron, ality, £10. to £15. Paris white, 60s. Emerald green, 10d. Derbyshire red 60s. Vermillionette, 5d. to 7d. per lb.

THE TYNE CHEMICAL REPORT.

TUESDAY.

hemical market here has been very excited for the last few rtly in consequence of the miners' strike affecting the coal of the Lancashire chemical manufacturers, and partly owing to e of the dock labourers at Liverpool, which has caused a good ders to come to the Tyne which would otherwise have gone to l. Caustic soda has advanced about 30s. per ton; bleach, 15s. soda ash about 10s. a ton; and soda crystals, which were y dropping, have recovered and improved about 2s. 6d. a ton e week. Sulphate of soda continues very firm at same price eek. Large American orders have come to hand for caustic, , bleach and crystals. prices are very irregular, but the may he taken as to-day's market quotations for prompt

nmediate shipment to the United States higher prices are

c soda, 77%, £12. per ton; bleaching powder, in softwood 5. per ton; in hardwood casks, £6. 2s. 6d; soda ash 48—52% per degree less 5%; soda crystals, 48s. 6d. per ton; sulphate ground, in casks, 42s. 6d. per ton, in bulk, 32s. 6d. per overed sulphur £4. 5s. per ton; chlorate of potash 5d. hyposulphite of soda, in 5—7 cwt. casks, £4. 5s. per ton; in egs, £4. 15s. per ton; silicate of soda, 140° Tw., £4. per ° Tw., £3. 7s. 6d. per ton; 75° Tw., £2. 10s. per ton; te sulphate of alumina, £4. 10s. per ton; blanc fixe, £7. 10s. ; chloride of barium, £8. per ton; nitrate of baryta, crystals ; ground, £19. per ton; sulphide of barium, £5. 10s.— Tyne or f.o.r. makers' works.

prices of manufacturing coals have not undergone any material out should the strike in the Midlands continue, prices may be to advance. Durham small is quoted about 10s. per ton, and berland steam small 8s. 6d. a ton.

WEST OF SCOTLAND CHEMICALS.

GLASGOW, Tuesday.

In the paraffin section there is an increased weakness to note as regards burning oil, which, nominally upheld by makers' minimum prices, is in reality very much of a drug on the magazines at the works, and in the hands of dealers who have been unfortunate enough to get into stock. American petroleum quotations have been unusually low, and that has reacted with damaging force on the Scotch article; but other influences have also been operating in the same adverse direction. Paraffin manufacturers are partly compensated by the increasing stability of scale and wax on the market, which are scarce and fetching values a good fraction in excess of the downward stop prices of the association. Lubricating oils also continue firm. There is perhaps a slight improvement to mark in the general chemicals market, although backwardness is still the feature in most lines. Bleach prices are slightly higher, but there is really little doing, and the article is unsettled to a degree. Caustic quotations rise steadily under prospects of a partial fuel famine, but dealers anticipate that at least a temporary level will be reached at an early date. Chlorate of potash is slightly better, but bichromates are all idle. Sulphate of ammonia remains but little changed. Chief prices current are:—Soda crystals, 46s. 6d. net Tyne; alum in lump £5., less 2½% Glasgow; borax, English refined £30., and boracic acid, £37. 10s. net Glasgow; soda ash, 1¼d. less 5% Tyne; caustic soda, white, 76°, £11. 10s.; 70/72°, £9. 15s., 60/62° £9, and 60/62° cream, £8. 2s. 6d., all less 2½% Liverpool; bicarbonate of soda, 5 cwt. casks, £5. 15s., and 1 cwt. casks, £6. net Tyne; refined alkali 48/52°, 1⅛d., less 2½% Tyne; saltcake, 30s. to 32s. 6d.; bleaching powder, £5. 10s. to £5. 15s. less 5% f.o.r. Glasgow; bichromate of potash, 4d., and of soda 3d., less 5 and 6% to Scotch and English buyers respectively; chlorate of potash, 5d., less 5% any port; nitrate of soda 8s. 1½d. to 8s. 6d.; sulphate of ammonia, spot, £11. 15s. f.o.b. Leith; salammoniac 1st and 2nd white, £37. and £35., less 2½% any port; sulphate of copper, £25. 10s. less 5% Liverpool; paraffin scale, hard and soft, 2⅛d. per lb.; paraffin wax, 120°, semi-refined, 3⅛d.; paraffin spirit (naphtha), 8½d.; paraffin oil (burning), 6¾d. to 7d. per gallon at Glasgow; ditto (lubricating), 865°, £5. 10s. to £6. per ton; 885°, £6. 10s.; and 890/895°, £7. 10s. to £8. Last week's imports of sugar at Greenock were 3,574 bags.

TAR AND AMMONIA PRODUCTS.

The benzol market has shown increasing signs of weakness, 90s. benzol not being worth more than 3s. 4d., and 50/90s. 2s. 6d. to 2s. 6½d. Solvent naphtha still continues to be well enquired for, and the same may be said of cresote; while crude carbolic acid is still under a cloud, and B quality of anthracene has fallen to 1s. 1d. per unit. The pitch market is exceedingly weak, and while 31s. f.o.b. has been paid in the Thames and Humber, 29s. is the price at Garston. The outlook is not assuring, but perhaps the finer weather and the long evenings may in the long-run help to re-establish prices.

The sulphate of ammonia market is practically without change, £11. 17s. 6d. down to £11. 16s. 3d. having been obtained f.o.b. Hull, the same prices ruling for London and Leith. At Liverpool the dock strike has prevented business, and quotations are therefore only nominal, as most of the business has been done from other ports.

MISCELLANEOUS CHEMICAL MARKET.

Prices of all products of the alkali trade have been very irregular during the past week. Several of the works in Lancashire have had to suspend operations owing to the short supply of fuel, and others will be in the same position during the coming week if the coal strike is prolonged. Caustic soda has been scarce and difficult to obtain, and 70's have realised £11. 10s f.o.b.; 60's, £10. 10s., and other strengths are advanced in equal degree. Owing to the same causes, soda ash at 1½ per deg. for ordinary kinds; saltcake at 35s. per ton; and soda crystals at £3. per ton, have been eagerly taken up by those who have been compelled to place orders. Bleaching powder has also benefited by the general rush to place orders, and is at £5. 10s. on rails for softwood and £6. f.o.b in hardwood casks. Chlorate of potash remains steady at 5d. per lb. Vitriol, muriatic acid, and sulphur are without change meantime. Sulphate of copper still active, spot price being £26. per ton, and forward deliveries £23. to £24. 10s. Manganese ore still scarce and realised 1s. 4½d. to 1s. 6d. per unit. Lead acetate is firmer, and foreign makers are quoting higher prices—white at £24. and brown £18. 10s. per ton c.i.f. Nitrate of lead quiet at £22. The demand for lead salts is still, however, below the average. Litharge of fine qualities, £15. 15s. to £16. net at usual points of delivery. Acetates of lime are still firm, but only moderate amount of business passing at last week's quotations. Arsenic firm and

THE

CHEMICAL TRADE JOURNAL.

Publishing Offices: 32, *BLACKFRIARS STREET, MANCHESTER.*

No. 148. SATURDAY, MARCH 22, 1890. Vol. V]

Contents.

Notices.

All communications for the *Chemical Trade Journal* should be addressed, and Cheques and Post Office Orders made payable to—

DAVIS BROS., 32, Blackfriars Street, MANCHESTER.

Our Registered Telegraphic Address is—

"Expert, Manchester."

The Terms of Subscription, commencing at any date, to the *Chemical Trade Journal*,—payable in advance,—including postage to any part of the world, are as follow:—

Yearly (52 numbers) .... **12s. 6d.**
Half-Yearly (26 numbers) .... **6s. 6d.**
Quarterly (13 numbers) .... **3s. 6d.**

Readers will oblige by making their remittances for subscriptions by Postal or Post Office Order, crossed.

Communications for the Editor, if intended for insertion in the current week's issue, should reach the office not later than **Tuesday Morning.**

Articles, reports, and correspondence on all matters of interest to the Chemical and allied industries, home and foreign, are solicited. Correspondents should condense their matter as much as possible, write on one side only of the paper, and in all cases give their names and addresses, not necessarily for publication. Sketches should be sent on separate sheets.

We cannot undertake to return rejected manuscripts or drawings, unless accompanied by a stamped directed envelope.

Readers are invited to forward items of intelligence, or cuttings from local newspapers, of interest to the trades concerned.

As it is one of the special features of the *Chemical Trade Journal* to give the earliest information respecting new processes, improvements, inventions, etc., bearing upon the Chemical and allied industries, or which may be of interest to our readers, the Editor invites particulars of such—when in working order—from the originators; and if the subject is deemed of sufficient importance, an expert will visit and report upon the same in the columns of the *Journal*. There is no fee required for visits of this kind.

We shall esteem it a favour if any of our readers, in making inquiries of, or opening accounts with advertisers in this paper, will kindly mention the *Chemical Trade Journal* as the source of their information.

Advertisements intended for insertion in the current week's issue, should reach the office by **Wednesday morning** at the latest.

Advertisements.

Prepaid Advertisements of Situations Vacant, Premises on Sale or To be Let, Miscellaneous Wants, and Specific Articles for Sale by Private Contract, are inserted in the *Chemical Trade Journal* at the following rates:—

30 Words and under.......... 2s. od. per insertion.
Each additional 10 words .... os. 6d. ,,

Advertisements of Situations Wanted are inserted at one-half the above rates when prepaid, viz:—

30 Words and under.......... 1s. od. per insertion.
Each additional 10 words .... os. 3d. ,,

Trade Advertisements, Announcements in the Directory Columns, and all Advertisements not prepaid, are charged at the Tariff rates, which will be forwarded on application.

FACTORY SMOKE.

DURING the past week or two, several correspondent the Manchester papers have exercised their though ness in endeavouring to persuade the public that b smoke is preventable, and that efforts should be made to sure its total abolition. Dr. Patterson, from Oldham, op the ball by suggesting the formation of an associatio prosecute the offenders. Mr. Fred Scott, the secretary o Manchester and Salford Sanitary Association, followe pointing out that his association had been working at the ject for some considerable time, and that, therefore, all ef towards Smoke Abolition should be centred in that Ass tion. These two letters led to another, by "Engineer," pointed out the possibility of preventing factory smoke, writing of an exhibition test, where 10 to 11 lbs. of water evaporated per lb. of fuel by one of the well-known mecha stoking appliances. The correspondence having rea this length, Mr. A. E. Fletcher, the chief inspe of alkali works for the Government, entered the l and adduced reasons why something should be done tow the solution of the problem. Another letter signed "Cle ness," and still another by Mr. J. W. Holden, conclude that has been said within the past few weeks, and it is ceedingly strange that all these correspondents have negle an enumeration of the evils arising from the combustio the sulphur of the coal. Why they should have so negle the sulphurous acid, which is the real damage-doing gredient, it is hard to conceive, but until this ev thoroughly appreciated by our legislators, as well as public, we are afraid the smoke problem will remain solved. It will be of but little use to the general publi see the carbon of the smoke totally consumed, and the still laden with those sulphurous vapours which destro time, almost anything with which they come in contact. correspondent suggests that moderate fines have done good, and by this we presume he would go in for a whole slaughter with heavy fines, and a daily penalty so long the smoke continued. Nothing could be more errone than the belief that such method would lead to good res There are so many smoke producers in the country that one with any sense would dare to tackle them all at one ti and any wholesale system of fining would certainly defea own object in the long run.

There is no doubt but that as the consumption of increases, so will the devastation produced from coal sm also increase, and this may be seen plainly enough in suburbs of all our large towns. Even within the past tw years this is only too patent around Manchester, and have also heard our friends make the same remarks regard to London, Birmingham, Leeds, Nottingham, Leicester. None of the correspondents we have alr quoted as writing to the Manchester papers have even hi what they would have us do with the sulphurous acid of coal smoke. We remember some time ago having our at tion called to an invention for washing coal smoke with l by means of which the sulphurous acid was extracted;

IMPORTS OF CHEMICAL PRODUCTS AT THE PRINCIPAL PORTS OF THE UNITED KINGDOM.

NDON.

nding Mar. 11th.

Ore—
5 t. C. J. Fox & Co.
1 Pillow, Jones, & Co.
H. Rogers, Sons, & Co.
25 pkgs. A. & M. Zimmermann
1 Arnati & Harrison
oride—
40 Petri Bros.
80 t. C. C. Bryden & Co.
76 Gonne, Croft, & Co.
10 „
2½ G. Davis & Co.
t. F. Parbury & Co.
17 c. G. Boor & Co.
2 F. Smith
J. Kitchin
B. Jacob & Sons
Johnson & Hooper
W. C. Bacon & Co
25 t. Flack, Chandler, & Co.
2½ „
00 Brandram, Bros. & Co.
cks. Leach & Co.
Magee, Taylor, & Co.
pkgs. Leach & Co.
ck. Brandram, Bros., & Co.
W. Harrison & Co.
D. Storer & Sons
3 Props. Scott's Wf.
2 J. L Lyon & Co.
199 t. Charlesworth & Co.
1 Goad, Rigg, & Co.
1½ Redfern, Alexander, & Co.
c—
26¾ c. L. & I. D. Jt. Co.
37 Elkan & Co.
7½ Clarke & Smith
13 Wrightson & Son
1½ Belize Estate & Prod. Co.
21 L. & I. D. Jt. Co.
43 Forbes, Forbes, & Co.
33 Henderson Bros.
118 R. & J. Henderson
12¾ Wallace Bros.
11 L. & I. D. Jt. Co.
198 W. Brandt, Sons, & Co.
20 Matheson & Co.
15 Wallace Bros.
9 L. & I. D. Jt. Co.
90 Kebbel, Son, & Co.
47½ „
14 D. C. Thomas & Son
59 L. & I. D. Jt. Co.
tar—
4 pkgs. E. W. Carling & Co.
16 B. & F. Wf. Co.
5 Middleton & Co.
5 W. C. Bacon & Co.
5 J. Puddy & Co.
5 B. & F. Wf. Co.
5 W. C. Bacon & Co.
5 B. & F. Wf. Co.
8 T. Ronaldson & Co.
—
58 t. L. & I. D. Jt. Co.
(otherwise undescribed)
£60 A. & M. Zimmermann
8 T. H. Lee
20 W. Harrison & Co.
21 W. G. Jacobs
12 L & I. D. Jt. Co.
806 A. & M. Zimmermann
69 Craven & Co.
11 Hotzapfel
—
50 pkgs. Bailey & Leetham
80 T. H. Lee
79 Levinstein & Sons
95 „
1 G. Meyer & Co

Holland, 63 Burt, Boulton & Co.
Germany, 2 L. & I. D. Jt. Co.
„ 2 Fulwood & Bland
Austria, 700 A. J. Humphrey
Canada, 40 „
Saffron
Spain, 1 pkg. C Brumlen
Sumac
Italy, 530 pkgs. W. France & Co.
„ 950 J. Kitchin
„ 500 W. Shelcott
„ 50 Larkins, Malcolmson, & Co.
Cutch
E. Indies, 500 pkgs. Wallace Bros.
„ 1,000 Hoare, Wilson, & Co
„ 40 W. Isted
Capula
E. Indies, 12 pkgs. Wallace Bros
Turmeric
E. Indies, 100 pkgs. Wallace Bros.
Gambier
E. Indies, 417 pkgs. Johnson, Rolls, & Co.
„ 283 L. & I. D. Jt. Co.
„ 168 P. & S. Evans & Co.
„ 195 Elkan & Co.
„ 1,030 E. Boustead & Co.
„ 276 J. H. & G. Scovell
„ 347 Beresford & Co.
„ 845 Kleinwort, Sons, & Co.
Myrabolams
E. Indies, 2,314 pkgs. L. & I. D. Jt. Co.
„ 90 „
„ 334 „
„ 667 D. Sassoon & Co.
„ 173 Lewis & Peat
„ 2,668 Marshall & French
„ 330 Baxter & Hoare
„ 334 L & I. D. Jt. Co.
„ 667 „
Tanners' Bark
Belgium, 50 t. J. Graves
U. S., 30 H. Kohnstamn
Victoria, 316 Beresford & Co.
„ 127½ „
„ 58¾ Hicks, Nash, & Co.
Belgium. 12 S. Bevington & Sons
Indigo
E. Indies, 121 chts. Elkan & Co.
„ 514 L. & I. D. Jt. Co.
„ 54 Patry & Pasteur
„ 16 F. H. Allen & Co.
„ 9 Reynolds & Sellers
„ 3 W. Brandt, Sons, & Co.
„ 15 Phillipps & Graves
„ 22 Parsons & Keith
„ 108 Ernsthausen & Co.
„ 9 Stansbury & Co.
„ 10 Benecke, Souchay, & Co.
„ 30 J. Owen
„ 17 Langstaff & Co.
„ 10 cs. Gt. N. Ry. Co.
„ 324 chts. L. & I. D. Jt. Co.
„ 102 J. A. Payne
„ 88 Arbuthnot, Latham, & Co.
„ 41 Temperleys & Co.
„ 16 Frauling & Fisher
„ 10 Kaul & Haenlein
„ 28 F. W. Heilgers & Co.
„ 96 Ernsthausen & Co.
„ 18 Hatton, Hall, & Co.
„ 5 F. Stahlschmidt & Co.
„ 34 Wallace Bros
Germany, 12 F. W. Heilgers & Co.
E. Indies, 4 Elkan & Co.
„ 5 F. Huth & Co.
„ 4 C. J. Hambro & Son
„ 5 T. Barlow & Bros.
„ 20 H. Johnson & Sons
„ 196 Ernsthausen & Co.
„ 166 L. & I. D. Jt. Co.
„ 19 Ralli Bros.
Indigo
E. Indies, 364 chts. L. & I. D. Jt. Co.
„ 40 „
„ 12 F. Stahlschmidt & Co.
„ 15 Ralli Bros.
„ 16 Patry & Pasteur
„ 10 G. Dards
„ 20 J. Forsey
„ 5 J. Owen
„ 65 Phillipps & Graves
„ 95 1 bx. Elkan & Co.
„ 25 Sawer, Mead, & Co.
„ 89 pkgs. T. Ronaldson & Co.
„ 6 bxs. Henderson Bros.
„ 10 chts. Seton, Laing, & Co.
„ 10 2 cs. L. & I. D. Jt. Co.

„ 49 F. W. Heilgers & Co
„ 6 Oldemeyer & Hadenfeldt
„ 29 Stansbury & Co.
„ 115 Ralli Bros.
„ 58 Fruhling & Goschen
„ 24 Ernsthausen & Co.
„ 69 Elkan & Co.
„ 5 Kaul & Haenlein
„ 11 Phillipps & Graves
„ 5 F. Huth & Co.
„ 2 G Dards
Cochineal
Canaries, 17 pkgs. Bruce & White
„ 18 J. Goddard & Co.
Orchella
Germany, 82 pkgs. Kebbel, Son, & Co.
Italy, 8 Wrightson & Son
Argols
France, 50 Skilbeck Bros.
Italy, 531 B. Jacob & Sons
„ 200 D. Magnus
Gutta Percha—
E. Indies, 1,160 c. H. W. Jewesbury & Co.
„ 483 Kleinwort, Sons, & Co
„ 250 J. H. & G. Scovell
„ 80 G. Ward & Sons
„ 545 H. W. Jewesbury & Co.
„ 249 Kaltenbach & Schmitz
„ 1,069 R. & J. Henderson
„ 245 Kaltenbach & Schmtz
Glucose—
Germrny, 110 pkgs. 110 owts. J. Barber & Co.
„ 100 100 Hyde, Nash & Co.
„ 340 520 Barrett, Tagant, & Co.
„ 25 240 J. H. Epstein
U. States, 40 212 L. Sutro & Co.
France, 220 220 De Paiva, Norman, & Co.
„ 80 80 Seaward Bros.
Germany, 25 140 J. H. Epstein
„ 115 181 Prop. Chamberlain's Wf.
„ 32 34 F. Stalhschmidt & Co
U. S., 49 345 T. M. Duche & Sons
Glycerine—
Holland, £50. F. W. Heilgers & Co.
N. Zealand, 83½ J. Connell & Co.
Germany, 160 Beresford & Co.
France, 480 Spies Bros. & Co.
„ 95 F. W. Heilgers & Co.
Germany, 125 Beresford & Co.
N. Zealand, 126 „
Germany, 83 Craven & Co.
Isinglass—
E. Indies, 14 pkgs.
„ 14 L. & I. D. Jt. Co.
Germany, 2 T. H. Lee
E. Indies, 8 Kaltenbach & Schmitz
„ 36 Clarke & Smith
„ 20 L. & I. D. Jt. Co.
„ 3 Hale & Son
China, 2 W. M. Smith & Sons
E Indies, 19 L. & I. D. Jt. Co.
Magnesia —
Holland, £250. Beresford & Co.
Magnesium —
Germany, £40. Phillipps & Graves
Manure—
Phosphate Rock
France, 180 t. A. Hunter & Co
B. W. Indies, 350 t. Pickford & Winkfield
Belgium, 170 Lawes, Manure Co.
U. S., 2,732 Wylie & Gordon
Manure *(otherwise undescribed*
France, 100 t. J. Burnett & Sons
U. S, 2 t. C. & A. H. Nichols
Naphtha—
Holland, 4 drs. Lister & Biggs
Belgium, 19 Stein Bros.
Potassium Sulphate—
Norway, 57 pkgs. B. & F. Wf. Co.
Petroleum—
S. Russia, 895,000 gals Smith & Charles
U. S., 4,300 brls. J. H. Usmar & Co.
„ 30 M. Dk. Co.
Holland, 34 drms Burt, Boulton, & Co.
U. S, 50 brls. Barrett, Tagant, & Co.
Belgium, 140 T. Steam Tug Co.
Pearl Ash—
France, 40 c. F. W. Berk & Co.
Pyrites—
Spain, 987 t. G. Ward & So

Paraffin Wax—
U. S., 75 brls. W. H. Mitchell
„ 75 G. H. Frank
„ 3,659 J H. Usmar & Co.
„ 200 Rose, Wilson, & Rose
„ 50 H. Hill & Sons
„ 25 W. H. Mitchell
Pitch—
France, 70 brls. Berlandina Bros. & Co.
Plumbago—
Ceylon, 80 cks. Doulton & Co.
„ 15 Anderson Bros.
„ 105 Hoare, Wilson, & Co.
„ 82 H. Lambert
„ 3 Marshall & French
„ 391 Thredder, Son, & Co.
„ 15 L. & I. D. Jt. Co.
„ 187 Prop. Chamberlain's Wf.
Holland, 124 Brown & Elmslie
Saltpetre—
Germany, 70 cks. J. Hall & Son
„ 80 kgs. P Hecker & Co.
E. Indies, 236 bgs. C. Wimble & Co.
„ 480 Clift, Nicholson, & Co.
„ 1,223 C. Wimble
„ 495 C. Wimble & Co.
Germany, 8 cks. Craven & Co.
Stearine—
Belgium, 109 bgs. H. Hill & Sons
Sugar of Lead—
Germany, 12 pkgs. W. J. Crook
„ 14 Hernu, Peron, & Co.
Sodium Acetate—
France, £78. W. G. Blagden
Tartars—
Italy, 11 pkgs Fellows & Co.
„ 31 Thames S. Tug Co.
Tartaric Acid—
Holland, 12 pkgs. B. & F. Wf. Co.
„ 16 „
Tartar Salts—
France, 40 c. F. W. Berk & Co.
Ultramarine—
Germany, 25 pkgs. G. Steinhoff
„ 40 „
Holland, 9 Hernu, Peron, & Co.
„ 10 G. Rahn & Co.
Verdigris—
France, £42. Skilbeck Bros.
„ 40 C. F. Gerhardt
Zinc Oxide—
Holland, 130 brls. M. Ashby
„ 97 Beresford & Co

LIVERPOOL.

Week ending Mar. 13th.

Antimony Ore—
Smyrna, 975 bgs.
Acetic Acid—
Rouen, 2 cks. R. J. Francis & Co.
Acids *(otherwise undescribed)*—
Hamburg, 4 cks.
Bones—
Constantinople, 200 bgs.
Calcutta, 588
Smyrna, 210 t.
Bone Meal—
Calcutta, 2,666 bgs.
Barytes—
Rotterdam, 12 cks.
Caoutchouc—
Accra, 8 brls. F. & A. Swanzy
„ 2 cs. Davies, Robbins, & Co.
Salt Pond, 2 1 brl. Radcliffe & Durant
„ 4 4 Pickering & Berthoud
„ 3 W. Griffiths & Co.
„ 1 Millersen & Co.
„ 4 Miller, Bros., & Co.
„ 2 1 S. & C. Nordlinger
„ 1 2 Edwards Bros.
Cape Coast, 7 cs. Millerson & Co.
Salt Pond, 1 brl.
„ 6 Ihlers & Bell
„ Havard & Co.
„ 4 1 cs. C. L. Clare & Co.
„ 5 Fletcher & Fraser
„ 1 W. Duff & Co.
„ 1 Pickering & Berthoud
„ 1 T. E. Tomlinson & Co.
„ 1 1 cs. A. Reis
Sierra Leone, 5 cks. Paterson & Zachonis
„ 27 brls.
„ 9 Pickering & Berthoud

Bathurst, 3 cks. Cie Francaise
Old Calabar, 2 W. Dodd & Co.
" 4 Pickering & Berthoud
" 3
Accra, 2 brls. Davies, Robbins, & Co.
" 2 Morgan & Co.
Sierra Leone, 10 Pickering & Berthoud
" 1 keg. 4 cks. Paterson, Zachonis, & Co.
" 4 brls. N Waterhouse & Sons
" 17 Cie Francaise
" 67 33 cks
New York, 4 cs.
Para, 479 R. Singlehurst & Co
" 1 Thin & Sinclair
" 59 Bieber & Co
" 1,662
" 4
Cameroons, 1 ck. Rider, Son, & Co.
" 2 Lucas Bros.
" 2 J. Holt & Co.
" 2 A. Herschell
" 22 R. & W. King
Victoria, 1 cs. E. Ost & Co.
Old Calabar, 4 J. Holt & Co.
New York, 30 bgs. C. Macintosh & Co.

Cream of Tartar—
Barcelona, 6 hds.
Rotterdam, 20 kegs.

Chemicals *(otherwise undescribed)*
Dunkirk, 29 cks.
Antwerp, 3 3 cs.

Dyestuffs—

Cochineal
Teneriffe, 2 bgs. Beriro & Co.

Sumac
Barcelona, 100 bgs. A. Ruffer & Sons

Gambier
Singapore, 432 bls. E. Boustead & Co
" 581 395 bgs.

Argols
Oporto, 22 cks. German Bank of London

Indigo
Calcutta, 275 chts.

Annatto
Bordeaux, 13 cks.

Myrabolams
Bombay, 2,013 bgs.
" 866 Ralli Bros.
" 4,703 D. Sassoon & Co.
" 496 Beytes, Craig, & Co.

Extracts
Boston, 10 brls.
Havre, 27 cks. Cunard S. S. Co
Rouen, 299 Co-op. W'sale Socy., Ltd.
" 100 "
New York, 100 bxs.
Boston, 150

Valonia
Smyrna, 350 bgs. S. Vivante & Son

Orchella
Lisbon, 22 bgs. Leeds & Liverpool Canal Co.

Dyestuffs *(otherwise undescribed)*
New York, 9 cs. Snow Hill E. Co.

Dried Blood—
Monte Video, 109 sks. J. Burdan

Glucose—
New York, 90 brls.
Philadelphia, 98
Hamburg, 15 cks. 220 bgs.

Glycerine—
Rotterdam, 14 dms.
" 3 crbys. 5 cs.

Isinglass—
Para, 17 cs. Bieber & Co.
" 14 Firth, Sands & Co.
Maranham, 4 pkgs. H. Rogers, Sons & Co.

Manganese Ore—
Oporto, 200 t. 5 brls.

Oxalic Acid—
Dunkirk, 4 cks.

Paraffin Wax—
Rangoon, 488 bxs.

Potassium Bromide—
New York, 20 cs.

Potash—
Rouen, 90 cks. 21 dms. Co-op. W'sale Society, Limited
Dunkirk, 20 cks.
Rouen, 16 "
Portland, 25 brls. Makin & Bancroft

Phosphate—
Ghent, 450 t. The Phosphate Guano Co.

Pitch—
Rotterdam, 21 cks.
" 44

Pyrites—
Huelva, 934 t. Tennants & Co.
" 1,494 "

Sodium Nitrate—
Rotterdam, 5 cks.
Caleta Buena, 10,413 bgs.

Sugar of Lead—
Rouen, 38 cks. R. J. Francis & Co.

Sulphur Ore—
Huelva, 1,121 t. Matheson & Co.
" 1,362 "

Soda—
Hamburg, 17 cks.

Saltpetre—
Calcutta, 1,069 bgs.

Stearine—
Philadelphia, 25 hds.

Sodium Aluminate—
Hamburg, 1 ck.

Tartar—
Bordeaux, 10 cks.
Havre, 7 Cunard S.S. Co.

Tartar Emetic—
Hamburg, 4 cks.

Tartaric Acid—
Rotterdam, 33 cks.
" 6

Ultramarine—
Rotterdam, 40 brls. 16 cks. 5 cs.
" 20

Verdigris—
Bordeaux, 2 cks.

Waste Salt—
Hamburg, 102 bgs.

Zinc Oxide—
Rotterdam, 100 cks. O. & H. Marcus
Antwerp, 50 brls.

Zinc Ashes—
Rotterdam, 17 cks.
Boston, 22 brls.
Hamburg, 98 bgs.

HULL.

Week ending March 13th.

Acetic Acid—
Rotterdam, 17 galls. Hutchinson & Son

Acid *(otherwise undescribed)*
Bremen, 1 ck. Hill & Co.
Rotterdam, 2 Hutchinson & Son

Alumina—
Rotterdam, 42 cks. W. & C. L. Ringrose

Barytes—
Rotterdam, 11 cks.
Bremen, 23 Tudor & Sons
" 10

Chemicals *(otherwise undescribed)*
Dunkirk, 90 pks. Wilson, Sons & Co.
Antwerp, 100 bgs. 103 pkgs. "
Bari, 37 cks. "
Siomemunde, 4 "
Bari, 22 "
Rouen, 39 Rawson & Robinson
Rotterdam, 2 3 cs. Hutchinson & Son
Dunkirk, 24 4 Wilson, Sons & Co.

Caoutchouc—
Hamburg, 2 cs.

Dyestuffs—

Alizarine
Rotterdam, 7 pks. W. & C. L. Ringrose
" 67 cks. Hutchinson & Son
" 27 pkgs. W. & C. L. Ringrose
" 62 cks. Hutchinson & Son
" 78 W. & C. L. Ringrose

Extracts
Dunkirk, 46 cks. Wilson, Sons & Co.
Trieste, 39 brls. "
" 20 "
Fiume, 450 "
Rouen, 74 cks. Rawson & Robinson
Bordeaux, 50 "
Dunkirk, 38 bls. Wilson, Sons & Co.

Madder
Rotterdam, 3 cks. J. Pyefinch & Co.

Valonia
Smyrna, 332 bgs. 55 tns.

Sumac
Trieste, 72 bgs. Wilson, Sons & Co.
Palermo 800 "
" 200 "

Glucose—
Hamburg, 30 cks. Wilson, Sons & Co.
" 15 Woodhouse & Co.
" 100 bgs.
New York, 180 brls. Wilson, Sons & Co.
Stettin, 34 cks. 180 bgs. Wilson, Sons & Co
" 1,200 C. M. Lofthouse & Co.
" 100
Dunkirk, 100 bgs. Wilson, Sons & Co.

Glycerine—
Hamburg, 1 drum.

Naphthol—
Rotterdam, 5 brls. Wilson, Sons & Co.

Pitch—
Antwerp, 10 cks. Wilson, Sons & Co.

Phosphorus—
Hamburg, 24 cs.

Phosphate—
Dunkirk, 200 bgs. Wilson, Sons & Co.
Ghent, 260 t. "
Dunkirk, 1,500 bgs. "

Pyrites—
Huelva, 1,445 t. J. Dalton, Holmes & Co.

Soda—
Rotterdam, 8 cks. Hutchinson & Son
" 75 G. Lawson & Sons

Sodium Nitrate—
Iquique, 5,674 bgs.

Saltpetre—
Hamburg, 22 brls.

Sulphur—
Catania, 1,494 pks. Wilson, Sons & Co.
" 130 t 135 pkgs.

Tartar—
Bordeaux, 5 cks. Rawson & Robinson

Tartaric Acid—
Rotterdam, 4 cks. G. Lawson & Sons

Tar—
Danzig. 57 cks. Wilson, Sons & Co.
Dunkirk, 3 cs. "

TYNE.

Week ending Mar. 13th.

Barytes—
Rotterdam. 46 cks. Tyne S. S. Co.

Copper Precipitate—
Huelva, 6,958 bgs. Scott Bros.

Glucose—
Hamburg, 9 cks. Tyne S. S. Co.

Manganese Ore—
Batoum, 650 t.

Phosphate of Lime—
Antwerp, 600 t. 2,000 bgs. Langdale's Manure Co.

Petroleum—
Hamburg, 49 brls. Tyne S. S. Co.

Saltpetre—
Hamburg, 10 kgs. Tyne S. S. Co.

Sulphur Ore—
Drontheim, 470 t. R. S. Story & Co.
Pomaron, 950 Mason & Barry

Tartaric Acid—
Rotterdam, 2 cks.

Zinc Oxide—
Antwerp, 25 brls. Tyne S. S. Co.

GLASGOW.

Week ending March 13th.

Aluminium—
Antwerp, 2 cs.
" 1

Alum—
Rotterdam, 7 cks. J. Rankine & Son

Barytes—
Bremerhaven, 20 cks.
Rotterdam, 71 bgs. J Rankine & Son

Copper Ore—
Huelva, 1,960 t. Th
" 1,723

Chrome Ore—
Macri, 1,941 t.

Dyestuffs—

Alizarine
Rotterdam, 58 cks.
" 36

Extracts
Rouen, 9 cks.
Antwerp, 8
Belize, 465 t. McArthur, Scc
Rotterdam, 3 cks. J. Ranki

Glucose—
Philadelphia, 120 brls.
New York, 150
Rotterdam, 16 cks. J Ranki

Plumbago—
Leghorn, 10 bgs.

Phosphate—
Antwerp, 100 t.

Pitch—
—— 9 cks.

Soda—
Antwerp, 19 cks.

Saltpetre—
Rotterdam, 7 cks. J. Ranki

Tartar—
Oporto, 6 cks. J. & P. H

Tartaric Acid—
Rotterdam, 1 ck. J. Ranki

Ultramarine—
Rotterdam, 5 cs. "

Waste Salt—
Hamburg, 400 t.

Zinc Ashes—
Antwerp, 8 cks.

Zinc Oxide—
Antwerp, 50 cks.
—— 75

GRIMSBY

Week ending Mar. 8t.

Albumen—
Dieppe, 15 pkgs.

Caoutchouc—
Hamburg, 5 cs.

Chemicals *(otherwise unde*
Antwerp, 4 pkgs.

Glucose—
Hamburg, 26 cs.
" 15

Zinc Ashes—
Antwerp, 16 cks.

GOOLE.

Week ending March 5

Alumina Sulphate—
Antwerp, 40 brls.

Caoutchouc—
Boulogne, 8 cks.

Dyestuffs—

Alizarine
Rotterdam, 39 cks.

Extracts
Rotterdam, 2 cks.

Dyestuffs *(otherwise unde*
Boulogne, 28 cks. 24 cs.

Glycerine—
Hamburg, 3 cs. 2 cyb

Glucose—
Hamburg, 40 bgs.
Dunkirk, 220

Naphthol—
Rotterdam, 5 cks.

Phosphate—
Ghent, 800 bgs.

Potash—
Dunkirk, 16 cks.
Calais, 22

Plumbago—
Boulogne, 3 cks.

Saltpetre—
Rotterdam, 325 bgs.
Antwerp, 100

Potassium Muriate—
Rotterdam, 11 cks.

Sugar of Lead—
Hamburg, 33 cks.

Waste Salt—
Hamburg, 55 cks.

EXPORTS OF CHEMICAL PRODUCTS

LONDON.

…k ending March 8th.

…Acid—
…rne, 6 cks. £21
…ury, 15 cs. 27
…ium Sulphate—
…d, 89 t. 2 c. £1,225
…rg, 2,296 1,361
20 230
…ra, 25 250
…, 100 18 1,160
…rg, 50 14 605
…k, 20 8 245
36 6 425
…rg, 20 200
…ra, 23 3 234
…rg, 50 14 605
…, 10 5 120
…ium Muriate—
…, 1 t. 2 c. £29
…ake—
…a, 27 t. 10 c. £100
…ium Carbonate—
…ma, 3 t. £112
… (otherwise undescribed)
100 cs. £250
…l, 5 c. 42
…e—
…rne, 5 t. £79
…c Acid—
…rg, 2 c. £33
…sh—
…lles, 10 t. 13 c. £190
…roducts—
…htha
…urne, 15 drs. £15
…ls, 2 2
…urne, 63 146
…zole
…, 19 pns. £375
10 cks. 206
…urg, 19 pns. 290
…dam, 30 437
…alla, 4 drs. 109
…dam, 25 19 cks 905
…urg, 20 343
…h
…rk, 140 t. £215
…, bulk. 406
…rk, 371 631
225 383
…, 80 brls. 31
1,093 1,913
…y, 50 16
51 22
…, bulk. 1,669
…dria, 120 36
…, 250 £150
…, 100 runlets 61
…o, 60 36
…i, 100 52
…y, 200 110
…i, 200 53
…, 4 cks. 3
…e, 30 drs. 8
…d, 15 brls. 20
100 runlets 55
10 cks. 74
50 drs. 4
…ancisco, 500 31
…Kong, 20 cks. 15
…y. 68 runlets 51
…idine Bases
…rg, 12 cks. £17
…ral Naphtha
…rne, 25 drms. £17
…sote
…ux, 2,200 brls. £1,320
…olic Acid
…rg, 5 cks. £32
…racene
…lorf, 210 c. £348
…e, 200 340
…lorf, 480 cks. 3,308
…lam, 565 3,906
63 340
…thaline
…a, 2 c. £19
3 2
…ork, 201 95
…lam, 10 cks. 10
…il
…les, 100 brls. £75
…olic Acid Crystals
…les, 120 cks. 1,680
…ork, 10 cs. 65
…c Solid—
…ma, 2,000 lbs. £173
…c Acid—
…am, 49 cks. £285
…Bay, 20 drms. 18
…ma, 2,000 lbs. 144

Caustic Soda—
Oporto, 4 t. £40
Auckland, 10 87
Brisbane, 5 12 c. 48
Cream of Tartar—
Wellington, 1 t. 3 c. £120
Auckland, 1 105
Dunedin, 15 100
Auckland, 1 105
Melbourne, 5 cs. 31
Adelaide, 15 84
Chemicals *(otherwise undescribed)*
Amsterdam, 4 c. £40
Barcelona, 7 cs. 95
Dunedin, 26 pkgs. 55
Callao, 10 16
" 14 40
New York, 7 t. 17 c. 800
" 10 cs. 140
Copper Sulphate—
Treport, 7 t. 2 c. £173
Venice, 100 cks. 634
Trieste, 20 127
Venice, 5 5 c. 135
Genoa, 61 14 1,792
Venice, 21 16 553
Trieste, 5 1 128
Treport, 5 129
Citric Acid—
Libau, 20 cks. £677
Sydney, 5 c. 40
Baltimore, 1 t. 5 185
Oporto, 3 cks. 22
Genoa, 25 187
Paris, 1 c. 7
Hamburg, 3 kegs 22
Marseilles, 121 t. 4 c. 559
Adelaide, 14 lbs. 1
Cologne, 5 c. 35
New York, 2 t. 265
Santander, 6½ c. 39
Disinfectants—
Madras, 21 c. £30
Melbourne, 2 t. 10 c. 40
Paris, 368 gls. 30
Halifax, 1 ck. 10
Dried Blood—
Oporto, 6 t. 3 c. £50
Fluoric Acid—
Melbourne, 540 gls. £33
Guano—
Trinidad, 89 t. 2 c. £841
Demerara, 75 t. 750
St. Kitts, 25 250
Hydrochloric Acid—
Rosario, 10 cs. £11
Hard Water Crystals—
Halifax, 2 cs. £22
Insect Powder—
Pt. Natal, 6 c. £52
Manure—
Fertilizers
Brisbane, 20 t £270
New York, 3 45
Stettin, 226 2 c. 940
Gandia, 641 11 5,457
Hamburg, 320 1,245
Gothenburg 322 12 1,308
Rotterdam, 9 15 42
" 29 6 88
St. Lucia, 20 220
Antigua, 25 250
Rotterdam, 15 135
Demerara, 52 250
Rotterdam, 121 4 559
" 6 18 24
Dunkirk, 15 60
Singapore, 14 5
Cologne, 39 7 185
Stettin, 49 6 214
Bordeaux, 25 100
Rouen, 122 10 600
Demerara, 540 bgs. 516
" 80 800
Oxalic Acid—
Antwerp, 5 t. 11 c. £200
" 6 10
Melbourne, 10 22
Potassium Chlorate—
Philadelphia, 1 t. 5 c. £60
Picric Acid—
Melbourne, 1 cs. £12
Superphosphate—
Rotterdam, 31 t. 14 c. £160
Saltpetre—
Rio Janeiro, 2 t. 10 c. £56
Porto Allegre, 17 21
Rio Janeiro, 1 11 35
Sulphur Flowers—
Algoa Bay, 2 t. 5 c. £27
Lisbon, 250 brls. 193

Sodium Sulphate—
Ghent, 40 t. £40
Sulphuric Acid—
Port Natal, 60 cs. £75
Trinidad, 1 pkg. 1
Rosario, 10 cs. 10
Karachi, 100 drs. 132
Sodium Silicate—
Yokohama, 6 t. 18 c. £29
Sheep Dip Powder—
Algoa Bay, 80 cs. £98
M. Video, 110 drs 75
Algoa Bay, 305 875
E. London, 150 725 597
Algoa Bay, 1 t. 42
Sydney, 49 gls. 12
Sulphur—
E. London, 160 cks. £141
Calcutta, 7 t. 19 c. 38
Soda Crystals—
Yokohama, 5 t. 11 c. £23
Gibraltar, 10 2 30
Sodium Bicarbonate—
Auckland 2 t. £87
Sodium Acetate—
Melbourne, 2 cks. £17
Superphosphate—
Penang, 2 t. £12
Cherbourg, 185 555
Salammoniac—
Smyrna, 11 c. £21
Salonica, 1 t 16 c. 60
Tartaric Acid—
Wellington, 5 c. £33
Auckland, 5 34
Dunedin, 7 58
Hamburg, 10 6

LIVERPOOL

Week ending Mar. 12th.

Antimony—
Madras, 6 c. £21
Alum Cake—
Calcutta, 12 t. 12 c. £54
Alum—
Galatz, 5 t. 5 c. 3 q. £26
Oporto, 5 14 1 28
Halifax, 2 12 13
Smyrna, 2 4 1 13
Galatz 7 11 39
Bombay, 8 3 42
Piræus, 4 10 22
Talcahuano, 1 5
Rio de Janeiro, 4 1 12
Ammonium Sulphate—
Antwerp, 1 t. 10 c. £18
Hamburg, 79 948
Bordeaux, 25 300
Demerara, 15 11 202
" 119 9 2 q. 1,497
Hamburg, 90 17 3 1,030
Las Palmas, 5 2 61
Hamburg, 29 7 336
Bordeaux, 20 240
Ammonium Carbonate—
Ancona, 5 c. 2 q. £17
Rotterdam, 2 t. 5 84
Genoa, 1 10 49
Philadelphia, 10 19 2 q. 494
Leghorn, 1 33
Valparaiso, 1 5 44
Hamburg, 5 3 7
Havana, 6 9
Bleaching Powder—
Bordeaux, 27 cks.
Boston, 252
Genoa, 26 pps.
Odessa, 6
Rio Janeiro, 10 cs.
Rotterdam, 26
Rouen, 196
San Sebastian, 18 brls.
Santos, 1
Bilbao, 54
Boston, 192
Chicago, 125 pkgs.
Hamburg, 29 55 tcs.
New York, 351
Vera Cruz, 8
Boracic Acid—
New Orleans, 6 c. £11
Bisulphide of Carbon—
Maranham, 5 c. £8
Borax Salts—
Havana, 1 t. 5 c. £68
Caustic Soda—
Adelaide, 150 dms.
Algiers, 6
Amsterdam, 14
Ancona, 15
Bahia, 40 dms.
Bordeaux, 17
Boston, 200
Corunna, 20 brls.
Corfu, 40
E. London, 2 bxs.
Fiume, 15
Havana, 10 40 kgs.
La Guayra, 5
Malta, 18
Mexico City, 300
Nantes, 69
New York, 658
Oporto, 64
Portland, 200
Rio Grande do Sul, 30 dms.
Rouen, 20
San Sebastian, 10 brls.
Santander, 3
Venice, 72
Philadelphia, 100
Genoa, 30
Alicante, 8
Amsterdam, 175
Ancona, 10
Barcelona, 100
Beyrout, 5
Chicago, 80 bxs.
Galatz, 10 50 kgs.
Genoa, 12
Hamburg, 55
Havana, 10
Havre, 18
Leghorn, 51
Madeira, 70
Mauritius, 10
Mazatlan, 100
New Orleans, 56
New York, 925 10 brls.
Odessa, 30
Parahiba, 20
Philadelphia, 86
Rotterdam, 47
San Francisco, 100
St. John, 25
Seville, 24
Singapore, 3
Tarragona, 40
Valencia, 25 cks.
Vera Cruz, 10
Carbolic Solid—
St. Louis, 5 t. 2 c. £464
Hamburg, 12 3 q. 50
Cobalt Oxide—
Gijon, 1 c. £45
Valencia, 2 q. 50
Chlorate—
Vigo, [illegible] c. £18
Chloride of Lime—
New York, 51 [illegible] 5 c. £295
Bilbao, 8 1 53
Carbon Sulphide—
Rio de Janeiro, 3 [illegible] c. £5
Charcoal—
Hamburg, 113 t. 13 c. 3 q. £595
Chemicals *(otherwise undescribed)*
Halifax, 1 t. 3 c. 1 q. £35
Havre, 1 4 2 140
Copper Sulphate—
Bilbao, 20 t. 1 q. £399
Marseilles, 8 3 c. 1 204
Barcelona, 9 18 1 244
Leghorn, 1 25
St. John's, 14 21
Tarragona, 10 3 3 35
" 10 [illegible]
Barcelona, 15 [illegible]
Bordeaux, 2 10 35
Ancona, 4 15 2 60
Genoa, 10 11 96
Leghorn, 5 1 195
Venice, 9 6 94
Barcelona, 25 8 2 q. 738
Bordeaux, 22 3 2 738
Barcelona, 4 411
Nantes, 10 5 40
St. John's, 4 3 216
Tarragona, 30 3 2 6
Barcelona, 18 2 674
Genoa, 126 10 410
" 3 1 2,396
Havre, 5 35
Naples, 3 11 1 125
Bordeaux, 182 19 3 67
Leghorn, 1 8 2 4,555
Lisbon, 1 8 3 25
Malta, 4 18 32
Venice, 34 2 2 114
635
Carbolic Acid—
New York, 3 t. 15 c. £233
Hamburg, 8 4 741
Havre, 19 10 239
Calcutta, 9 6 2 q. 140
Cadiz, 1 3 12
Carthagena 1 3 12
Boston, 230

Citric Acid—
- Piræus, 5 c. £37
- Carril, 4 — 28
- Portland, 2 — 1 q. 16

Copperas—
- Valparaiso, 7 t. 4 c. 1 q. £16
- Alexandria, 14 3 3 — 29

Caustic Potash—
- Chicago, 4 t. 7 c. £87
- Shanghai, 1 — 1 q. 21

Epsom Salts—
- Malaga, 14 brls.

Guano—
- Malaga, 29 t. 17 c. 3 q. £291

Glycerine—
- Vera Cruz, 2 drs. £5

Iodine—
- New York, 2 t. £1,200

Magnesium Calcide—
- Genoa, 84 lbs. £100

Magnesium Chloride—
- Bombay, 23 t. 17 c. £74

Manure—
- Havre, 150 t. 2 c. 3 q. £476

Magnesia—
- Portland, 5 cs.
- Rangoon, 10 kegs.
- Rio Janeiro, 10
- Genoa, 4
- Gijon, 1

Oxalic Acid—
- San Francisco, 2 t. 11 c. 2 q. £75

Pitch—
- Talcahuano, 6 t. 5 c.
- Bilbao, 11 t.

Potassium Oxalate—
- Madras, 5 c. 3 q. £9

Potassium Chlorate—
- Barcelona, 1 t. 4 c. £50
- Corunna, 10 — 22
- New York, 10 — 448
- Trieste, 2 10 — 130
- Yokohama, 5 — 210
- Ancona, 6 — 14
- Carril, 10 — 22
- Barcelona, 4 16 — 185
- Boston, 5 — 198

Phosphate—
- Hamburg, 25 t. 14 c. £130

Potassium Carbonate—
- Chicago, 5 t. 2 c £91
- New York, 3 15 — 68

Phosphorus—
- Bilbao, 2 c. £32
- Mexico, 18 — 152

Potash—
- Buenos Ayres, 15 c. £40

Potassium Bichromate—
- New York, 18 t. £490
- Havre, 10 — 310
- Seville, 7 c. 2 q. 14

Salt—
- Ambriz, 5 t.
- Cameroons, 44¾
- Durban, 1¾
- Iquique, 10 c.
- Kingston, 2
- Landana, 7
- Loanga, 15
- Manaos, 19 13
- Sydney, 300
- Vera Cruz, 12 10
- Antwerp, 304
- Belize, 222
- Benin, 22½
- Bonny, 18
- Buenos Ayres, 50
- Calcutta, 3,205
- Cape Coast, 6¼
- Gothenburg, 7¾
- Halifax, 800
- Iceland, 90
- Manaos, 7
- Mandal, 40
- Maryborough, 35
- Melbourne, 50
- St John's 10
- Santos, 100 cs.

Sodium Nitrate—
- Christiania, 4 t. 1 c 1 q. £37
- Santander, 2 — 2 18
- Barbadoes, 10 — 86
- Oporto, 1 6 1 11
- Shanghai, 5 — 3 43
- Valencia 16 — 7

Sulphur—
- Algoa Bay, 4 t. 9 c. 1 q. £38
- Pernambuco, 2 3 1 16
- Iquique, 5 — 24
- St. John's, 6 — 38
- Madeira, 2 10 — 22
- St. John, 15 — 5
- Iquique, 10 — 50

Salammoniac—
- Mersyne, 1 t. 1 q. £35
- Naples, 1 1 34
- Ancona, 5 c. 9
- Trieste, 1 11 3 56
- Alexandria, 1 — 35
- Odessa, 6 1 2 201
- Philadelphia, 4 2 141
- Piræus, 3 — 5
- New York, 4 19 2 167
- Syra, 3 3 7
- Hamburg, 5 — 9
- Bourgas, 13 — 25
- Genoa, 1 5 1 51

Superphosphate—
- Hamburg, 69 t. 6 c. 3 q. £100
- " 227 11 3 628

Saltpetre—
- Ancona, 3 t. £60
- St. Catherina, 1 — 22
- Syra, 2 1 c. 1 q. 44
- St. John. 1 10 — 32
- St. Louis, 2 — 5

Saltcake—
- Bari, 300 t. 1 c. £413

Sheep Dip—
- Valparaiso, 25 drs. £6

Sulphuric Acid—
- Santos, 19 crbys. £25

Sodium Biborate—
- Antwerp, 5 t. 7 c. 3 q. £161
- Rotterdam, 10 2 2 116

Sodium Chlorate—
- Vera Cruz, 1 ck. £9

Soda Ash—
- Baltimore, 68 cks. 112 tcs.
- Boston, 500 bgs. 305
- Corfu, 100 dms.
- Ghent, 43 cks.
- Madeira, 2 tcs.
- Maranham, 30 brls.
- Melbourne, 302 cks.
- Naples, 33
- Sherbrooke, 17 tcs.
- Smyrna, 200 brls.
- Alicante, 30 cks.
- Ancona, 3
- Baltimore, 492 132 tcs.
- Boston, 466 1,520 bgs.
- Demerara, 40
- Madrid, 28 brls.
- Monte Video, 100
- New Orleans, 7 cks. 39 brls.
- New York, 10 960 tcs.
- Portland, M., 67
- Rio Grande do Sul, 10 brls.
- Syra, 10

Soda Crystals—
- Alexandria, 34 brls.
- Boston, 280 cks,
- Constantinople, 50 brls.
- San Sebastian, 30
- St. John, N.B., 200
- Rosario, 100
- Smyrna, 35
- Valencia, 8 cks.
- Malta, 50 brls.
- St. John, 100 cks.

Sodium Bicarbonate—
- Antwerp, 135 kgs.
- Hamburg, 90 6 cks.
- Rouen, 65
- Stettin, 40
- Antwerp, 10 cks.
- Brisbane, 200 kgs.
- New York, 200

Sodium Silicate—
- Barcelona, 16 cks.
- Carthagena, 6
- Havre, 4
- Leghorn, 4 brls.
- Madeira, 10 cks.
- Savanilla, 5
- Seville, 35 brls.
- Venice, 32 cks.
- Odessa, 75
- Malta, 10 brls.

Tartaric Acid—
- Carril, 2 c. £11
- Portland, 17 — 85

Wood Acid—
- Vera Cruz, 6 cs. £16

Zinc Muriate—
- Halifax, 2 t. 8 c. 1 q. £47
- Bombay, 5 17 — 88

TYNE.

Week ending Mar. 13th.

Alum Cake—
- Copenhagen, 12 t. 16 c.

Ammonium Sulphate—
- Rotterdam, 20 t. 3 c.

Alkali—
- Hamburg, 25 t. 19 c.
- Antwerp, 21 1
- Amsterdam, 37 8
- Rotterdam, 15 2

Bleaching Powder—
- Messina, 1 t. 14 c.
- Hamburg, 109 18
- Gothenburg, 5 4
- Antwerp, 48 19
- Christiania, 20 1
- Konigsberg, 8 16
- Amsterdam, 119
- Rotterdam, 49 18
- Copenhagen, 28 4
- Dunkirk, 34 18

Caustic Soda—
- Hamburg, 5 c.
- Antwerp, 150 t.
- Amsterdam, 13 14
- Valencia, 10 3

Dyestuffs (*otherwise undescribed*)
- Malmo, 1 q.

Ferro Manganese—
- Bilbao, 40 t.

Litharge—
- Hamburg, 2 t. 12 c.

Magnesia—
- Hamburg, 11 c.
- Antwerp, 1

Potassium Chlorate—
- Hamburg, 3 t. 6 c.

Sodium—
- Hamburg, 1 cs.

Soda—
- Malmo, 5 t. 18 c.
- Randers, 24 10
- Aarhuus, 17 5
- Rotterdam, 36 7

Sodium Chlorate—
- Hamburg, 1 t.
- Rotterdam, 1

Sodium Bicarbonate—
- Amsterdam, 2 t. 2 c.

Soda Salts—
- Amsterdam, 5 t.

Soda Ash—
- Christiania, 38 t. 5 c.
- Konigsberg, 85 7

Sodium Sulphate—
- Gothenburg, 6 t. 2 c.

Superphosphate—
- Aarhuus, 5 t. 4 c.

Tar—
- Hamburg, 20 t.
- Copenhagen, 4

Manure—
- Antwerp, 228 bgs.
- Drontheim, 425
- Ghent, 602
- Hamburg, 264
- Rotterdam, 171

Naphtha—
- Bergen, 2 cks.

Pitch—
- Antwerp, 341 t.
- Dunkirk, 134
- Genoa, 208
- Leghorn, 61

Superhosphate—
- Drontheim, 1,200 bgs.

Zinc Muriate—
- Kjoge, 190 cks.

HULL.

Week ending Mar. 11th.

Alkali—
- Drontheim, 41 drms.
- Harlingen, 86 cks.

Bleaching Powder—
- Copenhagen, 1 ck.
- Stettin, 1 66

Bone Size—
- Bremen, 10 cks.
- Copenhagen, 2
- Harlingen, 10

Caustic Soda—
- Hango, 89 drms.
- Libau, 69
- Rotterdam, 37

Chemicals (*otherwise undescribed*)
- Amsterdam, 20 pkgs
- Antwerp, 26 cs. 10 cks.
- Boston, 125
- Christiania, 10
- Christiansand, 10
- Copenhagen, 12
- Dunkirk, 15
- Genoa, 102 4
- Gothenburg, 31 325 slabs
- Hamburg, 74 27
- Libau, 37 1 12 165
- Leghorn, 10 16
- Marseilles, 55 15
- Rotterdam, 1 67
- Rouen, 48
- Stettin, 4

Dyestuffs (*otherwise undescribed*)
- Rotterdam, 68 bgs. 4 cks.
- Reval, 3 cs.

GLASGOW.

Week ending Mar. 13th.

Ammonium Sulphate—
- Trinidad, 69 t. 9½ c.

Bleaching Powder—
- Sydney, 10 t.

Caustic Soda—
- Trinidad, 3 t. 10½ c. £27.

Chemicals (*otherwise undescri*
- Trinidad,

Epsom Salts—
- Sydney, 2¾ c. 3 lbs.

Manure—
- Trinidad, 311 t. 9¾ c. £

Oxalic Acid—
- Sydney,

Paraffin Wax—
- Bombay, 27¾ c.
- Rouen, 10 t 7¾ c.

Potassium Bichromate—
- Christiania, 2 cks.
- Bergen, 1
- Bombay, 2 t. 3¾ c.
- Rouen, 29 2¾
- Rotterdam, 64 cks.

Pitch—
- Amsterdam, 107 t.

Sodium Bichromate—
- Boston, 322 c.
- Bombay, 20
- Rotterdam, 2 cks.
- Rouen, 13 t. 4¾ c.

Sulphuric Acid—
- Rangoon, 12 t. 7¾ c.
- Bombay, 56 5

Superphosphate—
- Christiania, 100 t.

Tar—
- Madras,

Waste Salt—
- Calcutta, 110 c.

GOOLE.

Week ending Mar. 5th.

Benzole—
- Antwerp, 56 cks.
- Hamburg, 16 drums.
- Rotterdam, 58

Bleaching Preparations—
- Dunkirk, 97 cks.

Chemicals (*otherwise undescri*
- Dunkirk, 36 cks.

Coal Products—
- Boulogne, 2 cks.

Dyestuffs (*otherwise undescri*
- Boulogne, 4 cks.
- Ghent, 6 1 cs.
- Rotterdam, 3

Manure—
- Boulogne, 67 bgs.
- Dunkirk, 210
- Ghent, 222

Pitch—
- Antwerp, 240 t.

GRIMSBY.

Week ending March 8th.

Chemicals (*otherwise undescri*
- Antwerp, 10 cks.
- Dieppe, 54 pkgs.
- Hamburg, 16
- Rotterdam, 20 cs.

Coal Products—
- Dieppe, 2 cs. 110 cks. 22 pkgs.

PRICES CURRENT.

WEDNESDAY, March 19, 1890.

Prepared by Higginbottom and Co., 116, Portland Street, Manchester.

…ues stated are F.O.R. at maker's works, or at usual ports of shipment in U.K. The price in different localities may vary.

		£	s.	d.
25 °/₀ and 40 °/₀	per cwt	7/6 & 12/9		
Glacial	,,	3	5	0
, S.G., 2000°	,,	0	12	9
c 82 °/₀	nett per lb.	0	0	6½
	,,	0	0	3¾
ic (Tower Salts), 30° Tw.	per bottle	0	0	8
(Cylinder), 30° Tw.	,,	0	2	11
80° Tw.	per lb.	0	0	2
	,,	0	0	1¾
	nett ,,	0	0	3¼
	,,	0	1	4
ric (fuming 50 °/₀)	per ton	15	10	0
(monohydrate)	,,	5	10	0
(Pyrites, 168°)	,,	3	2	6
(,, 150°)	,,	1	8	0
(free from arsenic, 140/145°)	,,	1	10	0
rous (solution)	,,	2	15	0
(pure)	per lb.	0	1	0½
c	,,	0	1	2½
white powdered	per ton	13	10	0
se lump)	,,	4	15	0
Sulphate (pure)	,,	5	0	0
,, (medium qualities)	,,	4	10	0
m	per lb.	0	15	0
, ·880=28°	per lb.	0	0	3
=24°	,,	0	0	1⅞
Carbonate	nett ,,	0	0	3½
Muriate	per ton	23	10	0
,, (sal-ammoniac) 1st & 2nd	per cwt.	37/-& 35/-		
Nitrate	per ton	42	0	0
Phosphate	per lb.	0	0	10½
Sulphate (grey), London	per ton	11	16	3
,, (grey), Hull	,,	11	17	6
il (pure)	per lb.	0	0	11
alt ,,	,,	0	0	9½
	per ton	75	0	0
(tartar emetic)	per lb.	0	1	1
(golden sulphide)	,,	0	0	10
hloride	per ton	8	0	0
arbonate (native)	,,	3	15	0
ulphate (native levigated)	,,	45/- to 75/-		
Powder, 35 %	,,	5	10	0
Liquor, 7 %	,,	2	10	0
e of Carbon	,,	13	0	0
1 Acetate (crystal	per lb.	0	0	6
Chloride	per ton	2	2	6
y (at Runcorn) in bulk	,,	17/6 to 35/-		
Dyes :—				
e (artificial) 20 %	per lb.	0	0	9
a	,,	0	3	9
Products				
ene, 30 % A, f.o.b. London	per unit per cwt.	0	1	5
90 % nominal	per gallon	0	3	4
50/90 ,,	,,	0	2	6
c Acid (crystallised 35°)	per lb.	0	0	10½
,, (crude 60°)	per gallon	0	2	10
e (ordinary)	,,	0	0	2½
(filtered for Lucigen light)	,,	0	0	3
Naphtha 30 % @ 120° C.	,,	0	1	5
Oils, 22° Tw.	per ton	3	0	0
.o.b. Liverpool or Garston	,,	1	9	0
Naphtha, 90 % at 160°	per gallon	0	1	11
1 Oils (crude)	,,	0	0	2½
Chili Bars)	per ton	47	7	6
lphate	,,	26	0	0
ide (copper scales)	,,	51	0	0
(crude)	,,	30	0	0
(distilled S.G. 1250°)	,,	55	0	0
	nett, per oz.	8½d. to 9d.		
hate (copperas)	per ton	1	10	0
hide (antimony slag)	,,	2	0	0
et) for cash	,,	14	0	0
1arge Flake (ex ship)	,,	15	10	0
tate (white ,,)	,,	23	15	0
, (brown ,,)	,,	18	10	0
bonate (white lead) pure	,,	19	0	0
ate	,,	22	0	0
Lime, Acetate (brown)	,,	8	5	0
,, ,, (grey)	per ton	14	0	0
Magnesium (ribbon and wire)	per oz.	0	2	3
,, Chloride (ex ship)	per ton	2	6	3
,, Carbonate	per cwt.	1	17	6
,, Hydrate	,,	0	10	0
,, Sulphate (Epsom Salts)	per ton	3	0	0
Manganese, Sulphate	,,	18	0	0
,, Borate (1st and 2nd)	per cwt.	60/- & 42/6		
,, Ore, 70 %	per ton	4	10	0
Methylated Spirit, 61° O.P.	per gallon	0	2	2
Naphtha (Wood), Solvent	,,	0	3	11
,, ,, Miscible, 60° O.P.	,,	0	4	9
Oils :—				
Cotton-seed	per ton	22	10	0
Linseed	,,	23	0	0
Lubricating, Scotch, 890°—895°	,,	7	5	0
Petroleum, Russian	per gallon	0	0	5¼
,, American	,,	0	0	5⅜
Potassium (metal)	per oz.	0	3	10
,, Bichromate	per lb.	0	0	4
,, Carbonate, 90% (ex ship)	per ton	18	0	0
,, Chlorate	per lb.	0	0	5
,, Cyanide, 98%	,,	0	2	0
,, Hydrate (Caustic Potash) 80/85 %	per ton	22	10	0
,, ,, (Caustic Potash) 75/80 %	,,	21	10	0
,, ,, (Caustic Potash) 70/75 %	,,	20	15	0
,, Nitrate (refined)	per cwt.	1	3	6
,, Permanganate	,,	4	5	0
Prussiate Yellow	per lb.	0	0	9½
,, Sulphate, 90 %	per ton	9	10	0
,, Muriate, 80 %	,,	7	15	0
Silver (metal)	per oz.	0	3	7¾
,, Nitrate	,,	0	2	5½
Sodium (metal)	per lb.	0	4	0
,, Carb. (refined Soda-ash) 48 %	per ton	6	5	0
,, ,, (Caustic Soda-ash) 48 %	,,	5	10	0
,, ,, (Carb. Soda-ash) 48%	,,	5	12	6
,, ,, (Carb. Soda-ash) 58 %	,,	6	15	0
,, ,, (Soda Crystals)	,,	3	0	0
,, Acetate (ex-ship)	,,	15	15	0
,, Arseniate, 45 %	,,	10	10	0
,, Chlorate	per lb.	0	0	6¼
,, Borate (Borax)	nett, per ton.	30	0	0
,, Bichromate	per lb.	0	0	3
,, Hydrate (77 % Caustic Soda)	per ton.	13	0	0
,, ,, (74 % Caustic Soda)	,,	12	10	0
,, ,, (70 % Caustic Soda)	,,	11	10	0
,, ,, (60 % Caustic Soda, white)	,,	10	10	0
,, ,, (60 % Caustic Soda, cream)	,,	10	0	0
,, Bicarbonate	,,	6	0	0
,, Hyposulphite	,,	5	5	0
,, Manganate, 25%	,,	9	0	0
,, Nitrate (95 %) ex-ship Liverpool	per cwt.	0	8	1½
,, Nitrite, 98 %	per ton	27	10	0
,, Phosphate	,,	15	15	0
,, Prussiate	per lb.	0	0	7¾
,, Silicate (glass)	per ton	5	7	6
,, ,, (liquid, 100° Tw.)	,,	3	17	6
,, Stannate, 40 %	per cwt.	2	0	0
,, Sulphate (Salt-cake)	per ton.	1	15	0
,, ,, (Glauber's Salts)	,,	1	10	0
,, Sulphide	,,	7	15	0
,, Sulphite	,,	5	5	0
Strontium Hydrate, 98 %	,,	8	0	0
Sulphocyanide Ammonium, 95 %	per lb.	0	0	8
,, Barium, 95 %	,	0	0	5¼
,, Potassium	,,	0	0	8
Sulphur (Flowers)	per ton.	7	10	0
,, (Roll Brimstone)	,,	6	0	0
,, Brimstone : Best Quality	,,	4	7	6
Superphosphate of Lime (26 %)	,,	2	10	0
Tallow	,,	25	10	0
Tin (English Ingots)	,,	95	0	0
,, Crystals	per lb.	0	0	6⅜
Zinc (Spelter)	per ton.	21	15	0
,, Chloride (solution, 96° Tw.	,,	6	0	0

THE

CHEMICAL TRADE JOURNAL.

Publishing Offices: 32, *BLACKFRIARS STREET, MANCHESTER.*

No. 149. **SATURDAY, MARCH 29, 1890.** **Vol. VI.**

Contents.

Notices.

All communications for the *Chemical Trade Journal* should be addressed, and Cheques and Post Office Orders made payable to—

DAVIS BROS., 32, Blackfriars Street, MANCHESTER.

Our Registered Telegraphic Address is—

"Expert, Manchester."

The Terms of Subscription, commencing at any date, to the *Chemical Trade Journal*,—payable in advance,—including postage to any part of the world, are as follow:—

Yearly (52 numbers) .... **12s. 6d.**
Half-Yearly (26 numbers) .... **6s. 6d.**
Quarterly (13 numbers) .... **3s. 6d.**

Readers will oblige by making their remittances for subscriptions by Postal or Post Office Order, crossed.

Communications for the Editor, if intended for insertion in the current week's issue, should reach the office not later than **Tuesday Morning.**

Articles, reports, and correspondence on all matters of interest to the Chemical and allied industries, home and foreign, are solicited. Correspondents should condense their matter as much as possible, write on one side only of the paper, and in all cases give their names and addresses, not necessarily for publication. Sketches should be sent on separate sheets.

We cannot undertake to return rejected manuscripts or drawings, unless accompanied by a stamped directed envelope.

Readers are invited to forward items of intelligence, or cuttings from local newspapers, of interest to the trades concerned.

As it is one of the special features of the *Chemical Trade Journal* to give the earliest information respecting new processes, improvements, inventions, etc., bearing upon the Chemical and allied industries, or which may be of interest to our readers, the Editor invites particulars of such—when in working order—from the originators; and if the subject is deemed of sufficient importance, an expert will visit and report upon the same in the columns of the *Journal*. There is no fee required for visits of this kind.

We shall esteem it a favour if any of our readers, in making inquiries of, or opening accounts with advertisers in this paper, will kindly mention the *Chemical Trade Journal* as the source of their information.

Advertisements intended for insertion in the current week's issue, should reach the office by **Wednesday morning** at the latest.

Advertisements.

Prepaid Advertisements of Situations Vacant, Premises on Sale or To be Let, Miscellaneous Wants, and Specific Articles for Sale by Private Contract, are inserted in the *Chemical Trade Journal* at the following rates:—

30 Words and under .... 2s. 0d. per insertion.
Each additional 10 words .... 0s. 6d. „

Advertisements of Situations Wanted are inserted at one-half the above rates when prepaid, viz:—

30 Words and under .... 1s. 0d. per insertion.
Each additional 10 words .... 0s. 3d. „

Trade Advertisements, Announcements in the Directory Columns, and all Advertisements not prepaid, are charged at the Tariff rates which will be forwarded on application.

THE PATERNAL GOVERNMENT AGAIN.

TO say in one breath that "comparisons are odio[...] and in the next that "all things are comparati[...] would seem to imply either that we have risen to [...] height of paradox, or that we have lost our love [...] things in general. Perhaps at times we might blush [...] each of the soft impeachments, but at the present mom[...] we are too much concerned with something else, to blus[...] anything. We had drifted into what we may call [...] comparative line of thought, after a perusal of a rec[...] number of the *Times*. In one column was treated [...] case of Madame Tshebrikova, and, incidentally, [...] gigantic abuses of power, and the cruelties of the des[...] ism, under which Russia even at the present day [...] groaning and travailing. And here came in the "odi[...] comparison." Verily, we thought, our land is not as t[...] land; our rulers are not as those rulers; England bask[...] the sunshine of justice, while Russia simply writhes un[...] the administration of the law. And going on in this [...] congratulatory strain, we had almost suceeded in c[...] vincing ourselves that we dwelt in the midst of perfecti[...] and that the Utopia of Sir Thomas More's aspirati[...] was no longer a myth but a reality. And then, our e[...] wandering over the pages before us, fell on the word "gol[...] and that word (we admit it shamelessly), rivetted [...] attention immediately. It was not entirely unkn[...] to us. We were acquainted with the chemical sym[...] of the substantive to which it referred, and had e[...] studied some of its salient properties—and we knew, t[...] though usually classed among the fixed elements, it is m[...] often endowed with a volatility that is simply irrepressi[...] We were, therefore, not surprised to find that the circ[...] stances under which it appeared in the *Times* had refere[...] to a difficulty in its production. Sir Pritchard Morgan, [...] seems, after having, in company with other speculat[...] pockets, invested almost untold wealth in the Cambrian g[...] mines, and after having proved the exceeding richness of [...] deposits, has been rudely awakened from his golden dre[...] by an unappreciative Government, which has stepped in [...] a vital moment with its paternal claims, and caused the s[...] pension of operations.

We are not entirely at one with Sir Pritchard Morgan [...] garding all that he says about this gold mine, nor are [...] entirely at variance with him. He regards it as a field [...] labour. So do we. But we may differ somewhat in [...] estimate of its productiveness, for we have reliable adv[...] that it is *not* one of the richest mines in the world, and is [...] likely ever to be a second Mount Morgan in anything but [...] name. But, be this as it may, we have profound sympa[...] with Sir Pritchard Morgan as a victim of red-tape int[...] ference or departmental stupidity. And this brings us b[...] to the second of our original propositions—that "all thi[...] are comparative," for even in this land of progress, where [...] machinery of the State is supposed to be so highly organis[...] it is only too painfully evident that there is a huge want [...] ordinary intelligence in high places, and that the busin[...]

affairs of the nation are too often conducted upon the most unsound of business principles. We hope that Sir Pritchard Morgan's letter to Lord Salisbury will procure him an early opportunity of gathering in his golden harvest, and that it may be the means of stimulating the powers that be to a more intelligent administration of the functions committed to their charge.

BRINE PUMPING IN CHESHIRE.

A NEW SCHEME.

THE inhabitants of the salt district are for once making common cause with their enemies the brine pumpers. A new pumping company has been formed which, it is believed, will seriously affect the interests of all parties in the salt district. With their houses tumbling about their ears, the people of Northwich are, of course, against an extension of operations which they believe have brought about this state of things. But there is a good deal more in the matter than this. The new company proposes to tap the Northwich brine, and take it elsewhere for manufacture. A shaft is to be sunk at Wincham, which is just outside Northwich, and the brine is to be sent in pipes to Widnes, some ten miles off, where it will be converted into salt or used in the manufacture of chemicals. Well, how would this affect the interests of the Northwich district? Take first the case of the public. Great damage has been done to property by subsidences of land due, there can be little doubt, to the pumping of brine. No compensation has been given, and the brine pumpers do not admit liability. But the district receives a kind of indirect compensation. The salt industry and the industries associated with it find employment for the bulk of the people, and the works are large contributors to the rates. Moreover, so much salt is carried down the Weaver that until lately the Weaver Trustees have been able out of surplus revenue to contribute £14,000 or £15,000 in relief of the county rates. But it is said that if the new company sends its brine to Widnes, which is in another county, the manufacturing industries will follow, and that Northwich and Winsford, robbed in a manner of their daily bread, would continue to suffer from subsidences without hope of compensation. The bill promoted by the new company makes some references to damage which may be caused by brine pumping, but these are regarded by the opponents of the scheme as "hypothetical and illusory."

As to the manufacturers' side of the case, they say that if salt and alkali are manufactured on the banks of the Mersey they will, in self-defence, have to establish works in the same locality and adopt new means of sending their brine there. It is clear that a company manufacturing salt and chemicals at Widnes would have at a disadvantage rivals who had to pay the Weaver dues; and it is also clear that if the manufacture of salt and alkali is removed to the seaside, it will put the Weaver Trustees, with their large debt, in a very serious position. The Northwich and Winsford manufacturers argue that a competition such as is contemplated, is wholly uncalled for in the public interest. In a petition against the bill they suggest that the new company is promoted exclusively in the interest of chemical manufactures at Widnes, who seek to obtain brine for use in their works at a somewhat less cost than at present is paid for common salt, and at the same time to establish a salt trade wholly foreign to the locality. It is also pointed out that Parliament has hitherto protected the inhabitants and the great industry of the district by making statutory provisions to prevent brine leaving the district, and has uniformly rejected all attempts in this direction. We do propose here to discuss the rights and wrongs of this important matter, which will shortly come before Parliament. The new company, which bears the name of the "Widnes Brine Supply Company," have lodged their bill in the House of Lords, and it will probably come before a committee about the end of next month. It will be opposed by the Local Boards of Northwich, Winsford, Middlewich, and Sandbach, by the Salt Union, and by Messrs. Brunner, Mond and Co., Limited, and other local manufacturers. Professor Boyd Dawkins made a visit to the district a day or two ago, and we understand he will be called as a witness against the Bill. We are informed that he entertains no doubt that the subsidences at Northwich are due to the pumping of brine—a theory which has long been maintained by those more conversant with the district.

CRYSTALS OF GYPSUM.—Very interesting specimens of crystals of gypsum are being found in the drift clay through which a sewer is being formed at the Gateshead end of the Redheugh Bridge. The sewer is 30 ft. deep, and crystals of gypsum, or sulphate of calcium, in the very perfect and varied rhomboidal forms, are frequently being found. Specimens from various formations may be seen in the excellent geological museum.

CHEMICAL ENGINEERING*

XXIV.

TRANSPORTING GASES.

WE may follow the same course in dealing with gase we have already done with liquids; but with gases problem is much simpler in character, owing to the fact gases must be entirely closed in the conduit in which t are flowing, so that if we confine ourselves to a study of question it is nearly all we have to consider from a theore standpoint.

Head or pressure must be taken into our calculation the same manner as in dealing with liquids, but if we fo the formula given by Professor Pole for the discharge of through pipes we must measure that head or pressur terms of inches of water. The tables already given enable the student to convert pounds per square i or inches of mercury pressure into inches of water pres and by the aid of the following formula there be no temptation to use the method of trial and error settling the size of pipe necessary to accommodate a cer flow of gas. One remark, however, may be necessary, a which the importance has forced itself upon me by experi —if a calculation for the diameter of a pipe, works out t inch and fractions, it is safer, especially if the pipe is not c pletely straight, to take the required size as the next even i to allow for friction—thus, if the calculation said 3·72 inc we should infer a 4·0 inch pipe to be necessary.

And here, again, may be noticed that a very consider difference in practical effect is obtained, whether the hea pressure is measured at the fountain head, such as in the vessel of the supply to a Weldon oxidiser, or in the holder of a gas-works, or whether it be measured in the conduit near to the point of exit of the gases. For all prac work the head should be measured *in the pipe*, and if th not practicable, a calculation must be made of the h necessary to put the fluid in motion.

If Q = cubic feet of gas per hour,
l = length of pipe in yards,
h = inches of water pressure,
s = Sp. gr. of gas, air = 1·0,
d = diameter of pipe in inches,

then when h is measured in the pipe or conduit:—

$$Q = 1350\, d^2 \sqrt{\frac{h\,d}{s\,l}}$$

$$h = \frac{Q^2\, s\, l}{(1350)^2\, d^5}$$

$$d = \sqrt[5]{\frac{Q^2\, s\, l}{(1350)^2\, h}}$$

From the above formulæ it will be seen that *cæteris par* the application of a fourfold pressure, will only resul doubling the quantity of discharge; and that when the le of a pipe is quadrupled the discharge will be only one-ha the original quantity. Again, the quantity of discharge be doubled by reducing the length of the pipe to one-fo

The disturbing causes, operating against the realisati the theoretical, or undisturbed flow of fluids in pipes, been fully treated of in Dr. Pole's paper before the R Society of 1852, the influence of sudden enlargem tending to produce eddies, the influence of shape o entrance, and the effect of bends and irregularities in pipe, all of which tend to reduce the maximum flow, be taken into account; and I cannot too strongly emph Dr. Pole's remarks when he says:—"Every prudent, prac man knows the necessity of leaving a margin in his calc tions for incidental or unforeseen disturbing causes." " simple rule will often attract a practical man to its use the complicated one will frighten him away from it gether. For the same reason, it is desirable, in co-effici to employ round numbers, if possible, instead of tiring patience with a useless array of decimals."

We have already seen how the friction of bends ma approximately estimated in the case of liquids. Dr.

* A course of lectures delivered by Mr. George E. Davis at the Man Technical School.

suggests that in the present state of our knowledge we can only assume that the head necessary to overcome the resistance of a bend may vary as the sp. gr. of the fluids. If such argument is admissible, the following formula will apply to all fluids, whether they be liquid or gaseous:—

$$p = \frac{S v^2}{670} \frac{\lambda}{\rho} \sqrt{\frac{d}{\rho}}$$

Where S = weight of cb. ft. in lbs.
v = velocity in ft. per second.
ρ = radius of curve of bend.
λ = length of curve of bend.
h = head of water in feet.
p = pressure in lbs. per square foot.

The foregoing formulæ have proved very useful to me on several occasions. At one time it was desired to pass 25,000 cubic feet of gas per hour through 26 narrow tubes under a pressure of 3·25 inches of water; the calculated orifices were carefully turned up on the lathe to a gauge, when it was found that 25 of the tubes passed the required quantity, or an error of 4 per cent., which was remedied by slightly reducing the pressure. Of course, it must be remembered that very narrow tubes, especially if they happen to be of any considerable length, will not pass the quantity of gas shown by the formulæ for large pipes. In some experiments which I made several years ago showed that when every care was taken in the calculation to ensure 1,370 cubic feet per hour passing through a pipe, 5-8th of an inch in diameter and five feet in length, in practice only 960 cubic feet were delivered, and it was this circumstance which led me to discover, what I afterwards found was well known, that the pressure should be taken in the pipe itself and due allowance made for friction. I think nothing shows better the effect of friction so well as the bellows aspirators used to withdraw samples of chimney gases for testing purposes. Let the student connect these bellows with a length of 3 feet of half-inch glass tubing, the bellows will open easily; let the half-inch tube be now replaced by one having a bore of 1-8th of an inch, when it will be found a very difficult task indeed to open the bellows quickly. This lesson once learned is not easily forgotten.

The friction of narrow tubes may be approximately calculated by the following formula:—

$$h = \frac{\left(\frac{Q}{60}\right)^2 l}{(3·7\, d)^5}$$

which will sufficiently show how the friction increases as the pipe lengthens, or is narrowed or overworked.

Before concluding the present chapter I would wish to call attention to the properties of gaseous bodies and vapours. This study can scarcely be included in a dissertation on Chemical Engineering, but every student should possess a good knowledge of these things before the study of Chemical Engineering is entered upon. The expansion of gases, the saturation of gases with vapours, heat developed by compression of gases and vapours, and the cold produced by dilatation, are all well set forth in our standard treatises on Physics, and it is to these that the attention of the student is directed.

(To be continued.)

Detection of Wood in Paper.—MM. R. Godeffroy and M. Coulon have worked out a method for estimating the amount of woody tissue in paper, based on the fact that cellulose which has been extracted with water, alcohol, and ether does not reduce a solution of gold chloride, whereas wood pulp quickly reduces such a solution on warming to metallic gold. They find that 100 grms. of well-dried sawdust are capable of separating 14·285 grms. of gold from its solution. In examining paper by this method they remove size by treating the material with cold and hot water, and then extract alumina by washing with an alcoholic solution of tartaric acid. The paper after this treatment is tried and extracted with alcohol and ether, and then treated with the gold solution.

Legal.

THE ACTION AGAINST CALICO PRINTERS F POLLUTION: THE CASE SETTLED.

SCHWABE *v.* CHADWICK.

THE plaintiffs in this action are Messrs. Salis Schwabe and (who carry on the business of calico printers and bleacher Middleton, in this county, and the defendants are Messrs. Chadv and Co., who are also in the same line of business at Chadderton, r Oldham. The action was before the court at its last sittings in M chester, and stood part heard. The plaintiffs brought the actior restrain the defendants from polluting a stream of water, which is u by them for the purpose of their bleaching business, by throwing it refuse water from their works higher up the stream. Evidence given at the trial that it was the custom of the defendants to pollute stream with dye refuse, and that the plaintiffs had suffered consi able injury by the polluted water getting into their reservoirs, : interfering with their business. Just before the rising of the court terday, Mr. Maberly stated that the parties had been able to com terms in the action, and he desired to mention it now in order th might be removed from the paper. He was glad to say that they been able to come to an arrangement, which it was proposed to emb in an order. The parties were *sui juris*, and unless the court des it he did not know that it was necessary to mention the terms of agreement at which they had arrived.

The Vice-Chancellor said he did not think it necessary that the te should be mentioned if the parties were agreed. They were all *sui ju* This was one of the cases in which he thought the parties had d very wisely in coming to an agreement, and he did not think he n ask counsel to go into details unless they desired him to know th He presumed that the terms would be such terms as would be wi the order of the court, and beyond that he did not wish to know a thing more about it.

Mr. Clare said that at the trial of the action his Honour threw a suggestion which the parties had taken advantage of, and which thought was a very reasonable settlement of the dispute.

The Vice-Chancellor said he was sure the parties were indebted b to the counsel and the solicitors for great assistance in the matter, : he was glad that they had taken a reasonable course and settled dispute.

The terms upon which the action has been settled are that the fendants undertake to construct settling tanks to the satisfaction c surveyor, to be mutually agreed upon between the parties.

THE STRAIGHTENING OF CHIMNEY SHAFT!

BY C. MOLYNEUX AND J. M. WOOD, A.M.INST.C.E.

THE deviation of a chimney shaft from the vertical is often to traced to the settlement or giving way of the foundation in cor quence of too great a load being placed upon it. In some cases i due to the rapid rate at which the chimney has been built, whereby mortar in the joints is unable to carry the superincumbent weight; the chimney may have been canted from the vertical by the force of wind whilst the mortar is still soft and yielding. There are also ca recorded in which one side of a chimney was exposed to a continu period of wet weather, which kept the mortar soft and allowed shaft to settle on one side.

The work of straightening a shaft is a work of considerable difficu besides entailing serious risks; in fact, it must be looked upon a bold piece of engineering, and one in which the utmost caution sho be used. All the conditions under which the chimney was built sho be known before any work of this character is undertaken; and it often be found advisable rather to rebuild a shaft than to attemp straighten it. Under no circumstances should an attempt be made restore the structure to the vertical position, by sawing out the joints, removing courses of brickwork—*i.e.*, "cutting back," as it is te nically called—unless the shell of the stalk is built solid as a whole, one material, and is absolutely homogeneous, the brickwork or maso being of the highest character throughout. Many existing *brick* chimn have been successfully cut back in the hands of skilful and cautious m but, on the other hand, manv have received fatal cuts, and have eit fallen under the operation or a short time afterwards, among which r be mentioned the noted Newlands Mill chimney. The cutting b of *stone* shafts is seldom attempted, there being only a few succes cases recorded. When it has been decided to straighten a chimn the aid of a specialist or professional "chimney doctor" should called in: for his experience and staff of workmen, &c., are essentia the successful treatment of the case.

rst operation is to determine exactly how much the shaft from the vertical, and from this a careful calculation is to be find the necessary gap or depth of the cut. The cut should wedge shape, with a maximum opening at the outer circum- apering away to nothing at the centre. Allowance has to be settlement, so that the chimney shall not afterwards incline pposite direction. When the deviation is considerable, it is make several cuts—*i.e.*, at several joints; but in no case hey be close together. If the shaft deviates, as a whole, from nt of the foundation, the cut is made as low as possible. It borne in mind that when once a chimney has been partly cut through, its stability depends entirely upon its own weight, sion at the bed joint being destroyed; it is, therefore, advis- keep the cuts as low down as is practicable. When a brick is of considerable diameter, the method sometimes adopted out either one or two courses of bricks half-way round the on the side opposite to that towards which the chimney nd replace the same by a thinner course of brick or hard stone hape form. In executing this operation the brickwork is cut segments of, say, about 2 ft., and is replaced by a thinner The first incision is made directly opposite to the direction h the shaft leans; the next incision is made to the right or left, on alternately until one-half of the chimney has been cut . As this work proceeds wedges are driven in to keep the cut The cutting out can be carried out from both the inside and at the same time by the aid of long chisels. As soon as the been made half way round, and the wedges slackened, a loud is almost invariably heard, which is caused by the breaking of t of the joint. The chimney begins to move, and by slight ions slowly settles down on the thinner inserted course. A haft in Prussia, 331 ft. in height, was successfully cut back by ethod.

usual plan, however, now adopted is to make an incision into the shaft, as above described, and insert, as the brickwork away, strong steel screwjacks, whose function is to support the cumbent mass above as the masonry in the cut is removed, jacks are from 7in. to 10in. in length; above and below them ced strong wrought iron plates to distribute the load and obtain bearing. When the cut has been completed, and the jacks all d at their proper places, the next operation is to gradually and nly unscrew them, till the shaft has nearly regained its vertical n. The space between the jacks, and finally that occupied by is filled in with solid brickwork. It is usual to commence g good the cut before the chimney has quite returned to the ver- osition, otherwise the slight settlement of the new work will he chimney in the opposite direction.

Townsend Shaft, Glasgow, was straightened by sawing out the brick from the inside; no less than 12 cuts were made in the shaft, t at a height of 41ft, and the last 326ft. from the ground line. this operation, the shaft was stated to be 7ft. 9in. out of the l: it came back in a slightly oscillating manner, and the men ell when it was returning by the saws tightening (See Ban- "On Chimneys.") A few years ago a very massive chimney, in height, at the Royal Arsenal Gasworks, was successfully tened. The foundation settled, and the shaft inclined 3ft. 6in. the vertical. Notwithstanding that calculations were made showed that it could remain more than 3ft. out of the perpen- without over-reaching the limit of safety, it was deemed advi- o cut it back, and this was accordingly done.

onclusion, it may not be unnecessary to warn proprietors of y shafts against unscrupulous practitioners claiming to be ney doctors," but who are quacks and incompetent men, and vays ready to undertake any sort of work of this description. *ustries*.

HE CONCENTRATION OF LABORATORY REAGENTS.

recent number of the *Berlin und Deuts Chemical Ges.* (1890, 31), R. Blochmann calls attention to a matter which, we know xperience, is too often overlooked, though an important element systematic conduct of a laboratory—that is, the concentration reagents employed. It only needs to be pointed out to be at cognised by those who may not have considered this, even as ement of the routine of chemical work, that the ordinary opera- analysis would gain very much in precision by the adoption of a plan in the preparation of reagents; and the expediency plan being granted, preference would instinctively be to that based upon chemical equivalents. There is, of nothing essentially novel in the proposal to extend to hole of the reagents of a laboratory, the plan of d solutions, adopted in strictly volumetric analysis; indeed, we believe that in most technical laboratories a systematic pre- paration of reagents, those at least which are in constant use, is com- monly adopted. Further, we believe we are correct in stating, as a historical instance, that John Mercer carried out the plan of working with re-agents in the form of equivalent solutions, not only in the laboratory but in the colour-shop. The suggestion, therefore, though not more novel than the non-adoption of the plan may have made it, obviously commends itself to the attention of chemists, more especially, perhaps, to those who have the conduct of educational laboratories. The advantage of inculcating automatically the first principles of the science with the handling of every reagent by the student need scarcely be insisted upon.

If Shakespeare could be eloquent respecting "Sermons in stones," how much more modern chemical students, anent "professors in bottles!"

We need make no apology to the author of the paper in question for reproducing the tables which he gives of the composition of the principal reagents made up, for the most part, on the plan of normal solutions, as follows:—

I.—CONCENTRATED ACIDS.

	Sp. gr.	Wt. p.c.	1 litre containing	
Hydrochloric acid ..	1·160	31 8	369 gr. approx.	10 HCl.
Nitric ,, ..	1·305	48·1	628 ,, ,,	10 HNO_3
Sulphuric ,, ..	1·840	96·0	1,767 ,, ,,	36 $\frac{H_2SO_4}{2}$

II.—NORMAL SOLUTIONS.

(a) $\frac{2}{1}$ normal	wt. p.c.	(b) $\frac{1}{1}$ normal	wt. p.c.
Hydrochloric acid	7·1	Barium chloride (crystal)	11·2
Nitric acid	11·8	Calcium chloride	10·5
Sulphuric acid..........	9·2	Iron chloride (Fe_2 Cl_6)	5·2
Acetic acid	11·8	Potassium sulphate	8·1
Oxalic acid	12·3	Magnesium sulphate (crystal)	11·6
Tartaric acid	14·1	Copper sulphate (crystal)....	11·6
Potassium hydrate	10·3	Sodium disphosphate (crystal)	11·4
Sodium hydrate	7·4	Lead acetate (crystal)	16·9
Ammonia..............	3·5	Potassium chromate........	9·0
Ammonia sulphide	6·8	Potassium ferrocyanide	10·0
		(c) ½ normal.	
Ammonia chloride	10.4	Platinum chloride..........	8·0
Ammonia carbonate	9·4	Silver nitrate..............	8·0
Sodium carbonate (ant.)	9·6	Mercuric chloride	6·4
Sodium acetate (crystal) ..	25·2	Sodium nitrate	6·2

III.—OXIDISING AND REDUCING AGENTS.
(1 litre $= \pm \frac{O}{2} =$ 8gms. oxygen.)

	wt. p.c.	1 litre contain.	
Potassium bichromate	4·7	⅙ $K_2Cr_2O_7$	= 49·00%
Sodium hypochlorite	3·7	½ NaClO	= 37·2%
Potassium nitrite	4·2	½ KNO_2	= 42·5%
Stannous chloride (crystal)	10·5	½ $SnCl_2·2Ag$	= 112·5%

IV.—SATURATED SOLUTIONS.

	wt. p. ct. at 15°		Ratio to normal solution.
Hydrogen Sulphide ..	0·48	H_2S	1 : 3·5
Barium hydrate	5·95·	$Ba(OH)_2 8H_2O$	1 : 2·6
Calcium hydrate......	0·13	$Ca(HO)_2$	1 : 21·5
Calcium sulphate	0·26·	$CaSO_4 2H_2O$	1 : 33·0
Bromine (water)......	3·23	Br.	1 : 2·5 (?)

It is unnecessary to enter into the details of carrying out this plan. It is sufficient to have reproduced for our readers the outline of a very practical and instructive suggestion.

THE TECHNICAL INSTRUCTION ACT.

At the last meeting of the Manchester School Board, the Chairman made a statement with regard to the position of the Board in reference to this matter. He said that on the 14th of February a deputation representing the principal bodies in Manchester having control of science and art instruction waited upon the City Council with a memorial urging the adoption in Manchester of the Technical Instruction Act, 1889. At a recent meeting of the Council a special committee was appointed to report upon this matter. As it was very desirable that the Board should be in a position to supply the City Council with all necessary particulars in case it was decided to give a grant under the new Act in aid of science and art classes, he had thought it well to bring before the meeting the question of the appointment of a special committee to supply any information that might be needed by the City Council relative to the Board's science and art and commercial evening classes. For the information of the Board he had taken out certain particulars relative to science and art instruction in Manchester. These were the official figures published by the Science and Art Department in their twenty-sixth report, issued in May of last year. First, as to science classes, the number under instruction in Manchester School Board classes was 2,404, and the amount of the grant £2,739. 12s.; in classes conducted by other bodies in the city, 2,128, grant, £1,134. Then as to art classes, the number under instruction in Manchester School Board classes was 1,402, and the grant £400. 12s; in classes conducted by other bodies in the city, 2,105, and the amount of grant £1,316. 9s. Taking for the basis of comparison the Government grants earned by the science and art classes in Manchester, it appeared that the total sum paid by the Government during the year 1887-8 was £5,590. 13s. Of this amount the sum of £3,140. 4s. was earned in classes conducted by the Board. In view of the assistance which it was hoped that the City Council would give to science and art and commercial instruction, under the the Technical Instruction Act, 1889, the Board in November last adopted a comprehensive scheme for the better organisation of their evening classes. According to this scheme there would next winter be opened science and art schools at the following five centres:—Waterloo-road Board School, Cheetham; Birley-street Board School, Beswick; Bangor-street Board School, Hulme; St. Matthew's Board School, Ardwick; Central Board School, Deansgate. It was proposed that there should be a systematic course of science and art instruction at the foregoing centres, and as the experiment of the commercial evening school at the Central School, Deansgate, had been a great success this session, it had been resolved to open four commercial evening schools next autumn, namely:—Ducie-avenue Board School, Greenheys; Waterloo-road Board School, Cheetham; St. Matthew's Board School, Ardwick; Central Board School, Deansgate. The object of the Board was to group their evening instruction in science and art and commercial subjects at the above-named centres. As it would be necessary very soon to complete the arrangements for the various evening schools to be opened by the Board next session, it was desirable to provide the City Council with full particulars regarding the work of the School Board in providing science and art and commercial evening instruction, in case it was decided to adopt the powers of the new Act of Parliament. With the approval of the Board the matter might well be referred to the sub-committee which organised the evening schools and classes last winter, with the addition of such other members as it might be thought well to add.

Mr. Schou said the Council would probably impose a halfpenny rate for the purposes of the Technical Instruction Act. He understood that other educational bodies in Manchester intended to make large demands, and the School Board must take care that it received its due share.

Mr. Nunn asked if the commercial classes were self-supporting.

The Chairman said they were, but the fees were too high for many of those who desired to attend.

A sub-committee was appointed, in accordance with the suggestion of the Chairman.

COLOURING WORKING DRAWINGS.—A "Draftsman" writes to the *Builder and Woodworker* as follows:—"The colours used for representing wood, iron, and other materials, are: For soft pine, a very pale tint of sienna; for hard pine, burnt sienna, with a little carmine added; for oak, a mixture of burnt sienna and yellow ochre is used. Mahogany is represented by burnt sienna and a portion of dragon's blood. For walnut, dragon's blood and burnt umber are used. For bricks, burnt sienna and carmine make a good colour. Gray stones are represented by a mixture of black and white, with a little of prussian blue and carmine added—pale ink alone is sometimes used for stone work. Brown freestone is represented by burnt sienna, carmine and ink. Wrought iron is represented by a light tint of prussian blue, and cast iron by a gray tint composed of black, white, and a little indigo. Brass is tinted with gamboge. Gamboge, slightly mixed with vermilion, makes a good colour for copper. Silver is represented by an almost invisible blue. Make use of the best colours only. Do not mix with too little water. If the first coat is not dark enough, wait till dry, and give another coat. Make up your mind what portion you are going to colour before applying a drop of paint. Do not stop in the middle of a wash, but when once the brush touches the paper, go straight through with the portion you begin. If obliged to leave the job for a minute, paint up to a line. A dotted line will do if there is not a "full one handy; this will hide the join between the two patches of colour. Do not let your brush be too wet nor too dry, a few trials will soon show the right amount of colour to take up. Use the best English drawing paper; if you then find any trouble, a little prepared ox-gall mixed with the colour will do wonders. Clouded drawings, as a rule, are caused by letting the work dry and going over the edges again when starting afresh. No piece of colouring should be left until finished.

EXTRACTS FROM THE BOARD OF TRAD JOURNAL.

NEW CUSTOMS TARIFF OF VICTORIA.

The following rates of import duty are now levied under the r Customs tariff of Victoria, dated the 4th November last:—

Classification of Articles.	Rates of Duty.	
Acid, acetic, containing not more than 30 per cent. acidity	Pint or lb.	0 c
Do., do., for every extra 10 per cent. or part of 10 per cent. above 30 per cent.	,,	0 c
Do., muriatic	Cwt.	0 5
Do, nitric	,,	0 5
Do., sulphuric	,,	0 5
Ammonia, carbonate of	Pint or lb.	0 c
Do., liquid	,,	0 c
Chlorodine	25% *ad. val.*	
Cocculus Indicus	Lb.	0 1
Gelatine	,,	0 c
Gycerine, pure	,,	0 c
Do, crude	,,	0 c
Morphia	Oz.	0 1
Nitrate of silver	,,	0 c
Nux Vomica	Lb.	0 0
Strychnine	Oz.	0 1

CUSTOMS TARIFF OF JAMAICA.

The following is a statement of the rates of import duty now le under the Customs tariff of Jamaica:—

Classification of Articles.	Rates of duty no levied.		
Ale, beer, and porter	Gall.	0	0
Candles, composition	lb	0	0
Gunpowder	,,	0	1
Indigo	,,	0	0
Lard	,,	0	0
Matches, lucifers and others, per gross of 12 dozen boxes, each box to contain 100 sticks, and boxes containing any greater or lesser quantity to be charged in proportion		0	5
Salt	100 lbs.	0	1
Soap	,,	0	5
Spirits: Brandy	Gall.	0	10
,, Gin	,,	0	10
,, Rum, the produce of and imported from British possessions	,,	0	10
,, Whiskey	,,	0	10
Spirits of wine, alcohol and all other spirits, cordials, or spirituous compounds	,,	0	10
Sugar, refined	lb	0	0
,, unrefined	100 lbs.	0	10

THE EFFICIENCY OF CHIMNEYS.

THE "*Journal du Gaz et de l'Electricité*" quotes from a Ge source some experiments upon works' chimneys. An old chim 67 feet high, with internal diameter of 19·6 to 13·8 inches, and total passage, from fire to chimney top, of 98 feet, was taken down a new chimney, with an intended total draught of 95 feel ar minimum internal diameter of 25·5 inches, was planned out. V the chimney had gone up 39 feet it was tried; already there w great improvement on the old chimney; again at 46 feet, still be and at 52½ feet the draught was excellent, the smoke issued c without soot; and there was an economy of from 15 to 20 per cen fuel; so the chimney was finished of at that height. Herr Huth tl chimneys are usually made too narrow, and the mischief is aggra by increasing their height; so fuel escapes unburned. Herr Ram of Gotha, confirms this, and recommends a uniform internal dia as being more rational, and as protecting the brickwork from th and rapid axial stream. The cross-section of the chimney shoul from one-fourth to one-eighth the grate-area; and the height, not than 50 feet, should not exceed 100 to 120 feet (the diameter being to suit) unless the chimney is at a distance, in which case it m 160 to 200 feet, the diameter being regulated according to the am of soot which escapes.

CALIFORNIA QUICKSILVER.

l report to the *Engineering and Mining Journal*, Mr. J. pl, of San Francisco, gives the following table, showing the f the several California mines for the past six years :—

nes.	1884.		1885.		1886.
lmaden	20,000	..	21,400	..	18,000
....	2,931	..	1,309	..	3,478
'onsolidated	1,316	..	2,197	..	1,769
Western	3,292	..	3,469	..	1,949
r Bank..........	890	..	1,296	..	1,449
lria	1,025	..	1,144	..	1,406
Eastern..........	332	..	446	..	735
ton	881	..	385	..	409
lupe	1,179	..	35	..	..
rd Consolidated..	..	..	..	..	..
s	7	..	392	..	786
otal flasks........	31,913	..	32,073	..	29,981
	dols.		dols.		dols.
t price per flask ..	26·00	..	28·50	..	32·00
st price per flask..	35·00	..	32·00	..	39·00
ge per flask	30·50	..	30·25	..	35·50
al value at average price........	975,000	..	970,000		1,060,000
	1887.		1888.		1889.
Almaden	20,000	..	18,000	..	13,100
....	2,694	..	950	..	..
Consolidated	2,880	..	4,065	..	4,500
Western	1,446	..	625	..	550
ur Bank..........	1,890	..	2,164	..	2,150
Idria	1,490	..	1,320	..	1,000
Eastern..........	689	..	1,151	..	1,350
gton	673	..	126	..	800
alupe	..	..	..	..	..
ford Consolidated ..	1,543	..	3,848	..	1,700
ous	455	..	992	..	500
Total flasks........	33,760		33,250		25.650*
	dols.		dols.		dols.
est price per flask...	36·50	..	37·00	..	40·00
est price per flask ..	48·00	..	48·00	..	50·00
age per flask	42·25	..	42·50	..	45·00

alue at average price, $1,425,000, $1,415,000, $1.154,000.

otal production for 1889 is a near approximation.

ing to Mr. Randol, the monthly production and highest and ices prevailing during the past year have been as follows :—

Month.	Monthly production. Flask.		Highest price per flask.		Lowest price per flask.
ry..................	2,270	..	$43·00	..	$41·50
ary	1,740	..	42·00	..	41·50
h	2,125	..	41·50	..	40·00
....	2,134	..	41·00	..	40·00
....	1,840	..	45·00	..	41·00
....	2,225	..	50·00	..	46·50
....	2.021	..	47·50	..	46·00
st	2,060	..	47·50	..	46·00
mber................	2,030	..	47·50	..	46·00
er..................	2,440	..	47·00	..	46·50
mber................	2,460	..	48·00	..	46·00
mber................	2,705*	..	48·00	..	47·00

ber product estimated.

ing, Mr. Randol says: "The total production for 1889, sks, compared with the previous year, shows a decrease of s, and is the smallest quantity in any year since 1873, when ction was 27,642 flasks. New Almaden's production shows ,900 flasks, and is its lowest yield since 1874, when its pro- as 9,084 flasks. Napa Consolidated retains its position of ghest producer, and increased its output to 4,500 flasks, a 35 flasks. Ætna was dropped off the list. Bradford, the e rank last year, produced only 1,700 flasks, a loss of 2,148 reat Western produced 550 flasks, a loss of 75 flasks. Sul- k also shows a slight decrease, 2,150 against 2,164. New a like misfortune, 1,000 against 1,320. Great Eastern, an at increase, 1,350 against 1,320. Redington, in a last expiring effort, turned out 800 against 126. And various odds and ends of mines gathered 500 against 992 in 1888.

"This decrease all a long the line (except Napa Consolidated) emphasises the poverty of the mines; the higher price of quicksilver has failed to arrest the decline in production, and the future outlook is far from hopeful. Still higher prices must prevail in 1890; and this industry must be protected by a liberal duty—at least ten cents per pound—otherwise we may look for a further decline in production, to a point where the output will be insufficient to pay costs; and then—extinction."

REPORT ON THE USE OF THE MIXED INDIGO AND INDOPHENOL VAT.

By M. Galland.

(Chemist at Loenach Mulhouse Industrial Society).

THIS short note deals with the mixed vat (indigo and indophenol), the idea of which is due to the firm of Durand, Huguenin and Co., and I should like to call your attention, not only to its evident economy, but also to the fact that the two colours are taken up at the same time.

The reducing agent best adapted for use with the mixture of these two colours is sodium hyposulphite, which was discovered by M. Schutzenburger, and applied by him and M. de Lalande to the reduction of indigo. The vat is made up as follows :—Into a cask of 500 litres capacity, 30 litres of ground indigo, to which 3·5 kilos. of indophenol have been previously added, are put

Ground indigo.... { 10 kilos. indigo,
30 litres of water,
2 litres of soda at 35°.

These are left in contact for twelve hours and then ground up for six hours.

To this mixture is added 48 litres of sodium bisulphite solution of 40° B., and then slowly, so as to avoid too great a rise of temperature, 9 kilos. of zinc dust suspended in 10 litres of water. The whole is well stirred for half-an-hour, and 30 litres of caustic soda at 38°B added. This done, it is made up with water to 500 litres and allowed to stand for two or three days. The dyeing vat has a capacity of 5,000 litres. To make it up the contents of two casks of the above mixture are poured into 4,000 litres of water, to which sufficient hyphosulphite has previously been added to. absorb the dissolved oxygen. The amounts required by our vats are :—

2 kilos. of zinc dust,
12·5 litres of bisulphite at 40° B.
25 litres water,
8 litres caustic soda at 38°.

A vat thus made up imparts the dark shade (corresponding to 450-500 grammes of indigo by the old process) to 30 pieces of cotton passed through three times at such a rate that the fabric remains two minutes in the liquid, care being taken to keep the vat up to strength, and to reduce at the end of each operation, and to subsequently chrome the cloth in a cold solution containing 2 grammes of bichromate of potash per litre. To restore the vat, 123 litres of the mother liquid are added for every passage of 30 pieces, corresponding to 2·5 kilos. of indigo.

Working in this manner, each piece requires for the shade mentioned above only 250 grammes of indigo.

We have never exceeded this amount since we commenced to work with a mixed vat reduced by hyposulphite.

When a vat has been worked it is well to ascertain by a dyeing test on a small scale its state of reduction, and to add the amount of hyposulphite requisite to reduce it to its former condition.

This method of procedure is very simple and easy of application. The bath is much clearer than in the other process, does not so soon become muddy, and gives much more uniform shades. The treatment with chromate is quite as satisfactory as when indigo alone is used.

In proof of the statement made at the commencement of this article, namely, that both the colours are taken up at the same time, it may be mentioned that when a portion of the fabric is treated with alcohol, the indophenol dissolves even in the cold, leaving the insoluble indigo on the fibre. If the above instructions be rigidly adhered to, no difficulties will be experienced in the working of the mixed vat, the economy of which is obvious.—*Moniteur de la Teinture.*

Chemicals and Strikes.—The inter-relationships of manufactures are illustrated by the fact that the strikes in Lancashire have given activity and animation to the chemical trade of Durham. The explanation is simple: users abroad find a difficulty in obtaining chemicals in Lancashire because of the stoppage of fuel supply; and the orders for Germany and for America are coming to this side. The increased trade may be a temporary one—probably it will be—but in the meantime the chemical makers here benefit, as well as some shipowners.

A NEW METHOD OF PYRITES ANALYSIS.

THE sulphur in pyrites can be accurately estimated in a comparatively short time by the following method, which is more especially advantageous when the sample contains substances which render the application of solvents more difficult, but allow of the expulsion of the sulphur in the dry way.

The finely divided pyrites is placed in a small boat and introduced into a combustion tube which can be heated in a small furnace. A current of air (about 150—200 bubbles per minute), dried by means of sulphuric acid and saturated with nitric acid vapor by passing through a Drechsel's flask containing about 54 c.c. of fuming nitric acid, is passed through the combustion tube into a tubulated receiver containing 100 c.c. of bromine water, to which is attached a Peligot's tube containing 40 c.c. of bromine water, and finally through a cylinder of distilled water.

The pyrites is gradually heated, commencing at the hinder portions of the tube, the front parts being also warm to prevent the condensation of the sulphuric acid. In about three-quarters of an hour, the process being at an end, the Dreschel's flask containing nitric acid is removed and the boat is allowed to cool slowly in a stream of air. The sublimation of the sulphur by too rapid heating is to be avoided.

The contents of the receivers are washed into a beaker, the excess of bromine evaporated, the solution mixed with 1 c.c. of concentrated hydrochloric acid, heated to boiling, and the sulphuric acid precipitated with barium chloride; the precipitate must be carefully washed with boiling dilute hydrochloric acid, to free it from barium nitrate, but any admixture with this salt can be avoided by evaporating completely in a porcelain basin before adding the barium chloride.

The boat containing the ferric oxide, which is quite free from unchanged pyrites, is warmed with concentrated hydrochloric acid, the solution evaporated almost to dryness and the insoluble residue estimated in the usual manner; the iron solution is completely free from sulphuric acid.

Two analyses by this method gave the following results:—

	Per cent.		Per cent.
Sulphur	52·59	..	52·67
Iron	45·67	..	45·52
SiO_2, &c.	1·08	..	1·12
	99·34		99·31

Ex.

ALUMINOUS GLASS.

IT is well know that certain varieties of glass when exposed to the action of fire, even for a short time, become opaque and the surface roughened.

Certain products of Thuringia, on the contrary, enjoy a well merited reputation, because glass coming from the manufactories of that country supports without deterioration fusion, remelting, reblowing, etc. This property has been attributed to the introduction of a certain sand obtained from a neighbourhood of the village of Martinsroda.

It is interesting to understand the chemical composition of this sand.

Schott has made the following analysis of the sand, and side by side are the results of an analysis of the glass made from it.

	Sand from Martinsroda.	Glass made with sand from Martinsroda.
Silica	91·38	67·74
Alumina	3·66	3·00
Oxide of Iron	0·47	0·42
Lime	0·31	7·38
Magnesia	none	0·26
Oxide of Manganese	traces	0·52
Potash	2·99	3·38
Arsenious acid	none	0·24
Soda	0·50	16·01
Total	99·31	98·95

The extremely constant proportion of alumina in both sand and glass s remarkable, and calls for attention.

In order to arrive at the exact significance of the presence of this base, various experiments were undertaken with glass, some samples being made with pure silica, and others were used into the composition of which a little alumina had been introduced in quantity about equivalent to that contained in the sand of Martinsroda.

In submitting these different glasses thus prepared to the operation of remelting down into window glass, it was evident that it was more easily effected in those samples which contained the alumin

In short, these experiments, which were very thorough, proved the admixture of feldspar or alumina communicated to the glass gr strength, and made it more easy to work.

M. Schott attributes these advantages to the property which alu possesses of lessening the volatilisation of their alkalies in the s ficial layers, and preventing that crystalline structure which al characterises an easily broken glass, and which is due to the tend of the double silicate of calcium and sodium to take the form of cry. —*Scientific American.*

AN ADVANCING MARKET FOR ALKALIES AMERICA.

SINCE the beginning of the year there has been a steady th gradual advance in the market prices for soda ash and caustic and there is no indication yet that the top of the market has reached. In fact, if present indications hold good, prices will be higher during the balance of 1890 than they have been at any within many years. Apart from the fact that the manufacturer tired of losing money in unprofitable competition, there are several reasons for the upward movement of prices, and but for the fact that are two distinct processes of manufacture, between the exponer which competition is extremely active, the rise would no doul greater and would have come sooner.

The chief reason for the advance is found in the increased co salt, owing to the control exercised over that commodity by the Union of Great Britain. To some of the larger alkali makers th itself, would have been of no consequence since they are indepe of the combination having salt wells of their own, but to the sn concerns, whose product in the aggregates makes up the bulk of the ply of these chemicals, the restriction placed upon them by the adv in the cost of salt is a very serious matter. The old process men fee more keenly than their competitors who use the newer method of n facture, but to either class of producers the increased cost of raw ma is a most potent reason why they should get more money for their duct. A second reason is found in the advance in the price of and a final reason to some of the makers and will probably soon all, is the higher cost of labour.

If there are so many good reasons for the advance, the opport to make it with the best results is not lacking. For the past thr four months, but particularly since the beginning of the current yea demand for soda ash and caustic soda has been unprecedentedly The consumption all over the world seems to have increased enorm In this country, although the receipts have increased over ten or t per cent. during the past two months, and the supply has been incr by the added output of domestic factories, it is almost impossit secure anything for prompt or early delivery, and many of the n facturers have their product for several months to come under tract. At the present rate of consumption it seems likely that ther not be enough stock of soda ash to supply all requirements by the spring weather arrives, and caustic soda is in almost the same pos —*Oil, Paint, and Drug Reporter.*

PRICE'S PATENT CANDLE COMPANY (LIMITED), reports th business in 1889 has been satisfactory, although, as the shareh were informed by last year's report would be the case, excep expenditure took place to meet growing requirements in some bra of the trade. Adopting the usual comparison for three years c ordinary expenditure on replacements and repairs and on new macl and plant, the yearly cost has been:—Replacements and repairs— £26,248.; 1888, £22,863.; 1889, £30,140.; new machinery and —1887, £4,197.; 1888, £3,972.; 1889, 6,203. The ordinar extraordinary expenditure on the two factories amounted in 18 £37,133., as against a corresponding expenditure of £28,836. in and the company has obtained full value for the outlay. The pr 1889 amounted to £70,289. 6s. 11d., to which has to be £5,603. 17s. 10d. carried forward from the previous year, m together £75,893. 4s. 9d. Deducting the dividend paid in Sept last, £18,750., £10,000. for depreciation of plant, and, as the dir recommend a like sum in reduction of the goodwill account, ther remain an available sum of £37,143. 4s. 9d. Out of that su directors recommed the payment on the 18th March of a divid 10s. and a bonus of 5s. per share, making with the September div a distribution of 25s. per share for the year. Such a paymen absorb £28,125., and will leave £9,018. 4s. 9d. undivided. In e of their statutory power, the directors unanimously appointe Calderwood, who had been the company's manager since March, to the seat at their Board, rendered vacant by the death of G Brownrigg, C.B., with the title of Managing Director, and th confident that the appointment meets, and will continue to mee the hearty approval of their fellow-shareholders.

ERIMENTS WITH OXYGEN CYLINDERS.

RIES of experiments were recently conducted at Stevenston the Scotch and Irish Oxygen Company, Limited (Brin's pro- the purpose of demonstrating the absolute safety of their for containing compressed gases. It will be remembered atal accident occurred to the foreman of their works, at Pol- g the bursting of a cylinder some six weeks ago, and the im- was conveyed to the public that the explosion was due either lefect in the cylinder, or to its strength being overtaxed by urged to an excessive pressure. A thorough investigation was, made into the cause of the accident, and the facts disclosed oubt that it occurred through a mistake on the part of the un- man himself. It was clear that he had introduced oxygen linder which was already partly charged with hydrogen, these ming an explosive mixture when brought together. One fact y which pointed at once to this conclusion was that n portion of the exploded cylinder, found almost on the ere the accident occurred, was almost too hot to be ; whereas, the portions of cylinders which have since been in- ly burst by being subjected to excessively severe treatment nd to be perfectly cold. In consequence of this erroneous n it was thought advisable to subject the cylinders to an ex- series of tests, so as to prove beyond dispute that they are ily adapted for the purpose of carrying the gases, and capable anding the most extreme amount of rough usage to which ld possibly be subjected during transit from place to place. nders used are of different sizes, varying from 1ft. long by 3½ meter to 6ft. 6in. long by 5⅛in. diameter. They are made of ought steel ¼ of an inch in thickness, and before being used inder is subjected to a hydraulic test of at least twice its work- sure, and is afterwards stamped with the pressure to which it jected, the date of the test, and the test mark. The cylinders sted periodically, and for the purposes of safety the custom of pany has been to paint those for the different gases in distinc- ours, so as to avoid the danger of mixing. As a further pre- every cylinder which is brought into the works after been in use is at once emptied. To make the recurrence of isfortune as the recent accident absolutely impossible, it has cided to adopt a left-handed thread for the valves on the n and coal gas cylinders, so that by no possible inadvertency y be filled at the oxygen pump. A number of experiments ade about ten days ago which gave excellent results. A cylinder, weighing about 1 cwt., was twice raised to the f 35 ft. and dropped horizontally upon a solid iron block 12in. and weighing 3½ cwt., each blow bending it to the extent of ree-quarters of an inch. It was then dropped vertically on to d end, having a clear fall of 31 ft., when it was found that the had only flattened a part of about the size of a pennypiece. ext placed across the iron block, and an iron weight of dropped on to its centre from a height of 35 ft., the blow in the side to the extent of ⅞ of an inch. The cylinder was ently placed on two iron blocks, set 4 ft. 1 in. apart, so as to the ends, and the same weight again let fall upon it from the eight, with the result that it was bent 4½ in. from the but did not explode. Another cylinder was afterwards the same manner, with the exception of the crushing blow, his case even a more satisfactory result was obtained, as it was the extent of 7⅛in. by the bending blow and still remained A smaller cylinder, measuring 31in. long by 5⅛in. diameter, ng 17lb. liquified carbonic acid gas, was also dropped cross- d vertically from the same height, and was afterwards flattened extent of 1¾in. by dropping the 6¼cwt. upon it, without it otherwise than in shape. Each of these tested cylinders bsequently found to contain the full quantity of gas, and to be y sound. Yesterday's experiments were of a similar character, re equally satisfactory. A 6ft. 6in. cylinder, weighing 107lb., og the contents, was dropped four times across the iron block eight of 35ft., these trials producing a bend of 2⅛in. It was owed to fall on its end, with little perceptible result. A cylinder was treated in the same manner, and sustained no injury than a few dents. From these particulars it will be seen tests were eminently satisfactory, for it is inconceivable that inders during transit could undergo anything like the severe nt to which they were subjected, while it was evident that they sessed a considerable reserve of strength.

WIDNES SALT SCHEME.—We understand that a number of turers have withdrawn from the scheme. The bed of rock ld through is 60 feet thick, and it was met with at about 250 1 the surface, but there are some physical difficulties in the way ing it satisfactorily.

PRIZE COMPETITION.

A LITTLE more than twelve months ago, one of our friends suggested that it might be well to invite articles for competition, on some subject of special interest to our readers; giving a prize for the best contribution, he at the same time offering to defray the cost of the first prize, and suggesting a subject in which he was personally interested.

At that date we were totally unprepared for such a suggestion, but on thinking over the matter at intervals of leisure, we have come to think that the elaboration of such a scheme would not only prove of benefit to the manufacturer, but would go some way towards defraying the holiday expenses of the prize winner.

We are therefore prepared to receive competitions in the following subjects:—

SUBJECT I.

The best method of pumping or otherwise lifting or forcing warm aqueous hydrochloric acid of 30 deg. Tw.

SUBJECT II.

The best method of separating or determining the relative quantities of tin and antimony when present together in commercial samples.

On each subject, the following prizes are offered:—

One first prize of five pounds, one second prize of one pound, five additional prizes to those next in order of merit, consisting of a free copy of the *Chemical Trade Journal* for twelve months.

The competition is open to all nationalities residing in any part of the world. Essays must reach us on or before April 30th, for Subject II, and on or before May 31st, in Subject I. The prizes will be announced in our issue of June 28th.

We reserve to ourselves the right of publishing the contributions of any competitor.

Essays may be written in English, German, French, Spanish or Italian.

Trade Notes.

Another Trust.—The large smelting organisations of the United States have formed a trust, with a capital of £5,000,000. All but five of the smelting and refining companies in the United States are reported to be in it. Of the capital stock £3,000,000. is to be Common stock, and the remaining £2,000,000. Preferred stock. The chief object of the smelters, it is said, is to place their interests beyond the absolute control of the Lead Trust.

What May Happen.—As underground conductors make way in London, we may occasionally expect something like this in our dailies, unless precautions, in the nature of those which we have already indicated, are provided. A New York paper of recent date says: "A manhole plate of ordinary size covers a manhole in the bend of Minetta-street. Yesterday that plate sailed skyward for a dozen feet, followed by a volume of flame beautiful to behold, and attended by a crash of glass from near-by windows, and a reverberating report."—*Electrical Review.*

Ammonin.—This is a product made at Heidelberg, and said to be of great value for cleansing, prior to bleaching, not only cotton but all kinds of vegetable fibres. The method of manufacture is kept secret. It is a grey powder, colourless, and partially soluble in water; an analysis made by Dr. Zirnite shows it to contain 27 per cent. of soluble matters, 21 per cent. of which was carbonate of soda; 30·8 per cent. consisted of silica and oxides of iron and alumina; there was 34 per cent. of lime, with small quantities of sulphide of lime. Where the cleansing properties come in one fails to see; better results could be got by using plain soda-ash.

The Adulteration of Tallow.—Tallow is sometimes adulterated with stearic acid, and to detect the fraud H. Jaffe proposes to estimate how much standard alkali the sample absorbs. He warms 5 grammes of the tallow with 10 grammes of olive oil; adds a little turmeric solution as indicator, then runs in the standard alkali, keeping the mixture at 20° to 30° C. Olive oil should not show more than 2·5 per cent. and tallow 3·6 per cent. free acid. Stearic acid is equivalent to 100 per cent., so that a little of it throws off the balance of the tallow considerably.

An Automatic Evaporator.—Mr. F. E. Ray describes, in the *Pharmaceutical Record*, a very ingenious means of checking the evaporation of a solution at the right movement. He rigs up a kind of a balance of which one pan is the empty evaporating dish, the other a tin of sand. He makes this balance; then he lets the evaporating dish down into the sand-bath, and puts so much sand into the opposing tin as corresponds to the quantity he desires to leave in the evaporating dish. When the evaporation has gone to the desired point the dish rises out of the sand-bath, and a string connected with the beam shuts off the gas.

Glass Enamelled Steel Casks.—Glass enamelled steel casks are being made in the States for use as filters in glucose and sugar refineries, evaporating tanks for salt works, and other purposes, and to take the place of casks in breweries. They are said to be the finest specimens of enamelled steel work yet produced. The body of the cask is composed of a number of welded steel rings a quarter of an inch thick with right angle flanges at each edge. The heads are stamped from single sheets of steel in a powerful hydraulic press, and the inside is coated with a glass enamel melted into the steel at a high heat. The sections and heads are bolted together with half inch bolts two inches apart, and the flanges are reinforced by continuous steel washers. The casks can be drilled at any point without chipping the enamel, which shows the tenacious union effected between the steel and the enamel.

Weekly Wages.—A Bill introduced by Mr. Fenwick, M.P., requires the wages of every workman engaged by time to be paid in full at least once in each week. When a workman is engaged otherwise than by time, his wages are directed to be paid in full at the expiration of not more than 14 days from the commencement of the employment or from the last payment in full. If a workman thus engaged gives one day's previous notice to the employer that he requires it, at least 75 per cent. of the wages earned by him are to be paid at the expiration of not more than seven days from the commencement of the employment or from the last payment in full. Moreover, on application at the pay office or some convenient place not later than the day previous to the day on which under the Bill his wages are payable, a workman is to be entitled to receive from his employer a "pay note" showing the amount of wages earned. A form is given in the Bill as to the general effect of the contents of a pay note. For contravening the provisions of the Bill an employer is to forfeit from £5. to £10. on the first occasion, and from £10 to £20 for the second offence; and for the third offence, which is regarded as a misdemeanour, the penalty may go up as high as £100. In the term "workman" is included any person who is a labourer, servant in husbandry, journeyman, artificer, handicraftsman, miner, railway servant, or is otherwise engaged in manual labour.

The Production of Ivory.—There are annually killed in A a minimum of 65,000 elephants, yielding a production of a quant raw ivory, the selling price of which is some £850,000. This qu is shipped to various parts of the world—to the American. European, and the Asian markets. A large quantity is, however, by the native princes of Africa, who are very fond of—and, as a very good judges of—ivory. The production out of Africa is insignificant, and India, Ceylon, and Sumatra together produce some 20,000 kilogs. per year. India is the largest consumer of i and China is also a good market.

The Subsidences in Cheshire.—At the Northwich Local meeting, on the 19th inst., on the question of the opposition given to the Widnes Brine Bill coming up for discussion, it was s that Professor Boyd Dawkins, of Manchester, had visited North and had inspected the town, as well as the site of the pro pumping station at Wincham, with a view to giving a report t Parliamentary Committee. The result so far had been highly factory, and the Committee considered that the report would pr great value. It was reported that the Bill would not be befo Committee until the end of April or the beginning of May. O motion of Mr. Maddocks it was resolved to forward all details c subsidence question to members of Parliament and to the Presid the Local Government Board, showing how the people of the d had suffered and the position in which they stood at present asking for a Royal Commission to inquire fully into the subject.

The Cheshire Salt Supply.—The new shaft which has sunk at the Northwich Salt Company's new works to a depth of eighty odd feet, and which is thirteen feet square (being one o largest in the district), has, it is believed, struck a main flow of according to experts who have been to the place. The compan some difficulty in sinking the shaft, owing to meeting with a quic twenty feet deep, the result of which was that the shaft will have continued with a tube, but at the bottom of this quicksand the came through the "metal" in immense volumes, and has spru the shaft a height of forty feet. A powerful pump, with quick a has been inserted, and Mr. Clarke's intention is to get down as possible with cylinders. The flow of brine is such that a great s is expected. A number of houses have been built for the work and more are being erected, so that the place has all the appeara successful enterprise.

The Fullers' Earth Union, Limited.—The capital c new undertaking is £105,125., in 105,000 ordinary shares of £1. and 125 Founders' shares of £1. each. The directorate is a one, with Mr. Alfred G. D. Moger, J.P., banker, of Bath, as man Among the members of the board is the Right Hono Lord Kilmorey. The object of the company is to acquire and c date the Fullers' Earth Works in the only two places in whic earth is found of a character and under conditions which allow being profitably utilised for commercial purposes, namely, R Surrey, and Bath, Somerset. It has been ascertained on enqu the office of the Geological Survey that these are the only two in which works for digging and preparing the earth are in ope Fullers' earth is largely used for washing rugs and blankets, carp worsted yarns, woollen and worsted cloths and silk, and other f in the course of manufacture and dyeing. In addition to this use is a great demand for the earth (in a highly powdered and refine dition) for chemical and toilet purposes, and also in various sta refining h oils, tallows, fats and wax, and for numerous other pur Although the earth is largely used, not only in the centres of the w trade in Great Britain and Ireland, but also in France, Ho Germany, the United States, India, and the Colonies, prac speaking the supplies are drawn exclusively from the sources ment No less than eleven works will be amalgamated, and these will tically form a monopoly in their particular line. As will be se the perusal of the details, there can be but little doubt as to the p bility of the concern as newly organised and developed.

Market Reports.

TAR AND AMMONIA PRODUCTS.

The benzol market is again a shade weaker than reported last and prices, both for 90s and 50/90s may be called 1d. less per Solvent naphtha still continues in good request, and the home m are being supplied with 90% at 160° at 1s. 7d. per gallon, guar equal to sample. Creosote is moving off at good paying prices, anthracene is still again weaker. Pitch remains as quoted week, 29s, Garston, and 30s. East coast ports.

In sulphate of ammonia there has been a fair amount of bu sellers having met the market at the ruling rates. The lower of nitrate do not appear so far to have affected sulphate. Th ttle quieter feeling to-day. London quotations have kept £11. 12s. 6d. being the price at Beckton, and £11. 15s. ondon makes. At the other ports business has been fairly at £11. 15s. to £11. 16s. 3d.; Leith and Liverpool to £11. 15s.

METAL MARKET REPORT.

	Last week.				This week.		
....	51/0			....	51/4		
....	£90	7	6	....	£90	7	6
per	47	2	6	....	48	17	6
ter	21	15	0	....	21	15	0
d	12	7	6	....	12	12	6
er	43 3/4 d.			....	43 3/4		

COPPER MINING SHARES.

	Last week.			This week.	
Tinto	15 1/2	15 5/8	....	16	16 1/8
on & Barry	6 5/16	6 5/16	....	6 5/16	6 7/16
rsis	4 7/16	4 9/16	....	4 1/2	4 5/8
e Copper Co. ..	3 7/16	3 9/16	....	3 1/2	3 5/8
naqua	1 7/8	2 1/8	....	1 7/8	2 1/8
iapo	2 7/16	2 9/16	....	2 7/16	2 9/16
ulcillo	1	1 1/8	....	15/16	1 1/16
w Quebrada	5/8	3/4	....	5/8	3/4
iola	2 7/16	2 9/16	....	2 7/16	2 9/16
copilla	1/3	1/9	....	1/3	1/9
gentella	3d.	9d.	....	3d.	6d.

SCELLANEOUS CHEMICAL MARKET.

has been considerable disturbance in all branches of the trade, caused by the irregularities in supply of fuel during the ation. Prices of all products in the alkali trade have been by short supplies, and have not yet given way to any appreci- nt. Caustic soda 70's, for early deliveries is at £11. to per ton f.o.b., and 74 % £12. 5s. per ton. Soda ash is still d at 1 3/8d. to 1 1/2d. per deg. for ordinary grades. Soda £3. per ton. Salt cake 32s. 6d. to 35s. per ton in bulk. powder is at £5. 7s. 6d. to £5. 10s. per ton on rails for veries, but can be bought at easier rates for forward supplies. of potash is steady at 5d. Vitriol and muriatic acid remain late prices. Sulphur weak, and business done in recovered r ton at works. Sicilian 3rds £3. 10s. c.i.f. Sulphate of 26. for prompt delivery. Tin crystals 6 3/4d. per lb. Lead t, and prices nominally as last week, though re-sales are done Acetates of lime and acetic acid are quiet, and prices of the shade easier. Potash caustic and carbonate unaltered.

REPORT ON MANURE MATERIAL.

has been a moderate business doing, but mostly for early or e delivery and for limited quantities, manure makers being to concern themselves about forward supplies. Prices of us material continue to droop, and phosphate values are barely ed.

is very little doing in mineral phosphates, except in retail s for early or immediate deliveries, and quotations all round experienced any change during the past few weeks.

ve no fresh business to report in cargoes of River Plate bones. value we should quote £5. 7s. 6d. for December sailing. We ar of any fresh business in crushed bones, but £5. 5s. would be accepted for shipments shortly due. Further sales of East one meal have been made at £5. 10s. ex quay, and that price o for 200 tons Calcutta arrived at Liverpool. Common bones are worth £5., and £5. 2s. 6d. per ton for immediate ex quay Liverpool, but ahead they can be bought at . per ton less, easier freights being obtainable, and upplies being on the way. For next season we do not ny transactions so far. Nitrate continues very weak in all Spot price cannot be quoted over 8s. to 8s. 1 1/2d. per l due cargoes would not fetch over 8s. Prices asked for ship- 3d. to 4 1/2d. per cwt. more, but they don't result in business. arket for dried blood is very dull. There are sellers now of e prepared under 11s. per unit, and of River Plate at 10s. 3d. of ammonia, but these prices don't attract buyers in the uch cheaper nitrate soda and sulphate ammonia. Ammonite ther scarce, and quotation remains 9s. 6d. per unit, in bags, ghts, delivered f.o.r. at works.

hosphates are rather irregular in price, there being con- quantities in second hands offering. Quotation is nominally 46s. 3d. per ton in bulk, f.o.r. at works, but 1s. 3d. to 2s. 6d. less would be business for a good line.

The strike in Liverpool is still seriously interfering with deliveries of goods from the quay, but there seems now a fair prospect of adjustment of differences.

WEST OF SCOTLAND CHEMICALS.

GLASGOW, Tuesday.

The general chemicals market has been considerably excited and quotations feverishly on the rise; but there has been no corresponding bulk of business passed, or anything like it, and it is expected that in the result this excitement will be shown to have been unwarranted. Buyers are only buying what they must buy in the meantime, and with the labour troubles settling down the high level of quotation cannot be long maintained. Sulphate of ammonia has not shared in the prevailing inflation, but on the contrary is a little flatter than it has been, as low as £11. 13s. 9d. having been taken, and no more than £11. 15s. got for some days, despite allegations to the contrary. With the exception of burning oil paraffins are all healthy, and for good scale and wax in the hands of dealers rather more than the association fixtures, as given below, is to be obtained. Calico printing in Scotland is not so badly off for working orders, but the rest of the Scotch dyeing departments are all more or less idle, and have been so since January. Bichromates continue in very poor demand, and dyes and drysalteries generally are as a rule depressed. Scotch bleach quotations are irregular, and ruling lower than the figures lately current for Tyneside deliveries. Chief prices current are:—Soda crystals, 50s. to 52s. 6d., less 2 1/2% Glasgow; alum in lump £5., less 2 1/2% Glasgow; borax, English refined £30., and boracic acid, £37. 10s. net Glasgow; soda ash, 1 1/8d. net Tyne; caustic soda, white, 76°, £13., 70/72°, £11., 60/62° £10., and 60/62° cream, £9., all less 2 1/2% Liverpool; bicarbonate of soda, 5 cwt. casks, £6. 2s. 6d, and 1 cwt. casks, £6. 10s. net Tyne; refined alkali 48/52°, 1 1/4d., net Tyne; saltcake, 30s. to 32s. 6d.; bleaching powder, £5. 12s. to £5. 15s. less 5% f.o.r. Glasgow; bichromate of potash, 4d., and of soda 3d., less 5 and 6% to Scotch and English buyers respectively; chlorate of potash, 5d., less 5% any port; nitrate of soda 8s. 1 1/2d. to 8s. 3d.; sulphate of ammonia, spot, £11. 13s. 9d. to £11. 15s. f.o.b. Leith; sal-ammoniac, 1st and 2nd white, £37. and £35., less 2 1/2% any port; sulphate of copper, £26., less 5% Liverpool; paraffin scale, hard and soft, 2 1/2d. per lb.; paraffin wax, 120°, semi-refined, 3 5/16d.; paraffin spirit (naphtha), 8 1/2d. a gallon; paraffin oil (burning), 6 3/4d. to 7d. at Glasgow; ditto (lubricating), 865°, £5. 10s. to £6. per ton; 885°, £6. 10s.; and 890/895°, £7. 10s. to £8. Last week's imports of sugar at Greenock were 41,816 bags and 2,908 baskets.

THE LIVERPOOL MINERAL MARKET.

The advance previously reported in our market has been more than continued. Manganese: Arrivals are slow and meagre, and sales are being made for prompt and forward delivery at considerably advanced figures. Magnesite: Stocks of raw lump are very large, both at this side and at the port of shipment; large reductions in price, are, therefore, anticipated; raw ground, £6. 10s.; and calcined ground, £10. to £11. Bauxite (Irish Hill brand): Supplies are scarcely able to compete with the demand; prices, therefore, are very firm—Lump, 20s.; seconds, 16s.; thirds, 12s.; ground, 35s. Dolomite, 7s. 6d. per ton at the mine. French Chalk: Arrivals this week have been more than the average; but, in consequence of the strike, difficulties have been experienced in delivery, and sales have been made above last quotations, especially for G.G.B. "Angel-White" brand—90s. to 95s. medium, 100s. to 105s. superfine. Barytes (carbonate) continues easier, chiefly owing to useless competition; selected crystal lump scarce at £6.; No. 1 lumps, 90s.; best, 80s.; seconds and good nuts, 70s.; smalls, 50s.; best ground, £6.; and selected crystal ground, £8. Sulphate has somewhat improved, especially best brands, on account of the increased demand and short supplies; best lump, 35s. 6d.; good medium, 30s.; medium, 25s. 6d. to 27s. 6d.; common, 18s. 6d. to 20s.; ground best white, G.G.B. brand, 65s.; common, 45s.; grey, 32s. 6d. to 40s. Pumicestone easy; ground at £10., and specially selected lump, finest quality, £13. Iron ore in demand at full prices. Bilbao and Santander, 9s. to 10s. 6d. f.o.b.; Irish, 11s. to 12s. 6d.; Cumberland in strong demand at 18s. to 24s. Purple ore steady at last quotations. Spanish manganiferous ore selling freely at full prices. Emery-stone: Best brands scarce both for spot and forward delivery; prices stronger. No. 1 lump is quoted at £5. 10s. to £6., and smalls £5. to £5. 10s. Fullers' earth steady; 45s. to 50s. for best blue and yellow; fine impalpable ground, £7. Scheelite, wolfram, tungstate of soda, and tungsten metal inquired for. Chrome metal, 5s. 6d. per lb. Tungsten alloys 2s. per lb. Chrome ore: More offering, high grades bringing full prices.

Antimony ore and metal have further advanced. Uranium oxide, 24s. to 26s. Asbestos: Best rock, £17. to £18.; brown grades, £14. to £15. Potter's lead ore: Prices are firm; smalls, £14. to £15.; selected lump, £16. to £17. Calamine: Best qualities scarce—60s. to 80s. Strontia steady; sulphate (celestine) steady, 16s. 6d. to 17s. Carbonate (native), £15. to £16.; powdered (manufactured), £11. to £12. Limespar: English manufactured, old G.G.B. brand in demand, and brings full prices; 50s. for ground English. Felspar, 40s. to 50s.; fluorspar, 20s. to £6. Bog ore firm at 22s. to 25s. Plumbago: steady; Spanish, £6.; best Ceylon lump at last quotations; Italian and Bohemian, £4. to £12. per ton; founders, £5. to £6.; Blackwell's "Mineraline," £10. French sand, in cargoes, continues scarce on spot—20s. to 22s. 6d. Ferro-manganese and silicon spiegel in good demand at stronger figures. Chrome iron, 20 per cent., £24. to £25. Ground mica, £50. China clay freely offering—common, 18s. 6d.; good medium 22s. 6d. to 25s.; best, 30s. to 35s. (at Runcorn).

THE LIVERPOOL COLOUR MARKET.

COLOURS.—Steady, prices unchanged. Ochres: Oxfordshire quoted at £10., £12., £14., and £16.; Derbyshire, 50s. to 55s.; Welsh, best, 50s. to 55s.; seconds, 47s. 6d.; and common, 18s.; Irish, Devonshire, 40s. to 45s.; French, J.C., 55s., 45s. to 60s.; M.C., 65s. to 67s. 6d. Umber: Turkish, cargoes to arrive, 40s. to 50s.; Devonshire, 50s. to 55s. White lead, £21. 10s. to £22. Red lead, £18. Oxide of zinc: V.M. No. 1, £25.; V.M. No. 2, £23. Venetian red, £6. 10s. Cobalt: Prepared oxide, 10s. 6d.; black, 9s. 9d.; blue, 6s. 6d. Zaffres: No. 1, 3s. 6d.; No. 2, 2s. 6d. Terra Alba: Finest white, 60s.; good, 40s. to 50s. Rouge: Best, £24.; ditto for jewellers', 9d. per lb. Drop black, 25s. to 28s. 6d. Oxide of iron, prime quality, £10. to £15. Paris white, 60s. Emerald green, 10d. per lb. Derbyshire red, 60s. Vermillionette, 5d. to 7d. per lb.

THE COPPER MARKET.

Messrs. Harrington and Co.'s last fortnightly copper report, states:—Chili copper charters for first half of March were advised yesterday as 300 tons fine. Price of bars nominal, and exchange 24⅝d. The total charters since 1st January have been 5,200 tons, against 6,200 tons for the same time last year. Since the issue of our last the market has been quiet, and only a moderate business done, prices for g.m.b.'s varying between £46. 10s. and £47. 15s. cash, and £47. 2s. 6d. to £48. three months. Deliveries have been impeded owing to the strike of dock labourers. Smelters complain of want of orders from home and abroad, and are consequently buying sparingly. We close firm at £47. 6s. 3d. for cash, and £47. 12s. 6d. for three months prompt. The total decrease in stocks during the fortnight is 2,217 tons. Refined and manufactured sorts remain quiet, at £52. 10s. to £53. 10s. for tough, £54. to £55. for b.s., £61. to £62. for strong sheets, £58. to £59. for India sheets, and 6d. per lb. for yellow metal sheets. The present stock of English g.m.b.'s in warehouse, Liverpool and Swansea, is 6,837 tons, against 7,040 tons on the 28th ult., showing a decrease of 203 tons. The sales of furnace materials comprise—At Liverpool: 120/135 tons argentiferous Chili regulus, on private terms; 1,675 tons argentiferous Anaconda matte, on private terms; 250 tons Boston Montana matte, at 10s.; 100 tons argentiferous Montana matte, on private terms; and 200 tons Spanish ore, at 8s. 3d. per unit. At Swansea: 300 tons Spanish ore, to arrive, at 9s. 3d. per unit.

THE TYNE CHEMICAL REPORT.

TUESDAY.

Owing to the settlement of the coal dispute, and the resumption of shipments at Liverpool, the chemical market here is settling down after the past week's excitement. Caustic soda and soda ash have advanced 20s. and 15s. per ton respectively since last week's report. Crystals are slightly higher in price, but bleach is not quite so firm. Quotations are more or less nominal, and very little enquiry for forward business. To-days prices are:—

Caustic soda, 77%, £13. per ton, prompt delivery; bleaching powder, in softwood casks, £5. 15s. to £6. per ton, with 2s. 6d. per ton extra for hardwood casks; soda crystals, £2. 10s. per ton; soda ash 48—52% 1½d. per degree less 2½%; sulphate of soda, ground, in casks, 42s. 6d. per ton, sulphate in bulk, 32s. 6d. per ton; chlorate of potash 5d. per lb.; recovered sulphur, £4. 5s. per ton; silicate of soda, 140° Tw., £4. per ton; 100° Tw., £3. 7s. 6d. per ton; 75° Tw., £2. 10s. per ton; hyposulphite of soda, in 5—7 cwt. casks, £4. 5s. per ton; in 1 cwt. kegs, £4. 15s. per ton; pure white sulphate of alumina, £4. 10s. per ton; blanc fixe, £7. 10s. per ton; chloride of barium, £8. per ton; nitrate of baryta, cr £18. 10s.; ground, £19. per ton; sulphide of barium, £5. per ton—all f.o.b. Tyne or f.o.r. makers' works.

Dr. R. Spence Watson's award, in connection with the wages di between Messrs. C. Tennant and Partners, Limited, Hebburn, their workmen, was made known on Saturday last. The wages tion affected the men employed in the alkali and bleach departi only, numbering about 240, and of this number about one-fifth been awarded a slight advance on their old rates, the wages c others remaining unaltered.

There is no change in the price of manufacturing coals this v Northumberland steam small is quoted 8s. to 8s. 6d. per ton Durham small 9s. 6d. to 10s. per ton.

Gazette Notices.

First Meetings and Public Examinations.

HARRIS, LEON, Olive-terrace, Plashet-lane, Upper Park-east, London, varni colour manufacturer, and oil and glass merchant. April 10th, 33, Carey- Lincoln's Inn-fields; April 24th, 34, Lincoln's Inn-fields, London.

Notices of Dividends.

BISHOP, ALBERT BICKLEY, Atherstone, hat manufacturer. First and final di of 2s. 10½d., March 20th, 120, Colmore-row, Birmingham.

EASTWOOD, FRED, Accrington, late of Church, near Accrington, mineral an duce broker, and patent size manufacturer. First and final dividend of 3 March 29th, Official Receiver's Office, Preston.

New Companies.

ANGLO-CALIFORNIAN PETROLEUM GROUND, LIMITED.—This company registered on the 15th inst., with a capital of £250,000., in £1. shares, to a and work the petroleum wells of Messrs. Lacey and Rowland at Los A California, U.S.A., and there carried on under style of the "Puente Oil Com The subscribers are:— Sh

W. Buttle, Marden-park, solicitor ....
C. Watkins, 32, Shardeloes-road, New-cross, clerk ....
P S. Tempest, 9, Westeria-road, Lewisham ....
A. Nisbet, 12, Manor-road, Twickenham, clerk of works ....
H. Woods, 35, Queen Victoria-street, engineer..........
A. W. Woods, 16, St. George's-road, Regent's-park, secretary to a company ....
G. A. Ollard, 15, Bedford-row, solicitor ....

DURAND ELECTRIC PETROLEUM AND GAS ENGINE AND MANUFACT COMPANY, LIMITED.—This company was registered on the 15th inst., with a of £100,000., in £5. shares, to acquire the business now carried on at 163, A Victor Hugo, Paris, by Eugene Durand, with all the patents, invention contracts connected therewith. The nature of the business to be acquired stated. The first subscribers are:— Sh

J. D. Cahill, 6, Great George-street, Cork..........
T. B. Wells, 6, Pembroke-square, W. ....
F. Davies, 22, Davies-street, Berkeley-square ....
J. Russell, 175, Strand ....
F. Street, 16, Victory-square, S.E..........
C. Furtado, 172, Oakley-square, N.W..........
R. E. Pritchett, Rhiewport, Hornsey ....

The Patent List.

This list is compiled from Official sources in the Manchester Tec Laboratory, under the immediate supervision of George E. Davi Alfred R. Davis.

APPLICATIONS FOR LETTERS PATENT.

Brick-making Machinery. R. Parry. 3,744. March 10.
Incandescent Gas Lamps. A. Wenzel. 3,767. March 10.
Gas Motor Engines. C. D. Abel. 3,774. March 10.
Photographic Apparatus. W. Langden-Davies. 3,775. March 10.
Improvements relating to Basic Furnaces.—(Complete specification Psaczolka. 3,782. March 10.
Manufacture of Litharge and of Red Lead. G. Larrouy. 3,786. Ma
Treatment of Dye-woods and their Extracts. M. G. Lindemann. March 10.
Electrical Terminals. W. Brierley. 3,805. March 11.
Refrigerating Apparatus. F. N. Mackay. 3,806. March 11.
The reduction of Sodium or other Metals. G. A. Jarvis. 3,829. Mar
Apparatus for Controlling the Admission of Air to Heating Appara (Complete specification.) W. P. Thompson. 3,831. March 11.
Water-heating Apparatus. R. A. Haggard. 3,834. March 11.
Regulation of Alternating Generators.—(Complete specification.) S. Currie. 3,840. March 11.
Automatic Apparatus for Compensating Variations of Current in tric Conductors. Siemens Bros. and Co., Limited. 3,841. March 11.
Circuit Breaker for Secondary Currents. Siemens Bros. and Co., L 3,842. March 11.
The Shrinking and Fulling of Felt.—(Complete specification.) H. H. 3,846. March 11.

H. H. Lake. 3,852. March 11.
d Damper Frames. R. Cole and C. Simmonds. 3,868.
Photographic and other purposes. A. Watt and C. Symes.
h 12.
nsulation and Cooling. L. Pyke and H. T. Barnett. 3,900.
Apparatus. R. Harvey. 3,907. March 12.
ors. M. E. McLennan. 3,908. March 12.
urrent Transformers. G. Kapp, W. C. Johnson, and S. E.
913. March 12.
Apparatus. J. Schwager. 3,915. March 12.
s. M. Daelstaen. 3,921. March 12.
tteries.—(Complete specification.) H. H. Lake. 3,924. March 12.
Earthenware and Metallic Pipes. B. G. Smith. 3,931.
h Lanterns. W. S. Rawson and C. S. Snell. 3,937. March 13.
d Pipes. G. H. Smith and B. Cooper. 3,942. March 13.
for Electric or other Lamps. J. Clegg. 3,947. March 13.
achinery. E. Martin. 3,937. March 13.
Switches. W. Scott. 3,963. March 13
pparatus. W. de Morgan. 3,966. March 13.

Elements for Secondary Batteries. J. S. Stevenson. 3,967. March 13.
Apparatus for Cleansing Wool. H. A. A. Dombrain. 3,971. March 13.
Brick Kilns. G. Möller. 3,985. March 13.
Sulphurous Acid and Sulphites. G. Horsley and A. C. Wilson. 3,992. March 14.
Electric Regulators. H. Pieper, Fils. 4,000. March 14.
A Continuous Mechanical Filter. H. Yeomans. 4,005. March 14.
Manufacture of Alkaline Silicate. A. J. Boult. 4,021. March 14.
Evaporating and Distilling Apparatus. J. Foster. 4,024. March 14.
Electric Incandescent Lamps. A. A. Goldston. 4,025. March 14.
Gas Storage Apparatus. W. T. Walker. 4,027. March 14.
Pressure Gauges. A. T. Clarkson and J. B. Spurge. 4,037. March 14.
Blowers and Exhausters. T. P. Richards 4,044. March 15.
Safety Devices for High-pressure Electrical Currents. B. M. Drake and J. M. Gorham. 4,053. March 15.
Flushing Apparatus. J. Archer. 4,080. March 15.
Coupling and Insulating Apparatus for Dynamos. C. J. Barley and H. Stevenson. 4,086. March 15.
Secondary Voltaic Batteries. H. T. Cheswright. 4,087 March 15.
Hydro-atmospheric Motor. C. Burn. 4,093. March 15.
The Production of Cyanides from Ferro-cyanides. E. Bergmann. 4,095. March 15.

IMPORTS OF CHEMICAL PRODUCTS

AT

THE PRINCIPAL PORTS OF THE UNITED KINGDOM.

ONDON.

ending March 19th.

therwise undescribed)
17 c. T. H. Lee
£300. Ohlenschlager Bros.
y—
y, 94 t. Allatini Bros.
71 H. R. Merton & Co.
y Ore—
14 t. Bank of N. S. Wales
114 H. Emanuel
—
£208. Walsh, Lovett, & Co.
of Lime—
£119 W. Thatcher
100 Lister & Biggs
120 "
196¼ Johnson & Hooper
cid—
200 pkgs. Beresford & Co.
28 Leach & Co.
97
—
50 cks. Leach & Co.
11 J. L. Lyon & Co.
43 Hemmingway & Co.
18 W. Harrison & Co.
y, 23 Constable Henderson, & Co.
23 O'Hara & Hoar
110 J. Matton
600 A. Lumbeck & Co.
y, 7 G. Steinhoff
23 O'Hara & Hoar
24 Pillow, Jones & Co.
23 W. Harrison & Co.
28 J. L. Lyon & Co
22 W. Harrison & Co.
ne—
25 t. J. Kitchin
y, 36 cks. Craven & Co.
27 t 18 c G. Boor & Co.
50 W C. Bacon & Co.
29 7 G. Boor & Co.
27 t. A. Shaw & Co.
les, 41 17 c. Goad, Rigg, & Co.
10 Culverwell, Brooks, & Co.
a, 12 t. L. & I. D. Jt. Co.
alia, 1¼ c. Flack, Chandler, & Co.
ales, 9 2 Goad, Rigg, & Co.
and, 2 3 Flack, Chandler, & Co.

Chloride—
y, £40. Petri Bros.
als *(otherwise undescribed)*
y, £66. T. H. C. Bade
165 A. & M. Zimmermann
240 "
25 B. & F. Wf. Co.
20 J. Owen
20 Phillipps & Graves
6 C. Brumlen
y, 8 T. H. Lee
207 Phillipps & Graves
566 A. & M. Zimmermann
112 T. H. Lee.
" 20 L. & I. D. Jt Co.
" 27 A. & M. Zimmermann
" 12 Phillipps & Graves

Caoutchouc—
E. Africa, 5 c. Messageries, Maritimes de France
Madagascar, 59 W. Johnson & Co.
" 12¼ Kebbel, Son, & Co.
" 6¾ Matheson & Co.
France, 80 W. Balchin
Madagascar, 45 Arbuthnot, Ewart, & Co.
Spain, 14 Mildred, Goyeneche, & Co.
U. S., 14 L. & I. D. Jt. Co.
" 23 Mildred, Goyeneche & Co.
Madagascar, 22 L. & I. D. Jt. Co.
Cape, 40 Hammond & Co.
Madagascar, 32 A. Frey & Co.
" 127 Kebbel, Son, & Co.
" 60 Union Lighterage Co.
Columbia, 22 Stiebel Bros.
E. Africa, 15 Kebbel, Son, & Co.
Holland, 8 Lewis & Peat
E. Africa, 1 L. & I. D. Jt. Co.
Madagascar, 67 W. Johnston & Co.
Holland, 33 Helibut, Symons, & Co.

Caustic Soda—
Holland, 115 c. R. W Greeff & Co.

Carbonic Acid—
Holland, 12 pkgs. G. Rahn & Co.

Cream Tartar—
France, 2 pkgs. B. & F. Wf. Co.
" 12 "
" 5 M. Ashby
" 5 W. C. Bacon & Co.
" 10 J. Puddy & Co.
" 5 B. & F. Wf. Co.

Cobalt Ore—
Sydney, 48 t. Fellows, Morton, & Co.

Dyestuffs—

Madder
Holland, 12 pkgs. Williams, Torrey, & Co.

Rennet
Denmark, 5 cks. Bailey & Leetham

Orchella
Ceylon, 1 pkg. J. Porter & Co.

Argols
Cape, 38 pkgs. Hale & Son
" 2 Brown & Elmslie
S. Australia, 2 L. & I. D. Jt Co.
Italy, 804 pkgs. B. Jacob & Sons
Cape, 3 Brown & Elmslie

Extracts
Germany, 2 pkgs. Langstaff & Co.
U. S., 100 M. D. Co.
" 40 W. A. Bowditch
Belgium, 50 Best, Ryley, & Co.
Denmark, 1 Fullwood & Bland
U. S., 50 W. Burton & Sons
Denmark, 13 Bailey & Leetham
Germany, 5 Boutcher & Co.
U. S., 413 T. Ronaldson & Co.

Sumac
Italy, 250 Brown & Elmslie
" 350 W. France & Co.
" 300 J. Kitchen
" 250 "

Saffron
Mauritius, 1 R McAndrew & Co.
Spain, 1 Burgoyne, Burbidge & Co.

Cochineal
Canaries, 2 t. Swanston & Co.
" 9 pkgs. "
" 3 Kuhner, Henderson & Co.
" 15 A. Jimiez & Sons
" 30 H. R. Toby & Co.
" 22 Forwood Bros. & Co.

Myrabolams
E. Indies, 1,520 pkgs. Marshall & French
" 2,001 L. & I. D. Jt. Co.

Indigo
E. Indies, 5 chts. L. & I. D. Jt. Co.
" 91 "
" 9 F. W. Heilgers & Co.
" 16 "
" 6 Gt. N. Ry. Co.
" 110 L. & I. D. Jt. Co.
" 39 Elkan & Co.
" 5 Kleinwort, Sons, & Co.
" 205 P. & O. Co.
" 8 H. Grey, jun.
" 227 L. & I. D. Jt. Co.
" 205 Indigo Co.
" 29 F. W. Heilgers & Co.
" 2 bxs. Elkan & Co.
" 532 L. & I. D. Jt. Co.
" 9 F Stahlschmidt
" 34 Parsons & Keith
" 10 W. Brandt, Sons, & Co.
" 6 Stansbury & Co.
" 7 Ernsthausen & Co.
" 17 Arbuthnot, Latham, & Co.
" 14 F. W. Heilgers & Co.
" 10 Lond. & Hanseatic Bank
" 6 Elkan & Co
" 2 W. Isted
" 16 Langstaff & Co.
" 12 Stansbury & Co.
" 92 L. & I. D. Jt. Co.
" 5 Parsons & Keith

Valonia
A Turkey, 25 t. A. & W. Nesbitt
" 23¼ Adam Bros.
" 15 c. F. Hush & Co.

Tanners' Bark
Victoria, 31¼ t. Leach & Co.
Belgium, 40 J. J. Williamson & Sons
U. S., 15 Lindsay, Bird, & Co.
Victoria, 10 c. T. J. & T. Powell
——— 70½ t. Baxter & Hoare
Belgium, 10¼ Leach & Co.
Cape, 29½ Baxter & Hoare

Annatto
Ceylon. 12 pkgs. Hoare, Wilson, & Co.

Gambier
E. Indies, 429 pkgs. L. & I. D. Jt. Co.
" 845 Cox, Paterson, & Co.

Dyestuffs *(otherwise undescribed)*
Germany, 5 pkgs. L. & I. D. Jt. Co.

Glycerine—
Holland, £500 Spies, Bros., & Co.
Germany, 190 A. & M. Zimmermann
Holland, 60 L. & I. D. Jt. Co.
" 105 F. W. Heilgers & Co.
Germany 226 Craven & Co.
Holland, 120 F. W. Heilgers & Co.
Germany, 150 Beresford & Co.

Glucose—
U. S., 80 pkgs. 440 c. C. Southwell & Co.
France, 300 300 De Paiva, Norman, & Co.
" 80 80 Seaward Bros.
Germany, 40 325 L. Sutro & Co.
U. S., 80 400 T. M. Duche & Sons
Germany, 12 115 Seaward Bros.
" 20 180 Deutschmann & Co.
" 100 100 Fellows, Morton, & Co.
" 20 140 Dutrulle, Solomon, & Co.
" 300 300 J. Barber & Co.
" 200 200 Barrett, Tagant, & Co.
U. S., 300 1,881 L. Sutro & Co.
" 100 560 Props. Scott's Whf.
" 50 300 Pillow, Jones, & Co.
" 165 825 T. M. Duche & Sons
" 95 500 J. Keiller & Son
" 40 300 Page, Son, & East
Germany, 240 240 L. & I. D. Jt. Co.
U. S., 50 300 Pillow, Jones, & Co.

Guano—
Chile, 940 t. Ang. Cont. Guano Wks. Agency

Gutta Percha—
Holland, 34 c. Kleinwort, Sons, & Co.
Germany, 1 Lavington Bros.
France, 5 Flageollet & Co.

Horn Piths—
N. S. Wales, 4 t. 2 c. Goad, Rigg, & Co.

Isinglass—
China, 5 pkgs. W. M. Smith & Sons
Brazil, 1 Lewis & Peat
E. Indies, 6 pkgs. L. & I. D. Jt. Co
Brazil, 1 "
Denmark, 10 "
E. Indies, 3 "
Germany, 3 E. Davis & Co.
Cape, 2 Lewis & Peat

Manure—

Phosphate Rock
U. S., 1,661 t. Fox, Roy, & Co.
Aruba, 240 Lawes Manure Co.
Belgium, 130 D. de Pass & Co.
France, 135 Anglo Contl. Guano Works
U. S., 660 Odams, Ltd.
France, 105 A. Hunter & Co.
" 30 London Manure Co.
Belgium, 145 Schofield, White & Co.
Guernsey, 10 P. Hecker & Co.

Nitre—
Chile, 7 962 cubic bgs. A. Ford & Co.
" 13,183 W. Montgomery & Co.
" 14,613 "
" 14,613 F. Huth & Co.
" 9,395 Ang. Contl. Guano Wks. Agency
Holland, 7 cks. J. Owen

Naphtha—
Germany, 21 brls. Stein Bros.

Potassium Oxymuriate—
Belgium, £230 G. Boor & Co.

Pyrites—
Spain, 1,398 t Societie Commerciale
Pomaron, 900 Mason & Barry
Spain, 350 F. C. Hills & Co.

Potash—
Holland, £75 R Morrison & Co.

Plumbago—
E. Indies, 681 cks. L & I. D. Jt Co
Ceylon, 25 brls. Doulton & Co.
" 231 Thredder, Son, & Co.
Cape, 22 bgs. W. Dunn & Co.
Ceylon, 145 cks. Prop Chamb. Whf.
E. Indies, 302 cks. J. Forsey
Germany, 14 E & G. Hirsch & Co.
Ceylon, 155 H. J. Ison
" 296 L. & I. D. Jt. Co.
Holland, 17 Brown & Elmslie
Ceylon, 175 R. G. Hall & Co.
" 85 Hoare, Wilson, & Co.
" 128 Marshall & France
" 201 L. & I. D. Jt. Co.

Paraffin Wax—
U.S., 78 brls. G. H. Frank
" 67 Mordant Bros.
" 419 H. Hill & Sons
" 1,000 J. H. Usmann & Co.
" 112 cs. H. Hill & Sons

Petroleum—
Russia, 2 brls. L. & I. D. Jt Co.
U. S., 30 Mwl. Dk. Co.
S. Russia, 407,273 galls. Aschenheim & Fuhrken
U. S., 250 brls. H. Hill & Sons
Germany, 25 T. H. Lee
U. S., 186 Rose, Wilson, & Co.

Pearl Ash—
France, 100 c. F. W. Berk & Co.

Potassium Carbonate—
France, 8 pkgs. J. A. Reid

Potassium Bicarbonate—
Germany, £45 A & M. Zimmermann
Holland, 35 "

Potassium Sulphate—
Germany, 61 pkgs. H. Lambert
" 508 Anglo Contl. Guano Works
" 10 H. Lambert
Fredrikstad, 30 B. & F. Wf. Co.

Potassium Prussiate—
Germany, 11 pkgs. A. & M. Zimmermann

Sodium Acetate—
France, £133 A. Zumbeck & Co.
" 75 Prop. Hay's Wf.

Stearine—
Holland, 22 cks. Perkins & Horner
Germany, 48 W. Balchin
" 74 "
U. S., 5 Beresford & Co.

Sodium Nitrate—
Holland, £74 Kaltenbach & Schmitz

Saltpetre—
Germany, 72 cks. P. Hecker & Co.
E. Indies, 579 bgs. W. W. & R. Johnson & Sons
Germany, 79 cks. J. Hall & Son

Sugar of Lead—
Germany, 33 pkgs. C. J Capes & Co.

Tartaric Acid—
Holland, 27 pkgs. B. & F. Wf. Co
" 9 "
Germany, 6 O. Andrea & Co.
" 12 "

Tartars—
Italy, 31 pkgs. B. Jacob & Sons
" 10 Thames S. Tug Co.
" 610 Saeger, White, & Co.

Tar—
Germany, 2 brls. Langstaff, Ehrenberg, & Co.

Ultramarine—
Belgium, 4 cks. Leach & Co.
" 3 "
" 4 "
Holland, 5 Ohlenschlager Bros.
Germany, 1 G. Steinhoff
" 5 cs. H. Brooks & Co.
Holland, 4 Seaward Bros.
" 4 Spies Bros

Zinc Oxide—
Holland, 25 cks. Burrell & Co.
" 105 M. Ashby
" 300 Soundy & Son
U. S., 100 M. Ashby
Germany, 80 J. Matton

Zinc Sulphate—
Germany, £62 Beresford & Co.

LIVERPOOL.

Week ending Mar. 20th.

Antimony Ore—
Hamburg, 110 bgs.

Barytes—
Hamburg, 14 cks.

Bones—
Boston, 3 tcs.
Colon, 1 bx. 1,046 bgs. A. Dobell & Co.
Jaffa, 17 t. J. Jowett & Son
Beyrout, 201
Constantinople, 50 t.

Brimstone—
Catania, a quantity.

Borax—
Antofagasta, 780 bgs.
Leghorn, 100 cs.

Boracic Acid—
Leghorn, 30 cks. Shropshire Union Co.
" 12

Caoutchouc—
Banana, 14 bgs. Daumas & Co.
Old Calabar, 5 cks. Pickering & Berthoud
Addah, 2 cs. 11 W. J. Radatz & Co
Accra, 1 F. & A. Swanzy
Winnebah, 3 C. Lane & Co.
Salt Pond, 2 2 Pickering & Berthoud
" 1 6 Edwards Bros.
" 1 Shaeffer & Co.
" 2 1 A. Millerson & Co.
" 1 2 Radcliffe & Durant
" 3 T. M. Flood
" 1 4 A. Miller, Bros., & Co.
" 1 5 W. Griffiths & Co.
" 5 W. B. M'Iver & Co.
Cape Coast, 3 Manson & Co.
" 1 L. Hart & Co.
" 5 W. B M'Iver & Co.
" 1 Havard & Co.
" 2 F. & A. Swanzy
" 1 3 Radcliffe & Durant
" 1 Hobson & Co.
" 3 W Griffiths & Co.
" 4 J. Smith
" 2 Koresch & Siolzthy
" 1 W. Lucanor & Co.
" 1 T. E. Tomlinson & Co.
" 4 Edwards Bros.
" 3 W. Duff & Co.
" 2 Bennett Brotherton
" 4 6 Grimwade Ridley & Co.
" 1 9 F. & A Swanzy
" 1 Davies, Robbins, & Co.
Grand Bassam, 1 brl. 1 bx. Grimwade, Ridley, & Co.
" 10 2 bgs. Edwards Bros.
" 2 2 W. D. Woodin
" 1 M. L. Levin
" 7 M. Ridyard
" 2 Applemheimer Bros
" 2 S. C. Nordlinger
" 1 J. Sladelmann
" 1 cs. 1 5
Sherbro, 2 Cie Francaise
Sierra Leone. 1 pn. 43 brls. Pickering & Berthoud
" 61 110 3 bts.
Accra, 1 cs. 1 ck. E. G. Gunnell
" 3 Hutton & Co
" 4 M. Herschell & Co.
" 9 Edwards Bros.
" 27 J. P. Werner
" 17 J. J. Fischer & Co.
Akassa, 399 bgs.
Bay Beach, 2 cks. 2 cs. 3 brls. Pickering & Berthoud
" 11 pns. 6 brls. 3 bxs. 1½ hd. F. & A. Swanzy
" 3 cks. C. Rockmann
Quittah, 8
Winnebah. 1 brl. C. Lane & Co.

Cream of Tartar—
Newhaven, 2 cks.
Genoa, 8

Chemicals *(otherwise undescribed)*
Gothenburg, 5 bxs. Bahr, Behrend, & Ross

Dyestuffs—
Extract
Havre, 100 cs 18 cks. Cunard S. S Co, Ltd.

Orchella
Loanda, 30 brls. A Barbosa & Co.
Lisbon, 87 "
" 123 "
Loanda, 15 pckts.

Logwood.
Boston, 80 bxs. B. Wilkinson & Co.

Valonia
Constantinople 579 bgs

Cochineal
Teneriffe 4 bgs. Bruce & Still

Sumac
Palermo, 150 bgs.

Divi Divi
Maracaibo, 143 t.

Fustic
Maracaibo, 77 t.
Savanilla, 2,030 bgs. G. H. Muller & Co

Saffron
Valencia, 1 cs.

Glucose—
Hamburg, 12 cks.

Iodine—
Valparaiso, 192 brls. A. Gibbs & Sons

Insect Powder—
Trieste, 6 brls. G. Pototsching
" 75

Manganese Ore—
Fiume, 295 t. H. Borner

Pitch—
Amsterdam, 13 cks.

Pyrites—
Huelva, 1,595 t. Matheson & Co.
" 1,220 Tennants & Co.
" 1,430 Matheson & Co.
" 1,412 "

Potassium Bicarbonate—
Rotterdam, 8 cks.

Phosphate—
St. Valery, 297 t.

Paraffin Wax—
New York, 100 brls. Makin & Bancroft

Potash—
Boston, 30 brls. Makin & Bancroft
New York, 29 "

Saltpetre—
Hamburg, 51 cks.
" 52

Stearine—
Boston, 119 tcs.

Silica—
Hamburg, 200 bgs.

Sodium Nitrate—
Junin, 8,985 bgs.

Sulphur—
Catania, 300 bgs. 50 brls.

Tar—
Wilmington, 1,250 brls.
" 2,733

Tartar—
Naples, 2 cks.
Messina, 8

Ultramarine—
Trieste, 50 cs.

Verdigris—
Bordeaux, 9 cks.

GLASGOW.

Week ending March 20th.

Alum—
Rotterdam, 41 cks.

Bleaching Powder—
New York, 100 bxs McGradie & Christ
" 110 Hugh Moore & Co.

Dyestuffs—
Alizarine
Rotterdam, 88 cks. J. Rankine & Son
" 40

Logwood
Jamaica, 629½ t. Wm. Counal & Co.

Extracts
Bordeaux, 100 cs.
Rouen. 1 ck.
Rotterdam, 2 J. Rankine & Son

Madder
Rotterdam, 2 butts "

Sumac
Palermo, 5 bls. 1,155 bgs.

Dyestuffs *(otherwise undescribed)*
New York, 6 cs

Pyrites—
Huelva, 1,660 Tharsis Co.

Potash—
Boston, 32 brls. Rowley & Dick

Sodium Nitrate—
Pisagua, 1,130 t. Thomson Aikmann

Saltpetre—
Rotterdam, 7 cks. J. Rankine

Sulphur—
Palermo, 206 bgs.
Messina, 50

Tartar Emetic—
Hamburg, 2 cs.

Tartaric Acid—
Rotterdam, 7 cks. J. Rankine

Tartar—
Bordeaux, 3 cks. J & P. Hutc

Ultramarine—
Rotterdam, 8 cks. J. Rankine

HULL.

Week ending Mar. 20th.

Acids *(otherwise undescribed)*
Hamburg, 1 ck. Bailey & Le
Rotterdam, 4 G. Lawson &

Barytes—
Antwerp, 43 cks. Bailey & Le
Bremen, 22 T. W. Flint
" 42

Borax—
Danzig, 9 bgs. Wilson, Sons,

Bone Meal—
Kurrachee, 300 bgs.

Bones—
Kurrachee, 1,334 bgs.

Chemicals *(otherwise undescr*
Hamburg, 10 cks. Wilson, Sons,
Antwerp, 15 pks. 53 brls. "
Rotterdam, 9 cks. W. & C. L. Ri
New York, 50 Wilson, Sons,
Naples, 5 "
Genoa, 10 "
Bremen, 2 Veltmann
Danzig, 2

Dyestuffs—
Alizarine
Rotterdam, 3 cks. W. & C. L. Ri
" 3 "
" 65 pkgs. Hutchinson
" 21
" 75 cks. W. & C. L. Ri

Extracts
New York, 25 brls.
Rouen, 21 cks. Rawson &

Madder
Amsterdam, 1 ck. W. & C. L. Ri

Rennet
Rotterdam, 9 "
Copenhagen, 8 pks. Wilson, Sons,

Glucose—
Hamburg, 15 cks Woodhouse
" 10 Rawson & Ro
" 25
" 12 Wilson, Sons,
Stettin, 100 bgs. "
" 45 cks.
Hamburg, 18

Glycerine—
Hamburg, 5 drums.
" 5 pkgs.

Pitch—
Amsterdam, 13 cks. H. J.

Plumbago—
Rotterdam, 3 cks. Johnson
" 2 cks.

Phosphate—
Ghent, 1,204 bgs. 110 t. Wilson
" 160 t. "

Potash—
Rotterdam, 40 cks. W. & C. L.

Sulphate—
Christiania, 100 brls. Wilson, S

Stearine—
Rotterdam, 40 bgs. Wilson, S

Tar—
Marseilles, 24 cks.
Libau, 131

Tartaric Acid—
Rotterdam, 41 cks. W. & C. L.

TYNE.

Week ending Mar. 20th

Glucose—
Hamburg, 12 cks. Tyne S.
New York, 40 brls.

Phosphate—
Antwerp, 190 t. Langdale's Cl
" 220 "

Barcelona, 12 2 255
Bordeaux, 25 16 2 q. 634
Valencia, 5 5 115
Tarragona, 15 347
Almeria, 2 57
Rouen, 12 300
Tarragona, 5 64
Alicante, 3 18 3 91
Barcelona, 5 1 2 94
Bilbao, 41 1 798
Bordeaux, 20 9 3 420
Bilbao, 9 18 1 200
Bordeaux, 168 4 4,200
Nantes, 60 1 1,530
Rouen, 147 9 3,771
Valencia, 1 2 20
Hong Kong, 1 30
Genoa, 2 10 1 45

Carbolic Acid—
Bilbao, 13 c. £65
New York, 1 t. 5 75
Rouen, 1 2 121
Seville, 1 3 q. 12
Hamburg, 19 3 112

Chloride of Lime—
Alicante, 24 t. £150
Iquique, 12 c. 10

Calcium Chloride—
Rouen, 5 t. 7 c. £9

Chemicals *(otherwise undescribed)*
Guayaquil, 1 c. 3 q. £5
New York, 10 t. 9 64
Havre, 1 9
Colon, 1 30

Caustic Potash—
Tallerus, 3 c. £5

Citric Acid—
Batoum, 2 c. £15

Chlorate—
Lisbon, 3 c. £8
Genoa, 1 t. 56

Dried Blood—
Barcelona, 5 t. 2 c. 1 q. £41

Epsom Salts—
Rio Janeiro, 100 kegs.

Iron Oxide—
Callao, 30 t. £90
Gijon, 1 4 c. 3 q. 7
Leghorn, 2 1 1 15
Philadelphia, 8 15 148

Leather Waste—
Havre, 9 t. £72

Manure—
Bordeaux, 250 t. 8 c. £896
St. Nazaire, 350 2 3 q. 1,069
Valencia, 75 820
Cullera, 148 3 1,080

Magnesia—
Porto Alegre, 11 cs.
Carril, 1
Rio Janeiro, 10

Magnesium Chloride—
Bombay, 8 t. £29

Muriatic Acid—
Savanilla, 4 c. 2 q. £2

Oxalic Acid—
San Francisco, 1 t. £38

Potassium Chlorate—
New York, 12 t. 10 c. £583
Shanghai, 1 10 77
Yokohama, 5 216
Barcelona, 1 43
Hiogo, 2 10 112
Vera Cruz, 1 10 70
Havana, 15 15
Hiogo, 2 5 3 q. 111
Genoa, 1 44
Gothenberg, 5 200
Antwerp, 1 47
Hiogo, 7 10 300
Rotterdam, 1 10 59

Potash—
Monte Video, 1 t. 15 c. 2 q. £55

Phosphate—
Hamburg, 22 t. £100

Phosphorous—
Havana, 7 c. 2 q. £115
Vera Cruz, 3 3 40
" 1 t. 3 3 115

Potassium Chromate—
Havre, 6 t. £185

Potassium Bichromate—
Marseilles, 1 t. 7 c. £49

Sodium Chlorate—
Antwerp, 10 c. £28

Sodium Bicarbonate—
Batoum, 60 kegs.
Constantinople, 20
Odessa, 200
Seville, 20 25 brls.
Valparaiso, 8

Antwerp, 16 cks.
Barcelona, 100
Bremerhaven, 35
Rotterdam, 20
Seville, 4
Sydney, 120

Sodium Silicate—
Seville, 50 brls.
Valencia, 35
Barcelona, 32
Savanilla, 15 brls.

Soda Ash—
Baltimore, 282 cks.
Boston, 115 1,800 bgs.
Corunna, 2
Melbourne, 451
New York, 149 731 tcs.
Santander, 21 14 brls.
Sydney, 64
Windsor Mills, 103 tcs.
Barcelona, 90 brls.
Buenos Ayres, 175 200
Hamilton, 90
Kingsey, 15
Lisbon, 40
Monte Video, 50
New York, 46 260
Philadelphia. 90
Porto Allegre, 10
Rio Grande do Sul, 50
Salonica 50
Sherbrooke, 16
Tarragona, 20

Salt—
Adelaide, 125 t.
Barbadoes, 50
Cape Town, 13½
Cette Cama, 10
New York, 65
Para, 107
Calcutta, 2,710
Cape Coast, 7½
New York, 4 9 c.
Opobo, 50
San Francisco. 200
Santa Catherina, 15 3
Port Natal. 45

Soda Crystals—
Honolulu, 15 brls.
Malta, 50
New York, 140 cks.
St. John's, N B, 20
Valparaiso, 25
Constantinople, 50
Philadelphia, 167
Santa Catherina, 30 kegs.
Callao, 50

Sulphur—
Iquique, 5 t. 7 c. 1 q. £32
Rouen, 15 65
Porto Allegre, 2 t. 10 c. £15

Superphosphate—
Las Palmas, 5 t £18
" 8 28

Sodium Biborate—
Montreal, 1 t. 2 c. 2 q. £34
Yokohama, 3 11 2 107

Sugar of Lead—
Calcutta, 5 c. £6

Spent Oxide—
Dunkirk, 304 t. 17 c. £245

Salammoniac—
Marseilles, 11 c. 1 q £18
Mersyne, 8 14
New York, 4 t. 16 163
Gijon, 1 1 3 41
Syra, 3 5

Sheep Dip—
Cape Town, 20 cs. £40

Salt Nitrate—
Havana, 13 c. £15

Tar—
Honolulu, 70 brls.

Tartaric Acid—
Havana, 6 c. 1 q. £50

Verdigris—
Rio Janeiro, 1 c. 3 q. £11
New York, 3 cks. 58

Zinc Chloride—
Bombay, 4 t. 12 c. £96

TYNE.

Week ending Mar. 20th.

Alum—
Copenhagen, 10 c.

Alkali—
Naples, 1 t. 3 c.
Antwerp, 10 11
Barcelona, 14 1
Copenhagen, 5 8

Arsenic—
Barcelona, 1 t. 5 c.

Bleaching Powder—
Newfairwater, 14 t. 7 c.
Baltimore, 49 19
Malmo, 1 16
Antwerp, 27 17
Hamburg, 38 22
Christiania, 4 7
Aarhuus, 1 2
Philadelphia, 25 4
Rotterdam, 28 12
Copenhagen, 7 16
Libau, 56 15

Barytes Carbonate—
Hamburg, 20 t.

Caustic Soda—
Newfairwater, 10 t. 3 c.
Baltimore, 203 2
Antwerp, 4 19
Philadelphia, 72 10
Barcelona, 5 16
Copenhagen, 2 6
Libau, 29

Copperas—
Oporto, 10 t. 10 c.

Litharge—
Antwerp, 1 t. 14 c
Hamburg, 2 1

Magnesia—
Hamburg, 15 c.
Barcelona, 11 t 11

Naphthalene—
Libau, 25 t. 5 c.

Paraffin Wax—
Barcelona, 2 t. 1 c.

Phosphorus—
Malmo, 6 c.
Barcelona, 16

Potassium Chlorate—
Philadelphia, 2 t. 10 c.
Rotterdam, 7 9

Quinine—
Barcelona, 7 c.

Sodium Sulphate—
Baltimore, 261 t. 14 c.
Copenhagen, 5 5

Soda Ash—
Odense, 17 t. 4 c.
Gothenburg, 45 14
Hamburg, 18 11
Christiania, 10
Newfairwater, 9 19
Savona, 5 19
Naples, 1 2
Lisbon, 12 7
Copenhagen, 13 18

Sodium Silicate—
Barcelona, 16 t. 9 c.

Sodium Chlorate—
Hamburg, 2 t. 10 c.
Rotterdam, 1 2

Soda—
Svendborg, 5 t.
Antwerp, 2 15 c.
Aalborg, 18 11
Odense, 10 13
Aarhuus, 15 5
Philadelphia, 301 7
Kolding, 4 15

Sodium—
Hamburg, 1 cs.

Tar—
Newfairwater, 33 t.

HULL.

Week ending March 18th.

Alkali—
Venice, 50 drums.

Bleaching Powder—
Trieste, 2 cks.
Venice, 39

Bone Size—
Harlingen, 30 cks.

Benzol—
Hamburg, 14 drms.

Caustic Soda—
Bari, 20 cks.
Libau, 18 drums.

Chemicals *(otherwise undescribed)*
Amsterdam, 40 cks.
Bremen, 2 cs.
Bergen, 20 cks. 14 pkgs.
Bordeaux, 20 cs.
Christiania, 14
Copenhagen, 1
Gothenburg, 40 13 105 pkgs.
328 slabs.
Hamburg, 60 333 11 drums.
Libau, 5 100 pkgs.
New York, 59 75 cs.
Rouen, 30
Rotterdam, 42 35 pkg
Stettin, 10
Trieste, 2
Venice, 17

Dyestuffs *(otherwise undescri*
Bergen, 2 cks.
Hamburg, 7 1 keg.

Manure—
Dunkirk, 150 bgs.
Esbjerg, 800
Hamburg, 966
Rouen, 11

Pitch—
Antwerp, 130 t
Dunkirk, 217

Tar—
Christiania, 10 cks
Copenhagen, 11
Hamburg, 1

GLASGOW.

Week ending Mar. 20th.

Ammonia—
Trinidad, 7 t. 3¾ c.

Ammonium Sulphate—
Demerara, 68 t. 9 c.

Bone Ash—
Morlaix,

Chemicals *(otherwise undescr*
Melbourne, 24 c.

Dyestuffs—
Extracts
New York, 14½ c.

Guano—
Trinidad, 3 t.

Manganese Oxide—
New York, 206½ c.
" 199¾

Potassium Bichromate—
Rotterdam, 19 cks.
Antwerp, 309 c.

Potassium Prussiate—
New York,

Sodium Bichromate—
New York, 325 c.
Antwerp, 107 t.

Sodium Phosphate—
New York, 6 t. 6 c.

Soda Ash—
Baltimore, 1,317¾ c.
New York, 214¾

Salt—
Jamaica, 18 t. 18 c.

Tar—
Constantinople, 1,013 t. 2 c.

GOOLE.

Week ending March 12t

Benzol—
Antwerp, 26 cks.
Ghent, 16 drums.

Coal Products—
Antwerp, 23 cks.
Calais, 25
Hamburg, 50 8 drums.
Rotterdam, 170

Chemicals *(otherwise undescri*
Antwerp. 19 drums.
Boulogne, 1 ck.
Ghent, 1
Hamburg, 44
Rotterdam, 1

Dyestuffs *(otherwise undescri*
Ghent, 7 cks.

Manure—
Bruges, 1,441 bgs.
Dunikrk. 984
Ghent, 665

Pitch—
Antwerp, 193 t.

GRIMSBY.

Week ending March 15th.

Alkali—
Antwerp, 8 cks.

Chemicals *(otherwise undescri*
Antwerp, 99 cks.
Dieppe, 2 124 pkgs.
Hamburg, 8

PRICES CURRENT.

WEDNESDAY, March 26, 1890.

Prepared by Higginbottom and Co., 116, Portland Street, Manchester.

es stated are F.O.R. at maker's works, or at usual ports of shipment in U.K. The price in different localities may vary.

			£	s.	d.
25 °/₀ and 40 °/₀		per cwt	7/6 & 12/9		
Glacial		,,	3	5	0
S.G., 2000°		,,	0	12	9
82 °/₀	nett	per lb.	0	0	6½
		,,	0	0	3¾
(Tower Salts), 30° Tw.		per bottle	0	0	8
(Cylinder), 30° Tw.		,,	0	2	11
0° Tw.		per lb.	0	0	2
		,,	0	0	1¼
	nett	,,	0	0	3¼
		,,	0	1	4
ic (fuming 50 °/₀)		per ton	15	10	0
(monohydrate)		,,	5	10	0
(Pyrites, 168°)		,,	3	2	6
(,, 150°)		,,	1	8	0
(free from arsenic, 140/145°)		,,	1	10	0
ous (solution)		,,	2	15	0
(pure)		per lb.	0	1	0½
		,,	0	1	2½
hite powdered		per ton	13	10	0
se lump)		,,	4	15	0
Sulphate (pure)		,,	5	0	0
,, (medium qualities)		,,	4	10	0
m		per lb.	0	15	0
, ·880=28°		per lb.	0	0	3
=24°		,,	0	0	1¾
Carbonate	nett	,,	0	0	3½
Muriate		per ton	23	10	0
,, (sal-ammoniac) 1st & 2nd		per cwt.	37/- & 35/-		
Nitrate		per ton	42	0	0
Phosphate		per lb.	0	0	10½
Sulphate (grey), London		per ton	11	16	3
,, (grey), Hull		,,	11	15	0
Oil (pure)		per lb.	0	0	11¼
alt ,,		,,	0	0	9
		per ton	75	0	0
(tartar emetic)		per lb.	0	1	1
(golden sulphide)		,,	0	0	10
Chloride		per ton	8	0	0
Carbonate (native)		,,	3	15	0
ulphate (native levigated)		,,	45/- to 75/-		
Powder, 35 %		,,	5	7	6
Liquor, 7 %		,,	2	10	0
e of Carbon		,,	13	0	0
n Acetate (crystal		per lb.	0	0	6
Chloride		per ton	2	2	6
ay (at Runcorn) in bulk		,,	17/6 to 35/-		
Dyes:—					
ne (artificial) 20 %		per lb.	0	0	9
a		,,	0	3	9
Products					
cene, 30 % A, f.o.b. London		per unit per cwt.	0	1	5
90 % nominal		per gallon	0	3	3
50/90 ,,		,,	0	2	4
c Acid (crystallised 35°)		per lb.	0	0	10½
,, (crude 60°)		per gallon	0	2	9
e (ordinary)		,,	0	0	2½
(filtered for Lucigen light)		,,	0	0	3
Naphtha 30 % @ 120° C.		,,	0	1	5
Oils, 22° Tw.		per ton	3	0	0
.o.b. Liverpool or Garston		,,	1	9	0
Naphtha, 90 % at 160°		per gallon	0	1	11
n Oils (crude)		,,	0	0	2½
Chili Bars)		per ton	48	10	0
lphate		,,	26	0	0
ide (copper scales)		,,	51	0	0
(crude)		,,	30	0	0
(distilled S.G. 1250°)		,,	55	0	0
		nett, per oz.	8½d. to 9d.		
hate (copperas)		per ton	1	10	0
hide (antimony slag)		,,	2	0	0
et) for cash		,,	14	0	0
harge Flake (ex ship)		,,	15	10	0
tate (white ,,)		,,	23	15	0
,, (brown ,,)		,,	18	10	0
bonate (white lead) pure		,,	19	0	0
rate		,,	22	0	0

		£	s.	d.
Lime, Acetate (brown)	,,	8	2	6
,, ,, (grey)	per ton	14	0	0
Magnesium (ribbon and wire)	per oz.	0	2	3
,, Chloride (ex ship)	per ton	2	6	3
,, Carbonate	per cwt.	1	17	6
,, Hydrate	,,	0	10	0
,, Sulphate (Epsom Salts)	per ton	3	0	0
Manganese, Sulphate	,,	18	0	0
,, Borate (1st and 2nd)	per cwt.	60/- & 42/6		
,, Ore, 70 %	per ton	4	10	0
Methylated Spirit, 61° O.P.	per gallon	0	2	2
Naphtha (Wood), Solvent	,,	0	3	11
,, ,, Miscible, 60° O.P.	,,	0	4	7½
Oils:—				
Cotton-seed	per ton	22	10	0
Linseed	,,	24	10	0
Lubricating, Scotch, 890°—895°	,,	7	5	0
Petroleum, Russian	per gallon	0	0	5¼
,, American	,,	0	0	5¾
Potassium (metal)	per oz.	0	3	10
,, Bichromate	per lb.	0	0	4
,, Carbonate, 90% (ex ship)	per ton	18	0	0
,, Chlorate	per lb.	0	0	5
,, Cyanide, 98%	,,	0	2	0
,, Hydrate (Caustic Potash) 80/85 %	per ton	22	10	0
,, ,, (Caustic Potash) 75/80 %	,,	21	10	0
,, ,, (Caustic Potash) 70/75 %	,,	20	15	0
,, Nitrate (refined)	per cwt.	1	3	6
,, Permanganate	,,	4	5	0
Prussiate Yellow	per lb.	0	0	9½
,, Sulphate, 90 %	per ton	9	10	0
,, Muriate, 80 %	,,	7	15	0
Silver (metal)	per oz.	0	3	7¾
,, Nitrate	,,	0	2	5½
Sodium (metal)	per lb.	0	4	0
,, Carb. (refined Soda-ash) 48 %	per ton	6	2	6
,, ,, (Caustic Soda-ash) 48 %	,,	5	5	0
,, ,, (Carb. Soda-ash) 48%	,,	5	10	0
,, ,, (Carb. Soda-ash) 58 %	,,	6	15	0
,, ,, (Soda Crystals)	,,	3	0	0
,, Acetate (ex-ship)	,,	15	15	0
,, Arseniate, 45 %	,,	10	10	0
,, Chlorate	per lb.	0	0	6¼
,, Borate (Borax)	nett, per ton.	30	0	0
,, Bichromate	per lb.	0	0	3
,, Hydrate (77 % Caustic Soda)	per ton.	13	0	0
,, ,, (74 % Caustic Soda)	,,	12	0	0
,, ,, (70 % Caustic Soda)	,,	11	2	6
,, ,, (60 % Caustic Soda, white)	,,	10	2	6
,, ,, (60 % Caustic Soda, cream)	,,	9	17	6
,, Bicarbonate	,,	6	0	0
,, Hyposulphite	,,	5	5	0
,, Manganate, 25%	,,	9	0	0
,, Nitrate (95 %) ex-ship Liverpool	per cwt.	0	8	1½
,, Nitrite, 98 %	per ton	27	10	0
,, Phosphate	,,	15	15	0
,, Prussiate	per lb.	0	0	7¾
,, Silicate (glass)	per ton	5	7	6
,, ,, (liquid, 100° Tw.)	,,	3	17	6
,, Stannate, 40 %	per cwt.	2	0	0
,, Sulphate (Salt-cake)	per ton.	1	12	6
,, ,, (Glauber's Salts)	,,	1	10	0
,, Sulphide	,,	7	15	0
,, Sulphite	,,	5	5	0
Strontium Hydrate, 98 %	,,	8	0	0
Sulphocyanide Ammonium, 95 %	per lb.	0	0	8
,, Barium, 95 %	,	0	0	5¼
,, Potassium	,,	0	0	8
Sulphur (Flowers)	per ton.	7	10	0
,, (Roll Brimstone)	,,	6	0	0
,, Brimstone: Best Quality	,,	4	5	0
Superphosphate of Lime (26 %)	,,	2	10	0
Tallow	,,	25	10	0
Tin (English Ingots)	,,	94	10	0
,, Crystals	per lb.	0	0	6¼
Zinc (Spelter)	per ton.	21	12	6
,, Chloride (solution, 96° Tw.	,,	6	0	0

THE

CHEMICAL TRADE JOURNAI

Publishing Offices: 32, BLACKFRIARS STREET, MANCHESTER.

No. 150. **SATURDAY, APRIL 5, 1890.** **Vol.**

Contents.

Notices.

All communications for the *Chemical Trade Journal* should be addressed, and Cheques and Post Office Orders made payable to—

DAVIS BROS., 32, Blackfriars Street, MANCHESTER.

Our Registered Telegraphic Address is—

"Expert, Manchester."

The Terms of Subscription, commencing at any date, to the *Chemical Trade Journal,*—payable in advance,—including postage to any part of the world, are as follow:—

Yearly (52 numbers) .... **12s. 6d.**
Half-Yearly (26 numbers) .... **6s. 6d.**
Quarterly (13 numbers) .... **3s. 6d.**

Readers will oblige by making their remittances for subscriptions by Postal or Post Office Order, crossed.

Communications for the Editor, if intended for insertion in the current week's issue, should reach the office not later than **Tuesday Morning.**

Articles, reports, and correspondence on all matters of interest to the Chemical and allied industries, home and foreign, are solicited. Correspondents should condense their matter as much as possible, write on one side only of the paper, and in all cases give their names and addresses, not necessarily for publication. Sketches should be sent on separate sheets.

We cannot undertake to return rejected manuscripts or drawings, unless accompanied by a stamped directed envelope.

Readers are invited to forward items of intelligence, or cuttings from local newspapers, of interest to the trades concerned.

As it is one of the special features of the *Chemical Trade Journal* to give the earliest information respecting new processes, improvements, inventions, etc., bearing upon the Chemical and allied industries, or which may be of interest to our readers, the Editor invites particulars of such—when in working order—from the originators; and if the subject is deemed of sufficient importance, an expert will visit and report upon the same in the columns of the *Journal.* There is no fee required for visits of this kind.

We shall esteem it a favour if any of our readers, in making inquiries of, or opening accounts with advertisers in this paper, will kindly mention the *Chemical Trade Journal* as the source of their information.

Advertisements intended for insertion in the current week's issue, should reach the office by **Wednesday morning** at the latest.

Advertisements.

Prepaid Advertisements of Situations Vacant, Premises on Sale or To be Let, Miscellaneous Wants, and Specific Articles for Sale by Private Contract, are inserted in the *Chemical Trade Journal* at the following rates:—

30 Words and under.......... 2s. 0d. per insertion.
Each additional 10 words .... 0s. 6d. ,,

Advertisements of Situations Wanted are inserted at one-half the above rates when prepaid, viz:—

30 Words and under.......... 1s. 0d. per insertion.
Each additional 10 words .... 0s. 3d. ,,

Trade Advertisements, Announcements in the Directory Columns, and all Advertisements not prepaid, are charged at the Tariff rates which will be forwarded on application.

THE MORAL OF THE GAS FRAUD

THE TIPPING SYSTEM.

A FEW weeks ago the public conscience received shock from sundry revelations shewing how a noted asbestos company had been accustomed to its business. The said conscience will scarcely ha soothed by what transpired at the Leeds Assizes la A more glaring case of commercial immorality has scarc been brought to light, and though Mr. LEVER received what severe pecuniary chastisement, one cannot help t that, if he has got his deserts, then his sometime co HUNTER, is getting a greater punishment than is h But we are not trying to enlist sympathy for eithe men whose iniquities have been so completely bro light. We simply use the case to point a moral which i to be the tendency of the day to overlook. We are ingly loth to breathe a whisper which could be interpr a reflection upon the fair fame of the Corporation of S but we cannot help thinking it exceedingly strange that the years covered by the Gas Frauds, they should hav so lamentably ignorant on the subject of the price One is tempted to wonder if they made their own b contracts in the same happy-go-lucky fashion, and if t lowed their own employés the same latitude that b about the ruin of the Gas Manager. Is it not high-tin Municipal honours should be associated with a little responsibility than has hitherto been the case? Is much to ask that the men who do us the honour of gov us, should pledge themselves in a something more tha tioneering twaddle to consider our interests as they co their own? Had some greater measure of respons attached to the Corporation of Salford there would hav far less of that peculation, bribery, and corruption th terminated so disastrously for those who indulged in it. thoughtless shopkeeper who exposes his goods within of the hungry waif and the shoeless stray is generall justly, censured; but we are not yet educated up to the of extending our censure to those who deserve it much —to those who voluntarily accept offices which they c or will not, efficiently perform, though they know pe well that their negligence means the wasting of the money, and the encouragement of trickery and fraud public service.

It may be urged that our suggestion, if acted upon, tend to keep many a useful citizen from a seat in the C Chamber, and deprive his fellows of the benefit of his a istrative talent. Perhaps so; but its deterrent action be most strongly marked in the case of those whom i the interest of the community to keep out of office, ar public would gain in the long run, even at the cost of t clusion of a few who would have faithfully discharg duties of the coveted position.

CHEMICAL ENGINEERING*

XXV.

EXHAUSTING AND COMPRESSING GASES.

GASES are pumped and moved much in the same way as liquids, and there is scarcely any appliance used for pumping or otherwise moving a liquid, that cannot also be employed (with suitable variation on account of difference of density) for aiding the change of motion of a gas. The fan and the centrifugal pump; the pressure blower and the rotary pump; the air compressor and the ordinary force pump; the water injector and the steam jet exhauster are all parallels and the rules that apply to one will also apply to the other.

I think I should commence by describing the ordinary valve compression pump such as may be seen in almost every chemical works, used either for storing up force, for raising fluids, such as vitriol, to the top of denitrating or absorbing columns, or for the supply of air for purely chemical operations, such as the oxidation of manganese protoxide. These are well-known operations in the heavy chemical trade; but there are a multitude of uses to which small compressors are and may be put, in other branches of the industry; not merely as compressors, but as exhausters. And as we shall have to consider this later on, in connection with refrigeration and evaporation it will be as well to thoroughly understand this subject now.

Whatever operation is performed in connection with chemical processes, some attempt should be made to aim at efficiency, and if this can be attained at a reasonable cost there is no reason why it should not be done. There is not much *nous* in selecting an air compressor simply on account of its cheapness, any non-technical man could do that; but it is far more difficult to give one's reason for any definite choice.

PISTON COMPRESSORS.

In pumps of this order there are several points worthy of considerable attention, the most important of which is that the dead space or clearance between the cylinder ends and the piston should be restricted to the narrowest limits. It will be at once seen that in these pumps, the suction valve only lifts when the pressure behind the piston is less than the pressure of the air, and the compressed air in the dead space expands while the piston is receding, and so diminishes the effective capacity of the cylinder. Under ordinary pressures of, say, one atmosphere, or 15 lbs. per square inch, the influence of dead space is not so marked as at pressures of five to ten atmospheres, where the evil effect of clearance becomes very serious.

The next points to consider, are the forms of valves and their proper proportions to the pressures that have to pass through them—I shall not attempt to defend or uphold any of the ordinary valves I have seen working in my practical career; by doing so, I feel sure, I should meet with as many opponents to my views as supporters, and so will content myself with advising every one who finds himself in favour of any particular form, to take an indicator diagram from the compression cylinder, when perhaps he will find his baby is not the most beautiful in the world. The compression of a gas gives rise to the liberation of heat, so that an important feature of most air compressors is the cooling apparatus, and some of my hearers will no doubt have noticed that water is often injected into the compression cylinder to absorb this heat. But in some applications the air is required dry, so that some other system of cooling must be adopted. Most of you will be familiar with the blowing engines made by the Alkali work engineers, Messrs. English, The Widnes Foundry, Messrs. W. Neill and Sons, Messrs. Robinson Cooks, Messrs. Walker Bros., of Wigan, [illegible] am sorry I cannot show you a drawing of one of them. [illegible] however, exhibits the compressor made by Messrs. [illegible] and Son, of London. The crank shaft, actu-[illegible] belt or engine direct, raises the plunger and [illegible] through the balanced valve beneath, until it reaches the upper extremity of its stroke, [illegible] and—the bottom valve being now closed—[illegible] cylinder is gradually compressed until [illegible] chamber of the plunger, when, on the next [illegible] delivers the charge now in the inner chamber [illegible] valve in the bottom of the fixed pipe, which also [illegible] guide, into the receiver or air vessel; now, it will [illegible] be apparent that, by proportioning the sizes of the [illegible] and the inner chamber of the plunger, an equal [illegible] put on the up and down strokes, each load for [illegible] pressures being very moderate; for example, take the [illegible] der to be 6 in. diameter, and the area = 28 in., [illegible] diameter of the chamber to be 2¼ = 4 in. area, or [illegible] the capacity of the large chambers, the air would [illegible] pressed in it to 53 lb. per square inch, which would [illegible] pressure exerted at the end of the stroke on the plunger, [illegible] load on it would be 28 × 53 = 1484 lbs., and the [illegible] would deliver air through the valve in the pipe at 370 lbs. [illegible] square inch, with an equal load; it will be noticed [illegible] between the outer thickness of plunger and the wall of [illegible] chamber there is an annular space, which is kept suppl[illegible] with water by two pipes, one for admission and the other [illegible] exit, which is a most effectual means of keeping down the [illegible] created during compression.

FIG. 1.

A very important feature of this compressor is the abs[illegible] of clearance already alluded to—a clearance space under [illegible] plunger or piston of ⅛ inch causes a loss of one inch of [illegible] travel when working to four atmospheres pressure. Profe[illegible] Wellner, of Brunne, first pointed out the evil eff[illegible] of clearance space, and he endeavoured to overco[illegible] the difficulty, but was not successful, still he led [illegible] way for Mssrs. Burckhardt and Weiss' improveme[illegible] and still later the system of Wegelin and Hubner. [illegible] compression and evacuation has been very [illegible] studied by the foreign constructors of sugar boiling [illegible]

---

* [illegible] delivered by Mr. George E. Davis at the Manchester [illegible]

tus and this is a very good illustration of one industry helping another, and no one can have seen these appliances working abroad without coming to the conclusion that our chemical engineering industry would have been leading the van to-day if the engineer had not so often played schoolmaster to the scientist.

Messrs. Wegelin and Hubner's system of building compressors, lies in abolishing the effect of clearance in admitting the air to the cylinder by means of a slide valve, and adding a secondary slide or equalizer which allows the air that is compressed in front of the piston to find its way to the back (suction side) where it is drawn in with the incoming air. When the piston recedes from the suction side, the air pressure is immediately lowered, so that the piston produces suction from the commencement of the stroke.

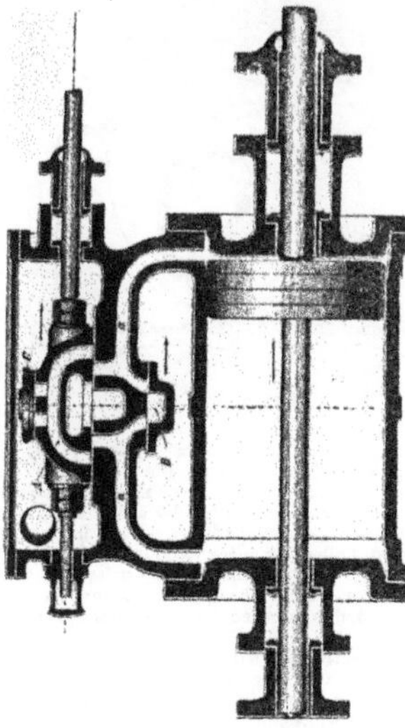

FIG. 2.

The accompanying illustration (Fig. 2.) shows the details of Messrs. Wegelin and Hübner's System. A is the distributing slide valve, for the purpose of letting the air in and out of the cylinder, while B is the equalizing valve uniting the two air passages *a a* just in the right moment, when the piston has arrived at the dead points at the end of each stroke.

If N = effective capacity of cylinder,
v = pressure of mean suction,
p = pressure of compression,
a = proportionate capacity of dead space,

then by the ordinary system, N will equal:—

$$N = \frac{1 - \frac{p}{v} a}{1 - a}$$

and with the equalizer:—

$$N = \frac{1 - \frac{p}{3} \times a^2}{1 - a^2}$$

From these formulæ it will be seen that if the dead space occupies a volume equal to 5 per cent. of the total capacity of the cylinder, the effective capacity with the old-fashioned valves will only be 80 per cent. at pressures of 60 lbs. to the square inch over the atmosphere, while with the new valves the effective capacity is 98 per cent. The difference is more striking at higher pressures—at 15 atmospheres of absolute pressure the older patterns could only yield a maximum efficiency of 26 per cent. of the total cylinder capacity, while with equalizers the yield would be 96 per cent.

CENTRIFUGAL MACHINES.

Leaving piston pumps, where a moderate volume of gas is required to be delivered at moderate or high pressures, we may turn our attention to those machines where a large volume is required at low pressures, and in this direction I cannot point to a better specimen than the Blackman air propeller, Fig. 3, in which every 10,000 cubic feet of air delivered per minute consumes about one H.P.

FIG. 3.

In this form of fan a very large volume of air can be set in motion, but there must not be much resistance, still it is a most useful piece of machinery for many chemical purposes. In constructing drying-ovens, stoves, and such like appliances, it is too often forgotten that the rapidity of drying depends in a great measure upon the quick removal of a moisture-laden atmosphere, which can in most cases be done more economically with a fan, than by allowing the moist atmosphere to find its way out of the chamber by simple diffusion. Many instances can be cited where a temperature of 100° C., with a frequent change of atmosphere, would do the work quicker and better than with the apparatus in common use. As an instance, there is the drying of precipitated phosphate of lime; if this substance is overheated, there is a considerable percentage rendered insoluble in citrate of ammonia, and damage of quality sustained thereby, but if the precipitate is dried at 100° C. in a strong current of air, there is scarcely any insoluble in the citrate of ammonia solution.

As before stated, the foregoing form of fan is unsuited for blowing against resistance or pressure; the blades merely impart to the particles of air a momentum corresponding to their velocity, but when resistance occurs the volume delivered is in ratio to that resistance, till a point is reached when the momentum and resistance are equal, when no air whatever is discharged, the fan will continue running absorbing much power, but producing no practical effect.

PRESSURE BLOWERS.

If it is necessary to blow against pressure, other forms must be chosen, these are called pressure blowers. Fig. 4 shows a variety in common use, called Root's blower, in which the two figure of 8 rollers both exhaust and compress the air within the casing during their revolution. They are made both as exhausters and compressors, and are capable of delivering a large volume of air at low pressures. They are employed in chemical works for various purposes, such as exhausting the dilute hydrochloric acid gas from the "Har salt-cake process, and drawing the gases through hing powder chambers in the "Deacon" process. also used for blowing cupolas where pressures of ounces per square inch only are required. Some e made by a committee of the Franklin Institute in ch will give an idea of the working of this pressure

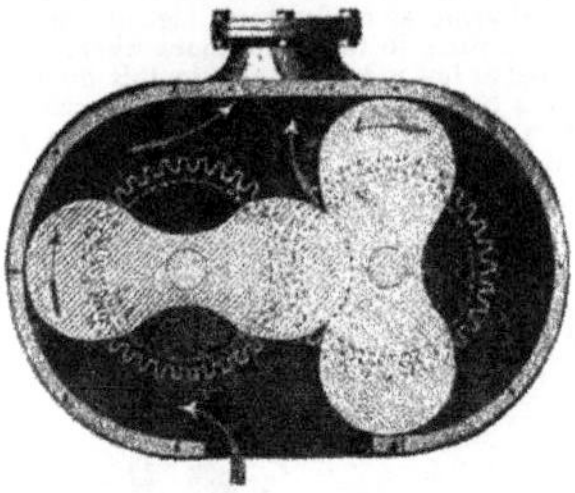

FIG. 4.

:—A Root blower rated to discharge 13·5 cubic feet olution was run at 180 revolutions per minute, and ed 17 ounces pressure per square inch on the guage, no time exceeded 18 ounces, sometimes falling as low ounces, while the pulsation was so great that the became invisible from the rapidity of its movement. ower consumed was 8·43 H.P.

illustration shown in Fig. 5. is a section of Baker's rotary re blower, much used in blowing cupolas in smelting It will supply a blast equal to 2 or 3 lbs. per square r produce a corresponding vacuum. The largest size hine yet delivered, I believe, forces 600,000 cubic feet er hour against a pressure of 3 lbs. per square inch, unning at 100 revolutions per minute.

FIG. 5.

influence of resistance, or pressure against even this f blower, may be seen in the experiments carried out Committee of the Franklin Institute. Working under sure of 6 inches of water (say 4 ounces per square inch) er exhauster of 13 cubic feet displacement capacity s 4½ H.P. to drive it, but in the trials already alluded blast of 12 ounces per square inch consumed 8·13 H.P. le's exhauster is too well known in the gas industry to description; in its simplest form, it consists of a er revolving eccentrically within a cast-iron case, and ed with sliding plates placed so that the ends are always tact with the inner surface of the case. The several makers of this appliance have departed somewhat from the original design, and Fig. 6 shows a section of the modification made by Messrs. Waller and Co., and, with whose workmanship, I have on several occasions been satisfied. The latest improvement is a revolving drum with four blades, which enter and extend themselves from the periphery during a complete revolution, which may be gathered from an examination of the illustration. I cannot say, off-hand, to what pressure these "exhausters" will work effectively or economically, but I know from experience they will work well to 60 inches of water.

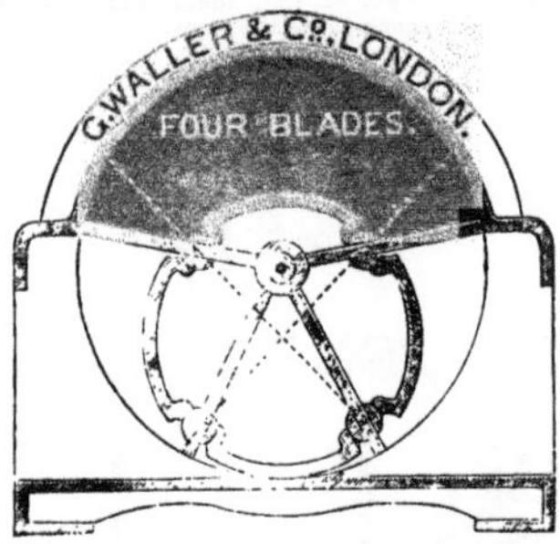

FIG. 6.

Some interesting experiments were made by Mr. C. Woodall, and published in 1877, in the Journal of Gas Lighting, wherein a Beale's exhauster, made by Messrs. Bryan, Donkin and Co., was driven by a horizontal non-condensing engine, with an expenditure of 2·11 lbs. of coke burned in a Cornish boiler, for every 1,000 cubic feet of gas passed. The vacuum on the inlet to the exhauster was 3 inches, and the pressure on the outlet, 38 inches. The indicated H.P. exerted, amounted to 0·206 per 1,000 cubic feet.

All the foregoing methods of exhausting or compressing, or forcing gases require the intervention of a motor of some description. The steam jet apparatus, working after the manner of a Giffard injector, requires nothing but a steam pipe; but, of course, the apparatus is not available where the steam has to be kept apart from the gas that has to be moved. To work effectively, they require to be made for each special operation, according to the nature of the gas to be moved, the volume to be passed, the pressure to be overcome, and the working steam pressure; the work to be done may be all exhaustion or forcing only, or a combination of the two; in other words, the business of the apparatus is to overcome the difference of pressure between the inlet and the outlet; and in order to do this effectively, the areas of both steam nozzle and air channel must be very correctly apportioned. When required to produce a vacuum, these appliances should always be designed for the specific conditions: with a working steam pressure of 50 lbs., a vacuum of 24 inches of mercury can easily be obtained, 26 inches being nearly the limit; but, with a higher steam pressure, a larger vacuum may be obtained.

These "ejectors" are largely used for a variety of purposes, such as agitating liquids by drawing air through a perforated pipe or false bottom, exhausting filters, oxidising liquids, exhausting creosote tanks, for vacuum pans in dye-wood extracting, and many other purposes.

Figs. 7, 8, and 9 are different forms of this article made by Messrs. Meldrum Bros., of Manchester. Fig. 7 is an ordinary ejector and Fig. 9 an air compressor. When the work to be done is not known within moderate limits, or varies considerably, the apparatus is supplied with the regulating spindle shown in Fig. 8.

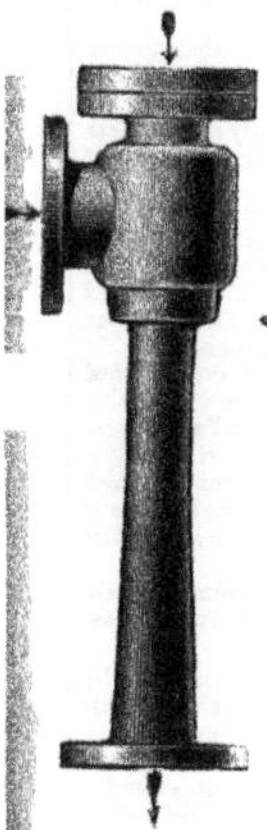

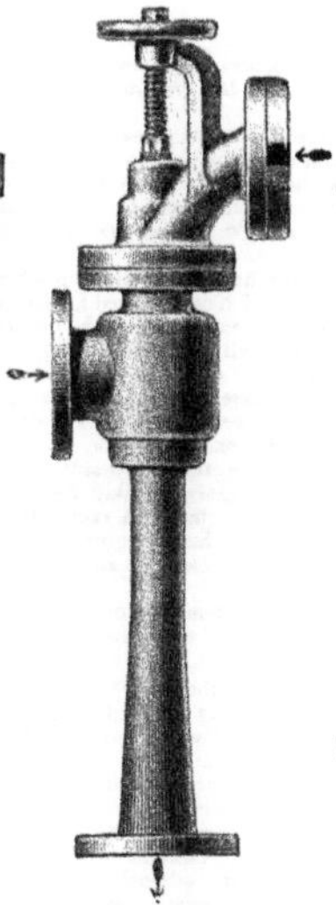

FIG. 7. FIG. 8. FIG. 9.

It often becomes necessary to know how much steam passes with the gas during the working of these steam-jet apparatus. Of course, much will depend upon their efficiency; but, with a well-made injector, the following rule will be very near the mark:—

N = cb. ft. gas per hour.
p = counter pressure in ins. water.
S = lbs. steam required.

$$S = \frac{N\sqrt[7]{p}}{55}$$

Which means that 20 lbs. of steam are required for every 1,000 cubic feet of air delivered, when the difference of pressure between inlet and outlet does not exceed two inches of water.

(To be continued)

Our Book Shelf.

LIQUID FUEL FOR MECHANICAL AND INDUSTRIAL PURPOSES, by E. A. BRAYLEY HODGETTS, p.p. 127 and 106 figures in the text. London: E. and F. N. Spow, 1890.

We have here a valuable addition to the literature of fuel. The book does not pretend to bring much that is new before the reader, but rather to gather together the facts that are already known, and to place side by side the results of the most important investigations on the subject. The author gives the average evaporative efficiency of petroleum residual—the particular liquid full mainly referred to—as 7 compared with 4 for coal, but as the difference in price is much more than proportionate we are compelled to agree with him that "not until new oilfields have been discovered, cheaper methods of transport introduced, and oil residuals brought down in price to about double that of coal, will it be possible for them to compete with the latter in Western Europe." The value of the work would be considerably enhanced by a careful revision of the numerical data as we notice several errors have crept in here.

GLASS COLOURS.

IN the *Diamant*, of Leipzig, an account is given of R. Zsigmo experiments in colouring glass with metallic sulphides, suc molybdenite, and sulphurets of antimony, copper, bismuth, and ni Tests made with batches of 10-20 kg., and with a heat not too g gave good results, as follows:—

Sand 65 dekagrammes, potash 15, soda 5, lime 9, molybdenit sulphide of sodium 2, gave a dark reddish-brown ruby glass. thinner layers this glass appeared light brownish-yellow. Flashed opal, it became a smutty black-brown.

Sand 50 dg., potash 15, soda 5, lime 9, molybdenite 1, sulphur sodium 2, gave a yellow glass.

Sand 10 parts, potash 3·3, soda 0·27, lime 1·64, molybdenite gave a reddish-yellow glass with a fine tinge of red.

Sand 100 parts, potash 26, soda 1·8, lime 12, sulphuret of co 1·7, sulphide of sodium 2·3, gave a dark-brown colour, varying sepia to sienna. In thick layers it was no longer transparent, but clear and unclouded. When heated, this glass became smutty bl brown and clouded.

A fine copper-red ruby was obtained from a batch of 10 lb. s potash 3 lb., lime 1·2 lb., soda 0·25 lb., sulphuret of copper 7·5 sulphide of sodium 10·5 kg., borax 9·5 kg.

The attempts to colour with sulphurets of antimony and bism failed. The sulphurets were volatilised, and the resulting glass almost colourless. But the addition of 5 per cent. of sulphur nickel to an ordinary batch gave a glass of fine amethyst violet col

NEW ANILINE BLACKS.

UNDER the names of NAPHTHOL BLACK 3B and NAPH LAMINE BLACK D the well-known firm of Cassella and Co Frankfort, are offering two new specialities.

For dyeing with the former the following process is recommende

LIGHT WOOLLEN FACRICS AND YARN: Boil first the goods du half to one hour with 5 to 10% acetic acid of 7 to 8° Tw. or 4 to of tartar, then add the well-dissolved dyestuff and boil from one to hours according to the quality of the wool.

Shortly before finishing the dyeing operation add 1 to 2% acetic in order to exhaust the dyebath. If feasable lower the temperatu the dyebath to 140—150° F. before adding the dyestuff, and gradu raise the temperature again to the boil.

COMMON WOOLS may be mordanted with either 2 to 3% mur or 1% sulphuric acid, in place of acetic acid, but acetic acid is al to be used for exhausting the bath.

CLOTH, FELT and such goods generally as are difficult to throughout, boil 1 to 2 hours with 5 to 10% acetic acid—muriat sulphuric acid ought *not* to be used—and then dye as above indica

SHADING: For redder shades add Naphtol-Black 4R, bril scarlets, amaranth or milling-reds, for blue shades add acid-gr naphtol-green or a soluble blue; for deep to dead blacks add In yellow G.

The contact with metals, ought, wherever this is feasible, to avoided.

Although the mordanting and dyeing in one bath offers no diffic it might be found economical to separate the two operations; mordanting bath can then continually be used, and the dyeing can be started at a low temperature without loss of fuel. In this neutralise the alkali in the dyestuff by adding to the dyebath the s weight of acetic acid as of naphtol-black; water containing lime likewise be neutralised with acid.

In the case of Naphtylamine-Black D the best method of applica is the following:—

WOOL can be dyed in neutral and in slightly acidulated bath. recommend for dyeing piece-goods and yarn:—For light sha addition of about 10 grs. of common salt per litre (14 oz. pe gallons) water, or 20 grs. of crystallized Glauber's salt; and for shades, addition of 5 to 7½ lbs. of acetic acid per 100 lbs. of w Loose wool dye adding 5 lbs. of alum and 15 lbs. of calcined Glau salt per 100 lbs. of wool. In all cases dye in boiling bath. 3 to 4 of Naphtylamine-Black D produce a full deep shade, which, adding some green, may be turned into a dead black. The dye are very fast to milling.

SILK dye in boiling bath with addition of some acetic acid or a 7% of dyestuff produce a fine deep black which resists to the actic water and even strong soaping.

MIXED FABRICS (wool and silk) dye in boiling bath with add of about 6 lbs. acetic acid per 100 lbs. of material. If alum is use place of acetic acid a deep dead black will be obtained.

NAPHTYLAMINE-BLACK D does not rub off; it dyes in neutral quite evenly; it is fast to light and milling; these good prope ought to secure for the new dyestuff a large consumption.

Industrial Celebrities.

CHRISTIAN ALLHUSEN.

VIII.

STIAN ALLHUSEN died on the 13th January of the present r (1890), and his will has been proved during the last few upwards of £1,126,000. personal estate. His father was a n of fortune, a Schleswig-Holsteiner ; he resided at Kiel. the armies of Napoleon Buonaparte over-ran Europe they the department of Elbmündungen, and General Davoust le its Governor. The hardness and cruelty of this man's had earned for him the name, "Butcher." He occupied the of the Allhusens as his head-quarters. Dispossessed of their d wealth the family was scattered, and the sons had to seek lihood by engaging in business.

youth Christian Allhusen obtained an appointment in the f one of the first firms in the grain trade on the Continent, Koch and Sons, of Rostock. When about nineteen years of ame across to England to Newcastle-on-Tyne, whither two of ers had preceded him, and had obtained employment with Campbell and Reveley, who also were grain merchants. Allhusen was admitted into the same office, and remained e firm until his brothers left to e business as corn merchants on n account.

very short time elapsed before one others went to London, the other ained in Newcastle took Christian tnership, although he was only ne years of age. But this arrange- l not last long, the senior partner t of the business and leaving Eng- ettle in Nova Scotia.

Christian Allhusen was at Rostock, ssrs. Koch & Sons, the senior clerk office was Mr. Bolckow ; between men an intimate friendship sprang now when Allhusen was left alone, eded in getting Bolckow to join him. ung men did not confine themselves ain trade, but struck out as general ts, also creating for themselves a iness as ship and insurance brokers, of the firm was "Christian Allhusen After a time Allhusen retired from t personal management of the grain ping business, although retaining an in it ; in later years the firm was from Christian Allhusen & Co., in instance to Boldman, Borries & Co., rwards to Borries, Craig, & Co.

as a young man, Allhusen had imself endowed with a spirit of enter- d had shewn that he possessed the faculties of a financier and usiness.

CHRISTIAN ALLHUSEN.

the year 1840 Allhusen discerned the opportunity which the ure of alkali presented, and although no chemist, and with no ce as a manufacturer, he entered on the industry before which great future was opening. He and Bolckow severed their on ; the talents of Bolckow were directed to the manufacture of became the financial partner in the firm of Bolckow, Vaughan ; ughan, who possessed the technical knowledge and ability, e practical manager of the works, he having gained his ce in the employ of Messrs. Losh, Wilson, and Bell, at Walker. same year that Henry Bolckow, John Vaughan, and Joseph eld their memorable meeting in Pilgrim-street, Newcastle, y formed the resolution to smelt the Cleveland ore at Middles- Christian Allhusen became the possessor of the small chemical p works which Charles Attwood and Co. started in the year This firm failed to make them a success during the six years re in operation. From 1840 to 1846 the business was con- under the title of Allhusen, Turner, and Co. ; then it was to C. Allhusen and Co , and afterwards to C. Allhusen and

the Limited Liability Acts were passed, and when the chemi- e was in a prosperous state, this private firm was converted oint-stock company, entitled the Newcastle Chemical Works y, Limited. It was registered on the 30th December, 1871— tal was £510,000 , in 60,000 shares of £8. 10s. each ; but in 83, it was reconstructed, the capital being reduced to £300,000., £240,000. being in 60,000 shares of £4. each, and £60,000. in 60,000 preference shares of £1. each, *the dividend of the preference shares being cumulative.*

This concern has suffered from the general depression which for so many years has rested on the Leblanc soda trade; and, although Christian Allhusen has been the chairman of the company, yet with all his astuteness he has not been able to earn for the ordinary shareholders any dividend since the year 1880, and in that year only 1⅓ per cent. was paid.

The policy which this firm has pursued in the conduct of their works has been the prompt introduction of improved appliances and new processes ; and Christian Allhusen always acted on the principle that whatever was worth doing at all was worth doing well. He prided himself on the perfection with which the plant was constructed ; he believed that a niggardly management was an extravagant one.

Surrounded by great engineering industries in which every method that science can suggest to economise labour, and to prevent loss by imperfect combustion is adopted, these Tyneside chemical works were similarly conducted. The engineering skill for which Newcastle is so distinguished, not only contributes to create the position which Newcastle engineers themselves hold, but other industries are greatly assisted and advanced by the scientific and progressive mechanical engineering of that district.

In Allhusen's works the system of working mechanical salt-cake furnaces was first wrought out by Messrs. Jones and Walsh, whose furnace was patented in 1875.

At the present day a very large outlay is being incurred to carry out thoroughly Mr. Chance's process of sulphur recovery from vat waste. Then again, these works have been specially laid out on an extensive scale for the production of caustic soda of high strength. As far as we can judge, this business has not been signalised by any important original discoveries or inventions, but whenever anything new seemed capable of profitable employment it was promptly adopted and effectively carried out. This must be regarded as the principle feature of Allhusen's policy as a manufacturer, to be ever on the lookout for improvements, and never to risk the chance of failure by unwise parsimony in construction.

Although Christian Allhusen remained chairman of the Newcastle Chemical Company until the close of his life, he did not continue to reside in Northumberland, but removed to Stoke Court, Slough, Buckinghamshire. His spirit of enterprise led him to accept positions in connection with undertakings that had their origin in London.

At one time, we find, he was a director of the Royal Aquarium, Westminster, but he does not appear to have found this an object worth sacrificing time and attention to, and his name is only on the list for a short period.

In the north, he connected himself with projects for the supply of water to Newcastle and Gateshead. He promoted the Whittle Dene Water Co. and became a director of the company into which this developed, the Newcastle and Gateshead Water Company ; but in London he became a director of the "Brazil Great Southern Railway Co ," "The British Land Mortgage Co. of America," "The International Bank of London," "The New Oriental Bank Corporation," and of the "Grand Hotel," the "Hotel Metropole," and others.

The ability that fitted him to attain to the position of director of these various companies enabled him also to utilise the opportunities that were thus afforded him ; he was able to get on the course that led him to the golden goal which he finally attained. It was mainly his financial ability, not his success as a chemical manufacturer, that enabled him to amass the fortune which he left behind him. He appears to have had the genius to discern the right moment and the proper sphere in which to act, and his calmness of judgment and promptness of action earned for him the reputation that everything he touched he turned to gold.

But Christian Allhusen not only had the prescience to discern the flowing of the tide that led on to fortune, he had grasped the principles of political economy. The prime of his life was spent in an active political sphere, for Newcastle has always been a school of strong, vigorous, independent, political opinion. Its sons have been stirring politicians, and have seriously studied with earnest purpose not only home affairs, but the wide field of international and foreign politics. At one time David Urquhart made Newcastle the scene of his agitation ; then its voice was heard in the manly, independent utterances of Joseph Cowen, and to-day it is represented by the man who, whilst he is an accomplished scholar and a famous man of letters, is also the embodiment of philosophical Radicalism.

But not only the place, the time also during which Allhusen lived —especially in his earlier days—was of exceptional interest; it was the day of agitation in favour of the abolition of the Corn Laws, and he was associated with the trade which most of all would be directly affected by any fiscal changes.

He was the contemporary of those great popular agitators, Richard Cobden and John Bright, with whom he was personally acquainted, and he mixed amidst that group of illustrious characters whose names are for ever associated with the great struggle and triumph of Free Trade.

Christian Allhusen became a sound, thorough Free Trader, his clear vision in business enabled him to discern the policy that secures the wealth of nations.

When, in the year 1860, the Cobden Treaty was being negotiated with France the alkali trade of the north was represented by Christian Allhusen, that of Lancashire having selected Mr. E. K. Muspratt to be associated with him.

Mr. Muspratt, recalling those events, says, "We met frequently from 1860 to 1870, when the various commercial treaties were being negotiated, and were together in Vienna, in 1865. I was specially impresssed with his great energy and commercial ability, and with the thorough grasp he had of Free Trade principles. He was a thorough Free Trader, and I think we made some impression on the French Government, as the suggested 30 per cent. duty was finally reduced to 10 per cent. He took great interest in every movement in favour of freer commercial intercourse with the Continent of Europe."

As a merchant, who was recognised as a most astute and shrewd man of business, his influence would be considerable, and his opinion carry great weight.

We learn from the *Newcastle Chronicle* that:—"Mr. Allhusen was a large shareholder in the Northumberland and Durham District Bank. When it failed (1857) he propounded a project for certain shareholders to take over the responsibilities of the Consett Iron Works, upon the liquidators of the Bank giving them a guarantee that no call would be made upon the shareholders who did this. The reason for this arrangement was that the iron works had to be carried on, but the liquidators were not in a position to become traders or manufacturers. This daring project, as it was then esteemed, met with a good deal of opposition, but Mr. Allhusen succeeded in inducing the late Mr. R. P. Philipson, Mr. John Benson, and others, to interest themselves in it; and a number of the Bank shareholders took over the works on the condition named. The result justified to the letter all the foresight and business acuteness of Mr. Allhusen. Since its reconstruction there is no enterprise that has been so conspicuously successful in this neighbourhood as the Consett Iron Works."

Were we permitted to enter more minutely into the story of this transaction, we may say it would be a very entertaining and instructive one, and one that would bring out the remarkable astuteness of Christian Allhusen, how well he knew "what he was about."

The following facts may indicate the magnitude and prosperity of this great concern with which Allhusen had so much to do. The Consett Iron Co., Limited, was registered in 1864. The original capital of the Company was £400,000., but it has since been increased at various times (viz. in 1866, 1872, 1880, and 1886) to its present nominal amount of £1,000,000. On the old shares the following dividends have been paid:—

1880-1	..	26⅔ per cent.
1881-2	..	20 ,,
1882-3	..	18⅓ ,,
1883-4	.	18⅓ ,,
1884-5	..	10
1885-6	..	10
1886-7	..	11⅓
1887-8	..	11⅓ ,,
1888-9	..	20 ,,

The company is connected with another company, the Consett Spanish Ore Co., Limited. The Directorate is almost the same in both companies. The capital of this mining company is £55,200., and the dividends that have been divided are:—

1880-1	..	10 per cent.
1881-2	..	15 ,,
1882-3	..	15
1883-4	..	32½
1884-5	..	36¼
1885-6	..	42½
1886-7	..	40
1887-8	..	38¾ ,,
1888-9	..	37½ ,,

Christian Allhusen was at one time president of the Newcastle Chamber of Commerce, he was for some years a member of the Gates-head Town Council, and was one of the members of the Tyne mission, but his specialty was finance, and commercial enterpri for him an absorbing interest.

During his life in Newcastle, at least from the year 1842, he r at Elswick Hall. We are indebted to the *Newcastle Journal* f following information:—"When Allhusen took up his resider Buckinghamshire, Elswick Hall came into the market, and great were made to induce the Corporation to purchase it, and conve grounds into a park. At that time the Corporation were not pre to do anything of the sort, and, in order to prevent the land f into the hands of the builders, several public-spirited gentlemen b it, as they put it, in trust for the people, being sure that sooner or the Town Council would perceive the advisability of a west end and would be convinced of the eminently suitable position c Elswick Hall grounds. In the course of time the Council were edu up to the point, and finally acquired the place, and converte grounds into the beautiful park which is now the delight of west-er and they found in the hall a capital resting place for the Lougl Noble models."

We could have wished that it were a part of our story to tel by the munificence of the successful merchant and manufacture people of Newcastle would remember his remarkable career by manent benefaction, that would have for generations have pror their health and happiness; but such expressions of a desire to b the people do not appear to have been consonant with his views he is not singular in the opinion, that the man who in pursuit of hi personal enrichment, and who with broad and intelligent concep of the science of business seeks his own interest, is a benefactor o people. His prosperity affords occupation to hundreds of families wise expenditure of capital fosters invention; his business becor nucleus whence other trades and occupations spring, and a fine w erected on a site, that, by its suitability, attracts other similar w gives a character to a population and tends to promote the institu that elevate and enrich social life. Undoubtedly, there is much said for such a view, at the same time there is a brightness in a benevolence which are prompted by those elementary virtu sympathy, compassion and goodwill.

It was only as politics were connected with trade that Chr Allhusen took much interest in them publicly. He was the chief moter of a banquet to Mr. Gladstone, at Newcastle, in the year but this was to celebrate the successful conclusion of the Trea Commerce with France, a negotiation in which, as we have al noticed, Allhusen had been one of the delegates of the alkali trad

In his latter years, when the Irish Question had come to the fo very naturally sided with the opinions of those with whom, b wealth and recognised ability, he had succeeded in becoming ciated.

County gentlemen and London financiers, whether plebeia patrician, have little sympathy with the popular movements o present day; they dread the socialistic spirit which in one for other is manifesting itself, and he joined the ranks of the Unionis opposition to the views and policy of Mr. Gladstone.

We can hardly speak of Christian Allhusen as an "Indu Celebrity;" he has little or nothing in common with Gossag Deacon, Kurtz or Gamble, Spence or Shanks; these men had themselves acquainted with chemical science, and they devoted lives to the application of their knowledge to manufactures; they inventors, they were engineers, they were trained to direct and m men, but Allhusen was eminently the man of business, early plac the counting-house of a great mercantile firm, which proved a scene genial to the earliest development of his gifts and talents: called a before he was of age to direct an independent business, thrown his own resources, quickened by the spirit of the times and the c tions by which he was surrounded, and the men with whom he ciated to great and active enterprise in the pursuit of riches; flo his vessel upon the rising tide of a new and lucrative industry; res from ruin a business in which he saw possibilities of great prospe and then seizing the most favourable moment to spread all his s catch the favouring winds of commercial progress and financial dev ment, he reached the haven of opulence and commanding status.

Born in 1806, making money whilst he was in his teens, he favoured with a career of exceptional length and opportunity, a his eighty-fourth year he died a millionaire.

The Best Known Temperatures at the Present Ti Ice melting point, 0° C.; water boiling point, 100° C.; naphth boiling point, 218° C.; then, with possible errors of 2°, we mercury boiling point, 360° C., and sulphur boiling point, 448° C.; with possible errors of 25° C., we have aluminium melting point, C.; and gold melting point, 1,045°C.; and lastly, with possible of 50° C., we have palladium melting point, 1,500° C., and plat melting point, 1,775° C.

IMPORTANT NOTICE.

SPECIAL EXHIBITION NUMBER

OF THE

"*CHEMICAL TRADE JOURNAL.*"

The issue for April 19th will appear as a Special Exhibition Number with a largely increased circulation.

Each Exhibit will receive a detailed notice, descriptive of the goods shewn.

As this number will possess unusual advantages as an advertising medium, it is particularly advisable that application for space should be made as early as possible in order to avoid disappointment.

GLASS BOTTLE INDUSTRY OF THE WORLD.

WHILE it is well known that the manufacture of glass bottles forms an important industry throughout the civilised world its full extent is not generally realised. Following is an estimate of the output per day of glass bottles :—

	Gross per day.
Great Britain and Ireland	6,206
Sweden	960
Norway	600
Denmark	360
Germany and Belgium	30,039
France	100
Austria	7,000
United States	840
Canada	120
Australia	207
Total	46,432

Taking the year at 300 working days, this would make a total output of 13,929,600 gross per year, showing how wide is the field open for this industry. It will be seen that Germany is far in advance of Great Britain in its output, and a considerable portion of the bottles used in England are imported from Germany. It is stated that during the last ten years, prices of bottles have declined on an average about 25 per cent. It is difficult to furnish separate particulars of the imports and exports of glass bottles to and from the United Kingdom, but the following figures from the Board of Trade Returns will give an idea of the importance of the glass trade :

Imports of Glass.	1887. £	1888. £	1889. £
Window and German sheet, including shades and cylinders	497,313	499,946	453,094
Plate, silvered or not	159,481	200,975	239,562
Manufactures, unenumerated	1,022,112	1,205,404	1,092,667

In the case of the imports the value of the bottles imported is not separated from other articles, but it may safely be assumed that a large proportion of the item of unenumerated manufactures consists of bottles. The Board of Trade returns furnish the following figures of the exports from the United Kingdom :—

Exports of Glass.	1887. Value. £	1888. Value. £	1889. Value. £
Plate, rough or silvered glass	262,680	260,867	243,991
Flint, of all kinds, and manufactures [illegible]	236,689	283,674	274,940
[illegible] manufactures of green or [illegible]	390,585	405,166	464,228
[illegible], unenumerated	131,075	159,634	164,553

brought in a fresh set of producers, who made money as fast as they could without reference to combinations. This did away with the enforced limit of production. The high price of copper, on the other hand, gave employment to labourers, and there was a dearth of shipping facilities. The following note gives the yearly production of nitrate from 1884 to 1888 inclusive :—

	1884.		1885.		1886.		1887.		1888.	
Production......	550,000		350,000		450,000		700,000		800,000	
Average price for cargoes arrived	s.	d.	s.	d.	s.	d.	s.	d.	s.	d.
March 31.	9	3	9	6	11	1½	10	3	10	3
Do. distant cargoes	9	3	9	9	10	0	8	3	9	0
Freights March 31	32	6	37	6	25	0	32	6	32	6

Among other details furnished by Consul Newman it is stated that the price of a quintal of nitrate on June 3 last, including cost of placing it in bulk, bagging and loading in cars, carriage by rail, export duty of 2s. 4d., port expenses, and shipping clear of surf, was $2. 46c., against a selling value of $2. 60c. This calculation takes no account of interest or depreciation. "The business of nitrate production," writes Consul Newman, "is a good honest business, with a sound natural basis, but even for a good thing one may pay too much; besides it has risks inherent to itself beyond the question of supply and demand. The disappointing demand for nitrate on the Continent has, it is believed, been partly the result of the collapse of the Copper Syndicate and Panama Canal Company in France. This tends to hamper banking facilities." There was a demand last year in France and Belgium for nitrate as a fertiliser, especially to beet, and if the Continent took 240,000 tons in 1881-2, and 527,000 tons in 1888-9, there is no reason why the consumption should not go on increasing. Consul Newman throws cold water on companies and over production, but his views have been challenged. The quantities imported into the United Kingdom have not fallen off.

IMPORTS.

GREAT BRITAIN AND IRELAND.

	1887.	1888.	1889.
	cwts.	cwts.	cwts.
Nitre Cubic (Nitrate of Soda)	1,736,802	2,053,282	2,351,310

We need only say that British capital has largely been invested in nitrate companies, and consequently we trust that the demand will be sustained. It will be observed that the imports into the United Kingdom in 1889 were 117,565 tons, or considerably less than the imports of France.—*Liverpool Journal of Commerce.*

THE NITRATE MARKET.

COLONEL NORTH presided a few days ago at Winchester House, London, over an ordinary general meeting of the Paccha and Jaz-Pampas Nitrate Company. He stated that the meeting was held in order to comply with the requirements of the law, but they had no report or accounts to submit. He was, however, pleased to see them, because for some time past there had been what was called a panic in nitrate—that was, more nitrate had been produced than it was thought could be consumed. They found now, however, that that was not the case, for over 80,000 tons had been consumed in January and February last in excess of the consumption in the same months of 1889. If the consumption increased at that rate the companies would hardly be able to meet the demand. The low price of nitrate had been to a certain extent the cause of this increased consumption, but he was sure that once a farmer used nitrate he would not care for any other manure. The nitrate companies had been competing against each other in making nitrate and giving it away to the consumer. This had been attended with beneficial results, but the nitrate makers had had enough of it, and all the large companies had formed a company to buy all the nitrate at the price of about 2dol. 70c., instead of 2dol. 25c. or 2dol. 30c. He believed that the arrangement connected with this company would be shortly completed.

DEATH AND BURIAL OF A BANK NOTE.—There is a certain ceremony which attends the death and burial of a Bank of England note. It is only three days after its cancelling that it is carried to its last home in the Bank-note Library. Its first dark day of nothingness is spent in the inspector's office, where severe judges sit in judgment on its virtue. During its second day it and its thirty or forty thousand fellows, done up in parcels, are counted and sorted; that is to say, each parcel is dealt out like a pack of cards, according to dates and denominations of value. The third day they are posted in ledgers, which are kept as indexes to the paid notes; and then, on the evening of their last day in the upper regions of light and air, they are carried down with scant ceremony in huge bags to the Bank-note Library.

BESSEMER PIG IRON IN ENGLAND AND T UNITED STATES.

THE remarkable situation in the English hematite or Besseme iron market that we pointed out in an editorial last week s to be even more notable than the figures then at hand justified u stating. Our English Exchanges of two weeks ago show that war were at least 19s. lower than makers' prices. Probably never ir history of the English market has there been such a margin bet makers' prices and the price of warrants, warrants being 63s. makers' prices 80s. to 82s. This price for warrants is certainly 1: 15s. less than the actual cost of production, based on the ruling of raw materials in England.

To produce a ton of pig iron in the West Coast of England requ say, 35 cwt. of ore and 22 cwt. of coke. West Coast hematite worth at the mine from 16 to 20 shillings, and Spanish ore is quot 16s. 6d. at the port of entry. Taking the average cost at the fur at 18s. a ton, and coke delivered at 35s. a ton, the cost of Bess pig iron at a Cumberland furnace would be about as follows :—

	s.	d.
Ore, 1¾ tons at 18s....	31	6
Coke, 1 1-10 tons at 35s.	38	6
Lime	3	6
Wages, incidentals, etc.	8	6
	82	0

This would make the cost of a ton of pig iron in the West Coast or, on the basis of 24.2 cents to the shilling, 19 dols. 84 cents a tc

Now, how does this compare with the cost of Bessemer pig irc the United States at the present time? Taking as a basis pr costs of ore and coke, which are those we have taken for England assuming that the cost per unit of iron to make a ton of ore is 12 c and that 94 units of iron and 1 1-10 tons of coke are required to a ton of pig, we would have the following as the cost of a ton c iron at Pittsburgh at the present time :—

	Dols.
Ore, 94 units at 12 cents	11.28
Coke, 1 1-10 tons at 2 dols. 95 cents................	3.24
Lime	.60
Labour	1.35
Relining	1.25
Incidentals..	1.25
	17·97

This makes the cost of a ton of Bessemer iron in Pittsburgh at present time, without counting anything for interest, plant, &c., tically 18 dols., about 1·87 dols. less than the price in England.

Analysing the different elements of cost, it will be seen that the of ore to a ton of pig in the United States is about 40 per cent than in England, while the cost of coke is only about one-third wl is at English furnaces; lime in the United States is less, while la in the United States is considerably more than it is in England, incidentals are less.

The cost of English ores and coke are abnormally high, and they may be maintained for a short time, it is more than can be pected by even the most sanguine that they will long remain as are. The action of the speculators in unloading their warran prices far below the present price of production, based on the pr prices of raw material, indicates that they at least do not believe the present rates can be permanently maintained. It is also stated there have not been many orders placed recently for the constructi ships in Great Britain, ship plates being one of the articles into w these hematite irons enter most largely. If this statement is tr seems evident that there must be in the near future a decline eve the price of makers' irons in Great Britain.—*American Manufact.*

Market Reports.

MISCELLANEOUS CHEMICAL MARKET.

Business is still in an unsettled state, and prices all round are irregular. Caustic soda is still comparatively scarce for early delive makers, as a rule, being well filled up with orders. Nearest values are for 70's £10. 10s.—60's £9. 10s. Cream, 60%, £9 ton—all f.o.b. Liverpool. Bleaching powder, £5. 10s. per tor rails. Soda ash is at 1¼d. to 1⅜d. per deg. for ordinary kind rails; high test ash is still very firm at 1½d. per deg. f.o.b. La shire and Tyne; soda crystals, £2. 15s. per ton in Lancasl ɔotash stationary at 5d. per lb.; sulphur in all grades ıge; chloride of calcium, 42s. 6d. per ton; vitriol and ds firm at current quotations; chromic acid, 6½ per lb; copper still very scarce for early deliveries, and spot ıe £26. to £26. 10s. per ton. Forward deliveries easier. ıate of lime and other fertilizers are now receiving more usual in view of spring demand. Nitrate of soda, 8s. per te of potash, £7. 15s. to £8 per ton on 80% c.i.f.; lime have been moving more readily at rather easier rates, brown and £13. 15s. for grey, at usual centres of con- Acetates of lead and nitrate of lead continue firm on the ıarge, £15. 15s. net per ton. In potash, caustic, and ıere is more demand and a stronger market. Oxalic acid ∫d. to 3⅝d. per lb.; iodine, 8½d. per oz.; white powdered ing freely at £12. 10s. f.o.b. The demand for chemicals e trades is still far from satisfactory, and has a depressing prices of many of the finer chemicals, colours, &c.

THE TYNE CHEMICAL REPORT.

TUESDAY.

mical market continues very firm, with good demand for of manufacture. Soda ash has declined in price, but this ıt might be expected after the sharp advance, in consequence sual temporary demand. Bleaching powder and caustic n unchanged at last week's quotations. Soda crystals very d have advanced 2s. 6d. per ton during the week. f soda in good demand, and prices raised half-a-crown a osulphite of soda has been advanced in price by English ı manufacturers. Quotations are now 20s. a ton above late

h and Russian shipments are expected to be made in about days for the first time this season. Quotations to-day llow:—Caustic soda, 77%, £13. per ton; bleaching ı softwood casks, £5. 15s. to £6. per ton; soda ash, ⅜d. per degree less 2½%; soda crystals, £2. 12s. 6d. ulphate of soda, in bulk, 35s. per ton; ground, in casks, on; chlorate of potash, 5d. per lb.; recovered sulphur, r ton; silicate of soda, 140° Tw., £4. per ton; 100° Tw., l. per ton; 75° Tw., £2. 10s. per ton; hyposulphite of 7 cwt. casks, £5. 5s. per ton; in 1 cwt. kegs, £5. 15s. pure white sulphate of alumina, £4. 10s. per ton; blanc 10s. per ton; chloride of barium, £8. per ton; nitrate of stals, £18. 10s. per ton; ground, £19. per ton; sulphide of 5. 10s. per ton—all f.o.b. Tyne or f.o.r. makers' works. ham Paper Pulp Company, Limited, which took over the John Smalley, West Hartlepool, for the manufacture of ·ood pulp, expect to recommence operations in the course ıys. The Marsden Chemical Company, Limited, are now ·ing sulphite wood pulp. Their works are at Marsden, tween South Shields and Sunderland, and their offices at ·reet, Newcastle-on-Tyne.

no change to report in the price of manufacturing coals. ıall is quoted 9s. 6d. to 10s. per ton, and Northumberland ı. to 8s. 6d. per ton.

ꞁEPORT ON MANURE MATERIAL.

ıs been a good business passing during the week for all kinds , for prompt or early delivery, the exceptionally fine weather ıght orders freely to the manure makers, and they being thus gauge more exactly the balance of their spring requirements

ot hear of any important transactions in mineral phosphates, ɔns for Charleston are somewhat easier, and a concession of ınit would probably have to be made in order to induce much Quotations for Canadian seem to be quite out of reach at present, and we think they will have to be considerably ·efore, at any rate, the United Kingdom will be tempted to

. nothing to report in cargoes of River Plate bones, there ing offering but a December sailing now fully due, and for 7s. 6d. would probably be accepted. There is nothing ıshed bones to arrive to report, but several parcels arrived at vill be offered when landed there, and these will probably be at £5. 5s. ex quay. From the quay 200 tons Calcutta bone ɜd, sold at £5. 7s. 6d. per ton, which price has repeatedly d for Bombay now in dock at Liverpool. £5. 10s. per ton ‹ quay, and sales have been made at £5. 12s. 6d. per ton, Common grinding bones are worth about £5. per ton, and less may have to be taken, considerable shipments having gether at Liverpool.

Nitrate of soda continues without change. Spot price is 8s. per cwt. for fine quality, and due cargoes cannot be quoted as worth more than 7s. 10½d. per cwt, though 8s. is generally asked.

We have nothing fresh to report in home prepared dried blood, but sales of River Plate on spot and to arrive have been made at 10s. 3d. per unit, which price would be accepted for further parcels arrived in Liverpool.

There is more ammonite now offering, and probably something under 9s. 6d. per unit, in bags, gross weights, delivered f.o.r. at works would be accepted.

The somewhat easier market for superphosphates, indicated in our last report, continues, and value cannot be quoted now at over £2. 3s. 9d. to £2. 5s. per ton, in bulk, f.o.r. at works, there being a considerable quantity in second-hands offering on the market.

Kainit is offering at 38s. to 40s. per ton, according to quantity, net cash, in bags, delivered f.o.r. Birkenhead or Garston.

METAL MARKET REPORT.

	Last week.		This week.
Iron	51/4	....	48/11
Tin	£90 7 6	....	£90 5 0
Copper	48 17 6	....	47 12 6
Spelter	21 5 0	....	21 0 0
Lead	12 12 0	....	12 10 6
Silver	43¾d.	....	43⅞

COPPER MINING SHARES.

	Last week.			This week.	
Rio Tinto	16	16⅛	....	15⅞	16
Mason & Barry	6 5/16	6 7/16	....	6 5/16	6 7/16
Tharsis	4½	4⅝	....	4½	4⅝
Cape Copper Co.	3½	3⅝	....	3⅜	3½ x.d.
Namaqua	1⅞	2⅛	....	1¾	2
Copiapo	2 7/16	2 9/16	....	2⅜	2½ x.d.
Panulcillo	15/16	1 1/16	....	1	1⅛
New Quebrada	⅝	¾	....	⅞	1
Libiola	2 7/16	2 9/16	....	2⅜	2⅝
Tocopilla	1/3	1/9	....	1/-	1/6
Argentella	3d.	6d.	....	3d.	9d.

WEST OF SCOTLAND CHEMICALS.

GLASGOW, Tuesday.

Matters are settling down slowly after recent uncertainty and excitement, but evidences of a disturbed balance in the market are still plentiful enough. High values in some seem to be giving way; in others they are as yet fully maintained; caustic soda for example having even registered a further slight rise in one of the strengths. Nitrate of soda continues on a very low level, and business has been done without difficulty at 8s., although the nominal quotation is a fraction higher. Sulphate of copper is being done this week 10s. down, and there has also been a further fall in borax, while soda crystals, refined alkali, and soda ash are all easing off a fraction. Sulphate of ammonia also is weaker, a few holders reconciling themselves yesterday to a selling price of £11. 12s. 6d. spot Leith, while others offer freely at £11. 13s. 9d. For forward delivery sellers are asking £11. 15s. without inducing more than a very moderate show of business. Paraffins (with the exception of burning oil) are still firm, and for scale and wax the agreement prices are freely obtainable under the new contracts now being entered into. The calls for bichromate of potash and bichromate of soda, of late rather unsatisfactory, have somewhat improved during the week. Chief prices current are:—Soda crystals, 50s. to 52s. 6d., less 2½% Glasgow; alum in lump £5., less 2½% Glasgow; borax, English refined £29., and boracic acid, £37. 10s. net Glasgow; soda ash, 1⅜d. net Tyne; caustic soda, white, 76°, £13., 70/72°, £11., 60/62° £10., and 60/62° cream, £9. 5s., all less 2½% Liverpool; bicarbonate of soda, 5 cwt. casks, £6. 5s., and 1 cwt. casks, £6. 10s. net Tyne; refined alkali 48/52°, 1⅜d., net Tyne; saltcake, 30s.; bleaching powder, £5. 10s. to £5. 15s., less 5% f.o.r. Glasgow; bichromate of potash, 4d., and of soda 3d., less 5 and 6% to Scotch and English buyers respectively; chlorate of potash, 5d., less 5% any port; nitrate of soda 8s. to 8s. 3d.; sulphate of ammonia, spot, £11. 12s. 6d. f.o.b. Leith; salammoniac, 1st and 2nd white, £37. and £35., less 2½% any port; sulphate of copper, £25. 10s., less 5% Liverpool; paraffin scale, hard and soft, 2½d. per lb.; paraffin wax, 120°, semi-refined, 3⅝d.; paraffin spirit (naphtha), 8d. a gallon; paraffin oil (burning), 6¾d. to 7d. at Glasgow for Scotch consumption (English orders done at about 1d. a gallon lower); ditto (lubricating), 865°, £5. 10s. to £6. per ton; 885°, £6. 10s.; and 890/895°, £7. 10s. to £8. Week's imports of sugar at Greenock were 30,255 bags.

£14. to £15. Potter's lead ore: Prices are firm: smalls, £14. to £15.; selected lump, £16. to £17. Calamine: Best qualities scarce—60s. to 80s. Strontia steady; sulphate (celestine) steady, 16s. 6d. to 17s. Carbonate (native), £15. to £16.; powdered (manufactured), £11. to £12. Limespar: English manufactured, old G G.B. brand in demand, and brings full prices; 50s. for ground English. Felspar, 40s. to 50s.; fluorspar, 20s. to £6. Bog ore firm at 22s. to 25s. Plumbago: steady; Spanish, £6.; best Ceylon lump at last quotations; Italian and Bohemian, £4. to £12. per ton; founders, £5. to £6.; Blackwell's "Mineraline," £10. French sand, in cargoes, continues scarce on spot—20s. to 22s. 6d. Ferro-manganese and silicon spiegel in good demand at stronger figures. Chrome iron, 20 per cent., £24. to £25. Ground mica, £50. China clay freely offering—common, 18s. 6d.; good medium 22s. 6d. to 25s.; best, 30s. to 35s. (at Runcorn).

Gazette Notices.

Partnerships Dissolved.

W. JOHNSON AND R. JOHNSON, Leek, chemists and druggists.

Adjudications.

ARCHIBALD LITTLE, Wandsworth and Mortlake, brick, lime, and cement merchant

New Companies.

FULLERS' EARTH UNION, LIMITED.—This company was registered on the 21st ult, with a capital of £105,125, in £1 shares, to acquire certain Fullers' earth mines and Fullers' earth works, situate at Nutfield, Surrey, and at Bath, and the property held therewith, and certain contracts relating to the supply of Fullers' earth. The first seven subscribers are:— Founders' Shares.

	Founders' Shares.
A. G. Moger, Bath, solicitor	1
W. Blewitt, Wanstead, solicitor	1
Earl Kilmorey, Carlton Club	1
E. G. Butler, J.P., Combe-hay, Bath	1
C. W. Cawley, Nutfield, managing director	1
Austin J. King, 13, Queen-square, Bath, solicitor	1
R. Le Brasseur, 12, New-court, Lincoln's-inn, solicitor	1

READ, HOLLIDAY AND SONS, LIMITED.—This company was registered on the 24th ult., with a capital of £200,000, in £10 shares, of which 9,000 are 5 per cent. cumulative preference shares, to acquire the business, undertaking, and assets of Read, Holliday and Sons, at Huddersfield, Manchester, and Glasgow, and in New York, Brooklyn Philadelphia, and Boston, U.S.A., dyers and manufacturers of dyes, colours, chemical substances, &c. An unregistered agreement of the 22nd inst. will be adopted. The subscribers are:— Shares.

	Shares.
Thos. Holliday, Huddersfield, chemical manufacturer	1
R. Holliday, Huddersfield, chemical manufacturer	1
W. Heppenstall, Huddersfield, bookkeeper	1
J. Pogson, Huddersfield, traveller	1
Mrs R. Holliday, Huddersfield	1
G. P. Norton, C A., Huddersfield	1
E. Cockshaw, Huddersfield, cashier	1

THOMAS AND GREEN, LIMITED.—This company was registered on the 25th ult., with a capital of £39,000, in 200 ordinary and 190 6 per cent preference shares of £100 each, to take over the business of paper manufacturers, carried on at Wooburn, Bucks, by John Thomas and Roland Green, and in conjunction with such business to manufacture gas and electric power for lighting and other purposes. The subscribers are:— Shares.

	Shares.
R. Green, Maidenhead, paper manufacturer	1
Mrs. Green, Maidenhead	1
James Thomas, Bisham, Berks, paper manufacturer	1
Miss E. Thomas, Great Marlow	1
John Thomas, Wooburn, Bucks	1
Mrs. Thomas, Wooburn, Bucks	1
Laurence Green, Maidstone, paper manufacturer	1
Herbert Green, Maidstone, paper manufacturer	1

The Patent List.

This list is compiled from Official sources in the Manchester Technical Laboratory, under the immediate supervision of George E. Davis and Alfred R. Davis.

APPLICATIONS FOR LETTERS PATENT.

Electric Arc Lamps. J. Campbell. 4,103. March 17.
Lubricators. A. Bradshaw. 4,107. March 17.
Maturing Spirits and Apparatus therefor. J. McKinless. 4,108. Mar. 17.
Magneto Call Bell Generators. F. J. Young. 4,116. March 17.
Smoke Prevention. J. Shaw. 4,118. March 17.
Rock-boring Apparatus. P. De Baère. 4,119. March 17.
Destructive Distillation of Mineral Oils. J. Laing. 4,120. March 17.
Water Gauges for Steam Boilers. J. McEwen. 4,131. March 17.
Grinding and Amalgamating Pans.—(Complete Specification). W. Roberts and H. R. Belden. 4,133. March 17.
Ozone Oil.—(Complete Specification). F. Piekenbrock. 4,137. March 17.
Water Gauge. H. Gläser. 4,139. March 17.
Treatment of Sewage. F P. Candy 4,151. March 19.
Grinding and Mixing Machine. M. H. Simonet. 4,153. March 17.
Apparatus for Concentrating Sulphuric Acid. C. S. Negrier. 4,171. Mar. 17.
Dust Separator. G. H. Oliver. 4,180. March 17.
Ball Socket Electric Switch. F. Suter. 4,189. March 18.
Generating Heat, Steam, and Light by Electricity.—(Complete Specification). T. D. Farrall. 4,191. March 18.

Cobalt Ore—
New Caledonia, 124 t. Newcomb & Son

Caoutchouc—
E. Indies, 33 c. Ross & Deering
E. Africa, 50 Hammond & Co.
" 6¼ Lewis & Peat
E. Indies, 11 L. & I. D. Jt. Co.
Madagascar, 7 "
" 4 Kebbel, Son, & Co.
France, 21½ Flageollet & Co.
France, 118 Kleinwort, Sons, & Co.
E. Africa, 52 L. & I. D. Jt Co.
Germany, 13¾ Heilbut, Symons, & Co.

Chemicals (*otherwise undescribed*
Germany, £9. T. H. Lee
" 414 Phillipps & Graves
" 14 L. & I. D. Jt. Co.
" 97 T. L. Lee
" 1,130 A. & M. Zimmermann
" 26 T. H Lee
" 15 Phillipps & Graves
" 12½ G. Boor & Co.
" 705 A. & M. Zimmermann
" 10 Burgoyne, Burbidge, & Co.
Holland, 10 W. Boyce

Dyestuffs—
Annatto
Ceylon, 10 pkgs. Eastern Prod. & Est. Co.
St. Vincent, 3 bgs. C. Neilson & Sons
Cutch
Holland, 20 pkgs. Brinkmann & Co.
Extracts
France, 70 pks. Levinstein & Sons
Holland, 55 Burt, Boulton, & Co.
U. S., 12 T. Ronaldson & Co.
Denmark, 7 Bailey & Leetham
France, 25 T. H Lee
" 251 Prop. Chamble's Wf.
U. S., 120 A. J. Humphrey
Orchella
France, 17 pkgs Hoare, Wilson, & Co.
Indigo
E. Indies, 16 chts. Langstaff & Co.
" 23 5 bxs. Ross & Deering
" 68 G. Yule & Co.
E. Indies, 75 chts. Stansbury & Co.
Greytown, 8 Cotesworth & Powell
" 118 L. & I. D. Jt. Co.
" 27 "
" 1 Cayzer, Irvine, & Co.
" 20 F. W Heilgers & Co.
" 5 Lewis & Peat
" 160 L. & I. D. Jt. Co.
" 11 H. Johnson & Sons
Tanners' Bark
Belgium, 14 t. 17 c. H. Henle Succrs
" 12 S. Barrow & Bro.
Germany, 10 Baxter & Hoare
" 20 Bevington & Sons
" 50 J. Graves
Yellow Berries
France, 30 pkgs. L. & I. D. Jt. Co.
Argols
Italy, 332 pkgs. Thames S. T. Co.
France, 4 pkgs. D. Magnus
Sumac
Italy, 590 pkgs. J. Kitchin
" 300 Hicks, Nash. & Co.
" 200 W. France
Extracts
France, 310 pkgs. Levinstein & Sons

Gutta Percha—
E. Indies, 223 c. Kaltenbach & Schmitz

Guano—
Chile, 1,120 t. Ang. Contl. Guano Wks.

Glycerine—
Holland, £42 F. W. Heilgers & Co.
Germany, 140 Beresford & Co.
Holland, 10½ Geelkerken & Co.
Germany, 130 T. H. Lee
" 500 Spies, Bros., & Co.

Glucose—
Germany, 400 pkgs. 400 c L. & I. D. Jt. Co.
" 10 91 J. H. Epstein
U. S., 75 450 L. Sutro & Co.
Belgium, 206 68 J. T. Morton
Germany, 21 100 Horne & Co.
U. S., 50 278 Props. Scott's Whf.

U. S., 50 brls. Gellatly & Co.
" 25 J. G. Rollins & Co.
S. Russia, 650 Grosseurth & Lubboldt

Paraffin Wax—
U. S., 100 brls. G. H. Frank
" 39 H. Hill & Sons

Pearl Ash—
France, 120 c. Petri Bros.

Plumbago—
Holland, 51 cks. Brown & Elmslie

Potassium Permanganate—
Germany, 6 pkg. Spies, Bros., & Co

Sodium Hyposulphate—
Germany, £38 Beresford & Co.

Sodium Acetate—
France, £100 Hunter, Waters, & Co.
Belgiam, 84 . Leach & Co.

Saltpetre—
Germany, 79 cks. J. Hall & Son
" 118 P. Hecker & Co.

Stearine—
Germany, 186 cks. W. Balchin
Belgium, 154 bgs. H. Hill & Sons
France, 10 cks. Van Geelkerken & Co.

Sugar of Lead—
Germany, 1 pkg. C. J. Capes & Co.

Tartaric Acid—
Holland, 7 pkgs B. & F. Wf. Co.
" 4 A. Finsler
Germany, 25 W. C. Bacon & Co.

Tartars—
Italy, 200 pkgs. Seager, White, & Co.

Ultramarine—
Holland, 12 cs. Ohlenschlager Bros.
" 24 pkgs. Hernu, Peron, & Co
Belgium, 4 cks. G Steinhoff
Germany, 2 "
Germany, 4 "
Holland, 10 pks. Haeffner, Hilpert, & Co

Zinc Oxide—
Holland, 20 brls. Beresford & Co.
" 50 M. Ashby
" 300 pks. Soundy & Son
" 140 Hernu, Peron, & Co.

LIVERPOOL

Week ending Mar. 27th.

Antimony Ore—
Smyrna, 754 bgs.

Boracic Acid—
Leghorn, 76 cks.

Bone Meal—
Bombay, 968 bgs. Okell & Owen
" 417
Calcutta, 91 Dalzeil & Co.
" 2,986

Bones—
Smyrna, 160 t.
Coquimbo, 6 R. R. Steel & Co.
Constantinople, 45 t. Barff & Co.

Brimstone—
Catania, 300 t. R. Robertson & Co.

Borate of Lime—
Antofagasta, 719 bgs. Cockbain, Allardice, & Co.

Copper Regulus—
Carrizal, 402 t. Balfour, Williamson, & Co.

Cream of Tartar—
Bordeaux, 5 cks.
Genoa, 10
Barcelona, 1 F. Reid & Co.
" 5
Tarragona, 5

Caoutchouc—
Ceara, 2 bls
" 68 Rosing, Bros., & Co
Salt Pond, 1 brl.
" 1 W. Griffiths & Co.
Cape Coast, 4 cks. 1 cs. Fletcher & Fraser
" 2 2 A. Millerson & Co.
" 1 Havard & Co.
" 3 T. E. Tomlinson & Co.
" 2 L. Hart & Co.
" 1 Davies, Robbin, & Co.
" 1 F. J. Eaton
" 1 W. Duff & Co.
Axim, 3 bgs. Radcliffe & Durant
" 2 Edwards Bros.
" 2 brls. F. & A. Swanzy
Grand Bassam, 3 cks. A. Verdier
" 3 bgs. Edwards Bros.
" 1 J. Stradelmann
" 2 brls.
" 2 Cie de l'Afrique
Sulymah, 3 Palma Tradg. Co.
Sierra Leone, 3
" 11 Pickering & Berthoud
" 2 Hutton & Co.
" 6 Paterson, Zachonis, & Co.
Bathurst, 4 A. Reis
" 3 chts. A. Cros & Co.
Colon, 29 brls. T. M. Williams

Dyestuffs—

Fustic
Vera Cruz, 1,323 ps. R. B. Watson & Co.
Tampico, 8,184 A. Ugarte

Sumac
Palermo, 350 bgs. 100 bls.
Catania, 150
Palermo, 326

Extract
New York, 100 cs.
" 6 brls.
" 42 J. Singleton
St. Nazaire, 150 cks.
New York, 50 bxs. Armour & Co.
Antwerp, 7 cs. J. T. Fletcher & Co.

Valonia
Patras, 625 bgs. C. J. Papayanni & Jeremias
Smyrna, 460 S. Vivante & Son
" 20 S. Spartali & Son
" 155½ t.
—— 205 Barry Bros.
Marothonisi, 107 t. C. G. Papayanni & Jeremias

Cochineal
Grand Canary, 7 bgs. Lathbury & Co.
" 30 Swanston & Co

Logwood
New York, 50 brls.
Belize. 169 t. P. Leckie & Co.
Boston, 100 brls. 10 cs. J. Kenyon & Son

Myrabolams
Bombay, 100 bgs. W. & R. Graham & Co.

Dried Blood—
Monte Video, 465 bgs. Merchants Banking Co., London

Glucose—
Philadelphia, 178 brls.
New York, 200
" 20
Rouen, 5 cks. Co-op. W'sale Socy., Ltd.
Stettin, 37 cks.
Hamburg, 5

Glycerine—
St. Nazaire, 29 cks. 4 dms.
Venice, 20

Manure—

Phosphate Rock
Coosaw, 900 t.

Nitrate—
Iquique, 12,344 bgs.

Phosphate—
Terneuren, 380 t 200 bgs. Phospho Guano Co.

Pitch—
Rouen, 15 cks. Co-op. W'sale Socy., Lmtd.

Potash—
Rouen, 34 cks. Co op W'sale Socy., Lmtd.

Pyrites—
Huelva, 1,664 t. Tennants & Co.

Stearine—
New York, 50 hds. 25 brls.
Antwerp, 13 bgs. J. T. Fletcher & Co.
" 60 10 cks

Sodium Nitrate—
Iquique, 9,970 bgs. W. & J. Lockett
Caleta Buena, 8,334 bgs.

Sulphur Ore—
Huelva, 1,639 t. Matheson & Co.

Saltpetre—
Calcutta, 665 bgs
Hamburg, 83 cks.

Salt—
Philadelphia, 65 tcs.
Hamburg, 102 bgs.

Sulphur—
Catania, 30 bgs.

Tartar—
Naples, 4 cks.

Zinc Ashes—
Boston, 11 brls.

Zinc Oxide—
St. Nazaire, 25 brls.

GLASGOW.

Week ending Mar. 27th.

Alum—
Rotterdam, 42 cks.

Acetate of Lime—
New York, 760 bgs.

Cream of Tartar—
Marseilles, 5 brls. 1 cs.

Chemicals *(otherwise undescribed)*
Hamburg, 1 cs.

Dyestuffs—

Cutch
Rangoon, 3,800 cs. P. Henderson & Co

Sumac
Marseilles, 150 bgs.

Madder
Marseilles, 6 brls.
Rotterdam, 2 cks. 1 box.

Alizarine
Rotterdam, 109 cks.
" 139

Extracts
Antwerp. 2 cks.

Logwood
Old Harbour, 448 t. M'Arthur, Scott & Co.
Belize, 100 t. 7 c. Allison, Cousland & Co.

Myrabolams
Bombay, 667 pkgs.

Glycerine—
Marseilles, 10 brls. A. Steven & Son

Glucose—
Philadelphia, 180 brls.

Gutta Percha—
Amsterdam, 7 bgs.

Paraffin Wax—
Rangoon, 364 cs. P. Henderson & Co.

Potash—
Rotterdam, 19 cks. J. Rankine & Son

Potassium Carbonate—
Hamburg, 55 cks.

Phosphate—
Antwerp, 450 t.

Saltpetre—
Hamburg, 182 cks. J. Poynter & Son

Tartaric Acid—
Bilbao, 16 brls.

Tar—
Marseilles, 69 brls.

Tartar—
Bordeaux, 4 cks. J. & P. Hutchinson
" 6

Ultramarine—
Rotterdam, 3 cs. 4 cks. J. Rankine & Son

Zinc Oxide—
Marseilles, 368 bgs.

HULL.

Week ending Mar. 27th.

Acetic Acid—
Rotterdam, 1 ck. Hutchinson & Son
" 1 Furley & Co.
" 1 cs. Hutchinson & Son
" 45 balloons

Barytes—
Ghent, 22 cks. Johnson Bros.
Bremen, 23 Tudor & Son
" 61 Tudor & Son

Chemicals *(otherwise undescribed)*
Hamburg, 6 cs. Malcolm & Son
Antwerp, 28 brls. 31 pks. Wilson, Sons, & Co.
Dunkirk, 326
Rotterdam, 1 ck. G. Lawson & Son
" 2 cs. Hutchinson & Son
Ghent, 20 cks. W. S. Merrikin
" 39 Sissons Bros. & Co.
Bremen, 5 crbys. Veltmann & Co.
Dunkirk, 30 drms. Wilson, Sons, & Co.
" 9 cks.
Antwerp, 57 pkgs. "

Dyestuffs—

Rennet
Copenhagen, 1 ck. 3 bxs. Wilson, Sons, & Co.

Extracts
Rouen, 171 cks. Rawson & Robinson
" 70 Wilson, Sons, & Co.

Alizarine
Rotterdam, 13 cks. W. & (
" 14 pks.
" 76 Hutc
" 10 W. & C

Myrabolams
Bombay, 9,161 bgs.

Glucose—
Hamburg, 600 bgs.
" 100 A. C. Ho
" 100
Copenhagen, 1,200 bgs.
Hamburg, 15 cks. Woo
" 12
Dunkirk, 40 bgs.

Horn Piths—
Lisbon, a quantity Laml

Manganese Borate—
Bremen, 1 kg. J

Manure—

Kainit
Hamburg, 1,016 bgs. G.
" 660 E

Potash—
Rotterdam, 32 cks.

Phosphate—
Dunkirk, 499 bgs.

Petroleum—
New York, 11,834 brls.

Pitch—
Libau, 1 brl.

Stearine—
Antwerp, 12 bgs. Wilson,

Turpentine—
Wilmington, 250 brls.

Tar—
Wilmington, 1,750 brls.

TYNE.

Week ending Mar.

Glucose—
New York, 25 brls.
Rotterdam, 20 bgs. T

Magnesia—
Hamburg, 3 cs. T

GRIMSB

Week ending Mar.

Chemicals *(otherwise u.*
Rotterdam, 5 cks.

Glucose—
Hamburg, 8 cks. 160 bg

GOOLE

Week ending Mar.

Alum—
Rotterdam, 27 cks.

Chrome Oxide—
Rotterdam, 2 brls.

Caoutchouc—
Boulogne, 8 cks.

Dyestuffs—

Alizarine
Rotterdam, 78 cks.

Dyestuffs *(otherwise u*
Boulogne, 44 cs. 56 cks.

Glucose—
Calais, 100 bgs.

Manure—

Phosphate Rock
Ghent, 800 bgs.

Muriate—
Rotterdam, 38 cks.

Phosphate—
Antwerp, 500 bgs.

Phosphate of Lime—
Antwerp, 250 t.

Potash—
Hamburg, 8 bgs.
Dunkirk, 2 cks. 17 dru

Sugar of Lead—
Hamburg, 1 ck.

Soda—
Rotterdam, 7 cks.

Sodium Silicate—
Rotterdam, 1 ck.

Sodium Nitrate—
Rotterdam, 5 cks.

Saltpetre—
Rotterdam, 385 bgs.

Waste Salt—
Hamburg, 1,217 bgs.

Chemicals (*otherwise undescribed*)
Hamburg, 36 pkgs. £77
Sydney, 46 207
Yokohama, 29 cs. 305
" 75 pks. 395
Madras, 44 170
Limon, 31 cks. 138
Yokohama, 100 kgs. 74

Condy's Fluid—
Sydney, 23 cs. £67

Citric Acid—
Cadiz, 5 c. £40
Brussels, 10 74
Hiogo, 3 24
Yokohama, 3 20
Madras, 1 8
Seville, 3 20
Amsterdam, 3 kegs. 22
Hamburg, 2 15

Disinfectants—
Natal, £71
Melbourne, 50 drs. 30 cks. 42
" 10 pkgs. 26
Sydney, 60 cs. 23
Hamburg, 40 cks. 200

Guano—
Mauritius, 31 t. 4 c. 298

Hyposulphite of Lime—
Rotterdam, 20 cs. £120

Manure—
Cherbourg, 70 t. £460
Malaga, 35 18 c. 300
Rotterdam, 31 160
Stettin, 204 8 800
Cologne, 19 12 122
Barbadoes, 3 31
Hamburg, 200 2,400
Townsville, 4 3 75
Trinidad, 90 657
St. Kitts, 5 39
Barbadoes, 2 9 17
Rochefort, 100 350
Cologne, 29 10 205
Mauritius, 84 15 404
Melbourne, 70 245
Sydney, 45 600
Harlingen, 78 1 345
Rotterdam, 15 135
Newfairwater, 680 4 2,840
Genoa, 23 120

Muriatic Acid—
Malaga, 4 t. 2 c. £30

Oxalic Acid—
Yokohama, 20 kgs. £32

Phosphorus—
Kobé, 1 t. 7 c. £361
Yokohama, 34 cs. £472
" 25 258
Kobé, 6 75
Yokohama, 40 477

Potassium Chlorate—
Melbourne, 1 t. £44

Potash—
Gothenburg, 10 c. £24

Picric Acid—
B. Ayres, 1 cs. £11

Potassium Sulphate—
Mauritius, 26 t. £260

Potassium Cyanide—
Madras, 1 t. £52
Algoa Bay, 1 10 c. 112

Soda—
Sydney, 2 t. 1 c. £13

Sheep Dip—
Algoa Bay, 5 t. £269
New York, 22 6 c. 1,125
Natal, 3 8 145
" 300 drs. 101
" 105 37
Algoa Bay, 17 cs. 48

Sulphur—
Madeira, 14 t. 14 c. £154
" 60 17 638
Calcutta, 15 18 83

Saltcake—
Antwerp, 130 t. £227

Salammoniac—
Wellington, 12 c. £20
B. Ayres, 4 cks. 41

Saltpetre—
Sydney, 2 t. £45
Limon, 5 112
Greytown, 10 c. 12
Bilbao, 2 10 55
Wellington, 1 10 34

Sulphur Flowers—
Madeira, 5 t. 5 c. £50
Melbourne, 4 19 45
Lisbon, 950 brls. 734
Madeira, 480 370
Wellington, 2 t. 14

Superphosphate—
Charente, 11 t. 16 c. £34
Mauritius, 21 14 175
Cherbourg, 185 t. 555
Rotterdam, 37 12 c. 150
Almeria, 4 19 28
Bordeaux, 29 17 90

Sodium Tartrate—
Melbourne, 5 c. £22

Sulphuric Acid—
Natal, 3 t. 13 c. £26

Sodium Nitrate—
Kobé, 34 t. 17 c. £318

Salicylic Acid—
Melbourne, 3 cs. £20
" 1 c. 29

Sodium Acetate—
New York, 5 t. 11 c. 80

Sodium Sulphate—
Ghent, 50 t. £30

Sheep Wash—
Kingston, Ja., 60 drs. £20

Tartaric Acid—
Sydney, 3 t. 3 c. £434
Wellington, 5 40
Fremantle, 4 40
Piraeus, 1 ck. 30
Melbourne, 10 c. £63
" 2 t. 266
Sydney, 1 129
" 5 34
Wellington, 3 21

LIVERPOOL.

Week ending Mar. 26th.

Alum—
Calcutta, 15 t. 1 c. £37
Oporto, 5 1 20
Calcutta, 3 1 2 q. 9
Genoa, 2 2 1 13

Acetic Acid—
Vera Cruz, 2 cs. 20 cks. £105

Alum Cake—
Calcutta, 8 t. 8 c. £40

Alum Sulphate—
Vera Cruz, 2 cks. £20

Ammonium Muriate—
Havre, 1 t. 6 c. £21
Gothenburg, 13 3 q. 15

Ammonium Sulphate—
Antwerp, 31 t. 10 c. £378
Rio Janeiro, 19 2 q. 13
Bremerhaven, 10 120
Gothenburg, 53 636
Gr. Canary, 6 3 1 30
Hamburg, 20 240
Las Palmas, 4 3 47
Antwerp, 31 10 378
Barbadoes, 50 10 606
Valencia, 5 2 2 59
" 5 3 2 58

Ammonium Carbonate—
Corunna, 5 c. £8
Leghorn, 10 3 q. 18
Barcelona, 18 2 20
Genoa, 17 26
Calcutta, 1 t. 10 52

Bleaching Powder—
Antwerp, 25 brls.
Bilbao, 96 cks.
Boston, 131 165 tcs.
Calcutta, 81 cs.
Hamburg, 27
Naples, 13
New York, 25
Philadelphia, 60
Rouen, 103
Boston, 214 93 tc
New York, 145
Rouen, 14
Vera Cruz, 30 brls.

Boracic Acid—
Palermo, 2 c. 3 q. £5

Brimstone—
Calcutta, 7 t. 19 c. 1 q. £37

Borax Crystals—
Callao, 1 t. £31

Iron Oxide—
Valparaiso, 59 t. 10 c. 3 q. £149
B. Ayres, 37 185

Magnesium Citrate—
Corunna, 1 c. £17

Manure—
Antwerp, 19 t 4 c. 1 q. £100

Pitch—
Genoa, 1,179 t.
Syra, 20 brls.

Potassium Carbonate—
Madrid, 4 c. £5

Potassium Chlorate—
Hamburg, 10 c. £22
Vigo, 2 11
Gothenburg, 34 t. 1,467
New York, 5 222

Phosphate—
Hamburg, 32 t. £120

Potassium Bichromate—
Havana, 8 c. 3 q. £16
Alexandria, 12 23

Sodium Bicarbonate—
Alexandretta, 8 kgs
Barcelona, 110
Bilbao, 100
Bordeaux, 60
Boston, 110
Marseilles, 240
Nantes, 30
Palermo, 26
St. Johns, Nfd. 40
San Sebastian, 10
Sydney, 30 67 cks.
Yokohama, 1,000
Antwerp, 32 6
Canada, 100 29
Hamburg, 22
Rotterdam, 23

Salt—
Calcutta, 3,681¼ t.
Edina, 25
Fredericia, 485
Isles de Los, 25
New Orleans, 70
Sierra Leone, 50
Valparaiso, 51
Brass, 50
Calcutta, 3,171
Chicago, 65
Conakry, 50
Coquimbo, 50
Faroe Islands, 70
Ibrail, 544
Lavannah, 50
Madeira, 4
Manoh, 122 7 c.
Salt Pond, 6¼
Sydney, 250

Soda Ash—
Boston, 220 tcs. 64 cks. 1,696 bgs.
B. Ayres, 20 brls.
Calcutta, 6
Havana, 4
New York, 133
Philadelphia, 153 355 tcs.
Rio Janeiro, 25
Sydney, 64
Yokohama, 120
Beyrout, 47
Boston, 303
Canada, 122
Malta, 40
New York, 422
Puerto Cabelo, 4
San Francisco, 68
Smyrna, 50
Syra, 15
Philadelphia, 415
Smyrna, 90

Sodium Silicate—
Barcelona, 20 cks.
Lisbon, 16
Santander, 32
Seville, 20
Varna, 10 brls.

Soda Crystals—
Philadelphia, 440 cks.
Rotterdam, 20 brls.
Halifax, 20

Sulphuric Acid—
Iquique, 1 t. £11

Sheep Dip—
Algoa Bay, 3 t. £100

Saltpetre—
Paranagua, 3 t. 10 c. 2 q. £80
Parahyba, 1 14 1 39
Piræus, 3 15 82
Callao, 1 5 27
Bahia, 1 2 28
Pernambuco, 5 14 128

Soda Salts—
Montreal, 1 t. 1 q. £28

Salammoniac—
Alexandretta, 17 c. 2 q. £31
Alexandria, 6 11
Antwerp, 10 3 19
Palermo, 4 7
Patras, 6 1 11
Smyrna, 1 t. 35
Acajutla, 10 20
Galatz, 3 1 5

Sulphur—
Iquique, 34 t. 18 c. £279
Calcutta, 15 90
St. John, N.B., 4 t. 1 c. 2 q. 26
Boston, 75 2 338
" 50 2 225
" 100 400
Madras, 5 5 30

Superphosphate—
Hamburg, 20 t. £75

Sodium Nitrate—
Balcelona, 5 t. 1 c. £48

Saltcake—
Baltimore, 100 t. 7 c. £241
Philadelphia, 37 16 80

Tar—
Bilbao, 30 brls.

Tin Oxalate—
Vera Cruz, 3 brls. £35

Tin Oxide—
Canada, 6 c. 1 q. £43

Zinc Chlorate—
Huelva, 1 t. 19 c. £28

Zinc Muriate—
Bombay, 2 t. 19 c. 2 q. £47

GLASGOW.

Week ending March 27th.

Ammonium Sulphate—
Demerara, 46 t. 1¼ c. £579

Bleaching Powder—
Dunedin, 2 t. 10 c. £20

Copper Sulphate—
Venice. 5 t. 3 c. £102
Trieste, 28 14½ 583
Bordeaux, 16 19½ 362

Dyestuffs—
Indigo
New York, 72 c. £80
Logwood
Dunedin, 40 t. 16¼ c. £449

Manure—
Trinidad. 314 t. 9 c. £3,449
Demerara, 320 17½ 2,422

Oxalic Acid—
Dunedin, 2 t. 15½ c. £116

Paraffin Wax—
Naples, 20 c. £33
Bombay, 1 t. ½ 32
Dunedin, 47¾ 70
Genoa, 16 16 388

Potassium Bichromate—
Bombay, 17¼ c. £31
Rotterdam, 19 cks. 267
Dunedin, 2 t. 13¾ 97
Venice, 25½ 45
Bordeaux, 5 12¾ 180
Genoa, 269½ 597
" 83¼ 141
Amsterdam, 3 cks. 25
Christiania, 1 1

Pitch—
Chili, £192
Amsterdam, 250 t. £250

Potassium Prussiate—
New York, 29 c. £114

Potassium Carbonate—
Bombay, 1 t. 16¼ c. £33. 11s.

Sulphuric Acid—
Madras, 8 t. 7½ c. £40

Soda Ash—
Dunedin, 3 t. 17¼ c. £24

Superphosphate—
Trinidad, 169 t. 6 c. £550

Sodium Bichromate—
Nantes, 5 t. 5¾ c. £160
Bombay, 1 13½ 44
Genoa, 10 t. 285

Sugar of Lead—
Bombay, 3 t. 2¼ c. £59

Tar—
Colombo, 12 t. £26
Madras, 16 34

TYNE.

Week ending Mar. 27th.

Antimony—
Antwerp, 1 t.
Batoum, 3 3 c.
Copenhagen, 6
New York, 17 t.
Rouen, 5

Alkali—
Newfairwater, 31 t. 1 c.
Gothenburg, 8 6
Antwerp, 15 7
Copenhagen, 1 4
Stettin, 4 16
Rotterdam, 19
New York, 176 1
Philadelphia, 200 10

Alumina Sulphate—
Rouen, 7 t 6 c.

Ammonium Carbonate—
Rotterdam, 2 t. 5 c.

Barytes Nitrate—
Rouen, 4 t. 19 c.

Barytes Carbonate—
Antwerp, 11 t.
Rouen, 49 2 c.
New York, 50

Bleaching Powder—
Newfairwater, 42 t. 7 c.
Gothenburg, 5 12
Antwerp, 50 9
Hamburg, 183 14
" 52 15
Genoa, 10
Libau, 186 17
Christiania, 20 3
Aalborg, 10
Malmo, 13 4
Rotterdam, 12 19
" 19 2
Stettin, 262 11
New York, 176 10
Philadelphia, 100 8
Rouen, 5 1

Copperas—
Randers, 1 t. 2 c.
Aalborg, 2 2
Copenhagen, 5 14
Stettin, 3 2

Chloride of Lime—
New York, 140 t. 9 c.

Caustic Soda—
Newfairwater, 14 t. 3 c.

Caustic Soda—
Antwerp, 6 t. 12 c.
Odessa, 70 4
Christiania, 6 14
Dunkirk, 21 16
New York, 158 2
Philadelphia, 29
Rouen, 3 3

Gypsum—
New York, 39 t. 5 c.

Litharge—
Gothenburg, 5 c.
Lisbon, 17
Copenhagen, 5
Riga, 4 t. 18
New York, 10 8
Rouen, 2 17

Manure—
Rotterdam, 40 t.

Magnesia—
Hamburg, 9 c.
Genoa, 1 t. 3

Phosphate—
Drontheim, 10 t.

Potassium Chlorate—
Randers, 1 t.

Soda Ash—
Antwerp, 10 t.
Hamburg, 9 18 c.
Genoa, 10 8
Libau, 20 5
Stettin, 10 6
Newfairwater, 111 4
New York, 166 3

Soda—
Gothenburg, 2 t. 17 c.
Genoa, 8 18
Aarhuus, 11 5
Esbjerg, 3 4
Rotterdam, 3 4
New York, 277 10
" 27
Philadelphia, 471 17

Sodium—
Hamburg, 1 cs.

Sodium Hyposulphate—
Rouen, 7 t.

Sodium Sulphate—
Malmo, 341 t.

Sodium Bicarbonate—
Aalborg, 12 c.

Superphosphate—
Korsoer, 25 t.

Sodium Chlorate—
Hamburg, 1 t. 2 c.

Tar—
Riga, 45 t.

HULL.

Week ending March 25th.

Benzol—
Hamburg, 14 drums.

Bone Size—
Harlingen, 1 ck.

Caustic Soda—
Antwerp, 1 ck.
Christiansund, 29 drms.
Libau, 306

Chemicals (*otherwise undescri*
Antwerp, 4 cks.
Amsterdam, 5
Bremen, 99
Bombay, 1,308
Boston, 70 50 pks.
Christiania, 12 164 slbs.
Copenhagen, 11
Danzig, 41 40
Drontheim, 10
Dunkirk, 20
Gothenburg, 11 15 39
Hamburg, 12 80 120
Hango, 10
Libau, 10 cs. 18 cks. 45 pks.
Lisbon, 13 31
Reval, 2 28
Rotterdam, 42 49 40

Dyestuffs (*otherwise undescrib*
Antwerp, 1 kg.
Bergen, 1 cs.
Boston, 30 cks.
Hamburg, 2 kg. 1 cs. 3 dru
Reval, 36 cks.
Rotterdam, 2 cs. 3 cks. 1
Rouen, 1

Manure—
Drontheim, 335 bgs.
Ghent, 1,638
Hamburg, 687
Stettin, 250

Pitch—
Antwerp, 426 t.
Drontheim, 230

Tar—
Hamburg, 5 cks.

GRIMSBY.

Week ending March 22nd.

Coal Products—
Dieppe, 35 cks. 65 pkgs.

Chemicals (*otherwise undescri*
Antwerp, 45 cks.
Dieppe, 170
Hamburg, 15
Rotterdam, 2

GOOLE.

Week ending March 19th.

Alkali—
Boulogne, 12 cks.

Bleaching Powder—
Rotterdam, 28 cks.

Benzol—
Dunkirk, 46 cks.
Rotterdam, 48

Coal Products—
Boulogne, 1 ck. 2 cs.
Rotterdam, 83 cks.

Chemicals (*otherwise undescri*
Ghent, 15 cks.
Hamburg, 114
Rotterdam, 1 pkg.

Dyestuffs (*otherwise undescrib*
Dunkirk, 4 cks.
Ghent, 4 drums.

Manure—
Antwerp, 580 bgs.
Hamburg, 438
Rotterdam, 2 cks.

Pitch—
Antwerp, 489 t.

PRICES CURRENT.

WEDNESDAY, April 2, 1890.

Prepared by Higginbottom and Co., 116, Portland Street, Manchester.

...es stated are F.O.R. *at maker's works, or at usual ports of shipment in U.K. The price in different localities may vary.*

		£	s.	d.
25 °/ₒ and 40 °/ₒ	per cwt	7/6 & 12/9		
Glacial..	,,	67/6		
, S.G., 2000°	,,	0	12	9
: 82 °/ₒ .. .. nett	per lb.	0	0	6½
.. ..	,,	0	0	3¼
c (Tower Salts), 30° Tw.	per bottle	0	0	8
(Cylinder), 30° Tw.	,,	0	2	11
o° Tw.	per lb.	0	0	2
.. ..	,,	0	0	1¼
. .. . nett	,,	0	0	3½
.. ..	,,	0	1	4
ic (fuming 50 °/ₒ)	per ton	15	10	0
(monohydrate)	,,	5	10	0
(Pyrites, 168°)	,,	3	2	6
(,, 150°)	,,	1	8	0
(free from arsenic, 140/145°)	,,	1	10	0
ous (solution)..	,,	2	15	0
(pure)	per lb.	0	1	0½
:	,,	0	1	2½
hite powdered..	per ton	13	0	0
se lump)	,,	4	15	0
Sulphate (pure)	,,	5	0	0
,, (medium qualities)	,,	4	10	0
n	per lb.	0	15	0
, ·880=28°	per lb.	0	0	3
=24°	,,	0	0	1¼
Carbonate .. .. nett	,,	0	0	3½
Muriate..	per ton	23	0	0
,, (sal-ammoniac) 1st & 2nd ..	per cwt.	37/- & 35/-		
Nitrate	per ton	41	10	0
Phosphate	per lb.	0	0	10½
Sulphate (grey), London	per ton	11	13	9
,, (grey), Hull	,,	11	12	6
il (pure)	per lb.	0	0	11½
ılt ,, ...	,,	0	0	8½
.. ..	per ton	75	0	0
(tartar emetic)	per lb.	0	1	1
(golden sulphide)..	,,	0	0	10
hloride	per ton	8	0	0
arbonate (native)	,,	3	15	0
ulphate (native levigated)	,,	45/- to 75/-		
Powder, 35 %	,,	5	10	0
Liquor, 7 %	,,	2	10	0
e of Carbon	,,	13	0	0
ı Acetate (crystal	per lb.	0	0	6
Chloride	per ton	2	2	6
y (at Runcorn) in bulk	,,	17/6 to 35/-		
Dyes :—				
e (artificial) 20 %	per lb.	0	0	9
a	,,	0	3	9
Products				
ene, 30 % A, f.o.b. London ..	per unit per cwt.	0	1	5
90 % nominal	per gallon	0	3	4
50/90 ,, ...	,,	0	2	7
: Acid (crystallised 35°)	per lb.	0	0	10¼
,, (crude 60°)	per gallon	0	2	8
e (ordinary)..	,,	0	0	2½
(filtered for Lucigen light)	,,	0	0	3
Naphtha 30 % @ 120° C.	,,	0	1	4
Oils, 22° Tw.	per ton	3	0	0
.o.b. Liverpool or Garston..	,,	1	8	0
Naphtha, 90 % at 160°	per gallon	0	1	9
ı Oils (crude)	,,	0	0	2½
Chili Bars)	per ton	47	12	6
lphate	,,	26	0	0
:ide (copper scales)	,,	51	0	0
(crude)..	,,	30	0	0
(distilled S.G. 1250°)..	,,	55	0	0
.. ..	nett, per oz.	8½d. to 9d.		
hate (copperas)	per ton	1	10	0
hide (antimony slag)	,,	2	0	0
et) for cash	,,	14	0	0
ıarge Flake (ex ship)	,,	15	10	0
tate (white ,,)	,,	23	10	0
, (brown ,,)	,,	18	5	0
bonate (white lead) pure	,,	19	0	0
ate	,,	22	0	0

		£	s.	d.
Lime, Acetate (brown)	,,	8	0	0
,, ,, (grey)	per ton	13	15	0
Magnesium (ribbon and wire)	per oz.	0	2	3
,, Chloride (ex ship)	per ton	2	6	3
,, Carbonate	per cwt.	1	17	6
,, Hydrate	,,	0	10	0
,, Sulphate (Epsom Salts)	per ton	3	0	0
Manganese, Sulphate	,,	18	0	0
,, Borate (1st and 2nd)	per cwt.	60/- & 42/6		
,, Ore, 70 %	per ton	4	10	0
Methylated Spirit, 61° O.P.	per gallon	0	2	2
Naphtha (Wood), Solvent	,,	0	3	10
,, ,, Miscible, 60° O.P.	,,	0	4	6
Oils :—				
Cotton-seed..	per ton	22	10	0
Linseed	,,	24	10	0
Lubricating, Scotch, 890°—895°	,,	7	5	0
Petroleum, Russian	per gallon	0	0	5¼
,, American..	,,	0	0	5¼
Potassium (metal)	per oz.	0	3	10
,, Bichromate	per lb.	0	0	4
,, Carbonate, 90% (ex ship)	per ton	18	10	0
,, Chlorate	per lb.	0	0	5
,, Cyanide, 98%	,,	0	2	0
,, Hydrate (Caustic Potash) 80/85 % ..	per ton	22	10	0
,, ,, (Caustic Potash) 75/80 % ..	,,	21	10	0
,, ,, (Caustic Potash) 70/75 % ..	,,	20	15	0
,, Nitrate (refined)	per cwt.	1	3	6
,, Permanganate	,,	4	5	0
Prussiate Yellow	per lb.	0	0	9½
,, Sulphate, 90 %	per ton	9	10	0
,, Muriate, 80 %	,,	7	15	0
Silver (metal)	per oz.	0	3	7⅞
,, Nitrate..	,,	0	2	5½
Sodium (metal)	per lb.	0	4	0
,, Carb. (refined Soda-ash) 48 %	per ton	6	2	6
,, ,, (Caustic Soda-ash) 48 %	,,	5	2	6
,, ,, (Carb. Soda-ash) 48%	,,	5	10	0
,, ,, (Carb. Soda-ash) 58 %	,,	6	15	0
,, ,, (Soda Crystals)	,,	2	17	6
,, Acetate (ex-ship)	,,	15	15	0
,, Arseniate, 45 %	,,	11	0	0
,, Chlorate	per lb.	0	0	6¼
,, Borate (Borax)	nett, per ton.	30	0	0
,, Bichromate	per lb.	0	0	3
,, Hydrate (77 % Caustic Soda)	per ton.	13	0	0
,, ,, (74 % Caustic Soda)	,,	11	15	0
,, ,, (70 % Caustic Soda)	,,	10	10	0
,, ,, (60 % Caustic Soda, white) ..	,,	9	10	0
,, ,, (60 % Caustic Soda, cream) ..	,,	9	0	0
,, Bicarbonate	,,	6	0	0
,, Hyposulphite	,,	5	10	0
,, Manganate, 25%	,,	9	0	0
,, Nitrate (95 %) ex-ship Liverpool ..	per cwt.	0	8	0
,, Nitrite, 98 %	per ton	27	10	0
,, Phosphate	,,	15	15	0
,, Prussiate	per lb.	0	0	7¼
,, Silicate (glass)	per ton	5	7	6
,, ,, (liquid, 100° Tw.)	,,	3	17	6
,, Stannate, 40 %	per cwt.	2	0	0
,, Sulphate (Salt-cake)	per ton.	1	12	6
,, ,, (Glauber's Salts)	,,	1	10	0
,, Sulphide	,,	7	15	0
,, Sulphite	,,	5	5	0
Strontium Hydrate, 98 %	,,	8	0	0
Sulphocyanide Ammonium, 95 %	per lb.	0	0	8
,, Barium, 95 %	,	0	0	5¼
,, Potassium	,,	0	0	8
Sulphur (Flowers)	per ton.	7	15	0
,, (Roll Brimstone)	,,	6	0	0
,, Brimstone : Best Quality	,,	4	5	0
Superphosphate of Lime (26 %)	,,	2	10	0
Tallow	,,	25	10	0
Tin (English Ingots)	,,	94	0	0
,, Crystals	per lb.	0	0	6¼
Zinc (Spelter)	per ton.	21	0	0
,, Chloride (solution, 96° Tw.	,,	6	0	0

No. 151. **SATURDAY, APRIL 12, 1890.** **Vol. VI**

Contents.

Notices.

All communications for the *Chemical Trade Journal* should be addressed, and Cheques and Post Office Orders made payable to—

DAVIS BROS., 32, Blackfriars Street, MANCHESTER.

Our Registered Telegraphic Address is—

"Expert, Manchester."

The Terms of Subscription, commencing at any date, to the *Chemical Trade Journal*,—payable in advance,—including postage to any part of the world, are as follow :—

Yearly (52 numbers) .... **12s. 6d.**
Half-Yearly (26 numbers) .... **6s. 6d.**
Quarterly (13 numbers) .... **3s. 6d.**

Readers will oblige by making their remittances for subscriptions by Postal or Post Office Order, crossed.

Communications for the Editor, if intended for insertion in the current week's issue, should reach the office not later than **Tuesday Morning.**

Articles, reports, and correspondence on all matters of interest to the Chemical and allied industries, home and foreign, are solicited. Correspondents should condense their matter as much as possible, write on one side only of the paper, and in all cases give their names and addresses, not necessarily for publication. Sketches should be sent on separate sheets.

We cannot undertake to return rejected manuscripts or drawings, unless accompanied by a stamped directed envelope.

Readers are invited to forward items of intelligence, or cuttings from local newspapers, of interest to the trades concerned.

As it is one of the special features of the *Chemical Trade Journal* to give the earliest information respecting new processes, improvements, inventions, etc., bearing upon the Chemical and allied industries, or which may be of interest to our readers, the Editor invites particulars of such—when in working order—from the originators; and if the subject is deemed of sufficient importance, an expert will visit and report upon the same in the columns of the *Journal*. There is no fee required for visits of this kind.

We shall esteem it a favour if any of our readers, in making inquiries of, or opening accounts with advertisers in this paper, will kindly mention the *Chemical Trade Journal* as the source of their information.

Advertisements intended for insertion in the current week's issue, should reach the office by **Wednesday morning** at the latest.

THE PROPERTIES OF ALUMINUM, WITH SO INFORMATION RELATING TO THE METAL.*

By Alfred E. Hunt, John W. Langley, and Charles Hall.

PRACTICAL HINTS.

DIPPING AND PICKLING.—Remove the dirt and grease from plates by dipping in benzine. To whiten the metal, leavin the surface a beautiful white mat, the sheet should be first dipped strong solution of caustic potash. This solution should then be dip in a mixture of concentrated acids, two parts nitric acid; then solution of undiluted nitric acid; then in a mixture of vinegar water, equal parts; then washed thoroughly in water and dried as u in hot sawdust.

TO POLISH.—Use a fine white polishing composition or rouge, tripoli, and a sheep-skin or chamois-skin buff, although it is of polished with an ordinary rag buff.

For fine work, to polish aluminum, use a mixture of equal parts, weight of olive oil and rum, made into an emulsion by being well shak together in a bottle. The polishing stone is dipped in this liquid, a the metal is polished, without using, however, too much pressure.

Aluminum may be easily ground by using olive oil and pumice.

The surface of aluminum treated with a varnish of four parts oil turpentine to one of stearic acid, or with a mixture of olive oil a rum shaken into an emulsion, allows an engraving tool to work aluminum as on pure copper.

FOR BURNISHING.—Use a blood-stone or steel burnisher. F hand burnishing, use either kerosene oil or a solution composed of t tablespoonfuls of ground borax dissolved in about a quart of hot wat with a few drops of ammonia added.

FOR LATHE WORK.—The burnisher should wear upon the finge of his left hand a piece of canton flannel, keeping it soaked with ker sene, and bringing it in contact with the metal, supplying a consta lubricant.

Very fine effects can be produced by first burnishing or polishing t metal, and then stamping it in polished dies, showing unpolish figures in relief.

SCRATCH BRUSHING.—Polish or burnish the surface and then u a fine steel scratch brush. A very fine finish is attained by rubbi with ground pumice stone and water.

In spinning aluminum, plenty of oil should be used to prevent t clogging of the tool and to make it cut smooth in the turning and assist in the spinning.

TO SOLDER THE METAL.—Soldering the metal in large surfaces h not been successfully accomplished up to the present. Small surfac of the metal can be readily soldered by the use of pure zinc and Ven tian turpentine. Place the solder upon the metal, with the Veneti turpentine, and heat gently with a blow pipe until the solder is melt It will then be found to have fixed itself firmly to the aluminum. T trouble with this, as with other solders, is that it will not flow on t metal. Therefore, large surfaces are not easily soldered.

In cold rolling aluminum, upon rolls designed for cold rolling ha crucible steel, it has been found possible to reduce aluminum throu the same sections as hard steel; the aluminum required, on t average, five annealings, where the steel required three to satisfactor withstand the same work.

ALLOYS OF ALUMINUM AND COPPER.—Ten per cent. aluminu with 90 per cent. copper (called 10 per cent. aluminum bronze), roll into plates, has an elastic limit of 70,000 to 80,000 pounds per square in a tensile strength of from 100,000 to 120,000 pounds per square inch reduction of area of from 20 to 40 per cent., with an elongation of fr 5 to 10 per cent. in 8 inches. The metal is a beautiful yellow colo and is susceptible of taking a fine polish. One great advantage of t metal is its freedom from corrosion from the action of the air, eit moist or dry, or water, upon it. Its specific gravity, in castings, is ab 7·84, and in rolled sheets about 7·89. Its modulus of elasticity is ab 18,000,000 pounds. In castings, it has a tensile strength of betwe

*Paper read at the meeting of the American Institute Mining Engineers, F 1890.

80·000 pounds per square inch, with a reduction of area of
r cent.
ion tests upon 10 per cent. aluminum bronze ¼-inch
d 2 inches long, gave an ultimate compressive strength of
unds per square inch, the specimens being shortened by
A similar piece of 5 per cent. bronze was shortened to
nd gave an ultimate compressive strength of 153,000 pounds
inch.
cent. aluminum bronze in tension has an elastic limit of
00 pounds per square inch; a tensile strength of about
nds per square inch; a reduction of area of from 30 to 50
Its specific gravity is from 8·20 to 8·30. Two and one-
nt. aluminum bronze has a specific gravity of 8·6.
lting point of 10 per cent. bronze is about 1700° Fahrenheit
higher than that of ordinary brass. The metal shrinks a little
¼" to the foot, or a little less than ordinary brass. It solidi-
apidly from the molten condition, and it is necessary to pour
ickly. The feed gates should be made large enough to pre-
metal freezing. Hot baked sand molds should be used for
Precautions should be taken also to prevent oxidisation of
, for without it, the oxide is carried into the metal, which
ts rolling into sheets. It is well also to bottom pour the
the mold—that is, to cast the metal into a hot ladle, having
n the bottom in direct connection with the gate of the casting the metal to settle, so that the oxide and dross shall come
face, in this way preventing its entering into the casting
The surface of the molten bath shall be kept covered with
charcoal. It is also advantageous to keep the bath covered
x in some cases, although the disadvantage of this is that the
t to cut the sides of the pot and add silicon to the metal. It is
p the metal in an inert atmosphere (casting in a cloud of smoke
e), to prevent the oxidation from the air in the mold attacking
l.

num bronze is an extremely dense, close metal. It can be
t a bright red heat as easily as can wrought iron. In this
differs from all other forms of bronze, which are red-short at
t. The fact that aluminum bronze is malleable at a red-heat,
ds this temperature without change, makes it especially adapt-
blast furnace tuyeres. The metal can be hardened to a
able extent by working without annealing.
neal aluminum bronze, heat to a dull red heat and permit it
gradually.
lloy of aluminum and copper does not volatilise at any ordinary
tures used in fusing it, and, consequently, it can be frequently
without any appreciable change in the chemical constituents
netal. This has great advantages in the economic use of the
s the scrap in castings or rolling can be readily remelted into
f the same quality metal.
inum bronze can be brazed as well as any other metal, using
der :—Zinc, 50 per cent.; copper, 50 per cent.; using ¼ of the
nd ¼ borax and cryolite in equal parts.
pure aluminum, as now manufactured by the Pittsburgh
ion Company, very pure aluminum bronze alloys can readily be
he impurities in the aluminum being reduced to one-tenth their
on being diluted with pure electrolytic lake copper.
following are some of the analyses of aluminum bronzes lately
y the Pittsburgh Reduction Company and by the Scovill Manu-
g Company :—

ind of Alloy.	Per ct. aluminum.	Per cent. copper.	Per ct. graphitoidal silicon.	Per ct. non-graphitoidal silicon.	Per ct. iron.	Specific gravity.
cent. bronze cast-made December 1889	9.20	90.00	0.117	0.370	0.077	7.690
cent. bronze cast-made December 1889	4·70	94.84	0.080	0.320	0.060	8.25
cent. bronze cast-made December 1889	2.35	97.29	0.05	0.26	0.050	8.61
cent. bronze cast-made at New-n	6.32	91.98	0.09	1.09	0.48	8.01

ALLOYS OF ALUMINUM AND IRON.

INUM IN WROUGHT IRON.—The influence of aluminum in wrought iron fluid has been taken advantage of in the well-Mitis process of making castings of wrought iron. Aluminum furnished by the Pittsburgh Reduction Company has been found to be very advantageous, and is largely used in the manufacture of Mitis metal. Aluminum will also increase the tensile strength of wrought iron and improve the fibre, if added either as pure metal or in the form of ferro-aluminum to the molten bath, just before the metal comes to nature in the puddling furnace.

ALUMINUM IN CAST IRON.—The influence of aluminum in cast iron is to turn the combined carbon to graphite—that is, to make the white iron gray, and also to close the texture of the metal (W. J. Keep.) It makes the metal ordinarily more fluid, and it also makes it susceptible of taking a better polish and retaining it free from oxidation. Aluminum will also increase the tensile strength of many grades of cast iron, and aids in obtaining sound castings free from blow-holes.

It has been used in the preparations from one-tenth of one per cent. to two per cent. with good results, with various grades of iron.

ALUMINUM IN STEEL.—The influence of aluminum in steel of high carbon, is to turn the carbon combined into a graphite, and destroys the hardening action of the carbon in tool steel. Aluminum in this sense softens steel. In structural steel of 20 per cent. carbon, a small amount of aluminum, up to 1 per cent. increases the tensile strength without to any great degree decreasing the ductility. By its aid, a higher tensile strength can be obtained in thick sections of steel which have not been subjected to cut little work, than can be otherwise obtained. Although aluminum, with considerable quantities of graphitoidal silicon have been added to steel, no graphitoidal silicon has been found in the steel afterwards, it being all found in the amorphorous or combined state in the resulting steel. The influence of aluminum also is to lower the melting point of the steel, and in this way make it more fluid. Its influence also is to make the ingots of steel more solid and more free from blow-holes. It can be most advantageously used in proportions of from one-tenth of 1 per cent. up to 3 per cent. of aluminum.

TABLE OF TENSILE TESTS OF IRON AND STEEL CONTAINING ALUMINUM.

CHARACTER OF MATERIAL TESTED.	Elastic limit per sq. in.	Tensile strength per sq. in.	Per cent. elongation in 8 in.	Per cent. reduction of area.	Character of fracture.	Analysis. Per ct. aluminum.	Per ct. carbon.	Per ct. manganese.	Per ct. silicon.	Per ct. phosphorus.
All iron muck bar rolled, Union Mills of C., P. and Co., Limited	27,300	46,500	28·00	36·08	Fibrous					
Three parts all iron with two parts aluminum muck..................	26,800	49,000	29·25	40·32	"					
All aluminum muck bar	29,500	55,300	14·00	22·56	"					
Four parts all iron with one part aluminum muck..................	28,510	49,190	29·50	37·34	"					
Open-hearth steel, with 1-10 of 1 per cent. aluminum, ¼-in. thick, plates..........	40,250	58,590	30·10	63·82	Silky cupped	0·04	0·12	0·41	0·02	0·05
Open-hearth steel, with 1-10 of 1 per cent. aluminum, 1-in. thick, plates	47,950	68,360	25·30	52·10	Silky	0·07	0·16	0·48	0·03	0·03
Open-hearth steel, with 1-10 of 1 per cent. aluminum, 2-in. thick, bar steel	46,260	63,760	26·10	47·90	Silky	0·03	0·16	0·39	0·08	0·05
Open-hearth steel, with 1-10 of 1 per cent. aluminum, 2-in. thick, bar steel	45,980	67,560	29·20	54·80	Silky	0·06	0·15	0·55	0·04	0·05

METHODS OF ANALYSIS OF ALUMINUM AND ITS ALLOYS.

As most of the properties of aluminum are very materially affected by the purity of the metal, and as the literature of methods of analyses of aluminum metals and its alloys is very scant, we have considered it appropriate to append to the paper the methods used by the Pittsburgh Testing Laboratory, chemists for the Pittsburgh Reduction Company, in their various analyses referred to, and which have been largely used as a basis of information in the conclusions drawn in this paper.

SCHEME FOR ANALYSIS OF METALLIC ALUMINUM.

FOR IRON.—Dissolve from 1 to 5 grammes of the metal in hydrochloric acid, and reduce with stannic chloride, taking up the excess of stannic chloride with bi-chloride of mercury. Titrate with a standard solution of bi-chromate of potash.

FOR SILICON.—Dissolve one gramme of the metal in aquaregia, composed of equal parts of nitric and hydrochloric acids; evaporate to dryness; take into solution with 5 cubic centimetres of strong hydrochloric acid; add water, boil and filter off the grey residue; wash with dilute hydrochloric acid; ignite and weigh as silicon, plus silica; fuse with carbonate of soda; dissolve in dilute muriatic acid; evaporate to dryness; re-dissolve in dilute hydrochloric acid and filter off total silica. The difference between the weight of the silicon and the silica being the oxygen, which is united with the graphitoidal silicon, from which the amount of graphitoidal silicon can be calculated.

COPPER AND LEAD.—Dilute the filtrate from the silica till free hydrochloric acid is only about 1 per cent. of the solution. Pass sulphuretted hydrogen through the hot solution, and filter off sulphate of copper and lead. Dissolve in 1·2 nitric acid and filter of sulphur. Add 5 cc. sulphuric acid and evaporate to sulphur fumes. Dilute, and in a short time filter from sulphate of lead into a platinum dish. The lead sulphate should be weighed and lead calculated from it. Nearly neutralise with ammonia, leaving about two drops of sulphuric acid (dilute) and precipitate copper by battery.

ALUMINUM.—Receive filtrate from sulphate of copper and lead in 500 cc. flask and make up to the 500 cc. mark with water. Take 50 cc. of the solution (equal to 1-10 gramme of the metal), boil off the free sulphuretted hydrogen; oxidize the iron in the solution with a little bromine water, and add sufficient ammonia to precipitate the sesqui-oxide of iron and alumina, which, after careful washing, ignite and weigh. Subtract the sesqui-oxide of iron calculated from the iron found by titration and the weight of any phosphoric acid which may be present, and calculate the remainder as Al^2O^3, from which the percentage of aluminum can be readily reckoned. (We have not found more than a trace of phosphorus in any of the samples of aluminum manufactured by the Pittsburgh Reduction Company.)

CALCIUM.—A trace may be found in the filtrates from iron and alumina, and is separated in the usual way. The filtrates from the calcium can be readily examined for magnesium. (We do not find any magnesium present in the metal manufactured by the Pittsburgh Reduction Company.)

ALKALIES.—Dissolve one gramme in hydrochloric acid and evaporate in a platinum dish to separate the silica, and proceed as usual until the sesqui-oxides of alumina and iron are precipitated with ammonia. Dissolve this, after washing, in nitric acid, in platinum and evaporate to dryness on the water-bath. Finally, heat over a small flame until the nitrate of alumina is all decomposed. This condition is indicated when the residue is white and friable, and no acid fumes can be detected by ammonia. Treat residue with hot water and get alkali nitrates in solution. This, together with the filtrate from alumina, contains all the alkali present, which should be evaporated, and after separating contained portions of lime and magnesia, weighed as chlorides. (No sodium or potassium is found in the metal manufactured by the Pittsburgh Reduction Company.)

METHODS OF ANALYSIS OF ALUMINUM IRON AND STEEL.

The accurate determination of aluminum when present in minute proportions, as less than ·10 per cent., in iron or steel, has been a very difficult problem, requiring very skilled work to obtain accurate results. The method given by Mr. Andrew A. Blair, in his work upon the Chemical Analysis of Iron, is undoubtedly a correct one, and is to be recommended. It is, however, tedious and requires several days to make the determination. The method of Mr. John E. Stead for the determination of minute quantities of aluminum in iron and steel, has given very satisfactory results, and is as follows:—

Weigh off 11 or 22 grammes of the iron or steel to be examined; the smaller quantity is preferable unless less than ·01 per cent. of aluminum is expected. Dissolve the 11 grammes in 44 cc., strong hydrochloric acid, or the 22 grammes in 88 cc., on the sand bath in a 600 cc., beaker. When dissolved, evaporate to dryness, and redissolve in hydrochloric acid; filter and wash the silica. Let the total bulk of the filtrate and washings not exceed 200 cc.

To separate the alumina from the main bulk of the iron, add 3 cc. of a saturated solution of sodium or ammonium phosphate (of course, free from alumina); then add dilute ammonia till the free acid is neutralized. This point is readily known by a small quantity of phosphate of iron and alumina remaining insoluble after repeatedly shaking the solution. Add hydrochloric acid, drop by drop, till the solution is clear, and then set the beaker on the bath to boil. Add 50 cc. of sodium hyposulphite (of course, carefully tested to be free from alumina) and continue the boiling until the solution does not give off any more sulphurous acid. If there is any doubt whether or not there is excess of hyposulphite, add a few more cc., and if this does not give a precipitate of sulphur, sufficient hyposulphite has been added. The smell Sulphur must not be mistaken for sulphurous acid, as the former always present. One hour's boiling will eliminate all the sulphu acid. Filter on washed filter paper (washed with hydrochloric aci and wash with boiling water to free it from all soluble iron and oth salts. Dissolve all the soluble matter out of the precipitate on t filter by pouring over it 5 cc. of hydrochloric acid and 5 cc. of boili water, allowing the solution to collect in the beaker in which t alumina was originally precipitated. Wash out of the filter all solub matter with as little wash water as possible. The matter left on t filter should be all sulphur and may be thrown away.

The solution must now be transferred to a platinum dish from t glass beaker, and evaporated to dryness over a beaker of boilin water. When dry, add two grammes of pure sodium hydrate, fr from silica (great care should be taken with reference to this re-age to see that it is perfectly free from alumina), place the sodium on t bottom of the dish, then add about 1 cc. of boiling water to partial dissolve the hydrate of soda. Place the dish over a flame of a Fletche rose burner until the mass is in a state of tranquil fusion. Allow cool, and add 50 cc. of boiling water, and place the dish over a Buns burner and boil for five minutes. More water must be added so th the total bulk measures exactly 110 cc. Filter off the insolub contents through a dry filter paper, which has been previously washe with acid to remove all alumina. The first portions filtered will b unclear, and must therefore be refiltered.

When 100 cc. (equivalent to 10 grammes of the 11 of steel taken i the first place) have passed through, measure off that quantity exact and reject the insoluble matter and the remainder of the solutio Neutralise the caustic soda with hydrochloric acid until the solutic reddens blue litmus paper. Add 3 cc. of sodium phosphate and the hyposulphite of soda in large excess. Boil till all sulphurous acid h been expelled, and add 3 cc. ammonium acetate, boiling for a fe minutes longer and filter through a washed filter. Wash well wi boiling water until the washings are free from chlorides, and ignite an weigh.

The precipitate is found to consist of $AlPO_4$, containing 22·36 pe cent. of aluminum.

Determinations can readily be made by this method in 12 hour time. Very accurate results can be obtained by this method, described in the Journal of the Society of Chemical Industry, date December 31, 1889.

Especial care must be taken that all re-agents are pure, and that t filter papers are washed free from alumina.

METHODS OF PREPARATION OF ALUMINA—THE ORE OF ALUMINU

As considerable inquiry has been made as to the ore from whi aluminum is made, it may be well to state here that aluminum is no being manufactured from the oxide, alumina, which is purified chem cally from silica and iron, from the native bauxite mineral. Bauxite found in considerable quantities and fully as pure in quality as the be foreign mineral in the states of North Carolina and Georgia, and the are vast deposits of it in Ireland and Northern France. The avera composition of bauxite is about as follows:—

	Al^2O^3	SiO^2	Fe^2O^3	H^2O	TiO^2	CaC
White from Beaux, France	58·10	21·70	3·00	14·00	3·30	trac
Brownish red from Revest, France	57·60	2·80	25·30	10·80	3·10	0·4
Oolitie from Allaveh, France	55·40	4·80	24·80	11·60	3·20	0·2
White from Georgia	58·00	5·20	2·30	30·10	3·25	0·3
Redish brown from Georgia	58·70	4·80	20·50	13·20	2·10	0·4

*Authority: Dana.
‡Authority: Hunt and Clapp.

The two methods of purification of bauxite are as follow:—

Bauxite, or a rich clay, chosen as free from iron as possible, roasted at a low red heat, and afterwards is treated with sulphu acid, which combines with the alumina present, forming sulphate alumina. This is readily dissolved by water, leaving the great bulk silica and iron behind. The solution of sulphate of alumina allowed to settle, the supernatant liquid syphoned off into an evapo ating tank, and evaporated to dryness. The dry sulphate of alumi is calcined at a red heat, driving off the sulphuric acid, leaving, as residue, anhydrous alumina. This calcination seems to be as easy the calcination of alumina hydrate, and there appears to be no dif culty in condensing the volatilised sulphuric acid, which can be us over again. This process is easier, on a laboratory scale, than t soda carbonate method, which is about as follows:—

Bauxite is fused with carbonate of soda, in a reverberatory furna The fused mass is lixiviated with water, which dissolves aluminate soda, which is decanted off. The solution of aluminate of soda decomposed by carbonic acid gas, which forms carbonate of sod which remains in solution, and the alumina hydrate is precipitate This alumina hydrate is afterwards washed repeatedly with wat dried, and calcined at a red heat for a considerable time, which for anhydrous alumina.

CHEMICAL ENGINEERING*

XXVI.

WEIGHING AND MEASURING.

HING also includes measuring, for what is weight a measure of the gravitation of bodies towards the gravity. Weight, so far as it relates to the force of n, is absolute, but weights and measures, as we d them, are both comparative. It is not with the neasuring, however, that we are just now interested, r a description of standards and appliances adapted uirements of works' management and necessarily to Engineering.

stated, all measurements are comparative and have to some standard or settled unit; atmospheric (*a*) is based on the pressure of a column of liquid; ure (*b*) is measured by comparison with the lineal n of a fluid; time (*c*) is reckoned by the fractional he mean solar day, such as the hour, minute, or measures of extension (*d*) whether linear, square, or e based on one fixed unit and standard of length d at a fixed temperature. The density of a body is contained in a unit volume, and the specific gravity relative density compared with some standard sub- adopted as a unit, and measured at some fixed ture and pressure. Then we have the combined s of extension, weight, time, and temperature, such quired in the measurement of mechanical work.

nly standards or units we need consider are the or metric, and the English or feet and pounds. The ystem has completely driven the English system from ratory, but we fear it will be a long time before it is in ordinary works' routine. It is, however, pleasing that a very simple relationship exists between the and French systems when large weights and volumes be considered. Roughly, 1,000 kilogrammes make lish ton, and 100 kilos. or centner, or metric quintal, e hundredweight; it is therefore very easy to compare s in this measure with workings in English measures. strange coincidence, grammes per litre of the French ds is equivalent to ounces per cubic foot in English e, so that experiments in the laboratory may be compared with large scale working without any tion whatever. On the English standard, the cubic water at 62° F., with the barometer at 30 inches of y pressure, weighs 252·458 grains, when weighed brass weights, so that the gallon contains 277·274 nches, and weighs 10lbs. The cubic foot of water ently weighs 62·321lbs. For all practical purposes, ic foot of water is taken at 1,000 ounces. The correct in air is 997·136 ounces, and it may be thought that rection would upset the comparison with the metric d we have already made. This is not the case, how- e kilogramme is a cubic decimetre of water, measured mperature of maximum density, 39° F. When, how- e water is at 62° F., and weighed against brass weights he weight is 998·717 grammes.

ing back to our standards or units, we have (*a*) the heric pressure measured by the height of a column of upported; that liquid is generally mercury, though nd glycerine have also been employed. Measure- re given in millimetres on the French method, and in and tenths on the English system, so that besides table for easy conversion of one into the other, more is required.

HE BAROMETER IN INCHES AND MILLIMETRES.

a.		Inches.	Ins.		Mm.
[illegible]	....	0·03937	1	....	25·4
[illegible]	....	0·07874	2	....	50·8
[illegible]	....	0·11811	3	....	76·2
[illegible]	....	0·15748	4	....	101·6
[illegible]	....	0·19685	5	....	127·0
[illegible]	....	0·23622	6	....	152·4
7	....	0·27559	7	....	177·8
8	....	0·31496	8	....	203·2
9	....	0·35433	9	....	228·6
10	....	0·39370	10	....	254·0
50	....	1·96852	20	....	508·0
700	....	27·55930	30	....	762·0

*e of lectures delivered by Mr. George E. Davis at the Manchester School. *All rights reserved.*

(*b*.) Temperature is differently reckoned on the French and English systems; the measure is made by comparison with the dilatation of a column of mercury or any other convenient fluid, but the graduations of the scale are fixed by the freezing and the boiling points of water, or rather, the boiling point and the melting of ice. In the Fahrenheit instrument the distance traversed by the expansion of a column of mercury between these two points is divided into 180 parts, while in the Centigrade thermometer the distance is divided into 100 parts.

It is a great pity the Fahrenheit thermometer cannot be weeded out from all our manufacturing establishments, but until it is, the following table will enable comparisons to be made between the two systems:—

TEMPERATURE TABLE.

To convert ° Fahrenheit into ° Centigrade, subtract 32 and find the number corresponding to the residue.			To convert ° Centigrade into ° Fahrenheit, find the number and add 32.		
1° F	....	0·5555	1° C	....	1·8
2°—	....	1·1110	2°—	....	3·6
3°—	....	1·6665	3°—	....	5·4
4°—	....	2·2220	4°—	....	7·2
5°—	....	2·7775	5°—	....	9·0
6°—	....	3·3330	6°—	....	10·8
7°—	....	3·8885	7°—	....	12·6
8°—	....	4·4440	8°—	....	14·4
9°—	....	4·9995	9°—	....	16·2
10°—	....	5·5555	10°—	....	18·0

(*c*.) The French system, included in its inception a centesimal measurement of time, and centesimal divisions of the circle; the day was to have been divided into 40 hours, each hour into 100 minutes, and each minute into 100 seconds, while the circle was divided into 400 grades, each grade into 100 minutes, and each minute into 100 seconds. The world, however, did not take kindly to this innovation, and the method of division derived from the old Arabian astronomers and navigators, is to-day the time and circle standard: twenty four hours to the day, while the hour and the minute each has 60 divisions. It is not necessary to give any tables for the conversion of centesimal time or divisions of the circle, to sexagesimal time or angle, and perhaps it would have been unnecessary to notice the system, were it not for the fact that the centesimal division of the circle is still explained in some elementary mathematical treatises.

(*d*.) Measures of extension are too firmly rooted amongst us to undergo much alteration now; the foot and the metre form the basis of nearly all the commercial transactions on the surface of the globe, the square foot and the square metre measure surfaces, while the cubic foot and the cubic metre measure solids and gases. In the manufacturing processes of this country, the measures of capacity and weight are the gallon, cubic foot, and pound; on the French system they are the litre, the hectolitre and the kilogramme, so that it will be necessary to give a few tables for the ready conversion of the one into the other, in order that metric measurements may readily be converted into English, and *vice versa*.

METRES AND MILLIMETRES TO FEET AND INCHES.

Metres.		Feet.	Millimetres.		Inches.
1	....	3·281	1	....	0·039
2	....	6·562	2	....	0·078
3	....	9·843	3	....	0·118
4	....	13·124	4	....	0·157
5	....	16·404	5	....	0·196
6	....	19·685	6	....	0·236
7	....	22·966	7	....	0·275
8	....	26·247	8	....	0·315
9	....	29·528	9	....	0·354
10	....	32·808	10	....	0·393

FEET AND INCHES TO METRES AND MILLIMETRES.

Feet.	Metres.	Inches.	Millimetres.
1	0·3048	1	25·4
2	0·6096	2	50·8
3	0·9144	3	76·2
4	1·2192	4	101·6
5	1·5240	5	127·0
6	1·8288	6	152·4
7	2·1336	7	177·8
8	2·4384	8	203·2
9	2·7432	9	228·6
10	3·0480	10	254·0

FRACTIONS OF AN INCH TO MILLIMETRES.

Inch.	Millimetres.	Inch.	Millimetres.
$\frac{1}{16}$	1·58	$\frac{9}{16}$	14·29
$\frac{1}{8}$	3·17	$\frac{5}{8}$	15·87
$\frac{3}{16}$	4·76	$\frac{11}{16}$	17·46
$\frac{1}{4}$	6·35	$\frac{3}{4}$	19·05
$\frac{5}{16}$	7·94	$\frac{13}{16}$	20·64
$\frac{3}{8}$	9·52	$\frac{7}{8}$	22·22
$\frac{7}{16}$	11·11	$\frac{15}{16}$	23·81
$\frac{1}{2}$	12·70	1·0	25·40

The foregoing tables will enable plans and specifications, drawn on either of the two systems under consideration, to be compared with the utmost facility. It may be advisable now to add several tables of square measure, as square metres (m^2) and square feet are very common dimensions.

METRIC SQUARE TO ENGLISH SQUARE.

Sq. Metres.	Sq. Feet.	Sq. Yards.
1	10·764	1·19
2	21·528	2·38
3	32·292	3·57
4	43·056	4·76
5	53·820	5·95
6	64·584	7·14
7	75·348	8·33
8	86·112	9·52
9	96·876	10·71
10	107·640	11·90

ENGLISH SQUARE TO METRIC SQUARE.

Sq. feet.	Sq. metres.	Sq. yards.	Sq. metres.
1	0·0929	1	0·8361
2	0·1858	2	1·6722
3	0·2787	3	2·5083
4	0·3716	4	3·3444
5	0·4645	5	4·1805
6	0·5574	6	5·0166
7	0·6503	7	5·8527
8	0·7432	8	6·6888
9	0·8361	9	7·5249
10	0·9290	10	8·3610

The HECTARE is 10,000 square metres and is equal to 2·47 acres.

The English measures of capacity and solidity are the gallon and the cubic foot, and though the grain measure is used in the laboratory in some instances, yet for all scientific purposes the cubic centimetre and the litre are employed. On the metric system the large scale measure of capacity is the hectolitre and the measure of solidity the cubic metre, it will, therefore, be necessary to have a table for the ready computation of these in both systems.

CUBIC METRES AND CUBIC FEET.

Cubic metres.	Cubic feet.	Cubic feet.	Cubic metres.
1	35·317	1	0·02831
2	70·634	2	0·05663
3	105·951	3	0·08494
4	141·268	4	0·11326
5	176·585	5	0·14157
6	211·902	6	0·16989
7	247·219	7	0·19820
8	282·536	8	0·22652
9	317·853	9	0·25483
10	353·170	10	0·28315

LITRES, HECTOLITRES, AND GALLONS.

Hectolitres.	Gallons.	Cubic feet.	Bushels.	Litres.	Gallons.	Cubic inches.	Gallons.	Hectolitres.	Litres.
1	22·009	3·53	2·7512	1	0·2200	61·028	1	0·0454	4·541
2	44·018	7·06	5·5024	2	0·4401	122·056	2	0·0908	9·082
3	66·027	10·59	8·2536	3	0·6602	183·084	3	0·1362	13·623
4	88·036	14·12	11·0048	4	0·8803	244·112	4	0·1816	18·164
5	110·045	17·65	13·7560	5	1·1004	305·140	5	0·2270	22·705
6	132·054	21·18	16·5072	6	1·3205	366·168	6	0·2724	27·246
7	154·063	24·71	19·2584	7	1·5406	427·196	7	0·3178	31·787
8	176·072	28·24	22·0096	8	1·7607	488·224	8	0·3632	36·328
9	198·081	31·77	24·7608	9	1·9808	549·252	9	0·4086	40·869
10	220·097	35·30	27·512	10	2·2009	610·280	10	0·4541	45·410

[A] SIMPLE STORAGE BATTERY.

ıalf-round porous cups and a round glass jar large enough two porous cups to stand in upright. Get two plates of e-sixteenth of an inch thick, wide enough to fit the half-. the porous cups and deep enough to come an inch or so ıp edge of the cups and jar. Solder a stout copper wire or to each lead plate at the top. Place the lead plates in the the cups nearly full with a paste made of red lead mixed ion of sulphate of soda thin enough to run like a cement. ır containing the two cups should be filled to within half an p of cups with sulphuric acid and water, about one d eight parts of water. One plate should be marked in charging, the currents will be correctly connected. y be charged by attaching to a series of a dozen sulpper cells for twenty-four hours, or from a dynamo. It ıys be charged in same direction, and it will improve by ıargings. A wooden cover may be fitted to the glass jar, ation of the fluid should be replenished by adding water. re cells of this battery will work small motors, lamps, and oils, and if thoroughly charged will retain a large volume of for considerable time. After once being well charged, four of sulphate of copper battery will recharge it.—*Journal of* *ıph.*

AGRICULTURAL CHEMISTRY.

THE ESSEX FIELD EXPERIMENTS.

ome years past the Essex Agricultural Society have been ring out experiments for the purpose of determining the value al manures for special crops, these being in the charge of Mr. Dyer, F.C.S., and Mr. Edward Rosling, of Chelmsford. The those carried out in 1889 is now being issued to members. two sets were carried through, the first consisting of the oats on the plots occupied by the experimental mangold 1888, and the second of a new set of experiments on the of mangolds. In each case the results are very instructive.

oat experiments, as in 1887 and 1888, no manures at all were o the land for the oat crop itself, the object of the experiments ot to ascertain the best mode of manuring oats, but to test the y effect upon this crop of the course of manuring followed for golds which occupied the ground during the previous year. as been afloat among farmers a general impression that the anuring of mangolds with artificials, although it may yield a p, does so at the expense of the condition of the soil, and that crop suffers in consequence. The experiments on this point and 1888 showed very clearly that such was not the case, t, generally, the heaviest mangolds were followed (withher manuring) by the heaviest oat crops, and that even ressings of nitrate of soda on mangolds told well on the ng oats, either in virtue of a portion of the nitrate remaining soil unexhausted or in virtue of the organic "dressing" l by the increased mass of decaying rootlets and tops in the or nitrate-fed, mangold plots, or from both these causes. From age results it will be seen that the use of dung alone gave an of less than half a ton of mangolds (owing largely to partial nd consequent irregularity of plant) and an increase of 12½ of oats and 6 cwt. of straw. The use of artificials without f soda, in addition to dung, gave nearly three tons of maner dung alone, but not an appreciably larger oat crop. Dressh including 2 cwt. of nitrate of soda, in addition to dung, gave ur tons mangolds more than dung alone, three bushels more of l 4 cwt. more straw. When the artificial dressing added to included 4cwt of nitrate of soda per acre, the increase over ne was on the average nearly 6½ tons of mangolds and 12 of oats. The plots without dung, but with artificial dressings, luding 4 cwt. nitrate of soda per acre, gave on an average 3½ nangolds, about 10¼ bushels of oats, and 5 cwt. of straw per re than the average of the unmanured plots. Although these re not so satisfactory to regard as those of the previous year, the irregularity of the mangold crop, yet they clearly conexperience already arrived at that the heavy manuring of manth artificials does not injuriously affect the land for the next crop, he contrary, enriches it.

ollowing table shows the cost of the manures, taking dung at ton; also the value of the increased yield of mangolds, l at 10s. per ton, of oats at 2s. per bushel, and of straw at £2.

Plot.	Manure per Acre.	Cost. £	s.	d.	Value of increase in crops as compared with no manure. £	s.	d.
*F.	12 tons dung, no artificials	3	0	0	2	1	10
J.	12 tons dung, 3 cwt. superphosphate	3	7	6	3	7	0
D.	12 tons dung, 2 cwt. nitrate of soda	4	0	0	4	10	9
C.	12 tons dung, 3 cwt. superphosphate, 2 cwt. nitrate of soda	4	7	6	4	12	6
M.	12 tons dung, 4 cwt. nitrate of soda	5	0	0	7	1	0
L.	12 tons dung, 3 cwt. superphosphate, 4 cwt. nitrate of soda	5	7	6	7	0	0
K.	12 tons dung, 4 cwt. basic cinder, 4 cwt. nitrate of soda	5	7	0	7	7	3
B.	12 tons dung, 2 cwt. guano	3	14	6	6	19	0
E.	12 tons dung, 2 cwt. guano, 2 cwt. nitrate of soda	4	14	6	5	8	0
*G.	12 tons dung, 4 cwt. guano	4	9	0	1	5	8
*H.	12 tons dung, 4 cwt. guano, 2 cwt. nitrate of soda	5	9	0	5	3	0
I.	12 tons dung, 4 cwt. guano, 4 cwt. nitrate of soda	6	9	0	5	14	0
R.	No dung, 4 cwt. nitrate of soda	2	0	0	3	3	6
N.	No dung, 4 cwt. basic cinder, 4 cwt. nitrate of soda	2	7	0	4	2	6
*P.	No dung, 3 cwt. superphosphate, 4 cwt. nitrate of soda	2	7	6	1	13	6
Q.	No dung, 6 cwt. guano, 4 cwt. nitrate of soda	4	3	6	4	5	10

*On these plots the mangold plant in 1888 partially failed.

It will be seen that on the two crops dung alone made a loss, and dung with superphosphate about paid expenses. Dung with 4 cwt. of nitrate of soda made a profit of £2. 1s. per acre, and when to the dung and 4 cwt. of nitrate 3 cwt. of superphosphate were added the profit fell to £1. 12s. 6d. per acre. Superphosphate did not answer quite so well as basic cinder. Dung and 2 cwt. of guano gave a profit of £3. 4s. 6d. per acre. The heavy dressing of 4 cwt. of guano and 4 cwt., of nitrate, in addition to dung, gave a loss on the two crops. On the no-dung plots a better result was obtained when basic cinder was coupled with nitrate—viz., £1. 15s. 6d. per acre—than when nitrate only was used, which gave a total profit of £1. 3s. 6d. Whether superphosphate or basic cinder was the better could not be told, as the superphosphate plot was not reliable, owing to the failure of the plant. In the report on the 1887 and 1888 experiments was a table of the dunged plots, in which from the cost of manuring was deducted the cost of the dung, and from the value of increase in crops was deducted the value of the increase produced by dung alone. In this way the profit on the use of the artificials was made more clear and striking. A similar table is given for this series of experiments, but it is right again to remind the reader that the plot which had dung alone was a partial failure as regards the mangold crop, so that its yield does not properly show with how much manuring power the dung should throughout be credited:—

Plot.	Manure per acre.	Cost of artificials. £	s.	d.	Value of increase in crops as compared with dung alone. £	s.	d.
J.	Dung and superphosphate	0	7	6	1	4	5
D.	Dung and 2 cwt. nitrate of soda	1	0	0	2	9	8
C.	Dung, 3 cwt. superphosphate, 2 cwt. nitrate of soda	1	7	6	2	11	0
M.	Dung, 4 cwt. nitrate of soda	2	0	0	4	19	1
L.	Dung, 3 cwt. superphosphate, 4 cwt. nitrate of soda	2	7	6	4	18	1
K.	Dung, 4 cwt. basic cinder, 4 cwt. nitrate of soda	2	7	0	5	5	4
B.	Dung and 2 cwt. guano	0	14	6	4	17	0
E.	Dung, 2 cwt. guano, 2 cwt. nitrate of soda	1	14	6	3	6	1
*G.	Dung, 4 cwt. guano	1	9	0	decrease		
*H.	Dung, 4 cwt. guano, 2 cwt. nitrate of soda	2	9	0	3	1	2
I.	Dung, 4 cwt. guano, 4 cwt. nitrate of soda	3	9	0	3	12	2

* On these plots the mangold plant in 1888 partially failed.

A new scheme of mangold experiments was devised in 1889. The conclusions had in earlier experiments been weakened by partial failures of plant, owing to irregularities in the land, which caused drought to affect some parts of the field more seriously than others. To diminish the effect of discrepancies from this cause, it was resolved to have three plots under each kind of treatment to be adopted. At the same time this necessarily curtailed the scope of the experiments in other directions, for otherwise the result of the multiplication of plots would have been so great as to render them unmanageable. The land chosen was in another part of the field, in which the experimental mangolds of 1887 were grown. The roots were grown on the "ridge" system, as in previous years, the manure being sown in furrows, which, split by the plough, yielded new alternate ridges, enclosing the manure, on which ridges the seed was drilled. The sowing took place between May 6th and May 13th. The manures used were, in addition to farm-yard dung, Peruvian guano, at £7. per ton, containing nitrogen equal to 4·50 %, ammonia and phosphoric acid equal to 50·8 %, phosphate of lime, and nitrate of soda of high commercial quality, costing £9. 10s. per ton. The guano was in each case applied just before sowing. Where 4 cwt. per acre of nitrate of soda was used, the first cwt. was sown also at seed-time, the remainder being put on in successive top dressings of 1 cwt. each. Where only 2 cwt. of nitrate was used both cwts. were put on as top dressings. The dates of top dressing with nitrate were, for the 2 cwt. plots, July 20th and August 19th, and for the 4 cwt. plots, July 20th, August 19th, and September 5th. There was a partial failure of plant in the upper part of the field, owing to an attack of grub. There were three plots under each manurial treatment adopted. The following table gives the average results:—

Manure per acre.	Mangolds. Yield per acre.		
	tons.	cwts.	qrs.
No manure (mean yield of D, J, and P.)............	9	14	0
12 tons dung (mean yield of A, G, and M)..........	10	10	2
12 tons dung and 2 cwt. nitrate of soda (mean yield of B, H, and N.)...............................	14	0	3
12 tons dung and 4 cwt. nitrate of soda (mean yield of C, I, and O.)	15	8	3
No dung, 6 cwt. guano, and 2 cwt. nitrate of soda (mean of plots E, K, and Q)....................	14	10	1
No dung, 6 cwt. guano, and 4 cwt. nitrate of soda (mean of plots F, L, and R.)	15	13	0

In the next table is given—(1) The average increase of crop under each treatment, obtained by deducting the mean yield of unmanured plots from that of each set of manured plots. (2) The cost of manure, taking dung, as in previous reports, at 5s. per ton; and (3) the value of the grain in mangolds at 10s. per ton:—

Manure.	Increase in mangold crop.			Cost of Manure.			Value of increase.		
	tons	cwts.	qrs.	£	s.	d.	£	s.	d.
12 tons dung only	0	16	2	3	0	0	0	8	3
12 tons dung and 2 cwt. nitrate..................	4	6	3	3	19	0	2	3	4
12 tons dung add 4 cwt. nitrate..................	5	14	3	4	18	0	2	17	4
No dung, 6 cwt. guano, and 2 cwt. nitrate...........	4	16	1	3	1	0	2	8	1
No dung, 6 cwt. guano, and 4 cwt. nitrate...........	5	19	0	4	0	0	2	19	6

In each case the manure shows a greater cost than was returned in the crop, at the value taken, though the previous experience renders it safe to predict that the difference will, in most cases, be more than realised in the increase in the following grain crop.

SCHMIEDBARENGUSS—A NEW METAL.—This is a composition of pig iron, wrought iron, copper, and aluminum, bronze alloy and a flux (for which several letters-patent are now pending). It is produced direct in the cupola without annealing. It is a weldable, malleable, soft steel casting to all appearances. It has remarkable ductility and strength. It is produced at less cost than malleable iron or steel castings. It is free from detrimental impurities. It is very tenacious and homogeneous. It is distinguished from all other castings, and has remarkable characteristics. The following is a copy of test, already made by the United States Government officials:—Size of test bar ·225 × ·512, subjected to tensil strain of 26,200, equal to 200,766 lbs. per square inch *without breaking the test bar.* Transverse ultimate strength broke at 7,232 lbs. per square inch; length between supports, 12 inches. 29 test bars that were broken showed a reduction of area of 28 to 67 per cent.; elongation per centum 52·10; resilience 37 to 45 per cent.; compressive strength per square inch, 200,853 pounds; specific gravity, 7·16.

EXPERIMENTS ON A STEAM ENGINE, THE CYLINDER OF WHICH IS HEATED FROM THE EXTERIOR BY GAS FLAMES.

BY BRYAN DONKIN.

THE object of the following experiments is to ascertain to what degree the condensation of steam in the cylinder can be avoided, and what economy of steam may be effected by heating the cylinder externally by means of gas flames, so as to raise the temperature of its walls considerably.

The engine employed was horizontal, and with a single cylinder. The diameter of the cylinder was 0·214 metres, the length of stroke 0·405 metres. The rod of the piston was directly united to a rotary exhauster employed in compressing air so as to use up the force supplied by the motor.

One series of experiments was made without condensing the steam at the exit of the cylinder, and a second with condensation. The amounts of water vapourised were carefully measured in all cases, and indicator diagrams frequently taken at either end of the cylinder.

The arrangement of the gas jets was a little defective, but the necessary precautions were taken throughout the experiments to obtain exact results.

Many experiments were made successively with and without flames. When the latter were employed, the amount of gas consumed was varied in different experiments. The Bunsen burners were placed very close to the exterior of the cylinder, and the flames were in contact with about one-half of its walls. A piece of asbestos cardboard, arranged concentrically with the wall of the cylinder, compelled the gases to circulate round it, and prevented loss of heat by radiation. The expansion of the steam was kept the same throughout the experiments.

RESULTS OF THE EXPERIMENTS.

The results are recorded in the following tables. It will be seen that the temperature of the cylinder, when gas flames were employed, was much above that of the steam admitted.

A considerable economy of steam is effected by the use of exterior gas heating, but the value of the gas consumed must, of course, be deducted, so that the final result, from a financial point of view, will vary from place to place, according to the prices of fuel and of gas.

Since the completion of these experiments a better arrangement of gas-burners for heating the cylinder has been adopted. Professor Dwelshauver Dery, of the University of Liege, has carefully calculated the results obtained by this new way of heating the cylinder, and has compared them with those given by two other arrangements intended to reduce condensation in the cylinder. This was attempted by Willan by increasing the speed of the piston, whilst the same object was aimed at in Hirn's experiments by superheating the steam. The comparison of the efficacy of these three methods is interesting.

PRINCIPAL RESULTS OF THE EXPERIMENTS.

TABLE I.

Results of experiments made on a steam engine of 6 h.p., *without condensation*, with and without external heating of the cylinder by means of gas-burners, made at Bermondsey in 1888:—

	Without Gas Flames.	With Gas Flames.		
Experiments..........	3	6	9	8
Dates in 1888	26/7	28/7	31/7	31/7
Duration of experiment..	4 hours.	4 hours.	—	3 hours.
Water evaporated—per H.P., per hour, in kilos.	20·139	15·421	15·337	15·337
Horse power—indicated on the diagrams	6·25	6·17	6·33	6·31
London gas—burned at Bunsen burners—litres per hour	0	1386	1132	991
Temperatures—at the exterior of the cylinder in degrees C.	120	212	—	168·8
Pressure of steam above that of the atmosphere:				
In the boiler, in kilos.	2·50	2·46	2·50	2·42
In the cylinder ,, ...	2·23	2·45	2·45	2·39
Velocity—revolutions per minute	90·73	90·79	89·7	90·1
Total revolutions	21,776	21,790	16,146	16,218
Steam used per H.P. from diagram, with 8/10 admission ..	15·874	15·600	15·457	—
Pressure utilised of steam introduced	80	101	101¼	

TABLE I.—The engine, without condensation, with and without gas.—A comparison of experiments 3 and 9 shows that there is a saving of 24 % in the amount of water vaporised per horse power per hour, when the cylinder is heated, and that this result is obtained with the same pressure at the boilers, the same expansion, and the same rate of working. It will be observed that the exterior temperature of the wall of the cylinder is much greater when the burners are lighted than when they are extinguished. The initial pressure of steam in the cylinder indicated by the diagrams is also about 10 % higher with gas than without; the weight of the steam utilised, according to the diagram, with 8/10 expansion, is only 80 per cent. of the total quantity admitted, so that it follows from the experiments that, without gas, 20% of the steam is condensed, while with gas, the diagrams show that all the water vaporised acts in the cylinder, and there is no loss.

TABLE II.

Results of experiments on a steam engine of 6 h.p. *with condensation*, with and without external heating of the cylinder by gas burners, made at Bermondsey in 1888:—

	Without flames.	With flames.	
Experiments	4	3	2
Dates, 1888	9/8	9/8	8/8
Duration	3 hours	3 hours	3 hours
Water evaporated—kilos. per h.p. per hour	17·461	13·924	13·743
H.P. shown by diagrams	5·61	6·05	5·47
London gas—burned by Bunsen burners in litres per hour	0	990	792
Temperatures—at the exterior of cylinder in degrees C.	101°	114°	115°
Pressure of steam above that of air in boiler, in kilos.	2·46	2·42	2·42
Pressure of steam above that of air in cylinder, in kilos.	1·03	1·07	0·915
Velocity—revolutions per minute	92·27	93·66	90·16
Total revolutions	16,680	16,680	16,230
Steam utilised—steam utilised per h.p. according to diagram with 8/10 admission	12·699	12·382	12·382
Steam utilised—percentage utilised of total steam admitted	73½	89	90

Table II.—Engine with condensation, with and without gas.

Experiments 2 and 4 show an economy of 21 per cent on the amount of water vaporised per h p. and per hour, when the cylinder is heated, other conditions being practically the same. The external temperature of the cylinder is also higher with gas.

The initial pressure in the cylinder is greater with gas. Calculating from the diagram it is found that only 73 per cent. of the steam admitted is utilised when the cylinder is not heated, whilst 90 per cent. is utilised when heat is applied, a saving of 17 per cent. being thus effected. A comparison of Tables I. and II. shows that the useful effect of heat applied to the cylinder is greater without than with condensation. Experiments in which the delivery of the burners employed was increased, show that after the cylinder has reached a certain temperature, increase of heat does not increase the economy. It is probable that the excess of heat is radiated to the exterior, and is partially absorbed by the extractor connected with the cylinder.

Some experiments were made at a rate of 45 revolutions, which also showed an economy when heat was applied, but to a much less extent than when the engine was working at 90 revolutions.

Messrs. Woodhouse & Rawson, Limited, have decided to open a branch warehouse and show room in Johannesburg, which will be under the management of Mr. R. Lewis Cousens, M.I.E.E., late Electrician to the Kimberley Town Council. Special attention will be paid to the carrying out of such work as the lighting of cities, hotels, mines, docks, railways, etc., the transmission of power for running stamps, tramways, light railways, telpher transport, etc., and the depot itself will be stocked with a most complete assortment of engineering and electrical requisites of the latest types.

INCREASING PRODUCTION OF SALT IN AMERICA.—Kansas promises to develop into an important competitor of New York, Michigan, Ohio and West Virginia, as a producer of salt. According to the *Engineering and Mining Journal* but one small plant in Kingman County, Kan., produced any salt in 1888, while in the following year seventeen companies were in operation, the total output for 1889 being 19,056 tons in bulk and 547,254 barrels. This was the production of four counties. The rapid development of such beds as these is a strong argument against the possibility of controlling the production of salt in America on the principle that has proved so successful in England. By the way, we believe the projectors of the "great American salt company" promised that it would be in successful operation by the beginning of this year. Nearly three months of the year have already gone and we have heard nothing to indicate that the scheme is to be revived.

SOCIETY OF CHEMICAL INDUSTRY.

GLASGOW AND SCOTTISH SECTION.

THE annual general meeting of this section took place on the 1st inst., in the rooms of the Society, Bath-street, Glasgow. The chairman, Mr. R. R. Tatlock, presided. Being the occasion of the annual re-adjustment of the official list of the section, the chemical papers arranged for presentation were briefer than is customary. Mr. Tatlock read a short reference to some experiments he had been making on the fatty acids of olive and other oils, with a special application to soap manufacture, regarding which the interim chairman, Mr. Stanford, expressed a hope that, as the speculations arising out of these partial experiments were highly interesting in character, Mr. Tatlock might at some early date present to the Society an exhaustive exposition of the subject. Thereafter, Mr. James Hope read a paper on "The estimation of cobalt and nickel," which was accepted by the meeting as a valuable contribution so far, but this also proved to be of the nature of an instalment, Mr. Hope holding himself pledged to resume the subject and carry it farther in a future paper. The proposals of the Committee as to the re-arrangement of the official list were then put to the meeting by the Chairman, and carried unanimously, the new order not to come in force, however, till the end of June, according to the custom of these appointments. Mr. E. C. C. Stanford becomes chairman, *vice* Mr. R. R. Tatlock, retiring; Prof. Dittmar, Mr. G. G. Henderson, and Mr. W. J. Chrystal, continue severally in the vice-chairmanship, the hon. secretaryship, and the hon. treasurership; and the new members of committee are Messrs. W. J. A. Donald, C. A. Fawcitt, J. Falconer King, T. P. Miller, J. Pattison, and R. R. Tatlock. Proceedings were brought to a close by the reading of the annual report of the secretary on the year's record and present condition of the section. Removals by death, change of residence, &c., have been very heavy, amounting to 28 in all; but the new nominations have been pretty numerous also, 23 in number, so that the membership remains at about the old strength, namely, 240. During the session 14 papers were presented, and there was an average membership attendance of 30.

MANCHESTER SECTION.

The annual meeting of the Manchester section of the Society of Chemical Industry was held on the evening of April 2nd, in the Victoria Hotel. Mr. I. Levinstein presided. After the election of the committee for the ensuing year, Dr. Bailey described a new method of analysis, and Mr. H. Grimshaw made a communication on the decomposition of chlorides of magnesium, calcium, and zinc by heat—three substances which, he said, were extensively used for processes which involved their application to cotton and woollen fabrics. His experiments had been directed to ascertain at what temperature and how low a temperature decomposition took place. The result of his experiments was to show that the only one of the three chlorides which suffered decomposition was the chloride of magnesium, which parted with a portion of its acid, thus affecting the colour and strength of the cotton and woollen fibres.

TANNING WITH EXTRACTS.

MONS. VILLON has the following remarks on tanning extracts in his treatise on "The Manufacture of Leather":—"Extracts, as we have previously said, are tanning materials from which the aqueous matter has been extracted. The real utility of extracts is for materials containing less than 10 per cent. of tannin; but some manufacturers for export use substances which are richer in tanning. The extracts of commerce come to the consumer under three forms: 20° Baumé, 30°, and solid. We reject entirely the employment of solid extracts, which are of no advantage, and give too much colour to the leather. We do not advise the employment of 30° Baumé, except in very special cases; it is better to use from 20 to 25° Baumé, which are the least coloured and the least falsified. The quality of extracts is also presented in three forms to the tanner: ordinary extracts, which are made without any manipulation, and which precipitate abundantly in cold; clarified extracts, from which the matters soluble in cold have been removed; the uncoloured and clarified extracts, which seldom yield the results that one expects from them generally. We should then choose ordinary extracts, which are well made, without too much colour, and concentrated in a vacuum.

"The dissolving of commercial extracts ought to be done in a pit heated to 30° of temperature, with a circular copper steam-tube. The liquid is pumped through a cask containing new tan on a double bottom, which acts as a filter. The clear liquor is received in a reservoir. Do this until the liquor registers 10° Baumé. The deposit which is formed ought to be removed, because it remains completely inert in the operations of tanning; while, of course, when tanning, every part of the extract should be dissolved, and the temperature in the pelt should never exceed 25°, so it would not be possible to dissolve matters which have been put into contact with water at 30°. Always, with solid extracts, one is obliged to dissolve by ebullition; but, we repeat it, don't use solid extracts, however well manufactured."

Trade Notes.

THE LARGEST NICKEL MINE IN THE WORLD.—On a little branch of the Canadian Pacific road, near Sudbury, Canada, is a nickel mine that produces more nickel, it is said, than the entire market of the world calls for. It is found at a depth of about 300 feet below the surface, in a layer of oxydised Laurentian rock characteristic of that region. Immediately the mineral is hoisted from the mine it is broken up and calcined for the purpose of eliminating the sulphur it contains. When this process is completed the residuum is conveyed to the smelter. After the dross of the molten metal flows off, the nearly pure nickel and copper are blended together, forming an alloy, 70 per cent. of which is nickel, and 30 per cent. copper, which it draws off at the base of the furnace and allowed to cool. When cold, the product is shipped to Swansea, Wales, and Germany, where the constituent metals are separated and refined by secret process, known only to the manufacturers, and jealously guarded. The present output of the mine is stated at 4,000 tons of nickel annually.

CHROMIUM IN DYEING.—The quantity of chromium fixed on the fibre is (says a writer in a contemporary) a most important matter in connection with the fastness to light of logwood blues and blacks. If more than a certain amount is present, the black, in a short time, is most likely to turn green. In all cases of logwood blacks that have turned green, which the writer has examined, an excess of chrome has been found. It is difficult to state a limit, since much depends upon the condition in which it exists, but not more than 1·5 per cent. of bichromate of potash should be fixed upon the fibre. Where the goods are chromed after dyeing, as well as before, with the object of making the wool clean, there is no doubt that part of the chromium is often fixed on the fibre as chromate or chromic acid; and such being the conditions, the black is very liable to turn green. It is most probable that in after-chroming a finishing by the use of too much bichromate of potash, the logwood black is rendered liable to turn green. If chromic acid is deposited on the fibre in the first chroming, it is all probably reduced to chromic oxide when it comes into the logwood.

Market Reports.

MISCELLANEOUS CHEMICAL MARKET.

Business in the alkali trade continues very active—manufacturers are very fully engaged with orders, and prices all round are satisfactory. Bleaching powder remains nominally at £5. 7s. 6d. per ton, on rails and makers generally are firm, notwithstanding that some re-sales have been made below this figure. Caustic soda has been somewhat irregular but for 70's the general quotation is £10. 5s. f.o.b., and for 60's £9. 7s. 6d. to £9. 10s. Recent stoppages at works have reduced stocks, and the prospect of an early demand from continental markets strengthens the position. The demand for soda ash continues good all round, and ordinary carbolic and caustic ash cannot be obtained under 1⅜d. per deg. f.o.b. Carbonate 58% ash, strong at 1 15-32d. Soda crystals £2. 17s. 6d. f.o.r. Saltcake 27s. 6d. per ton, in bulk. Sulphur without change. Chlorate of potash steady at 5d. Superphosphate of lime 26/28%, ruling at £2. 10s. to £2. 12 6d. at works and makers are readily filling up with orders at these prices. Sulphate of copper still scarce for early deliveries, and £26. to £26. 10s. paid this week. The future of this article remains uncertain, and forward quotations range from £20. to £24. per ton, according to period of delivery. Lead acetates and nitrate are slow of sale, but prices remain firm. Acetates of lime are more freely offered at £8 and £13. 15s. per ton respectively, for brown and grey. Other acetates and acetic acids quiet and prices stationary. Carbolic acids are dull, with a tendency to lower prices. Aniline oil and salt are firmer again, in sympathy with an upward move in benzol. Oxalic acid is difficult to find at 3⅜d. net. Prussiate of potash quiet, but price firm at 9⅜d. Bichromates and chromic acid without change. Muriate of ammonia stagnant in consequence of reduced consumption but an early resumption of normal demand is expected. Potash caustic, and carbonate are meeting with ready sale, and prices are firm. Green copperas is in active demand, and supplies are fully taken up. General business in chemicals, and especially in the fine goods of the trade is quiet, and prices are unsatisfactory as a consequence.

METAL MARKET REPORT.

	Last week.	This week.
....	48/11	48/9
....	£90 5 0	£90 0 6
r	47 12 6	48 0 0
r	21 0 0	20 12 6
....	12 10 6	12 10 0
....	43⅞d.	44d.

COPPER MINING SHARES.

	Last week.		This week.	
'into	15⅞	16	16⅛	16¼
a & Barry	6 5/16	6 7/16	6 5/16	6 5/16
is	4½	4⅝	4½	4⅝
Copper Co. ..	3⅜	3½x.d.	3½	3⅝
iqua	1¾	2	1¾	2
po	2⅜	2½x.d.	2 5/16	2 7/16
cillo	1	1⅛	1	1⅛
Quebrada	⅞	1	⅝	¾
la	2⅛	2⅜	2⅛	2⅜
pilla	1/-	1/6	1/-	1/6
atella........	3d.	9d.	3d.	6d.

REPORT OF MANURE MATERIAL.

been a good prompt business passing, notwithstanding , but prices have been the turn in favour of buyers, both tic and nitrogenous material.

shippers continue to invite offers for Charleston River ock under 11d. per unit, cost, freight, and insurance, for United Kingdom, but the syndicate, being sold a good way n to be firm in their quotation of 11d. For Charleston b½d. would be accepted, and probably ¼d. less for good- s, freights being somewhat easier. We have no alteration values of either Belgian or Somme phosphates. Quota- nadian continue without change, but they are quite above ited Kingdom buyers, though sales of the higher qualities are resumably for the Continent. Prices for Florida phos- hardly yet to be fixed, and most of what is coming forward ributed among buyers mainly for trial purposes.

still nothing to report in River Plate cargoes of bones to gdom, the December cargo referred to in our last report not arrived at port of call, and buyers being indisposed to bid ime. Sales of crushed East India bones have been made l at £5. 3s. 9d. per ton, ex quay, up to £5. 5s., the latter required for further quantity. No business reported to tons Bombay bone meal sold at £5. 10s. per ton, ex quay at which price further quantity may be had ex quay. lowest ex store. There have been sales made of common nes at £4. 17s. 6d. up to £5., the lower price being, per- st value for fair ordinary quality, considerable quantities e way or having arrived.

f soda continues without change, though on the Continent has made some advance. Spot price is 8s. per cwt. Due worth 7s. 10½d. to 8s., according to size and quality. For ment considerably more money is required, the very large far this year encouraging the hope that the position of this be considerably improved before next season.

els of roughish River Plate dried blood sold at 9s. 6d. per re Liverpool: 10s. 3d. required for fine parcels, and s so far accepted for fine quality. For home prepared 10s. 9d. per unit, nearest value, with a somewhat slower per unit, in bags, gross weights, delivered f.o.r. at works cepted for ground hoof.

nains without change at 38s. to 40s. per ton, in bags, e to rails, according to quantity.

sphates are somewhat better, most of the parcels in second- ng on the market, having been sold at 46s. 3d. per ton, bulk, delivered f.o.r. at works, is again the nearest value.

E LIVERPOOL MINERAL MARKET.

ket has continued strong, and in some minerals a still nce has been experienced. Manganese: Arrivals are les continue to be made at advanced figures. Magnesite: w lump continue large, and prices are in buyers' favour. , £6. 10s.; and calcined ground, £10. to £11. Bauxite (Irish Hill brand): The strong demand continues, and prices remain very firm—Lump, 20s.; seconds, 16s.; thirds, 12s.; ground, 35s. Dolomite, 7s. 6d. per ton at the mine. French Chalk: Arrivals this week have increased, all of which, however, have, gone direct into consumption, and prices remain firm, especially for G.G.B. "Angel-White" brand—90s. to 95s. medium, 100s. to 105s. superfine. Barytes (carbonate) continues easier, selected crystal lump scarce at £6.; No. 1 lumps, 90s.; best, 80s.; seconds and good nuts, 70s.; smalls, 50s.; best ground, £6.; and selected crystal ground, £8. Sulphate has somewhat improved, best lump, 35s. 6d.; good medium, 30s.; medium, 25s. 6d. to 27s. 6d.; common, 18s. 6d. to 20s.; ground best white, G.G.B. brand, 65s.; common, 45s.; grey, 32s. 6d. to 40s. Pumicestone quiet; ground at £10., and specially selected lump, finest quality, £13. Iron ore in demand at full prices. Bilbao and Santander, 9s. to 10s. 6d. f.o.b.; Irish, 11s. to 12s. 6d.; Cumberland in strong demand at 18s. to 24s. Purple ore steady at last quotations. Spanish manganiferous ore selling freely at good prices. Emery-stone: Best brands continue scarce both for spot and forward delivery; prices firmer. No. 1 lump is quoted at £5. 10s. to £6., and smalls £5. to £5. 10s. Fullers' earth steady; 45s. to 50s. for best blue and yellow; fine impalpable ground, £7. Scheelite, wolfram, tungstate of soda, and tungsten metal inquired for. Chrome metal, 5s. 6d. per lb. Tungsten alloys 2s. per lb. Chrome ore: High grades are inquired for, and bring full prices. Antimony ore and metal have further advanced. Uranium oxide, 24s. to 26s. Asbestos: Best rock, £17. to £18.; brown grades, £14. to £15. Potter's lead ore: Prices are firm; smalls, £14. to £15.; selected lump, £16. to £17. Calamine: Best qualities scarce—60s. to 80s. Strontia steady; sulphate (celestine) steady, 16s. 6d. to 17s. Carbonate (native), £15. to £16.; powdered (manufactured), £11. to £12. Limespar: English manufactured, old G.G.B. brand in demand, and brings full prices; 50s. for ground English. Felspar, 40s. to 50s.; fluorspar, 20s. to £6. Bog ore firm at 22s. to 25s. Plumbago: steady; Spanish, £6.; best Ceylon lump at last quotations; Italian and Bohemian, £4. to £12. per ton; founders, £5. to £6.; Blackwell's "Mineraline," £10. French sand, in cargoes, continues scarce on spot—20s. to 22s. 6d. Ferro-manganese and silicon spiegel in good demand at stronger figures. Chrome iron, 20 per cent., £24. to £25. Ground mica, £50. China clay freely offering—common, 18s. 6d.; good medium 22s. 6d. to 25s.; best, 30s. to 35s. (at Runcorn).

THE LIVERPOOL COLOUR MARKET.

COLOURS.—Steady, prices unchanged. Ochres: Oxfordshire quoted at £10., £12., £14., and £16.; Derbyshire, 50s. to 55s.; Welsh, best, 50s. to 55s.; seconds, 47s. 6d.; and common, 18s.; Irish, Devonshire, 40s. to 45s.; French, J.C., 55s., 45s. to 60s.; M.C., 65s. to 67s. 6d. Umber: Turkish, cargoes to arrive, 40s. to 50s.; Devonshire, 50s. to 55s. White lead, £21. 10s. to £22. Red lead, £18. Oxide of zinc: V.M. No. 1, £25.; V.M. No. 2, £23., Venetian red, £6. 10s. Cobalt: Prepared oxide, 10s. 6d.; black, 9s. 9d.; blue, 6s. 6d. Zaffres: No. 1, 3s. 6d.; No. 2, 2s. 6d. Terra Alba: Finest white, 60s.; good, 40s. to 50s. Rouge: Best, £24.; ditto for jewellers', 9d. per lb. Drop black, 25s. to 28s. 6d. Oxide of iron, prime quality, £10. to £15. Paris white, 60s. Emerald green, 10d. per lb. Derbyshire red, 60s. Vermillionette, 5d. to 7d. per lb.

WEST OF SCOTLAND CHEMICALS.

GLASGOW, Tuesday.

All paraffin products, with the continued exception of burning oil alone, incline to the strengthening side, and scale and wax in free hands fetch prices slightly higher than those at which makers are now delivering under their lately signed time contracts. Were it not for the backward position of burning oil, matters would be looking brighter for the Scotch oil companies; but the burning article in a sense is the one by which their undertakings stand or fall, and when it continues to droop uninterruptedly for a season, there is sure trouble ahead. Dividend declarations for the year are due within a week or two, and it is certain that some of the concerns will this time have nothing to declare but loss. Two or three of the better equipped ones will come out of the ordeal fairly well; but in no case will the affluent dividends of former days be repeated, or even approached. The general chemicals market has been on the down grade during the week, but partly owing to the stoppage of transactions for Monday's holiday, there is no increase of turnover to report. The easier prices ought, however, to influence the bulk of the business done between now and next week. Buying and selling in sulphate of ammonia have been somewhat disorganised owing to the closed days, but there is sufficient to indicate a further weakening of the article, and to-day's price is no more than

£11. 10s. Makers have been offering for forward delivery at £11. 15s., but most likely are ready to accept less. Chief prices current are :— Soda crystals, 52s. 6d., less 2½% Glasgow; alum in lump £5., less 2½% Glasgow; borax, English refined £29., and boracic acid, £37. 10s. net Glasgow; soda ash, 1⅛d. net Tyne; caustic soda, white, 76°, £13., 70/72°, £10. 7s. 6d., 60/62° £9. 10s., and 60/62° cream, £9., all less 2½% Liverpool; bicarbonate of soda, 5 cwt. casks, £6. 5s., and 1 cwt. casks, £6. 10s. net Tyne; refined alkali 48/52°, 1⅛d., net Tyne; saltcake, 30s.; bleaching powder, £5. 10s. to £5. 15s., less 5% f.o.r. Glasgow; bichromate of potash, 4d., and of soda 3d., less 5 and 6% to Scotch and English buyers respectively; chlorate of potash, 5d., less 5% any port; nitrate of soda 8s. to 8s. 3d.; sulphate of ammonia, spot, £11. 10s., f.o.b. Leith; salammoniac, 1st and 2nd white, £37. and £35., less 2½% any port; sulphate of copper, £25. 10s., less 5% Liverpool; paraffin scale, hard and soft, 2½d. per lb.; paraffin wax, 120°, semi-refined, 3$\frac{5}{32}$d.; paraffin spirit (naphtha), 8d. a gallon; paraffin oil (burning), 6¾d. to 7d. at Glasgow for Scotch consumption (English orders done at about 1d. a gallon lower); ditto (lubricating), 865°, £5. 10s. to £6. per ton; 885°, £6. 10s.; and 890/895°, £7. 10s. to £8. Week's imports of sugar at Greenock were 70,272 bags.

TAR AND AMMONIA PRODUCTS.

The benzol market retains its firmness, and 50/90's have been sold at 2s. 7½d., and 90's stand as quoted last week. There is every expectation of this market retaining its present position, as there are no disturbing elements in the horizon. Solvent naphtha is exceedingly firm, as also is creosote, while anthracene is still a very gloomy picture. Pitch remains weak, and 28s. on West Coast, and 29s. to 29s. 6d. on East Coast ports may be called to-day's prices.

The sulphate of ammonia market after experiencing some irregularity appears now to be settling down to a more steady condition, and parcels which have been pressing have mostly been cleared out, and a better feeling is therefore noticeable. It has been stated that £11. 10s. f.o.b. Hull, was accepted last week, but this could only have been an exceptional transaction. To-day's values are £11. 12s. 6d., Hull, £11. 10s. to £11. 12s. 6d., Leith, and £11. 10s. f.o.b. Liverpool. There is no doubt this drop has been caused by the Leith makers losing their heads at a time when additional firmness was required. There have also been transactions at a low price in Liverpool by dealers who should have known the state of the market better, and it is a well-known fact that dealers have offered parcels on the Continent at 5s. per ton below prices which they have offered producers on this side.

THE TYNE CHEMICAL REPORT.

TUESDAY.

The chemical market has been very quiet this week, with little business doing on 'Change owing to the holidays. Prices remain much about the same as last week. Nitrate of baryta has been advanced 5s. a ton. Caustic soda 77 %, £13 per ton; bleaching powder, £5. 15s. to £6 per ton, softwood casks; soda ash, 48—52%, 1⅛d. per degree, less 2½%; soda crystals, £2. 12s. 6d. per ton; sulphate of soda, in bulk, 35s. per ton; ground, in casks, 45s. per ton; recovered sulphur, £4. 5s. per ton; chlorate of potash, 5d. per lb.; silicate of soda, 75° Tw., £2. 10s. per ton; 100° Tw., £3. 7s. 6d. per ton; 140° Tw., £4. per ton; hyposulphite of soda, in 1 cwt. kegs, £5. 15s. per ton, in 5—7 cwt. casks, £5. 5s. per ton; pure white sulphate of alumina, £4. 10s. per ton; blanc fixe, £7. 10s. per ton; chloride of barium, £8. per ton; nitrate of baryta, crystals, £18. 15s. per ton; ground, £19. 5s. per ton; sulphide of barium, £5. 10s. per ton—all f.o.r. makers' works or f.o.b. Tyne.

Small coals continue in good demand, and prices remain very firm. Northumberland steam small quoted 8s. to 8s. 6d. per ton, and Durham small, 10s. per ton.

The exports of chemicals from the Tyne for the first three months of this year show a considerable increase in tonnage over the exports for the same period of last year.

	1889. Tons.	1890. Tons.
Alkali and soda ash	2,702	3,704
Bicarb. soda	26	24
Bleach powder	3,911	6,424
Manure	1,963	2,955
Soda crystals	2,989	3,303
Sulphate of soda	187	1,107
Other chemicals	3,761	4,911
	15,539	22,428

Gazette Notices.

Partnerships Dissolved.

HILTON, ANDERSON AND CO., Faversham, Upnor, and Halling, and U[p] Thames-street, London, manufacturers of Portland, Roman, and other ceme[nts] as far as regards P. Hilton.

P. OVENDEN, C. RHIND, J. FOSTER, and R. M. DOUGLAS, under the styl[e] the Malago Vale Ochre and Colour Company, Bristol, ochre and col[our] manufacturers and merchants.

D. Y. CLIFF, C. CLIFF, P. CLIFF, and B. CLIFF, under the style of J. Cliff [&] Sons, Runcorn and Leeds, chemical apparatus manufacturers and fireb[rick] merchants, as far as regards P. Cliff and B. Cliff.

The Bankruptcy Act, 1883.

Receiving Orders.

WALTER SLOCOMBE, Romford, oil and colourman.

New Companies.

INTERNATIONAL EXHIBITION OF MINING AND METALLURGY.—This comp[any] was registered on the 31st ult., as a company limited by guarantee to £1. e[ach] member to establish and maintain an exhibition, in or near London, of mining [and] metallurgy. The word "limited" is omitted from the title by Board of Tr[ade] licence. The subscribers are:—

H. C. Gray, 139, Lothian-road, N., railway clerk.
Hy. Eve, 18, Finch-lane, business manager, *Mining Journal*.
G. A. Ferguson, 18, Finch-lane, editor *Mining Journal*.
W. Pritchard Morgan, M.P., Dolgelly.
Hy. Cribb, Bishop's Stortford.
T. R. R. Jordan, 15, George-street, E C., engineer.
E. C. Bennett, ME, 20, Bucklersbury.

OTTOMAN PAPER MANUFACTURING COMPANY, LIMITED.—This company registered on the 27th ult., with a capital of £325,000., divided into 16,250 prefer[ence] shares, 16,150 ordinary, and 100 founders' shares of £10. each, to acquire all or of the shares in the capital stock of a Société Anonyme constituted in the Otto[man] Empire, for acquiring a concession from the Sultan of Turkey for the construc[tion] and working of a paper manufactory in His Majesty's dominions. There ar[e] signatories, who take amongst them 23 founders' shares, the first seven being :—

Sh[ares]

The Hon E. W. Douglas, Christchurch ....
Lord Clinton, 32, Bruton-street ....
J. C. Thynne, The Cloisters, Westminster ....
H. W. Cobb, Salisbury ....
J. Judd, J.P., St. Andrew's Hill ....
Herbert Spicer, 165, Queen Victoria-street ....
W. H. Richardson, J.P., Jarrow ....

VICTORIA PETROLEUM COMPANY, LIMITED.—This company was registered the 28th ult., with a capital of £10,000. in £10. shares, to import, export, and ma[nu]facture, and deal in mineral, vegetable, or animal oils. The subscribers are :—

Sh[ares]

R. Johnston, Cardiff, merchant ....
J. Milligan, Liscard, accountant ....
S. Woodhead, Oxton, freight manager ....
A. H. Woodhead, Oxton, cashier ....
J. Caesar, Walton, accountant ....
A E. Wills, Seaforth, freight manager ....
E. J. Corney, 9, Tynwald-hill, Liverpool, bookkeeper ....

The Patent List.

This list is compiled from Official sources in the Manchester Techn[ical] Laboratory, under the immediate supervision of George E. Davis [and] Alfred R. Davis.

APPLICATIONS FOR LETTERS PATENT.

Briquettes or Block Fuel. T. C. Fawcett and J. B. Swallow. 4,542. March
Dynamos.—(Complete Specification). F. V. Andersen and J. O. Girdlest[one]. 4,557. March 24.
Tin Refining. T. Teague. 4,560. March 24.
Apparatus for Extracting Juices from Fruits. J. Coppard. 4,562. March
Gas and Petroleum Motors.—(Complete Specification). E. Kaselowsky. 4, March 24.
Grey Colouring Matters. O. Imray. 4,577. March 24.
Manufacture of Carbonate of Magnesium and Double Carbonate Magnesium and Potassium. H. Precht. 4,588. March 24.
Improved Steam Boiler. W. Hornsby and R. Edwards. 4,589. March 24
Kilns for Firing Glass and Enamel Ware. H. T. Parfitt. 4,592. March
Separating Liquids or Solids from Gases. E. Jones. 4,594. March 24.
Manufacture of Sulphonated Thionines. S. Pitt. 4,596. March 24.
Manufacture of Gas. W. Gray. 4,598. March 24.
Electric Switches. A. A. C. Swinton. 4,605. March 25.
Transparent Flexible Films for Photography. J. Leslie. 4,606. March
Filtering Oil. W. E. Crane. 4,607. March 25.
Smoke Consumption. J. Dean and H. Powell. 4,613. March 25.
Drying Apparatus. M. Guthrie. 4,614. March 25.
Photometers. F. Greene and F. H. Varley. 4,622. March 25.
Water Heaters and Steam Condensers. C. E. Masterman. 4,643. March
Dynamos. P. F. Degn. 4,648. March 25.
Purifying Brine.—(Complete Specification). J. C. Mewburn. 4,652. March
Electrical Soldering.—(Complete Specification). H. H. Lake. 4,657. March
Electrical Heating Devices for Forging Machines.—(Complete Specificat[ion]). O. Imray. 4,670. March 25.
The Welding of Metals.—(Complete Specification). R. J. Tieford. 4, March 25.
The Dyeing of Yarns.—(Complete Specification). A. J. Boult. 4,683. March
Concentrating Magnetic Iron Ore.—(Complete Specification). H. J. Had[dan]. 4,690. March 25.
Production of Salts from Brine. E. G. Lawrance. 4,694. March 25.
Apparatus for Drying Salt. E. G. Lawrance. 4,695. March 25.
Closing Vessels to promote Sterilisation. E. Sonstadt. 4,701. March 2
Measuring Apparatus for Liquids. J. W. White. 4,707. March 26.
Smoke Consumption. R. M. Whitaker. 4,708. March 26.

LONDON.

Week ending April 1st.

Alum—
Holland, £30 Ohlenschlager Bros.
" 25 "

Antimony Ore—
France, 10 t. Pillow, Jones, & Co.

Acetic Acid—
Holland, 72 pks. Dunn & Co.
Germany, 16 Lister & Biggs

Arsenic—
Germany, £125 Burgoyne, Burbidge, & Co.

Bromide of Iron—
Germany, £400 A. & M. Zimmermann
" 400 "

Bones—
Victoria, 8 t. Anning & Cobb
U. S., 10 A. & W. Nesbitt
B. W. I., 1 Culverwell, Brooks, & Co
Victoria, 3 L. & I. D. Jt. Co.
E. Indies, 100 J. Mitchell
" 600 Gonne, Croft, & Co.

Barium Chloride—
Germany, £100 J. Knill & Co.

Barytes—
Germany, 59 cks. H. Lambert
Belgium, 131 A. Zumbeck & Co.
Holland, 69 A. Boltz
" 23 Props. Scott's Whf.
" 44 W. Harrison & Co.
Germany, 19 Litchfield & Co.

Caoutchouc—
E. Indies, 50 c. N. S. S. Co., Ltd.
Belgium, 31 Lewis & Peat
E. Africa, 1¼ "
U. S., 20 Kebbel, Son, & Co.
E. Indies, 33 Arbuthnot, Ewart, & Co.
" 18 Caldwell, Watson, & Co.
France, 10 Flageollet & Co.
Germany, 10 H. Grey, jun.
E. Indies, 200 Huttenbach & Co.
" 4½ Kebbel, Son, & Co.
Holland, 49 Heilbut, Symons, & Co.
E. Africa, 1¼ L. & I. D. Jt. Co.

Cream of Tartar—
Spain, 5 pks. B. & F. Wf. Co.
" 16 W. C. Bacon & Co.
" 6 Christopherson & Co.
France, 5 pks. J. Puddy & Co.
Portugal, 3 B. & F. Wf. Co.
Holland, 8 "
Spain, 13 W. C. Bacon & Co.
Portugal, 2 B. & F. Wf. Co.
France, 5 M. Ashby
" 8 E. W. Carling & Co.
" 5 B. & F. Wf. Co.
" 43 N. S. S. Co., Ltd.

Copper Ore—
Sydney, 101 t. W. H. Pridham & Co.

Caustic Potash—
Germany, 16 pkgs. W. Balchin

Caustic Soda—
Belgium, 255 c. Beresford & Co.

Chemicals *(otherwise undescribed)*
Germany, £45 T. H. Lee
" 25 Phillipps & Graves
" 30 L. & I. D. Jt. Co.
France, 12 C. Brumlen
Germany, 97 T. H. Lee
" 198 A. & M. Zimmermann
Holland, 6 Beck & Pollitzer
" 10 Phillipps & Graves
Germany, 5 L. & I. D. Jt. Co.
" 256 A. & M. Zimmermann
" 406 Phillipps & Graves
" 12½ Becker & Ulrich
France, 15 C. Brumlen
Germany, 23 F. Stahlschmidt & Co.

Dyestuffs—
Yellow Berries
France, 30 pks. Arnati & Harrison
Indigo
E. Indies, 17 chts. Elkan & Co.
" 69 Patry & Pasteur
" 17 Stahlschmidt & Co.
" 7 L. & I. D. Jt. Co.
" 89 "
" 32 N. S. S. Co. Ltd.
" 50 Benecke, Souchay, & Co.
" 79 Arbuthnot, Latham, & Co.
" 49 W. Brandt, Sons, & Co.
" 25 Elkan & Co.
U. S., 3 Patry & Pasteur
E. Indies, 47 L. & I. D. Jt. Co.
" 12 Elkan & Co.
" 30 Benecke, Souchay, & Co.
" 33 Stansbury & Co.
" 342 L. & I. D. Jt. Co.
" 122 Arbuthnot, Latham, & Co.
" 22 Parsons & Keith
" 11 Walker, Munsie, & Co.
" 9 F. H. Allen & Co.
" 10 chts. W. Isted.
" 5 Ernsthausen & Co.
" 15 L. & I. D. Jt. Co.
" 4 W. Isted
" 9 Frauling & Goschen
Germany, 11 Chalmers, Guthrie, & Co.
E. Indies, 23 L. & I. D. Jt. Co.
" 7 F. Stahlschmidt & Co.
" 151 W. Balchin
" 205 Indigo Co.
Annatto
Norway, 1 pkg. Johnsen & Jorgensen
Ceylon, 2 Hoare, Wilson, & Co.
Extracts
France, 250 pks. Levinstein & Sons
Holland, 73 J. Forsey
" 26 W. Balchin
France, 50 B. & F. Wf. Co.
" 95 L. J. Levenstein & Sons
Belgium, 100 B. Kuhn
U. S., 1 Laughlin & Co.
Austria, 800 A. J. Humphrey
U. S., 60 T. Ronaldson & Co.
Denmark, 1 Petty & Co.
" 4 Fulwood & Bland
Norway, 3 Johnsen & Jorgensen
Aniline
Holland, 5 pks. R. Baals & Co.
Belgium, 10 Union Lighterage Co.
Sumac
Italy, 100 pks. F. Roberts
" 500 Brown & Elmslie
Saffron
France, 1 pkg. Major & Field
Safflower
E. Indies, 32 pks. Lewis & Peat
Myrabolams
E. Indies, 500 pks. A. Hagan
" 550 L. & I. D. Jt. Co.
Cutch
E. Indies, 500 pks. Middleton's S. S. Whf.
" 917 Wrightson & Son
Gambier
E. Indies, 556 pks. S. Barrow & Bro.
" 820 Beresford & Co.
Germany, 274 Vonder, Meden, Bros., & Co.
Divi Divi
B. W. Indies, 316 pks. G. T. Benton & Sons
Tanners' Bark
Belgium, 10 t. C. S. Lovell
" 50 Boutcher & Co.
" 10 H. Henle's succrs.
" 31½ J. J. Williamson & Sons
" 10½ Leach & Co.
Victoria, 26 Baxter & Hoare
Orchella
Ceylon, 25 pks. E. Barker & Co.
E. Indies, 3 C. C. Bryden & Co.

Dyestuffs *(otherwise undescribed)*
A. Turkey, 125 pks. [illegible]

Gutta Percha—
Germany, 17 c. [illegible]
E. Indies, 100 [illegible]
" 397 [illegible]
B. Guiana, 30 [illegible]
E. Indies, [illegible] [illegible]
Holland, [illegible] [illegible]
E. Indies, 640 J. [illegible]

Glucose—
Germany, 10 pks. 20 c. [illegible]
" 200 [illegible] [illegible]
" 200 200 J. [illegible]
" 800 800 [illegible]
" 3 24 [illegible]
" 180 190 [illegible]
U. S., 25 130 [illegible]
France, 300 300 De [illegible]
Germany, 25 [illegible] [illegible]
" 5 25 [illegible]
" 400 400 [illegible]
" 25 200 [illegible]
" 100 [illegible] L. & I. [illegible]
" 45 490 [illegible]
U. S., 240 240 [illegible]
" 87 534 R. G. [illegible]

Guano—
Belgium, 30 t. Ang. [illegible]

Glycerine—
Victoria, £173 J. [illegible]
Germany, 114 [illegible]
" 63
Holland, 450 [illegible]

Isinglass—
B. W. I., 1 pkg. L. & [illegible]
China, 9 pks. W. [illegible]

Magnesia—
Holland, £40 [illegible]

Manure—
Germany, 29 t. [illegible]
" 30 [illegible]
Phosphate Rock
France, [illegible] t. [illegible]
Belgium, 260 [illegible]
" 200 [illegible]
" [illegible] [illegible]
France, 570 [illegible]

Manure *(otherwise undescribed)*
Germany, 30 t. [illegible]
U.S. 4½ C. [illegible]

Manganese Borate—
Germany, £25 Petri Brs.
Nitre—
Chile, 13,761 cubic bgs. Ang. Contl. Guano Wks. Agency
Naphtha—
Germany, 2 drs. H. Lorenz
" 2 brls. L. & I. D Jt. Co.
Belgium, 31 Domeier & Co.
Potassium Oxymuriate—
Belgium, 100 pks. G. Boor & Co.
Potassium Sulphate—
Germany, 500 pks. Hunter, Waters & Co.
France, 14 Carey & Sons
Germany, 63 H. Lambert
Paraffin Wax—
U.S., 200 brls. G. H. Frank
" 6,193 J. H. Usmar & Co.
" 236 H. Hill & Sons
" 140 112 cs. "
Potassium Carbonate—
Germany, 11 pks. Northcott & Sons
Pearl Ash—
France, 200 c. F. W. Berk & Co.
Pitch—
Holland, 8 brls. Poth, Hille & Co.
Germany, 70 Berlandina Brs. & Co.
Plumbago—
Holland, 37 cks. Brown & Elmslie
" 16 cs. J. Barber & Co.
Pyrites—
Spain, 2,193 t. C. Tennant, Sons & Co.
Pomaron, 1,000 Mason & Barry, Ld.
Spain, 1,909 C. Tennant, Sons & Co.
Saltpetre—
Germany, 43 cks. Craven & Co.
E. Indies, 327 L. & I. D. Jt. Co.
" 387 Clift, Nicholson & Co.
" 1,092 C. Wimble & Co.
" 327 L. & I. D. Jt. Co.
A. Turkey, 17 cks. Brown & Elmslie
Salammoniac—
Holland, £50 Barr, Moering & Co.
Sodium Sulphate—
Holland, £50 Prop. Dowgate Dk.
Sodium Acetate—
France, £75 S. F. Waters & Co.
" 146½ A. Zumbeck & Co.
Stearine—
Germany, 408 cks. Ross & Deering
Tartars—
Italy, 43 pks. B. Jacob & Sons
Tartaric Acid—
Holland, 12 pks. B. & F. Wf. Co.
" 4 Augspurg, Hopf & Co.
France, 8 B. & F. Wf. Co.
Holland, 23 "
" 4 Moering & Co.
Germany, 12 C. Andrae & Co.
Tincal—
E. Indies, £120 J. Puddy & Co.
Ultramarine—
Germany, 19 cks. G. Steinhoff
" 4 "
Holland, 2 Hernu, Peron & Co.
Belgium, 6 cs. W. Harrison & Co.
Holland, 11 pks. Hernu, Peron & Co.
" 4 cks. Ohlenschlager Brs.
Zinc Oxide—
U.S., 12 pks. W. Harrison & Co.
Holland, 130 M. Ashby
" 165 "
Germany, 100 cks. D. Storer & Son
" 98 M. Ashby
Zinc Sulphate—
Germany, £65 F. Smith

GLASGOW.

Week ending April 3rd.

Albumen—
Rotterdam, 3 brls. J. Rankine & Son
Barytes—
Rotterdam, 114 cks.
Dyestuffs—
Alizarine
Rotterdam, 80 cks.
" 185
Sumac
Rouen, 100 bgs.
Glucose—
Hamburg, 13 cks.
New York, 100 brls.
Glycerine—
Rotterdam, 8 cs. J. Rankine & Son
Magnesium Silicate—
New York, 480 bgs. £250
Potash—
New York, 2 cks. Hurlet and Campsie Alum Co.
Rotterdam, 20 J. Rankine & Son
Dunkirk, 9 brls
Potassium Carbonate—
Hamburg, 12 cks.
Pyrites—
Seville 1,081 t. Tharsis Co.
Huelva, 1,600 "
Sodium Nitrate—
Junin, 9,158 bgs. Thomson Aikman
Salt—
Hamburg, 1,060 bgs.
Sodium Acetate—
Rouen, 10 cks.
Sulphur Ore—
Drontheim, 220 t.
Saltpetre—
Rotterdam, 7 cks. J. Rankine & Son
Tar—
Wilmington, 1,500 brls.
Ultramarine—
Rotterdam, 10 cs. J. Rankine & Son
Zinc Oxide—
Rotterdam, 20 cks. J. Rankine & Son

LIVERPOOL

Week ending April 2nd.

Antimony—
Havre, 1 cs.
Antimony Ore—
Smyrna, 50 bgs.
Marseilles, 1,010 bgs.
Acetate of Lime—
New York, 2,488 bgs.
Acids (*otherwise undescribed*)—
Leghorn, 30 cks.
Bones—
Bombay, a quantity.
Smyrna, 246 t. J. Jowett & Sons
Constantinople, 65 bls. Jivan, Tekian, & Co.
Smyrna, 208 t.
Bombay, a quantity
Kurrachee, 8,813 bgs. Ralli Bros.
" 504 W. & R. Graham & Co.
Bone Meal—
Bombay, 1,200 bgs.
" a quantity.
Kurrachee, 1,000 bgs. Ralli Bros.
" 122 W. & R. Graham & Co.
Borax—
Havre, 135 cks.
Carbolic Acid—
Havre, 10 bskts.
Caoutchouc—
Manaos, 1,683 cs.
Akassa, 722 bgs. Royal Niger Co.
Accra, 11 cks. J. P. Werner
" 12 C. Lane & Co.
" 5 Davies, Robbin, & Co.
" 1 1 cs. Edwards Bros.
" 4 F. & A. Swanzy
Salt Pond, 2 cs. 3 brls. 2 pks. W. Griffiths & Co.
" 1 S. & C. Nordlinger
" 3 5 Edwards Bros.
" 4 Pickering & Berthoud
" 1 Radcliffe & Durant
" 1 Davies, Robbins, & Co.
" 1 brl. Tomlinson & Co.
" 3 W. B. McIver & Co.
" 4 Lintott Bros. & Co.
" 7 A. Miller Bros. & Co.
Cape Coast, 3 cks. S. Henrichson & Co.
" 3 L. Hart & Co.
" 4 W. Griffiths & Co.
" 2 A. Millerson & Co.
" Davies, Robbins, & Co.
" A. Miller & Co.
" 6 C. Friere
" 3 Ihlers & Bell
" 10 W. Duff & Co.
" 5 J. Smith & Co.
" 1 Tomlinson & Co.
" 2 H. B. W. Russell
" W. B. McIver & Co.
" 2 1 cs.
" 3 1 C. L. Clare & Co.
" 1 H. C. Nordlinger
" 6 Fletcher & Fraser
Dixcove, 1 brl. 1 cs. F. & A. Swanzy
Axim, 3 bgs. Edwards Bros.
" 2 Radcliffe & Durant
" 1 A. Miller & Co.
" 1 M. L. Levin
" 2 Pickering & Berthoud
Assinee, 3 brls. Millward, Bradbury, & Co.
" 1 cs. W. D. Woodin
" 1 A. Verdier
Grand Bassam, 2 bgs. T. E. Tomlinson
" 1 A. Miller, Son, & Co.
" 1 Grimwade, Ridley, & Co.
" 3 Pickering & Berthoud
" 1 W. D. Woodin
" 2 Openheimer Bros.
" 1 M. Ridyard
" 1 W. Comel Bank
" 1 bt. 1 brl. F. & A. Swanzy
" 1 cs. 1
Sierra Leone, 23 brls. Pickering & Berthoud
" 52 pns. 24 brls. Comp. Francaise
" 12 3 Broadhurst & Co.
" 44 pks. Paterson, Zachonis, & Co.
" 1 ck.
Conakry, 3 pks. 3 brls. A. Cros & Co.
" 12 pns. 2 Cie Francaise
Ceara, 39 bls. Leech, Harrison, & Co.
Cobalt Ore—
Valparaiso, 127 sks. A Gibbs & Sons
Copper Ore—
Valparaiso, 17 bgs. Balfour, Williamson, & Co.
" 727 Brownells, Lewis, & Co.
Cream of Tartar—
Barcelona, 5 hds. Wright, Crossley, & Co.
" 10
Tarragona, 3 cks.
" 5
Barcelona, 14 hds.
Chemicals (*otherwise undescribed*)
Rotterdam, 19 cks.
Dyestuffs—
Sumac
Palermo, 100 bgs. H. Rooke, Sons, & Co.
" 300
Barcelona, 100 A. Ruffer & Son
" 50
Palermo, 770 50 bls.
Annatto
New York, 3 brls.
Divi Divi
New York, 50 bgs. W. Clucas
" 220 D. Midgley & Son
Juan Guigo, 33 t. Muhl Pardo & Co.
Logwood
Jamaica, 476¾ t.
Valonia
Smyrna, 97 t.
Saffron
Valencia, 1 cs.
Myrabolams
Bombay, 408 bgs.
" 2,749 D. Sassoon & Co.
" 316 B. Craig & Co.
" 200
" 3,876 D. Sassoon & Co.
" 1,133 Ralli Bros.
Indigo
Bombay, 4 pks.
Extracts
Rouen, 35 cks. R. J. Francis & Co.
New York, 50 brls.
Boston, 340 bxs.
Havre, 1 ck.
Argols
Bordeaux, 101 cks.
Oporto, 22 cks. G. Bank of London
Fustic
Barcelona, 1,989 ps. Muhl Pardo & Co.
Orchella
Lisbon, 7 bls. A. Barbosa & Co.
Alizarine
Rotterdam, 2 cks.
Cochineal
Grand Canary, 29 bgs. Lathbury & Co.
Glycerine—
Rotterdam, 10 crbys. 8 dms.
Glucose—
New York, 90 brls.
" 549
Iodine—
Valparaiso, 9 brls. W. & J. Lockett
Manganese Ore—
Bordeaux, 6 cks.
Oxalic Acid—
Rotterdam, 4 cks.
Phosphate—
Ghent, 700 bgs.
Pyrites—
Huelva, 1,490 t. Tennar
" 1,620 Mathes
" 1,379 ,
" 2,046 ,
Phosphoric Acid—
Bremerhaven. 6 cs.
Potassium Bicarbonate—
Rotterdam, 8 cks.
Potash—
Rouen, 21 dms. Co-op. W'sale
Portland, 30 brls.
Sodium Silicate—
Rotterdam, 2 cks.
Sodium Nitrate—
Iquique, 11,037 bgs. Graham,
Tartaric Acid—
Rotterdam, 13 cks.
Ultramarine—
Rotterdam, 8 cks. 4 cs. 1 bx.
Zinc Oxide—
Antwerp, 25 brls.
Marseilles, 97 bgs. 8 brls.
Rotterdam, 55 cks. O. & H.

HULL.

Week ending April 3r

Acid (*otherwise undescribe*
Rotterdam, 4 cks. Hutchinso
" 10 bkts. G. Lawso
" 1 ck. Furle
Albumen—
Hamburg, 14 cks. Wilson, Son
Genoa, 8 cs. "
Ammonia—
Hamburg, 15 brls. "
Barytes—
Rotterdam, 6 cks.
Antwerp, 29 Bailey &
Bremen, 19 cks. Blundell, Spenc
" 23 Hanger, Watson, &
" 46
Bones—
Messina, 36 t. Wilson, Son
Chloral Hydrate—
Rotterdam, 1 cs. Hutchinso
Chemicals (*otherwise undes*
Fiume, 80 cks. Wilson, Son
Genoa, 15 "
Rotterdam, 5 cks. W. & C. L. F
" 3
Cobalt Oxide—
Hamburg, 4 cs. G. Malcol
Dyestuffs—
Rennet
Hamburg, 1 ck. G. Lawson
Rotterdam, 6 ck. Hutchinso
Argols
Bordeaux, 18 bgs. Rawson & F
Alizarine
Rotterdam, 48 pkgs. W. & C.
" 31 Hutchinso
" 20 cks. "
" 45 "
" 5 7 cs. 6 brls. "
" 11 W. & C. L.
Extracts
Trieste, 99 brls. Wilson, Son
Fiume, 450 "
Bordeaux, 50 cks. Rawson & F
Rouen, 78 cks. 20 cs. Rawson &
New York, 200 bxs. Wilson, So
Logwood
—— 25 brls.
Fustic
Boston, 40 bxs. Wilson, So
Sumac
Palermo, 705 bgs. Wilson, So
" 200
Glucose—
Hamburg, 25 cks.
" 100 bgs. Th McK
Stettin, 1,235 cks. C. M. I
" 73
Hamburg, 6
New York, 40 brls. Wilson,
Manure—
Phosphate Rock
Charleston, 1,940 t. Hammon
Naphthol—
Rotterdam, 2 brls. Hutchinso

LONDON.

Week ending March 29th.

Acids (*otherwise undescribed*)
Cologne, 300 brls. £457
Reval, 100 cs. 290
Demerara, 6 crbys. £33

Alum—
Salonica, 20 t. £118

Alum Cake—
Calcutta, 172 cks. £190

Arsenic—
B. Ayres, 1 t. 10 c. £48

Ammonium Sulphate—
Hamburg, 150 t. £1,800
Ghent, 35 410
" 30 11 c. 350
Hamburg, 60 19 720
Ghent, 15 1,050
Demerara, 5 17 73
Berbice, 3 37
Demerara, 20 245

Ammonia—
B. Ayres, 30 cs. £39

Acetic Acid—
Cape Town, 5 c. £15

Ammonium Carbonate—
Halifax, 20 cs. £45

Copper Sulphate—
Lisbon, 5 t. 11 c. £126
Genoa, 10 t. 246
Treport, 7 8 c. 180
Bordeaux, 3 3 75
Malta, 2 56
Salonica, 1 23

Clarifiants—
Bordeaux, 35 cks. £65

Caustic Soda—
Lisbon, 5 t. £40
Natal, 10 c. 17
Algoa Bay, 1 5 88
B. Ayres, 5 16 62

Citric Acid—
Rotterdam, 3 cs. £21
Hamburg, 2 cks. 70
Copenhagen, 10 kgs. 70
Brussels, 10 c. 74
Libau, 8½ 79
B. Ayres, ½ 4
Genoa, 5 37
Melbourne, 10 kgs. 70
Bombay, 1 cs. 5

Chemicals (*otherwise undescribed*)
Melbourne, 15 pks. £129
Yokohama, 40 cs. 227

Carbolic Acid—
Bombay, 5 t. £30

Coal Products—

Tar
St. John's, N'fld. 100 brls. £30
Sydney, 51 17
Dunkirk, bulk 4,000
Havre, bulk 212
Dunkirk, bulk 265
Havre, 35 cks. 101
Mauritius, 10 brls. 8
Calcutta, 24 rults. 10
St. John's, N'fld. 83 37
Natal, 30 drms. 6
Mauritius, 10 brls. 14
Sydney, 20 21
Bombay, 96 rults. 77
Calcutta, 100 48

Anthracene
Cologne, 252 cks. 1,591
Rotterdam, 450 2,600

Benzol
Hamburg, 20 drs. 285
Rotterdam, 24 300

Naphthalene
New York, 11 cks. £90
Marseilles, 12 10
Hamburg, 9 19
Treport, 55 53

Disinfectants—
Paris, 154 gls. £10

Guano—
Demerara, 50 t. £500
" 12 6 c. 129
Berbice, 6 7 67
Hamburg, 10 4 58

Manure—
Gothenburg, 186 t. 2 c. 944
Dunkirk, 11 10 48
Hamburg, 49 5 170
" 100 1,200
Rouen, 27 10 115
Bordeaux, 60 250
Barbadoes, 50 600
Antigua, 30 300
Guadeloupe, 15 420
Ghent, 130 1,560
Rotterdam, 58 11 365
Hamburg, 21 19 90

Nitric Acid—
Algoa Bay, 3 cs. £5
Cape Town, 5 10

Oxalic Acid—
New York, 3 t. 18 c. £190
Lisbon, 13 15
Algoa Bay, 10 cs. 29
Antwerp, 5 t. 16 c. 222
Yokohama, 1,282 lbs. 17

Superphosphate—
Cherbourg, 130 t. 990

Salicylic Acid—
Brisbane 1 cs. £29

Strontium Nitrate—
New York, 17 c. £26
" 3 t. 18 118

Sulphuric Acid—
Durban, 1 t. 10 c. £7
Penang, 1 17 90

Sheep Dip—
Algoa Bay, 28 pks. £35
E. London, 50 cs. 140
" 340 gls. 90
Algoa Bay, 50 cs. 1,750 drs. £567

Sulphur Flowers—
Lisbon, 250 brls. £193
Adelaide, 5 t. 1 c. 45

Salammoniac—
Salonica, 1 t. 19 c. £60
Gothenburg, 1 5 41
Shanghai, 5 10

Sulphur—
Brisbane, 6 t. 10 c. £37
Madeira, 4 4 33

Soda Crystals—
Hong Kong, 3 t. 10 c. £12
Sydney, 10 14 28

Sodium Bicarbonate—
Calcutta, 3 t. £18
Wellington, 2 10 c. 15

Tartaric Acid—
Auckland, 5 c. £34
Wellington, 5 32
Sydney, 5 32
Townsville, 5 61

LIVERPOOL.

Week ending April 1st.

Alum Cake—
Philadelphia, 6 t. 16 c. 1 q. £20

Antimony Regulus—
Alexandria, 2 c. £8
Hamburg, 1 t. 5 c. £84

Ammonium Carbonate—
Hamburg, 1 t. £30
Toronto, 15 c. 3 q. 26
New York, 14 6 3 480
Bombay, 2 5 44

Alum—
Bombay, 9 t. 4 c. £48

Ammonium Muriate—
Philadelphia, 21 t. 8 c. £504
Christiania, 7 3 q. 800

Bleaching Powder—
Amsterdam, 11 cks.
Barcelona, 10
Bombay, [illegible]
Boston, 400
Ghent, 130
Havana, 30
Rotterdam, 2
Santander, 50
Barcelona, 20
Boston, 121
Genoa, 24
Havre, 25
Montreal, 5 [illegible]
San Sebastian, 11

Boracic Acid—
Genoa, 1 t. [illegible]

Borax—
Christiania, 3 t. 11 c.
Rouen, 24 4
Ibrail, 1 10 3
Adelaide, 1 3

Borax Salts—
Havana, 1 t. 4 c.

Caustic Soda—
Antwerp, 20 drms.
Boston, 200
Cape Town, 30
Constantinople, 34
Coquimbo, 60
Galatz, 30
Havana,
Ibrail, 30
Malaga, 25
Marseilles, 28
Matanzas, 70
New York, 700
Odessa, 200
Vigo, 20
Corunna,
Amsterdam, 142 [illegible]
Antwerp,
Barcelona, 350
Bilbao, 20
Boston, 200
Cartagena, 25 40 kgs.
Ceara, 50
Genoa, 55
Hamburg, 15
La Guayra, 30
Madrid, 100
Montreal, 30
New Orleans, 50
New York, 523
Oporto, 54
Philadelphia, 100
Rio Janeiro, [illegible]
San Sebastian, [illegible]

Carbolic Acid—
Genoa, 9 c. 3 q. £65

Copper Sulphate—
Genoa, 2 t. 13 c £47
Adelaide, 20 574
Marseilles, 7 12 190
Barcelona, 17 1 3 q. 336
Bilbao, 4 19 129
Bordeaux, 6 8 2 155
Santos, 1 19 1 51
Valencia, 10 1 240
Barcelona, 4 18 3 100
Pasages, 10 260
Trieste, 5 19 1 130
Venice, 45 9 1 886
Genoa, 5 1 125
Tarragona, 20 2 1 400
Barcelona, 10 2 178
Bordeaux, 9 4 189
Venice, 7 17 1 167

Calcium Chloride—
New York, 25 t. 10 c. £64
" 4 19 10

Chemicals (*otherwise undescribed*)
Honolulu, 11 cs. £7
Baltimore, 30 t. 14 c 1 q. 130
New York, 10 30

Disinfectants—
Baltimore, 10 cs. £57
Calcutta, 5 c. 10

Epsom Salts—
Seville, 20 brls.
" 36

Guano—
Havana, 499 t. 5 c 2 q. £4,241

Iron Oxide—
Philadelphia, 12 t. 18 c. £310

Magnesium Chloride—
Bombay, 10 t. 3 c. £27

Manure—
Havre, 200 t. 4 c. 3 q. £660

Picric Crystals—
New York, 2 t. 5 c. £311

Phosphorous—
Shanghai, 100 lbs. £7

Potassium Bichromate—
Naples, 1 t. 2 c. £43
Yarmouth, 4 3 q. 9

Potassium Chromate—
Genoa, 5 t 2 c. £135

Potassium Carbonate —
Bombay, 10 c. 1 q. £14

Potassium Chlorate—
Hamburg, 1 t. 5 c. £52
Oporto, 2 10 93
Toronto, 7 16
New York, 12 10 564
Hamburg, 2 84
Havana, 2 93
Kobe, 25 1,200
Naples, 10 25
Rotterdam, 16 34
Stockholm, 5 233
Vera Cruz, 4 10
Amsterdam. 5 12

Phosphate—
Cape Town, 5 t. 1 c. £12

Sodium Bicarbonate—
Amsterdam, 41 kgs.
Barcelona, 100 150 cks 50 brls.
Boston, 100
Hamburg, 10
Havana, 20
Malaga, 20 1
Hamburg, 74
Kobe, 500
Mayaguez, 20
St John's, 50
Seville, 50

Soda Crystals—
New York, 747 kgs.
St. John's, Nfld., 95 kgs. 20 brls.
Boston, 280
Seville, 20

Soda Ash—
Baltimore, 330 cks. 255 tcs.
Boston, 200
New York, 4 419
Oporto, 16
Philadelphia, 146 666
Boston, 28
Ceara, 5 brls.
Genoa, 192
New York, 146
Seville, 6
Venice, 21 21 cks.

Sodium Silicate—
Bilbao, 20 brls.
Cartagena, 65
Oporto, 25
Seville, 100

Salt—
Boston, 116 t.
Callao, 2
Old Calabar, 150 4 c.
Pernambuco, 15 cs.
Portland M., 120
Sierra Leone, 50
Sydney, 103½
Banana, 50
Bonny, 18
Iquique, 50 cs.
Melbourne, 103 12
Boma, 14
Cameroons, 95

Saltpetre—
Ceara, 8 c. £10
Hamilton, 2 t. 5 49
Mexico City, 8 3 q. 10
Almeria, 1 t. 11 3 34

Sulphur—
Belize, 5 brls £5
Boston, 50 t. 2 c. 200
Vera Cruz, 10 2 q. 12
Port Alegre, 2 19 16

Saltcake —
Philadelphia, 37 t. 18 c. £81
Baltimore, 105 1 184

Sheep Dip—
Algoa Bay, 5t. £200

Salammoniac—
Aleppo, 5 c. £10
Naples, 10 2 q. 18
Philadelphia, 3 19 139
Galatz, 1 t. 4 32
" 9 15
Leghorn, 1 4 3 41

Strontia—
Rouen, 67 t. 3 c. £694

Sodium Biborate—
Hamburg, 9 t. 15 c. 1 q. £263
Rio Janeiro, 5 3 9

Tar—
Honolulu, 150 brls.

Zinc Chlorate—
Canada, 1 t. 3 c. 3 q. £25

Zinc Sulphate—
Hamburg, 4 t. 18 c. £75

TYNE.

Week ending April 3rd.

Alkali—
Amsterdam, 5 t. 4 c.
New York, 50 11
Gothenburg, 2 3
Christiania, 4 18
Genoa, 4
Leghorn, 8 5
Rotterdam, 9
Copenhagen, 2 14

Antimony—
New York, 15 t.
Christiania, 6 c.
Rotterdam, 2 10
New York, 15

Arsenic—
Venice, 10 t. 7 c.

Ammonium Sulphate—
Hamburg, 50 t.
Nykjobing, 5

Barium Chloride—
Genoa, 1 t. 1 c.
Leghorn, 7

Barytes—
Antwerp, 3 t.

Barytes Nitrate—
Genoa, 18 c.

Bleaching Powder—
Amsterdam, 36 t.
New York, 206 2 c.
Gothenburg, 1 7
Antwerp, 61 19
Hamburg, 26 3
Christiania, 12 12
" 10 8
Bergen, 5
Rotterdam, 29 6
New York, 174 16
Copenhagen, 13 11
Newfairwater, 81 19
Dansig, 3 1

Caustic Soda—
Amsterdam, 2 t. 18 c.
New York, 101 11
Gothenburg, 21 17
Antwerp, 8 2
Genoa, 18 17
" 19 9
Barcelona, 30

Gypsum—
New York, 35 t. 11 c.

Hydrochloric Acid—
Huelva, 78 lbs.

Litharge—
Amsterdam, 3 t. 15 c.
Hamburg, 1 1
Rotterdam, 1 19

Magnesia—
New York, 2 t. 4 c.
Hamburg, 7
Barcelona, 4 12

Potassium Bicarbonate—
Huelva, 10 lbs.
New York, 1 t. 17 c.

Potassium Bichromate—
Rotterdam, 1 t. 1 c.

Potassinm Chlorate—
Antwerp, 1 t.
New York, 10

Phosphorus—
Barcelona, 1 t. 12 c.

Soda—
Leghorn, 9 t. 6 c.
" 5 2
Naples, 7 18
Nakskov, 12
Nykjobing, 37 9
Amsterdam, 5 7
New York, 486 19
" 269 2
" 150 9
" 103 19
Copenhagen, 20 5

Soda Ash—
Genoa, 3 t. 18 c.
New York, 702 3
Venice, 20 15
New York, 98 17
Newfairwater, 40 3

Sodium Hyposulphate—
New York, 14 t. 1 c.

Sodium Hyposulphite—
Genoa, 1 t. 9 c.

Salammoniac—
Hamburg, 1 t. 5 c.

Sodium Nitrate—
Gothenburg, 1 t.

Sodium Bicarbonate—
Amsterdam, 1 t. 10 c.

Sulphuric Acid—
Huelva, 120 lbs.

Sodium Sulphate—
New York, 110 t. 15 c.

Sodium Bichromate—
New York, 9 t. 10 c.

Tar—
Hamburg, 5 t.
Malmo, 1 q.

GLASGOW.

Week ending April 3rd.

Ammonium Sulphate—
Nantes, 5 t. 3½ c. £60

Bleaching Powder—
Sydney, 121 c. £40
New York, 978 25 lbs. 405
Rouen, 106 38
New York, 1,070½ 390

Copper Sulphate—
Rouen, 383 c. £475
Bergen, 12½ 15

Caustic Soda—
Rouen, 295 c. £107

Iodine—
Philadelphia, 6,702 lbs. £4,022

Litharge—
Halifax, 15 t. 3½ c. £251
Boston, 15 10½ 225

Potassium Bichromate—
Gothenburg, 210¾ c. £365
Rouen, 31 t. 17¾ 1,039
Gothenburg, 7 ½ 227
Halifax, 83 155
Venice, 43¾ 90
Rotterdam, 2 cks. 30
Amsterdam. 13 190
New York, 501 c. 936
Antwerp, 460 669

Paraffin Wax—
Sydney, 29 t. 16 c. £804
Rouen, 7 8½ 180
Nantes, 9 9 245

Pitch—
Santander, 1,020 t.

Potassium Prussiate—
Halifax, 12½ c. £24

Soda Ash—
Baltimore,
Buenos Ayres, 100 cks.

Sodium Bichromate—
Rouen, 12 t. 4 c. £
Antwerp, 26

Sodium Silicate—
Sydney, 5 t. 6 c.

Superhosphate—
Barbadoes, 49 t. 16 c. £

Sulphuric Acid—
Bombay, 177 t. 12 c. £1,

HULL.

Week ending April 1st.

Chemicals (*otherwise undescribe*
Antwerp, 24 cs. 9 cks.
Amsterdam, 300 pks.
Bremen, 11
Christiania, 161 slbs.
Dunkirk, 30
Genoa, 4 156 cks.
Gothenburg, 15 17 15 pk
Hamburg, 22 47
Harlingen, 9
Konigsberg, 25 25
Leghorn, 2
Marseilles, 20
New York, 625
Odessa, 1 205
Rotterdam, 10 5 70
Rouen, 30
Stettin, 5 20

Caustic Soda—
Dunkirk, 18 drums.

Dyestuffs (*otherwise undescribe*
Antwerp, 5 cks.
Hamburg, 1
New York, 48 kegs.
Reval, 1 cs.

Glycerine—
Hamburg, 3 drums.

Manure—
Antwerp, 492 bgs.
Dunkirk, 836
Esbjerg, 100
Gothenburg, 306
Stettin, 226

Pitch—
Antwerp, 192 t.
Dunkirk, 221
Genoa, 277
Venice, 324

Superphosphate—
Esbjerg, 300 bgs.

Tar—
Antwerp, 2 cks.
Copenhagen, 6
Christiania, 5
Fiume, 10
Gothenburg, 150

GRIMSBY.

Week ending Mar. 29th.

Coal Products—
Dieppe, 4 cs. 35 cks.

Chemicals (*otherwise undescribe*
Dieppe, 79 cs.
Hamburg, 4
Rotterdam, 9

GOOLE.

Week ending March 26th.

Bleaching Powder—
Dunkirk, 72 cks.
Hamburg, 131

Benzol—
Antwerp, 66 cks.
Dunkirk, 46

Coal Products—
Boulogne, 3 cks.
Calais, 30
Rotterdam, 27

Copper Precipitate—
Rotterdam, 263 cks.

Chemicals (*otherwise undescribe*
Hamburg, 140 drums.

Dyestuffs (*otherwise undescribe*
Ghent, 1 cs. 22 cks.

Manure—
Dunkirk, 89 bgs.
Boulogne, 21
Ghent, 807

Pitch—
Antwerp, 519 t.

PRICES CURRENT.

WEDNESDAY, April 9, 1890.

Prepared by Higginbottom and Co., 116, Portland Street, Manchester.

…lues stated are F.O.R. *at maker's works, or at usual ports of shipment in U.K. The price in different localities may vary.*

		£	s.	d.
—				
…, 25 °/o and 40 °/o	per cwt	7/6 & 12/9		
Glacial	„	67/6		
…ic, S.G., 2000°	„	0	12	9
…nic 82 °/o	nett per lb.	0	0	6¾
…ic	„	0	0	3¾
…atic (Tower Salts), 30° Tw.	per bottle	0	0	8
(Cylinder), 30° Tw.	„	0	2	11
… 80° Tw.	per lb.	0	0	2
…us	„	0	0	1¾
…c	nett „	0	0	3¾
…	„	0	1	4
…uric (fuming 50 °/o)	per ton	15	10	0
„ (monohydrate)	„	5	10	0
„ (Pyrites, 168°)	„	3	2	6
„ („ 150°)	„	1	8	0
„ (free from arsenic, 140/145°)	„	1	10	0
…urous (solution)	„	2	15	0
…ic (pure)	per lb.	0	1	0½
…ric	„	0	1	2½
… white powdered	per ton	13	0	0
…oose lump)	„	4	15	0
… Sulphate (pure)	„	5	0	0
„ (medium qualities)	„	4	10	0
…um	per lb.	0	15	0
…ia, ·880=28°	per lb.	0	0	3
=24°	„	0	0	1¾
Carbonate	nett „	0	0	3½
Muriate	per ton	23	0	0
„ (sal-ammoniac) 1st & 2nd	per cwt.	37/- & 35/-		
Nitrate	per ton	41	10	0
Phosphate	per lb.	0	0	10½
Sulphate (grey), London	per ton	11	12	6
„ (grey), Hull	„	11	12	6
… Oil (pure)	per lb.	0	0	11¾
… Salt „	„	0	0	9½
…y	per ton	75	0	0
(tartar emetic)	per lb.	0	1	1
(golden sulphide)	„	0	0	10
… Chloride	per ton	8	0	0
… Carbonate (native)	„	3	15	0
… Sulphate (native levigated)	„	45/- to 75/-		
…g Powder, 35 %	„	5	7	6
Liquor, 7 %	„	2	10	0
…de of Carbon	„	13	0	0
…m Acetate (crystal	per lb.	0	0	6
Chloride	per ton	2	2	6
…lay (at Runcorn) in bulk	„	17/6 to 35/-		
…r Dyes:—				
…ine (artificial) 20 %	per lb.	0	0	9
…ta	„	0	3	9
…r Products				
…acene, 30 % A, f.o.b. London	per unit per cwt.	0	1	5
…l, 90 % nominal	per gallon	0	3	4
50/90 „	„	0	2	7½
…lic Acid (crystallised 35°)	per lb.	0	0	10¾
„ (crude 60°)	per gallon	0	2	8
…ote (ordinary)	„	0	0	2½
(filtered for Lucigen light)	„	0	0	3
Naphtha 30 % @ 120° C.	„	0	1	4
… Oils, 22° Tw.	per ton	3	0	0
f.o.b. Liverpool or Garston	„	1	8	0
…t Naphtha, 90 % at 160°	per gallon	0	1	9
…en Oils (crude)	„	0	0	2½
… Chili Bars)	per ton	48	7	6
…ulphate	„	26	0	0
…xide (copper scales)	„	51	0	0
…e (crude)	„	30	0	0
(distilled S.G. 1250°)	„	55	0	0
…	nett, per oz.	8½d. to 9d.		
…phate (copperas)	per ton	1	10	0
…phide (antimony slag)	„	2	0	0
…eet) for cash	„	14	0	0
…harge Flake (ex ship)	„	15	10	0
…etate (white „)	„	23	10	0
„ (brown „)	„	18	5	0
…rbonate (white lead) pure	„	19	0	0
…trate	„	22	0	0

		£	s.	d.
Lime, Acetate (brown)	„	8	0	0
„ „ (grey)	per ton	13	15	0
Magnesium (ribbon and wire)	per oz.	0	2	3
„ Chloride (ex ship)	per ton	2	5	0
„ Carbonate	per cwt.	1	17	6
„ Hydrate	„	0	10	0
„ Sulphate (Epsom Salts)	per ton	3	0	0
Manganese, Sulphate	„	18	0	0
„ Borate (1st and 2nd)	per cwt.	60/- & 42/6		
„ Ore, 70 %	per ton	4	10	0
Methylated Spirit, 61° O.P.	per gallon	0	2	2
Naphtha (Wood), Solvent	„	0	3	10
„ „ Miscible, 60° O.P.	„	0	4	6
Oils:—				
Cotton-seed	per ton	22	10	0
Linseed	„	24	10	0
Lubricating, Scotch, 890°—895°	„	7	5	0
Petroleum, Russian	per gallon	0	0	5¾
„ American	„	0	0	5¾
Potassium (metal)	per oz.	0	3	10
„ Bichromate	per lb.	0	0	4
„ Carbonate, 90% (ex ship)	per ton	19	0	0
„ Chlorate	per lb.	0	0	5
„ Cyanide, 98%	„	0	2	0
„ Hydrate (Caustic Potash) 80/85 %	per ton	22	10	0
„ „ (Caustic Potash) 75/80 %	„	21	10	0
„ „ (Caustic Potash) 70/75 %	„	20	15	0
„ Nitrate (refined)	per cwt.	1	3	6
„ Permanganate	„	4	2	6
Prussiate Yellow	per lb.	0	0	9½
„ Sulphate, 90 %	per ton	9	10	0
„ Muriate, 80 %	„	7	15	0
Silver (metal)	per oz.	0	3	7½
„ Nitrate	„	0	2	5½
Sodium (metal)	per lb.	0	4	0
„ Carb. (refined Soda-ash) 48 %	per ton	6	2	6
„ „ (Caustic Soda-ash) 48 %	„	5	2	6
„ „ (Carb. Soda-ash) 48%	„	5	10	0
„ „ (Carb. Soda-ash) 58 %	„	6	15	0
„ „ (Soda Crystals)	„	2	17	6
„ Acetate (ex-ship)	„	15	15	0
„ Arseniate, 45 %	„	11	0	0
„ Chlorate	per lb.	0	0	6¾
„ Borate (Borax)	nett, per ton.	29	10	0
„ Bichromate	per lb.	0	0	3
„ Hydrate (77 % Caustic Soda)	per ton.	12	15	0
„ „ (74 % Caustic Soda)	„	11	15	0
„ „ (70 % Caustic Soda)	„	10	5	0
„ „ (60 % Caustic Soda, white)	„	9	10	0
„ „ (60 % Caustic Soda, cream)	„	9	0	0
„ Bicarbonate	„	6	0	0
„ Hyposulphite	„	5	10	0
„ Manganate, 25%	„	9	0	0
„ Nitrate (95 %) ex-ship Liverpool	per cwt.	0	8	0
„ Nitrite, 98 %	per ton	27	10	0
„ Phosphate	„	15	15	0
„ Prussiate	per lb.	0	0	7¾
„ Silicate (glass)	per ton	5	7	6
„ „ (liquid, 100° Tw.)	„	3	17	6
„ Stannate, 40 %	per cwt.	2	0	0
„ Sulphate (Salt-cake)	per ton.	1	7	6
„ „ (Glauber's Salts)	„	1	10	0
„ Sulphide	„	7	15	0
„ Sulphite	„	5	5	0
Strontium Hydrate, 98 %	„	8	0	0
Sulphocyanide Ammonium, 95 %	per lb.	0	0	8
„ Barium, 95 %	„	0	0	5¾
„ Potassium	„	0	0	8
Sulphur (Flowers)	per ton.	7	15	0
„ (Roll Brimstone)	„	6	0	0
„ Brimstone: Best Quality	„	4	5	0
Superphosphate of Lime (26 %)	„	2	10	0
Tallow	„	25	10	0
Tin (English Ingots)	„	94	0	0
„ Crystals	per lb.	0	0	6¾
Zinc (Spelter)	per ton.	20	15	0
„ Chloride (solution, 96° Tw.	„	6	0	0

THE

CHEMICAL TRADE JOURNAL.

Publishing Offices: 32, BLACKFRIARS STREET, MANCHESTER.

No. 152. **SATURDAY, APRIL 19, 1890.** **Vol. VI.**

Contents.

Notices.

All communications for the *Chemical Trade Journal* should be addressed, and Cheques and Post Office Orders made payable to—

DAVIS BROS., 32, Blackfriars Street, MANCHESTER.

Our Registered Telegraphic Address is—
"Expert, Manchester."

The Terms of Subscription, commencing at any date, to the *Chemical Trade Journal*,—payable in advance,—including postage to any part of the world, are as follow:—

Yearly (52 numbers)	**12s. 6d.**
Half-Yearly (26 numbers)	**6s. 6d.**
Quarterly (13 numbers)	**3s. 6d.**

Readers will oblige by making their remittances for subscriptions by Postal or Post Office Order, crossed.

Communications for the Editor, if intended for insertion in the current week's issue, should reach the office not later than **Tuesday Morning.**

Articles, reports, and correspondence on all matters of interest to the Chemical and allied industries, home and foreign, are solicited. Correspondents should condense their matter as much as possible, write on one side only of the paper, and in all cases give their names and addresses, not necessarily for publication. Sketches should be sent on separate sheets.

Advertisements.

OPENING OF THE PERMANENT CHEMICA[L] EXHIBITION.

ON Tuesday evening last, the Permanent Chemical E[x]hibition was opened in presence of a large and repr[e]sentative gathering of gentlemen connected with t[he] chemical and allied industries. The opening address w[as] delivered by Mr. Ivan Levinstein, chairman of the Manche[s]ter Section of the Society of Chemical Industry, and chai[r]man of the Chemical Committee of the Manchest[er] Chamber of Commerce.

In a short opening address he said that when he was aske[d] by his friend, Mr. George E. Davis, to open the handsom[e] and well-arranged chemical exhibition which they had ju[st] had the pleasure of inspecting, he consented with very gre[at] pleasure, because he looked upon the establishment of th[e] exhibition as another sign of the activity and increase[d] enterprise in the chemical trade of this neighbourhood, an[d] he believed that such an exhibition must be useful in th[e] development of our industries.

That chemical exhibitions could be made instructive as we[ll] as attractive was proved by the popularity of the chemic[al] section of the Manchester Jubilee Exhibition, and it was n[ot] at all unlikely that its success first suggested the idea [of] perpetuating such an exhibition in this city. However th[at] might be, he thought that the promoters deserved all cred[it] for the manner in which they had carried out their scheme.

The object of the exhibition was not only the illustration by specimens and otherwise, of the results of chemic[al] operations and kindred processes, but also the illustration [of] the appliances necessary for arriving at these results in th[e] best and most economical manner. In addition to bringin[g] together the various exhibits of chemicals and products of [a] similar character and the necessary machinery for their pr[o]duction, the promoters had set aside in the building for th[e] convenience of visitors writing-rooms, a smokeroom, lavator[y] telephones, &c., with a view to facilitate the transaction [of] business and ultimately make use of this part of the buildin[g] as a kind of chemical exchange—an idea which he hope[d] might be realised before long.

Other useful attractions, such as a very comprehensiv[e] technical library, including a number of periodicals and trad[e] catalogues, ought to assist in making the exhibition buildin[g] a favourite meeting place of many who were interested i[n] the chemical trade. There was no doubt in his mind th[at] such an exhibition, under good management, must be usef[ul] in many ways. Take, for example, the large number [of] appliances intended for saving labour in chemical works, an[d] which was constantly being increased by the introduction [of] new and improved machinery and plant. Now even if [it] must be admitted that many of these appliances did n[ot] fulfil all that was claimed by the makers of them, on th[e] other hand there were many which would be welcomed an[d] introduced by manufacturers if they only knew their merit[s]. The difficulty, however, was to differentiate their value, an[d] this was not at all surprising, as those daily engaged i[n] practical work could ill afford the time and money necessa[ry] to experiment and test all of them.

ng-felt want" could be successfully met, then, by a
ent exhibition, showing by drawings and, where at all
ble, by working models, the advantages of the
ve appliances, the working of which could be ex-
by the attendants in the exhibition, or by reference
nanager.

ight be allowed, perhaps, to suggest that if space per-
there should be included in the exhibition smoke-
ing appliances, of which we had lately heard a great
d this mostly from people who knew very little of
actical working. Many of those who were engaged
hemical trade had tried smoke-preventing apparatus
form or another, and almost as many, he dared say,
lled them out again, either on account of their in-
cy or the greatly increased expenditure, and in the
hat thus far their experience had taught them that
l the best and cheapest remedy was plenty of boiler
nd slow and regular steady firing. If, however, work-
dels could be seen which could prove to their satis-
that any real advantage would result from their use,
no doubt that they would be only too glad to adopt
st suitable and best appliances, notwithstanding the
experience which some of them might hitherto have

her advantage which he anticipated from the exhibi-
as the collection of specimens of various chemicals
cognised makers—not mere specimens specially got
exhibition purposes. Such a collection might be found
seful for reference in disputes, which sometimes
n interpreting the words used in chemical contracts,
ary good commercial quality."

ould like also to suggest that in addition to periodicals,
s of a technical nature, newspapers, &c., statistical
nsular reports should be kept with reference to the
of chemicals, &c., and that samples and specimens
micals and similar products intended for export,
as well as English, should be exhibited, in order to
ur merchants and manufacturers the different methods
ing up in this and other countries. That this was
le, he thought, would be admitted when it was remem-
hat it had often occurred that foreign goods, especially
n goods, had been preferred to ours, for no other
than that their labels or general make-up was nicer in
ance, although the quality of the goods was in no
superior to the quality of ours.

idering all the advantages offered by the promoters to
lic, he heartily wished success to this new enterprise,
now declared the exhibition open. (Cheers.)

llowing descriptions represent the bulk of the exhibits now *in*
many others are now ready to come in.

The Blackman Ventilating Co.

specimen of their well-known Air Propeller, which may be
to many purposes in connection with the chemical industries.
ixed against the wall, it shows the application of this appliance
ntilating and equalisation of air in factories. Mechanical ap-
up to within the last three or four years, have not found a
ting for these purposes; but the ease with which the Black-
propeller can be applied in nearly every case and in every
changed the condition of things as well as the atmosphere of
our factories and workshops. In a woollen factory in the
England a 48-inch Blackman Air Propeller was fixed for dry-
on the existing tables, having about 500 square feet of wire
overed with it, and through this the Air Propeller draws or
air at the rate of 20,000 cubic feet per minute. All this air
from a very large adjoining room, so that this room is venti-
le the fan is doing work in another direction. Until the last
machines for moving air were designed as a rule for forcing
ntities against considerable resistance, as for blowing smiths'
cupolas. For these purposes the form and capabilities of the
l fan are well suited, but for dealing with freer air in quantity
othing like the Blackman Air Propeller. The scoop shape of
with its peripheral flange has given this appliance several
t advantages, and when we come to consider that an 84-inch
, making 300 revolutions a minute, will move not less than
100,000 cubic feet of air in the same time, with an expenditure of six H.P., the advantages of such a machine need no further demonstration.

Messrs. G. G. Blackwell and Co.,

the mineral merchants, of Liverpool, show a very handsome case, which they had formerly on view at the Irish Exhibition in London. There are some fine specimens of manganese of various strengths, and the summit of the pile of specimens inside the case is topped with one of the finest samples of manganese ore we have ever seen. Amongst the others stand out cryolite, French chalk, or steatite, of which Mr. Blackwell makes a speciality, and which, besides being in powder and lump, is sawn into small blocks or pencils for the use of engineers and others, for marking upon iron. Barytes, both in the form of carbonate and sulphate, is another good show of Mr. Blackwell's, as well as bauxite, of which he is, perhaps, the largest provider in this country, for alum making, alum cake, aluminoferric, and many other purposes. The case also contains some specimens of sulphate of alumina, of great interest to the technologist, and especially is this so of that made by means of Newland's process, and which is of very good quality. There is also a specimen of Weldon mud, the artificial oxide of manganese, on which so much turned in the celebrated trial between Messrs. Spence and Kurtz, showing that to-day it is a merchantable article, and can be bought on application to the proper quarter. A sample of water, purified by the A.B.C. process of the Native Guano Company, finds also a space in Mr. Blackwell's collection.

Messrs. Bracher and Co., of Wincanton,

have two appliances in the Exhibition; one of their Patent Mixer, which, though small in size, is able to get through a large quantity of work, and is much used by the smaller chemical manufacturers for such purposes as mixing dry paints, powders, such as baking powder, and other operations, which, though very important, are often performed in a very slip-shod manner, owing to the want of some such ready appliance as this.

Bracher's Automatic Still, in copper, is also an exhibit. An account of this still has already been given in the *Chemical Trade Journal*, and it may be again stated here that its peculiarity consists in its capability of doing the maximum quantity of work when the condensing water is kept nearly at the boiling point. There are many uses to which this still can be put, and we feel sure that no one would prefer to purchase distilled water when a small and continuously working still of this character can be obtained. Messrs. Bracher have several other appliances of extreme utility.

Messrs. Brunner, Mond and Co., Limited,

of Northwich, who are, we suppose, the largest makers of soda ash in the world, and that by the ammonia process, have an exhibit which has returned from the Melbourne Exhibition. The case has not suffered by its perambulations, which argues well for the care with which our common carriers perform their duties. In six large handsome glasses we find specimens of Messrs. Brunner Mond and Co.'s productions. There is the ordinary bi-carbonate of soda, which, on being calcined, furnishes their high strength soda ash of commerce, which is so well-known throughout the entire world. Then there is the purified bi-carbonate of soda, which we believe to be quite as good (at a less price) as some of the more highly prized qualities to be found in the market at much higher figures. We also find samples of sulphate of ammonia, muriate of ammonia, sesqui-carbonate of ammonia and soda crystals, which this firm now produce in the ordinary way of trade. These samples should be inspected by all paper makers, dyers, calico printers, and users of alkali products generally, for there is no doubt that the purchase of a high-class alkali such as this is, is far cheaper in the long-run than the use of such abominations as caustic bottoms and weak soda ash, which lead to all manner of difficulty when they are employed in industries where the presence of impurities is apt to do so much damage.

The Buckley Brick and Tile Co., Limited,

of Buckley, near Chester, have a stand showing a great many samples of their well-known firebrick ware. There is exhibited, for instance, a ring of well-made blocks which this firm supplies as a speciality for forming the arches in Glover and Gay Lussac towers. Anyone who values the permanence of his structure would do well to consult this form of building. It is far more lasting and far more solid than any arch built with ordinary bricks. There are tiles such as are used in the Chance-Claus kiln; elevators employed in the wet copper extraction process, spreading roses for saltcake pots, stoneware packing for absorbing towers, and revolver breaker blocks. We do not suppose that this list is an exhaustive one. The Buckley Brick and Tile Co. are an enterprising firm, and if they see their way to business in any particular direction, they are ever ready to consult the desires of their customers, and produce anything in fireclay ware that is wanted.

Mr. W. F. Clay,

the chemical bookseller, of Edinburgh, has a fine collection of chemical literature. Mr. Clay is one of those men always on the look out for something new, and even the old does not escape his notice, so that anyone requiring chemical literature, whether it be old or new, is sure to find it in Mr. Clay's list. We have just taken up a catalogue from beneath this exhibit, and there we find many works on the chemical and allied industries—brewing, dyeing and calico printing, astronomy, botany, electricity, engineering, mechanics, geology and mineralogy, meteorology, metallurgy, natural history, medicine, and surgery. In looking over the list we find that Mr. Clay does not confine his attention to English literature, American and German works forming quite as important a section with him as British, so that visitors to this Exhibition, if they are in want of any technical information, cannot do better than consult Mr. Clay's catalogue and exhibit.

The Clayton Aniline Co.

of Clayton, near Manchester, have a splendid show case, which was formerly in the Paris Exhibition.

The articles exhibited by the Company are:—Crude naphtha, light oil, pure benzole, pure toluole, pure xylole, nitrobenzole, nitrotoluole, nitroxylole, nitrobenzole for explosives; binitrobenzole, binitrotoluole, essence of mirbane, aniline, toluidine, xylidine, orthotoluidine, paratoluidine, aniline salt, metaxylidine, orthonitrotoluole, paranitrotoluole, azobenzole, azotoluole, benzidine, toluidine sulphate, carnotine sulphate, and antifebrine. Rouge drap Clayton, Clayton yellow, aurotine, Stanley red for silk, muriatic acid, nitric acid, sulphuric acid recovered from waste acid, &c., &c. The above colours, except carnotine, being patented by the Company.

The Company was formed in May, 1876; the works had to be built, and they began to manufacture in March, 1877. At first the works were fitted up for a production of only 250/300 tons of aniline per annum. They produced their own nitric acid, but no sulphuric acid, nor the pure muriatic acid required for the manufacture of aniline salt, and they bought their benzole.

Since then the Company, under the management of Dr. Dreyfus, has increased their power of production considerably. They hold the position of being the largest producers of aniline in this country.

The Company manufacture their own nitric acid, and the cylinder acid (pure hydrochloric acid), required for the manufacture of aniline salt; they make besides very large quantities of pure benzole, pure toluole, pure xylole from crude naphtha and light oil, which is used for their own manufacture. The quantities of crude naphtha, light oil, and benzole worked by the Company are very great.

A great improvement has been made in these works in the recovery of nitric acid, nitrobenzole, and sulphuric acid from the so-called waste acid from the nitrobenzole, nitrotoluole, and binitro manufacture.

In nitrifying benzole or toluole, a mixture of nitric acid and sulphuric acid at 168° is employed, the proportion being—

About 300 lbs. mixed acids for 100 lbs. pure benzole.
,, 260 ,, ,, toluole.

This gives nitrobenzole, and a weaker acid having a specific gravity of about 1,640 = 128°, which is called waste acid. As every 100 lbs. benzole nitrified, give about 230 lbs of this waste acid, and as the Company nitrify, on an average, 9,000 lbs. pure benzole per day, the quantity of waste acid produced is about 20,700 lbs. per day.

This large quantity of waste acid is too weak to be used as such for the production of nitric acid, and was, and is still, a very great burden to aniline manufacturers. This company are the first aniline manufacturers, who, not having a vitriol works at their disposal, have succeeded in making from this waste acid the ordinary rectified vitriol, such rectifi vitriol being in every respect as good as the ordinary rectified vitri direct from the vitriol works, and the Company are using such r covered vitriol for nitrifying and other purposes.

This method and apparatus for the recovery of the waste ac has been communicated to a friendly firm in France, Messrs. Duran Huguenin and Cie., of St. Fons, near Lyons, who have also exhibite at Paris, and who will bear out the statement as to the advantag of such recovery for aniline works. Another improvement made i the recovery of useful products from waste acid is the following:— The waste acid contains always about 10% of nitric acid and ½% nitrobenzole or nitrotoluole. Dr. Dreyfus has succeeded, at a sma cost only, in extracting these valuable substances from the waste ac and he believes he is the only aniline manufacturer doing this. H also utilizes the waste from the manufacture of binitro-benzole an binitro-toluole, as well as the nitric acid and binitro-benzole toluole that is left in such waste acid. The quality of the articl manufactured by the Company bears a high reputation for its purit The pure benzole, pure toluole, and pure xylole, are of the highe standard of purity for commercial products, as are the nitro produc and the binitro products therefrom. The pure aniline has an averag specific gravity at 15° Centigrade of 1·0270, and boils within degree. The ordinary toluidine contains over 35% of para toluidine the para toluidine contains under 1% of ortho, and the ortho toluidi under 0·5 para. The aniline salt is as nearly chemically pure possible.

The binitro-benzole has a solidifying point of 83°.
,, toluole ,, ,, 65°,

and they are both free from acid and moisture.

The other articles are at present used for the colour works of th Company. These works were begun in 1888, with the manufacture carnotine, a colour in every respect the same as primuline. Sin then the production has been extended by the manufacture of aurotin a colour patented by the Company, and which dyes wool a fa yellow, standing fulling, not bleeding, and very fast to light and air also Clayton cloth red, another colour patented by the Company, whic dyes wool in fast shades, standing the action of air and light, fullin and not bleeding. This latter colour has the remarkable property dyeing a red on silk, which does not bleed; it can be used also f producing pinks and reds on calico by printing. These colours hav been protected by letters patent.

Clayton yellow is another patented colour of the Company, whic dyes cotton without a mordant, and is useful for printing; it dyes sil also.

As distillers of naphtha, the Company supply solvent naphtha india-rubber manufacturers, and Dr. Dreyfus has recently taken o letters patent in France and England and elsewhere for the productic of a purified solvent naphtha, with which india-rubber goods, free fro the well-known obnoxious smell, can be produced.

Mr. J. Cortin, of Newcastle-on-Tyne,

has a very fine exhibit of non-rotative acid valves, which have bee devised to fulfil a want long felt by chemical manufacturers and othe of a safe, durable, and efficient means for stopping or regulating th flow of acids or other corrosive fluids. The ordinary cocks and oth means hitherto employed have many drawbacks, such as plugs liftin and flying out, scalding attendants, causing accidents, leakage, an the great wear and tear by friction of these soft metals cause them become useless. The body and valve of these appliances are made i a special mixture of regulus, and withstand the action of acids an other corrosive fluids for a very much longer period than any oth metal. The peculiarity lies in the plug being non-rotative, rising falling into its seat out of a mitred setting without turning round, an is therefore free from friction in working, and the wear and tear thus reduced to the smallest possible amount. Most chemical factori of any repute use these valves, and we are sure the exhibit will b looked at by all visitors with great interest. Mr. Cortin also show his combination of stone and antimony cocks, and his all-antimon bib-cocks. He also exhibits his steam tap for the regulation of stea into vitriol chambers, which is graduated so that the attendant ma be able to know whether the steam has been altered or not in h

There is also the usual hydrogen apparatus for the auto- ldering of lead, with bellows, jets, and nozzles. Altogether, it is a very interesting one, especially from a chemical engi- ɔoint of view.

Messrs. Davis Bros.,

g chemists and chemical engineers, have exhibited some of ucts from processes in which they are interested. The pro- istilling coal for the sake of the products, as now carried on sole licensees, Messrs. Newton, Chambers, and Co., near , is illustrated by samples of pure benzene and pure toluene in the operation, as well as samples of 90 per cent. benzol ent naphtha. The process is also applicable to the gases from ns, and samples are also shown of solvent naphtha, burning, icating oils, obtained from that source. These identical were first exhibited at the Inventions Exhibition in London, and are now in as good a condition as when sent from the the first instance.

roprietors also show a series of products obtained from the t of phosphate of alumina with sand, salt cake, and coal. By this mixture an insoluble double silicate of alumina and soda l, together with tribasic phosphate of soda, known in com- "Tripsa." The exhibit contains samples of the phosphate a deposits from Alta Vela and from the islands of Redonda Roques.

sa" has been employed as a preventive of boiler incrustation, for the purification of waste waters from mills and factories, mparative high price has hindered its becoming generally he application of "Tripsa" to the purification of sewage, waste waters from mills and factories, is shown by a series of and in connection therewith, is another series of the dried sludge obtained in the process.

orion evaporator, as applied to the recovery of soda from aper mill liquor, and to the recovery of potash from wool- liquor also is shown by an interesting collection of samples.

Messrs. Galloways, Limited,

exhibition two large photographs, one of their boiler shed, the largest in the world, being 420 feet long, and 180 feet ile another photograph is of three of their large steel boilers, ong and 8 feet in diameter, which are now driving 1,500 H.P. otographs themselves are very good specimens of the photo- rt, and give the observer a very good idea of the extent of the oducing industry of the present day.

Messrs. Grimshaw Bros., Limited.,

on, near Manchester, exhibit a very good collection of the eous chemicals used, not only in the india-rubber trade, but e industry of sizing and weaving. We find here, specimens rubber and india-rubber solution, starch and sizing materials nds; sizing tallow, sizing grease, French chalk, all of which special bearing upon the the textile industries. But the ipt running through Messrs. Grimshaw's exhibit, is the ation of the use of many zinc compounds. Chloride of de of zinc, carbonate of zinc, sulphate of zinc, zinc in zinc ferro-cyanide, zinc phosphate, zinc silicate, and other ons of zinc, are here exhibited. All these compounds are connection with some industry or another, and Messrs- w have for many years made a speciality of this kind of

Messrs. Jewsbury and Brown,

hester, show an excellent collection of ærated waters. This has become of immense importance during the past few d a very large demand has sprung up for non-intoxicating s. It is to provide for this that such firms as Messrs. and Brown have built up their businesses, and in the case nd in the exhibition, may be seen the ingenuity of many brains for the requirements of the public. Potash and soda waters sented, as well as Seltzer, Vichy, Carrara, and lime water, have a goodly collection of compound beverages, such as d beer, champagne cider, ginger beer, &c., &c. It may be y some that this is hardly a chemical industry, but if ill only reflect, they will understand that it is essentially a chemical industry. Carbonic acid is required and manufactured. This is forced through water under strong pressure. The various flavourings and ingredients forming the various beverages have to be manufactured, so that if this is not a chemical industry, it at least fosters and encourages a great many branches of it.

Messrs. C. R. Lindsey and Co,

of Clayton, near Manchester, are makers and manufacturers of pyro- ligneous products, as well as a variety of other chemicals. The case they have in the Exhibition contains a large sample of commercial grey sugar of lead, the crude material before refining, while the bottom shelf of their case contains a good assortment of sulphate of copper crystals. Commercial nitric acid and nitrate of lead may also be found exhibited, and there is a large sample of peroxide of lead. All these products are used in the industries of which Manchester is the centre, and the excellent quality of the products made by Messrs. Lindsey has obtained for them a good reputation.

Messrs. Charles Lowe and Co.,

of Reddish, near Stockport, have no exhibit of their products, but they have placed a large mural tablet on the Exhibition walls to inform visitors that they are makers of carbolic crystals, melting at 40°, at 35°, and also at 29°, and that they are makers of cresylic acid, of picric acid, and aurine, of phenates and cresylates, and of the various metallic salts of sulphophenic acid. They are also makers of sheep dips and disinfecting powders. Perhaps Messrs. Chas. Lowe and Co. are of opinion that "good wine needs no bush," and therefore it is unneces- sary to exhibit the actual substances produced in their extensive works at Reddish, but it would be much more interesting to intending pur- chasers to see the actual substances they are likely to require.

The Manchester Aniline Co.

have sent the case they had on view at the Manchester Exhibition, but the exhibits have been re-arranged, and the patterns, etc., dyed with the latest colours. Upon a shelf at the back of the stand may be seen a series of products from crude benzol right up to aniline, which is the base of most of the colours exhibited. There are also to be found dyed patterns of cotton yarns, raw cotton and prints, to illustrate the uses and applications of aniline black, and these are blended with a few fancy colours on jute, cotton yarns, satteens, prints, etc. There is also a very good collection of miscellaneous chemical products, such as acetate of chrome, and phthalic acid, in some of the best crystals it is possible to conceive. A collection of materials for softening and finish- ing cotton goods concludes the list which this firm have placed for visitors to see.

Messrs. Meldrum Bros.,

of Cathedral-yard, Manchester, have collected round one of the central pillars of the establishment a very complete show of their well-known steam ejectors, exhausters, silent boiling jets, and air compressors, so that visitors may be familiar with the style of work this firm is capable of turning out. There is to be found a large steam blower capable of delivering 50,000 cubic feet of air per hour, under a 2in. or 3in. water pressure; a boiling jet or silent heater, to boil 800 gallons of water per hour; a gas producer blower with inlet air valve, to gasify up to 5cwts. of coal per hour against a counter-pressure of 12in. or upwards; an air compressor for 35,000 cubic feet per hour, to a pressure of 5lbs. per square inch and upwards; an ejector or exhauster for 15,000 cubic feet per hour, which, with a 50lb. steam pressure, will give a vacuum of 26in. of mercury; and a 1in. acid elevator with five loose flanges for lifting 400 gallons of acid per hour. It may be well known to most of our readers that Messrs. Meldrum Bros. have fitted up a large number of under-grate blowers in con- nection with their patent for burning breeze, ashes, and refuse coal. Some day Messrs. Meldrum may possibly see their way to fit up one of these appliances in this Exhibition, anyhow, the matter deserves special attention on the part of coal consumers, who have to face the difficulty of a higher coal bill than they have had before for many years. A drawing of this patent may be seen in the Exhibition.

Messrs. E. D. Milnes and Brother,

of Bury, exhibit a very fine case of dyewoods and dyewood extracts. The show they have made has produced a good effect, and was much appreciated. In the centre of the case is a large and well-executed photograph of their Bury Works, around which is draped goods which have been dyed with their extracts. It is a very interesting sight, in this Exhibition, to be able to compare the old colours produced from natural products with the newer ones produced artificially from perfectly non-coloured raw materials, and perhaps if Messrs. Milnes could carry us back to the days when even extracts were unknown, the raw woods and materials being employed, we should see that a great advance in tinctorial power has resulted from the application of science to this industry.

The case contains specimens of almost all the natural products employed in the tinctorial arts, logwood, barwood, fustic, myrabolams, turmeric, va'onia, sanders wood, camwood, Brazil wood, sapan wood, larch bark, sumac, Persian berries, quercitron, quebracho, madder roots, lima wood, cutch, galls, divi divi, and cube gambier. Nicely polished pieces of the various woods accompany the glasses containing the specimens ready prepared for the dyer.

The Newcastle Chemical Works Co., Limited,

of Newcastle-on-Tyne, exhibit some of their well-known products, solid caustic soda of 77% strength, both solid and ground, and some of the Chance recovered sulphur, which is now being so largely produced by several of Mr. Chance's licensees. This exhibit is a parallel to that of Messrs. Brunner, Mond, & Co.; the latter deal exclusively with high strength carbonate of soda, the former deal almost exclusively with high strength caustic soda, and it is, no doubt, being perceived by all users of these products that the high strength is the cheaper in the long run. We know of many firms, who at one time, were perfectly satisfied with the results obtained by the use of 48% soda ash, and 60% caustic soda, that have now taken to use these high strength products almost exclusively, and the results obtained have been so much superior to the old *regime* that it is not likely to be departed from. The Newcastle Chemical Works Company have certainly produced an article in the shape of ground caustic, which is destined to come much more into use than even it is at present. The form itself is so handy compared with the old-fashioned way of breaking up a drum of solid caustic, that anyone who has used the ground article will at once appreciate its advantages.

Mr. Alfred Smith,

of Clayton, near Manchester, has on view a number of the products of which he makes a speciality, shellac and bleached lac, india-rubber substitute and india-rubber solution; chemicals employed in the rubber trade, such as hyposulphite of lead (black hypo), zinc oxide, iron oxide, light surface-black, sulphide of antimony, vermillion, and sublimed sulphur. There are also specimens of chloride of sulphur, bisulphide of carbon, solvent naphtha, and liquid ammonia. Mr. Alfred Smith is also the maker of the "Excelsior" disinfecting fluid which is exhibited in his case.

Messrs. Robinson, Cooks, and Co.

of the Atlas Foundry, St. Helens, exhibit two large drawings of their gas compressor and vacuum pump, fitted with Pilkington and Forrest's patent valves. These valves are so much superior to the old make of valves, that all who employ mechanical compressors should see these drawings. Messrs. Robinson, Cooks, and Co., have promised to add further details of this form of construction.

The Worthington Pumping Engine Co.

have on view one of their well-known "Duplex" pumps. It may be unknown to our readers that the Worthington pump was introduced into this country in 1885 after having been, for many years, in successful operation in the United States of America. There are now about 60,000 of these pumps at work, and at present they are being produced at the rate of 600 per month. Wherever they have been exhibited, either at work or at rest, they have secured the highest awards, and at the recent Paris Exposition they obtained the only "Grand prix" awarded to pumping machinery.

The pump is designed so as to obtain a positive steam valve motion without the aid of tappets, crankshaft, fly-wheel, connecting or eccentric rods. This is effected by arranging two steam pumps alongside each other, and causing the piston rod of the one to actuate the valve rod of the other by means of a swing lever and rocking-shaft. At the end of the stroke each pump rests till its steam valve is opened by the travel of its neighbour, and thus all slamming of water valves, and loss through "slip" are avoided. Before one pump l rest, the other has commenced to travel, and thus a cont uniform discharge of water is maintained.

This feature of the pump is of the greatest value in transp volumes of liquor to great distances, as in transport of by pipe lines, or as in the proposed transport of salt brin Cheshire saltfields to the place of manufacture; and has exclusive adoption in the National Transit Company's petr lines in America, after other pumps had been tried and fa pumps are self-contained, and require little foundation, a work done occupy very little space.

They are now in extensive use in all industries in this c water supply, boiler-feed, liquor transmission, hydraulic pre pumps, and, in short, for every purpose for which pumps can One guarantee of the success of the pump is the number of which have been put into the market.

Messrs. Musgrave and Co,

the well-known heating engineers of Belfast, have heated th with several of their slow combustion stoves. The exhibiti a large one, and the stove fixed close to the entrance has s capable of more than heating the place very uniformly. In room a much smaller stove is fixed, and the experience attended the lecture was that this size of stove is capable o room of twice the cubic capacity. The laboratories connect establishment are also heated with these slow combustion is a great pity that the excellence of these heating arrangen more widely known. Coke is the fuel employed, the absolutely smokeless, and their economy in the matter of fu marvellous.

Mr. J. Royle

has exhibited one of his steam-reducing valves, which Messrs. Benson, Pease and Company use in connection with the they have on exhibition. The general experience of ste that reducing valves are too delicate to be relied upon, but valve is a notable exception to this general rule, as it is reliable, and may be used with very great advantage whe pressure of steam is desired, such as, for instance, in the vitriol chambers, and stills of all descriptions.

The Chadderton Ironworks Company

have on view one of their patent steam dryers of large size, of which they have placed one of the Chadderton steam tra McDougall's Patent). The exhibit itself shows at once ho to fix either of these appliances to ordinary steam mains, and one and all were much struck with the simplicity of the arr There is no doubt the McDougall steam trap is one of the l *the* best—steam trap in the market, and those who have gi other kinds as being unreliable will no doubt be pleased wit ing of this one.

Messrs. R. and J. Dempster,

the engineers, of Manchester, have placed a large series photographs, showing many forms of work they execu annular condensers, telescopic gas holders, gas purifi exhausters, the Livesey patent washers and scrubbers; a very comprehensive picture of what they are able to exec works at Newton Heath.

Messrs. Joseph Wright and Company,

of Tipton, have placed two photographs illustrating the ap the Berryman feed-water heater, and they have also reser for the exhibition of one of these appliances. There is dc the high price of fuel now prevailing will be an incentive t many things tending to the econonmy of fuel, and much c that is being allowed to escape into the open air will in th utilised by appliances of this kind.

Messrs. Doulton and Co., Lambeth.

To praise the wares turned out by this celebrated firm wou superfluous, but we cannot refrain from remarking th admiration that was expressed at the beautifully finished s chemical stone ware exhibited. The collection consists mixing and boiling pots, stills with still-heads and conde acid receivers, taps, perforated tiles, and store jars for acid this last class recently delivered deserves a special word of mention It was of 600 gals. capacity, and a special kiln had to be constructed for firing it. In reference to the tiny taps displayed in the case we heard many visitors speculating as to the possibility of obtaining these for laboratory purposes—and it might be worth the while of the exhibitors to make a special push in this direction.

Messrs. R. Daglish and Co., St. Helens,

are represented in shadow rather than in substance, their exhibit consisting of two handsomely mounted collections of photographs, shewing typical machinery of their construction—blowing engines, a sludge pump, Brock and Minton's revolving filter press, etc. Perhaps the most striking feature of the collection is the photograph of an improved revolving black ash furnace, 30' 6" long with a diameter of 12' 6".

Mr. Joseph Aird, Greatbridge,

has made a very successful attempt to bring his speciality prominently before the eyes of visitors. The highly coloured tablet, four feet square, occupying a central position at one end of the exhibition, will allow no one to go away in a state of ignorance as to where iron tubing is to be obtained, or at least no one can attribute such ignorance to the fault of the enterprising manufacturer.

Messrs. Ashmore, Benson, Pease, and Co., Stockton,

are represented by three interesting exhibits. The patent ammonia still, of which they make a speciality, is illustrated by a large-sized model, completely fitted up, and capable of turning out a ton of sulphate of ammonia per week. These stills are either working or in course of erection in many places throughout the country—from Sunderland to Hastings—and the manufacturers have no less than four now in hand for immediate delivery. No better testimony to their efficiency can be desired than the fact that one well-known sulphate maker is at the present moment adding a fifth to his installation of four.

No less interesting is the small working model of their approved gasholder, constructed on what is now known as the wire-rope principle. The vertical guiding columns are superseded by a system of wire ropes, so arranged as to effectually distribute all lateral pressure, and obviate the necessity for outside support. Two accompanying photographs show a triple lift holder, fitted with the wire rope arrangement at Rochdale-road, Manchester. Another holder, on the same principle, is in successful operation at Haslingden, and a third is in course of erection at Newburn, near Newcastle.

This firm is also exhibiting one of Green's canvas screens—an arrangement in which canvas sheets are substituted for boards for gas scrubbing purposes. The main advantage consists in the possibility of obtaining twice the scrubbing surface in the same space.

Mr. Ernst Fahrig, St. Helen's Ozone Works, Plaistow.

We have here one of the most novel features of the Exhibition—a collection of samples of ozonised products. It must require a most courageous spirit to attempt to give practical life and being to a body so little known as ozone, but we think we can congratulate Mr. Fahrig upon the possession of this most essential quality. Samples I. and II. show esparto pulp bleached by ozone. Where this agent is employed there is said to be absolutely no "going back;" in fact, an imperfectly bleached material will become whiter by standing, as though some residual ozone were slowly spending itself, and thereby gradually bleaching the fibres. No. III. is ozonised water, suitable for killing microbes, and for sterilising purposes generally. No. IV. is a phial of ozonised oil. This is available either for medical or manufacturing purposes. In the latter case it may be utilised in the preparation of floorcloth, linoleum, &c., such as is shown in Sample VII. No. V. shows what is called "ozone ammoniated lime," the peculiarity being that a considerable quantity of nitrogen is said to be fixed in combination with the lime. No. VI. is "ozone oxidised mangan," a high oxide of manganese, formed by the action of ozone on a lower oxide. No. VIII. shows a bleached solution of sugar. Before treatment with [illegible] this liquor was jet black. The bleaching may be performed [illegible] or after boiling. It is also applicable to dry sugar of all

It is quite certain that neither of these processes could meet the requirements of the chlorine manufacture of the present day.

Nine years afterwards (March 16, 1847, No. 11,624), the well-known Dunlop process was patented. Dunlop describes his invention as follows :—"My invention consists of an improved method of producing chlorine fit for manufacturing purposes, by effecting mutual decompositions between the following substances :—Muriate of soda, or any other muriate, nitrate of soda, or any other nitrate, muriatic acid, nitric acid. The assistance of sulphuric acid is in some cases required. I generally prefer employing it, in order to obtain for residuum, sulphate of soda, suitable for the manufacture of soda, &c. The above substances may be employed either altogether, or only two of them ; as, for instance, nitrate with a muriate (in which case, of course, sulphuric acid must be employed), or a muriate with nitric acid, or a nitrate with muriatic acid, or muriatic and nitric acids may be used together. For the latter cases sulphuric acid is employed, according to the results wished to be obtained. On the application of heat, chlorine, an oxide of azote, and muriatic acid are evolved. These gases are then passed through a suitable condenser charged with sulphuric acid of a strength sufficient to absorb the oxide of azote. The chlorine and muriatic acid are then separated by means of water." In dealing with the foregoing nitro-sulphuric acid Dunlop says, "The process I usually follow is to introduce the nitrous sulphuric acid into a suitable vessel when, by the addition of water and heat, the oxide of azote is disengaged. This latter is caused to traverse a suitable condenser along with a sufficient quantity of air and steam, or water, and is by this means all transformed into nitric acid, which can be used again in the manufacture of chlorine, and again recovered in the manner described, and so on."

In the next year, 1848, McDougall and Rawson (No. 12,333, Nov. 21st) filed an omnibus specification in which the collection and oxidation of nitrous fumes from any source is described. They pass the lower oxides of nitrogen alternatively into water, or an aqueous solution of nitric acid, with air containing oxygen, and this patent was actually granted in face of the one granted the year before to Dunlop.

Swindells and Nicholson, in 1852 (No. 390), patented a process for the production of chlorine, by adding nitric acid to hydrochloric acid, and heating the same so as to produce chlorine and nitrochloric acid, and this being brought into contact with oxygen gas, with the vapour of water, nitric acid is produced, and the chlorine conveyed where required.

Higgin, in 1854 (No. 766), in describing his invention for a mode of separating metals from each other states that one of the fluids he prefers to employ is a mixture of muriatic acid and nitrate of soda, which yields chlorine in solution ; but it is not even hinted at as being a process for the industrial manufacture of chlorine gas. Gatty, in 1857 (No. 2,230), proposed to employ nitrates in the manufacture of chlorine, but only as an oxidiser for the manganese, but he patented the application of the nitrous fumes for use in the manufacture of sulphuric acid, either "at once," or after they had passed through condensers.

In the year 1858, Messrs. Roberts and Dale patented a process (No. 2,242) identical with that previously protected in 1847 by Dunlop. To be more correct, perhaps I may say identical up to a point. Messrs. Roberts and Dale pursued absolutely the same methods as Dunlop for the production of their chlorine ; but when the nitrogen oxides were driven off from the nitrosulphuric acid, instead of oxidising them for employment again, Messrs. Roberts and Dale conducted them into a solution of alkali.

We may now pass to the year 1864 and turn to Specification No. 2,313, which is that of Messrs. Baggs and Simpson. This specification only received provisional protection—whether it was discovered that it ran on lines so similar to that of Dunlop as to be a very doubtful possession, I know not, but it is a noteworthy fact that it was abandoned. There is, however, one point of interest that concerns us now. It is that the inventors, operating upon nitric acid with hydrochloric acid, in a separate vessel, think it necessary to specify that a gentle heat should be applied. Chloronitric and chloronitrous gas with free chlorine are given off. Although Messrs. Baggs and Simpson abandoned the foregoing specification, yet it is doubtful whether it left their minds altogether, as Isham Baggs in 1866 (No. 3,296), obtained letters patent for improvements in the manufacture and treatment of hydrochloric and nitric acids, which seem to me to bear somewhat upon the processes at which they had previously been working. These improvements were the treatment of liquid nitric or muriatic acid with sulphuric acid in order to secure the absorption of the water by the latter.

Up to this time we seem to have no method differing essentially from the 1847 process of Dunlop ; it is true that Baggs and Simpson preferred to bring the hydrochloric acid in contact with the nitric acid in a separate vessel, while Dunlop, as also did Roberts and Dale, preferred to mix the nitrate with the salt, and to decompose both with sulphuric acid, but in all three specifications there is a total absence of the why and wherefore of the various processes, and the claims are so wide that had the processes been successful, they might have been intended to include anything. They are just the kind of specifications it should be the aim of the legislature to reject, as such patents are not only of no use to their possessors, but they stand in the way of industrial improvement to a much greater extent than is generally imagined.

We may now pass over a period of twenty years, rather a long time in chemical history, but a period in which the chlorine maker's attention was almost entirely rivetted upon the problem of manganese recovery from the older chlorine process. On the 1st of October, 1884, John Taylor applied for provisional protection for a new process for the preparation of chlorine gas, and on the 12th day of June, 1885, he filed his complete specification, from which I gather what he intended to be his *modus operandi*. This specification is an improvement on all that have gone before. It certainly bears a nearer resemblance to that of Baggs and Simpson than to that of Dunlop, but, in my opinion, it is not identical with either. I have before you a model of the Taylor plant, constructed from my own ideas gathered from his specification. Taylor says :—" In carrying out my invention gaseous-hydrochloric acid, obtained directly from the salt-cake furnace, or in any other convenient way, is passed through a tower or tube, or other suitable apparatus containing fragments of pumice stone, coke, or other suitable inert material, through which cold strong nitric acid is allowed to flow. The action of the nitric acid causes the formation of chlorine, nitrosyl chloride, water, and small quantities of nitrogen tri- and tetroxides, about two-thirds of the chlorine of the hydrochloric acid being set free at once. Taylor does not give us the details of the reaction, probably owing to the complications introduced by the presence of air, which always exists in very large quantity in furnace gases ; but the main reaction is probably as follows :—

$$2HCl+2HNO_3=2Cl+N_2O_4+2H_2O.$$

But besides this, varying amounts of nitrosyl chloride are formed by subsequent combinations that take place between the free chlorine and nitrogen oxides. The water formed during the reaction, and also that, I presume, coming from the vitriol used for decomposing the salt, is stated to be periodically removed from the bottom of the nitric acid tower. Taylor then states that the mixed gases, chlorine, nitrosyl chloride, and nitrogen tri- and tetroxides are passed into strong sulphuric acid, which absorbs the nitrous fumes, permits the chlorine to pass on unchanged, and decomposes nitrosyl chloride with the formation of hydrochloric acid and nitrosyl sulphate.

The resulting gaseous mixture of chlorine with a certain proportion hydrochloric acid is submitted a second time to nitric acid, and afterwards to sulphuric acid. The resulting gases are put through the processes a third time if necessary. The two alternations will produce a chlorine containing 90 per cent. of the hydrochloric acid as free chlorine, while three alternations produce about 96 per cent. In actual practice this chlorine is further washed with water to dissolve out the last traces of HCl.

In order to regenerate the nitric acid employed, Taylor places the nitro-sulphuric acid in a suitable vessel, and dilutes it by the gradual addition of water, while a rapid current of air is passed through. The nitrous vapours evolved are passed through water in a series of Woolf's bottles, whereby nitric acid of sufficient strength for use over again is obtained.

Let us now leave this picture and experiment and carry ourselves to the year 1887. In this year William Donald obtained letters patent for an invention " to obtain chlorine from hydrochloric acid, and to recover the bye-products formed during the operation." Hydrochloric acid gas is first dried by passing it through a suitable receptacle charged with sulphuric acid. The dried gas is then passed into a vessel charged with a mixture of strong nitric and sulphuric acids, where the hydrochloric acid gas undergoes decomposition. Donald represents the reaction that takes place as follows :—

$$2HCl + 2HNO_3 = 2H_2O + N_2O_4 + Cl_2,$$

and states that the vessel containing the nitric and sulphuric acids should be kept at a temperature about 0°C., while that of the gases ought not to exceed 30°C. The gaseous mixture resulting from the foregoing operation is now passed into a vessel containing cold dilute nitric acid, whereby nearly the whole of the oxides of nitrogen become fixed or absorbed, the reaction being expressed as follows :—

$$N_2O_4 + H_2O = HNO_3 + HNO_2.$$

The chlorine, still containing small quantities of nitrogen compound, is now passed into a second absorbing vessel containing cold nitric acid, and finally through a scrubber containing sulphuric acid. This completes the elimination of the nitrogen compounds, and leaves the chlorine to pass on for use.

The nitric acid is regenerated by warming the nitrous liquors from the absorbers, and blowing a steady current of air through.

At a suitable temperature the nitrous acid very readily absorbs oxygen and becomes converted into nitric acid. The current of air requires very careful regulation, the object being, of course, to obtain a maximum of oxidation with a minimum of lower nitrogen compounds carried forward. Under any circumstances, some nitrous fumes will escape from the warmed liquor, but their hydration and oxidation will be effected as they pass (together with the excess of air that must always be employed) in contact with the moistened surfaces exposed in the scrubber above. If this regenerative process be properly conducted, and the plant not worked beyond its capacity, only the merest traces of oxides of nitrogen will pass the scrubbers, and these can be recovered by absorption with strong sulphuric acid, so that the residual nitrogen and the excess of oxygen finally pass away from the apparatus absolutely free. It will be seen at once that this is a most important feature in a process involving the use of large quantities of so costly a material as nitric acid.

The apparatus you see built up before you has been designed as nearly as possible from Donald's specification, and, I think, will show you fairly well how the inventor, when framing this, intended to work the process on the large scale.

The sulphuric acid that has done duty in the absorber is mixed with nitric acid to form the decomposing mixture above mentioned. When the action of the hydrochloric acid upon this has removed the nitric acid the sulphuric acid remains behind, weakened, it is true, by the absorption of the water formed in the process, but still sufficiently strong to act as a drier upon the pot or roaster gases. It is then evaporated to say 150°Tw. and used for decomposing a subsequent batch of salt. Thus you will notice that it makes its entrance, so to say, at the end of the series of operations, and moves steadily forward, stage by stage, to its exit at the beginning. You will also notice that the nitric acid moves in a sphere equally important and equally well defined. It is perpetually being pulled to pieces and put together again. By its decomposition with hydrochloric acid gas it is robbed of its water and a portion of its oxygen; by absorption in dilute nitric acid, and subsequent treatment with air, this water and oxygen are restored, and it returns to the scene of its first labours as efficient as ever.

I am sorry Mr. Donald is unable to be here, as he could have gone more deeply into the matter than I can possibly do. Standing as I do in the position of having been consulted by both Mr. Donald and the Taylor Syndicate, I am only at liberty to draw upon such materials for this paper as are accessible to everyone that may be interested in this subject, but I think I have said sufficient to shew you that the nitric acid chlorine process has passed the fairy-tale stage of its history—that it is no longer confined to the realm of myth and legend, but has entered upon the field of actual, and, I may say, startling possibilities. I hope that before long we may meet together again these rooms to hear news of further development, and in the meantime I would advise those most interested in the matter to consider what will be the position of affairs when possibilities have become realities, and the Weldon Process a thing of the past.

FRENCH AGRICULTURE IN 1789 AND 1889.*

THE area of land capable of cultivation in France is about 47½ million hectares, one-third of which is taken up with cereals (15,440,000 hectares), this being an increase of about two million hectares since 1789 (one hectare corresponds to about 2½ acres). At that date 10 million hectares of land were lying fallow, whilst to-day no more than one-third of that area (3,644,000 hectares) is in the same condition. A hundred years ago there were 7,600,000 hectares of moorland, which have now been reduced to 3,880,000 hectares. In other words, about one-fifth of the cultivatable land has been actually brought under cultivation in the hundred years, and now bears crops. The land laid out in gardens and orchards has remained practically the same, while vineyards have increased almost one-twentieth.

The extension of cultivation, due to the reclamation of moorland and the reduction of the amount of fallow land, has taken effect principally in the formation of natural and artificial pastures, and the increased production of fodder and potatoes, as is shown by the following numbers:—

	1789.	1889.
	Hectares.	
Potatoes	4,300	1,488,000
Artificial meadows (land sown with grass or clover)	1,000,000	3,253,000
Root crops	100,000	1,397,000
Meadows and pastures	3,000,000	5,827,000

This increase in the cultivation of plants intended for feeding cattle naturally indicates that the number of domestic animals has undergone a large increase during the century; the number of horses in 1789 was 2,400,000, and has risen to 2,900,000, an increase of about one-fifth. The number of cattle, however, has been almost doubled, rising from 7,655,000 in 1789 to 13,400,000 in 1889. There are fewer sheep in France at the present time than there were a hundred years ago, 23 millions now, against 27 in 1789. As to pigs, the statistics of 1789 are [illegible], but their numbers decreased from 23 millions in 1840 to 7 millions at the present day, the difference being more than covered by [illegible]ation.

*From an article by Mr. Grandeau in the *Temps*.

The production and consumption of meat have proceeded on p
lines to the alteration in the number of cattle which we have jus
tioned. The total consumption of meat in France in 1789 w
million kilogrammes (1 kilogramme = 2 lbs., or 50 kilos. = 1
corresponding to a mean annual consumption of 17 kilos. per h
the inhabitants, and a total value of 203 million francs. The
consumption at present is 30·360 kilos. per head, representing
weight of 1,200 million kilos., and a value of 1,632 million
These numbers, which have changed so largely during a century
how small the consumption of meat still remains among us, and
an enormous gap remains to be filled by home production in t
mal food of our population. Lavoisier estimated the produc
wheat in 1789 at 31 million hectolitres (1 hectolitre = 2¾ bushel
present production is 110 millions, 15 of which are required f
crop of the following year, and 4 million are employed for ind
purposes, leaving 90-92 million hectolitres disposable for food,
about 104 millions are required. The consumption of corn in the
about 1834 was 1·64 hectolitres per inhabitant, whilst it is now
hectolitres. The area devoted to wheat amounted in 1789 to only
one-third of the whole amount employed for cereals—say 4 or
millions; to-day it amounts to 7 millions, and makes up one
the whole cereal producing area. A hundred years ago the
class scarcely knew of the use of wheaten bread, whilst to-day,
greater part of France, wheaten bread is gradually replacing tha
with rye, oats, and buckwheat. But here too much remains
done. Our mean harvest is seven times the amount sown, and a
to 15 hectolitres per hectare. It will have to increase by 2 or 3
litres if the use of wheaten bread is everywhere to replace that
inferior cereals. This result, as our readers are aware, is
attain.

Let us now inquire into the increase of the value of agricultu
duce in France. Lavoisier estimated the total value of the p
obtained in 1789 from arable land, meadows, and vineya
2,000,750,000 francs. It had increased by 1840 to 5,000,
francs, amounted to 7,664,000,000 in 1862, and now stands
figure of 8,000 million francs (£320,000,000).

It is beyond doubt that, although this very marked increase
value of agricultural produce in France is largely due to the
and industry of our worthy agricultural population, yet the inc
action of public authorities in aid of education and the introduc
scientific methods of treating the soil and of handling cattle ma
claim a share in the progress which has been made.

Almost completely deprived of all encouragement—financ
otherwise—agriculture was formerly given up to routine an
without schools, without teaching of any description, was, as i
bound to a stationary condition which it was beyond the power
individual efforts to appreciably modify. The Government
Republic has, however, rightly understood the function of th
with regard to agriculture, its proper duty being, before all, t
velopment of instruction and the propagation by all possible m
applications of chemical and biological science to the cultiva
plants and the rearing of animals. The rapid increase of adva
accorded to agriculture and the development of agricultural inst
of all kinds, which has distinguished the last twenty years, ha
rendered visible to the visitors to the Paris Exhibition by an in
arrangement of gilded cubes, which represent, by their increasin
the annual votes made by the State for general agricultural p
during the century.

In 1789, the expenditure under this heading amounted to
francs; in 1829 to 297,828 francs; to 898,000 francs in 1849,
as much as a quarter of a million francs in 1869. It has now r
8,339,000 francs, a sum which is still small enough, if we judge
what has already been effected, the progress which yet remains
made in increasing the wealth of the country by large grants fo
cultural education in all its forms. It appears even less consi
when compared with the amounts expended on this object by
nations. But whatever remains still to be desired, the agricultur
France can never, without ingratitude, refuse to appreciate the
which public authorities are showing on their behalf, and to ac
ledge the steady improvement caused by their liberality in the co
of the various agricultural stations. Although small in comp
with many foreign budgets, the actual amount paid by the Min
Agriculture leaves far behind the wretched sums which were de
to agriculture by the Governments anterior to 1870. The la
years have been particularly fertile in useful innovations. The
witnessed the restoration of the Agricultural Institution, the
schools of practice and of the departmental chairs of agricultu
reorganization of our great national veterinary schools, the e
ment of experimental plots for teaching purposes, and, finall
increase of our agricultural stations and laboratories.

All these institutions were worthily represented at the Paris E
tion, and gave evidence of their vitality by the large amo
scientific and practical work they had produced during the ten y
their existence.

A QUICK FREEZING REGENERATIVE ICE-MAKING APPARATUS.

THIS machine, made by the Pulsometer Engineering Company, of London, is designed to produce ice in blocks of about 1½" to 3" thick continuously, after the brine has been reduced to the proper temperature. The illustration shows a machine which will make about 40 lbs. of ice per hour. The apparatus is of the simplest construction and very cheap. It comprises a steam cylinder driving a compression pump direct. The ice-moulds and refrigerator are all contained in one tank, and the arrangement is such that the refrigerator cannot frozen up. The condenser is contained in a circular tank, fixed on same base-plate that carries the engine, and therefore the machin self-contained, and no expensive fixing or foundations are requir The apparatus is so arranged that it can also be used for cooling liqu or refrigerating rooms to any desired temperature.

Its functions are controlled by a single valve, and the mechanis so simple that it can be operated by persons quite unfamiliar w machinery, and for this reason the machine is specially applicable cooling liquids, gases, &c., and in all other trades where skil mechanics are not employed.

SULPHUR COMPOUNDS IN CRUDE PETROLEUM.

IN a recent number of the *Bericht der deutschen chemischen Gesellschaft*, Charles F. Mabery and Albert W. Smith have contributed a note on the results of their investigation on the sulphur compounds of the crude petroleum from the Ohio Oil Field. They find that on fractionating the crude oil, the sulphur compounds collect in the portion boiling between 200° and 300° C., and that they can be easily separated from this fraction by the addition of concentrated sulphuric acid. From this acid solution, after neutralising with lead carbonate or lime, the compounds separate on warming. They appear to be very unstable, but when the aqueous solution is distilled, a yellow oil collects in the distillate, which contains 14·97 %. of sulphur. When they attempted to distil the oil, it decomposed, evolving sulphuretted hydrogen. Under diminished pressure the oil was fractionated into portions which, on being tested, contained no thiophene or mercaptan compounds, but consisted of various organic sulphides. The portions having the lowest boiling point yielded on distillation a hydrocarbon which formed a bromine addition compound of the formula $C_{17}H_{14}Br_{4}$. By the addition of alcoholic mercuric chloride they also succeeded in isolating the double compounds of methyl, ethyl, and propyl sulphides, and obtained indications of the presence of butyl sulphide.

SOME EXPERIENCES WITH ZINC.

ZINC is often used in boilers and hot water tanks to prevent the c rosive action of the water on the metal of which the tank or bo is composed. The action appears to be an electrical one, the iron be one pole of the battery and the zinc being the other. Under the act of the current of electricity so produced, the water in the tank is slo decomposed into its elements, oxygen and hydrogen. The hydro is deposited on the iron shell, where it remains. It will not unite w iron to form a new compound, but if any iron rust (known to chemists as oxides of iron) is present, it will remove the oxygen fr this and deposit the metallic iron on the plates. The oxygen of water that is decomposed, instead of going to the iron, goes to the z and forms oxide of zinc, and in the course of time the zinc will be fou to be almost entirely converted into oxide, only a small fraction of original metal being left.—*Invention.*

"FROM COAL TO CALICO."—A lecture under this title was gi by the Rev. W. G. Whittam last Friday before the Eastbourne Natu History Society. Dealing with the aniline colours, the lecturer e mated the annual saving to the country by their use, instead of the madder-root preparations, at £4,000,000.

PRIZE COMPETITION.

E more than twelve months ago, one of our friends ited that it might be well to invite articles for , on some subject of special interest to our ring a prize for the best contribution, he at the offering to defray the cost of the first prize, and a subject in which he was personally interested. date we were totally unprepared for such a sug- t on thinking over the matter at intervals of have come to think that the elaboration of such a ld not only prove of benefit to the manufacturer, go some way towards defraying the holiday the prize winner.

herefore prepared to receive competitions in the ubjects :—

SUBJECT I.

method of pumping or otherwise lifting or forcing ous hydrochloric acid of 30 deg. Tw.

SUBJECT II.

method of separating or determining the relative f tin and antimony when present together in com- nples.

subject, the following prizes are offered :—

prize of five pounds, one second prize of one additional prizes to those next in order of merit, of a free copy of the *Chemical Trade Journal* for ths.

petition is open to all nationalities residing in any world. Essays must reach us on or before April ibject II, and on or before May 31st, in Subject I. will be announced in our issue of June 28th.

rve to ourselves the right of publishing the con- f any competitor.

ay be written in English, German, French, Spanish

Trade Notes.

OF SALT EXPORTS IN CHESHIRE.—The export list n the Custom-house returns and the home trade show that quantity of salt exported from Cheshire was only 35,531 74,161 tons for March of last year. The decrease is r example, the United States took only 2,367 tons, against rch of last year; British North and South America took gainst 16,233 last year; Africa, 2,155 tons, against 3,417; mah, and other Asian ports, 15,501 tons, against 40,384; s parts of Europe, 2,916 tons, against 4,148. The amount January and February is a little in favour of this year's a total of 119,267 tons, against 106,293. The falling off ttributed to the agitation over the coal trade and to a rise *Times.*

LOCOMOTIVE COAL.—As the time approaches for enter- usual yearly contracts for the supply of locomotive fuel, companies in particular are now showing a good deal of e matter, as there does not appear to be any question that or the ensuing year will be much higher than those now in 1889, most of the companies were supplied with the best aire "hards" at about 6s. per ton; in July, 1889, the price rom that date was raised to 8s. 6d., and in the early part t year the tender for the North-Eastern was 10s. 6d. per tter rate, at least, is expected to be the one that will be e end of June. This will make the increase in the price fuel equal to 75% in less than two years. How it will way companies will be gathered from the fact that before in the price of steam coal the London and North-Western o. a year for coal and coke for locomotives, the Midland 000., the North-Eastern, £200,000., and the Great ething like £190,000. The South Yorkshire Steam Coal- ciation, who supply several railway companies with steam neeting at Barnsley on Wednesday evening, when it was e prices now obtained for steam coal ranged from 10s. 3d. n, whilst, with the increase in the price of coal, there was ng advance in the cost of nearly every colliery require-

Market Reports.

MISCELLANEOUS CHEMICAL MARKET.

A steady business continues to be done in soda ash and caustic soda, and prices are well maintained for early deliveries. Ordinary grades of soda ash at 1⅜d. to 1½d.; refined, 1¾d. Caustic soda, 70%, £9. 17s. 6d. per ton on rails; 60% at £8. 17s. 6d., and cream, 60%, at £8. 12s. 6d. Saltcake, 25s. per ton in bulk. Bleaching powder has had a distinct decline in value, and offers for spot delivery at £5. per ton on rails, and for April-June at lower figures. Chlorate of potash 5d. per lb., makers quoting forward 5¼d. per lb. Soda crystals, £2. 15s. to £3. per ton in bags and casks respectively. Vitriol and muriatic acid continue to find a ready outlet, and prices are firm. Sulphur in all grades is quiet and without change in value. Sulphate of copper, £25. per ton nearest spot value; £24. per ton for May-June; and for June-December, £20. to £21. per ton. Oxalic acid scarce and now quoted 4d. per lb. Aniline salt and oil firm, and in rather better demand. Carbolic acids, crude and crystal, distinctly weaker, and offered within the week at considerable concessions on prices lately ruling: crystals 34° now at 9¼d. to 9½d. per lb. Pale liquid unchanged at 1s. to 1s. 1d. per gallon. Acetates of lime quiet, but the market on the whole is firm at £8. for brown, though some parcels have been changing hands at lower figures; grey, £13. 15s. per ton. Acetate of soda, £16. 10s. per ton. Acetic acids generally are quiet and without change in value. Lead salts: both nitrate and acetate are quiet; prices unaltered. Tin crystals rather lower at 6⅜d. per lb. Yellow prussiate of potash 9½d. per lb., but some odd parcels offering a shade under this. Potash, carbonate, and caustic maintain a firm position, and are commanding rather higher prices. All sundry chemicals for textile trades continue dull, and there is very little movement in prices.

METAL MARKET REPORT.

	Last week.		This week.
Iron	48/9	....	46/0
Tin	£90 0 6	....	£89 7 6
Copper	48 0 0	....	48 10 0
Spelter	20 12 6	....	20 7 6
Lead	12 12 0	....	12 10 0
Silver	44d.	....	45⅜d.

COPPER MINING SHARES.

	Last week.			This week.	
Rio Tinto	16⅛	16¼	....	16¼	16⅜
Mason & Barry	6 5/16	6 5/16	....	6¼	6⅜
Tharsis	4½	4⅝	....	4⅞	5
Cape Copper Co.	3½	3⅝	....	3½	3⅝
Namaqua	1¾	2	....	1 15/16	2 1/16
Copiapo	2 5/16	2 7/16	....	2⅜	2½
Panulcillo	1	1⅛	....	15/16	1 1/16
New Quebrada	⅝	¾	....	½	⅝
Libiola	2⅛	2⅜	....	2¼	2⅜
Tocopilla	1/-	1/6	....	1/-	1/6
Argentella	3d.	6d.	....	3d.	9d.

THE COPPER MARKET.

From Messrs. Harrington and Co.'s report of the 16th we note that the total Chili charters since January 1st have been only 7,600 tons, as against 9,900 tons for the same time last year. The actual stocks of copper in English and French warehouses are 87,770 tons, as compared with 118,503 tons at same date last year. The Tharsis Sulphur and Copper Co. will pay next month a dividend of 20% for 1889, being the same as paid for 1888.

WEST OF SCOTLAND CHEMICALS.

GLASGOW, Tuesday.

There has been rather a better turnover in the general market, as was expected, the break of prices having given a much needed impetus to buying. The drop is pretty general here, as elsewhere, but with us bleaching powder perhaps shows slightly stiffer than either in Lancashire or on Tyne, and last week's quotations are about maintained.

Refined alkali, chlorate of potash and sulphate of copper are all lower, and caustic soda has made a further drop all along the line, to the distinct promotion of business, the bulk of orders placed being much larger than for the past few weeks. Borax has gone up 10s., this forming the only exception of note. Sulphate of ammonia in this market, for whatever reason, has gone worse and worse, and to-day it has passed hands certainly at as low a figure as £11. 6s. 3d., and probably at £11. 5s. The price, spot, Leith must stand, therefore, as £11. 5s. to £11. 7s. 6d. on even the most favourable interpretation. Forward business has come down correspondingly, and delivery two or three months ahead has been fixed at £11. 10s if not lower. Chief prices current are :—Soda crystals, 52s. 6d, less 2½% Glasgow; alum in lump £5., less 2½% Glasgow; borax, English refined £29., and boracic acid, £37. 10s. net Glasgow; soda ash, 1⅛d. net Tyne; caustic soda, white, 76°, £12. 10s., 70/72°, £10. 2s. 6d., 60/62° £9. 5s, and cream, 60/62° £8. 15s., all less 2½% Liverpool; bicarbonate of soda, 5 cwt. casks, £6. 5s., and 1 cwt. casks, £6. 10s. net Tyne; refined alkali 48/52°, 1½d., net Tyne; saltcake, 30s.; bleaching powder, £5. 10s. to £5. 15s., less 5% f.o.r. Glasgow; bichromate of potash, 4d., and of soda 3d., less 5 and 6% to Scotch and English buyers respectively; chlorate of potash, 4⅞d., less 5% any port; nitrate of soda 8s. to 8s. 3d.; sulphate of ammonia, spot, £11. 5s. to £11. 7s. 6d. f.o.b. Leith; salammoniac, 1st and 2nd white, £37. and £35., less 2½% any port; sulphate of copper, £25., less 5% Liverpool; paraffin scale, hard and soft, 2½d. per lb.; paraffin wax, 120°, semi-refined, 3⅜d.; paraffin spirit (naphtha), 8d. a gallon; paraffin oil (burning), 6¾d. to 7d. at Glasgow for Scotch consumption (English orders done at about 1d. a gallon lower); ditto (lubricating), 865°, £5. 10s. to £6. per ton; 885°, £6. 10s.; and 890/895°, £7. 10s. to £8. Week's imports of sugar at Greenock were 13,404 bags.

REPORT OF MANURE MATERIAL.

There continues to be a good prompt business passing at some decline for nitrogenous, but at steady prices for phosphatic material. The position of mineral phosphates is without change, there being nothing in prices quoted to induce purchasing ahead, and manure manufacturers being too busily engaged getting out their manures to have much time for considering their next season's requirements. Prices last quoted must, therefore, be taken as nominal values of to-day.

We don't hear of any business in cargoes of bones, and anything arriving at port of call will be now almost too late for the present season. Further sales of East India crushed bones have been made at £5. 3s. 9d. per ton ex quay Liverpool, which is nearest closing value. Sales of Bombay bone meal have been made at £5. 10s. per ton, to arrive, ex quay, and this price has also been accepted for a good line ex store Liverpool. Several parcels of common grinding bones arrived at Liverpool have been sold at £4. 10s. for a rough and dirty Turkish parcel, up to £4. 18s. 9d. for clean, dry parcels, and the good demand for this class of bones has kept the market steady in face of rather large arrivals.

Nitrate of soda is just steady. Spot price remains 8s. per cwt. Due cargoes are worth about 7s. 10½d. to 8s., according to size and quality. Buyers do not seem much disposed to give the higher prices demanded for autumn arrival, and consequently we do not hear of any important business in that position.

Further sales of River Plate dried blood have been made at 10s. per unit of ammonia, ex-store Liverpool, which price must be quoted as the nearest present value. For home prepared, 10s. 6d. to 10s. 9d. per unit required, according to position of the works in relation to market.

Nine shillings per unit of ammonia f.o.r. at works, is quoted for ground hoof, in bags, gross weights.

Kainit remains as last quoted, at 38s. to 40s. per ton, in bags, delivered free to rails, according to quantity.

Superphosphates remain as quoted a week ago, at 46s. 3d. per ton, net cash, in bulk, delivered f.o.r. at works, the bulk of the stuff in second hands having been placed, and at lower prices.

THE TYNE CHEMICAL REPORT.

TUESDAY.

The Chemical Market continues quiet with little business doing, and prices seem to be tending to the normal condition they were in previous to the late advance. Caustic soda has dropped 30s. a ton since last report. Bleaching powder is 5s. to 10s. a ton lower; soda ash about 5s. per ton in favour of buyers. Crystals are very firm at late quotations, and stocks said to be very light; sulphate of soda quiet, but no change in price. To-day's quotations are :—Caustic soda 77%, £11. 10s. per ton; pure ground caustic in 3 and 4 cwt. barrels, £15. per ton; bleach powder, in softwood casks, £5. 10s. per ton; sod ash, 1⅛d. per degree, less 7½%; soda crystals, in casks, 52s. 6 to 55s. per ton; in 2 cwt. bags, £2. 15s. per ton, net weight; in 1 cwt bags, £2. 17s. 6d. per ton, net weight; sulphate of soda, ground, i casks, £2. 5s. per ton; in bulk, £1. 15s. per ton; chlorate of potash 5d. per lb.; recovered sulphur, £4. 5s. per ton; silicate of soda 75° Tw., £2. 10s. per ton; 100° Tw., £3. 7s. 6d. per ton; 14 Tw., £4 per ton; hyposulphite of soda, 1 cwt. kegs, £5. 15s. p ton, in 5—7 cwt. casks, £5. 5s. per ton; pure white sulphate alumina, £4. 10s. per ton; blanc fixe, £7. 10s. per ton; chlorid of barium, £8. per ton; nitrate of baryta, crystals, £18. 15s. per ton ground, £19. 5s. per ton; sulphide of barium, £5. 10s. per ton—a f.o.b. Tyne, or f.o.r. makers' works. Small coals continue in goo demand. Manufacturers have great difficulty in getting deliverie owing to supplies being curtailed in consequence of the short tim worked by the miners last week.

Northumberland steam small quoted at 8s. to 8s. 6d. per ton, an Durham small at 10s. per ton.

The Durham miners have made application, through their Associa tion, for a further advance of 15 per cent. The owners have deferre the consideration of the matter until the selling price has been asce tained for the three months ending 31st March.

TAR AND AMMONIA PRODUCTS.

The Tar products market remains in much the same condition as r ported last week. Benzoles, both 90's and 50/90's selling at old rate though we hear various accounts of the prices at which actual busine has been done. Solvent naphtha is still in good demand, and creoso remains a good market. Crude carbolic acid and anthracene are bo very dull, and prices do not seem inclined to mend. Pitch is also wea with but little demand.

In sulphate of ammonia, the lower prices have induced a fa amount of business, and, on the whole, the position has now assume a steadier appearance. The business for Hull delivery has bee limited, prices varying from £11. 7s. 6d. to £11. 10s. The bulk the transactions has been negotiated for delivery f.o.b. Leith, princ pally in forward contracts at £11. 7s 6d. Quotations f.o.b. Live pool stand at £11. 7s. 6d. Beckton stands nominally at £11. 10s which price is also the value for outside makes.

THE LIVERPOOL MINERAL MARKET.

Our market has continued strong, and the late advance has bee well maintained. Manganese: Arrivals are proportionately small prices unaltered. Magnesite: Stocks of raw lump continue large, a prices are very easy. Raw ground, £6. 10s.; and calcined groun £10. to £11. Bauxite (Irish Hill brand) in increased demand very strong prices—Lump, 20s.; seconds, 16s.; thirds, 12s.; groun 35s. Dolomite, 7s. 6d. per ton at the mine. French Chalk: Arriva this week have increased, but all have practically gone into co sumption; prices, therefore, are firm at last figures, especially f G.G.B. "Angel-White" brand—90s. to 95s. medium, 100s. to 105 superfine. Barytes (carbonate) continues easier; selected cryst lump scarce at £6.; No. 1 lumps, 90s.; best, 80s.; seconds and good nu 70s.; smalls, 50s.; best ground, £6.; and selected crystal groun £8. Sulphate has somewhat improved, best lump, 35s. 6d.; go medium, 30s.; medium, 25s. 6d. to 27s. 6d.; common, 18s. 6d. to 20s ground best white, G.G.B. brand, 65s.; common, 45s.; grey, 32s. 6 to 40s. Pumicestone quiet; ground at £10., and specially select lump, finest quality, £13. Iron ore unaltered. Bilbao and Santa der, 9s. to 10s. 6d. f.o.b.; Irish, 11s. to 12s. 6d.; Cumberland strong demand at 18s. to 24s. Purple ore steady at last quotation Spanish manganiferous ore selling freely at good prices. Emery-stone Best brands still scarce both for spot and forward delivery; pric firmer. No. 1 lump is quoted at £5. 10s. to £6., and smalls £5. £5. 10s. Fullers' earth steady; 45s. to 50s. for best blue and yello fine impalpable ground, £7. Scheelite, wolfram, tungstate of sod and tungsten metal inquired for. Chrome metal, 5s. 6d. per l Tungsten alloys 2s. per lb. Chrome ore: The present demand h been supplied by fairly large arrivals, prices, however, are unaltere Antimony ore and metal have further advanced. Uranium oxid 24s. to 26s. Asbestos: Best rock, £17. to £18.; brown grade £14. to £15. Potter's lead ore: Prices are firm; smalls, £14. £15.; selected lump, £16. to £17. Calamine: Best qualities scarce 60s. to 80s. Strontia steady; sulphate (celestine) steady, 16s. 6 to 17s. Carbonate (native), £15. to £16.; powdered (ma factured), £11. to £12. Limespar: English manufactured, G.G.B. brand in demand, and brings full prices; 50s. for grou English. Felspar, 40s. to 50s.; fluorspar, 20s. to £6. Bog ore demand, firm at 22s. to 25s. Plumbago: steady; Spanish, £6 best Ceylon lump at last quotations; Italian and Bohemian, £4.

on; founders, £5. to £6.; Blackwell's "Mineraline," ich sand, in cargoes, continues scarce on spot—20s. . Ferro-manganese and silicon spiegel in good demand figures. Chrome iron, 20 per cent., £24. to £25. a, £50. China clay freely offering—common, 18s. 6d.; m 22s. 6d. to 25s.; best, 30s. to 35s. (at Runcorn).

IE LIVERPOOL COLOUR MARKET.

.—Unaltered. Ochres: Oxfordshire quoted at £10., 14., and £16.; Derbyshire, 50s. to 55s.; Welsh, to 55s.; seconds, 47s. 6d.; and common, 18s.; Irish, , 40s. to 45s.; French, J.C., 55s., 45s. to 60s.; M.C., 6d. Umber: Turkish, cargoes to arrive, 40s. to 50s.; , 50s. to 55s. White lead, £21. 10s. to £22. Red lead, xide of zinc: V.M. No. 1, £25.; V.M. No. 2, £23., ed, £6. 10s. Cobalt: Prepared oxide, 10s. 6d.; black, ue, 6s. 6d. Zaffres: No. 1, 3s. 6d.; No. 2, 2s. 6d. Terra est white, 60s.; good, 40s. to 50s. Rouge: Best, £24.; ditto rs', 9d. per lb. Drop black, 25s. to 28s. 6d. Oxide of iron, ty, £10. to £15. Paris white, 60s. Emerald green, 10d. erbyshire red, 60s. Vermillionette, 5d. to 7d. per lb.

New Companies.

AND PREECE, LIMITED.—This company was registered on the 1st; capital of £25,000. in £10. shares, to take over the business of electrica rried on by Thomas Reginald Andrews and Thomas Preece, at the lls, Bradford, York. 300 of the shares are deferred shares, to rank for t six per cent. has been paid on the ordinary shares The subscribers

	Shares.
rews, Bradford	1
reece, Bradford	1
nson, Thornleigh, Appleby, electrician	1
gatroyd, Halifax	1
h Brayshaw, Manningham, draper	1
glas, Bradford, electrical engineer	1
f, Bradford, accountant	1

PHOSPHATE COMPANY, LIMITED.—This company was registered on , with a capital of £5,000., in £1. shares, to acquire and work mining hts, lands, and property in Canada or elsewhere. The subscribers

	Shares.
on, Sylphide-villas, Forest-hill, clerk	1
ry, 10, Dusany-road, West Kensington, clerk	1
nett, Mackenzie-road, Beckenham, clerk	1
er, 4, Fenchurch-avenue, merchant	1
, 4, Fenchurch-avenue chemical broker	1
, 11, Queen Victoria-street, clerk	1
ney, 11, Queen Victoria-street, clerk	1

PORTLAND CEMENT COMPANY, LIMITED.—This company was regis- and inst., with a capital of £25,000., in £5. shares, to purchase from n, of Millwood-house, Grays, Essex, cement manufacturer, certain ., for the manufacture of concrete and metal telegraph poles, signal and nd other articles. The subscribers are:—

	Shares.
'allet, 15, Billiter-street, merchant	1
1, West India-road, E., merchant	1
unr., 14, St. Mary-axe, ship-owner	1
son, Grays, Essex, cement manufacturer	1
105, Leadenhall-street, insurance broker	1
ove, 54, Comerford-road, Brockley	1
h, Hainhault-road, Leytonstone, clerk	1

Gazette Notices.

Partnerships Dissolved.

S. A. BISHOP and C. FOREMAN, under the style of Brittan and Foreman Hermes-street, Pentonville, London, telegraphic instrument makers.

G. TAYLOR, H. MATTINSON, and J. TAYLOR, under the style of A. and G. Taylor, Bradford, Leeds, Huddersfield, and Dewsbury, photographers, as far as regards J. Taylor.

Adjudications.

JOSEPH ABRAHAM, Bristol, soapmaker.
WALTER SLOCOMBE, Romford, oil and colourman.

The Patent List.

This list is compiled from Official sources in the Manchester Technica. Laboratory, under the immediate supervision of George E. Davis ana Alfred R. Davis.

APPLICATIONS FOR LETTERS PATENT.

Sterilising Liquids by Magnetism. A. de Méritens. 5,048. April 1.
" " " **Electricity.** " " 5,049. April 1.
Softening Brittle Paper Pulp. C. Kellner. 5,053. April 1.
Bleaching Fibrous Material by Electricity. C. Kellner. 5,054. April 1.
Apparatus for Manufacturing Sulphuric Acid. E. Delplace and J. Delplace. 5,058. April 1.
Steam Boilers.—(Complete Specification.) W. H. Smith and W. A. Knapp 5,063. April 1.
Steam Boilers.—(Complete Specification.) N. P. Towne. 5,064. April 1.
Treatment of Phosphatic Minerals. L. Mond. 5,072. April 1.
Inclined Gas Retorts. C. Hunt. 5,078. April 1.
Electric Cables.—(Complete Specification.) H. H. Lake. 5,087. April 1.
Steam Boiler Furnaces. W. H. Wilson and J. Welsh. 5,117. April 2.
Modifying the Properties of Fibres for Paper Making. C. Kellner. 5,128. April 2.
Water Tube Steam Boilers. T. W. Lapworth and A. W. Hayes. 5,130 April 2.
Coating Surfaces with Metal. W. A. Thoms. 5,140. April 2.
Cold Air Refrigerating Machines. W. Garden. 5,152. April 2.
Manufacture of Colouring Matters. The Clayton Aniline Co., Limited and J. Hall. 5,155. April 2.
Charging and Discharging Gas Retorts. S. J Woodhouse. 5,163. April 3.
Brickmaking Machinery. G. H. Shanks, P. R. Tardew and J. Andrews. 5,175. April 3.
Brick and Tile-making Machinery. S. Witt, jun. 5,180. April 3.
Brickmaking Machinery. J. Dixon and R. T. Dixon. 5,188. April 3.
Colour Printing on Textiles. W. Dorrington. 5,190. April 3.
Gas and Petroleum Motors. A. G. Melhuish. 5,192. April 3.
Treatment of Coke. G. Waller. 5,194. April 3.
Treatment of Copper Nickel Matte. Electric Construction Corporation Limited, T. Parker. and A. E. Robinson. 5,199. April 3.
Sulphate of Ammonia Saturators. T. Wilton. 5,205. April 3.
Explosives. H. H. Lake. 5,209. April 3.
Instantaneous Production of Water Gas. J. Blum. 5,239. April 5.
Mixing Apparatus. J. Rowley and J. W. Rowley. 5,241. April 5.
Apparatus for Carburetting Water Gas. H. Fourness. 5,243. April 5.
Multiple Effect Evaporating Apparatus.—(Complete Specification.) A. Chapman. 5,267. April 5.
Reducing Valves. J. Kirkaldy. 5,277. April 5.
Steam Boilers. J. Kirkaldy. 5,278. April 5.
Washing and Cleaning Raw Sugar. G. F. Redfern. 5,282. April 5.
Manufacture of Sugar. G. F. Redfern. 5,283. April 5.
Bleaching Vegetable Fibres. C. Kellner. 5,285. April 5.
Manufacture of Phosphates. L. J. de Graef. 5,287. April 5.

IMPORTS OF CHEMICAL PRODUCTS

AT

THE PRINCIPAL PORTS OF THE UNITED KINGDOM.

ONDON.

ending April 8th.

	£50.		Beresford & Co.
de—			
	37.		Taylor, Sommerville, & Co.
Ore—			
	30 t.		Pillow, Jones, & Co.
	52		H. R. Merton & Co.
	70		Pillow, Jones, & Co.
ld—			
	30 pks.		Fuerst Bros.
	15 pks.		A. & M. Zimmermann
Acetate of Lime—			
U. S.,	£220.		Lister & Biggs
"	220		"
Ammonia—			
Holland,	£150.		H. Lorenz
Bones—			
N. S. Wales,	3 t.	8 c.	Leach & Co.
Holland,	200 t.		Ang.-Cont. G. Works
E. Indies,	100		C. C. Bryden & Co.
Bone Meal—			
E. Indies,	250 t.		J. Mitchell
Brimstone—			
Italy,	10 t.		F. & L. Sitzler
"	29	18 c.	G. Boor & Co.
"	21		H. S. Smith & Co.
"	5		A. Hughes
"	25		A. Zumbeck & Co.
"	32		Pigon, Wilks, & Co.
Barytes—			
Germany,	38 cks.		Burrell & Co.
"	6		H Lambert
"	23		W. Harrison & Co.
Belgium,	38		"
"	20		W. A. Rose & Co.
"	41		Magee, Taylor, & Co.
"	20		Leach & Co.
"	20		W. A. Rose & Co.
"	44		Magee, Taylor, & Co.
Germany,	23		W. Harrison & Co.
Cream of Tartar—			
Holland,	20 pks.		Middleton & Co.
France,	4		B. & F. Wf. Co.
"	6		W. E. Pantin
France,	5 pks.		M. Ashby
Spain,	35		W. C. Bacon & Co.
"	6		F. Smith
Copperas—			
Germany,	£75.		Charles & Fox
Chemicals *(otherwise undescribed)*			
Holland,	£8.		A. & M. Zimmermann
Germany,	30		T. H. Lee
"	408		A. & M. Zimmermann
E. Africa,	213		"
"	120		T. H. Lee
Germany,	169		A. & M. Zimmermann
"	14		Haeffner & Co.

[illegible] 25 c. — Ross & Deering
[illegible] Henderson Bros.
[illegible] 90 L. & I. D. Jt. Co.
776 Kebbel, Son, & Co.
T. H. Lee
[illegible] 56 R. & J. Henderson
200 Kaltenbach & Schmitz
168 Kleinwort, Sons, & Co.
[illegible] 10 Phillipps & Graves
97 Kebbel, Son, & Co.
39 W. Johnston & Co., Ld.
[illegible] Settlements, 109 c. E. Boustead & Co.

Potash—
[illegible] 2 pks. Beresford & Co.
92 F. W. Berk & Co.

[illegible]
[illegible] -12 pks. Bruce & White
[illegible] 1 pk. Lavington Bros.
[illegible] 273 pks. L. & I. D. Jt. Co.
467 Marshall & French
690 "
607 Baxter & Hoare
[illegible] 342 pks. W. H. Cole & Co.
25 Hoare, Wilson, & Co.
175 Elkan & Co.
427 Kleinwort, Sons, & Co.
73 S. Barrow & Bro.
205 Beresford & Co.
568 R. & J. Henderson
213 L. & I. D. Jt. Co.
179 Brinkmann & Co.
214 L. & I. D. Jt. Co.
[illegible] 37 chts. "
73 Benecke, Souchay, & Co.
22 W. Brandt, Sons, & Co.
66 A. Harvey
60 Ernsthausen & Co.
1 crn. L. & I. D. Jt. Co.
91 "
[illegible] 5 chts. Langstaff & Co.
149 Henderson Bros.
25 H. J. Pearlbach & Co.
105 L. & I. D. Jt. Co.
12 Arbuthnot, Latham, & Co.
0 F. W. Heilgers & Co.
4 Parsons & Keith
5 Elkan & Co.
4 Patry & Pasteur
3 Stansbury & Co.
34 L. & I. D. Jt. Co.
69 Ernsthausen & Co.
91 L. & I. D. Jt. Co.

[illegible]-ore Bark
[illegible] 10 t. H. Heule's Succrs.
20½ Prop. Scott's Wf.
20 Fardell & Co.
90 Boutcher, Mortimer, & Co.
90 Hicks, Nash, & Co.
[illegible] 50 t. J. Vicary & Sons
426 pks. B. Jacob & Sons
347 "
903 "
750 Brown & Elmslie
290 L. & N. W. Ry. Co.
600 J. Kitchen
90 W. France & Co.
209 A. Laming & Co.
210 Beresford & Co.
90 H. Johnson & Sons
[illegible] 39 pks. Burt, Boulton, & Co.
8 T. H. Lee
6 Kaltenbach & Schmitz
50 L. J. Levinstein & Sons
25 J. Graves
80 L. J. Levinstein & Sons
[illegible] 55 pks. L. & I. D. Jt. Co.
[illegible] 40⅜ t. J. Graves
[illegible] 31 Hicks, Nash, & Co.
[illegible] 200 J. Graves
[illegible] 1 pks. Lee, Crerar, & Co.
[illegible] 107 pks. Anderson, Weber, & Smith
[illegible] 690 T. H. Lee
73 A. & M. Zimmermann
23 Beresford & Co.
9 H. Lambert
400 Spies, Bros., & Co.
68 Beresford & Co.
[illegible] 785 t. Jensen & Co.
283 "

Glucose—
Germany, 200 pks. 200 c. J. Barber & Co.
" 200 400 "
" 65 524 L. Sutro & Co.
" 40 317 Pillow, Jones, & Co.
U. S., 50 250 A. J. Humphrey
" 75 375 T. M. Duche & Sons
" 35 210 Hyde, Nash, & Co.
" 800 800 A. Dawson
" 25 125 M. D. Co.
Germany, 115 219 "
" 1 7 Stein Bros.
" 800 900 Dutrulle, Solomon, & Co.
France, 100 100 "
" 500 500 Carey & Sons
" 400 400 Trier, Meyer, & Co.
" 220 220 J. Barber & Co.
Germany, 15 122 T. H. Lee
" 25 275 R. G. Hall & Co.
" 74 571 J. Knill & Co.
U. S., 60 500 P. W. Hore
" 800 800 A. Dawson
" 45 275 M. D. Co.
" 100 500 P. Windmuller
" 200 1,140 C. Czarnikow

Gutta Percha—
E. Indies, 1,085 c. Kaltenbach & Schmitz
" 2,265 R. J. Henderson
" 430 H. W. Jewesbury & Co.
" 76 Kleinwort, Sons, & Co.

Isinglass—
Str. Settlements, 70 pks. L. & I. D. Jt. Co.
E. Indies, 9 "
" 10 Hale & Sons

Manure—
Germany, 50 t. F. W. Berke & Co.
Phosphate Rock
Belgium, 260 t. De de Pass & Co.
" 227 Andrew Hunter & Co.
" 205 Miller & Johnson
" 130 Lawes Manure Co.
" 130 Anglo Contl. Guano Wks. Agency

Manganese—
France, 2 t. R. Wolff & Co.

Naphtha—
Germany, 16 drs. Stein Bros.

Potassium Carbonate—
France, 12 pks. Beresford & Co.
" 10 J. A. Reid
" 18 Charles & Fox
Holland, 1 H. Boyce

Potassium Permanganate—
Holland, 10 pks. H. Boyce

Potassium Silicate—
Holland, 20 pks. Beresford & Co.

Potassium Sulphate—
Holland, 65 pks. H. Lambert
France, 43 J. Knight & Son

Pearl Ash—
France, 64 c. Petri Bros.

Plumbago—
Holland, 105 cks. Brown & Elmslie
Ceylon, 32 Doulton & Co.
" 302 Thredder, Son, & Co.
France, 2 Hernu, Peron, & Co.
Ceylon, 108 Prop. Chambs. Whf.
Germany, 337 cs. Becker & Ulrich

Paraffin Wax—
U. S., 127 brls. H. Hill & Sons
" 500 "
" 220 G. H. Frank

Pyrites—
La Jaja, 400 t. Forbes, Abbott, & Co.
" 500 G. Ward & Sons
" 440 T. S. Whitehead & Chambers

Saltpetre—
Germany, 100 cks. T. Merry & Son
Belgium, 135 bgs. G. Meyer & Co.
E. Indies, 1,639 Clift, Nicholson, & Co.
Germany, 27 cks. P. Hecker & Co.
" 2 Craven & Co.
" 79 J. Hall & Son

Stearine—
France, 60 bgs. H. Hill & Sons
" 51 "
U. S., 20 hds. Beresford & Co.
Holland, 13 bgs. 13 sks. Perkins & Homer

Sugar of Lead—
Germany, 19 pks. Petri Bros.

Tartar—
Italy, 183 pks. Seagar, White, & Co.

Tartaric Acid—
Holland, 100 pks. N. Smith
" 50 B. & F. Wf. Co.
Germany, 64 O. Andrea & Co.
Holland, 16 B. & F. Wf. Co.

Ultramarine—
Holland, 12 pks. Becker & Ulrich
" 26 cs. G. Steinhoff
Germany, 14 pks. "
" 8 Blundell, Spence, & Co.
Belgium, 11 Hernu, Peron, & Co.

Zinc Oxide—
Germany, 9 cks. H. Lambert
Holland, 500 Soundy & Son
" 40 M. Ashby
" 200 "

LIVERPOOL.

Week ending April 10th.

Acids (*otherwise undescribed*)
Leghorn, 30 cks. Shropshire Union Co.

Barytes—
Antwerp, 26 cks.

Bones—
Boston, 6 tcs.
Natal, 20 t. H. Burk & Co.
Karachi, 4,941 bgs. W. & R. Graham & Co.
Smyrna, 51 t. O. Gantes & Co.
Colon, 382 bgs. A. Dobell & Co.

Bone Meal—
Boston, 20 brls. "
Karachi, 49 bgs. W. & R. Graham & Co.

Caoutchouc—
Maranham, 3 bxs. H. Evans
Colon, 19 bls. J. M. Williams
Para, 2 cs. A. Christiansen
" 119
" 17 4 srns. A. Booth & Co.
Lisbon, 341 bgs.
New York, 75

Cream of Tartar—
Patras, 8 brls. C. Raftopulo & Co.
Tarragona, 5 cks.

Carbolic Acid—
Havre, 39 btls.

Chemicals (*otherwise undescribed*)
Rotterdam, 24 cks.

Dried Blood—
Monte Video, 226 bgs. J. Purdon

Dyestuffs—
Extracts
Boston, 25 brls. L. Levinstein & Sons
Rouen, 411 cks.
Baltimore, 50
Fiume, 139 Boutcher Mortimer, & Co.
" 121 Millett Tanning Extract
Colon, 50 cs. E. Chesney & Co.
Fustic
Colon, 246 ps. A. Dobell & Co.
" 1,581 "
Cochineal
Las Palmas, 4 bgs. J. Goddard & Co.
" 6 Swanston & Co.
Orchella
Lisbon, 12 bgs. Leeds & Liverpool Canal Co.
Indigo
Calcutta, 45 cs. 1 bx. Margan & Co.
" 23 Baring Bros.
" 45
Divi Divi
Hamburg, 100 bgs.
—— 204
Amsterdam, 100
Alizarine
Rotterdam, 4 brls.
Valonia
Patras, 447 bgs. Papayanni & Jeremias
Smyrna, 50 t. S. Vinzenti & Son
" 450 bgs.
Argols
Naples, 4 cks.
Brindisi, 313 sks. 27 brls.

Glucose—
Hamburg, 50 cks.
Stettin, 24 Bostock & Co.
" 200 bgs.

Glycerine—
Amsterdam, 5 crbys. 3 cs.
Rotterdam, 5 5
Marseilles, 6 dms. Francis & Co.

Isinglass—
Para, 12 cs. Hooson & Yates
" 21 Bieber & Co.
Maranham, 2 cks. Gunston, Son, & Co.
" 1 Schill, Mofers, & Co.

Manganese—
Coquimbo, 1,000 t.

Nitrate—
Pisagua, 2 389 bgs. Hainsworth, Wasson, & Co.

Paraffin Wax—
Baltimore, 100 brls.

Pitch—
Hamburg, 15 bgs.
Amsterdam, 13 cks.
Rotterdam, 45

Potassium Prussiate—
Rotterdam, 4 cks.

Picric Acid—
Rotterdam, 2 cks.

Phosphate of Lime—
Dunkirk, 4,200 bgs.

Pyrites—
Huelva, 2,713 t. Matheson & Co.
" 1,613 "
" 1,504 Beauford & Co.
" 1,183 Matheson & Co.

Stearine—
Boston, 250 tcs.
New York, 262
Antwerp, 38 bgs.

Sulphur Ore—
Huelva, 1,129 t. Matheson & Co.

Salt—
Rotterdam, 1 ck.

Saltpetre—
Hamburg, 150 cks.
Calcutta, 461 bgs.
Karachi, 109 W. & R. Graham & Co.

Tartaric Acid—
Rotterdam, 2 cks.

Tartar—
Bordeaux, 11 cks.

Tincal—
Calcutta, 551 bgs.
" 223

Ultramarine—
Rotterdam, 4 cs. 28 cks.

Verdigris—
Bordeaux, 1 ck.

Zinc Oxide—
Antwerp, 10 brls. J. T. Fletcher & Co.
" 243

GLASGOW.

Week ending April 10th.

Barytes—
Bremerhaven, 40 cks.

Barytes Sulphate—
Antwerp, 130 bgs.

Bone Meal—
Rouen, 117 bgs. J. & P. Hutchinson

Charcoal—
Gothenburg, 20 bgs.
Antwerp, 107

Chemicals (*otherwise undescribed*)
Hamburg, 3 cs.

Chrome Ore—
Glemlek, 1,890 t. Stevenson, Carlisle & Co.

Dyestuffs—
Alizarine
Rotterdam, 42 cks.
" 21 J. Rankine & Son
Extracts
Bilbao, 25 brls.
Halifax, 120 cks.
Bordeaux, 223 cks.
Rouen, 1 J. & P. Hutchinson
Antwerp, 1 bx.

Glycerine—
Rouen, 2 cs. J. & P. Hutchinson
Hamburg, 20 drms.

Gutta Percha—
New York, 6 cs.

Glucose—
New York, 120 brls.

Manure—
Phosphate Rock
Aruba, 381 t. A. Cross & Son

Oxalic Acid—
Rotterdam, 5 cks.

Potash—
Hamburg, 80 cks.

Phosphate—
Treport, 1,865 sks.
Salt—
Hamburg, 101 t.
" 508 bgs.
Saltpetre—
Hamburg, 48 cks. J. Poynter & Co.
Tartar—
Bordeaux, 33 cks.
Tartaric Acid—
Rotterdam, 4 cks.
Zinc Oxide—
Antwerp, 30 cks.

HULL.

Week ending April 10th.

Acid (*otherwise undescribed*)
Amsterdam, 30 cks. W. & C. L. Ringrose
Borax—
Boston, 1 brl. Wilson, Sons, & Co.
Barytes—
Antwerp, 200 bgs. H. F. Pudsey
Bremen, 22 cks. Tudor & Sons
" 40
Bone Meal—
Bombay, 770 bgs.
Chemicals (*otherwise undescribed*)
Rotterdam, 18 cs. G. Lawson & Sons
Dunkirk, 150 34 pks. Wilson, Sons, & Co.
Antwerp, 24 "
Hamburg, 25 "
Rotterdam, 2 G. Lawson & Sons
" 10 cks.
" 54 pks. W. & C. L. Ringrose
Bremen, 3 pgs. Veltmann & Co.
" 2 cs. Wilson, Sons, & Co.
Dyestuffs—
Myrabolams
Bombay, 2,466 bgs.

Madder
Rotterdam, 1 ck. G. Lawson & Sons
Amsterdam, 2 cks. W. & C. L. Ringrose
Rennet
Copenhagen, 33 cks. Wilson, Sons, & Co.
Rotterdam, 9
Alizarine
Rotterdam, 4 pks. W. & C. L. Ringrose
" 45 Hutchinson & Son
" 2 cks. W. & C. L. Ringrose
Extracts
Rouen, 86 cks. 190 cs. Rawson & Robinson
Boston, 50 brls.
New York, 4 Wilson, Sons, & Co.
Glucose—
Hamburg, 21 cks. Wilson, Sons, & Co.
" 10 Rawson & Robinson
" 20 Bailey & Leetham
" 13
" 13
Stettin, 200 bgs. 5 cks. Wilson, Sons, & Co.
" 100 11
Hamburg, 14
New York, 40 brls. Wilson, Sons, & Co.
Manure—
Dunkirk, 100 bgs. Wilson, Sons, & Co.
Rouen, 71 cks. Rawson & Robinson
Naphthol—
Rotterdam, 10 pks. Wilson, Sons, & Co.
Stearine—
Rotterdam, 15 bgs. Wilson, Sons, & Co.
Soda—
Rotterdam, 3 cks. G. Lawson & Sons
Salt—
Hamburg, 205 bgs. Farley & Co.
Tar—
Dunkirk, 3 pks. Wilson, Sons, & Co.
Libau, 40 brls.

TYNE.

Week ending Apr. 10th.

Barytes—
Antwerp, 21 cks. Tyne S. S. Co.
Rotterdam, 12 "
Copper Precipitate—
Huelva, 7,237 bgs. Scott Bros.
Glycerine—
Rotterdam, 5 pks. Tyne S. S. Co.
Pyrites—
Huelva, 490 t. A. Guthrie
" 1,099 Scott Bros.
Phosphate—
Antwerp, 280 t. Langdale's Manure Co.
" 440 1,000 bgs. "
Salt—
Hamburg, 102 t.
Saltcake—
Hamburg, 278 t.
" 404
Sodium Nitrate—
Iquique, 12,550 bgs. J. C. Swan
Sulphur Ore—
Pomaron, 1,201 t. Mason & Barry
Zinc Oxide—
Antwerp, 25 brls. Tyne S. S. Co.

GOOLE.

Week ending April 2nd.

Antimony Ore—
Boulogne, 8 cks.
Caoutchouc—
Boulogne, 16 cks.
Dyestuffs—
Alizarine
Rotterdam, 57 cks.
Madder
Rotterdam, 15 cks.
Dyestuffs (*otherwise undescribed*)
Boulogne, 4 cs. 86 cks. 4 pks.

Glucose—
Hamburg, 290 bgs. 1 ck.
" 90
Manure—
Phosphate Rock
Ghent, 800 bgs.
Naphthalamine—
Rotterdam, 1 ck.
Naphthol—
Rotterdam, 5 cks.
Potash—
Dunkirk, 80 drms.
Calais, 112
Phosphate—
Ghent, 800 bgs.
Soda—
Rotterdam, 7 cks.
Sodium Silicate—
Rotterdam, 2 cks.
Saltpetre—
Rotterdam, 285 bgs.
Salt—
Hamburg, 1,018 bgs.

GRIMSB[Y]

Week ending Apr[il]

Caoutchouc—
Hamburg, 2 pks.
Chemicals (*otherwise un[described]*)
Hamburg, 31 cs. 6 pks.
Dyestuffs—
Extract
Dieppe, 19 cks.
Dyestuffs (*otherwise un[described]*)
Antwerp, 8 cs.
Glucose—
Hamburg, 17 cks.
Zinc Ashes—
Antwerp, 23 cks.
" 9 cks.

EXPORTS OF CHEMICAL PRODUCTS

OF

THE PRINCIPAL PORTS OF THE UNITED KINGDOM.

LONDON.

Week ending April 5th.

Ammonium Carbonate—
Libau, 34 kgs. 30 cks. £178
Reval, 7 t. 10 c. 285
Ammonia—
Gothenburg, 10 c. £18
Arsenic—
Rangoon, 1 t. 5 c. £17
Ammonium Sulphate—
Colombo, 2 t. 17 c. £38
Carbolic Solid—
Hamburg, 4 t. 19 c. £405
Copper Sulphate—
Salonica, 1 t. 8 c. £25
Canterbury, 1 20
Caustic Potash—
Otago, 1 t. 18 c. £40
Citric Acid—
Halifax, 5 c. £35
New York, 2 t. 10 384
Copenhagen, 5 kgs. 31
Hamburg, 1 t. 10 c. 10 kgs. 208
Stettin, 1 7
Coal Products—
Naphthalene
Hamburg, 87 cks. £175
" 6 cs. 12
Benzol
Hamburg, 1 drm. £15
Anthracene
Dusseldorf, 88 cks. £475
Tar
Karachi, 100 rults. £50
Calcutta, 204 275
Pitch
Boston, 200 cks. £127
Antwerp, 340 t. 470
Chemicals (*otherwise undescribed*)
Melbourne, 15 kgs. £105
Auckland, 41 pks. 87
Bahia, 13 36
[D]isinfectants—
[Co]lombo, 1 t. 10 c. £37

Manure—
Hamburg, 29 t. 11 c. £290
" 175 13 703
Nakskov, 450 t. 4,070
Auckland, 60 292
Gothenburg, 243 t. 12 c. 1,316
Hamburg, 58 5 273
Auckland, 25 807
Potassium Chlorate—
Sydney, 5 t. £211
Sulphur—
Calcutta, 19 t. 10 c. £102
Sydney, 100 275
Sodium Tartrate—
Toronto, 1 ck. £11
Superphosphate—
Malmo, 25 t. £300
Soda Crystals—
N. Orleans, 26 t. 5 c. £100
Canterbury, 4 18 16
Soda Ash—
Auckland, 10 t. £65
Saltpetre—
Adelaide, 1 t. £21
Halifax, 1 12 c. 33
Tartaric Acid—
Adelaide, 1 t. £133
Calcutta, 300 lbs. 21
Toronto, 5 cks. 130
Canterbury, 5 c. 40
Auckland, 5 40

LIVERPOOL

Week ending April 9th.

Ammonia—
Valparaiso, 4 c. £7
Ammonium Sulphate—
Antwerp, 50 t. £605
Valencia, 50 600
Antimony Regulus—
Algiers, 1 t. 19 c. 1 q. £131
Bordeaux, 7 11 1 905

Ammonium Carbonate—
Oporto, 10 c. 1 q. £17
Portland, 1 t. 35
Ceara, 3 3 9
Genoa, 1 3 44
Stockholm, 7 5
Christiania, 7 3 8
Alum—
Callao, 2 t. 9 c. £24
Galata, 1 16 9
Ibrail, 9 8 47
Ancona, 4 3 22
Portland, 3 13 18
Bilbao, 5 25
Santos, 10 5
Syra, 4 3 22
Valparaiso, 1 10 1 q. 7
Ammonium Muriate—
New York, 10 t. 9 c. £290
Antwerp, 6 3 q. 9
Philadelphia, 16 9 1 369
Brimstone—
Calcutta, 8 t. 1 q. £38
Boston, 25 106
Borax—
Monte Video, 2 t. 12 c. £47
Christiania, 5 18 123
Vera Cruz, 1 1 1 q. 33
Havana, 5 2 8
Boracic Acid—
Stockholm, 24 cks. £141
Bone Waste—
Rouen, 2 t. 15 c. £27
Barbadoes, 20 214
Bleaching Powder—
Barcelona, 20 brls.
Boston, 184 cks.
Genoa, 27
Gothenburg, 4
Naples, 74
Philadelphia, 30
Rouen, 15
Trieste, 20
Boston, 537 10 cs. 168 tcs.
Hamburg, 28
Philadelphia, 122
Rotterdam, 8
Rouen, 59

Caustic Soda—
Seville, 100 drms.
Ancona, 10
Barcelona, 211 35 [illegible]
Bilbao, 180 13
Catania,
Corunna, 24
Genoa, 47
Gothenburg, 3
Hamburg, 18
Lisbon, 226
Naples, 5
New York, 395
Portland, 200
Rio Grande do Sul, 66 drm[s.]
Rouen, 40
Santos, 70
Alicante, 50
Anersia, 77
Antwerp, 28
Bahia, 21
Barcelona, 278
Boston, 765
Naples, 30
New York, 877
Pasages, 100
Ponce, 4
Portland, 100
Rouen, 30
Santos, 15
Ibrail, 30
Carbolic Acid—
Rotterdam, 24 t. 5 c.
Hamburg, 7
Havre, 19
Mayaguez, 3
Portland, 10
Paris, 2 19
Cobalt Oxide—
Gijon, 1 c.
Charcoal—
Hamburg, 37 t. 11 c.
Havre, 17 16
Citric Acid—
Barcelona, 1 t. 2 c.
Copper Sulphate—
Genoa, 5 t. 1 c.
" 34 4

Leghorn, 9 17 3 q. 225
Marseilles, 202 8 3 4,217
Leghorn, 5 3 120
Nantes, 29 18 3 679
Alexandria, 1 2 23
Pasages, 9 18 2 253
Bilbao, 5 1 90
Bordeaux, 4 19 1 125
St Nazaire, 5 3 1 145
San Sebastian, 10 220
Barcelona, 35 850
,, 10 19
Tarragona, 1 2 3 22
,, 4 19 1 100
Trieste, 20 1 1 455
Venice, 15 8 3 285
Pasages, 21 5 443
San Sebastian, 27 16 695
Syra, 8 2 11
Trieste, 4 10 110
Valencia, 8 19 2 180
Bordeaux, 5 1 1 110
Havre, 5 120
Rouen, 40 12 1,013
Vera Cruz, 5 cks. 13
Pasages, 10 2 3 233
Bilbao, 10 1 3 232
,, 6 106
Genoa, 5 6
Bordeaux, 5 1 107
Barcelona, 4 18 1 112
Tarragona, 5 1 2 110

Copperas—
Piræus, 8 t. 18 c. 2 q. £20
Galatz, 4 9
,, 24 15 59
Alexandria, 13 14 27

Cream of Tartar—
Savanilla, 5 c. 1 q. £28
St. John's, N. F., 1 t. 11 3 159

Caustic Potash—
St. John's, N. F, 3 c. £6
Hamburg, 7 8
San Francisco. 10 12
New York, 14 16

Chemicals (*otherwise undescribed*)
New York, 1 t. 10 c. £77

Chalk Precipitate—
Philadelphia, 1 t. 19 c. £27

Chloride of Lime—
New York, 50 t. £280

Epsom Salts—
Rio Janeiro, 50 kgs.
Rouen, 6 cks.

Gypsum—
Canada, 5 t. 10 c. £23

Iron Oxide—
Gijon, 1 t. 4 c. 2 q. £7

Iodine—
New York, 8 t £4,800

Leather Waste—
Havre, 1 t. 4 c. £9

Manure—
Antwerp, 14 t. 12 c. £66

Magnesia—
Antwerp, 2 cs.
Genoa, 1

Napthaline—
Antwerp, 202 t. 6 c. £161

Potassium Carbonate—
Philadelphia, 6 c. £10
Boston, 13 t. 168

Potassium Chromate—
Havre, 2 t. £70

Phosphorus—
Shanghai, 8 c. £48
Monte Video, 10 cs. 165
Buenos Ayres, 10 165

Potassium Chlorate—
Corunna, 10 c. £23
New York, 5 t. 200
Stockholm, 4 168
Yokohama, 5 276
Genoa, 1 47
Bari, 10 20

Phosphate—
Hamburg, 10 t. 9 c. £55

Potassium Silicate—
Amsterdam, 1 t. 2 c. 3 q. £10

Potassium Bichromate—
Marseilles, 1 t. 7 c. 2 q. £61

Pitch—
Ancona, 2,208 t.
Brindisi, 1,715

Sodium Silicate—
Algoa Bay, 10 cks.
Cartagena, 8
Lisbon, 8
Odessa, 100
Oporto, 8
Barcelona, 10
Bilbao, 16

Sodium Bicarbonate—
Amapala, 20 kgs.
Antwerp, 60 8 cks.
Bombay, 250
Calcutta, 500
Hamburg, 80 4
Ibrail, 60
Marseilles, 50
Rotterdam, 50
Santander, 20
Valparaiso, 100
Antwerp. 12 pks. 42 cks.
Barcelona, 300 kgs.
Bari, 60
Bordeaux, 100
Genoa, 100
Hamburg, 309
Leghorn, 70
New York, 200
Rotterdam, 10
Rouen, 260 25
St. John's, 100
Salonica, 20
Santander, 20
Ancona, 82

Superhosphate—
Gr. Canary, 5 t. £18

Saltcake—
Philadelphia, 37 t. 18 c. £81
Boston, 100 213

Saltpetre—
Piræus, 1 t. 7 c. 1 q. £30
Portland. 3 7 3 75
Syra, 3 2 2 69

Sulphur—
Rotterdam, 2 t. 18 c 3 q. £20
Boston, 49 19 210
Portland, 5 18 3 37
Rouen, 20 90
Sydney, 3 1 29
Pernambuco, 1 8 2 10
Calcutta, 15 90
Madras, 3 16 1 24
Boston, 50 2 220

Salammoniac—
Genoa, 2 t. 4 c. 3 q. £72
Naples. 9 17
New York, 5 170
Bari, 10 19
Piræus, 2 3 5
Genoa, 10 2 18
Philadelphia, 5 3 1 181
Marseilles, 1 2 36
Mersyne, 1 2 3 75
,, 8 14

Sodium Biborate—
Stettin, 10 t. 3 c. 2 q. £306
Stockholm, 10 3 2 306
Antwerp, 26 2 781
Ibrail, 6 2 10

Sodium Sulphate—
Philadelphia, 13 t. 6 c. £69
Rio de Janeiro, 1 14 1 q. 6
B. Ayres, 2 4 3 18

Soda Ash—
Baltimore, 119 cks.
Barcelona, 10
Boston, 373 214 bgs. 262 tcs.
Cadiz, 6
Kingsley, 16
Montreal, 41
New York, 1,311
Philadelphia, 142 231
Syra, 30
Valparaiso, 10 brls.
Baltimore, 65 288
Barcelona, 100 brls.
Beyrout, 50
New York, 968
Patras, 20
Philadelphia, 309

Soda Crystals—
New York, 140 cks.
Philadelphia, 140
New York, 140

Salt—
Barbadoes, 18 t.
Cameroons, 49
Chicago, 195
Old Calabar, 100
Portland, 160
Santos, 20 cs.
Algoa Bay, 6 t.
Bonny, 10
Brass, 50
B. Ayres, 482
Cape Lopez, 25
Cape Town, 11
E. London, 10 6 c.
Gr. Bassam, 9 18
Halifax, 599 13
Lagos, 100
Penang, 1 16
Rouen, 10
Sierra Leone, 50

Tartaric Acid—
Savanilla, 1 c. 1 q. £9

Tar—
Cabenda, 9 brls.
St. John's, 50

Zinc Chloride—
Bombay, 2 t. 12 c. £44

TYNE.

Week ending April 10th.

Alkali—
Stettin, 2 t. 12 c.
Rotterdam, 10
Antwerp, 18 11

Antimony—
Rotterdam, 5 t.
,, 5

Barytes Carbonate—
Hamburg, 5 t. 5 c.
Stettin, 4 15

Bleaching Powder—
Antwerp, 103 t. 14 c.
Brake, 1 8
Hamburg, 169 7
Christiania, 1 12
Odense, 10 17
Hamburg, 22 4
Riga, 40 4
Stettin, 215 4
Gothenburg, 4 16
Konigsberg, 91 11
Rotterdam, 33

Copperas—
Christiania, 9 t.

Caustic Soda—
Gothenburg, 29 t. 17 c.
Antwerp, 145 9
Brake, 7
Rotterdam, 1 4

Litharge—
Antwerp, 11 c.
Hamburg, 5 t.
Christiania, 1 7
Riga, 1 8
Stettin, 4 2

Manure—
Christiania, 25 t.
Esbjerg, 31 8 c.

Magnesia—
Antwerp, 3 q.
Hamburg, 14 c.

Oxalic Acid—
Odense, 3 t. 4 c.

Sodium Hyposulphite—
Antwerp, 5 t. 4 c.

Sodium Bicarbonate—
Brake, 10 c.

Soda—
Kallundborg, 6 t. 5 c.
Aalborg, 27 7
Odense 9 13
Randers, 39 7
Aarhuus, 14 4
Grenaa, 15 1
Rotterdam, 28 10

Soda Ash—
Brake, 10 c.
Dunkirk, 7 t. 4
Gothenburg. 11 18
Stettin, 1 19
Gothenburg, 9
Konigsberg, 80 3

Sodium Chlorate—
Hamburg, 6 t.

HULL.

Week ending April 8th.

Alkali—
Bremen, 2 cks.
Copenhagen 10
Drontheim, 17 drums.

Bleaching Powder—
Drontheim, 1 ck.

Chemicals (*otherwise undescribed*)
Amsterdam, 5 cks.
Bordeaux, 20 cs.
Bergen, 23
Boston, 50
Copenhagen, 17 pks.
Christiania, 8 7
Drontheim, 72
Danzig, 42
Ghent, 4
Gothenburg, 7 5 cs. 10 318 slt
Hamburg, 22 128 6
Hango, 1
Konigsberg, 26
Libau, 22
Melbourne, 2
New York, 2
Riga, 2
Rotterdam, 26 30 26 cks.
Reval, 2
Rouen, 30
Stockholm, 7 3

Caustic Soda—
Danzig, 35 drums.
Konigsberg, 21
Libau, 75

Dyestuffs (*otherwise undescribed*)
Antwerp, 2 cks
Boston, 5
Hamburg, 102 c.
Hango, 5 pks.
Rotterdam, 1 11 kgs. 5 cs.

Manure—
Aalborg, 208 t.
Dunkirk, 672 bgs.
Gothenburg, 50
Hamburg, 111

Pitch—
Antwerp, 230 t.

Tar—
Danzig, 18 cks.
Gothenburg, 150
Hamburg, 12
Konigsberg, 6

GLASGOW.

Week ending April 10th.

Chloride of Tin—
Japan, 8¾ c. £[illegible]

Dyestuffs—
Logwood
Halifax, 88¾ c. £11[illegible]

Magnesium Silicate—
Penang, 2 t. 18 ¾ c. £[illegible]

Manure—
Mauritius, 70 t. 9¾ c. £7[illegible]
,, 105 8[illegible]

Magnesium Chloride—
Penang, 2 t. 11¾ c. £2[illegible]

Paraffin Wax—
Rouen, 14 t. 17¾ c. £4[illegible]
Japan, 135¾ 1[illegible]

Potassium Bichromate—
Boston. 332 c. £6[illegible]
Amsterdam, 13 cks. 1[illegible]
Barcelona, 331½ c. 18 lbs. £6[illegible]
Rotterdam, 63 cks. 4[illegible]

Pitch—
Caen, 720 t.

Sodium Bichromate—
Halifax, 2 t. 17¾ c. £[illegible]
Boston, 355 4[illegible]
Rouen, 16 9¾ 4[illegible]

Soda Ash—
Baltimore, 605¾ c. £2[illegible]

GOOLE.

Week ending April 2nd.

Benzole—
Ghent, 23 drums

Coal Products—
Boulogne, 10 cks.
Hamburg, 50
Rotterdam, 80

Chemicals (*otherwise undescribed*)
Calais, 10 cks.

Dyestuffs (*otherwise undescribed*)
Boulogne, 16 cks 2 drums.
Ghent, 8 6 10 tins.

Manure—
Boulogne, 87 bgs.
Dunkirk, 100
Ghent, 743
Hamburg, 422

Pitch—
Antwerp, 170 t.

Sulphur Waste—
Rotterdam, 100 t.

GRIMSBY.

Week ending April 5th.

Coal Products—
Dieppe, 27 cks.

PRICES CURRENT.

WEDNESDAY, APRIL 16, 1890.

PREPARED BY HIGGINBOTTOM AND CO., 116, PORTLAND STREET, MANCHESTER.

es stated are F.O.R. at maker's works, or at usual ports of shipment in U.K. The price in different localities may vary.

		£	s.	d.
25 %, and 40 %	per cwt	7/6 & 12/9		
Glacial	,,	67/6		
, S.G., 2000°	,,	0	12	9
: 82 %	nett per lb.	0	0	6½
	,,	0	0	3¾
c (Tower Salts), 30° Tw.	per bottle	0	0	8
(Cylinder), 30° Tw.	,,	0	2	11
o° Tw.	per lb.	0	0	2
	,,	0	0	1¾
	nett ,,	0	0	4
	,,	0	1	2½
ic (fuming 50 %)	per ton	15	10	0
(monohydrate)	,,	5	10	0
(Pyrites, 168°)	,,	3	2	6
(,, 150°)	,,	1	8	0
(free from arsenic, 140/145°)	,,	1	10	0
ous (solution)	,,	2	15	0
(pure)	per lb.	0	1	0½
:	,,	0	1	2½
hite powdered	per ton	13	0	0
e lump)	,,	4	15	0
ulphate (pure)	,,	5	0	0
,, (medium qualities)	,,	4	10	0
1	per lb.	0	15	0
·880=28°	per lb.	0	0	3
=24°	,,	0	0	1¾
Carbonate	nett ,,	0	0	3½
Muriate	per ton	23	0	0
,, (sal-ammoniac) 1st & 2nd	per cwt.	37/- & 35/-		
Nitrate	per ton	40	0	0
Phosphate	per lb.	0	0	10½
Sulphate (grey), London	per ton	11	10	0
,, (grey), Hull	,,	11	7	6
l (pure)	per lb.	0	0	11
lt ,,	,,	0	0	9
	per ton	75	0	0
(tartar emetic)	per lb.	0	1	1
(golden sulphide)	,,	0	0	10
loride	per ton	8	0	0
rbonate (native)	,,	3	15	0
lphate (native levigated)	,,	45/- to 75/-		
Powder, 35 %	,,	5	0	0
Liquor, 7 %	,,	2	10	0
of Carbon	,,	13	0	0
Acetate (crystal	per lb.	0	0	6
hloride	per ton	2	2	6
(at Runcorn) in bulk	,,	17/6 to 35/-		
Dyes :—				
(artificial) 20 %	per lb.	0	0	9
	,,	0	3	9
Products				
ene, 30 % A, f.o.b. London	per unit per cwt.	0	1	5
90 % nominal	per gallon	0	3	4
50/90 ,,	,,	0	2	7½
Acid (crystallised 35°)	per lb.	0	0	9¼
,, (crude 60°)	per gallon	0	2	4
(ordinary)	,,	0	0	2½
(filtered for Lucigen light)	,,	0	0	3
aphtha 30 % @ 120° C.	,,	0	1	3
Oils, 22° Tw.	per ton	3	0	0
o.b. Liverpool or Garston	,,	1	8	0
Naphtha, 90 % at 160°	per gallon	0	1	9
Oils (crude)	,,	0	0	2½
ili Bars)	per ton	48	7	6
phate	,,	25	0	0
de (copper scales)	,,	51	0	0
crude)	,,	30	0	0
distilled S.G. 1250°)	,,	55	0	0
	nett, per oz.	8½d. to 9d.		
ate (copperas)	per ton	1	10	0
ide (antimony slag)	,,	2	0	0
t) for cash	,,	14	0	0
rge Flake (ex ship)	,,	15	10	0
ate (white ,,)	,,	23	10	0
(brown ,,)	,,	18	5	0
onate (white lead) pure	,,	19	0	0
ate	,,	22	0	0

		£	s.	d.
Lime, Acetate (brown)	,,	7	17	6
,, ,, (grey)	per ton	13	15	0
Magnesium (ribbon and wire)	per oz.	0	2	3
,, Chloride (ex ship)	per ton	2	5	0
,, Carbonate	per cwt.	1	17	6
,, Hydrate	,,	0	10	0
,, Sulphate (Epsom Salts)	per ton	3	0	0
Manganese, Sulphate	,,	18	0	0
,, Borate (1st and 2nd)	per cwt.	60/- & 42/6		
,, Ore, 70 %	per ton	4	10	0
Methylated Spirit, 61° O.P.	per gallon	0	2	2
Naphtha (Wood), Solvent	,,	0	3	10
,, ,, Miscible, 60° O.P.	,,	0	4	6
Oils :—				
Cotton-seed	per ton	22	10	0
Linseed	,,	24	10	0
Lubricating, Scotch, 890°—895°	,,	7	5	0
Petroleum, Russian	per gallon	0	0	5¼
,, American	,,	0	0	5¾
Potassium (metal)	per oz.	0	3	10
,, Bichromate	per lb.	0	0	4
,, Carbonate, 90% (ex ship)	per ton	19	0	0
,, Chlorate	per lb.	0	0	5
,, Cyanide, 98%	,,	0	2	0
,, Hydrate (Caustic Potash) 80/85 %	per ton	22	10	0
,, ,, (Caustic Potash) 75/80 %	,,	21	10	0
,, ,, (Caustic Potash) 70/75 %	,,	20	15	0
,, Nitrate (refined)	per cwt.	1	2	6
,, Permanganate	,,	4	2	6
Prussiate Yellow	per lb.	0	0	9½
,, Sulphate, 90 %	per ton	9	10	0
,, Muriate, 80 %	,,	7	15	0
Silver (metal)	per oz.	0	3	7¾
,, Nitrate	,,	0	2	5¾
Sodium (metal)	per lb.	0	4	0
,, Carb. (refined Soda-ash) 48 %	per ton	6	5	0
,, ,, (Caustic Soda-ash) 48 %	,,	5	5	0
,, ,, (Carb. Soda-ash) 48%	,,	5	12	6
,, ,, (Carb. Soda-ash) 58 %	,,	6	15	0
,, ,, (Soda Crystals)	,,	2	17	6
,, Acetate (ex-ship)	,,	15	15	0
,, Arseniate, 45 %	,,	11	0	0
,, Chlorate	per lb.	0	0	6¼
,, Borate (Borax)	nett, per ton.	29	0	0
,, Bichromate	per lb.	0	0	3
,, Hydrate (77 % Caustic Soda)	per ton.	12	15	0
,, ,, (74 % Caustic Soda)	,,	11	10	0
,, ,, (70 % Caustic Soda)	,,	9	17	6
,, ,, (60 % Caustic Soda, white)	,,	8	15	0
,, ,, (60 % Caustic Soda, cream)	,,	8	12	6
,, Bicarbonate	,,	6	0	0
,, Hyposulphite	,,	5	10	0
,, Manganate, 25%	,,	9	0	0
,, Nitrate (95 %) ex-ship Liverpool	per cwt.	0	8	0
,, Nitrite, 98 %	per ton	28	0	0
,, Phosphate	,,	15	15	0
,, Prussiate	per lb.	0	0	7¾
,, Silicate (glass)	per ton	5	7	6
,, ,, (liquid, 100° Tw.)	,,	3	17	6
,, Stannate, 40 %	per cwt.	2	0	0
,, Sulphate (Salt-cake)	per ton.	1	5	0
,, ,, (Glauber's Salts)	,,	1	10	0
,, Sulphide	,,	7	15	0
,, Sulphite	,,	5	5	0
Strontium Hydrate, 98 %	,,	8	0	0
Sulphocyanide Ammonium, 95 %	per lb.	0	0	8
,, Barium, 95 %	,,	0	0	5¼
,, Potassium	,,	0	0	8
Sulphur (Flowers)	per ton.	7	15	0
,, (Roll Brimstone)	,,	6	0	0
,, Brimstone : Best Quality	,,	4	5	0
Superphosphate of Lime (26 %)	,,	2	10	0
Tallow	,,	25	10	0
Tin (English Ingots)	,,	93	0	0
,, Crystals	per lb.	0	0	6¾
Zinc (Spelter)	per ton.	20	10	0
,, Chloride (solution, 96° Tw.	,,	6	0	0

No. 153. SATURDAY, APRIL 26, 1890. Vol. VI.

Contents.

Notices.

All communications for the *Chemical Trade Journal* should be addressed, and Cheques and Post Office Orders made payable to—

DAVIS BROS., 32, Blackfriars Street, MANCHESTER.

Our Registered Telegraphic Address is—
"Expert, Manchester."

The Terms of Subscription, commencing at any date, to the *Chemical Trade Journal*,—payable in advance,—including postage to any part of the world, are as follow :—

Yearly (52 numbers) .... **12s. 6d.**
Half-Yearly (26 numbers) .... **6s. 6d.**
Quarterly (13 numbers) .... **3s. 6d.**

Readers will oblige by making their remittances for subscriptions by Postal or Post Office Order, crossed.

Communications for the Editor, if intended for insertion in the current week's issue, should reach the office not later than **Tuesday Morning.**

Articles, reports, and correspondence on all matters of interest to the Chemical and allied industries, home and foreign, are solicited. Correspondents should condense their matter as much as possible, write on one side only of the paper, and in all cases give their names and addresses, not necessarily for publication. Sketches should be sent on separate sheets.

We cannot undertake to return rejected manuscripts or drawings, unless accompanied by a stamped directed envelope.

Readers are invited to forward items of intelligence, or cuttings from local newspapers, of interest to the trades concerned.

As it is one of the special features of the *Chemical Trade Journal* to give the earliest information respecting new processes, improvements, inventions, etc., bearing upon the Chemical and allied industries, or which may be of interest to our readers, the Editor invites particulars of such—when in working order—from the originators; and if the subject is deemed of sufficient importance, an expert will visit and report upon the same in the columns of the *Journal*. There is no fee required for visits of this kind.

We shall esteem it a favour if any of our readers, in making inquiries of, or opening accounts with advertisers in this paper, will kindly mention the *Chemical Trade Journal* as the source of their information.

Advertisements intended for insertion in the current week's issue, should reach the office by **Wednesday morning** at the latest.

Advertisements.

Prepaid Advertisements of Situations Vacant, Premises on Sale or To be Let, Miscellaneous Wants, and Specific Articles for Sale by Private Contract, are inserted in the *Chemical Trade Journal* at the following rates :—

30 Words and under.......... 2s. 0d. per insertion.
Each additional 10 words .... 0s. 6d. „

Advertisements of Situations Wanted are inserted at one-half the above rates when prepaid, viz :—

30 Words and under.......... 1s. 0d. per insertion.
Each additional 10 words .... 0s. 3d. „

Trade Advertisements, Announcements in the Directory Columns, and all Advertisements not prepaid, are charged at the Tariff rates which will be forwarded on application.

THE COPPER MARKET.

A Review of the Present Position.

WHEN the operations of what will probably for a lon time to come be known as *the* Copper Syndicate, cam to such a disastrous end in the beginning of 1889, the mos gloomy apprehensions were rife as to the future of th Copper Market. As the huge combination was seen to b tottering to its fall, and vainly struggling, like a fly in spider's web, to free itself from the toils that enveloped i the attention of almost every thinking individual was draw to the subject, and hundreds of thousands of people who ha previously associated the metal only with pence and hal pence, were then made aware for the first time of the im portance of the *rôle* that copper plays in the economy of th world. The general opinion was that when the inevitabl crash came, the stock that had for sixteen months been in creasing with a fatal rapidity, would so flood the market tha for years to come its organisation would be gone, and ever man's hand would be against that of his neighbour's in hi eagerness to sell. Then the crash came, and for a short tim the fears of the pessimists seemed to be only too wel founded. A fortnight saw G.M.B.S. fall from £77. 15s. to £35. and the holders of stocks were ruined. But gradually th state of things improved, and the position to-day shews tha if the speculators of 1887 and 1888 had been content to rais Chili bars to say £55., the Syndicate might still have been in existence and have realised considerable profits. Of course the collapse of the combination removed the great incentiv to production, while at the same time consumption receive a tremendous stimulus. Copper returned to the place tha nature appears to have indicated for it in the arts and manu factures, and the search for substitutes ceased to be re munerative. How prices gradually improved to the neigh bourhood of £50. is now a matter of history, and it reflect no small credit on those to whom fell the manipulation o the huge stocks left by the Syndicate, that these prices hav been fairly well maintained up to the present. It is no difficult to account for the state of affairs that has made th task of the holders a much easier one than seemed probabl some twelve months since. That period has seen an immens activity in most of the industries where copper plays a part In the single field of electricity its use has increased by leap and bounds, and the number of works lately erected, or now in course of erection, for the manufacture of sulphate o copper, testifies to the growing consumption of that article In addition to this, it must be remembered that the stock o old copper, which proved such an awkward factor in th calculations of M. Secretan and his colleagues, has bee pretty well exhausted, and that many mines which were onl called into a state of production by the abnormal prices o 1887-8, have been relegated to obscurity. On duly weighin these considerations it is easy to see why the total stocks in English and French warehouses have fallen from 118,50 tons to 87,770 within the last twelve months, or a total de crease of nearly 31,000 tons. But if we take the figures o the last three and a half months—from the 1st of January t the 15th of April—we note that the decrease in stocks is onl

s, shewing that the diminution has fallen off con-
in rapidity, and that supply and demand are coming
d closer together. Probably this is to be accounted
o ways. In the first place a lull has come over trade
, and the copper industries participate in the general
nd in the second place it must be remembered that
is still a very remunerative price to mineowners who
able to produce when C.B.S. stood at £40., and
cement is certain to operate in favour of an increased
It is one of the signs of the times that Messrs.
nd Barry, Limited, expect to produce twice as much
in 1890 as they dealt with in 1889. If only trade
good and absorb the output, we see no reason why
ould not be maintained for a long time to come, but
either of the causes above-mentioned, the supply
xceed the demand, even to a very slight extent, we
id the effect upon the market would be most
us.

E SPONTANEOUS IGNITION OF COAL CARGOES.

OFESSOR VIVIAN B. LEWES, ASSOCIATE, F.C.S., F.I.C.

the Thirty-first Session of the Institution of Naval Architects,
rch 28th, 1890, the Earl of Ravensworth in the chair.]

EN years ago the loss of life and property caused by the spon-
eous ignition of coal cargoes became so serious that the Board of
ting in unison with the Committee of Lloyds', urged upon the
ent the necessity for appointing a Royal Commission to inquire
eport upon the possibility of preventing a class of disaster as
in nature as it is destructive in result. The Commission,
s appointed in April, 1875, had the good fortune to be a aided
quiry by the scientific knowledge of Dr. Percy and Mr., now
rick, Abel, and after collecting and collating all the evidence
ld be obtained, they in the following year published their
on the subject. In this report much valuable evidence is
review, and much sound advice is given for the guidance of
loyed in the shipment of coals; but whether it is that the
blue book scares the ordinary reader, or that the circulation
terature, often of the most valuable character, is of necessity
ed, the fact remains that the conditions at present existing
as bad as they were before the publication of the report.
nine years immediately following the report, viz., 1875 to
coal-laden vessels are known to have been lost from spon.
nition of their cargoes, whilst during the same period 328
ing from unknown causes, a large percentage of these losses
oubtedly due to the same cause; and these, again, form but a
l percentage of the cases in which cargoes have heated and
in which the vessel has been saved; and now, within the
ears, the statement has gained ground that with the general
f temperature in steamships, due to the introduction of triple
engines and high pressure boilers, spontaneous firing in the
ers and coal cargoes, is very much increased, and that many
arise from this cause.
these circumstances, Mr. Martell suggested to me some time
n inquiry into the causes and possible prevention of this very
il would be work, not only likely to be acceptable to the
of this Institute, but also one that was needed in the Service,
in the mercantile marine; and I have now much pleasure in
before you the results obtained in a long series of experiments,
, taken in conjunction with the work done by others on the
hrows a somewhat clearer light upon the causes of this particu.
of phenomenon, and enables suggestions to be made for its
n.
a substance of purely vegetable origin, formed out of contact
y long exposure to heat and pressure, from the woody fibre
ous constituents of a monster vegetation which flourished long
earth was inhabited by man; and coal may therefore be
on as a form of charcoal, which having been formed at a tem.
ower than that of the charcoal burner's heap and under great
is very dense, and still retains a quantity of those constituents,
the latter case, are driven off, as tar, wood naphtha, &c.;
bodies consist essentially of compounds containing carbon
gen, together with a little oxygen and nitrogen, and form the
atter and hydrocarbons of the coal.
the carbon and hydrocarbons, coal also contains certain
odies which were mostly present in the sap and fibre of the original vegetation, and which give the ash which is left behind when the coal is burnt.

These substances consist chiefly of sulphate of lime or gypsum, silica and alumina, whilst in nearly all kinds of coal is to be found a substance called disulphide of iron, coal brasses or pyrites, which has been formed by the gradual reduction of the sulphates by carbonaceous matter in the presence of iron salts, and which, during the combustion of of the coal, is decomposed, giving off sulphur compounds and leaving behind oxide of iron, which gives the reddish brown colour to the ash left by many kinds of coal.

Of these constituents of coal, the only ones which play no part in the phenomena attending heating and spontaneous ignition, are the mineral constituents other than the pyrites, and we have therefore to deal with the chemical actions which take place when the carbon, hydrocarbons, and brasses contained in newly-won coal come in contact with air and moisture.

(a) THE INFLUENCE OF CARBON IN PRODUCING HEATING.

Carbon is one of those substances which possess to an extraordinary degree the power of attracting and condensing gases upon their surface, this power varying with the state of division and density of the particular form of carbon used. The charcoal obtained from dense forms of wood, such as box, exhibit this property to a high degree, one cubic inch of such charcoal absorbing.* :—

Ammonia gas	90	cubic inches.
Sulphuretted hydrogen	55	"
Carbon dioxide	35	"
Ethylene (olefiant gas)	35	"
Oxygen	9·25	"
Nitrogen	6·5	"

whilst certain kinds of coal also exhibit the same power, although to a less degree.

The absorptive power of newly-won coal due to this surface attraction varies, but the least absorbent will take up 1¼ times its own value of oxygen, whilst in some coals more than three times their volume of the gas is absorbed. This absorption is very rapid at first, but gradually decreases, and is, moreover, influenced very much by temperature, for reasons which will be explained later.

The absorption is at first purely mechanical, and itself causes a rise of temperature, which in the case of charcoal formed in closed retorts, as in preparing alder, willow, and logwood charcoal for powder making, would produce spontaneous ignition, if it were not placed in sealed cooling vessels for some days before exposure to air.

The rate of absorption varies with the amount of surface exposed, and therefore able to take part in this condensing action, so that when coal or charcoal is finely powdered, the exposed surface being much greater, absorption becomes more rapid, and rise of temperature at once takes place. If charcoal is kept for a day after it has been made, out of contact with air, and is then ground down into a powder, it will frequently fire after exposure to the air for thirty-eight hours; whilst a heap of charcoal powder, of 100 bushels or more, will always ignite. It is for this reason that, in making the charcoal for powder, it is always kept, after burning, for three or four days in air-tight cylinders before picking over, and ten days to a fortnight before it is ground.

In the case of coal, this rise in temperature all tends to increase the rate of the action which is going on; but is rarely sufficient to bring about spontaneous ignition, as only about one-third the amount of oxygen being absorbed by coal that is taken up by charcoal, and the action being much slower, tends to prevent the temperature reaching the high ignition point of the coal. Air-dry coal absorbs oxygen more quickly than wet coal.

(b) THE ACTION OF THE BITUMINOUS CONSTITUENTS OF THE COAL IN SPONTANEOUS IGNITION.

All coal contains a certain percentage of hydrogen, which is in combination with some of the carbon, and also with the nitrogen and oxygen, and forms with them the volatile matter in the coal, and the amount present in this condition varies very largely, being very small in anthracite and very great in cannel and shale. When the carbon of the coal absorbs oxygen, the compressed gas becomes very chemically active, and very soon commences to combine with the carbon and hydrogen of the bituminous portions, converting them into carbon, dioxide, and water vapour. This chemical activity increases rapidly with rise of temperature, so that the heat generated by the absorption of the oxygen causes it to rapidly enter into chemical combination. Chemical combination of this kind—*i.e.*, oxidation—is always accompanied by evolution of heat, and this further rise of temperature again increases rapidity of oxidation, so that a steady rise of temperature is set up, and this taking place in the centre of a heap of small coal, which, from the air and other gases enclosed in its interstices, is an admirable non-conductor of heat, will often cause such heating of the mass that if air can percolate slowly into the heap in sufficient quantity to supply the necessary percentage of oxygen for

* Saussure.

the continuance of the action, the igniting point of the coal would be soon reached.

The effect of rise of temperature in increasing the rapidity of chemical actions of this kind can be realised from the effect which it has in the spontaneous ignition of oily waste or rag.

If a substance like cotton waste be rendered oily with anything except the mineral oils, it acquires the power of taking up oxygen from the air, and this oxidising the oil gives rise to heat. At ordinary temperatures this oxidation is slow, and, consequently, it may be days before the rise in temperature becomes sensible, but when this point is reached the oxidation proceeds with remarkable rapidity, and in a few hours the point of ignition is reached and the mass bursts into flame, whilst if the oily waste be placed in a warm place at first, spontaneous ignition is only a question of hours or sometimes even minutes.

Galletley found that oily cotton at ordinary temperatures took some days to heat and ignite, whilst placed in a chamber warmed to 130° to 170° F. (54° to 76° C.), the cotton greasy with boiled linseed ignited in one hour fifteen minutes, and olive oil on cotton in five hours, and in a chamber heated to 180° to 200° F. (82° to 93° C.) olive oil on cotton ignited in two hours.

It has been suggested that very bituminous coal, such as cannel, shale, and coals containing schist, is liable to spontaneous ignition from the fact that a rise in temperature would cause heavy oils to exude from them, which, by undergoing oxidation, might cause rapid heating. But experiment not only shows that this is not the case, but that the heavy mineral oils have a remarkable influence in retarding heating cotton waste, oily with easily oxidisable oils mixed with 20 per cent. of heavy mineral oil, being exempt from heating.

(c) THE ACTION OF IRON DISULPHIDE, PYRITES, OR COAL BRASSES IN PROMOTING SPONTANEOUS IGNITION.

Ever since Berzelius first expressed the opinion that the heat given out by the oxidation of iron disulphide into sulphates of iron might have an important bearing on the heating and ignition of coal, it has been adopted as the popular explanation of that phenomenon; and although the work of Dr. Richters clearly proves this not to be the case, the old explanation is still given, a notable exception, however, being in the case of our great metallurgist, Dr. Percy, who, as early as 1864, pointed out that probably oxidation of the coal had also something to do with spontaneous combustion, a prediction amply verified by Dr. Richters' researches some six years later.

This disulphide of iron is found in coal in several different forms, sometimes as a dark powder distributed throughout the mass of the coal, and scarcely to be distinguished from coal itself. In larger quantities it is often found forming thin golden-looking layers in the cleavage of the coal, whilst it sometimes occurs as large masses and veins, often an inch to two inches in thickness, but inasmuch as these masses of pyrites are very heavy, they rarely find their way into the screened coal for shipment, many hundreds of tons of these "brasses" being annually picked out from the coal at the pit's mouth, and utilised in various manufacturing processes. If the air is dry the pyrites undergo but little change at ordinary temperatures; but in moist air they rapidly oxidise when in a finely-divided condition, the first action being the formation of ferrous sulphate and sulphur dioxide, together with the liberation of sulphur, the relative amounts of the two latter being regulated by the temperature and the supply of air, whilst longer contact with moist air converts the ferrous sulphate in a basic ferric sulphate generally termed "misy."

It is during this process of oxidation that the heat supposed to cause the ignition is evolved. But when it is considered that some of the coals most prone to spontaneous combustion contain only eight-tenths of a per cent. of iron pyrites, and rarely more than 1¼ per cent., the absurdity of imagining this to be the only cause of ignition becomes manifest. If 100 lbs. of coal were taken, and the whole of the pyrites in it concentrated in one spot and rapidly oxidised to sulphate, the temperature would barely be raised to 100° C., if all loss of heat could be avoided. Besides which, in certain manufactures pure iron pyrites are largely used, and, when free from carbonaceous matter, may be kept in a state approaching to powder in heaps containing many hundred tons; and although undergoing continual oxidation, I have been unable to trace a single case of heating, much less a rise of temperature which would approach the igniting point of coal. When, however, it is mixed with finely divided carbonaceous matter, then heating and ignition is a frequent occurrence in even moderate sized heaps.

I have carefully determined the igniting point of various kinds of coal, and find, that—

Cannel coal	ignites at	698 degrees F.	=	370	degrees C.
Hartlepool coal	,,	766 ,,	=	408	,,
Lignite	,,	842 ,,	=	450	,,
Welsh steam coal	,,	870·5 ,,	=	477	,,

So that no stretch of imagination could endow the small trace of pyrites scattered through a large mass of coal, and undergoing slow oxidation, with the power of reaching the needful temperature.

Dr. Richters fully realises this point, and discards the idea of the pyrites doing anything more than adding their mite to the caus which bring about rise of temperature. In this, however, I thi he is mistaken, my experiments, which will be publish when complete, pointing to the fact that they may increase t liability to ignition when present in large quantities, and do so liberating sulphur under certain conditions. Now, sulphur h an igniting point of 482 F. or 250 C., so that the presen of free sulphur would lower the ignition point of tl coal by considerably over 100 degrees Centigrade, the sulphur in tl case playing exactly the same part that it does in gunpowder, which it lowers the point of ignition, and increases the rapidity combustion. A still more important part played by the pyrites is th as they become oxidised to ferrous sulphate they swell in size, and tend to split up the coal into small pieces, and by exposing a lar extent of fresh surface to the air cause increase of temperature ai energetic chemical action.

We can now trace the actions which cumulate in ignition. Tl newly-won coal is brought to the mouth of the pit, and at once cor mences by virtue of its surface action to absorb oxygen from the ai but unless piled in unusually large heaps, and a good deal broken, does not, as a rule, show signs of heating, as the exposed surface comparatively small, and the air finding its way freely between tl lumps keeps down the temperature. The coal is now screened, ar the obtrusively large lumps of brasses picked out; it is then put in tl trucks and enjoys the disintegrating processes of joltings and shuntin innumerable, every jar adding to the percentage of small coal preser and a corresponding increase in the size of the surface exposed to tl air. Arrived at the docks it has to be transferred from the truck to tl ship, which is done by one of the numerous forms of tips, shoots, spouts employed for the purpose, and it is during this operation th more harm is done than at any other period. The coal first shot in the vessel by reason of the distance which it has to fall is broken down in small lumps, and having to bear the impact of the succeeding load fallii upon it from a height, rapidly becomes powdered into slack, whilst t succeeding loads falling in on the cone so formed get more or less brok down, so that by the time the cargo is all taken in, a dense mass small coal is to be found under the hatchway, and it is invariably this point that heating takes place, as the large surface exposed fresh the air by the breaking down of the coal causes rapid absorption oxygen, and consequent rise of temperature. This sets up chemic combination between the oxygen absorbed by the coal and the hydr carbons and the coal brasses.

The combination of the brasses with oxygen causes the swelling the oxidised mass and splitting up of the coal; fresh surfaces are e posed, and more absorption of oxygen takes place, and the igniti point of the sulphur vapour and sulphur compounds distilled out the pyrites is reached, and rapidly raises the temperature to the igniti point of the coal. It is only in cases where large quantities of den coal brasses are present that this action can take place, as in t ordinary case, where 1 or 2 per cent. only of pyrites are present, t sulphur vapour distilled out from the pyrites is oxidised to sulphur dioxi at temperatures far below the point of ignition of sulphur vapour; a in such cases the heat of absorption and oxidation of the bitumino portions of the coal is amply sufficient to raise the temperature the requisite 752° to 932° Fahr. (400° to 500° C.)

(To be continued.)

THE THARSIS SULPHUR AND COPPER COMPAN

ON Friday last the ordinary general meeting of the Tharsis Sulph and Copper Company was held in Glasgow. Sir Charles Tennai Bart., who presided, moved the adoption of the report. He said th total output of refined copper for 1889 had exceeded that of 1888 2,256 tons, and, owing to the pressing wants of the trade, they we enabled to invoice their sales very closely up. They had writt £20,000. off the mines (which now stood at £280,000.), £20,000. off t railways and pier, £98,950. off works, buildings, machinery, and pla account in Spain, and £6,411. off the works in Britain. As to tl Société des Metaux, the amount of their claim at the date of the 31 December last was £263,915., which they had carried to a suspen account. This amount would of course be largely increased by the e of the current year, at which date their three years' contract would e pire. What might be the result of the legal proceedings they had stituted, and which were now pending, it was difficult to say. Tl were hopeful of being allowed to rank as creditors for the total amou of their claim, in which case they should probably, or, it was perha safer to say they might, receive some dividend. The total sum rema ing at the credit of profit and loss account was £249,169, which enabl the directors to pay a dividend of 20 per cent. and to carry forwa £14,237. 1s. 6d. to the credit of 1890. With reference to the pro pects of the company for the current year, the Chairman said their sa of pyrites were the same as last year's, both as to quantity and pric The whole of their production of iron ore had been placed at sligh higher prices. The report was adopted.

[R]EGULATION OF THE SALE OF FERTILIZERS IN GERMANY.

. the liberal government which France has chosen for
; it has been thought necessary to enact special laws regu-
sale of manures.
'atic Germany the State has not yet turned its attention to
n, and it is through the private action of the individuals
-both buyers and sellers—that proposals have been made
ulation of transactions in fertilizing materials.
thesis is sufficiently striking to give interest to some de-
the resolutions adopted by a committee composed of
and agriculturists.
Jnited States, in France, and in Belgium, says the
andwirthschaftliche Presse, from which we borrow these
buses have given rise to legislation against the delinquents.
any, where there is just as much to complain of, it is be-
be thought necessary to protect those who are the victims,
ished to effect this without having recourse to State
1. It would be grand, worthy of the German spirit,
chultz, to attain the end desired by a common effort,
by the private action of manufacturers, merchants, agri-
and representatives of science, associated together with
f putting the trade in fertilizing materials on a secure
oughout Germany. To succeed in such a movement, and
independent of the public authorities, this would be a
ar for the industry and trade in fertilizing materials for
ientific men in particular, for the Society of Agriculture,
ver, adds he, for the German nation, and for the lofty
spirit which animates it.
usiastic appeal of M. Schultz has not fallen unheeded upon
. The section dealing with the subject at the Magdeburg
ppointed a special committee upon it, consisting of nine
three representing agriculture, three the manufacturing
d three the agricultural stations. Meeting at Berlin on the
er last, under the presidency of Dr. Maercker, the eminent
t Hallé, this committee has decided upon the following
lations:—
essary that all sacks containing commercial fertilizers offered
uld bear upon them in printed characters the exact designa-
manufacturer, the name of the fertilizer, and the amount
of valuable substances contained in it. It is also desirable
ks should always be of uniform weight, which might be
75, or 100 kilos.; the sacks should be completely filled.
ds the name and composition of the fertilizer, the sub-com-
arrived at the following conclusions:—
se of *superphosphate*, without any qualification, the amount
phosphoric acid should not be indicated.
e superphosphate contains phosphoric acid partly soluble in
partly insoluble, the value of which must therefore be ex-
e quantity of each of these two forms of phosphoric acid
e stated on the sack.
precipitate should bear a statement of the total phosphoric
hose of *finely powdered Thomas slag*, both of phosphoric
e dust.
bone meal, *fish guano*, *meat dust*, *Peru guano*, *wool dust*,
to be labelled with their contents of phosphoric acid and of

acal superphosphate, obtained from superphosphate and
nium sulphate should tell the amount of soluble phosphoric
nitrogen.
se of mixtures of *nitrogenous substances with superphos-*
superphosphate, associated with organic nitrogenous matter
ather, dust, wool, &c.), the amount of total nitrogen, and
ely the proportion of organic matter to ammoniacal
ould be stated.
uperphosphates, mixtures of superphosphates with Chili
uld be marked with the amounts of nitrogen and soluble
acid.
of *nitrate of soda*, *sulphate of ammonia*, and *superphosphate*
belled with their percentage of phosphoric acid, soluble in
te, and ammoniacal nitrogen.
od, *bone* and other *nitrogenous manures* should have their
percentage of nitrogen indicated. In all mixtures of potash
should be stated the percentage of potash together with that
valuable substances present. It would be well to arrange
ns concerned to give the percentage of potash in kainite and
salts instead of, as at present, the amount of sulphate of
ained in them.
ling should take place immediately upon receipt of the
should be done in the presence of an agent of the seller or
ial and competent person.
committee of the 15th October, has decided to put these
o practice until the autumn of 1890. It has also expressed
a desire to see a uniform method of examining manures adopted, and has given it to be understood that a conference of delegates from the experimental stations, with representatives of the manufacturers, would not only contribute towards this result, but also clear up the different questions raised on this subject.

Such are the commercial regulations which the German Society of Agriculture has just adopted, with the hope of ensuring integrity in the manure trade and thus rendering it more active. Small dealers will be the first to profit by them; small farmers, so often exposed to fraud against which they are unable to defend themselves, will regain confidence, and an excellent effect will shortly be produced on the progress of agriculture.

NEW DYES.

A series of chameleon colours, including red, purple, pink, brown, scarlet, maroon, orange, and yellow.

(MANCHESTER ANILINE COMPANY).

For 100lbs. Cotton (Yarn or Cloth).

1ST BATH.—Dissolve in 4 gallons of boiling water 5lbs. chameleon, 20lbs. common salt, and pour the solution into the dye-bath containing about 200 gallons water. Enter the cotton at boiling point, boil for 30 minutes, give several turns, lift and wash. (The bath can be renewed to its original strength by the addition of 2½lbs. chameleon.)

2ND BATH—cold—1lb. of nitrite of soda, 2lbs. sulphuric acid, 168° Tw. Turn continually for 15 minutes, take out, rinse in cold water, and enter at once into the

3RD BATH OR DEVELOPING BATH, which is made up, according to the colour required, as described below:—

Work the cotton in this bath cold for 30 minutes. Lift out, wash and dry, and the dyeing is completed.

RED.—Dissolve 1½lbs. red developer with equal weight of liquid caustic soda, 150° Tw., in 2 gallons of hot water.

SCARLET AND ORANGE.—These developers are dissolved in the same proportions as for red.

PINK.—Dissolve 1lb. developer in cold water with sufficient soda crystals to neutralise the bath. To brighten the shade, wash in boiling water, then in hot soap, and again in boiling water. Wring and dry. In dyeing this colour proportionately less quantities of chameleon, nitrate of soda, and sulphuric acid are required than for the red.

MAROON.—Dissolve 15 to 20lbs. developer in hot water and neutralise with soda ash.

PURPLES.—Dissolve 1½lbs. developer in hot water and proceed as for red.

YELLOW AND BROWN.—10lbs. of each of these developers are required for medium shades. With the brown a little hydrochloric acid should be used.

MEM.—The first and third bath can be kept for further use, but the second bath must be made up afresh every day. The dyeing processes must be carried through in quick succession.

DIAMINE BLACK R O (Leopold Cassella and Co.).—The great value of this dye is for mixing purposes. It can be combined with any other substantive dyes, especially our diamine colours, viz.: Diamine red N O, diamine yellow, diamine blue B and 3 R, cotton brown A and N, and thioflavine S, in any required proportions, thus producing almost any shade on cotton in one bath and at a very small cost.

COTTON.—Dye boiling, adding 2 oz. of crystallised Glaubers' salt per gallon of water to the dye-bath; common salt, soda, potash, phosphate of soda, borax, or soap may likewise be used. By topping with brilliant green in a cold bath without any further addition a dead black can be obtained, diamine black acting as a mordant for basic dye-stuffs, such as brilliant green, solid green, safranine, magenta, and thioflavine T. Dyeings done with diamine black R o are extremely fast to washing and light. Acid renders the shade bluer and brighter.

SILK.—Dye with addition of some acetic acid.

SILK AND COTTON (mixed) FABRICS dye like pure cotton. By using soap in the place of Glaubers' salt, the cotton takes up the dye-stuffs, the silk remaining undyed.

WOOL AND COTTON (mixed) FABRICS can be dyed in one bath by addition of Glaubers' salt, using diamine black R o for the cotton and our naphtylamine black D pat. for the wool in such proportions as the fabric contains more or less of the two materials. If the dyeing is done in copper or tin vessels care must be taken that the bath does not turn alkaline, owing to the simultaneous use of naphtylamine black D.

PRINTING AND DISCHARGING.—Zinc dust will discharge diamine black R o. For calico-printing, special directions are issued.

THE GLOBE CHEMICAL WORKS, WIDNES.—G. H. Scott and Co., of New Mills, near Stockport, have taken over the above for the manufacture of India-rubber chemicals—chloride of sulphur, bisulphide of carbon, and golden sulphuret of antimony, etc. They are reconstructing the premises, and almost entirely renewing the plant.

THE ECONOMICS OF ILLUMINATING AGENTS.

THE question as to the actual cost to the consumer, everything being included, of the light produced by the various illuminating agents now in use has not yet received a clear and definite answer. Careful attention should therefore be paid to any evidence which is based upon a serious study of the question, so that at least some idea of the truth may be gained. It is for this reason that we reproduce below extracts and tables, taken from a paper presented by M. C. Rolland, engineer at Mons, to the Society of Engineers of the Liege Institute, and published in the *Revue Universelle des Mines*.

According to this author, "it follows from the examination of these tables that lighting by gas, and even by petroleum, is by no means on the point of being replaced by the electric light." These tables establish the fact that under the most fovourable conditions for the production of the electric light, that is to say, where spare motive power can be applied to its production, this system of lighting still costs 0·0016 francs per candle power per hour (Table VII.), while gas, costing 0·15 francs per cubic metre, burnt in the recuperative lamps of Siemens, or still better, of Wenham, gives a light which only costs 0·001011 francs, and even only 0·000698 francs per candle power per hour. (Table II.)

PETROLEUM.

The luminous intensity given by a petroleum lamp varies considerably with the quality of the oil employed, and also with the quality, the state of cleanliness, and the cut of the wick. A badly-trimmed wick, or one not cut level, gives, for example, a flame which on one side is too long, and slightly yellow or brown, because the supply of air at that point in insufficient, and on the other side is perhaps too small, the air supply being there in excess. Now, it is well known that under both of these conditions the illuminating power suffers.

The management and regulation of wicks is delicate work, and certainly ought to be included in an estimate of the expense of a lamp.

Another source of expense in using this class of illuminant is that arising from the use of the lamp—repairs, consumption of wicks and chimneys, losses of oil, and waste used for cleaning (column 6 of Table I.).

It may be estimated that these sources of expense amount per lamp and hour to—

Expense of lamp........................	0·0100 franc.
Wicks, chimneys, &c..........	0·0045 „
	0·0145 franc.

Finally, to include the additional trouble connected with the use of this illuminant, such as purchasing the oil, storing oil, risk of fire, etc., we think that the theoretical price ought to be increased by 25 per cent. It will be observed that the luminous intensity diminishes considerably when the photometric measurement is made at an angle of 45°.

According to the experiments of M. Heim, the loss of illuminating power is proportional to the diameter of the burner. It is 20 per cent. for burners with an ordinary long flame; about 35 per cent. for the 30 mm. burners used in lamps which give an intense round flame, and amounts to 5 per cent. with 60 mm. burners. It will be possible to compare the luminous intensities of the various systems of lamps and burners at this angle of 45°. The one class giving the maximum intensity at 0° (horizontal rays), the other at 90° (vertical rays).

Moreover, lamps are usually employed under such conditions that their rays are at this angle.

ILLUMINATING GAS.

Column 7 of Table II. gives the expenses of putting down apparatus and mains for candle power per hour, calculated on the following grounds:—

In Belgium the cost of establishing a public burner may be estimated at 15 francs (for private illumination, stations, rooms, etc., where more elegant appliances are used, the cost may rise to 25 or 35 francs per burner), and maintenance and interest may be taken at 10 per cent. of this amount, say 150 francs per annum. This cost must be spread over the entire consumption, depending on the number of hours during which it is in use.

Assuming a minimum of 700 hours per annum, and a consumption of 250 litres per hour, it will be found that at 15 centimes per cubic metre, interest and maintenance must be valued at about 6 per cent. of the value of the gas.

The corresponding expenses for a regenerative Siemens or Wenham lamp, which costs at least 60 francs for a consumption of 250 litres per hour, may be calculated in the following manner, a saving of 50 % of gas being supposed:—

I.

2 ordinary burners ....value	3 fr.
2 secondary tubes..........................	4 fr.
Piping and main tube (30 fr., less 7 fr.)	23 fr.
	30 fr.

II.

One complete regenerative burner, giving the same light as the two burners above	60 fr.
Piping and principal tube	23 fr.
Secondary tube..........................	2 fr.
	85 fr.

The expense x per cent. of maintenance and interest is then foun by the following proportion:—

$$30 : 6 :: 85 : x$$

$$x = \frac{6 \times 85}{30} = 17\ \%$$

We have, therefore, calculated at this rate the expenses tabulated i column 7, table II., for the illumination given by Siemens and Wer ham lamps.

For Clamond and Aver burners we have adopted 10 % of the ga consumed as representing the corresponding expenses. The rate of 6 remains for ordinary burners, the cost of which does not exceed francs.

It must be borne in mind that these numbers should vary inversel with the number of hours during which the burners are employed since the sources of expense remain almost constant whatever be th consumption of gas.

Before passing to the examination of some other modes of illumina tion, a word may be said on the subjects of recuperation and intens burners.

The luminous intensity of a flame increases very rapidly with it temperature, and may be approximately represented by an expressio of the form $I = a^t$, when I is the intensity of the flame, and t its tem perature. The great increase of intensity obtained by superheating th air may thus be conceived, when it becomes sufficient to sensibly in crease the temperature of the flame and so produce a more efficaciou combustion. In the Siemens and Wenham lamps the air may be heate to 400° or 600°, but the apparatus soon wears out if the latter tempera ture be habitually employed.

To render the recuperation rational and efficacious it is essential tha the apparatus be so arranged that the air is heated by the products o combustion, and *not simply by the flame of the burner*, for this ca only give up part of its heat at the expense of its illuminating power.

It has not been found advantageous to strongly heat the gas itself, o the contrary, deposits are thus formed which rapidly block up th orifices by which it escapes from the burner.

For a given consumption of gas there is a definite supply of ai which corresponds to a maximum of luminous intensity. Th determination of this quantity is of special importance when ordinar burners with cold air are employed, but even when hot air is supplied it i still important to ascertain exactly the draught which correspond to a maximum of illuminating power.

Everyone can convince himself of the effect of the draught o volume of air brought into contact with luminous flame on its illumi nating power by simply placing a glass chimney on the top of the glas of an Argand burner, so that the height of the chimney is doubled. I will be observed that the flame immediately becomes lower and throw a less powerful light on surrounding objects, although it may itsel become whiter. Inversely, the intensity of a flame, which is burnin with an excess of air, may be increased by diminishing the supply o the latter. This increase continues until the flame becomes brow towards the top, after which further diminution of the air supply cause a rapid diminution in the brightness of the flame, the combustion the becoming incomplete.

This experiment justifies the conclusion that for burners with col air the maximum of light corresponds to the minimum of air whic permits of complete combustion, and that the yellow flame is mor economical than a white flame obtained by means of an excess o cold air.

When the air arrives at the burner heated to 500°, the influence o the draught will, of course, not be so great, but it must be remembere that the temperature of the flame expressed in degrees is twice a great as this.

The economy effected by these intense flames, that is to say, flame produced by a large amount of gas, may be explained by the facts that in the first place, the amount of air immediately surrounding the flam is smaller for an equal volume of flame than in the ordinary burner and that the loss of heat is thus also rendered smaller; and that in th second place the surfaces and volumes of neighbouring parts of th apparatus so situated as to be capable of absorbing heat from the flam are also smaller in proportion to the gas consumed; finally, a wide flam consists of an interior cone of gas, surrounded by a thicker incandescer layer (which must be traversed by the air) than is the case in a narro flame burning the same volume of gas.

On account of these various circumstances the temperature of a larg flame is appreciably higher than that of a small one, and hence aris its greater illuminating power. This luminous intensity, therefor results from a more effective combustion, which produces an increa in the light radiated from the flame.

TABLE I.—PETROLEUM.

	Chief Types of Lamp. Angle with Horizon. 0°.	Angle 45°.	Actual Expense, 25 in excess of calculated.
Candle power	40	29·3	
Consumption of petroleum of best quality, 800 grammes per litre	90gr.	90	
Consumption per candle power per hour	2·25	3·037	3·796
Expense per candle per hour, at 0·15 francs. per litre		0·000508	0·000710

	Chief Types of Lamp. Angle with Horizon. 0°	Angle 45°.	Actual Expense, 25% in excess of calculated.
Working expenses per candle per hour	0·	0·00049	0·00049
Total expenses per candle per hour		0·001058	0·00130
Total expenses per lamp per hour, 29 candle power at 45°		0·0313	0·03545

TABLE II.—ILLUMINATING GAS.

Type of burner.	Angle with horizon.	Candle power	Gas per hour in cubic metres.	Consumption of gas per candle power per hour in litres.	Value at 0·15 francs per cubic metre.	Working expenses.	Total per candle power per hour.
Split burner	0 degs.	16·9	0·251	14·8			
	45 ,,	17·2	0·256	14·9	0·002235 francs	0·000134	0·002369
Argand burner	0 ,,	21·9	0·239	10·9			
	45 ,,	19·4	0·241	12·4	0·001860	0·000111	0·001971
New Clamond burner	45 ,,	21·1	0·190	9	0·00135	0·00013	0·001480
Aver or de Pintsch burner	0 ,,	14·4	1·0951	6·60			
	45 ,,	10·5	0·1037	9·88	0·001482	0·000148	0·001630
Cardinal or de Brauer burner.	45 ,,	21·9	0·219	10	0·00150	0·000095	0·00159
Siemens regenerator	0 ,,	65·3	0·460	7·05			
No. 3	45 ,,	46·9	0·456	9·75	0·001462	0·000248	0·001710
Wenham, No. 2	0 ,,	28·4	0·249	8·77			
	45 ,,	44·5	0·257	5·77	0·000864	0·000147	0·001011
	90 ,,	45·8	0·256	5·58			
Wenham, No. 4	0 ,,	99·0	0·285	6·92			
	25 ,,	152·0	0·686	4·51			
	45 ,,	170·0	0·677	3·98	0·000597	0·000101	0·000968
	65 ,,	200·0	0·685	3·42			
	90 ,,	202·0	0·671	3·33			

TABLE III.—WATER GAS.

Photometric tests of Some's burner.

Consumption per hour. Cubic feet.	Litres.	Pressure at the burner. Inches.	Mm.	Candle power.	Candle power per cubic foot.
9·66	272·4	2·25	57·15	12·85	1·33
8·31	234·3	2·37	60·19	10·88	1·31
7·90	222·7	2·50	63·50	12·24	1·55
6·70	188·9	1·75	44·45	8·48	1·26
6·70	188·9	1·00	25·40	8·41	1·25
5·58	157·3	3·25	82·55	9·94	1·78
5·10	143·8	4·50	38·10	6·85	1·34
3·96	111·6	2·00	50·80	5·47	1·38
53·91	1519·9			75·14	

Consumption per candle power per hour ...... 20·20 litres.
Price of gas at Frankfort per cubic metre...... 0·075 francs.
Cost per candle power per hour ....0·0015 ,,
Working expenses, 10 % ....0·00015 ,,
Total cost per candle power per hour..........0·00165 ,,

TABLE IV.—MAGNESIUM LAMPS.

No. of ribbons	Illuminating power. With reflector.	Without reflector.	Candle power per ribbon.	Consumption per hour per ribbon.	Consumption per hour per candle.	Total cost per candle Magnesium at 37·5fs. per kilo.
1	150	3,200	150	16·7 grms.	0·1114 grms.	0·0044
2	237	5,880	118·7	16·7 ,,	0·1410 ,,	
4	450	8,000	112·5	16·7 ,,	0·1480 ,,	
6	700	11,300	117	16·7 ,,	0·1415 ,,	
8	950	17,000	117	16·7 ,,	0·1430 ,,	

TABLE V.—ELECTRIC LIGHT—ARC LAMPS.

(M. HEIM.)

	Circuit lamp of Pieper.		Piette Krizich (Schuckert) Differential Lamp.		Siemen's and Halske Differential Lamp.	
Diameter of Carbons	6·7m.m.	5·0	1·0		14	
Length of Arc	2 ,,	2	4	4		4·5
Angle with horizon	0°	45°	0°	45°	0°	45°
Candle power	126	377	220	1,420	575	3,830
Electric work in volts ampires	160	153	414	410	918	912
Volts ampires per candle power	1·27	0·405	1·88	0·291	1·60	0·238
Candle power per horse power	433	1,360	293	1,890	344	2,310

TABLE VI. INCANDESCENT LAMPS.

(After Heim.)

Types.	Candle power.	Electric work in volts ampire.	Volts Ampires per candle.	Candle per H. P.	Lamps per H. P.
Edison lamp, old model	16	72	4·50	122	7·6
,, new ,,	16	60	3·75	147	9·2
Swan lamp, old ,,	16	66	4·13	133	8·3
,, new ,,	16	56	3·50	157	9·8
Siemens and Halske	16	52	3·25	169	10·6
Bustien (Cannstadt)	16	56	3·50	157	9·8

TABLE VII. INCANDESCENT LAMPS.

Kind of Installation.	Price per candle power per hour.
Private installation for 200 lamps at least, with special motive power	0·0031 fr.
When a part of the labour is on hand and an excess of power can be utilised for the production of the light	0·0016 fr.
Special installation for private lighting	0·0047 —0·0062 fr.

SOLUBILITY OF GLASS IN WATER.—(1) Water-glass is decomposed by water into free alkali and silicic acid, a certain proportion (varying with the time of action, concentration, and temperature) of the latter becoming hydrated and dissolved. (2) Potash-glasses are far less soluble than soda-glasses, but the difference decreases with increase of the proportion of lime present. (3) Soda and potash are united in glass both to the silica and the lime. The resistance of glass towards the action of water is dependent on the presence of double silicates of soda or potash and lime. (4) Of all sorts of glass, the plumbiferous flint-glasses are least soluble in boiling water. (5) The relative resistance of glasses is different towards hot and cold water.—*Mylius and Foerster.*

DYNAMITE EXPLOSION.—An explosion occurred at Nobel's Dynamite Works, Stevenston, at one o'clock on the 17th of April, in the refuse acid sheds, doing considerable damage to property and slightly injuring one man. It appears that the refuse-acid had been run into the apparatus in which it was treated, and that the men had ceased to draw off acid from the still, when an explosion occurred in the pipe which brings the acid to the still from the outside tank. The explosion was a severe one, the pipe being torn to atoms, about 50 panes of glass in the roof being broken, some slates torn off, and one of the doors blown out. The man in charge was severely cut about the face, but after being treated at the works he was able to walk home. The cause of the accident is supposed to have been the presence of nitro-glycerine in the acid pipe. The refuse-acid is the acid which remains after manufacture of nitro-glycerine. It is taken to the acid shed to be treated, and it sometimes contains a little of the explosive.—*Times.*

SOCIETY OF CHEMICAL INDUSTRY.

LONDON SECTION.

THE extra meeting on April 21st, was held as usual at the Chemical Society's rooms, at Burlington House, Mr. David Howard in the chair. After the minutes of the last meeting had been read and confimed, the secretary, Mr. Tyrer, read a paper by Dr. Moritz and Professor Meldola, entitled, "Note on the expulsion of ammoniacal compounds from sulphuric acid used in Kjeldahl determinations."

The authors have conducted a series of experiments on the process advanced by Professor Meldola some time back, and reported in the Journal of the Society (1884, p. 63), based on the use of a nitrite which was adversely criticised by Lunge, on the ground that the nitrite added was not completely eliminated. These show that, provided heating be continued for at least two hours, and the temperature attained be high, no such danger exists.

In the discussion which followed it was pointed out that acid of the necessary purity could now be bought, and that, therefore, the need of any process of purification was less marked than formerly. Mr. Foster, commenting on the general applicability of the Kjeldahl process, stated that it had failed in his hands when used for gas coke and petroleum, even when Nordhausen acid was employed. Mr. Blount remarked that the attainment of a high temperature could be insured, thus securing the expulsion of nitrous acid and favouring the complete decomposition of the substance analysed, by the addition of potassium sulphate, a device due to Mr. Chattaway. After a few words from Mr. Tyrer, thanking the Section for the consideration shown him as secretary, and introducing his successor, Mr. Mumford, Mr. Boverton Redwood read his paper on "The petroleum fields of India," which he had substituted for the joint paper by himself and Professor Dewar on the conversion of heavy hydrocarbons with illuminating oils, on account of the latter being not quite completed. The oil bearing districts of India are four in number, petroleum being found in Burmah, Assam, the Punjaub, and Beloochistan. In Lower Burmah wells exist in the Aracan Islands and on the mainland at Myangung. The industry is ancient, but has been of small importance, chiefly from the crude native methods in use. The oil is light in colour, and some samples yield as much as 66 per cent. of kerosene. Mention of petroleum in Upper Burmah is found in the diary of Captain Cox, dated 1797, in which he compares it with the product then recently obtained by Lord Dundonald by the distillation of coal. In a report of the Indian Geological Survey, a return is made of the wells in the two oil fields of Twingoung and Berne, in Upper Burmah, according to which there are at the former 166 worthless wells, 89 giving a comparatively small quantity, and 120 having a good yield. In these wells the most primitive methods were in vogue, the digging being done by hand, in spite of the high temperature, the impossibility of using a light on account of the inflammability of the oil vapour, and the asphyxiating effect on the workmen, which limited the shifts to 290 seconds as a maximum. To economise time the digger was blindfolded before descending, so that he might begin work at once, and not need to wait until his eyes were accustomed to the feeble light. Messrs. Findlay, Fleming, and Co. have imported transatlantic drillers, properly equipped, and though obtaining at present only about 3-15 barrels of oil per well per day, are confident of securing a larger output. Many difficulties have to be encountered, one of the most serious being scarcity of water, but this has been met by laying a pipe line from the river. With regard to the quality and composition of these oils, wide variations occur, but two distinct kinds can be traced. As a type of the first may be taken an oil with a sp. gr. of ·937, flashing point 150° F., solidifying point below 0° F., and of high viscosity; while the second is represented by one with a sp. gr. of ·887, flashing point 110° F., solidifying point 82° F., and of low viscosity. The quantity of burning oil that can be obtained by distillation of the crude oil is small, and it has been with a view to increase this to even 70-80 per cent. that Professor Dewar and the author have been working. As may be supposed from its solidifying point, the second oil mentioned above yields much paraffin, most of which is exported, the hot climate of Burmah rendering stearin candles preferable to those made of it.

The oil fields in Assam are mentioned in an official report as early as 1825, and in 1865 the Indian Geological Survey reported that drillings were then being begun. A well 275 ft. deep has been sunk at Digboy and at first yielded well, but the oil struck appeared to be only a pocket and soon ceased. A district some 60 miles square may be reckoned as oil-bearing. From a sample examined 89 % of lubricating oil was obtained, but no kerosene. In the Punjaub, Nobel has begun exploiting the district known to contain oil, and thinks well of the enterprise. At the Khatan oil field, in Beloochistan, petroleum containing a good deal of asphaltum in solution is obtained, which has proved good and economical as fuel for locomotives.

In the discussion which followed, Mr. Charles Marvin, who has done much to direct public attention to these petroleum deposits, pointed out the political and commercial importance of their use locomotive fuel. Mr. Watson Smith inquired whether coal found Burmah was anthracite, like that in Pennsylvania, situated near th oil wells. Mr. J. B. White stated that samples of oil examined England had often been collected from surface pits, and thus havi lost their more volatile constituents by evaporation, gave a small percentage of kerosene than if taken direct from a flowing wel with regard to Mr. Watson Smith's question, the coal found w not anthracite but a bituminous coal much like that from Sou Wales collieries. Mr. Boverton Redwood did not think that parallel could be drawn between the occurrence of coal and oil t gether in Burmah and Pennsylvania, as their geological positic differed materially. At the conclusion of the paper, which was exce lently illustrated by specimens, diagrams, photographs, and a map the district, the meeting was adjourned to May 5th.

NOTES FROM KUHLOW.

TESTING CAST IRON.—In the case of those foundries which obta their pig directly from blast furnaces the testing of cast iron is esp cially important, as charcoal blast furnaces are very sensitive to ar accidental change in the mixture. The metal, which is taken from tl furnace by means of a ladle, the matter floating on the surface bei removed, is poured into an open sand mould in the form of a cavity about twenty centimetres in diameter and seven or eight centimetres depth. Iron which is rich in silicium and carbon becomes rapid coated on the surface with a dull glowing cover of oxide formation These dull formations also indicate an iron too rich in graphite. Brig and long-lasting formations distinguish the iron best adapted for cas ing purposes. If the iron in a little time becomes rapidly blistered, if it throws off hissing sparks, it is a proof that it is poor in silicium ar hard. A practised eye will readily perceive the peculiarities in tl nature of the iron by carefully observing the formations.

THE WEISSWASSER paper and cellulose manufacturers have ju introduced into the market under the names of uni-coloured and tw coloured watertight cellulose papers a cellulose material that can l applied to the most varied purposes. The cellulose paper can be us for book backs, table cloths, and as a temporary covering for roofs, well as for packing goods. It can be laid on damp walls and as a co ing for maps, in short its applicability is extraordinarily manifold. Th cellulose paper is far cheaper than parchment. It does not becon sticky through heat, nor does it crack from the cold, as is the case wi oil cloth. The disagreeable asphalte odour is not perceptible.

SPIRIT PRODUCTION IN THE GERMAN EMPIRE.—In the peri from the 1st of October, 1889, to the end of March 2,193,922 hect litres of pure alcohol were produced in the distilleries in the Germ Empire, and introduced into free traffic on payment of the consumpti imposts, viz., 50 pfgs. on 799,392 hectolitres, and 70 pfgs. on 412,8 hectos.

FLUID MIXTURE FOR TOYS.—Fine ground argillaceous slate per cent., rag paper paste 20 per cent., and 30 per cent. of bur plaster are mixed with the necessary volume of water to form a past which is then cast in moulds, the moulds having been previous daubed with finely ground slate, powdered plaster, or fat. A sufficient thick crust will form in a few minutes, when the residuum of tl mixture must be poured out of the mould. The mixture, which unbreakable, hardens very rapidly. The castings thus produced m be immersed in paraffin or stearine, or they can be japaned. In tl latter case it is desirable, so as not to consume too much paint, to fi apply a coat of quick drying boiled oil, and when the oil has becor hard the article is to be painted.

NOTE ON MANURE DRYERS.—This being the busy time at man works, says Vincent Edwards, F.C.S., in the *Chemical News*, a f words in connection with the above may be of interest. It is, course, necessary to add to some manures substances which will only increase bulk, but at the same time absorb moisture. I ha come to the conclusion that there is a good deal of danger of seri error in this matter; in fact, of undoing what has been done at so mu expense by adding "dryers" containing iron and alumina; in fa some manufacturers may be deluded by unscrupulous dealers, as it almost certain that many of these stuffs are worthless. I have ma some experiments with ashes from furnaces, and find that some cont large quantities of iron, which, notwithstanding any new ideas to contrary, is a very dangerous ingredient to supply without discri nation to plants. It is true that some may be benefitted by the use iron as a manure, but the advantages of its general use are more tl doubtful. Manufacturers should see that all substances used for dry are nearly free from iron. No superphosphate or bone manure sho contain more than 0·7 per cent. of iron (Fe), which is poison to plant and injury to the manure.

Correspondence.

ANSWERS TO CORRESPONDENTS.

VICTOR.—Soda crystals are often employed in works to make up standard solutions: the "points" as the large crystals are called are always employed, as they do not vary much in composition.

X. J. V. E. Q.—1. 20 per cent. of sulphuric anhydride dissolved in the H_2SO_4. 2. 146°/145° is the density on Twaddells' hydrometer. Pyrites 168 degs. is pyrites, rectified oil, pyrites 150 degs. is pyrites brown oil. 2. We could not, the labour would be too great. 3. We cannot tell. 4. We note your remarks, and will see what can be done.

W. O. T.—The correct formula is $Ca\,SO_4\,2H_2O$, and this is the state in which it is found; when burnt, however, it becomes $CaSO_4$ or 136 parts becomes 100 You are correct, it is 21 per cent. nearly.

E. A.—We should like a copy of your paper when it appears.

"ONLY BORN."—We really cannot advise you. There are really so many chemists now in the market that we fear yours is not a solitary case. Your best course is personal application. It was very unwise of you to resign your appointment until you saw at least the shadow of another.

ADVERTISER.—You cannot surely expect one advertisement to do you any good. Has it ever struck you that you have competitors.

BETA.—We forwarded your letter in due course.

MYRABOLAMS.—Your query is only suitable for our advertising columns.

PRIZE COMPETITORS.—Your essays have been duly received, but will remain unopened until the time specified in the notice.

MACHINERY.—You can do no better than consult our second-hand machinery columns. (See advt.)

EXHIBITOR.—We accept your apologies for not being present at the opening of the exhibition; we shall be pleased to see you over at any time.

NEMO.—We are sending you an exhibition prospectus.

ALKALI.—Please address the query to us again. We fear your last communication has been mislaid.

C.D.—Thanks for the correction, it has been put right in this issue.

SUGAR MANUFACTURE IN INDIA.

THE new *Kew Bulletin* contains a despatch from the Government of India to Lord Cross respecting sugar manufacture in India. The improvement of sugar production and manufacture in India (the despatch says) has been the subject of attention both of the authorities and of capitalists since the beginning of the century, and various attempts have been made to establish factories, none of which appear to have been attended with any permanent success unless supplemented by the sale of rum and liquors. Sugar refining alone has not proved sufficiently profitable to maintain a factory. Some of the main difficulties against which the industry has to contend are that the cultivation of sugar-cane is limited by the supply, not only of water for irrigation, but also of manure; it is confined to small farms or holdings, and each cultivator who is able to grow the crop at all can only find manure enough for a small area, generally less than half an acre, of sugar-cane. The plots of sugar-cane are, therefore, greatly scattered even in a canal-irrigated tract. A central factory has accordingly to bring in its supplies of cane in small quantities over varying and sometimes considerable distances. The carriage of canes over a long distance is detrimental to the juice for the purpose of sugar making, especially in India, where the canes ripen at the season when the atmosphere is driest, and suffer, therefore, the *maximum* of injury. The amount of cane which can be grown, limited as it is by the supply of water and manure, barely suffices for the wants of the Indian population. It seems to be at present as profitable to produce coarse sugar for their use as highly refined sugar for export. There is, therefore, no sufficient inducement to capital to embark on the more difficult and expensive system. A further obstacle to sugar refining in India is the high differential rate which the conditions of the excise system require to be placed upon spirits made on the European method as compared with that levied on spirits manufactured by the native process. The sugar refiner in India is thus placed at a disadvantage in respect to the utilisation of his molasses in the form of spirits. Hence the Government is unable to support to project of State model sugar factories, but is inclined to look to the gradual improvement of the ryot's method of manufacture rather than the introduction of more expensive and centralising systems.—*Times*.

ACTION OF SULPHURIC ACID UPON ALUMINIUM.—A. Ditte.—Cold dilute sulphuric acid seems to have no action upon aluminium, [illegible] as the formation of aluminium hydroxide evolves 195·8 calories, [illegible] should at ordinary temperatures decompose water and *a* [illegible] acids. The author in this paper demonstrates that such [illegible] that if a plate of aluminium immersed in dilute sulphuric [illegible] be attacked, the cause is that it becomes coated with [illegible] hydrogen which prevents all direct contact with [illegible], ex., No. 11.

Trade Notes.

Alcohol from Chestnuts.—The *Journal de la Chambre de Commerce de Constantinople* says that a company has recently been formed in Paris for the working of products derived from chestnuts, and chiefly the production of alcohol from chestnuts according to improved processes.

A High Chimney.—The Fall River Iron Company's 105 metre high chimney will soon be surpassed in dimensions by one about to be erected in the Imperial foundry at Halsbruch, near Freiburg, Saxony. According to Mr. Huppner's plans it will be 138m. high, with an interior diameter of 4·80m. The pedestal is 12m., the shaft 8·70m., the chimney proper 117·30m., the total length of the flues 984m. The structure is to be built on an elevation of the ground 79m. above the workshops, and thus towering over these 217m. The height is intended to neutralise the poisonous effects of the gases.

Bricks from Coke.—The use of coke, coke dust, or graphite from gas-retorts in the manufacture of refractory bricks for lining iron furnaces seems like a contradiction of nature; but it appears from several communications to a recent meeting of the Society of German Iron Manufacturers that an industry in the manufacture of such bricks for ironworks is actually established, and is growing. Hitherto nothing has been found capable of withstanding the corrosive action of blast-furnace slag, which is alternately acid and basic, and carries away the lining of the hearths of the furnaces as though it possessed no resistance, although, as a matter of fact, everything is done to prevent this action. The best refractory materials, if placed in the way of a current of slag, will completely melt away in an hour or two. The observation that slag runs best in a channel of coke or coal ash turned attention to this material for lining furnaces; and Mr. F. Burgess, of Gelsenkirchen, states that in his first experiments, in 1883, he tried a combination of coal, coke dust, graphite, and clay, moulded in the form of bricks. Unfortunately, in the process of burning these carbon bricks, the carbon largely burnt out; but even so they gave satisfactory results. The process could not be patented because it is on record that furnaces in the Hartz Mountains have been lined with a similar combination of coke, dust, and clay. It appears also, from a paper by M. Purcel, that in a certain district of France the hearths and bottoms of furnaces have for some years been lined with graphite brick. The raw material of these bricks was gas-retort graphite, ground and mixed with tar and then calcined. Part of the tar is coked, and binds the graphite into hard and durable bricks. Coke, poor in ash, treated in the same way, yields good results. These bricks give satisfaction in furnaces which are severely pushed. The cost is about £5., per ton in Germany.—*Journal of Gas Lighting.*

Market Reports.

MISCELLANEOUS CHEMICAL MARKET.

Caustic soda has been gradually returning to its normal level of value in relation to soda ash and as the pressure of orders brought about by the recent curtailments of production has now been overcome, the quotations for spot and early deliveries are for 70%, £9. 5s. to £9. 7s. 6d. f.o.b.; and for 60% white, £8. 15s. per ton, and cream, 60%, £8. 12s. 6d. per ton. The higher strengths are obtainable in the Tyne district at £11. 10s. net, for 77%. In Lancashire, 74%, £11. per ton, f.o.b. Soda ash maintains a firm position, and makers are well booked with orders. Prices on the spot 1½d. per degree for ordinary carb ash, 48% to 58%, and for caustic ash, 1⅜ per degree. Soda crystals firm at £3. Chlorate of potash quiet, but without change, at 5d. per lb. Sulphur meets with little fresh inquiry, but makers are fairly well engaged for the present output. Bleaching powder quiet and prices for early deliveries, £4. 17s. 6d. to £5. per ton for softwood, on rails. Hardwood casks for export £5 7s. 6d. f.o.b., Liverpool, and £5. 10s. f.o.b., Tyne. Sulphate of copper is still the subject of much speculation as to future prospect. Spot quotations £24. per ton, and for latter part of 1889, £18. 10s. to £19. per ton. Acetates of lime are quiet, but prices steady for prompt and forward deliveries at £7. 17s. 6d. for brown, and £13. 15s. for grey, at usual centres of consumption. Carbolic and picric acids nominally without change, but with a weaker tendency. Lead salts quiet in all grades, but firmly held at current quotations. Green copperas in active demand and prices firmer. White powdered arsenic is moving off freely at £13. per ton, at usual centres of distribution. Aniline oil and salt are firm on the spot and more business doing at the advanced prices. Oxalic acid still scarce, and nominally 4d.; citric and tartaric acids quiet. Chromic acid and bichromates, prussiate of potash, tin and zinc salts, sulphocyanides, and miscellaneous goods for textile trades are extremely quiet in consequence of prevailing dullness of trade.

METAL MARKET REPORT.

	Last week.		This week
Iron	46/-	....	44/11
Tin	£89 7 6	....	£90 5 0
Copper	48 10 0	....	48 7 6
Spelter	20 7 6	....	20 15 0
Lead	12 10 0	....	12 17 6
Silver	45⅜d.	....	46d.

COPPER MINING SHARES.

	Last week.			This week.	
Rio Tinto	16¼	16⅜	....	16¼	16 5/16
Mason & Barry	6¼	6⅜	....	6 5/16	6 7/16
Tharsis	4⅞	5	....	4 15/16	5 1/16
Cape Copper Co.	3½	3⅝	....	3½	3⅝
Namaqua	1 15/16	2 1/16	....	2	2⅛
Copiapo	2⅜	2½	....	2 5/16	2 7/16
Panulcillo	15/16	1 1/16	....	15/16	1 1/16
New Quebrada	½	⅝	....	½	⅝
Libiola	2¼	2⅜	....	2⅛	2¼
Tocopilla	1/-	1/6	....	1/-	1/6
Argentella	3d.	9d.	....	3d.	9d.

REPORT ON MANURE MATERIAL.

The active business for prompt and early delivery experienced since Easter has been continued during the past week, but prices for phosphatic material still favour buyers, as also do those for nitrogenous material, with the exception of nitrate, which closes somewhat firmer.

Outside shippers of Charleston phosphates, favoured by easier freights, are still inviting offers for cargoes for shipment during the summer months, and buyers have been asked to bid even as low as 10d., though whether business would result at the price seems doubtful. No doubt the prospective shipments of Florida rock will, to some extent, affect the market; and as the present manure season has probably scarcely fulfilled the expectations of manufacturers, they seem at present inclined to defer their purchases for their next season's manures. Prices for other descriptions of rock continue nominally without change, but they are too high to induce United Kingdom buyers to come to terms.

A mixed cargo of River Plate bones and ash—December sailing—has arrived at port of call, and will have to be sold—£5. 7s. 6d. and £4. 17s. 6d. for bones and ash respectively—would probably be business. Further sales of East India crushed bones have been made at Liverpool at £5. 2s. 6d. ex-quay, and some business has been done in meal at £5. 10s., but the quay has now been almost cleared, and buyers do not seem much inclined to purchase afloat so late in the season. Closing values are £5. 2s. 6d. ex-quay for crushed bones, and £5. 10s. ex-quay or store for bone meal. Several parcels of Mediterranean and Brazil bones arrived at Liverpool have been sold at £4. 10s. for sugar stained, up to £4. 17s. 6d. for good clean hard bones, suitable for charring.

There has been a good business passing in nitrate of soda, both on spot and for cargoes, and the market closes firmer. Spot price is 8s. 1½d. Cargoes at coast have been sold at 8s. June and July sailings are worth 8s. 4½d., and for later sailings still higher prices are required, very heavy deliveries still continuing; and it being reported that several of the works in Chili have been closed on account of the recent low prices.

There is nothing fresh to report in dried blood, except that small sales of home-prepared have been made at 10s. 6d. per unit, free to rails at works.

Kainit remains as last quoted, 38s. to 40s. in bags, delivered free to rails.

Superphosphates are quiet, but prices remain as last quoted, 46s. 3d. per ton in bulk, delivered free to rails at works.

TAR AND AMMONIA PRODUCTS.

The benzol market is slightly firmer than it was at this time last week; 2s. 7½d. for 50/90's and 3s 5½d. for 90's are prices which we learn have been actually paid. Solvent naphtha is still in good demand, and especially that quality known as "Odorless solvent." Creosote maintains its price notwithstanding the approach of light evenings, the bulk of it going for creosoting timber, as against the winter use for lighting purposes. Anthracene and carbolic acid remain without change. The pitch market is still depressed, and lower values are mentioned as being very probable.

›hate of ammonia market is very dull indeed, prices having since our last report. London is now quoted as £11. 6s. 3d., on at various prices from £11. 7s. 6d. to £11. 5s., with but ess doing at these low rates. Hull is spoken of as £11. 6s. downward tendency, while Leith is reported at £11. 5s. ding buyers at this price. We do not believe this drop occasioned by the laws of supply and demand, as at the year we shall probably find there has been as much f ammonia exported as ever.

HE LIVERPOOL COLOUR MARKET.

s.—Unaltered. Ochres: Oxfordshire quoted at £10., 14., and £16.; Derbyshire, 50s. to 55s.; Welsh, to 55s.; seconds, 47s. 6d.; and common, 18s.; Irish, e, 40s. to 45s.; French, J.C., 55s., 45s. to 60s.; M.C., 6d. Umber: Turkish, cargoes to arrive, 40s. to 50s.; e, 50s. to 55s. White lead, £21. 10s. to £22. Red lead, xide of zinc: V.M. No. 1, £25.; V.M. No. 2, £23., ed, £6. 10s. Cobalt: Prepared oxide, 10s. 6d.; black, lue, 6s. 6d. Zaffres: No. 1, 3s. 6d.; No. 2, 2s. 6d. Terra est white, 60s.; good, 40s. to 50s. Rouge: Best, £24.; ditto rs', 9d. per lb. Drop black, 25s. to 28s. 6d. Oxide of iron, ity, £10. to £15. Paris white, 60s. Emerald green, 10d. erbyshire red, 60s. Vermillionette, 5d. to 7d. per lb.

IE LIVERPOOL MINERAL MARKET.

rket has been somewhat easier this week. Manganese: ive somewhat improved; prices firm. Magnesite: Stocks p continue large, and prices are still easier. Raw ground, and calcined ground, £10. to £11. Bauxite (Irish Hill strong demand at full prices—Lump, 20s.; seconds, 16s.; ; ground, 35s. Dolomite, 7s. 6d. per ton at the mine. alk: Arrivals this week have been larger, but the majority nto consumption; prices, therefore, are firm at last figures, for G.G.B. "Angel-White" brand—90s. to 95s. medium, 05s. superfine. Barytes (carbonate) easy; selected crystal e at £6.; No. 1 lumps, 90s.; best, 80s.; seconds and good smalls, 50s.; best ground, £6.; and selected crystal ground, hate has somewhat improved, best lump, 35s. 6d.; good os.; medium, 25s. 6d. to 27s. 6d.; common, 18s. 6d. to 20s.; st white, G.G.B. brand, 65s.; common, 45s.; grey, 32s. 6d. umicestone quiet; ground at £10., and specially selected st quality, £13. Iron ore unchanged. Bilbao and Santan- 10s. 6d. f.o.b.; Irish, 11s. to 12s. 6d.; Cumberland in nand at 18s. to 24s. Purple ore steady at last quotations. anganiferous ore selling freely at good prices. Emery-stone: ls still continue scarce, especially for spot delivery; prices No. 1 lump is quoted at £5. 10s. to £6., and smalls £5. to Fullers' earth steady; 45s. to 50s. for best blue and yellow; able ground, £7. Scheelite, wolfram, tungstate of soda, sten metal inquired for. Chrome metal, 5s. 6d. per lb. alloys 2s. per lb. Chrome ore: More arrivals, but prices Antimony ore and metal have further advanced. Uranium to 26s. Asbestos: Best rock, £17. to £18.; brown grades, 5. Potter's lead ore: Prices are firm; smalls, £14. to cted lump, £16. to £17. Calamine: Best qualities scarce— os. Strontia steady; sulphate (celestine) steady, 16s. 6d. Carbonate (native), £15. to £16.; powdered (manu- £11. to £12. Limespar: English manufactured, old and in demand, and brings full prices; 50s. for ground Felspar, 40s. to 50s.; fluorspar, 20s. to £6. Bog ore in rm at 22s. to 25s. Plumbago: steady; Spanish, £6.; n lump at last quotations; Italian and Bohemian, £4. to ton; founders, £5. to £6.; Blackwell's "Mineraline," nch sand, in cargoes, continues scarce on spot—20s. l. Ferro-manganese and silicon spiegel in good demand er figures. Chrome iron, 20 per cent., £24. to £25. ca, £50. China clay freely offering—common, 18s. 6d.; um 22s. 6d. to 25s.; best, 30s. to 35s. (at Runcorn).

WEST OF SCOTLAND CHEMICALS.

GLASGOW, Tuesday.

enerally have continued to give way, but even with this in- buyers have not been offering on any scale beyond that of requirements, and their calculation no doubt is that will be easier still within a few days. Consequently, the nanimate, except in one or two unimportant sections of the forward supply department; in these, some contracts of rather improved bulk have been carried through. Caustic soda is dropping steadily day by day, and the position of bleach is also less satisfactory, although the tendency to fluctuation here seems not so prominent as it is elsewhere. Last week after report, the Scotch sulphate of ammonia market stiffened under some rather bulky, immediate orders for Glasgow shipment, and as high as £11. 10s. was got by sellers, thanks to this stimulant; but the pressure is off again, and to-day's price is from £11. 7s. 6d., to £11. 8s. 9d., spot Leith. Forward movement feeble. The paraffin section shows hardly any change, and scale and wax are still healthy, and burning oil still very much otherwise. Lubricating oils firm. One of the Oil Companies has announced results for the year ending March 31st. This is the Clippens Company. It has nothing to divide. The Broxburn Oil Company, it is believed, will again pay 15 per cent. The Tharris Sulphur and Copper Company repeat their 20 per cent. this year, to the surprise of many, as copper has come down. Chief prices current are:—Soda crystals, 51s., net Tyne; alum in lump £5., less 2½% Glasgow; borax, English refined £29., and boracic acid, £37. 10s. net Glasgow; soda ash, 1¼d. net Tyne; caustic soda, white, 76°, £12., 70/72°, £9. 12s. 6d., 60/62° £8. 15s., and cream, 60/62° £8. 10s., all less 2½% Liverpool; bicarbonate of soda, 5 cwt. casks, £6., and 1 cwt. casks, £6. 5s. net Tyne; refined alkali 48/52°, 1½d., net Tyne; saltcake, 28s.; bleaching powder, £5. 7s. 6d. to £5. 12s. 6d., less 5% f.o.r. Glasgow; bichromate of potash, 4d., and of soda 3d., less 5 and 6% to Scotch and English buyers respectively; chlorate of potash, 4⅞d., less 5% any port; nitrate of soda 8s. to 8s. 1½d.; sulphate of ammonia, spot, £11. 7s. 6d. to £11. 8s. 9d. f.o.b. Leith; salammoniac, 1st and 2nd white, £37. and £35., less 2½% any port; sulphate of copper, £25., less 5% Liverpool; paraffin scale, hard and soft, 2½d. per lb.; paraffin wax, 120°, semi-refined, 3⅝d.; paraffin spirit (naphtha), 6½d. a gallon; paraffin oil (burning), 6¾d. to 7d. at Glasgow for Scotch consumption only (English orders done at about 1d. a gallon lower); ditto (lubricating), 865°, £5. 10s. to £6. per ton; 885°, £6. 10s.; and 890/895°, £7. 10s. to £8. Week's imports of sugar at Greenock were 33,859 bags.

THE TYNE CHEMICAL REPORT.

TUESDAY.

After a dull week, the Chemical Market to-day shows signs of more activity and prices are a little firmer. Quotations are still in favour of buyers. Caustic soda has had a further drop of 10s. a ton. Soda crystals are 1s. 6d per ton less than last week. Soda ash is slightly easier, and bleaching powder half-a-crown a ton lower; sulphate of soda remains unchanged. To-day's prices are as follows:—Bleaching powder, in softwood casks, £5. 7s. 6d. per ton; caustic soda 77% £11. per ton; soda-ash, 1¼d. per degree; soda crystals, in casks, £2. 11s., per ton, gross weight; in 1 cwt. bags, £2. 13s. 6d. per ton, net weight; in 2 cwt. bags, £2. 11s. per ton, net weight; sulphate of soda, ground, in casks, £2. 5s. per ton; in bulk, £1. 15s. per ton; recovered sulphur, £4. 5s. per ton; chlorate of potash, 4⅜d. per lb.; silicate of soda, 75° Tw., £2. 10s. per ton; 100° Tw., £3. 7s. 6d. per ton; 140° Tw., £4. per ton; hyposulphite of soda, 1 cwt. kegs, £5. 15s. per ton, in 5—7 cwt. casks, £5. 5s. per ton; pure white sulphate of alumina, £4. 10s. per ton; blanc fixe, £7. 10s. per ton; chloride of barium, £8. per ton; nitrate of baryta, crystals, £18. 15s. per ton; ground, £19. 5s. per ton; sulphide of barium, £5 10s. per ton—all f.o.b. Tyne, or f.o.r. makers' works.

Manufacturing coals continue firm in price, principally owing to the heavy foreign demand. Durham small is quoted 10s. per ton, and Northumberland steam small 8s. to 8s. 6d. per ton.

The Durham miners' application for a further advance in wages is still in abeyance, as the County average selling price for the last three months is not yet known.

The Patent List.

This list is compiled from Official sources in the Manchester Technical Laboratory, under the immediate supervision of George E. Davis and Alfred R. Davis.

APPLICATIONS FOR LETTERS PATENT.

Manufacture of Steel. J. Lewthwaite. 5,319. April 8.
Dynamometers.—(Complete Specification.) H. H. Lake. 5,344. April 8.
Treatment of Uranium Compounds. T. G. Martyn. 5,348. April 8.
Manufacture of Artificial Building Material. C. O. Weber and G. F. Freeman. 5,349. April 8.
Manufacture of Nitro-Cellulose. J. R. France. 5,364. April 8.
Gualacol Ether. O. Imray. 5,366. April 8.
Refining of Camphor.—(Complete Specification.) H. H. Lake. 5,367. April 8.
Oil Refining.—(Complete Specification) E. Noppel. B. Grosche and J. Bigler. 5,375. April 8.

Nitration and Denitration of Cellulose. H. de Chardonnet. 5,376. April 8.
Asbestos Filters.—(Complete Specification.) F. Breyer. 5,377. April 8.
Resins and Varnishes. A. F. St. George. 5,380. April 9.
Improved Anti-Corrosive Compound.—(Complete Specification.) D. Fulton. 5,387. April 9.
Reduction of Zinc Ores. O. Lumaghi. 5,407. April 9.
Steam Boiler and other Furnaces. T. G. Lishman. 5,410. April 9.
Wool Cleaning. W. P. Thompson. 5,413. April 9.
Brick Machines. W. W. Horn. 5,415. April 9.
Treatment of Organic Matters.—(Complete Specification.) J. Guillaume. 5,419. April 9.
Manufacture of Cellulose by Electricity. C. Kellner. 5,420. April 9.
Air Compressors. W. Tattersall. 5,429. April 10.
Manufacture of Water Gas. H. Williams 5,434. April 10.
Improved Antiseptic Deodorant and Germicide. J. Wheeler. 5,438. April 10.
Manufacture of Nitrate of Ammonium. C. A. Burghardt. 5,442. April 10.
Water Filtration.—(Complete Specification.) J. A. Crocker. 5,445. April 10.
Smoke Combustion. W. Ackroyd, T. H. Ackroyd, and J. Willoughby. 5,453. April 10.
Application of Ceramic Colouring Matters to India-rubber. J. Mills. 5,455. April 10.
Manufacture of Sulphuric Acid. R. H. Wilson. 5,469. April 10.
Wool Scouring. G. W. Arnott, P. A. Olivier, and G. Seagrave. 5,482. April 10.
Improvements in Combustion of Fuel. J. Westray. April 11.
Steam Boilers and their Furnaces.—(Complete Specification.) E. J. Duff. 5,487. April 11.
Manufacture of Chlorine. W. Donald. 5,488. April 11.
Steam Boiler Furnaces. J. Riley. 5,489. April 11.
Filters. C. E. Gittens 5,495. April 11.
Smoke Consumption. G. Hewitt. 5,501. April 11.
Automatic Measuring Apparatus for Liquids. J. Leloup. 5,514. April 11.
Improved Material for Building Purposes. P. M. Justice. 5,515. April 11.
Steam Generators.—(Complete Specification.) W. Chambers. 5,517. April 11.
Asphalt Paving. A. McLean. 5,533. April 11.
Furnaces for Burning Liquid Fuel. E. Manbre. 5,543. April 12.
Manufacture of Iron and Steel. A. Turner, A. Baird, and M. B. Baird. 5,545. April 12.
Manufacture of Wire. B. Mountain. 5,546. April 12.
Dyeing and Apparatus Therefor. B. Haigh. 5,547. April 12.
Manufacture of Chlorine. A. Campbell and W. Boyd. 5,571. April 12.
Cement Kilns. G. H. Skelsey. 5,581. April 12.

Gazette Notices.

Partnerships Dissolved.

WADDINGTON and RAMSBOTTOM, Manchester, paper merchants.

Adjudications.

FREDERICK ARTHUR PECK, Cleethorpes and Great Grimsby, chemists.
MATTHEW HENRY NEALE, East Greenwich, chemist.

New Companies.

AUSTIN CRAVEN LIMITED.—This company was registered on the 12th inst., with a capital of £8,000., in £5. shares, 400 of which are £6. per cent. cumulative preference shares, to take over the business of Austin Craven, of Manchester, mineral water manufacturer, etc. The subscribers are:—

	Shares.
H. Cardwell, Hulme, Manchester, brewer	1
J. Thackeray, Hulme. Manchester, licensed victualler	1
Austin Craven, Hulme, Manchester, mineral water manufacturer	1
J. Pollitt, Church, hotel proprietor	1
P. R. Davies, Greenheys, licensed victualler	1
F. J. Ashbury, Manchester, chartered accountant	1
J. Winstanley, Moss Side, Manchester, cashier	1

CAM PORTLAND CEMENT COMPANY, LIMITED.—This company was registered on the 16th inst., with a capital of £20,000., in £1. shares, to purchase from T. Westrope Bowman, of Meldreth, Cambs., 7⅞ acres of land at Meldreth at the price of £200. per acre, with an option to purchase the adjoining 26½ acres within two years at the same price per acre, and to carry on business as cement, lime, and plaster-makers. The subscribers are:—

	Shares.
A. Russell, Huntingdon, draper	800
T. W. Bowman, Meldreth farmer	2,000
G. Worboy, Bassingham, builder	250
W. E. Cocks, Bassingham, engineer	250
J. Bowman, Lithington, farmer	200
Sarah D. Bowman, Meldreth	100
J. Westnights, Cambridge, analyst	20

DUNTON GREEN BRICK AND TILE WORKS.—This company was registered on the 11th inst., with a capital of £20,000., in £1. shares, to take over the business of brick and tile manufacturer, carried on by William James Thompson, at Dunton Green, near Sevenoaks, Kent. The subscribers are:—

	Shares.
J. Wescomb, Dunton Green, brickmaker	20
W. E. Scott, Dunton Green, tile maker	10
J. W. Breething, Dunton Green, brickmaker	5
W. J. Breething, Riverhead, Kent, brickmaker	5
A. E. Partridge, Dunton Green, clerk	2
R. Inkpen, Dunton Green, brickmaker	1
T. Collins, Dunton Green, brickmaker	1

ROBOTTOM BORACIC ACID SYNDICATE, LIMITED.—This company was registered on the 14th inst., with a capital of £5,000., in £20. shares, whereof 100 are founders' shares, to carry on in the United Kingdom, South America, or elsewhere, the business of miners, refiners, distillers, and manufacturers of boracic acid, borate of lime, borax, and other similar ore, mineral substance, or product. An unregistered agreement with Arthur Robottom will be adopted. The subscribers are:—

	Shares.
A. Robottom, 3, St. Alban's-villas, Highgate-road, merchant	1
M. L. Ingram, 18, Mincing-lane, merchant	1
L. H. Gillham, 69, Lombard-street, iron agent	1
W. H. Graybrook, Borax Works, Liverpool	1
Hodgson Watt, 12, Commercial Sale Rooms, broker	1
R. W. Shaw, 9, Fenchurch-street, shipowner	1
W. Shaw, 9, Fenchurch-street, shipowner	1

IMPORTS OF CHEMICAL PRODUCTS

AT

THE PRINCIPAL PORTS OF THE UNITED KINGDOM.

al
113 pks. W. H. Cole & Co.
15 W. M. Smith & Sons
20 Swanston & Co.
42 J. Goddard & Co.
19 W. M. Smith & Sons
4 Swanston & Co.
9 Kuhner, Hendschel Co.
8 Johnson, Rolls, & Co.
Bark
25 pks. Dalgety & Co.
l
74 pks. Burt, Boulton, & Co.
15 Humphrey & Co.
2 R. J. Fullwood & Bland
1 pkg. M. D. Co.
6 pks. A. Laming & Co.
o E. Austin & Co.
o F. Roberts
o cs, Wrightson, & Son
2 Brinkmann & Co.
6 pks. Brinkmann & Co.
4 Beresford & Co.
9 Elkan & Co.
2 L. & I. D. Jt Co.
7 Brinkmann & Co.
16 pks. W. H. Cole & Co.
Bark
8 t. S. Barrow & Bros.
16¾ A. & W. Nesbitt
50 t. Devitt & Hett
112½ Baxter & Hoare
15½ G. B. Mackareth
115 Baxter & Hoare
16 G. B. Mackeith
816 pks. B. Jacob & Sons
188 "
6 pks. L. & I. D. Jt. Co.
6 pks. Hurford & Middleton
2 Hoare, Wilson & Co.
s (*otherwise undescribed*)
1 pks. Burt, Boulton, & Co.
–
£150 Beresford & Co.
35 "
17 "
70 T. H. Lee
ha—
134 c. Hultenbach & Co.
940 H. W. Jewesbury & Co.
80 pks. 450 c. Horne & Co.
100 550 F J. Warren
40 200 T. M. Duche & Sons
50 400 "
1 190 J. Knill & Co.
228 1,824 Anderson, Weber, & Co.
15 90 M. D. Co.
8 74 L. & I. D. Jt. Co.
25 200 L. Sutro & Co.
160 468 "
140 840 Barrett, Tagant, & Co.
40 217 Page, Son, & East
60 326 Pillow, Jones, & Co
21 190 J. Knill & Co.
180 180 J. Barker & Co.
6 10 Spies, Bros., & Co.
100 550 P. Windmuller
13 110 J. Knill & Co.
00 100 Union Lighterage Co.
29 236 J. Knill & Co.
00 200 J. F. Ohlmann
50 100 C. Atkins & Co.
00 400 Baxter & Hoare
00 110 Harnett & Co.
65 354 T. M. Duche & Sons
60 660 J. Barber & Co.
50 100 T. H. Lee
3 25 C. Atkins & Co.
25 220 H. Lorenz
09 1,020 Anderson, Weber, & Smith

France, 200 200 Baxter & Hoare
U. S., 110 550 A. Ashby
" 2,000 2,000 Union Lighterage Co.
" 500 500 M. D Co

Guano—
Sweden, 50 t. Anglo Swedish Co.

Isinglass—
Sts. Settlements, 4 pks. L. & I. D Jt. Co.
E. Indies, 13 pkgs. "
Singapore, 3 "
Penang, 34 pks. "
China, 1 pkg. E. Boustead & Co.
Germany, 5 pks. Atkinson, Bethe, & Co.
Rio Janeiro, 1 pkg. L. & I. D. Jt. Co.
E. Indies, 10 "
Singapore, 2 pks. Hale & Son

Manure—
Phosphate Rock
U. S., 2,000 t. Fox, Roy, & Co.
St. Valery, 167 Lawes Manure Co., Limited
France, 305 Anglo Contl. Guano Wks.
" 5 J. Burnett & Sons
U. S., 2 C A. & H. Nichols
Germany, 50 F. W. Berk & Co.
Sydney, 11 Goad, Rigg, & Co.
Germany, 50 Perkins & Homer
E. Indies, 14 t. G. H. Wilson & Sons
Germany, 50 F. W. Berk & Co.
" 60 Adley, Tolkin, & Co.
E. Indies, 500 Gonne, Croft, & Co.

Manganese—
Germany, £30 Burgoyne, Burbidge, & Co.

Manganese Ore—
Adelaide, 201 t. F. Manders

Naphtha—
Holland, 73 drs. Burt, Boulten, & Co.

Potash—
Germany, 18 cks. W. Balchin
U S., 111 c. Charles & Fox

Plumbago—
E. Indies, 138 cks. H. W. Ison
" 174 Barron & Gibson
" 67 H. J. Perlbach & Co.
" 183 J. Thredder, Son, & Co.
" 61 Prcp. Chamb's Whf.
Germany, 18 B. Gallaway
" 4 T. Jordan & Co.

Pitch—
U. S., 48 brls. W. Isted
France, 92 Berlandina Bros.

Pearl Ash—
U. S., 52 c. Charles & Fox
France, 20 Petri Bros.

Paraffin Wax—
U. S., 82 G. H. Frank
" 120 Rose, Wilson, & Sons
" 48 "
" 200 E. H. Holdsworth

Red Arsenic—
Germany, £36

Saltpetre—
E. Indies, 1,629 bgs. C. Wimble & Co.
" 13,964 Brown & Elmslie
" 742 Clift, Nicholson, & & Co.
" 610 C. Wimble & Co
" 1,173 Ralli Bros.

Stearine—
U. S., 56 cks. Rose, Wilson, & Rose
Melbourne, 339 Ross & Deering
Germany, 393 W. Balchin

Sodium Acetate—
Belgium, £95 Soundy & Son
" 116 c. A. Zumbeck & Co.
France, 214 "

Sugar of Lead—
Germany, 28 pks. Beresford & Co

Sulphur Oil—
Italy, £520 A. Laming & Co

Tartaric Acid—
Holland, 19 pks. I. Sauer
" 19 A. Hughes
France, 18 B. & F. Wf. Co.
Holland, £350 Middleton & Co.
" 650 W. C. Bacon & Co.
" 1,485 B & F. Wf. Co.
" 475 C. Christopherson & Co.
" 120 Hernu, Peron, & Co.

Tartars—
Italy, 53 pks. B. Jacob & Sons

Ultramarine—
Holland, 4 cks. Spies, Bros., & Co.
" 22 cs. Hernu, Peron, & Co.
" 8 pks. Ohlenschlager Bros.
" 11 Haeffner, Hilpert, & Co.
Belgium, 10 Leach & Co.
Germany, 10 G Steinhoff

Zinc Oxide—
Holland, 170 cks. M. Ashby
U. S., 100 brls. "
Germany, 50 cks. D. Storer & Sons, Limited
Holland, 100 brls. N. Smith
Germany, 10 cks. Beresford & Co.
Holland, 434 Soundy & Son

Zinc Sulphate—
Germany, £80 I. Matton

LIVERPOOL

Week ending April 17th.

Antimony Ore—
Marseilles, 967 bgs.
Hamburg, 415

Acetate of Lime—
St. Nazaire, 84 cks.
New York, 1,202 bgs.

Bones—
Boston, 162 sks. J. Gordon & Co.
New York, 407 bgs.
Constantinople, 15 t.
Ceara, 173 bgs. Leech, Harrison & Co
" 785 R. Singlehurst & Co.
" 100

Bone Ash—
Rosario, 220 bgs.

Barytes—
Rotterdam, 17 cks.

Boracic Acid—
Hamburg, 5 cks.

Caoutchouc—
St. Nazaire, 4 bls.
Ceara, 2
" 5 Leech, Harrison & Co.
" 2 Stadebauer & Co.
Bayin, 5 bgs. A. Millerson & Co.
" 3 Halsan & Co.
" 5 Edwards Brs.
" 14 3 brls. Pickering & Berthoud
Half Jack, 2 M. L. Levin
Three Towns, 1 ck.
Grand Bassam, 1 brl. H. S. Attia Brs.
Sierra Leone, 76 Pickering & Berthoud
" 18 pns. 4 brls. Patersen, Zachonis & Co.
" 18 brls. Cie Francaise Co.
Bathurst, 22 cks. "
" 1 Radcliffe & Durant
" 4 A. Cros & Co.
Salt Pond, 2 pks. A. Reis
Cape Coast. 1 brl. Edwards Brs.
" 1 1 cs. Davies, Robbins & Co.
" 1 Emil Grund
" 5 Mordaunt Brs.
" 9 5 A. Miller, Brs. & Co.
" 6 L. Hart & Co.
" 2 W. B. McIver & Co.
" 11
Axim, 4 cks. 1 cs. Pickering & Berthoud
" 1 bg. Radcliffe & Durant
Attaboe, 6 Edwards Brs.
" 28 Pickering & Berthoud
Old Calabar, 5 cks. Pickering & Berthoud
Forcados, 1 brl. J. Chorlton & Co.
Addah, 13 W. J. Rodatz & Co.
Accra, 2 cs. 4 cks. G. W. Christie & Co.
" 3 pns. 1 brl. 1 cs. F. & A. Swanzy
" 13 Hutton & Co.
" 3 1 Davies, Robbins & Co.
" 10 2 3 J. J. Fischer & Co.
Salt Pond, 2 cks. H. B. McIver & Co.
" 2 bxs. Pickering & Berthoud
" 4 brls. 2 cs. Miller Brs.
" 1 T. Millington & Co.
" 2 Edwards Brs.
Axim, 2 bgs. "
Grand Bassam, 19 cs. A. Verdier
" 1 Radcliffe & Durant
" 1 brl Latchford & Co.
" 1 S. & C. Nordlinger
" 5 bgs. W. D. Woodin
" 4 A. Reis
" 1 Edwards Brs.
" 1 Millward, Bradbury & Co.
" 1 F. & A. Swanzy
Sierra Leone, 42 cks. Fisher & Randall
" 17 Pickering & Berthoud
" 7 Broadhurst, Sons & Co.
Accra, 2 brls. 1 cs. Hutton & Co.
" 1 Elder, Dempster & Co.
" 1 Pitt Brs. & Co.
" 1 Radcliffe & Durant
C.C. Castle, 18 cks. 1 cs. Ihlers & Bell
" 2 F. & A. Swanzy
" 3 brls. C. L. Clare & Co.
" 1 1 cs. E. Grime
" 1 Pitt Brs. & Co.
Axim, 3 Millward, Bradbury & Co.
" 1 bg. Radcliffe & Durant
" 1 Pickering & Berthoud
Akassa, 16 cs. 20 bgs. Royal Niger Co.
Quittah, 14 brls. 2 pns. F. & A. Swanzy
Accra, 13 cks 1 cs. Pickering & Berthoud
" 1 F. Mosher
" 38 J. P. Werner
" 8 4 J. J. Fischer & Co.
" 6 1 F. & A. Swanzy
" 4 M. Herschell & Co.
" 1 Davies, Robbins & Co.
" 11 C. Lane & Co
" 7 brls. Edwards Brs.

Chemicals (*otherwise undescribed*)
Rotterdam, 2 cks.
" 1

Cream of Tartar—
Bordeaux, 3 cks.
Genoa, 12
Barcelona, 12
Rotterdam, 10

Chloral Hydrate—
Rotterdam, 3 cs.

Copper Ore—
Galveston, 3,050 sks.
New York, 3,959

Dyestuffs—
Sumac
Catania, 10 bls.
Palermo, 860 bgs.
Cutch
Rangoon, 3,105 bxs.
Logwood
Laguna, 326 t. R. B. Watson & Co.
Cochineal
Grand Canary, 4 bgs. Widow, Duranty & Son
Teneriffe, 25 Bruce & White
" 16 H. R. Toby & Co.
Grand Canary, 30 Lathbury & Co.
Divi Divi
Hamburg, 207 bgs.
Extracts
St. Nazaire, 350 cks.
Havre. 25 cs. 15 cks. Cunard S.S. Co.
Bordeaux, 1 G. G. Blackwell
Gambier
Singapore, 3,591 bls.

Glucose—
Hamburg, 25 cks.
" 10 50 bgs.
Philadelphia, 88 brls.

Glycerine—
Rotterdam, 20 cs.

Horn Piths—
Antwerp, a quantity.

Iodine—
Valparaiso, 22 brls. Gibbs & Sons
" 52 A. Gibbs & Sons

Litharge—
Hamburg, 40 cks.
Manganese Ore—
Gothenburg, 160 t. MacQueen Brs.
Manganese Powder—
Hamburg, 1 ck.
Potassium Muriate—
Hamburg, 203 bgs.
Pyrites—
Huelva, 2,055 t. Matheson & Co.
" 1,420 "
" 1,386 Tennants & Co.
" 1,234 "
Paraffin Wax—
Rangoon, 525 bxs.
Baltimore, 100 brls.
Phosphate—
Bruges, 1,200 bgs.
St. Valery, 210 t.
Potash—
Rotterdam, 29 cks.
Salt—
Hamburg, 100 bgs.
Saltpetre—
Hamburg, 32 cks.
" 20
Antwerp, 135 bgs. J. T. Fletcher & Co.
Stearine—
Rotterdam, 20 bls.
Antwerp, 39 bgs.
Sulphate—
Antwerp, 6 cks.
Sulphur—
Catania, 140 bgs. 50 brls.
Palermo, 60
Sodium Nitrate—
Rotterdam, 9 cks.
Tar—
Wilmington, 3,300 brls.
Tartaric Acid—
Rotterdam, 48 cks.
" 15
Tartars—
Bari, 5 cks.
Naples, 13
Rotterdam, 1
Ultramarine—
Rotterdam, 4 cks. 50 brls.
" 14 cs. 19 cks.
Zinc Oxide—
Hamburg, 10 cks.
Rotterdam. 30
New York, 50 brls.
Antwerp, 20
" 5

GLASGOW.

Week ending April 17th.

Alkali—
Rotterdam, 200 bgs. J. Rankine & Son
Chemicals *(otherwise undescribed)*
Hamburg, 1 cs.
Dyestuffs—
Alizarine
Rotterdam, 44 cks. J. Rankine & Son
" 65
Extracts
Antwerp, 6 cks.
Cutch
Rangoon, 1,684 bxs. P. Henderson & Co.
Fustic
Jamaica, 42 t. 8 c. Wallace, Wilkie, & Co.
Logwood
Jamaica, 20 t. 18 c. Wallace, Wilkie, & Co
Sumac
New York, 40 brls.
Gutta Percha—
Amsterdam, 5 bgs.
Hamburg, 3 bskts. 1 bg.
Glucose—
New York, 105 brls.
" 300
Rotterdam, 10 cks. J. Rankine & Son
Saltpetre—
Rotterdam, 7 cks. J. Rankine & Son
Sulphur—
Catania, 1,200 t.
Sulphur Ore—
Drontheim, 250 t.
Salt—
Hamburg, 143 cks. 508 bgs. 50 t.
Stearine—
New York, 11 brls.
" 14
Tar—
Wilmington, 1,500 brls.
Ultramarine—
Rotterdam, 3 cs. J. Rankine & Son
" 2 cks. "
Antwerp, 2 cs.
Zinc Ashes—
Antwerp, 5 cks.
Zinc Oxide—
Antwerp, 100 cks.
New York, 150 brls.
Rotterdam, 100

HULL.

Week ending Apr. 17th.

Albumen—
Reval, 30 cks. Wilson, Sons, & Co.
Acids *(otherwise undescribed)*
Rotterdam, 2 cks. Hutchinson & Son
Antimony—
Reval, 1 ck.
Barytes—
Hamburg, 68 cks. Wilson, Sons, & Co.
Rotterdam, 17 brls.
Antwerp, 25 Bailey & Leetham
Chemicals *(otherwise undescribed)*
Dunkirk, 30 pks. Wilson, Sons, & Co.
Antwerp, 290 "
Bremen, 20 kgs. 3 cs. Veltmann & Co.
Dyestuffs—
Alizarine
Rotterdam, 10 cks. Hutchinson & Son
" 35 "
" 20 "
" 15
" 12 pks. G. Lawson & Sons
Extracts
Trieste, 20 brls. Wilson, Sons, & Co.
Fiume, 900 "
Rouen 85 cks. Rawson & Robinson
Valonia
Smyrna, 478 bgs.
Trieste, 76 Wilson, Sons, & Co.
Sumac
Trieste, 75 bls. Wilson, Sons, & Co.
Palermo, 300 bgs. Kupers, Ostler, & Scott
Madder
Rotterdam, 2 cks. G. Lawson & Sons
Glycerine—
Hamburg, 5 drums.
Glucose—
Hamburg, 200 bgs 38 cks. Wilson, Sons, & Co.
" 22 cks. Bailey & Leetham
" 5
" 49
Stettin, 40 bgs. Wilson, Sons, & Co.
Rotterdam, 50 Hutchinson & Son
Manure—
Kainit
Hamburg, 500 bgs. E. Kipps & Co.
" 50 t.
Ghent, 160 Wilson, Son, & Co.
Pitch—
Hamburg, 65 brls.
Phosphate—
Antwerp, 90 t. Bailey & Leetham
Potash—
Rotterdam, 11 cks. G. Lawson & Sons
Potassium Muriate—
Rotterdam, 2 cks. Hutchinson & Son
Salt—
Hamburg, 508 bgs. Bailey & Leetham
" 508 C. M. Lofthouse & Co.
Ultramarine—
Rotterdam, 4 cks.

TYNE.

Week ending April 17th.

Alum Earth—
Hamburg, 19 cks. Tyne S. S. Co.
Copper Ore—
Huelva, 318 t. Bede Metal Co.
Copper Precipitate—
La Jaja, 1,418 bgs. Bede Metal Co.
Huelva, 2,725
" 23 cks. 4,425 bgs. H. Scholfield & Son
Glucose—
Hamburg, 5 cks. Tyne S. S. Co
Manure—
Phosphate Rock
Coosaw, 1,820 t. Langdale's Manure Co
Manganese Ore—
Hamburg, 30 t. Tyne S. S. Co
Huelva. 200 t. Scott Bros
Pyrites—
Huelva, 1,110 t. Scott Bros
La Jaja, 250 t. Bede Metal Co
Phosphate—
Aruba, 450 t. Langdale's Manure Co
Rennet—
Copenhagen, 2 cks. Wilson, Sons, & Co
Soda—
Rotterdam, 15 cks. Tyne S. S. Co
Zinc Oxide—
Antwerp, 25 brls. Tyne S. S. Co

GOOLE.

Week ending April 9th.

Alum—
Antwerp, 39 cks.
Chrome Chlorate—
Rotterdam, 11 cks.
Dyestuffs—
Divi Divi
Rio Hacha, 217 t.
Madder
Rotterdam, 6 cks.
Alizarine
Rotterdam, 23 cks.
Dyestuffs *(otherwise undescribed)*
Boulogne, 1 cs. 54 cks.
Glycerine—
Hamburg, 2 drums. 3 cs.
Rotterdam, 10
Glucose—
Hamburg, 53 cks. 200 bgs.
Dunkirk, 120
Potash—
Rotterdam, 5 cks.
Soda—
Rotterdam, 8 cks.
Salt—
Hamburg, 554 t. 227 cks. 400 bgs.
Saltpetre—
Rotterdam, 225 bgs.
Tartar—
Rotterdam, 4 cks.

GRIMSBY.

Week ending Apr. 12th.

Albumen—
Hamburg, 5 cks.
" 14
Dieppe, 4
Chemicals *(otherwise undescribed)*
Hamburg, 13 pks.
Glucose—
Hamburg, 12 cks.
Zinc Ashes—
Antwerp, 16 cks.

Coal Products—

Solvent Naphtha
Calais, 24 cks. £87

Toluole
Rotterdam, 18 t. 17 c. £455

Naphthalene
Philadelphia, 79 cks. £100
Treport, 39 20
Rotterdam, 27 brls. 37
Hamburg, 11 t. 5 c. 187
Stettin, 5 2 45

Anthracene
Rotterdam, 252 cks. £6,200
" 126 1,275
" 195 737
Dusseldorf, 68 328
Rotterdam, 300 2,800

Benzol
Rotterdam, 47 cks. £340
" 50 drms. 988
Antwerp, 19 puns. 340
Treport, 11 270
Rotterdam, 54 cks. 635
" 24 25 drms. 751
Calais, 19 417
Rotterdam, 20 puns. 328

Tar
Sulina, 20 brls. £35
Bombay, 145 rdlts. 120
" 37 44
Algoa Bay, 300 drms. 30
Calcutta, 200 rlts. 169

Bone Pitch
Rotterdam, 18 cks. £25
Rangoon, 30 pkgs. 12
Sydney, 20 cks. 19
Madras, 50 rlts. 26
Cette, 1,165 t. 2,372

Pitch
Sulina, 15 brls. £10
Bombay, 100 57
New York, 4 cks. 10
Antwerp, 796 t. 1,290
Dunkirk, 406 652
Amsterdam, 938 1,649
Antwerp, 419 625
Genoa, 1,814 3,500
Rotterdam, 12 cks. 25
Rangoon, 30 pkgs. 12
Sydney, 20 cks. 19
Madras, 50 rdlts. 26
Cette, 1,165 t. 2,372
Sulina, 15 brls. 10
Bombay, 100 57
New York, 4 cks. 10
Antwerp, 380 t. 600
" 416 630
Dunkirk, 296 492
" 150 230
Amsterdam, 938 1,642
Antwerp, 419 605
Genoa, 1,824 3,500

Chemicals (*otherwise undescribed*)
Brisbane, 67 pkgs. £83
Genoa, 22 cks. 70
Dunedin, 25 199
Rio Janeiro, 7 78
Auckland, 8 40
" 3 38

Disinfectants—
Paris, 420 gls. £42
Hamburg, 33 cks. 215
Melbourne, 2 cs. 26
Calcutta, 30 drs. 30
Melbourne, 42 pkgs. 53
Brisbane, 15 cs. 24
Lyttelton, 4 13
Sydney, 6 33

Fluoric Acid—
Melbourne, 540 gls. 33

Guano—
Antigua, 16 t. 113

Manure—
Rouen, 48 t. 4 c. £200
Gandia, 530 19 3,390
Valencia, 110 17 840
Rotterdam, 17 5 103
" 72 8 576
" 19 16 166
Tarragona, 145 1,160
Alicante, 75 600
Valencia, 350 2,800
Malaga, 50 400
St. Lucia, 38 380
Demerara, 20 200
Stettin, 68 1 q. 390
Rotterdam, 29 5 153
Gothenburg, 283 11 1,335
Genoa, 10 100
Valencia, 70 11 600
Lyttelton, 130 2,025
Gothenburg, 100 16 839
Melbourne, 80 1,080

Nitric Acid—
Auckland, 4 cs. £18

Oxalic Acid—
Rotterdam, 5 t. 7 c. £250
Antwerp, 170 4 2,085
Lisbon, 1 4 34

Potassium Chlorate—
Hiogo, 5 t. £216
Hong Kong, 2 5 c. 96
" 1 46
New York, 5 222
" 2 88
Yokohama, 10 22

Potassium Carbonate—
Santos, 3 cs. £12

Potassium Bicarbonate—
Salonica, 7 c. £13

Potassium Bichromate—
Hong Kong, 4 c. £11

Phosphorus—
Kobe, 1 t. 2 c. £201
Yokohama, 20 cs. 119
Hiogo, 30 415

Potash—
Calcutta, 8 cs. £48

Potash Crystals—
Sydney, 13 pkgs. £35

Potash Solution—
Sydney, 3 cs. £13

Sheep Dip—
Port Elizabeth, 3 t. 8 c. £190
Algoa Bay, 2 10 140
" 1 47
" 25 cs. 60

Sodium Bicarbonate—
Auckland, 5 t. 8 c. £27

Sodium Silicate—
Auckland, 2 t. 5 c. £10

Sulphur—
Madras, 4 t. 19 c. £38
Amsterdam, 1 10 14

Sulphuric Acid—
Karrachi, 30 drs. 66

Sulphur Flowers—
Oporto, 19 t. £173
" 20 250
E. London, 100 pkgs. 57
Lisbon, 1,260 brls. 973

Salammoniac—
Salonica, 2 t. 2 c. £64
" 7 brls. 30
Galatz, 1 t. 1 c. 35
Paris, 1 8 49
Yokohama, 10 cks. 85
Samsoun, 3 c. 6

Soda Crystals—
Rockhampton, 30 t. £93
Hong Kong, 8 2 c. 24

Saltpetre—
Melbourne, 2 £43
Rio Janeiro, 2 10 c. 56
Oporto, 7 8 158
Rio Janeiro, 2 7 53
" 2 7 53
Adelaide, 1 10 35
Gibraltar, 25 528
Rockhampton, 10 17

Sodium Bisulphide—
Philadelphia, 5 c. £21

Sodium Carbonate—
Rockhampton, 1 t. 19 c. £10

Sodium Sulphate—
Antwerp, 140 t. £175

Sanitas Powder—
Bombay, 12 t. 1 c. £114

Saltcake—
Antwerp, 125 t. £220

Soda and Lime Salts—
Halifax, 10 pkgs. £115

Superphosphate—
Martinique, 267 t. 1 c. £1,060

Tartaric Acid—
Rockhampton, 10 £72
Brisbane, 5 33
Adelaide, 11 68
Lyttelton, 6 cs. 41
Melbourne, 20 137
Cossack, 3 25
Canterbury, 5 35

LIVERPOOL.

Week ending April 16th.

Ammonium Sulphate—
Gr. Canary, 10 t. 2 c. £120
Valencia, 51 1 1 q. 670

Ammonium Muriate—
New York, 5 t. 2 c. 2 q. 187

Acetic Acid—
Porto Alegre, 3 2 £8
Vera Cruz, 1 t. 4 19

Albumen—
New York, 2 t. 7 c. 20

Antimony Regulus—
New York, 2 t. 15 c. 220

Alum Cake—
Gothenburg, 10 t. 5 c. 30

Arsenic Acid—
New York, 15 t. 7 c. 2 q. 218

Ammonium Carbonate—
Genoa, 16 c. £30
Naples, 9 3 q. 18
Genoa, 1 t. 15 3 49
Venice, 1 3 2 22
St. Louis, 5 1 3 24

Alum—
Cape Town, 2 t. 14 c. £24
Galatz, 2 17 1 q. 14
St. John's, N. B., 2 t. 18 c. 12
Calcutta, 45 4 220
Galatz, 3 1 16
Havana, 8 12 43
Melbourne, 3 7 17
Portland, 7 15 39
Montreal, 5 6 24
Vera Cruz, 9 7
Baltimore, 27 5 2 q. 120
Bombay, 1 1 30
Melbourne, 4 [illegible]
Calcutta, 10 30

Bleaching Powder—
Antwerp, 79 cks.
Boston, 394
Genoa, 19
Hamburg, 80
Ghent, 107
Antwerp, 30
Bari, 6
Bombay, 10 cs.
Calcutta, 100
New York, 41 pkgs. 100 brls. 85 tcs.

Borax—
Yokohama, 3 t. 8 c. £83
Rouen, 3 4 87
St. John's, N. B., 3 3 q. 6
Santander, 6 1 13
Malaga, 6 1 10

Borax Salts—
Havana, 1 t. £100

Boracic Acid—
Marseilles, 10 c. £20

Chalk Precipitate—
New York, 3 t. 10 c. 3 q. £63

Chromic Acid—
Hamburg, 4 c. 2 q. £11

Citric Acid—
Genoa, 10 3 71

Caustic Soda—
Ancona, 20 drms.
Barcelona, 200
Bilbao, 100
Boston, 490 10 brls.
Carril, 10
Ceara, 50
Chicago, 100
Genoa, 40
Hamburg, 58
Jaffa, 16
Naples, 32
Nantes, 72
New York, 450
Philadelphia, 100 270
Rio de Janeiro, 15
Rotterdam, 10
San Francisco, 90 25 brs.
Seville, 100
Valencia, 16
Vigo, 10 brls.
Antwerp, 34
Charlotte Town, 10
Copenhagen, 43
Genoa, 65
Hamburg, 10
Honolulu, 20
Maceio, 30
Matanzas, 96
New York, 474 110
Pernambuco, 50
Philadelphia, 150 brls.
Portland, 204 drms.
Rio Grande du Sol, 30 drms.
Rotterdam, 20
Rouen, 8
Santos, 10
Seville, 30 brls.
Singapore, 3
Stettin, 60
Sydney, 35
Trieste, 15
Venice, 30

Copper Sulphate—
Nantes, 254 t. 9 c. £5,800
Barcelona, 10 2 2 q. 401
Philadelphia, 15 12

Bilbao,
Barcelona,
Genoa,
Havre,
Marseilles
Venice,
Leghorn,
Trieste,
Genoa,
Venice,
Barcelona
Tarragona
Bordeaux
Passages,
Rouen,
Genoa,
Halifax,
Lisbon,
Tarragona
Barcelona
Trieste,
Venice,
Ancona,
Naples,
Passages,
Barcelona
Bari,
Bordeaux
Genoa,
Halifax,
Hamburg
Tarragona
Leghorn,
Venice,
Bilbao,
Barcelona
Catania,
Tarragona
Catania,
Barcelona
Bilbao,
Vigo,
Calcutta,
Bilbao,
Bordeaux
Nantes,
Barcelona

Calcium
Buenos A
New Yor
"
Boston,

Carbolic
Montreal

Carbolic
New Yor
Philadelp
Rouen,
Hambur
New Yor
Antwerp
New Yor
Philadelp
Havre,

Charcoal
Hong K

Chloride
New Yor
Boston,
New Yor
Canada,
Rotterda

Caustic
Sydney,

Copperas
St. John
Velo,

Chemical
Boston,
Vera Cr

Disinfec
Bombay

Epsom
New Yo

Glycerin
Hambur
Colon,
Parim,

Ground
Hambur

Hydro
New Yo

Iron Ox
Gijon,
Philade

Iron Sil
New Yo

Lead S
Rouen,

Lead A
Vera Cr

Manure—
Valencia, 60 t. £140
Havre, 100 3 c. 2 q. 299
St. Nazaire, 23 12 3 171
Bordeaux, 250 5 772
Antwerp, 9 13 71
St. Malo, 273 10 3 1,398
St. Nazaire, 176 9 2 880

Magnesia—
Leghorn, 9 cs.
London, 12
Mexico City, 1

Magnesium Chloride—
Nantes, 5 t. 8 c £27
Rouen, 12 10 3 q. 27
Montreal, 2 10 7

Phosphorus—
Canton, 1 c. 3 q. £77
Kingston, 3 11
Vera Cruz, 1 t. 14 1 210
Monte Video, 9 54

Potassium Carbonate—
New York, 33 t. 8 c. £397

Potash—
Lisbon, 3 c. 2 q. £25
Amsterdam, 1 t. 8 53
Havana, 5 1 11

Potassium Chlorate—
New York, 3 t. 15 c. £200
Boston, 2 10 130
Leghorn, 15 35
Boston, 2 10 134
Newfairwater, 2 10 105
Copenhagen, 3 137
Hamburg, 2 10 105
Santos, 2 93
New York, 15 595
Monte Video, 1 1 3 q. 48
Hamburg, 1 47
Santos, 3 135

Pitch—
Carril, 60 brls.

Sodium Nitrate—
Almeria, 3 t. 14 c 1 q. £31
Shanghai, 5 1 1 42
Trieste, 42 10 2 356

Soda Ash—
Baltimore, 133 cks.
Boston, 122 tcs. 82 2,462 bgs
Canada, 11
Ceara, 40 brls.
Jaffa, 23
Maranham, 30
New York, 645 478 tcs.
Paysandu, 100
Philadelphia, 430
Rio Janeiro, 4
Valparaiso, 10
Varna, 30
Baltimore, 709 tcs. 473 cks.
Boston, 57 1,800 bgs 300
Bruges, 8
Buenos Ayres, 200 brls.
Calcutta, 7
Constantinople, 200
Malta, 20
New York, 1,833 tcs. 203
Philadelphia, 887
Portland M., 133
Portland Or., 2
Puerto Cabello, 32 brls.
Rio Janeiro, 315
Rosario, 50
Santos, 45

Sodium Bicarbonate—
Amapala, 20 kgs.
Callao, 25
Constantinople, 30
Leghorn, 20
Nagasaki, 200
Rotterdam, 20
Stettin, 45 2 cks.
Trieste, 300 20
Bari, 50
Antwerp, 18
Bremen, 25
Buenos Ayres, 100
Calcutta, 580
Genoa, 51
Hamburg, 20
Rotterdam, 20
Santander, 20
Seville, 200
Stettin, 320 pks.

Salt—
Algoa Bay, 5 t.
Calcutta, 4,630
Ghent, 139
Grand Bassa, 7 3 c.
Kingston, 9
New York, 250
Para, 288 13
Paspebiac. 389
Portland, 5
Port Natal, 15
Sierra Leone, 110
Sinoe, 10
Boston, 66
Brisbane, 50
Baltimore, 100
Amsterdam, 250
Black Point, 5
Boston, 30
New York, 25
Bruges, 205
Calcutta, 12,464
Cavally, 11½
E. London, 24¾
Grand Bassa, 25
Halifax, 290
Hamburg, 193
Manaos, 19 13
Milwaukee, 65
Monte Video, 11 3
Oporto, 50
Quebec, 400
St. John's, 25 cs.
Salt Pond, 25
San Francisco, 93
Shediac, 400
Sierra Leone, 100

Soda Crystals—
Boston, 100 kgs. 672 brls.
Calcutta, 100
Constantinople, 100
Gibraltar, 30
New York, 522 140 cks.
Philadelphia, 280
St. John, 80 kgs.
St. John's, 50
Philadelphia, 50

Sheep Wash—
Belsize, 2 cs. £5
Buenos Ayres, 3 t. 14 c. 74

Saltcake—
Baltimore, 50 t. 6 c. £88
" 369 8 727
Philadelphia, 28 81

Sheep Dip—
Algoa Bay, 2 t. 10 c. £135

Saltpetre—
Pernambuco, 5 t. 14 c. 1 q. £128
Corrinto, 1 6 2 40
Monte Video, 5 2 6
Porto Alegre, 10 11
Para, 10 11

Sulphur—
St. John's, N. B., 2 t. 10 c. £18
" 1 8
Boston, 50 2 200
" 150 9 656
Halifax, 50 2 220
New York, 100 1 500
Boston, 100 2 442
Calcutta, 29 18 3 q. 150

Sodium Biborate—
Antwerp, 9 t. 4 c 2 q. £277
Rotterdam, 14 2 22

Salammoniac—
Samsoun, 6 c. £11
Malta, 6 10
Larnaco, 4 1 q. 7
Trieste, 1 t. 1 35
New York, 4 18 2 166
Samsoun, 6 1 11

Sodium Chloride—
Hamburg, 10 c. £27
New York, 2 t. 10 137

Strontia—
Havre, 65 t. 6 c £459

Sodium Nitrate—
Bari, 3 t. 2 c. £35
Alicante, 1 6 2 q. 10
Almeria, 1 4 3 10

Tin Phosphate—
Calcutta, 5 c. £33

Tartaric Acid—
Montreal, 18 c. £123

Tar—
St. John's. 50 brls.

Zinc Muriate—
Bombay, 1 t. 2 c. 2 q. £22
" 16 cks. 139

Zinc Chlorate—
Rotterdam, 5 t. 8 c. £36

TYNE.

Week ending April 17th.

Alkali—
Amsterdam, 21 t. 7 c.
Hamburg, 4 17

Antimony—
Rotterdam, 10 t.

Alum—
Pernau, 1 t. 14 c.

Arsenic—
Barcelona, 5 t. 5 c.

Bleaching Powder—
Baltimore, 50 t. 5 c.
Copenhagen, 103 14
Christiania, 24 15
Amsterdam, 59 6
Gothenburg, 10 3
Lisbon, 10
Antwerp, 17 18
Hamburg, 36 8
Rotterdam, 1 9
Copenhagen, 39 17

Barytes—
Vienna, 5 c.

Caustic Soda—
Baltimore, 87 t. 1 c.
Copenhagen, 2 7
Barcelona, 6
" 18 8

Copperas—
Vienna, 17 c.

Hydrochloric Acid—
Pomaron, 17 c.

Litharge—
Antwerp, 5 c.
Hamburg, 3 t. 1
Copenhagen, 16
" 1 1

Manure—
Kainit
Frederikshavn, 9 t. 16 c.

Magnesia—
Vienna, 1 c.
Hamburg, 1 t. 7

Nitric Acid—
Pomaron, 4 c.

Pitch—
Antwerp, 100 t.

Soda—
Svendborg, 21 t. 18 c.
Antwerp, 6 17
Randers, 50 17
Copenhagen, 2 1

Soda Ash—
Christiania, 10 t.
Gothenburg, 67 15 c.
Baltimore, 50 7
Copenhagen, 9 16
Rotterdam, 13 14

Sulphuric Acid—
Barcelona, 2 c.

Sodium Sulphate—
Copenhagen, 5 t. 8 c.
Lisbon, 5 7
Copenhagen, 5 13

Superphosphate—
Esbjerg, 27 t. 10 c.
Frederikshavn, 206 5
Reval, 249 16

Sodium Bicarbonate—
Rotterdam, 6 c.

Tar—
Pernau, 12 t. 4 c.

HULL.

Week ending April 15th.

Bleaching Powder—
Rouen, 6 cks.
Stettin, 56 60 kgs.

Chemicals *(otherwise undescribed)*
Antwerp, 26 cs. 54 pks.
Alexandria, 12
Bombay, 130 81 cks.
Bergen, 25
Bremen, 2 20
Copenhagen, 6
Christiania, 6 7
Gothenburg, 53 5 320 slbs.
Hamburg, 138 9 45 cks.
Konigsberg, 9
Rouen, 30
Stettin, 81 80 40

Dyestuffs *(otherwise undescribed)*
Dunkirk, 100 bgs.
Hamburg, 4 drms. 8 cks.
Reval, 2

Manure—
Antwerp, 489 bgs.
Dunkirk, 56
Hamburg, 180 t.
Stettin, 570 bgs.
Ghent, 391

Naphtha—
Melbourne, 35 drums.

Pitch—
Antwerp, 205 t.

Tar—
Christiania, 10 cks.
Hamburg, 12

GLASGOW.

Week ending April 17th.

Ammonium Sulphate—
Trinidad, 141 t. 16¾ c. £1,806

Chloride of Lime—
New York, 53 t. 16¾ c. £391
" 1,100 399

Copper Sulphate—
Bordeaux, 23 t. 17 c. £440

Dyestuffs—
Logwood
Lisbon, £244
" 227
Fustic
Boston, 23 c. £47

Epsom Salts—
Christiania, 100 bgs. £38

Glucose—
Cape Colony, 5 t. 15 c. £110
Bombay, 12 14¾ 205

Manure—
Lisbon, 60 t. £190
Trinidad, 51 19¾ c. 353

Pitch—
Christiania, 80 brls. £15
Amsterdam, 212 t. 212
Bordeaux, 161 6 c. 160

Potassium Prussiate—
New York, 19¾ c. £76

Potassium Bichromate—
Rotterdam, 12 cks. £174
New York, 165 c. 309
Lisbon, 102¾ 193

Paraffin Wax—
Cape Colony, 3 t. 7½ c. £120

Soda Ash—
New York, 50 t. ¼ c. £339

Sulphuric Acid—
Bombay, 7½ t. £75
Colombo, 8 15 c. 38
Madras, 4 13 £22 17s.
Bombay, 11 5 112

Salt—
Normanton, 100 t. 3½ c. £107

Sodium Sulphate—
Lisbon, £27

Sodium Bichromate—
New York, 172 c. £240

Tar—
St. John's, £35
Calcutta, 112

GRIMSBY.

Week ending April 12th.

Coal Products—
Dieppe, 12 cks.

Chemicals *(otherwise undescribed)*
Antwerp, 6 cks.
Dieppe, 2 130 cs.
Hamburg, 30

GOOLE.

Week ending April 9th.

Benzole—
Antwerp, 122 cks.
Dunkirk, 50
Ghent, 17 drums.
Hamburg, 16
Rotterdam, 96 cks. 17 drums.

Bleaching Powder—
Hamburg, 30 cks.

Bleaching Preparations—
Rotterdam, 15 cks.

Copper Precipitate—
Rotterdam, 251 cks.

Coal Products—
Dunkirk, 25 cks.
Hamburg, 50
Rotterdam, 96

Chemicals *(otherwise undescribed)*
Dunkirk, 30 cks.
Hamburg, 44

Dyestuffs *(otherwise undescribed)*
Hamburg, 200 cks. 3 drums.

Manure—
Boulogne, 89 bgs.
Ghent, 454
Hamburg, 90

Pitch—
Antwerp, 249 t.
Ghent, 280

PRICES CURRENT.

WEDNESDAY, April 23, 1890.

Prepared by Higginbottom and Co., 116, Portland Street, Manchester.

…s stated are F.O.R. at maker's works, or at usual ports of shipment in U.K. The price in different localities may vary.

		£	s.	d.
5 %, and 40 %	per cwt	7/6 & 12/9		
ilacial	,,		67/6	
S.G., 2000°	,,	0	12	9
82 %	nett per lb.	0	0	6½
	,,	0	0	3¼
(Tower Salts), 30° Tw.	per bottle	0	0	8
(Cylinder), 30° Tw.	,,	0	2	11
° Tw.	per lb.	0	0	2
	,,	0	0	1¼
	nett ,,	0	0	4
	,,	0	1	2½
c (fuming 50 %)	per ton	15	10	0
(monohydrate)	,,	5	10	0
(Pyrites, 168°)	,,	3	2	6
(,, 150°)	,,	1	8	0
(free from arsenic, 140/145°)	,,	1	10	0
ous (solution)	,,	2	15	0
pure)	per lb.	0	1	0½
	,,	0	1	2½
hite powdered	per ton	13	0	0
e lump)	,,	4	15	0
ulphate (pure)	,,	5	0	0
,, (medium qualities)	,,	4	10	0
	per lb.	0	15	0
·880=28°	per lb.	0	0	3
=24°	,,	0	0	1¼
Carbonate	nett ,,	0	0	3½
Muriate	per ton	23	0	0
,, (sal-ammoniac) 1st & 2nd	per cwt.	37/-& 35/-		
Nitrate	per ton	40	0	0
Phosphate	per lb.	0	0	10½
Sulphate (grey), London	per ton	11	7	6
,, (grey), Hull	,,	11	5	0
l (pure)	per lb.	0	0	11
lt ,,	,,	0	0	9
	per ton	75	0	0
(tartar emetic)	per lb.	0	1	1
(golden sulphide)	,,	0	0	10
loride	per ton	8	0	0
rbonate (native)	,,	3	15	0
lphate (native levigated)	,,	45/- to 75/-		
Powder, 35 %	,,	5	0	0
Liquor, 7 %	,,	2	10	0
of Carbon	,,	13	0	0
Acetate (crystal	per lb.	0	0	6
hloride	per ton	2	2	6
(at Runcorn) in bulk	,,	17/6 to 35/-		
Dyes:—				
(artificial) 20 %	per lb.	0	0	9
	,,	0	3	9
Products				
ene, 30 % A, f.o.b. London	per unit per cwt.	0	1	5
90 % nominal	per gallon	0	3	5½
50/90 ,,	,,	0	2	7½
Acid (crystallised 35°)	per lb.	0	0	9¼
,, (crude 60°)	per gallon	0	2	3
(ordinary)	,,	0	0	2½
(filtered for Lucigen light)	,,	0	0	3
aphtha 30 % @ 120° C.	,,	0	1	3
ils, 22° Tw.	per ton	3	0	0
.b. Liverpool or Garston	,,	1	8	0
Naphtha, 90 % at 160°	per gallon	0	1	9
Oils (crude)	,,	0	0	2½
ili Bars)	per ton	48	7	6
phate	,,	24	0	0
de (copper scales)	,,	51	0	0
crude)	,,	30	0	0
distilled S.G. 1250°)	,,	55	0	0
	nett, per oz.	8½d. to 9d.		
ate (copperas)	per ton	1	10	0
ide (antimony slag)	,,	2	0	0
t) for cash	,,	14	0	0
rge Flake (ex ship)	,,	15	10	0
ate (white ,,)	,,	23	10	0
(brown ,,)	,,	18	5	0
onate (white lead) pure	,,	19	0	0
te	,,	22	0	0

		£	s.	d.
Lime, Acetate (brown)	,,	7	17	6
,, ,, (grey)	per ton	13	15	0
Magnesium (ribbon and wire)	per oz.	0	2	3
,, Chloride (ex ship)	per ton	2	5	0
,, Carbonate	per cwt.	1	17	6
,, Hydrate	,,	0	10	0
,, Sulphate (Epsom Salts)	per ton	3	0	0
Manganese, Sulphate	,,	18	0	0
,, Borate (1st and 2nd)	per cwt.	60/- & 42/6		
,, Ore, 70 %	per ton	4	10	0
Methylated Spirit, 61° O.P.	per gallon	0	2	2
Naphtha (Wood), Solvent	,,	0	3	10
,, ,, Miscible, 60° O.P.	,,	0	4	7½
Oils:—				
Cotton-seed	per ton	22	10	0
Linseed	,,	24	10	0
Lubricating, Scotch, 890°—895°	,,	7	5	0
Petroleum, Russian	per gallon	0	0	5¼
,, American	,,	0	0	5¼
Potassium (metal)	per oz.	0	3	10
,, Bichromate	per lb.	0	0	4
,, Carbonate, 90% (ex ship)	per ton	19	0	0
,, Chlorate	per lb.	0	0	5
,, Cyanide, 98%	,,	0	2	0
,, Hydrate (Caustic Potash) 80/85 %	per ton	22	10	0
,, ,, (Caustic Potash) 75/80 %	,,	21	10	0
,, ,, (Caustic Potash) 70/75 %	,,	20	15	0
,, Nitrate (refined)	per cwt.	1	2	6
,, Permanganate	,,	4	2	6
Prussiate Yellow	per lb.	0	0	9½
,, Sulphate, 90 %	per ton	9	10	0
,, Muriate, 80 %	,,	7	15	0
Silver (metal)	per oz.	0	3	7½
,, Nitrate	,,	0	2	5½
Sodium (metal)	per lb.	0	4	0
,, Carb. (refined Soda-ash) 48 %	per ton	6	7	6
,, ,, (Caustic Soda-ash) 48 %	,,	5	2	6
,, ,, (Carb. Soda-ash) 48%	,,	5	12	6
,, ,, (Carb. Soda-ash) 58 %	,,	6	15	0
,, ,, (Soda Crystals)	,,	2	17	6
,, Acetate (ex-ship)	,,	15	15	0
,, Arseniate, 45 %	,,	11	0	0
,, Chlorate	per lb.	0	0	6¼
,, Borate (Borax)	nett, per ton.	30	0	0
,, Bichromate	per lb.	0	0	3
,, Hydrate (77 % Caustic Soda)	per ton.	12	0	0
,, ,, (74 % Caustic Soda)	,,	11	0	0
,, ,, (70 % Caustic Soda)	,,	9	7	6
,, ,, (60 % Caustic Soda, white)	,,	8	12	6
,, ,, (60 % Caustic Soda, cream)	,,	8	10	0
,, Bicarbonate	,,	6	0	0
,, Hyposulphite	,,	5	10	0
,, Manganate, 25%	,,	9	0	0
,, Nitrate (95 %) ex-ship Liverpool	per cwt.	0	8	0
,, Nitrite, 98 %	per ton	28	0	0
,, Phosphate	,,	15	15	0
,, Prussiate	per lb.	0	0	7¼
,, Silicate (glass)	per ton	5	7	6
,, ,, (liquid, 100° Tw.)	,,	3	17	6
,, Stannate, 40 %	per cwt.	2	0	0
,, Sulphate (Salt-cake)	per ton.	1	5	0
,, ,, (Glauber's Salts)	,,	1	10	0
,, Sulphide	,,	7	15	0
,, Sulphite	,,	5	5	0
Strontium Hydrate, 98 %	,,	8	0	0
Sulphocyanide Ammonium, 95 %	per lb.	0	0	8
,, Barium, 95 %	,,	0	0	5¼
,, Potassium	,,	0	0	8
Sulphur (Flowers)	per ton.	7	15	0
,, (Roll Brimstone)	,,	6	0	0
,, Brimstone: Best Quality	,,	4	5	0
Superphosphate of Lime (26 %)	,,	2	10	0
Tallow	,,	25	10	0
Tin (English Ingots)	,,	93	0	0
,, Crystals	per lb.	0	0	6¼
Zinc (Spelter)	per ton.	20	0	0
,, Chloride (solution, 96° Tw.	,,	6	0	0

No. 154. SATURDAY, MAY 3, 1890. Vol. VI.

Contents.

Notices.

All communications for the *Chemical Trade Journal* should be addressed, and Cheques and Post Office Orders made payable to—

DAVIS BROS., 32, Blackfriars Street, MANCHESTER.

Our Registered Telegraphic Address is—
"Expert, Manchester."

The Terms of Subscription, commencing at any date, to the *Chemical Trade Journal*,—payable in advance,—including postage to any part of the world, are as follow :—

Yearly (52 numbers) .... **12s. 6d.**
Half-Yearly (26 numbers) .... **6s. 6d.**
Quarterly (13 numbers) .... **3s. 6d.**

Readers will oblige by making their remittances for subscriptions by Postal or Post Office Order, crossed.

Communications for the Editor, if intended for insertion in the current week's issue, should reach the office not later than **Tuesday Morning.**

Articles, reports, and correspondence on all matters of interest to the Chemical and allied industries, home and foreign, are solicited. Correspondents should condense their matter as much as possible, write on one side only of the paper, and in all cases give their names and addresses, not necessarily for publication. Sketches should be sent on separate sheets.

We cannot undertake to return rejected manuscripts or drawings, unless accompanied by a stamped directed envelope.

Readers are invited to forward items of intelligence, or cuttings from local newspapers, of interest to the trades concerned.

As it is one of the special features of the *Chemical Trade Journal* to give the earliest information respecting new processes, improvements, inventions, etc., bearing upon the Chemical and allied industries, or which may be of interest to our readers, the Editor invites particulars of such—when in working order—from the originators; and if the subject is deemed of sufficient importance, an expert will visit and report upon the same in the columns of the *Journal*. There is no fee required for visits of this kind.

We shall esteem it a favour if any of our readers, in making inquiries of, or opening accounts with advertisers in this paper, will kindly mention the *Chemical Trade Journal* as the source of their information.

Advertisements intended for insertion in the current week's issue, should reach the office by **Wednesday morning** at the latest.

Advertisements.

Prepaid Advertisements of Situations Vacant, Premises on Sale or To be Let, Miscellaneous Wants, and Specific Articles for Sale by Private Contract, are inserted in the *Chemical Trade Journal* at the following rates :—

30 Words and under.......... 2s. 0d. per insertion.
Each additional 10 words .... 0s. 6d. „

Advertisements of Situations Wanted are inserted at one-half the above rates when prepaid, viz :—

30 Words and under.......... 1s. 0d. per insertion.
Each additional 10 words .... 0s. 3d. „

Trade Advertisements, Announcements in the Directory Columns, and all Advertisements not prepaid, are charged at the Tariff rates which will be forwarded on application.

ON GULLIBILITY.

WE hasten to take an opportunity of exchanging ideas wit our readers on this subject, because childlike fait and simple credulousness are so rapidly dying out under th exterminative influence of the School Board, that gullibilit will soon become as extinct as the dodo, and dishonesty ceas to exist for the simple reason that mankind will be too muc on the alert to fall a prey to its wiles. But these halcyo days have not yet actually arrived. They may not be far of but the proceedings in our law courts shew us occasionall that specimens of the *genus* gull still survive. A case recentl tried in the Stockport County Court, and reported elsewher in these columns, throws some little light upon the progres of the evolutionary operation which is daily making ma sharper than he used to be. As has often been the cas before, the fertilizing phospho-guano only called into bein the apple of discord, and as Messrs. S. and T. Rogerson, o Cheadle, had expected a crop of a different nature the sought a solatium for their disappointment from the adminis trator of the law. Perhaps this last proceeding is the on that most fully justifies these present remarks, but we prefe to leave it for the moment and deal with some other aspect of the case. It appears that the Manchester Phosguan Company sold the plaintiffs at the rate of 35s. per ton som of the residual matter left after bones had been treated fo the extraction of phosphoric acid. Messrs. Rogerson, a their counsel explained to the court, were very chary over th purchase, as they knew that in the manure trade things ar not always "what they seem." However, their knowledg availed them little, for the eloquence of the vendor secure an order. *After* having made the bargain the buyers, in thei exceeding wisdom, had the stuff analysed, and their peace o mind was not promoted by a report from Dr. Voelcker tha the manure was worthless. Its effect, too, upon the crop t which it was applied was absolutely *nil*. The scientific evi dence was conflicting, and it appears to us that the truth la somewhere between the opinions expressed by the well-know analysts who gave evidence. If this was so, probably Judg Hughes was not far wrong in coming to the conclusion tha the market value of the manure was 17s. per ton.

But why did Messrs. Rogerson buy without guarantee a article which they had evidently regarded with a certai amount of suspicion? Or why did they not have it analyse *before* making the bargain? They belong to an Association one of the main objects of which is to protect their interest under such circumstances as these, but although the bes scientific aid was to their hand, they did not trouble to avai themselves of it. One cannot altogether blame the manu facturer for trying to get the highest price possible for hi wares, and there is nothing to shew that in this case he di not honestly think he was giving a *quid pro quo*. The "goo old days" of the manure maker are over. He can no longe grind up broken bricks and sell them at £10. per ton. H has to be circumspect in his ways and walk on the same foot path as other men.

Then another question arises. When buyer and selle ound themselves at variance upon the value of the materi bought and sold, would it not have been a less costly and more satisfactory proceeding to have submitted the case to competent arbitration? If the services of an impartial expert had been called in—one in whom both parties had complete confidence—and he had been requested to give a report upon the value of the manure, based upon its chemical analysis, we fail to see where there could have been any necessity for taking the matter to the County Court at all.

CHESHIRE COUNTY COUNCIL.

POLLUTION OF THE IRWELL AND THE SHIP CANAL.

ONE of the chief items of business before the Cheshire County Council is the consideration of correspondence that has passed between the clerk and the Local Government Board with reference to the pollution of the rivers Irwell and Mersey and of the Manchester Ship Canal. The Local Government Board having requested the County Council to furnish them with the reasons on which they based their application for the creation of a joint committee representing all the administrative counties in the watershed of the Irwell and Mersey for enforcing the provisions of the Rivers Pollution Prevention Act, a draft reply has been prepared by the clerk to the County Council setting forth that the immediate reason for the petition is the imperative need that the water in the Manchester Ship Canal should be reasonably pure. The Cheshire County Council, in order to secure this, proposed that the area constituting the district should be the watershed of the Mersey and Irwell as far as the tideway at Warrington, this area including all the water flowing into the Ship Canal. The proposal of the Lancashire County Council dealt with the whole watershed of the Mersey and Irwell, which included not only the waters flowing into the Ship Canal, but also the tidal waters of the estuary of the Mersey; and for the purpose of preventing the pollution of the Ship Canal it was not necessary that these tidal waters of the Mersey should be included in the proposed district, as tidal waters might be safely left to themselves. The drainage of Warrington and of all places lower down the Mersey would not fall into the Canal but into the tidal waters of the Mersey. The area given in the Cheshire County Council map was a compact district, and, with the Ribble watershed which the Lancashire County Council was petitioning to have placed under another joint committee, would, roughly speaking, cover the whole of the cotton district. This cotton district it was of the utmost importance to treat in a uniform manner. The area proposed by the Lancashire County Council was unwieldy and unnecessary, including, for instance, the River Weaver, which drained a district apart from the salt trade purely agricultural. Birkenhead could not have identical interests with the cotton manufacturing districts; and, further, if the tidal waters of the Mersey were included within the proposed district, the joint committee exercising authority thereover might be brought into conflict with the Mersey Conservancy Commissioners and the River Weaver Trustees. The extension of the area, as proposed by the Lancashire County Council, did not seem to have any advantages, but it did seem to have great disadvantages. It was difficult to understand why the West Riding of Yorkshire should be included in the Lancashire scheme; and, further, if the River Weaver was included within such district, then Staffordshire must also be covered by the same authority. The Cheshire County Council desired to observe that for the purpose of securing the prevention of the pollution of the Manchester Ship Canal, the foregoing, among other reasons, pointed to the area shown upon the Cheshire map as being sufficient, and that there was no necessity for the enlarged district proposed by the Lancashire County Council. The area proposed by the Cheshire authority was the same as that described in the Mersey, &c., Protection Act, 1862, which prohibited the throwing of solid or bulky substances into the Mersey or Irwell, and their tributaries above Warrington bridge, without qualification as to preventing the flow or polluting the waters thereof.

A NEW USE OF GRAPHITE.—As a hint to blast furnace managers, the following account is given by *The Mechanical News* of a new use for graphite. An experiment was tried at the Crown Point Iron Company's Plant some time ago, on the occasion of their re-lining and starting in blast one of their large furnaces. After the fire-bricks were in place, a cheap article of graphite or plumbago was bought, reduced to a paste [illegible] and the interior of the furnace washed with the plumbago [illegible] gave a slippery glaze—incidental to the lubricating quality [illegible]bago—to the fire-brick lining, which lessened the time [illegible] heat quite a percentage. The slippery surface to be [illegible] or other refuse and the charge passed down in less time [illegible] free and clear. The Crown Point managers claim [illegible] scheme. The expense of the trail was very [illegible] value of the result. The graphite used came [illegible]ucible Company, Jersey City, N. J.

Finally, we may add that fluorine and hydrogen combine when cold and in darkness. This is the only example of two simple gaseous matters directly combining without the intervention of a foreign energy. Chlorine and hydrogen require light; hydrogen, and oxygen require an electric spark or a flame; hydrogen and fluorine combine directly.

Moreover, this chemical activity has been very demonstrated by Messrs. Berthelot and Moissan who have determined the heat of combination of hydrogen and fluorine to be 37·6 calories, that is to say it is greater than that of the hydroacids for iodine, bromine, and chlorine. Thus the fluorine is the most active element known at present, and on account of this very property, we maintain that it will be called upon to furnish chemists the most interesting reactions.—*La Nature.*

THE SPONTANEOUS IGNITION OF COAL CARGOES.

BY PROFESSOR VIVIAN B. LEWES, ASSOCIATE, F.C.S., F.I.C.

[Read at the Thirty-first Session of the Institution of Naval Architects, March 28th, 1890, the Earl of Ravensworth in the chair.]

(Continued from page 267.)

ON examining the evidence to be obtained as to the conditions under which spontaneous ignition of coal in ships usually takes place, it is found that liability to ignition increases with:—

(1) The increase in tonnage of cargoes.

Thus, in cargoes of under 500 tons the cases reported amount to a little under ½% for shipments out of Europe; from 500 to 1,000 tons, to over 1%; from 1,000 to 1,500 tons, to 3·5%; 1,500 to 2,000 tons, to 4·5%; and over 2,000 tons, to no less than 9%.

The evidence demonstrating this very remarkable result is to be found in the Report of the Royal Commission for 1875, p. viii., and clearly shows the influence of mass upon this action which acts in two ways:—

(*a*) The larger the cargo, the more non-conducting material will there be between the spot at which heating is taking place and the cooling influence of the outer air.

(*b*) The larger the cargo the greater will be the breaking-down action of the impact of coal coming down the shoot upon the portions first loaded into the ship, and the larger, therefore, the fresh surface exposed to the action of the air.

(2) The ports to which shipments are made, 26,631 shipments to European ports in 1873 only resulting in 10 casualties, whilst 4,485 shipments to Asia, Africa, and America gave no less than 60.

This startling result is partly due to the length of time the cargo is in the vessel, the absorption and oxidation being a comparatively long action, but a far more active cause is the increase in the action brought about by the increase of temperature in the tropics, which converts a slow action into a rapid one, and if statistics had been taken, most of the ships would have been found to have developed active combustion somewhere about the neighbourhood of the Cape, the active action developed in the tropics having raised the temperature to the igniting point of the coal by that time.

(3) The kind of coal of which the cargo consists, some coals being specially liable to spontaneous heating and ignition.

This is a point on which great diversity of opinion exists, but I think it will be pretty generally admitted that cases of heating and ignition are more frequent in coals shipped from East Coast ports than in shipments of the South Wales coals. As has been pointed out, however, so much depends on the amount of small coal present that a well-loaded cargo of any coal would be safer than a cargo of Welsh steam coal in which a quantity of dust had been produced during loading.

The idea that the percentage of pyrites present is any indication of the liability to spontaneous combustion must be entirely discarded, as experiment shows that many coals poor in pyrites frequently ignite, whilst others rich in them are perfectly safe.

A much surer guide is to be found in the quantity of moisture present in an air-dried sample of coal, which is a sure index to the absorptive power, the higher the amount of moisture held by the coal after exposure for some time to dry air, the greater will be its power of absorption for oxygen, and the greater therefore its liability to spontaneous heating and ignition.

This is beautifully shown by the table on opposite page, in which the percentage of pyrites and moisture present in some coals are contrasted with their liability to self-ignition.

(4) The size of the coal, small coal being much more liable to spontaneous ignition than large.

This, as has been pointed out, being entirely due to the increase in active absorbent surface exposed to the air, a fact which is verified by the experience of large consumers of coal on land, gas managers recognising the fact that coal which has been stamped down or shaken down during storage being more liable to heat than if it has been more tenderly handled, the extra breakage causing the extra risk.

(5) Shipping coals rich in pyrites whilst wet.

Liability to Spontaneous Ignition.	Pyrites per cent.		Moisture pe[r] cent.
Very slight	1.13	..	2·54
	1·01 to 3·04	..	2·75
	1·51	..	3·90
Medium	1·20	..	4·50
	1·08	..	4·55
	1·15	..	4·75
Great	1·12	..	4·85
	0·83	..	5·30
	0·84	..	5·52
	1·00	..	9·01

The effect of external wetting on coal is to retard at first the absor[p]tion of oxygen and so to check the action; but it also increases t[he] rate of oxidation of the pyrites, and so causes disintegration of t[he] coal, with consequent crumbling and heating due to exposure of fre[sh] dry surfaces.

(6) Ventilation of the cargo.

The so-called ventilation, which has from time to time been intro[]duced into coal ships, is undoubtedly one of the most prolific causes [of] spontaneous ignition.

For ventilation to do any good cool air would have to sweep con[]tinuously and freely through every part of the cargo, a condition im[]possible to attain, whilst anything short of that only increases th[e] danger, the ordinary methods of ventilation supplying just about th[e] right amount of air to create the maximum amount of heating. Th[e] reason of this is clear. A steam coal absorbs about twice its ow[n] volume of oxygen, and takes about ten days to do it under favourabl[e] conditions, and it is this oxygen which in the next phase of the actio[n] enters into chemical combination and causes the serious heating.

A ton of steam coal occupies 42 to 43 cubic feet, and if properl[y] loaded contains between the lumps as nearly as possible 12 cubic fee[t] of air space, that is to say, of the 42 cubic feet, 12 cubic feet is air an[d] 30 cubic feet is coal.

Thirty cubic feet of coal, with its fresh absorbing surfaces laid bar[e] by the crushing incidental to loading, will, in the first ten days afte[r] being taken on board, absorb 60 cubic feet of oxygen, if it can get it. Now, air contains only, roughly, one-fifth of its volume of oxygen, s[o] that 60 cubic feet represent 300 cubic feet of air, or twenty-five time[s] as much as is present; so that it is evident that if air could be ex[]cluded there would be only one-twenty-fifth the quantity of oxyge[n] present which is needed for complete action, and any heating would in consequence, be very slight; whilst to produce the greatest heatin[g] it would be necessary to change the entire air in the cargo twenty-fiv[e] times in the first ten days, and this is just about what the ordinar[y] method of taking a box shaft along the keelson with Ventian lattic[e] upshafts from it would give.

The most forcible illustration of the evil of such ventilation is to b[e] found in the case of the four colliers, "Euxine," "Oliver Cromwell," "Calcutta," and "Corah," which were loaded at Newcastle under th[e] same tips, at the same time, with the same coal, from the same seam. The first three were bound for Aden, and were all ventilated. Th[e] "Corah" was bound for Bombay, and was not ventilated. The thre[e] thoroughly ventilated ships were totally lost from spontaneous ignitio[n] of their cargo, while the "Corah" reached Bombay in perfect safety.

(7) Rise in temperature in steam colliers due to the introduction o[f] triple expansion engines and high-pressure boilers.

It has been fully pointed out that anything which tends to increas[e] of initial temperature increases the rapidity of chemical action. Stea[m] at 80 lbs. boiler pressure has a temperature of 324° F. (162° C.), and [a] common stoke-hold temperature with boilers worked at this pressure i[s] 100° to 130° F. (or 38° to 54° C.). Steam at a boiler pressure o[f] 155 lbs. has a temperature of 368° F. or 186° C., and gives a corres[]ponding increase of temperature in the stoke-hold and other adjacen[t] portions of the vessel, the temperature in the stoke-hold under thes[e] conditions being from 110° F. (43·5° C.) to 140° F. (60°), an increas[e] of about 10° F.

It is, however, difficult in the mercantile marine to get a direc[t] comparison of the increase in temperature due to this cause, but in th[e] service some of the troopships have been from time to time refitted and in the case of the "Malabar" and "Crocodile," the tempera[]tures existing with the old simple low-pressure engines, the old com[]pound engines, and the new triple expansion engines can be contrasted and will be found in the accompanying table, for which I am indebte[d] to Mr. White.

From this it will be seen that, taking the voyage through, th[e] average increase of temperature in the stoke-hold from the use of th[e] triple expansion engines is about 5°. This table is of great interest, a[s] the temperatures taken during the outward voyage of the "Crocodile," December, 1883, point to the coals in the bunkers commencing to heat and also illustrate where and at what temperature it commenced.

TEMPERATURE RECORDS. H.M.S. "MALABAR" AND "CROCODILE."

Ship	Locality.	Date.	Deck.	Coal Bunkers.	Engine Room.	Stoke-hold.
"CROCODILE," Original simple low-pressure engines.		1874				
	Channel and Bay	Jan. 15	52	72	98	90
		„ 18	58	80	99	92
	Mediterranean	„ 23	65	86	98	92
		„ 27	66	84	100	92
	Red Sea	Feb. 6	68	92	108	106
		„ 10	81	100*	114	110
		1875				
	Channel and Bay	Feb. 11	46	66	90	105
		„ 12	54	70	94	102
	Mediterranean	„ 19	53	74	92	89
		„ 26	66	74	98	104
	Red Sea	Mar. 4	77	90	104	115
		„ 6	82	94	98	120
	Indian Ocean	„ 12	79	98*	104	114

Ship	Locality.	Date.	Deck.	Coal Bunkers.	Engine Room.	Stoke-hold.
"MALABAR," Old compound engines.		1885				
	Channel and Bay	Oct. 3	59	85	82	89
		„ 6	63	86	86	110
	Mediterranean	„ 8	72	90	98	94
		„ 9	73	90	96	100
		„ 10	70	90	90	110
	Red Sea	„ 19	81	92	95	107
		„ 20	84	100	96	108
		„ 21	88	104*	104	119
		„ 23	87	102	97	112
"CROCODILE," Compound engines.		1888				
	Channel and Bay	Dec. 4	55	78	72	90
		„ 6	54	85	80	84
		„ 8	60	92	82	88
	Mediterranean	„ 12	64	97	80	100
		„ 15	60	95	78	90
		„ 17	62	112	83	94
	Red Sea	„ 22	74	113	92	100
		„ 23	80	115	96	103
		„ 24	82	116*	98	110
	Indian Ocean	„ 31	78	99	97	118

Ship	Locality.	Date.	Deck.	Coal Bunkers.	Engine Room.	Stoke-hold.
"MALABAR," New triple engines.		1888				
	Channel and Bay	Dec. 13	49	77	78	72
		„ 14	56	86	86	88
		„ 16	62	80	84	90
	Mediterranean	„ 18	63	88	90	94
		„ 19	62	88	90	94
		„ 22	68	87	90	100
		„ 23	74	89	90	98
	Red Sea	„ 28	79	96	102	110
		„ 29	82	100	100	110
		„ 30	84	105*	100	108
		„ 31	83	104	104	110
		1889				
	Channel and Bay	Oct. 5	65	84	86	94
		„ 7	67	84	90	102
	Mediterranean	„ 10	76	100	92	102
		„ 11	80	100	93	99
		„ 15	79	104	96	100
		„ 16	80	106	96	106
	Red Sea	„ 19	83	106	103	102
		„ 20	86	110	102	110
		„ 21	89	115	108	108
		„ 22	92	120*	113	122

* Maximum temperature during voyages from Portsmouth to Bombay.

NOTE.—The Registers of the original engines of "Malabar" are destroyed; temperature records of "Crocodile" are given in lieu for a period when she was fitted with old low-pressure engines.

Crossing the bay, the stokehold and bunker temperatures agree fairly well, but entering the Mediterranean, the stokehold having become heated to 100° F. on the 12th, action commenced in the bunkers, and this rapidly raised their temperature higher than that of the stokehold, the action, however, not being violent, and the bunker temperature rapidly falling below that of the stokehold as the chemical action died away.

Having now discussed the chemical and physical conditions which lead to the phenomenon known as "spontaneous ignition," we can formulate precautions which will tend to prevent such disasters.

(1) THE CHOICE OF COAL FOR SHIPMENT TO DISTANT PORTS.

The coal should be as large as possible, free from dust, and with as little "smalls" as can be helped.

It is better as free from pyrites as possible, in order to prevent disintegration after shipment, *and it should contain when air-dried not more than 3 per cent. of moisture.*

(2) PRECAUTIONS TO BE TAKEN DURING SHIPMENT.

No coal should be shipped to distant ports until at least a month has elapsed since it was brought to the surface at the pit's mouth. Every precaution should be taken to prevent breaking up of the coal whilst being taken on board, and on no account must any accumulation of fine coal be allowed under the hatchways.

When possible the coal should be shipped dry, as external wet, by producing oxidation of the pyrites, causes disintegration.

(3) PRECAUTIONS TO BE TAKEN ON BOARD COAL-LADEN SHIPS.

This phase of the question is undoubtedly the most important, and in order to ensure any successful treatment of the coal cargo at sea to prevent undue heating and ignition, the means adopted must be as nearly automatic in their working as possible, as it is useless to expect the master or any officer on board a collier during rough weather, &c., to comply with any instructions, such as daily taking the temperatures in various parts of the cargo, and so on.

The coal compartments should be made gas tight, as far as the bulkheads separating them from the rest of the ship is concerned; and as no difficulty is found in doing this when provision for forced draught is being ... difficulty should be found in this case.

... coal has all been taken in, it should be battened down, ... should not be again opened until the vessel reaches ..., the only ventilation allowable being a 2 inch pipe ... the crown of each coal compartment, and led 12 ft. ... the top being left open. This would be quite ... egress to any gases evolved by the coals, but ... access of air.

Into the body of the coal cargo itself would be screwed, at regular intervals of about 6 ft., iron pipes, closed at the bottom, and containing alarm thermometers, constructed in the following way:—A long bulb of glass, containing mercury, has an insulated wire inserted into the quicksilver, and making contact with it, whilst the stem attached to the bulb has a second wire in it, so arranged that when a rise of temperature causes expansion of the mercury, in rising in the tube it makes contact, and the wires from these tubes are in connection with an electric bell, index board, and battery in the captain's room, so that the moment the temperature is reached to which the thermometers have been set, the bell rings, and will continue to ring until the temperature again sinks, the spot in which heating is taking place being indicated by the index board.

In the evidence given before the Commissioners in 1875, Mr. J. Glover strongly advocated the use of carbon dioxide, or carbonic acid gas as it is more usually termed, for extinguishing ignition when it had broken out in a coal cargo, and for stopping heating when it had reached a dangerous pitch. His proposal was to generate the gas by the action of hydrochloric acid upon chalk, and to lead it by gas pipes to the compartment affected, and this gas, being heavier than air and a non-supporter of combustion, was to displace the air and its contained oxygen, and so to prevent further action by surrounding the coal with an atmosphere which could not carry on combustion. The idea was a good one, but there were many difficulties in the way of carrying it out, one being that for every thousand tons of coal carried, 80cwt. of hydrochloric would have had to be shipped, also the gas could not have been driven down into the hold if any serious heating had taken place, as an up current would have been formed and would have carried it away, whilst in this state of gas it fails to give any great cooling effect and so would have exercised but little influence upon the mass of red hot fuel. These objections weighed so strongly with the Commissioners, that in their final report we find the following sentences:—

"Several methods for generating carbonic acid gas and applying it to the ignited portion of a coal cargo have been proposed for our consideration. We consider, however, that although this gas might be useful by excluding atmospheric air (which is essential to support combustion), yet it will not, as water does, exert any very sensible cooling effect, *which is a point of vital importance in the case of a mass of ignited coal.* We are of opinion that water and steam are the only agents practically available for the purpose of extinguishing fire in coal cargoes."

Applied in the way which was suggested, there is no doubt but that the carbonic acid gas would have been practically useless; but there is another way in which it could be used, which would make it a most powerful cooling agent, an instantaneous quencher of fire, and would prevent any further tendency to heat on the part of the coal treated with it.

If carbonic acid gas is compressed under a pressure of 36 atmospheres at a temperature of 32° F. (0° C.), it is condensed to the liquid state, and can be obtained in steel vessels, closed with screw valves. On opening the valve some of the liquid is ejected into the air, and on coming into the ordinary atmospheric pressure, is in a moment converted into a large volume of gas. Conversion from the liquid to the gaseous state means the absorption of a large amount of heat, and so great is this that everything near the stream of new-born gas is cooled down, and some of the escaping liquid is frozen to a solid, having a temperature of—78° C., or − 108·4 F.

This liquid carbonic acid gas is now extensively manufactured, and is used abroad to a large extent for aërating waters, driving torpedoes, and for freezing machines; and I should suggest its use in the following way for the checking of ignition in the coal cargo:—

The nozzle attached to the screw-valve on the bottle of condensed gas would have a short metal-nose-piece screwed on to it, the tube in which would be cast in solid, with an alloy of tin, lead, bismuth, and cadmium, which can be so made as to melt at exactly 200° F. (93° C.). The valve would then be opened, and the steel bottle buried in the coal during the process of loading. The temperature at which the fusible metal plug will melt is well above the temperature which could be reached by any legitimate cause, and would mean that active heating was going on in the coal; and under these conditions the pressure in the steel cylinder would have reached something like 1700 lbs., and the moment the plug melted the whole contents of the bottle would be blown out of it into the surrounding coal, producing a large zone of intense cold, and cooling the whole of the surrounding mass to a comparatively low temperature. The action, moreover, would not stop here, as the cold, heavy gas would remain for some time in contact with the coal, diffusion taking place but slowly through the small exit pipe.

When coal has absorbed as much oxygen as it can, it still retains the the power of absorbing a considerable volume of carbonic acid gas; and when coal has heated, and then been rapidly quenched, the amount of gas so absorbed is very large indeed, and the inert gas so taken up remains in the power of the coal, and prevents any further tendency to heating; indeed, a coal which has once heated, if only to a slight degree, and has then cooled down, is perfectly harmless, and will not heat a second time. It is not by any means necessary to replace the whole of the air in the interstices of the coal with the gas, as a long series of experiments show that 60 per cent. of carbonic acid gas prevents the ignition of the most pyrophoric substances.

One hundred cubic feet of gas can be condensed in the liquid state in a steel cylinder 1 foot long and 3 inches diameter, and it has been shown that a ton of coal contains air spaces equal to about 12 cubic feet; therefore, one of these cylinders would have to be put in for every 8 tons of coal, and these would be distributed evenly throughout the cargo, and near the alarm thermometers, which would be set to ring a degree or two below the point at which the fusible plug would melt.

The bell ringing in the captain's room would warn him that heating was taking place, and the bell would continue to ring until the cylinder had discharged its contents, and had cooled down to a safe degree, so that the whole arrangement would be purely automatic, and yet the officers would know if everything was safe.

This liquid is now being made at a comparatively cheap rate, and with any demand for it machinery could be put up at the principal coaling ports to charge empty cylinders at a very low rate, so that the initial cost of the steel cylinders once got over the expenses would not be worth considering, more especially as one, or two at most, would be likely to go off.

If the precautions advocated were taken no danger could arise until the arrival of the ship at her destination, and the commonest precautions would then suffice. On removing the hatches, no naked light must be allowed near them, and no one must be allowed to descend into the hold until all the gases have had time to diffuse out into air. If the cylinders have gone off there will be but little fear of explosion, as a high percentage of the carbonic acid gas lowers the explosive power which the mixture of marsh gas (given off from some coals) and air possess; but the carbonic acid gas would overcome and suffocate a man descending into an atmosphere containing any considerable percentage of it. When a safety lamp, lowered into the hold continues to burn as brightly as it did in the open air, then it is perfectly safe to descend.

When once coal in a cargo has fired, pumping in water is of practically no use, as the fire is, as a rule, near the bottom of the mass of coal, and the flow of water is so impeded, that in percolating through the interstices of the heated coal, it is converted into steam before it can reach the seat of combustion. The most effective way to apply water would be to have four 3-inch pipes laid along the floor of the coal compartments, about 6 feet apart, these tubes having a ¼-inch hole bored in the upper side every foot or so, and each pair of pipes coming through the bulk-head, and connecting one to two 6-inch pipes passing through the side of the vessel, the sea water being prevented from entering by means of screw valves. As soon as the alarm thermometer gave notice that heating had reached a dangerous point, these valves could be opened and the lower portion of the cargo drenched with salt water. This, evaporating rapidly, would give large volumes of water vapour, which, passing up through the heated coal, would lower its temperature, but would not be nearly as effective as the method before advocated. It might, however, be used in conjunction with that method, and would, in many cases, save the carbonic acid gas.

The question of preventing the heating and ignition of stores of coal on land and ready for use in bunkers, cannot be met so well by the use of the liquid gas, and, in these cases, it would be found beneficial to dress the coals with a little tar or tar oil, which would close the pores, and to a great extent prevent oxidation. I believe this was advocated by Lachman about 1870.

Crude petroleum in small quantities for this purpose would also be found valuable, for, as already shown, it has no tendency to oxidise itself, and lowers the tendency in other bodies, besides coating them, and so preventing access of oxygen.

The labour trouble in the coal trade is hampering all branches of industry, and, with increasing difficulty in obtaining coal, all sorts of rubbish will be shipped, and many a cargo of coal will go out during the next few months which, under normal conditions, would never have been allowed on board; and we may expect a heavy increase in the loss of life and property from spontaneous ignition, This will probably result in a rise in the rate of premium, and further check to our export coal trade; and I sincerely trust that, should these gloomy forebodings be fulfilled, the suggestions made in this paper, and based on experimental facts, may be found of value.

In conclusion, I wish to acknowledge my indebtedness to the researches of Dr. Richters, whose papers on the weathering of coal are to be found in Dingler's *Polytechnisches Journal* for 1870; to the report of the Royal Commissioners on this question in 1876, and finally, to Mr. Martell for suggesting the subject of this paper.

PHOSGUANO.

COUNTY COURT ACTION.

HIS Honour Judge Hughes was engaged last Friday at the Stockport County Court for several hours hearing a case, in the result of which the farmers of the district appear to take a lively interest, as evidenced by the large number who patiently waited through its hearing. It was one in which Messrs. S. and T. Rogerson farmers, of Heathside, Cheadle, sued George Henry Holden, trading as the Manchester Phosguano Company, of 41, Corporation-street, Manchester, manure manufacturers, for £4. 4s. 9d., as damages for breach of contract. Mr. R. Brown, solicitor, Stockport, appeared for the plaintiffs, and Mr. T. R. Sutton, barrister, Manchester, for the defendant.

Mr. Brown said in the spring of last year Messrs. Rogerson were waited upon by a Mr. Smith, who was a representative of the defendant He informed them that he was in a position to supply them with a quantity of bone phosphate manure at a very cheap rate. The conversation took this form. Smith represented that his employer had a very large quantity of bone phosphate manure which was lying at their works somewhere in the neighbourhood of Manchester, which had been condemned by the local authority. This they were bound to get rid of, and were prepared to sell the manure at a great sacrifice. Mr Rogerson, as a practical farmer, knowing very well that there were all kinds of worthless materials in the market, and that people went round to farmers with it, was very careful in his enquiries as to what this material was. Smith represented that he had already sold to Mr. Booth, who was one of the largest farmers in the district, a quantity of the same stuff and that it had proved successful. Mr. Rogerson subsequently said, "I generally use nitrate of soda, but upon your representation I will buy some of your manure." Smith added that the material was fairly worth £3 10s. per ton, but he would sell it to Mr. Rogerson at 35s per ton. Mr. Rogerson, thinking it would be valuable and suitable for that year's crop, gave an order for three tons. It was laid at the same time as was a quantity of nitrate of soda, but there were places left where nothing was sown at all. The result of the bone phosphate was nil, while the nitrate of soda produced excellent results. Mr Rogerson, seeing the results, went over to the offices of the firm on May 3rd, and saw one of the representatives of the defendant there He told him the condition of the manure, and also said he was going to have a portion of it analysed, and intended sending it to Dr Voelcker, of London, that gentleman being the analyst to the Royal Agricultural Society, of which society Mr. Rogerson happened to be member. All members of the society were entitled to send two samples of manure, &c., to him for analysis at a very small cost. The reply of the analyst was to the effect that the phosphate was unfit fo

vas refuse material which he would not have advised
Mr. -Rogerson, on May 27th, sent a letter to the
orming him what the reply of the doctor was, and
what he intended doing in the matter. In
aid he should certainly insist upon payment.
n wrote back to the effect that whereas the
a which had been applied produced very satisfactory
ne manure was not successful in the least. Continuing,
id the tale made out by Smith that it was necessary to get
phate in a few days because it had been condemned by the
es, was not exactly true. The material was manufactured
ilman, who was engaged in extracting phosphoric acids
He got all the phosphoric acids or phosphate which he
the bones, after which he threw away the latter as use-
this very heap of Esilman's which was purchased by a
on behalf of the defendants at 6d. per ton. They were
buy it at 6d. per ton and sell it at 35s. if they liked.
ving been sown and the results being nil, Mr. Rogerson
by the defendants for payment, but he was never sued.
used this fertilizer for that year's crop, but the result was
n the land where this manure was sown as the other upon
ad been sown. He trusted to Smith's warranty that it
rtilizer for crops, but as it proved not to be the claim for
ich he had sustained thereby constituted the present
conclusion, Mr. Brown said what was really sold to
n was gypsum, which was nothing more nor less than
Brown then called a number of witnesses to support
t, the first being Mr. James Rogerson, who said the
er 100 acres in extent. In reply to Mr. Sutton he said he
. per ton for nitrate of soda, but where he put 6 cwt. of
ate on an acre of land he only used 1½ cwt. of nitrate of
same space. The latter cost about 12s. and the former
Before the court adjourned for luncheon Mr. Brown said
our cases. In the present one neither plaintiff nor dean analyst; but in the second the sample had been
It was therefore desirable that they should get on to the
that afternoon. Mr. Sutton thought it was a pity that
ase had not been taken first. The Judge: I will sit at
e second case is finished. I suppose that will be about
The other witnesses called were Mr. Rogerson, farmer, of
Henry Booth, of Dairy House Farm, Handforth; John
osphoric acid manufacturer, of Bradford, near Man-
Thompson, of Manchester, chemist and public analyst for
who said that the material was not phosphate at all; it
of lime, and was not worth putting on land, because the
contained could be found in almost any soil. There
blue in it, which was a poison to plants.
Sutton had addressed the Court at some length for the
. Charles Estcourt, Mr. Philip Escourt, and Mr. Spencer,
Manchester and district, were called. All of these had
ples of the manure to analyse, and found it very good as a

djournment his honour said that, carefully weighing the
the analytical chemists on both sides, he came to the con-
the value of the manure was 17s. per ton, as against 35s.
arged, and he found for the plaintiffs for £2. 19s. 3d.
breach of contract, with costs. As other actions had been
were awaiting the issue of this suit he would grant an
e parties could agree upon a case.
a number of farmers interested in the case, viz:—Thos.
n, of Old Hall Farm, Woodford, who sues the Phosguano
r £7. 18s. 6d.; Isaac Leah, farmer, of Gill Bent, Cheadle
sues for £5. 19s.; and Isaac D. Burrows, also of Cheadle
sues for £6. 6s.

NSULATOR.—We understand that Dr. Purcell Taylor has,
nsiderable trouble, succeeded in obtaining a new insulating
ing a higher di-electric resistance than gutta-percha, at about
st of the latter substance. The new meterial may be made
r, either rigid or flexible, and to melt at any desired tem-
is claimed that it is very tough and tenacious.—*Invention.*

NAGEMENT OF STEAM ENGINES AND BOILERS.—A Bill
roduced into Parliament by the following labour represen-
essrs. Fenwick, Pickard, Broadhurst, Wm. Crawford, and
vide for certificates to persons in charge of steam engines,
on land. Every person who before January 1st, 1891, has
tical control or management for a period of not less than
a steam-engine exceeding five-horse power, or steam
y closed vessel for the generation of steam for such engine
, railway or manufactory in the United Kingdom, or on
coasting steamer of like power or pressure, shall be entitled
te of service to be delivered to him by the Board of Trade
ory proof that he has had such experience, and is a person
nd general good conduct.

USES AND VALUE OF ASBESTOS.

IN a recent paper Prof. J. Donald, M.A., of Montreal, states that the asbestos of commerce is the product of two widely separated countries—Italy and Canada. The Italian article was first in the market, but the Canadian product soon made for itself a place and a name, and the mineral is now shipped from Canada to Italy; while toward the close of 1889 the United Asbestos Company, Limited, of London, which controls the mines of Italy, acquired property in the Canadian field, and is equipping the same with a complete plant preparatory to operations on a large scale. It is very evident, then, that the Canadian fibre is, to say the least, no mean factor in the asbestos industry.

The various names by which this mineral is known are as follows:—The French Canadians call it *pierre a coton*—i.e., cotton-stone; the Germans speak of it as *steinflachs*—stone-flax; *amianto*, the Italian name, indicates that which is undefiled, in allusion to the fact that it may be cleansed by fire; and asbestos, the name by which it is generally known, is a Greek word, signifying endless, ceaseless, and points to its fire-resisting properties.

Prof. Donald's recent analysis of some Canadian asbestos showed the following results:—

Magnesia	40·07
Silica	39·05
Water	14·48
Alumina	3·69
Oxide of Iron	2·41
Undetermined	·30
Total	100·00

Most, if not all, the asbestos of commerce is classified as "fibrous serpentine," from the fact that the mineral occurs principally in a fibrous form, the fibres being so fine and flexible that they may be spun and woven as cotton and flax are; and, moreover, the fabric so obtained is capable of resisting a very high temperature. Some varieties are said to have resisted a temperature of 5,000° F. It is noted, however, that although asbestos is incapable of being fused, except at extraordinarily great heat, its fibres lose their flexibility and become brittle at a temperature only sufficiently high to deprive it of the water that is found in its composition. Altogether, it is a singular mineral, but one that is gradually coming into common use in a variety of forms suited to a great many purposes.

"Not only does our mineral resist high temperatures," says Prof. Donald, "but it is also proof against the action of the majority of chemicals. It therefore forms a very valuable substance for use in filtering apparatus, especially where acid and alkaline liquids, which corrode ordinary filtering paper and cloth, have to be dealt with. As a filtering medium it is used not only in chemical laboratories, but in manufacturing establishments as well."

At Salt Lake City, Utah, the Wyoming asbestos has been sold for about 80 dols. per ton. When the assorted mineral is woven into cloth its market value is about 300 dols. a ton.

Its uses are numerous. For roofing it is deemed the best known material. It is used also extensively in steam packing, while for boiler felting it has no equal. In printing and writing paper its indestructible qualities make it invaluable for documents that need to be preserved. As wall paper it aids in rendering a building fire-proof. In the form of rope it becomes available as fire-escapes and fire-resistant supports. Stokers and furnace-men use it for serviceable gloves, and full suits of clothing, including stockings, are already made of it, as are salvage blankets for rescuing goods from burning buildings. It can, indeed, be almost universally adopted; for mail bags, theatre curtains, drapery and dresses. Its non-conducting properties also commend it highly for application in connection with steam engines and boilers, for packing pistons, flange joints, hot-air joints, cylinder heads, and similar purposes—in the form of yarn, or rope, or felt—the latter making a non-conducting covering for steam pipes, and furnished in sections to fit any size of pipe and into rolls and sheets for large surfaces.

A lower grade of asbestos is less fibrous, more resembling gypsum, and this, when ground finely, becomes the basis of an excellent fire-proof paint.—*Exchange.*

THE ELECTRICAL DEMAND FOR COPPER.—One of the principal items entering into the manufacture of electrical apparatus and goods is copper. Perhaps a better appreciation than generally obtains of this fact may be formed from the statement which we heard last week, and is true, that one of the large companies had just placed an order for 1,000,000 pounds of copper, and had at that time an order in for 250,000 pounds more. Enormous as this quantity may seem, it is understood that it will all be used up by July 1st. The activity prevailing may readily be inferred from these figures, which are those of a single concern, though of course it is one of the most prominent.—*Electrical Engineer.*

BI-CARBONATE OF SODA FOR CARBONIC ACID GAS.

A WELL-KNOWN Liverpool firm of aerated water manufacturers states, as a result of their experience, that the use of bi-carbonate of soda is far preferable to and very much cheaper than the whiting process, says *The Mineral Water Trades' Recorder*. The firm has given much attention to the production of carbonic acid gas, which has hitherto been almost invariably made from carbonate of lime or whiting. As a result of perfectly reliable and frequently repeated experiments it has, however, been found that by the bi-carbonate of soda process a cheaper and better carbonic acid gas is obtained—a gas chemically pure and rich, as is seen in champagne wine. The difference in quality of the two gases may be demonstrated by a simple experiment. Put some bi-carbonate of soda in one glass, and some whiting in another; pour into each a small quantity of water; add a few drops of sulphuric acid, when effervescence will at once take place and carbonic acid gas will be generated. On smelling at each in turn it will be found that the gas in the vessel containing whiting emits an offensive smell, while that generated in the bi-carbonate of soda will be found to be perfectly inodorous. The purity of the gas made from bi-carbonate of soda tells in its favour as compared with that made from whiting. Then as to the cost, the firm in question have clearly demonstrated to our representative, when he called upon them, that the carbonic acid gas manufactured by means of bi-carbonate of soda is not only beyond comparison purer, but it is also very much cheaper. Experiments have been made as to the cost of one ton of carbonic acid gas, the prices of sulphuric acid being £6. per ton, of bi-carbonate of soda £5. per ton, and of whiting £1. 10s. per ton. By the bi-carbonate process the cost would be £16. 12s., made up of two tons of bi-carbonate of soda and one ton two cwts. of sulphuric acid. By the whiting process the cost would be £22. 10s., made up of three tons of whiting and three tons of sulphuric acid. The difference in favour of the former process is thus seen to be no less than £5. 18s. per ton. In addition to this, by the bi-carbonate process glauber-salts to the value of £4. 19s. are obtained. Deducting the cost of manufacture and packing—£1. 17s. 6d., there is a balance remaining of £2. 16s. 6d., which, added to the difference in favour of bi-carbonate process—£5. 18s., represents a total saving for the ton of carbonic acid gas manufactured of £8. 14s. 6d. This will, indeed, be a matter of great consideration in the future to mineral water manufacturers.

INCREASING CONSUMPTION OF NAPHTHALINE.

THE extreme scarcity and high price of camphor has raised into prominence an industry which only a few years ago was one of the most uncertain and unimportant known in the chemical world. About four years ago all the refined naphthaline used in this country (and the quantity was insignificant), was imported from Europe. The field presented by the consumption of this product of modern chemistry was so limited as to be almost overlooked by American enterprise. Two or three firms saw enough in it to add the production of naphthaline to their other business, and from this unimportant beginning has grown up an extensive and valuable industry. While there has been no increase in the number of producers since the output of this chemical began to assume important proportions in this country, the production has reached what may be considered very large proportions. It is stated that only three years ago the annual consumption in the United States did not exceed twenty tons, while for the current year the consumption, it is estimated, will reach at least five hundred tons.

The sudden and unprecedented growth of demand is due largely if not entirely to the extraordinary advance in the price of camphor and the impossibility of obtaining adequate supplies of that gum. The manufacturers were not prepared for this enlargement of the demand, and they were caught with insufficient facilities for turning out naphthaline as fast as wanted by the trade. The result is that the three producers have booked more orders than they can fill for the next three months and as the demand in Europe has increased in the same ratio there is little or nothing there that can be spared for this market. Naphthaline is produced in several forms, the more saleable being balls, tablets, scales, and granulated. The first named is more popular than the other kinds and the demand for this description has been so large that at the present time there is none to be had either here or abroad, except possibly from second hands at enhanced prices.

It is only a question of enlarging present producing facilities to furnish a practically unlimited supply of all grades, but the industry is yet so young, and its future so uncertain, that those engaged in it are not prepared to increase their expenses to the extent necessary to accommodate the output to the requirements of consumption.

The effect of the increased consumption is shown by the advance [in] market values. Before the present movement began, three and [a] quarter cents per pound was considered a fair average price. At [the] present time the average quotation for the cheapest grade is five ce[nts] per pound, and the market has a decided upward tendency, owi[ng] to the small available supply here or abroad. The difficulty is [to] supply the immediate wants of consumers. The manufacturers c[an] offer plenty of goods for delivery after June, but as by that time [the] consuming season would have ended, buyers do not care to make c[on]tracts so far ahead, particularly as they would, in carrying the st[ock] over to the next season, have to stand a loss in weight from evapo[ra]tion, and deterioration in the colour of the stock from atmosphe[ric] influences that might render it unsaleable.

So far this country has not been able to supply a sufficient quant[ity] of crude naphthaline to meet the needs of consumption, partly beca[use] enough attention has not been given to the subject by those who are [in] a position to most profit by producing the chemical, but more partic[u]larly because our coal is not as rich in naphthaline as the Europe[an] varieties. Consequently for some time to come at least market pric[es] here will be influenced to a considerable extent by the conditions p[re]vailing in Europe from whence the bulk of our supply of the cru[de] article will still be drawn.

While naphthaline has come into prominence during the past year [as] a substitute for camphor in the preservation of woollens, furs, and ot[her] articles from the destructiveness of insects it has other uses. Am[ong] these may be mentioned its recent application to carburetting gas, [for] disinfecting purposes, and its more ancient use in medicine. In all these fields its usefulness is extending, and there is every reaso[n to] suppose that it has become a fixed important article of commerce, [and] that as such its market value will become established on a highly p[ro]fitable basis.—*Oil, Paint, and Drug Reporter.*

NEW METHOD OF TREATING ORES.

MR. R. J. W. POUND has taken out protection in Victoria f[or a] method of separating metals from their matrices, as well a[s] raising the ores from mines. A sensational experiment was ma[de at] the offices of the Australian Ventilating Company, Little Bourke-st[reet], Melbourne, on December 9th, when, in the presence of a numb[er of] experts and others interested in mining pursuits, Mr. Pound sh[owed] the results of his method of separation. A piece of stone, conta[ining] (by analysis) gold, silver, and copper, was broken into several pi[eces], one of which was pulverised in a mortar, and the stuff produced [taken] by Mr. Pound into an airt-tight room, his imperative instructions [being] to liberate him in seven and a half minutes. At the expiration o[f the] time the door was opened, when Mr. Pound was discovered in [a con]dition of insensibility, but he quickly recovered on being take[n to] one of the cooling rooms of the establishment. On regaining [con]sciousness he immediately returned to the now comparatively ven[tilated] apartment in which he had conducted his operations, and in the [course] of a few minutes reappeared with a small quantity of each of th[e] metals that had been contained in the stone. The process ha[d been] carried out under conditions inimical to animal life, but with [proper] mechanical arrangements no one need be in the chemically v[itiated] atmosphere. But Mr. Pound's scheme includes also the raising [of] ores by means of containers upon vertical railways. When thu[s raised] the stone will be crushed three times. After the second crushing [it will] be carried by compressed air action and delivered between th[e ...] whence it will be caught by two intersecting steel wheels, t[hus en]gendering heat to about 80 or 85°. After this the chemical [process] comes into operation. The crushed and heated ore is pushed [through] a steel chamber into a battery, where all the "binding propert[ies" are] neutralised, allowing the gold, silver, and other metallic sub[stances] together with the residuum, to be passed on in semi-separate c[ondition] through a copper chamber to a battery for final treatment. [The] gold, silver, and copper are to be extracted and delivered in [...] containers, the balance of the ore substance passing off through [...] Mr. Pound claims that the ore can be treated at the cost (i[ncluding] raising from the slopes) of about 3s. per ton.—*American Manu[facturer.]*

RESTRICTING THE USE OF SACCHARIN.—The Portuguese [Govern]ment has decreed that saccharin, whether alone or mixed [with any] other product, shall be sold by chemists only on the prescript[ion of a] legally qualified medical man. Every contravention of this en[actment] as well as the employment of saccharin in the manufacture [of] meats and drinks, is made punishable by definite penalties.

A NEW VINE MANURE.—M. Ville, a professor of che[mistry at] Paris, states that he has discovered a new chemical manure [...] and almost miraculous in its effects on the vine. It con[sists of a] mixture of phosphate of lime, carbonate of potash, and s[ulphate of] lime, which, if placed round vine-growths, will enable them [to resist the] onslaughts of the phylloxera.

NT REGULATING BUNSEN BURNER.

ier, which is applied by John J. Griffin & Sons, Limited, unique, as by a novel and ingenious arrangement, with the central gas jet as used in all other ers, and by one single movement regulates simultaneously and the air. The gas passes into the burner tube through the side of the tube, which is therefore open from top to consequently cannot become choked.

: consists of 3 pieces only, which can be very easily and apart or put together, without the use of tools, even if ving to prolonged use, as the various parts fit together vs.

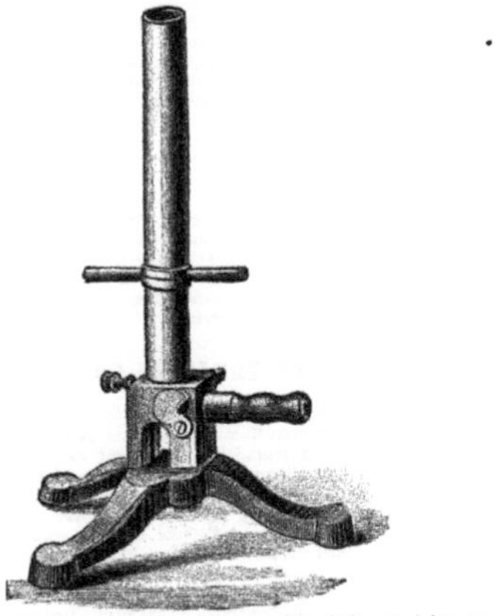

able discs are provided at the side of the air inlets, to inflow of air under varying gas pressures.

advantage is that as the tube is open from top to bottom hance of the jet becoming choked. The burner contains power of adjustment to suit all pressures, and is supplied at ost.

IMETRIC METHOD OF ESTIMATING TANNIN IN BARKS, ETC.

BY SAMUEL J. HINSDALE, FAYETTEVILLE.

VE 0·04 gramme potassic ferryicyanide in 500 cubic centis of water, and add to it 1·5 cubic centimetres (about 22 : ferri chloridi. Call this iron mixture.

·04 gramme "pure" tannin (galotannic acid), which has t 212°F., in 500 cubic centimetres of water. Call this on.

·8 gramme oak bark with boiling water, and make it up to ntimetres with cold water.

2-ounce clear glass tumblers (or beaker glasses) on a white in one of them, with a dropping pipette (about 4 inches inch wide) about half filled, put five drops of the infusion in the others, with the same pipette (after rinsing), put , seven, and eight drops of the "tannin solution" (the infusion and of the tannin solution must be uniform. The ne pipette, about half filled, insures that).

to each 5 cubic centimetres of "iron mixture," and in inute add to each tumbler about 20 cubic centimetres of vithin three minutes observe the shades of colour. The drops of "tannin solution" used in the tumbler which in shade of colour to the tumbler containing the infusion cates the percentage of tannin in the bark, *i.e.*, if it is the seven drops were placed, the tannin strength of the bark

o observe the shades of colour horizontally rather than d to hold up the infusion tumbler, with the one which most nearly corresponds, opposite to a white wall, with your back to the light.

The above is written for oak bark, but the same process will answer for any substance containing less than 10 % of tannin. The results are necessarily in terms for commercial gallotannic acid, and not in those of pure tannin or of the particular tannin in the material assayed.

For substances containing between about 10 and 20 % it is best to dilute the infusion with an equal part of water and proceed as above, using five drops of the dilute infusion, and for the answer double the result. Thus, if the diluted infusion of tea required eight drops "tannin solution" to correspond, call the percentage 16.

For substances containing less than 1 or 1½ %, exhaust 8 grammes instead of 0·8 gramme, and take one-tenth of the result for the answer. For substances containing more than 20 %, as galls, sumach, catechu, etc., you may dilute the infusion with two, three, or more times its bulk with water, and calculate as above (as with tea), or you may use one, two, three, or four drops of the undiluted infusion in the first glass and make the calculation thus, *i.e.* : As the number of drops of infusion used is to the number of drops "tannin solution" used (to correspond), so is five to the answer, thus : Suppose two drops infusion were used and the corresponding tumbler contained 15 drops tannin solution—2 : 15 : 5, answer 37·5 %.

The object in diluting the infusions is because the infusion glass may be of too deep a blue shade. It is better that it should just produce a light blue.

The tumblers must be perfectly clear and clean.

The "iron mixture," "tannin solution," and infusion must be freshly prepared and not exposed to the rays of the sun.

The water used must be free of iron and tannin.

Brit. and Col. Druggist.

PRIZE COMPETITION.

A LITTLE more than twelve months ago, one of our friends suggested that it might be well to invite articles for competition, on some subject of special interest to our readers; giving a prize for the best contribution, he at the same time offering to defray the cost of the first prize, and suggesting a subject in which he was personally interested.

At that date we were totally unprepared for such a suggestion, but on thinking over the matter at intervals of leisure, we have come to think that the elaboration of such a scheme would not only prove of benefit to the manufacturer, but would go some way towards defraying the holiday expenses of the prize winner.

We are therefore prepared to receive competitions in the following subjects :—

SUBJECT I.

The best method of pumping or otherwise lifting or forcing warm aqueous hydrochloric acid of 30 deg. Tw.

SUBJECT II.

The best method of separating or determining the relative quantities of tin and antimony when present together in commercial samples.

On each subject, the following prizes are offered :—

One first prize of five pounds, one second prize of one pound, five additional prizes to those next in order of merit, consisting of a free copy of the *Chemical Trade Journal* for twelve months.

The competition is open to all nationalities residing in any part of the world. Essays must reach us on or before April 30th, for Subject II, and on or before May 31st, in Subject I. The prizes will be announced in our issue of June 28th.

We reserve to ourselves the right of publishing the contributions of any competitor.

Essays may be written in English, German, French, Spanish or Italian.

A NEW LUBRICANT.—A new mineral oil, as thick as butter, has been introduced as a lubricator by the Compagnie Francaise de Graisses Minerals Consistantes. It is stated to be free from acid, resin, or drying oils, and does not alter with exposure to the air. Its melting point is 84° C., and it does not inflame at a lower temperature than 220° C. While it resembles butter in colour it is odourless, and has no chemical action on metals.

Trade Notes.

ANTIPYRIN.—Dr. Knorr, the discoverer and patentee of antipyrin, is said to have cleared more than £200,000. during the late epidemic of influenza.

SULPHUR COMPANY, LIMITED.—Mr. Samuel Harrison, for many years formerly connected with Kurtz's, has been appointed, we understand, manager to this company.—*Liverpool Journal of Commerce.*

THE LEYTON SEWAGE DISPOSAL WORKS.—At the meeting of the Junior Engineering Society on the 11th inst., Mr. J. G. Browning read a paper on the "Leyton Sewage Disposal Works," describing very fully the methods there adopted for treating the sewage. Up till 1883 the sewage of the town was dealt with by simple filtration through straw and charcoal, at which date treatment by sulphate of alumina was introduced, but abandoned twelve months later on account of its expense. Mr. Hanson's chemical process was then substituted, and has since been worked with satisfactory results down to the present time. The chemicals used in this process are lime and a powerfully-oxidising powder prepared from alkali waste containing both hyposulphites and sulphate of lime. About 16 cwt. of lime and 3 cwt. of the powder are used per million gallons of sewage; and the resulting effluent is free of offensive odour, and is discharged into the River Lea. The quantity treated per 24 hours averages about 1½ million gallons, the main part of which flows into the mixing and settling tanks without requiring to be pumped. The sludge obtained, amounting to 110 tons per week, is pressed into cakes in the usual way, and is used as manure, though farmers do not consider it sufficiently valuable to cart away to any considerable distance.

THE FIXATION OF ATMOSPHERIC NITROGEN IN GAS PURIFIERS.—The subject of the fixation of atmospheric nitrogen in the forms of cyanogen and ammonia compounds is likely to attract an increasing amount of attention in connection with the employment of air to assist in the purification of coal gas. Many attempts have been made to produce cyanides directly from atmospheric nitrogen, as well as to obtain ammonia from the same source; but all manufacturing processes especially designed for this purpose have failed commercially, although some have attained experimental success. Yet cyanides are constantly being formed incidentally to furnace operations, and where alkalized carbon at a high heat is exposed to air and steam. Under such conditions, at a low heat, the nitrogen may be fixed as ammonia, and at a high heat as a cyanide. It is a question how far the reactions that go on inside an oxide of iron gas purifier include the fixation of atmospheric nitrogen; but cyanides and ferro-cyanides are being manufactured in Germany from the spent oxide of gas-works with which probably air has not been used. There are gas works in the United Kingdom where 3 per cent. of air and upwards is admitted into oxide purifiers for the sake of revivifying the purifying material *in situ.* It would be interesting to know whether, and if so under what conditions, any of the nitrogen of this air is fixed in the material as cyanides or as ammonia.—*Journal of Gas Lighting.*

Market Reports.

METAL MARKET REPORT.

	Last week.		This week
Iron	44/11	....	44/8½
Tin	£90 5 0	....	£92 0 0
Copper	48 7 6	....	49 0 0
Spelter	20 15 0	....	21 7 6
Lead	12 17 6	....	12 7 6
Silver	46d.	....	47d.

COPPER MINING SHARES.

	Last week.			This week.	
Rio Tinto	16¼	16 5/16	....	17	17⅛
Mason & Barry	6 3/16	6 7/16	....	6½	6⅝
Tharsis	4 15/16	5 1/16	....	4 15/16	5 1/16
Cape Copper Co.	3½	3⅝	....	3⅞	4
Namaqua	2	2⅛	....	2 7/16	2 9/16
Copiapo	2 5/16	2 7/16	....	2⅜	2½
Panulcillo	15/16	1 1/16	....	15/16	1 1/16
Libiola	2⅛	2¼	....	2¼	2⅜
New Quebrada	½	⅝	....	½	⅝
Tocopilla	1/-	1/6	....	1/-	1/6
Argentella	3d.	9d.	....	3d.	9d.

TAR AND AMMONIA PRODUCTS.

The Tar products market remains in about the same condition
reported last week; 90's Benzol is selling for 3s. 6d., and 2s. 7d.
2s. 7½d. for 50/90's. Solvent naphtha and creosote still in go
demand. The future of crude carbolic acid is still uncertain. Th
are indications that so soon as the demand reaches the supply we sh
have better prices again, as all that is at present being produced
finding a sale without any further reduction in price. Anthracene s
remains at old rates, the markets is lifeless and buyers are not anxi
to operate. Pitch remains at about 28s., West Coast ports.
The sulphate of ammonia market is without any new feature, and l
little disposition is shown to purchase for near delivery, buyers doli
out their orders only when they have immediate wants. Makers, ho
ever, are not now quite so pressing, and if they continue in this cou
and only offer to meet the demand, the position will right itself in
time. Values are: Beckton, £11. 5s., London outside mak
£11. 6s 3d., Hull, £11. 2s. 6d. to £11. 3s. 9d., according to quali
The price at Leith varies from £11. 1s. 3d. to £11. 2s. 6d.

MISCELLANEOUS CHEMICAL MARKET.

There has been more firmness in bleaching powder during the p
week, and prices remain at £5. per ton on rails and £5. 5s.
£5. 7s. 6d. f.o.b. The caustic soda reaction has not yet ceased, a
values on the spot are 7s. 6d. to 10s. below the quotations of a w
ago. 70's are offering at £8. 17s. 6d. to £9., and 74 % at £10. 1
f.o.b. Liverpool. White and cream, 60 % less abundant in supp
stand at £8. 7s. 6d. and £8. 5s. per ton respectively at makers' wor
Soda ash still very firm for ordinary carb, 48 % to 58 %. at 1½d.
deg. f.o.b.; caustic ash at 1⅜d. per deg. f.o.b. Soda cryst
£2. 17s. 6d. in casks on rails. Salt cake in bulk, 25s. per ton on ra
Chlorate of potash has given way, and can be bought from makers a
re-sellers at 4¾d. to 4⅞d. per lb. Vitriol and muriatic acids conti
in good demand, and prices are unaltered. Sulphur quiet, and pri
without change. Sulphate of copper somewhat irregular, and realis
from £24. 10s. to £25. 10s. for prompt delivery. The position f
ward is unchanged, acetates of lime moving slowly at £7. 17s. 6d.
£8. for brown and £13. 15s. for grey. Acetate of soda rather firm
Acetic acids quiet. Acetates of lead dull, but makers are firm, a
good white realises £23. 5s. to £23. 10s. per ton c.i.f. at usual por
brown, £18. 5s. to £18. 10s. Nitrate of lead also firm at £22.
£22. 10s. Nitrate of soda still offering at 8s. per cwt. Potash cau
and carbonate scarce, and firm for early deliveries. Arsenic in g
demand, and makers fully booked with orders. Carbolic acids d
and prices continue to droop. Tartaric acid, 1s. 3d. per lb. Th
are yet no signs of improvement in demand from the general shipp
or home trade.

REPORT ON MANURE MATERIAL.

The business during the past week has been very much of the sa
character as that throughout April—mainly for prompt or ea
delivery, and with falling prices, except in the case of nitrate of so
which closes firm at a shade above values a month ago. Min
phosphates continue without much alteration in the situation,
attitude of manure manufacturers being that of waiting for easier pri
for material while still busy getting out their manures. The mi
cargo of River Plate bones and bone ash, December sailing, name
week ago as having arrived at port of call, has been sold on priv
terms, but understood to be a shade under £5. 7s. 6d. for bones,
£4. 17s. 6d. or 70% for the bone ash.
East India Bone Meal is somewhat easier, £5. 7s. 6d. having b
accepted ex-store, Liverpool, and a shipment arrived subseque
hanging fire at the price. To arrive near at hand less would proba
do, and for shipment ahead £5. would be cabled out for chance
business. Some further sales of crushed Kurrachee bones have b
made at £5. 2s. 6d. per ton, ex-quay Liverpool, and that price wo
be accepted for 200 tons now in dock—perhaps £5. 1s. 3d. Sev
parcels of common grinding bones have been sold during the we
at £4. 5s. for a dirty parcel of Turkish, up to £4. 16s. 3d. for a lot
clean dry hard Brazils, there being still a demand for really good cl
bones, both for charring and grinding. Nitrate of soda is decide
firmer on the week, and large sales have been made both *on spot, at co*
and *for shipment*, at advanced prices. Spot is firm at 8s. 1½d.
good ordinary quality, and at 8s. 3d. for refined. 8s. 1½d. has b
paid for a refined cargo at port of call for the Mersey, and 8s. 0¾d
an ordinary cargo for U.K. Business has been done at from 8s. 4
for June sailing up to 8s. 9d. for September-October. Home p
pared dried blood continues to offer at 10s. 6d. per unit of ammo
free rails at works, and this price asked for River Plate in sec

·ound hoof is still to be had at 9s. per unit, f.o.r. at works, ·uld probably do for quantity.

getting scarcer, and 40s. now generally required.

·sphates are somewhat firmer, stocks having been much the continued drain upon them. Manufacturers seem in- ·mand higher prices for any orders outside their regular con-

·WEST OF SCOTLAND CHEMICALS.

GLASGOW, Tuesday.

·is not much of a front of improvement to show, we are at least ·off than we were a week ago. Caustic soda has continued the ·nd is now down to about the figures of the beginning of March ·buyers are once more showing some life and a very fair bulk ·assing. Orders for bichromate of potash and bichromate of ·ave been coming in more freely during the week, and the ·of the previous ten days has been partly made up for. In ·urning oil is, if anything, lower of vitality still, with Scotch ·cially upheld, and values to English consumers a halfpenny ·wer; but scale and wax are both very strong, and there is ·prospect of prices being further advanced by the combined ·of Scotland and America. This, if realised, will be followed ·candles, on the part of the associated English and Scotch ·In addition to caustic, soda ash also is the turning of the scale ·bleach with us is pretty firm as things go, and quotations ·ist maintained. Sulphate of ammonia has again weakened ·ptibly, and buyers may to-day supply themselves freely at ·spot Leith; while now also for forward, contracts are being ·w down as £11. 7s. 6d. Chief prices current are:—Soda ·os. 6d. net Tyne; alum in lump £5., less 2½% Glasgow; ·glish refined £29., and boracic acid, £37. 10s. net Glasgow; 1¼d. less 2½%, Tyne; caustic soda, white, 76°, 70/72°, £9., 60/62° £8. 5s., and cream, 60/62° £8., ·% Liverpool; bicarbonate of soda, 5 cwt. casks, £6. 5s., ·. casks, £6. 10s. net Tyne; refined alkali 48/52°, ·t Tyne; saltcake, 25s. 6d., to 27s.; bleaching powder, ·o £5. 15s. less 5% f.o.r. Glasgow; bichromate of potash, ·of soda 3d., less 5 and 6% to Scotch and English buyers ·y; chlorate of potash, 4⅞d., less 5% any port; nitrate of ·to 8s. 3d.; sulphate of ammonia, spot, £11. 5s. f.o.b. ·lammoniac, 1st and 2nd white, £37. and £35., less ·port; sulphate of copper, £25., less 5% Liverpool; ·cale, hard and soft, 2½d. per lb.; paraffin wax, 120°, ·d, 3⅝d.; paraffin spirit (naphtha), 8d. a gallon; paraffin ·g), 6¾d. to 7d. on rails, Glasgow for Scotch consumption ·lish orders done at about 1¼d. a gallon lower); ditto ·g), 865°, £5. 10s. to £6. per ton; 885°, £6. 10s.;—and ·£7. 15s. to £8. Week's imports of sugar at Greenock ·o bags.

·HE LIVERPOOL MINERAL MARKET.

·little change to report in our market this week. Manganese: ·ave rather fallen off, and prices continue firm. Magnesite: ·n to the previous stocks large arrivals have taken place and ·quite in hands of buyers. Raw ground, £6. 10s.; and ·round, £10. to £11. Bauxite (Irish Hill brand) in excep- ·rong demand at full prices—Lump, 20s.; seconds, 16s.; ·.; ground, 35s. Dolomite, 7s. 6d. per ton at the mine. ·alk: Arrivals this week are very small, and prices un- ·specially for G.G.B. "Angel-White" brand—90s. to 95s. ·00s. to 105s. superfine. Barytes (carbonate) easy; selected ·ip scarce at £6.; No. 1 lumps, 90s.; best, 80s.; seconds and ·70s.; smalls, 50s.; best ground, £6.; and selected crystal ·. Sulphate has somewhat improved, best lump, 35s. 6d.; good ·os.; medium, 25s. 6d. to 27s. 6d.; common, 18s. 6d. to 20s.; ·st white, G.G.B. brand, 65s.; common, 45s.; grey, 32s. 6d. ·umicestone quiet; ground at £10., and specially selected ·st quality, £13. Iron ore unchanged. Bilbao and Santan- ·10s. 6d. f.o.b.; Irish, 11s. to 12s. 6d.; Cumberland in ·hand at 18s. to 24s. Purple ore as last quoted. Spanish ·ous ore selling freely at good prices. Emery-stone: ·s still asked for, and business is being done at full prices, ·for prompt. No 1 lump is quoted at £5. 10s. to £6., and ·to £5. 10s. Fullers' earth steady; 45s. to 50s. for best blue ·; fine impalpable ground, £7. Scheelite, wolfram, tungstate ·d tungsten metal inquired for. Chrome metal, 5s. 6d. per lb. ·lloys 2s. per lb. Chrome ore: Still further arrivals, prices ·ntimony ore and metal have further advanced. Uranium ·. to 26s. Asbestos scarce, bringing increased prices. ·ad ore firm; smalls, £14. to £15.; selected lump, £16. to £17. Calamine: Best qualities scarce—60s. to 80s. Strontia steady; sulphate (celestine) steady, 16s. 6d. to 17s. Carbonate (native), £15. to £16.; powdered (manufactured), £11. to £12. Limespar: English manufactured, old G.G.B. brand in demand, and brings full prices; 50s. for ground English. Felspar, 40s. to 50s.; fluorspar, 20s. to £6. Bog ore in demand, firm at 22s. to 25s. Plumbago: steady; Spanish, £6.; best Ceylon lump at last quotations; Italian and Bohemian, £4 to £12. per ton; founders, £5. to £6.; Blackwell's "Mineraline," £10. French sand, in cargoes, continues scarce on spot—20s. to 22s. 6d. Ferro-manganese and silicon spiegel in good demand at stronger figures. Chrome iron, 20 per cent., £24. to £25. Ground mica, £50. China clay freely offering—common, 18s. 6d.; good medium 22s. 6d. to 25s.; best, 30s. to 35s. (at Runcorn).

THE LIVERPOOL COLOUR MARKET.

COLOURS.—Unaltered. Ochres: Oxfordshire quoted at £10., £12., £14., and £16.; Derbyshire, 50s. to 55s.; Welsh, best, 50s. to 55s.; seconds, 47s. 6d.; and common, 18s.; Irish, Devonshire, 40s. to 45s.; French, J.C., 55s., 45s. to 60s.; M.C., 65s. to 67s. 6d. Umber: Turkish, cargoes to arrive, 40s. to 50s.; Devonshire, 50s. to 55s. White lead, £21. 10s. to £22. Red lead, £18. Oxide of zinc: V.M. No. 1, £25.; V.M. No. 2, £23., Venetian red, £6. 10s. Cobalt: Prepared oxide, 10s. 6d.; black, 9s. 9d.; blue, 6s. 6d. Zaffres: No. 1, 3s. 6d.; No. 2, 2s. 6d. Terra Alba: Finest white, 60s.; good, 40s. to 50s. Rouge: Best, £24.; ditto for jewellers', 9d. per lb. Drop black, 25s. to 28s. 6d. Oxide of iron, prime quality, £10. to £15. Paris white, 60s. Emerald green, 10d. per lb. Derbyshire red, 60s. Vermillionette, 5d. to 7d. per lb.

THE TYNE CHEMICAL REPORT.

TUESDAY.

The Chemical Market has been decidedly dull during the past week, and prices generally have been gradually dropping every day. The chief feature has been the heavy fall in the price of caustic soda, brought about, principally by the collapse of the Liverpool market for this article. To-day's price is 20s. a ton under that of a week ago. Bleaching powder has dropped half-a-crown a ton, and continues quiet with little business doing. Soda ash is slightly easier in price by about 10s. 6d. per ton. Sulphate of soda is little enquired after, and prices are merely nominal. Soda crystals are the only exception to the general state of dulness. There is an excellent demand for them, and last week's quotations are fully maintained. The following are to-day's market prices:—Bleaching powder, in softwood casks, £5. 5s. per ton; caustic soda 77%. £10. per ton, ground in 3-4 cwt. barrels, £14. per ton; soda ash, 1¼d. per degree less 2½%; soda crystals, in 1 cwt. bags, net weight, £2. 13s. 6d. per ton, in 2 cwt. bags, net weight, £2. 11s. per ton; in casks, gross weight, £2. 11s. per ton; sulphate of soda, in bulk, £1. 15s. per ton; ground, in casks, £2. 5s. per ton; recovered sulphur, £4. 5s. per ton; chlorate of potash, 4¾d. per lb.; silicate of soda, 75° Tw., £2. 10s. per ton; 100° Tw., £3. 7s. 6d. per ton; 140° Tw., £4. per ton; hyposulphite of soda, 1 cwt. kegs, £5. 15s. per ton, in 5—7 cwt. casks, £5. 5s. per ton; pure white sulphate of alumina, £4. 10s. per ton; blanc fixe, £7. 10s. per ton; chloride of barium, £8. per ton; nitrate of baryta, crystals, £18. 15s. per ton; ground, £19. 5s. per ton; sulphide of barium, £5. 10s. per ton—all f.o.b. Tyne, or f.o.r. makers' works.

Manufacturing coals continue in strong demand, and prices remain very firm. Northumberland steam small has been advanced sixpence per ton during the past week, and are now quoted from 8s. 6d. to 8s. 9d. per ton. Durham small is quoted 10s. per ton.

Mr. James M'Culloch, manager of the Sulphuric Acid and Decomposing Departments of Messrs. Charles Tennant and Partners, Limited, Hebburn, has been appointed resident manager of the chemical works of Messrs. C. Tennant and Co., Limited, of St. Rollox, Glasgow.

New Companies.

KELSEY'S PATENT ADJUSTABLE HAT SYNDICATE, LIMITED.—This syndicate was registered on the 18th inst., with a capital of £6,000., in £1. shares, to acquire and work the letters patent granted to John Robinson Kelsey for the manufacture of soft fitting hats, and for the preparation and employment of ingredients for the purpose of such manufacture. The subscribers are:—

	Shares.
L. F. Marsh, Bristol, manufacturer	1
W. S. Halbert, 16, Ashburn-place, S.W.	
J. H. Wallery, Bristol, chartered accountant	1
A. S. Trotman, Bristol, solicitor	1
W. H. Wall, Bristol, traveller	1
R. H. N. G. F. Lambert, Bristol, merchant	1
C. C. Hinshelwood, 5, Portland-mansions, Oxford-street	1

VENEZUELA TELEPHONE AND ELECTRICAL APPLIANCES COMPANY, LIMITED.—This company was registered on the 22nd inst., with a capital of £70,000., in £1.

shares, to carry on in Venezuela the business of a telephone, telegraph, and electric company. The subscribers are :—

	Shares.
Sir Douglas Fox, Kingston-on-Thames	250
J. W. Phillips, M.P., 24, Queen Anne's-gate	250
L. Franke, Winchester-house	2,000
W. P. Van Laar, 41, Rue de Petites Ecurier, Paris, merchant	250
D. Antaiminson, Amsterdam, banker	500
J. Simpson, Bushey-view, Hampton Wick, merchant	100
H. P. Gilbert, 7, Poplar-grove, Shepherd's Bush-road	250

YACOBE PETROLEUM SYNDICATE, LIMITED.—This company was registered on the 18th inst., with a capital of £30,000., in £5. shares, to acquire and work petroleum properties in any part of Russia or elsewhere, and for such purposes to adopt an unregistered agreement of 27th ult., made with Sir Peter Tait. The subscribers are :—

	Shares.
Wm. Ellison Macartney, M.P., Palace-chambers, Westminster	1
F. Falden, M.P., Carlton Club	1
E. Hugovin, 12, Queen-street, E.C.	1
A. J. Miell, Claremont-road, Highgate	1
W. Whiten, 56, Plimsoll-road, Finsbury-park	1
G. S. Williams, 46, Huxley-road, Queen's-park	1
W. S. Newland, 32, Medora-road, Forest-hill	1

THE CHEMICAL SALT CO., LIMITED.—This company has been registered at Edinburgh, with a capital of £40,000., divided into 4,000 £10. shares, to take over and purchase from the firm of Charles Tennant and Co., St. Rollox, Limited, the lease of certain mines, pits, veins, and seams of salt and salt rock, and to carry on in the United Kingdom or elsewhere the trade or business of salt miners and masters, salt and chemical manufacturers and merchants. The first subscribers are :—

	Shares.
Charles Tennant, Bart	
Hugh Brown	
William McLure	
James King, Bart.	
Jonathan Thompson	
Francis John Tennant	
Thomas Alexander	

The Patent List.

This list is compiled from Official sources in the Manchester Technical Laboratory, under the immediate supervision of George E. Davis and Alfred R. Davis.

APPLICATIONS FOR LETTERS PATENT.

Mechanical Stokers. J. Proctor. 5,603. April 14.
Tin-Smelting Furnaces.—(Complete Specification). J. Letcher. 5,608. Ap. 14.
Furnace Bars and Grates. W. Fraser. 5,616. April 14.
Manufacture of Washing Soap. J. Bowden and S. Hern. 5,620. April 14.
Brewing. R. Haddan. 5,629. April 14.
Steam Generators. J. M. Stratton. 5,633. April 14.
Tin Smelting. T. Teague. 5,638. April 14.
Revolving Boilers used in Paper Making. H. J. Rogers and J. Paramore. 5,661. April 15.
Fermentation. J. Salomon. 5,673. April 15.
Automatic Regulation of Electric Currents. J. Kalb. 5,678. April 15.
Producing Low Tension Alternating Electric Currents. E. Manville and W. L. Madgen. 5,716. April 15.
Manufacture of Cement. W. R. Taylor. 5,719. April 15.
Colouring Matters. J. Hall. 5,721. April 15.
Diquinolyline Derivatives. B. Willcox. 5,722. April 15.
Phenacetine Derivatives. B. Willcox. 5,723. April 15.
Promoting Circulation in Water and Steam Boilers. O. Imray. 5, April 15.
Removable Fire-Bricks for Boiler Furnaces. O Imray. 5,734. Apri
Azo Colouring Matters Derived from Azoxyamines. J. Imray. 5 April 15.
Colouring Matters Derived from Fluores Ceine. J. Imray. 5,737. Apri
Testing Galvanometers. A. Jamieson. 5,748; April 16
Armatures for Dynamos. R. E. B. Crompton. 5,750. April 16.
Linseed Presses. H. King and W. Marshall. 5,765. April 16.
Steam Boilers. G. Ellams 5,766. April 16.
Grey Basic Colouring Matters. O. Imray. 5,777. April 16.
Black Colouring Matters. O. Imray. 5,780. April 16.
Heating and Cooling Liquids. H. H. Lake. 5,785. April 16.
Galvanic Batteries. F. L. Rawson. 5,810. April 17.
Substitute for Glass. S. Raudnitz. 5,815. April 17.
Wool Extracting. C. Womach. 5,828. April 17.
Refining Cotton and other Oils. G. Tall. 5,833. April 17.
Automatic Electric Controlling Apparatus. J. Radcliffe. 5,836. April
Distillation of Mineral Oils. C. M. Pielsticker. 5,838. April 17.
Rods of Parchmentized Fibre.—(Complete Specification). R. P. Frist. 5, April 17.
Generating Steam and Economising Fuel. W. B. Leachman. 5, April 17.
Manufacture of Slag Fibre. C. Wood. 5,847. April 17.
Utilising Exhaust Steam in Ammonia Refrigerators. J. A. Osenbri 5,855. April 17.
Regulating Supply of Oxygen and Hydrogen Gases to Burners. E. Presbrey. 5,861. April 17.
Evaporating Salt Water for Marine Boilers. J. Kerr. 5,891. April 19
Steam Generators. A. G. Brown. 5,897. April 19.
Insulating Electrical Conductors. A. J. Jarman. 5,910. April 19.
Polishing Compounds. J. Hill, senior, J. Hill, junior, and A. R. Hill. 5, April 19.
Secondary Batteries. P. Schoop. 5,920. April 19.
Steam Boilers. P. Oriolle. 5,922. April 19.
Smelting of Metals. W. C. Loe. 5,937. April 19.
Secondary Batteries. K. E. Boettcher. 5,938. April 19.
Extracting and Purifying Oils and Perfumes.—(Complete Specificati H. H. Lake. 5,940. April 19.
Manufacture of Ceramic Ware. H. M. Ashley. 5,970. April 19.
Distillation of Oils. J. Dewar and B. Redwood. 5,971. April 19.

Gazette Notices.

PARTNERSHIPS DISSOLVED.

ADELAIDE HART AND B. HART, under the style of Albert Hart, Houndsdi London, felt hat and cap manufacturers.
EDWARDS AND GEORGE, Newport, Mon., brewers.

THE BANKRUPTCY ACT, 1883.

Receiving Orders.

PERCY JOHN OTTEY, Burton-on-Trent, Chemist and Druggist.

IMPORTS OF CHEMICAL PRODUCTS

AT

THE PRINCIPAL PORTS OF THE UNITED KINGDOM.

LONDON.

Week ending April 22nd.

Acetone—		
Holland,	£50.	Beresford & Co.
Acetic Acid—		
Holland,	95 pks.	A. & M. Zimmermann
Arsenic—		
Germany,	£60	L. & I. D. Jt. Co.
Acetate of Lime—		
U.S.,	£205	W. Thatcher
Antimony—		
Japan,	35 t.	H. Rogers, Sons & Co.
Antimony Ore—		
A. Turkey,	5 t.	J. Ford & Co.
Spain,	10	H. Bath & Son
"	9	J. W. Cater, Sons & Co.
Austria,	61	H. R. Merton & Co.
A. Turkey,	5	French & Smith
Portugal,	50	H. Emanuel
Sydney,	9	H. Bath & Sons
A. Turkey,	38	H. Emanuel
Bones—		
Holland,	58 t.	E. Willson
Sydney,	10	Coml. Bkg. Co., of Sydney
U. S.,	15	Hicks, Nash & Co.
Morocco,	12	T. Clark
Trebizonde,	19	Culverwell, Brooks & Co.
Bone Meal—		
E. Indies,	250 t.	Gonne, Croft & Co.
Barytes—		
Belgium,	117 cks.	A. Zumbeck & Co.
"	11	J. L. Lyon & Co.
Germany,	23	O'Hara & Hoar
"	66	D. Storer & Sons, Ld.
Holland,	22	H. Harrison & Co.
"	3	O'Hara & Hoar
Carbonic Acid—		
Holland,	12 pks.	J. G. Fischer
Caoutchouc—		
France,	34 c.	L. & I. D. Jt. Co.
E. Indies,	111	Ross & Deering
"	74	Henderson Brs.
Colon,	4	Lewis & Peat
"	2	L. & I. D. Jt. Co.
Cape,	7	W. M. Smith & Sons
S. Africa,	6	J. Owen
"	10	Hammond & Co.
Mauritius,	107	L. & I. D. Jt. Co.
France,	4	Kebbel, Son & Co.
Inhambane,	16	"
Tamatane,	150	
"	718	Union Lighterage Co.
Germany	2	L. & I. D. Jt. Co
Cream of Tartar—		
Italy,	10 pks.	B. & F. Wf. Co.
"	5	F. Smith
Spain,	2	B. & F. Wf. Co.
"	20	W. C. Bacon & Co.
Copper Ore—		
Natal,	9 t.	H. Bath & Son
Chemicals *(otherwise undescribed)*		
Germany,	£135	A. & M. Zimmermann
U.S.,	296	Lister & Biggs
Germany,	15	F. Stahlschmidt & Co.
"	178	W. Balchin
"	5	H. J. Perlbach & Co.
"	70	T. H. Lee
"	382	A. & M. Zimmermann
France,	15	Cartnutts & Co.
Germany,	43	W. Burton & Sons
"	30	G. Steinhoff
Holland,	5	Phillipps & Graves
Dyestuffs—		
Orchella		
France,	27 bls.	Kebbel, Son & Co.
Valonia		
A. Turkey,	158 t.	A. & W. Nesbitt
"	167	Hicks, Nash & Co.
"	220	J. Graves
"	42	Anning & Cobb
"	100	Boutcher, Mortimore & Co.
"	50	Adam I
"	75	Hicks, Nash &
Fustic		
Belize,	5 t.	A. Ha
Logwood		
Mexico,	25 t.	C. Leary &
Cochineal		
Canaries,	49 pks.	Zinton &
"	22	W. M. Smith & S
Myrabolams		
E. Indies,	317 pks.	Marsh F
"	300	Arbuthnot, Ewa
"	700	Ide & Chri
Dyestuffs *(otherwise undescribed*		
Germany,	19 pks.	Fleming's Oi Chemical Co., Limi
France,	5	C. Atkins & Nis
"	150	Hemingway &
Denmark,	1	R. J. Fullwooc Bl
Bahia,	40 t.	L. & I. D. Jt.
E. Indies,	120	Benekendorff &
"	210	W. A. Bowdi
"	80	Falkenburg & He
Indigo		
E. Indies,	64 chts.	L. & I. D. Jt.
"	4	F. Huth &
"	3	Patry & Past
"	122	L. & I. D. Jt.
"	45	W. Brandt, Sons &

9 srns. L. & I. D. Jt. Co.
18 Patry & Pasteur
18 A. Juninez & Sons
11 chts. L. & I. D. Jt Co.
23 "
90 E. H. Gauntlett
10 L. & I. D. Jt. Co.
1 srn. "
96 chts. "
11 Patry & Pasteur
5 G. N. Ry. Co.
10 Phillipps & Graves

162 pks. J. H. & G Scovell
257 S. Barrow & Bros.

Bark
293 t. Leach & Co.
10 H. Henle's Sucrs.
50 J. Vicary & Sons
y, 227 Baxter & Hoare
40 A. & W. Nesbitt
28 Devitt & Helt

30 pks. T. Ronaldson & Co.

360 "
103 Best, Ryley, & Co.
200 A. J. Humphrey
10 Elkan & Co.
365 L. J. Levinstein & Sons

pks. Swanston & Co.

ks. Thames S. Tug Co.
B. Jacob & Sons

37 pks. Brown & Elmslie

400 pks. 400 c. J. Barber & Co.
15 90 M. D. Co.
100 800 T. M. Duche & Son
25 200 L. Sutro & Co.
00 600 "
50 300 Prop. Scott's Whf
00 840 J. H. Epstein
25 200 L. Sutro & Co.
45 354 "
12 110 L. & I. D. Jt. Co.
25 301 Props. Chambs. Whf.
20 160 T. M. Duche &
00 200 J. Barber & Co.
33 1,333 W. R. Spence
00 500 Barrett, Tagant, & Co.
48 380 J. Knill & Co.
95 2,650 Anderson, Weber, & Smith
25 220 J. H. Epstein
88 846 Seaward Bros.
25 204 Pillow, Jones, & Co.

£115 Beresford & Co.
520 Spies, Bros., & Co.

ha—
315 c. Kaltenbach & Schmitz
158 Kleinwort, Sons, & Co.
380 J. H. & G. Scovell
1 Lavington Bros.

2 t. Anglo Contl. Guano Wks.

pks. L. & I. D. Jt Co.
Lewis & Peat
L. & I. D. Jt. Co.
Props. Hay's Whf.

osphate
02 t. Lawes Manure Co., Lmtd.
530 t. J. Burnett & Sons
50 A. Hunter & Co.
199 Ang. Cont. Guano Wks.

,281 cu. bgs. A. Ford & Co.

Bicarbonate—
pks. A. & M. Zimmermann

Sulphate—
cks. Weatherley, Mead, & Co.
102 t. Anglo Contl. Guano Wks.

Plumbago—
E. Indies, 37 cks. H. Lambert
" 80 brls. Doulton & Co.
" 160 Sawer Mead, & Co.
" 139 T. Jordan & Co.
" 193 J. Thredder, Son, & Co.
" 348 cks. L. & I. D. Jt. Co.
" 384 brls. J. Forsey
Germany, 37 cks. B. Gallaway

Paraffin Wax—
U. S., 565 brls. H. Hill & Sons
" 165 G. H. Frank
Germany, 44 cs. Grossmith & Luboldt
U. S., 1,350 brls. J. H. Usmar & Co.

Potassium Bromide—
Germany, £95 L. & I. D. Jt. Co.

Pearl Ash—
France, 200 c. F. W. Berk & Co.
" 140 Petri Bros.

Potassium Permanganate—
Holland, £56 Spies, Bros., & Co.

Pyrites—
Spain, 2,041 t. C. Tennant, Sons, & Co.
Portugal, 1,150 Mason & Barry, Lmtd.

Saltpetre—
Germany, 915 cks. P. Hecker & Co.
" 100 T. Merry & Son
E. Indies, 796 bgs. Henckell du Buisson & Co.

Stearine—
Germany, 145 cks. W. Balchin
France, 55 Beresford & Co.
" 55 sks. J. Goddard & Co.

Tartaric Acid—
Italy, 57 pks. B. Jacob & Sons
Holland, 20 B. & F. Wf. Co.
France, 11 "
Holland, 18 "
" 5 Northcott & Sons

Tartars—
Italy, £2,380 Thames S. Tug Co.
" 1,009 Seager, White, & Co.

Ultramarine—
Germany, 2 cs. H. Brooks & Co.
Holland, 8 Becker & Ulrich
Belgium, 6 Leach & Co.

Verdigris—
France, £78 Skilbeck Bros.

Zinc Oxide—
Germany, 50 cks. W. A. Rose & Co.

LIVERPOOL

Week ending Apr. 24th.

Albumen—
Genoa, 10 cs. H. Fehr & Co.

Acetic Acid—
Rouen, 2 cks. R. J. Francis & Co.
" 10 "
Rotterdam, 6 bskts.

Alum—
Rotterdam, 40 cks.
" 6

Antimony Ore—
Marseilles, 668 bgs.

Borate of Lime—
Antofagasta, 1,530 bgs. Cockbain, Allardice, & Co.

Bleaching Powder—
Ancona, 1 brl.

Borax—
Leghorn, 127 cks.

Bones—
Syra, 2,096 bgs. C. Nicolaidis
Constantinople, 110 bgs.
Smyrna, 35 t.
Colon, 418 bgs. A. Dobell & Co.
Sidon, 180
Tripoli, 85 t.
Valparaiso, 673 sks. Jones & Roberts
Constantinople, 12 bgs. Jivan, Tekian, & Co.
" 249 Jivan Tekeian
" 25 t.

Barytes—
Antwerp, 41 bgs.

Caoutchouc—
Ambriz, 13 cks. Taylor, Laughland, & Co.
Ambrizette, 9 "
Muculla, 47 37 bgs. "
Kinembo, 5 "
" 5 T. Millington
" 3 Samson & de Liarge
Mussera, 4 "
Ambrizette, 77 "
Muculla, 99 "
Loango, 10 Edward. Bros.
Salt Pond, 3 brls. Lintott, Bros., & Co.
" 1 N. P. Nathan & Son
" 1 W. Griffiths & Co.
" 1 1 cs. F. & A. Swanzy
" 2 1 F. Schaeffer & Co.
" 4 W. B. M'Iver & Co.
" 2 3 Edwards Bros.
Addah, 2 cks. Pitt, Bros., & Co.
" 1 Radcliffe & Durant
" 22 pns. J. P. Werner
Cape Coast, 4 Davies, Robbins, & Co.
" 1 A. Breslaur & Co.
" 1 T. E. Tomlinson
" 2 F. J. Eaton & Sons
" 8 A. Miller, Son, & Co.
" 2 J. H. Rayner & Co.
" 14
" 1 J. Swift
" 1 Edwards Bros.
" 2 3 cks 2 cts. W. B. M'Iver & Co.
" 2 Radcliffe & Durant
" 1 pn. 14 3 cts. Fletcher & Fraser
" 6 1 A. Miller, Bros., & Co.
Assinee, 13 bgs. 1 brl. Millward, Bradbury, & Co.
Sierra Leone, 24 cks. Paterson, Zachonis, & Co.
" 48 Pickering & Berthoud
" 1 bg. W. C. Taylor
" 29 Cie Francaise
Pernambuco, 2 bls. 14 brls.
C.C. Castle, 5 pks. W. B. M'Iver & Co.
" 2 brls. A. Millerson & Co.
" 2 Fletcher & Fraser
" 4 pks. A. Miller, Bros., & Co.
" 2 cks. 1 cs. C. L. Clare & Co.
" 2 2 A. Reis
" 6
Grand Bassam, 1 Grimwade, Ridley, & Co.
" 1 Schaeffer & Co.
" 3 A. Reis
Sierra Leone, 12 Cie Francaise
" 4 Paterson, Zachonis, & Co.
" 15 Pickering & Berthoud
Bathurst, 11 Cie Francaise
" 5 A. Cros & Co.
" 3 A. Reis
Cameroons, 12 R. & W. King
" 3 Rider, Son, & Co.
" 4 A. Herschell
" 2 pns. Lucas, Bros., & Co.
Addah, 1 cs. 2 cks. 3 pns. W. J. Rodatz
" 1 bg. 23 Radcliffe & Durant
" 23 J. P. Werner
Accra, 1 brl. 3 pns. F. & A. Swanzy
Salt Pond, 2 brls. 1 cks. W. B. M'Iver & Co.
" 20 Miller, Bros., & Co.
" 2 Radcliffe & Durant
" 3 cs. W. Griffiths & Co.
" 1 ck. F. & A. Swanzy
Colon, 3 brls. A. Dobell & Co.
" 1 J. M. Williams
Mussera, 4 bgs. Samson & de Liagre
Ambrizette, 4 cks. Taylor, Laughland, & Co.
" 3 J. Holt & Co.
" 29 bgs.
Muculla, 14 cks. Taylor, Laughland, & Co.
" 44 bgs. Sampson & de Liagre
Quinzao 152 "
Banana, 4 brls. Daumas & Co.
Black Point, 2 cks. 1 cs. Pinto Leite, & Nephew
Loango, 4 Edwards Bros.
Sette Cama, 23 J. Holt & Co.
" 4 1 brl. R. W. Roulston
Cape Lopez, 38 cks. J. Holt & Co.
Gaboon, 6 "
Elobey, 16 "
Gaboon, 7 Daumas & Co.
" 21 A. V. Heyder
Old Calabar, 26 African Association
" 5 brls. E. Grand
Accra, 1 1 cs. A. Reis
" 13 cks. 2 brls. D. P. Pearce
" 8 J. F. Morton
" 2 Davies, Robbins, & Co.
Salt Pond, 2 1 cks. Pickering & Berthoud
" 3 1 W. B. M'Iver & Co.
Cape Coast, 2 brls. 1 cs. W. D. Woodin
" 1 1 Kronig & Liegler
" 1 1
" 8 Iblers & Bell
" 2 A. Millerson & Co.
" 1 Davies, Robbins, & Co.
Secondee, 3 pns. Paterson, Zachonis, & Co.
" 26 Broklehurst, Son, & Co.
" 27 brls. Pickering & Berthoud
" 13 2 pks. 7 pns.
Dixcove, 1 1 cs. "
" 1 bg 1 Radcliffe & Durant
" 1 brl. 2 F. & A. Swanzy
Barcelona, 6 bls. Arkle & Mounsden

Charcoal—
Rouen, 40 bgs. Co-op. W'sale Socy, Lmtd.

Cream Tartar—
Barcelona, 5 hds. F. Reid & Co.

Chemicals (*otherwise undescribed*)
Rouen, 9 cks. Co-op. W'sale Socy, Lmtd.
Rotterdam, 12 cks.

Dyestuffs—

Argols
Bordeaux, 24 sks.

Sumac
Barcelona, 100 bgs. A. Ruffer & Son
" 50
Palermo, 450 bgs.

Logwood
Boston, 250 bxs.
Belize, 340 Brownbill & Co.
Jamaica, 40 t.

Valonia
Syra, 35 bgs. Bucura & Economi
Smyrna, 407 t. Barry Bros.
" 100 Morsi N. Frauses
Patras, 225½ t. C. G. Papayanni & Jeremias
Marathonisi, 116 t. Papayanni & Jeremias
Syra, 956 bgs. "
Smyrna, 30 bgs. 200 t.

Extracts
Nantes, 30 cks. R. J. Francis & Co.
Havre, 100 20 cs.
New York, 10 bxs. 11 brls.
Rouen, 540 cks. Co-op. W'sale Soc., Ld.
Marseilles, 65 brls.
Philadelphia, 75 175 bxs.

Myrabolams
Bombay, 667 bgs.
" 340 W. & R. Graham & Co.
" 500

Fustic
Carthagena, 1,553 ps. Mildred, Goyenecbe
Colon, 1,337 A. Dobell & Co.
" 941 "

Logwood
New York, 10 brls.
" 42

Glucose—
Hamburg, 200 bgs. 25 cks.

Horn Piths—
Hamburg, 25 t.

Manganese—
Carthagena, 1,950 t. Wigan Coal & Iron Co.

Manure—
Bruges, 30 bgs.

Manganese Ore—
Coquimbo, 40 t.
Carrizal, 940

Nitrate—
Calais, 566 t.

Phosphoric Acid—
Bremerhaven, 16 cs.

Potash—
Rouen, 16 cks. Co-op. W'sale Society, Limited
Dunkirk, 9

Phosphates—
Bruges, 510 bgs. J. T. Fletcher & Co.
" 510

Pitch—
Rotterdam, 28 cks.
" 10

Paraffin Wax—
Baltimore, 100 brls.
Sugar of Lead—
Rouen, 34 cks. R. J. Francis & Co.
Sodium Nitrate—
Iquique, 8,220 bgs.
Pisagua, 11,580 F. Lacioz
Stearine—
Antwerp, 38 bgs.
New York, 50 hds.
Tartar—
Bordeaux, 2 cks.
" 3
Tartaric Acid—
Rotterdam, 12 cks.
" 11
Marseilles, 1 brl.
Ultramarine—
Rotterdam, 6 cs.
Waste Salt—
Hamburg, 350 t.
Zinc Ashes—
Boston, 19 brls.
Zinc Oxide—
Antwerp, 5 brls. J. T. Fletcher & Co.
" 85

HULL.

Week ending April 24th.

Aluminium Sulphate—
Rotterdam, 40 cks. W. & C. L. Ringrose
Acid *(otherwise undescribed)*
Rotterdam, 12 cks. W. & C. L. Ringrose
" 1 G. Lawson & Sons
Albumen—
Hamburg, 10 cks. Wilson, Sons, & Co.
Arsenic—
Hamburg, 100 kgs. "
Barytes—
Antwerp, 100 bgs. Wilson, Sons, & Co.
Rotterdam, 23 cks.
Ghent, 40 H. F. Pudsey
Bremen, 19 cks. Blundell, Spence, & Son
" 23 Tudor & Son
Bones—
Trieste, 3 pks. Wilson, Sons, & Co.
Chemicals *(otherwise undescribed)*
Genoa, 10 cks. Wilson, Sons, & Co.
Stettin, 42 "
Bremen, 3 2 cs. Veltmann & Co.
Dyestuffs—
Sumac
Palermo, 1,110 bgs. Wilson, Sons, & Co.
" 28
Extracts
Bordeaux, 50 cks. Rawson & Robinson
Hamburg, 80
Trieste, 40 brls. Wilson, Sons, & Co.
Fiume, 260 "
Rennet
Rotterdam, 8 cks. G. Lawsen & Sons
Orchella
Lisbon, 9 bgs. Bailey & Leetham
Alizarine
Rotterdam, 24 pks. W. & C. L. Ringrose
" 31 "
" 25 cks. Hutchinson & Son
" 6 "
" 8 W. & C. L. Ringrose
Glucose—
Dunkirk, 40 bgs. Wilson, Sons, & Co.
Hamburg, 21 cks. Woodhouse & Co.
" 21 "
" 100 bgs. G. Lawson & Sons
" 200 J. Cowburn
" 5 cks.
Stettin, 5 Wilson, Sons, & Co
Hamburg, 17
" 5
Stettin, 31
Hamburg, 100 cks. C. M. Lofthouse
" 15 Woodhouse & Co.
Horn Piths—
Stockholm, 1 bkt.
Manure—
Kainit
Hamburg, 508 bgs. J. Dalton, Holmes, & Co.
" 500 A. E. Peacock
Naphthol—
Rotterdam, 3 brls. Hutchinson & Son
" 1 bx. Wilson, Sons, & Co.
Naphtha—
Hamburg, 3 tns. 5 kgs. R. T. Bruce & Co.
Oxide—
Rotterdam, 5 cs. Hutchinson & Son
Hamburg, 8 G. Malcolm & Son
Pitch—
Rotterdam, 24 cks. Hutchinson & Son
Danzig, 46 Wilson, Sons, & Co.
Phosphate—
Dunkirk, 200 bgs. "
Ghent, 320 t. 204 bgs. "
Phosphorus—
Rotterdam, 8 drms. T. F. Bell & Co.
Rouen, 150 cs. Wilson, Sons, & Co.
Potash—
Rotterdam, 34 cks W. & C. L. Ringrose
Dunkirk, 54 pks. 51 drms. Wilson, Sons, & Co.
Antwerp, 12 cks. "
Pyrites—
Huelva, 1,504 t. J. Dalton, Holmes, & Co.
Plumbago—
Genoa, 16 cks. Wilson, Sons, & Co.
Sulphate—
Antwerp, 25 cks. H. F. Pudsey
Silicate—
Antwerp, 3 cks. Wilson, Sons, & Co
Sulphur—
Marseilles, 20 bgs.
Catania, 207 brls. 400 bgs. 1,709 pks.
Saltpetre—
Rotterdam, 170 bgs. Wilson, Sons, & Co.
Sodium Nitrate—
Iquique, 6,034 bgs.
Tartar—
Bordeaux, 2 cks. Rawson & Robinson
Tar—
Dunkirk, 7 cs. Wilson, Sons, & Co.
Stockholm, 10 brls. "
Ultramarine—
Rotterdam, 3 cks.
Waste Salt—
Hamburg, 105 bgs. C. M. Lofthouse & Co.

GLASGOW.

Week ending April 24th.

Albumen—
Rotterdam, 7 cks. J. Rankine & Son
Barytes—
Rotterdam, 46 cks.
Barytes Sulphate—
Antwerp, 60 bgs. 6 cks.
Chloral Hydrate—
Rotterdam, 1 cs. J. Rankine & Son
Dyestuffs—
Logwood
Rotterdam, 1 ck. "
Alizarine
Rotterdam, 86 cks.
Extracts
Nantes, 90 cks.
Antwerp, 3
Madders
Bordeaux, 5 J. & P. Hutchinson
Glycerine—
Hamburg, 40 drums.
Glucose—
Hamburg, 100 cs.
Manure—
Phosphate Rock
Antwerp, 100 t.
Potash—
Rotterdam, 11 cks. J. Rankine & Son
" 20
" 5 "
Potassium Carbonate—
Hamburg, 45 cks.
Sodium Silicate—
Rotterdam, 1 cs. J. Rankine & Son
Sodium Sulphate—
Antwerp, 100 cks.
Tartar—
Bordeaux, 2 cks. J. & P. Hutchinson
" 2
" 10 "
" 5
Tartaric Acid—
Rotterdam, 3 cks. J. Rankine & Son
Ultramarine—
Rotterdam, 4 cks. 20 cs. J. Rankine & Son
Waste Salt—
Hamburg, 1,100 t. Lesler, Beck, & C
" 508 cks.
Zinc Oxide—
Antwerp, 50 cks.

TYNE.

Week ending April 24th.

Chemicals *(otherwise undescribe*
Rotterdam, 1 ck. Tyne S. S. C
Glycerine—
Hamburg, 5 cs. Tyne S. S. C
Glucose—
Rotterdam, 1 ck. Tyne S. S. C
Saltpetre—
Hamburg, 7 cks. Tyne S. S. C
Salt—
Hamburg, 132 t. 6 bgs. "
Sulphur Ore—
Pomaron, 1,500 t. Mason & Ba
Drontheim, 670 R. S. Story & C

GOOLE.

Week ending April 16th.

Alum—
Rotterdam, 16 cks.
Caoutchouc—
Boulogne, 2 cks.
Dyestuffs—
Alizarine
Rotterdam, 1 ck.
Logwood
Hayti, 320 t.
Dyestuffs *(otherwise undescribed,*
Boulogne, 96 cs. 47 cks. 3 pks.
Glycerine—
Rotterdam, 3 pks.
Potash—
Calais, 20 cks.
Dunkirk, 2
Potassium Muriate—
Hamburg, 184 bgs.
Waste Salt—
Hamburg, 132 t. 508 bgs.

GRIMSBY.

Week ending April 19th.

Albumen—
Antwerp, 30 cs.
Chemicals *(otherwise undescribe*
Dieppe, 10 cks.
Glucose—
Hamburg, 28 cks.
Manure—
Hamburg, 500 bgs.

EXPORTS OF CHEMICAL PRODUCTS

OF

THE PRINCIPAL PORTS OF THE UNITED KINGDOM.

LONDON.

Week ending April 19th.

Ammonium Carbonate—
Yokohama, 10 c. £20
B. Ayres, 35
" 1 t. 40
St. Petersburg, 300 kgs. 15 cks. 531
Gutujewski, 40 30 190
Ammonium Sulphate—
Dunkirk, 20 t. £230
Cologne, 15 173
Rotterdam, 60 720
Trinidad, 15 6 c. 194
Martinique, 15 4 220
Demerara, 25 3 307
Bordeaux, 29 6 360
Arsenic—
Wellington, 1 t. 10 c. £20
Benzole—
Rotterdam, 19 c. £767
Benzoic Acid—
New York, £33
Carbolic Solid—
Rotterdam, 5 t. £397
Cream of Tartar—
E. London, 4 c. £26
Melbourne, 1 t. 110
Adelaide, 10 50
Melbourne, 20 kgs. 102
Adelaide, 10 57
Brisbane, 12 70
Copper Sulphate—
Naples, 5 t. £122
Genoa, 25 1 c. 635
Treport, 10 215
Genoa, 30 666
Naples, 5 119
Hobart, 2 44
Genoa, 5 120
" 5 120
Treport, 7 1 175
Carbolic Acid—
Rotterdam, 25 cks. £80
Le Treport, 2 t. 5 c. 18
Carbolic Solid—
Valencia, 14 c. £100
Barcelona, 1 t. 140
Citric Acid—
Amsterdam, 2 kgs. £15
Hamburg, 15 c. 112
Cape Town, 1½ 12
Hamburg, 2 15
Monte Video, 1 ck. 36
Marseilles, 10 101
Melbourne, 10 80
Brussels, 10 74
Hamburg, 4 140
Melbourne, 20 kgs. 41
Hong Kong, ½ 4
Melbourne, 20 138
Montreal, 2 15
Riga, 1 8
Cologne, 5½ 30
Caustic Soda—
New Orleans, 120 t. 3 c. £
Algoa Bay, 9
Sydney, 2 2
Caustic Potash—
Sydney, 1 t. 13 c.
Coal Products—
New York, 51 rults.
Cette, 1,008 t. 1,
Dunkirk, 672 1,
" 250
" 150
Caen, 146
Cette, 979 1,
Dunkirk, 483
" 642 1,
Rouen, 15 brls.
Odessa, 200
Konigsberg, 135 t.
Cette, 1,199 2,
Tar
Karachi, 100 rnlts.
Stockholm, 25 cks.
Bombay, 61 rnlts.

Benzole
Rotterdam, 62 cks. £969
" 30 pns. 436
" 39 cks. 767
" 47 drms. 30 cks. 1,148

Anthracene
Dusseldorf, 93 cks. £445
Calais, 40 737
Hamburg, 16 drms. 319
Dusseldorf, 134 cks. 743
Rotterdam, 130 718

Naphthalene
New York, 54 pks. £180
" 26 cks. 33
Treport, 39 50 bgs. 90
Christiania, 1 cs. 12

Solvent Naphtha
Ghent, 8 cks. £21

Carbolic Acid
Rotterdam, 40 cks. £397
Treport, 10 28
Stettin, 25 165

Toluole
Rotterdam, 19 cks. £433

Aniline
Melbourne, 1 cs. £13

Creosote Salts
Philadelphia, 150 t. £320

Chemicals (*otherwise undescribed*)
Melbourne, 10 cs. 1 pk. £121
Calcutta, 25 brls. 150
Demerara, 1 cs. 90
Adelaide, 15 kgs. 25
Rio Janeiro, 20 pks. 45

Disinfectants—
Paris, 10 cks. £50
Antwerp, 25

Guano—
Trinidad, 200 t. £1,000
Morlaix, 160 1,600
Trinidad, 2 18 c. 29
Jamaica, 10 8 174

Hydrochloric Acid—
Trinidad, 1 t. 10 c. £6

Manure—
Riga, 35 t. 10 c. £285
Cullera, 363 11 3,630
Martinique, 24 12 245
Singapore, 20 130
St. Vincent, 50 500
Hamburg, 49 5 170
Bordeaux, 20 75
Dantzig, 19 14 110
Rouen, 60 805
" 15 80
Bordeaux, 35 7 262
Barbadoes, 750 1,480
St. Lucia, 32 264
Berbice, 6 63
Demerara, 18 189
Trinidad, 25 1 62

Nitric Acid—
Jersey, 2 t. 8 c. £43

Oxalic Acid—
Antwerp, 5 t. 3 t. £175
New York, 3 1 100
Sydney, 5 92

Potassium Chlorate—
Gothenburg, 3 t. £205
Sydney, 5 415
New York, 3 130
Oporto, 10 c. 25

Superphosphate—
Riga, 50 t. £150
Bordeaux, 100 401

Sulphur—
Lisbon, 250 brls. £193

Saltpetre—
Azores, 12 c. £13
Melbourne, 2 t. 10 60
Corunna, 2 44

Salammoniac—
Antwerp, 1 t. £34
Hamburg, 1 2 c. £45
Varna, 1 1 35
Salonica, 1 1 35
Bourgas, 1 4 40

Saccharum Cake—
Dunkirk, 10 t. 4 c. £25

Sheep Dip—
E. London, 225 drs. £63
" 75 28
" 10 c. 96
Natal, 278 pks. 126
Pt. Elizabeth, 1 t. 19 c. 70
Algoa Bay, 42 cs. £90
" 200 gls. 41
" 250 58

Saltcake—
Antwerp, 125 t. £220

Sodium Sulphate—
Ghent, 65 t. £65
Norrkoping, 100 200

Sodium Hyposulphite—
Melbourne, 3 t. £24

Sulphur Flowers—
Algoa Bay, 6 t. 15 c. £59

Sodium Bicarbonate—
Melbourne, 2 t. 6 c. £21

Sulphuric Acid—
Hamburg, 1 t. £20

Tartaric Acid—
Brisbane, 10 c. £75
Melbourne, 30 kgs. 171
Adelaide, 20 86
Montreal, 4 33
Dunedin, 5 27
Melbourne, 5 38
" 35 pks. 173
Adelaide, 15 56

LIVERPOOL.

Week ending April 23rd.

Alkali—
Dantzic, 18 cks.

Antimony Regulus—
Rotterdam, 3 t. 4 c. £360

Ammonium Carbonate—
Bahia, 9 c. £26
Bilbao, 10 2 q. 48
San Sebastian, 4 8
Genoa, 1 t. 10 49
Amsterdam, 4 7
Odessa, 1 1 36
Portland, 19 1 27

Alum Cake—
Montreal, 20 t. 5 c. £44

Ammonium Muriate—
Antwerp, 1 c. 1 q. £9

Asphaltum—
Boston, 7 t. 16 c. £40

Ammonium Sulphate—
Valencia, 66 t. £738
" 30 3 c. 1 q. 339
" 50 4 2 650

Alum—
Odessa, 2 t. 1 c. £11
Salonica, 4 25
Melbourne, 1 4
Madras, 15 17 1 q. 80
Smyrna, 7 2 3 35
Wellington, 1 1 5
Cape Town, 4 16 47
Montreal, 1 19 8
Oporto, 5 2 24
Saloura, 1 2 6
Volo, 1 8 2 8

Bleaching Powder—
Antwerp, 32 cks.
Baltimore, 130
Barcelona, 10
Boston, 692
Calcutta, 130 c.
Chicago, 37
Dunkirk, 18
Genoa, 402
Ghent, 163
Hamburg, 170
Montreal, 35
New York, 605
Vera Cruz, 90
Antwerp, 25 cs.
Bombay, 5
Bordeaux, 35
Boston, 597
Hamburg, 29
Lisbon, 35
Malaga, 10
Montreal, 70 pks.
Newfairwater, 57
New York, 233 20
Philadelphia, 90 tcs.
Rotterdam, 14 30
Rouen, 94
Trieste, 30

Bone Waste—
Rouen, 2 t. £15

Boracic Acid—
Rotterdam, 1 t. 2 c. 2 q. £98

Bone Ash—
Stettin, 174 t. 1 c. 1 q. £870

Brimstone—
New York, 99 t. 16 c. £415
Rotterdam, 11 18 3 q. 80

Borax—
San Sebastian, 5 c. 3 q. £9
Stettin, 19 3 29
Rouen, 29 t. 6 816
Malaga, 4 2 7
Christiania, 5 18 115

Caustic Soda—
Antwerp, 10 drms. 29 brls.
Barcelona, 250 40
Bilbao, 18
Calcutta, 106
Carthagena, 20
Hamburg, 35
Brazil, 20
Lisbon, 14
Malaga, 20
Malta, 10
Maranhão, 25
New York, 649
Oporto, 43
Philadelphia, 136
Portland, 50
Port Natal, 25 bxs.
Seville, 20
Valencia, 25
Vigo, 10
Montreal, 83
Callao, 20
Carthagena, U. S. 69
" Spain, 47 65 pks.
Constantinople, 11 brls.
Genoa, 183
Halifax, 15
Havana, 50
La Guayra, 14
Lisbon, 19
Malta, 100
Marseilles, 20
Melbourne, 26 pks.
Montreal, 136
Para, 11
Pernambuco, 5
Philadelphia, 130 450 brls.
Santander, 20
Santos, 20
Stettin, 25 bxs. 20 10 pks.
Yokohama, 8

Caustic Potash—
Montreal, 3 t. 13 c. £53
New York, 7 15 132
Sydney, 1 12 30
Cadiz, 11 12

Charcoal—
Hamburg, 150 t. £750
Rotterdam, 50 3 c. 200
Hamburg, 55 9 280

Chlorate—
New York, 3 t. £163

Carbolic Solid—
Amsterdam, 8 c. £30
New York, 5 t. 1 q. 425

Copperas—
Santos, 1 t. 9 c. 2 q. £3

Calcium Chloride—
New York, 4 t. 19 c. £10
Sydney, 20 22

Copper Sulphate—
Nantes, 62 t. 2 c. 1 q. £1,401
Ancona, 6 138
Barcelona, 20 3 400
Bilbao, 5 3 93
" 20 3 430
Constantinople, 12 2 30
Bordeaux, 10 17 1 288
" 15 13 3 320
Dunkirk, 4 19 3 280
Genoa, 17 2 342
Leghorn, 2 18 30
Marseilles, 10 3 176
Genoa, 19 4 1 380
Trieste, 5 2 106
Barcelona, 21 3 3 411
Tarragona, 13 1 2 265
Leghorn, 14 2 14
Genoa, 33 3 4 644
" 8 10 151
Vera Cruz, 1 18 47
Barcelona, 10 9 1 421
Alicante, 1 18
Almeria, 1 15 34
Tarragona, 2 36
Barcelona, 25 5 2 610
Bilbao, 10 1 240
Bordeaux, 123 7 1 2,560
Brindisi, 5 18 105
Genoa, 153 2 2,850
Leghorn, 8 7 2 130
Venice, 33 10 2 635
Passages, 10 2 193
Bilbao, 5 1 100
Nantes, 53 6 1,395
Rouen, 41 13 905
Valencia, 15 8 1 351
Genoa, 58 15 2 1,350
Venice, 23 15 3 390

Citric Acid—
Barcelona, 1 t. £143

Carbolic Acid—
New York, 5 t. 10 c. £100
Boston, 2 193
New York, 1 18 1 q. 26
" 2 206

Disinfectants—
Calcutta, 30 cks. 37

Pernambuco, 10 cs.
Portland, 180
Rotterdam, 472
Santos, 175
Sierra Leone, 190
Sydney, 12
Valparaiso, 50
Prince Edward Island, 152 t. 6 c.
Quebec, 1,110 t. 9 c.
St. John, N.B., 125
San Francisco, 59
Shediac, 229½
Sherbro, 99½
Sierra Leone, 279½
Trinidad, 42
Tuborg, 698
Valparaiso, 100

Soda Ash—
Baltimore, 288 cks. 1,114 tcs.
Barcelona, 40
Boston, 531 273
B. Ayres, 200 brls.
Constantinople, 100
Coquimbo, 16
Corfu, 89
Genoa, 52
Ibrail, 20
Lisbon, 106
Madrid, 28
Oporto, 25 tcs.
Philadelphia, 245 cks.
Santos, 10
Sydney, 40 brls.
Alicante, 2
Baltimore, 232
Boston, 1,157
Canada, 350 tcs.
Corfu, 11
Genoa, 35
Leghorn, 102
Monte Video, 20
New York, 808
Philadelphia, 44
Rio Janeiro, 50
Smyrna, 100
Varna, 100

Sodium Silicate—
Alexandria, 10 cks.
Bilbao, 16
Carthagena, 6 65 pks.
Odessa, 28
Para, 5
Venice, 16

Soda Crystals—
Boston, 750 cks.
Constantinople, 50 brls.
Hamburg, 100 bgs.
Malta, 50
New York, 750 cks.
St. John's, N.B., 225 brls.
St. John's, Nfld., 45 kgs.
Alexandria, 40 brls.
Boston, 500 kgs.
Callao, 60
Charlotte Town, 15 cks.
Constantinople, 50
East London, 30
New York, 420
Rotterdam, 20
St. John's, Nfld., 50
Wellington, 10

Sodium Bicarbonate—
Antwerp, 20 kgs.
Astoria, 100
Auckland, 32
Copenhagen, 25 2 cks.
Danzic, 24 brls.
Dunedin, 100
Fairwater, 240
Genoa, 60
Gothenburg, 4 cks.
Hamburg, 20
Havre, 14
Lisbon, 30
Melbourne, 74
Rotterdam, 30
St. John's, N.B., 50
Stettin, 28
Wellington, 170
Antwerp, 1
Canda, 775 35
Charlotte Town, 50
Melbourne, 200
Rouen, 5 t. 8 c.
Trieste, 70

Soda Salts—
Montreal, 6 t. 18 c. £195

Saltcake—
Baltimore, 31 t. 1 c. £128
Ghent, 100 140
Boston, 100 6 213
Philadelphia, 37 t. 15 c. 80

Sodium Biborate—
Antwerp, 2 c. 1 q. £5
Palermo, 4 3 8
Rotterdam, 6 1 9

Sodium Sulphate—
Lyttleton, 5 t. 1 c. £37
Wellington, 7 14 54

Sulphur—
Boston, 50 t. £217
" 50 2 c. 225
" 25 1 106
San Luis, 13 t. 10 c. 3 q. 87
Lisbon, 7 7 1 59
Boston, 50 2 225
" 50 2 225

Sulphuric Acid—
New York, 1 t. 2 c. 2 q. £65

Sal Acetos—
Trieste, 4 c. 3 q. 16

Salammoniac—
Rouen, 1 t. 10 c. £23
Melbourne, 17 3 q. 33
Montreal, 1 35
Alexandria, 10 17
Syra, 4 3 9
Beyrout, 1 1 38
Bilbao, 8 14
Boston, 5 1 172
Calcutta, 3 19 1 136
Trieste, 1 1 36
Valparaiso, 6 1 11
Alexandretta, 4 19

Sheep Dip—
East London, 1 t. 10 c. £80
Cape Town, 2 10 135
Natal, 1 10 80

Saltpetre—
Almeria, 12 c. £14
Barcelona, 3 t. 11 1 q. 65
Montreal, 5 5
Syra, 1 18 2 43
Portland, 4 7 3 98
Piræus, 7 10 163
Pernambuco, 2 45
Halifax, 10 11
Montreal, 1 21

Sodium Nitrate—
Odessa, 2 t. 11 c. 22
Fiume, 6 6 3 61
Alicante, 3 16 1 32
Malaga, 6 7 50
Valencia, 3 2 3 25

Sheep Wash—
New York, 2 t. 10 c. £85
Monte Video, 7 11
Mossel Bay, 15 40

Tannic Acid—
Rio de Janeiro, 1 c. £8

Tin Crystals—
Barbadoes, 17 c. 1 q. £80

Zinc Chloride—
Boston, 1 t. 5 c. 1 q. £27

TYNE.

Week ending April 24th.

Alkali—
Christiania, 5 t. 8 c.
New York, 12 17
Rotterdam, 1 19
Carthagena, 16 14
New York, 100 8

Alum—
Carthagena, 5 t. 9 c.

Antimony—
New York, 30 t.

Barytes Carbonate—
Antwerp, 10 t.
Stettin, 164 5 c.
Copenhagen, 2

Baryta Chlorate—
Hamburg, 5 c.

Bleaching Powder—
Gothenburg, 10 t. 18 c.
Antwerp, 9
Hamburg, 36 9
Dunkirk, 4 18
Pasages, 20 4
New York, 140 6
Rotterdam, 22 2
Stettin, 222 15
Riga, 11 4
Quebec, 18 16
Aarhuus, 9 19
Horsens, 1 2
Carthagena, 3 4
Copenhagen, 19 2
Libau, 20 5

Caustic Soda—
Gothenburg, 2 t. 4 c.
Antwerp, 5 3
Pasages, 29
New York, 100 2
Quebec, 14 10
New York, 205 4

Gypsum—
New York, 72 t. 7 c.
" 36 11

Litharge—
Stettin, 7 t. 19 c.
Quebec, 35 4

Magnesia—
Hamburg, 9 c.
Genoa, 15

Napthalene—
Riga, 25 t. 7 c.

Potassium Chlorate—
New York, 10 t.

Pitch—
Copenhagen, 5 t.

Phosphorus—
Malmo, 4 c.

Sulphur—
Hamburg, 1 t.

Sodium Sulphate—
Gothenburg, 10 t. 4 c.
Quebec, 1
Libau, 20 11

Soda Ash—
Gothenburg, 5 t. 4 c.
Antwerp, 6 12
Hamburg, 13
Venice, 10 14
New York, 115 1
Stettin, 10
Riga, 3 4
New York, 100 11

Soda—
Aarhuus, 7 t. 6 c.
Frederikshavn, 12 15
Gothenburg, 4 18
Christiania, 5 16
" 5 16
New York, 413
Rotterdam, 5 1
Quebec, 341 17
Copenhagen, 16 2
New York, 154

Sodium Chlorate—
Rotterdam, 1 t. 2 c.

Sodium Hyposulphite—
Quebec, 2 t. 5 c.

HULL.

Week ending Apr. 22nd.

Alkali—
Konigsberg, 11 drums.
New York, 191 cks.
Trieste, 19

Barytes—
Marseilles, 25 t.

Bleaching Powder—
Christiania, 34 cks.
Stettin, 135

Chemicals *(otherwise undescribed)*
Amsterdam, 20 cks. 32 pks.
Bergen, 3
Bari, 50
Christiania, 5 1 cs.
Copenhagen, 6 1
Constantinople, 50
Dunkirk, 454
Danzig, 50 1
Gothenburg, 5 36
Ghent, 2
Genoa, 5 22 26
Helsingfors, 6
Hamburg, 20 11 159
Konigsberg, 10
Lisbon, 1 5
Libau, 5
Leghorn, 5
Marseilles, 50
New York, 54 35 33
Odessa, 384 80
Rotterdam, 202 40 70
Reval, 5
Stettin, 9 5
Stockholm, 1
Trieste, 31 50

Caustic Soda—
Danzig, 5 drms.
Libau, 70

Dyestuffs *(otherwise undescribed)*
Antwerp, 1 ck.
Hamburg, 1 1 cs. 2 drums.
Rotterdam 7 1 1
Reval, 3

Manure—
Ghent, 647 bgs.
Hamburg, 479
Rotterdam 734

Pitch—
Genoa 162 t.

Salt—
Bergen, 1 ck.

Tar—
Amsterdam, 10 cks.
Bergen, 2
Drontheim, 10
Gothenburg, 155
Hamburg, 5
Stettin, 6
Odessa, 200

GLASGOW.

Week ending April 24th.

Ammonium Sulphate—
Trinidad, 114 t. 18 c. £1,024
Demerara, 10 19 150
" 12 13 137

Bleaching Powder—
Brisbane, 13 t. 1 c. £96

Cream of Tartar—
Brisbane, 22 c. £29

Caustic Soda—
New York, 204 c. £58

Chemicals *(otherwise undescribed)*
Penang, 3 t 8¾ c. £36. 18s.

Dyestuffs—
Logwood
Dunedin, 2 t. £26. 10s.

Magnesium Silicate—
Penang, 2 t. 1 c. £31. 10s.

Potassium Bichromate—
Italy, 18 t. 244¾ c. £985
Rotterdam, 13 cks. 184
" 1 16
New York, 496 c. 926
Antwerp, 322¾ 495
Stockholm, 2 cks 22
Karachi, 1 t. 6 c. £46

Pitch—
Sable D'Ohnne, 1,000 t.
St. Nazaire, 900

Paraffin Wax—
Italy, 14½ c. £20
Dunedin, 48¾ 11 lbs. 100

Soda Ash—
Baltimore, 1,817 c. £430

Sodium Silicate—
Sydney, £33

Sodium Bichromate—
Naples, 15 t. 5 c. £435
New York, 506 £709

Sulphuric Acid—
Bombay, 4 t. 10 c. £48
Karachi, 167

Tartaric Acid—
Brisbane, £38

Tar—
St. John's, N. F., £23

GRIMSBY.

Week ending April 19th.

Alkali—
Dieppe, 6 cks.

Coal Products—
Dieppe, 30 cks.
Malius, 25

Chemicals *(otherwise undescribed)*
Antwerp, 2 cks.
Dieppe, 3 cs. 30 153 pks.
Hamburg, 10

Manure—
Phosphate Rock
Malius, 25 t.

GOOLE.

Week ending April 16th.

Benzole—
Hamburg, 15 drums.

Chemicals *(otherwise undescribed)*
Ghent, 3 cks.
Hamburg, 67

Coal Products—
Hamburg, 38 cks.
Rotterdam, 74

Dyestuffs *(otherwise undescribed)*
Boulogne, 2 cks. 8 drums.
Ghent, 10 8
Hamburg, 1 cs.

Manure—
Boulogne, 43 bgs.
Rotterdam, 490

Pitch—
Antwerp, 146 t.
Ghent, 146

PRICES CURRENT.

WEDNESDAY, April 30, 1890.

Prepared by Higginbottom and Co., 116, Portland Street, Manchester.

…s stated are F.O.R. at maker's works, or at usual ports of shipment in U.K. The price in different localties may vary.

		£	s.	d.
…5 %, and 40 %	per cwt	7/3 & 12/6		
…lacial	,,	67/6		
S.G., 2000°	,,	0	12	9
82 %	nett per lb.	0	0	6½
	,,	0	0	3¾
… (Tower Salts), 30° Tw.	per bottle	0	0	8
(Cylinder), 30° Tw.	,,	0	2	11
…° Tw.	per lb.	0	0	2
	,,	0	0	1¾
	nett ,,	0	0	4
	,,	0	1	2½
…ic (fuming 50 %)	per ton	15	10	0
(monohydrate)	,,	5	10	0
(Pyrites, 168°)	,,	3	2	6
(,, 150°)	,,	1	8	0
(free from arsenic, 140/145°)	,,	1	10	0
…ous (solution)	,,	2	15	0
(pure)	per lb.	0	1	0½
	,,	0	1	3
…hite powdered	per ton	13	0	0
…se lump)	,,	4	15	0
…Sulphate (pure)	,,	5	0	0
,, (medium qualities)	,,	4	10	0
…m	per lb.	0	15	0
…, ·880=28°	per lb.	0	0	3
=24°	,,	0	0	1¾
Carbonate	nett ,,	0	0	3½
Muriate	per ton	23	0	0
,, (sal-ammoniac) 1st & 2nd	per cwt.	37/- & 35/-		
Nitrate	per ton	40	0	0
Phosphate	per lb.	0	0	10½
Sulphate (grey), London	per ton	11	6	3
,, (grey), Hull	,,	11	2	6
…il (pure)	per lb.	0	0	11
…alt ,,	,,	0	0	9
…	per ton	75	0	0
(tartar emetic)	per lb.	0	1	1
(golden sulphide)	,,	0	0	10
…hloride	per ton	8	0	0
…arbonate (native)	,,	3	15	0
…ulphate (native levigated)	,,	45/- to 75/-		
… Powder, 35 %	,,	5	0	0
Liquor, 7 %	,,	2	10	0
…le of Carbon	,,	13	0	0
…n Acetate (crystal	per lb.	0	0	6
Chloride	per ton	2	2	6
…ay (at Runcorn) in bulk	,,	17/6 to 35/-		
… Dyes :—				
…ne (artificial) 20 %	per lb.	0	0	9
…ta	,,	0	3	9
…r Products				
…cene, 30 % A, f.o.b. London	per unit per cwt.	0	1	5
…, 90 % nominal	per gallon	0	3	6
50/90 ,,	,,	0	2	7
…c Acid (crystallised 35°)	per lb.	0	0	9
,, (crude 60°)	per gallon	0	2	3
…te (ordinary)	,,	0	0	2½
(filtered for Lucigen light)	,,	0	0	3
Naphtha 30 % @ 120° C.	,,	0	1	3
Oils, 22° Tw.	per ton	3	0	0
f.o.b. Liverpool or Garston	,,	1	8	0
…t Naphtha, 90 % at 160°	per gallon	0	1	9
…n Oils (crude)	,,	0	0	2½
…hili Bars)	per ton	49	0	0
…lphate	,,	25	0	0
…xide (copper scales)	,,	52	10	0
… (crude)	,,	30	0	0
(distilled S.G. 1250°)	,,	55	0	0
	nett, per oz.	8¾d. to 9d.		
…hate (copperas)	per ton	1	10	0
…hide (antimony slag)	,,	2	0	0
…eet) for cash	,,	14	0	0
…harge Flake (ex ship)	,,	15	10	
…etate (white ,,)	,,	23	5	
,, (brown ,,)	,,	18	5	
…rbonate (white lead) pure	,,	19	0	
…rate	,,	22	5	8

		£	s.	d.
Lime, Acetate (brown)	,,	7	17	6
,, ,, (grey)	per ton	13	15	0
Magnesium (ribbon and wire)	per oz.	0	2	3
,, Chloride (ex ship)	per ton	2	5	0
,, Carbonate	per cwt.	1	17	6
,, Hydrate	,,	0	10	0
,, Sulphate (Epsom Salts)	per ton	3	0	0
Manganese, Sulphate	,,	18	0	0
,, Borate (1st and 2nd)	per cwt.	60/- & 42/6		
,, Ore, 70 %	per ton	4	10	0
Methylated Spirit, 61° O.P.	per gallon	0	2	2
Naphtha (Wood), Solvent	,,	0	3	10
,, ,, Miscible, 60° O.P.	,,	0	4	7½
Oils :—				
Cotton-seed	per ton	22	10	0
Linseed	,,	25	0	0
Lubricating, Scotch, 890°—895°	,,	7	5	0
Petroleum, Russian	per gallon	0	0	5⅛
,, American	,,	0	0	5⅛
Potassium (metal)	per oz.	0	3	10
,, Bichromate	per lb.	0	0	4
,, Carbonate, 90% (ex ship)	per ton	19	0	0
,, Chlorate	per lb.	0	0	4⅜
,, Cyanide, 98%	,,	0	2	0
,, Hydrate (Caustic Potash) 80/85 %	per ton	22	10	0
,, ,, (Caustic Potash) 75/80 %	,,	21	10	0
,, ,, (Caustic Potash) 70/75 %	,,	20	15	0
,, Nitrate (refined)	per cwt.	1	2	6
,, Permanganate	,,	4	2	6
Prussiate Yellow	per lb.	0	0	9½
,, Sulphate, 90 %	per ton	9	10	0
,, Muriate, 80 %	,,	7	15	0
Silver (metal)	per oz.	0	3	10⅛
,, Nitrate	,,	0	2	7
Sodium (metal)	per lb.	0	4	0
,, Carb. (refined Soda-ash) 48 %	per ton	6	10	0
,, ,, (Caustic Soda-ash) 48 %	,,	5	5	0
,, ,, (Carb. Soda-ash) 48%	,,	5	15	0
,, ,, (Carb. Soda-ash) 58 %	,,	6	17	6
,, ,, (Soda Crystals)	,,	2	15	6
,, Acetate (ex-ship)	,,	15	15	0
,, Arseniate, 45 %	,,	11	0	0
,, Chlorate	per lb.	0	0	6¾
,, Borate (Borax)	nett, per ton.	30	0	0
,, Bichromate	per lb.	0	0	3
,, Hydrate (77 % Caustic Soda)	per ton.	11	10	0
,, ,, (74 % Caustic Soda)	,,	10	10	0
,, ,, (70 % Caustic Soda)	,,	9	0	0
,, ,, (60 % Caustic Soda, white)	,,	8	10	0
,, ,, (60 % Caustic Soda, cream)	,,	8	5	0
,, Bicarbonate	,,	6	0	0
,, Hyposulphite	,,	5	10	0
,, Manganate, 25%	,,	9		0
,, Nitrate (95 %) ex-ship Liverpool	per cwt.	0	8	0
,, Nitrite, 98 %	per ton	28	0	0
,, Phosphate	,,	15	15	0
,, Prussiate	per lb.	0	0	7¾
,, Silicate (glass)	per ton	5	7	6
,, ,, (liquid, 100° Tw.)	,,	3	17	6
,, Stannate, 40 %	per cwt.	2	0	0
,, Sulphate (Salt-cake)	per ton.	1	12	0
,, ,, (Glauber's Salts)	,,	1	10	0
,, Sulphide	,,	7	15	0
,, Sulphite	,,	5	5	0
Strontium Hydrate, 98 %	,,	8	0	0
Sulphocyanide Ammonium, 95 %	per lb.	0	0	8
,, Barium, 95 %	,	0	0	5¾
,, Potassium	,,	0	0	8
Sulphur (Flowers)	per ton.	7	15	0
,, (Roll Brimstone)	,,	6	0	0
,, Brimstone : Best Quality	,,	4	5	0
Superphosphate of Lime (26 %)	,,	2	10	0
Tallow	,,	25	10	0
Tin (English Ingots)	,,	96	10	0
,, Crystals	per lb.	0	0	6¾
Zinc (Spelter)	per ton.	21	0	0
,, Chloride (solution, 96° Tw.	,,	6	0	0

THE

CHEMICAL TRADE JOURNAL.

Publishing Offices: 32, *BLACKFRIARS STREET, MANCHESTER.*

No. 155. **SATURDAY, MAY 10, 1890.** **Vol. VI**

Contents.

Notices.

All communications for the *Chemical Trade Journal* should be addressed, and Cheques and Post Office Orders made payable to—

DAVIS BROS., 32, Blackfriars Street, MANCHESTER.

Our Registered Telegraphic Address is—

"Expert, Manchester."

The Terms of Subscription, commencing at any date, to the *Chemical Trade Journal,*—payable in advance,—including postage to any part of the world, are as follow:—

Yearly (52 numbers) .... **12s. 6d.**
Half-Yearly (26 numbers) .... **6s. 6d.**
Quarterly (13 numbers) .... **3s. 6d.**

Readers will oblige by making their remittances for subscriptions by Postal or Post Office Order, crossed.

Communications for the Editor, if intended for insertion in the current week's issue, should reach the office not later than **Tuesday Morning.**

Articles, reports, and correspondence on all matters of interest to the Chemical and allied industries, home and foreign, are solicited. Correspondents should condense their matter as much as possible, write on one side only of the paper, and in all cases give their names and addresses, not necessarily for publication. Sketches should be sent on separate sheets.

We cannot undertake to return rejected manuscripts or drawings, unless accompanied by a stamped directed envelope.

Readers are invited to forward items of intelligence, or cuttings from local newspapers, of interest to the trades concerned.

As it is one of the special features of the *Chemical Trade Journal* to give the earliest information respecting new processes, improvements, inventions, etc., bearing upon the Chemical and allied industries, or which may be of interest to our readers, the Editor invites particulars of such—when in working order—from the originators; and if the subject is deemed of sufficient importance, an expert will visit and report upon the same in the columns of the *Journal.* There is no fee required for visits of this kind.

We shall esteem it a favour if any of our readers, in making inquiries of, or opening accounts with advertisers in this paper, will kindly mention the *Chemical Trade Journal* as the source of their information.

Advertisements intended for insertion in the current week's issue, should reach the office by **Wednesday morning** at the latest.

Advertisements.

SOCIETY OF CHEMICAL INDUSTRY.

LONDON SECTION.

AN ordinary meeting of the section was held on Monday, May 5tl at Burlington House, Mr. David Howard in the chair. M William Webster's paper on the electrical treatment of sewage havin been postponed *sine die.* Mr. Watson Smith first communciated a no on a reaction for distinguishing gallic from tannic acid that had recentl appeared in Liebig's Annalen. Treatment with phenylhydrazine an caustic soda develops a blue colour with tannic acid, and a golde yellow with gallic acid. The former fades to a yellowish tint on stanc ing. The reaction having been shown satisfactorily, excited som comment, and in reply to questions Mr. Watson Smith said that present it was a purely qualitative test, and that the precise chang occurring was not yet worked out. The same author then read hi paper on the "Chemical Aspects of the Dinsmore Process." The prc cess was described, and its advantages in the manufacture of coal-g pointed out in a paper read before the Liverpool section of the Societ in 1889, and may be briefly described as a method of enriching the g at the expense of the tar. The idea is old, a patent of which th main object appears to have been the disposal of an objectionabl by-product, viz., tar, having been taken out in 1830; since then others have been granted, but have not achieved success. Even i the original Dinsmore process before the modifications due t Mr. Carr and Mr. Pritchard were introduced, the plan of runnin the tar direct into the carbonising retort was productive c troublesome stoppages. The improved method now worked con sists in passing the crude gas before the tar accompanying i has been condensed, with a heated retort, thus gasifying muc of the tar and enriching the gas in illuminants. While ordinar town gas containing one third of Dinsmore gas has a composition c 50.60 % of H, 33.39 % of CH_4, and 4.37 % of olefines, pure Dinsmor gas contains 43.98 % H, 44.34 % of CH_4, and 6.76 % of olefines. Th latter is also free from carbon dioxide, and gains in illuminating powe from this cause also. Distinguishing the Dinsmore gas made at th Widnes gasworks as A, and the town gas containing one third c Dinsmore gas as B, the following contrasts are observed:—

(1). Composition of the tars obtained.

	A.		B.
Water	1.1 %	..	7.1 %
Light oils	1.3 ,,	..	5.4 ,,
Creosote oils	16.5 ,,	..	17.8 ,,
Anthracene oils....	12 1 ,,	..	8.6 ,,
Pitch	69.0 ,,	..	61.1 ,,
Sp. gr.......	1.157 ,,	..	1.150 ,,

The small quantity of light oils and the fact that the creosote oi yielded 2.5 % of tar acids in the case of A, while B gave 5.4 % ar very significant. From A, 0 79 % of real anthracene, while from B onl 0.32 % can be obtained. The pitch from A had a softening point c 67·5°C., and a melting point of 94° C., while the figures for B were 68°C and 96°C. respectively; the tensile strength of that from A was 150lbs and from B 132 lbs. per sq. in. The difference in the quality and yiel of gas is very considerable—9,000 c. ft. of 15 c. p. for B, as again 9,800 c. ft. of 20-21 c. p. for A. The industrial importance of th process appears when the growing scarcity of cannel is considerec The increase in illuminating power may be ascribed to the breakin down of benzene into acetylene, and the reduction of phenols to the corresponding aromatic hydrocarbons. Bethelot has proved th former reaction to occur, and the latter can even be used to prepare benzene, free from thiophen, and is practicable as a lectur experiment.

In opening the discussion, the Chairman pointed out the serio result to colour makers of the scarcity of aromatic hydrocarbons th the general adoption of the process would entail, but welcomed it as being a step in the direction of allowing gas engineers and chemists to regulate the quality and output of gas and its by-products, according to the current market prices of the several commodities. Professor Foster, having studied the process chiefly from the point of view of the increased luminosity of the gas, gave the results of his experience, and ascribed the gain in illuminating power to the breaking up of hydrocarbons of the paraffin series into those of the olefine series and free hydrogen. A tangible advantage was that cheaper coal could be used, even such as was not generally looked upon as of gas-making quality. Mr. Scudder pointed out that the greater illuminating power of Dinsmore gas might well be due to its freedom from CO^2 and N, and the probable presence of benzene vapour. Dr. C. Alder Wright remarked on the discrepency between the sp. gr. of the gas as given by Prof. Foster and its composition. Prof. Foster, in replying, touched on the weak point in the comparison instituted between the two kinds of gas, as the coal used in their production was not identical.

The hour being late, the paper on the chemistry of hypochlorite Bleaching, by Messrs. Cross and Bevan was postponed, and Mr. Tyrer read a note by an American member, Mr. Martin L. Griffin, on "Moisture in Paper Pulp." The author having commented on the extremely vague and unsatisfactory nature of the term "air-dried," suggested as a standard resulting from very numerous observations during all sorts of weather, the allowance of 6 % of water on the undried pulp, *i.e.* 6.38% on the dried pulp, and states that it has been largely accepted by manufacturers. The difficulty of making a fair deduction, based on the determination of moisture in a sample drawn from the outer portion of a bale packed dry, and then exposed to damp air, thus absorbing water in its exterior layers, having been pointed out by Dr. Wright, the meeting was adjourned.

MANCHESTER SECTION.

The sixth meeting of the present session of this Society was held last evening at the Victoria Hotel, Dr. Bailey (Owens College) presiding. Dr. Liebmann read a paper on pure aniline and its homologues. He said that since the discovery of aniline in 1826 it had formed the basis of much investigation, but there were always new facts connected with it cropping up. On the strength of scientific research he had devised a method for the analysis of these products, which he hoped to further elucidate. A brief discussion took place upon the paper, and the thanks of the meeting were accorded to Dr. Liebmann. Mr. J. Carter Bell followed with a paper on the liability of certain oils to spontaneous combustion. He said that for the past four months he had been making a great many experiments upon oils, to see whether some definite conclusion could be arrived at as to the liability of certain oils to show the phenomenon of spontaneous combustion. He was sorry to say that the results were not very concordant. Mr. Bell described in detail the experiments he had made with various oils and wools, and gave the results obtained in each. There was, he said, amongst some manufacturers an opinion that white wools when greased were more liable to heat than dyed wools. There was some reason to think that that might be so, but his experiments proved exactly the reverse. They were quite opposed to the theory that undyed wools would heat sooner than dyed wools, and light wools sooner than dark. He had never been able to get wool to burst into a flame, it always smouldered away. He had made inquiries as to whether people had ever seen a case of spontaneous combustion—the whole thing bursting into a flame—and he did not think that such cases had really been seen in mills. The experiments showed, however, the great importance of thoroughly sweeping up and clearing all places where wool was kept. A discussion followed, and the meeting concluded with the usual votes of thanks.

SCOTTISH SECTION.

The usual monthly meeting of this section was held in the rooms of the Philosophical Institution, Edinburgh, on the evening of Tuesday, Mr. J. H. Beilby in the chair. The papers of the occasion con- [illegible] of the second instalment of Dr. J. B. Readman's exposition of [illegible] ufacture of phosphorus, in the first part of which the process [illegible] cterised as a costly one, both from wear and tear of plant and [illegible] loss of material at various stages, and an account of [illegible] and other explosives by Mr. G. MacRoberts, which [illegible]. Dr. Readman's second reading entered into details of some of the experiments carried [illegible] endeavouring to simplify the process of [illegible] experiments have spread over a period of [illegible] the possibility of superseding the employment [illegible] altogether, by the use of silica and abandoning [illegible] Specimens of waste slags illustrating the [illegible] exhibited, and typical results were given in [illegible] every case The patents taken out for the [illegible] facture were discussed as far as they bore on the [illegible] the paper. Mr. MacRoberts entered at considerable [illegible] comparative energies of the now numerous and various [illegible] blasting gelatine, the most powerful of all, down to [illegible] of gunpowder. The superiority of the new explosives [illegible] than that of energy, was dwelt upon, emphasis being [illegible] parative powerlessness (in some cases) of water or [illegible] the explosive faculty, and on the comparative [illegible] nitro glycerine compounds, commonly governs the [illegible] itself.

ENERGY OF EXPLOSIVES.

QUANTITY OF EXPLOSIVE USED IN MORTAR—[illegible]

Name of Explosive.	Ingredients.
1. No. 1 Dynamite	Kieselguhr .... Nitro-Glycerine ....
2. No. 1 Blasting Gelatine	Nitro-Glycerine .... Nitro-Cotton ....
3. Nitro-Glycerine	Nitro-Glycerine ....
4. Gelatine Dynamite....	Nitro-Glycerine, 92.3% Nitro-Cotton, 7.7% Nitrate of Potash .... Wood Meal ....
5. Gelignite	Nitro-Glycerine, 94.8% Nitro-Cotton, 5.2% Nitrate of Potash Wood Meal ....
6. Detonator Mixture....	Chlorate of Potash........ Fulminate of Mercury ....
7. Fulminate of Mercury..	Fulminate of Mercury ....
8. Ammonia Powder	Nitrate of Ammonia........ Wood Meal .... Sulphur........
9. Nitrate of Ammonia Powder	Nitrate of Ammonia........ Picric Acid ....
10. Nitrate of Potash and Picric Powder	Nitrate of Potash .... Picric Acid ....
11. Nitrate of Soda and Picric Powder......	Nitrate of Soda .... Picric Acid ....
12. Nitrate of Soda and Picric Powder	Nitrate of Soda .... Picric Acid ....
13. Nitrate of Soda and Picric Powder	Nitrate of Soda .... Picric Acid ....
14. Nitrate of Ammonia and Ferrocyanide of Potassium	Nitrate of Ammonia .... Ferrocyanide of Potassium........
15. "Securite"	Nitrate of Ammonia........ Dinitrobenzol ....
16. Nitrate of Potash and Dinitrobenzol	Nitrate of Potash .... Dinitrobenzol ....
17. Chlorate of Potash and Dinitrobenzol	Chlorate of Potash........ Dinitrobenzol ....
18. Chlorate of Potash and Paraffin	Chlorate of Potash........ Paraffin ....
19. Chlorate of Potash and Dinitrocotton	Chlorate of Potash........ Dinitrocotton ....
20. Dinitrocotton	Dinitrocotton ....
21. Maxim's Powder......	Nitrate of Potash .... Sulphur .... Paraffin ....
22. Gunpowder	Gunpowder, RLG4 A ....
23. Gunpowder	Gunpowder, RLG4 D ....
24. Gunpowder	Gunpowder, RLG4 F ....
25. Gunpowder	Gunpowder, Brown, C H ....
26. Vul Powder, Blasting..	Ferrocyanide of Potassium .... Nitrate of Potash .... Solid Paraffin .... Peroxide of Iron........ Charcoal ....
27. "Roburite"	Nitrate of Ammonia .... Dinitrobenzol .... Moisture, &c. ....
28. Tonite..............	Gun Cotton .... Nitrate of Barium .... Nitrate of Potash .... Carbonate of Soda .... Moisture ....
29. "Potentite"	Gun Cotton .... Nitrate of Potash .... Moisture ....
30 Gunpowder (for cannon)	Gunpowder ....

At the close, thanks were awarded to both [illegible]

THE USE OF NITRIC ACID FOR THE PREPARATION OF CHLORINE.

FROM time to time the chlorine manufacturer has been assailed with rumours of new processes that were to revolutionise the trade. One of these rumours is now current, and, if it is to be believed, the problem of the economical use of nitric acid in the preparation of chlorine has at length been solved by Messrs. Davis Bros., of Manchester, who have just taken out a batch of patents on the subject. The various reactions that take place between nitric and hydrochloric acids have long been known, and the student of Gmelin is surprised when he considers how little has been added in recent years to our knowledge in this direction. The difficulties in the way of a nitric-acid chlorine process are mainly difficulties of engineering, and are such as the mere chemist is not likely to overcome. Since the days of Dunlop no very serious attempt was made to surmount these obstacles, till Taylor in 1844 took out his patent for the preparation of chlorine gas, and the syndicate which supported him erected an experimental plant at Bristol. That this was not an unqualified success will surprise no one who has studied the matter. Taylor had forgotten to deal with the water formed during the reactions, and, presumably, had forgotten, like Dunlop, to calculate how much vitriol would be required to absorb the oxides of nitrogen produced by the decomposition. Moreover, his experimental plant was erected in such a manner that the very elements of chemical construction were conspicuous only by their absence.

The process patented by Donald in 1887 has perhaps seen scarcely so much light as even that of Taylor. The patentee has made two steps in the right direction—by effecting the decomposition of the hydrochloric acid by nitric acid in the presence of sulphuric acid in order to eliminate the water of the reaction, and in using cold dilute nitric acid as an absorbent for the nitrogen oxides. It is rather curious that in their specifications both Taylor and Donald omit all recognition of the air in the furnace gases, which must needs be an important factor in a process which is essentially one of oxidation and reduction. Donald's process was worked for some months on a small scale at Widnes, but although it had the resources and the energy of the house of Muspratt behind it, the results were not such as to give either it or its inventor a permanent abiding place in that salubrious town. The single fact that Donald proposes to work his gases through columns of liquid equal in the aggregate to some 40 feet of water pressure, will provide much food for the reflective mind that appreciates the difficulty of constructing and connecting vessels to withstand the corrosive action of the various gases and liquids with which they will have to come into contact. Messrs. Davis see their way to avoid this stumbling-block, and also declare the possibility of dispensing with the refrigerating apparatus that plays a prominent part in the Donald process. It remains to be seen what will be their measure of success.

THE BAKU PETROLEUM WELLS.

WRITING from Batoum, on March 8th last, Mr. D. R. Peacock, the British Consul at that port, deals with the statements made about the alleged threatened exhaustion of the Baku petroleum wells. To the multitudes of people capitalists and labourers employed the exhaustion of the wells would be an irreparable calamity. However, so far as the yearly quantities of crude oil yielded during the last decade may serve as a criterion, the Baku petroleum fields cannot be said to show signs of exhaustion. What is presumably meant by people asserting the contrary is that, on the whole, within the comparatively limited area of the territory thoroughly explored and actually yielding oil, in order to reach the oil-bearing strata wells now-a-days have to be sunk to a considerably greater depth than formerly. No local well owner denies this. The question of exhaustion might thus be resolved on technical and commercial grounds respectively into the question of increased difficulties encountered in boring operations, and that of the profitableness or otherwise of carrying on such operations. The technical difficulties at present are not insurmountable, and, as to the commercial side of the matter, it must be admitted that the production of crude oil, owing to the rise in the price of that article, proved a more lucrative business in 1889 than previously. Occasional deficiencies of supply at Batoum, or the rise of prices for petroleum exports, may be correctly ascribed to the greater demand for these products. The quantity of crude oil produced in 1888 was 2,580,000 tons; in 1889 it was 3,306,000 tons. Shipments from Baku and Batoum were in 1888 1,741,958 tons; in 1889, 2,413,170 tons, or an increase in favour of 1889 in the production of crude oil of above 28 %, and in the total shipments of above 37 %. At present there is, therefore, no ground for alarm, and should the limited territory of the Apsheron, where borings are actually being carried on, prove insufficient, there are other petroleum fields along the Caspian as yet not explored, because there has not yet been necessity for so doing. When it does come capitalists will readily turn their attention to the new fields. The crisis is n likely to come soon. At the present time, besides the many still pr ductive pumping wells near Baku, there are three fountains at the o place spouting with unabated force, and yielding daily above 5,0 tons. With reference to shipping, Mr. Peacock says that the numb of tank boats engaged in carrying petroleum in bulk at Batoum num bered in 1889 28, of which 22 were under the British flag, 3 Germa 2 under Russian, and 1 under Belgian. The United States have, it said, about 30 boats actually engaged in the trade in America, so it evident that the chances of the Caucasian petroleum industry in com peting with that of America must materially diminish.

A FERTILIZER DECISION IN PHILADELPHIA

THE case of Heller et al. vs. Cadwalader (Collector of the port was tried before Judge McKenna in the United States Ci cuit Court of Philadelphia on April 14th and 15th, and resulted i a verdict for the plaintiff.

The suit was brought by the plaintiffs against the Collector the port of Philadelphia to recover the amount of duty paid by them under protest, on an importation of 50 tons manure salt importe by them on the ship Cuba in May, 1888. The case is in al respects similar to the one brought by the plaintiffs against th collector of the port of New York, and which was tried befor Judge Lacombe on May 22nd, 1889, and which also resulted in verdict for the plaintiffs.

The plaintiffs contend that the manure salt imported by them and which contains 90 to 98% sulphate of potash is entitled to fre entry under the provisions of the free list for "guanos, manure and all substances expressly used for manures." The defenda claimed that under the Treasury ruling of August 2nd, 1870 (S 715), concerning "sulphate of potash," this importation of manu salt was dutiable at 20% ad valorem The Treasury agents hav been assiduously questioning all the importers of sulphate of potash and some of the chemical manufacturers who, during the past fe years, have bought this article for use as a raw material in th manufacture of bichromate of potash, alum, and other chemicals.

Not one of the Government witnesses could prove why sulphat of potash should be protected, why the farmer should be deprive of it, or why the chemical manufacturer who desired to use it as raw material should be compelled to pay an extra price for it o account of a 20% duty. It was proved at the trial that every poun of this importation was used for fertilizer purposes, that there a many kinds of sulphate of potash manufactured from various source that the article in question is a crude or raw material, and not crystal line, and that it was impossible to use it for chemical or medicinal pu poses without being purified and refined, which is a very expensiv process. Evidence to this effect was given by Drs. Williams an Miller, of Philadelphia, Mr. C. M. Stillwell, of Stillwell and Gla ding, New York, Mr. Albert U. Andrus, of Lazell, Dalley and Co. and Mr. Horace M. Olmstead, of E. Merck, and after a very abl charge by Judge McKenna the jury brought in a verdict for th importers.

THE PAPER TRADE OF PORTUGAL.

Judging from recent reports, the paper trade of Portugal is not present in a particularly flourishing condition. The consumption of pap in the country, is, it is true, considerable, but the greater part of this is im ported, the Portuguese paper mills being few in number and mostl unimportant. Importation from other countries is encouraged by lo tariff duties, which vary from 15 to 100 reis per kilo,* the higher rat being imposed on writing papers, scarcely any being produced in Po tugal itself; the lowest rate applies to wrappings and printings, whil cardboards pay about 7½ reis per kilo. The imported paper com principally from Germany, Belgium, and France, England being a ba fourth. A few years ago the annual arrivals of paper were valued about 450,000 pesetas (about a third being cardboards, strawboard &c.), and this total has since been largely exceeded, printings comin over in larger quantities every year. The principal importers in Li bon are Messrs. Rodrigues and Rodrigues, Machado and Co., Virw Macieva and filhos, and Bonaventura da Costa Marques; besides the there are several smaller firms engaged in the trade As regards th materials for home production, most of the caustic soda comes fro England, the colours from Germany, alum from France, and star and farina from Belgium; esparto is also used to some extent. Pa ments are generally made by three months' bills discounted at banks i the usual way.—*Paper Trade Review.*

* 1,000 reis (milreis) = 4/5½ English. 1 kilogramme nearly 2⅕lbs. English. 1 peseta = about 8d. English.

LEGED FRAUDS ON LISTER AND CO., AT BRADFORD.

day, at the Borough Court, Bradford, Francis Stubbs, (44), Heaton, and Henry Varley, (32), drysalter, East-street, ndered to answer the charge of having conspired to cheat Lister and Co., Limited, Manningham Mills, of various ey. The magistrates on the bench were Mr. Skidmore), Ald. Smith Feather (Mayor), Ald. John Hill, Ald. Priestman, Ald. Thomas Hill, Mr. W. Oddy, Mr. T. A. :. R. Kell, Mr. Abraham Mitchell, and Mr. H. Waud. as crowded.

gh (instructed by Mr. A. Neill) appeared for the prosecution; w (instructed by Mr. J. Freeman) represented Stubbs; and ı (Messrs. Ford and Warren, Leeds) defended Varley.

haw complained of the inadequacy of the particulars of the ıished by the prosecution, contending that they did not ıformation required to enable counsel to pick out the line of o prepare for cross-examination.

ren took a similar objection, remarking that the information ı the particulars was so meagre as to lead to the conclusi on l for the defence were being misled either intentionally or ally.

gh replied that if his friends made a grievance of this matter e willing to have a further adjournment.

for the defence intimated that they did not desire a further ıt.

cussing the point futher, the Stipendiary ruled that the ob- ed was not applicable to a preliminary investigation, though l it would be fair on an indictment. He, however, was cquiesce in a suggestion of Mr. Warren to allow the cross- n of witnesses to be deferred, if counsel thought it desirable. was then proceeded with, all the witnesses, on the appli- Ir. Warren, being requested to leave the court.

ıgh, in stating the case for the prosecution, said the prisoners ed with conspiracy, and also with obtaining money by false rom Lister and Co., Limited. Stubbs had been employed osecuting company under an agreement made the 3rd of , 1889, as their foreman dyer. The company in the course siness carried on extensive dyeing operations at their mills, ım, and the salary Stubbs received as foreman was £550. per t was part of his duties, as provided by the agreement, to nd diligently discharge his duties to the best of his skill . Stubbs had also to carry out to the best of his skill such ıe company might from time to time devise. Stubbs was a ıtly trusted by his employers, and they had confidence in his and integrity, and in the course of his duties Stubbs was give orders for such dye materials as should be required by The custom was that from time to time a clerk would be to order goods from various people, but chiefly from Mr. arley, drysalter, of Leeds, in whose employment the prisoner . According to the custom adopted, the quantities were ın to the clerk by Stubbs, and he would write out the orders ley, no price being inserted. The prices were left entirely etion of Stubbs, and he fixed them in accordance with the in from time to time by Varley, and which had been agreed en these invoices were checked to see that the goods which red corresponded with the quantities contained in them, and as so, then the invoices were passed. Of course the com- mployed Stubbs assumed that he had to the best of his n that the goods which were delivered were of the ılity, and that the prices he had arranged for were fair ices. His judgment and integrity were relied upon act, absolute confidence was placed in him. If he ın not to be trusted, he had every opportunity to defraud ers by agreeing to prices which were not fair prices, and the person who supplied the goods to obtain a larger profit, practically robbing the purchaser. He would be able to lench that from time to time the prisoner Varley came over ham Mills apparently for the purpose of arranging prices He did not know whether the prisoner Varley was a partner of that name or not, but he was authorised to draw cheques greements with respect to the supply of goods on behalf of ıd had power to make contracts for the sale of articles. which the prosecution alleged against the prisoners was it was agreed between the two men that sums in excess et prices should be charged for goods, and that the plunder ıined by the firm of Richard Varley from Lister and Co. ivided in certain proportions agreed upon between Richard the prisoner Stubbs. He would be able to show with he whole of the goods that were supplied by Varley s were made, and that a share of the profit was paid by prisoner Stubbs. There was only one exception made in the transaction, and that was in the price charged for fullers' earth. The prosecution had selected five of the principal articles out of the whole of those supplied by Varley to Lister and Co., and he proposed to take those five articles and show the style in which the business of the firm was conducted by the prisoners. The prosecution had confined their action to the past twelve months, because from that time the business was incorporated in a limited liability company, and it would not be possible to go back any further without preferring another indictment. He might say, however, that the system had gone on for years, the only difference being that with increasing confidence in their system as time went on the prisoners increased the amount they put into their pockets. An article called sumac was one of those supplied to Lister and Co. The market price of that never exceeded 13s. 3d. per cwt, but Lister and Co. had the privilege of paying Mr. Varley the sum of 15s. per cwt., the difference of 1s. 9d. being divided between the prisoners, Stubbs receiving 6d. per cwt. and Varley the remaining 1s. 3d., those sums being over and above what Lister and Co. should have paid if they had only paid the fair market price. There was a total sum on that article alone of £105 17s. 11d., which had been paid in excess owing to the system pursued by the prisoners, and Stubbs' share of it was £30. 4s. 9d., and the balance of £74. 13s 2d. was retained by Varley. There was another article called cutch supplied, the regular market price of which was 27s. 6d. per cwt; but in one instance, and in one instance alone, namely, to a firm called Messrs. Young and Co., had Mr. Varley charged 31s. per cwt. Although the whole of the other customers of the firm had only paid 27s. 6d. per cwt., he would accept the highest figure of 31s. per cwt. as the market price. That charged Lister and Co. was 37s. per cwt, being 9s. 6d. more than was charged other people, and 6s. a cwt. more than was paid by Young and Co. Stubbs' allowance of that excess price was 1s. a cwt. There was a total amount of £84. 15s. paid in excess on that item, and of that Stubbs received £14. 2s. 6d., and Varley £70. 12s. 6d. The highest price paid for fustic in the whole of Yorkshire was 6s. 6d. per cwt., and the price charged to Lister was 9s. 6d., Stubbs receiving 3d. per cwt. and Varley 2s. 9d. The total excess paid was £98. 16s. 3d., Stubbs receiving £8. 4s. 8d., Varley £90. 11s. 7d. There was a dye called fast red. He would be able to prove that Varley purchased that at 3s. per lb., and allowing a fair profit of 9d. per lb., 3s. 9d. would be a fair price to pay. The price charged Messrs. Lister was 7s. 9d. per lb., and of the excess of 4s. Stubbs received 1s. 6d. per lb. and Varley 2s. 6d. £636 on this item alone was paid in excess, Stubbs receiving £238. 10s., and Varley £397. 10s. Encouraged, as it seems, by these prices, the prisoners went further. There was a dye called "Blue 00 x," manufactured in Germany by a firm who had an agent in Bradford named Kilner. Varley had the privilege of supplying this article to Lister and Co., and it could be supplied at from 5s. 6d. to 5s. 9d. per lb. It was supplied in tins on which there was a certain mark in ink, which would be rubbed off at Varley's, and then the material was forwarded to Lister and Co. Taking as the top market price 6s. 6d. per lb., it would be shown that the price charged to Lister and Co. was 18s. 6d. per lb. The excess paid on this item was £1,038. The share paid to Stubbs was £346, and to Varley £692. The total amounts thus received in excess during the 12 months was £1,961, and of that Stubbs received £637. 1s. 11d., Varley the balance of the plunder. After Mr. Waugh had proceeded to explain how the transactions had been carried on by referring to books that had been siezed at Varley's place of business,

Abijah Taylor, chemist and sample dyer in the velvet department at Manningham Mills, said he wrote out orders for drugs required according to instructions given by Stubbs, who was foreman of the department. Stubbs also fixed upon the drysalter to whom the order was to be given. He had told him to send orders for certain goods to Mr. Richard Varley, Leeds, and the invoices now produced relating to such goods had been initialled by him and F. Hull. The initialling meant that the weights and prices were correct Harry Varley was in the habit of calling at Manningham Mills once or twice a week to see Stubbs, who would then ask him (witness) if anything was required. After making out an order he handed it to Stubbs. At the latter's request he regularly prepared a report as to the quantity of drugs and dyewares received each week from every firm of drysalters. In reply to questions from the bench, witness added that the initialling of the invoices by himself and F. Hull meant that the weights and prices were correct. He got the prices from a book which was formerly in the weaving office.

It now transpired that the book witness was quoting from was only a copy, and his examination was adjourned for the production of the original book.

Benjamin Thomas Gibbins, a director of Lister and Co., and a manager of the velvet department at Manningham Mills, gave evidence that Stubbs was foreman dyer in the same department. His duties were to see to the dyeing of goods and buy all material that was necessary. It was also the duty of Stubbs to fix the prices at which the goods were to be bought.

Abijah Taylor recalled, produced the original book, which he said he had received five years since from another person in the department; but he was unable to say in whose handwriting the entries were made. He had at various times altered the prices in accordance with those of later invoices.

In cross-examination by Mr. Kershaw, he said that the drysalters who called did not enter the dyeing department but waited at the lodge. From 1879 till 1884 the orders for dyewares were given by Freeman Hull; and after it became Stubbs duty to give the orders he (witness) was told to check the invoice prices with the entries for similar goods in a book. After Stubbs gave this direction he accordingly put down in a book the price of "fast red" when it fell from 8s. 9d. to 8s. In the same way the figures were afterwards altered from 8s. to 7s. 9d. The price of blue oox was originally entered at 24s 6d. per lb., but it was subsequently reduced at various times to 23s., 21s., and 20s. 6d. The latter alteration was made in the time of Stubbs giving the orders, and subsequently it was altered to 18s. 6d. Pure blue, which was considered nearly equal to oox, had been obtained from Messrs. Staley and Co., and the price put down for pure blue xx was 27s. per lb. Since then the price had been reduced, and it now stood at 21s., the alteration being made by order of Stubbs. To retain uniformity in shades it was important that the wares should be obtained from the same drysalter. Cutch was an article which varied in quality, and Stubbs had gone to several drysalters for it, the major portion of the quantity used being obtained from firms other than Varley's. Its price from all drysalters was 37s per cwt. Fustic fluctuated much in value. In 1885 the price put down for it was 9s. 6d. per cwt. It was reduced to 9s., but the last entry for it was again 9s. 6d. It was a wood dye, used to produce a fine yellow, and for the purposes of Lister and Co. the best quality was required. Sumac was supplied sometimes in the leaf and sometimes ground. It came from Sicily, and was sold under different brands, which varied in quality.

At this point the Court adjourned till Monday. On reopening, Mr. Warren asked for an order to be allowed to inspect the stock-book of dyewares in the velvet department, and the books relative to purchases from other drysalters than Varley's, as he wanted to ascertain whether or not such persons had charged the same price, or less or more, than those paid to Varley's.

Mr. Waugh objected to the application, and said that if the defence would adopt the ordinary course by serving a notice to produce, the matter could be considered. But there were so many books that he must insist upon certain books being specified, otherwise they might have to bring a waggon-load.

The Stipendiary (Mr. Skidmore) said if the prosecution had nothing to fear from the inspection of the books, he did not see why they should object. Of course, if they stood upon their technical right that notice had not been served, there was nothing further to be said.

Mr. Warren intimated that he would serve a notice to produce.

Mr. Kershaw complained that the prosecution objected to the defence copying any portion except two pages of a book which was produced by the witness Abijah Taylor on Saturday.

Mr. Waugh said his answer was that the book was produced and particular pages referred to as being those by which the prices in Richard Varley's list were checked. The rest of the book was not put in as evidence, and the defence would have objected if he had attempted to do it.

Mr. Kershaw replied that the book was put in as a list of prices, which might have been a long sheet, and not a book; and he could not understand the difficulties which were thrown in the way of the defence.

The Stipendiary said he thought the defence were entitled to it.

The book was accordingly handed to Mr. Kershaw.

Abijah Taylor, chemist and sample dyer in the velvet department at Manningham Mills, was again called and cross-examined by Mr. Kershaw. He said the business had greatly extended since 1879, and to a great extent the dying of particular shades had caused the success of the plush trade. The shades were due to the experiments conducted by Stubbs. Probably it would be dangerous to Lister and Co.'s trade to change the dyer. He knew that Stubbs had given notice to leave the service of Lister and Co, it being the common talk at the mill. He believed that at Saltaire Mills the manufacture and dyeing of plushes were carried on. He did not know anybody who had the same reputation as Stubbs for dyeing plushes. Supposing Stubbs went to Saltaire Mills in the same capacity, and obtained dyewares from the same drysalters, there would be a danger of some of Lister and Co.'s business being transferred. As to the ordering of goods, if Subbs was away when the drysalter called, the order would be given by himself, if any were required. He had frequently given orders under such circumstances. In certain drugs Stubbs had told him to distribute the orders as equally as possible, among different drysalters; for instance, he had been told to distribute the orders for aniline dyes between Varley's and Staley's. Blue entered into about two-thirds of the different shades of plush, which were 150 in number. He had known cases in which Stubbs asked him the price which was being paid for certain dyewares, and to answer him it was necessary to consult the book which I (witness) kept. Stubbs knew that this book was kept as a check on tl prices. Goods for other departments were also obtained from Varley' and the invoices for such goods would be initialled and checked by son one else. He had heard Stubbs complain of the quality of the yai which he had to dye. During the time that F. Hull made out tl orders for goods some experiments were made with aniline dyes. Tl result was shown to Stubbs, who was allowed to exercise his discretio in the matter.

Cross-examined by Mr. Warren, witness said a book was kept i which were entered the names of persons from whom goods were pui chased. There was another book called the velvet dyer's purchasin book, in which the prices of goods received were entered.

Mr. Warren asked for the production of these books, or for an op portunity of inspecting them.

The Stipendiary said he thought the defence were entitled to see th books indicated.

Mr. Waugh replied that he would confer with his advisers in refer ence to the most convenient time for the inspection of the books.

Benjamin T. Gibbins, a director of Lister and Co. Limited, an manager of the velvet department at Manningham Mills, whose evi dence was given on Saturday, was recalled for the purpose of cros examination. Questioned by Mr. Kershaw, he admitted that till th engagement of Stubbs the firm had not been able to obtain a satisfac tory dyer for plush. The success of the firm in the plush trade ha been partially due to the shades of the dyeing, and to the fact that th shades were fast to light. Of course, there was not really such a thin as "fast to light," although the tickets on the pieces stated that the were "fast colours." The firm did not guarantee them as fast colours the term was merely a trade one, meaning that the colours wer durable. He had no doubt that he had told Stubbs repeatedly that i was important to have the goods dyed with fast colours. Between 187 and 1884 Stubbs was allowed to do as he liked with regard to seein the drysalters and selecting the dyewares required. He remembere that on one occasion he heard that Stubbs was going to leave. Tha was in 1884, and he was told that Stubbs had arranged to go to th firm at Saltaire. He considered it important that the services of Stubb should be retained, and arrangements were made for him to remain It came to his knowledge that Stubbs at that time complained that h had not been allowed to see the drysalters or to order goods; but h did not remember that Stubbs gave that as a reason for desiring t leave. He did not remember Stubbs saying that he had not been properl treated by Mr. Hull, who had the right to purchase the drugs. H said he had made an absolute contract with Messrs. Salt, and showe the agreement to himself and Mr. Reixach, the general manager. I was arranged that he should continue in the service of Messrs. Liste and Co., by giving him an indemnity against any proceedings b Messrs. Salt. He could not recollect that Stubbs then said that a there had been complaints about the colours not being fast, he wishe to see the drysalters. So far as he knew, Stubbs had always seen them at the lodge, none of them being allowed in the works. Mr Hull acted as clerk to Stubbs, and also as clerk in the weavin department.

In reply to Mr. Warren, witness said he could not recollec whether Abijah Taylor was in the department before 1884, nor ha he any recollection of Mr. Reixach having stated to him that Stubb was continually complaining that he was not allowed to order his dye wares. As salesman of the department, he (witness) had never sol goods at three times the value of their cost. He did not think i would be good policy to buy in the cheapest market and sell at th highest price; he should do his best, but on the principle of doin unto others as he expected to be done unto.

Freeman Hull, clerk in the velvet-weaving department, said hi initialling of the invoices had nothing to do with the prices, but merel referred to the distribution of the items in the respective departments. Formerly he gave the orders to the drysalters when told by Stubb what was required, his instructions by the latter being to the effec that he was to see that the prices were as nearly as possible to thos entered in the book as the price previously given. The previous price had probably been fixed in the same way. He had not fixed any o the original prices without having the sanction of Stubbs. In cross examination, witness said this arrangement had existed for severa years prior to 1885, when it ceased. He believed the book of origina prices was in the hand-writing of a clerk named Woodcock, employe in the weaving department; but he could not say that he gave Wood cock instructions to enter the prices. He had seen Stubbs at th lodge with Varley before 1885, but he could not say how often. H had never heard that Varley was told that he must not supply red c blue dyewares to any other firm than Lister's. He could not recolle giving Varley any samples to match, or telling him the prices paid t other people.

Edward Woodhead, clerk to Messrs Brown and Co., 6, Chape street, Liverpool, was called to prove sales of cutch to Mr. Richai Varley, Leeds, in August last; but as he only produced a press cop

, which was not in his own handwriting,
objected to his evidence, on the ground that he was not could answer the questions which would have to be put nation.
liary upheld the objection.
iteman, clerk with Messrs. John Fletcher, jun., and Co., rpool, proved that during January, 1889, his firm sold ard Varley, of Leeds; and in cross-examination he stated of cutch varied from 15s. to 34s. per cwt. During last s scarce, and the price mentioned in the invoice of sale to ty was 28s. per cwt. His firm did not pay the carriage. ons of the firm of Varley with his firm had always been orward character.
hen adjourned till Tuesday.

ME MORDANTS IN WOOL DYEING.

as a mordant for polygenetic colours in wool dyeing is employed in the form of potassium or sodium bichromate. strong oxidising agent, may under certain circumstances ool or the colouring matter employed, and more or less Hence the desirability of a chromium salt not posessing In view of this, those chromium salts in which the rms the base, such as chromium sulphate, chrome alum, urally turned to. The last-named salt has long been proefficient substitute for potassium bichromate. As a bye manufacture of alizarine, it it comparatively cheap, and ave expected that it would frequently be employed in prac-, no doubt, would be the case if it had been found superior te of potash. It is well known that chromium sulphate, are not satisfactorily fixed on cotton by the methods which h the corresponding aluminium salts, and it would not be chromium and aluminium salts showed similar differences s mordants for wool. From experiments made by Gardner at chromea'um may be applied in the same way as common n addition of cream of tartar, but whereas with one molenon alum four molecules of cream of tartar are sufficient, lecule of chrome alum it is necessary to use from twelve to lecules of cream of tartar. The great expense of this large cream of tartar is probably the reason why chrome alum is lly employed.
of chromium (CrF_34H_2O) has recently been put upon the ler the name of "fluorchrome." This substance, like n and oxalic acid, is a non-oxidising mordant, and also een (Cr_2O_3) mordanted wool, but attempts to obtain it right full shades as with the use of potassium bichromate acid were not successful. It appears also to give somen shades, but possibly these defects may be overcome on s. The defects noticed are possibly due to the hydrofluoric d in the mordanting operation. "Fluorchrome" is conore expensive than chrome alum and oxalic acid.
d with other metals chromium appears to be the most eful as a mordant for wool, since it gives both red, yellow, lours with different colouring matters, so that, by using a olouring matters, a very large variety of compound shades ained. Of the various chromium salts which have been chromates are the best for general use, being cheap, easily l apparently not materially affecting the fibre, although the much questioned. The shades produced by chromium e a rule dark, brilliant, and very fast to light and alkalies. *Mercury.*

ATED OIL GAS.—Dr. Thorne, of the Brin's Oxygen Comcently carried out a series of experiments with the view of dvantages of the Tatham patent for the manufacture of an ry high illuminating power. It will be in the recollection rs that Mr. Valon has shown, at Ramsgate, that oxygen with advantage for the purification of coal gas, and that it in his process the addition of oxygen to the gas caused in the illuminating power. The same is now found to be en oxygen is added to oil gas, and it has been shewn that as per cent. of oxygen can be added to the oil gas without any ng an explosive mixture. The process at present used at orks at Westminster consists in first heating American a moderate temperature in an iron retort, and then adding and vapours produced about 20 per cent. of oxygen, which subsequent condensation of the vapours and at the same usly increases the illuminating power. The necessary very simple description, and, as far as one can judge from nts which have been carried out, there seems every ground that the process is one which will be found to work well cale and be capable of producing an illuminating gas of high a comparatively cheap rate —*Industries.*

PERMANENT CHEMICAL EXHIBITION.

JOSEPH AIRD, GREATBRIDGE.—Iron tubes and coils of all kinds.

ASHMORE, BENSON, PEASE AND CO., STOCKTON-ON-TEES.—Sulphate of Ammonia Stills, Green's Patent Scrubber, Gasometers, and Gas Plant generally.

BLACKMAN VENTILATING CO., LONDON. — Fans, Air Propellers, Ventilating Machinery.

G. G. BLACKWELL AND CO, LIVERPOOL.—Manganese Ores, Bauxite, French Chalk. Importers of minerals of every description.

BRACHER AND CO., WINCANTON.—Automatic Stills, and Patent Mixing Machinery for Dry Paints, Powders, &c.

BRUNNER, MOND AND CO, NORTHWICH.—Bicarbonate of Soda, Soda Ash, Soda Crystals, Muriate of Ammonia, Sulphate of Ammonia, Sesqui-Carbonate of Ammonia.

BUCKLEY BRICK AND TILE CO, BUCKLEY.—Fireclay ware of all kinds—Slabs, Blocks, Bricks, Tiles, "Metalline," &c.

CHADDERTON IRON WORKS CO., CHADDERTON.—Steam Driers and Steam Traps (McDougall's Patent).

W. F. CLAY, EDINBURGH.—Scientific Literature—English, French, German, American. Works on Chemistry a speciality.

CLAYTON ANILINE CO., CLAYTON.—Aniline Colours, Aniline Salt, Benzole, Toluole, Xylole, and Nitro-compounds of all kinds.

J. CORTIN, NEWCASTLE-ON-TYNE.—Regulus and Brass Taps and Valves, "Non-rotative Acid Valves," Lead Burning Apparatus.

R. DAGLISH AND CO., ST. HELENS.—Photographs of Chemical Plant —Blowing Engines, Filter Presses, Sludge Pumps, &c.

DAVIS BROS., MANCHESTER.—Samples of Products from various chemical products—Coal Distilling, Evaporation of Paper-lyes, Treatment of waste liquors from mills, &c.

R. AND J. DEMPSTER, MANCHESTER.—Photographs of Gas Plants, Holders, Condensers, Purifiers, &c.

DOULTON AND CO., LAMBETH.—Specimens of Chemical Stoneware, Stills, Condensers, Receivers, Boiling-pots, Store-jars, &c.

E. FAHRIG, PLAISTOW, ESSEX. — Ozonised Products. Ozone Bleached Esparto-Pulp, Ozonised Oil, Ozone-Ammoniated Lime, &c.

GALLOWAYS, LIMITED, MANCHESTER. — Photographs illustrating Boiler factory, and an installation of 1,500-h.p.

GRIMSHAW BROS., LIMITED, CLAYTON.—Zinc Compounds. Sizing Materials, India-rubber Chemicals.

JEWSBURY AND BROWN, MANCHESTER.—Samples of Aerated Waters.

JOSEPH KERSHAW AND CO., HOLLINWOOD.—Soaps, Greases, and Varnishes of various kinds to suit all requirements.

C. R. LINDSEY AND CO., CLAYTON. — Pyroligneous Products. Acetate of Lead, Sulphate of Copper, &c.

CHAS. LOWE AND CO., REDDISH.—Mural Tablet-makers of Carbolic Crystals, Cresylic and Picric Acids, Sheep Dip, Disinfectants, &c.

MANCHESTER ANILINE CO., MANCHESTER. — Aniline Colours, Samples of Dyed Goods and Miscellaneous Chemicals, both organic and inorganic.

MELDRUM BROS., MANCHESTER. — Steam Ejectors, Exhausters, Silent Boiling Jets, Air Compressors, and Acid Lifters.

E. D. MILNES AND BROTHER, BURY.—Dyewoods and Dyewood Extracts. Also samples of dyed fabrics.

MUSGRAVE AND CO., BELFAST.—Slow Combustion Stoves. Makers of all kinds of heating appliances.

NEWCASTLE CHEMICAL WORKS COMPANY, LIMITED, NEWCASTLE-ON-TYNE.—Caustic Soda (ground and solid, Soda Ash, Recovered Sulphur, etc.

ROBINSON, COOKS, AND COMPANY, ST. HELENS.—Drawings illustrating their Gas Compressors and Vacuum Pumps, fitted with Pilkington and Forrest's patent valves.

J. ROYLE, MANCHESTER.—Steam Reducing Valves.

A. SMITH, CLAYTON.—India-rubber Chemicals, Rubber Substitute, Bisulphide of Carbon, Solvent Naphtha, Liquid Ammonia, and Disinfecting Fluids.

WORTHINGTON PUMPING ENGINE COMPANY, LONDON.—Pumping Machinery. Speciality, their "Duplex" Pump.

JOSEPH WRIGHT AND COMPANY, TIPTON.—Berryman Feed-water Heater. Makers also of Multiple Effect Stills, and Water-softening Apparatus.

FLAMES.*

I WILL begin my remarks with a paradox. Flames are of no importance, yet the subject is a very important one indeed, and has not received the amount of attention which it deserves, either from a commercial or a scientific point of view. Flame is really nothing but a sign of an incomplete or transition state of chemical combination. Its presence during combustion is not always necessary, and I think I am within the actual facts when I say that its presence, under practical conditions in commercial use, indicates always a loss of work.

The appearance of flame is misleading, and the greater the flame the smaller the work done, other things being equal. I have been asked by a well-known engineer if I could explain why certain boilers gave such an exceedingly small duty for the fuel consumed when the flues were, as he said, "filled from end to end with magnificent flame;" the fact was that his so-called magnificent flame was a delusion, hollow and cold inside, and not coming into contact with his boiler at all. When the same fuel was burnt with a very small flame, hardly visible over the bridge, the duty increased some thirty per cent.

I will now give you some practical demonstrations of the various characters of flames and their delusive appearances. A cotton handkerchief, as you no doubt are aware, will burn readily to ashes, and I will prove this to you by burning one. I have here another, precisely the same, which I will saturate with proof spirits of wine, and, as you see, the flame, although apparently a fierce one, is not hot enough to ignite the handkerchief, which comes out of the fiery test without a singe or mark. The fact is that the flame is not only comparatively a cold and wet one, but it is also hollow, and does not come in contact with the handkerchief at all, the space between the handkerchief and the flame being filled with cold vapour, which only burns when it comes in contact with the surrounding air. You may perhaps smile at the mention of a "wet" flame, but the term is perfectly correct. All flames, in which hydrogen is burnt along with oxygen, contain water, and, in addition to this, water formed by the process of combustion. The proof spirit contains also a large quantity. A great proportion of the energy of combustion is absorbed in converting this water into steam, the temperature being reduced to the lowest point at which flame can exist; in fact, as the spirit is consumed more rapidly than it has power to dissipate the water, the temperature of the flame gradually lowers until it actually becomes too cold to maintain its existence. To make the internal space in a flame visible to you it is necessary to experiment in a different way. I now take a burner eight inches in diameter, supplied with a mixture of coal gas and air, the air being in a quantity sufficient to increase the bulk of the flame, but not enough to enable the gas to burn, except on the outer surface where the mixture comes in contact with the surrounding air, with which it combines. This great flame is of very little use for any purpose, and I will now proceed to prove how great a delusion it is by placing a ball of paper inside it, then some loose gunpowder on an open paper, and again a ball of guncotton, all of which remain untouched. The outer film of flame is hot enough to burn my hand if left in it; but if the body of my hand is protected I can, as you see, put my naked fingers inside the flame without discomfort, and pick the paper of gun cotton out of the centre of the flame.

I will prove the presence of unburnt gas in the centre of the flame by leading some of it out with a tube and igniting it in a separate flame; and I will again prove the presence of unburnt gas in the centre of this second flame, by leading some of it again and producing a third separate flame from the gas abstracted from the centre of the original flame.

By an alteration of the burner, admitting sufficient air with the gas to form an explosive mixture, which will burn without assistance from the external air, the flame instantly becomes solid, much smaller, less visible, and at once explodes the gunpower. To reduce the flame to a still smaller size, a different form of burner is necessary, with a supply of air under pressure, and you now see the same quantity of gas and air burning, but the space taken up by the combustion, instead of being as at first eight inches wide, and 18 inches high, is less than one hundredth part the size; and instead of being able to put my bare hand in, it will fuse wrought iron instantly. It is well known that the available duty of any source of heat is, other things being equal, in direct proportion to the difference of temperature between this source and the object to be heated; and, therefore, we shall get a much larger amount of work from our small high temperature flame than from the large and colder one. I will now dispense with flame altogether, and show you the same quantity of gas and air burning as before, but in the most perfect form, the combination taken place without any flame and on the surface of the substance to be heated. To show this so that all can see, the mixture of gas and air is directed on a large ball of iron wire, flame being used at first to heat the wire to the necessary temperature to continue the combustion. By stopping the gas supply for an instant the flame is extinguished, and the combustion is now continued without any flam[e] but with an enormous increase in the heat obtained. This invisible [or] flameless combustion is only possible under certain conditions, and o[ne] essential point is that the combustible mixture shall come in absolu[te] contact with a substance at a high temperature which is capable of a[b]sorbing the heat as it is generated. I will now heat this small furna[ce] to a temperature sufficient to cause combustion without flame, and wi[ll] then remove the side, showing you the interior of the furnace with [a] crucible being kept at a white heat by blowing a cold mixture of g[as] and air into it. In the absence of a solid substance at a high temperature, it is, so far as is known, impossible to cause combustion withou[t] flame; and when a flame is used it is also impossible to make th[e] flame touch a cold surface.

Many of you will imagine that if a solid body is surrounded by [a] flame the flame touches it; this is altogether a mistake. There is a spac[e] between the two which it is impossible to pass, a cold and flamele[ss] zone which surrounds the cold surface, which is quite impassable t[o] flame under any conditions, and which most seriously obstructs the wor[k] of heating. To enable you to see that this impassable cold zon[e] exists beyond any doubt, I have here a copper vessel containing wate[r] and on the side of this vessel I have pasted a thin paper label. O[n] this I will direct the powerful flame, which you have seen will fu[se] wrought iron instantly, and the paper remains untouched, without [a] trace of singeing. The full force of the flame, urged by a heavy bla[st] of air, may be directed on this paper for any length of time without th[e] slightest effect, so long as the vessel contains any water. You will, n[o] doubt, imagine that the heat is absorbed so quickly that the paper ha[s] not time to get hot, but it is very easy to prove that this is not the cas[e] by heating a wire in the same flame and touching the paper with it[,] causing instant charring. The cause of this extraordinary result ha[s] never yet been fully explained, but I believe it to be that all substance[s] have an adherent film of air which resists the passage of any flame, bu[t] which is, of course, instantly removed by the application of any soli[d] substance. This theory has one weak point, that the cold zone is im[]passable also to radiant heat; and in the face of this fact I must sa[y] that a really satisfactory explanation is yet to be found which will agre[e] with the present accepted theories of heat. The action of flame o[r] heated matter on moist surfaces is much more easily explained. It [is] known that a moist hand or stick can be passed through molten iro[n] without burning, owing to the film or steam evolved, which pre[]vents contact with the metal. The same reason accounts fo[r] the fact that I can burn gun cotton on my hand without feelin[g] any heat, the moisture present absorbing the heat as fast a[s] it is evolved. In the case of the paper label on the metal vesse[l] there is no moisture, the label having been carefully drie[d] to prevent the sudden formation of steam lifting it away fro[m] the metal surface; and I think that the peculiar resistance to flam[e] contact with a cold surface requires some further explanation than th[e] present theories can account for. In connection with the subject o[f] flames, I may refer, as a curiosity, to the enormous volume of soun[d] of different tones which is produced by placing various sizes o[f] chimneys on a gauze burner consuming a mixture of gas and air. Th[e] sound is as powerful, but certainly not so pleasing, as that of a fog[]horn. In the combustion of iron, magnesium, and other substance[s] of which the product is a solid, and not a vapour of gas, flam[e] proper never exists, although in the combustion of magnesium an[d] zinc it is apparently present. The brilliant incandescence of th[e] particles of oxide thrown off at a very high temperature causes [a] deceptive appearance.

(To be continued.)

* Corn Exchange, Oxford, April 25th, 1890. Lecture delivered by Thoma[s] Fletcher, F.C.S.

SULPHITE PULP IN AMERICA.

THE following abstract is from a letter on the above subject by M[r.] Waldemar Thilmany, Kaukauna, Wisconsin, U.S.A., on Marc[h] 13th, to the *Papier Zeitung*, No. 35, 1st May, 1890.

At the Thilmany Pulp and Paper Works, the Mitcherlich Sulphite pr[o]cess is at work. The first mill using this process was started at Alpen[a,] Michigan, in November, 1886, with a daily output of 20 tons of 2,000lb per ton.

It interested him much to find out how the different kinds of woo[d] would behave in the sulphite process, and he therefore allowed a boil[er] full of each of the following different woods to be digested, viz.; Beec[h] *(Fagus Sylvatica L)*, birch *(Betula Alba)*, maple *(Acer Pseudopl[a]tanus* and *Acer Platanoides L)*, American larch *(Pinus Larix L[)]* aspen *(Populus tremula L)*, black poplar *(Populus Nigra L)*, red-[pine] *(Pinus Picea)*, spruce or white pine *(Pinus Abies)*, yellow pi[ne] *(Pinus Sylvestris)*.

The wood was freshly felled, and had a diam. ivarying from five [to] eight inches. It was cut into discs of 1¼ inches thick, and then digest[ed] whole with a solution of 5½° Baumé by the Mitcherlich metho[d] voods with the exception of the *Pinis Sylvestris* (which d in the centre), were well-boiled and completely soft. ch, and maple yielded snow-white sulphite pulp, whose r, was considerably shorter than that from the red or Pinus Picea and Pinus abies), and consequently they did strong a paper.

, some boilers full of larchwood ("Tamrack") were d the pulp run off on a cylinder machine in the same way o the paper mills. The fibre appeared good and strong, r was mottled, *i.e.*, it varied from dark brown to citron

t the seven mills in America, working the Mitcherlich luce daily about 70 tons pulp, and use exclusively spruce or rood (*Pinus Abies*).

er mills, situated in the Fox River Valley, with eight ach 14 feet diameter and 42 feet long, produce daily 20 ; whilst another mill at Madison, Maine, has 20 digesters size, equal to a daily output of 20 tons pulp.

pulp prepared by the Mitcherlich process is clean if the perly sorted. There is apparently a difference in the the fibre obtained from American spruce by the sulphite compared with imported sulphite pulps, even those made herlich system in Europe. Tests of raw American-made ewspaper made from it, in a paper mill belonging to Mr. brother, revealed the fact that from 10 to 15% more im- had to be used to obtain the same results as regards

s imported to America are frequently mixed, so that the ter is not always provided with the same make. How the importer only knows. This he keeps secret, and ay untrue assertions regarding it. The papermaker who o made good newspaper from 20% sulphite and 80% mech- od pulp, suddenly finds that his paper is much weaker, quently many complaints pour in upon him. He then his pulps, and finds among them the most diverse

hite pulp made in America and manufactured into news- sed for this purpose in the unbleached state.

PRIZE COMPETITION.

TLE more than twelve months ago, one of our friends gested that it might be well to invite articles for ion, on some subject of special interest to our giving a prize for the best contribution, he at the ne offering to defray the cost of the first prize, and ng a subject in which he was personally interested.

t date we were totally unprepared for such a sug- but on thinking over the matter at intervals of we have come to think that the elaboration of such a would not only prove of benefit to the manufacturer, ld go some way towards defraying the holiday of the prize winner.

e therefore prepared to receive competitions in the ; subjects :—

SUBJECT I.

st method of pumping or otherwise lifting or forcing ueous hydrochloric acid of 30 deg. Tw.

SUBJECT II.

st method of separating or determining the relative s of tin and antimony when present together in com- samples.

h subject, the following prizes are offered :—

rst prize of five pounds, one second prize of one e additional prizes to those next in order of merit, of a free copy of the *Chemical Trade Journal* for onths.

mpetition is open to all nationalities residing in any e world. Essays must reach us on or before April Subject II, and on or before May 31st, in Subject I. s will be announced in our issue of June 28th.

serve to ourselves the right of publishing the con- of any competitor.

nay be written in English, German, French, Spanish

THE VALUE OF CORN AS A FUEL, AND AS A FERTILISER, AND BOTH.

Under the above somewhat sensational heading, the *American Economist* draws attention to the fact that a superabundant production of corn has led to the necessity for finding a new outlet. The outlet proposed may at first sight startle our readers. "While," the writer goes on to say, "this condition is very distressing it is not without its palliating circumstances, which are :—

1st, its value for fuel, the same being tested by the cost of coal to farmers, and 2nd, the value of its ashes as a fertiliser after burning, as tested by an analysis of the same.

The first point is approximately fixed by a communication from an esteemed friend, who is a Kansas farmer, thus :

Coal at the mines in Franklin and Osage Counties, 80 lbs. to the bushel, sells at 8 to 10 cents per bushel, or 25 bushels to the ton, equal to $2 to $2 50 per ton.

With corn at the same price per bushel which is just now unfortunately a not uncommon rate that amounts practically to a exchange of a bushel of corn for a bushel of coal.

True, the weight of a bushel of coal (70lbs) is a quarter more than that of a bushel of corn (56lbs), but that is probably fully made up by the more perfect combustion of corn, which leaves no residuum but the ashes, while coal leaves a great quantity of almost useless residuum.

True, the farmer in this instance has to haul the coal *from* the mines, but that is only equivalent on the average to carting corn *to* the railroads.

In Franklin County, 70 miles from Kansas City, corn has ruled as 15 cents per bushel ; at 100 miles west from this county 10 cents has been the price.

In the middle and western counties of Kansas there is no coal and very little timber ; consequently coal is imported from the East and from Colorado. The cost, delivered, is 16 to 18 cents per bushel, equal to $4 to $4.50 per ton.

The price of coal in M'Pherson County, one of the central counties of our State, is $5 per ton.

The latter showing is not so good as the former. If he gets no more for his corn than in the first quotation he practically swaps off two (2) bushels of corn for one (1) bushel of coal.

A bad showing enough, certainly, but it has the mitigating and consoling feature that he largely preserves the plant food, used by the corn in its growth, in the ashes retained after burning, which, thanks to the labours of agricultural chemists, can be ascertained with entire accuracy.

Among such authorities Colonel George E. Feraring is conspicuous, and in his valuable book entitled "Elements of Agriculture," published by the New York Tribune Association, when Horace Greeley administered it, is a table showing the amount of such plant food in pounds for *ten bushels of corn*, thus :

Potash pounds	2·78
Lime	0·12
Magnesia	1·52
Phosphoric acid	4·52
Silica (sand)	0·06

This is corroborated by another analysis of the ashes of corn, showing in every 1,000 parts of potash 261 parts ; phosphorus, 449 parts.

Disregarding the value of all elements but potash and phosphoric acid, and applying the values per pound of the same at the prices fixed at the New Jersey Experimental Station, the following results appear for every ten bushels burned :—

Potash, 278 lbs. at 6 cents, 16 68, say	17 cents
Phosphoric acid, 452 lbs. at 8 cents, 36·16	36 „
Value of ashes from 10 bushels corn	53 cents

or five and three-tenths cents per bushel on all corn fed or burned on the farm, which is lost when the corn is sent away to market.

This, of course, should be noted when prices are considered, as it nearly doubles the minimum value of such corn as is used for fuel."

PREPARATION OF IODINE.—It is found that in every 100 kilograms of impure nitrate there are 50 grams of iodine, and that in the crystallisation of the nitrate the water used, called, technically, "agua vieja," holds it in solution. In order to extract it, this old water is drawn off into a separate tank and charged with sulphite of soda, forming the iodide of sodium, precipitating the iodine, which, containing more or less impurities, is refined by sublimation and condensation.

During the ten years ending December 31st, 1888, the exports of iodine amounted to 1,588,074 kilograms, with a total value of $19,333,757, upon which the Government collected in export duties $1,172,576.

GOLD ORES REDUCTION.

IN response to an invitation from the Gold Ores Reduction Company (Limited), a number of gentlemen attended at the company's works, Hackney Wick, a few days ago, to witness the process of gold extraction under their patents. This company has, for some time past, been treating refractory ores under methods invented by Mr. H. Hutchins, and they claim to have secured a degree of success hitherto unknown. The whole system was demonstrated at the works for the benefit of the visitors, and the inventor gave detailed explanations of the process. A large crusher was filled with lumps of rock, which it champed and then ejected in small pieces. From this crusher the ore was passed to an adjacent mill, where it was ground as fine as snuff, and then worked through the hair-meshes of an oscillating sifter; thence it was transferred to the furnace, where the powder was subjected to a thorough roasting by hot air. It is here that the special method of the patent takes effect, it being claimed that a complete oxi lisation of the intractable sulphurets and arsenical ores takes place, and by artificial means the gold is freed. The whole system, from the crushing to the panning, is computed at about an hour. In the condition of a fine rouge-like product the ore is passed from the furnace to the amalgamators, where the powder is ground with water until the gold has flown to its natural amalgam, where it is drawn off and is ready for retort distillation in the laboratory. Two or three hours, it is stated, are sufficient to treat a ton of minerals, and the plant at present erected on the works has a capacity to treat 100 tons a week; but it is intended to supplant this by others. The moderate cost of the process is represented as one of its chief attractions; a furnace can be erected for about £200., and adapted to any existing works, being added between the stamps and the amalgamator. A feature of the system is that flower of quicksilver is entirely obviated by the new process.

Market Reports.

METAL MARKET REPORT.

	Last week.		This week
Iron	44/8½	....	44/5
Tin	£92 0 0	....	£92 0 0
Copper	49 0 0	....	49 13 9
Spelter	21 7 6	....	21 5 0
Lead	12 7 6	....	12 17 6
Silver	47d.	....	46d.

COPPER MINING SHARES.

	Last week.			This week.	
Rio Tinto	17	17 1/8	....	17 1/8	17 1/2
Mason & Barry	6 1/2	6 5/8	....	6 3/4	6 7/8
Tharsis	4 15/16	5 1/16	....	4 13/16	5 1/16
Cape Copper Co.	3 7/8	4	....	3 7/8	4
Namaqua	2 7/16	2 9/16	....	2 3/8	2 1/2
Copiapo	2 3/8	2 1/2	....	2 7/16	2 9/16
Panulcillo	15/16	1 1/16	....	1	1 1/8
Libiola	2 1/4	2 3/8	....	2 3/8	2 5/8
New Quebrada	1/2	5/8	....	1/2	3/4
Tocopilla	1/-	1/6	....	1/-	1/6
Argentella	3d.	9d.	....	3d.	9d.

REPORT ON MANURE MATERIAL.

There has been during the past week a continued active retail business in both phosphatic and nitrogenous material for immediate delivery.

The manure manufacturers, too, are very busy, and as they see their stocks being rapidly reduced, they seem more inclined to consider the purchase of their next season's material.

The easier freights obtainable for Charleston phosphate rock permit of lower prices than were current a month or two ago, and it is quite likely that about 10d. per unit might now be business for land rock—perhaps also for River—in cargoes, c.i.f. to United Kingdom. We also think that some concession will have to be made for Somme and Canadian to induce any important business in these descriptions.

A large portion of the River Plate production of bone ash is reported to have changed hands recently at £5. 2s. 6d. up to £5 7s. 6d., on 70%, partly for Germany and partly for United Kingdom, indicating on the part of buyers a strong opinion as to the future

There is nothing to report in cargoes of River Plate bones, it being now almost too late for present United Kingdom season, and the United States not so far having come forward as buyers for shipment ahead, as they have done for some years back. Common grinding bones are selling at Liverpool at £4. 3s 9d. for rough and dirty parcels, up to £4. 15s. for good, clean, dry lots.

East Indian bone meal is selling at £5. 10s ex quay Liverpool, and same price required to arrive. We do not hear of any forwar business.

Two hundred tons crushed Kurrachee bones arrived, Liverpool, sol at £5, ex quay, which price is asked for shipment ahead. Nitrate c soda without much change in values, but market quieter. Spot pric 8s. for ordinary; 8s 1½d. for refined. Due cargoes are worth abou 8s., and for shipment ahead 8s. 4½d. to 8s. 9d. would have to be paid according to position.

Some considerable sales of home prepared dried blood have bee made, and makers seem inclined to hold now for higher prices, bu buyers don't, so far, follow. Ten shillings and sixpence to 11s., according to position of works, now asked.

Superphosphates are more enquired for, and 47s. 6d. up to 50s., ne cash, in bulk, f.o.r., at works, now asked, but in some quarters less i being taken.

TAR AND AMMONIA PRODUCTS.

The benzol market is decidedly firmer than it was at this time las week, and 3s. 9d. may be quoted for 90's benzol, and 2s. 9d. fo 50/90's. Solvent naphtha is hardened likewise, and prices may be re garded as 1d. dearer than in our last report. Creosote is still movin off freely, while crude carbolic and anthracene are still neglected. Th pitch market is very dull, and no sales of any moment have taken place 28s. on East coast ports is about to-day's nominal value, but we canno hear of any business whatever in the meantime.

The sulphate of ammonia market is still in a peculiar position. A fair amount of business has been passing at current rates, and in som quarters a steadier tone is reported. There is not so much eagernes on behalf of makers to realise, and provided discretion in offerin parcels is now exercised no further shrinkage of value should tak place. On the other hand, there are some who believe that prices wil go still lower, but that we shall see a reaction for forward deliver before the end of this month. Prices to-day stand at £11. 2s. 6d. f o l Hull and Leith, while Beckton, as well as outside London makes, ar selling for £11 6s. 3d.

MISCELLANEOUS CHEMICAL MARKET.

There has been an absence of activity in nearly all branches of th trade, and few changes of any importance in values during the pas week. In the alkali trade prices all round remain steady. Causti soda recovered slightly, and any further decline in value is not looke upon as probable. F.o b. Liverpool quotations are for 74%, £10. pe ton; 70%, £9 to £9. 5s.; 60%, £8. 5s. Soda ash still in goo demand at 1½d. per deg at Liverpool for certain brands. 1⅜d. pe deg. for caustic ash. Soda crystals £2. 17s. 6d. at makers' works i casks. Chlorate of potash 4¾d to 5d. per lb. Bleaching powder i fair request and steady at £5. per ton, on rails, and at £5. 7s. 6d. f.o.b Recovered sulphur in all grades rather slow of sale, and prices withou alteration. Sulphate of copper, prompt delivery, £25. to £26. pe ton, and scarce; £23. to £24. for June, and lower quotations forward White powdered arsenic in good demand and firm at £13. to £13. 10s net at usual points Acetates of lime more freely offered; brown a £7. 15s and in plentiful supply; grey in limited supply at £13 15s. t £14. per ton. Acetate of soda dearer on the spot. Acetic acid without change. Acetate of lead and nitrate of lead firm, and rathe more inquiry. Flake litharge, £15. 15. per ton. Potash, causti and carbonate meeting with ready sale at current rates. Yellov prussiate of potash continues to hold a firm position at 9½d. t 9¾d. for British makes, and 9¼d. for foreign. Bichromate and chromic acid unaltered in price. Tartaric acid firm at 1s. 3d. pe lb. Carbolic acid crystals have still a downward tendency, owing t the falling off in demand. Aniline salt and oil are firm in prices, an meeting with a ready sale; pure oil at 10¾d to 11d. per lb., and pur salt at 9d. to 9½d. per lb. Ammonium salts all round are dull, an prices stationary.

THE COPPER MARKET.

Messrs. Harrington and Co.'s copper report, dated Liverpool, 1s May, states: Chili copper charters for second half of April are advise to-day as 900 tons fine. The total charters since 1st January have bee 8,500 tons, against 9,900 tons for same time last year. The trad have participated pretty freely, as also in furnace material; bu speculation, we think, has been interfered with by the excited an unsettled state of the silver market. The Association for yellov metal on the 17th ult., reduced its price to 5½d., but have to-da advanced same to 5¾d. per lb. It is reported that the French banke have just sold the greater portion of their American stock of ingots an slab copper, and that the quantity they now control there does no exceed 6,000 tons. The present stock of English g.m.b.'s in ware house, Liverpool and Swansea, is 6,092 tons, against 6,401 tons o wing a decrease of 309 tons for the fortnight and 494 tons ı. The sales of furnace stuffs comprise—Liverpool, 130 ore, at 9s. 9d.; 126 tons Italian ore (10%) at 8s.; 200 ore (15 to 20%) at 9s.; 26 tons (35%) at 10s.; 300 tons ½d.; 583 tons ordinary Anaconda Matte (second hands) tons at 10s.; 212 tons at 10s. 1d.; and 1,000 tons Argenti- ivate terms. Swansea—1,100 tons Namaqua ore (to arrive) , and 765 tons at 9s. 3d.; also 1,250 tons Spanish ore er and rich in sulphur) at 8s. 6d. Precipitate: 400 tons Mora (re-sale) at 9s. 6d.; 60 tons Seville at 10s. 4½d.; Miguel, at 10s.; 219 tons Mason's (old syndicate stock) at 164 tons at 9s. 7½d.; 100 tons English, at 10s. 3d. per

THE TYNE CHEMICAL REPORT.

TUESDAY.

nical market is much steadier this week, although there is ss doing. The general feeling is that prices have about bottom for the present. Bleaching powder keeps quiet are 2s. 6d. to 5s. per ton lower than last week. Caustic y steady at same price as quoted a week ago. Soda crystals brisk demand and prices are unchanged. Soda ash is just asier, and sulphate of soda 1s. 6d. per ton under last

To-day's quotations are as follows:—Bleaching powder, l casks, £5. to £5. 2s. 6d. per ton; caustic soda 77%. £10. ound in 3·4 cwt. barrels, £13. per ton; soda ash, 48/52%, degree less 3½%; soda crystals, in 1cwt. bags, net weight, d. per ton, in 2 cwt. bags, net weight, £2. 11s. per ton; ross weight, £2. 11s. per ton; sulphate of soda, in bulk, er ton; ground, in casks, 43s. 6d. per ton; recovered 4. 5s. per ton; chlorate of potash, 4¾d. per lb.; silicate ° Tw., £2. 10s. per ton; 100° Tw., £3. 7s. 6d. per ton; 140° per ton; hyposulphite of soda, in 1 cwt. kegs, £5. 15s. per —7 cwt. casks, £5. 5s. per ton; pure white sulphate of 4. 10s. per ton; blanc fixe, £7. 10s. per ton; chloride £8. per ton; nitrate of baryta, crystals, £18. 15s. per ton; 19. 5s. per ton; sulphide of barium, £5. 10s. per ton— Tyne, or f.o.r. makers' works.

s no change in the price of small coal for manufacturing Northumberland steam quoted 8s. 6d. to 8s. 9d. per ton, am small 10s. per ton. The representatives of the Durham ers' and Miners' Associations met together in Newcastle last d discussed the question of a further advance of wages, but unable to come to a decision; the meeting was therefore for another week.

oduction of magnesia will shortly be recommenced in a ing chemical works. The same company are preparing to re pearl hardening for paper making purposes.

lowing are the exports of chemicals from the Tyne for the April, and the corresponding month last year:—

	April 1890. Tons.	April 1889. Tons.
kali and soda ash	1,175	1,590
carbonate soda	10	8
eaching powder	3,278	2,035
nure	579	816
la crystals	2,701	2,482
lphate of soda	87	6
her chemicals	1,657	2,877
	9,487	9,814

WEST OF SCOTLAND CHEMICALS.

GLASGOW, Tuesday.

nates (both potash and soda) which were in rather better re. week before last, have again become comparatively neglected cotch makers are now sending out very restricted quantities eaching powder also is decidedly weaker, and prices have a trifle. Caustic soda is lower than it was a week ago, but have touched bottom, and the feeling is now one of recovery, ry more active and quotations again rather on the mount. tals are easier, also chlorate of potash, but nitrate of soda is e least little bit better off than it has been for the past two ith actual quotations, however, not much changed. Sul. ammonia has declined persistently, although at a very ate, and at present the value, prompt Leith, does not exceed d. It is asserted on the market here that sellers have, in a nces, parted at a lower figure still; but this has not sfactorily verified, and the probability is that the above resents the extremity for the time. Paraffins unchanged. iners are again talking of a further restriction of the output, as a measure for forcing a rise of wages. Chief prices current are:—Soda crystals, 50s. net Tyne; alum in lump £5., less 2½% Glasgow; borax, English refined £29., and boracic acid, £37. 10s. net Glasgow; soda ash, 1¼d. less 2½%, Tyne; caustic soda, white, 76°, £11., 70/72°, £8. 15s., 60/62° £8., and cream, 60/62° £7. 15s, all less 2½% Liverpool; bicarbonate of soda, 5 cwt. casks, £6. 5s., and 1 cwt. casks, £6. 10s. net Tyne; refined alkali 48/52°, 1½d., net Tyne; saltcake, 25s. to 26s.; bleaching powder, £5. 5s. to £5. 10s. less 5% f.o.r. Glasgow; bichromate of potash, 4d., and of soda 3d., less 5 and 6% to Scotch and English buyers respectively; chlorate of potash, 4⅝d., less 5% any port; nitrate of soda 8s. to 8s. 3d.; sulphate of ammonia, spot, £11. 5s. f.o.b. Leith; salammoniac, 1st and 2nd white, £37. and £35., less 2½% any port; sulphate of copper, £25., less 5% Liverpool; paraffin scale, hard and soft, 2½d. per lb.; paraffin wax, 120°, semi-refined, 3⅝d.; paraffin spirit (naphtha), 8d. a gallon; paraffin oil (burning), 6¾d. to 7d. on rails, Glasgow to all Scotch buyers (fixed) (English orders negotiable about 1¼d. a gallon lower); ditto (lubricating), 865°, £5. 10s. to £6. per ton; 885°, £6. 10s.; and 890/895°, £7. 15s. to £8. Week's imports of sugar at Greenock were 32,793 bags.

THE LIVERPOOL COLOUR MARKET.

COLOURS are unchanged. Ochres: Oxfordshire quoted at £10., £12., £14., and £16.; Derbyshire, 50s. to 55s.; Welsh, best, 50s. to 55s.; seconds, 47s. 6d.; and common, 18s.; Irish, Devonshire, 40s. to 45s.; French, J.C., 55s., 45s. to 60s.; M.C., 65s. to 67s. 6d. Umber: Turkish, cargoes to arrive, 40s. to 50s.; Devonshire, 50s. to 55s. White lead, £21. 10s. to £22. Red lead, £18. Oxide of zinc: V.M. No. 1, £25.; V.M. No. 2, £23., Venetian red, £6. 10s. Cobalt: Prepared oxide, 10s. 6d.; black, 9s. 9d.; blue, 6s. 6d. Zaffres: No. 1, 3s. 6d.; No. 2, 2s. 6d. Terra Alba: Finest white, 60s.; good, 40s. to 50s. Rouge: Best, £24.; ditto for jewellers', 9d. per lb. Drop black, 25s. to 28s. 6d. Oxide of iron, prime quality, £10. to £15. Paris white, 60s. Emerald green, 10d. per lb. Derbyshire red, 60s. Vermillionette, 5d. to 7d. per lb.

THE LIVERPOOL MINERAL MARKET.

Our market has continued to rule steady this week. Manganese: Arrivals are very much easier, and prices continue firm. Magnesite: Additional arrivals have told upon the tenor of the market, and prices for raw lump are flat. Raw ground, £6. 10s.; and calcined ground, £10. to £11. Bauxite (Irish Hill brand) still continues in exceptionally strong demand, bringing top prices—Lump, 20s.; seconds, 16s.; thirds, 12s.; ground, 35s. Dolomite, 7s. 6d. per ton at the mine. French Chalk: Stocks have been reduced to nil, and prices have advanced, especially in view of small arrivals. G.G.B. "Angel-White" brand—95s. to 100s. medium, 105s. to 110s. superfine. Barytes (carbonate), easy; selected crystal lump scarce at £6.; No. 1 lumps, 90s.; best, 80s.; seconds and good nuts, 70s.; smalls, 50s.; best ground, £6.; and selected crystal ground, £8. Sulphate has somewhat improved, best lump, 35s. 6d.; good medium, 30s.; medium, 25s. 6d. to 27s. 6d.; common, 18s. 6d. to 20s.; ground best white, G.G.B. brand, 65s.; common, 45s.; grey, 32s. 6d. to 40s. Pumicestone quiet; ground at £10., and specially selected lump, finest quality, £13. Iron ore steady. Bilbao and Santander, 9s. to 10s. 6d. f.o.b.; Irish, 11s. to 12s. 6d.; Cumberland in strong demand at 18s. to 24s. Purple ore as last quoted. Spanish manganiferous ore continues in fair demand. Emery-stone: Best brands still scarce, and inquired for at full prices. No 1 lump is quoted at £5. 10s. to £6., and smalls £5. to £5. 10s. Fullers' earth steady; 45s. to 50s. for best blue and yellow; fine impalpable ground, £7. Scheelite, wolfram, tungstate of soda, and tungsten metal further inquired for. Chrome metal, 5s. 6d. per lb. Tungsten alloys 2s. per lb. Chrome ore: High percentages selling at full prices. Antimony ore and metal have further advanced. Uranium oxide, 24s. to 26s. Asbestos scarce, bringing increased prices. Potter's lead ore firm; smalls, £14. to £15.; selected lump, £16. to £17. Calamine: Best qualities scarce—60s. to 80s. Strontia steady; sulphate (celestine) steady, 16s. 6d. to 17s. Carbonate (native), £15. to £16.; powdered (manufactured), £11. to £12. Limespar: English manufactured, old G.G.B. brand in demand; and brings full prices; 50s. for ground English. Felspar, 40s. to 50s.; fluorspar, 20s. to £6. Bog ore in demand, firm at 22s. to 25s. Plumbago steady; Spanish, £6.; best Ceylon lump at last quotations; Italian and Bohemian, £4. to £12. per ton; founders, £5. to £6.; Blackwell's "Mineraline," £10. French sand, in cargoes, continues scarce on spot—20s. to 22s. 6d. Ferro-manganese and silicon spiegel in good demand at stronger figures. Chrome iron, 20 per cent., £24. to £25. Ground mica, £50. China clay freely offering—common, 18s. 6d.; good medium 22s. 6d. to 25s.; best, 30s. to 35s. (at Runcorn).

Gazette Notices.

Partnerships Dissolved.

NUTHALL, THORNLEY, and BOOTH, Hyde, hat manufacturers, as far as regards W. Nuthall.

C. F. Hole, and C. Maybury, under the style of the Nottingham Chemical Company, Nottingham, chemical manufacturers.

Adjudications.

EDWARD PRIME, Barrington, Cambridgeshire, cement manufacturer.

The Patent List.

This list is compiled from Official sources in the Manchester Technical Laboratory, under the immediate supervision of George E. Davis and Alfred R. Davis.

APPLICATIONS FOR LETTERS PATENT.

Mechanical Stokers. J. W. Claridge. 5,991. April 21.
Fluid-tight Joints for Pipes. J. Newton and D. A. Quiggin. 5,992 April 21.
Electric Heating Apparatus.—(Complete Specification). M. W. Dewey. 6,032. April 21.
Utilisation of Electrical Energy.—(Complete Specification). M. W. Dewey. 6,041. April 22.
Syphons.—(Complete Specification). E. A. Meyer. 6,043. April 22.
Manufacture of Alkaline Silicates and Aluminates. J. Pointon. 6,046. April 22.
Manufacture of Alkaline Aluminates. J. Pointon. 6,047. April 22.
Manufacture of Alkaline Silicates. J. Pointon. 6,048. April 22.
Steam Generators. R. Cunliffe. 6,054. April 22.
Apparatus for Smoke Prevention. E. Cockerill and Son, and T. Cockerill. 6,055. April 22.
Steam Feed Water Apparatus. A. MacLaine. 6,059 April 22.
Photographic Developers. B. Jumeaux. 6,066. April 22.
Water Filters. J. W. Ward and C. Griffith. 6,080. April 22.
Kilns. E. Brook. 6,088. April 22.
Evaporation of Brine.—(Complete Specification). J. C. Mewburn. 6,107. April 22.
Pulverizing Machinery. G. J. Smith and T. H. Tregoning. 6,129. April 22.
Condensing and Cooling Apparatus.—(Complete Specification). J. Klein. 6,136. April 22.
Water Gauges. J. A. Hopkinson and J. Hopkinson. 6,141. April 22.
Electrical Switch. G. Schultz. 6,156. April 23.
Linings for Pulp Boilers. R. N. Redmayne. 6,161. April 23.
Secondary Batteries. G. Barker. 6,177. April 23.
Blue Colouring Matters for Cotton. S. S. Bromhead. 6,195. April 23.
Glazed Bricks and Tiles. W. R. May. 6,196. April 23.
Washing Fluid.—(Complete Specification). A. W. Curtis. 6,199. April 23.
Electrolytic Bleaching. S. Stepanow. 6,202. April 23.
Extraction of Gold and Platinum. C. T. J. Vautin. 6,216. April 23.
Production of Combustible Gas. S. Griffin. 6,217. April 23
Galvanic Batteries. M. Ancizar. 6,223. April 23.
Ovens or Kilns for Firing Pottery.—(Complete Specification). A Fielding. 6,231. April 24.
Tertiary Electric Battery.—(Complete Specification). A. J. Jarman. 6,236. April 24.
Supplying Water to Steam Boilers. C. Fouracre. 6,243. April 24.
Treatment of Sewage C. G. Moor. 6,245. April 24.
Prevention of Boiler Incrustation. C. R. Bonné. 6,246. April 24.
Filtering or Purifying Liquids. J. Longshaw. 6,252. April 24.
Application of Anthracite Gas for Metallurgical and other purposes. T. D. Rock. 6,256. April 24.
Electrolysis. J. Marx. 6,266. April 24.
Electrolytic Apparatus. J. Marx. 6,267. April 24.
Apparatus for Controlling and Measuring Liquids.—(Complete Specification). H. Börner. 6,273. April 24.
Production of Blue-Grey Colouring Matters. O. Imray. 6,274. April 24.
Recovering Sludge of Washed Coals or Ores. H. Simon. 6,275. April 24
Charging Coke Ovens. R. de Soldenhoff. 6,282. April 24.
Rag-Cleaning Machinery. W. Gillespie. 6,297. April 25.
Washing, Scouring, and Bleaching Fibrous Materials. R. Harrison 6,315. April 25.
Open Hearth System of Manufacturing Steel. S. Fox. 6,316. April 25.
Smoke Consumption. T. T. Vernon. 6,323. April 25.
Electric Meters. E. Batault. 6,326. April 25.
Production of Magnesic Oxychloride. F. M. Lyte. 6,333. April 25.
Improvements in Malting. W. Paterson. 6,337. April 25.
Dyestuffs. The British Alizarine Co., Limited, and D. C. Bendix. 6,346. Ap. 25.
Manufacture of Coloured Paper.—(Complete Specification). H. H. Lake. 6,348. April 25.
Chimneys and Chimney Tops—(Complete Specification). T. A. Meggeson and J. Hadfield. 6,358. April 26.
Machinery for Cleaning Cotton Seed. G. H. Croker. 6,368. April 26.
Improved Lining for Pump Boilers.—(Complete Specification). C. Kellner. 6,369. April 26.
Smoke Prevention. S Bamforth. 6,371. April 26.
Storing and Expelling Liquid under Pressure. R. Haddan. 6,382. Ap. 26.
Electric Motors. A. Blattner. 6,389. April 26.
Production of Caramel-Malt. L. Ramsel. 6,394. April 26.
Treatment of Sewage. F. B Hill. 6,397. April 26.
Improved Steam Generator.—(Complete Specification). G. Dürr. 6,398. April 26.
Reverberatory Furnaces. J. von Langer and L. Cooper. 6,399. April 26.
Utilizing Waste Heat of Furnaces. J. von Langer and L. Cooper. 6,400. April 26.

New Companies.

J. A. EGESTORFF AND COMPANY, LIMITED.—This company was registered on the 24th ult., with a capital of £10,000., in £5. shares, to acquire the business of manufacturers of ultramarine. lime blue, wash blue paste, red oxide of iron, and metal polishing paste, carried on by Messrs. Egestorff and Son, of Backbarrow. The subscribers are:—

	Shares.
J. A. Egestorff, Backbarrow, near Ulverston, colour merchant	1
J. C Egestorff, Backbarrow	1
F. O. Fell, Ulverston, contractor	1
C. B. Daniell, Ulverston, solicitor	1
A. H. Thorn, Cartinel, solicitor	
J. Atkinson, Ulverston, stationer	1
S. E. Mayor, Barrow-in-Furness, solicitor	1

ST. ASAPH IRON, LEAD, AND BARYTA COMPANY, LIMITED.—This company was registered on the 29th ult., with a capital of £30,000., in £1. shares, to acquire land, and to carry on business as miners, smelters, and reducers of ores and minerals. The subscribers are:—

	Shares.
E. R. Jameson, 111, Bushey-hill-road, Camberwell, accountant	1
A. R. Hewitt, 15, Shepperton-road, N., clerk	1
W. F. Gardner, Rushey-green, Catford	1
J. M. Niven, 13, Blomfield-road, Shepherd's-bush	1
Frank Fuller, Esk-villa, Leytonstone, secretary	1
W. C. Castle, 123, Plimsoll-road, N., clerk	1
T. H. Baxter, 29, Hannibal-road, Mile-end, clerk	1

UNIVERSAL GREASE AND OIL COMPANY, LIMITED.—This company was registered on the 24th ult., with a capital of £5,000., in £1. shares, to manufacture and trade in oils, fat, chemicals, and other similar articles. The subscribers are:—

	Shares.
A. W. Kerley, 14, Great Winchester-street, solicitor	1
G. S. Pitt, East Dene, Willesden-green	1
T. H. Briden, 9, Lorrimore-square, S.E.	1
Hy. Verden, 14, Great Winchester-street, clerk	1
H. W. Broadhurst, 137, Fenchurch-street, grease manufacturer	1
Alex. Kerly, 14, Great Winchester-street, solicitor	1
A. G. Thiselton, 20, Queen's-terrace, Peckham, clerk	1

IMPORTS OF CHEMICAL PRODUCTS AT THE PRINCIPAL PORTS OF THE UNITED KINGDOM.

LONDON.

Week ending April 29th.

Alum—		
Holland,	£25	Hernu, Peron & Co.
"	325	Ohlenschlager Brs.
Alkali—		
Holland,	300 c.	E. Ogett & Sons
"	200	Taylor, Sommerville & Co.
"	208	Beresford & Co.
"	200 bgs.	H. Grey, jun.
"	200	T. H. Lee
Acetone—		
Holland,	£53	Beresford & Co.
Acetic Acid—		
Holland,	101 pks.	A. & M. Zimmermann
France,	1	Pokorny, Fielder & Co.
Acetate of Lime—		
U.S.,	£204	Johnson & Hooper
Halifax,	100	Bryce, Junor & Co.
Antimony Ore—		
Sydney,	9 t.	L. & I. D. Jt. Co.
Portugal,	10	H. Emanuel
Spain,	10	
S. Russia,	1	M. " D. Co.
Bones—		
Brisbane,	2 t.	L. & I. D. Jt. Co.
U.S.,	34	Hicks, Nash & Co.
Auckland,	2	Flack, Chandler & Co.
Launceston,	1	J. Terry & Co.
E. Indies,	207	T. H. Allen & Co.
Sydney,	7	Flack, Chandler & Co.
Barytes—		
Germany,	22 cks.	J. W. Harrison & Co.
Belgium,	130	A Zumbeck & Co.
Holland,	91	A. Boltz
"	45	W. Harrison & Co.
Germany,	47	"
Barium Chloride—		
Germany,	£80	Petri Brs.
Cream of Tartar—		
Italy,	20 pks.	C. Christopherson & Co.
France,	3	T. Ronaldson & Co.
Copper Ore—		
Sydney,	72 t.	L. & I. D. Jt. Co.
Cossack,	27	W. Marden
Adelaide,	7 c.	H. Bath & Son
"	12 t.	Harrold Brs.
Caoutchouc—		
E. Indies,	91 c.	Wrightson & Son
"	105	Ralli Brs.
U.S.,	126	Kebbel, Son & Co.
"	31	Mildred, Goyeneche & Co.
Mozambique,	23	Elkan & Co.
Zanzibar,	52	A. Frey & Co.
"	13	Wrightson & Son
Natal,	9	J. T. Rennie, Son & Co.
Cape,	17	Lewis & Peat
France,	15	Stiebel Brs.
Zanzibar,	37	L. & I. D. Jt. Co.
Quilimane,	24	L. & I. D. Jt. Co.
Carbonic Acid—		
Holland,	12 pks.	J. G. Fisher

s—
30 Phillipps & Graves
otherwise undescribed)
73 A & M. Zimmermann
4 pks. J. A. Reid
10 Petri Brs.
50 T. H. Lee
04 A. & M. Zimmermann
78 W. Burton & Sons
50 W. Nicolson & Co.

20 pks B. & F. Wf. Co.
2 C. Atkins & Nisbet
25 Pickford & Co
1 R. J. Fullwood & Bland
13 G. S. N. Co.
20 J. Graves
50 L. J. Levinstein & Sons
50 "
11 G. S. N. Co.
46 G. B. Mackereth
3 Tegner Price & Co.
9 Boutcher, Mortimore & Co.

Bark
4 t. Arnati & Harrison
25 Dunlop Brs. & Co.
207 Baxter & Hoare
d, 19 W. H. Boult
54 Baxter & Hoare
11 Leach & Co.
40 J Graves
10 H. Henlis' Succrs.
95 Baxter & Hoare

82 chts. L. & I. D. Jt. Co.
10 W. Isted
24 Ross & Deering
58 L. & I. D. Jt. Co.
25 "
275 T. H. Allan & Co.
110 L. & I. D. Jt. Co.
16 F. W. Heilgers & Co.
6 F. Huth & Co.
57 L & I. D. Jt. Co.
31 Benecke, Souchay & Co.

ams
700 pks. Idle & Christie
,974 L. & I. D. Jt. Co.
350 Baxter & Hoare
,732 Beresford & Co.

120 pks. S. Barrow & Brs.
93 Elkan & Co.
401 Beresford & Co.
167 P. & S. Evans & Co.
678 Elkan & Co.
216 H. Lambert

451 pks. B Jacob & Sons

6 pks. W. Balchin

114 pks. Hoare, Wilson & Co.

600 pks. Wrightson & Son
500 Middleton's S.S. Wf. Co.
710 Anderson, Weber & Smith
250 Middleton's S.S. Wf. Co.

ne
51 pks. Burt, Boulton & Co.

l
16 pks. Swanston & Co.
9 Kuhner, Hendschel, & Co.

8 bls. P. Macfadyen & Co.
5 H. Rooke, Sons, & Co.
5 J. T. Rennie, Sons, & Co.
(otherwise undescribed)
3 cks. C. Atkins & Nisbet
2 A. Henry & Co.

15. Beresford & Co.
110 T. H. Lee
300 Spies Bros. & Co.
98 Hoare, Wilson, & Co
100 Beresford & Co.
75 F. W. Heilgers & Co.
70 H. Lambert

138 pk. 650 cwts. T. M. Duche & Sons
30 150 U. Lighterage Co., Ld.
00 100 J. Barber & Co.
5 45 Perkins & Homer
00 400 Horne & Co.

" 25 240 Fellows, Morton, & Co.
" 24 210 Horne & Co.
" 37 325 L. & D. Jt. Co.
" 50 50 J. Barber & Co.
" 100 108 J Forsey
" 300 300 J. Barber & Co.
" 200 200 Barrett, Tagant, & Co.
U. S., 500 500 "
" 100 1,100 U. Lighterage Co., Ld.
" 800 801 A. Dawson
" 50 303 Proprs. Scotts' Whf.
" 160 1,200 L. Sutro & Co.
Germany, 24 198 J Cooper
" 9 82 J. Knill & Co.
" 25 200 T. M. Duche & Sons
" 25 235 J. Cooper
" 55 434 L. Sutro & Co
" 200 200 J. Greig
France, 100 100 J. Barber & Co.
" 200 200 De Paiva Norman & Co.
U. S, 100 600 U. Lighterage Co. Ld.
Germany, 800 800R E. Drummond & Co.
Holland, 1 6 C. B de Witt

Guano—
Peru, 1,220 t Anglo Contl. Guano Wks.
Sweden, 45 Anglo Swedish Co

Gutta Percha—
Singapore, 604 c. Kaltenbach & Schmitz
Berbice, 10 D. Baird & Sons
France, 33 Kleinwort, Sons, & Co.

Hyposulphite—
Germany, £75. Beresford & Co.

Kainit—
Germany, £403 Perkins & Homer

Manganese—
Germany, £30. Petri Bros.

Manganese Ore—
Adelaide, 9. t. Harrold Bros.

Manure—
Germany, 11 t. W. & C Paintin
" 10 F. W Berk & Co
Abberville, 20 Lawes Manure Co., Ld
Germany, 50 F. W. Berk & Co.
" 10 Petri Bros

Phosphate Rock
Aruba, 705 t. G. M. Bauer

Naphtha—
Holland, 8 drs. Burt, Boulton, & Co
" 13 Domeier & Co.

Nitre—
Caleta Buena, 14,147 bgs. W Montgomery & Co.

Potassium Hydrate
France, 4 pks. Spies, Bros. & Co.

Potassium Sulphate—
Germany, 203 pks. T. Farmer & Co.
France, 4 Carey & Sons

Potassium Prussiate—
Germany, £850.

Potassium Carbonate—
Germany, 91 cks. Beresford & Co.
France, 10 J. A Reid
Germany, 35 W. Balchin

Potassium Muriate—
Germany, 1 pkg. Beresford & Co.

Potassium Oxymuriate—
Holland, £230 G. Boor & Co.

Potassium Bicarbonate—
Holland, £80. A. & M. Zimmermann

Plumbago—
Germany, 17 cks. Fellows, Morton, & Co.
Holland, 72 Browne & Elmslie
Germany, 17 brls. Beresford & Co.
Italy, 300 J. Thredder, Son, & Co.
" 160 bgs. B. Gallaway
E Indies, 59 brls. L. & I D. Jt. Co.
" 40 Hoare, Wilson, & Co.
Germany. 1 ck. B. Gallaway
E Indies, 82 J. Thredder, Son, & Co.

Paraffin Wax—
U. S., 420 brls. H. Hill & Sons
" 200 G. H. Frank
" 7,150 J H. Usmar & Co.
Batoum, 600 Grosscurth & Luboldt
Germany, 3 cs. "

Pearl Ash—
France. 40 c. Petri Bros.

Pitch—
France, 63 cks. H. Hill & Sons
E. Indies, 1,244 bgs. C. Wimble & Co.
" 241 bils. Elkan & Co.
" 592 Louis, Achard & Co.

Germany, 70 P. Hecker & Co.
36 cks. Craven & Co.
" 127 J. Hall & Son
" 72 P. Hecker & Co
" 80 "
" 100 T. Merry & Son

Sugar of Lead—
France, 10 pks. Prop Hay's Whf

Sodium Acetate—
Holland, £95. Soundy & Son

Sodium Nitrate—
Holland, £50. J. Owen

Sulphuric Acid—
France, 41 pks. Langstaff & Co.

Soda Ashes—
Holland, 205 c Henderson, Craig, & Co

Tartaric Acid—
Holland, 6 pks. Major & Field
" 7 B. & F. Wf. Co.
" 18 "
" 9 Augspurg, Hopf, & Co.

Ultramarine—
Holland, 1 ck. Ohlenschlager Bros.
Germany, 20 H. Lambert
" 2 J. Kitchen Ld.
Holland, 1 Glaeser & Vogeler
Belgium, 4 W. Harrison & Co
Holland, 11 Seaward Bros
Belgium, 8 H. Rohnstamn
Germany, 50 cs. G. Steinhoff
" 4 cks. Hernu, Peron, & Co.
Holland, 4 H. Kohnstamn
" 2 Ohlenschlager Bros.
Belgium, 7 Leach & Co.

Zinc Oxide—
U. S., 100 brls. M. Ashby, Ld.
Holland, 100 "
" 25 G. Dards
Germany, 300 cks Soundy & Son
Holland, 15 brls. Beresford & Co.

LIVERPOOL.

Week ending May 1st.

Albumen—
Antwerp, 3 cks. J. T. Fletcher & Co.

Borax—
Hamburg, 1 brl.

Boracic Acid—
Leghorn, 130 cks.

Bones—
Kurrachee. 4,000 bgs Ralli Bros.
" 216 W. & R. Graham & Co.
Constantinople, 185 bgs.

Bone Ash—
Rio Grande, 107 t. 5,200 kilos. J. Beckwith & Co.

Bone Meal—
Bombay, 2,000 bgs.
Kurrachee, 2,000 Ralli Bros.

Barytes—
Antwerp, 28 cks.

Caoutchouc—
Akassa, 186 cks 280 bgs. Royal Niger Co.
Quittah, 3 brls. F. & A. Swanzy
Lamie, 7 pns. "
Accra, 1 bg. A. Reiss
" 8 cs. Pickering & Berthoud
Salt Pond, 13 brls. C. Rattmann
" 2 Radcliffe & Durant
C. C. Castle, 1 1 cs. C. L. Clare & Co.
" 2 W. D. Woodin
" 11 pks. A. Miller Bros.
" 2 brls. E. Grund
" 4 cks. W. B. M'Iver & Co.
" 2 brls. 1 cs.
Dixcove, 1 cs. F. & A. Swanzy
Axim, 6 bgs. Universal Co Bk.
" 1 A. Millerson & Co.
" 5 Pickering & Berthoud
" 1 brl. Lintott, Bros, & Co.
Assinee, 1 ck. Millward, Bradbury, & Co.
" 1 F. & A. Swanzy
" 1 cs. Pitt, Bros, & Co.
" 1 F. Schaeffer
" 2 bgs. Powell & Sing
Grand Bassam, 1 brl. 2 cks. S. Henrickson & Co.
" 2 bgs. W. D. Woodin
" 1 Radcliffe & Durant
Manaos, 20 cs. Bieber & Co.
" 25 W Carden & Co.
" 75 J. Lilley & Son
" 64 R Vietz
" 40 Bard & Co.
" 1 077
Para, 211 F. Snarez & Co.
" 573

Grand Bassam, 5 bgs. Edwards Bros.
" 3 A. Reis
" 6
" 2
" 7 cs. A Verdier
Sierra Leone, 10 pns. Paterson, Zachonis, & Co.
Rotterdam, 12 cks 209 bgs. 5 cs.

Copper Ore—
Lisbon, 21 t. 55 cks. J Harrington & Co.
Rosario, 153 bgs. A. Gibbs & Sons

Cream of Tartar—
Barcelona, 2 cks.
Bordeaux, 2
Tarragona, 5 brls Sachse & Klemn

Charcoal—
St. Nazaire, 700 sks.

Dyestuffs—

Cochineal
Las Palmas, 35 bgs. Lathbury & Co.
Teneriffe, 25 M. Pardo & Co.
Hamburg, 1 cs.

Extracts
Philadelphia, 200 bxs 6 brls
Havre, 10 cs. Cunard S. S. Co., Ltd.
" 120 20 cks "
Messina, 34 cs.
New York, 1 bx. A. Holt
Rouen, 25 cks. R. I. Francis & Co.

Myrabolams
Bombay, 230 bgs. D. Sassoon & Co
" 4,712 "
Calcutta, 1,656 pks.

Sumac
Palermo, 100 bgs. J. Glynn & Son
" 100 J. Kitchen, Ltd.
" 375

Valonia
Syra, 200 bgs. Andronlis Bros.

Argols
Oporto, 22 cks. F. Leyland & Co.
Bordeaux, 14

Orchella
Lisbon, 32 bls. 17 bgs.

Fustic
Havre, 489 bgs. Cunard S. S. Co., Ltd.
Havana, 549 ps.

Divi Divi
Havre, 404 bgs. Cunard S.S. Co., Ld.

Logwood
Hamburg, 10 cs.
Belize, 360 t. E. Brownbill & Co.

Guano—
Rio Grande, 1,155 bgs. J. Beckwith & Co.

Glucose—
Hamburg, 44 cks.
" 40 bgs.
New York, 132 bls. Schoell, Rope & Co.
" 50 brls. G. E. Bastol
" 80

Glycerine—
Monte Video, 100 cs. Mariano Yaro

Horn Piths—
Rotterdam, 24 bls. and a quantity.
Rio Grande, 5,500 J. Beckwith & Co.

Iodine—
Valparaiso, 107 brls. A. Gibbs & Sons

Manure—

Phosphate Rock
Savannah, 730 t.
Beaufort, 980

Pyrites—
Huelva, 1,420 t. Matheson & Co.
" 1,270 Tennants & Co.
" 1,080

Potash—
Hamburg, 1 cs.

Potassium Muriate—
Hamburg, 100 bgs.

Saltpetre—
Hamburg, 21 cks.
" 54
Calcutta, 688 bgs. Ralli Brs.
" 1,035

Stearine—
Rotterdam, 10 cks.
Antwerp, 37 bgs.

Sodium Nitrate—
Pisagua, 9,933 bgs. W. & J. Lockett

Sulphur Ore—
Huelva, 1,452 t. Matheson & Co.
" 1,657 "

Tar—
Wilmington, 1,000 brls.

Tartar—
Bordeaux, 69 cks.
Naples, 3 R. Powles
"

Tartar Salts—
Rotterdam, 10 cks.

Tartaric Acid—
Rotterdam, 14 cks.

Ultramarine—
Hamburg, 1 cs.
Rotterdam, 30 brls. 7 cs. 24 cks.
" 6 2

Verdigris—
Bordeaux, 1 ck.

Waste Salt—
Hamburg, 1,304 bgs.

Zinc Oxide—
Antwerp, 25 cks. 53 brls 1 cs.

GLASGOW.

Week ending May 1st.

Antimony Regulus—
Marseilles, 38 brls. J. W. Quincy & Co.

Barytes—
Rotterdam, 46 cks.
" 9c

Chemicals (*otherwise undescribed*)
Hamburg, 3 cs.

Cream Tartar—
Marseilles, 5 brls.

Chrome Alum—
Rotterdam, 26 cks. J. Rankine & Son

Dyestuffs—

Alizarine
Rotterdam, 199 cks.
" 61

Extracts
Antwerp, 10 cks.
Rouen, 14

Sumac
New York, 30 bgs.
Marseilles, 100

Indigo
Marseilles, 4 cs

Glycerine—
Marseilles, 10 brls. A. Stewart & Son

Glucose—
Amsterdam, 5 cks

Pyrites—
Huelva, 1,592 t. Tharsis Co.

Phosphate of Lime—
Antwerp, 100 t.

Sulphur Ore—
Bergen, 180 t.

Tartaric Acid—
Marseilles, 10 brls.
Rotterdam, 8 cks.

Ultramarine—
Antwerp, 5 cs.
Rotterdam, 1 cs. 4 cks. J Rankine & Son

Waste Salt—
Hamburg, 203 bgs.

Zinc Oxide—
Rotterdam, 50 cks.
" 200

HULL.

Week ending May 1st.

Acids (*otherwise undescribed*)
Rotterdam, 3 cks. G. Lawson & Sons
" 1
" 20 "
" 37 Wilson, Sons, & Co.

Ammonia—
Antwerp, 1 ck. Wilson, Sons, & Co.

Barytes—
Rotterdam, 23 cks. Hutchinson & Son
Bremen, 10 cks Veltmann & Co.
" 19 Blundell Spence, & Co.
" 60
Antwerp, 20 Bailey & Leetham

Bones—
Riga, 363 bgs. Wilson, Sons, & Co.

Bone Meal—
Bombay, 4228 bgs

Chemicals (*otherwise undescribed*)
Hamburg, 5 cs. Wilson, Sons, & Co.
Antwerp, 38 pks. "
Hamburg, 5 cks.
Rotterdam, 4 W. & C. L. Ringrose

Dyestuffs—

Alizarine
Rotterdam, 1 brl. G. Lawson & Sons
" 10 cks. Hutchinson & Son
" 1 W. & C. L. Ringrose
" 59
" 36 pks. Hutchinson & Son

Valonia
Smyrna, 218 t. 990 bgs. Brown, Atkinson, & Co.
" 530 t.

Extract
Rouen, 48 cks. Rawson & Robinson

Sumac
Catania, 50 bgs. Wilson, Sons, & Co.
Palermo, 1,535 pks

Myrabolams
Bombay, 10,254 bgs.

Rennet
Rotterdam, 2 cks. W. & C. L. Ringrose

Glycerine—
Amsterdam, 1 drum.

Glucose—
Hamburg, 15 cks. Woodhouse & Co.
Stettin, 57 Wilson, Sons, & Co.
" 40
" 800 bgs. C. M. Lofthouse & Co.
New York, 95 brls.

Manure—
Ghent, 50 t. Hammond, Emes, & Co

Phosphate Rock
Coosaw, 1 930 t.

Naphthalene—
Rotterdam, 19 cks. G. Lawson & Son

Potash—
Antwerp, 26 cks. Wilson, Sons, & Co.

Pitch—
Antwerp, 11 cks. H. F. Pudsey
Dunkirk, 25 Wilson, Sons, & Co.

Plumbago—
Hamburg, 10 brls. Wilson, Sons, & Co.

Soda—
Rotterdam 50 cks. G. Lawson & Sons

Sodium Nitrate—
Iquique, 7,345 bgs

Sulphur—
Catania, 455 pks Wilson, Sons, & Co.

Sulphur Ore—
Seville, 1,230 t J. Dalton, Holmes, & Co.

Sulphate—
Christiansand, 100 pks. Wilson, Sons, & Co.

Stearine—
Rotterdam, 75 bgs Wilson, Sons, & Co.
Antwerp, 6 "

Silicate—
Antwerp, 10 cks. Wilson, Sons, & Co.

Saltpetre—
Hamburg, 100 bgs. Wilson, Sons, & Co

GOOLE.

Week ending April 23rd.

Chemicals (*otherwise undescribed*)
Rotterdam, 1 ck.

Dyestuffs—

Alizarine
Rotterdam, 30 cks. 1 bx.

Logwood
Jamaica, 500 t.
Monte Christi, 391

Dyestuffs (*otherwise undescribed*)
Boulogne, 56 cks. 4 cs 3 pks.

Glucose—
Hamburg 40 cks. 20 bgs.
Dunkirk, 120

Potash—
Hamburg, 12 cks.

Potassium Muriate—
Rotterdam, 20 cks.
Hamburg, 1,000 bgs.

Soda—
Rotterdam, 5 cks.

Saltpetre—
Rotterdam, 285 bgs.

Waste Salt—
Hamburg, 498 bgs.

TYNE.

Week ending May 1st.

Copper Precipitate—
Pomaron, 3 633 bgs. Mason & Barry

Glucose—
Hamburg, 50 bgs. Tyne S. S. Co.
New York, 25 brls. B. J. Sutherland & Co.

Manure—
Hamburg, 870 t J. C. Swan & Co

Pyrites—
Huelva, 1,373 t. Scott Bros.

Sulphur Ore—
Pomaron, 1,050 t. Mason & Barry

Salt—
Hamburg, 101 bgs. Tyne S. S. Co.

Saltcake—
Hamburg, 160 t.

Zinc Oxide—
Antwerp 50 brls. Tyne S. S. Co.

GRIMSBY.

Week ending April 26th.

Chemicals (*otherwise undescribed*)
Hamburg, 33 cks. 10 pks.

Dyestuffs—

Extracts
Dieppe, 16 pks.

Dyestuffs (*otherwise undescribed*)
Hamburg, 2 cks.

Glycerine—
Dieppe, 10 cs.

Zinc Ashes—
Antwerp, 12 cks.

EXPORTS OF CHEMICAL PRODUCTS

OF

THE PRINCIPAL PORTS OF THE UNITED KINGDOM.

79 t. 16 c. £640
30 00
40 400
phosphate—
39 t. 5 c. 115
11 t. 5 c. £100
19 70
39 419
224 3 845
100 800
573 16 4,806
67 11 220
10 1 85
12 6 55
120 bgs. 26
30 9 210
48 6 299
34 3 170
119 450
115 1,552
9 15 50
2 22
4 45
36 19 290
82 13 912
8 89
17 2 188
25 250
10 30
50 600
25 250
—
4 c. £14
l—
1 t. 8 c. £55
6 12
1 3 40
Chlorate—
12 c. £27
1 t. 45
5 210
5 210
phate—
50 t. £380
en, 206 5 c. 547
—
Bay, 20 cs. £50
17 pks. 28
22 t. 16 c. 1,120
10 26
120 brls. £92
250 193
735 pks. 606
lphate—
115 t. £115
-
7 t. 7 c. £154
5 109
2 10 55
iro, 1 12 36
4 18 106
230 t. £322
als—
10 t. 1 c. £30
32 10 93
t., 7 t. 12 c. £39
5 8 30
carbonate—
r., 5 t. £31
cid—
1 t. £128
5 34
1 18 245

ERPOOL

nding Apr. 30th.

Muriate—
12 c. 1 q. £15
21 t. 8 504
5 1 3 126
l—
3 q. £7
Regulus—
4 c. 2 q. £16
12 43
556
Carbonate—
2 t. 2 q. £66
1 6 c. 3 44
1 2 9
1 3 5
2 2 3 70
3 t. 15 c. 1 q. £17
8 9 1 40
2 18 14

Salvora, 1 t. 5
Sydney, 1 5
Pasages, 10 12 c. 52
Montreal, 25 9 118
Baltimore, 18 14 1 q. 85
Boston, 2 16 2 18
Valencia, 1 3 1 6
Ammonium Sulphate—
Valencia, 50 t. £625
Bordeaux, 10 2 c. 1 q. 125
Valencia, 50 6 3 640
Acetic Acid—
Santos, 11 c. 2 q. £13
Borax—
Matanzas, 9 c. 1 q. £14
Samsoun, 16 1 30
London (Ont.), 11 1 17
Odessa, 2 t. 3 2 62
Rotterdam, 4 6
Blood Albumen—
Santos, 11 c. 3 q. £57
Brimstone—
Pernambuco, 1 t. 13 c. £13
Ponce, 4 5 1 q. 27
Boston, 25 106
Bleaching Powder—
Baltimore, 52 cks.
Barcelona, 40
Bordeaux, 33
Boston, 774 172 tcs.
Calcutta, 75 cs.
Copenhagen, 14
Ghent, 160
Hamburg, 56
Malaga, 30 brls.
Montreal, 56 tcs.
New York, 339
Odessa, 31
Palermo, 4
Philadelphia, 53
Rotterdam, 29
Terneuzen, 25
Amsterdam, 6
Antwerp, 62
Bordeaux, 25
Boston, 58
Genoa, 240
Hamburg, 29
Montreal 202 pks.
Nantes, 14
New York, 5 bxs.
Oporto, 100 kgs.
Tarragona, 6 brls.
Caustic Potash—
Montreal, 8 t. 1 c. £170
Napier, 1 20
New York, 10 10
Cobalt Oxide—
New York, 8 c. 3 q. £425
Valencia, 1 55
Copper Precipitate—
Naples, 20 t. 1 c. £725
Charcoal—
Rotterdam, 50 t. £210
Hamburg, 51 255
Carbolic Acid—
Hamburg, 4 t. 15 c. 1 q. £67
Philadelphia, 1 3 90
Coquimbo, 1 6
Yokohama, 2 17 176
New York, 3 19 244
Hamburg, 6 16 1 778
" 4 460
Havre, 1 4 2 140
Rotterdam, 4 10 620
Copper Sulphate—
Barcelona, 38 t. 18 c. 2 q. £956
" 2 4 53
Bordeaux, 120 18 2,831
Odessa, 2 53
Tarragona, 3 15 89
Valencia, 5 1 120
Barcelona, 5 2 100
B. Ayres, 1 15 47
Marseilles, 5 3 1 103
Piræus, 9 12
B. Ayres, 1 14 51
Rio de Janeiro, 3 6
Tarragona, 10 246
Valencia, 10 3 254
Bilbao, 10 2 240
Genoa, 8 2 190
Nantes, 28 410
Venice, 10 3 236
Barcelona, 10 3 120
Bordeaux, 79 9 1,835
Galatz, 2 9 2 55
Genoa, 105 2 1 1,950
Havre, 5 1 95
Genoa, 31 12 3 568
Marseilles, 192 12 1 4,570
Trieste, 1 13 2 40
Venice, 20 4 370
Valencia, 14 19 1 365

Almeria, 1 16
Bordeaux, 5 3 110
Barcelona, 3 90
Pasages, 10 2 3 200
Genoa, 23 18 3 505
Leghorn, 8 12 2 182
Naples, 10 200
Trieste, 17 358
Venice, 4 5 90
Calcutta, 5 102
Barcelona, 10 212
Bordeaux, 8 17 216
Rouen, 59 11 4,487
Marseilles, 55 15 1,365
Genoa, 5 1 3 90
Venice, 10 2 202
Nantes, 5 3 134
Calcium Chloride—
Wellington, 4 t. £8
" 2 6
Brisbane, 2 11
Chrome Acetate—
Naples, 4 c. 2 q. £11
Copperas—
Smyrna, 3 t. 6 c. 3 q. £10
Montreal, 3 3 1 7
Calcutta, 11 14 3 27
Constantinople, 3 t. 12 c. 2 q. 8
Quebec, 3 7 1 2
Rio Janeiro, 2 8 1 5
Rotterdam, 20 19 2 44
Salonica, 4 6 10
Smyrna, 6 10 3 14
Syra, 6 11 3 15
" 3 19 9
Adelaide, 6 14
Chemicals *(otherwise undescribed)*
Antwerp, 2 t. 16 c. 3 q. £78
Hamburg, 17 12 431
Constantinople, 3 13 2 107
Ibrail, 1 10 1 52
Galatz, 8 14
Rio Janeiro, 2 5 22
Vera Cruz, 2 13 71
Caustic Soda—
Adelaide, 5 pks.
Amsterdam, 55 dms.
Ancona, 10
Antonina, 3 cks.
" 53
Arica, 20 bxs.
Boston, 300
Catania, 35
Dunedin, 15
E. London, 2
Genoa, 55
Hamburg, 6
La Guayra, 10
Leghorn, 50
Maceio, 30
Malaga, 10
Marseilles, 10
Montreal, 50
Algiers, 5
Amsterdam, 60
Antwerp, 49
Barcelona, 90 36 cks.
Calcutta, 106
Carthagena, 25
Copenhagen, 6 brls.
Dunedin, 16
Fairwater, 50
Galatz, 50
Havre, 150
Kurrachee, 36
Lisbon, 20
Maceio, 80
Malaga, 10 2
Malta, 60
Marseilles, 15
Matanzas, 8
Monte Video, 25
Montreal, 150
Naples, 20
New Orleans, 100
" 125
New York, 1,538
Oporto, 75
Pasages, 50
Pernambuco, 10
Petrolia, 10
Philadelphia, 100
Rotterdam, 20
Rouen, 65
San Sebastian, 4
Santander, 24
Seville, 174
Sydney, 11
Valencia, 16
Vigo, 3
Wellington, 32
Nantes, 35
Napier, 10 bxs.
Naples, 32
New York, 100
" 143 105 pks.
Odessa, 350

Oporto, 16
Philadelphia, 50
Piræus, 15
Portland Or., 75
Rotterdam, 58
Santander 6
Seville, 165
Dried Blood—
Barbadoes, 50 t. £450
Disinfectants—
Madras, 6 t. 11 c. 3 q. £66
New York, 5 cks. 45
Baltimore, 20 cs. 90
Epsom Salts—
Constantinople, 10 cks.
Montreal, 40 brls.
Rio Janeiro, 159 kgs.
Guano—
Hamburg, 19 t. £70
Havana, 401 14 c. £3,408
Glycerine—
Coquimbo, 2 c. 2 q. £10
St. John's, N. F., 1 t. 3 c. 120
Ground Spar—
Havre, 96 t. 9 c. £802
Gypsum—
New York, 10 t. 1 c. £30
Iron Oxide—
Callao, 5 t. £15
Lime Superphosphate—
Montreal, 25 t. 3 c. 2 q. £87
Lime Hypophosphate—
Calcutta, 1 cs. £25
Lead Nitrate—
Lisbon, 3 c. 3 q. £5
Magnesia—
Corunna, 1 cs.
Magnesium Citrate—
Corunna, 2 q. £9
Manure—
Trinidad, 1 t. 12 c. £9
Bordeaux, 2 10 20
Magnesium Sulphate—
New York, 4 t. 7 c. £30
Oxalic Acid—
Barcelona, 5 t. 1 c. 3 q. £127
New York, 3 2 1 102
Barcelona, 1 7 2 43
New York, 9 14 320
Pitch—
Cette, 1,162 t.
Genoa, 300
Havre, 1 ck.
Rouen, 49 18 c.
Potassium Prussiate—
Rio de Janeiro, 10 c. £45
Potassium Carbonate—
Montreal, 5 t. 11 c. £78
New York, 9 6 131
Potassium Chlorate—
Boston, 5 t. £198
Brindisi, 1 10 c. 66
Coquimbo, 1 6
Montreal, 1 10 68
Genoa, 7 47
B. Ayres, 1 10 76
Hamburg, 2 5
Yokohama, 12 10 685
New York, 5 222
Hiogo, 7 10 300
Stockholm, 10 400
Mexico, 8 3 q. 20
Potassium Bichromate—
Rio de Janeiro, 1 t. 4 c. 1 q. £45
Calcutta, 11 21
Phosphorous—
Shanghai, 15 c. 3 q. £118
Picric Crystals—
Sydney, 2 c. 2 q. £16
Soda Ash—
Sydney, 321 brls.
Syra, 100
Talavera, 20
Toronto, 140
Torrijos, 20
Volo, 20
Wellington, 32
Yokohama, 127
Aivali, 50
Alicante, 10 cks.
Auckland, 22
Baltimore, 219 tcs. 326
Barcelona, 60
Boston, 160 219
B. Ayres, 150
Constantinople, 47
Dunedin, 60
Hiogo, 40
Leghorn, 78
Maceio, 15

Madrid, 85 brls.
Napier, 20
Oporto, 5 36 cks.
Philadelphia, 113 1,321 tcs.
Rio de Janeiro, 179
Rosario, 50
Seville, 50
Montreal, 54
Adelaide, 50
Baltimore, 66 44
Barcelona, 18
Boston, 139
B. Ayres, 200
Genoa, 15
Lisbon, 15
Montreal, 128
New York, 311
Odessa, 100
Philadelphia, 566
Rio de Janeiro, 360
Rosario, 50
Smyrna, 100
Vigo, 2

Soda Crystals—
Malta, 10 brls.
Montreal, 764
Napier, 34
New York, 140 cks.
Philadelphia, 280
Portland Or., 50
Singapore, 12
Wellington, 68
Calcutta, 50
Freemantle, 15 34
Philadelphia, 340
Portland Or., 335

Salt—
Aarhuus, 878 t.
Ahgwey, 10
Ambriz, 10
Archangel, 400
Barbadoes, 67
Benin, 25
Brisbane, 50
Cameroons, 100
Dunedin, 15
Gaboon, 50
Kallundborg, 156
Landana 7
Loango, 25
Montreal, 423 2 c.
New York, 1,487½
Old Calabar, 60¼
Philadelphia, 850
Quebec, 1,093 14
Sierra Leone, 200
Antwerp, 105
Baltimore, 200
Boston, 270
Brass, 50
Bruges, 198
Chicago, 80
Copenhagen, 649
Freemantle, 5
Hamburg, 312¾
Monrovia, 10
Montreal, 521 6
North Sydney, 522 6
Picton, 498 6
Sierra Leone, 71
Sinoe, 75
Trinidad, 35 17
Ukaka, 48
Cape Coast, 6¼
Old Calabar, 70

Sheep Dip—
E. London, 1 t. 10 c. £80
Cape Town, 15 40
Algoa Bay, 1 10 80

Sheep Wash—
Montreal, 11 c £30

Sodium Silicate—
Venice, 16 cks.
Genoa, 173 brls.
Messina, 10
Naples, 100
Oporto, 6
Piræus, 15
Santander, 3

Sodium Bicarbonate—
Batoum, 100 kgs.
B. Ayres, 50 cks.
Chalcis, 20
Lyttletown, 20
Nantes, 4
Rotterdam, 25
Seville, 100 33 brls.
Sydney, 100
Wellington, 50
Adelaide, 190
Alexandria, 1 ck.
Algiers, 20
Antwerp, 5 pks.
Barcelona, 250
Bombay, 250
Boston, 50
Calcutta, 500
Genoa, 40 cks.
Montreal, 1,070 pks.
New York, 20 kgs.
Portland, 150
Rotterdam, 20
St. Petersburg, 400

Sodium Bisulphide—
Wellington, 15 c. £9

Saltcake—
Baltimore, 200 t. 7 c. £443

Sodium Nitrate—
Ancona, 7 t 12 c. 1 q. £63
Bari, 3 1 3 25
Seville, 12 12 104
Odessa, 10 1 87
Trieste, 13 17 42

Sodium Bichromate—
Syra, 10 c. £13

Sodium Biborate—
Alexandria, 5 c. 3 q. £9
Rotterdam, 2 t. 10 3 77
Montreal, 5 9 2 164

Sodium Chlorate—
Rotterdam, 1 t. £54

Salammoniac—
Genoa, 10 c. 2 q. £19
Hamburg, 9 3 18
New York, 5 t. 2 172
Oporto, 5 2 10
Palermo, 5 3 11
Syra, 9 1 16
Philadelphia, 4 2 1 144
Maceio, 5 10
Hiogo, 2 12 2 91
Colastine, 2 cs. 8

Saltpetre—
Palermo, 10 c. £11
Piræus, 15 t. 326
Ibrail, 5 6
Maceio, 3 1 62
Pernambuco, 8 6 3 q. 192
" 5 7 120
Maceio, 1 2 3 27

Sulphur—
Maceio, 7 c. 2 q. £8
Montreal, 1 t. 5 11
Boston, 50 217
Quebec, 1 12 9
Boston, 50 2 225
Rouen, 30 13 153

Tar—
Loanda, 20 brls.
Manilla, 9 cks.

Tartaric Acid—
Coquimbo, 1 c. £10

Tin Oxymuriate—
Rio Janeiro, 2 c. 2 q. £8

Tannic Acid—
Rouen, 7 c. £42

Zinc Chloride—
Bombay, 2 t. 1 c. 3 q. £42

TYNE.

Week ending May 1st.

Antimony—
Hamburg, 10 c.

Anthracene—
Rotterdam, 11 t. 19 c.

Alkali—
Hamburg, 4 t. 6 c.
" 5 13
Christiania, 5 8
Amsterdam, 22 15
Rotterdam, 5 11

Bleaching Powder—
Newfairwater, 9 t. 12 c.
Norrkoping, 54 7
Gothenburg, 12 3
San Francisco, 86 9
Hamburg, 24 11
Antwerp, 47 15
Dunkirk, 30 1
Amsterdam, 6 1
Copenhagen, 5 18
Rotterdam, 21 7
St. Petersburg, 879 13
Libau, 10 2

Caustic Soda—
Odessa, 6 t. 19 c.
Gothenburg, 10 4

Ferro Manganese—
Dahlsbruk, 5 t. 5 c.

Glycerine—
Rotterdam, 6 t 11 c.

Litharge—
Malta, 2 t. 4 c.
Constantinople, 1 5
Hamburg, 10
Genoa, 1 1
Gothenburg, 2 t. 14 c.
Copenhagen, 6
Rotterdam, 10
St. Petersburg, 2 14
" 8 10

Magnesia—
Hamburg, 16 c.
Antwerp, 2
St. Petersburg, 5

Muriatic Acid—
Monte Video, 2 t.

Napthalene—
Antwerp, 306 bgs.

Pitch—
Copenhagen, 20 t.

Potassium Chlorate—
Newfairwater, 6 t. 15 c.

Soda—
Svendborg, 20 t. 3 c.
Stockholm, 199 19
Odense, 23 14
Svendborg, 25 1
Ahus, 107 16
Drontheim, 36 18
Norrkoping, 47 19
Amsterdam, 25
Copenhagen, 29 4
Rotterdam, 11 1

Soda Ash—
San Francisco, 25 t. 7 c.
Hamburg, 3 13
Gothenburg, 45 2
Newfairwater, 20 13
Odessa, 6 11
Copenhagen, 15 15
Libau, 10 5

Sodium Chlorate—
Rotterdam, 1 t. 2 c.

Salammoniac—
St. Petersburg, 30 t. 5 c.
" 3 4
" 12 2

Sodium Sulphate—
St. Petersburg, 60 t. 13 c.

Superphosphate—
Esbjerg, 11 t. 15 c.

Sodium Hyposulphite—
San Francisco, 10 t.

Sodium Silicate—
Malta, 10 t. 4 c.

Sulphur—
Gothenburg, 40 t.

Sulphuric Acid—
Monte Video, 50

Tar—
Odessa, 12 t. 13 c.

HULL.

Week ending April 29th.

Alum—
San Francisco, 181 cks.

Alkali—
New York, 191 cks.

Benzole—
Hamburg, 74 drms.

Bone Size—
Bremen, 10 cks.

Bleaching Powder—
Antwerp, 24 cks.
Christiania, 28

Chemicals *(otherwise undescribed)*
Antwerp, 32 cs. 4 cks.
Abo, 18
Bombay, 209 100 pks.
Bordeaux, 20
Bremen, 51
Bergen, 12 40 10
Copenhagen, 12 cs. 10
Christiania, 66
Dunkirk, 22 30
Gothenburg, 5
Hamburg, 2 19 76
Konigsberg, 120
Norrkoping, 20
Rouen, 30
Riga, 4 36 320
Rotterdam, 30 13 300
Stettin, 26 85
Stockholm, 5 13

Dyestuffs *(otherwise undescribed)*
Antwerp, 1 ck. 3 kgs.
Boston, 14
Hamburg, 4
New York, 21 pks.

Manure—
Dunkirk, 738 bgs.
Riga, 2 t.
Rotterdam, 21

Pitch—
Antwerp, 400 t.
Dunkirk, 146

Tar—
Copenhagen, 20 cks.
Hamburg, 10
Stettin, 3

GLASGOW.

Week ending May 1st.

Ammonia—
Penang, 2 t. 16 c. £70

Ammonium Sulphate—
Valencia, 159 t. 18¾ c. £1,840
Antigua, 2,472¼ 1,525

Alumina Sulphate—
Montreal, 10 t. 18 c. £40

Benzol—
Rotterdam, 13 drums. £298

Copper Sulphate—
Rouen, 4 t. 19¼ c. £109
Rangoon, 12 130
Calcutta, 57½ 20

Dyestuffs—
Logwood
Quebec, 612½ c. 222
Montreal, 45¼ 58
Fustic
Montreal, 68 19
Dyestuffs *(otherwise undescribed)*
Rangoon, 100 lbs. 22

Epsom Salts—
Montreal, 15 t. 72
Christiania, 50 bgs. £40

Litharge—
Montreal, 9 t. 2 c. £150
Quebec, 5 85

Manure—
Antigua, 279½ c. 145

Potassium Sulphate—
Antigua, 84¾ c. 32

Potassium Bichromate—
Antwerp, 103¼ c. 175
Rouen, 49 t. 14¾ 1,551
Gothenburg, 5 7¾ 175
Riga, 24½ 44

Paraffin Wax—
Bordeaux, 20 t. ½ c. 567

Pitch—
Gothenburg, 245 t. 16 c. £600

Salt—
Antigua, 44 t. 15½ c. £63

Superphosphate—
Christiania, 50½ t. 252

Sulphuric Acid—
Antigua, 5,040 lbs. 46

Sodium Bichromate—
Antwerp, 100 c. £125
Rouen, 17 t. 1¼ 468

GOOLE.

Week ending April 23rd.

Bleaching Preparations—
Antwerp, 15 cks.

Benzole—
Antwerp, 79 cks.
Rotterdam, 146

Coal Products—
Rotterdam, 163 cks.

Chemicals *(otherwise undescribed)*
Ghent, 1 ck. 13 drms.
Hamburg, 50
Rotterdam, 20

Dyestuffs *(otherwise undescribed)*
Boulogne, 2 cs. 18 cks.
Ghent, 2 drms. 1 3 10 tins

Manure—
Antwerp, 580 bgs.
Boulogne, 165 35 shts.
Rotterdam, 198

Pitch—
Antwerp, 60 t.
Ghent, 170
Boulogne, 15 cks.

GRIMSBY.

Week ending April 26th.

Coal Products—
Dieppe, 55 cks.

Chemicals *(otherwise undescribed)*
Dieppe, 6 cks.

PRICES CURRENT.

WEDNESDAY, MAY 7, 1890.

PREPARED BY HIGGINBOTTOM AND CO., 116, PORTLAND STREET, MANCHESTER.

stated are F.O.R. at maker's works, or at usual ports of shipment in U.K. The price in different localities may vary.

		£	s.	d.
%, and 40 %	per cwt	7/3 & 12/6		
acial..	,,	67/6		
.G., 2000°	,,	0	12	9
2 % .. .. nett	per lb.	0	0	6½
.. ..	,,	0	0	3¾
Tower Salts), 30° Tw.	per bottle	0	0	8
Cylinder), 30° Tw.	,,	0	2	11
Tw.	per lb.	0	0	2
.. ..	,,	0	0	1¾
.. .. nett	,,	0	0	4
.. ..	,,	0	1	2½
(fuming 50 %)	per ton	15	10	0
(monohydrate)	,,	5	10	0
(Pyrites, 168°)	,,	3	2	6
(,, 150°)	,,	1	8	0
(free from arsenic, 140/145°)	,,	1	10	0
ns (solution)..	,,	2	15	0
ure)	per lb.	0	1	0½
.. ..	,,	0	1	3
ite powdered..	per ton	13	0	0
lump)	,,	4	17	6
lphate (pure)	,,	5	0	0
,, (medium qualities)	,,	4	10	0
.. ..	per lb.	0	15	0
·880 = 28°	per lb.	0	0	3
= 24°	,,	0	0	1¾
Carbonate .. .. nett	,,	0	0	3½
Muriate..	per ton	23	0	0
,, (sal-ammoniac) 1st & 2nd ..	per cwt.	37/- & 35/-		
Nitrate	per ton	40	0	0
Phosphate	per lb.	0	0	10½
Sulphate (grey), London	per ton	11	6	3
,, (grey), Hull..	,,	11	2	6
(pure)	per lb.	0	0	10¾
lt ,, ...	,,	0	0	9½
.. ..	per ton	75	0	0
(tartar emetic)	per lb.	0	1	1
(golden sulphide)..	,,	0	0	10
loride	per ton	8	0	0
rbonate (native)	,,	3	15	0
lphate (native levigated)	,,	45/- to 75/-		
Powder, 35 %	,,	5	0	0
Liquor, 7 %	,,	2	10	0
of Carbon	,,	13	0	0
Acetate (crystal	per lb.	0	0	6
hloride	per ton	2	2	6
(at Runcorn) in bulk	,,	17/6 to 35/-		
Dyes:—				
(artificial) 20 %	per lb.	0	0	9
.. ..	,,	0	3	9
Products				
ne, 30 % A, f.o.b. London ..	per unit per cwt.	0	1	5
0 % nominal	per gallon	0	3	9
50/90 ,, ...	,,	0	2	9
Acid (crystallised 35°)	per lb.	0	0	8¾
,, (crude 60°)	per gallon	0	2	3
(ordinary)..	,,	0	0	2½
(filtered for Lucigen light)	,,	0	0	3
aphtha 30 % @ 120° C.	,,	0	1	3
Oils, 22° Tw.	per ton	3	0	0
o.b. Liverpool or Garston..	,,	1	8	0
Naphtha, 90 % at 160°	per gallon	0	1	9
Oils (crude)	,,	0	0	2½
ili Bars)	per ton	49	12	6
hate	,,	25	0	0
de (copper scales)	,,	52	10	0
crude)	,,	30	0	0
listilled S.G. 1250°)..	,,	55	0	0
.. ..	nett, per oz.	0	0	9
ate (copperas)	per ton	1	10	0
de (antimony slag)	,,	2	0	0
) for cash	,,	14	10	0
rge Flake (ex ship)	,,	15	10	0
ate (white ,,)	,,	23	5	0
(brown ,,)	,,	18	5	0
onate (white lead) pure	,,	19	0	0
te	,,	22	5	0

		£	s.	d.
Lime, Acetate (brown)	,,	7	15	0
,, ,, (grey)	per ton	13	15	0
Magnesium (ribbon and wire)	per oz.	0	2	3
,, Chloride (ex ship)	per ton	2	5	0
,, Carbonate	per cwt.	1	17	6
,, Hydrate	,,	0	10	0
,, Sulphate (Epsom Salts)	per ton	3	0	0
Manganese, Sulphate	,,	18	0	0
,, Borate (1st and 2nd)	per cwt.	60/- & 42/6		
,, Ore, 70 %	per ton	4	10	0
Methylated Spirit, 61° O.P.	per gallon	0	2	2
Naphtha (Wood), Solvent	,,	0	3	10
,, ,, Miscible, 60° O.P.	,,	0	4	7½
Oils:—				
Cotton-seed..	per ton	22	10	0
Linseed	,,	26	0	0
Lubricating, Scotch, 890°—895°	,,	7	5	0
Petroleum, Russian	per gallon	0	0	5½
,, American..	,,	0	0	5½
Potassium (metal)	per oz.	0	3	10
,, Bichromate	per lb.	0	0	4
,, Carbonate, 90% (ex ship)	per ton	19	0	0
,, Chlorate	per lb.	0	0	4½
,, Cyanide, 98%	,,	0	2	0
,, Hydrate (Caustic Potash) 80/85 % ..	per ton	22	10	0
,, ,, (Caustic Potash) 75/80 % ..	,,	21	10	0
,, ,, (Caustic Potash) 70/75 % ..	,,	20	15	0
,, Nitrate (refined)	per cwt.	1	2	6
,, Permanganate	,,	4	2	6
Prussiate Yellow	per lb.	0	0	9½
,, Sulphate, 90 %	per ton	9	10	0
,, Muriate, 80 %	,,	7	15	0
Silver (metal)	per oz.	0	3	10
,, Nitrate..	,,	0	2	7
Sodium (metal)	per lb.	0	4	0
,, Carb. (refined Soda-ash) 48 %	per ton	6	7	6
,, ,, (Caustic Soda-ash) 48 %	,,	5	5	0
,, ,, (Carb. Soda-ash) 48%	,,	5	15	0
,, ,, (Carb. Soda-ash) 58 %	,,	6	17	6
,, ,, (Soda Crystals)	,,	2	17	6
,, Acetate (ex-ship)	,,	15	15	0
,, Arseniate, 45 %	,,	11	0	0
,, Chlorate	per lb.	0	0	6¼
,, Borate (Borax)	nett, per ton.	29	0	0
,, Bichromate..	per lb.	0	0	3
,, Hydrate (77 % Caustic Soda)	per ton.	11	0	0
,, ,, (74 % Caustic Soda)	,,	9	17	6
,, ,, (70 % Caustic Soda)	,,	9	0	0
,, ,, (60 % Caustic Soda, white) ..	,,	8	5	0
,, ,, (60 % Caustic Soda, cream) ..	,,	8	2	6
,, Bicarbonate	,,	6	0	0
,, Hyposulphite	,,	5	10	0
,, Manganate, 25%	,,	9		0
,, Nitrate (95 %) ex-ship Liverpool ..	per cwt.	0	8	0
,, Nitrite, 98 %	per ton	28	0	0
,, Phosphate	,,	15	15	0
,, Prussiate	per lb.	0	0	7¾
,, Silicate (glass)	per ton	5	7	6
,, ,, (liquid, 100° Tw.)	,,	3	17	6
,, Stannate, 40 %	per cwt.	2	0	0
,, Sulphate (Salt-cake)	per ton.	1	6	6
,, ,, (Glauber's Salts)	,,	1	10	0
,, Sulphide	,,	7	15	0
,, Sulphite	,,	5	5	0
Strontium Hydrate, 98 %	,,	8	0	0
Sulphocyanide Ammonium, 95 %	per lb.	0	0	8
,, Barium, 95 %	,,	0	0	5¼
,, Potassium	,,	0	0	8
Sulphur (Flowers)	per ton.	7	15	0
,, (Roll Brimstone)	,,	6	0	0
,, Brimstone : Best Quality	,,	4	7	6
Superphosphate of Lime (26 %)	,,	2	10	0
Tallow	,,	25	10	0
Tin (English Ingots)	,,	95	10	0
,, Crystals	per lb.	0	0	6¼
Zinc (Spelter)	per ton.	21	10	0
,, Chloride (solution, 96° Tw.	,,	6	0	0

No. 156. SATURDAY, MAY 17, 1890. Vol. VI

Contents.

Notices.

All communications for the *Chemical Trade Journal* should be addressed, and Cheques and Post Office Orders made payable to—

DAVIS BROS., 32, Blackfriars Street, MANCHESTER.

Our Registered Telegraphic Address is—

"Expert, Manchester."

The Terms of Subscription, commencing at any date, to the *Chemical Trade Journal*,—payable in advance,—including postage to any part of the world, are as follow :—

Yearly (52 numbers) .... **12s. 6d.**
Half-Yearly (26 numbers) .... **6s. 6d.**
Quarterly (13 numbers) .... **3s. 6d.**

Readers will oblige by making their remittances for subscriptions by Postal or Post Office Order, crossed.

Communications for the Editor, if intended for insertion in the current week's issue, should reach the office not later than **Tuesday Morning.**

Articles, reports, and correspondence on all matters of interest to the Chemical and allied industries, home and foreign, are solicited. Correspondents should condense their matter as much as possible, write on one side only of the paper, and in all cases give their names and addresses, not necessarily for publication. Sketches should be sent on separate sheets.

We cannot undertake to return rejected manuscripts or drawings, unless accompanied by a stamped directed envelope.

Readers are invited to forward items of intelligence, or cuttings from local newspapers, of interest to the trades concerned.

As it is one of the special features of the *Chemical Trade Journal* to give the earliest information respecting new processes, improvements, inventions, etc., bearing upon the Chemical and allied industries, or which may be of interest to our readers, the Editor invites particulars of such—when in working order—from the originators; and if the subject is deemed of sufficient importance, an expert will visit and report upon the same in the columns of the *Journal*. There is no fee required for visits of this kind.

We shall esteem it a favour if any of our readers, in making inquiries of, or opening accounts with advertisers in this paper, will kindly mention the *Chemical Trade Journal* as the source of their information.

Advertisements intended for insertion in the current week's issue, should reach the office by **Wednesday morning** at the latest.

Advertisements.

Prepaid Advertisements of Situations Vacant, Premises on Sale or To be Let, Miscellaneous Wants, and Specific Articles for Sale by Private Contract, are inserted in the *Chemical Trade Journal* at the following rates :—

30 Words and under.......... 2s. 0d. per insertion.
Each additional 10 words .... 0s. 6d. ,,

Advertisements of Situations Wanted are inserted at one-half the above rates when prepaid, viz :—

30 Words and under.......... 1s. 0d. per insertion.
Each additional 10 words .... 0s. 3d. ,,

Trade Advertisements, Announcements in the Directory Columns, and all Advertisements not prepaid, are charged at the Tariff rates which will be forwarded on application.

THE WIDNES BRINE SUPPLY BILL.

THE Bill to incorporate a company with £100,000. share capital in £10. shares, and £25,000. in debentures, for the construction of a pumping station for brine at Wincham, with a service reservoir at Keckwich, and lines of connecting pipes to take the brine to Widnes has just come before Lord Basing's Select Committee of the House of Lords. Mr. Littler, Q.C., Mr. Pembroke Stephens, Q.C., and Mr T. F. Squarey, appeared for the promoters of the Bill. Petitions in opposition to the Bill were deposited by the Cheshire County Council the Manchester Ship Canal Company, the Northwich Local Board the Widnes Local Board, the Northwich Guardians of the Poor, the trustees of the Weaver Navigation, property Owners in Wincham, the London and North-Western Railway, the Great Western and London and North-Western Railways, and the Bucklow District Highway Board. Sir Richard Brooke's petition against was withdrawn. Mr Littler, in his opening statement, said the proposal was an entirely new one. Widnes contained some of the largest alkali works in the world, and had the reputation of being the greatest vegetable killing place in the kingdom. Alkali, of course, could not be made without salt, but recently the old expensive Le Blanc process had been superseded by the ammonia-soda process, which had enabled Brunner, Mond, and Company to pay 30 per cent. dividends. The Bill was earnestly desired by the people of Widnes, and something must be done in the matter or Widnes would be ruined. The ammonia-soda process, in addition to its cheapness, had the advantage of causing no noxious gases or abominable waste. Salt could formerly be purchased in Widnes at 6s. 6d., but some ingenious gentlemen had devised the Salt Union, and, without any increase in railway charges put the price up to 9s. 6d. to customers who bound themselves for five years, to 12s. 6d. for those who bound themselves for two years, and to 14s. 6d. to those who would not bind themselves for any period at all. This increase in price had reduced the export trade in salt from 650,000 tons to 400,000 tons a year. The proposal was to pump the brine out of a farm of 75 acres out of a great bed of 2,500 acres. Lancashire ought no more to be debarred from taking brine from Cheshire than Cheshire ought to be debarred from taking coal from Lancashire. The whole interested opposition, the learned counsel concluded by saying, was that of the Salt Union.

At a later sitting, Mr. Littler intimated, on behalf of the promoters, that they were willing to be taxed to the amount of a penny per thousand gallons of brine pumped towards the county rates. Mr. Balfour Brown opened the case for the Nantwich Guardians, and contended that the Widnes trade would prosper in spite of the ammonia-soda process. The real purpose of the Bill was to break up the Salt Union. Mr. Thomas Ward, the Cheshire manager of the Salt Union, was then examined against the scheme, as was also Mr. James Mactear, who shewed that the evidence given in support of the Bill by Mr. A. E. Fletcher, the Chief Inspector of Alkali Works, was fallacious, and that, in fact, that so far from being ruined, the trade of Widnes had even up to the present day been expanding. He also stated that the establishment of ammonia soda works in Widnes would mean simply the addition of another trade to those already in existence. The continuance of the Le Blanc process was necessary for the production of bleaching powder of which the world required at least 140,000 tons annually. Mr. George E. Davis was also called by the opponents to shew that any decrease in the trade of Widnes had been brought about by artificial means, and he substantially corroborated Mr. Mactear's evidence.

The committee subsequently decided that the preamble was not proved, and the Bill was, therefore, thrown out.

EGED FRAUDS ON LISTER AND CO., LIMITED, AT BRADFORD.

(Continued from page 300.)

inst., at the Borough Court, Bradford, Francis Stubbs, Heaton, and Henry Varley, drysalter, East-street, Leeds, dered to answer the charge of having conspired by false defraud Lister and Co., Limited, at Manningham Mills, ams of money. The magistrates on the bench were Mr. Stipendiary), Ald. Smith-Feather (Mayor), Mr. R. Kell, bler, and Mr. H. Waud.

(in the absence of Mr. Waugh, who was engaged else- ared for the prosecution; Mr. Kershaw (instructed by Mr. represented Stubbs; and Mr. Warren (Ford and Warren, nded Varley.

ter Henry Stancomb, assistant cashier at Manningham uced a bundle of receipts and invoices relating to transac- Richard Varley between March, 1889, and February, 1890.

ard Woodhead, clerk with Messrs. Brownhill and Co., 6, et, Liverpool, produced a copy of his firm's invoice to rley of the sale of 250 boxes of cutch at 25s. 6d. per cwt., 25th, 1889; and of 10 boxes at 26s. per cwt., on October oss-examination by Mr. Warren, he said the price of cutch siderably—from 15s. to 34s. per cwt. The terms of sale n 14 days, the buyer paying carriage from Liverpool, and ds would be from 9d. to 1s. per cwt. He was not aware as of sale to Lister and Co. It was a common practice for o buy large quantities of cutch for storage. Some qualities ried greatly in storage, and becoming less in weight, it com- higher price. He had been with Messrs. Brownhill and Co. years, and their transactions with Richard Varley had always straightforward character.

lerick Maton, market clerk with Messrs. Dalton and Young, g-lane, London, produced copy of an invoice of cutch sold Varley on January 10th, 1889.

rren objected to this as being too remote in date to be he charge being that the accused conspired between March, February, 1890.

ll said he thought he was justified in putting in evidence as in January.

pendiary Magistrate upheld the objection, and the witness llowed to proceed.

eph Hindley, a partner in the firm of Buckle and Company, eworks, Horton, stated that his firm in July last bought from Varley 50 boxes of cutch at 27s. 6d. per cwt.; in February was purchased at 27s. 6d. per cwt.; and in March and April , per cwt. The carriage was paid by the seller. In cross- on by Mr. Warren, he said the cutch supplied to his firm was "bull cutch," and it was bought for furure delivery as without any alteration in price. He was aware that cutch of y fluctuated in value, and that at present it was dearer. He that his firm had got a good bargain in purchasing cutch at named.

illiam Storey, clerk to Messrs. W. and A. Bowditch, 43, ane, London, proved that in July last his firm sold to Richard tons of fustic roots at £3. per ton, the buyer paying the —In cross-examination, he said roots were the lowest quality being 20s. or 25s. cheaper than the best Jamacia fustic. The £3. per ton was the lowest price he had known during his e for thirty years. "Selected fustic" was the portion picked the bulk.—In re-examination by Mr. Neill, he said that the icked fustic was £5. to £5. 10s. per ton.

alter Fenton, cashier to Messrs. Sands, Wilson and Company, manufacturers, Birstall, proved the purchase from Richard November last and March last of rasped dried fustic at 6s. 3d. and in January last of chip fustic at 6s. 3d. per cwt.—In nination by Mr. Warren, he said the fustic purchased by his not the best quality.

orge Wood, dyer, Birstall, stated that in July and August rchased from Richard Varley rasped fustic at 6s. per cwt. xamination, he said he had no knowledge of plush dyeing, fore he was unable to say whether the use of this quality of ld be prejudicial to plush.

omas R. Kaye, manager for Messrs. Young and Co., dyers ers, Valley Mills, Bradford, proved that in February last his ased from Richard Varley a quantity of bull cutch at 31s. carriage paid. In cross-examination, he said that at the aking the contract, which was twelve months before the cutch was at the lowest price he had ever known it. The otation during the past fortnight was 38s. for a small quan. could not get a quotation for a large quantity. He knew about plush finishing, having been for four years with ister and Co., as plush finisher. At the beginning of 1884, the orders for dye-wares were given by F. Hull, who certainly saw the prices, if he did not fix them. Stubbs told A. Taylor what articles were required, and the list, after being made out by the latter, was handed to Hull, by whom it was passed on to the drysalters. While he was finisher, Messrs. Lister and Co. reduced the quality of their yarn, which rendered the work of dyeing and finishing more difficult. He had occasion to complain of this, and repeatedly asked Stubbs if he had changed his drugs. Mr. Reixach, the manager, put a similar question to him; and, when he answered that he had not, he was told that he must keep up the quality, and not get cheap drugs. In cross-examination by Mr. Warren, witness said that the contract for cutch to be supplied at 31s. per cwt. extended over two years, and he considered the bargain was a good one. In re-examination by Mr. Neill, he said it was a matter of common knowledge at Manningham Mills that F. Hull saw the drysalters at the lodge. After 1884 Stubbs saw them, and he had the fixing of the prices.

Mr. James McMullen, import manager for Messrs. D. Midgley and Co., general merchants, Princess-street, Manchester, proved that his firm on the 12th February last sold to Richard Varley a quantity of maracaibo for fustic at £4 per ton, carriage being paid by the buyer. In reply to Mr. Warren, he said this was the commonest quality of fustic.

Mr. Ferdinand Roussey, bookkeeper and correspondent with Messrs. Lloyd and Co., general merchants, Gracechurch-street, London, stated that his firm sold to Richard Varley, on July 8th, 1889, a quantity of sumac at 9s. per cwt., on October 23rd at 9s. 6d. per cwt., and on February 25th, 1890, at 9s. 6d. per cwt. The terms were cash within 14 days, and 1¼ per cent. discount, the buyer paying the carriage from Palermo. In cross-examination by Mr. Warren, he said that in March, 1889, there was a sale of star brand sumac to Richard Varley at 10s. 3d. per cwt., but he had not brought a copy of that invoice, because it was not asked for the Chief Constable. There was a higher-priced sumac than that supplied to Varleys, and it was sold at from 6d. to 9d. per cwt. more.

Mr. Owen Elias, clerk with Messrs. Rock and Co., wool, leather, and sumac merchants, London, proved the sale to Richard Varley during last year of various quantities of sumac at 9s. 6d. per cwt., the carriage being paid by the buyer. In reply to Mr. Warren, he said there was no other price for sumac at that time.

Mr. Moses Gaunt, tanner, Armley, said for the past eighteen years he had dealt with Richard Varley, and during the past twelve months he had purchased sumac from him. In December last he bought a quantity at 13s. 3d. per cwt., delivered without charge. In February last he made another purchase on the same terms. Cross-examined by Mr. Warren, he said there were many qualities of sumac, and he had paid as much as 16s. 6d. per cwt. to Varley's and others for it. Tanners used large quantities of sumac, and therefore bought large quantities at a time. He had always found Varley's firm strictly honourable in all their transactions.

Mr. James A. Swan, clerk with Messrs. Schott, Segner, and Co., aniline dye merchants, Manchester, proved that between July, 1889, and March 3rd, 1890, there had been about twelve sales of acid magenta (fast red) by his firm to Richard Varley, the price in each instance being 3s. per lb., carriage to Leeds paid, and a discount of 2½ per cent.

Mr. William Kilner, drysalter, 42, Brook-street, Bradford, said that for 20 years he had held an agency for Messrs. Kalle and Co., Biebrich, Germany. For several years he had sold to Richard Varley a blue known as OOX; but at first it was described as "Four B." At Varley's request it was altered in shade, a deeper blue being required. The price in April, 1884, was 8s. Afterwards a mixture of two qualities was made to match a sample supplied by Varley's, as Kalle and Co.'s "Five B" was too strong in colour. The article known as Blue OOX was supplied in tins, which he forwarded to Varley's in the same state as they were delivered to him. During last year and up to February of the present year he had supplied quantities of Blue OOX to Varley's, but to no other drysalters. He had, however, tried to sell to dyers, and had called at Manningham Mills on eight or ten occasions during the past year to endeavour to get an order. He had several times seen Stubbs there, but had not been able to get an order for blue. During the past fortnight he had obtained an order for blue from Lister and Co., and he had sold it at 6s. per lb. It was the same quality of blue as he had supplied to Varley's at 5s. 9d. per lb., and which they invoiced to Lister and Co. as blue OOX.

Cross-examined by Mr. Kershaw.—He several times saw Stubbs when he called at Manningham Mills; but he was not able to tell Stubbs that the blue he was offering was identical with the blue OOX which was being supplied by Varley's. On one occasion Stubbs said he should like to do business with him, but it would not do to change their drugs then, because their shade cards had been sent out, and those would have to be matched in the goods. Stubbs had never said if he (witness) could produce something new it might be put on the shade card, and he would get the benefit of it.

Cross-examined by Mr. Warren.—The marks on the tins had reference to the tare of the empty tins, and if they were not rubbed off there would be nothing to indicate to Lister and Co. where Varley's had obtained them. On the 22nd ult. he received a message requesting him to go to Manningham Mills; and after being shown the tins supplied by Varley's, he identified them as having been obtained from him. He was then asked to quote a price, and he mentioned 7s., but as he found that they had ascertained the price at which it had been sold to Varley's, he agreed to supply it at 6s.

Mr. Frederick John Hicks, cashier at the Leeds and County Bank, Limited, said he appeared under a subpœna calling upon him to produce all cheques on the bank issued by Richard Varley on the said bank, and cashed at such bank or any of its branches, between February, 1889, and February, 1890. He did not produce them, because they were not under his control.

Mr. Neill said he saw the bank manager, and told him exactly what was wanted. It was at that gentleman's suggestion that he served the subpœna on Mr. Hicks. He must now serve the notice on the manager.

Mr. Warren said this was another "fishing" expedition on the part of the prosecution, although the higher Court had already decided against a similar application.

The Court then adjourned till Wednesday morning.

The first witness called by Mr. Waugh was George Whitford, who has already been referred to in the course of the case as the person who gave the information in the first instance to Messrs. Lister and Company, which caused investigations to be made, and ultimately proceedings to be taken. The witness deposed that he was by business a clerk, and resided at 6, Heed Terrace, Burmantofts, Leeds. Formerly for about nine years he was employed by Mr. Richard Varley, drysalter, of East-street, Leeds, as bookkeeper, but was discharged on March 21st or 22nd. He was told at the time that the reason of his dismissal was because he went late, and Mr. Varley also told him that he would not be allowed to do as he liked at his next place. After witness had given evidence at considerable length as to the mode or ordering goods by Varley and the persons from whom they were ordered, Mr. Waugh questioned him with regard to the supply of goods by Varley to Listers. Mr. Richard Varley had supplied cutch to Listers during the 12 months ending March 1st, 1890, in large quantities. They had also supplied cutch during the same period to Messrs. Young and Company, Bradford, Messrs. J. Buckle and Company, Great Horton, and Messrs. Salt, of Saltaire. Salts had different brands of the cutch, but Listers had always had "Cock" brand except on the last occasion, when "Bull" cutch was sent. The cutch was sent from Liverpool by rail in boxes, or from London by steamer, to Varleys'. The boxes were not opened or anything done to the cutch before it was sent to Listers'. The cost of cutch to Varley was from 25s. to 28s. per cwt. Mr. Waugh was asking further questions with reference to the prices at which the goods were sold to customers, but Mr. Warren objected, unless books or invoices were produced. The Stipendiary upheld the objection. Further questions were asked from witness as to the supply to Varley of "Star" sumac, such as was supplied to Listers by them. The sumac cost 9s. 6d. per cwt. in Palermo. The carriage to London was 12s. 6d. per ton, and from London to Leeds 15s. The sumac purchased from Rock and Co. cost 9s. 6d. in Liverpool. It was named "Basso" brand, and the carriage from Liverpool to Leeds payable by Varley's was 15s. per ton. Listers were charged 15s. per cwt. for both brands, while small quantities of it were supplied to Mr. Moses Gaunt, tanner, of Leeds, at 13s. 3d. per cwt. Salt and Co. had "Basso" brand supplied to them at 11s. per cwt. Dealing next with fustic, witness detailed the firms from which Varley's made purchases, the price paid varying from £3. to £4. 17s. 6d. per ton delivered at Goole. The carriage from Goole to Leeds was 5s. per ton. It came in logs, and was either "chipped" or "rasped" by machinery, which would have the effect of making it into a powder. Varleys charged 30s. per ton for doing this work for other people, and after "rasping" water was added, which would increase the weight from three to five cwt. to the ton. The quality sold to Listers was composed of best and common mixed, and contained fustic roots, Maracaibo, Corimbo, and Jamaica. Listers were charged 9s. 6d. per cwt. for it, and other firms were charged from 6s. to 6s. 6d. for the same kind. Varley's bought "acid magenta" from Schott, Segner and Co., Manchester. The tins in which the dye was sent were removed by Varley's, and Varley's labels put on in their place. The cost price was 3s. per pound delivered in Leeds, and it was invoiced to Lister and Co. at 7s. 9d. per lb. They did not supply it to anyone else. Varley's also supplied Lister and Company with a dye called Blue OOX. It was really an alkaline blue known to the trade as BBBBB, and was bought from Mr. William Kilner, of Brook-street, Bradford, the price paid being 5s. 9d. per pound. Prior to the past twelve months Varley had bought similar blue to that referred to from Messrs. Offenbach, of Bradford, at the same price they paid to Kilner, and had sent tins of it to Listers along with the tins of Kilner's. Listers paid 18s. 6d. per lb. for the blue delivered, Varley's paying the carriage, which would not amount to ¼d. per lb. Witness kept some of the books of Richard Varley, and knew Harry Varley kept an account with the prisoner Stubbs. The book in which it was kept was kept in his private drawer. Two books were produced at this stage, and witness said they were principally in Harry Varley's handwriting. Every month Harry Varley took the weights of the goods supplied to Listers from the sales day book and entered them in one of the books produced. Witness had assisted him by calling out the weights from the day book. After the account was made up for the month, Harry Varley used, as a rule, to make a cheque out to himself for the amount which was entered in the books as due to F. Stubbs. There were two enteries in the book already referred to for July 21st and October 21st, 1889, respectively. The first said, "Paid for this month, £75. 3s.," and the second one, "Paid on account, £100. Due F. S. £25. 14s. 9d.' The entries were in the handwriting of Harry Varley. He had cashed the cheques for Harry Varley. Telegrams had passed between the two prisoners when the cheques were drawn. He had heard conversations between Harry Varley and Richard Varley. About two months ago Harry Varley and Richard Varley were talking about Stubbs, and Harry Varley said Stubbs could not leave under six months notice. Witness was 22 years of age. From time to time he had received money for Varley's a few times—perhaps a score or two; he could not say how many. On the 6th of March last witness received £5. 6s. 8d. from Messrs. Richardson and Company at Varley's office. It was not his duty to enter into the ledger the receipt of money unless specially told to do so. He had made entries into the ledger of moneys purporting to have been received. Answering further questions by Mr. Warren, the witness said that when he received money he entered the amount on the counterfoil of the receipt book. Three or four days after he left he went to Varley's and was asked where the sum of £5 6s. 8d. was, and he afterwards told Varley that he had paid it to him at the office. Varley replied that he did not remember it. He first gave information in this case on Easter Monday. He was discharged on the 21st March. He was not discharged because he was always behind with his work. He made an extract from the books before they were taken possession of by the police. He refused to say where he made the extract at first, but upon being pressed he stated that he made the extract at Listers' office. He took the book to Listers' offices, having previously obtained it from Varley's on the Thursday morning of Easter week. He had then left Varley's employment for over a week. He got the book about half-past eight in the morning. He did not expect any of the Varley's would be there then. There were some of Varley's men present. The book was in a locked drawer, and he got it by taking out an unlocked drawer which was above. He did not steal the book, he only borrowed it. Mr. Lee, secretary to Lister and Company, said he was to bring the book, but not how he was to get it. He had told Mr. Lee he could get it. Mr. Reixach was present. Witness made the copies of the accounts partly in the presence of Mr. Lee and Mr. Reixach. The witness went on to explain how he returned the book to Messrs. Varley's office, stating that just as he was putting it into the drawer someone came into the room and asked him what he was doing in the room. Witness made the excuse that he had come for a coat which he had left some time previously.

The examination of this witness was resumed on the following day.

In reply to questions, witness said that cutch was purchased by Varley and Company from a great many persons and sold again to a lot of people. Sumac was purchased also from a lot of people and stored. He had had no communication verbally or otherwise with anyone about the matter prior to the writing of the letter to Mr. Lister.—Cross-examined by Mr. Kershaw: He never saw Stubbs at Varleys' works.—Re-examined by Mr. Waugh: He had never known any blue to be returned from Listers to Varley. There was no truth in the suggestion that he retained the sum of £5. 8s. 9d. about which he was cross-examined yesterday.—Mr. Hermann Offenbach, drysalter, of St. John-street, Bradford, gave evidence as to the supply of alkaline blue to Richard Varley, of Leeds. Cross-examined by Mr. Kershaw: He had been at Listers, had seen Mr. Stubbs there, and had got orders for other kinds of dyes. He had supplied Indian yellow, and Mr. Stubbs had got the price reduced from 6s. 6d. per lb. to 3s. 6d.—Mr. Sidney Smith, sub-manager of the Bradford District Bank, deposed that Stubbs had had a banking account there since 1881. From July to October 8, 1889, Stubbs paid in to his account £600, and there was only one withdrawal—for £80. He paid in various £100 and £50 Bank of England notes.—Mr. Alfred John Holliday, chief cashier to the firm of Lister and Company, said he had paid Stubbs's salary, which had been £550 per annum since December 1, 1888, when Lister and Company became a limited company. He never paid to Stubbs a £50 or £100 note.—Inspector Edwin Hey gave evidence as to the arrest of the prisoner Varley, at Leeds, on April 15th. Varley's father and brother came up, and they asked to be allowed to see their solicitor, Mr. Warren. This was complied with, Mr. Warren being seen in the Great Northern Hotel. While waiting for Mr. Warren, Mr. Richard

o the prisoner: "I don't see we have anything to fear in don't see we have done anything wrong. We have simply Stubbs commission for orders we received from him. If we e so somebody else in the trade would have done so. We ne this loosely, as you will see from the books you have ave kept a proper account of what we have paid Stubbs. r Varley then remarked, "I have done nothing no more than paying him his commission. If I had not done so, in the same trade would have done so and got the trade." Sidney then remarked, "If we were not to do this, we ll put the shutters up." Detective Butterworth said that of April he went with the chief constable to Manningham ere arrested the prisoner Stubbs. The Chief Constable rrant, and in reply the prisoner said, "I have nothing to shall have to see my solicitor. I don't know anything y's only what the firm does. This closed the case for the
After some conversation with the counsel engaged, the said he did not think he should be justified in withdrawing m a jury. Counsel for the prisoners then intimated that reserve their defence.

COMMERCIAL TURPENTINE OIL.

By Arthur Wilson.

ection of the adulterants of turpentine oil, especially when ring in small quantities, present difficulties even to the l operator. They may be arranged in two classes, viz.,—1. f turpentine with petroleum spirit, shale spirit, coal tar osin oil, &c. 2. Mixture of two classes of turpentine such as d American.
ell-known fact that Russian turpentine oil is wholly unsuit- st of the purposes to which turpentine oil is applied, so that nportant to be able to detect such an addition, and it may e, that it is a work of no ordinary difficulty. According to hat is known as the turpentine "fake" is practised, which bringing up the density of an adulterated oil of turpentine to genuine article, by the addition of rosin oil of high density. his be the truth or not, in a large number of samples I have failed to detect its presence. The examination of oil at the present time relies very much on the behaviour of on distillation, together with the Halogen absorption, but perfect scheme includes

1. Ordinary and fractional distillation.
2. Halogen absorption.
3. Behaviour with sulphuric acid.
4. Optical activity.
5. Certain simple tests.
6. Specific gravity.

cific gravity of commercial oil of turpentine varies between and any samples above or below these must be regarded cion. Of 20 samples recently examined, the highest was e lowest 865·01. Specific gravity is best determined by the balance. The optical activity of commercial turpentine oil ery much value, but if an oil possesses a high value for S. D. possibly be highly adulterated. The values of S. for the D. number of samples, I have recently examined were

	Maximum.	Minimum.
$\{S\}D=$	+ 15·29	+ 12·05

—A Laurent instrument was used for these determinations.
f the above values were for oils sold at small shops as well upplied to large consumers. The optical activity is of no istinguish between American and Russian oil of turpentine, se in the case of French oil on account of the laevorotatory the latter. Certain simple tests have been proposed, but they y little value, in mixtures of turpentine and its adulterants. of sulphur has been proposed by Warren, but my experience ts favour, although as a test for the fixed oils and fats it is of e; the same remarks apply also to Valenta's test with glacial l. Rosin oil is usually tested for by means of stannic chloride , stannic bromide as proposed by Mr. Allen.
t deal depends upon how the stannic bromide is prepared, so y case it is better to substitute the test with acetic anhydride uric acid, in which case a beautiful red or reddish violet n is produced if rosin oil be present. The test can be applied idue left on distillation by steam. Warren's test with alcoholic in my experience, of very little use.
logen absorbtion of turpentine is of much value if properly l, in all cases a large excess of the reagent being present, his be bromine or iodine.
um spirit and shale spirit absorb 30 and 70% respectively of bromine. I have usually found the bromine value to average 215, and the iodine value (by Hubbs' process) 331.

The distillation test, when conducted in its simplest form, is an excellent distinguishing test for turpentine and turpentine adulterants, when alone, and even when present in moderate proportion in a mixture, but when small percentages are present its indications are not of much use. The following table shows the average results, when 100 c.c. of the sample were distilled in a 16 oz. flask with a side tube, the thermometer bulb being placed in the usual position near the side tube:—

Percentage distilling at T. temperature.	T.
40·1	148·5°C.
62·0	149·0
78·0	150·0
90·0	153·0

The greatest range I have met with in genuine turpentine was in the case of a sample of boiling point 157·5C., giving 20% at 158·5C., and 92% at 184·0C.

Fractional distillation, when carried out by the aid of a dephlegmator such as Le Bel, Glynsk, or Hempel, are of much more value than the above simple distillation test. The undermentioned results were obtained by a Glynsk dephlegmator in a similar manner to the analysis of benzols. The specific gravity of the distillate is taken by a Sprengel tube:—

Temperature C.°	Volume distillate.	Specific gravity of distillate.
156°	20%	862·9
156·25	40%	864·28
156·7	60%	863·1
159·0	80%	864·54
Tailings		873·21

Difference between density of tailings and 1st fraction 10·4
Difference between original density and tailings ...... 6·61

Fractional distillation is much more likely to detect such adulterants as shale and petroleum spirit, than is a simple distillation. It may be interesting to show the difference when fractional distillation is applied to a sample, as against a simple test in a flask with a side tube.

Fractional Distillation. Temperature, C.°	Vol. Distillate.	Simple Distillation. Temperature, C.°	Vol. Distillate.
156°	20 %	160°	23 %
156·25	40 ,,	161	45 ,,
156·75	60 ,,	162	57·5 ,,
159·0	80 ,,	163	65·5 ,,
..	..	164	71·0 ,,
..	..	165	78·0 ,,
..	..	166	86·0 ,,

A method of analysis, due to Professor Armstrong depends on the polymerization of turpentine oil, whilst petroleum naphtha is very little affected. On the other hand, rosin spirit is very much affected, and cannot safely be detected. 250 c.c. of the sample are treated with 75 c.c. of a cooled mixture of 2 parts of H_2SO_4 and 1 of water, in a globular separator. The acid should be added about 20 c.c. at a time, agitating well and avoiding much rise of temperature. The acid layer is then tapped off, and the oily liquid steam-distilled till nothing more comes over. The water is separated from the oily portion of the distillate, and the latter measured, and treated again with half its volume of 4/1 acid, then steam-distilled as before.

Genuine oil of turpentine usually gives a second distillate, averaging from 3 to 6% of the original liquid taken, according as the operation has been more or less successful. Any excess over 4% may be reckoned as due to a foreign oil. I have found that by carefully conducting the operation, and using 3/1 acid in the first treatment, the second will not be more than 4% by volume. Should it measure more it is advisable to treat it again with 4/1 acid. The amount of heat developed in the acid treatment varies, but it is not advisable to add the whole of the acid at one time.

In conclusion, the oleo-refraction has been proposed to detect turpentine adulterants, but very little work appears to have been done with it, at least in this country.

The dinner given the other day in Baltimore, where, according to the *Electrical World*, waiters were superseded by an electric car that darted out of the pantry and served the courses with neatness and dispatch, opens a magnificent vista of future peace. Such a servitor would be a blessed relief from the polyglot importunity of the animate things that we endure to-day. It would demand only its regular wages of current strictly in advance, and would never extend a wheel or switch for tips. Better than all, it is dumb and of unequalled sobriety. Hasten the day when we shall find it in every household.

FLAMES.*

(Concluded from page 303.)

FLAME never exists except during the process of combination of two or more gases or vapours. As a familiar instance of both forms of combustion, there is no more striking example than coke or charcoal, which if burnt at once to carbonic acid burns entirely without flame. If the supply of air is deficient, carbonic oxide is formed, which, being a combustible gas, burns with a flame. I have here a curious experiment showing that the so-called air gas, which is composed of air saturated with gasoline or benzoline vapour, is neither a gas nor a perfect mixture or combination. The flame, produced under the conditions I show you, proves beyond question that the volatile gasoline is unequally diffused through the air, and that it exists in small centres. You will see that each of these becomes a separate and distinct centre of combustion, the result being, perhaps, one of the most beautiful effects ever produced by flames.

It is a very common thing to hear of white flames, although, so far as I am aware, white flames do not exist. We can get them red, blue, green and yellow, shading off to very pale lemon colour, but a really white flame I have never seen, nor have I seen a flame of any pure colour, all are so mixed as to be useless for optical experiments on colours. The sources of light which are really white are, to the best of my knowledge, all flameless, or, to speak more correctly, the flame is not the source of the light. I may refer, as examples, to the electric arc, the lime light, and the magnesium light, all of which are caused by solid bodies intensely heated. You will see that when I burn this magnesium wire the gas flame which is usually called white is really a dirty yellow. When a flame is coloured it imparts its colour to all the surfaces it falls on, and if the flame could be made a pure colour, without any white in it, all other colours are invisible in its light.

In these experiments ordinary air has been used to combine with the combustible materials, air being composed of four parts of nitrogen and one of oxygen. The nitrogen is quite inert, having no power to combine; in fact, it does nothing except to dilute the oxygen and reduce the temperature obtained. If we remove the nitrogen, and use oxygen alone, the combustion is far more rapid and intense; and I will now show you some of the wonderful effects which can be produced with ordinary coal gas and the compressed oxygen prepared by the Brin Oxygen Co. by a very simple and cheap process. The blowpipe I shall use is a very noisy one, but as it was only designed as a cheap arrangement for workshop emergencies, this is not a matter of great importance, as, by using a more expensive form of blowpipe and a slightly different gas arrangement, the apparatus can be made absolutely silent. I have here a plate of steel ¼ inch thick, which, when hardened and tempered, is absolutely drill proof, and I will now proceed to fuse a slot in it, which, as you see, can be done with the greatest ease at the rate of something like three inches per minute. I now take a heavy ordinary engineer's file, and a clear hole is fused through in less than a minute; and, using the same blowpipe, I will direct it on to a block of lime, and obtain a light which is strong enough to throw shadows of ordinary gas flames.

You will remember the heat obtained in the small furnace by the flameless combustion of air and gas. I will now take the same furnace and use the air deprived of its inert nitrogen. In the furnace is a crucible filled with scraps of steel; the crucible is one which will stand perfectly the fusion of pure wrought iron at a clear blinding white heat, and yet you will see that both crucible and steel are fused with the greatest ease.

You will no doubt observe that in the production of these enormously high temperatures a little flame is visible, but it is so trifling as to be of no importance; part of it being caused by the combustion of the dust in the air drawn towards the outer surface of the zone of combustion. If it were possible in practice to consume the mixture entirely without flame, the results would undoubtedly be still greater than you now see.

When speaking of the properties of flame I am on the borders of the unknown, and my experiments and knowledge of this subject are very incomplete. The study is to myself one of both business and pleasure combined, and I am still studying experimentally the cold zone or space which exists between all flames and cold substances to which they are applied. If this cold zone can be passed in practice, and the flames can be applied in direct contact with the vessels to be heated, we shall then obtain something approaching the full theoretical duty of the fuel consumed, and our waste of fuel will drop to a very small fraction. I have succeeded in obtaining this direct flame contact in certain forms of apparatus for boiling water by the addition of solid projecting studs to the surface of the vessel; these studs become sufficiently hot to admit of contact with the flame, and by a series of careful experiments I have come to the conclusion that the effective value of this surface is about six times as great as that of surfaces with which water is in direct contact, *i.e.*, one square foot of surface on a projecting stud is equivalent in effective work to six square feet of ordinary boiler or water tube surface. In a paper which I had the honour of reading before the Iron and Steel Institute, I referred to one problem in heating which, if solved, would reduce our waste of fuel to zero, *i.e.*, the conversion of a large bulk of heat of low intensity to a smaller bulk of heat of high intensity. This conversion is possible with all other natural forces, such as light, electricity, &c., and I believe it to be possible with heat also, the only objection being, so far as I can see, that we do not know how to do it at present; but I have no doubt that the time will come when this problem will be solved by someone who will be rewarded by both fame and fortune.

* Corn Exchange, Oxford, April 25th, 1890. Lecture delivered by Thomas Fletcher, F.C.S.

Trade Notes.

THE TELEPHONE IN MANCHESTER.—We are glad to notice that there is a prospect of a great reduction in telephonic rates, which at present are so excessive as to greatly restrict the use of this feature of modern progress. Under the name of the Mutual Telephone Co., Limited, a company has been registered which proposes to provide shareholders and others within a radius of one mile from the Manchester Exchange, with an exchange wire at a cost of £5. and £6. per annum respectively. The "Mann" system of switching will be employed, and the wires be laid on the metallic circuit principle.

THE EXPORTATION OF SALT FROM THE MERSEY.—The exports of salt to foreign countries from Liverpool during April amounted to 59,537 tons, against 48,536 tons during April last year, an increase of 11,001 tons. The shipments during the first five or six months of last year were equal to the average of previous years, the quantity being assisted very much by the balances of contracts at cheap prices. As the total output for the first three months of this year exceeded the output of the same period of last year, the maintenance of the foreign exports at a good average must be very reassuring to those interested in the salt trade, and who believed our English production of the article would be taken as usual abroad at prices that must leave good profits to the producers.

CAVES containing deposits of earth with from 4 to 30 per cent. of calcium nitrate, and 5 to 60 per cent. of calcium phosphate, says the *American Druggist*, are common in Venezuela, not only in the littoral mountain chains, but also on the flanks of the Cordillera of the Andes. In these deposits are imbedded remains of mammalian bones, preserving their form, but so friable as to fall to powder when they are extracted. They consist solely of calcium phosphate; the gelatin has been nitrified and dissolved out, and the calcium carbonate of the bones has been used up in neutralising the nitric acid produced. The nitric ferment is found in abundance throughout the deposits in a very well developed form. Some of these deposits are ten metres thick.

METROPOLITAN WATER SUPPLY.—Dr. Frankland reports to the Registrar-General the results of the chemical analyses of the waters supplied to the inner, and portions of the outer, circle of the metropolis during the month of March. Taking the average amount of organic impurity contained in a given volume of the Kent Company's water during the nine years ended December, 1876, as unity, he finds that the proportional amount contained in an equal volume of water supplied by each of the metropolitan water companies, and by the Tottenham Local Board of Health, was:—Kent, 0·5; New River, 1·3; Colne Valley, 1·4; Tottenham, 1·7; Grand Junction, 2·1; Southwark and Lambeth, 2·4; East London, 2·6; West Middlesex, 2·9; and Chelsea, 3·5. The water drawn from the Thames by the Chelsea, West Middlesex, Southwark, Grand Junction, and Lambeth Companies again exhibited a considerable reduction in the proportion of organic matter, the Grand Junction Company's water taking the lead in this respect. The water was in every case efficiently filtered. The water obtained chiefly from the River Lea by the New River and East London Companies showed a corresponding improvement as compared with the supply in February. It was efficiently filtered. The deep-well waters of the Kent and Colne Valley Companies and of the Tottenham Local Board of Health were, as usual, of a high degree of organic purity, but only the supply of the Kent Company excelled that of the New River Company in this respect. The Tottenham water was slightly turbid owing to suspended particles of rust of iron. The Kent and Colne Valley waters were clear and bright. That of the Colne Valley Company was softened before delivery, and was thus rendered suitable for washing. Seen through a stratum two feet deep, the waters presented the following appearances:—Kent, Colne Valley, and New River, clear and colourless; Tottenham slightly turbid and colourless; the remaining waters clear and very pale yellow.

MEETINGS OF COMPANIES.

N OIL COMPANY, LIMITED.—The report for the year l 2nd, states that the amount written off for depreciation is cing 5 per cent. upon the capital expenditure. The profit :luding £793. brought forward, is £42,250., and the :commend a dividend at the rate of 6 per cent on the and one at the rate of 15 per cent. on the ordinary shares, e in equal proportions on June 4th, and December 4th, less The sum of £4,000. is also to be placed to the reserve will then amount to £27,467., £2,287. being carried forward.

TO COMPANY.—The ordinary general meeting was held on oder the presidency of Mr. H. M. Matheson. The Chair-ving the adoption of the report, said that after the experi-8, the board did not believe even if the prices were higher sent there could be for years to come any marked effect upon ion. On the other hand, with the new uses to which this :tal was being applied, it might safely be said that the in a healthy condition. Their deliveries of pyrites; the n in England, Germany, and elsewhere under contracts for ir and copper, the former at fixed contract prices and the : price of the month for delivering had amounted to 395,081 average copper contents had been 2·595 per cent. dry ig, 9,416 tons of copper. These figures were the highest d by the company. The copper made at the mines amounted ons, and they had brought to the market and realised 17,667 ng in all with the pyrites copper, 27,083 tons. The stocks were large, but the great bulk was in the form of "matt" ipitate," and there were also 890 tons of refined, all of which :arried over in the books at cost price. Owing to the un-success of sulphate of copper in checking the ravages of vines a large demand for this copper product has arisen, and had taken the means to profit by it. Mr. T. C. Bruce The adoption of the report was carried.

AND BARRY, LIMITED.—The 12th annual general meeting reholders was held on Tuesday at the City Terminus Hotel, ınder the presidency of Mr. F. T. Barry, M.P. The Chair-tted that the accounts did not permit the payment of a higher idend for the year than 6s. per share, after placing £21,000. l funds. Having alluded with satisfaction to the increase year in the market value of copper, he stated that their liabilities, exclusive of the balance of profit for the year, £30,571. as compared with £50,034. at the end of 1888. moderate reduction had been made in the valuation of the s at the mine, the valuation of the various stocks of ore and cipitate, after allowing for shipment per Cornelia, valued at stood at a lower figure than in the previous balance-sheet. assets, after allowing for the difference between the balance it the end of 1889 and at the close of 1888, amounted to 000. more than at the 31st of December, 1888. As stated, in the balance-sheet, one cash item embraced the sum of the payment of which they were still endeavouring to en-gal proceedings. Turning to the profit and loss account, he they had received payment of their claim for over-assessment -tax, alluded to by him at their two last annual meetings. ecided to carry forward £44,203. in consequence of the unsettled hich he would presently more fully refer. The quantity of and raised at the mine during 1889 gave a total of 176,228 is large reduction was made in consequence of the breakage the current make of copper precipitate having been under sideration by the directors at a board meeting, held on the , 1889, when the quotation for G.M.B. copper was under ton. At the board meeting in question they approved the of copper precipitate being reduced to the equivalent of ine copper, but as the price of copper improved during the latter year they took immediate steps for increasing the produc-eactual make amounted to 4,592 tons of fine copper. For the r they estimated the quantity of ore broken and raised would 000 tons. The annual average for the eight years, 1882-89, o 326,743 tons. Naturally, after the extraction of such a the present workings were at a much lower depth in the ney were in 1882, and the ore they were now breaking did uite so high a per centage of copper as the ore broken and the higher workings. The production of copper precipitate is estimated at 5,000 to 5,500 tons fine copper. At the 9 they had under treatment for the extraction of copper nillions of tons of mineral, and the whole of their workings proved that there was ample ore in sight ture breakage. The legal proceedings they had been institute against La Société Industrielle et Commerciale of Paris, and the Comptoir d'Escompte de Paris, for yment of the value of shipments of copper precipitate delivered under the company's contract during January, February, and March, 1889, were still pending, but he hoped that their action against the Comptoir d'Escompte would come on for trial during next month. The amount of the Company's claim for copper precipitate delivered was £99,351., subject to a credit for resale of March delivery, amounting to £24,595., thus reducing their claim to £74,756. Their company ranked, of course, as creditors against the estate of La Société des Métaux in liquidation, but, in addition, they had secured, under an attachment order, a payment into Court of £9,486. They now awaited the decision of the Court of Queen's Bench as to whether this sum should be paid over to this company, or whether the official liquidator could claim it on behalf of the general creditors of La Société des Métaux. They had also a second attachment order running against copper, assumed to be the property of that company, in the hands of third parties. He thought the prospects of the copper market very satisfactory. The large accumulated stock of copper held at this date last year by those who were associated in financing the French speculation was reported to have been reduced from 179,000 tons to about 89,000 tons, and should the consumption of copper continue upon the scale of the last six months, he thought they might reasonably expect that the present price of copper would gradually advance. The recent rise in the value of silver had also a direct tendency to strengthen the copper market by increasing the demand for copper in India, and by reducing the shipments of copper from Chili. He concluded by moving the adoption of the report and the payment of a final dividend of 4s. a share, making 3 per cent. for the year. Mr. Francis Ricardo seconded the motion. In answer to questions, the Chairman expressed the hope, in regard to the question of dividend, that they might meet next year under more favourable auspices. It was, however, necessary to keep a large sum in hand pending the result of the litigation referred to. The motion was then adopted.

NOBEL-DYNAMITE TRUST COMPANY, LIMITED.—The following is the report by the directors for the past year:—Annexed are the accounts for the year ending 30th April, 1890: The net profit for this period, after payment of income tax, amounts to £142,306. 1s. 5d., and, including the sum of £10,153. 5s. 6d. brought forward from last account, there is a total available profit of £162,359. 6s. 11d., which the directors propose to be applied in the following manner, viz.:—In extinction of preliminary expenses, £4,500.; in payment of a dividend of 8½ per cent., free of income tax, on 160,397 shares, £136,337. 9s.; carrying forward to next account, £11,521. 17s. 11d.—total, £152,359. 6s. 11d. The dividend on the remaining 15,000 new shares recently issued will be paid out of an amount received in special consideration for their admission to participation, as hereinafter explained. While in previous years the profits of the subsidiary companies, whose dividends constitute your revenues, had been almost exclusively derived from the production and sale of blasting agents, and were consequently nearly entirely dependent on this one source of income, your directors have now to report a series of occurrences which have effected a considerable change in this respect, enlarging the sphere of your interests, and placing them upon a materially broadened basis. Nitro-glycerine compounds, although pre-eminent among explosives as blasting agents for mining purposes, had not hitherto been found capable of satisfactory employment as propelling agents for military purposes. The difficulties in the way of rendering them adapted for such purposes were very great, and were of a description to make the very highest demands upon the skill of the inventor. It must, under these circumstances, be regarded as an event of the first importance, that your honorary president, Mr. Alfred Nobel, has been enabled to announce the invention of military powders on a nitro-glycerine basis the practical manufacture of which on a large scale has since been perfected by our subsidiary companies. The directors were enabled, and it at once became their duty, to secure for the subsidiary companies the benefits of this new invention so far as regards its employment in the United Kingdom, the British Colonies, the German and Austrian Empires, and certain other countries. An arrangement with this object has accordingly been made with Mr. Nobel on the basis of pre-existing general arrangements, and on terms considered by the directors to be equitable and fair. The new invention as perfected may now confidently be asserted to fulfil the highest ballistic requirements of a smokeless powder. Concurrently with this acquisition, through which the sphere of the company's interests was extended by the inclusion of military powders among the products to be manufactured by the subsidiary companies, a somewhat analogous process of development was in operation in the business of the leading Continental powder manufacturers, tending to bring them into keen competition, not only with your prospective interests in the powder trade, but also in connection with the production of high explosives generally. The manufacturers more particularly referred to are the United Rhenish-Westphalian Powder Factories, of Cologne, and the Rottweil-Hamburgh Powder Factory, of Rottweil, acting in combination with Messrs. Cramer and Bucholz, of Ronsahl, and Messrs. Wolff and Co., of Walsrode. This group had long occupied a position of exceptional eminence at the head of their trade as the manufacturers of the well-known cocoa and other powders, and were at this time themselves commencing the production of chemical smokeless powders. The directors must ascribe it chiefly to the efforts of their colleagues, Mr. Max A. Philipp and Mr. J. N. Hiedemann (whose exertions were attended with the special advantage due to his exceptional connection with both parties), that it has been possible to arrive at an understanding with these powder factories, whereby a complete identity of interests is secured for the future between the two groups. The new agreement provides that for a period of about 35 years ending 31st December, 1925, the profits of the two groups are to be pooled and re-apportioned in certain fixed percentages, which have been determined with due regard to the previous revenues and present prospects of the respective groups. The entire arrangement has been the subject of very anxious and careful consideration at the hands of your directors, and they feel satisfied that it will be found to operate to the advantage of both parties. Irrespective of the avoidance of a mutually injurious competition, the new combination may be confidently expected to secure that which the shareholders will doubtless join with the directors in regarding as of the highest importance, viz., a greater immunity to both parties from irregularity of annual results, whether arising from fluctuations of trade or from the uneven incidence of the risks of explosion. It also offers every probability of enabling considerable economies to be effected, both in manufacturing costs and capital outlay, it being now possible for both groups to arrange to produce at a common source various materials, which would otherwise have had to be separately manufactured by each. It remains to be added that the agreement with the Powder Group came into operation as from 1st July last. In January of this year a new issue of 15,000 shares was made to provide funds for the purchase by the Dynamit Actien Gesellschaft, of Hamburg, of a controlling interest in the important Chemical Company, the "Chemische-Fabrik-in-Billwärder-vorm-Hell-and-Sthamer, A. G.," and for the anticipated requirements of the subsidiary companies in connection with the erection of works for the manufacture of smokeless powders, and for other purposes. Of these shares 5,900 were issued for cash, and the remaining 9,100 were allotted in conversion of a new issue, made with the sanction of the board, of M1,000,000 (about £50,000) of shares in the Dynamit Actien Gesellschaft. The sum of M2,000,000 (about £100,000) was applied by that company out of the proceeds of these shares in purchase of the interest in the Chemische-Fabrik-in-Billwärder, consisting likewise of M2,000,000 in shares. This investment has been made in concurrence with the Powder Group for the promotion of the common interest, and is confidently anticipated to yield a satisfactory rate of profit. The balance of the proceeds of the new capital of the Dynamit Actien Gesellschaft, as well as of the 5,900 shares issued for cash, is intended for the other purposes above referred to. Both the 5,900 shares allotted for cash and the 9,100 shares issued in conversion of the new capital of the Dynamit Action Gesellschaft were disposed of at a premium of £5. per share. The total premiums on the new issues will be treated as reserves, the amount accruing to this company being £29,500., as shown in the balance-sheet, while the addition to the reserves of the Dynamit Action Gesellschaft from this source amounts to £45,500. In addition to the premium which has been mentioned, a special payment of £1. per share was received in express consideration for the admission of the new shares to participation in the forthcoming dividend, the proceeds of these shares not having been available for contribution to revenue during the past year. Of this amount £12,750. has been carried to profit and loss, being the equivalent of the proposed dividend of 8½ per cent., the remaining £2,250. being dealt with as premium. The general business of the past year in the dynamite branch, although subject to some fluctuations, may on the whole be characterised as satisfactory. The business of the Powder Group has temporarily been adversedly affected owing to its state of transition from the old products to the new smokeless powders. From this cause, one of the principal powder factories was for several months lying idle, during extensive alterations, but work has now been actively resumed on an increased scale. The directors regret to announce the resignation during the past year of their colleague, Mr. John Taylor, occasioned by the pressure of other engagements upon his time and attention. The vacancy thus created at the board has been filled by the appointment of the Right Hon. Lord Ribblesdale. Lord Ribblesdale, Mr. Philipp, and Mr. Reid, are the directors retiring at the forthcoming general meeting, and, being eligible, are cordially recommended for re-election, as are also Messrs. Cooper Brothers and Co., for the office of auditors to the company.

SHIPBUILDING on the Tyne is now becoming less brisk. From the Walker yard of Sir W. G. Armstrong, Mitchell, and Co., hands are being paid off. This firm have now on the stocks three petroleum steamers unsold. Large numbers of workmen are also being discharged from the Hebburn yard of Messrs. Hawthorn, Leslie, and Co.

PERMANENT CHEMICAL EXHIBITION.

THE proprietors wish to remind subscribers and their fr generally that there is no charge for admission to the Exhib Visitors are requested to leave their cards, and will confer a favo making any suggestions that may occur to them in the directi promoting the usefulness of the Institution.

JOSEPH AIRD, GREATBRIDGE.—Iron tubes and coils of all kinds.

ASHMORE, BENSON, PEASE AND CO., STOCKTON-ON-TEES.—Sul of Ammonia Stills, Green's Patent Scrubber, Gasometers, an Plant generally.

BLACKMAN VENTILATING CO., LONDON. — Fans, Air Prope Ventilating Machinery.

G. G. BLACKWELL AND CO., LIVERPOOL.—Manganese Ores, Ba French Chalk. Importers of minerals of every description.

BRACHER AND CO., WINCANTON.—Automatic Stills, and P Mixing Machinery for Dry Paints, Powders, &c.

BRUNNER, MOND AND CO, NORTHWICH.—Bicarbonate of S Soda Ash, Soda Crystals, Muriate of Ammonia, Sulphat Ammonia, Sesqui-Carbonate of Ammonia.

BUCKLEY BRICK AND TILE CO., BUCKLEY.—Fireclay ware kinds—Slabs, Blocks, Bricks, Tiles, "Metalline," &c.

CHADDERTON IRON WORKS CO., CHADDERTON.—Steam Drier Steam Traps (McDougall's Patent).

W. F. CLAY, EDINBURGH.—Scientific Literature—English, Fr German, American. Works on Chemistry a speciality.

CLAYTON ANILINE CO., CLAYTON.—Aniline Colours, Aniline Benzole, Toluole, Xylole, and Nitro-compounds of all kinds.

J. CORTIN, NEWCASTLE-ON-TYNE.—Regulus and Brass Taps Valves, "Non-rotative Acid Valves," Lead Burning Apparat

R. DAGLISH AND CO., ST. HELENS.—Photographs of Chemical —Blowing Engines, Filter Presses, Sludge Pumps, &c.

DAVIS BROS., MANCHESTER.—Samples of Products from va chemical products—Coal Distilling, Evaporation of Paper-Treatment of waste liquors from mills, &c.

R. & J. DEMPSTER, MANCHESTER.—Photographs of Gas Pl Holders, Condensers, Purifiers, &c.

DOULTON AND CO., LAMBETH.—Specimens of Chemical Stone Stills, Condensers, Receivers, Boiling-pots, Store-jars, &c.

E. FAHRIG, PLAISTOW, ESSEX.—Ozonised Products. O Bleached Esparto-Pulp, Ozonised Oil, Ozone-Ammon Lime, &c.

GALLOWAYS, LIMITED, MANCHESTER.—Photographs illustr Boiler factory, and an installation of 1,500-h.p.

GRIMSHAW BROS., LIMITED, CLAYTON.—Zinc Compounds. S Materials, India-rubber Chemicals.

JEWSBURY AND BROWN, MANCHESTER.—Samples of Aerated W

JOSEPH KERSHAW AND CO., HOLLINWOOD.—Soaps, Greases, Varnishes of various kinds to suit all requirements.

C. R. LINDSEY AND CO., CLAYTON. Lead Salts (Acetate, Ni etc.) Sulphate of Copper, etc.

CHAS. LOWE AND CO., REDDISH.—Mural Tablet-makers of Car Crystals, Cresylic and Picric Acids, Sheep Dip, Disinfectants

MANCHESTER ANILINE CO., MANCHESTER. — Aniline Col Samples of Dyed Goods and Miscellaneous Chemicals, organic and inorganic.

MELDRUM BROS., MANCHESTER. — Steam Ejectors, Exhau Silent Boiling Jets, Air Compressors, and Acid Lifters.

E. D. MILNES and BROTHER, BURY.—Dyewoods and Dye Extracts. Also samples of dyed fabrics.

MUSGRAVE AND CO., BELFAST.—Slow Combustion Stoves. M of all kinds of heating appliances.

NEWCASTLE CHEMICAL WORKS COMPANY, LIMITED, NEWCAS ON-TYNE.—Caustic Soda (ground and solid), Soda Ash, covered Sulphur, etc.

ROBINSON, COOKS, AND COMPANY, ST. HELENS.—Drawings, trating their Gas Compressors and Vacuum Pumps, fitted Pilkington and Forrest's patent valves.

J. ROYLE, MANCHESTER.—Steam Reducing Valves.

A. SMITH, CLAYTON.—India-rubber Chemicals, Rubber Subst Bisulphide of Carbon, Solvent Naphtha, Liquid Ammonia, Disinfecting Fluids.

WORTHINGTON PUMPING ENGINE COMPANY, LONDON.—Pun Machinery. Speciality, their "Duplex" Pump.

JOSEPH WRIGHT AND COMPANY, TIPTON.—Berryman Feed-Heater. Makers also of Multiple Effect Stills, and W softening Apparatus.

Correspondence.

ANSWERS TO CORRESPONDENTS.

W.V.C.—Sorry to have overlooked your enquiry. You will find the information you require in *Electrolysis*—by Fontaine and Berly—published by Messrs. E. and F. N. Spon, London.

J. A. W.—We insert your query below.

K. P.—Wait a little while. We shall have more to tell you soon.

R. D.—Yes. The cost sheet may look well now, but what is to happen if you increase production as you propose doing.

DELTA.—An advertisement in this issue gives what you want.

QUERY.—Has the Emulseme (for raising liquids as described by Mactear Journal Soc. Chemical Industry, vol. 6, 1887), met with any success in Great Britain, and would not the air, liberated at the delivery pipe carry away considerable quantities of acid fumes (in the case of raising strong hydrochloric acids), unless provided with some small washing arrangement?—J. A. W.

*** We do not hold ourselves responsible for the opinions expressed by our correspondents.*

To the Editor of the Chemical Trade Journal.

SIR,—I notice in your issue of March 1st, 1890, a letter on the determination of "moisture in wood pulp," and beg to say that a number of years' extensive experience in England, America, and Continental Europe has taught me that the air-dried basis is most unsatisfactory to both manufacturer and consumer, and last—but by no means least—to the dealer or importer. I thoroughly agree with Mr. John Pattinson's remarks and claims that selling and buying on "dry weight in water jacketed oven" (as a contract I recently heard read), would avoid much trouble with all pulps.

Further, the method of calculation adopted ought to be more definitely established between seller and buyer, and specified in the contracts.

As the matter now stands, and with existing trade customs, there is far less trouble in the pulp trade from anything connected with the tests than there is from the fact that the average book-keeper in a paper mill is often "way off" in his method of calculation.—Yours respectfully,

H. A. RADEMACHER, C.E.

Bridgeport, Conn., U.S.A.

PRIZE COMPETITION.

A LITTLE more than twelve months ago, one of our friends suggested that it might be well to invite articles for competition, on some subject of special interest to our readers; giving a prize for the best contribution, he at the same time offering to defray the cost of the first prize, and suggesting a subject in which he was personally interested.

At that date we were totally unprepared for such a suggestion, but on thinking over the matter at intervals of leisure, we have come to think that the elaboration of such a scheme would not only prove of benefit to the manufacturer, but would go some way towards defraying the holiday expenses of the prize winner.

We are therefore prepared to receive competitions in the following subjects:—

SUBJECT I.

The best method of pumping or otherwise lifting or forcing warm aqueous hydrochloric acid of 30 deg. Tw.

SUBJECT II.

The best method of separating or determining the relative quantities of tin and antimony when present together in commercial samples.

On each subject, the following prizes are offered:—

One first prize of five pounds, one second prize of one pound, five additional prizes to those next in order of merit, consisting of a free copy of the *Chemical Trade Journal* for twelve months.

The competition is open to all nationalities residing in any part of the world. Essays must reach us on or before April 30th, for Subject II, and on or before May 31st, in Subject I. The prizes will be announced in our issue of June 28th.

We reserve to ourselves the right of publishing the contributions of any competitor.

Essays may be written in English, German, French, Spanish or Italian.

clipped by the rubber ring, the joint made good by the clamps, and filtration effected through a plane surface instead of the usual cone. Professor McLeod showed a diagram of his apparatus for evaporation, and the drying of solids in vacuo, which he has proposed as possibly adapted for the method of water analysis of Frankland and Armstrong. Dr. Plympton had two excellent exhibits: (1) An apparatus for estimating sulphur and the halogens in organic compounds, the principle of which is the same as that employed in determining sulphur in coal-gas; a small flame of coal gas (or hydrogen in the case of sulphur determinations) burns under a lamp chimney in a brisk current of air aspirated by a water pump, and into it is pushed a little platinum basket containing the substance to be analysed, the products of combustion being caught in a tower filled with glass beads moistened with caustic soda. In the case of a volatile liquid, the gas is simply passed through it and burnt along with the vapour it takes up. (2) A set of Gooch crucibles and some modifications of his devising.

Dr. Norman Collie showed several clever contrivances useful to those making many combustions, notably a tube for collecting the water, which, externally resembling an ordinary U tube, contained in the near limb a little tube receptacle to catch the bulk of the water and to allow the rated passage of the gas to be seen, and a drying tube of pumice soaked in sulphuric acid which could be renewed without disconnection. Mr. S. U. Pickering showed the apparatus used in his researches on the nature of solution, much of which had a value besides that given it be its scientific merits, which roused covetous feelings, excusable when the almost fabulous price which platinum has now reached is considered.

That the exhibition was a complete success there can be no doubt, and its repetition may well be effected after a reasonable interval.

THE LEAD-POISONING QUESTION AT SHEFFIELD.—In accordance with the recommendation which followed the investigation last February, by a Special Committee of the Sheffield Corporation, into the prevalence of lead-poisoning in the borough and surrounding district, "a small but sufficient quantity of chalk in a state of minute division" is being added to the Redmires water before its distribution to the town, by way of experiment. Mr. E. M. Eaton, Assoc. M. Inst. C.E., the engineer of the water department, first made experiments with the water in the Hadfield dam, at Crookes, but, experiencing a difficulty in securing a good mixture, he turned his attention to Redmires; and for the last month or so, it is stated, there has been an uninterrupted addition of chalk to the water, and that in a way which secures a perfect admixture. The culvert through which the water flows into the town commences in a little wood by the side of the third or lowest dam at Redmires. As the water leaves the dam, it runs into a miniature reservoir, and then falls into the culvert some feet below. It is here that the chalk is applied. The means adopted are both simple and inexpensive. In a shed close to the culvert there is a large cask constantly supplied with water, into which is put a prescribed quantity of chalk at stated intervals. A churn-like apparatus fixed into the cask secures the perfect mixture of the chalk with the water. A lead pipe runs from the cask to the head of the culvert, and from it there drops into the water, just as it is tumbling into the culvert, a continuous supply of milk or cream of chalk. The water, of course, becomes instantly coloured; but it is said that by the time it reaches Sandygate, it is as clear as it was before it entered the conduit. That this plan secures a thorough admixture of the chalk with the water there can be no question; and a further advantage connected with it is, that the supply of one to the other can be regulated by turning a tap at the head of the pipe. The tests, of course, have not been carried on sufficiently long to warrant the statement that a perfect remedy has been found, but Mr. Eaton is said to be satisfied with the results which he has obtained so far.

Market Reports.

REPORT ON MANURE MATERIAL.

There has been a fair business passing during the week for immediate delivery, but, as we approach the end of the spring season, the orders are becoming more retail in their character, buyers being unwilling to saddle themselves with anything they may be obliged to hold over the current season. Prices on the whole, both for phosphatic and nitrogenous material remain without much alteration, but perhaps a shade in favour of buyers.

The situation as regards Charleston phosphate rock continues without much alteration, outside sellers still inviting bids of 10d. per unit for cargoes of river rock c.i.f. to United Kingdom, and buyers being in no hurry to make their purchases for next season, having still hopes of being able to do better by waiting. Quotations for Canadian and high class Somme remain without alteration, but they do not admit of business, and sellers will have to modify their ideas before they can expect to attract United Kingdom orders. There is nothing special to note in Belgium phosphates, and as regards Florida, the results of exper ments of trial shipments are awaited before buyers can entertain larg purchases of this description of rock.

We do not hear of any transactions in River Plate cargoes of bone and as yet it seems too early to determine how the market is likely t rule for next season, the American demand for shipment up to the en of the year which may be expected having so far not made its appea ance. We do not hear of any fresh business in bone ash, but th market is very firm. Sales of East India bone meal have been mad at £5. 10s. ex quay and ex store, Liverpool, and £5. 10s. is closin value on spot. For shipment ahead £5. 5s. would no doubt be bus ness, and in some cases buyers are invited to bid as low as £5., but w doubt whether the latter price would be business. The market ha been cleared of crushed East Indian bones on spot, but business i doing for shipment ahead at very full prices, and sellers do not seer inclined to commit themselves to any very large sales as they did la year to their cost. Nitrate of soda is a shade easier on the week, spo value being 8s. per cwt. for ordinary up to 8s. 1½d. for refined. Du cargoes are worth scarcely 8s. per cwt., and for shipment ahead ther would be sellers at 8s. 4½d., June-July shipment, and at 8s. 8¼d. fo September-October-November.

The market for dried blood remains firm at 10s. 6d. to 11s. per uni for home prepared, delivered free to rails at works. There is no Rive Plate offering at the moment.

Superphosphates are getting scarce, and 47s. 6d., net cash, in bulk f.o.r. at works, now lowest, 50s. being asked in some instances.

METAL MARKET REPORT.

	Last week.		This week
Iron	44/5	....	44/6
Tin	£92 0 0	....	£93 5 0
Copper	49 13 9	....	52 5 0
Spelter	21 5 0	....	23 0 0
Lead	12 17 6	....	12 17 6
Silver	46d.	....	47¼d.

COPPER MINING SHARES.

	Last week.			This week.	
Rio Tinto	$17\frac{3}{8}$	$17\frac{1}{2}$	....	$18\frac{11}{16}$	$18\frac{13}{16}$
Mason & Barry	$6\frac{3}{4}$	$6\frac{7}{8}$	....	$7\frac{7}{16}$	$7\frac{9}{16}$
Tharsis	$4\frac{15}{16}$	$5\frac{1}{16}$	....	$5\frac{5}{16}$	$5\frac{7}{16}$
Cape Copper Co.	$3\frac{7}{8}$	4	....	$4\frac{7}{16}$	$4\frac{9}{16}$
Namaqua	$2\frac{3}{8}$	$2\frac{1}{2}$	....	$2\frac{7}{8}$	3
Copiapo	$2\frac{7}{16}$	$2\frac{9}{16}$	....	$2\frac{13}{16}$	$2\frac{15}{16}$
Panulcillo	1	$1\frac{1}{8}$	....	1	$1\frac{1}{8}$
New Quebrada	$\frac{1}{2}$	$\frac{3}{4}$	....	$\frac{5}{8}$	$\frac{3}{4}$
Libiola	$2\frac{3}{8}$	$2\frac{5}{8}$	....	$2\frac{13}{16}$	$2\frac{15}{16}$
Tocopilla	1/-	1/6	....	1/-	1/6
Argentella	3d.	9d.	....	3d.	9d.

MISCELLANEOUS CHEMICAL MARKET.

There has been a further decline in value of caustic soda during th past week, 70% having been done at £8. 5s. per ton. Other strength are lower in a corresponding degree; 77% f.o.b. Tyne, £9. 15s. to £10. 74% f.o.b. Liverpool at £9. per ton Values all round are now withi 5s. to 7s. 6d. per ton of the lowest points touched at the end of Janua last, and prior to the strong and sudden advances which commence from that time. With soda ash at 1¼d. per deg., we now have th anomaly of caustic soda at a lower price, viz 1$\frac{3}{16}$d. per deg.; certai grades of soda ash are, however, distinctly lower in value and can b got from 1⅛ per deg. Bleaching powder quiet and unchanged i value at £5. per ton on rails. Soda crystals, £2. 17s. 6d. per to Vitriol more plentiful, and rather lower in price. Muriatic aci without change. Sulphate of copper is firmer in sympathy with th copper market, and spot prices are £25. to £26. per ton. Lime £23. 10s. to £24. Lead salts are firmly maintained in prices, an white acetate reported firmer in some quarters, but no general advanc has been established. White lead, nitrate of lead, litharge, etc., hav all rather a higher tendency. Firmness in the metal markets has als influenced salts of tin, zinc, copper, and silver. Acetates of lime ar in about the same position as last reported, and prices nominally th same. There is, however, only a small amount of business passing i these or other acetates or acetic acids. For oxalic acid there has bee considerable enquiry, speculative and resale transactions being re corded at 3¾d. to 4d. per lb. Tartaric and citric acids are witho quotable change. Potash prussiate and bichromate are quiet. Potas carbonate and caustic firm, and fair amount of business being done a current rates, both for early and forward deliveries. Carbolic crysta still declining in value, and quotations somewhat irregular—at 7½ to 8¼d. for 34° (in quantity). Other miscellaneous goods witho special change in position, and trade generally is reported quiet in mo branches of chemical manufacture.

TAR AND AMMONIA PRODUCTS.

:e quoted 3s. 9d. and 2s. 9d. for 90's and 50/90's
but the prices are merely nominal and there is so little
ı their continuance that forward contracts could probably be
easier terms. The heavy demand for solvent naphtha is
ices stand fairly firm at 1s. 8d. to 1s. 9d.. Creosote
to 2½d. according to quality, while the crude carbolic
ınsettled and irregular, buyers offering 2s., and sellers
ss than 2s. 2d. Anthracene and pitch are in practically
ɔsition as last week.
of ammonia has improved considerably since last week.
es have been accorded at all the ports; and the present
ıins reassuring. At Beckton the quotation has risen to
l.—£11. 15s., but no business is recorded at these prices.
don makes may be taken as £11. 8s. 9d., and £11. 7s. 6d.
or Leith.

E LIVERPOOL MINERAL MARKET.

on the whole have ruled steadier during the past week.
Arrivals small; prices firm but unaltered. Magnesite:
ivals have taken place, and prices for raw lump are in
buyers. Raw ground, £6. 10s.; and calcined ground,
l. Bauxite (Irish Hill brand) continues in exceptionally
and, bringing full prices—Lump, 20s.; seconds, 16s.;
ground, 35s. Dolomite, 7s. 6d. per ton at the mine.
ılk firm at last quotations, especially G.G.B. "Angel-
ınd—95s. to 100s. medium, 105s. to 110s. superfine.
:bonate), easy; selected crystal lump scarce at £6.;
ps, 90s.; best, 80s.; seconds and good nuts, 70s.;
; best ground, £6.; and selected crystal ground, £8.
s somewhat improved, best lump, 35s. 6d.; good medium,
m, 25s. 6d. to 27s. 6d.; common, 18s. 6d. to 20s.;
white, G.G.B. brand, 65s.; common, 45s.; grey, 32s. 6d.
aicestone quiet; ground at £10., and specially selected
t quality, £13. Iron ore steady. Bilbao and Santan-
ıos. 6d. f.o.b.; Irish, 11s. to 12s. 6d.; Cumberland in
ınd at 18s. to 24s. Purple ore as last quoted. Spanish
us ore continues in fair demand. Emery-stone: Best
scarce, and inquired for at full prices No 1 lump
at £5. 10s. to £6., and smalls £5. to £5. 10s.
th steady; 45s. to 50s. for best blue and yellow;
able ground, £7. Scheelite, wolfram, tungstate of soda,
ı metal further inquired for. Chrome metal, 5s. 6d. per lb.
lloys 2s. per lb. Chrome ore: steadier, prices un-
ntimony ore and metal have further advanced. Uranium
to 26s. Asbestos scarce, bringing increased prices.
l ore firm; smalls, £14. to £15.; selected lump, £16. to
ne: Best qualities scarce—60s. to 80s. Strontia steady;
estine) steady, 16s. 6d. to 17s. Carbonate (native), £15.
wdered (manufactured), £11. to £12. Limespar: English
d, old G.G.B. brand in demand, and brings full prices;
ınd English. Felspar, 40s. to 50s.; fluorspar, 20s. to £6.
emand, firm at 22s. to 25s. Plumbago steady; Spanish,
'eylon lump at last quotations; Italian and Bohemian, £4.
ton; founders, £5. to £6.; Blackwell's "Mineraline,"
ch sand, in cargoes, continues scarce on spot—20s.
Ferro-manganese and silicon spiegel in good demand
figures. Chrome iron, 20 per cent., £24. to £25.
ı, £50. China clay freely offering—common, 18s. 6d.;
m 22s. 6d. to 25s.; best, 30s. to 35s. (at Runcorn).

ŒEST OF SCOTLAND CHEMICALS.

GLASGOW, Tuesday.

al market in this quarter has further lost in tone since the
st week, and caustic soda, at that time giving what were
ns of recovery, has gone back again, and to-day is obtain-
noney in all the strengths, and yet not very much doing.
ɔwder also is easier, and the other market chemicals
ıe to the fall, with the exception, perhaps, of sulphate of
Sulphate took a change for the better towards the close of
ıd to-day it is very firm, especially for forward delivery,
ally declining to name any prices on this basis at present.
are plentiful at recovery up to £11. 5s., but sellers are
this, and a further rise is probable. Deliveries of bichro-
ported as mending slightly, but inaction is still the pre-
ure. Paraffins unchanged on the market, but makers are
ened with trouble on the part of the miners. The
announcement in the Scotch morning papers of the new Davis discoveries, threatening to revolutionise the alkali manufacture, and particularly the chlorine department, has been received with much interest in chemical circles, and developments are watched for. Chief prices current are:—Soda crystals, 50s. 6d. net Tyne; alum in lump £5., less 2½% Glasgow; borax, English refined £29., and boracic acid, £37. 10s. net Glasgow; soda ash, 1¼d. less 5%, Tyne; caustic soda, white, 76°, £10., 70/72°, £8. 10s., 60/62° £7. 15s., and cream, 60/62° £7. 10s., all less 2½% Liverpool; bicarbonate of soda, 5 cwt. casks, £6. 5s., and 1 cwt. casks, £6. 10s. net Tyne; refined alkali 48/52°, 1½d., net Tyne; saltcake, 25s. to 26s.; bleaching powder, £5. 5s. to £5. 7s. 6d. less 5% f.o.r. Glasgow; bichromate of potash, 4d., and of soda 3d., less 5 and 6% to Scotch and English buyers respectively; chlorate of potash, 4⅝d., less 5% any port; nitrate of soda 8s. 1½d. to 8s. 3d.; sulphate of ammonia, spot, £11. 5s. to £11. 6s 3d. f.o.b. Leith; salammoniac, 1st and 2nd white, £37. and £35., less 2½% anyport; sulphate of copper, £24. 10s., less 5% Liverpool; paraffin scale, hard and soft, 2½d. per lb.; paraffin wax, 120°, semi-refined, 3⅛d.; paraffin spirit (naphtha), 8d. a gallon; paraffin oil (burning), 6¾d. to 7d. on rails Glasgow to all Scotch buyers (English orders negotiable about 1¼d. a gallon lower); ditto (lubricating), 865°, £5. 10s. to £6. per ton; 885°, £6. 10s.; and 890/895°, £7. 15s. to £8. Week's imports of sugar at Greenock were 25,549 bags.

THE LIVERPOOL COLOUR MARKET.

COLOURS are unaltered. Ochres: Oxfordshire quoted at £10., £12., £14., and £16.; Derbyshire, 50s. to 55s.; Welsh, best, 50s. to 55s.; seconds, 47s. 6d.; and common, 18s.; Irish, Devonshire, 40s. to 45s.; French, J.C., 55s., 45s. to 60s.; M.C., 65s. to 67s. 6d. Umber: Turkish, cargoes to arrive, 40s. to 50s.; Devonshire, 50s. to 55s. White lead, £21. 10s. to £22. Red lead, £18. Oxide of zinc: V.M. No. 1, £25.; V.M. No. 2, £23., Venetian red, £6. 10s. Cobalt: Prepared oxide, 10s. 6d.; black, 9s. 9d.; blue, 6s. 6d. Zaffres: No. 1, 3s. 6d.; No. 2, 2s. 6d. Terra Alba: Finest white, 60s.; good, 40s. to 50s. Rouge: Best, £24.; ditto for jewellers', 9d. per lb. Drop black, 25s. to 28s. 6d. Oxide of iron, prime quality, £10. to £15. Paris white, 60s. Emerald green, 10d. per lb. Derbyshire red, 60s. Vermillionette, 5d. to 7d. per lb.

THE TYNE CHEMICAL REPORT.

TUESDAY.

The state of the chemical market is much about the same as reported last week. Business keeps quiet, but there is no material alteration in prices except for caustic soda, which is 10s. per ton under last quotation. This reduction makes a total decrease in price of £3. 10s. per ton during the last month. Bleaching powder is steady at last week's rates, with a little more enquiry for prompt shipment. Soda crystals are well sustained, and prices are stiffening. Soda ash has fallen off by one and a half per cent., and sulphate of soda quiet and easier in price by 1s. 6d. per ton. The production of caustic soda on an extensive scale by a new process will shortly be commenced on Tyneside. The Wallsend Chemical Company, Wallsend, who gave up balling their sulphate about three years ago, and disposed of it to the Jarrow Chemical Co., Tyne Dock, have recommenced the manufacture of soda ash and crystals. The market prices current are:—Bleaching powder, in softwood casks, £5. to £5. 2s. 6d. per ton; caustic soda, 77%., in drums, £9. 10s. per ton, ground in 3-4 cwt. barrels, £13. per ton; soda ash, 48/52%, 1¼d. per degree less 5½; soda crystals, in 1 cwt. bags, net weight, £2. 14s. per ton, in 2 cwt. bags, net weight, £2. 11s. 6d. per ton; in casks, gross weight, £2. 11s. 6d. per ton; sulphate of soda, in bulk, 32s. per ton; ground and packed in casks, 42s. per ton; recovered sulphur, £4. 5s. per ton; chlorate of potash, 4⅝d. per lb.; silicate of soda, 75° Tw., £2. 10s. per ton; 100° Tw., £3. 7s. 6d. per ton; 140° Tw., £4. per ton; hyposulphite of soda, in 1 cwt. kegs, £5. 15s. per ton, in 5—7 cwt. casks, £5. 5s. per ton; pure white sulphate of alumina, £4. 10s. per ton; blanc fixe, £7. 10s. per ton; chloride of barium, £8. per ton; nitrate of baryta, crystals, £18. 15s. per ton; ground, £19. 5s. per ton; sulphide of barium, £5. 10s. per ton—all f.o.b. Tyne, or f.o.r. makers' works.

Small coals for manufacturing purposes are slightly easier in price, owing to the lessened demand, and consequently more coal being thrown on the market. Durham small quoted 9s. 6d. to 10s. per ton, and Northumberland small, 8s. to 8s. 6d. per ton.

At the adjourned meeting of the Durham coalowners and the miners' representatives held on Friday last, the coalowners adopted the following resolution:—"After considering the request of the Federation Board for a further advance in wages, we are not prepared at present to give any advance, but in a month's time the tendency of trade may be made more clear, and we will then meet and discuss the matter further with the Federation Board."

Gazette Notices.

THE BANKRUPTCY ACT, 1883.

Partnerships Dissolved.

HOWLETT BROS., Hanley, brick and tile manufacturers.
G. W. HAWKSLEY, J. WILD, and D. HAIGH, under the style of G. W. Hawksley and Co., or Hawksley, Wild and Co., Sheffield, boiler manufacturers, also under the style of the Belle Isle Dye Works Company, Wakefield, dyers.
WATKINS and SOULBY, St. Helens, oil and colour manufacturers.

New Companies.

BANNER, SON AND COMPANY, LIMITED.—This company was registered on the 2nd inst., with a capital of £1,000., in £1. shares, to carry on in Liverpool and elsewhere the business of chemists, druggists, &c. Registered office, 179, London-road, Liverpool.

KEFLEY BRICK COMPANY, LIMITED.—This company was registered on the 3rd inst., with a capital of £10,000., in £10. shares, to carry on business as brick and tile makers. The subscribers are :—

	Shares.
J. Hassal, Ashbo de la Zouch, earthenware manufacturer	1
W. Hassall, Woodville, earthenware manufacturer	1
E. Ison, Ashby de la Zouch, metal trade valuer	1
G. Williams, Sheffield, brass founder	1
W. T. Skelding, Stourbridge, brick manufacturer	1
T. Spencer, Bradford, contractor	1
C. Skelding, Stourbridge, brick manufacturer	1
J. Jordan, Wollaston, Stourbridge, solicitor	1
E. Webster, Harborne, brickworks manager	1

MASON'S PAPER MILL COMPANY, LIMITED.—This company was registered on the 7th inst., with a capital of £25,000., in £10. shares, to carry on the business of paper manufacturers, and for such purposes to acquire the leasehold ground known as Tovell's Wharf, situate in the Parish of St. Peter, Ipswich, and to enter into an agreement with Frank Wm. Mason. The subscribers are :—

	Shares.
G. C. Mason, Ipswich, merchant	1
F. W. Mason, St. Nicholas, Ipswich, merchant	1
H. W. Mason, Ipswich, merchant	1
G. Mason, Ipswich	1
Rev. S. W. Stagg, Colchester	1
Miss G. E. Mason, Ipswich	1
Miss C. M. G. Mason, Ipswich	1

The Patent List.

This list is compiled from Official sources in the Manchester Technical Laboratory, under the immediate supervision of George E. Davis and Alfred R. Davis.

APPLICATIONS FOR LETTERS PATENT.

Manufacture of Chlorine. G. E. Davis and A. R. Davis. 6,416. April 28.
Mechanical Stoker.—(Complete Specification). C. Tschoepe. 6,424. April 28.
Chemical Furnaces. G. E. Davis and A. R. Davis. 6,433. April 28.
Smelting. H. Napier and T. Carr. 6,462. April 28.
Preparation of Oxygen.—(Complete Specification). E. Neave. 6,463. Ap. 28.
Distillation of Hydro-carbons and other Oils. C. M. Pielsticker. 6,465. April 28.
Distillation of Hydro-carbons and other Oils. C. M. Pielsticker. 6,466. April 28.
Electric Batteries. H. T. Eagar and R. P. Milburn. 6,476. April 28.
Preventing Incrustation in Boilers.—(Complete Specification). J. Efrem. 6,478. April 28.
Evaporating Pans. G. Fletcher. 6,484. April 28.
Nitrated Colouring Matters. J. Imray. 6,486. April 28.
Low Water Alarm for Steel Boilers.—(Complete Specification). O. Imr 6,489. April 28.
Electric Refrigeration.—(Complete Specification). M. W. Dewey. 6,4 April 28.
Feed-Water Heaters.—(Complete Specification). W. P. Thompson. 6,5 April 28.
Apparatus for Burning Hydro-carbon Oil. H. H. Lake. 6,502. April
Dyeing. W. H. Booth. 6,506. April 28.
Electro Deposition. W. Ryland. 6,524. April 29.
Hardening Steel.—(Complete Specification). J. H. Bramwell. 6,529. April
Reversing Valves for Gas Furnaces. S. Fox and T. Williamson. 6,5 April 29.
Production of Hydrochloric Acid and Ammonia. J. Plummer, jun. 6,5 April 29.
Preparation of Oxygen.—(Complete Specification). A. Longsdon. 6,5 April 29.
Insulating and Fire-proofing Material. G. Brewer. 6,555. April 29.
Electrical Measuring Instrument.—(Complete Specification). H. H. La 6,568. April 29.
Electrical Measuring Instrument.—(Complete Specification). H. H. La 6,569. April 29.
Pressure Gauges.—(Complete Specification). J. Jackson. 6,573. April 29.
Manufacture of Gas. J. H. W. Stringfellow. 6,575. April 29.
Dessicating Apparatus.—(Complete Specification). W. P. Thompson. 6,5 April 29.
Repairing Carbon Filaments of Electric Lamps.—(Complete Specificatic H. H. Lake. 6,581. April 29.
Gas Producers.—(Complete Specification). J. J. Shedlock and T. Den 6,584. April 29.
Distillation and Rectification of Alcohol.—(Complete Specification). W. Whiteman. 6,587. April 29.
Bleaching. R. T. Webb, care of The Ards Weaving Co., Limited, Newtowna Down. 6,612. April 30.
Manufacture of Perfumes from Hydro-carbons. F. Valentiner. 6,6 April 30.
Electrical Switches. J. Warden and G. Smith. 6,623. April 30.
Steam Boiler Feeding, Regulating, and Indicating Apparatus. Murrie. 6,625. April 30.
Treatment of Paper-making Fibre Materials. J. Johnston and G. Johnst 6,644. April 30.
Manufacture of Iron and Steel. J. H. Lancaster and M. R. Cowley. 6,6 April 30.
Furnaces for Destroying Organic Matter. M. M. Brophy, of the firm J. Slater and Co., 4, South-street, Finsbury, London. 6,660. April 30.
Electric Meter. A. Frager. 6,667. April 30.
Deposition of Alloys. S. O. Cowper-Coles. 6,670. April 30.
Smoke Prevention. T. Thornley. 6,692. May 1.
Manufacture of Chlorine. G. E. Davis and A. R. Davis. 6,698. May 1.
Wool Scouring. C. W. Kimmins and T. Craig. 6,706. May 1.
Marine Cement. F. C. Goodall. 6,711. May 1.
Steam Boiler and other Furnaces. H. C. Ashlin. 6,718. May 1.
Drying Kilns. J. Kirkby. 6,722. May 1.
Colouring Matters. C. Dreyfus. 6,729. May 1.
Production of Coloured Rubber Goods. C. Dreyfus. 6,730. May 1.
Generation of Carbonic Acid. J. McEwen. 6,758. May 2.
Secondary Batteries. H. Edmunds. 6,795. May 2.
Illuminating or Heating Gas.—(Complete Specification). W. C. Andre 6,800. May 2.
Purification of Illuminating Gas.—(Complete Specification). W. C. Andre 6,804. May 2.
Manufacture of Copper. Sir H. H. Vivian. 6,821. May 2.
Preparation of Chlorine. G. E. Davis and A. R. Davis. 6,831. May 3.
Treatment of Soap Makers' Leys. J. Taylor. 6,834. May 3.
Steam Boiler Furnaces. A. Sharratt. 6,842. May 3.
White Pigment. E. Parry and H. Foskett. 6,850. May 3.
Refrigerating, Evaporating, and Dessicating Apparatus. A. G. Sout and F. D. Blyth. 6,851. May 3.
Portland Cement.—(Complete Specification). G. Higham. 6,855. May 3.
Gas Producers for Metallurgical and other Purposes.—(Complete Spe cation). J. J. Shedlock and J. J. Meldrum. 6,864. May 3.
Asbestos Fuel.—(Complete Specification). H. C. Turner. 6,868. May 3.
Substantive Colouring Matters. J. Y. Johnson. 6,874. May 3.
Colouring Matters. J. Y. Johnson. 6,875. May 3.
Smoke Prevention. J. Ashworth and W. Kneen. 6,876. May 3.
Combustible Coal-Brick.—(Complete Specification). J. J. Hiertz and A Garnett. 6,877. May 3.
Colouring Matters. H. H. Lake. 6,879. May 3.
Galvanic Element with Constant Current.—(Complete Specification). van der Poppenburg. 6,803. May 3.

IMPORTS OF CHEMICAL PRODUCTS

AT

THE PRINCIPAL PORTS OF THE UNITED KINGDOM.

LONDON.

Week ending May 6th.

Alkali—

Holland, 400 c. R. W. Greeff & Co.
France, 210 Arnati & Harrison
Holland, 9 G. Rahn & Co.

Acetic Acid—

Holland, 52 pks. A. & M. Zimmermann
Germany, 6 G. Meyer & Co.
Holland, 25 pks. Beresford & Co.
" 26 A. & M. Zimmermann
" 170 Phillipps & Graves

Acetate of Lime—

U. S., £150. Gillespie Bros. & Co.
" 85 Lister & Biggs

Antimony Ore—

Melbourne, 14 t. G. Bore & Co.
Spain, 20 Pillow, Jones, & Co.
E. Turkey, 10 c. L. & I. D. Jt. Co.
Germany, 10 O. Scholzig

Acetone—

Holland, £100. Beresford & Co.

Ammonia—

Holland, £25. Weatherley, Mead, & Co.

Bromide—

Germany, £50. Howard & Son

Boracic Acid—

Italy, 20 pks. J. Puddy & Co.

Barytes—

Germany, 14 cks. O'Hara & Hoa
Germany, 23 cks. Constable, Hende &
" 24 W. Harrison &
" 103 J. Ma
Belgium, 10 J. J. Lyon &

Bones—

E. Indies, 301 t. Gonne, Croft, &
" 25 Tullock &
" 25 Culverwell, Brooks, &
U. S., 37 H. A. Lane &
Zanzibar, 10 L. & I. D. Jt.
Sydney, 8 Goad, Rigg, &
" 14 Hicks, Nash, &

G. Boor & Co.
C. J. Capes & Co.
F. Smith
Hyslop & Symons
A. Hughes
W C. Bacon & Co.
F. Parbury & Co.
Union Lighterage Co., Limited
2. L. & I. D. Jt. Co.
Adamson, Gilfillan, & Co.
Kebbel, Son, & Co.
L. & I. D. Jt. Co.
C. Atkins & Nisbet
Hoare, Wilson & Co.
Knight, Hayman, & Co.
Kebbel, Son, & Co.
L. & I. D. Jt. Co. ,,
J. E. Wilson
L. & I. D. Jt. Co.
Kebbel, Son, & Co.
Thames S. T. Co.
Anderson, Weber, & Smith

tar—
pks. B. & F. Wf. Co.
C. F. Gerhardt ,,
Major & Field
W. C. Bacon & Co.
5 C. Christopherson & Co.
5 W. C. Bacon & Co.

sh—
4. Petri Bros.
4 t. Harrold Bros.
3 H. Bath & Son
0 W. H. Pridham & Co.

therwise undescribed)
0. A. & M. Zimmermann
5 Page, Son, & Co.
50 A. & M. Zimmermann
15 Phillipps & Graves
11 Hernu, Peron, & Co.
18 T. H. Lee
16 Phillipps & Graves
00 A. & M. Zimmermann

Bark
8 t. Von der Meden & Co.
07 Baxter & Hoare
10 H. Henle
10 G. S N. Co.
95 Dalgety & Co.
38 Baxter & Hoare
20 J. Vicary & Sons
20 Devitt & Hett
10 W. A. Bowditch
28 Magee, Taylor, & Co.
50 J. J. Williamson & Son
pks. T. Ronaldson & Co.
L. J. Levinstein & Sons
Bailey & Leetham ,,
Union Lighterage Co., Limited.
W. France & Co.
G. Scruton & Co
A. J. Humphrey
W. Barton & Sons
Boutcher, Mortimore, & Co.

s
pks. A. Everingham & Co.
Benecke, Souchay, & Co.
J. P. Alpe & Co.
Forbes, Forbes, & Co.
A. & W. Nesbitt
Baxter & Hoare
L. & I. D. Jt. Co.
B. Jacob & Sons
Sollas & Sons
Blundell, Spence, & Co.
J. Kitchen, Ld.
pks. Burt, Boulton, & Co.
pks. Swanston & Co. ,,
t. M. D. Co.
9 chts. Patry & Pasteur
1 L. & I. D. Jt. Co.
8 ,,
5 Elkan & Co.
9 L. & I. D. Jt. Co.
2 F. Stahlschmidt & Co.
4 L. & I. D. Jt. Co.

Gambier
Singapore, 378 L & I. D. Jt. Co.
,, 415 R. & J. Henderson
,, 1,347 W. H. Cole & Co.
,, 10 t. Elkan & Co.
,, 214 L. & I. D Jt. Co.
,, 128 pks. H. Lambert
,, 208 T. J. & T. Powell
,, 86 L. & I. D. Jt. Co.

Alizarine
Holland, 57 J. Ferry & Co.

Cutch
E Indies, 4,000 Hoare, Wilson, & Co.
,, 1,232 H. A. Lichfield & Co.

Dyestuffs *(otherwise undescribed)*
Germany, 19 pks. Flemings' Oil and Chemical Co.

Glycerine—
Holland, £365 Spies, Bros., & Co.
,, 69 Knight & Morris
Germany, 245 Beresford & Co
,, 170 ,,
Holland, 64 F. W. Heilgers & Co.

Glucose—
Germany, 600 pks 600 c. J. Barber & Co.
,, 25 799 L. Sutro & Co.
,, 300 300 Dutrulle, Solomon & Co.
,, 130 1,015 F. J. Warren
U. S, 100 500 R. G. Hall & Co.
Germany, 11 180 J. Knill & Co.
U. S., 440 1,240 T. M. Duche & Sons
,, 25 125 Hyde, Nash, & Co.
Germany, 20 160 C. Hapke
,, 35 280 T. M. Duche & Sons
U. S, 1,600 1,600 A. Dawson
,, 120 120 T. M. Duche & Sons
,, 100 560 Proprts. Scott's Whf.
Germany, 24 202 J. Knill & Co.
,, 200 200 J. F. Ohlmann

Gutta Percha—
Singapore, 90 c. N. S. S. Co.
,, 40 Jackson & Till
,, 485 H. W. Jewesbury & Co
France. 20 Flageollet & Co.
Singapore, 934 Kaltenbach & Schmitz

Guano—
Lobos, 1,180 t. Anglo Cont Guano. Wks

Isinglass—
Singapore, 4 pks. S. Barrow & Bros.
E. Indies, 4 Clarke & Smith
,, 5 L. & I. D. Jt. Co.
Singapore, 24 ,,
Demerara, 6 ,,
Singapore, 3 ,,
Brettesnoe, 53 J. Jensen & Co., Ltd.
Demerara, 6 Lewis & Peat

Manure—
Belgium, 170 t. A. Hunter & Co.
Germany, 50 ,,
,, 5 Anglo Contl. Guano Wks.
U. S., 8 ,,
Las Palmas, 69 N. W African Co., Lmtd.
Holland, 318 Lawes Manure Co.
France, 120 T. Farmer & Co.
,, 150 S. F. Waters & Co.
Holland, 370 A. Hunter & Co.
Germany, 40 F. W. Berk & Co.

Manganese Ore—
Adelaide, 171 t. Harrold Bros

Naphthalene—
Holland, £20 R. Morrison & Co.

Naptha—
Germany, 27 drs. Stein Bros.
Holland, 1 ,,

Nitre—
Iquique, 12,565 bgs. W. Montgomery & Co.
,, 12,565 H. Bath & Son
Peru, 8,650 cu. bgs. A. Ford & Co.

Pearl Ash—
France, 132 c. Petri Bros.

Potassium Sulphate—
Germany, 68 pks. Bessler, Wachter, & Co.
,, 30 H. Lambert
,, 21 cks. Burgoyne, Burbidge, & Co.
France, 21 pks. F. W. Berk & Co.

Potassium Hydrate—
Germany, 3 pks. Hernu, Peron, & Co.

Potassium Muriate—
Germany, 100 pks. F. W. Berk & Co.

Potassium Carbonate—
France, 29 pks. Beresford & Co.
,, 14 ,,

Potassium Bichromate—
France, 21 pks. W. H. Pridham & Co.

Paraffin Wax—
U. S., 830 brls. 84 cs. H. Hill & Sons
,, 325 ,,
,, 75 Prop. Chambs. Whf.
,, 41 L. & I. D. Jt. Co.
,, 120 Rose, Wilson, & Rose
,, 100 G. H. Frank
Rangoon, 290 brls. H. Hill & Sons

Pyrites—
Spain, 200 t. Anderson, Anderson, & Co.

Plumbago—
Holland, 4 cks. Brown & Elmslie
E. Indies, 87 brls. Hoare, Wilson, & Co.
Germany, 147 cks. Brown & Elmslie
Holland, 15 ,,
E. Indies, 183 brls. J. Thredder, Son, & Co.
,, 122 J. Owen
,, 55 Doulton & Co.
Holland, 20 cs. J. Barber & Co.

Pyrites—
Spain, 1,340 t. Societie Commerciale

Sodium Acetate—
Treport, £85 Prop. Hay's Wf.

Saltpetre—
E. Indies, 225 bgs. Clift, Nicholson, & Co.
Germany, 11 cks. Craven & Co.
,, 118 P. Hecker & Co.
,, 250 bgs. C. Wimble & Co.
E. Indies, 1,435 ,,
,, 100 L. & I. D. Jt. Co.
,, 1,822 ,,

Stearine—
France, 9 sks. Van Geelkerken, & Co.
Holland, 25 bgs. Perkins & Homer
Belgium, 62 H. Hill & Sons

Tartaric Acid—
Holland, 40 pks. B. & F. Wf. Co.
,, 12 ,,
,, 5 Northcott & Sons

Tartars—
Italy, £1 400 B. Jacob & Sons
,, 407 Fellows, Morton, & Co.
,, 1,400 Thames S. Tug Co.
,, 1,000 Howards & Sons
,, 605 Seager, White, & Co.

Tar—
Germany, 200 brls. Linck, Moeller. & Co.
France, 21,000 g. Burt, Boulton, & Co,

Ultramarine—
Holland, 15 pks. Haeffner, Hilpert, & Co.
Germany, 15 cs G. Steinhoff
Belgium, 9 pks. W. Harrison & Co.

Verdigris—
France, £29 R. Womersley & Son

Zinc Sulphate—
Germany, £60 Howards & Son

Zinc Oxide—
U. S., 100 brls. M. Ashby, Ltd.
Holland, 110 M. Ashby
,, 15 Beresford & Co.
,, 50 cks. M. Ashby, Ltd.
Germany, 25 Burrell & Co

LIVERPOOL.

Week ending May 8th.

Albumen—
Bruges, 20 cs. J. T. Fletcher & Co.

Acetate of Lime—
New York, 762 bgs.
,, 861
,, 713

Borax—
New York, 1 brl.

Bones—
Boston, 154 bgs.
Cadiz, 1,120 Strumpel and M. S. Brown

Barytes—
Antwerp, 9 cks. 30 bgs.

Bone Ash—
Rio Grande, 81 t. Redfern, Alexander, & Co.

Bone Meal—
Bombay, 3,133 bgs.
,, 4,000

Caustic Potash—
Dunkirk, 21 dms.

Copper Regulus—
Huelva, 94 t.

Caoutchouc—
New York, 17 pks. W. Symington & Co.
,, 3 cs. Farnworth & Jardine
Addah, 22 J. P. Werner
,, 9 cks. F. & A. Swanzy
Sierra Leone, 10 brls. Pickering & Berthoud
Rangoon, 656 bgs.
Pram Pram, 45 cks. J. P. Werner
,, 2 Kronig & Siegler
,, 1 Edwards Bros.
Salt Pond, 2 5 cs. F. & A. Swanzy
,, 1 F Schaeffer & Co.
,, 1 H. Havard & Co.
,, 1 N. P. Nathan & Son
,, 3 cs. 1 Davies, Robbins, & Co.
,, 2 1 Pickering & Berthoud
,, 1 28 A. Miller, Bros., & Co.
,, 3 R. Barbour & Bros.
,, 1 2 R. Muller
,, 1 Pitt Bros.
Cape Coast, 1 3 cs. Miller, Bros., & Co.
,, 5 L. Hart & Co.
,, 1 J. Bowden & Co.
,, 1 cs. 11 Ihlers & Bell
,, 7 ,,
,, 1 W. A. M'Iver & Co
,, 3 Fletcher & Fraser
,, 1 Mordaunt Bros.
,, 2 Davies, Robbins, & Co.
Assinee, 1 W. D. Woodin
,, 1 Pickering & Berthoud
,, 1 A. Verdier
,, 1 F. Schaeffer & Co.
,, 1 Edwards Bros.
,, 1 Radcliffe & Durant
,, 1 bg. ,,
,, 3 A. Herschel
,, 2 Millward, Bradbury, & Co.
Grand Bassa, 1 brl. Cie Francaise
Sierra Leone, 15 cks. Paterson, Zachonis, & Co.
,, 2 Ashton, Kinder, & Co.
,, 13 Cie Francaise
,, 29
,, 136 Pickering & Berthoud
,, 4 A. Herschell
Bathurst, 6 Cie Francaise
,, 14 A. Reis
,, 7 A. Cross & Co.
,, 1 2 cs.
Goree. 3 A. Cross & Co.
,, 1 Cie Francaise

Copper Precipitate—
Villa Real, 17,000 bgs. Mason & Barry
Huelva, 1,519

Chemicals *(otherwise undescribed)*
Rotterdam, 10 cks.

Dyestuffs—

Divi Divi
New York, 30 bgs. W. Clucas

Fustic
Jamaica, 148 t.
Carthagena, 788 ps. Mildred, Goyenecke, & Co.

Extract
New York, 50 bxs.
Havre, 16 cks. Cunard S. S. Co., Lmtd.
Rouen, 161 100 cs. Co-op. W'sale Socy., Lmtd.

Gambier
Singapore, 1,674 bls.

Cutch
Rangoon, 2,088 bgs. 20,574 bxs.

Cochineal
Teneriffe, 40 bgs. Kuhner, Henderson, & Co.
,, 13 H. R. Toby & Co.
,, 3 ,,
Gr. Canary, 20 sks. Swanston & Co.
,, 67 F. A. Woygt & Co.

Myrabolams
Bombay, 72 bgs. Beyts, Craig, & Co.
,, 512 ,,
,, 660 D. Sassoon & Co.
,, 2,814 ,,
,, 310

Logwood
Savannah-la-Mar, 104 t. E. Brownbill & Co
Jamaica, 434

Dyestuffs (*otherwise undescribed*)
New York, 5 cs. Hatrick & Co.

Glucose—
New York, 200 brls.
Hamburg, 47 cks.
New York, 68
" 50

Horn Piths—
Rio Grande, 10,000 Redfern, Alexander, & Co

Isinglass—
Maranham, 5 cks. Gunston, Sons, & Co.
" 1 R. Singlehurst & Co.

Manganese Ore—
Gothenburg, 150 t. Macqueen Bros.

Manure—
Jersey, 109 bgs. 1 pt-bg. G. Hadfield & Co.

Naphtha—
New York, 3,700 bgs. Meade-King, Robinson, & Co.

Oxalic Acid—
Rotterdam, 6 cks.

Paraffin Wax—
New York, 130 brls. Richardson, Spence, & Co

Phosphate of Lime—
Antwerp, 380 t.

Potash Salts—
Hamburg, 1,700 bgs.

Pyrites—
Huelva, 939 t. Tennants & Co.
" 1,046 Mathesen & Co
" 1,500 "
" 1,586 "
" 1,528 Tennants & Co.

Soda—
Rotterdam, 65 brls.

Sodium Nitrate—
Iquique, 12,990 bgs.
" 9,200 J. T. North

Saltpetre—
Hamburg, 104 cks.
" 12

Tar—
Barcelona, 1 cs. • H. Tozeret
Wilmington, 2,889 brls.

Tartar—
Bordeaux, 52 cks.

Tartaric Acid—
Rotterdam, 12 cks.
" 3
Antwerp, 1

Ultramarine—
Rotterdam, 4 cks. 5 cs. 40 brls.

Waste Salt—
Hamburg, 162 t.
" 1,016 bgs.

Zinc Oxide—
Rotterdam, 100 cks. O. H. Marcus
Antwerp, 95 brls. J. T. Fletcher & Co.
" 115
Hamburg, 96

GLASGOW.

Week ending May 8th.

Chemicals (*otherwise undescribed*)
Hamburg, 3 cks. 1 cs.

Alum—
Hamburg, 240 cks.

Dyestuffs—
Madder
Rotterdam, 7 cks. J. Rankine & Son
Alizarine
Rotterdam, 85 cks.
" 99
Extracts
New York, 10 cs.
Nantes, 549 cks.

Glycerine—
Hamburg, 20 drms.

Muriatic Acid—
Hamburg, 1 cs.

Nitric Acid—
Hamburg, 1 cs.

Paraffin Wax—
New York, 505 brls. H. Hill & Son
" 370

Saltpetre—
Rotterdam, 6 cks. J. Rankine & Son
Hamburg, 100 kgs. J. Poynter & Son, Limited

Sulphuric Acid—
Hamburg, 1 drum.

Tartaric Acid—
Rotterdam, 3 cks.

Waste Salt—
Hamburg, 1,000 bgs.
" 223 t.

TYNE.

Week ending May 8th.

Barytes—
Antwerp, 50 bgs. Tyne S. S. Co.

Copper Precipitate—
Huelva, 256 t.

Copper Ore—
Bergen, 41 cks.

Chemicals (*otherwise undescribed*)
Hamburg, 1 ck. Tyne S. S. Co.

Glucose—
Hamburg, 20 cks. Tyne S. S. Co.
Rotterdam, 20 bgs. "

Guano—
Bole, 2,162 bgs.

Pyrites—
Huelva, 1,046 t.

Phosphate—
Antwerp, 185 t. Langdale's Chemical Manure Co.

Sodium Silicate—
Rotterdam, 1 ck. Tyne S. S. Co.

Sulphur Ore—
Pomaron, 1,200 t. Mason & Barry

Waste Salt—
Hamburg, 165 t.

HULL.

Week ending May 8th.

Acetate of Lime—
St. Malmo, 29 bgs. Hutchinson & Son

Ammonia—
Rotterdam, 1 pkg. W. & C. L. Ringrose

Alum—
Rotterdam, 16 cks.

Alumina Sulphate—
Rotterdam, 80 cks. W. & C. L. Ringrose

Acids (*otherwise undescribed*)—
Hamburg, 1 cs. Wilson, Sons, & Co.
Rotterdam, 6 cks. W. & C. L. Ringrose
Rotterdam, 2 cks. Hutchinson & Son

Barytes—
Bremen, 23 cks. Wilson, Sons, & Co.
Antwerp, 80 H. F. Pudsey
Strauss, 23 Tudor & Son

Benzole—
Danzig, 4 pks. Wilson, Sons, & Co.

Copperas—
Strauss, 31 cks.

Chemicals (*otherwise undescribed*)
Antwerp, 34 pkgs. Wilson, Sons, & Co.

Dyestuffs—
Alizarine
Rotterdam, 20 cks. Hutchinson & Son
" 6 W & C. L. Ringrose
" 16 "
" 25 Hutchinson & Son
" 60 3 cs. "
Extracts
Rouen, 85 20 Rawson & Robinson
" 40 Wilson, Sons, & Co.
Rennet
Rotterdam, 2 cks.

Ferro Chrome—
Rotterdam, 1 cs. Hutchinson & Son

Glycerine—
Amsterdam, 10 pks. W. & C. L. Ringrose

Glucose—
Hamburg, 10 cks. Rawson & Robinson
" 12 brls Bailey & Leetham
" 40 100 bgs.
" 10
" 13
Stettin, 117 100 Wilson, Sons, & Co.

Naphthaline—
Rotterdam, 8 cks. G. Lawsen & Sons

Naphthol—
Rotterdam, 3 brls. Wilson, Sons, & Co.

Oxide—
Rotterdam, 12 cks. Hutchinson & Son

Potash—
Rotterdam, 45 cks.

Phosphate—
Ghent, 160 t. Wilson, Sons, & C

Pyrites—
Huelva, 930 t J. Dalton, Holmes, & C

Sulphur—
Antwerp, 15 bgs. Wilson, Sons, & C
Rotterdam, 4 cs. G. Lawson & So

Turpentine—
Libau, 128 brls. 1 cs.

GRIMSBY.

Week ending May 3rd.

Chemicals (*otherwise undescribe*
Hamburg. 5 cks.

Dyestuffs (*otherwise undescribe*
Antwerp, 6 pks.

Manure—
Kainit
Hamburg, 500 bgs.

Zinc Ashes—
Antwerp, 25 cks.

GOOLE.

Week ending April 30th.

Alkali—
Rotterdam, 100 bgs.

Aluminium Sulphate—
Antwerp, 43 brls.

Antimony—
Boulogne, 4 cks.

Antimony Ore—
Boulogne, 1 ck.

Chrome Acetate—
Rotterdam, 11 cks.

Dyestuffs—
Alizarine
Rotterdam, 48 cks.
Logwood
Jamaica, 251 t.
Dyestuffs (*otherwise undescribed*)
Boulogne, 20 cks. 4 cs. 2 crbys.

Glucose—
Hamburg, 18 cks. 100 bgs.

Manure—
Phosphate Rock
Ghent, 800 bgs.

Potassium Muriate—
Rotterdam, 10 cks.

Plumbago—
Boulogne, 1 cs. 13 cks.

Potash—
Dunkirk, 37 drums. 57 cks.
Calais, 8 brls.

Saltpetre—
Rotterdam, 100 bgs.
Hamburg, 88 cks.

Tartars—
Rotterdam, 5 cks.

Waste Salt—
Hamburg, 1,391 cks. 355 t.

Pitch
Bombay, 50 brls. £15
Antwerp, 135 t. 250
Galatz, 50 15
Gijon, 612 912
Caen, 517 827
Dunkirk, 319 510
Melbourne, 22 pks. 4
Constantinople, 150 brls. 45
Paris, 13 cks. 50
Bordeaux, 200 t. 1,300
Dinsburg, 125 187

Disinfectants—
Hamburg, 25 cks. £150
" 19 114
Sydney, 48 pks. 31
Havre, 5 10
Natal, 12 44
Paris, 55 cks. 80
Sydney, 20 32

Guano—
Guadiloupe, 190 t. £1,900
Antigua, 6 2 c. 74
Trinidad, 41 17 440

Hyposulphite of Lime—
Rotterdam, 1,600 lbs. £180

Hydrochloric Acid—
Bombay, 5 pks. £14

Manure—
Malaga, 50 t. £400
Nevis, 35 380
Antigua, 10 100
Dominica, 5 50
St. Kitts, 55 550
Barbadoes, 60 9 c. 600
St. Vincent, 10 100
Barbadoes, 19 11 390
Danzig, 15 85
Amsterdam, 160 1,280
Rotterdam, 60 480
Valencia, 168 11 1,380
Gothenburg, 97 9 367
Pt. Chalmers, 100 1,350
Trinidad, 180 1,348
Barbadoes, 50 650
St. Kitts, 6 69
Antigua, 1 5 15
Havre, 20 220
Rouen, 10 180

Oxalic Acid—
Rotterdam, 3 t. 9 c. £100

Potassium Prussiate—
Lisbon, 8 c. £30

Potash—
Lisbon, 14 c. £19
Calcutta, 20 cs. 115
Gothenburg, 10 27

Phosphorus—
Yokohama, 900 lbs. £62
Hong Kong, 3 c. 37
Santa Cruz, 6 c. 15

Potassium Chlorate—
Rotterdam, 1 t. 41
Melbourne, 1 45
Algoa Bay, 1 c. 7
Melbourne, 1 45

Potassium Permanganate—
Yokohama, 900 lbs. £25

Potassium Iodide
Yokohama, 900 lbs. £180

Saltpetre—
Sydney, 1 t. £24
Madeira, 15 c. 22
Brussels, 2 1 42
Rosario, 4 92

Soda Salts—
Montreal, 4 cs. £22

Sodium Carbonate—
Marseilles, 10 t. £66

Superphosphate—
Canaria, 9 t. £46
Halifax, 27 110
Trinidad, 10 52

Sodium Bicarbonate
Sydney, 19 c. £11
Yokohama, 10 pks. 25

Salammoniac—
Paris, 1 t. 8 c. £50

[illegible]ic Acid
[illegible] Town, 2 t. 5 c. £21
1 66
£140
30

Soda Crystals—
Newcastle, 9 t. 15 c. £29

Tartaric Acid—
Sydney, 3 £20
Wellington, 2 14
Sydney, 10 cks. 67
" 20 kgs. 114

LIVERPOOL
Week ending May 7th.

Ammonium Carbonate—
Genoa, 1 t. 10 c. £55
Hamburg, 1 2 2 q. 28

Albumen—
Boston, 1 t. 15 c. 3 q. £130

Alum Cake—
Calcutta, 4 t. 5 c. £13

Alum—
Melbourne, 5 t. 10 c. £28
Montreal, 12 5 149
Volo, 2 3 11
Bombay, 8 42
Cape Town, 4 17
Baltimore, 15 6 3 q. 71
Bari, 1 7 6

Ammonia—
Valparaiso, 6 c. £8

Ammonium Muriate—
Boston, 3 t. 9 c. £87
Philadelphia, 5 16 3 q. 133

Ammonium Sulphate—
Antwerp, 10 t. £120
Bremerhaven, 19 228
Hamburg, 20 240
Valencia, 127 1,524
" 50 600
Nantes, 30 11 c. 410
Valencia, 101 17 1,190
Hamburg, 5 3 q. 56
Rotterdam, 30 360
Bordeaux, 15 1 224
St. Nazaire, 66 19 3 204
Salonica, 10 12
Valencia, 87 1,044

Bleaching Powder—
Ancona, 10 pps.
Boston, 1,120 cks.
Bruges, 25
Genoa, 78
Ghent, 29
Hamburg, 291
Leghorn, 11
Mexico City, 60 brls.
Montreal, 44 95 tcs.
Naples, 30 pps.
New York, 253
Philadelphia, 118
Rio Janeiro, 10 cs.
Rouen, 196
San Sebastian, 29
Trieste, 36
Vera Cruz, 10
Ancona, 12
Antwerp, 40
Ghent, 14
Hamburg, 50
New York, 98
Odessa, 61
Oporto, 7
Pernambuco, 3
St. Louis, 35 pks. 25 15 bxs.

Boracic Acid—
Antwerp, 2 t. 12 c. 1 q. £95
Stockholm, 4 12 138

Borate of Lime—
Rotterdam, 9 t. 18 c. 3 q. £120

Bone Waste—
Rouen, 29 t. £760
Hamburg, 29 130

Brimstone—
Philadelphia, 6 t. £164

Borax—
Montreal, 1 t. £29
Corunna, 9 c. 2 q. 15
Rouen, 32 8 972
Stockholm, 3 cks. 13
Montreal, 1 bl. 6 57

Charcoal—
Hamburg, 31 t. £225

Cobalt Oxide—
Bordeaux, 14 c. £6

Citric Acid—
Corunna, 2 c. £14

Copperas—
Rotterdam, 1 t. 16 c. £40
Constantinople, 8 6 2 q. 19
Sydney, 5 12 13
Montreal, 1 10 5

[illegible] of Tartar—
[illegible], 10 c. 3 q. £57

Caustic Soda—
Adelaide, 20 dms.
Batoum, 176
Copenhagen, 20
Genoa, 30
" 20
Hamburg, 240 kgs.
La Guayra, 10
Malta, 72
Montreal, 300
New York, 300
Oporto, 15
Pernambuco, 30
Philadelphia, 50
Rio Grande du Sol, 50
Rouen, 36
St. Petersburg, 79
Sydney, 40
Ancona, 5
Antwerp, 53
Bahia, 10
Barcelona, 26
Calcutta, 3 brls.
Chicago, 50
Corunna, 20
Dunedin, 51
Galatz, 10
Hamburg, 6
Maranham, 5
Montreal, 225
Nagasaki, 50
Naples, 25
New York, 50
Odessa, 225
Philadelphia, 135
Rio Grande du Sol, 75
Rotterdam, 4
St. Louis, 38 25 bxs.
San Sebastian, 10
Santander, 24
Seville, 100
Sydney, 45

Copper Sulphate—
Bordeaux, 34 t. 8 c. £1,043
Sydney, 2 47
Iquique, 10 8 244
Marseilles, 5 3 119
Odessa, 5 3 120
Tarragona, 15 2 348
Bilbao, 5 1 3 q. 90
Bordeaux, 10 15 242
Havre, 1 19 33
Malaga, 19 18 2 195
Antwerp, 10 2 1 223
Bordeaux, 84 13 1,841
Venice, 10 1 240
Vigo, 9 12
Genoa, 33 19 1 624
Leghorn, 2 3 38
Barcelona, 5 3 1 129
Bari, 18 3 20
Marseilles, 50 18 3 1,315
Nantes, 32 19 795
Tarragona, 10 248
Pasages, 10 3 195
Nantes, 10 220
Santos, 6 6
Venice, 19 19 3 399
Nantes, 39 13 975
Bordeaux, 20 16 3 431
Salonica, 10 12
Genoa, 19 11 3 328
Marseilles, 9 17 2 208
Rouen, 27 17 825
Bordeaux, 40 5 1 830
" 30 757
Leghorn, 10 2 2 253
Marseilles, 23 4 3 481

Carbolic Acid—
Rotterdam, 14 t. 5 c. £1,710
New York, 1 10 145
Paris, 2 9 255
Genoa, 19 2 q. 100
Leghorn, 5 3 36
St. Louis, 5 1 459
New Orleans, 11 3 93
Genoa, 9 3 65

Chemicals *(otherwise undescribed)*
Trinidad, 175 cs. £260
Vera Cruz, 5 t. 16 c. £56
" 1 cs. 12 brls. 84
Philadelphia, 11 13

Disinfectants—
Calcutta, 495 galls. £39

Dried Blood—
Montreal, 1 t. £15

Glycerine—
Hamburg, 8 t. 3 c. £196

Guano—
Havana, 265 t. 13 c. 2 q. £2,268

Ground Spar—
Havre, 49 t. 6 c. £355

Gypsum—
Boston, 10 t. £30
New York, 10 2 c. 30
Boston, 1 1 3 q. 10

Iron Oxide—
Gijon, 2 t. 10 c. 1 q. £2

Iodine—
New York, 4 t. £2,400
Genoa, 10 c. 6[illegible]

Lime Sulphate—
Santander, 5 t. £2

Magnesia—
Rio Janeiro, 5 cs.
Genoa, 1

Magnesium Calcide—
Genoa, 1 cs. £3

Manure—
Hamburg, 24 t. 5 c. 3 q. £7
Barbadoes, 50 2 14
Rochefort, 200 8 62
Havre, 50 15

Magnesium Chlorate—
Genoa, 16 t. 1 q. £1[illegible]

Nitric Acid—
Santos, 5 carboys. £10

Potassium Chromate—
Genoa, 1 t. £23

Potassium Bicarbonate—
Montreal, 5 c. £7

Potash—
Palermo, 9 t. 15 c. £210

Phosphorus—
Shanghai, 10 cs. £96
Santander, 3 c. 3 q. 33
Singapore, 15 70
Havana, 10 1 177
Mexico, 1 t. 219

Pitch—
Galatz, 100 brls.
Genoa, 1,842 t.

Potassium Manganate—
Valparaiso, 2 c. 1 q. £11

Potassium Stannate—
Bombay, 12 c. 2 £40

Potassium Bichromate—
Salonica, 1 t. 7 c. £47
Santos, 1 3 50

Potassium Chlorate—
Catania, 10 c. £23
Gothenburg, 2 t. 10 168
St. Petersburg, 2 10 111
San Sebastian, 3 3 q. 13
Santander, 10 21
B. Ayres, 3 10 180
Shanghai, 1 2 63
Stockholm, 4 168
Sydney, 10 450
Montreal, 1 10 66
Philadelphia, 5 [illegible]
San Francisco, 2 93
Santander, 10 23
Oporto, 1 40
Shanghai, 3 15 170
Vera Cruz, 6 11
Gothenburg, 15 640
Hiogo, 3 13 1 q. 140
Constantinople, 2 10 116
Hamburg, 6 10 185
San Francisco, 1 5 55

Potassium Carbonate—
Philadelphia, 9 c. 3 q. £13

Sodium Chlorate—
Hamburg, 1 t. £55

Sodium Silicate—
Ancona, 100 cks.
Salonica, 15
Oporto, 6
Seville, 50

Soda Ash—
Baltimore, 683 cks. 573 tcs.
Barcelona, 32 brls.
Boston, 131 58 1,700 bgs.
Callao, 20 dms.
Canada, 5
Genoa, 19 brls.
Lisbon, 16
Melbourne, 60
Monte Video, 50
Montreal, 34 33
Newcastle, 20
New Glasgow, 30
New York, 970
Oporto, 31
Philadelphia, 395 546
Porto Alegre, 100
Rio Janeiro, 121
Santos, 2
Seville, 50
Sherbrooke, 18
Smyrna, 36
Baltimore, 60
Barcelona, 18
Maranham, 80
Monte Video, 50
Montreal, 67
New York, 130

Rio Grande du Sol, 50
Rio Janeiro, 70
Santanter, 21 cks.
Syra, 250

Salt—
Amsterdam, 200 t.
Bathurst, 401
Belize, 10
Bonny, 75
Boston, 350
Brisbane, 100
Boctouche, 175
Buguma, 25
Charlotte Town, 35
Korsoer, 390
Lagos, 100
Montreal, 875 13 c.
New Orleans, 4 3
Old Calabar, 125 6
Paspebiac, 375
Quebec, 1,031 13
Rotterdam, 343
Salonica, 6¼
Shediac, 200
Sierra Leone, 100
Bakana, 10
Baltimore, 140
Bruges, 150
Calcutta, 4,597
Cape Palmas, 5½
Dunedin, 38½
East London, 22¼
Milwaukee, 65
Montreal, 7 16
Newcastle, 520
New York, 65
Philadelphia, 65
San Francisco, 66 10
Santos, 4¼

Sodium Bicarbonate—
Alexandria, 50 brls.
Ancona, 40 kgs.
Antwerp, 20
Bari, 40
Bordeaux, 15
Boston, 68 cks.
Hamburg, 12 pks. 8
Kingston, 16
London, Ont., 50
Marseilles, 20
Montreal, 25
Naples, 100
New Orleans, 200
Rouen, 120
Seville, 125 brls.
Toronto, 100
Vera Cruz, 40
Danzig, 100
Stettin, 80 10

Soda Crystals—
Bilbao, 34 brls.
Singapore, 17 cks.
Alexandria, 20
Gibraltar, 40
London, O., 10 kgs.
Malta, 100
New York, 50
Quebec, 25

Sodium Bichromate—
Piræus, 6 c. £125
Havre, 4 t. 15 140

Sodium Biborate—
Hamburg, 7 t. 17 c. 1 q. £220
S. Petersburg, 9 19 3 300
Stettin, 12 3 3 366
Stockholm, 12 15 2 326

Sodium Arseniate—
Boston, 3 t. 11 c. 1 q. £40

Sulphuric Acid—
Pernambuco, 4 c. 2 q. £10

Superphosphate—
Gr. Canary, 20 t. 2 c. 1 q. £148

Sulphur—
Lisbon, 11 t. 3 q. £89
Boston, 50 2 c. 207
Vigo, 10 1 58
Boston, 49 19 240
Calcutta, 22 10 115

Sodium Nitrate—
Santander, 1 t. 17 c. 1 q. £16

Saltcake—
Baltimore, 79 t. 3 c. 2 q. £183
Philadelphia, 49 7 114
Baltimore, 95 18 185
Ghent, 100 140

Sugar of Lead—
Naples, 2 t. £45
Calcutta, 4 c. 3 q. £6

Saltpetre—
Para, 1 t. 1 c. 1 q. £2
Santos, 2 19 7
Syra, 1 17 2 4
Pernambuco, 2 18 2 7

Salammoniac—
Beyrout, 11 c. 1 q. £20
Constantinople, 1 2 3 40
Malta, 4 7
Alexandria, 11 20
Barcelona, 12 cks. 40
Constantinople, 2 68
Odessa, 1 8 1 48

Soda Salts—
Montreal, 5 £130

Sheep Dip—
Natal, 1 t. 10 c. £80
E. London, 1 10 80
Algoa Bay, 1 10 80
Natal, 1 10

Tar—
Curacoa, 50 dms.
Sierra Leone, 17 brls.
Belize, 50
San Francisco, 50

Tartaric Acid—
Talcahuano, 2 c. 3 q. £10
Corunna, 2 14

Zinc Chlorate—
Bombay, 2 t. 17 c. £60
" 5 2 q. 110
" 2 6 1 33

Zinc Oxide—
Piræus, 10 c. 2 q. £13

TYNE.

Week ending May 8th.

Alumina—
Rouen, 9 t. 19 c.

Alkali—
Hamburg, 4 t. 16 c.
Christiania, 37 7
Stockholm, 2 12
Rotterdam, 1 2

Arsenic—
Helsingfors, 1 t. 1 c.

Antimony—
New York, 30 t.
Rotterdam, 4 10 c.

Ammonium Carbonate—
Rotterdam, 2 t. 4 c.

Barytes Carbonate—
Hamburg, 5 t.

Barytes—
Rouen, 15 t.

Barium Chloride—
Rouen, 5 t. 3 c.

Bleaching Powder—
Ancona, 16 t. 7 c.
New York, 50 4
Lisbon, 9 17
Malmo, 9 5
Christiania, 1 17
Rotterdam, 5 8
" 3 7
Odense, 24 17
Antwerp, 79 4
Hamburg, 70 11
Stockholm, 3 17
Helsingfors, 12 14
Rouen, 25 14
St. Petersburg, 113 t. 18 c.

Borax—
Bilbao, 1 c.

Carbolic Acid—
Hamburg, 25 brls.

Caustic Soda—
New York, 100 t. 2 c.
Philadelphia, 154 1
Antwerp, 100 2
Hamburg, 5 12
Gothenburg, 9 4
Abo, 2 2
Stockholm, 20
" 6 14
Wasa, 2
Abo, 9
Rouen, 11 17

Epsom Salts—
Gothenburg, 1 3

Gypsum—
New York, 72 t. 13 c.

Litharge—
Stockholm, 3 t. 1 c.
Antwerp, 11
Gothenburg, 1 19
Stockholm, 1 18
Rouen, 10 16

Magnesia—
Hamburg, 5
Antwerp, 3 q.
Rouen, 5

Manure—
Rotterdam, 50 8

Napthalene—
Antwerp, 14 t. 10 c.

Potassium Chlorate—
Antwerp, 5 t. 5 c.
Rotterdam, 3 15

Pitch—
Antwerp, 100

Sodium Chlorate—
Philadelphia, 3 t. 15 c.
Hamburg, 5

Soda—
Philadelphia, 230 t. 1 c.
Antwerp, 25 10
Rotterdam, 1 5
" 30
Esbjerg, 18
Aalborg, 18 11
Aarhuus, 17 6
Nakskov, 3 2
Stockholm, 131 16
Wasa, 3 14
Abo, 9 14
Helsingfors, 4 16
New York, 150 9
Stavanger, 5 3
Malmo, 5 18

Soda Ash—
Gothenburg, 25 t. 10 c.
Christiania, 8 5
Aalborg, 3 4
Philadelphia, 100 5
Antwerp, 10 6
Hamburg, 7 19
Ancona, 1 5
New York, 165 8
Stockholm, 50 11
Wasa, 26 5

Sodium Hyposulphate—
Christiania, 1 t. 2 c.

Saltcake—
New York, 49 t. 3 c.

HULL.

Week ending May 6th.

Alkali—
Drontheim, 17 drums.

Borax—
Hango, 1 ck.

Bleaching Powder—
Drontheim, 59 cks.
St. Petersburg, 50

Caustic Soda—
Antwerp, 18 drums. 3 cks.
Dunkirk, 19
Danzig, 71
Helsingfors, 125
Hango, 14
Konigsberg, 34
Libau, 56
Petersburg, 500

Chemicals *(otherwise undescribed)*
Antwerp, 9 cs. 3 cks. 2 pks.
Amsterdam, 8
Abo, 14 100
Bremen, 2 cs. 4
Bergen, 10
Boston, 7
Copenhagen, 119
Christiania, 55 6
Danzig, 44 27 154 slbs.
Gothenburg, 3 11
Genoa, 76 41
Hamburg, 79 8 553
Harlingen, 4
Helsingfors, 20
Hango, 16
Konigsberg, 18 30
Marseilles, 25
Naples, 40
New York, 30
Rouen, 33
Rotterdam, 21 24 104
Stettin, 18
St. Petersburg, 394 140
Libau, 9 50

Dyestuffs *(otherwise undescribed)*
Hamburg, 1 cs. 11 drs.
Rotterdam, 2 2 cks. 2 pks.
Reval, 46

Manure—
Drontheim, 30 bgs.
Hamburg, 343
Rouen, 50
Rotterdam, 500
Stettin, 479

Naphtha—
Dunkirk, 23 cks.

Pitch—
Antwerp, 200 t.
Genoa, 302

Tar—
Abo, 150 cks.
Bremen, 100
Gothenburg, 150
Hamburg, 4
Wasa, 10

GLASGOW.

Week ending May 8th.

Alum—
Quebec, 214½ c.

Bones—
Rouen, 4 t. 6 c. £

Boracic Acid—
Stockholm, 4 cks. £1

Dyestuffs—
Madders
Melbourne, 12¼ c. £
Extracts
Quebec, £1
Melbourne, 44¼
Logwood
Quebec, 2 t. 11 c. £
Melbourne, 358¼ 1

Glycerine—
Rouen, 12 t. 5 c. £2

Magnesium Sulphate—
Melbourne, 11 t. 3¼ c. £

Naphtha—
Rouen, £1

Oxalic Acid—
Barcelona, 17¼ c. £

Pitch—
Chili, £1

Potassium Bichromate—
Rouen, 10 t. 13¾ c. £3
Barcelona, 108¾ 2
" 105¾ 1
Gothenburg, 5 6¾ 1
Stockholm, 1 ck.
Oporto, 109½ 2
Malta, 36¾
Rotterdam, 125 8

Paraffin Wax—
Trieste, 62 c. £
Rouen, 7,380 lbs. 1
Barcelona, 11 t. 13¼ 3
Stockholm, 50 cks.
Malta, 5 7½ 1

Potassium Prussiate—
Rouen, £3
Amsterdam, 4 cks.

Salt—
Chili, 4 t. 10¼ c. £

Sodium Bichromate—
Quebec, 3 t. 6 c. £1
Rouen, 7 17¾ 1

Tartar—
Quebec, 5½ c. £

GRIMSBY.

Week ending May 3rd.

Alkali—
Antwerp, 2 cks.

Coal Products—
Antwerp, 30 cks.
Dieppe, 79

Chemicals *(otherwise undescribed)*
Antwerp, 8 cks. 4 pks.
Rotterdam, 5

GOOLE.

Week ending Apr. 30th.

Alkali—
Boulogne, 6 cks.

Benzole—
Antwerp, 24 cks.
Ghent, 15 drs.
Hamburg, 32
Rotterdam, 25 cks.

Bleaching Preparations—
Rotterdam, 28 cks.

Coal Products—
Ghent, 2 cs.
Rotterdam, 64 cks.

Chemicals *(otherwise undescribed)*
Dunkirk, 2 cks.

Dyestuffs *(otherwise undescribed)*
Boulogne, 2 cks.

Manure—
Boulogne, 38 bgs.
Bruges, 49
Ghent, 199

Pitch—
Ghent, 180 t.

PRICES CURRENT.

WEDNESDAY, MAY 14, 1890.

PREPARED BY HIGGINBOTTOM AND CO., 116, PORTLAND STREET, MANCHESTER.

tated are F.O.R. at maker's works, or at usual ports of shipment in U.K. The price in different localities may vary.

			£	s.	d.
, and 40 °/。		per cwt.	7/- & 12/6		
ial		,,	67/6		
., 2000°		,,	0	12	9
°/。	nett	per lb.	0	0	6¼
		,,	0	0	3¾
ower Salts), 30° Tw.		per bottle	0	0	8
linder), 30° Tw.		,,	0	2	11
w.		per lb.	0	0	2
		,,	0	0	1¾
	nett	,,	0	0	4
		,,	0	1	2
uming 50 °/。)		per ton	15	10	0
onohydrate)		,,	5	10	0
yrites, 168°)		,,	3	2	6
,, 150°)		,,	1	6	0
ee from arsenic, 140/145°)		,,	1	7	6
(solution)		,,	2	15	0
e)		per lb.	0	1	0½
		,,	0	1	3
powdered		per ton	13	0	0
mp)		,,	4	17	6
hate (pure)		,,	5	0	0
, (medium qualities)		,,	4	10	0
		per lb.	0	15	0
o=28°		per lb.	0	0	3
=24°		,,	0	0	1¾
rbonate	nett	,,	0	0	3½
riate		per ton	23	0	0
,, (sal-ammoniac) 1st & 2nd		per cwt.	37/-& 35/-		
trate		per ton	40	0	0
osphate		per lb.	0	0	10½
phate (grey), London		per ton	11	8	9
,, (grey), Hull		,,	11	7	6
ure)		per lb.	0	0	10¾
,,		,,	0	0	9½
		per ton	75	0	0
rtar emetic)		per lb.	0	1	1
lden sulphide)		,,	0	0	10
ide		per ton	8	5	0
nate (native)		,,	3	15	0
ate (native levigated)		,,	45/- to 75/-		
wder, 35 %		,,	5	0	0
uor, 7 %		,,	2	10	0
Carbon		,,	13	0	0
etate (crystal		per lb.	0	0	6
ride		per ton	2	2	6
t Runcorn) in bulk		,,	17/6 to 35/-		
s :—					
rtificial) 20 %		per lb.		0	9
		,,	8	3	9
oducts					
, 30 % A, f.o.b. London		per unit per cwt.	0	1	5
% nominal		per gallon	0	3	9
90 ,,		,,	0	2	9
id (crystallised 35°)		per lb.	0	0	8
(crude 60°)		per gallon	0	2	0
rdinary)		,,	0	0	2½
ltered for Lucigen light)		,,	0	0	3
tha 30 % @ 120° C.		,,	0	1	3
, 22° Tw.		per ton	3	0	0
Liverpool or Garston		,,	1	8	0
htha, 90 % at 160°		per gallon	0	1	8
s (crude)		,,	0	0	2½
Bars)		per ton	52	0	0
te		,,	25	15	0
(copper scales)		,,	54	0	0
de)		,,	30	0	0
illed S.G. 1250°)		,,	55	0	0
		nett, per oz.	0	0	9
(copperas)		per ton	1	10	0
(antimony slag)		,,	2	0	0
r cash		,,	14	10	0
Flake (ex ship)		,,	15	10	0
(white ,,)		,,	23	10	0
(brown ,,)		,,	18	5	0
te (white lead) pure		,,	19	0	0
		,,	22	5	0

		£	s.	d.
Lime, Acetate (brown)	,,	7	15	0
,, ,, (grey)	per ton	13	15	0
Magnesium (ribbon and wire)	per oz.	0	2	3
,, Chloride (ex ship)	per ton	2	5	0
,, Carbonate	per cwt.	1	17	6
,, Hydrate	,,	0	10	0
,, Sulphate (Epsom Salts)	per ton	3	0	0
Manganese, Sulphate	,,	18	0	0
,, Borate (1st and 2nd)	per cwt.	60/- & 42/6		
,, Ore, 70 %	per ton	4	10	0
Methylated Spirit, 61° O.P.	per gallon	0	2	2
Naphtha (Wood), Solvent	,,	0	3	10
,, ,, Miscible, 60° O.P.	,,	0	4	6
Oils :—				
Cotton-seed	per ton	22	10	0
Linseed	,,	26	0	0
Lubricating, Scotch, 890°—895°	,,	7	5	0
Petroleum, Russian	per gallon	0	0	5¼
,, American	,,	0	0	5¾
Potassium (metal)	per oz.	0	3	10
,, Bichromate	per lb.	0	0	4
,, Carbonate, 90% (ex ship)	per ton	19	10	0
,, Chlorate	per lb.	0	0	4¾
,, Cyanide, 98%	,,	0	2	0
,, Hydrate (Caustic Potash) 80/85 %	per ton	22	10	0
,, ,, (Caustic Potash) 75/80 %	,,	21	10	0
,, ,, (Caustic Potash) 70/75 %	,,	20	15	0
,, Nitrate (refined)	per cwt.	1	2	6
,, Permanganate	,,	4	2	6
Prussiate Yellow	per lb.	0	0	9½
,, Sulphate, 90 %	per ton	9	10	0
,, Muriate, 80 %	,,	7	15	0
Silver (metal)	per oz.	0	3	11¼
,, Nitrate	,,	0	2	8
Sodium (metal)	per lb.	0	4	0
,, Carb. (refined Soda-ash) 48 %	per ton	6	7	6
,, ,, (Caustic Soda-ash) 48 %	,,	5	5	0
,, ,, (Carb. Soda-ash) 48%	,,	5	15	0
,, ,, (Carb. Soda-ash) 58 %	,,	6	17	6
,, ,, (Soda Crystals)	,,	2	17	6
,, Acetate (ex-ship)	,,	16	10	0
,, Arseniate, 45 %	,,	11	0	0
,, Chlorate	per lb.	0	0	6¼
,, Borate (Borax)	nett, per ton.	29	0	0
,, Bichromate	per lb.	0	0	3
,, Hydrate (77 % Caustic Soda)	per ton.	10	0	0
,, ,, (74 % Caustic Soda)	,,	9	0	0
,, ,, (70 % Caustic Soda)	,,	8	7	6
,, ,, (60 % Caustic Soda, white)	,,	8	0	0
,, ,, (60 % Caustic Soda, cream)	,,	7	15	0
,, Bicarbonate	,,	6	0	0
,, Hyposulphite	,,	5	10	0
,, Manganate, 25%	,,	9		0
,, Nitrate (95 %) ex-ship Liverpool	per cwt.	0	8	1½
,, Nitrite, 98 %	per ton	28	0	0
,, Phosphate	,,	15	15	0
,, Prussiate	per lb.	0	0	7¾
,, Silicate (glass)	per ton	5	7	6
,, ,, (liquid, 100° Tw.)	,,	3	17	6
,, Stannate, 40 %	per cwt.	2	0	0
,, Sulphate (Salt-cake)	per ton.	1	5	0
,, ,, (Glauber's Salts)	,,	1	10	0
,, Sulphide	,,	7	15	0
,, Sulphite	,,	5	5	0
Strontium Hydrate, 98 %	,,	8	0	0
Sulphocyanide Ammonium, 95 %	per lb.	0	0	7¾
,, Barium, 95 %	,	0	0	5¾
,, Potassium	,,	0	0	8
Sulphur (Flowers)	per ton.	7	15	0
,, (Roll Brimstone)	,,	6	0	0
,, Brimstone : Best Quality	,,	4	7	6
Superphosphate of Lime (26 %)	,,	2	12	6
Tallow	,,	25	10	0
Tin (English Ingots)	,,	98	0	0
,, Crystals	per lb.	0	0	6¼
Zinc (Spelter)	per ton.	23	0	0
,, Chloride (solution, 96° Tw.	,,	6	0	0

No. 157. SATURDAY, MAY 24, 1890. Vol. VI.

Contents.

Notices.

All communications for the *Chemical Trade Journal* should be addressed, and Cheques and Post Office Orders made payable to—

DAVIS BROS., 32, Blackfriars Street, MANCHESTER.

Our Registered Telegraphic Address is—

"Expert, Manchester."

The Terms of Subscription, commencing at any date, to the *Chemical Trade Journal*,—payable in advance,—including postage to any part of the world, are as follow:—

Yearly (52 numbers) .... **12s. 6d.**
Half-Yearly (26 numbers) .... **6s. 6d.**
Quarterly (13 numbers) .... **3s. 6d.**

Readers will oblige by making their remittances for subscriptions by Postal or Post Office Order, crossed.

Communications for the Editor, if intended for insertion in the current week's issue, should reach the office not later than **Tuesday Morning.**

Articles, reports, and correspondence on all matters of interest to the Chemical and allied industries, home and foreign, are solicited. Correspondents should condense their matter as much as possible, write on one side only of the paper, and in all cases give their names and addresses, not necessarily for publication. Sketches should be sent on separate sheets.

We cannot undertake to return rejected manuscripts or drawings, unless accompanied by a stamped directed envelope.

Readers are invited to forward items of intelligence, or cuttings from local newspapers, of interest to the trades concerned.

As it is one of the special features of the *Chemical Trade Journal* to give the earliest information respecting new processes, improvements, inventions, etc., bearing upon the Chemical and allied industries, or which may be of interest to our readers, the Editor invites particulars of such—when in working order—from the originators; and if the subject is deemed of sufficient importance, an expert will visit and report upon the same in the columns of the *Journal*. There is no fee required for visits of this kind.

We shall esteem it a favour if any of our readers, in making nquiries of, or opening accounts with advertisers in this paper, will kindly mention the *Chemical Trade Journal* as the source of their information.

Advertisements intended for insertion in the current week's issue, should reach the office by **Wednesday morning** at the latest.

Advertisements.

Prepaid Advertisements of Situations Vacant, Premises on Sale or To be Let, Miscellaneous Wants, and Specific Articles for Sale by Private Contract, are inserted in the *Chemical Trade Journal* at the following rates:—

30 Words and under.......... 2s. od. per insertion.
Each additional 10 words .... os. 6d. ,,

Advertisements of Situations Wanted are inserted at one-half the above rates when prepaid, viz:—

30 Words and under.......... 1s. od. per insertion.
Each additional 10 words .... os. 3d. ,,

Trade Advertisements, Announcements in the Directory Columns, and all Advertisements not prepaid, are charged at the Tariff rates, which will be forwarded on application.

THE PATENT LAWS AND THEIR ABUSE.

IT is a very generally recognised principle of our econom that the man who makes an invention is entitled to reward proportionate to the benefit that he confers upon th community at large. We say, advisedly, "entitled," becau it by no means follows that he obtains his deserts. In fac the lot of the inventor is generally a hard one, as the histori of most of our great discoverers will testify. The wor soothes its conscience by recognising a man's claims—ar lets him starve in the meantime.

The Patent Laws have done much in the direction of justi by securing to an inventor the opportunity of recoupir himself for his expenditure of time, thought, or money; ar so long as only real inventors, and real inventors are question, there can be no doubt as to the wholesomeness their operations. But, like all good and useful institution the Patent Laws are open to abuse, and, as they at prese stand, may be made to offer a premium to a class of m whose extermination, rather than encouragement, should l the object of the State. We refer to those men who rush to the Patent Office with vaguely worded specifications, cor prising an abundance of indefinite claims, and drawn up such a manner as to include as many as possible of the futu allied inventions or improvements which the crafty patente though unable to discover or define himself, is yet astu enough to see looming in the vista before him. A ma for example, patents a method of dealing with th waste acid chloride of iron liquors from pickling. H process involves concentration, and this operation l performs in "a suitable vessel." A suitable vess forsooth! The real invention would be the discovery of th "suitable vessel," but, as the Patent Laws now stand, th reward of such invention would probably fall to the origin individual, whose speciously worded specification was a sna so to say, cunningly set for the capture of this "suitab vessel," *when someone else should discover it.* Or, to ta another example, it is well known, that the constitution certain series of chemical compounds is arranged upon w defined lines, so that from the composition and ch acteristics of known members of the series, very corre inferences can be drawn with regard to the members th are, as yet, unknown. Here is a very fertile field for hi who would reap where he has not sown. What is mo likely to turn out a lucrative investment than a few guine spent in patenting the use or manufacture of some of the "missing links?"

As our readers may gather from our opening remarks, have great respect for the Patent Laws, but we wo subject them to a very vigorous amendment. When a m applies for a patent, we would compel him to state, in pl and unmistakable language, what he demands a patent f Ambiguous claims we would reject entirely, and we wo make it impossible for the ignorance of one inventor, invalidate the discoveries of another, as is often the case no We understand that strong efforts are being made to sec a very radical reform, and we wish the promotors of t movement every success.

matter of the trimming is. The plaintiff has made arrangements with the captains of vessels, brokers, or captains or owners of vessels that have been employed to take this pitch. He has entered into arrangements with them, that he shall have the perquisite or benefit of trimming cargoes. Now, the actual cost of trimming the cargoes is proved by Jones to be 2d. It is said there were some baskets and shovels or pickaxes or something that had to be provided, but substantially we can put it at 2d. a ton. It is said that Beckton does it for 2½d. When Beckton does it for 2½d., I suppose there is a profit on it. So far, the price of Beckton would seem to support the story of Jones, that the man who does it gets 2d. a ton for doing it. That would be taken to be the proper cost of the labour employed in trimming the cargo. Beckton does it for 2½d. Mr. Leonard, who is a person, if I may say so, in the same position as that of Beckton—he is a seller of pitch—evidently wished to have the advantage. It is more convenient, and there is more profit in doing it yourself, and therefore he was desirous of having it done by himself. They do not care to have it done by other people if they can get the advantage and have the convenience of doing it themselves. We can suppose the feeling exercised by Mr. Leonard would be the feeling generally entertained by sellers of these goods. The plaintiff, who was a clerk of Mr. Dasnieres', and who was bound to give him all his time and do the best he possibly could for his interests—without consulting Mr. Dasnieres or mentioning the subject to him, seems for some years to have undertaken this duty upon himself, not charging 2d., so far as I understand, not always charging 2½d., but sometimes 3d. and 3½d., and one lad who was called says he knows of one case where 4d. was charged, and between 4d. and 2½d., the Beckton price, there is what I may call a considerable proportion. It is a large proportion of the total, 1½d. or 2½d. is a considerable difference on that small sum of money, and I daresay on a thousand tons that is worthy of consideration. It is stated, and it is for you to judge, that this is against the interests of Mr. Dasnieres.

There is one thing that is perfectly clear, it is not done with the knowledge of Mr. Dasnieres. He is not told of it. There is no entry in his books. There is no mention of it to him, although it may have been known to two or three lads who were in his office, but he does not know it until he is told by Mr. Leonard, who makes a ground of complaint about it.

It is stated that this is prejudicial to the interests of Mr. Dasnieres, because it may affect his freights. Mr. Dasnieres was entitled to rely upon the plaintiff to get the freights as low as possible. It is all very well to say that Mr. Dasnieres, having been in London, ought to know what freights are, but freights constantly vary, and freights are not precisely certain. Freights are what can be obtained by bargaining and haggling on one side and the other, and if a captain knows that "I have got to allow to the clerk in the office a bigger "price for trimming the cargo than I should have to pay for getting it "done myself by the Beckton people, or Leonard, the people who sell "the cargo, or whoever the people may be"—if he thought he would have to pay more in dealing with Dasnieres, it is natural that he might be a little stiffer on the question of freight, and that would prejudice Dasnieres.

On the other hand, it is said—"Oh, no, the freights are communi"cated to Dasnieres, and he need not accept them." That is very true. Dasnieres is in Paris; he cannot tell to what extent further coaxing might get 2d. or 2½d., or whatever it is per ton off the freight of this stuff. He cannot tell exactly. He is entitled to rely on his manager, and if he does not suspect his manager is getting anything—I will not say out of the freight, but out of trimming the cargo—he naturally trusts him, and thinks the manager has done the best he can, and therefore he may accept freights which might have been pressed lower down if the captain had known that he was to get the trimming done at a lower price.

It does certainly seem to be inconsistent with the duty of a servant, without the knowledge of his master, to be taking advantages of this description, and that it may more or less trouble him in dealing with captains and obtaining the best freights that he ought to obtain.

(*To be concluded.*)

A NEW PROCESS IN GLASS BOTTLE BLOWING.

THE Manchester Glass Bottle Company have introduced a new process for the manufacture of bottles, which, it is claimed, not only dispenses with the bottle finisher, but enables 25 per cent. more work to be done with the reduced gang, and at the same time renders breakages less frequent owing to dispensing with the handling occasioned by a boy passing the bottles from the blowers to the finishers. The mould or stamp of the old process is taken pretty much as it stands, and to it is fitted a collar with a knife cutting-edge at the extreme top. The service performed by the collar in conjunction with a corresponding plug with a similar knife cutting-edge immediately above it is to execute the work formerly done by the finisher. The operation is performed by pedal pressure with the utmost accuracy and neatness, and the bottle is then ready for what is called in the new process glazing, and which imparts a higher finish to the lips of the bottle than was possible under the old system. The apparatus includes a sort of invertible iron box or cage, which holds the bottle mouth downwards over a Bunsen burner while a smooth gloss is being given to that part. The manual work entailed is performed by a boy. One great difficulty the enterprising inventor has had to encounter was to simplify the parts of the machine, so that on changing at a few minutes' notice from making one class of bottle to another, the bottle blower may adjust the machine by the screwing and unscrewing of two or three nuts. In this he seems to have arrived at a point of perfection. Owing to the cheapness of Continental labour, the bottle-making firms on the Continent have had pretty much of a monopoly all over the world, the United Kingdom included, in certain of the smaller kinds of bottles used for patent medicines and the like. The Manchester Glass Bottle Company assert that when once they have got thoroughly to work this will be the case no longer, as they will be able to turn out their wares at a price which the Continental makers will not be able to touch.

HOW PUBLIC ANALYSTS ARE APPOINTED.

THERE was considerable interest shown in the appointment of a public analyst by the Vestry of Clerkenwell at their meeting on May 8—a good whip up of members below, and a gallery full of evidently professional gentlemen. Mr. F. G. Scheib (churchwarden) was in the chair, and, in introducing the subject, called upon the clerk (Mr. R. Paget) to read the following letter :—

2, Fisher Street, Red Lion Square,
R. Paget, Esq., Vestry Clerk. April 29, 1890.

DEAR SIR,—When you informed me that your Vestry had not re-elected me as their public analyst, but were opening the appointment to other gentlemen, I, of course, considered that my name would be submitted as an intending candidate for the vacant position. Understanding that it was omitted at the last meeting, I now ask that it may be placed upon the list, as I am willing to continue to serve the Vestry in this very important and responsible position. T. REDWOOD.

Mr. Kelly said the name of Dr. Redwood was before the Vestry at the last meeting. Mr. Ross: And he is now too late in his application. Let's have a young man. Mr. Walton: Surely you will try to be fair to a man who has served you well in years gone by. I shall move that Dr. Redwood's name be added to the list. Mr. Putterill seconded, on the ground of Dr. Redwood's claims as an old officer of the board. Mr. Ross: Professions are overstocked unfortunately; and, when old men have held appointments for a long time, they should make way for young ones. Mr. Atkinson submitted that the Vestry would be doing a very unjust thing if they did not allow Dr. Redwood to be a candidate simply because he had not sent formal application in time. A vote was taken, with a large majority against Dr. Redwood. Mr. Ross: Now let's vote for a Clerkenwell man. Mr. Walton: We don't want Clerkenwell men. We don't want an analyst close to our doors. Mr. Ross (ironically): Of course I didn't look at it in that way. I thought we were going to have an honest man. The following names were then submitted to the vote :—

Mr. Wm. Chattaway, F.I.C., A.S.P.A., member of Pharmaceutical Society, and member of Society of Chemical Industry.

Mr. J. K. Colwell, Ass. I.C, A.S P.A., Public Analyst for Bedford; is in charge of St. Marylebone Laboratory.

Mr. C. H. Cribb, B.Sc. (Lond.), F.C.S., F.I.C., M.S.P.A., Public Analyst for Strand district.

Mr. W. Johnson, F.I.C., F.C.S, F.G.S., M.S.P.A., and Society of Chemical Industry.

Mr. J. P. Laws, F.I.C, F.C.S.

Mr. A. P. Luff, M.D., B.Sc. (Lond.) M.R.C.P. M.R.C.S., L.S.A., F.I.C. F C.S

Mr. A. W. Stokes, F.I.C., F.C.S., M.S.P.A., Public Analyst for Paddington and St. Luke's.

Mr. J. Wilkie, F.C.S., M.C.S.

There were 31 votes in favour of Mr. Colwell, and not half-a-dozen each recorded in favour of the other candidates. Mr. Colwell was consequently declared elected, and called before the board to receive his appointment. Mr. Colwell said he thanked the board for the honour, and would endeavour to do his duty as well as he possibly could. He was a young man, and therefore should use all his exertions to merit the confidence bestowed. With regard to the word "jobbery," which he had heard a member apply to the matter—Mr. Walton: I protest! This gentleman is not a member of the board. Mr. Ross: No; he was in the ratepayers' gallery, but is supposed not to have heard what was said in debate. (Laughter.) Mr. Kelly: I am told that Mr. Walton says it is a planned job. I never saw Mr. Colwell in my life until last Tuesday; but he is a capital fellow, and I am glad he has the appointment. Mr. Ross: I voted for him because he is a Clerkenweller, and has all the qualifications for the appointment. Mr. Colwell then retired, and the next business was proceeded with.

IZATION OF WASTE IN GLUE MANUFACTURE.

facture of leather, in these days of keen competition, *Oil, Paint, and Drug Reporter*, cannot be too strictly the principles of economy; first, because the process, watched and well regulated right the way through, loss to the tanner of no inconsiderable amount, which, d add to his profits a large item. There are so many his industry of leather manufacture by which loss may behoves every tanner and those connected with the trade means to improve upon their present methods, and to r present resources. For it must be borne in mind that ole of the process, from beginning to end, is chiefly of a ure; and, considering that this view has only recently we cannot perhaps be discontented with the advances been already made. However, some of the principal occur, and which ought to have our special attention, are processes, the extracting of the tannin, &c., from the stances, the necessity of removing the lime from the hides them to the pit to be tanned (as the lime always destroys centage of tannin), the prevention of fermentation both in the tannin liquors, and the utilization of waste. This last nyone who will undertake the trouble of studying it, the manufacture of glue, gelatine, and similar substances, an important position in our industries, and the demand asing, both for the arts and for other purposes.

tanner is the very man who should utilize his resources, ing that he has most of the material at his command, why ot turn them to his own account, instead of sending or em away? No man is in a much better position than he. already at hand innumerable scraps of hide, the skin, and sheep, and very often skins that are fit for no other purpose tock, besides the skins of such animals as the rabbit its fur, scraps of parchment, pieces and roundings from id leather, and many other parts, which often have the of being worse than useless. All these he may turn to his own advantage. Now, we have it from the best that glue does not pre-exist in the hides and skins of nless under abnormal conditions, as when suffering from e, but is brought about by a variety of circumstances, and product of one simple process, as many suppose.

en, is the matter or substance derived from various animal , when such substances are boiled with water, and it is ying of the hide, the boiling of it down, and then in finally jelly obtained, where the art lies, and all the operations they are conducted, tend greatly to alter the quantity, the d value of the glue which is obtained. For instance, the ned from a fresh green hide will be very different from that btained from hides which have been both limed and dried, ly which is obtained after boiling differs entirely from the is obtained when that jelly is dried, as well as differing m a glue solution. So that the process of the manufacture he result of a combination of conditions.

material, or the animal skin from which we obtain glue, two parts—the upper layer, termed the epidermis, which is glue-making material, and the under layer, or dermis, titutes the true skin, and it is this we require for glue manu- he portions of fat and flesh which often adhere to this, ned the adipose tissue, are also entirely useless, and must rid of before we proceed. All the animal trimmings are as the heads, the ragged edges, the ears, the tails, feet pieces. It must also be borne in mind that the older the better will be the quality of the glue obtained. It is there- ble to keep the pieces of ox by themselves, as well as the heep by themselves, and the pieces of hog, hare, and rabbit hemselves, and to give a separate boiling to each lot, as the animal, the less solid the glue. Thus different qualities ed, which renders them more especially fit for certain and for certain classes of trade. The size is used in the e of paper, whilst the glue may be used by carpenters, d in numerous other manufactures.

tion of glue.—Glue is composed of :—

	Per cent.
arbon	49·1
ydrogen	6·5
itrogen	18·3
xygen and sulphur	26·1

position of it makes it nearly analogous to that skin; in nd its modifications of isinglass, jelly, crude glue, &c., are y many to really have the same chemical composition, but ysical characters, just the same as with the starches and

PERMANENT CHEMICAL EXHIBITION.

THE proprietors wish to remind subscribers and their friends generally that there is no charge for admission to the Exhibition. Visitors are requested to leave their cards, and will confer a favour by making any suggestions that may occur to them in the direction of promoting the usefulness of the Institution.

JOSEPH AIRD, GREATBRIDGE.—Iron tubes and coils of all kinds.

ASHMORE, BENSON, PEASE AND CO., STOCKTON-ON-TEES.—Sulphate of Ammonia Stills, Green's Patent Scrubber, Gasometers, and Gas Plant generally.

BLACKMAN VENTILATING CO., LONDON. — Fans, Air Propellers, Ventilating Machinery.

G. G. BLACKWELL AND CO, LIVERPOOL.—Manganese Ores, Bauxite, French Chalk. Importers of minerals of every description.

BRACHER AND CO., WINCANTON.—Automatic Stills, and Patent Mixing Machinery for Dry Paints, Powders, &c.

BRUNNER, MOND AND CO, NORTHWICH.—Bicarbonate of Soda, Soda Ash, Soda Crystals, Muriate of Ammonia, Sulphate of Ammonia, Sesqui-Carbonate of Ammonia.

BUCKLEY BRICK AND TILE CO, BUCKLEY.—Fireclay ware of all kinds—Slabs, Blocks, Bricks, Tiles, "Metalline," &c.

CHADDERTON IRON WORKS CO., CHADDERTON.—Steam Driers and Steam Traps (McDougall's Patent).

W. F. CLAY, EDINBURGH.—Scientific Literature—English, French, German, American. Works on Chemistry a speciality.

CLAYTON ANILINE CO., CLAYTON.—Aniline Colours, Aniline Salt, Benzole, Toluole, Xylole, and Nitro-compounds of all kinds.

J. CORTIN, NEWCASTLE-ON-TYNE.—Regulus and Brass Taps and Valves, "Non-rotative Acid Valves," Lead Burning Apparatus

R. DAGLISH AND CO., ST. HELENS.—Photographs of Chemical Plant —Blowing Engines, Filter Presses, Sludge Pumps, &c.

DAVIS BROS., MANCHESTER.—Samples of Products from various chemical products—Coal Distilling, Evaporation of Paper-lyes, Treatment of waste liquors from mills, &c.

R. & J. DEMPSTER, MANCHESTER —Photographs of Gas Plants, Holders, Condensers, Purifiers, &c.

DOULTON AND CO., LAMBETH.—Specimens of Chemical Stoneware. Stills, Condensers, Receivers, Boiling-pots, Store-jars, &c.

E. FAHRIG, PLAISTOW, ESSEX.—Ozonised Products. Ozone-Bleached Esparto-Pulp, Ozonised Oil, Ozone-Ammoniated Lime, &c.

GALLOWAYS, LIMITED, MANCHESTER.—Photographs illustrating Boiler factory, and an installation of 1,500-h.p.

GRIMSHAW BROS., LIMITED, CLAYTON.—Zinc Compounds. Sizing Materials, India-rubber Chemicals.

JEWSBURY AND BROWN, MANCHESTER.—Samples of Aerated Waters.

JOSEPH KERSHAW AND CO, HOLLINWOOD.—Soaps, Greases, and Varnishes of various kinds to suit all requirements.

C. R. LINDSEY AND CO, CLAYTON.—Lead Salts, (Acetate, Nitrate, etc.) Sulphate of Copper, etc.

CHAS. LOW AND CO., REDDISH.—Mural Tablet-makers of Carbolic Crystals, Cresylic and Picric Acids, Sheep Dip, Disinfectants, &c.

MANCHESTER ANILINE CO., MANCHESTER.—Aniline Colours. Samples of Dyed Goods and Miscellaneous Chemicals, both organic and inorganic.

MELDRUM BROS., MANCHESTER.—Steam Ejectors, Exhausters, Silent Boiling Jets, Air Compressors, and Acid Lifters.

E. D. MILNES AND BROTHER, BURY.—Dyewoods and Dyewood Extracts. Also samples of dyed fabrics.

MUSGRAVE AND CO., BELFAST.—Slow Combustion Stoves. Makers of all kinds of heating appliances.

NEWCASTLE CHEMICAL WORKS COMPANY, LIMITED, NEWCASTLE-ON-TYNE.—Caustic Soda (ground and solid), Soda Ash, Recovered Sulphur, etc.

ROBINSON, COOKS, AND COMPANY, ST. HELENS.—Drawings, illustrating their Gas Compressors and Vacuum Pumps, fitted with Pilkington and Forrest's patent Valves.

J. ROYLE, MANCHESTER.—Steam Reducing Valves.

A. SMITH, CLAYTON.—India-rubber Chemicals, Rubber Substitute, Bisulphide of Carbon, Solvent Naphtha, Liquid Ammonia, and Disinfecting Fluids.

WORTHINGTON PUMPING ENGINE COMPANY, LONDON.—Pumping Machinery. Speciality, their "Duplex" Pump.

JOSEPH WRIGHT AND COMPANY, TIPTON —Berryman Feed-water Heater. Makers also of Multiple Effect Stills, and Water-Softening Apparatus.

PRIZE COMPETITION.

A LITTLE more than twelve months ago, one of our friends suggested that it might be well to invite articles for competition, on some subject of special interest to our readers; giving a prize for the best contribution, he at the same time offering to defray the cost of the first prize, and suggesting a subject in which he was personally interested.

At that date we were totally unprepared for such a suggestion, but on thinking over the matter at intervals of leisure, we have come to think that the elaboration of such a scheme would not only prove of benefit to the manufacturer, but would go some way towards defraying the holiday expenses of the prize winner.

We are therefore prepared to receive competitions in the following subjects:—

SUBJECT I.

The best method of pumping or otherwise lifting or forcing warm aqueous hydrochloric acid of 30 deg. Tw.

SUBJECT II.

The best method of separating or determining the relative quantities of tin and antimony when present together in commercial samples.

On each subject, the following prizes are offered:—

One first prize of five pounds, one second prize of one pound, five additional prizes to those next in order of merit, consisting of a free copy of the *Chemical Trade Journal* for twelve months.

The competition is open to all nationalities residing in any part of the world. Essays must reach us on or before April 30th, for Subject II, and on or before May 31st, in Subject I. The prizes will be announced in our issue of June 28th.

We reserve to ourselves the right of publishing the contributions of any competitor.

Essays may be written in English, German, French, Spanish or Italian.

MANUFACTURE OF "COMPRESSED YEAST."

IN a thesis presented to the school of pharmacy of the University of Wisconsin, Mr. Alfred J. M. Lasche describes how compressed yeast is made in various parts of the United States. The thesis is printed in the *Pharmaceutische Rundschau* of New York. In regard to the preparation of the mash, it is stated that 3,130 lb. of ground corn are mixed with 4,500 gallons of water. This mixture is heated to 190° F. (to swell the starch, and thereby facilitate its inversion) and subsequently cooled to 154° F., then 1,920 lb. of ground rye and 550 lb. of ground malt are added, the malt being specially employed for the amount of diastase it contains, and is indispensable in the converting process. This mixture is then allowed to stand one hour, and is finally cooled to 80° F. The proportions of the different grains are of course largely a matter of opinion, and the various yeast manufacturers have different working formulas.

When the mash has cooled to 80° F. it is drawn off into another tub, and one gallon of concentrated sulphuric acid is added, in order to dissolve all remaining starch, dextrin, and glutinous matter, and to convert them into grape sugar. Finally, a quantity of compressed yeast is added to start the fermentation. This yeast settles to the bottom of the tub, but as soon as fermentation has started (usually in half an hour), and carbonic acid is being generated, the current of the latter gradually carries the yeast to the top of the liquid. It remains there, covered by a layer of the chaffy parts of the grain, until the yeast has accumulated in a sufficiently large quantity, and the current of carbonic acid has become strong enough, when it eventually breaks this film of chaffy particles, and collects on top of it in the form of foam. This goes on until all the nutritive matter has been assimilated. The foam, containing all the yeast, rises about two feet above the top of the liquid, dependent on the size of the tub, and when no more effervescence is noticeable, fermentation is complete.

Immediately after fermentation has ceased the foam is drawn off by means of troughs, and run, together with a fresh supply of water, into a revolving, six-sided and declining cylinder, lined with a sufficiently fine strainer. During this step of the process nearly all the chaffy remnants of the grain are separated, and the liquid, containing the yeast plant in suspension, is allowed to flow into a basin, whence, by means of a trough, it finally flows into a large tub.

The product in this tub is prevented from further fermentation by the addition of a sufficient quantity of ice. The yeast is now allowed to settle, the supernatant liquid drawn off, and the residue repe washed to free it from all mechanical impurities.

When sufficiently cleansed, it is run into a press by means of a pump. The press is constructed of a column of iron frames, sides of each frame being covered with a very fine straining cloth all the parts fitting tightly into each other. The yeast having pumped into such a press, the water is separated from it by mea the strainer, and carried off through a waste pipe.

The yeast, now compressed, is taken out in the form of large c and in this condition it is brought into commerce.

ANOTHER YEAR WITH FUEL GAS.*

ONE year ago it was my privilege to present to this associati paper on "Fuel Gas," and to-day I am called upon by worthy secretary to renew this subject.

Many of you, no doubt, have heard at the various association n ings, or read in our gas journals, a great many papers and article this subject during the past year. Some of them have been of ar couraging nature, while others—and I might say the most of the have been decidedly the other extreme. But if you stop to consic moment you will remember that the words "assume," "if," "theoretically" occur at frequent intervals. No doubt the wi have given the subject much careful thought and study, but who an you can tell by theorizing what will occur in practice?

I could monopolize your time for hours on the theoretical possibi of fuel gas, but that does not satisfy. Facts are what we want—demonstrated by actual results; and until such facts can be prese beyond all possibility of doubt, few gas managers would care to ven in a business way, into the unknown; few would care to run the of success or failure judging of the conditions from a theore standpoint.

There is no longer any necessity to theorize. There is no lo any reason to doubt, for seventy miles from this city there is a fue plant in operation, doing a business that ought to convince the sceptical. A plant that is not of the mushroom order, but is bui stay, and is second to none of its size in the country in mecha appliances for reducing labour to the minimum.

It is somewhat alarming to contemplate the erection of a fuel plant, looking at it from the standpoint taken by Mr. John Youn his recent paper read before the American Association. But a estimates are taken and his conclusions drawn from his experienc the sale and delivery of natural gas, I am not surprised at the fig In fact, I can easily credit the statements since I made a persona vestigation of the appliance in use at Alleghany. No attempts been made, that I could ascertain, to use the gas economically. coal stoves and furnaces filled with broken brick and the gas atta thereto were in universal use. Under such conditions, is it to be v dered at that such vast volumes of gas are required, and that fluctuations in demand should exist? As an illustration of the d ences that proper appliances make in the consumption of gas, I give you the figures of gas consumed by two houses, one fitted with a gas furnace and the other with the gas attached to an ordi coal furnace filled with brick. The former is a 14-roomed house, was warmed throughout during the winter months with an ave consumption of 39,000 cubic feet per month. The latter w 10-room house, and required a consumption of 125,000 cubic feet month.

In the manufacture and sale of illuminating gas the whole distribu occurs in practically four hours of each day, and necessitates a sto capacity at least equal to the output, or a manufacturing capacity e to three times the output if the storage be reduced one-half. mains also have to be of sufficient size to supply the entire daily ou in practically four hours.

Is it not plain to be seen that certain mains supplying a given qua of illuminating gas, would, if used to supply fuel gas, where they a use thirteen hours each day, distribute four and one-half times as n as the latter? Is it not consistent to claim a decrease in holder capa in the same ratio? There is equally as wide a difference in the wi and summer output of illuminating gas, as could possibly exist ir supply of fuel gas. Mr. Young must certainly have had some stan of comparison in mind when he said that the storage capacity of a gas plant would have to be simply erroneous. But he certainly c not have used the supply of illuminating gas as a comparison to d his conclusions from. My daily record shows no such erratic den as Mr. Young assumes must exist. He is generous enough to c the possibility of getting 30,000 cubic feet of fuel gas from a to bituminous coal, while the facts are, we are getting 50,000 cubic and have every reason to expect 50,000 from improvements no process of construction.

* Paper read by Charles H. Evans, of Jackson, Ohio, at the Toledo Meeti the Ohio Gas Light Association.

tutes the success of any business if it is not the selling of The fact that people buy shows that there are certain gas, else why should they continue to use it? Why should ontinue to grow and our receipts to increase if the supply not a success? Is it necessary the gas should be in every son before it is entitled to the claim of success? Such ie impression of some gas engineers.

ision a short time ago to read a number of letters from managers, which were in answer to inquiries of a certain posed to introduce fuel gas into the city where he resided. y surprised to learn the opinions of some of the success plant, especially so in this case, because I am positive ieers in question never were in Jackson to investigate the but were only too willing to cry down something they about.

is to be a popular fallacy existing among a great many gas illuminating gas has two and a half times the practical of fuel water gas. This is probably due to the assump- heat of combustion of a composition of gases is the same elements. This fact is usually the reverse. Thus, marsh hich forms more than one-third of the bulk of coal gas, is three parts of weight of carbon and one part by weight In one pound of marsh gas the heat of combus- ements, when burned separately, is for carbon 10,914 hydrogen 15,381, making a total of 26,295 units as the t which one pound of marsh gas ought to yield. But by nents the heat evolved is only 23,616 units; hence marsh g evolves 2,672 units per lb. less than its elements, or a 12 per cent. Another serious drawback to the practical the heat in the combustion of illuminating gas is due to e of such vast volumes of nitrogen, which of necessity n in to obtain the proper amount of oxygen. This nitrogen rbs the heat and detracts largely from the flame tempera- s, but by its insidious presence it casts off into the atmos- umed gases the moment a cold vessel is placed in contact e. In fact, I can safely assert that the air dilution neces- complete combustion of illuminating gas, detracts from its lue fully 50 per cent. This would not appear to be the ecific heat of the nitrogen, which enters into the compo- stracted from the theoretical heat evolved. But this is r illustration of the difference between theory and practice. gas requires but 2·47 cubic feet of air for the perfect com- ubic foot, it is slightly affected, comparatively speaking, ution.

have demonstrated the fact that illuminating gas used for heating is worth but 30 per cent. more than fuel water gas o and one-half times as much.

ew the oft-trodden ground and see if there is not some ssibility of making the general sale of fuel gas a success. rtainly elements of advantage in its use that ought eferable even at a higher cost than a solid carbonaceous

e cost, labour, and inconvenience of handling a heavy oided, the fuel being capable of easy distribution.

t is in a form also free from those material impurities d a large residual waste, besides impairing combustion. is also free (if it be a purely combustible gas) from those hich, in the presents methods of heating, involve even n the cause last mentioned.

is in precisely the candition to unite perfectly and in- with the oxygen of the air, thus securing a thorough Hence, it gives an immediate and uniform result, and erature is constant.

e intense and steady heat of the flame just mentioned e and money, by presenting an even fire surface ready at f ignition.

is a fire, capable of concentration upon the precise point ult is desired, and one that is thoroughly under control, a valve, starting, graduating or stopping the combustion

The general cleanliness of the system, no dirt or residuals

he decided advantage from a sanitary standpoint of g combustible gases in our dwellings, instead of attempt- them as well, by means of the imperfect gas machines In the one case, the only risk arises from the possibility lily detected by the sense, and having simply mechanical the other, it is a much more serious risk because the emical one, consequent upon imperfect combustion, and f poison into the atmosphere is likely to be frequent and ay nothing of the deoxidation of the air by contact with urfaces.

apear to be overwhelming arguments in favour of a against a gross form, and it is a matter of considerable

surprise to me that every family on our line of main is not a consumer. But there is one class of people that always wait to see what their neighbour is going to do, and another class that is afraid of everything new. But, nevertheless, there are between eight and nine hundred families consuming our gas, and nine-tenths of all the business places in the city are lighted by our method of illumination. To this last feature I want to draw your special attention, for I do not believe a fuel gas business can be perfect and complete without an incandescent burner. The only one we have in use at Jackson (the Fahnehjelm) has, during the past year made such rapid strides towards perfection that to one who remembers it, when first brought before the public, the change seems marvellous. The comb has now double the life it had a year ago, and emits a beautiful straw-coloured light, that is very attractive to the public, as the fate of the now defunct Jackson Gas Light Company will testify.

A series of photometrical observations which I have made from time to time with the Fahnehjelm burner, since its improvement shows an average of 5 candle power per cubic foot of gas burned for the first 150 hours burning, and a gradual diminution, until the expiration of 250 hours, it has decreased to a shade less than three candles per cubic foot of gas consumed.

I removed four burners from a chandelier in my house that had been in use for 38 days, in the months of December and January, and had been burned at least 190 hours. A photometrical observation of the burners showed an illumination equal to 3·9 candles per cubic foot of gas burned.

One burner which costs the consumer four cents practically represents the oil that is used to carburate a thousand feet of water gas, as well as the additional amount of fuel expended in bringing to proper tem- perature the surface used to gasify the oil. In many places the cost of carburation by oil reaches from 25 cents to 30 cents per 1,000, and in no instance has the cost reached so low a figure as is represented by our Fahnehjelm burner.

That fuel gas has established itself in the field to stay I do not believe there is any longer need to doubt.

In answer to question following the reading of his paper, Mr. Evans said: The coal used in the gas plant at Jackson is Youghiogheny nut, the coal mined at Jackson containing considerable sulphur and being little used. The furnace used in Jackson is one made specially for the gas, so that the products of combustion will go out of the chimney at the very lowest possible degree of temperature.

It costs about $4 to heat a thousand cubic feet of space under general conditions. The cost of gas for a month, for a family of six, cooking on a four-hole stove, will be about $1.30 if used economically. This is on a stove made for use of gas. An ordinary stove fitted for gas would consume three times as much.

The lowest rate for which Jackson fuel gas is sold is 38 cents per thousand. Mr. Evans finds by experiment and actual measurement that for room heating 20,000 cubic feet of gas is equal to one ton of coal; the difference in cost, therefore, is not great.

CROXLEY PAPER MILLS.

IN the House of Commons, last Friday, Mr. Lawson asked the President of the Local Government Board, whether he would inquire into the cause of the poisoning and pollution of the river Colne and its tributaries by the Croxley Paper Mills, near Watford, on Wednesday, May 7th, whereby private property to the value of many hundreds of pounds was destroyed; and whether he had any power to insist upon the provision of necessary apparatus whereby poisonous, noxious, or polluting liquid or matter could be dealt with so as to prevent any such injury or damage for the future; and, if so, whether he would exercise such powers in the interests of the owners, tenants, and rate- payers of the river and fisheries in the neighbourhood.

In reply, Mr. Ritchie said:—Since the notice of this question was given by the hon. member, I have communicated with the sanitary authority of the Watford Union, and I learn from their clerk that no complaint has been made to the authority, as to the occurrence on May 7th, at the Croxley Paper Mills. I am informed, however, by communications which I have received, that a tank at the mills which was charged with chloride of lime burst, the result being to poison great quantities of fish, and to affect the water for a long distance below the mills. I have no authority to insist upon the provision of necessary apparatus whereby poisonous, noxious, or polluting liquid or matter can be dealt with, so as to prevent any such injury or damage for the future. If, as I infer is the case, it is considered that the provisions of the Rivers Pollution Prevention Act are contravened, the sanitary authority are empowered, with the consent of the Local Government Board, to institute proceedings for the enforcement of that Act. If any such consent is applied for, the matter will receive my prompt consideratian. In the meantime, I will bring the matter specially under the attention of the sanitary authority, and request them to investigate the facts, and report the result to the Board.

SMOKE ABATEMENT.

ON Wednesday, a public meeting was held at the Mansion-house "to promote the national work undertaken by the committee for testing smoke-preventing appliances." The Lord Mayor presided, supported by Lord Derby, Lord Howard, of Glossop, Earl Fitzwilliam, Sir F. Abel, Sir H. Roscoe, M.P., and Sir Douglas Galton; and letters, regretting their inability to attend, were received from the Duke of Westminster, Lord Rayleigh, Lord Rosebery, Lord Cross, Lord Percy, Lord Egerton of Tatton (the president of the committee), and the Mayors of Manchester, Salford, Sheffield, Nottingham, and Leicester. The chairman, in opening the proceedings, said he thought that every one must appreciate the importance of, and the necessity for the purest air we could possibly obtain. Mr. A. E. Fletcher, her Majesty's chief inspector of alkali works, and chairman of the executive of the Committee for Testing Smoke-Preventing Appliances, said the committee did not wish to recommend any particular appliance, nor did they stand forward as inventors. Lord Derby then proposed a resolution approving the objects of the committee, as stated. He thought that the diminution of smoke and its necessary accompaniment dirt, was a matter which concerned every one, except those who were fortunate enough to live away from great towns. Indifference was the real difficulty which they had to encounter, but in England, anything which came to be recognized as a want, was eventually supplied. The expenditure of fuel in creating dirt—for that was what it came to—was a waste of fuel itself, and the injury caused to property was not inconsiderable. He believed that more than three-fourths—he would say something like nine-tenths—of the smoke from collieries and factories was absolutely preventable, though some trouble and outlay would be required. Possibly more stringent legislation would be needed, but let them first try the experiment of enforcing the laws which they already had. Lord Howard of Glossop seconded, and Professor Chandler Roberts-Austen supported the resolution, which was carried unanimously. On the motion of Sir Henry Roscoe, M.P., seconded by Earl Fitzwilliam, and supported by Alderman Bowes (Salford), a resolution was next passed in favour of raising a fund to meet the expenses of the work. Mr. Ernest Hart, the Mayor of Rochdale, and Mr. T. C. Horsfall afterwards addressed the meeting, and a vote of thanks to the Lord Mayor closed the proceedings.

THE EXTENSIVE FRAUDS AT BRADFORD.

ON Wednesday, at the Crown Court, Leeds, before Mr. Justice Vaughan Williams, was concluded the case of Francis Stubbs (foreman dyer) and Harry Varley (drysalters' traveller), who were indicted for conspiring to defraud S. C. Lister and others, and also for obtaining money by false pretences. This case, which raised questions similar to those which are familiar to the public in connection with the Salford Gas frauds, created great interest, and there was a large attendance.

Mr. Lockwood, Q.C., Mr. Asquith, Q.C., and Mr. Waugh prosecuted; Mr. Waddy, Q.C., Mr. C. Matthews, and Mr. Kershaw defended Stubbs; and Sir Charles Russell, Q.C., Mr. E. T. Atkinson, Q.C., and Mr. Manisty represented Varley.

The prosecution was instituted by Messrs. Lister and Co. (Limited), silk spinners and manufacturers, Manningham, Bradford. The defendant Stubbs was for many years the foreman dyer in the velvet department of the business, and in such capacity was almost entirely responsible for the purchase of dye wares for his department, for the selection of the drysalters from whom they were procured, and it was also alleged, for the prices that were paid for them; and since 1889, at which period he had stipulated for the right to see the drysalters alone and without any interference, he had exercised absolute control. Prior to the beginning of these proceedings he had given notice of his intention to leave the employment of the prosecutors. Richard Varley, a drysalter in Leeds, supplied dye wares in considerable quantity to the velvet department, and the defendant Varley acted on behalf of his father, Richard, in the transaction of the business, and was in frequent communication with Stubbs for that purpose.

It was alleged on behalf of the prosecutors that between March 1889, and February, 1890, dye wares had been sold to the prosecuting company for their velvet department by Richard Varley, and that by arrangement between the two defendants the prices paid for them had been largely in excess of what was fair and reasonable, and that the defendant Stubbs had received large payments in consideration of obtaining such excessive prices from his employers. The total amount so received during that period in excess of what was fair and reasonable was alleged to be £1,961, and of this Stubbs got £637. The evidence was confined to allegations with regard to certain dyewares—viz., cutch, fustic, sumac, fast red, and an aniline blue known in the as blue oox.

SIR CHARLES RUSSELL, in addressing the jury on behalf defendant Varley, did not deny that Stubbs had received payme way of commission upon the business done by his department wi firm his client represented, and did not defend this practice, but viewing the evidence he contended that the prosecutors had fai prove that which was essential to the conviction of the defend viz., that these payments had been made and received with a frau intention. He also justified the large profit derived upon the fast red and aniline blue upon the ground that the evidence estab that there was no market price for these dyes, that the blue was standard article but a mixture invented by Richard Varley himse that the sale of both blue and red was exclusively confined by Messrs. Lister. He further suggested that it was fair to infer fro evidence and from the failure to call Mr. Reixach, the manager, witness, that Messrs. Lister had winked at the payments now to be condemned as criminal.

Mr. WADDY followed, on behalf of Stubbs, and whilst claimir benefit of all that had been urged for Varley, pointed out to the that the prisoners were entitled to be acquitted unless the jury satisfied that the parties understood that the payments made were out of the pockets of Messrs Lister and not out of the pock Richard Varley.

The learned JUDGE, in summing up the case to the jury, di them that it was not necessary for them, in order to find the int to defraud proved, to be satisfied that the defendants themselves nized the dishonesty of the act done, if the nature of the act wa as to have the effect of defrauding.

The prisoners having been found guilty, Stubbs was senten nine months' hard labour, and Varley three months' hard labour.

Trade Notes.

PARAFFIN AND PETROLEUM IN SERVIA.—The Servian S schina has passed a Bill granting to an Anglo-German syndicate a poly for the working of paraffin and petroleum in Servia. Th oil deposits in the neighbourhood of Alexinatz will be the first tapped.

THE FIRST LOADED PETROLEUM TANK-SHIP ON THE R —On Saturday, the 3rd inst., a loaded tank vessel passed the havens, being the first ship of the kind seen on the Rhine. A prepared means of transport for petroleum on the Rhine thus be an accomplished fact. The honour, which otherwise would fallen to Duisburg, is due to Mannheim. A tank vessel ha constructed at Duisburg, but owing to the objections of the authorities and of an adjoining chemical manufacturer, the vess to be taken to pieces to be constructed again in another place ' against fire." The petroleum tank vessel now at work is call "Petrolia," and was built at the Berninghaus Yard, Duisbu the account of the great petroleum import firm of Horstman Co., Rotterdam. The cargo is about 8,500 cwts. petroleum, Rotterdam to Mannheim. The petroleum is imported direct fro United States in tank steamers, and either pumped from them into the Rhenish tank vessels or into reservoirs at Rotte At its destination (Duisburg or Mannheim) the tank vessel disch the petroleum into iron reservoirs, to which filling apparatus fo are attached. This form of petroleum transport will much inju railways, on which the petroleum traffic is now very large. *(Ku*

LOCOMOTIVE COAL.—The existing contracts for the supply of motive fuel to several of the leading railway companies will exp the end of next month, and a meeting of coalowners on the subj the new charges for 1890 has just been held in the South Yor district, which supplies steam coal by contract to several comp As regards that district, it appears that up to the end of 1888 th tract price for locomotive coal was 6s. per ton. At the beginn 1889 a short-time contract was conceded at 8s. per ton, and in last the contracts all round were raised to 8s. 6d. The Eastern, contracting from the beginning to the end of each ye January last was charged 11s. per ton for the South Yor "hards," and that is now understood to be the rate at whic contracts beginning in July next will be taken, although in som tricts it is likely to be even more. Taking the all-round price as 11s. per ton, the London and North-Western, it is estimated have to pay about £210,000. more for its coal in 1890-91 than in 1888, the Midland upwards of £180,000., the Great-Weste less than £150,000., and the North-Eastern (which consumes 50,000 tons of coal monthly) fully £160,000.

TRADE.—A conference representing employers and em-
orth of England iron and steel district was held at
Tyne, on the 20th inst., to consider a proposal by the
educe wages by 10 per cent., particularly in application
steel-smelting departments. After a long discussion it
at the reduction should begin on June 2.

UNION AND THE SALT WORKERS.—The Amalgamated
and Alkali Workers, Northwich, recently sent a list of
e Salt Union for increase of pay and amelioration of
These demands have been considered by the directors,
cessions made, but in most instances the proposals have
. The men have requested an interview with the
h will shortly take place.

HELENS COPPER COMPANY, LIMITED.—In the Chancery
Saturday, a petition was presented for the reduction of
f the St. Helens Copper Company, Limited. The
formed with a capital of £100,000. in £1. shares, all of
sued, and £80,000. was paid up. The company had
sperous, and did not require £30,000. of its capital. It
posed to return that amount to its shareholders, and
of all further liability on their shares, thus leaving the
,000. It had no debts whatever. Mr. Justice Chitty,
d that it was quite refreshing to find a company like this,
er.

LUCIUS, AND BRUNING.—At the general meeting of the
of the above-named company, held last Saturday, under
ship of Dr. Eugen Lucius, the proposals of the directors
lication of the profits were ratified. The proposal of the
put the whole of the shares of the first and second issues
of their proprietors, and, with that object to issue new share
to introduce a corresponding change in the statutes, was
as stamp exemption for the new script was not to be
M. C. F. W. Meister, one of the founders, and a member
torate, having been induced through advanced age to
fice, Mr. Walter vom Rath was elected in his stead. In
g his post Mr. Meister has decided to give to the workmen
k of his recognition, by placing in the hands of the
sum of 100,000 marks. (£5,000.) to be applied in the
workmen's home. *(Kuhlow.)*

FOR THE OIL RIVERS DISTRICT.—At the quarterly
the Belfast Chamber of Commerce, held on Thursday last,
of supporting the action of the Liverpool Chamber in its
to the granting of the above-named charter was fully
In the general interest of the mercantile community it was
hat the Belfast Chamber is of opinion that the granting of
ch as that sought for by the African Association is liable to
ical to public interests by the creation of monopolies; and
he Council to take such steps as they think proper to sup-
emorial of the Liverpool Chamber in opposition to the
the charter." The authorities of that Chamber have thereto the Prime Minister in the sense of the resolution, and
a circular to members of Parliament and others, requesting
their influence against the granting of the charter sought
African Association, and to advocate the formation of the
a Crown colony, either in connection with or apart from
f Lagos.

LITAN WATER SUPPLY.—Dr. Frankland reports to the
eneral the results of the chemical analyses of the waters
he inner and portions of the outer circle of the metropolis
month of April. Taking the average amount of organic im-
ined in a given volume of the Kent Company's water during
rs ending December, 1876, as unity he finds that the propor-
it contained in an equal volume of water supplied by each
politan water companies, and by the Tottenham Local
alth, was—Kent, 0·6; New River, 0·7; Tottenham, 1·0;
, 1·6; Lambeth, 1·8; Grand Junction, 2·0; Chelsea, 3·3;
sex and Southwark, 2·4; and East London, 2·5. The water
the Thames by the Chealsea, West Middlesex, Southwark,
ion, and Lambeth Companies was, in every case, efficiently
exhibited for river water, a high degree of organic purity.
obtained chiefly from the river Lea by the New River
ndon Companies was also efficiently filtered. That of the
Company was of a very high degree of organic purity,
ast London Company's supply was slightly inferior in this
l the Thames-derived waters. The deep-well waters of the
olne Valley Companies and of the Tottenham Local Board
ere, as usual, of a high degree of organic purity. The
Colne Valley waters were clear and bright, but the
water was again slightly turbid owing to suspended particles
on. The supply of the Colne Valley Company was the
ter delivered in London. Seen through a stratum 2ft.
aters presented the following appearances:—New River,
Colne Valley, clear and colourless; Tottenham, slightly
colourless; the remaining waters clear and very pale

THE LIABILITY OF DIRECTORS.—The Grand Committee on Trade sat at the House of Commons yesterday, under the presidency of Mr. Arthur O'Connor, and proceeded to consider the Directors Liability Bill. On clause 1 an amendment changing the title of the Acts from "The Companies Prospectuses" to "The Directors' Liability" Act was on a division carried by 18 votes to 3. Clauses 3, 4, and 5, embodying the main features of the bill, were struck out in favour of a new clause proposed by Mr. Warmington, to the effect that every director of a company, and every other person who has authorised the issue of a prospectus or notice, shall be liable to pay compensation for the loss or damage sustained by reason of any inaccurate or misleading statement in the prospectus or notice, unless he proves that he made careful and reasonable inquiry and examination into the statement made and had reasonable grounds to believe that the statement was true, or unless he proves that he had not consented to become a director of the company, or, having consented, he withdrew his consent before the issue of the prospectus or notice, and that the prospectus or notice was issued without his authority or consent. The date of the commencement of the Act was altered from the 1st September, 1890, to the 1st January, 1891. The bill was ordered to be reported with amendments to the House.

THE FATALITIES AT THE LEEDS FORGE.—The report of Dr. Stevenson, who, at the request of the Home Secretary, attended the inquest in November last on the bodies of two men who met their death by water-gas poisoning at the Leeds Forge, has been issued. He states that it is now nearly half a century since Leblanc first announced the poisonous character of carbonic oxide gas, which is the poisonous constituent of a great number of mixed and commercial gases, such as charcoal vapours or fumes, the vapour of burning coke, the combustible gases from blast-furnaces, producer gas, forge gas, gas from regenerator furnaces, gas leaking from stove pipes, and the so-called water-gas. After describing the manufacture of water-gas, Dr. Stevenson says that it consists approximately of 40 per cent. by volume of carbonic oxide and 50 per cent. of hydrogen, the remainder being diluent gases. It is practically odourless, hence leakages of the gas cannot be detected by the sense of smell. He goes on to state that in the eight years ending 1887 the fatal cases of water-gas poisoning in the four American cities of New York, Boston, Baltimore, and Chicago were 295. The manufacture of water-gas had, it was stated, been prohibited in the States of New Jersey and Massachusetts and in Paris. The alleged odorisation at the Leeds Forge was ineffectual. "It must be borne in mind," says Dr. Stevenson, "that safety is conferred not by simply imparting a decided and unmistakable odour to the gas, but by rendering it so far odorous that it is readily detected by smell when mixed with at least 5,000 volumes of air. Anything short of this confers only fancied safety, so poisonous is carbonic oxide gas." In dealing with the Leeds case he simply gives a detailed account of the accident and the sensations experienced by the medical men at the *post mortem* examination, and then the inquest.

DYEING.

THE AGEING OF LOGWOOD.

MESSRS. Clarkson and Macfarlane have contributed to the Franklin Institute an interesting account of some experiments they have been making on ageing logwood. This ageing is done at present by stacking the rasped wood into heaps, and sprinkling it with weak alkaline liquors, the process being one of oxidation, resulting in the conversion of the hæmatein into hæmatoxyline. It has been observed before that chlorine has a similar action, but no advantage was taken of this, inasmuch as generally the action was too great, and the hæmatoxyline was practically destroyed and rendered useless for dyeing purposes. The authors named, in the course of some experiments, found that by regulating the quantity of chlorine used the hæmatein could be converted into hæmatoxyline without any destructive effect; and the dyeing power of the wood was considerably increased, as much as 150 per cent., without any deterioration in the quality and tone of colour produced. One interesting part of their experiments tends to show that hæmatein and hæmatoxyline have special dyeing properties of their own; the one requires to be dyed in a special way to develop all its power, while the other must be used somewhat differently. Thus, for dyeing wool mordanted with bichrome and tartar, they found the oxdised product to yield by far the best results, the shades obtained being twice as deep as from the unoxidised wood. On the other hand, when used for dyeing in the cold with slightly alkaline liquors, the unoxidised product gave the best results; from which they infer that for wool dyeing the cured wood would give the best results, while for cotton the dry cut wood should be used, as this is generally dyed cold and with alkaline solutions.

THE ADVANTAGES OF THE APPLICATION OF LIME TO SOILS.

ALFRED DRIEBERG.

FOR improving certain kinds of soils, lime is esteemed one of the best means we possess. It is required for the growth of all kinds of cultivated plants, especially those belonging to the natural order *Leguminosæ*. Therefore it is absolutely necessary that lime should be present to some degree in all culivated soils. While lime is always present in soils that admit of cultivation, the quantity contained in them is very often insufficient for the healthy growth of certain crops. Therefore if lime be applied to such soils it naturally increases their fertility. On soils of this kind the striking effects of lime is best seen when the soil contains in abundance all the other essential elements of fertility with the exception of lime. Lime not only acts directly as a manure, but increases the other materials necessary for the growth of crops. In most cases the beneficial influence of lime is due to its chemical action in the soil. Lime preserves clay in an "open" condition, thus making heavy soils friable and pervious to water. It also promotes the decomposition of vegetable matter and the formation of nitrates. A soil whose fertility has been impaired by an excess of organic matter can be rendered fertile by a large dose of quicklime. In some soils that are infested with insects an application of lime destroys them entirely at little expense. When applied in large quantities to clay land, it opens and loosens the clay and gives it a certain amount of porosity, and as a consequence, it brings about further improvement by exposing a larger extent of surface to the action of the atmosphere. The quantity of lime applied to the soil varies with the purpose it is intended to serve. If lime is applied to destroy an excess of organic matter, a large dose will be necessary, but when a soil is naturally deficient, a far smaller dose will be sufficient. For obtaining the fullest effects of lime, small doses at short intervals are very effective. Where the opposite course is adopted there is considerable waste and a gradually diminishing effect, as the natural tendency of lime is to sink down into the subsoil. A certain quantity of lime is dissolved and removed by drainage water, and the remainder, in a few years, sinks below the cultivated depth, or chemical changes take place which render it effete. On arable land, the plough for a season or two brings it back to the surface, but after a time it gets beyond the depth of the plough and is lost. This strong tendancy of lime to sink into the subsoil, shows us that when liming, we should not plough the lime in, but keep it as near the surface as possible. The land should be ploughed first, then the lime spread and simply harrowed in. Burnt lime is much more powerful in its action on vegetable matter than chalk or marl, it should be used with discrimination, lest the humus of the soil be unduly diminished. Heavy clays or soils rich in humus are those most benefited by burnt lime. In reclaiming peat-bogs, lime is of the highest value. The acid humic matter of the peat is neutralised by the lime, and the conditions are made suitable for the oxidation of the nitrogenous organic matter, and the formation of ammonia and nitrates. The general effect of lime is to render available the plant food already in the soil without itself supplying any significant amount. Vegetable remains under peculiar circumstances refuse to decay, and accumulate to an injurious extent. This kind of vegetable matter is generally found in undrained or badly drained land. To remove this sour humus, lime is generally employed, which by acting upon the insoluble vegetable matter, hastens its decay, and "sweetens the land," for by decay these materials furnish carbonic acid and other useful food-materials for plants. The lime thus converts a noxious ingredient into a source of fertility. Lime economises the use of potash; for certain crops, where potash is not abundant in the soil, have, to some extent, the power of utilizing lime in its place. Lime also improves the quality of grain, grasses, and other crops, the finer grasses on certain lands refusing to grow until the land has been limed. It hastens the maturity of crops and checks the growth of moss and weeds in the soil. The effect of lime on the mechanical texture of many soils is also great. It pulverizes and lightens strong soils, at once improving their drainage and rendering them more easily tilled. It also improves the texture of light soils – provided an overdose be not applied—even when they contain but little organic matter. The avidity of lime for moisture, added to the chemical changes brought about by it, have the effect of increasing the absorptive and retentive power of soils to a considerable degree. A deep soil requires a heavier dressing of lime than a shallow one, and deep tillage will call for larger applications than where the cultivation is shallower. A sandy soil requires less than a heavy clay, and soils poor in vegetable matter require less than soils which are rich in organic matter. A small quantity of lime will have greater effect on drained land than a larger dose on wet or undrained land. Lime slakes best and quickest when laid down in small heaps and slightly covered with fine soil. This saves refilling and recarting. The heaps should be put down at equal distances apart, so that when the lime is slaked, it could be spread out easily. There should not be to much magnesium carbonate present in a limestone, as it is consid[illegible] less valuable for agricultural purposes. As I have said before, best to apply lime in small doses at short intervals rather than l[illegible] doses at long periods, as Darwin has shown us that the action of ea[illegible] worms tends to bury it. The weight of lime per bushel varies f[illegible] 75 lb. to nearly 1 cwt. according to the particular kind. The b[illegible] it is burnt the lighter it is comparatively. Pure varieties of limes[illegible] yield a little over 11 cwt. of burnt lime per ton Lime shoul[illegible] applied *as a rule* to soils containing much clay or humus; not in c[illegible] contact with nitrogenous manures such as dung or guano, as it sets the ammonia, which is liable to escape into the air. But lime, s[illegible] it sets free plant food, tends to exhaust the soil, and "lime wit[illegible] manure, will make both farm and farmer poor."

The following are the chemical changes which lime passes i[illegible] application to land:—

1st $Ca\,C\,O_3$. Pure limestone rock, calcium carbonate before burr[illegible]

2nd CaO. Lime, calcium oxide, quicklime, caustic lime, shells, as it comes from the kiln after burning.

3rd CaH_2O_2. Slaked lime, calcium hydrate, fallen shells, water has been put on it, or it has absorbed water from the atmosph[illegible] still somewhat caustic, and usually spread on the land in this state.

5th $CaCO_3$. Mild lime, the state to which it eventually return[illegible] long exposure; has yielded up its water and absorbed carbon dic[illegible] (CO_2.) In this condition it is equivalent to finely powd[illegible] limestone or chalk, as it is identical in chemical composition them.

Effect of Lime on Soils.

1. Acts with felspar or clay, setting free potash, or other alkali[illegible]
2. Acts on vegetable matter, setting free ammonia, water, r[illegible] acid, and carbon dioxide (which it unites with), tending to de[illegible] excess of humus in the soil.
3. Neutralises organic acids—humic, ulmic, geic, &c.,—sweetening the soil.
4. Takes up nitric acid as formed by the nitrifying bacteria.
5. Is a plant food in itself.
6. Renders harmless injurious salts of copper, iron, &c.
7. Opens up clay soil from the "curdling" effect it has on molecules of that substance.

Soils which contain more than four per cent. of lime (carbo[illegible] should not have any applied as a rule.

Loamy and clay soil contain one to three per cent. of calcium [illegible]bonate, and defective soils less than one per cent.—*Tropical [illegible]culturalist.*

Market Reports.

REPORT ON MANURE MATERIAL.

There has been a moderate business passing, both *on spot* [illegible] *forward*, at higher prices for phosphates, but without much chang[illegible] nitrogenous material.

The position as regards mineral phosphates is still unchanged, large shippers being well sold a long way ahead, will not accept [illegible] 11d. per unit, C. F. & I. to U. K., for cargoes of hot-air-dried [illegible] Rock, but outside shippers are still inviting bids at less money. [illegible] are sellers of Charleston Land Rock at 10d. per unit, and seller[illegible] inviting offers under that price. We do not hear of any impo[illegible] transactions in other descriptions, but now that we are nearing th[illegible] of the manure season, manufacturers seem more at liberty and inclined to consider their next season's supply of material.

There are now several cargoes of River Plate bones on the ma[illegible] February and March sailing, but we do not hear of any transac[illegible] £5 7s. 6d. asked, delivered U.K. ports, but something less [illegible] probably be business. It is reported that large sales of crushed Indian bones have been made for direct shipment to U.K. port[illegible] very full prices, and at the moment there are buyers at £4. 17s. 6d [illegible]quay, Liverpool, May and June steamer shipment, but no selle[illegible] the price. Spot has been cleared, and there are many enquirie[illegible] rather dirty parcel of uncrushed Brazils, sold on Wednesd[illegible] £4. 12s. 6d., ex-quay Liverpool, which is quite 5s. up from the l[illegible] point touched. Sales of East Indian bone meal continue to be [illegible] at £5. 10s., ex-quay or store Liverpool, and we understand th[illegible] shipment ahead, £5. 5s. has been bid and refused for quantity. Shi[illegible] having been caught last year, seem very shy about committing [illegible]selves to large sales this year, unless at prices quite out of rea[illegible] buyers.

Nitrate of soda is quiet at 8s per cwt. on spot. Due cargo[illegible] worth barely 8s.; May and June sailings are worth about 8s. 3[illegible] and September, October, and November, about 8s. 7½d. to 8s.

Dried blood is more enquired for, without alteration in valu[illegible]

·ed ; 60 tons of rough River Plate sold at 10s. per unit ex-·pool.

sellers of ground hoof and horn at 9s. per unit, delivered ·ks, in bags, gross weights, but not much demand just now ·le.

quiet at 40s.

·phates are again a shade easier, and 46s. 3d. in bulk for-·ks, may be taken as nearest value of 26%.

TAR AND AMMONIA PRODUCTS.

·'s prices stand good for benzols to-day, and there is an air ·bout the market that was wanting when we wrote last ·rt. Solvent naphtha is still quoted at 1s. 9d. Anthracene ·ite nominal at 1s. 5d. ; 30% B can be bought at 1s.

·ate of ammonia market is somewhat quieter, and prices ·l slightly. £11. 5s. to £11 6s. 3d. may be taken as the ·ndon, Hull, and Leith. The Beckton quotation is £11. 10s.

·CELLANEOUS CHEMICAL MARKET.

·ut little change to report in the position of the alkali trade past week. Caustic soda has slightly recovered from the ·ression, and there has been more business doing in 70% at ·l. to £8. f.o.b. Liverpool ; 60% at £7. 2s. 6d.; cream per ton For the higher strengths : 74% is value for ·l. f.o.b. Liverpool ; while 77% on the Tyne is at £9. 15s. ·b. at that port. Soda ash is maintained in price at 1½ per ·ling brands, but there is less strength in the position for ·ries. According to quantity and period of delivery prices ·d. to 1½ for carb. ash 48% to 58% ; and for caustic 1 5/16d. per deg. Chlorate of potash is slightly better at ·. Soda crystals are well sustained at £3. to £3 5s., · package. Saltcake in bulk 25s. to 26s. per ton. Vitriol ·ful, and prices easier. Muriatic acid without change. ·owder remains steady, at £5. per ton on rails and £5. ·. in export casks. Sulphate of copper firmer, with recent ·netal, and speculative business for forward deliveries has ·ed to some extent from this cause. Prompt price is £26. ·l June to September £23 10s. per ton. White powdered ·inding ready outlet, and prices are firmly maintained. ·d acetic acids stationary in value, and business in this the whole, is dull. Carbolic acid crystals are rather ·¾d. to 8d. for 34° for early deliveries ; but the outlook afford any distinct encouragement to makers. Lead salts ·l there is more inquiry. Muriate of ammonia is slightly ·n home and export trades. Potash, caustic, and carbonate with fair demand, and are firm in prices both for early deliveries. Citric and tartaric acids remain steady in ·lic acid 3¾d. to 4d. per lb. There is no improvement the demand for chemicals from the textile manufacturing

THE TYNE CHEMICAL REPORT.

TUESDAY.

, on the whole, are much steadier this week. There has advance in the price of soda ash and soda crystals, the two and a half per cent., and the latter, one shilling per ·an last week. Bleaching powder and sulphate of soda ·t, without any chahge in prices. Caustic soda is ·ve shillings per ton, but the market closes steady, better demand. To-days market prices are:—Bleaching softwood casks, £5. to £5. 2s. 6d. per ton; 77%., in drums, £9. 5s. per ton, ground in 3-4 cwt. ; soda ash, 48/52%, 1¼d. per degree less 2½% ; soda ·cwt. bags, net weight, £2. 15s. per ton, in 2 cwt. bags, ·2. 12s. 6d. per ton ; in casks, gross weight, £2. 12s. 6d. ·hate of soda, in bulk, 32s. per ton ; ground and packed in ·r ton ; recovered sulphur, £4. 5s. per ton ; chlorate of per lb. ; silicate of soda, 75° Tw., £2. 10s. per ton ; 100° ·d. per ton ; 140° Tw., £4. per ton ; hyposulphite of soda, ·, £5. 15s. per ton, in 5-7 cwt. casks, £5. 5s. per ·ute sulphate of alumina, £4. 10s. per ton ; blanc fixe, ton ; chloride of barium, £8. per ton ; nitrate of baryta, crystals, £18. 15s. per ton ; ground, £19. 5s. per ton ; sulphide of barium, £5. 10s. per ton—all f.o.b. Tyne, or f.o.r. makers' works.

There is no change to report in the price of small coals for manufacturing purposes. Northumberland steam in good demand, at prices ranging from 8s. to 8s. 6d. per ton. Durham small, quoted, 9s. 6d. to 10s. per ton.

METAL MARKET REPORT.

	Last week.		This week
Iron	44/6	....	44/2½
Tin	£93 5 0	....	£94 12 6
Copper	52 5 0	....	52 17 6
Spelter	23 0 0	....	23 5 0
Lead	12 17 6	....	12 17 6
Silver	47¼d.	....	47 7/16d.

COPPER MINING SHARES.

	Last week.			This week.	
Rio Tinto	18 11/16	18 13/16	....	19	19⅛
Mason & Barry	7 7/16	7 9/16	....	7 5/16	7 7/16
Tharsis	5 5/16	5 7/16	....	5 5/16	5 5/16
Cape Copper Co.	4 7/16	4 9/16	....	4⅜	4½
Namaqua	2⅞	3	....	2⅞	3
Copiapo	2 13/16	2 15/16	....	2¾	2⅞
Panulcillo	1	1⅛	....	15/16	1 1/16
Libiola	2⅜	2⅝	....	2⅜	2¾
New Quebrada	½	¾	....		
Tocopilla	1/-	1/6	....		
Argentella	3d.	9d.	....	3d.	9d.

THE LIVERPOOL COLOUR MARKET.

COLOURS without change. Ochres: Oxfordshire quoted at £10., £12., £14., and £16.; Derbyshire, 50s. to 55s.; Welsh, best, 50s. to 55s.; seconds, 47s. 6d.; and common, 18s.; Irish, Devonshire, 40s. to 45s. ; French, J.C., 55s., 45s. to 60s. ; M.C., 65s. to 67s. 6d. Umber: Turkish, cargoes to arrive, 40s. to 50s. ; Devonshire, 50s. to 55s. White lead, £21. 10s. to £22. Red lead, £18. Oxide of zinc: V.M. No. 1, £25.; V.M. No. 2, £23. Venetian red, £6. 10s. Cobalt: Prepared oxide, 10s. 6d. ; black, 9s. 9d.; blue, 6s. 6d. Zaffres: No. 1, 3s. 6d. ; No. 2, 2s. 6d. Terra Alba: Finest white, 60s. ; good, 40s. to 50s. Rouge: Best, £24.; ditto for jewellers', 9d. per lb. Drop black, 25s. to 28s. 6d. Oxide of iron, prime quality, £10. to £15. Paris white, 60s. Emerald green, 10d. per lb. Derbyshire red, 60s. Vermillionette, 5d. to 7d. per lb.

WEST OF SCOTLAND CHEMICALS.

GLASGOW, Tuesday.

Sulphate of ammonia, which was on the rising grade last week at a good £11. 5s., stiffened steadily, for spot delivery, till £11. 7s. 6d. was topped at the close of last week and the beginning of this. To-day it feels as if the edge was just off this ascent, and probably the full £11. 7s. 6d. is no longer readily procurable ; but, the article is, nevertheless, healthy, and no one as yet looks for much of a drop, while another and immediate rally would not excite surprise. For delivery forward, over July to September, £11. 12s. 6d. is said to have been done ; but this is doubtful, and the forward market is essentially barren of authenticated results for the moment, owing to the independent attitude of makers. Chemicals generally still incline to the fall here, and caustics have encountered a further considerable drop in all strengths except the highest. Bleach also is a little cheaper, but with a good show of put through business for the week ; and chlorate of potash is another eighth down. As an exception, soda crystals rise, and now quote 52s., if not more. Bichromates are in very weak demand, but paraffins (except burning oil) retain positions well. Wax and scale in dealers hands are fetching a fraction over stop prices as below, but makers, for the most part, work on old unexhausted contracts concluded on these terms. Chief prices current are :—Soda crystals, 52s. net Tyne ; alum in lump £5., less 2½% Glasgow ; borax, English refined, £29., and boracic acid, £37. 10s. net Glasgow ; soda ash, 1¼d. less 5%, Tyne ; caustic soda, white, 76°, £10., 70/72°, £7. 15s., 60/62° £7. 2s. 6d., and cream, 60/62° £6. 17s. 6d., all less 2½% Liverpool ; bicarbonate of soda, 5 cwt. casks, £6. 5s., and 1 cwt. casks, £6. 10s. net Tyne ; refined alkali, 48/52°, 1½d. less 1¼%, Tyne ; saltcake, 25s. to 26s. ; bleaching powder, £5. 2s. 6d. to £5. 5s. less 5% f.o.r. Glasgow ; bichromate of potash, 4d., and of soda 3d., less 5 and 6% to Scotch and English buyers respectively; chlorate of potash, 4½d., less 5% any port; nitrate of soda 8s. to 8s. 1½d.; sulphate of ammonia, spot, £11. 6s. 3d. to £11. 7s. 6d. f.o.b. Leith; salammoniac, 1st and 2nd white, £37. and £35., less 2½% any port; sulphate of copper, £25. less 5% Liverpool; paraffin scale, hard and soft, 2½d. per lb.; paraffin wax, 120°, semi-refined, 3⅝d.; paraffin spirit (naphtha), 8d. a gallon; paraffin oil (burning), 6¾d. to 7d. on rails Glasgow to all Scotch buyers (English orders negotiable about 1¼d. a gallon lower); ditto (lubricating), 865°, £5. 10s. to £6. per ton; 885°, £6. 10s.; and 890/895°, £7. 15s. to £8. Week's imports of sugar at Greenock were 18,101 bags.

THE LIVERPOOL MINERAL MARKET.

Our market has ruled steady during the past week. Manganese: Arrivals very small; prices continue firm. Magnesite: There is little demand for raw lump, whilst the production has very much increased; prices, therefore, are in the hands of buyers, and sales can be made at lower prices than any time previously. Raw ground, £6. 10s.; and calcined ground, £10. to £11. Bauxite (Irish Hill brand) continues in increasingly strong demand, bringing tip-top prices—Lump, 20s.; seconds, 16s.; thirds, 12s.; ground, 35s. Dolomite, 7s. 6d. per ton at the mine. French Chalk: The only arrivals this week have been of G.G.B. "Angel-White" brand, which sells freely at full prices *i.e.*, 95s. to 100s. medium, 105s. to 110s. superfine. Barytes (carbonate), easy; selected crystal lump scarce at £6.; No. 1 lumps, 90s.; best, 80s.; seconds and good nuts, 70s.; smalls, 50s.; best ground, £6.; and selected crystal ground, £8. Sulphate has somewhat improved, best lump, 35s. 6d.; good medium, 30s.; medium, 25s. 6d. to 27s. 6d.; common, 18s. 6d. to 20s.; ground best white, G.G.B. brand, 65s.; common, 45s.; grey, 32s. 6d. to 40s. Pumicestone quiet; ground at £10., and specially selected lump, finest quality, £13. Iron ore steady. Bilbao and Santander, 9s. to 10s. 6d. f.o.b.; Irish, 11s. to 12s. 6d.; Cumberland easier, 14s. to 18s. Purple ore quiet. Spanish manganiferous ore continues in fair demand. Emery-stone: Best brands still scarce, especially on spot: prices unaltered No 1 lump is quoted at £5. 10s. to £6., and smalls £5. to £5. 10s. Fullers' earth steady; 45s. to 50s. for best blue and yellow; fine impalpable ground, £7. Scheelite, wolfram, tungstate of soda, and tungsten metal further inquired for. Chrome metal, 5s. 6d. per lb. Tungsten alloys 2s. per lb. Chrome ore: steadier, prices unaltered. Antimony ore and metal have further advanced. Uranium oxide, 24s. to 26s. Asbestos scarce, bringing increased prices. Potter's lead ore continues firm; smalls, £14. to £15.; selected lump, £16. to £17. Calamine: Best qualities scarce—60s. to 80s. Strontia steady; sulphate (celestine) steady, 16s. 6d. to 17s. Carbonate (native), £15. to £16.; powdered (manufactured), £11. to £12. Limespar: English manufactured, old G.G.B. brand in demand, and brings full prices; 50s. for ground English. Felspar, 40s. to 50s.; fluorspar, 20s. to £6. Bog ore in demand, firm at 22s. to 25s. Plumbago steady; Spanish, £6.; best Ceylon lump at last quotations; Italian and Bohemian, £4. to £12. per ton; founders, £5. to £6.; Blackwell's "Mineraline," £10. French sand, in cargoes, continues scarce on spot—20s. to 22s. 6d. Ferro-manganese and silicon spiegel easier. Chrome iron, 20 per cent., £24. to £25. Ground mica, £50. China clay freely offering—common, 18s. 6d.; good medium 22s. 6d. to 25s.; best, 30s. to 35s. (at Runcorn).

New Companies.

African Petroleum Exploration Syndicate, Limited.—This syndicate was registered on the 10th inst., with a capital of £10,000., in £1. shares, to acquire lands, concessions, mining and oil rights, and real and personal property in South Africa, and to carry on business as oil merchants. The subscribers are:—

	Shares.
R. Gudridge, 19, Sussex-street, Poplar, clerk	1
F. Vigo, 93, Earl's-court-road	1
A. Burnie, 165, Fenchurch-street, solicitor	1
J. J. Allen, 6, Stafford-road, Brixton	1
H. T. Bull, 12, Effingham-street	1
Wm. Young, 7, Vicarage-road, Camberwell	1
A. W. Heron Maxwell, 56, St. James's street, S.W., secretary to a company	1

Metropolitan Cement and Brick Works, Limited.—This company was registered on the 13th inst., with a capital of £10,000., in £1. shares, to carry on business as cement and brick manufacturers. The subscribers are—

	Shares.
H. Tapp, M.E., 20, Bucklersbury	1
F. Mortlemann, 71, Fenwick-road, East Dulwich, clerk	1
G. H. Taperell, Harewood, Chingford	1
F. Wingrove, 8, St. Paul's crescent, N., commission agent	1
H. Emery Fuller, 13 and 15, Leadenhall-buildings	1
W. Browne, 16, Craigmillar-road, Edinburgh, medical practitioner	1
G. R. Poole, 17, Brooke-road, Wood-green	1

Monarch Soap Company, Limited.—This company was registered 13th inst., with a capital of £2,000., in £1. shares, to trade as soap, chemica varnish manufacturers. The subscribors are:— Sh

- W. Minchin, 38, Great St Helens, soap maker
- A. A. Coster, Junior Carlton Club
- H. H. Hughes, 44, The Grove, Hammersmith
- F. Minchin, 14, Grace-street, Bromley-by-Bow
- E. E. Johnson, 110, Cannon-street, accountant
- R. Kurtz, Carshalton, Surrey
- F. W. Heather, 5, Pump-court, Temple, barrister

San Francisco Breweries, Limited.—This company was registered 13th inst., with a capital of £1,000,000., in £10. shares, to trade as br maltsters, etc., and for such purposes to adopt an unregistered agreement w F. Lenon and C. E. Ertz. The subscribers are— Sh

- T. Willerson, 7, Nightingale-road, Clapton, accountant
- F. W. Hull, 68, Disraeli-road, Putney, accountant
- R. Gordon, New Malden
- H. J. Bass, 10, Rodney-terrace, Putney, factory manager
- E. E. Friend, 8, Larcom-street, S.E., merchant
- J. Kelly, 8, Gladstone-terrace, S.W., merchant
- G. A. Buckler, Pembroke-road, South Tottenham

Gazette Notices.

Partnerships Dissolved.

H. Moss and T. M. Broughton, under the style of Moss, Broughton & Lawson-street, Great Dover-street, Southwark, London, Mechanical Elecrical Engineers.

S. S. Fewings and O. P. Traherne, under the style of S. S. Fewings an Seaton, Mineral Water Manufacturers.

The Patent List.

This list is compiled from Official sources in the Manchester Tech Laboratory, under the immediate supervision of George E. Davi Alfred R. Davis.

APPLICATIONS FOR LETTERS PATENT.

Fire-Lighting and Fuel. W. Gordon and G. A. Newton. 6,900. May
Non-Conducting Composition. F. W. Moore. 6,905 May 5.
Brick Making Machinery. T. C. Fawcett. 6,914. May 5.
Fuel-Compressing Machinery. H. G. Holborow. 6,919. May 5.
Electric Batteries. H. T. Eagar and R. P. Milburn. 6,924. May 5.
Air Compressors. S. E. Martyn. 6,928. May 5.
Manufacture of Colouring Matters. C. D. Abel. 6,931. May 5.
do. do. do. 6,932. May 5.
Extraction of Gold from Ores. G. Bamberg. 6,936. May 5.
Electrical Switch. J. Appleton. 6,945. May 5.
Basic Blue Colouring Matters. S. Pitt. 6,946. May 5.
Lining of Pulp Boilers. C. Kellner. 6,951. May 6.
Coating Metals with Cement. C. Kellner. 6,952. May 6.
Preventing Incrustation in Steam Boilers.—(Complete Specification Draper. A. Holmgren, H. R. Mount, and J. Barnes. 6,955. May 6.
Carbonising. J. Horne. 6,966. May 6.
Manufacture of Paper Pulp.—(Complete Specification). C. Kellner. May 6.
Disintegrating Apparatus. P. U. Askham and W. Wilson. 7,009. May
Manufacture of Indigo. W. Cole. 7,018. May 6.
Manufacture of Soap. C. A. Serra. 7,019. May 6.
Making Fibrous Material Impervious to Water.—(Complete Specific A. J. Boult. 7,024. May 6.
Carburetors.—(Complete Specification). J. J. Cooper. 7,040. May 6.
Treatment of Ores Containing Copper, Nickel, and Cobalt. A. Si 7,055. May 6.
Liquefying and Storing Chlorine. J. Y. John. 7,058. May 6.
Corburetted Water-Gas. H. Fourness. 7,063. May 7.
Filtering Apparatus. B. Hunt and W. McD. Mackey. 7,075. May 7.
Fermentation of Amylaceous Matter. A. J. Boult. 7,098. May 7.
Brick Kilns. B. F. McTear. 7,099. May 7.
Manufacture of Soap. H. Scott. 7,117. May 7.
Electric Meters. O. Romanze, W. Weise, and F. W. White. 7,125. May
Reduction of Iron from its Ores. J. D. Danton. 7,129. May 7.
Water Softening Apparatus. W. Lawrence. 7,138. May 7.
Employment of Silicic Acid for Decolourising Oils. H. Stern. 7,142.
Smokeless Fire Lighters. T. Williams. 7,144. May 7.
Glass Bottle Making Machinery. D. Rylands. 7,145. May 8.
Smoke Prevention. T. G. Hardie. 7,162. May 8.
Manufacture of Paper. E. Partington. 7,163. May 8.
Drying Sludge. F. N. Mackay. 7,165. May 8.
Feed-Water Heaters. D. B. Moirson. 7,167. May 8.
Manufacture of Gas Retorts and other Clay Ware. C. H. Edwards. May 8.
Welding Metals by Electricity.—(Complete Specification). W. P. Thomp 7,185. May 8.
Welding Metals by Electricity.—(Complete Specification). W. P. Thon 7,186. May 8.
Bleaching Textile Fibres. W. H. Spencer. 7,198. May 8.
Dyeing Wool. P. Cavailles. 7,202. May 8.
Electric Meter. J. Perry. 7,219. May 9.
Enamelling Metals. J. King. 7,220. May 9.
Oil Filter.—(Complete Specification). C. A. Kœllner. 7,229. May 9.
Rust Prevention Composition. L.E.A.F.D. de Liebhaler. 7,238. May
Manufacture of Ammonia. H. Baudouin and P. F. Escarpit. 7,247. M
Converting Iron into Steel. W Hodge. 7,253. May 9.
Obtaining Pigments from Salts of Iron. M. N. d'Andria. 7,260. Ma
Manufacture of Gas. G. M. Cruikshank. 7,277. May 10.
Manufacture of Explosives. C O. Lundbolm and G. H. Horsie. May 10.
Drainage and Sewerage. J. P. Bayly. 7,318. May 10.
Pots for Melting Glass. F. O. Thompson. 7,327. May 10.

IMPORTS OF CHEMICAL PRODUCTS

AT

THE PRINCIPAL PORTS OF THE UNITED KINGDOM.

NDON.

ding May 13th.

Ohlenschlager Bros.
—
ks. Beresford & Co.
pks A. & M. Zimmermann
Phillips & Graves
Beresford & Co.
Lime—
gs. Lister & Briggs

M. D. Co.
re—
c. Pillow, Jones & Co.
L. & I. D. Jt. Co.
H. R. Merton & Co.
H. Bath & Son
t H. Emanuel
French & Smith
Pillow, Jones & Co.
H. Emanuel
Quirk, Barton & Co.

c. Olyett & Sons

4 t. M. D. Co.
o G. Soanes & Co.
o Clift, Nicholson & Co.
10 Goad, Rigg & Co.
3 "
—
o t. A. Hunter & Co.
52 Gonne, Croft & Co.
50 C. C. Bryden & Co.
50 Anglo-Con. G. Wks.
ld—
J. Puddy & Co.
—
o t. W. C. Bacon & Co
00 J. Besant
00 Seager, White & Co.

45 cks. A. Boltz
12 Blundell, Spence & Co. Ltd.
9 W. Harrison & Co.
22 "
42 "
34 Magee, Taylor & Co.
31 50 bgs. Leach & Co.
19 cks. P. Jantzen
47 D. Storer & Sons, Ltd.
58 O'Hara & Hoar
32 A. Lumbeck & Co.
83 W. Harrison & Co.
47 "
oride—
75 Petri Bros.

185 T. Morson & Sons
—
2½ t. H. Bath & Son
87 "
artar—
5 pks. M. Ashby, Ltd.
5 B. & F. Wf Co.
10 cks. "
4 pks. T. Ronaldson & Co.
2 Burgoyne, Burbidge & Co.
10 W. C. Bacon & Co.
—
10 c. L. & I. D. Jt. Co.
3 "
9 Anderson, Weber & Smith
5 Clarke & Smith
27 "
25 L. & I. D. Jt. Co.
3 Matheson & Co.
95 Kleinwort, Sons & Co.
12 Wallace Bros.
5 L. & I. D. Jt. Co.
22 "
20 Brown & Elmslie
ash—
00 c. J. Knight & Sons

Chemicals *(otherwise undescribed)*
Holland, £54 Kaltenbach & Schmitz
Germany, 59 T. H. Lee
" 305 A. & M. Zimmermann
" 68 Spies Bros & Co.
France, 550 W. C. Bacon & Co
Germany, 12 Elkan & Co.
" 500 A. & M. Zimmermann
" 140 T. H. Lee
" 278 Craven & Co.
Holland, 45 G. Rahn & Co.
Germany, 5 L. & I. D. Jt. Co.
" 112 W. Burton & Sons
" 65 Spies Bros. & Co.

Dyestuffs—
Argols
Italy, 720 pks. B. Jacob & Sons
Orchella
E. Indies, 15 E. Barber & Co.
Sumac
Italy, 400 W. France & Co
" 500 W. Shelcott
Tanners' Bark
Holland, 11 t. G. S. N. Co.
" 20 T. H. Lee
Adelaide, 77 Devitt & Hett
" 77 Baxter & Hoare
Belgium, 10 Leach & Co.
Logwood
Jamaica, 5 E. D. & F. Man
Holland, 13 Chilworth Gunpowder Co
Valonia
A. Turkey, 170 t. Hicks, Nash & Co.
" 51 A. & W. Nesbitt
Indigo
E. Indies, 26 chts. L. & I. D. Jt. Co.
" 23 Benecke, Souchay & Co.
" 21 G. Croshaw & Co.
" 10 Langstaff & Co.
U. S., 13 L. & I. D. Jt. Co.
E. Indies, 29 L. & I. D. Jt. Co.
" 7 W. Isted
" 4 G. Croshaw & Co.
" 35 Elkan & Co.
" 30 pks. Ross & Deering
" 61 chts. L. & I. D. Jt. Co.
" 72 W. Brandt Sons & Co.
Myrabolams
E. Indies, 350 pks. Tod, Durant & Co.
" 400 Marshall & French
" 550 Beresford & Co.
" 6,913 Arbuthnot, Latham & Co.
" 1,021 Hoare, Wilson & Co.
" 761 A. Everingham & Co.
" 1,971 Benecke, Souchay & Co.
Saffron
Spain, 1 pks. T. Allen
" 2 R Quincey & Son
Gambier
Singapore, 637 pks. L. & I. D. Jt. Co.
" 24 Brinkmann & Co.
E. Indies, 825 bls. L. & I. D. Jt. Co
" 186 bgs. Elkan & Co.
Singapore, 29 pks. L. & I. D. Jt. Co.
Cutch
E. Indies, 400 pks. Wrightson & Son
Holland, 50 Brinkmann & Co.
Rangoon, 1,212 Brown and Elmslie
Extracts
U. S., 10 pks. A. F White & Co.
France, 4 R J. Fullwood & Bland
Germany, 1 Domeier & Co.
U. S., 200 T. Ronaldson & Co.
Italy, 4½ cs. L. & N. W. Ry. Co.
Madders
Holland, 12 pks. W. Balchin
Dyestuffs *(otherwise undescribed)*
E. Indies, 551 pks. L. & I. D. Jt. Co.
" 220 Props. Butler's Wf.
Holland, 76 Burt, Boulton & Co
E. Indies, 300 L. & I. D. Jt. Co

Guano—
France, 2,112 t. Anglo Contl. G. Wks.

Glycerine—
France, £65 Major & Field
Holland, 480 Spies, Bros. & Co.
" 135 F. W. Heilgers & Co.
Germany, 130 Beresford & Co.

Glucose—
U. S., 25 pks. 200 cwts. Union Lighterage Co., Ltd.
" 118 650 cwts. C. Southwell & Co.

Glucose—*continued*
Germany, 25 pkgs. 200 cwts. L. Sutro & Co.
" 100 100 M. D Co.
" 100 100 J. Barber & Co.
" 330 1 000 Baxter & Hoare
" 75 600 J Knill & Co.
" 10 75 J. Cooper
" 30 40 Botolph & Nichs. Whvs. Co.
" 50 400 L. Sutro & Co.
" 150 1,260 J. H Epstein
" 159 604 Props. Chamberlain's Whf
U. S., 50 265 Pillow, Jones & Co
" 175 875 T. M. Duche & Sons
" 240 240 "
" 240 240 Howell & Co.
" 400 400 Barrett, Tagant & Co.
" 25 135 M. D. Co.
Germany, 130 1,015 F. J. Warren
" 25 197 Pillow, Jones & Co
France, 100 100 J Barber & Co.
" 340 340 Trier, Meyer & Co
U. S., 30 165 Pillow, Jones & Co
" 40 219 Page, Son & East
Germany, 10 90 J. Knill & Co.
U. S., 100 600 Union Lighterage Co., Ltd.

Gutta Percha—
Singapore, 605 c. Soundy & Son
" 345 J. H. & G. Scovell
U. States, 68 Kleinwort, Sons & Co

Isinglass—
E. Indies, 12 pks. Hale & Son
" 9 L. & I. D. Jt. Co.
Portugal, 3 "
E. Indies, 15 Clarke & Smith

Manure—
Holland, 50 t Anglo-Contl. G Wks.
Belgium, 100 T. Farmer & Co.
U. States, 2 C. A. H. Nielio & Co.
Germany, 72 t. Union Lighterage Co., Ld.
Sydney, 3 t. Goad, Rigg & Co.
Holland, 1 G. Rahn & Co.
" 280 Anglo-Contl. G. Wks
Phosphate Rock
Belgium, 120 t. Miller & Johnson
" 130 London Manure Co., Ltd.
U States, 2,100 Wyllie & Gordon
Holland, 45 G. Rahn & Co.
Gypsum
France, 330 t. J. Burnett & Sons

Nitre—
Holland, 7 cks. J. Owen

Naphtha—
Holland, 10 drs. Domeier & Co
" 35 Burt, Boulton & Co
U. States, 953 brls. C. F. Bowring & Co.
Holland, 40 drs. Burt, Boulton & Co.
Belgium, 7 cks. Stein Bros.
Holland, 28 drs. Domeier & Co.

Potassium Sulphate—
Germany, £10 H. Lambert

Potash—
Holland, £3 Spies Bros. & Co.

Potassium Carbonate—
France, 10 cks. J. A. Reid
" 20 Charles & Fox
" 10 Beresford & Co

Pearl Ash—
France, 110 c. F. W. Berk & Co.

Plumbago—
Germany, 40 pks. S. H. Lee
" 20 cks. Le Sueur & Co.
E. Indies, 471 L. & I. D. Jt. Co
Holland, 19 Brown & Elmslie
E Indies, 91 brls. Price, Hickman & Co.

Pyrites—
Spain, 1,914 t. C. Tennant, Sons & Co.

Paraffin Wax—
U. States, 100 brls. G. H Frank
" 238 H. Hill & Sons
" 75 Mordaunt Bros.
" 603 J. H. Usmar & Co.
" 110 G W. Frank
" 414 pks. H. Hill & Sons

Potassium Bicarbonate—
Holland, £55 A. & M. Zimmermann

Sodium Acetate—
France, £57 Barr, Moering & Co.

Sodium Sulphate—
Holland, £45 J. Owen
" 25 H. Lorenz

Salammoniac—
Holland, £133 Barr, Moering & Co.

Saltpetre—
Germany, 11 cks. J. A. Reid
" 35 P. Hecker & Co.
E. Indies, 333 bgs. C. Wimble & Co.
" 321 Clift, Nicholson & Co.

Stearine—
Belgium, 130 cks. H. Hill & Sons
Germany, 15 cs. T. H. Lee
France, 51 bgs. H. Hill & Sons

Tartaric Acid—
Holland, 22 pks. B. & F. Wf. Co.
France, 8 "
Holland, 20 "
" 2 Middleton & Co.
" 2 B. & F. Wf. Co.
" 13 C. Christopherson & Co.

Tartar—
Italy, £925 Thames S. Tug Co.

Ultramarine—
Germany, 2 cs. G. Steinhoff
Belgium, 9 cks. Leach & Co.
" 4 W. Harrison & Co.
Holland, 5 Ohlenschlager

Verdigris—
France, 2 cks. N. Steinberg

Zinc Sulphate—
Germany, £60 Howards & Son

Zinc Oxide—
Germany, 5 cks. Beresford & Co.
" 50 brls. M. Ashby, Ld.
Holland, 75 "

LIVERPOOL.

Week ending May 15th.

Ammonia—
Bordeaux, 1 cs.

Acetate of Lime—
St. Nazaire, 84 cks.

Albumen—
Bruges, 20 cs. J. T. Fletcher & Co.

Barytes—
Antwerp, 45 bgs. 12 cks.

Cream of Tartar—
Bordeaux, 4 cks.
" 1

Caoutchouc—
Lagos, 1 cs. Pitt Bros.
Quittah, 1 ck. J. H. Rayner & Co.
" 2 brls. W. J. Rodatz & Co.
Accra, 3 Hutton & Co.
" 1 Davis, Robbins & Co.
" 2 Edwards Bros.
" 5 M. Herschell & Co.
" 8 Pickering & Berthoud
" 12 J. J. Fischer & Co.
Cape Coast, 1 pn. 1 bg. Edwards Bro.
" 3 brls. A. M. Waite & Co.
" 5 1 cs. Fletcher & Fraser
" 1 Symington & Co.
" 1 pn. S. & C. Nordlinger
" 1 ck. W. B. McIver
" 1 brl. C. L. Clare & Co.
" 1 Havard & Co.
" 1 S. Henrichsen
" 2 Mordaunt Bros.
" 5 20 pks. A. Miller, Bro. & Co.
" 3 cs. 2 cks. 4 pns. F. & A. Swanzy
" 3 brls.
Axim, 1 bg. Pickering & Berthoud
" 1 brl. Edwards Bro.
Grand Bassam, 6 bgs. 1 cks. "

Caoutchouc—*continued.*
Grand Bassam, 8 bgs. 5 cks. W. D. Woodin
" 2 1 E. Wayland
" 5 F.&A Swanzy
" 3 A. Miller, Son & Co
" 1 Hobson & Co.
" 1
" 1 1 ck. S. & C. Nordlinger
" 2 cs Grimwade Ridley & Co
Sierra Leone, 1 brl T. Christy & Co.
" 49 Pickering & Berthoud
" 2 pns. 19 brls. Paterson Zachonis & Co
" 20 Broadhurst, Sons & Co
" 9 hds. 39 brls.
Isles de Los, 415 F. Colin & Co.
" 25 pns. 2 Cie Francaise
Chemicals *(otherwise undescribed)*
Rotterdam, 4 cks.
Dyestuffs—
Fustic
Savanilla, 2,832 lgs. G. Muller & Co.
Carthagena, 451 ps. Mildred Goyeneche
Logwood
Boston, 200 bxs.
Belize 60¼ t.
Divi Divi
Carthagena, 8 bgs. Mildred Goyeneche
Havre, 202 bgs Cunard S.S. Co, Limited
Extracts
Havre, 64 cks. Cunard S.S. Co., Limited
Bordeaux, 70
New York, 5 brls.
Philadelphia, 50
Argols
Oporto, 23 cks F. Leyland & Co.
" 23 Ger. Bank, London
Brindisi, 568 bgs.
Myrabolams
Bombay, 410 bgs. Forbes, Forbes & Co.
Glucose—
New York, 390 brls
Dried Blood—
B Ayres 967 bgs.
Iodine —
Valparaiso, 162 brls A. Gibbs & Sons
" 22 W. & J. Lockett
Manure—
Bruges, 22 bgs.
Magnesium Sulphate—
Bordeaux, 1 ck.
Pitch—
Rotterdam, 65 cks.
Pyrites—
Huelva, 1,617 t. Matheson & Co.
" 971 Tennants & Co.
" 1,420 Matheson & Co.
Potassium Muriate—
Hamburg, 2,954 bgs.
Phosphate—
St. Valery, 150 t.
Antwerp, 600 bgs.
Sodium Silicate—
Rotterdam, 1 ck.
Stearine—
Antwerp, 51 bgs.
Saltpetre—
Antwerp, 135 bgs. J. T. Fletcher & Co.
Sulphur Ore—
Huelva, 1,116 t. Matheson & Co.
Tartar—
Havre, 7 Cunard S.S. Co, Ltd.
Bari, 2 cks.
Tartaric Acid—
Rotterdam, 24 cks.
Rotterdam, 5
Turpentine—
Wilmington, 400 brls.
Ultramarine—
Rotterdam, 5 cs. 5 brls.
Rotterdam, 12 cks.
Verdigris—
Bordeaux, 2 cks.
Zinc Oxide—
Antwerp, 50 cks.

TYNE.

Week ending May 15th.

Barytes—
Rotterdam, 22 cks. Tyne S. S. Co.
Copper Precipitate—
Huelva, 5,832 bgs. Scott Bros
Copper Ore—
Huelva, 168 t Rio Tinto Co.
Christiania, 1 bx.
Huelva, 1,056 Scott Bros.
Ferro Chrome—
Antwerp, 20 cs. Tyne S. S. Co.
Glucose—
Hamburg, 100 bgs Tyne S. S. Co.
" 3 cks. "
Saltpetre—
Hamburg, 2 cks. Tyne S. S. Co
Saltcake—
Hamburg, 220 t.
Waste Salt—
Hamburg, 140 t.

HULL.

Week ending May 15th.

Acids *(otherwise undescribed)*
Rotterdam, 12 cks. Hutchinson & Son
Barytes—
Rotterdam, 22 cks.
Antwerp, 12
Bremen, 19 Blundell, Spence & Co.
Chemicals *(otherwise undescribed)*
Genoa, 17 cks. Wilson, Sons & Co.
Dunkirk, 139 pks. "
Antwerp, 77 "
Rotterdam, 3 cs. G. Lawson & Sons
Stettin, 29 cks. Wilson, Sons & Co.
Bremen, 2 Veltmann & Co.
Dyestuffs—
Rennet
Rotterdam, 4 cks. W & C. L. Ringrose
Copenhagen, 3 cks. Wilson, Sons & Co.
Extracts
Rouen, 54 cks Rawson & Robinson
Alizarine
Rotterdam, 1 brl. G. Lawson & Sons
" 38 pks. W. & C. L. Ringrose
" 53 cks. Hutchinson & Son
" 3 W. & C. L. Ringrose
" 14 "
Madders
Rotterdam, 2 cks.
Glucose—
Dunkirk, 60 bgs. Wilson, Sons & Co.
Hamburg. 300 "
" 37 cks. Woodhouse & Co
" 10 Rawson & Robinson
New York, 120 brls.
Hamburg, 9 cks. G. Lawson & Sons
" 100 bgs. Thompson, McKay and Co.
Stettin, 250 12 cks Wilson, Sons and Co.
Glycerine—
Rouen, 1 ck. Rawson & Robinson
Manure—
Ghent, 190 t. Wilson, Sons & Co.
Oxide—
Hamburg, 6 cs. G. Malcolm & Son
Phosphate—
Ghent, 100 t. 1,713 bgs Wilson, Sons and Co.
Stearine—
Rotterdam, 50 cks. G. Lawson & Sons
Tartar—
Bordeaux, 1 ck.
Tartaric Acid—
Rotterdam, 4 cks
" 4
Ultramarine—
Bremen, 1 ck. Furley & Co.

GRIMSBY.

Week ending May 10th.

Glucose—
Hamburg, 55 cks.
Zinc Ashes—
Antwerp, 10 cks.

GLASGOW.

Week ending May 15th.

Aluminium—
Antwerp, 3 cs.
Barytes Sulphate—
Antwerp, 200 bgs.
Barytes—
Rotterdam, 116 cks.
Dyestuffs—
Sumac
Palermo, 1,250 bgs. 50 bls.
Extracts
Bordeaux, 100 cks. J. & P. Hutchinson
Rouen, 2
Antwerp, 9 1 cs.
Dunkirk, 5
Madder
Bordeaux, 13 cks.
Gutta Percha—
Rotterdam, 20 bkts. J. Rankine
Glycerine—
Rouen, 10 cks.
Rotterdam, 10 cs. J. Rankine
Glucose—
New York, 80 brls.
Nitrate—
Iquique, 9,110 bgs. Thomson, Ail
Oxalic Acid—
Rotterdam, 5 cks.
Pyrites—
Huelva, 1,800 t. Crossley, Tilburn
Sulphur—
Palermo, 260 bgs.
Saltpetre—
Rotterdam, 7 cks. J Rankine
Tartar—
Bordeaux, 6 cks. J. & P. Hutcl
" 11
Tartaric Acid—
Rotterdam, 4 cks. J. Rankine
Zinc Oxide—
Antwerp, 10 cks.
" 125

GOOLE.

Week ending May 7th.

Antimony—
Boulonge, 4 cks.
Caoutchouc—
Boulogne, 11 cks.
Dyestuffs—
Alizarine
Rotterdam, 22 cks. 71 pks.
Logwood
Kingston, 500 t.
Jamaica, 373
Dyestuffs *(otherwise undescr*
Boulogne, 95 cks. 61 cs.
Glucose—
Hamburg, 140 bgs. 18 cks.
Dunkirk, 120
Plumbago—
Boulogne, 180 cks.
Picric Acid—
Antwerp, 11 bgs.
Pyrites—
Huelva, 1350 t.
Saltpetre—
Antwerp, 100 bgs.
Rotterdam, 240 bgs.
Soda—
Rotterdam, 6 cks.
Tartaric Acid—
Rotterdam, 2 cks.
Ultramarine—
Rotterdam, 5 cs.
Hamburg, 30
Waste Salt—
Hamburg, 708 bgs. 665 t.
" 711 457

Tar
Sydney, 24 pkgs. £38
Natal, 850 drms. 34
Cape Town, 25 17
[illegible], 200 rlts. 126
[illegible], 328 209
Abo, 600 287
[illegible], 200 brls. 120
Danzig, 250 190
Abo, 500 [illegible]
Gothenburg, 100 68
[illegible], 336 154
[illegible]
[illegible], 19 cks. £430
[illegible], 30 drms. [illegible]
Rotterdam, 53 981
Hamburg, 20 pns. 350
Stettin, 1,250 rlts. 672
Rotterdam, 30 50 1,780
" 40 659
Hamburg, 1 drm. 2
[illegible], 50 737
Anthracene
[illegible], 250 cks. 950
Dunkirk, 67 765
Naphthaline
Le Treport, 5 cks. 6
Montreal, 2 2 cs. 7
Stockholm, 2 17
Hamburg, 7 55
Stockholm, 58 7 151
Hamburg, 20 57
Alizarine
Rouen, 20 12 kgs. 384
Naphtha
Brussels, 2 cks. £10
Brisbane, 20 drs. 24
Benzole
Rotterdam, 19 pns. £237
Creosote
Cologne, 994 cks. 696 bgs. £307
Cherbourg, 605 brls. 130
Bayonne, 1,026 575

Copperas—
Rotterdam, 100 kgs. £20

Chemicals (*otherwise undescribed*)
Cape Town, 330 cks. £290
Calcutta, 20 cs. 252
Adelaide, 7 pks. 119
Toronto, 5 150
Madrid, 15 95
Sydney, 50 179

Disinfectants—
Brisbane, 5 cs. £41
" 13 pks. 24
Hamburg, 15 cks. 90
Le Treport, 12 15
Calcutta, 2 t. 12

Guano—
Landerneau, 191 t. £1,900
Berbice, 5 13 c. 60
Nevis, 25 250
Barbados, 5 3 24

Hydro Carbon—
Brussels, 2 t. 2 c. £38

Manure—
Christiania, 5 t. 8 c. £35
Berbice, 25 256
Malaga, 32 12 320
" 25 3 211
Havre, 30 120
Dominica, 29 12 280
Cherbourg, 25 275
Nevis, 15 190
Demerara, 50 500
Treport, 300 bgs. 210
Lyttletown, 100 1,350
Malaga, 50 400
Valencia, 170 11 c. 1,364
" 652 4 6,155
Trinidad, 110 1,040
Barbadoes, 200 2,275
" 100 1,000
Nevis, 2 20
Baltimore, 20 63
Rouen, 34 10 180
Bordeaux, 35 200
Hobart, 52 708
Launceston, 72 1,013

Muriatic Acid—
Natal, 2 t. 5 c. £21

Nitric Acid—
Natal, 10 c. £15
Jersey, 4,280 lbs. 53

Oxalic Acid—
Rotterdam, 15 c. £00

Potassium Chlorate—
Hong Kong, 2 t. 5 c. £99

[illegible]phoric Acid—
[illegible]ane, 15 t. £250
[illegible]rara, 2 32

[illegible] Citrate—
[illegible] 7 c. £45

Sodium Sulphate—
Norkoping, 200 t. £090

Sheep Dip—
Algoa Bay, 7 t. 4 c. £290
Natal, 3 8 130
Cape Town, 2 35
Karachi, 200 drs. 24

Saltcake—
Antwerp, 115 t. £200

Sulphur—
Lisbon, 8 t. 2 c. £164

Sulphuric Acid—
Natal, 5 t. 5 c. 25
Algoa Bay, 20 cs. 13

Sulphur Flowers—
Cape Town, 35 cks. 40 kgs. £38

Salammoniac—
Cape Town, 1 t. 1 c. £36
Libau, 2 cks. 20
Galatz, 13 23
Constantinople, 1 10 50
Galatz, 6 27
Hamburg, 1 2 61

Sodium Silicate—
Vancouver, 5 t. 17 c. £22
Auckland, 9 2 67

Sodium Bicarbonate
Auckland, 2 t. £14

Superphosphate—
Cherbourg, 160 t. £480

Saltpetre—
Bordeaux, 1 t. 10 c. £40
Newcastle, 2 20
Syra, 6 5 130
Oporto, 9 8 190
Melbourne, 2 10 25
Oporto, 11 5 250
Rio Janeiro, 1 18 43

Soda Crystals—
Hong Kong, 15 t. £44

Tartaric Acid—
Brisbane, 2 c. £13
" 4 28
Dunedin, 5 44
Canterbury, 5 40
Adelaide, 1 t. 153
Auckland, 5 45
Melbourne, 10 78
Sydney, 40 cs. 200

LIVERPOOL.

Week ending May 14th.

Ammonium Carbonate—
Bilbao, 6 c. £10
San Francisco, 1 t. 6 1 q. 41
Leghorn, 15 1 27
Montreal, 1 28
Genoa, 11 3 16
" 6 8

Acetate of Lime—
Hamburg, 10 t. £80

Ammonium Sulphate—
Barbadoes, 30 t. 9 c. 3 q. £371
Valencia, 50 7 1 600

Antimony Regulus—
Alicante, 10 c. £35

Ammonium Muriate—
Havre, 1 t. 2 c. £20
New York, 8 199

Alum—
Galatz, 9 t. 1 c. 2 q. £47
Madras, 8 2 38
Melbourne, 1 5
Madras, 4 9 29
Malta, 1 1 5

Antimony Sulphide—
Barcelona, 2 c. £13

Acetic Acid—
Rio Grande du Sol, 1 c. 3 q. £5

Bleaching Powder—
Bari, 40 kgs.
Bordeaux, 31 cks.
Genoa, 27
Montreal, 60
New York, 84 tcs.
Pasages, 7
Baltimore, 145
Barcelona, 15
Batoum, 20
Boston, 410
Calcutta, 150 cs.
Genoa, 381
Hong Kong, 30
Leghorn, 10
Montreal, 161
New York, 258
Oporto, 27
Philadelphia, 117
San Sebastian, 29

Borax—
Oporto, 18 c. 2 q. £[illegible]
Antwerp, [illegible] [illegible] [illegible]
Melbourne, [illegible] [illegible]
London (Ont.), 8 3 [illegible]

Boracic Acid—
Havana, 3 c. £[illegible]
Zante, 2 1 q. 71

Borate of Lime—
Rotterdam, 10 t. 3 c. 2 q. £105

Carbolic Acid—
Hamburg, 6 t. 2 c. 1 q. £443
Calcutta, 1 2 2 [illegible]
Madras, 1 18 3 [illegible]
Calcutta, 11 2 12
Genoa, 1 3 15

Charcoal—
Hamburg, 50 t. £[illegible]
" 151 7 c. 2 q. [illegible]

Copperas—
Callao, 3 t. 10 c. £8

Caustic Soda—
Alicante, 10 drms.
Amsterdam, 24
Antwerp, 200
Barcelona, 120
Bari, 69
Bilbao, 16
Bombay, 18
Bordeaux, 200
Calcutta, 160
Carthagena, 15
Ceara, 60
Demerara, 2
Fiume, 5
Galatz, 44
Genoa, 60
Halifax, 5
Hamburg, 26
Havana, 90
Hong Kong, 16
Malaga, 74 18 brls. 30 kgs.
Montreal, 105
Newfairwater, 108
New Orleans, 100
New York, 200
Pasages, 100
Rotterdam, 2 brls.
St. Petersburg, 50 drms.
Seville, 89
Sydney, 30
Trieste, 5
Venice, 15
Philadelphia, 50
Genoa, 180
Hamburg, 225
Leghorn, 32
Madrid, 100
Matanzas, 12
" 3 10 kgs.
Monte Video, 100
Montreal, 37
Nantes, 67
Naples, 40
New Orleans, 100
New York, 500
Odessa, 25
Alicante, 20
Amsterdam, 40
Antwerp, 55
Bahia, 25 30 kgs.
Barcelona, 80
Bari, 401
Bordeaux, 10
Boston, 750
Bruges, 29
B. Ayres, 10
Calcutta, 170
Carthagena, 24
Corfu, 40
Parahiba, 5
Rio Janeiro, 15
Rotterdam, 32
Rouen, 58
St. Petersburg, 1,028
Santos, 35
Seville, 90 30 brls.
Sydney, 23
Valencia, 45
Venice, 89

Calcium Chloride—
Melbourne, 5 t. £19

Copper Sulphate—
Naples, 7 t. 6 c. 2 q. £148
Pasages, 10 1 [illegible]
Alicante, 10 [illegible]
Ancona, 49 1 [illegible]
Bari, 5 1 1 [illegible]
Bilbao, 30 1 3 [illegible]
Bordeaux, 2 13 [illegible]
Brindisi, 5 1 2 102
Genoa, 20 9 1 [illegible]
" 43 [illegible] 0 [illegible]
Trieste, 7 [illegible] [illegible]
Venice, 50 26 1 1,0[illegible]
Havre, 4 17 3 [illegible]

Salt—*continued.*
Blackpoint, 15
Boston, 150
Brisbane, 1
Bruges, 150
B. Ayres, 4
Cape Coast. 6¾
Copenhagen, 753
Demerara, 17 17 c.
E. London, 17 18
Eloby, 7½
Halmstat, 250
Hamburg, 50
Montreal, 295
New York, 80
Portland, 550 10
Port Natal, 32
Sydney, 450
Trinidad, 8 1

Sodium Bicarbonate—
Bremen, 4 kgs.
Calcutta, 50
Genoa, 50
Hamburg, 70 15 brls.
Nantes, 26 10 cks.
Newcastle. 40
Port Said,
Rosario, 100
St. John, 250
Sydney, 350
Valencia, 125
Valparaiso, 11
Boston, 100 cks.
Copenhagen, 17 12 kgs.
Genoa, 27
Konigsberg, 28
Naples, 80
Odessa, 185
Stettin, 15
Sydney, 54 20
Victoria, 20

Soda Ash—
Baltimore, 84 tcs. 122 cks.
Barcelona, 6 35 brls.
Bombay, 25
Boston, 210
Calcutta, 6
Ceara, 50
Ghent, 42
New York, 288 382
Parahiba, 25
Philadelphia, 460
Piræus, 50
Rio Janeiro, 245
San Francisco, 209
Santos, 35 brls.
Sydney, 14 cks.
Baltimore, 166 tcs.
Bombay, 3 cks.
Boston, 1,040 bgs
B. Ayres, 200 brls.
Callao, 30
Havana, 2 cks.
Hong Kong, 4 brls.
New York, 463 tcs.
Philadelphia, 311
Rio Janeiro, 75 brls.
Hamburg, 5

Sodium Silicate—
Galatz, 10 brls.
Seville, 17

Soda Crystals—
Alexandria, 20 brls.
B. Ayres, 34 kgs.
Calcutta, 100
Gibraltar, 10
Malta, 50
Boston, 280

Salammoniac—
Naples, 15 c. £26
Ancona, 10 17
Genoa, 1 t. 7 2 q. 48
Alexandretta, 2 70
Smyrna, 1 35
Leghorn, 14 1 26
Malta, 4 7
Mersyne, 3 5
New York, 5 1 170
San Francisco, 10 3 19
Salaera, 3 6
Genoa, 1 14 1 59

Saltpetre—
Montreal, 10 c. £11
Piræus, 3 t. 15 82
Pernambuco, 2 10 64
Ceara, 9 2 q. 11
Syra, 3 2 2 69
Vera Cruz, 1 15 2 44

Sulphur—
Boston, 50 t. £220
" 50 217
Lisbon, 19 13 c. 1 q. 158
Vera Cruz, 17 3 6

Superphosphate—
Gr. Canary, 11 t. 3 c. 1 q. £30
Hamburg, 20 65
Las Palmas, 10 9 1 102

Sodium Bichromate—
New York, 8 t. 10 c. £230

Sodium Biborate—
Antwerp, 6 t. 1 c. 3 q. £183
Genoa, 1 3 2 45
Hamburg, 1 1 2 32
Montreal, 4 16 1 145

Saltcake—
Dunedin, 10 t. 11 c. 3 q. £26
New York, 50 1 125

Sodium Nitrate—
Valencia, 5 t. 19 c. 2 q. £50
Trieste, 9 17 2 82

Tar—
Grand Canary, 10 brls.

Zinc Chlorate—
Bombay, 17 c. 3 q. £15

GLASGOW.

Week ending May 15th.

Ammonia—
Trinidad, 4 t. 7 c. £46
Demerara, 283 1 3,049

Bleaching Powder—
Sydney, £35

Copperas—
Karachi, £348

Copper Sulphate—
Rouen, 2 t. 16¾ c. £56
Italy, 110 c. 125

Dyestuffs—
Extracts
Quebec, £107

Epsom Salts—
Sydney, £41

Gutta-Percha—
Sydney, £54

Iodine—
Philadelphia, 60 c. £3,840

Manure—
Trinidad, 47 t. £374
Demerara, 25 £243. 15s.
" 30 250
" 12 2 c. 142
Trinidad, 150 630

Paraffin Wax—
Hong Kong, 2 t. 1¼ c £57
Japan, 4,191 lbs. 66
Italy, 32 18 c. 932

Potassium Prussiate—
Quebec, 20 c. £88
Italy, 16¼ £45
New York, 68½ 267
Gothenburg, 19 t. 10 c. 44

Potassium Bichromate—
Boston, 332 c. £620
Quebec, 4 t. 8 c. 155
New York, 165 c. 309
Rouen, 31 t. 13½ 1,115
Gothenburg, 5 7 145
Antwerp, 309¼ c. £435
Bergen, 1 ck. 14

Phosphate—
Demerara, 66 t. £230

Pitch—
Quebec, 23 t. 10 c. £53
Gothenburg, 244 13 330
Granville, 100 160

Superphosphate—
Demerara, 10 t. 6 c. £40
Christiania, 20 100

Sulphuric Acid—
Karachi, £56
Bombay, 562

Sodium Bichromate—
Rouen, 11 t. 11¼ c. £247
Rotterdam, 62 cks. 530
Italy, 238
" 79¼ c. 110

Tar—
Calcutta, £63
Karachi, 58

HULL.

Week ending May 13th.

Alkali—
New York, 239 cks.
Trieste, 8

Borax—
Helsingfors, 1 ck.

Bleaching Powder—
Trieste, 36 cks.
Venice, 25

Caustic Soda—
Abo, 30 drums.
Helsingfors, 40
" 111
Konigsberg, 111
Libau, 36
Reval, 20 cks.
Wyburg, 87 drums.

Chemicals (*otherwise undescribed*)
Antwerp, 48 cks. 5 pks.
Abo, 1
Bremen, 50 2 cs.
Bari, 10 50 kgs. 17
Christiania, 100 10 161 slbs.
Copenhagen, 14
Gothenburg, 5 2 313
Helsingfors, 1
Hamburg, 5 236
Helsingfors, 3
Libau, 10 pks.
Messina, 5
New York, 50
Odessa, 207 52
Palermo, 1 ck.
Rouen, 36
Rotterdam, 32 32 cs.
Stettin, 1
St. Petersburg, 2 2 20 pks.

Dyestuffs (*otherwise undescribed*)
Antwerp, 1 drum.
Hamburg, 9 2 cks.
Rotterdam, 3 3 cs. 5 kgs.

Manure—
Antwerp, 576 bgs.
Hamburg, 21 bls.

Pitch—
Antwerp, 485 t.
Venice, 200

Salt—
Hango, 1 cs.

Tar—
Amsterdam, 6 cks.
Abo, 150
Bergen, 16
Copenhagen, 16
Hamburg, 2
Rotterdam, 11

Turpentine—
Stettin, 15 cks.

TYNE.

Week ending May 15th.

Antimony—
Hamburg, 1 t. 10 c.
New York, 30

Alum—
Rotterdam, 10 t. 2 c.

Anthracene—
Rotterdam, 25 brls.

Alkali—
Gothenburg, 21 t. 7 c.
Antwerp, 4 16
Genoa, 5 10
Stettin, 5 11
Amsterdam, 15 18
Rotterdam, 11

Ammonium Sulphate—
Valencia, 121 t. 3 c.
Stettin, 10

Bleaching Powder—
Pasages, 91 t. 3 c.
Gothenburg, 7
St. Petersburg, 122 t. 13
Hamburg, 43 t. 1
Antwerp, 50 4
Christiania, 20 1
Dunkirk, 25 1
Stettin, 78 7
Riga, 58 2
Amsterdam, 15 8
New York, 50 8
Rotterdam, 31 19
San Francisco, 65 t. 5

Barytes Carbonate—
Hamburg, 10 t.
Riga, 157 11 c.

Calcined Magnesia—
Genoa, 7 c.

Caustic Soda—
Genoa, 3 c.
Stettin, 10 t. 3
Riga, 16 5
New York, 229 2
Wasa, 14 10
Valencia, 14 10
Gothenburg, 2 10

Copperas—
Pasages, 14 c.
Copenhagen, 2 t. 8

Gypsum—
New York, 72 t. 10 c.

Litharge—
Gothenburg, 1 t. 6 c.
" 1 2
Hamburg, 2 8
Amsterdam, 5 c.
Oporto, 3 t. 19
New York, 2 10
Rotterdam, 11
Stettin, 3 6
Riga, 4 4
Copenhagen, 11

Magnesia—
Valencia, 5 c.
Hamburg, 1 t. 5
New York, 1 2

Oxalic Acid—
New York, 11 t. 4 c.

Potassium Chlorate—
Antwerp, 2 t. 2 c.
Stettin, 4 10
Rotterdam, 1

Pitch—
Elsinore, 46 t. 10 c.

Salt—
Esbjerg, 2 t.

Sodium Sulphate—
St. Petersburg, 12 t. 3 c.
Copenhagen, 15 t. 14

Soda Ash—
Pasages, 12 c.
Gothenburg, 6 t. 14
Gothenburg, 30
Hamburg, 13 10
Stettin, 11 11
New York, 50 8
San Francisco, 100 t. 7
Copenhagen, 41 14

Soda—
Skjelskor, 18 t. 5 c.
Genoa, 5
Esbjerg, 5 19
Aarhuus, 13
Frederickshavn, 1 t. 17
Antwerp, 15 11
New York, 108 7
Rotterdam, 5
Wasa, 3 5

Sodium Hyposulphite—
Rotterdam, 1 t 11 c.

Sodium Chlorate—
Hamburg. 6 t. 12 c.

Sodium—
Hamburg, 2 cs.

Superphosphate—
Valencia, 5 t.

Sulphur—
Gothenburg, 15 t.

Tar—
Hamburg, 110 brls.
Elsinore, 847
Memel, 1 t. 5 c.
Wasa, 1 4

GRIMSBY.

Week ending May 10th.

Coal Products—
Dieppe, 30 cks.

Chemicals (*otherwise undescri*
Dieppe, 65 cks. 65 cs.
Hamburg, 2

Dyestuffs (*otherwise undescril*
Antwerp, 2 cks.

GOOLE.

Week ending May 7th.

Benzole—
Antwerp, 25 cks.
Ghent, 16 drs.
Rotterdam, 217 cks.

Coal Products—
Rotterdam, 118 cks.

Chemicals (*otherwise undescri*
Dunkirk, 30 cks.

Dyestuffs (*otherwise undescri*
Boulogne, 3 cks.
Hamburg, 142

Manure—
Antwerp, 83 bgs.
Boulogne, 24
Bruges, 84
Calais, 130
Rotterdam, 354

Pitch—
Antwerp, 250 t.
Ghent, 18

PRICES CURRENT.

WEDNESDAY, MAY 22, 1890.

PREPARED BY HIGGINBOTTOM AND CO., 116, PORTLAND STREET, MANCHESTER.

The values stated are F.O.R. at maker's works, or at usual ports of shipment in U.K. The price in different localities may vary.

Article	Unit	£	s.	d.
Acids:—				
Acetic, 25 °/o and 40 °/o	per cwt.	7/- & 12/6		
,, Glacial	,,	65/-		
Arsenic, S.G., 2000°	,,	0	12	9
Chromic 82 °/o .. nett	per lb.	0	0	6½
Fluoric	,,	0	0	3¼
Muriatic (Tower Salts), 30° Tw.	per bottle	0	0	8
,, (Cylinder), 30° Tw.	,,	0	2	11
Nitric 80° Tw.	per lb.	0	0	2
Nitrous	,,	0	0	1¼
Oxalic .. nett	,,	3¼d to 4d.		
Picric	,,	0	1	2
Sulphuric (fuming 50 °/o)	per ton	15	10	0
,, (monohydrate)	,,	5	10	0
,, (Pyrites, 168°)	,,	3	2	6
,, (,, 150°)	,,	1	5	0
,, (free from arsenic, 140/145°)	,,	1	7	6
Sulphurous (solution)	,,	2	15	0
Tannic (pure)	per lb.	0	1	0½
Tartaric	,,	0	1	3
Arsenic, white powdered	per ton	13	0	0
Alum (loose lump)	,,	4	17	6
Alumina Sulphate (pure)	,,	5	0	0
,, ,, (medium qualities)	,,	4	10	0
Aluminium	per lb.	0	15	0
Ammonia, ·880=28°	per lb.	0	0	3
,, =24°	,,	0	0	1¼
,, Carbonate .. nett	,,	0	0	3½
,, Muriate	per ton	23	0	0
,, ,, (sal-ammoniac) 1st & 2nd	per cwt.	37/- & 35/-		
,, Nitrate	per ton	40	0	0
,, Phosphate	per lb.	0	0	10½
,, Sulphate (grey), London	per ton	11	6	3
,, ,, (grey), Hull	,,	11	5	0
Aniline Oil (pure)	per lb.	0	0	10½
Aniline Salt ,,	,,	0	0	9¼
Antimony	per ton	75	0	0
,, (tartar emetic)	per lb.	0	1	1
,, (golden sulphide)	,,	0	0	10
Barium Chloride	per ton	8	5	0
,, Carbonate (native)	,,	3	15	0
,, Sulphate (native levigated)	,,	45/- to 75/-		
Bleaching Powder, 35 %	,,	5	0	0
,, Liquor, 7 %	,,	2	10	0
Bisulphide of Carbon	,,	13	0	0
Chromium Acetate (crystal	per lb.	0	0	6
Calcium Chloride	per ton	2	2	6
China Clay (at Runcorn) in bulk	,,	17/6 to 35/-		
Coal Tar Dyes:—				
Alizarine (artificial) 20 %	per lb.	0	0	9
Magenta	,,	0	3	9
Coal Tar Products				
Anthracene, 30 % A, f.o.b. London	per unit per cwt.	0	1	5
Benzol, 90 % nominal	per gallon	0	3	9
,, 50/90 ,,	,,	0	2	9
Carbolic Acid (crystallised 35°)	per lb.	0	0	8
,, (crude 60°)	per gallon	0	2	2
Creosote (ordinary)	,,	0	0	2½
,, (filtered for Lucigen light)	,,	0	0	3
Crude Naphtha 30 % @ 120° C.	,,	0	1	4
Grease Oils, 22° Tw.	per ton	3	0	0
Pitch, f.o.b. Liverpool or Garston	,,	1	8	0
Solvent Naphtha, 90 % at 160°	per gallon	0	1	9
Coke-oven Oils (crude)	,,	0	0	2½
Copper (Chili Bars)	per ton	52	17	6
,, Sulphate	,,	26	0	0
,, Oxide (copper scales)	,,	55	0	0
Glycerine (crude)	,,	30	0	0
,, (distilled S.G. 1250°)	,,	60	0	0
Iodine	nett, per oz.	0	0	9
Iron Sulphate (copperas)	per ton	1	10	0
,, Sulphide (antimony slag)	,,	2	0	0
[Le]ad (sheet) for cash	,,	14	10	0
Litharge Flake (ex ship)	,,	15	10	0
Acetate (white ,,)	,,	24	0	0
.. (brown ,,)	,,	18	0	0
[Carbo]nate (white lead) pure	,,	19	0	0
[illegible]	,,	22	5	0

Article	Unit	£	s.	d.
Lime, Acetate (brown)	,,	7	15	0
,, ,, (grey)	per ton	13	10	0
Magnesium (ribbon and wire)	per oz.	0	2	3
,, Chloride (ex ship)	per ton	2	5	0
,, Carbonate	per cwt.	1	17	6
,, Hydrate	,,	0	10	0
,, Sulphate (Epsom Salts)	per ton	3	0	0
Manganese, Sulphate	,,	17	10	0
,, Borate (1st and 2nd)	per cwt.	60/- & 42/6		
,, Ore, 70 %	per ton	4	10	0
Methylated Spirit, 61° O.P.	per gallon	0	2	2
Naphtha (Wood), Solvent	,,	0	3	10
,, ,, Miscible, 60° O.P.	,,	0	4	6
Oils:—				
Cotton-seed	per ton	22	10	0
Linseed	,,	26	0	0
Lubricating, Scotch, 890°—895°	,,	7	5	0
Petroleum, Russian	per gallon	0	0	5¼
,, American	,,	0	0	5¼
Potassium (metal)	per oz.	0	3	10
,, Bichromate	per lb.	0	0	4
,, Carbonate, 90% (ex ship)	per ton	19	10	0
,, Chlorate	per lb.	0	0	4¼
,, Cyanide, 98%	,,	0	2	0
,, Hydrate (Caustic Potash) 80/85 %	per ton	22	10	0
,, ,, (Caustic Potash) 75/80 %	,,	21	10	0
,, ,, (Caustic Potash) 70/75 %	,,	20	15	0
,, Nitrate (refined)	per cwt.	1	2	6
,, Permanganate	,,	4	2	6
Prussiate Yellow	per lb.	0	0	9½
,, Sulphate, 90 %	per ton	9	10	0
,, Muriate, 80 %	,,	7	15	0
Silver (metal)	per oz.	0	3	11½
,, Nitrate	,,	0	2	8
Sodium (metal)	per lb.	0	4	0
,, Carb. (refined Soda-ash) 48 %	per ton	6	5	0
,, ,, (Caustic Soda-ash) 48 %	,,	5	5	0
,, ,, (Carb. Soda-ash) 48%	,,	5	12	6
,, ,, (Carb. Soda-ash) 58 %	,,	6	15	0
,, ,, (Soda Crystals)	,,	2	17	6
,, Acetate (ex-ship)	,,	16	0	0
,, Arseniate, 45 %	,,	11	0	0
,, Chlorate	per lb.	0	0	6¼
,, Borate (Borax)	nett, per ton.	29	0	0
,, Bichromate	per lb.	0	0	3
,, Hydrate (77 % Caustic Soda)	per ton.	9	15	0
,, ,, (74 % Caustic Soda)	,,	8	15	0
,, ,, (70 % Caustic Soda)	,,	7	17	6
,, ,, (60 % Caustic Soda, white)	,,	6	17	6
,, ,, (60 % Caustic Soda, cream)	,,	6	15	0
,, Bicarbonate	,,	6	0	0
,, Hyposulphite	,,	5	10	0
,, Manganate, 25%	,,	9		0
,, Nitrate (95 %) ex-ship Liverpool	per cwt.	0	8	1½
,, Nitrite, 98 %	per ton	28	0	0
,, Phosphate	,,	15	15	0
,, Prussiate	per lb.	0	0	7¾
,, Silicate (glass)	per ton	5	7	6
,, ,, (liquid, 100° Tw.)	,,	3	17	6
,, Stannate, 40 %	per cwt.	2	0	0
,, Sulphate (Salt-cake)	per ton.	1	5	0
,, ,, (Glauber's Salts)	,,	1	10	0
,, Sulphide	,,	7	15	0
,, Sulphite	,,	5	5	0
Strontium Hydrate, 98 %	,,	8	0	0
Sulphocyanide Ammonium, 95 %	per lb.	0	0	7¼
,, Barium, 95 %	,,	0	0	7¼
,, Potassium	,,	0	0	[illegible]
Sulphur (Flowers)	per ton.	7	10	0
,, (Roll Brimstone)	,,	6	0	6
,, Brimstone: Best Quality	,,	4	7	6
Superphosphate of Lime (26 %)	,,	2	12	6
Tallow	,,	25	10	0
Tin (English Ingots)	,,	98	0	0
,, Crystals	per lb.	0	0	6¼
Zinc (Spelter)	per ton.	23	0	0
,, Chloride (solution, 96° Tw.	,,	6	0	0

THE

CHEMICAL TRADE JOURNAL.

Publishing Offices: 32, *BLACKFRIARS STREET, MANCHESTER.*

No. 158. **SATURDAY, MAY 31, 1890.** **Vol. VI**

Contents.

Notices.

All communications for the *Chemical Trade Journal* should be addressed, and Cheques and Post Office Orders made payable to—

DAVIS BROS., 32, Blackfriars Street, MANCHESTER.

Our Registered Telegraphic Address is—
"Expert, Manchester."

The Terms of Subscription, commencing at any date, to the *Chemical Trade Journal*,—payable in advance,—including postage to any part of the world, are as follow:—

Yearly (52 numbers)	**12s. 6d.**
Half-Yearly (26 numbers)	**6s. 6d.**
Quarterly (13 numbers)	**3s. 6d.**

Readers will oblige by making their remittances for subscriptions by Postal or Post Office Order, crossed.

Communications for the Editor, if intended for insertion in the current week's issue, should reach the office not later than **Tuesday Morning.**

Articles, reports, and correspondence on all matters of interest to the Chemical and allied industries, home and foreign, are solicited. Correspondents should condense their matter as much as possible, write on one side only of the paper, and in all cases give their names and addresses, not necessarily for publication. Sketches should be sent on separate sheets.

We cannot undertake to return rejected manuscripts or drawings, unless accompanied by a stamped directed envelope.

Readers are invited to forward items of intelligence, or cuttings from local newspapers, of interest to the trades concerned.

As it is one of the special features of the *Chemical Trade Journal* to give the earliest information respecting new processes, improvements, inventions, etc., bearing upon the Chemical and allied industries, or which may be of interest to our readers, the Editor invites particulars of such—when in working order—from the originators; and if the subject is deemed of sufficient importance, an expert will visit and report upon the same in the columns of the *Journal*. There is no fee required for visits of this kind.

We shall esteem it a favour if any of our readers, in making enquiries of, or opening accounts with advertisers in this paper, will kindly mention the *Chemical Trade Journal* as the source of their information.

Advertisements intended for insertion in the current week's issue, should reach the office by **Wednesday morning** at the latest.

Advertisements.

Prepaid Advertisements of Situations Vacant, Premises on Sale or To be Let, Miscellaneous Wants, and Specific Articles for Sale by Private Contract, are inserted in the *Chemical Trade Journal* at the following rates:—

30 Words and under	2s.	0d.	per insertion.
Each additional 10 words	0s.	6d.	,,

Advertisements of Situations Wanted are inserted at one-half the above rates when prepaid, viz:—

30 Words and under	1s.	0d.	per insertion.
Each additional 10 words	0s.	3d.	,,

Trade Advertisements, Announcements in the Directory Columns, and all Advertisements not prepaid, are charged at the Tariff rates, which will be forwarded on application.

Legal.

HIGH COURT OF JUSTICE—QUEEN'S BENCH DIVISIO

TUESDAY, APRIL 29.

(Before Mr. Justice DAY *and a Special Jury.)*

STOCKMAN v. DASNIERES—A TRANSACTION IN PIT

(Continued from page 331.)

SHORTHAND WRITER'S NOTES OF JUDGE'S SUMMING UP.

IT is a thing, no doubt, that a lot of people have been getting a g
deal used to this system of perquisites, and of persons, if I ma
so, warping their own judgments and diminishing their own effici
by taking perquisites. It is done in households, no doubt, to a
large extent, and I very little doubt that masters who employ serv
are not benefited by the fact that your servants get perquisites
allowances from various tradesmen with whom you deal. But it is
because it is generally done by, we will say, domestic servants, th
is lawful or proper, or if I use another expression, looking at it str
honest, because you have no business to receive money by way of
quisites which must be charged against your master in one wa
the other.

That is the thing that is complained of on the part of the defend
when he says, my servant has been taking trimming without my kn
ledge, not telling me about it or asking me whether I approve of it,
he has been doing this, and doing it systematically, and he has b
charging prices such as we have had proved in evidence before
Certainly the prices we have now were higher than the 2d. pai
Jones, and, no doubt, more than Becton, who used, as I gather,
said before, to get 3d., and at the most 3½d., while in one insta
we know the plaintiff has obtained as much as 4d.

It might be inconvenient with reference to Becton people or pe
like Mr. Leonard, who complained of this. Mr. Leonard m
object and might dislike it, and as Mr. Dasnieres says—"It is not c
venient for me to make things unpleasant to the persons with who
deal. It is inconvenient in my business to do things, or for my serv
rather to do things, that are inconvenient to the seller of the goods.

He naturally wishes to be on good terms with the person with wh
he deals, and Mr. Leonard does object to this system, and made
objection at this time on the occasion of the dinner in August, 18
Then, it is said that there is no harm done to Mr. Dasnieres, beca
Mr. Dasnieres would not have done it himself, but why would not
Dasnieres have done it? He says—"It is not my business, an
would not do it, I wish to please my customers." The servant is
justified by saying—"My master would not do it, and therefore I
it myself." If he had consulted his master, very likely Mr. Dasni
would have done it himself, Mr. Dasnieres says it is inconvenient for
to. I will not say be put on unfriendly terms, but to do a thing wh
is obnoxious to Mr. Leonard. This thing has been going on for ye
in secrecy. It has been communicated to him, but done without
consent. He says he could not possibly have thought of such a th
going on. It is for you to say whether such a course of conduct does
justify the dismissal, not of an ordinary servant, but of a manager, y
alter ego, the person who is managing the things for you, and wh
you have to trust to a considerable extent.

Then, gentlemen, the substantive ground of justification is the
with reference to the transaction in pitch. You have to consider t
transaction in November. There is evidence before you of a somew
serious character upon that transaction. It was clearly the duty
Stockman, whatever he thought about his master, I should say, to co
municate to him offers; and if 4,000 tons were offered, it was not
should have thought, for him to determine whether his master sho
buy upon a rising market or not, it was for him to communicate to

oo tons were offered. Mr. Stockman seems to have he said that his impression was that he had com- his employer. When he is asked—"You could only ted it by letter or by telegram," he says—

The one was in Paris and the other was in London. ugh the telegrams you do not find any communication gram to the effect that there were 4,000 tons offered. ion for you to consider, it seems to me at least, whether was over-bought or was not over-bought. We are servant. It was for the servant to tell his master, oo tons offered, will you have them or not?" But to be found in any of the letters or correspondence eived; and it is not to be found in the copy letter the telegrams, and I think you will probably come to hat Stockman did fail to communicate that offer to his n what is done? who does buy it? Stockman buys ou will probably be of opinion that not having com- his employer that there was this 1,000 tons in the an buys it. Why does Stockman buy it? It cannot was under-bought or over bought, because he had no him except the open market. He bought for a rise— stuff was likely to rise. Surely it was his duty, as a is master have the benefit of an opportunity of making ,000 tons. He does not communicate with his master, in the probable rise, he goes and buys a 1,000 tons. me to be the gravaman of the whole charge. It is not ud, as it has been put to be. It is a failure on the part to communicate to his master an offer of goods which y appearance of involving satisfactory results, because it himself. He would not have bought it if he had at this sort of stuff was going to rise. He does not the opportunity of buying it, but being here to look 's interest, he buys it himself. What next took place? wanted, and on the 13th November, eight days after on the 5th by Stockman. What does he do? He does ell his master. "I can let you have some at a slight have got a lot—1,000 tons—which I can sell to you." ing to his master at all, but actually realises his profit and exposes his master to have to pay Blagden com- purchase by Dasnieres; because, when he realises his agden, by closing his account with him, and receiving he sells to Blagden as though Blagden were principal, lerstanding that it is to be handed over to Dasnieres at price which he receives the benefit of. I can only say be no question in point of law that that profit, at any or which the servant is accountable to the master. Of be no question. But there is still the question, whether ismissal. That is a matter which is for you. It is done secrecy. No trace of it is communicated to Mr. he goods are bought without any communication to Mr. t they were in the market. They were offered to the Ir. Dasnieres under circumstances of the market which kman to think it was a desirable thing for him to and, without any communication to Mr. Dasnieres, he ered to his master for himself, upon which he makes a ich goods he afterwards sells to his master, without any 1 of any sort or kind. That is a matter which certainly very serious consideration. It is for you to say what ourselves would take, or what you think a prudent mer- ind that the manager, when he lives in another country, oys to manage his affairs, does act in that manner.

nsider that, and consider the question of the trimming. that it is necessary to trouble you with any observations ous other matters, to which attention has been called. tters seem to be the more serious features in the case. It ion necessarily of fraud and dishonesty; but it is a ether this man has so conducted himself in his employment, reasonable for a master to say: "You shall be manager nger; I dismiss you at once," if, in your judgment, it reasonable master in acting in that way, if that is the think a reasonable employer would take, to dismiss, t, the justification is made out, and you will find your defendant, upon the question of justification.

d be of another opinion, and think that these measures Mr. Dasnieres in dismissing the plaintiff, then you have at damages Mr. Stockman is entitled to for being the service of Mr. Dasnieres, practically, three months e, when he could be dismissed. As I have said, being e three months notice, he would be now, practically, at the ee months notice, and he would be entitled to com- that remaining period. His salary, as you know, was ar, and he was also entitled to commissions, the arned up to the time of his leaving the service. You le about the others, because those will, in any event, be provided for. You have to consider what you will allow in respect of loss of salary, and also commissions for the three months, what would be a reasonable compensation for not having three months notice, three months during which he might have earned the fourth part of £300 a year, £75, and might have earned commissions. It is for you to say what would be a reasonable compensation for being deprived, for that time, altogether of the situation, in which he might have made the emoluments to which I have referred. That will be in the event of your finding that justification is not made out. If justification is made out, you will simply find your verdict for the defendant. The earned commission, as I say, to be payed by the defendant to the plaintiff, will be ascertained out of court, and you need not trouble yourselves with reference to it.

There is one other question which has been raised, and that involves simply a question of law, viz., the question of the right of the defendant to an account with reference to profits made in respect of the trimming, and also profit made in respect of the transaction in pitch. There are questions which seem to me to be questions of law, and no question of fact really arises, the substantial question for you is, whether the justification has been made out or not. Consider your verdict, gentlemen, and say how you find.

The Jury consulted together for a short time.

The ASSOCIATE: Gentlemen, are you all agreed?

The FOREMAN: Yes, we are all agreed.

The ASSOCIATE: Do you find for the plaintiff, or for the defendant?

The FOREMAN: The defendant.

Mr. JELF: I should ask your Lordship, then, for judgment for the defendant, upon the cause, so far as it is an action for wrongful dismissal, and with regard to the other part of the case—

Mr. JUSTICE DAY: The defendant is entitled to an account,

Mr. JELF: I agree, I was only going to say that the costs of fighting this action should be ours.

Mr. JUSTICE DAY: Of course, the costs of trial will be the costs of the defendant, but there will be some costs for the plaintiff. I mean the Writ, and there may be the Statement of Claim. There must be some costs.

Mr. DICKENS: The usual costs of the issue upon which the plaintiff has succeeded and the usual costs of the issue upon which the defendant has succeeded.

Mr. JUSTICE DAY: It is merely a question of taxation. I direct an account. The master will arrange it.

Mr. JELF: That will depend upon what happens with regard to the contra account.

Mr. JUSTICE DAY: Then I had better not determine it now.

Mr. JELF: If your Lordship will give us liberty to apply.

Mr. JUSTICE DAY: Certainly.

Mr. DICKENS: With reference to the £56., they are entitled to that. There is no question about that. With reference to the trimmings I deny that they are entitled to that.

Mr. JELF: We do not ask for the whole trimmings. It is so much of the trimmings.

Mr. JUSTICE DAY: That is a matter which you had better reserve.

Mr. JELF: I was thinking that my friend and I might consider these things and set off one against the other.

Mr. JUSTICE DAY: You can have it brought before me in Middlesex Juries Chambers, or Bankruptcy, or elsewhere. You can have it put before me.

Mr. JELF: We had better take the verdict of the Jury and your Lordship orders the costs of the trial to be ours.

Mr. JUSTICE DAY: If you fix any day convenient, I daresay I shall be sitting here for weeks. I do not know what I may be ordered to do to-morrow, but so far as I know I shall be here and any day which is convenient to you to come before me, you can without having a special appointment. You can come before me to-morrow morning, or any time when it is convenient to yourselves, if you two can agree any time convenient to yourselves.

Mr. JELF: May we mention a day?

Mr. JUSTICE DAY: No, instead of mentioning a day, come and argue the matter. Come in when it is convenient to argue the question at any time and I will take it.

Mr. JELF: If your Lordship pleases.

SUNLIGHT FOR MATURING WINES.—Experiments have recently been made in Spain, on the action of sunlight in maturing wines. Layers of new wines in bottles of coloured glass have been exposed to the direct rays of the sun, with the result, that both the flavour and quality have been improved. In the South of Europe, there has been a practice of ripening cognac, by exposing the bottle on the roof for years.

GRANTS TO AGRICULTURAL AND DAIRY SCHOOLS.

A PARLIAMENTARY Paper issued by the Board of Agriculture contains a report on and a list of the distribution of the Parliamentary grants in aid of agricultural and dairy schools, and for agricultural experiments, in the year ended March 31, 1890. The Board of Agriculture, on its formation on the 9th of September, 1889, was intrusted with the duty, formerly devolving on the Lords of the Committee of Council on Agriculture, of awarding grants for the promotion of technical instruction in agriculture. The present system is only an experiment following the policy of the Privy Council in the previous year. The following table shows that 25 separate institutions received assistance from the grants, and also the amount and the purpose in each case :—

Institution or School.	Dairy Instruction.	General Agricultural Instruction.	Experiments.	Forestry.	Total.
England and Wales.	£	£	£	£	£
Bath and West of England Society and Southern Counties Association (Dairy Schools and Experiments)	300	—	150	—	450
Aspatria Agricultural College (General Agricultural Instruction and Dairying)	150	250	—	—	400
University College of North Wales, Bangor (General Agricultural Instruction and Dairying) ..	105	275	20	—	400
Leicestershire Education Committee (Daily Lectures) ..	250	—	—	—	250
Eastern Counties Dairy Institute (Dairy Classes and Local Lectures)	250	—	—	—	250
Cheshire County Dairy Institute	200	—	—	—	200
British Dairy Farmers' Institute, Aylesbury	200	—	—	—	200
Norfolk Chamber of Agriculture (Experiments)	—	—	150	—	150
Sussex Association for the Improvement of Agriculture (Experiments)	—	—	150	—	150
Gloucester Dairy School	75	—	—	—	75
Royal Manchester, Liverpool, and North Lancashire Society (Experiments)	—	—	35	—	35
Hereford and Ross Dairy Lectures	25	—	—	—	25
Swanley Horticultural and Technical College (Fruit-growing and other Lectures)	—	25	—	—	25
Scotland.					
University of Edinburgh (Teachers' Class and Forestry Lectures) ..	—	400	—	100	500
Glasgow and West of Scotland Technical College (General Agricultural Instruction) ..	—	350	—	—	350
Scottish Dairy Institute, Kilmarnock (Cheese-making and General Dairy Work)	250	—	—	—	250
University of Aberdeen, Fordyce Trust (General Agricultural Instruction)	—	200	—	—	200
Heriot Watt College, Edinburgh (General Agricultural Instruction)	—	150	—	—	150
Aberdeenshire Agricultural Research Association (Experiments)	—	—	150	—	150
Institute of Scottish Teachers of Agriculture (Providing Schoolmasters with Agricultural Information)	—	100	—	—	100
Wigtownshire Dairy Association	100	—	—	—	100
Stewartry of Kirkcudbright Dairy Association	70	—	—	—	70
Angus and Mearns Dairy School	50	—	—	—	50
Dumfriesshire Dairy Association	30	—	—	—	30
Dounby Science School, Orkney (Experiments)	—	—	25	—	25
Total Great Britain ..	2055	1750	680	100	4585

PERMANENT CHEMICAL EXHIBITION.

THE proprietors wish to remind subscribers and their fr generally that there is no charge for admission to the Exhib Visitors are requested to leave their cards, and will confer a favo making any suggestions that may occur to them in the directi promoting the usefulness of the Institution.

JOSEPH AIRD, GREATBRIDGE.—Iron tubes and coils of all kinds

ASHMORE, BENSON, PEASE AND CO., STOCKTON-ON-TEES.—Sul of Ammonia Stills, Green's Patent Scrubber, Gasometers, anc Plant generally.

BLACKMAN VENTILATING CO., LONDON. — Fans, Air Prope Ventilating Machinery.

G. G. BLACKWELL, LIVERPOOL. — Manganese Ores, Bau French Chalk. Importers of minerals of every description.

BRACHER AND CO., WINCANTON.—Automatic Stills, and P Mixing Machinery for Dry Paints, Powders, &c.

BRUNNER, MOND AND CO., NORTHWICH.—Bicarbonate of Soda Ash, Soda Crystals, Muriate of Ammonia, Sulphat Ammonia, Sesqui-Carbonate of Ammonia.

BUCKLEY BRICK AND TILE CO, BUCKLEY.—Fireclay ware kinds—Slabs, Blocks, Bricks, Tiles, "Metalline," &c.

CHADDERTON IRON WORKS CO., CHADDERTON.—Steam Drier Steam Traps (McDougall's Patent).

W. F. CLAY, EDINBURGH.—Scientific Literature—English, Fr German, American. Works on Chemistry a speciality.

CLAYTON ANILINE CO., CLAYTON.—Aniline Colours, Aniline Benzole, Toluole, Xylole, and Nitro-compounds of all kinds.

J. CORTIN, NEWCASTLE-ON-TYNE.—Regulus and Brass Taps Valves, "Non-rotative Acid Valves," Lead Burning Apparat

R. DAGLISH AND CO., ST. HELENS.—Photographs of Chemical —Blowing Engines, Filter Presses, Sludge Pumps, &c.

DAVIS BROS., MANCHESTER.—Samples of Products from va chemical products—Coal Distilling, Evaporation of Paper Treatment of waste liquors from mills, &c.

R. & J. DEMPSTER, MANCHESTER.—Photographs of Gas Pl Holders, Condensers, Purifiers, &c.

DOULTON AND CO., LAMBETH.—Specimens of Chemical Stone Stills, Condensers, Receivers, Boiling-pots, Store-jars, &c.

E. FAHRIG, PLAISTOW, ESSEX. — Ozonised Products. O Bleached Esparto - Pulp, Ozonised Oil, Ozone - Ammon Lime, &c.

GALLOWAYS, LIMITED, MANCHESTER.—Photographs illustr Boiler factory, and an installation of 1,500-h.p.

GRIMSHAW BROS., LIMITED, CLAYTON.—Zinc Compounds. S Materials, India-rubber Chemicals.

JEWSBURY AND BROWN, MANCHESTER.—Samples of Aerated Wa

JOSEPH KERSHAW AND CO, HOLLINWOOD.—Soaps, Greases, Varnishes of various kinds to suit all requirements.

C. R. LINDSEY AND CO., CLAYTON.—Lead Salts, (Acetate, Nit etc.) Sulphate of Copper, etc.

CHAS. LOW AND CO., REDDISH.—Mural Tablet-makers of Car Crystals, Cresylic and Picric Acids, Sheep Dip, Disinfectants,

MANCHESTER ANILINE CO., MANCHESTER.—Aniline Col Samples of Dyed Goods and Miscellaneous Chemicals, organic and inorganic.

MELDRUM BROS., MANCHESTER.—Steam Ejectors, Exhau Silent Boiling Jets, Air Compressors, and Acid Lifters.

E. D. MILNES AND BROTHER, BURY.—Dyewoods and Dye Extracts. Also samples of dyed fabrics.

MUSGRAVE AND CO., BELFAST.—Slow Combustion Stoves. Ma of all kinds of heating appliances.

NEWCASTLE CHEMICAL WORKS COMPANY, LIMITED, NEWCAS ON-TYNE.—Caustic Soda (ground and solid), Soda Ash, Reco Sulphur, etc.

ROBINSON, COOKS, AND COMPANY, ST. HELENS.—Drawings, trating their Gas Compressors and Vacuum Pumps, fitted Pilkington and Forrest's patent Valves.

J. ROYLE, MANCHESTER.—Steam Reducing Valves.

A. SMITH, CLAYTON.—India-rubber Chemicals, Rubber Substi Bisulphide of Carbon, Solvent Naphtha, Liquid Ammonia, Disinfecting Fluids.

WORTHINGTON PUMPING ENGINE COMPANY, LONDON.—Pum Machinery. Speciality, their "Duplex" Pump.

JOSEPH WRIGHT AND COMPANY, TIPTON.—Berryman Feed- Heater. Makers also of Multiple Effect Stills, and W Softening Apparatus.

ELECTRICITY AS A HASTENER IN THE TANNING OF HIDES.

IT is somewhat remarkable that so little application has been made of electricity to industrial processes in which chemical operations are involved. If the metallurgical operations of electro-depositing and electric smelting be excepted, the Cassel gold chlorination, Hermite's bleaching, Webster's sewage, and Figge's unsavoury process for "de-gumming" rhea fibre, have alone been tried on any large scale. Regarding an electric current as a mode of conveying energy, it has to compete with heat and the various forms of chemical energy, the applications of which entail, generally, a comparatively small outlay for plant; but where a distinct local effect, or where the action of so-called nascent elements is required, it would seem that electricity might frequently be employed with economy.

Many of the most important chemical processes of the day, have been worked out in the laboratory by the patient application of chemical principles at the hands of skilled chemists, and in this respect, the electric method of tanning is singular, since we have not been able to hear of even the existence of a chemical hypothesis to account for its results. The British Tanning Company have recently started, at Rothsay Street, Bermondsey, London,, S.E., a practical method of tanning "by electricity." The company was formed about a year ago, to work in this country, the patents of Worms and Ball, who, after many experiments, managed, in Paris, to successfully overcome the difficulties which Goulard and the earlier investigators in this direction had met with. It will be within the recollection of many of our readers that, at the Paris Exhibition, several specimens of leather, tanned in France by this process, were shown in the saddlery department. The English company at first started with a small plant of French manufacture, but they have recently erected a new series of drums, which have been at work for some weeks past, yielding very satisfactory results. In the present process, the hides, after the usual liming and depilating process, are transferred to large wooden drums, nearly 12ft. in diameter, and as many feet in length. These are mounted on a horizontal axis so arranged as to permit of being slowly rotated by steam power. Each drum is then charged with about 500 gallons of ordinary tanning liquor, of about 20° strength, made from either oak bark or chestnut, and, after the addition of a small quantity of turpentine, is hermetically sealed. Thus far, the method differs but little from the various processes which have been tried, with small success, to substitute for the hand labour involved in daily turning the hides, a mechanical movement to the tanning extract, and suspended hides. In the electrical process, which we had an opportunity of seeing at work last week, a few feet of copper wire, forming the electrodes for a current which enters at one end and leaves at the other, are fastened within the flat sides of the rotating drums. The copper electrodes are provided on the shafts outside, with rubber contacts, consisting of ordinary dynamo brushes, to which the leads from a small dynamo are attached. At the Bermondsey works, there are five such drums, and a current of twelve amperes, at a pressure of 70 volts, is used, or about 14 volts for each drum. The joint effect of this mechanical agitation of the pelts and liquor, and the passage of this low density current, is stated to reduce the time required to successfully tan half a ton of hides per drum from several months, to four or five days. After working for some little time a slight heating of the liquor is noticed, and the electrolytic gases produced exert a small pressure within the drum, which is controlled by a spring valve, but no data seem to be forthcoming as to what influence this increase of temperature and pressure have upon the result. The former effect might, of course, be obtained more economically, while the latter is in this case small, and has been applied much more elaborately in the hydraulic process of Spilsbury in 1823, and the vacuum process of Knowles and Duesbury, both of which failed many years ago.

We are informed that the simple mechanical agitation has been found to reduce the time necessary for tanning to one-half, the rest of the great economy being attained by the inexplicable electric action. It is obvious that it is extremely difficult even to devise experiments which would be successful in differentiating the part played by the various forms of energy which apparently influence the result, but until this is done a full explanation of the increased rapidity of the tannage is impossible.

A considerable stock of finished butts and skins were on view at the time of our visit; and, in the opinion of experts, the tanning process was stated to be fairly complete and well struck through, but the butts were generally more supple than is considered advisable for boot soles. The continual mechanical agitation in bulk would naturally open the pores of the hide; and it is suggested, somewhat vaguely, that the electric current has a similar effect, presumably in a manner akin to the action of a current on osmosis and dialysis, though in this case there is no definite direction, as the hides are being continually churned in the liquor. Turnbull, in 1845, patented a process in which

More than forty years ago Peter Barlow wrote:—"Of late years considerable changes have taken place in the modes of tanning, so that this operation, which a few years ago took several months to complete, can now be done in as many days;" but he goes on to say,—"the leather, however, made by this hastened process is considered to be less perfect than by the old method, an opinion which is supported by many eminent tanners and scientific men, among whom may be mentioned Sir Humphrey Davy." While the former part of this quotation exactly expresses the result of the electric methods of tanning, we hope that history will not repeat itself, but that we have in these processes the beginning of a new and important industry. In the application of electricity to any industry, considerable time must necessarily elapse before "the trade" will give way its prejudices against a new process, and when, as in this case, there are alternative schemes before the public, the delay is likely to be prolonged by litigation. We believe that both processes are capable of yielding a valuable commercial leather, and that the manufacturers who adopt the new method will find that there are many advantages both in time, labour, and space over the older methods of tanning.—*Industries.*

THE PREPARATION OF CRYSTALLIZED EGG ALBUMEN.*

By F. HOFMEISTER.

IN the course of investigation upon the behaviour of albuminoid bodies towards saline solutions, the author has repeatedly made observations indicating that the animal albuminoid substances which have not hitherto been obtained crystalline are not devoid of crystallizability. As the result of experiments directed especially to that object, he reports that he has worked out a method for obtaining egg albumen in a crystalline condition.

Fresh white of egg, free from yolk, is beaten into a fine froth with a good whisk, and allowed to stand for twenty-four hours. The nearly clear thin liquid solution of white of egg that collects at the bottom is then poured off from the froth, and treated with a cold saturated neutral solution of ammonium sulphate, in order to separate globulin. The resulting precipitate is filtered off, and the perfectly clear liquid, containing saline matter, is left to evaporate at the ordinary temperature in a wide flat-bottomed dish. After some days there is deposited at the bottom a layer of a finely granular white, sometimes yellowish or reddish coloured precipitate, which, under the microscope appears to be composed of tolerably large, transparent light refracting spheres or spheroid aggregates, without radiating or stratified structure (globulites). The thin membrane that covers the surface consists of the same form elements. When all increase of the precipitate upon further evaporation ceases to be perceptible, it is filtered off, and in this way almost the whole of the albumen is obtained in the form of a coarsely granular, pure white or slightly-coloured mass, completely soluble in water, which can be pretty well freed from mother-liquor by pressure. In order to purify this further it is dissolved in a semi-saturated solution of ammonium sulphate, and again left to evaporate spontaneously, and the operation is repeated as long as the albumen separates in globulites.

Usually in the third or fourth separation, by microscopical examination, fine needles may be seen mixed with the globulites, and upon further standing these rapidly increase, partly at the expense of the globulites present. The needles appear sometimes isolated, and sometimes in stellate groups. Frequently it is observed that the globulites, out of a crystalline nucleus contained in them, are gradually changed into a stellate group.

If when the beginning of crystallization becomes perceptible the shallow dish be covered with a glass plate, so as to moderate the evaporation, the greater part, or sometimes the entire mass of the albumen obtained gradually in the form of needles or thin oblique-angled plates. Mostly, however, a portion of the albumen that has separated in globulities resists this change. Neither is this object obtained by re-crystallization in the manner described, probably because the evaporation, even at the ordinary temperatare, always goes on too rapidly, and especially too unequally. When this stage, however, has once been reached, a complete conversion into crystals is easily produced if the filtered and pressed precipitate, redissolved in a semi-saturated solution of ammonium sulphate, be filled into a tube made of parchment paper, well closed at both ends, and laid in a dish containing a semisaturated solution of ammonium sulphate, so that it is everywhere surrounded by the liquid. The increase in concentration of the external liquid withdraws water very gradually and uniformly from the albumen solution contained in the tubes. If the albumen be already very pure it separates them directly in tables; or if there be first a formation of globulites these are soon completely rep needles or fine plates.

For the purpose of more rapid purification it would appear on the one hand to filter off and remove the portion that sep globulities, because in this way some difficultly soluble impur separated; and on the other hand in the crystallization not to a evaporation to go so far that the separation of crystalline am sulphate begins, because, according to the author's experie portion last to separate possesses only a small disposition to cr The separated crystalline mass can be advantageously freed mother liquor by washing it with a solution of ammonium sul the same density as the liquor. In order to avoid loss, howe washing should not be too prolonged. The albumenoid body pared proved upon close examination to be identical with th egg albumen, as prepared by Starke. This was only what w expected, because in the preparation other white of egg subst the character of albumen were not separated The condition th of egg—apart from the easily removed globulin—contains egg as the single albumenoid body present in considerable quantity looked upon as very favourable to its crystallization.

Unfortunately the author did not succeed in bringing to crysta a solution of the crystallized white of egg freed from salt by d This and the closely connected question whether the crystalli prepared consisted simply of albumen or, what he considers to probable, a compound of it with ammonium sulphate, he h compelled by circumstances to leave undetermined for a time.

* *Zeitschrift fur Physiologische Chemie*, vol. xiv., p. 651.

CROP AND STOCK PROSPECTS.

PROFESSOR MALDEN, in a letter to the *Times*, s "Last year at this season there was a very luxuriant g grass and corn in most districts; as wet weather continued t became over-rank, and the wonderful promise of May was tained, for the grass, though plentiful, was poor, and cattle d little good on it, while that which was mown for hay has gener proved so valuable for feeding as usual. Except in a few f districts, the yield of corn was considerably below the co average before harvest. Last year's growth was unhealthily lu while, so far as I have been able to see from visits during week or two to the gravels of Berks, the heavy loams of No fens of Lincolnshire, the mixed soils of Beds, and the chalks o and Hants, this year's growth is most healthy. Good plants are to be met with everywhere, and, with average weather, bid yield better than they have for years. The grass is not so h last year, but the cattle are doing so well that, though the stor been bought in at high prices, they will probably clear the when they are fat, which is more than looked probable at on There is an abundance of sheep-keep everywhere, and sto hardly fall in price in the face of it. Mangolds are coming up

LAYING PIPES UNDER WATER.

MR. F. S. Pecke, a civil engineer at Watertown, U.S.A. accomplished in a very simple, cheap, and expeditio what is usually a difficult and expensive operation—the laying o line of pipe in deep water. He had occasion to lay nearly 1,c of suction pipe at Rouse's Point. The water was needed for m turing purposes, and as it was found that water near the sh more or less impure, it was necessary to place the inlet a consi distance out in the lake. He purchased for the purpose pressure pipe of 8 in. diameter, manufactured by the Spiral Wel Co., at East Orange, N.J., and used for couplings cast iron weighing, with bolts and gaskets, about 65 lb. to the pair. P the end of the first length, he pushed it out on the surface c Champlain, and connected the second length, pushing this out until the whole line was coupled. It then presented the spectacle of a line of 8 in. pressure pipe nearly 1,000 ft. long, with a displacement of only 3½ in. of its diameter. When th site length had been connected the line was towed to positi plug at the end removed, and the pipe sank easily in 16½ ft. c without breaking a joint or receiving any injury. No buoys c were used in the operation, and no apparatus of any kind. T is now in use as the suction of a steam pump, and gives perfec faction. Work of this kind usually involves the use of expens troublesome flexible joints, and Mr. Pecke's neat and in expedient is worthy of record and of imitation under like condit

It is obvious, says *Engineering News*, that this could hard been done with cast iron pipe, on account of its rigidity and to fracture.

I OF PLATES OF STEAM BOILERS.

ιY THOMAS T. P. BRUCE WARREN.

orrosion of the iron plates of a steam boiler is due at
wo causes; in one case the corrosion, indicated by
lace above the water line, and in the other case by
ırface covered by incrustation or deposit. In both
marked absence of uniformity of corrosion, some
›usly affected, whilst others in the same boiler and in
›f the boiler remain intact. This has been explained
ıeous character of the plates, due either to chemical
fferences.
ıquires no great amount of chemical skill to deter-
ese factors operate in shortening the life of a boiler.
appears to me to have escaped the attention of writers
deserves mention. It is well known that a rolled
a "skin" or surface which resists oxidation. If we
scratching or filing, the new surface is rapidly coated
:posure to damp air. If we protect with a varnish
:s corrosion is not perceived. If a freshly-rolled plate
ce remains intact, whereas the edges rust in a very
. piece of sheet-iron be immersed in dilute acid, the
with what is known as the natural skin, resists the
less, whilst the recently-cut surfaces are speedily

›ubt, due to a difference in the electrical relation of
one being much more electro-positive than the other.
of pitting, which is not only local, but, in some parts
e, serious, can be explained in the same way.
ction of a boiler there are several operations which
formity of surface; the rivets, the holes, whether
ɜd, all introduce an individual function in the wear
oiler. The natural skin of an iron plate, if partly
; a surface which is electro-positive to the other parts,
l in a corrosive liquid, the parts of the plate which
ve actually assist in the destruction of the parts which
egative; hence it is that pitting is promoted when it

edy seems to be the entire removal of the surface after
ht then ensure an uniform surface, when alteration
l be less likely to lead to local chemical action.
the magnet for determining inequalities in a boiler
e very problematical, but the difference of potential
are immersed in the same liquid, as might be observed
:ter, would reveal the existence of chemical or
rences of structure which would be favourable to

›posed to use zinc as a protection against corrosion;
by rendering the boiler surface electro-negative, we
rrosion, but, unfortunately, the instructions given for
ding. In some cases pieces of zinc are simply thrown
long as they remain *clean* and are in contact with a
, the preservative action of the zinc is ensured, but
ith mud or boiler deposit, they are shielded from
If the zinc be suspended or attached *metallically* to
the water line, we are more likely to prevent corrosion
ls, the iron being electro-negative to zinc.
of the zinc being greatest at the points of suspension,
:ar away and fall into the deposit, where it becomes,
mical sense, lost, but so long as it remains suspended,
r at the points of attachment keep the contacts clean.
ber that the greatest loss of zinc is where the metallic
that by increasing the area of the zinc plate, so long
ntact is constant, we are doing little or no good. The
: areas of contact of the two metals, so as to ensure
dvantage, when once determined for any particular
most rigidly enforced. When a boiler is cleaned out
to "put the zinc back again;" clean contacts, or
es in the zinc plate if worn, should never be neglected.
ıe plate should be entirely renewed. It is well-known
should be allowed to introduce an electro-positive
›oiler-plates. Spelter or sheet-zinc frequently contains
of lead; in time the zinc dissolves away, leaving a
lead; we are then in a fair way of doing more harm
boiler plates, in spite of the solubility of chloride of

. recommended for keeping a boiler clean, and here, I
:al of carelessness has been introduced by writers on
ne-half of a boiler is in contact with a hot solution of
other half not; the condition of the two parts of the
, as regards liability to corrosion, must be too evident
t.

The unequally heated parts of a boiler will require a difference of potential proportional to the energy of the liquid, and here we have probably an explanation for the increased activity of alkaline solutions in removing scale.

From these considerations it will appear that if we can keep a boiler electro-negative, by an independent source of electrification, to any destructive agency which might arise in the boiler, we might reasonably hope to prolong its life.—*Chemical News.*

THE TENDERING OF TISSUES BY IRON MORDANTS.

THIS was the subject of a recent communication to the Société Industrielle de Mulhouse, by M. Jeanmaire. Every cotton tissue, on being immersed in iron mordant, is weakened after a shorter or longer contact. The degree of the change increases with the concentration of the mordant and the temperature. The influence of an acid pre-existing on the tissue is nil, provided it has no oxidising action. If the tissue shows no trace of weakening before the mordanting, the degree of weakening by the iron will not be greater than that produced on neutral tissue, whatever be the nature or quantity of the acid or the manner of drying. In certain cases the tissue is protected by the mordant itself from a subsequent change which the acid might eventually produce. If the tissue is weakened more or less already before the application of the mordant, the total weakening will be equal to the one originally observed, increased by that produced by the mordant on unimpaired tissue. An alkaline discharge increases the extent to which the fibre is attacked by the iron mordant. —*Textile Mercury.*

NOTES FROM KUHLOW.

SULPHUR.—This product is undoubtedly one of the most important articles in the chemical trade. In Germany in recent years it has mainly been extracted by the regeneration of sulphur forming the residuum in the manufacture of Leblanc soda, which is accumulated in enormous heaps at almost all works. The Schafer process rendered it possible to regain the sulphur from the residuum, and in this way chemical manufacturers were enabled to make a profit where it would otherwise have been impossible. An occurence of some interest from a commercial point of view thereupon took place. In the course of a few years the prices of raw sulphur in Sicily declined so heavily, and freights were reduced to that extent, that it became cheaper to bring Silician sulphur to Germany than to regenerate the sulphur left over from the Leblanc process. In consequence, Silician sulphur is almost alone found in trade. Extensive experiments are being conducted in Upper Silesia with the object of extracting sulphur, flowers of sulphur and liquid sulphuric acid from the torrified gases escaping from sulphuric acid in the manufacture of zinc.

THE LIEBENWERDA BRIQUETTE WORKS.—The lignite used for the briquettes is brought to the works by means of a wire rope railway. The speed is 1½ metres per second. Some seventy-five tons of lignite are brought into the works inside twenty-four hours. Each truck contains about seven cwts. After the lignite has passed through several wood and iron bars it falls into a rolling mill, where it is completely pulverised. The lignite, which is then in dust form and contains 50 per cent. of water, is thereupon transferred to a drying furnace (steam drying machinery), where it must pass through nineteen divisions lying one above the other, in order that the moisture may as far as possible be evaporated. It is then placed in the pressing machine which has a horizontal motion, and turns out about 120 tons of briquettes daily. The briquettes are sold on the spot at about £4. per ten tons. The coarse lumps of coal that cannot pass through the bars are collected and sold as fuel at about 25s. per 10 tons at the works. An electric light machine provides the necessary illumination for night work. The wire rope railway connecting the works with the mines is 6,150 metres long, and the journey lasts about an hour.

AN IMPORTANT WIND MOTOR, with pumping works, has just been constructed for the Griefswald waterworks by Messrs. Friedr. Filler & Hinsch, of Eimsbüttel, proving, in a convincing manner, the applicability of such motors for the purpose in question. The motor drives four pumps, and serves for supplying the town with water. A very favourable impression is made by the gigantic dimensions and elegant construction of the whole. The motor itself, is 12·20 metres high, and is fixed on an iron tower 20 metres in height. With a speed of wind of 4·5 metres, it developes a power of 25 horse-power, and raises in an hour, 162,000 litres of water about 15 metres high. The work is a credit to Hamburg industry, in general, and to Messrs.

Friedr. Filler & Hinsch in particular. It may be mentioned that the firm has carried off no less than 24 awards.

FARBSTOFF-IMPORT-ACTIEN GESELLSCHAFT, VORMALS BÖSENBERG & JANKE, HAMBURG.—A company has been formed at Hamburg, under the above title, with a capital of 500,000 mks., for the purpose of acquiring the import business in tanning and dyeing materials of the firm of Bösenburg & Janke, and the quebracho and dyewood mills, situated at Blankenese, on the Lower Elbe, and belonging to the same firm. The management of the business will, in the first instance, be carried on by Messrs. Messrs. H. Bösenberg, and A. Janke. The prosperity of the new enterprise appears assured, from the fact that the company has joined the Sales Syndicate of the Hamburg Quebracho Mills.

BLECHWALZWERK SCHULZ KNAUDT ACTIEN GESELLSCHAFT, ESSEN.—As appears from a notice of the Royal *Amtsgericht* at Essen, the above-named company has empowered Mr. Carl Ohly to sign, per procurationem, such signature, to be accompanied by that of a director or of another procurator.

SEPARATION OF THE THREE XYLENES OF COAL TAR.

SULPHURIC acid containing of 80 per cent. of H_2SO_4 still acts on meta-xylene, whilst diluted to contain about 84 per cent., its action on para- and ortho-xylene ceases. Ordinary sulphuric acid (93 to 95 per cent.) acts, therefore, on meta-xylene, until in the mixture the ratio of sulphuric acid to water is about 80 : 20. In para- and ortho-xylene the action will have already ceased when the ratio attains 80 : 16. This happens, for instance, as soon as 202 grms. of sulphuric acid (93 per cent.) have dissolved 100 grms. of meta-xylene, or as soon as 233 grms. of the same acid have dissolved 100 grms. of para- or ortho-xylene. Hence it follows that to a sulphuric acid saturated with meta-xylene, about one-sixth of its original weight of new acid may be added without the mixture being rendered inoperative for dissolving para- and ortho-xylene.

The crude xylene worked with contained generally more than 60 per cent. of meta-xylene. The treatment with sulphuric acid was carried out in a cast-iron cylinder with a stirring apparatus, or for analytical purposes, in glass flasks with good shaking apparatus. The temperature during the operation rose as high as 80°. Under these conditions the sulphuric acid, if added in sufficient quantity, attacks meta-xylene only. In the selection of the raw material care should be taken to select a product which does not begin to distil under 136°, otherwise the toluene will afterwards render the para- and ortho-xylene impure. Good results have been obtained with products that began to distil at 136°, and of which 90 per cent. had passed over at 145°. The raw material should, before separation, be repeatedly washed with small quantities of sulphuric acid (which is to be kept cold) till it becomes not more than slightly brown.

Meta-xylene.—It follows from these remarks that the total amount of meta-xylene can be separated from the raw xylene as sulphonic acid in one operation, by means of sulphuric acid, the quantity of which is determined by a preliminary trial.

The complete removal of the meta-xylene is the object of a subsequent preparation of pure ortho-xylene. One kilo. of washed raw xylene is put into a (graduated) flask, which can be mechanically shaken during several hours, while the quantity of sulphuric acid, sufficient to dissolve 60 per cent. of meta-xylene, calculated as shown above, is gradually added. After allowing it to settle it will not often be found that too little of the substance has been dissolved—that is, that the meta-xylene present was less than 60 per cent. If the solution was normal, a further addition of one-sixth of the sulphuric acid employed will, in most cases, dissolve a few more per cents. If by this the decrease of hydrocarbon is made greater by nearly 60 per cent., the addition of sulphuric acid is, of course, to be repeated. The necessary amount of sulphuric acid, roughly estimated in this manner should be controlled by repeating the experiment.

One-third of the quantity of acid thus determined is added to the raw xylene in a suitable stirring apparatus; the remainder and about one-sixth of the acid in excess is added after some time. The influx is to be so regulated that the temperature during the reaction does not exceed 80°. Under these conditions all the meta-xylene passes into solution exclusively as 1, 3, 4 meta-xylenesulphonic acid. After allowing it to settle for some time the acid is drawn off. It still contains 2 to 3 per cent. of hydrocarbons mechanically dissolved; most of them separate when the acid is diluted. The mixture of acids is, by the addition of water, brought to a specific gravity of 1·4. A considerable rise of the temperature above 100° is of direct advantage, as it prevents a too early crystallisation of the sulphonic acid, so that the mechanically dissolved hydrocarbons can separate at the surface. At the same time the greater part of the sulphonic acids of the olefines of the raw xylene, which are present in small quantities, is decom[p] The sulphonic acids of the xylenes begin to decompose onl[y] slightly at 150°.

On cooling slowly, the whole mass consolidates to a firm c[ake of] crystals of the α-meta-xylenesulphonic acid. The mother-[liquor] obtained by treating the crystals in a centrifugal machine, sep[arates] into two layers, a lower one containing nearly pure sulphuric acid [sp.] gr. 1·57-1·60, and above it a layer of sulphonic acid, for the most [part] still in a liquid state. From this, repeated crystallizations of α-[meta-]xylenesulphonic acid are obtained after destruction of the olefi[ne sul]phonic acid still present by heating, and subsequent addition o[f sul]phuric acid.

If the pure hydrocarbon is to be regenerated from the sulp[honic] acid, the latter may be decomposed, quantitatively, with wate[r in a] boiler at 220°; under pressure. In the case of meta-xyl[ene a] carefully-directed dry distillation of the ammonium salt is also pos[sible,] but not in the case of para-and ortho-zylene. The meta-xylene i[s] acidulated a little with sulphuric acid, and will then give a [better] yield than by dry distillation of the free acid, without any other seco[ndary] decomposition, except a partial carbonization. The yield by [dis]tillation of the ammouium salt, amounted to 80-90 per cent.

Para-xylene.—The remainder of hydrocarbon, separated fro[m the] meta-xylenesulphonic acid is, in the same stirring apparatus, a[t] treated with an excess of sulphuric acid. The elevation of temperat[ure] if necessary, supported by indirect steam, and is to be kept for [some] time at 80°; the whole substance, except about 5-10 per cent. then dissolve. The mixture of acids is drawn of, and again dilut[ed] in the case of meta-xylene, and left to crystallize in a cool place. [The] mass at once separates into two layers, of which the lower, conta[ining] the sulphuric acid, considerably exceeds in quantity that of the [meta-]xylene, but also contains no appreciable quantity of dissolved h[ydro]carbons. The upper layer coagulates to a pulp of needles of the [para-]xylenesulphonic acid, in the still liquid ortho-xylenesulphonic acid

The para-compound crystallizes completely, and may be obtain[ed] pure, by means of a good centrifugal machine that the hydro-ca[rbon] isolated from it, melts at 5°-7°. As the sulphonic acid is deliques[cent] a thick and thorough separation of these crystals is essential fo[r the] subsequent preparation of pure ortho-xylene.

Toluenesulphonic acid behaves like para-xylenesulphonic acid, [is not] easily distinguishable from it externally, and, in the above opera[tion] renders the para-xylenesulphonic acid impure, if the amount of to[luene] in the raw xylene be considerable.

If a greater purity in the para-xylene be desired, the sulphonic [acid] freed from mother-liquor by the centrifugal machine, is to be diss[olved] in water, and re-crystallized by addition of sulphuric acid. [The] barium salts have not rendered any particularly good service i[n the] separation of the xylenes. For the preparation of absolutely [pure] para-xylene the beautifully crystallizing sodium salt of the sulp[honic] acid, obtained from the calcium salt, is most preferable.

The only way of isolating the para-xylene is by decomposing it [with] water in the pressure boiler, where decomposition will take place [quan]titatively; the sodium salt must previously be acidulated. A [dry] distillation of the acid, or of the acidulated ammonium salt, is [inad]missible, scarcely 20 per cent. of the theoretical produce [being] obtainable.

Ortho-xylene.—The mother-liquors of the para-xylenesulphoni[c acid] contain, indeed, all the ortho-xylene, but do not, by simple de[com]position, produce any hydro-carbon of sufficient purity. It is b[est to] form first the lime salts, and to prepare the sodium salts, whic[h, in] ortho-xylene, are particularly easy to obtain in fine crystals, ne[ed] not even to be re-crystallized in order to furnish, after decompos[ition,] a pure hydrocarbon with a constant boiling point between 143° [and] 145°. In experimental decompositions the sodium salt behave[s like] the para-compound. With dilute sulphuric acid or hydrochloric [acid] the decomposition in the pressure-boiler takes place quantitati[vely.] Dry distillation produces even worse results than with para-xylene

The portion of raw xylene which remained undissolved in sulp[huric] acid consists of paraffins, and not polymerized terpenes. A h[ydro]carbon, with its boiling point between para- and ortho-xylene, i[s not] present, according to the author's experiments.

After removal of the meta-xylene from larger quantities of [the] mixture by means of sulphuric acid, the two hydrocarbons might, [with] advantage, be obtained sufficiently pure for most purposes by fract[ional] distillation with a difference of 7° in the boiling point. The sulphonic acids become turbid on the addition of water, not on[ly on] account of the separation of the excess of sulphuric acid, but because, besides a few aromatic hydrocarbons, they contain mec[hani]cally dissolved olefines and the like, which separate out on diluti[on, as] mentioned under meta-xylene. If the operation were conducted [with] sulphonic acid only, the mechanical separation from these ac[com]panying bodies is to be effected, if necessary, by dilution or hea[ting,] before the preparation of the hydrocarbons can take place. If sa[lts of] sulphonic acids are used, these impurities are not taken into acc[ount.]
—*Chem. Zeitung.*

CARBURETTED WATER GAS.

remembered that a few months ago two workmen lost
s at the Leeds Forge in consequence of inhaling water
employed at these works for welding purposes. The
naturally given rise to a strong prejudice against the use
nd various efforts have been made to overcome so serious
to its use for manufacturing or other purposes. Among
I. Fourness, of Manchester, has been carrying out experi-
is direction; and he appears to have been successful in
object in view, inasmuch as he has patented a process for
the gas, which he claims not only destroys all its poison-
es, but produces a gas of high illuminating power.
s is well known, is produced by steam coming in contact
scent coke or other fuel, the result being the giving off of
e gas. With the water gas thus produced Mr. Fourness
latile oil while the gas is in a heated state—a thorough
the two being thus secured, and the water gas effectually
By this process, as already stated, the poisonous car-
gas is destroyed, and to the water gas is imparted high
properties; gas manufactured by this process having been
22-candle power. Mr. Fourness claims that gas which is
iceable for welding and heating purposes as heretofore,
n be employed both for lighting and domestic purposes,
factured by his process at a cost not exceeding 6d. per
feet, as compared with 1s. 5d. per 1,000 cubic feet, which
be the cheapest rate at which coal gas is produced. The
gas thus manufactured by the Fourness process is very
t of ordinary coal gas, so that any escape can be readily
hile another distinguishing feature is that the water gas
is rendered permanent, and may be stored in holders for
le period without decomposing. So far Mr. Fourness has
ctured his gas in small experimental apparatus; but we
opportunity of inspecting the product, and certainly, so
nating properties are concerned, it compares very favour-
dinary coal gas. If Mr. Fourness's calculations as to
ect, it certainly promises to be an important step in pro-
w and cheap illuminant.—*Journal of Gas Lighting.*

SS ENAMELLED STEEL CASKS FOR RS AND MANUFACTURING CHEMISTS.

namelled steel casks are now extensively made in America
in breweries and processes of a similar nature. They are
use as filters in glucose and sugar refineries, evaporating
t works and other purposes, and to replace wooden tubs
breweries, and have been widely adopted by the latter in
ith the well-known vacuum system of ageing and ripening
body of the cask is composed of a series of welded steel
in thickness, with right angle flanges at each edge. The
mped from single sheets of steel ⅝ in. in thickness in a
raulic press, and the inside is coated with a glass enamel
the steel at a high heat. These sections and heads are
ly together with ⅝ in. bolts, two inches apart, and the
-enforced by continuous steel washers ⅜ in. thick. Be-
ints a very thin asbestos and plumbago packing, odorless
, is used. The height of each ring is 30 in., and the
in. The heads are dished 10 in. and given a smooth,
beautiful form in the hydraulic press. Each ring is made
ous sheet of the best homogeneous steel, 35 in. wide,
circular form, when the ends are welded together by a
ss, making a continuous weld 35 in. long of ¼ in. steel.
e then placed in a flanging machine, and a flange is spun
right angles 2½ in. wide. The welding and flanging of
re beautifully done, and are triumphs of fine work in steel.
together with four rings or sections, they make a cask
t 110 barrels, having the strength of steel outside and the
tinuous glass inside, and are odourless, tasteless, acid-proof,
lly indestructible. They are also air-tight under both
internal pressure. They have cast iron legs made adjust-
e inequalities of the floor, and are fitted with convenient
other openings. For the purposes designed, they are both
nequalled. Each section, including the heads, is enamelled
d revolves while heating in a furnace specially constructed
se, by which the enamel is burned on uniformly at all
casks can be drilled at any point without chipping the
ch shows a union between the steel and enamel of ex-
enacity, which is not found in the case of enamelled cast

THE HORSE-POWER OF BOILERS.

IT should be remembered that the horse-power for boilers is largely a matter of arrangement and varies according to the nature of the work to be done by the steam generated. As the horse-power of an engine is usually empirically figured from the area of the cylinder, so is the power of boilers not unfrequently estimated from their size merely, without regard to those other conditions which in practice are all important—such as setting, size of grate, heating surface, rate of combustion, etc. To meet some of these difficulties in the way of fairly estimating, in comprehensive fashion, the comparative horse-power of different boilers, it is considered best, on the whole, to reckon the power of a boiler by its capacity of evaporating water under given conditions. It is found that with a moderately good engine 30 pounds of water (raised to its equivalent in steam) will develope one horse-power (indicated) per hour. Bearing this fact of practice in mind the United States experts have agreed to regard the nominal horse-power for a boiler as an evaporation of 30 pounds of water (about one-half cubic foot) per hour into steam at 70 pounds steam pressure, the water being fed into the boiler at 200 degrees Fahrenheit.

As will be seen, the above figuring is confined to the boilers. When we convey steam from the boiler to the engine and seek to convert it into power, under control and useful, we are confronted with a new set of problems. These are too numerous and complex to be considered here. Briefly, as illustrating what we have to say, it is found that the ordinary slide-valve engine requires the evaporation of between 50 and 60 pounds of water per hour to produce one indicated horse-power, whilst large condensing engines require about 20 pounds, and compound condensing even less. Turning to another and important aspect of the subject, we are indebted to Jas. E. Denton, M.E., professor of experimental mechanics at the Stevens Institute of Technology, Hoboken, N. J., for some useful particulars upon the subject of boilers in connection with heating purposes, and as his words have the merit of being at once simple and practical, we produce them here:

"For heating purposes the horse power of a boiler is measured by the cubic feet of space it will heat; the cubic feet of space is given a radiating surface in the proportion of 1 square foot of radiating surface to about 100 cubic feet of space, and the boiler is given heating surface in the proportion of 12 square feet of radiating surface to 1 square foot of heating surface in the boiler, assuming that each square foot of the latter can evaporate 3 pounds of steam per hour at about 70 lbs. pressure, and finally the boiler maker sells the boiler on the assumption that each 15 square feet of heating surface is a horse power, which, evidently, is equivalent to $15 \times 3 = 45$ pounds of steam for a horse power. For steam heating practice, therefore, a 100 horse power is practically expected to be one that will evaporate 4,500 pounds of water per hour, and should a boiler be tested and rated at 100 horse power on the 30 pound basis, independently of the kind of work it was to do, that boiler, if used for steam heating, might fail to heat as large a space as steam heaters had been accustomed to heat with a 100 horse power boiler."

From what we have said it will be seen that the horse power of a boiler is subject to considerable variations, depending largely upon the purposes for which it is required, as also upon the engines it has to feed.

THE RECKLESS USE OF EXPLOSIVES.

ONE of the most fruitful sources of accident in the use of dynamite, says *The Times*, is the thawing of that explosive in order to render it plastic and fit for use. It is a peculiarity of dynamite that it solidifies at the comparatively high temperature of about 40° Fahrenheit, when it becomes inert. To thaw it tin warming pans are, or ought to be, used, although they are not always provided, as they undoubtedly should be. But provided or not, it not unfrequently enters the heads of labouring miners to thaw the cartridges of dynamite in their own wilful way. And they are very ingenious in devising methods of doing this in the most criminally unsafe manner possible. A reference to the reports of her Majesty's inspectors of explosives reveals the fact that from the winter of 1871-2 to February 18 last no fewer than 63 accidents have occured in the United Kingdom through the improper thawing of dynamite. As a result of these accidents 50 lives have been sacrificed, whilst in addition 76 persons have been more or less seriously injured. These reports also describe the methods by which these inevitable accidents are brought about, and which are either due to ignorance or to sheer recklessness. As a matter of fact, instances are recorded of workmen frying, boiling, toasting, and baking dynamite with the view of thawing it. Up to February last the idea of steaming it like a potato does not appear to have occured to any one. This method, however, was adopted at Colwill quarry, near Egg Buckland, Devonshire, with the result that two men were killed. In this case the practice was to use an old paint can half filled with water and placed on a sledge hammer head over the smithy fire. Over the can was placed a piece of canvas, varied with an old straw hat, and on the canvas or in the hat the cartridges were laid to thaw. The process of thawing was being carried on in this way when an explosion occured which killed the two men. What had happened was that in previous heatings of cartridges some of the nitro-glycerine had exuded from the dynamite, and had filtered through the sacking or the straw hat into the pot. Being heavier than the water the nitro-glycerine sank to the bottom, and upon the temperature reaching the exploding point (350° to 400° Fahrenheit) the catastrophe occurred. In the evidence given at the inquest it was shown the agent who supplied the dynamite took no steps to acquaint the purchaser of the proper method of thawing dynamite nor to inquire as to his knowledge of the process. Hence, Major Cundill, R.E., in his report on this accident does not hold this dealer blameless. With carelessness on one hand and recklessness on the other it is not easy to see how these malpractices are to be stopped. A salutary lesson might, however, be taught if some responsible person, not an unlucky, ignorant labourer, were tried for manslaughter.

PRIZE COMPETITION.

A LITTLE more than twelve months ago, one of our friends suggested that it might be well to invite articles for competition, on some subject of special interest to our readers; giving a prize for the best contribution, he at the same time offering to defray the cost of the first prize, and suggesting a subject in which he was personally interested.

At that date we were totally unprepared for such a suggestion, but on thinking over the matter at intervals of leisure, we have come to think that the elaboration of such a scheme would not only prove of benefit to the manufacturer, but would go some way towards defraying the holiday expenses of the prize winner.

We are therefore prepared to receive competitions in the following subjects:—

SUBJECT I.

The best method of pumping or otherwise lifting or forcing warm aqueous hydrochloric acid of 30 deg. Tw.

SUBJECT II.

The best method of separating or determining the relative quantities of tin and antimony when present together in commercial samples.

On each subject, the following prizes are offered:—

One first prize of five pounds, one second prize of one pound, five additional prizes to those next in order of merit, consisting of a free copy of the *Chemical Trade Journal* for twelve months.

The competition is open to all nationalities residing in any part of the world. Essays must reach us on or before April 30th, for Subject II, and on or before May 31st, in Subject I. The prizes will be announced in our issue of June 28th.

We reserve to ourselves the right of publishing the contributions of any competitor.

Essays may be written in English, German, French, Spanish or Italian.

WASTE ACIDS FROM GALVANISING WORKS.

ALL of our readers who are engaged in the iron trade are aware how difficult it is for the manufacturers of galvanised iron to deal with the spent acids which have to be got rid of during the process. In various parts of the kingdom this difficulty has brought manufacturers of galvanised iron into collision with the local authorities in connection with the contamination of streams and rivers, and simple though the subject may at first sight appear, it is undoubtedly one of the greatest evils with which manufacturers in this class of business have to contend. We have, therefore, pleasure in calling attention to the result of experiments which have for some months past been carried on by Mr. Thomas Turner, lecturer on metallurgy at Mason College, Birmingham. There are several systems in use at present, one of the best of which is the lime-purifying system. Under Mr. Turner's new process the waste liquor is merely boiled down to dryness, and the solid residue heated to low redness. Oxide of iron remains in the furnace, while free hydrochloric acid distils off, is condensed, and can be used over and over again. The new process differs from all preceding methods, in that it requires no chemical agents, that nothing whatever is thrown away, that no nuisance is made, and that a clear profit is realised. As a result of experience it is found that the fu will work three months without stoppage, that they only use 4 fuel to treat completely a ton of waste liquor, that the acid rec is perfectly suitable for using over and over again, and that the of iron recovered has a value which goes a long way towards for the fuel used. The experiments have been conducted ar system perfected at the works of Messrs. Walker Bros., of Wals Netherton, one of the largest firms of galvanisers in Staffor There is great need for some efficient system, otherwise the galva and sheet iron trades, which have immensely developed in the Mi during the last ten or fifteen years, and which at present employ sands of operatives and hundreds of thousands of capital, will ce continue—as they have already begun to do—to drift to Lanc Bristol, the banks of the Thames below London, and other place the coast, where the waste acids can be readily disposed of by allowed to flow out to sea with the tide.

[We extract the above from the *Engineer*. The account process would have been more perfect if a description had been of the vessel in which the waste liquor is to be boiled down to dr —Ed. *C. T. J.*]

A RISE IN BRIMSTONE IN AMERICA.

THERE has been a disposition for some time past, amon buyers of Sicily brimstone, to refrain from purchasing, ow a belief on their part that values were on a higher basis than th ditions of the market and the statistical position of stocks warr and the theory has been advanced at various periods, that prices sooner or later show a decline, in which event they would suppl wants. Dealers, on the other hand, have as strenuously cont that there was nothing in the outlook to indicate lower value they have in many instances warned buyers against "holding off long, and the tenor of late cable advices has caused the dealers t more than ordinary faith in their predictions of higher figures. which were received here on Saturday, proved conclusively that have waited a little too long before effecting their purchases, and result of their policy, they will be forced to pay higher pric market having materially advanced, since the date of our last and the cables prognosticate a still further advance. The chief r assigned for the course of the market, are the high asking r freight, and the lack of supplies at the Sicily ports, there bein said, a decrease of some 50,000 tons since January first, whi been brought about by circumstances previously noted in these col The circulars, it is asserted, regarding the stocks of Sicily, have incorrect and misleading, and according to dealers, there is prospect of a "squeeze," for lots up to August shipment, parc this position being quoted a trifle less than those for shipment that date, while the amount in store, of Sicily, being infinitesim relief from this quarter may be expected. True, there is some brimstone here, but this is also held at high figures and the qu is not exorbitant. Taken altogether, the outlook for buyers appear rather blue, as the market has every indication of rulin for a considerable time to come, unless something unforeseen occur, which seems hardly probable at the moment.

Trade Notes.

THE BRITISH ASSOCIATION: LEEDS MEETING.—The provi arrangements have been made for the sixtieth meeting of the Association, which will be held in Leeds during September next Monday, September 1st, the Reception-room will be opened Victoria Hall. On Wednesday, the 3rd, at eight o'clock, the r president will resign the chair in favour of Sir Frederick Augustus C.B., F.R.S., who will deliver the inaugural address. On Thu at eight o'clock, a soirée; on Friday, at half-past eight, a lecture Saturday, a lecture to the operative classes, by Professor Perry, F on Monday, at half-past eight, a lecture; and on Tuesday, at ei soirée. The lectures will be given by C. V. Boys, F.R.S., and Poulton, M.A., F.R.S. The sections will sit daily under their r tive presidents. A.—Mathematical and Physical Science, J. Glaisher, D.Sc., F.R.S., president. B.—Chemical Science, Pr E. Thorpe, F.R.S., president. C.—Geology, Professor A. H. M.A., F.R.S., president. D.—Biology, Prof. A. Milnes Ma M.A., F.R.S., president. E.—Geography, Lieut.-Colonel Lambert Playfair, K.C.M.G., president. F.—Economic Scienc Statistics, Prof. Alfred Marshall, M.A., president. G.—Mech Science, Captain A. Noble, C.B., F.R.S., president. H.—A pology, John Evans, D.C.L., Treas. R.S., president. The afte of Saturday, September 6th, and Thursday, September 11th, v devoted to visits and excursions to places of interest in Leeds a neighbourhood.

NDER GREAT PRESSURES.—In the coal mines at Kla-ia, there are located two pairs of compound pumping orm a notable plant. They drive double acting plunger in. stroke for one engine and 3 foot stroke for the other, beed of from 40 to 72 revolutions per minute. The ated 1700 feet below the surface of the ground, and they inst this whole head, doing the work with ease and . The pumps are the invention of Professor Ricldcr, nic Institute at Berlin, and the design has given remark-erever used.

IN BEER.—There is a revival of the rumour that Nux-e being used to give bitterness to beer, as a substitute it can be none other than an idle rumour, because no e foolish enough to run the risk. He needs not to run drop, in a gallon of beer, of picric acid, the bitterest igs, would make a bung-hole twist again, to say nothing er's mouth. If they would only keep out of his beer hamomile flowers, colombo root, burnt sugar, liquorice, lt, the consumer needs not to concern himself about the When the beer is hard or sour, when it is thick and ter is the place for it. When the beer is cloudy, there oducts of a second fermentation and God knows what. ay that when the beer is clear the beer is pure?

E.—It is not a new story this about the use of Nux-manufacture of bitter beer. It doubtless had its origin ne bitter-making properties of the drug (1 part in parts of water is still perceptible), and in the greatly rtation of it from India. But the latter is explained by ise of it as a poison for vermin, especially in Australia. eeds are yielded by the Koochla tree (*Strychnos Nux-* h flourishes in Ceylon, on the Coromandel coast, and India. Strychnine, so popular with our homœopathists ative Asiatic practitioners, is the deadly poison in Nux--the same poison that is got from St Ignatius's beans, angostura bark. It is the dread arrow poison—the outh American Indians.

E OF FRENCH COCHIN-CHINA.—Mr. Tremlett, the at Saigon, in his report on the trade of Cochin-China ear, observes that crops were short and much suffering ig the natives in consequence. This is not so apparent here a large number of people find employment; but ance from the capital the poorer classes have long been abour for food. The rice exports last year were only or 220,000 tons less than those of the previous year. e Straits, and the Philippines are the chief markets for ongkong taking more than half the total export. Last rade with Japan in raw cotton began, which is likely to nportance in view of the development of the cotton t country. All the trade statistics for the year show a -off, owing to the inferior rice crop and partly to the eavy duties levied upon foreign imports, which French manufactures do not replace. The tariff continues to pon foreign goods, particularly piecegoods, which still ur with the natives, upon whom falls the burden of duty. factures offer little relief, not being suitable either in e; the tax upon foreign tissues was increased upon July othing more has been heard of a projected loan; it has, into oblivion, like so many others. The Consul is er, that relief of some kind must be found for the lget.

URING IN NATAL.—The Government of Natal, has es and regulations for the law to make provision for y rewards, colonial manufacturing industries. Its two it features are, the appointment of a Commissioner of l the offer of rewards, varying in amount from £100. to he establishment of new industries, of which, twenty are e amount of rewards extends from one-eighth to two-lue of the out-put. The amount of the capital to be in-. from £500. to £25,000. The aggregate amount of d, is £25,000. The principal industries which the ishes to build up are leather, woollen, and iron. The industries mentioned in the schedule, embraces woollen blankets, and rugs from colonial wool. The value of be produced in one year and a half, is about £8,000., on d or bonus of £3,000. will be paid, while the amount of d in plant must be £15,000. Class 5 embraces —tweeds, serges, and flannels from colonial wool. The ital to be invested is £7,500., and the value of the out-ar and a half, must equal £5,000., on which a bonus of e paid by the Government. Class 6 embraces yarns om colonial wool. The amount of capital to be invested, 500., and on the output of a value of £1,000. per annum, it will pay a bonus of £300. Natal is determined, if she p home manufactures.

A GLASS COMPANY'S OFFER.—The Sumner Glass Company, of Steubenville, has made a proposition to the business men of that place for an enlargement of the works which will double their capacity and semi-monthly pay roll. The company proposes to the citizens that if the latter will raise $4,000. and present to them, they will immediately erect another furnace 85 feet high, holding 12 large pots, with accompanying lears, etc. This would give employment to 100 or 120 hands, and make the pay roll not less than $1,500. per week. The entire cost of the improvement is about $15,000., which the company is not able to raise from its own funds, but it can be accomplished with a little help from citizens. The money has been about all raised.

A GUSHER IN WEST VIRGINIA.—A Parkersburg, W. Va., press telegram announces that "the Johnson and Brockmier well, on the river bank at Belmont, which came in recently as a 750-barrel a day well, and which has been flowing about 500 barrels a day ever since, is now doing over 1,000 barrels a day. When it first came in, the drill had just merely touched the oil sand, and yesterday the drill was again dropped in the well and the sand slightly agitated. As a consequence the oily stream began spurting out at the rate of over 1,000 barrels a day, and is keeping it up. Nine other wells in the immediate vicinity are expected in at any time. Some of them are on the top of the sand."

METAL THAT MELTS EASILY.—A metal that will melt at such a low temperature as 150° is certainly a curiosity, but John E. White, of Syracuse, N.Y., has succeeded in producing it. It is an alloy composed of lead, tin, bismuth, and calcium, and, in weight, hardness, and colour, resembles type-metal. So easily does it melt that, if you place it on a comparatively cool part of the stove with a piece of paper under it, it will melt without the paper being scorched. Another peculiarity about it is, that it will not retain heat, and becomes cold the moment it melts. It is used in the manufacture of the little automatic fire alarms for hotels, and which give an electric alarm when the metal melts, owing to the rising of the temperature by fire.

INDIAN FORESTS.—According to the last official return on the subject, there were in India at the close of 1889 54,917 square miles of forest demarcated and reserved by the State. The area has increased especially since 1877-8. In that year it was only 17,705 square miles; in the following year it amounted to 40,425 square miles, in consequence of the energetio operations carried on in the central provinces. These latter have now the largest area of reserved forests of any province in India. It amounts to 19,712 square miles, Bombay coming next with 10,236. The areas elsewhere are:—In Lower Burmah, 5,111, Bengal, 4,988, Madras, 3,727, North-Western Provinces and Oude, 3,727, Assam, 3,447, the Punjab, 1,535, and Berar, 1,059 square miles.

NEW COMPANY IN CANADA.—Gas and oil, natural gas and petroleum oil, are what the Wentworth Gas and Oil Company is principally after. This company, of which Messrs. Lewis Springer, Barton, R. R. Waddell, J. V. Teetzel, J. N. Waddell, P. C. Brown, and F. R. Waddell, Hamilton, are the provisional directors, want incorporation to acquire land, sink wells, and construct machinery thereon for the purpose of obtaining natural gas, and to lay pipes conducting the same to the city of Hamilton, also for purchasing refineries, plant and machinery, and buying, selling, and producing salt, crude petroleum, oil, etc. Their proposed capital is $45,000., and they expect to buy a lot of property in a direct line with the "Well and Flew," whatever that is.—*Petrolea Topic.*

AN INDIAN NATIVE MINT.—Captain Temple, in an article on the coins of modern Punjaub chiefs in the new *Indian Antiquary*, describes the Patiala Mint and the methods of minting practised there. The Mint, he says, is an ordinary Punjaub courtyard, about 20 ft. square in the open part, entered by a gateway leading into a small apartment doing duty as an entrance-hall, the remainder of the courtyard being surrounded by low sheds opening into it. These buildings, which look like the "rooms" of a *serai*, are the workshops. The method of coining in this very primitive "Mint" is described as follows: "The silver after being assayed is cast into small bars by being run into grooved iron moulds. The melting is done in the courtyard in very small quantities in little furnaces improvised for each occasion. The thickness of the bars is about the diameter of the rupee, and when cold they are cut up by a hammer and chisel by guess work, into small weights and weighed in small balances as accurately as hand weighing will permit. These are afterwards heated and rounded by hammering into discs, and again weighed by hand and corrected by small additions of silver hammered in cold or by scraping. After this the disc is handed over to the professional weigher, who finally weighs it by hand and passes it. It is then stamped by hammering, being put between two iron dies placed in a strong wooden frame. These dies are very much larger than the coins, so that only a portion of the legend can come off, and the coiners are not at all careful as to how much appears on the coin. The only thing they do is to try and make the particular mark of the reigning chief appear. If they do not succeed it does not matter much."

LONDON WATER SUPPLY.—The Vestry of St. James', Westminster, have invited the other vestries and district boards of the metropolis to appoint delegates to a conference shortly to be held for the purpose of considering the whole subject of metropolitan water supply, and especially the desirableness of asking the Government to introduce a Bill (1) either to confer on the County Council power to acquire the present undertakings or to establish a competing supply; and (2) to require the water companies to supply water by meter at a fixed tariff.

CONVENTION OF GERMAN SODA FACTORIES.—The central office of the ring is in Bernburg, and the object of the convention is the regulation of the output and sale of ammonia soda and soda crystals throughout Germany. The new firm, which comprises eight works—viz., the German Solvay Works, Bernburg; Chemical Works, Buckan; R. Sourmondt & Co., Montwy; the Nüremberg Soda Works; Engelcke & Krause, Trotha; Rothenfelder Company, Rothenfelde; G. Egestorff's Salt Works, Linden; and the Chemical Factory, Schöningen—will be known as the "Syndicate of German Soda-works," and is under the management of Dr. Karl Wessel, in Bernburg. The Leblanc Works are outside the combination.

Market Reports.

TAR AND AMMONIA PRODUCTS.

The holidays have not interfered much with the course of business, as little Benzol has been changing hands, and, in the meantime, prices remain as quoted in our last week's report; 90s. benzol being value for 3/9, and 50/90s. for 2/9. Crude carbolic acid is still under a cloud, though there is no doubt that large quantities are still finding a ready outlet. Pitch remains weak, chiefly on account of the season being over and the hot weather that has arrived. Anthracene may be quoted at 1s. per unit for good quality B.

The sulphate of ammonia market remains practically unchanged, and, though quiet, it cannot be called, by any means, flat. At Hull £12. 5s. is to-day's value, and the same may be quoted for Leith. In London, things remain *in statu quo*, though prices of Beckton are a trifle lower than at this time last week.

MISCELLANEOUS CHEMICAL MARKET.

Business during the week has been interfered with by the Whitsuntide holidays, which are very generally observed in the large manufacturing centres. In the alkali trade transactions have been small owing to this cause, and prices are nominally as they stood at the end of the past week—viz., for 70%, £8. 5s. to £8. 10s.; 60%, £7. 5s.; and cream 60%, £7. 2s. 6d. f.o.b. Liverpool. For 74% and 77% prices are £9. and £9. 15s per ton respectively, f.o.b. Liverpool and Tyne. Soda ash varies between 1⅜ and 1½ per deg. for carbonated 48% to 58%; but makers are fairly well engaged with orders for some little time forward, and buyers do not find it easy to obtain early supplies at low prices. Soda crystals are fairly strong, and prices unaltered. Chlorate of potash and soda 4¾d. and 6d. per lb. respectively. Vitriol and muriatic acids are quiet. Bleaching powder steady at £5. per ton rails; £5. 7s. 6d. f.o.b. Sulphur without change. Sulphate of copper steady at £25. to £25. 10s. for early delivery. Salammoniac and muriate of ammonia are moving off more readily. Other ammonia salts still quiet, and prices unaltered. Acetate of lime: Brown without change, and only a small amount of business being done. Acetic acids of all kinds dull. Potash: Caustic and carbonate are moving readily, and prices are well maintained. Oxalic acid has again settled down, and is quiet at 3¾d. per lb. Prussiate of potash moving slowing at 9½d. per lb. Sulphocyanides firm in price, but little or no business doing. Tartaric and citric acids steady at 1s. 2¾d. and 1s. 4d. per lb. Chromic acid quiet at 6½d. per lb. Bichromates without change.

REPORT ON MANURE MATERIAL.

During the past week, the holidays have interfered with business, and there is very little to report.

Quotations for mineral phosphates remain without alteration, but there is a greater disposition on the part of manure makers to be covering part of their requirements ahead, and we shall expect before long to hear of a considerable business doing. It is still uncertain how far the Florida phosphate may affect prices generally, but trial shipm made seem to be satisfactory, and this new source of supply cannot think, fail to influence the market in buyers' favour, though it take some considerable time for it to have its full effect.

There are buyers of cargoes of River Plate bones, spring shipm at £5. 5s. per ton, delivered U.K. ports, but sellers want 2s. more. There are buyers of crushed Kurrachee bones, May- shipments, at £4. 17s. 6d. ex quay, Liverpool, but shippers have advanced their price to £5. 2s. 6d., which figure of course ent precludes further business at present. East Indian bone meal conti to sell at £5. 10s. ex quay and store, and we understand that £5 has been bid and refused for a large line, Summer-Autumn shipm Shippers seem afraid to commit themselves to very extensive sales, knowing what opposition they may have to contend with in buyin the other side to cover their sales here. There is a demand for pa of uncrushed bones, and £4. 12s. 6d. would readily be paid for g clean bones—perhaps £4. 15s. for anything specially good.

Nitrate of soda remains very quiet at 8s. per cwt. on spot. cargoes are worth barely 8s. May-June sailing may be quoted al 8s. 3¾d., and September-October-November about 8s. 7½d. per

We have no alteration to note in dried blood or ammonite.

There is a quiet market for kainit at 40s..

There is a less demand for superphosphates, and slightly e prices would be accepted—say 45s. in bulk, nett cash, delivere rails at seller's station.

THE TYNE CHEMICAL REPORT.

MONDA

Owing to it being Bank Holiday to-day, there has been no busi done on 'Change. Chemicals since last report have continued steady, with no material alteration in prices. Soda ash, crystals, caustic are in good demand, and last week's rates are fully maintai Bleaching powder is rather neglected at the moment, but quotat remain unchanged. Sulphate of soda quiet and little do The following may be taken as the prices current:—Bleac powder, in softwood casks, £5. to £5. 2s. 6d. per t caustic soda, 77%., in drums, £9. 5s. per ton, ground, in 3-4 barrels, £12.; soda ash, 48/52%, 1¼d. nett to 1¼d. less 2½%; crystals, in 1 cwt. bags, net weight, £2. 15s. per ton, in 2 cwt. b net weight, £2. 12s. 6d. per ton; in casks, gross weight, £2. 12s. per ton; sulphate of soda, in bulk, 32s. per ton; ground and packe casks, 42s. per ton; recovered sulphur, £4. 5s. per ton; chlorat potash, 4⅝d. per lb.; silicate of soda, 75° Tw., £2. 10s. per ton; Tw., £3. 7s. 6d. per ton; 140° Tw., £4. per ton; hyposulphite of s in 1 cwt. kegs, £5. 15s. per ton, in 5-7 cwt. casks, £5. 5s. ton; pure white sulphate of alumina, £4. 10s. per ton; blanc £7. 10s. per ton; chloride of barium, £8. per ton; nitrate of bar crystals, £18. 15s. per ton; ground, £19. 5s. per ton; sulphid barium, £5 10s. per ton—all f.o.b. Tyne, or f.o.r. makers' works.

Small coals for manufacturing purposes are in good demand, prices rule very steady. Durham small, quoted, 9s. 6d. to 10s. ton. Northumberland steam small from 8s. to 8s. 6d. per ton.

WEST OF SCOTLAND CHEMICALS.

GLASGOW, Tuesda

Since last report, sulphate of ammonia has failed to recover f from the depression then becoming discernable, but on the cont has weakened persistently, until to day the highest value procura spot delivery, is £11. 2s. 6d. only; some buyers holding off for lower terms. Makers, however, have elected to play the waiting ga stocks not being out of the way for bulk, and have drawn the lin the figure named. Some makers profess themselves decided no take that, and with buying and selling opinions so much at varia there is scarcely any business doing. The aspect of the forward sec is unchanged, and there seems hardly anything carried through. the general market the chief feature is a recovery in caustics, pri pally noticeable in the 70-72 strength. Bleaching powder has m another slight declension, and the late idleness in both classes of Scotch bichromate manufacture is, if anything, accentuated. Para are unchanged, a healthy feeling prevailing all through, except in b ing oil, which now looks forward to the fall of the year for the ear chance of amendment. Most of the Scotch Mineral Oil Compa have declared working results for the past year, and, with the excep of the Burntisland Oil Company, which confesses a loss, these considerably better than had been generally anticipated. Chief pr current are:—Soda crystals, 52s. 6d. net Tyne; alum in lump £5.,

% Glasgow; borax, English refined, £29. 10s., and boracic acid, '. 10s. net Glasgow; soda ash, 1¼d. less 3½%, Tyne; caustic soda, te, 76°, £10., 70/72°, £8. 2s. 6d., 60/62° £7. 5s., and cream, 60/62° , all less 2½% Liverpool; bicarbonate of soda, 5 cwt. ks, £6. 5s., and 1 cwt. casks, £6. 10s. net Tyne; refined ali, 48/52°, 1½d. less 1¼%, Tyne; saltcake, 25s. to 26s.; aching powder, £5. 2s. 6d. to £5. 5s. less 5% f.o.r. Glasgow; bichro- .e of potash, 4d., and of soda 3d., less 5 and 6% to Scotch English buyers respectively; chlorate of potash, 4½d., less 5% any ; nitrate of soda 8s. to 8s. 1½d.; sulphate of ammonia, spot, . 2s. 6d. f.o.b. Leith; salammoniac, 1st and 2nd white, £37. £35., less 2½% any port; sulphate of copper, £25. less 5% Liver- l; paraffin scale, hard and soft, 2¾d. per lb.; paraffin wax, 120°, i-refined, 3¼d.; paraffin spirit (naphtha), 7d. a gallon; paraffin burning), 6¼d. to 7d. on rails Glasgow to all Scotch buyers glish orders negotiable about 1¼d. a gallon lower); ditto ricating), 865°, £5. 10s. to £6. per ton; 885°, £6. 10s.; and /895°, £7. 15s. to £8. Week's imports of sugar at Greenock e 16,906 bags.

THE LIVERPOOL COLOUR MARKET.

OLOURS are quieter. Ochres: Oxfordshire quoted at £10., ., £14., and £16.; Derbyshire, 50s. to 55s.; Welsh, , 50s. to 55s.; seconds, 47s. 6d.; and common, 18s.; Irish, 'onshire, 40s. to 45s.; French, J.C., 55s., 45s. to 60s.; M.C., to 67s. 6d. Umber: Turkish, cargoes to arrive, 40s. to 50s.; 'onshire, 50s. to 55s. White lead, £21. 10s. to £22. Red lead, . Oxide of zinc: V.M. No. 1, £25.; V.M. No. 2, £23., etian red, £6. 10s. Cobalt: Prepared oxide, 10s. 6d.; black, 9d.; blue, 6s. 6d. Zaffres: No. 1, 3s. 6d.; No. 2, 2s. 6d. Terra a: Finest white, 60s.; good, 40s. to 50s. Rouge: Best, £24.; ditto jewellers', 9d. per lb. Drop black, 25s. to 28s. 6d. Oxide of iron, ne quality, £10. to £15. Paris white, 60s. Emerald green, 10d. lb. Derbyshire red, 60s. Vermillionette, 5d. to 7d. per lb.

THE LIVERPOOL MINERAL MARKET.

ur market has continued to rule fairly steady during the past week. ganese: Arrivals practically nil; prices firm, without alteration. gnesite: No arrivals except a cargo at an outport, which has been very low, and sales, especially for forward delivery, are in the ds of buyers. Raw ground £6. 10s., and calcined ground £10. to . Bauxite (Irish Hill brand): The production is scarcely able to pace with the demand; prices, therefore, are very steady. Lump, ; seconds, 16s.; thirds, 12s.; ground, 35s. Dolomite, 7s. 6d. per at the mine. French Chalk: Arrivals comparatively small; prices —*i.e.*, 95s. to 100s. medium, 105s. to 110s. superfine. Barytes (car- ate), easy; selected crystal lump easier at £6.; No. 1 lumps, 90s.; , 80s.; seconds and good nuts, 70s.; smalls, 50s.; best ground £6.; selected crystal ground, £8. Sulphate: Improvement continues, prices ; best lump, 35s. 6d.; good medium, 30s.; medium, 25s. 6d. to 27s. common, 18s. 6d. to 20s.; ground, best white, G.G.B. brand, 65s.; mon, 45s.; grey, 32s. 6d. to 40s. Pumicestone quiet; ground at £10., specially selected lump, finest quality, £13. Iron ore somewhat easier. ao and Santander, 9s. to 10s. 6d. f.o.b.; Irish, 11s. to 12s. 6d.; nberland easier, 14s. to 18s. Purple ore quiet. Spanish ganiferous ore continues in fair demand. Emery-stone: Best nds are still scarce on spot: prices firm. No 1 lump quoted at £5. 10s. to £6., and smalls £5. to £5. 10s. ers' earth steady; 45s. to 50s. for best blue and yellow; impalpable ground, £7. Scheelite, wolfram, tungstate of soda, tungsten metal in fair request. Chrome metal, 5s. 6d. per lb. gsten alloys 2s. per lb. Chrome ore: steadier, prices unaltered. imony ore and metal have further advanced. Uranium oxide, 24s. 6s. Asbestos continues scarce, bringing increased prices. Potter's ore continues firm; smalls, £14. to £15.; selected lump, £16. to . Calamine: Best qualities scarce—60s. to 80s. Strontia steady; hate (celestine) steady, 16s. 6d. to 17s. Carbonate (native), £15. 16.; powdered (manufactured), £11. to £12. Limespar: English ufactured, old G.G.B. brand in demand, and brings full prices; for ground English. Felspar, 40s. to 50s.; fluorspar, 20s. to £6. ore in demand, firm at 22s. to 25s. Plumbago steady; Spanish, ; best Ceylon lump at last quotations; Italian and Bohemian, £4. 12' per ton; founders, £5. to £6.; Blackwell's "Mineraline," . French sand, in cargoes, continues scarce on spot—20s. 2s. 6d. Ferro-manganese and silicon spiegel easier. Chrome iron, per cent., £24. to £25. Ground mica, £50. China clay freely ing—common, 18s. 6d.; good medium 22s. 6d. to 25s.; best, 30s. 5s. (at Runcorn).

New Companies.

BIRMINGHAM OXYGEN COMPANY, LIMITED.—This company was registered on the 15th inst., with a capital of £50,000., in £5. shares, to carry on business as manufacturers of oxygen, and also of aerated waters, and for such purposes to adopt an unregistered agreement made respectively with Robert Dempster and John Dempster, James E. Longsen and R. A. Sacré. The subscribers are:—

	Shares.
R. A. Sacré, Knowle, Birmingham	200
W. J. Sacré, Higher Broughton, solicitor	5
J. E. Langdon, Royal Exchange, Manchester, merchant	1
G. W. Hulme, Rochdale, cotton spinner	1
T. Bardsley, Didsbury, yarn salesman	1
A. Melvin, Higher Broughton, clerk	1
G. Baron, Davenport, Cheshire, cashier	1

IRRIGATION SYNDICATE, LIMITED.—This company was registered on the 17th inst., with a capital of £10,000., in £200. shares, to acquire concessions for the establishment of irrigation, drainage, and water-works, from any State, Government, municipal, or other officers. The subscribers are:—

	Shares.
M. Epstein, 93, Albert-street, Regent's-park	1
D. Singer, 52, Leinster-square, Bayswater	1
G. Bergel, 17, Pembridge-crescent, W.	1
C. E. Lester, 180, East India-road, clerk	1
S. C. L. Brown, 3, Church-lane, Hornsey, stockbroker	1
H. Ansbacher, 7, Upper Bedford-place, stockbroker	1
C. P. Kirkham, 18, Willes-road, Kentish Town	1

MEXICAN PORTLAND CEMENT COMPANY, LIMITED.—This company was registered on the 15th inst., with a capital of £20,000., in £1. shares, 100 being founders' shares, to acquire upon terms of an agreement of the 14th, certain patent rights of J. C. Gibbon and H. W. Gibbon, of Mexico, relating to the manufacture of Portland cement and artificial stone in Mexico. The purchase consideration is 100 founders' shares and 4,000 ordinary shares, all credited as fully paid up. The subscribers are:—

	Shares.
H. K. Gow, 36 and 37, Leadenhall-street, solicitor	1
C. Baker, 6, St. Andrew's-villas, Gravesend, cement expert	1
C. W. Woods Curtis, 15, Ramberg-road, Upper Tooting engineer	1
A. E. Carey, C.E., 9, Dean's-yard, Westminster	1
E. C. Friend, Hayes-common, Kent, paint manufacturer	1
W. H. P. Stevens, Buscot, Barnet, secretary to a company	1
A. C. Hartley, 21, Medina-villas, West Brighton	1

The Patent List.

This list is compiled from Official sources in the Manchester Technical Laboratory, under the immediate supervision of George E. Davis and Alfred R. Davis.

APPLICATIONS FOR LETTERS PATENT.

Filtering Machinery. E. Martin. 7,361. May 12.
Electric Switches. G. S. Hooker. 7,373. May 12.
Do. do. A. Brüll. 7,378. May 12.
Electric Battery.—(Complete Specification). J. N. Levsen. 7,383. May 12.
Injectors.—(Complete Specification). W. P. Thompson. 7,388. May 12.
Decortication of Fibrous Vegetable Stems.—(Complete Specification). J. Longmore and W. L. Watson. 7,389. May 12.
Decortication of Fibrous Vegetable Stems.—(Complete Specification). J. Longmore and W. L. Watson. 7,390. May 12.
Injectors.—(Complete Specification). W. P. Thompson. 7,391. May 12.
Electric Heating. R. Kennedy. 7,399. May 13.
Steam Gauges. E. Outram. 7,412. May 13.
Refuse Destructors. W. Walkington. 7,419. May 13.
Production of Hydroxy-Quinone. R Holliday. 7,421. May 13.
Vacuum Pumps.—(Complete Specification). H. H. Lake. 7,426. May 13.
Brick-making Machinery.—(Complete Specification). R. Knickerbocker. 7,427. May 13.
Electric Switches. B. Dukes. 7,436. May 13.
Annealing of Metals.—(Complete Specification). H. H. Lake. 7,461. May 13.
Amalgamating Machinery. W. J. Smuts. 7,464. May 13.
Refuse Destructors.—(Complete Specification). L. Hesse. 7,500 May 14.
Unhairing Skins. F. R. Maggs 7,509. May 14.
Improved Brush for Dynamos. M. Immisch. 7,511. May 14.
Means for Preventing Polarisation in Primary Batteries. A. Walker and T. J. D. Rawlins. 7,518. May 14.
Liquid Metres. R M. L. G. Renaud. 7,519. May 14.
Improvements in Connection with Indigo Dyeing. W. Elbers. 7,522. May 14.
Improvements in Connection with the Ammonia Soda Process. J Vivian and G. Bell. 7,527. May 14.
Liquid-measuring Apparatus. J. Leloup. 7,537. May 14.
Filters. J. P. Bayly. 7,546. May 14.
Ammeters and Voltmeters. C. R. G Smythe and E. Payne. 7,560 May 15.
Preparing Animal Fibres for Spinning. K. T. Sutherland and G. Esdaile. 7,568. May 15.
Producing Air Gas. T. Le Poidevin. 7,569. May 15.
The Production of New Amido Bases. A. G. Green and T. A. Lawson. 7,575. May 15.
Electro Disposition of Heavy Metals. W. A. Thoms. 7,881. May 15.
Registering Apparatus for Paper-making Machines. C. J. Richardson. 7,598. May 15.
Improvements in Gas Holder Guides. E. L. Pease. 7,616. May 16.
Smoke Consumption. W. D. Grimshaw. 7,625. May 16.
Re-tinning Copper Utensils. C. Knight and W. F. Sharp. 7,647. May 16.
Pressure Pumps. C. Kingsford. 7,667. May 16.
Electric Meters. C. D. Abel. 7,670. May 16.
Treatment of Sewage. E L. Mayer. 7,673. May 16.
Manufacture of Soap. J. A. Clough. 7,706. May 17.
Measuring Oxygen or Air into Gas Purifier. C. B. Newton. 7,714. May 17.
Treating Chemically Softened Waters for Boiler Use. L. Archbutt and R. M. Deeley. 7,719. May 17.
Preventing the Access of Air to Iron and Steel during Annealing.—(Complete Specification). H. H. Lake. 7,733. May 17.
Manufacture of Water Gas. J. Von Langer and L. Cooper. 7,739. May 17.

Gazette Notices.

Partnerships Dissolved.

B. PIFFARD and W. H. CRANSTONE, under the style of the Patent Silicate Manure Co., Hemel Hempstead, manufacturers of artificial manures.

CLARKE & COOBAN, Blaenau, Festiniog, mineral water manufacturers.

THE BANKRUPTCY ACT, 1883.

Receiving Orders.

THOMAS WILLIAMS and JOHN HENRY THOMAS (trading as Grindley Chester, chemists and druggists.

LAWRENCE HAYWARD COOK, Prospect Hill, Walthamstow, and Amh Hackney, London, drysalter.

Adjudications.

GEORGE ELLIS MOSES, Manchester, drysalter.

IMPORTS OF CHEMICAL PRODUCTS

AT

THE PRINCIPAL PORTS OF THE UNITED KINGDOM.

LONDON.

Period ending May 17th.

Acetic Acid—
Holland, 73 pks. A. & M. Zimmermann
" 1 Barr Moering & Co.

Ammonia—
Holland, £28 W. H. Cole & Co.

Antimony—
Belgium, 3 t. Girard Bros. & Co.
Austria, 42 t. H. R. Merton & Co.

Antimony Ore—
New Zealand, 42 t. N. Z. Antimony Co.
France, 73 Smith Sundius & Co.

Alkali—
Holland, 200 c. Taylor Sommerville & Co.

Bones—
E. Indies, 200 t. Gonne Croft & Co.
Trinidad, 24 Beresford & Co.
U. S. 25 A. W. Nesbitt
E. Indies, 298 Gonne Croft & Co.

Boracic Acid—
Italy, 50 pks. J. Bath & Co.

Brimstone—
Italy, 20 t. G. Boor & Co.
" 5 F. Smith

Barium Chloride—
Germany, £80 Petri Bros.

Barytes—
Germany, 24 cks. O'Hara & Hoar

Carbonic Gas—
Holland, £15 Phillipps & Graves

Copper Ore—
Belgium, 30 t. A. Gibbs & Sons
Cossack, 20 t. W. Marden
Belgium, 10 F. Manders

Cobalt Ore—
France, 210 t. Newcomb & Son

Cream of Tartar—
Holland, 10 pks. Middleton & Co.
Italy, 20 W. C. Bacon & Co.

Caustic Potash—
France, £120 F. W. Berk & Co.

Caoutchouc—
Zanzibar, 64 c. L. & I. D. Jt. Co.
U. S. 160 J. E. Wilson
France, 71 c. Hernu Peron & Co.
E. Indies, 7 L. & I. D. Jt. Co.
" 90 G. Ward & Sons
" 67 Ross & Deering
U. S. 22 L. & I. D. Jt. Co.
Portugal, 400 c. Kleinwort Sons & Co.

Copper Sulphate—
Holland, £220 Middleton & Co.

Chemicals (*otherwise undescribed*)
Germany £820 A. & M. Zimmermann
" 16 Phillipps & Graves
" 60 Beresford & Co.
" 10 H. J. Pearlbach & Co.
" 113 A. & M. Zimmermann
Holland, 16 T. H. Lee

Dyestuffs—

Gambier
Singapore, 131 pks. Beresford & Co.
" 858 L. & I. D. Jt. Co.
" 1,182 R. & J. Henderson

Saffron
Spain, 1 pks. C. Brumlen

Cutch
E. Indies, 20 pks. Mee Billing & Co.
" 1,500 H. A. Litchfield & Co.
" 859 L. & I. D. Jt. Co.
" 200 Brinkermann & Co.

Indigo
E. Indies, 10 chts. G. S. N. Co.
" 13 Benecke Souchay & Co.
" 5 L. & I. D. Jt. Co.
" 26 pks. Ross & Deering
" 9 chts. Elkan & Co.
" 46 L. & I. D. Jt. Co.
" 11 J. Owen
" 8 Thomasset & Co.
" 62 L. & I. D. Jt. Co.
" 7 W. Isted
" 11 Parsons & Keith
" 2 Benecke Souchay & Co
" 10 L. & I. D. Jt. Co.
" 56 Arbuthnot Latham & Co.

Extracts
France, 100 pkgs. L. J. Levinstein & Son
Austria, 817 A. J. Humphrey
Holland, 155 Union L. Co., Ld.
" 221 Prop. Chmbln's. Wf.

Tanners Bark
Holland, 37 t. J. J. Williamson & Co.
" 50 J. Graves
Italy, 9 t. A. & W. Nesbitt
Holland, 9 S. Bevington & Sons
" 15 Dalgety & Co.
Melbourne, 37 t. Baxter & Hoare
Adelaide, 7 t. Devitt & Hett

Myrabolams
E. Indies, 3,391 pks. Marshall & French
Holland, 200 "

Sumac
Italy, 800 pks. J. Kitchin, Ltd.

Argols
Italy, 1,649 pks. B. Jacob & Sons
" 311 Thames S. Tug Co.

Alizarine
Durban, 12 pks. J. Forsey

Glucose—
Germany, 25 pks. 221 c. Craven & Co.
U. S., 100 650 L Sutro & Co.
" 20 120 T. H. Lee
France, 1 7 Carey & Sons
U. S., 200 1,215 T. M. Duche & Sons
" 112 600 Harne & Co.

Gutta-Percha—
Singapore, 266 c. Jackson & Zill
" 164 L. & I. D. Jt. Co.
" 516 Kleinwort, Sons & Co.
" 177 Kaltenbach & Schmitz

Isinglass—
E. Indies, 14 pks. Clarke & Smith

Manure—
Holland, 175 t. Lawes Manure Co.
" 195 Ang.-Contl. G. Wks.
" 195 A. Hunter & Co.
Germany, 50 "
" 11 Beresford & Co.
U. S., 5 C. A. & H. Nichols

Naphtha—
Holland, 41 drs. Burt, Boulton & Haywood

Potassium Carbonate—
Germany, 10 pks. F. J. Putz & Co.

Paraffin Wax—
U. S., 296 brls. 64 bgs. H. Hill & Sons
" 242 "

Pearl Ashes—
France, 100 c. F. W. Berk & Co.
Holland, 400 "

Potassium Muriate—
Holland, 4 cks. Spies Bros. & Co
" 26 H. Johnson & Sons

Potassium Hydrate—
France, 4 pks. Spies Bros & Co.

Potassium Sulphate—
Germany, 112 pks. Howards & Son
" 284 Bessler, Waechter & Co.

Plumbago—
E. Indies, 179 brls. W. H. Ison
" 91 Arbuthnot, Latham & Co.
Germany, 25 cs. T. H. Lee
" 63 "
E. Indies, 56 brls. J. Thredder, Son & Co.
Germany, 60 cs. T. H. Lee
E. Indies, 80 brls. Doulton & Co.
" 159 Props. Chamblns. Wf.

Pitch—
France, 95 brls. Berlandrina Bros. & Co.

Saccharine—
Portugal, £370 J. Hall Jun. & Co.

Sugar of Lead—
Holland, 14 pks. Beresford & Co.

Saltpetre—
E. Indies, 593 bgs. 200 cks. Henschell, du Buisson & Co.

Tartaric Acid—
Holland, 4 pks. Middleton & Co.
" 37 B. & F. Wf. Co.
France, 8 "

Tartar—
Italy, £1,400 B. Jacob & Sons
" 1,050 Seager, White & Co.

Ultramarine—
Holland, 3 pks. Ohlenschlager Bros.
Germany, 7 H. Brooks & Co.
Holland, 2 cks. Phillipps & Graves
Germany, 4 J. Kitchin, Ltd.
" 4 G. Steinhoff

Zinc Oxide—
Germany, 50 brls. M. Ashby, Ltd.
U. S., 200 "
Holland, 42 cks. Beresford & Co.
" 65 brls. M. Ashby, Ltd.
" 250 cks. Soundy & Son
" 200 brls. M. Ashby, Ltd.

LIVERPOOL.

Period ending May 20th.

Acetate of Lime—
New York, 818 bgs.
" 1,350

Albumen—
Havre, 15 cs. Cunard S. S. Co. Ld.

Acids (*otherwise undescribed*)
Leghorn, 53 cks.

Arsenic—
Santander, 169 cs.

Bones—
Boston, 5 tcs.
Ceara, 118 bgs. R Singlehurst & Co.
" 1,429 Leach, Harrison, & Co.

Bone Meal—
Kurrachee, 1,000 bgs. Ralli Bros.

Borax—
Leghorn, 150 cs.

Caoutchouc—
Brazil, 531 cs. R. Singlehurst & Co.
Ceara, 1 1 bl. "
Pernambuco, 19
Accra, 7 Davies, Robbins, & Co.
Salt Pond, 1 cs. Radcliffe & Durrant
" 1 ck. Edwards Bros.
" 2 A. Millerson & Co.
" 1 W. Griffiths & Co.
" 1 cs. A. & C. Nordlinger
" 2 ck. Pitt Bros. & Co.
" 2 8 cs. Pickering & Berthoud
" 1 E. G. Gunnel

Caoutchouc—*continued.*
Salt Pond, 8 cks. Miller B
" 3 cs Davie
Cape Coast, 8 2 Miller, I
" 1 1 Ed
" 3 W.
" 3 H. B.
" 7 Davies,
" 1 1 F. J. E
" 3 W. B.
" 1 S He
" 1 3 Fletche
" 7 Th
" 8 1
Axim, 3 bgs. 1 brl. Pickeri
" 5 1 Edw
" 1 A. Mille
" 4 Radcliffe
" 1 Millward, Bradb
Goree, 8 brls. S Johns
Bathurst, 4 cks. Ci

Copper Ore—
Leghorn, 26 t.
Sestri Levante, 1,025 t. H. B

Cream of Tartar—
Rotterdam, 10 cks.

Chemicals (*otherwise un*
Hamburg, 9 cs.
Rouen, 88 cks. Co-op. W

Dyestuffs—

Sumac
Palermo, 400 bgs. J. Kit
" 550

Valonia
Alexandria, 150 bgs. C
Patras, 840 C G. P

Fustic
Savanilla, 2,616 logs. G. H. M
Colon, 11 t. Melchers, Ru
Ceara, 5½ Leach, Harr
Corfu, 4 P. V. Marri

Argols
Bordeaux, 76 cks. 95 bgs.
Oporto, 23 Ge

Extracts
Rouen, 406 cks. Co-op. W'sal

Cutch
Calcutta, 400 cs.

Myrabolams
Bombay, 2,847 bgs. D. Sas
" 1,526
" 296 Beyts, C
" 1,335 D. Sas
" 986
" 505
Calcutta, 702 bgs.
Bombay, 80 R. C
" 41 Tod, Du
" 2,105 D. Sas
" 338

Glucose—
Hamburg, 24 cks. 200 bgs.
New York, 20 brls.

Horn Piths—
Brazil, 5 t. Gillies
Ceara, 73 bgs. R. Singlel

Manganese Ore—
Oporto, 110 t.

Nitrate—
Pisagua, 9,371 bgs.

Phosphoric Acid—
Bremenhaven, 3 cs. D. Cu

Potash Salts—
Bremerhaven, 14 cks. D. C

Phosphate—
Bruges, 600 bgs.
Potash—
Montreal, 10 brls. Makin & Bancroft
" 20 Haines & Co.
" 15
Potassium Prussiate—
Rotterdam, 4 cks.
Pyrites—
Huelva, 1,559 t. Matheson & Co.
" 1,340 Tenants & Co.
" 1,238 "
Paraffin Wax—
New York, 100 bdls. Meade, King, & Co.
Sulphur—
Palermo, 440 bgs.
Saltpetre—
Calcutta, 1,190
Kurrachee, 109 W. & R. Graham & Co.
Stearine—
Rotterdam, 3 cks. 20 bgs.
Salt—
Madeira, 140 bgs.
Tartaric Acid—
Bilbao, 30 brls.
Tartar—
Naples, 2 cks.
Bordeaux, 38
Rotterdam, 1
Waste Salt—
Hamburg, 494 bgs.
" 640

GLASGOW.

Week ending May 22nd.

Alum—
Rotterdam, 45 cks.
Barytes—
Antwerp, 100 bgs.
Chrome Ore—
Rotterdam, 2,100 bgs.
Dyestuffs—
Alizarine
Rotterdam, 89 cks. J. Rankine & Son
Madders
Rotterdam, 1 ck J. Rankine & Son
Extracts
New York, 27 cs.
Antwerp, 1 ck. 1 cs.
Fustic
Jamaica, 9 t. 15 c. Wallace Wilkie & Co.
Glycerine—
Rotterdam, 5 cs.
Glucose—
New York, 208 brls.
Manganese Chlorate—
Rotterdam, 1 ck.
Pyrites—
Huelva, 1,193 t. Tharsis Co.
Stearine—
Antwerp, 12 bgs.
Saltpetre—
Hamburg, 48 cks.
Sulphur Ore—
Bergen, 300 t.
Tartaric Acid—
Rotterdam, 4 cks. J. Rankine & Son
Ultramarine—
Rotterdam, 3 cs. J. Rankine & Son
Waste Salt—
Hamburg, 320 t.
" 271 cks.

TYNE.

Period ending May 20th.

Barytes—
Antwerp, 50 bgs. Tyne S. S. Co
Glucose—
Hamburg, 90 bgs. Tyne S. S. Co.
" 12 cs.
Rock Salt—
Hamburg, 130 t.
Salt—
Harburg, 130 t.
Tartaric Acid—
Rotterdam, 4 cks. Tyne S. S. Co.
Zinc Oxide—
Antwerp, 20 brls. Tyne S. S. Co

HULL.

Week ending May 22nd.

Acetic Acid—
Rotterdam, 10 pks. Hutchinson & Son
Acids (*otherwise undescribed*)
Rotterdam, 1 ck. W. & C. L. Ringrose
" 4 "
Barytes—
Bremen, 60 cks.
Antwerp, 38 brls. H. F. Pudsey
" 18 cks.
Bremen, 23 Tudor & Son
" 23 T. W. Flint & Co.
" 23 J. W. Davis & Son
" 18
Bones—
Trieste, 2 pks. Wilson, Sons & Co.
Chemicals (*otherwise undescribed*)
Dunkirk, 70 pks. Wilson, Sons & Co.
Antwerp, 138 "
Hamburg, 245 "
Bremen, 1 cs. 21 kgs. Veltmann & Co.
Genoa, 16 cks. Wilson, Sons & Co.
Marseilles, 5
Rotterdam, 1 W. & C. L. Ringrose
Dunkirk, 22 pks. Wilson, Sons & Co.
Dyestuffs—
Cutch, 1 bl. G. Lawson & Sons
Argols
Bari, 24 cks. Wilson, Sons & Co.
Extracts
Rouen, 58 cks. Rawson & Robinson
Fiume, 700 bls. Wilson, Sons & Co.
Dunkirk, 23 cks. "
Sumac
Bordeaux, 100 bls. Rawson & Robinson
Trieste, 20 bgs. Wilson, Sons & Co.
Palermo, 376 "
" 350
Alizarine
Rotterdam, 78 cks. W. & C. L. Ringrose
" 10 G. Lawson & Sons
" 5 W. & C. L. Ringrose
" 20 pks. "
" 21 cks. Hutchinson & Son
Rennet
Copenhagen, 8 cks. Wilson, Sons & Co.
Rotterdam, 5
Fustic
Hamburg, 317 pcs. Wilson, Sons & Co.
Glucose—
Hamburg, 6 cks. Wilson, Sons & Co.
" 8 C. M. Lofthouse & Co.
Stettin, 210 bgs. Wilson, Sons & Co.
" 11 cks. M. Whitfield & Sons
" 23 Hull Dock Co
" 200 bgs. Thompson, McKay & Co.
New York, 80 cks.
Hamburg 12 Bailey & Leetham
" 200 bgs.
" 12 cks. Bailey & Leetham
" 17
Kainit—
Hamburg, 300 bgs. G. Lawson & Sons
Manure—
Ghent, 160 t.
Rouen, 4 cks. Rawson & Robinson
Naphthaline—
Hamburg, 1 brl. Wilson, Sons & Co.
Oxide—
Rotterdam, 4 cks. Hutchinson & Son
Pyrites—
Huelva, 1,518 t. J. Dalton, Holmes & Co.
Pitch—
Dunkirk, [illegible]
Plumbago—
Hamburg, 46 cks.
" 21 bls.
Sulphur—
Catania, 240 pks. [illegible]
Turpentine—
Lisbon, 30 bls.
Tar—
Marseilles, 25 cks.

GRIMSBY.

Week ending [illegible]

Glucose—
Hamburg, 14 cks.
Kainit—
Hamburg, 500 bgs.

GOOLE.

Week ending [illegible]

Alum—
Rotterdam, 21 cks.
Barytes—
Antwerp, 30 bgs.
Chrome Alum—
Rotterdam, 30 cks.
Dyestuffs—
Alizarine
Rotterdam, 61 pks.
Madders
Rotterdam, 5 cks.
Logwood
Jamaica, 150 t.
Dyestuffs (*otherwise* [illegible])
Antwerp, 5 cs. [illegible]
Glucose—
Hamburg, 48 cks. [illegible]
Calais, [illegible]
Manure—
Phosphate Rock
Ghent, 800 bgs.
Potash—
Hamburg, 100 bgs.
Calais, 117 drs.
Dunkirk, 58 cks.
Saltpetre—
Hamburg, 20 cks.
Sodium Silicate—
Rotterdam, 1 ck.
Waste Salt—
Hamburg, 377 t. [illegible]

EXPORTS OF CHEMICAL PRODUCTS

OF

THE PRINCIPAL PORTS OF THE UNITED KINGDOM.

LONDON.

Period ending May 14th.

Ammonium Sulphate—
Rotterdam, 85 t. £1,020
Emmerich, 125 1,500
Arsenic—
Sydney, 1 t. £15
Ammonium Carbonate—
Sydney, 15 cs. £77
Benzoic Acid—
New York, 8 cs. £67
Boracic Acid—
Melbourne, 18 c. £36
Copper Sulphate—
Sydney, 2 t. £49
Genoa, 5 125
Citric Acid—
Libau, 1 c. £8
Bombay, 2 14
Yokohama, 5 cks. 38
Progreso, 1 cs. 20
Toronto, 6 43
St. Petersburg, 1 t. 149
Hamburg, 2 kgs. 43
B. Ayres, 5 brls. 105
Clarifiants—
Bordeaux, 45 cks. £90
Carbolic Solid—
Barcelona, 10 c. £70
Yokohama, 1,000 lbs. 57
Carbolic Acid—
Algoa Bay, 3 cs. £39
Calomel—
New York, 20 cs. £164
Copperas—
Melbourne, 5 t. 13 c. £13
Disinfectants—
Sydney, 43 cs. £91
Penang, 200 gls. £32
Guano—
Barbadoes, 30 t. £300
" 320 3,200
Manure—
Berbice, 20 t. £105
Rotterdam, 60 510
Antigua, 25 4 c. 306
Riga, 50 150
Nitric Acid—
Jersey, 2 t. 7 c. £43
Potash—
Melbourne, 5 c. £22
Gothenburg, 10 22
Potassium Cyanide—
B. Ayres, 2 cs. £20
Phosphoric Acid—
Demerara, 3 t. 10 c. £86
Potassium Chlorate—
Sydney, 5 t. £211
New York, 10 c. 28
Sheep Dip—
Algoa Bay, 1 t. 5 c. £66
Pt. Elizabeth, 6 15 300
Salammoniac—
Aden, 1 t. 1 c. £37
Sydney, 13 24
Salicylic Acid—
Singapore, 2 c. £17
Zinc Ashes—
Antwerp, 21 t. 16 c. £400

LIVERPOOL.

Period ending May 17th.

Bleaching Powder—
Barcelona, 70 cks.
Boston, 421 84 tcs.
Calcutta, 25 cs.
Genoa, 67
Halifax, 2
Hamburg, 27 20 kgs.
Lisbon, 29
Montreal, 32 tcs.
New Fairwater, 57 cks.

Bleaching Powder—*continued*.
New York, 452 cks. [illegible]
Rotterdam, 29
Vera Cruz,
Antwerp, 27
Bilbao, [illegible]
Boston, 188
Carthagena, [illegible]
Genoa, 58
Ghent, 23
Hamburg, 28
Havanna, 8
Leghorn, 10
Malaga, 5
Naples,
New York, 298
Philadelphia, 48
Rio Janeiro,
Rouen, 44
Caustic Soda—
Stettin, 6 drms.
Trieste, 73
Valparaiso, 70
Stockholm, 23
Adelaide, 32
Algiers, 5
Amsterdam, 23
Antwerp, 76 14
Batoum, 138
Bilbao, 20
Bordeaux, 37

Caustic Soda—*continued.*
Calcutta, 10 dms.
Dunkirk, 6
Galatz, 100
Genoa, 116
Gothenburg, 5
Hamburg, 498
La Guayra, 46
Leghorn, 28
Lisbon, 18
Madeira, 5
Madrid, 10 20 brls.
Marseilles, 45
Melbourne, 34
Messina, 80
Montreal, 483
Naples, 20
New Fairwater, 113 dms.
New York, 1300
Odessa, 250
Oporto, 30 8 cks.
Rotterdam, 26
Santos, 7
Bahia, 10
Barcelona, 300 80 20 brls.
Bilbao, 126 35
Bruges, 6 4
Genoa, 182
" 15
Gothenburgh, 4
Hamburg, 460
Hiogo, 70
Leghorn, 85
Lisbon, 94 11 4
Malaga, 10
Manilda, 50
MonteVideo, 200
Montreal, 100
New Fairwater, 29
New York, 250 80 brls.
Philadelphia, 135
Piræus, 85
Puerto Cabello, 11 dms.
Rouen, 52
San Sebastian, 40
Santos, 11
Seville, 25
Stettin, 27
Valencia, 41
Yokohama, 30
Gibraltar, 20
Rotterdam, 54

Epsom Salts—
Batoum, 13 cks.
Maranham, 8
RioJaneiro, 100 kgs.
Maceio, 1

Magnesia—
Antwerp, 1 cs.
Rio Janeiro, 10
Vera Cruz, 2 cks.

Salt—
New York, 25 t.
Old Calabar, 154¾ t.
Opobo, 45 9 c.
Para, 177¾
Philadelphia, 80
Port Mulgrave, 250 t.
Rangoon, 25
Rotterdam, 203
St. John, N.B., 800
Sandheads, 2145
Sydney, 50
Vadsol, 465
Winnebah, 10
Abonnema, 43
Ambriz, 5
Amsterdam, 261
Antwerp, 42
Benin, 22
Boston, 156
Calcutta, 1320
Cameroons, 4
Campbeltown, 150
Halifax, 1419¾
Hamilton, 6¾
Loango, 12½
Miramichi, 180
Montreal, 301 10 c.
Baltimore, 100
Benin, 23
Bonny, 36
Boston, 100 2 cks.
Bruges, 200
Chicago, 299
Coquimbo, 50
Elobey, 5
Grand Bassam, 12 t.
Hamburg, 50
Manaos, 52¾
Montreal, 223
Newcastle, 100
Old Calabar, 115¾
Rouen, 10
Shediac, 10
Sydney, 175
Victoria, 150
Vigo, 2

Sodium Bicarbonate—
Bombay, 10 kgs.
Calcutta, 500
Genoa, 15 cks.
Halifax, 25
Hamburg, 50
Havana, 25
Miramichi, 15
Montreal, 400 19
Nagasaki, 200
Odessa, 100
Palermo, 20
Lante, 2 brls.
Amsterdam, 20
Calcutta, 500
Hamburg, 20 dms.
Santander, 40
Smyrna, 15

Sodium Silicate—
Barcelona, 32 cks.
Bilbao, 32
Galatz, 15 brls.
La Guayra, 10
Lisbon, 32
Madrid, 10
Montreal, 42
Seville, 25
VictoriaB.C., 32

Soda Ash—
Oporto, 10 pks.
Philadelphia, 427 cks. 28 tcs.
Piræus, 36 brls.
Sherbrooke, 16 tcs.
Smyrna, 5
Varna, 60
Windsor Mills, 100
Yokohama, 222
Bahia, 20
San Francisco, 109 cks.
Adelaide, 79 brls.
Baltimore, 509
Bilbao, 50
Boston, 397 130 tcs.
Constantinople, 40 brls.
Genoa, 7 cks.
Hamburg, 4
Hamilton, 120
Kingsey, 14
Leghorn, 42
Madeira, 5
Madrid, 30
Montreal, 69 17
New York, 306 1055
Baltimore, 378 tcs. 134
Boston, 10
Corfu, 50 brls.
Ghent, 12
Lisbon, 13
Monte Video, 10
Philadelphia, 685
Puerto Cabello, 10
Rio Janeiro, 150
Valparaiso, 15

Soda Crystals—
Bilbao, 17 brls. 6 cks.
Boston, 280
Calcutta, 10
Malta, 53
New York, 280
Singapore, 9
VictoriaB.C., 20
Monte Video, 67 brls.
New York, 56
Portland, 100
Quebec, 50
Singapore, 12 cks.

Tar—
Rouen, 59 t. 14 c.
Calcutta, 24 brls.

GLASGOW.

Week ending May 22nd.

Ammonia—
Hong Kong, 2 t. 18 c. £70. 10s.

Charcoal—
Antwerp, 118 c. £27

Caustic Soda—
New York, 200¼ c. £56

Copper Sulphate—
Rouen, 2 t. 16½ c. £57

Chloride of Lime—
San Francisco, 306¼ c. £96

Citric Acid—
Lisbon, 5 c. 37

Dyestuffs—
Logwood
Quebec, 8 t. 13½ c. £69

Epsom Salts—
Christiana, 50 bgs. £33

Manganese Oxide—
New York, 200¼ c. £93

Manure—
Christiania, 2 c. £1

Paraffin Wax—
Barcelona, 10 t. 3½ c. £280
Oporto, 10 4 250
" 2 16½ 79
Bordeaux, 5 132
Fremantle, 1 15 67
Bilbao, 46 3 45
Antwerp, 103 155

Potassium Bichromate—
Antwerp, 130¼ c. £194
New York, 493 921
Rouen, 16 t. 18½ 559
Barcelona, 109 200
Antwerp 103¾ 150

Potassium Prussiate—
Rouen, 116¾ £305

Pitch—
La Rochelle, 628 t. 3 c. £785
Bilbao, 563 12½ 1,005

Soda Ash—
San Francisco, 310¼ c. £86

Sodium Bichromate—
New York, 495 c. 694
Rouen, 12 t. 18¼ c. 292
Barcelona, 43 60

Superphosphate—
Christiania, 20 t. £100

HULL.

Week ending May 20th.

Alkali—
Harlingen, 94 cks.

Bone Size—
Amsterdam, 10 cks.

Bleaching Powder—
Drontheim, 30 cks.
St. Petersburg, 50

Borax—
Libau, 7 cks.

Caustic Soda—
Libau, 154 drums.
Riga, 158
St. Petersburg, 350

Chemicals (*otherwise undescribed*)
Antwerp, 29 cs. 2 cks.
Abo, 9
Bombay, 24
Christiania, 43 20 pks.
Copenhagen, 6
Drontheim, 1
Danzig, 7
Gothenburg, 35 40 5 334 slbs.
Hamburg, 15
Harlingen, 4
Lisbon, 5
Libau, 14
Norrkoping, 1
New York, 20
Rotterdam, 41 28
Riga, 20 100
Rouen, 31
S. Petersburg, 7 188
Stettin, 6
Wasa, 14 1 pkg.

Dyestuffs (*otherwise undescribed*)
Antwerp, 1 ck.
Boston, 5
Hamburg, 13
Rotterdam, 1 cs. 5
Bombay, 43

Manure—
Antwerp, 490 bgs.
Drontheim, 40
Rotterdam, 491 111 t.

Pitch—
Antwerp, 499 t.
Dunkirk, 113

Salt—
Lisbon, 1 cs.

Tar—
Christiania, 6 cks.
Copenhagen, 13
Gothenburg, 150
Hamburg, 14
Stettin, 1
Wasa, 40

GRIMSBY.

Week ending May 17th.

Chemicals (*otherwise undescribed*)
Hamburg, 12 pks.
Rotterdam, 10 cs.

TYNE.

Period ending May

Ammonia—
Barcelona, 5 t. 17 c.

Antimony—
Barcelona, 2 t. 4 c.

Alkali—
Hamburg, 3 t. 10 c.
Aarhuus, 2 18
Rotterdam, 3 11

Baryta Hydrate—
Danzig, 5 t. 7 c.

Barytes Carbonate—
Antwerp, 10 t. 16 c.

Baryta Nitrate—
Antwerp, 6 c.

Bleaching Powder—
Norrkoping, 1 t.
Antwerp, 19 1 c.
" 20 14
Hamburg, 20 18
Aarhuus, 2 16
Rotterdam, 36 10
Danzig, 2 18

Copperas—
Rotterdam, 2 t. 19 c.

Caustic Soda—
Antwerp, 5 t. 3 c.
Christiania, 7 6

Iron Oxide—
Malmo, 6 c.

Litharge—
Gothenburg, 9 c.
Antwerp, 11
Danzig, 5

Manure—
Christiania, 15 t.

Magnesia—
Barcelona, 12 t. 3 c.

Potassium Chlorate—
Hamburg, 5 c.

Potassium Bichromate
Norrkoping, 4 t. 18 c.

Phosphorus—
Malmo, 6 c.

Pitch—
Malmo, 5 t.

Sulphur—
Gothenburg, 10 t.

Sodium Chlorate—
Rotterdam, 1 t. 12 c.

Sodium Sulphate—
Danzig, 1 t. 10 c.

Soda Ash—
Christiania, 10 t. 5 c.
Newfairwater, 19 t. 12

Soda—
Antwerp, 5 t. 7 c.
Frederickshavn. 6 t. 7
Huelva, 2 t. 3
Horsens, 11 8
Rotterdam, 11 8
Odense, 10 13

Superphosphate—
Christiania, 10 c.

Tar—
Malmo, 1 t. 5 c.

Turpentine—
Antwerp, 1 c.

GOOLE.

Week ending May

Benzole—
Antwerp, 24 cks.
Rotterdam, 48

Bleaching Preparation
Hamburg, 18 cks.

Copper Precipitate—
Rotterdam, 242 cks.

Coal Products—
Rotterdam, 163 cks.

Chemicals (*otherwise un*
Dunkirk, 15 cks.
Hamburg, 97
Rotterdam, 25

Dyestuffs (*otherwise un*
Ghent, 21 cks.

Manure—
Boulogne, 153 bgs.
Dunkirk, 50
Rotterdam, 550

Pitch—
Antwerp, 531 t.
Ghent, 106

PRICES CURRENT.

TUESDAY, MAY 27, 1890.

PREPARED BY HIGGINBOTTOM AND CO., 116, PORTLAND STREET, MANCHESTER.

The values stated are F.O.R. *at maker's works, or at usual ports of shipment in U.K. The price in different localities may vary.*

		£	s.	d.
Acids :—				
Acetic, 25 °/ₒ and 40 °/ₒ	per cwt.	7/- & 12/6		
,, Glacial	,,	65/-		
Arsenic, S.G., 2000°	,,	0	12	9
Chromic 82 °/ₒ .. .. nett	per lb.	0	0	6½
Fluoric	,,	0	0	3¾
Muriatic (Tower Salts), 30° Tw.	per bottle	0	0	8
,, (Cylinder), 30° Tw.	,,	0	2	11
Nitric 80° Tw.	per lb.	0	0	2
Nitrous	,,	0	0	1¼
Oxalic .. .. nett	,,	0	0	3¾
Picric	,,	0	1	2
Sulphuric (fuming 50 °/ₒ)	per ton	15	10	0
,, (monohydrate)	,,	5	10	0
,, (Pyrites, 168°)	,,	3	2	6
,, (,, 150°)	,,	1	4	6
,, (free from arsenic, 140/145°)	,,	1	7	6
Sulphurous (solution)	,,	2	15	0
Tannic (pure)	per lb.	0	1	0½
Tartaric	,,	0	1	3
Arsenic, white powdered	per ton	13	0	0
Alum (loose lump)	,,	4	17	6
Alumina Sulphate (pure)	,,	5	0	0
,, ,, (medium qualities)	,,	4	10	0
Aluminium	per lb.	0	15	0
Ammonia, ·880=28°	per lb.	0	0	3
,, =24°	,,	0	0	1¼
,, Carbonate .. .. nett	,,	0	0	3½
,, Muriate	per ton	23	0	0
,, ,, (sal-ammoniac) 1st & 2nd ..	per cwt.	37/- & 35/-		
,, Nitrate	per ton	40	0	0
,, Phosphate	per lb.	0	0	10½
,, Sulphate (grey), London	per ton	11	6	3
,, ,, (grey), Hull	,,	11	5	0
Aniline Oil (pure)	per lb.	0	0	10½
Aniline Salt ,, ...	,,	0	0	9¼
Antimony	per ton	75	0	0
,, (tartar emetic)	per lb.	0	1	1
,, (golden sulphide)	,,	0	0	10
Barium Chloride	per ton	8	5	0
,, Carbonate (native)	,,	3	15	0
,, Sulphate (native levigated)	,,	45/- to 75/-		
Bleaching Powder, 35 %	,,	5	0	0
,, Liquor, 7 %	,,	2	10	0
Bisulphide of Carbon	,,	13	0	0
Chromium Acetate (crystal	per lb.	0	0	6
Calcium Chloride	per ton	2	2	6
China Clay (at Runcorn) in bulk	,,	17/6 to 35/-		
Coal Tar Dyes :—				
Alizarine (artificial) 20 %	per lb.	0	0	9
Magenta	,,	0	3	9
Coal Tar Products				
Anthracene, 30 % A, f.o.b. London ..	per unit per cwt.	0	1	5
Benzol, 90 % nominal	per gallon	0	3	9
,, 50/90 ,, ...	,,	0	2	9
Carbolic Acid (crystallised 35°)	per lb.	0	0	8
,, (crude 60°)	per gallon	0	2	2
Creosote (ordinary)	,,	0	0	2½
,, (filtered for Lucigen light)	,,	0	0	3
Crude Naphtha 30 % @ 120° C.	,,	0	1	4
Grease Oils, 22° Tw.	per ton	3	0	0
Pitch, f.o.b. Liverpool or Garston	,,	1	8	0
Solvent Naphtha, 90 % at 160°	per gallon	0	1	9
Coke-oven Oils (crude)	,,	0	0	2½
Copper (Chili Bars)	per ton	53	0	0
,, Sulphate	,,	25	10	0
,, Oxide (copper scales)	,,	55	0	0
Glycerine (crude)	,,	30	0	0
,, (distilled S.G. 1250°)	,,	60	0	0
Iodine	nett, per oz.	0	0	9
Iron Sulphate (copperas)	per ton	1	10	0
,, Sulphide (antimony slag)	,,	2	0	0
Lead (sheet) for cash	,,	14	10	0
,, Litharge Flake (ex ship)	,,	15	10	0
,, Acetate (white ,,)	,,	24	0	0
,, ,, (brown ,,)	,,	18	0	0
,, Carbonate (white lead) pure	,,	19	0	0
,, Nitrate	,,	22	5	0
Lime, Acetate (brown)	,,	7	15	0
,, ,, (grey)	per ton	13	10	0
Magnesium (ribbon and wire)	per oz.	0	2	3
,, Chloride (ex ship)	per ton	2	5	0
,, Carbonate	per cwt.	1	17	6
,, Hydrate	,,	0	10	0
,, Sulphate (Epsom Salts)	per ton	3	0	0
Manganese, Sulphate	,,	17	10	0
,, Borate (1st and 2nd)	per cwt.	60/- & 42/6		
,, Ore, 70 %	per ton	4	10	0
Methylated Spirit, 61° O.P.	per gallon	0	2	2
Naphtha (Wood), Solvent	,,	0	3	10
,, ,, Miscible, 60° O.P.	,,	0	4	6
Oils :—				
Cotton-seed	per ton	22	10	0
Linseed	,,	26	0	0
Lubricating, Scotch, 890°—895°	,,	7	5	0
Petroleum, Russian	per gallon	0	0	5¼
,, American	,,	0	0	5¼
Potassium (metal)	per oz.	0	3	10
,, Bichromate	per lb.	0	0	4
,, Carbonate, 90% (ex ship)	per ton	19	10	0
,, Chlorate	per lb.	0	0	4¼
,, Cyanide, 98%	,,	0	2	0
,, Hydrate (Caustic Potash) 80/85 % ..	per ton	22	10	0
,, ,, (Caustic Potash) 75/80 % ..	,,	21	10	0
,, ,, (Caustic Potash) 70/75 % ..	,,	20	15	0
,, Nitrate (refined)	per cwt.	1	2	6
,, Permanganate	,,	4	2	6
Prussiate Yellow	per lb.	0	0	9½
,, Sulphate, 90 %	per ton	9	10	0
,, Muriate, 80 %	,,	7	15	0
Silver (metal)	per oz.	0	3	11¼
,, Nitrate	,,	0	2	8
Sodium (metal)	per lb.	0	4	0
,, Carb. (refined Soda-ash) 48 %	per ton	6	5	0
,, ,, (Caustic Soda-ash) 48 %	,,	5	5	0
,, ,, (Carb. Soda-ash) 48%	,,	5	12	6
,, ,, (Carb. Soda-ash) 58 %	,,	6	15	0
,, ,, (Soda Crystals)	,,	2	17	6
,, Acetate (ex-ship)	,,	16	0	0
,, Arseniate, 45 %	,,	11	0	0
,, Chlorate	per lb.	0	0	6¼
,, Borate (Borax)	nett, per ton.	29	0	0
,, Bichromate	per lb.	0	0	3
,, Hydrate (77 % Caustic Soda)	per ton.	9	15	0
,, ,, (74 % Caustic Soda)	,,	9	0	0
,, ,, (70 % Caustic Soda)	,,	8	5	0
,, ,, (60 % Caustic Soda, white) ..	,,	7	5	0
,, ,, (60 % Caustic Soda, cream) ..	,,	7	2	6
,, Bicarbonate	,,	6	0	0
,, Hyposulphite	,,	5	10	0
,, Manganate, 25%	,,	9		0
,, Nitrate (95 %) ex-ship Liverpool ..	per cwt.	0	8	1½
,, Nitrite, 98 %	per ton	28	0	0
,, Phosphate	,,	15	15	0
,, Prussiate	per lb.	0	0	7¼
,, Silicate (glass)	per ton	5	7	6
,, ,, (liquid, 100° Tw.)	,,	3	17	6
,, Stannate, 40 %	per cwt.	2	0	0
,, Sulphate (Salt-cake)	per ton.	1	5	0
,, ,, (Glauber's Salts)	,,	1	10	0
,, Sulphide	,,	7	15	0
,, Sulphite	,,	5	5	0
Strontium Hydrate, 98 %	,,	8	0	0
Sulphocyanide Ammonium, 95 %	per lb.	0	0	7¼
,, Barium, 95 %	,,	0	0	5¼
,, Potassium	,,	0	0	8
Sulphur (Flowers)	per ton.	7	10	0
,, (Roll Brimstone)	,,	6	0	0
,, Brimstone : Best Quality	,,	4	7	6
Superphosphate of Lime (26 %)	,,	2	12	6
Tallow	,,	25	10	0
Tin (English Ingots)	,,	98	0	0
,, Crystals	per lb.	0	0	6¼
Zinc (Spelter)	per ton.	23	0	0
,, Chloride (solution, 96 Tw.	,,	6	0	0

THE
CHEMICAL TRADE JOURNAI

Publishing Offices: 32, *BLACKFRIARS STREET, MANCHESTER.*

No. 159. **SATURDAY, JUNE 7, 1890.** **Vol.**

Contents.

Notices.

All communications for the *Chemical Trade Journal* should be addressed, and Cheques and Post Office Orders made payable to—

DAVIS BROS., 32, Blackfriars Street, MANCHESTER.

Our Registered Telegraphic Address is—
"Expert, Manchester."

The Terms of Subscription, commencing at any date, to the *Chemical Trade Journal*,—payable in advance,—including postage to any part of the world, are as follow:—

Yearly (52 numbers) .... **12s. 6d.**
Half-Yearly (26 numbers) .... **6s. 6d.**
Quarterly (13 numbers) .... **3s. 6d.**

Readers will oblige by making their remittances for subscriptions by Postal or Post Office Order, crossed.

Communications for the Editor, if intended for insertion in the current week's issue, should reach the office not later than **Tuesday Morning.**

Articles, reports, and correspondence on all matters of interest to the Chemical and allied industries, home and foreign, are solicited. Correspondents should condense their matter as much as possible, write on one side only of the paper, and in all cases give their names and addresses, not necessarily for publication. Sketches should be sent on separate sheets.

We cannot undertake to return rejected manuscripts or drawings, unless accompanied by a stamped directed envelope.

Readers are invited to forward items of intelligence, or cuttings from local newspapers, of interest to the trades concerned.

As it is one of the special features of the *Chemical Trade Journal* to give the earliest information respecting new processes, improvements, inventions, etc., bearing upon the Chemical and allied industries, or which may be of interest to our readers, the Editor invites particulars of such—when in working order—from the originators; and if the subject is deemed of sufficient importance, an expert will visit and report upon the same in the columns of the *Journal*. There is no fee required for visits of this kind.

We shall esteem it a favour if any of our readers, in making inquiries of, or opening accounts with advertisers in this paper, will kindly mention the *Chemical Trade Journal* as the source of their information.

Advertisements intended for insertion in the current week's issue, should reach the office by **Wednesday morning** at the latest.

Advertisements.

COAL SMOKE.

To strain at a gnat and swallow the proverbial
an operation that has aforetime been consid
worthy of imitation. Once in a way, the perform
be considered instructive, if it be only to demons
folly of agitating our minds over small but visible
the total exclusion of the invisible ones of magnitud

Nothing can illustrate these remarks so well as th
agitation respecting coal smoke. It is, however, t
to be perceived in the Metropolis that unless some
taken to promote legislation against the present n
burning coal in private houses no improvement can
for.

If this be true with regard to visible smoke, it t
what we have preached for so long a time. At
delivered in Manchester in 1882, the writer said: "
believe that were we to banish all the factories for
round you would have the so-called fogs quite as ir
ever. If we wish to clear the atmosphere our first
is with domestic chimneys. It has hitherto been co
a subject too vast to be dealt with, and our sanitary le
have perhaps been wise in their generation by pr
letting it alone."

An American expert says: "Factory furnaces, h
are not to be charged with all the smoke. If they wer
of the English cities would be practically smokele
vast amount of coal consumed for domestic purposes
for the only difficulty they have in clearing the atm
for in London smoke from factory chimneys is
cally prevented, while from those in Paris volumes o
rivalling Cincinnati, are constantly emitted, yet the
atmosphere is smoky while that of Paris is clear. In
vast quantities of bituminous coal are consumed for c
purposes, in Paris, none. We must therefore concl
the difference in the atmosphere of the two capitals i
a difference in domestic conditions."

It is strange that in the discussion of the smoke
the evils due to sulphurous acid should be so systen
neglected. No doubt the cry for the suppression
smoke is well enough in its way, but it resembles
no less than the sulphurous acid does the camel.

The *British Medical Journal* has undertaken the ta
forming its readers "that smoke from factories or
any kind no longer plagues London." The Augea
has been cleansed, we are informed, from the info
afforded by the long and costly trials carried out in
tion with the Smoke Abatement Exhibition, and thr
agency of the Police and the Home Office, as well as
the energetic action of the smoke inspector, London
practically freed from factory smoke. All this no dou
very well, but what about the sulphurous acid, whi
well known to the authorities as to be exempted by
mention in the 29th Section of the Alkali, &c., Worl
lation Act of 1881: has there been any attempt to
evolution?

No doubt some of our correspondents will be askin
question—what do you propose, to limit the evils aris

evolution of sulphurous acid from the combustion of coal? ere are many replies that could be given to such a ques-h, but we are afraid improvement is impossible until we e up the practice of legislating for the suppression of each lividual evil, instead of endeavouring to suppress each class evils, in the abstract.

Ve were pleased to find the bill introduced into the House Lords, for the purpose of giving power to local authorities require that any new house erected should be fitted with thods by which bituminous coal could be burned without the oduction of smoke, was thrown out after the second read-;, and Lord Wemyss is to be congratulated on the success his opposition, though we do not agree with the arguments ich he and others employed to gain their object. When argument is used that the forcible suppression of a public sance is an invasion of the private rights of the builder l householder, we are not at one with his lordship. There no doubt that a man may do what he pleases with his own operty provided he does not infringe the laws of the realm, l what is more, that he does not injure his neighbour by result of any act indulged in for his own enjoyment.

Leaving smoke production and the fogs they engender on one side, we still have the question before us: what will done with the sulphurous acid? In this connection we y mention that we have lately seen a fuel, which on burn-evolves no sulphurous acid. It is well known to chemists t coal dust mixed with lime or chalk furnishes a fuel giving ictically no volatile sulphur. The fuel brought under our ice was in the form of briquettes, giving a smokeless ne, the sulphur remaining in the ash.

Ve fancy we hear some of our readers exclaiming, "Surely are not going to advocate compulsion to burn a block l?" Why not? The Legislature has compelled chemical nufacturers to build costly condensers to suppress the ape of acid vapours, which do not form a tithe of that it forth into the air by ordinary coal consumers; then why uld not all folk (to use the words of the Alkali Act) "use the t practicable means for preventing the discharge into the nosphere of all noxious gases?"

f the sulphurous acid of coal smoke could be suppressed, getation would flourish once more in our towns, open ces would be cheering to the eye, instead of the desolate l grassless squares which they are now in many places; vn residences would be more attractive and more eagerly ight after, rents would not suffer depreciation in the burbs, and buildings would suffer less from premature ay than they unfortunately do at the present time.

The man who solves this question will deserve the best nks of his countrymen, who no doubt will reward him as y have done all other public benefactors, viz., starvation a solatium and a molten image to perpetuate his mory.

WATER GAS.*

HE Board, since its last Report, has continued its investigations on the subject of water gas. It is understood that the parties nterest do not desire to offer further evidence in the cases pending ne close of the year. The Board believes that some modification of present law is desirable, and that a report of the work of the Com-sioners, in this branch of their duties, ought to be made to the islature at the present time.

n the course of its inquiries the Board has endeavoured to ascer-what commercial advantages, if any, might accrue to the com-ies from an extension of the present statutory limit of carbonic e, believing that in any such advantages the consumers would ultimately share, and that the companies ought to be permitted to avail themselves of them, if they could do so without prejudice to the public interest. For this purpose information has been obtained by personal inspection and correspondence from more than a hundred companies. The Board desires to gratefully recognise the uniform courtesy which has been freely extended it by the engineers in charge of water-gas works in other States, and to acknowledge its obligations for the information so freely furnished.

It is clear that the cost of gas, whether coal or water, is not an inflexible figure, the same for all times and places. An advance of fifty cents per ton in coal, or of one cent per gallon in oil, making an increase of ten per cent. in the cost of coal and water gas respectively. In certain large works, where both kinds of gas have been made, the advantage in cost of production has been sometimes with one, sometimes with the other, kind of gas. There are more exclusively water-gas works in the anthracite coal region in Pennsylvania, and within easy reach of it, than in any other portion of the country of the same area: because there the price of anthracite coal has been low, and that of gas coal relatively very high. The particular location of a gas works, and its arrangement, the quality of the materials, and kind of apparatus used, and the kind of labour and superintendence available, may make even larger variations in the cost of either gas.

The expenses of distribution, management, taxes, &c., are not likely to vary greatly, whether coal or water gas is the product; so that the fact of value in estimating the cost of the two gases is the cost to manufacture or cost in the holder, as distinguished from cost at the meter.

The principal items of cost in the holder are for materials, coal and oil, labour and repairs. From the cost of materials in coal gas is to be deducted the very considerable receipts from the sale of coke, tar, ammoniacal liquor. Under the best conditions in this State these receipts have reached 50 per cent. of the cost of coal, and in some companies more favourably situated they have reached a much higher percentage.

A ton of good gas coal will yield approximately ten to eleven thousand feet of gas. In order to secure the candle-power usually supplied in this State there must be added about ten per cent. of cannel, or something more than five gallons of oil, equivalent to more than one-half gallon per thousand feet.

In the production of illuminating water gas there is commonly used more than fifty pounds of hard anthracite coal, and from five to six and one-half gallons of oil for every thousand feet of gas, although in the very best works, and under very favourable conditions, these quantities may be slightly reduced. These two items nearly make up the cost of water gas in the holder.

In coal gas the cost of labour is a most important element. It often nearly equals one-half the gross cost of materials, and in many companies does not vary much from the net cost of coal. In water gas the cost of labour is small.

It has been very difficult for the Board to obtain definite and reliable information as to the cost of repairs in water-gas works. It is not a very considerable item in coal gas, and is probably about one-half as much in water gas.

The following figures show the range of cost in certain companies, each item being calculated independently of the other, and, it is believed, fairly indicate the difference of cost in the holder of the two gases—

MATERIALS.

Coal Gas.

Gas coal	40 to 43	cents.
Oil or cannel	3 to 7	,,
Residuals	13 to 23	,,
Nett, for materials	20 to 30	,,
Labour	14 to 20	,,
Repairs	5 to 8	,,
Total	50 to 57	,,

Water Gas.

Coal	11 to 15	cents.
Crude oil	13 to 18	,,
Naphtha	24 to 33	,,
Materials	26 to 33	,, *
	38 to 45	,, †
Labour	5 to 11	,,
Repairs	2 to 3	,,
Total	46 to 55	,,

The figures relating to coal gas are taken from the returns of certain companies in this State, as made to the Board. The figures relating to water gas are estimates based upon information procured without the State, and intended to show what the same companies might do at current rates for materials, if they were making this instead of coal gas. It may be noted that the companies now making

* From the Fifth Annual Report of the Board of Gas and Electric Light missioners of the Commonwealth of Massachusetts. 1890.

* Crude oil. † Naphtha.

water gas in this State have not yet been able to equal the figures given. The relations shown above may be greatly changed at any time. The figures given for naphtha and crude oil are suggestive. A few years ago, when the number of water-gas works was small, the price of naphtha was about the same as that of crude oil to-day. As the demand for naphtha for gas-making purposes increased, the price advanced to a point which nearly neutralised the advantages of its use. Recently a crude oil, from the Ohio fields, difficult to refine, has come across the market at a low price, and an apparatus specially adapted for it has again reduced the cost of water gas to the low point reached when naphtha was first introduced. There are now signs of an advance in its price.

There have been great improvements in water-gas apparatus, and there is much activity in this direction now. It is now claimed that cheap bituminous coal and slack can be made to do the work of expensive anthracite; and, if the expectations of the inventors of this new process are realised, the cost of producing water-gas may be greatly lessened.

In companies whose output is below a certain limit, the peculiar conditions under which they make gas cause the cost to vary greatly from the figures given above. In these the cost per thousand feet for both coal and labour rises rapidly where coal gas is made. There is little or nothing obtainable from residuals, as all or nearly all the coke is used to carbonise the coal, and the production of gas per man is much reduced, since, of necessity, the gas maker must be idle much of the time. In a properly-constructed water-gas works of very small output, while the cost of coal may be easily doubled by the necessity for frequently cooling and reheating the apparatus, the cost of labour is not materially different from the same item in large works. In a single day one man may make and store water gas sufficient for several days' consumption, during which the works need no attention, and the gas maker can devote himself to the numerous other details of the business. Instances have been observed by the Board, in other States, where small works have been carried on in this manner with apparently good results. For reasons not easy to enumerate, the attempts by small companies in this State to pursue a similar course have not heretofore been entirely successful.

Although in the larger works, except in favoured localities, the commercial advantages of manufacturing exclusively coal or water gas may not be considerable, with the methods and apparatus now most in use, the advantage to company and consumer in the combination of the two gases is now generally recognised by the larger companies, and nearly one-third of all the companies in the country manufacture some water gas. The development of electric light, combined with other causes, has greatly increased the public demand for light; a gas of fifteen or sixteen candles is no longer acceptable, and from eighteen to twenty candle gas is as low as companies in the larger towns and cities think it wise to distribute. Gas of this quality cannot be produced with ordinary coal alone, but requires the addition of cannel coal or of oil, which latter, from its low cost and the facilities for handling, has been most generally adopted. For this use of oil the ordinary appliances of a coal-gas works are but poorly adapted. When oil is used in them, it has shown a decided tendency to pass with the coal gas as a vapour, much of which is subsequently deposited in the mains or fixtures in a liquid form, or, by causing a hard substance to form at the tip, obstructs the flow of gas, and causes the forked and irregular flame so common where such gas is used; and the gas often manifests an uncontrollable tendency to smoke. To a high candle-power secured in this way is often due the the blackened spots in the ceilings of dwelling-houses where such gas is used.

A water-gas apparatus affords the means for avoiding these difficulties. In this the heats best adapted to breaking up the oily particles into a fixed gas may readily be attained, and gas of a very high candle-power be made. When coal gas is made at the same works, and the two are mixed in the proper proportions and best manner, the unfortunate results mentioned as due to oil should not occur. The gas is composed of the same constituents, but in different proportions, and reaches the burner with that combination of heating and illuminating elements calculated to produce a clear, white flame, free from smoke.

Numerous other reasons suggested by local conditions have induced companies to add to coal-gas works apparatus for making water gas. It enables a company to easily meet a sudden demand for an unusual amount of gas. It can be made ready for gas-making at an hour's notice, then allowed to cool when the demand is over, and this course repeated indefinitely without injury, while coal-gas benches once treated in this manner, would probably need rebuilding. An unexpected dark day may thus be readily provided for; and the considerable changes in consumption, where street lights are burned by moon schedule, but are unexpectedly called for on stormy nights. In some places the increased consumption has made necessary some increase in the capacity of the works, which has been secured with less expense for extension by the introduction of water-gas apparatus, with a consequent saving in capital account. A brief period of extraordina sumption in midwinter is sometimes provided for in this wa many works where both kinds of gas are made, coke produced works is substituted for anthracite coal, furnishing a profitable surplus coke, and preventing a reduction of its price.

These are some of the indirect advantages resulting from a m ture of both kinds of gas in the same works. They ar independent of whether one gas made alone costs a little more than the other. To secure them requires the distribution of a g taining somewhat more than 10 per cent. of carbonic oxide.

The question of the comparative safety of water and coal g very perplexing and difficult one to solve satisfactorily. One o members of the Board have personally examined the premises w large number of deaths have occurred from asphyxiation by i illuminating gas in Brooklyn, New York City, Baltimore, an cities. A few of these deaths were caused by coal gas, probably, by a mixed gas; but the larger number were due to gas. Some valuable suggestions have been obtained from inquiries, although great difficulty was experienced in securin or definite information. It was impossible, for instance, in mo to ascertain the length of time the gas was discharged into the and in many cases the size of the burner, and whether parti wholly open, were only matters of conjecture. Most of the ac investigated occurred at the cheaper boarding and lodging hous was found in one city—and the same facts also applied to the in a less degree—that a very low price for gas, great activity part of competing companies in canvassing for customers, a very favourable terms that were made for piping buildings a nishing fixtures, had led to the introduction of gas into th cheapest lodging and boarding houses, that are frequented dissolute and ignorant classes. In several instances it was that the inhaling of gas alone would not have produced deat caused it by aggravating diseases existing at the time of the ac Some of the rooms contained less than 500 cubic feet of spac the average of all the rooms was about 1,100 cubic feet. The amount of gas was 75 cubic feet, as nearly as could be ascertai a room containing 748 cubic feet of space; but there was a transom at the time of the accident, and the person lived twelv after being discovered. The smallest amount of gas was 1 feet, in a room containing 806 cubic feet of space; and the pers dead when found. The average amount of gas in the rooms wa 38 cubic feet. But, as before intimated, the information in re the amount of gas in the rooms where deaths occurred is unsatis In three instances where a man and a woman occupied rooms to the men died and the women recovered.

CORNERING.

THIS subject has of late years and recent months receive attention in the 'Change world. It has had the considera the most acute lawyers and the sharpest speculators, to say not the attention of those being ruined or on the road to ruin b may aptly be called gambling in stocks and produce. It ha suggested in the States that the system of "futures" should b illegal. By this is meant that A shall not sell to B that whic yet his or within his power, or perhaps is not yet in existen next crop's cotton. In England there is an almost forgotten A in existence, passed in the sixteenth century, prohibiting what w in the days of the sixth Edward known as "forestalling." This stalling" consisted in buying or contracting for any merchandis coming to market, or persuading persons to enhance the pric such goods or food produce was on the market, either of which p would be calculated to make the market dearer. According to t —were it hauled out of its dusty recess, where it has lain dorm so many generations—a modern "corner" in corn in Liverpool be illegal, as the word used in 5 and 6 Edward VI., cap. "victual." So would a corner in cotton for the word "mercha is used. But we do not suppose that anyone would ever dr enforcing the letter of the law. Some ingenious means would be to escape the provisions of such legislation. There are men to- able as even the great orator of the past, who remarked that the not an Act of Parliament on the statute book through which h not drive a carriage and four. Of course, "cornering" is no mercially moral. None will argue that; and there is some sati in knowing that whilst corners ruin the innocent and put a bar way of remunerative labour, they generally bring down a well-n reward on the heads of their organisers. The French Corre Tribunal has recently made an example of the leaders in the corner, of which we have recently heard so much. The charges the originators, Mon. Secretan and Mons. Laveissiére were— serious in British commercial eyes, and the other not so serious. our very high standpoint of principle we look upon the declara

ious dividends as certainly bad, if discovered. Generally speaking, sin is not so great if it does not come to light; and yet this tious dividend facturing is done in the open daylight every day of year. The system obtains in concerns in the directorate list of ch rank the names of men high in standing, and to all appearances ve suspicion. These Frenchmen at the head of the Society de aux are now in goal, because they created apparent profits—profits llen to more than normal extent by placing on certain material an ficial and exaggerated value. We are not a bit sorry for them. To a commonplace phrase, they went the whole hog. They wanted to opolise an entire department of trade. They thought there was a nce to succeed. They ran the risk, knowing it was a risk, and they d not who fell if they did not fall, and thought not of the ruin they ld wreak on those who fell when they were unable to meet their agements. But this creating an apparent profit, by setting on given k an exaggerated and artificial value, is no worse than we see done very great centre day by day. Do we not see bogus companies ted—a large capital offered for subscription for the purpose of ing up a business that is not worth one tithe of the amount inally asked? The one case is every bit as bad as the other, and would indeed be a happy day for commercial England if some drastic slative action were taken which would make a directorate directly onsible for the statements in prospectuses, and ensure that inflated itals should not be raised upon flimsy concerns which, when they e posed on paper before a gullable public, lead to disappointment l subsequent bankruptcy.—*Liverpool Journal of Commerce.*

Our Book Shelf.

IE TWENTY-FIFTH AND TWENTY-SIXTH REPORTS of the Chief Inspector of Alkali, etc, Works; being his proceedings during the years 1888 and 1889. London, Eyre & Spottiswoode:

Extreme pressure of business, at this time last year, compelled us to t the former of these two reports aside, until a more convenient son for detailed notice. A year has passed and gone, and though may be charged with procrastination, we may urge, in defence, that has given the opportunity for a comparison of two year's work, hin the precincts of the same notice.

But *revenons a nos montons*, the 25th report tells us, that in England l Ireland there were 926 works registered at the close of 1888, ile, from the 26th report, we gather that only 903 works were gistered in 1889. As a total, there were 1,057 registered works in United Kingdom at the end of 1888, but only 1,032 at the close 1889.

Passing over the details of "visits to works, and tests made" in the h report, we may make passing reference to the statement, that in the case of hydrochloric acid, "the average of the whole amount aping, is 0·089 grain of hydrochloric acid per cubic foot of air; an ount slightly less than that of previous years.

The 26th report, in referring to the escape of hydrochloric acid, says: he limit laid down in the Alkali Act is 0·20, or two-tenths of a grain. is satisfactory to note that the average for the whole year is distinctly low the half of this, or only 88, in the place of the 200. This is most identical with the figure 89, reached in the year 1888.

While giving due praise to the inspectors for the energies with ich they have carried on their investigations, every one who has ecial knowledge of the alkali trade must agree that the cause of the minution of escapes of acid gases is not to be found in Government pection. We may go even farther than this, and say that the alkali de has suffered to an immense extent, measured, perhaps, in millions pounds sterling, by the indirect action of the Alkali Acts. We are ite aware that several leaders of the trade have expressed the inion that inspectors are blessings in disguise, but human nature s, all the world over, a tendency to greater activity in the tongue, in to foresight in the brain.

If, then, the diminution of the escape of acid gases be not due, in the in, to Government inspection, to what is it due? To reply is easy. fore the year 1863, in the majority of instances, the alkali manu- cture was carried on in a most crude manner. If we are to credit the tistics published in the first, second, and third reports of the first Chief spector (Dr. Angus Smith), it was no uncommon thing to find 30–40 r cent. of the hydrochloric acid wasted; and, in some cases, *the iole of it.* These were the halcyon days of the soda industry, days ien soda making yielded a handsome profit, without the necessity of oking to the hydrochloric acid as a source of income, but to-day the tter is very different, the ammonia soda process must be looked upon, a time at least, as the universal provider of carbonate of soda, or at is commercially known as soda ash, while the Le Blanc process ist supply the world with caustic soda, and if it cannot do so at a price, at least equal to the causticising of ammonia soda, then it must cease to exist.

But to return to our point, soda ash is now the bye-product, and hydrochloric, the milch cow of the Le Blanc process, therefore it is to the manufacturers interest to secure as much hydrochloric acid as possible. If we take the average escape as 0·089 for 1888 and 0·088 for 1889, we may infer (the difference is but one-thousandth of a grain per cubic foot) that the condensation of hydrochloric acid gas is now as complete as it can be made practically. This, as we have said before, cannot be attributed solely to Government inspection, the various economic changes that have taken place in the allied industries, have had much to do with the question. We have already endeavoured to show that Le Blanc soda ash has had a formidable rival in ammonia soda, but this, so to speak, internal rivalry, is as nought when compared to external effects. At one time, the whole of the soda employed in the paper industry was run away as black liquor into the nearest stream, and as there are considerably more than three hundred paper mills in the United Kingdom, the recovery of this soda, or 70% of it, must be a factor to be reckoned with. There are many mills now recovering from 24 to 30 tons of soda ash weekly, thus securing a greater degree of purity for the streams in the neighbourhood than was heretofore possible. But soda pulp has had a powerful rival in what is known as sulphite pulp, so that the soda trade has not received a corresponding impetus from the growth of the paper trade. The tendency has, therefore, been to carefully nurse the bleaching powder industry and to take advantage of all the knowledge imparted by earlier investigators, first, in increasing the yield of acid from the salt decomposed, and secondly, in wasting less of it in the chlorine manufacture than was done formerly.

If manufacturers have thus benefited, how is it possible for us to establish the proposition, that by reason of inspection "the alkali trade has suffered to an immense extent?" Many of the earlier manufacturers prospered like the green bay tree; the outside world did not know the organisation required in, and the cares and troubles of a chemical business; it was a new occupation for adventurers, a calling in which years of enforced apprenticeship were unnecessary. One cannot wonder, then, that the chemical industry became a centre of attraction, even drawing men from other well-established businesses and walks in life. But it is a notorious fact that many men, enticed from a mechanical calling to one strictly physical and chemical, were totally ignorant of chemical facts. While profits were ample they did not bother their heads in studying manufacturing economy, and chemical economy was out of the question. Many of these works would have gone out of existence, and competition been less keen to-day than it is, but inspection came in and said: "You must stop your extravagant methods, you must economise your waste products, you must poison the land no longer;" and it was so. To the ill-conditioned, ill-constructed, and badly managed works, no doubt inspection is a "distinct benefit to the owners of such works,"* but it is very much open to question whether the Act is a "benefit" to the scientific worker, to the man who can work and be guided by the results of his own investigations, rather than by what he is able to glean from the doings of his neighbours. Judging from the number of prosecutions under the Act, all the works seem to be behaving themselves exceedingly well, so that one is almost tempted to ask is there need for inspection any longer. In the 25th report, nine prosecutions are recorded, but seven of these were for non-registration, and the heinousness of this crime may be gathered from the fact that a firm in Norfolk was fined £30. and costs for non-registration, while another firm was only fined £20. and costs for failing to use the best practicable means for preventing the escape of acid gas from the manure plant. The 26th report states, that there has been only one prosecution during the year 1889, and that for failing to adopt the best practicable means for preventing the escape of sulphuretted hydrogen during the manufacture of sulphate of ammonia. Penalty £10. and costs, while a manure works at Chelmsford the year before were fined £17. for non-registration. Better to register and make a smell, than not to register!

From the 25th report we learn that 590,312 tons of salt were decomposed by the Le Blanc process during 1888, and the 26th report gives us the information that 584,203 are the figures for 1889.

In the 25th report, on page 13, when recording the progress of sulphur recovery from alkali waste, the chief inspector says "great care will be needed in preventing all escape of the sulphuretted hydrogen, a highly poisonous and offensive gas. It will be the duty of the inspectors under the Alkali Act to see that none of it passes into the atmosphere, but that the last traces are uniformly retained." Have the inspectors done this? The 26th report is most unsatisfactory on this point. It does seem to us a case of special pleading when a small producer of sulphuretted hydrogen is fined £10. and costs, while a hundred times the quantity is let loose in Widnes, but there, Widnes has a special inspector to itself.

* We borrow these words from Mr. A. M. Chance, which may be found on page 10 of the 25th report of the Chief Inspector of Alkali Works.

Considerable space is taken up in the 25th report with a description of plant, used for suppressing the nuisance arising from the escape of sulphuretted hydrogen. In a multitude of counsellors there is wisdom, but he will be a clever man who can tell *the best practicable means* from the digest of the 117 cases therein contained.

The production of salt seems to have fallen from 2,039,867 tons in 1888, to 1,792,790 tons in 1889.

From the report of Dr. Affleck, on the Widnes district, we gather that "the nitric acid process, patented by Mr. Donald (for the production of chlorine), and which was alluded to last year, has not yet proved to be a success on an industrial scale and has been abandoned in the Widnes works, where it was being tried during the past 18 months."

The reports of all the district inspectors are very interesting reading, and especially so to ourselves, but we would give a hint that it may be well to regard. Assuming that a register number is necessary to conceal the identity of the works, it would be better to curtail the list of "additions and improvements" thus, on page 36 it is easy from the information there given to lift the curtain of concealment, even by an outsider, and to residents in the vicinity there is no concealment at all.

THE FIFTH ANNUAL REPORT of the Board of Gas and Electric Light Commissioners of the Commonwealth of Massachusetts, Boston. Wright and Potter. 1890.

This is a public document of 208 octavo pages, from which a great deal of useful information may be gleaned, especially that relating to water-gas and to the employment of electric lighting in the State.

In another column we publish some of the remarks upon water-gas, which may prove of service to our readers, and below is given a portion of the report of the Gas Inspector which goes to show generally what is done with regard to these matters in America.

"The tests of the gas of the larger companies were always made at some distance from the works, and usually at the companies' offices. The gas of the smaller companies was tested at hotels and town halls, as well as the companies' offices. The inspections have been made at rather irregular intervals, and no notice has ever been given when they were to occur, so as to avoid giving the companies any opportunity of preparing a special quality of gas, if they had any such desire. More inspections have been made in the fall and winter months than in the spring and summer months, so as to follow the production of gas to a certain extent. At Chicopee, Dedham, Fitchburg, Gloucester, Manufacturers' of Fall River, Marlborough, Plymouth, and Webster, the inspections were made at the works, as there seemed to be no other place available.

The results of the tests made during the year by the inspectors, and given in the following table, were furnished to the Board of Gas and Electric Light Commissioners at their request.

Number of Inspections made.	NAME OF PLACE OR COMPANY.	CANDLE-POWER.			GRAINS PER 100 FEET OF GAS OF	
		Average.	Highest.	Lowest.	Iphur.	Ammonia.
52	Boston	17·91	19·3	16 5	10·59	5·86
10	Brookline	18·31	18·7	17·8	9·57	4·80
20	Cambridge	17·29	18·3	16·1	11·44	2·16
14	Charlestown	17·17	18·1	16 0	11·59	1 —
7	Chelsea	18·67	19·2	17 6	9 61	1·20
10	Dorchester	17·01	18·3	15·6	10·71	2·85
8	East Boston	18·04	19·0	17·4	10 22	4·71
9	Fall River	18·94	22·1	17 2	9·03	1·74
9	Haverhill	18·32	19·5	17·7	9·96	6·80
9	Holyoke	17·86	18·6	16·1	8·14	3·62
7	Jamaica Plain	17·57	18·0	17 0	9·84	2·80
14	Lawrence	17·21	18·2	15·7	10·34	1 —
33	Lowell	18·11	18·8	17·4	10 98	1·63
9	Lynn	18·18	19·1	17·5	12·43	1 —
7	Malden	17·33	18·2	16·4	10·11	3·14
9	New Bedford	17·26	18·4	16·5	8 88	1·14
9	Newton	17·58	18·3	17·1	12·16	1 —
25	Roxbury	17·97	19·8	17·0	10·39	6·09
8	Salem	18·00	18·6	17·3	7·25	3·02
12	South Boston	17·80	18·5	17·0	8·98	8·23
14	Springfield	18·58	19·4	17·5	9·72	2·10
7	Taunton	17·69	18·1	17·3	8·29	1·11
7	Waltham	17·74	18·5	16 9	6·06	4 53
18	Worcester	17·97	19·3	15·3	11·54	1 —
	Average	17·85	—	—	9·91	2·98

Number of Sample.	Candle-power. — Coal Gas.	PLACE OR COMPANY.	Illuminants.	Marsh Gas.	Hydrogen.	Carbonic Oxide.	Nitrogen.	Oxygen.	Carbonic Acid.
1	18.5	Boston	6.92	35.85	48.13	6.89	1.78	0	0.43
2	17.6	Lowell	4.87	36.46	50.12	6.10	1.18	0	1.27
3	16.4	Lawrence..	4.90	37 21	48.91	6.35	1.11	0	1.52
4	17.8	Charlestown	6.61	36.06	47.57	7.30	1.00	0	1.46
5	17.2	"	5.43	34.87	49.82	6 86	1.08	0	1.94
6	18.0	Worcester	7.17	36.55	44.87	5.46	4.57	0	1.38
7	—	Lynn	4.74	34.19	51.24	6.70	2 00	0	1 13
8	19.0	South Boston	5.68	38.72	47.70	6.99	0.85	.06	0
	Mixtures of Coal Gas, Petroleum Gas, and Water Gas.								
9	19.8	Roxbury	8.53	31.07	44.79	14.50	0.88	0	.23
10	18.8	"	8.80	34.78	46.35	9.15	0.92	0	0
11	17.5	"	6.33	33.98	43.85	9.33	4.77	.69	1.05
12	17.4	"	5.61	34.24	48.43	7.11	3.24	.13	1.24
13	20.6	Fall River	6.38	30.79	48.52	12.67	.92	0	.52
14	18.0	"	5.63	34.15	50 26	8.77	.84	.09	.26
15	17.2	"	4.44	35 41	51.20	6.77	1.11	0	1.07
16	17.5	"	6 08	28.21	52.41	8.26	3.74	0	1.30
17	17.9	"	7.25	25.14	42.44	21.90	1.57	0	1.70
18	20 7	"	11 93	29.87	36 78	18.62	1 81	0	.99
19	18.5	"	5 78	35.69	49.15	7.29	1.08	.03	.98
20	17.2	Dorchester	7.87	30.59	43.13	12.37	4 04	.34	1 66
21	18.0	"	7.00	30 99	46.24	10.97	3.15	.32	1 33
22	20.8	Spencer	9.48	31.73	36.90	14.76	3.12	.12	3.89
23	16.7	"	7.93	29.49	42.20	13.51	4.29	.13	2.45
24	20.7	"	9.60	34.87	35.42	14 26	3.65	.15	2 05
	Carburetted Water Gas.								
25	14.3	Athol	7.07	20.66	[illegible]	[illegible]	[illegible]	[illegible]	[illegible]

PETROLEUM NOTES.

RUSSIA'S GAIN FROM OIL.

WHILE the presence of oil in New Zealand encourages th that England may some day become the petroleum po the South Pacific, the opening up of Persia to British trade attention to a quarter too long neglected by this country. Moi once I have expressed the opinion that the development of pet at Baku has done more to expand and consolidate Russia's posi the Kaspian region than all the military successes against the Tur and the annexation of Merv and Penjdeh. When the Nobels com their oil operations at Baku ten years ago, that corner of the (was far less developed than Burma is to-day. The town was unknown, no railway connected it with the Black Sea, and the (had the reputation of being almost as barbarous and backw commercial activity as the Sea of Aral. In the interval, the pop of Baku has risen from 15,000 to 75,000 people. Its wells prod million gallons of crude oil every year. More than 100 ref stretching along the coast of Baku Bay, turn out 200 million gal refined oil annually. From the forty piers are exported million tons of refuse oil to provide fuel for 1000 steamers, locor and factory engines in various parts of Russia. Over 1000 oil st are incessantly running from Baku to the Volga. From the 10,000 tank cars are employed in conveying the oil by rail various parts of the empire. Hundreds of iron reservoirs ere different centres, provide accommodation for 120,000,000 gallon The Trans-Caspian railway, nearly 100 miles long, uses fr extremity to the other oil fuel, and no other. On the Volga, liq is replacing wood and saving the forests from ruthless destructio Batoum, thanks to the millions of gallons of oil pouring through ope, keeping a fleet of tank steamers incessantly employed, a ble Turkish town has developed in a few years into a powerful rosperous mercantile community, exporting to India alone last 7,000,000 gallons of petroleum, of which a million gallons went very Burma where our eastern oil fields are situated. Finally, a xcise tax on kerosene, newly imposed last year, yielded nearly quarters of a million sterling, and enabled the Minister of Finance lare the best budget for thirteen years.—*Marvin's* "*Our Un- iated Petroleum Empire.*"

RAILWAY TRANSMISSION OF PETROLEUM.

Liverpool Mercury has the following:—"Nature is bounteous provision she has made for the supply of petroleum oil on the nents both of Europe and America, and the results are an ious development in the trade, and in the requirements neces- for the storage and transportation of that commodity. Tank ers are rapidly replacing sailing vessels, and comprehensive gements have been or are being made for storing the oil in There are, indeed, several private establishments and a er of public wharves where storage tanks have been erected relieve the distributor of investing capital in expensive installa- as they are now required, and place at the disposal of shippers oil-producing countries additional means of assisting them in trade, which cannot fail to be of the greatest advantage. In Britain and Ireland there is storage capacity, approximately, 9,500 barrels. Liverpool and Birkenhead alone have accommo- for 236,500 barrels, and Barrow, which is a distributing for the Liverpool market, for 100,000 more. But the problem ns to be solved as to the carrying of the oil in bulk from the by railway into the interior, and this is a question which requires s consideration. The storage and carriage of petroleum are by ans unattended with danger, and it is obvious that for the safety the traffic must be conducted under very strict regula- —*Liverpool Journal of Commerce.*

THE PETROLEUM FIELDS OF INDIA.

India petroleum occurs in Upper and Lower Burmah (including rakan Islands), in Assam, in the Punjaub, and in Beloochistan. rakan the most productive fields are believed to be those of i and the eastern Baranga Island, in both of which localities eum of a very high quality, in some cases sufficiently pure to able of being burned in ordinary lamps in its crude state has been ed. In India, as in Russia and Galicia, owing to the dis- l character of the strata in which the petroleum is found, drilling nded with difficulties. The most abundant supplies of oil were obtained from the wells at Khatan, any one of the five wells ly drilled being capable of furnishing the entire supply of 50,000 s of oil a year, which is estimated to be the amount required for ind-Pishin section of the North-Western Railway. From the of view of the kerosene manufacturer, the Arakan oil was the but, on the other hand, much of the Yonangyoung oil yielded a percentage of paraffin, which was a valuable product; and the of kerosene from the latter oil was capable of being largely sed by the newly-adopted processes. Too little was at present n about the oil fields of India to admit of a confident prediction their future. Unless the necessary operations were carried out skilful, energetic, and experienced management, disappointment liscouragement would probably ensue, and the development of l fields of India would be retarded. The action of the Indian rnment in proposing to sell the oil-producing territories in sections, d of granting prospecting licences, is likely to afford some guarantee icient development.

SALFORD SEWAGE EXPERIMENTS.

HT millions of gallons of sewage per day, and how to dispose of it? is the problem which—owing to the progress of the Man- r Ship Canal—the Salford Corporation finds itself compelled to Under ordinary circumstances the addition of such an effluent river Irwell obviously raises many serious questions, but, when nsider that the flowing, if not silvery, Irwell will, in effect, be by- ye converted into a series of docks or settling ponds, more or less nt, it will at once be seen that some process which, after treating wage matters, gives an inoffensive and stable effluent, becomes solute necessity. Up to the present, Salford has adopted the process of precipitation, which, as is well known, will not suffice et the more stringent requirements of the future. Therefore, a onths ago, the Salford Corporation invited promoters and tees of various systems of sewage purification to institute experi- l trials at Mode Wheel, so as to show the results which could ly be obtained from Salford sewage. e of these experiments have now been undertaken by the Inter- al Water and Sewage Purification Company, Limited. This company's system is already in operation at several places in the neighbourhood of London, and several provincial authorities have decided upon its adoption. In this system a patented reagent styled "Ferozone" is used as a precipitant. Without commenting upon the nomenclature, we may say this compound as been analysed, and stated to contain principally sulphate of iron and magnetic oxide of iron, besides salts of alumina and magnesia. It is stated that the quantity required for the complete precipitation of all impurities ranges from five to eight grains per gallon, or seven to ten cwts. per million gallons of sewage. As this is only one-half the precipitant added in the lime process, and less than one-tenth of the weight added in some other processes, the statement has important bearings on the quantity of sludge which ultimately has to be disposed of. All precipitants naturally add to the amount of sludge to be eventually dealt with, and, of course, from this standpoint, the system which adds the least is the best. At Salford sewage works, about three grains of "ferozone" are added per gallon of sewage, which is then allowed to flow into a settling tank of some ten thousand gallons capacity, where it is allowed to become perfectly quiescent while an alternative and similar tank is being filled. After settling for two or three hours, the supernatent liquid is run into one of two alternate filters of some forty square yards area each, which consist of: first, a top layer of sand nine inches deep, then twelve inches mixed "polarite" and sand, under which are further layers of sand, peagravel, and shingle, in which are embedded the pipes for carrying away the effluent. It is said that effective filtration can be carried on at the rate of one thousand gallons per square yard per twenty-four hours. From an analysis of "polarite"—which is the active ingredient of the filter—it appears to consist of something like fifty-three per cent. of magnetic oxide of iron, twenty-five per cent. of silica, and some lime and magnesia The magnetic oxide of iron is familiar to most of us as blacksmiths' scales, and is a mineral well known to metallurgists; but its use as a filtering medium is somewhat novel, and, so far—in towns where it has been tried, as at Acton, Tottenham, Hendon, &c.—it appears to have given satisfaction. For the purposes of the "International" system, the natural ore is sub- mitted to a special treatment before use as a filtering medium. It might be thought that after continuous use the filters would be liable to clogging by arrested impurities, but, on this point, it is stated that after the filter beds at Acton had been in use for fourteen months, the only cleansing had been the removal of the surface layers of sand. The Borough Engineer of Salford reports that the amount of sludge by the "International" system is less than half that produced by the ordinary lime treatment. Moreover, after the sludge has been pressed into bricks, burnt and ground up, it is said to be sold at Acton for manure at thirty shillings per ton. As sludge has generally been looked upon as a dead loss, such a result in Salford would be a valuable feature if it could be disposed of at anything like this price.

Electrolysis is to be applied to the sewage by the Electrical Purifica- tion Association, Limited, a company which has been formed to work the system of sewage treatment patented by Mr. William Webster, F.C.S., whose experiments at Crossness have recently attracted atten- tion. No chemicals are added, and, after electrolysis, the impurities separate by gravitation, and the effluent is said to be quite free from any tendency to putrefaction. At the experimental works, the crude sewage is passed through a channel of brickwork ninety feet long, four feet nine inches broad, and about fourteen inches deep. In this channel are suspended three hundred and sixty-four common cast iron electrodes, four feet deep, two feet eight inches wide, and half an inch thick. These are, for electrical reasons, divided into twenty-eight sections of thirteen plates each, and connected by copper strips to a sixty-five volt dynamo. The plates are hung longitudinally in the channel, so as to present their edges to the flow of the sewage, and as they are only five-eighths of an inch apart, every particle of sewage is subjected to the action of the electric current. Some of the impurities separate themselves as a flocculent mass buoyed up by the liberated gases at first, but, after remaining in a settling tank for a time, they fall to the bottom, and the supernatant liquid is the purified effluent. As the dissolved iron is the only addition, and as this only amounts to two grains of iron per gallon of sewage treated, or two cwts. of cast iron per eight million gallons per day, it is claimed that the amount of sludge formed is less than by any other process. Chemists who have reported on the trials of this process state that none of the effluents showed any signs of putrefaction, and also, that they showed a marked absence of sulphuretted hydrogen, and any filtration was superfluous.

The results of these experiments will be awaited with considerable interest; for while the question is a very pressing one in Salford, it is of vital importance to many other towns in the United Kingdom. We understand that among other systems to be experimentally tried will be the fish brine or "Amines" process.

A HORN PITH SYNDICATE.—We learn that a syndicate contem- plating the purchase of all the horn pith works in this country.

PERMANENT CHEMICAL EXHIBITION.

THE proprietors wish to remind subscribers and their friends generally that there is no charge for admission to the Exhibition. Visitors are requested to leave their cards, and will confer a favour by making any suggestions that may occur to them in the direction of promoting the usefulness of the Institution.

JOSEPH AIRD, GREATBRIDGE.—Iron tubes and coils of all kinds.

ASHMORE, BENSON, PEASE AND CO., STOCKTON-ON-TEES.—Sulphate of Ammonia Stills, Green's Patent Scrubber, Gasometers, and Gas Plant generally.

BLACKMAN VENTILATING CO., LONDON. — Fans, Air Propellers, Ventilating Machinery.

GEO. G. BLACKWELL, LIVERPOOL.—Manganese Ores, Bauxite, French Chalk. Importers of minerals of every description.

BRACHER AND CO., WINCANTON.—Automatic Stills, and Patent Mixing Machinery for Dry Paints, Powders, &c.

BRUNNER, MOND AND CO, NORTHWICH.—Bicarbonate of Soda, Soda Ash, Soda Crystals, Muriate of Ammonia, Sulphate of Ammonia, Sesqui-Carbonate of Ammonia.

BUCKLEY BRICK AND TILE CO, BUCKLEY.—Fireclay ware of all kinds—Slabs, Blocks, Bricks, Tiles, "Metalline," &c.

CHADDERTON IRON WORKS CO., CHADDERTON.—Steam Driers and Steam Traps (McDougall's Patent).

W. F. CLAY, EDINBURGH.—Scientific Literature—English, French, German, American. Works on Chemistry a speciality.

CLAYTON ANILINE CO., CLAYTON.—Aniline Colours, Aniline Salt, Benzole, Toluole, Xylole, and Nitro-compounds of all kinds.

J. CORTIN, NEWCASTLE-ON-TYNE.—Regulus and Brass Taps and Valves, "Non-rotative Acid Valves," Lead Burning Apparatus.

R. DAGLISH AND CO., ST. HELENS.—Photographs of Chemical Plant—Blowing Engines, Filter Presses, Sludge Pumps, &c.

DAVIS BROS., MANCHESTER.—Samples of Products from various chemical products—Coal Distilling, Evaporation of Paper-lyes, Treatment of waste liquors from mills, &c.

R. & J. DEMPSTER, MANCHESTER—Photographs of Gas Plants, Holders, Condensers, Purifiers, &c.

DOULTON AND CO., LAMBETH.—Specimens of Chemical Stoneware, Stills, Condensers, Receivers, Boiling-pots, Store-jars, &c.

E. FAIRRIG, PLAISTOW, ESSEX.—Ozonised Products. Ozone-Bleached Esparto-Pulp, Ozonised Oil, Ozone-Ammoniated Lime, &c.

GALLOWAYS, LIMITED, MANCHESTER.—Photographs illustrating Boiler factory, and an installation of 1,500-h.p.

GRIMSHAW BROS., LIMITED, CLAYTON.—Zinc Compounds. Sizin Materials, India-rubber Chemicals.

JEWSBURY AND BROWN, MANCHESTER.—Samples of Aerated Waters.

JOSEPH KERSHAW AND CO, HOLLINWOOD.—Soaps, Greases, and Varnishes of various kinds to suit all requirements.

C. R. LINDSEY AND CO, CLAYTON.—Lead Salts, (Acetate, Nitrate, etc.) Sulphate of Copper, etc.

CHAS. LOWE AND CO., REDDISH.—Mural Tablet-makers of Carbolic Crystals, Cresylic and Picric Acids, Sheep Dip, Disinfectants, &c.

MANCHESTER ANILINE CO., MANCHESTER.—Aniline Colours. Samples of Dyed Goods and Miscellaneous Chemicals, both organic and inorganic.

MELDRUM BROS., MANCHESTER.—Steam Ejectors, Exhausters, Silent Boiling Jets, Air Compressors, and Acid Lifters.

E. D. MILNES AND BROTHER, BURY.—Dyewoods and Dyewood Extracts. Also samples of dyed fabrics.

MUSGRAVE AND CO., BELFAST.—Slow Combustion Stoves. Makers of all kinds of heating appliances.

NEWCASTLE CHEMICAL WORKS COMPANY, LIMITED, NEWCASTLE-ON-TYNE.—Caustic Soda (ground and solid), Soda Ash, Recovered Sulphur, etc.

ROBINSON, COOKS, AND COMPANY, ST. HELENS.—Drawings, illustrating their Gas Compressors and Vacuum Pumps, fitted with Pilkington and Forrest's patent Valves.

J. ROYLE, MANCHESTER.—Steam Reducing Valves.

A. SMITH, CLAYTON.—India-rubber Chemicals, Rubber Substitute, Bisulphide of Carbon, Solvent Naphtha, Liquid Ammonia, and Disinfecting Fluids.

WORTHINGTON PUMPING ENGINE COMPANY, LONDON.—Pumping Machinery. Speciality, their "Duplex" Pump.

JOSEPH WRIGHT AND COMPANY, TIPTON.—Berryman Feed-water Heater. Makers also of Multiple Effect Stills, and Water-Softening Apparatus.

SOCIETY OF CHEMICAL INDUSTR

LONDON SECTION.

THE last ordinary meeting of the session was held at House on the 2nd inst., Mr. David Howard in the cl Secretary, Mr. Tyrer, having read the minutes of the last m announced the arrangements made for the annual genera which will be held at Nottingham, Mr. Cross, before re paper of the evening, exhibited and described an ingenious ac devised by Mr. Watkins, of Hereford, for determining ph exposures, the excellent work resulting from its use being den by a series of photographs taken under most diverse conditi Cross then read the paper by himself and Mr. Bevan, entitl considerations on the chemistry of Hypochlorite Bleachir authors confined themselves to bleaching by means of hypocl this instance, and discussed various points, the elucidation was desirable. A fact which their work had brought into pr and which was generally overlooked, was the chlorination of that commonly occurred. The degree to which this took pended on many circumstances, notably on the particular hy used, it being greater in the case of calcium hypochlorite tha of magnesium hypochlorite prepared by double decomposi bleaching powder, while this again caused more chlorinatio bleaching solution got by the electrolysis of magnesium Moreover, the amount of chlorination varied according to th bleaching desired, material such as linen subjected to bleaching" suffering less than that which was only "wh bleached, because the former had had its fibres freed from e except pure cellulose by previous drastic treatment with whereas the impurities contained in the latter were capable of chlorine to a considerable extent, rendering the consumption relatively high. That chlorination of the fibre actually c proved in the case of esparto pulp, both by analysis and by from it an organic oily body which contained chlorine. This tion at once differentiates hypochlorite bleaching from ble means of pure oxidants, such as permanganate and hydrogen and the cause assigned for it is the presence and activity c oxygen in the fibre.

It being necessary in the course of the investigation to est free base in hypochlorite solutions, the authors, after an exam the methods already in use, devised the following process, a mend it as preferable to all of them. To the solution to hydrogen peroxide of known acidity is added to destroy chlorite, after which the free base can be titrated with standa the ordinary way. It was found incidentally that the quanti lime per unit volume in filtered solutions of calcium hypochl approximately constant whatever the strength of the hypochlo other words, the proportion of lime referred to the availabl varied inversely as the concentration of the liquor.

With regard to the question of bleaching efficiency, experi cates that of the several factors of importance a high *ten* though increasing the speed, is wasteful of bleach, and tends t ate the fibre, while the presence of much *free base* is d tendering the fibre and giving a poor colour. The whole su needs much patient and intelligent study.

In opening the discussion the Chairman pointed out that li and bleached in the days of our grandmothers could still favourably in quality with modern productions, and congrat authors on their contribution to our knowledge. Dr. Thorne from his own experience many of the authors' statements, p as to the formation of organic chlorinated compounds, some were volatile and could be detected by the sense of smell d process of bleaching. It was to the loss of such substanc away by the current of inert nitrogen that he attributed the air to facilitate bleaching when used in conjunction with bleach, although the use of oxygen undiluted proved effec also commented on the tendency of pulp bleached hot, to ret colour after 10—12 hours. After a brief reply by Mr. Cros Thorp moved a vote of thanks to Mr. David Howard for his services as chairman for so long a period, the motion being se Mr. Samuel Hall, and carried with acclamation. Mr. E replying said that he hoped to be present frequently, thoug ally, at future meetings. Dr. Thorne then proposed a vote to the honorary secretary of the section, Mr. Tyrer, now re his accession to the chair, a motion which found a hearty se Mr. Howard, and was carried unanimously.

Mr. Watson Smith having exhibited a new form of buns the chief feature of which was a side slit for the admission of of the usual vertical tube, together with some modification the meeting adjourned to Nottingham in July.

ate—
600 bgs.
—
al, 10 brls. Makin & Bancroft
20 Haines & Co.
15
um Prussiate—
am, 4 cks.
—
1,559 t. Matheson & Co.
1,340 Tenants & Co.
1,238 ,,
Wax—
ork, 100 bdls. Meade, King, & Co.
—
, 440 bgs.
re—
, 1,199
iee, 109 W. & R. Graham & Co.
e—
am, 3 cks. 20 bgs.
, 140 bgs.
c Acid—
32 brls.
—
2 cks.
ix, 38
lam, 1
Salt—
rg, 494 bgs.
640

LASGOW.

ek ending May 22nd.

am, 45 cks.
—
, 100 bgs.
Ore—
am, 2,100 bgs.
s—
rine
am, 89 cks. J. Rankine & Son
lers
am, 1 ck J. Rankine & Son
acts
ork, 27 cs.
, 1 ck. 1 cs.
c
. 9 t. 15 c, Wallace Wilkie & Co.
ne—
am, 5 cs.

Glucose—
New York, 208 brls.
Manganese Chlorate—
Rotterdam, 1 ck.
Pyrites—
Huelva, 1,195 t. Tharsis Co.
Stearine—
Antwerp, 12 bgs.
Saltpetre—
Hamburg, 48 cks.
Sulphur Ore—
Bergen, 300 t.
Tartaric Acid—
Rotterdam, 4 cks. J. Rankine & Son
Ultramarine—
Rotterdam, 5 cs. J. Rankine & Son
Waste Salt—
Hamburg, 320 t.
,, 271 cks.

TYNE.

Period ending May 20th.

Barytes—
Antwerp, 50 bgs. Tyne S. S. Co
Glucose—
Hamburg, 50 bgs. Tyne S. S. Co.
,, 12 cs.
Rock Salt—
Hamburg, 150 t.
Salt—
Harburg, 150 t.
Tartaric Acid—
Rotterdam, 4 cks. Tyne S. S. Co.
Zinc Oxide—
Antwerp, 20 brls. Tyne S. S. Co.

HULL.

Week ending May 22nd.

Acetic Acid—
Rotterdam, 10 pks. Hutchinson & Son
Acids *(otherwise undescribed)*
Rotterdam, 1 ck. W. & C. L. Ringrose
,, 4 ,,
Barytes—
Bremen, 62 cks.
Antwerp, 38 brls. H. F. Pudsey
,, 18 cks. ,,
Bremen, 23 Tudor & Son
,, 23 T. W. Flint & Co.
,, 23 J. W. Davis & Son
,, 18
Bones—
Trieste, 2 pks. Wilson, Sons & Co.
Chemicals *(otherwise undescribed)*
Dunkirk, 70 pks. Wilson, Sons & Co.
Antwerp, 138 ,,
Hamburg, 245 ,,
Bremen, 1 cs. 11 kgs. Veltmann & Co.
Genoa, 16 cks. Wilson, Sons & Co.
Marseilles, 5 ,,
Rotterdam, 1 W. & C. L. Ringrose
Dunkirk, 22 pks. Wilson, Sons & Co.
Dyestuffs—
Cutch 1 bl. G. Lawson & Sons
Argols
Bari, 24 cks. Wilson, Sons & Co.
Extracts
Rouen, 58 cks. Rawson & Robinson
Fiume, 700 bls. Wilson, Sons & Co.
Dunkirk, 25 cks. ,,
Sumac
Bordeaux, 100 bls. Rawson & Robinson
Trieste, 20 bgs. Wilson, Sons & Co.
Palermo, 376 ,,
,, 350
Alizarine
Rotterdam, 78 cks. W. & C. L. Ringrose
,, 20 G. Lawson & Sons
,, 5 W. & C. L. Ringrose
,, 20 pks. ,,
,, 21 cks. Hutchinson & Son
Rennet
Copenhagen, 8 cks. Wilson, Sons & Co.
Rotterdam, 5
Fustic
Hamburg, 317 pcs. Wilson, Sons & Co.
Glucose—
Hamburg, 6 cks. Wilson, Sons & Co.
,, 8 C. M. Lofthouse & Co.
Stettin, 210 bgs. Wilson, Sons & Co.
,, 11 cks. M. Whitfield & Sons
,, 23 Hull Dock Co
,, 200 bgs. Thompson, McKay & Co.
New York, 80 cks.
Hamburg. 12 Bailey & Leetham
,, 200 bgs.
,, 12 cks. Bailey & Leetham
,, 17
Kainit—
Hamburg, 500 bgs. G. Lawson & Sons
Manure—
Ghent, 160 t.
Rouen, 4 cks. Rawson & Robinson
Naphthaline—
Hamburg, 1 brl. Wilson, Sons & Co.
Oxide—
Rotterdam, 4 cks. Hutchinson & Son
Pyrites—
Huelva, 1,518 t. J. Dalton, Holmes & Co.
Pitch—
Dunkirk, 56 cks. Wilson, Sons & Co.
Plumbago—
Hamburg, 42 cks.
,, 21 bls. Wilson, Sons & Co.
Sulphur—
Catania, 868 pks. Wilson, Sons & Co.
Turpentine—
Libau, 50 bls.
Tar—
Marseilles, 25 cks.

GRIMSBY.

Week ending May 17th.

Glucose—
Hamburg, 14 cks.
Kainit—
Hamburg, 500 bgs.

GOOLE.

Week ending May 14th.

Alum—
Rotterdam, 21 cks.
Barytes—
Antwerp, 50 bgs.
Chrome Alum—
Rotterdam, 30 cks.
Dyestuffs—
Alizarine
Rotterdam, 61 pks.
Madders
Rotterdam, 5 cks.
Logwood
Jamaica, 559 t.
Dyestuffs *(otherwise undescribed)*
Antwerp, 5 cs. 97 cks.
Glucose—
Hamburg, 48 cks. 200 bgs.
Calais, 100
Manure—
Phosphate Rock
Ghent, 800 bgs.
Potash—
Hamburg, 100 bgs.
Calais, 117 drs.
Dunkirk, 58 cks.
Saltpetre—
Hamburg, 20 cks.
Sodium Silicate—
Rotterdam, 1 ck.
Waste Salt—
Hamburg, 537 t. 1,015 bgs. 662 cks.

EXPORTS OF CHEMICAL PRODUCTS

OF

THE PRINCIPAL PORTS OF THE UNITED KINGDOM.

Caustic Soda—*continued.*
Calcutta, 10 dms.
Dunkirk, 6
Galatz, 100
Genoa, 116
Gothenburg, 5
Hamburg, 498
La Guayra, 46
Leghorn, 28
Lisbon, 18
Madeira, 5
Madrid, 10 20 brls.
Marseilles, 45
Melbourne, 34
Messina, 80
Montreal, 483
Naples, 20
New Fairwater, 113 dms.
New York, 1300
Odessa, 250
Oporto, 30 8 cks.
Rotterdam, 26
Santos, 7
Bahia, 10
Barcelona, 300 80 20 brls.
Bilbao, 126 35
Bruges, 6 4
Genoa, 182
" 15
Gothenburgh, 4
Hamburg, 460
Hiogo, 70
Leghorn, 85
Lisbon, 94 11 4
Malaga, 10
Manilda, 50
Monte Video, 200
Montreal, 100
New Fairwater, 29
New York, 250 80 brls.
Philadelphia, 135
Piræus, 85
Puerto Cabello, 11 dms.
Rouen, 52
San Sebastian, 40
Santos, 11
Seville, 25
Stettin, 27
Valencia, 41
Yokohama, 30
Gibraltar, 20
Rotterdam, 54

Epsom Salts—
Batoum, 13 cks.
Maranham, 8
Rio Janeiro, 100 kgs.
Maceio, 1

Magnesia—
Antwerp, 1 cs.
Rio Janeiro, 10
Vera Cruz, 2 cks.

Salt—
New York, 25 t.
Old Calabar, 154¾ t.
Opobo, 45 9 c.
Para, 177¾
Philadelphia, 80
Port Mulgrave, 250 t.
Rangoon, 25
Rotterdam, 203
St. John, N.B., 800
Sandheads, 2145
Sydney, 50
Vadsol, 465
Winnebah, 10
Abonnema, 43
Ambriz, 5
Amsterdam, 261
Antwerp, 42
Benin, 22
Boston, 156
Calcutta, 1320
Cameroons, 4
Campbeltown, 150
Halifax, 1419¾
Hamilton, 6¾
Loango, 12½
Miramichi, 180
Montreal, 301 10 c.
Baltimore, 100
Benin, 23
Bonny, 36
Boston, 100 2 cks.
Bruges, 200
Chicago, 299
Coquimbo, 50
Elobey, 5
Grand Bassam, 12 t.
Hamburg, 50
Manaos, 52¾
Montreal, 223
Newcastle, 100
Old Calabar, 115¾
Rouen, 10
Shediac, 10
Sydney, 175
Victoria, 150
Vigo, 2

Sodium Bicarbonate—
Bombay, 10 kgs.
Calcutta, 500
Genoa, 15 cks.
Halifax, 25
Hamburg, 50
Havana, 25
Miramichi, 15
Montreal, 400 19
Nagasaki, 200
Odessa, 100
Palermo, 20
Lante, 2 brls.
Amsterdam, 20
Calcutta, 500
Hamburg, 20 dms.
Santander, 40
Smyrna, 15

Sodium Silicate—
Barcelona, 32 cks.
Bilbao, 32
Galatz, 15 brls.
La Guayra, 10
Lisbon, 32
Madrid, 10
Montreal, 42
Seville, 25
Victoria B.C., 32

Soda Ash—
Oporto, 10 pks.
Philadelphia, 427 cks. 28 tcs.
Piræus, 36 brls.
Sherbrooke, 16 tcs.
Smyrna, 5
Varna, 60
Windsor Mills, 100
Yokohama, 222
Bahia, 20
San Francisco, 109 cks.
Adelaide, 79 brls.
Baltimore, 509
Bilbao, 50
Boston, 397 130 tcs.
Constantinople, 40 brls.
Genoa, 7 cks.
Hamburg, 4
Hamilton, 120
Kingsey, 14
Leghorn, 42
Madeira, 5
Madrid, 30
Montreal, 69 17
New York, 306 1055
Baltimore, 378 tcs. 134
Boston, 10
Corfu, 50 brls.
Ghent, 12
Lisbon, 13
Monte Video, 10
Philadelphia, 685
Puerto Cabello, 10
Rio Janeiro, 150
Valparaiso, 15

Soda Crystals—
Bilbao, 17 brls. 6 cks.
Boston, 280
Calcutta, 10
Malta, 53
New York, 280
Singapore, 9
Victoria B.C., 20
Monte Video, 67 brls.
New York, 56
Portland, 100
Quebec, 50
Singapore, 12 cks.

Tar—
Rouen, 59 t. 14 c.
Calcutta, 24 brls.

GLASGOW.

Week ending May 22nd.

Ammonia—
Hong Kong, 2 t. 18 c. £70. 10s.

Charcoal—
Antwerp, 118 c. £27

Caustic Soda—
New York, 200¾ c. £56

Copper Sulphate—
Rouen, 2 t. 16½ c. £57

Chloride of Lime—
San Francisco, 306¾ c. £96

Citric Acid—
Lisbon, 5 c. 37

Dyestuffs—
Logwood
Quebec, 8 t. 13½ c. £69

Epsom Salts—
Christiana, 50 bgs. £33

Manganese Oxide—
New York, 200¾ c. £93

Manure—
Christiania, 2 c. £1

Paraffin Wax—
Barcelona, 10 t. 3½ c. £280
Oporto, 10 4 250
" 2 16½ 79
Bordeaux, 5 132
Fremantle, 1 15 67
Bilbao, 46 3 45
Antwerp, 103 155

Potassium Bichromate—
Antwerp, 130¾ c. £194
New York, 493 921
Rouen, 16 t. 18½ 559
Barcelona, 109 200
Antwerp 103¾ 150

Potassium Prussiate—
Rouen, 116¾ £305

Pitch—
La Rochelle, 628 t. 3 c. £785
Bilbao, 563 12½ 1,005

Soda Ash—
San Francisco, 310¼ c. £86

Sodium Bichromate—
New York, 495 c. 694
Rouen, 12 t. 18¾ c. 292
Barcelona, 43 60

Superphosphate—
Christiania, 20 t. £100

HULL.

Week ending May 20th.

Alkali—
Harlingen, 94 cks.

Bone Size—
Amsterdam, 10 cks.

Bleaching Powder—
Drontheim, 30 cks.
St. Petersburg, 50

Borax—
Libau, 7 cks.

Caustic Soda—
Libau, 154 drums.
Riga, 158
St. Petersburg, 350

Chemicals (*otherwise undescribed*)
Antwerp, 29 cs. 2 cks.
Abo, 9
Bombay, 24
Christiania, 43 20 pks.
Copenhagen, 6
Drontheim, 1
Danzig, 7
Gothenburg, 35 40 5 334 lbs.
Hamburg, 15
Harlingen, 4
Lisbon, 5
Libau, 14
Norrkoping, 1
New York, 20
Rotterdam, 41 28
Riga, 20 100
Rouen, 31
S. Petersburg, 7 188
Stettin, 6
Wasa, 14 1 pkg.

Dyestuffs (*otherwise undescribed*)
Antwerp, 1 ck.
Boston, 5
Hamburg, 13
Rotterdam, 1 cs. 5
Bombay, 43

Manure—
Antwerp, 490 bgs.
Drontheim, 40
Rotterdam, 491 111 t.

Pitch—
Antwerp, 499 t.
Dunkirk, 113

Salt—
Lisbon, 1 cs.

Tar—
Christiania, 6 cks.
Copenhagen, 13
Gothenburg, 150
Hamburg, 14
Stettin, 1
Wasa, 40

GRIMSBY.

Week ending May 17th.

Chemicals (*otherwise undescribed*)
Hamburg, 12 pks.
Rotterdam, 10 cs.

TYNE.

Period ending May

Ammonia—
Barcelona, 5 t. 17 c.

Antimony—
Barcelona, 2 t. 4 c.

Alkali—
Hamburg, 3 t. 10 c.
Aarhuus, 2 18
Rotterdam, 3 11

Baryta Hydrate—
Danzig, 5 t. 7 c.

Barytes Carbonate—
Antwerp, 10 t. 16 c.

Baryta Nitrate—
Antwerp, 6 c.

Bleaching Powder—
Norrkoping, 1 t.
Antwerp, 19 1 c.
" 20 14
Hamburg, 20 18
Aarhuus, 2 16
Rotterdam, 36 10
Danzig, 2 18

Copperas—
Rotterdam, 2 t. 19 c.

Caustic Soda—
Antwerp, 5 t. 3 c.
Christiania, 7 6

Iron Oxide—
Malmo, 6 c.

Litharge—
Gothenburg, 9 c.
Antwerp, 11
Danzig, 5

Manure—
Christiania, 15 t.

Magnesia—
Barcelona, 12 t. 3 c.

Potassium Chlorate—
Hamburg, 5 c.

Potassium Bichromate
Norrkoping, 4 t. 18 c.

Phosphorus—
Malmo, 6 c.

Pitch—
Malmo, 5 t.

Sulphur—
Gothenburg, 10 t.

Sodium Chlorate—
Rotterdam, 1 t. 12 c.

Sodium Sulphate—
Danzig, 1 t. 10 c.

Soda Ash—
Christiania, 10 t. 5 c.
Newfairwater, 19 t. 12

Soda—
Antwerp, 5 t. 7 c.
Frederickshavn, 6 t. 7
Huelva, 2 t. 3
Horsens, 11 8
Rotterdam, 11 8
Odense, 10 13

Superphosphate—
Christiania, 10 c.

Tar—
Malmo, 1 t. 5 c.

Turpentine—
Antwerp, 1 c.

GOOLE

Week ending May

Benzole—
Antwerp, 24 cks.
Rotterdam, 48

Bleaching Preparatio
Hamburg, 18 cks.

Copper Precipitate—
Rotterdam, 242 cks.

Coal Products—
Rotterdam, 163 cks.

Chemicals (*otherwise u*
Dunkirk, 15 cks.
Hamburg, 97
Rotterdam, 25

Dyestuffs (*otherwise un*
Ghent, 21 cks.

Manure—
Boulogne, 153 bgs
Dunkirk, 50
Rotterdam, 550

Pitch—
Antwerp, 531 t.
Ghent, 106

THE TYNE CHEMICAL REPORT.

TUESDAY.

narket for chemicals has been very quiet during the past week. es all round are easier, and in favour of buyers. Bleaching is decidedly dull, and price has dropped half-a-crown a ton st report. Caustic soda is also lower by 5s. per ton. Soda lengthened its discount by 2½ per cent., and crystals are 6d. to 1s. a ton. The quotation for recovered sulphur is nominal, as, owing to a breakdown in the machinery, the process has been stopped for a short time. Makers are also ld, and are not likely to be in the market for this article few weeks yet. To-day's quotations are:—Bleaching in softwood casks, £5. per ton; caustic soda, 77%, in £9. per ton; ground, in 3-4 cwt. barrels, £12. per ton; sh, 48/52%, 1¼d. per degree, less 5%; soda crystals, in bags, net weight, £2. 14s. 6d. per ton, in 2 cwt. bags, net £2. 12s. per ton; in casks, gross weight, £2. 12s. per ton; of soda, in bulk, 32s. per ton; ground and packed in 12s. per ton; recovered sulphur, £4. 5s. per ton; chlorate of 4½d per lb., less 5%; silicate of soda, 75° Tw., £2. 10s. ; 100° Tw., £3. 7s. 6d. per ton; 140° Tw., £4. per ton; phite of soda, in 1 cwt. kegs, £5. 15s. per ton, in 5-7 cwt. £5. 5s. per ton; pure white sulphate of alumina, £4. 10s. ; blanc fixe, £7. 10s. per ton; chloride of barium, £8. per trate of baryta, crystals, £18. 15s. per ton; ground, £19. 5s. ; sulphide of barium, £5 10s. per ton—all f.o.b. Tyne, or akers' works.

s of manufacturing coals continue very firm, although all other of coal are lower. There is an exceedingly heavy foreign for smalls, which more than compensates for the great falling e home consumption.

umberland steam small is 8s. to 8s. 6d per ton, and Durham . 6d. to 10s. per ton.

bulk of the workpeople employed by the Tyne Plate Glass ny, Limited, South Shields, terminated their engagements on y last, after working a fortnight's notice.

ems that, a few months ago, the men in the various departments glassworks applied for an advance of wages. It was mutually o refer the matter to arbitration, and Dr. Spence Watson was as arbitrator. Three weeks ago the evidence of both sides mpleted before the arbitrator, and it was decided that the should remain unaltered until his award was made known. ult has not yet been published The women employed by the llowing the example of the men, applied for an advance, and oulated that an improved method of working in the women's ent, which the firm intended to bring into operation, should introduced. This demand the company would not entertain, he stoppage of the works. It is expected that some arrange- ill be arrived at on the arrival of the chairman of the directors, M. Palmer, who is expected in a few days. The number of mployed is about 600.

MISCELLANEOUS CHEMICAL MARKET.

e in alkalies is fairly good, and caustic soda is firm all round at ek's prices, with a tendency to further advance, viz.: 74% £8. 17s. 6d. to £9., 70%, £8. 5s.; 60%, white and cream, and £7. respectively. Soda ash in steady demand, and prices 1½d. per deg., according to brand. Soda crystals in good , and prices firmly maintained at £3. to £3. 5. f.o.b. Liver- d Tyne. Bicarbonate slightly easier, at £5. 15s. per ton. e of potash dull at 4½d. per lb. Sulphur has been finding utlet for export to the U.S., and prices are steady. Muriatic d vitriol in less active demand, and prices in buyers' favour. ng powder reported rather firmer for future delivery, but spot doing at £4. 17s. 6d. on rails, and £5. for July forward, and rt at £5. 5 f.o.b. Liverpool. Sulphate of copper steady at s. to £25. for June delivery, and makers well booked with Lead acetates and nitrate continue neglected, but prices steady, and there is no pressure of stock. Acetates of lime entiful, and prices the turn easier—brown at £7. 12s. 6d. to .; grey at £13. to £13. 5s. at usual points. Acetic acids of es quiet, and unchanged in price. Acetate of soda £16. 10s.

Arsenic in good demand, and white-powdered firm in price nett. Carbolic acids are without material change in position, y little business passing meantime. Aniline oil and salt are ly unchanged in prices. Potash, caustic, and carbonate are price, and in good demand both for early and forward delivery.

REPORT ON MANURE MATERIAL.

The manure season being now almost over, business in material for prompt delivery has during the past week been on a very limited scale, but for forward a large business has been done in mineral phosphates. There are sellers of hot-air-dried bull River Rock at 10½d. per unit, in cargoes c.f. and i. to U.K. Bids of ½d. less are still invited by outside shippers, but we have not heard of any transactions under 10½d. Several cargoes of land rock have been placed at 9½d. per unit, c.i.f. to U.K., and there are still sellers at the price.

Quotations for Somme and Belgian descriptions are without material alteration. Considerable business is reported to have been done in the lower grades, but prices for the higher kinds still rather preclude buying. There is more disposition on the part of sellers of Canadian to meet buyers, and a larger business in this description of rock may now be anticipated Since the sale of three cargoes of River Plate bones at £5. 7s. 6d. about a week ago, we do not hear of any fresh transactions. Sellers are now asking £5. 10s., delivered good U.K. port, for summer shipment, but nearest value is probably £5 7s. 6d, the last price paid for March sailing. For River Plate bone ash about the same price would have to be paid, basis 70% phosphate of lime, a large proportion of the production having been sold at something near that figure.

For the present shippers of East Indian crushed bones and bone meal seem afraid of offering for shipment ahead, and there is no business reported. £4 18s. 9d. bid and refused for crushed Kurrachee bones, and £5 5s for Bombay bone meal, for shipment, delivered ex-quay Liverpool On spot small sales of Bombay meal have been made at £5. 5s. ex-quay, but practically the price is £5. 10s., ex-store for finer quality and £5. 5s. for irregular. There are no crushed bones offering. Value of common grinding bones has improved from 5s. to 7s. 6d. per ton from the lowest point touched, and ordinary parcels would now fetch £4. 10s. to £4. 12s. 6d.; fine, clean, dry lots would realize £4. 15s. to £4. 17s. 6d. per ton ex-quay.

The nitrate of soda market remains very dull in all positions. Spot value is 8s. for fine quality. Due cargoes are worth barely 8s. May-June sailings, ordinary quality, might fetch about 8s. 3d. per cwt., and September-October shipments 8s. 6d. to 8s. 7½d. per cwt. In other ammoniacal material there is nothing of interest to note, prices for dried blood, ammonite, etc., being without alteration. Kainit, scarce at 40s. per ton. Supers, quiet at 46s. 3d. per ton, in bulk, f.o.r., at Works for 26%.

TAR AND AMMONIA PRODUCTS.

The benzol market has retained its firmness, no great change, however, having taken place since our last report. 90s. benzol stands at 3/9, and 50/90s. at 2/9 to 2/10 Carbolic acid is quiet, while creosote, anthracene, and pitch retain their old prices.

Since this time last week there has been a very quiet market in sulphate of ammonia, but the steady tone is, however, maintained, and buyers offer in vain £11. 2s. 6d., most sellers being firm at £11. 3s. 9d. to £11. 5s. f.o.b. Hull, at which price a few transactions have taken place. The value at Leith remains at £11. 3s. 9d. for prompt parcels. The Hull stocks were 450 tons at the end of last week, but no doubt the bulk of it will be cleared off in the usual course. In any case there is no reason for any lowering of prices on this account, although dealers may make use of the fact to depress values.

THE LIVERPOOL COLOUR MARKET.

COLOURS.—There is little change to report in figures this week. Ochres: Oxfordshire quoted at £10., £12., £14., and £16.; Derbyshire, 50s. to 55s.; Welsh, best, 50s. to 55s.; seconds, 47s. 6d.; and common, 18s.; Irish, Devonshire, 40s. to 45s.; French, J.C., 55s., 45s. to 60s.; M.C., 65s. to 67s. 6d. Umber: Turkish, cargoes to arrive, 40s. to 50s.; Devonshire, 50s. to 55s. White lead, £21. 10s. to £22. Red lead, £18. Oxide of zinc: V.M. No. 1, £25.; V.M. No. 2, £23., Venetian red, £6. 10s. Cobalt: Prepared oxide, 10s. 6d.; black, 9s. 9d.; blue, 6s. 6d. Zaffres: No. 1, 3s. 6d.; No. 2, 2s. 6d. Terra Alba: Finest white, 60s.; good, 40s. to 50s. Rouge: Best, £24.; ditto for jewellers, 9d. per lb. Drop black, 25s. to 28s. 6d. Oxide of iron, prime quality, £10. to £15. Paris white, 60s. Emerald green, 10d. per lb. Derbyshire red, 60s. Vermillionette, 5d. to 7d. per lb.

THE

CHEMICAL TRADE JOURNAI

Publishing Offices: 32, *BLACKFRIARS STREET, MANCHESTER.*

No. 159. **SATURDAY, JUNE 7, 1890.** **Vol.**

Contents.

Notices.

All communications for the *Chemical Trade Journal* should be addressed, and Cheques and Post Office Orders made payable to—

DAVIS BROS., 32, Blackfriars Street, MANCHESTER.

Our Registered Telegraphic Address is—

"Expert, Manchester."

The Terms of Subscription, commencing at any date, to the *Chemical Trade Journal*,—payable in advance,—including postage to any part of the world, are as follow:—

Yearly (52 numbers) .... **12s. 6d.**
Half-Yearly (26 numbers) .... **6s. 6d.**
Quarterly (13 numbers) .... **3s. 6d.**

Readers will oblige by making their remittances for subscriptions by Postal or Post Office Order, crossed.

Communications for the Editor, if intended for insertion in the current week's issue, should reach the office not later than **Tuesday Morning.**

Articles, reports, and correspondence on all matters of interest to the Chemical and allied industries, home and foreign, are solicited. Correspondents should condense their matter as much as possible, write on one side only of the paper, and in all cases give their names and addresses, not necessarily for publication. Sketches should be sent on separate sheets.

We cannot undertake to return rejected manuscripts or drawings, unless accompanied by a stamped directed envelope.

Readers are invited to forward items of intelligence, or cuttings from local newspapers, of interest to the trades concerned.

As it is one of the special features of the *Chemical Trade Journal* to give the earliest information respecting new processes, improvements, inventions, etc., bearing upon the Chemical and allied industries, or which may be of interest to our readers, the Editor invites particulars of such—when in working order—from the originators; and if the subject is deemed of sufficient importance, an expert will visit and report upon the same in the columns of the *Journal*. There is no fee required for visits of this kind.

We shall esteem it a favour if any of our readers, in making inquiries of, or opening accounts with advertisers in this paper, will kindly mention the *Chemical Trade Journal* as the source of their information.

Advertisements intended for insertion in the current week's issue, should reach the office by **Wednesday morning** at the latest.

Advertisements.

Prepaid Advertisements of Situations Vacant, Premises on Sale or To be Let, Miscellaneous Wants, and Specific Articles for Sale by Private Contract, are inserted in the *Chemical Trade Journal* at the following rates:—

30 Words and under.......... 2s. 0d. per insertion.
Each additional 10 words .... 0s. 6d. "

Advertisements of Situations Wanted are inserted at one-half the above rates when prepaid, viz:—

30 Words and under.......... 1s. 0d. per insertion.
Each additional 10 words .... 0s. 3d. "

Trade Advertisements, Announcements in the Directory Columns, and all Advertisements not prepaid, are charged at the Tariff rates, which will be forwarded on application.

COAL SMOKE.

To strain at a gnat and swallow the proverbial c an operation that has aforetime been conside worthy of imitation. Once in a way, the performa be considered instructive, if it be only to demonst folly of agitating our minds over small but visible tl the total exclusion of the invisible ones of magnitude

Nothing can illustrate these remarks so well as the agitation respecting coal smoke. It is, however, be to be perceived in the Metropolis that unless some s taken to promote legislation against the present m burning coal in private houses no improvement can b for.

If this be true with regard to visible smoke, it be what we have preached for so long a time. At a delivered in Manchester in 1882, the writer said: " believe that were we to banish all the factories for te round you would have the so-called fogs quite as int ever. If we wish to clear the atmosphere our first d is with domestic chimneys. It has hitherto been con a subject too vast to be dealt with, and our sanitary leg have perhaps been wise in their generation by pra letting it alone."

An American expert says: "Factory furnaces, h are not to be charged with all the smoke. If they were of the English cities would be practically smokeless vast amount of coal consumed for domestic purposes a for the only difficulty they have in clearing the atmo for in London smoke from factory chimneys is cally prevented, while from those in Paris volumes of rivalling Cincinnati, are constantly emitted, yet the I atmosphere is smoky while that of Paris is clear. In I vast quantities of bituminous coal are consumed for d purposes, in Paris, none. We must therefore conclu the difference in the atmosphere of the two capitals is a difference in domestic conditions."

It is strange that in the discussion of the smoke n the evils due to sulphurous acid should be so system neglected. No doubt the cry for the suppression o smoke is well enough in its way, but it resembles tl no less than the sulphurous acid does the camel.

The *British Medical Journal* has undertaken the tas forming its readers "that smoke from factories or w any kind no longer plagues London." The Augean has been cleansed, we are informed, from the infor afforded by the long and costly trials carried out in c tion with the Smoke Abatement Exhibition, and thro agency of the Police and the Home Office, as well as t the energetic action of the smoke inspector, London h practically freed from factory smoke. All this no dout very well, but what about the sulphurous acid, whic well known to the authorities as to be exempted by mention in the 29th Section of the Alkali, &c., Works lation Act of 1881: has there been any attempt to li evolution?

No doubt some of our correspondents will be asking question—what do you propose, to limit the evils arisi

…, Refining, and Casting Metals by Electricity—(Complete Specification). H. H. Lake. 7876. May 20.
…itation and Apparatus therefor. F. König. 7878. May 20.
…ements in Brewing. H. T. Brown, G. H. Morris, and E. R. Moritz. May 20.
…ion of Combustion in Furnaces. T. Norman and H. T. Simpson. May 22
…cture of Gas. W. Dyson. 7889. May 22.
…omposition for Tillage Purposes. W. Dyson. 7890. May 22.
…g Metal Plates and Sheets. F. T. Thomas. 7916. May 22
…ements in Bottles for Aerated Waters. J. Monteith. 7922. May 22.
…cture of Dioxynaphthaline Carbon Acids. J. Y. Johnson. 7926. …2.
…ls for Electrolytically Deposited Tubes F. E. Elmore. 7932. …2.
… Kilns. E. H. Edwards. 7942. May 22.
… Consumption. J. Lever, J. Holland and W. H. Todd. 7952. May 22
" " " " " 7953. May 22.
…tion of Slate for Portland Cement. R. Watkins. 7962. May 22.
…ent of Sewage. J. W. Lodge. 7963. May 22.
… Regulating Valves. J. Weir and G. Weir. 7967. May 22.
…g Vegetable Oils. A T. Hall. 7976. May 22.
…os and Electro Motors. E. Sechehaye. 7981. May 22.
…cture of Ammonium Nitrate. J. C. Butterfield, T. C. Batchelor, and …rley. 7983. May 22.
… Welding—(Complete Specification). C. L Coffin. 7988. May 22.
" " " " 7989. May 22.
" " " " 7990 May 22.
" " " " 7991. May 22.
" " " " 7992. May 22.
" " " " 7993. May 22
" " " " 7994. May 22.
…g Cotton Seed. W. Gray and W. P. Thompson. 7099. May 22.

An Improved Amalgamator—(Complete Specification). H. H. Lake. 8004. May 22.
Recovery of Gold and Silver by means of Chlorine. W. H. Dowland. 8032. May 22.
Preparation of Cement. G. H. Skelsey. 8033. May 22.
Improved Electric Conductor. R. W. Eddison. 8036. May 23.
Smoke Consumption. J. Torkington. 8050. May 23.
Treating Phosphatic Minerals. R. Fullarton 8064. May 23,
Hollow Furnace Fire Bars. J. B. Foxwell. 8067. May 23.
Manufacture of India-rubber Materials. T. Birnbaum. 8072. May 23.
Improved Composition of Soap—(Complete Specification). G. Allen. 8077 May 23.
Secondary Batteries. M. Bailey and J. Warnes. 8086. May 23.
Packing Washer for Flange Joints. A. Scott. 8088. May 23.
Manufacture of Bleaching Powder and Caustic Soda—(Complete Specification). J. D. Penock and J. A. Bradburn. 8090. May 23.
Concrete. F. Sarg. 8108 May 23.
Manufacture of Gas. E. Freund. 8109 May 23.
Electric Meters. G. Forbes. 8110. May 23.
Brick-making Machinery. W. Rushforth 8122. May 24.
Machinery for Dyeing, Scouring, and Drying Hanks. E. Sykes and D. Sykes. 8134. May 24.
Heating Fluids and Condensing Steam—(Complete Specification). J. Wright. 8145. May 24.
Improvements in Stills of the Coffey type. A. Chapman. 8147. May 24.
Treatment of Alkali Waste, and Manufacture of Sulphuretted Hydrogen and Sulphide of Ammonium. J. Leith. 8150. May 24.
Smoke Consumption and Fuel Economy. G. Moffatt and S. Stuttaford. 8167. May 24
Apparatus for delivering Air into Boiler Furnaces. " " 8168. May 24.
Manufacture of Illuminating Gas. W. Clarke. 8170. May 24.

IMPORTS OF CHEMICAL PRODUCTS
AT
THE PRINCIPAL PORTS OF THE UNITED KINGDOM.

LONDON.
…eek ending May 24th.

…—
…d, £30 Ohlenschlager Bros.
…Acid—
…d, 20 pks. Beresford & Co.
10 Petri Bros.
172 A. & M. Zimmermann
…e of Lime—
…tes, 432 pks. Lister & Briggs
—
…, 50 c. P. Hecker & Co.
…ony Ore—
… 1 t. Borne & Co., Ltd.
… 7 Vivian, Younger, & Co.
…gal, 23 H. Emanuel
…rkey, 52 Quirk, Barton, & Co.
…n's Bay, 118 t. G. Boor & Co.
…y, 17 t. Bank of N. S. Wales
14 " "
—
…ne, 4 t. Goad, Rigg, & Co.
…ampton, 7 t Goldsborough & Co.
43 t. D. de Pass & Co.
…ide, 2 Hicks, Nash, & Co.
…iffe, 7 Dyster, Nalder, & Co.
…tes, 31 H. A. Lane & Co.
…y, 8 Goad, Rigg, & Co.
…lies, 600 Gonne, Croft, & Co.
…tone—
6 t. Typke & King
400 Brandram Bros. & Co.
…s—
…d, 23 cks. W. Harrison & Co.
22 " "
…choue—
10 c. Hammond & Co.
…nbique. 56 c. Elkan & Co.
…lies, 41 c. Caldwell, Watson, & Co.
…ar, 17 L. & I. D. Jt. Co.
…bane, 36 " "
… 17 Kebbel, Son, & Co.
… 1 J. Owen
…tes, 5 Beresford & Co.
…nbique, 90 c. Kleniwort, Sons & Co.
…bane, 10 c. W. M. Smith & Sons.
…ar, 15 Clarke & Smith
45 Wrightson & Son
…lies, 152 E. Barber & Co.
50 Ralli Bros.
…e, 20 Flageollet & Co.
52 Stiebel Bros.
…pore, 228 Huttenbach & Co.
…e, 48 " "
6 Flageollet & Co.
79 Kebbel Son & Co.

Carbonic Acid—
Holland, 25 pks. J. G. Fisher
" 27 J. Owen
" 27 J. Owen
Cream of Tartar—
France, 5 pks. B. & F. Wf. Co.
" 6 " "
" 10 W. C. Bacon & Co.
" 1 T. Ronaldson & Co.
Spain, 14 B. & F. Wf. Co.
Holland, 5 Middleton & Co.
Italy, 10 B. & F. Wf. Co.
Chemicals *(otherwise undescribed)*
Germany, £18 Brasch & Rothenstein
" 168 A. & M. Zimmermann
" 72 Phillipps & Graves
Holland, 20 " "
Germany, 540 A & M. Zimmermann
" 30 T. H. Lee
" 15 L. & I. D. Jt. Co.
Norway, 60 A. Hughes
Germany, 36 Phillipps & Graves
Copperas—
Germany, £120 Charles & Fox
Caustic Potash—
Holland, 21 pks. Spies Bros. & Co.
Dyestuffs—
Tanners Bark
Holland, 9 t. S. Bevington & Sons
N. S. Wales, 75 t. L. & I. D. Jt. Co.
Melbourne, 150 t. Beresford & Co.
France, 112 L. & I. D. Jt Co.
Austria, 1 B. Kuhn
Durban, 11 A. & W. Nesbitt
Adelaide, 25 W. H. Boult
" 73 Goad, Rigg, & Co.
Natal, 24 Hicks, Nash, & Co.
Adelaide, 51 J. Vicary & Sons
Melbourne, 46 Devitt & Hett
Belgium, 9 Leach & Co.
Holland, 80 S. Barrow & Bro.
Melbourne, 200 Baxter & Hoare
Extracts
Holland, 21 pks. Soundy & Co.
Austria, 650 A. J. Humphery
U. States, 51 W. Burton & Sons
Indigo
Sweden, 1 cht. M. Erdmann & Co.
E. Indies, 41 L. & I. D. Jt. Co.
" 72 " "
" 24 G. Croshaw & Co.
" 9 Patri & Pasteur
Germany, 4 F. W. Heilgers
E Indies, 25 Elkan & Co.
" 10 F. Huth & Co.
" 39 Benecke, Souchay, & Co.
" 8 L. & I. D. Jt. Co.
" 3 Ross & Deering
" 10 H. J. Perlbach & Co.

Myrabolams
E. Indies, 412 pks. Ralli Bros.
" 900 Union L. Co. Ltd.
Orchella
E. Indies, 7 pks. Hoare, Wilson, & Co.
" 13 H. Gluck & Co.
Cochineal
Teneriffe, 10 pks. W. M. Smith & Sons
Canaries, 10 Kuhner Henderson & Co.
Teneriffe, 25 F. Huth & Co.
" 10 Bruce & White
Canaries, 4 W. M. Smith & Son
Saffron
Spain, 1 pkg. Pickford & Co.
Gambier
Singapore, 207 pks. W. H. Cole & Co.
" 176 Beresford & Co.
Argols
Cape, 5 pks. J. R. Thomson & Co
Italy, 541 R. Jacob & Sons
Sumac
Italy, 520 pks. J. Kichin, Ltd
" 200 W. France & Co.
" 100 " "
" 200 G. Lewis
Cutch
E Indies, 1170 pks. Anderson, Weber, & Smith
" 20 Johnson, Rolls, & Co.
Rangoon, 200 Brinkmann, & Co.
Annatto
E. Indies, 14 pks. Hoare, Wilson, & Co.
Dyestuffs—
Divi Divi
E. Indies, 88 pks. Beresford & Co.
" 88 " "
Valonia
Smyrna, 110 t. Hicks, Nash & Co
" 184 A. W. Nesbitt
" 231 J. Graves
Yellow Berries
Smyrna, 156 bgs. Major & Field
Tumeric
E. Indies, 200 pks. H. J. Perlbach & Co.
Dyestuffs *(otherwise undescribed)*
U. S., 5 pks. Bailey & Leetham
Holland, 33 Burt Boulton & Co.
Glycerine—
Holland, £45 F. W. Heilgers & Co.
Germany, 85 Beresford & Co.
" 65 " "
" 126 A & M. Zimmerman
Holland, 230 F. W. Heilgers & Co.

Gutta Percha—
Singapore, 2 c. Jackson & Till
" 470 H. W. Jewsbury & Co.
" 148 L. & I. D. Jt. Co.
" 236 Kaltenbach & Schmitz
Glucose—
Germany, 200 pks. 200 cwts. M. D. Co.
France, 100 100 J. Barber & Co.
U. S., 1,020 1,020 T. C. Howell
" 500 500 Barrett Tagant & Co.
" 200 1,100 C. Czarnikow & Co.
" 40 220 L. Sutro & Co.
Germany, 122 122 L. & J. D. Jt. Co.
" 418 418 " "
" 16 143 J. Knill & Co.
" 400 400 L. & I. D. Jt. Co.
" 327 1,777 L. Sutro & Co.
" 25 200 Pillow Jones & Co.
" 50 100 C. Atkins & Nisbet
" 100 100 J. Barber & Co
" 33 263 Proprs. Chmblns Wf.
U.S., 100 600 L. Sutro & Co.
" 20 110 Teede & Bishop
Isinglass—
Kobe, 3 pks. Hale & Son
N. Russia, 4 J. D. Hewitt & Co.
Manure—
Holland, 174 t. Anglo-Contl. G. Wks.
" 50 " " "
Belgium, 160 Lawes Manure Co., Ltd.
Holland, 180 Anglo-Contl. G. Wks.
France, 150 Lawes Manure Co. Ltd.
" 200 J. Burnett & Sons
Belgium, 50 Leach & Co.
Magnesia—
Holland, £65 A. & M. Zimmerman
Naphtha—
Holland, 20 drs. Burt Boulton & Co.
Oxalic Acid—
Norway, 29 pks. A. Hughes
" 16 C. Atkens & Nisbet
Pearl Ashes—
France, 80 cs. F. W. Berk & Co.
Montreal, 80 Charles & Fox
Potash—
Holland, £40 Spies Bro. & Co.
Potassium Carbonate—
France, 37 pks. F. J. Putz & Co.
Potassium Oxymuriate—
Germany, £233 G. Boor & Co.
Potassium Sulphate—
Germany, 67 pks. H. Lambert
" 59 Petri Bros

Paraffin Wax—
U. S., 200 brls. L. & I. D. Jt. Co.
" 150 J. H. Usmar & Co.
" 35 cs. 590 brls. H. Hill & Sons
" 200 brls. Rose, Wilson & Rose
Holland, 180 " "
" 110 H. Hill & Sons
" 1,718 J. H. Usmar & Co.
U. S. 99 G. H. Trank
" 592 cs. J H. Usmar & Co.

Pitch—
France, 70 cks. H. Hill & Sons

Plumbago—
Germany, 40 cks. Brown & Elmslie
E. Indies, 18 brls. Doulton & Co.
" 87 Arbuthnot Latham & Co.
Germany 19 cks. H. Henle, Succrs.
E. Indies, 96 brls. R. G. Hall & Co.
" 91 Price Hickman & Co.
" 19 Hoare, Wilson & Co.
Germany, 20 cks. Pokorny Fjelder & Co.
" 20 Le Sueur & Co.
" 19 brls. Prop. C. Chamberlains Wf.
E. Indies, 128 J. Thedder, Son & Co.

Pyrites—
Spain, 400 t. T. C. Hills & Co.

Stearine—
France, 200 bgs. G. Boor & Co.
Holland, 22 bls. J. H. Usmar & Co.
Belgium, 25 bgs. H. Hill & Sons
France, 60 " "

Sodium Acetate—
France, £82 W. G. Blagden

Sugar of Lead—
Holland, 2 pks. Beresford & Co.

Tar—
Germany, 200 brls. Linck Moeller & Co.

Tartaric Acid—
Holland, 4 pks. Middleton & Co.
" 18 W. J Crook
" 4 Northcott & Sons
" 8 Augspurg & Co.
" 2 B. & F. Wf. Co. Ltd.
France, 6 " " "

Ultramarine—
Belgium, 6 cks. Leach & Co.
Holland, 20 Peron & Co.
Germany, 64 cs. G. Steinhoff
Holland, 28 pks. Haeffner Hilpert & Co.
Belgium, 6 W. Harrison & Co.
" 5 Leach & Co.

Verdigris—
France, 2 pks. C. Brumlen

Zinc Oxide—
Holland, 50 brls. M. Ashby, Ltd.
Germany, 50 " "

LIVERPOOL.

Week ending May 27th.

Acetate of Lime—
St. Nazaire, 42 cks.

Borax—
Leghorn, 30 cks. Shropshire Union Ry. Co.
" 60

Bones—
Bombay, 4 bgs.
Para, 10 t. Gillies & Da Costa
Cadiz, 184 bgs. 8 brls. E. F. Callister & Co.
Pernambuco, 37 t.

Bone Meal—
Bombay, 2,133 bgs.

Chemicals *(otherwise undescribed)*
Rotterdam, 2 cks.

Cream of Tartar—
Rotterdam, 4 cks.
Bordeaux, 7
Genoa, 18

Copper Ore—
New Orleans, 269 sks. J. Lewis & Son

Charcoal—
St. Nazaire, 640 sks.

Caoutchouc—
Manaos, 9 cs. Blagden & Prince
" 210
Itacoatiara, 6
Para, 6 Bieber & Co.
" 225
Boma, 1 brl. Valle & Azevedo
Ambriz, 7 Taylor, Laughland & Co.
Mussera, 3 "
Kinsembo, 27 "
" 24 Samson & de Liagre

Caoutchouc—*continued.*
Ambrizette, 10 J. Holt & Co.
Muculla, 20 Taylor, Laughland & Co.
" 81 Samson & de Liagre
Banana, 13 Mender & Frinas
Cabenda, 57 bks. Hatton & Cookson
Black Point, 6 cs 5 cks. Pinto, Leite & Co.
" 1 cks. Edwards Bros.
Loango, 4 "
Sette Cama, 30 J. Holt & Co.
" 55 Hatton & Cookson
N'gove, 15 R. W. Roulston
Cape Lopez, 14 J. Holt & Co.
Gaboon, 21 A. V. Heyder
" 10 Daumas & Co.
Elobey, 8 J. Holt & Co.
Cameroons, 5 Rider, Sons & Co.
" 10 R. & W. King
" 14 J. Holt & Co.
" 5 Lucas Bros & Co.
" 13 A. Herschell
Victoria, 1 cs. 3 brls. Ambos Bay Co.
Bibundi, 1 cks. L. Hart & Co.
Old Calabar, 5 African Assen
" 1
" 2 cs. Pickering & Berthoud

Dyestuffs—
Argols
Oporto, 23 cks. Ger. Bank, London
Brindisi, 51 271 sks.
Divi Divi
Hamburg, 213 bgs.
Orchella
Hamburg, 3 bgs.
Extracts
New York, 11 kgs. A. J. White
Marseilles, 126 brls.
Antwerp, 7 cs. J. T. Fletcher & Co.
Philadelphia, 25 brls.
Hamburg, 1 cs.
Havre, 15 cks. Cunard S.S. Co., Ld.
Fustic
Pernambuco, 2,490 kilos. W. Blackburn

Glycerine—
Amsterdam, 6 crbys. 6 cs.
Marseilles, 2 brls. H. Appleby
Hamburg, 1 cs.

Glucose—
Hamburg, 5 cks. D. Currie & Co.
" 12
" 10
New York, 60 brls.
Philadelphia, 218

Horn Piths—
Para, 5 t. Gillies & Da Costa

Iodine—
Valparaiso, 110 brls. A. Gibbs & Sons

Manganese Ore—
Gothenburg 80 t. Masqueen Bros.
Batoum, 75 t.
" 435

Nitrate—
Valparaiso, 2 cs. W. & J. Lockett

Phosphates—
Antwerp, 1 pkg.
" 2,350 bgs.
Charleston, 2,000 t.

Pitch—
Amsterdam, 14 cks.
Rotterdam, 46

Pyrites—
Huelva, 1,390 t. Matheson & Co.
" 1,430 "
" 1,557 "
" 2,058 "

Potash—
Rotterdam, 30 cks.

Saltpetre—
Calcutta, 1,789 bgs.
Hamburg, 41 cks.

Sulphur Ore—
Huelva, 1,441 t. Matheson & Co.
" 1,600 "

Sodium Silicate—
Rotterdam, 1 ck.

Stearine—
Antwerp, 12 bgs. J. T Fletcher & Co.

Tartar—
Bordeaux, 5 cks.

Tartaric Acid—
Marseilles, 3 brls.

Ultramarine—
Rotterdam, 22 cks. 13 cs.

Verdigris—
Marseilles, 6 brls.
Bordeaux, 5 cks.

Waste Salt—
Hamburg, 607 bgs. 130 t. D. Currie & Co.

Zinc Oxide—
Antwerp, 175 brls.

GLASGOW.

Week ending May 29th.

Aluminium—
Antwerp, 15 cs.

Barytes—
Antwerp, 200 bgs.

Carbonic Acid—
Rotterdam, 12 cks. J. Rankine & Son

Cream of Tartar—
Bombay, 10 cks.

Dyestuffs—
Alizarine
Rotterdam, 9 cks. J. Rankine & Son
" 152
Extracts
Rouen, 40 cs. 5 cks.
Antwerp, 5 cks.
" 3 4 pks.
Rouen, 12
" 60 bgs.
Sumac
Bombay, 150 bgs.
Logwood
Jamaica, 492 t. W. Connal & Co.
Rotterdam, 2 cks.
Madders
Rotterdam, 7 cks.

Manure—
Phosphate Rock
Charlestown, 1,050 t. A. Cross & Sons
" 150

Pitch—
Antwerp, 17 cks.

Stearine—
Antwerp, 63 bgs.

Sodium Silicate—
Rotterdam, 2 cks.

Tartaric Acid—
Bombay, 21 brls.

Tartar—
Bordeaux, 9 cks. J. & P. Hutchinson
" 4

Ultramarine—
Rotterdam, 5 cks. J. Rankine & Son
Antwerp, 4 cs 2

Zinc Oxide—
Antwerp, 11 cks.

TYNE.

Week ending May 27th.

Copper Precipitate—
Huelva, 2,198 bgs. Scott Bros.
Pomaron, 3,454 Mason & Barry

Pyrites—
Huelva, 1,296 t. Scott Bros.

Rock Salt—
Hamburg, 155 t.

Sulphur Ore—
Pomaron, 1,046 t. Mason & Barry

Saltpetre—
Hamburg, 10 cks. Tyne S. S. Co.

Ultramarine—
Rotterdam, 10 cs. Tyne S S. Co.

Zinc Oxide—
Antwerp, 55 brls. Tyne S. S. Co.

HULL.

Week ending May 29th.

Alum—
Hamburg, 24 cks. Bailey & Leetham

Albumen—
St. Petersburg, 70 cs.

Acids *(otherwise undescribed)*
Rotterdam, 18 cks. G. Lawson & Sons

Bone Meal—
Bombay, 638 bgs.

Barytes—
Antwerp, 63 cks. H. F. Pudsey
" 38 A. Sanderson & Co.
" 5 Bailey & Leetham
" 49 H. F. Pudsey

Carbonic Acid—
Amsterdam, 30 cks. W. & C. L. Ringrose

Chemicals *(otherwise undescribed)*
Antwerp, 86 pks. Wilson, Sons & Co
Hamburg, 16 8 cks. "
Rouen, 25 cks. Rawson & Robinson
Danzig, 1
Stettin, 26 Wilson, Sons & Co.
Bremen, 2 J. Pyefinch & Co.
Hamburg, 15 Wilson, Sons & Co.

Dyestuffs—
Extracts
Rouen, 82 cks.
Rennet
Rotterdam, 2 cks. W. & C. L Ringrose

Dyestuffs—*continued.*
Alizarine
Rotterdam, 9 cks. W. & C.
" 5
" 8
" 77 Hutch
Myrabolams
Bombay, 6,764 bgs.
Valonia
Smyrna, 600 bgs.

Glucose—
Hamburg, 200 bgs. 25 cks.
" 15 cks. Rawson
" 10
" 10 Wood
" 100 bgs. C. M Loft
Stettin, 200 bgs. Wilson
Amsterdam, 200
Hamburg, 10 cks.
Stettin, 52 200 bgs.
" 1,000 bgs. C. M. Lof

Manure—
Kainit
Hamburg, 1,067 bgs. 70 t.

Naphthaline—
Hamburg, 9 cks. Wilson

Naphthol—
Rotterdam, 4 brls. Hutcl

Oxide—
Rotterdam, 2 cs. Hutch

Pitch—
Rotterdam, 20 cks. G. La

Phosphorus—
Rouen, 10 cs. Rawsor

Saltpetre—
Hamburg, 20 kgs. 7 cks.

Tar—
Wilmington, 2,200 brls.
Gothenburg, 4 brls. Wilsor

Ultramarine—
Rotterdam, 20 cs. W. & C

GRIMSB

Week ending Ma

Chemicals *(otherwise u*
Rotterdam, 10 ck

Manure—
Kainit
Hamburg, 200 bg

GOOLE

Week ending May

Acetic Acid—
Rotterdam, 7 brl

Antimony—
Boulogne, 1 ck.
" 2

Barytes—
Rotterdam, 6 ck

Caoutchouc—
Boulogne, 4 ck

Dyestuffs—
Logwood
Tagunade Terminos, Jamaica, 1090 t.
Madder
Rotterdam, 12 ck
Alizarine
Rotterdam, 53 ck
Orchella
Rotterdam, 30 pk

Dyestuffs *(otherwise u*
Hamburg, 1 bg.
Boulogne, 32 cs. 8 ck
" 2 25
" 7 14

Glucose—
Hamburg, 60 bgs. 11 ck
" 25
" 200 70

Muriate—
Rotterdam, 32 ck

Manure—
Phosphate Rock
Ghent, 800 b

Potash—
Dunkirk, 40 d

Soda—
Rotterdam, 5 c

Saltpetre—
Rotterdam, 200 b

Ultramarine—
Rotterdam, 3 c

Waste Salt—
Hamburg, 50 tns. 763 b
" 240 102 b
" 160 512

EXPORTS OF CHEMICAL PRODUCTS

OF

THE PRINCIPAL PORTS OF THE UNITED KINGDOM.

LONDON.

Week ending May 21st.

Ammonia—
Santos, 12 c. £20

Ammonium Sulphate—
Barbadoes, 100 t. £1,180
Sydney, 25 2 c. 307
Cologne, 19 10 220

Ammonium Carbonate—
Paris, 3 t. 14 c. £111
Bordeaux, 1 14 50

Ammonium Muriate—
Sydney, 11 c. £14

Arsenic—
Sydney, 1 t. £15

Boracic Acid—
Melbourne, 20 pks. £56
" 1 t. 2 c. 43

Brimstone—
Algoa Bay, 2 t. 1 c. £16

Carbolic Solid—
Paris, 1 t. 3 c. £77

Copper Sulphate—
Genoa, 5 t. 5 c. £120
Calcutta, 10 248
Genoa, 5 110
Bordeaux, 10 220
" 20 475

Cream of Tartar—
Nelson, 10 c. £59
Brisbane, 1 t. 10 100
Melbourne, 10 55
" 1 1 107
Lyttleton, 10 60
Newcastle, 3 cs. 17

Citric Acid—
Melbourne, 20 kgs. £140
Toronto, 20 140
Halifax, 2 c. 14
Rangoon, 1 cs. 5
Antwerp, 2 kgs. 15
Hamburg, 3 c. 21
Calcutta, ¼ 2
Nelson, 1 10
Barcelona, 1 t. 148
New York, 3 400
Hamburg, 4 cks. 59
Brussels, 6 kgs. 43

Clarifiants—
Bordeaux, 45 cks. £67

Caustic Soda—
Batoum, 10 t. 3 c. £100

Caustic Potash—
Melbourne, 1 t. £26

Chemicals *(otherwise undescribed)*
Adelaide, 100 cs. £120
Karrachi, 12 35
Sydney, 5 145
Barcelona, 7 50

Coal Products—

Naptha
Winsloe, 50 cks. £150

Carbolic Acid
Stettin, 25 cks. 1,024 gls. £163

Benzole
Rotterdam, 77 drms. £1,189
Terneuzen, 16 329

Tar
Delgoa Bay, 40 brls. £18
Norrkoping, 600 40
Sydney, 250 drs. 65

Creosote
Lisbon, 720 brls. £280

Anthracene
Dusseldorf, 88 cks. £518

Napthaline
Rotterdam, 572 bgs. £138

Pitch
Singapore, 176 t £250
Penang, 15 brls. 13
Alexandria, 100 28
Sables d'Olonne, 937 t. 1,430
Brest, 125 t. 190
Antwerp, 104 166
Gijon, 778 1,167

Condy's Fluid—
Sydney, 17 cs. £39

Disinfectants—
Sydney, 2 cs. £17
" 8 83
Melbourne, 52 400 drs. 320
Hamburg, 25 cks. 136

Hydro Carbon—
Brussels, 2 t. 2 c. £38

Manure—
Riga, 99 t. 10 c. £340
St Vincent, 10 101
Martinique, 393 6 2,310
" 100 1,000
Cherbourg, 30 526
Sydney, 10 135
Riga, 97 900

Nitric Acid—
Montreal, 10 cs. £22

Phosphorus—
Dunedin, 1,100 lbs £30

Potassium Bromide—
Sydney, 3 kgs. £23

Potash—
Sydney, 5 c. £22

Potassium Permanganate—
Melbourne, 5 c. £13

Potassium Chlorate—
Melbourne, 6 c. £13

Saltpetre—
Santos, 4 t. 13 c. £110
Oporto, 3 14 77
Vigo, 1 10 32
Rio de Janeiro, 1 9 43
Santos, 4 9 100

Sodium Bicarbonate—
Nelson, 2 t. £15

Sodium Sulphate—
Sydney, 3 t. 4 c. £11

Sulphur—
Lisbon, 200 brls. £154
Algoa Bay, 150 kgs. 69

Saltcake—
Hamburg, 127 t. 5 c. £178

Sheep Dip—
Algoa Bay, 25 cs. £70
" 50 140
New York, 13 t. 8 c. 075
Algoa Bay, 6 16 300
Wellington, 200 drs. 119

Sheep Wash—
B. Ayres, 100 drms. £140

Sulphuric Acid—
Rangoon, 3 t. 15 c. £30
Singapore, 1 13
Trinidad, 3 5 12

Superphosphate—
Martinique, 138 t. £332

Tartaric Acid—
Nelson, 5 c. £40
Lyttleton, 5 39
Newcastle, 2 cs. 11
Halifax, 3½ c. £24
Melbourne, 5 38

LIVERPOOL.

Week ending May 24th.

Ammonium Sulphate—
Grand Canary, 1 t. 15 c. 2 q. £21
Las Palmas, 15 9 1 100
Valencia, 127 1,512
Gr. Canary, 14 162
Rotterdam, 50 8 2 640

Antimony Regulus—
Montreal, 4 c. 2 q. £16

Ammonium Carbonate—
Hamburg, 7 c. £12
Genoa, 4 t. 12 2 q. 195
Vigo, 1 3 34

Alum—
Bombay, 7 t. £37
Rio de Janeiro, 1 2 c. 6
Ibrail, 10 5 2 q. 45
Lisbon, 10 10 1 54
Valparaiso, 3 16
Montreal, 10 6 50
Calcutta, 9 15 30
Montreal, 10 2 51
St. John, N.B., 12 10 1 63
Smyrna, 2 2 11
Valencia, 5 5 1 27
Baltimore, 15 4 1 80
Calcutta, 8 1 35

Ammonium Muriate—
Philadelphia, 3 t. 6 c. 1 q. £96
New York, 8 7 204
" 3 9 50

Bleaching Powder—
Ancona, 14 cks.
Baltimore, 107
Bordeaux, 10
Boston, 401
Calcutta, 81 cs.
Genoa, 293
Ghent, 28
Hamburg, 100
Havana, 21 brls.
Hong Kong, 60
Leghorn, 8
Montreal, 70 13 pks.
New York, 356
Oporto 52
Philadelphia, 57
San Sebastian, 17
Trieste, 27
Vera Cruz, 20
Antwerp, 11
Bilbao, 25
Boston, 855
Bruges, 25
B. Ayres,
Chicago, 27
Genoa, 178
Ghent, 130
Halifax, 2 cs.
Hamburg, 104
Montreal 83
Naples, 21
New York, 184
Rotterdam, 111
San Francisco, 44 96
Venice, 18

Borate of Lime—
Rotterdam, 10 t. 3 q. £120

Borate of Lime—
Havre, 23 c. 10 c. 1 q. £158

Borax—
Montreal, 2 t. 5 c. £68
Rotterdam, 5 4 145

Brimstone—
Boston, 35 £150
New York, 50 210

Caustic Soda—
Pernambuco, 59 dms.
Philadelphia, 30 cks.
St. John, N.B. 50
San Francisco, 75
Santander, 10 brls.
Santos, 20
Savanilla, 20 50 kgs.
Shanghai, 4
Singapore, 11
Talavera, 30
Tarragona, 112 20 cks.
Tunis, 10
Valencia, 55
Iloilo, 2
Leghorn, 60
Maceio, 30
Madrid, 70
Malaga, 190 pks.
Malta, 35 dms.
Melbourne, 30
Messina, 25
Montreal, 100
Nagasaki, 200
Nantes, 69
Naples, 28
New York, 1,355
Odessa, 300
Pasages, 55
Adelaide, 35
Alicante, 50
Amsterdam, 20
Bahia, 15 50 kgs.
Barcelona, 120
Bari, 51
Batoum, 161
Boston, 150
Bremen, 2
Cienfuegos, 20
Genoa, 183
Gothenburg, 176
Guantanamo, 10
Hamburg, 54
Algiers, 35
Ancona, 30
Antwerp, 74
Bahia, 15
Baltimore, 50
Bari, 17
Batoum, 232

Caustic Soda—*continued.*
Bilbao, 31 dms.
Bombay, 10
Boston, 800
Carthagena, 5
Cienfuegos, 20
Dunedin, 10
Dunkirk, 36
Nagasaki, 50
Nantes, 35
Naples, 105
New Orleans, 200 kgs.
Oporto, 15
Otago, 3
Palermo, 5
Penang, 90
Philadelphia, 140 84 brls.
Rotterdam, 34
St. John, N.B., 20
St. Louis, 100 15 15 bxs.
San Sebastian, 130
Fairwater, 70
Fiume, 18
Galatz, 90
Gefle, 5
Hamburg, 3
Ibrail, 133
Kurrachee, 20
Leghorn, 35
Malta, 5
Marseilles, 48
Melbourne, 144
Messina, 15
Monte Video, 180
Montreal, 218
Sfax, 10
Stettin, 65 10 cks.
Stockholm, 15
Trieste, 25
Vigo, 15

Citric Acid—
Rangoon, 1 cs. £8

Cobalt Oxide—
Gijon, 2 c. £90
Seville 2 q. 30

Chalk Precipitate—
New York, 2 t. 10 c. 3 q. £60

Carbolic Solid—
Rouen, 6 c. £30

Cream of Tarter—
Gibraltar, 2 c. £11

Copperas—
Constantinople, 7 t. 11 c. £13
Calcutta, 13 15 16
Smyrna, 7 7 3 15
Soerbaya, 20 5 37
Rotterdam, 19 17 2 42

Caustic Potash—
Boston, 1 t. 15 c. £35
New York, 8 13 173

Copper Sulphate—
Cape Town, 4 c. £6
Barcelona, 10 t. 1 1 q. 137
Bordeaux, 56 16 3 1,149
" 4 14 99
" 133 19 3,448
Genoa, 34 3 3 643
Naples, 4 18 3 89
Trieste, 2 1 37
Rouen, 51 8 1,090
Hamburg, 3 7 3 50
Algiers, 9 3 8
Alicante, 2 3 40
Bilbao, 15 2 3 310
Bordeaux, 148 15 1 1,513
Marseilles, 110 17 3 2,719
Naples, 3 10 64
Leghorn, 2 2 34
Genoa, 37 2 2 355
Nantes, 35 2 2 2,125
Naples, 4 19 3 100
Pernambuco, 10 14
Venice, 3 18 3 80
Algiers, 10 2 2 250
Almeria, 5 1 96
Genoa, 50 11 1,152
Iquique, 11 1 226
Valparaiso, 3 12 86
Pasages, 4 19 2 100
Tarragona, 15 385

Carbolic Acid—
Havre, 4 t. 1 c. £143
New York, 2 13 150
Rotterdam, 59 8 7,227
Hamburg, 4 460

Carbolic Acid—*continued.*
Kurrachee, 1 t. 17 c. £14
Paris, 2 5 19 25
Philadelphia, 11 9 3 200
Genoa, 1 105

Chemicals (*otherwise undescribed*)
Puerto Cabello, 1 cs. £6
Trinidad, 75 230

Disinfectants—
Demerara, 1 t. £18
Calcutta. 14 pks. 5

Dried Blood—
Montreal, 2 t. 1 c. 1 q. £25

Epsom Salts—
Calcutta, 10 cks.
Genoa, 6
Yokohama, 20 brls.

Glycerine—
Vera Cruz, 1 c. £5

Guano—
Havana, 246 t. 1 c. 2 q. £2,140
Valencia, 298 19 1 2,987

Gypsum—
New York, 10 t. 3 q. £30

Magnesia—
Talcahuano, 2 kgs

Magnesium Chlorate—
Singapore, 10 c. £6

Manure—
Demerara, 134 t. 13 c. £1,470

Oxalic Acid—
Barcelona, 5 t. 13 c. £142
New York, 3 8 2 q. 112

Potassium Sulphate—
Barbadoes, 20 t. 4 c. 3 q. £155

Picric Acid—
New York, 15 c. £150

Potassium Bichromate—
Santos, 6 brls. £40
Smyrna, 13 c. 25
Aleppo, 7 13

Pitch—
Antwerp, 304 t.
Catania, 166 cks.
Marseilles, 970
Rouen, 60 18 c.

Potassium Carbonate—
Boston, 12 t. 5 c. £186
New York, 3 14 50

Potashes—
Catania, 5 t. 17 c. £127

Potassium Prussiate—
Rio de Janeiro 1 cs. £9

Potassium Chlorate—
Shanghai, 1 t. 16 c. £77
Genoa, 1 43
Rotterdam, 1 42
New York, 2 10 128
Genoa, 1 5 65
Naples, 2 6 140
New York, 2 10 130
Odessa, 10 25
Carril, 1 48
Stockholm, 8 364
Santander, 1 42
Santos, 2 93
Gothenburg, 7 310

Sulphur Chloride—
Barcelona, 2 c. £6

Sodium Biborate—
Rotterdam, 12 t. 2 c. £17
Yokohama, 3 11 2 q. £107

Salt—
Benin, 45 t.
Bergen, 540
Bilbao, 2½
B. Ayres, 40
Calcutta, 7,630
Chicago, 100
Degama, 50
Dunedin, 10
E. London, 22
Fredericia, 725
Halifax, 25
Isles de Los, 25
Melbourne, 85
Montreal, 815
Opobo, 100
Algoa Bay, 6
Barbadoes, 20
Cape Town, 10
Dunedin, 30
Halifax, 883
Montreal, 210
Pugwash, 50
Quebec, 600
Shediac, 390
Para, 178½
Portland, 1,048½
St. John's, Nfld., 5
Stockholm, 150
Valparaiso, 4½
Victoria, 175

Soda Ash—
Baltimore, 140 tcs. 415 cks.
Bombay, 2
Boca, 41 brls.
Boston, 74 46 850 bgs.
B. Ayres, 200 brls.
Cape Town, 20
Constatinople, 10
Galatz, 42
Hong Kong, 12
Melbourne, 225
Para, 8
Paranagua, 18
Philadelphia 598 210
Rio Janeiro, 67
Seville, 7 45
Sherbrooke, 18
Varna, 20
Alicante, 30
Baltimore, 162 70
Boston, 134 191 900 bgs.
Corunna, 2
Ghent, 42
Monte Video. 100 brls.
New Orleans, 7
New York, 598 856
Philadelphia, 43
Piræus, 100
Puerto Cabello, 2
Rio Janeiro, 230
Santos, 10
Valparaiso, 5

Sodium Bicarbonate—
Antwerp, 19 cks.
Bilbao, 30 kgs
B. Ayres, 25
Calcutta, 800
Hamburg, 20
Mantanzas, 20
Monte Video. 200
Montreal, 350
Naples, 22
Rio Janeiro, 50
Seville, 14 brls. 20
Toronto, 50
Trieste, 115
Antwerp, 29 60 50 bgs.
Bombay, 250
Constantinople, 50
Gutujewsky, 100
Hamburg, 20
Las Palmas, 40
Melbourne, 50
Newfairwater, 90
Quebec, 50

Soda Crystals—
Boston, 280 brls.
Constantinople, 50
Iquique, 100 kgs.
Monte Video, 30 cks. 25 brls.
New York, 190
Boston, 140
Constantinople, 50
Malta, 200
Pernambuco, 100
Shanghai, 16

Sodium Silicate—
Cienfuegos 8 brls.
Bilbao, 58
Genoa, 150

Sulphuric Acid—
Vera Cruz, 4 c. 1 q. £5

Sodium Nitrate—
New York, 16 c. 3 q. £22
Demerara, 20 15 1 187

Sodium Chlorate—
Lisbon, 5 c. £14

Saltcake—
Baltimore, 60 t. 9 c. £106
Montreal, 100 11 251

Sulphur—
Rouen, 30 t. 4 c. £136
Rangoon, 2 9
Ceara, 16 3
St. John, N.B 2 15 19
New York, 100 4 1,124
Boston, 70 2 330

Saltpetre—
Ceara, 8 c. £9
Maceio, 15 17
Pernambuco, 1 t. 10 40

Salammoniac—
Calcutta, 15 c. £27
Galatz, 9 16
Genoa, 17 3 q. 31
Antwerp, 4 6
Trieste, 4 2 36
Smyrna, 2 t. 8 1 83
Antwerp, 19 2 38
Philadelphia, 7 15 1 272
Alexandria, 19 34
Montreal 1 1 1 36
Pernambuco, 2 10 65
Galatz, 11 3 19

GLASGOW.

Week ending May 29th.

Alumina Sulphate—
Melbourne 215 c. £34. 10s.

Ammonium Sulphate—
Trinidad, 221 t. ¼ c. £2734
Demerara, 67 6¾ 770
" 34 6 385

Benzole—
Rotterdam, 125 brls. 63 drms. £1945

Bleaching Powder—
Calcutta, 100 c. £30

Charcoal—
Rotterdam, 5 t. 4 c. £17

Chemicals (*otherwise undescribed*)
Quebec, 9 t. 10¼ c. £95
Chili, 11½ c. £36

Dyestuffs—
Extracts
Melbourne, 22¼ c. £48

Epsom Salts—
Quebec, 131 c £24

Lead Nitrate—
Montreal, 2 t. 2 c. £40

Manure—
Trinidad, 309 t. 14 c. £2,279
" 318 17½ 2,543
" 10 113

Paraffin Wax—
Melbourne, 238 c. £361

Potassium Bichromate—
Stockholm, 13 cks £135
Melbourne, 22 c. 40
Calcutta, 11½ 20

Potassium Prussiate—
New York, 117 c. £475
Antwerp, 407¾ 591

Pitch—
Amsterdam, 150 t. £78. 15s.
Madras, £29
Quebec, 58 t. 10 c. 102

Tar—
Madras, £31
Calcutta, £56. 10s

TYNE.

Week ending May 27th.

Antimony—
Hamburg, 3 t.

Alkali—
Hamburg, 3 t. 11 c.
Christiania, 12
St. Petersburg, 29 19
Montreal, 25 5

Alumina Sulphate—
Montreal, 5 t. 6 c.

Barytes Carbonate—
Genoa, 10 c.
Hamburg, 6 t.

Bleaching Powder—
Hamburg, 20 t. 19 c.
Gothenburg, 10 4
Antwerp, 33 19
Montreal, 12 16
St. Petersburgh, 238 t. 18 c.
Christiania, 30
Lisbon, 10 t.
Venice, 3 13 c.
Naples, 2 14
Oporto, 10 2
Flensburg, 16 15
Montreal, 15

Copperas—
Sundsvall, 14 t. 3 c.
Fredericia, 5 11
Montreal, 7 14

Caustic Soda—
Genoa, 42 t. 14 c.
Flensburg, 5
Montreal, 7 10
Gothenburg, 4 18
Hamburg, 8 19
Antwerp, 14
Montreal, 7 5

Litharge—
Hamburg, 3 t. 19 c.
Montreal, 5 10

Magnesia—
Hamburg, 2 c.
Antwerp, 1
Montreal, 3 t.

Pitch—
Antwerp, 100 t.
Montreal, 2 c.

Potassium Chlorate—
St. Petersburg, 5 t.

Sodium Sulphate—
Aarhuus, 10 t. 10 c.

Sodium Silicate—
Antwerp, 1 t.

Soda—
Aalborg, 35 t. 14 c.
Wasa, 17 18
Helsingfors, 1 1
Antwerp, 1
Sundsvall, 51 10
Fredericia, 40 1
Montreal, 139 15

Soda Ash—
Venice, 27 t. 15 c.
Leghorn, 4 8
Abo, 73 13
Gothenburg, 2 16
Antwerp, 1 6

Salammoniac—
St. Petersburgh, 57 t.

Sulphur—
Gothenburg, 10 t.

HULL.

Week ending May 27

Benzole—
Hamburg, 15 drms.

Caustic Soda—
Harlingen, 18 drms.
Reval, 459

Chemicals (*otherwise und*
Antwerp, 4 cks.
Amsterdam, 1
Bremen, 1
Bergen, 2
Bordeaux, 20 cs.
Christiania, 50
Copenhagen, 11 cks.
Dunkirk, 20 cks.
Gothenburg, 32 cs.
Hamburg, 50 63
Konigsberg, 16
New York, 170
Rotterdam, 4 10
Rouen, 30
St. Petersburg, 217 cks.
Stockholm, 1 ck. 2 cs.
Stettin, 12

Dyestuffs (*otherwise un*
Antwerp, 8 cks.
Hamburg, 2 1 cs.
Rotterdam, 9 4
St. Petersburg, 50 cks.

Manure—
Antwerp, 500 bgs.
Dunkirk, 530
Hamburg, 2216
Rotterdam, 470
Stettin, 190

Tar—
Antwerp, 16 cks.
Amsterdam, 10
Copenhagen, 6
Gothenburg, 300 brls.
Rotterdam, 3

GRIMSBY

Week ending May 2

Coal Products—
Dieppe, 47 cks.

Chemicals (*otherwise un*
Dieppe, 160 pks.
Hamburg, 19

GOOLE.

Week ending May

Benzole—
Antwerp, 25 cks.
Dunkirk, 49
Hamburg, 46 drms
Rotterdam, 194

Coal Products—
Ghent, 20 drms.

Chemicals (*otherwise un*
Antwerp, 18 drms.

Dyestuffs—
Logwood
Antwerp, 100 bgs.

Dyestuffs (*otherwise un*
Boulogne, 2 cs. 7 cks.
Hamburg, 157
" 1
Rotterdam, 3

Manure—
Ghent, 500 bgs.
Rotterdam, 741

Pitch—
Antwerp, 160 t.

PRICES CURRENT.

PREPARED BY HIGGINBOTTOM AND CO., 116, PORTLAND STREET, MANCHESTER.

The values stated are F.O.R. at maker's works, or at usual ports of shipment in U.K. The price in different localiti

		£	s.	d.
Acids:—				
Acetic, 25 °/ₒ and 40 °/ₒ	per cwt.	7/- & 12/6		
,, Glacial	,,		65/-	
Arsenic, S.G., 2000°	,,	0	12	9
Chromic 82 °/ₒ	nett per lb.	0	0	6¼
Fluoric	,,	0	0	3¾
Muriatic (Tower Salts), 30° Tw.	per bottle	0	0	8
,, (Cylinder), 30° Tw.	,,	0	2	11
Nitric 80° Tw.	per lb.	0	0	2½
Nitrous	,,	0	0	1¼
Oxalic	nett ,,	0	0	3¾
Picric	,,	0	1	1½
Sulphuric (fuming 50 °/ₒ)	per ton	15	10	0
,, (monohydrate)	,,	5	10	0
,, (Pyrites, 168°)	,,	3	2	6
,, (,, 150°)	,,	1	4	6
,, (free from arsenic, 140/145°)	,,	1	7	6
Sulphurous (solution)	,,	2	15	0
Tannic (pure)	per lb.	0	1	0½
Tartaric	,,	0	1	3
Arsenic, white powdered	per ton	13	0	0
Alum (loose lump)	,,	4	17	6
Alumina Sulphate (pure)	,,	5	0	0
,, ,, (medium qualities)	,,	4	10	0
Aluminium	per lb.	0	15	0
Ammonia, ·880 = 28°	per lb.	0	0	3
,, = 24°	,,	0	0	1¼
,, Carbonate	nett ,,	0	0	3½
,, Muriate	per ton	23	0	0
,, ,, (sal-ammoniac) 1st & 2nd	per cwt.	37/- & 35/-		
,, Nitrate	per ton	40	0	0
,, Phosphate	per lb.	0	0	10½
,, Sulphate (grey), London	per ton	11	0	3
,, ,, (grey), Hull	,,	11	5	0
Aniline Oil (pure)	per lb.	0	0	10½
Aniline Salt ,,	,,	0	0	9
Antimony	per ton	75	0	0
,, (tartar emetic)	per lb.	0	1	1
,, (golden sulphide)	,,	0	0	10
Barium Chloride	per ton	8	5	0
,, Carbonate (native)	,,	3	12	6
,, Sulphate (native levigated)	,,	45/- to 75/-		
Bleaching Powder, 35 %	,,	4	17	6
,, Liquor, 7 %	,,	2	10	0
Bisulphide of Carbon	,,	13	0	0
Chromium Acetate (crystal	per lb.	0	0	6
Calcium Chloride	per ton	2	2	6
China Clay (at Runcorn) in bulk	,,	17/6 to 35/-		
Coal Tar Dyes:—				
Alizarine (artificial) 20 %	per lb.	0	0	9
Magenta	,,	0	3	9
Coal Tar Products				
Anthracene, 30 % A, f.o.b. London	per unit per cwt.	0	1	5
Benzol, 90 % nominal	per gallon	0	3	9
,, 50/90	,,	0	2	9
Carbolic Acid (crystallised 35°)	per lb.	0	0	8
,, (crude 60°)	per gallon	0	2	2
Creosote (ordinary)	,,	0	0	2½
,, (filtered for Lucigen light)	,,	0	0	3
Crude Naphtha 30 % @ 120° C.	,,	0	1	4
Grease Oils, 22° Tw.	per ton	3	0	0
Pitch, f.o.b. Liverpool or Garston	,,	1	8	0
Solvent Naphtha, 90 % at 160°	per gallon	0	1	9
Coke-oven Oils (crude)	,,	0	0	2½
Copper (Chili Bars)	per ton	54	10	0
,, Sulphate	,,	25	0	0
,, Oxide (copper scales)	,,	55	0	0
Glycerine (crude)	,,	30	0	0
,, (distilled S.G. 1250°)	,,	60	0	0
Iodine	nett, per oz.	0	0	9
Iron Sulphate (copperas)	per ton	1	10	0
,, Sulphide (antimony slag)	,,	2	0	0
Lead (sheet) for cash	,,	14	10	0
,, Litharge Flake (ex ship)	,,	15	10	0
,, Acetate (white ,,)	,,	24	0	0
,, ,, (brown ,,)	,,	18	0	0
,, Carbonate (white lead) pure	,,	19	0	0
,, Nitrate	,,	22	0	0

Lime, Acetate (brown)	,,
,, ,, (grey)	per t
Magnesium (ribbon and wire)	per
,, Chloride (ex ship)	per t
,, Carbonate	per cw
,, Hydrate	,,
,, Sulphate (Epsom Salts)	per t
Manganese, Sulphate	,,
,, Borate (1st and 2nd)	per cwt.
,, Ore, 70 %	per t
Methylated Spirit, 61° O.P.	per gall
Naphtha (Wood), Solvent	,,
,, ,, Miscible, 60° O.P.	,,
Oils:—	
Cotton-seed	per t
Linseed	,,
Lubricating, Scotch, 890°—895°	,,
Petroleum, Russian	per gall
,, American	,,
Potassium (metal)	per
,, Bichromate	per
,, Carbonate, 90% (ex ship)	per t
,, Chlorate	per l
,, Cyanide, 98%	,,
,, Hydrate (Caustic Potash) 80/85 %	per
,, ,, (Caustic Potash) 75/80 %	,,
,, ,, (Caustic Potash) 70/75 %	,,
,, Nitrate (refined)	per cw
,, Permanganate	,,
Prussiate Yellow	per l
,, Sulphate, 90 %	per t
,, Muriate, 80 %	,,
Silver (metal)	per o
,, Nitrate	,,
Sodium (metal)	per l
,, Carb. (refined Soda-ash) 48 %	per t
,, ,, (Caustic Soda-ash) 48 %	,,
,, ,, (Carb. Soda-ash) 48%	,,
,, ,, (Carb. Soda-ash) 58 %	,,
,, ,, (Soda Crystals)	,,
,, Acetate (ex-ship)	,,
,, Arseniate, 45 %	,,
,, Chlorate	per l
,, Borate (Borax)	nett, per to
,, Bichromate	per l
,, Hydrate (77 % Caustic Soda) (f.o.b.)	per to
,, ,, (74 % Caustic Soda) ,,	,,
,, ,, (70 % Caustic Soda) ,,	,,
,, ,, (60 % Caustic Soda, white) (f.o.b.)	,,
,, ,, (60 % Caustic Soda, cream)	,,
,, Bicarbonate	,,
,, Hyposulphite	,,
,, Manganate, 25%	,,
,, Nitrate (95 %) ex-ship Liverpool	per cw
,, Nitrite, 98 %	per t
,, Phosphate	,,
,, Prussiate	per l
,, Silicate (glass)	per t
,, ,, (liquid, 100° Tw.)	,,
,, Stannate, 40 %	per cw
,, Sulphate (Salt-cake)	per to
,, ,, (Glauber's Salts)	,,
,, Sulphide	,,
,, Sulphite	,,
Strontium Hydrate, 98 %	,,
Sulphocyanide Ammonium, 95 %	per l
,, Barium, 95 %	,,
,, Potassium	,,
Sulphur (Flowers)	per to
,, (Roll Brimstone)	,,
,, Brimstone: Best Quality	,,
Superphosphate of Lime (26 %)	,,
Tallow	,,
Tin (English Ingots)	,,
,, Crystals	per l
Zinc (Spelter)	per to
,, Chloride (solution, 96° Tw.	,,

THE

CHEMICAL TRADE JOURNAI

Publishing Offices: 32, *BLACKFRIARS STREET, MANCHESTER.*

No. 160. **SATURDAY, JUNE 14, 1890.** **Vol.**

Contents.

Notices.

All communications for the *Chemical Trade Journal* should be addressed, and Cheques and Post Office Orders made payable to—

DAVIS BROS., 32, Blackfriars Street, MANCHESTER.

Our Registered Telegraphic Address is—

"Expert, Manchester."

The Terms of Subscription, commencing at any date, to the *Chemical Trade Journal*,—payable in advance,—including postage to any part of the world, are as follow:—

Yearly (52 numbers) .... **12s. 6d.**
Half-Yearly (26 numbers) .... **6s. 6d.**
Quarterly (13 numbers) .... **3s. 6d.**

Readers will oblige by making their remittances for subscriptions by Postal or Post Office Order, crossed.

THE LAW OF PATENTS.—SUGGESTI AMENDMENTS.

THE Chemical and Allied Trades Section of the Mı Chamber of Commerce has for some time had under cı tion the bearing and application of the Patents, Designs, aı Marks Acts of 1883-88.

The Committee of the Section has arrived at the conclu certain provisions of the Patent Act of 1883 are mischievous ı impracticable when applied to chemical patents. The Com also of the unanimous opinion that some provision ought to l in the law for distinguishing between mechanical and chemical for the reason that however applicable the existing statutes m the former, it frequently happens that its provisions are absolu practicable when they come to be applied to the latter.

This fact has already been fully recognised in the Germa Law by the enactment of special clauses and provisions administration of the law in the case of chemical patents. Tl mittee is further of opinion that if the principle were adc distinguishing between mechanical and chemical patents, th ability of litigation in the case of chemical patents would be c ably decreased, and greater justice would be done to litigants.

For the present the attention of the committee has been dir the anomalies which result from the operation of Section 18, for future deliberation other sections and provisions of the Pat of 1883 to 1888. This section gives power to a patentee frc to time to disclaim, correct, or explain his specification. ' practically no safeguard for the public against the far-reaching thus granted. It is true that the second clause of this sectic any person the right to oppose an application for leave to ame also that the eighth clause apparently limits the rights granted applicant by Clause 1; but, in fact, both these protective cla altogether ineffectual to prevent unscrupulous patentees from both the letter and spirit of the law.

According to the existing law a patentee invalidates his p amongst other reasons:—

1. If he describes in his specification a process that d produce the result claimed.
2. If he specifies one or more alternative processes for ob similar or equivalent results, and one of those process not produce the result claimed.
3. If, in his specification, he uses ambiguous, mislead evasive language.
4. If his description is insufficient to enable those skilled art to carry out the invention without other aid th afforded by the specification.

The Committee holds that these just and equitable principle been made absolutely inoperative by the application and ef Section 18, and that a patentee may now commit breaches of t principles first set forth, and may abuse a monopoly and p which, in their opinion, is already far too easily granted t without running the slightest risk of vitiating his patent, beca can confidently trust to his benevolent friend, Section 18, to him with all the protection of which he may stand in need, obtain for him condemnation of his former abuse of privileg thus enable him to make good a patent, to the possession of wh was never entitled.

A patentee may now specify and claim almost anything tl speculative and imaginative mind may suggest, without ever made a single experiment to ascertain the possibility of the rea s dreams, so long as he is clever enough to make his original fication sufficiently wide and comprehensive, and to formulate it ch a manner as to permit him, in accordance with Section 18, time to time, to disclaim, correct or explain, without in any way rging his claims, or, after such amendment, making the invention antially different from or larger than the original specification. oured by section 18, the speculative faculties of some patentees of late years been so marvellously extended, that it now frequently pens that a patentee specifies, and claims as part of his invention, nical products or substances which he has never even seen, much made or produced.

may, therefore, well be asked, what are the consequences of such nomalous state of things, in which we have, on the one hand, -established law and practice, tending to prevent the abuse of a ilege, and on the other, laws to make their operation ineffective.

n illustration will, perhaps, serve to bring more clearly before the d the ground on which these assertions are made, and the manner hich advantage may be and is taken by patentees of the provisions ection 18.

would-be patentee, for example, gets the idea into his head that ably a combination of aniline with napthol might produce a uring matter, and on trying this in the laboratory he really obtains loured reaction. Here, then, is the foundation for a patent, which t be applied for without loss of time, before anybody gets a similar . His chemical knowledge will suggest to him that aniline belongs class known to chemists as "amines," that there exist a number odies belonging to this class, and that napthol belongs to a group hemicals known as "phenols," also including a large number of lar bodies, and further, that these amines and p enols, again com- separately with other substances. The idea fulther occurs to him all these amines and phenols, and their complex resultant bodies, ht also, when combined, probably produce similar, or even much e valuable colouring matters than the simple combination of aniline napthol. Now here is a dilemma: for his chemical knowledge hes him at the same time that although it is probable that all these binations might produce colouring matters similar to the one he hastily detected when combining aniline with napthol, yet, the ability is by no means excluded that some of these might not luce such results.

'e will stop here for a moment to review the position. The gnised principles of law would, under these conditions, only permit to specify that which he has really discovered and ascertained by eriment to give a certain result, which would be the combination of ine with napthol; and if he did specify and claim more than he had overed, and it should turn out afterwards that one or several of e combinations will not work or give the specified result, then his nt would be null and void, and rightly so, for his patent has ally and actually been taken out under false pretences.

ut, to follow up the illustration, the would-be patentee sets the law aught by relying on section 18 to get him out of his difficulties if should happen to go against him, and it should ultimately turn that some of these combinations either give practically no result, ne different from the result claimed. He proceeds, after a very f consideration, and without wasting any more time in experiments, raw up his specification somewhat in the following manner:—

have discovered that new and valuable colouring matters, from ows to reds, may be produced by first converting amines, such as ine and its homologues and their sulpho-acids, into their ective diazo compounds, and combining these diazo compounds, in anner well known to chemists, with phenols, their sulpho and o acids, &c., &c." Then follows the example with aniline and hthol, in order to show the manner or process of practically carrying his invention.

ere we have a high-sounding and truly comprehensive patent for ingeniously-speculative would-be patentee. From a coloured tion produced by the combination of two chemicals, our inventor increased his invention to several hundred combinations. But the st of it is that this patent is absolutely good, and he will not only a monopoly for 14 years, and put a tax on our industries, but he also prevent other inventors from labouring in a special field of arch which he has wrongfully claimed as belonging to him, and ld anyone by laborious experiment discover amidst this labyrinth robabilities anything really valuable and useful, he will soon find a put to his inquisitive curiosity. Our patentee, however, well ws that to take action successfully means that he must have a good nt which will stand the scrutinising tests of the Courts, and he has his time about found out that some of his patented combinations e excellent in theory but would not work in practice. What must o then? He simply applies to the Comptroller for an amendment is patent under section 18, explaining that in the practical working is processes he has found that some of his results as claimed are not so ul as others, and that he is anxious to give the public the benefit of later experience. He then commences to apply the pruning knife ly, always conforming to clause 8, and without enlarging his patent he complies with it by narrowing down his claims by disclaimer; he strikes out the useless matters, corrects and explains away that which was false and misleading, until of a copious specification apparently only the mere skeleton is left.

But this change has converted a legally-valueless patent into a valuable one, and by virtue of clause 9, which provides that the amendment shall in all courts and for all purposes be deemed to form part of the specification, he is now safe against all opposition and interference on the part of the public. Surely the Comptroller cannot object to our patentee disclaiming, for instance, toluidin and xylidin and naphthylamine, which are amines, and disclaiming, for example, phenol, cresol, resorcinol, or other phenols, their sulpho and carbo acids, etc., because by so doing he actually narrows down his claim. In the eyes of the Comptroller, with the limited powers with which he is invested, the applicant has in every respect complied with the provisions of section 8, because the invention has not been enlarged or rendered substantially different. The invention and claims were for colouring matters obtained by combining amines and their sulpho-acids with phenols and their sulpho and carbo-acids, and it will remain such after the amendment has been granted.

By the foregoing illustration, which might be multiplied almost *ad libitum*, it has been attempted to show the manner in which great facilities are afforded by section 18 to unscrupulous patentees for evading the law, and the Committee trust they have succeeded in establishing the conviction that this section in its present form operates hurtfully and injuriously on our trades and industries. They now submit their propositions for preventing this abuse of privilege, without doing any injustice or restricting in any way the rights of honest inventors. These proposals are:—

1.—To add to part II., section 5, sub-section 4, of the Act of 1883, the following provisions, which shall be known as sub-section 4b:—"When generic terms are used in chemical patents, the patentee shall specify *by name or otherwise* each substance or compound which he intends to include within such term or terms, and he shall only be entitled to claim as his invention those substances or compounds which have been so specified by him."

2.—Sub-section 4c:—"There shall be deposited at the Patent Office, along with the complete specification of chemical patents, duplicate samples, sufficient for analysis or examination of the result or results claimed, and of all raw and intermediate products used in obtaining such result of results, unless such raw or intermediate products shall be ordinary articles of commerce easily obtainable in this country."

3.—Sub-section 4d:—"The complete specifications of all patents relating to coal-tar colours, in addition to the samples of the colouring matters themselves, and the intermediate substances used in obtaining them, shall be accompanied by dyed specimens of cotton-wool or silk, showing the results obtained with one per cent. and two per cent. respectively of the dyes claimed. A complete statement of the method employed in dyeing in each case shall also be given."

The Committee are further of opinion that the words in clause 1, section 18, "from time to time," shall be altered, in the case of chemical patents, into "two years from date of application."

The Committee recommends that a copy of this Report be sent to all Members of Parliament representing the Manchester district, with the request that they will take the first opportunity of bringing the questions of chemical patents and the working of section 18 before the House of Commons, and that the co-operation of other Chambers of Commerce be invited, especially of all those which are established in chemical centres, such as London, Liverpool, Glasgow, Newcastle, &c.

IVAN LEVINSTEIN,

Chairman of the Chemical and Allied Trades Sectional Committee.

RELATIVE COSTS OF TRANSMISSION OF POWER.—The following comparisons of cost of transmission of power by various methods appeared in the *Revue Universelle des Mines*:—(1) Comparative cost on 10-horse power transmitted 1093 yards: By cables, 1·77 per effective horse-power per hour; by electricity, 2·21; by hydraulics, 2·90; by compressed air, 2·98. (2) Comparative cost on 50-horse power transmitted 1093 yards: By cables, 1·35 per effective horse-power per hour; by hydraulics, 1·87; by electricity, 2·07; by compressed air, 2·29. (3) Comparative cost on 10 effective horse-power transmitted 5465 yards: By electricity 2·64 per effective horse-power per hour; by compressed air, 4·66; by cables, 4·69; by hydraulics, 5·29. (4) Comparative cost on 50 effective horse-power transmitted 5465 yards: By electricity, 2·37 per effective horse-power per hour; by cables, 2·65; by compressed air, 2·99; by hydraulics, 3·02. Steam was the prime mover used in each of the above instances, and it appears that for long distances electricity takes the lead in economy over all other systems. It has also a great advantage in the facility with which the power may be subdivided, and there appears to be no doubt that in future coal mining electricity will be much used for coal-cutting, tunnelling, hauling, pumping, etc., as well as for lighting.

Legal.

QUEEN'S BENCH DIVISION.

On June 3rd.

THE CAPE COPPER COMPANY v. THE COMPTOIR D'ESCOMPTE DE PARIS—THE SOCIETE INDUSTRIELLE ET COMMERCIALE DES METAUX.

(Before Mr. Justice Denman *and* Mr. Justice Charles.

THIS case had, with many others, arisen out of transactions alleged to have been entered into with the object of enhancing the prices of copper; and, indirectly, it raised a question as to whether combinations with that object are now illegal by the law of England since the Act of 1844 (7 and 8 Vict., c. 24), abolishing the offences of forestalling, regrating, and engrossing, and for repealing certain statutes passed in restraint of trade, the preamble of which was read to the Court in these terms, reciting that it was expedient that such statutes should be repealed, as it had been found by experience that the restraint laid by them upon the dealing in corn, meal, flour, cattle, and sundry other sorts of victuals, by preventing a free trade in those commodities, had a tendency to discourage the growth and enhance the price, &c. And this case also with the others raised a question as to the legality of combinations to fix prices so as to affect other traders injuriously, so as to create a monopoly, and prevent them from obtaining commodities at all, except at prohibitory prices—a question somewhat analogous to that involved in the case of the Mogul Steamship Company, lately decided in the Court of Appeal, and now before the House of Lords. The particular case, with others, had arisen out of a contract entered into in March, 1888, between the Cape Copper Mining Company (the plaintiffs) and the Société des Métaux, whereby the latter agreed with the former that the former should sell, and the latter buy, the copper to be obtained from the Copper Company's mines during three years, at the price of £70. a ton up to 5,750 tons, and for any excess beyond that quantity £42. 10. a ton, and a guarantee given in July by the Comptoir d'Escompte to the Copper Company. (There was another similar contract with a similar guarantee, in which latter document an English domicile in the City was adopted, and the law of England was to be applicable.) It appeared that in March, 1889, the Société des Métaux suspended payment, and a decree of liquidation was pronounced against them, and they, it was alleged, made default in their payments, and the plaintiffs (the Copper Company) sought to recover certain sums of £50,662. and £23,642 from the Comptoir d'Escompte upon their guarantee, and they had refused to pay, and then this action was brought, alleging that the Société des Métaux had failed to accept and pay for 2,000 tons of best select copper produced by the Copper Company at their works, and the plaintiffs alleged that they had sold for the Société 543 tons of the copper bought by the Société under the agreement, and prior to the sales had credited the Société with £44,822. in respect of such sales, which, however, had only realised £23,231., upon which the plaintiffs claimed to be entitled to the difference, and the total amount they claimed to recover was over £150,000, while in the other action the total amount involved above £600,000. The Comptoir d'Escompte had originally pleaded as their defence that the guarantee was *ultra vires*—that is, beyond their power to give; and they now sought, in consequence of recent disclosures, to add a defence, founded on the alleged illegality of the transaction, either by French or English law, as entered into with the object of raising the prices of the commodity and creating a monopoly, and tending to prevent other traders from obtaining it except at prohibitory prices, &c.

The Attorney-General (with Mr. Bigham, Q.C., and Mr. H. Tindal Atkinson) appeared for the Comptoir d'Escompte in support of their application to be allowed to raise this defence, urging that the transaction certainly was illegal by French law, and stating that the manager had actually been prosecuted and convicted in France for the offence, under section 419 of the French Code Pénal, of a combination to prevent other traders from obtaining the commodity. Then, as to illegality of the transaction by English law, he urged that it was still illegal according to our law, and, at all events, the defence should be allowed to be raised, especially as in the Mogul case the question was indirectly raised; and it was now in the House of Lords.

Mr. Finlay, Q.C. (with Mr. F. M. Abrahams) appeared for the plaintiffs (the Copper Company), and urged that the effect of allowing the defence would be to delay the trial. Moreover, he urged that the defence was not meritorious, for the Société des Métaux and the Comptoir d'Escompte had got an English copper company to enter into the transaction, and then set up that it was illegal. Further he denied the alleged illegality of the transaction, at all events by English law. He read the preamble of the Act of 1844 abolishing the old law as to "forestalling," and "regrating," and as to penalties on "badgers." [Mr. Justice Denman.—What are badgers?"] They wer who bought corn or other victuals in one place and carriec another to enable them to raise the price. The defence s was founded on English law was futile, and as to Fencl illegality, if it existed, must have been known long ago, and have been set up at the first. Now the only object of allowin be to delay the trial.

The Attorney-General, in reply, urged that it was by to be taken as settled that such transactions were not still i English law. For in the Mogul Steam Company case the l the Rolls dissented from the decision, and the case was n appeal to the Lords. Then, as to the illegality by French lav the law was known it was not known whether there had bee tion of it in this transaction until the prosecution of the manag conviction had taken place only ten days before the present a was first made.

The Court, in the course of the argument, had intimate doubt whether the new defence ought to be set up so far as on illegality by English law, but that it ought to be allowed legality by French law, and at the close of the arguments (wh lengthy) they gave judgment to that effect.

Mr. Justice Denman, in giving judgment, said he tho defence must be restricted to illegality by French law, as he no English law by which such a transaction would be illegal a It could only be so, he said, by virtue of the old law of "fore or "regrating," or "engrossing" (which latter was the te would embrace such transactions as this), and that law wa abolished by the Act of 1844, which had been referred defence that the transaction was *ultra vires* was already raised that the transaction was beyond the powers of the company. now sought to set up the defence of positive illegality bo English and French law, and as to the latter alone he tho defendants ought not to be precluded from setting it up, but n to the former—that is, the English law. As to the objection he thought the case might come on for trial in a reasona within the present sittings, if both parties did their best to s The defence, therefore, as to illegality by French law allowed.

Mr. Justice Charles concurred.

IN THE HIGH COURT OF JUSTICE.

(Before Lord Coleridge *and* Mr. Justice Wills.

LOCK v. WESTWOOD.

THIS was a case under the Employers' Liability Act, arisin circumstances rather singular. The plaintiff, the workn employed in a factory at Millwall, on the floor of which was a weighing two tons, which had been standing there six months being secured in any way except by its own weight. It wa and of iron, and it was about 8ft. high, standing on one of which were about 2ft square. Over the workshop where it sto "traveller," as it is called—a machine running on wheels o form, and with chains hanging from it to attach to any objects The plaintiff was engaged in painting the rafters of the roof, been told by the foreman to "shift the traveller when it wa way." He accordingly was so shifting it when, in some way not explain (and no one else was present), the machine sudd upon him and crushed him, doing him terrible injuries. He this action in the Bow County Court to recover compensation, Judge leaving the case to the jury, they found a verdict in his The employers appealed.

Mr. Tatlock, on their part, contended that there was no of any negligence on their part, and, even assuming the f direction "to shift the traveller" was wrong, there was no connect it with the accident.

Mr. Moyses, on the part of the plaintiff, contended tha machine had stood in the workshop six months and could fallen unless it had been moved, and the only thing that co moved it was the chains of the traveller, it must be assumed chains had come in contact with it, and, therefore, there was of negligence in leaving the machine unsecured in any way an the chains hanging down over it, and telling the man to " traveller."

The Court, however, came to the conclusion that there evidence of any negligence of the employers or their forem nected with or conducing to the accident; and, therefore, verdict could not stand.

Lord Coleridge, in giving judgment, said with every re the learned Judge, he ought not to have left the case to the there was no evidence whatever to sustain the action. It was to support such an action that there should be evidence negligence, and there was none. It would be a shocking abu Act if, under such circumstances, an action could be ma

rere too apt, however, in such cases to give verdicts in favour men injured without any evidence of negligence to affect the ers. Moreover, the plaintiff, and even the other workmen, aat the machine was there and the "traveller" above it, and if as any danger they knew of it, so that even if there had been gligence it would by no means follow that the plaintiff would tled to recover. But, in fact, there was no evidence of any such nce as would sustain the action, and therefore the judgment e entered for the defendants.

Justice Wills concurred.

ment accordingly entered for the defendants.

AMERICAN IMPORTS.

e year 1889 the total value of the imports of merchandise into e United States was 741½ millions of dollars, of which 484¾ s were dutiable and 256½ millions were free. On the dutiable im- e average *ad valorem* rate of duty collected was 45 per cent., the mount of duty collected having been 220½ millions of dollars, at 45 millions sterling. This corresponds to an average of about er head of the population, and is not, therefore, a very large sum f—not so much, indeed, as the amount of duty levied *per capita* e other countries, and not much more than the *per capita* ns' revenue collected in the United Kingdom. But the incidence American Customs Tariff is exceptional. The amounts of duty ed on the principal classes of imports in each of the years 1880 389 were as under :—

	1880.	1889.
	1=1,000 dols	1=1,000 dols.
stuffs	1,356	1,161
cals, drugs, &c.	4,079	5,017
manufactures	9,976	10,841
nware and china	2,356	3,694
articles, perfumery, &c.	2,258	2,781
and nuts	3,401	4,007
hemp, and jute	1,415	2,482
. .. .	2,811	4,526
ery and precious stones	707	1,232
er and its manufactures	3,411	3,417
, confectionery, &c...	42,210	55,995
. .. .	14,016	17,342
and its manufactures	29,238	41,355
and its manufactures	1,336	1,776
co	4,681	11,194
nd steel	23,244	16,909

hing could be clearer than the fact which this statement brings hat the American tariff does not have the effect that it is intended ig about. As a fiscal instrument intended to exclude foreign ce it is a failure. It is not, and does not claim to be, a revenue like that of the United Kingdom; for it yields so enormous a hat it has for years past been an embarrassment to American ians to know what to do with their surpluses. The balance in blic treasury has increased by "leaps and bounds," having more oubled between 1884 and 1889, and at the end of the latter year ounted to the unprecedented sum of 659½ million dollars, or 132 ns sterling. The National Debt has been liquidated with a swift- aat is unique in the history of modern nations. At the end of he outstanding principal of the debt, which was 1,919¼ millions lars at the end of 1880, had been reduced to 976 millions of s, or 194 millions sterling, less the cash in the Treasury. Put ther way, the debt of the United States, which was 69¼ dollars *pita* in 1867, and 41 dollars *per capita* in 1879, had fallen to 15 dollars in 1889, while the interest *per capita* of the debt had educed from 3·84 dollars in 1867 to about half a dollar in 1889. y, therefore, the National Debt of the United States is being uished at so rapid a rate that in another few years, given the conditions as now, no debt will remain, and a burden which was ed for the benefit of posterity, and of which posterity ought to s full share, will be discharged almost entirely by the generation ncurred it. Fiscal necessities, in these circumstances, do not call y increase of Customs duties, which already contribute more than lf of the total revenue of the country.

eign Languages.—In the House of Commons, Mr. H. Wilson that a better system of French teaching should be adopted, in to give the students a colloquial knowledge of the ge, which would be useful to those who might engage in com. l pursuits. Seeing the number of German clerks who come here, e of their superior knowledge of foreign languages, he did not hat the teaching of foreign languages could be considered to be a He regarded it as a very important matter. (Hear, hear).

PERMANENT CHEMICAL EXHIBITION.

THE proprietors wish to remind subscribers and their friends generally that there is no charge for admission to the Exhibition. Visitors are requested to leave their cards, and will confer a favour by making any suggestions that may occur to them in the direction of promoting the usefulness of the Institution.

Joseph Aird, Greatbridge.—Iron tubes and coils of all kinds.

Ashmore, Benson, Pease and Co., Stockton-on-Tees.—Sulphate of Ammonia Stills, Green's Patent Scrubber, Gasometers, and Gas Plant generally.

Blackman Ventilating Co., London. — Fans, Air Propellers, Ventilating Machinery.

Geo. G. Blackwell, Liverpool.—Manganese Ores, Bauxite, French Chalk. Importers of minerals of every description.

Bracher and Co., Wincanton.—Automatic Stills, and Patent Mixing Machinery for Dry Paints, Powders, &c.

Brunner, Mond and Co, Northwich.—Bicarbonate of Soda, Soda Ash, Soda Crystals, Muriate of Ammonia, Sulphate of Ammonia, Sesqui-Carbonate of Ammonia.

Buckley Brick and Tile Co., Buckley.—Fireclay ware of all kinds—Slabs, Blocks, Bricks, Tiles, "Metalline," &c.

Chadderton Iron Works Co., Chadderton.—Steam Driers and Steam Traps (McDougall's Patent).

W. F. Clay, Edinburgh.—Scientific Literature—English, French, German, American. Works on Chemistry a speciality.

Clayton Aniline Co., Clayton.—Aniline Colours, Aniline Salt, Benzole, Toluole, Xylole, and Nitro-compounds of all kinds.

J. Cortin, Newcastle-on-Tyne.—Regulus and Brass Taps and Valves, "Non-rotative Acid Valves," Lead Burning Apparatus.

R. Daglish and Co., St. Helens.—Photographs of Chemical Plant —Blowing Engines, Filter Presses, Sludge Pumps, &c.

Davis Bros., Manchester.—Samples of Products from various chemical processes—Coal Distilling, Evaporation of Paper-lyes, Treatment of waste liquors from mills, &c.

R. & J. Dempster, Manchester.—Photographs of Gas Plants, Holders, Condensers, Purifiers, &c.

Doulton and Co., Lambeth.—Specimens of Chemical Stoneware, Stills, Condensers, Receivers, Boiling-pots, Store-jars, &c.

E. Fahrig, Plaistow, Essex. — Ozonised Products. Ozone-Bleached Esparto-Pulp, Ozonised Oil, Ozone-Ammoniated Lime, &c.

Galloways, Limited, Manchester.—Photographs illustrating Boiler factory, and an installation of 1,500-h.p.

Grimshaw Bros., Limited, Clayton.—Zinc Compounds. Sizing Materials, India-rubber Chemicals.

Jewsbury and Brown, Manchester.—Samples of Aerated Waters.

Joseph Kershaw and Co, Hollinwood.—Soaps, Greases, and Varnishes of various kinds to suit all requirements.

C. R. Lindsey and Co., Clayton.—Lead Salts, (Acetate, Nitrate, etc.) Sulphate of Copper, etc.

Chas. Lowe and Co., Reddish.—Mural Tablet-makers of Carbolic Crystals, Cresylic and Picric Acids, Sheep Dip, Disinfectants, &c.

Manchester Aniline Co., Manchester.—Aniline Colours. Samples of Dyed Goods and Miscellaneous Chemicals, both organic and inorganic.

Meldrum Bros., Manchester.—Steam Ejectors, Exhausters, Silent Boiling Jets, Air Compressors, and Acid Lifters.

E. D. Milnes and Brother, Bury.—Dyewoods and Dyewood Extracts. Also samples of dyed fabrics.

Musgrave and Co., Belfast.—Slow Combustion Stoves. Makers of all kinds of heating appliances.

Newcastle Chemical Works Company, Limited, Newcastle-on-Tyne.—Caustic Soda (ground and solid), Soda Ash, Recovered Sulphur, etc.

Robinson, Cooks, and Company, St. Helens.—Drawings, illustrating their Gas Compressors and Vacuum Pumps, fitted with Pilkington and Forrest's patent Valves.

J. Royle, Manchester.—Steam Reducing Valves.

A. Smith, Clayton.—India-rubber Chemicals, Rubber Substitute, Bisulphide of Carbon, Solvent Naphtha, Liquid Ammonia, and Disinfecting Fluids.

Worthington Pumping Engine Company, London.—Pumping Machinery. Speciality, their "Duplex" Pump.

Joseph Wright and Company, Tipton.—Berryman Feed-water Heater. Makers also of Multiple Effect Stills, and Water-Softening Apparatus.

THE BURNELL WOOL-WASHING MACHINE.

THIS is the invention of Mr. George Burnell, of Leeds, which substitutes the solvent method for the emulsion method that is in general use at present. The machine, which was constructed by Messrs. Pullan, Tuke, and Gill, of Elland-road, to the order of the patentee, has been exhibited on several occasions before experts, and its work tested, and similar trials of the machine have also been made in South Australia, where Mr. Burnell has worked for many years, and where he invented the machine. As a result of these trials several important improvements were suggested and have now been carried out by Mr. Burnell. The original machine, it will be remembered, consisted of two V shaped tanks, each containing in the upper part a large central drum surrounded by a series of rollers. The first or larger tank was filled with water up to the level of the lowest roller, and above that a continuous flow of benzene was kept up, the wool to be treated passing through the benzene in a direction opposite to that of the flow, and receiving a succession of squeezes between the rollers and the drum. This removed the fat and a large portion of the dirt, and the wool on being passed to the second tank, which contained warm water, was relieved of the potash and other remaining impurities. Methods were also adopted for recovering the benzene for use again, and for recovering the potash, wool fat, and other residuals for commercial purposes. The improved machine will consist of four instead of two stages, and will, in fact, embrace four machines in one. In the first tank the benzene will remove the wool fat and the heaviest part of the dirt. The second, which will also contain pure benzene, will remove the last traces of grease, and as the benzene will flow from the second tank into the first, it will be seen that after having the major part of its impurities removed the wool is immersed in a purer bath of benzene, and thus has the fat effectually washed out of it. The third tank will contain hot water to wash out the potash; and the fourth, a solution of potash soap to give the wool a last rinse and impart to it the beautiful white colour which is the great desideratum of the wool-washer. Each process—that of removing the grease and that of removing the potash, &c.—has thus been doubled, and it must add to the effectiveness of the work. All four tanks are fitted with a simple device by which the dirt settling to the bottom can be removed. The whole machine will be much simplified by the improvements. All the rollers will be more easy of access, and any of them will be removable and replaceable without interference with the others. The rollers are made of brass, which will obviate all inconvenience or damage through rust. The cover, or hood, which is necessitated by the volatile and inflammable nature of the benzene, will be more substantial, and will be made in separate parts, each of which can be removed and replaced separately. It will also be fitted with plate-glass windows, through which the work may be inspected while in progress. A model machine, of one-fourth the dimensions of the original, embodying these and other improvements, has been constructed, and was publicly tested on Wednesday at the Neville Works, Elland-road, before an assemblage of manufacturers—wool-scourers and others, when, we are informed, satisfactory results were obtained. The principle of the machine is believed to be good, though, of course, it has yet to undergo the further test of everyday use in manufacture. It is intended to float a company to work the patent.

DYEING BLACK ON COTTON WOOL.

THERE are at present three methods generally used for dyeing black cotton wool.

THE CATECHU BLACK

is obtained by first steeping the wool in a boiling bath overnight containing 10-20°/₀ catechu, ½°/₀ $CuSO_4$ (copper sulphate). In the morning it is taken out and entered into a boiling bath of 1°/₀ bichromate of potash, after being worked in here for ½ hour, it is then well washed and a fresh bath is got up in which there is boiled out 100% logwood chips for 1 hour, the logwood being withdrawn, the cotton wool is then entered and boiled for 1 hour, by this means we have obtained a good black, but it still wants the logwood fixing by entering into a solution of iron liquor and working in this bath for 20 minutes longer. After softening off the cotton wool (to render it easier to work up into yarn) by means of soap and oil it is then ready to be dried.

THE AMERICAN OR LOGWOOD BLACK

is dyed in a bath of a strong solution of logwood extract to which has been added a slight quantity of copper carbonate.

This latter salt is made by precipitating copper sulphate with soda ash.

The cotton wool is entered into this bath and boiled for 1 hour, it is then lifted out and put into a wool waggon and covered up so that it keeps warm, as it becomes much heavier and darker colour.

Next morning the wool is again entered into the same bath, to which has been added about 1 lb. of soda ash. The bath is kept bo[…] 1 hour, the wool is again lifted and put into a wool waggon a[…] warm by covering up, next morning it may be washed off and into a bath containing 1 lb. chrome to fully oxidize the logwoo[…]

It only needs entering into a bath containing soap and o[…] minutes, when it is ready to be dried.

This yields a very good black, and it is fairly fast to milling.

BENZIDENE BLACK.

This black product is one of the benzidene series, simil[…] benzopurpurine and chrysamine colours.

It is dyed with simply adding the 6 % dyeware along wi[…] glaubers' salt. The great fault of this is that it gives very bro[…] red blacks, and it is sometimes topped with ½ % aniline gr[…] cold bath. This is merely to produce an optical illusion, as t[…] is altered into a blue black. The benzidene black is very ap[…] for dark slates and dark blues, and it is also being used[…] Continent as a bottom or ground for the dyeing of aniline blac[…]

At present the high price is keeping it from being ext[…] applied to the dyeing of cotton wool black.

ANILINE BLACK.

Aniline black, by its properties, its mode of application, completely from the other blacks and from other aniline colou[…] the latter are brought to the dyer already prepared for immedi[…] Aniline black, on the contrary, is formed only on the fibre itsel[…]

While most of the other aniline colours are fugitive, this contrary, is noted for its fastness to all reagents.

For a long time aniline black was only employed for[…] tissues. The price of bichromate of potash, aniline, and the [...] to which aniline blacks were liable, hindered it from being e[…] as a dyeing agent for the raw material trade.

Below is a recipe for dyeing aniline which is at once e[…] permanent. The cotton is first wet out, and it is then entere[…] cold bath composed of

Aniline oil 8 %,
Hydrochloric acid 16 %,
Sulphuric acid 16 %,
Bichromate of potash 16 %,
Ferrous sulphate 16 %.

The cotton wool is worked in this bath for 1½ hours. It is th[…] washed and boiled out in a bath containing

5 % Soap.

This softens the cotton, and yields a good blue black, fast t[…] milling, stoving, and does not turn green on exposure to the air.

NO NECESSITY FOR TALL CHIMNEYS.

A MECHANICAL engineer, recently, in a conversation regar[…] accident to the Clark Thread Company's chimney, express[…] self as believing that "the time will come when these tall c[…] would serve no more useful purpose than that of monument[…] folly of their builders." The fuel used in a manufactory costs to[…] to allow twenty-five per cent of it to escape up a brick shaft, to[…] up out of doors." It is true that the gases must leave the boi[…] comparatively high temperature, but there are factors enter[…] boiler at a much lower temperature to which this heat coul[…] advantage, be transferred. In the feed-water, and the supply o[…] the furnace, much of this heat may be directly returned to th[…] itself, while, by its application to the reheating of exhaust steam[…] it for heating and manufacturing processes, to the inter[…] cylinders of compound engines, and directly to purposes usefu[…] arts, the necessity for the direct combustion of an additional am[…] coal to supply heat for these purposes, is avoided. But, it[…] urged, it is in the very direction of allowing of a considerable co[…] the furnace gases that tall chimneys are built, for, to gain a s[…] draught with comparatively cool gases, the chimney must have e[…] height or area, or both. To such an observation, the ge[…] quoted, referred to the high rates of economy realized in mari[…] tice, where anything like an approach to a tall stack is ou[…] question. In his opinion, future development will be in the d[…] of regaining from the flue gases, every possible unit of heat, a[…] maintenance of the draft not by the maintenance of a tall co[…] hot gas, but directly by mechanical means. Mr. Hoadly has[…] that the heat derivable from the gases is more than pra[…] equivalent to the power required to operate a fan or blower to[…] air to the furnace. If this proves to be true in general applica[…] may dispense with the expensive, and often unsightly, tall ch[…] obtain a better control of the air supply to the furnace, and prev[…] tendency to leakage of cold air through the setting, by mainta[…] the furnace and combustion chamber a pressure greater in[…] less than that of the atmosphere.—*Steam Power.*

THE WATER SAFETY LAMP.

E attention of the Home Office authorities having been drawn by Her Majesty's inspectors of explosives to the fact that be- 300 and 400 deaths and about 2,000 cases of serious injury occur lly through the use of imperfect and dangerous mineral oil lamps, requested Professor Boverton Redwood, F.I.C., in conjunction Sir Frederick Abel, F.R.S., to advise them upon this subject. of the lamps submitted to and reported upon by Professor Red- is the water safety lamp, in which a reservoir of water, the ge of which is open to the flame of the lighted wick, is made to guish the latter immediately the lamp is overturned. The water tained in a casing or jacket placed directly over the oil chamber, aving an annular opening around the wick chamber. Upon the being overturned the water immediately rushes out of the jacket, passing through the perforated air-passage to the wick or combus- chamber, immediately extinguishes the flame. This ingenious but e device is the invention of Mr. Devoll, C.E., who has thus eded in superseding the various mechanical arrangements by 1 the automatic extinction of this class of lamp has hitherto been ed. The present invention embodies the requisite qualities of icity, certainty of action, and cheapness, and it is equally applicable ordinary cheap artisans' paraffin lamp and to those of the most kind. A demonstration was given with this lamp yesterday at avoy Hotel, when a number of lamps were instantly extinguished ing simply overturned or thrown against a wall or on the floor. icial test was the partial emptying of the water reservoir of a lamp, it up with paraffin, lighting the wick, and overturning the lamp, the light was at once extinguished.

THE COPPER TRADE.

DGING from present indications, it seems that the outlook in the copper market is most encouraging. During the first four months e present year the production of lake copper was apparently about r cent. more than for the corresponding period of 1889, yet in the of this increased production prices have advanced. On May 10 s reported that there were only 4,000,000 pounds of copper at Superior, and the bulk of this was sold. In former years there from 22,000,000 to 28,000,000 pounds on hand at the opening of gation. At the same time it is announced that the banker's ngs of copper, the remains of the French syndicate, have all been d out. This showing that the demand has increased more ly than the supply is taken by some as an indication that the et could have taken care of even a larger amount. The stocks of Lake Superior mines are almost entirely owned in Boston. The of the present stimulus upon the market value of those stocks, comparison of their present value with that existing under the of the French syndicate prior to its collapse, may be of interest to who watch trade developments. Of the mine dividend-paying er stocks on the Boston list, five are now selling higher than they under the syndicate stimulus. The lowest market value of the s of these companies, before the formation of the syndicate, shows erage of 46¾; the highest price under the syndicate shows an ge of 88⅝; the lowest since the syndicate, 45; and the highest recently touched shows an average of 82 for the same companies' s. This places the aggregate market value of Lake Superior ig stocks prior to the operations of the French syndicate at 170,000; under syndicate control, when ingot copper sold at 16½, the market value of these stocks was $55,712,500; after the ose in March, 1889, when lake ingot sold at 11 cents, the market of the stock declined to $29,756,250, while the value, with ingot cents, is $51,845,000.—*American Manufacturer.*

MANUFACTURING SALT "IN VACUO."

R. Perry F. Nursey, past president of the Society of Engineers, read a paper before the members of that body, at the Westminster Town on the subject of "Picks' System of Manufacturing Salt *in vacuo.*" Henry Adams, president of the Society, was in the chair.—In the e of his paper, Mr. Nursey observed that the manufacture of salt carried on in the present day mainly in the same way as it was by emote ancestors—namely, in open evaporating brine pans. He ibed an ordinary salt plant, pointing out the disadvantages of the nt system, including the constant expense in repairs and renewal pans, which sealed and buckled and at the best only lasted about years; the heavy cost of fuel; for land (the pans covering large) and for labour; and the production of noxious gases, which were deleterious to animal and vegetable life. He then proceeded to ibe Dr. Picks' vacuum process, which was based on the Rillieux or -effect system. Dr. Picks' apparatus, he explained, is made in duplicate sections, each consisting of four main parts, namely, the boiling chamber, the heating chamber, the collecting chamber, and the filtering chamber. The steam used enters the heating chamber of the first section and there heats the brine, and the steam given off from the brine enters the steam chamber of the second section and heats the brine there. The same process is repeated in the third section. The steam generated in the latter section from the brine, is drawn off by means of a vacuum pump condenser. The advantages, as demonstrated by the working of an apparatus put up by Mr. Nursey at a salt works in Staffordshire, were stated to be a saving in fuel, in labour, in area occupied, and the avoidance of the production of deleterious gases. The consumption of coal by the pan process was 12 cwt. per ton of fine white salt. By the new process a saving of 7 cwt. per ton was said to be effected where waste or exhaust steam was not available. The process the lecturer showed to be automatic, and to require no skilled labour in carrying it on. With regard to space, it was stated that an apparatus consisting of three sections, each 2¾ yards in diameter, would turn out 50 tons of salt per day, or 300 tons per week, while a salt pan 12 yards long by 2½ yards wide, exclusive of brickwork, only turned out about 40 tons per week. The paper was illustrated by diagrams o the old and new systems, and the author exhibited samples of salt made from the same brine and at the same works, under the old and new processes. The superiority of the salt produced *in vacuo* was declared manifest both as regards fineness and greater density.

THE FRENCH AND BRITISH WORKMAN.

MR. BERNAL, in his last consular report from Havre, makes a brief comparison of the condition of the French and British working-classes. He refers to a statement made by Mr. Potter, one of a deputation of British working men who attended the Paris Exhibition, that he thought, on the whole, French workmen are better off than the industrial classes in England, and that the deputation did not see a drunkard or beggar during their stay in Paris. He would see too many of the latter in Havre, and after a comparison of the prices of provisions, groceries, &c., in that town, with those at which similar articles can be bought in the United Kingdom, Mr. Potter would probably change the first part of his opinion. The advantage the French Workman has over his English colleague is that the former, even in the case of equally steady men, will get more amusement out of a less expenditure of money; that the greater number of French women can make very savoury soups or dishes out of comparatively nothing, which hardly any English working-class woman knows how to do. The dock labourers of Havre, who go in from all parts of the country, and especially from Brittany, receive—the foremen 5s., and the men 4s. 2d. a day. No man is taken for less than half a day, and if kept after noon is entitled to a day's pay. If wanted after knocking-off time, 7½d. an hour is paid him. Night work is paid as day work, four hours counting as half a day; but a whole night is paid as a day and a half. Working hours vary according to the season of the year. After deducting half an hour for breakfast and two hours—12 to 2—for dinner, they range from 7½ in the winter to 9½ in the summer months. The wages paid in the other trades are as follows:—Masons and bricklayers, 6d., their helps, 3½d. an hour; locksmiths, black smiths, painters, 6d.; joiners, 5½d. to 6d.; house carpenters. 6½d.; slaters, 7d.; workers in metal, 6d. to 7d. an hour; coopers, 4s. 10d., and ship carpenters, 6s. 5d. a day. The hours of labour are about ten in summer and nine in the winter months. The rents paid by workmen range from £6. to £10. a year. The wages of bakers are, for a first-class hand, from £3. 4s. to £3. 12s., and for a second-class man from £2. to £2. 8s a month, with board. The working hours are generally from 1 a.m. to 1 p.m., and even longer. The price of ordinary household bread is 10½d. the leaf of 6 lb. 9¾ oz.

THE TRADE OF INDIA.—The trade and navigation accounts o British India, made up to the close of the official year ending March 31st last, show a slight falling off in the value of imports, and a large increase, amounting to upwards of £4,295,000. sterling, in the value of exports. The total imports of merchandise and Government stores (excluding treasure), amounted in value to Rx.69,199,376 (Rx.1= ten rupees), being Rx.241,090 below the total for the previous year, but above the annual average of the previous five years by as much as Rx.7,683,100. The falling off in the value of imports as compared with the figures for the year before, is explained by the large decrease in the imports of cotton yarns and textile fabrics, which quite over-balanced the increase under the head of metals, chiefly copper, which plays such a large part in native industries. The exports of Indian produce and manufactures during the 12 months amounted in value to over Rx.99,088,333, being an increase of Rx.6,445,598, over the figures for the previous year. This increase is mainly due to greater activity in the export of raw cotton, jute, and rape-seed.—*Times.*

THE IRON AND STEEL INDUSTRIES OF THE UNITED STATES.

IF anything more were needed to illustrate the material progress made in recent years by the great Rebublic on the other side of the Atlantic, it is amply provided for by the Report—always of considerable interest, but more than usually attractive on the present occasion—on the American iron trade in 1889, which has just been presented by Mr. James M. Swank to the American Iron and Steel Association, the headquarters of which are at Philadelphia. Beginning with pig iron, concerning which reliable *data* are available, thanks to the efforts of the American Iron and Steel Association and the British Iron Trade Assaciation, we find that the output of crude iron during the last few years in the two countries was as follows :—

Years.	Great Britian.	United States.
	Tons.	Tons.
1882	8,586,680	4,623,323
1883	8,529,300	4,595,510
1884	7,811,727	4,097,868
1885	7,415,469	4,044,526
1886	7,009,754	5,683,329
1887	7,550,518	6,417,148
1888	7,998,909	6,489,738
1889	8,245,336	7,603,842

It will be seen that the British *maximum* production was attained in the year which begins the above table, while the United States are increasing their production of pig iron so steadily that they are evidently destined to become, at an early date, the leading pig-iron producing country in the world, possibly attaining this distinction in 1890. That the United States are using proportionately more pig iron than Great Britian is shown by the fact that the consumption of crude iron in this country rose from 7,052,291 tons in 1888 to 7,692,230 tons in 1889, while in the United States it jumped from 5,864,950 tons to 6,962,800 tons.

If we turn to the production of steel in the United States and Great Britain, the comparison is even more striking. We subjoin a table showing the production of Bessemer steel ingots (including Clapp-Griffiths steel ingots) and steel rails in Great Britain in the last 13 years, compared with the production of the United States in the same period. For the sake of strict accuracy, it should be observed that in the ingot tonnage for the United States for 1889 is also included the small quantity of Robert-Bessemer steel made in that year.

Years.	Great Britain.		United States	
	Ingots.	Rails.	Ingots.	Rails.
	Tons.	Tons.	Tons.	Tons.
1877 ..	750,000	508,400	500,524	385,865
1878 ..	807,527	622,390	653,773	491,427
1879 ..	834,511	520,231	829,439	610,682
1880 ..	1,044,382	732,910	1,074,262	852,196
1881 ..	1,441,719	1,023,740	1,374,247	1,187,770
1882 ..	1,673,649	1,235,785	1,514,687	1,284,067
1883 ..	1,553,380	1,097,174	1,477,345	1,148,709
1884 ..	1,299,676	784,968	1,375,531	996,983
1885 ..	1,304,127	706,583	1,519,430	959,471
1886 ..	1,570,520	730,343	2,269,190	1,574,703
1887 ..	2,089,403	1,021,847	2,936,033	2,101,904
1888 ..	2,032,794	979,083	2,511,161	1,386,277
1889 ..	2,140,793	943,048	2,930,204	1,510,057

The race between the two largest steel producers of the world has been keen, but the United States have now for some years eclipsed this country in the output of Bessemer steel ingots and rails. In the production of ingots, the race was won by America only in 1884, while with regard to Bessemer rails, the supremacy of the United States was asserted in 1879. Since that year, the construction of railways proceeded at a tremendous pace in the States, the mileage built reaching a total of 11,569 miles in 1882, after which, the rate of building fell off somewhat; but the total rose to 12,872 miles of new-track in 1887, the largest annual record ever made by any country.

If we consider all kinds of steel made by the two great steel-making countries, the United States were ahead of their European rival in 1886 and 1887. In 1888 and 1889, however, Great Britain again asserted her supremacy. She now excels as a steel producer, because of her large annual output of open-hearth steel, which is extensively used in ship-building in this country, a field in which the United States is idle, owing to its protective tariff having killed its mercantile marine.

The figures for the last six years' production of open-hear are :—

Years.	Great Britain.	United S
	Tons.	Ton
1884	475,250	117,5
1885	583,918	133,3
1886	694,150	218,9
1887	981,104	322,0
1888	1,292,742	314,3
1889	1,429,169	374,5

The production of steel ingots of all kinds in the two countries may be summarized as follows :—

Description.	Great Britain.	United S
	Tons.	Ton
Bessemer steel	2,140,793	2,930,
Open-hearth steel	1,429,069	374,
Crucible steel (about)	100,000	75,
Other steel	Nominal	5,
Totals	3,669,862	3,385,

The English statistics of finished iron do not cover the same gr those published by the American Iron and Steel Associatio therefore it is difficult to make comparisons. The output of p bars in Great Britain was as high as 2,841,534 gross tons (of 2,: in 1882; it fell to 1,616,701 tons in 1886, rising again to 1,; tons in 1887, 2,031,473 tons in 1888, and 2,253,756 tons in The production of rolled iron of all descriptions in the United in 1882 was 2,493,831 net tons (of 2,000 lb.); 1883, 2,348,87 1884, 1,957,307 tons; 1885, 1,804,526 tons; 1886, 2,283,622 1887, 2,588,500 tons; 1888, 2,411,654 tons; 1889, 2,586,38 In gross tons the quantities were 2,153,263 in 1888, and 2,309,2 in 1889.

Reference was made above to the large use of open-hearth s this country for shipbuilding purposes. When it is considered t shipyards of the United States constructed but 34,354 tons of ne ping in 1887, 36,719 tons in 1888, and 53,513 tons in 1889, wl totals for Great Britain in those years were 577,327, 904,32 1,288,251 tons respectively, the insignificance of the America building industry is obvious. The backward state of the Unite with regard to the manufacture of tinplates, which is still in its mental stage in that country, will probably not continue if the pro of the Tariff Bill now before Congress are retained when the finally passed. It is surprising that no efforts have been m "domicile" such an important industry in the United States, use of tinplates is larger there than in any other country, and t imports, from this country chiefly, are enormous. The British i of tinplates into the United States have steadily grown from 82,9 in 1871 to 331,311 tons in 1889; the total quantity of imports i 19 years being 3,293,404 tons, and their total foreign $283,671,246 (£56,724,249), irrespective of the freights and duti by the importers. Possibly the increase in the tinplate impor still continue, for the requirements of provision "canners" and pe "packers" are still growing, and they do not look with favour increase of duty now proposed.

DYED SUGAR.

AN important meeting of the Society of Public Analysts took on Wednesday evening, chiefly to hear Mr. Cassal's vi the subject of "Dyed Sugar." The interests of the trade, whic thought to be in some peril from the action of the author, w slashing controversialist, were well represented by the prese Messrs. Newlands and Heron. It appears that in Mr. C ordinary practice as a public analyst, he has recently come numerous samples of sugar coloured yellow with some foreign co matter, probably of the nature of methyl orange, and out of 15 s collected in Kensington, 10 were found to be thus treated. opinion this was nothing less than a fraud, and he had so rega in his certificates of analysis. The reactions of the dye-stuff vogue are pretty definite, and its detection therefore easy: trated hydochloric acid added direct to about an ounce of the su sample produces a purple pink colour, becoming reddish br standing; strong sulphuric acid has the same effect, but the rea marred by subsequent charring. The substance may be isolated by extracting the sugar with 90% alcohol, evaporat solution to dryness, taking up again in alcohol, and dyeing wool, slightly mordanted with aluminium acetate, in the soluti fast yellow colour on the skein is a sure indication of its presence sugar, thus prepared, is known in the trade as yellow cryst

12-13s. per cwt., as against 17-18s. for Demerara. Mr. ...ds, taking up the cudgels, revealed a state of things sufficiently ...g to the outsider, though familiar enough to the expert. It ... that practically all sugar, white included, is coloured; these colouring matters being used for one class, chloride of tin for ..., and ultramarine or an aniline blue for the white varieties. ...antities are, however, in all cases, small; *e.g.*, 1 part in 50,000 ...elin O O O, and a more tangible, but still harmless, quantity ...ride of tin, so that the matter seemed of little moment. The ...ject of colouring the sugar was merely to maintain a standard ... that any given sugar could be sold as a recognised grade of the ...

The actual figures ruling for Demerara were 15s. 6d. to 16s. 9d. ...t., and for "yellow crystals," 16s 3d. to 17s. 3d., so that the ...nce in price was actually in the reverse direction to that stated by ...assal. The method of testing proposed was fallacious, as many ... artificially coloured gave no reaction. Mr. Heron mainly en...red to allay alarm as to the alleged toxic action of these ...ing matters. He assured the meeting that nothing whose ...logical action was baneful was used. This brought Mr. Hehuer ...e fray, he declining to accept any assurance that was of so ...l a character. As to the notion that had been put forward that ...blic taste demanded coloured sugars, it had no weight with him

The popular taste used to demand beautiful green pickles ...red with copper), brilliantly red bloater paste, mustard reduced ...tency with flour, cocoa of the homœopathic variety in which ... predominated, and so forth, and it had been the duty of the ... analyst, and one which he had successfully performed, to ...mn such products without regard to "popular taste" based on ...r ignorance. Had manufacturers been content to educate the ... with the preferential consumption of colourless sugars, instead ...dering to their prejudice, none of this trouble would have arisen. ...pshot appears to be that the practice of colouring sugar is almost ...rsal, and will not readily be displaced, and provided nothing be ... which has not been proved innocuous, no real harm will result, ...h there is little doubt a technical offence will have been com...d under the Food and Drugs Act.

Trade Notes.

"Truth," last week, asked "Who is Joule?" We have pleasure ...swering that he was an earnest seeker after Truth.

...e two following announcements were inadvertently omitted from ...ast week's issue:—

...bes and Fittings. Joseph Aird, Wellington Tube Works, ...t Bridge, Staffordshire, announces a reduction in the price of ... and fittings, of 5% gross.

...essrs. Hildick & Hildick, of Walsall, Tube Manufacturers, ...announce that their discounts on plain and galvanized wrought iron ..., are increased 5% gross, as from May 20th.

...xalic Acid.—Passing along Liverpool Streets between 8 and ...n., the observer will note a number of porters and errand-boys ...ing the brass window-plates with solutions of oxalic acid. These ...lmost exclusively contained in wine or beer bottles; occasionally a ...r-beer bottle varies the monotony. It is rarely that a label is ...d on any of these bottles, notwithstanding the care taken by the ...gists in selling the acid at the hands of a qualified man, and in ...ing it "Poison." Reflecting on this, one must either believe ... a special providence watches over solutions of oxalic acid, or that ...d deal of nonsense is talked in judicial quarters about the law pro...ng the public.

...e Leon Cobalt and Copper Company.—A new company ...een formed in England under special circumstances to work ...s in Spain. This venture deserves some attention, as it marks ... beginning of a new era in mining in Spain. The Spaniards ...rally, after getting possession of a new mine, hand it over to ...n capitalists, who form companies of which the working capital ... foreign. The Spanish vendors receive shares as payment. The ... of the Leon Cobalt and Copper Company is a new departure. ...Spanish holders of these mines, which cover some 350 hectares, ...d them for sale in England, and a company was organised in ...lon with a capital of £50,000. The bankers, the Union Bank of ... and England, offered these shares for sale in Spain, with the ...ecedented result that the demands for shares at Madrid and Bilbao ... have been double the required capital. It is not yet known ... applications there have been at the branches of Barcelona, ...ncia, and Seville, nor in the mother establishment at London. ... explanation of this is that the Spanish capitalists, especially ... of Bilbao, have shown more confidence in the English manage... of mining business than in their own, and that they are more ...g to put their money with them than in a mining business of a ...y Spanish character.

Dr. Corsar Ewart is Professor of Natural History in Edinburgh University, having been appointed when Professor Ray Lankester refused the post. Professor Ewart has for some time past professed a decided knowledge of pisciculture, and recently published in a Scotch newspaper a series of articles on mussel culture. The other day a journalist residing in the neighbourhood of Edinburgh raised an action in the Court of Session against the Professor for remuneration for these articles, which he declares were written by him. Professor Ewart kept the matter out of Court, it seems, by paying the plaintiff £50 and the costs of the case. But I should like to know who really wrote the articles. If the Professor, why did he pay the £50? If the journalist, why is he not made the Professor of Natural History in the University?—*Truth*.

Cocoanut Butter.—Within the last few months, says the *American Druggist*, a new trade has risen in India and has attained extraordinary dimensions. About two years ago a German chemist, Dr. Schlunk, discovered that excellent butter could be made from cocoanut milk. It is, according to a Bombay newspaper, pleasant to taste and smell, of a clear whitish colour, singularly free from acids, easily digestible, and an imcomparably healthier and better article of diet than the cheap poor butters and oleomargarines of European markets. The manufacture is carried on in Germany, where one firm turns out from 3,000 to 4,000 kilogrammes daily. The cocoanuts required are imported from India, chiefly Bombay, in large and increasing numbers, and the trade seems likely to attain still greater importance.

A Pioneer of Industry.—Mr. Arthur Robottom, well known in Liverpool, Birmingham, and London, &c., as a pioneer of commerce and introducer of new trade products, has left Liverpool by the *John Elder* for Brazil, Chili, and Peru on a voyage of discovery. One of his principal objects is to inspect a mountain in the Andes that is reported to contain a variety of chemicals naturally formed, and also borate of lime and nitrate of soda deposits. Mr. Robottom was one of the first to sell nitrate of soda for the manufacture of nitric acid. He was engaged in the manufacture of paraffin oil, and was instrumental in the introduction of corozo nut or vegetable ivory for button making. He also introduced to the trade piassava from Brazil, the keltol from Ceylon, borax from California, and a variety of other articles that now find employment for thousands of people who are engaged in working them up.

Stockport Guardians and the Rating of Machinery Bill.—At the meeting of the Guardians of the Stockport Union, Mr. C. Earnshaw presiding, the question of taking further action in support of this measure was considered. The Guardians had petitioned Parliament in its favour.—Mr. R. Hammond, chairman of the Reddish Local Board, observed that amendments were threatened to the Rating of Machinery Bill, amendments which if passed would be a most serious thing to users of machinery in that neighbourhood. They should do all they could to strengthen the hands of the promoters of the bill.—Sir William Houldsworth asked that pressure might be put upon Mr. Ritchie before the third reading, which was appointed for the 18th instant; and as the Guardians were so largely interested in machinery and in the welfare of the district, they ought to do what they could to promote the passing of the bill as it now stood. The bill provided that for rating purposes the following machinery should be taken into consideration:—"(1) Outer wheels, steam, gas, air, and electric engines,steam boilers, and all other fixed motive powers, and the fixed appurtenances thereof; (2) shafts, wheels, drums, and other fixed power machinery which transmits the action of motive power to other machinery, fixed or loose (2a) save as in the last section provided, no machinery, whether attached to the tenement or premises or not, shall be taken into consideration in estimating such rateable value." Mr. E. T. Cunliffe (Handforth): That is the bill as amended and accepted by Sir William Houldsworth, Mr. Mowbray, and those who have charge of it. Lieutenant Colonel Wilkinson and others spoke in favour of the bill, and on the motion of Alderman William Leigh, the Board decided unainmously to forward to Mr. L. J. Jennings, M.P., a memorial in support of the measure for presentation to Mr. Ritchie.

Illegitimate German Enterprise.—In granting a perpetual injunction, with costs, last week against a German firm and their agents, restraining them from interfering with the rights of the proprietors of the liqueur manufactured at the famous Monastery of La Grande Chartreuse, Mr. Justice Kay made some severe remarks. The defendant's counsel had "wished to state" that in Germany it was competent for them to imitate the plaintiff's label, and that the bottles and labels in question had been sent over to this country under the impression that the English law would permit their use. The learned judge, who, the public will be pleased to hear, has quite recovered his health, answered this plea by declaring that "according to their own counsel's statement the defendants labelled their goods as being the product of La Grande Chartreuse well knowing they were not. They knew what they were doing was a lie intended to be imposed on people, and knowing it was a lie the contention was that what they did was done with a view of finding out whether it was legal in England." Illegitimate German enterprise finds no favour in the eyes of Mr. Justice Kay.

SINGKEP, THE NEW TIN ISLAND IN EAST INDIA.—The Singkep Tin Company, concessionaire of the Island of Singkep, has raised its capital from 1,250,000 to 1,500,000 guilders—say £125,000—to ensure the regular working of the tin mines. Singkep promises to become a formidable rival to its great tin-producing neighbours, Banca and Blitong. Its size is about 50,000 hectares—say, 123,550 acres—and it belongs to the Lingga Islands, a group off the east coast of Sumatra, and formerly belonging to Johore and Malacca. The tin mines of Singkep are supposed to be of great antiquity. They were hitherto worked by the native Sultan, who is a vassal of Holland. He exported the ore to Muntok, the capital of Banca, or to Singaporee, but Singkep tin was not known by itself in the European market, being sold for Banca or Blitong.

IRON MANUFACTURE IN INDIA.—The ironworks owned by the Bengal Iron and Steel Company (Limited), at Burakur, are a few miles distant from the junction of the East Indian and Bengal-Nagpore Railways, where arrangements have been made for the production of 30,000 tons of pig iron per annum from the ironstone and coal found in close proximity on the property. In the foundry preparations have been made for turning out cast iron water-pipes, sleepers, and other railway materials in very large quantities, the moulding being accomplished by hydraulic machines of the latest pattern, at rates which, it is expected, will render European competition difficult, if not impossible. The output of these works will not be confined to pig iron and casting, but as soon as practicable the production of wrought iron and steel, and its manufacture into beams and bars, will be commenced, and eventually the rolling of rails, girders, joists, and all sections of iron and steel in common use on railways and for building purposes.—*Indian Engineering.*

USES FOR OLD PAPER.—Most housekeepers know how invaluable newspapers are for packing away the winter clothing, the printing ink acting as defiance to the stoutest moth, some housewives think, as successfully as camphor or tar paper. For this reason newspapers are invaluable under the carpet, laid over the regular carpet paper. The most valuable quality of newspapers in the kitchen, however, is their ability to keep out the air. It is well known that ice, completely enveloped in newspapers so that all air is shut out, will keep a longer time than under other conditions; and that a pitcher of ice water laid in a newspaper, with the ends of the paper twisted together to exclude the air, will remain all night in any summer room with scarcely any perceptible melting of the ice. These facts should be utilized oftener than they are in the care of the sick at night. In freezing ice cream, when the ice is scarce, pack the freezer only three-quarters full of ice and salt, and finish with newspapers, and the difference in the time of freezing and quality of the cream is not perceptible from the result where the freezer is packed full of ice. After removing the dasher, it is better to cork up the cream and cover it tightly with a packing of newspapers than to use more ice. The newspapers retain the cold already in the ice better than a packing of cracked ice and salt, which must have crevices to admit the air.—*Scientific American.*

MINING IN DEVON AND CORNWALL.—The last annual report of Mr. Pinching, Inspector of Mines for Devon and Cornwall, has just been laid before Parliament. The total number of persons employed was 13,464, or 275 less than the previous year. The quantity of minerals wrought from the various mines of the two counties was 118,201 tons, of which 12,464 tons consisted of tin ore, 7,688 tons of arsenical pyrites, 7,618 tons of dressed copper ore, 4,758 tons of arsenic, 3,200 tons of iron ore, and 1,200 tons of barytes. Of the rest 81,051 tons consisted of clays, stone, and slate, and there were, in addition, small quantities of zinc ore, ochre, and copper. The number of deaths by accident in 1889 was 23, being one less than 1888, and the number per 1,000 was 1·70. The average for the past 17 years was 1·63. The inspector thinks that many of the accidents underground are due to the negligence of the miners themselves. There were four prosecutions, followed by the conviction of mine managers, during the year. Mr. Pinching thinks the proportion of accidents in connection with man engines is still larger than it should be. Tin mining, he observes, has progressed steadily during the year, in spite of market fluctuations. Amongst the several lodes which have shown a marked improvement the most important is that being worked at the Carn Brea mine, which has caused the shares to rise 500 per cent. value. Dolcoath mine adit, although sunk to a depth of over 400 fathoms, still continues to make large returns, and the lode at this great depth shows no sign of decreasing in value. Amongst other mines which have much improved during the year are Tincroft, Wheal Grenville, South Frances, West Kitty, and Wheal Kitty. During the year 24 mines, several of which were suspended last year, have been abandoned, 12 have been suspended, and 16 new mines have been started.

THE STORAGE OF GRAIN IN THE UNITED STATES.—The Consul at Baltimore in his last report describes a system of storing grain which, if generally employed, would revolutionize the present methods. Should it prove a success, the elevators now in general use wil obsolete, and every farmer will be supplied with a substitute he will be able to store his grain for years at little cost and The inventor maintains that he can manufacture steel tan required capacity, and at a cost averaging from, say, 2d. to each bushel of capacity, as against 1s. 8d. to 2s. now exp wooden elevators. The steel tanks will be filled with gr simple process. Whan the tank is full a percentage of air is e and a quantity of carbonic acid gas admitted. The valves closed and the grain will keep sound for years. Having e the oxygen there is no chance of fermentation, and, as a cons no decay or rot. At the same time, all animal life perishes grain is secure against the ravages of weevils, which are so de With a tank costing £100. the farmer has storage for 10,000 and can hold his crop against low prices until the market i Fire cannot burn it nor rain injure it. The process by which is manipulated is as simple as it is effective. It consists in an receiver, leading from which is a large pipe so arranged as t over the grain in the receiver it is desired to unload. By m suction-fan the air is exhausted in the receiver, and the rush t vacuum is sufficient, it is said, to draw the grain into the pipe, which it passes to the tank. Should the process prove a s will most eventually put an end to the proceedings of thos whose business appears to be the making of fictitious valu bread stuffs.—*Times.*

THE AFFAIRS OF JAMES BERGER SPENCE.—A meeting last week at the London Bankruptcy Court under the failur debtor, who formerly traded at Manchester as a chemical, m mineral merchant. He first acted as agent to his father in t chester business, and afterwards opened a branch in London. he started on his own account he had no capital, but was subs joined by various partners, who introduced a total sum of £1 capital. The partnerships have all been dissolved and duly He now states that he has debts amounting to about £10,000. was the promoter of the Kruis River Cobalt Company, and £2,000. for advertising that undertaking. He claims 2,500 Spence's Metal Company, the value of which is uncertain, bu not yet received the shares, as they have not gone to allotme further states that he receives them in consideration of his under another company of a similar nature now wound up. holds 50 shares in the North Wales Gold Exploration Compan Chairman stated that only two proofs had been lodged. Th had not yet filed his statement, although the proceedings stituted in February last. In explanation the debtor stated papers and documents had been in the hands of a former cle without him he could not prepare a proper statement. Mr. Pa he represented Mr. Robertson, the gentleman referred to by th and the only document held was a Swaziland concession. U Chairman inquiring if there was any offer to be submitted, th said he had 60 acres of land at Erith upon a building lease at acre, and there was every prospect of being able to sell at sufficient to pay the unsecured creditors in full. The C intimated that he should shortly apply to the Court for an adju No resolutions were passed, and the matter was left in the the Court to be wound up in the usual way.

A CHEMICAL WASTE.—It is not many years since sul ammonia was a bye product, which was wholly lost. In hi for the year 1889, Mr. A. E. Fletcher, the chief-inspector u Alkali, &c. Works Regulation Act, makes a statement regar production of ammonia in the United Kingdom. In 1889, g ironworks, shaleworks, and coke and carbonizing works pro less than 133,604 tons. This is a great increase in twelve m over 122,785 tons. In 1877 the production was only 113, and odd. It is estimated that the value of last year's produc about £1,500,000. Yet, great as the saving already realized it is small in comparison with what might be made. I ammonia were saved from coke ovens, ironworks, and other where the consumption of coal is large, the addition to the wealth would exceed ten times the amount now realized. little strange that so much of the sulphate of ammonia produ is exported to Germany. There it is used principally in the cu of beetroot. Seeing that the larger portion of this ammoni duced in the gasworks scattered through the country, it surprising that the British farmer does not see it to be to hi to buy that which may be had in his own market town rathe allow it to be carried away and shipped for the use of the f Germany, who has to pay the carriage and other charges in to the price for which our farmers at home could purchase it.

THE CROWN ALUMINIUM WORKS, late in the occupation of t Aluminium Company, at Yardley Wood, which were recentl by auction, have been sold by private contract to a firm manufacturers, but the price is not disclosed. The works, large residences, occupy about six acres of freehold land. intention of the purchasers to re-open the works shortly.

NG OF MACHINERY.—A conference between the members ed in the Rating of Machinery Bill, and the President of the overnment Board, the Attorney-General, and Mr. Walter Long, d at the House of Commons last week. The decision come to it the amendment agreed upon between the Attorney-General, part of the Government, and Sir Henry James, on the part of moters of the Bill, should be placed upon the paper by Sir James. The amendment proposes to make clause 1 run as : "In estimating for the purpose of any valuation list, or poor, local rate, the gross estimated rental or rateable value of any ment occupied for any trade, business or manufacturing purny increased value arising from machinery, which is machinery manufacturing process, and is only fixed to or sunk in the ment for the purpose of steadying it, and which can be removed injury to the hereditament, or to itself, and does not require cial construction or adaptation to the hereditament in which it is all be excluded."

Market Reports.

REPORT ON MANURE MATERIAL.

increased activity which we referred to in our last report has ontinued throughout the past week, and a large volume of s has been concluded.

e are sellers of cargoes of hot-air dried Bull River phosphate 10½d. per unit, cost, freight, and insurance to good U. K. nd bids of less than this are being invited. Cargoes of land ay be had at 9½d. per unit, cost, freight, and insurance to Quotations for the higher grades of Somme rather preclude s in large lines, 15¼d. c.i.f. asked for 70/75%, and 12d. for , 1-5th rise in each case. For 55/60% about 10d. would be ed. The price for high grade Florida rock has hardly been but probably about 15d. on 80% in the dry would be business. t values of Canadian are 1s. 0½d. for 70%, 1s. 2d. for 75%, and ½d. for 80%, full delivered terms, 1-5th rise in each case. ions for Belgian remain unchanged.

rther cargo of River Plate bones, loading, is reported to have old at £5. 7s. 6d. for U. K., and a bid of 15d. less for another cargo has not led to business. The ideas of holders of East bone meal are strengthened by the advancing rate of exchange, is also stopping business for shipment. It does not seem likely ent that anything under £5. 10s. would be accepted for shipand nothing less would do for fine quality on spot, ex store. is nothing fresh to report in crushed bones. Any parcels of on grinding bones would sell readily at £4. 12s. 6d. for common parcels, up to £4. 17s. 6d. for clean dry hard bones. For Plate bone ash £5. 7s. 6d. on 70%, nearest value.

hin the past week over 30,000 tons nitrate of soda arrived at or due, sold at 7s. 10½d. to 7s. 11¼d., since which transactions been paid for several further cargoes for U. K. The coast has een effectively cleared, and at the close the market is very firm advance. Spot cannot be bought under 8s. 1½d. for ordinary, 3d. for refined quality. Due cargoes are worth 8s , June-July ent about 8s. 4½d., September-October about 8s. 9d. per cwt., red good U. K. ports.

ed blood offers rather more freely, but there is nothing to be der 10s. 6d. per unit, f.o.r. at works, which price will probably uired for a 40-ton parcel of fine River Plate, now on the way, e in about ten days.

nonite is quiet at 9s. 6d. per unit, f.o.r. at works. Kainit is at 40s. per ton. Supers quiet at 46s. 3d. per ton in bulk, f.o.r. ks, for 26%.

MISCELLANEOUS CHEMICAL MARKET.

re have been no changes of importance during the past week in value of alkalies for early deliveries. Caustic soda remains firm e at last week's quotation, viz., £9. for 74%, and £8. 5s. for o.b.; lower grades in small demand, but prices nominally und. The high strength caustic soda from the Tyne works es to find ready outlet for export trade. Soda ash generally is and though most of the makers are well engaged for some little head, and can command 1⅜d. per degree for early deliveries, sition forward is more favourable to buyers, and quotations are the market at 1¼d. to 1$\frac{5}{16}$d. at maker's works. Soda crystals ged, and chlorate of potash without any improvement. Bleachwder steady at £4. 17s. 6d. rails for prompt, and £5. per ton, nd forward. Vitriol and muriatic acids without change. Saltcake, 27s. to 28s. in bulk. Sulphate of copper rather more plentiful, but for June delivery firm at £24. 10s.; the high value of copper preventing any downward movement for the present. For future deliveries there are various speculative quotations from £20. to £22. per ton, according to period of deliveries required. Lead salts of all kinds still very quiet in demand, but without alteration in value. Tin crystals dearer in sympathy with the metal. Zinc oxides and salts are all firm in price owing to the high spelter market. Acetate of lime, brown, distinctly easier and more plentiful at £7. 10s. to £7. 12s. 6d. for early and forward deliveries; grey scarce at £13. Acetic acids of all grades are without any material changes. Potash, caustic, and carbonate are selling more freely, and makers are well booked with orders. Superphosphates of lime are in good demand as usual at this season, but supply is equal to requirements, and prices remain steady. Prussiate of potash keeps steady at 9½d. per lb., though no large business is being done. Bichromates quiet. Tartaric and citric acids are steady at 1s. 3d. per lb. for both.

TAR AND AMMONIA PRODUCTS.

The benzol market still retains its firm position, and 3s. 9d. for 90's and 2s. 9d. for 50/90's may be quoted as to-day's values. Solvent naphtha is still in good request, and creosote is moving off very freely. Crude carbolic acid is still flat, but sellers seem to have made up their minds not to accept any lower offers, and although 2s. has been bid for good 60's, makers everywhere refuse less than 2s. 2d. Anthracene is still weak, and sales of B have been reported at 1s. There have been no transactions in A quality, and therefore price cannot be stated. Pitch is out of season, and weak.

In sulphate of ammonia values remain unchanged at £11. 2s. 6d. to £11. 3s. 9d. f.o.b. Hull, and £11. 1s. 3d. to £11. 2s. 6d. Leith and Liverpool. It was only to be expected that the recent action of speculators has had the effect of driving away orders from the market momentarily, in consequence but few transactions have taken place, sellers seemingly being unwilling, so far, to concede the low rates demanded by buyers. Beckton price is nominally £11. 5s., while outside London makes are quoted at £11. 2s. 6d.

WEST OF SCOTLAND CHEMICALS.

GLASGOW, Tuesday.

There are still awanting genuine symptoms of recovery as regards the chemical market here, and on the contrary, the position to-day is, on the whole, even less satisfactory than it was a week ago. The extreme top price for sulphate of ammonia, spot, Leith delivery, is to-day no more than £11. 2s. 6d., or barely that. It is on report that £11. 1s. 3d. has been taken, but the evidence is of doubtful character, and probably no dealing has as yet taken place at so low a dip. Nevertheless buyers do not pay £11. 2s. 6d. at all readily, and makers being disinclined to give way further, there is hardly anything doing, and the forward basis is also idle. The output from the bichromate factories shows a slight improvement, but still halts considerably short of what is desirable. Bleach is nominally unchanged, but the market is quite a listless one. Caustics continue dropping, and borax, sulphate of copper, and nitrate of soda are all easier; most of the other common chemicals only hold former quotations, and about the only instance of a veritable rise is found in saltcake, which this week quotes dearer. Paraffins unchanged, scale and wax retaining their firmness, and burning oil still poorly. The last of the Scotch mineral oil companies to declare working results is the West Lothian, with no dividend; but the general outcome of the industry is on the whole satisfactory for the year. Chief prices current are:—Soda crystals, 52s. net Tyne; alum in lump £5., less 2½% Glasgow; borax, English refined, £29., and boracic acid, £37. 10s. net Glasgow; soda ash, 1¼d. less 5%, Tyne; caustic soda, white, 76°, £9. 10s., 70/72°, £8., 60/62° £7. 1s. 3d., and cream, 60/62° £6. 17s. 6d., all less 2½% Liverpool; bicarbonate of soda, 5 cwt. casks, £5. 15s., and 1 cwt. casks, £6. net Tyne; refined alkali, 48/52°, 1½d., less 1¼%, Tyne; saltcake, 27s. to 29s.; bleaching powder, £5. to £5. 2s. 6d., less 5% f.o.r. Glasgow; bichromate of potash, 4d., and of soda 3d., less 5 and 6% to Scotch and English buyers respectively; chlorate of potash, 4½d., less 5% any port; nitrate of soda 8s.; sulphate of ammonia, spot, £11. 2s. 6d. f.o.b. Leith; salammoniac, 1st and 2nd white, £37. and £35., less 2½% any port; sulphate of copper, £24., less 5% Liverpool; paraffin scale, hard and soft, 2¾d. per lb.; paraffin wax, 120°, semi-refined, 3¼d.; paraffin spirit (naphtha), 7d. a gallon; paraffin oil (burning), 6¾d. to 7d. on rails Glasgow to all Scotch buyers (English orders negotiable about 1¼d. a gallon lower); ditto (lubricating), 865°, £5. 10s. to £6. per ton; 885°, £6. 10s.; and 890/895°, £7. 15s. to £8. Week's imports of sugar at Greenock were 33,868 bags.

THE LIVERPOOL COLOUR MARKET.

COLOURS are unchanged. Ochres: Oxfordshire quoted at £10., £12., £14., and £16.; Derbyshire, 50s. to 55s.; Welsh, best, 50s. to 55s.; seconds, 47s. 6d.; and common, 18s.; Irish, Devonshire, 40s. to 45s.; French, J.C., 55s., 45s. to 60s.; M.C., 65s. to 67s. 6d. Umber: Turkish, cargoes to arrive, 40s. to 50s.; Devonshire, 50s. to 55s. White lead, £21. 10s. to £22. Red lead, £18. Oxide of zinc: V.M. No. 1, £25.; V.M. No. 2, £23. Venetian red, £6. 10s. Cobalt: Prepared oxide, 10s. 6d.; black, 9s. 9d.; blue, 6s. 6d. Zaffres: No. 1, 3s. 6d.; No. 2, 2s. 6d. Terra Alba: Finest white, 60s.; good, 40s. to 50s. Rouge: Best, £24.; ditto for jewellers, 9d. per lb. Drop black, 25s. to 28s. 6d. Oxide of iron, prime quality, £10. to £15. Paris white, 60s. Emerald green, 10d. per lb. Derbyshire red, 60s. Vermillionette, 5d. to 7d. per lb.

THE LIVERPOOL MINERAL MARKET.

Our market has continued to rule fairly steady during the past week, and prices remain practically without alteration. Manganese: Arrivals have been somewhat larger, but as most of them have gone into consumption prices are unaltered. Magnesite: No alteration to report; prices in buyers' favour. Raw lump 26s., raw ground £6. 10s., and calcined ground £10. to £11. Bauxite (Irish Hill brand) in strong demand and prices firm—Lump, 20s.; seconds, 16s.; thirds, 12s.; ground, 35s. Dolomite, 7s. 6d. per ton at the mine. French Chalk: Arrivals have been larger, but the majority have gone direct into consumption, and prices are unchanged: 95s. to 100s. medium, 105s. to 110s. superfine. Barytes (carbonate) easier; selected crystal lump is scarce at £6.; No. 1 lumps, 90s.; best, 80s.; seconds and good nuts, 70s.; smalls, 50s.; best ground £6.: and selected crystal ground, £8. Sulphate: firm, and prices are better; best lump, 35s. 6d.; good medium, 30s.; medium, 25s. 6d. to 27s. 6d.; common, 18s. 6d. to 20s.; ground best white, G.G.B. brand, 65s.; common, 45s; grey, 32s. 6d. to 40s. Pumicestone unaltered; ground at £10., and specially selected lump, finest quality, £13. Iron ore somewhat easier. Bilbao and Santander, 9s. to 10s. 6d. f.o.b.; Irish, 11s. to 12s. 6d.; Cumberland, 13s. to 15s. Purple ore quiet. Spanish manganiferous ore continues in fair demand. Emery-stone in fair demand; prices unchanged. No. 1 lump is quoted at £5. 10s. to £6, and smalls £5. to £5. 10s. Fullers' earth steady; 45s. to 50s. for best blue and yellow; fine impalpable ground, £7. Scheelite, wolfram, tungstate of soda, and tungsten metal continue in fair request. Chrome metal, 5s. 6d. per lb. Tungsten alloys 2s. per lb. Chrome ore continues to offer pretty freely, and prices are easier. Antimony 24s. to 26s. Asbestos; the demand has improved; prices firmer. Potter's lead ore continues firm; smalls, £14. to £15.; selected lump, £16. to £17. Calamine: Best qualities scarce—60s. to 80s. Strontia steady; sulphate (celestine) steady, 16s. 6d. to 17s. Carbonate (native), £15. to £16.; powdered (manufactured), £11. to £12. Limespar: English manufactured, old G G.B. brand in demand, and brings full prices; 50s. for ground English. Felspar, 40s. to 50s.; fluorspar, 20s. to £6. Bog ore in demand, firm at 22s. to 25s. Plumbago steady; Spanish, £6.; best Ceylon lump at last quotations; Italian and Bohemian, £4. to £12. per ton; founders, £5. to £6.; Blackwell's "Mineraline," £10. French sand, in cargoes, continues scarce on spot—20s. to 22s. 6d. Ferro-manganese and silicon spiegel easier. Chrome iron, unaltered at last quotations. Ground mica, £50. China clay freely offering—common, 18s. 6d.; good medium 22s. 6d. to 25s.; best, 30s. to 35s. (at Runcorn).

THE TYNE CHEMICAL REPORT.

TUESDAY.

The Chemical market has been dull this week, and prices are again tending downwards. Transactions are few, and are chiefly for present wants. Bleaching powder has declined half-a-crown a ton, and soda crystals are one shilling per ton lower. Sulphate of soda unchanged, with little doing; soda ash and caustic soda are steady at last week's quotations. To-day's prices are as follows:—Bleaching powder, in softwood casks, £4. 17s. 6d. to £5. per ton; caustic soda, 77%, £9. per ton; ground, in 3-4 cwt. barrels, £12. per ton; soda ash, 48/52%, 1¼d. per degree, less 5%; soda crystals, in 1 cwt. bags, £2. 13s. 6d. per ton, net weight; 2 cwt. bags, £2. 11s. per ton, net weight; in casks, £2. 11s. per ton, gross weight; sulphate of soda, in bulk, 32s. per ton; ground and packed in casks, 42s. per ton; recovered sulphur, £4. 5s. per ton; chlorate of potash, 4½d. per lb.; s soda, 75° Tw., £2. 10s. per ton; 100° Tw., £3. 7s. 6d. 140° Tw., £4. per ton; hyposulphite of soda, in 1 cw £5. 15s. per ton, in 5-7 cwt. casks, £5. 5s. per ton; pu sulphate of alumina, £4. 10s. per ton; blanc fixe, £7. 10s. chloride of barium, £8. per ton; nitrate of baryta, crystals, £ per ton; ground, £19. 5s. per ton; sulphide of barium, per ton—all f.o.b. Tyne, or f.o.r. makers' works.

Small coals, for manufacturing purposes, are now showing easiness, although colliery owners are still holding out for figure for first-class smalls; inferior qualities have been reduc a shilling a ton during the week. Prices range from 7s. to 8s ton for Northumberland small, and 8s. 6d. to 9s. 6d. per Durham small.

The following are the exports of chemicals from the Tyn month of May, and corresponding month of last year:—

	May, 1889. tons.		May, 189 tons.
Alkali & Soda Ash........	1,349	......	1,582
Bicarbonate of Soda	10	......	8
Bleaching Powder	2,141	......	2,317
Manure	67	......	622
Soda Crystals	2,593	......	1,511
Sulphate of Soda	7	......	53
Caustic Soda.... }	(no separate return)....		1,240
Other Chemicals }	3,766	......	1,183
	9,933		8,516

The Patent List.

This list is compiled from Official sources in the Manchester Laboratory, under the immediate supervision of George E. L Alfred R. Davis.

APPLICATIONS FOR LETTERS PATENT.

Treatment of Paper. J. M. Campbell. 8,183. May 27.
Feed Tank for Water Pipes.—(Complete Specification). E. D. Ber May 27.
Apparatus for Heating, Purifying, and Distilling Liquids. G. F 8,192. May 27.
Production of Colouring Matter. J. Hall. 8,215. May 27.
Extraction of Gold from its Ores. W. A. A. Dowden. May 27.
Treatment of Sewage.—(Complete Specification). W. S. We May 27.
Electric Welding.—(Complete Specification). W. R. Lake. 8 242.
Manufacture of Colouring Matters. W. R. Lake. 8,243. May 2
Treatment of Waste India-Rubber. G. L. Hille. 8,250. May 27
Scouring and Wiping Tinned Sheet Iron. G. F. Redfern. 8,251
Manufacture of Manures. Sir E. R. Sullivan. 8,262. May 27.
Washing, Dyeing, and Treating Textile Materials. E. Sutcliffe Sutcliffe. 8,270. May 28.
Smoke Consumption. J. Moss. 8,275. May 28.
Dynamometers. E. Nixon and W. Millichamp. 8,298. May 28.
Solidifying Mineral Oils and their Application for Combusti Lagoutte, and G. de Velna. 8,204. May 28.
Refactory Linings for Converter Bottoms.—(Complete Sp B. Versen. 8,295. May 28.
Brick-making Machinery. C. Huelser. 8,301. May 28.
Alizarine Blue Colouring Matters. J. Y. Johnson. 8,303.
Manhole Lids. J. Birch, 8,307. May 27.
Smoke Combustion. J. C. Jopling. 8,319. May 29.
Recovery of Mercury after Use for Gold and Silver Extractio Wethered and A. E. Morgans. 8,324. May 29.
Treatment of Galvanisers' Waste Acids. A. G. Greenw May 29.
Concentrating or Separating Ores. W. McDermott. 8,352. May
Treatment of Fuller's Earth. W. F. Keevil. 8,355. May 29.
Coating Metallic Surfaces with Tin. S. O. Cowper-Coles. 8,357
Cleaning Tin Plates. M. R. Waddle. 8,364. May 30.
Manufacture of Colouring Matters.—(Complete Specification). R. Kramer, and W. Herking. 8,389. May 30.
Electric Switches. B. M. Drake, J. M. Jorham and P. J. Pri May 30.
Battery Compound. W. Wright. 8,394. May 30.
Purifying Feed Water. A. K. J. Krais. 8,395. May 30.
Treatment of Iron Ores. H. Aitken. 8,399. May 30.
Dealing with Distillery Wash. A. Walker. 8,408. May 30.
Production of Coumarin Colouring Matters. B. Willcox. 8,411
Manufacture of Artificial Stone. J. Hartnell. 8,413. May 30.
Smoke Prevention. S. K. Barnes. 8,418. May 30.
Apparatus for Manufacture of Wood Wool.—(Complete S O. Evenstad and O. Senstad. 8,422. May 30.
Separation of Textile Fibres. E Knecht. 8,452. May 31.
Solution of Calcium Phosphate in Carbonic Acid Water. T. I and W. H. Symons. 8,454. May 31.
Manufacture of Soap Combined with Mineral Oils J. B. Rol May 31.
Electric Switches. C. A. McEvoy. 8,474. May 31.
Machinery for Preparing Flax and other Fibre. A. T. L Lawson and S. Dear. 8,475. May 31.
Electric Switches. S. F. Beevor. 8,477. May 31.
Priming and Detonating Composition.—(Complete Specification) and E. M. Gregory. 8,481. May 31.
Purifying Oils and Fatty Substances. W. B. G. Hogg and J. 8,484. May 31

Copper Sulphate—*continued.*
Lisbon, 5 t. 1 c. £125
Constantinople, 2 t. 51
Genoa, 5 t. 110
Treport, 10 214
Bordeaux, 47 1,125

Coal Products—

Carbolic Acid
Stettin, 10 cs. £33

Naptha
Calais, 1 cask. £6
Copenhagen, 10 drms. 80 gals. 14

Anthracene
Rotterdam, 130 cks. £[illegible]
Dusseldorf, 252 2,[illegible]
Cologne, 252 1,[illegible]
Rotterdam, [illegible] [illegible]

Naphthaline
Hamburg, 9 cks. £71
Libau, 2 14
Hamburg, 6 55
Gothenburg, 1 10
Philadelphia, [illegible] 172
Christiania, 1 c. 7
New York, 7 [illegible]

Benzole
Rotterdam, [illegible] £[illegible]
" [illegible] 760
Antwerp, 20 pns. 4[illegible]

Tar
Danzig, [illegible] brls. £[illegible]
" [illegible] 110

Pitch
Dunkirk, 106 t £150
Rouen, 129 [illegible]
Constantinople, 200 brls. 6[illegible]
Alexandria, 100 [illegible]
New York, 10 cks. 15
" 4 20
Antwerp, 123 t. 178
Odessa, 950 brls. 247

Chemicals *(otherwise undescribed)*
Dunedin, 26 pks. £1[illegible]
Melbourne, 15 54

Disinfectants –
Calcutta, 105 cks. £67
Hong Kong, 75 drs. 54
Shanghai, [illegible] cks. 40
Colombo, 12 90
Jamaica, 81 pks. 38

Guano—
Dominica, 19 t. £429
Barbadoes, 20 200
Trinidad, 100 1,000

Manure—
Danzig, 9 t. 19 c. £55
St. Kitts, 3 31
Barbadoes, 26 276
Antigua, 36 360
Bordeaux, 21 10 150
Jamaica, 33 6 334
Barbadoes, 84 924
Guadeloupe, 49 5 550
Motril, 300 3,000
Malaga, 19 13 200
Dunedin, 100 1,350
Natal, 100 1,350
Demerara, 20 200
Barbadoes, 100 1,250
St. Lucia, 40 16 431
Barbadoes, 25 250
Jamaica, 120 bgs. 126
Salobrena, 194 1 1,649

Potassium Iodide—
Yokohama, 2½ c. £167

Potassium Chlorate—
Philadelphia, 5 t. £250

Sulphur Flowers—
E. London, 2 t. 4 c. £23

Sodium Sulphate—
Ghent, 65 t. £65

Sheep Dip—
E. London, 6 t. 2 c. £308

Sodium Silicate—
Lyttelton, 20 brls. £28

Sulphur—
Calcutta, 15 t. 18 c. £90
Adelaide, 25 204
Mauritius, 103 538

Saltpetre—
Melbourne, 1 t. £22
" 2 43

Soda Crystals –
Constantinople, 3 t. 7 c. £105
[illegible], 1 11 10

[illegible]ric Acid—
[illegible] 1760 lbs. £50
[illegible] 2000 12

[illegible] 11 c. £95

[illegible] [illegible]00 lbs. £71
[illegible] 2 cs. 4

LIVERPOOL.

Week ending May 31st.

Alum—
St. John, N.B., 2 t. 12 c. £13
Sydney, 3 15
Batoum, 10 58
Constantinople, 12 1 62
Galatz, 9 15 1 q. 50
Catania, 2 4 1 12
Calcutta, 52 2[illegible]

Alum Cake—
Ancona, 14 t. 7 c. £41

Ammonium Nitrate –
Port Elizabeth, 1 t. 12 c. 1 q.

Ammonium Carbonate –
Montreal, 1 t. [illegible] £28
Genoa, 9 15
Havana, 6 [illegible] q. 11
Barcelona, 1 4 30
Odessa, [illegible] 15

Ammonia—
Talcahuano, 1 [illegible] £5

Ammonium Muriate –
Bilbao, 6 t. 10 c. 1 q. £16[illegible]
Syra, 3 [illegible]
Antwerp, [illegible] [illegible]
New York, 1 1 7

Ammonium Sulphate—
Valencia, 107 t. 9 c. 2 q. £1,195
Malaga, 101 12 1,148
Nantes, 41 6 4[illegible]3

Antimony Regulus –
Montreal, [illegible] £[illegible]

Alum Sulphate—
Calcutta, 2 t. 1 q. £15

Borax—
Rouen, 57 t. 7 c. £1,462
Havana, 4 7
Norrkoping, 1 13 46

Brimstone—
Calcutta, 2[illegible] t. £1[illegible]

Borate of Lime –
Rotterdam, 10 t. 4 c. 2 q. £123
New York, 94 1,092

Bleaching Powder—
Antwerp, 30 cks. 7 brls.
Bordeaux, 29
Boston, 226
Calcutta, 70 cs.
Copenhagen, 55
Genoa, 30 [illegible]
Hamburg, 80
Leghorn, 153 [illegible]
Lisbon, [illegible]
Montreal, 22
New York, 154
Oporto, 7
Rouen 131
San Francisco, 190
Valparaiso, 79 dms.
Ancona, 120 cks.
Boston, 207
Genoa, 155
Ghent, 132
Hamburg, 27
Montreal, 22 tcs.
Rotterdam, 29
Venice, 40

Caustic Soda—
Amsterdam, 135 dms.
Ancona, 5
Antwerp, 17 26 brls.
Bahia, 5
Baltimore, 50
Barcelona, 25
Bari, 28
Batoum, 804
Bilbao, 40 17
Carthagena, 10
Constantinople, 31
Fairwater, 74
Genoa, 183
Ghent, 15
Hamburg, 10
Hong Kong, 28
Ibrail, 52
Leghorn, 61
Lisbon, 60
Malaga, 7
Marseilles, 20
Montreal, 50
Nantes, 104
Nagasaki, 50
Naples, 96
New York, 978
Oporto, 00
Philadelphia, 75 136
Rotterdam, 81
Rouen, 20

Caustic Soda—*continued.*
S. Petersburg, 1,726 dms.
San Francisco, 75
San Sebastian, 15 15 cks.
Santos, 20
Seville, 100
Talcahuano, [illegible]
Tarragona, 2[illegible] 30 brls.
Valparaiso, 1[illegible]
Venice, 7[illegible]
Algiers, 5
Algoa Bay, 23
Bari, [illegible]
Batoum, 100
Bilbao, 6
Carthagena, 55 25
Dunkirk, 18
E. London, 15
Genoa, 1[illegible]
Halifax, 16
Malaga, 126 13
Newfairwater, 2[illegible]
New Orleans, 50
New York, [illegible]
Rotterdam, 1[illegible]
Tunis, [illegible]
Yokohama, 10[illegible]

Calcium Chloride –
Calcutta, 2 t. 1 [illegible] £6

Citric Acid –
St. John's, N. F., 2 q. £[illegible]

Copperas –
Salaora, 1 t. 4 c. 2 q. £[illegible]

Charcoal—
Havre, 13 t. £47

Copper Sulphate—
Rouen, 91 t. 9 c. £1,[illegible]
Leghorn, 17 2 q. 1[illegible]
Genoa, 26 11 2 49[illegible]
Venice, 2[illegible] 2 2 411
Bordeaux, 50 5 1,165
Havre, 5 1 2 89
La Union, 2 3 70
Leghorn, 2 10 15
Venice, 10 16 2 [illegible]33
Genoa, 140 9 1 2,975
Ancona, 4 19 90
Antwerp, 4 19 1 95
Barcelona, 19 11 418
" 19 19 1 425
Bari, 6 110
Bilbao, 15 3 2 315
Bordeaux, 50 1 2 1,250
Leghorn, 27 4 525
Naples, 1 19 2 40
Tarragona, 4 19 110
Bilbao, 10 [illegible]40
Bordeaux, 22 14 521
Genoa, 45 11 1 1,[illegible]
Havre, 5 120
Monte Video, 2 4 51
Nantes, 44 16 1,105
Bordeaux, 59 12 1,334
Nantes, 10 200
Barcelona, 10 1 224
Bordeaux, 29 18 645
Genoa, 27 13 3 508
Leghorn, 5 2 2 112
Marseilles, 19 7 2 381

Carbolic Solid—
Hamburg, 2 c. 2 q. £10
St. Louis, 5 t. 2 465

Carbolic Acid—
New York, 3 t. £174
Paris, 1 4 c. 2 q. 120
Rotterdam, 59 13 7,253
New York, 4 10 450
Havre, 7 10 95
Madras, 1 12

Chemicals *(otherwise undescribed)*
Rio de Janeiro, 7 c. £13
Valparaiso, 6 35
Santos, 30 crbys. 34

Epsom Salts—
Bombay, 50 bgs.

Ground Spar—
Havre, 69 t 15 c. £500

Guano—
Bilbao, 1 t. 18 h. 2 q. £16

Gypsum—
New York, 10 t. 1 q. £30

Hypophosphate of Lime—
Calcutta, 1 cs. £23

Iodine—
Hamburg, 6 t. £1,600

Magnesia—
Manila, 2 cs.
Rio Janeiro, 5
Porto Alegre, 11

Manure—
St. Nazaire, 130 t. 1 c. £383
Barbadoes, 5 60
Gr. Canary, 24 180

Manure—*continued.*
Havre, 55 t. 4 c. 1 q.
Antwerp, 5 1
Gr. Canary, 52 15
Hamburg, 51

Oxalic Acid—
Malta, 6 c.

Potassium Bichromate—
Antwerp, 16 c. 1 q
Rio de Janeiro, 3

Phosphorus –
Vera Cruz, 12 c.
" 13 2 q.
Havana, 18
Kobe, 2

Pitch—
Montreal, 200 brls.

Potassium Carbonate—
Boston, 8 t. 15 c.

Potashes—
Monte Video, 12 c.

Potassium Chromate—
Havre, 1 t. 12 c.

Potassium Chlorate—
St. Petersburg, 12 t. 10 c.
Monte Video, 1
Genoa, 1 10
Hamburg, 1
Havana, 10
Bari, 1 10
" 12
Genoa, 2 10
Venice, 7
B. Ayres, 5
New York, 15
Pasages, 1 0
Philadelphia, 5
St. Louis, 3 16
Amsterdam, 1
Bari, 10
Genoa, 12
Naples, 2
New York, 12 10
Shanghai, 5

Potassium Oxalate—
Madras, 5 c. 1 q.

Sodium Bicarbonate—
Amsterdam, kgs.
Bremerhaven, 37 cks.
Canada, 300 10
Copenhagen, 20 50
Dunedin, 130
Kingston, 9 pks.
Konigsberg, 50
Kobe, 150
Libau, 90
Malaga, 20
Montreal, 600
Oran, 50
Randers, 30
Rotterdam, 4[illegible]
Rouen, 11[illegible]
Salonica, [illegible]
Sydney, 30
Yokohama, 7[illegible]
Alexandria, 50 brls.
Ancona, 10
Copenhagen, 113 cks.
Halifax, 200
Hamburg, 4[illegible]
Melbourne, 20
Stettin, 20
Valparaiso, 5[illegible]

Soda Ash—
Baltimore, 448 tcs.
Barcelona, 00 brls.
Bombay, 13
Boston, 199 640 pks. [illegible]
B. Ayres, 400
Calcutta, 7
Canada, 43
Ceara, 50
Corfu, 50
Genoa, 3
Havre, 42
Madras, 50 kgs.
Melbourne, 25
MonteVideo, 67 tcs.
N. York, 2,140
Oporto, 15
Philadelphia, 422 tcs.
Rio Grande du Sol, 50 brls.
Rio Janeiro, 230
Santos, 55
Sherbrooke, 16 tcs.
Wellington, 40
Baltimore, 150 cks. 40 tcs.
Boston, 76
Chicago, 67
Lisbon, 21
New York, 35
Philadelphia, 204
Rio Janeiro, 50 brls.
San Francisco 73

Ultramarine—
Holland, 4 cks. Ohlenschlager Bros.
" 20 Hernu Peron & Co.
Belgium, 5 Leach & Co.
Koln, 17 G. Steinhoft
Holland, 30 cs. "
" 13 Phillipps & Graves

Zinc Oxide—
Germany, 50 cks. J. Matton
Holland, 100 brls. M. Ashby Ld.
Germany 50 "

LIVERPOOL.

Week ending June 3rd.

Acetic Acid—
Rouen, 2 cks. R. J. Francis & Co.

Bones—
Monte Video, 450 bgs. A. Horny & Co.
" 700 "

Bone Meal—
Bombay, 323 bgs. A. V. Chutsey & Co.

Boracic Acid—
Leghorn, 42 cks.

Barytes—
Antwerp, 10 cks.

Borax—
Leghorn, 120 cs.

Cream of Tartar—
Barcelona, 5 hds. Sachse & Klemn
" 10 cks.

Copper Precipitate—
Huelva, 72 t.

Chemicals (*otherwise undescribed*)
Rouen, 19 cks. R. J. Francis & Co.

Dyestuffs—
Myrabolams
Bombay, 168 bgs. D. Sassoon & Co.
" 2107
Sumac
Palermo, 700 bgs.
" 600 260 bls.
Orchella
Lisbon, 100 bls.
" 12 A. Barbosa & Co.
Yellow Berries
Smyrna, 60 bgs.
Valonia
Smyrna, 208 t.
" 20 bgs. C. M. Elliadi
" 415 t. Barry Bros.
" 100 114 bgs.
Rennet
Rotterdam, 1 ck.
Extracts
Rouen, 2 cs. R. J. Francis & Co.
Philadelphia, 25 brls.
Logwood
Savanna la Mar, 462 t. 11 c.

Horn Piths—
Monte Video, 90 t. R. B. Watson & Co.

Manganese—
Odessa, 392 t.
Poti, 1850
Antwerp, 1 ck.

Manganese Ore—
Carrizal Bajo, 510 t.
Coquimbo, 490
Poti, 1850

Phosphate—
Montreal, 250 t.

Pyrites—
Huelva, 1470 t. Matheson & Co.

Permanganate—
Rotterdam, 5 cks.

Saltpetre—
Antwerp, 135 bgs. J. T. Fletcher & Co.

Stearine—
Rotterdam, 30 bgs.
Antwerp, 52

Sulphur—
Catania, 60 bgs. 6 cks.
Palermo, 160

Tartar—
Bari, 2 cks.
Naples, 2

Waste Salt—
Hamburg, 500 bgs. and a quantity.

Zinc Oxide—
Rotterdam, 100 cks. O. & H. Marcus
Antwerp, 85 brls.
" 135

Zinc Ashes—
Boston, 24 brls.

GRIMSBY.

Week ending May 31st.

Chemicals (*otherwise undescribed*)
Antwerp, 4 cks.
Rotterdam, 10 pks.

Glucose—
Hamburg, 200 bgs. 12 cks

Manure—
Kainit
Hamburg, 500 pks. 500 bgs.

Zinc—
Antwerp, 25 cks.

GOOLE.

Week ending May 28th.

Antimony—
Boulogne, 2 cks. 1 cs.

Caoutchouc—
Boulogne, 14 cks.

Dyestuffs—
Extracts
Rotterdam, 6 pks.
Alizarine
Rotterdam, 8 pks.
Logwood
Monte Christi, 366 t.
Dyestuffs (*otherwise undescribed*)
Boulogne, 60 cks. 8 cs. 2 crbys. 2 pks.

Glucose—
Hamburg, 33 cks.
" 25
Dunkirk, 100 bgs.

Muriate—
Rotterdam, 10 cks.

Potash—
Hamburg, 12 cks.
Calais, 54
Dunkirk, 6 21 drms.

Saltpetre—
Rotterdam, 100 bgs.

Waste Salt—
Hamburg, 650 cks. 320 t.

Zinc Oxide—
Antwerp, 20 brls.

HULL.

Week ending June 5th.

Arsenic—
Hamburg, 3 cks. C. M. Lofthouse & Co.

Acetate of Lime—
Helsingfors, 2 bgs. J. Good & Sons

Barytes—
Antwerp, 29 cks. H. F. Pudsey
" 19 Bailey & Leetham
" 33 H. F. Pudsey
Bremen, 23 Tudor & Son
" 66

Brimstone—
Catania, 878 pks. Wilson, Sons & Co.

Bone Meal—
Bombay, 862 bgs.

Chemicals (*otherwise undescribed*)
Antwerp, 10 cks. Wilson, Sons & Co.
Rotterdam, 2 cs. W. & C. L. Ringrose
Bremen, 1 kg. 2 cs. 1 ck. Veltmann & Co.
" 1 "
Bari, 35 cks. Wilson, Sons & Co.

Dyestuffs—
Logwood
Boston, 200 bxs. Wilson, Sons & Co.
Extracts
Rouen, 7 cks. Rawson & Robinson
Trieste, 20 brls. Wilson, Sons & Co.
Fiume, 245
Sumac
Trieste, 30 bls. Wilson, Sons & Co.
Palermo, 625 bgs. "
" 200
Valonia
Trieste, 86 bgs. Wilson, Sons & Co.
Madders
Rotterdam, 2 cks. Hutchinson & Son
Alizarine
Rotterdam, 5 cks. W. & C. L. Ringrose
" 103 Hutchinson & Son
" 3 cs. 14 cks. W. & C. L. Ringrose
" 8 cks.
Myrabolams
Bombay, 8050 bgs. Wilson, Sons & Co.

Glucose—
New York, 25 brls.
Hamburg, 100 bgs.
" 25 cks. Bailey & Leetham
Stettin, 1200 bgs. C. M. Lofthouse & Co.
" 400 37 cks.
Hamburg, 15 cks. Rawson & Robinson
Rotterdam, 40 bgs. Hutchinson & Son

Glycerine—
Hamburg, 10 cs. Lofthouse & Saltmer

Pitch—
Dunkirk, 20 cks. Wilson, Sons & Co.
Rouen, 23

Phosphorus—
Rouen, 150 cs. Wilson, Sons & Co.

Pyrites—
Huelva, 547 t. J. Dalton, Holmes & Co.
" 300

Sodium Nitrate—
Hamburg, 808 bgs. Wilson, Sons & Co.

Sulphur—
Catania, 25 t.

Turpentine—
Libau, 139 brl.
Charleston, 200

Tartaric Acid—
Rotterdam, 7 cks.

GLASGOW

Week ending June 5t

Alum—
Rotterdam, 15 cks. J. Rankir

Barytes—
Rotterdam, 46 cks.

Dyestuffs—
Extracts
Nantes, 75 cks.
New York, 6 brls.
Dunkirk, 13 cks.
Antwerp, 4
Alizarine
Rotterdam, 205 cks.
" 57

Glycerine—
Hamburg, 20 drms.

Paraffin Wax—
New York, 314 brls.

Pyrites—
Huelva, 1,646 t. Th

Sodium Silicate—
Rotterdam, 5 cks. J. Ranki

Tartaric Acid—
Rotterdam, 2 cks J. Ranki

Tartar—
Oporto, 8 cks.

Tartar Emetic—
Hamburg, 7 cks.

Ultramarine—
Rotterdam, 3 cs. J. Ranki

Waste Salt—
Hamburg, 296 bgs. J. Poynt
" 100 t. 508 bgs.

Zinc Oxide—
Antwerp, 75 cks.

TYNE.

Week ending June 3r

Barytes—
Rotterdam, 86 cks. Tyne

Copper Precipitate—
Huelva, 5,981 bgs. S

Glycerine—
Hamburg, 3 drums Tyne

Glucose—
New York, 49 brls. C

Paraffin Wax—
New York, 175 brls. C

Pyrites—
Huelva, 1,295 t. S

Saltpetre—
Hamburg, 7 cks. Tyne

Stearine—
New York, 25 hgds. C

Zinc Oxide—
Antwerp, 50 brls. Tyne

EXPORTS OF CHEMICAL PRODUCTS

OF

THE PRINCIPAL PORTS OF THE UNITED KINGDOM.

LONDON.

Week ending May 28th.

Acids (*otherwise undescribed*)
St. Petersburg, 100 cs. £250
Cologne, 400 pks. 636

Ammonium Sulphate—
Hamburg, 50 t. 14 c. £570
Rotterdam, 60 720
" 30 10 348
Cologne, 40 10 455
Demerara, 50 3 615

Anthracene—
Dusseldorf, £348

Arsenic—
Wellington, 2 t. £26
Yokohama, 1 10 c. 22

Alum Cake—
Calcutta, 250 cks. £171

Acetic Acid—
Melbourne, 3 t. 6 c. £82

Ammonium Carbonate—
Yokohama, 1 t. £38
Bremen, 56 pks. 157

Bone Ash—
Mauritius, 300 t. £1,354

Citric Acid—
Boston, 1 t. £135
Smyrna, 1 kg. 10
Odessa, 4 29
Hamburg, 2 c. 14
Brussels, 3 23
Yokohama, 10 68
Genoa, 10 74

Citric Acid—*continued.*
Leghorn, 10 c. 74
Naples, 2 15
Marseilles, 2 15
Melbourne, ½ 4
Seville, 15 111
Philadelphia, 10½ c. 74
M. Video, 2 16
B. Ayres, 4 32
St. Petersburg, 15 111

Cream of Tartar—
Dunedin, 1 t. £101
Adelaide, 10 c. 56
Rockhampton, 1 10 162
Sydney, 1 19 215
Melbourne, 5 24
M. Video, 4 44
Adelaide, 10 56
B. Ayres, 2 22

Carbolic Acid—
Genoa, 9 c.
Hawkes Bay, 50 drs.

Carbolic Solid—
Genoa, 18 c.
New York, 555 lbs.

Caustic Soda—
Sydney, 3 t. 2 c.
Bilbao, 13 13
Lyttleton, 15 4
Adelaide, 30

Copper Sulphate—
Marseilles, 2 t. 4 c.
Bordeaux, 25
Sydney, 2

PRICES CURRENT.

WEDNESDAY, June 11, 1890.

Prepared by Higginbottom and Co., 116, Portland Street, Manchester.

lues stated are F.O.R. *at maker's works, or at usual ports of shipment in U.K. The price in different localities may vary.*

		£	s.	d.
c, 25 °/₀ and 40 °/₀	per cwt.	7/- & 12/6		
Glacial	,,		65/-	
ic, S.G., 2000°	,,	0	12	9
nic 82 °/₀ .. nett	per lb.	0	0	6½
ic	,,	0	0	3¼
atic (Tower Salts), 30° Tw.	per bottle	0	0	8
(Cylinder), 30° Tw.	,,	0	2	11
80° Tw.	per lb.	0	0	2
us	,,	0	0	1¼
ic .. nett	,,	0	0	3¼
	,,	0	1	1½
huric (fuming 50 °/₀)	per ton	15	10	0
,, (monohydrate)	,,	5	10	0
,, (Pyrites, 168°)	,,	3	2	6
,, (,, 150°)	,,	1	4	6
,, (free from arsenic, 140/145°)	,,	1	7	6
hurous (solution)	,,	2	15	0
ic (pure)	per lb.	0	1	0½
aric	,,	0	1	3
, white powdered	per ton	13	0	0
loose lump)	,,	4	17	6
a Sulphate (pure)	,,	5	0	0
,, (medium qualities)	,,	4	10	0
ium	per lb.	0	15	0
nia, ·880=28°	per lb.	0	0	3
=24°	,,	0	0	1¼
Carbonate .. nett	,,	0	0	3½
Muriate	per ton	23	0	0
,, (sal-ammoniac) 1st & 2nd	per cwt.	37/- & 35/-		
Nitrate	per ton	40	0	0
Phosphate	per lb.	0	0	10½
Sulphate (grey), London	per ton	11	6	3
,, (grey), Hull	,,	11	5	0
Oil (pure)	per lb.	0	0	10½
Salt ,,	,,	0	0	9
ony	per ton	75	0	0
(tartar emetic)	per lb.	0	1	1
(golden sulphide)	,,	0	0	10
Chloride	per ton	8	5	0
Carbonate (native)	,,	3	12	6
Sulphate (native levigated)	,,	45/- to 75/-		
ing Powder, 35 %	,,	4	17	6
Liquor, 7 %	,,	2	10	0
ide of Carbon	,,	13	0	0
ium Acetate (crystal	per lb.	0	0	6
n Chloride	per ton	2	2	6
Clay (at Runcorn) in bulk	,,	17/6 to 35/-		
ar Dyes:—				
rine (artificial) 20 %	per lb.	0	0	9
enta	,,	0	3	9
Tar Products				
racene, 30 % A, f.o.b. London	per unit per cwt.	0	1	5
ol, 90 % nominal	per gallon	0	3	10
50/90 ,,	,,	0	2	10
olic Acid (crystallised 35°)	per lb.	0	0	7½
,, (crude 60°)	per gallon	0	2	2
sote (ordinary)	,,	0	0	2½
,, (filtered for Lucigen light)	,,	0	0	3
e Naphtha 30 % @ 120° C.	,,	0	1	4
se Oils, 22° Tw.	per ton	3	0	0
, f.o.b. Liverpool or Garston	,,	1	8	0
ent Naphtha, 90 % at 160°	per gallon	0	1	9
ven Oils (crude)	,,	0	0	2½
(Chili Bars)	per ton	58	5	0
Sulphate	,,	24	10	0
Oxide (copper scales)	,,	57	10	0
ne (crude)	,,	30	0	0
(distilled S.G. 1250°)	,,	60	0	0
	nett, per oz.	0	0	9
lphate (copperas)	per ton	1	10	0
lphide (antimony slag)	,,	2	0	0
sheet) for cash	,,	14	10	0
Litharge Flake (ex ship)	,,	15	10	0
Acetate (white ,,)	,,	24	0	0
,, (brown ,,)	,,	18	0	0
Carbonate (white lead) pure	,,	19	0	0
Nitrate	,,	22	0	0

		£	s.	d.
Lime, Acetate (brown)	,,	7	10	0
,, ,, (grey)	per ton	13	0	0
Magnesium (ribbon and wire)	per oz.	0	2	3
,, Chloride (ex ship)	per ton	2	5	0
,, Carbonate	per cwt.	1	17	6
,, Hydrate	,,	0	10	0
,, Sulphate (Epsom Salts)	per ton	3	0	0
Manganese, Sulphate	,,	17	10	0
,, Borate (1st and 2nd)	per cwt.	60/- & 42/6		
,, Ore, 70 %	per ton	4	10	0
Methylated Spirit, 61° O.P.	per gallon	0	2	2
Naphtha (Wood), Solvent	,,	0	3	11
,, ,, Miscible, 60° O.P.	,,	0	4	4½
Oils:—				
Cotton-seed	per ton	22	10	0
Linseed	,,	26	0	0
Lubricating, Scotch, 890°—895°	,,	7	5	0
Petroleum, Russian	per gallon	0	0	5⅛
,, American	,,	0	0	5¼
Potassium (metal)	per oz.	0	3	10
,, Bichromate	per lb.	0	0	4
,, Carbonate, 90% (ex ship)	per ton	19	10	0
,, Chlorate	per lb.	0	0	4½
,, Cyanide, 98%	,,	0	2	0
,, Hydrate (Caustic Potash) 80/85 %	per ton	22	10	0
,, ,, (Caustic Potash) 75/80 %	,,	21	10	0
,, ,, (Caustic Potash) 70/75 %	,,	20	15	0
,, Nitrate (refined)	per cwt.	1	2	6
,, Permanganate	,,	4	2	6
Prussiate Yellow	per lb.	0	0	9½
,, Sulphate, 90 %	per ton	9	10	0
,, Muriate, 80 %	,,	7	15	0
Silver (metal)	per oz.	0	4	0¼
,, Nitrate	,,	0	2	9
Sodium (metal)	per lb.	0	4	0
,, Carb. (refined Soda-ash) 48 %	per ton	6	2	6
,, ,, (Caustic Soda-ash) 48 %	,,	5	2	6
,, ,, (Carb. Soda-ash) 48%	,,	5	7	6
,, ,, (Carb. Soda-ash) 58 %	,,	6	10	0
,, ,, (Soda Crystals)	,,	2	17	6
,, Acetate (ex-ship)	,,	16	0	0
,, Arseniate, 45 %	,,	11	0	0
,, Chlorate	per lb.	0	0	6¼
,, Borate (Borax)	nett, per ton.	29	0	0
,, Bichromate	per lb.	0	0	3
,, Hydrate (77 % Caustic Soda) (f.o.b.)	per ton.	9	15	0
,, ,, (74 % Caustic Soda) ,,	,,	9	0	0
,, ,, (70 % Caustic Soda) ,,	,,	8	5	0
,, ,, (60 % Caustic Soda, white) (f.o.b.)	,,	7	5	0
,, ,, (60 % Caustic Soda, cream) ,,	,,	7	2	6
,, Bicarbonate	,,	5	15	0
,, Hyposulphite	,,	5	10	0
,, Manganate, 25%	,,	9		0
,, Nitrate (95 %) ex-ship Liverpool	per cwt.	0	8	0
,, Nitrite, 98 %	per ton	28	0	0
,, Phosphate	,,	15	15	0
,, Prussiate	per lb.	0	0	7¼
,, Silicate (glass)	per ton	5	7	6
,, ,, (liquid, 100° Tw.)	,,	3	17	6
,, Stannate, 40 %	per cwt.	2	2	6
,, Sulphate (Salt-cake)	per ton.	1	7	6
,, ,, (Glauber's Salts)	,,	1	10	0
,, Sulphide	,,	7	15	0
,, Sulphite	,,	5	5	0
Strontium Hydrate, 98 %	,,	8	0	0
Sulphocyanide Ammonium, 95 %	per lb.	0	0	7¼
,, Barium, 95 %	,	0	0	5¼
,, Potassium	,,	0	0	8
Sulphur (Flowers)	per ton.	7	10	0
,, (Roll Brimstone)	,,	6	0	0
,, Brimstone: Best Quality	,,	4	7	6
Superphosphate of Lime (26 %)	,,	2	12	6
Tallow	,,	25	10	0
Tin (English Ingots)	,,	100	0	0
,, Crystals	per lb.	0	0	6¾
Zinc (Spelter)	per ton.	23	0	0
,, Chloride (solution, 96° Tw.	,,	6	0	0

Salt—
Ambriz, 25 t.
Amsterdam, 100
Antwerp, 220
Barbadoes, 66½
Berufgord, 198
Black Point, 5
Boston, 385
Calcutta, 8,071
Cameroons, 40
Conakry, 50
Coquimbo, 377½
Gfle, 286
Ghent, 205
Halifax, 1,376
Hamburg, 130
Kingston, 8¾
Landana, 8
Melbourne, 200
Montreal, 1,939
Newfahrwaster, 463
Norrkoping, 291
Ofjord, 208
Old Calabar, 100½
Abonnema, 20
Montreal, 100
New York, 660
Para, 107¾
" 196 t. 8 c.
Quebec, 500
St. John N.B., 1,104
Shediac, 300 1c
Valparaiso, 91
Montego Bay, 346

Sodium Silicate—
Bari, 30 brls.
New York, 10

Soda Crystals—
Canada, 100 kgs.
Constantinople, 107 brls.
Coquimbo, 50
Malta, 50
Montreal, 50
Talcahuano, 20
Toronto, 100 kgs.
Valparaiso, 50
Boston, 280
Halifax, 40

Sodium Bichromate—
Genoa, 8 c. £11
Leghorn, 2 t. 10 70

Sodium Chlorate—
Stettin, 9 c. £25

Strontia Sulphate—
Valencia, 8 c. £20

Saltcake—
Baltimore, 202 t. 5 c. £462

Saltpetre—
Pernambuco, 5 t. 14 c. 1 q. £129
Maceio, 10 11
" 1 10 32
Pernambuco, 2 13 2 40
San Francisco, 5 115
Piræus, 3 15 82
Barbadoes, 19 2 14

Salammoniac—
Beyrout, 1 t. 19 c. £67
Malta, 3 5
Alexandria, 1 34
Bombay, 5 4 185
Montreal, 10 1 q. 18
Constantinople, 2 t. 68
Salonica, 1 34

Sheep Dip—
Algoa Bay, 4 t. 7 c. £175
Baltimore, 2 1 2 q. 50
E. London, 3 14 130
Montreal, 5 cks. 23
E. London, 5 18 200

Sodium Biborate—
Rotterdam, 4 t. 13 c. 1 q. £140
Hamburg, 16 3 482
Antwerp, 3 11 108
Hamburg, 10 4 2 307

Soda Salts—
Montreal, 3 t. £76

Sulphur—
Montreal, 25 t. 1 c. 1 q. £141
Nantes, 19 1 3 416
Porto Alegre, 1 3 2 8
Montreal, 4 31
" 4 2 1 27
New York, 50 2 220
Pernambuco, 1 1 2 7

Sodium Nitrate—
Demerara, 15 t. 9 c. 1 q. £126
Lisbon, 50 2 2 407
New York, 1 2 30
Ancona, 1 6 11

Sulphuric Acid—
Monte Video, 1 t. 10 c. £18

Tar—
Leghorn, 1 ck.
Rotterdam, 93 t. 1 c.
Quebec, 6 brls.
Sierra Leone, 20 dms.

Tartaric Acid—
Havana, 1 c. £8

Zinc Chloride—
New York, 2 t. 11 c. £51

TYNE.

Week ending June 3rd.

Antimony—
Hamburg, 10 t. 10 c.

Alkali—
Rotterdam, 3 t. 5 c.
New York, 116 16
Calmar, 28 14
Amsterdam, 2 10
Riga, 6 15
Horsens, 1 4
Helsingfors, 2 13

Alum—
Odense, 5 t. 7 c.

Ammonium Carbonate—
Rotterdam, 2 t. 4 c.
Barcelona, 11
Randers, 17

Ammonium Sulphate—
Rotterdam, 35 t.
Rouen, 10 2 c.
Valencia, 30

Barytes Carbonate—
Rouen, 25 t. 15 c.

Barium Chloride—
Hango, 1 t. 2 c.

Bleaching Powder—
Malmo, 10 t. 3 c.
Rotterdam, 13 14
Drontheim, 14
Copenhagen, 11 3
Liban, 4 5
Gothenburg, 16 15
New York, 185 9
Rouen, 5 10
Amsterdam, 44 10
Bergen, 5
Riga, 49 18
San Francisco, 106
Antwerp, 18 6
Hamburg, 35 5
Barcelona, 1 3
Gothenburg, 14
Christiania, 2 5
Horsens, 1 4
Odense, 3 12
Helsingfors, 30 4
Lisbon, 5 5

Barytes—
Hamburg, 1 t.

Baryta Nitrate—
Barcelona, 19 c.

Copperas—
Hango, 1 t. 1 c.

Caustic Soda—
Copenhagen, 2 t. 6 c.
New York, 252 9
Riga, 35 9
San Francisco, 14 10
Antwerp, 9 9
Hamburg, 10 19
Barcelona, 15
Horsens, 1 9
Abo, 2

Copperas—
Bergen, 1 t. 7 c.

Litharge—
Copenhagen, 1 t. 15 c.
Rouen, 3 2
Riga, 2 15
Barcelona, 10

Magnesia—
Libau, 1 c.
Hamburg, 10
Barcelona, 9 t. 12

Naphthaline—
Riga, 25 t. 9 c.
Antwerp, 19 19

Plumbago—
Bilbao, 12 c.

Potassium Chlorate—
Malmo, 1 t. 8 c.
Rotterdam, 3 16
Antwerp, 1 2
Hamburg, 1 10
Horsens, 1
Wyburg, 1 10

Pitch—
Antwerp, 100 t.

Paraffin Wax—
Barcelona, 8 t. 3 c.

Saltpetre—
Copenhagen, 1 t.

Sodium Bicarbonate—
Odense, 7 t. 19 c.

Soda—
Malmo, 5 t. 13 c.
Rotterdam, 11 17
Nakskov, 19 19
Drontheim, 11 11
Odense, 13 3
Aalborg, 7 10
New York, 100
Amsterdam, 11 17
Aarhuus, 7 2
Randers, 22
Christiania, 5 6
Horsens, 19 8
Odense, 21 1
Abo, 3 15
Hango, 1 7
Wyburg, 11

Superphosphate—
Riga, 220 t. 1 c.

Soda Ash—
Rotterdam, 6 t. 13 c.
Copenhagen, 30 8
Libau, 40 5
Gothenburg, 26 12
Aalborg, 3 6
New York, 129 2
Calmar, 76 16
San Francisco, 25
Antwerp, 2 13
Gothenburg, 25
Christiania, 2 10
" 2 9
Odense, 17 5
Hango, 3 10
Lisbon, 40 1

Sodium—
Hamburg, 1 box

Sodium Hyposulphate—
Barcelona, 12 c.

Sulphuric Acid—
Hango, 11 c.

Sodium Sulphate—
Riga, 12 t. 7 c.
Hango, 1 6

Zinc Sulphate—
Barcelona, 18 c.

HULL.

Week ending June 3rd.

Alum—
San Francisco, 94 cks.

Bleaching Powder—
Stettin, 316 cks.

Borax—
Libau, 1 ck.
Riga, 2

Bone Size—
Bremen, 10 cks.
Harlingen, 10

Caustic Soda—
Libau, 170 drums.
Marseilles, 120
Nicolaieff, 108
Riga, 528

Chemicals *(otherwise undescribed)*
Amsterdam, 30 cks.
Antwerp, 9 74 pks.
Alexandria, 12 cs.
Boston, 100
Copenhagen, 4 21
Danzig, 87 18
Guernsey, 20
Genoa, 30
Gothenburg, 5 12
Hamburg, 9 68 116
Konigsberg, 25
Libau, 98
Lisbon, 4
Marseilles, 4 14 25
New York, 65 6
Odessa, 235 4
Riga, 8 300
Rotterdam, 32 87 21
Stettin, 3 96
Stockholm, 1
St. Petersburg, 790 cks.

Dyestuffs *(otherwise undescribed)*
Amsterdam, 2 cs.
Hamburg, 4 cks.
Rotterdam, 5 cks. 3 cs. 1 drm. 24 bgs.
Rouen, 1 cs.
St. Petersburg, 56 cks.

Manure—
Hamburg, 306 bgs.
Rotterdam, 30 t.
Stettin, 461 bgs.

Pitch—
Antwerp, 473 t.

Tar—
Christiania, 5 cks.
Genoa, 8
Hamburg, 8
Stavanger, 3
Rotterdam, 7

Turpentine—
Stettin, 15 cks.

GLASGOW

Week ending June

Ammonium Carbonate—
Christiania, 3 cks. 13 kgs.

Alkali—
Little Bay,

Copper Sulphate—
Nantes, 20 t. 12¾ c.

Copperas—
Karachi, 15 t. 7½ c.

Epsom Salts—
Calcutta, 4 t. 15¼ c.
Karachi, 5

Glycerine—
Rouen, 7 t. 5 c.

Litharge—
Karachi, 31¼ c.

Muriatic Acid—
Karachi, 21 t. 3 c.

Naphtha—
Rouen, 90 cks.

Potassium Prussiate—
New York, 29 c.

Potassium Bichromate—
Odessa, 14 c.
Rotterdam, 34 cks.
Karachi, 2 t. 12½ c.
Bergen, 1 ck.
Christiania, 1
New York, 208
Rouen, 17 t. 15¾
Marseilles, 1 3¾

Pitch—
Gothenburg, 232 t. 17 c.
Amsterdam, 250
Barbadoes, 131

Paraffin Wax—
Nantes, 2420 lbs.
Sydney, 203 c.

Sulphuric Acid—
Bombay, 225 c.

Sodium Bichromate—
Karachi, 1 t. 19¼ c.

Salt—
Brisbane, 163 t.

Tar—
Karachi,
Little Bay,

GOOLE.

Week ending May 2

Benzole—
Antwerp, 52 cks.
Rotterdam, 25 18 drms

Bleaching Preparation
Antwerp, 37 cks.

Coal Products—
Ghent, 2 cs.
Rotterdam, 30 cks.

Chemicals *(otherwise un*
Rotterdam, 5 cks.
Rouen, 50 pks.
Ghent, 41

Dyestuffs *(otherwise un*
Boulogne, 5 cks.
Ghent, 1 3 drms
Rotterdam, 1
" 220 bgs.
Hamburg, 66

Manure—
Rotterdam, 296 bgs.

Pitch—
Antwerp, 98 t.

GRIMSB

Week ending May

Coal Products—
Dieppe, 87 cks.

Chemicals *(otherwise un*
Dieppe, 4 cs. 1 ck.
Hamburg, 10
Rotterdam, 26

PRICES CURRENT.

WEDNESDAY, June 11, 1890.

Prepared by Higginbottom and Co., 116, Portland Street, Manchester.

alues stated are F.O.R. at maker's works, or at usual ports of shipment in U.K. The price in different localities may vary.

Article	Unit	£	s.	d.
——				
ic, 25 °/₀ and 40 °/₀	per cwt.	7/- & 12/6		
Glacial	,,	65/-		
nic, S.G., 2000°	,,	0	12	9
mic 82 °/₀	nett per lb.	0	0	6½
ric	,,	0	0	3¾
iatic (Tower Salts), 30° Tw.	per bottle	0	0	8
(Cylinder), 30° Tw.	,,	0	2	11
ic 80° Tw.	per lb.	0	0	2
ous	,,	0	0	1¾
lic	nett ,,	0	0	3¾
ic	,,	0	1	1½
phuric (fuming 50 °/₀)	per ton	15	10	0
,, (monohydrate)	,,	5	10	0
,, (Pyrites, 168°)	,,	3	2	6
,, (,, 150°)	,,	1	4	6
,, (free from arsenic, 140/145°)	,,	1	7	6
phurous (solution)	,,	2	15	0
nic (pure)	per lb.	0	1	0½
taric	,,	0	1	3
ic, white powdered	per ton	13	0	0
(loose lump)	,,	4	17	6
na Sulphate (pure)	,,	5	0	0
,, (medium qualities)	,,	4	10	0
inium	per lb.	0	15	0
onia, ·880=28°	per lb.	0	0	3
=24°	,,	0	0	1¾
Carbonate	nett ,,	0	0	3½
Muriate	per ton	23	0	0
,, (sal-ammoniac) 1st & 2nd	per cwt.	37/-& 35/-		
Nitrate	per ton	40	0	0
Phosphate	per lb.	0	0	10½
Sulphate (grey), London	per ton	11	6	3
,, (grey), Hull	,,	11	5	0
ne Oil (pure)	per lb.	0	0	10½
ne Salt ,,	,,	0	0	9
nony	per ton	75	0	0
,, (tartar emetic)	per lb.	0	1	1
,, (golden sulphide)	,,	0	0	10
m Chloride	per ton	8	5	0
,, Carbonate (native)	,,	3	12	6
,, Sulphate (native levigated)	,,	45/- to 75/-		
hing Powder, 35 %	,,	4	17	6
,, Liquor, 7 %	,,	2	10	0
phide of Carbon	,,	13	0	0
mium Acetate (crystal	per lb.	0	0	6
um Chloride	per ton	2	2	6
Clay (at Runcorn) in bulk	,,	17/6 to 35/-		
Tar Dyes:—				
zarine (artificial) 20 %	per lb.	0	0	9
genta	,,	0	3	9
Tar Products				
thracene, 30 % A, f.o.b. London	per unit per cwt.	0	1	5
nzol, 90 % nominal	per gallon	0	3	10
,, 50/90 ,,	,,	0	2	10
rbolic Acid (crystallised 35°)	per lb.	0	0	7½
,, (crude 60°)	per gallon	0	2	2
eosote (ordinary)	,,	0	0	2½
,, (filtered for Lucigen light)	,,	0	0	3
ide Naphtha 30 % @ 120° C.	,,	0	1	4
ease Oils, 22° Tw.	per ton	3	0	0
ch, f.o.b. Liverpool or Garston	,,	1	8	0
lvent Naphtha, 90 % at 160°	per gallon	0	1	9
-oven Oils (crude)	,,	0	0	2½
er (Chili Bars)	per ton	58	5	0
Sulphate	,,	24	10	0
Oxide (copper scales)	,,	57	10	0
rine (crude)	,,	30	0	0
,, (distilled S.G. 1250°)	,,	60	0	0
e	nett, per oz.	0	0	9
Sulphate (copperas)	per ton	1	10	0
Sulphide (antimony slag)	,,	2	0	0
(sheet) for cash	,,	14	10	0
Litharge Flake (ex ship)	,,	15	10	0
Acetate (white ,,)	,,	24	0	0
,, (brown ,,)	,,	18	0	0
Carbonate (white lead) pure	,,	19	0	0
Nitrate	,,	22	0	0

Article	Unit	£	s.	d.
Lime, Acetate (brown)	,,	7	10	0
,, ,, (grey)	per ton	13	0	0
Magnesium (ribbon and wire)	per oz.	0	2	3
,, Chloride (ex ship)	per ton	2	5	0
,, Carbonate	per cwt.	1	17	6
,, Hydrate	,,	0	10	0
,, Sulphate (Epsom Salts)	per ton	3	0	0
Manganese, Sulphate	,,	17	10	0
,, Borate (1st and 2nd)	per cwt.	60/- & 42/6		
,, Ore, 70 %	per ton	4	10	0
Methylated Spirit, 61° O.P.	per gallon	0	2	2
Naphtha (Wood), Solvent	,,	0	3	11
,, ,, Miscible, 60° O.P.	,,	0	4	4½
Oils:—				
Cotton-seed	per ton	22	10	0
Linseed	,,	26	0	0
Lubricating, Scotch, 890°—895°	,,	7	5	0
Petroleum, Russian	per gallon	0	0	5⅜
,, American	,,	0	0	5¾
Potassium (metal)	per oz.	0	3	10
,, Bichromate	per lb.	0	0	4
,, Carbonate, 90% (ex ship)	per ton	19	10	0
,, Chlorate	per lb.	0	0	4½
,, Cyanide, 98%	,,	0	2	0
,, Hydrate (Caustic Potash) 80/85 %	per ton	22	10	0
,, ,, (Caustic Potash) 75/80 %	,,	21	10	0
,, ,, (Caustic Potash) 70/75 %	,,	20	15	0
,, Nitrate (refined)	per cwt.	1	2	6
,, Permanganate	,,	4	2	6
Prussiate Yellow	per lb.	0	0	9½
,, Sulphate, 90 %	per ton	9	10	0
,, Muriate, 80 %	,,	7	15	0
Silver (metal)	per oz.	0	4	0¾
,, Nitrate	,,	0	2	9
Sodium (metal)	per lb.	0	4	0
,, Carb. (refined Soda-ash) 48 %	per ton	6	2	6
,, ,, (Caustic Soda-ash) 48 %	,,	5	2	6
,, ,, (Carb. Soda-ash) 48%	,,	5	7	6
,, ,, (Carb. Soda-ash) 58 %	,,	6	10	0
,, ,, (Soda Crystals)	,,	2	17	6
,, Acetate (ex-ship)	,,	16	0	0
,, Arseniate, 45 %	,,	11	0	0
,, Chlorate	per lb.	0	0	6¼
,, Borate (Borax)	nett, per ton.	29	0	0
,, Bichromate	per lb.	0	0	3
,, Hydrate (77 % Caustic Soda) (f.o.b.)	per ton.	9	15	0
,, ,, (74 % Caustic Soda) ,,	,,	9	0	0
,, ,, (70 % Caustic Soda) ,,	,,	8	5	0
,, ,, (60 % Caustic Soda, white) (f.o.b.)	,,	7	5	0
,, ,, (60 % Caustic Soda, cream) ,,	,,	7	2	6
,, Bicarbonate	,,	5	15	0
,, Hyposulphite	,,	5	10	0
,, Manganate, 25%	,,	9		0
,, Nitrate (95 %) ex-ship Liverpool	per cwt.	0	8	0
,, Nitrite, 98 %	per ton	28	0	0
,, Phosphate	,,	15	15	0
,, Prussiate	per lb.	0	0	7¾
,, Silicate (glass)	per ton	5	7	6
,, ,, (liquid, 100° Tw.)	,,	3	17	6
,, Stannate, 40 %	per cwt.	2	2	6
,, Sulphate (Salt-cake)	per ton.	1	7	6
,, ,, (Glauber's Salts)	,,	1	10	0
,, Sulphide	,,	7	15	0
,, Sulphite	,,	5	5	0
Strontium Hydrate, 98 %	,,	8	0	0
Sulphocyanide Ammonium, 95 %	per lb.	0	0	7¾
,, Barium, 95 %	,,	0	0	5¾
,, Potassium	,,	0	0	8
Sulphur (Flowers)	per ton.	7	10	0
,, (Roll Brimstone)	,,	6	0	0
,, Brimstone: Best Quality	,,	4	7	6
Superphosphate of Lime (26 %)	,,	2	12	6
Tallow	,,	25	10	0
Tin (English Ingots)	,,	100	0	0
,, Crystals	per lb.	0	0	6¾
Zinc (Spelter)	per ton.	23	0	0
,, Chloride (solution, 96° Tw.	,,	6	0	0

THE

CHEMICAL TRADE JOURNAL

Publishing Offices: 32, *BLACKFRIARS STREET, MANCHESTER.*

No. 161. **SATURDAY, JUNE 21, 1890.** **Vol.**

Contents.

Notices.

All communications for the *Chemical Trade Journal* should be addressed, and Cheques and Post Office Orders made payable to—

DAVIS BROS., 32, Blackfriars Street, MANCHESTER.

Our Registered Telegraphic Address is—

"Expert, Manchester."

The Terms of Subscription, commencing at any date, to the *Chemical Trade Journal,*—payable in advance,—including postage to any part of the world, are as follow:—

Yearly (52 numbers)	**12s. 6d.**
Half-Yearly (26 numbers)	**6s. 6d.**
Quarterly (13 numbers)	**3s. 6d.**

Readers will oblige by making their remittances for subscriptions by Postal or Post Office Order, crossed.

Communications for the Editor, if intended for insertion in the current week's issue, should reach the office not later than **Tuesday Morning.**

Articles, reports, and correspondence on all matters of interest to the Chemical and allied industries, home and foreign, are solicited. Correspondents should condense their matter as much as possible, write on one side only of the paper, and in all cases give their names and addresses, not necessarily for publication. Sketches should be sent on separate sheets.

We cannot undertake to return rejected manuscripts or drawings, unless accompanied by a stamped directed envelope.

Readers are invited to forward items of intelligence, or cuttings from local newspapers, of interest to the trades concerned.

As it is one of the special features of the *Chemical Trade Journal* to give the earliest information respecting new processes, improvements, inventions, etc., bearing upon the Chemical and allied industries, or which may be of interest to our readers, the Editor invites particulars of such—when in working order—from the originators; and if the subject is deemed of sufficient importance, an expert will visit and report upon the same in the columns of the *Journal.* There is no fee required for visits of this kind.

We shall esteem it a favour if any of our readers, in making inquiries of, or opening accounts with advertisers in this paper, will kindly mention the *Chemical Trade Journal* as the source of their information.

Advertisements intended for insertion in the current week's issue, should reach the office by **Wednesday morning** at the latest.

Advertisements.

Prepaid Advertisements of Situations Vacant, Premises on Sale or To be Let, Miscellaneous Wants, and Specific Articles for Sale by Private Contract, are inserted in the *Chemical Trade Journal* at the following rates:—

30 Words and under	2s. 0d.	per insertion.
Each additional 10 words	0s. 6d.	,,

Advertisements of Situations Wanted are inserted at one-half the above rates when prepaid, viz:—

30 Words and under	1s. 0d	per insertion.
Each additional 10 words	0s. 3d.	per insertion.

Trade Advertisements, Announcements in the Directory Columns, and all Advertisements not prepaid, are charged at the Tariff rates, which will be forwarded on application.

SACCHARINE.

THIS now famous substitute for sugar is only ten years old,
been discovered a decade since by Mr. Fahlberg, a N
chemist. It is extracted by a rather protracted process from
formed in the dry distillation of coal. In a pure condition it is
amorphous powder, the sweetening capacity of which is three
times that of the finest refined sugar. When the discovery
announced it was naturally received with incredulity by perso
regarded it as one of those fictions in which the United Stat
fruitful. When, however, through the publication of the pr
was possible to produce saccharine in any laboratory, the fac
invention had to be admitted, but it was believed that the prod
only value as a pharmaceutical preparation. The fallacy of th
was speedily demonstrated, for, favoured by its comparative c
and some other circumstances, it soon acquired an important
in industrial life. It readily found employment in the manufa
preserves, sweets, chocolate, liqueurs, and even in the produ
wine and beer, in the place of sugar which till then had been e
ly employed. It was only natural that the Fahlberg discover
stimulate other chemists to fresh experiments in the same fi
among the competing articles arose the so-called methyl sac
patented by the proprietors of the well-known tar manufa
Ludwigshafen-on-Rhine.

The growing popularity of saccharine was the means of attra
attention of a number of governments, who were mainly co
with the question as to the influence the new product would
the revenue. Several authorities also affirmed that the use of sa
was not altogether conducive to health. In this relation the r
the experiments of a number of scientists, undertaken at the
of a Comité Consultatif d'hygiéne, of Paris, are of much imp
It was established that saccharine restrained the assimilation b
tive organs of starch and albuminous-containing foods taken at
time, in consequence of its antiseptic properties, and that it w
no food stuff, as after being eaten it underwent no alteration
expelled from the system in its original condition.

In view of these circumstances, the Dutch delegates at th
national Congress for the Abolition of Sugar Bounties, held in
in 1888, were induced to recommend the adoption of a comn
of action on the part of the governments against saccharine.
has, however, been jointly accomplished, but a number of gove
have since adopted measures of more or less severity with the
restricting the consumption of saccharine in place of sugar.
from England, in which country an act had been pass
hibiting the employment of saccharine in the produc
excisable articles, beer in particular, the Portuguese Gov
was the first to adopt decisive measures. In a royal decree is
9th of August, 1888, the importation into Portugal of saccha
saccharine-containing products was entirely prohibited, excep
medical purposes, for which special permission had to be obta
each occasion. The importation of saccharine into France and
was prohibited by a decree of the President of the Republic d
1st of December, 1888. In a decree dated the 3rd of April 18
Spanish Government not only prohibited the importation of sa
into Spain, but also made its employment in food stuffs as a su
for sugar a penal offence. In Belgium, the law of the 21st
1889, imposed an entrance duty of 140 frcs. per kilo. on saccha
on all articles containing down to ½ per cent. saccharine. In H
accharine has for long had to pay an ad valorem duty of 5 p
as a drug, but it is probable that the long-entertained intentio
pose a duty of 60 florins per kilo, will be carried into effect

AUTOMATIC GAS RETORTS.

e ordinary process of gas manufacture the coal to be carbonised ed into horizontal retorts at one end, the red-hot coke being rds raked out from the other end. This entails severe manual and is attended by other disadvantages in gas-making. The nt evils, especially those affecting the workmen, have been, to ttent, mitigated by what are known as mechanical stokers and contrivances, but these have by no means come into general from the earliest times of gas-making it has heen seen that if nciple of gravitation could be utilised, the severity of the labour ging and discharging the retorts would be greatly reduced, important economy in gas manufacture would be effected. To d Clegg, one of our earliest gas engineers, proposed, in 1804, to retorts at a considerable angle with the horizon, so as to charge raw them quickly, and with a minimum of manual labour. followed in the same direction, but without success, the proper iot having been hit upon, nor proper means taken to insure the ng of the coal evenly over the bottom of the retort. Mr. Coze, nch gas engineer, has, however, been the first to succeed in ucing the principle of gravitation successfully into practice. e has been enabled to do by adopting the exact angle of slope at coal will rest when delivered into the retort, and at which vill slide out upon being subjected to the slightest disturbance. ystem has been in use for some time past with every success at a rks in France, and has been visited and favourably reported upon glish gas engineers, who have further testified to the great ny insured by the system.

Coze system was soon introduced into this country, and a full of retorts has been put up at the Brentford Gasworks at ll. The charging apparatus and the general arrangements, er, proved to be somewhat cumbersome, and the details were not to lend themselves successfully to English gasmaking e. The system has, therefore, been adapted to the requirements country by Messrs. Morris and Van Vestraut, two English gas ers, and a full setting of retorts on their system has also been at the Brentford works. In this system the retorts are set at le of about 30 deg. with the horizon, and are charged by means oveable shoot, which is made telescopic so as to enable it to each tier of retorts. At the lower end of the shoot is an adjust- uide, by means of which the velocity with which the coal enters ort is controlled. If the coal is charged at too high a velocity it sh to the bottom of the inclined retort, if at too low a velocity lodge in the upper part of the retort. The guide, however, can to the exact slope required, and this, combined with the angle retort, causes the charge of coal to be evenly distributed over tom of the retort. The coke slips out from the lower end of lined retort at the slightest touch, so that the system is practi- utomatic. In order to test the system a single retort was first at Southall, and this was inspected and its working reported y Mr. Perry F. Nursey, C.E., a short time since. In his report ursey says:—

ie working retort was of the ordinary ⊍ section, 20 ft. long by wide and 13 in. deep, and was set at an angle of about 30° with izon. I first saw this retort drawn. The coke in it was laid in ctly even bed from end to end, and it came out readily on being loosened with a light slice. I then saw the retort charged cwt. of Pelaw Main coal, the usual charge being 6 cwt. for ed retort; but, all the irregularities being filled up, a heavier can be used. The retort was charged in the remarkably short of 4½ seconds."

advantages of the new system may be summed up as obviating luous work of retort charging and drawing; economising the g, skilled labour not being required; saving the wear and tear , and effecting an economy in construction. Upon this latter may be observed that a retort-house for the new system need half the width of one on the old system for the same number rts. Moreover, in converting an existing retort-house to the stem, two benches of retorts can be erected in the place of the is single bench, as stage work and stage coal are dispensed Another important advantage is that, skilled labour being ed with, the new system will give gas companies great command e carbonizing department of their works in the event of future difficulties. The system of inclined retorts is being adopted, ly by the Brentford Gas Company, but by the South Metro- Gas Company, the Gas Light and Coke Company, and by the al provincial gas companies.

monstration of the charging and drawing of a full setting of etorts on the Morris and Van Vestraut system at the Southall of the Brentford Gas Company took place a few days since in the e of a number of gas engineers and other gentlemen. In order the test comparative, a bench of seven 20 ft.-through retorts ordinary horizontal system was first drawn and afterwards charged by ordinary trained stokers. There were three men at each end of the setting, and the time occupied in drawing the coke was 4½ minutes per retort. The time occupied in charging each retort with 6 cwt. of coal was 50 seconds. In the case of the inclined system there was also a setting of seven through retorts of the same length and section as in the previous case, the work here being performed not by regularly trained stokers, but by yardmen. With one man at each end of the setting, the drawing of the red-hot coke from each retort was effected in 1¼ minutes, and the charging of each with 6 cwt. of coal in 15 seconds, and this notwithstanding that the new system is not yet complete in all its working details, more being left to be done by manual labour than will be the case when the coaling arrangements have been completed.

ADULTERATION IN THE UNITED STATES.

REPORTS of officials charged with the duty of food inspection and analysis almost invariably show, says *Bradstreet's*, that the exercise of such functions under governmental authority continues to be a necessity. The report recently published of the State Dairy and Food Commissioner of Minnesota shows that adulteration is still practised to a considerable extent in the territory within his jurisdiction. The Commissioner found a considerable amount of adulteration in such articles as lard, vinegar, and baking powder. Cotton-seed oil is used as an adulterant of lard. In a large number of instances baking powders are adulterated with alum, and coloured low wine vinegar is palmed off as cider vinegar. The report of the analyst in charge of the Minneapolis laboratory showed a large amount of adulteration in such articles as coffee, ground mustard, vinegar, and milk. Of the samples examined by him, coffee showed the largest amount of adulteration, 100 per cent.; ground mustard came next, with 82 per cent.; then baking powders, with 47 per cent of adulteration, and milk, with 42 per cent. It should be remarked, by the way, that the samples of coffee and mustard examined were few in number. It is encouraging to know, however, that according to the Commissioner's observations adulteration in food products is of less frequent occurrence than it has been.

TRUSTS IN AMERICA.

HOW PRICES ARE KEPT HIGH.

MR. DONALD, in the *Contemporary Review*, has a remarkable article on American Trusts. In the course of it he says:—Another powerful combination is the great Sugar Trust. Sugar presented an excellent opportunity for the trust makers. It is protected by a duty which averages about eighty per cent., and a bounty is paid by the Government on all sugar exported. Sugar is one of the necessaries of life, and is used in every household. The sugar refiners discovered in 1887 that too much sugar was being manufactured, so they consolidated to reduce the supply and raise the price. The real value of the property "trusted" was 15,000,000 dols., but "trust" certificates were issued which "watered" it up to 60,000,000 dols. The trust first depressed the price of raw sugar, and then raised the price of cut loaf and crushed sugar by one and a-half cents per lb., and granulated sugar by one cent per lb. A rise of one cent per lb. on the sugar consumed in the United States would mean an increased profit of 30,000,000 dols. Strong opposition has been made to this trust, but it still holds its own. English people have nothing to complain of in this matter. They ought to appreciate the friendly attitude of the United States Government, as it helps to pay for their sugar.

When the Steel Combination pressed on the western plough manufacturers they in turn organised a trust, and squeezed the farmers, who are now contemplating a similar course to resist the pressure. A Steel Rail Combination has been in existence since 1877. It is not formed on trust lines, but serves the same purpose. The "iron lords" and "steel lords" are bound together by the closest ties of self-interest in the American Iron and Steel Association. The Association keeps the prices as high as the tariff will allow, and does all it can by the circulation of pamphlets, by employing "lobbyists," and by resorting to other well-known methods, to maintain a feeling in favour of the continuance of a protective tariff on iron and steel.

The principal manufacturers of American whisky got up "pools" now and then between 1878 and 1887 to arrange prices. The "pools" were not quite so successful as the distillers desired, and in 1887 they discovered that the hitch arose because there was too much whisky. This discovery was worthy of temperance reformers, but the object of the distillers was not to help forward the prohibition movement or the temperance cause. Nor was their ultimate aim the limitation of whisky-drinking. They only wanted to temporarily limit the supply. They organised the Western Distillers and Cattle Feeders' Trust—a compound sort of trust. On its formation seventy distillers joined it, and the price of whisky was at once raised from 30 to 40 per cent. Fifty-seven distilleries were closed, and the remaining thirteen left to make profits for the time being for all the shareholders. The owners of the distilleries which were lying idle, therefore, did not lose anything. The wages of the men still left at work were cut down from 10 to 20 per cent. But the trust had been too grasping, and competition began to reappear. New distilleries were opened, and as these had to be crushed or absorbed, down went the price of whisky—lower than it had ever been before—until they succumbed. The trust now controls more than half the distilleries in the country. It also fixes the price for "mash" used for feeding cattle—hence its double-barrelled name. The duty on alcohol is 171·85 per cent., and the duty on spirits distilled from grain—such as the trust makes—rises to 396·43 per cent.

Selected Correspondence.

SULPHATES IN NITRIC ACID.

BY JAMES H. HUXLEY IN *Chemical News*.

I have recently been a good deal perplexed by finding sulphuric acid in steel and pig iron, in which I knew it was not likely to be present, and on careful examination of the acids employed for solution, I found that the nitric acid pur., 1·42 s. g., contained a considerable per centage of sulphuric acid. I have used this same make of acid for many years without having any reason to suspect its purity, though, only occasionally, for sulphur estimations. The makers have been communicated with by the persons who supply me, and I understand their explanation is that the acid is quite pure as it leaves their factory, and that the sulphuric acid is derived from the bottles in which it is sent out. These are Winchesters of greenish colour, and new ones are always used. The acid of which I am speaking is part of a batch which has been in bottle since last November, and the makers are not surprised at its having picked up sulphuric acid from the bottles. No doubt there is sulphate of soda in the material of which the glass is made, but I find that I can obtain a pure re-distilled nitric acid in bluish-coloured Winchesters which is perfectly free from sulphuric acid, though it has been in bottle about the same length of time as the other. I don't know whether any of your readers have had similar trouble, but I think it may be of value to warn users of so-called pure nitric acid of the necessity of frequently testing it for sulphates, and I should be glad to have information as to the action of nitric acid upon glass bottles in which it is stored, as the present state of things seems to be rather anomalous; viz, that greenish glass Winchesters sometimes contain sulphates capable of being dissolved out by nitric acid, but that bluish glass Winchesters have not been known to contaminate nitric acid in this way; at least, so far as I can ascertain.

WHAT IS CAST STEEL?

BY SIR HENRY BESSEMER IN *The Engineer*.

Your correspondent, "Crucible," writing under the above heading, asks, "Will Sir Henry kindly state, for the benefit of the trade generally, the present meaning of the term 'cast steel?'"

In reply I would remark that there are known in commerce several different kinds of steel, each bearing a name which more or less accurately indicates its mode of manufacture or its specific character. Thus we have blister steel, shear steel, double shear, German or natural steel, and puddled steel, in all of which the bars are produced by the union or welding together of solid pieces, a process which gives to them a more or less laminated structure, and consequently an admixture of scoria or other extraneous matters between the different layers of which the bar is composed.

Huntsman's great invention consisted simply in fusing such imperfect steel bars in a crucible, and, while in a fluid state, casting the molten steel into an ingot, whereby the laminated character of the steel bar is lost, and its mechanically mixed scoria is separated from it. The cast ingot so produced forms a homogeneous mass, which is crystalline in structure and perfectly free from lamination, admixture with scoria, or other extraneous solid matter, all of which advantages are derived from the fluid condition of the metal, and its being cast or run into a mould, and hence such steel is distinguished by the very correct and obvious name of cast steel. There is in this great transformation no particular virtue in the crucible itself, which is only a means to an end; that end, and all the advantages flowing from it, is simply derived from the fluidity of the metal, and the casting therefrom of a homogeneous mass.

Your correspondent very naïvely remarks that "it would simplify matters if the generic titles of 'crucible' and 'Bessemer' were kept separate, and not allowed to become confused in any way by the connection of the term cast steel." However much it may be desired by some persons to draw a hard-and-fast line that shall separate Bessemer steel from crucible steel, it is impossible to do so, for they are indissolubly united in the same category by one great distinctive character which is common to each process, viz., the fluid condition products, and the formation of this fluid into an ingot or hom mass by the act of casting. Nor is this all, for Bessemer st crucible steel, may be either mild or highly carburetted at th the manufacturer; it may, like crucible steel, be alloyed du process with any other metal, and, like crucible steel, it ma good or bad quality, dependent chiefly on the quality of material put into the crucible or the converter.

And further, it may be said that Bessemer steel, like crucib is crystalline in structure, and not laminated; it breaks like steel with a beautiful conchoidal fracture, and may, like fluid steel, be made into a vast variety of useful articles by fou casting in suitable moulds; and, in fact, Bessemer steel is with crucible steel in its molecular structure, its chemical cons and in its physical properties, and is in all respects as true cas that made in the crucible.

INDIAN INDUSTRIAL PRODUCTS.

The following information respecting raw cotton, dyes and materials, and lac are extracted from a memorandum on inland trade compiled in the Revenue and Agricultural Depar the Government of India, and which has recently been issued Government central printing office at Simla:—

COTTON (RAW).—Though the production and exportation o have never attained the dimensions which they reached when the cotton famine due to the American civil war raised t of exports to 37 millions sterling, it is still one of the great s the Indian export trade.

Cotton enters into the agriculture of almost every part of In its cultivation is concentrated in the region, embracing roughly square miles, occupied by the basaltic formation known geolog the Deccan trap, the decomposition of which has created commonly designated black cotton soil. The region comp provinces of Bombay and Berar, and the Western half of the India States, the Nizam's dominions and the Central Pr Within it, and particularly in the Bombay Presidency an several exotic and hybrid varieties of cotton are grown; it co the largest share to the export trade, and it still offers an field for the expansion and improvement of cotton cultivation.

The area under cotton, as well as the out-turn in any year, to large fluctuations. It is a crop easily affected by unse rains, and the extent of the sowing depends upon the fall favourable, and to a less extent upon rotation. Including the which it is grown as a mixed crop (a common practice in agriculture), the present normal acreage under cotton estimated at 14 million acres, and the out-turn of 129 lakhs of m

The subjoined figures gave the details of this estimate; but i as they do native States and other tracts for which reliable are not available, they must be accepted as open to wide correc

(00000 omitted.)

Block.	Normal Area.	Normal Out-turn Cleaned Cotton.	A E
	Acres.	Mds.	
Bombay	52	51	
Berar	20	14	
North-Western Provinces and Oudh	17	16	
Madras	17	10	
Nizam's Territory	10	7	
Rajputana and Central India States	8	11	
Punjab	8	13	
Central Provinces	5	3	
Bengal	2	2	
Sindh	1	2	
Total	140	129	

Cotton is sown at the commencement of the rains and pick October to December. In Southern India there is a second s less commercial importance, known as the late crop, which e September and is picked from February to April. A fair yiel whole was obtained from late crops in 1888. The out-turn of crops of that year was not above middling, though superior and, owing to the early cessation of the rains and the absenc timely falls during the ripening and picking season, better

*Lakh of maunds=73,475 cwt.

anticipations founded on a decreased acreage. The inland trade of the year stands at 89 lakhs of maunds, showing an increase of 6 per cent. As might be expected from its proximity to the great cotton-growing area, Bombay town took the lion's share of the exports, viz., 64 lakhs of maunds.

Among cotton producers the Bombay Presidency takes foremost rank. Of the five million acres which it cultivates with cotton, about one-half belong to native States within its boundaries. The late crop (grown principally in Khandesh) of the year 1888—89, equalled that of 1887; the early crop was 3½ per cent. below it. Exports to the port-town consequently show a fall from 34 to 30 lakhs of maunds, but this was more than made up by Berar, which stands next in importance to the Bombay Presidency as a cotton producer. An exceptionally favourable yield was secured in this province, and the despatches to Bombay town rose from 9 to 18 lakhs of maunds.

The Central Provinces also had a good crop, and exhibit an increase from 91,000 maunds to over 2 lakhs of maunds.

In Northern India the seasons have been favourable to cotton for the past two or three years. Fears were entertained that the Punjab out-turn would prove worse than the poor crop of 1887; in the end these fears were not realised, the area being 17 per cent. greater, but the yield does not appear to have been proportionately favourable, as the total exports show a falling off of 1 lakh of maunds.

A contrary result is to be inferred from the export figures for the North-Western Provinces and Oudh, where, in spite of a decrease of 8 per cent. in area, the exports remained at the level of the previous year (11½ lakhs of maunds).

Dyes and Tans comprise indigo, myrabolams, cutch, turmeric, aniline dyes, and "others"; the first is by far the most important and stands at 323 lakhs of rupees* in a total trade in dyes and tans valued at 442 lakhs of rupees.

Bengal, the North-Western Provinces and Oudh, and Madras, are the principal sources of commercial Indigo, and their combined exports during the year amounted to 289½ lakhs of rupees, viz.: Bengal, 187½ lakhs of rupees; North-Western Provinces and Oudh, 73½ lakhs of rupees; and Madras, 28½ lakhs of rupees. It is also grown rather extensively in the Punjab, but chiefly for local consumption. Elsewhere its cultivation is not unknown, but it is unimportant.

Indigo manufacture in Bengal has a long history marked by many vicissitudes of fortune; but notwithstanding some serious checks, and in later years the competition of aniline and other dyes, it continues to hold its place as one of the great industries of the province. The total area under indigo in Bengal is estimated to be 588,000 acres, and the manufacture is in the hands of European capitalists. The season was fairly prosperous and the exports amounted to 90,616 maunds.

In the North-Western Provinces and Oudh indigo is largely cultivated in the districts to the east of Allahabad, and in the central and western half of the tract lying between the Ganges and Jumna rivers, where canal irrigation has led to a considerable extension of the acreage under this crop. The total area now under indigo in the North-Western Provinces and Oudh averages about 337,000 acres. Unlike Bengal, the manufacture is, except in the eastern districts adjacent to Behar, in the hands of natives. The crop of 1888—89 was not a good one owing to heavy and frequent rainfall, and exports fell from 43,000 maunds in 1887—88 to 40,000 maunds.

Madras indigo is commercially less valuable than that of Northern India; its cultivation is confined to the Northern Circars, but is extending, and the total area is between 400,000 and 500,000 acres. The exports were 23,868 maunds against 26,000 maunds in the preceding year.

The indigo area in Punjab is returned as 138,000, of which 82,000 are comprised in the districts of Multan, Muzaffargarh, and Dera Ghazi Khan, where the indigo exported from the province is manufactured. The exports go to Sindh and Karachi. During 1888—89 amounted to 14,085 maunds, valued at 12·39 lakhs of rupees.

A noticeable peculiarity of indigo culture is the extent to which the cultivators of Behar depend on the North-Western Provinces and Oudh and the Punjab for seed. The exports of seed from these provinces average 1·5 lakhs of maunds a year.

Myrabolams, the fruit of a species of *Terminalia*, are exported principally from the forests of the Central Provinces and Bombay to Bombay town. During the year the exports of the Central Provinces amounted to 5·71 lakhs of rupees in value, and of the Bombay Presidency to 3·34 lakhs of rupees. With the opening of the railway from Nagpur to Bengal the trade may be expected to increase rapidly.

The Indian trade in cutch is insignificant compared with that of Burma, and [illegible] value was under 8 lakhs of rupees. Of this about 2·5 [illegible] rupees is the value of cutch imported through Calcutta [illegible] exports from Burma. Punjab and Bombay [illegible] from the North-Western Provinces and Oudh.

---

* [illegible] 1,00,000 rupees.

PERMANENT CHEMICAL EXHIBITION.

THE proprietors wish to remind subscribers and their friends generally that there is no charge for admission to the Exhibition. Visitors are requested to leave their cards, and will confer a favour by making any suggestions that may occur to them in the direction of promoting the usefulness of the Institution.

JOSEPH AIRD, GREATBRIDGE.—Iron tubes and coils of all kinds.

ASHMORE, BENSON, PEASE AND CO., STOCKTON-ON-TEES.—Sulphate of Ammonia Stills, Green's Patent Scrubber, Gasometers, and Gas Plant generally.

BLACKMAN VENTILATING CO., LONDON. — Fans, Air Propellers, Ventilating Machinery.

GEO. G. BLACKWELL, LIVERPOOL.—Manganese Ores, Bauxite, French Chalk. Importers of minerals of every description.

BRACHER AND CO., WINCANTON.—Automatic Stills, and Patent Mixing Machinery for Dry Paints, Powders, &c.

BRUNNER, MOND AND CO., NORTHWICH.—Bicarbonate of Soda, Soda Ash, Soda Crystals, Muriate of Ammonia, Sulphate of Ammonia, Sesqui-Carbonate of Ammonia.

BUCKLEY BRICK AND TILE CO., BUCKLEY.—Fireclay ware of all kinds—Slabs, Blocks, Bricks, Tiles, "Metalline," &c.

CHADDERTON IRON WORKS CO., CHADDERTON.—Steam Driers and Steam Traps (McDougall's Patent).

W. F. CLAY, EDINBURGH.—Scientific Literature—English, French, German, American. Works on Chemistry a speciality.

CLAYTON ANILINE CO., CLAYTON.—Aniline Colours, Aniline Salt, Benzole, Toluole, Xylole, and Nitro-compounds of all kinds.

J. CORTIN, NEWCASTLE-ON-TYNE.—Regulus and Brass Taps and Valves, "Non-rotative Acid Valves," Lead Burning Apparatus.

R. DAGLISH AND CO., ST. HELENS.—Photographs of Chemical Plant —Blowing Engines, Filter Presses, Sludge Pumps, &c.

DAVIS BROS., MANCHESTER.—Samples of Products from various chemical processes—Coal Distilling, Evaporation of Paper-lyes, Treatment of waste liquors from mills, &c.

R. & J. DEMPSTER, MANCHESTER.—Photographs of Gas Plants, Holders, Condensers, Purifiers, &c.

DOULTON AND CO., LAMBETH.—Specimens of Chemical Stoneware, Stills, Condensers, Receivers, Boiling-pots, Store-jars, &c.

E. FAHRIG, PLAISTOW, ESSEX. — Ozonised Products. Ozone-Bleached Esparto-Pulp, Ozonised Oil, Ozone-Ammoniated Lime, &c.

GALLOWAYS, LIMITED, MANCHESTER.—Photographs illustrating Boiler factory, and an installation of 1,500-h.p.

GRIMSHAW BROS., LIMITED, CLAYTON.—Zinc Compounds. Sizing Materials, India-rubber Chemicals.

JEWSBURY AND BROWN, MANCHESTER.—Samples of Aerated Waters.

JOSEPH KERSHAW AND CO., HOLLINWOOD.—Soaps, Greases, and Varnishes of various kinds to suit all requirements.

C. R. LINDSEY AND CO., CLAYTON.—Lead Salts, (Acetate, Nitrate, etc.) Sulphate of Copper, etc.

CHAS. LOWE AND CO., REDDISH.—Mural Tablet-makers of Carbolic Crystals, Cresylic and Picric Acids, Sheep Dip, Disinfectants, &c.

MANCHESTER ANILINE CO., MANCHESTER.—Aniline Colours. Samples of Dyed Goods and Miscellaneous Chemicals, both organic and inorganic.

MELDRUM BROS., MANCHESTER.—Steam Ejectors, Exhausters, Silent Boiling Jets, Air Compressors, and Acid Lifters.

E. D. MILNES AND BROTHER, BURY.—Dyewoods and Dyewood Extracts. Also samples of dyed fabrics.

MUSGRAVE AND CO., BELFAST.—Slow Combustion Stoves. Makers of all kinds of heating appliances.

NEWCASTLE CHEMICAL WORKS COMPANY, LIMITED, NEWCASTLE-ON-TYNE.—Caustic Soda (ground and solid), Soda Ash, Recovered Sulphur, etc.

ROBINSON, COOKS, AND COMPANY, ST. HELENS.—Drawings, illustrating their Gas Compressors and Vacuum Pumps, fitted with Pilkington and Forrest's patent Valves.

J. ROYLE, MANCHESTER.—Steam Reducing Valves.

A. SMITH, CLAYTON.—India-rubber Chemicals, Rubber Substitute, Bisulphide of Carbon, Solvent Naphtha, Liquid Ammonia, and Disinfecting Fluids.

WORTHINGTON PUMPING ENGINE COMPANY, LONDON.—Pumping Machinery. Speciality, their "Duplex" Pump.

JOSEPH WRIGHT AND COMPANY, TIPTON.—Berryman Feed-water Heater. Makers also of Multiple Effect Stil and Water-Softening Apparatus.

THE PURIFICATION OF WATER BY MEA METALLIC IRON.

THE purifying properties of the metal iron and its benefic in the natural soil are well known.

Dr. Medlock was probably the first who endeavoured practical use of metallic iron as a purifier of water. He t patent in 1857 for a process in which iron wires or plates w suspended in tanks through which the impure water was to 1867, Dr. Thomas Spencer brought out a material which h magnetic carbide, in which iron was the active reagent. Lit ever, was accomplished, on a practical scale, previous to the by Professor Gustav Bischof of the material known as spor The spongy iron is produced by heating hematite ore to a ten of a little below that of fusion, and thus rendering it porous in form. Dr. Bischof's material has long been utilized in filters, and the spongy iron filter is at the present time secon in its remarkable purifying properties, and in the permanen action during the whole period that the mass of the material porous.

In supplying the city of Antwerp with water from the Riv it was found necessary to adopt some artificial means for impr quality of the water. Spongy iron was selected. Three filters were constructed, each pair consisting of an upper b taining a mixture of spongy iron and gravel, three feet in t and a lower basin containing a bed of river sand two feet thi river water was pumped into the upper basin, flowing thr bed of spongy iron on to the sand filter where the oxide of retained. The chemical results and the great improvemen appearance of the water were all that could be desired, and f two years it seemed that a practical process of purifying foul a large scale had been found. After a time, however, as the for water increased in the city, and the filters were required to their calculated output, it was found that the mass of sp mixture was caking together and becoming daily less porous. at length became so serious that it was with the greatest diffi sufficient filtered water could be obtained to meet daily requ The spongy iron beds had to be dug over, by manual labour, loosen the material, and restore, in some degree, its porosity.

In order to overcome this difficulty, Mr. Wm. Anders many experiments, accomplished the desired end by the revolving purifier, wherein, instead of allowing t to flow downwards through a motionless mass of the material, the purifying material is showered down divided particles in a flowing stream of water. The consists of a horizontal cylinder capable of revolving in bearings. Attached to the internal periphery of the cylir series of short curved shelves, arranged either in hori diagonal rows at equal distances. A sixth row of curved replaced by a line of small square plates, which, by mea outside the cylinder, can be set at an angle with the a apparatus. By regulating the angle of these plates the show can be directed back to the inlet end of the purifier, and the of the flow of water to carry forward the purifying material acted. Gearing is provided for rotating the cylinder, and outlet pipes are fixed axially at either end. On being starte sufficient metallic iron to fill one-tenth of the cylinder is i through the manhole, the iron being in a suitable state of su The purifier is then filled with water through the sluice-cock apparatus is set in rotation. The effect of rotation is to sco iron particles and to shower them down through the flowi Contrary to expectations, it was found that a contact var three and a-half to five minutes was sufficient to effect the purification of the majority of waters. Further, it was that when used in the revolving purifier *spongy* iron had merit. Any form of waste iron—such as cast iron, bori punchings, etc., gave equally good chemical results, and wer preferable to spongy iron, the irregularity of whose form with the automatic and continuous renewal of the active surf purifying material by attrition. The theory of the process a be as follows:—

The action on the iron in the purifier is one of reductio bonic acid brought by the impure water dissolving a minute the metal and forming a protosalt of iron. On issuing cylinder into the open air this protosalt is gradually convert action of the atmospheric oxygen, into ferric hydrate, wh soluble. It is stated that nascent ferric hydrate destroys th matter; but there is no doubt that the coagulation of the ferri carries down a great proportion of suspended organic matt obvious that a further process is necessary to remove this p

* Abstract of a paper read by Mr. Easton Devonshire before th Institute.

This is accomplished by filtration through 18 inches of sand, whereby the water is freed from its suspended matters. "Sandfilters have been "at work at Antwerp for more than five years; they have never been "cleaned below the surface, and yet at the present time the water "issuing from them contains no free ammonia at all, and less than [illegible] of "a part in a million of albuminoid ammonia; the river water containing "on the average [illegible] of a part in a million of each of these forms of "ammonia."

The general results obtained are said to be as follows:—Firstly, all colour is removed from the water. Secondly, oxidisable organic matter as measured by its power of reducing permanganate of potash is reduced in proportions varying from forty-five to ninety per cent., according as the organic matter is principally of vegetable or of animal origin. Thirdly, free ammonia and nitrous acid are entirely removed. Fourthly, albuminoid ammonia is reduced from sixty to ninety per cent. Lastly, micro-organisms are entirely destroyed or removed by this process. Professors Blas, of Louvain; Jorissen, of Liege; Swarts and Van Ermengem, of Ghent, were appointed as a commission to examine and report upon the practical results of actual working by this process. Quarterly reports were made by the commission, giving the results of the weekly analyses. Each of these reports stated that the very impure and dirty water of the river Nethe was transformed by the process of purification into a liquid equal, from a hygienic point of view, to the purest and most healthy spring water.

Legal.

COURT OF APPEAL.

(Before the MASTER *of the* ROLLS *and Lords Justices* LINDLEY *and* LOPES.*)*

THE CAPE COPPER COMPANY (LIMITED) v. COMPTOIR D'ESCOMPTE DE PARIS.

THIS was an appeal by the defendants from an order of the Divisional Court (Mr. Justice Denman and Mr. Justice Charles). The action was brought to recover £600,000, and arose out of the French combination or ring to raise the price of copper. The Société des Métaux had agreed to purchase from all the chief copper companies, including the plaintiff company, all the copper produced for three years, at greatly enhanced prices. The defendants guaranteed the performance of this agreement by the Société des Métaux. The Société des Métaux and the defendants subsequently went into liquidation, and the French ring collapsed. The price of copper accordingly fell considerably, and the plaintiffs brought this action against the liquidator of the Comptoir d'Escompte, upon the guarantee, to recover damages for the loss sustained by them in consequence of the non-performance of the agreement by the Société des Métaux in respect of certain deliveries of copper. The pleadings having been closed in December, 1889, the defendants subsequently applied for leave to add a defence that the agreement was illegal and void according to the English law, as being in restraint of trade, and therefore against public policy. The matter having been referred to the Court from Chambers, the Divisional Court refused to allow the amendment, giving leave, however, to add a defence that the agreement was illegal and void according to French law, and ordering the trial to take place on June 23rd. The defendants appealed.

Counsel having been heard,

The Master of the Rolls said that in this action, involving so large an amount of money, the Divisional Court had, in his opinion, struck too soon. The defendants ought to be allowed to plead that these contracts were avoided by English law. That being so, it would be futile to allow the amendment, and at the same time to fix June 23rd for the trial. They had come to the conclusion that they ought not to fix any day for the trial, but they did not for a moment say that the case must be thrown over the long vacation. If the defendants did not prosecute the commission with due diligence, the plaintiffs could make a further application to the Court. The costs caused by the amendment must be paid by the defendants, and the costs of the appeal must abide the event of the action.

The Lords Justices concurred.

THE MINERAL PALACE OF COLORADO.—The Colorado *Western Manufacturer* states that a mineral palace is to be erected at Pueblo, Col., which will cost about $250,000. It will be of handsome design, the exterior being a series of square columns and beautifully polished stone. All parts of the building will be made of the products of Colorado's mines, the owners in all the counties in the State having sent in their choicest and richest specimens. In the interior will be seen every variety of mineral production, from stone and coal to pure gold, the value of which will be at least $750,000. It is intended to be a permanent exhibit. The building will be lighted by 3,000 incandescent electric lights.

Trade Not

A LARGE ORDER.—The whole of the
supplied by Galloways, Limited, Manchester
steam required in the Edinburgh Exhibition
by the North British Rubber Co., for their C

GREAT STOCK OF TOBACCO IN LIVERPOO
at the Liverpool Docks there are just
tobacco, as against 31,932 in those of Londo
months, upwards of 8,000 hogsheads have be
of 11,000 have been delivered. The tobacco
greater than that of all the other British ports

ONE WELL THAT SPOUTETH.—We learn fro
Commerce, that at Pittsburg, a poor little debt-
of Forest-grove, has become rich by a strike of
wells sunk now average an out-put of 1,000 barr
will amount to 90,000 dollars a year. The
this says, "the deacons are beside themselves
is more joy in that church over one well that sp
sinners who repent. We can well believe it.

DISCOVERY OF NICKEL DEPOSITS IN
nickel have been very recently discovered nea
by Herr Reitsch, a mining engineer. The
length of ten kilometres, and are one kilome
only a few metres from the surface. Accordi
papers, the mineral is free from copper, antim
averages about 5 per cent. of nickel, though
much as 13 per cent. The discovery has gi
Germany, where it is hoped the importati
Caledonia will in future be dispensed with,
easily smelted on the spot.

GERMANY AND THE MCKINLEY BILL.—
not entirely without resources to meet the
Negotiations are at this moment proceeding b
and German Governments with the object of
the prohibition on the import of American
considering whether the amelioration at lea
should not be made a condition of the throw
into this country of American meat. Very g
nected with the American meat trade, and
they would be exerted in favour of ameliorating
duties if it were seen that the removal of th
products were the result.—*Kuhlow*.

BRITISH TRADE WITH THE DANUBIAN
letter, referring to the opening of the new rai
port of Bourgas to Yamboli and the new faci
lish trade with the Danubian States, was r
Manchester Chamber of Commerce:—"Fore
Sir,—With reference to the letter from [redacted]
am directed by the Marquis of Salisbury to
formation of the Manchester Chamber of Com
Yamboli to Bourgas was formally opened
nouncing this fact to the Secretary of State,
Consul-General at Sofia again expresses th
steamship company will shortly establish dire
London and the port of Bourgas, for unless
is soon taken he is apprehensive that Ger
Hamburg on especially favourable conditio
advantageous terms with British trade that
siderable loss.—I am, &c., JAMES FERGUSSO

ADULTERATION OF NORMANDY BUTT
Consul at Caen, in his last report, says that,
adopted very recently against the fraudul
dealers and merchants in largely introducing
to mix with the pure butter, this fraud h
carried on extensively, and the exporters
been introducing the hitherto excellent N
English market largely adulterated with the
of the butter merchants of Northern France h
have recently issued an appeal to all the ho
endeavour to avert this disaster to the trade
the exportation to foreign markets, attributabl
the hitherto pure butter exported. The expo
1882 to the English markets alone amounted to
while in the year 1887 it had fallen to 58,
latter year, there has been a still further decr
poses, as a precautionary measure, that a new
oblige the makers of the compound to give it
other than the hue of butter in its unadulter
it will then be impossible to mix it wit
showing some trace of margarine. This
extensively carried on in Caen.—*Times*.

ESTABLISHMENT OF A COMMERCIAL MUSEUM AT ALENÇON.—The *Moniteur Officiel du Commerce* announces that the French Minister of Commerce, Industry, and the Colonies, has, by decision of the 11th April, approved the establishment of a commercial musuem by the Alençon Chamber of Commerce.

A RUSSIAN EMPLOYERS' LIABILITY BILL.—A Bill is before the Russian Council of the Empire which will throw heavy expenses upon the proprietors of manufacturing establishments in Russia. If a workman is injured in the manufactory the proprietor will have to pay all costs during the whole time he is unable to work and the whole of his ordinary wages; if the workman is killed on the spot, or if his death afterwards is the consequence of the injuries received, the proprietor must pay the funeral costs and annuities to his widow and his children. The widow receives an annuity during her life of 30 per cent. of her deceased husband's wages, and the children an annuity of 15 per cent. each of their father's wages until they reach their fifteenth year; if they have no mother the annuities are increased to 20 per cent. If a working child is killed or mutilated the parents receive annuities of 15 per cent. each. There is no doubt this Bill will be voted by the Council of the Empire, and will receive Imperial sanction.

SMOKELESS POWDER.—In some comparative experiments with the smokeless powder C/89 fired from cannon of various calibres, which took place on the shooting ground connected with the Grusonwerk, Buckau, near Magdeburg, it was proved that the new powder yielded an effect from three to four times greater per kilo. of the charge than the old sorts of powder. The powder C/89 threw off a thin brownish vapour which was so thin that another shot could immediately be fired, the butt being distinctly visible. Even in rainy weather the vapour was completely dissipated within three seconds, whilst the smoke resulting from black powder hung about the gun for a long time, making a quick firing impossible. The powder C/89 leaves so little residuum after combustion that the bore remains almost clean, and the heating of the barrel and cartridge holder is much less than with black powder.—*Kuhlow.*

KEROSENE IN STEAM BOILERS.—At the Cincinnati meeting of the American Institute of Mechanical Engineers, "Notes on Kerosene in Steam Boilers" was the title of a paper presented by Professor Carpenter. The boilers experimented on were of the ordinary tubular type, 12ft. in height, four being 4ft. in diameter, and the other two 5ft. in diameter. When the use of oil was begun the boilers were badly incrusted with a hard scale. The first application of oil was made by inserting about a gallon, then filling the boiler with water, heating to the boiling point and allowing the water to stand two or three weeks before removing. This succeeded in removing fully one-half the scale. This was found to be a more effective way than that of applying the oil in small quantities when the boiler was in use. Kerosene oil was used in this case. It was found that there was no advantage to be derived from the use of more than a certain quantity of oil. For boilers 4ft. in diameter and 12ft. long, the best results were obtained with two quarts of oil for each boiler per week. The boilers now have less scale in them than at any previous time within four years, and the small quantity remaining in them seems to be soft and gradually disappearing. In the discussion which followed, C. W. Nason said that in one case, with the use of crude petroleum in a boiler badly scaled, it was noticed that the scales came off in cakes, the oil seeming to enter between the scale and iron, and thereby separating the two. These cakes varied considerably in size, and in some cases were quite large.

THE BANKRUPTCY BILL.—The House of Commons Standing Committee on Trade gave further consideration on Thursday to the Bankruptcy Bill introduced by Sir A. Rollit, Mr. A. O'Connor presiding. On the motion of Sir A. Rollit, the Committee added the following sub-section to Clause 9, which lays down the conditions on which a bankrupt shall obtain his discharge:—"That the bankrupt has, within three months preceding the date of the receiving order, incurred liabilities in order to make his assets equal to 10s. in the pound on the amount of his unsecured liabilities." Mr. Dixon-Hartland proposed to omit from the clause the provision that on proof of any of the offences mentioned in the section, or in section 28 of the principal Act, the Court should have power to suspend the bankrupt's discharge for not less than five years. Mr. Chamberlain pointed out that if the amendment were carried it would be necessary to strike out some portion of the old Act. He hoped the amendment would be withdrawn and the *minimum* period reduced. The amendment was negatived without a division, and it was then resolved, by 23 to 10, on the motion of Mr. Gedge, to substitute two years for five years. Sir A. Rollit moved the following addition to the clause:—"Provided that, if at any time after the expiration of two years from the date of any order made under this section the bankrupt shall satisfy the Court that there is no reasonable probability of his being in a position to comply with the terms of such order, the Court may modify the terms of the order, or of any substituted order, in such manner and upon such conditions as it may think fit." The words were added to the Bill. The 4th and 5th sub-sections were struck out. The Committee adjourned.

REPORTED CLOSING OF HEIDELBERG UNIVERSITY.—A con able sensation has been created at Heidelberg by an alarming that the Government contemplates closing the University of tha It is difficult to conceive the object of so retrograde a step—w moreover, would be exceedingly unpopular in Germany.—*Truth.*

SUNDAY WORK WITHOUT PAY.—At Wakefield County before Judge Greenhow, Locke and Co., of St. John's Colliery, manton, were sued for £3. 12s., in lieu of notice, by an ex-d named Edwin Bennett. Plaintiff's case was that early in Janu was engaged as deputy by the certified manager, Mr. Wilson, nc having been said at the time about Sunday work. Finding th predecessor used to do Sunday duty, the plaintiff voluntarily sented to follow the example; but on the 19th April he was to the underground steward that there would be no work for hin day, as work was only available for a limited number. Plaintif then reminded that next day (Sunday) it would be his turn for whereupon the plaintiff rejoined that if he could not work for p Saturday he was not going to work for nothing on Sunday. O ensuing Monday morning, when he presented himself at the co he was sent about his business, and hence the action. His H gave a verdict for the plaintiff, with costs.

STANDARD TESTS AND METHODS OF TESTING OF MATER —A committee of the American South-Western Mining Assoc (says the *Engineering and Mining Journal*) was appointed abou years ago, and consisted of Messrs. Henry R. Towne, Gu Henning, R. H. Thurston, Chas. H. Morgan and Thomas Egl A preliminary report with a copious appendix is presented by Henning, the reporter of the committee, which shows that it has doing a great deal of work in the direction of establishing a un system of testing and reporting on materials of construction. practice of England, France, Belgium, and Germany has been stu and it is found that in Germany far more attention has been p scientific testing than in any other country. Duplicate tensil pieces of steel and iron were sent to twenty-four places having t machines, including iron, steel, and bridge works, technical co and testing laboratories. Sixteen of these have reported the resu full, which are published in the report, and the uniformity c results is stated by the committee to be satisfactory, showing that harmony exists among experts in the United States in rega methods pursued in testing and in making out reports. The comr make a number of recommendations in regard to standard methc testing. One of the most important is a new method of measurin reporting the elongation of the test piece, the old method having results not properly comparable, on account of the fracture t place sometimes near the middle of the piece and sometimes a ends. By the new method, the specimen is divided by gauge m before test, into inches and half inches. After fracture, measure are made as follows:—Having divided the standard length into 24 parts, measure from the point of fracture to cover all divisio to 12 to the right—or 12 parts—then measure from the fr towards the left to four, and add the elongation of parts 4 measured to right of fracture, or eight parts, to this; thu elongation of the standard 24 parts will be obtained as thoug fracture were located exactly at the middle division, for the elon of the divisions 4—12 on the right is, undoubtedly, practicall same as that of a similar original length would be if located o other side of the fracture.

Market Reports.

THE LIVERPOOL MINERAL MARKET.

During the past week our market has maintained its firmne prices, whilst a fair amount of business has been done. ganese: Arrivals have been somewhat small this week, and both for spot and forward delivery are firm. Magnesite offering, and prices in buyers' favour. Raw lump 26s., ground £6. 10s., and calcined ground £10. to £11. Ba (Irish Hill brand): The production is scarcely able to pace with the demand, and prices are very firm—Lump, seconds, 16s.; thirds, 12s.; ground, 35s. Dolomite, 7s. 6d. ton at the mine. French Chalk: There have been more ar this week, but the best part have gone direct into consumption prices are unchanged, especially for G. G. B. "Angel White" brand, 9 100s. medium, 105s. to 110s. superfine. Barytes (carbonate) more inc selected crystal lump is scarce at £6.; No. 1 lumps, 90s.; best, seconds and good nuts, 70s.; smalls, 50s.; best ground £6.; and sel crystal ground, £8. Sulphate firm, and prices are better; best l 35s. 6d.; good medium, 30s.; medium, 25s. 6d. to 27s. 6d.; com

6d. to 20s.; ground best white, G.G.B. brand, 65s.; common, grey, 32s. 6d. to 40s. Pumicestone unaltered; ground at £10., specially selected lump, finest quality, £13. Iron ore easier. io and Santander, 9s. to 10s. 6d. f.o.b.; Irish, 11s. to 12s. 6d.; berland, 10s. to 12s. 6d. Purple ore quiet. Spanish manganiferous continues in fair demand. Emery-stone: Finest lump sought and brings full prices. No. 1 lump is quoted at £5. 10s. to and smalls £5. to £5. 10s. Fullers' earth steady; to 50s. for best blue and yellow; fine impalpable nd, £7. Scheelite, wolfram, tungstate of soda, and tungsten l continue in fair demand. Chrome metal, 5s. 6d. per lb. me ore: Best and lower grades offered freely, and prices asier. Antimony oxide, 24s. to 26s. Asbestos in strong demand, cially rock; prices firmer. Potter's lead ore continues firm; ls, £14. to £15.; selected lump, £16. to £17. Calamine: Best ities scarce—60s. to 80s. Strontia steady; sulphate (celes- steady, 16s. 6d. to 17s. Carbonate (native), £15. to £16.; dered (manufactured), £11. to £12. Limespar: English ufactured, old G.G.B. brand in demand, and brings full prices; for ground English. Felspar, 40s. to 50s.; fluorspar, 20s. to £6. ore in demand, firm at 22s. to 25s. Plumbago steady; Spanish, ; best Ceylon lump at last quotations; Italian and Bohemian, £4. 12. per ton; founders', £5. to £6.; Blackwell's "Mineraline," . French sand, in cargoes, continues scarce on spot—20s. 2s. 6d. Ferro-manganese and silicon spiegel easier. Chrome iron, tered at last quotations. Ground mica, £50. China clay freely ing—common, 18s. 6d.; good medium 22s. 6d. to 25s.; best, 30s. 5s. (at Runcorn).

THE TYNE CHEMICAL REPORT.

TUESDAY.

'here is very little change to report in chemicals this week. Prices round, with the exception of bleaching powder, have kept fairly dy, and are much about the same as last week. Bleaching powder ery weak and has dropped half-a-crown a ton, with little doing at moment. Caustic soda is in better demand and market stronger. la ash, crystals, and sulphate of soda practically unchanged. Prices rent are as follows:—Bleaching powder, in softwood casks, £4. 15s. ton, nominal quotation; caustic soda, 77%, £9. per ton; ground packed in 3-4 cwt. barrels, £12. per ton; soda ash, 48/52%, d. per degree, less 5%; soda crystals, in casks, £2. 11s. per ton, s weight; in 2 cwt. bags, £2. 11s. per ton, net weight; in 1 cwt. s, £2. 13s. 6d. per ton, net weight; sulphate of soda, in bulk, 32s. ton; ground and packed in casks, 42s. per ton; recovered hur, £4. 5s. per ton; chlorate of potash, 4½d. per lb.; silicate of , 75° Tw., £2. 10s. per ton; 100° Tw., £3. 7s. 6d. per ton; Tw., £4. per ton; hyposulphite of soda, in 1 cwt. kegs, 15s. per ton, in 5-7 cwt. casks, £5. 5s. per ton; pure white hate of alumina, £4. 10s. per ton; blanc fixe, £7. 10s. per ton; ride of barium, £8. per ton; nitrate of baryta, crystals, £18. 15s. ton; ground, £19. 5s. per ton; sulphide of barium, £5. 10s. ton—all f.o.b. Tyne, or f.o.r. makers' works.

he price of small coals, for manufacturing purposes, is again lower week. The stoppage of the Tyne Plate Glass Works throws t 800 tons of small weekly into the market. Northumberland n is quoted from 7s. to 7s. 6d. per ton, and Durham small from d. to 8s. 6d. per ton. The Northumberland miners have sent in plication for an advance of wages, and it is expected the colliery rs will meet together about the end of the month to consider the cation.

THE COPPER MARKET.

m Messrs. Harrington and Co.'s report of June 17th we note Chili Copper Charters for first half of June are advised as 800 fine. Price of bars about 25·40 dols., and exchange 23¾d. charters since January 1st have been 11,300 tons, against o tons for same time last year. An extensive business has been in G.m.b.'s and G.o.b.'s, and with but few oscillations market y advanced from £54. 5s. to £59. 15s. cash, and £54. 15s. to or three months prompt; closed steady at £59. 5s. and £59. 15s. on respectively. The estimated decrease in visible supply, nd and France, is 6,700 tons fine.

present stock of English G.m.b.'s in warehouse, Liverpool and ea, is 5,286 tons, against 5,383 tons on 31st ult., showing a se of 97 tons in the fortnight. The feature of the fortnight has a winding-up of sales of the entire stock of Anaconda Matte in April was 23,461 tons) on behalf of French holders—ing with sales at 9s. 9d., they finally sold at 11s. 6d.

MISCELLANEOUS CHEMICAL MARKET.

There has been some slight improvement in the position of [illegible] soda during the week, and prices are rather higher. Quo[illegible] the spot are:—For 77%. on the Tyne, £9. 15s., and f.o.[illegible] 74%, £9 per ton; 70%, £8. 10s.; 60%, £7. 10s. S[illegible] more easily obtainable, and for prompt and early de[illegible] range from 1$\frac{1}{12}$d. to 1⅛d. per degree at makers [illegible] steady at £2. 18s. 9d. to £3 2s. 6d. at wo[illegible] Chlorate of potash dull at 4½d. per lb. Bleaching pow[illegible] to £5. per ton on rails, according to brand, makers being [illegible] with orders. Saltcake 27s. in bulk. Vitriol and muriatic ac[illegible] changed in values. Recovered sulphur well sold, and a firm [illegible] Demand for sulphate of copper has to some extent fallen off, [illegible] prices are a turn lower for near deliveries—on the spot £24 [illegible] £24. 10s., and July £22. 10s. Lead salts quiet, and prices [illegible] changed. Acetate of lime, brown, £7. 7s. 6d. to £7. 10s. per ton at usual points, and in plentiful supply. Acetic acids and acetate of soda without material change. In tartaric and citric acids there are small sales at 1s. 2½d. and 1s. 3d per lb. Oxalic acid quiet again at 3¼d. to 3¾d. per lb. Picric acid 1s. 1d. to 1s. 2d. per lb. For carbolic acid crystals the demand is still small, and prices consequently are at a low point. Prices during the week have been a shade easier, but any appearance of larger orders in the market will directly influence the value of this article, which at present is far from satisfactory to the producers. Potash, caustic and carbonate, are selling freely, and prices remain firm. Sulphocyanides are in small demand, and supplies have been curtailed. Aniline oil and salt are steadily maintained in price.

REPORT ON MANURE MATERIAL.

There has been a very fair amount of business doing during the past week, and prices all round are maintained, but there has scarcely been so much activity as there was at the beginning of the month.

We have nothing fresh to report in the position of mineral phosphates, nearest values for cargoes of Bull River rock being 10½d. per unit cost, freight, and insurance to United Kingdom, and of land rock 9½d. per unit. 70/75% Somme is still quoted 15¼d., and 60/65% 12d., one-fifth rise in each case. The quotations for Canadian and Belgian remain as quoted a week ago.

We do not hear of any business in cargoes of River Plate bones for the United Kingdom, but several sales have been made for the United States at the equivalent of £5. 7s. 6d. to United Kingdom. The advanced rate of exchange with India is stopping business in crushed bones and bone meal, and we do not hear of any sales having been made for shipment ahead. On spot fine Bombay meal is selling at £5. 10s. ex-store, and inferior quality at £5. 5s. There is no stock of crushed bones here, and nothing under £5. 5s. would be cabled *for shipment*. Parcels of common grinding bones still meet with a ready demand at £4. 15s. to £4. 17s. 6d. Nearest value of River Plate bone ash is £5. 7s. 6d. on 70%. Some further cargoes of nitrate of soda have been sold at 8s. per cwt. for ordinary quality, March shipment up to 8s. 4½d. for refined. Due cargoes are worth fully 8s. ordinary quality, and at present sellers will not accept that price. June-July shipments are worth 8s. 4½d., and September-October about 8s. 9d., delivered good United Kingdom ports.

The position of dried blood and ammonite remains unchanged, but there is decidedly more inquiry for both.

Kainit is scarce at 40s. per ton.

Supers are somewhat irregular, owing to parcels in second hands being offered for re-sale, and the nearest value of 26% is probably 45s. per ton, in bulk, f.o.r. at works.

TAR AND AMMONIA PRODUCTS.

Prices for benzol may be taken as the same as last week, 3s. 9d. for 90s and 2s. 9d. for 50/90s. The feature of the anthracene market is a confederation which has just been formed by some half-dozen of the largest tar distillers, with a view to maintaining the price of A at 1s. 6d., and B at 1s. 3d. The success of the movement is very problematical.

Sulphate of ammonia prices show a still further decline. After touching £11. there has, at the last moment, been a slight recovery to £11. 1s. 3d., London and country, while the Beckton price remains nominally £11. 5s.

THE LIVERPOOL COLOUR MARKET.

...URS unaltered. Ochres: Oxfordshire quoted at £10., £14., and £16.; Derbyshire, 50s. to 55s.; Welsh, best, 50s. ...; seconds, 47s. 6d.; and common, 18s.; Irish, Devonshire, ... 45s.; French, J.C., 55s., 45s. to 60s.; M.C., 65s. to ... Umber: Turkish, cargoes to arrive, 40s. to 50s.; Devon... 50s. to 55s. White lead, £21. 10s. to £22. Red lead, £18. ... of zinc: V.M. No. 1, £25.; V.M. No. 2, £23. Venetian ... £6. 10s. Cobalt: Prepared oxide, 10s. 6d.; black, ...d.; blue, 6s. 6d. Zaffres: No. 1, 3s. 6d.; No. 2, 2s. 6d. Terra ...: Finest white, 60s.; good, 40s. to 50s. Rouge: Best, £24.; ditto ...jewellers, 9d. per lb. Drop black, 25s. to 28s. 6d. Oxide of iron, ...e quality, £10. to £15. Paris white, 60s. Emerald green, 10d. ... lb. Derbyshire red, 60s. Vermillionette, 5d. to 7d. per lb.

WEST OF SCOTLAND CHEMICALS.

GLASGOW, Tuesday.

Although the position of caustic soda has improved in quite a measurable degree, the general market is still very flat, and prices are even easier in some of the lines. Nitrate of soda is firmer and dearer at the moment, but as yet its home competitor sulphate of ammonia shows no signs of taking the turn, rather the reverse, for last week's value has not been maintained, and Leith delivery has been done at £11. 1s. 3d. spot, not to say quite freely, for a few holders decline to part at that, but still in some volume. Indeed, one parcel went at £11., but this particular lot not having been up to the average mark in respect of quality, the transaction was not strictly within the regular market category, and the true representative value, therefore, has not yet quite descended to that low point. Enquiry about forward sulphate has been pretty active, without result, makers' notions being incompatible with the prevailing buying idea. Scotch bleach makers have come down very low to make business, and £4. 12s. 6d. to £4. 15s. has been done. Bichromates still in feeble demand. Paraffins unchanged, and pretty firm—all except burning oil, now entering the contract season against coming winter. Hardly anything has been booked yet for Scotch delivery, but some English orders (at lower prices) have been secured. Chief prices current are:—Soda crystals, 52s. net Tyne; alum in lump £5., less 2½% Glasgow; borax, English refined, £29. 10s., and boracic acid, £37. 10s. net Glasgow; soda ash, 1¼d. less 5% Tyne; caustic soda, white, 76°, £9. 10s., 70/72°, £8. 10s, 60/62° £7. 10s., and cream, 60/62° £7., all less 2½%. Liverpool; bicarbonate of soda, 5 cwt. casks, £5. 15s., and 1 cwt. casks, £6. net Tyne; refined alkali, 48/52°, 1½d., less 1¼%, Tyne; saltcake, 27s. to 29s.; bleaching powder, £4 12s. 6d., to £4. 15s., less 5% f.o.r. Glasgow; bichromate of potash, 4d., and of soda 3d., less 5 and 6% to Scotch and English buyers respectively; chlorate of potash, 4½d., less 5% any port; nitrate of soda 8s. 1½d. to 8s. 3d.; sulphate of ammonia, spot, £11. 1s. 3d. f.o.b. Leith; salammoniac, 1st and 2nd white, £37. and £35., less 2½% any port; sulphate of copper, £23. 10s., less 5% Liverpool; paraffin scale, hard and soft, 2¾d. per lb.; paraffin wax, 120°, semi-refined, 3¼d.; paraffin spirit (naphtha), 7d. a gallon; paraffin oil (burning), 6¾d. to 7d. on rails Glasgow to all Scotch buyers (English orders negotiable about 1¼d. a gallon lower); ditto (lubricating), 865°, £5. 10s. to £6. per ton; 885°, £6. 10s.; and 890/895°, £7. 15s. to £8. Week's imports of sugar at Greenock were 57,111 bags.

The Patent List.

This list is compiled from Official sources in the Manchester Technical Laboratory, under the immediate supervision of George E. Davis and Alfred R. Davis.

APPLICATIONS FOR LETTERS PATENT.

Steam Reducing Valves. T. Charlton and W. Pepper. 8,491. June 2.
Treatment of Sewage. J. Macfarlane and O. P. Macfarlane. 8,498. June 2.
Air Pumps.—(Complete Specification). H. Schulze-Berge and F. Schulze-Berge. 8,499. June 2.
Paper Finishing Apparatus. F. Butterfield and W. Renton. 8,504. June 2.
Yellow Colouring Matter. O. Imray. 8,506. June 2.
Heat Generating Apparatus. E. Broadhead, W. Beesley, and D. Evans. 8,510. June 2.
Boiler Gauge Glass Fittings. J. Burlinson. 8,511. June 2.
Manufacture of Cement. O. Prinz. 8,519. June 2.
Decorticating Textile Plants. P. A. Favier. 8,520. June 2.
Manufacture of Aerated Liquids. C. Lock. 8,524. June 2.
Azo Colours upon Fibres. B. Willcox. 8,530. June 2.
Edge Runner Grinding Mills. P. J. Neate. 8,531. June 2.
Scrapers for Edge Runner Grinding Mills. P. J. Neate. 8,532. June 2.
Secondary Batteries. N. de Binardos and Lloyd and Lloyd. 8,534. June 2.
Manufacture of Illuminating Gas. S. Pitt. 8,535. June 2.
Development of Colouring Matters in Textile Fabrics.—(Complete Specification). J. J. Hart. 8,541. June 3.
Filter Presses.—(Complete Specification). W. H. Johnson and C. C. Hut... 8,544. June 3.
Apparatus for Charging Retorts. C. Eitle. 8,554. June 3.
Brick-making Machinery.—(Complete Specification). B. C. White ... A Boyd. 8,563 June 3.
Galvanizing. T. L. Thomas and J. B. Hillman. 8,572. June 3.
Secondary Batteries.—(Complete Specification). J. Y. Johnson. 8,578.
Paper Winding Apparatus.—(Complete Specification). J. Y. Johnson ... June 3.
Manufacture of Building Blocks from Slag. S. G. T. C. Bryan. ... June 3.
Amalgamating Machinery. W. W. Fyfe. 8,609. June 3.
Filtering Apparatus. W. L. Wise. 8,614. June 4.
Stone-Breaking Machinery. S. B. Goodwin, W. Barsby, and R. ... 8,622. June 4.
Grinding and Crushing Mills. W. Barford and J. E. S. Perkins. ... June 5.
Mauufacture of India-rubber Rolled Fabrics. R. K. Bailey, of the Chas. Macintosh & Co., Ltd., and J. Dempsey. 8,683. June 5.
Treatment of Sewage. J. Chaffer and L. Simms, both of the British ... and Refuse Co. 8,685. June 5.
Furnace Grates. J. Keith. 8,688. June 5.
Pulverizers and Amalgamators. T. E. Bickle. 8,696. June 5.
Gas Furnaces. J. von Langer and L. Cooper. 8,707. June 5.
Electrolytic Treatment of Sulphite of Zinc. E. Edwards. 8,716. ...
Substitutes for Fulminate of Mercury. E. Blinkhorn. 8,717. June ...
New Derivatives of Alizarine and its Analogues. B. Willcox. ... June 5.
Manufacture of Naphthaline. F. Fenner and F. W. Colls. 8,728. J...
Manufacture of Carbonic Acid Gas. W. Bruce. 8,733. June 6.
Separating Oxygen and Nitrogen from Atmospheric Air. W. ... 8,752. June 6.
Manufacture of Iron and Steel. J. von Langer. 8,759. June 6.
Electric Motors.—(Complete Specification). H. Gloswith. 8,778. June ...
Liquid Meters. J. A. Müller. 8,784. June 6.
Manufacture of Fuming Sulphuric Acid. G. Léon. 8,786. June 6.
Electric Motors and Dynamos. A. Schanschieff. 8,787. June 6.
Feed Water Purifying Apparatus.—(Complete Specification). J. Y. J... 8,788. June 6.
Apparatus for Evaporating, Concentrating, and Distilling. W. R. ... and R. A. Robertson. 8,790. June 7.
Dyeing. A. North. 8,799. June 7.
Secondary Batteries. W. P. Thompson. 8,800. June 7.
Dyeing. W. L. Wise. 8,809. June 7.
Manufacture of Gunpowder. H. Kolf. 8,811. June 7.
Obtaining Alloys of Aluminium. J. Clark and G. W. Clark. 8,820. ...
Gold Extraction. J. Coombs. 8,826. June 7.
Measuring Liquids.—(Complete Specification). W. A. Sheppard. ... June 7.
Rendering Nitrate of Ammonium Non-Hygroscopic. E. Blinkhorn ... June 7.
Nitrifying Cotton and Similar Material.—(Complete Specificatio... Müller. 8,843. June 7.
Manufacture of Whisky and other Spirits. E. J. Taylor. 8,850. ...
Electric Meters. L. A. W. Desruelles and R. F. O. Chauvin. 8,854. J...

New Companies.

BROWN PAPER FIBRE COMPANY, LIMITED.—This company was regis... the 6th inst., with a capital of £2,000., in £1. shares, to manufacture fr... wood, earth, or other substances, a flock or material for manufacturing pa... for upholstery and other purposes. An unregistered agreement beetween ... London and Wm. Snowden Hedley will be adopted. The subscribers are ...

S

W. S. Hedley, M.D., Midford, near Bath ...
Mrs. Hedley, Midford, near Bath ...
A. S. King, 12, Queen-square, Bath, solicitor ...
C. E. Thomas, 13, Queens-square, Bath ...
J. A. London, 116, Sandyford-road, Newcastle, paper manufacturer ...
J. A. London, junr., 116, Sandyford-road, Newcastle, clerk ...
A. Robinson, 27, Perry-park-road, solicitor ...

JAMES BRIERLEY AND COMPANY, LIMITED.—This company was regis... the 10th inst., with a capital of £30,000., in £5. shares, to adopt an unre... agreement of 24th ult., with James Brierley, and to carry on at Heywood, ... where, the business of brewers. The subscribers are:—

S

J. Tetlow, Heywood, beerseller ...
R. Arnfield, Heywood, mill manager ...
J. R. Greenhalgh, Heywood, joiner and builder ...
J. Holt, Heywood ...
C. Brierley, Rochdale, beerseller ...
T. Jaques, Middleton, beerseller ...
W. H. Heaton, Bury, licensed victualler ...

Gazette Notices.

Partnerships Dissolved.

RADAMS BROS., Birmingham, paint and colour manufacturers.
HALLS AND PHILLIPS, Atherstone and Nuneaton, hat manufacturers.

THE BANKRUPTCY ACT, 1883.

Receiving Orders.

JAMES LEARMOUNT, South Shields, oil dealer and drysalter.

Adjudications.

ROBERT SNELL WILLIAMS AND WILLIAM MILLNER (trading as the Boo... Stores Company), Bootle, paint and oil dealers.

IMPORTS OF CHEMICAL PRODUCTS

AT

THE PRINCIPAL PORTS OF THE UNITED KINGDOM.

ONDON.

ending June 7th.

114 c. Arnati & Harrison
100 Charles & Fox

£60. Gilliard & Holt

Ore—
40 t. Bk. of N. S Wales
10 H. Emanuel
10 "
and, 12 t. R. T. Turnbull & Co.
, 16 H. Emanuel

Acid—
, 90 pks. A. & M. Zimmermann
21 Phillipps & Graves
150 Beresford & Co.

of Lime—
25 t. Johnson & Hooper

ales, 1 t. Dalgety & Co., Ld.
es, 4 L. & I. D. Jt. Co.
land, 1 R. Brooks

one—
11 t. R. Morrison & Co.
20 B. Jacob & Son
400 W. C. Bacon & Co.
400 Brandram, Bros., & Co.
200 Northcott, Sons, & Co

s—
m, 22 cks. Leach & Co.
ny, 19 H. A. Litchfield & Co.
19 "
22 W. Harrison & Co
m, 140 A. Zumbeck & Co.

Regulus—
y, 45 t. A. Gibbs & Sons
25 Vivian, Younger, & Co.

ras—
ny, £55. Burgoyne, Burbidge, & Co

c Potash—
e, 8 drs. J. A. Reid
26 F. W. Berk & Co.

ic Soda—
nd, 100 c. Beresford & Co.

chouc—
bar, 15 c Anderson, Weber, & Smith
tave, 40 W. Johnston & Co , Ld.
e, 13 Thames S. T. Co.
gascar, 160 c. L. & I. D. Jt. Co.
, 85 Kebbel Son & Co.
e, 21 "
tave, 48 "
323 A. Frey & Co., Ld.
, 10 Gallaty, Hankey, & Co.
, 66 Anderson, Weber, & Smith
Africa, 189 W. Johnston & Co., Limited
e, 60 L. & I. D. Jt. Co.
bar, 21 Clarke & Smith
, 47 Lewis & Peat
dies, 121 C. Scheeps & Co
20 Anderson, Weber, & Smith
bar, 230 L. & I. D. Jt. Co.

r Sulphate—
um, 32 cks. Middleton & Co
nd, £338. "

t Ore—
ey, 4 c. Mason, Bros., Ld.

n of Tartar—
e, 5 pks. B. & F. Wf Co., Ld.

icals *(otherwise undescribed)*
e, £96. E. W. Carling & Co.
any, 58 T. H. Lee
50 Craven & Co.
384 A. & M. Zimmermann
, 20 L. & I. D. Jt. Co.

Chemicals—*continued*
Holland, £10 H Boyce
" 15 Phillipps & Graves
" 85 R. Morrison & Co.
Germany, 14 F. Stahlschmidt & Co.
" 120 C. Faust & Co.
U. S., 16 G Boor & Co.
Germany, 19 J. James & Son
" 50 Phillipps & Graves
" 135 T. H. Lee
" 150 A. & M. Zimmermann
Holland, 14 R. Morrison & Co.
Germany, 60 T. H. Lee
" 900 A. & M. Zimmermann

Dyestuffs—

Anthracene
Holland, 33 pks. Little & Johnson

Valonia
A. Turkey, 56 t. A. & W. Nesbitt
" 20 J. Graves
" 60 Anning & Cobb

Extracts
U.S., 61 pks. G. B. Mackereth
" 55 T. Ronaldson & Son
France, 250 L.J. Levinstein & Sons

Gambier
E. Indies, 210 W. H. Cole & Co.
Singapore, 213 L. & I. D. Jt. Co.
Sts. Settlements, 294 brls. R. & J. Henderson
" 967 W. H. Cole & Co.
China, 648 L. & I. D. Jt. Co.

Cochineal
Singapore, 219 pks. "
Canaries, 4 Johnson, Rolls, & Co.
" 12 Schenker & Co.
" 21 Bruce & White

Alizarine
U. S., 5 British Alizarine Co.

Annatto
France, 4 Hurford & Middleton

Tanners' Bark
U. S., 27 t. J. Graves
Adelaide, 79 Devitt & Hett
" 17 Goad, Rigg, & Co.
Melbourne, 85 Baxter & Hoare
Sydney, 1 L. & I. D. Jt. Co.
Melbourne, 4 Anning & Cobb
Holland, 20 H Henle
" 50 Hicks, Nash, & Co.
" 5 Baxter & Hoare
Melbourne, 400 "
" 103 Beresford & Co.
Sydney, 10 Anning & Cobb
Launceston, 203 A. & W. Nesbitt
" 52 Culverwell, Brooks, & Co.
" 10 Hicks, Nash, & Co.
Holland, 24 G S. N. Co.
" 50 Boutcher, Mortimore & Co.
" 27 J. J. Williamson & Son
Melbourne, 102 A. & W. Nesbitt
Sydney, 25 W. H. Boult

Indigo
Holland, 2 chts. L. & I. D. Jt. Co.
Sweden, 1 "
E. Indies, 22 "
" 18 Parsons & Keith
" 10 W. Isted
E. Indies, 15 chts. Arbuthnot, Latham, & Co.
" 10 L. & I. D. Jt. Co.
" 83 "
" 203 Barlow & Bros.
" 57 L. & I. D. Jt. Co.
E. Indies, 86 "
" 22 J. Forsey
Germany, 1 box L. & I. D. Jt. Co.

Myrabolams
E. Indies, 4,098 pks. W. Shelcott
" 173 L. & I. D. Jt. Co.
" 173 "
" 1,107 "
" 1,340 Beresford & Co.
" 1,033 J. P. Alpe & Co.

Turmeric
E. Indies, 294 Newcomb & Son

Safflower
E. Indies, 25 bls L. & I. D. Jt. Co.

Argols
Italy, 808 pks. B. Jacob & Son

Sumac
Italy, 1,215 pks. J. Kitchen, Ld.
" 400 W France & Co.
" 400 W. Shelcott

Cutch
E. Indies, 900 Wrightson & Son
" 200 Middleton's S. S. Wf. Co.

Gutta Percha—
Singapore, 423 c. Kaltenbach & Schmitz

Glycerine—
Germany, £120 Beresford & Co.
" 32 T. H. Lee

Glucose—
U. States, 40 pks. 211 cwts. Page, Son, & East
" 25 148 Union Lighterage Co. Ltd.
" 120 668 L. Sutro & Co.
Germany, 57 405 "
" 70 680 L. & I. D. Jt. Co.
" 75 595 Pillow, Jones & Co.
" 1,000 1,000 R. E. Drummond & Co.
" 25 200 T. M. Duche & Sons
" 1,000 1,000 T. C Howell
" 310 300 J. Barber & Co.
" 100 100 Fellows, Morton & Co.
" 26 212 J. Knill & Co.
France, 100 100 Barrett, Tagant & Co.
U. States, 800 800 "
" 72 360 T. H. Lee
" 100 560 Union Lighterage Co. Ltd.
Germany, 15 135 J. Knill & Co.
U. States, 150 900 L. Sutro & Co.
" 38 190 "
Germany, 24 200 J. Knill & Co.
" 25 199 Pillow, Jones & Co.
" 30 236 Prprs. Chmbln's Whf.
U. States, 50 250 Allan Bros.
" 64 360 Union Lighterage Co Ltd.
Germany, 100 100 J. Barber & Co.
" 50 395 Pillow, Jones & Co.
" 40 349 Proprs. Chmbln's Whf.
U. States, 150 710 Union Lighterage Co. Ltd.

Isinglass—
Germany, 4 pks. L. & I. D. Jt. Co.
E. Indies, 18 "
" 17 cs. Clarke & Smith
France, 44 bgs. L & I D. Jt. Co.
Demerara, 2 brls. J. Hales, Caird & Co.
E. Indies, 1 pkg. Forbes, Forbes & Co.
Demerara, 2 L. & I. D. Jt. Co.
U. States, 10 pks. W. M. Smith & Sons
Germany, 17 L. & I. D. Jt. Co.

Manganese Ore—
Japan, 10 t. Matheson & Co.
Pt. Augusta, 54 F. Manders
" 38 Harrold Bros.

Manure—
Holland, 133 t. Miller & Johnson
Aruba, 290 J. A. Cockman
Holland, 205 t. Lawes Manure Co. Ltd.
France, 136 Anglo Cntnl. Guano Works
" 115 A. Hunter & Co.
B. Ayres, 6 L. & I. D. Jt. Co.

Phosphate Rock
Belgium, 138 t. A. Hunter & Co.

Naptha—
Belgium, 6 drs. Lister & Biggs
Holland, 19 Little & Johnston
Belgium, 44 Domeier & Co.

Paraffin Wax—
U. States, 75 brls. Prop. Chamberlain's Whf.
" 330 H. Hill & Sons
" 1,075 J. H. Usmar & Co.

Potash—
Holland, 14 pks. Spies Bros. & Co.

Potassium Bicarbonate—
Holland, 9 cks. A. & M. Zimmermann

Plumbago—
China, 76 brls. L. & I. D. Jt. Co.
Germany, 17 cks. H. J. Huth & Co.
E. Indies, 80 brls. Props. Chamberlain's Wharf
" 225 E. Barber & Co.
Germany, 335 cs. Props. Scott's Wharf
E. Indies 108 brls. J. Thredder, Son & Co.

Sodium Hyposulphate—
Germany, £90 J. James & Son, Germany
Holland, 130 Beresford & Co

Sugar of Lead—
Holland, £40 Beresford & Co.

Sodium Acetate—
Holland, £95 Soundy & Son

Saltpetre—
Germany, 8 cks. Craven & Co.
" 100 T. Merry & Son
Holland, 64 C. Wimble & Co.
E Indies, 1,330 bgs. Ralli Bros.
" 1,178 C. Wimble & Co.
Germany, 118 cks. P. Hecker & Co.

Soda Ashes—
Holland, 200 c. Taylor, Sommerville & Co.

Stearine—
France, 10 sks. Van Geelkerken & Co.
Belgium, 125 bgs. H. Hill & Sons
Holland, 54 "
France, 293 "
" 54 sks. Beresford & Co.

Tar—
France, 3 brls. Burt, Boulton & Co.

Tartaric Acid—
France, 24 pks. B. & F. Wf. Co. Ltd.
" 40 W. C. Bacon & Co.
Holland, 4 Northcott & Sons

Tartar—
Italy, £2,265 Thames S Tug Co.
" 550 Seager, White & Co.

Turpentine—
U.States, 2,652 brls. Nickoll & Knight
Brunswick, 1,500 brls. Nickoll & Knight

Ultramarine—
Holland, 4 cks. Ohlenschlager Bros.
Germany, 4 G. Steinhoff
" 16 cs. Hughes, Barrett & Co.

Zinc Oxide—
France, 19 cks. W. Harrison & Co.
Holland, 20 G. Dards
" 161 brls. M. Ashby

LIVERPOOL.

Week ending June 10th.

Albumen—
Hamburg, 1 ck.

Bones—
Talcahuano, 750 bgs. T. Berg
Parnahiba, 2 t. R. Singlehurst & Co.
Smyrna, 161 bgs. O. Gantes & Co.

Caustic Potash—
Dunkirk, 33 dms.

Cream of Tartar—
Tarragona, 5 cks.
Patras, 8 brls. Papayanni & Jeremias
Copper Ore—
St. John's, N'fd., 1 brl. J. J. Langley
New Orleans, 343 sks R. Bulman & Co
" 138 Cunningham & Shaw
Caoutchouc—
Para, 237 cs. 75 srs. F. Snarez & Co.
" 35 73 R. Singlehurst & Co.
" 39 104 "
Ceara, 5 bls. Leech, Harrison & Co.
Parnahiba, 2 cs. R Singlehurst & Co.
Accra, 14 cks. Pickering & Berthoud
" 13 Taylor, Laughland & Co.
" 14 J. J. Fischer & Co.
" 3 Hutton & Co.
" 10 Edwards Bros.
" 5 C. Lane & Co.
" 28 Davies, Robbins & Co.
" 5 J. H. Rayner & Co.
" 8 F. & A Swanzy
" 4 cs. 65 J. P. Werner
" 1 bg. Radcliffe & Durant
Sierra Leone, 154 brls. Pickering & Berthoud
" 104 cks.
Chemicals *(otherwise undescribed)*
Rotterdam, 7 cks.
Bremerhaven, 3 cs.
Dried Blood—
B. Ayres, 230 bgs.
Dyestuffs—
Argols
Brindisi, 285 sks.
Extracts
New York, 8 brls.
Divi Divi
Amsterdam, 844 bgs. Browne, Geveke & Co.
Hamburg, 208
Havre, 168 bgs.
Valonia
Smyrna, 1 bg. G. Galatti
" 2
Dyestuffs *(otherwise undescribed)*
New York, 25 cs.
Glucose—
Stettin, 300 bgs. 12 cks.
Dunkirk, 100
New York, 130 brls.
Hamburg, 40 bgs. 15
Philadelphia, 40 brls.
Horn Piths—
B. Ayres, 33,245 kilos. Williamson, Milligan & Co.
Rosario, 34,654
Paysandu, 36,000
B. Ayres, 39,540
Isinglass—
Para, 24 cs. Bieber & Co.
" 12 Frith, Sands & Co.
Manure—
Phosphate Rock
Oruba, 320 t. Phospho Guano Co.
Charleston, 825 "
Montreal, 200
Corsaw, 1,660
Manganese Ore—
Poti, 1,880 t.
Gothenburg 100 Macqueen Bros.
Coquimbo, 850
Nitrate—
Iquique, 2,859 bgs.
Potash—
Dunkirk, 11 cks.
Pyrites—
Huelva, 1,624 t. Matheson & Co.
Phosphoric Acid—
Bremerhaven, 15 cs.
Potash Salts—
Hamburg, 100 brls.
Potassium Bicarbonate—
Rotterdam, 4 cks.

Saltpetre—
Hamburg, 6 cks.
Sodium Silicate—
Rotterdam, 1 ck.
Sulphur Ore—
Huelva, 1,130 t. Matheson & Co.
Tartaric Acid—
Rotterdam, 2 cks.
Turpentine—
Brunswick, 50 brls.
Ultramarine—
Rotterdam, 5 cks. 40 brls.
Hamburg, 30 cs.
Waste Salt—
Hamburg, 130 t.
Zinc Ashes—
Trieste, 5 brls. Hughes, Chenery & Co.

GLASGOW.

Week ending June 12th.

Barytes—
Rotterdam, 48 cks.
Caoutchouc—
New York, 312 bgs. 18 cs. 14 brls.
Chemicals *(otherwise undescribed)*
Rotterdam, 1 cs. J. Rankine & Son
Hamburg, 5 4 cks.
" 1
Dyestuffs—
Logwood
Jamaica, 704 t.
Alizarine
Rotterdam, 32 cks.
Extracts
Bordeaux, 375 cks.
Rouen, 4
Annatto
Bordeaux,
Gutta Percha—
Rouen, 2 cs.
Gypsum—
Rouen, 1 lot
Oxalic Acid—
Rotterdam, 5 cks.
Paraffin Wax—
New York, 175 brls. H. Hill & Son
Pitch—
St. Nazaire, 95 cks.
Phosphate—
Montreal, 100 t.
Sulphur Ore—
Huelva, 1,702 t. Matheson & Co.
Bergen, 180 t.
Tartaric Acid—
Rotterdam, 1 ck. J. Rankine & Son
St. Nazaire, 2 brls. F. & L. Sitzler
Tartar—
Bordeaux, 13 cks.
Ultramarine—
Rotterdam, 2 cks. J. Rankine & Son
Waste Salt—
Hamburg, 130 t.
" 260 1,320 bgs.
Zinc Oxide—
Antwerp, 5 brls.

HULL.

Week ending June 12th.

Albumen—
St. Petersburg, 60 cs.
Acids *(otherwise undescribed)*
Rotterdam, 5 cks. Wilson, Sons & Co.
Barytes—
Antwerp, 49 cks. Bailey and Leetham
" 61 H. F. Pudsey
Bremen, 10 cks. Veltmann & Co
" 46 T. W. Flint & Co.

Caoutchouc—
Marseilles, 49 pks. Wilson, Sons & Co.
Chemicals *(otherwise undescribed)*
Genoa, 10 cks. Wilson Sons & Co.
Stettin, 31 "
Dunkirk, 250 "
Antwerp, 70 pks. "
Hamburg, 796 "
Dantzic, 1 pkg.
Dyestuffs—
Argols
Bordeaux, 97 bgs. Rawson & Robinson
Extracts
Dunkirk, 47 cks. Wilson, Sons & Co.
Rouen, 30 Rawson & Robinson
Rennet
Copenhagen, 1 cs. Wilson Sons & Co.
Rotterdam, 5 cks. G. Lawson & Sons
Alizarine
Rotterdam, 5 cks. Hutchinson & Son
" 10 Hutchinson & Son
" 1 cs. 68 cks. W. & C. L. Ringrose
Madders
Rotterdam, 1 ck. J. Pyefinch & Co.
Glycerine—
Amsterdam, 5 crbys.
Glucose—
Dunkirk, 120 bgs.
Stettin, 200 18 cks. Wilson Sons & Co.
" 1,200 C. M. Lofthouse & Co.
" 200 25 cks.
Isinglass—
Hamburg, 1 cs. Wilson, Sons & Co.
Oxide—
Rotterdam, 1 cs. Hutchinson & Son
Pitch—
Hamburg, 61 brls.
Dantzig, 13 cks.
Amsterdam, 13 cks. H. F. Pudsey
Plumbago—
Hamburg, 1 brl. Furley & Co.
" 23 F. B. Grotrian
" 1 cs. 15 cks.
Stearine—
Rotterdam, 53 bls. G. Lawson & Sons
" 35 bgs. Wilson, Sons & Co.
Soda—
Rotterdam, 8 cks. G. Lawson & Sons
Sulphur—
Marseilles, 15 cks.
Tartar—
Bordeaux, 7 cks. Rawson & Robinson
Turpentine—
Libau, 51 brls.
Savannah, 1,500 cks.
Ultramarine—
Rotterdam, 1 ck.
" 2

TYNE.

Week ending June 10th.

Alum—
Rotterdam, 31 cks. Tyne S. S. Co.
Alum Earth—
Hamburg, 29 cks. Tyne S. S. Co.
Barytes Sulphate—
Antwerp, 52 cks. Tyne S. S. Co.
Barytes—
Antwerp, 64 cks.
Rotterdam, 86 Tyne S. S. Co.
Copper Precipitate—
Huelva, 2,481 bgs. Bede Metal Chemical Co.
" 3,937 bgs. 2 cks. Henry Schofield & Son
" 14,754 bgs. A. Guthrie
" 2,819 Scott Bros.

Copper Ore—
Huelva, 375 t. Bede Metal C
Glucose—
Hamburg, 16 cks. Tyne S.
Pyrites—
Huelva, 2,185 t. Sco
" 1,756
" 300 t. A.
" 1,260 Scot
Phosphate—
Antwerp, 260 t. Langdale's C
Salt Cake—
Harburg, 150 t.
Saltpetre—
Harburg, 268 t. Clephans & W
Sulphur Ore—
Pomaron, 1,000 t. Mason &
Drontheim, 620
Ultramarine—
Rotterdam, 20 cs. Tyne S.
Waste Salt—
Harburg, 250 t. Jarrow Chemi

GOOLE.

Week ending June 4th

Alkali—
Ghent, 2 cks.
Rotterdam, 35
Alum—
Monte Christi, 40 brls.
Caoutchouc—
Boulogne, 5 cks.
" 4 1 cs.
Chemicals *(otherwise undesc*
Rotterdam, 1 ck.
Dyestuffs—
Divi Divi
Rio Hacha, 260 t.
Alizarine
Rotterdam, 21 cks.
Logwood
Monte Christi, 432 t.
Rotterdam, 691 pieces
Jamaica 478 t.
Monte Christi, 330
Cutch
Boulogne, 6 cs.
Dyestuffs *(otherwise undescri*
Boulogne, 1 cs. 2 pks. 44 c
" 2 9
" 2 2 51
Glucose—
Hamburg, 23 cks. 100 bgs.
" 12 100
Potash—
Hamburg, 4 cks.
Dunkirk, 22
Calais, 41 drums
Red Lead—
Hamburg, 6 cks.
Saltpetre—
Hamburg, 3 cks.
Waste Salt—
Hamburg, 152 t. 296 brls.
" 335

GRIMSBY.

Week ending June 7th.

Caoutchouc—
Hamburg, 5 cks.
Chemicals *(otherwise undesc*
Rotterdam, 1 ck.
Glucose—
Hamburg, 10 cks.

EXPORTS OF CHEMICAL PRODUCTS

OF

THE PRINCIPAL PORTS OF THE UNITED KINGDOM.

LONDON.

Week ending June 4th.

Ammonium Sulphate—

Brught,	139 t.		£1,668
Emmerick,	101		1,212

Ammonium Sulphate—*continued.*

Barbadoes,	100 t.		£1,228
Amsterdam,	50	15 c.	560
Berbice,	5		63
Rotterdam,	20	6	225

Ammonium Carbonate—

Bremen,	24 pks.	£79
Montreal,	40 kgs.	65
New York,	13 t. 8 c.	£415

Ammonium Muriate—

Auckland,	1 t. 15 c.

Arsenic—

Melbourne,	5 t.

Acid—
l, 50 pks. £120
ton, 4 t. 9 c. £101
Acid—
s, 2 cks. £18
rne, 1 t. 3 c. 45
Acid—
rg, 2 cs. £44
lphia, 1½ c. 39
Potash—
iver, 600 lbs. £23
Bisulphide—
lphia, 4 t. 12 cs. £59
s, 60 drs. 29
Sulphate—
ux, 4 t. £87
own, 10 c. 12
, 21 18 455
s, 2 60
c Acid—
ama, 10 cs. £60
ux, 10 pks. 30
c Solid—
ork, 30 cs. £105
lles, 120 cks. 1,647
rg, 4 t. 16 c. 400
e of Lime—
own, 1 t. 5 c. £12
n Chloride—
17 t. £48
Soda—
2 t. 18 c £30
10 20
ren, 10 1 100
of Tartar—
nd, 1 t. £99
stle, 10 c. 51
5 26
Acid—
eal, 5 c. £40
, 5 cks. 36
is, 10 74
2½ 20
tersburg, 1 t. 140
rdam, 2 14
roducts—
h
pore, 50 brls. £17
tta, 100 rlts. 43
ndria, 100 brls. 30
a, 1,976 t. 3.221
oping, 1,000 brls. £594
tta, 50 44
300 rlts. 171
asole
rdam, 50 rlts. 342
ama, 10 drs. 26
rdam, 43 840
, 20 cks. 337
hracene
rdam, 252 cks. £1,999
61 435
ldorf, 60 336
60 sks. 348
rdam, 150 cks. 950
od Naptha
tta, 10 dms. 50 gls. £14
cals *(otherwise undescribed)*
y, 22 pks. £194
, 2 26
26 46
ndon, 5 33
Blood—
itius, 18 t. 10 c £120
ectants—
o, 2 c. £25
gton, 84 pks. 50
urg, 20 cks. 100
y, 100 gls. 30
urg, 28 cks. 140
25 150
annite—
itius, 10 t. £300
100 cks. 150
ulphite of Lime—
York, 3 cs. £32
3 63
Sulphate—
urg, 11 t. 14 c. £55
tic Acid—
, 4 t. £72
re—
, 15 t. £100
incent, 10 101
urg, 30 9 c. 118
nica, 19 16 200
does, 25 250
urg, 3 1 30
does, 6 61
ica, 3 31
urg, 210 2,625

Manure—*continued.*
Madeira, 19 t. 17 c. 182
Gr. Canary, 120 1,080
Hamburg, 100 1,250
Oxalic Acid—
Antwerp, 8 cks. £60
Potassium Chlorate—
New York, 1 t. 10 c. £63
" 1 42
Phosphorus—
Cape Town, 9 c. £84
Hiogo, 1,600 lbs. 105
Yokohama, 500 60
M Video, 4 cs. 33
Dunedin, 600 lbs. 35
Sodium Hyposulphite—
New York, 5 cs. £53
" 2 32
Sodium Bisulphide—
Philadelphia, 7 c. £28
Sodium Acetate—
New York, 5 t. £79
Sulphur Flowers—
Cape Town, 450 kgs. 125 cks. £352
Madeira, 3 t. 14 c. 41
Sulphurous Acid—
Shanghai, 3 c. £35
Sheep Dip—
Algoa Bay, 15 t. £840
Fremantle, 2 16 c. 140
Natal, 1 9 70
Algoa Bay, 30 cs 84
E London, 1 t. 14 c. 75
Natal, 5 cs 55
E. London, 81
" 85 pks. 260
Algoa Bay, 10 cs. 27
Superphosphate—
Mauritius, 78 t. 19 c. £350
Lyttleton, 99 7 395
Sulphuric Acid—
Singapore, 3 t. 4 c. £13
Algoa Bay, 30 cs. 21
Trinidad, 3 t. 4 c. 12
Singapore, 1 3 20
Sulphur—
E. London, 60 pks. £35
Algoa Bay, 3 t. 7 c. 25
" 1 7 16
Lisbon, 250 brls. 193
Saltpetre—
Syra, 10 t 12 c. £220
Sydney, 2 43
Camocin, 10 225
Oporto, 3 15 76
Melbourne, 1 22
Gothenburg, 2 3 42
Salammoniac—
Hamburg, 19 c. £33
Gothenburg, 1 t. 4 41
Saltcake—
Antwerp, 115 t. £201
" 150 263
Tartaric Acid—
Wellington, 3 cs. £23
Newcastle, 3 c. 20
Bombay, 4 34
Cape Town, 5 39
Montreal, 15 110
Otago, 2 13

LIVERPOOL.

Week ending June 7th.

Ammonium Sulphate—
Antwerp, 30 t. £360
Barbadoes, 24 10 c. 288
Las Palmas, 37 612
Valencia, 77 924
Nantes, 29 13 332
Valencia, 76 1,032
Las Palmas, 10 120
Bordeaux, 8 2 3 q. 110
Demerara, 58 19 707
Valencia, 50 11 3 700
Antigua, 15 1 3 282
Antimony Regulus—
New York, 5 t. 14 c. £350
" 9 17 701
Acetate of Lead—
Colon, 9 c. 3 q. £14
Alum—
Havana, 4 t. 17 c. £24
Vigo, 2 1 11
Melbourne, 1 1 q. 5
Canea, 1 10 8
Calcutta, 25 125
Colon, 9 5
Cape Town, 2 3 11
Calcutta, 101 7 608
Rio Janeiro, 19 5

Acetic Acid—
Colon, 1 t. 3 c. 2 q. £19
Porto Alegre, 1 3 5
" 1 3 5
Albumen—
San Francisco, 15 c. £25
Colon, 5 22
Lisbon, 5 2 q. 20
Ammonium Muriate—
Philadelphia, 6 t. 10 c. £149
" 5 12 128
New York, 5 5 130
Philadelphia, 23 1 3 q. 550
Ammonium Carbonate—
Rotterdam, 3 c. £6
Venice, 4 1 q. 6
Genoa, 1 t. 33
" 6 1 10
Gothenburg, 5 3 7
Venice, 4 7
Bleaching Powder—
Antwerp, 27 cks.
Baltimore, 109
Boston, 156
Naples, 3
Vigo, 4
Barcelona, 50
Bombay, 6 cs.
Boston, 161
Havana, 21 brls.
Naples, 15 pps.
New York, 86 tcs.
Rio Janeiro, 10 cs.
Rotterdam, 41
Bone Waste—
Rouen, 9 t. 19 c. £75
Borate of Lime—
Rotterdam, 10 t. 6 c. 1 q. £124
New York, 106 3 1,234
Borax—
Savanilla, 1 c. £6
Antwerp, 9 t. 15 1 q. 263
Rotterdam, 166 cks. 200
Antwerp, 12 cs. 36
Rotterdam, 2 t. 4 66
Halifax, 1 2 1 32
Copperas—
San Francisco, 10 t. £21
Piræus, 3 t. 4 c. 2 q. 8
Caustic Soda—
Adelaide, 35 dms.
Bahia, 30 20 kgs.
Baltimore, 200
Barcelona, 20
Dunedin, 36 cks.
Genoa, 36
Havana, 10
Ibrail, 15
Malta, 5
Maranham, 10
Marseilles, 5
Melbourne, 35
Montreal, 170
New Orleans, 35
New York, 350
Palermo, 5
Rosario, 30
Rouen, 48
Stettin, 54
Sydney, 50
Tampico, 250 kgs.
Venice, 20
Alexandria, 36
Antwerp, 25
Bilbao, 50 brls.
Bombay, 10
Calcutta, 178
Corunna, 20
Danzig, 50
E. London, 10
Ensenda, 25
Fiume, 15
Galatz, 30
Ghent, 18
Hamburg, 6 5
Havana, 15
Ibrail, 32
La Guayra, 30
Malaga, 26 pks
Marseilles, 15
Melbourne, 25
Montreal, 130
Naples, 20
Newfairwater, 60
New York, 278
Oporto, 30
Pasages, 25 brls.
Penang, 100
Pernambuco, 15
Riga, 200
St. Petersburg, 50
Santander, 15
Seville, 322
Singapore, 12

Caustic Soda—*continued.*
Stettin, 95 dms.
Trieste, 30
Carbolic Solid—
Rotterdam, 1 c. 1 q. £9
Carbolic Acid—
Amsterdam, 5 c. £32
Philadelphia, 1 t. 10 81
Rio Janeiro, 2 6 66
Montreal, 11 30
Philadelphia, 5 7
Copper Precipitate—
Antwerp, 24 t. 6 c. £725
Copper Sulphate—
Rouen, 37 t. 3 c. £943
Bordeaux, 56 17 2,891
Nantes, 16 19 721
Madras, 10 240
Bordeaux, 23 18 1,178
Genoa, 2 8 40
Marseilles, 204 13 2 q. 4,725
Pasages, 10 233
Bordeaux, 5 7 134
Havre, 5 2 107
Genoa, 25 5 558
Bari, 1 1 18
Barcelona, 10 181
Sydney, 2 47
Genoa, 115 2,679
Havre, 2 45
Bordeaux, 58 3 1,397
Mersyne, 10 15
Venice, 15 9 2 273
Bordeaux, 38 15 814
Nantes, 49 5 1,221
Barcelona, 7 3 2 177
Bordeaux, 1 18 45
Barcelona, 35 2 2 875
Bordeaux, 110 9 2 2,750
Valencia, 4 19 125
La Union, 2 45
Nantes, 112 16 1 2,800
Calcium Chlorate—
Constantinople, 2 t. 14 c. £5
Charcoal—
Rouen, 20 t. £100
Hamburg, 30 150
Rouen, 15 75
Havre, 15 54
" 15 54
Chemicals *(otherwise undescribed)*
Colon, 4 c. 2 q. £9
Rio de Janeiro, 1 5
Taltal, 7 2 19
Disinfectants—
Kingston, 4 c. £6
Bombay, 13 t. 17 166
Baltimore, 20 cs. 88
Madras, 1 5
Calcutta, 2 39
Shanghai, 3 8
Epsom Salts—
Calcutta, 100 kgs.
Rio Janeiro, 50
Ground Spar—
Havre, 98 t. 13 c. £738
Rotterdam, 50 15 800
Guano—
Hamburg, 2 t. £10
Gypsum—
New York, 10 t. 2 c. 2 q. £30
Lead Nitrate—
Colon, 5 c. 1 q. £7
Magnesium Sulphate—
Seville, 1 t. 10 c. 3 q. £8
Magnesia—
Havana, 10 cs.
San Francisco, 11 pks.
Trieste, 2 cs.
Manure—
Valencia, 50 t. £140
St. Nazaire, 81 1 c. 2 q. 235
Gr. Canary, 19 12 158
Rouen, 50 3 460
Magnesium Chlorate—
Bombay, 5 t. 10 c. £20
Muriatic Acid—
Ensenada, 1 t. £13
Magnesium Calcide—
Trieste, 38 lbs. £45
Oxalic Acid—
Barcelona, 5 t. 2 c. 2 q. £165
Calcutta, 6 11
Potassium Carbonate—
Corunna, 4 c. 2 q. £5
Pitch—
Coquimbo, 100 brls.
Potashes—
Palermo, 3 t. 5 c. £59
Potassium Silicate—
Amsterdam, 1 t. 2 c. 1 q. £6

Phosphorus—
Vera Cruz, 1 t. 15 c. 2 q. £177
Havana, 2 26

Potassium Chlorate—
St. Petersburg, 2t. 10 c. £108
Hamburg, 3 136
Copenhagen, 7 10 350
New Fairwater, 2 t. 10 111
Riga, 5 t. 207
Corunna, 10 22
Gothenburg, 12 503
Hamburg, 10 22
Corunna, 1 56
St. Petersburg, 5 11
Havana, 4 8 190

Potassium Bichromate—
Yokohama, 2 t. £75
Hiogo, 3 15 c. 145
Havre, 4 16 135
New York, 8 2 q. 61

Sodium Biborate—
Antwerp, 5 t. 4 c. 1 q. £156
Rotterdam, 2 13 80
Yokohama, 3 11 2 107

Superphosphate—
Barbadoes, 14 t.

Saltpetre—
Matanzas, 7 c. 2 q. £9

Salt—
Antwerp, 136 t.
Archangel, 514
Boston, 200
Faroe Islands, 182
Ghent, 95
Laurvig, 300
Para, 35¾
Pernambuco, 2 cs.
Podpakta, 200 t.
Antwerp, 25
Barbadoes, 100
Bata, 10
Bonny, 50
Calcutta, 9,675
Cameroons, 140
Cape Coast, 6
Cape Town, 20
Ghent, 175
Iceland, 96
Montreal, 494
New York, 70
Para, 107 2 c.
Quebec, 494
Rotterdam. 324
St. John, N.B. 300
Salt Pond, 6¾
Savannah, 602
Trinidad, 13½

Sodium Bicarbonate—
Boca, 40 cks.
B. Ayres. 30
Calcutta, 1,000 kgs.
Champerico, 40
Hamburg, 22
Melbourne, 54 220
Montreal, 150
New York, 33
Rotterdam, 3 25
St. Petersburg, 240
San Francisco, 200
Havana, 20
Marseilles, 3

Soda Ash—
Baltimore, 245 cks.
Boston, 850 bgs
Monte Video, 7 brls.
Montreal, 26 tcs.
New York, 606
Para, 6
Philadelphia, 552
Rio Janeiro, 50
San Francisco, 11
Baltimore, 28 34 cks.
Bilbao, 50 brls.
Boston, 67
B. Ayres, 70
Buffalo, 60
Ghent, 8
Malta, 30
New York, 106
Patras, 30
Rio Janeiro, 180
Santos,
Sydney, 40
Oporto. 10

Soda Crystals—
Calcutta, 50 brls.
Gibraltar, 10
St. John's, Nfld. 20 kgs.

Sodium Silicate—
Carthagena, 7 cks.
Corunna, 17 brls.
Madras, 5 cks.
Santander, 32
Seville, 7
Montreal, 39

Sodium Silicate—*continued.*
Salonica, 15 cks.
Venice, 32

Tar—
Bahia, 40 brls.
Sierra Leone, 15

Sulphuric Acid—
Kingston, 15 galls. £7

Sugar of Lead—
Rio de Janeiro, 5 c. 1 q £9

Sulphur—
Rouen, 22 t. 19 c. £104
New York, 25 1 110
Boston, 75 3 356
Porto Alegre, 16 7
Halifax, 7 10 1 q. 45
Madras, 5 10 2 42
New York, 18 6 82
" 225 1,005
Boston, 45 181
Rangoon, 2 9
New York, 50 2 225

Sodium Chlorate—
Rotterdam, 1 t. £35

Sodium Bichromate—
Marseilles, 18 c. £25

Saltcake—
Montreal, 50 t. 5 c. £126
Philadelphia, 30 45

Salammoniac—
Alexandria, 10 c. £17
Beyrout, 12 21
Malta, 10 17
Constantinople, 3 5
Batoum, 2 t. 3 q. 70
New York, 4 18 168
Corunna, 5 9
Alexandria, 2 5 78
Corfu, 5 7
Philadelphia, 7 11 1 265
Gothenburg, 6 2 11
Syra, 17 1 31

Superphosphate of Lime—
Valencia, 50 t. £125

Sodium Sulphate—
Talcahuano, 2 t. 10 c. £32
Hamburg, 2 10
Wellington, 2 2 16

Sodium Nitrate—
Toronto, 2 t. £16
Ancona, 19 154
Bari, 7 2 c. 1 q. 60
Antigua, 5 1 1 39
Malaga, 2 1 16
Pasages, 6 64
Malaga, 3 6 1 28
Montreal, 2 16

Tartaric Acid—
Bahia, 8 c. £58
Havana, 1 t. 155

Tar—
Old Calabar, 5 brls

Zinc Muriate—
Kurrachi, 2 t. 16 c. £50
Bombay, 1 15 24

TYNE.

Week ending June 10th.

Antimony—
Copenhagen, 1 t.
Rotterdam, 1
St. Petersburg, 11 t.

Ammonium Carbonate—
Copenhagen, 6 c.

Ammonium Sulphate—
Rotterdam, 50 t.

Aluminium—
St. Petersburg, 305 lbs.

Alkali—
Rotterdam, 6 t. 7 c.
St. Petersburg, 107 t. 15
Gothenburg, 9 cks.

Barium Chloride—
St. Petersburg, 1 t. 2 c.

Bleaching Powder—
Copenhagen, 4 t. 1 c
Libau, 24 19
Rotterdam, 16 2
St. Petersburg, 10 t.
" 45
Gothenburg, 57 cks.
Konigsburg, 15 t. 6 c.
Hamburg, 67 13
Antwerp, 73 6
Malmo, 5
Christiania, 30 2

Copperas—
Sundswall, 187 t.
Tromsoe, 13 c.

Caustic Soda—
Rotterdam, 7 t. 5 c.
Hamburg, 10 3
Antwerp, 4 18
Christiania, 4 19

Copper Precipitate—
Hamburg, 7 c.

Litharge—
Antwerp, 1 t. 12 c.
Copenhagen, 1 3

Magnesia—
Hamburg, 19 c.

Magnesium Carbonate—
Libau, 19 c.

Potassium Chlorate—
Hamburg, 2 t. 4 c.
Antwerp, 2 4

Pitch—
Antwerp, 50 t.

Potassium Bichromate—
Malmo. 5 t. 5 c.

Sodium Chlorate—
Hamburg, 5 t.

Soda Ash—
Konigsberg 281 t.
Seville, 2 t. 18 c.
Hamburg, 9 3
Odense, 2 16

Soda—
Aarhuus, 20 t. 2 c.
Frederikshaven, 9 11
" 2 3
Malmo, 5 8
Copenhagen, 26 18
Esbjerg, 6 19
Groto, 1 1
Sundswall, 59 6
Kolding, 33
Tromsoe, 1 12
Wasa, 12
Helsingfors, 3 6
Svendborg, 19 16

Salt—
Esbjerg, 1 c.

Sodium Nitrate—
Huelva, 91 t. 14 c.

GLASGOW.

Week ending June 12th.

Alkali—
Dunedin, 3 t. 4 c. £24

Ammonium Sulphate—
Cullera, 55 t. 5¾ c. £611

Caustic Soda—
Dunedin, 103¾ c. £46

Copper Sulphate—
Italy, 23½ t. 508

Chemicals (*otherwise undescribed*)
Italy, 5 c. £65

Dyestuffs—
Turmeric
Montreal, 34 c. £23. 10s.
Fustic
Montreal, 13½ c. £27
Alizarine
Dunedin, 660 lbs. £44

Glucose—
Cape Town, 6 t. 14½ £120

Manure—
Antigua, 110 c. £39

Pearl Ash—
Dunedin, 29 c. £27

Phosphorus—
Dunedin, 1 t. 14½ c. £160

Potassium Bichromate—
Dunedin, 23¾ c. £42
" 3 t. ½ c. 24
Rotterdam, 69 cks. 620
Calcutta, 11 c. £20 15s.
Italy, 13 t. 13¾ £461
" 14 22
" 81½ 150
" 295 551

Paraffin Wax—
Dunedin, 10 t. 1¾ c. £500
Antwerp, 100 133
Italy, 3 9 87
Cape Town, 5 2 183

Soda Ash—
Dunedin, 7 t. 17¾ c. £53

Sulphuric Acid—
Madras, 9 t. 2½ c. £47 15s.

Sodium Bichromate—
Italy, 33 t. 11 c. £887
Quebec, 314 439

HULL.

Week ending June 10t[h.]

Alkali—
Bari, 2 cks.
Bergen, 12
Trieste, 61

Bleaching Powder—
Trieste, 4 cks.
Venice, 25

Borax—
Helsingfors, 1 ck.
Hango, 2

Bone Size—
Bremen, 10 cks.
Harlingen, 28
Rotterdam, 5

Benzole—
Dunkirk, 24 cks.

Caustic Soda—
Abo, 50 drms.
Helsingfors, 52
Hango, 26
Rotterdam, 20
Reval, 125

Chemicals (*otherwise undesc[ribed]*)
Antwerp, 36 cs. 2 cks.
Amsterdam, 16
Bremen, 1
Bari, 9 4
Christiania, 163 slbs. 2 24
Drontheim, 2
Dunkirk, 2 cs.
Gothenburg 639slbs 39 cs. 20 cks.
Helsingfors, 31 cks.
Hamburg, 266 cs. 41 3
Hango, 5
Libau, 25
Palermo, 3
Petersburg, 6 50
Rotterdam, 1
Trieste, 39
Uleaborg, 49
Wasa, 2
Wyburg, 14 1

Dyestuffs (*otherwise undesc[ribed]*)
Gothenburg, 1 ck.
Hamburg, 2 kgs. 2 ck.
Rotterdam, 4 ck. 2 cs.

Manure—
Hamburg, 1,901 bgs.

Pitch—
Antwerp, 434 t.
Dunkirk, 91

Salt—
Copenhagen, 1 t.

Tar—
Gothenburg, 25 cks.
Helsingfors, 350
Copenhagen, 3

GRIMSBY.

Week ending June 7t[h.]

Coal Products—
Dieppe, 39 cks.

Chemicals (*otherwise undes[cribed]*)
Antwerp, 6 drums
Dieppe, 102 pks.

GOOLE.

Week ending June 4[th.]

Benzole—
Ghent, 8 drums.
Hamburg, 15
Rotterdam, 49 cks.

Coal Products—
Ghent, 2 cs.
Rotterdam, 96 cks.

Caoutchouc—
Boulogne, 26 cks.

Chemicals (*otherwise unde[scribed]*)
Antwerp, 10 cks.
Hamburg, 10
Ghent, 1

Dyestuffs (*otherwise unde[scribed]*)
Boulogne, 5 cs. 6 cks.
Ghent, 29 tins 1
Rotterdam, 4
Ghent, 270 bgs.

Manure—
Hamburg, 935 bgs.
Rotterdam, 176 1 cks.

Pitch—
Antwerp, 280 t.
Boulogne, 15 cks.

PRICES CURRENT.

WEDNESDAY, June 18, 1890.

Prepared by Higginbottom and Co., 116, Portland Street, Manchester.

s stated are F.O.R. at maker's works, or at usual ports of shipment in U.K. The price in different localities may vary.

		£	s.	d.
5 °/₀ and 40 °/₀	per cwt.	7/-	&	12/3
lacial	,,		65/-	
S.G., 2000°	,,	0	12	9
82 °/₀	nett per lb.	0	0	6½
..	,,	0	0	3¾
(Tower Salts), 30° Tw.	per bottle	0	0	8
(Cylinder), 30° Tw.	,,	0	2	11
° Tw.	per lb.	0	0	2
..	,,	0	0	1¾
.	nett ,,	0	0	3¾
.	,,	0	1	1½
ic (fuming 50 °/₀)	per ton	15	10	0
(monohydrate)	,,	5	10	0
(Pyrites, 168°)	,,	3	2	6
(,, 150°)	,,	1	4	6
(free from arsenic, 140/145°)	,,	1	7	6
rous (solution)	,,	2	15	0
(pure)	per lb.	0	1	0½
c	,,	0	1	2½
white powdered	per ton	13	0	0
ose lump)	,,	4	17	6
Sulphate (pure)	,,	5	0	0
,, (medium qualities)	,,	4	10	0
m	per lb.	0	15	0
, ·880=28°	per lb.	0	0	3
=24°	,,	0	0	1¾
Carbonate	nett ,,	0	0	3½
Muriate	per ton	23	0	0
,, (sal-ammoniac) 1st & 2nd	per cwt.	37/-	&	35/-
Nitrate	per ton	40	0	0
Phosphate	per lb.	0	0	10½
Sulphate (grey), London	per ton	11	1	3
,, (grey), Hull	,,	11	1	3
Oil (pure)	per lb.	0	0	10½
Salt ,,	,,	0	0	9
y	per ton	75	0	0
(tartar emetic)	per lb.	0	1	1
(golden sulphide)	,,	0	0	10
Chloride	per ton	8	5	0
Carbonate (native)	,,	3	12	6
Sulphate (native levigated)	,,	45/-	to	75/-
ng Powder, 35 %	,,	4	17	6
Liquor, 7 %	,,	2	10	0
lde of Carbon	,,	13	0	0
um Acetate (crystal	per lb.	0	0	6
Chloride	per ton	2	2	6
Clay (at Runcorn) in bulk	,,	17/6	to	35/-
ar Dyes:—				
rine (artificial) 20 %	per lb.	0	0	9
nta	,,	0	3	9
ar Products				
racene, 30 % A, f.o.b. London	per unit per cwt.	0	1	5
ol, 90 % nominal	per gallon	0	3	9
50/90 ,,	,,	0	2	9
olic Acid (crystallised 35°)	per lb.	0	0	7¼
,, (crude 60°)	per gallon	0	2	2
ote (ordinary)	,,	0	0	2½
, (filtered for Lucigen light)	,,	0	0	3
Naphtha 30 % @ 120° C.	,,	0	1	4
e Oils, 22° Tw.	per ton	3	0	0
, f.o.b. Liverpool or Garston	,,	1	8	0
nt Naphtha, 90 % at 160°	per gallon	0	1	9
en Oils (crude)	,,	0	0	2½
(Chili Bars)	per ton	59	10	0
Sulphate	,,	22	0	0
Oxide (copper scales)	,,	57	10	0
ne (crude)	,,	30	0	0
(distilled S.G. 1250°)	,,	60	0	0
..	nett, per oz.	0	0	9
lphate (copperas)	per ton	1	10	0
lphide (antimony slag)	,,	2	0	0
heet) for cash	,,	14	10	0
itharge Flake (ex ship)	,,	15	10	0
cetate (white ,,)	,,	24	0	0
,, (brown ,,)	,,	18	0	0
arbonate (white lead) pure	,,	19	0	0
itrate	,,	22	0	0

		£	s.	d.
Lime, Acetate (brown)	,,	7	7	6
,, ,, (grey)	per ton	13	0	0
Magnesium (ribbon and wire)	per oz.	0	2	3
,, Chloride (ex ship)	per ton	2	5	0
,, Carbonate	per cwt.	1	17	6
,, Hydrate	,,	0	10	0
,, Sulphate (Epsom Salts)	per ton	3	0	0
Manganese, Sulphate	,,	17	10	0
,, Borate (1st and 2nd)	per cwt.	60/-	&	42/6
,, Ore, 70 %	per ton	4	10	0
Methylated Spirit, 61° O.P.	per gallon	0	2	2
Naphtha (Wood), Solvent	,,	0	3	11
,, ,, Miscible, 60° O.P.	,,	0	4	3
Oils:—				
Cotton-seed	per ton	22	10	0
Linseed	,,	26	0	0
Lubricating, Scotch, 890°—895°	,,	7	5	0
Petroleum, Russian	per gallon	0	0	5½
,, American	,,	0	0	5¾
Potassium (metal)	per oz.	0	3	10
,, Bichromate	per lb.	0	0	4
,, Carbonate, 90% (ex ship)	per ton	19	10	0
,, Chlorate	per lb.	0	0	4½
,, Cyanide, 98%	,,	0	2	0
,, Hydrate (Caustic Potash) 80/85 %	per ton	22	10	0
,, ,, (Caustic Potash) 75/80 %	,,	21	10	0
,, ,, (Caustic Potash) 70/75 %	,,	20	15	0
,, Nitrate (refined)	per cwt.	1	2	6
,, Permanganate	,,	4	2	6
Prussiate Yellow	per lb.	0	0	9½
,, Sulphate, 90 %	per ton	9	10	0
,, Muriate, 80 %	,,	7	15	0
Silver (metal)	per oz.	0	4	0½
,, Nitrate	,,	0	2	9
Sodium (metal)	per lb.	0	4	0
,, Carb. (refined Soda-ash) 48 %	per ton	6	0	0
,, ,, (Caustic Soda-ash) 48 %	,,	5	0	0
,, ,, (Carb. Soda-ash) 48%	,,	5	5	0
,, ,, (Carb. Soda-ash) 58 %	,,	6	5	0
,, ,, (Soda Crystals)	,,	2	17	6
,, Acetate (ex-ship)	,,	16	0	0
,, Arseniate, 45 %	,,	11	0	0
,, Chlorate	per lb.	0	0	6¾
,, Borate (Borax)	nett, per ton.	29	0	0
,, Bichromate	per lb.	0	0	3
,, Hydrate (77 % Caustic Soda) (f.o.b.)	per ton.	9	15	0
,, ,, (74 % Caustic Soda) ,,	,,	9	0	0
,, ,, (70 % Caustic Soda) ,,	,,	8	10	0
,, ,, (60 % Caustic Soda, white) (f.o.b.)	,,	7	10	0
,, ,, (60 % Caustic Soda, cream) ,,	,,	7	5	0
,, Bicarbonate	,,	5	15	0
,, Hyposulphite	,,	5	10	0
,, Manganate, 25%	,,	9		0
,, Nitrate (95 %) ex-ship Liverpool	per cwt.	0	8	1½
,, Nitrite, 98 %	per ton	28	0	0
,, Phosphate	,,	15	15	0
,, Prussiate	per lb.	0	0	7¾
,, Silicate (glass)	per ton	5	7	6
,, ,, (liquid, 100° Tw.)	,,	3	17	6
,, Stannate, 40 %	per cwt.	2	2	6
,, Sulphate (Salt-cake)	per ton.	1	7	6
,, ,, (Glauber's Salts)	,,	1	10	0
,, Sulphide	,,	7	15	0
,, Sulphite	,,	5	5	0
Strontium Hydrate, 98 %	,,	8	0	0
Sulphocyanide Ammonium, 95 %	per lb.	0	0	7¾
,, Barium, 95 %	,	0	0	5¾
,, Potassium	,,	0	0	8
Sulphur (Flowers)	per ton.	7	10	0
,, (Roll Brimstone)	,,	6	0	0
,, Brimstone: Best Quality	,,	4	7	6
Superphosphate of Lime (26 %)	,,	2	12	6
Tallow	,,	25	10	0
Tin (English Ingots)	,,	100	0	0
,, Crystals	per lb.	0	0	6⅞
Zinc (Spelter)	per ton.	23	10	0
,, Chloride (solution, 96° Tw.	,,	6	0	0

THE

CHEMICAL TRADE JOURNAL

Publishing Offices: 32, *BLACKFRIARS STREET, MANCHESTER.*

No. 162. **SATURDAY, JUNE 28, 1890.** **Vol. V**

Contents.

Notices.

All communications for the *Chemical Trade Journal* should be addressed, and Cheques and Post Office Orders made payable to—

DAVIS BROS., 32, Blackfriars Street, MANCHESTER.

Our Registered Telegraphic Address is—

"Expert, Manchester."

The Terms of Subscription, commencing at any date, to the *Chemical Trade Journal*,—payable in advance,—including postage to any part of the world, are as follow:—

Yearly (52 numbers) .... **12s. 6d.**
Half-Yearly (26 numbers) .... **6s. 6d.**
Quarterly (13 numbers) .... **3s. 6d.**

Readers will oblige by making their remittances for subscriptions by Postal or Post Office Order, crossed.

Communications for the Editor, if intended for insertion in the current week's issue, should reach the office not later than **Tuesday Morning.**

Articles, reports, and correspondence on all matters of interest to the Chemical and allied industries, home and foreign, are solicited. Correspondents should condense their matter as much as possible, write on one side only of the paper, and in all cases give their names and addresses, not necessarily for publication. Sketches should be sent on separate sheets.

We cannot undertake to return rejected manuscripts or drawings, unless accompanied by a stamped directed envelope.

Readers are invited to forward items of intelligence, or cuttings from local newspapers, of interest to the trades concerned.

As it is one of the special features of the *Chemical Trade Journal* to give the earliest information respecting new processes, improvements, inventions, etc., bearing upon the Chemical and allied industries, or which may be of interest to our readers, the Editor invites particulars of such—when in working order—from the originators; and if the subject is deemed of sufficient importance, an expert will visit and report upon the same in the columns of the *Journal*. There is no fee required for visits of this kind.

We shall esteem it a favour if any of our readers, in making inquiries of, or opening accounts with advertisers in this paper, will kindly mention the *Chemical Trade Journal* as the source of their information.

Advertisements intended for insertion in the current week's issue, should reach the office by **Wednesday morning** at the latest.

Advertisements.

Prepaid Advertisements of Situations Vacant, Premises on Sale or To be Let, Miscellaneous Wants, and Specific Articles for Sale by Private Contract, are inserted in the *Chemical Trade Journal* at the following rates:—

30 Words and under .... 2s. 0d. per insertion.
Each additional 10 words .... 0s. 6d. ,,

Advertisements of Situations Wanted are inserted at one-half the above rates when prepaid, viz:—

30 Words and under .... 1s. 0d per insertion.
Each additional 10 words .... 0s. 3d. per insertion.

Trade Advertisements, Announcements in the Directory Columns, and all Advertisements not prepaid, are charged at the Tariff rates, which will be forwarded on application.

THE LABOUR MARKET.

FOR the purposes of reasoning and argument, labou be considered in pretty much the same way a consider more tangible commodities. It is true that the nature of the case its treatment must in some res differ from that of these, but in the few remarks that i we regard labour, from a broad and general view, as n something of which there is always a supply correspo more or less approximately to an equally increasing dem something that is subject to the same fluctuations in as bleaching powder or soda crystals—something that for its artificial regulation the same inducements that called into being a Salt Union or a Whiskey Trust.

We are far from lacking sympathy for the working m rather, as we should call him here, the vendor of la We are strongly of opinion that his prosperity and prosperity of his country mean one and the same thing our sympathy does not prevent us from seeing his s comings, and lamenting the unbusinesslike manner in v he conducts the sale of that one and only article that he l dispose of. From the moment when the first faint stre the light of activity gave evidence last year of the co dawn of an improvement in trade, the mind of the la seller began to ferment within him, and the anxiety th has displayed ever since to raise the price of his ware to our thinking, been out of all proportion to what the cir stances of the case would justify. When a man, who ha monopoly of an article, refuses to sell except at a figure prohibits its use, it is manifest that he defeats his own e and this seems to us to be just the position that the la seller of to-day is occupying. His disputes with customers (*i.e.*, with his employers) have become monotc by reason of their multitude. Concessions have only s to increase his demands. Already the young bud of imp ment has received a severe check, and the probabiliti its coming to flower grow less day by day. To-day we the strike, to-morrow the concession, and the day afte stoppage of the mill or factory through inability to pro at the ruling prices. The political agitator—the glib-ton misleader of the people—has doubtless much of this to an for; but it is a great reflection upon the common sense o working men of the country that they so persistently refu see the things that have so much importance for them, and are so patent to everyone but themselves. Of what use i five, ten, or twenty per cent. rise of wages if it drive work away from the factory and sets the hands adrift? present seems to us a most ill-advised time for the B workman to go into the "striking" business. That bus has been over-done. The only outcome that is likely t from it is the hastening of the transfer of what is left o trade to the shores of our neighbours, and a return o distress that was so prevalent a short time since.

CANADIAN SUGAR.—Agents of an English syndicate at Mo are trying to buy the sugar interest of the Dominion.

URGH INTERNATIONAL EXHIBITION.

(By our Scotch Representative.)

Edinburgh, June 23rd.

Edinburgh's second "International," the first having taken so recently as 1886, preceded at short distances of time by 'orestries" and "Fisheries" of a minor and tentative These latter were popular and successful financially, and the national" also fully justified the greater scope of the under- ough the surplus, as rather too hastily announced at the lually dwindled, under a soberer and more leisurely balancing and adjusting accounts, to a comparatively educible minimum. Glasgow's great show followed realising an undoubted success in every way, with a dsome surplus, and no doubt enticing Edinburgh to effort which is now running its course. It was thrown he beginning of last month in a wofully unfinished and now, after nearly two months of struggling into order, short of the completing touches, and some of the leading are not yet under way. It was to have been an Exhibition al Appliances only, as at first schemed; but the promoters enlarged their views as the work was proceeding, and it is cumstance, and not to any shortcoming of personal manage- at the present somewhat unsatisfactory condition of the isplay is due. The Exhibition, which stands on ground a ance to the south-west of the city, contains many things well examination, but exhibitors and the examining public alike great disadvantages. Between the Machinery Hall and the n courts and annexes, there is an abundance of floor space; uite promiscuously apportioned, with stand numbers not even in nce at consecutiveness, and if a particular exhibit be wanted, or two may be expended in the search, the catalogue, with its show of order, proving rather a hindrance than otherwise. no classification, properly speaking, and all that the critic can take up a feature here and there, and get through his d task the best way he can. The purely chemical examples scanty indeed, but a number of the general exhibits are of terest to the chemical manufacturer, and these, or a part of ill be included in the mention which is to follow. reat English manufacturers of chemicals, so well represented Glasgow Exhibition two years ago, have practically put in no nce at all; and even of the Scotch contingent, about the only nous example is that of

YOUNG'S PARAFFIN AND MINERAL OIL COMPANY,

West Court, stand 633. This stands on the floor by itself, cased, and in general appearance and arrangement a replica of e firm's exhibit at Glasgow, which many visitors will remem- It provides a very complete illustration of the fecundity of the ising-looking shale block in the production of numerous and al paraffin derivatives in marketable form, including all kinds of phthas, scale, wax, solid burning paraffin, fruit and flowers in , candles, tapers, and vestas, also sulphate of ammonia and the other by-products of this great industry. These are grouped kind of composite trophy, surmounted by busts, in refined wax, of Sir Lyon Playfair and David Landale. Accessory articles exhibit are Young's lamps, stoves, and other apparatus for g, cooking, and heating by means of paraffin oil and solid , for the most part manufactured at the lamp works of the y in England. Adjoining the foregoing (Stand 634) is a case lar form and of remarkable effectiveness by

ICE'S PATENT CANDLE COMPANY, BATTERSEA, LONDON.

sists largely of paraffins in the marketable forms of candles, ghts, oils, &c.; but includes also specimens of the soap manu- in toilet, sanitary and household sorts, stearine samples &c., ms on the whole a very interesting and instructive grouping of cturing results in this growing department of industrial tion.

of the most striking of the features of the Exhibition, for the al eye at least, is to be found in the Boiler House, where five of ilers of Galloways Ltd., Manchester, are constantly at work, ng steam for the whole of the uses of the building. This is magnificent sight in its way, giving the onlooker at once an f the perfection to which this noted firm has carried their ty. These boilers, each measuring 30 feet by 8 ft. diameter, een acquired, as they stand, by the

NORTH BRITISH RUBBER COMPANY, EDINBURGH,

ir Castle Mills Works, to which they will be removed at the October. This firm are also exhibitors (West Court), showing india-rubber products for chemical, mechanical, and manufac- purposes, and including articles of clothing in great variety, the whole composing an illustration which will repay examination. Another illustration of india-rubber adaptability of a less general character is to be found in the

SPHINCTER GRIP ARMOURED HOSE COMPANY,

exhibit, Machinery Hall, consisting of their now well-known flexible india-rubber "indestructible" hose; a remarkable contrivance, and, to the eye, giving strong promise of satisfactory endurance, if not of absolute indestructibility.

The Machinery Hall contains the bulk of the electrical exhibits, a few of which may be noticed here.

WOODHOUSE AND RAWSON, UNITED, LTD.,

of London, Manchester, and Bradford, who, in addition, are the producers of the high candle-power incandescent lamps, illuminating the grand stand in the grounds, show electric lamps, lanterns, bells, switches, wire testers, railway and post office transmitters and receivers in great variety, including all electrical appliances within the very wide range embraced by the firm.

Messrs. Woodhouse and Rawson also exhibit Horn's Electrical Tachometer—for indicating directly the speed of a rotating shaft. The action of this is founded upon Arago's discovery, of the currents set up in a copper disc when rotating in a magnetic field. The disposition of the instrument is such that its calibration is not affected by a variation in the strength of the permanent magnet employed, while its sensitiveness is said to be superior to any form of centrifugal speed indicator. It is specially useful for marine and railway purposes, where it is subject to shocks which have practically no effect upon it. Its sensitiveness enables it to be specially useful in detecting slipping belts, which are a fertile source of loss of power in mills, etc.

KING, MENDHAM AND CO., BRISTOL,

have an excellent little case showing their "Wimshurst" influence electrical, glass-plate, and ebonite cylinder machines, and other electrical apparatus.

PRIESTMAN BROTHERS, HULL,

engineers, have, at stand 59a, an exhibit the purpose of which, in its entirety, is to demonstrate a complete electric light installation, as worked by the Priestman oil engine, suitable for mills, factories, stations, mansions, and other small aggregations of working or dwelling populations. The main feature is the engine itself, of 2 h.p., apparently a sound piece of workmanship, and, in form, similar to the gas engine type. One of these (9 h.p. nominal) is in a house of its own on the grounds, driving a dynamo for the experimental search light on the main entrance tower.

At stand 134, there are three samples of the Blackman air propeller, one of them combined with motor for driving, by electric current, a new patent of the proprietors, the

BLACKMAN AIR PROPELLER VENTILATING COMPANY, LONDON.

Another at work shows the facilities of the Blackman fan in exhausting and propelling air in buildings, and for cooling and drying purposes.

Close by, are found some fine specimens of machine, engine, and other beltings by the

ROSSENDALE BELTING COMPANY,

Mosley-street, Manchester, including woven hair belting with patent edge, sewn cotton belting, carrying belts for collieries, dredgers, factories, &c., and samples also of the patent horse tracing belts of the firm. These will be examined with interest by the visitor of manufacturing proclivities; and a portion of this passing attention is also paid to the stand adjoining, smaller, but including some belting specimens of merit by the proprietors, the

GANDY BELT MANUFACTURING COMPANY, LIVERPOOL.

Here and there are a few illustrations of products in paints, varnishes, &c., and also in practical painting and varnishing.

THE PATENT LEAD AND ZINC-WHITE COMPANY, CAMBERWELL,

London, have, in the North Court, six doors finished (including artistic decorations) with their patent non-poisonous paints. Each is in a colour of its own, although treated in the same design, and they form an exhibit which must be, and is, very interesting to the trade. The patentees assert that these coatings are proof against the most virulent of chemical and manufacturing gases.

THE YORKSHIRE VARNISH COMPANY, RIPON AND LONDON,

have a large and very attracting case (same court), containing all sorts of varnishes for railway, coach, house, ship, and cabinet making work; also some Yorkshire art and other enamels, and wood stains, such as naphthaline, &c.

Closely allied to the same line are

J. AND D. MACNAIR AND CO., GLASGOW AND MANCHESTER

(331a), with their fine samples of the various gums used in the manufacture of varnishes and polishes, also of wood naphthas and shellacs. This is a good case, but, like many others, sadly out of its proper order in the numbering, and very difficult of discovery by any one trusting to the offices of the catalogue alone.

WALKERS, PARKER AND CO., NEWCASTLE

(131), show some fine specimens of their products, including red, white, orange and other leads, litharge and lead in pigs, sheets and pipes.

STOREY AND CO., LANCASTER

(627), are also well to the front with their makes in colours, sizes, manganese dryers, size softeners, &c. Stand 148a

A. B. FLEMING AND CO., EDINBURGH,

is very interesting in its varied selections from the list of this well known firm, including inks of all kinds, colliery greases, solidified and other oils, lubricators, and the new mineral fibre "Azolete," rock and ground. At stand 625, West Court, there are good samples of borax and boracic acid, also borax crystals, by W. G. PURSELL AND CO., LEITH.

ASPINALL, ASPINALL AND CO., NEW CROSS, LONDON,

at 358, same Court, show, within a large and excellently-arranged case, some samples of enamel and water paint which are worth looking at. The above-named comprise about the whole of the gleanings within the chemical and allied category, and, except in general excellence of individual display, it is not by any means a strong muster; but the fault lies not with the exhibiting manufacturers of the country, and is to be found in the primary haste, subsequent indecision, and final rush and hurry, amid which this rather ramshackle thing in exhibitions has been got together. Probably attempting are petition so soon after 1886 and 1888 was taxing the complaisance of exhibitors and visitors overmuch; but the radical change, while in progress, from a mere departmental exhibition of electrical appliances and discoveries, to the International and Comprehensive Industrial scope, was a fatal vacillation, which introduced all the mischief, and has resulted in perhaps the most slipshod achievement of the kind ever put in evidence since the institution of the "Exhibition" was first invented. It was a case of swopping horses when crossing the stream, and the result forms the strongest confirmation of the soundness of the American adage, which forbids that very risky operation.

The foreign exhibitors are housed in the East Court, but their exhibits are of very miscellaneous description, with hardly anything applicable to our present purpose of special review. There is a railway annexe off the machinery hall, which is, perhaps, the most compact and most interesting part of the exhibition. Nowhere else has classification been successfully practised. It comprises samples of locomotives, wagons and carriages, from the earliest to the most recent times. Although some are old friends at past shows, they are welcome here as links in the line of illustration from primitive to perfect types. The grounds of the exhibition are extensive, with various electrical, ship, water-gliding, and telpherage miniature railways, all more or less backward, but the *raree-show* element is otherwise too abundant, making the area not unlike an ordinary fair ground, with its shouting of excited touters; and the visitor finds he can see little there unless willing to disburse unlimited extras.

STRIKING TO STRIKERS.

THE recent strike of stokers at the South Metropolitan Gas Company's works has developed a phase of the labour question which is not without importance. It is well known that when the retort houses were deserted by what we may call the trained stokers, many of the vacant places were filled by raw recruits from the country. These new men did not pretend to be skilled in the art of filling retorts, but they were willing to be taught; and under the South Metropolitan Company's engineers and foremen they learned their lesson thoroughly and quickly. The man who can handle a spade in the field does not require long to learn how to handle a shovel in the retort-house—a simple fact which the strike leaders thought fit to omit from their calculations; with dire consequences to the misguided men who quietly followed their advice. The agricultural labourer soon became a competent gas stoker—with the result that he has now got two strings to his bow, while the stoker who boasted that he was so skilled that no outsider could fill his place, has none at all. With the recurrence of the summer season the agricultural-stoker gives up stoking and betakes himself to agriculture; when the winter returns he will forsake the frozen fields for the warmer climate of the retort-house. Between agriculture and gas-making his whole time may be occupied—to his own advantage and to the confusion of the man who prefers to fill retorts for six months of the year and loaf about during the remaining six. Many of the men employed by the South Metropolitan Company during the past winter are now returning to their country labours, and each is provided with an addressed envelope, to be used should he desire to return to gas-stoking when winter comes round. The possibilities of a dual occupation such as is here indicated are, we think, susceptible of a development that cannot but result in good for the workman.—*The Gas World.*

PERMANENT CHEMICAL EXHIBITIO[N]

THE proprietors wish to remind subscribers and their [...] generally that there is no charge for admission to the Ex[...] Visitors are requested to leave their cards, and will confer a f[...] making any suggestions that may occur to them in the dir[...] promoting the usefulness of the Institution.

JOSEPH AIRD, GREATBRIDGE.—Iron tubes and coils of all ki[...]

ASHMORE, BENSON, PEASE AND CO., STOCKTON-ON-TEES.—[...] of Ammonia Stills, Green's Patent Scrubber, Gasometers, Plant generally.

BLACKMAN VENTILATING CO., LONDON. — Fans, Air Pr[...] Ventilating Machinery.

GEO. G. BLACKWELL, LIVERPOOL.—Manganese Ores, French Chalk. Importers of minerals of every descriptio[n]

BRACHER AND CO., WINCANTON.—Automatic Stills, and Mixing Machinery for Dry Paints, Powders, &c.

BRUNNER, MOND AND CO., NORTHWICH.—Bicarbonate [...] Soda Ash, Soda Crystals, Muriate of Ammonia, Sulp[...] Ammonia, Sesqui-Carbonate of Ammonia.

BUCKLEY BRICK AND TILE CO, BUCKLEY.—Fireclay wa[...] kinds—Slabs, Blocks, Bricks, Tiles, "Metalline," &c.

CHADDERTON IRON WORKS CO., CHADDERTON.—Steam D[...] Steam Traps (McDougall's Patent).

W. F. CLAY, EDINBURGH.—Scientific Literature—English, German, American. Works on Chemistry a speciality.

CLAYTON ANILINE CO., CLAYTON.—Aniline Colours, Anili[...] Benzole, Toluole, Xylole, and Nitro-compounds of all kir[...]

J. CORTIN, NEWCASTLE-ON-TYNE.—Regulus and Brass T[...] Valves, "Non-rotative Acid Valves," Lead Burning App[...]

R. DAGLISH AND CO., ST. HELENS.—Photographs of Chemic[...] —Blowing Engines, Filter Presses, Sludge Pumps, &c.

DAVIS BROS., MANCHESTER.—Samples of Products from chemical processes—Coal Distilling, Evaporation of Pa[...] Treatment of waste liquors from mills, &c.

R. & J. DEMPSTER, MANCHESTER.—Photographs of Gas Holders, Condensers, Purifiers, &c.

DOULTON AND CO., LAMBETH.—Specimens of Chemical St[...] Stills, Condensers, Receivers, Boiling-pots, Store-jars, &c.

E. FAHRIG, PLAISTOW, ESSEX. — Ozonised Products. Bleached Esparto-Pulp, Ozonised Oil, Ozone-Amn[...] Lime, &c.

GALLOWAYS, LIMITED, MANCHESTER.—Photographs ill[...] Boiler factory, and an installation of 1,500-h.p.

GRIMSHAW BROS., LIMITED, CLAYTON.—Zinc Compounds. Materials, India-rubber Chemicals.

JEWSBURY AND BROWN, MANCHESTER.—Samples of Aerated [...]

JOSEPH KERSHAW AND CO, HOLLINWOOD.—Soaps, Greas[...] Varnishes of various kinds to suit all requirements.

C. R. LINDSEY AND CO., CLAYTON.—Lead Salts, (Acetate, etc.) Sulphate of Copper, etc.

CHAS. LOWE AND CO., REDDISH.—Mural Tablet-makers of [...] Crystals, Cresylic and Picric Acids, Sheep Dip, Disinfecta[...]

MANCHESTER ANILINE CO., MANCHESTER.—Aniline Samples of Dyed Goods and Miscellaneous Chemica[...] organic and inorganic.

MELDRUM BROS., MANCHESTER.—Steam Ejectors, Ex[...] Silent Boiling Jets, Air Compressors, and Acid Lifters.

E. D. MILNES AND BROTHER, BURY.—Dyewoods and I[...] Extracts. Also samples of dyed fabrics.

MUSGRAVE AND CO., BELFAST.—Slow Combustion Stoves. of all kinds of heating appliances.

NEWCASTLE CHEMICAL WORKS COMPANY, LIMITED, NEW[CASTLE-]ON-TYNE.—Caustic Soda (ground and solid), Soda Ash, R[...] Sulphur, etc.

ROBINSON, COOKS, AND COMPANY, ST. HELENS.—Drawin[gs illus]trating their Gas Compressors and Vacuum Pumps, fitt[...] Pilkington and Forrest's patent Valves.

J. ROYLE, MANCHESTER.—Steam Reducing Valves.

A. SMITH, CLAYTON.—India-rubber Chemicals, Rubber Su[...] Bisulphide of Carbon, Solvent Naphtha, Liquid Ammo[...] Disinfecting Fluids.

WORTHINGTON PUMPING ENGINE COMPANY, LONDON.—[...] Machinery. Speciality, their "Duplex" Pump.

JOSEPH WRIGHT AND COMPANY, TIPTON.—Berryman Fe[...] Heater. Makers also of Multiple Effect Stills and Softening Apparatus.

JID VOLATILE NICKEL COMPOUND.

seems destined to startle the modern chemical world. being a comparative rarity, except on plated goods, it be- non laboratory material but a little while ago, then its nsidered an element was impugned, and now it is both re-instated and its vapour density, for the first time These results have followed from the researches of Mr. 'rs. Quincke, and Langer, which were recently made the paper before the Chemical Society. The investigation gin to the need for removing carbonic oxide from producer to use it in the gas battery described in Mr. Mond's address at the annual meeting of the Society of Chemical year, nickel and cobalt being found to effect this object. in this direction, it has been found that a direct compound l carbonic oxide, viz, $Ni (CO)_4$ exists, the new substance *urless liquid*, volatile at ordinary temperatures, boiling at ig a specific gravity of 1·3185 at 17° C, soluble in alcohol, id chloroform, and not acted upon by dilute acids and ts vapour explodes when heated to 60° C. Its vapour rmined by Victor Meyer's method is 6·01, instead of the value, 5·9 ; from this, the atomic weight of nickel is found to ie metal itself can be obtained from it in the form of etallic mirrors of such purity as to form a splendid raw r the re-determination of the atomic weight. The mean ned by reducing nickel oxide from this source by heating in f electrolytic hydrogen, was 58·61, corresponding closely previously accepted, viz., 58·52. This proves conclusively and Schmidt's assertion that the metal hitherto considered nickel is contaminated with another element, and that all rning it consequently need revision, cannot be sustained, es nickel as we have always known it among the elements. stitution of the compound $Ni(CO)_4$ is still the subject of the peculation, Mr. Mond declined to be "drawn" on this itting the temptation to represent it by some fascinating ring ut contenting himself at present with a statement of the h connection with this curious body, it is to be noted that no balt compound can be obtained, thus establishing another f differentiation and separation between it and its twin brother,

THE ARECA-NUT.

CENTLY published paper on "The Narcotics and Spices of the ast," which was read by Dr. Dymock before the Anthro- l Society of Bombay, contains, according to the Calcutta *man*, some interesting information about the areca-nut, which d *supari* by natives, and usually betel-nut by Europeans. h the nut is so well known, it has only been scientifically ated in comparatively recent years. The palm on which it s supposed to be indigenous in the Malayan peninsula and but is now only known in the cultivated state. Few persons y idea of the consumption of the nut in India ; but, as a matter in addition to the vast quantity locally produced (Dr. Dymock o,000,000 people eat it every day of the year), there is an annual of about 30,400,000 lbs. from Ceylon, the Straits Settlements, matra. On the other hand, there is a small annual export of an 500,000 lbs. for the use of Indians living in Zanzibar, ius, Aden, China, and other countries. It is well known to the s that the fresh nuts have intoxicating properties and produce ess, and that the nuts from certain trees possess these properties unusual extent, and even retain them when dry. These intoxi- properties are much diminished by heat, and as the nuts which s them are apt to be mixed up with the common sort, many us people decline to use any except the red nuts of commerce, have all undergone a process of cooking. Dr. Dymock inclines opinion that the original wild nut must have been an intoxicant, ally as the unripe nuts of the best trees produce slightly intoxi- effects. The betel-leaf or *pan*, with which natives eat the areca- highly esteemed, and its thirteen properties are enumerated in ncient books of the Hindus. Until very recently the nut was sed by European medical writers to be simply astringent, and the cating properties of the *bira* or *pan*, the universal native pledge of ship, were supposed to be due to the leaf, and to the spices which ime are put into the *pan*. But the process of organic chemistry d to the discovery of organic properties in the nut, the active ple of which, if injected under the skin of rabbits and cats, causes death in a few minutes. At the same time the essential oils of leaves have been found to be highly beneficial in catarrhal ons and throat inflammations.

THE ANALYSIS OF FATS.

THE free fatty acids are determined by extracting the total fats with petroleum ether, and phenolphthalein is used as the indicator in the titration with alkali ; the results are calculated as oleic acid. In order to judge of the rancidity of fats and oils of commerce by the amount of free acids they contain, it is necessary to know the percentage of free acids in the pure products. The results of experiments in this direction are shown in the following table, both for oils and oil cakes. The numbers given are the mean values found :

	100 parts contain		
	Free fatty acids.	Total fats.	Per cent. of free fatty acids on total fat.
I.—Oils.			
Rape (Brassica rapa)...........	0·42	37·75	1·10
Cabbage (Brassica campestris) ..	0·32	41·22	0·77
Poppy (Paparer somniferum)	3·20	46·90	6·66
Earth nut (Arachis hypogæa)			
(a) seed	1·91	46·09	4·15
(b) outside pale yellow husk	1·91	4·43	43·10
Sesame (Sesamum orientale)	2·21	51·59	4·59
Castor (Ricinus communis)	1·21	46·32	2·52
Palm nut (Elais guinensis) (with 6 per cent. husk)	4·19	49·16	8·53
Coprah (Cocos nucifera)	2·98	67·49	4·42
II.—Cakes.			
Rape	0·93	8·81	10·55
Poppy.......................	5·66	9·63	58·89
Earth nut	1·42	7·65	18·62
Sesame	6·15	15·44	40·29
Palm nut	1·48	10·39	14·28
Earth nut meal (extracted)	1·55	18·68	8·29
Cocoa.......................	1·31	13·11	10·51
Linseed	0·75	8·81	9·75
Castor......................	1·27	6·53	20·07

The oils obtained by pressing only contain a portion of the free fatty acids. The so-called "technical oils" which result from a second and third pressing of the seeds also contain free fatty acids, but the greater portion thereof remains behind in the cakes.

In the case of "extracted oils," as opposed to oils obtained by pressing, it is pointed out that the composition of the fat extracted is the same as that left behind in the seeds (*i.e.*, both contain the same percentage of free acids), since both glycerides and fats are equally dissolved by the solvents. The test for free acids, together with a microscopical examination, serves therefore to distinguish "extracted" from "pressed" meal ; and further, by testing the oils also for free acid "extracted" and "pressed" oils can be distinguished. This holds as long as the product is unadulterated. A number of analyses of palm-nut cake and of cocoa-nut cake are given which bear out this point.—*Oil, Paint, and Drug Reporter (Zeits. Anal. Chemie.).*

ANALYSTS AT VARIANCE.

AN extraordinary conflict of opinion between analysts occurred at the Portsmouth Police Court a few days since, in connection with an information against a grocer of that town for selling adulterated butter. Three separate certificates were handed in. The first, which was from the Borough Analyst (Dr. B. H. Mumby), showed that the article sold consisted entirely of margarine ; the second, signed by the County Analyst (Dr. Angell), to whom the defendant forwarded sample after breaking the seal, was to the effect that the butter w quite pure, though not of the highest quality ; and the third, signed b three of the official analysts at Somerset House, showed that of 85 pe cent. of fat in the butter, 15 per cent. only was foreign fat. Mr Matthews, barrister-at-law, who defended, claimed an acquittal, on the ground that the various analyses were in complete conflict. The magistrates, however, declined to give full weight to Dr. Angell's certificate, on the ground that since the seal had been broken, there was nothing to show that the sample forwarded to Dr. Angell was the same as that left with the defendant by the sanitary officer. They convicted on the Somerset House certificate, and fined the defendant £1. and costs. At a meeting of the Portsmouth Grocers' Association on Tuesday evening, attention was called to the extraordinary discrepancy between the certificate of Dr. Mumby and that of the officia chemists at Somerset House, and it was resolved to write to the Sanitary Council, requesting that the Borough Analyst should be called upon for an explanation.

PRIZE COMPETITION.

Owing to the number of competitors and the variety of schemes and processes submitted to us, we are compelled to postpone the prize award for two or three weeks, in order to allow us to try all the processes in our own laboratory.

WATER-GAS IN AMERICA.

A GERMAN-AMERICAN gas engineer, writing to the *Journal fur Gasbeleuchtung* on the gas industry in America, says most of the gas companies which supply the electric light are losing money in that department. About half the lighting gas now supplied is water-gas. There is no great difference between the cost of water-gas and that of coal-gas; in some cases the balance of pecuniary advantage is the one way, and sometimes the other. The advantages of water-gas are (1) a white, smokeless flame, (2) the process can be interrupted or forced on, which is convenient where the gasholders are small, (3) the apparatus is soon ready for working, (4) fewer hands are required, 10 men as against 25 for 1,750,000 feet per day, and no special call for strength and endurance, which is important in view of strikes. Most works which make coal-gas have now water-gas apparatus ready in case of need. On the other hand (1) unenriched water-gas contains 50 per cent. of carbonic oxide, enriched (through the oxidising action of steam used along with the naphtha) contains, as in New York, only 25 per cent.; still it has introduced a new danger (three cases of death in the first half of February, 1890, in New York alone), and many cases of suicide are due to its now well-known property of ensuring an easy and painless death; formerly gas escapes generally induced unconsciousness and then a racking headache; now they are generally fatal; (2) the tar produced is as thin as ink, has a powerful smell, and is very difficult to get rid of; when thrown away it results in actions for damages; (3) there is more condensation, and that more oily, in the street pipes; (4) this condensation destroys the leather of dry meters and causes an increasingly heavy outlay for repairs. The use of gas for cooking and heating is rapidly extending; but hopes of replacing solid fuel by water-gas or Siemens gas are not being realised; Mr. Westinghouse has used every effort to devise good gaseous fuel, and has abandoned all the forms, returning to coal-gas as the best for fuel; but even coal-gas is not cheap enough for general use as a fuel. Electricity is pushing so hard that the good times, when the gas manager stayed in his office while his customers came to him, are now over; he must now bestir himself like any other man of business or the undertaking goes backwards.

CHROME MORDANTS FOR ALIZARINE COLOURS.

ALTHOUGH the alizarine colours, as they are called, can be dyed or printed on wool, cotton, and silk, with alumina and iron mordants, yet the chrome mordants are by far the best for the purpose. This arises from the fact that chrome mordants on the whole fix these colours better and therefore give faster shades than the other mordants; further, they yield purer shades, do not alter them as much, for instance, as iron mordants do; and they are more generally applicable. The chrome mordants most generally used are the bichromates of potash and soda, chrome alum, and acetate of chrome. Chrome alum is the double sulphate of chromium and potassium, which crystallises with 24 molecules of water of crystallisation in the form of octahedral crystals, and is obtained as a by-product in the manufacture of alizarine. It is well suited for the mordanting of wool, although it is rarely used, which is due to the fact that it takes longer boiling to bring about the mordanting of the wool unless larger quantities of cream of tartar are used with it; this is not admissible on account of the increased expense.

The bichromate of potash is the chrome mordant mostly used in the wool dyeing. The operation of this salt in mordanting is based upon the fact that the chromic acid is reduced to oxide of chromium, and the latter is, as it were, deposited on the wool fibre in a nascent condition, thus bringing about an unusually intimate and firm fixation of the colouring matter. This reduction of the chromic matter takes place when reducible substances, such as organic acids or salts, are added, at the expense of the latter, or if these be not used, or non-reducible or inorganic acids or salts be added to the mordanting bath, the chromic acid is converted into oxide, but to a less degree, by the reducing action of the fibre itself.

The former method, the addition of reducible organic substances, is preferable, in order to preserve the strength, etc., of the wool fibre. The reducing agents mostly used are tartar (bitartrate of potash), and oxalic acid. Sometimes small quantities of organic or inorganic acids or acid salts are added when the water used contains a large of lime; acetic or oxalic acids give the best results in this ca mordanting bath must be slightly acid, and, lime having a neu effect, these acid additions must be made to correct this acti these mordant assistants (tartar, oxalic acid) are rather ex cheap substitutes are often offered, but these give very inferior both as regards thorough exhaustion of the dyebath, and bright solidity of the shades obtained.

When bichromate of potash and organic reducing agents a the mordanted wool has a green colour, due to the oxide of ch which is deposited on the fibre; while, if the bichromate be use or with sulphuric acid or inorganic salts, the wool has a yellow due to its taking up chromic acid. The former condition suitable for taking up such dyestuffs as alizarine, and the latter wood dyeing. As the result of experience the proportions mordanting agents that give the best results have been found per cent. of bichromate of potash and 2½ per cent. of tartar, cent. bichromate and 1 per cent. sulphuric acid The mordanti must be about 20 or 30 times the weight of the wool, the sma proportion is, the slower, and consequently the less perfect wil precipation of the chromium oxide on the wool. Wher quantities of wool are to be mordanted in large vats, it is advi increase this proportion. Each ingredient is dissolved separate added to the bath. The wool is then entered, and the wh rapidly boiling for from 1½ to 2 hours. For very light requiring little dyestuff, it is best to diminish the proportions mordant.—*Kuhlow.*

INDAMINE BLUE.

THIS dyestuff, which is sold in the form of a paste, is the chloride of a base of the induline series. Like other men the same group, it is used for dyeing shades of blue, resembling on the various fibres. On wool it may be applied either in a or slightly acid bath. It gives very even shades, but the possess the disadvantage of rubbing off. Cotton can be dyed in ways. The brightest shades are obtained by mordanting with acid, fixing with a salt of antimony, dyeing in a bath of the d and then immediately, without rinsing, working the dyed mater bath containing either 15 per cent. of the weight of the c potassium bichromate or 2·5 per cent. of potassium bichromate w per cent. of sulphuric acid, at a temperature of 50°—60° for 10 n For dyeing a full shade, the cotton must be mordanted with 10 per cent. of its weight of tannic acid.

Another plan consists in applying the dyestuff on unmo cotton in a bath containing sodium acetate; 10—20 per cent dyestuff used remains in the dye-bath. The cotton is, without chromed with 15 per cent. of its weight of bichromate of pot better still, with a mixture of 2 per cent. of this salt, and 3·5 p of copper sulphate, these being the proportions required t copper chromate. As a certain amount of metallic oxide fixes cotton, in combination with the insoluble colour formed, the go be topped with logwood, &c., without further mordanting. It i ever, preferable, when such colouring matters are to be subse applied, to increase the quantity of copper chromate in the ox bath to, say, 6 per cent. of potassium bichromate and 10 per copper sulphate, and to boil the goods in such a solution for an

For calico printing, the paste only requires mixing with a thic containing a little acetic acid, and steaming, in order to give fa colours, whose fastness is further increased by chroming The and tartrate of the colour-base give the best results. The c actions of an aqueous solution of indamine blue R. are as sodium hydrate precipitates the base of a dark red violet tannic acid completely precipitates; concentrated sulphuric a solves the dyestuff with a blue colour; stannous chloride pro slight precipitate; zinc dust and ammonia decolourize.—*Chem.*

ALLSOPPS' SHAREHOLDERS.—A meeting of the sharehol Allsopp and Sons was held the other day to receive the report Committee of Investigation, appointed at a meeting in Mar The Chairman of the Committee read the statement which h presented to them by the accountants. The latter had examin books, and found that there had been no misrepresentation prospectus, but the accounts showed that during the last thre there had been a decrease of 23½ per cent. in the profits. Th mittee submitted that the vendors should waive their dividend ordinary stock until the other holders had received a dividend per cent. A resolution was proposed calling upon the ver refund the sum of half a miilion, as being the excess of the gc and this was unanimously adopted.

COAL MINING STATISTICS.

rt of Mr. Joseph Dickinson, H.M. Inspector of Mines, Manchester and Ireland district, contains many facts and any matters of interest. It deserves to be better known fidential intimations of apprehended danger or actual ent to the inspectors are attended to, whether anonymous his district only 17 were received in 1889, and in one of nymous warner had forgotten to mention the name of the ch he desired to call attention. What is more to the pur-case, where the complaint was investigated it was well-is suggested that some plan like that in force in Germany, esentative committee of workmen have periodical meetings ine managers for the purpose of considering all reports of proposals of safeguards, might usefully be adopted also in this Every reasonable precaution that can be devised ought to be , and the progress that already has been made should be an to further exertion in this direction. When the figures s and fatalities in North and East Lancashire are examined, that since the passing of the Act of 1850, "the ratio of has diminished nearly one-half, and the deaths more than in proportion to the number of miners employed." Mr. mentions that of the 10,265,221 tons of coal wrought in ,876 came from a depth of less than 100 yards, 7,910,463 en 100 and 500 yards, and 1,129,882 from between 500 yards. In the Ashton Moss Colliery, the coal is worked ,020 yards vertical depth, and in the Pendleton Colliery at o yards. Taking the whole of the United Kingdom, there o persons employed in and about mines. The production is d in tons: Coal, 176,916,724; fireclay, 2,192,346; iron-70,542; oil shale, 2,014,860; minor minerals, 239,184; a 9,633,656 tons. The fatal accidents in 1889, numbered 912, ed in the death of 1,128 persons. The importance of our dustry is not likely to be under-estimated, and it is at once and good policy to safeguard as far as it is humanly prac-lives of those who are engaged in the arduous task of win-the depths of the earth the raw material on which depends of the industry and wealth of the country.

THE GOLD DEPOSITS OF PERU.

harles Mansfield, the British Minister at Lima, has recently t to the Foreign Office an elaborate report on the auriferous of Peru, drawn from the report of the Lima School of Mining. ountain ranges of Peru, in the vicinity of the seaboard, when-rocks are of a crystalline character, gold is found in veins of hich have been intruded into the granite and syenite. In this st all the spurs of the Andes are of these formations, and the s quartz is almost invariably accompanied by oxide of iron and The proportion of oxide of iron varies considerably. Every form is met with, from white quartz, permeated with small ous spots, to a reddish rock so charged with oxide of iron that almost the whole of the auriferous mineral, the quartz g, as it were, in an accidental manner. The quartz in the s minerals of the coast varies considerably in appearance, being ystallized in prisms, in semi-crystallized grains, agglomerated united by oxide of iron, or in amorphous masses more or less , or friable, with the appearance of scoria. In this district the s quartz is often associated with other minerals, such as a white substance, smooth to the touch, with a silky, almost silvery with flakes of carbonate of lime of a laminar structure, and of a resinous appearance. Lastly, on the cost of Peru gold etimes be discerned in copper minerals, as well as in those ith chalk, copper pyrites, malachite, alacamite, and silicate of In the upland districts, where the formations exhibit the r of aqueous deposits, veins of gold are not only found in e earths, but also in metamorphic rocks, such as quartzites schist, intruding themselves into the sedimentary and eruptive ns. In these veins the gold is sometimes in a pure state, as n pyrites, sulphuret of iron, or accompanied by other metallic ts more or less auriferous, copper pyrites, panabase, bournonite, jamesonite, &c. Gold in the mountain range is found in veins ads, and in the alluvial districts of the same in flakes and In the Cordillera Oriental, in the district called the Montana, sually found in quartz veins injected into talc and clay slate by l of crystalline rocks. The quartz which accompanies the gold istrict is white, and occasionally exhibits marks of oxide of the latter is never found in the same abundance as in the s minerals of the seaboard. In this part of Peru there are t beds of auriferous soil, and it is from here that the greater of nuggets have been extracted. The report then passes in review all the districts of the country where, as far as is known at present, the existence of gold has been authenticated, following the departments in geographical succession from north to south. The information is given in great detail, and refers to all the mines, worked and unworked, and gold deposits known to exist in Peru, with details respecting the quantity of the metal obtainable. This portion of the report occupies 26 pages.

MIXTURES FOR CLEANING THE HANDS.

IN chemical works, it is not an uncommon occurrence for one's hands to become so soiled with the various well defined and separate nastinesses to be found there, as to be quite insusceptible of cleansing by ordinary soaps or soap powders. One or two chemists of our acquaintance use a mixture under these circumstances, which we publish for the benefit of those of our readers who care to try it. Take about two or three grammes of bleaching powder, the same quantity of soda ash, and about twice their bulk of sawdust. Completely saturate these with caustic soda liquor, say 10 or 15 Twaddell, and quickly rub over the hands. As soon as the desired effect is produced, rinse the hands with water. It is occasionally necessary to repeat the process, but, as a rule, one application suffices to make the hands perfectly clean. There is an odour of bleaching powder perceptible from hands thus treated, to which some may object, but this may be destroyed, and the appearance of the hands still further improved by rubbing them over with a little sulphurous acid solution, or by rubbing first with a solution of sulphite or hyposulphite of soda, or sulphite of ammonia, and, while still wet from this, rubbing over with very dilute hydrochloric or sulphuric acid. The hands should be well washed with water, and a little ten per cent. glycerine rubbed in to keep them soft.

Nitric acid stains on the hands still appear to defy all comers, except pumice stone, but inks, and organic stains may—in the absence of bleaching powder - be generally removed by a mixture of chlorate of potash, and hydrochloric acid.

AFRICAN INDIGO.

THE production of indigo in West Africa, says the *Deutsche Wollen Gewerbe*, is almost entirely in the charge of women, and its extent depends upon the manufacture of cotton goods by the natives. How important this industry is can be judged from the fact that millions of metres of cotton fabrics are annually manufactured upon the primitive hand-looms of the country for the domestic consumption and for export. Especially extensive is the export of these goods to Brazil, where they have become very fashionable, and are particularly used for decorative purposes. The most popular colour for these fabrics is the blue derived from indigo. A commission which, in 1886, was sent by the Government from Lagos to Yoruba, to report on the culture of indigo, stated that in the city of Ibadan, with a population of about 150,000, nearly everybody is clothed in blue stuffs. Upon the banks of the Gambia River this industry is carried on very extensively. The indigo is there known under various names, as "Carro" in Mandingo, "N'Gangba" in Volof, "Elu" in Yoruba, "Suini" or "Luni" in Houssa, while the plant is called "Baba." In the valley of the Niger River the pure precipitate is produced, in which form alone the indigo has a market value. In Gambia and Yoruba it is found in the form of balls of rotten leaves; mostly mixed with cowdung, and without commercial value outside the country. The process of extracting the indigo is as follows:—In an earthen vessel of about 60 quarts capacity the leaves are steeped, and thereby an extract produced which is fermented; then the liquid in poured off and exposed to the action of the air. When the precipitation takes place, and all the dyestuff has settled to the bottom of the vessel, the supernatant liquor is poured off, the pulverulent pecipitate mixed with a little gum, and formed into small balls, &c. The materials to be dyed are steeped in the extract before exposing it to the air, and dried in the open air, which operation is repeated until the desired shade is obtained. For the production of stripes or of patterns in different shades of colour, the material is sewed together where a lighter shade is desired, whereby the intensity of the blue is diminished.

STRIKE OF ST. HELENS GLASS WORKERS.—At one o'clock, on Saturday, the casting hall men at Messrs. Pilkington's top works, St. Helens, who are agitating for an advance of wages, left work in accordance with their resolution. In other departments, such as grinding, smoothing, and polishing, the machinery is stopped, and about 1,500 hands will be thrown idle.

THE NEW FERTILISER LAW.

THE following act relating to the fertiliser trade in the State of New York, passed the Legislature of this State at its recent session, and in order to give information to all interested in the business, the New York Fertiliser and Chemical Exchange decided to publish a copy of the bill for gratuitous circulation.

SECTION 1. All commercial fertilisers which shall be offered for sale, to be used in this state, shall be accompanied by an analysis stating the per centages contained therein of nitrogen or its equivalent of ammonia, of soluble and available phosphoric acid, the available phosphoric acid either to be soluble in water or in a neutral solution of citrate of ammonia, as determined by the methods agreed upon by the American Society of Agricultural Chemists, and of potash soluble in distilled water. A legible statement of the analysis of the goods shall be printed on, or attached to each package of fertilisers offered for sale for use in this state; and where fertilisers are sold in bulk, to be used in this state, an analysis shall accompany the same, with an affidavit that it is a true representation of the contents of article or articles.

SEC. 2. Manufacturers residing in this state and agents or sellers of fertilisers made by persons residing outside the limits of this state, shall, between the first and twentieth days of July in each year, furnish to the Director of the New York State Agricultural Experiment Station at Geneva a list of the commercial fertilisers they manufacture or offer for sale for use in this state, with the names or brands by which they are known on the market, and the several per centages of nitrogen or its equivalent of ammonia, of phosphoric acid soluble and available, and of potash, either single or combined, contained in said fertiliser, as called for in section one of this act. Whenever any fertiliser, or fertilising ingredients, are shipped or sold in bulk for use by farmers in this state, a statement must be sent to the Director of the New York State Agricultural Experiment Station, at Geneva, giving the name of the goods so shipped, and accompanied with an affidavit from the seller, giving the analysis of such per centage guaranteed.

SEC. 3. Whenever a correct chemical analysis of any fertiliser offered for sale in this state shall show a deficiency of not more than one-third of one per cent. of nitrogen or its equivalent of phosphoric acid, or one-half of one per cent. of ammonia soluble or available, and one-half of one per cent. of potash soluble in distilled water, such statements shall not be deemed false within the meaning of this act. This act shall apply to all articles of fertilisers offered or exposed for sale for use in the State of New York, the selling price of which is ten dollars per ton or higher, and of which they are part or parcel, and of any element into which they enter as fertilising material, among which may be enumerated nitrate of soda, sulphate of ammonia, dissolved bone black and bone black undissolved, any phosphate rock, treated or untreated with sulphuric or other acids, ashes from whatever source obtained, potash salts of all kinds, fish scrap, dried or undried, also all combinations of phosphoric acid, nitrogen or potash, from whatever source obtained, as well as all and every article that is or may be combined for fertilising purposes.

SEC. 4. All manufacturers or dealers exposing or offering for sale in this state fertilisers containing roasted leather or any other form of inert nitrogenous matter, shall, in legible print, state the fact on the packages in which the fertilisers are offered or exposed for sale.

SEC. 5. Every person, firm, or corporation violating any of the provisions of this act shall, upon conviction thereof, for the first offence be punished by a fine not less than 50 dollars and not more than 200 dollars, and for the second offence by double the amount in the discretion of the court; such fines to be paid to the officer whose duty it is to enforce the provisions of this act, to be used by him for that purpose, and to be accounted for to the comptroller.

SEC. 6. The Director of the New York State Agricultural Experiment Station at Geneva is charged with the enforcement of the provisions of this act, and shall prosecute in the name of the people, for violations thereof; and for that purpose he may employ agents, counsel, chemists, and experts, and the Court of Special Sessions shall have concurrent jurisdiction to hear and determine charges for violating the provisions of this act committed in their respective counties, subject to the power of removal provided in chapter one of title six, of the code of criminal procedure.

SEC. 7. And the said Director of the New York State Agricultural Experiment Station at Geneva, or his duly authorised agents, shall have full access, egress and ingress to all places of business, factories, buildings, cars, vessels, or other places where any manufactured fertiliser is sold, offered for sale, or manufactured. Such Director shall also have power to open any package, barrel, or other thing containing manufactured fertiliser, and may take therefrom sufficient samples; and whenever any such fertiliser is so taken for samples, it may be divided into different portions, and one or more portions sealed [illegible] a way that it cannot be opened without upon examination evidence of having been opened to the person sealing the sam[illegible] delivered to the person from whom said example is taken, or an[illegible] person that may be agreed upon, by the said director or his agen[illegible] takes the same, and the person from whom it is taken, which po[illegible] delivered may, upon consent of the parties, be delivered to a c[illegible] for the purpose of being analysed other than the chemist emplo[illegible] said Director.

SEC. 8. The sum of 20,000 dollars, or so much thereof as r[illegible] necessary, is hereby appropriated out of any money in the treasu[illegible] otherwise appropriated, to be used by said Director of the New [illegible] State Agricultural Experiment Station at Geneva, as shall be aut[illegible] by the board of control thereof, in enforcing the provisions of th[illegible] Said sum shall be paid to said Director by the treasurer up[illegible] warrant of the Comptroller, upon vouchers to be approved [illegible] Comptroller, in such sums and at such times as said Directo[illegible] require, who shall file a statement for what purposes he desi[illegible] same.

SEC. 9. Agents, representatives, or sellers of manufactured fer[illegible] or fertilising material made or owned by parties outside of this [illegible] and offered for sale for use in this State, shall conform to the pro[illegible] of this Act, and shall be subject to its penalties, and in all parti[illegible] shall take the place of their non-resident principal.

SEC. 10. Chapter 222 of the laws of 1878 is hereby repealed.

SEC. 11. This act shall take effect immediately.

—*Oil Paint and Drug Repo[illegible]*

THE IMPENDING FALL IN COAL PRICE[illegible]

THERE are already indications that coal prices are giving in several distinct parts of the kingdom; and, therefore, gas companies that have held out against the combination of [illegible] owners and the inflated state of the market, will, in all probal[illegible] soon have the advantage of such of their compeers as have made [illegible] tracts at top prices. Of course, a desperate struggle will be ma[illegible] a few weeks to check the inevitable tumble, or, at least, to pos[illegible] it until most of the large gas coal purchases for the year have bee[illegible] cluded; but such efforts must be futile. The reason for this c[illegible] tion lies on the surface. It is universally admitted that the iron m[illegible] has gone to pieces. After a phenomenal gamble, the reaction h[illegible] in; and iron manufacturers are now accumulating stocks [illegible] extremely uncomfortable rate for themselves. The price of oven [illegible] has gone down to one-half the highest rate of the past winter, an[illegible] probably fall lower. In these circumstances, how is the value o[illegible] to be maintained? If the great coal and iron companies cannot [illegible] making and selling iron, they must part with their coal; and [illegible] matter of fact, several sellers who a week or so ago were most ex[illegible] over the prospect of having a good turn at last, are approa[illegible] buyers and offering supplies at a marked reduction. They c[illegible] help themselves. In a normal condition of trade, at least 40 per [illegible] of the gross coal output of the country goes in the iron manufa[illegible] When the iron trade is very brisk, as it has lately been, a se[illegible] additional quantity of coal is drafted into the iron-works, which n[illegible] coal scarce elsewhere. The slightest check in the demand for ir[illegible] any of its merchantable forms is quickly felt in a reduced deman[illegible] fuel; and so the wheels go round. No combination among [illegible] owners, who are so often iron-masters also, can hold good for [illegible] after the market turns against them. They will talk very big [illegible] their rights, and their determination to enforce them; but this [illegible] fidence only lasts while the market is rising, and when any combin[illegible] that can hold together for a week or two must make matters bett[illegible] them. So soon as things begin to appear queer, each looks ou[illegible] evidence of the other giving way; and so, at the first disqui[illegible] rumour, the portentous combination dissolves into a swarm of mut[illegible] destructive individuals. When the coal owners touch bottom a[illegible] nobody will pity them. Some, who have managed to secure [illegible] contracts while the "boom" has lasted, will not look for commi[illegible] tion. What has specially enraged purchasers of coal during the [illegible] few months is the unreasonableness of the advances asked and ins[illegible] upon in some deplorable instances. A moderate increase of pri[illegible] we have all along contended, would have been paid without [illegible] grumbling; but the coal owners want to recoup all their past loss[illegible] one stroke, and have over-reached themselves in the attempt. If [illegible] gas companies had not hurried forward their contracts, th[illegible] playing into the hands of the coal-owners, the downward move[illegible] would have set in earlier; but it will now in all probability be ge[illegible] and irresistible, if not rapid.—*Journal of Gas Lighting.*

Correspondence.

not hold ourselves responsible for the opinions expressed by our correspondents.

SULPHATES IN NITRIC ACID.

To the Editor of the Chemical Trade Journal.

In your issue of the 21st inst, under *Selected Correspondence*, I interest a letter by Mr. J. H. Huxley from the *Chemical News*, at he had been recently a good deal perplexed by finding ic acid" (sic) in steel and pig iron, in which he knew it was to be present; further, that he found the nitric acid S.g. 1·42 "a considerable per centage of sulphuric acid," and, further- t the makers assert that the sulphuric acid is derived from sh glass Winchesters in which it was supplied. estion affects a large section of your readers, viz.: Manu- of pure nitric acid, and works' chemists, besides analysts of inations. I have therefore written to you as being a trade wide circulation to give you my own experience. For five more I had to make daily analyses of steel of the greatest al purity, especially as regards sulphur (not sulphuric acid), this it was necessary to be absolutely certain that the acids of the highest attainable purity. I have purchased for my s of green glass Winchesters filled with nitric acid S.g. 1·42 ll-known provincial firm, and I have *never* found a quantity ric acid impurity in the nitric acid which, on the quantity capable of influencing by ·001 % the per centage of sulphur eels. I am one of those heretics who believe that ·0001 % rth arguing; and in nearly every case I have found the nitric lutely free from H_2SO_4 after most careful testing of samples most cases had been three months in bottle in my own labora- is by no means infrequent to find chemists complaining of the which are supplied to them by dealers, and it is to be regretted any cases the complaint is justified by the facts. Now, sir, in it would perhaps be not too much to ask whether sufficient been taken to be sure that this "considerable per centage" ric acid was considerable enough to vitiate the results for n a first-class steel. In other words, we of the trade would ore definite statement as to the facts, particularly as to the r litre of H_2SO_4 by $BaSO_4$, the amount of dilution, and the test. I have controlled the manufacture of considerable s of nitric acid, and I am sure I should have been very pleased *definite* complaint against my product. But I think that if d that H_2SO_4 in my acid was accounted for to any appreciable the greenness of the bottles, I should have done so under a eservation involving a greenness of a totally different kind. ologies, I am, F. I. C.

Trade Notes.

ICAL PATENT CASE.—In the Chancery division, on Monday, Mr. Justice Kekewich, the case of Rawes v. Chance Brothers, for hearing. With Sir Richard Webster, Q.C., M.P., there l for the plaintiffs Mr. Theodore Aston, Q.C., and Mr. Moulton, Q.C., M.P., with whom were Mr. Lawson and Mr and for the defendents, Sir Horace Davey, Q.C., M.P., and mington, Q.C., M.P., with whom was Mr. Carpmael. There aidable array of chemical experts. As the case was proceeding e went to press, we withhold a fuller report.

T ARE FRENCH WINES MADE OF?—Last year France pro- 3,000,000 hectolitres of wine, and herself consumed 45,000,000 es, to say nothing of the exportation. How, then, was the he "wine" produced? Clearly it was artificially produced v. But how? From Levantine raisins, some say. But the atistics show that less than two million hectolitres were made year from dried grapes It may, therefore, not unfairly be l that the fifty or sixty million hectolitres which remain to be l are made from even more illegitimate materials. It is s that good French wine is not to be obtained in France. e of the grape is sent abroad, while the juice of other things oots, for aught we know—is sold to the Frenchman, who imself upon recognising a good glass of wine when he tastes oss the Channel this matter of dried grapes has become quite M. Tirard's Cabinet went out on it three months ago, and nt Government is proposing to adopt a little more protection ng wine made from foreign grapes with a duty of three francs olitre.

MIXING OF AMERICAN AND RUSSIAN OILS.—*Bradstreet's* qu a statement of a Philadelphia journal to the effect that 10,000, gallons of crude petroleum have been purchased for export to mixed with Russion oil, furnishing "a startling instance of a practi that has been in vogue in Europe for a long time, to the great injury o the American oil export trade" Austrian refiners have been buying crude American oil, mixing it with the much inferior petroleum from the Russian wells, and selling the mixture as genuine American kerosene. It is added that "under Amercian brands this bogus oil has been sold in the foreign markets wherever American oil is in demand, and the inferior quality of the spurious kerosene has given a bad name to all oils bearing American brands."

THE ROYAL SCHOOL OF MINES.—An important regulation has been decided upon, affecting students at the Normal School of Science and Royal School of Mines. As a result, no student will, after the session 1890-91, be entered for the associateship course, except under very special circumstances, unless he has passed in the first class of the elementary stage of mathematics, chemistry, and physics, and can show to the satisfaction of the Council that he possesses the necessary elementary knowledge of those subjects. The Council has been compelled to take this step by the fact that without some pre- liminary knowledge of mathematics, chemistry, and physics it is im- possible for students to follow the course with advantage.

INDIA-RUBBER BALLS.—It is reported that the annual production of india-rubber toy balls in different countries—in dozens—is: Germany, 2,850,000; France, 800,000; Russia, 750,000; England, 630,000; Austria, 520,000; America, 500,000; Italy, 450,000; total, say, 6,500,000 dozen. The demand on the German manufacturers is in creasing yearly. Germany alone exports to England about 850,000 dozen; to America about 900,000 dozen; keeping for home trade about 550,000 dozen; and sending the balance to Holland, France, Spain, Portugal, Norway, Sweden, and the Australian Colonies. The "Presents from Southport" and "Presents from Blackpool" balls will all be found to have inscribed on one side, "manufactured in Germany."

EFFECT OF THE MCKINLEY TARIFF BILL.—*Kuhlows* states that the most careful estimate furnished by a leading firm importing seal plushes, who have taken the trouble to determine the exact effect of the tariff on fine standard qualities of their lines, first-choice goods, is as follows:—

	24"	24"	50"	50"	50"
No.	1	2	3	4	5
Whether silk or cotton predominant.	Cotton.	Cotton.	Cotton.	Silk.	Silk.
Present Price,	$1·30	$1·85	$4·50	$5·50	$7·50
Proposed *Additional Duty*	·49	·52	1·29	3·60	4·59
McKinley prices,	$1.79	$2·37	$5·79	$9·10	$12·09

and when it is remembered that the *present* duty is 50%, it will be seen how largely the duty is to be increased. Besides this, an additional 5% is levied by the Customs Administrative Bill, which puts a duty on packages. This shows, pretty clearly, brother Jonathan's ideas on the subject of protection.

THE DECIMAL SYSTEM.—The Vice-President of the Council (Sir W. Hart Dyke) last week received a deputation at the Education Department from the Associated Chambers of Commerce, which urged that the study of the decimal system of coinage, weights, and measures should be made a compulsory subject in all elementary schools. The deputa- tion consisted of Colonel Hill, M.P. (President of the Associated Chambers), Sir Henry Roscoe, M.P., Mr. S. Montagu, M.P., Mr. Thomas Shaw, M.P., Sir Philip Magnus, Sir V. K. Barrington, Colonel Robson, Mr. James Hole, Mr. E. W. Fithian, Mr. J. D. Leader, Major Hughes, and others. The Vice-President, in reply, said that personally he had long been in favour of the adoption of a decimal currency by this country, as he believed the gain, commercial and otherwise, would be enormous. He cordially sympathized with the deputation, but so vast and considerable a change could only be adopted probably after a long discussion and debate in Parliament, and on the responsibility and suggestion of the Government. The duty of the Department was to secure the best elementary education they could for the large mass of the children, and they thought they were performing their duty by training every child in the elementary schools in such a manner that he might immediately take advantage of that training when he went forth into life. The difficult question, therefore, arose, whether until this system had been adopted in this country they were justified in carrying out to the full the suggestions made by the deputation. If he could in any degree forward the ob- jects they supported he should be glad, but the Department must have in view that it was their duty to give not only the best education, but the one immediately suitable to the welfare, progress, and success in life of the children.

MINERAL INDIA RUBBER ASPHALTE.—Another article formerly considered worthless has been added to the useful products, and is known and is called mineral indiarubber asphalte. It is produced during the process of refining tar by sulphuric acid, and forms a black material very much like ordinary asphalte, and elastic like india-rubber. When heated so that the slimy matter is reduced to about 60 per cent. of its former size, a substance is produced hard like ebony. It can be dissolved in naphtha, and is an excellent non-conductor of electricity, and therefore valuable for covering telegraph wires and other purposes where a non-conducting substance is needed. Dissolved, the mineral indiarubber produces a good waterproof varnish. The manufacture of the material is very profitable, and pays the inventor 400 or 500 per cent.—*Kuhlow.*

THE MINERAL WEALTH OF NEWFOUNDLAND.—In their June circular Messrs. Seward and Co., of 7, Draper's-gardens, London, E.C., point out that the recent political dissatisfaction in Newfoundland instead of doing harm ought to have the effect of directing attention to the island and to its mineral wealth and capabilities. Undoubtedly, there is copper in abundance, and now that the price of this metal has so considerably advanced money should be readily forthcoming to develop the country. The Cape Copper Company are leading the way, and have made arrangements to take over and work the Tilt Cove Mine. Copper has been found in many of the Newfoundland Colonisation and Mining Company's Grants. The manager is at present directing his efforts chiefly to developing the La Manche Lead Mine, which is opening out satisfactorily, and it is probable that before long the Colonisation and Mining Company will be able to profitably dispose of the mine. The completion of the railway system in the island will at once put considerable value on the 100 square miles of land held by the Colonisation and Mining Company.

THE COLOUR OF WATER.—Like the gases oxygen and hydrogen, of which it is composed, pure water has no taste or smell, and, like air, it appears to be colourless when in thin layers, but when looked at in large masses, as in the sea, and in deep lakes, it is blue. Pure water, especially sea water for example, is limpid, clear, and transparent. It absorbs all the prismatic colours, except that of ultramarine, which being reflected in every direction, imparts a hue approaching the azure of the sky. The true tint of water when not exposed to atmospheric influence is always uniform, but it changes its colour in certain localities from the presence of infusoria, vegetable substances, and minute particles of matter. Water, in a natural state, is said to be never wholly pure. Drawn from a shallow well, it may look bright and sparkling, and yet be full of deadly poison for the animal and human system. Speaking generally, pure water has the bluish hue, yet some has a strong brown or yellowish tint from peat or iron, and yet is free from impurity. The water of Loch Katrine, which supplies Glasgow, is thus coloured, and some of the purest from artesian wells. The blue tint of water may be discerned in the following manner:—Let down into water a metal tube (open at the top and closed with a clean glass plate at the bottom) near to a white object 20 feet below the surface. The object when looked at through the tube has a most beautiful blue colour. It would have appeared to be yellow, if its colour was due to the light reflected by extremely small particles of matter suspended in the water.

THE ST. HELENS CHEMICAL WORKERS' ASSOCIATION.—A well-attended meeting of the members of this association was held last week in the coach house, Cotham-street. Mr. P. J. King addressed the meeting at some length, he said that some weeks ago he told them they were organising a campaign against the cowards, blacklegs, and knob-sticks. The Union in St. Helens was now strong enough to assert itself, and he hoped that during the election of the various yard committees that was to take place, anyone connected with the previous committee who was afraid to assert himself, was ashamed of his connection with the Union, and frightened to speak to the managers or foremen, would be relegated into private life. They did not want irrational, imprudent, or hot-headed men, or fiery enthusiasts connected with the committee. They wanted reasonable, rational men who would look at everything put before them from a business standpoint, and deal with it accordingly. He reminded his hearers that he had now got an office at 32, Claughton-street. It was only a temporary one, and he hoped in a short time to get a more imposing structure, where the chemical workers could meet and discuss their grievances. It would be open from 9 to 11 in the forenoon and from 5 to 7 in the afternoon, and during those hours they would find someone who would take down their grievances. If they thought there was anything wrong at the works, they ought to come there and tell him, and he would not be slow to take action. He particularly wished to ask them, if any accident occurred either night or day, to let him know about it and he would get up and go to the spot, to be able to give evidence when the case came before the judge. If they refused him admission to the works it would not tell very favourably; it would show there was something wrong. He instructed them as to several of the rules, and then said he wanted to give a finishing touch to the organisation that month, and if they would help him he would undertake that in the next month would not be a man, connected with a chemical works in St. He who would be outside the Union.

CAST STEEL.—Mr. B. H. Thwaite writes to the *Enginee* follows:—"There is as great a relative difference between crucibl cast steel and Bessemer cast steel as there is between varying b of pig iron. Crucible or pot steel, as generally understood, is m factured from raw materials of a refined character, which from fi last is steel—the process merely converts the integral parts int whole—removes the slag and converts the oxide of the scale of integral parts of the charge. The metal thus undergoes a fu process of purification, and its structural uniformity is also increase think that a simple way of defining the respective qualities of castings produced from (1) blister steel bars by the pot or cru process, (2) by the Martin process is an open-hearth furnace, (3) b pneumatic or Bessemer process in a converter, may be found by a ing a technical brand nomenclature as follows:—(1) Cast steel brand, P B brand mark; (2) ditto ditto, open-hearth brand, O brand mark; (3) Bessemer or converter brand, C B or B B t mark. If annealed the letter A may be added. The excellen the quality of the steel will be found to be in the order given, an cost of production will be in inverse ratio. If makers of the P B O H B quality of steel castings would adopt it, this system of defir would soon be universally accepted."

THE CONDITION OF THE RIVER MERSEY.—At the fortni meeting of the Altrincham Union Rural Sanitary Authority, he Knutsford on Wednesday, Dr. Fox, medical officer for Mid-Ches reported on a death from typhoid fever which had occurred i township of Partington under peculiar circumstances. He said i one they must regard most seriously. A man, 21 years of a bread-winner, the support of his mother, was carried off by a ventible disease. The cottages in one of which the young man abut upon the river Mersey, which here, as elsewhere, has no appearance than that of an open sewer. After a long investiga the conclusion he had arrived at was that the putrid emanations the river, and especially from the banks left exposed when the was low, were the cause of the fever in this particular case. young man came down a stranger from the county of Kent to li the very brink of this sewage-laden stream. He only worked or Ship Canal a fortnight, which was the usual period of incubatio typhoid fever, after which time he took to his bed, and in an fortnight the disease had ended fatally. Would it be possible, h regard to the vast interests of local industries, to bring the river i state of decency? If so, by what body or combination of b should the gigantic task be undertaken? He had thought it rig draw attention to this exceptional death from typhoid fever as sho the injurious influence upon a stranger of an atmosphere followin course of the river, laden with every kind of abominable odo which those who had been born and bred in the neighbour seemed to be insensible and indifferent, possessing the imm which was often associated with acclimatisation.—The Guardia the township confirmed the Medical Officer's report, but no resol was come to on the subject.

Market Reports.

TAR AND AMMONIA PRODUCTS.

There is little change in these; the prices for benzol may be t as unchanged, 3s 9d for 90s. and 2s. 9d. for 50/90s. Anthrac still stationary at last week's prices: 1s. 5d. per unit for 30% A f. London; carbolic acid crystals without change, current quota being 7¼d. to 7½d. per lb. for 34/35° and 8½d. to 9d. for 40°.

Sulphate of ammonia: Hull price at the beginning of the wee to £11, but a recovery has taken place, and prices may be tak £11. 2s. 6d. Leith is quoted as £11. 1s. 3d., some sellers a £11. 2s. 6d., but purchasers are not forthcoming The Beckton is still reported £11. 5s. 0d., but no business is reported. A ten to improvement in the state of the market is anticipated.

REPORT ON MANURE MATERIAL.

Market very quiet, and not much doing. Mineral phosphates r firmer, the Florida not coming into the market so largely as expected, and there being a large falling off in the productio Somme, prices without material change.

Nothing to report in bone cargoes. A small parcel of bone r rather inferior, sold at £5. 5s. per ton ex quay. Nitrate ca quiet, nothing doing.

The position of other articles unchanged.

[MI]SCELLANEOUS CHEMICAL MARKET.

business continues to be done in caustic soda both for prompt [an]d deliveries, and prices are firmer all round. Current quo[tations]:—For 77%, £9 10s. to £9 15s., f.o.b. Tyne; 74%, £9. , £8. 15s.; and 60°, £7. 15s, all f.o.b. Liverpool. Soda ash ¾d. to 1⅞d. per degree at works. Soda Crystals, £2. 17s. 6d. ton. Chlorate of potash, dull and small sales at 4½d. per lb. powder is more firmly held for forward deliveries, transactions delivery are recorded from £4. 12s. 6d. to £4. 15s. per ton, t works during past week, but for future deliveries makers posed to book under £5. per ton, and more has been obtained. ement on foot to establish the union of Leblanc soda works, is o advance values of the product for forward deliveries, makers being indisposed to commit themselves to contracts. Vitriol t 26s. to 28s. for brown, and £3. for rectified at works. of copper easier on the spot at £22 to £22. 10s. per ton, and for ecember £21. Acetates of lime without change. Acetates of lead firm at current rates. Brown sugar of lead £18s. 10s. Nitrate of lead still quiet and a little business being done at ton. Potash, caustic and carbonate, in good demand and prices firm.

THE METAL MARKETS.

LONDON.

[I]RON market has a firmer tone, and although the end of the is usually a dull time, prices of warrants have improved, and a little more doing in manufactured.

[COPPE]R.—Chili and G.M.B. bars continue firm, and closed £59. ers and £59. 15s. three months. Sulphate of copper is flat, e of spot £21. per ton.

[TIN.]—Foreign is steady and dearer at £95. 10s. cash, with a g tendency. English ingots, £99.

[TIN-]PLATES firm, with a better demand all round, though prices at show only a slight advance.

is firm, and other metals steady.

[Co]pper shares Rio Tinto rose ⅜ to 23 and eleven-sixteenths, and Barry five-sixteenths to 8 and nine-sixteenths, and Cape [Co]piapo one-sixteenth each to 4⅞ and 3 and one-sixteenth. Rio t one time touched 24⅛, but this led to sales being made for account.

GLASGOW.

market strong, with a fair business. Scotch done at 44s. 10d. 1½d. cash, also at 44s. 11d. to 45s. 3d. one month, closing 45s. 1d., cash sellers ask ½d. more. Middlesbrough done at . to 41s. 10½d. cash, also at 41s. 10½d. to 42s. 0½d. one closing buyers at 41s. 11d. cash, sellers ask 42s. Hematite 51s. to 51s. 1½d. cash, also at 51s. 0½d. to 51s. 2½d. one closing buyers 51s. 1½d. cash, sellers at 51s. 2d.

WOLVERHAMPTON.

[So]mewhat more cheerful tone was noticeable to-day, 'Change [f]avourably influenced by better reports from Glasgow, Middles[broug]h, and Barrow. Marked bars are in good request at £9. 10., [Pl]ates and angles also moved more freely. Common bars [so]ld at £6. 17s. 6d. Sheets weak and irregular. Tredegar tin [b]ars delivered in Staffordshire were quoted £5. 15s. Derbyshire [fo]rge pig is selling at 42s. 3d., delivered at Staffordshire works. [B]ar hematite pig delivered in Staffordshire was quoted 65s. for [a]nd 55s for No. 4.

WEST OF SCOTLAND CHEMICALS.

GLASGOW, Tuesday.

ers have been sending out in increased quantity, but there is no ement to report in current values, except in the case of caustic soda, s again firmer, and fetching better prices all round. The local f soda crystals is pretty well taken up for the time being, and it on report that some makers have had to buy in the southern s in order to meet past made engagements, maturing at the t in excess of production. Glasgow price on rails is £2. 15s. 6d. %. Bleaching powder has descended a further notch, and is at est point over a period of years. Saltcake is about two shillings and sulphate of copper continues gradually on the slide, being ow at £22. or less. The three Scotch makers of bichromate of and bichromate of soda concur in describing the present stream ery to order is very slack, with much too wide a margin between l and works capacity of production. Prices of both as formerly under articles of agreement. The position of sulphate of ia has improved slightly, perhaps, but still leaves a lot of space onable recovery. It was done towards close of week at £11., but there have been symptoms of a rally, and quotation is now £11. 1s. 3d., to £11. 2s. 6d., Leith, as under. Chief prices current are:—Soda crystals, 51s. net, Tyne; alum in lump £5., less 2½% Glasgow; borax, English refined, £29. 10s, and boracic acid, £37. 10s. net Glasgow; soda ash, 1⅛d. less 4%, Tyne; caustic soda, white, 76°, £9. 15s., 70/72°, £8. 13s. 9d., 60/62°, £7. 12s. 6d., and cream, 60/62°, £7. 5s., all less 2½% Liverpool; bicarbonate of soda, 5 cwt. casks, £5. 15s, and 1 cwt. casks, £6. net Tyne; refined alkali, 48/52°, 1⅛d., less 1¼%, Tyne; saltcake, 26s. to 28s.; bleaching powder, £4. 10s. to £4. 12s. 6d., less 5% f.o.r. Glasgow; bichromate of potash, 4d., and of soda, 3d., less 5 and 6% to Scotch and English buyers respectively; chlorate of potash, 4½d., less 5% any port; nitrate of soda 8s. 1½d. to 8s. 3d.; sulphate of ammonia, spot, £11. 1s. 3d. to £11. 2s. 6d., f.o.b. Leith; salammoniac, 1st and 2nd white, £37. and £35., less 2½% any port; sulphate of copper, £22., less 5% Liverpool; paraffin scale, hard and soft, 2¼d. per lb.; paraffin wax, 120°, semi-refined, 3¼d.; paraffin spirit (naphtha), 7d. a gallon; paraffin oil (burning), 6¾d, to 7d. on rails Glasgow to all Scotch buyers (English orders negotiable about 1¼d a gallon lower); ditto (lubricating), 865°, £5. 10s. to £6. per ton; 885°, £6. 10s.; and 890/895°, £7. 15s. to £8. Week's import of sugar at Greenock were 33,350 bags.

THE TYNE CHEMICAL REPORT.

TUESDAY.

Chemicals during the last few days have undergone a decided improvement in this market. Bleaching powder, which last week was neglected and in a weak state, made a further drop in price of half-a-crown a ton, but has now recovered and is in good demand at the moment. Makers are now asking from £4. 15s. to £5. per ton, according to brand. Caustic soda is very strong and price has been advanced ten shillings per ton since last report. Soda ash, crystals, and sulphate of soda steady, and practically unchanged in price.

This being the Newcastle race week, most of the chemical workers here have from two to seven days holiday. Manufacturers, as is usual at this time, take advantage of the stoppage to get all necessary repairs done and flues cleaned.

Messrs. Bell Bros., Limited, of the ammonia soda works, Middlesbrough, are now manufacturing bicarbonate of soda, suitable for making carbonic acid gas, and they recommend it to mineral water manufacturers as being much superior to whiting on the following grounds: It largely increases the production of gas, as the same generator will give three times as much gas from bicarbonate as from whiting. The gas produced is absolutely free from impurities. The waste product, instead of being a worthless nuisance like lime mud from whiting, yields a saleable article, viz., glauber salts. It is much cheaper, as a ton of gas from whiting costs £19. 2s. 6d, and from bicarbonate £15. 10s, exclusive of the value of the glauber salts, if they are recovered.

Chemicals to-day are quoted, Bleaching powder, £4. 15s. per ton, in softwood casks; caustic soda, 77%, £9. 10s per ton; ground and packed in 3-4 cwt. barrels, £12. 10s. per ton; soda ash, 48/52%, 1¼d. per degree, less 5%; soda crystals, £2. 10s 6d. per ton, in casks, gross weight; £2. 10s. 6d. per ton, in 2 cwt. bags. net weight; and £2. 13s per ton, in 1 cwt. bags, net weight; sulphate of soda, in bulk, 32s. per ton; ground and packed in casks, 42s. per ton; recovered sulphur, £4. 5s. per ton; chlorate of potash, 4½d. per lb.; silicate of soda, 75° Tw., £2. 10s. per ton; 100° Tw., £3 7s. 6d. per ton; 140° Tw., £4. per ton; hyposulphite of soda, in 1 cwt kegs, £5. 15s. per ton; in 5-7 cwt. casks. £5. 5s. per ton; pure white sulphate of alumina, £4 10s. per ton; blanc fixe, £7. 10s. per ton; chloride of barium, £8. per ton; nitrate of baryta, crystals, £18. 15s. per ton; ground, £19. 5s. per ton; sulphide of barium, £5. 10s. per ton—all f.o.b. Tyne, or f.o.r. makers' works.

There is a feeling of easiness about the price of small coals for manufacturing purposes, but no appreciable reduction in quotations.

Durham smalls range from 7s. 6d. to 8s. 6d. per ton, and Northumberland steam from 7s. to 7s. 6d. per ton.

THE LIVERPOOL MINERAL MARKET.

Minerals have maintained their firmness, whilst some ores have further increased in value. Manganese arrivals increased, but the largest portion has gone into consumption, and prices are well maintained. Magnesite freely offering, and prices totally in buyers' favour. Raw lump, 26s., raw ground £6. 10s., and calcined ground £10. to £11. Bauxite (Irish Hill Brand) in strong demand, bringing full prices. Lump, 20s.; seconds, 16s.; thirds, 12s.; ground, 35s. Dolomite, 7s. 6d. per ton at the mine. French Chalk: Arrivals well maintained, demand brisk, and prices are unaltered, especially for G.G.B. "Angel White" brand, 95s. to 100s. medium, 105s. to

110s. superfine. Barytes (carbonate) more inquiry; selected crystal lump is scarce at £6.; No. 1 lumps, 90s.; best, 80s.; seconds and good nuts, 70s.; smalls, 50s.; best ground £6.; and selected crystal ground, £8. Sulphate firm, and prices are better; best lump, 35s. 6d.; good medium, 30s.; medium, 25s. 6d. to 27s. 6d.; common, 18s. 6d. to 20s.; ground best white, G.G.B. brand, 65s.; common, 45s.; grey, 32s. 6d. to 40s. Pumicestone in fair request; ground at £10., and specially selected lump, finest quality, £13. Iron ore easier. Bilbao and Santander, 9s. to 10s. 6d. f.o.b.; Irish, 11s. to 12s. 6d.; Cumberland, 10s. to 12s. 6d. Purple ore quiet. Spanish manganiferous ore continues in fair demand. Emery-stone: Best brands inquired for, and bring full prices. No. 1 lump is quoted £5. 10s. to £6., and smalls £5. to £5. 10s. Fullers' earth steady; 45s. to 50s. for best blue and yellow; fine impalpable ground, £7. Scheelite, wolfram, tungstate of soda, and tungsten metal continue in fair demand. Chrome metal, 5s. 6d. per lb. Chrome ore offering more freely, and prices are easier. Antimony oxide, 24s. to 26s. Asbestos is in strong demand, especially rock; prices firmer. Potter's lead ore continues firm; smalls, £14. to £15.; selected lump, £16. to £17. Calamine: Best qualities are scarce—60s. to 80s. Strontia steady; sulphate (celestine) steady, 16s. 6d. to 17s. Carbonate (native), £15. to £16.; powdered (manufactured), £11. to £12. Limespar: English manufactured, old G.G.B. brand in demand, and brings full prices; 50s. for ground English. Felspar, 40s. to 50s.; fluorspar, 20s. to £6. Bog ore in demand, firm at 22s. to 25s. Plumbago steady; Spanish, £6.; best Ceylon lump at last quotations; Italian and Bohemian, £4. to £12. per. ton; founders', £5. to £6.; Blackwell's "Mineraline," £10. French sand, in cargoes, continues scarce on spot—20s. to 22s. 6d. Ferro-manganese and silicon spiegel easier. Chrome iron, unaltered at last quotations. Ground mica, £50. China clay freely offering—common, 18s. 6d.; good medium 22s. 6d. to 25s.; best, 30s. to 35s. (at Runcorn).

THE LIVERPOOL COLOUR MARKET.

COLOURS unaltered. Ochres: Oxfordshire quoted at £10., £12., £14., and £16.; Derbyshire, 50s. to 55s.; Welsh, best, 50s. to 55s.; seconds, 47s. 6d.; and common, 18s.; Irish, Devonshire, 40s. to 45s.; French, J.C., 55s., 45s. to 60s.; M.C., 65s. to 67s. 6d. Umber: Turkish, cargoes to arrive, 40s. to 50s.; Devonshire, 50s. to 55s. White lead, £21. 10s. to £22. Red lead, £18. Oxide of zinc: V.M. No. 1, £25.; V.M. No. 2, £23. Venetian red, £6. 10s. Cobalt: Prepared oxide, 10s. 6d.; black, 9s. 9d.; blue, 6s. 6d. Zaffres: No. 1, 3s. 6d.; No. 2, 2s. 6d. Terra Alba: Finest white, 60s.; good, 40s. to 50s. Rouge: Best, £24.; ditto for jewellers, 9d. per lb. Drop black, 25s. to 28s. 6d. Oxide of iron, prime quality, £10. to £15. Paris white, 60s. Emerald green, 10d. per lb. Derbyshire red, 60s. Vermillionette, 5d. to 7d. per lb.

Gazette Notices.

Partnerships Dissolved.

T. MADEW, M. HENSHALL, E. GATER, S. OAKES, AND J. HENSHALL, under the style of the Sun Street Brick and Marl Co., Hanley, brick manufacturers.

New Companies.

CHESHIRE SALT CORPORATION, LIMITED.—This company was registered on the 16th inst., with a capital of £250,000., in £5. shares, to acquire, upon terms of an unregistered agreement of 3rd inst. with Thos. Barrow, certain freehold salt properties at Winsford, Cheshire, and upon terms of an unregistered agreement of 9th inst., to acquire the letters patent of Sigismund Pick, for the manufacture of salt, dated April 6, 1887, No. 5,124. The subscribers are:—

	Shares.
H. E. Smith, 26, Park-place, Leyton	1
W. Parker, 10, Rosslyn-hill, Hampstead	1
J. Bostock, 25, Lebert-road, Forest-gate, insurance agent	1
W. Danney, 6, Lexham gardens, W.	1
B. Hill, 22, Parkhurst-hill-road, N.W., solicitor	1
F. Spooner, 83, Rendlesham-road, N.E., insurance agent	1
J. L. Standen, 60, Moray-road, Tollington-park	1

CHEMISTS' SUPPLY, LIMITED.—This company was registered on the 12th inst., with a capital of £10,000., in £5. shares, to trade as chemists, druggists, soap boilers, drysalters, oil and colourmen, and to adopt an unregistered agreement between F. G. Treharne and H. E. Sparks. The subscribers are:—

	Shares.
F. G. Treharne, 86 and 88, Leadenhall-street, chemist	1
H. E. Sparks, 6, Shaftesbury-terrace, Harringay, accountant	1
R. M. Sopwith, 6, Hall-street, City-road, druggist	1
J. E. Moxey, Enfield, coal merchant	1
Mrs. Moxey, Enfield	1
Miss F. L. Treharne, Enfield	1
Miss Treharne, Enfield	1

GENERAL PHOSPHATE CORPORATION, LIMITED.—This company was regis[tered] on the 13th inst., with a capital of £1,000,000 in £10. shares, 500 being fou[nders'] shares the owners of which are each to apply for 50 ordinary shares of the first [...] and to pay £30. in respect of each founders' share, towards the preliminary exp[enses]. After provision has been made for a reserve fund, a preference non-cum[ulative] dividend of £10 per cent. per annum is to be paid on the ordinary shares, one m[oiety] of the residue of the net profits to be divided amongst the holders of the fou[nders'] shares, and the other moiety amongst the ordinary shareholders. The object o[f the] company is to acquire lands and mines in Canada, the United States, the [...] Indies, Norway, Spain, France, Belgium, and elsewhere, which may contain, [or be] supposed to contain, phosphates of lime, or other phosphates. An agreement of [...] inst., between Knud Sando and H. Mallaby Deeley, provides for payment of [pre]liminary expenses, and the issue of the founders' shares. The subscribers are:—

	Founders' Shares.	Ord[inary] Shar[es].
Lord Stalbridge, 12, Upper Brook-street	1	5[0]
Sir James Whitehead, Bart., Highfield House, Catford-bridge	1	5[0]
Sir Jacob Wilson, 5, Great George-street, S.W.	1	5[0]
The Hon. C. P. Parker, Eccleston, Chester, land agent	1	5[0]
Sir G. S. Baden Powell 8, St. George's-place, Hyde-park	1	5[0]
Sampson S. Lloyd, 2, Cornwall-gardens, S.W.	1	5[0]
H. Wallaby Deeley, B.A., LL.B., Curson-park, Chester	1	5[0]

THE LIVERPOOL PATENT SOAP COMPANY, LIMITED has been registered w[ith a] capital of £50,000., divided into 4,940 ordinary (7 per cent. cumulative prefere[nce]) of £10. each, and 600 deferred shares of £1. each. The company proposes to [carry] on business as soap manufacturers and merchants.

The Patent List.

This list is compiled from Official sources in the Manchester Tech[nical] Laboratory, under the immediate supervision of George E. Davis [and] Alfred R. Davis.

APPLICATIONS FOR LETTERS PATENT.

Grinding Machinery.—(Complete Specification). T. Breakell. 8,864. Ju[ne ...]
Extraction of Gold. J. W. Macfarlane. 8,884. June 9.
Manufacture of Azo Colours. Read Holliday & Sons, Ltd., and T. Holli[day]. 8,895. June 9.
Treatment of Cotton Dyed with Azo Colours. Read Holliday & Sons, [...] and T. Holliday. 8,896. June 9.
Treatment of Sulphurised Ores. J. C. Butterfield. 8,900. June 9.
Treatment of Alkali Waste for the Production of Sulphuretted Hydro[gen]. R. H. Davidson, and R. H. Davis. 8,901 June 9.
Extraction of Gold from Refractory Ores. F. G. Jordan. 8,908. June [...]
Charging Inclined Gas Retorts H Woodall. 8,909. June 9.
The Production of Ferro-ferric and Ferric Oxides. A. Crossley and J[...] Jones. 8,914 June 9.
Condensation of Nitric Acid and other Distillation Products.—(Com[plete] Specification). O. Guttmann. 8,915. June 9.
Extraction of Lead. F. Ellershausen. 8,916. June 9.
Manufacture of Cement. D. Wilson. 8,919. June 9.
Treatment of Iron Residues resulting from reduction of Organic N[itro] compounds. T. Peters. 8,922. June 9.
Refrigerating Apparatus.—(Complete Specification). H. J. Allison. [...] June 10.
Coating Metallic Surfaces with Alloys S O. Cooper-Coles. 8,929. Jun[e ...]
Charging and Drawing Gas Retorts. L. S. d'Iszoro. 8,930. June 10.
Electro-deposition.—(Complete Specification). G. H. Felt. 8,933. June [...]
Manufacture of Sulphuretted Hydrogen, Sulphide of Ammonium [...] Alkali. J. Leith. 8,942. June 10.
Agglutinants for Coal Blocks. W. P. Thompson. 8,943. June 10.
Furnaces for use in Manufacture of Tin and Terne Plates. J. Ha[...] 8,954. June 10.
Electro Motors. R. Kennedy. 8,964. June 10.
Prevention of Oxidation of Copper during Annealing.—(Complete Sp[ecifi]cation). H. H. Lake. 8,989. June 10.
Mauufacture of Iron and Steel.—(Complete Specification). H. H. Lake. [...] June 10.
Reduction of Ore.—(Complete Specification). H W. Lash, and J. Johnson. [...] June 10.
Production of Colouring Matters from Nitroso Compounds. B. Wil[...] 9,001. June 11.
Apparatus for the Manufacture of Aerated Liquids. L. G. Chinnery [...] S M. Chinnery. 9,048. June 11.
Electrical Meters. S. Z de Ferranti and A. Wright. 9,061. June 11.
Separation of Fats from Emulsions. C. D. Heustrom 9,062. June [...]
Manufacture of Phenolether and Oxydiphenyl. J. Dawson and R. Hi[...] 9,080 June 12.
Manufacture of Caustic Soda. F. Ellerhausen. 9,112. June 12.
Extraction of Precious Metals. R. D. Bowman and H. J Anderson. [...] June 12.
Manufacture of Acid Phosphates for Manures.—(Complete Specifica[tion]). H. H. Lake. 9,129. June 12.
Manufacture of Paper. J. Craig. June 13.
Smoke Consumption. J Lever, J. Holland, and W. H. Todd. 9,139. Ju[ne ...]
Electric Switches. F. M. Newton and T. Hawkins. 9,163. June 13.
Do. do. do. do. 9,164. June 13.
Do. do. do. do. 9,165. June 13.
Do. do. do. do. 9,166. June 13.
Treatment of Waste Animal Matter and Extraction of Fat theref[rom]. H. H. Lake. 9,176. June 13.
Extraction of Gold. C. R. Western. 9,182. June 13.
Recovery of Gaseous Products given off during Fermentation. C. [...] Tichborne, A. E Darley, M. F. Purcell, and S. Geogheagan. 9,183. Jun[e ...]
Utilization of Waste Potassium Salts. W. L. Wise. 9,198. June 13.
Artificial Fuel. J. Bowing. 9,199. June 13.
Manufacture of Ultramarine. R. W. E. MacIvor. 9,200. June 13.
Electro-Magnetic Separator. J. Roncrewski 9,246. June 14.
Desilverising Lead. E. Edwards 9,247. June 14.
Electrolytic Deposition of Aluminium S. Wohle. 9,257. June 14.
Production of Colouring Matters. O. Imray. 9,258. June 14.

IMPORTS OF CHEMICAL PRODUCTS

ONDON.

k ending June 14th.

400 c. H. Wallace & Co.
170 P Hecker & Co.

16 pks. Ohlenschlager Bros.

—
£37. Weatherley, M[illegible] & Co.
herwise undes[illegible])
, 16 pks. O. Andrae & Co

of Lime—
688 pks. Lister & Biggs
408 Bryce, Junr., & White

cid—
28 pks. Spies, Bros., & Co.
13 C. Christopherson & Co.
57 Beresford & Co

y Ore—
29 t. Pillow, Jones, & Co.
20 H. R. Merton & Co
95 E. J. Hudaberdoglie
las, 47 t. W Balchin
11 c. L. & I. D Jt. Co.
ıd, 30 N. Z. Antimony Co.
8 H Bath & Son
30 H. Emanuel
las, 3 F. Manders
1½ Rottman, Strome, & Co.

—
Isles, 12 t. Falkland Islands Co

—
, £100. A. & M. Zimmermann

, 3 t. L. & I. D. Jt. Co.
9 Beresford & Co
10 Goad, Rigg, & Co.
10 "

ne—
1 t. R. Morrison & Co.

—
, 38 cks Burrell & Co.
13 O'Hara & Hoar
22 "
50 pks. Magee, Taylor, & Co.
66 T Rend
, 21 W. A. Rose & Co.
41 W. Harrison & Co.
10 cks. Coste & Co.
, 22 W. Harrison & Co.

ore—
40 t. Royal Mail S. P. Co.

f Tartar—
2 pks. L. & I. D. Jt. Co.
2 Evans, Lescher, & Co.
10 W. C. Bacon & Co.
5 B. & F. Wf. Co.
5 "
10 W. C. Bacon & Co.
2 Evans, Lescher, & Co.

ouc—
13 c. Kleinwort, Sons, & Co
, 8 Stiebel Bros.
11 L. & I. D. Jt. Co.
137 Cundall & Co
, 27 Kebbel, Son, & Co.
5 Hammond & Co
ne, 30 Major & Field
50 Pickford & Co.
, 11 Kebbel, Son, & Co.
10 Elkan & Co
s, L. & I. D. Jt. Co.
que, 114 c. "
3 G. H. Penny & Co.
18 L. & I. D. Jt. Co
6 "
, 50 "
, 2 "
40 J. Hale & Son

Potash—
11 pks. Petri Bros.

ls (otherwise undescribed)
£87. H. Boyce
, 8 T. H. Lee
700 A. & M. Zimmermann
22 F. Stahlschmidt & Co.
8 Kaul & Haenlein
, 20 L. & I. D. Jt. Co.
13 Pot. & M'ch's. Wvs. Co.
9 T. H. Lee

s—
to
7 pks. Hoare, Wilson, & Co.

la
, 38 pks. Anderson, Weber, & Smith

ood
51 t. G. T. Benton & Son

Cutch
Holland, 30 pks. J Barber & Co.

Extracts
Adelaide, 100 pks. A. J. Humphrey
Holland, 1 G Meyer & Co.
France, 25 T. H. Lee
" 120 L. J. Levinstein & Sons
U. S., 40 W. A. Bowditch
Halifax, 55 T. Ronaldson & Co.

Argols
Cape, 213 W. H. Cole & Co.
" 9 J. R. Thomson & Co.

Turmeric
E. Indies, 75 pks. L. & I. D. Jt Co.
" 200 J. Graves
Cochin, 132 L. & I. D. Jt. Co
" 448 J. P. Alpe & Co.

Fustic
Spain, 58 t. Churchill & Sim
Colon, 33 "

Indigo
E. Indies, 50 chts. L & I. D. Jt. Co.
" 7 A. Harvey
" 1 T. H. Allan & Co.
" 20 F W Heilgers & Co.
" 7 W. Brandt, Sons, & Co.
" 47 L. & I. D Jt. Co.
Holland, 2 "
W. Indies, 7 srns. Chalmers, Guthrie, & Co.
E. Indies, 7 chts. Darling Bros.
" 2 L. & I. D. Jt. Co.

Myrabolams
E. Indies, 1,366 pks. "
" 1,558 W. Shelcott
" 327 L. & I. D Jt. Co.
" 4,833 Hoare Wilson & Co
" 3,431 Benecke, Souchay, & Co.
" 2,029 Beresford & Co.
" 550 Lewis & Peat
" 6,860 Beresford & Co.
" 1,366 J. P. Alpe & Co.

Valonia
Smyrna, 35 t. A. & W. Nesbitt
" 63 J. Graves

Tanners' Bark
Melbourne, 31 t. Beresford & Co
Durban, 25 A. & W. Nesbitt
Adelaide, 52 Baxter & Hoare
Natal, 21 Hicks, Nash, & Co.
Sydney, 41 Dalgety & Co.
Holland, 10 L. Henle
" 11 G. S. N. Co.
" 50 J Graves
" 50 J. J. Williamson & Sons
New Zealand 8 Von der Meden & Co.
Adelaide, 120 Baxter & Hoare

Gambier
Singapore, 437 bls. Brinkmann & Co.
" 152 pks. Elkan & Co.
" 194 H. Lambert
" 353 S. Barrow & Bros.
" 213 W. H Cole & Co.
" 109 Elkan & Co.
" 158 Lewis & Peat
" 292 J. H & G. Scovell
" 422 Cox, Paterson, & Co
" 432 R. & J. Henderson
" 84 T. J. & T. Powell

Dyestuffs *(otherwise undescribed)*
U. S., 7 pks. Katrick & Co
Canaries, 10 Kuhner & Henderson

Guano—
Brettesnoes, 200 t J. Jenson & Co. Limited
Pt. Stanley, 20 W. Horton & Yates
Lobos, 1,500 Anglo Contl. G. Wks.
" 1,882 "

Gutta-Percha—
Singapore, 110 c Jackson & Till
" 106 Soundy & Co.

Glycerine—
Holland, £130. Beresford & Co.
Germany, 65 H. Lambert
Holland, 128 Burgoyne, Burdidge, & Co.
" 112 F. W. Heilgers & Co.
Germany, 180 Beresford & Co.
Holland, £500. Spies, Bros. & Co.

Glucose—
Germany, 800 pks. 800 cwts. J. Barber & Co.
" 50 420 Union Lighterage Co., Ld.
" 400 400 J. Barber & Co.
" 24 240 Fellows, Morton & Co.
U. States, 500 500 L & I. D. Jt. Co.
" 183 1,000 Horne & Co.
" 36 188 Union Lighterage Co, Ld.
" 290 490 T. M Duche & Son

Glucose—*continued.*
Germany, 105 pks. 105 cwts. J Barber & Co.
" 105 100 Hyde, Nash, & Co.
" 700 700 J. Barber & Co.
" 25 117 J. Fink & Co.
" 400 400 Baxter & Hoare
" 200 200 Barrett, Tagant, & Co
" 140 328 Props Chamberlain's Whf.
" 58 404 Sutro & Co.
" 26 208 T. M Duche & Son

Isinglass—
E. Indies, 2 pks. Clarke & Smith
" 4 "
Penang, 4 L. & I. D. Jt Co

Manure—
Phosphate Rock
U. States, 2,486 t Wyllie & Gordon
Montreal, 300 A. Hunter & Co.

Manure—
Belgium, 140 t. Odams' Chemical Manure Co.
" 155 A. Hunter & Co.
Holland, 155 Couper, Millar, & Co.
France, 260 Anglo Contl. G. Works
Holland, 224 "
France, 112 T Farmer & Co.
" 220 Lawes' Manure Co., Ld.
Holland, 244 A. Hunter & Co
U. States, 218 C A. & H. Nichols

Nitre—
Germany, 3 cks. Craven & Co.
Peru, 10 505 bgs. W. Montgomery & Co.

Naphtha—
Belgium, 8 drs. Domeier & Co.
France, 79 cks. Burt, Boulton, & Co.
Holland, 20 drs. "

Oxalic Acid—
Norway, 76 pks. J. Kitchen, Ld.
" 21 C. Atkins & Nisbet

Potash—
Germany, £225. G. Boor & Co.

Potashes—
Montreal, 150 c. Charles & Fox
" 235 S. Ward & Co

Potassium Carbonate—
Germany, 41 cks. E. Cook & Co.

Pitch—
Holland, 12 cks L. Fischel
Belgium, 20 Berlindina Bros. & Co.

Pyrites—
Lajaja, 500 t. Forbes, Abott, & Co.
" 430 G. Ward & Sons
Spain. 1,428 Societe Commerciale, etc.

Potassium Sulphate—
Germany, 508 pks. Anglo Contl. Guano Works

Pearl Ashes—
France, 300 c. Petri Bros.
" 163 "

Paraffin Wax—
U. States, 120 brls. H. Hill & Sons
" 75 M. Dk. Co.
" 109 G. H. Frank
" 200 cs. H. Hill & Sons
" 141 G. H. Frank
" 100 Rose, Wilson, & Rose
" 148 H Hill & Sons

Plumbago—
E. Indies, 172 brls. J. Thredder Son, & Co.
" 57 H. W. Ison
" 58 Hoare, Wilson, & Co.
" 50 J. Forsey
" 349 J. Thredder, Son, & Co.
" 723 H. W. Ison
Holland, 11 cs. G. Rahn & Co.
E Indies, 4 J. Forsey
" 378 L. & I. D. Jt. Co.
" 108 Craven & Co.
" 53 L. & I. D. Jt. Co.
" 8 H. Johnson & Sons
" 171 T. Jordan & Co,
Germany, 50 cs. Beresford & Co.

Sodium Nitrate—
Germany, £30. C. Faust & Co.

Saltpetre—
Germany, 101 cks. T. Merry & Son
" 80 P. Hecker & Co.
" 8 "
" 3 Craven & Co.
" 62 J. Hall & Son

Soda—
Holland, £70 R. Morrison & Co.

Sodium Silicate—
Holland, £35 Spies, Bros., & Co
Sodium Sulphide—
Holland, £120 W. Balchin
Sodium Acetate—
France. 10 pks. Prop Hays Wf

Stearine—
U. States, 41 hds. T. Ronaldson & Co
" 100 brls. H. Hill & Sons
Holland, 15 cks. 10 bgs. Perkins & Homer

Tartaric Acid—
Holland. 78 pks. B. & F. Wf. Co., Ld.
, 42 "

Ultramarine—
Holland, 4 cks. H. Hodson & Co.
" 4 Spies, Bros., & Co.
" 1 Ohlenschlager Bros
" 16 pks. Haeffner & Co.
France, 27 Flageollet & Co.
Holland, 4 cks. H. Lambert
" 5 G. Steinhoff

Zinc Oxide—
Holland, 25 brls. M. Ashby, Ld.
U. S., 50 "
Holland, 200 "
Germany, 50 cks. J. Matton
Holland, 15 Beresford & Co.
" 50 M. Ashby, Ld

LIVERPOOL.

Week ending June 17th.

Antimony Ore—
Genoa, 1,019 bgs.

Acetate of Lime—
New York, 786 bgs.
" 2,567

Bones—
Colon, 636 bgs. A. Dobell & Co.
Philadelphia, 80

Bone Ash—
Rio Grande du Sol, 160 t. Haycroft & Pethick

Bone Meal—
Bombay, 344 bgs. Okell & Owen
" 816 A. Vinayik, Chates & Co.
" 1,287

Barytes—
Rotterdam, 5 cks.

Borax—
Rouen, 19 cks. Co-op. W'sale Soc.
Leghorn, 10 cs.

Boracic Acid—
Leghorn, 30 cs. Shropshire Union Co.
" 87 cks.

Caoutchouc—
Old Calabar, 13 cks. African Association, Ld.
" 7 brls.
C. C. Castle, 6 cs. 3 brls. W. Duff & Co.
" 4 C. Freese
" 72 Miller Brs. & Co.
" 7 Ihlers & Bell
" 2 Radcliffe & Durant
" 1 L. Hart & Co.
" 4 Ellis & Co.
" 1 E. Wayland
" 1 A. Millerson & Co.
" 1 Hobson & Co.
" 5 Davies, Robbins & Co.
" 1 Havard & Co.
" 4 1 F J. Eaton & Son
" 2 S. Henrichsen & Co.
Rangoon, 691 bgs.
C. C. Castle, 1 brl. Mordaunt Brs.
" 1 cs 6 Edwards Bros.
" 6 W. B. McIver & Co.
" 3 J. Smith & Co.
" 3 W. Griffiths & Co.
" 11 19 Fletcher & Fraser
" 4 F. & A. Swanzy
" 2 A. Reis
" 30

Cream of Tartar—
Bordeaux, 7 cks. 1 brl.
Barcelona, 37 hds.

Charcoal—
Rouen, 650 bgs. Co-op. W'sale Soc.
Rotterdam, 216

Copper Ore—
Leghorn, 126 t.
" 126

Chemicals *(otherwise undescribed)*
Rouen, 45 cks. Co-op. W'sale Soc.

Dyestuffs—
Divi Divi
Hamburg, 183 bgs.
Maracaibo, 139 t. D. Midgley & Sons
Logwood
Hamburg, 6939 ps. E. Brownbill & Co
Myrabolams
Bombay, 1337 D. Sassoon & Co.
" 667 Beyts, Craig & Co.
" 1217
" 3128 bgs. D. Sassoon & Co
" 599
Argols
Oporto, 23 cks. German Bank of London
Gambier
Singapore, 633 bls. National S S. Co. Ld.
" 3225
Valonia
Marathonisi, 54 sks. Robinson & Hadwen
Sumac
Palermo, 100 bgs.
" 125
Cutch
Rangoon, 1245 bgs. 990 cs.
Bombay, 1809 bxs
Rangoon, 4981
Turmeric
Calcutta, 10 bgs.
Fustic
Savanilla, 1478 lgs. G. H. Muller
Colon, 1009 ps.
Brazil, 143
Extracts
Baltimore, 52 brls. Mucklow & Co.
" 20
Dyestuffs *(otherwise undescribed)*
Colon, 25 bxs. Chesney & Co.

Glucose—
New York, 100 brls. Richardson, Spence & Co.
" 49
Hamburg, 33 cks.
New York, 112 brls.

Glycerine—
Rotterdam, 1 dm.

Horn Piths—
Rouen, a quantity, Co-op. W'sale Soc.
Rio Grande du Sol, 15,000 Haycroft & Pethick

Paraffin Wax—
Rangoon, 538 bxs.
New York, 100 brls Meade-King & Robinson

Phosphate—
Montreal, 200 t.
" 348 Phosphate of Lime Co.
" 100

Pyrites—
Huelva, 1,624 t. Matheson & Co.
" 858 Tennants & Co.
" 1,407 Matheson & Co.
" 1,134 "

Potash—
Rouen, 42 dms. 59 cks. Co-op. W'sale Society
Rotterdam, 38

Pitch—
Rouen, 25 cks. Co-op. W'sale Soc.
Rotterdam, 40
" 25

Plumbago—
Genoa, 80 bgs.

Stearine—
New York, 25 hds.
Bordeaux, 6 cs.
Antwerp, 39 bgs.

Saltpetre—
Calcutta, 171 bgs. Ralli Bros.
Antwerp, 135 J. T. Fletcher & Co.
" 135 bgs. "
Hamburg, 100 cks. 30 brls.

Sodium Acetate—
Rouen, 9 cks. Co-op. W'sale Soc.

Sulphur—
Catania, 150 bgs.

Sulphur Ore—
Huelva, 1,504 t. Matheson & Co.

Tartar—
Naples, 6 cks. 1 cs.
Bordeaux, 5

Tartaric Acid—
Rotterdam, 7 cks.

Tartar Emetic—
Hamburg, 12 cks.

Turpentine—
Savannah, 2,422 brls.

Ultramarine—
Hamburg, 30 cs.
Rotterdam, 4 cks.

Verdigris—
Bordeaux, 2 cks.

Zinc Oxide—
Antwerp, 50 cks. J. T. Fletcher & Co.
" 167
" 75 brls. J T. Fletcher & Co.
" 125

Zinc Sulphate—
Rotterdam, 7 cks.

GOOLE.

Week ending June 11th.

Alkali—
Rotterdam, 1 ck.

Antimony—
Boulogne, 2 cks.

Barytes—
Rotterdam, 3 cks.

Caoutchouc—
Boulogne, 3 cks.

Dyestuffs—
Alizarine
Rotterdam, 5 brls.
Logwood
Rotterdam, 3 cks.
Monte Christi, 110 t.
Belize, 354
Monte Christi, 384

Dyestuffs *(otherwise undescribed)*
Boulogne, 71 cks. 5 cs. 3 pks.
" 2

Glucose—
Hamburg, 140 bgs. 36 cks.
" 20 24

Potash—
Rotterdam, 20 cks.
Hamburg, 5 8 bgs.
Calais, 3 15 drms.
Dunkirk, 13 9

Saltpetre—
Rotterdam, 100 bgs.

Sodium Nitrate—
Rotterdam, 5 cks.

Waste Salt—
Hamburg, 110 t. 1,014 bgs.
" 254
" 203
" 220 418 cks.

HULL.

Week ending June 19th.

Albumen—
Hamburg, 20 cks Wilson Sons & Co.

Acid—
Rotterdam, 4 cks. Hutchinson & Sons

Barytes—
Antwerp, 21 cks. H. F. Pudsey
Bremen, 22 Wilson Sons & Co.
" 9 Storry Smithson & Co.
" 19 Blundell Spence & Co.

Chemicals *(otherwise undescribed)*
Hamburg, 39 pks. 5 cs. Wilson Sons & Co.
Antwerp, 219 "
Dunkirk, 259 "
Christiania, 4 cks. "
Bremen, 2 J. Pyefinch & Co.
Rotterdam, 1 cs. W. & C. L Ringrose
Dunkirk, 306 pks. Wilson Sons & Co.

Dyestuffs—
Alizarine
Rotterdam, 17 cks. W. & C. L. Ringrose
" 10 Hutchinson & Son
" 13 W. & C. L. Ringrose
" 37 pks. W. & C. L. Ringrose
" 52 Hutchinson & Son
Extracts
Rouen, 22 cks. Rawson & Robinson
Rennet
Rotterdam, 1 ck. G. Lawson & Sons
Copenhagen, 10 Wilson Sons & Co.
Logwood
Boston, 50 brls.

Glucose—
Hamburg, 21 cks. Rawson & Robinson
" 15 Woodhouse & Co.
New York, 50 brls.
Hamburg, 23 cks. C. M. Lofthouse & Co.
" 34 cks.
Stettin, 1,200 bgs. C. M. Lofthouse & Co.
" 14 200 Wilson Sons & Co.
" 200

Pitch—
Hamburg, 6 cks.
Dunkirk, 64

Phosphorus—
Rouen, 10 cs. Rawson & Robinson

Saltpetre—
Hamburg, 2 cks. Wilson Sons & Co.
" 10 Bailey & Leetham

Stearine—
Rotterdam, 51 bgs. Hornby & Needler

Tartaric Acid—
Rotterdam, 4 cks. W. & C. L

Tar—
Konigsberg, 25 cks.

Turpentine—
Libau, 50 cks.

TYNE.

Week ending June 17th

Phosphate—
Antwerp, 200 t. 1,000 bgs. Lau Chemi

Waste Salt—
Hamburg, 130 t.

GRIMSBY.

Week ending June 14t

Chemicals *(otherwise undesc.*
Hamburg, 7 pks. 3 cs.
Rotterdam, 2

Glucose—
Hamburg, 12 cks.

GLASGOW.

Week ending June 19th

Charcoal—
Montreal, 210 bgs.

Dyestuffs—
Madders
Rotterdam, 2 cks. J. Rankine
Marseilles, 5 brls. Rob M'Intosh
Rotterdam, 4 cks. J. Rankine
Sumac
Marseilles, 50 bgs.
Extracts
Antwerp, 2 cks.
Marseilles, 57 brls
New York, 100 cs.
Alizarine
Rotterdam, 22 cks.
" 63 J. Rankine

Glucose—
New York, 125 brls.

Gutta Percha—
Rotterdam, 4 bskts. J. Rankine

Manure—
Phosphate Rock
Bilbao, 2,250 t. Coltness I
Aruba, 330 A. Cross

Stearine—
St. John's, N.F., 9 cks. Th &

Tartaric Acid—
Marseilles, 10 brls.

Ultramarine—
Antwerp, 13 cs.

Zinc Oxide—
Montreal, 100 cks.
New York, 20 brls.
Rotterdam, 125 cks. J. Rankine

EXPORTS OF CHEMICAL PRODUCTS

OF

THE PRINCIPAL PORTS OF THE UNITED KINGDOM.

LONDON.

Week ending June 11th.

Acids *(otherwise undescribed)*
Lisbon, 11 pks. £39

Ammonium Sulphate—
Cologne, 37 t. 15 c. £425
Antwerp, 130 1,560
Rotterdam, 11 13 130
Antwerp, 10 10 115

Ammonium Carbonate—
Oporto, 11 c. £25
Yokohama, 1 t. 10 57

Arsenic—
Melbourne, 13 t. £199
Yokohama, 1 10 c. 22

Bone Ash—
Mauritius, 250 £1,115

Benzoic Acid—
New York, 6 cs. £50

Carbon Bisulphide—
Rotterdam, 5 t. 6 c £63

Caustic Soda—
Algoa Bay, 1 t. 1 c. £58
Seville, 18 220

Copper Sulphate—
Oporto, 4 t. 3 c. £100
Bordeaux, 23 470
" 5 110
Genoa, 5 114
Bordeaux, 4 87
Genoa, 5 120
Lisbon, 5 18 120

Cream of Tartar—
Sydney, 10 c. £57
Normanton, 10 45
Townsville, 11 59
Adelaide, 10 51
Melbourne, 2 t. 210
" 10 51
Brisbane, 2 12
Christchurch, 5 27

Citric Acid—
Hamburg, 6 cks. £156
Santander, 1 c. 8
Adelaide, 10 70
Genoa, 1 t. 145
Rotterdam, 2 15
" 1 keg 8
Genoa, 2 150
Boulogne, 3 22
Rotterdam, 11 75
Christiania, 5½ 35
Hamburg, 2 14
Sydney, 1 139
New York, 3 14 490
Brisbane, 1 8
Hamburg, 5 40
St. Petersburg, 10 kegs 74

Carbolic Solid—
Barcelona, 8 c. £23
New York, 16 cs. 58

Copperas—
Varna, 16 t. 14 c. £35

Carbolic Acid—
Rotterdam, 2 t.

Clarifiants—
Bordeaux, 30 cks.

Coal Products—
Creosote Salts
Cologne, 407 bgs. 123 cks.
Naphthaline
Hamburg, 25 cks.
Philadelphia, 99
Barcelona, 5
Le Treport, 2
Anthracene
Rotterdam, 20 cks.
Dusseldorf, 146
" 63
Benzole
Rotterdam, 20 cks.
" 24
" 12 drs.
" 32
Terneuzen, 32 dms.
Rotterdam, 20 pns.

25 rlts. £22
, 52 34
, 74 42
164 91
100 brls. 49
500 245
16 cks. 9
on, 140 dms. 11

ia, 300 kgs. £35
, 81 bls 64
50 15
175 t. 767
, 140 216
25 cks. 60
, 1062 1,593
100 brls. 26
160 232

ls (otherwise undescribed)
22 cs. £240
lo, 20 pks. 60
4 c. 69
, 69 259
20 54
d, 3 cs. 24
, 15 140
33 74

tants—
20 cks. £32
, 7 t. 10 c. 200 gls. 25
32 pks. 51

a, 7 t. £70
eau, 191 1 c. 1,910
2 18 32

—
ue, 20 t. £200
100 1,250
am, 100 1,250
s, 99 19 c. 1,195
, 50 500
, 3 30
170 7 1,443
700 1 5,870
69 300
d, 80 1,080
31 10 320
, 338 11 2,708

cid—
, 1 t. £30
ne, 1 6 c. 42

m Chlorate—
ong, 2 t. 5 c. £93

rus—
, 2,300 lbs. £168
18 c. 225
ong, 3 36
na, 1,500 lbs. 172

—
1 t. 3 c. £28

m Cyanide—
i, 7 c. £22

Crystals—
ne, 1 t. £43

m Iodide—
urg, 2 c. £64

m Prussiate—
14 c. £58

c Acid—
ong, 10 cs. £12

e—
1 t. 10 c. £32
g, 2 1 44
, 1 21
3 65
1 4 27

osphate—
ue, 20 t. £70
220 781
30 105

Bicarbonate—
12 c. £12

Sulphate—
ing, 150 t. £150

ystals—
le, 10 t. 18 c. £33

Flowers—
wn, 250 brls. £146

ip—
500 gls. £59

oniac—
, 1 t. 17 c. £70
1 3 40
13 22
16 26

Silicate—
d, 5 t. 1 c. £27

Acid—
ton, 5 c. £35
, 1 t. 133
, 2 16

LIVERPOOL.

Week ending June 14th.

Ammonium Muriate—
Antwerp, 7 c. 3 q. £9

Alum Cake—
Dunedin, 2 t. 1 c. 3 q. £7

Alum—
Oran, 1 t. 12 c. £8
Galatz, 5 15 29
Havana, 2 3 3 q. 10
Leghorn, 5 6 25
Madras, 8 16 55
Malta, 1 5
Melbourne, 5 25
Montreal, 3 16 20
Baltimore, 36 16 2 171
Calcutta, 4 4 1 12
Constantinople, 2 17 1 14
Wellington, 1 1 5
Sydney, 1 1 5
Ibrail, 63 12 1 319
Volo, 1 8 7
Smyrna, 9 16 2 51
Odessa, 5 1 1 26
Kustendje, 1 5
Halifax, 5 15 1 28
Baltimore, 39 12 449

Ammonium Carbonate—
Vigo, 10 c. £23
Oporto, 1 t. 42
New York, 1 13 3 q. 35
Gothenburg, 5 9
Barcelona, 1 33

Ammonium Sulphate—
Rotterdam, 50 t. 11 c. 3 q. £570
Valencia, 103 2 2 1,218
" 50 4 590
" 50 10 585

Bleaching Powder—
Galatz, 29 cks.
New York, 83 25 bxs.
Rouen, 94
Trieste, 17
Baltimore. 91 cks.
Barcelona, 40 brls.
Bordeaux, 33
Boston, 702
Corunna, 16 dms.
Genoa, 209
Ghent, 277
Hamburg, 8
Havre, 15
Lisbon, 15
Montreal, 88 tcs.
Oporto, 100
Philadelphia, 114

Borate of Lime—
Rotterdam, 10 t. 4 c. 3 q. £120

Bone Waste—
Hamburg, 15 t. £50

Borax—
Rio de Janeiro, 12 c. £19
Valparaiso, 1 t. 5 38

Brimstone—
Otago, 51 t. 14 c. £214

Caustic Soda—
Alexandria, 16 dms.
Antwerp, 20
Batoum, 191
Boston, 150
Galatz, 29
Ibrail, 30
Maceio, 39
Malaga, 50 kgs.
Marseilles, 28
Monte Video, 25
Montreal, 40
Motril, 320
New York, 25 bxs.
Philadelphia, 135 brls.
Porto Cabello, 30
Rouen, 72
St. Petersburg, 3,428 dms.
Santander, 16 18 brls.
Seville, 65
Singapore, 45
Stettin, 163
Syra, 5
Valencia, 32
Vigo, 5
Alexandria, 26
Algiers, 20
Algoa Bay, 75 bxs.
Amsterdam, 70
Ancona, 25
Antwerp, 25
Barcelona, 50
Boston, 350
Calcutta, 3 cks.
Carthagena, 20
Ceara, 10
Coruna, 16

Caustic Soda—*continued.*
Dunedin, 11 dms.
Galatz, 6
Genoa, 45
Hamburg, 33
Ibrail, 105
Lisbon, 90
Malaga, 100
Monte Video, 100
Montreal, 200
Napier, 3
Naples, 82
" 28
New York, 775
Odessa, 280
Oporto, 82
Philadelphia. 150 brls.
Rio Grande du Sol, 10 dms.
Rio Janeiro, 15
Rotterdam, 10
St. Petersburg, 708
San Francisco, 20 bxs. 100 dms.
Seville, 25 kgs. 62
Stettin, 16
Tampico, 20
Wellington, 15
Yokohama, 100

Chlorate—
Lisbon, 2 brls. £17

Caustic Potash—
San Francisco, 1 t. 13 c. £35
Naples, 10 25

Copperas—
Monte Video, 2 t. 18 c. 2 q. £6
Syra, 2 19 2 6

Calcium Chloride—
New York, 20 t. 5 c. £42

Carbolic Acid—
New York, 3 t. 14 c. £294
Rotterdam, 5 28
Hamburg, 13 3 q. 71
Genoa, 11 3 57
Rotterdam, 6 36

Copper Sulphate—
Alicante, 10 t. 2 c. 2 q. £113
Bilbao, 5 1 3 112
Rouen, 79 3 75
Barcelona, 19 15 2 429
Patras, 13 18
Bordeaux, 24 17 3 555
" 4 16 116
Genoa, 9 18 243
Gijon, 9 18 240
Trieste, 1 2 27
Bordeaux, 55 4 1 1,338
Nantes, 14 7 967
Bordeaux, 7 18 197
Santos 10 15
Amsterdam, 8 2 9
Barcelona, 10 212
Bilbao, 14 7 3 357
Pasages, 30 910
Bordeaux, 27 8 1 597
Vigo, 10 3 13
Tarragona, 20 3 1 472
Trieste, 1 1 22
Genoa, 7 7 2 155
Iquique, 20 2 444
Ibrail, 5 5
Leghorn, 9 10 218
Sydney, 1 1 3 26
Barcelona, 3 3 80
Bilbao, 20 2 438
Bordeaux, 19 19 3 420
Melbourne, 12 5
Monte Video, 1 1 22
Calcutta, 2 10 51
Sydney, 1 24
Genoa, 20 2 420

Citric Acid—
Vigo, 3 c. £35
Calcutta, 7 2 q. 28

Chemicals *(otherwise undescribed)*
New York, 7 c. 2 q. £165

Disinfectants—
Kurrachee, 23 pks. £9
Calcutta, 3 cs. 6

Epsom Salts—
Rio Janeiro, 50 kgs.
Seville, 10 brls.
Valparaiso, 10 cks.

Glycerine—
Hamburg, 7 t. 17 c. £182

Ground Spar—
Hamburg, 30 t. 16 c. £250

Guano—
Barcelona, 20 t. 7 c. 1 q. £198

Gypsum—
New York, 10 t. 1 c. £30

Magnesia—
Valparaiso, 50 cs.
Constantinople, 2
Vera Cruz, 2 cks. 3

Magnesium Chlorate—
Bombay, 2 t. 4 c. 3 q. £6
" 4 19 13
Montreal, 13 34

Oxalic Acid—
Rouen. 2 t. 3 c. £75
Mexico City, 2 16

Pitch—
Sydney, 6 brls.
Demerara, 14

Potash—
B. Ayres, 1 t. 2 c. £22

Potassium Nitrate—
Carthagena, 4 t. 17 c. 2 q. £90

Potassium Carbonate—
Boston, 13 t. 15 c. £186

Potassium Cyanide—
Halifax, 1 cs. £8

Phosphorus—
B. Ayres, 4 c. 2 q. £335
Vera Cruz, 2 16

Phosphates—
Hamburg, 20 t. 2 c. £100
" 22 2 110

Phosphate of Lime—
Philadelphia, 10 c. £140

Potassium Chlorate—
Shanghai, 4 t. 10 c. £189
Mexico City, 10 21
Palermo, 10 25
Odessa, 1 50
Philadelphia, 2 10 105
St. Petersburg, 15 750
Shanghai, 2 10 125
Vigo, 1 42
Naples, 10 25
" 10 25

Soda Ash—
Baltimore, 307 cks. 344 tcs.
Barcelona, 60 brls.
Boston, 217 500 bgs. 155
B. Ayres, 600 brls.
Leghorn, 76
Melbourne, 36
Montreal, 149
New York, 35
Odessa, 5
Philadelphia, 60
Rio Janeiro, 140
Smyrna, 70
Baltimore, 378 cks.
Boston, 46
Monte Video, 200
Santander, 22
Philadelphia, 100

Sodium Bicarbonate—
Bombay, 350 kgs.
Christiania, 75
Constantinople, 20
Dantzig, 4 cks.
Halifax, 120
Hamburg, 22
Konigsberg, 30
Melbourne, 36 tcs.
Odessa, 660
Riga, 30
Rotterdam, 20
St. John, N.B., 250
St. Petersburg, 100
Stettin, 40
Boston, 100 cks.
Rouen, 155
St. Petersburg, 300
Vera Cruz, 13 cks.

Soda Crystals—
Boston, 250 cks.
Calcutta, 20 brls.
Constantinople, 30
Malta, 150
Alexandria, 20
Coquimbo, 20
Callao, 10
St. John, N.B., 200

Salt—
Abonnema, 37 t.
Amsterdam, 54
Antwerp. 25
Baltimore, 300
Barbadoes, 45¾
Benin, 22
Bonny, 18
Calcutta, 6723
Cameroons, 45
Chicago, 318
E. London, 9
Faroe Islands, 150
Grand Bassa, 14
Iceland, 150
Manaos, 18
Montreal, 400
New York, 675
Old Calabar, 100
Opobo, 50
Para, 230 10 c.

Salt—*continued.*
Port Natal, 188 t.
St. John's, Nfld., 4
Bilbao, 5 c.
Calcutta, 2,556
Demerara, 27
Ghent 150
Iceland, 479
Manaos, 96
Melbourne, 102
Monte Video, 28
Montreal, 313
New York, 405
Para, 507¾
Seville, 30 cs.
Trinidad, 36 3
Valparaiso, 8 9
Wilmington, 320
Port Natal, 75
Quebec, 500
St. John's, Nfld., 20 cs.
San Francisco, 199¾
Sherbro, 100 9 c.
Talcahuano, 120
Trinidad, 37
Valparaiso, 301
Wellington, 518
Quebec, 416

Sodium Silicate—
Zante, 5 brls.
Galatz, 15
Genoa, 17

Sugar of Lead—
Colon, 4 c. £5

Salt Cake—
Baltimore, 101 t. 10 c. £177
" 204 355

Sodium Biborate—
Hamburg, 17 t. 18 c. 2 q. £539
Rotterdam, 1 1 2 33
St. Petersburg, 15 10 1 466
Stettin, 1 15 1 53

Sodium Chlorate—
Mexico City, 5 c. £14

Salammoniac—
Antwerp, 5 c. 3 q. £10
Alexandria, 1 t. 1 28
Patras, 9 17
Larnaca, 1 2 38
Malta, 8 15
Patras, 5 2 9
Naples, 9 1 16
Salaora, 6 10
Volo, 11 19
Genoa, 8 14

heep Dip—
Algoa Bay, 2 t. 5 c. £125
Natal, 1 10 80
B. Ayres, 3 1 q. 8
New York, 4 4 60
Algoa Bay, 1 9 57
E. London, 15 40
Monte Video, 300 drs. 106
B. Ayres, 12 20

Saltpetre—
Rio Grandé du Sol, 6 c. 7
Toronto, 6 6

Sulphur—
Maceio, 1 t. 10 c. £10
Madras, 4 15 1 q. 29
New York, 11 8 51
Oporto, 181 733
Boston, 75 3 315
Rangoon, 5 28
Boston, 25 109

Tartar—
Bordeaux, 6 t. £570

Tar—
Santander,

Zinc Chlorate—
Rotterdam, 5 t. 8 c. £36

TYNE.

Week ending June 17th.

Alumina Sulphate—
Pasages, 21 t. 14 c.

Anthracene—
Rotterdam, 9 t. 19 c.

Ammonium Sulphate—
Hamburg, 50 t.
Odense, 5

Alkali—
Copenhagen, 6 t. 7 c.
Rotterdam, 4 16
Stockholm, 2 18
Gefle, 1 2
Leghorn, 11 10
S. Petersburg, 51 8
Christiania, 6 8
Gothenburg, 18 9
Rotterdam, 5 6
Riga 16 11

Alum—
Pasages, 25 t. 2 c.

Bleaching Powder—
Copenhagen, 16 t. 2 c.
Rotterdam, 27 3
" 9
Danzig, 1 4
Newfairwater, 56 9
Lisbon, 4 19
New York, 100 3
Stockholm, 3
Genoa, 15
S. Petersburg 191 18
Bilbao, 20 1
Abo, 24 1
Antwerp, 17 11
Genoa, 10 1
Hamburg, 12 14
" 21 2
Pasages, 10
Montreal, 25 17
Christiania, 5 2
Esbjerg, 1 3
Odense, 1 15
Rotterdam, 6 1
" 9 11
" 10 9
St. Petersburg, 105 2
Riga, 17 4

Barytes—
Rotterdam, 2 t. 4 c.

Copperas—
Genoa, 5 t. 5 c.
Vienna, 3 19
Esbjerg, 6 2

Caustic Soda—
Copenhagen, 4 t. 7 c.
Rotterdam, 2 12
Reval, 5 16
Newfairwater, 20
Gefle, 4
Genoa, 34 16
Naples, 17
St. Petersburg, 37 7
Antwerp, 8 3
Hamburg, 3 14
Rotterdam, 4 18
Riga, 20 6

Ferro Manganese—
Bilbao, 44 t. 14 c.

Gypsum—
New York, 38 16 c.

Litharge—
Copenhagen, 1 t. 9 c.
Newfairwater, 1
Riga, 11 2
Hamburg, 2 15

Magnesia—
Newfairwater, 12 c.
Hamburg, 4

Oxalic Acid—
New York, 10 t. 6 c.

Potassium Chlorate—
Genoa, 1 5
St. Petersburg, 2 10
Rotterdam, 6 c.

Phosphorus—
Malmo, 4 c.

Salammoniac—
St. Petersburg, 17 t. 15 c.

Sulphur—
Rotterdam, 8 t. 16 c.

Soda—
Copenhagen, 26 t. 18 c.
Libau, 25 6
Rotterdam, 10 17
Stockholm, 18 12
Gefle, 35 3
St. Petersburg, 4 19
Montreal, 40 11
Christiania, 10 3
Antwerp, 5
Aarhuus, 24 15
" 23
Esbjerg, 12 4
Odense, 22 10
Rotterdam, 29 7
" 11 10
Horsens, 8 3
Svendborg, 4 5
Randers, 22 10

Soda Ash—
Libau, 20 t. 6 c.
Newfairwater, 70 19
Lisbon, 5 17
Genoa, 9 19
Abo, 2 19
Christiania, 9 19
Hamburg, 7 12
Gothenburg, 98 11

Sodium Sulphate—
Aarhuus, 21 t. 11 c.

Sodium Chlorate—
Rotterdam, 1 t 16 c.

Superphosphate—
Libau, 770 t. 1 c.
Riga, 200 6
" 339 1

Sodium Hyposulphate—
Antwerp, 1 t. 6 c.

Zinc Ashes—
Copenhagen, 10 c.

GLASGOW.

Week ending June 19th.

Ammonium Sulphate—
Trinidad, 7 t. 3½ c. £82
Demerara, 5 17¾ 62
" 50 13¾ 571
Mauritius, 9 15¾ 100
Nantes, 70 8½ 875

Benzole—
Dieppe, 185 brls. 15 pns. £1,045

Bleaching Powder—
Sydney, 6 t. 3 c. £38

Caoutchouc—
Rangoon, 34 c. £232

Copperas—
Karachi, 11 t. 5 c. £22

Copper Sulphate—
Nantes, 21 t. 12¾ c. £387

Chemicals (*otherwise undescribed*)
Rangoon, 18½ c. £20

Dyestuffs—
Extracts
Quebec, 98½ c. £146
Oporto, 5 t. 15¾ 45
Montreal, 161¼ 250
" 87 98
Argols
Quebec, 22¾ c. £21
Logwood
Quebec, 443½ c. £165
Montreal, 573½ 207

Epsom Salts—
Christiania, 50 bgs. £40
Calcutta, 6 t. 1½ c. 46

Guano—
Demerara, 10 t. £105

Litharge—
Quebec, 4 t. 9¾ c. £72
Karachi, 2 2¼ 31

Manure—
Trinidad, 127 t. 1½ c. £1,205
" 140 6 1,114
" 10 80
Demerara, 10 95
" 54 9 536
Mauritius, 3 ½ 80
" 200 10 1,554
Penang, 200 13 c. 2,100

Manganese Oxide—
New York, 228 c. £93

Nitric Acid—
Bombay, 1 t. 7 c. £43 15s.

Naptha—
Rouen, 50 cks. £450

Potassium Bichromate—
Quebec, 9 t. 9¾ c. £318
" 4 3 154
Oporto, 1 1½ 36
Rouen, 16 6½ 549
Antwerp, 510
Karachi, 3 9¼ 123

Potassium Carbonate—
Karachi, 1 t. 18 c. £33

Paraffin Wax—
Hong Kong, 1 t. 18½ c. £53
Nantes, 18 16½ 480
Japan, 4½ 6

Potassium Chloride—
Pasages, 5 t. ¾ c. £50

Potassium Prussiate—
New York, 5 t. 9 c. £361

Sulphuric Acid—
Rangoon, 12 t. 2¼ c. £132
Bombay, 20 12 103

Soda Ash—
New York, 176½ £49

Sodium Bichromate—
Quebec, 5 t. 16¼ c. £154
Karachi, 33¼ 44

Verdigris.—
New York, £200

HULL.

Week ending June 17t.

Bleaching Powder—
Bremen, 10 cks.
Drontheim, 59
Rotterdam, 7

Barytes—
Stettin, 150 t.

Caustic Soda—
Amsterdam, 20 drms.
Bergen, 4
Gothenburg, 248
Konigsberg, 105
Libau, 145
Riga, 465
St. Petersburg, 450

Chemicals (*otherwise undes.*
Antwerp, 1 cs. 2 cks. 48
Amsterdam, 1 1 36
Boston, 20
Bombay, 542 235
Bordeaux, 20 50
Bremen, 2 6
Christiania, 5 27 82
Copenhagen, 11
Danzig, 15 36
Gothenburg, 14 44 3
Ghent, 4 40
Hamburg, 25 8
Helsingfors, 1
Konigsberg, 29 20
Lisbon, 1
Libau, 85
New York, 25 6
Rotterdam, 31 1
Rouen, 36
St. Petersburg, 5 2 20
Stettin, 24 123
Stavanger, 1
Stockholm, 1 10
Wyburg, 1

Dyestuffs (*otherwise undesc.*
Antwerp, 8 cks.
Amsterdam, 2
Boston, 4
Hamburg, 16 1 drum
Reval, 3 1 cs.
Rotterdam, 33 4 kgs.
St. Petersburg, 2 cs.
Stettin, 300 bgs.
Melbourne, 2 cks.

Manure—
Stettin, 484 bgs.

Pitch—
Antwerp, 446 t.

Tar—
Amsterdam, 5 cks.
Hamburg, 2
Stockholm, 5

GOOLE.

Week ending June 11t.

Benzole—
Calais, 25 cks.
Rotterdam, 48

Chemicals (*otherwise undes.*
Dunkirk, 36 cks.
Ghent, 2

Caoutchouc—
Boulogne, 4 cks.

Dyestuffs (*otherwise undesc.*
Boulogne, 3 cs. 5 cks.
Ghent, 30 bgs. 12 tins. 1 ck. 12
Rotterdam, 66 bgs.
" 85

Manure—
Rotterdam, 176 bgs.

Pitch—
Antwerp, 737 t.

GRIMSBY.

Week ending June 14.

Coal Products—
Dieppe, 98 cks.

Chemicals (*otherwise undes.*
Antwerp, 49 pks.
Dieppe, 60
Hamburg, 10 cks. 2 cs.

Dyestuffs (*otherwise undes.*
Dieppe, 3 cks.

Pitch—
Antwerp, £328.

PRICES CURRENT.

WEDNESDAY, JUNE 25, 1890.

PREPARED BY HIGGINBOTTOM AND CO., 116, PORTLAND STREET, MANCHESTER.

The values stated are F.O.R. at maker's works, or at usual ports of shipment in U.K. The price in different localities may vary.

Article		£	s.	d.
Acids:—				
Acetic, 25 °/o and 40 °/o	per cwt.	7/- & 12/3		
" Glacial	"	64/-		
Arsenic, S.G., 2000°	"	0	12	9
Chromic 82 °/o .. nett	per lb.	0	0	6½
Fluoric	"	0	0	3¼
Muriatic (Tower Salts), 30° Tw.	per bottle	0	0	8
" (Cylinder), 30° Tw.	"	0	2	11
Nitric 80° Tw.	per lb.	0	0	2
Nitrous	"	0	0	1¼
Oxalic .. nett	"	0	0	3¾
Picric	"	0	1	1
Sulphuric (fuming 50 °/o)	per ton	15	10	0
" (monohydrate)	"	5	10	0
" (Pyrites, 168°)	"	3	2	6
" (" 150°)	"	1	4	6
" (free from arsenic, 140/145°)	"	1	7	6
Sulphurous (solution)	"	2	15	0
Tannic (pure)	per lb.	0	1	1
Tartaric	"	0	1	2½
Arsenic, white powdered	per ton	13	0	0
Alum (loose lump)	"	4	17	6
Alumina Sulphate (pure)	"	5	0	0
" " (medium qualities)	"	4	10	0
Aluminium	per lb.	0	15	0
Ammonia, ·880=28°	per lb.	0	0	3
" =24°	"	0	0	1¾
" Carbonate .. nett	"	0	0	3½
" Muriate	per ton	23	0	0
" " (sal-ammoniac) 1st & 2nd	per cwt.	37/- & 35/-		
" Nitrate	per ton	40	0	0
" Phosphate	per lb.	0	0	10½
" Sulphate (grey), London	per ton	11	2	6
" " (grey), Hull	"	11	2	6
Aniline Oil (pure)	per lb.	0	0	10½
Aniline Salt "	"	0	0	9
Antimony	per ton	75	0	0
" (tartar emetic)	per lb.	0	1	1
" (golden sulphide)	"	0	0	10
Barium Chloride	per ton	8	7	6
" Carbonate (native)	"	3	12	6
" Sulphate (native levigated)	"	45/- to 75/-		
Bleaching Powder, 35%	"	4	15	0
" Liquor, 7%	"	2	10	0
Bisulphide of Carbon	"	13	0	0
Chromium Acetate (crystal	per lb.	0	0	6
Calcium Chloride	per ton	2	2	6
China Clay (at Runcorn) in bulk	"	17/6 to 35/-		
Coal Tar Dyes:—				
Alizarine (artificial) 20%	per lb.	0	0	9
Magenta	"	0	3	9
Coal Tar Products				
Anthracene, 30% A, f.o.b. London	per unit per cwt.	0	1	5
Benzol, 90% nominal	per gallon	0	3	9
" 50/90 "	"	0	2	9
Carbolic Acid (crystallised 35°)	per lb.	0	0	7¼
" (crude 60°)	per gallon	0	2	2
Creosote (ordinary)	"	0	0	2½
" (filtered for Lucigen light)	"	0	0	3
Crude Naphtha 30% @ 120° C.	"	0	1	4
Grease Oils, 22° Tw.	per ton	3	0	0
Pitch, f.o.b. Liverpool or Garston	"	1	8	0
Solvent Naphtha, 90% at 160°	per gallon	0	1	9
Coke-oven Oils (crude)	"	0	0	2½
Copper (Chili Bars)	per ton	59	5	0
" Sulphate	"	22	0	0
" Oxide (copper scales)	"	57	10	0
Glycerine (crude)	"	30	0	0
" (distilled S.G. 1250°)	"	60	0	0
Iodine	nett, per oz.	0	0	9
Iron Sulphate (copperas)	per ton	1	10	0
" Sulphide (antimony slag)	"	2	0	0
Lead (sheet) for cash	"	14	10	0
" Litharge Flake (ex ship)	"	15	15	0
" Acetate (white ")	"	24	0	0
" " (brown ")	"	18	5	0
" Carbonate (white lead) pure	"	19	0	0
Nitrate	"	22	0	0
Lime, Acetate (brown)	"	7	7	6
" " (grey)	per ton	13	0	0
Magnesium (ribbon and wire)	per oz.	0	2	3
" Chloride (ex ship)	per ton	2	5	0
" Carbonate	per cwt.	1	17	6
" Hydrate	"	0	10	0
" Sulphate (Epsom Salts)	per ton	3	0	0
Manganese, Sulphate	"	17	10	0
" Borate (1st and 2nd)	per cwt.	60/- & 42/6		
" Ore, 70%	per ton	4	10	0
Methylated Spirit, 61° O.P.	per gallon	0	2	2
Naphtha (Wood), Solvent	"	0	3	10
" " Miscible, 60° O.P.	"	0	4	3
Oils:—				
Cotton-seed	per ton	22	10	0
Linseed	"	26	0	0
Lubricating, Scotch, 890°—895°	"	7	5	0
Petroleum, Russian	per gallon	0	0	5¼
" American	"	0	0	5¾
Potassium (metal)	per oz.	0	3	10
" Bichromate	per lb.	0	0	4
" Carbonate, 90% (ex ship)	per ton	19	10	0
" Chlorate	per lb.	0	0	4½
" Cyanide, 98%	"	0	2	0
" Hydrate (Caustic Potash) 80/85%	per ton	22	10	0
" " (Caustic Potash) 75/80%	"	21	10	0
" " (Caustic Potash) 70/75%	"	20	15	0
" Nitrate (refined)	per cwt.	1	2	6
" Permanganate	"	4	2	6
Prussiate Yellow	per lb.	0	0	9¾
" Sulphate, 90%	per ton	9	10	0
" Muriate, 80%	"	7	15	0
Silver (metal)	per oz.	0	3	11¾
" Nitrate	"	0	2	8
Sodium (metal)	per lb.	0	4	0
" Carb. (refined Soda-ash) 48%	per ton	6	0	0
" " (Caustic Soda-ash) 48%	"	5	0	0
" " (Carb. Soda-ash) 48%	"	5	5	0
" " (Carb. Soda-ash) 58%	"	6	5	0
" " (Soda Crystals)	"	2	17	6
" Acetate (ex-ship)	"	16	0	0
" Arseniate, 45%	"	11	0	0
" Chlorate	per lb.	0	0	6¾
" Borate (Borax)	nett, per ton.	29	0	0
" Bichromate	per lb.	0	0	3
" Hydrate (77% Caustic Soda) (f.o.b.)	per ton.	9	15	0
" " (74% Caustic Soda) "	"	9	5	0
" " (70% Caustic Soda) "	"	8	15	0
" " (60% Caustic Soda, white) (f.o.b.)	"	7	15	0
" " (60% Caustic Soda, cream) "	"	7	10	0
" Bicarbonate	"	5	15	0
" Hyposulphite	"	5	10	0
" Manganate, 25%	"	9		0
" Nitrate (95%) ex-ship Liverpool	per cwt.	0	8	1½
" Nitrite, 98%	per ton	28	0	0
" Phosphate	"	15	15	0
" Prussiate	per lb.	0	0	7¼
" Silicate (glass)	per ton	5	7	6
" " (liquid, 100° Tw.)	"	3	17	6
" Stannate, 40%	per cwt.	2	2	6
" Sulphate (Salt-cake)	per ton.	1	7	6
" " (Glauber's Salts)	"	1	10	0
" Sulphide	"	7	15	0
" Sulphite	"	5	5	0
Strontium Hydrate, 98%	"	8	0	0
Sulphocyanide Ammonium, 95%	per lb.	0	0	7¾
" Barium, 95%	"	0	0	5¾
" Potassium	"	0	0	8
Sulphur (Flowers)	per ton.	7	10	0
" (Roll Brimstone)	"	6	0	0
" Brimstone: Best Quality	"	4	7	6
Superphosphate of Lime (26%)	"	2	7	6
Tallow	"	25	10	0
Tin (English Ingots)	"	99	0	0
" Crystals	per lb.	0	0	6¾
Zinc (Spelter)	per ton.	23	10	0
" Chloride (solution, 96° Tw.	"	6	0	0

INDEX.

VOL. VI.

Salt—*continued.*
Port Natal, 188 t.
St. John's, Nfld., 4
Bilbao, 5 c.
Calcutta, 2,556
Demerara, 27
Ghent 150
Iceland, 479
Manaos, 96
Melbourne, 102
Monte Video, 28
Montreal, 313
New York, 405
Para, 507¾
Seville, 30 cs.
Trinidad, 36 3
Valparaiso, 8 9
Wilmington, 320
Port Natal, 75
Quebec, 500
St. John's, Nfld., 20 cs.
San Francisco, 199¾
Sherbro, 100 9 c.
Talcahuano, 120
Trinidad, 37
Valparaiso, 301
Wellington, 518
Quebec, 416

Sodium Silicate—
Zante, 5 brls.
Galatz, 15
Genoa, 17

Sugar of Lead—
Colon, 4 c. £5

Salt Cake—
Baltimore, 101 t. 10 c. £177
" 204 355

Sodium Biborate—
Hamburg, 17 t. 18 c. 2 q. £539
Rotterdam, 1 1 2 33
St. Petersburg, 15 10 1 466
Stettin, 1 15 1 53

Sodium Chlorate—
Mexico City, 5 c. £14

Salammoniac—
Antwerp, 5 c. 3 q. £10
Alexandria, 1 t. 1 28
Patras, 9 17
Larnaca, 1 2 38
Malta, 8 15
Patras, 5 2 9
Naples, 9 1 16
Salaora, 6 10
Volo, 11 19
Genoa, 8 14

heep Dip—
Algoa Bay, 2 t. 5 c. £125
Natal, 1 10 80
B. Ayres, 3 1 q. 8
New York, 4 4 60
Algoa Bay, 1 9 57
E. London, 15 40
Monte Video, 300 drs. 106
B. Ayres, 12 20

Saltpetre—
Rio Grandé du Sol, 6 c. 7
Toronto, 6 6

Sulphur—
Maceio, 1 t. 10 c. £10
Madras, 4 15 1 q. 29
New York, 11 8 51
Oporto, 181 733
Boston, 75 3 315
Rangoon, 5 28
Boston, 25 109

Tartar—
Bordeaux, 6 t. £570

Tar—
Santander,

Zinc Chlorate—
Rotterdam, 5 t. 8 c. £36

TYNE.

Week ending June 17th.

Alumina Sulphate—
Pasages, 21 t. 14 c.

Anthracene—
Rotterdam, 9 t. 19 c.

Ammonium Sulphate—
Hamburg, 50 t.
Odense, 5

Alkali—
Copenhagen, 6 t. 7 c.
Rotterdam, 4 16
Stockholm, 2 18
Gefle, 1 2
Leghorn, 11 10
S. Petersburg, 51 8
Christiania, 6 8
Gothenburg, 18 9
Rotterdam, 5 6
Riga 16 11

Alum—
Pasages, 25 t. 2 c.

Bleaching Powder—
Copenhagen, 16 t. 2 c.
Rotterdam, 27 3
" 9
Danzig, 1 4
Newfairwater, 56 9
Lisbon, 4 19
New York, 100 3
Stockholm, 3
Genoa, 15
S. Petersburg 191 18
Bilbao, 20 1
Abo, 24 1
Antwerp, 17 11
Genoa, 10 1
Hamburg, 12 14
" 21 2
Pasages, 10
Montreal, 25 17
Christiania, 5 2
Esbjerg, 1 3
Odense, 1 15
Rotterdam, 6 1
" 9 11
" 10 9
St. Petersburg, 105 2
Riga, 17 4

Barytes—
Rotterdam, 2 t. 4 c.

Copperas—
Genoa, 5 t. 5 c.
Vienna, 3 19
Esbjerg, 6 2

Caustic Soda—
Copenhagen, 4 t. 7 c.
Rotterdam, 2 12
Reval, 5 16
Newfairwater, 20
Gefle, 4
Genoa, 34 16
Naples, 17
St. Petersburg, 37 7
Antwerp, 8 3
Hamburg, 3 14
Rotterdam, 4 18
Riga, 20 6

Ferro Manganese—
Bilbao, 44 t. 14 c.

Gypsum—
New York, 38 16 c.

Litharge—
Copenhagen, 1 t. 9 c.
Newfairwater, 1
Riga, 11 2
Hamburg, 2 15

Magnesia—
Newfairwater, 12 c.
Hamburg, 4

Oxalic Acid—
New York, 10 t. 6 c.

Potassium Chlorate—
Genoa, 1 5
St. Petersburg, 2 10
Rotterdam, 6 c.

Phosphorus—
Malmo, 4 c.

Salammoniac—
St. Petersburg, 17 t. 15 c.

Sulphur—
Rotterdam, 8 t. 16 c.

Soda—
Copenhagen, 26 t. 18 c.
Libau, 25 6
Rotterdam, 10 17
Stockholm, 18 12
Gefle, 35 3
St. Petersburg, 4 19
Montreal, 40 11
Christiania, 10 3
Antwerp, 5
Aarhuus, 24 15
" 23
Esbjerg, 12 4
Odense, 22 10
Rotterdam, 29 7
" 11 10
Horsens, 8 3
Svendborg, 4 5
Randers, 22 10

Soda Ash—
Libau, 20 t. 6 c.
Newfairwater, 70 19
Lisbon, 5 17
Genoa, 9 19
Abo, 2 13
Christiania, 9 9
Hamburg, 7 12
Gothenburg, 98 11

Sodium Sulphate—
Aarhuus, 21 t. 11 c.

Sodium Chlorate—
Rotterdam, 1 t 16 c.

Superphosphate—
Libau, 770 t. 1 c.
Riga, 200 6
" 339 1

Sodium Hyposulphate—
Antwerp, 1 t. 6 c.

Zinc Ashes—
Copenhagen, 10 c.

GLASGOW.

Week ending June 19th.

Ammonium Sulphate—
Trinidad, 7 t. 3½ c. £82
Demerara, 5 17¾ 62
" 50 13¾ 571
Mauritius, 9 15¾ 100
Nantes, 70 8½ 875

Benzole—
Dieppe, 185 brls. 15 pns. £1,045

Bleaching Powder—
Sydney, 6 t. 3 c. £38

Caoutchouc—
Rangoon, 34 c. £232

Copperas—
Karachi, 11 t. 5 c. £22

Copper Sulphate—
Nantes, 21 t. 12¾ c. £387

Chemicals (*otherwise undescribed*)
Rangoon, 18½ c. £20

Dyestuffs—
Extracts
Quebec, 98½ c. £146
Oporto, 5 t. 15¾ 45
Montreal, 161¼ 250
" 87 98
Argols
Quebec, 22¾ c. £21
Logwood
Quebec, 443½ c. £165
Montreal, 573½ 207

Epsom Salts—
Christiania, 50 bgs. £40
Calcutta, 6 t. 1½ c. 16

Guano—
Demerara, 10 t. £105

Litharge—
Quebec, 4 t. 9¼ c. £72
Karachi, 2 2¼ 31

Manure—
Trinidad, 127 t. 1½ c. £1,205
" 140 6 1,114
" 10 80
Demerara, 10 95
" 54 9 536
Mauritius, 3 ½ 80
" 200 10 1,554
Penang, 200 13 c. 2,100

Manganese Oxide—
New York, 228 c. £93

Nitric Acid—
Bombay, 1 t. 7 c. £43 15s.

Naptha—
Rouen, 50 cks. £450

Potassium Bichromate—
Quebec, 9 t. 9¾ c. £318
" 4 3 154
Oporto, 1 1½ 36
Rouen, 16 6½ 549
Antwerp, 510
Karachi, 3 9¼ 123

Potassium Carbonate—
Karachi, 1 t. 18 c. £33

Paraffin Wax—
Hong Kong, 1 t. 18½ c. £53
Nantes, 18 16¼ 480
Japan, 4½ 6

Potassium Chloride—
Pasages, 5 t. ¾ c. £50

Potassium Prussiate—
New York, 5 t. 9 c. £361

Sulphuric Acid—
Rangoon, 12 t. 2¼ c. £132
Bombay, 20 12 103

Soda Ash—
New York, 176½ £49

Sodium Bichromate—
Quebec, 5 t. 16¾ c. £154
Karachi, 33¼ 44

Verdigris.—
New York, £200

HULL.

Week ending June 17t.

Bleaching Powder—
Bremen, 10 cks.
Drontheim, 59
Rotterdam, 7

Barytes—
Stettin, 150 t.

Caustic Soda—
Amsterdam, 20 drms.
Bergen, 4
Gothenburg, 248
Konigsberg, 105
Libau, 145
Riga, 465
St. Petersburg, 450

Chemicals (*otherwise undes*
Antwerp, 1 cs. 2 cks. 48
Amsterdam, 1 1 36
Boston, 20
Bombay, 542 235
Bordeaux, 20 50
Bremen, 2 6
Christiania, 5 27 82
Copenhagen, 11
Danzig, 15 36
Gothenburg, 14 44 3
Ghent, 4 40
Hamburg, 25 8
Helsingfors, 1
Konigsberg, 29 20
Lisbon, 1
Libau, 85
New York, 25 6
Rotterdam, 31 1
Rouen, 36
St. Petersburg, 5 2 20
Stettin, 24 123
Stavanger, 1
Stockholm, 1 10
Wyburg, 1

Dyestuffs (*otherwise undesc.*
Antwerp, 8 cks.
Amsterdam, 2
Boston, 4
Hamburg, 16 1 drum
Reval, 3 1 cs.
Rotterdam, 33 4 kgs.
St. Petersburg, 2 cs.
Stettin, 300 bgs.
Melbourne, 2 cks.

Manure—
Stettin, 484 bgs.

Pitch—
Antwerp, 446 t.

Tar—
Amsterdam, 5 cks.
Hamburg, 2
Stockholm, 5

GOOLE.

Week ending June 11t

Benzole—
Calais, 25 cks.
Rotterdam, 48

Chemicals (*otherwise undes*
Dunkirk, 36 cks.
Ghent, 2

Caoutchouc—
Boulogne, 4 cks.

Dyestuffs (*otherwise undesc*
Boulogne, 3 cs. 5 cks.
Ghent, 30 bgs. 12 tins. 1 ck. 12
Rotterdam, 66 bgs.
" 85

Manure—
Rotterdam, 176 bgs.

Pitch—
Antwerp, 737 t.

GRIMSBY.

Week ending June 14

Coal Products—
Dieppe, 98 cks.

Chemicals (*otherwise undes*
Antwerp, 49 pks.
Dieppe, 60
Hamburg, 10 cks. 2 cs.

Dyestuffs (*otherwise undes*
Dieppe, 3 cks.

Pitch—
Antwerp, £328.

WALKER BROS.,

ENGINEERS, PAGEFIELD IRONWORKS,

WIGAN.

WALKER'S PATENT

AIR COMPRESSING ENGINES.
GAS COMPRESSING ENGINES.

70,000 I.H.P. ALREADY AT WORK.

WALKER'S PATENT [a6774246

EXHAUST AND BLAST FANS.

SPENCE'S
LUMP ALUM,
POTASH, OR AMMONIA

IS THE FREEST FROM IRON. [822154

PETER SPENCE & SONS, Manchester Alum Works, MANCHESTER.

Also manufacturers of **SULPHATE OF ALUMINA & ALUMINOFERRIC.**

GEORGE SCOTT & SON,

Engineers, Ironfounders, & Boilermakers, since 1834,

MAKERS OF EVERY DESCRIPTION OF CHEMICAL PLANT AND MACHINERY,

White Lead, Oil, Colour, and Soap Makers' Machinery and Plant.

Air Compressors. Apparatus for Liquefying Gases. Edge Runners. Blowing Engines for Mixing and Agitation.	Vacuum Pans. Vacuum Pumps. Filter Presses, with or without absolute Extraction. Special Acid Filter Presses. Charcoal Filters.	Hydro-Extractors for Sulphate of Ammonia and other Chemical Products. Hydraulic Presses. Roller and Horizontal Mills.

Telegrams—"Thirty-four, London." Telephone 4,370. [a0617122

44 and 46, CHRISTIAN STREET, LONDON. E.

WINCE PULLEYS.

RUSCOE'S PATENT GROOVED PULLEYS ENAMELLED ARE THE BEST.

They withstand the acid, are well balanced, and move freely on the shaft; little wear and tear.

Once used, always continued. [a670136

References, Prices, and other information, apply to the Patentee and Sole Maker:

JOHN RUSCOE, HYDE, near MANCHESTER.

PURE SULPHUROUS ACID,

SULPHITES AND BISULPHITES OF LIME, SODA, &c.

FOR EXPORT. [970154

S. KEYS & CO. Limited, WEST BROMWICH.

Tenders

BOROUGH OF MIDDLESBR

WANTED, an Experie
gineer and Manager to take cha
Works. He must have a thorough scienti
knowledge of the Manufacture and Dist
and the Manufacture of Sulphate of Amm
also understand Regenerative Heating,
practical acquaintance with Main and Serv
must be competent to design and prepare
carry out all constructive works and alter
be required. His whole time must be
duties of his office. He must not be unde
than 45 years of age.

The Office will be held subject to three m
either side.

The make of Gas is about 300 million fee
Applications, stating age, qualificatio
ployment, and amount of salary required,
not more than three recent testimonials,
the undersigned not later than the 22nd of
endorsed "Gas Manager."

Any applicant canvassing will be disquali

GEORGE BAINBRIDGE

Municipal Buildings, Middlesbrough,
Dec 24, 1889.

THE Corporation of Midd
prepared to receive TENDERS for
the SULPHURIC ACID required at their
the Fifteen months ending the 31st of Marc

Further particulars may be obtained from
Gas Manager, Middleton.

Tenders, addressed to the Chairman of
mittee, to be delivered at my Office, on or
day, the 15th of January, 1890, endorse
Sulphuric Acid."

The Corporation do not bind themselve
lowest or any tender.

FREDERICK ENTWISTLE

Town Hall, Middleton, near Manchester,
Dec. 24, 1889.

HALIFAX CORPORATION GAS.

THE Halifax Corporation
pared to receive TBNDERS (*a*) for
SULPHURIC ACID, and (*b*) for the purc
OXIDE OF IRON for the year endin
December, 1890.

Full particulars and form of Tender may
application to Mr. Thomas Holgate, F.C.S
Halifax.

Sealed tenders, endorsed "Sulphuric Aci
Oxide," as the case may be, to be sent to t
not later than Twelve o'clock, on the 15
1890.

The Corporation do not bind themselves
lowest, highest, or any tender.—By order,

KEIGHLEY WALTON

Town Hall, Halifax, Dec. 27, 1889.

Miscellaneous.

MANAGER, (Foreman)
wanted for small chemical works, m
understand the manufacture of Brimstone,
and Muriatic Acids, Tin Crystals, Lac Spi
Iron, Glauber Salts, and other dyers chemi
perience, age, and wages expected. Address
Trade Journal Office.

TARTARIC ACID.—W
Chemist thoroughly competent to
manufacture of this article, liberal terms t
party.—Address, "Tartaric," *Chemical*
Office.

WANTED, second-ha
Disintegrator For sale, No. 3 Ca
rator, perfect order.—W. M. Fuller, Ju
Chambers, Wolverhampton.

TO BE SOLD, the LANC
MANURE COMPANY'S WORK
with Machinery and 24,490 square yards of Fr
Address J. Hosking, 16, Fenwick Street, Li

FOR SALE, a Works, in
bourhood of Manchester, suitable fo
uses: comprising one acre of land, building
power, stills, tanks, store capacity 1,500 t
Appleby & Wood, 47, Spring Gardens, Ma

CHARCOA

FILTERING, POWDERED an

WOOD NAPHTHA, ACETATE

TAR, LAMPBLACK, FOUNDERS' BLA

Lydbrook Chemica

183, CANNON ST., LONDON,

Works and Mills, LYDBROOK, For

D CHEMICAL BOOKS.

ndition and sent Carriage Paid.

rd to Arts and Manufactures. By ence (including Schorlemmer, Crookes, *f Machinery, etc.*, 8 vols., £4, for 45s.
emmer's Chemistry. Complete set; rganic; 8 vols., £5 18s. 6d.
Makers' Pocket Book, 1889. 6s. 3d.
ary of Chemistry and the Allied mplete and unabridged edition; 9 vols., 15., for £8 8s.
ION in 4 vols., re-written and revised id MUIR. Vols. I. and II., complete, ols., half morocco, 42s., for 32s. each.
cts of the Alkali Trade, 8s
echt's Coal-Tar Colours, 5s. 9d.
Book of Chemistry. By H. WATTS. Organic and Inorganic, with Index, 19), scarce, £20, for £7 10s.
r and Ammonia. *New Edition*, 27s.
ition (S.H.) for 12s.
; and Calico Printing, 40s., for 21s.
Watts' Chemical Technology of ufactures. 5 vols., £5 5s., for 30s.
ALIES, AND SALTS, 3 vols., £4, for 15s.
Journal, 1878-87. 20 vols., £8 8s.

AM F. CLAY,

nical Bookseller,

PLACE, EDINBURGH.

ited for any books. Odd Vols. or Nos. e, and will receive prompt attention.
se "Journal of Society of Chemical ny other Chemical Literature, for nged at Highest Market Value.
[986169

TUBES

AND FITTINGS for Gas, Water, Steam, Liquid aulic Work, galvanised white enamelled Dr. Angus Smith's Composition.
ting, super-heating, distilling, and es, to 500 feet long without joints. Iron valves. [933173

N SPENCER,
s, Wednesbury, and 14, Great St. stle, Cannon-st., London, E.C.

HED 1869.

ANESE

ACKWELL,
TER & MANUFACTURER.
ks: GARSTON.
FFICES:
l Chambers, W.,
POOL. [994135

Bauxite,
Barytes,
Fluor Spar,
Magnesite,
Chrome Ore,
Emery Stone,
China Clay,
French Chalk,
Oxides,
Ochres,
Umbers,
Bog Ore,
and all
Minerals.

---

f the Chemical Trade Journal.

"CHEMICAL TRADE JOURNAL" nber, 1887), with Title-page and in cloth and lettered, may be ice, delivered by parcel post. e United Kingdom, or 11s. 6d.

ary to June, 1888), Volume III cember, 1888), Volume IV. June, 1889), and Volume V. nber, 1889). Price, delivered 9s. per Volume for the United 0s. abroad. Apply to

9, Blackfriars-st., Manchester.

PHILIP HARRIS & CO. LIMITED,

BULL RING, BIRMINGHAM,

HAVE JUST ISSUED

A CATALOGUE OF CHEMICAL APPARATUS,

Containing about 1,200 illustrations, giving list of apparatus and reagents for use in the analysis of Iron, Steel, Fuel, &c., by various methods; also particulars of Bunte's, Hempel's, and other Gas analysis apparatus; **Harris's Polariscope, Microscope, &c., &c.** (See *Chemical Trade Journal*, February 16th, page 102.)

GLUE, GELATINE,

AND

PREPARED HORN PITHS.

THOMSON & SON, Manufacturers,

TRANMERE, BIRKENHEAD.

CUBLEY & PRESTON,

4, High Street, SHEFFIELD.

PURE CHEMICALS FOR ANALYSIS.

Chemical and Scientific Apparatus.

[696136

Special—ARNOLD'S APPARATUS for Estimation of Sulphur, in Steel, Iron, etc., £4 4s. complete.

IMPORTANT TO CHEMICAL MANUFACTURERS

CORTIN'S NEW PATENT

Non-Rotative Acid Valve,

The invention of this Valve has been the means of dispensing with all other Taps, hitherto used, owing to its great safety in working with Acids.

Highest Testimonials from all parts of the globe. No compensation bills to pay since the invention of this Valve.

Apply

J. CORTIN,

Chemical Plant Manufacturer, Newcastle-on-Tyne.

CHARLES BIRCHALL,

32, CASTLE STREET,
(FIRST FLOOR.)

Advertisements inserted in every Newspaper in the World.

Liverpool: Exchange Chambers
lasgow Charles-st., St. Rollox

CANNINGTON, SHAW, & CO.

CARBOYS.

SHERDLEY GLASSWORKS, ST. HELENS.

Tenders.

MANCHESTER CORPOR
GASWORKS.
TO TAR DISTILLERS AND OTH
The Gas Committee of the Corporation of
are prepared to RECEIVE TENDERS for
CHASE and REMOVAL of the GAS TAR
duced at their several stations, during a period
or three years, commencing as follows, viz.:—
Bradford Road Station, from noon on the 31st M
Rochdale Road Station, from noon on the 24th
Gaythorn Station, from noon on the 1st Augu
The Committee do not bind themselves to
highest or any tender, and reserve the right to
tenders as they may consider advisable. Seal
addressed to the Chairman of the Gas Com
endorsed "Tender for Tar," must be delivered a
of the Gas Department, Town Hall, on or bef
day the 13th of February next. Forms of
other particulars may be obtained on applicatio
to Mr. Charles Nickson, Superintendent of the D
—By order of the Gas Committee.
WM. HENRY TALBOT, To
Town Hall, Manchester, January 17, 1890.

Miscellaneous.

THE Valuable and C
CHEMICAL MANURE WORKS, kn
Silvertown Chemical Works, Silvertown, for SA
premises have been constructed at considerable
purpose of a very large trade, and the machinery
in every respect for the immediate carrying
business; there is a wharf with a frontage of
feet to the River Thames, so that all goods can
and unloaded direct from the wharf, also a rail
connecting the works with the various railway
the United Kingdom; the premises compri
manufacturing, storage, and packing works, pyri
acid chambers capable of producing 5000 tons c
per annum, office, laboratory, and caretaker's
plant is capable of producing 20,000 tons of
annum, which could at a small outlay be greatl
all parts of works are connected by tram lines
ticulars apply to Walter W. Feast, Esq., charter
ant, St. George's House, Eastcheap, London,
Messrs. Hatchett, Jones, and Co., Solicitors,
lane, London, E.C.

ALKALI MAKERS' PLANT AND MAC
FOR SALE, by Private Tr
the Jarrow Chemical Company's Tyne D
South Shields, a large quantity of ACID, AL
CAUSTIC SODA PLANT, in good condition,
be viewed at any time.—Apply to the Jarrow Ch
Limited, Tyne Dock, South Shields

A YOUNG MAN, with Lab
experience, seeks situation with a Chem
facturing or Merchant Firm, either in Office or
Works, or as Assistant Chemist. Matriculated
University, holds certificate of Finsbury Technic
knows French and part German, and writes
Can give first-class references. A.C., *Chem*
Journal Office.

A CHEMIST, doing busi
Broker and Agent in Paris, able to co
English, seeks the representation of Englis
Manufacturing firms.—Apply VIII, *Chemic*
Journal.

WANTED, a Situation, by a
thoroughly experienced in the Man
Sulphuric Acid, would also be capable of tak
other branch of Chemical Industry.—Appl
Chemical Trade Journal.

FOR SALE BY PRIVATE TREAT

A VALUABLE UMBER

Thoroughly proved and in good working ord
Cargoes tested by manufacturers; Color an
power very fine and equal to best Foreign Umber
position to manipulate and prepare for mark
Patent "double-current" levigator just out by
Bate, Gorton; can ship direct; capital safe
There is also on the estate well defined Lo
Copper and Hematite Iron Ore.—For further
and samples, apply direct, J. T. Foulis, M
Ramsey, I.O.M.

STANDARD CHEMICAL BOOKS.

All in good condition and sent Carriage Paid.

Chemistry applied to Arts and Manufactures. By writers of eminence (including Schorlemmer, Crookes, etc.): *Plates of Machinery, etc.*, 8 vols., £4., for 45s.
Roscoe & Schorlemmer's Chemistry. Complete set; organic and inorganic; 8 vols., £5 18s. 6d.
Lunge's Alkali-Makers' Pocket Book, 1889. 6s. 3d.
Watts' Dictionary of Chemistry and the Allied Sciences. Complete and unabridged edition; 9 vols., cloth, (new), £15., for £8 8s.
Ditto, New Edition in 4 vols., re-written and revised by Morley and Muir. Vols. I. and II., complete, *now ready*, 2 vols., half morocco, 42s., for 32s. each.
Kingzett's Products of the Alkali Trade, 8s.
Benedikt and Knecht's Coal-Tar Colours, 5s. 9d.
Gmelin's Hand-Book of Chemistry. By H. Watts. Complete set, Organic and Inorganic, with Index, 19 vols., 8vo. cloth, scarce, £20, for £7 10s.
Lunge's Coal Tar and Ammonia. *New Edition*, 27s.
Ditto, Previous Edition (S.H.) for 12s.
Crookes's Dyeing and Calico Printing, 42s., for 21s.
Richardson and Watts' Chemical Technology of Arts and Manufactures. 5 vols., £5. 5s., for 30s.
Ditto, Acids, Alkalies, and Salts, 3 vols., £4, for 15s.
Chemical Society Journal, 1878-87. 20 vols., £8 8s.

WILLIAM F. CLAY,

Chemical Bookseller,

2, TEVIOT PLACE, EDINBURGH.

Communications invited for any books. Odd Vols. or Nos. wanted or for sale, and will receive prompt attention.
Wanted to purchase "Journal of Society of Chemical Industry" and any other Chemical Literature, for Cash, or Exchanged at Highest Market Value. [986169

TUBES

IRON TUBES AND FITTINGS for Gas, Water, Steam, Liquid Ammonia, and Hydraulic Work, galvanised white enamelled inside, or coated by Dr. Angus Smith's Composition.
COILS for heating, super-heating, distilling, and refrigerating purposes, to 500 feet long without joints. Iron and brass cocks and valves. [933173

JOHN SPENCER,
Globe Tube Works, Wednesbury, and 14, Great St. Thomas Apostle, Cannon-st., London, E.C.

ALEX. ROWAT & CO.,

Wire Work & Wire Cloth Manufacturers,

AND

BOILER TUBE BRUSH MAKERS,

25, CANDLERIGGS, GLASGOW.

Established 1867. [m880159

Back Volumes of the Chemical Trade Journal.

Volume I. of the "Chemical Trade Journal" (May to December, 1887), with Title-page and Index, bound in cloth and lettered, may be obtained. Price, delivered by parcel post, 10s. 6d. for the United Kingdom, or 11s. 6d. abroad; also

Volume II. (January to June, 1888), Volume III (July to December, 1888), Volume IV. (January to June, 1889), and Volume V. (July to December, 1889). Price, delivered by parcel post, 9s. per Volume for the United Kingdom, or 10s. abroad. Apply to

---

PHILIP HARRIS & CO. LIMITED,

BULL RING, BIRMINGHAM,

HAVE JUST ISSUED

A CATALOGUE OF CHEMICAL APPARATUS,

Containing about 1,200 illustrations, giving list of apparatus and reagents for use in the analysis of Iron, Steel, Fuel, &c., by various methods; also particulars of Bunte's, Hempels, and other Gas analysis apparatus; **Harris's Polariscope, Microscope, &c., &c.** (See *Chemical Trade Journal*, February 16th, page 102.)

RETORTS.

PLAIN AND TUBULATED.

SAMUEL RALPHS,

PRUSSIA STREET FLINT GLASS WORKS,

Oldham Road, MANCHESTER.

Established 1869.

MANGANESE

GEO. G. BLACKWELL,
Mine Proprietor, Importer & Manufacturer.
Works and Mills: Garston.
Chief Offices:
26, 26, & 27, Irwell Chambers, W., LIVERPOOL. [994135

Bauxite,
Barytes,
Fluor Spar,
Magnesite,
Chrome Ore,
Emery Stone,
China Clay,
French Chalk,
Oxides,
Ochres,
Umbers,
Bog Ore,
and all Minerals.

IMPORTANT TO CHEMICAL MANUFACTURERS

CORTIN'S NEW PATENT

Non-Rotative Acid Valve,

The invention of this Valve has been the means of dispensing with all other Taps, hitherto used, owing to its great safety in working with Acids.
Highest Testimonials from all parts of the globe. No compensation bills to pay since the invention of this Valve.
Apply [d9ed

J. CORTIN,
Chemical Plant Manufacturer, Newcastle-on-Tyne.

OILS, SOAPS, GREASES.

London Oil Refining Co.,

GARDEN LANE,

SALFORD.

Strap and Boiler Compositions. Hot Neck Grease, &c.

NOW READY.
A Second Edition of

SIZING AND MILDEW IN COTTON GOODS.

By GEORGE E. DAVIS, F.I.C., F.R.M.S.; and CHARLES DREYFUS, Ph.D.

PRICE TEN SHILLINGS. [68x

Manchester: Palmer and Howe, Princess Street.
London: Simpkin Marshall, and Co.

LAMP BLACK

And other BLACKS
Loose and in Packets
for Home and Export.
Instantaneous dried Blue Varnish.

JAMES DRINKWATER,
Springfield Lane Chemical Works,
SALFORD,
MANCHESTER.

---

rpool: Exchange Chambers
gow: Charles-st., St. Rollox

CANNINGTON, SHAW, & CO.

CARBOYS.

SHERDLEY GLASSWORKS, ST. HELENS.

CHARLES BIRCHALL,

32, Castle Street.
(FIRST FLOOR.)

Advertisements inserted in every Newspaper in the World.

SPENCE'S
LUMP ALUM,
POTASH, OR AMMONIA
IS THE FREEST FROM IRON. [821154

PETER SPENCE & SONS, Manchester Alum Works, MANCHESTER.
Also manufacturers of **SULPHATE OF ALUMINA & ALUMINOFERRIC.**

COILS OF ALL DESCRIPTIONS.
[856157

CHARCOAL:
FILTERING, POWDERED and LUMP.
WOOD NAPHTHA, ACETATE OF LIME,
TAR, LAMPBLACK, FOUNDERS' BLACKINGS.
Lydbrook Chemical Co.,
153, CANNON ST., LONDON, E.C. 601
Works and Mills, LYDBROOK, Forest of Dean.

WINCE PULLEYS.

RUSCOE'S PATENT GROOVED PULLEYS ENAMELLED ARE THE BEST.

They withstand the acid, are well balanced, and move freely on the shaft; little wear and tear.

Once used, always continued. [

References, Prices, and other information, apply to the Patentee and Sole Maker:

JOHN RUSCOE, HYDE, near MANCHESTER.

PURE SULPHUROUS ACID,
SULPHITES AND BISULPHITES OF LIME, SODA, &c.
FOR EXPORT. [970154
S. KEYS & CO. Limited, WEST BROMWICH.

Tenders.

MANCHESTER CORPOR
GASWORKS.
TO TAR DISTILLERS AND OTH
The Gas Committee of the Corporation of are prepared to RECEIVE TENDERS for CHASE and REMOVAL of the GAS TAR duced at their several stations, during a period or three years, commencing as follows, viz.:—
Bradford Road Station, from noon on the 31st
Rochdale Road Station, from noon on the 24th
Gaythorn Station, from noon on the 1st Augu
The Committee do not bind themselves to highest or any tender, and reserve the right to tenders as they may consider advisable. Seal addressed to the Chairman of the Gas Comr endorsed "Tender for Tar," must be delivered a of the Gas Department, Town Hall, on or be day the 13th of February next. Forms of other particulars may be obtained on applicatio to Mr. Charles Nickson, Superintendent of the D —By order of the Gas Committee.
WM. HENRY TALBOT, To
Town Hall, Manchester, January 17, 1890.

NORTH EASTERN RAII
CREOSOTE OIL.

The Directors are prepared to receive t the supply of 200,000 to 300,000 gallons thereof) of CREOSOTE OIL of the very b free from all impurities, and delivered at this Creosote Works, North Dock, Monkwear Greenland, West Hartlepool, as may be require
Specification and form of tender may be application to Mr. J. S. Mitford, Permanent keeper, Central Station, Newcastle-on-Tyne.
Sealed tenders, marked "Tender for Creosote be sent to the Secretary, at York, not later th Saturday, 15th February, 1890.
The Directors do not bind themselves to lowest or any tender.
C. N. WILKINSON, S
York, 29th January, 1890.

Miscellaneous.

Replies to Advertisements addres "Chemical Trade Journal" office are for Advertisers, post free, on the day of their

---

THE Valuable and C
CHEMICAL MANURE WORKS, kr Silvertown Chemical Works, Silvertown, for S. premises have been constructed at considerable purpose of a very large trade, and the machinery in every respect for the immediate carrying business; there is a wharf with a frontage o feet to the River Thames, so that all goods car and unloaded direct from the wharf, also a rai connecting the works with the various railway the United Kingdom; the premises compris manufacturing, storage, and packing works, pyri acid chambers capable of producing 5000 tons cl per annum, office, laboratory, and caretaker's plant is capable of producing 20,000 tons of annum, which could at a small outlay be greatl all parts of works are connected by tram lines ticulars apply to Walter W. Feast, Esq., charter ant, St. George's House, Eastcheap, London, Messrs. Hatchett, Jones, and Co., Solicitors, lane, London, E.C.

A CHEMIST, doing busi
Broker and Agent in Paris, able to co English, seeks the representation of Englis Manufacturing firms.—Apply V111, *Chemi Journal.*

AGENT (traveller) wanted i
chester, calling on shipping houses with and on chemical manufacturers to sell Phos makers. Commission.—Apply Deacons, P314, Street, London.

NICKEL REFINERY.—V
an experienced Manager to superinte and working of a Plant for the Wet Process H. Ball, Castle Street, Birmingham.

TO BE SOLD, the LANCAS
MANURE COMPANY'S WORKS with Machinery and 24,420 square yards of Freeh Address J. Hosking, 16, Fenwick Street, Liverp

---

SITUATION wanted, by
practical man, 10 years' experience as T and Sulphate of Ammonia Manufacturer; able take management of plant.—Address Williams, 6, Tunstall Street, Beswick, Manchester.

FOR SALE, a Works, in the
bourhood of Manchester, suitable for a uses: comprising one acre of land, buildings, off power, stills, tanks, store capacity 1,500 tons. Appleby & Wood, 47, Spring Gardens, Manche

WALKER BROS.,

ENGINEERS, PAGEFIELD IRONWORKS,

WIGAN.

WALKER'S PATENT

AIR COMPRESSING ENGINES.
GAS COMPRESSING ENGINES.

70,000 I.H.P. ALREADY AT WORK.

WALKER'S PATENT [a0774148

EXHAUST AND BLAST FANS.

SPENCE'S
LUMP ALUM,

POTASH, OR AMMONIA

IS THE FREEST FROM IRON. [801154

PETER SPENCE & SONS, Manchester Alum Works, MANCHESTER.

Also manufacturers of **SULPHATE OF ALUMINA & ALUMINOFERRIC.**

GEORGE SCOTT & SON,

Engineers, Ironfounders, & Boilermakers, since 1834,

MAKERS OF EVERY DESCRIPTION OF CHEMICAL PLANT AND MACHINERY,

White Lead, Oil, Colour, and Soap Makers' Machinery and Plant.

Air Compressors. Apparatus for Liquefying Gases. Edge Runners. Blowing Engines for Mixing and Agitation.	Vacuum Pans. Vacuum Pumps. Filter Presses, with or without absolute Extraction. Special Acid Filter Presses. Charcoal Filters.	Hydro-Extractors for Sulphate of Ammonia and other Chemical Products. Hydraulic Presses. Roller and Horizontal Mills.

Telegrams—"Thirty-four, London." Telephone 4,870.

44 and 46, CHRISTIAN STREET, LONDON, E. [v94P

PATENT TWISTED TAPS, DIES, AND CUTTERS.

Thousands in constant work, giving universal satisfaction.

Work Easier, Less Friction, Last Longer, and do Better Work than straight grooves.

MAKER OF

SCREWING MACHINES, STOCKS & DIES;

LATHES, DRILLING AND SHAPING MACHINES, &c.

Enquiries and Trial Orders Solicited. Everything guaranteed. Apply to

JOHN RUSCOE, [108d

ALBION WORKS, HYDE, MANCHESTER.

PURE SULPHUROUS ACID,

SULPHITES AND BISULPHITES OF LIME, SODA, &c.

FOR EXPORT. [970154

S. KEYS & CO. Limited, WEST BROMWICH.

Tenders.

MANCHESTER CORPOR
GASWORKS.
TO TAR DISTILLERS AND OTH

The Gas Committee of the Corporation of are prepared to RECEIVE TENDERS for CHASE and REMOVAL of the GAS TAR duced at their several stations, during a period or three years, commencing as follows, viz.:—

Bradford Road Station, from noon on the 31st
Rochdale Road Station, from noon on the 24th
Gaythorn Station, from noon on the 1st Augu

The Committee do not bind themselves to highest or any tender, and reserve the right to tenders as they may consider advisable. Seal addressed to the Chairman of the Gas Com endorsed "Tender for Tar," must be delivered of the Gas Department, Town Hall, on or be day the 13th of February next. Forms of other particulars may be obtained on applicatio to Mr. Charles Nickson, Superintendent of the —By order of the Gas Committee.

WM. HENRY TALBOT, To

Town Hall, Manchester, January 17, 1890.

NORTH EASTERN RAI
CREOSOTE OIL.

The Directors are prepared to receive the supply of 200,000 to 300,000 gallons thereof) of CREOSOTE OIL of the very free from all impurities, and delivered at this Creosote Works, North Dock, Monkwear Greenland, West Hartlepool, as may be require

Specification and form of tender may be application to Mr. J. S. Mitford, Permanent keeper, Central Station, Newcastle-on-Tyne.

Sealed tenders, marked "Tender for Creosot be sent to the Secretary, at York, not later th Saturday, 15th February, 1890.

The Directors do not bind themselves to lowest or any tender.

C. N. WILKINSON, S

York, 29th January, 1890.

Miscellaneous.

Replies to Advertisements addres "Chemical Trade Journal" office are for Advertisers, post free, on the day of their

---

THE Valuable and C
CHEMICAL MANURE WORKS, kn Silvertown Chemical Works, Silvertown, for S premises have been constructed at considerable purpose of a very large trade, and the machinery in every respect for the immediate carrying business; there is a wharf with a frontage of feet to the River Thames, so that all goods can and unloaded direct from the wharf, also a rail connecting the works with the various railway the United Kingdom; the premises compris manufacturing, storage, and packing works, pyrit acid chambers capable of producing 5000 tons ch per annum, office, laboratory, and caretaker's plant is capable of producing 20,000 tons of annum, which could at a small outlay be greatly all parts of works are connected by tram lines. ticulars apply to Walter W. Feast, Esq., charter ant, St. George's House, Eastcheap, London, Messrs. Hatchett, Jones, and Co., Solicitors, lane, London, E.C.

GRINDING MILL, suitat
grinding hard Coke to pass No. 20 Siev least wear and tear.—Address "Mill," *Chemi Journal* Office.

WANTED, Man thoroughl
versant with the distillation of Wood N the making of Iron Liquor.—Apply with fullest p M113, *Chemical Trade Journal* Office.

MANAGER or Sub-Mana
Practical Chemist, 29, with ten years' in management and construction of Sulphuric Superphosphate Plants, is open to engagem guarantee economical work.—Address H.B.G., *Trade Journal*, Blackfriars Bridge, Manchester.

NICKEL REFINERY.—W
an experienced Manager to superinten and working of a Plant for the Wet Process. H. Ball, Castle Street, Birmingham.

TO BE SOLD, the LANCAS
MANURE COMPANY'S WORKS a with Machinery and 24,420 square yards of Freeho Address J. Hosking, 16, Fenwick Street, Liverp

---

SITUATION wanted, by
practical man, 10 years' experience as Ta and Sulphate of Ammonia Manufacturer; able take management of plant.—Address Williams, c 6, Tunstall Street, Beswick, Manchester.

Miscellaneous.

FOR SALE, a Works, in the neighbourhood of Manchester, suitable for a variety of uses: comprising one acre of land, buildings, offices, steam power, stills, tanks, store capacity 1,500 tons. Address Appleby & Wood, 47, Spring Gardens, Manchester.

FOR SALE BY PRIVATE TREATY

A VALUABLE UMBER MINE.

Thoroughly proved and in good working order; Sample Cargoes tested by manufacturers; Color and Staining power very fine and equal to best Foreign Umbers; splendid position to manipulate and prepare for market, by New Patent "double-current" levigator just out by Follows and Bate, Gorton; can ship direct; capital safe investment. There is also on the estate well defined Lodes of rich Copper and Hematite Iron Ore.—For further particulars and samples, apply direct, J. T. Foulis, Mine Agent, Ramsey, I.O.M.

ADVERTISER, aged 20, holding high certificates in Chemistry, nearly four years' training in chemical laboratory, desires situation as Assistant Analyst or Assayer; references.—H. Hember, Queen Charlton, Bristol.

CEMENT WORKS MANAGER, with thorough Chemical, and general practical knowledge of the process of Cement Manufacture from limestone, required to proceed to China to erect there a Cement Mill, Kilns, &c., and superintend the making of the Cement. A THOROUGHLY PRACTICAL FOREMAN, and also a BURNER, accustomed to the making of Cement from limestone will also be required.—Address with fullest details as to past experience, age, &c., to "Cement," c/o J. W. Vickers, 5, Nicholas Lane, London, E.C.

HIGH CLASS Chemist wanted for Yorkshire Tar Distillery; must be thoroughly experienced; please furnish full particulars.—Address "Tar."

CHARCOAL:

FILTERING, POWDERED and LUMP.

WOOD NAPHTHA, ACETATE OF LIME,

TAR, LAMPBLACK, FOUNDERS' BLACKINGS.

Lydbrook Chemical Co.,

143, CANNON ST., LONDON, E.C. 601

Works and Mills, LYDBROOK, Forest of Dean.

TUBES

IRON TUBES AND FITTINGS for Gas, Water, Steam, Liquid Ammonia, and Hydraulic Work, galvanised white enamelled inside, or coated by Dr. Angus Smith's Composition.

COILS for heating, super-heating, distilling, and refrigerating purposes, to 500 feet long without joints. Iron and brass cocks and valves. [933:73]

JOHN SPENCER,
Globe Tube Works, Wednesbury, and 14, Great St. Thomas Apostle, Cannon-st., London, E.C.

ESTABLISHED 1869. Bauxite,

PHILIP HARRIS & CO. LIMITED
BULL RING, BIRMINGHAM,
HAVE JUST ISSUED
A CATALOGUE OF CHEMICAL APPARATUS,

Containing about 1,200 illustrations, giving list of apparatus and reagents for use in the analysis of Iron, Steel, Fuel, &c., by various methods; also particulars of Bunte's, Hempel's, and other analysis apparatus; **Harris's Polariscope, Microscope, &c., &c.** (See *Chemical Trade Journal*, February 16th, page 102.)

MATHIESON'S
PURE AMMONIA ALKALI.

Commercial Test 59° equal to 99 °/₀ of Carbonate of Soda.

The purest and most economical article for the use of Paper Makers, Bleachers, Dyers, Calico Printers, Glass Manufacturers, Soap Makers, Wool Scourers, etc.

Samples sent free on application to

W. H. COWBURN,
3, Clarence Street, Albert Square, MANCHESTER
AGENT FOR ENGLISH CONSUMERS.

Over 5000 in use.

THE
CYCLONE DUST COLLECTOR

The most simple and effective device for cleaning the dust laden air of

CHEMICAL, CEMENT, and MANURE WORKS;
COTTON, WOOLLEN, PAPER, and BONE MILLS, &c.

NO FIRE RISK. NO MOTIVE POWER.

Full particulars and prices on application to

HENRY SIMON,
20, Mount Street, MANCHESTER.

OILS, SOAPS, GREASES.

London Oil Refining Co.,
GARDEN LANE,
SALFORD.

Strap and Boiler Compositions, Hot Neck Grease, &c.

NOW READY.
A Second Edition of

LAMP BLACK And other BLACKS

Tenders.

TAR.

THE Directors of the Rea
Company are prepared to receive T
purchase of their TAR for one year, from t
next.

Specifications for the contracts can be see
or will be sent on application.

Rail and water communication direct to th
Tenders to be sent not later than Monda
March, to

ROBT. BRA

Office: 22, Market Place, Reading.
February 25th, 1890.

MIRFIELD GAS COMPAN

THE Directors of this Con
prepared to receive Tenders for the
and AMMONIACAL LIQUOR produced
works for the year, from the 1st day of April
For Tar, state price at the Company's W
thorpe.

For Ammoniacal Liquor, state price per
strength from 8 to 12 ounces, tested by Wills's
either at the Company's Works or into Cont
Tenders to be sent in, endorsed "Tend
Ammoniacal Liquor," not later than Mon
day of March, 1890, addressed to

SAMUEL DA

Ravensthorpe, *via* Normanton.
March 12th.

Miscellaneous.

Replies to Advertisements addr "Chemical Trade Journal" office are f Advertisers, post free, on the day of th

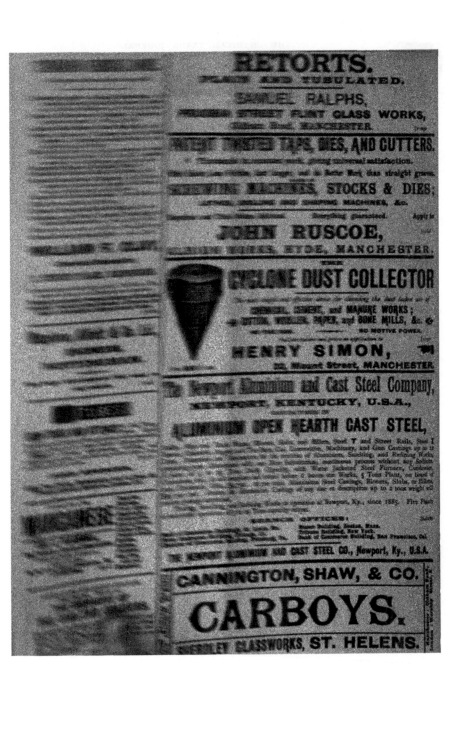

Tender.

THE SCARBOROUGH GAS CO

TAR AND AMMONIACAL LIQ

THE Directors of the Sca Gas Company are prepared to receive for the purchase of the AMMONIACAL I surplus TAR produced at their works for the 18th day of April next.

Conditions of tender and necessary parti supplied on application to the undersign tenders, in accordance with such conditions, endorsed, must be delivered not later than noo the 7th day of April next.

By Order,
WILLIAM J. MOON, Secretary
Gas Offices, Scarborough,
19th March, 1890.

Miscellaneous.

Replies to Advertisements addr "Chemical Trade Journal" office are f Advertisers, post free, on the day of the

TO BE SOLD, the LANC MANURE COMPANY'S WORKS with Machinery and 24,420 square yards of Fre Address J. Hosking, 16, Fenwick Street, Live

BICHROMATES OF POTASH AN PRACTICAL CHEMIST, connected with an American Chrome he has been engaged for over five years, and viously had seven years' experience in a Works in Scotland, is open for engagement sub-manager.—Apply to H. G. S., 35, Leir Hyde Park, London.

FOR SALE, a Works, in t bourhood of Manchester, suitable for uses: comprising one acre of land, buildings, power, stills, tanks, store capacity 1,500 to Appleby & Wood, 47, Spring Gardens, Manc

LIXIVIATING TANKS. WANTED, a number of once, must be in good condition.—A Post Office, Middlesbrough.

YOUNG MAN, 22, B.Sc. who has worked for three years in (Glasgow and Edin.), desires situation as Assistant Chemist in Works.—Address "Alp *Trade Journal*, Blackfriars Street, Manchest

WANTED, as Foreman, who thoroughly understands the m Sulphuric Acid from Brimstone and Pyrite premises. A married man without fami Address "Acid," 34, Union Street, Birmingh

AN ANALYST and Te Chemistry, of much experience, has a Resident Pupil. Facilities can be given for th of other studies concurrently if desired. Mc Comfortable home in one of the healthiest sea England.—G. R. T., *Chemical Trade Jo* Blackfriars Bridge, Manchester.

ON SALE or TO LET, Lane Works, Radcliffe. Railway sid Apply to Watsons', Solicitors, Bury; or F Solicitor, 55, Cross Street, Manchester.

Now Ready, with more than 600 Illustrations 30s.

CHEMICAL TECHNOL OR, CHEMISTRY IN ITS APPLICA ARTS AND MANUFACTURE Edited by CHARLES EDWARD GRO Editor of the "Journal of the Chemical So WILLIAM THORP, B.Sc.

Vol. I., FUEL AND ITS APPLICA By E. J. MILLS, D.Sc., F.R.S., & F. J. R

"We have here a book not merely admirable, from the patience and erudition authors in collecting the materials, as well judgment in elaborating the vast mass of has been in their hands. Nor must we number, the appropriateness, and good exe illustrations. . . . To competent men t carry with it its own urgent recommendation *News*.

London:
J. & A. CHURCHILL, 11, New Burling

Tender.

Miscellaneous.

WALKER BROS.,

ENGINEERS, PAGEFIELD IRONWORKS,

WIGAN.

WALKER'S PATENT

AIR COMPRESSING ENGINES.
GAS COMPRESSING ENGINES.

70,000 I.H.P. ALREADY AT WORK.

WALKER'S PATENT [v198a

EXHAUST AND BLAST FANS.

SPENCE'S

LUMP ALUM,

POTASH, OR AMMONIA

IS THE FREEST FROM IRON. [821154

PETER SPENCE & SONS, Manchester Alum Works, MANCHESTER.

Also manufacturers of **SULPHATE OF ALUMINA & ALUMINOFERRIC.**

J. GREGORY & CO.,

Deepcar Ganister & Firebrick Works,

Near SHEFFIELD.

SPECIALITIES—

Silica Bricks and Fire Bricks for Siemens Furnaces, Glass and Chemical Works, and other intense continuous or intermittent heats. Special Bricks for Gas Works and Coke Ovens.

The Highest Grades and Quality Guaranteed. Moderate Prices.

114J] Telegrams:—"GREGORY, DEEPCAR."

Acetate of Lime. Wood Tar and Pitch.

CHARCOAL.

Best and Common Lump, Filtering, Powdered, &c.

Lampblack, Founders' Blackings, Plumbago, etc.

LYDBROOK CHEMICAL CO.,

195A] **143, CANNON STREET, LONDON, E.C.**

Works and Mills: LYDBROOK, FOREST OF DEAN; AND OAKWOOD, PARKEND, GLOS.

MATHIESON'S

PURE AMMONIA ALKALI.

Commercial Test 59° equal to 99 °/₀ of Carbonate of Soda.

The purest and most economical article for the use of Paper Makers, Bleachers, Dyers, Calico Printers, Glass Manufacturers, Soap Makers, Wool Scourers, etc.

Samples sent free on application to

W. H. COWBURN,

70, MARKET STREET, MANCHESTER.

AGENT FOR ENGLISH CONSUMERS. [112H

Tenders.

TO TAR DISTILLERS.

THE Directors of the Ashto[n] Company invite Tenders for the purcha[se of the] surplus TAR that will be produced at their wor[ks for] a period of twelve months, commencing the fir[st] August next.

Any further information required can be ob[tained on] application to the undersigned, to whom se[aled] endorsed Tenders must be delivered not later [than] o'clock at noon, on Monday, the 19th May next.

DAVID CL[...]

Gas Offices,
Ashton-under-Lyne,
30th April, 1890.

TO TAR DISTILLERS AND MANUFACT[URING] CHEMISTS.

THE Directors of the Welli[ngton] (Salop) Coal and Gas Light Company invit[e] for the purchase of the surplus TAR to be pro[duced at] their works for one year, from 1st June next.

Particulars may be obtained from the unders[igned, to] whom Tenders should be sent on or before May [...]

JNO. C. BELTON[,] M[...]

Gas Office, Market-st,
Wellington, Salop, April 30th, 1890.

Miscellaneous.

☞ Replies to Advertisements addressed [to the] "Chemical Trade Journal" office are forwa[rded to] Advertisers, post free, on the day of their re[ceipt].

TO BE SOLD, the LANCAS[HIRE] MANURE COMPANY'S WORKS at [...] with Machinery and 24,420 square yards of Freehold [...] Address J. Hosking, 16, Fenwick Street, Liverpool[.]

A THOROUGHLY practical [...]tical Chemist, F.C.S., and Accountant, is o[pen to] engagement in the Manure or other business; h[...] years' experience in all branches of the Manure [...] at present acting as assistant secretary and Chem[ist ...] complete private laboratory; highest testimonials.—Box 50, *Chemical Trade Journal*, Manchester.

AMMONIUM Nitrate wanted [...] ton lots; free from Sulphate and Chlorid[e ...] 14 lbs. sample (which will be paid for), in air tight [...] with lowest price, to J. C. Butterfield, 13, Victori[a ...] Westminster, London, S.W.

WANTED, to place a Young [...] with a firm of Chemical Manufacturers, [...] business; premium.—Address H. 130, *Chemica[l Trade] Journal*, Blackfriars-street, Manchester.

FOR SALE. Twenty-six stron[g] Casks, tinned inside; capacity, forty gallo[ns ...] fitted with interchangeable screw bungs.—Addres[s ...] *Chemical Trade Journal*, Blackfriars-street, Man[chester].

FOR SALE, a Works, in the [neigh]bourhood of Manchester, suitable for a v[ariety of] uses: comprising one acre of land, buildings, office[s ...] power, stills, tanks, store capacity 1,500 tons. [...] Appleby & Wood, 47, Spring Gardens, Mancheste[r].

ADVERTISER (age 20) d[esires] situation as Assistant Analyst, Works Che[mist ...] Demonstrator; nearly six years' experience; no [...] to going abroad; some experience in Portland [...] qualified teacher.—Apply A. E. Turner, We[...] Ditchling, Sussex.

ENGINEER, residing in Ant[werp ...] late Briquettes Manufactory and Tar [...] Manager, desires to represent an English fi[rm ...] Distiller or Pitch Merchant) for the sale of [...] Belgium.—Address Henry Del, Ingénieur, Rue [...] Antwerp.

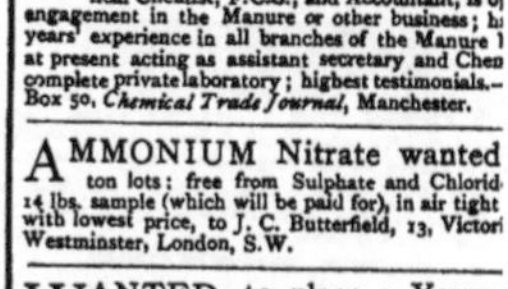

COILS OF ALL DESCRIPTIO[N]

PERMANENT CHEMICAL EXHIBITION.

A permanent chemical exhibition was opened on the evening of the 15th April, in the premises of Messrs. Davis Brothers, proprietors of the *Chemical Trade Journal*, Blackfriars Street, Manchester. Large rooms have been set apart for the purposes of the exhibition. Here it is proposed to establish a chemical exchange for the city, in which there will be permanently on view a collection of chemical products, apparatus, samples, &c., such as will be of great interest to those engaged in the chemical trade. One or two of the exhibits which were in the Chemical Section of the Manchester Jubilee Exhibition are to be seen here. It is hoped that a benefit will be conferred upon those who at present, in coming to Manchester, have either to drag their samples about with them or have to do their business without any samples at all. Such samples may in future be deposited in this new institution in Blackfriars Street, where business may be done by those who wish to use it. There will be a writing and reading room attached to the place, furnished with all the leading chemical papers and the market reports. The exhibition was declared open by Mr. Ivan Levinstein, chairman of the Manchester section of the Society of Chemical Industry, in presence of a large company. He spoke of the usefulness of the exhibition and the great advantages to be derived from the proposed exchange, which, he said, would fill a long felt want. Those engaged in the chemical trade might go there and see working models, by which they could judge of the merit of appliances recommended to them, and see whether they possessed the advantages claimed. He hoped the institution might become a favourite meeting place for chemical men. He would suggest that there might be seen there a collection of smoke-preventing appliances, for in Manchester that was a subject about which a great deal was heard. They were all desirous of preventing any smoke nuisance, and though he had no doubt many of them had tried smoke-preventing appliances he was afraid many of them had to put them out again, either on account of the great expenditure or on account of inefficiency. (Hear, hear.) An exhibition of smoke-preventing appliances would help them considerably to judge of their efficacy. He also hoped the proprietors would see their way to add to the library such valuable literature as consular reports and statistical accounts with reference to the subject of chemicals. In that respect a collection of the chemical products exported from this country to transatlantic countries and to our colonies would be most valuable, as well as such exhibits as might enable them to see how the Germans or the French were making up their goods for the Indian or Chinese markets. (Hear, hear.) He heartily wished success to the exhibition.—*Manchester Examiner*, April 16, 1890.

Tenders.

MANCHESTER CORPORATION GAS \
TAR AND AMMONIACAL LIQUC

THE Gas Committee of th
chester Corporation are prepared to
TENDERS for the PURCHASE and REM
the TAR and AMMONIACAL LIQUOR to b
at their Droylsden station during a period of o
years from the 24th June next.

The Committee do not bind themselves to
highest or any tender.

Sealed tenders, addressed to the Chairman
Committee, and endorsed "Tender for Tar ar
Droylsden," must be delivered at the offices
Department, Town Hall, on or before Thursda
day of June next.

Forms of tender and further particulars may b
on application in writing to Mr. Charles Nick
intendent of the Gas Department.

By order of the Gas Committee,
WM. HENRY TALBOT
Town Hall, 9th May, 1890. Tow

MANCHESTER CORPORATION GAS \
SPENT OXIDE.

THE Gas Committee of the
ation of Manchester are prepared to
TENDERS for the PURCHASE and REM
the SPENT OXIDE to be produced at their
Rochdale-road, and Bradford-road Works duri
of one, two, or three years from the 30th June n

The Committee do not bind themselves to
highest or any tender.

Sealed tenders, addressed to the Chairman
Committee, and endorsed "Tender for Spent O
be delivered at the offices of the Gas Departm
Hall, on or before Thursday, the 5th day of Jun

Forms of tender and further particulars may b
on application in writing to Mr. Charles Nick
intendent of the Gas Department.

By order of the Gas Committee,
WM. HENRY TALB
Town Hall, 9th May, 1890, Tow

TAR AND AMMONIA WATEI

THE Directors of the Loughb
Gas Company are prepared to receive T
for the surplus TAR and AMMONIA WAT
produced at their Works for One, Two, or Th
from the 30th day of June, 1890.

The estimated quantity will be about 350 to
and 900 tons of Ammonia water.

The Tar and Liquor to be delivered into th
tors' Boats at the Wharf on the Derby Roa
borough.

The strength of the Liquor to be tested by
Hydrometer (temperature 60°); and tenders to
price per ton to be paid for every ½° of strength

The Boats for the removal of the Tar and
Water to be sent at such times as may be require

Payments monthly.

Further information may be obtained on app
the undersigned, to whom tenders, in accord
this advertisement, must be sent not later than t
May, 1890.

The Directors do not bind themselves to
highest or any tender.

J. B. BALL, C.E., M
Loughborough, April 30th, 1890.

Miscellaneous.

☞ Replies to Advertisements addresse "Chemical Trade Journal" office are forw Advertisers, post free, on the day of their r

TO BE SOLD, the LANCAS
MANURE COMPANY'S WORKS a
with Machinery and 24,420 square yards of Freehol
Address J. Hosking, 16, Fenwick Street, Liverpo

AMMONIUM Nitrate wante
ton lots; free from Sulphate and Chlori
14 lbs. sample (which will be paid for), in air tigh
with lowest price, to J. C. Butterfield, 13, Victo
Westminster, London, S.W.

A YOUNG MAN who has s
Chemistry, theoretically, desires Situation
cal Works. Address, H 130, *Chemical Trade*
Blackfriars Street, Manchester.

FOR SALE. Twenty-six stron
Casks, tinned inside; capacity, forty gall
fitted with interchangeable screw bungs.—Addr
Chemical Trade Journal, Blackfriars-street, Ma

GEORGE SCOTT & SON,

Engineers, Ironfounders & Boilermakers since 1834.

Makers of every description of Chemical Plant and Machinery,

WHITE LEAD, OIL, COLOUR, and SOAP MAKERS MACHINERY and PLANT,

And Sole Makers of Plant for a New Process for Making Acetic Acid.

- Air Compressors
- Apparatus for Liquefying Gases
- Edge Runners
- Blowing Engines for Mixing and Agitation
- Vacuum Pans, Charcoal Filters
- Special Acid Filter Presses
- Hydraulic Presses
- Roller and Horizontal Mills
- Hydro-Extractors for Sulphate of Ammonia and other Chemical Products.

Filter Presses with or without absolute extraction. Vacuum Filters.

v174P] Vacuum Pumps to give an almost absolute vacuum.

Telegrams: "Thirtyfour, London." Telephone 4390.

44 & 46, Christian Street, London, E., England.

SPENCE'S
LUMP ALUM,
POTASH, OR AMMONIA

IS THE FREEST FROM IRON. [206

PETER SPENCE & SONS, Manchester Alum Works, MANCHESTER.

Also manufacturers of **SULPHATE OF ALUMINA & ALUMINOFERRIC.**

J. GREGORY & CO.,
Deepcar Ganister & Firebrick Works,
Near SHEFFIELD.

SPECIALITIES—
Silica Bricks and Fire Bricks for Siemens Furnaces, Glass and Chemical Works, and other intense continuous or intermittent heats. Special Bricks for Gas Works and Coke Ovens.

The Highest Grades and Quality Guaranteed. Moderate Prices.

114J] Telegrams:—"GREGORY, DEEPCAR."

Acetate of Lime. Wood Tar and Pitch.
CHARCOAL.
Best and Common Lump, Filtering, Powdered, &c.

Lampblack, Founders' Blackings, Plumbago, etc.

LYDBROOK CHEMICAL CO.,

1956] **143, CANNON STREET, LONDON, E.C.**

Works and Mills: LYDBROOK, FOREST OF DEAN; AND OAKWOOD, PARKEND, GLOS.

MATHIESON'S
PURE AMMONIA ALKALI.

Commercial Test 59° equal to 99 °/₀ of Carbonate of Soda.

The purest and most economical article for the use of Paper Makers, Bleachers, Dyers, Calico Printers, Glass Manufacturers, Soap Makers, Wool Scourers, etc.

Samples sent free on application to

W. H. COWBURN,
70, MARKET STREET, MANCHESTER.
AGENT FOR ENGLISH CONSUMERS. [112H

Tenders.

**MANCHESTER CORPORATION GAS WORKS.
TAR AND AMMONIACAL LIQUOR.**

THE Gas Committee of the Manchester Corporation are prepared to RECEIVE TENDERS for the PURCHASE and REMOVAL of the TAR and AMMONIACAL LIQUOR to be produced at their Droylsden station during a period of one or more years from the 24th June next.

The Committee do not bind themselves to accept the highest or any tender.

Sealed tenders, addressed to the Chairman of the Gas Committee, and endorsed "Tender for Tar and Liquor, Droylsden," must be delivered at the offices of the Gas Department, Town Hall, on or before Thursday, the 5th day of June next.

Forms of tender and further particulars may be obtained on application in writing to Mr. Charles Nickson, superintendent of the Gas Department.

By order of the Gas Committee,
WM. HENRY TALBOT,
Town Hall, 9th May, 1890. Town Clerk.

**MANCHESTER CORPORATION GAS WORKS.
SPENT OXIDE.**

THE Gas Committee of the Corporation of Manchester are prepared to RECEIVE TENDERS for the PURCHASE and REMOVAL of the SPENT OXIDE to be produced at their Gaythorn, Rochdale-road, and Bradford-road Works during a period of one, two, or three years from the 30th June next.

The Committee do not bind themselves to accept the highest or any tender.

Sealed tenders, addressed to the Chairman of the Gas Committee, and endorsed "Tender for Spent Oxide," must be delivered at the offices of the Gas Department, Town Hall, on or before Thursday, the 5th day of June next.

Forms of tender and further particulars may be obtained on application in writing to Mr. Charles Nickson, superintendent of the Gas Department.

By order of the Gas Committee,
WM. HENRY TALBOT,
Town Hall, 9th May, 1890. Town Clerk.

**TAUNTON GAS LIGHT AND COKE COMPANY.
TENDER FOR TAR.**

THE Taunton Gas Light and Coke Company are prepared to receive TENDERS for the purchase of the surplus TAR produced at their works over one year, commencing 1st July next.

Tenders, stating price offered per ton, put into Contractor's Tanks at Taunton Railway Station, to be delivered to the undersigned not later than 1st June next.

ALFRED EDWARDS,
May 14th, 1890. Secretary and Manager.

SURPLUS TAR AND LIQUOR.

THE Gas Committee of the Bingley Improvement Commissioners invite competent firms to TENDER for the purchase of the SURPLUS TAR and LIQUOR as proposed to be made at their Gas Works, Bingley, during the 12 months ending June 30th, 1891. Form of Tender to be had on application to the undersigned.

Tenders to be addressed to James Gatecliff, Esq., Chairman of the Gas Committee, Commissioners' Offices, Main-street, Bingley, endorsed "Tender for Tar and Liquor," on or before the 9th day of June, 1890.

By Order,
GEORGE DUNBAR MALAM,
May 20th, 1890. Engineer.

TO MANUFACTURING CHEMISTS & OTHERS.

THE BRIGHOUSE LOCAL BOARD are prepered to receive TENDERS for the PURCHASE OF THE SURPLUS TAR produced at their Gas Works, for a term of One, Two, or Three Years, from the 1st of July next.

The quantity of Tar made is about 350 tons per annum. The Tar can be loaded into Contractors' carts on the works, or into Barges on the Aire and Calder Canal (which runs alongside) by pipe direct from the works. Any further particulars can be obtained on application to the undersigned.

Sealed Tenders, endorsed on the outside "Tender for Tar," stating price per ton at the Gas Works, must be sent in on or before Thursday, the 5th day of June, 1890, addressed to Richard Kershaw, Esq., Chairman of the Local Board, Public Offices, Brighouse.

The Board do not bind themselves to accept the highest or any Tender.

By order,
JAMES PARKINSON,
Brighouse, May 20th, 1890 Manager.

**BOROUGH OF ST. HELENS GAS DEPARTMENT.
TO TAR DISTILLERS.**

THE Gas Committee of the St. Helens Corporation invite TENDERS for the purchase of the SURPLUS TAR produced at their Works; estimated quantity about 1100 tons.

Forms of Tender and all particulars may be obtained on application to the undersigned.

The Tenders, properly endorsed, are to be delivered on or before 10th June next, addressed to the Chairman of the Gas Committee, Warrington Old Road, St. Helens.

Gas Works, Warrington Old Road,
May, 1890

Tenders.

MANCHESTER CORPORATION GAS
TAR AND AMMONIACAL LIQU

THE Gas Committee of th chester Corporation are prepared to TENDERS for the PURCHASE and RE the TAR and AMMONIACAL LIQUOR to b at their Droylsden station during a period of years from the 24th June next.

The Committee do not bind themselves to highest or any tender.

Sealed tenders, addressed to the Chairman Committee, and endorsed "Tender for Tar a Droylsden," must be delivered at the offices Department, Town Hall, on or before Thursd day of June next.

Forms of tender and further particulars may on application in writing to Mr. Charles Nick intendent of the Gas Department.

By order of the Gas Committee,
WM. HENRY TALBO

Town Hall, 9th May, 1890. To

MANCHESTER CORPORATION GAS
SPENT OXIDE.

THE Gas Committee of the ation of Manchester are prepared to TENDERS for the PURCHASE and RE the SPENT OXIDE to be produced at thei Rochdale-road, and Bradford-road Works duri of one, two, or three years from the 30th June

The Committee do not bind themselves to highest or any tender.

Sealed tenders, addressed to the Chairman Committee, and endorsed "Tender for Spent O be delivered at the offices of the Gas Depart Hall, on or before Thursday, the 5th day of Ju

Forms of tender and further particulars may on application in writing to Mr. Charles Nic intendent of the Gas Department.

By order of the Gas Committee,
WM. HENRY TAL

Town Hall, 9th May, 1890. To

TAUNTON GAS LIGHT AND COKE C
TENDER FOR TAR.

THE Taunton Gas Light ar Company are prepared to receive TEND purchase of the surplus TAR produced at their one year, commencing 1st July next.

Tenders, stating price offered per ton, pu tractor's Tanks at Taunton Railway Station, to to the undersigned not later than 1st June next.

ALFRED EDWARDS,
May 14th, 1890. Secretary and

SURPLUS TAR AND LIQUOR.

THE Gas Committee of the Improvement Commissioners invite com to TENDER for the purchase of the SUR and LIQUOR as proposed to be made at their Bingley, during the 12 months ending June Form of Tender to be had on application to signed.

Tenders to be addressed to James Gatecliff, man of the Gas Committee, Commissioners' Of street, Bingley, endorsed "Tender for Tar an on or before the 9th day of June, 1890.

By Order,
GEORGE DUNBAR
May 20th, 1890. Engineer

TO MANUFACTURING CHEMISTS &

THE BRIGHOUSE L BOARD are prepered to receive TEN the PURCHASE OF THE SURPLUS TA at their Gas Works, for a term of One, Two Years, from the 1st of July next.

The quantity of Tar made is about 350 tons The Tar can be loaded into Contractors' carts or or into Barges on the Aire and Calder Canal (alongside) by pipe direct from the works. particulars can be obtained on application to signed.

Sealed Tenders, endorsed on the outside " Tar," stating price per ton at the Gas Works, in on or before Thursday, the 5th day of addressed to Richard Kershaw, Esq., Chair Local Board, Public Offices, Brighouse.

The Board do not bind themselves to accept or any Tender.

By order,
JAMES PARKIN
Brighouse, May 20th, 1890

BOROUGH OF ST. HELENS GAS DEPA
TO TAR DISTILLERS.

THE Gas Committee of the St. Corporation invite TENDERS for the the SURPLUS TAR produced at their Works quantity about 1100 tons.

Forms of Tender and all particulars may be application to the undersigned.

The Tenders, properly endorsed, are to be or before 10th June next, addressed to the the Gas Committee, Warrington Old Road, St

Gas Works, S. GLOVER,
Warrington Old Road, Engineer a
May, 1890

Tenders.

COUNTY BOROUGH OF STOCKPORT.
(GAS DEPARTMENT.)
RESIDUAL PRODUCTS FOR SALE.

THE Gas Committee are prepared to receive TENDERS for the purchase of the surplus TAR and AMMONIACAL LIQUOR produced at their Works for a period of One, Two or Three years, commencing on the 1st of July next.

Tenders to state the price per ton for Tar, and price per ton for Liquor 6° Twaddel.

Tenders to be delivered, addressed to the Chairman of the Gas Committee, on or before June 12th, 1890, endorsed "Tender for Tar, &c."

Particulars as to quantity and conditions of contract may be obtained on application to the Engineer.

The Committee do not bind themselves to accept the highest or any tender.

JAMES JACQUES,
Engineer and Manager.

Gas Works, May 26, 1890.

COUNTY BOROUGH OF HALIFAX.
TO MANUFACTURING CHEMISTS AND TAR DISTILLERS.

THE Gasworks Committee of the Halifax Corporation are prepared to receive TENDERS for the purchase of the surplus TAR—about 3,500 tons—produced at their Gasworks during the year ending 30th June, 1891. Delivery may take place from the Railway Siding of the Corporation Gasworks, or from the Canal Wharf of the Calder and Hebble Navigation, with which a Tar pipe communicates. Full particulars may be obtained on application to the Engineer, Mr. Thos. Holgate, F.C.S., Gasworks, Halifax. Sealed tenders endorsed "Tar," stating the price offered per ton, and the mode of removal, to be sent to the undersigned not later than 11 a.m., on Friday the 13th June, 1890. The highest or any tender will not necessarily be accepted.

By Order,
KEIGHLEY WALTON,
Town Clerk.

Town Hall, Halifax,
19th May, 1890.

Miscellaneous.

TO BE SOLD, the LANCASHIRE MANURE COMPANY'S WORKS at Widnes with Machinery and 24,430 square yards of Freehold Land.—Address J. Hosking, 16, Fenwick Street, Liverpool. [169h

---

MANAGER Wanted, to go abroad to erect and work a Wood Distilling Plant.—Write, giving full particulars, to F 134, *Chemical Trade Journal*, Blackfriars-street, Manchester.

WANTED, by a Young Man, situation as Chemist or Assistant Chemist; has had considerable experience in Wood Distilling, Acetates and Acetic Acid making.—Address S 135, *Chemical Trade Journal*, Blackfriars-street, Manchester.

WANTED, in a provincial Laboratory, Senior Assistant, thoroughly conversant with analytical and technical processes, and possessing a sound knowledge of physics.—Apply with full particulars to No. 137, *Chemical Trade Journal*, Blackfriars-street, Manchester.

PATENTS AND TRADEMARKS
Secured, Bought & Sold.

Charges for Securing Patents:
Great Britain provisional, £2 10s.; complete, £5 10s.
America, £12 10s.
Special attention given to Chemical Inventions. [117L

The International Patent Agency,
27, Martin's Lane, Cannon Street, London, E.C.

TUBES

IRON TUBES AND FITTINGS for Gas, Water, Steam, Liquid Ammonia, and Hydraulic Work, galvanized white enamelled inside, or coated by Dr. Angus Smith's Composition.

COILS for heating, super-heating, distilling, and refrigerating purposes, to 500 feet long without joints. Iron and brass cocks and valves. [933173

JOHN SPENCER,
Globe Tube Works, Wednesbury, and 14, Great St. Thomas Apostle, Cannon-st., London, E.C.

PHILIP HARRIS & CO. LIMITED,
BULL RING, BIRMINGHAM,
HAVE JUST ISSUED
A CATALOGUE OF CHEMICAL APPARATUS,

Containing about 1,200 illustrations, giving list of apparatus and reagents for use in the analysis of Iron, Steel, Fuel, &c., by various methods; also particulars of Bunte's, Hempels, and other Gas analysis apparatus; **Harris's Polariscope, Microscope, &c., &c.** (See *Chemical Trade Journal*, February 16th, page 102.) [15F

WINCE PULLEYS.

RUSCOE'S PATENT GROOVED PULLEYS ENAMELLED ARE THE BEST.

They withstand the acid, are well balanced, and move freely on the shaft; little wear and tear.

Once used, always continued.

References, Prices, and other information, apply to the Patentee and Sole Maker:

JOHN RUSCOE, HYDE, near MANCHESTER.

RETORTS.
PLAIN AND TUBULATED.

SAMUEL RALPHS,
PRUSSIA STREET FLINT GLASS WORKS,
Oldham Road, MANCHESTER. [11w

IRON CARBOY HAMPERS.
All Qualities. Lowest Prices.

H. LEIGH & SONS
ORLANDO MILL,
BOLTON.

PURE SULPHUROUS ACID,
SULPHITES AND BISULPHITES OF LIME, SODA, &c.
FOR EXPORT.
S. KEYS & CO. Limited, WEST BROMWICH.

Liverpool Exchange Chambers
Glasgow Charles-st., St. Rollox

CANNINGTON, SHAW, & CO.
CARBOYS.
SHERDLEY GLASSWORKS, ST. HELENS.

GEORGE SCOTT & SON,

Engineers, Ironfounders & Boilermakers since 1834.

Makers of every description of Chemical Plant and Machinery,

WHITE LEAD, OIL, COLOUR, and SOAP MAKERS MACHINERY and PLANT,

And Sole Makers of Plant for a New Process for Making Acetic Acid.

Air Compressors	Special Acid Filter Presses
Apparatus for Liquefying Gases	Hydraulic Presses
Edge Runners	Roller and Horizontal Mills
Blowing Engines for Mixing and Agitation	Hydro-Extractors for Sulphate of Ammonia
Vacuum Pans, Charcoal Filters	and other Chemical Products.

Filter Presses with or without absolute extraction. Vacuum Filters.

V174P] Vacuum Pumps to give an almost absolute vacuum.

Telegrams: "Thirtyfour, London." Telephone 4390.

44 & 46, Christian Street, London, E., England.

SPENCE'S
LUMP ALUM,
POTASH, OR AMMONIA

IS THE FREEST FROM IRON. [206v

PETER SPENCE & SONS, Manchester Alum Works, MANCHESTER.

Also manufacturers of **SULPHATE OF ALUMINA & ALUMINOFERRIC.**

J. GREGORY & CO.,
Deepcar Ganister & Firebrick Works,
Near SHEFFIELD.

SPECIALITIES—
Silica Bricks and Fire Bricks for Siemens Furnaces, Glass and Chemical Works, and other intense continuous or intermittent heats. Special Bricks for Gas Works and Coke Ovens.

The Highest Grades and Quality Guaranteed. Moderate Prices.

114J] Telegrams:—"GREGORY, DEEPCAR."

Acetate of Lime. Wood Tar and Pitch.
CHARCOAL.
Best and Common Lump, Filtering, Powdered, &c.

Lampblack, Founders' Blackings, Plumbago, etc.

LYDBROOK CHEMICAL CO.,

1954] 143, CANNON STREET, LONDON, E.C.

Works and Mills: LYDBROOK, FOREST OF DEAN; AND OAKWOOD, PARKEND, GLOS.

MATHIESON'S
PURE AMMONIA ALKALI.

Commercial Test 59° equal to 99 °/o of Carbonate of Soda.

The purest and most economical article for the use of Paper Makers, Bleachers, Dyers, Calico Printers, Glass Manufacturers, Soap Makers, Wool Scourers, etc.

Samples sent free on application to

W. H. COWBURN,
70, MARKET STREET, MANCHESTER.

AGENT FOR ENGLISH CONSUMERS. [112H

Tenders.

SURPLUS TAR AND LIQUOR.

THE Gas Committee of the Improvement Commissioners invite com to TENDER for the purchase of the SURP and LIQUOR as proposed to be made at their Bingley, during the 12 months ending June Form of Tender to be had on application to signed.

Tenders to be addressed to James Gatecliff, E man of the Gas Committee, Commissioners' Offi street, Bingley, endorsed "Tender for Tar an on or before the 9th day of June, 1890.

By Order,
GEORGE DUNBAR M
May 20th, 1890. Engineer.

COUNTY BOROUGH OF STOCKPO
(GAS DEPARTMENT.)
RESIDUAL PRODUCTS FOR SAL

THE Gas Committee are prep receive TENDERS for the purchase of t TAR and AMMONIACAL LIQUOR produc Works for a period of One, Two or Three years, ing on the 1st of July next.

Tenders to state the price per ton for Tar, an ton for Liquor 6° Twaddel.

Tenders to be delivered, addressed to the Cl the Gas Committee, on or before June 12th, 1890 "Tender for Tar, &c."

Particulars as to quantity and conditions o may be obtained on application to the Engineer.

The Committee do not bind themselves to highest or any tender.

JAMES JACQUE
Gas Works, May 16, 1890. Engineer and

THE SHREWSBURY GAS LIGHT COM
DISPOSAL OF SURPLUS TAR.

THE Directors of the above Co are desirous of receiving TENDERS f chase of the surplus TAR made at their works year commencing July 1st, 1890.

The estimated quantity for disposal is about per annum.

Tenders must be made on Forms (containi particulars) which may be obtained on applicat Company's Works, or by post, and must be sen undersigned not later than the 20th day of June,

The Directors do not bind themselves to a highest or any Tender.

By Order,
WM. BELTON, A.M.I.C.
Secretary and M
Gas Works, Shrewsbury,
29th May, 1890.

Miscellaneous.

☞ Replies to Advertisements addresse "Chemical Trade Journal" office are forw Advertisers, post free, on the day of their r

BY ORDER OF THE TRUSTEE OF TH OF JAMES FARMER, ESQ., DECEASE TON BROOK CHEMICAL WORKS, MANCHESTER.

WHEATLEY KIRK, PRI GOULTY are instructed to offer FOR PUBLIC AUCTION, at the Mitre Hotel, Gates, Manchester, on Thursday, the 19th day 1890, at Seven o'clock prompt in the evening, conditions of Sale to be then produced, the wh LAND, BUILDINGS, MOTIVE POWE PLANT, comprising the Gorton Brook Chemic situate in Brook-street, West Gorton (near Station), which are now in work under the ma of the Trustee.

The property is, as to an area of 1,796 squa freehold, subject to a low perpetual chief rent, an area of 5,976 square yards held on a 999 years a low rent. The works can produce about 3 Vitriol per week, and Spirits of Salt, Caustic Sod or Printers' Solutions in proportion, and othe Drysaltery Goods.

The WORKS are STILL IN OPERATION continue so whether SOLD or NOT.

Plans and particulars may be had on applicati Auctioneers, Albert-square, Manchester, and 5 Victoria-street, London, E.C.; or from Mr. J BECK, Trustee, on the premises; or from Mes C. CHEW & SON, Solicitors, 10, Kennedy-stre chester.

TO BE SOLD, the LANCAS MANURE COMPANY'S WORKS at with Machinery and 24,420 square yards of Freehol Address J. Hosking, 16, Fenwick Street, Liverpo

Tenders.

NEW TENDER.

TAR AND AMMONIA LIQUOR

THE Penrith Local Board are prepared to receive TENDERS for the PURCHASE of the SURPLUS TAR and AMMONIACAL LIQUOR produced at their works during the year ending June 30th, 1891.

Full particulars may be obtained on application to Mr. E. Shaul, Gas Manager.

Sealed tenders endorsed "Tender for Tar, &c.," to be delivered to the undersigned not later than Monday, June 30th, 1890.

G. WAINWRIGHT,
Clerk to Local Board,
Penrith.

Public Offices,
June 12th, 1890.

COUNTY BOROUGH OF SALFORD.

GAS DEPARTMENT.

TAR.

THE Gas Committee invite Tenders for the purchase and removal of the surplus TAR to be produced at their Bloom-street, Regent-road, and Liverpool-street Works in 12 months from July 13th next.

The approximate quantity will be 4,500 tons.

Forms of Tender and Conditions of Contract may be obtained on application to the Gas Engineer, Gas Offices, Bloom-street, Salford.

Sealed Tenders, endorsed "Tender for Tar," addressed to the Chairman of the Gas Committee, must be delivered to me on or before Thursday the 10th July next, at five o'clock in the evening.

By Order.
SAM'L BROWN,
Town Clerk.

Town Hall, Salford,
12th June, 1890.

GEORGE SCOTT & SON,

Engineers, Ironfounders & Boilermakers since 1834.

Makers of every description of Chemical Plant and Machinery,

WHITE LEAD, OIL, COLOUR, and SOAP MAKERS MACHINERY and PLANT,

And Sole Makers of Plant for a New Process for Making Acetic Acid.

Air Compressors
Apparatus for Liquefying Gases
Edge Runners
Blowing Engines for Mixing and Agitation
Vacuum Pans, Charcoal Filters
Special Acid Filter Presses
Hydraulic Presses
Roller and Horizontal Mills
Hydro-Extractors for Sulphate of Ammonia and other Chemical Products.
Filter Presses with or without absolute extraction. Vacuum Filters.
Vacuum Pumps to give an almost absolute vacuum.

v174P]

Telegrams: "Thirtyfour, London." Telephone 4390.

44 & 46, Christian Street, London, E., England.

SPENCE'S
LUMP ALUM,
POTASH, OR AMMONIA

IS THE FREEST FROM IRON. [206v

PETER SPENCE & SONS, Manchester Alum Works, MANCHESTER.

Also manufacturers of **SULPHATE OF ALUMINA & ALUMINOFERRIC.**

J. GREGORY & CO.,
Deepcar Ganister & Firebrick Works,
Near SHEFFIELD.

SPECIALITIES—

Silica Bricks and Fire Bricks for Siemens Furnaces, Glass and Chemical Works, and other intense continuous or intermittent heats. Special Bricks for Gas Works and Coke Ovens.

The Highest Grades and Quality Guaranteed. Moderate Prices.

114J] Telegrams:—"GREGORY, DEEPCAR."

MATHIESON'S
PURE AMMONIA ALKALI.

Commercial Test 59° equal to 99 °/。 of Carbonate of Soda.

The purest and most economical article for the use of Paper Makers, Bleachers, Dyers, Calico Printers, Glass Manufacturers, Soap Makers, Wool Scourers, etc.

Samples sent free on application to

W. H. COWBURN,
70, MARKET STREET, MANCHESTER.
AGENT FOR ENGLISH CONSUMERS. [112H

Tenders.

CORPORATION OF MOSSLI

THE Gas Committee of th Corporation are prepared to receive Te surplus TAR that may be produced at their Roaches, for 12 months from the 30th June,

Particulars and form of Tender may be ob the manager, Mr. R. MERRELL, at the Gas

Tenders endorsed "Tender for Tar," to be noon on Monday, July 7th, at the Town C Mossley.

The Committee do not bind themselves t highest or any Tender.

JOSEPH HYDE, To

Miscellaneous.

THE BANKHALL OIL AND CHEMICAL LIMITED (formerly Thacker & Co.), LIV

FOR SALE by PRIVATE T as a going concern, the above old-establis known Oil and Chemical Works, including t the Machinery and Plant, and the goodwill of The works stand on about 15,530 square ya which is held on lease for 75 years from Marc ground rent of £700, per annum. In additio good private road on the east side of the wor occupied jointly with the Lancashire and Yor way Co. The situation of the works is eve could be desired. They are bounded on the east by the Lancashire and Yorkshire Rail west by the Leeds and Liverpool Canal, and by Bankhall-st. There are sidings into the wo Lancashire and Yorkshire Railway, and there access to the canal. The works are also in clo to the docks, and to the London and No Midland, and Manchester Sheffield and Linc way stations, and are only about 1½ mil Liverpool Exchange. An important feature i with these works is that they are not liable to be prosecuted for emitting noxious odours, etc. T comprise extensive Oil and Cake Mills, with and Tallow Refineries, Oil Stores, Tar and Na leries, Soapery, large Warehouses with ver fireproof vaults and steam hoist, Cooperage, the usual Offices. The machinery and plant s tively new, in excellent condition, and capab a very extensive manufacturing business. T connection is a most valuable one. With a view realisation, a very moderate price will b

For orders to inspect the works and to tre the Liquidator, T. THEODORE ROGERS John-street, Liverpool; or to WM. F. MOR CO., Solicitors, 5, Castle-street, Liverpool.

TO BE SOLD, the LANCA MANURE COMPANY'S WORKS with Machinery and 24,420 square yards of Freeh Address J. Hosking, 16, Fenwick Street, Liver

WANTED, from Septemb 1890, to March 31st, 1891, the hire Travelling Tanks for conveying Tar and Liq with price and particulars "X," *Chemical Tra* Blackfriars street, Manchester.

TO PHOSPHORIC ACID MA Wanted, a Foreman having had good e the manufacture of this article. Liberal suitable man. All communications will be strict confidence by "Phosphoric," c/o Deac tising Offices, Leadenhall Street, London, E.C.

WANTED, a small exper Vacuum Pan.—Address George A. Cl Needham Market.

CHEMICAL MANURE W for Sale; at side of Great Eastern Railwa ford, near Ipswich, as now carried on by Chapman & Co.; to be sold in consequence of of the managing partner; freehold; offers capitalist an opportunity of securing a good go —For particulars apply to Messrs. Garrod, Son, Auctioneers, Ipswich.

TO Analytical Chemists.— for one or two months to act as *locum* thoroughly qualified Chemist, not under 30 ye applicants must have had complete training an experience in brewery analytical work, and forward also copies of testimonials.—Apply by to Gravity, Box No. 4, Post-office, Fiveways, Bi

PATENTS AND TRADE
Secured, Bought

Charges for Securing Patents:

Great Britain provisional, £2 10s.; comple America, £12 10s.

Special attention given to Chemical Inven

The International Patent
27, Martin's Lane, Cannon Street, Lond

Tenders.

COUNTY BOROUGH OF SAL[...]

GAS DEPARTMENT.

TAR.

THE Gas Committee invite[...] for the purchase and removal of the sur[...] be produced at their Bloom-street, Rege[...] Liverpool-street Works in 12 months from Jul[...]

The approximate quantity will be 4,500 tons[...]

Forms of Tender and Conditions of Cont[...] obtained on application to the Gas Engineer, Bloom-street, Salford.

Sealed Tenders, endorsed "Tender for Ta[...] to the Chairman of the Gas Committee, must[...] to me on or before Thursday the 10th July[...] o'clock in the evening.

By Order.

SAM'L BR[...]

T[...]

Town Hall, Salford,
12th June, 1890.

CORPORATION OF MOSSL[...]

THE Gas Committee of th[...] Corporation are prepared to receive Te[...] surplus TAR that may be produced at their[...] Roaches, for 12 months from the 30th June.

Particulars and form of Tender may be o[...] the manager, Mr. R. MERRELL, at the Ga[...]

Tenders endorsed "Tender for Tar," to be[...] noon on Monday, July 7th, at the Town C[...] Mossley.

The Committee do not bind themselves [...] highest or any Tender.

JOSEPH HYDE, T[...]

Miscellaneous.

THE BANKHALL OIL AND CHEMICA[...] LIMITED (formerly Thacker & Co.), LIV[...]

FOR SALE by PRIVATE T[...] as a going concern, the above old-establi[...] known Oil and Chemical Works, including [...] the Machinery and Plant, and the goodwill of[...] The works stand on about 15,530 square y[...] which is held on lease for 75 years from April[...] a ground rent of £700. per annum. In additi[...] good private road on the east side of the wo[...] occupied jointly with the Lancashire and Yo[...] way Co. The situation of the works is eve[...] could be desired. They are bounded on th[...] east by the Lancashire and Yorkshire Rai[...] west by the Leeds and Liverpool Canal, and [...] by Bankhall-st. There are sidings into the wo[...] Lancashire and Yorkshire Railway, and there[...] access to the canal. The works are also in clo[...] to the docks, and to the London and No[...] Midland, and Manchester Sheffield and Linc[...] way stations, and are only about 1¼ mil[...] Liverpool Exchange. An important feature [...] with these works is that they are not liable to be[...] prosecuted for emitting noxious odours, etc. T[...] comprise extensive Oil and Cake Mills, with[...] and Tallow Refineries, Oil Stores, Tar and Na[...] leries, Soapery, large Warehouses with ve[...] fireproof vaults and steam hoist, Cooperage, [...] the usual Offices. The machinery and plant [...] tively new, in excellent condition, and capal[...] a very extensive manufacturing business. T[...] connection is a most valuable one. With a view[...] realisation, a very moderate price will [...]

For orders to inspect the works and to tr[...] the Liquidator, T. THEODORE ROGERS[...] John-street, Liverpool; or to WM. F. MOR[...] CO., Solicitors, 5, Castle-street, Liverpool.

TO BE SOLD, the LANCA[...] MANURE COMPANY'S WORKS [...] with Machinery and 24,420 square yards of Free[...] Address J. Hosking, 16, Fenwick Street, Liver[...]

WANTED, by a Chemist ([...] Geneva University and Polytechn[...] Austria), situation in Dyeworks, Printworks, [...] other industrial establishment.—Address "C[...] W. H. Smith & Son, Blackfriars-street, Manc[...]

WANTED, thoroughly [...] man, for the manufacture of Lemon[...] Chromes.—Address C142, *Chemical Tra[...]* Blackfriars-street, Manchester.

STANDARD CHEMICAL BOOKS.

All in good condition and sent Carriage Paid.

Chemistry applied to Arts and Manufactures. By writers of eminence (including Schorlemmer, Crookes, etc.): *Plates of Machinery, etc.*, 3 vols., £4, for 45s.
Roscoe & Schorlemmer's Chemistry Complete set: organic and inorganic; 8 vols., £5 18s. 6d.
Lunge's Alkali-Makers' Pocket Book, 1889. 6s. 3d.
Watts' Dictionary of Chemistry and the Allied Sciences. Complete and unabridged edition; 9 vols., cloth, (new), £15., for £8 8s.
Ditto, New Edition in 4 vols., re-written and revised by Morley and Muir. Vols. I. and II., complete, *now ready*, 2 vols., half morocco, 42s., for 12s. each.
Kingzett's Products of the Alkali Trade, 6s.
Benedikt and Knecht's Coal-Tar Colours, 5s. 9d.
Gmelin's Hand-Book of Chemistry. By H. Watts. Complete set, Organic and Inorganic, with Index, 19 vols., 8vo. cloth, scarce, £20, for £7 10s.
Lunge's Coal Tar and Ammonia. *New Edition*, 27s
Ditto, Previous Edition (S.H.) for 12s.
Crookes's Dyeing and Calico Printing, 42s., for 21s.
Richardson and Watts' Chemical Technology of Arts and Manufactures. 5 vols., £5. 5s., for 30s.
Ditto, Acids, Alkalies, and Salts, 3 vols., £4. for 15s.
Chemical Society Journal, 1878-87. 20 vols., £8 8s.
Thorpe's Dictionary of Applied Chemistry (Technical). 3 vols., forming supplement to Watts'. Vol. 1 *now ready*, 42s. for 32s.

WILLIAM F. CLAY,

Chemical Bookseller,

2, TEVIOT PLACE, EDINBURGH.

Communications invited for any books. Odd Vols. or Nos. wanted or for sale, and will receive prompt attention.
Wanted to purchase "Journal of Society of Chemical Industry" and any other Chemical Literature, for Cash, or Exchanged at Highest Market Value. [9861-

---

REVISED AND ENLARGED EDITION,
Small Post 8vo. Price 6s 6d.

THE CHEMISTRY OF THE COAL-TAR COLOURS.

Translated from the German of Dr R. BENEDIKT, and Edited, with Additions, by E. KNECHT, Ph.D. (Head Master of the Dyeing Department of the Technical College, Bradford), Editor of the *Journal of the Society of Dyers and Colourists.*

1170] LONDON:
GEORGE BELL & SONS, York Street, Covent Garden.

TUBES

IRON TUBES AND FITTINGS for Gas, Water, Steam, Liquid Ammonia, and Hydraulic Work, galvanized white enamelled inside, or coated by Dr. Angus Smith's Composition.

COILS for heating, super-heating, distilling, and refrigerating purposes, to 500 feet long without joints. Iron and brass cocks and valves. [933173

JOHN SPENCER,

Globe Tube Works, Wednesbury, and 14, Great St. Thomas Apostle, Cannon-st., London, E.C.

Established 1869.

MANGANESE

Bauxite, Barytes, Fluor Spar, Magnesite, Chrome Ore, Emery Stone, China Clay, French Chalk, Oxides, Ochres, Umbers, Bog Ore, and all Minerals

GEO. G. BLACKWELL,
Proprietor, Importer & Manufacturer.
Works and Mills: Garston.
CHIEF OFFICES:
26, 26, & 27, Irwell Chambers, W., LIVERPOOL. [1636

ALEX. ROWAT & CO.,

Wire Work & Wire Cloth Manufacturers,

AND

BOILER TUBE BRUSH MAKERS,

5, CANDLERIGGS, GLASGOW.

[Estab]lished 1837. [m880159

PHILIP HARRIS & CO. LIMITED,

BULL RING, BIRMINGHAM,

HAVE JUST ISSUED

A CATALOGUE OF CHEMICAL APPARATUS,

Containing about 1,200 illustrations, giving list of apparatus and reagents for use in the analysis of Iron, Steel, Fuel, &c., by various methods; also particulars of Bunte's, Hempels, and other Gas analysis apparatus; **Harris's Polariscope, Microscope, &c., &c.** (See *Chemical Trade Journal*, February 16th, page 102.) [[illegible]

WINCE PULLEYS.

RUSCOE'S PATENT GROOVED PULLEYS ENAMELLED ARE THE BEST.

They withstand the acid, are well balanced, and move freely on the shaft; little wear and tear.

Once used, always continued. [[illegible]

References, Prices, and other information, apply to the Patentee and Sole Maker:

JOHN RUSCOE, HYDE, near MANCHESTER.

ACCURATE STANDARD SOLUTIONS.

Mr. Francis Sutton, F.I.C., F.C.S., has a Special Laboratory for the Manufacture of these Solutions for every variety of Analysis by Volumetric Methods, and on any System of Measurement; also Indicators, Test Papers, and all requisites.
Special Prices quoted to Large Consumers. Price List on application to

FRANCIS SUTTON & CO., NORWICH.

IRON CARBOY HAMPERS.

All Qualities. Lowest Prices.

H. LEIGH & SONS

ORLANDO MILL,

BOLTON. [1671

PURE SULPHUROUS ACID,

SULPHITES AND BISULPHITES OF LIME, SODA, &c.

FOR EXPORT. [180

S. KEYS & CO. Limited, WEST BROMWICH.

Liverpool: Exchange Chambers
Glasgow: Charles-st., St. Rollox

CANNINGTON, SHAW, & CO.

CARBOYS.

SHERDLEY GLASSWORKS, ST. HELENS.

Manchester: Oldham Road.
London: Worship Street, E.

[1980

Lightning Source UK Ltd.
Milton Keynes UK
UKHW020155260219
337978UK00012B/1399/P

9 780331 035094